ARENE CHEMISTRY

ARENE CHEMISTRY

Reaction Mechanisms and Methods for Aromatic Compounds

Edited by

JACQUES MORTIER

WILEY

Published by John Wiley & Sons, Inc., Hoboken, New Jersey
Published simultaneously in Canada

Library of Congress Cataloging-in-Publication Data:

Arene chemistry : reaction mechanisms and methods for aromatic compounds / edited by Jacques Mortier.
pages cm
Includes index.
ISBN 978-1-118-75201-2 (cloth)
1. Aromatic compounds. 2. Chemistry, Organic. I. Mortier, Jacques, 1959– editor.
QD331.A74 2016
547′.61–dc23

2015024766

Set in 9/11pt Times by SPi Global, Pondicherry, India

10 9 8 7 6 5 4 3 2 1

1 2016

CONTENTS

LIST OF CONTRIBUTORS

Igor V. Alabugin Department of Chemistry & Biochemistry, Florida State University, Tallahassee, FL, USA

Marco Bandini Dipartimento di Chimica "G. Ciamician", Alma Mater Studiorum, Università di Bologna, Bologna, Italy

Gonzalo Blay Departament de Química Orgànica, Facultat de Química, Universitat de València, Burjassot (València), Spain

Sergio M. Bonesi CIHIDECAR CONICET, Departamento de Química Orgánica, Facultad de Ciencias Exactas y Naturales, Universidad de Buenos Aires, Buenos Aires, Argentina

Anne Boussonnière Institut des Molécules et Matériaux du Mans, Faculté des Sciences et Techniques, UMR CNRS 6283, Université du Maine and CNRS, Le Mans Cedex, France

Tore Brinck Applied Physical Chemistry, KTH Royal Institute of Technology, Stockholm, Sweden

María E. Budén INFIQC, Departamento de Química Orgánica, Facultad de Ciencias Químicas, Universidad Nacional de Córdoba, Córdoba, Argentina

Paola R. Campodónico Centro de Química Médica, Facultad de Medicina, Clínica Alemana Universidad del Desarrollo, Santiago, Chile

Anne-Sophie Castanet Institut des Molécules et Matériaux du Mans, Faculté des Sciences et Techniques, UMR CNRS 6283, Université du Maine and CNRS, Le Mans Cedex, France

Floris Chevallier Chimie et Photonique Moléculaires, UMR 6226 CNRS-Université de Rennes 1, Rennes, France

Renato Contreras Departamento de Química, Facultad de Ciencias, Universidad de Chile, Chile

Michael R. Crampton Department of Chemistry, University of Durham, Durham, UK

Charles B. de Koning School of Chemistry, University of the Witwatersrand, Johannesburg, South Africa

Marion H. Emmert Department of Chemistry and Biochemistry, Worcester Polytechnic Institute, Worcester, MA, USA

Maurizio Fagnoni PhotoGreen Lab, Department of Chemistry, University of Pavia, Pavia, Italy

Francisco Foubelo Departamento de Química Orgánica, Facultad de Ciencias, Universidad de Alicante, Alicante, Spain

Francisco Galindo Departamento de Química Inorgánica y Orgánica, Universitat Jaume I de Castellón, Castellón de la Plana, Spain

Javier F. Guastavino INFIQC, Departamento de Química Orgánica, Facultad de Ciencias Químicas, Universidad Nacional de Córdoba, Córdoba, Argentina

Norbert Hoffmann Institut de Chimie Moléculaire de Reims, UMR 6229 CNRS et Université de Reims Champagne-Ardenne, UFR Sciences, Reims, France

M. Consuelo Jiménez Departamento de Química/Instituto de Tecnología Química UPV-CSIC, Universitat Politècnica de València, València, Spain

Anny Jutand Ecole Normale Supérieure-PSL Research University, Département de Chimie, Sorbonne Universités, UPMC Univ Paris 06, CNRS UMR 8640 PASTEUR, Paris, France

Dipannita Kalyani St. Olaf College, Northfield, MN, USA

Masanam Kannan Department of Chemistry, Indian Institute of Technology Guwahati, Guwahati, India

Kemal Kaya Department of Chemistry & Biochemistry, Florida State University, Tallahassee, FL, USA

Oxana A. Kholdeeva Boreskov Institute of Catalysis, Novosibirsk State University, Novosibirsk, Russia

Douglas A. Klumpp Department of Chemistry and Biochemistry, Northern Illinois University, DeKalb, IL, USA

Guillaume Lefèvre CEA–Saclay, IRAMIS Institute, SIS2M/LCCEF, Gif-sur-Yvette, Cedex, France

Christopher J. Legacy Department of Chemistry and Biochemistry, Worcester Polytechnic Institute, Worcester, MA, USA

Frédéric R. Leroux Laboratoire de Chimie Moléculaire, CNRS and University of Strasbourg, UMR CNRS 7509, ECPM, Strasbourg Cedex 2, France

Simon E. Lewis Department of Chemistry, University of Bath, Bath, UK

Magnus Liljenberg Applied Physical Chemistry, KTH Royal Institute of Technology, Stockholm, Sweden

Raimondo Maggi "Clean Synthetic Methodology Group", Dipartimento di Chimica dell'Università, Università degli Studi di Parma, Parma, Italy

Mieczysław Mąkosza Institute of Organic Chemistry, Polish Academy of Sciences, Warsaw, Poland

Elodie E. Marlier St. Olaf College, Northfield, MN, USA

Miguel Angel Miranda Departamento de Química/Instituto de Tecnología Química UPV-CSIC, Universitat Politècnica de València, València, Spain

Rana K. Mohamed Department of Chemistry & Biochemistry, Florida State University, Tallahassee, FL, USA

Florence Mongin Chimie et Photonique Moléculaires, UMR 6226 CNRS-Université de Rennes 1, Rennes, France

Marc Montesinos-Magraner Departament de Química Orgànica, Facultat de Química, Universitat de València, Burjassot (València), Spain

Jacques Mortier Institut des Molécules et Matériaux du Mans, Faculté des Sciences et Techniques, UMR CNRS 6283, Université du Maine and CNRS, Le Mans Cedex, France

Rodrigo Ormazábal-Toledo Departamento de Fisica, Facultad de Ciencias, Universidad de Chile, Chile

Armen Panossian Laboratoire de Chimie Moléculaire, CNRS and University of Strasbourg, UMR CNRS 7509, ECPM, Strasbourg Cedex 2, France

José R. Pedro Departament de Química Orgànica, Facultat de Química, Universitat de València, Burjassot (València), Spain

F. Christopher Pigge Department of Chemistry, University of Iowa, Iowa City, IA, USA

Amalia Isabel Poblador-Bahamonde Department of Organic Chemistry, University of Geneva, Geneva, Switzerland

Tharmalingam Punniyamurthy Department of Chemistry, Indian Institute of Technology Guwahati, Guwahati, India

Emmanuel Riguet Institut de Chimie Moléculaire de Reims, UMR 6229 CNRS et Université de Reims Champagne-Ardenne, UFR Sciences, Reims, France

Roberto A. Rossi INFIQC, Departamento de Química Orgánica, Facultad de Ciencias Químicas, Universidad Nacional de Córdoba, Córdoba, Argentina

Sethuraman Sankararaman Department of Chemistry, Indian Institute of Technology Madras, Chennai, India

Veronica Santacroce "Clean Synthetic Methodology Group", Dipartimento di Chimica dell'Università, Università degli Studi di Parma, Parma, Italy

Roberto Sanz Área de Química Orgánica, Facultad de Ciencias, Universidad de Burgos, Burgos, Spain

Giovanni Sartori "Clean Synthetic Methodology Group", Dipartimento di Chimica dell'Università, Università degli Studi di Parma, Parma, Italy

Mani Sengoden Department of Chemistry, Indian Institute of Technology Guwahati, Guwahati, India

Timothy J. Snape School of Pharmacy and Biomedical Sciences, University of Central Lancashire, Preston, UK

Anisley Suárez Área de Química Orgánica, Facultad de Ciencias, Universidad de Burgos, Burgos, Spain

Ryo Takita RIKEN Center for Sustainable Resource Science, Wako-shi, Saitama, Japan and Graduate School of Pharmaceutical Sciences, The University of Tokyo, Bunkyo-ku, Tokyo, Japan

Ken Tanaka Department of Applied Chemistry, Graduate School of Science and Engineering, Tokyo Institute of Technology, Meguro-ku, Tokyo, Japan

Masanobu Uchiyama RIKEN Center for Sustainable Resource Science, Wako-shi, Saitama, Japan and Graduate School of Pharmaceutical Sciences, The University of Tokyo, Bunkyo-ku, Tokyo, Japan

Willem A. L. van Otterlo Department of Chemistry and Polymer Science, Stellenbosch University, Matieland, South Africa

Miguel Yus Departamento de Química Orgánica, Facultad de Ciencias, Universidad de Alicante, Alicante, Spain

PREFACE

Benzenoid aromatic compounds or arenes have tremendous importance in academic and industrial chemical applications. Of the circa 10 million compounds that are known today, about three millions are arenes. Reactions involving arenes represent key steps in fundamental synthesis, especially in the pharma, agrochemical, and polymer fields. Arene compounds are also widely used as starting materials to obtain dyes, perfumes, explosives, preservatives, etc. New applications include sectors such as functional materials, organic electronics, and molecular machines.

The success of these industries is, in large part, due to the towering achievements of arene chemistry, a mature discipline that emerged well over 150 years ago. Without a doubt, arene chemistry research is now in its golden age, and its knowledge is indispensable for any synthetic chemists. Despite these extraordinary academic and commercial implications, there are, as yet, no books focusing on mechanisms and strategies in this continuing developing field, with a comprehensive coverage of classical and more recent reactions.

To date, the commonly accepted books on arene chemistry are either out of date or only deal with specific reaction types. For instance, *Modern Arene Chemistry* by Astruc (Wiley-VCH, Weinheim, 2002) is overly involved in the materials science end of the chemistry covered, while *Aromatic Chemistry* by Hepworth, Waring, and Waring (Royal Society Cambridge, 2002), which is intended specifically for basic-level chemistry students, is only 168 pages, of which the last 20 are answers to problems.

Arene chemistry is growing so rapidly that one cannot keep up with progress, and to get information on aromatic reactions, one needs to consult many different books. Although there are already many books on the market about nucleophilic aromatic substitution (including *Modern Nucleophilic Aromatic Substitution*, by Terrier, Wiley-VCH, Weinheim, 2013), aromatic rearrangements, reductions, oxidations, dearomatization reactions, and photochemical and biochemical transformations, it is quite difficult to get an overview of the significant impact of each topic. On the other hand, electrophilic aromatic substitution, aryne chemistry, and directed aromatic metalation have advanced dramatically in understanding over the years but rarely received appropriate attention. Moreover, metal-catalyzed cross-coupling and CH-functionalization reactions, which have known a recent booming development widely covered by an extremely abundant literature, deserve to be summarized and commented to meet the needs of a broader readership.

Arene Chemistry: Reaction Mechanisms and Methods for Aromatic Compounds is the first book of its kind that furnishes a complete overview and a guide and collects in a single opera all the topics related to the field. This compendium connects methodology and reactivity of aromatic compounds with mechanism, at the interface of synthesis and physical organic chemistry. It is organized according to reaction classes, so that someone who would like to run their first aromatic oxidation can flip open to the corresponding chapter to learn the basics quickly. This book also establishes the interesting connectivity between the different subjects. Because the presentation of the material organized according to reaction mechanisms is of central significance to students of organic chemistry, I feel fairly confident that the pedagogical approach followed will render the content readily comprehensible. In addition, the text grouped on reaction mechanism type as opposed to the reaction products should be much more intuitive to aid a deeper understanding of the area.

Considering that arene chemistry is a field that evolves in parallel in laboratories throughout the world, I sought to select younger active colleagues and leading senior experts who were not only authoritative but also as geographically distributed as the field itself. The contributors' expertise allows them to frame the literature contextually for the audience while providing a critical view of the state of the art in terms of potential for growth, future outlook, and limitations. In a rather limited space, each chapter is organized to understand and expand on aromatic reactions covered in foundation courses to the latest understanding and to apply them in a practical context by designing syntheses.

In building the project, 32 topics divided into 10 parts have been identified as deserving a special coverage. There is detailed contents from which I believe it will be possible to track down most points. Each chapter covers basics as well as most recent areas of interest to give a complete picture to both teach and bridge the primary literature. Each chapter should also lead the reader to consult the secondary literature sources cited by the authors including reviews, books, and monographs, in order to understand the subject in a more comprehensive manner. This book is organized with the intention of providing a platform for scientists from different disciplines to generate new ideas and thoughts by inspiring each other.

As aromatic compounds are ubiquitous, this book should be especially relevant to a large audience, which covers advanced undergraduates through postgraduates and right up to academic faculty, and the chemical industry, that is, almost the entire organic chemistry community. The work published on heteroaromatic chemistry is so extensive that it was impractical to attempt to review progress in this area at the same time. If so, this would have doubled or tripled the content. However, synthetic applications described in the different sections can be related to the preparation of carbocyclic aromatic (benzene) rings embedded in a heterocycle. Typical examples of industrial applications of the relevant technologies are appropriately illustrated throughout the text.

To sum up, the coverage presents the most significant results and the underlying principles that are emerging in arene chemistry. Since this book directly addresses arenes and encapsulates most important synthetic applications, it should be an easy choice for people looking for information on aromatic reactions, both from mechanistic and synthetic viewpoints.

At the start of the project about 3 years ago, I was aware of the immensity of the task and the difficulty of covering such a broad area. I hope that this book reflects recent changing trends in research so that it will cater for the maximum possible range of interests. I accept the entire responsibility for any significant omission. I heartily encourage those who read and use this book to contact me directly with comments, errors, and publications that might be appropriate for eventual future editions.

My email address is jacques.mortier@univ-lemans.fr, my blog is http://jmortier.unblog.fr

As an editor, it has been a very exciting experience to collaborate with acknowledged experts from all over the world. I wish to express my profound gratitude for the time and effort that they have dedicated to this process. This work would not have been accomplished without the acknowledged experts (over 150) including most of the contributors of this book who agreed to read the chapters and contributed to improving the quality of the book.

I like to extend my warm thanks to all of my students, postdoctoral researchers, and colleagues from the university and the industry for their intellectual contribution and dedication.

I thank the publisher Wiley, especially Jonathan Rose for contacting me to write this book and for his understanding and help in preparing the book. I am also grateful to François Pascal Raj of SPi Global who led the copyediting process with great skill.

It would not have been possible to put the book in its final form without the support, encouragement, love, and patience of my wife, Marie-Jeanne, and my two sons, Rik and Jan. They are tenderly acknowledged.

I also think of my mom, Marie-Thérèse Bernier, whom I miss so much. This book is dedicated to her memory.

JACQUES MORTIER

PART I

ELECTROPHILIC AROMATIC SUBSTITUTION

1

ELECTROPHILIC AROMATIC SUBSTITUTION: MECHANISM

DOUGLAS A. KLUMPP

Department of Chemistry and Biochemistry, Northern Illinois University, DeKalb, IL, USA

1.1 INTRODUCTION

Electrophilic aromatic substitution (S_EAr) is one of the most important synthetic organic reactions [1]. Since its discovery in the 1870s by Charles Friedel and James Crafts [2], it has become a general route to functionalized aromatic compounds. The chemistry is used extensively in the chemical industry, providing millions of tons of aromatic products annually for chemical feedstock, commodity chemicals, and consumer applications. For example, detergents (i.e., **1**, Scheme 1.1) are commonly prepared using two S_EAr reactions: alkylation and sulfonation. The antibacterial agent *sulfadiazine* (**2**) is prepared using nitration and chlorosulfonation reactions during the course of its synthesis, while the disperse dye (**3**) is prepared using an azo coupling reaction. Several important polymers, such as thermosetting phenol-formaldehyde resins, are also prepared via S_EAr reaction steps. In other applications, the chemistry is commonly used in natural product and target-directed syntheses [1].

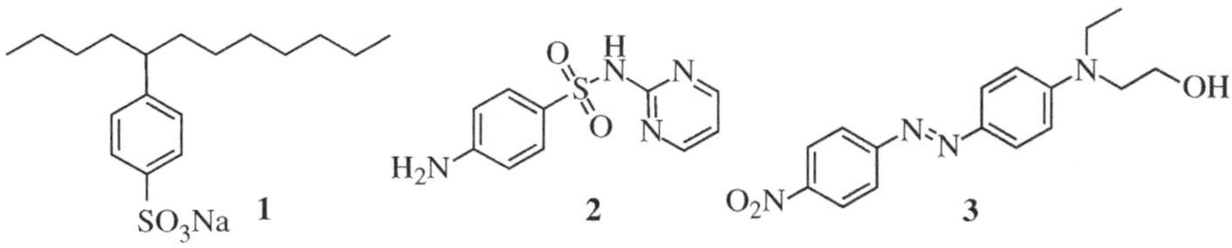

SCHEME 1.1 Products from S_EAr reactions.

Arene Chemistry: Reaction Mechanisms and Methods for Aromatic Compounds, First Edition. Edited by Jacques Mortier.

In addition to its application in synthetic chemistry, S_EAr has one of the most thoroughly studied mechanisms among organic reactions. These studies have paralleled the development of chemistry itself—from the understanding of ions in chemistry and aromaticity in π-systems to the development of high-level theoretical calculations and ultrafast spectroscopic methods. Our understanding of these mechanisms has evolved steadily since the time when the chemistry was first described. This area continues to be an active area of research, and these studies provide new insights into the mechanisms of these valuable organic transformations. The importance of this mechanistic understanding cannot be overstated. Because the chemistry has significant economic value, mechanistic understanding is crucial for chemists to maximize reaction yields, reduce costs, and minimize the environmental impacts of these synthetic processes. In the following chapter, I will provide an overview of the S_EAr reaction mechanism, discussing the salient features of these processes and efforts to understand them.

1.2 GENERAL ASPECTS

The S_EAr reactions involve more than 20 distinctly different types of substitutions, yet these transformations have similar overall mechanisms. The commonly proposed mechanism involves interaction of an electrophilic species with the π-system of an arene (Scheme 1.2) [3]. The electrophile (E+) itself is often a cationic species (*vide infra*), but S_EAr reactions may also be initiated by dipolar

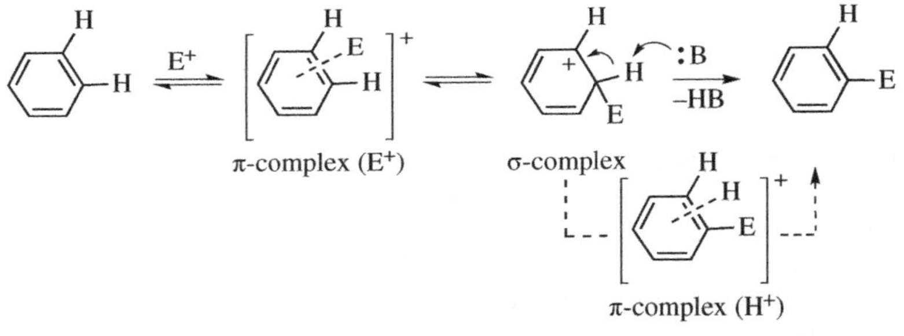

SCHEME 1.2 Proposed mechanism for the S_EAr reaction.

groups or molecules. The initial interaction may lead to the formation of a π-complex or an encounter complex. The π-complex often forms the σ-complex intermediate, also known as the Wheland complex. In the final step, a base removes the *ipso* proton and the substitution product is obtained. This mechanistic interpretation also allows for the formation of a second π-complex from the σ-complex intermediate, where the proton is loosely bound to the π-system. With the regeneration of the aromatic π-system, product stability typically leads to a fast reaction in the final step.

There are several variations of this mechanism. For example, in nitrations, there is considerable evidence to suggest single electron transfer between the nitronium cation (NO_2^+) and the arene (*vide infra*), followed by coupling of the product radicals to give a σ-complex intermediate [4]. There are also examples known involving addition of radical species into the arene (such as $\cdot NO_2$) as a route to substitution products [5]. Moreover, there are examples of S_EAr reactions in which (cationic) groups other than H^+ leave the final reaction step [6].

1.3 ELECTROPHILES

The electrophiles in S_EAr reactions may be divided into two basic categories: those with fully formed cationic charge centers and those having reactive, polarized bonds. For example, Friedel–Crafts alkylation often occurs through the involvement of discrete carbocation intermediates (see Chapter 2).

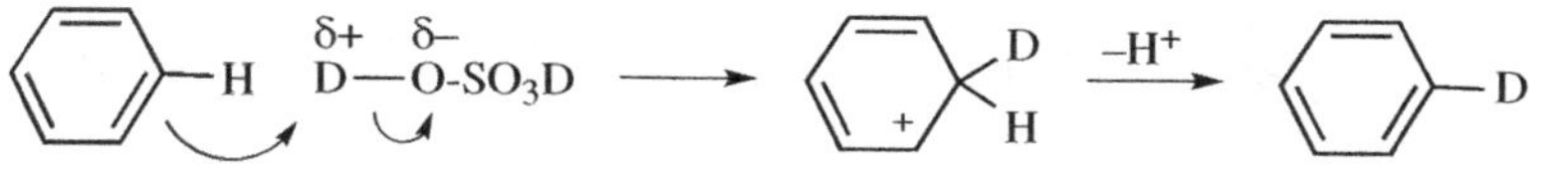

(1.1)

This may be contrasted with the D_2SO_4-promoted hydrogen–deuterium exchange at an arene (Eq. 1.1). In this case, the electrophilic chemistry occurs at the polarized deuterium–oxygen bond, where the deuterium atom carries a significant positive charge. Although the various S_EAr synthetic reactions do share a common basic mechanism (Scheme 1.2), they often differ considerably in the means or mechanisms by which the electrophiles are generated. Several of the common mechanistic types are described below.

Many electrophilic species are generated by the action of Lewis acid catalysts. For example, Friedel–Crafts acylation may occur through the involvement of the acylium ion (i.e., **4**) often generated by Lewis acid-promoted halide abstraction (Eq. 1.2) [7]. Similar Lewis acid-promoted reactions may be used to give carbocationic species from alkyl halides, carboxonium ions from acetals and related precursors, iminium ions from α-haloalkylamines, and others.

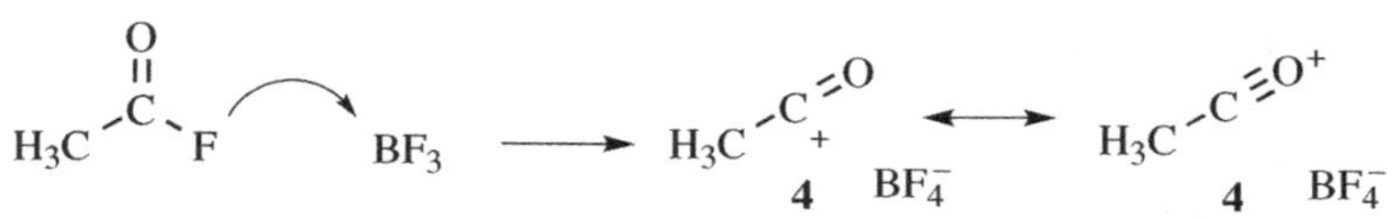

(1.2)

While discrete cationic species may be formed by Lewis acid reactions, highly polarized species may also be the active electrophiles in the transformations. In the case of brominations (Scheme 1.3), Br_2 itself may develop a small dipole (**5**) with approach to an electron-rich arene (such as phenol). Interaction with the Lewis acid may increase the degree of polarization (**6**) or, in the limiting case, give the bromonium ion (**7**). The exact nature of the reacting electrophile depends on several factors, including the reactivity of the arene nucleophile, temperature, strength of the Lewis acid, or solvent ionizing power.

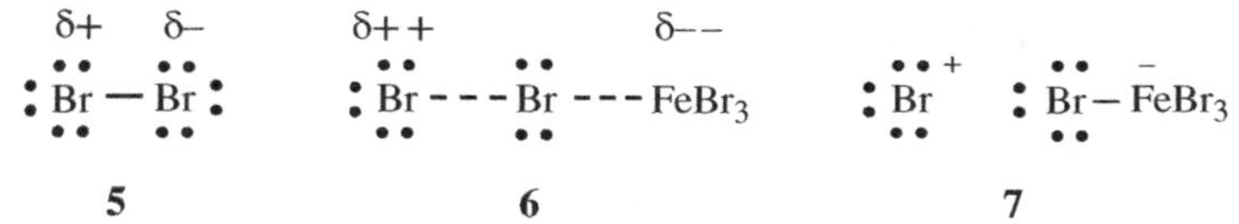

5 **6** **7**

SCHEME 1.3 The development of cationic charge on bromine.

In the case of Brønsted acid catalysts, cationic electrophiles may be generated by the direct protonation of a functional group (Fig. 1.1). This type of chemistry is especially important in the S_EAr reactions of carbonyl compounds and olefins. The carboxonium ions (**8** and **9**) and nitrilium ion (**10**) are formed by protonation at a nonbonding electron pair, while protonation at the olefinic π-bond gives the carbocation (**11**). Both solid (i.e., zeolites) and liquid Brønsted acids may generate electrophiles by this chemistry.

In many types of S_EAr reactions, cationic electrophile formation requires one or more steps after functional group protonation or activation (Fig. 1.2). Alcohols and related functional groups are protonated, and with subsequent cleavage of C—O bond, the carbocation electrophile (**11**) is formed. In a similar respect, a common method of nitration involves the use of HNO_3 with H_2SO_4. The nitronium ion electrophile (NO_2^+, **12**) is formed by protonation of nitric acid and subsequent loss of water by cleavage of the N—O bond [8]. The nitrosonium ion electrophile (NO^+) may be generated by an analogous transformation from nitrous acid, HNO_2 [9]. Likewise, *N*-acyliminium ion electrophiles (i.e., **13**) may be formed by ionization of *N*-hydroxymethylamides [10].

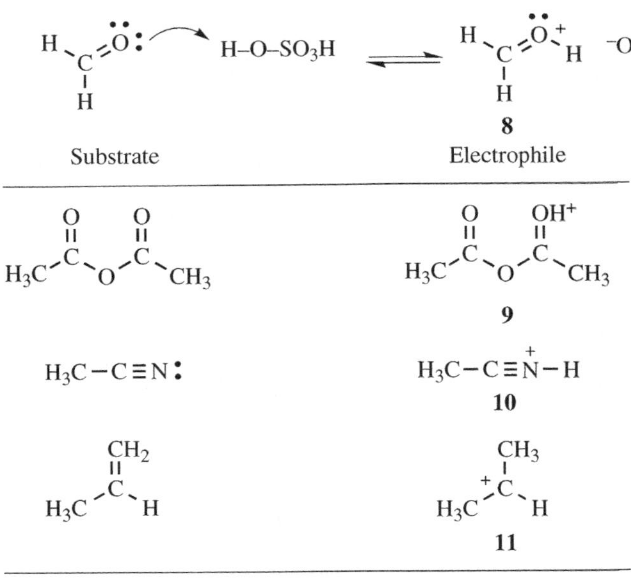

FIGURE 1.1 Examples of electrophiles formed by direct protonation.

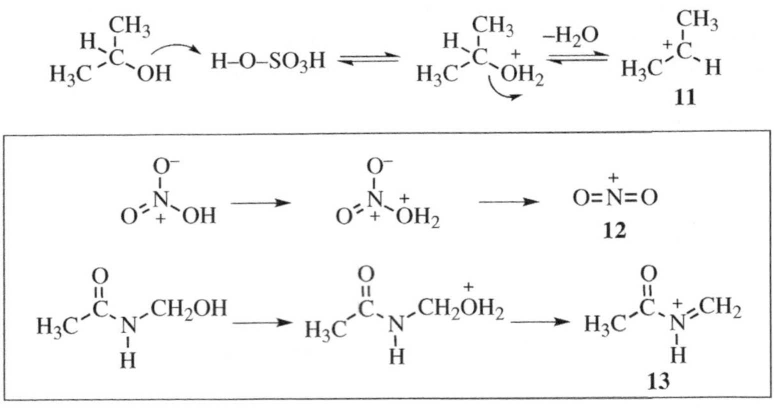

FIGURE 1.2 Electrophiles generated from Brønsted acids.

There are many examples of Brønsted acid-promoted reactions where highly polarized functional groups are the active electrophiles. For example, Olah and coworkers reported *N*-chlorosuccinimide to be a powerful chlorinating agent with superacidic BF_3–H_2O [11]. The active electrophile is likely the diprotonated (**14**) or triprotonated species (**15**, Eq. 1.3),

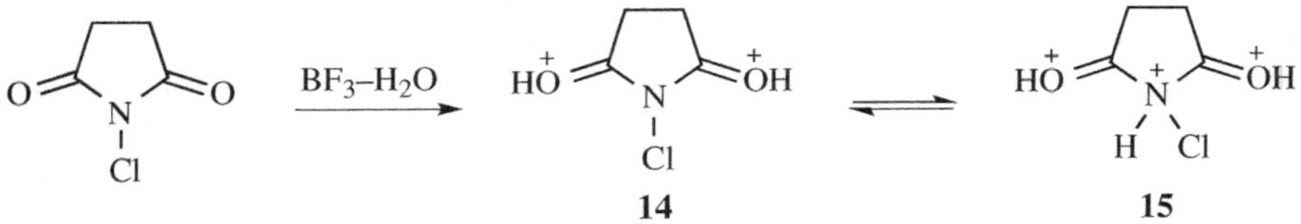

(1.3)

which transfers Cl^+ directly to the arene nucleophile. This system is capable of chlorinating nitrobenzene—a strongly deactivated arene—in 69% yield. In S_EAr reactions with epoxides, the C—O bond may undergo nucleophilic ring opening following protonation or strong hydrogen bonding at the oxygen. Thus, the epoxide substrate (**16**) provides the cyclialkylation product (**17**) in quantitative yield by the action of 1,1,1,3,3,3-hexafluoroisopropanol (Scheme 1.4) [12]. It is suggested that the epoxide is protonated (or coordinated through hydrogen bonding), leading to a nucleophilic ring opening of the epoxide.

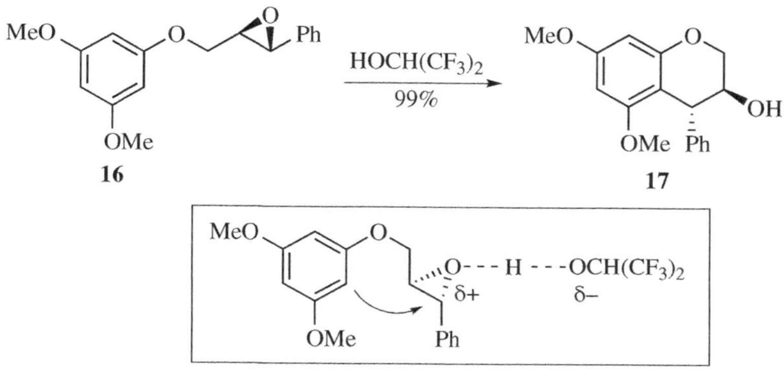

SCHEME 1.4 An intramolecular S_EAr reaction with epoxide **16**.

There are numerous multistep processes that generate electrophiles. As examples of these types of reactions, we will consider the diazotization of anilines and the formation of chloroiminium ions in the Vilsmeier–Haack reaction. Aryl diazonium ions are useful in the modification of arenes by the Sandmeyer reaction and as electrophilic intermediates in diazonium coupling reactions for the synthesis of dyes and pigments. Several types of synthetic methods have been developed for this chemistry, and the mechanism varies depending on the methodology [13]. Under some conditions, the nitrosonium ion (**18**) initiates the process (Scheme 1.5). N_2O_5 and NOCl have also been proposed as intermediates in diazotization—both are considered as nitrosonium ion carriers. The aniline reacts

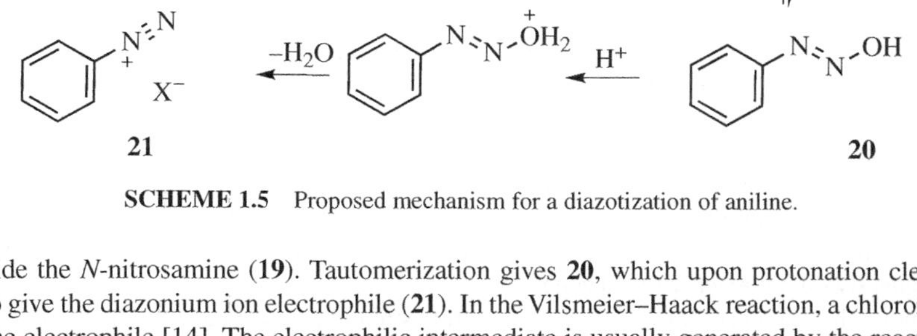

SCHEME 1.5 Proposed mechanism for a diazotization of aniline.

to provide the *N*-nitrosamine (**19**). Tautomerization gives **20**, which upon protonation cleaves off water to give the diazonium ion electrophile (**21**). In the Vilsmeier–Haack reaction, a chloroiminium ion is the electrophile [14]. The electrophilic intermediate is usually generated by the reaction of a formamide with $POCl_3$ (Scheme 1.6). Thus, phosphorous oxychloride reacts at the carbonyl group to provide the addition compound **22**. Cleavage of the C—O bond then provides the electrophilic chloroiminium ion (**23**). Following the S_EAr reaction, hydrolysis gives the formyl group.

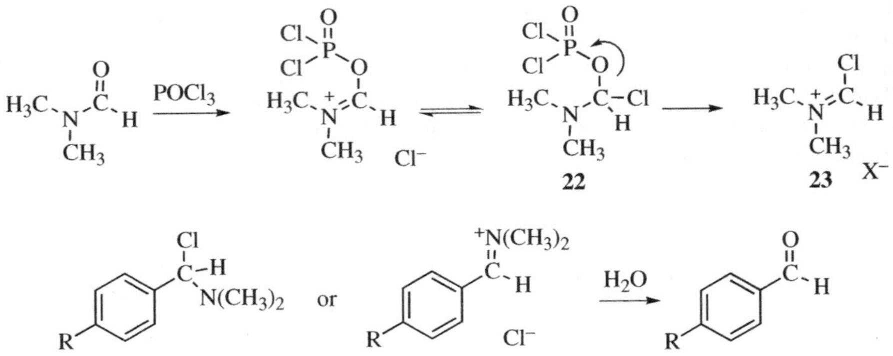

SCHEME 1.6 Proposed mechanism for the Vilsmeier–Haack reaction.

A number of S_EAr reactions have also been developed in which the electrophile is generated by an oxidative process. For example, She and coworkers recently used a (diacetoxyiodo)benzene-promoted phenol oxidation to generate the cationic species (**24**) that undergoes cyclization to the natural product—gymnothelignan N (**25**, Scheme 1.7) [15]. Other oxidative synthetic methods have been developed for electrophilic halogenation [16], aminations [17], and nitrations [18].

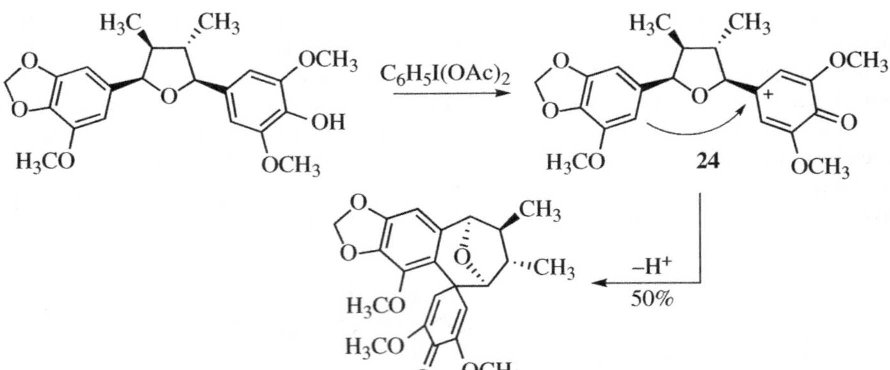

SCHEME 1.7 Oxidative route to electrophile and the cyclization to gymnothelignan N (**25**).

In addition to the mechanisms of electrophile formation, another critical consideration relates to electrophile strength. There has been a vast amount of work done to characterize electrophile strengths [19]. Although much of the work relates to chemistry with *n*-type nucleophiles and nonaromatic π-nucleophiles, some studies have sought to estimate electrophile strengths in S_EAr reactions. Relative electrophile strengths became apparent as the synthetic S_EAr reactions were developed. While the nitronium ion (NO_2^+) salts react with benzene under mild conditions, carboxonium ions such as protonated formaldehyde (CH_2=OH+) are weaker electrophiles and consequently do not react with benzene.

Among the methods for evaluating electrophile strength, a useful approach involves comparing relative reaction rates with benzene and toluene [20]. More reactive electrophiles are expected to be less selective in competition reactions between the two arenes. As noted by Stock and Brown [21],

TABLE 1.1 Relative Yields and k_T/k_B for Competitive S_EAr Reactions with Toluene and Benzene

	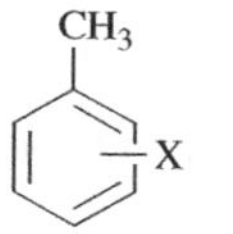	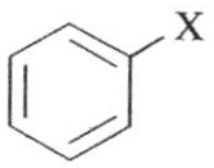	k_T/k_B
NO_2BF_4 sulfolane, 25°C	62.5%	37.5%	1.67
	($X = NO_2$)		
Cl_2, CH_3CN, 25°C	99.9%	0.1%	1650
Cl_2, CH_3CO_2H, 24°C	99.7%	0.3%	353
Cl_2, CH_3NO_2, 25°C $FeCl_3$	93.1%	6.9%	13.5
	(X = Cl)		

an exceptionally reactive electrophile should exhibit no selectivity and give relative rates, k_T/k_B, of approaching 5/6 (reflecting the number of arene C—H positions) or 0.833. Conversely, less reactive electrophiles are expected to be more selective in the reaction with toluene (the more nucleophilic arene) and provide larger values for k_T/k_B. Many electrophilic systems have been studied using this approach, and the data is generally consistent with observations from synthetic chemistry. For example, nitronium tetrafluoroborate is considered a very strong electrophilic system, and it gives nitration products with k_T/k_B of 1.67 (Table 1.1) [22].

Without a Lewis acid catalyst, molecular chlorine is a rather weak electrophile, and this is reflected in its highly selective reaction with toluene (k_T/k_B 1650) [21]. When the reaction is done in acetic acid (a strong hydrogen bonding solvent and Brønsted acid catalyst), the chlorine is somewhat polarized. The more reactive Cl_2 electrophile exhibits slightly lower selectivity with k_T/k_B 353. With the use of a strong Lewis acid, a highly electrophilic system is formed and the chlorination shows only modest selectivity with k_T/k_B of 13.5. Extensive studies of k_T/k_B rates have been presented in the literature [21].

Other approaches of evaluating electrophile strengths have been developed, including comparisons of the regioselectivities of electrophilic attacks. The assumption is that more reactive electrophiles should exhibit less positional selectivity in chemistry with substituted arenes. In a method developed by Brown and associates [21], electrophiles were compared by the relative reactivities of the *meta* and *para* positions of toluene, where more reactive electrophiles give increasing proportions of *meta* substitution. While this approach has some general applicability, problematic cases are known. For example, Olah has described a series of very reactive electrophiles exhibiting low substrate selectivity but high positional selectivity [23]. To explain this, it was suggested that the highest-energy transition state resembles the σ-complex in some S_EAr reactions and the π-complex in other S_EAr reactions (*vide infra*). Correlating electrophilic reactivities with positional regioselectivity seems to work best in the former case.

The Mayr group has applied Equation 1.4 to determine the electrophilicity parameters [19b], *E*, in reactions between electrophiles and nucleophiles, where *N* is the

$$\log k(20°C) = s\,(N + E) \qquad (1.4)$$

nucleophilicity parameter and *s* is the nucleophile-dependent slope parameter. By analyzing pseudo-first-order rate constants from reactions with various types of nucleophiles, the electrophilicities of numerous cationic species have been established. Several of the characterized electrophiles are known to be involved in S_EAr reactions [24], such as carbocations **26** and **27**, the iminium ion **28**, and carboxonium ion **29** (Scheme 1.8). For comparison, toluene and anisole have

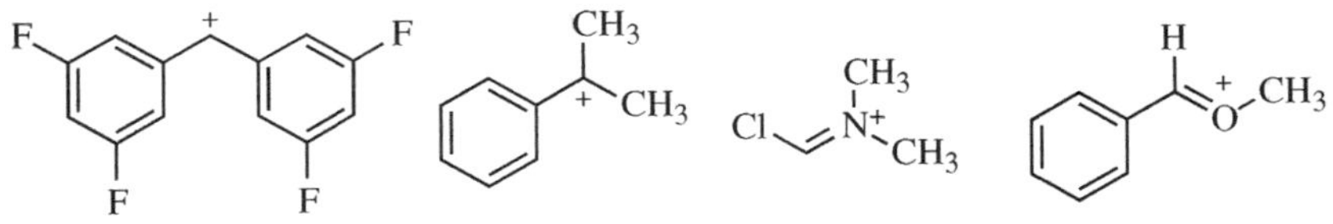

	26	**27**	**28**	**29**
E parameter:	8.02	5.73	–5.77	2.97

SCHEME 1.8 Electrophilicity parameters, *E*, for electrophiles **26–29**.

respective nucleophilicity parameters, *N*, of −4.67 (S_N 1.77) and −1.18 (S_N 1.20). Recently, this approach was used to predict the rate constant for the S_EAr reactions involving iminium ion salts and pyrroles (for the rate-determining C—C bond-forming step) [25].

Among the most reactive electrophiles, an informal benchmark of reactivity has been often cited—the S_EAr reaction with nitrobenzene. As a strongly deactivated arene (*vide infra*), nitrobenzene only reacts with the most active electrophiles. Moreover, since many S_EAr reactions are done in highly acidic media, the nitro group may itself be completely protonated (pK_a −11.3), thereby enhancing the deactivation of the aromatic ring. Examples of electrophiles capable of reacting with nitrobenzene are iodine (I) trifluoromethane sulfonate **30** [26], nitronium salts **31** and **32**, and the carboxonium ions **33** and **34** (Scheme 1.9) [27–29]. Dications **32–34** are examples of superelectrophilic species, a class of reactive intermediates usually formed by (multi-)protonation equilibria or multidentate Lewis acid

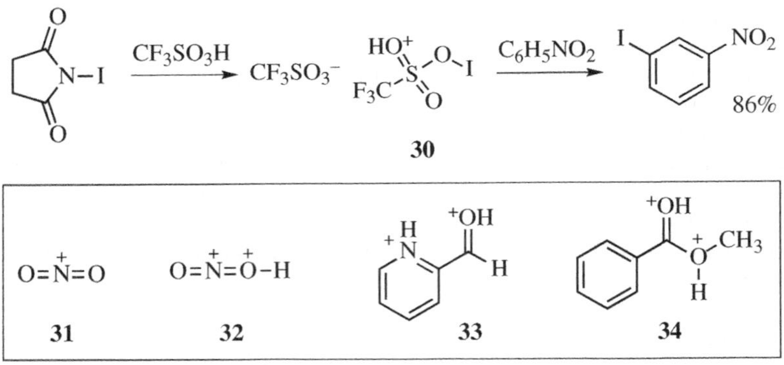

SCHEME 1.9 Examples of electrophiles shown to react with nitrobenzene.

interactions. Superelectrophiles are often capable of reacting with the weakest nucleophiles [30]. For example, the protio-nitronium dication (**32**) has been shown to nitrate even 1,3-dinitrobenzene.

A very active area of research in S_EAr chemistry is in the field of asymmetric synthesis. This chemistry involves a unique set of electrophiles—those in which a chiral environment must exist near the electrophilic reaction site. In most cases, these asymmetric synthetic reactions are accomplished with a chiral electrophile or a chiral catalyst (or counterion) in tight coordination to the electrophile. For example, Stadler and Bach used a chiral electrophile (**36**) in an S_EAr reaction leading to the natural product (−)-podophyllotoxin **38** (Scheme 1.10) [31]. With planar sp^2 carbocation centers, facial selectivity may be controlled by neighboring groups, in this case the adjacent vinyl group on the lactone **36**. The Friedel–Crafts chemistry provides intermediate **37**, which is then converted to (−)-podophyllotoxin (**38**) as a single enantiomer.

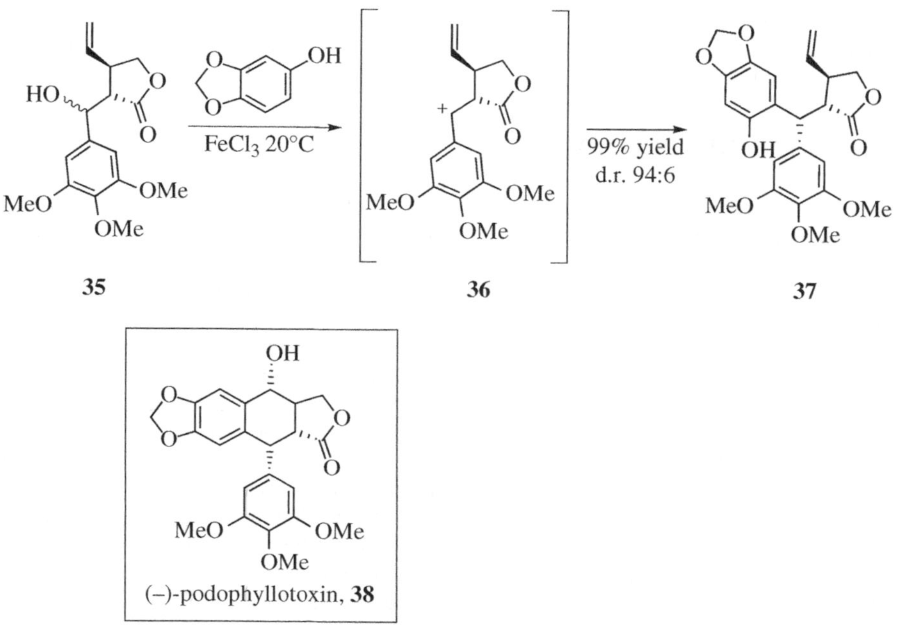

SCHEME 1.10 S_EAr reaction with a chiral electrophile.

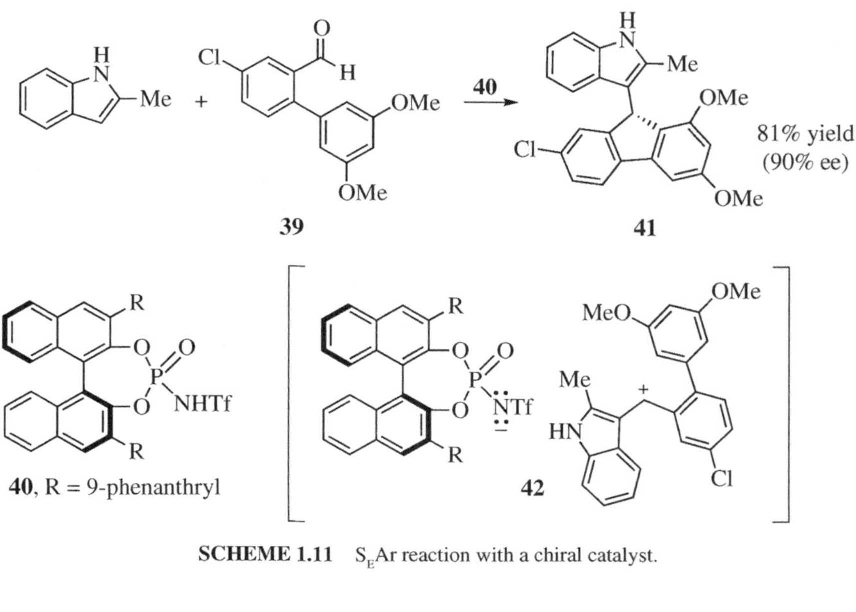

SCHEME 1.11 S_EAr reaction with a chiral catalyst.

Both chiral Brønsted and Lewis acids have been useful in asymmetric Friedel–Crafts reactions. For example, the chiral Brønsted acid **40** was used in the asymmetric synthesis of chiral fluorenes from an achiral indole and the biarylaldehyde **39** (Scheme 1.11) [32]. Initial steps in the conversion lead to the ion pair **42**. Through ion pairing with the electrophilic carbocation, the chiral anion directs ring closure to provide the S_EAr product **41** in reasonable enantioselectivity. Owing to the value of this type of synthetic methodology, asymmetric S_EAr reactions have been the subject of several recent reviews [33].

1.4 ARENE NUCLEOPHILES

As described earlier, the S_EAr involves the reaction of an electrophilic species with an arene nucleophile. There are several types of arenes common to the S_EAr reactions: substituted benzenes, polycyclic aromatic compounds, and heterocyclic compounds. Substituent effects largely control the chemistry of substituted benzenes and related compounds. This includes both activating and directing effects of substituents on the S_EAr reaction.

Functionalized arenes are generally compared to benzene with respect to their relative reactivities. Substituents are described as activating groups if they increase the S_EAr reaction rates compared to benzene, and they are described as deactivating groups if they decrease the relative reaction rate. These substituent effects may be understood in terms of the Hammond postulate. In many cases, the σ-complex is the highest-energy intermediate in the S_EAr reaction, and the transition state leading to it resembles the σ-complex. Any substituent capable of stabilizing the cationic σ-complex should also stabilize the corresponding transition state, lowering the energy barrier and increasing the relative reaction rate (Fig. 1.3). Thus, σ-complex **43** may be stabilized by an electron-donating group (i.e., EDG=$-OCH_3$), while benzene or a derivative with a deactivating group (i.e., EWG=$-CF_3$) leads to the less stable σ-complex **44.** The respective transition states ($\ddagger_1$ and $\ddagger_2$) are raised or lowered accordingly, and this affects the relative reactivities or reaction rates.

As a carbocationic species, the σ-complex is stabilized by electron-donating groups. These interactions may involve resonance and/or inductive effects. With substituents such as the methoxy group, σ-complex stabilization may occur through resonance interaction with the *n*-electrons (Eq. 1.5). Hyperconjugation is also an important stabilizing effect for σ-complexes (Eq. 1.6), an interaction that explains the activating effects of alkyl groups. Similar interactions are important

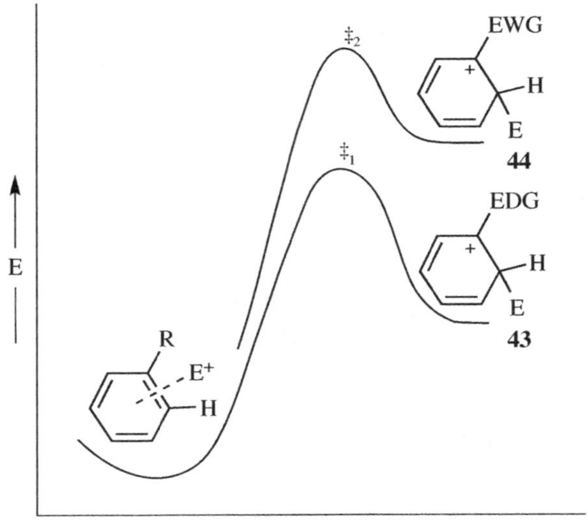

FIGURE 1.3 The effects of substituents on σ-complex stability and energy barriers. EDG, electron-donating group; EWG, electron-withdrawing group.

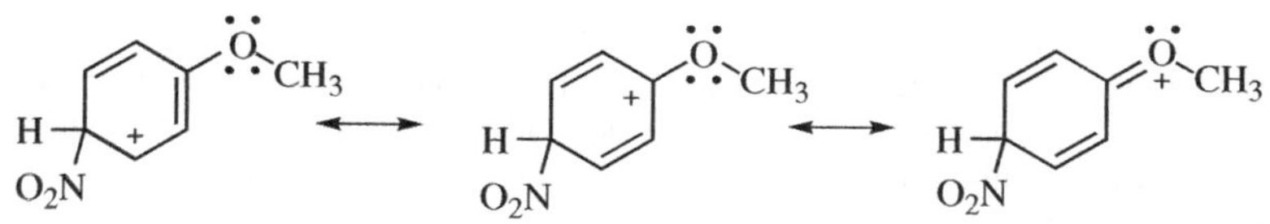

(1.5)

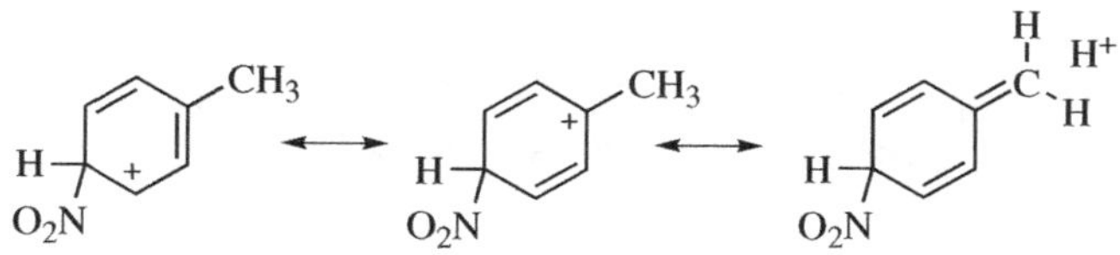

(1.6)

with *para* attack and σ-complex formation. In contrast to the stabilizing effects of electron-donating groups, the σ-complex is destabilized by electron-withdrawing groups (Scheme 1.12). Among these structure types, there are groups that destabilize the σ-complex primarily by inductive effects (**45**) and by unfavorable charge–charge repulsive interactions (**46** and **47**). The electron-withdrawing groups are deactivating in S_EAr reactions.

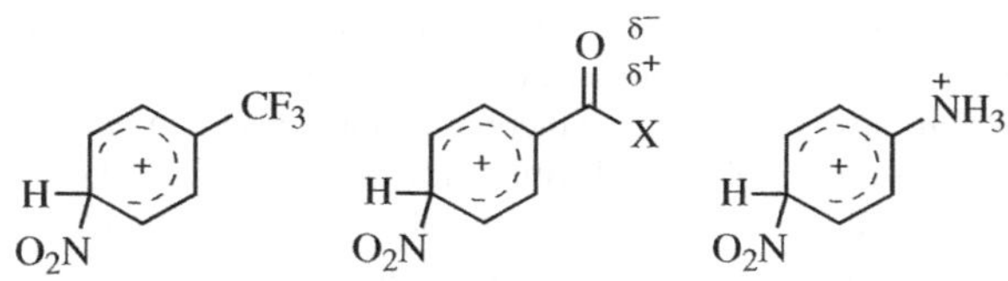

45 **46** **47**

SCHEME 1.12 Destabilized σ-complexes.

The activating or deactivated effects of the substituents may be quantitatively described through the use of partial rate factors [34]. These values estimate the reactivities of positions on an arene relative to the carbons of benzene. For example, partial rate factors may be calculated for a monosubstituted benzene (Ar) at the *ortho*, *meta*, and *para* positions. For a given electrophilic system, the partial rate factor, *f*, is calculated using the relative rates of reactions ($k_{benzene}$ and k_{Ar}) and the fraction or percent of each regioisomer (Eq. 1.7). The first term is a statistical factor that accounts for the

Partial rate factor at the *ortho* position of arene Ar $$f_o^{Ar} = \left(\frac{6}{2}\right)\left(\frac{k_{Ar}}{k_{benzene}}\right)\left(\begin{matrix}\%\ ortho\\ \text{isomer}\end{matrix}\right) \quad (1.7)$$

number of reactive positions on benzene versus those on the substituted arene—6/2 for the *ortho* and *meta* and 6/1 for the *para*. The second term describes the ratio of the overall rate difference between reactions at the arene (k_{Ar}) and benzene ($k_{benzene}$). For sulfonation (H_2SO_4–H_2O, 25°C) of toluene, the $k_{toluene}/k_{benzene}$ ratio is found to be 31.0 and the product sulfonic acids are formed in an isomer ratio 36:5:59 (*ortho*:*meta*:*para*) [21]. Solving for the partial rate factor at the *ortho* position, $f = (6/2)(31.0)(0.36) = 34.0$. For the sulfonation, partial rate factors at the *meta* and *para* positions are, respectively, 4.3 and 112. These values not only reflect the relative reactivity of toluene compared to benzene—large values indicate a substantially more reactive arene—but they also indicate the relative reactivities of the positions on the substituted arene. The higher reactivities of the *ortho* and *para* positions of toluene are clearly apparent by the respective partial rate factors.

As expected from the previous discussion on $k_{toluene}/k_{benzene}$, the partial rate factors are quite sensitive to the nature of the electrophilic system [35]. For toluene, strong electrophiles (i.e., NO_2BF_4) have $k_{toluene}/k_{benzene}$ approaching unity, while weaker electrophiles (i.e., Br_2) show more selectivity. These trends are seen in the partial rate factors for the reactions (Scheme 1.13). The less reactive nitrating system (HNO_3) shows increasing $k_{toluene}/k_{benzene}$ selectivity, compared to the nitronium salt, and the partial rate factors are correspondingly higher.

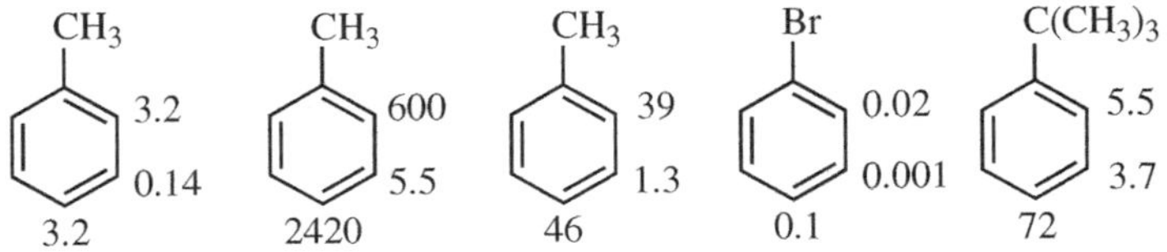

Electrophilic system:	NO_2BF_4 sulfolane	Br_2, AcOH	HNO_3 CH_3NO_2	HNO_3 CH_3NO_2	HNO_3 CH_3NO_2

SCHEME 1.13 Partial rate factors for S_EAr reactions.

Partial rate factors are also strongly influenced by the nature of substituents on the arene, where electronic and steric effects are both found to be significant. Thus, halogens are found to be deactivating substituents, as seen in the low values for the partial rate factors in nitration. With *tert*-butylbenzene, the partial rate factor for the *ortho* position is significantly lower than the *ortho* partial rate factor for toluene. This results from the steric effects from the *tert*-butyl group.

Several research groups have also used theoretical methods in an effort to understand the activating and deactivating effects of the substituents in S_EAr reactions. For example, Galabov and coworkers have developed a computational approach for determining *electrophile affinity*, Eα, as a measure to determine arene reactivity and positional selectivity in S_EAr reactions [36]. Other recent approaches to this problem include the development of reactive hybrid orbital analysis [37], the topological analysis of electron localization function [38], the calculations of electrostatic potentials at the arene carbons [39], and several other methods. A comprehensive summary of this area is beyond the scope of this chapter; however, the interested reader may consult one of the recent reviews of this topic [40].

A critically important consideration also involves the directing effects of the arene substituents. As seen in the calculation of partial rate factors (*vide supra*), arene substituents may tend to favor, or direct, reactions at specific sites on the ring. Among the substituents that activate the arenes toward S_EAr reactions, many of these are *ortho*/*para* directing groups (Table 1.2). Deactivating substituents are often *meta* directing groups.

The halogens are an exception to this trend, as they are mildly deactivating but *ortho*/*para* directing. Alkyl groups tend to be activating and *ortho*/*para* directing; however, electronegative atoms/groups on the alkyl group may change this. For example, the trifluoromethyl group is deactivating and *meta* directing, due to the inductive effects of the fluorine atoms. Nitration of α,α,α-trifluorotoluene provides the *meta* substitution product in good yield (Eq. 1.8) [41].

TABLE 1.2 Relative Activating and Directing Effects of Substituents

	Activating Groups	Deactivating Groups
Ortho/*para* directing	Amino, hydroxy (strongly) alkyl, alkoxy, amido, aryl (weakly)	Halogens
Meta directing	None	Nitro, ammonium, sulfonium (strongly) aldehydes, ketones, acids, (weakly) esters, carboxamides

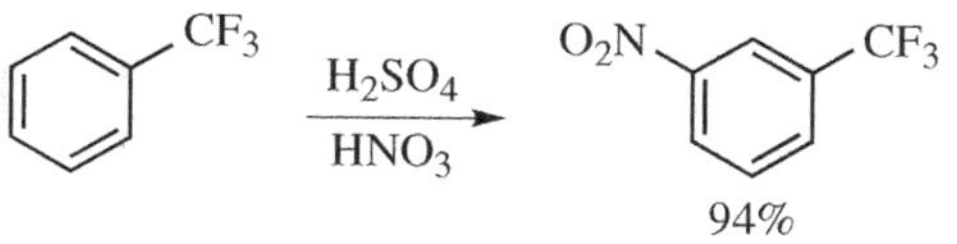

(1.8)

The directing effects may again be understood by a stabilization or destabilization of the σ-complex. Since most S_EAr reactions are kinetically controlled processes, product distributions are generally controlled by the relative stabilities of the transition states leading to the σ-complex. As noted earlier, any substituent that stabilizes or destabilizes the σ-complex will likely have a similar effect on the transition state. The σ-complex, and to some extent the preceding transition state, is an example of the cyclohexadienyl cation-type structure (Scheme 1.14). Using resonance structures, the cyclohexadienyl cation is characterized by positive charge centers at the *ortho* and *para*

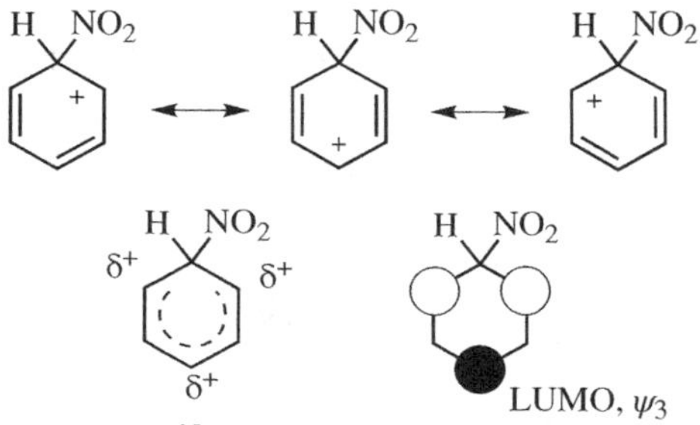

48 **49**

SCHEME 1.14 Charge distribution in a σ-complex and the orbital coefficients for the LUMO, ψ_3.

positions to the attacking electrophile. This means that the most significant positive charge is located at the *ortho* and *para* positions (**48**). In considering the Hückel-type molecular orbitals of this system, the lowest unoccupied molecular orbital (LUMO, ψ_3) has large coefficients at the *ortho* and *para* positions (**49**). Thus, electron-donating substituents stabilize the σ-complex most effectively when they are *ortho* and *para* to the position of electrophilic attack. This also tends to lower the energies of the transition states leading to the *ortho* and *para* σ-complex intermediates and increases the relative rates of these reaction paths.

As an example of these directing effects, Qin and coworkers utilized Friedel–Crafts chemistry to generate functionalized pyrroloindolines [43], a class of structures known for their biological activities (Scheme 1.15). The benzylic carbocation **50** was generated by silver-promoted halide abstraction, and this species was reacted with toluene. The *ortho* and *para* directing effect of the alkyl group is apparent in resonance structure **51b** where the cationic charge center is adjacent to the methyl group. As noted previously, the σ-complex is stabilized by the hyperconjugative resonance with the methyl group. Following rearomatization of the tolyl group, the functionalized pyrroloindoline (**52**) is obtained in good stereoselectivity and yield.

Electron-donating groups at the *meta* position of the σ-complex have a much smaller effect on the stability of the σ-complex, as these groups are unable to directly interact with the cationic charge center (alternatively, the *meta* position is a nodal point in the LUMO, ψ_3, and this prevents the groups from donating electron density into the LUMO). This is seen, for example, in the σ-complex resulting from bromination of anisole (Eq. 1.9). With less stable σ-complex intermediates, *meta* attack tends to be disfavored in these types of systems.

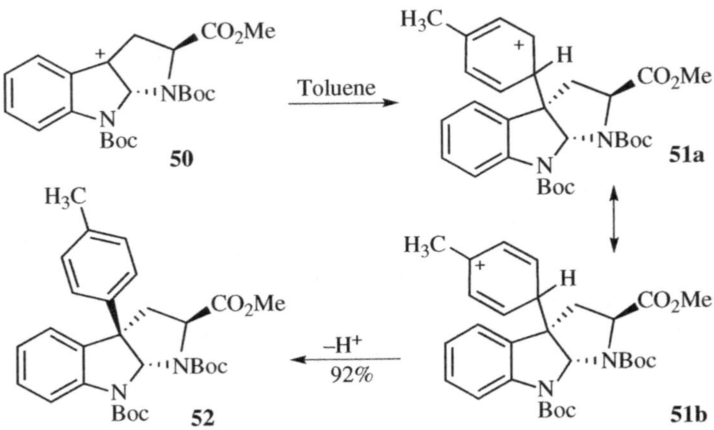

SCHEME 1.15 *Ortho* and *para* directing effect with toluene.

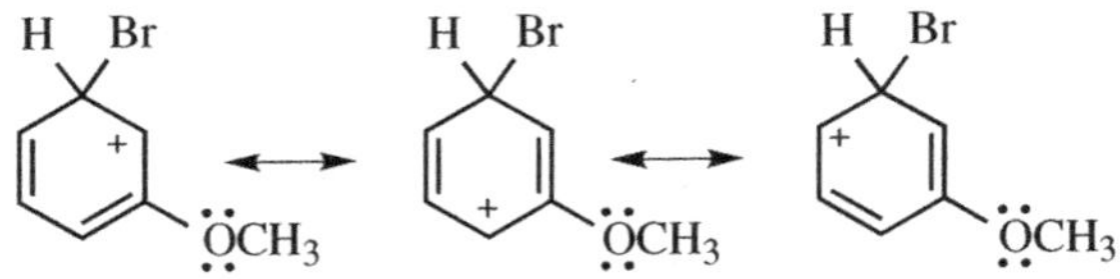

(1.9)

In order to explain the *meta* directing effects of electron-withdrawing groups, it is also useful to consider the position of the cationic charge center in the σ-complex (Scheme 1.16). Nitration of aniline has been accomplished in acidic media to give *meta* nitroaniline as the major product with the *para* isomer as a minor product. Comparing the respective σ-complexes, *para* attack leads to **54** and *meta* attack leads to **56**. Intermediate **54** is destabilized by unfavorable electrostatic interactions, as the ammonium cation is adjacent to a charge center in the σ-complex. Moreover, the ammonium cation—an electron-withdrawing group—is bonded to a ring carbon having a

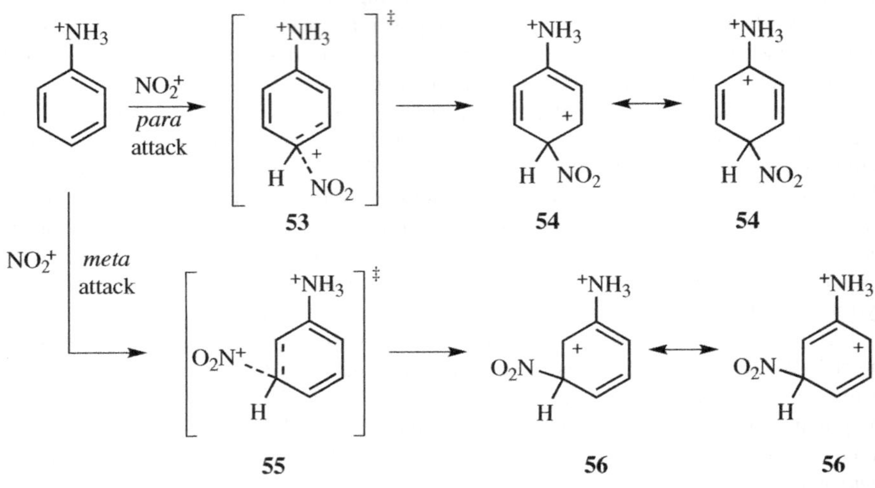

SCHEME 1.16 Transition states (**53** and **55**) and σ-complexes (**54** and **56**) for nitration of the anilinium ion.

large LUMO coefficient. In the case of **56**, the cyclohexadienyl cation system does not generate a positive charge center adjacent to the ammonium cation, and this gives the σ-complex **56** a small measure of stability. The respective transition states (**53** and **55**) are also affected by the destabilizing/stabilizing interactions. This leads to more rapid formation of the *meta* regioisomeric product.

An approach has also examined directing effects from the perspective of the arene starting material. For example, theoretical calculations have been used to estimate the electron densities at the ring carbons of substituted benzenes [46]. Electron-donating groups such as the ethyl, hydroxyl, and amine groups all exhibit greatest π-electron populations at the *ortho* and *para* positions (Scheme 1.17). Conversely, electron-withdrawing groups exhibit the greatest π-electron populations at the *meta* position. These observations are consistent with the general

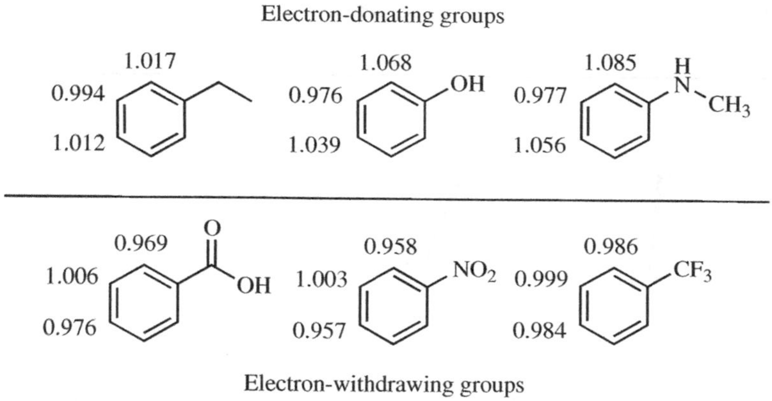

SCHEME 1.17 Calculated (HF/STO-3G level) π-electron populations at ring positions of substituted benzenes.

directing effects of substituents, but the electronic structure of the arene starting material may also be more important in cases where an early transition occurs (i.e., with highly reactive electrophiles).

1.5 π-COMPLEX INTERMEDIATES

In the initial stages of an S_EAr reaction, a π-complex often forms between the electrophile and the arene. Similar donor–acceptor complexes have been long known from solution-phase studies. These complexes were observed to be nonconducting, colored solutions formed from mixing an aromatic compound with a π-acceptor, such as HCl, Ag salts, or I_2.

The involvement of π-complexes in S_EAr reactions was first proposed by Dewar to explain relative reaction rates for some conversions [48]. For example, the relative stabilities of arene π-complexes (with HCl) have been shown to correlate with the relative rates of nitration (Table 1.3) [49]. The π-complex for *m*-xylene is estimated to be only about twice as stable as that for benzene. The relative rates of nitration for these two arenes are similar, suggesting a role of the π-complex in the rate-determining step of the nitration. In contrast, chlorination exhibits a markedly greater rate of reaction with *m*-xylene compared to benzene. This suggests that the rate-determining step for chlorination involves a transition state resembling the σ-complex. Thus, the importance of π-complexes varies among different S_EAr reactions.

TABLE 1.3 Relative Stabilities of HCl–Arene π- and σ-Complexes and Relative Rates of Reactions

	Relative stabilities			Relative reaction rates	
	π-complex	σ-complex		Nitration	Chlorination
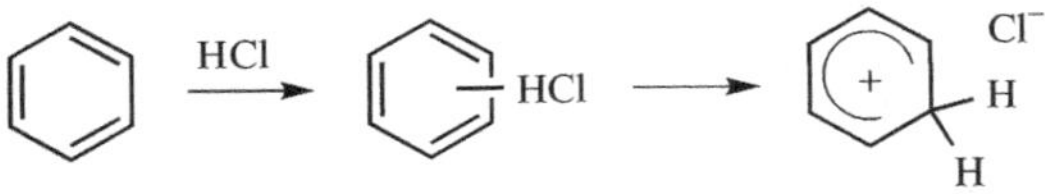			Benzene	0.51	0.0005
	0.61	0.09			
	1.26	26			
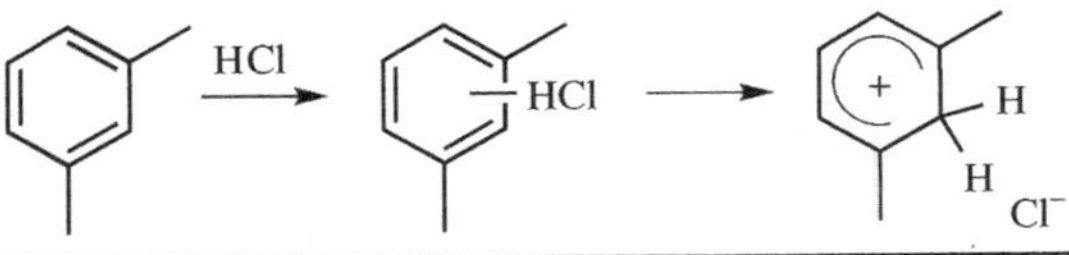			*m*-Xylene	0.84	200

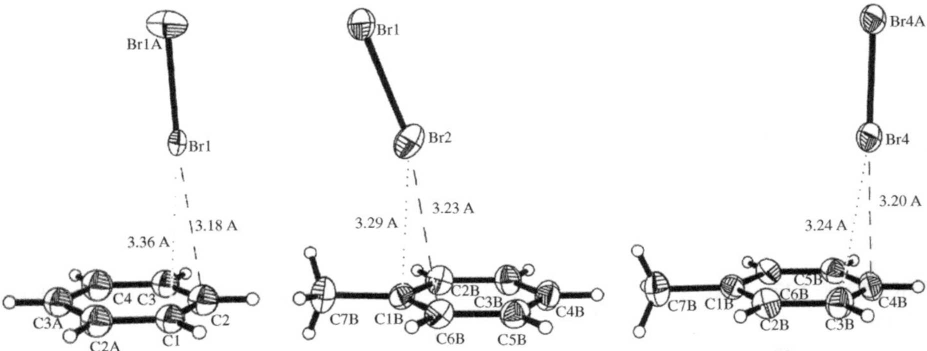

57 **58** **59**

FIGURE 1.4 X-ray crystal structures of the [Br_2, C_6H_6] π-complex (**57**) and the [Br_2, $C_6H_5CH_3$] π-complexes (**58** and **59**). From Vasilyev *et al.* [42]—Reproduced with permission from the Royal Society of Chemistry.

The Kochi group has conducted several comprehensive studies of π-complexes in S_EAr reactions [50]. Most notably, this group has obtained accurate structural parameters of several π-complexes from low-temperature X-ray diffraction studies of stable π-complex crystals (Fig. 1.4). When Br_2 and C_6H_6 are sealed in a capillary and cooled to −150°C, the π-complex (**57**) is obtained [42]. This structure confirms the prediction from previous theoretical calculations showing that η^2 hapticity is favored over η^1 or η^6 with the [Br_2, C_6H_6] π-complex [51]. The ring-Br_2 bond distance is found to be less than the van der Waals radii, consistent with a weak donor–acceptor interaction within the π-complex. Analysis also shows there is little Br—Br bond elongation observed and the C—H bonds remain within the ring plane. Upon warming the π-complex crystals to −78°C, HBr is released and bromobenzene is formed quantitatively [42].

A similar π-complex was obtained with Br_2 and toluene at −150°C [42]. Two structures were observed from the crystalline products: a π-complex with Br_2 located near the *ortho* carbon (**58**) and a π-complex with Br_2 located near the *para* carbon (**59**). Both structures show the Br_2 perpendicular to the plane of the ring with η^2 hapticity. Interestingly, electrophiles tend to react at the *ortho* and *para* positions of toluene, and these π-complexes are consistent with the observed regiochemistry. This is also consistent with calculated electron densities and charges at the ring carbons [23, 51], where the *ortho* and *para* carbons are shown to have the highest degree of electron density. As an activated arene, toluene is considered a better electron donor compared to benzene. This is seen in the respective π-complexes, as Br_2 is located slightly closer to the aromatic ring with toluene compared to benzene.

Another study by Kochi *et al.* examined the chemistry of nitrosation [52]. Several [NO, arene]$^+$ π-complexes were characterized by UV–visible, NMR, and IR spectroscopy. X-ray quality crystals were also obtained at −78°C with mesitylene, hexamethylbenzene, and other arenes. The highly colored nitroso π-complexes were prepared directly from nitrosonium salts (i.e., NO^+ $SbCl_6^-$) and the aromatic compound. Structural studies revealed the distance between the ring and the electrophile is roughly 1 Å less than the sum of the van der Waals radii and the N—O bond length is significantly lengthened as a result of strong donor–acceptor interaction.

In the [NO, hexamethylbenzene]$^+$ complex, the N—O bond length is measured as 1.108 Å, increasing from 1.06 Å found in the uncomplexed nitrosonium ion (NO^+) [53]. The effects of inner-sphere electron donation are also observed in the infrared N—O stretching frequencies. With the uncomplexed nitrosonium ion, N—O stretch is observed at 2272 cm^{-1}, while the [NO, hexamethylbenzene]$^+$ π-complex exhibits a stretch at 1885 cm^{-1} [52]. For comparison, nitric oxide (·NO) has an estimated bond length of 1.15 Å and an N—O stretch frequency of 1876 cm^{-1}. These data indicate a significant degree of electron transfer in the π-complex; however, the complexes are ESR silent, suggesting the nitric oxide and hexamethylbenzene radical cation are not fully formed species.

The [NO, hexamethylbenzene]$^+$ π-complex has also been studied by UV–Vis and NMR spectroscopy [54]. In ^{13}C NMR spectroscopy, hexamethylbenzene exhibits two resonances at δ, 17.0 and 133.2, while the [NO, hexamethylbenzene]$^+$ π-complex shows signals at δ, 17.8 and 150.8. The significant downfield shift of the ring carbons is consistent with charge transfer in the donor–acceptor complex. The same π-complex exhibits UV–Vis absorption bands at 337 nm (strong) and 500 nm (weak). The low-energy absorption band has been shown to be closely related to the E°_{ox} (electron donor strength) of the arene.

Several π-complexes have also been characterized using theoretical methods [55]. For example, the potential energy surfaces of the nitrosonium ion/benzene and nitronium ion/benzene reactions have been studied using ab initio molecular orbital calculations (Scheme 1.18) [53]. The first minimum for nitrosation is the π-complex (**60**), which is found to be 36 kcal/mol below the starting materials. Calculations indicate there is no barrier to formation of the π-complex. The subsequently formed σ-complex is characterized as a transition state structure (**61**), a stationary point that is located at a saddle point (+23 kcal/mol). Rearomatization of the ring leads to the final *N*-protonated product in a deep potential energy minimum (−42 kcal/mol).

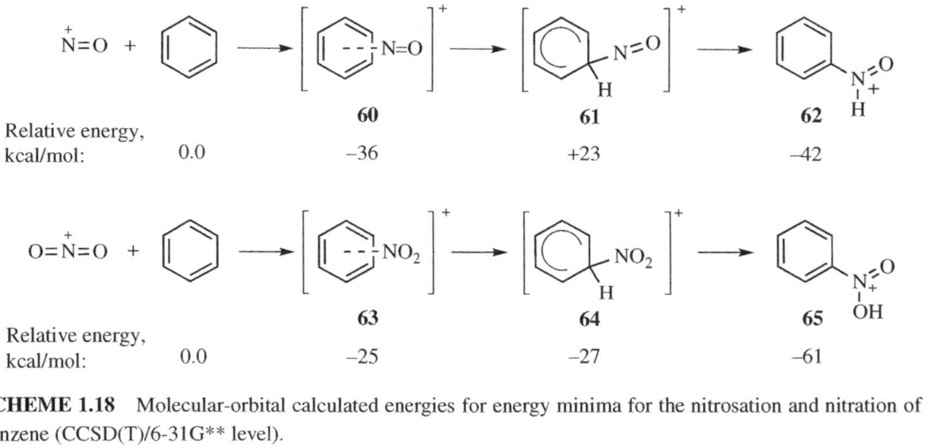

SCHEME 1.18 Molecular-orbital calculated energies for energy minima for the nitrosation and nitration of benzene (CCSD(T)/6-31G** level).

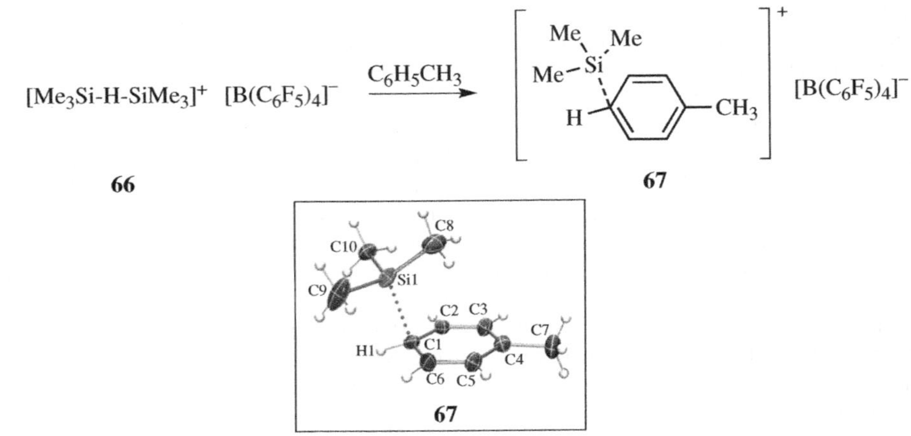

FIGURE 1.5 Preparation of silylium ion complex **67** and its crystal structure. Adapted with permission from Ibad *et al.* [44]. © (2011) American Chemical Society.

In the nitration, both π-complex (**63**) and σ-complex (**64**) are located at energy minima. The π-complex is located 25 kcal/mol below the starting materials, and it is characterized by a significant degree of bending of the nitronium ion and donor–acceptor bonding above the rim of the benzene ring. Despite the obvious transfer of π-electron density toward NO_2^+, there is little distortion of the C—H bond angle. Formation of the σ-complex (**64**, −27 kcal/mol) is followed by rearomatization and product formation. These results confirmed the earlier suggestion by Olah and coworkers [23] that the σ-complex is not necessarily associated with the highest-energy species on the reaction coordinate. As such, the π-complex is expected (in some cases) to strongly influence positional selectivity in S_EAr reactions. Interestingly, the molecular orbital calculated energy profiles of nitrosation and nitration have been shown to coincide with a theoretical treatment using Marcus–Hush theory [53].

Although not commonly involved in S_EAr reactions, there has been extensive work related to π-complexes and σ-complexes involving silylium ions (R_3Si^+) and arene donors. Several crystal structures have been obtained for these donor–acceptor complexes, beginning with Lambert's $[Et_3Si]^+$ $[B(C_6F_5)_4]^-\cdot C_6H_5CH_3$ isolated in 1993 [56]. A recent study examined several complexes of trimethylsilylium cations with arenes [44], some of which were isolated as crystalline solids (Fig. 1.5). For example, a relatively stable silylium ion donor–acceptor complex (**67**) was prepared from the hydride-bridged silane adduct cation (**66**). Complex **67** is found to be stable to 80°C, but at higher temperatures, the crystalline material decomposes. A major product of the decomposition is Me_3SiF, suggesting fluoride abstraction from the borate anion is favored over the S_EAr reaction. Similar crystalline solids were isolated from benzene, ethylbenzene, *n*-propyl and *i*-propylbenzene, xylenes, and trimethylbenzenes. Each was characterized by Raman and IR spectroscopy as well as X-ray crystallography.

Analysis of the structural parameters reveals some interesting trends. As can be seen in structure **67**, the geometry around silicon atom deviates considerably from the planar structure expected from the uncomplexed silylium ion. The sum of the C—Si—C bond angles in **67** is found to be about 341°, compared to 360° for the silylium cation (Me_3Si^+) and 328.4° for a tetrahedral structure. The coordination is also clearly a η^1-type interaction. This raises an obvious question: is it a π-complex or a σ-complex? There is no definitive point at which a structure becomes a σ-complex. As noted by Sidorkin, the ideal π-complex has a 90° angle between the ring and the electrophilic center [57], while the ideal σ-complex has a bonding angle of about 125° (Scheme 1.19). A continuum of structures is also suggested, which includes a mixed π/σ-type complex. In complex **67**, the α-bond angle is 102.4°. The measured C_1—Si bond length in **67** is 2.135 Å. This is longer than the sum of the C–Si covalent radii (1.91 Å) but much shorter than the sum of the van der Waals radii (3.8 Å). These data suggest a structure of mixed π/σ type for the $[Me_3Si]^+$ $[B(C_6F_5)_4]^-\cdot C_6H_5CH_3$ complex (**67**).

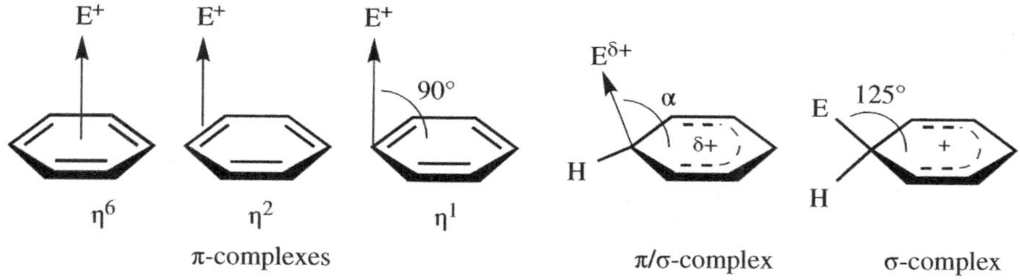

SCHEME 1.19 Structure types for intermediates in the S_EAr reaction.

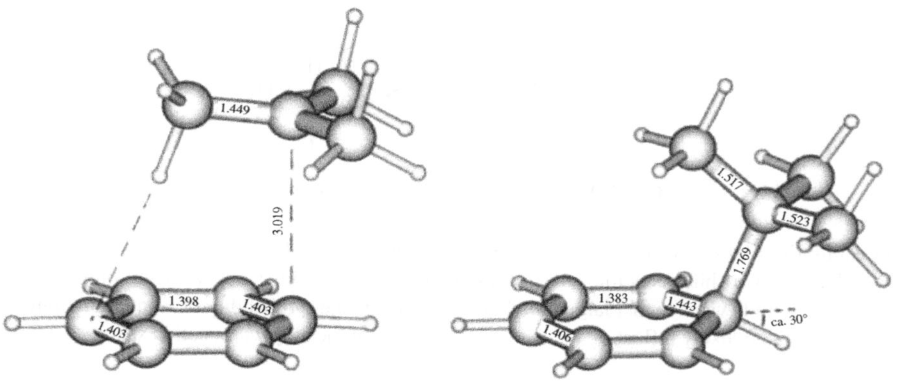

FIGURE 1.6 MP2/6-31+G**(fc) calculated π-complex and σ-complex for the alkylation of benzene. From Heidrich [45]—Reproduced with permission from WILEY-VCH Verlag GmbH & Co. KGaA, Weinheim.

Notably, the progression from a π-complex to a π/σ-complex to a σ-complex involves an increasing amount of charge transfer from the arene to the electrophile. A fully formed σ-complex thus has a cyclohexadienyl cation structure. Based on the data from crystallographic studies, the extent of σ-complex formation depends on the nature of both electrophile and arene. The silylium cation studies showed that increasing the nucleophilic character of the arene leads to increasing amounts of σ-complex character. This was evident from measurement of the α-bond angles and C_1—Si bond lengths.

Similar trends were observed in recent computational study that examined the potential energy surface for the Friedel–Crafts alkylation of benzene [45]. The MP2/6-31+G**(fc) calculations studied the reactions of benzene with the methyl cation, the isopropyl cation, and the *tert*-butyl cation. A stable π-complex was only found in the case of the *tert*-butyl cation reacting with benzene (Fig. 1.6). Both the η^1 and η^2 hapticity π-complex structures were found to be comparable in energy and about 10 kcal/mol more stable than the gas-phase starting materials. The planar structure of the *tert*-butyl cation suggests only a minimal amount of electron density has been transferred from benzene. The stable σ-complex was also located, and it was estimated to be approximately 4 kcal/mol more stable than the π-complex. Interestingly, the π-complexes could not be located for either the reaction of isopropyl cation or the methyl cation, but the reaction proceeds directly to the σ-complex. This reflects an increasing electron demand at the 2° isopropyl cation and methyl cation centers. It is also a further indication that the π-complex intermediate may not be involved, or have a much diminished lifetime, in some S_EAr reactions.

Gas-phase ion chemistry has also been used as an interesting approach in the study of π- and σ-complexes. For example, an isotopically labeled sample of *sec*-butylbenzene was subjected to chemical ionization with the CD_5^+ ion and studied by collisionally activated dissociation and metastable dissociations using tandem mass spectroscopy (MS/MS) [58]. The chemical ionization with methane is thought to be a clean method of preparing gas-phase arenium ions (Scheme 1.20). Examination of the fragments suggests that isomerization of alkyl group takes place in the gas phase. This occurs through

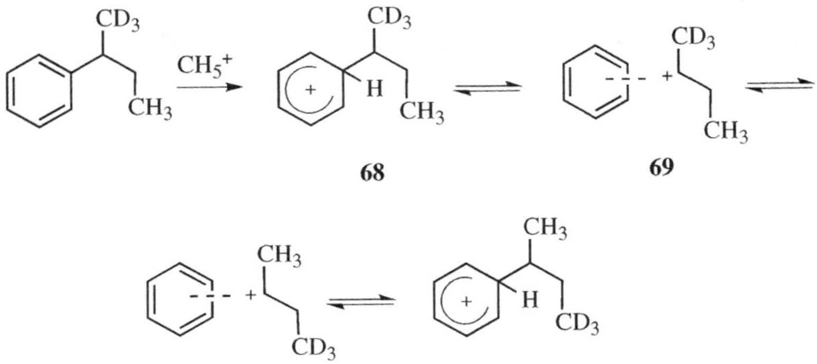

68 69 70 71

SCHEME 1.20 Gas-phase isomerization via π-complex.

an equilibrium between the initially formed σ-complex (**68**) and a subsequently formed π-complex (**69**). 1,2-Hydride shift gives the isomeric π-complex (**70**), and this leads to the σ-complex (**71**). The study also found evidence that larger alkyl groups had a greater probability of isomerizing through the π-complex route, likely reflecting the increasing stability of the alkyl carbocation species.

1.6 σ-COMPLEX OR WHELAND INTERMEDIATES

As described earlier, the S_EAr reaction mechanism generally involves the formation of a σ-complex intermediate. This species is formally a cyclohexadienyl cation, and it has also been called the Wheland intermediate, Pfeiffer–Wizinger complex, arenium ion, benzonium ion, and benzenium ion. Since it was first proposed as an intermediate in Friedel–Crafts reactions, there has been a considerable amount of evidence for its involvement in the reactions. As described previously, the involvement of the σ-complex provides a good basis for the understanding of some activating and directing effects in S_EAr reactions.

Early studies of acid–base chemistry suggested a role of cyclohexadienyl cations in S_EAr reactions. While arenes and HCl provide nonconducting solution, use of the Brønsted–Lewis acid conjugate HCl–$AlCl_3$ leads to colored solutions that conduct electric current [59]. This was interpreted as formation of the cyclohexadienyl cation and $AlCl_4^-$ ion pair, as the HCl–$AlCl_3$ is an exceptionally strong proton donor. Thus, the proton serves as the electrophile to generate the cyclohexadienyl cation. Brown and associates conducted a number of studies demonstrating a linear relationship between σ-complex stability (determined by protonation equilibria) and the rates of a variety of S_EAr reactions [21, 60]. These results are considered strong evidence for the σ-complex as the key intermediate in these reactions.

The final step in the S_EAr reaction mechanism involves deprotonation of the σ-complex intermediate to regenerate the aromatic π-system, and this is expected to be a very fast step. Since the C—H bond is not being broken in a rate-determining step, there is usually little or no detectable kinetic isotope effect (KIE) for S_EAr reactions [61]. Thus, studies of KIEs are also consistent with the involvement of the σ-complex. Larger KIEs have been observed in conversions involving weak electrophiles, such as nitrosations and diazonium coupling reactions [61c].

Several examples have been reported of S_EAr reactions providing different products under thermodynamic and kinetic control. For example, sulfonation of naphthalene at 80°C gives predominantly the α-isomer (the kinetic product), while reaction at an elevated temperature provides the β-isomer (the more stable thermodynamic product) as the major product (Scheme 1.21) [62, 63]. These results may be understood by considering the stabilities of the respective σ-complex intermediates. The σ-complex-α is a more stable intermediate than σ-complex-β, as it benefits from favorable resonance stabilization. This

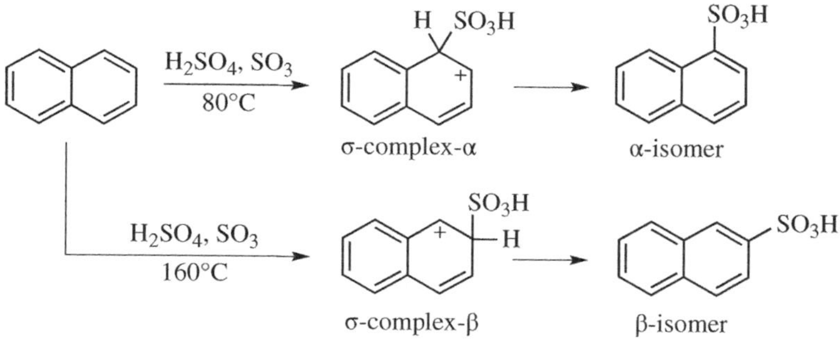

SCHEME 1.21 Sulfonation of naphthalene.

leads to formation of the α-isomer under kinetically controlled conditions. However, the α-isomer itself is less stable than the β-isomer, due to steric effects involving the sulfonic acid group and the peri position hydrogen. Under thermodynamic control, the reversible S_EAr reaction gives the more stable β-isomer.

Numerous studies have sought to directly observe σ-complexes using spectroscopic methods. This has not been an easy task, as rapid deprotonation of the σ-complex often limits the lifetime of this reactive intermediate. Under normal S_EAr reaction conditions, even rapid spectroscopic methods—such as time-resolved UV–visible spectroscopy—generally do not detect the σ-complexes. A common method of stabilizing and observing the σ-complex species has been through the use of hexamethylbenzene (and other hexa-substituted benzenes). This arene produces a σ-complex that cannot undergo deprotonation to give a stable aromatic substitution product. Doering and Saunders first used this method in 1958 to characterize the methylation of σ-complex intermediate using ^{1}H NMR (Eq. 1.10) [64]. As expected, the resulting σ-complex (**72**) gives four ^{1}H signals in a 3:6:6:6 ratio.

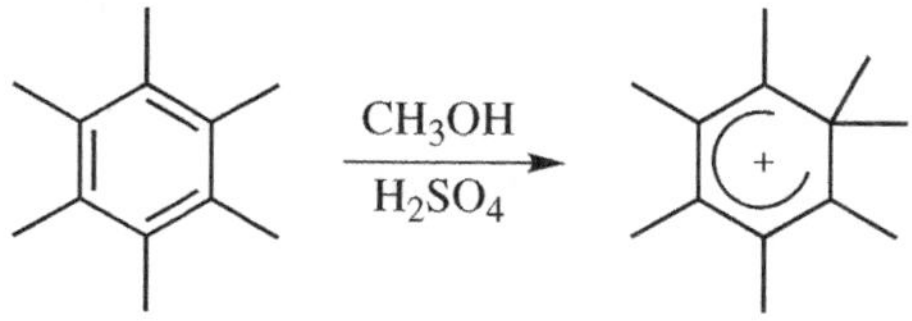

72 (1.10)

Subsequent studies have provided detailed structural information related to these σ-complexes [50, 65], including species from halogenation, nitration, sulfonation, acylation, alkylation, and other methods (Table 1.4). Importantly, the NMR characterizations of these structures showed clear differences from the related π-complexes. For example, nitrosonium cation (NO^+) salts provide the π-complex with hexamethylbenzene giving rise to a ^{13}C spectrum with two signals (*vide supra*) [50]. Nitronium cation (NO_2^+) salts provide the stable σ-complex (**74**), giving a ^{13}C spectrum with eight signals (four ring carbons and four methyl carbons). The ring carbons are observed at δ, 98.4 (C_1), 141.8 ($C_{3,5}$), 181.5 ($C_{2,6}$), and 204.7 (C_4) [50]. The downfield signals at the $C_{2,6}$ and C_4 are consistent with the cyclohexadienyl cation structure.

Several σ-complex structures have also been studied using X-ray crystallography. The crystalline adducts **73**, **75**, **76**, **81**, and **83** from hexamethylbenzene have all been characterized by X-ray diffraction [65f–j]. In complex **75**, the C—Cl bond distance is measured to be 1.81 Å, a value typical

TABLE 1.4 Observed σ-Complexes from Hexamethylbenzene

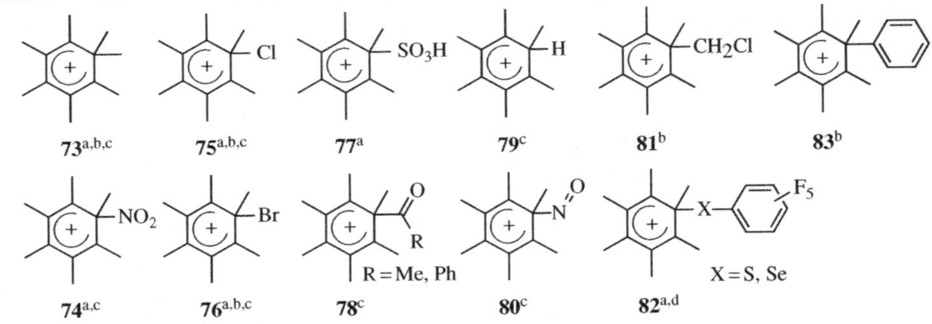

(a) ^1H and ^{13}C NMR (b) X-ray crystal structure (c) UV–vis (d) ^{19}F NMR

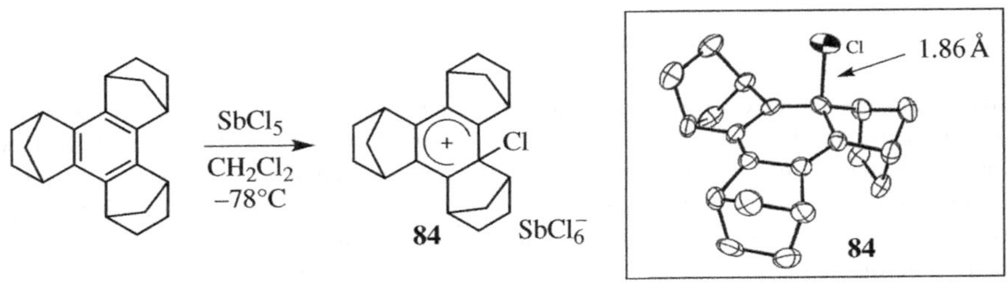

FIGURE 1.7 Preparation of σ-complex **84**. Adapted with permission from Rathore *et al.* [47]. © (1994) American Chemical Society.

for a C—Cl single bond [65j]. The C_2—C_1—Cl bond is about 106°, which is near the expected value for an sp^3 carbon center. Kochi and coworkers have used the sterically crowded aromatic donor to prepare the relatively stable σ-complex **84** from electrophilic chlorination (Fig. 1.7) [47]. Analysis of the crystal structure reveals the C—Cl bond distance to be 1.86 Å. Reed *et al.* have also used low-coordinating anion chemistry to prepare several σ-complexes as protonated arenes [66], for example, obtaining X-ray crystal structures of protonated *m*-xylene, mesitylene, pentamethylbenzene, and hexamethylbenzene. Both steric and electronic stabilizing effects may be used to stabilize σ-complexes, as the crystalline salt **85** was isolated from 1,3,5-pyrrolidinobenzene (Scheme 1.22) [67]. The electron-donating properties of the pyrrolidine nitrogens are clearly apparent by shortening of the C—N bonds to the cyclohexadienyl ring upon formation of the σ-complex.

Another valuable method of studying the σ-complexes involves the use of superacidic and stable ion conditions pioneered by Olah and colleagues [68]. These solutions are nonnucleophilic and

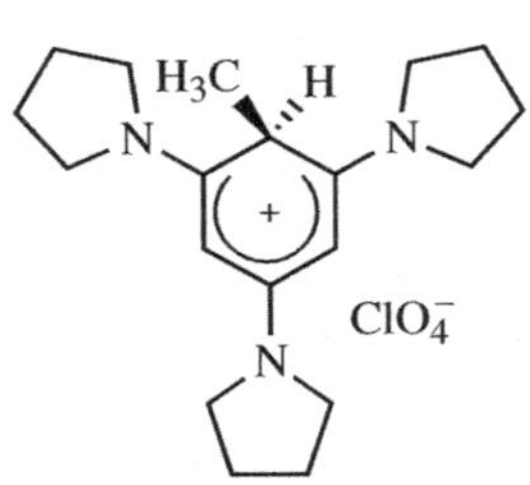

85

SCHEME 1.22 Stable σ-complex **85**.

nonbasic, enabling cationic species to be long-lived (at low temperature) and amenable for study. For example, this chemistry has been used to generate the parent σ-complex, the benzenium ion (**86**) from protonation of benzene in solutions of HF—SbF_5, FSO_3H—SbF_5, and carborane superacid (Scheme 1.23) [68]. Many other aromatic hydrocarbons have been studied using this technique, leading to stable σ-complexes such as **87** and **88** [68, 69]. Although the cationic species such as **86–88** possess very high acidity, the superacidic solution prevents deprotonation equilibria.

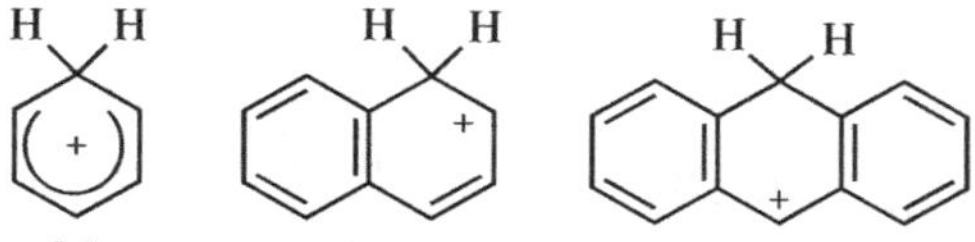

86 **87** **88**

SCHEME 1.23 Protonated aromatic hydrocarbons.

Conventional Friedel–Crafts σ-complexes have likewise been generated under stable ion conditions. Olah and Kuhn generated alkylation and formylation of σ-complexes [70], both isolated as solids with well-defined ionic character (conductance studies), and heating of the solids produced the S_EAr products (Scheme 1.24). A subsequent NMR study involving the low-temperature ethylation of 1,3,5-triethylbenzene also revealed the presence of σ-complex intermediates [65a]; however, isomeric species were formed rapidly by hydride/alkyl shifts. Indeed, this aspect of cyclohexadienyl cation chemistry has made the study of σ-complexes difficult, as the barrier for isomerization is often quite low.

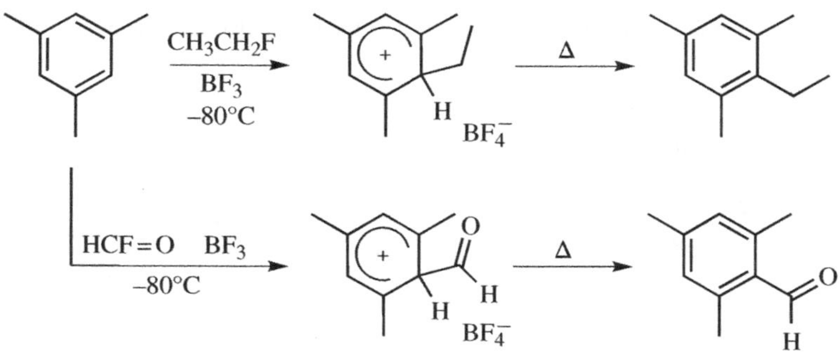

SCHEME 1.24 Intermediate σ-complexes and the conversion to Friedel–Crafts products.

Likewise, the isomerization of the σ-complex **83** was observed in solution phase [71]. While the crystalline complex **83** is stable to 70°C, dissolving the salt in solvent leads to rapid isomerization at 25°C (Eq. 1.11). Moreover, the rates of isomerization have been shown to vary with differing counterions, such as $AlCl_4^-$ and BPh_4^-. Regarding σ-complex isomerization, it has also been shown that an

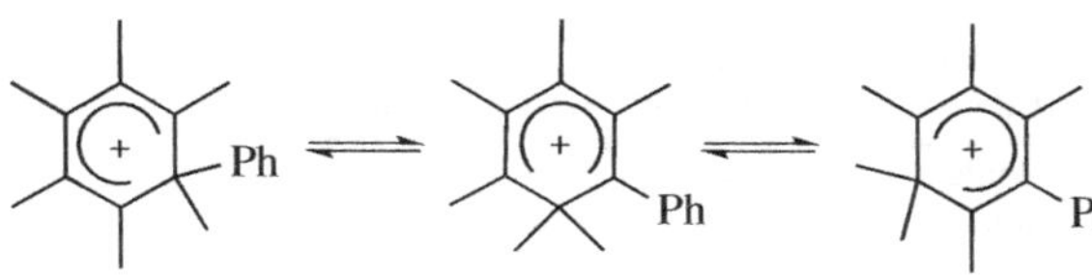

83

(1.11)

Al_2O_3 surface can accelerate alkyl group migrations in some σ-complexes [72]. The 1,2-alkyl shifts are also known to retain their configuration at the migrating carbon [73], in accordance with Woodward–Hoffmann rules for the sigmatropic rearrangement.

The isomerization of σ-complexes does have significant commercial importance. For example, this chemistry is used to prepare isomeric xylenes—important feedstock chemicals for aromatic dicarboxylic acids [74]. The proportion of xylene isomers depends on the conditions used, as strongly acidic conditions (i.e., HF—BF_3) are employed to favor the *meta* isomer and shape-selective zeolites are used to favor the *para* isomer (the isomer with the smallest cross-sectional area). The thermodynamically most stable σ-complex for the protonated xylenes is the arenium ion **89**, which leads to the *meta* xylene product (Scheme 1.25).

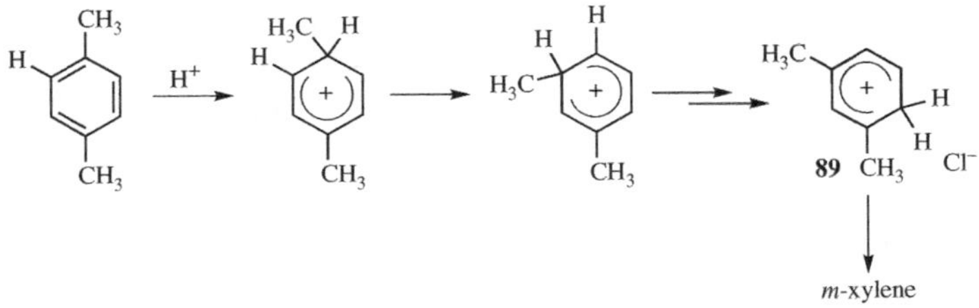

SCHEME 1.25 Isomerization of *p*-xylene to *m*-xylene.

The chemistry of the σ-complex is also important in transalkylations and related reactions. For example, monoalkylbenzenes undergo disproportionation reactions by transalkylation, which in the case of cumene provides diisopropylbenzene and benzene (Eq. 1.12) [75]. In this case, cumene serves as a nucleophile, while the isopropyl cation is the likely electrophile. Shape-selective zeolite catalysts

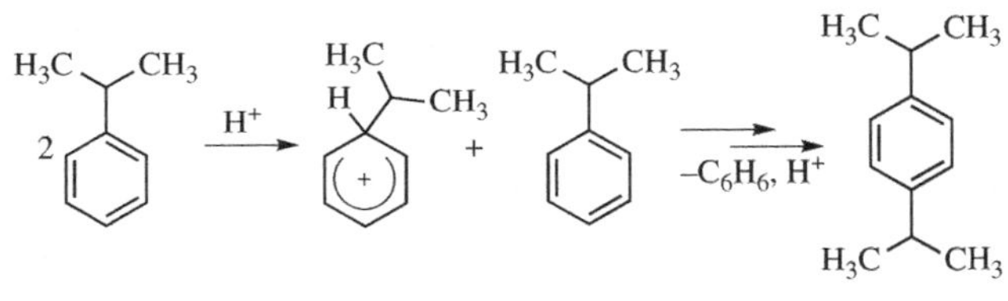

(1.12)

may be used to preferentially give the *para* isomer [75b]. The tendency for groups to undergo transalkylation follows the well-known trends of carbocation stability (isopropyl > ethyl > methyl) [76]. Transacylation was demonstrated with acetylmesitylene [77], where the acetyl cation is transferred from the incipient σ-complex (**90**) to the arene nucleophile (Eq. 1.13). Other examples of deformylation, desilation, dealkylation, and similar reactions of the σ-complex intermediates are known [78].

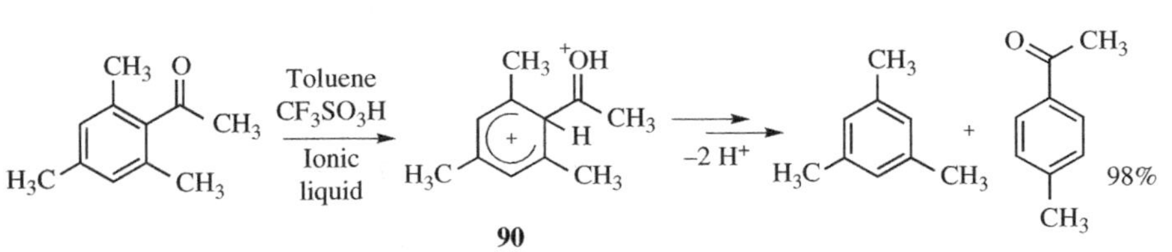

(1.13)

Several examples of dearylation reactions have also been reported [79]. These are thought to proceed through *ipso* protonation of aryl groups [80], with formation of a σ-complex (**91**) and cleavage of the aryl group (Scheme 1.26). This chemistry has been utilized as a general route to aza-polycyclic aromatic compounds [81]. A similar reaction pathway has been proposed for the acid-promoted depolymerization of coals [82].

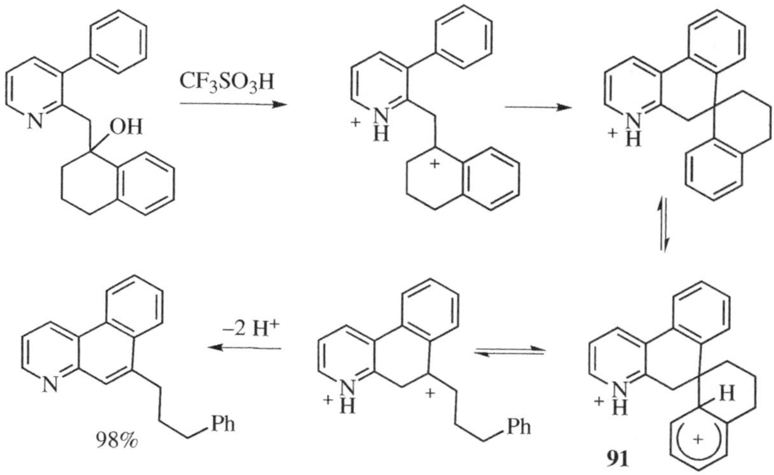

SCHEME 1.26 Dearylation through *ipso* protonation.

1.7 SUMMARY AND OUTLOOK

Mechanistic considerations of the S_EAr reaction began shortly after Friedel and Crafts reported their interesting conversions. These studies have continued for nearly 140 years, as chemists investigated the mechanisms of these valuable reactions. The studies have involved—and in some cases contributed to the development of—major areas of organic chemistry including the theory of aromaticity, the role of reactive ionic intermediates, the concept of resonance interactions, linear free energy relationships, the transition state theory, and reaction kinetics. The S_EAr reaction mechanism has been studied by most spectroscopic methods, and recently, intermediates have been examined by X-ray crystallography. Theoretical approaches have also provided many useful insights. With these considerations, the S_EAr may be the most thoroughly studied reaction mechanism in organic chemistry.

Despite the extensive mechanistic studies, there continues to be the need for more work. Many questions remain to be answered. For example, how do asymmetric environments affect the S_EAr reaction? What factors provide the highest regioselectivity in S_EAr reaction? When does the π-complex form and how does it affect the S_EAr reaction? How does σ-complex stability affect the outcome of these synthetic conversions? Can electrophiles be generated that show unusual reactivities? How do environmentally friendly catalysts affect the S_EAr reaction mechanism? These and other questions will certainly be addressed by research chemists in the decades ahead.

ABBREVIATIONS

Å Ångströms
B Base
E Electrophilicity parameter
E+ Electrophile

EDG	Electron-donating group
ESR	Electron spin resonance
EWG	Electron-withdrawing group
IR	Infrared
k	Rate constant
KIE	Kinetic isotope effect
k_T/k_B	Relative rates of reaction with toluene (k_T) and benzene (k_B)
LUMO	Lowest unoccupied molecular orbital
MS	Mass spectroscopy
N	Nucleophilicity parameter
NMR	Nuclear magnetic resonance
OMe	Methoxy
Ph	Phenyl
S_EAr	Electrophilic aromatic substitution
UV	Ultraviolet

REFERENCES

[1] (a) Olah, G. A., Praksh, G. K. S. (1991) *Comprehensive Organic Synthesis*, Vol. **3** (ed.: Trost, B. M.), Pergamon Press, Oxford, UK, pp. 293–335. (b) Sartori, G., Maggi, R. (2009) *Advances in Friedel-Crafts Acylation Reactions Catalytic and Green Processes*, CRC Press, Boca Raton, FL. (c) Bandini, M., Umani-Ronchi, A. (eds.) (2009) *Catalytic Asymmetric Friedel-Crafts Alkylations*, Wiley-VHC, New York.

[2] Friedel, C., Crafts, J. M. (1877) *Compt. Rend.*, **84**, 1392; 1450.

[3] Carey, F. A., Sundberg, R. J. (2007) *Advanced Organic Chemistry*, 5th Ed., Springer, New York, pp. 771–814.

[4] (a) Johnson, J. F., Ridd, J. H., Sandall, J. P. B. (1989) *J. Chem. Soc. Chem. Commun.*, 244. (b) Esteves, P. M., de M. Carneiro, J. W., Cardoso, S. P., Barbosa, A. G. H., Laali, K. K., Rasul, G. G., Prakash, G. K. S., Olah, G. A. (2003) *J. Am. Chem. Soc.*, **125**, 4836.

[5] (a) Squadrito, G. L., Fronczek, F. R., Church, D. F., Pryor, W. A. (1990) *J. Org. Chem.*, **55**, 2616. (b) Shiri, M., Zolfigol, M. A., Kruger, H. G., Tanbakouchian, Z. (2010) *Tetrahedron*, **66**, 9077.

[6] (a) Olah, G. A., Kuhn, S. J. (1964) *J. Am. Chem. Soc.*, **86**, 1067. (b) Bartlett, P. D., Roha, M., Stiles, R. M. (1954) *J. Am. Chem. Soc.*, **76**, 2349. (c) Deans, F. B., Eaborn, C. (1959) *J. Chem. Soc.*, 2299. (d) Wilson, S. R., Jacob, L. A. (1986) *J. Org. Chem.*, **51**, 4833.

[7] (a) Davieva, M. G., Lindeman, S. V., Neretin, I. S., Kochi, J. K. (2005) *J. Org. Chem.*, **70**, 4013. (b) Seel, F. Z. (1943) *Z. Anorg. Allg. Chem.*, **250**, 331; **252**, 24. (c) Boer, F. P. (1966) *J. Am. Chem. Soc.*, **88**, 1572. (d) Sato, Y., Yato, M., Ohwada, T., Saito, S., Shudo, K. (1995) *J. Am. Chem. Soc.*, **117**, 3037.

[8] Ingold, C. K. (1969) *Structure and Mechanism in Organic Chemistry*, 2nd Ed., Cornell University Press, Ithaca, NY, p. 330.

[9] Williams, D. L. H. (1988) *Nitrosation*, Cambridge University Press, Cambridge, UK, pp. 58–76.

[10] (a) Maryanoff, B. E., Zhang, H.-C., Cohen, J. H., Turchi, I. J., Maryanoff, C. A. (2004) *Chem. Rev.*, **104**, 1431. (b) Speckamp, W. N., Moolenaar, M. J., (2000) *Tetrahedron*, **56**, 3817. (c) Zhang, Y., DeSchepper, D. J., Gilbert, T. M., Sai, K. K. S., Klumpp, D. A. (2007) *Chem. Commun.*, 4032.

[11] Prakash, G. K. S., Mathew, T., Hoole D., Esteves, P. M., Wang, Q., Rasul, G., Olah, G. A. (2004) *J. Am. Chem. Soc.*, **126**, 15770.

[12] Li, G.-X., Qu, J. (2010) *Chem. Commun.*, **46**, 2653.

[13] (a) Hegarty, A. F. (1978) *The Chemistry of Diazonium and Diazo groups* (ed.: Patai, S.), John Wiley & Sons, Ltd, Chichester, UK. (b) March, J. (1992) *Advanced Organic Chemistry*, 4th Ed., John Wiley & Sons, Inc., New York, pp. 635–637.

[14] Just, C. (1976) *Iminium Salts in Organic Chemistry*, Vol. **9** (eds.: Bohme, H., Viehe, H. G.), in Advances in Organic Chemistry: Methods and Results, Wiley-InterScience, Hoboken, NJ, pp. 225–342.

[15] Li, H., Zhang, Y., Xie, X., Ma, H., Zhao, C., Zhao, G., She, X. (2014) *Org. Lett.*, **16**, 4440.

[16] (a) Suzuki, H., Nonoyama, N. (1998) *Tetrahedron Lett.*, **39**, 4533. (b) Doyle, M. P., Van Lente, M. A., Mowat, R., Fobare, W. F. (1980) *J. Org. Chem.*, **45**, 2570.

[17] Chow, Y. L. (1980) *Reactive Intermediates*, vol. **I** (ed.:Abramovitch, R. A.), Plenum Press, New York, p. 151.

[18] Suzuki, H., Tomaru, J.-i., Murashima, T. (1994) *J. Chem. Soc. Perkin Trans.* **1**, 2413.

[19] (a) Swain, C. G., Scott, C. B. (1953) *J. Am. Chem. Soc.*, **75**, 141. (b) Lucius, R., Loos, R., Mayr, H. (2002) *Angew. Chem. Int. Ed.*, **41**, 92, and references cited therein.

[20] Brown, H. C., Nelson, K. L. (1953) *J. Am. Chem. Soc.*, **75**, 6292.

[21] Stock, L. M., Brown, H. C. (1963) *Adv. Phys. Org. Chem.*, **1**, 35.

[22] Olah, G. A., Kuhn, S. J., Flood, S. (1961) *J. Am. Chem. Soc.*, **83**, 4571.

[23] Olah, G. A. (1971) *Acc. Chem. Res.*, **4**, 240.

[24] Mayr, H., Kempf, B., Oflal, A. R. (2003) *Acc. Chem. Res.*, **36**, 66.

[25] Lakhdar, S.; Mayr, H. (2011) *Chem. Commun.*, **47**, 1866.

[26] Olah, G. A., Wang, Q., Sandford, G., Prakash, G. K. S. (1993) *J. Org. Chem.*, **58**, 3194.

[27] Olah, G. A., Germain, A., Lin, H. C., Forsyth, D. (1975) *J. Am. Chem. Soc.*, **97**, 2928.

[28] Klumpp, D. A., Lau, S. (1999) *J. Org. Chem.*, **64**, 7309.

[29] Hwang, J. P., Prakash, G. K. S., Olah, G. A. (2000) *Tetrahedron*, **56**, 7199.

[30] Olah, G. A., Klumpp, D. A. (2008) *Superelectrophiles and Their Chemistry*, John Wiley & Sons, Inc., New York.

[31] Stadler, D., Bach, T. (2008) *Angew. Chem. Int. Ed.*, **47**, 7557.

[32] Wang, S.-G., Han, L., Zeng, M., Sun, F.-L., Zhang, W., You, S.-L. (2012) *Org. Biomol. Chem.*, **10**, 3202.

[33] (a) Lu, H. H., Tan, F., Xiao, W.-J. (2011) *Curr. Org. Chem.*, **15**, 4022. (b) Poulsen, T. B., Jørgensen, K. A. (2008) *Chem. Rev.*, **108**, 2903. (c) Naredla, R. R., Klumpp, D. A. (2013) *Chem. Rev.*, **113**, 6905. (d) Parmar, D., Sugiono, E., Raja, S., Rueping, M. (2014) *Chem. Rev.*, **114**, 9047.

[34] Taylor, R. (1990) *Electrophilic Aromatic Substitution*, John Wiley & Sons, Inc., New York, p. 40.

[35] Stock, L. M. (1976) *Prog. Phys. Org. Chem.*, **12**, 21.

[36] Koleva, G., Galabov, B., Wu, J. I., Schaefer, H. F., Schleyer, P. v. R. (2009) *J. Am. Chem. Soc.*, **131**, 14722.

[37] Hirao, H., Ohwada, T. (2003) *J. Phys. Chem. A*, **107**, 2875.

[38] Fuster, F., Sevin, A., Silvi, B. (2000) *J. Phys. Chem. A*, **104**, 852.

[39] Politzer, P., Abrahmsen, L., Sjoberg, P. (1984) *J. Am. Chem. Soc.*, **106**, 855.

[40] (a) Galabov, B., Ilieva, S., Koleva, G., Allen, W. D., Schaefer, H. F., Schleyer, P. v. R. (2013) *Wiley Interdiscip. Rev. Comput. Mol. Sci.*, **3**, 37. (b) Bachrach, S. M., (2012) *Annu. Rep. Prog. Chem. Sect. B Org. Chem.*, **108**, 334.

[41] Ghaffarzadeh, M., Rahbar, S. (2014) *J. Chem. Res.*, **38**, 200.

[42] Vasilyev, A. V., Lindeman, S. V., Kochi, J. K. (2001) *Chem. Commun.*, 909.

[43] Wang, Y., Kong, C., Du, Y., Song, H., Zhang, D., Qin, Y. (2012) *Org. Biomol. Chem.*, **10**, 2793.

[44] Ibad, M. F., Langer, P., Schulz, A., Villinger, A. (2011) *J. Am. Chem. Soc.*, **133**, 21016.

[45] Heidrich, D. (2002) *Angew. Chem. Int. Ed.*, **41**, 3208.

[46] Hehre, W. J., Radom, L., Pople, J. A (1972) *J. Am. Chem. Soc.*, **94**, 1496.

[47] Rathore, R., Loyd, S. H., Kochi, J. K. (1994) *J. Am. Chem. Soc.*, **116**, 8414.

[48] Dewar, M. J. S. (1954) *J. Chem. Soc.*, **406**, 777.

[49] (a) Brown, H. C., Brady, J. D. (1952) *J. Am. Chem. Soc.*, **74**, 3570. (b) Kilpatrick, M., Luborsky, F. E. (1953) *J. Am. Chem. Soc.*, **75**, 577. (c) Olah, G. A., Kuhn, S. J., Flood, S. H. (1961) *J. Am. Chem. Soc.*, **83**, 4571.

[50] Hubig, S. M., Kochi, J. K. (2000) *J. Org. Chem.*, **65**, 6807.

[51] Lu, Y.-X., Zou, J.-W., Wang, Y. H., Yu, Q.-S. (2007) *Int. J. Quantum Chem.*, **107**, 1479.

[52] Rosokha, S. V., Kochi, J. K. (2001) *J. Am. Chem. Soc.*, **123**, 8985.

[53] Gwaltney, S. R., Rosokha, S. V., Head-Gordon, M., Kochi, J. K. (2003) *J. Am. Chem. Soc.*, **125**, 3273.

[54] (a) Kim, E. K., Kochi, J. K. (1991) *J. Am. Chem. Soc.*, **113**, 4962. (b) Borodkin, G. I., Shubin, V. G. (2001) *Russ. Chem. Rev.*, **70**, 211.

[55] (a) Grozema, F. C., Zijlstra, R. W. J., Swart, M., Van DuiJhen, P. Th. (1999) *Int. J. Quantum Chem.*, **72**, 709. (b) Mebel, A. M., Lin, H. L., Lin, S. H. (1999) *Int. J. Quantum Chem.*, **72**, 307. (c) Ammal, S. S. C., Ananthavel, S. P., Venuvanalingam, P., Hegde, M. S. (1998) *J. Phys. Chem. A*, **102**, 532. (d) Matsuzawa, A., Osamura, Y. (1997) *Bull. Chem. Soc. Jpn.*, **70**, 1531.

[56] Lambert, J. B., Zhang, S., Stern, C. L., Huffman, J. C. (1993) *Science*, **260**, 1917.

[57] Sidorkin, V. F., Doronina, E. P., Belogolova, E. F. (2012) *Organometallics*, **31**, 7511.

[58] Holman, R. W., Gross, M. L. (1989) *J. Am. Chem. Soc.*, **111**, 3560.

[59] Smith, M. B., March, J. (2001) *March's Advanced Organic Chemistry*, 5th Ed., John Wiley & Sons, Inc., New York, p. 679.

[60] (a) Brown, H. C., Brady, J. D. (1952) *J. Am. Chem. Soc.*, **74**, 3570.

[61] (a) Melander, L. (1949) *Acta Chem. Scand.*, **3**, 95. (b) Bonner, T. G., Bower, F., Williams, G. (1953) *J. Chem. Soc.*, 2650. (c) Zollinger, H. (1964) *Adv. Phys. Org. Chem.*, **2**, 163.

[62] Nowicki, L., Zarzycki, R. (1987) *J. Chem. Technol. Biotechnol.*, **39**, 149.

[63] For a closely related conversion involving thiophenes, see: Arai, Y., Nakazaki, J., Segawa, H. (2008) *Tetrahedron Lett.*, **49**, 5810.

[64] Doering, W. v. E., Saunders, M., Boyton, H. G., Earhart, H. W., Wadley, E. F., Edwards, W. R., Laber, G. (1958) *Tetrahedron*, **4**, 178.

[65] (a) Olah, G. A., Spear, R. J., Messina, G., Westerman, P. W. (1975) *J. Am. Chem. Soc.*, **97**, 4051. (b) Olah, G. A., Lin, H. C., Mo, Y. K. (1972) *J. Am. Chem. Soc.*, **94**, 3667. (c) Mamatyuk, V. I., Rezvukhin, A. I., Detsina, A. N., Buraev, V. I., Isaev, I. S., Koptyug, V. A. (1973) *Zh. Org. Khim.*, **9**, 2429. (d) Koptyug, V. A. (1984) *Top. Curr. Chem.*, **122**, 247 pp. (e) Nugent, W. A. (1980) *J. Org. Chem.*, **45**, 4534. (f) Borodkin, G. I., Nagy, S. M., Gatilov, Yu. V., Shakirov, M. M., Rybalvo, T. V., Shubin, V. G. (1992) *Zh. Org. Khim.*, **28**, 1806. (g) Borodkin, G. I., Gatilov, Yu. V., Nagy, S. M., Shubin, V. G. (1996) *Zh. Org. Khim.*, **37**, 534. (h) Borodkin, G. I., Chernyak, E. I., Shakirov, M. M., Shubin, V. G. (1994) *Zh. Org. Khim.*, **37**, 397. (i) Borodkin, G. I., Nagy, S. M., Mamatyuk, V. I., Shakirov, M. M., Shubin, V. G. (1996) *Zh. Org. Khim.*, **37**, 534. (j) Rathore, R., Hecht, J., Kochi, J. K. (1998) *J. Am. Chem. Soc.*, **120**, 13278.

[66] Reed, C. A., Fackler, N. L. P., Kim, K.-C., Stasko, D., Evans, D. Ŕ., Boyd, P. D. W., Rickard, C. E. F. (1999) *J. Am. Chem. Soc.*, **121**, 6314.

[67] Effenberger, F. (1989) *Acc. Chem. Res.*, **22**, 27.

[68] Olah, G. A., Prakash, G. K. S., Molnar, A., Sommer, J. (2009) *Superacid Chemistry*, 2nd Ed., John Wiley & Sons, Inc., New York.

[69] Shubin, V. G., Borodkin, G. I. (2004) *Carbocation Chemistry*, (eds.: Olah, G. A., Prakash, G. K. S.), John Wiley & Sons, Inc., New York.

[70] Olah, G. A., Kuhn, S. J. (1958) *J. Am. Chem. Soc.*, **80**, 6541.

[71] Borodkin, G. I., Nagy, S. M., Mamatyuk, V. I., Shakirov, M. M., Shubin, V. G. (1984) *Zh. Org. Khim.*, **20**, 552.

[72] Olah, G. A., Prakash, G. K. S., Molnar, A., Sommer, J. (2009) *Superacid Chemistry*, 2nd Ed., John Wiley & Sons, Inc., New York, p. 136.

[73] Borodkin, G. I., Panova, Y. B., Shakirov, M. M., Shubin, V. G. (1979) *J. Chem. Soc. Chem. Commun.*, 354.

[74] Olah, G. A., Molnar, A. (1995) *Hydrocarbon Chemistry* John Wiley & Sons, Inc., New York, pp. 112–115.

[75] (a) Ordomsky, V. V. Ivanova, I. I. Knyazeva, E. E. Yuschenko, V. V., Zaikovskii, V. I. (2012) *J. Catal.*, **295**, 207. (b) Kaeding, W. W. (1989) *J. Catal.*, **120**, 409.

[76] Beltrame, P., Beltrame, P. L., Carniti, P., Nespoli, G. (1978) *Gazz. Chim. Ital.*, **108**, 651.

[77] Sarca, V. D., Laali, K. K. (2004) *Green Chem.*, **6**, 245.

[78] Reed, C. A., Fackler, N. L. P., Kim, K.-C., Stasko, D., Evans, D. R., Boyd, P. D. W., Rickard, C. E. F. (1999) *J. Am. Chem. Soc.*, **121**, pp. 615–617.

[79] (a) Koltunov, K.-Yu., Prakash, G. K. S., Rasul, G., Olah, G. A. (2004) *Heterocycles*, **62**, 757. (b) Necula, A., Racoveanu-Schiketanz, A., Gheorghiu, M. D., Scott, L. T. (1995) *J. Org. Chem.*, **60**, 3448. (c) Fonken, G. J. (1963) *J. Org. Chem.*, **28**, 1909.

[80] Kethe, A., Li, A., Klumpp, D. A. (2012) *Tetrahedron*, **68**, 3357.

[81] (a) Li, A., DeSchepper, D., Klumpp, D. A. (2009) *Tetrahedron Lett.*, **50**, 1924. (b) Li, A., Gilbert, T. M., Klumpp, D. A. (2008) *J. Org. Chem.*, **73**, 3654. (c) Li, A., Kindelin, P. J., Klumpp, D. A. (2006) *Org. Lett.*, **8**, 1233.

[82] Shimizu, K., Ikuo, I. (1998) *Energy Fuel*, **12**, 115.

2

FRIEDEL–CRAFTS ALKYLATION OF ARENES IN TOTAL SYNTHESIS

GONZALO BLAY, MARC MONTESINOS-MAGRANER, AND JOSÉ R. PEDRO

Departament de Química Orgànica, Facultat de Química, Universitat de València, Burjassot (València), Spain

2.1 INTRODUCTION

With more than 130 years of history, Friedel–Crafts (FC) alkylation of aromatic compounds has become one of the most important tools in organic synthesis. Nowadays, the FC alkylation represents an essential synthetic step in a number of academic syntheses as well as in commercial processes in the bulk chemical industry. In this chapter, the authors provide a survey of the application of this reaction in the total synthesis of natural products and biologically significant compounds. An exhaustive review of the literature is not possible, and it is not necessary, within the limited space of this chapter. We have tried to keep a balanced and representative number of examples featuring different electrophiles, stoichiometric and catalytic procedures, and substrate structures. With few exceptions, the syntheses have been selected from the most recent literature within 2000–2014. The chapter is organized according to the characteristics of the FC reaction. The first part shows total syntheses involving intermolecular FC alkylation as a means of introducing alkyl substituents on the arene ring. Simple alkylations have not been considered. Instead, we have chosen examples where a substantial part of the molecule or functionalized groups, which allow further modification on the alkyl chain, is introduced. The construction of cyclic structures via intramolecular FC alkylation is considered next. Reactions are grouped according to the homocyclic or heterocyclic character of the newly formed ring. Non-enantioselective and enantioselective cascade and tandem reactions

Arene Chemistry: Reaction Mechanisms and Methods for Aromatic Compounds,
First Edition. Edited by Jacques Mortier.

involving an FC alkylation are dealt with in a separate part as they allow great structural modification with a reduced number of operations. Finally, an example of total synthesis involving a less common *ipso*-FC alkylation is presented.

2.2 TOTAL SYNTHESIS INVOLVING INTERMOLECULAR FC ALKYLATIONS

2.2.1 Synthesis of Coenzyme Q_{10}

Coenzyme Q_{10} (**6**) is important for energy production in higher animals and also has an effect on the treatment of heart-related diseases. Although mass production by fermentation currently offers economical advantage, different chemical syntheses have been reported. The key issue in these chemical syntheses is the coupling between the quinone core and the polyprenyl side chain. However, problems associated with cyclization within the polyprenyl chain and stereochemical integrity of polyprenol, which is used as electrophile, constitute serious drawbacks in these procedures. An improved synthesis of coenzyme Q_{10} developed by Koo *et al.* [1] is outlined in Scheme 2.1. The key step is the FC allylation of commercial tetramethoxytoluene (**1**) with a benzenesulfonyl substituted C5 allylic chloride **2**. This coupling with compound **2a** was initially performed with 1.2 equiv. of $ZnCl_2$ at 80°C to give **3a** in 69% yield (*E/Z* = 10:1). Since the *E* configuration of **3a** deteriorated at high temperatures or with prolonged reaction times, compounds **3b–d** with substituted arylsulfonyl moieties were prepared and tested. Best results for the FC reaction were obtained with 30 mol-% of $FeCl_3$ at 40°C. Noteworthy, the substituted benzenesulfonyl groups facilitated the activation of the C5 allylic chloride regardless of their electronic nature. Transformation of **3b** involved oxidation of the quinone with CAN, reduction back to the dihydroquinone, and protection as MOM ether (the MOM group facilitates the oxidation of the dihydroquinone in a further stage). This was followed by coupling of **4b** with solanesyl bromide. Removal of the sulfone with palladium and $LiEt_3BH$ and deprotection and oxidation with $FeCl_3$ and HCl gave coenzyme Q_{10} (**6**).

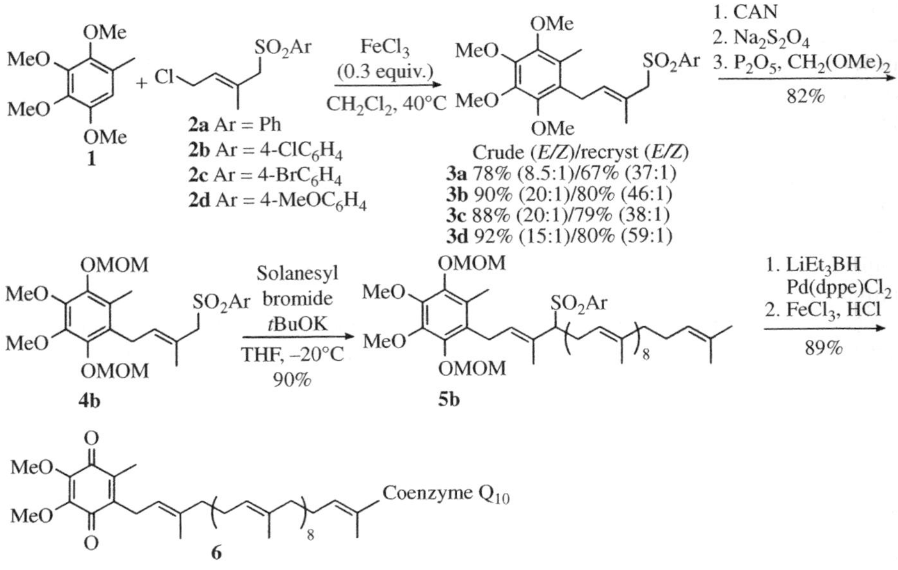

SCHEME 2.1 Synthesis of coenzyme Q_{10}.

2.2.2 Total Synthesis of (±)-Brasiliquinone B

An FC alkylation with bromolactone **11** is the key step in the first total synthesis of the angucycline antibiotic (±)-brasiliquinone B (Scheme 2.2) [2]. The key tetralin derivative **10** was prepared from 7-methoxytetralone (**7**) in four steps. The FC reaction between **10** and **11** was carried out in the presence of excess $SnCl_4$ to afford regioselectively lactone **12** [3], which was reductively opened to give acid **13**. Cyclization of **13** with TFAA followed by oxidation with CrO_3 gave brasiliquinone B dimethyl ether (**15**) in good yield. Reaction of **15** with $AlCl_3$ brought about demethylation to afford (±)-brasiliquinone B (**16**).

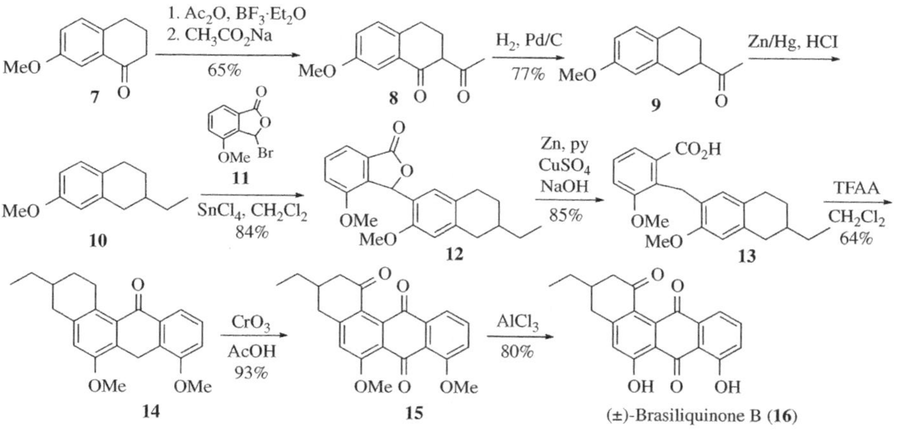

SCHEME 2.2 Synthesis of (±)-brasiliquinone B.

2.2.3 Synthesis of (−)-Podophyllotoxin

As it is known, besides halides, alcohols, especially allylic and benzylic ones, can be used to generate electrophiles in FC alkylations. An example is provided with the synthesis of the cytotoxic compound (−)-podophyllotoxin (Scheme 2.3) [4]. The key step in this synthesis is the reaction between sesamol (**20**) and benzyl alcohol **19**, which is prepared by aldol condensation between aldehyde **17** and enantiomerically pure Taniguchi lactone (**18**). The use of a catalytic amount of $FeCl_3$ (5 mol-%) outperformed other Lewis acid catalysts, $Bi(OTf)_3$ or $AuCl_3$, as well as stoichiometric HBF_4, allowing the obtention of compound **21** with high diastereoselectivity (94:6) in almost quantitative yield. Heck coupling and cleavage of the terminal alkene moiety in compound **23** completed the synthesis of (−)-podophyllotoxin (**24**).

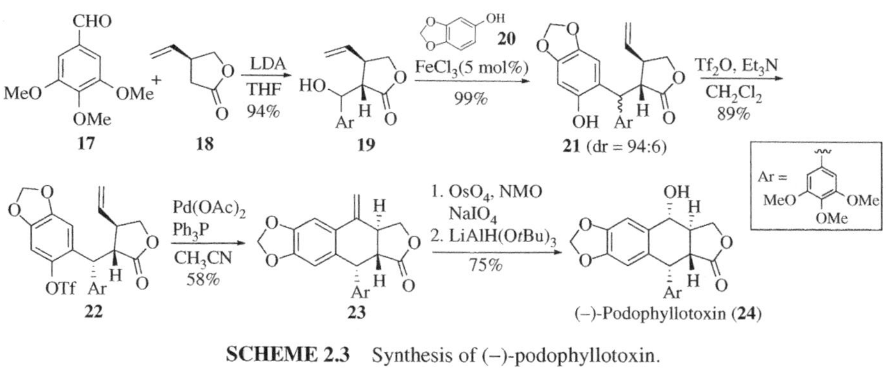

SCHEME 2.3 Synthesis of (−)-podophyllotoxin.

2.2.4 Synthesis of Puupehenol and Related Compounds

Alvarez-Manzaneda *et al.* [5] have developed a synthetic strategy toward the tetracyclic marine metabolite puupehenones involving a FC reaction with enone **25**, a common enantiopure nordrimane synthon utilized in terpenoid synthesis, as the electrophilic partner (Scheme 2.4). The reaction is carried out in the presence of the cationic resin Amberlyst A15 to afford ketone **27** in high yield and with complete diastereoselectivity. The reaction only works with highly activated arenes such as **26**, as neither anisole nor catechol dimethyl ether reacted under identical conditions. Ketone **27** can be straightforwardly transformed into triflate **30**, which is an intermediate in the synthesis of puupehenone-related metabolites such as puupehenol (**31**).

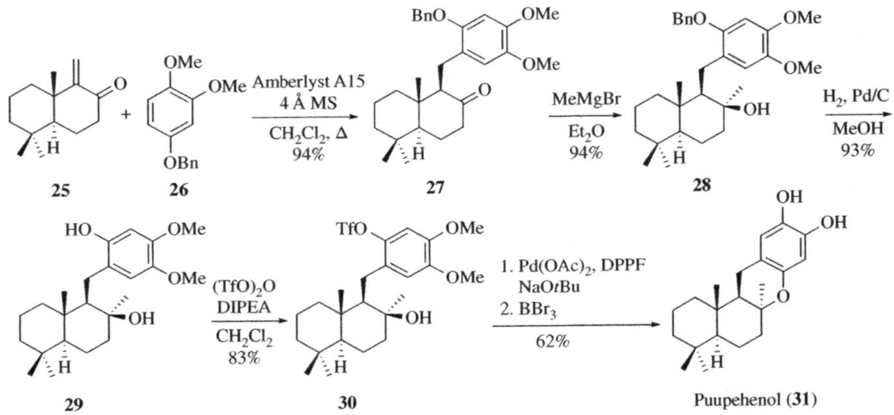

SCHEME 2.4 Synthesis of puupehenol.

2.2.5 Synthesis of (−)-Talaumidin

An intermolecular FC hydroxyalkylation on a carboxonium ion is the key step in the enantioselective synthesis of the neurotrophic lignan (−)-talaumidin (Scheme 2.5) [6]. The carboxonium precursor **36** was prepared from alcohol **33** (obtained in eight steps from aldehyde **32** with high diastereo- and enantioselectivity). The crucial diastereoselective FC step was achieved upon treatment of **36** with 1,2-methylenedioxybenzene (**37**) and $SnCl_4$, which afforded straightforwardly the desired product **39** in 89% yield through cation **38**. Debenzylation of **39** with $Pd(OH)_2$ furnished (−)-(2*S*,3*S*,4*S*,5*S*)-talaumidin (**40**) in 77% yield.

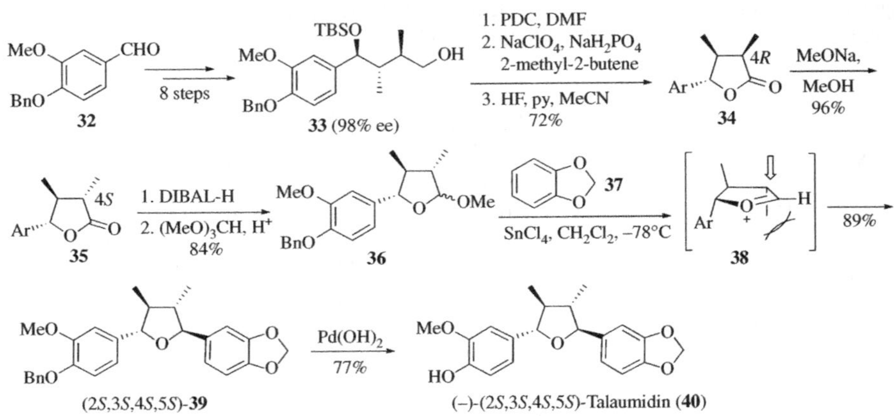

SCHEME 2.5 Synthesis of (−)-talaumidin.

2.2.6 Total Synthesis of (±)-Schefferine

Related with the previous synthesis, an FC aminoalkylation through a tosyliminium ion allowed the total synthesis of (±)-schefferine (Scheme 2.6) [7], which is a phenolic alkaloid isolated from *Schefferomitra subaequalis*, a liana climber found on rain forest trees. The known amidal **41** was treated at −78°C with 4-allyl-1,2-dimethoxybenzene (**42**) and $BF_3{\cdot}Et_2O$ to regioselectively give **43** in 88% yield. The allylic chain of the aromatic ring was then modified to furnish alcohol **44**, which after deprotection of the tosylamide was cyclized via a Mitsunobu reaction to give (±)-schefferine (**46**).

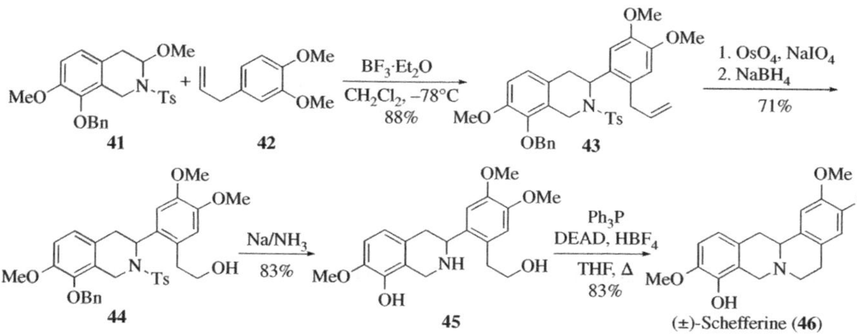

SCHEME 2.6 Synthesis of (±)-schefferine.

2.3 TOTAL SYNTHESIS INVOLVING INTRAMOLECULAR FC ALKYLATIONS

Intramolecular FC alkylation is a powerful tool for the construction of cyclic structures. Depending on the length and composition of the linker between the arene and the electrophilic group and the reaction conditions, a large variety of saturated or unsaturated homocyclic or heterocyclic compounds can be obtained.

2.3.1 C—C Bond Formation Leading to Homocyclic Rings

2.3.1.1 Synthesis of Dihydrexidine Dihydrexidine has been designed as a dopamine D1 agonist and examined for the treatment of Parkinson's disease and schizophrenia. This compound shows a high level of enantiospecificity in its interaction with the D1 receptor, and hence, methods for its synthesis in enantiomeric pure form have been devised, some of them through aminotetralin **50** (Scheme 2.7) [8]. A retrosynthetic analysis of compound **50** reveals that it can be obtained by an FC cyclization of 1,2-aminoalcohol **49**. Enantiomerically pure *syn*-**49** was prepared in six steps from Garner's aldehyde **47** (a D-serine derivative). The FC cyclization of compound **49** was best carried out with neat TFA to give **50** in 70% yield and excellent diastereoselectivity (dr>99:1). Treatment of *N*-nosyl amide **51** with sodium hydride and MOMCl gave compound **52**, which was cyclized with TMSOTf to give tetracyclic **53** in 40–52% yield depending on the quality of the reagent. Deprotection of the nosyl group and demethylation with BBr_3 afforded dihydrexidine hydrobromide (**55**). Finally, successive treatment with $NaHCO_3$ and HCl provided dihydrexidine hydrochloride (**56**) (ee>99%).

In a shorter synthesis of dihydrexidine [9], synthetic intermediate **51** was obtained via an FC reaction of a chiral aziridine intermediate **60** (Scheme 2.8). Styrene **58** was prepared from a Wittig reaction of aldehyde **57** as a mixture of diastereomers (*E/Z*=78:22) and treated with PhINNs in the presence of the BOX ligand **59** and $Cu(OTf)_2$. It was found that $Cu(OTf)_2$ was a dual catalyst for the enantioselective aziridination and FC reaction. The *E/Z* mixture of styrenes **58** provided exclusively *trans*-2-amino-1-phenyltetralin **51** in 87% ee, which could be increased up to 99% after crystallization. The *cis*-cyclized product was not detected.

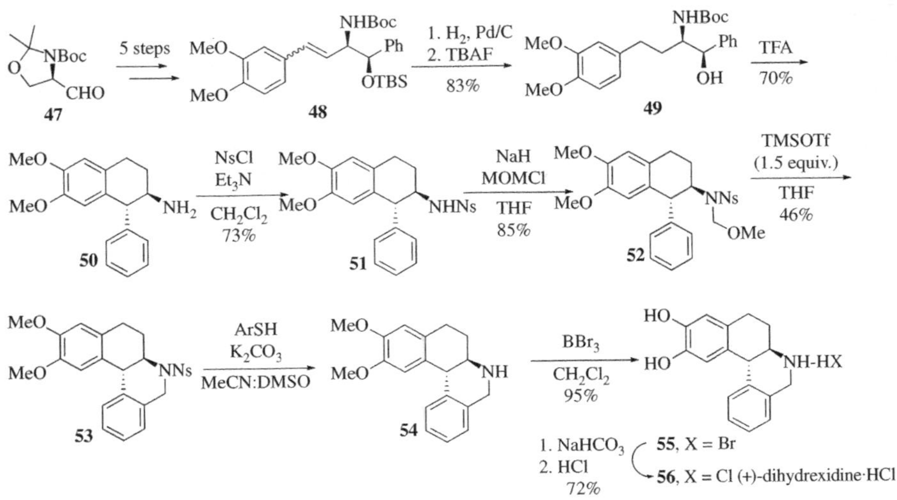

SCHEME 2.7 Synthesis of dihydrexidine hydrochloride via an FC reaction of aminoalcohol **49**.

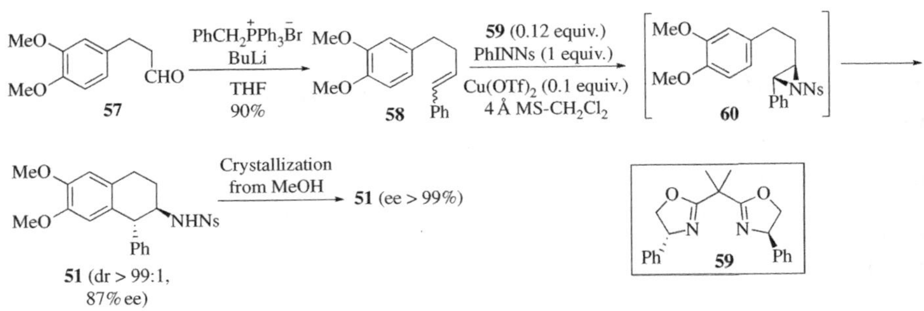

SCHEME 2.8 Formal synthesis of dihydrexidine via an aziridination-FC pathway.

2.3.1.2 Synthesis of (−)-Aphanorphine An FC cyclization of compound **66** has been employed by the groups of Zhai [10] and Yus [11] for the construction of ring B of aphanorphine, an alkaloid isolated from the blue-green alga *Aphanizomenon flos-aquae*, which incorporates a 3-benzazepine scaffold that resembles that of benzomorphan analgesics. While the synthesis by Zhai is based on the use of chiral 4-hydroxyproline as starting material, Yus employs *tert*-butylsulfinylimide (**62**) as chiral auxiliary (Scheme 2.9), which makes this procedure more adequate for the preparation of both enantiomers of aphanorphine. Accordingly, the asymmetric indium-mediated α-aminoallylation of (*p*-methoxyphenyl)acetaldehyde (**61**) was carried out to give homoallylic amine **63**, which after epoxidation and regioselective opening of the epoxide gave 3-pyrrolidinol **65**. All attempts to carry out the FC cyclization with compound **65** using Lewis or Brønsted acids gave rise to decomposition of the starting material. The amine protecting group was then changed to benzoyl group, to give the key intermediate **66**. Treatment of the diastereomeric mixture of **66** with $AlCl_3$ gave the tricyclic compound **67** with complete diastereoselectivity, which was transformed into aphanorphine (**70**) after debenzoylation and methylation of the *N* atom.

2.3.1.3 Synthesis of (−)-8-Demethoxyrunanine and (−)-Cepharatine D An intramolecular FC reaction on dihydroindolone **73** is the key step in the enantioselective syntheses of the hasubanan

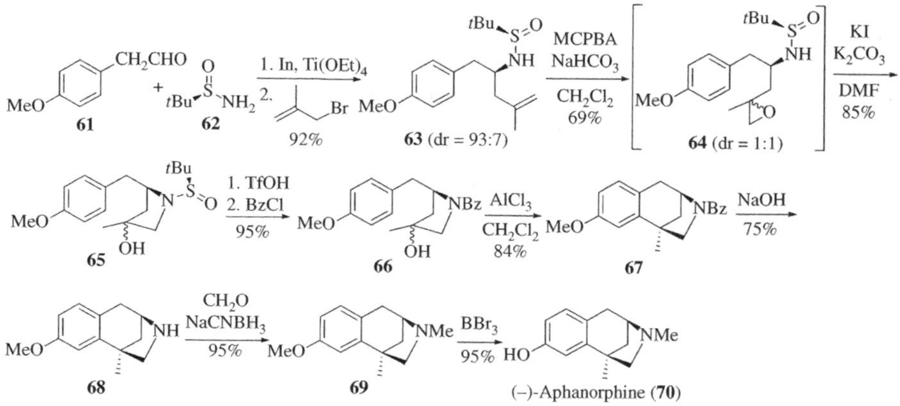

SCHEME 2.9 Synthesis of tricyclic alkaloid (−)-aphanorphine.

alkaloid (−)-8-demethoxyrunanine, which features an aza[4.4.3]propellane core, and the structurally related cepharatines A, C, and D, isolated from *Stephania cepharantha* (only the synthesis of (−)-cepharatine D is shown in Scheme 2.10) [12]. Dihydroindolone **73** was prepared in enantiomerically pure form from 2-bromo-4-methoxyphenol (**71**) and chiral sulfinylimide **62**, through dienone imine **72**. Cyclization of dienone **73** was carried out with excess of the strong Brønsted acid TfOH in dichloromethane, which performed better than other Lewis acid tested, to give propellane **74** in 97% yield. Epoxidation of compound **74** followed by epoxide cleavage with tetrabutylammonium methoxide gave (−)-8-demethoxyrunanine (**76**). On the other hand, when epoxidation of compound **74** was followed by long exposure to silica gel, a rearranged aminal **78** was obtained in 76% yield via enol **77**. Dehydrogenation of aminal **78** with excess KHMDS and *N-tert*-butylbenzenesulfinimidoyl chloride provided cepharatine D (**79**) in 22% overall yield from **71**.

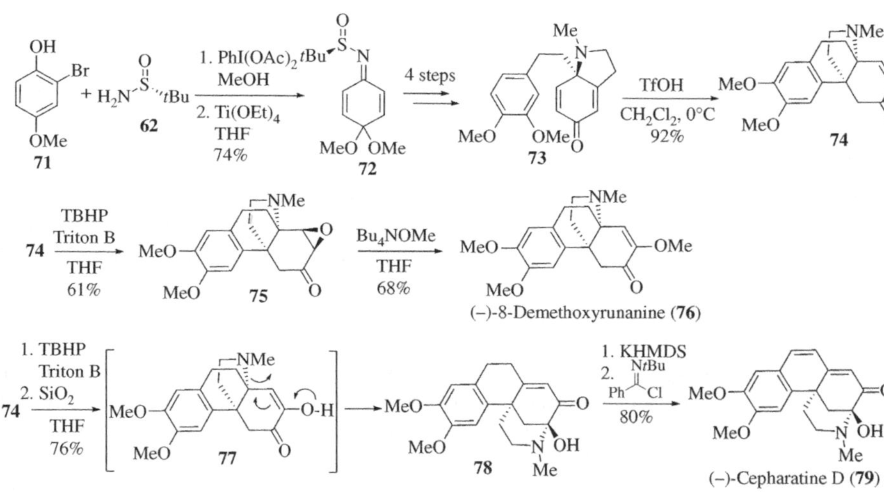

SCHEME 2.10 Synthesis of (−)-8-demethoxyrunanine and (−)-cepharatine D.

2.3.1.4 *Total Synthesis of (−)-Plicatic Acid*

The enantioselective synthesis of (−)-plicatic acid, which has been identified as the causative agent of occupational asthma, features an interesting FC cyclization of a chiral epoxide **81** leading to an α-hydroxy ketone bearing a quaternary stereocenter **82** (Scheme 2.11) [13]. The chiral epoxide precursor **81** was prepared by asymmetric epoxidation of the Knoevenagel adduct **80** with the chiral hydroperoxide TADOOH. An FC reaction of epoxide **81** was best performed with a catalytic amount of TfOH to give α-hydroxyketone **82** as a 4:1 diastereomeric mixture favoring the desired diastereomer, which was separated by chromatography. Notably, the FC reaction also proceeded in a highly regioselective manner as the other regioisomer was not detected. Compound **82** was silylated with $ClSi(Me)_2CH_2Br$ to give **83**, which underwent a SmI_2-mediated intramolecular Barbier reaction to afford hydroxysilane **84**, which was subjected to a Fleming–Tamao–Kumada oxidation to furnish triol ester **85**. Finally, ester hydrolysis with sodium propanethiolate and deprotection of the Bn groups afforded (−)-plicatic acid (**87**).

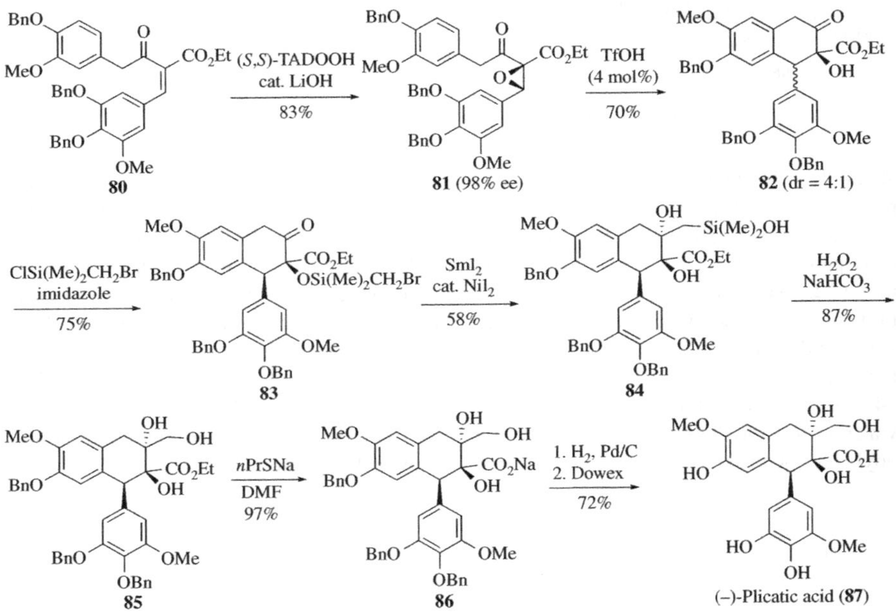

SCHEME 2.11 Enantioselective synthesis of (−)-plicatic acid.

2.3.1.5 Total Synthesis of the Proposed Structure for Kealiinine C The following example features the formation of a fully unsaturated 6-member ring during the synthesis of the proposed structure of kealiinine C (Scheme 2.12) [14], belonging to a family of 2-aminoimidazole alkaloids isolated from sponges of the *Leucetta* family. Metalation of diiodo-1-methylimidazole **88** followed by reaction with 3,4,5-trimethoxybenzaldehyde gave alcohol **89**. Again, metalation followed by treatment with *N*-methyl formanilide gave aldehyde **90**, which was reacted with *p*-methoxyphenyl magnesium bromide to give diol **91**. Treatment of unpurified **91** with HCl resulted in an intramolecular FC cyclization–dehydration sequence to provide naphthimidazole **92**. C2-metalation with BuLi followed by exposure to $TrisN_3$ produced the azido compound **93**, which upon reduction of the azide to amine provided compound **94**, whose structure was confirmed by X-ray. However, the spectroscopic features of the synthetic material showed discrepancies with those described for the natural product, which were attributed to the existence of a tautomeric equilibrium.

2.3.1.6 Total Syntheses of (±)-Isopaucifloral F, (±)-Quadrangularin A, and (±)-Pallidol Isopaucifloral F, quadrangularin A, and pallidol are natural polyphenols that belong to the resveratrol family and possess a variety of biological activities. In particular, pallidol shows estrogen-like

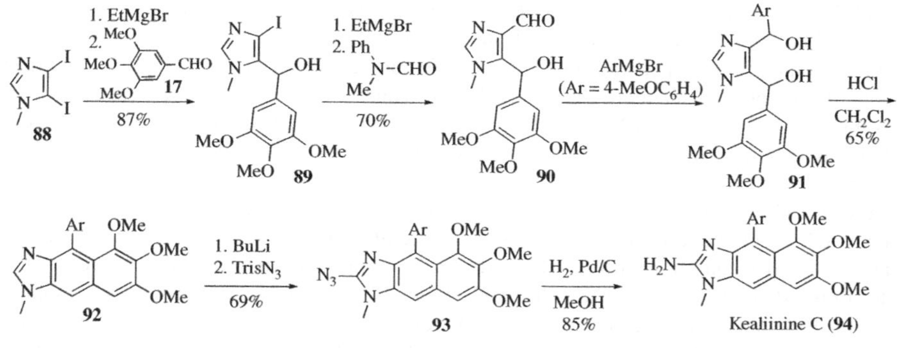

SCHEME 2.12 Total synthesis of the proposed structure for kealiinine C.

activity. A recent synthesis of these compounds [15] features the formation of two five-member rings via FC-type cyclizations (Scheme 2.13). Compound **97** (diastereomer mixture) was readily prepared from 3,5-dimethoxybenzoic acid (**95**) by radical dimerization and Wittig–Horner olefination. Upon treatment of enone **97** with $BF_3 \cdot Et_2O$, a Nazarov cyclization produced compound **98a** with both aromatic rings in *trans* disposition, from which (±)-isopaucifloral F (**98b**) was obtained after deprotection of the phenol groups. Reduction of the ketone in compound **98a** followed by Ramberg–Backlund olefination gave compound **100a**, which after demethylation afforded (±)-quadrangularin A (**100b**). Attempts to carry out intramolecular FC cyclization with alkene **100a** with different Brønsted or Lewis acids failed. Therefore, compound **100a** was regioselectively converted into alcohol **101** via standard hydroboration/oxidation procedure. An FC cyclization of **101** was accomplished with $AlCl_3$ to give 68% yield of compound **102a** having two fused cyclopentane rings, which was readily converted into (±)-pallidol (**102b**) with BBr_3.

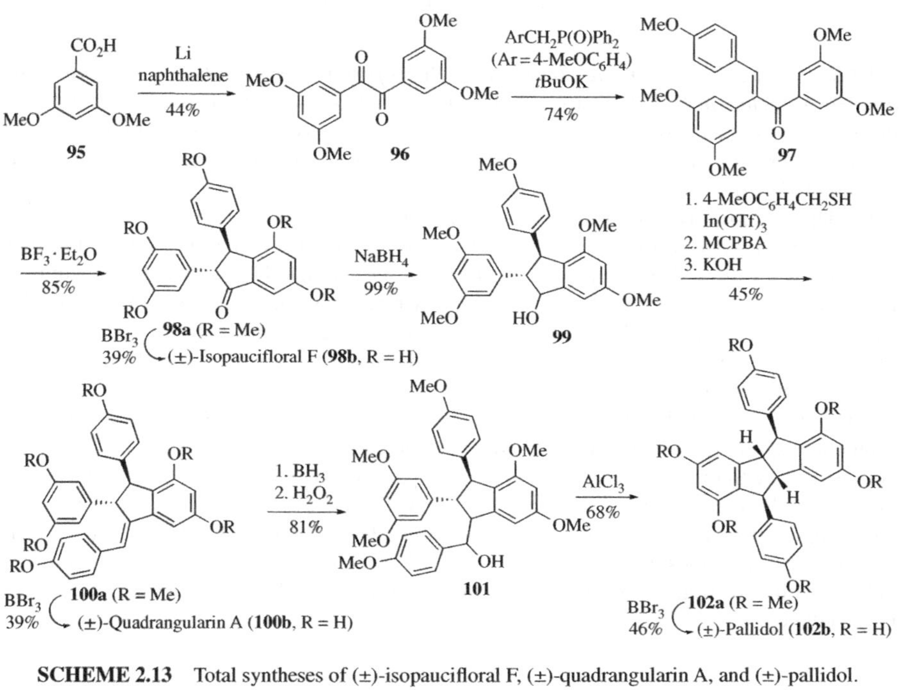

SCHEME 2.13 Total syntheses of (±)-isopaucifloral F, (±)-quadrangularin A, and (±)-pallidol.

2.3.1.7 Synthesis of (±)-Microstegiol A Nicholas reaction and an FC cyclization reaction are the two key steps in the construction of the cyclohepta[*d,e*]naphthalene framework of microstegiol (Scheme 2.14) [16], a rearranged abietane featuring a seven-member ring, isolated from the roots of a number of plants of the genus *Salvia*, which presents antileukemic activity. The 1,6-disubstituted naphthalene **104**, with the two arene rings differently activated, was subjected to a Nicholas reaction with the alkyne dicobalt complex **103**. The reaction was carried out in the presence of an excess amount of $BF_3{\cdot}Et_2O$ and followed by oxidative decomplexation of Co with I_2 to give **105** in good yield. Hydrogenation of the triple bond was accompanied by removal of the benzyl ether, and the resulting naphthol unit was readily converted into its triflate **106**. Palladium-catalyzed cross-coupling with DABAL-Me_3 gave compound **107**, which after treatment with MeLi provided the key tertiary alcohol **108**. When exposed to one drop of H_2SO_4, a solution of alcohol **108** underwent ring closure in conjunction with tautomerization of the naphthol function to a cyclohexenone ring giving **109** in 81% yield. Finally, oxidation of **109** with oxygen gave (±)-microstegiol (**110**) in 64% yield.

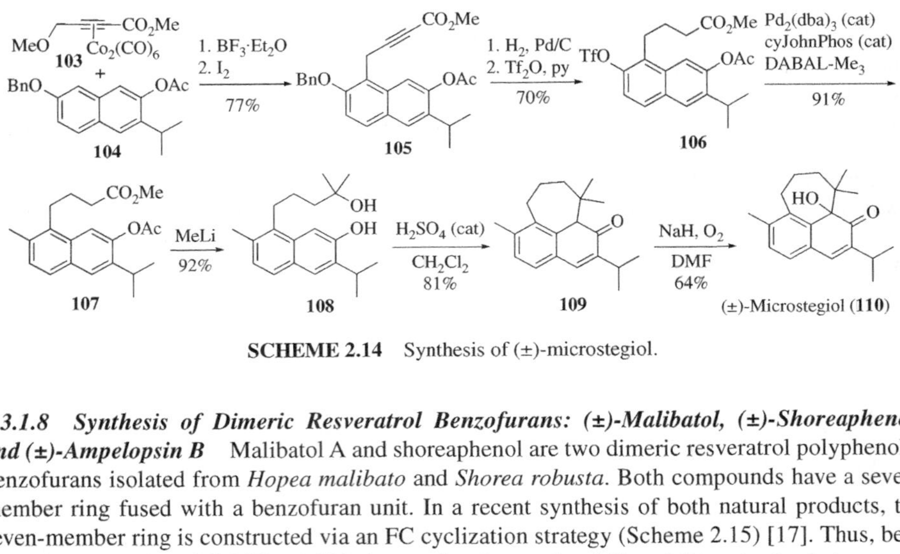

SCHEME 2.14 Synthesis of (±)-microstegiol.

2.3.1.8 Synthesis of Dimeric Resveratrol Benzofurans: (±)-Malibatol, (±)-Shoreaphenol, and (±)-Ampelopsin B Malibatol A and shoreaphenol are two dimeric resveratrol polyphenolic benzofurans isolated from *Hopea malibato* and *Shorea robusta*. Both compounds have a seven-member ring fused with a benzofuran unit. In a recent synthesis of both natural products, the seven-member ring is constructed via an FC cyclization strategy (Scheme 2.15) [17]. Thus, benzylic deprotonation (LiTMP) of **111**, intramolecular nucleophilic addition to the ketone, and dehydration of the resulting alcohol gave **112**. Epoxidation of stilbene **112** with the bromohydrin protocol was followed by treatment of the resulting epoxide **113** with BBr_3 (12 equiv.). This resulted in the FC cyclization to a seven-member ring, presumably through the intermediacy of **114**, and concomitant global demethylation giving racemic malibatol A (**115**) as a single diastereomer, although in low yield (20%). Oxidation of malibatol A with PDC then afforded (±)-shoreaphenol (**116**).

On the other hand, the group of Yang has disclosed a convenient Tf_2O-catalyzed FC alkylation protocol to synthesize the dibenzo[*a,d*]cycloheptene core of this kind of resveratrol dimers, which is illustrated with the concise synthesis of (±)-ampelopsin B (Scheme 2.16) [18]. Thus, dimerization by oxidative coupling of resveratrol (**117**) with $FeCl_3$ gave benzofuran ε-viniferin (**118**), which upon treatment with a catalytic amount (10 mol-%) of Tf_2O in toluene at 100°C provided (±)-ampelopsin B (**119**) in 41% yield. The reaction is thought to be initiated by reaction of Tf_2O with trace water to release a proton that coordinates with the alkene to form a carbocation intermediate.

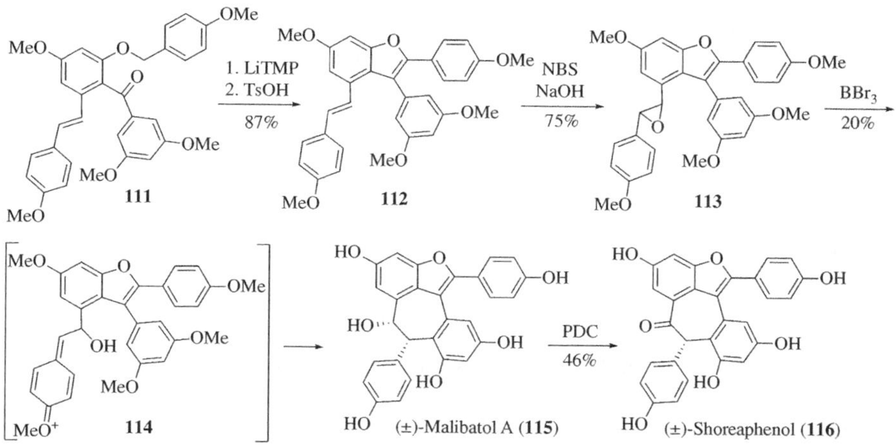

SCHEME 2.15 Synthesis of (±)-malibatol A and (±)-shoreaphenol.

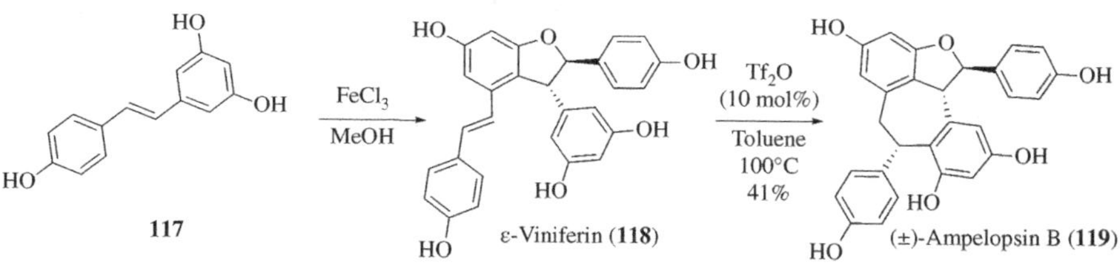

SCHEME 2.16 Synthesis of (±)-ampelopsin B.

2.3.2 C–C Bond Formation Leading to Oxygen-Containing Rings

2.3.2.1 Synthesis of Lamellarin G Trimethyl Ether Lamellarin G belongs to a group of marine natural products that contain the 5-oxa-6*b*-aza-dibenzo-[*a,i*]fluoren-6-one skeleton. A recent synthesis [19] of this compound involved an intramolecular FC cyclization of an aryl cinnamate **124** as a key step to the creation of the chroman-2-one moiety present in the molecule (Scheme 2.17). The synthesis involved an FC acylation of 1,2-dimethoxybenzene (**120**) with maleic anhydride (**121**) and esterification with 3,4-dimethoxyphenol (**123**) to give the FC substrate **124**. Upon treatment with NBS (1.1 equiv.) and catalytic $Sm(OTf)_3$ (10 mol-%), compound **124** underwent an intramolecular bromoarylation reaction, probably through a bromonium ion intermediate, to give a brominated chroman-2-one **125** in 93% yield. Finally, coupling of **125** with 6,7-dimethoxy-1,2,3,4-tetrahydroisoquinoline hydrochloride (**126**) in the presence of K_2CO_3 gave lamellarin G trimethyl ether (**127**) in 63% yield.

2.3.2.2 Total Synthesis of Cacalol Kedrowski and Hoppe [20] have reported a synthesis of the natural sesquiterpene cacalol that emphasizes the synthetic utility of the FC reaction in forming arene–carbon bonds. The short synthesis features up to three different FC-type reactions, one of them leading to the formation of a heterocyclic benzofuran ring (Scheme 2.18). The synthesis began with the *ortho*-lithiation of 4-methylanisole (**128**) and alkylation with 5-iodopent-1-ene to give alkene **129**. This molecule was then cyclized in an intramolecular FC alkylation by treatment with $AlCl_3$ to give tetralin **130**, which is a key intermediate in previous syntheses of cacalol. An FC

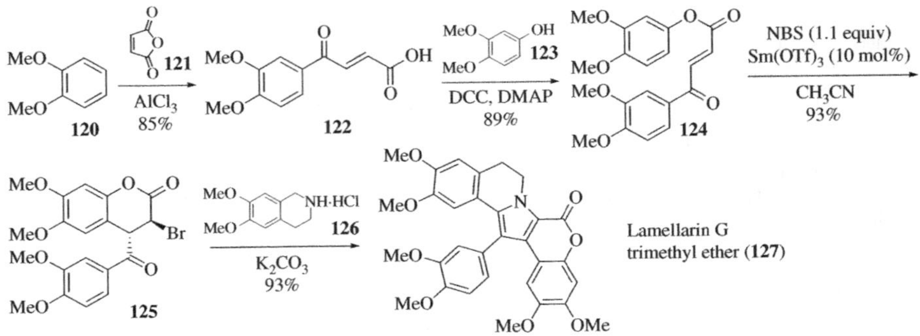

SCHEME 2.17 Synthesis of lamellarin G trimethyl ether.

formylation with dichloromethyl methyl ether and $TiCl_4$ yielded 96% of a mixture of two regioisomeric aldehydes **131** and **132** in a 14:1 ratio that was converted into phenol **133** after oxidation with MCPBA, hydrolysis, and separation. The phenol **133** was alkylated with chloroacetone. An FC cyclodehydration in concentrated sulfuric acid formed a furan ring to give cacalol methyl ether (**135**), which was finally demethylated to cacalol (**136**) with BBr_3.

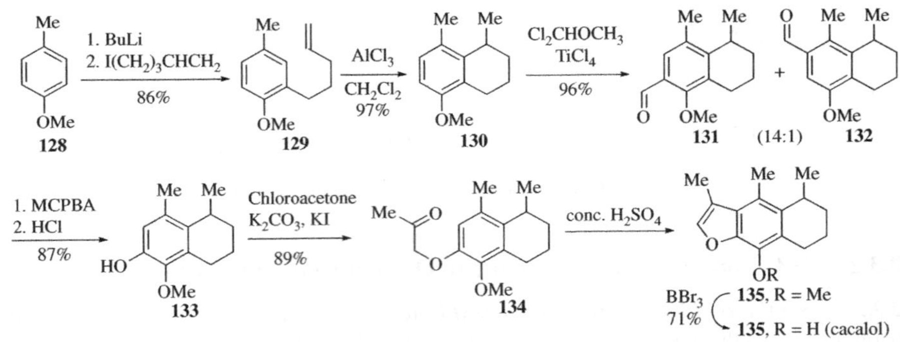

SCHEME 2.18 Total synthesis of cacalol.

2.3.2.3 Total Synthesis of the Macrocyclic Antibiotic Kendomycin Although intramolecular FC alkylations are mostly used for the construction of medium-sized rings, one can find in the literature examples where larger cycles are formed using this reaction. The synthesis of the kendomycin, an ansamycin antibiotic isolated from various *Streptomyces* species, is a representative example of formation of a macrocycle via an FC alkylation (Scheme 2.19) [21]. The elaborated compound **137** was converted into acetal **138**. Attempts to cyclize **138** under a variety of FC conditions were unsuccessful, mainly leading to hydrolysis of the anomeric acetate. However, the reaction employing phenol **139** as substrate with $SnCl_4$ smoothly afforded the desired macrocycle **141** as a single diastereomer in 40–70% yield. The reaction is thought to proceed through the formation of an *O*-glycoside intermediate **140**. After exchange of protecting groups, oxidative modification of the aryl core with IBX and removal of the TES group afforded kendomycin (**144**).

2.3.3 C—C Bond Formation Leading to Nitrogen-Containing Rings

2.3.3.1 Total Synthesis of Hunanamycin A Hunanamycin A is a natural product isolated from a marine-derived *Bacillus humanensis*. This compound presents a nitrogenated tricyclic structure related to riboflavin degradation products and shows antimicrobial activity against *Salmonella*

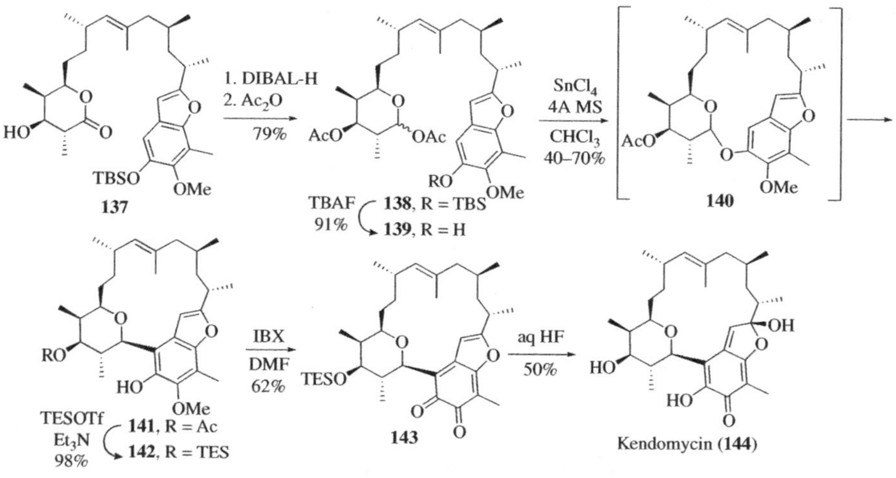

SCHEME 2.19 Total synthesis of kendomycin.

bacteria. Its first reported total synthesis (Scheme 2.20) [22] began with the monoalkylation of quinoxalindione **145** with prenyl bromide. Then, the intramolecular FC alkylation carried out with $AlCl_3$ in chlorobenzene as the solvent allowed obtaining the tricyclic aglycon **147**. The authors reported the reaction to be explosive if carried out in nitromethane. The *N*-alkylation of **147** with allylic bromide **148** (prepared from D-mannitol) and diastereoselective dihydroxylation of the *Z*-olefin **149** gave **150**, which on deprotection with aqueous acetic acid furnished hunanamycin (**151**).

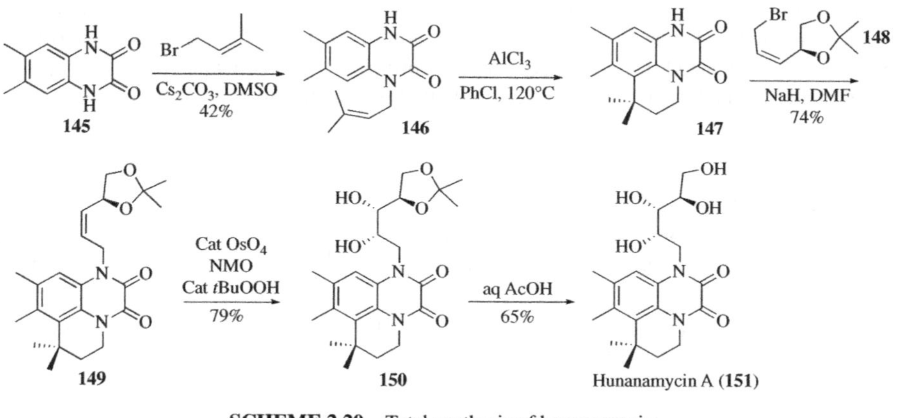

SCHEME 2.20 Total synthesis of hunanamycin.

*2.3.3.2 **Total Synthesis of Murrayazoline*** An intramolecular FC alkylation was a key step in the total synthesis of murrayazoline, a carbazole alkaloid isolated as a racemic or an optically active compound from medicinal plants of the genus *Murraya* [23]. Murrayazoline features a characteristic hexaheterocyclic structure, which was constructed by a combination of FC-type Michael addition and Pd-catalyzed C—O coupling reaction (Scheme 2.21). A Pd-catalyzed double N-arylation of amine **153** with dibromodiphenyl derivative **152** provided the FC precursor **154**. The treatment of **154** with 50 mol% of $Sc(OTf)_3$ in dichloroethane and H_2O induced deprotection of the ethylene ketal group as well as the intramolecular FC-type Michael addition and deprotection of the *O*-MOM group to give the pentacyclic ketone **156** in 73% yield. Treatment of triflate **157** with

MeMgBr gave the tertiary alcohol **158**, which upon a Buchwald–Hartwig C—O coupling reaction with stoichiometric $Pd(OAc)_2$ provided murrayazoline (**159**).

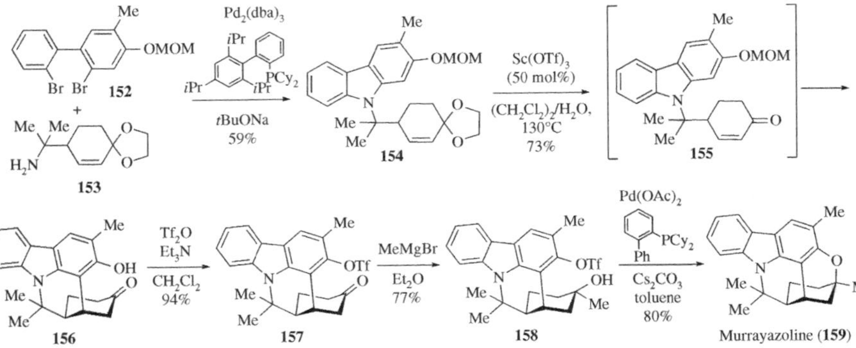

SCHEME 2.21 Total synthesis of murrayazoline.

*2.3.3.3 **Formal Total Synthesis of (−)-Cephalotaxine*** The polycyclic alkaloid (−)-cephalotaxine is a fascinating target for total synthesis because its structure contains a number of interesting features, including an embedded 1-azaspiro[4,4]nonane ring system. A formal synthesis of (−)-cephalotaxine has been reported recently by Hayes (Scheme 2.22) [24]. Spiropyrrolidine derivative **161** prepared from proline was subjected to epoxidation followed by Lewis acid-mediated intramolecular epoxide opening to give tricyclic compound **163**, which after elimination and carbamate hydrolysis provided spirocyclic amino alcohol **164**. Alkylation of compound **164** with nosylate **165** produced the FC cyclization precursor **166**. In a previously attempted synthesis of (−)-cephalotaxine [25], a diastereomer of compound **166** failed to undergo the FC cyclization when treated with polyphosphoric acid, probably due to sensitivity of the methylenedioxy moiety to harsh acidic conditions. However, upon treatment of allylic alcohol **166** with $SnCl_4$ in CH_2Cl_2/nitromethane, a nitrogenated seven-member ring was formed, delivering the pentacyclic compound **167** (46% yield) that had been previously converted into (−)-cephalotaxine [25].

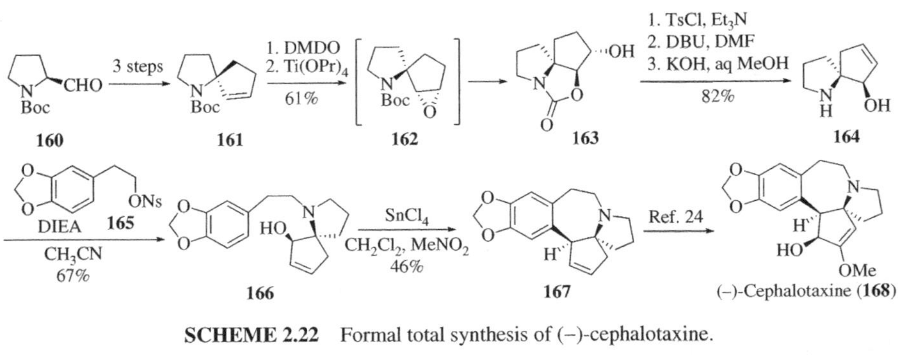

SCHEME 2.22 Formal total synthesis of (−)-cephalotaxine.

2.4 TOTAL SYNTHESIS THROUGH TANDEM AND CASCADE PROCESSES INVOLVING FC REACTIONS

2.4.1 C—C Bond Formation Leading to Homocyclic Rings

*2.4.1.1 **Synthesis of Taiwaniaquinol Diterpenoids*** Taiwaniaquinol diterpenoids present a 6,5,6-ABC tricyclic skeleton relatively uncommon in natural products. Their intriguing structures and their biologically significant activities have attracted the attention of organic chemists. She's group [26] has reported an acid-promoted domino FC reaction, leading to the construction of the core 6,5,6-ABC tricyclic skeleton of these diterpenoids (Scheme 2.23). Starting from 1,3-dimethoxy-2-isopropylbenzene (**169**) and cyclogeranic acid (**170**), the key domino reaction proceeded smoothly, using neat methanesulfonic acid in the presence of P_2O_5 (20:1), to give the thermodynamically more favorable tricyclic *cis*-indanone **171** in good yield. The regioselectivity of this reaction was controlled by the *ortho* directing effect of the MeO groups. Then, the total synthesis of taiwaniaquinol B (**173**) was accomplished through selective demethylation, CAN oxidation, and sodium dithionite reduction.

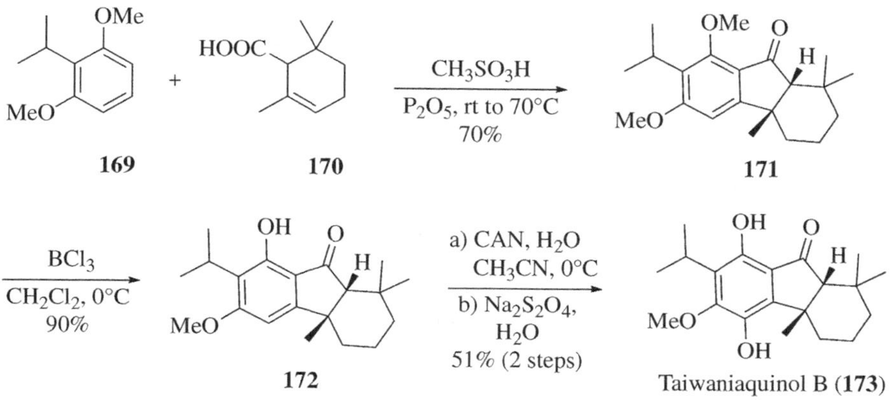

SCHEME 2.23 Total synthesis of taiwaniaquinol B.

2.4.1.2 Synthesis of Clavilactone D Clavilactone D (**181**) is a new member of the tyrosine kinase inhibitors, which has attracted considerable attention owing to its intriguing structure and biological significance. The aldol reaction of aldehyde **174** with chiral lactone **175** allowed to obtain the substrate **176** in good yield (92%), although moderate diastereoselectivity (**176a**/**176b**/**176c**=3:1:2) (Scheme 2.24) [27]. Cyclization of alkenyl alcohol **176a** with three equivalents of iodine provided fused polycyclic compound **179** in moderate yield (27%) along with other nonmacrocyclic compounds. The reaction was considered to take place through the initial generation of an iodonium intermediate that underwent tetrahydrofuran formation. The tetrahydrofuran-tethered system served as a molecular scaffold that increased the proximity of the aromatic ring to the carboxonium center generated at the lactone acetal, both being suitably positioned for their connection through an FC cyclization. The polycyclic lactone **179** was subsequently transformed into the final product **180** with the clavilactone D core.

2.4.1.3 Synthesis of Hainanensine Hainanensine (**188**) is a structurally unique minor alkaloid component identified from the antileukemia plant *Cephalotaxus hainanensis*. Li *et al.* [28] have achieved a facile total synthesis of this alkaloid via an unusually effective rearrangement–annulation cascade as key step (Scheme 2.25). From readily available 3,4-dihydroisoquinoline **182**, the aldehyde derivative **185** was prepared in 70% overall yield. This aldehyde was transformed into enone **186** by Horner–Wadsworth–Emmons olefination. Upon refluxing in formic acid for 6 h, enone **186** underwent an acid-promoted ring expansion process to intermediate (i), followed by tautomerization to (ii) and an FC-type cyclization–dehydration leading to **187** with a benzazepine ring system. A five-step sequence of functional group transformations led to hainanensine (**188**).

2.4.1.4 Synthesis of (−)-Haouamine A Haouamine A (**193**) was isolated in 2003 from a marine ascidian (*Aplidium haouarianum*) and displayed potent and selective cytotoxic activity against the HT-29 human colon carcinoma cell line. Aubé *et al.* [29] have culminated a formal

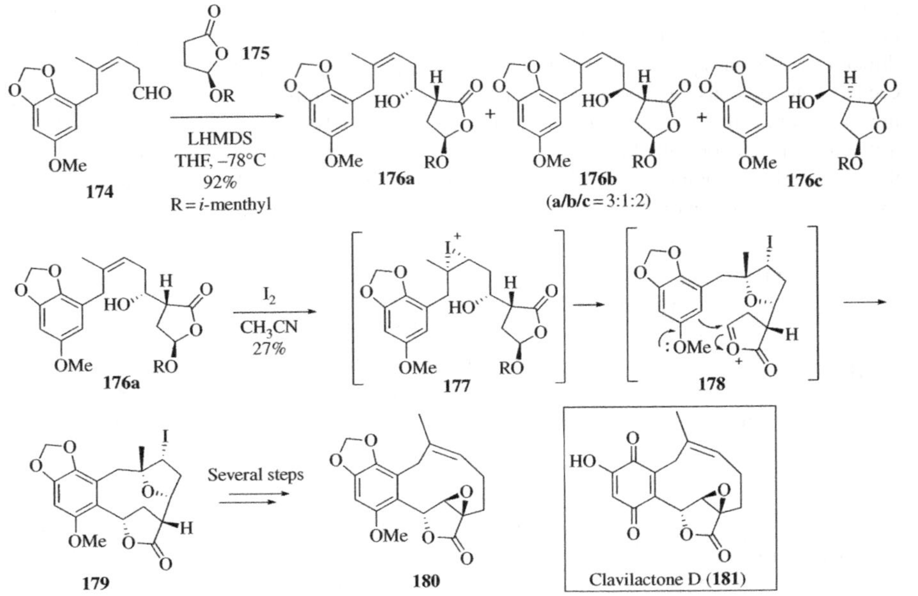

SCHEME 2.24 Total synthesis of clavilactone D.

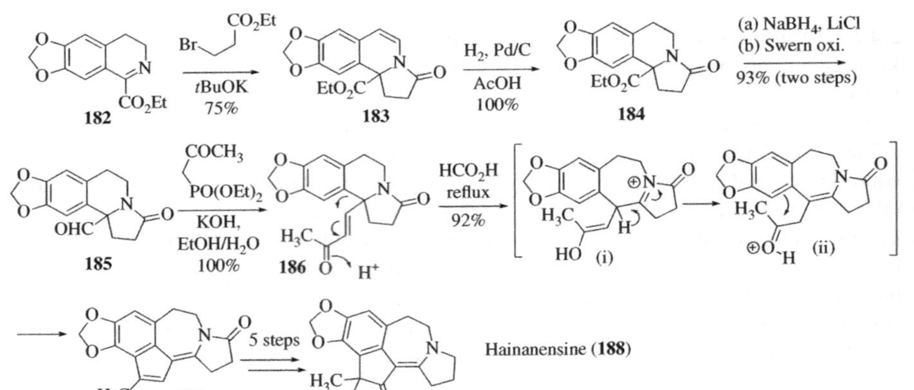

SCHEME 2.25 Total synthesis of hainanensine.

synthesis of (−)-haouamine A in which the tricyclic indeno-tetrahydropyridine core was constructed in 13 steps from serine, a tandem Prins/FC reaction being the key step (Scheme 2.26). This tandem reaction was carried out with enal **189** as a substrate using Lewis acids bearing "non-nucleophilic" anions, BF_3 or $Al(OTf)_3$, in nitromethane, leading exclusively to the tricyclic compound **190** (65% yield) as a mixture of diastereomers at C2 (1.5:1). This mixture was oxidized to ketone **191**, and the Cbz protecting group was removed by hydrogenolysis with concomitant debromination to produce the haouamine tricyclic core in compound **192**, which is a synthetic intermediate to (−)-haouamine A (**193**).

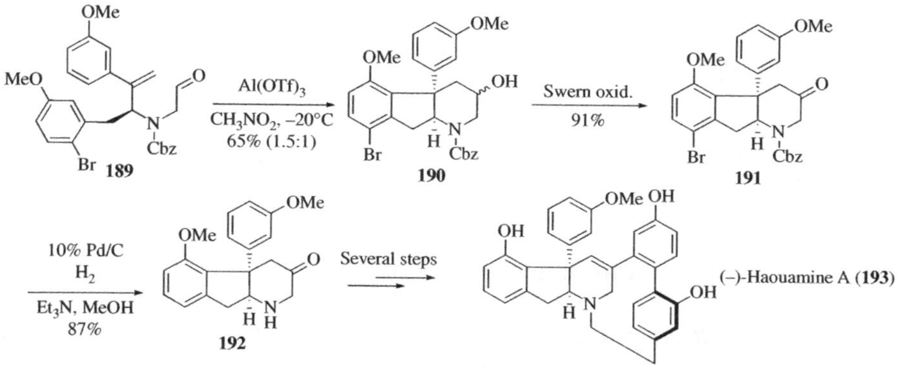

SCHEME 2.26 Synthesis of (−)-haouamine A.

2.4.2 C—C Bond Formation Leading to Oxygen-Containing Rings

2.4.2.1 ***Synthetic Studies Toward Fumimycin*** Fumimycin (**200**) is a novel nonpeptidic metabolite isolated from the fermentation broth of *Aspergillus fumisynnematus F746*, which exhibits bacterial peptide deformylase inhibitor activity. Zhou *et al.* [30] have carried out the synthesis of an advanced intermediate **199**. This intermediate has been synthesized in 18% total yield from vanillin as the starting material and a tandem FC alkylation/lactonization as the key reaction step (Scheme 2.27). First, vanillin (**194**) was converted to its allyl ether **195** that underwent a Dakin oxidation to give phenol **196**. An FC alkylation with methyl 2-acetamidoacrylate followed by tandem lactonization gave rise to the benzofuranone **197** with 85% yield. The Claisen rearrangement of allyl ether in **197** smoothly provided **198**. Isomerization of the terminal double bond was carried out in ethanol at reflux in the presence of $RhCl_3{\cdot}3H_2O$. Surprisingly, the olefin isomerization gave the *Z* configuration product **199** exclusively, while in the natural fumimycin, the double bond presents the *E*-configuration.

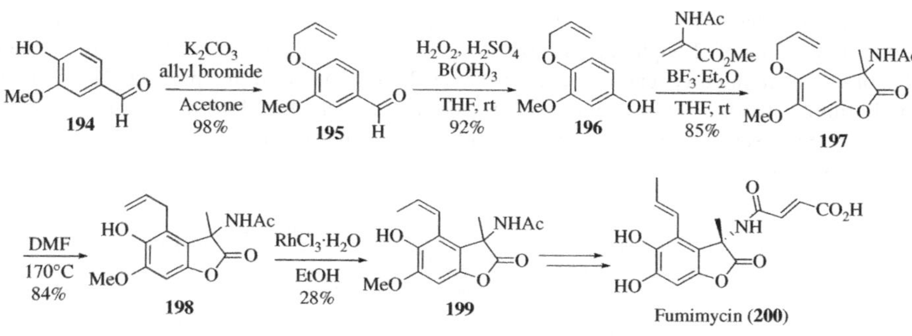

SCHEME 2.27 Synthesis of an advanced intermediate to fumimycin.

2.4.2.2 ***Synthetic Route to Norneolignans*** Seo *et al.* [31] have accomplished the total synthesis of several norneolignans in a synthetic sequence that involves the one-pot reaction of 4-hydroxyphenylacetone (**201**) with 2-chloro-2-(methylthio)acetophenone derivatives **202** under FC conditions to give compound **203** with a 2-arylbenzofuran ring, as the key step (Scheme 2.28). Reduction of the ketone carbonyl group, desulfurization of the carbon–sulfur bond, methanesulfonate formation, and elimination of this group completed the synthetic sequence.

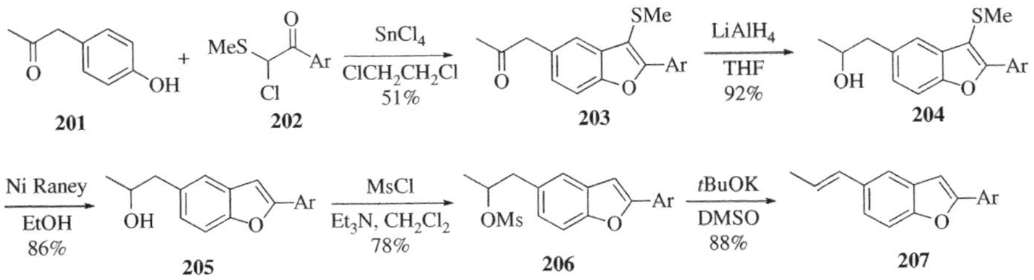

SCHEME 2.28 Synthetic route to norneolignans.

2.4.2.3 Synthetic Routes to Chromanes Chromanes are important structural motifs in organic synthesis and have been found as core structural elements commonly present in many bioactive compounds. Jørgensen *et al.* [32] have developed a tandem reaction that involves a Lewis acid-catalyzed oxa-Michael addition of phenols **208** to β,γ-unsaturated α-ketoesters **209**, followed by an intramolecular FC alkylation to form chromanes **211** (Scheme 2.29). The reaction proceeds under the influence of a Mg-BOX (**210**) catalyst to give diastereomerically pure chromanes with enantioselectivities up to 81% and excellent yields. The best results were obtained with *m*-methoxyphenol, while *m*-*N,N*-dimethylaminophenol afforded the corresponding chromane as single diastereomer in excellent yield but with low enantioselectivity (<20% ee).

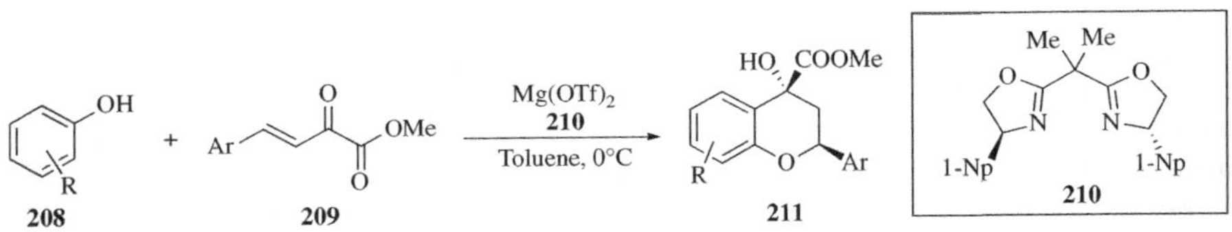

SCHEME 2.29 Enantioselective tandem oxa-Michael addition FC alkylation of phenols with β,γ-unsaturated α-ketoesters.

On the other hand, Yang *et al.* [33] have developed an organocatalyzed enantioselective FC-type addition reaction of 2-naphthol **212** with β,γ-unsaturated α-ketoesters **209** using a cinchona alkaloid-derived thiourea catalyst **213** (Scheme 2.30). The resulting product **214** is in rapid equilibrium with the cyclic hemiketal **215**, which was dehydrated with a catalytic amount of concentrated H_2SO_4 in a one-pot fashion, providing the naphthopyran derivatives **216** with moderate to good yields (51–91%) and enantioselectivities (57–90% ee).

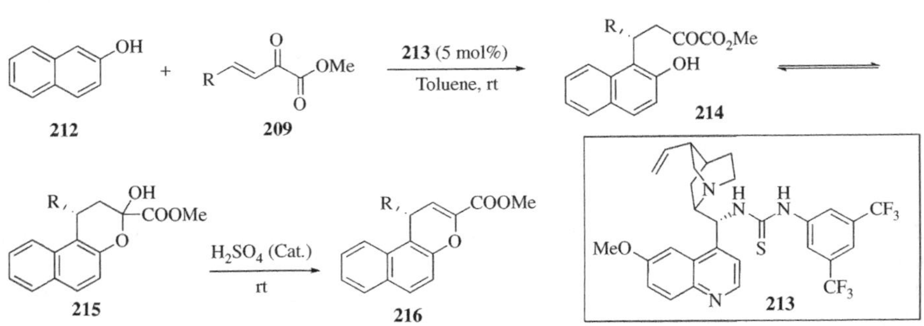

SCHEME 2.30 Organocatalytic asymmetric FC alkylation/cyclization cascade reaction of 2-naphthols with β,γ-unsaturated α-ketoesters.

Wang *et al.* [34] have also described an organocatalyzed enantioselective FC alkylation/cyclization cascade reaction of 1-naphthols **217** and α,β-unsaturated aldehydes **218** promoted by a diphenylprolinol ether **219** as a catalyst (Scheme 2.31). This method affords one-pot access to chiral and synthetically useful chromanes and dihydrobenzopyranes **220** in high yields and enantioselectivities.

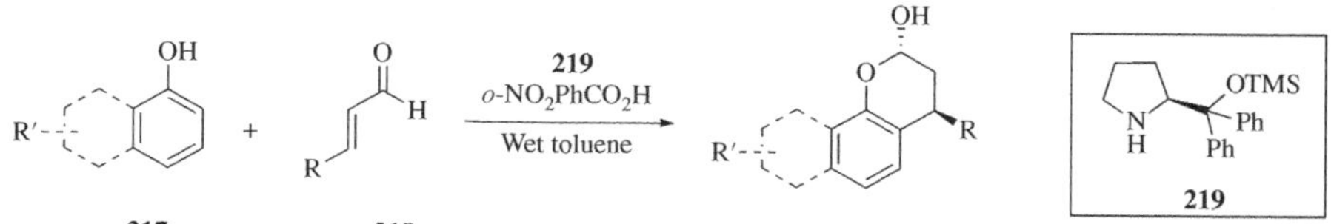

SCHEME 2.31 Organocatalytic asymmetric FC alkylation/cyclization cascade reaction of 1-naphthols with α,β-unsaturated aldehydes.

Later, the same group [35] disclosed the synthesis of various potential bioactive chiral functionalized chromanes **222** with high levels of enantio- and diastereoselectivity (up to 76% ee and >20:1 *dr*) via the rosin-derived tertiary amine-thiourea **221** catalyzed enantioselective FC alkylation/cyclization cascade of 1-naphthol **217** with a variety of β,γ-unsaturated α-ketoesters **209** (Scheme 2.32).

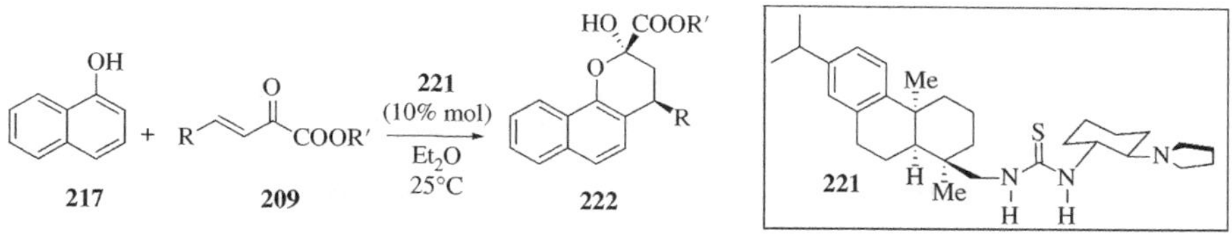

SCHEME 2.32 Synthesis of modified chromanes.

Finally, Feng and Zhang [36] have described the organocatalytic domino Michael-type FC alkylation/cyclization of 2-naphthols **212** with alkylidene Meldrum's acids **223** obtaining a variety of enantioenriched β-arylsplitomicins **225** with moderate to good yields (up to 99%) and moderate enantioselectivities (up to 79% ee). Screening of chiral bifunctional organocatalysts revealed the thiourea-tertiary amine catalyst **224** as the optimal catalyst in terms of yield and enantioselectivity. Cyclohexyl alkylidene Meldrum's acids were the best electrophiles in this reaction (Scheme 2.33).

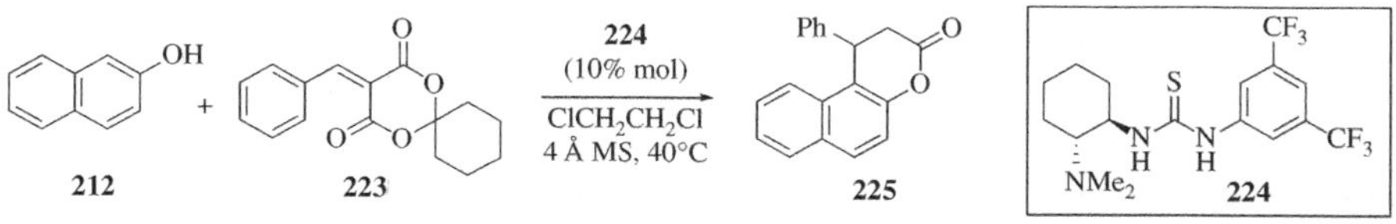

SCHEME 2.33 Enantioselective domino Michael-type FC alkylation/cyclization of 2-naphthols with alkylidene Meldrum's acids.

2.4.2.4 Synthesis of Heimiol A and Hopeahainol D Heimiol A and its epimer hopeahainol D are two unique resveratrol-derived dimers with a [3.2.2] bicycle system. The key step in their synthesis was a iodolactonization/intramolecular FC cascade reaction of the carboxylic acid **226** promoted by the iodonium source $[(Et_2SI)_2Cl_2]SbCl_6$ (IDSI) in MeCN at 25°C (Scheme 2.34) [37]. This reaction provided the entire [3.2.2] bicycle with the required all-*syn* stereochemistry present in the target molecules. Then, the methyl ethers were removed (BBr_3), the new phenolic hydroxyls were protected as benzyl ethers, and the lactone moiety reacted with 4-benzyloxybromobenzene/*n*-BuLi to give compound **228** as a single stereoisomer. The tertiary alcohol was reduced with $Et_3SiH/BF_3{\cdot}OEt_2$ giving protected hopeahainol D **229**, which upon hydrogenation over Pd/C afforded hopeahainol D (**230**). Finally, treatment of hopeahainol D with a mixture of $BF_3{\cdot}OEt_2$ and BCl_3 in MeOH at 25°C resulted in complete epimerization to give heimiol A (**231**).

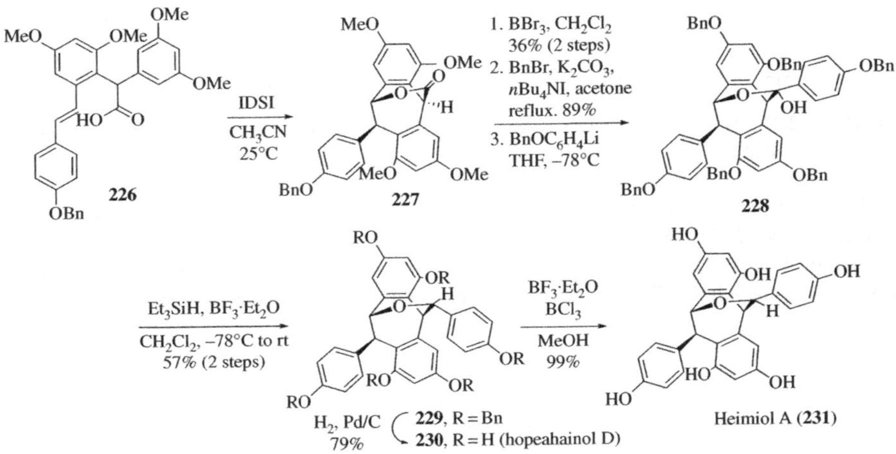

SCHEME 2.34 Total synthesis of heimiol A and hopeahainol D.

2.4.3 C—C Bond Formation Leading to Nitrogen-Containing Rings

2.4.3.1 Synthesis of Tetrahydroquinoline Alkaloids The pyranoquinoline core structure is the backbone of a wide range of alkaloids with biological activities. Yadav *et al.* [38] have described a synthesis of optically active pyranotetrahydroquinolines **234** in a one-pot operation using arylamines **232** and sugar-derived δ-hydroxy-α,β-unsaturated aldehydes **233**, which undergo a tandem Michael and intramolecular FC-type cyclization in the presence of Lewis acids such as $InCl_3$, $Yt(OTf)_3$, or $Bi(OTf)_3$ (Scheme 2.35).

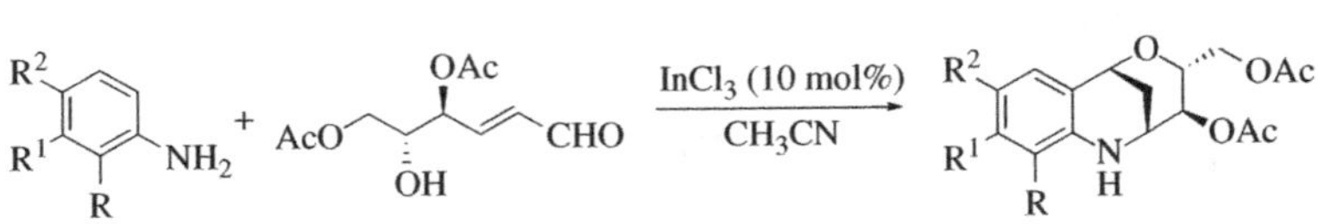

232 233 234

SCHEME 2.35 Synthesis of pyranotetrahydroquinolines from arylamines and δ-hydroxy-α,β-unsaturated aldehydes.

2.4.3.2 Synthetic Studies Toward Saframycin B and Ecteinascidin 743 (Yondelis) Tetrahydroisoquinoline alkaloids are a broad family of biologically active natural products. Takemoto *et al.* [39] have carried out a concise synthesis of the CDE ring system present in this kind of alkaloids (Scheme 2.36). Both Au(I)-catalyzed intramolecular hydroamination of alkynyl amide and NBS-mediated oxidative FC cyclization were utilized as key reactions. By reductive alkylation of α-amino acid derivative **235** with aldehyde **236**, the coupling product **237** was obtained in 71% yield. After carbamoylation of **237**, the key hydroamidation of **238** was carried out in the presence of 10 mol-% of $AuCl(PPh_3)/AgNTf_2$. The reaction occurred with excellent regio- and stereoselectivity to give the desired cyclized adduct **239** in 85% yield as a single isomer. Finally, the oxidative FC cyclization of the benzyl derivative **240** was carried with NBS, leading to compound **241** in 70% yield. The formation of compound **241** from **240** most probably involves oxidation by NBS to give an allylic bromide that eliminates an iminium cation, which is the electrophile in the FC reaction. Isomerization of the enamide moiety from *Z*- to *E*-configuration was observed in this last step. The synthesis of saframycin B (**242**) from **241** has been achieved by Kubo's group [40].

Following a similar strategy, the same group has also developed an efficient route to compound **243** bearing the pentacyclic segment present in ecteinascidin 743 (Et-743, **244**). Et-743 has been recently approved and sold as Yondelis for the treatment of advanced soft tissue sarcoma.

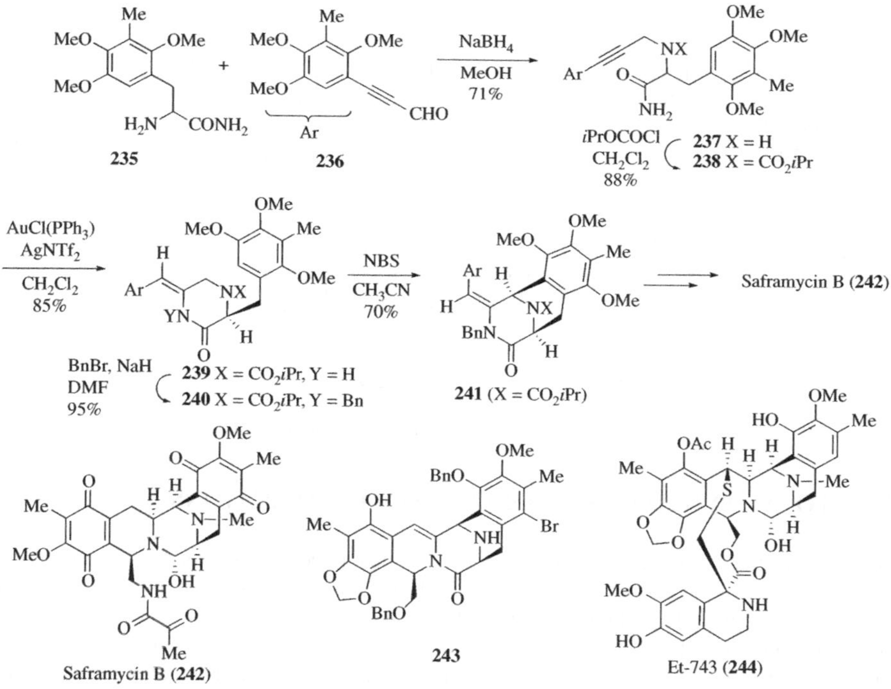

SCHEME 2.36 Synthetic studies toward saframycin B and ecteinascidin 743 (Yondelis).

2.4.3.3 ***Synthesis of Ammosamide B*** Recently, Nagasawa's group has devised a tandem FC-based method for the construction of the tricyclic pyrroloquinoline skeleton [41], which was applied to the synthesis of ammosamide B, an alkaloid isolated from marine *Streptomyces* (Scheme 2.37). The strategy involved applying sequential FC reactions via an indole intermediate **246**, starting from a symmetric tetrasubstituted benzene **245** (prepared from *m*-dichlorobenzene in two steps). A variety of Lewis acids were tested to promote the FC reaction, most of them giving rise to the formation of insoluble oligomers. Only $BF_3{\cdot}Et_2O$ yielded indole **246** (52%) when the reaction was carried out in dichloromethane, while tricyclic **247** was obtained when it was carried out in toluene, albeit in low yield (35%) due to insolubility of **247** during work-up and purification. In fact, compound **248** could be obtained in 46% yield from **245** by excluding purification processes. Hydrogenation of the nitro groups in **248** and trifluoracetylation of the resulting amino groups gave compound **250**. Completion of the synthesis of ammosamide B involved carbamoylation of C4 via a palladium-catalyzed methoxycarbonylation of chloride **252** and chlorination of the resulting product **253** with NCS to give **254**, which after simultaneous removal of the trifluoroacetyl groups and formation of the amide with ammonia in methanol provided ammosamide B (**255**).

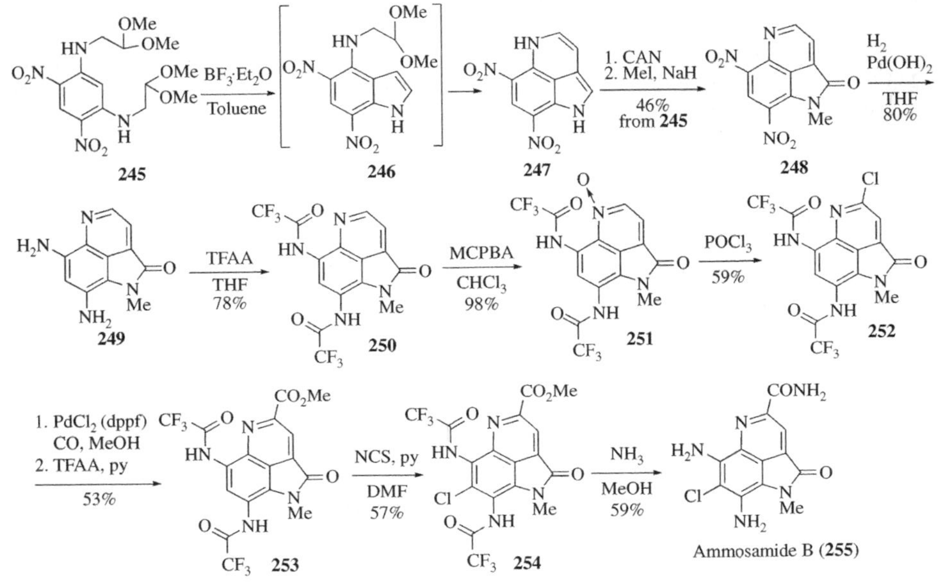

SCHEME 2.37 Synthesis of ammosamide B.

2.5 TOTAL SYNTHESIS INVOLVING *ipso*-FC REACTIONS

Reaction with electrophiles can also take place on substituted positions of the arene ring. In some cases, recovery of the aromaticity from the resulting cationic intermediate involves elimination of the initial substituent to yield a new substituted arene. These reactions are known as electrophilic *ipso*-substitutions and are often observed with silyl and sulfonyl arenes. On the other hand, very electron-rich arenes, such as phenols, can undergo alkylation reactions on alkyl-substituted *ortho* and *para* positions, promoted by either acidic or basic conditions. These reactions are reported in the literature as *ipso*-FC reactions despite the fact that aromaticity is not recovered because of the poor leaving group ability of the alkyl groups. Such processes are used for the formation of cyclohexanedienones bearing an all-carbon quaternary center.

2.5.1 Synthesis of (*S*)-(−)-Xylopinine

The group of García-Ruano has described the use of a sulfinyl group as an *ipso* director in aromatic electrophilic substitution during a concise enantioselective synthesis of (*S*)-xylopinine (**262**), an alkaloid of the protoberberine family isolated from *Xylopia discreta* (Scheme 2.38) [42], which features a tetracyclic ring skeleton. Condensation of the carbanion derived from chiral sulfoxide (*S*)-**256** with sulfinylimine (*S,E*)-**257** gave a 2:1 mixture of tetrahydroisoquinolines **258a** and **258b**, differing only in configuration at sulfur. *N*-desulfinylation of this mixture gave the diastereomeric sulfoxides **259a** and **259b**, which, without separation, were converted into (*S*)-(−)-xylopinine under Pictet–Spengler conditions. This reaction presumably occurs by electrophilic attack of the iminium species on the sulfur-bearing carbon atom of the electron-rich aromatic ring followed by expulsion of the arylsulfinyl moiety. This reaction represents a unique example of participation of an arylsulfinyl group in an electrophilic *ipso* aromatic substitution process.

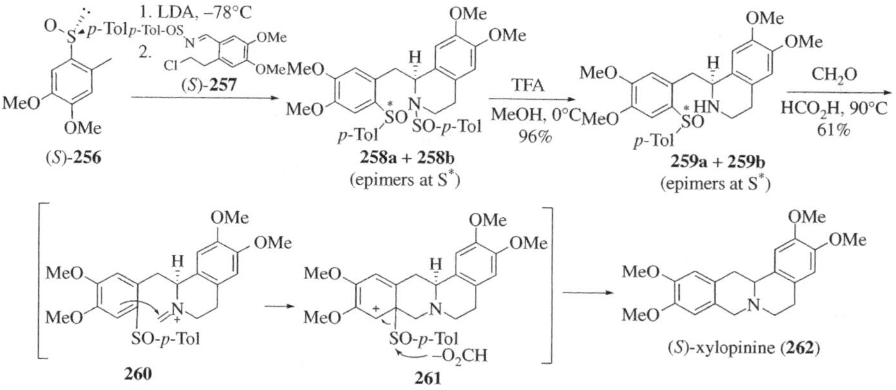

SCHEME 2.38 Synthesis of (*S*)-(−)-xylopinine.

2.5.2 Synthesis of Garcibracteatone

Garcibracteatone (**269**) is a polycyclic polyprenylated acylphloroglucinol isolated from the bark of *Garcinia bracteata*. Its structure features a highly compact polycyclic ring containing seven stereocenters, five of which are quaternary. A recent biomimetic synthesis of garcibracteatone [43] involves an *ipso*-FC alkylation in the construction of one of these quaternary centers (Scheme 2.39). The synthesis started from phloroglucinol (**263**), which after an FC acylation with benzoyl chloride followed by FC alkylation with prenyl bromide (**265**) gave symmetric benzenetriol **266** in 34% yield. Upon treatment with NaH and lavandulyl iodide (**267**), the *ipso*-alkylation produced compound **268** as an inseparable mixture of two diastereomers, with each diastereomer existing as a mixture of two tautomers. The

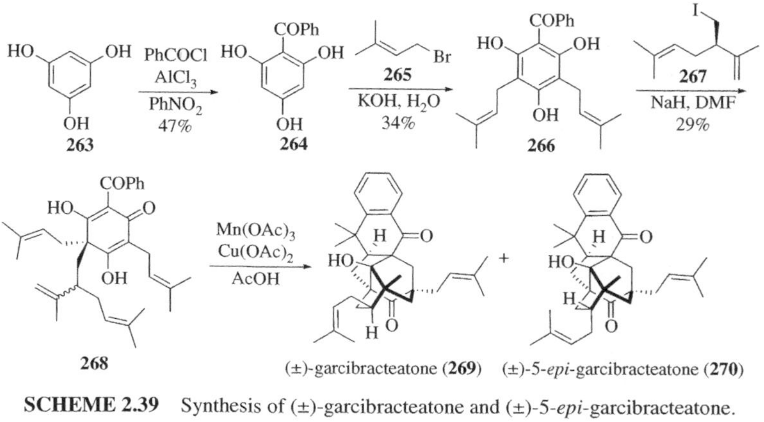

SCHEME 2.39 Synthesis of (±)-garcibracteatone and (±)-5-*epi*-garcibracteatone.

modest yield (29%, 50% based on recovered starting material) reflected the sterically crowded nature of the all-carbon quaternary center formed, as well as the low reactivity of the alkyl iodide. Finally, oxidation of **268** with $Mn(OAc)_3$ and $Cu(OAc)_2$ promoted a radical cascade sequence that furnished a separable mixture of (±)-garcibracteatone (**269**, 14%) and (±)-5-*epi*-garcibracteatone (**270**, 8%).

2.6 SUMMARY AND OUTLOOK

With more than 140 years of history, the FC alkylation has become an essential synthetic tool for the formation of arene–carbon bonds that has been used in countless syntheses of natural products and bioactive compounds with structures ranging from the simple to the very complex. Intermolecular FC reactions have allowed the introduction of simple to highly functionalized alkyl fragments on the aromatic core, while intramolecular FC alkylations have been used to construct the polycyclic skeleton of different synthetic targets, including terpenoids, alkaloids, antibiotics, and others. Most synthetic limitations are related with functional group compatibility under the harsh conditions sometimes required to generate the electrophilic intermediate. New acid catalysts and electrophile precursors that could work under milder conditions would be very desirable. The design and application of new cascade processes involving an FC alkylation step are sought as a methodology to create multiple covalent bonds and molecular complexity in a simple step, thus increasing the synthetic efficiency, one of the goals of the current organic synthesis. Finally, a number of methodologies for asymmetric FC alkylations have already been developed. However, its application in total synthesis has serious limitations in terms of substrate scope, especially with electron-neutral or electron-deficient arenes. Progress in this area is desirable for the synthesis of complex molecular targets.

2.7 ACKNOWLEDGMENT

Financial support from the MINECO (Gobierno de España) and FEDER (European Union) (CTQ2009-13083 and CTQ2013-47494-P) and from Generalitat Valenciana (ISIC2012/001) is gratefully acknowledged. M. M-M thanks the Universitat de València for a predoctoral grant (Atracció de Talent).

ABBREVIATIONS

Ac	Acetyl
Bn	Benzyl
Bz	Benzoyl
CAN	Cerium ammonium nitrate
DABAL-Me_3	1,4-Diazabicyclo[2.2.2]octane-trimethylaluminum adduct
DBU	1,8-Diazabicyclo[5.4.0]undec-7-ene
DEAD	Diethyl azodicarboxylate
DIBAL-H	Diisobutylaluminium hydride
DIPEA	Diisopropylethylamine
DMDO	Dimethyldioxirane
DPPF	1,1′-Bis(diphenylphosphino)ferrocene
dr	Diastereomeric ratio
ee	Enantiomeric excess
HMDS	Hexamethyldisilylamide
IBX	2-Iodoxybenzoic acid

LDA	Lithium diisopropylamide
LiTMP	Lithium tetramethylpiperidide
MCPBA	*m*-Chloroperoxybenzoic acid
MOM	Methoxymethyl
Ms	Methanesulfonyl (Mesyl)
MS	Molecular sieves
NBS	*N*-Bromosuccinimide
NCS	*N*-Chlorosuccinimide
NMO	*N*-Methylmorpholine *N*-oxide
Np	Naphthyl
Ns	*p*-Nitrophenylsulfonyl (Nosyl)
PDC	Pyridinium dichromate
py	Pyridine
TADOOH	5-[(Hydroperoxydiphenyl)methyl]-2,2-dimethyl-1,3-dioxolan-4-yl] diphenylmethanol)
TBAF	Tetrabutylammonium fluoride
TBHP	*tert*-Butylhydroperoxide
TES	Triethylsilyl
Tf	Trifluoromethanesulfonyl
TFA	Trifluoroacetic acid
TFAA	Trifluoroacetic anhydride
TMS	Trimethylsilyl
$TrisN_3$	2,4,6-Triisopropylbenzenesulfonyl azide

REFERENCES

[1] (a) Min, J.-H., Lee, J.-S., Yang, J.-D., and Koo, S. (2003) *J. Org. Chem.*, **68**, 7925–7927; (b) Oh, E.-T., Kim, H. J., Oh, J. T., Su, L., Yun, I., Nam, K., Min, J.-H., Kim, J. W., and Koo, S. (2012) *Eur. J. Org. Chem.*, 4954–4962.

[2] Patil, M. L., Borate, H. B., Ponde, D. E., Bhawal, B. M., and Deshpande, V. H. (1999) *Tetrahedron Lett.*, **40**, 4437–4438.

[3] During a synthesis of the cytotoxin dynemicin A, a related FC alkylation was carried out using AgOTf, because of the sensitivity of the reaction substrates to strong Lewis acids: Taunton, J., Wood, J. L., and Schreiber, S. L. (1993) *J. Am. Chem. Soc.*, **115**, 10378–10379.

[4] Stadler, D. and Bach, T. (2008) *Angew. Chem. Int. Ed.*, **47**, 7557–7559.

[5] Alvarez-Manzaneda, E., Chahbou, R., Cabrera, E., Alvarez, E., Haidour, A., Ramos, J. M., Alvarez-Manzaneda, R., Tapia, R., Es-Samti, H., Fernández, A., and Barranco, I. (2009) *Eur. J. Org. Chem.*, 1139–1143.

[6] Esumi, T., Hojyo, D., Zhai, H., and Fukuyama, Y. (2006) *Tetrahedron Lett.*, **47**, 3979–3983.

[7] Bianchi, D. and Kaufman, T. S. (2000) *Can. J. Chem.*, **78**, 1165–1169.

[8] (a) Malhotra, R., Ghosh, A. Ghosh, R., Chakrabarti, S., Dutta, S., Dey, T. K., Roy, S., Basu, S., and Hajra, S. (2011) *Tetrahedron: Asymmetry*, **22**, 1522–1529; (b) Malhotra, R., Ghosh, A., Dutta, S., Dey, T. K., Basu, S., and Hajra S. (2014) *RSC Adv.*, **4**, 48344–48347; (c) Malhotra, R., Chakrabarti, S., Dey, T. K., Dutta, S., Alapati, K. B., Dutta, S., Roy, S., Basu, S., and Hajra, S. (2013) *Tetrahedron*, **69**, 8994–9000.

[9] Hajra, S. and Bar, S. (2011) *Tetrahedron: Asymmetry*, **22**, 775–779.

[10] Ma, Z., Hu, H., Xiong, W., and Zhai, H. (2007) *Tetrahedron*, **63**, 7523–7531.

[11] Medjahdi, M., Gonzalez-Gomez, J. C., Foubelo, F., and Yus, M. (2011) *Eur. J. Or. Chem.*, 2230–2234.

[12] Chuang, K. V., Navarro, R., and Reisman, S. E. (2011) *Angew. Chem. Int. Ed.*, **50**, 9447–9451.

[13] Sun, B.-F., Hong, R., Kang, Y.-B., and Deng, L. (2009) *J. Am. Chem. Soc.*, **131**, 10384–10385.

[14] Das, J., Koswatta, P. B., Jones, J. D., Yousufuddin, M., and Lovely, C. J. (2012) *Org. Lett.*, **14**, 6210–6213.

[15] Zhong, C., Zhu, J., Chang, J., and Sun, X. (2011) *Tetrahedron Lett.*, **52**, 2815–2817.

[16] Taj, R. A. and Green, J. R. (2010) *J. Org. Chem.*, **75**, 8258–8270.

[17] Chen, D. Y. K. Kang, K., and Wu, R. T. (2010) *Molecules*, **15**, 5909–5927.

[18] Wang, G.-W., Wang, H.-L., Capretto, D. A., Han, Q., Hu, R.-B., and Yang, S.-D. (2012) *Tetrahedron*, **68**, 5216–5222.

[19] Yadav, J. S., Gayathri, K. U., Subba Redy, B. V., and Prasad, A. R. (2009) *Synlett*, 43–46.

[20] Kedrowski, B. L. and Hoppe, R. W. (2008) *J. Org. Chem.*, **73**, 5177–5179.

[21] Yuan, Y., Men, H., and Lee, C. (2004) *J. Am. Chem. Soc.*, **126**, 14720–14721.

[22] Shingare, R. D., Velayudham, R., Gawade, J. R., and Reddy, D. S. (2013) *Org. Lett.*, **15**, 4556–4569.

[23] Ueno, A., Kitawaki, T., and Chida, N. (2008) *Org. Lett.*, **10**, 1999–2002.

[24] Iosno, N. and Mori, M. (1995) *J. Org. Chem.*, **60**, 115–119.

[25] Hameed, A., Blake, A. J., Hayes, C. J. (2008) *J. Org. Chem.*, **73**, 8045–8048.

[26] Tang, S., Xu Y., He, J., He, Y., Zheng, J., Pan, X., and She, X. (2008) *Org. Lett.*, **10**, 1855–1858.

[27] Yoshimitsu, T., Nojima, S., Hashimoto, M., Tsukamoto, K., and Tanaka, T. (2009) *Synthesis*, **17**, 2963–2969.

[28] Zhang, Z.-W. and Li, W.-D. Z. (2010) *Org. Lett.*, **12**, 1649–1651.

[29] Fenster, E., Fehl, C., and Aubé, J. (2011) *Org. Lett.*, **13**, 2614–2617.

[30] Zhou, Z.-W., Li, W.-C., Hu, Y., Wang, B., Ren, G., and Feng, L.-H. (2013) *Res. Chem. Intermed.*, **39**, 3049–3054.

[31] (a) Seo, P.-J., Choi, H.-D., and Son, B.-W. (2004) *Arch. Pharm. Res.*, **27**, 1189–1193; (b) Choi, H.-D., Seo, P.-J., and Son, B.-W. (2002) *Arch. Pharm. Res.*, **25**, 786–879; (c) Choi, H.-D., Ha, M.-C., Seo, P.-J., Son, B.-W., and Song, J.-C. (2000) *Arch. Pharm. Res.*, **23**, 438–440.

[32] van Lingen, H. L., Zhuang, W., Hansen, T., Rutjes, F. P. J. T., and Jørgensen, K. A. (2003) *Org. Biomol. Chem.*, **1**, 1953–1958.

[33] Wang, X.-S., Zheng, C.-W., Zhao, S.-L., Chai, Z., Zhao, G., and Yang, G.-S. (2008) *Tetrahedron Asymmetry*, **19**, 2699–2704.

[34] Hong, L., Wang, L., Sun, W., Wong, K., and Wang, R. (2009) *J. Org. Chem.*, **74**, 6881–6884.

[35] Jiang, X., Wu, L., Xing, Y., Wang, L., Wang, S., Chen, Z., and Wang, R. (2012) *Chem. Commun.*, **48**, 446–448.

[36] Wang, J.-Y., Zhang, H., Liao, Y.-H., Yuan, W.-C., Feng, Y.-J., and Zhang, X.-M. (2012) *Synlett*, **23**, 796–800.

[37] Snyder, S. A., Wright, N. E., Pflueger, J. J., and Breazzano S. P. (2011) *Angew. Chem. Int. Ed.*, **50**, 8629–8633.

[38] (a) Yadav, J. S., Reddy, B. V. S., and Padmavani, B. (2004) *Synthesis*, 405–408; (b) Yadav, J. S., Reddy, B. V. S., Parimala, G., and Raju, A. K. (2004) *Tetrahedron Lett.*, **45**, 1543–1546.

[39] (a) Obika, S., Yasui, Y., Yanada, R., and Takemoto, Y. (2008) *J. Org. Chem.*, **73**, 5206–5209; (b) Enomoto, T., Yasui, Y., and Takemoto, Y. (2010) *J. Org. Chem.*, **75**, 4876–4879.

[40] Kubo, A., Saito, N., Yamato, H., Masubuchi, K., Nakamura, M. J. (1988) *J. Org. Chem.*, **53**, 4295–4310.

[41] Takayama, Y., Yamada, T., Tatekabe, S., and Nagasawa, K. (2013) *Chem. Commun.*, **49**, 6519–6521.

[42] Mastranzo, V. M., Yuste, F., Ortiz, B., Sánchez-Obregón, R., Toscano, R. A., and García Ruano, J. L. (2011) *J. Org. Chem.*, **76**, 5036–5041.

[43] Pepper, H. P., Lam, H. C., Bloch, W. M., George, J. H. (2012) *Org. Lett.*, **14**, 5162–5164.

3

CATALYTIC FRIEDEL–CRAFTS ACYLATION REACTIONS

Giovanni Sartori, Raimondo Maggi, and Veronica Santacroce
"Clean Synthetic Methodology Group", Dipartimento di Chimica dell'Università, Università degli Studi di Parma, Parma, Italy

3.1 INTRODUCTION AND HISTORICAL BACKGROUND

The Friedel–Crafts acylation reaction generates aromatic ketone products from the combination of an aromatic substrate with an acyl component typically in the presence of a catalyst [1]. Interest in this electrophilic process and the optimization of preparative methods is generated by the considerable practical value of the aromatic ketone products. In fact, these compounds constitute fundamental intermediates or target products in the pharmaceutical, fragrance, flavor, dye, and agrochemical industries [2–4].

The Friedel–Crafts acylation reaction can be performed without any catalyst, but very harsh conditions are necessary [5]. Conventionally, and more efficiently, this electrophilic acylation using acyl chlorides or anhydrides is catalyzed by Lewis acids (such as $ZnCl_2$, $AlCl_3$, $FeCl_3$, $SnCl_4$, and $TiCl_4$); when carboxylic acids are directly utilized as acylating reagents, strong protic acid catalysts (such as H_2SO_4 and HF) are needed. The simplified mechanism of the reaction involving an acyl chloride and $AlCl_3$ is depicted in Scheme 3.1.

Depending on the strength of the Lewis acid and on the structure of the R^1 substituent, the actual acylating agent may be the acid–base complex **1** or the acyl cation **2**. The active reagents **1** or **2** or even a mixture of both undergo the electrophilic aromatic substitution affording the final ketone product **3** [1].

Friedel–Crafts acylation follows the established "activation–orientation rules" of electrophilic aromatic substitution. However, the acylation of some highly activated aromatic compounds such as phenols and aromatic amines preferentially occurs at the substituent heteroatom, affording esters or amides instead of the more valuable aromatic ketones. These *ortho*-hydroxy- or

Arene Chemistry: Reaction Mechanisms and Methods for Aromatic Compounds, First Edition. Edited by Jacques Mortier.

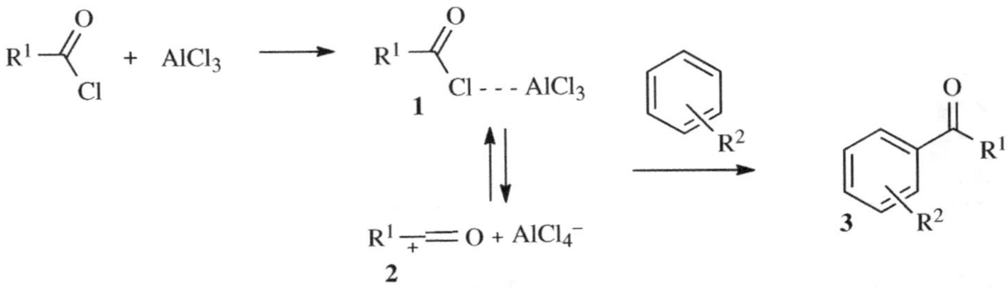

SCHEME 3.1 $AlCl_3$-promoted mechanism of the Friedel–Crafts acylation reaction.

ortho-amino-ketones can instead be produced using alternate procedures (see Section 3.4 of this chapter and Chapter 31).

To minimize waste production and lower the economical impact of a given synthetic process, replacement of homogeneous Lewis or protic acid catalysts with potentially recyclable heterogeneous solid acid catalysts is an attractive proposition. Additionally, the direct use of carboxylic acids as the acylating agents in these heterogeneous acid-catalyzed systems generates water as the only stoichiometric by-product. A further advantage connected with this strategy is the possible application of solid catalysts in continuous flow reactors that can improve the reaction selectivity and ease of product purification.

This chapter focuses on the advantages achieved with different procedures and makes comparative analyses of the synthetic results. Furthermore, despite utility at small scale, the use of Lewis acid/acyl chlorides reagents on an industrial scale is frequently discouraged due to the large amounts of waste produced. Practically, the use of homogeneous Lewis acids in combination with acyl chlorides or anhydrides poses a serious problem for the efficient recovery and disposal of metal oxides and protic inorganic acid by-products.

The remaining discussion is organized in three sections dealing, respectively, with the use of homogeneous catalysts, heterogeneous catalysts, and direct phenol acylation. Each section contains various subtopics based on the nature and role of each class of catalysts.

3.2 CATALYTIC HOMOGENEOUS ACYLATIONS

The use of metal halides (e.g., $AlCl_3$) as catalysts in the acylation of aromatic compounds can sometimes cause problems due to the strong complex formed between the ketone product and the metal halide itself. This can necessitate the use of more than stoichiometric amounts of catalyst, and the work-up commonly requires hydrolysis of the complex, leading to the loss of the catalyst and giving large amounts of corrosive waste streams. For these reasons, during the past decade, the development of more environmentally friendly Friedel–Crafts acylation reactions has become a goal of the general "green revolution" that has spread in all fields of synthetic chemistry [6]. A key objective of this work is the identification of new and increasingly efficient catalysts [7]. Great efforts have been made by different research groups to achieve the goal of making the Lewis acid role catalytic in Friedel–Crafts acylation [8].

3.2.1 Metal Halides

It is known that the benzoylation of 2-methoxynaphthalene (**4**) with an equimolar amount of benzoyl chloride in the presence of aluminum chloride (5% mol) gives products **5**, **6**, and **7** in 51% yield and 63:25:12 ratio (Scheme 3.2) [9].

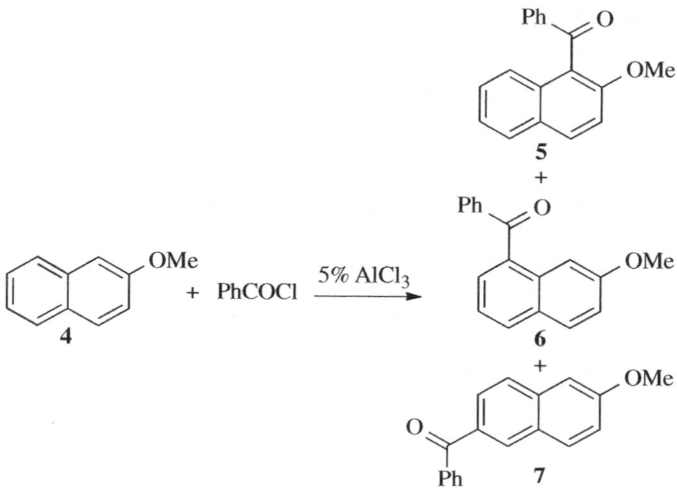

SCHEME 3.2 $AlCl_3$-promoted benzoylation of 2-methoxynaphthalene.

The regiochemistry of the reaction is strongly influenced by the identity of the Lewis acid catalyst employed as well as by the reaction temperature. For example, with indium chloride in place of aluminum chloride, compound **5** is obtained in higher selectivity (total yield 58%, **5**:**6**:**7** = 88:9:3). On increasing the temperature in the reaction with indium chloride, the product composition is changed and compound **6** is the predominant one (total yield 57%, **5**:**6**:**7** = 0:84:16). Quite similar results are achieved with iron trichloride, tin tetrachloride, and zinc chloride, while in the case of antimony pentachloride and titanium tetrachloride, compound **5** is the major product.

Hard Lewis acid chloroaluminate ionic liquids show catalytic activity in the Friedel–Crafts acylation reaction; however, they suffer from the same issues as aluminum chloride [10]. Of particular interest of these reactions is the replacing of aluminum chloride by indium trichloride to form chloroindate(III) ionic liquids [11]. The advantage of using indium trichloride compared with aluminum chloride is represented by its hydrolytic stability and reduced oxophilicity. Chloroindate(III) ionic liquids are synthesized by mixing 1-butyl-3-methylimidazolium chloride with anhydrous indium trichloride at 80°C [12]. In the benzoylation of anisole with benzoic anhydride at 80°C, a 79% yield of 4-methoxybenzophenone is achieved when indium trichloride (5% mol) is used in conjunction with ionic liquids, either as a solution in 1-butyl-3-methylimidazolium bis(trifluoromethylsulfonyl) imide or as chloroindate(III) binary ionic liquid, for 3 h (*ortho*:*para* = 6:94). The chloroindate(III) catalyst can be recycled five times with only a loss in activity from 79 to 62% yield, and no change in selectivity. Benzene, isobutylbenzene, toluene, and naphthalene give good yields with benzoyl chloride (81–96%), whereas with benzoic anhydride, lower yields are obtained (22–87%). Worth noting is the fact that naphthalene gives the less favored two-substituted product in the reaction with the more bulky benzoic anhydride with respect to the benzoyl chloride [10].

Catalytic Friedel–Crafts acylation can be greatly improved by using a combination of Lewis acid with silver or lithium perchlorates. Good results can be achieved by using a combination of antimony pentachloride (4% mol) and lithium perchlorate (100% mol) in refluxing methylene chloride [13]. An interesting observation for this particular system is that anhydrides give better yields than the corresponding chlorides. For example, anisole reacts with valeric anhydride to give *para*-methoxyphenyl butyl ketone in 85% yield, while 11% yield of the same product is obtained in the reaction with valeric chloride under the same reaction conditions.

3.2.2 Perfluoroalkanoic Acids, Perfluorosulfonic Acids, and Their (Metal) Derivatives

Metal triflates that can be easily prepared from metal halides and triflic acid at −78°C [14] show several unique properties compared with the corresponding metal halides. The use of bismuth(III) triflate allows for acylation of both activated and deactivated aromatic compounds with anhydrides and acyl chlorides [15]. Thus, the acylation of deactivated aromatics such as trifluoromethoxybenzene, fluorobenzene, and chlorobenzene can be achieved in high yields with benzoyl chloride in the presence of bismuth(III) triflate (10% mol) without solvent. The *para*-acylation product is the most abundant in all cases (trifluoromethoxybenzene: 87% yield, *ortho*:*para* 4:96; fluorobenzene: 86% yield, *ortho*:*para* 0:100; chlorobenzene: 89% yield, *ortho*:*para* 13:87).

Gallium nonafluorobutanesulfonate (nonaflate) [Ga(ONf)3] (5% mol) shows a very high activity in the acylation of aromatic compounds with acyl chlorides without solvent [16]. Benzene and deactivated benzenes such as fluorobenzene, chlorobenzene, and even dichlorobenzene react smoothly to afford the corresponding aromatic ketones in 71–98% yield (Scheme 3.3).

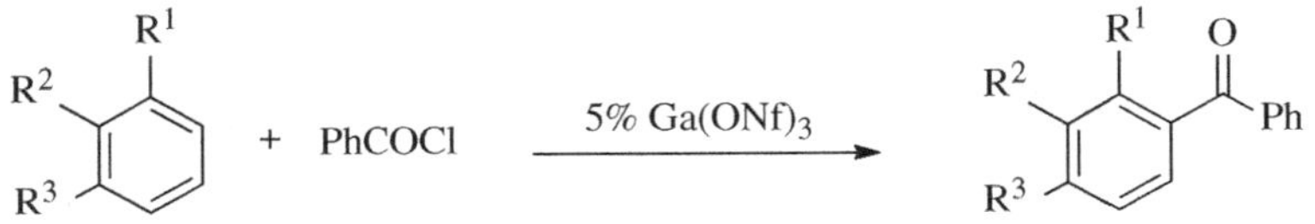

8 **9**

R^1 = H, Me; R^2 = H, Cl; R^3 = H, Me, F, Cl

SCHEME 3.3 Benzoylation of aromatics promoted by gallium nonaflate.

Mixed anhydrides derived from acyl chlorides and triflic acid (Scheme 3.4) can undergo efficient electrophilic acylation reactions [17].

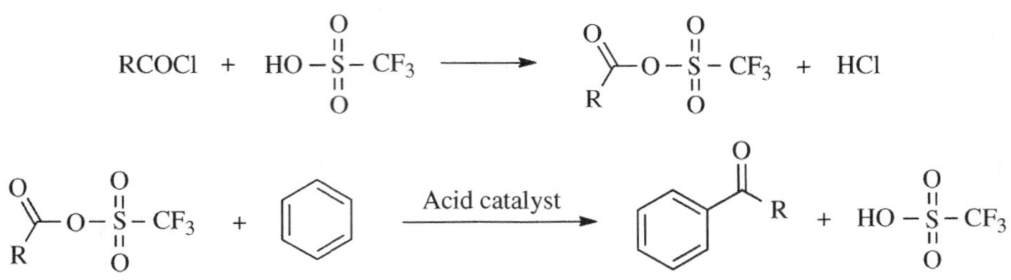

R = alkyl, aryl

SCHEME 3.4 Preparation and use of mixed anhydrides.

This acylation method is advantageous in terms of the mild conditions employed and the easy availability of acyl chlorides. While in the benzoylation of benzene with benzoyl chloride in the presence of hafnium triflate (5% mol) or triflic acid (5% mol) only 5–10% yields in benzophenone are obtained, with a combination of Lewis and protic acids (5% mol hafnium triflate and 5% mol triflic acid) benzophenone is isolated in 77% yield [18]. The yield is improved to 82% when hafnium triflate and triflic acid are used in higher quantities (10% mol each); a further increase of the catalyst amount does not improve the yield. Aromatic substrates such as benzene, toluene, chlorobenzene, and fluorobenzene react smoothly under these conditions with both aliphatic and aromatic acid chlorides, affording the corresponding *para*-arylketones **10** in 60–83% yield and 15:85–1:99 isomer distribution (Scheme 3.5).

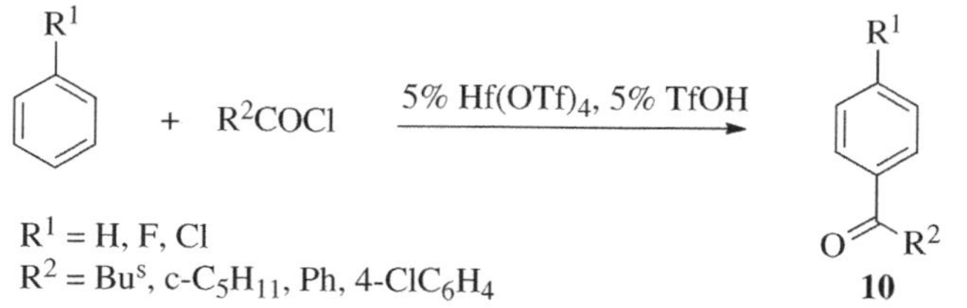

SCHEME 3.5 Functionalization of deactivated aromatic substrates promoted by hafnium triflate and triflic acid.

More efficiently, the benzoylation of activated and deactivated aromatic compounds is achieved by using bismuth triflate (10% mol) and carboxylic acids as acylating agents in the presence of trifluoroacetic anhydride or heptafluorobutyric anhydride [19]. *meta*-Xylene undergoes acetylation with 100% yield by using acetic acid in the presence of trifluoroacetic anhydride under bismuth or scandium triflate catalysis (10% mol). The bismuth triflate-catalyzed reaction can be extended to different aromatics as well as to aliphatic and aromatic carboxylic acids, giving ketones in nearly quantitative yield. Benzene and chlorobenzene are benzoylated with a benzoic acid/heptafluorobutyric anhydride mixture in the presence of bismuth triflate (10% mol), giving the corresponding benzophenones in 90 and 65% yield (*ortho*:*para* 6:94), respectively. The catalyst can be recovered by precipitation with hexane from the final reaction mixture and reused in the model benzoylation of *meta*-xylene, giving the product with similar yield (~95%) for four successive runs.

Ionic liquids can provide an ideal medium for reactions that involve reactive ionic species due to their ability to stabilize charged intermediates such as acyl cations [20]. The Friedel–Crafts benzoylation of aromatic compounds with benzoyl chloride to obtain the corresponding benzophenone derivatives using bismuth triflate in the presence of 1-butyl-3-methylimidazolium trifluoromethanesulfonate is carried out under microwave activation [21]. This catalytic system promotes the reaction in good yields and short reaction times. Under microwave irradiation, the reaction at 120°C for 30 min provides a 98% conversion of anisole into 4-methoxybenzophenone (96% yield). No reaction is observed in the absence of bismuth triflate, and only 77% yield of the product is obtained with bismuth triflate alone. Benzoylation of electron-rich arenes, including naphthalene, proceeds efficiently, and the corresponding products are obtained in good yields (70–95%). In addition, the catalytic system bismuth triflate/1-butyl-3-methylimidazolium trifluoromethanesulfonate is stable to moisture and can be easily recovered and reused without any significant loss of the catalytic activity.

Ionic liquids in combination with supercritical fluids are a versatile tool for the immobilization and recycling of homogeneous catalysts, allowing continuous Friedel–Crafts acylation reactions to be realized. The acylation of anisole with acetic anhydride is carried out in a flow system using a metal triflate immobilized in the ionic liquid 1-butyl-4-methylpyridinium bis(trifluoromethylsulfonyl)imide as catalyst and $scCO_2$ as continuous extraction phase [22]. Different metal triflates are utilized under continuous flow conditions using high pressure: yttrium triflate possesses the best balance between sufficient acidity for catalytic activity and softness to release the product and so permits a good catalyst reuse (TONs up to 190).

3.2.3 Miscellaneous

Iodine (2% mol) can be used as a catalyst for the acetylation of electron-rich aromatic compounds with aliphatic and aromatic acyl chlorides or anhydrides in 25–93% yields [8]. In successful acylations, the violet-colored refluxing mixture disappears after 15–30 min. Heterocyclic compounds such as furan and thiophene derivatives undergo easy acylation in the presence of variable amounts of iodine. The process is of particular synthetic interest since these heterocycle compounds are

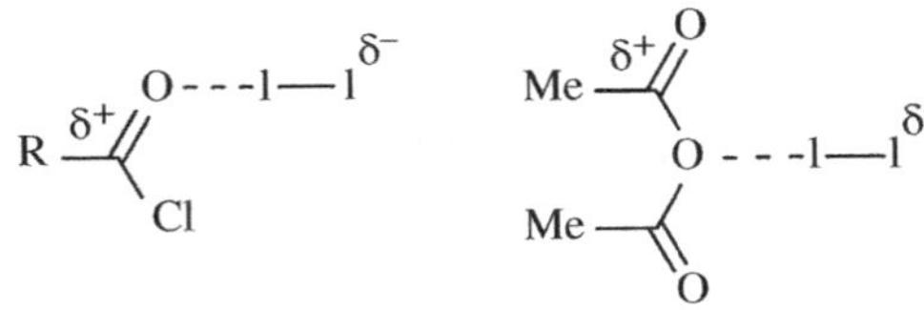

FIGURE 3.1 Iodine-promoted activated intermediates for Friedel–Crafts acylation reactions.

particularly sensitive to acid treatment and give rise to telomerization in classical Friedel–Crafts acylations. Catalysis by iodine is due to the possibility of iodine to act as an electron acceptor and form polarized complexes, giving activated intermediates of the type reported in Figure 3.1 [23].

3.3 CATALYTIC HETEROGENEOUS ACYLATIONS

Even though the catalytic systems described in the preceding sections provide greater practicality than stoichiometric systems, their application in industrial synthesis is not widespread. However, in more recent times, maximum effort has been directed toward the use of solid acid catalysts. In fact, heterogeneous catalysts can be easily separated from the reaction mixture and reused; they are generally not corrosive and do not produce problematic side products. Different classes of materials have been studied and utilized as heterogeneous catalysts for Friedel–Crafts acylations: these include zeolites, (acid-treated) metal oxides, and heteropoly acids (HPAs) already utilized in the hydrocarbon reactions [24]. Moreover, the application of clays, perfluorinated resin-sulfonic acids, and supported (fluoro) sulfonic acids, mainly exploited in the production of fine chemicals, is the subject of intensive studies in this area.

3.3.1 Zeolites

Zeolite-based molecular sieves are crystalline aluminosilicates with a two- or three-dimensional framework consisting of SiO_4 and AlO_4 tetrahedra that are connected by bridging oxygen atoms [25]. Because of the lower valence of aluminum compared to silicon, the number of AlO_4 tetrahedra controls the negative charge on the zeolite framework. This charge is compensated by cations or protons. The protons represent the Brønsted acid sites and participate in the acid-catalyzed transformations of organic molecules. Moreover, dehydroxylation of bridging hydroxyl groups leads to the formation of unsaturated aluminum species that act as Lewis acid sites. Surface acidity, as well as hydrophobicity of zeolites, is usually related to their composition, in particular the silica/alumina ratio (SAR). In the present description, this crucial parameter will be reported between brackets at the end of the zeolite-type code [es: HBEA(26)] [26].

Crystalline structures of the zeolite containing tetrahedral silicon, aluminum, and phosphorus, as well as transition metals and many group elements with the valence ranging from I to V, have also been synthesized with the generic name of zeotypes, including $AlPO_4$-, SAPO-, MeAPO-, and MeAPSO-type molecular sieves [27, 28]. Zeolites are conventionally defined as ultralarge (>12-), large (12-), medium (10-), or small (8-membered ring) pore materials corresponding to a channel diameter between 20 and 5 Å (Table 3.1).

The hydrothermally prepared HBEA zeolites, using amorphous silica as the silicon source, give a very selective synthesis of 4-methylacetophenone **11** in the acylation of toluene with acetic anhydride. In particular, it is shown that HBEA(32) gives product **11** with 80% yield and 98% selectivity (Scheme 3.6) [31].

The effect of the concentration of the acid sites on the catalyst surface has been evaluated in the acylation of toluene with isobutyryl chloride. By decreasing the concentration of acid sites

TABLE 3.1 Pore Type and Dimension of the Zeolites Utilized

Zeolite	Pore Type	Dimension
Beta (HBEA) [29]	Interconnected channels	7.6×6.4Å channel 5.5×5.5Å channel
Y (HFAU) [29]	Interconnected spheres	7.4Å diameter pore 11.8Å diameter cavity
ZSM-5 (HMFI) [29]	Interconnected channels	5.5×5.6Å channel 5.1×5.5Å channel
Mordenite (HMOR) [29]	Interconnected channels	6.5×7.0Å channel 2.6×5.7Å channel
Nu-10 (HTON) [30]	One-dimensional channels	5.5×4.5Å channel
X (HFAU) [29]	Interconnected spheres	7.4Å diameter pore 11.8Å diameter cavity

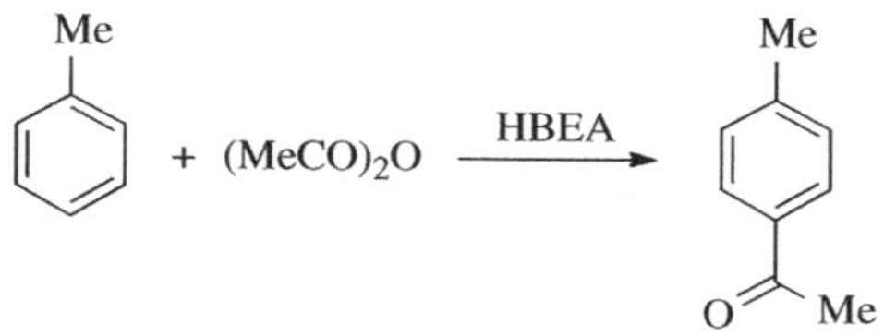

11

SCHEME 3.6 HBEA-promoted toluene acetylation with acetic anhydride.

[HBEA(75) and HBEA(140) series], only a slight decrease of isobutyryl chloride conversion is observed (56 and 53%, respectively). However, the higher turnover frequency (TOF) observed with HBEA(140) (3.11 min^{-1}) underscores the importance of the hydrophobicity of zeolites in the reaction between polar and nonpolar compounds. The selectivity toward the isopropyl *para*-tolylketone for HBEA(25), HBEA(75), and HBEA(140) is 75, 78, and 80%, respectively [32].

A further interesting application of HBEA zeolite is illustrated by the direct benzoylation of arenes with benzoic acids. The synthetic method is the subject of a patent and shows very interesting properties from a green chemistry point of view [33]. As an example, a mixture of toluene, *para*-chlorobenzoic acid, and a HBEA previously calcined at 400°C (150 mg/mmol) pressurized with nitrogen to 2×10^5 Pa, heated to 200°C, and stirred for 4 h gives 4-chloro-4′-methylbenzophenone in 84% yield.

An important example of the application of these solid catalysts is represented by the comparison of the activities of HBEA and HBEA-supported In_2O_3 (In_2O_3-BEA) (20 wt % loading) in the acylation of benzene with benzoyl chloride [34]. In_2O_3-BEA shows higher activity (80% benzoyl chloride conversion in 1.5 h vs. 50% benzoyl chloride conversion with HBEA in 3 h). This catalytic activity enhancement can be explained by considering the activation of benzoyl chloride by both protonic acid and redox sites present in In_2O_3-BEA. In fact, rare earth metal oxide species are known for their redox properties, which are expected to play a significant role in activating the benzoyl chloride according to the pathway depicted in Scheme 3.7.

$$
\begin{aligned}
m\,\mathrm{PhCOCl} + \mathrm{M}^{n+} &\longrightarrow m\,\mathrm{PhCOCl}^{+\cdot} + \mathrm{M}^{(n-m)+}\\
m\,\mathrm{PhCOCl}^{+\cdot} &\longrightarrow m\,\mathrm{PhCO}^{+} + m\,\mathrm{Cl}^{\cdot}\\
m\,\mathrm{Cl}^{\cdot} + \mathrm{M}^{(n-m)+} &\longrightarrow \mathrm{M}^{n+} + m\,\mathrm{Cl}^{-}\\
\mathrm{PhCO}^{+} + \mathrm{ArH} &\longrightarrow \mathrm{PhCOAr} + \mathrm{H}^{+}\\
\mathrm{H}^{+} + \mathrm{Cl}^{-} &\longrightarrow \mathrm{HCl}
\end{aligned}
$$

SCHEME 3.7 Redox process for the production of the benzoyl cation.

Thus, the benzoyl cation produced according to this redox process undergoes the acylation reaction.

A quite interesting behavior was observed by evaluating the activity of CeY in the acylation of toluene with linear carboxylic acids of increasing chain length (from acetic [C_2] to behenic acid [C_{22}]) [35]. The reaction shows an extraordinary high *para* shape-selectivity effect, and in all cases, the yield of *para* isomer is at least 94%. For the different linear chain carboxylic acids, the yield increases linearly with the number of carbon atoms, according to the increasing lipophilicity of the carboxylic acid that can strongly interact with the apolar porous network of this zeolite. Thus, for example, butyric acid (C_4) gives the acylated products in 20% yield, whereas lauric acid (C_{12}) affords the product in 96% yield.

The acylation of naphthalene **12** with benzoyl chloride and acetic anhydride is carried out in the presence of different zeolites, including HBEA, in comparison with $AlCl_3$ (Scheme 3.8) [36–38].

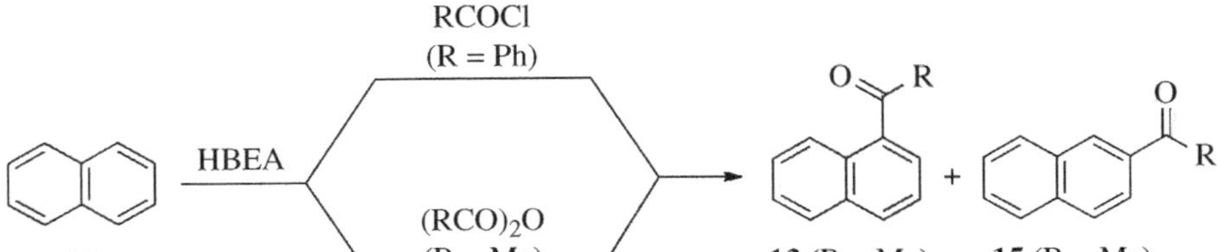

12

13 (R = Me)
14 (R = Ph)

15 (R = Me)
16 (R = Ph)

SCHEME 3.8 HBEA-promoted naphthalene acylation.

HBEA is more efficient and selective with respect to the production of 2-benzoylnaphthalene (**16**) [**16**:**14**=81:19], in comparison with non-shape-selective $AlCl_3$, which preferentially affords compound **14** [**16**:**14**=18:92].

Significant differences in the acetylation of naphthalene with acetic anhydride (2:1 molar ratio) over HBEA are observed with decalin or sulfolane as solvents; the different hydrophilicities of these solvents dramatically influence the resulting naphthalene conversion. The hydrophilic sulfolane interacts more strongly with the zeolite surface, thus blocking the acid sites that are less available for the acylation reaction (naphthalene conversion=14%); on the contrary, the hydrophobic decalin enables the adsorption of acetic anhydride and increases the rate of acylation reaction (naphthalene conversion=25%). Due to the defined structure of the HBEA, the selective formation of isomer **15** is probably achieved via a restricted transition state selectivity (**15**:**13**=81:19 at 35% naphthalene conversion). It must be underlined that different secondary products are, in general, produced on the catalyst surface due to consecutive reaction of the products.

Fairly good results are achieved in the acylation of 2-methylnaphthalene with butyric anhydride in the presence of HBEA(47) [38]. Under optimized reaction conditions, the 2-methyl-6-butyrylnaphthalene and 2-methyl-7-butyrylnaphthalene are isolated in 78% yield and in 58:42 isomer distribution.

The Friedel–Crafts acylation of aromatic ethers has attracted considerable interest in organic synthesis and in industrial chemistry because of the widespread application of the corresponding ketones as valuable intermediates in fine chemistry [39–43]. An example is the selective acetylation of 2-methoxynaphthalene at the carbon in position 6 owing to the great interest in 2-methoxy-6-acetylnaphthalene, an intermediate in the preparation of the anti-inflammatory drug (*S*)-Naproxen [44, 45].

Synthetically valuable methods for the efficient acylation of aryl ethers with anhydrides over different HBEA zeolites under batch conditions are also reported [43–46]. The acetylation of anisole can be performed without solvent, and the catalyst can be reused with unchanged activity for three further cycles after recovery and regenerated by heating in air at 550°C for 3 h [43]. The catalyst shows a partial deactivation [47–49] due, to a large extent, to the inhibition effect of the product. However, when the reaction is performed in a fixed-bed reactor, the catalyst deactivation is much slower, particularly when an anisole-rich mixture is used, with the *para*-product being obtained in high selectivity (>98%). The conclusion is that the use of an excess of anisole enhances the catalyst stability and limits both retention of product and formation of polyacetylated by-products, which deactivate the catalyst even if, as previously underlined, the reaction product continued to be the major poison of the catalyst [39].

An efficient method for the continuous flow selective acylation of aromatics, including ethers, over HBEA zeolite is reported [41, 42]. The process is performed in a fixed-bed tubular reactor charged with HBEA(25) zeolite deposited on a small layer of glass pellets or directly packed without any support or mixed with alumina. The aromatic substrate and acetic anhydride are passed through the reactor heated at 90–150°C for a period of 6–7 h. Substituted acetophenones are so obtained in 63–100% yield.

An interesting method is reported in which catalytically active and selective HBEA coatings can be prepared on ceramic monoliths constituted either of pure silica or of cordierite ($2Al_2O_3{\cdot}5SiO_2{\cdot}2MgO$) and on metallic wire gauze packings [50]. The average loading values of the HBEA coatings are the following: silica monolith, 6.6 wt %; cordierite monolith, 5.4 wt %; and silica precoated wire gauze packing, 5.6 wt %. The activity of the coated structures was measured (batch-wise) in the acylation of anisole utilized as solvent–reagent with linear carboxylic acids (Scheme 3.9).

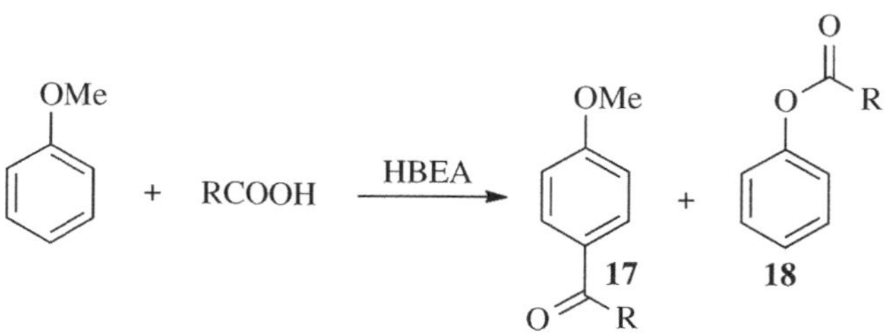

R = Me, Et, Pr, Bu, C_5H_{11}, C_6H_{13}, C_7H_{15}

SCHEME 3.9 HBEA-promoted anisole acylation.

For comparison, experiments were also performed under continuous flow conditions using octanoic acid: the liquid reaction mixture is passed through the monoliths using a nitrogen flow. The silica monolith shows the highest activity ($k = 0.05\ l/g_{cat}{\cdot}h$), which was only 10% lower than that of HBEA particles in slurry. The activities of the coated cordierite monolith and the metal wire gauze packing are much lower than that of the HBEA particles in slurry. The selectivity is maintained as all monoliths give a similar selectivity of about 77–85% toward the main *para* isomer **17** ($R = C_7H_{15}$) compared to the 84% selectivity of the original HBEA tested in slurry. In the solid acid-catalyzed acylation with carboxylic acids, inhibition of the active sites occurs due to water adsorption. A monolithic structured reactor can make it possible to remove water *in situ* by a stripping operation using a gas flow through the reactor. Under these conditions, the HBEA-coated monoliths can be reused after regeneration by washing with acetone, drying at 120°C (16 h), and calcining at 450°C for 4 h.

Acylation of anisole with carboxylic acids, in particular with octanoic acid, is described to be carried out over variously activated HBEA zeolites [51, 52]. Activation of the zeolite by treatment with oxalic acid after steaming results in a significant decrease of the surface area (from 670 to 500 m^2/g) and an increase of the bulk SAR (from 13 to 51). This provides an increase in activity and selectivity: in fact, HBEA itself exhibits an activity, defined as the initial apparent first-order constant *k*, of $0.03\ l/g_{cat}{\cdot}h$ and a selectivity toward compound **17** ($R = C_7H_{15}$) of 80% at 50% conversion, whereas a great activity increase ($k = 0.12\ l/g_{cat}{\cdot}h$) and a selectivity improvement (up to 95%) are observed with the acid-treated catalyst. The activity and selectivity enhancement can be due to the increased accessibility or participation of active sites, rather than the formation of additional active species. Indeed, removal of extra framework aluminum and, in general, of "site-blocking" species results in an increase of the number of active sites that are accessible.

HBEA zeolite supported on macroscopic SiC material (HBEA/SiC) has been investigated as a high performance catalyst for benzoylation of anisole with benzoyl chloride (anisole/benzoyl chloride ratio = 1.5) [53]. HBEA zeolite is synthesized and *in situ* deposited on SiC extrudates previously prepared by a reported technology [54, 55]. This composite catalyst shows relatively high activity with a total 71% conversion of benzoyl chloride after 24 h on stream (95% selectivity with respect to *para*-methoxybenzophenone). The catalyst can be reused after washing with methylene chloride, showing quite similar results for three tests. Comparative experiments carried out with unsupported more highly acidic HBEA show that in the first cycle similar activity and ketone selectivity are achieved, but significant catalyst deactivation is observed during the second cycle.

As already underlined, the acetylation of 2-methoxynaphthalene with acetic anhydride is particularly investigated due to the use of 2-acetyl-6-methoxynaphthalene (**21**) as an intermediate for the synthesis of (*S*)-Naproxen, a nonsteroidal anti-inflammatory drug (Scheme 3.10).

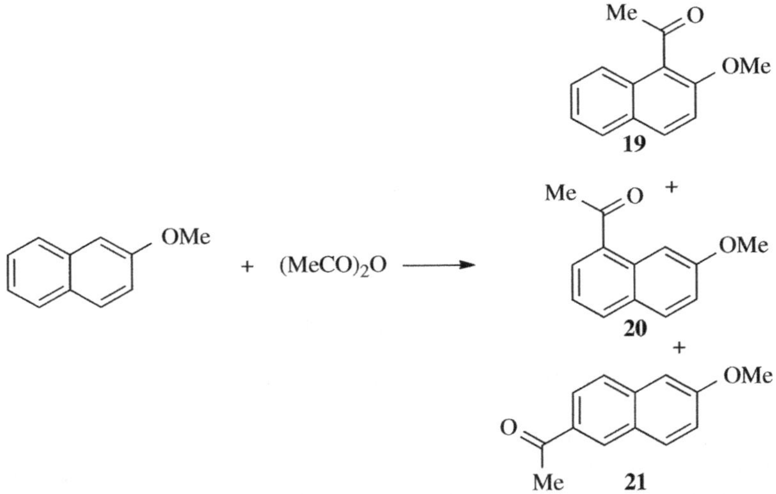

SCHEME 3.10 Acetylation of 2-methoxynaphthalene.

The desired product 2-acetyl-6-methoxynaphthalene (**21**) is obtained in 80% yield in the presence of a polar solvent such as nitrobenzene and zeolite HBEA as catalyst, because it has a more restricted pore structure that allows the preferential formation of isomer **21** less hindered with respect to isomers **19** and **20** [56].

The SAR value is expected to play a major role in the distribution of isomers **19**, **20**, and **21** in the reaction depicted in Scheme 3.10 (2-methoxynaphthalene/acetic anhydride ratio = 0.5). In a comparative study with three HY zeolites, namely, HY(15), HY(40), and HY(100), an increase of the initial activity (TOF 1.72, 8.5, and 9.3 m^{-1}, respectively) is observed with the increase of the SAR value and a consequent lowering of the surface acidity (0.65, 0.15, and 0.062 meq H^+/g, respectively) [57]. These results confirm that the activity does not stem from an increased acid strength of the catalyst [58] and suggest, in agreement with some conclusions by Corma [59], that the catalytic activity can be better related to a more hydrophobic character of dealuminated HY zeolites. In all cases, the reaction leads to the formation of compound **19** as the major product (95% yield), **21** (obtained only up to 4% yield), and traces of **20**.

3.3.2 Clays

Clays consist of crystalline sheets of aluminosilicate that are negatively charged. Indeed, as already shown for zeolites, random substitutions for the aluminum and silicon result in a net negative charge within the sheets; typical substitutions include Al^{3+} for Si^{4+} and Mg^{2+} for Al^{3+}. To balance this negative charge, cations such as Na^{+}, K^{+}, or Ca^{2+} are located in an amorphous interlayer between the sheets. The interlayer also includes water molecules, some of which are complexed to the interlayer cations. Clay particles are composed of alternating sheets and interlayers. Different clay materials (natural, acid-treated, cation-exchanged [60], and pillared clays [61]) are effective catalysts for a wide variety of organic reactions [62], including Friedel–Crafts acylation; both Brønsted and Lewis acidities can play a role in the catalytic activity.

Two groups of materials are studied in catalytic reactions, namely, clays modified through exchange of the interlayer cations or through impregnation of metal chlorides [63–66]. The acid Fe-K10, obtained by exchanging K10 montmorillonite clay with Fe(III) ions, gives outstanding results in the acylation of mesitylene (Scheme 3.11) and *para*-xylene with benzoyl chloride or benzoic anhydride (arene/acylating agent ratio = 7) affording the products in 98–100% yield.

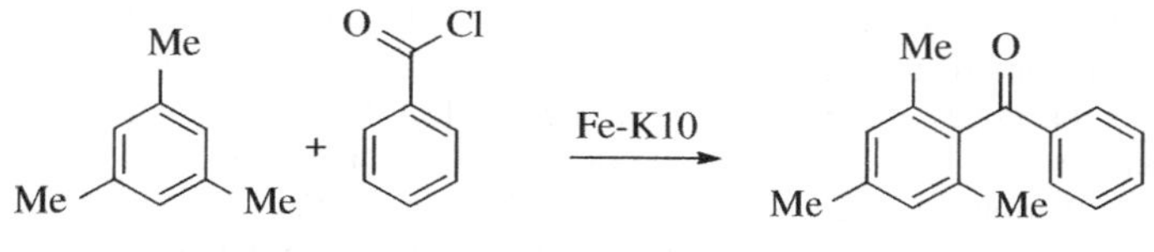

SCHEME 3.11 Fe-K10-promoted benzoylation of mesitylene.

Cation-exchanged montmorillonites are utilized by Geneste *et al.* [67, 68] in the acylation of aromatics with carboxylic acids. Best results are achieved with Al^{3+}-exchanged montmorillonite (60% yield of the *ortho*, *meta*, and *para* isomers). The influence of the chain length of different carboxylic acids on the yield of acylation appears to be similar on clays and on zeolites [35]. Indeed, acylation of toluene with $CH_3(CH_2)_nCOOH$ over Al^{3+}-montmorillonite gives 12% yield for $n=1$, 45% for $n=6$, and 80% for $n=14$.

Fe^{3+}- and Zn^{2+}-exchanged clays promote the acylation of 2-methoxynaphthalene with acetic anhydride affording the kinetically favored 1-acetyl-2-methoxynaphthalene in 82% yield [69–71].

A quite interesting result in the acylation of benzo-15-crown-5 (**22**) with acetyl chloride is reported by using Sn-K10 catalyst [72]. The monoacylated product **23** is obtained in 90% yield. The redox mechanism previously depicted in Scheme 3.7 can account for the higher activity of Sn-K10 in comparison to K10 montmorillonite itself (Scheme 3.12).

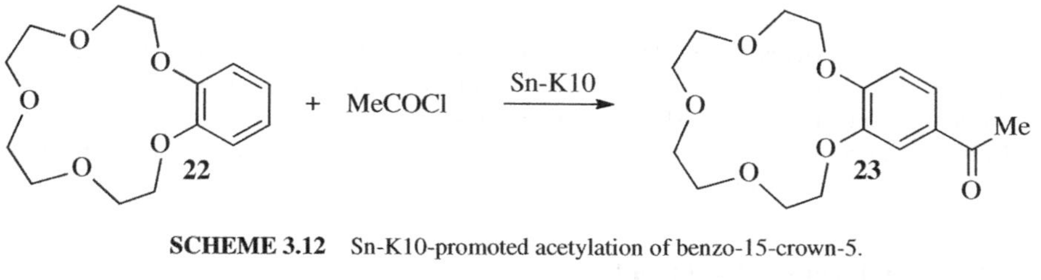

SCHEME 3.12 Sn-K10-promoted acetylation of benzo-15-crown-5.

Some interesting results are observed in the study of bentonite-supported polytrifluoromethanesulfosiloxane (B-PTMSS) as a catalyst in the acylation of ferrocene with different acyl chlorides [73] (ferrocene/acyl chloride ratio = 0.5) (Scheme 3.13).

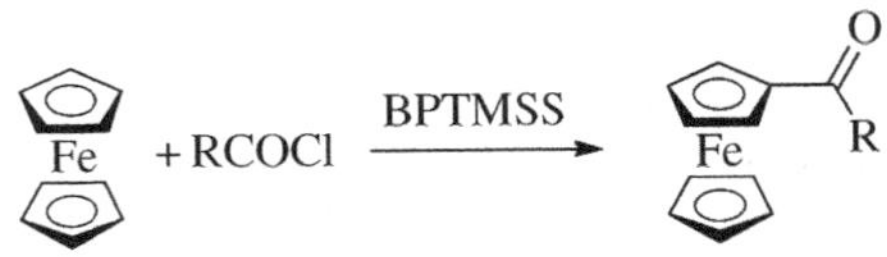

R = Et, Pr, Bu, C_6H_{13}, C_7H_{15}, $Cl(CH_2)_3$, Ph

SCHEME 3.13 BPTMSS-promoted ferrocene acylation.

The conversions of acyl chlorides increase in the following order: aromatic acyl chlorides (~20%) < chlorobutyryl chloride (~30%) < long chain acyl chlorides (70–80%). The catalyst can be utilized for at least three cycles in the acylation of ferrocene with butyryl chloride, showing similar activity (81, 73, and 86% ferrocene conversion, respectively).

3.3.3 Metal Oxides

Metal oxides are usually utilized in their pure form or in different mixtures as solid catalysts in fine chemical preparation since they are commercially available or easily synthesized, they are stable toward moisture, and their properties can be tailored by doping with convenient metal ions [74]. Dimethylbenzophenones can be obtained in high yield (88–97%) by reaction between benzoyl chloride and the three isomeric xylenes (xylene/benzoyl chloride ratio = 1.5) in the presence of iron(III) oxide. Various researchers have observed leaching of $FeCl_3$ (produced by reaction of iron oxide with HCl) into solution, suggesting a possible catalytic contribution from this homogeneous iron-based Lewis acid.

Zinc oxide, an inexpensive and commercially available inorganic solid, can be utilized as an efficient catalyst in the Friedel–Crafts acylation of activated and deactivated aromatic compounds with acyl chlorides at room temperature for 5–120 min giving products in 87–98% yield [75]. Acylation is claimed to occur exclusively at the *para* position of the monosubstituted aromatic compounds. The catalyst can be recovered and reused, after washing with methylene chloride, for at least two further cycles, showing quite similar high yield (~90%) in the model benzoylation of anisole. Again, zinc chloride is implicated as a possible true catalyst, generated *in situ* by the reaction of zinc oxide with hydrogen chloride.

An efficient and eco-friendly procedure for the small-scale acylation of ferrocene with carboxylic acids is based on the *in situ* production of the mixed carboxylic-trifluoroacetyl anhydride [76]. The reaction is simply performed by adsorbing ferrocene on the surface of activated alumina (preheated at 150°C for 3 h) and adding a mixture of carboxylic acid and trifluoroacetic anhydride at room temperature for a selected time. Products are recovered in 55–98% yield simply by elution with diethyl ether.

The regioselective acylation of aromatic ethers with carboxylic acids (aromatic ether/carboxylic acid ratio = 1) can be performed with an equimolecular mixture of trifluoroacetic anhydride adsorbed on the surface of alumina without any solvent [77]. The process can be applied, with nearly quantitative yields, to anisole and the three isomeric dimethoxybenzenes by using carboxylic acids. The authors outline that in the case of anisole, the acylation selectively occurs at the *para* position to the methoxy group. The reaction requires a large amount of alumina and trifluoroacetic anhydride, and consequently, it can only be applied at the laboratory scale. The intervention of a mixed carboxylic acid/trifluoroacetic acid mixed anhydride intermediate is presumed.

3.3.4 Acid-Treated Metal Oxides

The acylation of toluene with anhydrides is performed in the presence of solid superacids based on zirconia. The catalyst is prepared from $ZrOCl_2{\cdot}8H_2O$ treated with aqueous ammonia and washed with water and a solution of chlorosulfonic acid in ethylene dichloride followed by calcination at 650°C.

In a model reaction, the solid catalyst named UDCaT-5 gives a 62% propionic anhydride conversion and 67% 4-methylacetophenone selectivity by performing the reaction at 180°C for 3 h [78].

Sulfated metal oxide catalysts represent a class of extremely attractive strong solid acids showing widespread application to different chemical transformations [79]. Sulfated zirconia (SZ), prepared by the treatment of zirconia with sulfuric acid or ammonium sulfate, was reported to exhibit extremely strong acidity [80, 81].

SZ can be applied to the efficient acylation of anisole, with a variety of anhydrides, affording the *para* isomer in 96–99% yield, accompanied by a small amount of the *ortho* isomer [82, 83]. In addition, electron-rich and electron-poor aromatic compounds react with benzoic anhydride, and generally, high yields of the desired aromatic ketones are obtained (70–92%) [83, 84]. By using chiral anhydrides such as (*S*)-2-methylbutyric anhydride, the pure (*S*)-1-(4-methoxyphenyl)-2-methylbutan-1-one is isolated in 95% yield [82]. Similar good synthetic results are achieved by using acyl chlorides as acylating agents. In this case, too, the *para* isomers are obtained in 67–92% yield and 96–97% selectivity, the *ortho* isomers being obtained in lower amounts.

A double metal oxide sulfate solid superacid, namely, alumina-zirconia/persulfate, can be prepared by treatment of a mixture of aluminum hydroxide and zirconium(IV) hydroxide with an aqueous solution of ammonium persulfate, followed by calcination at 650°C [85]. This catalyst can be utilized in the benzoylation of arenes with benzoyl and *para*-nitrobenzoyl chloride giving benzophenones in 60–89% yield. Even if 1 g of catalyst is needed for 40 mmol of acyl chloride, the process seems to be quite useful because the catalyst can be readily regenerated by heating after washing with acetone and diethyl ether and reused four times.

Gallium(III)- and iron(III)-promoted SZ can be supported on mesoporous materials such as MCM-41 silica [86]. The catalysts, named GaSZ/MCM-41 and FeSZ/MCM-41, respectively, are prepared by incipient wetness impregnation using zirconium(IV) sulfate as the precursor in combination with gallium(III) nitrate or ferric nitrate and calcined at 700°C for 3 h. These catalysts were studied in the acylation of veratrole with acetic anhydride; the reaction gives only 3,4-dimethoxyacetophenone. The most active catalyst is GaSZ/MCM-41 (78% yield, 100% selectivity), followed by SZ/MCM-41 alone (68% yield, 100% selectivity) in the reaction carried out at 80°C for 3 h (Scheme 3.14).

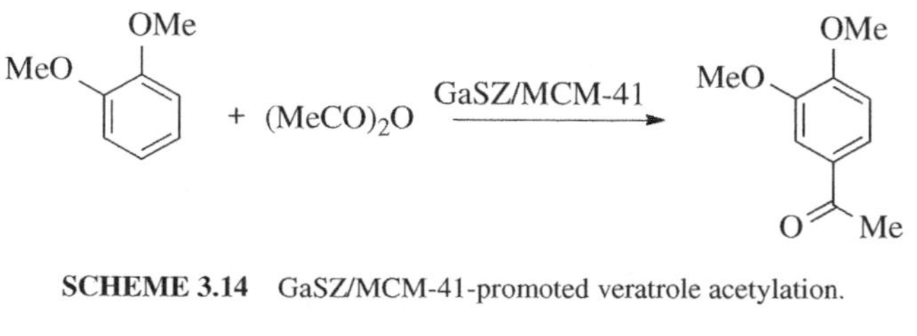

SCHEME 3.14 GaSZ/MCM-41-promoted veratrole acetylation.

IR studies confirm that not only Brønsted acid sites but also Lewis acid sites are effective in the activation of the acylating agent. In fact, the FT-IR C═O adsorption analysis shows that Lewis acidity is present in the case of GaSZ/MCM-41 and SZ/MCM-41 catalysts, whereas it is almost absent in the FeSZ/MCM-41 sample. The best catalyst GaSZ/MCM-41 can be recycled three times after washing and calcination in air at 450°C for 90 min, giving 78, 76, and 70% yield in the three cycles, respectively.

3.3.5 Heteropoly Acids (HPAs)

HPAs are Brønsted acids composed of heteropoly anions and protons as countercations; the most commonly utilized HPA is the phosphotungstic acid $H_3PW_{12}O_{40}$ (PW). HPAs are stronger than many conventional solid acids such as mixed oxides and zeolites. One important advantage of these catalysts is that they can be utilized both homogeneously and heterogeneously. The homogeneous reactions occur in polar media at approximately 100°C; on the other hand, when using nonpolar solvents, the reactions proceed heterogeneously. The HPA catalysts are easily separated from the homogeneous systems by extraction with water and from the heterogeneous reaction mixture by filtration [87, 88].

The acylation of anisole with acetic anhydride is carried out in the presence of silica-supported PW [89–91]. The supported catalysts are prepared by impregnating silica (surface area: 300 m^2/g) with a methanol solution of PW. The acylations are carried out in batch phase at 110°C in a glass reactor charged with aromatic substrate and acetic anhydride, the substrate taken in excess over the acylating agent; no solvent is used. At 110°C, the *para*-acylation process dominates (98% yield), with only a small percentage of the *ortho*-acylation product (~2%) when 40% PW/SiO_2 is utilized. The reaction appears to be truly heterogeneous; the catalyst is found to be reusable, although gradual decline of activity is observed. Better results are obtained when, after the first run, the catalyst is filtered off and washed with methylene chloride; apparently, this treatment removes tars more efficiently from the catalyst surface. Such a procedure allows 82% yield of the *para*-methoxyacetophenone to be obtained in the second run. Coking may cause partial deactivation of the catalyst, which is evident from the dark brown color of the catalyst, initially a white powder. A similar catalyst prepared by impregnation of PW onto mesoporous MCM-41 silica and utilized in the same model reaction performed at 120°C for 1 h gives *para*-methoxyacetophenone with 100% acetic anhydride conversion and 100% selectivity [92].

The acylation of toluene and anisole with C_2–C_{12} aliphatic carboxylic acids can be carried out with $Cs_{2.5}H_{0.5}[PW_{12}O_{40}]$, affording the corresponding products in 41–71% yield [93]. These solid acids are superior in activity to the conventional acid catalysts such as sulfuric acid and zeolites and can be reused after a simple work-up, albeit with reduced activity. As already underlined, the activity of this kind of catalyst can be improved by carrying out the reaction in the presence of trifluoroacetic anhydride [94]. The acylation of anisole is conducted by acetic or benzoic acid in the presence of trifluoroacetic anhydride catalyzed by $AlPW_{12}O_{40}$ with excellent yields (94–96%). Under the same reaction conditions, 2-methyl- and 4-methylanisoles are also acylated, giving the corresponding ketones in 90–97% yield.

A hybrid material based on aluminum tungstophosphate can be utilized in the acylation of anisole with acetic anhydride [95]. The catalyst is prepared by immobilizing aluminum tungstophosphate in a polymeric blend formed by polyvinyl alcohol and polyethylene glycol by the freeze-thawing method. The salt retains the Keggin structure of the heteropolyanion during the immobilization procedure. *para*-Methoxyacetophenone is obtained in 90% yield in 30 min at 100°C accompanied by 10% of *ortho*-methoxyacetophenone.

3.3.6 Nafion

Nafion, a perfluorinated polymer containing pendant sulfonic acid groups, is generally considered to be a solid superacid whose pK_a ranges from −5 to −9. It represents an efficient catalyst for promoting various organic reactions such as alkylation, isomerization, disproportionation, transalkylation, acylation, nitration, hydration, rearrangement, etc. [96, 97].

Preparation of fluorenone and related cyclic and heterocyclic ketones **25** from the corresponding compounds **24** or the appropriate benzoic acid derivatives can be obtained under relatively mild conditions [98, 99]. The reaction is carried out by heating a mixture of the carboxylic acid derivatives **24** and the solid Nafion in 1,2-dichlorobenzene at about 180°C, affording products **25** in 82–95% yields (Scheme 3.15).

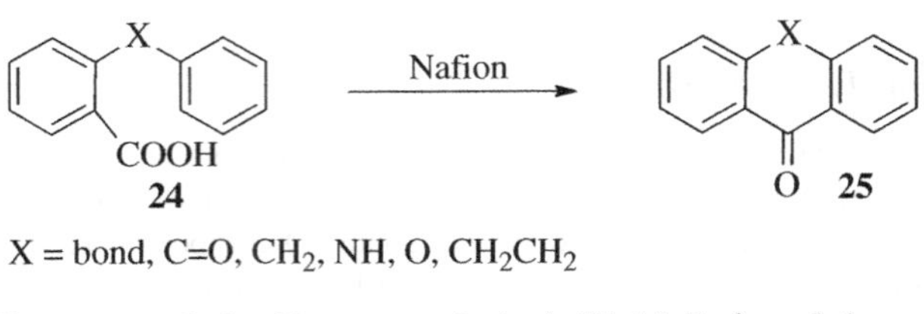

SCHEME 3.15 Fluorenone and related ketones synthesis via Friedel–Crafts acylation reaction with Nafion.

The reaction can also be performed in refluxing *para*-xylene in the case of the carboxylic acid (100% yield with diphenylethane-2-carboxylic acid after 12 h) and in refluxing benzene with the corresponding acyl chloride, giving similar good results. Moreover, 4-arylbutanoic acids cyclize in refluxing *para*-xylene in the presence of Nafion, giving 1-tetralone analogs in nearly quantitative yields.

SAC materials are Nafion/silica composites prepared by an *in situ* sol-gel technique, in which suspensions of Nafion resin nanoparticles are mixed with soluble silicon sources to form a gel that is dried to the final glass-like material [100]. These materials can be ion exchanged by stirring them in a saturated aqueous solution of the selected metal salt. The solid catalyst SAC-13 (silica containing 13 wt % of pure Nafion) promotes the acetylation of 2-methoxynaphthalene with acetic anhydride giving 1-acetyl-2-methoxynaphthalene (**19**—Scheme 3.10) with 94% acetic anhydride conversion and 90% selectivity, whereas SAC-80 ion exchanged with $AgNO_3$ in the same reaction gives mostly 2-acetyl-6-methoxynaphthalene (**21**—Scheme 3.10) (96% acetic anhydride conversion, 69% selectivity).

3.3.7 Miscellaneous

Graphite can promote the Friedel–Crafts acylation reaction of active aromatic compounds such as anisoles and polymethylbenzenes with acyl halides to afford the corresponding aromatic ketones [101]. In a typical experiment, graphite is added to a mixture of anisole and benzoyl bromide in benzene, and the mixture is heated under reflux for 8 h to afford *para*-methoxybenzophenone in 80% yield. Different acyl halides and several anisoles and polymethylbenzenes are utilized to give the corresponding products in high yields (60–92%).

To overcome the problem of the high amount of graphite utilized and mainly to avoid the use of the expensive and problematic acyl halides, graphite can be coupled with *para*-toluenesulfonic acid and utilized to activate the more eco-compatible carboxylic acids toward electrophilic acylation [102]. With this catalyst, a solvent-free process can be operated and high yields (84–96%) can be obtained even with deactivated aromatic substrates. The graphite can be reused after simple washing with ethyl acetate and water, but the *para*-toluenesulfonic acid, which is not adsorbed on the graphite during the reaction, must be added again for the successive runs.

3.4 DIRECT PHENOL ACYLATION

Hydroxylated aromatic ketones represent valuable intermediate compounds in the synthesis of important fragrances and pharmaceuticals. For example, the resorcinol acylation products are employed for the production of valuable fine chemicals such as 4-*O*-octyl-2-hydroxybenzophenone (UV light absorbent for polymers) and ipriflavone (antiosteopenic drug). *o*-Hydroxyacetophenone (*o*-HAP) and *p*-hydroxyacetophenone (*p*-HAP) are widely used for the synthesis of aspirin and paracetamol (4-acetaminophenol), respectively [103]. *o*-HAP represents also a key intermediate for the production of 4-hydroxycoumarin and warfarin, which are both used as anticoagulant drugs in the therapy of thrombotic disease [104], and it is also employed for the synthesis of flavonones [105, 106].

The direct acylation of phenols represents an extremely complex reaction. Indeed, phenol is a typical ambident system and reacts with acylating reagents such as acetyl chloride in the presence of convenient homogeneous or heterogeneous acid catalysts affording phenyl acetate by O-acetylation as well as *o*-HAP and *p*-HAP by C-acetylation of the activated aromatic ring. Furthermore, phenyl acetate can undergo Fries rearrangement (see Chapter 31), thus complicating the entire process. The O-acylation process is much more rapid than the C-acylation one, and in general, the *o*-HAP is prevalent with respect to *p*-HAP [107]. In our opinion, this is probably due to the *ortho*-directing effect of the hydroxyl group, which can stabilize the transition state by hydrogen bonding leading to the *ortho* attack in a way resembling the so-called complex-induced proximity effect (CIPE) (Scheme 3.16) [108].

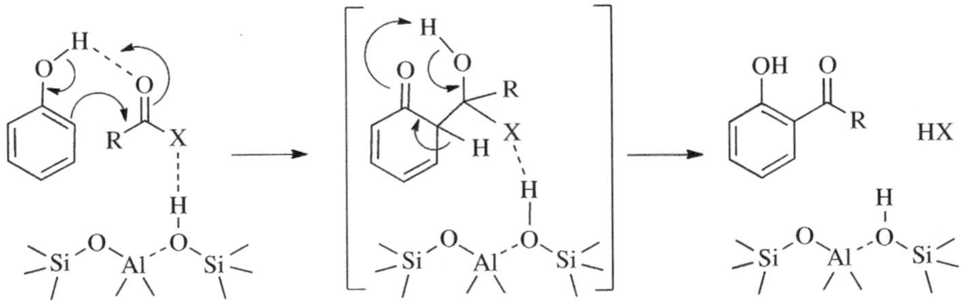

SCHEME 3.16 Hypothesized mechanism for the *ortho*-functionalization of phenol promoted by a heterogeneous catalyst.

It must also be observed that the mode of *p*-HAP formation is probably different form that of *o*-HAP: the *ortho* isomer is a primary product, while the *para*-one seems to be a secondary product. Of course, other ways for the formation of *o*-HAP can result from the acylation of phenol with phenyl acetate, which is a better acylating agent than acetic acid.

Mechanistic differentiation between direct phenol C-acylation and O-acylation followed by Fries rearrangement is often difficult. Therefore, in this section are included synthetic processes where both reagents, the phenol substrate and the acylating agent, and the catalyst are mixed together in the starting reaction mixture aside from the specific reaction mechanism.

Metal phenolates are utilized for the preparation of *ortho*-hydroxyarylketones. The selectivity of the process strongly depends on the coordinating ability of the metal: for example, the reaction of aluminum tri(2-*tert*-butyl)phenolate with monochloroacetyl chloride carried out in toluene exclusively affords the direct *ortho*-acylation product (71% yield, 98% selectivity) [109]. On the contrary, by using bromomagnesium 2-*tert*-butylphenolate and different acyl chlorides (RCOCl), the phenol oxygen competes with the *ortho*-carbon affording a mixture of esters and *ortho*-ketones. Interestingly, an increase in the electron-withdrawing power of R in RCOCl results in a gradual increase in the *ortho*-C/O reactivity ratio. This effect is observed in spite of the obvious increase in size and increase in the steric requirements of the reagents involved; a reasonable correlation between the logarithm of the *ortho*-C/O reactivity ratio and the polar substituent constant σ_I (the true measure of the inductive effect of the substituent R group) is reported (Fig. 3.2) [110].

The metal-template effect is also exploited as a powerful tool to achieve highly selective phenol *ortho*-acylation with trichloroacetonitrile according to a typical Houben–Hoesch reaction [111]. The best catalyst is boron chloride, which, by reacting with phenols in 1,2-dichloroethane, produces dichlorophenoxyboron derivatives, the reactive species responsible for the activation of trichloroacetonitrile as well as for the *ortho*-regioselective attack. By using zinc chloride and acetic anhydride without any solvent, an equimolecular mixture of 2- and 4-hydroxyacetophenone (36 and 40% yield, respectively) is produced, excluding any remarkable metal-template effect in the $ZnCl_2$-based system.

A very selective and effective Friedel–Crafts acylation of phenols with acyl chlorides is represented by the use of concentrated triflic acid as a catalyst [112]. The reaction carried out with phenols and naphthols under solventless conditions for 1 h affords the corresponding ketones with very high yields and excellent selectivities. It is worth noting that the use of 1% triflic acid/acetonitrile results in the exclusive production of the corresponding O-acylation products.

Metal triflates can also be efficiently utilized as reusable catalysts in the acylation of phenols [104, 113, 114]. For example, with scandium triflate (5% mol), a complete regioselectivity is

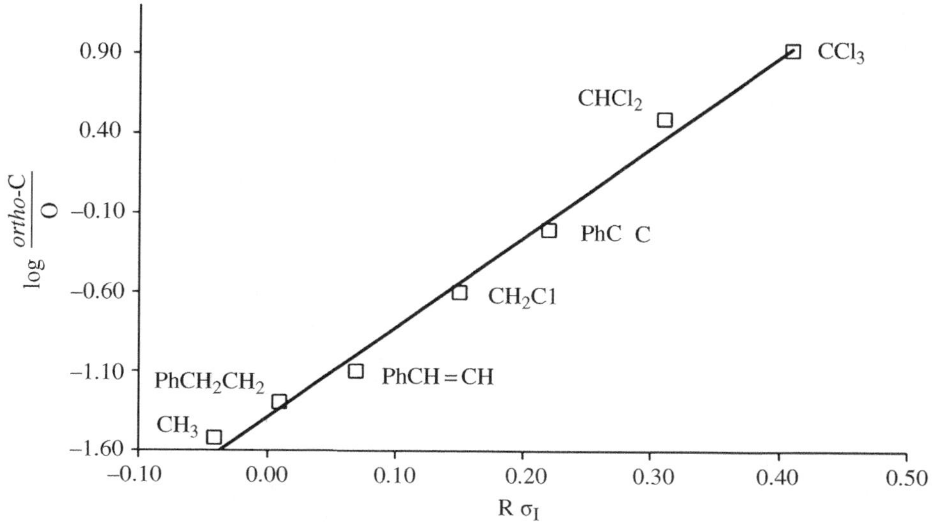

FIGURE 3.2 Electron-withdrawing power effect of R group in RCOCl on the log *ortho* C/O reactivity.

observed with *meta*-cresol and 1-naphthol derivatives, which afford the products deriving from the attack at the less hindered position. The high catalytic activity of scandium triflate depends on the fact that it is not trapped by the free hydroxyl groups of the phenols as well as the carbonyl oxygens of the products in a different way from what happens with other classic Lewis acids (i.e., aluminum chloride).

Solid acid catalysts such as clays and zeolites are also utilized for phenol acylation; however, these processes suffer from catalyst deactivation problems and lack C-selectivity. In the acylation of phenol with acetic anhydride, HZSM-5 zeolite shows a very high *ortho*-selectivity (48% *o*-HAP yield, <1% *p*-HAP yield), although phenyl acetate is isolated in only approximately 20% yield [115]. The SAR value has a remarkable influence on the selectivity of the process: when the reaction is carried out in the presence of HZSM-5(30), HZSM-5(150), and HZSM-5(280) zeolites, the *o*-HAP yields are 42, 40, and 15%, respectively, whereas the O-acylation is noticeably increased. These results mean that C-acylation requires higher Brønsted acidity and that lower acidity leads to phenyl acetate formation. It must be noted that the reaction performed with an amorphous aluminosilicate acid catalyst gives mostly phenyl acetate without isomer selectivity. These results suggest that the C-acylation of phenol occurs in the channels of zeolites and not on the external surface.

The same trend is observed in the gas-phase reaction of phenol with acetic acid, a greener acylating agent. *o*-HAP still represents the major isomer [116]; by using HZSM-5(83.6), the *o*:*p* isomer ratio is 87:13, and it becomes 93:7 in the presence of a HZSM-5(84.8) [117].

The HBEA(20) deactivation in the reaction between phenol and phenyl acetate has also been studied [118]. In this case, the organic material trapped in the zeolite can be recovered following two methods: (i) Soxhlet extraction of the zeolite [Ext] and (ii) extraction of coke by dissolution of the zeolite itself in a 40% solution of hydrofluoric acid [Coke]. The acylation reactions are carried out in two classical solvents, dodecane and sulfolane, and in both cases, a significant lowering of the rate of HAPs' formation with time is observed. This deactivation is faster in dodecane (~1 h) than in sulfolane (~2 h).Whatever the solvent, the two reactants are the main components of the material retained in the catalyst; nevertheless, in the case of sulfolane, their contents in Ext and Coke are similar to that of the reaction mixture, whereas when the less polar solvent dodecane is employed, their contents in Ext and Coke are greater than that in the reaction mixture. In addition, sulfolane constitutes a very significant part of Ext and Coke, whereas practically no dodecane is detected. These differences can be related to the difference in solvent polarity: the polar sulfolane enters the pores of this zeolite, while dodecane cannot enter them. For the same reason, the amount of phenyl acetate found in Coke is very small compared to those of the more polar phenol and sulfolane. The catalyst deactivation can be attributed to the product inhibition: indeed, the reaction rate is very low when phenyl acetate is added to the zeolite impregnated with phenol, with the higher rate being obtained when phenol is added to the zeolite impregnated with the less polar phenyl acetate. The conclusion is that the decrease in catalyst activity is not necessarily due to formation of the heavy secondary products but most likely to a limitation of the access of phenyl acetate to the zeolite pores occupied by the very polar phenol.

Mesoporous Al-MCM-41 also promotes the reaction between phenol and acetic anhydride or, better, acetic acid [119]. Phenyl acetate represents the major product, and in all experiments, *o*-HAP results the sole C-acylated product, without any detectable amount of *p*-HAP. Different behaviors are found with increasing temperature with the two acylating agents. When acetic anhydride is utilized, the overall conversion of the phenol decreases with the temperature increase; this can be ascribed to the decomposition of the phenyl acetate back to phenol, which results in an apparent conversion decline (from 99 to 52%). At the same time, the selectivity for *o*-HAP increases from <1 to 21%, indicating that at lower temperatures the O-acylation is the predominant process, but as the temperature increased, O-acylation is gradually replaced by C-acylation. On the contrary, with acetic acid, the overall conversion is much lower, but it increases at higher temperatures from 33 to 47% together with the *o*-HAP selectivity (from <1 to 24%).

Other kinds of heterogeneous catalysts are also employed in the acylation of phenols. To mention just an example, graphite in methanesulfonic acid shows very good activity in the direct acetylation of phenol with benzoic acid [120]. After 3 h at 120°C, the *ortho* isomer can be isolated in 81% yield; on the contrary, in the absence of graphite, the yield drops to 20%, and with graphite alone, the yield is negligible. By using different substituted benzoic acids, the process can be applied to a variety of phenol and naphthol derivatives, obtaining high yields and high levels of *ortho*-regioselectivity. Concerning the reusability of the catalyst, the yield of the *ortho*-product for three successive runs does not decrease, but as methanesulfonic acid is not adsorbed onto graphite, it is necessary to add it again for the successive runs.

As underlined in the introduction of this section, the resorcinol acylation represents a very important process. The benzoylation of resorcinol to produce 2,4-dihydroxybenzophenone, previously performed with benzotrichloride (which implies the coproduction of 3 mol of hydrochloric acid and consequently a large amounts of acid waste) [43, 48, 121], is carried out by using benzoic acid in the presence of HBEA zeolite (Scheme 3.17) [32, 46].

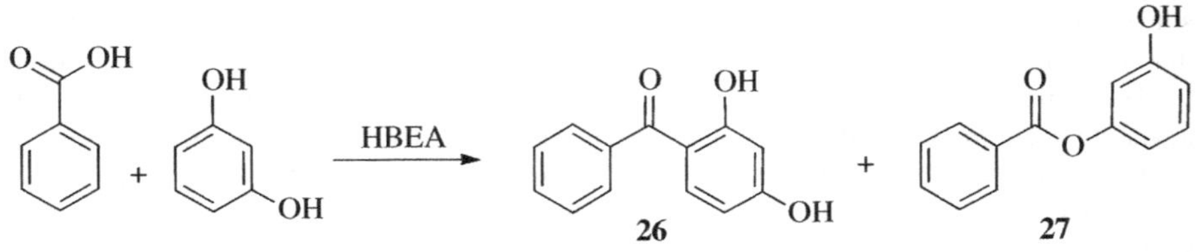

SCHEME 3.17 HBEA-promoted benzoylation of resorcinol.

With *para*-chlorotoluene as solvent, after 18 h 2,4-dihydroxybenzophenone (**26**) is isolated in 70% yield, together with 20% of resorcinol monobenzoate (**27**) and 3% of resorcinol dibenzoate.

Different montmorillonites are also employed to react resorcinol with phenylacetic acid chloride [122, 123]. A high resorcinol conversion is reached (65–80%), but the yield of the desired ketone is poor and only with KS montmorillonite is an appreciable yield of 25% achieved, the ester being nevertheless the main product. The ketone/ester ratio remains practically constant during the reactions with every catalyst, and in all cases, the ester concentration shows no maximum, which means that the ketone is not formed by a rearrangement of the ester but directly by ring acylation. Fe-K10 and Fe-KS clays are slightly more active than the undoped ones (i.e., 25 and 42% yield for KS and Fe-KS, respectively). In the presence of Fe-K10 without solvent in a melted phase, although the resorcinol conversion is slightly lower (~70%) than that in dichloroethane (80%), the ketone is isolated in 60% preparative yield; the solvent-free reaction has some other advantages: in fact, while the catalysts used in dichloroethane are generally deactivated after one cycle, the solvent-free reaction shows a satisfactory catalyst reusability after washing and drying.

In a more environmentally benign way, benzoic acid is employed in equimolecular amount as a resorcinol acylating agent in the presence of montmorillonite clay [124]. Resorcinol monobenzoate **27** and 2,4-dihydroxybenzophenone **26** are the sole products: the former is claimed by the authors to be a primary product, whereas the latter is a secondary one, being exclusively formed by Fries rearrangement. These results are not completely consistent with mechanistic conclusions reported: this is probably due to the use of different acylating agents [122]. On the basis of these data, the authors conclude that resorcinol monobenzoate first diffuses out of the catalyst pores and then, successively, is activated and transformed into 2,4-dihydroxybenzophenone. One possible explanation is that the high hydrophilic character of the clay surface, especially in samples having higher alumina content, makes the pores filled with more polar substances, such as benzoic acid and resorcinol, while less polar compounds rapidly diffuse out of the clay porosity and undergo consecutive transformation at the active sites located at the external particle surface. The reaction between resorcinol and benzoic acid to yield resorcinol monobenzoate is fully reversible, and the hydrolysis of resorcinol monobenzoate to yield resorcinol and benzoic acid occurs with water retained in clays; the removal of water from the bulk liquid by azeotropic distillation makes it possible to considerably increase the conversion, which reached 80% since the esterification reaction is favored; in this case, 2,4-dihydroxybenzophenone is isolated in 62% selectivity.

3.5 SUMMARY AND OUTLOOK

It is currently recognized that catalytic processes are key for sustainable development in the chemical industry. They allow synthesis of products in a resource-sparing way, with less consumption of energy and with minimum formation of waste [125].

One of the salient features of the original Friedel–Crafts acylation reaction is the requirement of one or more than one equivalent of Lewis acid catalyst that cannot be recovered and reused [1]. The possibility to carry out the reaction with only catalytic amounts of Lewis acids (such as iron trichloride, zinc chloride, iron, and iodine, or, better, with metal triflates such as lanthanide triflates) represents a significant development [113]. Metal triflates, developed in particular by Kobayashi, are very active (1–5% mol with respect to the acylating agent) and insensitive to the water content of the reaction medium and can be reused [126].

The aforementioned results, however, describe examples of preparative methods where the catalyst still represents the most expensive component. Therefore, effective recovery of the said catalysts is an important objective. In certain cases, separation of the target product from homogeneous catalysts can be problematic and hamper efforts to achieve the most economical process (not to mention potentially lowering the isolated product quality). The use of heterogeneous catalysts offers a way to avoid this situation by ensuring simple product/catalyst separation via simple filtration.

Various classes of solid catalysts have been utilized for Friedel–Crafts acylation reactions; in some cases, the process can be efficiently industrialized, showing that many of the problems of reactivity, catalyst deactivation, and engineering can be solved by using zeolites as catalysts [41, 127, 128].

An additional great advantage of heterogeneous catalysts is represented by the possibility to carry out the process with fixed-bed reactors in continuous flow mode. For this purpose, zeolites have been supported onto ceramic materials giving macroscopic shapings utilizable in fixed-bed reactors [129, 130] for the acylation of anisole with carboxylic acids [52, 131].

However, the most important industrial catalytic processes were developed by purely empirical methods and countless screening experiments; the complexity of the solid catalysts still represents a serious obstacle to the understanding of the structure–reactivity relationship [132]. Recent studies where *in situ* analysis was applied have driven great improvements in the comprehension of the mode of catalyst operation in these reactions. However, only a fraction of mechanistic analyses can be conducted under *in situ* conditions so there remains an opportunity in this area.

Objectively, reports of the surface acidity–catalytic activity correlation remain somewhat conflicting: differences usually arise from the use of varying reaction conditions and approaches to prepare or modify the catalysts and also from a poor characterization of the materials employed. Indeed, the detailed physicochemical characterization of the catalytic materials as well as the study of their interaction with reagents and products still represents an unsolved problem for the use of heterogeneous catalysis for organic syntheses. To address this issue, further research into standardized methods for evaluation of solid catalyst efficiency, reusability [133], and "true heterogeneity" (possible leaching phenomena) are required [134].

Finally, although in the last decade a great number of studies on both homogeneous and heterogeneous catalytic Friedel–Crafts acylations have been conducted, a general method for the efficient and environmentally acceptable large-scale production of aromatic ketones is still missing. Future development of new sustainable synthetic routes to aromatic ketones will surely rely upon a synergistic cooperation between catalysis, physicochemical characterization, and engineering groups, at both industrial and academic levels.

ABBREVIATIONS

B-PTMSS	Bentonite-supported polytrifluoromethanesulfosiloxane
CIPE	Complex-induced proximity effect
HPAs	Heteropoly acids
o-HAP	*o*-Hydroxyacetophenone
p-HAP	*p*-Hydroxyacetophenone
SAR	Silica/alumina ratio
SZ	Sulfated zirconia
TOF	Turnover frequency

REFERENCES

[1] Gore, P. H. (1964) *Friedel-Crafts and Related Reactions*, (ed. Olah, G. A.), John Wiley & Sons Inc., London, Vol. **III**, Part 1, p 1.

[2] Franck, H. G. (1988) *Industrial Aromatic Chemistry*, Springer, Berlin.

[3] Horsely, J. A. (1997) *ChemTech* **27**, 45.

[4] Bauer, K., Garbe, D., and Surberg, H. (1990) *Common Fragrance and Flavor Materials*, WHC Vertagsgesellschaft, Weinheim, p 83.

[5] Milto, V. I., Mironov, G. S., and Kopelkin, V. V. (1989) *Zh. Org. Khim.* **25**, 2372–2374.

[6] Sheldon, R. A. and Downing, R. S. (1999) *Appl. Catal. A Gen.* **189**, 163–183.

[7] Anastas, P. T., Bartlett, L. B., Kirchhoff, M. M., and Williamson, T. C. (2000) *Catal. Today* **55**, 11–22.

[8] Pearson, D. E. and Buehler, C. A. (1972) *Synthesis* 533–542.

[9] Pivsa-Art, S., Okuro, K., Miura, M., Murata, S., and Nomura, M. (1994) *J. Chem. Soc. Perkin Trans.* **1**, 1703–1707.

[10] Adams, C. J., Earle, M. J., Roberts, G., and Seddon, K. R. (1998) *Chem. Commun.* 2097–2098.

[11] da Silveira Neto, B. A., Ebeling, G., Gonçalves, R. S., Gozzo, F. C., Eberlin, M. N., and Dupont, J. (2004) *Synthesis* 1155–1158.

[12] Earle, M. J., Hakala, U., Hardacre, C., Karkkainen, J., McAuley, B. J., Rooney, D. W., Seddon, K. R., Thompson, J. M., and Wähälä, K. (2005) *Chem. Commun.* 903–905.

[13] Mukaiyama, T., Suzuki, K., Han, J. S., and Kobayashi, S. (1992) *Chem. Lett.* 435–438.

[14] Olah, G. A., Farooq, O., Farnia, S. M. F., and Olah, J. A. (1988) *J. Am. Chem. Soc.* **110**, 2560–2565.

[15] Desmurs, J. R., Labrouillère, M., Le Roux, C., Gaspard, H., Laporterie, A., and Dubac, J. (1997) *Tetrahedron Lett.* **38**, 8871–8874.

[16] Matsuo, J. I., Odashima, K., and Kobayashi, S. (2000) *Synlett* **3**, 403–405.

[17] Effenberger, F., Eberhard, J. K., and Maier, A. H. (1996) *J. Am. Chem. Soc.* **118**, 12572–12579.

[18] Kobayashi, S. and Iwamoto, S. (1998) *Tetrahedron Lett.* **39**, 4697–4700.

[19] Matsushita, Y.-I., Sugamoto, K., and Matsui, T. (2004) *Tetrahedron Lett.* **45**, 4723–4727.

[20] Wasserscheid, P. and Keim, W. (2000) *Angew. Chem. Int. Ed.* **39**, 3772–3789.

[21] Tran, P. H., Le, D. N. B., and Le, T. N. (2014) *Tetrahedron Lett.* **55**, 205–208.

[22] Zayed, F., Greiner, L., Schulz, P. S., Lapkin, A., and Leitner, W. (2008) *Chem. Commun.* 79–81.

[23] Olah, G. A. (1973) *Friedel–Crafts Chemistry*, Wiley-Interscience, New York.

[24] Corma, A. (1995) *Chem. Rev.* **95**, 559–614.

[25] Čejka, J. and Wichterlová, B. (2002) *Catal. Rev.* **44**, 375–421.

[26] Meier, W. M. and Olson, D. H. (1992) *Zeolites* **12**, 449–656.

[27] Wilson, S. T., Lok, B. M., Messina, C. A., Cannan, T. R., and Flanigen, E. M. (1982) *J. Am. Chem. Soc.* **104**, 1146–1147.

[28] Lok, B. M., Messina, C. A., Patton, R. L., Gajek, R. T., Cannan, T. R., and Flanigen, E. M. (1984) *J. Am. Chem. Soc.* **106**, 6092–6093.

[29] Meier, W. M. and Olson, D. H. (1992) *Zeolites* **12**, issue 5 - special issue "Atlas of Zeolites Structure Type".

[30] Kokotalio, G. T., Schlenker, J. L., Dwyer, F. G., and Walyocsik, E. W. (1985) *Zeolites* **5**, 349–351.

[31] Botella, P., Corma, A., López-Nieto, J. M., Valencia, S., and Jacquot, R. (2000) *J. Catal.* **195**, 161–168.

[32] Klisáková, J., Cervený, L., and Čejka, J. (2004) *Appl. Catal. A Gen.* **272**, 79–86.

[33] Bourgogne, J.-P., Aspisi, C., Ou, K., Geneste, P., Durand, R., and Mseddi, S. (1992) FR 2,667,063.

[34] Choudhary, V. R., Jana, S. K., Patil, N. S., and Bhargava, S. K. (2003) *Microporous Mesoporous Mater.* **57**, 21–35.

[35] Chiche, B., Finiels, A., Gauthier, C., and Geneste, P. (1986) *J. Org. Chem.* **51**, 2128–2130.

[36] Bhattacharya, D., Sharma, S., and Singh, A. P. (1997) *Appl. Catal. A Gen.* **150**, 53–62.

[37] Cerveny, L., Mikulcová, K., and Čejka, J. (2002) *Appl. Catal. A Gen.* **223**, 65–72.

[38] Yuan, B., Li, Z., Liu, Y., and Zhang, S. (2008) *J. Mol. Catal. A Chem.* **280**, 210–218.

[39] Rohan, D., Canaff, C., Fromentin, E., and Guisnet, M. (1998) *J. Catal.* **177**, 296–305.

[40] Jaimol, T., Moreau, P., Finiels, A., Ramaswamy, A. V., and Singh, A. P. (2001) *Appl. Catal. A Gen.* **214**, 1–10.

[41] Spagnol, M., Gilbert, L., Guillot, H., and Tirel, P.-J. (1997) WO 9748665; (1998) *Chem. Abstr.* **128**, 90668.

[42] Spagnol, M., Gilbert, L., Benazzi, E., and Marcilly, C. (1996) WO 9635656; (1997) *Chem. Abstr.* **126**, 46970.

[43] Smith, K., Zhenhua, Z., and Hodgson, P. K. G. (1998) *J. Mol. Catal. A Chem.* **134**, 121–128.

[44] Badri, R. and Tavakoli, L. J. (2003) *J Incl. Phenom. Macrocycl. Chem.* **45**, 41–43.

[45] Zoeller, J. R. and Sumner, C. E. (1990) *J. Org. Chem.* **55**, 319–324.

[46] Kantam, M. L., Ranganath, K. V. S., Sateesh, M., Kumar, K. B. S., and Choudary, B. M. (2005) *J. Mol. Catal. A Chem.* **225**, 15–20.

[47] Derouane, E. G., Dillon, C. J., Bethell, D., and Derouane-Abd Hamid, S. B. (1999) *J. Catal.* **187**, 209–218.

[48] Derouane, E. G., Crehan, G., Dillon, C. J., Bethell, D., He, H., and Derouane-Abd Hamid, S. B. (2000) *J. Catal.* **194**, 410–423.

[49] Derouane, E. G., Schmidt, I., Lachas, H., and Christensen, C. J. H. (2004) *Catal. Lett.* **95**, 13–17.

[50] Beers, A. E. W., Nijhuis, T. A., Aalders, N., Kapteijn, F., and Moulijn, J. A. (2003) *Appl. Catal. A Gen.* **243**, 237–250.

[51] Beers, A. E. W., van Bokhoven, J. A., de Lathouder, K. M., Kapteijn, F., and Moulijn, J. A. (2003) *J. Catal.* **218**, 239–248.

[52] Beers, A. E. W., Nijhuis, T. A., Kapteijn, F., and Moulijn, J. A. (2001) *Microporous Mesoporous Mater.* **48**, 279–284.

[53] Winé, G., Tessonnier, J. P., Pham-Huu, C., and Ledoux, M. J. (2002) *Chem. Commun.* 2418–2419.

[54] Ledoux, M. J. and Pham-Huu, C. (2001) *CATTECH* **5**, 226–246.

[55] Winé, G., Tessonnier, J.-P., Rigolet, S., Marichal, C., Ledoux, M.-J., and Pham-Huu, C. (2006) *J. Mol. Catal. A Chem.* **248**, 113–120.

[56] Fromentin, E., Coustard, J.-M., and Guisnet, M. (2000) *J. Mol. Catal. A Chem.* **159**, 377–388.

[57] Meric, P., Finiels, A., and Moreau, P. (2002) *J. Mol. Catal. A Chem.* **189**, 251–262.

[58] Barthomeuf, D. (1987) *Mater. Chem. Phys.* **17**, 49–71.

[59] Corma, A., Climent, M. J., García, H., and Primo, J. (1989) *Appl. Catal.* **49**, 109–123.

[60] Alberti, G. and Costantino, U. (1996) *Comprehensive Supramolecular Chemistry*, (eds. Atwood, J. L., Davies, J. E. P., MacNicol, D. D., and Vögtle, F.), Pergamon Press, Oxford, Vol. **7**, pp 1–23.

[61] Gil, A., Gandía, L. M., and Vicente, M. A. (2000) *Catal. Rev. Sci. Eng.* **42**, 145–212.

[62] Balogh, M. and Laszlo, P. (1993) *Organic Chemistry Using Clays*, Springer-Verlag, New York.

[63] Cornélis, A., Gerstmans, A., Laszlo, P., Mathy, A., and Zieba, I. (1990) *Catal. Lett.* **6**, 103–110.

[64] Laszlo, P. and Montaufier, M.-T. (1991) *Tetrahedron Lett.* **32**, 1561–1564.

[65] Cornélis, A., Laszlo, P., and Wang, S. (1993) *Tetrahedron Lett.* **34**, 3849–3852.

[66] Cornélis, A., Laszlo, P., and Wang, S.-F. (1993) *Catal. Lett.* **17**, 63–69.

[67] Chiche, B., Finiels, A., Gauthier, C., Geneste, P., Graille, J., and Pioch, D. (1987) *J. Mol. Catal.* **42**, 229–235.

[68] Chiche, B. H., Geneste, P., Gauthier, C., Figueras, F., Fajula, F., Finiels, A., Graille, J., and Pioch, D. (1987) FR 2599275; (1988) *Chem. Abstr.* **109**, 172531.

[69] Choudary, V. R., Jana, S. K., and Narkhede, V. S. (2002) *Appl. Catal. A Gen.* **235**, 207–215.

[70] Choudary, V. R., Jana, S. K., and Mandale, A. B. (2001) *Catal. Lett.* **74**, 95–98.

[71] Choudary, B. M., Sateesh, M., Kantam, M. L., and Ram Prasad, K. V. (1998) *Appl. Catal. A Gen.* **171**, 155–160.

[72] Biró, K., Békássy, S., Ágai, B., and Figueras, F. (2000) *J. Mol. Catal. A Chem.* **151**, 179–184.

[73] Hu, R.-J. and Li, B.-G. (2004) *Catal. Lett.* **98**, 43–47.

[74] Tauster, S. J. (1987) *Acc. Chem. Res.* **20**, 389–394.

[75] Sarvari, M. H. and Sharghi, H. (2004) *J. Org. Chem.* **69**, 6953–6956.

[76] Ranu, B. C., Jana, U., and Majee, A. (1999) *Green Chem.* **1**, 33–34.

[77] Ranu, B. C., Ghosh, K., and Jana, U. (1996) *J. Org. Chem.* **61**, 9546–9547.

[78] Yadav, G.D. and Kamble, S. B. (2012) *Appl. Catal. A Gen.* **433–434**, 265–274.

[79] Song, X. and Sayari, A. (1996) *Catal. Rev. Sci. Eng.* **38**, 329–412.

[80] Hino, M., Kobayashi, S., and Arata, K. (1979) *J. Am. Chem. Soc.* **101**, 6439–6441.

[81] Hino, M. and Arata, K. (1980) *J. Chem. Soc. Chem. Commun.* 851–852.

[82] Deutsch, J., Trunschke, A., Müller, D., Quaschning, V., Kemnitz, E., and Lieske, H. (2003) *Catal. Lett.* **88**, 9–15.

[83] Deutsch, J., Trunschke, A., Müller, D., Quaschning, V., Kemnitz, E., and Lieske, H. (2004) *J. Mol. Catal. A Chem.* **207**, 51–57.

[84] Deutsch, J., Quaschning, V., Kemnitz, E., Auroux, A., Ehwald, H., and Lieske, H. (2000) *Top. Catal.* **13**, 281–285.

[85] Jin, T.-S., Yang, M.-N., Feng, G.-L., and Li, T.-S. (2004) *Synth. Commun.* **34**, 479–485.

[86] Breda, A., Signoretto, M., Ghedini, E., Pinna, F., and Cruciani, G. (2006) *Appl. Catal. A Gen.* **308**, 216–222.

[87] Kozhevnikov, I. V. (1998) *Chem. Rev.* **98**, 171–198.

[88] Kozhevnikov, I. V. (2003) *Appl. Catal. A Gen.* **256**, 3–18.

[89] Kaur, J., Griffin, K., Harrison, B., and Kozhevnikov, I. V. (2002) *J. Catal.* **208**, 448–455.

[90] Kaur, J., Kozhevnikova, E. F., Griffin, K., Harrison, B., and Kozhevnikov, I. V. (2003) *Kinet. Catal.* **44**, 175–182.

[91] Cardoso, L. A. M., Alves Jr., W., Gonzaga, A. R. E., Aguiar, L. M. G., and Andrade, H. M. C. (2004) *J. Mol. Catal. A Chem.* **209**, 189–197.

[92] Khder, A. S., Hassan, H. M. A., and El-Shall, M. S. (2012) *Appl. Catal. A Gen.* **411–412**, 77–86.

[93] Kaur, J. and Kozhevnikov, I. V. (2002) *Chem. Commun.* 2508–2509.

[94] Firouzabadi, H., Iranpoor, N., and Nowrouzi, F. (2003) *Tetrahedron Lett.* **44**, 5343–5345.

[95] Fuchs, V. M., Pizzio, L. R., and Blanco, M.B. (2008) *Catal. Today* **133–135**, 181–186.

[96] Olah, G. A., Prakash, G. K. S., and Sommer, J. (1985) *Superacids*, John Wiley & Sons, Inc., New York.

[97] Tanabe, K., Hattori, H., and Yamaguchi, T. (1990) *Crit. Rev. Surf. Chem.* **1**, 1–25.

[98] Olah, G. A., Mathew, T., Farnia, M., and Prakash, G. K. S. (1999) *Synlett* 1067–1068.

[99] Yamato, T., Hideshima, C., Prakash, G. K. S., and Olah, G. A. (1991) *J. Org. Chem.* **56**, 3955–3957.

[100] Schuster, H. and Hölderich, W. F. (2008) *Appl. Catal. A Gen.* **350**, 1–5.

[101] Kodomari, M., Suzuki, Y., and Yoshida, K. (1997) *Chem. Commun.* 1567–1568.

[102] Sarvari, M. H. and Sharghi, H. (2005) *Helv. Chim. Acta* **88**, 2282–2287.

[103] Fritch, J. R., Fruchey, O. S., and Horlenko, T. (1990) US 4 954 652; (1991) *Chem. Abstr.* **114**, 81258.

[104] Uwaydah, I. M., Aslam, M., Brown, C. H., Fitzhenry, S. R., and Mcdonough, J. A. (1997) US 5696274; (1997) *Chem. Abstr.* **127**, 135720.

[105] Climent, M. J., Corma, A., Iborra, S., and Primo, J. (1995) *J. Catal.* **151**, 60–66.

[106] Drexler, M. T. and Amiridis, M. D. (2003) *J. Catal.* **214**, 136–145.

[107] Neves, I., Jayat, F., Magnoux, P., Pérot, G., Ribeiro, F. R., Gubelmann, M., and Guisnet, M. (1994) *J. Mol. Catal.* **93**, 169–179.

[108] Jackman, L. M., Petrei, M. M., and Smith, B. D. (1991) *J. Am. Chem. Soc.* **113**, 3451–3458.

[109] Sartori, G., Casnati, G., Bigi, F., and Predieri, G. (1990) *J. Org. Chem.* **55**, 4371–4377.

[110] Chapman, N. B. and Shorter, J. (1978) *Correlation Analysis in Chemistry*, Plenum Press, New York, pp 357–396.

[111] Bigi, F., Maggi, R., Sartori, G., and Casnati, G. (1992) *Gazz. Chim. Ital.* **122**, 283–289.

[112] Murashige, R., Hayash, Y., Ohmori, S., Torii, A., Aizu, Y., Muto, Y., Murai, Y., Oda, Y., and Hashimoto, M. (2011) *Tetrahedron* **67**, 641–649.

[113] Kawada, A., Mitamura, S., and Kobayashi, S. (1993) *J. Chem. Soc. Chem. Commun.* 1157–1158.

[114] Kobayashi, S. and Nagayama, S. (1998) *J. Am. Chem. Soc.* **120**, 2985–2986.

[115] Subba Rao, Y. V., Kulkarni, S. J., Subrahmanyam, M., and Rama Rao, A. V. (1995) *Appl. Catal. A Gen.* **133**, L1–L6.

[116] Wang, Q. L., Torrealba, M., Giannetto, G., Guisnet, M., Pérot, G., Cahoreau, M., and Casso, J. (1990) *Zeolites* **10**, 703–706.

[117] Neves, I., Jayat, F., Magnoux, P., Pérot, G., Ribeiro, F. R., Gubelmann, M., and Guisnet, M. (1994) *J. Chem. Soc. Chem. Commun.* 717–718.

[118] Rohan, D., Canaff, C., Magnoux, P., and Guisnet, M. (1998) *J. Mol. Catal. A Chem.* **129**, 69–78.

[119] Bhattacharyya, K. G., Talukdar, A. K., Das, P., and Sivasanker, S. (2001) *Catal. Commun.* **2**, 105–111.

[120] Sharghi, H., Hosseini-Sarvari, M., and Eskandari, R. (2006) *Synthesis* 2047–2052.

[121] Myata, A., Matsunaga, K., and Ishikawa, M. (1995) JP 07089894; (1995) *Chem. Abstr.* **123**, 82946.

[122] Farkas, J., Békássy, S., Ágai, B., Hegedûs, M., and Figueras, F. (2000) *Synth. Commun.* **30**, 2479–2485.

[123] Békássy, S., Farkas, J., Ágai, B., and Figueras, F. (2000) *Top. Catal.* **13**, 287–290.

[124] Bolognini, M., Cavani, F., Cimini, M., Dal Pozzo, L., Maselli, L., Venerito, D., Pizzoli, F., and Veronesi, G. C. (2004) *R. Chim.* **7**, 143–150.

[125] Sheldon, R. A. (1997) *Chem. Ind.* 12–15.

[126] Kobayashi, S. (1994) *Synlett* 689–701.

[127] Sheldon, R. A. and Van Bekkum, H. (2001) *Fine Chemical Through Heterogeneous Catalysis*, Wiley-VCH, Weinheim.

[128] Vogt, A. and Pfenninger, A. (1996) EP 0,701,987.

[129] Xu, M., Cheng, M., and Bao, X. (2000) *Chem. Commun.* 1873–1874.

[130] García-Martínez, J., Cazorla-Amorós, D., Linares-Solano, A., and Lin, Y. S. (2001) *Microporous Mesoporous Mater.* **42**, 255–268.

[131] Nijhuis, T. A., Beers, A. E. W., Vergunst, T., Hoek, I., Kapteijn, F., and Moulijn, J. A. (2001) *Catal. Rev.* **43**, 345–380.

[132] Schlögl, R. (1993) *Angew. Chem. Int. Ed. Engl.* **32**, 381–383.

[133] Gladysz, J. A. (2001) *Pure Appl. Chem.* **73**, 1319–1324.

[134] Lempers H. E. B. and Sheldon, R. A. (1998) *J. Catal.* **175**, 62–69.

4

THE USE OF QUANTUM CHEMISTRY FOR MECHANISTIC ANALYSES OF S_EAr REACTIONS

Tore Brinck and Magnus Liljenberg
Applied Physical Chemistry, KTH Royal Institute of Technology, Stockholm, Sweden

4.1 INTRODUCTION

4.1.1 Historical Overview of Early Quantum Chemistry Work

Heitler and London made the first true quantum chemical calculation on the hydrogen molecule (H_2) in 1927 [1]. Their approach was later extended by Slater and Pauling to become what today is known as the valence bond (VB) method [2–5]. It is based on representing the total molecular wavefunction by functions that describe the pairwise interaction between atoms and thus mirrors the intuitive concept of a chemical bond. An extension of this method for many-atom systems has been developed by Goddard and coworkers and is commonly referred to as the generalized valence bond (GVB) method [6]. Although the VB method is conceptually well suited for describing bond breaking and bond formation processes, it has been applied rather sparsely to chemical problems. The reason can be traced to the difficulty of transforming the VB formalism into equations that can be solved efficiently by modern computers. Another approach, originally developed by Mulliken, Hund, Slater, and Lennard-Jones and usually called the molecular orbital (MO) method, has instead been the dominant ansatz in computational quantum chemistry [7–11]. In the MO method, the electrons are described by one-electron functions (orbitals) delocalized over the entire molecule. MOs appear as eigenfunctions to the Fock operator when the Hartree–Fock equations are applied to molecular systems [11, 12]. When the MOs are described as linear combinations of atom-centered basis functions, the Hartree–Fock equations can be solved efficiently on a computer using Roothaan's equations [13].

Arene Chemistry: Reaction Mechanisms and Methods for Aromatic Compounds,
First Edition. Edited by Jacques Mortier.

The downside of the Hartree–Fock method lies in the neglect of the instantaneous correlation of the motion of electrons, which in many instances result in erroneous potential energy surfaces (PES) of chemical reactions. The traditional approach to remedy the lack of electron correlation in Hartree–Fock has been to expand the wavefunction as a linear combination of antisymmetric wavefunctions (configuration functions) obtained by utilizing different combinations of the occupied and virtual Hartree–Fock MOs [14]. These so-called post-Hartree–Fock methods are in principle capable of producing the exact solution but are computationally very demanding for larger molecules. In some cases, such as open-shell singlets, the Hartree–Fock orbitals are not well suited for describing the dominating configuration functions, and an alternative approach, multiconfigurational SCF (MC-SCF) theory, is often used. In this approach, the orbitals are optimized to minimize the energy of the configuration interaction wavefunction [14–16]. The MC-SCF theory is computationally very demanding for large systems and has to be supplemented with additional expansions of the wavefunction to include dynamic electron correlation effects.

During the last 20 years, another MO method, Kohn–Sham density functional theory (KS-DFT), has emerged as the dominating approach for computational analysis of chemical reactions [17]. KS-DFT is unlike Hartree–Fock not based on traditional wavefunction theory, and the MOs are merely used for computing the kinetic energy and for representing the electron density. The theoretical foundation of KS-DFT comes from the Hohenberg–Kohn theorem, which states that the energy and all other ground state properties of a molecular system are uniquely defined by the electron density [18]. The big advantage of KS-DFT is that electron correlation effects are considered, even though the computational procedure and cost are very similar to Hartree–Fock. However, like Hartree–Fock theory, KS-DFT has problems describing systems that are not well represented by a single configuration function.

The early attempts to analyze aromatic molecules and their reactivity utilized for obvious reasons approximate MO theory, particularly the Hückel method and its variants [19]. The Hückel method can be seen as based on Hartree–Fock theory, but only the π-electrons are considered and most integrals are neglected. The final energy expression is only dependent on two types of integrals, which are not calculated but given empirical estimates. Most early attempts to calculate the orientation in aromatic substitutions, both electrophilic and nucleophilic, were focused on determining the net atomic charges at the possible points of substitution. Wheland was the first to use MO theory to find the energies of the structures contributing to the activated complex [20]. He associated the rate of reaction to the loss of π-electron binding energy when the carbon undergoing attack changes its hybridization to sp^3 and is removed from conjugation. In this way, he could treat electrophilic, nucleophilic, and radical reagents in a unified way. Even though he used a slight variation of the Hückel method and the few adjustable parameters that appeared were assigned in a rather arbitrary manner, the calculated effects were of correct order of magnitude both regarding the rates of substitution and the resonance moments. Therefore, the work of Wheland has to be regarded as a groundbreaking contribution to the theoretical analysis of electrophilic aromatic substitution (S_EAr) reactions.

Any historical expose like this would be inadequate without considering the work of Michael Dewar. In the early 1950s, he extended and generalized the original quantum mechanical treatment of Hückel by using resonance and perturbation theory. He showed that the use of resonance theory in the study of many reaction types, including S_EAr reactions, gave qualitatively correct predictions of reaction rates and orientation for all the investigated substituent types and annular heteroatoms [21–23]. One example is his study of the carbocyclic aromatics—benzene, naphthalene, phenanthrene, anthracene, naphthacene, and pentacene [22]. It is fair to say that this early work of Dewar was the starting point of the modern era of computational organic chemistry.

Later, he was instrumental in the development and use of semiempirical quantum chemistry methods, like MINDO, MNDO, and AM1, for analysis of organic reactions [24]. Semiempirical methods are generally based on the Hartree–Fock formalism, but the computational effort is reduced by various approximations to the two-electron and overlap integrals that appear in Hartree–Fock theory [25]. Integrals are completely omitted or approximated by simpler expressions, and Dewar pioneered their fitting to experimental data to obtain accurate reaction energetics. In 1965, Dewar and Thompson calculated electrophilic localization energies for a set of 21 nonequivalent positions in 12 aromatic hydrocarbons and showed that a semiempirical SCF-MO method gave better correlation than the Hückel method with experimental partial rate factors [26].

Since the advent of KS-DFT into computational quantum chemistry in the beginning of the 1990s, semiempirical methods have largely lost their importance to the field. A large number of theoretical studies on S_EAr have appeared in the literature during the last decades. Although KS-DFT has been the most commonly applied method, ab initio quantum chemical methods have also been used to a significant extent. Before we dwell into the results of these studies, we will summarize the knowledge acquired from experiment.

4.1.2 Current Mechanistic Understanding Based on Kinetic and Spectroscopic Studies

Let us, as a short introduction, repeat the putative mechanism for S_EAr, with nitration as model example, outlined in Scheme 4.1. It starts with the usually rapid and reversible complexation of the active electrophile with the π-system of the aromatic ring, a species commonly referred to as the π-complex. This species is often considered to be unoriented, that is, no positional selectivity is associated with it. In order for the substitution process to proceed, the π-complex must react to form another reaction intermediate, the σ-complex, a species also known as the Wheland intermediate or the arenium ion. In this intermediate, the cyclic conjugation of the aromatic system is broken, and the carbon at the site of the substitution is tetravalent and bonded via σ-bonds to both the nitrogen of the electrophile and the leaving group (H^+). Most S_EAr reactions are kinetically controlled, as the formation of the σ-complex is essentially irreversible and a proton is eliminated in the last fast step, giving the product. The formation of the σ-complex is generally considered the rate-limiting step.

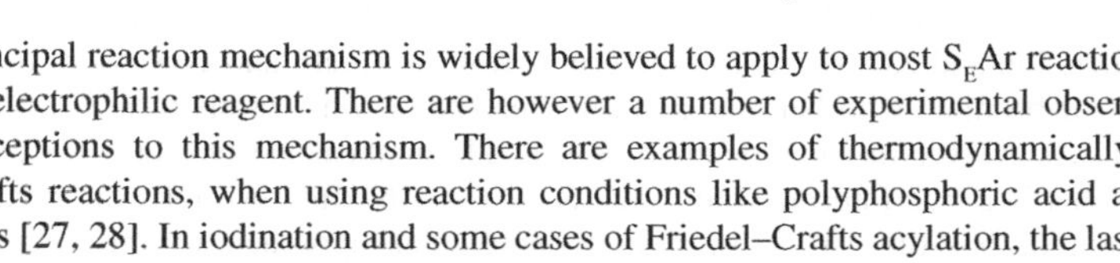

SCHEME 4.1 The putative mechanism for S_EAr nitrations.

This principal reaction mechanism is widely believed to apply to most S_EAr reactions irrespective of the electrophilic reagent. There are however a number of experimental observations that indicate exceptions to this mechanism. There are examples of thermodynamically controlled Friedel–Crafts reactions, when using reaction conditions like polyphosphoric acid and elevated temperatures [27, 28]. In iodination and some cases of Friedel–Crafts acylation, the last step of the reaction, the proton abstraction, has been shown to have a substantial kinetic isotope effect, which indicates that this step is at least partially rate limiting [29–31]. There are also still open questions regarding the exact nature of the reaction intermediates, and we will focus on these issues in the remaining part of the chapter.

Our current understanding of the S_EAr mechanism is largely based on the interpretation of kinetic data. The existence of a σ-complex (or Wheland) intermediate was first proposed by Pfeiffer and Winzinger in 1928. Ingold and Hughes deduced the general S_EAr mechanism of nitrations in 1946 based on kinetic measurements and showed that the active electrophile is the nitronium ion (NO_2^+). The work of Melander from 1949 was instrumental for establishing the S_EAr mechanism as multistep with a fast deprotonation step that is subsequent of the rate-determining step [32, 33]. Melander was able to demonstrate that there is only a small tritium isotope effect ($k_H/k_T < 1.3$) in nitrations and brominations of benzene derivatives. Brown and Stock later found good correlations between relative rates of halogenation and relative stabilities of σ-complexes, which indicated that the rate-determining transition state of S_EAr is similar to the σ-complex [34]. The generality of this interpretation was challenged by Olah based on the low substrate selectivity coupled with high positional selectivity for nitrations of activated aromatics [35]. Olah argued that these observations could be explained if the formation of a π-complex is the rate-determining step for strong electrophiles, such as NO_2^+. The existence of a π-complex had first been suggested by Dewar in 1946 [36]. Olah originally proposed that the π-complex must be oriented to allow for the observed positional selectivity [37]. However, he later revised this interpretation and claimed that the formation of the σ-complex determines the positional selectivity even for those S_EAr reactions where the rate-determining transition state resembles the π-complex [35].

Perrin in 1977 suggested that a one-electron transfer followed by the formation of a radical–radical cation complex is a more plausible mechanism for nitration of substrates more activated than toluene [38]. He showed that the electrochemical formation of the naphthalene radical cation in the presence of NO_2 gives the same product ratio as that formed in nitration. Kochi and coworkers followed up on this hypothesis and substantiated it further through the study of electron donor–acceptor complexes involved in the photochemical as well as in thermal activation of aromatic nitration [39, 40]. Even though the importance of the single-electron transfer (SET) mechanism in the case of nitration of activated aromatics has received a large acceptance, the details of the mechanism are still under debate and have been the focus of numerous theoretical studies. We will return to this issue in the next section of this chapter. There are also other mechanistic details of S_EAr reactions that have been difficult to understand from kinetic measurements. The kinetics of halogenations is, for example, frequently complex [41]. Also for Friedel–Crafts reactions, the kinetic data are often difficult to interpret, partly because promoters (often Lewis acids, like $AlCl_3$) form complexes with reactants, solvents, and products and these complexes interconvert in an often unknown manner during the course of the reaction [42].

In addition to kinetic measurements, the isolation and characterization of reaction intermediates have played key roles for reaching our current understanding of the S_EAr reaction. The π-complexes formed from the interaction of hexamethylbenzene and nitrosonium and iodonium ions have been observed by NMR and UV-Vis spectroscopy [43]. In the former case, it could also be demonstrated by low-temperature X-ray crystallography that the nitrosonium cation is located in a central position above the planar arene ring [43]. More recently, the gas-phase π-complexes of nitrosonium ion and benzene have been analyzed by IR multiple-photon dissociation (IRMPD) spectroscopy, and this study provided support of earlier theoretical studies indicating that S_EAr with the nitrosonium as electrophile differs in mechanism from general S_EAr (vid. infra) [44]. Of greater relevance were the crystallization and X-ray structure determination by Kochi and coworkers of metastable π-complexes of bromine with benzene and toluene. The brown crystals formed from bromine and benzene slowly evolved into hydrogen bromide and bromobenzene even at –78°C, suggesting that the π-complex is a true intermediate of the S_EAr [45, 46]. Interestingly, the π-complexes were not unoriented, but the Br_2 showed a perpendicular approach to the ring with C-atom coordination. In the case of the complexes with toluene, preferences for coordination to the *ortho* and *para* positions were found.

The σ-complexes present in the S_EAr mechanism are even more difficult to detect than π-complexes due to their ultrafast deprotonation. This problem can be circumvented by the preparation of σ-complexes without aromatic hydrogens from reactions between electrophiles and hexasubstituted benzenes. The first such example was the preparation of the heptamethylbenzenium cation, which was characterized by NMR [47]. Using the same principle, σ-complex intermediates of nitration and chlorinations have been prepared and subsequently characterized by NMR and UV-VIS spectroscopy [48]. The σ-complexes formed from the interaction of hexamethylbenzene and the cations silyl, methyl, bromonium, and chloronium were crystallized and analyzed by X-ray diffraction [43].

In 2000, Hubig and Kochi claimed to have made the first observation of a true Wheland intermediate of an S_EAr when studying nitrosation of methyl-substituted benzenes by time-resolved spectroscopy [49]. Irradiation of a suspension of $NO^+BF_4^-$ in dichloromethane containing mesitylene with a laser pulse of 355 nm generated a transient species, which was identified as the radical ion pair. This species rapidly converted to another transient, which was assigned as the Wheland intermediate. However, theoretical studies have indicated that the reaction does not proceed via a radical ion pair and that the Wheland structure is a transition state rather than an intermediate of the nitrosation reaction [50]. This interpretation also found support in the analysis of the π-complex of nitrosonium ion and benzene by IRMPD [44].

4.2 THE S_EAr MECHANISM: QUANTUM CHEMICAL CHARACTERIZATION IN GAS PHASE AND SOLUTION

The main body of this section is devoted to nitration, which is the S_EAr that has been most extensively studied by computational quantum chemistry. The comparisons and overview are greatly facilitated by the fact that the active electrophile commonly is believed to be NO_2^+ and this species has been considered in the majority of the studies.

4.2.1 Nitration and Nitrosation

A large number of quantum chemical studies have been devoted to this topic. However, as we will see, there are still open questions regarding the mechanistic details of the nitration reaction. We will start this survey with two seminal articles from 2003 that were written independently of each other.

The first study is by Esteves, Olah, and coworkers [51]. In this work, the detailed PES for the nitration of benzene by the NO_2^+ ion in the gas phase was investigated by restricted KS-DFT with the B3LYP functional. An important finding is that the formation of the σ-complex is preceded by the passing of two intermolecular complexes on the PES. The first complex is of C_{6v} – symmetry with a perpendicular coordination of the NO_2 group toward the center of the aromatic ring. The structure of this complex is well in line with Olah's earlier proposal of an unoriented π-complex as a key species in the nitration of arenes. On the gas-phase PES, the π-complex lies 19.2 kcal/mol (enthalpy) below the free reactants (NO_2^+ and benzene) and 14.3 kcal/mol above the σ-complex. In the second complex, the N-atom of the NO_2 group is coordinated to one of the C-atoms of benzene in a structure similar to the σ-complex. However, the N—C distance is 2.0 Å, indicating that the interaction has only a limited covalent character. This is in contrast to the σ-complex where the C—N bond is 1.51 Å and where the hydrogen coordinated to the same carbon is significantly bent out of the plane. The second complex is 13.6 kcal/mol lower in enthalpy than the first, and thus, it is only 0.7 kcal/mol above the σ-complex. Esteves *et al.* interpret the formation of the second complex as an SET and the complex itself as a radical–radical cation pair. The main justification for this interpretation is the similarity between the geometries of the NO_2 group and the free NO_2 radical. In contrast to NO_2^+, which is linear, both these structures are bent with an O—N—O angle close to 135° (137° and 134°, respectively). The validity of this interpretation can be questioned on methodological grounds. First, based on data from the article, it is clear that the theoretical level used by the authors significantly overestimates the driving force for electron transfer in this system. According to reliable gas-phase experimental data, the SET between the separated reactants, that is, the reaction according to Equation 4.1, is exothermic by 7.4 kcal/mol [51]:

$$NO_2^+ + C_6H_6 \rightarrow NO_2^{\cdot} + [C_6H_6]^{+\cdot} \qquad (4.1)$$

However, at the B3LYP/6-311++G(d,p) level, the exothermicity of the same reaction is 21.8 kcal/mol. Consequently, the driving force for SET is overestimated by more than 14 kcal/mol. The second problem is that restricted KS-DFT is not capable of describing an open-shell electronic structure, such as a complex between two radicals. A proper description will require multiconfigurational wavefunction theory, such as MC-SCF. The authors were clearly aware of this problem and performed a GVB(3)-CASSCF(6,6) single point calculation at the B3LYP-optimized geometry. They claim a significant multiconfigurational character of the total wavefunction but provide no quantitative data on the contribution from the $|NO_2,C_6H_6^+\rangle$ configuration. Thus, the true character of the second complex is difficult to determine based on the article.

The second important article from 2003 is by Head-Gordon, Kochi, and coworkers [50]. They used coupled-cluster theory at the CCSD(T)/6-31G(d,p) level to analyze the PES for nitration of benzene. The CCSD(T) method has the advantages compared to B3LYP that it provides more accurate energetics and a better description of structures with a substantial multiconfigurational character. However, the relatively small basis set used in the study is likely to reduce the accuracy of the relative energies. In this work, only one noncovalent complex prior to the formation of the σ-complex was found. This complex is very similar in structure to the second complex described by Esteves *et al.* [51]. However, the CCSD(T) method gives a slightly longer N—C distance of 2.16 Å. The energy is only 2 kcal/mol higher than for the σ-complex, and it lies 25 kcal/mol below the separated reactants. Also, the σ-complex is similar in structure to that obtained by Esteves *et al.*, but it has a slightly longer N—C bond of 1.55 Å. The authors advocate that the reaction proceeds via an SET mechanism, but with the significant difference from Esteves *et al.* that the electron transfer takes place upon the formation of the σ-complex. Thus, the π-complex is not considered to be a radical–radical cation complex, but rather a charge-transfer complex between two closed-shell molecules. The authors provide a semiquantitative analysis of the PES in terms of the Mulliken charge-transfer formalism and Marcus–Hush theory for electron transfer. Due to the large reorganization energy associated with the reduction of NO_2^+, the PES has two minima, the π-complex and the σ-complex, which correspond to the electron configurations $|NO_2^+,C_6H_6\rangle$ and $|NO_2,C_6H_6^+\rangle$, respectively. Interestingly, Kochi had prior to this article advocated the same type of analysis to justify the formation of an SET complex before the σ-complex [52]. Clearly, the results from the quantum chemical calculations spawned the alternative interpretation.

The same study also analyzed the nitrosation reaction of benzene [50]. Although the NO^+ has a similar reduction potential as NO_2^+, this reaction proceeds at a rate that is 10^4 times slower than the corresponding nitration. A π-complex with a nearly perpendicular approach of NO, and the N-atom coordinated to the center of the aromatic ring, was identified. In agreement with an earlier study by Skokov and Wheeler, it was found that the nitrosonium reaction features no stable σ-complex; the σ-bonded Wheland structure is rather a transition state of the reaction [53]. This feature could also be explained by the Marcus–Hush theory [50]; the low reorganization energy of NO^+, about half compared to NO_2^+, gives a stronger coupling between the configurations $|NO^+,C_6H_6\rangle$ and $|NO,C_6H_6^+\rangle$, and this results in only one broad minimum on the PES, the π-complex. The results show that the nitrosonium reaction has a different mechanism compared to nitration and that care should be taken when interpreting other S_EAr reactions based on structures of complexes between arenes and NO^+.

In their 2006 study, Esteves and coworkers analyzed the initial intermediates—that is, the ring-centered coordinated π-complex, the C-atom coordinated π-complexes, and the σ-complexes—in the nitration of a range of substituted benzenes [54]. They suggest that the deactivated benzenes follow a traditional Ingold–Hughes S_EAr mechanism, whereas the activated ones follow an SET mechanism. However, according to their interpretation, there is no sharp transition between the two mechanisms but rather a continuous transformation toward the SET as the benzenes become more activated. They were not able to locate the ring-centered coordinated π-complex for the activated benzenes, and this was seen as indication of the SET mechanism. The C-atom coordinated π-complexes were, in line with their earlier study, considered to be radical–radical cation complex formed from the SET. However, the authors again used restricted KS-DFT (B3LYP), and thus, their study provides no theoretical analysis of the open-shell character of the C-atom coordinated π-complexes. To gain further insight into the importance of SET, results from mass spectrometry studies on the gas-phase reactions between NO_2^+ and selected aromatics were analyzed. The reaction between isolated NO_2^+ and benzene yields ionized benzene as the dominant product due to the energy dissipation problem. However, monosolvated NO_2^+ yields a mixture between ionized benzene, the $[NO_2^{+}\cdot C_6H_6]$ adduct, and ionized phenol. The relative amount of the ionized aromatics increases with the activation tendency of the substituent. Thus, the anisole spectrum has a large peak due to the ionized aromatic, whereas the corresponding peak is missing in the nitrobenzene mass spectrum. This was interpreted as increasing importance for SET in activated aromatics [54].

The next important study is by Haas and coworkers from 2010 [55]. They investigated the S_EAr mechanism and the importance of SET for a number of different nucleophiles using MC-SCF theory. The advantage with MC-SCF theory is that it can describe closed- and open-shell systems, for example, biradical systems, with equal accuracy. In addition, it is possible to analyze low-lying excited states and their crossings with the ground state in a consistent manner. The authors analyzed the gas-phase reaction of NO_2^+ and benzene at the CASSCF(10,9)/cc-*p*VDZ level. Although this system lies on an excited state surface for separated reactants, the authors find that the interacting system is on the ground state surface. This is attributed to the small difference in ionization potential between NO_2 and benzene and the large reorganization energy involved in the reduction of NO_2^+. The interaction of NO_2^+ and benzene leads to the formation of a C-atom coordinated π-complex that collapses into the σ-complex with only a small barrier (not determined). The π-complex and σ-complex lie 12.5 and 42 kcal/mol, respectively, below the separated reactants. Starting from the π-complex, the $|NO_2^+,C_6H_6\rangle$ state can cross with the $|NO_2,C_6H_6^+\rangle$ state to form radical pair that has an oxygen atom of the NO_2 group coordinated to the benzene ring. This open-shell intermediate is only 0.7 kcal/mol above the π-complex, and it readily converts to an oxygen-bonded σ-complex in a process that is exothermic by 59 kcal/mol. The authors further conclude, but without the support of MC-SCF calculations, that a SET S_EAr mechanism is favored for more activated aromatics. However, they emphasize that the SET mechanism is less important in solution.

Haas and coworkers also investigated the nitrosation of benzene and aniline [55]. For benzene, their results are in line with results of Head-Gordon and Kochi [50], that is, the low-lying π-complex is on the $|NO^+,C_6H_6\rangle$ surface, and there is no radical pair involved in the reaction. There is a large barrier for the conversion of the π-complex into the N-bonded nitroso derivative via σ-complex. However, in contrast to Head-Gordon and Kochi, they find the σ-complex to be a high-energy minimum rather than a transition state on the PES. In the case of aniline, the driving force for electron transfer is larger, and two stable radical pairs have been optimized. However, there are several electronic states that are close in energy, and it does not seem that either of the radical pairs is an intermediate in the formation of a σ-complex.

The main criticism that can be drawn upon the work of Haas and coworkers is that the theoretical level they use does not account for the effects of dynamic electron correlation. This can explain some of the discrepancies in the relative energies of different intermediates when compared to the results of, for example, the Head-Gordon study [50]. Thus, care should be taken when interpreting the detailed energetics of S_EAr from the Haas study. However, it should provide a good qualitative interpretation of the mechanistic details of the reaction. To go beyond this, a methodology that can combine an exact treatment of static correlation effects with dynamic electron correlation, such as MR-CI or CAS-PT2, would be needed. Such a treatment is computationally very demanding and has to our knowledge not yet been realized for the S_EAr.

At this point, it may be appropriate to summarize our current understanding of the mechanism for gas-phase nitration of aromatics. The reaction can proceed via the formation of two intermediates before the formation of the σ-complex. The first is a ring-centered coordinated π-complex that is intermediate in energy between the separated reactants and the σ-complex. The second is the C-atom coordinated π-complex that is only slightly higher in energy than the σ-complex.

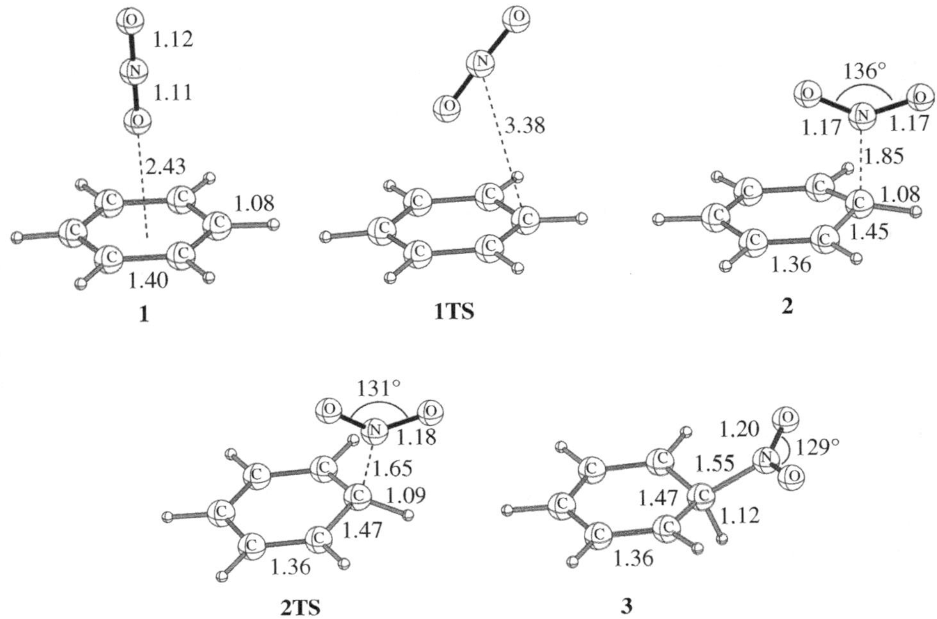

FIGURE 4.1 Structures of stationary points in the gas-phase nitration of benzene optimized at the M06-2X/6-311G(d,p) level. Bond lengths in angstroms and bond angles in degrees.

Whereas the second type of complex seems to be omnipresent in all nitrations, the first has not been found in all theoretical studies, and its stability generally decreases with increasing activation of the aromatic nucleophile. There seems to be a consensus that some degree of SET is present in nitrations and that this is most prominent for activated aromatics. However, there is little theoretical support for the interpretation of the C-atom coordinated π-complex as an SET complex.

We will now present recent, and yet unpublished, state-of-the-art computations on the nitration of benzene in the gas phase and in solution. Figures 4.1 and 4.2 show the structures of the stationary points and the free energy surface, respectively, for the nitration of benzene computed at the M06-2X/6-311G(2df,2p)//M06-2X/6-311G(d,p) level. The M06-2X functional generally provides more accurate energetics than the common B3LYP functional. This is particularly true for noncovalent complexes where the inclusion of dispersion effects in M06-2X is important. The overall driving force for SET, that is, the difference in ionization enthalpy between the NO_2 radical and benzene, is 13.0 kcal/mol. Thus, it is slightly overestimated compared to experiment, but to a much smaller extent than with the B3LYP functional. The initial step is the formation of the ring-centered π-complex (**1**), which lies 16 and 6 kcal/mol below the reactants in enthalpy and free energy, respectively. This intermediate is converted to the C-atom coordinated π-complex in a nearly barrierless process that is exergonic by 11 kcal/mol. In the C-atom coordinated π-complex (**2**), the NO_2 group has a geometry similar to the free NO_2 radical. However, the restricted KS determinant is stable toward becoming unrestricted. Considering that the stability of open-shell species generally is overestimated when represented by an unrestricted determinant, this strongly indicates that the C-atom coordinated π-complex is not a radical–radical cation complex resulting from an SET. The π-complex is converted to the σ-complex (**3**) without a significant barrier in a process that is exergonic by 4 kcal/mol. Thus, in agreement with earlier theoretical studies, the C-atom coordinated π-complex and the σ-complex are close in energy. The σ-complex lies 22 and 32 kcal/mol below the separated reactants in free energy and enthalpy, respectively.

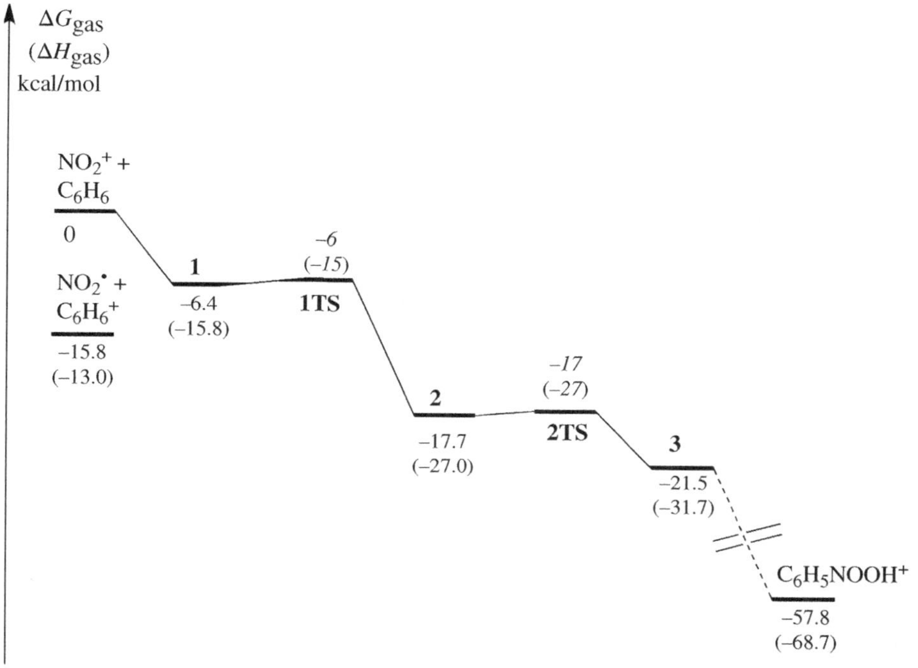

FIGURE 4.2 The free energy surface in kcal/mol for gas-phase nitration of benzene computed at the M06-2X/6-311G(2df,2p)//M06-2X/6-311G(d,p) level. Relative enthalpies of stationary points are in parentheses. Due to uncertainties in vibrational contributions, the relative standard free energies and enthalpies of transition states are given as integers (in italics).

The overall PES for nitration in the gas phase is very different from that in solution. In the gas phase, formation of the σ-complex from the free reactants is a very exothermic process with a zero overall barrier. However, compared to solution, the deprotonation of the σ-complex is a slower process. Different pathways for intramolecular proton transfer have been studied, including transfer to an oxygen as well as initial transfer to a nearby carbon [51, 55, 56]. We will not go into the details of these reactions, since they are of little relevance for solution chemistry.

Relatively few theoretical studies have been devoted to the nitration of aromatics in solution. The main reason stems from the difficulty of treating solute–solvent interactions within a quantum chemical framework. Although it is in principle possible to represent the solvent explicitly, such a procedure is generally not applicable, since a very large number of solvent molecules are needed to get an accurate representation of the PES for a solution reaction. This is particularly true for reactions involving ions or zwitterions. The most common approach for representing the solvent in quantum chemical calculations is instead by a dielectric continuum approach, such as the polarizable continuum model (PCM) [57]. This type of approach can be combined with a few explicit solvent molecules in cases where the solvent is expected to play an active role in bond breaking or bond formation [58].

Figure 4.3 shows the free energy surface for the nitration of benzene computed at the M06-2X/6-311G(2df,2p)/M06-2X/6-311G(d,p) level with PCM representation of the aqueous solution. The structures of the corresponding stationary states are depicted in Figure 4.4. A comparison of Figures 4.1 and 4.4 shows that essentially the same types of stationary points are found in gas phase and in solution, even though their detailed geometries are different. In contrast,

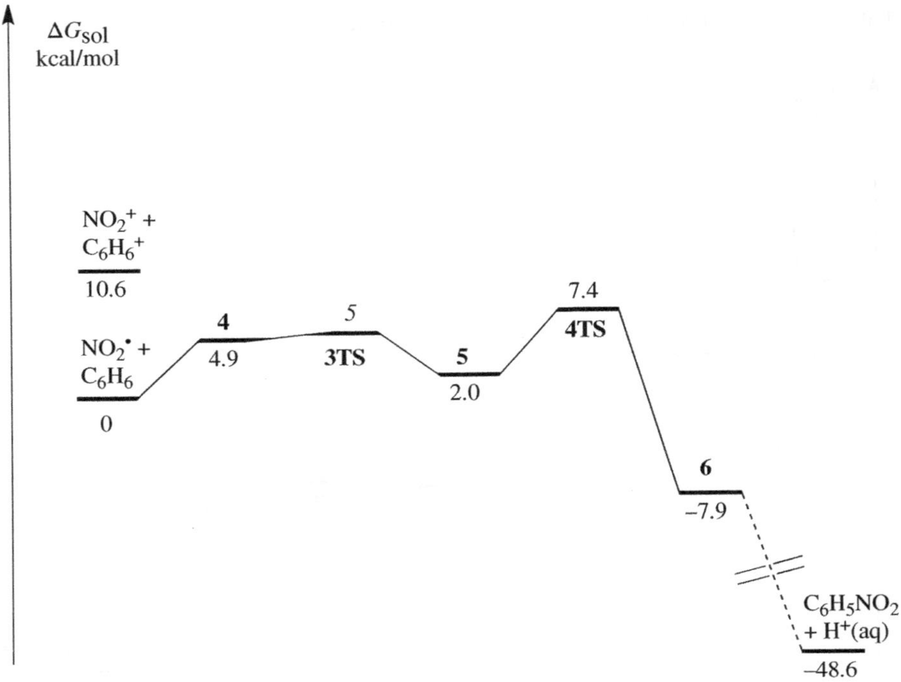

FIGURE 4.3 The free energy surface in kcal/mol for nitration of benzene in aqueous solution computed at the M06-2X/6-311G(2df,2p)//M06-2X/6-311G(d,p) level using SMD-PCM to account for solvation effects. Due to uncertainties in vibrational contributions, the relative standard free energy of **3TS** is given as an integer (in italics).

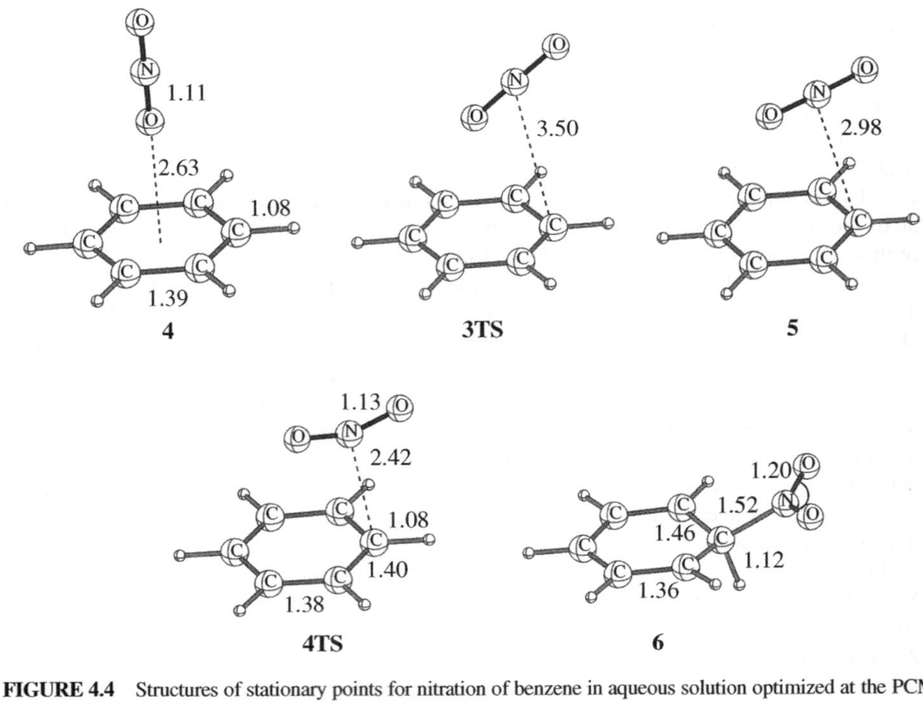

FIGURE 4.4 Structures of stationary points for nitration of benzene in aqueous solution optimized at the PCM-M06-2X/6-311G(d,p) level. Bond lengths in angstroms and bond angles in degrees.

the relative free energies of these points differ greatly between the two media. In aqueous solution, the formation of the first π-complex (**4**) is an endergonic process by 5 kcal/mol, which can be compared to exergonicity of 6 kcal/mol for the same process in gas phase. The N—C distance is slightly larger in solution indicating a reduced extent of charge transfer. However, the difference in structure between solution and gas phase is much larger for the second π-complex (**5**). Here, the solution complex can be viewed as a weak cation–molecule complex trapped in a solvent cage. The structure of the gas-phase complex (**2**), on the other hand, indicates a strong interaction and a significant degree of charge transfer. In terms of the structure, the transition state (**4TS**) for forming the σ-complex in solution is actually more weakly bonded than the second π-complex (**2**) in gas phase. Thus, the rate-determining transition state for nitration in solution is early and shows very little resemblance to the σ-complex. The energy of the transition state is also much closer to that of the π-complex than the σ-complex. This observation is in good agreement with the early proposal of Olah based on kinetic data of nitrations and stability data for π- and σ-complexes of arenes [35]. The structures of the σ-complex, though, are very similar in solution and gas phase.

We have also investigated the deprotonation of the σ-complex in solution by calculations including one or more explicit solvent molecules combined with PCM. This study will be discussed in more detail in a future publication. However, our results indicate that a water molecule can act as a base and deprotonate the σ-complex of a deactivated benzene in a nearly barrierless process. In the case of activated aromatics, such as phenol, we have been able to locate a transition state and estimate the barrier to a few kilocalories per mole. We can therefore conclude that even in the presence of only a weak base, such as water, the deprotonation is a faster process than the formation of the σ-complex.

4.2.2 Halogenation

The putative mechanism for aromatic halogenations is similar to that of nitrations, but unlike nitrations there is no commonly identified active electrophile. Molecular chlorine is believed to be the active electrophile in uncatalyzed chlorination, and the reaction follows the expected second-order kinetics [41]. The reaction is slow in nonpolar solvents, and under such conditions, it is catalyzed by the addition of acid. It has been suggested that the heterolytic cleavage of the Cl—Cl bond during the initial attack of the aromatics is promoted by protonation. Lewis acids, such as $AlCl_3$ or $FeCl_3$, are used to catalyze chlorinations. These reactions are overall third order and first order with respect to the Lewis acid [59]. The kinetics is consistent with the Cl^+ ion as the active electrophile, but there is little experimental or theoretical support for the formation of this species [60]. A more realistic hypothesis is the formation of a Lewis acid–Cl_2 complex that weakens the Cl—Cl bond and promotes the formation of the σ-complex. Hypochlorous acid is a relatively weak chlorinating agent that is activated in acidic solution. Acetyl hypochlorite is another effective chlorinating agent.

Similarly to chlorinations, molecular bromine is believed to be the electrophile of uncatalyzed brominations. In acidic solutions, the formation of the σ-complex seems to be reversible, and overall rate law depends on the bromide concentration [61]. Both the reaction rate and positional selectivity of bromination are heavily substituent dependent, which has been seen as an indication of a late transition state resembling the σ-complex in energy and structure [60]. Bromination is catalyzed by Lewis acid in a similar manner as chlorinations. Acetyl hypobromites are known to be very effective bromination agents, and the reactivity is enhanced by electron-withdrawing groups on the leaving group, that is, the activity is much higher for CF_3CO_2Br than CH_3CO_2Br [62]. Molecular iodine is a much weaker halogenating agent than bromine [60]. Fluorine, on the other hand, is too reactive and unselective to be of practical importance as electrophilic reagent.

Halogenation has been investigated in a number of theoretical studies. Most of these have focused on Cl_2 or Cl^+ as the active electrophile. Considering that there is little evidence for the formation of free Cl^+ in solution, we will focus on the former type studies.

In 2002, in a combined experimental and theoretical study, Zhang and Lund investigated the chlorination of toluene by Cl_2 in different solvents [63]. Computation was performed at the B3LYP/cc-pVTZ(-f) level with implicit consideration of solvent effects using a dielectric continuum model. They found two different reaction mechanisms depending upon the solvent polarity. In a low-polarity solvent and gas phase, they were not able to locate a stable arenium ion and instead predicted a concerted reaction mechanism with a transition state resembling a σ-complex but with a very long Cl—Cl distance. In solvent with a dielectric constant greater than 10, they found the σ-complex to be stable and the formation of the arenium ion and Cl^- to be an exothermic process. However, no transition state for the formation of the σ-complex was identified.

The subsequent year, Ben-Daniel *et al.* investigated the chlorination of benzene and its catalysis by alcohols [64]. The theoretical level was similar to that of Zhang and Lund, but it seems that solvent calculations were only performed a posteriori on optimized gas-phase structures. In the case of the uncatalyzed reaction, two stationary points prior to the σ-complex were characterized. The reaction begins with the formation of a π-complex, which has an almost perpendicular approach of Cl_2 toward the aromatic ring, that is, the C—Cl—Cl angle is close to 180°. This is a weak complex with a Cl—C distance of 4.4 Å. The next point, which is the transition state for forming the σ-complex, has very different structure, with a C—Cl—Cl angle close to 90° and the outermost Cl-atom pointing away from the ring. The Cl_2 unit is almost dissociated (Cl—Cl distance of 2.9 Å), and the Cl—C bond is relatively short (1.91 Å). Considering the very different geometries of these structures, it can be anticipated that there are additional stationary points that lie between them on the PES. The authors located also the σ-complex in which the Cl—Cl bond is further dissociated and the outermost Cl-atom (Cl^-) prepositioned for abstracting a proton. They found that the PES is extremely flat around this point and that the σ-complex easily collapses to the addition or substitution product. The activation energy for transformation from the π-complex to the σ-complex was computed to 41 kcal/mol. This is reduced to 12 kcal/mol in a solvent with a dielectric constant of 34. The effects of explicit alcohol catalysts, ethanol and trifluoroethanol, were further explored. In the high dielectric solution, the effect of ethanol was almost negligible, whereas trifluoroethanol reduced the barrier by an additional 6 kcal/mol. It was found that the catalyst provides a complementary charge template to the transition state.

In 2011, Filimonov *et al.* reported an analysis on the reaction mechanism for chlorination and iodination of substituted benzenes [65]. The computational level and the treatment of solvation effects are similar to that used in the study of Ben-Daniel *et al.* In the case of chlorination, the optimized stationary points are also similar in the two studies. It appears though that the structure of the rate-determining state has a slight dissimilarity in that the Cl_2 group in Filimonov's study is bent over the ring, rather than pointing away from the ring. This could also be the reason that they get a higher activation energy going from π-complex to the σ-complex in solution. Compared to the free reactants, the free energies in methanol for the π-complex, the transition state, and the σ-complex are −2, 23, and 22 kcal/mol, respectively. Filimonov *et al.* also reported a transition state for the deprotonation of the σ-complex. However, it is clear that there is a discontinuity in the PES since this latter transition state lies 13 kcal/mol (16 kcal/mol in gas phase) lower in free energy than the σ-complex.

In a recent study, Sakic and Vrcek investigated prereactive complexes in the chlorination (with Cl_2) of benzene, triazine, and tetrazine both in gas phase and in an apolar solvent (CCl_4) [66]. Using a stochastic search methodology, they were able to systematically investigate the configurational space by which Cl_2 can approach the aromatic π-system of benzene. With this methodology, they also find a number of new addition products and transition state structures. They could however not find the σ-complex of this system, neither in gas phase nor in CCl_4, without a counterion, like BF_4^- or $AlCl_4^-$, in close contact.

We have recently investigated the chlorination of substituted benzenes in aqueous solution at a corresponding level to that used for nitration (to be published). Figure 4.5 shows the structures of the π-complex (**7**), the rate-determining transition state (**5TS**), and the σ-complex (**8**). The π-complex is similar in geometry to earlier gas-phase structures. This is not surprising considering that it is a complex between two neutral molecules. There is a larger difference between gas phase and solution for the transition state; the transition state in solution has a shorter Cl—Cl bond of only 2.6 Å. In the σ-complex, the Cl—Cl

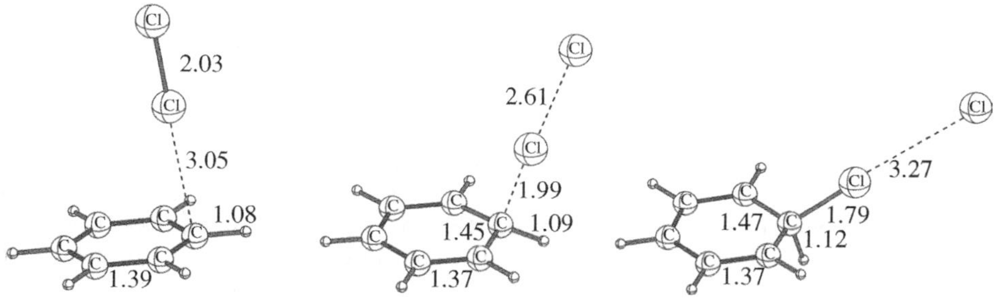

FIGURE 4.5 Structures of stationary points for chlorination of benzene in aqueous solution optimized at the PCM-M06-2X/6-31+G(d,p) level. Bond lengths in angstroms and bond angles in degrees.

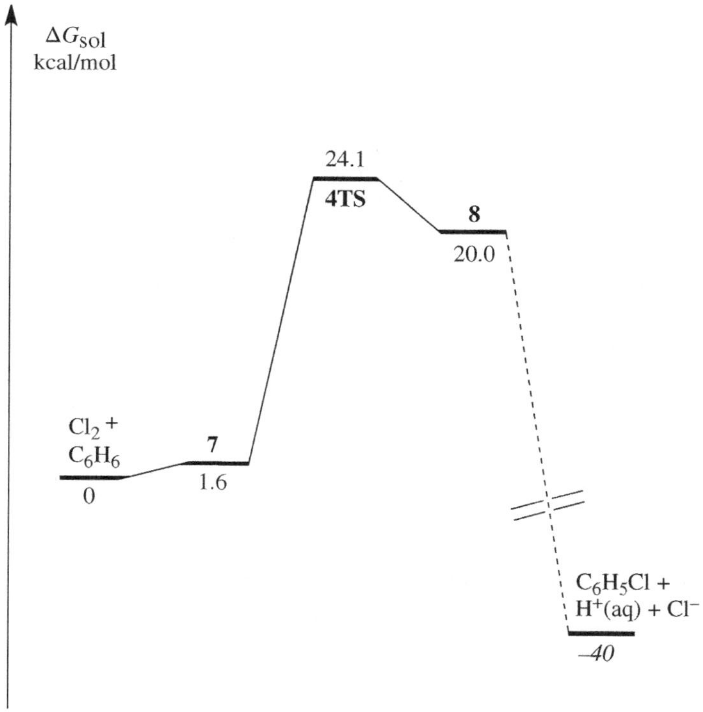

FIGURE 4.6 The free energy surface in kcal/mol for chlorination of benzene in aqueous solution computed at the M06-2X/6-311++G(2d,2p)//M06-2X/6-31+G(d,p) with a PCM representation of the solvent.

distance is increased to 3.3 Å, and the structure can essentially be viewed as an arenium ion with a Cl^- coordinated to the Cl substituent. We were not able to locate the corresponding arenium ion where the Cl^- is coordinated to leaving H^+. All attempts to optimize such a structure resulted in proton abstraction and the formation of HCl that dissociated from the chlorobenzene. We were also unsuccessful in optimizing a transition state that connects the dissociation **8** with the HCl formation. Considering the structure of **8**, it does not seem likely that the Cl^- formed from the attacking Cl_2 will serve as base in the deprotonation of the arenium ion. The free energy surface shown in Figure 4.6 confirms that this is a relatively slow reaction, where the formation of the σ-complex is the rate-determining step.

It is interesting to compare the rate-determining transition state of the nitration reaction in Figure 4.4 with the corresponding transition state for chlorination in Figure 4.5. It is clear that whereas the nitration transition state resembles the oriented π-complex, the chlorination structure is much closer to the σ-complex. Thus, Olah's original hypothesis is essentially confirmed [35].

Recently, Kong *et al.* published a manuscript that challenged the S_EAr mechanistic pathway for the reaction between Br_2 and different polybenzoid hydrocarbons, including benzene [67]. Based on computed PES in PCM solvent model (CCl_4) at the restricted B2-PLYP/6-311+G(2d,2p) level of theory, they showed that an addition–elimination mechanistic route can compete successfully with the standard S_EAr textbook mechanism in apolar solvents or in the gas phase. However, in order to obtain a reasonable reaction rate, most brominations are run in polar solvents, and this favors the S_EAr pathway as interactions with the solvent stabilize the σ-complex intermediate. Even though the classical S_EAr mechanism involving the σ-complex intermediate is not universally applicable, it is the most relevant mechanism when considering preparative or industrial applications.

4.2.3 Sulfonation

In aromatic sulfonation, the active electrophile is generally considered to be SO_3, either as a free species or complexed to a carrier, but detailed mechanistic studies have unfolded cases with many complexities like reversibility and different actual electrophiles [68]. In a recent article, Koleva *et al.* studied this reaction type with quantum computational methods at the M06-2X/6-311+G(2d,2p) and SCS-MP2/6-311+G(2d,2p) levels, both in gas phase and in solution [69]. Their results indicate that the mechanism does not involve a single SO_3 electrophile but that two SO_3 units participate in the reaction. They further found that the reaction proceeds via a single transition state through a concerted mechanism in gas phase and in apolar solvents like CCl_4 or $CFCl_3$. However, polar, higher dielectric SO_3-complexing media favor a classical S_EAr mechanism via a σ-complex arene-$(SO_3)_2$ dimer-type intermediate. The picture is also supported by experimental kinetic results showing that this reaction is second order in SO_3 in polar, complexing solvents, like CH_3NO_2 [70].

4.2.4 Friedel–Crafts Alkylations and Acylations

Apart from the arene substrate and an alkyl halide or acyl halide/acid anhydride, Friedel–Crafts alkylations and acylations most often involve a Lewis acid promoter in the form of a metal halide, like $AlCl_3$. Several species may function as the active electrophile, and both acyl halide/metal halide complexes and acylium ions have been observed experimentally. The complex forming ability of these metal halides complicates all mechanistic evaluations, and Friedel–Crafts reactions have rarely been the subject of quantum chemical mechanistic studies.

One exception is the study of Volkov *et al.*, who performed calculations in gas phase at the MP2/LANLZDZ(d)+ level on the $C_6H_7^+$ $M_2Cl_7^-$ system [71]. They found both for M = Al and M = Ga that the metal halide dimeric form of the σ-complex intermediate was considerably more stable than the corresponding monomeric form.

In another paper, Stashenko *et al.* studied the intramolecular Friedel–Crafts alkylation of *ortho*-allyl-*N*-benzylaniline, located all the stationary points on the PES with unrestricted Hartree–Fock theory, and computed energies using restricted MP2 theory [72]. However, this specific reaction was catalyzed with H^+ (sulfuric acid) and not with a metal halide.

4.3 PREDICTION OF RELATIVE REACTIVITY AND REGIOSELECTIVITY BASED ON QUANTUM CHEMICAL DESCRIPTORS

Historically, much of the theoretical analysis of the S_EAr reaction has been focused on the use of reactivity indices or, as they are today commonly referred to, reactivity descriptors. The basic assumption behind such studies is that the reactivity for a given substrate toward a particular chemical reaction can be correlated to ground state properties of the substrate. Taking the S_EAr reaction as an example, one would typically calculate properties of one or more aromatic substrates that reflect the strength of the most important interactions that take place upon forming the rate-determining transition state. Such an analysis can provide an indication of the reactivity at different positions of the aromatic nucleophile, that is, the regioselectivity for the reaction, but can also be used to compare the relative reactivity of different nucleophiles. It should be noted that reactivity indices work best for analyzing reactions with early transition states, where there are only minor perturbations in the electronic and nuclear structures of the interacting species upon forming the transition state. In addition, this type of approach is best suited for comparing substrates that have a similar chemical structure. It should further be noted that using reactivity indices as quantitative measures of relative reactivity generally requires calibration and validation against kinetic data obtained from experiments or accurate transition state calculations.

Although there are many drawbacks of using reactivity descriptors compared to quantum chemicals calculations of PES, the methodology has certain advantages. First of all, it is computationally inexpensive compared to rigorous transition state calculations. It is basically sufficient to perform one ground state calculation for each aromatic substrate rather than a labor- and computer-intensive probing of the PES to locate the rate-determining transition state. The computational simplicity of the descriptor approach allows automated scanning of a large number of substrates in a limited time. This type of approach can therefore, for example, be useful as a guide to synthesis planning. Another advantage is that reaction indices can provide information about the nature of the interaction that determines the variation of the activation energy among different substrates, and thus provide physical insight that can be difficult to obtain from a direct transition state calculation.

The early theoretical studies of the S_EAr reaction largely focused on correlating reactivity with partial atomic charges. This was noted already in the seminal article by Wheland from 1942 [20]. These analyses met rather limited success with respect to predicting regioselectivity and relative reactivity among congeners of substrates. One reason can be traced to the limited accuracy of the charge distributions obtained with the HMO method; for example, as noted by Fukui in 1952, the HMO predicts a uniformly zero charge for all unsaturated carbon atoms in a nonsubstituted aromatic hydrocarbon [73]. However, also more recent studies based on accurate electronic structure calculations have reached the conclusion that there is no clear correlation between S_EAr reactivity and ring carbon charges [69]. The physical basis for expecting such a correlation would be that the initial interaction leading to the formation of the transition state is largely electrostatic in nature. The limited success of charge analysis of S_EAr can thus be taken as an indication that electrostatics is not the most important interaction type for determining the activation energy. However, it should also be noted that atomic charge is not a rigorously defined physical property and that the entire charge distribution of the molecule must be considered for calculating the electrostatic interaction energy. A much better descriptor for analyzing electrostatic interactions is the molecular electrostatic potential ($V(\mathbf{r})$), which is defined by (Eq. 4.2)

$$V(\mathbf{r}) = \sum_A \frac{Z_A}{|\mathbf{R}_A - \mathbf{r}|} - \int \frac{\rho(\mathbf{r}')d\mathbf{r}'}{|\mathbf{r}' - \mathbf{r}|} \tag{4.2}$$

where Z_A is the charge on nucleus A, located at $\mathbf{R}_A$, and $\rho(\mathbf{r})$ is the electron density function. The electrostatic potential is a real physical property, which can be determined from experimental as well as theoretical electron density distributions. $V(\mathbf{r})$ gives the interaction energy between a point charge located at the position $\mathbf{r}$ and the unperturbed charge distribution of the molecule. To analyze chemical reactivity, $V(\mathbf{r})$ should preferably be computed on a molecular surface defined by a constant contour of electron density. The regions of negative $V(\mathbf{r})$ indicate positions where an electrophile is likely to attack, and the relative magnitudes of $V(\mathbf{r})$ at the most negative positions ($V_{S,min}$) are indicative of relative interaction strengths [74]. Although $V(\mathbf{r})$ is a much better measure of electrostatic interactions than atomic charges, using $V(\mathbf{r})$ as a sole descriptor for analyzing S_EAr has not proven to be a particularly effective approach. As noted by Murray and Politzer, $V(\mathbf{r})$ often varies slowly over the aromatic ring and does not always provide a clear indication of positional selectivity [75]. This indicates that other interactions, such as charge transfer and polarization, are likely to be important for determining activation energies of S_EAr.

Fukui pioneered the development of frontier molecular orbital (FMO) theory and its use for analyzing S_EAr [73, 76]. In contrast to atomic charges, which depend on all occupied orbitals, the FMO theory considers only the highest occupied molecular orbital (HOMO) and the lowest unoccupied molecular orbital (LUMO). The basis for using FMO for S_EAr is that the reaction has significant charge-transfer character and that electrons, according to Koopmans' theorem [77], are most easily transferred from the HOMO of the nucleophile to the LUMO of the electrophile. Thus, the density of the HOMO at the different aromatic carbons is expected to be indicative of positional selectivity for S_EAr. The HOMO density has also proven to be a more effective descriptor than atomic charges for predicting regioselectivity. However, this method fails to predict the correct ordering of carbon positions for some seemingly trivial cases, such as cyanobenzene and benzaldehyde [78]. This failure can be traced to the involvement of additional π-orbitals in the reaction.

On the basis of density functional theory, Parr and Yang defined a local property that they called the Fukui function $f(\mathbf{r})$ [79]. It is defined as the partial derivative of the electron density with respect to the total number of electrons (N) and taken at constant external potential $\upsilon(\mathbf{r})$ (Eq. 4.3):

$$f(\mathbf{r}) = \left(\frac{\partial \rho(\mathbf{r})}{\partial N}\right)_{\upsilon(\mathbf{r})} \tag{4.3}$$

Due to the discontinuity of the derivative at the value N, Parr and Yang proposed two reaction indices, $f^-(\mathbf{r})$ and $f^+(\mathbf{r})$, that they associated with the susceptibility toward the attack of an electrophile and a nucleophile, respectively. These functions are in the finite difference approximation given by (Eqs. 4.4 and 4.5)

$$f^-(\mathbf{r}) \approx \rho_N(\mathbf{r}) - \rho_{N-1}(\mathbf{r}) \tag{4.4}$$

$$f^+(\mathbf{r}) \approx \rho_{N+1}(\mathbf{r}) - \rho_N(\mathbf{r}) \tag{4.5}$$

In the further approximation of neglecting orbital relaxation upon oxidation and reduction, these functions are reduced to the HOMO and LUMO density, respectively, thus converging to the FMO theory. Condensed versions of the functions have also been defined in terms of atomic charges, q_k, of the ground state and the ions (Eqs. 4.6 and 4.7) [80]:

$$f_k^- = q_k(N) - q_k(N-1) \tag{4.6}$$

$$f_k^+ = q_k(N+1) - q_k(N) \tag{4.7}$$

The $f^-(\mathbf{r})$ performs generally much better than HOMO density for predicting positional selectivity of S_EAr [78]. However, it fails in certain cases, such as nitrobenzene where it underestimates the high selectivity for the *meta* position.

Ayers and coworkers derived a general reactivity index by combining $V(\mathbf{r})$ computed from atomic charges with an approximate Fukui function, f^-_{app}, *computed* from the condensed Fukui functions [81]. In the case of electrophilic attack, the reactivity index takes the form (Eq. 4.8)

$$\Xi^{\kappa}_{\Delta N\leq 0}=(\kappa+1)V_{\text{app}}(\mathbf{r})-\Delta N(\kappa-1)f^-_{\text{app}}(\mathbf{r}) \tag{4.8}$$

The κ parameter describes the character of the interaction. In the case of strong electrostatic control, κ has a value greater than 1, and in the case of a strong charge-transfer interaction, it is more negative than −1. However, most reactions lie in the intermediate region. The amount of charge transfer ΔN can be estimated from the ionization potentials and electron affinities of the electrophile and nucleophile. Using this general equation, Ayers and coworkers were able to describe the regioselectivity of S_EAr for the challenging borazarophenanthrenes, which due to their ionic character cannot be described as dominantly a charge-transfer interaction.

The average local ionization energy $I(\mathbf{r})$ is probably the most effective single reaction index for analyzing regioselectivity and relative reactivity of S_EAr. The $I(\mathbf{r})$ function was first proposed by Sjoberg *et al.* in 1990 [82] and is rigorously defined within the frameworks of Hartree–Fock theory and KS-DFT by (Eq. 4.9)

$$I(\mathbf{r})=\sum_i^{\text{occ}}\frac{-\varepsilon_i\rho_i(\mathbf{r})}{\rho(\mathbf{r})} \tag{4.9}$$

where ε_i is the energy of orbital i and $\rho_i(\mathbf{r})$ is the density of the orbital. According to Koopmans' theorem [77], the absolute values of the orbital energies are approximations to the ionization energies, and $I(\mathbf{r})$ can be interpreted as the average energy needed to ionize an electron at a point $\mathbf{r}$ in the space of a molecule or atom.

The $I(\mathbf{r})$ computed on a molecular surface defined by a contour of constant electron density is an effective tool for analyzing reactivity toward electrophiles [83]. The positions on the molecular surface where $I(\mathbf{r})$ has its lowest values, the local surface minima ($I_{S,\min}$), are indicative of the positions with the least tightly bound electrons and thus the sites that are most prone to electrophilic attack. In contrast to $V(\mathbf{r})$, $I(\mathbf{r})$ not only reflects a molecule's tendency for electrostatic interaction but also its ability for charge transfer and polarization interactions [84]. $I(\mathbf{r})$ has been shown to be particularly well suited to describe reactivity toward electrophiles in solution where electrostatic interactions are shielded and their importance is reduced [85].

Already, in the original study, it was demonstrated that $I(\mathbf{r})$ can predict regioselectivity for S_EAr. It was shown for benzene and eight substituted benzenes that the lowest $I_{S,\min}$ are found at the positions that are preferred for nitration and halogenation, for example, in toluene, $I_{S,\min}$ of similar magnitude are found over the *ortho* and *para* positions, whereas in nitrobenzene the lowest $I_{S,\min}$ is found at the *meta* position [82]. Even anomalous systems, such as protonated aniline ($C_6H_5NH_3^+$), which is a *meta*/*para* director, were correctly predicted. Furthermore, it was demonstrated that there is a linear correlation between the $I_{S,\min}$ at the *para* and/or *meta* positions for each system and the corresponding Hammett constant (σ_p or σ_m). Thus, the magnitude of the $I_{S,\min}$ reflects the relative reactivity of the different systems toward S_EAr. Later studies showed that $I(\mathbf{r})$ reflects positional selectivity and relative reactivity also in heteroaromatic systems [75, 86, 87].

In a more recent study, we investigated the capacity of $I(\mathbf{r})$ for quantitative prediction of regioselectivity in different types of S_EAr, including halogenation, nitration, and Friedel–Crafts acylation, and for more than 20 different aromatic substrates [85]. On the basis of the observed (experimental) positional selectivity, we computed the relative free energies of the transition states. In general, it was found that the differences in $I_{S,\min}$ between different positions were around three times larger than the differences in free energy between the corresponding transition states. A simple rationalization for this discrepancy in the magnitude of the energy differences is that only a fraction of an electron is transferred to the electrophile at the transition state and thus only a fraction of the local ionization energy is likely to contribute to the barrier. When the $I_{S,min}$ values were scaled by 1/3, that is, $I_{S,min}/3$, we were able to get good quantitative agreement with experimental regioselectivity for halogenations. The mean absolute deviation (MAD) between relative $I_{S,min}/3$ and relative transition state free energies was 1.1 kcal/mol [85]. The agreement with the experiment was slightly worse for nitrations where the MAD was 1.6 kcal/mol. In particular, $I(\mathbf{r})$ failed to predict the most reactive site for two systems with bulky substituents where the positions that have the lowest $I_{S,min}$ values are sterically hindered. The $I(\mathbf{r})$ approach did not perform very well for Friedel–Crafts acylations and only in a few cases where the lowest $I_{S,min}$ was found at the position corresponding to the dominating isomer. However, also other methods have problems with Friedel–Crafts acylations, and the regioselectivity seems to be highly dependent on reaction conditions, for example, thermodynamic rather than kinetic control of the reaction is not uncommon [85]. In addition, the reduced performance of the $I(\mathbf{r})$ approach for this type of reaction may be correlated to the bulkiness of the electrophiles, which renders steric effects more important.

The utility of $I(\mathbf{r})$ for predicting S_EAr has been further explored by Brown and Cockroft in a recent study [88]. They found a general correlation with a linear correlation coefficient of $R^2 = 0.86$ between partial rate constants for a large number of S_EAr reactions including nitration, bromination, chlorination, mercuration, ethylation, and benzylation reactions. The quality of the correlations improved significantly ($R^2 = 0.95 - 0.99$) when subsets of data from individual studies were selected. These correlations were of similar quality as those obtained with the computationally more elaborate electrophile affinity method, in which the activation energy is estimated from the energy difference between the σ-complex and the reactants (vid. infra). Brown and Cockroft further explored the capacity of the $I(\mathbf{r})$ approach for predicting regioselectivity and came to the conclusion that the method is much more reliable than traditional approaches, such as partial atomic charges, the electrostatic potential, FMO, and the Fukui function. It was found to predict the right isomer distribution also for problematic cases, such as the nitration of 1-nitronaphthalene and electrophilic substitution on borazarophenanthrenes, where the traditional reactivity indices fail.

4.4 QUANTUM CHEMICAL REACTIVITY PREDICTION BASED ON MODELING OF TRANSITION STATES AND INTERMEDIATES

4.4.1 Transition State Modeling

The most direct method for computing the reactivity is to characterize all the stationary points along the PES and to estimate the rate constant from the free energy difference between the rate-determining transition state and the reactants. This requires a rather elaborate computational procedure, which is difficult to automate. The computations are complicated by the fact that most S_EAr reactions take place in solution and a proper representation of solvation effects is crucial in order to obtain accurate kinetic data. Furthermore, estimation of entropy effects can be difficult, especially estimating the entropy of the free reactants in solution. In this context, it is not surprising that this full procedure has mostly been used to study the mechanisms of S_EAr reactions and not so frequently for computing relative reactivity in congeneric series of aromatics or for estimating regioselectivity.

One exception is the work of Arrieta and Cossio, which investigated the nitration of benzene, phenol, and benzonitrile at the B3LYP/6-31G(d) level within the PCM solvent model (where they used diethyl ether as model solvent), with special focus on the loss of aromaticity [89]. They identified the structure of the σ-complex and the transition state preceding it, but their use of the energies of this transition state gave a qualitatively wrong prediction of the isomer distribution of benzonitrile, as the *meta* isomer was not predicted to be the most stable one. This result did not change at higher levels of theory, for example, G3B3. The results for phenol were in good agreement with experiment when it comes to the *ortho*/*para* directing tendency.

4.4.2 The Reaction Intermediate or Sigma-Complex Approach

Since transition state calculations are still quite demanding, other means may be employed to estimate the transition state energy. In reactions that proceed via a two- or multiple-step mechanism, the intermediate formed in the rate-determining step can sometimes be used as an indicator of the transition state energy. The activation energy is estimated as the energy difference between the intermediate and the free reactants. In the case of regioselectivity prediction, it is sufficient to compute the energy of the intermediates as the different reaction pathways have the same starting point. This type of approach has frequently been used to study the S_EAr reaction, and in most studies, the σ-complex energy has been used to estimate the transition state energy. We will refer to this as the σ-complex approach.

Compared to reaction index calculations, the σ-complex approach is computationally more demanding. The computational requirements when it comes to the level of theory, the need of considering solvation effects, and the problem with entropy estimation are similar to direct transition state calculations. However, this approach has the advantage that it avoids difficult transition state optimizations and replaces them with optimizations to local minima. As such optimizations normally have rather large convergence radii for default starting structures, this approach lends itself to automated procedures in a way that transition state-based procedures cannot yet offer. One such application, presently under development, is in the virtual reactivity screening in pharmaceutical research and development, where the computations are performed automatically as "black box" calculations with the input being large substance libraries. The σ-complex approach can potentially be very accurate and sometimes rival the accuracy of direct transition state calculations. However, this requires that the reaction has a late transition state with a structure similar to that of the intermediate.

Historically, there have been a larger number of applications where the reactivity has been correlated to the stability of the σ-complex. For example, the traditional approach for determining regioselectivity based on resonance theory compares the different resonance forms of the σ-complex [90]. Stock and Brown argued already in the 1960s that the transition state of halogenations resembles the σ-complex based on correlations with experimental stabilities of non-S_EAr σ-complexes [34]. There have also been a number of studies correlating quantum chemical σ-complex energies with reactivity. We will not survey all these studies but only give some relevant examples of more recent studies along this approach.

Pankratov in 2000 computed relative σ-complex energies for nitration, which he referred to as cationic localization energies, for a large number of monosubstituted benzenes by the use of the semiempirical PM3 method [91]. Good linear relationships were found for predicting the positional selectivity from the σ-complex energies. However, different scaling factors were needed for the different positions, that is, the *ortho*, *meta*, and *para* positions.

Koleva *et al.* in 2009 also computed the relative energy for forming the σ-complex of S_EAr reactions from separated reactants but referred to this energy as the *electrophile affinity* [92]. In this study, energies and structures were computed at the B3LYP/6-311+G(2d,2p) level. A very good linear correlation ($R=0.99$) was found between the σ-complex energies and partial rate factors for chlorination. Good correlations ($R=0.97$) were also obtained for nitration of substituted benzenes and for benzylation of benzene and halobenzenes. As already mentioned, the correlations of Koleva *et al.* were of similar quality as those found by Brown and Cockroft using the $I(\mathbf{r})$ approach [88].

We have also in a previous article reported on the use of the σ-complex approach for predicting positional selectivity for halogenations, nitrations, and Friedel–Crafts acylations [85]. In this study, we compared the relative unscaled σ-complex energies directly with relative isomeric transition state energies obtained from experimental regioselectivities. This approach gave quantitative accuracy for halogenations, that is, the MAD value for those substrate combinations that had more than one observed isomer was only 0.4 kcal/mol. In contrast, the method failed for the nitration of monosubstituted benzenes. More specifically, we found a too high energy for the *ortho* isomer relative to the *para/meta* isomers in all investigated cases of monosubstituted benzenes, resulting in the amount of the *ortho* isomer being underestimated. The method also predicted a too high σ-complex energy for the *meta* isomer in the cases of *ortho/para* directing substituents. Considering that the rate-determining transition state of nitrations in solution more closely resembles the C-coordinated π-complex than the σ-complex, these results are not surprising. We have recently tested to predict the regioselectivity of nitrations based on π-complex energies. However, this approach performed even worse than the σ-complex approach. In the case of Friedel–Crafts acylations, the results of the σ-complex approach were erratic, and the agreement with experiment depended on the reaction conditions [85]. However, the method was qualitatively correct, that is, it predicts the dominating isomers correctly, for experiments in which the Lewis acid concentration was high.

4.5 SUMMARY AND CONCLUSIONS

In this chapter, we have reviewed the current mechanistic understanding of the S_EAr reaction based on experimental and theoretical studies. The nitration reaction often proceeds via the formation of two consecutive π-complexes before the appearance of the σ-complex. The first complex involves the coordination of an oxygen to the aromatic ring center, whereas the second complex, which seems to be present in all nitrations, has the N-atom of the nitronium ion coordinated to one of the ring C-atoms. In the gas phase, the nitration may have a contribution from a SET, and this contribution increases with the activation tendency of the aromatic substrate. The solution reaction lacks a driving force for SET, and the rate-determining step is the transformation of the C-atom coordinated π-complex into the σ-complex. This step has a very early transition state corresponding to a loosely bound π-complex. The deprotonation of the σ-complex is fast in the presence of water.

The uncatalyzed halogenation with molecular chlorine as the electrophile proceeds via a C-atom coordinated π-complex before the formation of the σ-complex. In solution, the latter has a Cl—Cl bond that is nearly dissociated. The rate-determining transition state is similar in structure to the σ-complex. Deprotonation of the σ-complex is easily facilitated even by a weak base, but the process requires the leaving of the Cl^- formed from the electrophile to occur before the abstraction of the proton.

Other S_EAr reactions are less well studied by theoretical methods, and kinetic measurements indicate that the mechanisms often are complex. In the case of sulfonation, both theory and experiments show that the reaction is second order in the electrophile and first order in the nucleophile in a nonpolar environment, whereas a polar solvent promotes a traditional second-order S_EAr mechanism.

Reaction indices have been frequently used to analyze and predict positional selectivity of S_EAr reactions. The average local ionization energy, $I(\mathbf{r})$, seems to have the best predictive power for this reaction type among the commonly used descriptors. Local minima in $I(\mathbf{r})$ reflect both the positional selectivity for S_EAr and the relative reactivity in aromatic and heteroaromatic systems. The $I(\mathbf{r})$ approach generally has quantitative predictive capacity for nitrations and halogenation, but it has problems for systems where steric hindrance is of key importance for the regioselectivity.

The σ-complex approach has emerged as an alternative to reaction indices for predicting regioselectivity in S_EAr. It is based on the assumption that the relative energies of the σ-complexes are similar to the corresponding transition state energies and thus reflect the positional selectivity for an aromatic substrate when reacting with a particular electrophile. The method provides quantitative predictions for halogenations, where it surpasses the $I(\mathbf{r})$ approach in accuracy. However, the σ-complex approach fails for nitrations. This is not surprising considering that the rate-determining transition state in nitrations is more similar in structure and energy to the π-complex than the σ-complex. The performance is varying for Friedel–Crafts acylations and depends on the reaction conditions.

ABBREVIATIONS

B3LYP	Becke's three-parameter exchange functional together with the Lee–Yang–Parr correlation functional
CAS	Complete active space
CCSD(T)	Coupled cluster with singles, doubles, and noniterative triples excitations
DFT	Density functional theory
FMO	Frontier molecular orbital
GVB	Generalized valence bond
HOMO	Highest occupied molecular orbital
KS	Kohn–Sham
LUMO	Lowest unoccupied molecular orbital
M06-2X	Minnesota exchange correlation functional published 2006 with double amount of exact exchange
MC-SCF	Multiconfigurational self-consistent field
MO	Molecular orbital
MP2	Second-order Møller–Plesset perturbation theory
MR-CI	Multireference configuration interaction
PCM	Polarizable continuum model
PES	Potential energy surface
S_EAr	Electrophilic aromatic substitution
SET	Single-electron transfer
TS	Transition state
VB	Valence bond

REFERENCES

[1] Heitler, W. and London, F. (1927) *Z. Phys.*, **44**, 455–472.

[2] Slater, J.C. (1931) *Phys. Rev.*, **37**, 481.

[3] Slater, J.C. (1931) *Phys. Rev.*, **38**, 1109.

[4] Pauling, L. (1931) *J. Am. Chem. Soc.*, **53**, 1367–1400.

[5] Pauling, L. (1931) *J. Am. Chem. Soc.*, **53**, 3225–3237.

[6] Goddard, W.A., Dunning, T.H., Hunt, W.J., and Hay, P.J. (1973) *Acc. Chem. Res.*, **6**, 368–376.

[7] Mulliken, R.S. (1927) *Phys. Rev.*, **29**, 637.

[8] Hund, F. (1928) *Z. Phys.*, **51**, 759–795.

[9] Lennard-Jones, J.E. (1929) *Trans. Faraday Soc.*, **25**, 668–686.

[10] Slater, J.C. (1930) *Phys. Rev.*, **36**, 57.

[11] Slater, J.C. (1930) *Phys. Rev.*, **35**, 210.

[12] Fock, V. (1930) *Z. Phys.*, **61**, 126–148.

[13] Roothaan, C.C.J. (1951) *Rev. Mod. Phys.*, **23**, 69–89.

[14] Szabo, A. and Ostlund, N.A. (1996) *Modern Quantum Chemistry: Introduction to Advanced Electronic Structure Theory*, Dover Publications, Inc, Mineola, NY.

[15] Hinze, J. and Roothaan, C.C.J. (1967) *Prog. Theor. Phys. Suppl.*, **40**, 37–51.

[16] Roos, B.O. (1987) *Advances in Chemical Physics: Ab Initio Methods in Quantum Chemistry Part 2, Volume 69* (ed. K.P. Lawley), John Wiley & Sons, Ltd, Chichester, pp. 399–445.

[17] Kohn, W. and Sham, L.J. (1965) *Phys. Rev. A*, **140**, 1133–1138.

[18] Hohenberg, P. and Kohn, W. (1964) *Phys. Rev. B*, **136**, 864–871.

[19] Hückel, E. (1931) *Z. Phys.*, **70**, 204–286.
[20] Wheland, G.W. (1942) *J. Am. Chem. Soc.*, **64**, 900–908.
[21] Dewar, M.J.S. (1952) *J. Am. Chem. Soc.*, **74**, 3341–3345.
[22] Dewar, M.J.S. (1952) *J. Am. Chem. Soc.*, **74**, 3357–3363.
[23] Dewar, M.J.S. and Longuet-Higgins, H.C. (1952) *Proc. R. Soc. Lond. A Math. Phys. Sci.*, **214**, 482–493.
[24] Dewar, M.J.S., Zoebisch, E.G., Healy, E.F., and Stewart, J.J.P. (1985) *J. Am. Chem. Soc.*, **107**, 3902–3909.
[25] Pople, J.A., Santry, D.P., and Segal, G.A. (1965) *J. Chem. Phys.*, **43**, S129–S135.
[26] Dewar, M.J.S. and Thompson Jr, C.C. (1965) *J. Am. Chem. Soc.*, **87**, 4414–4423.
[27] Agranat, I., Shih, Y.S., and Bentor, Y. (1974) *J. Am. Chem. Soc.*, **96**, 1259–1260.
[28] Levy, L., Pogodin, S., Cohen, S., and Agranat, I. (2007) *Lett. Org. Chem.*, **4**, 314–318.
[29] Grovenstein, E. and Aprahamian, N.S. (1962) *J. Am. Chem. Soc.*, **84**, 212–220.
[30] Miller, L.L. and Watkins, B.F. (1976) *J. Am. Chem. Soc.*, **98**, 1515–1519.
[31] Olah, G.A., Lukas, J., and Lukas, E. (1969) *J. Am. Chem. Soc.*, **91**, 5319–5323.
[32] Melander, L. (1949) *Acta Chem. Scand.*, **3**, 95.
[33] Melander, L. (1949) *Nature*, **163**, 599.
[34] Stock, L.M. and Brown, H.C. (1963) *Adv. Phys. Org. Chem.*, **1**, 35.
[35] Olah, G.A. (1970) *Acc. Chem. Res.*, **4**, 240–248.
[36] Dewar, M.J.S. (1946) *J. Chem. Soc.*, **406**, 777.
[37] Olah, G.A., Kuhn, S.J., and Flood, S.H. (1961) *J. Am. Chem. Soc.*, **83**, 4571–4580.
[38] Perrin, C.L. (1977) *J. Am. Chem. Soc.*, **99**, 5516–5518.
[39] Kochi, J.K. (1992) *Acc. Chem. Res.*, **25**, 39–47.
[40] Kim, E.K., Bockman, T.M., and Kochi, J.K. (1993) *J. Am. Chem. Soc.*, **115**, 3091–3104.
[41] Stock, L.M. and Baker, F.W. (1962) *J. Am. Chem. Soc.*, **84**, 1661–1667.
[42] DeHaan, F.P., Covey, W.D., Delker, G.L., Baker, N.J., Feigon, J.F., Ono, D., Miller, K.D., and Stelter, E.D. (1984) *J. Org. Chem.*, **49**, 3959–3963.
[43] Hubig, S.M. and Kochi, J.K. (2000) *J. Org. Chem.*, 6807–6818.
[44] Chiavarino, B., Crestoni, M.E., Fornarini, S., Lemaire, J., Maitre, P., and MacAleese, L. (2006) *J. Am. Chem. Soc.*, **128**, 12553–12561.
[45] Vasilyev, A.V., Lindeman, S.V., and Kochi, J.K. (2001) *Chem. Commun.*, 909–910.
[46] Vasilyev, A.V., Lindeman, S.V., and Kochi, J.K. (2002) *New J. Chem.*, **26**, 582–592.
[47] Doering, W.v.E., Saunders, M., Boyton, H.G., Hearhart, H.W., Wadley, E.F., Edwards, W.R., and Laber, G. (1958) *Tetrahedron*, **4**, 178–185.
[48] Olah, G.A., Lin, H.C., and Mo, Y.K. (1972) *J. Am. Chem. Soc.*, **94**, 3667–3669.
[49] Hubig, S.M. and Kochi, J.K. (2000) *J. Am. Chem. Soc.*, **122**, 8279–8288.
[50] Gwaltney, S.R., Rosokha, S.V., Head-Gordon, M., and Kochi, J.K. (2003) *J. Am. Chem. Soc.*, **125**, 3273–3283.
[51] Esteves, P.M., De M Carneiro, J.W., Cardoso, S.P., Barbosa, A.G., Laali, K.K., Rasul, G., Prakash, G.K., and Olah, G.A. (2003) *J. Am. Chem. Soc.*, **125**, 4836–4849.
[52] Rosokha, S.V. and Kochi, J.K. (2002) *J. Org. Chem.*, **67**, 1727–1737.
[53] Skokov, S. and Wheeler, R.A. (1999) *J. Phys. Chem. A*, **103**, 4261–4269.
[54] de Queiroz, J.F., Carneiro, J.W., Sabino, A.A., Sparrapan, R., Eberlin, M.N., and Esteves, P.M. (2006) *J. Org. Chem.*, **71**, 6192–6203.
[55] Xu, X.F., Zilberg, S., and Haas, Y. (2010) *J. Phys. Chem. A*, **114**, 4924–4933.
[56] Parker, V.D., Kar, T., and Bethell, D. (2013) *J. Org. Chem.*, **78**, 9522–9525.
[57] Cossi, M., Barone, V., Cammi, R., and Tomasi, J. (1996) *Chem. Phys. Lett.*, **255**, 327–335.
[58] Hu, C.H. and Brinck, T. (1999) *J. Phys. Chem. A*, **103**, 5379–5386.
[59] Caille, S.Y. and Corriu, R.J.P. (1969) *Tetrahedron*, **25**, 2005–2022.

[60] Carey, F.A. and Sundberg, R.J. (2007) *Advanced Organic Chemistry*. 5th Ed, Springer Science, New York, pp. 800–804.

[61] Brown, H.C. and Wirkkala, R.A. (1966) *J. Am. Chem. Soc.*, **88**, 1447–1452.

[62] Barnett, J.R., Andrews, L.J., and Keefer, R.M. (1972) *J. Am. Chem. Soc.*, **94**, 6129–6134.

[63] Zhang, M. and Lund, C.R.F. (2002) *J. Phys. Chem. A*, **106**, 10294–10301.

[64] Ben-Daniel, R., de Visser, S.P., Shaik, S., and Neumann, R. (2003) *J. Am. Chem. Soc.*, **125**, 12116–12117.

[65] Filimonov, V.D., Poleshchuk, O.K., Krasnokutskaya, E.A., and Frenking, G. (2011) *J. Mol. Model.*, **17**, 2759–2771.

[66] Sakic, D. and Vrcek, V. (2012) *J. Phys. Chem. A*, **116**, 1298–1306.

[67] Kong, J., Galabov, B., Koleva, G., Zou, J.-J., Schaefer, H.F., and Schleyer, P.v.R. (2011) *Angew. Chem. Int. Ed.*, **50**, 6809.

[68] Smith, M.B. and March, J. (2007) *March's Advanced Organic Chemistry: Reactions, Mechanisms, and Structure*. 6th Ed, John Wiley & Sons, Inc., Hoboken, NJ, pp. 695–697.

[69] Koleva, G., Galabov, B., Kong, J., Schaefer, H.F., and Schleyer, P.v.R. (2011) *J. Am. Chem. Soc.*, **133**, 19094–19101.

[70] Bosscher, J.K. and Cerfontain, H. (1968) *Recl. Trav. Chim. Pays-Bas.*, **87**, 873–887.

[71] Volkov, N.A., Timoshkin, A.Y., and Suvorov, A.V. (2004) *Int. J. Quantum Chem.*, **100**, 412–418.

[72] Stashenko, E.E., Martínez, J.R., Tafurt-García, G., Palma, A., and Bofill, J.M. (2008) *Tetrahedron*, **64**, 7407–7418.

[73] Fukui, K., Yonezawa, T., and Shingu, H. (1952) *J. Chem. Phys.*, **20**, 722.

[74] Brinck, T. (1998) *Theoretical and Computational Chemistry Vol 5*. (ed. C. Parkanyi), Elsevier, Amsterdam, pp. 51–94.

[75] Murray, J.S. and Politzer, P. (1998) *Theoretical and Computational Chemistry Vol 5*. (ed. C. Parkanyi), Elsevier, Amsterdam, pp. 189–203.

[76] Fukui, K., Yonezawa, T., Nagata, C., and Shingu, H. (1954) *J. Chem. Phys.*, **22**, 1433.

[77] Koopman, T.A. (1933) *Physica*, **1**, 104.

[78] Langenaeker, W., Demel, K., and Geerlings, P. (1991) *J. Mol. Struct. THEOCHEM*, **234**, 329–342.

[79] Parr, R.G. and Yang, W. (1984) *J. Am. Chem. Soc.*, **106**, 4049–4050.

[80] Yang, W. and Mortier, W.J. (1986) *J. Am. Chem. Soc.*, **108**, 5708.

[81] Anderson, J.S.M., Melin, J., and Ayers, P.W. (2007) *J. Chem. Theory Comput.*, **3**, 375–389.

[82] Sjoberg, P., Murray, J.S., Brinck, T., and Politzer, P. (1990) *Can. J. Chem.*, **68**, 1440–1443.

[83] Politzer, P., Murray, J.S., and Bulat, F.A. (2010) *J. Mol. Model.*, **16**, 1731–1742.

[84] Brinck, T., Murray, J.S., and Politzer, P. (1993) *Int. J. Quantum. Chem.*, **48**, 73–88.

[85] Liljenberg, M., Brinck, T., Herschend, B., Rein, T., Rockwell, G., and Svensson, M. (2010) *J. Org. Chem.*, **75**, 4696–4705.

[86] Brinck, T., Murray, J.S., Politzer, P., and Carter, R.E. (1991) *J. Org. Chem.*, **56**, 2934–2936.

[87] Murray, J.S., Brinck, T., and Politzer, P. (1992) *J. Mol. Struct. THEOCHEM*, **255**, 271–281.

[88] Brown, J.J. and Cockroft, S.L. (2013) *Chem. Sci.*, **4**, 1772.

[89] Arrieta, A. and Cossio, F.P. (2007) *J. Mol. Struct. THEOCHEM*, **811**, 19–26.

[90] Pauling, L. and Wheland, G.W. (1933) *J. Chem. Phys.*, **1**, 362–374.

[91] Pankratov, A.N. (2000) *J. Mol. Struct. THEOCHEM*, **507**, 239–244.

[92] Koleva, G., Galabov, B., Wu, J.I., Schaefer, H.F., and Schleyer, P.v.R. (2009) *J. Am. Chem. Soc.*, **131**, 14722–14727.

5

CATALYTIC ENANTIOSELECTIVE ELECTROPHILIC AROMATIC SUBSTITUTIONS

Marco Bandini

Dipartimento di Chimica "G. Ciamician", Alma Mater Studiorum, Università di Bologna, Bologna, Italy

5.1 INTRODUCTION AND HISTORICAL BACKGROUND

Benzylic stereocenters represent a structural motif widely distributed in naturally occurring compounds and biologically active organic structures [1]. In pharmaceutical and agrochemical compounds, stereoisomeric and particularly enantiomeric forms of aromatic structures frequently display different or even complementary activity toward biosystems. As a consequence, it is currently recorded as a growing demand for sustainable (chemical and economical) synthetic methodologies enabling the differentiation between stereochemical pathways of organic transformations. In this regard, stereoselective catalysis represents the ultimate chemical frontier for providing reliable solutions to these interrogatives both in small- and medium-scale productions.

Due to ubiquity of arenes in added value compounds, the stereochemical manipulation of aromatic species is rapidly becoming a "hot spot" in organic synthesis with developments both in metal-based and metal-free catalysis.

In the realm of arene functionalizations, electrophilic substitutions, also referred to as Friedel–Crafts-type alkylation (FCA) reactions [2], represent a pillar for the "decoration" of the arene periphery by the assistance of σ-Lewis acids (LAs) [3]. Due to its predominant role in the field, synthetic chemists historically addressed their efforts toward the reinvention of asymmetric FCA reactions by replacing common achiral LAs and Brønsted acids (i.e., $AlCl_3$, $FeCl_3$, BF_3, *p*TSA, TfOH) with chiral organometallic and metal-free species. Focusing on the electrophilic partners, the modern variants of enantioselective Friedel–Crafts-type alkylations generally employ prochiral organic reagents capable of generating new stereogenic centers upon aromatic electrophilic

Arene Chemistry: Reaction Mechanisms and Methods for Aromatic Compounds,
First Edition. Edited by Jacques Mortier.

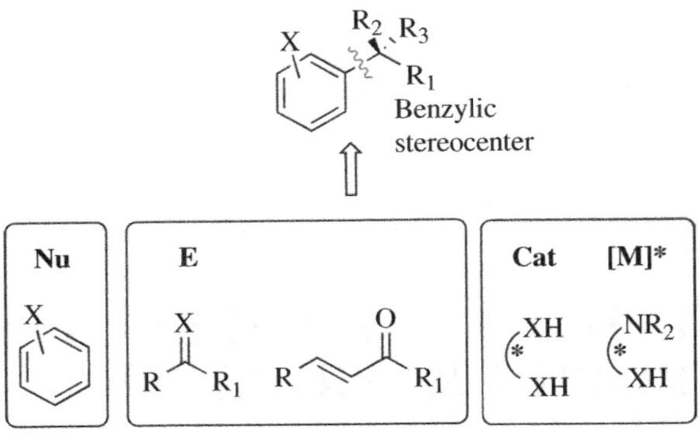

FIGURE 5.1 Pictorial representation of targeted benzylic stereocenters and common partners in catalytic asymmetric Friedel–Crafts-type alkylations.

substitution. Among the others, the following reaction partners deserve a particular mention: carbonyl compounds (aldehydes, ketones, and imines) and Michael acceptors.

From a mechanistic viewpoint, asymmetric Friedel–Crafts alkylation (AFCA) reactions are similar to the historical version, with the chiral additives responsible for triggering and controlling the overall aromatic C—H replacement. Interestingly, covalent (organocatalysis) or noncovalent (metal catalysis) interactions can take place between catalysts and electrophiles during the enantiodiscriminating step of the reaction course (Fig. 5.1).

Moreover, it should be mentioned that the nature of the aromatic hydrocarbons that found concrete applications in the field of AFCAs still suffers from severe limitations. Mild conditions (i.e., low reaction temperatures) and low catalyst loadings are experimental conditions frequently adopted in order to realize high regio-, chemo-, and stereocontrols. This aspect in combination with the use of electrophilic agents featuring lower reactivity with respect to classic alkyl halides accounts for the narrow substrate scope. In particular, only "electron-rich" benzene rings carrying electron-donating groups (EDGs) displayed sufficient nucleophilic character for actively taking part into asymmetric electrophilic aromatic substitution reactions. Finally, symmetrically substituted arenes are commonly utilized in order to circumnavigate regiochemistry issues during the C—H replacement.

Historically, one of the first examples of enantioselective arene (i.e., phenols) "decoration" was reported by an Italian team (Casiraghi and coworkers) [4] that documented on the efficiency of (−)-menthoxy(ethyl)aluminum chloride **3** as a chiral stoichiometric LA, in promoting the condensation of a range of phenols **1** with chloral **2**. *Enantiomeric excesses* up to 80% were obtained (Scheme 5.1a). Shortly after, the first catalytic variant was realized by Erker and coworkers that described the stereoselective addition (ee up to 84%) of 1-hydroxynaphthalene (**5a**) to methyl or ethylpyruvate (**6a**,**b**) in the presence of chiral zirconocene **7** (Scheme 5.1b) [5].

The authors justified the recorded preferential *ortho*-regiochemistry by considering a simultaneous coordination mode of the phenol/naphthol unit and the electrophile by the aluminum and zirconium centers, during the creation of the new stereogenic center (inset Scheme 5.1).

These processes emphasized for the first time the potential of chiral LAs in controlling the stereochemical profiles in AFCA reactions. In particular, the catalytic methodology elected electrophilic transition metal (TM) species as valuable promoters of arene functionalizations due to their intrinsic lower oxophilicity with respect to conventional σ-LAs. Despite seminal works lack in providing

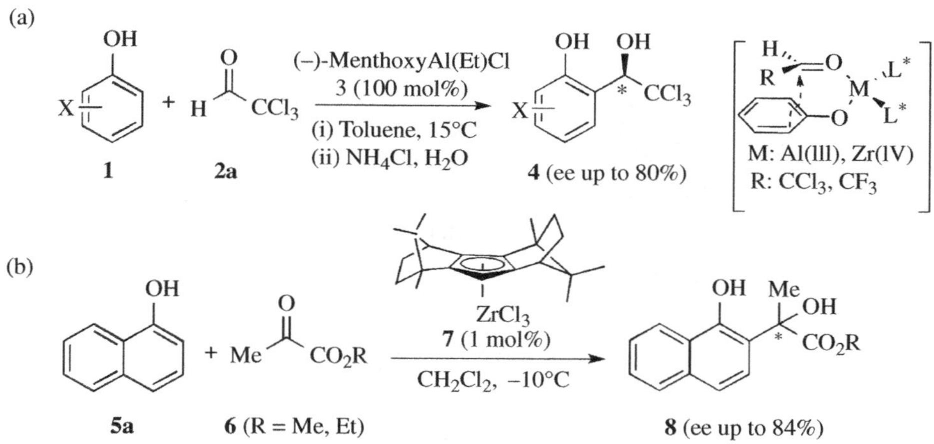

SCHEME 5.1 From stoichiometric (a) to catalytic (b) asymmetric Friedel–Crafts alkylation of hydroxyarenes.

detailed mechanistic proofs, it is reasonable to consider the electrophilic activation of the carbonyl compounds by the metal species as the key step in determining both kinetic and stereochemistry of the reaction machinery.

New horizons in the field have been traced

The new millennium signed a landmark in the segment of AFCAs with the documentation of always more performing metal-based and organocatalytic systems. Moreover, very recently, the synergetic use of metal and metal-free catalysis started gaining credit in the chemical community and applications in AFCAs. These classifications will be adopted also in organization of the chapter providing insights into the mechanism details when available from the original reports [6].

5.2 METAL-CATALYZED AFCA OF AROMATIC HYDROCARBONS

5.2.1 Introduction

The Friedel–Crafts alkylation reaction represents one of the oldest LA-assisted reactions in organic synthesis. As a consequence, the replacement of classic achiral metal catalysts with chiral analogues should represent a natural evolution in the field. However, important issues such as poisoning of the metal species by the reaction partners/products, side reactions triggered by Brønsted acidity generated during the reaction course, and scarce regioselectivity dramatically delayed the development of efficient catalytic methodologies for the alkylation of arenes in stereodefined manner. In the Chart 5.1, a list of metal and chiral "privileged" ligands [7] utilized in FC alkylations is reported.

Mechanistically, it is widely accepted that these procedures involve the coordination of the electrophilic agent by the metal species with the subsequent nucleophilic attack by the less hindered enantiotopic face (outer sphere mechanism). In some cases, a two-site binding mode of the catalyst has been also postulated with both partners of the process (inner sphere mechanism), simultaneously interacting with the catalytic unit. The latter arrangement will concur to define a positive proximity effect during the C–C bond-forming step.

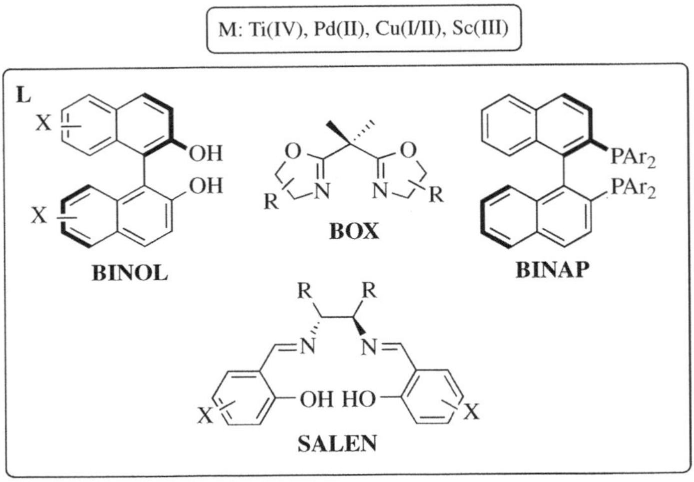

CHART 5.1 Chiral "privileged" ligand and metal sources commonly employed in the catalytic enantioselective alkylation of electron-rich arenes.

5.2.2 Metal-Catalyzed Condensation of Arenes with Carbonyl Compounds and Their Nitrogen Derivatives

The stereoselective synthesis of 1-aryl-2,2,2-trifluoroethanol derivatives **11** was documented by Mikami and coworkers, as one of the first examples of asymmetric FC alkylation of aromatic ethers **9** [8a,b]. The combined use of a catalytic amount (10 mol%) of enantiomerically pure (*R*)-6,6′-Br_2-BINOL-Ti(O^iPr)$_2$ (**10**) and (*R*)-6,6′-Br_2-BINOL (1:1 molar ratio with the metal catalyst) assisted the stereoselective addition of fluoral **2b** with aromatic ethers **9a–c**. The corresponding fluorinated secondary alcohols (**11a–c**) were isolated in good yields and ee up to 90% (Scheme 5.2). Interestingly, the isolation of the *para*-regioisomers as the major reaction products ruled out the simultaneous coordination to the metal center by the alkylating agent and the oxygen atom of the ether moiety [4]. Even if the real nature of the catalytically active species is still unclear, the addition of an equimolar aliquot of free ligand is thought to

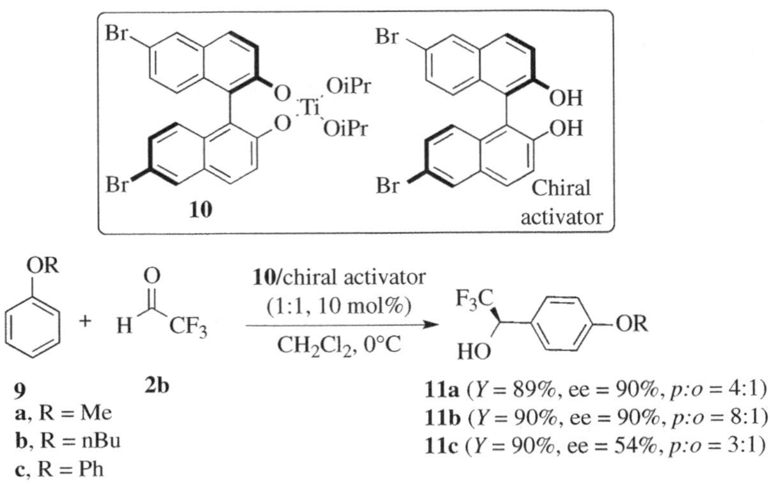

SCHEME 5.2 Synthesis of enantiomerically enriched 1-aryl-2,2,2-trifluoroethanols **11a–c** via titanium-catalyzed electrophilic substitution (**11c** was obtained in the absence of chiral activation).

increase the available amount in solution of active species of general formula [L_2Ti(IV)], deriving from the spontaneously formed oligomeric aggregates originated by mixing BINOL and Ti(O*i*Pr)$_4$ in 1:1 ratio [8c].

Trifluoromethyl ethyl pyruvate **6a** was also efficiently coupled with electron-rich benzenes such as 1,3-benzenediol **9d** in the presence of chiral copper as well as palladium-based complexes (**12** and **13**). In the Friedel–Crafts-type condensation, chiral bis-oxazoline-[Cu(OTf)$_2$] [9a] and BINAP-[Pd(SbF$_6$)$_2$] [9b] catalysts provided the corresponding tertiary α-trifluoromethyl-carbinol **14a** with ee = 86 and 89%, respectively (Scheme 5.3). Heterogeneous catalysis was also investigated for the first time in the enantioselective alkylation of electron-rich arenes (i.e., 1,3-dimethoxybenzene, **9d**) with trifluoromethyl ethyl pyruvate **6a**. In particular, the tethering of chiral bis-oxazoline ligand to silica or mesoporous MCM-41 inert matrix provided a reliable chiral support to undergo [Cu(II)]-catalyzed AFCAs with high levels of stereocontrol and reusability [9c].

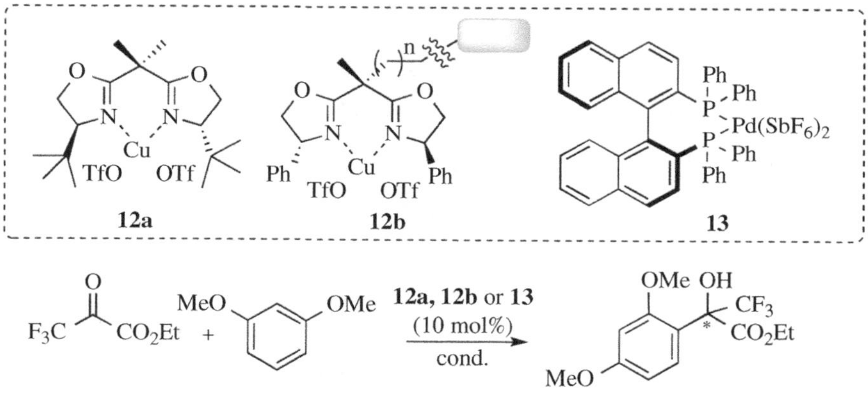

6a **9d** **14a** (**12a**: *Y*: 56%, ee: 86%)
(**12b**: *Y*: 90%, ee: 92%)
(**13**: *Y*: 69%, ee: 89%)

SCHEME 5.3 Regio- and stereoselective alkylation of resorcinol **9d** with trifluoromethyl ethyl pyruvate **6a**. Cond.: **12a**: Et_2O, 0°C → rt; **12b** (SiO_2): CH_3CN, rt; **13**: CH_2Cl_2, 30°C.

The intrinsic limitation of the previous methods (i.e., exclusive use of 1,3-dimethoxybenzene **9d**) was efficiently addressed by Chen under solvent-free conditions [10]. Besides increased reactivity, the absence of solvent positively affected also the stereoselectivity in the FC alkylation of substituted aryl ethers with trifluoromethyl ethyl pyruvate. A range of quaternary benzylic stereogenic centers was obtained in stereocontrolled manner (ee up to 93%, 99% after recrystallization). Interestingly, under solvent-free conditions, the loading of the catalyst **12c** was reduced up to 0.01 mol%, still maintaining a respectable isolated yield (70%) and enantiomeric excess for **14b** (85%; Scheme 5.4).

Jørgensen and coworkers elegantly documented on the suitability of α-imino esters **15** as electrophilic agents in asymmetric alkylation of electron-rich arenes [11]. Interestingly, the electrophiles were efficiently obtained *in situ* via an aza-Wittig reaction (*N*-carboalkoxyiminophosphane and alkyl glyoxylates) and then subjected to aromatic substitution without further purification. Best reaction conditions employed catalytic amounts of (*R*)-Tol-BINAP-CuPF$_6$ (**16a**) in toluene/CH_2Cl_2 (19:1) as the reaction mixture (Scheme 5.5). Interestingly, a range of nonnatural aromatic *R*-amino acids **17** was conveniently accessible in mild conditions and high stereocontrol (ee: 67–98%), by means of several amino- and alkoxy-substituted benzenes **9**.

Finally, polydentate chiral calixarene-like salen ligand **20** was demonstrated to be superior to privileged ligands salen **19a** and **b** in the titanium-catalyzed condensation of variously substituted aromatic amines with glyoxylate **2c** under mild conditions (catalyst loading = 5 mol%, Et_2O, 0°C; Scheme 5.6) [12]. Although the authors did not provide a conclusive response on the real structure of the [**20**-Ti] complex, an

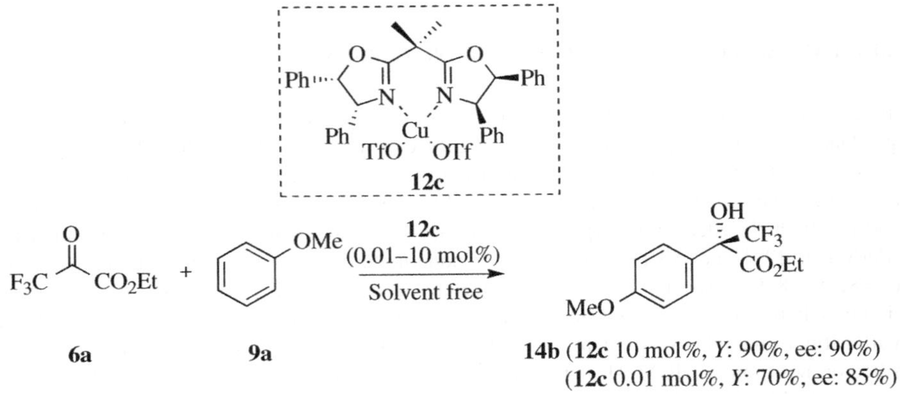

SCHEME 5.4 Solvent-free version of the asymmetric alkylation of aromatic ethers **9** with trifluoroethylpyruvate in the presence of chiral BOX-[$Cu(OTf)_2$] complex.

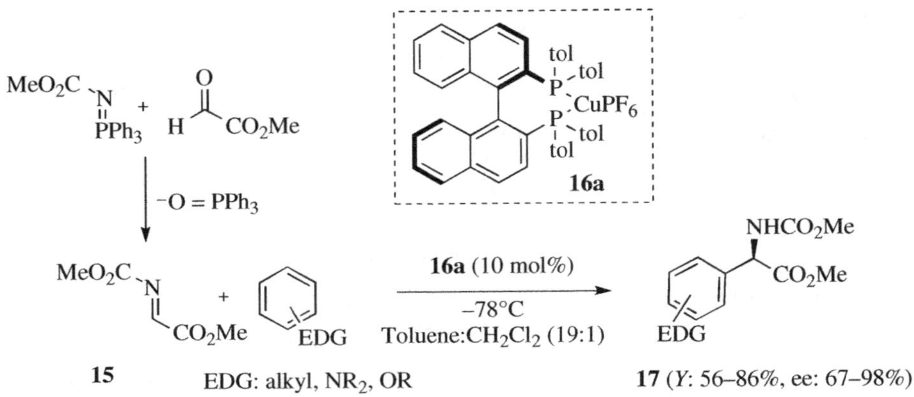

SCHEME 5.5 Ready access to nonnatural aromatic α-amino acids (**17**) via copper-catalyzed asymmetric FC alkylation of electron-rich benzenes.

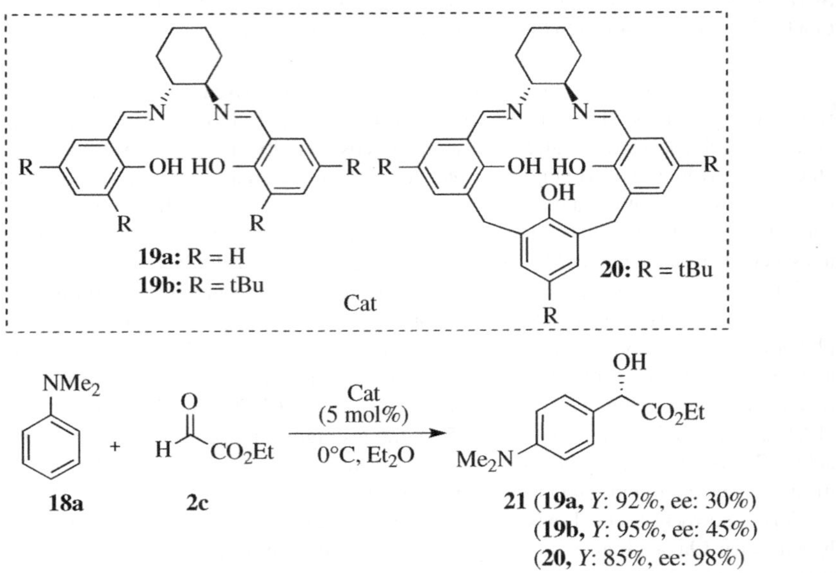

SCHEME 5.6 Chiral calixarene–salen-type ligand in asymmetric FC alkylation of aromatic amines **18a**.

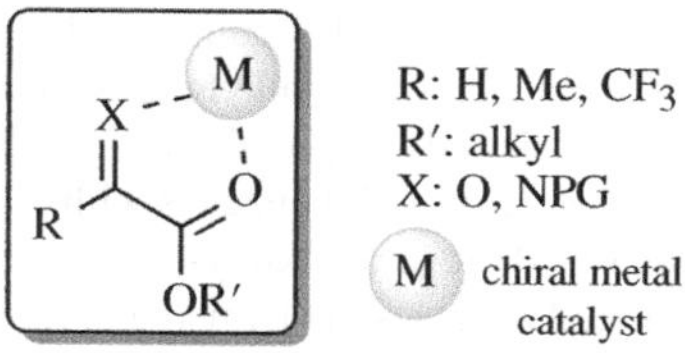

FIGURE 5.2 Chelating coordination mode between chiral metal complexes and pyruvate or glyoxylate derivatives.

efficient cavity recognition of the substrate by the catalyst was assumed to account for the higher stereoselectivity obtained with **20** in comparison to acyclic congeners **19a** and **b**.

At this stage, it should be emphasized that glyoxylate and pyruvate derivatives are potential two-site binding substrates in combination with metal complexes featuring at least two available coordination sites (*syn*). The five-membered ring aggregate, resulting from the chelation coordination mode (Fig. 5.2), generally displays high rigidity with consequent reduced number of possible conformers in the presence of C_2-symmetric chiral ligands. Rigidity and high symmetry in the coordination between the chiral metal species and reaction partners are valuable guidelines to predict high stereo-translations in organic synthesis [13].

In the chelating alkylating agents for catalytic enantioselective FC alkylation reaction of benzenes, it should be also highlighted the efficiency of *N*Ts-aldoimine (**22a**) in combination with chiral binuclear binol-based vanadium complex (**23**), for the enantioselective functionalization of 2-naphthol **5b**. The heterocoupling worked in a straightforward manner in chlorobenzene as the solvent at 10°C in the presence of 5 mol% of **23**. The corresponding aminophenol **24** was isolated in 70% yield and *enantiomeric excess* up to 90% (Scheme 5.7) [14].

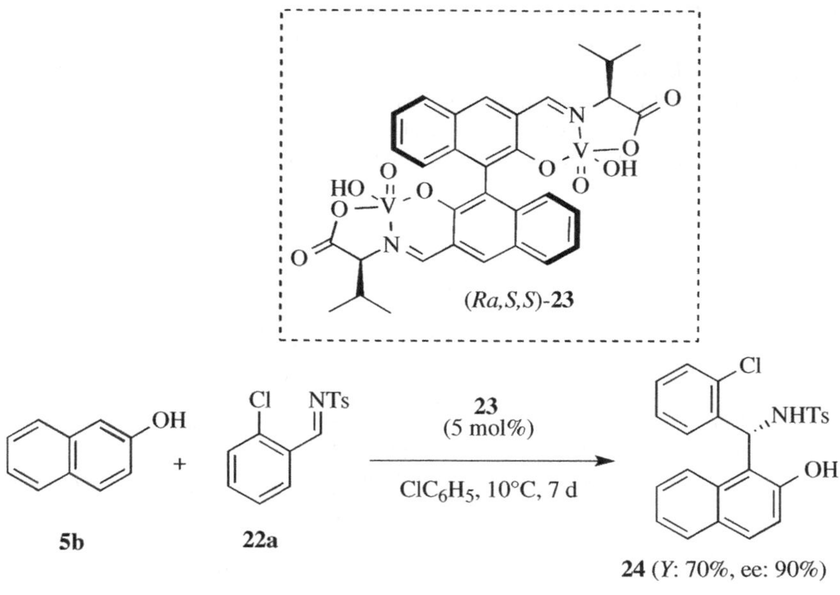

SCHEME 5.7 Chiral dinuclear vanadium Lewis acids for the enantioselective Friedel–Crafts alkylation of naphthols with *N*Ts-imines.

5.2.2.1 ***Michael Addition*** Enantioselective 1,4-conjugated addition reactions have been largely employed as surrogates of classic Friedel–Crafts alkylations for the functionalization of arenes. In this context, Desimoni's group envisioned the possibility to apply chiral cationic pyBOX-Sc(III) LAs in the conjugated addition of electron-rich benzenes with chelating Michael acceptors such as (*E*)-2-oxo-4-aryl-3-butenoates **25** [15]. Several alkoxy- and amino-substituted benzenes were efficiently coupled with β,γ-unsaturated α-ketoesters by using chiral TIPS-pybox ligand **26** (10 mol%; Scheme 5.8).

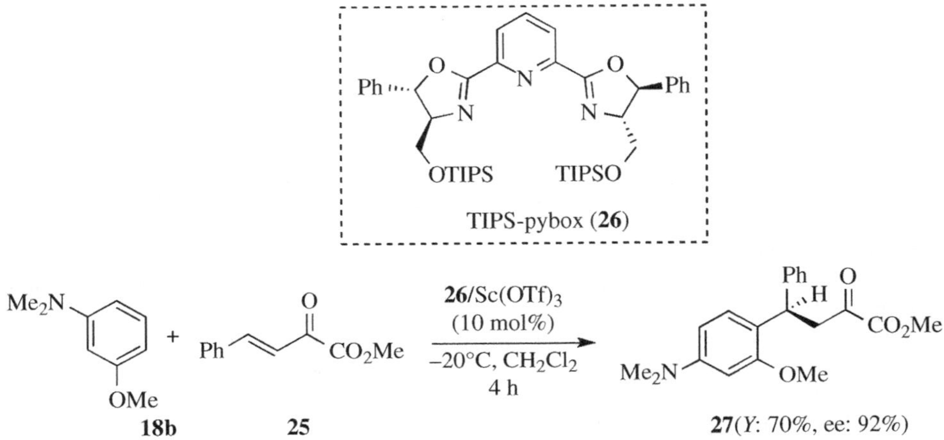

SCHEME 5.8 Michael-type Friedel–Crafts alkylation reaction mediated by chiral TIPS-pybox-[Sc(III)] Lewis acid.

Subsequently, Feng and coworkers documented on the LA-mediated Michael addition of phenols (**1**) to 4-oxo-4-arylbutenoates **28** in the presence of chiral *N,N'*-dioxide-[Sc(III)] complexes [16]. Sterically hindered groups at the amide nitrogen atom of the ligand **29** (i.e., 1-adamantyl) played a major role in the enantioselection of the process, delivering the 1,4-addition adducts **30** in high *enantiomeric excesses* (Scheme 5.9). Several alkoxyphenols were utilized displaying a remarkable selectivity toward C-alkylation with respect to the oxa-Michael counterpart.

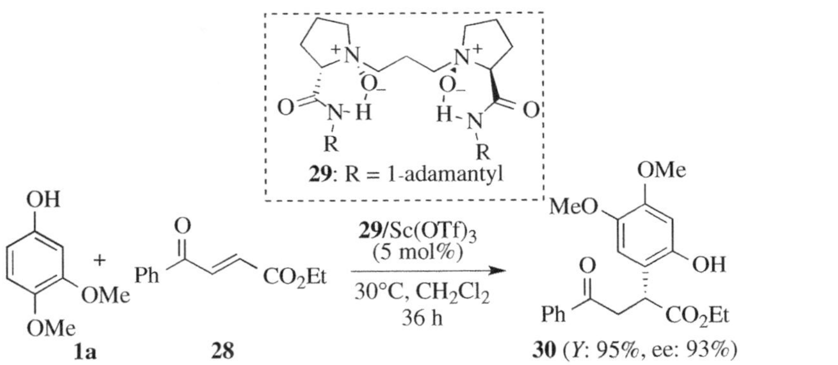

SCHEME 5.9 Michael-type Friedel–Crafts alkylation reaction mediated by chiral *N,N'*-dioxide-[Sc(III)] Lewis acid.

5.2.2.2 ***Asymmetric Allylic Alkylation*** Over the past two decades, asymmetric allylic alkylations (AAAs) have gained growing popularity in the organic chemistry scenario toward the direct realization of synthetically flexible organic building blocks in stereochemically defined manner [17]. Despite functional group tolerance and wide scope, electron-rich aromatic compounds emerged only recently as suitable soft nucleophilic species to be combined in the metal-assisted Friedel–Crafts-type nucleophilic allylic substitutions. In this direction, Bandini and Umani-Ronchi pioneered the field by presenting the direct enantioselective Pd-catalyzed allylic functionalization of indoles with allyl carbonates, delivering functionalized tetrahydro-β-carbolines in good yields and high enantiomeric excess [18, 19].

Focusing on benzene-like arenes, Shu-Li You described an efficient intramolecular Ir-catalyzed nucleophilic allylic substitution of phenols, addressing *meta*-substituted hydroxyarene with allylic carbonate side chains **31** as acyclic precursors [20]. A range of tetrahydroisoquinolines **33** was isolated in good yields (up to 90%) and ee up to 96% in the presence of chiral phosphoramidite ligand (S,S,S_a)-**32** (4 mol%) and $[IrCl(COD)]_2$ as the metal source (Scheme 5.10). The use of DMAP as an additive (200 mol%) in THF (rt to 50°C) allowed the regiochemical *ortho* aromatic C—H replacement (**33** vs. **33′**) prevalently.

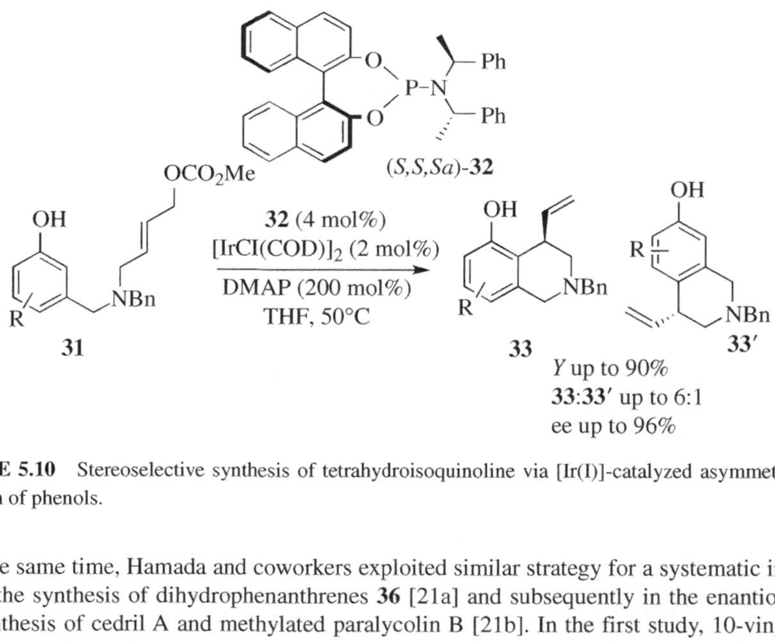

SCHEME 5.10 Stereoselective synthesis of tetrahydroisoquinoline via [Ir(I)]-catalyzed asymmetric allylic alkylation of phenols.

At the same time, Hamada and coworkers exploited similar strategy for a systematic investigation on the synthesis of dihydrophenanthrenes **36** [21a] and subsequently in the enantioselective total synthesis of cedril A and methylated paralycolin B [21b]. In the first study, 10-vinyl or 10-isopropenyl chiral dihydrophenanthrenes were obtained via Pd-catalyzed asymmetric allylic substitution under Friedel–Crafts-type alkylation conditions. In particular, the chiral complex obtained *in situ* from (*R*,*R*)-DACH-phenyl Trost ligand **35** (6 mol%) and $Pd(dba)_2$ (5 mol%) promoted the intramolecular allylic alkylation of the functionalized biaryl compound and trisubstituted alkene **34a**, delivering 10-isoprenyl derivative **36a** in 96% yield and 93% ee (Scheme 5.11a). Identical catalytic system displayed moderate activity in promoting the Friedel–Crafts alkylation of the acyclic precursor **34b**, in the presence of KOAc as an additive. Despite the synthetically unsustainable stereochemical control (*enantiomeric excess* = 66%), the tricyclic compound **36b** represented a valuable building block in the total synthesis of enantiomerically enriched natural compound (−)-cedril A (Scheme 5.11b).

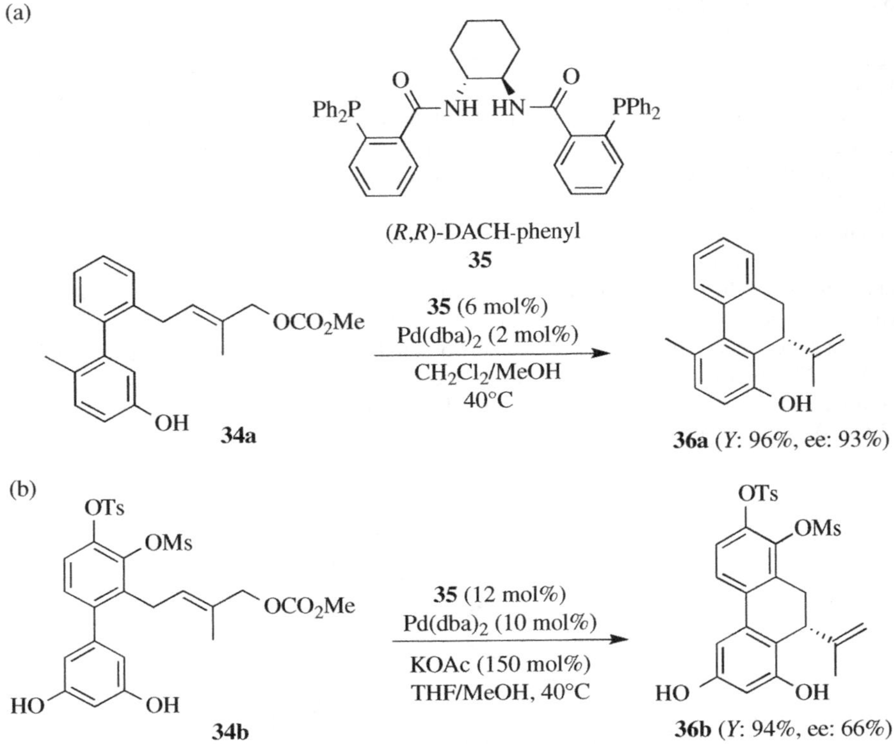

SCHEME 5.11 (a) Tsuji–Trost asymmetric allylic alkylation applied to the total synthesis of vinyl or isopropenyl chiral dihydrophenanthrenes. (b) Stereoselective synthesis of the (–)-cedril A precursor **36b**.

5.3 ORGANOCATALYZED AFCA OF AROMATIC HYDROCARBONS

5.3.1 Introduction

Asymmetric organocatalysis deals with the realization of enantioselective transformations by means of small chiral organic molecules [14]. In this realm, chiral acid catalysis and chiral aminocatalysis represent two well-established research fields featuring a rich portfolio of applications in asymmetric C—C and C—X bond-forming processes and reductions.

The first example of catalytic asymmetric functionalization of arenes under metal-free conditions was documented by MacMillan and coworkers in 2002 under aminocatalysis regime [1]. The combination of chiral imidazolidinone **38a** with enals **37** resulted into the formation of covalent aggregates (i.e., iminium ion **39**) featuring a lower LUMO (lowest unoccupied molecular orbital) energy with respect to the aldehydic precursor. Therefore, the chirally modified electrophile **39** can enter the catalytic cycle characterized by a diastereoselective nucleophilic attack of electron-rich benzenes followed by the restoring of the catalytically active secondary amine via hydrolysis of the immonium salt **40** (Scheme 5.12).

This work represents a landmark in the area of stereoselective metal-free (i.e., aminocatalysis) alkylation of benzenes based on Michael-type condensation via covalent catalyst–substrate interaction [22]. Subsequently, asymmetric acid catalysis based on hydrogen bond catalyst–substrate recognitions has found elegant applications in 1,4-conjugated additions and direct condensation of arenes with carbonyl compounds. The following sections will be organized based on the reactivity exploited in the arene functionalization.

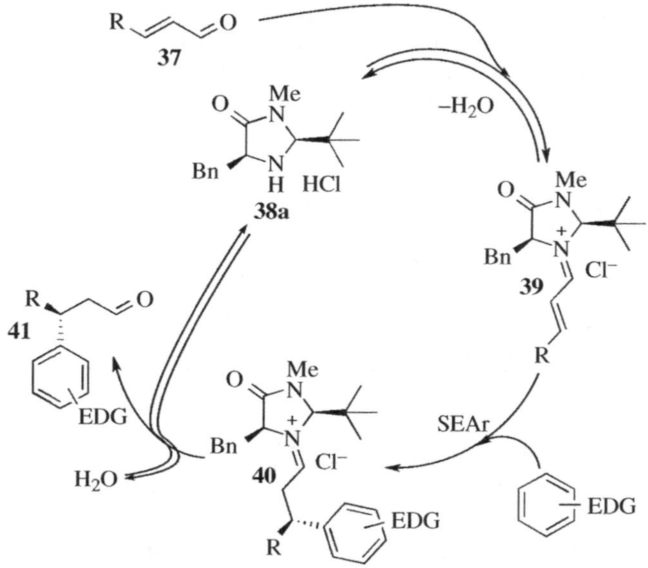

SCHEME 5.12 Reaction machinery of the asymmetric amino-catalyzed alkylation of benzenes.

5.3.2 Asymmetric Organocatalyzed Condensation of Arenes with Carbonyl Compounds and Their Nitrogen Derivatives

The organocatalyzed condensation of arenes (i.e., phenols) with a model carbonyl compound, 3,3,3-trifluoropyruvate **6a**, was first described in 2008 in the presence of *Cinchona* alkaloid **42a** (10 mol%). Crucial aspect to guarantee high levels of stereoinductions (ee up to 94%) lied on blocking the C9—OH with a hindered group (i.e., phenanthrene, PHEN) and leaving unprotected the C6′—OH function (Scheme 5.13) [23].

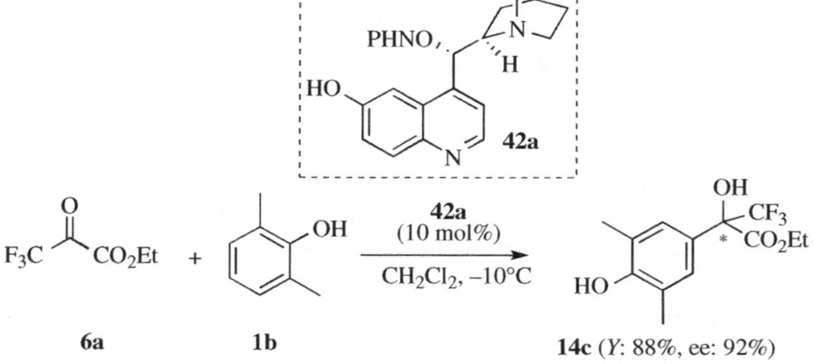

SCHEME 5.13 Asymmetric alkylation of phenols with carbonyl compounds in the presence of chiral *Cinchona* alkaloids (PH: 9-phenanthryl).

Accordingly, the fine tunability of *Cinchona* alkaloids allowed to optimize the sterical/electronic features of the organocatalysts in order to extend the previous methodology also to the aza-Friedel–Crafts reaction of naphthols **5b** [24a] and sesamols **5c** [24b] (i.e., *N*-sulfonyl aldimines **22b**, as FC electrophiles; Scheme 5.14). Interestingly, also a leucine-derived bifunctional catalyst **43** demonstrated efficiency in the enantioselective aza-Friedel–Crafts reaction with phenols **1**, providing the corresponding aminophenols **46** in excellent enantiomeric excesses [24c].

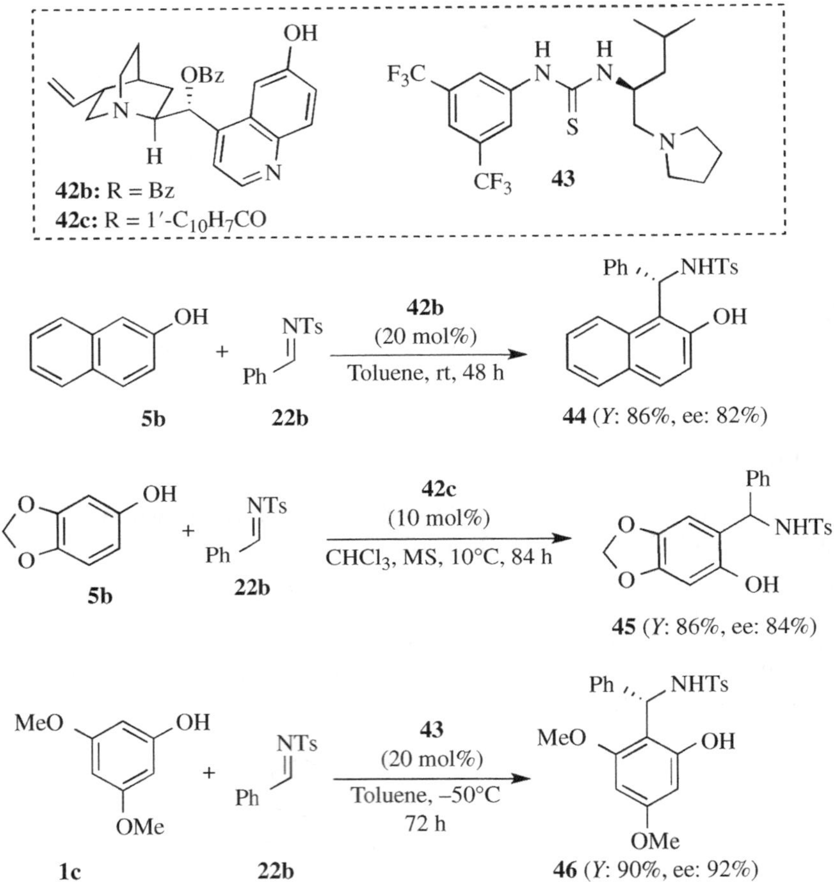

SCHEME 5.14 Asymmetric organocatalyzed aza-Friedel–Crafts alkylations of phenols, naphthols, and sesamols.

A dual activation mode based on a network of hydrogen bond interactions is assumed being active in all these methodologies. In particular, the C6′—OH group could activate the carbonyl derivatives, while the quinuclidine tertiary nitrogen atom could modulate the reactivity of the hydroxybenzene (the case of *Cinchona* alkaloids is reported in Fig. 5.3).

5.3.3 Asymmetric Organocatalyzed Alkylations of Arenes via Michael Additions

Pioneered by MacMillan's work (Scheme 5.12) [1], a range of catalytic methodologies has been subsequently reported. Here, enals, nitro olefins, and enylphosphonates proved efficient in the regio- and stereoselective functionalization of arenes in a straightforward manner.

In particular, second-generation MacMillan's catalyst **38b** (20 mol%) was found promoting the conjugate addition of the *N,N′*-dimethylanilines **18** with dimethyl 3-oxoprop-1-enylphosphonate **47** in 75–77% ee and 75–90% isolated yields (Scheme 5.15) [25]. Interestingly, this approach provided a direct entry to one member of the fosmidomycin antimalarial agent (**47**: R=H) [26] in good stereoinduction.

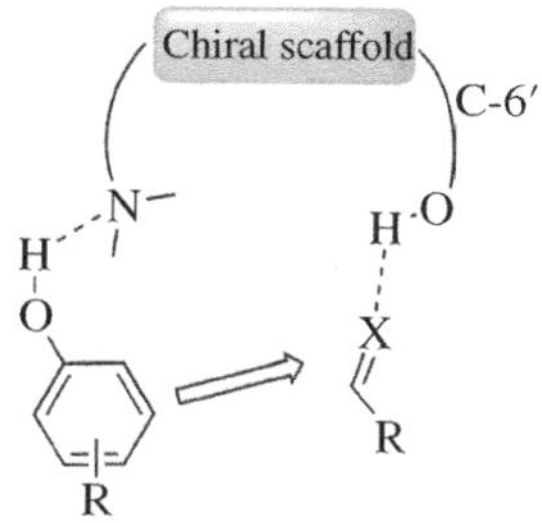

FIGURE 5.3 Dual mode activation of the *Cinchona* alkaloids in the aza-FC alkylation of electron-rich arenes.

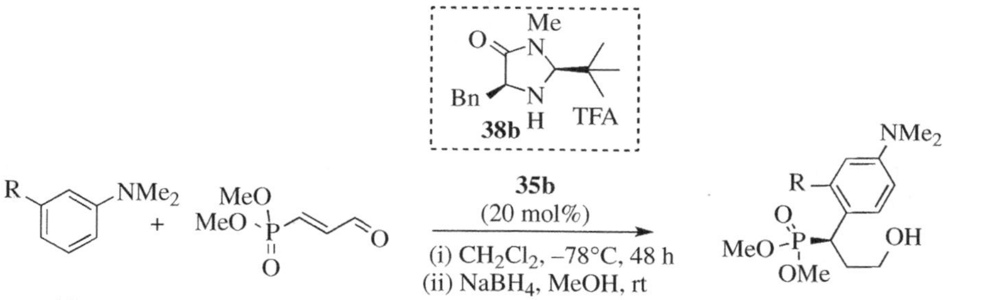

SCHEME 5.15 Asymmetric organocatalyzed synthesis of α-aryl phosphonates **48**.

Nitroalkenes represent a valuable synthetic building block in organic synthesis due to their high attitude in taking part in Michael addition as acceptors and due to the wide synthetic flexibility of the resulting nitroalkanes [27]. In this segment, Chen and coworkers first reported on the possibility to couple 2-naphthols and nitroalkenes **51** under the assistance of chiral bifunctional *Cinchonine*-based thiourea **49a** (10 mol%; Scheme 5.16a) [28]. 1-Naphthol **5a** was also tested providing the corresponding 2-alkylated arene in moderated ee (80%). Shortly after, Nagasawa and Sohtome extended the applicability of the method also to phenols **1** by means of conformationally flexible acyclic bis-thiourea organocatalyst **50** [29]. The authors suggested that a long chiral spacer between the catalytic centers of **50**, guanidinium cation and thiourea moiety, could positively impact on the stereochemical profile of the process that was demonstrated to occur under entropy control (Scheme 5.16b).

A synthetically powerful variant of the asymmetric FCA reaction via Michael addition involves cascade processes in which multiple catalytic events occur subsequentially. The approach has been efficiently employed in the synthesis of chromanes and polycyclic analogues via enantioselective annulation reactions.

Naphthols and phenols can be considered as "bifunctional" nucleophiles since besides the intrinsic carbon nucleophilicity displayed in the FC reactions, the hydroxyl group can still exert nucleophilicity if bifunctional electrophiles are employed.

Organocatalysis was utilized to emphasize this feature due to the high degree of chemoselectivity of asymmetric transformations based on aminocatalysis.

Wang elegantly demonstrated the potentiality of chiral diarylprolinol ether **54** in the synthesis of chromanes **56** via enantioselective Michael-type Friedel–Crafts alkylation/cyclization cascade synthetic sequence between **5a** and α,β-unsaturated aldehydes **37a** [30a]. Under optimal conditions, moderate diastereoselectivity and high enantioselectivity were obtained. Differently, phenol was found unreactive (Scheme 5.17a). The same team years later documented also the activity of a rosin-derived tertiary amine-thiourea **55** in similar process involving 1- and 2-naphthols and β,γ-unsaturated α-ketoesters **25** (Scheme 5.17b) [30b]. A proposed model of the enantiodiscriminating step of the reaction is also provided by the authors (**58**).

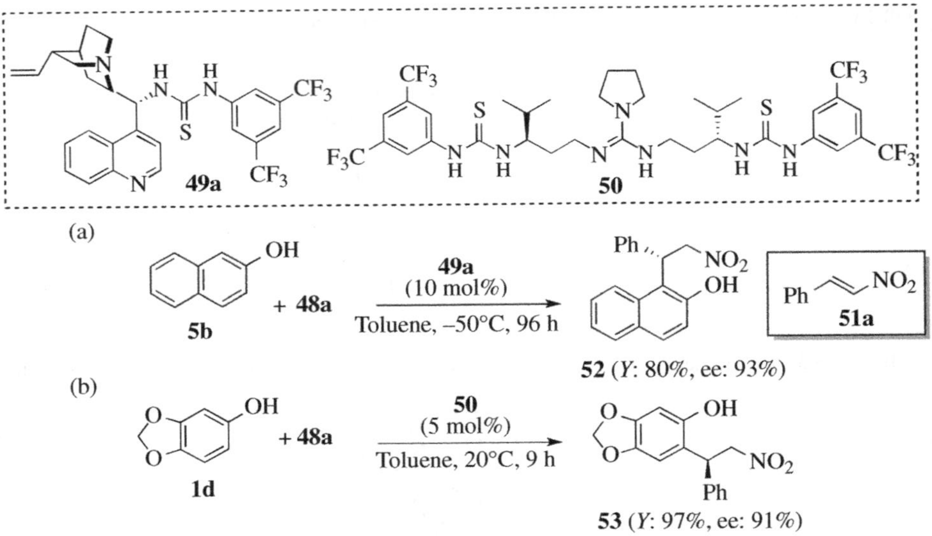

SCHEME 5.16 Chiral bifunctional organocatalysis in the Michael addition of hydroxybenzenes (a: naphthol and b; alkoxyphenol) to nitro olefins **51**.

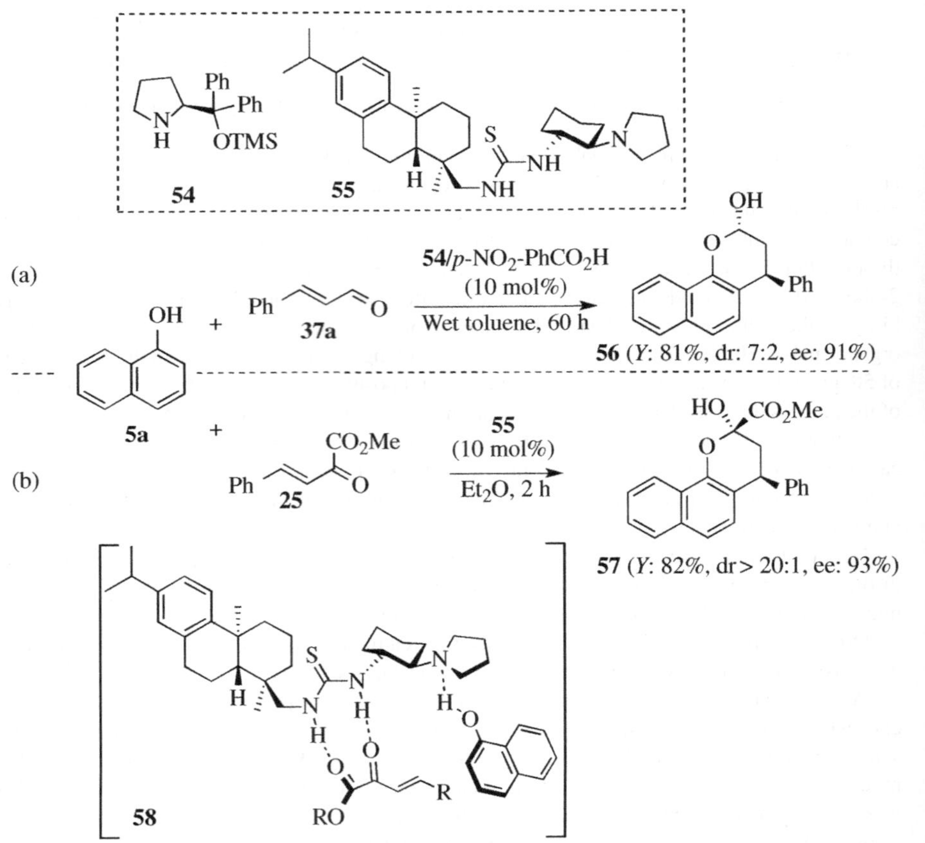

SCHEME 5.17 Diastereo- and enantioselective synthesis of chromanes via organocatalyzed cascade reactions. (a) With α,β-unsaturated aldehyde and (b) with β–γ-unsaturated β-ketoesters.

Similar chemical outcome was also obtained through the use of an emerging class of bielectrophilic synthons: (*Z*)-bromonitroalkenes **51b**. In particular, it was envisioned that Michael-type FCA followed by the nucleophilic substitution of the bromine atom by the hydroxy naphthol group could provide chiral *trans*-dihydroarylfuran derivatives **60** directly [31]. Extensive survey of reaction conditions led the authors to optimize the catalytic system with the squaramide catalyst **59** (10 mol%) and an external base (i.e., NaOAc) in order to neutralize the stoichiometric amount of HBr delivered by the catalytic cycle (Scheme 5.18). Unfortunately, the method appeared limited to naphthols. As a matter of fact, when substituted phenols were tested, the cascade process became sluggish; despite still good stereochemical profiles were recorded.

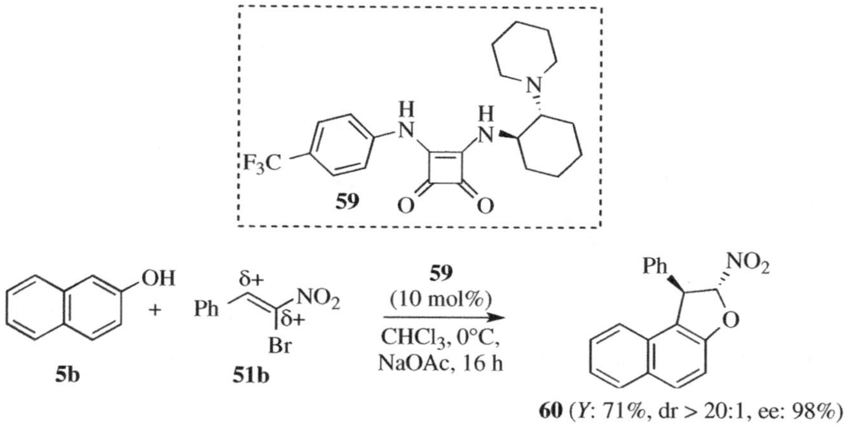

SCHEME 5.18 Stereoselective synthesis of *trans*-dihydroarylfurans **60** via organocatalyzed cascade reactions involving bromonitroalkenes **51b**.

Alkylideneindolenines **62** found important applications as Michael-type acceptors. Seminal works by Petrini's group [32] documented on the synthesis of **62** via base-assisted decomposition of arenesulfonylalkylindoles **61** (Fig. 5.4a).

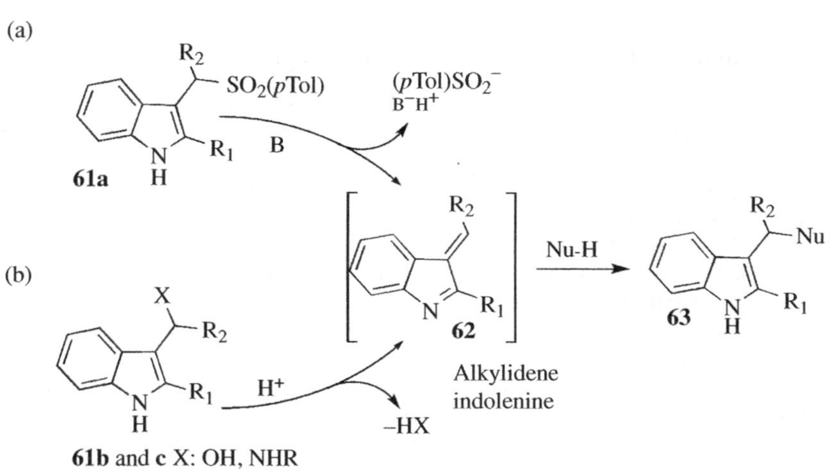

FIGURE 5.4 *In situ* generation of alkylidene indolenines (**62**): valuable Michael acceptors for organic transformations. (a) Base catalysis. (b) Acid catalysis.

This class of intriguing electrophilic agents has been very recently protagonist in the organocatalyzed enantioselective functionalization of variously substituted 2-naphthols under noncovalent activation conditions by Jing and coworkers [33]. A range of sulfonylalkylindoles **61a** was subjected to condensation with variously substituted 2-naphthols in the presence of chiral thiourea **49b** (10 mol%). The corresponding alkylated naphthols **64** were isolated with variable isolated yields (64–96%) and *enantiomeric excesses* markedly dependent from the nature of the substituent R_2. In particular, while very high stereoinductions were obtained for R_2 = aryl, disappointing results were collected when aliphatic counterparts were utilized (Scheme 5.19).

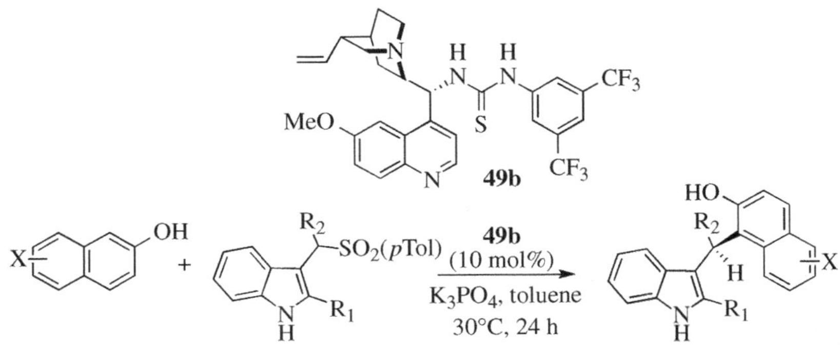

5b **61a** **64** (*Y*: 64–96%, ee: 0–98%)

SCHEME 5.19 Organocatalyzed enantioselective Michael additions of 2-naphthols to *in situ* generated alkylideneindolenines.

Similarly, C-3 benzyl alcohols and amino derivatives **61b** and **c** (Fig. 5.4b) are known to deliver highly reactive indolinium derivatives also when exposed to acid conditions.

In this segment, You and coworkers documented the suitability of the *in situ* generated electrophilic indolinium in the chiral Brønsted acid-mediated enantioselective decoration of electron-rich aromatic hydrocarbons [34]. Both alcohols and NTs amide compounds proved suitability for the enantioselective alkylation of alkoxy-substituted benzenes in inter- and intramolecular fashion. Contact ion pairing between the chiral BA anion and the indolinium intermediate was accounted as active species during the enantiodiscriminating event of the alkylation reaction (Scheme 5.20).

5.3.4 Organo-SOMO-Catalyzed Asymmetric Alkylations of Arenes

Within the wide scenario of organocatalysis, SOMO (singly occupied molecular orbital) catalysis is becoming extremely popular for the assembling of complex organic architectures through the use of weak nucleophilic reagents that proved incompatibility with classic methods [35]. Key aspect of the methodology involves the selective oxidation of electron-rich *in situ* formed enamines with the consequent formation of highly reactive 3π-radical cation species **70** (Fig. 5.5). Chemo- (O_2, TEMPO, CAN, …) as well as as well as photochemical enantioselective organocatalytic oxidation of aldo- and ketoenamines have been efficiently documented in the of carbonyl compounds with the benefit of umpolung the intrinsic nucleophilic α-position of carbonyl compounds.

Very recently, this methodology was utilized by both Nicolaou [36] and MacMillan [37] groups in the selective α-arylation of aldehydes by using not prefunctionalized benzene-like rings. The process could also be conveniently conceptualized also in terms of catalytic asymmetric FC-type alkylation reaction and, therefore, deserves particular mention.

Nicolaou's team envisioned the possibility to apply the asymmetric SOMO catalysis in the key step of the total synthesis of calamenene. Basically, the strategy concerns the use of acyclic aldehydic compounds **71** carrying electron-rich arene as model substrates and CAN (i.e., cerium(IV)

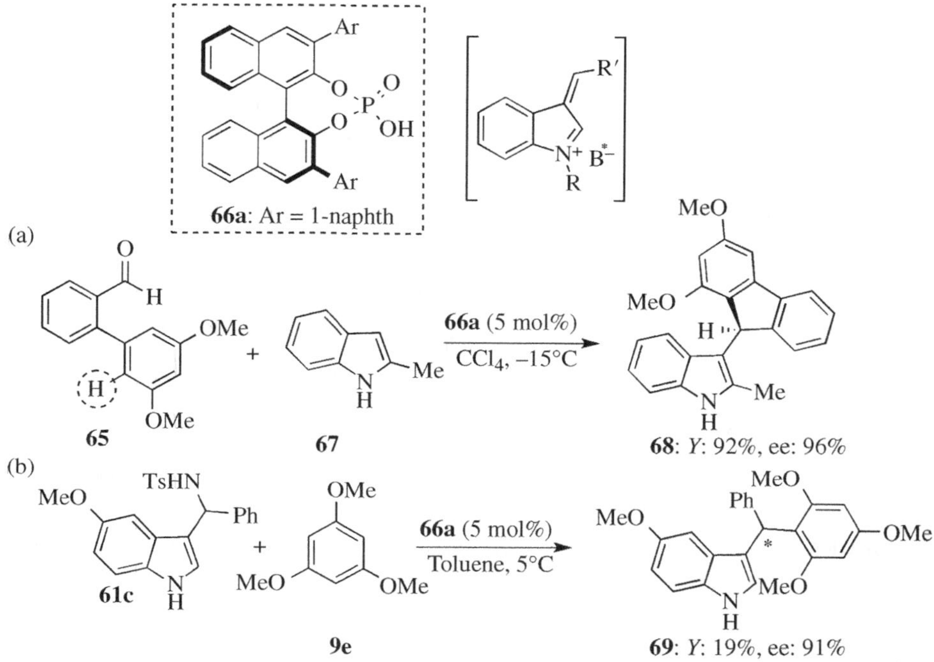

SCHEME 5.20 Brønsted acid (BA)-catalyzed alkylation of aromatic hydrocarbons. (a) Synthesis of indenes and (b) enantioselective alkylation of alkoxybenzene via electrophilic indolinium intermediates.

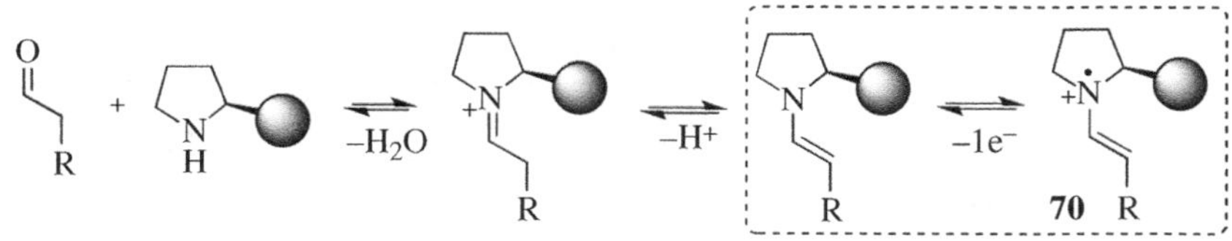

FIGURE 5.5 Single electron transfer (SET) oxidation of an *in situ* formed enamine to obtain electrophilic radical cations.

ammonium nitrate) as the stoichiometric oxidant. Enantiomerically enriched 2-*tert*-butyl-3-methyl-5-benzyl-4-imidazolidinone **38b** showed particular competence in delivering a range of tetrahydro naphthyl skeletons **73** in good yields (up to 80%) and very high enantiomeric excess (up to 97%). Remarkably, the bicyclic compound **73** obtained from selected aldehyde was employed to deliver enantiomerically pure demethyl calamenene **74** in a straightforward manner (Scheme 5.21).

Analogous process was investigated also by MacMillan's group that provided clear evidence on the efficiency of chiral imidazolidinone (2*R*,5*R*)-**38b** in controlling stereochemical profile of radical alkylation of aromatic electron-rich arenes [37a]. Here, $[Fe(phen)_3]\cdot(PF_6)_3$ was utilized as the stoichiometric single electron oxidant. Subsequently, DFT calculations allowed the unambiguous characterization of radical species involved in the catalytic cycle and provided a concrete rationale for the observed regiochemistry of the ring-closing step (i.e., *ortho*-selectivity for 1,3-disubstituted aromatics; see Scheme 5.22) [37b].

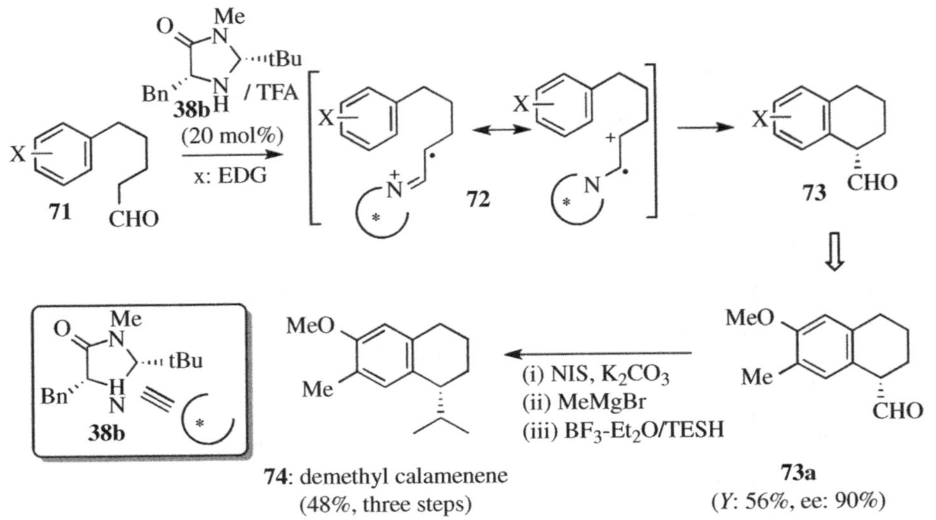

SCHEME 5.21 Application of the enantioselective SOMO catalysis to the synthesis of calamenene derivative **74** via formal FC-type alkylation reaction.

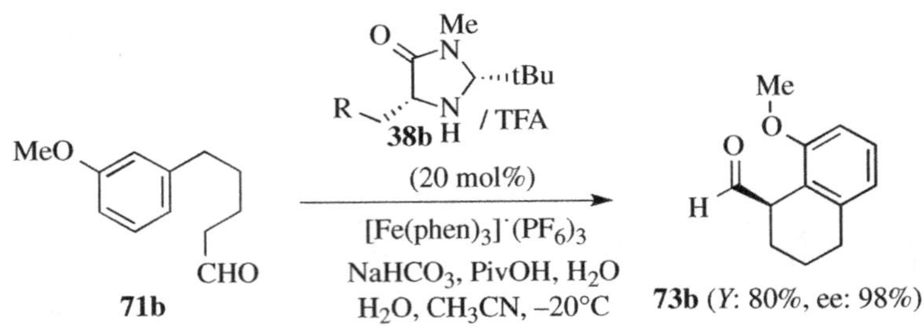

SCHEME 5.22 Asymmetric α-arylation of aldehydes via SOMO catalysis.

5.3.5 Miscellaneous in Asymmetric Organocatalyzed Alkylations of Arenes

The stereoselective Friedel–Crafts amination reaction has been also reported by the Jørgensen group via an asymmetric organocatalyzed reaction [38]. Interestingly, the chiral amination of 2-naphthols leads to non-biaryl atropoisomeric mixtures **75** obtained in highly stereocontrolled manner. The approach relies on the well-known condensation of naphthalenes and azodicarboxylates that was reported for the first time in 1921 [39]. In particular, the combined use of chiral tertiary amines and hydroxyarenes resulted in the formation of chiral naphtholate intermediate that smoothly underwent condensation with the diazo compound **74** in a highly regio- and stereoselective manner. Aminated naphthols **75** feature a rotation barrier around the C—N bond sufficiently high to be easily separated in the two enantiomeric forms via chiral HPLC analysis (Scheme 5.23).

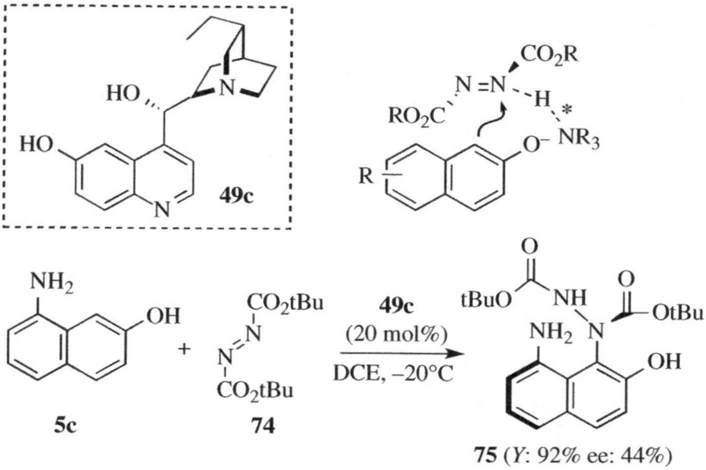

SCHEME 5.23 Asymmetric amination of naphthols via organocatalysis.

5.4 MERGING ASYMMETRIC METAL AND ORGANOCATALYSIS IN FRIEDEL–CRAFTS ALKYLATIONS

The merging of metal catalysis and organocatalysis has recently emerged as a powerful tool to realize unprecedented multistep reaction sequences, one pot. Remarkable applications were also found in asymmetric synthesis [40].

In the Friedel–Crafts realm, Lou and coworkers reported on the activity of Mg-phosphoric acid-based binary catalyst in the alkylation of free phenols via Michael addition of β,γ-unsaturated α-ketoesters **25** [41]. Despite the real structure of the binary organometallic catalytic species is still unknown, excellent levels of chemical and optical outcomes were achieved in the titled process (Scheme 5.24). To be mentioned that neither the BA nor MgF_2 alone could promote the model reaction at any extents, proving the formation of a concertedly activated catalytic aggregate between the two acids.

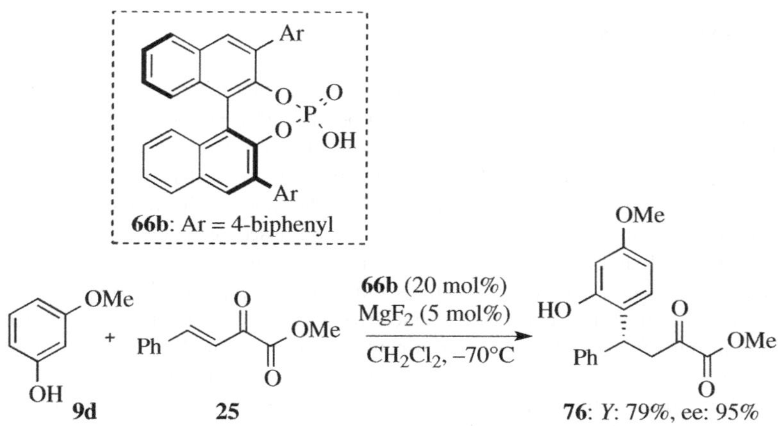

SCHEME 5.24 Binary catalytic system in the enantioselective alkylation of free phenols.

Finally, an elegant example of relay catalysis was applied by Terada and coworker in the asymmetric Pictet–Spengler condensation involving a ternary reaction sequence [42]. In particular, substituted tetrahydroisoquinolines **78** were readily accessible in moderate ee (up to 53%) starting from crotylamides **77** by exploiting simultaneously [Ru(II)] catalysis and BINOL-phosphoric acid catalysis (**66**c). The whole process comprises the following reaction sequence: isomerization of allylamide to enamide **79** (Ru-H), generation of the iminium intermediate **80** by protonation of the enamide (chiral BA), and *endo-trig* Pictet–Spengler cyclization under chiral anion control (Scheme 5.25).

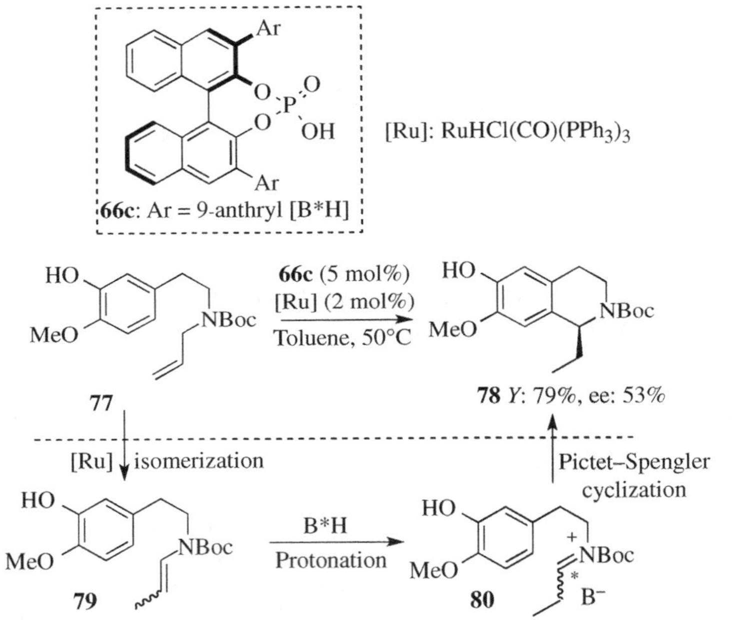

SCHEME 5.25 Asymmetric Pictet–Spengler cyclization under "relay" binary catalysis.

5.5 SUMMARY AND OUTLOOK

Benzylic stereogenic centers are widely diffused molecular motifs within naturally occurring compounds. The direct replacement of aromatic C—H bonds with C—C as well as C—X chemical connections represents one of the most valuable synthetic alternatives for the stereochemical decoration of arene peripheries. Over the past decades, catalytic enantioselective Friedel–Crafts-type alkylation reactions emerged as a potent synthetic tool for the functionalization of electron-rich arenes in the presence of substoichiometric amounts of metal-based or metal-free chiral entities. This scenario impacted dramatically the overall sustainability of the process (i.e., economical and environmental point of view), with respect to the seminal multistep protocols. However, severe limitations in terms of substrate scope still characterize these transformations that generally face a significant drop in efficiency when "electron-neutral" (i.e., benzene-like) or electron-deficient aromatics are involved. As a matter of fact, the mild reaction conditions routinely adopted in enantioselective catalytic aromatic functionalizations do not satisfy the demanding energy profiles of reaction courses involving deactivated arenes.

However, very recently remarkable progresses have been recorded in this segment allowing the realization of complex and stereochemically defined aromatic structures in a highly efficient mode, and a few of the most salient catalytic enantioselective FC alkylations of benzenes are described herein.

The field is facing a continuous expansion and together with the somehow complementary late transition metal-mediated aromatic C—H insertion reactions will contribute substantially in drawing new synthetic shortcut toward the realization of complex molecular targets.

ABBREVIATIONS

AFCA	Asymmetric Friedel–Crafts alkylation
BA	Brønsted acid
BINAP	2,2′-Bis(diphenylphosphino)-1,1′binaphthalene
BINOL	1,1′-Bi-2-naphthol
BOX	Bisoxazoline
CAN	Cerium ammonium nitrate
COD	1,5-Cyclooctadiene
DACH	*trans*-1,2-diaminocyclohexane
DMAP	4-Dimethylaminopyridine
EDG	Electron-donating group
EE	Enantiomeric excess
EWG	Electron-withdrawing group
FCA	Friedel–Crafts alkylation
LA	Lewis acid
LTM	Late transition metal
LUMO	Lowest unoccupied molecular orbital
MCM	Mobil composition of matter
PHEN	Phenanthrene
*p*TsOH	*p*-Toluenesulfonic acid.
SOMO	Singly occupied molecular orbital
TEMPO	2,2,6,6-Tetramethylpiperidine 1-oxyl
TfOH	Trifluoromethanesulfonic acid

REFERENCES

[1] Paras, N. A. and MacMillan, D. W. C. (2002) *J. Am. Chem. Soc.* **124**, 7894–7895.

[2] (a) Friedel, C. and Crafts, J. M. (1877) *Compt. Rend.* **84**, 1392–1395; (b) Friedel, C. and Crafts, J. M. (1877) *Compt. Rend.* **84**, 1450–1454.

[3] Bandini, M. (2009) *Catalytic Asymmetric Friedel-Crafts Alkylations* (eds. M. Bandini, A. and Umani-Ronchi), Wiley-VCH, Weinheim.

[4] Bigi, F., Casiraghi, C., Casnati, G., and Sartori G. (1985) *J. Org. Chem.* **50**, 5018–5022.

[5] Erker, G. and van der Zeijden, A. A. H. (1990) *Angew. Chem. Int. Ed. Engl.* **29**, 512–514.

[6] For general reviews of catalytic asymmetric methodologies for the alkylation of aromatics see: (a) Jørgensen K. A. (2003) *Synthesis* 1117–1125; (b) Bandini, M., Melloni, A., and Umani-Ronchi, A. (2004) *Angew. Chem. Int. Ed.* **43**, 550–556. (c) Bandini, M., Melloni, A., Tommasi, S., and Umani-Ronchi, A. (2005) *Synlett* 1199–1222; (d) Poulsen, T. B. and Jørgensen, K. A. (2008) *Chem. Rev.* **108**, 2903–2915; (e) Marqués-López, E., Diez-Martinez, A., Merino, P., and Herrera, R. P. (2009) *Curr. Org. Chem.* **13**, 1585–1609; (f) Rueping M. and Nachtsheim, B. J. (2010) *Beilstein J. Org. Chem.*, doi:10.3762/bjoc.6.6; (g) You, S.-L., Cai Q., and Zeng, M. (2009) *Chem. Soc. Rev.* **38**, 2190–2201; (h) Bartoli, G., Bencivenni, G.,

and Dalpozzo, R. (2010) *Chem. Soc. Rev.* **39**, 4449–4465; (i) Terrasson, V., de Figueiredo, R. M., and Campagne, J. M. (2010) *Eur. J. Org. Chem.* 2635–2655; (j) Zen, M. and You, S.-L. (2010) *Synlett* 1289–1301; (k) Chauhan, P. and Chimni, S. S. (2012) *RCS Adv.* **2**, 6117–6137.

[7] Soon, T. P. and Jacobsen, E. N. (2003) *Science* **299**, 1691–1693.

[8] (a) Ishii, A. and Mikami, K. (1999) *J. Fluor. Chem.* **97**, 51–55; (b) Ishii, A., Soloshonok, V. A., and Mikami, K. (2000) *J. Org. Chem.* **65**, 1597–1599; (c) Mikami, K., Terada, M., Korenaga, T., Matsumoto, Y., Ueki, M., and Angelaud, R. (2000) *Angew. Chem. Int. Ed.* **112**, 3532–3556.

[9] (a) Zhuang, W., Gathergood, N., Hazell, R. G., and Jørgensen, K. A. (2001) *J. Org. Chem.* **66**, 1009–1013; (b) Mikami, K., Aikawa, K., Kainuma, S., Kawakami, Y., Saito, T., Sayo, N., and Kumobayashi, H., (2004) *Tetrahedron Asymmetry* **15**, 3885–3889; (c) Corma, A., García, H., Moussaif, A., Sabater, M. J. R., Zniber, R., and Redouane, A. (2002) *Chem. Commun.* 1058–1059.

[10] Zhao, J.-L., Liu, L., Sui, Y., Liu, Y.-L., Wang, D., and Chen, Y.-J. (2006) *Org. Lett.* **8**, 6127–6130.

[11] (a) Saaby, S., Fang, X., Gathergood, N., and Jørgensen, K. A. (2000) *Angew. Chem. Int. Ed.* **39**, 4114–4116; (b) Saaby, S., Bayón, P., Aburel, P. S., and Jørgensen K. A. (2002) *J. Org. Chem.* **67**, 4352–4361.

[12] Zhu, C., Yuan, C., and Lv Y. (2006) *Synlett* 1221–1224.

[13] Kuwano, R. and Ito, Y. (1999) *Comprehensive Asymmetric Catalysis* (eds. E. N. Jacobsen, A. Pfaltz, and H. Yamamoto), Springer-Verlag, Berlin.

[14] Takizawa, S., Arteaga Arteaga, F., Yoshida, Y., Kodera, J. Nagata, Y., and Sasai, H. (2013) *Dalton Trans.* **42**, 11787–11790.

[15] Faita, G., Mella, M., Toscanini, M., and Desimoni, G. (2010) *Tetrahedron* **66**, 3024–3029.

[16] Bai, S., Liu, X., Wang, Z., Cao, W., Lin, L., and Feng, X. (2012) *Adv. Synth. Catal.* **354**, 2096–2100.

[17] (a) Trost, B. M. and Van Vranken, D. L. (1996) *Chem. Rev.* **96**, 395–422; (b) Trost, B. M. and Crawley, M. L. (2003) *Chem. Rev.* **103**, 2921–2944.

[18] Bandini, M., Melloni, A., Piccinelli, F., Sinisi, R., Tommasi, S., and Umani-Ronchi, A. (2006) *J. Am. Chem. Soc.* **128**, 1424–1425.

[19] For an example of Pd catalyzed C-allylation of phenols via an *O*-allylation–Claisen rearrangement sequence, see: Trost, B. M. and Toste, F. D. (1998) *J. Am. Chem. Soc.* **120**, 815–816.

[20] Xu, Q.-L., Dai, L.-X., and You, S.-L. (2012) *Org. Lett.* **14**, 2579–2581.

[21] (a) Suzuki, Y., Nemoto, T., Kakugawa, K., Hamajima, A., and Hamada, Y. (2012) *Org. Lett.* **14**, 2350–2353; (b) Suzuki, Y., Matsuo, N., Nemoto, T., and Hamada, Y. (2013) *Tetrahedron* **69**, 5913–5919.

[22] Lu, H.-H., Tan, F., and Xiao, W.-J. (2013) *Curr. Org. Chem.* **15**, 4022–4045.

[23] (a) Zhao J.-L., Liu L., Gu C.-L., Wang D., and Chen Y.-J. (2008). *Tetrahedron Lett.* **49**, 1476–1479; (b) Liu, G., Zhang, S., Li, H., Zhang, T., and Wang, W. (2011) *Org. Lett.* **13**, 828–831.

[24] (a) Chauhan P. and Singh Chimni, S. (2011) *Eur. J. Org. Chem.* 1636–1640; (b) Chauhan P. and Singh Chimni, S. (2013) *Tetrahedron Lett.* **54**, 4613–4616; (c) Li, G.-X. and Qu, J. (2012) *Chem. Commun.* **48**, 5518–5520.

[25] Guo, Y.-C., Li, D.-P., Li, Y.-L., Wang, H.-M., and Xiao, W.-J. (2009) *Chirality*, **21**, 777–785.

[26] Haemers, T., Wiesner, J., Busson, R., Jomaa, H., and Calenbergh, S. V. (2006) *Eur. J. Org. Chem.* 3856.

[27] (a) Tsogoeva S. B. (2007) *Eur. J. Org. Chem.* 1701–1716; (b) Almasi, D., Alonso, D. A., and Nájera, C. (2007) *Tetrahedron Asymmetry* **18**, 299–365; (c) Vicario, J. L., Badia, D., and Carrillo, L. (2007) *Synthesis* 2065–2092.

[28] Liu, T.-Y., Cui, H.-L., Chai, Q., Long, J., Li, B.-J., Wu, Y., Ding, L.-S., and Chen, Y.-C. (2007) *Chem. Commun.* 2228–2230.

[29] (a) Sohtome, Y., Shin, B., Horitsugi, N., Takagi, R., Noguchi, K., and Nagasawa K. (2010) *Angew. Chem. Int. Ed.* **49**, 7299–7303; (b) Sohtome, Y., Shin, B., Horitsugi, N., Noguchi, K., and Nagasawa, K. (2011) *Chem. Asian J.* **6**, 2463–2470.

[30] (a) Hong, L., Wang, L., Sun, W., Wong, K., and Wang, R. (2009) *J. Org. Chem.* **74**, 6881–6884; (b) Jiang, X., Wu, L., Xing, Y., Wang, L., Wang, S., Chen, Z., and Wang R. (2012) *Chem. Commun.* **48**, 446–448.

[31] Jarava-Barrera, C., Esteban, F., Navarro-Ranninger, C., Parra, A., and Alemán J. (2013) *Chem. Commun.* **49**, 2001–2003.

[32] (a) Ballini, R., Palmieri, A., Petrini, M., and Torregiani, E. (2006) *Org. Lett.* **8**, 4093–4096; (b) Martinelli, F., Palmieri, A., and Petrini M. (2011) *Phosphorus Sulfur Silicon* **186**, 1032–1045.

[33] Yu, L., Xie, X., Wu, S., Wang, R., He, W., Qin, D., Liu, Q., and Jing L. (2013) *Tetrahedron Lett.* **54**, 3675–3678.

[34] (a) Sun, F.-L., Zeng, M., Gu, Q., and You S.-L. (2009) *Chem. Eur. J.* **15**, 8709–8712; (b) Sun, F.-L., Zheng, X.-J., Gu, Q., He, Q.-L., and You S.-L. (2010) *Eur. J. Org. Chem.* 47–50.

[35] MacMillan D. W. C. (2008) *Nature* **455**, 304-308.

[36] Nicolaou, K. C., Reingruber, R., Sarlah, D., and Bräse S. (2009) *J. Am. Chem. Soc.* **131**, 11640–11641.

[37] (a) Conrad, J. C., Kong, J., Laforteza, B. N., and MacMillan D. W. C. (2009) *J. Am. Chem. Soc.* **131**, 11640–11641; (b) Um, J. M., Gutierrez, O., Schoenebeck, F., Houk, K. N., and MacMillan D. W. C. (2010) *J. Am. Chem. Soc.* **132**, 6001–6005.

[38] Brandes, S., Bella, M., Kjærsgaard, A., and Jørgensen K. A. (2006) *Angew. Chem. Int. Ed.* **45**, 1147–1151.

[39] (a) Diels, O. and Back I. (1921) *Chem. Ber.* **54**, 213–226; (b) Mitchell, H. and Leblanc Y. (1994) *J. Org. Chem.* **59**, 682–687.

[40] (a) Park, Y. J., Park, J.-W., Jun, C.-H. (2008) *Acc. Chem. Res.* **41**, 222–234; (b) Shao, Z. and Zhang H. (2009) *Chem. Soc. Rev.* **38**, 2745–2755; (c) Zhou J. (2010) *Chem. Asian J.* **5**, 422–434.

[41] (a) Lv, J., Li, X., Zhong, L., Luo, S., and Cheng J.-P. (2010) *Org. Lett.* **12**, 1096–1099; (b) Lv, J. and Luo S. (2013) *Chem. Commun.* **49**, 847–858.

[42] Tod, M. and Terada M. (2013) *Synlett* **24**, 75–7562.

PART II

NUCLEOPHILIC AROMATIC SUBSTITUTION

6

NUCLEOPHILIC AROMATIC SUBSTITUTION: AN UPDATE OVERVIEW

MICHAEL R. CRAMPTON
Department of Chemistry, University of Durham, Durham, UK

6.1 INTRODUCTION

When Bunnett and Zahler [1] published their landmark review in 1951, only two mechanisms of nucleophilic aromatic substitution were known. These were the unimolecular S_N1 process, typically observed with arenediazonium salts, as in Scheme 6.1, and the bimolecular S_NAr pathway, which is shown in Scheme 6.2 involving substitution of a halide ion by an anionic nucleophile and involving an anionic intermediate (**1**). As in aliphatic substitutions, both unimolecular and bimolecular pathways are possible.

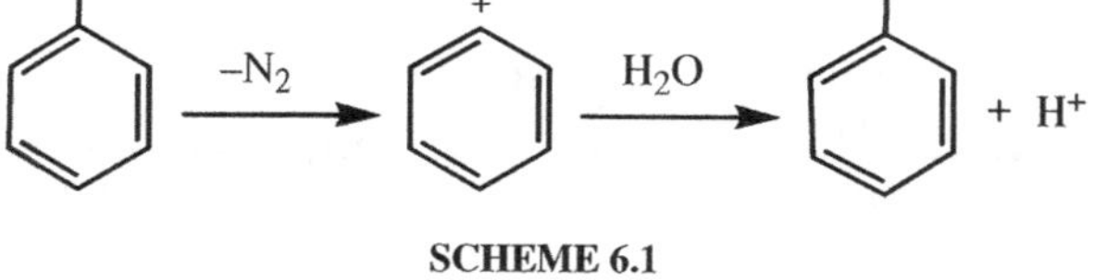

SCHEME 6.1

The S_NAr pathway involves nucleophilic addition followed by elimination of a leaving group, the nucleofuge, and requires the presence of at least one strongly electron-withdrawing substituent in the ring to stabilize the intermediate. The possibility of an elimination process followed by addition was established in the 1950s when it was shown [2, 3] that chlorobenzene labeled with ^{14}C at the 1-position reacted with potassium amide in liquid ammonia to give almost equal amounts of aniline labeled with ^{14}C at the 1- and 2-positions, as shown in Scheme 6.3. The discovery of the

Arene Chemistry: Reaction Mechanisms and Methods for Aromatic Compounds,
First Edition. Edited by Jacques Mortier.

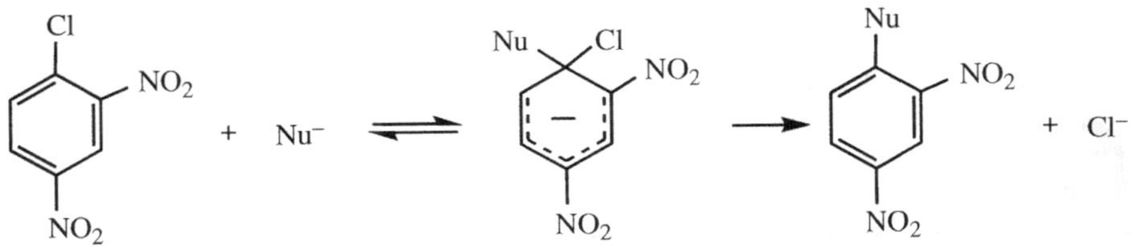

1

SCHEME 6.2

benzyne intermediate, (**2**), allows substitution in systems not activated by strongly electron-withdrawing groups (EWGs), such as the nitro group. This has opened up a whole new field of chemistry, which is detailed in Chapter 12.

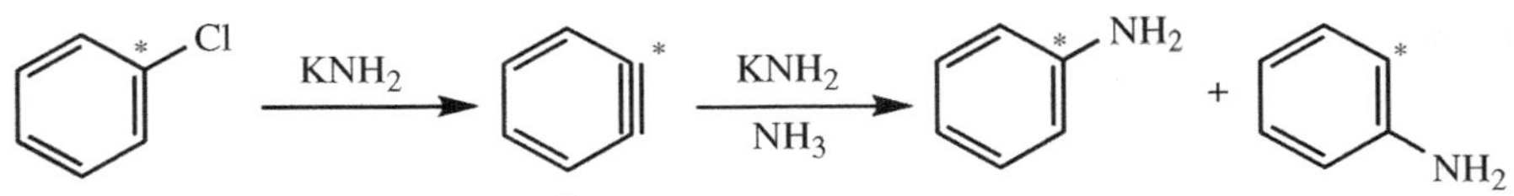

2

SCHEME 6.3

Early on, it was demonstrated that aromatic nitro compounds may form radical anions in alkaline solutions [4], with the possibilities of photochemical reactions [5]. There followed the development of the radical chain $S_{RN}1$ mechanism [6]. An interesting early demonstration of reaction by this mechanism was in the reaction of *ortho*-halogenoanisoles with potassium amide in liquid ammonia [7]. Reaction by the benzyne mechanism gives predominantly the *meta*-substituted product due to the electronic influence of the methoxy group. However, with an excess of potassium metal, which promotes electron transfer, the $S_{RN}1$ pathway predominates yielding *ortho*-anisidine, as shown in Scheme 6.4. The $S_{RN}1$ mechanism now forms an important synthetic pathway, and this and other homolytic processes are covered in Chapters 9 and 10.

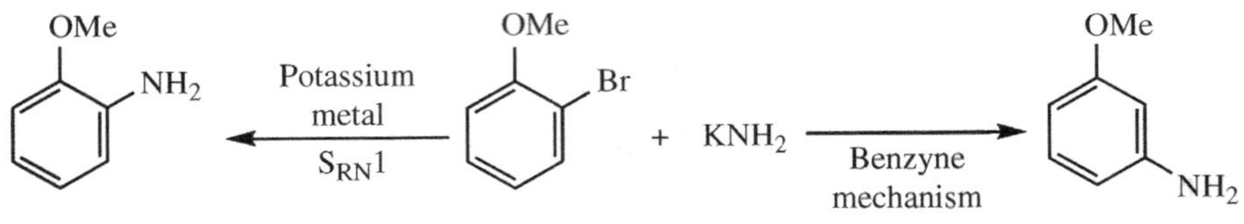

SCHEME 6.4

The mechanisms mentioned so far all require the presence in the arene of a nucleofugic group, often a halogen, nitro group, or alkoxy or phenoxy group. The substitution of ring hydrogen by these mechanisms is not usual since the potential leaving group, the hydride ion, has low stability. Nevertheless, it is known that nucleophilic attack at ring carbon atoms carrying hydrogen is usually a faster process than at similarly activated ring positions carrying substituents such as halogen or an alkoxy group [8, 9]. This is illustrated in Scheme 6.5 where (**3**) although formed more rapidly than (**4**) does not lead to substitution. Both *ortho* and *para* positions are activated by the nitro group, and both types of adduct may be present in very low concentrations. However, the irreversible loss of chloride from (**4**) eventually leads to the substituted product.

Nevertheless, processes are available allowing the conversion of adducts such as (**3**) to substituted products. These are covered in Chapter 11. Briefly, these may involve oxidation of

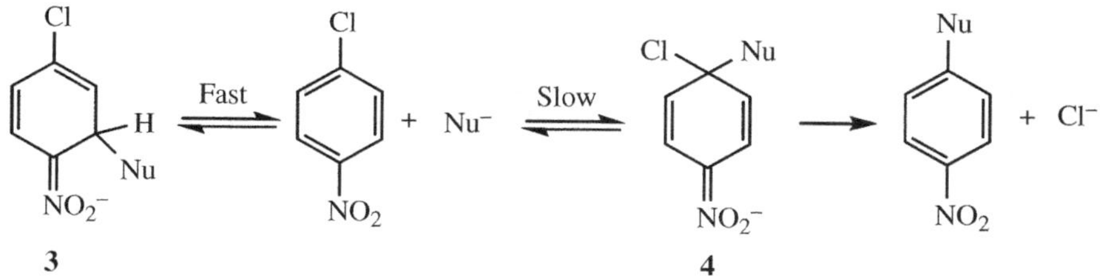

SCHEME 6.5

(**3**), either by external oxidants or by electron transfer processes with unreacted substrate [10] or by an important mechanism known as vicarious substitution [11]. Here, the nucleophile, such as the anion of chloromethyl phenyl sulfone, carries a leaving group, such as chlorine, which together with the ring hydrogen may undergo base-catalyzed β-elimination from the intermediate (**3**).

The possibilities for the formation of carbon–carbon bonds involving arenes have been dramatically increased in recent years by the use of transition metal catalysis. Copper-mediated reactions to couple aryl halides in Ullmann-type reactions [12, 13] have been known for many years, and copper still remains an important catalyst [14, 15]. However, the use of metals such as palladium [16, 17] to effect substitution has led to such an explosion of research that in 2011 transition metal-catalyzed processes comprised more than half of the reactions classified as aromatic substitutions in *Organic Reaction Mechanisms* [18]. The reactions often involve a sequence outlined in Scheme 6.6 where Ln represents ligand(s) for the palladium. Oxidative addition of the aryl halide to the palladium catalyst is followed by transmetalation with an aryl or alkyl derivative and by reductive elimination to give the coupled product and regenerate the catalyst. Part 6 of this book elaborates these and related processes.

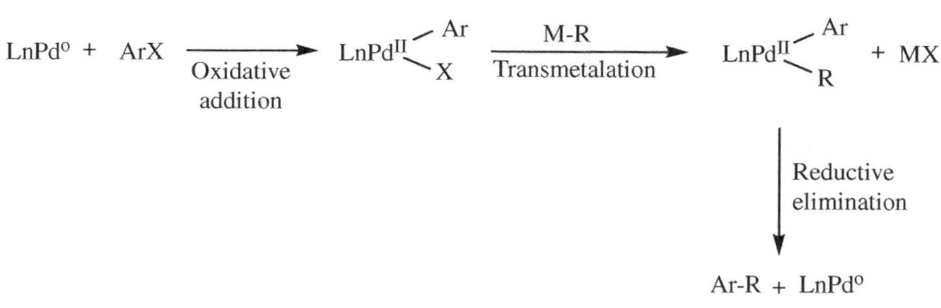

SCHEME 6.6

In a chapter of this length, it is not possible to cover the subject area comprehensively, and in the sections that follow, examples are given that hopefully will illustrate some of the main principles involved. For a more complete coverage of the literature and subject area, the readers are referred to the excellent book by F. Terrier [9].

6.2 THE S_NAr MECHANISM

The bimolecular mechanism of substitution generally involves two steps: nucleophilic attack at a ring carbon atom and expulsion of the leaving group. These are shown in Scheme 6.7, where EWG represents an EWG. The presence of at least one such group is necessary to give stability to the anionic intermediate. Nucleophilic attack will be rate determining if the energy barrier for formation of the intermediate is higher than that for its conversion to the substituted product as shown in case (a) in Figure 6.1. The reverse situation of rate-determining leaving group expulsion is case (b). Figure 6.1 also shows, in case (c), the possibility of a concerted process via a single transition state. The latter is the situation that normally prevails in bimolecular S_N2 reactions in aliphatic systems where the lack of an aromatic ring to delocalize negative charge makes an intermediate unviable.

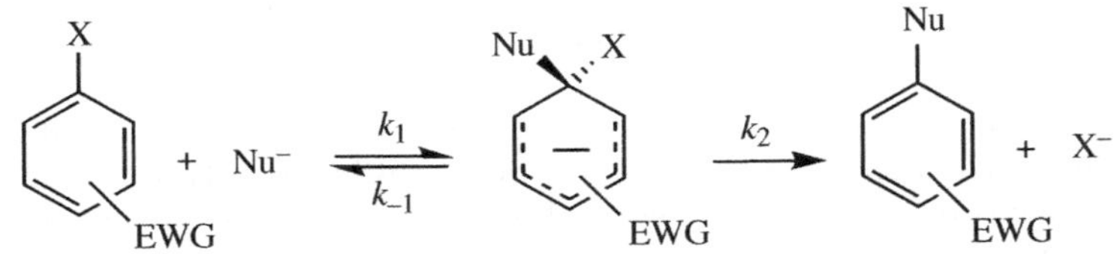

SCHEME 6.7

There are several pieces of experimental evidence for the formation of a discrete intermediate in aromatic systems. One of the earliest was Meisenheimer's observation [19] that addition of potassium methoxide to 2,4,6-trinitrophenetole or addition of potassium ethoxide to 2,4,6-trinitroanisole produced identical compounds, (**5**), which on acidification produced in each case a mixture of methyl and ethyl picrates. Meisenheimer complexes, which are anionic σ-adducts, are considered later in this chapter.

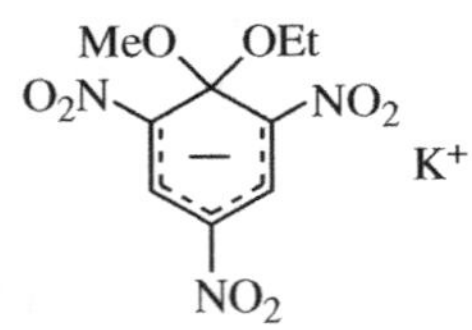

5

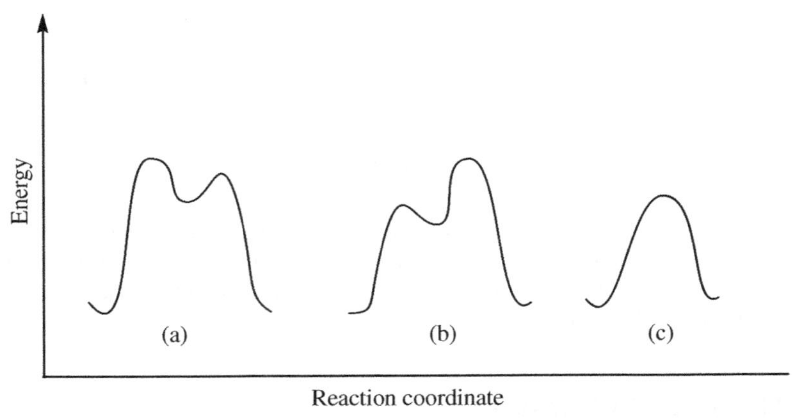

FIGURE 6.1 Energy diagrams for bimolecular substitution; (a) and (b) involve an intermediate with, respectively, nucleophilic attack and leaving group expulsion rate limiting; (c) represents a concerted mechanism with a single transition state.

Further evidence comes from the effects of different leaving groups on the overall reactivity. If the reactions were concerted, then breaking of the C—X bond, in Scheme 6.7, would be at least partially involved in the rate-determining step with the expectation that, because of the strength of the C—F bond, fluoro compounds would generally react more slowly than the other halogeno-compounds. However, in aromatic substitutions, fluorine is often a much better leaving group than the other halogens. For example, the reaction of 1-halogeno-2,4-dinitrobenzenes with hydroxide ions in water [20] gives the reactivity order F≫Cl, Br>I. This is to be expected when nucleophilic attack, the first step in Scheme 6.7, is rate determining. The two-step mechanism can also accommodate the reactivity order Br>Cl>F observed in the reaction of 1-halogeno-2,4-dinitrobenzenes with *N*-methylaniline in ethanol [21]. Here, leaving group expulsion, the second step in Scheme 6.7, is the slow step.

The observation of general base catalysis in some substitutions by amines is additional evidence for the presence of an intermediate. It is particularly significant that plots of the second-order rate constant versus the concentration of added base may be curved. Examples reported early on are in the reactions of 4-nitrophenylphosphate with secondary amines in water [22] and of 2,4-dinitrophenyl ether with piperidine [23]. At low concentrations of the base catalyst, the base-assisted second step is rate determining so that increases in catalyst concentration result in rate increases. However, a point is reached at high base concentrations, when the second step is sufficiently rapid, that the first step becomes rate limiting.

There is now considerable evidence that the vast majority of S_NAr reactions involve the two-step pathway with an intermediate sitting in a potential energy well [9, 24]. Nevertheless, evidence for a concerted pathway involving a single transition state was found [25, 26] in the displacement of 4-nitrophenoxide ions from 2-(4-nitrophenoxy)-4,6-dimethoxy-1,3,5-triazine by substituted phenoxide ions, shown in Scheme 6.8. A linear Brønsted plot of log *k*, where *k* represents the second-order rate constant, versus pK_a values was observed. The pK_a values are for the parent phenols from which the nucleophiles are formed. The linearity of the plot extended to pK_a values both higher and lower than that of the leaving group, 4-nitrophenol, and this linearity is evidence for a concerted process. A change in slope when $\Delta pK_a = 0$ (ΔpK_a being the difference in pK_a values of the entering and leaving groups) would be expected for a reaction by a two-step mechanism. A distinct change in slope of such a plot has been observed recently [27] in the related reaction of substituted phenoxide ions with 2,4,6-trinitrophenyl phenyl ethers indicative of a two-step pathway here.

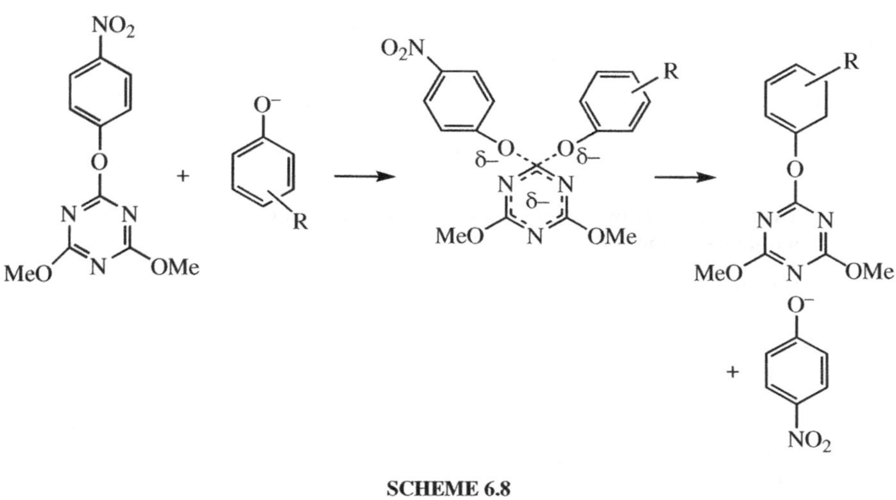

SCHEME 6.8

Some theoretical studies of gas-phase substitutions in weakly activated systems suggest the possibility of a reaction by a concerted process [28–30].

6.2.1 Effects of Activating Groups

The presence of the π-electron charge cloud inhibits the approach of nucleophiles to the aromatic ring, and EWGs are necessary to help dissipate negative charge, either inductively or through resonance interaction. The nitro group is particularly effective when located *ortho* and/or *para* to the position of attack as indicated by the resonance forms shown in Scheme 6.9. Theoretical calculations [31] of the stabilities of cyclohexadienyl anions give the values shown in Table 6.1. Values are quoted relative to the hydride adducts (**6**). The results show that for EWGs, activation decreases in the order *para*>*ortho*≫*meta*. For π-donor substituent, such as methoxy, there is a compromise between π-donation and σ-withdrawal leading to a reactivity order $m>o>p$. When groups capable of strong inductive withdrawal are bonded at the *ipso*-position, strong stabilization results. This accounts in part for the good leaving group ability of groups such as NO_2, OMe, and F. Kinetic data are available [32] for the reaction of a series of 4-substituted 1-chloro-2-nitrobenzenes with methoxide ions in methanol and with amines. Reactivity increases in the following order as the 4-substituent is varied: NH_2<OMe<Me<H<CO_2^-<Cl<$CONH_2$<CF_3<CO_2Me<COMe<SO_2Me<CN<NO_2<SO_2CF_3<NO. A Hammett plot requires the use of σ^- values, rather than σ values, to obtain good linearity, with $\rho=3.9$, signifying the importance of resonance interaction with the *para*-substituent. The trifluoromethylsulfonyl group [33] is more strongly activating than the nitro group, and more recently, the extreme activating powers of the $CF_3S(O){=}NSO_2CF_3$ group have been recognized [34].

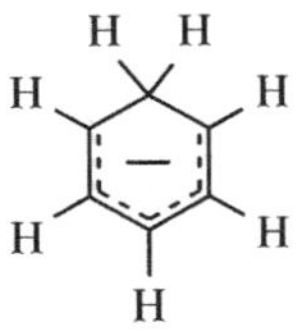

6

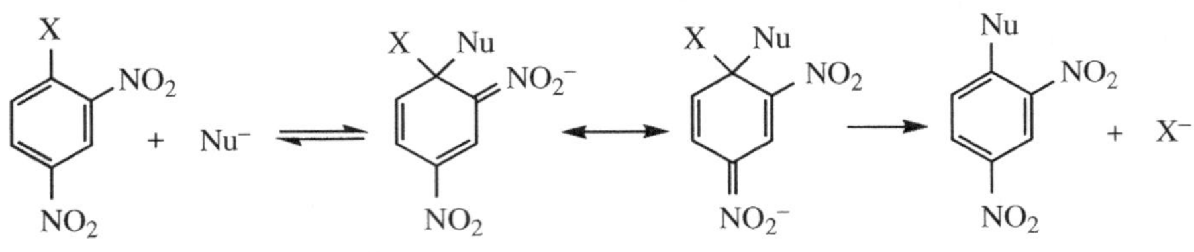

SCHEME 6.9

TABLE 6.1 Relative Stabilization Energies (kJ mol⁻¹) of Cyclohexadienyl Anions for Monosubstituted Benzenes

Substituent	*ortho*	*meta*	*para*	*ipso*
CN	125.8	69.6	148.6	5.4
NO_2	178.9	87.5	201.8	129.9
CO_2H			120.2	
F	10.4	27.9	−6.6	29
Me	−5.7	−4.6	−3.5	−5.3
OMe	−12.5	10	−31.2	43.3
NH_2	−47.6	−9.5	−78.8	15.8

Adapted with permission from Birch *et al.* [31]. © (1980) American Chemical Society.

In agreement with the stabilization energies reported in Table 6.1, *ortho*-nitro groups are generally found to be slightly less activating than their *para* isomers [9, 32]. For example, the *ortho/para* reactivity ratio is 0.3 for the reaction of 1-chloro-2-nitro- and 1-chloro-4-nitro-benzene with methoxide ions in methanol [32]. Nevertheless, in some cases, *ortho/para* ratios greater than unity are observed. Thus, in the reaction of the chloronitrobenzenes with potassium *t*-butoxide in *t*-butanol the *ortho/para* ratio is circa 100, ascribed [35] to specific stabilization of the transition state by association with the cation, as shown in (**7**). The difference between the two solvents, methanol and *t*-butanol, may lie in their difference in dielectric constant, so that potassium methoxide in methanol is dissociated while potassium *t*-butoxide in the parent alcohol is largely ion paired. For reactions involving amine nucleophiles, there is the possibility for favorable hydrogen bonding interaction, as shown in (**8**), with nitro groups that are *ortho* to the reaction center. This has been termed "built-in" solvation [1, 36] and leads to *ortho/para* ratios greater than unity. The effect becomes more important as the solvating properties of the solvent decrease; for example, for reaction of chloronitrobenzenes with piperidine, the ratio increases from 1.4 to 46 as the solvent is changed from water to benzene [37].

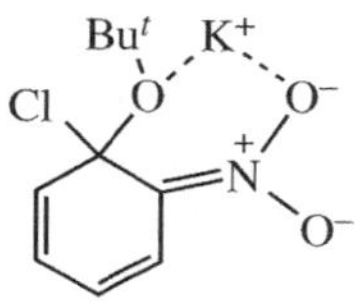

7

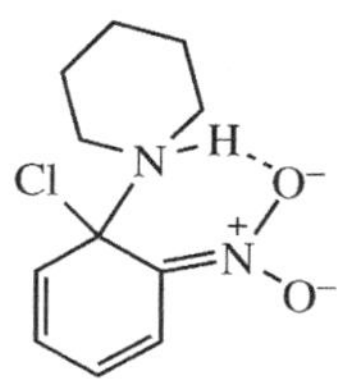

8

Naturally, the presence of increasing numbers of activating groups increases reactivity. Hence, picryl chloride (1-chloro-2,4,6-trinitrobenzene) is slowly solvolyzed in methanol or water without the need for an added base [32, 38]. The availability of polyfluoroaromatics has stimulated interest in their reactions with nucleophiles [39–41], and the activating effect of five fluorine atoms has been found to be roughly comparable with that of an *ortho*- or *para*-nitro group [42]. Similarly, the presence of the ring nitrogen in pyridine results in activation only slightly less than that from a nitro group [32]. However, substitutions in heterocyclic systems are beyond the scope of this book.

Annelation of a second benzene ring allows for the greater dispersion of negative charge so that naphthalene derivatives are generally more reactive than the corresponding benzene derivatives. Thus, 1-chloro-2,4-dinitronaphthalene is roughly an order of magnitude more reactive than 1-chloro-2,4-dinitrobenzene in nucleophilic substitutions [9]. A much more dramatic effect is observed when a furazan or furoxan ring is annelated onto a benzene ring. Thus, 7-chloro-4-nitrobenzofurazan (**9**) shows reactivity similar to that of picryl chloride [43]. It is noteworthy that in both these cases, nucleophilic attack at an unsubstituted ring position, C5 in the case of (**9**), precedes substitution of chloride but is unproductive [8]. There is some evidence that in the reaction of (**9**) with aniline derivatives in water/DMSO solvent nucleophilic attack, which is rate limiting, may involve an SET mechanism. Here, fast electron transfer from the aniline donor to (**9**) is followed by slower coupling of the cation and anion radicals in the solvent cage [44]. Similarly, in the reaction of 2,6-bis(trifluoromethylsulfonyl)-4-nitroanisole with substituted anilines, the results suggest that the intermediate may be formed by direct attack of aniline or by an SET process depending on the aniline basicity [45].

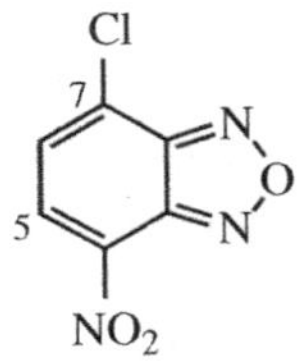

9

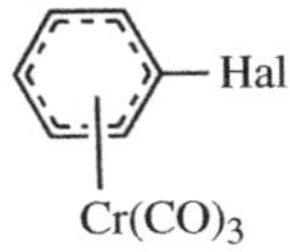

10

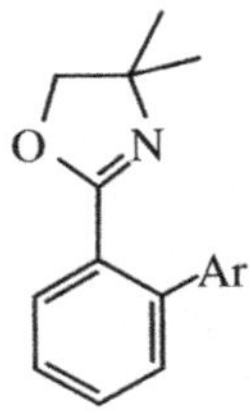

11

Although halogenoarenes without activating ring substituents do not normally undergo substitutions by the S_NAr mechanism, it has been shown that their π-complexes, such as (**10**), with transition metal compounds may do so. Kinetic studies of the reactions of halogenobenzenes with methoxide ions in methanol show that reactivity increases in the series $Cr(CO)_3 < Mo(CO)_3 < (\eta^5\text{-}C_5H_5)Fe^+ < Mn^+(CO)_3$. The halogen mobility order F>Cl indicates that nucleophilic attack is rate limiting here [46].

In the Meyers reaction, an oxazoline group that would not normally be considered to be strongly electron withdrawing activates an *ortho*-fluoro or *ortho*-alkoxy group to substitution by organolithium and organomagnesium reagents [47]. Alkyl or aryl derivatives, such as (**11**), may be formed, and reaction with lithium salts of primary and secondary amines yields aminated derivatives. Chelation-assisted substitutions of *ortho*-fluoro or *ortho*-methoxy groups by organolithium and Grignard reagents have been reported in naphthoic acids [48] and in 2-sulfonyl-1-methoxynaphthalenes [49].

A summary of ring activation would not be complete without mention of steric effects. Since nucleophilic attack occurs in a plane perpendicular to that of the aromatic ring, substitutions are not usually very sensitive to the bulk of *ortho*-substituents. However, steric hindrance by an *ortho*-methyl substituent may be observed as the steric requirements of the nucleophile increase. Thus, the introduction of a 6-methyl group in 1-chloro-2,4-dinitrobenzene causes rate decreases by factors of 14, 22, and 276 in reactions with methoxide, aniline, and piperidine, respectively. The first two figures are attributable to the electronic effect of the methyl substituent, but with piperidine, an additional steric effect is apparent [50].

Many activating groups, such as the nitro group, need to be coplanar with the aromatic ring to exert their maximum electron-withdrawing influence. Secondary steric effects may be observed when such groups are forced from the ring plane by *ortho*-substituents. For example, the presence of a 3-methyl group in 1-chloro-2,4-dinitrobenzene slows reaction with piperidine by nearly three orders of magnitude [50]. When substitution of a nitro group occurs, then steric effects may change the regioselectivity of reaction. Thus, in the reaction of 2,4-dinitrotoluene with ethanethiolate ions, the nitro group displaced is that at the 2-position, which is forced from the ring plane by the *ortho*-methyl group [51].

6.2.2 Leaving Group Effects

Partly due to their availability, most work has featured variations in reactivity in halogenoarenes, where the order of reactivity most commonly found is $F \gg Cl \sim Br > I$ [1, 9, 32]. This is shown by the results in Table 6.2 for reaction of 1-halogeno-4-nitrobenzenes with methoxide ions in methanol. The high reactivity of aryl fluorides is readily explained by the two-step S_NAr mechanism with the first step, nucleophilic attack, being rate limiting. The strong polarization of the $C^{\delta+}\text{–}F^{\delta-}$ bond compared to other carbon–halogen bonds is an important factor, complemented by the stabilizing effect of an *ipso*-fluorine on the reaction intermediate as shown in Table 6.1. Nevertheless, even when nucleophilic attack is rate limiting, the halogen reactivity ratios are strongly dependent on the nature of the nucleophile and of the solvent. This is illustrated in Table 6.2 by comparison of the

TABLE 6.2 Relative Reactivities of Methoxide and Thiophenoxide Ions Toward 1-Halogeno-4-nitrobenzenes in Methanol

Nucleofuge	$k(MeO^-)^a$	$k(PhS^-)^b$	$k(PhS^-)/k(MeO^-)$
F	1300	3.5	1.3
Cl	3	0.3	55
Br	2	0.8	200
I	1	1	480

Adapted with permission from Bartoli and Todesco [52]. © (1977) American Chemical Society.
[a] Reactivity relative to reaction of the iodo-compound with methoxide.
[b] Reactivity relative to reaction of the iodo-compound with thiophenoxide.

relative reactivities of methoxide and thiophenoxide. The polarizability of the leaving group and nucleophile have been found to be important factors [52], iodide being relatively easily replaced by the polarizable thiophenoxide nucleophile.

In fact, a large number of functional groups bound to the ring by oxygen, sulfur, and nitrogen atoms can act as leaving groups [1, 32]. It is not possible to define a unique order of leaving group abilities, but most experimental studies give the following order in reactions when nucleophilic attack is rate limiting: $F > NO_2 > OSO_2Ph > Cl$, $Br > I > SOAlk > SO_2Alk > OAlk \sim OPh > SAlk \sim SPh > SO_3^- > N(Alk)_2 > H$.

The high nucleofugality of the nitro group is noteworthy, due in part to its ability to stabilize the cyclohexadienyl intermediate (Table 6.1). Its high polarizability makes it particularly susceptible to replacement by polarizable nucleophiles, such as thiolate ions. Positively charged substituents, NMe_3^+ and SMe_2^+, are also readily replaced [53], as are protonated heteroarenes such as imidazolium cation [44].

It is worth noting that within a closely related series, good linear free energy relationships have been observed. Thus, in the reactions of ring-substituted 1-phenoxy-2,4-dinitrobenzenes and 2-phenoxy-5-nitropyridines with methoxide and hydroxide ions, linear Hammett plots of log *k* versus the pK_a value of the departing substituted phenoxide ions have been observed [1, 54].

When the second step of the reaction pathway, nucleofuge departure, is rate limiting, then the reactivity order is very different. Then, due to the strength of the carbon–fluorine bond, the halogen mobility order is reversed. For example, in the reaction with thiocyanate ions in methanol, 1-iodo-2,4-dinitrobenzene is circa 1000 times more reactive than the 1-fluoro analogue [32]. Examples where nucleophilic attack is not rate limiting have often been observed in reactions involving amine nucleophiles [36]. Here, nucleofuge departure is more likely to be rate limiting when the reaction involves sterically hindered amines, such as *N*-methylaniline or piperidine, and in nonpolar solvents. For example, it has been shown, using fluorine kinetic isotope effects, that in the reaction of 1-fluoro-2,4-dinitrobenzene with piperidine, the rate-determining step changes from nucleophilic attack to nucleofuge departure as the solvent is changed from acetonitrile to tetrahydrofuran [55].

6.2.3 The Attacking Nucleophile

The results in Table 6.2 suggest that no unique order of reactivity applying in all reactions can be given. However, an approximate order of decreasing reactivity [1, 32] that applies to most situations is $NH_2^- > SO_3^{2-} > RS^- > AlkO^- > PhO^- >$ piperidine $> OH^- >$ aniline $>$ ammonia $>$ halide ions $>$ water. There is some dependence on the basicity of the nucleophile, measured by the pK_a value of its corresponding acid particularly in series of structurally related nucleophiles. Examples are in the reactions of *m*- and *p*-substituted anilines with 1-chloro-2,4-dinitrobenzene [1] and of substituted phenoxide ions with halogenonitrobenzenes [9]. Similarly, the reactivities of *m*- and *p*-substituted thiophenoxide ions with 1-chloro-2,4-dinitrobenzene closely parallel their basicities [56].

The reduced reactivity here of *o*-methyl thiophenoxide may be ascribed to steric factors. In these and other examples, relatively good correlations with the Hammett and Brønsted equations may be observed, and it has been noted [57] that when nucleophilic attack is rate limiting, the Brønsted β_{Nuc} value is usually in the range 0.5–0.7. Studies of the substitution reactions of halogenonitrobenzenes with nucleophiles, including substituted fluorenide carbanions, such as (**12**), and phenothiazine nitranions (**13**) in DMSO solvent [53] indicated that for anions of the same basicity, the nucleophilic reactivity was in the order $S^- >> C^- > O^- > N^-$. The high reactivity of thianions is most likely due to their high polarizability.

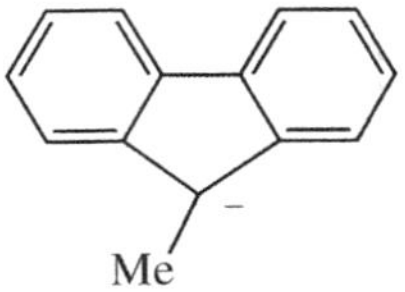

12

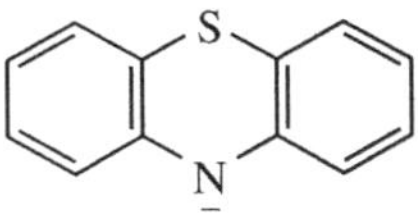

13

It should also be noted that there is the possibility of complications arising from reaction of the nucleophile with the solvent to give the lyate anions. For example, phenoxides may react with alcohols to form alkoxide ions, so that reaction of 4-nitrofluorobenzene with sodium phenoxide in methanol was found to give 99% of 4-nitroanisole and only 1% of the nitrophenyl phenyl ether [58].

Attempts to obtain general measures of nucleophilic reactivity include that of Ritchie [59]. He initially measured rate constants for combination of anionic nucleophiles with organic cations, such as triarylmethyl and tropylium cations, and was able to obtain the relation given in Equation 6.1 for nucleophilicity. Here, k and k_{H_2O} are respectively the rate constants for reaction of the cation with the nucleophile and with water.

$$N_+ = \log\left(\frac{k}{k_{H_2O}}\right) \tag{6.1}$$

Interestingly, the nucleophilicity parameters were shown to apply fairly well for reactions with the four 1-halogeno-2,4-dinitrobenzenes in water and in methanol. An exception was for reaction with the azide ion, which showed lower than expected reactivity in the substitution reaction [20]. Inclusion of an extra parameter increases the applicability of the equation so that Mayr [60] has shown that Equation 6.2, where S is a nucleophile specific slope parameter, allows the reactivities of very many nucleophiles

$$\log k = S(N + E) \tag{6.2}$$

and electrophiles to be correlated. Here, N and E are nucleophilicity and electrophilicity parameters, which have unique values for particular nucleophiles and electrophiles. Mayr's equation has been useful in describing the reactivities of strongly activated electrophiles, such as picryl chloride and 7-chloro-4,6-dinitrobenzofuroxan with π-excessive carbon nucleophiles, such as indoles [61].

Reactions involving amine nucleophiles have played an important part in the study of aromatic substitutions, partly because the observation of base catalysis in these reactions provides evidence for the two-step nature of the mechanism. Early work [1] on the reactions of aliphatic amines with 1-chloro-2,4-dinitrobenzene in ethanol, where nucleophilic attack is rate limiting, showed the importance of steric effects so that, for example, *iso*-propylamine is roughly 10 times less reactive than *n*-propylamine. The general pathway for these reactions is shown in Scheme 6.10. Here, the base B may be excess of the nucleophile or added bases such as DABCO or hydroxide ions. The

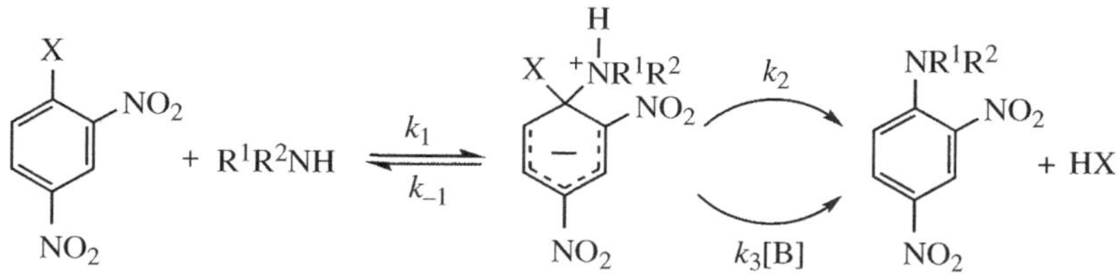

14

SCHEME 6.10

second-order rate constant for the reaction, k_A, is given by Equation (6.3). If the condition of Equation (6.4) applies, then conversion of the zwitterionic intermediate (**14**) to products is faster than its reversion to reactants, so that nucleophilic attack will be rate limiting. Base catalysis may be observed when the conversion to products is slower than reversion to reactants, represented by Equation (6.5):

$$k_A = \frac{\text{rate}}{[\text{ArX}][\text{NHR}^1\text{R}^2]} = \frac{k_1(k_2 + k_3[\text{B}])}{k_{-1} + k_2 + k_3[\text{B}]} \tag{6.3}$$

$$k_{-1} < k_2 + k_3[\text{B}] \tag{6.4}$$

$$k_{-1} > k_2 + k_3[\text{B}] \tag{6.5}$$

The results of many studies indicate that base catalysis is more likely to be observed with secondary than with primary amines and with poorer leaving groups, such as F, OR, and SR, rather than with better leaving groups, such as Cl, Br, and I. In nonpolar solvents such as cyclohexane and toluene, base catalysis is more likely than in polar hydroxylic solvents such as water and alcohols [55].

General, rather than specific, base catalysis is observed indicating that the rate-determining step involves proton transfer. The two likely possibilities, shown in Scheme 6.11, are proton transfer from the zwitterionic intermediate to base, the k_3 step, or from the conjugate acid of the base to the leaving group in the anionic intermediate (**15**). This is the k_4 step in Scheme 6.11. These possibilities are not distinguishable from the kinetics alone since in each case, the transition state contains the elements of the parent, the amine and the catalyzing base. Initially, the base catalysis was assumed to be due to rate-limiting deprotonation of the zwitterionic intermediate followed by rapid leaving group departure [62]. This is the rate-limiting proton transfer (RLPT) mechanism. However, when it became known that proton transfers between normal acids are very fast [63], this mechanism lost favor and was replaced by the specific base–general acid (SB–GA) mechanisms in which the k_4 step involving leaving group departure is involved in the rate-determining step. Strong evidence for the latter mechanism was obtained in the reaction of 1-ethoxy-2,4-dinitronaphthalene with *n*-butylamine in DMSO where the intermediate (**16**) was observed by UV–visible spectroscopy, and the kinetics of its conversion to products, catalyzed by butylammonium ions, could be measured [64]. The intermediate (**17**) has been identified using NMR-flow methods in the reaction of 2,4,6-trinitroanisole with butylamine, and there is good evidence for the SB–GA mechanism in reactions of strongly activated alkyl ethers with amines in DMSO [65]. One interesting observation is that in the reaction of 1-ethoxy-2,4-dinitronaphthalene with piperidine and pyrrolidine [66], the k_4 step is 1100 times slower for piperidine than pyrrolidine. This is attributed to a stereoelectronic effect related to the difficulty in rotating the piperidine function into the plane of the aromatic ring.

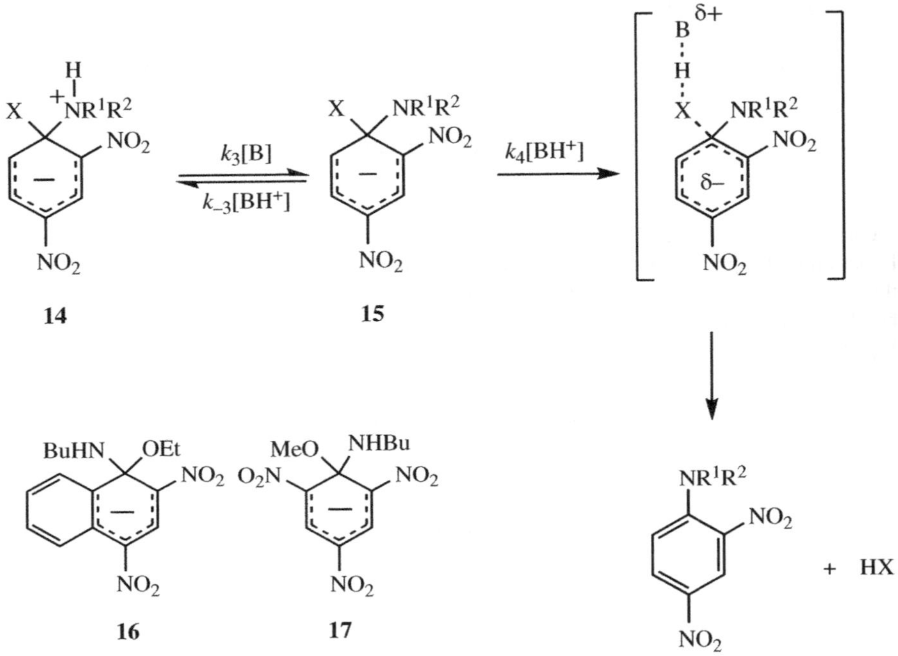

SCHEME 6.11

Nevertheless, in water and other hydroxylic solvents, there is good evidence that the RLPT mechanism is often favored. The availability of relaxation techniques, such as the temperature jump method, has allowed rate measurements on intermediates such as (**18**) shown in Scheme 6.12. These show that amine expulsion may be extremely rapid so that even though proton transfer from (**18**) to base approaches the diffusion-controlled limit, it may become rate limiting. Hence, the condition $k_{-1} > k_3$[B] necessary to observe base catalysis may apply [62]. Other results show that in the context of Scheme 6.11, the condition $k_4 > k_{-3}$ is likely to hold in many cases so that leaving group expulsion is unlikely to be rate limiting. There is also good evidence that the RLPT mechanism holds in reactions involving the displacement of phenoxide ions [67], aryl sulfides [68], and ethylsulfide [69] by aliphatic amines in DMSO and of phenoxide ions by aliphatic amines in DMSO and of phenoxide ions by amines in acetonitrile [70].

There is a greater likelihood of observing base catalysis in reactions involving secondary rather than primary amines. One possibility here is the lower rate of proton transfer from zwitterionic intermediates (**14**) to the catalyzing amine observed with secondary than with primary amines [71, 72]. It is known that as steric hindrance at the reaction center increases on changing from primary to secondary amines, rate constants for the proton transfer step, k_3 in Scheme 6.10, decrease so that the condition $k_{-1} > k_3$[B] leading to the observation of base catalysis is more likely. A further possibility is that in reactions with secondary amines, the amino proton may be hydrogen bonded to an *ortho*-nitro group, as shown in (**8**), making its transfer more difficult.

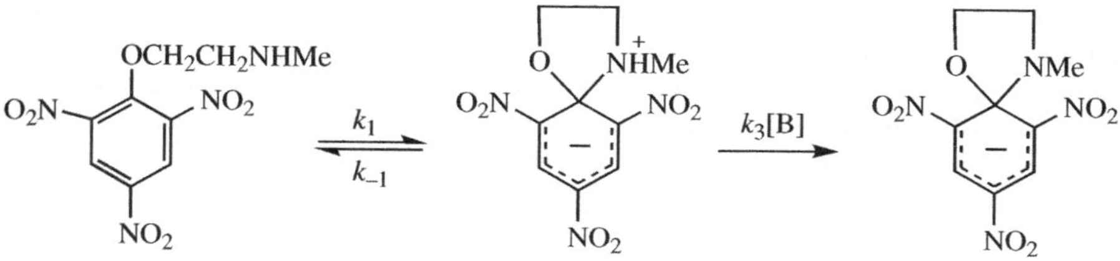

18

SCHEME 6.12

In reactions involving primary amines, there will be a nonhydrogen-bonded amino proton available for transfer.

Intramolecular hydrogen bonding is more likely in nonpolar solvents, such as cyclohexane and toluene, than in polar solvents, such as water and DMSO, where hydrogen bonding of amino protons to solvent is likely to be preferred. In these nonpolar solvents, reactions may show a greater than second-order dependence on the concentration of amine [73–75]. Additives capable of forming hydrogen-bonded aggregates with the amine may also accelerate reactions [76, 77]. Explanations for this behavior include reaction of the substrate with dimers of the amine in the formation of the intermediate [78, 79] and the formation of electron donor–acceptor complexes between the substrate and the amine or additive [80].

Regarding the uncatalyzed conversion of the zwitterionic intermediate to products, the k_2 step in Scheme 6.10, the most likely possibility is intramolecular transfer of a proton from the nitrogen atom to the leaving group [24]. In solvents with good hydrogen bonding ability, this may occur via a solvent molecule [24, 70], as indicated in (**19**) and (**20**).

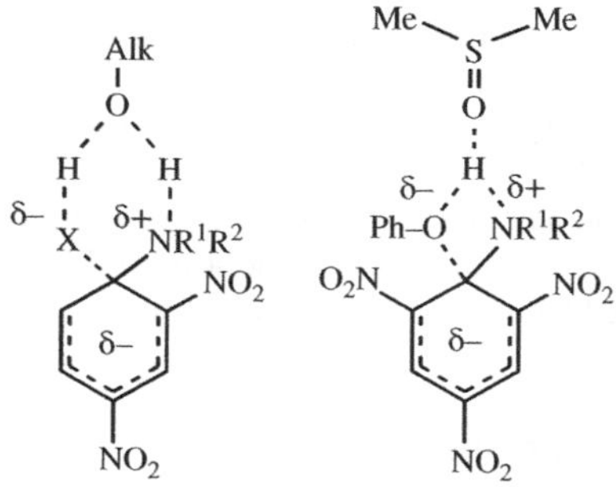

19 **20**

6.2.4 Solvent Effects

Substitutions involving anionic nucleophiles generally proceed much more rapidly in dipolar aprotic solvents, such as DMSO and DMF, than in protic solvents, such as water and methanol. For example, the substitution of fluoride ions by azide ions in 4-nitrofluorobenzene is roughly four orders of magnitude faster in DMSO than in methanol [81]. When nucleophilic attack is rate limiting, the solvation of the reactants and of the transition state for formation of the intermediate are important. Small anions with high charge density, such as hydroxide and methoxide, are well solvated in protic solvents by hydrogen bonding interactions but are poorly solvated in dipolar aprotic solvents. However, transition states, such as (**21**), where the negative charge is dispersed are well solvated by the dipolar aprotics. Hence, the energy barrier for reaction is decreased in the latter solvents. The effect becomes much smaller when larger polarizable nucleophiles, such as phenoxide and thiophenoxide, are involved [82–86].

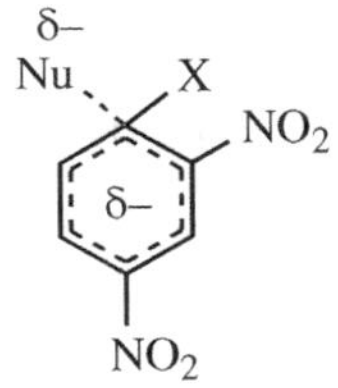

21

Solvation may also affect leaving group ability so that, for example, fluoride ions are much better solvated in water than in dipolar aprotic solvents. This can affect the nature of the rate-determining step in the overall reaction. Hence, in the reaction of 1-fluoro-2,4-dinitrobenzene with secondary amines, nucleophilic attack is rate limiting in water, but in acetonitrile, base catalysis is observed with the second step rate determining [87, 88].

The fact that when the fluoride ion is poorly solvated it is a good nucleophile but poor leaving groups may be put to good effect in the preparation of aryl fluorides. Thus, halide exchange reactions and displacement of nitro groups by fluoride can be successful in DMSO [89]. In such solvents, and even in benzene, the use of crown ethers may be beneficial. These solubilize salts such as potassium fluoride by complexing the potassium ions, leaving the fluoride ions in a reactive state [90]. Crown ethers have also proved effective in the phase-transfer catalyzed reaction of thiolate ions with relatively unactivated arenes such as dichlorobenzene [91].

Cationic micelles, such as those formed from cetyltrimethylammonium bromide, may also accelerate reaction. Studies involving reactions with alkoxide ions indicate that factors involved include the concentration of reagents within the micelle, desolvation of the alkoxide ions, and stabilization of the transition state [92]. There have also been studies of the effects of oil–water emulsions [93] and vesicles [94] on reactivity, and ionic liquids have been shown to accelerate the ethanolysis of 1-fluoro-2,4-dinitrobenzene [95].

Recently, the use of liquid ammonia as a solvent for substitutions has been advocated [96, 97], and the displacement of fluoride from 2-nitro- and 4-nitro-fluorobenzene by alkoxide and phenoxide ions has been achieved without competing solvolysis.

6.2.5 Intramolecular Rearrangements

Smiles and his coworkers [98], in the 1930s, were able to show that in alkaline conditions, *ortho*-hydroxysulfones (**22**) may undergo rearrangement to give sulfinic acids (**24**). Such processes, shown in Scheme 6.13, are known as Smiles rearrangements [1]. Generally, at least one strongly EWG, such as NO_2, is required to stabilize the intermediate (**23**) and allow the rearrangement to take place. Early on, it was shown [99, 100] that such reactions are subject to strong steric effects. Thus, the introduction of a methyl group at the 6-position in (**22**) speeds the reaction by more than five orders of magnitude. This was attributed to the conformation (**25**) required for ring formation being favored. It was also shown that such reactions may be reversed [101]. Hence, in weakly acidic solution, pH 2–6, where the sulfino, but not the hydroxyl, groups are ionized, the sulfone (**22**) may be reformed.

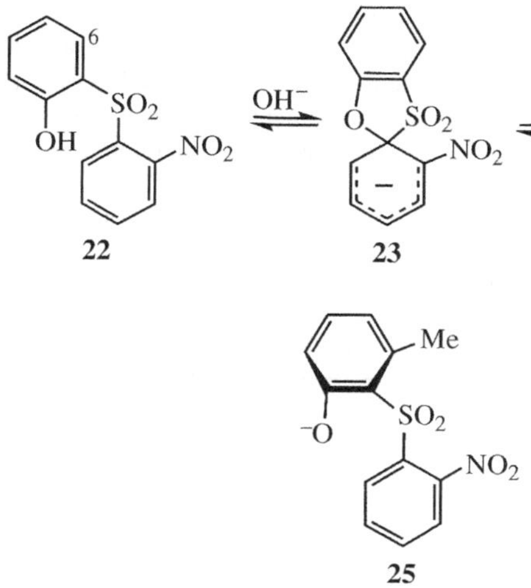

SCHEME 6.13

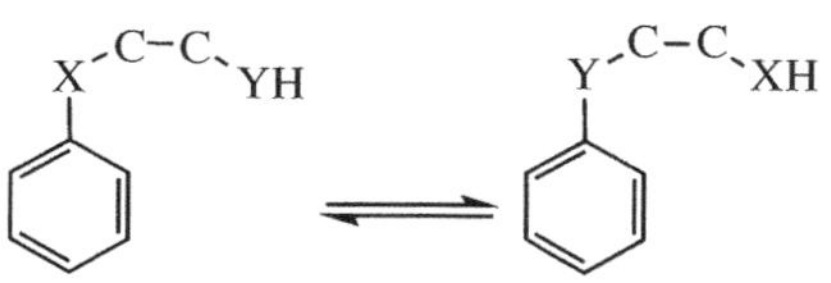

SCHEME 6.14

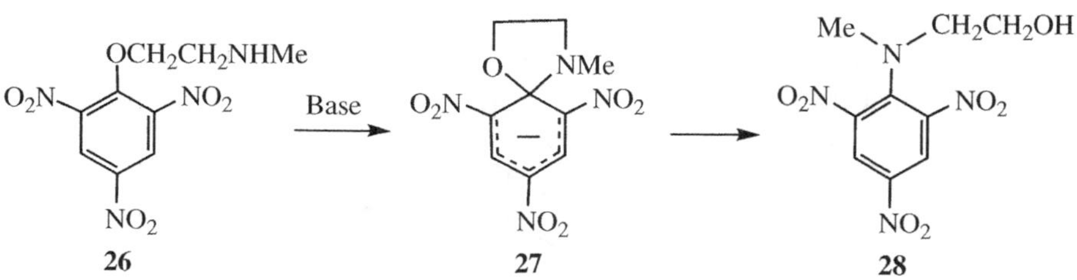

SCHEME 6.15

Such reactions may be generalized as shown in Scheme 6.14 with X typically O, S, SO_2, SO, and CO_2 and Y O, S, and NR. Reactions where X = O and Y = NR are designed as O→N rearrangements, and there are many examples in the literature [9]. For example, *N*-methyl-β-aminoethyl aryl ethers (**26**) very readily rearrange to the corresponding arylamines (**28**) as shown in Scheme 6.15. With strong activation in the ring, spiro Meisenheimer adducts such as (**27**) may be isolated [102]. The ease of this rearrangement may make it difficult to synthesize the aryl ethers such as (**26**). However, Bernasconi [102] has shown that rapid acidification of intermediates such as (**27**) results in rapid cleavage of the C—N bond allowing isolation of (**26**) as its ammonium salt. In a less activated system, it has been shown that the 4-nitro equivalent of (**26**) may be formed by reaction of 4-nitrochlorobenzene with the 2-aminoalcohol acting as an oxygen nucleophile in DMSO, where the rearrangements is slow [103]. Two similar rearrangements have been shown to be involved in the conversion, Scheme 6.16, of benzenesulfonamides into aniline derivatives [104].

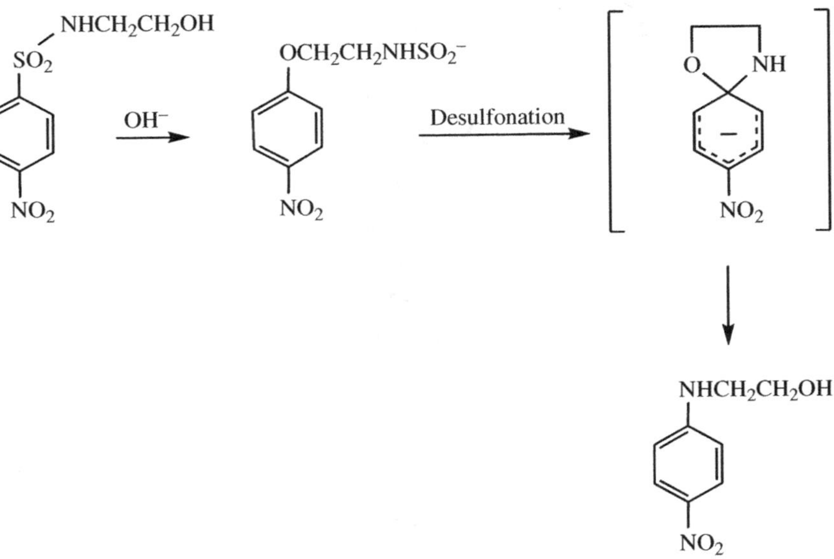

SCHEME 6.16

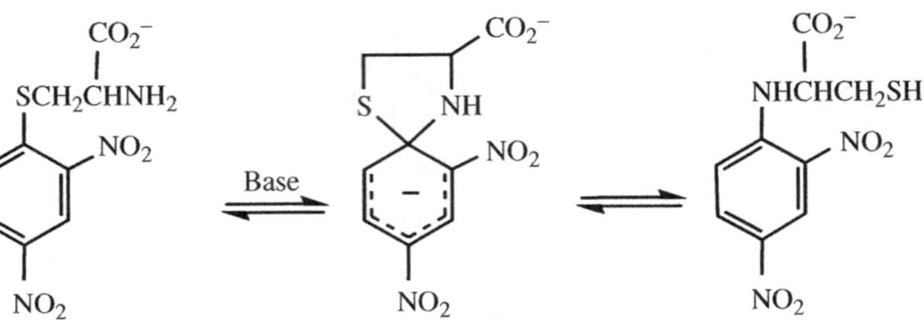

29 **30** **31**

SCHEME 6.17

An S→N rearrangement has been reported in the cysteine derivative (**29**), shown in Scheme 6.17. In methanol, the addition of a base such as Dabco is required, while in DMF or DMSO, the reaction is much faster and does not require added base [105]. Interestingly, in the presence of added base in DMF, the anionic intermediate (**30**) is observable spectroscopically, and due to ionization of the thiol group in (**31**), the reverse N→S rearrangement occurs to give a mixture of (**29**) and (**31**).

An O→Se rearrangement has been used in the conversion of phenols into selenols [106]. Schemes 6.18 and 6.19 show examples of an N→N rearrangement where a weakly basic amide group is replaced by an amine group [107] and of an O→O rearrangement during a Passerini-type reaction involving a phenol derivative, an aldehyde, and cyclohexyl (Cy) isocyanide [108].

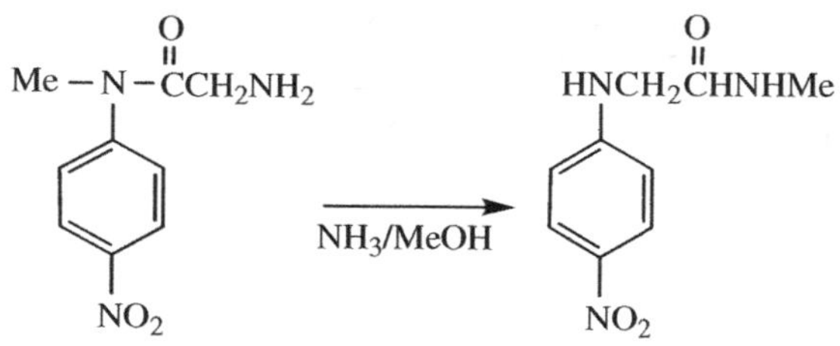

SCHEME 6.18

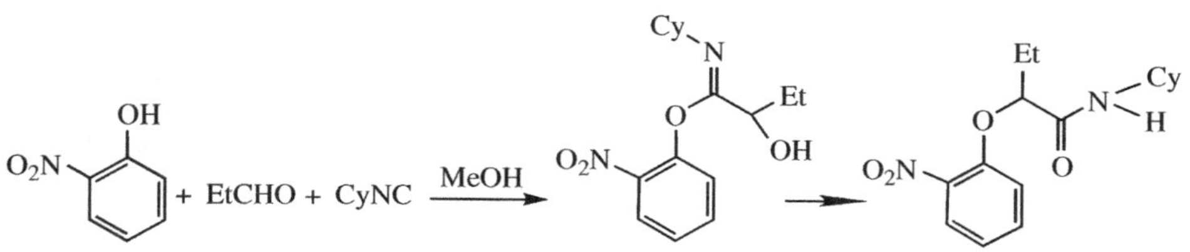

SCHEME 6.19

The Truce–Smiles rearrangement involves reaction of an internal carbon nucleophile [109] and differs from the Smiles reaction in that an activating group, such as a nitro group, may not be necessary. An early example, shown in Scheme 6.20, was the rearrangement of a phenyl sulfone to a sulfinic acid and involves carbanion formation by deprotonation of a methyl group by butyl lithium [110].

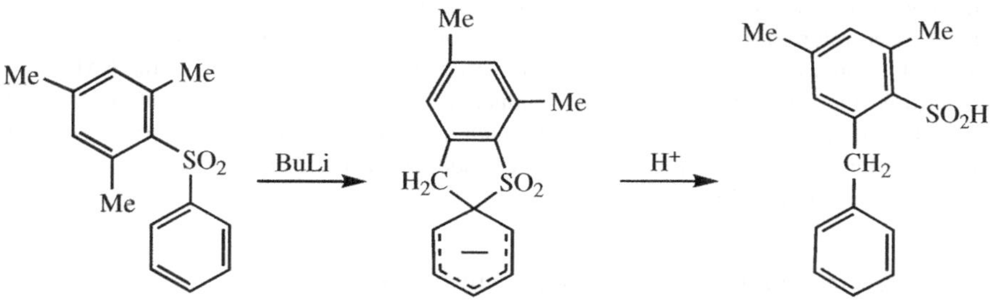

SCHEME 6.20

The reaction of the 2-bromophenylperfluoropyridyl ether (**32**) with butyl lithium may result in the formation of 4-(2-hydroxyphenyl)tetrafluoropyridine (**33**) by the sequence shown in Scheme 6.21 and also involves carbon–carbon bond formation [111].

The product (**33**) may undergo an oxydefluorination reaction to give the benzofuran derivative (**34**). Such intramolecular substitutions provide a useful synthetic methodology and often involve displacement of, or reaction with, a nitro group [112]. Some synthetic aspects are summarized in Section 6.5.

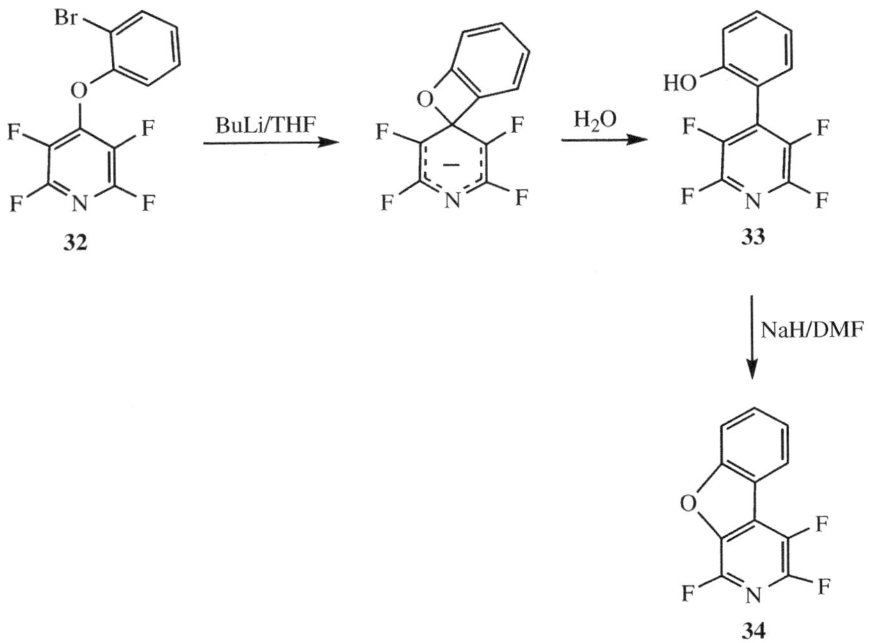

SCHEME 6.21

6.3 MEISENHEIMER ADDUCTS

Nucleophilic attack at a ring position activated with EWGs but which carries a poor leaving group may result in the isolation, or spectroscopic detection, of intermediates sitting in the potential energy well shown in Figure 6.1. Meisenheimer's early characterization [19] of the adduct (**5**) has already been referred to. There followed the use of spectroscopic techniques in the infrared and UV–visible ranges to investigate these species. However, the development of the NMR technique has allowed the structures of very many adducts to be determined. This technique, importantly, allows identification of the ring position at which nucleophilic attack has occurred. These species known as Meisenheimer complexes or anionic σ-adducts have been the subject of a book [113] and several reviews [114–117].

6.3.1 Spectroscopic and Crystallographic Studies

The infrared spectra of adducts such as (**5**) formed between alkyl picrates and alkoxides show changes relative to the parent compounds that are consistent with a fully covalent structure. A series of strong bands between 1225 and 1040 cm^{-1} is typical of ketals [118, 119]. Many sigma adducts are intensely colored, and attempts have been made to correlate UV–visible spectra with structural features [120]. However, these spectra do not generally provide such precise information as do NMR studies. Nevertheless, the absorption intensity usually obeys the Beer–Lambert law, so that absorption is directly proportional to concentration, and this is very useful in the determination of concentrations in kinetic and equilibrium studies.

The ^1H NMR spectra of sigma adducts were first reported in the 1960s [121, 122], and since then, the technique has proved invaluable in structural determinations. The method is particularly strong in that nucleophilic attack at a ring carbon atom results in a change from sp^2 to sp^3 hybridization, so that a pronounced shift in position at the resonance due to the atom or group at that position occurs. Hence, NMR spectroscopy allows the position(s) of attack to be defined. This was first illustrated in the case of trinitrobenzene (TNB) where the parent molecule in DMSO solvent shows a singlet at δ 9.2 ppm. The addition of sodium methoxide results in bands at δ 6.14(t) and δ 8.42(d) ppm with relative intensities 1:2 from the spin-coupled bands of adduct (**35**, R = OMe).

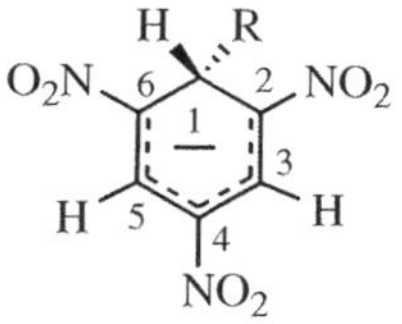

35

At high methoxide concentrations, a diadduct is formed involving the addition of methoxide ions at two ring positions and bands at δ 6.2 and δ 8.6 ppm having relative intensities 2:1 are present. Currently, ^{13}C NMR spectroscopy is widely used together with ^1H data to provide structural information and to give information about charge distribution in adducts. Some representative data, mostly obtained in DMSO-d_6 solvent, for adducts from TNB are given in Table 6.3. There is a loose correlation between the shift of the ring hydrogen at the point of attack and the electronegativity of the attacking atom, with shifts decreasing in the order O > N ~ S > C > H. Spin coupling, J ≃ 1H, is often observed between H-1 and H-3,5, the *meta* hydrogens.

The UV–visible spectra of these adducts typically show two absorption maxima, and data are included in Table 6.3. At higher methoxide concentrations, conversion to adducts with 2Nu:1 TNB stoichiometry results in a broad adsorption with maximum circa 500 nm [84].

A strength of the NMR method is that with ambident nucleophiles changes with time in the adducts present can be readily detected. Thus, with acetophenone anions [131], the oxygen-bonded

TABLE 6.3 Representative Data for Adducts with Structure (35) from Trinitrobenzene

	Chem Shifts[a]			UV–Vis;	
R	H-1	H-3,5	Reference	Max/nm (solvent)	Reference
OMe	6.14	8.42	[9]	425; 495 (MeOH)	[123]
OH	6.19	8.34	[114]	435; 500 (H_2O)	[124]
OEt	6.2	8.4	[9]	424; 497 (EtOH)	[125]
OPh	6.86[b]	8.34[b]	[114]	—	
SEt	5.75	8.30	[112]	460; 550 (MeOH)	[84]
SO_3^-	6.0	8.3	[117]	474; 550 (H_2O/DMSO)	[126]
NH_2	5.5	8.3	[118]	454; 542 (DMSO)	[127]
NEt_2	5.7	8.5	[118]	448; 526 (DMSO)	[127]
Pyrrolide ion	6.89	8.51	[119]	—	
CH_3	4.6[c]	8.2[c]	[120]	470; 572 (CH_3CN)	[128]
$CH_2(CH_2)_2CH_3$	4.8[c]	8.4[c]	[120]	474; 568 (CH_3CN)	[128]
CH_2COCH_3	5.1	8.4	[121]	465; 552 (DMSO)	[129]
Phenoxide (C-adduct)	5.51	8.31	[114]	468; 570 (DMSO)	[130]
Pyrrolide (C-adduct)	5.71	8.31	[119]	—	

[a] In DMSO-d_6 unless stated.
[b] In CD_3N/glyme.
[c] In CD_3CN.

enolate adduct may be observed initially followed by conversion to the thermodynamically more stable carbon-bonded adduct as shown in Scheme 6.22.

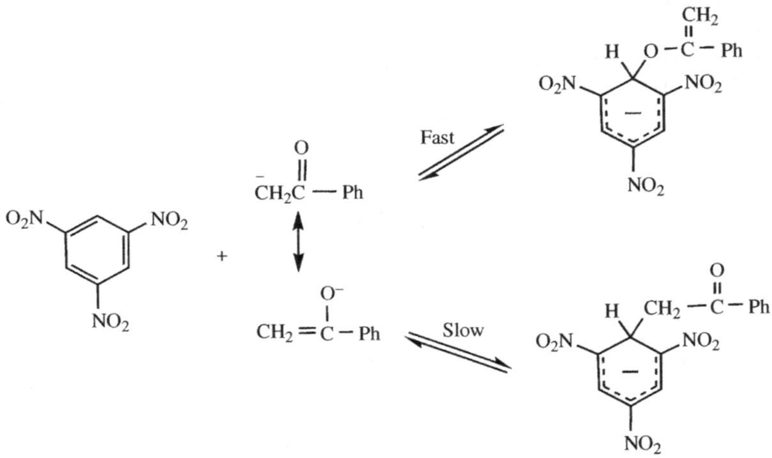

SCHEME 6.22

Similarly, with phenoxide ions, oxygen attack is kinetically favored, while attack by the *para*- or *ortho*-ring carbons gives thermodynamically more stable adducts [130]. It is worth noting that earlier work was hampered by the competition with phenoxide of attack by hydroxide ions formed from small amounts of water in the solvent. Ambident attack by pyrrolide anions may involve reaction at nitrogen or carbon centers [132]. DMSO usually acts as an inert solvent for these measurements, but attack by its anion, the dimsyl ion, may involve successive reaction at oxygen, sulfur, and carbon centers [133].

X-ray crystallographic study of the methoxide adduct (**36**) of 2,4,6-trinitroanisole [134] and its diethoxy equivalent [135] confirmed the essential equivalence of the two alkoxy groups at the 1-position. The results showing double-bond localization indicated that, in the crystal, the negative charge is concentrated at the *para*-nitro group as shown in the resonance form (**36**). The X-ray structure of the methoxide adduct (**35**) of TNB [136] indicates that the ring is somewhat distorted into a boat-like structure. The X-ray structure of the dioxolane spiro-adduct (**37**) has also been reported [137]. Here, the association of the potassium counterion with the *para*-nitro may overemphasize the importance of this resonance form in the crystal as theoretical calculations indicate nearly equal contributions from resonance forms involving all three nitro groups. Structures of the zwitterionic spiro-adducts (**38**) and (**39**) containing a tropylium cation [138, 139] also indicate nearly equal contributions from resonance forms involving *ortho*- and *para*-nitro groups.

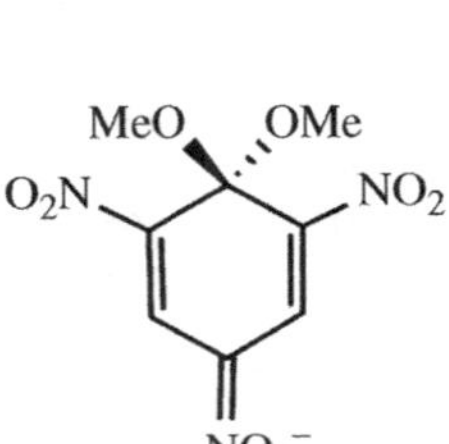

36

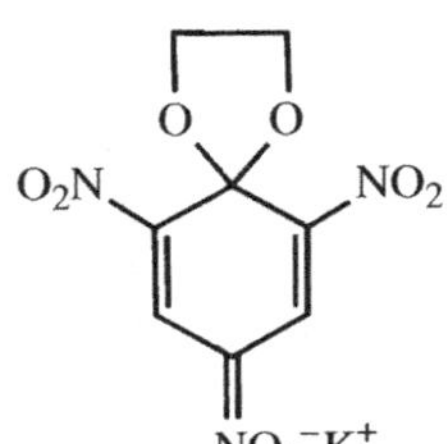

37

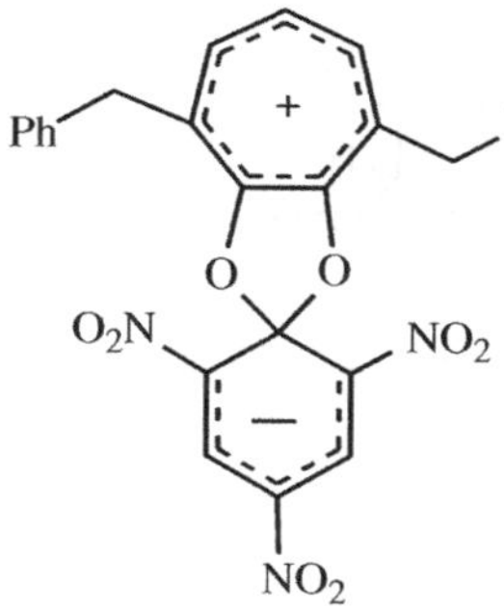

38

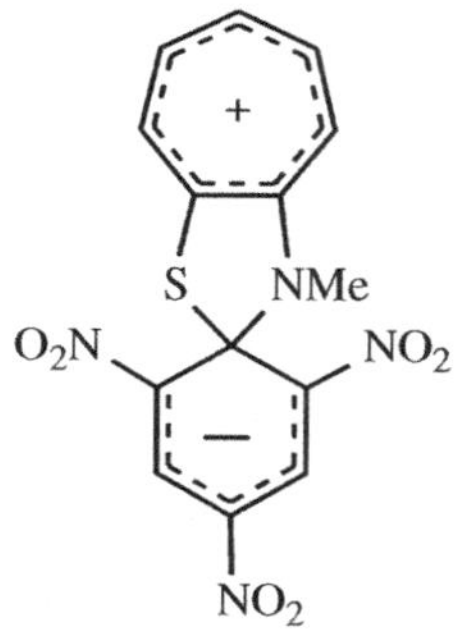

39

The structure of the spiro-adduct (**40**) formed by condensation of picric acid and diisopropylcarbodiimide has been confirmed by X-ray crystallography. Interestingly, it shows fluorescent properties [140]. Also of interest is the spiro-adduct (**41**) in which ring activation is provided by trifluoromethylsulfonyl groups [141]. The evidence, given in the next section, shows that this group has considerably greater stabilizing power than the nitro group.

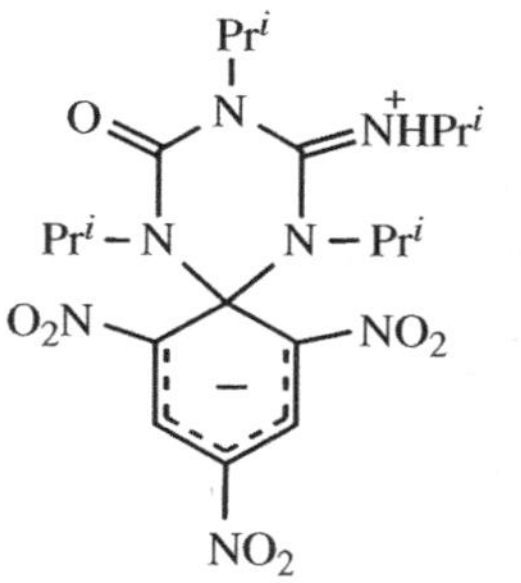

40

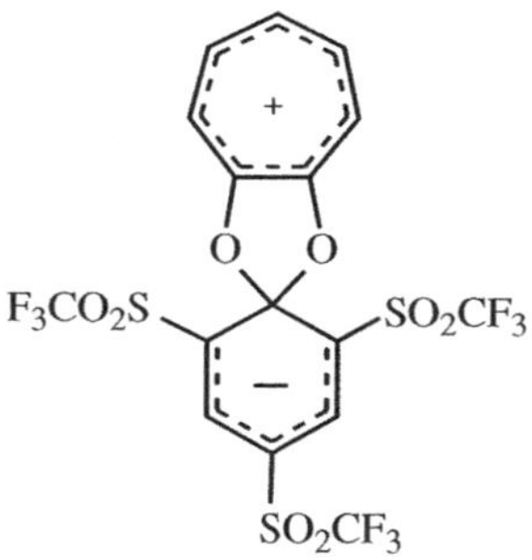

41

6.3.2 Range and Variety of Substrates and Nucleophiles

Meisenheimer's original work [19] involved the reaction of alkoxide ions with 2,4,6-trinitrophenyl ethers where thermodynamically stable adducts, such as (**43**) from 2,4,6-trinitroanisole and methoxide, are formed by attack at the 1-position. However, use of NMR spectroscopy in DMSO/methanol solvent showed [142, 143] that the initial attack is at the 3-position to give the isomeric adduct (**42**). Kinetic studies in methanol using fast reaction techniques [144] confirmed the kinetic preference for formation of (**42**), while the equilibrium constant for formation of (**43**) is circa four orders of magnitude higher than for formation of (**42**). This has been attributed to the relief of steric strain between the methoxy group at the 1-position and the nitro groups at the 2- and 6-positions on formation of (**43**) and also to the stabilizing effect of two *gem* methoxy groups on the sp^3-hybridized carbon atom [9, 113–116]. The faster attack at the 3-position is likely to be due to steric hindrance to approach of the nucleophile to the hindered 1-position, known as F-strain, and to the loss of resonance interaction, shown in (**45**), when attack occurs at the 1-position. The generally accepted mechanism [9, 113–116] for formation of the adduct (**43**) involves reversal of the initial step forming (**42**) and attack by methoxide on the parent anisole. However, a recent kinetic analysis [145] has suggested the possibility that formation of (**43**) involves methoxide attack at the 1-position of (**42**) to give a diadduct (**44**) followed by loss of methoxide from the 3-position, as indicated by path (a) in Scheme 6.23. That diadducts such as (**44**) may be formed in this system is known from NMR studies [142].

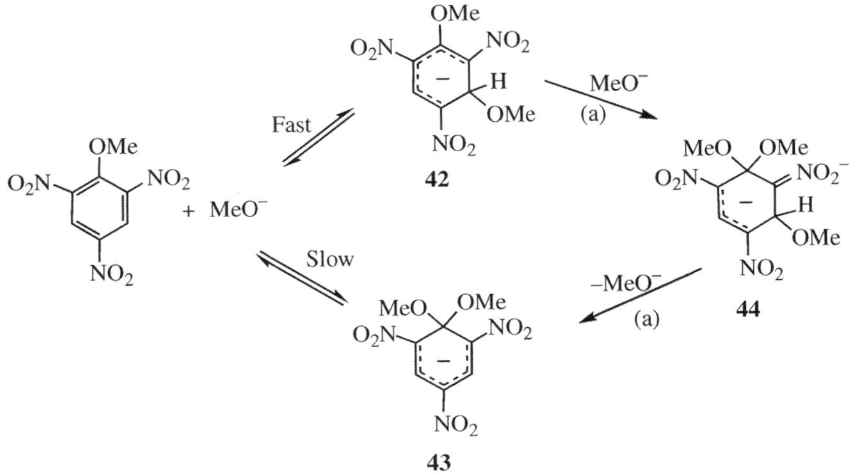

SCHEME 6.23

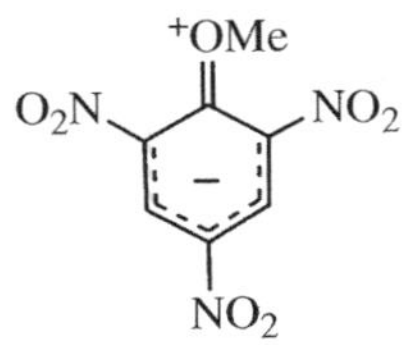

45

Studies have also been reported in less activated ring systems such as alkyl 4-substituted-2,6-dinitrophenyl ethers (**46**) and alkyl 2-substituted-4,6-dinitrophenyl ethers (**47**). These show, generally, that alkoxide attack occurs more rapidly at unsubstituted ring positions but that the thermodynamically preferred adducts are formed by attack at the alkoxy-substituted 1-position [9, 113–116]. Equilibrium constants decrease in the sequence of substituent, Z or Y: $NO_2 > CN > CHO \sim SO_2Me > CO_2Me \sim CF_3 > Cl > F > H$. The results also show that a nitro group *para* to the position of attack is more effective than an *ortho*-nitro group. For example, comparison of compounds where Z and Y = Cl shows that the 1,1-dimethoxy adduct of (**47**) has a stability three orders of magnitude higher than that of the 1,1-dimethoxy adduct of (**46**).

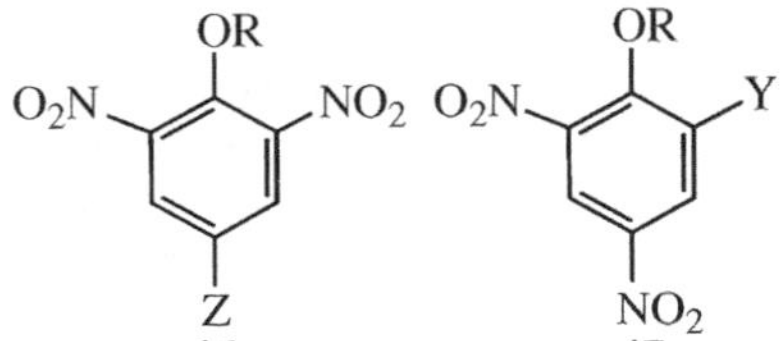

46 **47**

The extremely powerful stabilizing effect of the trifluoromethylsulfonyl group must be noted. Thus, the 1,1-dimethoxy adduct (**48**) is around four orders of magnitude more stable than the trinitro-analogue (**43**) and forms spontaneously in methanol without the need for an added base [146].

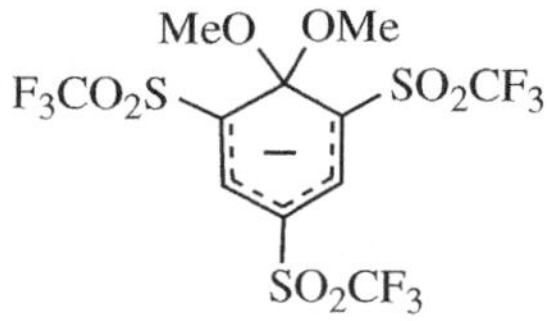

48

Annelation of a benzene ring to give naphthalene derivatives increases the possibilities of charge delocalization. Thus, the adduct (**50**) from dinitronaphthalene is four orders of magnitude more stable than the adduct (**49**) from dinitrobenzene but is still 200 times less stable than the methoxide adduct (**35**, R = OMe) from TNB [147, 148]. In anthracene, a single nitro group is sufficient to allow the detection of the adduct (**51**) in DMSO.

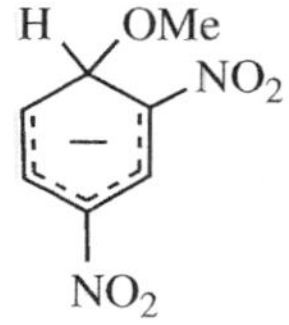

49

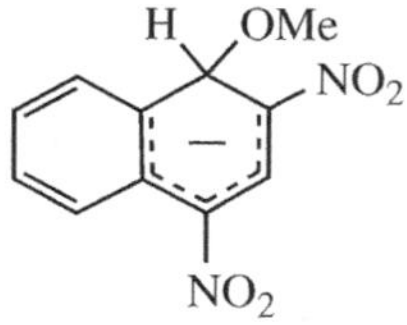

50

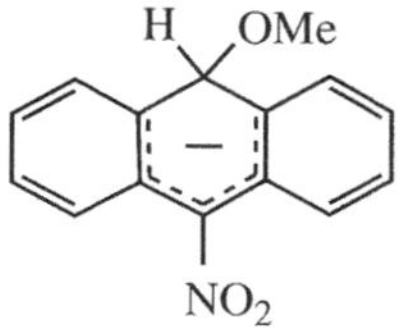

51

It should be noted that, as with nucleophilic substitutions considered in Section 6.2, the stabilities of adducts are greatly affected by the solvent used. Thus, changing from a hydroxylic solvent such as methanol, where methoxide ions are well solvated, to a dipolar aprotic solvent such as DMSO, where methoxide is desolvated but the large polarizable adducts are well solvated, increases adduct stability by many orders of magnitude [84].

Although this book is not concerned with heterocyclic systems, it should be mentioned that many adducts have been described from nitrogen, oxygen, and sulfur heterocycles [9, 113–116]. In pyridine derivatives, ring nitrogen has an activating effect only slightly lower than that of a nitro group and without the steric consequences [149, 150].

The reaction of TNB with primary or secondary aliphatic amines in DMSO [127] gives the anionic adducts (**53**) as shown in Scheme 6.24. Kinetic studies in DMSO [71] or water [151] show that proton transfer from the initially formed zwitterion (**52**) to a second amine molecule may be rate determining.

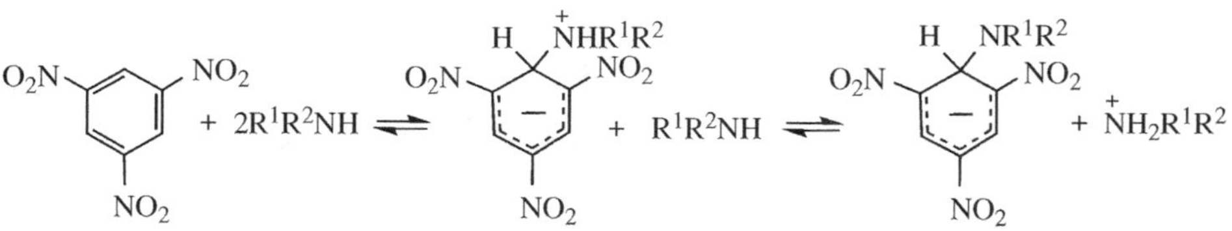

SCHEME 6.24

For steric reasons, the proton transfer is considerably slower when reaction involves a secondary amine, such as piperidine, than with a primary amine. Amide adducts such as (**53**, $R^1 = R^2 = H$) may also be formed in liquid ammonia [96, 152]. Formation of the anilide adduct (**53**, $R^1 = Ph$, $R^2 = H$) in DMSO requires the presence of a strong base, such as DABCO, to effect the proton transfer [153]. TNB does not form zwitterionic adducts from tertiary amines such as triethylamine. However, there is evidence for the formation of species such as (**54**) where the positive charge is delocalized [153]. Mention has already been made in Section 6.2.3 of adducts (**16**) and (**17**) formed during substitution reactions of alkyl ethers with amines. The reaction of dialkylamino-2,4-dinitronaphthalenes with primary amines in DMSO may result [154] in amine exchange via intermediates such as (**55**). Similar reactions have been observed in dialkylaminotrinitrobenzene derivatives [155], although here the initially formed, and observable, species such as (**56**) result from amine attack at an unsubstituted ring position [156].

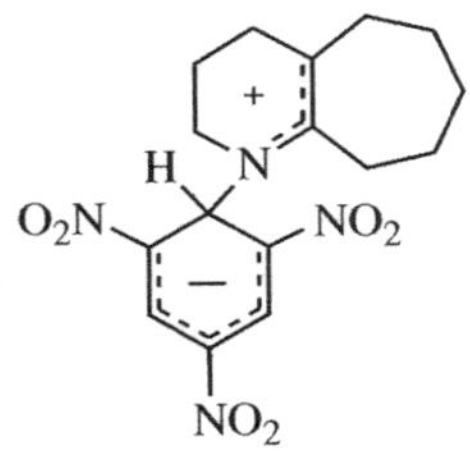

54

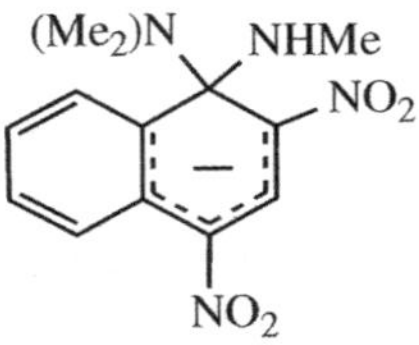

55

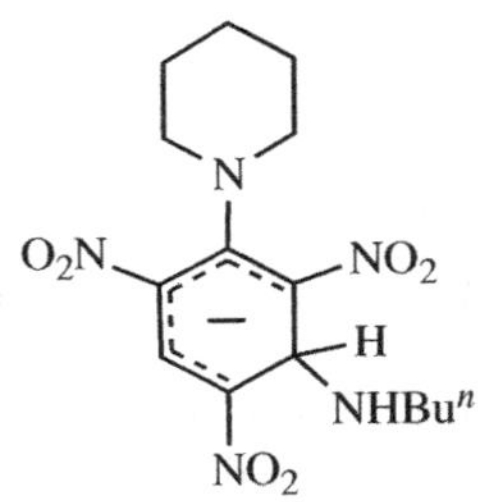

56

There have also been examples in trinitro-activated systems of adducts (**57**) formed from the addition of hydride [157], (**58**) from the addition of fluoride [117], and (**59**) from addition of phosphorus [158].

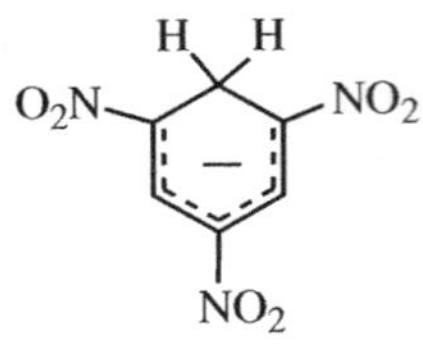

57

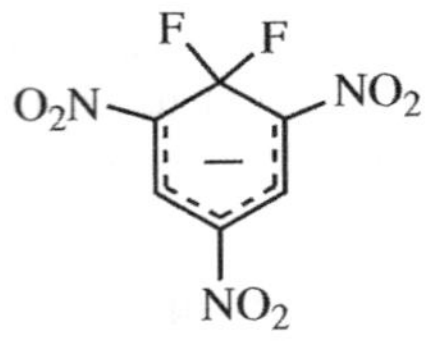

58

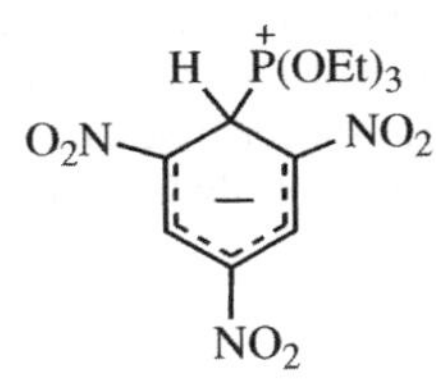

59

The reaction of sodium thioethoxide or sodium thiophenoxide with TNB in protic solvents or in DMSO results in the formation of 1:1 and 2:1 adducts by addition at 1 or 2 ring positions, respectively [84, 159]. With ethylthiopicrate thioethoxide, ions initially yield a mixture of the isomeric adducts (**60**) and (**61**), and this is followed by substitution of the *para*-nitro group to give 1,4-bisethylthio-2,6-dinitrobenzene [160].

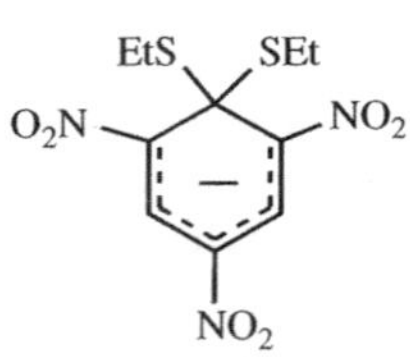

60

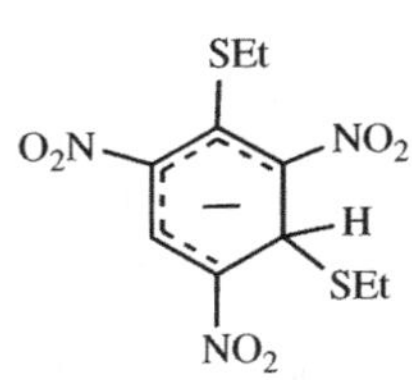

61

There have been several studies reporting equilibrium and rate measurements for reactions of TNB with thiolate ions. Values of the equilibrium constants measure the thermodynamic affinity of the nucleophile for the carbon center, a quantity usually known as the carbon basicity. This is distinct from nucleophilicity, which reflects the rate constant for nucleophilic attack. The results show that in comparison with oxygen bases, the polarizable sulfur bases have higher carbon basicities and nucleophilicities than expected from the pK_a values of the parent thiols [84, 161, 162].

Sulfite ions, too, show high reactivity in adduct-forming reactions. With TNB, 2,4,6-trinitroanisole and *N*-substituted picramide adducts with 1:1 and 2:1 stoichiometries may be formed by reaction at unsubstituted ring positions. Measurements in water and DMSO show that formation of 2:1 adducts (**62**) is favored in water that will solvate the localized negative charges [126].

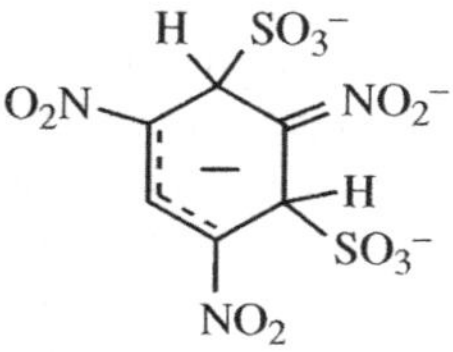

62

The possibility of the formation of *cis*/*trans* isomers in (**62**) has been confirmed both by kinetic [163] and NMR studies [164, 165].

Reaction of carbanions with TNB may result in the formation of new carbon–carbon bonds. Examples of adducts formed are (**63**) from ketones [166], (**64**) from aliphatic nitro compounds [167], and (**65**) from malononitrile [168]. There are various possibilities for further reaction of these adducts. These include the formation of diadducts or of *meta*-bridged adducts, such as (**66**), from suitable ketones [169, 170] or from amidines, or of the ionization of the exocyclic hydrogen in (**65**) to give a dianion.

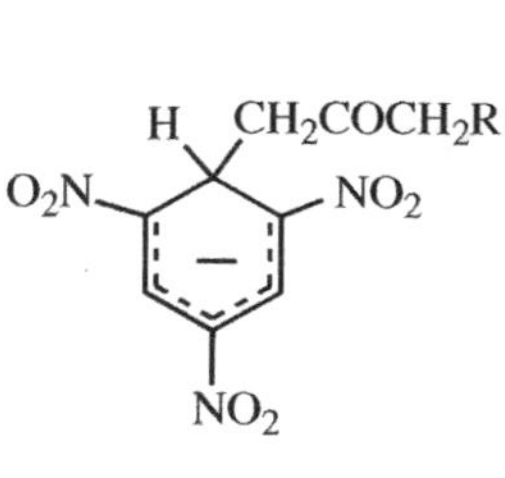

63

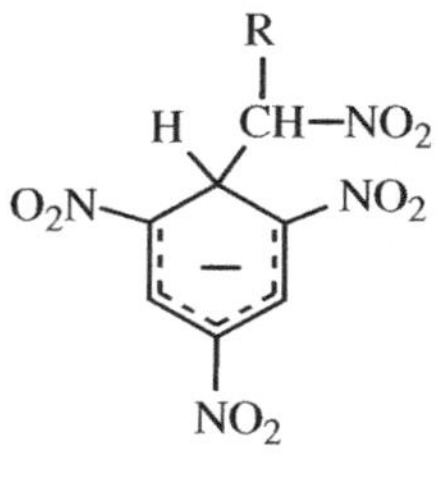

64

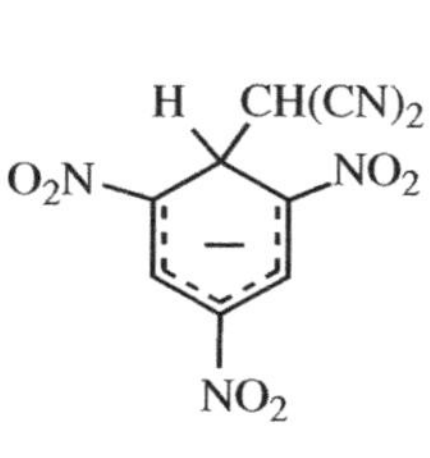

65

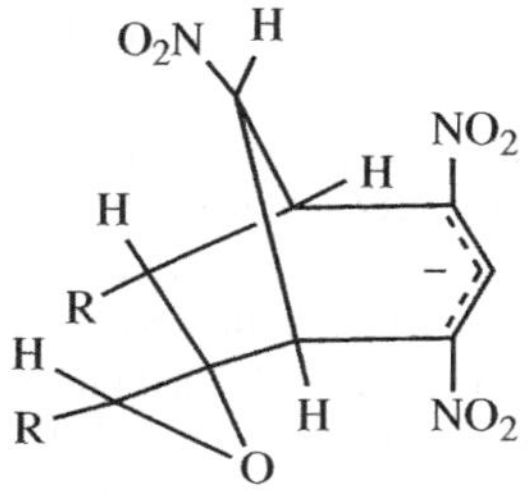

66

Reaction of the malonitrile anion with picryl chloride (1-chloro-2,4,6-trinitrobenzene) shows the usual reactivity pattern in that addition is faster at unsubstituted ring positions, while slower attack at the chloro-substituted position yields the substituted product [168]. It should be noted that kinetic studies show that comparisons of rate and equilibrium constants, for example, comparing carbanions and alkoxide ions, require the application of the concept of "intrinsic reactivity" [171, 172]. This involves the consideration of the electronic-structural and solvational reorganization during reaction. Some carbon nucleophiles may show ambident behavior. Examples have been given in Section 6.3.1 of ketones and phenols where reaction may involve oxygen or carbon centers and pyrrolide ions where reaction at nitrogen or carbon is possible.

In less activated systems containing a single nitro group, carbanion addition may occur at ring carbons carrying hydrogen to yield adducts. Although present in low concentrations, these adducts may lead to substituted products containing new carbon–carbon bonds through oxidative or vicarious pathways. These are dealt with in Chapter 11.

As noted in Section 6.2.5, spiro-adducts such as (**27**) may be isolated during intramolecular Smiles rearrangements. Cyclization of 2-hydroxyethoxy-trinitrobenzene gives the dioxolane adduct (**37**). Kinetic and equilibrium studies show that due to stereoelectronic effects, both the cyclization and ring-opening processes are much more rapid than the corresponding processes in dialkoxy adducts, and overall the spiro-adducts have much higher stabilities [173–175]. Increasing the size of the dioxolane ring from five to six or seven members results in large decreases in complex stability [174, 176]. The dithiolane (**67**) and oxathiolane (**68**) adducts have also been reported [177]. Interestingly, while ring-opening of the dioxalane ring in (**37**) is subject to general acid catalysis [177], opening of the dithiolane ring in (**67**) is catalyzed by mercury(II) ions but not by proton acids [178]. The fluorescent properties of the spiro-adduct (**40**) were referred to earlier. Here, deprotonation in the presence of base results in the loss of fluorescence. The trinitrophenyl spiro-complex of adenosine has also been shown to fluoresce when encapsulated in γ-cyclodextrin [179].

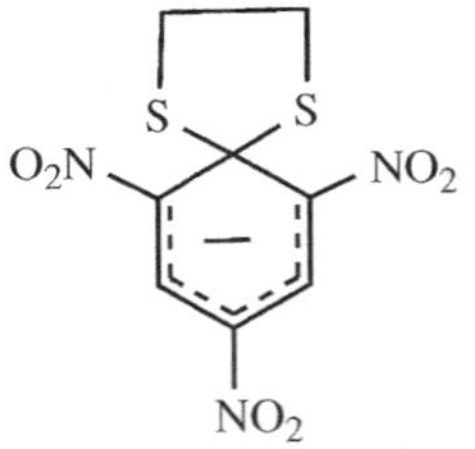

67

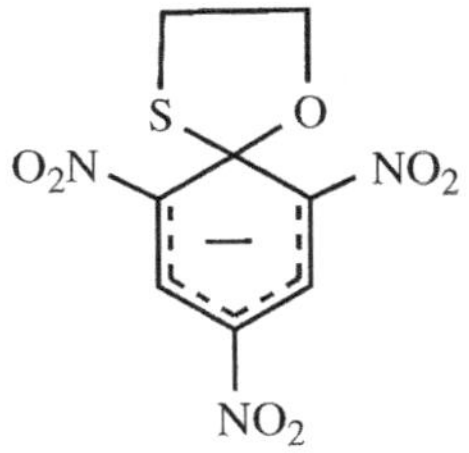

68

6.3.3 Superelectrophilic Systems

The very highly electrophilic character of compounds such as 4,6-dinitrobenzofuroxan (DNBF) and 4,6-dinitrobenzofurazan (DNBZ) has afforded them the status of "superelectrophiles." Characteristics include the ability to react with water or methanol in the absence of added base to form adducts, such as (**69**) from DNBZ and water [9, 180]. They will also react with weak carbon nucleophiles such as pyrroles and indoles to give adducts such as (**70**), formed from DNBF and indole. The zwitterionic (**71**) is formed from DNBF and tris(dialkylamino)benzene [181]. With aniline, the nitrogen-bonded adduct (**72**) is formed under kinetic control, while the carbon-bonded adduct (**73**) has higher thermodynamic stability [182, 183]. The superelectrophiles may be ranked in terms of their affinities for water by measuring the pK_a values for formation of adducts such as (**69**) with the liberation of a proton. Here, the value for DNBF is about 10 pK_a units lower than that for the reaction of TNB with water. The only benzene derivative that can approach this order of reactivity is 1,3,5-tristrifluoromethylsulfonyl benzene [9].

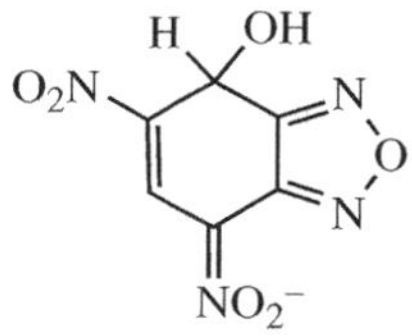

69

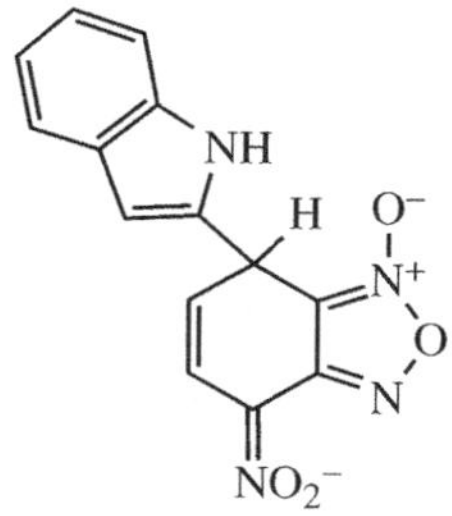

70

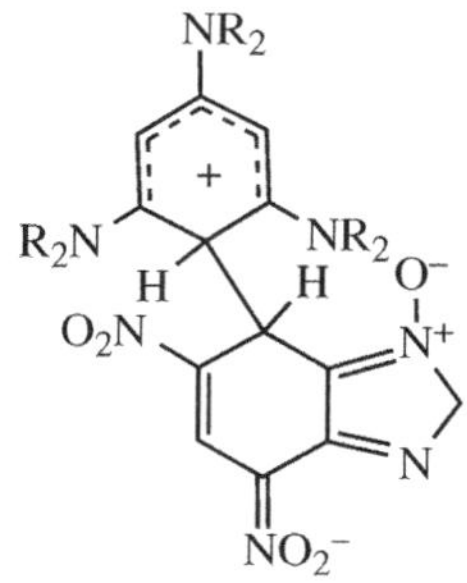

71

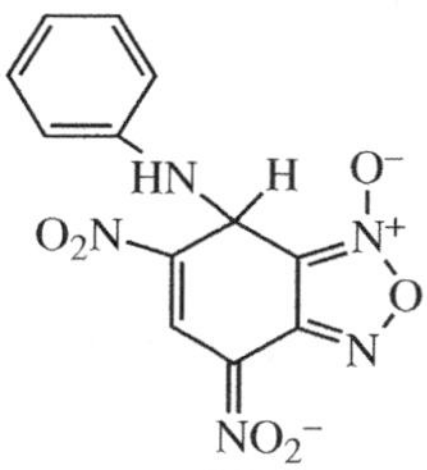

72

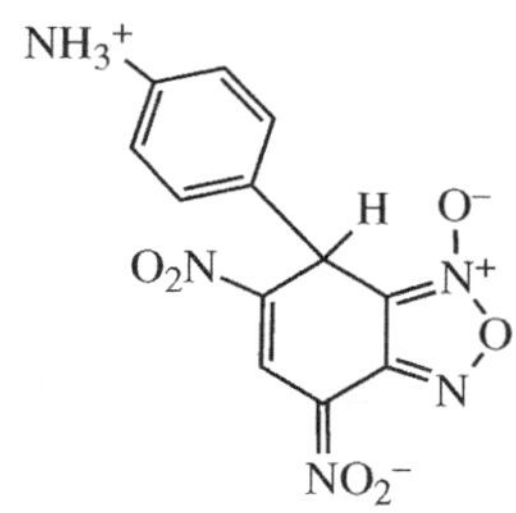

73

The electrophilicities of the superelectrophiles may also be compared using rate constants for their reactions with indoles, pyrroles, and enamines allowing them to be positioned on Mayr's electrophilicity scale [60, 61]. Also of interest is the ability of superelectrophiles to take part in Diels–Alder-type cycloaddition reactions to produce adducts with dienes, such as butadiene, and with ethyl vinyl ether [184].

6.4 THE S_N1 MECHANISM

6.4.1 Heterolytic and Homolytic Pathways

Due to the limited possibilities for the stabilization of aryl cations, the S_N1 mechanism is far less common in aromatic than in aliphatic systems [32]. However, it is well established for arenediazonium ions where the leaving group is a nitrogen molecule (see Scheme 6.1). As required for a rate-determining dissociation, the reactions are first order in the diazonium ion, and the rate of decomposition is independent of the nature or concentration of added metal halide salts. However, since the aryl cations produced are extremely reactive, they form products with any available nucleophile. Hence, in aqueous sodium chloride solutions, benzenediazonium ions will produce phenol and chlorobenzene with the proportion of the latter increasing with increasing salt concentration [185]. The reaction rate is retarded by electron-withdrawing *meta-* and *para-*substituents, such as Cl and NO_2, which destabilize the aryl cations. Electron-releasing *meta-*substituents, such as Me, accelerate reaction, but perhaps surprisingly, electron-releasing *para-*substituents slow the reaction rate. This has been attributed to contributions of resonance forms such as (**74**), which strengthen the carbon–nitrogen bond that has to be broken [58].

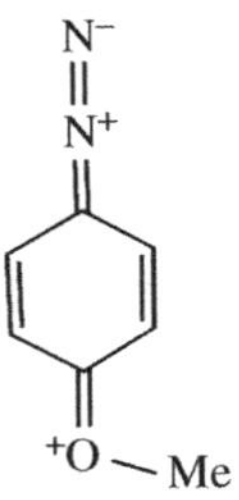

74

There is evidence, particularly in poorly solvating solvents such as trifluoroethanol, that the initial dissociation of the arenediazonium ions may be at least partially reversible. Thus, when the starting material is selectively labeled with ^{15}N, the rearrangement shown in Scheme 6.25 takes place. Also under an atmosphere of 300 atm of unlabeled nitrogen, there is some incorporation of the external unlabeled nitrogen into the unreacted reagent [186]. Recently, theoretical studies have considered the possibility of dediazoniation by an S_N2 pathway [187].

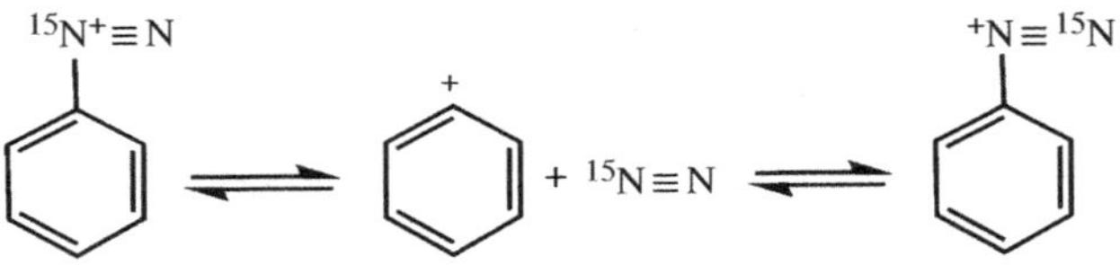

SCHEME 6.25

Dediazoniation may also involve the formation of reduction products, and this is indicative of the presence of pathways involving free radicals. For example, in a recent study [188] of the decomposition of 4-methylbenzenediazonium ions in ethanol–water mixtures, it was found that in acidic solutions with pH <2, heterolysis gave the aryl cation leading to the formation of 4-phenetole and 4-cresol. However, in less acidic solutions, the formation of toluene indicated a homolytic pathway. It is well known that in solutions of methoxide ions in methanol and ethoxide ions in ethanol diazonium ions may form (*Z*)- and (*E*)-arylazo alkyl ethers. The former may be involved in the formation of aryl radicals, which can abstract a hydrogen atom from the solvent to produce the reduction products [189, 190]. It has been found that in general electron-releasing ring substituents tend to favor the ionic decomposition pathway, while EWGs favor the radical pathway [191].

The formation of aryl chlorides from arenediazonium ions on reaction with copper(I) chloride is known as the Sandmeyer reaction and also involves the intermediacy of aryl radicals [192]. Copper salts are similarly involved in the reactions of diazonium ions with cyanide, nitrite, and sulfur dioxide. The uses of arenediazonium salts in synthetic reactions forming carbon–carbon, carbon–sulfur, and carbon–boron bonds have been summarized recently [193].

It is worth noting that the N_2^+ group may also act as a strongly electron-withdrawing substituent in S_NAr reactions, leading, for example, to substitution of a nitro group by a hydroxyl group during the diazotization of dinitroaniline [1, 192].

6.5 SYNTHETIC APPLICATIONS

Due to their availability, haloarenes are frequently used for synthetic reactions, and when nucleophilic attack is rate limiting, as is usually the case, the reactivity order is F≫Cl, Br>I. Although reaction can occur in unactivated systems, this requires the use of high temperatures and/or pressures and may be aided by metal catalysts. Thus, reaction of chlorobenzene with aqueous sodium hydroxide or ammonia can yield phenol and aniline, respectively. In systems containing one or more EWGs, reaction with alkoxides or phenoxides can be used to form alkyl aryl ethers and diphenyl ethers, respectively [1, 32]. Reaction with ammonia or amines may yield aniline derivatives, and it has been found that substitution of 4-nitrohalobenzenes by amines in THF is favored by the use of high pressures [194]. The use of microwave or ultrasonic radiation may also be beneficial in the reaction of 4-activated halogenobenzenes with imidazole [195]. Here, the benefit of using the fluoro derivative is shown by the formation of the substituted product from 4-fluorobenzaldehyde, as shown in Scheme 6.26, while the chloro derivative shows very limited reactivity.

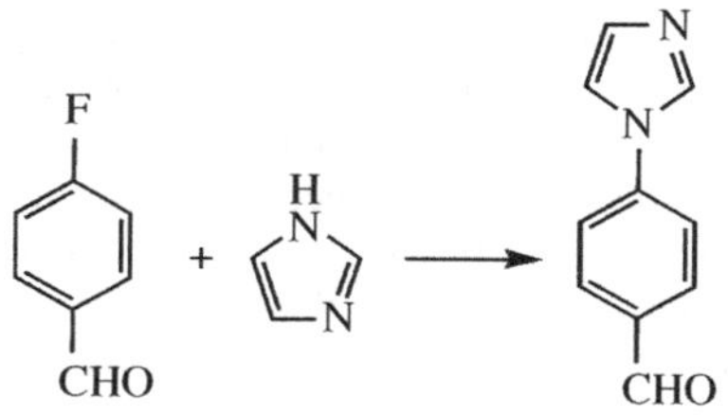

SCHEME 6.26

The aryl ether linkage is important in many natural products and pharmaceuticals, including the antibiotic vancomycin [196]. Diaryl ethers have been efficiently prepared by the coupling of fluorobenzonitriles with a series of phenols in DMSO using potassium carbonate as a base catalyst and with microwave irradiation [197]. A DFT study of the reaction of alkali metal phenoxides with fluorobenzenes indicated that the role of the metal cation is to aid the binding of the aryl halide and to facilitate fluoride displacement [198].

Ortho-selectivity is generally observed in the reactions of 2,4-dichloro- and 2,4-difluoronitrobenzene with alkoxide and thiophenoxide ions [199]. Also in less activated systems, nucleophiles generated from phenols and thiophenols with potassium fluoride-alumina and 18-crown-6-polyether will react in DMSO with cyano- or nitro-substituted fluoro- or chloro-benzenes [200]. Interestingly, the reaction of difluorobenzenes with two different alcohols can occur sequentially. Introduction of the first ether function deactivates the ring, and use of more forcing conditions allows substitution of the second fluorine [201]. Consecutive displacements of fluorine and nitro groups have been observed in the reaction of *ortho*-fluoronitrobenzene with chiral acyl bicyclic lactones in a highly enantioselective synthesis of spirooxindoles [202].

Alkyl- and aryl-thiolates are generally more reactive than their oxygen equivalents and will readily replace halogens under mild conditions to give thioethers. Hence, the use of an alcohol solvent rather than DMSO may be possible [203].

Aniline derivatives are widely used in areas such as pharmaceutical chemistry and the dyestuff industry and may be prepared by halogen displacement from arenes activated by a single EWG. Fluorine displacement occurs more readily than loss of chlorine, bromine, or iodine and may be achieved by heating with an amine in DMSO or by use of microwave or ultrasound methods. The reaction generates acid that can be scavenged by excess of the reacting amine, if it is inexpensive, or by a nonnucleophilic base such as carbonate or tertiary amines. Simply heating nitro- or cyanobenzenes carrying a halogen, alkoxy, or nitro leaving group at the 4-position in hexamethylphosphoramide (HMPA) [204, 205] may be sufficient to give 4-dimethylamino derivatives such as (**75**). Here, the dimethylamino group is derived from the solvent.

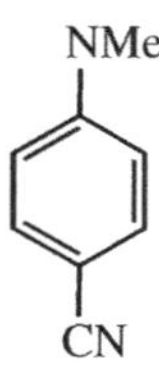

75

It should also be noted that copper-catalyzed Ullmann-type coupling of aryl halides with amines yields substituted products [206], and reaction with diphenylamine has been used to form triarylamines [207]. Triarylamines may also be formed in a variation of the Meyers reaction [47] by displacement by lithium amides of fluoro- or methoxy-substituents activated by an *ortho*-ester function [208]. The oxidation of σ-adducts is discussed in Chapter 11, but it should also be mentioned that aminated products may also be produced by the oxidation of adducts formed by the addition of amide or alkylamide ions at ring carbon atoms carrying hydrogen [209].

Halogen exchange reactions are often carried out conveniently in dipolar aprotic solvents and may be used, for example, in the reaction of 2,4-dinitrochlorobenzene with potassium fluoride to give 2,4-dinitrofluorobenzene [90]. Polyhalosubstitution activates the ring, and it has been shown that the activating effects of fluorine with respect to the point of attack decrease in the order *meta*>*ortho*>*para* [210]. Selective substitution is often possible; examples are the reaction of 2,3,4,6-tetrafluoronitrobenzene with ammonia that gives the 2-amino derivative [211] and the reaction of 2,4-difluoroiodobenzene with imidazole that gives the product of substitution of the 2-fluorine atom, or with copper catalysis the iodo-substituted product [212]. The reaction of octafluoronaphthalene with secondary amines in DMF occurs selectively at β-ring positions—the reaction with piperidine yielding the 2,3,6,7-substituted derivative [213]. The reaction with lithium amides may result in the substitution of up to seven ring fluorines, the reaction with dimethylamide giving (**76**) that acts as a strongly basic "proton sponge" [214]. All of the six fluorines in hexafluorobenzene may be replaced by reaction with pyrrolylsodium in DMF to give a propeller-shaped molecule [215]. The alkylation and arylation of carbon–fluorine bonds by Grignard reagents [216] occurs selectively at *ortho* positions to give disubstituted products such as (**77**).

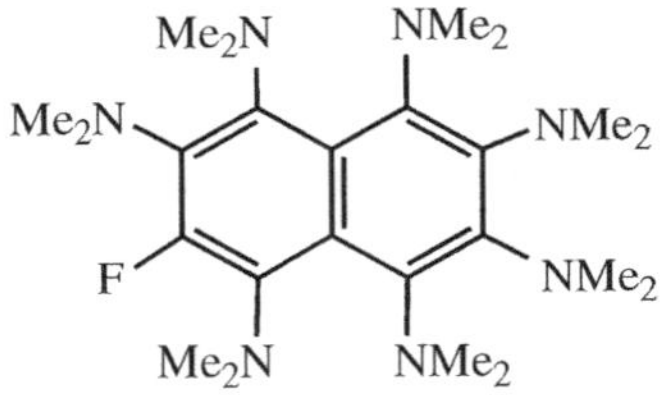

76

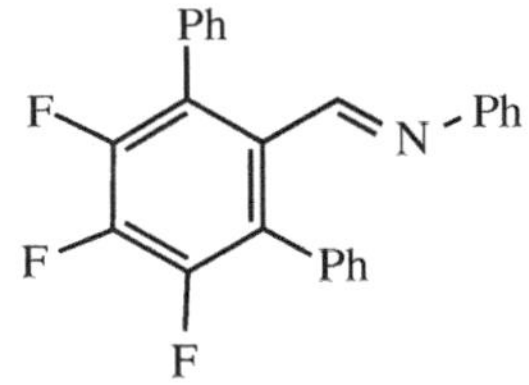

77

Substitution of halogens by carbanions may also be used synthetically although the competition with attack at unsubstituted positions is well documented (Chapter 11). The reaction of 2,4,-dinitrochlorobenzene with dimethyl malonate ions [217, 218] may lead to the formation of (**78**). Carbanions generated by the use of strong organic bases in dipolar aprotic solvents have been used to produce products such as (**79**) from 2-cyano- and 2-nitro-fluorobenzenes [219]. Chelation-assisted Meyers-type reactions are also useful in the formation of alkyl or aryl derivatives by substitution of fluoro or alkoxy groups by organolithium or Grignard reagents [47–49].

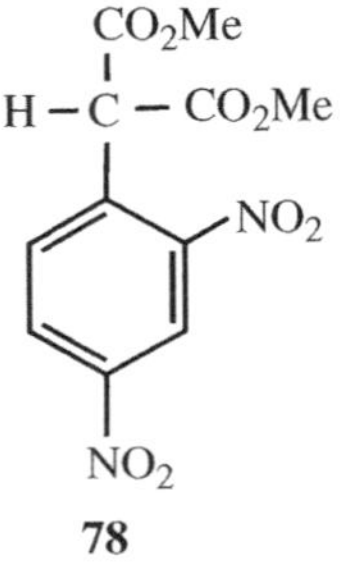

78

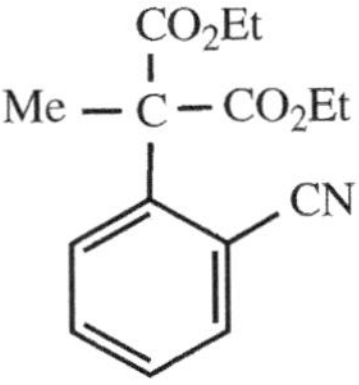

79

The replacement in activated systems of alkoxy and aryloxy groups by nucleophiles such as amines has been used synthetically. Many other groups can also be displaced, with positively charged substituents such as NMe_3^+ and SMe_2^+ rivalling the halogens for ease of displacement [6, 58]. Triflate (OSO_2CF_3) is also an excellent leaving group and may be readily substituted in mono-activated rings [220]. However, the nitro group as well as being strongly activating may also feature as a very good nucleofuge (leaving group).

Kornblum showed that in HMPA as solvent *o*-, *m*-, and *p*-dinitrobenzenes may suffer nitro group displacements by sulfur, oxygen, and carbon nucleophiles—examples being the formation of (**80**) with thiophenoxide (**81**) with methoxide and (**82**) with nitropropane anions [221]. Activation by a single electron-withdrawing substituent, such as PhCO, CN, CF_3, and CO_2Me, in *o*- or *p*-positions may also be sufficient to enable nitro group substitution by oxygen and sulfur nucleophiles in HMPA [221] or in DMSO [222, 223]. Fluoroalkoxide derivatives, such as (**83**), are readily prepared by this method [224]. Recently, the presence of the pentaflurosulfonyl group in (**84**), and in its *meta*-isomer, has been shown to activate the nitro group displacements by alkoxides and thiolates in DMF [225].

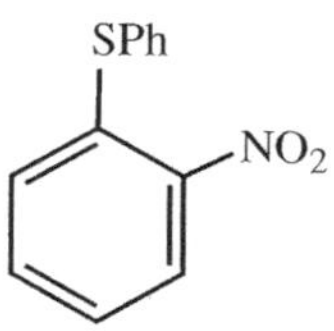

80

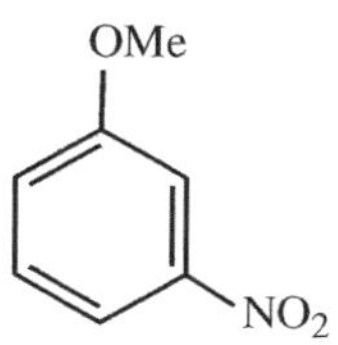

81

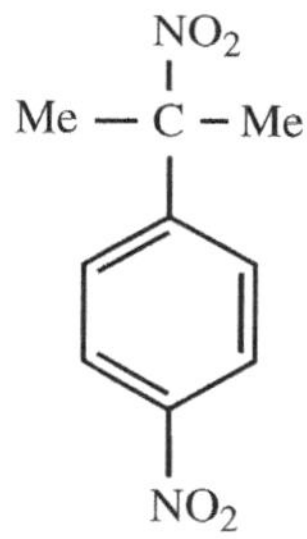

82

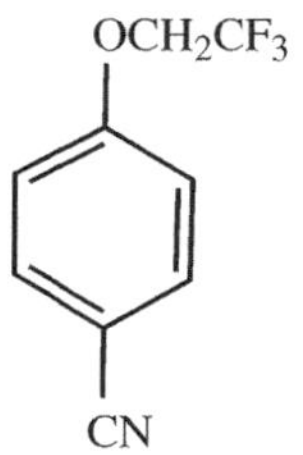

83

84

Mechanisms involving radicals have been proposed for the nitro group substitution of *p*-dinitrobenzene by 2,6-disubstituted phenoxide to give diphenyl ethers [226] and for the reactions of dinitrobenzenes with polyfluoro-alkoxides and -thioalkoxides [227].

It is noteworthy that an *o*- or *p*-alkylthio function is capable of activating the displacement of a nitro group. Thus, the reactions of dinitrobenzenes, 2,4-dinitrochlorobenzene, and picryl chloride with the sodium salt of isopropanethiol in HMPA may result in the complete substitution of chloro and nitro groups as illustrated in Scheme 6.27 [228]. The multiple substitution of nitro groups by methanethiolate anions in DMF has also been observed in dinitrobenzenes carrying CF_3 or $CONH_2$ substituents [229]. The reaction of 2,3-dinitrotoluene with ethanethiolate ions in HMPA results in selective displacement of the 2-nitro substituent. Generally, nitro groups adjacent to alkyl groups are most readily displaced. The explanation here is that steric strain results in twisting of the *ortho*-nitro group from the ring plane, allowing easier substitution [51]. *Ortho*-selectivity is also shown in

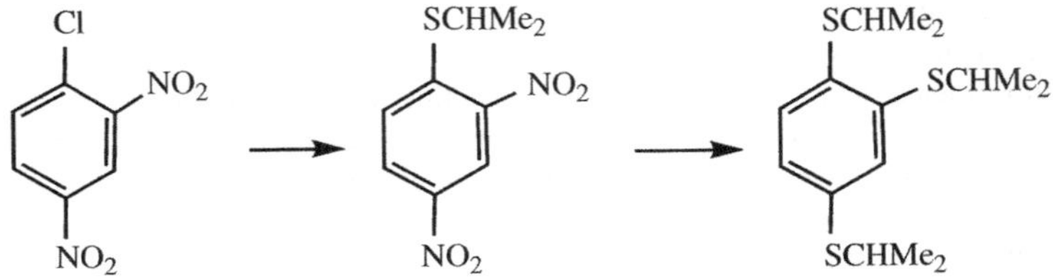

SCHEME 6.27

the reactions of 2,4,6-trinitrobenzamide, (**85**), with oxygen and sulfur nucleophiles in acetonitrile or DMF [230]. However, increasing the bulk of the 1-substituent too much as in the *N,N*-diethyl derivative of (**85**) results in substitution of the *para*-nitro group. Reaction of ethanethiolate ions with the thioether (**86**) in DMSO initially yields Meisenheimer adducts but is followed by substitution of the nitro group *para* to the ethylthio group [231].

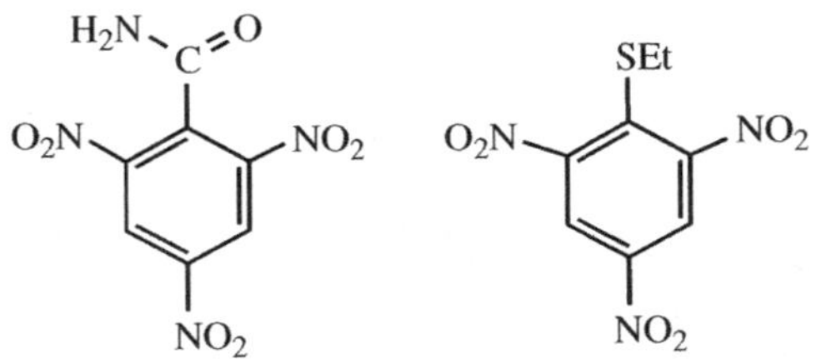

85 **86**

The reaction of *p*-dinitrobenzene with carbon acids in liquid ammonia in the presence of potassium *t*-butoxide [232] gives substituted products such as (**87**), and reaction with carbanions derived from nitroalkanes in DMSO [233] may also result in substitution to give products such as (**82**).

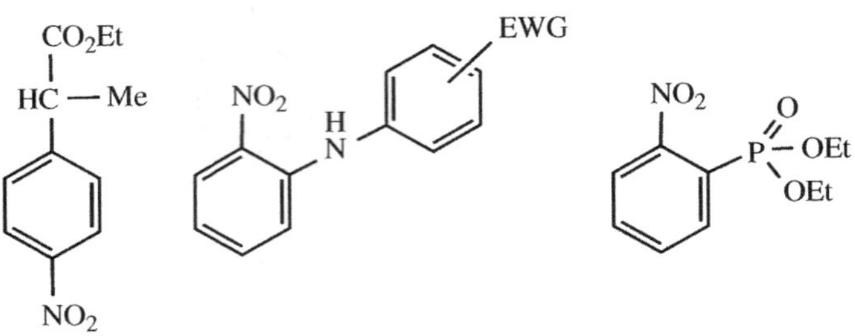

87 **88** **89**

Diphenylamines may be prepared by nitro group displacement from *o*- *or p*-dinitrobenzenes or from cyano-substituted nitrobenzenes. Thus, reaction with acidic anilines containing electron-withdrawing substituents in DMSO in the presence of potassium carbonate yields [234] products such as (**88**). *Ortho*-dinitrobenzene will also react with tervalent phosphorus compounds, such as trialkylphosphites, in acetonitrile to give substituted products; (**89**) is formed from triethylphosphite [235].

Fluoroarenes may be prepared by fluorodenitration reactions of activated substrates by reaction with potassium fluoride in sulfolane. Here, it is necessary to add a nitrite trap so that the liberated nitrite ion does not attack the aromatic ring [236].

Aniline derivatives are normally unreactive toward hydrolysis. However, since triaminobenzene may exist partially as its imine tautomer (**90**), it can be hydrolyzed to give the trihydroxy derivative, phloroglucinol, in a reaction that is accelerated by microwave radiation [237].

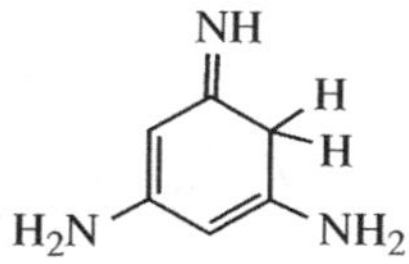

90

Intramolecular substitutions are useful synthetically particularly in the formation of heterocyclic compounds. Note that intramolecular rearrangements, such as the Smiles rearrangement, were described in Section 6.2.5. As observed with intermolecular processes, intramolecular substitutions usually involve the displacement of halogens or the nitro group by oxygen, nitrogen, sulfur, or carbon nucleophiles. The reaction of *o*-chloronitrobenzenes with *o*-aminophenols [112] may result in two substitutions leading to phenoxazine derivatives as shown in Scheme 6.28. Related reactions with pyrocatechol and thiopyrocatechol derivatives can produce phenodioxins and phenoxathiins, respectively. Displacements of a nitro group [238] and of fluorine [239] have been used, respectively, in the synthesis of dibenzo-oxazepine derivative (**91**) and dibenzo-oxazocine derivatives (**92**). The double displacement of chloride and a nitro group by dimethyldithiocarbamic acid in DMSO can lead [240] to the production of benzodithiol-2-ones as shown in Scheme 6.29, and the condensation of 2′-haloacetophenone and dithioesters, involving a thiodehalogenation process, produces thiochromene derivatives [241] as shown in Scheme 6.30.

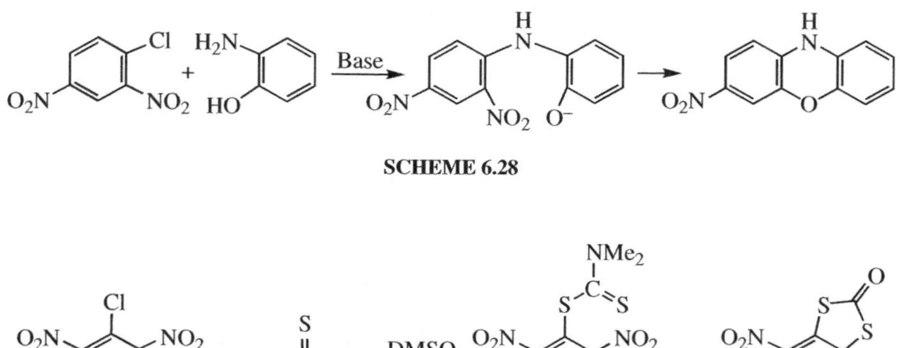

SCHEME 6.28

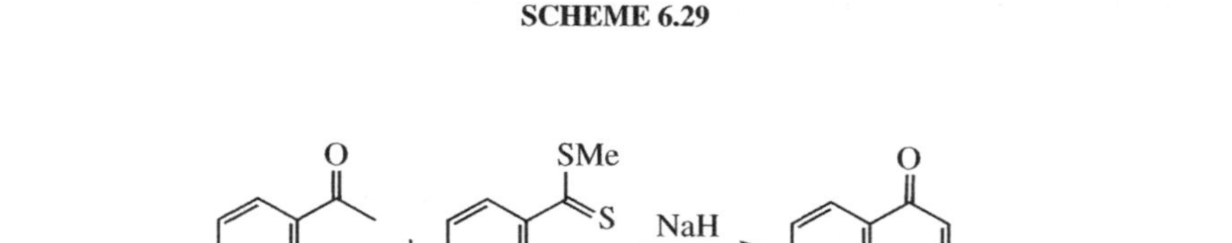

SCHEME 6.29

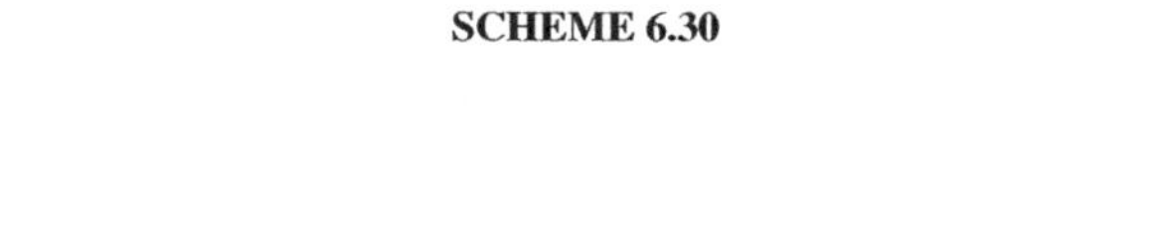

SCHEME 6.30

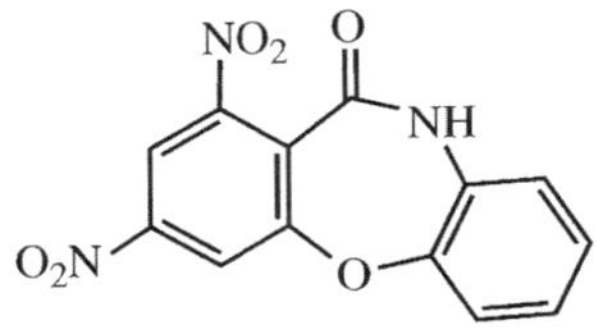

91

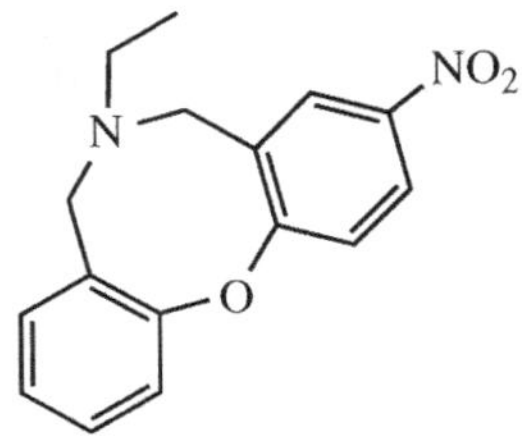

92

Internal amine nucleophiles have also been used to displace a nitro group forming nitrophenazines [112] in a process similar to that shown in Scheme 6.28. The amino-debromination pathway shown in Scheme 6.31 may produce benzimidazol-2-one derivatives [242]. An aminodefluorination process is involved in the reaction shown in Scheme 6.32 producing 2-arylindole derivatives. Here, the acetylene function acts as an EWG as well as a reagent [243].

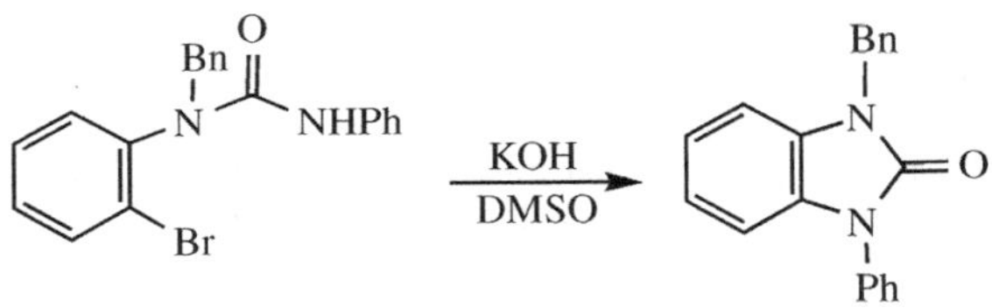

SCHEME 6.31

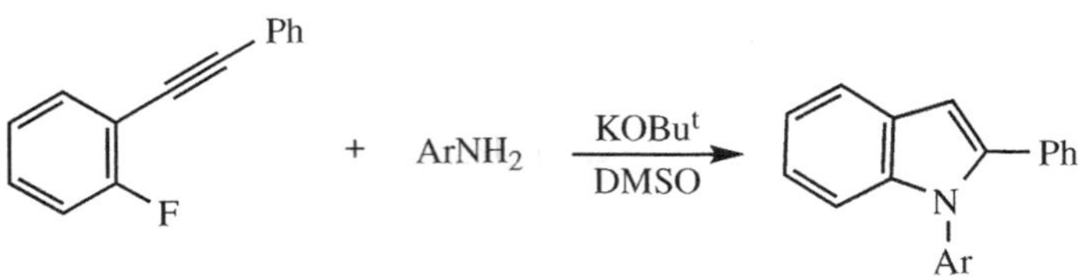

SCHEME 6.32

An unusual internal substitution of a dimethylamino group by an internal amine leads to indazole derivatives as shown in Scheme 6.33. Here, the ability of the parent to act as a "proton sponge" increases the potential for nucleophilic attack [244]. An example of the displacement of a nitro group by an internal carbanion is the formation of (**94**) from (**93**) in the presence of base [245].

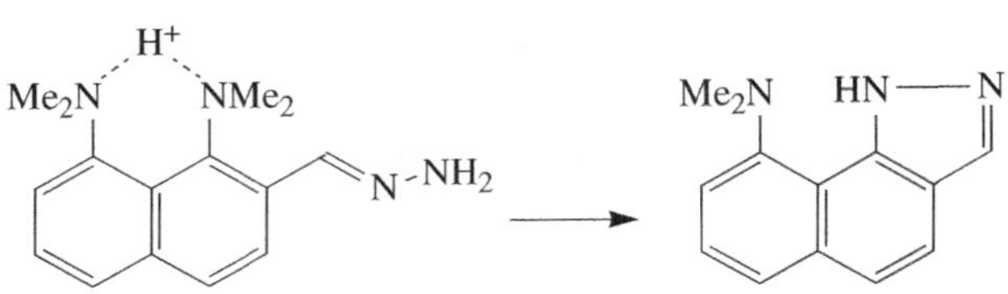

SCHEME 6.33

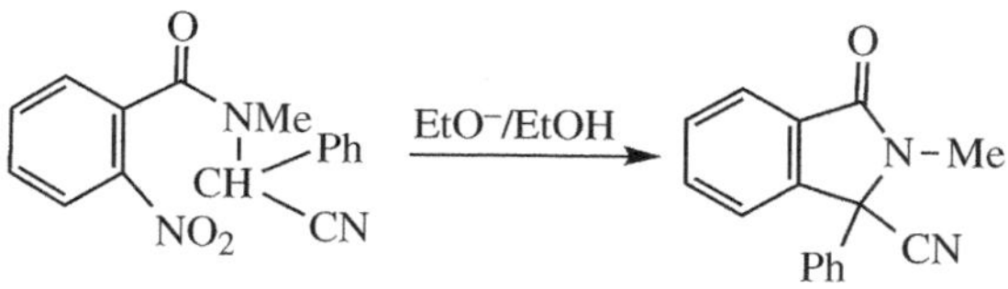

93 **94**

In a short survey, such as this, of synthetic applications, it is not possible to give a comprehensive coverage of the subject so, as elsewhere in the chapter, examples are given that hopefully illustrate the main points. A useful survey of practical applications of the use of S_NAr reactions in synthesis is given in Ref. 246. It should also be noted that in many cases, synthetic methods, dealt with in Part 6, using metal catalysts such as copper and palladium are available.

ABBREVIATIONS

CD_3CN	Trideuteroacetonitrile
DABCO	1,4-Diazabicyclo[2.2.2.]octane
DFT	Density functional theory
DMF	Dimethylformamide
DMSO	Dimethylsulfoxide
EWG	Electron-withdrawing group
SET	Single-electron transfer
TNB	1,3,5-Trinitrobenzene
UV	Ultraviolet

REFERENCES

[1] Bunnett, J.F. and Zahler, R.E. (1951) *Chem. Rev.*, **49**, 275.

[2] Roberts, J.D., Simmons, H.E., Carlsmith, L.A., and Vaughan, C.W. (1953) *J. Am. Chem. Soc.*, **75**, 3290.

[3] Roberts, J.D., Semenow, D.A., Simmons, H.E., and Carlsmith, L.A. (1956) *J. Am. Chem. Soc.*, **78**, 601.

[4] Russell, G.A. and Janzen, E.G. (1962) *J. Am. Chem. Soc.*, **84**, 4153.

[5] Havinga, E., de Jongh, R.O., and Dorst, W. (1956) *Rec. Trav. Chim.*, **75**, 378.

[6] Bunnett, J.F. (1978) *Acc. Chem. Res.*, **11**, 413.

[7] Kim, J.K. and Bunnett, J.F. (1970) *J. Am. Chem. Soc.*, **92**, 7464.

[8] Crampton, M.R., El Ghariani, M.A., and Khan, H.A. (1971) *Chem. Commun.*, 834.

[9] Terrier, F. (2013) *Modern Nucleophilic Aromatic Substitution*, Wiley-VCH, Weinheim.

[10] Chupakhin, O.N., Charushin, V.N., and van der Plas, H.C. (1994) *Nucleophilic Aromatic Substitution of Hydrogen*, Academic Press, San-Diego.

[11] Makosza, M. and Winiarski, J. (1987) *Acc. Chem. Res.*, **20**, 282.

[12] Ullmann, F. (1903) *Ber. Dtsch. Chem. Ges.*, **36**, 2382.

[13] Fanta, P.E. (1964) *Chem. Rev.*, **64**, 613.

[14] Hassan, J., Sevignon, M., Gozzi, C., Schulz, E., and Lemaire, M. (2002) *Chem. Rev.*, **102**, 1359.

[15] Ley, S.V. and Thomas, A.W. (2003) *Angew. Chem. Int. Ed.*, **42**, 5400.

[16] Wu, X.-F., Ambarasan, P., Neumann, H., and Beller, M. (2010) *Angew. Chem. Int. Ed.*, **49**, 9047.

[17] Seechurn, C.C.C.F., Kitching, M.O., Colacot, T.J., and Snieckus, V. (2012), *Angew. Chem. Int. Ed.*, **51**, 5062.

[18] Crampton, M.R. (2014) *Organic Reaction Mechanisms*, (Ed. A.C. Knipe), Ch. 5 (**2011**) John Wiley & Sons, Ltd, Chichester.

[19] Meisenheimer, J. (1902) *Ann. Chem.*, **323**, 205.

[20] Ritchie, C.D. and Sawada, M. (1977) *J. Am. Chem. Soc.*, **99**, 3754.

[21] Bunnett, J.F. and Randall, J.J. (1958) *J. Am. Chem. Soc.*, **80**, 6025.

[22] Kirby, A.J. and Jencks, W.P. (1965) *J. Am. Chem. Soc.*, **87**, 3217.

[23] Bunnett, J.F. and Garst, R.H. (1965) *J. Am. Chem. Soc.*, **87**, 3879.

[24] Bernasconi, C.F. (1973) *MTP Int. Rev. Sci. Org. Ser.* **1**, 3, 33.

[25] Renfrew, A.H.M., Taylor, J.A., Whitmore, M.J., and Williams, A. (1993) *J. Chem. Soc., Perkin Trans.* 2, 1703.

[26] Renfrew, A.H.M., Rettura, D., Taylor, J.A., Whitmore, J.M.J., and Williams, A. (1995) *J. Am. Chem. Soc.*, **117**, 5484.

[27] Crampton, M.R. and Robotham, I.A. (2013) *J. Phys. Org. Chem.*, **26**, 1084.

[28] Glukhovtsev, M.N., Bach, D.R., and Laiter, S. (1997) *J. Org. Chem.*, **62**, 4036.

[29] Giroldo, T., Xavier, L.A., and Riveros, J.M. (2004) *Angew. Chem. Int. Ed.*, **43**, 3588.

[30] Fernandez, I., Frenking, G., and Uggerud, E. (2010) *J. Org. Chem.*, **75**, 2971.

[31] Birch, A.J., Hinde, A.L., and Radom, L. (1980) *J. Am. Chem. Soc.*, **102**, 6430.

[32] Miller, J. (1968) *Aromatic Nucleophilic Substitution*, Elsevier, Amsterdam.

[33] Shein, S.M., Ignatov, V.A., Kozorev, L.A., and Chervatyuk, L.F. (1967) *Zh. Obshch. Khim.*, **37**, 114.

[34] Boiko, V.N., Kirii, N.V., and Yagupolskii, L.M. (1994) *J. Fluor. Chem.*, **67**, 119.

[35] Pietra, F., Vitali, D., Del Cima, F., and Cardiali, G. (1970) *J. Chem. Soc., B*, 1659.

[36] Terrier, F. (1991) *Nucleophilic Aromatic Displacement*, Wiley-VCH, Weinheim.

[37] Bunnett, J.F. and Morath, R.J. (1955) *J. Am. Chem. Soc.*, **77**, 5051.

[38] Bevan, C.W.L. and Hirst, J. (1956) *J. Chem. Soc.*, 254.

[39] Brooke, G.M. (1997) *J. Fluor. Chem.*, **86**, 1.

[40] Chambers, R.D. (2004) *Fluorine in Organic Chemistry*, Blackwell, Oxford.

[41] Baker, J and Muir, M. (2010) *Can. J. Chem.*, **88**, 588.

[42] Cacace, F., Speranza, M., Wolf, A.P., and Macgregor, R.R, (1982) *J. Fluor. Chem.*, **21**, 145.

[43] Di Nunno, L., Florio, S., and Todesco, P. (1975) *J. Chem. Soc., Perkin Trans.* 2, 1469.

[44] Terrier, F., Mokhtari, M., Goumont, R., Halle, J.-C., and Buncel, E. (2003) *Org. Biomol. Chem.*, **1**, 1757.

[45] El Guesmi, N., Berionni, G., and Asghar, B.H. (2013) *Monatsh. Chem.*, **144**, 1537.

[46] Knipe, A.C., McGuiness, S.J., and Watts, W.E. (1981) *J. Chem. Soc., Perkin Trans.* 2, 193.

[47] Mortier, J. (2011) *Curr. Org. Chem.*, **15**, 2413.

[48] Aissaoui, R., Nourry, A., Coquel, A., Dao, T.T.H., Derdour, A., Helesbeux, J.-J., Duval, O, Castanet, A.-S., and Mortier, J. (2011) *J. Org. Chem.*, **77**, 718.

[49] Hattori, T., Suzuki, M., Tomita, N., Takeda, A., and Miyano, S. (1997) *J. Chem. Soc., Perkin Trans.* 1, 1117.

[50] Capon, B. and Chapman, N.B. (1957) *J. Chem. Soc.*, 600.

[51] Benedetti, F., Marshall, D.R., Stirling, C.J.M., and Leng, J. (1982) *J. Chem. Soc., Chem. Commun.*, 918.

[52] Bartoli, G. and Todesco, P.E. (1977) *Acc. Chem. Res.*, **10**, 125.

[53] Bolto, B.A. and Miller, J. (1956) *Aust. J. Chem.*, **9**, 74.

[54] Bowden, K., Prasannan, S., and Ranson, R.J. (1987) *J. Chem. Soc., Perkin Trans.* 2, 181.

[55] Persson, J., Axelman, S., and Matsson, O. (1996) *J. Am. Chem. Soc.*, **118**, 20.

[56] Crampton, M.R. and Willison, M.J. (1974) *J. Chem. Soc., Perkin Trans.* 2, 238.

[57] Bordwell, F.G. and Hughes, D.L. (1986) *J. Am. Chem. Soc.*, **108**, 5991.

[58] Bunnett, J.F. (1958) *Chem. Soc., Quart. Rev.*, **12**, 1.

[59] Ritchie, C.D. (1972) *Acc. Chem. Res.*, **5**, 348.

[60] Mayr, H., Kempf, F., and Ofial, A.R. (2003) *Acc. Chem. Res.*, **36**, 66.

[61] Rodriguez-Dafonte, P., Terrier, F., Lakhdar, S., Kurbatov, S., and Goumont, R. (2009) *J. Org. Chem.*, **74**, 3305.
[62] Bernasconi, C.F. (1978) *Acc. Chem. Res.*, **11**, 147.
[63] Eigen, M. (1964) *Angew. Chem. Int. Ed.*, **3**, 1.
[64] Orvik, J.A. and Bunnett, J.F. (1970) *J. Am. Chem. Soc.*, **92**, 2417,
[65] Fyfe, C.A., Damji, S.W.H., and Koll, A. (1979) *J. Am. Chem. Soc.*, **101**, 951.
[66] Bunnett, J.F., Sekuguchi, S., and Smith L.A. (1981) *J. Am. Chem. Soc.*, **103**, 4865.
[67] Chamberlin, R.A. and Crampton, M.R. (1995) *J. Chem. Soc., Perkin Trans.* 2, 1831.
[68] Chamberlin, R.A. and Crampton, M.R. (1994) *J. Chem. Soc., Perkin Trans.* 2, 425.
[69] Chamberlin, R.A. and Crampton, M.R. (1993) *J. Chem. Soc., Perkin Trans.* 2, 75.
[70] Isanbor, C., Emokpae, T.A., and Crampton, M.R. (2002) *J. Chem. Soc., Perkin Trans.* 2, 2019.
[71] Crampton, M.R. and Gibson, B. (1981) *J. Chem. Soc., Perkin Trans.* 2, 533.
[72] Crampton, M.R. and Routledge, P. (1984) *J. Chem. Soc., Perkin Trans.* 2, 573.
[73] Hirst, J. (1994) *J. Phys. Org. Chem.*, **7**, 68.
[74] Nudelman, N.S., Alvaro, C.E.S., and Yankelevich, J.S. (1997) *J. Chem. Soc, Perkin Trans.* 2, 2125.
[75] Alvaro, C.E.S. and Nudelman, N.S. (2011) *Trends Org. Chem.*, **15**, 95.
[76] Onuoha, G.N., Onyido, I., and Hirst, J. (1988) *J. Chem. Soc., Perkin Trans.* 2, 971.
[77] Cattana, R.I., Singh, J.O., Anunziata, J.D., and Silber, J.J. (1987) *J. Chem. Soc., Perkin Trans.* 2, 79.
[78] Alvaro, C.E.S. and Nudelman, N.S. (2005) *J. Phys. Org. Chem.*, **18**, 880.
[79] Alvaro, C.E.S., Ayala, A., and Nudelman, N.S. (2011) *J. Phys. Org. Chem.*, **24**, 101.
[80] Forlani, L. (1999) *J. Phys. Org. Chem.*, **12**, 417.
[81] Miller, J. and Parker, A.J. (1961) *J. Am. Chem. Soc.*, **83**, 117.
[82] Cox, B.G. and Parker A.J. (1973) *J. Am. Chem. Soc.*, **95**, 408.
[83] Acenedo, O. and Jorgensen, W.L. (2004) *Org. Lett.*, **6**, 2881.
[84] Crampton, M.R. (1968) *J. Chem. Soc., B*, 1208.
[85] Fendler, J.H. and Larsen, J.W. (1972) *J. Org. Chem.*, **37**, 2608.
[86] Larsen, J.W., Amin, K., and Fendler, J.H. (1971) *J. Am. Chem. Soc.*, **93**, 2910.
[87] Um, I.-H., Min, S.-W., and Dust, J.M. (2007) *J. Org. Chem.*, **72**, 8797.
[88] Um, I.-H., Im, L.R., Kang, J.-S., Bursey, S.S., and Dust, J.M. (2012) *J. Org. Chem.*, **77**, 9738.
[89] Attina, M., Cacace, F., and Wolf, A.P. (1983) *J. Chem. Soc. Chem. Commun.*, 108.
[90] Adams, D.J. and Clark, J.H. (1999) *Chem. Soc. Rev.*, **28**, 225.
[91] Landini, D., Montunari, F., and Rolla, F. (1983) *J. Org. Chem.*, **48**, 604.
[92] Bunton, C.A. and Diaz, S. (1976) *J. Am. Chem. Soc.*, **98**, 5663.
[93] Athanassakis, V., Bunton, C.A., and de Buzzaccarini, F. (1982) *J. Phys. Chem.*, **86**, 5002.
[94] Ning, M.S., Price, S.E., Ta, J., and Davies, K.M. (2009) *J. Phys. Org. Chem.*, **23**, 2009.
[95] Tanner, E.E.L., Hawker, R.R., Yan, H.M., Croft, A.K., and Harper, J.B. (2013) *Org. Biomol. Chem.*, **11**, 7516.
[96] Ji, P., Atherton, J.H., and Page, M.I. (2011) *J. Org. Chem.*, **76**, 3286.
[97] Atherton, J.H., Page, M.I., and Sun, H. (2013) *J. Phys. Org. Chem.*, **26**, 1038.
[98] Levy, A., Rains, H.C., and Smiles, S. (1931) *J. Chem. Soc.*, 3264.
[99] McClement, C.S. and Smiles, S. (1937) *J. Chem. Soc.*, 1016.
[100] Bunnett, J.F. and Okamoto, T. (1956) *J. Am. Chem. Soc.*, **78**, 5363.
[101] Coats, R.R. and Gibson, D.T. (1940) *J. Chem. Soc.*, 442.
[102] Bernasconi, C.F., Gehriger, C.L., and de Rossi, R.H. (1973) *J. Org. Chem.*, **98**, 2838.
[103] Knipe, A.C., Sridhar, N., and Lound-Keast, T. (1977) *J. Chem. Soc., Perkin Trans.* 1, 581.
[104] Kleb, K.G. (1968) *Angew. Chem. Int. Ed.*, **7**, 291.
[105] Kondo, H., Moriuchi, F., and Sunamoto, J. (1981) *J. Org. Chem.*, **46**, 1333.

[106] Sorensen, A., Rasmussen, B., Agarwal, S., Schau-Magnussen, M., Solling, T.I., and Pittelkow, M. (2013) *Angew. Chem. Int. Ed.*, **52**, 12346.

[107] Gilman, N.W., Levitan, P., and Sternbach, L.H. (1973) *J. Org. Chem.*, **38**, 373.

[108] El Kaim, L., Gizolme, M., and Grimaud, L. (2006) *Org. Lett.*, **8**, 5021.

[109] Snape, T.J. (2008) *Chem. Soc. Rev.*, **37**, 2452.

[110] Truce, W.E., Ray, W.J., Norman, O.L., and Eickemeyer, D.B. (1958) *J. Am. Chem. Soc.*, **80**, 3625.

[111] Gonzalez, J.P., Edgar, M., Elsegood, M.R.J., and Weaver, G.W. (2011) *Org. Biomol. Chem.*, **9**, 2294.

[112] Preston, P.N. and Tennant, G. (1972), *Chem. Rev.*, **72**, 627.

[113] Buncel, E., Crampton, M.R., Strauss, M.J., and Terrier, F. (1984) *Electron-Deficient Aromatic- and Heteroaromatic-Base Interactions*, Elsevier, Amsterdam.

[114] Crampton, M.R. (1969) *Adv. Phys. Org. Chem.*, **7**, 211.

[115] Strauss, M.J. (1970) *Chem. Rev.*, **70**, 667.

[116] Terrier, F. (1982) *Chem. Rev.*, **82**, 77.

[117] Artamkina, G.A., Egorov, M.P., and Beletskaya, I.P. (1982) *Chem. Rev.*, **82**, 427.

[118] Foster, R. and Hammick, D.L. (1954) *J. Chem. Soc.*, 2153.

[119] Dyall, L.K. (1960) *J. Chem. Soc.*, 5160.

[120] Abe, T. (1966) *Bull. Chem. Soc. Jpn.*, **39**, 627.

[121] Crampton, M.R. and Gold, V. (1964) *J. Chem. Soc.*, 4293.

[122] Foster, R.A. and Fyfe, C.A. (1965) *Tetrahedron*, **21**, 3363.

[123] Gold, V. and Rochester, C.H. (1964) *J. Chem. Soc.*, 1692.

[124] Symons, E.A. and Buncel, E. (1972) *Can. J. Chem.*, **50**, 1729.

[125] Gan, L.H. and Norris, A.R. (1971) *Can. J. Chem.*, **49**, 2490.

[126] Crampton, M.R. (1967) *J. Chem. Soc., B*, 1341.

[127] Crampton, M.R. and Gold, V. (1967) *J. Chem. Soc., B*, 23.

[128] Taylor, R.P. (1970) *J. Org. Chem.*, **35**, 3578.

[129] Foster, R.A. and Fyfe, C.A. (1966) *J. Chem. Soc., B*, 53.

[130] Buncel, E. and Manderville, R. (1993) *J. Phys. Org. Chem.*, **6**, 71.

[131] Buncel, E., Dust, J.M., and Manderville, R.A. (1996) *J. Am. Chem. Soc.*, **118**, 6072.

[132] Halle, J.C., Terrier, F., Pouet, M.J., Simonnin, M.P. and Debleds, F. (1982) *Can. J. Chem.*, **60**, 1988.

[133] Buncel, E., Park, K.-T., Dust, J.M., and Manderville, R.A. (2003) *J. Am. Chem. Soc.*, **125**, 5388.

[134] Ueda, H., Sakabe, N., Tanaka, J., and Furusaki, A. (1968) *Bull. Chem. Soc. Jpn.*, **41**, 2866.

[135] Destro, R., Grammacioli, C., and Simonetta, M. (1968) *Acta Crystallogr.*, **B24**, 1369.

[136] Destro, R., Pilati, T., and Simonetta, M. (1979) *Acta Crystallogr.*, **B35**, 733.

[137] Borbulevych, O.Ya., Shishkin, O.V., and Antipin, M.Yu. (2002) *J. Phys. Chem. A*, **106**, 8109.

[138] Borbulevych, O.Ya., Antipin, M.Yu., and Olekhnovich, L.P. (1999) *Acta Crystallogr.*, **C55**, 2177.

[139] Borbulevych, O.Ya., Shishkin, O.V., Budarina, Z.N., Antipin, M.Yu., and Olekhnovich, L.P. (1999) *Acta Crystallogr.*, **C55**, 1915.

[140] Al-Kaysi, R.O., Guirado, G., and Valente, E.J. (2004) *Eur. J. Org. Chem.*, 3408.

[141] Morozov, P.G., Kurbatov, S.V., Dolgustrin, F.M., Antipin, M.Yu., and Olekhnovich, L.P. (2004) *Russ. Chem. Bull., Int. Ed.*, **53**, 2075,

[142] Servis, K.L. (1965) *J. Am. Chem. Soc.*, **87**, 5495.

[143] Crampton, M.R. and Gold, V. (1966) *J. Chem. Soc., B*, 893.

[144] Bernasconi, C.F. (1971) *J. Am. Chem. Soc.*, **93**, 6975.

[145] Parker, V.D., Handoo, K.L. Hao, W., and Cheng, J.-P. (2011) *J. Org. Chem.* **76**, 1250.

[146] El Guesmi, N., Boubaker, T., Goumont, R., and Terrier, F. (2008) *Org. Biomol. Chem.*, **6**, 4041.

[147] Hinze, W.L., Liu, L.-J., and Fendler, J.H. (1975) *J. Chem. Soc., Perkin Trans.* 2, 1751.

[148] Crampton, M.R. and Khan, H.A. (1973) *J. Chem. Soc., Perkin Trans.* 2, 710.

[149] Terrier, F., Millot, F., and Simonnin, M.P. (1971) *Tetrahedron Lett.*, **12**, 2933.

[150] Illuminati, G. and Stegel, F. (1983), *Adv. Heterocycl. Chem.*, **34**, 306.
[151] Bernasconi, C.F., Muller, M.C., and Schmid, P. (1979) *J. Org. Chem.*, **44**, 3189.
[152] Chudek, J.A., Ellingham, R.A., and Foster, R. (1985) *J. Chem. Soc., Perkin Trans.* 2, 1477.
[153] Boga, C. and Forlani, L. (1998) *J. Chem. Soc., Perkin Trans.* 2, 2155.
[154] Sekiguchi, S., Suzuki, T., and Hosokawa, M. (1989) *J. Chem. Soc., Perkin Trans.* 2, 1983.
[155] Sekiguchi, S., Ishikura, H., Hirosawa, Y., and Ono, N. (1990) *Tetrahedron*, **46**, 5567.
[156] Chamberlin, R.A. and Crampton, M.R. (1993) *J. Chem. Res. Synop*, 106.
[157] Taylor, R.P. (1970) *Chem. Commun.*, 1462.
[158] Onys'ko, P.P. and Gololobov, Yu.G. (1977) *J. Gen. Chem. USSR*, **47**, 2267.
[159] Crampton, M.R. and El Ghariani, M.A. (1971) *J. Chem. Soc., B*, 1043.
[160] Biggi, G. and Pietra, F. (1973) *J. Chem. Soc., Perkin Trans.* 1, 1980.
[161] Crampton, M.R. (1971) *J. Chem. Soc., B*, 2112
[162] Crampton, M.R. and Stevens, J.A. (1989) *J. Chem. Soc., Perkin Trans.* 2, 925.
[163] Bernasconi, C.F. and Bergstrom, R.G. (1973) *J. Am. Chem. Soc.*, **95**, 3603.
[164] Crampton, M.R. and Willison, M.J. (1973) *J. Chem. Soc., Chem. Commun.*, 215.
[165] Strauss, M.J. and Taylor, S.P.B. (1973) *J. Am. Chem. Soc.*, **95**, 3813.
[166] Renfrow, R.A., Strauss, M.J., and Terrier, F. (1980) *J. Org. Chem.*, **45**, 471.
[167] Fyfe, C.A. (1968) *Can. J. Chem.*, **46**, 3047.
[168] Crampton, M.R., Kee, T.P., and Wilcock, J.R. (1986) *Can. J. Chem.*, **64**, 1714.
[169] Strauss, M.J. (1974) *Acc. Chem. Res.*, **7**, 181.
[170] Bard, R.R. and Strauss, M.J. (1975) *J. Am. Chem. Soc.*, **97**, 3789.
[171] Bernasconi, C.F. (1982) *Pure Appl. Chem.*, **54**, 2335.
[172] Cox, J.P.L., Crampton, M.R., and Wight, P. (1988) *J. Chem. Soc., Perkin Trans.* 2, 25.
[173] Crampton, M.R. and Willison, M.J. (1974) *J. Chem. Soc., Perkin Trans.* 2, 1681.
[174] Bernasconi, C.F. and Gandler, J.R. (1977) *J. Org. Chem.*, **42**, 3387.
[175] Bernasconi, C.F. and Howard, K. (1982) *J. Am. Chem. Soc.*, **104**, 7248.
[176] Crampton, M.R. and Willison, M.J. (1976) *J. Chem. Soc., Perkin Trans.* 2, 155.
[177] Crampton, M.R. and Willison, M.J. (1974), *J. Chem. Soc., Perkin Trans.* 2, 1686.
[178] Crampton, M.R. and Willison, M.J. (1976) *J. Chem. Soc., Perkin Trans.* 2, 901.
[179] Green, T.K., Denoroy, L., and Parrot, S. (2010) *J. Org. Chem.*, **75**, 4048.
[180] Buncel, E. and Terrier, F. (2010) *Org. Biomol. Chem.*, **8**, 2288.
[181] Forlani, L., Boga, C., Mazzanti, A., and Zanna, N. (2012) *Eur. J. Org. Chem.*, 1123.
[182] Buncel, E., Renfrow, R.A., and Strauss, M.J. (1987) *J. Org. Chem.*, **52**, 488.
[183] Crampton, M.R., Rabbitt, L.C., and Terrier, F. (1999) *Can. J. Chem.*, **77**, 639.
[184] Halle, J.C., Vichard, D., Pouet, M.-J., and Terrier, F. (1997) *J. Org. Chem.*, **62**, 7178.
[185] Ambroz, H.B. and Kemp, T.J. (1979) *Chem. Soc. Rev.*, **8**, 353.
[186] Bergstrom, R.G., Landells, R.G.M., Wohl, G.H., and Zollinger, H. (1976) *J. Am. Chem. Soc.*, **98**, 3301.
[187] Martinez, A.G., Cerero, S.D.L.M., Barcina, J.O., Jimenez, F.M., and Maroto, B.L. (2013), *Eur. J. Org. Chem.*, 6098.
[188] Fernando Alonso, A. and Bravo-Diaz, C. (2010) *Helv. Chim. Acta*, **93**, 877.
[189] Broxton, T.J. and McLeish, M.J. (1982) *J. Org. Chem.*, **47**, 3673.
[190] Broxton, T.J. and McLeish, M.J. (1983), *Aust. J. Chem.*, **36**, 103.
[191] Broxton, T.J., Bunnett, J.F., and Paik, C.H. (1977) *J. Org. Chem.*, **42**, 643.
[192] Zollinger, H. (1973) *Acc. Chem. Res.*, **6**, 335.
[193] Mo, F., Dong, G., Zhang, Y., and Wang, J. (2013) *Org. Biomol. Chem.*, **11**, 1582.
[194] Ibata, T., Isogami, Y., and Toyoda, J. (1987) *Chem. Lett.*, **16**, 1187.
[195] Meciarova, M., Podlesna, J., and Toma, S. (2004) *Monatsh. Chem.*, **135**, 419.

[196] Pitsinos, E.N., Vidali, V.P., and Couladouros, E.A. (2011) *Eur. J. Org. Chem.*, 1207.

[197] Li, F., Meng, Q., Chen, H., Li, Z., Wang, Q., and Tao, F. (2005) *Synthesis*, 1305.

[198] Jones, G.O., Somaa, A.A., O'Brien, J.M., Albishi, H., Al-Megren, H.A., Abdulrahman, A.M., Alsewailem, F.D., Hedrick, J.L., Rice, J.E., and Horn, H.W. (2013) *J. Org. Chem.*, **78**, 5436.

[199] Wendt, M.P. and Kunzer, A.R. (2010) *Tetrahedron Lett.*, **51**, 3041.

[200] Sawyer, J.S., Schmittling, E.A., Palkowitz, J.A., and Smith, W.J. (1998) *J. Org. Chem.*, **63**, 6338.

[201] Kim, A., Powers, J.D., and Toczko, J.F. (2006) *J. Org. Chem.*, **71**, 2170.

[202] Sen, S., Potti, V.R., Surakanti, R., Murthy, Y.L.N., and Pallepogu, R. (2011) *Org. Biomol. Chem.*, **9**, 358.

[203] Li, Z., Yang, Q., and Qian, X. (2005) *Tetrahedron*, **61**, 8711.

[204] Pedersen, E.B., Perregard, J., and Lawesson, S.-O. (1973) *Tetrahedron*, **29**, 4211.

[205] Gupton, J.T., Idoux, J.P., Baker, G., Colon, C., Crews, A.D., Jurss, C.D., and Rampi, R.C. (1983), *J. Org. Chem.*, **48**, 2933.

[206] Chen, D., Yang, K., Xiang, H., and Jiang, S. (2012) *Tetrahedron Lett.*, **53**, 7121.

[207] Qian, X., Xu, S., Zong, Q., and Fang, D. (2012) *Chin. J. Chem.*, **30**, 2389.

[208] Hattori, T., Satoh, T., and Miyano, S. (1996) *Synthesis*, 514.

[209] Gulevskaya, A.V. and Pozharskii, A.F. (2008) *Russ. Chem. Bull., Int. Ed.*, **57**, 913.

[210] Chambers, R.D., Musgrave, W.K.R., Waterhouse, J.S., Williams, D.L.H., Burdon, J., Hollyhead, W.B., and Tatlow, J.C. (1974) *J. Chem. Soc., Chem. Commun.*, 239.

[211] Burdon, J., Fischer, D., Parsons, I.W., and Tatlow, J.C. (1981) *J. Fluor. Chem.*, **18**, 507.

[212] Jitchati, R., Batsanov, A.S., and Bryce, M.R. (2009) *Tetrahedron*, **65**, 855.

[213] Sorokin, V.I., Ozeryanskii, V.A., and Pozharskii, A.F. (2004) *Eur. J. Org. Chem.*, 766.

[214] Sorokin, V.I., Ozeryanskii, V.A. Pozharskii, A.F., and Starikova, Z.A. (2004) *Mendeleev Commun.*, **14**, 14.

[215] Biemans, H.A., Zhang, C., Smith, P., Kooijman, H., Smeets, W.J.J., Spek, A.L., and Meijer, E.W. (1996) *J. Org. Chem.*, **61**, 9012.

[216] Lu, F., Sun, H., Du, A., Feng, L., and Li, X. (2014) *Org. Lett.*, **16**, 772.

[217] Kavalek, J., Machacek, V., Pastrnek, M., and Sterba, V. (1977) Collect. *Czech. Chem. Commun.*, **42**, 2928.

[218] Kavalek, J., Pastrnek, M., and Sterba, V. (1978) *Collect. Czech. Chem. Commun.*, **43**, 1401.

[219] Ueno, M., Yonemoto, M., Hashimoto, M., Wheatley, A.E.H., Naka, H., and Kondo, Y. (2007) *Chem. Commun.*, 2264.

[220] Neuville, L., Bigot, A., Dau, M.E.T.H., and Zhu, J. (1999) *J. Org. Chem.*, **64**, 7638.

[221] Kornblum, N., Cheng, L., Kerber, R.C., Kestner, M.M., Newton, B.N., Pinnick, H.W., Smith, R.G., and Wade, P.A. (1976) *J. Org. Chem.*, **41**, 1560.

[222] Logue, M.W. and Ham, B.H. (1981) *J. Org. Chem.*, **46**, 1638.

[223] Kondoh, A., Yorimitsu, H., and Oshima, K. (2006) *Tetrahedron*, **62**, 2357.

[224] Idoux, J.P., Madenwald, M.L., Garcia, B.S., Chu, D.-L., and Gupton, J.T. (1985) *J. Org. Chem.*, **50**, 1876.

[225] Beier, P., Pastyrikova, T., Vida, N., and Iakobson, G. (2011) *Org. Lett.*, **13**, 1466.

[226] Sammes, P.G., Thetford, D., and Voyle, M. (1988) *J. Chem. Soc., Perkin Trans.* 1, 3229.

[227] Tejero, I., Huertas, I., Gonzalez-Lafont, A., Lluch, J.M., and Marquet, J. (2005) *J. Org. Chem.*, **70**, 1718.

[228] Cogalli, P., Testaferri, L., Tingoli, M., and Tiecco, M. (1979) *J. Org. Chem.*, **44**, 2636.

[229] Beck, J.R. and Yahner, J.A. (1978) *J. Org. Chem.*, **43**, 2048.

[230] Shkinyova, T., Dalinger, I.L., Molotov, S.I., and Shevelev, S.A. (2000) *Tetrahedron Lett.*, **41**, 4973.

[231] Biggi, G. and Pietra, F. (1973) *J. Chem. Soc., Perkin Trans.* 1, 1980.

[232] Iwasaki, G., Saeki, S., and Hamana, M. (1986) *Chem. Lett.*, **15**, 31.

[233] Norris, R.K. and Randles, D. (1979) *Aust. J. Chem.*, **32**, 2413.

[234] Gorvin, J.H. (1985) *J. Chem. Soc., Chem. Commun.*, **238**.

[235] Cadogan, J.I.G., Sears, D.I., and Smith, D.M. (1969) *J. Chem. Soc., C*, 1314.

[236] Maggini, M., Passudetti, M., Gonzales-Trueba, G., Prato, M., Quintily, U., and Scorrano, G. (1991) *J. Org. Chem.*, **56**, 6406.

[237] Lee, W.H., Kim, J.S., and Kim, S.C. (2009), *Bull. Kor. Chem. Soc.*, **30**, 3105.

[238] Samet, A.V., Marshalkin, V.N., Kislyi, K.A., Chenysheva, N.B., Strelenko, Y.A., and Senenov, V.V. (2005) *J. Org. Chem.*, **70**, 9371.

[239] Ouyang, X., Chen, Z., Liu, L., Dominguez, C., and Kiselyov, A.S. (2000) *Tetrahedron*, **56**, 2369.

[240] Rasheed, K. and Warkentin, J.D. (1977) *J. Org. Chem.*, **42**, 1265.

[241] Vijay, T.A.J., Nandeesh, K.N., Raghavendra, G.M., Rangappa, K.S., and Mantelinga, K. (2013) *Tetrahedron Lett.*, **54**, 6533.

[242] Beyer, A., Reucher, C.M.M., and Bolm, C. (2011) *Org. Lett.*, **13**, 2876.

[243] Bizier, N.P., Wackerly, J.W., Braunstein, E.D., Zhang, M., Nodder, S.T., Carlin, S.M., and Katz, J.L. (2013) *J. Org. Chem.*, **78**, 5987.

[244] Povalyakhina, M.A., Antonov, A.S., Dyablo, O.V., Ozeryanskii, V.A., and Pozharskii, A.F. (2011) *J. Org. Chem.*, **76**, 7157.

[245] Spence, T.W.M. and Tennant, G. (1972) *J. Chem. Soc., Perkin Trans.* 1, 835.

[246] Carron, S. and Ghosh, A. (2011) *Practical Synthetic Organic Chemistry: Reactions, Principles and Techniques*, John Wiley & Sons, Inc., Hoboken, 237.

7

THEORETICAL AND EXPERIMENTAL METHODS FOR THE ANALYSIS OF REACTION MECHANISMS IN S_NAr PROCESSES: FUGALITY, PHILICITY, AND SOLVENT EFFECTS[1]

Renato Contreras[1], Paola R. Campodónico[2], and Rodrigo Ormazábal-Toledo[3]

[1]*Departamento de Química, Facultad de Ciencias, Universidad de Chile, Chile*
[2]*Centro de Química Médica, Facultad de Medicina, Clínica Alemana Universidad del Desarrollo, Santiago, Chile*
[3]*Departamento de Fisica, Facultad de Ciencias, Universidad de Chile, Chile*

7.1 INTRODUCTION

The generally accepted nucleophilic aromatic substitution (S_NAr) mechanism occurs in activated aromatic compounds bearing good leaving groups (LG) through an addition–elimination route [1–8]. The first step is the nucleophilic attack toward an activated aromatic ring, leading to the formation of an anionic σ-adduct named Meisenheimer complex (MC) [6, 9, 10]. In a second step, the LG detaches after an intramolecular proton transfer from the nucleophile [11–14]. This last step may or may not proceed via a catalyzed pathway promoted by a second nucleophile molecule (see Scheme 7.1) [15–17]. This chapter is concerned with the theoretical and computational

[1]During the revision process of this chapter, we received a note from the editorial board concerning the Nucleophilic Substitution of Hydrogen in Electron-Deficient Arenes, where some mechanistic aspects significantly differ from the one addressed herein; mainly in those issues related to the experimental conditions. We recommend the reader to consult Chapter 11 by Prof. Mieczysław Mąkosza for a deeper discussion about this important point.

Arene Chemistry: Reaction Mechanisms and Methods for Aromatic Compounds,
First Edition. Edited by Jacques Mortier.

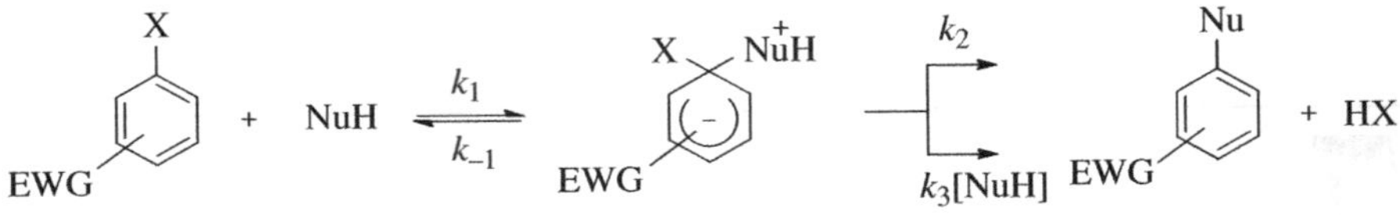

SCHEME 7.1 Generally accepted mechanism for nucleophilic aromatic substitution with neutral nucleophiles.

models that may assist in the rationalization and analysis of reaction mechanism in S_NAr processes. There exist a significant number of articles dealing with the theoretical description of S_NAr reactions [18–23]. They are mostly based on the complete exploration of the potential energy surface (PES), including the characterization of transition state (TS) structures and reaction intermediates. In this work, we describe a theoretical model that uses PES calculations as an input to perform a deeper analysis based on global, local, and nonlocal response functions defined in the frame of the conceptual density functional theory (DFT) of Parr and Yang [24].

The chapter is organized summarizing the main achievements on aromatic substrates that react through an S_NAr mechanism. Theoretical and experimental procedures commonly used to predict their mechanism and understand the reactivity of these systems are also described. The road map of the chapter integrates experiment and theory by using the experimental data to validate the different reactivity indices–based approach to describe philicity and fugality patterns and their reliability to explain the different stages of S_NAr processes, including solvation effects.

7.2 CONCEPTUAL DFT: GLOBAL, REGIONAL, AND NONLOCAL REACTIVITY INDICES

The DFT of atoms and molecules formulated by Parr and Yang [24] has emerged as a powerful tool to convert classical chemical concepts like electronegativity, softness, and hardness [25–28] and related reactivity indices like electrophilicity (ω^+) [26, 29, 30], nucleophilicity (ω^-) [31–33], electrofugality (ν^+) [34], and nucleofugality (ν^-) [35–37] into quantitative scales of reactivity. Their local or regional counterparts are readily obtained by projecting the global property using the Fukui function of the system. From this formalism, site reactivity describing regioselectivity can also be used to establish a hierarchy of site reactivity within a molecule [38–41]. Site activation, on the other hand, enters into this phenomenological reactivity theory model, by following the variations in local or regional properties along a well-defined reaction coordinate [42–46]. In this section, a brief account of this set of reactivity indices is presented.

One of the basic quantities defined in the DFT of Parr and Yang is the electronic chemical potential of the system μ, defined as the derivative of the total energy E with respect to the number of electrons N at constant external potential (i.e., the electric potential due to the nuclei in the system) $\upsilon(\mathbf{r})$. The operative expression given by Parr and Yang is as follows [24–26, 47, 48]:

$$\mu = \left[\frac{\partial E}{\partial N}\right]_{\upsilon(\mathbf{r})} \approx -\left(\frac{I+A}{2}\right) \approx \left(\frac{\varepsilon_{\mathrm{H}}+\varepsilon_{\mathrm{L}}}{2}\right) \tag{7.1}$$

In Equation 7.1, I and A are the vertical ionization potential and electron affinity, respectively. A closer look at the second equality reveals that the finite difference expression of μ is exactly the negative of Mulliken electronegativity [27, 49, 50]. This result is relevant, for it describes the electronic chemical potential as a fugacity term, that is, the propensity of electrons to abandon the atom or molecule. In this sense, the direction of the electronic charge flux is determined by the difference in electronic chemical potential; electrons will flow from the region of high electronic chemical potential of an interacting pair toward the zone where the electronic chemical potential is lower, until equilibrium is reached, and the electronic chemical potential equalizes when a precursor intermolecular complex is formed. The third equality in Equation 7.1 is the most used approximation that expresses the electronic chemical potential in terms of the one-electron energies of the frontier orbitals HOMO (H) and LUMO (L), which results after applying Koopmans theorem [51, 52]. Another pertinent global quantity is the chemical hardness η [27, 53–55]:

$$\eta = \left(\frac{\partial^2 E}{\partial N^2}\right)_{\upsilon(\mathbf{r})} \approx I - A \approx \varepsilon_L - \varepsilon_H \tag{7.2}$$

Chemical hardness represents the resistance of the system to exchange electrons with a specified or unspecified environment. Chemical softness S is simply defined as the inverse of chemical hardness: $S = 1/\eta$.

A significant progress was made, when the global electrophilicity was introduced by Parr *et al.* [29]. Formerly proposed by Maynard [56], the global electrophilicity index was given the following working expression in terms of the μ and η indices:

$$\omega^+ = \frac{\mu^2}{2\eta} \tag{7.3}$$

The global electrophilicity index describes an electron acceptor as a species displaying high electronegativity and low hardness (or high softness). The first application of this reactivity index allowed a successful classification of dienes as nucleophiles and dienophiles and dipolarophiles as electrophiles [38]. The set of global reactivity indices that is pertinent to this chapter is completed with the definition of the global nucleophilicity [31–33]. Among the different approaches proposed to describe the electron-releasing ability of atoms and molecules, we will rely on that proposal based on terms of the vertical ionization potential I [33]. The working formula is given by

$$\omega^- = -I \approx \varepsilon_H \tag{7.4}$$

where Koopmans theorem has been used again [51, 52].

Besides the global descriptors of reactivity, there are a number of local indices that describe the active sites within a molecule. Among them, of special interest is the Fukui function [48, 53, 57–60]. This index, besides being a local descriptor of reactivity by itself (i.e., regioselectivity), may act as a convenient distribution function for global quantities. The original definition is as follows [59, 61]:

$$f(\mathbf{r}) = \left[\frac{\partial \rho(\mathbf{r})}{\partial N}\right]_{\upsilon(\mathbf{r})} \tag{7.5}$$

Note that $f(\mathbf{r})$ describes the change in electron density at point $\mathbf{r}$ in space with respect to the variation in the total number of electrons N, at constant external potential (i.e., at fixed molecular geometry). There are several approximate formulas to evaluate the condensed to atom Fukui function. A three-point finite difference approach is normally used [24, 61], albeit it may be also readily obtained by single-point calculations, using the frozen core approximation. The former approach will be used herein. Using the normalization to unity condition [57, 58]:

$$\int d\mathbf{r}\, f(\mathbf{r}) = 1 \tag{7.6}$$

it is easy to show that for any well-defined global property P, say, its local counterpart may be written as

$$P^{\pm}(\mathbf{r}) = P f^{\pm}(\mathbf{r}) \tag{7.7}$$

where + and – refer to an electron-accepting or electron-releasing property, respectively. Using the three-point finite difference approach, we get

$$f^+ = \rho(N+1) - \rho(N) \tag{7.8}$$

$$f^- = \rho(N) - \rho(N-1) \tag{7.9}$$

where $\rho(N)$, $\rho(N+1)$, and $\rho(N-1)$ correspond to the densities of the system with N, $N+1$, and $N-1$ electrons. In particular, the following definitions are useful:

$$\omega^+(\mathbf{r}) = \omega f^+(\mathbf{r}) \tag{7.10}$$

$$\omega^-(\mathbf{r}) = \omega f^-(\mathbf{r}) \tag{7.11}$$

$$s^\pm(\mathbf{r}) = S f^\pm(\mathbf{r}) \tag{7.12}$$

They describe local electrophilicity, local nucleophilicity, and local softness, respectively.

However, in chemistry, one is more interested in properties of atoms or (functional) groups instead of properties of points in space. A suitable regional integration converts these local quantities into regional or condensed to atom ones:

$$\omega_k^+ = \omega f_k^+ \tag{7.13}$$

$$\omega_k^- = \omega f_k^- \tag{7.14}$$

$$s_k^\pm(r) = S f_k^\pm(r) \tag{7.15}$$

In Equations 7.13–7.15, f_k^+ and f_k^- are the electrophilic and nucleophilic Fukui functions, respectively, which can be easily obtained by a regional model reported elsewhere [57, 58].

Further significant progress has been reached by empirically defining nucleofugality and electrofugality indices by associating them with group nucleophilicity and group electrophilicity, respectively [34–37]. Fugality indices may be institutively understood using the simple Scheme 7.2.

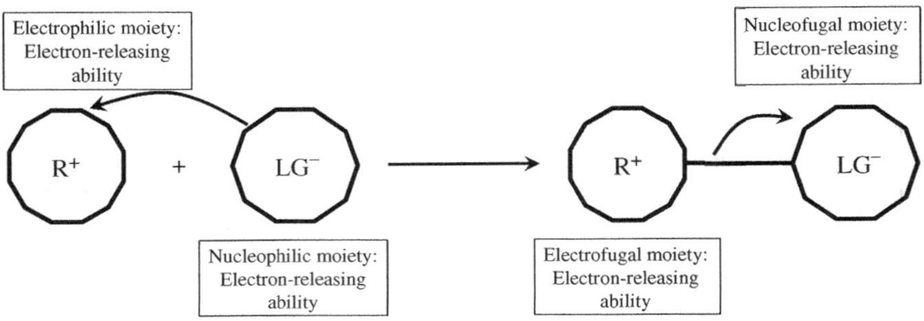

SCHEME 7.2 General heterolytic bond forming process defining the nucleofuge (leaving group, LG) and the electrofuge (permanent group, R). Adapted with permission from Ormazábal-Toledo *et al.* [34]. © (2011) American Chemical Society.

According to this scheme, the best electrofuge will be the fragment displaying the highest group nucleophilicity because during the process of heterolytic bond cleavage, the electrofuge is expected to act as an electron donor moiety. At the same time, the nucleofuge must display a high group electrophilicity to depart with the bonding electron pair [62]. In other words, while electrophilicity and nucleophilicity are properties that refer to a molecule as a whole, electrofugality and nucleofugality are essentially group properties [34]. Other models of fugacity indices have been reported [63–67]. The problem of experimentally evaluating regional properties is somehow hard to perform, and it is at this point that theory may help to assess them from theoretical models validated against observed quantities.

The definition of fugality indices entails the need for simple mathematical expressions to assess the group electrophilicity and group nucleophilicity. These definitions are based on the additivity rule for global softness S:

$$S=\int d\mathbf{r}s(\mathbf{r}) \tag{7.16}$$

Or its condensed to atom version:

$$S=\sum_k s_k^{\pm} \tag{7.17}$$

Using these properties, the definition of nucleofugality ν_G^+ and electrofugality ν_G^- of a group or molecular fragment G, say, may simply be expressed as [34, 37]

$$\nu_G^+=\sum_{k\in G}\omega_k^+ \tag{7.18}$$

and

$$\nu_G^-=\sum_{k\in G}\omega_k^- \tag{7.19}$$

The nucleofugality index has been successfully applied to rank functional groups as LG in nucleophilic substitution and elimination reactions [35–37]. Electrofugality, on the other hand, has been less experimentally studied. Mayr *et al.* defined a hierarchy of electrofuges in the study of phenylsufinate anions with a benzhydrylium cation [68]. This database was revisited to demonstrate that fugality quantities are essentially group properties and that the hierarchy of electrofuges experimentally established by Mayr's group displays a linear relationship with group nucleophilicity [68]. The resulting empirical relationship between electrofugality and group nucleophilicity is linear and reflects the inductive substituent effect, as given by Hammett's substituent constant σ [69]. A similar result was consistently obtained for the theoretical scale for nucleofuges [37]. Nucleofugality and electrofugality patterns in S_NAr reactions will be reevaluated in this chapter to quantitatively scale the LG ability of atoms or fragments for some selected examples.

7.3 PRACTICAL APPLICATIONS OF CONCEPTUAL DFT DESCRIPTORS

The direct approach for the scrutiny of reaction mechanisms is the full exploration of the PES. Within this approach, relative stability of ground states, TS, and intermediate structures can be assessed and the activation and reaction energies be quantitatively established. In the present approach, the PES is used as input information for the evaluation of reactivity indices to assess reactivity and site activation along a reaction coordinate.

In the addition–elimination mechanism, the nucleophile transfers some of its electronic charge to a vacant π^* orbital in the substrate. This interaction permits the addition of the nucleophile forming metastable species—the MC intermediate. If the attack occurs at a position of the aromatic ring substituted by a potential LG, the MC may break down yielding the substituted product, where the LG is expelled. The order of LG reactivity is still a matter of controversy, but the normally accepted trend is F>Cl>Br>I, for the halides family with different nucleophiles in protic solvents [8, 11–13]. In the same way, it is difficult to assign an empirical scale of nucleophilicity for S_NAr reactions [21, 22, 70, 71]. Nucleophilicity depends on the intrinsic capability to donate electronic charge of some bases. Additionally, nucleophilicity also depends on the solvent, the substrate, and the interactions displayed between these three bodies. The nucleophiles normally used in S_NAr reactions are alkoxides, phenoxides, sulfides, fluorides, thiocyanates, and amines.

7.3.1 Nucleophilicity and LG Scales

Nucleophilicity and LG abilities are quite difficult to assign. The study of the reactivity of several amines has been discussed from experimental and theoretical approaches. In this sense, the reaction between 1-fluoro-2,4-dinitrobenzene (FDNB) with amines of different nature in water is a suitable way to understand the main interactions between these nucleophiles in S_NAr reactions [21, 71]. The kinetic study provides a key body of data to deduce the reaction mechanism and the rate-determining step. Theoretical tools described earlier provide the complementary information to understand how a chemical reaction does occur. The analysis starts by considering three different amines, namely, propylamine, piperidine, and quinuclidine as representative neutral nucleophiles of primary, secondary, and tertiary amines, respectively. The target interaction is the potential hydrogen bond that can be formed [19, 21]. Quinuclidine is a tertiary amine and cannot establish this type of interaction. The kinetic data point out to the formation of the MC as the rate-determining step (see Section 7.4 for a deep discussion on this point) [72–74]. The reactivity trend is piperidine>propylamine>quinuclidine (or more general secondary amines>primary amines>tertiary amines). To rationalize the trends observed, a local analysis based on electrophilicity and nucleophilicity quantities was performed. A fragmentation model [18, 21, 34] (Scheme 7.3) was proposed to readily condense these indices using Equations 7.13 and 7.14. The condensed results are presented in Table 7.1.

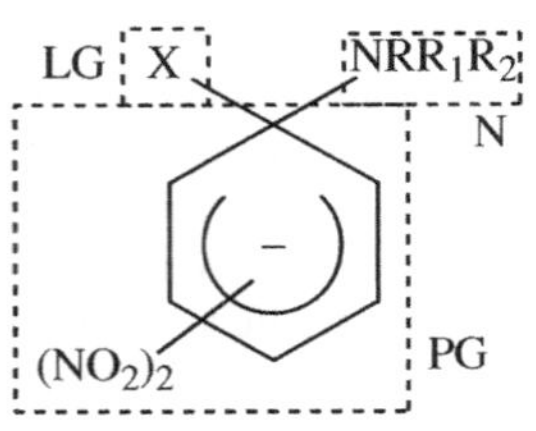

SCHEME 7.3 General fragmentation model of the electrophile–nucleophile pair. LG, PG, and N stand for leaving group, permanent group, and nucleophile. Adapted with permission from Ormazábal-Toledo *et al.* [18]. © (2013) American Chemical Society.

TABLE 7.1 Electrophilicity and Nucleophilicity Condensed into Fragments from Scheme 7.3 Evaluated at the Transition State on the PES[a]

Amine	$\omega^+(N)$	$\omega^+(LG)$	$\omega^+(PG)$	$\omega^-(N)$	$\omega^-(LG)$	$\omega^-(PG)$
Propylamine	0.00	0.01	1.09	5.02	0.11	5.19
Piperidine	0.01	0.01	1.02	7.41	0.03	2.29
Quinuclidine	0.01	0.01	0.93	7.64	0.02	1.83

[a]All values are expressed in eV and taken from Ormazabal-Toledo *et al.* [21]—Reproduced with permission from the Royal Society of Chemistry.

As shown in Table 7.1, the electrophilicity of the three fragments remains constant for all the nucleophiles in the interaction with FDNB considered. This descriptor is mainly located at the permanent group (PG) moiety showing that for all amines the value is almost the same, thereby suggesting that the activation promoted by (and toward) the amine is comparable in the three situations. On the other hand, the nucleophilicity patterns are more resolved for the series of three amines. The property condensed at the amine moiety shows that propylamine is less nucleophilic than secondary and tertiary amines [71]. Furthermore, the nucleophilicity condensed at the PG is interpreted here as the capability of the PG to transfer the electronic charge from the nucleophile to the LG, through the *ipso* carbon in the aromatic ring. Note that, in all cases, the PG is 2,4-dinitrobenzene group, but depending on the amine considered, the nucleophilicity (its capability to transfer electronic charge) may be modulated. The hypothesis is that this change in reactivity promoted by the nucleophile is conducted by the formation of an intermolecular hydrogen bond. When we tried to assess the differences between a secondary and a tertiary amine, the analysis was not sufficient because all values are closer to each other (see Table 7.1). However, the formation of the hydrogen bond promotes the charge transfer from the nucleophile to the PG. In this sense, hydrogen bond formed by the secondary amine facilitates the charge transfer to the PG. On the other hand, for the tertiary amine, there is no such hydrogen bond, thereby implying a lesser charge transfer. The usefulness of other tools like reactivity profiles and orbital interactions will be discussed in Section 7.5.1 [18, 19, 21].

The LG abilities in S_NAr are difficult to analyze using the method presented earlier. The reason is simple: according to Stirling's principle, the study of a local property must be made at the stage of reaction where the LG participates [62]. In round words, the LG ability may be observed in an advanced step of the reaction prior to the formation of the MC. To do this, a quasistatic analysis is presented in the next section.

7.3.2 Activation Properties: Reactivity Indices Profiles

The LG abilities in S_NAr has been discussed experimentally by Senger *et al.* [11] and Um *et al.* [11, 71, 72] in the reaction between 1-halo-2,4-dinitrobenzenes toward morpholine in water [11, 71, 72]. The main result shows that the LG abilities follow the order F>Cl>Br>I. To explain this outcome, reactivity profiles were obtained for the set of reactions. Reactivity profiles display the evolution of philicity and fugality patterns along a reaction path (Fig. 7.1) [18]. These profiles are useful tools to describe the charge transfer and bond-breaking/formation processes. To follow the reaction path, it is necessary to use an arbitrary fragmentation model, similar to that presented in Scheme 7.3. For a better visualization of the different stages of the reaction, Figure 7.2 displays the structures of the optimized reactants, TS, and MC. The reactivity profiles show that the electrophilicity pattern is mainly concentrated at the PG moiety (the 2,4-dinitrobenzene group). This result may be related to the capability of the ring to accept and stabilize electronic charge (Fig. 7.1a). This stabilization leads to the formation of an intermediate complex. Additionally, the electrophilicity index condensed at the LG fragment (the halogen group) reveals that in the case of iodine, the electrophilicity is dramatically enhanced (Fig. 7.1a). According to Um's proposal, iodine may detach as LG in an early stage of the reaction, prior to the formation of the MC [71]. From Equation 7.18, nucleofugality is defined as the group electrophilicity of the LG. Therefore, iodine is consistently predicted as nucleofuge at a stage prior to the intramolecular proton transfer of the second step of the reaction.

On the other hand, nucleophilicity profiles centered at the morpholine moiety (Fig. 7.1b) reveals that at the beginning of the reaction, morpholine interacts in different ways with the substrate (this point is discussed deeply in Section 7.5). Moreover, in this case, nucleophilicity reaches a minimum near the MC; for PG and LG moieties, the nucleophilicity reaches a maximum, thereby revealing

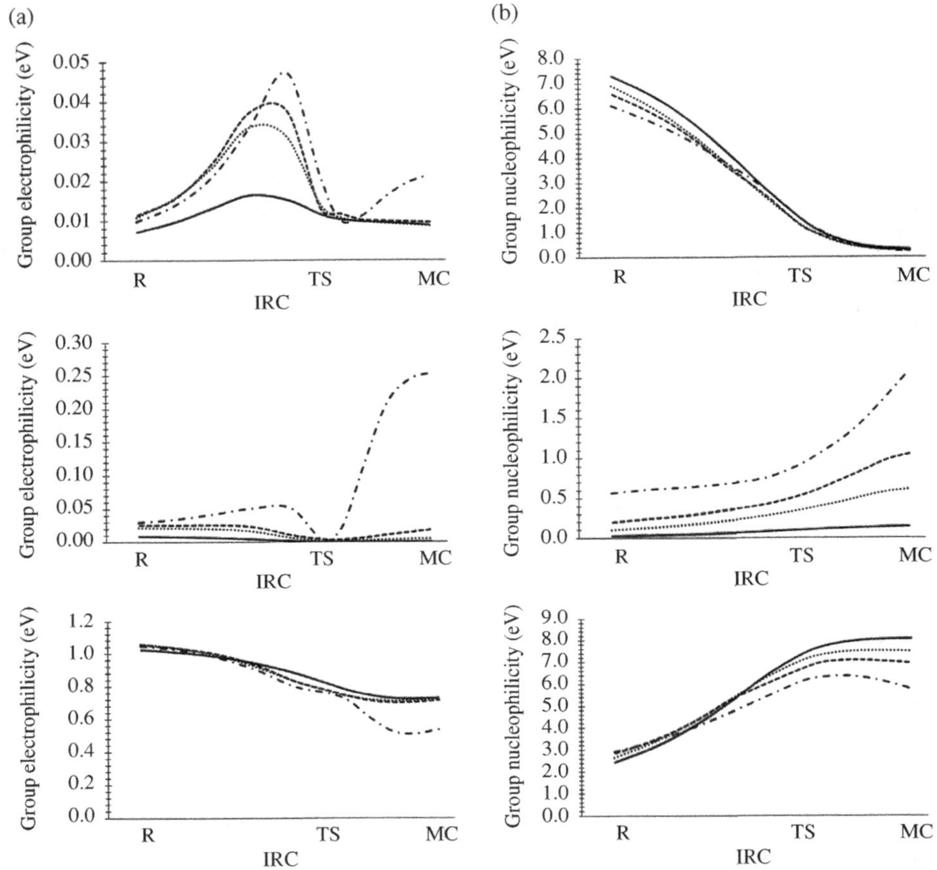

FIGURE 7.1 Reactivity profiles for the reaction between 1-halo-2,4-dinitrobenzene toward morpholine. (a) Electrophilicity profiles and (b) nucleophilicity profiles. Adapted with permission from Ormazábal-Toledo *et al.* [18]. © (2013) American Chemical Society.

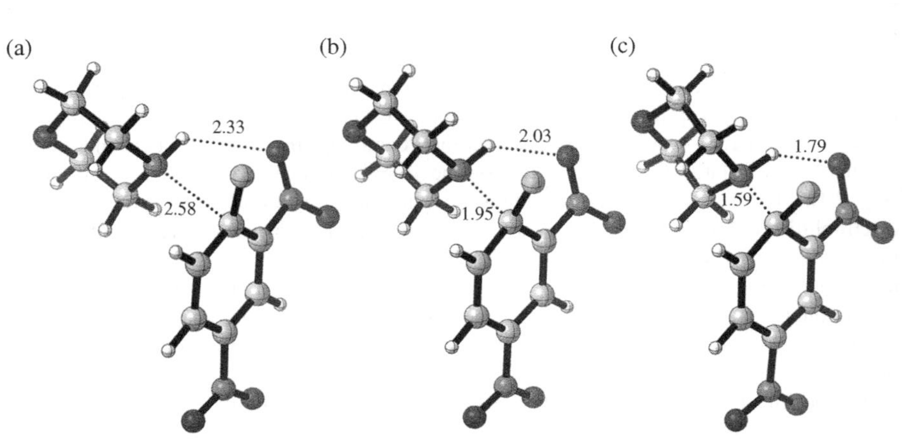

FIGURE 7.2 Optimized structures of (a) reactants, (b) transition states, and (c) Meisenheimer complex stages from the reactivity profiles in Figure 7.1. The numbers represent the distance in angstrom for the case of X=F.

the charge transfer process: the electronic charge is transferred from the nucleophile to the aromatic ring where it becomes stabilized, which is transferred to the LG assisted by the intramolecular proton transfer at the second stage of the reaction [72, 74].

7.4 S_NAr REACTION MECHANISM

Scheme 7.1 displays the commonly accepted mechanism of an S_NAr reaction that proceeds through an addition–elimination route [8, 72, 75]. In the first step, the nucleophilic attack to an activated aromatic ring leads to the formation of an anionic σ-adduct (MC). When the nucleophile is neutral, the second step of the mechanism may occur through two routes. The first one is the detachment of the LG from MC (k_2 in Scheme 7.1), followed by a fast proton transfer, yielding the substitution product. The other pathway is that the proton transfer from the MC to the LG be catalyzed by a second molecule of the neutral nucleophile (NuH), followed by a fast departure of the LG (k_3[NuH] in Scheme 7.1). It has been suggested that other channels may be operative, but in the experimental conditions considered in this chapter, they are normally discarded [72, 76]. Applying the steady-state condition to the MC intermediate in Scheme 7.1 leads to the rate law

$$v = k_N[S][\mathrm{NuH}] = k_{\mathrm{obs}}[S] \tag{7.20}$$

where [S] and [NuH] are the concentrations of the aromatic substrate and the nucleophile, respectively, and

$$k_N = \frac{k_{\mathrm{obs}}}{[\mathrm{NuH}]} = \frac{k_1(k_2 + k_3[\mathrm{NuH}])}{k_{-1} + k_2 + k_3[\mathrm{NuH}]} \tag{7.21}$$

is the rate coefficient for the nucleophilic attack.

It is worth emphasizing at this point that the general reaction mechanism depicted in Scheme 7.1 is not unequivocal, but several other reaction routes leading to the same reaction product in Scheme 7.1 can be operative—for example, it may be a concerted nucleophilic attack followed by deprotonation of the nucleophile or proton transfer assisted by the solvent and others [77]. However, for the purpose of illustrating the reliability of the reactivity indexes approach to describe reaction mechanisms, the analysis will be performed on the basis of Scheme 7.1. Additionally, Scheme 7.1 is valid only for neutral nucleophiles. It has been proposed, from computational studies, that the use of charged nucleophiles (anions) may lead to concerted routes [78], but for practical purposes, the discussion in this chapter will consider only neutral nucleophiles.

7.4.1 Kinetic Measurements

In the reaction between FDNB toward different amines in water, the kinetic data point out to the formation of the MC as the rate-determining step [21, 71, 72]. If this is the case, it follows that $k_2 + k_3[\mathrm{NuH}] >> k_{-1}$ is a result suggesting that the product formation through channels k_2 and k_3[NuH] is a faster process than the decomposition of MC to the reactants (k_{-1} channel). Replacing in Equation 7.21 yields

$$k_N = k_1 \tag{7.22}$$

and

$$k_{\mathrm{obs}} = k_1[\mathrm{NuH}] \tag{7.23}$$

TABLE 7.2 Summary of PK_a of Amines in Water and their Second-Order Rate Constants (k_1) for the Aminolysis of FDNB[a]

Nucleophiles	pK_a	k_1 (M^{-1} s^{-1})
Primary amines		
Propylamine	10.89	0.378
Glycine	9.76	0.142
Ethanolamine	9.50	0.098
Benzylamine	9.34	0.238
Hydrazine	8.10	0.411
Glycine ethyl ester	7.68	0.0267
Trifluoroethylamine	5.70	0.0017
Secondary amines		
Piperidine	11.22	9.37
Piperazine	9.82	5.48
1-(2-Hydroxyethyl)piperazine	9.38	1.43
Morpholine	8.36	1.07
1-Formylpiperazine	7.98	0.444
Piperazinium ion	5.68	0.0163
Tertiary amine		
Quinuclidine	11.4	0.0873

[a]Data taken from Ormazabal-Toledo *et al.* [21]—Reproduced with permission from the Royal Society of Chemistry.

where k_1 may be obtained from the slope of a graph between the rate coefficients of the reaction considering different amounts of nucleophile. The rate constants for the reaction considered are quoted in Table 7.2 [21].

Inspection of Table 7.2 reveals that the general trend in reactivity is SAA>PA>quinuclidine. The origin of this reactivity trends was discussed in Section 7.3.1 [21, 71, 72]. Table 7.2 includes the α-nucleophile hydrazine. This type of nucleophiles presents an enhanced reactivity than other isobasic molecules, induced by the presence of nonbonding electron pair vicinal to the nucleophilic center [79–81]. Finally, quinuclidine is the unique tertiary amine in the family considered, and it has the lowest rate coefficient, probably due to the absence of an acidic hydrogen atom capable to establish a hydrogen bond to the electrophile or to the solvent [19, 70].

In the reaction of FDNB toward SAA in MeCN, the kinetic evidence suggests that $k_2 + k_3[\mathrm{NuH}] \ll k_{-1}$ and Equation 7.21 may be simplified to

$$k_N = \frac{k_{\mathrm{obs}}}{[\mathrm{NuH}]} = \frac{k_1\left(k_2 + k_3[\mathrm{NuH}]\right)}{k_{-1}} = Kk_2 + Kk_3[\mathrm{NuH}] \quad (7.24)$$

and the observed rate coefficient becomes [72]

$$k_{\mathrm{obs}} = Kk_2[\mathrm{NuH}] + Kk_3[\mathrm{NuH}]^2 \quad (7.25)$$

where $K = k_1/k_{-1}$.

The rate coefficients are presented in Table 7.3.

As shown in Table 7.3, the rate-determining step is a combination between the catalyzed and the noncatalyzed pathways. In these systems, the catalyzed route exceeds in almost 200 times the non-catalyzed channel. This result points out to a most favored expulsion of the LG catalyzed by a second molecule of nucleophile [11, 71, 72].

TABLE 7.3 Summary of Microscopic Rate Constants for the Reactions of FDNB with SAA in MeCN at 25°C[a]

Amine	pK_a	Kk_2	Kk_3
Piperidine	18.8	111	19,000
Piperazine	18.2	54.0	16,700
1-(2-Hydroxyethyl)piperazine	17.6	7.63	1,340
1-Formylpiperazine	17.0	2.00	359
Morpholine	16.0	0.712	102

[a] Reprinted with permission from Um *et al.* [72]. © (2007) American Chemical Society.

7.4.2 Nucleophilicity, LG, and PG Abilities

7.4.2.1 Nucleophilicity and LG Abilities The reaction between FDNB and secondary amines in MeCN proceeds through a mechanism where the expulsion of the LG assisted by a second amine molecule is rate determining [72]. On the other hand, the use of primary amines in the reaction changes the slow step to the formation of the MC intermediate [71]. This result highlights the importance of the correct choice of the amine (primary or secondary) and the solvent, which determines the rate-determining step [72]. The study of other reactions in water using FDNB suggests that both families of nucleophiles have the same mechanism, being the nucleophilic attack the rate-determining step [82–84]. The differences between one solvent or another (MeCN or water) is related here to their quality as hydrogen bond donor. Water may donate a hydrogen bond to the LG, thereby assisting its expulsion more efficiently than the other amine [72]. On the other hand, primary amines fulfill this assistance in MeCN probably due to its less steric hindrance. It has been proposed that the LG may interact via hydrogen bonding to the nucleophile (see Scheme 7.4a) [71], but computational results discard this explanation [21]. In fact, the preferred interactions are those established between the nucleophile and the *ortho*-substituent (NO_2 in the present case; see Scheme 7.4b). Then, the more reactive channels promoted by primary amines would be those where the presence of an additional acidic hydrogen atom could help to considerably decrease the energy barrier of the k_3 pathway in the proposed mechanism.

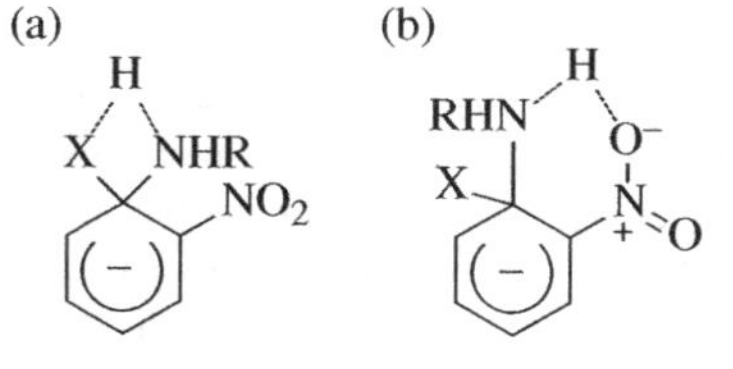

SCHEME 7.4 Two possible interactions displayed at the MC stage between the nucleophile and (a) the LG and (b) the o-NO_2 group. Reprinted with permission from Um *et al.* [71]. © (2012) American Chemical Society.

Using Mayr's equation, which relates nucleophilicity (N) and electrophilicity (E) by $\log k = s(E+N)$ using kinetic coefficients [68, 85], it may be possible to obtain an electrophilicity order for XDNB derivatives. The empirical electrophilicity numbers obtained are $E = -14.1$ (X=F), −17.6 (X=Cl and Br), and −18.3 (X=I) for the whole family of halides [11, 71]. The theoretical electrophilicity numbers obtained from Equation 7.3 are as follows: $\omega = 1.31$, 1.33, 1.30, and 1.28 for X=F, Cl, Br, and I, respectively. These results show that the electrophilicity index is not in line with the experimental figure even qualitatively. The explanation for this outcome may be traced to the absence of solvation effects that are incorporated in the solvent sensitivity parameter (s) in

Mayr's equation. Note that theoretical electrophilicities are obtained from the isolated ground state of molecules. In the next section, we consider this point by examining the electrophile–nucleophile interaction using a natural bond order population analysis.

7.4.2.2 PG Effect In nucleophilic aromatic substitution described in Scheme 7.1, the aromatic substrates need to bear good LG (normally halides or phenoxy groups). Additionally, they must be strongly activated by electron-withdrawing groups (EWG) as NO_2, CF_3, and CN in *ortho* and/or *para* positions, with respect to the LG [12, 86–88]. The presence of heteroatoms in the aromatic ring has been proposed as even more reactive substrate in different experimental conditions, namely, using water as solvents [89]. The reaction between some chlorobenzenes and chloropyridines has been studied in MeCN at 25°C (see Scheme 7.5). In all cases, the rate-determining step has been reported as the formation of the MC intermediate [12, 88].

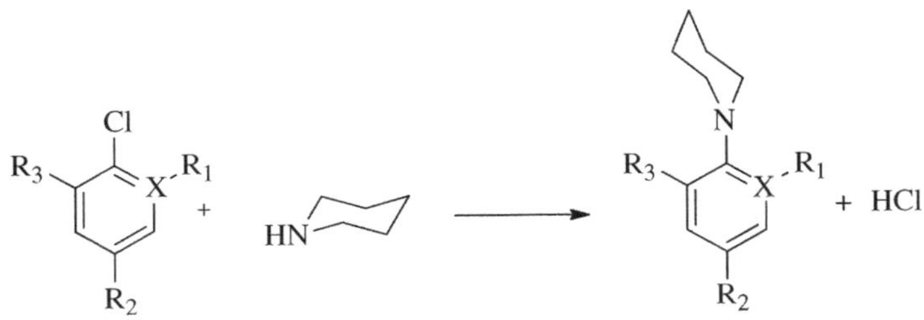

1 X = C	$R_1 = NO_2$	$R_2 = NO_2$	$R_3 = H$	$k_1 = 0.52\ M^{-1}s^{-1}$
2 X = C	$R_1 = NO_2$	$R_2 = H$	$R_3 = NO_2$	$k_1 = 0.0023\ M^{-1}s^{-1}$
3 X = N		$R_2 = NO_2$	$R_3 = H$	$k_1 = 0.0033\ M^{-1}s^{-1}$

SCHEME 7.5 Reaction of chlorobenzenes and chloropyridine derivatives toward piperidine in MeCN at 25°C. k_1 is the rate coefficient for each reaction. Data taken from Crampton *et al.* [12] and [88]—Reproduced with permission from WILEY-VCH Verlag GmbH & Co. KGaA, Weinheim.

For the reaction between 1-chloro-2,4-dinitrobenzene (**1**) and 1-chloro-2,6-dinitrobenzene (**2**), the reported rate coefficients show that the first substrate is about 200 times more reactive [12, 88]. The authors suggest that the different reactivity of both systems may be attributed to steric hindrance effects. During the nucleophilic attack, the geometry of substrate **2** will change during the approach of the nucleophile. In this sense, the resonance induced by the NO_2 in position 6 of the aromatic ring will be diminished. This effect proposed by Hammett is called steric inhibition of resonance and may be defined as the decrease of the ability of an EWG (or electro-releasing group). This effect has been traced to the loss of coplanarity of their orbitals, with respect to the π-system of the aromatic ring [90]. Both substrates show the deviation of coplanarity of the 2-NO_2 group induced by the formation of a hydrogen bond. In this case, the loss of coplanarity is compensated by the interaction of electrophile–nucleophile that facilitates the nucleophilic attack and stabilizes the MC intermediate. In the case of substrate **1**, the 4-NO_2 group is perfectly coplanar, thereby enhancing its electron-withdrawing efficiency, whereas for the 6-NO_2 group, the resonance and the electrophilic capacity of the substrate are lost.

In the case of substrates **1** and **3**, the comparison provides insights about the different reactivity with respect to a pyridine derivative [12, 88]. The reported reaction rates in MeCN reveal that when there is an NO_2 group in *ortho* position, the reaction is 20 times faster than the pyridine derivative. It is very important to stress at this point the role of the solvent as hydrogen bond donor–acceptor. The presence of a *o*-NO_2 group in the substrate inhibits the formation of hydrogen bonds between the solvent and the substrate, due to the *built-in solvation* effect [19, 86, 87]. Nevertheless, in pyridines and other heterocycles, it may be possible to observe these interactions [89]. This result may be attributed to hydrogen bond formation between the substrate and the nucleophile mediated by the solvent. In the reactions in Scheme 7.5, the rate coefficients were obtained in MeCN. In this case, the hydrogen bonding will be less efficient because the reaction media do not provide the necessary hydrogen to establish the interaction. This observation may be used to better understand the intrinsic reactivity of pyridines in S_NAr reactions.

7.5 INTEGRATED EXPERIMENTAL AND THEORETICAL MODELS

The studies summarized at this point reveal that a static analysis that considers just the electronic properties of the aromatic substrates or the nucleophile as infinitely separated molecules does not give substantial information about the global and local reactivity and their impact in the reaction mechanism. For this reason, more sophisticated analysis has been presented along the intrinsic reaction coordinate that explains in some cases the reactivity of aromatic substrates toward neutral nucleophiles. An additional proposal is presented in this section, which includes the orbital interaction analysis in selected cases between the electrophile and the nucleophile.

7.5.1 Hydrogen Bonding Effects

It has been reported that the presence of a NO_2 group in *ortho* position with respect to LG in aromatic substrates leads to more reactive systems in S_NAr reactions [12, 86–88, 91]. The electrophile–nucleophile interactions may be easily addressed by using the second-order perturbation theory [92, 93]. In this approach, the interaction between donor and acceptor Lewis-type orbitals measures the deviation of an idealized Lewis structure may be quantified. Normally, the interactions between a nonbonding Lewis orbital and an antibonding orbital are the most important contributions to the nucleophile/electrophile pair interaction. For instance, in the reaction between XDNB and morpholine, two interactions may be responsible for the high reactivity. The interactions are better represented in Scheme 7.6 and quantified in Table 7.4.

Table 7.4 shows that in all cases, the main differences are established at the MC stage because the distance between the nucleophile and the electrophile allows a high overlap between the orbitals. The interaction between the oxygen lone pairs in the o-NO_2 and the N—H antibonding orbital outweighs the interaction with the LG lone pairs. Even though in the four cases the interaction is the same, the values are dependent on the LG present in the substrate. The $E^{(2)}$ value diminishes downward the family of halides, a result attributed to a different stabilization along the reaction path [18]. Additionally, the interaction between the LG and the N—H antibonding orbital is marginal (Scheme 7.6c), thereby refuting the proposal of Um about the LG abilities [11, 71]. The LG ability may be explained on the basis of the stabilizing effect induced by the hydrogen bond at the TS and MC stages [18, 19, 70].

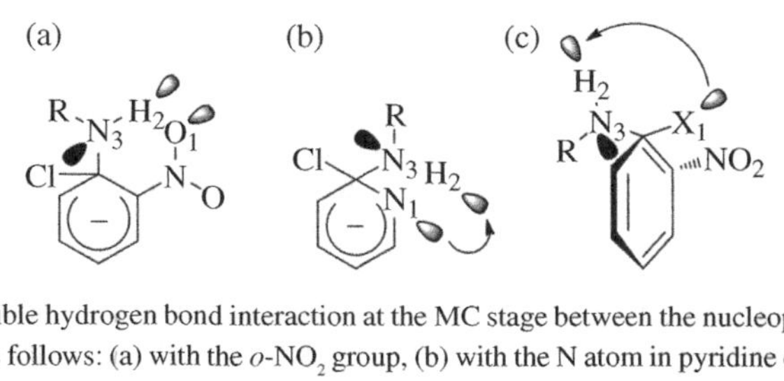

SCHEME 7.6 Possible hydrogen bond interaction at the MC stage between the nucleophile and the substrate. The interactions are as follows: (a) with the o-NO_2 group, (b) with the N atom in pyridine derivative, and (c) with the LG. Atom numbering follows the order of interaction quoted in Table 7.4. Adapted with permission from Ormazábal-Toledo *et al.* [18], © (2013) American Chemical Society; and Ormazábal-Toledo *et al.* [21], reproduced with permission from the Royal Society of Chemistry.

TABLE 7.4 Second-Order Perturbation Theory Energies for the Substrates in Scheme 7.5 at the Reactants, TS and MC Intermediate Stages of the Reaction toward Piperidine

			$E^{(2)}$ (kcal/mol)		
Substrate	Donor	Acceptor	R	TS	MC
1	*LP O1*	$BD^* N_3\text{–}H_2$	0.8	4.5	7.6
2	*LP O1*	$BD^* N_3\text{–}H_2$	1.7	8.6	12.3
3	*LP N1*	$BD^* N_3\text{–}H_2$	~0	~0	~0

From Ormazabal-Toledo *et al.* [21]—Reproduced with permission from the Royal Society of Chemistry.

In the reaction between 2-chloro-5-nitropyridine, the observed reactivity may be also explained using the perturbation theory approach. The interaction between the lone pair orbital in the N of pyridine with the nucleophile is less than the interaction with the *o*-NO_2 group (see Scheme 7.6a and 7.6c), thereby reinforcing the preceding discussion. This result suggests that the interaction between the N of pyridine and the amine does not assist the nucleophilic attack in MeCN [12, 88]. On the other hand, in the ClDNB substrate, the interaction is almost 5 kcal/mol greater, suggesting that the presence of an NO_2 group may enhance the reactivity of the system by forming a more efficient hydrogen bonding between the nucleophile and the substrate.

7.6 SOLVENT EFFECTS IN CONVENTIONAL SOLVENTS AND IONIC LIQUIDS

7.6.1 Preferential Solvation

Preferential solvation may be defined as the different composition of the internal and external shell of solvent interacting with a solute, with respect to different components of the solvent (usually binary mixtures) [94–98]. The external shell is commonly treated as the "bulk of the solvent" and may be described using long-range forces described by dielectric model. The effect of the bulk may be simply addressed using the classic theories of Kirkwood–Onsager or more sophisticated approaches as quantum mechanical models of solvation based on the reaction field theory [99–104]. In this sense, the preferential solvation may be approached as a solute–solvent interaction that represents local interactions associated with the first shell of solvation. Preferential solvation of 1-halo-2,4-dinitrobenzenes has been studied by Mancini and coworkers in dichloromethane with different polar protic and aprotic cosolvents [82, 83, 105–109]. In this section, we present a simple model to analyze preferential solvation in the reaction between phenyl-2,4,6-trinitrophenylether with morpholine and piperidine as representative SAA in water/ethanol mixtures [110]. It has been found that only piperidine is sensitive to preferential solvation because at high proportions of water, the rate coefficients increase, thereby suggesting an enhanced stabilization of the MC intermediate. On the other hand, other amines have the same rate constants for the whole series of ethanol/water mixtures. The solvent effect observed in the kinetic of the reaction with piperidine may be attributed to a different molecular interaction between the piperidine and the solvent molecules [111]. These results suggest that the MC environment changes for the set of mixtures. Nevertheless, for the remaining amines, the first shell is similar for varying compositions of the reaction media. This result may be due to the polar nature of the substituent at position 4 of the family of amines. This result suggests that the kinetic responses of the family of amines, but piperidine, depend only on the bulk induced by the solvent. In the case of piperidine, the kinetic behavior depends on the preferential hydrogen bonds displayed by the presence of more water molecules in the first shell.

7.6.2 Ionic Liquids and Catalysis

Ionic liquid (ILs) are fluids formed just by ions, with melting points below the melting point of water [112, 113]. They were considered as "green solvents" due to their properties of low vapor pressure, thermal stability, and recyclability [114, 115]. Nevertheless, in recent years, their greenery is under controversy due the presence of toxic groups or atoms as -CN, -F, -SCN, etc. [116]. ILs are normally composed by bulky organic cations (imidazolium or pyridinium derivatives) substituted with alkyl chains and an organic or inorganic counterion. As the combinatory of potential cations and anions is virtually infinite, ILs are considered as designer solvents because appropriate IL for specific reactions (task specific solvents) may be designed [112, 114, 117]. Welton and coworkers have studied the reaction between *p*-fluoronitrobenzene with *p*-anisidine in conventional solvents and in ILs [118]. The use of conventional solvents and the common ILs as $[C_4C_1Im][BF_4]$, $[C_4C_1Im][OTf]$, $[C_4C_1Im][NTf_2]$, and others based on anions from poor Brønsted acids leads to low yields (in general, <5%). The design of an IL capable to improve the reaction was done based on the fact that nucleophiles in S_NAr must be able to donate its acidic hydrogen at the nucleophilic atom more easily. If this is the case, the reaction will be faster because the LG may detach faster, rearomatizing the MC. Based on this hypothesis, four ILs based on the anions CF_3COO^- and $CH_3SO_3^-$ were synthesized resulting in a yield of the reaction near to 70%. That is, the design of an adequate IL allows increasing yields above the best conventional solvent [118].

Solvation effects promoted by ILs were recently studied by Gazitua *et al.* [119]. The reaction of 2,4-dinitrophenylsulfonyl chloride (DNBSCl) with piperidine was kinetically investigated for a set of 21 conventional solvents (including MeCN, benzene, acetone, water, ethanol, etc.) and 17 ILs. It was found that solvent polarity modulates the reaction path differently. The study developed in conventional solvents revealed that the ability of the solvent to establish hydrogen bonds drives the S_NAr process following a pathway where the formation of the MC is rate determining (see Scheme 7.1). The use of some solvents, as water and formamide, leads to a nucleophilic activation at the nitrogen atom in piperidine, as discussed in similar cases by other authors [70, 75, 120]. Reaction rates in ILs largely exceed various conventional solvents, even water. The IL $[C_2C_1IM]$ [DCN] based on the dicyanamide anion presents the best performance for S_NAr, a result probably due to the high polarizability of dicyanamide anion, as suggested by Welton [118, 121].

7.7 SUMMARY AND OUTLOOK

Theoretical aspects of the S_NAr mechanism have been described including solvent effects. The set of global and regional reactivity indices defined within the frame of conceptual DFT provide a complete body of concepts suitable to rationalize and to predict different aspects of S_NAr reactions. Nucleophiles and electrophiles are successfully described with the aid of the global electrophilicity and nucleophilicity indices, allowing their intrinsic reactivity to be described on quantitative basis. Their regional counterparts in turn help in quantitatively describing the electrofugality and nucleofugality patterns of PG and LG, respectively. Solvent effects are introduced using explicit solvent molecules, and their roles on mechanistic aspects of S_NAr reactions are highlighted using hydrogen bond acidity and basicity. This approach reveals a third ingredient of chemical reactivity—namely, site activation–deactivation responses. The electronic molecular responses are accounted for, using NBO techniques including second-order perturbation theory analysis. The future research in this field will be significantly devoted to specific solvation by the new generation of ILs, because hydrogen bonding has been shown here to participate by increasing the electrophilicity and nucleophilicity of reagents. New ILs may then be designed and prepared to improve the reactivity in S_NAr processes by modulating their respective Lewis hydrogen bond basicity and acidity.

ABBREVIATIONS

$[C_4C_1Im][BF_4]$	1-Butyl-3-methylimidazolium tetrafluoroborate
$[C_4C_1Im][NTf_2]$	1-Butyl-3-methylimidazolium bis((trifluoromethyl)sulfonyl)imide
$[C_4C_1Im][OTf]$	1-Butyl-3-methylimidazolium trifluoromethanesulfonate
BrDNB	1-Bromo-2,4-dinitrobenzene
ClDNB	1-Chloro-2,4-dinitrobenzene
DFT	Density functional theory
DNBSCl	2,4-Dinitrophenylsulfonyl chloride
EWG	Electron-withdrawing group
FDNB	1-Fluoro-2,4-dinitrobenzene
HOMO	Highest occupied molecular orbital
IDNB	1-Iodo-2,4-dinitrobenzene
IL	Ionic liquid
LG	Leaving group
LUMO	Lowest unoccupied molecular orbital
MC	Meisenheimer complex
MeCN	Acetonitrile
NBO	Natural bond orbital
NuH	Neutral nucleophile
PES	Potential energy surface
PG	Permanent group
SAA	Secondary alicyclic amines
S_NAr	Nucleophilic aromatic substitution
TS	Transition state
XDNB	1-Halo-2,4-dinitrobenzene

REFERENCES

[1] Buncel, E., Norris, A. R., and Russell, K. E. (1968) *Q. Rev. Chem. Soc.*, **22**, 123–146.

[2] Buck, P. (1969) *Angew. Chem. Int. Ed.*, **8**, 120–131.

[3] Bunnett, J. F. (1993) *Tetrahedron*, **49**, 4477–4484.

[4] Zoltewicz, J. (1975) *Topics in Current Chemistry, Organic Syntheses*; Springer: Berlin Heidelberg: Vol. **59**.

[5] Miller, J. (1968) *Aromatic Nucleophilic Substitution*; Elsevier Pub. Co.: Amsterdam.

[6] Bernasconi, C. F. (1978) *Acc. Chem. Res.*, **11**, 147–152.

[7] Bunnett, J. F. (1974) *J. Chem. Educ.*, **51**, 312–315.

[8] Bunnett, J. F. and Zahler, R. E. (1951) *Chem. Rev.*, **49**, 273–412.

[9] Artamkina, G. A., Egorov, M. P., and Beletskaya, I. P. (1982) *Chem. Rev.*, **82**, 427–459.

[10] Bernasconi, C. F. and De Rossi, R. H. (1976) *J. Org. Chem.*, **41**, 44–49.

[11] Senger, N. A., Bo, B., Cheng, Q., Keeffe, J. R., Gronert, S., and Wu, W. (2012) *J. Org. Chem.*, **77**, 9535–9540.

[12] Crampton, M. R., Emokpae, T. A., and Isanbor, C. (2007) *Eur. J. Org. Chem.*, **2007**, 1378–1383.

[13] Crampton, M. R., Emokpae, T. A., Howard, J. A. K., Isanbor, C., and Mondal, R. (2004) *J. Phys. Org. Chem.*, **17**, 65–70.

[14] Beck, J. R. (1978) *Tetrahedron*, **34**, 2057–2068.

[15] Emokpae, T. A. and Atasie, N. V. (2005) *Int. J. Chem. Kinet.*, **37**, 744–750.

[16] Mancini, P. M., Fortunato, G. G., and Vottero, L. R. (2004) *J. Phys. Org. Chem.*, **17**, 138–147.

[17] Bernasconi, C. F., De Rossi, R. H., and Schmid, P. (1977) *J. Am. Chem. Soc.*, **99**, 4090–4101.

[18] Ormazábal-Toledo, R., Contreras, R., and Campodónico, P. R. (2013) *J. Org. Chem.*, **78**, 1091–1097.

[19] Chéron, N., El Kaïm, L., Grimaud, L., and Fleurat-Lessard, P. (2011) *Chem. Eur. J.*, **17**, 14929–14934.

[20] Jacobsson, M., Oxgaard, J., Abrahamsson, C. O., Norrby, P. O., Goddard Iii, W. A., and Ellervik, U. (2008) *Chem. Eur. J.*, **14**, 3954–3960.

[21] Ormazabal-Toledo, R., Contreras, R., Tapia, R. A., and Campodonico, P. R. (2013) *Org. Biomol. Chem.*, **11**, 2302–2309.

[22] Fernández, I., Frenking, G., and Uggerud, E. (2010) *J. Org. Chem.*, **75**, 2971–2980.

[23] Acevedo, O. and Jorgensen, W. L. (2004) *Org. Lett.*, **6**, 2881–2884.

[24] Parr, R. G. and Yang, W. (1994) *Density-Functional Theory of Atoms and Molecules*; Oxford University Press: New York.

[25] Parr, R. G., Donnelly, R. A., Levy, M., and Palke, W. E. (1977) *J. Chem. Phys.*, **68**, 3801–3807.

[26] Geerlings, P., De Proft, F., and Langenaeker, W. (2003) *Chem. Rev.*, **103**, 1793–1873.

[27] Parr, R. G. and Pearson, R. G. (1983) *J. Am. Chem. Soc.*, **105**, 7512–7516.

[28] Pearson, R. G. (1963) *J. Am. Chem. Soc.*, **85**, 3533–3539.

[29] Parr, R. G., Szentpály, L. V., and Liu, S. (1999) *J. Am. Chem. Soc.*, **121**, 1922–1924.

[30] Chattaraj, P. K., Sarkar, U., and Roy, D. R. (2006) *Chem. Rev.*, **106**, 2065–2091.

[31] Cedillo, A., Contreras, R., Galván, M., Aizman, A., Andrés, J., and Safont, V. S. (2007) *J. Phys. Chem. A*, **111**, 2442–2447.

[32] Uggerud, E. (2006) *Chem. Eur. J.*, **12**, 1127–1136.

[33] Contreras, R., Andres, J., Safont, V. S., Campodonico, P., and Santos, J. G. (2003) *J. Phys. Chem. A*, **107**, 5588–5593.

[34] Ormazábal-Toledo, R., Campodónico, P. R., and Contreras, R. (2011) *Org. Lett.*, **13**, 822–824.

[35] Campodónico, P. R., Ormazábal-Toledo, R., Aizman, A., and Contreras, R. (2010) *Chem. Phys. Lett.*, **498**, 221–225.

[36] Campodónico, P. R., Andrés, J., Aizman, A., and Contreras, R. (2007) *Chem. Phys. Lett.*, **439**, 177–182.

[37] Campodonico, P. R., Aizman, A., and Contreras, R. (2006) *Chem. Phys. Lett.*, **422**, 340–344.

[38] Aurell, M. J., Domingo, L. R., Pérez, P., and Contreras, R. (2004) *Tetrahedron*, **60**, 11503–11509.

[39] Domingo, L. R. (2000) *Eur. J. Org. Chem.* 2265–2272.

[40] Contreras, R., Domingo, L. R., Andrés, J., Pérez, P., and Tapia, O. (1999) *J. Phys. Chem. A*, **103**, 1367–1375.

[41] Domingo, L. R., Arnó, M., and Andrés, J. (1999) *J. Org. Chem.*, **64**, 5867–5875.

[42] Elango, M., Parthasarathi, R., Subramanian, V., Sarkar, U., and Chattaraj, P. K. (2005) *J. Mol. Struct. THEOCHEM.*, **723**, 43–52.

[43] Ormazábal-Toledo, R., Contreras, R., and Campodónico, P. R. (2013) *J. Org. Chem.*, **78**, 1091–1097.

[44] Chattaraj, P. K., Gutiérrez-Oliva, S., Jaque, P., and Toro-Labbé, A. (2003) *Mol. Phys.*, **101**, 2841–2853.

[45] Kar, T., Scheiner, S., and Sannigrahi, A. B. (1998) *J. Phys. Chem. A*, **102**, 5967–5973.

[46] Kar, T. and Scheiner, S. (1995) *J. Phys. Chem.*, **99**, 8121–8124.

[47] Ayers, P. W. and Levy, M. (2000) *Theor. Chem. Accounts*, **103**, 353–360.

[48] Fuentealba, P. and Cedillo, A. (1999) *J. Chem. Phys.*, **110**, 9807–9811.

[49] Mulliken, R. S. (1935) *J. Chem. Phys.*, **3**, 573–585.

[50] Mulliken, R. S. (1934) *J. Chem. Phys.*, **2**, 782–793.

[51] Koopmans, T. (1934) *Physica*, **1**, 104–113.

[52] Janak, J. F. (1978) *Phys. Rev. B*, **18**, 7165–7168.

[53] Yang, W. and Parr, R. G. (1985) *Proc. Natl. Acad. Sci. U. S. A.*, **82**, 6723–6726.

[54] Parr, R. G. and Chattaraj, P. K. (1991) *J. Am. Chem. Soc.*, **113**, 1854–1855.

[55] Pearson, R. G. (1998) *Chemical Hardness: Applications from Molecules to Solids*; John Wiley & Sons: Weinheim.

[56] Maynard, A. T., Huang, M., Rice, W. G., and Covell, D. G. (1998) *Proc. Natl. Acad. Sci. U. S. A.*, **95**, 11578–11583.

[57] Fuentealba, P., Pérez, P., and Contreras, R. (2000) *J. Chem. Phys.*, **113**, 2544–2551.

[58] Contreras, R. R., Fuentealba, P., Galván, M., and Pérez, P. (1999) *Chem. Phys. Lett.*, **304**, 405–413.

[59] Li, Y. and Evans, J. N. S. (1995) *J. Am. Chem. Soc.*, **117**, 7756–7759.

[60] Fukui, K., Yonezawa, T., and Shingu, H. (1952) *J. Chem. Phys.*, **20**, 722–725.

[61] Yang, W. and Mortier, W. J. (1986) *J. Am. Chem. Soc.*, **108**, 5708–5711.

[62] Stirling, C. J. M. (1979) *Acc. Chem. Res.*, **12**, 198–203.

[63] Anderson, J. S. M., Liu, Y., Thomson, J. W., and Ayers, P. W. (2010) *J. Mol. Struct. THEOCHEM.*, **943**, 168–177.

[64] Ayers, P. W., Anderson, J. S. M., Rodriguez, J. I., and Jawed, Z. (2005) *Phys. Chem. Chem. Phys.*, **7**, 1918–1925.

[65] Gineityte, V. (2007) *J. Mol. Struct. THEOCHEM.*, **810**, 91–101.

[66] Jaramillo, P., Domingo, L. R., and Pérez, P. (2006) *Chem. Phys. Lett.*, **420**, 95–99.

[67] Streidl, N., Denegri, B., Kronja, O., and Mayr, H. (2010) *Acc. Chem. Res.*, **43**, 1537–1549.

[68] Mayr, H. and Ofial, A. R. (2005) *Pure Appl. Chem.*, **77**, 1807–1821.

[69] Hansch, C., Leo, A., and Taft, R. W. (1991) *Chem. Rev.*, **91**, 165–195.

[70] Alvaro, C. E. S., Ayala, A. D., and Nudelman, N. S. (2011) *J. Phys. Org. Chem.*, **24**, 101–109.

[71] Um, I. H., Im, L. R., Kang, J. S., Bursey, S. S., and Dust, J. M. (2012) *J. Org. Chem.*, **77**, 9738–9746.

[72] Um, I. H., Min, S. W., and Dust, J. M. (2007) *J. Org. Chem.*, **72**, 8797–8803.

[73] Isanbor, C. and Babatunde, A. I. (2009) *J. Phys. Org. Chem.*, **22**, 1078–1085.

[74] Mancini, P. M., Fortunate, G. G., and Vottero, L. R. (2005) *J. Phys. Org. Chem.*, **18**, 336–346.

[75] Banjoko, O. and Babatunde, I. A. (2004) *Tetrahedron*, **60**, 4645–4654.

[76] Bunnett, J. F. (1958) *Q. Rev. Chem. Soc.*, **12**, 1–16.

[77] Renfrew, A. H. M., Taylor, J. A., Whitmore, J. M. J., and Williams, A. (1993) *J. Chem. Soc. Perkin Trans.* **2**, 1703–1704.

[78] Liljenberg, M., Brinck, T., Herschend, B., Rein, T., Tomasi, S., and Svensson, M. (2012) *J. Org. Chem.*, **77**, 3262–3269.

[79] Edwards, J. O. and Pearson, R. G. (1962) *J. Am. Chem. Soc.*, **84**, 16–24.

[80] Ren, Y. and Yamataka, H. (2007) *J. Org. Chem.*, **72**, 5660–5667.

[81] Buncel, E. and Um, I. H. (2004) *Tetrahedron*, **60**, 7801–7825.

[82] Mancini, P. M. E., Terenzani, A., Adam, C., Pérez, A., and Vottero, L. R. (1999) *J. Phys. Org. Chem.*, **12**, 207–220.

[83] Mancini, P. M. E., Terenzani, A., Adam, C., and Vottero, L. R. (1999) *J. Phys. Org. Chem.*, **12**, 430–440.

[84] Mancini, P. M., Fortunato, G., Adam, C., Vottero, L. R., and Terenzani, A. J. (2002) *J. Phys. Org. Chem.*, **15**, 258–269.

[85] Mayr, H. and Patz, M. (1994) *Angew. Chem. Int. Ed.*, **33**, 938–957.

[86] Bunnett, J. F., Morath, R. J., and Okamoto, T. (1955) *J. Am. Chem. Soc.*, **77**, 5055–5057.

[87] Bunnett, J. F. and Morath, R. J. (1955) *J. Am. Chem. Soc.*, **77**, 5051–5055.

[88] Crampton, M. R., Emokpae, T. A., Isanbor, C., Batsanov, A. S., Howard, J. A. K., and Mondal, R. (2006) *Eur. J. Org. Chem.*, 1222–1230.

[89] Joule, J. A. and Mills, K. (2010) *Heterocyclic Chemistry*; John Wiley & Sons, Ltd: Chichester.

[90] Hammett, L. P. and Paul, M. A. (1934) *J. Am. Chem. Soc.*, **56**, 827–829.

[91] D'Anna, F., Frenna, V., Noto, R., Pace, V., and Spinelli, D. (2006) *J. Org. Chem.*, **71**, 5144–5150.

[92] Reed, A. E., Curtiss, L. A., and Weinhold, F. (1988) *Chem. Rev.*, **88**, 899–926.

[93] Reed, A. E. and Weinhold, F. (1985) *J. Chem. Phys.*, **83**, 1736–1740.

[94] Ben-Naim, A. (1990) *Pure Appl. Chem.*, **62**, 25–34.

[95] Covington, A. K. and Newman, K. E. (1979) *Pure Appl. Chem.*, **51**, 2041–2058.

[96] Langford, C. H. and Tong, J. P. K. (1977) *Acc. Chem. Res.*, **10**, 258–264.

[97] Petrov, N. K., Wiessner, A., and Staerk, H. (2001) *Chem. Phys. Lett.*, **349**, 517–520.

[98] Salari, H., Khodadadi-Moghaddam, M., Harifi-Mood, A. R., and Gholami, M. R. (2010) *J. Phys. Chem. B*, **114**, 9586–9593.

[99] Kirkwood, J. G. (1934) *J. Chem. Phys.*, **2**, 351–361.

[100] Onsager, L. (1936) *J. Am. Chem. Soc.*, **58**, 1486–1493.

[101] Tapia, O. and Goscinski, O. (1975) *Mol. Phys.*, **29**, 1653–1661.

[102] Tomasi, J., Mennucci, B., and Cammi, R. (2005) *Chem. Rev.*, **105**, 2999–3093.

[103] Miertš, S., Scrocco, E., and Tomasi, J. (1981) *Chem. Phys.*, **55**, 117–129.

[104] Klamt, A., Mennucci, B., Tomasi, J., Barone, V., Curutchet, C., Orozco, M., and Luque, F. J. (2009) *Acc. Chem. Res.*, **42**, 489–492.

[105] Mancini, P. M. E., Terenzani, A., Adam, C., Pérez, A. D. C., and Vottero, L. R. (1999) *J. Phys. Org. Chem.*, **12**, 713–724.

[106] Mancini, P. M. E., Terenzani, A., Adam, C., and Vottero, L. R. (1997) *J. Phys. Org. Chem.*, **10**, 849–860.

[107] Mancini, P. M. E., Terenzani, A., Gasparri, M. G., and Vottero, L. R. (1996) *J. Phys. Org. Chem.*, **9**, 459–470.

[108] Martinez, R. D., Mancini, P. M. E., Vottero, L. R., and Nudelman, N. S. (1986) *J. Chem. Soc. Perkin Trans.* **2**, 1427–1431.

[109] Nudelman, N. S., Mancini, P. M. E., Martinez, R. D., and Vottero, L. R. (1987) *J. Chem. Soc. Perkin Trans.* **2**, 951–954.

[110] Ormazabal-Toledo, R., Santos, J. G., Ríos, P., Castro, E. A., Campodónico, P. R., and Contreras, R. (2013) *J. Phys. Chem. B*, **117**, 5908–5915.

[111] McClelland, R. A., Kanagasabapathy, V. M., Banait, N. S., and Steenken, S. (1992) *J. Am. Chem. Soc.*, **114**, 1816–1823.

[112] Hallett, J. P. and Welton, T. (2011) *Chem. Rev.*, **111**, 3508–3576.

[113] Welton, T. (1999) *Chem. Rev.*, **99**, 2071–2083.

[114] Earle, M. J. and Seddon, K. R. (2000) *Pure Appl. Chem.*, **72**, 1391–1398.

[115] Rogers, R. D. and Seddon, K. R. (2003) *Science*, **302**, 792–793.

[116] Swatloski, R. P., Holbrey, J. D., and Rogers, R. D. (2003) *Green Chem.*, **5**, 361–363.

[117] Seddon, K. R. (1997) *J. Chem. Technol. Biotechnol.*, **68**, 351–356.

[118] Newington, I., Perez-Arlandis, J. M., and Welton, T. (2007) *Org. Lett.*, **9**, 5247–5250.

[119] Gazitua, M., Tapia, R. A., Contreras, R., and Campodonico, P. R. (2014) *New J. Chem.*, **38**, 2611–2618.

[120] Ji, P., Atherton, J. H., and Page, M. I. (2011) *J. Org. Chem.*, **76**, 3286–3295.

[121] Cerda-Monje, A., Ormazábal-Toledo, R., Cárdenas, C., Fuentealba, P., and Contreras, R. (2014) *J. Phys. Chem. B*, **118**, 3696–3701.

8

ASYMMETRIC NUCLEOPHILIC AROMATIC SUBSTITUTION

Anne-Sophie Castanet, Anne Boussonnière, and Jacques Mortier
Institut des Molécules et Matériaux du Mans, Faculté des Sciences et Techniques, UMR CNRS 6283, Université du Maine and CNRS, Le Mans Cedex, France

8.1 INTRODUCTION

Although the nucleophilic aromatic substitution of electron-deficient arenes with nucleophiles has been known for more than 100 years, to the best of our knowledge, there is no review on the asymmetric version of this reaction [1, 2]. Within the context of this chapter, we have classified the main approaches involving asymmetric nucleophilic aromatic substitution reactions (S_NAr^*). We detail the current mechanistic understanding based on experimental and computational studies, and we summarize a selection of representative synthetic applications.

In an S_NAr^* process, a nucleophile reacts stereoselectively with an electron-deficient arene activated by one or more EWGs located *ortho* and/or *para* to a leaving group (L), creating a new C—C, C—O, or C—N bond (Fig. 8.1). The reaction provides access to a variety of aromatic derivatives with axial, central, and planar chirality.

Atropisomerism is significant because it introduces an element of chirality in the absence of stereogenic atoms. Axial chirality is observed with stereoisomers (or atropisomers) that result from hindered rotation about a single C—C or C—N bond. The barrier of rotation between atropisomers must be high enough to allow for their isolation. A minimum of three *ortho*-substituents are generally required for an axially chiral biphenyl to have substantial stability toward racemization at room temperature. For general definitions and descriptions, see references [3–5].

Within the realm of natural products and bioactive compounds, the tri- or tetra-*ortho*-substituted biaryl subunit is by far the most common form of atropisomerism. The S_NAr^* reaction allows the formation of the backbone of important drugs with diverse biological activities such as (−)-steganone (**1**), a dibenzocyclooctadienyl lignan lactone exhibiting antileukemic activity [6]; spirooxoindolines

Arene Chemistry: Reaction Mechanisms and Methods for Aromatic Compounds, First Edition. Edited by Jacques Mortier.

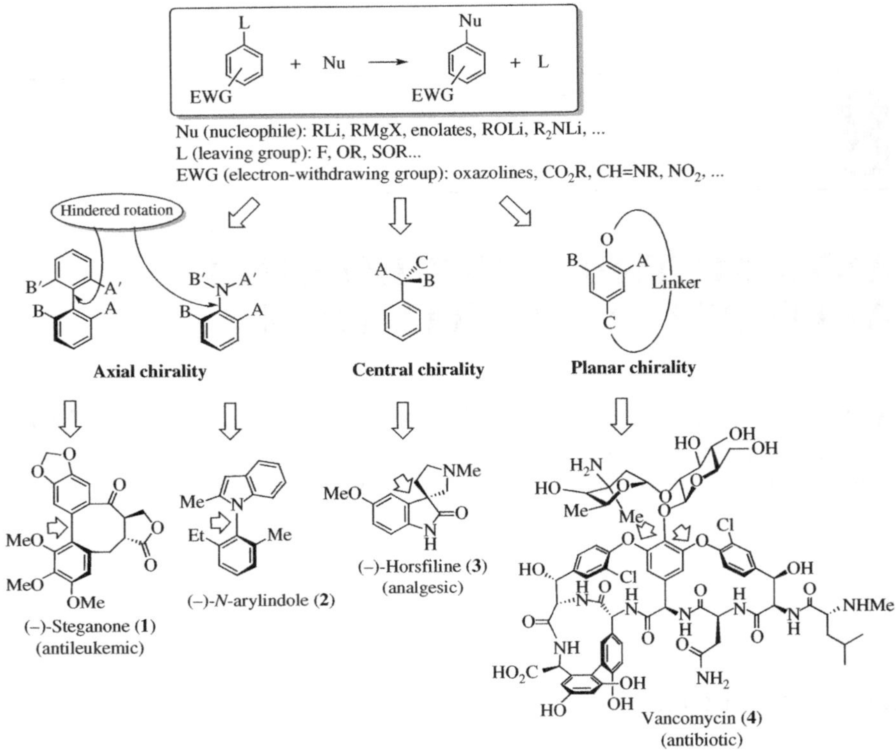

FIGURE 8.1 The S_NAr^* process.

derived from (–)-horsfiline (**3**), an alkaloid found in traditional herbal medicine [7, 8]; and cyclophane vancomycin (**4**), an exceedingly complex glycopeptide antibiotic consisting of three interlocking cyclic tripeptides having a conformationally rigid cup-shaped structure [9]. Enantiomerically pure compounds having an axially chiral C—N bond, like *N*-arylindole **2**, are of potential importance as chiral ligands and auxiliaries in asymmetric reactions and as intermediates for the preparation of natural products [10, 11].

The S_NAr^* reaction is circumscribed by the major mechanisms of nucleophilic aromatic substitution: the most common S_NAr addition–elimination (Chapter 6), the oxidative nucleophilic aromatic substitution of hydrogen (ONSH; Chapter 11), the benzyne mechanism (Chapter 12), and the radical-nucleophilic aromatic substitution $S_{RN}1$ (Chapter 10). The chiral inductor can be bonded (Section 8.2) or nonbonded to the reactants (Section 8.3) (Scheme 8.1).

The most popular S_NAr^* approach described so far relies on the use of chiral electron-withdrawing groups (EWGs*) (Section 8.2.1). The major breakthrough in this area is due to Meyers who developed a general access to axially chiral biaryls through the use of chiral aryloxazolines (the Meyers reaction). The nucleophile, usually a Grignard reagent, displaces the achiral leaving group L, which is generally an alkoxy group. Chiral esters and sulfonamides can take the place of the oxazolinyl. Stereocontrol can also result from the use of chiral alkoxide and sulfoxide leaving groups L* (Section 8.2.2). Furthermore, diastereoselective S_NAr^* can be performed starting from chiral planar substrates (Section 8.2.3), especially η^6-arene-$Cr(CO)_3$ complexes and cyclophanes. A chiral tether can be used for the construction of chiral planar macrocycles (Section 8.2.4). Alternatively, chiral nucleophiles (Nu*) such as enolates are able to induce stereocontrolled processes (Section 8.2.5).

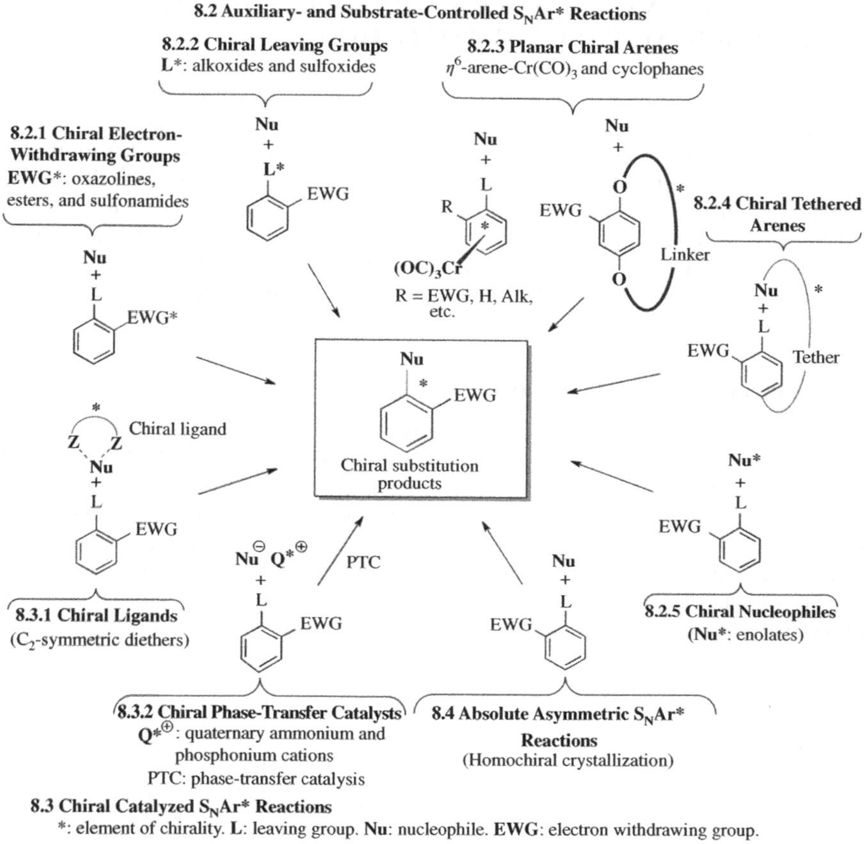

SCHEME 8.1 The S_NAr^* reaction: major mechanisms.

In the cases discussed in subsections 8.2.1, 8.2.2, 8.2.3, and 8.2.5, the chiral auxiliary is removed during or after the reaction is completed to provide the desired chiral substitution product: the reaction is said to be *auxiliary controlled.* When the formation of a new axial, central, or planar stereogenic unit is induced by a preexisting stereogenic element in the substrate or in the entering nucleophile, which is not removed at the end of the process (such as in Section 8.2.4), the reaction is said to be *chiral substrate controlled.*

In auxiliary- and substrate-controlled S_NAr^* reactions, stoichiometric amounts of enantiomerically pure compounds are required. In recent years, major progress has come from the development of chiral catalyzed reactions (Section 8.3). In the approach conceptualized by Tomioka (Section 8.3.1), the formation of a new stereogenic unit is induced by substoichiometric or catalytic amounts of a chiral neutral ligand able to chelate the nucleophile. Jørgensen and Maruoka have shown that chiral quaternary ammonium and phosphonium cations $Q^{*\oplus}$ can induce asymmetry during the S_NAr^* process by acting as chiral phase-transfer catalysts (Section 8.3.2). Finally, recent advances in absolute asymmetric S_NAr^* rely on new developments in homochiral crystallization (Section 8.4).

Partial aspects of the chemistry discussed in this chapter have been reviewed earlier: the Meyers reaction [12, 13], transition metal η^6-arene complex reactions [14], and atroposelective approaches to axially chiral biaryls [15]. The S_NAr^* approach is only briefly presented in François Terrier's monograph (3 pages over a 460-page book) [2].

8.2 AUXILIARY- AND SUBSTRATE-CONTROLLED ASYMMETRIC NUCLEOPHILIC AROMATIC SUBSTITUTION

In S_NAr^* reactions, substrate and reagent combine to form a diastereomeric transition state. In the case of auxiliary-controlled reactions, the asymmetric induction is promoted by a chiral element temporarily linked to the arene or the nucleophile. The ideal chiral auxiliary has to fulfill several requirements: (i) it must be easily available in both enantiomeric forms to permit selective synthesis of both enantiomers, (ii) it must induce good stereoselectivity, (iii) the diastereomeric products must be easily separated, and (iv) cleavage of the chiral auxiliary must provide the requisite enantiomer in high yield without racemization. Additionally, an efficient work-up to allow easy recovery of expensive chiral auxiliaries is highly desirable. Most chiral auxiliaries are either natural products (alcohols, amino acids, carbohydrates, etc.) or derived from natural products.

8.2.1 Chiral Electron-Withdrawing Groups

8.2.1.1 Chiral Oxazolines: The Meyers Reaction A general synthesis of biaryls of high optical purity through the use of chiral oxazolines EWGs* was devised by Meyers [12, 13]. Configurationally stable chiral biaryls **7** can be prepared by asymmetric S_NAr^* reaction of substituted Grignard reagents **6** with aryloxazolines **5**, readily accessible from benzoic acids and chiral aminoalcohols (Scheme 8.2) [16, 17]. The use of aryllithiums results in little or no selectivity, presumably due to the rapid rate of reaction [18]. In this process, the ligand displacement proceeds efficiently in dry ether or tetrahydrofuran (THF), and the diastereomeric products are usually separated by chromatography prior to hydrolytic deprotection of the oxazoline.

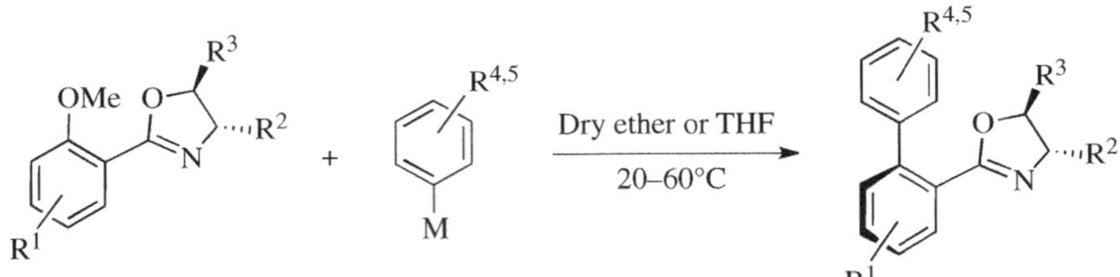

5 R^1 = alkyl, OMe
R^2 = *i*-Pr, *t*-Bu, CH_2OMe
R^3 = H, Ph

6 M = Li, MgBr
$R^{4,5}$ = alkyl, OMe, etc.

7

SCHEME 8.2 The Meyers reaction (1).

The best atropodiastereoselectivities are obtained with Grignard reagents **9** having an *ortho*-methoxy substituent (Scheme 8.3) [12, 19]. High selectivities in favor of atropisomer **10** are observed if the second substituent of **9** exhibits a weaker electron-donating effect (R = Me, CH_2OSiMe_2t-Bu). The sense of chiral induction is reversed if R is more electron donating than OMe (R = 1,3-dioxolan-2-yl group). Consequently, the atropodiastereoselectivity observed is poor if OMe and R display comparable electron-donating ability (R = CH_2OMe).

These observations can be rationalized by a model in which an addition–elimination pathway is operative, with the overall stereoselectivity determined by the two steps (Scheme 8.4). The first step presumably involves initial complexation of organomagnesium **9** with oxazoline **8**. The nitrogen lone pair would lie in the plane of the oxazoline and coordinate Mg. Furthermore, Mg is also coordinated to the methoxy leaving group. Although the mechanism of the Meyers reaction has not yet been elucidated computationally [20], it is assumed that chelate **B** is favored sterically over chelate **A**. The bulky isopropyl substituent favors attack of the nucleophile from the opposite face, leading to azaenolate **12**. The oxazoline moiety not only induces chirality but also favors addition of the

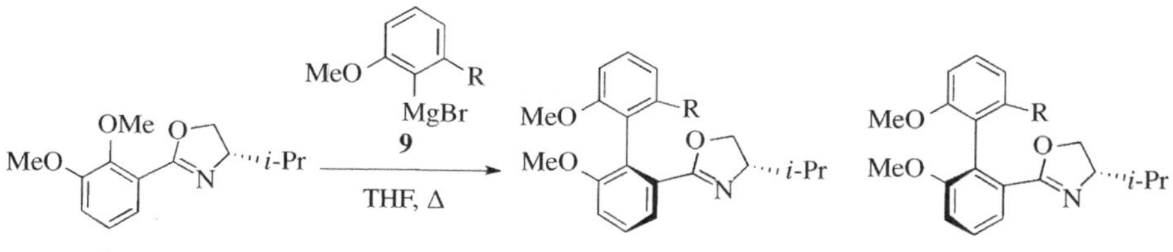

8 10 11

Electron-donating ability ↑

R	Yield (%)	de (%)	Major diastereomer
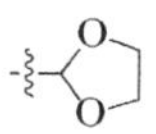	90	60	**11**
CH_2OMe	75	20	**11**
Me	79	80	**10**
CH_2OSiMe_2t-Bu	73	86	**10**

SCHEME 8.3 The Meyers reaction (2).

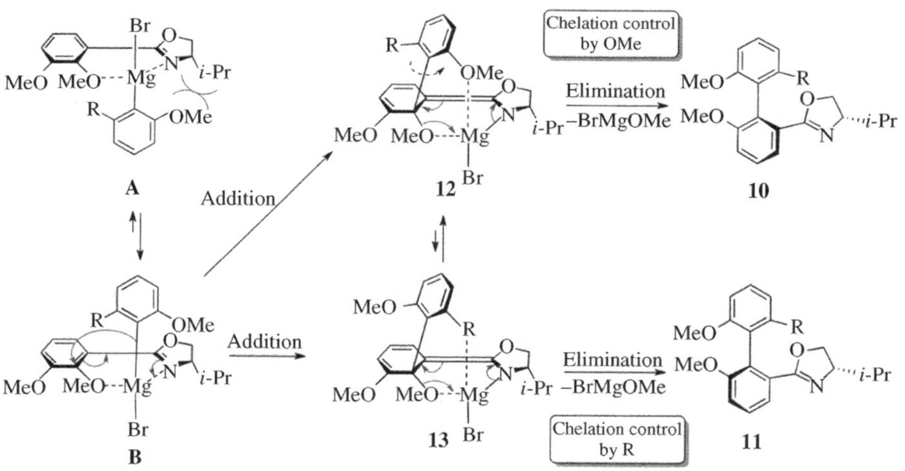

SCHEME 8.4 The Meyers reaction: mechanism.

nucleophile by stabilizing the negative charge. Meisenheimer complex **12** is a σ-complex that opens the possibility for rotation of the aromatic ring about the newly formed σ bond to provide the alternate conformation **13**. Complexation of the 2-methoxy group with magnesium in complex **12** would favor a rearrangement that produces diastereoisomer **10**. If the R group is suitable for coordination to the magnesium, the alternate pathway by which complex **13** gives atropisomer **11** becomes competitive.

The oxazoline ring of **10** and **11** can be transformed into a variety of functionalities: hydrolysis yields a carboxylic acid and regenerates the chiral auxiliary that can be recycled [12]. Alternatively, sequential N-methylation/reduction provides an aldehyde suitable for further transformation. Mild cleavage conditions must be selected to prevent racemization about the Csp^2—Csp^2 bond.

Although the substitution pattern of the nucleophile is limited by the presence of an *ortho* chelating substituent (OMe, 1,3-dioxan-2-yl group, etc.), a wide range of natural products and analogs have been synthesized using this methodology. The stereochemically decisive step in the synthesis of antileukemic dibenzocyclooctadiene lignan (–)-steganone (**1**) is the coupling reaction of chiral (tetramethoxyphenyl)oxazoline **14** with Grignard **15**, which leads to the atropisomer **16** in 65% yield and 75% de (Scheme 8.5) [6].

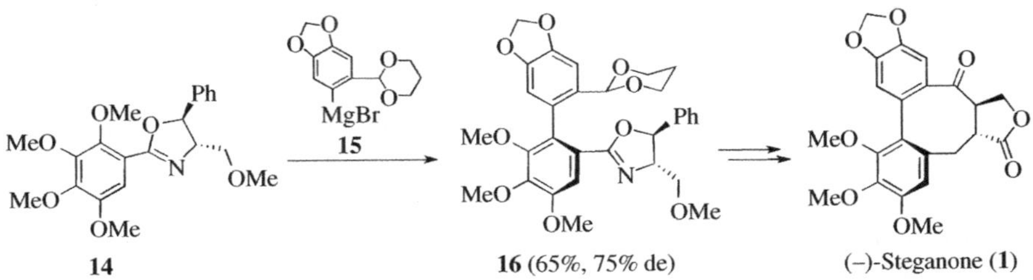

SCHEME 8.5 Synthesis of antileukemic lignan (–)-steganone.

The versatility of the Meyers reaction for the construction of the dibenzocyclooctadiene backbone is amply documented (Fig. 8.2). In particular, lignans (–)-interiotherin A (**17**) [21], (–)-schizandrin (**18**), (–)-isoschizandrin (**19**) [22], and (–)-gomisin E (**20**) [21] were prepared by this method. The synthetic utility of the Meyers reaction to create the axially chiral biaryl skeletal framework was further displayed by the atroposelective syntheses of natural naphthylisoquinoline alkaloids including (–)-*O*-methylancistrocladine (**21**) [23] and (–)-*O*-methylhamatine (**22**) [24], among others [25, 26], and by the synthesis of enantiomerically pure C_2-symmetric biphenyl ligand **23** and binaphthyl porphyrin **24** (Fig. 8.3), which have been used as intermediates for the preparation of chiral catalysts for asymmetric reduction and epoxidation [27, 28].

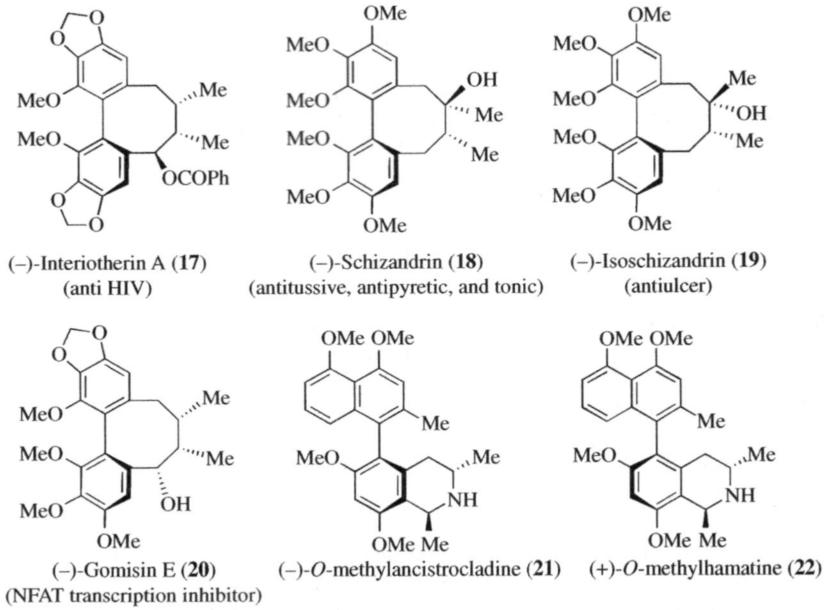

FIGURE 8.2 Synthesis of lignans **17–20** and naphthylisoquinoline alkaloids **21** and **22**.

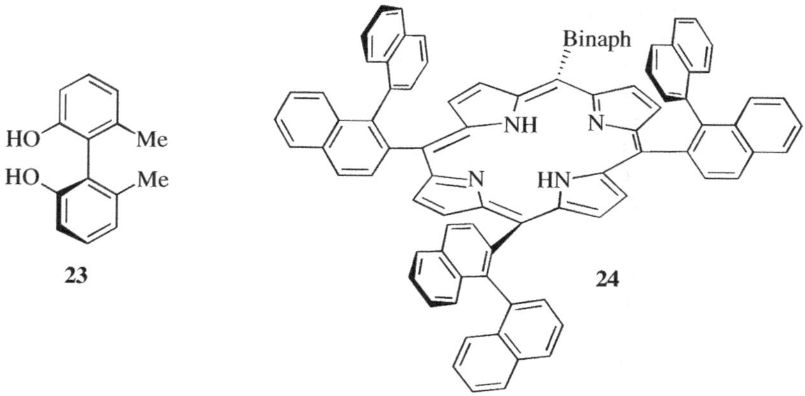

FIGURE 8.3 Synthesis of C_2-symmetric biphenyl ligand **23** and binaphthyl porphyrin **24**.

8.2.1.2 ***Chiral Esters. The Miyano Reaction. Chiral Sulfonamides*** Miyano showed that a chiral ester group can take the place of the oxazolinyl EWG to give similar chelation-assisted S_NAr^* reactions; however, the diastereoselectivity is usually less satisfactory (Scheme 8.6 and Table 8.1) [29, 30]. Displacement of the methoxy group of **25** and **26** possessing a (−)-menthyl and a (*S*)-1-phenylethyl ester activator leads to 1,1′-binaphthalene products **33/34** with low diastereomeric excesses (entries 1–4). This is presumably due to the fact that the chiral center is remote from the reaction site. The preferential binaphthyl axis is governed by the Grignard reagent used: the chiral axis induced with **31** (R^2=H) is opposite to that obtained with **32** (R^2=OMe). The chiral alkoxy leaving group determines configurational bias rather than the chiral ester group, and high levels of asymmetric induction (up to 98%) are achieved with naphthalenes **27–29** possessing a chiral 1-menthoxy leaving group (entries 5–10; see also the discussion in the next subsection).

Although the sulfonamide group exerts a stronger electron-withdrawing effect, it is a less effective activator (entries 11 and 12); however, the axial chirality is induced with higher fidelity [31, 32]. The direction of twist of the binaphthyl axis is again reversed by R^2 (H↔OMe) permutation.

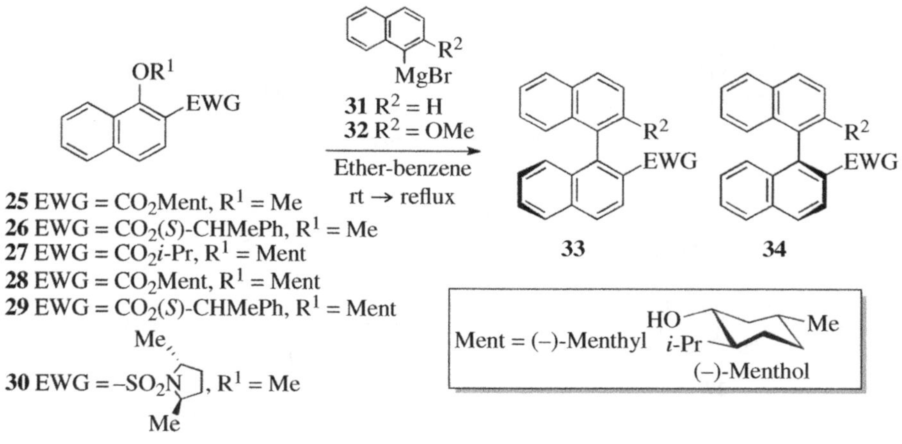

SCHEME 8.6 Chiral esters and sulfonamides.

TABLE 8.1 Chiral Esters and Sulfonamides

Entry	Naphthalene	Grignard	Yield (%)	de (%) [a]/ee (%) [b]
1	**25**	**31**	86	4 (**34**)
2	**26**	**31**	96	28 (**34**)
3	**25**	**32**	88	9.5 (**33**)
4	**26**	**32**	92	51 (**33**)
5	**27**	**31**	95	80 (**33**)
6	**27**	**32**	92	97 (**34**)
7	**28**	**31**	93	76 (**33**)
8	**28**	**32**	81	98 (**34**)
9	**29**	**31**	74	91 (**33**)
10	**29**	**32**	85	91 (**34**)
11	**30**	**31**	42	59 (**34**)
12	**30**	**32**	48	80 (**34**)

[a] Entries 1–4 and 7–12.
[b] Entries 5 and 6.

8.2.2 Chiral Leaving Groups

8.2.2.1 Chiral Alkoxy Leaving Groups The Meyers and Miyano strategies were extended by promoting the departure of a chiral menthoxy leaving group (Scheme 8.7). The prerequisite 1-(−)-menthoxy naphthalenes **27**, **35**, and **36** are prepared by nucleophilic displacement of the corresponding 1-methoxy or 1-bromonaphthalene derivatives with sodium (−)-menthoxide in DMF [33]. Reaction of Grignard **32** gives good atropostereoselectivities and moderate to good yields of coupling products **37–39** [34, 35]. To foster chelation between the substrate and the organomagnesium, the reaction is carried out in poorly coordinating solvents (benzene or ether-benzene). Useful levels of enantioselectivity (up to 67%) are reported by the coupling of

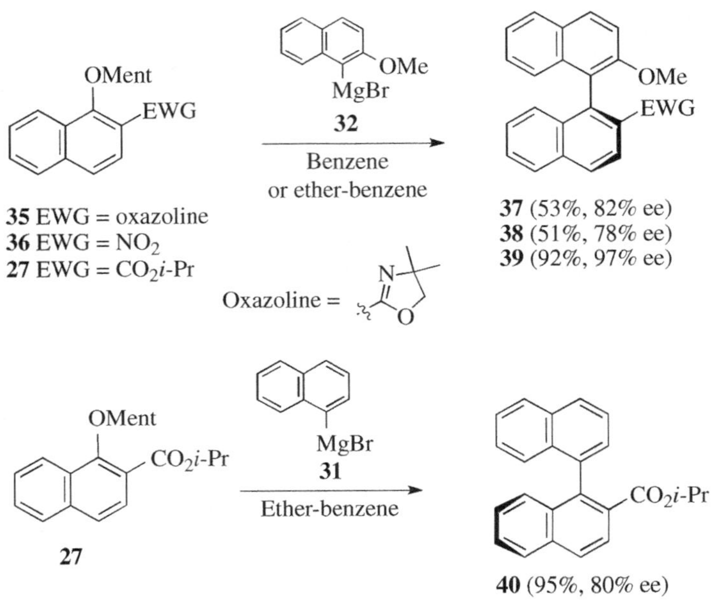

SCHEME 8.7 Chiral alkoxy leaving groups.

aryllithium reagents with naphthyloxazoline **35** in THF. The direction of chiral induction and level of atropenantioselectivity are dependent on the *ortho*-substituents of the nucleophile. By reaction with **31** and **32**, isopropyl 1-(−)-menthoxy-2-naphthoic ester **27** provides access to coupling ester products **39** and **40** of opposite conformation with high chemical and atropisomeric yields [29]. It is important to note that EWG* groups, that is, chiral oxazolines in the Meyers process and chiral esters in the Miyano reaction, lead to diastereoisomeric products that are in principle separable by chromatography or recrystallization, whereas naphthalenes **35**, **36**, and **27** bearing an achiral EWG give enantiomers. Naturally occurring fenchol, quinine, isoborneol, and quinidine are less efficient chiral alkoxy leaving groups than menthol in terms of chemical yield and/or stereoselectivity [33, 35].

In the biphenyl series, axially chiral tri- and tetra-*ortho*-substituted biphenyls **43** are successfully prepared with excellent optical yields from 2-*t*-butylphenyl 2-(−)-menthoxybenzoic esters **41** and aryl Grignard reagents **42** (Scheme 8.8) [36]. The steric bulk of the ester group prevents the undesired nucleophilic attack of the carbonyl by the aryl Grignard.

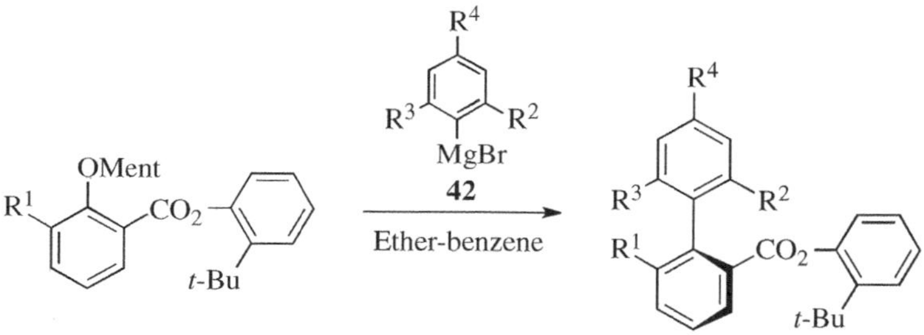

41 R^1 = Me, OMe
R^{2-4} = H, Me, OMe, CH_2OMOM

43 (up to 94% ee)

SCHEME 8.8 Synthesis of axially chiral tri- and tetra-*ortho*-substituted biphenyls.

The reaction of ester **27** with Grignards **31** and **32** was suggested to proceed by a chelation-assisted conjugate addition of the Grignard carbanion, followed by elimination of the alkoxide ion, a process closely related to the oxazoline-mediated Meyers reaction (Scheme 8.9) [30, 36]. The configuration of the chiral carbon of the (−)-menthoxy (marked with a star) presumably controls the stereoselectivity of the addition step. Based on steric consideration, chelate **B** would be favored over chelate **A**. Steric effects and electron-donating abilities of the substituents around the newly formed C—C bond in σ-complexes **44** and **45** are responsible for the stereoselectivity observed in the elimination step. In the case of **31** (R = H), complex **45** avoids steric repulsion between the peri hydrogen atom and the naphthalene ring undergoing substitution. The conformation affords binaphthalene **40** with good chemical yield and enantiomeric excess (ee) values after elimination of menthoxy magnesium bromide. With **32** (R = OMe), strong intramolecular chelation between the methoxy group and Mg in **44** overrides the destabilizing steric interactions between the aromatic rings and leads to the biaryl **39**. Since the direction of the axial twist induced by the oxazoline and the nitro is the same as that observed with the isopropoxycarbonyl activating group, it is assumed that their reactions follow similar pathways.

As a practical example of application of this method, the Miyano reaction proved useful for the preparation of chiral binaphthyl Hoveyda–Grubbs-type catalyst **47**, which finds utility in asymmetric metathesis (Scheme 8.10) [37]. This chiral complex possessing a stereogenic Ru center can be prepared in greater than 98% enantiomeric purity without resolution from optically pure binaphthalene **46** prepared by the Miyano method. However, in contrast to the Meyers reaction, the Miyano reaction has so far not been employed for the total synthesis of natural products.

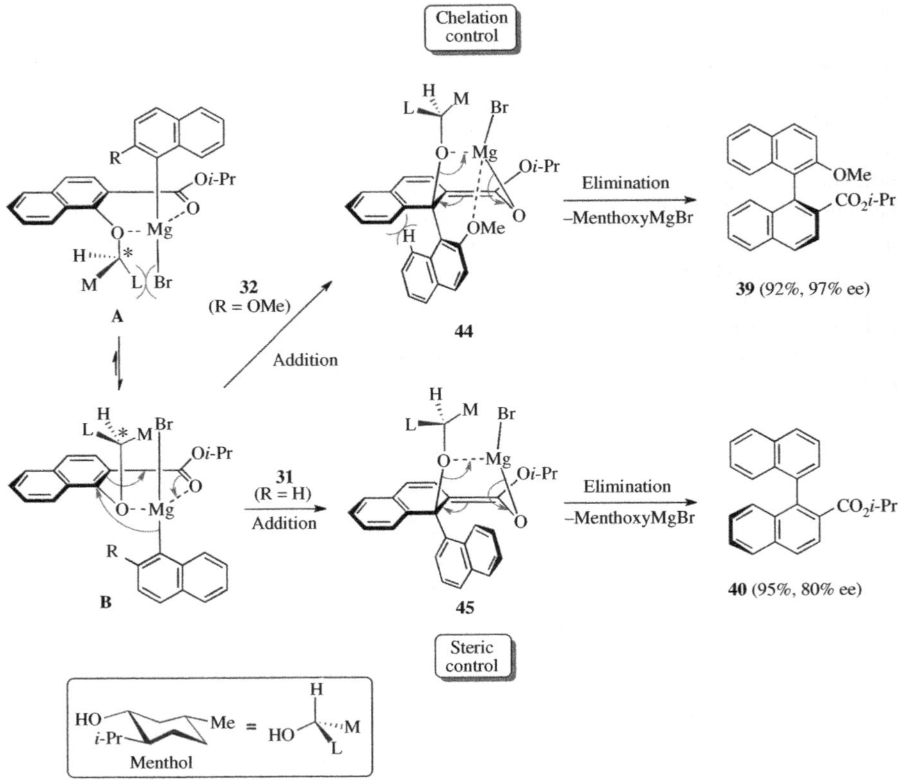

SCHEME 8.9 The Miyano reaction: mechanism.

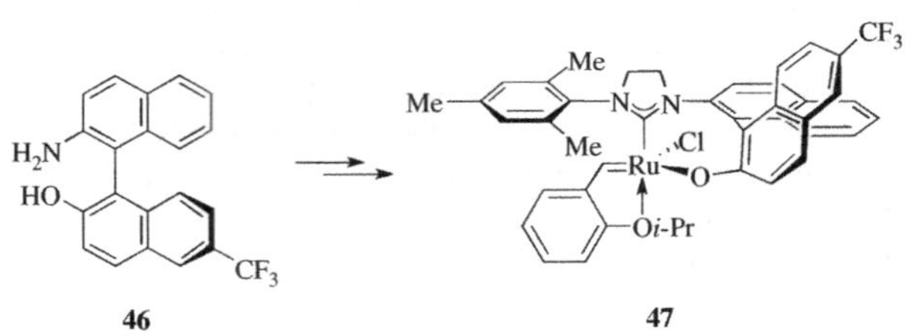

SCHEME 8.10 Preparation of chiral binaphthyl Hoveyda–Grubbs-type catalyst **47**.

8.2.2.2 ***Chiral Sulfoxide Leaving Groups*** In the ester-mediated binaphthyl coupling reaction of *p*-tolylsulfinyl naphthalene **48** with the 1-naphthyl Grignard **31** [38], displacement of a chiral sulfinyl leaving group can induce axial chirality in high optical yield (Scheme 8.11) [39, 40]. The chirality of **48** stems from the stable tetrahedral configuration of the sulfur atom. 1,1′-Binaphthalene product **40** does not result from a direct S_NAr route. According to the mechanism proposed by Baker and Sargent, the initial attack of the Grignard reagent onto the sulfoxide occurs axially from the side opposite to the oxygen ligand, leading to hypervalent pentacoordinate σ-sulfurane **49** [41]. The equatorial ester group and the *p*-tolyl are oriented anti for steric considerations. The orientation of the axial 1-naphthyl is such that nonbonded interactions are minimized. Depending on the nature of the substituents around

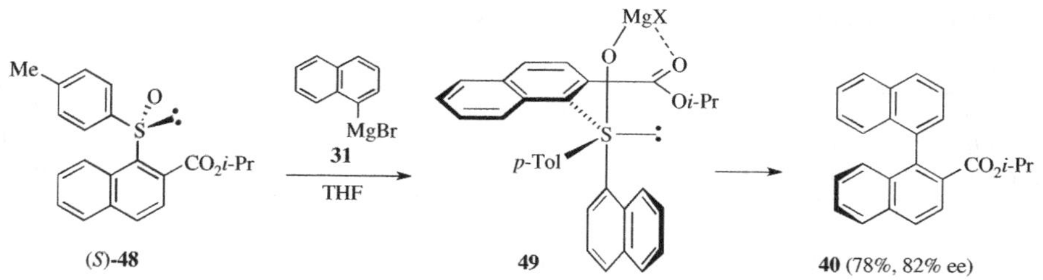

SCHEME 8.11 Chiral sulfoxide leaving groups.

the sulfur atom, sulfurane **49** can evolve in different manners. The reaction of (*S*)-**48** with **31** proceeds mainly by ligand coupling in a concerted manner, leading to atropisomer **40**. Side products arising out of ligand exchange are also isolated. Sulfinyl naphthylamides and sulfinyl naphthyloxazolines are also efficient electron-deficient arenes for S_NAr^* reactions [40]. However, these sulfinyl naphthalenes fail to couple with the more hindered 2-methoxy-l-naphthylmagnesium bromide.

Planar chiral cyclopentadienylmetal complexes (M=Zr, Ti, Co, Fe, Mo) are stereoselective catalysts, which have received attention in a number of asymmetric synthetic applications [42]. The coupling of 1-(*p*-tolyl-sulfinyl)naphthalene-2-carboxylate ester **50** with indenyl lithium **51** and fluorenyl lithium **52** provides access to 1-(2′-methyl-3′-indenyl)naphthalene (**53**) and 9-(1′-naphthyl)fluorene (**54**), respectively, as single rotamers (Scheme 8.12) [43–45]. The atropisomerism arises from hindered rotation about the bond between the cyclopentadienyl ring and the naphthalene moiety. Ligands **53** and **54** are building blocks for the enantiospecific synthesis of planar chiral bidentate ansa zirconocene complexes **55** and **56** [46, 47].

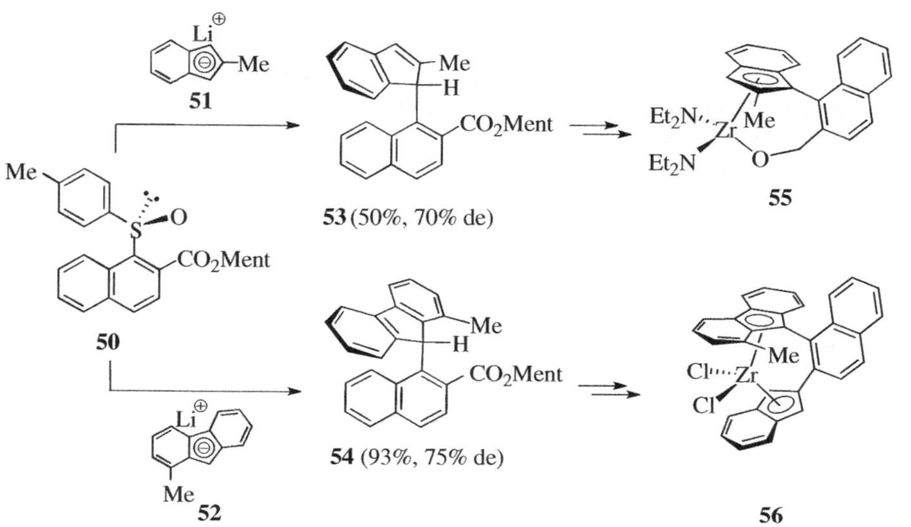

SCHEME 8.12 Synthesis of 1-(2′-methyl-3′-indenyl)naphthalene (**53**) and 9-(1′-naphthyl)fluorene (**54**).

8.2.3 Planar Chiral Arenes

8.2.3.1 Planar Chiral η^6-Arene-Cr(CO)$_3$ Complexes *N*-Aryl indoles and related compounds having a chiral C–N bond have received recent attention as novel atropisomeric intermediates for asymmetric synthesis [10, 48]. *N*-Aryl indoles **59**,**60** can be prepared by stereoselective

TABLE 8.2 Planar Chiral η^6-Arene-Cr(CO)$_3$ Complexes

57,58	R¹	R²	R³	Yield (%)	de (%)	Major dia.
57a,58a	H	Me	Et	94	>96	**60a,a**
57b,58b	$SiEt_3$	Me	H	90	>96	**59b,b**
57c,58b	Me	Me	H	87	66	**59c,b**

nucleophilic aromatic substitution of substituted indoles **57** with planar chiral arenetricarbonylchromium complexes **58** (Scheme 8.13 and Table 8.2) [11, 14, 49, 50]. In this approach, the Cr(CO)$_3$ moiety carries the chiral information by shielding one π-face of the arene ring and due to its strong electron-withdrawing nature facilitates the nucleophilic attack. The sense of asymmetric induction is strongly dependent on the steric demand of the R^1 substituent of indole **57**. When R^1 = H, *N*-aryl indole chromium complex of configuration **60** is isolated with high diastereomeric excess, whereas the opposite configuration **59** is preferentially obtained when R^1,R^2 ≠ H and R^3 = H. The tricarbonylchromium fragment is usually eliminated by photooxidative demetalation.

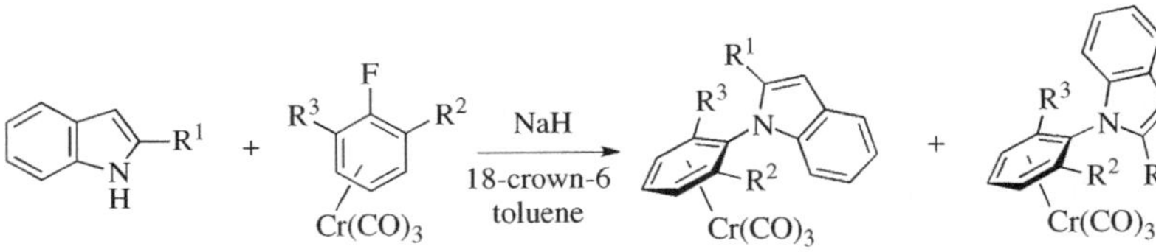

57 58 59 60

SCHEME 8.13 Planar chiral η^6-arene-Cr(CO)$_3$ complexes.

To rationalize these results, geometry optimizations for the coupling of **57c** with **58b** were carried out computationally with the RHF level of theory and 3-21G(d) basis sets for all atoms [49]. The reaction presumably proceeds via the generally accepted addition–elimination mechanism involving a Meisenheimer complex intermediate (Scheme 8.14). In the addition step, to avoid steric repulsion, the 2-methylindole approaches the arenechromium complex from the less sterically

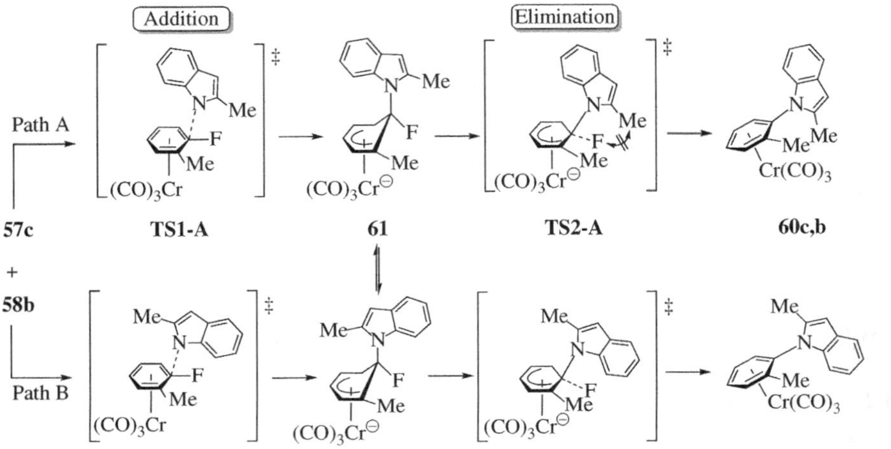

SCHEME 8.14 Planar chiral η^6-arene-Cr(CO)$_3$ complexes: mechanism.

hindered face, opposite to the $Cr(CO)_3$ fragment. Two pathways A and B leading, respectively, to the Meisenheimer complexes **61** and **62** can be operative. The determining step was found to be the elimination step (**TS2**) that has the highest-energy barrier and not the initial addition step (**TS1**). Thus, the stereochemistry of the products must be determined by the conformation at the second step. The transition state **TS2-B** is more stable than **TS2-A** because the methyl group of the indole in **TS2-A** interferes to a greater extent with the fluoride anion nucleofuge than the flat benzene ring of the indole in **TS2-B**. As a result, the formation of **59c,b** is favored over **60c,b**, in agreement with experimental results. Planar chiral (arene)$Cr(CO)_3$ complexes have also been used for the synthesis of axially chiral biaryls [51].

*8.2.3.2 **Planar Chiral Cyclophanes*** Planar chirality is also found in ansa compounds **63**, which are benzene derivatives bridged by a polymethylene linker (Scheme 8.15) [52]. Nucleophilic attack of the naphthyl Grignard reagent **32** followed by cleavage of the ansa chain delivers binaphthalene **65** with good chemical yield and high enantiomeric purity [53]. As the ansa chain effectively shields the upper face of the naphthalene ring, the nucleophile is forced to approach from the less sterically hindered face. The stereochemical outcome in Scheme 8.15 is rationalized by reference to intermediate **64**. This reaction represents the first example of facial chirality transfer from a cyclophane to a chiral axis in binaphthalene products with very high stereospecificity.

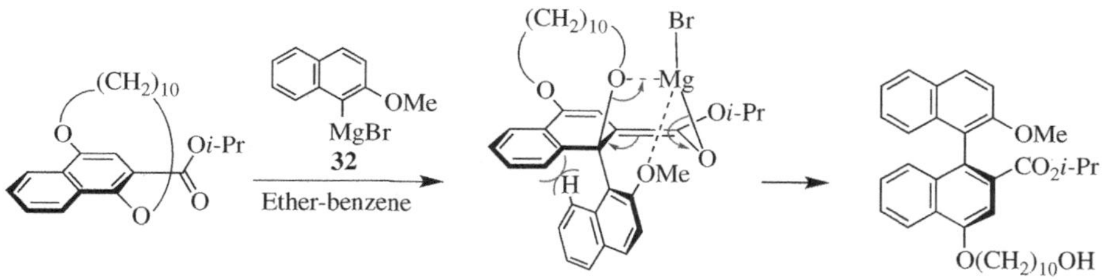

63 **64** **65** (82%, > 99% ee)

SCHEME 8.15 Planar chiral cyclophanes.

8.2.4 Chiral Tethered Arenes

Intramolecular displacement of a fluoride in an electron-deficient arene by a phenolic oxygen is a reliable method for the construction of macrocyclic diaryl ethers (**66**→**67**) and a useful alternative to the copper-catalyzed Ullmann ether synthesis (Scheme 8.16) [54, 55]. The high efficiency of the macrocyclization can be attributed to the tether in the cyclization precursor **66**, which does not predominantly exist in an extended conformation but preferentially adopts a bent conformation that is stabilized by

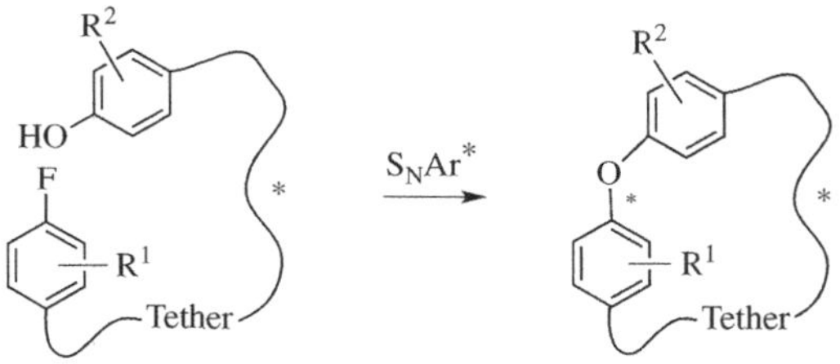

66 **67**

SCHEME 8.16 Chiral tethered arenes.

noncovalent interactions (π–π stacking, intramolecular hydrogen bonding, anion-π interactions, etc.) [56–58]. This preorganization brings the phenolic oxygen and the arylfluoride in close proximity. Macrocyclic biaryl ether **67** can exist as separable atropisomers provided that the free rotation around the aryl-oxygen bond is prevented by the tether. Although high diastereoselectivities can be achieved, the atroposelectivity is very sensitive to subtle structural modifications and it is difficult to predict [59].

The major role played by intramolecular S_NAr^* reactions in macrocyclization methodology was demonstrated by Evans in his total synthesis of vancomycin aglycon **74** (Scheme 8.17) [9, 60]. In the C-O-D and D-O-E oxydibenzene fragments, hindered rotations of the aromatic rings induce planar chirality. Construction of the C-O-D fragment is achieved by a substrate-controlled diastereoselective S_NAr^* by reacting macrocycle **68** with Na_2CO_3 in DMSO at room temperature. The nitro-activated S_NAr^* gives the desired product **69** as a mixture of atropodiastereomers (66% de, step a). Noteworthy, the arylglycine subunit of the tether does not racemize under these mild conditions. The EWG nitro group can be removed by sequential reduction/diazotation/reduction (steps b and c). The second macrocyclization provides **72** in excellent yield (95%) and good diastereomeric excess (83%). This reaction is conducted by adding CsF to a DMSO solution of **71** (step d) [55]. The diastereoselectivity of this cycloetherification is controlled by the conformation of the AB biaryl [61]. Reduction of the nitro group of **72** followed by Sandmeyer reaction ($NH_2 \rightarrow Cl$) provides **73** possessing a chloro substituent on the E ring (steps e and f). Deprotection steps deliver vancomycin aglycon **74**.

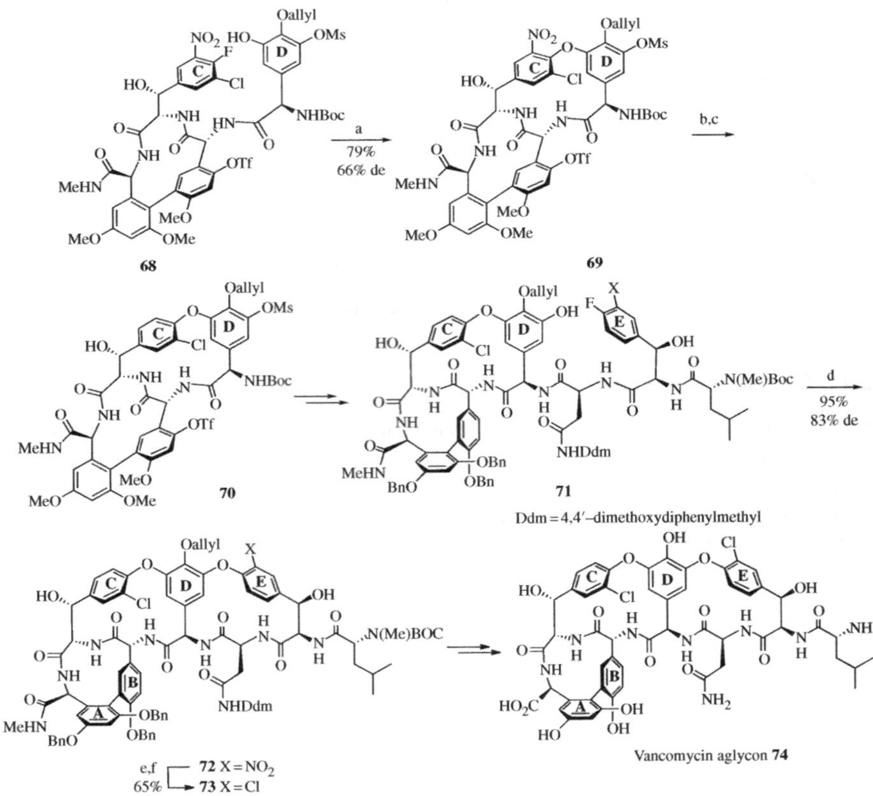

a. Na_2CO_3, DMSO, rt, then Tf_2NPh.
b. Zn, AcOH, EtOH, 40°C.
c. $NaNO_2$, H_3PO_2, cat. Cu_2O, THF/H_2O, 0°C
d. CsF, DMSO, rt.
e. Zn, AcOH, EtOH, 40°C.
f. HBF_4, *t*-BuONO, MeCN, then CuCl, $CuCl_2$, H_2O

SCHEME 8.17 Total synthesis of vancomycin aglycon **74**.

8.2.5 Chiral Nucleophiles

The preceding reactions dealt with the use of chiral auxiliaries linked to the electrophilic arene partner. The entering nucleophile can also serve as a chiral controller in diastereoselective S_NAr^* reactions. This approach was successfully employed for the arylation of enolates derived from amino acids. To illustrate the potential of the method, two examples have been selected. Arylation of Schöllkopf's bislactim ether **75** with aryne **77** as electrophilic arylation reagent was demonstrated by Barrett to provide substitution product **81** with good yield (Scheme 8.18) [62, 63]. Aryne **77** arises from the *ortho*-lithiation of **76** between the methoxy and the chlorine atom followed by elimination of LiCl. Nucleophilic attack of **77** by the lithiated species **78** occurs by the opposite face to that carrying the *i*-Pr substituent. Inter- or intramolecular proton transfer at the α-face of the newly formed carbanion **79** affords the anionic species **80**. Subsequent diastereoselective reprotonation with the bulky weak acid 2,6-di-*t*-butyl-4-methylphenol (BHT) at the less hindered face provides the *syn* product **81**. Hydrolysis and *N*-Boc protection give the unnatural arylated amino acid **82**. The proposed mechanism is supported by a deuterium-labeling experiment. Unnatural arylated amino acids have found application as intermediates for the construction of pharmaceutically important products such as peptidomimetics, enzyme inhibitors, etc. [64, 65].

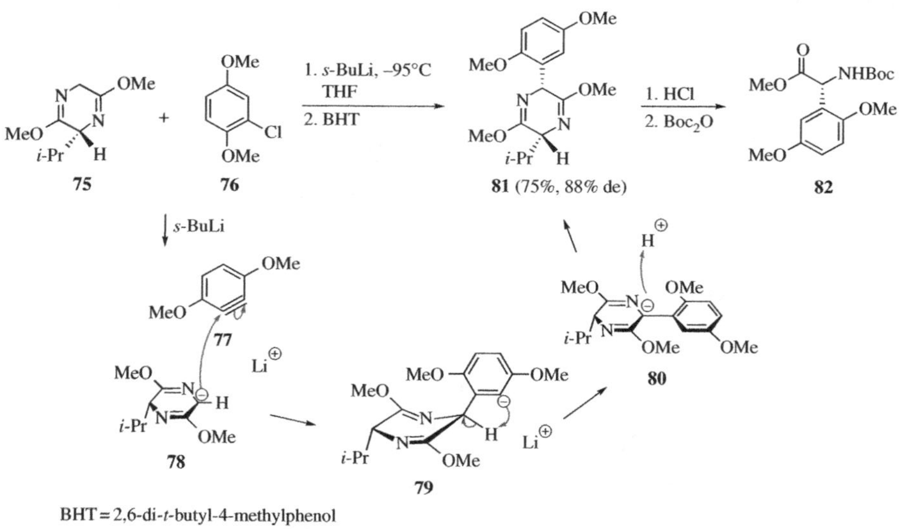

SCHEME 8.18 Arylation of Schöllkopf's bislactim ether **75** with aryne **77**.

Mąkosza demonstrated that optically pure (*R*)-4-nitroarylprolines **87** can be prepared by ONSH of nitroarenes (Scheme 8.19) [66, 67] using Seebach's procedure of self-reproduction of chirality [68]. The trimethylsilyl ester of *N*-trimethylsilyl-L-proline (**83**) is first condensed with pivaldehyde to afford oxazolidinone **84** in enantiomerically pure form. The chiral enolate obtained by treatment of **84** with potassium hexamethyldisilazane (KHMDS) is then allowed to react with various nitroarenes to yield adducts **85**, where the incoming nitroarene approaches by the side occupied by the *t*-butyl. The addition takes place regioselectively in the *para* position to the nitro group. Oxidation with 2,3-dichloro-5,6-dicyanobenzoquinone (DDQ) followed by acid hydrolysis gives (*R*)-4-nitroarylproline **87** as a single detectable stereoisomer. Unnatural α-substituted proline derivatives are generally used as chiral building blocks and organocatalysts.

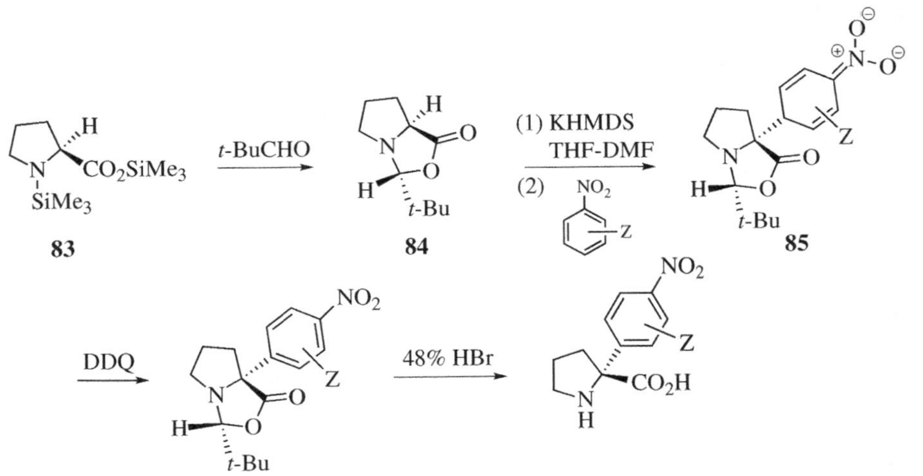

86 (15–72%, 100% de) (*R*)-**87**

SCHEME 8.19 Synthesis of (*R*)-4-nitroarylprolines **87** by oxidative nucleophilic substitution of hydrogen (ONSH) of nitroarenes.

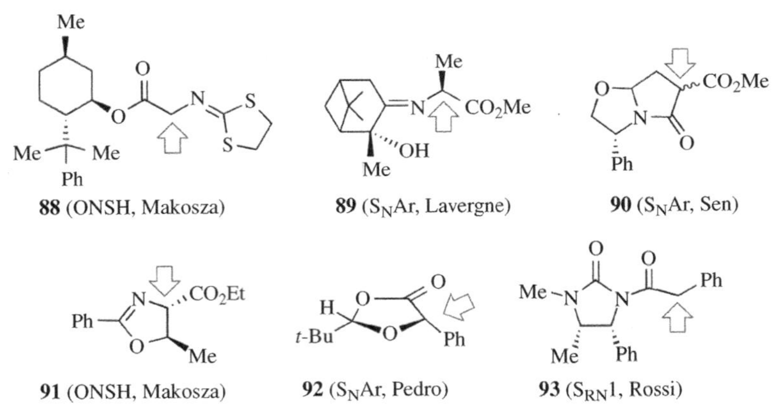

FIGURE 8.4 Other chiral nucleophiles.

Good to excellent diastereoselectivities are also obtained with chiral enolates derived from menthol ester **88** [69], Schiff base of 2-hydroxypinan-3-one **89** [70], bicyclic lactam **90** [71], 2-phenyloxazoline **91** [72], (2*S*,5*S*)-*cis*-1,3-dioxolan-4-one **92** [73], and imidazolidinone **93** [74, 75] (Fig. 8.4)

8.3 CHIRAL CATALYZED ASYMMETRIC NUCLEOPHILIC AROMATIC SUBSTITUTION

There are only few reports of catalytic regio- and enantioselective S_NAr^* reactions in which a chiral ligand or a chiral organocatalyst is responsible of the asymmetric induction.

8.3.1 Chiral Ligands

In his groundbreaking work in the field, Tomioka reported an efficient asymmetric conjugate addition–elimination of 1-fluoro-2-naphthaldehyde (2,6-diisopropylphenyl)imine (**94**) with 1-naphthyllithium **95** giving chiral 1,1′-binaphthalene **98** with good ee and high chemical yield by using the external chiral diether ligand **96** in catalytic amount at –45°C (Scheme 8.20) [76, 77]. In this reaction, the central chirality of the Meisenheimer complex **97** is converted to axial chirality in the product **98** after elimination of LiF. Regeneration of naphthyllithium-diether complex **A** from LiF-diether complex **B** through ligand exchange is crucial for propagation of the catalytic asymmetric process [78]. This regeneration is sluggish with a methoxy leaving group. Because THF competes with the chiral ligand to form a complex with the organolithium, toluene is the solvent of choice. Organolithiums are also less reactive in toluene because of their higher aggregation state in this solvent. The use of LiBr-free naphthyllithium is crucial in order to form a tight chiral lithium–ligand nucleophilic complex that gives high stereoselection. Under similar conditions, sterically hindered naphthoic acid esters give less satisfactory results [79].

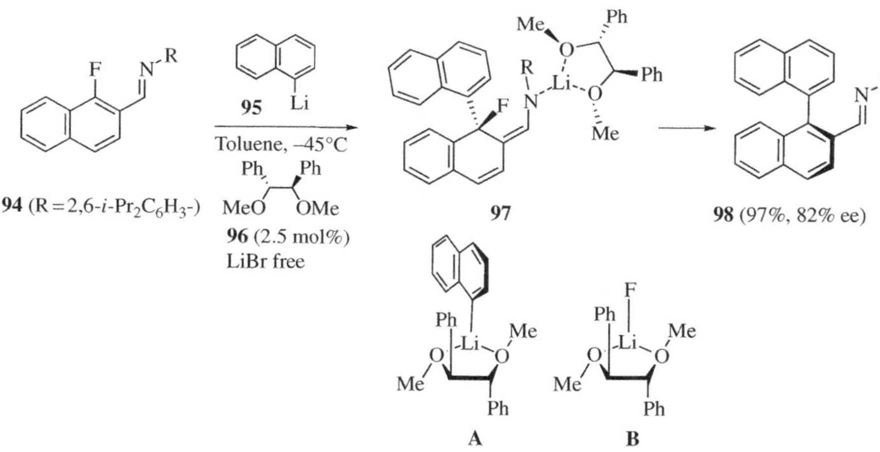

SCHEME 8.20 Chiral catalyzed asymmetric nucleophilic aromatic substitution.

8.3.2 Chiral Phase Transfer Catalysts

Jørgensen devised a simple and efficient organocatalytic approach for the synthesis of spiro-pyrrolidone-3,3′-oxoindoles with ee's up to 85%, by regio- and enantioselective S_NAr^* reaction of activated aromatic compounds with enolates derived from 1,3-dicarbonyl compounds, using phase-transfer catalysis (PTC) conditions (Scheme 8.21) [80, 81]. KOH initially removes the acidic proton of the β-ketoester **99** generating an ambident nucleophilic enolate, which interacts with the cinchona alkaloid-derived catalyst **101** forming a chiral ion pair [82, 83]. The tight binding creates a "chiral pocket" around the nucleophile so that the arylation agent **100** approaches by the less hindered face. S_NAr adducts **102** are solids and the ee can be increased significantly (up to 99%) by recrystallization. A dramatic improvement of both the regio- (C vs. O-arylation) and enantioselectivity is realized by simply replacing a benzyl alcohol by a benzoate (marked with a star) in the catalyst **101**. Less reactive arenes (like 1-fluoro-4-nitrobenzene) give only traces of the desired adducts. By treatment with Pd/C in trifluoroacetic acid (TFA), the optically active product **102** bearing a quaternary stereocenter is converted to the spiro-oxyindole **103**, a motif found in numerous natural substances [84].

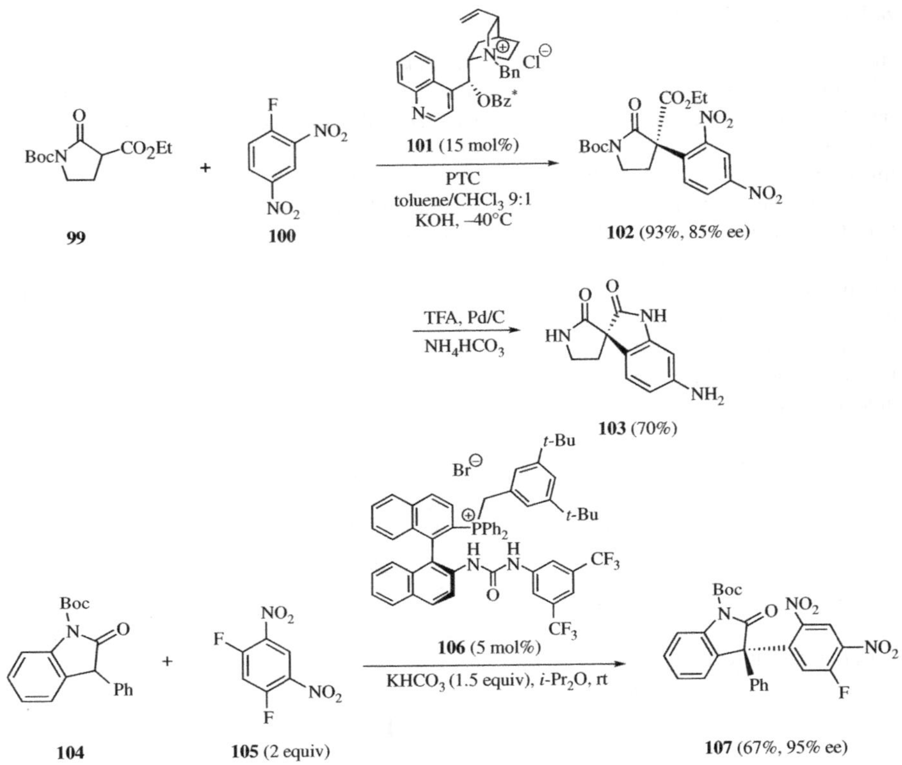

SCHEME 8.21 Chiral organocatalysts (1).

Under similar PTC conditions, Maruoka described the catalytic asymmetric synthesis of triarylmethanes **107**, possessing a chiral all-carbon quaternary center with excellent ee's (up to 95%), by reaction of trisubstituted enolates derived from 3-aryloxindoles **104** with activated fluorobenzenes **105** in the presence of chiral bifunctional quaternary phosphonium bromide **106** (5% mol) [85].

Recently, asymmetric PTC using chiral quaternary ammonium salt **110** has proven to be an effective method for the enantioselective α-arylation of α-imino acid derivatives **108** via asymmetric nucleophilic aromatic substitution, to give α,α-disubstituted α-amino acids **111** in good to high enantiomeric purity (Scheme 8.22) [86].

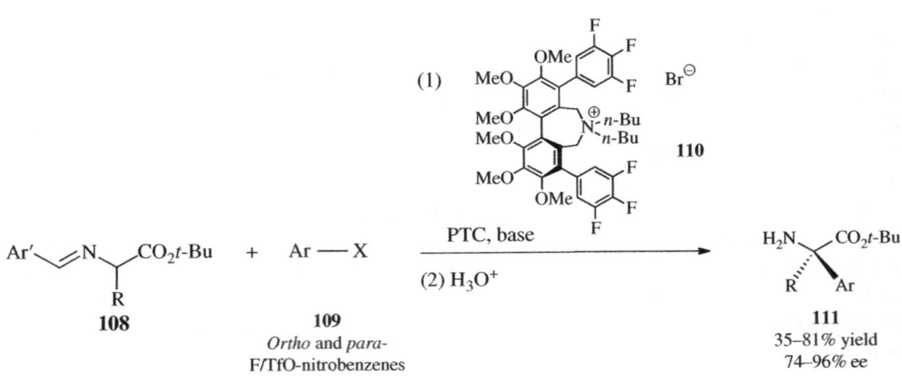

SCHEME 8.22 Chiral organocatalysts (2).

An enantioselective method in which a chiral catalyst is responsible of a desymmetrizing nucleophilic aromatic substitution through discrimination in the displacement of an enantiotopic leaving group has been reported very recently in the heterocyclic series. Under PTC conditions, the chiral counterion **113** (10% mol) directs the substitution of prochiral dichloropyrimidine **112** by PhSK by a tandem desymmetrization/kinetic resolution mechanism, leading to the chiral product **114** (Scheme 8.23) [87].

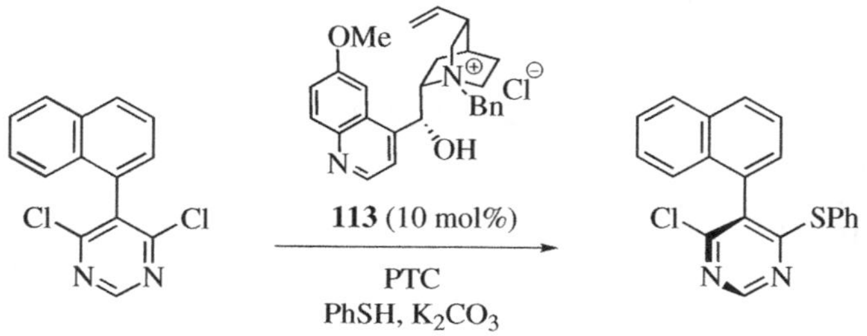

112 **114** (93%, 94% ee)

SCHEME 8.23 Chiral organocatalysts (3).

In addition, Snyder demonstrated that the use of a lipophilic quaternary proline-derived ammonium salt chiral selector to transfer the nucleophile (a hydroxide ion) from the aqueous layer to the organic layer containing an electron-deficient arene allows the spectroscopic characterization of stable chiral Meisenheimer adducts [88].

8.4 ABSOLUTE ASYMMETRIC NUCLEOPHILIC AROMATIC SUBSTITUTION

The synthesis of enantioenriched compounds from achiral starting materials in the absence of chiral reagents or catalysts—the so-called absolute asymmetric synthesis—has long been an intriguing challenge to chemists [89, 90]. The two most common methods rely on the irradiation with circularly polarized light and on spontaneous production of optically pure enantiomorphic crystals from achiral compounds [91, 92]. Crystallization of naphthamides **115/116** yields spontaneously homochiral crystals (Scheme 8.24). A large quantity of the desired crystals can be prepared by seeding the desired crystal during the crystallization process from the melt. Single X-ray crystallographic analyses reveal that both naphthamides have axial chirality with an orthogonal conformation for the naphthalene plane and the amide. Whereas chiral crystals of **115/116** dissolved in THF rapidly racemize at room temperature, the axial chiral integrity is retained by dissolving the enantiomorphic crystals in THF or toluene at −80°C [93].

Atropisomerization of the C-amide bond can be suppressed by introducing a bulky *t*-butyl group in the naphthalene by an S_NAr^* reaction. Addition of crystals of (+)-**115/116** to *t*-BuLi in cold toluene provides with good ee's the naphthalene derivatives **117/118** of undetermined absolute configuration. Kinetic resolutions of racemic amines were also performed using the provisional chiral molecular conformation derived from chiral crystals [94]. Despite the attractive features of these examples, planning of absolute asymmetric S_NAr^* transformation remains a difficult task since only about 10% of achiral substrates crystallize in a chiral fashion.

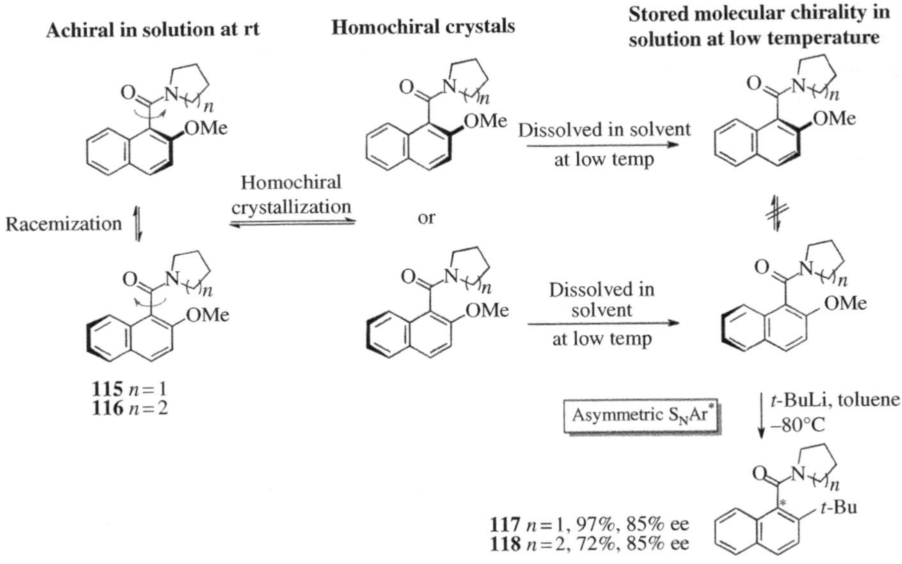

SCHEME 8.24 Absolute asymmetric nucleophilic aromatic substitution.

8.5 SUMMARY AND OUTLOOK

Over the past two decades, S_NAr^* has emerged as an efficient synthetic tool for the construction of complex and stereochemically defined arene scaffolds. To provide a comprehensive and useful understanding, we have classified these reactions into three main categories (auxiliary- and substrate-controlled S_NAr^*, chiral catalyzed S_NAr^*, and absolute S_NAr^*), and we have focused our presentation on the detailed current mechanistic understanding based on experimental and computational studies. Although not exhaustive in literature review, examples have been chosen to highlight the potential of these reactions in the area of pharmaceuticals and organic materials.

Most current S_NAr^* reactions take advantage of chiral features in the substrate or the nucleophile. Grignard reagents, organolithiums, enolates, amide bases, and alkoxide bases are able to react as nucleophiles with arenes that are usually activated by one or more EWGs. The selectivity is primarily dictated by the steric and electronic nature of the arene substrate, and the process provides access to a wide variety of aromatic derivatives with axial, central, and planar chirality. While the Meyers reaction has received broad attention, the synthetic potential of coupling reactions with ester EWGs still remains to be evaluated for more elaborated compounds.

Ligand- and organocatalyst-mediated atroposelective S_NAr^* methodologies that have emerged in the recent years should open up new synthetic perspectives. Additional experimental and theoretical efforts are needed for guiding the future development of this field.

ABBREVIATIONS

BHT	2,6-Di-*t*-butyl-4-methylphenol
Boc	*t*-Butoxycarbonyl
DDQ	2,3-Dichloro-5,6-dicyanobenzoquinone
ee	Enantiomeric excess

EWG	Electron-withdrawing group
EWG*	Chiral electron-withdrawing group
KHMDS	Potassium hexamethyldisilazane
L	Leaving group
L*	Chiral leaving group
Ment	Menthyl
MOM	Methyl methoxy
Nu	Nucleophile
Nu*	Chiral nucleophile
ONSH	Oxidative nucleophilic aromatic substitution of hydrogen
PTC	Phase transfer catalysis
S_NAr^*	Asymmetric nucleophilic aromatic substitution
TFA	Trifluoroacetic acid
THF	Tetrahydrofuran

REFERENCES

[1] Smith, M. B. (2013) in *March's Advanced Organic Chemistry: Reactions, Mechanisms, and Structure*, 7th ed., John Wiley & Sons, Inc., Hoboken, p. 732.

[2] Terrier, F. (2013) in *Modern Nucleophilic Aromatic Substitution*, Wiley-VCH Verlag, Weinheim/Hoboken, pp. 73–76.

[3] Testa, B. (2013) *Helv. Chim. Acta*, **96**, 351–374.

[4] Ōki, M. (1983) in *Topics in Stereochemistry*, vol. **14** (eds.: Allinger, N. L., Eliel, E. L., Wilen, S. H.), John Wiley & Sons, Inc., Hoboken, pp. 1–81.

[5] Wolf, C. (2008) in *Dynamic Stereochemistry of Chiral Compounds: Principles and Applications*. Royal Society of Chemistry, Cambridge, pp. 84–104.

[6] Meyers, A. I., Flisak, J. R., Aitken, R. A. (1987) *J. Am. Chem. Soc.*, **109**, 5446–5452.

[7] Damerla, V. S. B., Tulluri, C., Gundla, R., Naviri, L., Adepally, U., Iyer, P. S., Murthy, Y. L. N., Prabhakar, N., Sen, S. (2012) *Chem. Asian J.*, **7**, 2351–2360.

[8] Hong, S., Jung, M., Park, Y., Ha, M. W., Park, C., Lee, M., Park, H.-G. (2013) *Chem. Eur J.*, **19**, 9599–9605.

[9] Kahne, D., Leimkuhler, C., Lu, W., Walsh, C. (2005) *Chem. Rev.*, **105**, 425–448.

[10] Clayden, J., Moran, W. J., Edwards, P. J., LaPlante, S. R. (2009) *Angew. Chem. Int. Ed.*, **48**, 6398–6401.

[11] Kamikawa, K., Kinoshita, S., Matsuzaka, H., Uemura, M. (2006) *Org. Lett.*, **8**, 1097–1100.

[12] Meyers, A. I., Nelson, T. D., Moorlag, H., Rawson, D. J., Meier, A. (2004) *Tetrahedron*, **60**, 4459–4473.

[13] Mortier, J. (2011) *Curr. Org. Chem.*, **15**, 2413–2437.

[14] *(Arene)Cr(CO)$_3$ Complexes: Aromatic Nucleophilic Substitution* by Semmelhack, M. F., Chlenov, A. (2004) in *Topics in Organometallic Chemistry*, vol. **7** (ed.: Kündig, E. P.), Springer Verlag, Berlin, pp. 43–69.

[15] Bringmann, G., Price Mortimer, A. J., Keller, P. A., Gresser, M. J., Garner, J., Breuning, M. (2005) *Angew. Chem. Int. Ed.*, **44**, 5384–5427.

[16] Meyers, A. I., Meier, A., Rawson, D. J. (1992) *Tetrahedron Lett.*, **33**, 853–856.

[17] Meyers, A. I. (2005) *J. Org. Chem.*, **70**, 6137–6151.

[18] Meyers, A. I., Himmelsbach, R. J. (1985) *J. Am. Chem. Soc.*, **107**, 682–685.

[19] Meyers, A. I., Lutomski, K. A. (1982) *J. Am. Chem. Soc.*, **104**, 879–881.

[20] Chen, Y.-M., Chass, G. A., Fang, D.-C. (2014) *Phys. Chem. Chem. Phys.*, **16**, 1078–1083.

[21] Singidi, R. R., RajanBabu, T. V. (2008) *Org. Lett.*, **10**, 3351–3354.

[22] Warshawsky, A. M., Meyers, A. I. (1990) *J. Am. Chem. Soc.*, **112**, 8090–8099.

[23] Leighton, B. N., Rizzacasa, M. A. (1995) *J. Org. Chem.*, **60**, 5702–5705.

[24] Rizzacasa, M. A., Sargent, M. V. (1990) *J. Chem. Soc. Chem. Comm.*, 894–896.

[25] Chau, P., Czuba, I. R., Rizzacasa, M. A., Bringmann, G., Gulden, K.-P., Schäffer, M. (1996) *J. Org. Chem.*, **61**, 7101–7105.

[26] Baker, R. W., Kyasnoor, R. V., Sargent, M. V., Skelton, B. W., White, A. H. (2000) *Aust. J. Chem.*, **53**, 487–506.

[27] Moorlag, H., Meyers, A. I. (1993) *Tetrahedron Lett.*, **34**, 6993–6996.

[28] O'Malley, S., Kodadek, T. (1989) *J. Am. Chem. Soc.*, **111**, 9116–9117.

[29] Suzuki, T., Hotta, H., Hattori, T., Miyano, S. (1990) *Chem. Lett.*, **19**, 807–810.

[30] Hattori, T., Hotta, H., Suzuki, T., Miyano, S. (1993) *Bull. Chem. Soc. Jpn.*, **66**, 613–622.

[31] Hattori, T., Suzuki, M., Komuro, Y., Miyano, S. (1995) *J. Chem. Soc., Perkin Trans.* **1**, 1473–1474.

[32] Hattori, T., Suzuki, M., Tomita, N., Takeda, A., Miyano, S. (1997) *J. Chem. Soc., Perkin Trans.* **1**, 1117–1124.

[33] Wilson, J. M., Cram, D. J. (1984) *J. Org. Chem.*, **49**, 4930–4943.

[34] Hattori, T., Takeda, A., Yamabe, O., Miyano, S. (2002) *Tetrahedron*, **58**, 233–238.

[35] Wilson, J. M., Cram, D. J. (1982) *J. Am. Chem. Soc.*, **104**, 881–884.

[36] Hattori, T., Koike, N., Miyano, S. (1994) *J. Chem. Soc., Perkin Trans.* **1**, 2273–2282.

[37] Van Veldhuizen, J. J., Gillingham, D. G., Garber, S. B., Kataoka, O., Hoveyda, A. H. (2003) *J. Am. Chem. Soc.*, **125**, 12502–12508.

[38] Chiral *p*-tolylsulfinyl naphthalene **48** is readily prepared by treating optically active (−)-menthyl (−)-arenesulfinates with arylmagnesium bromides: Andersen, K. K., Gaffield, W., Papanikolaou, N. E., Foley, J. W., Perkins, R. I. (1964) *J. Am. Chem. Soc.*, **86**, 5637–5646.

[39] Baker, R. W., Pocock, G. R., Sargent, M. V., Twiss, E. (1993) *Tetrahedron Asymmetry*, **4**, 2423–2426.

[40] Baker, R. W., Sargent, M. V. (1994) *Pure Appl. Chem.*, **66**, 2143–2150.

[41] Finet, J.-P. (1998) *Ligand Coupling Reactions with Heteroatomic Compounds*, Elsevier Science, Oxford.

[42] Halterman, R. L. (1992) *Chem. Rev.*, **92**, 965–994.

[43] Baker, R. W., Foulkes, M. A., Taylor, J. A. (1998) *J. Chem. Soc., Perkin Trans.* **1**, 1047–1058.

[44] Baker, R. W., Hambley, T. W., Turner, P. (1995) *J. Chem. Soc. Chem. Commun.*, 2509–2510.

[45] Baker, R. W., Hambley, T. W., Turner, P., Wallace, B. J. (1996) *Chem. Commun.*, 2571–2572.

[46] Baker, R. W., Wallace, B. J. (1999) *Chem. Commun.*, 1405–1406.

[47] Baker, R. W., Foulkes, M. A., Turner, P. (2000) *J. Chem. Soc. Dalton Trans.*, 431–433.

[48] Bringmann, G., Gulder, T., Reichert, M., Meyer, F. (2006) *Org. Lett.*, **8**, 1037–1040.

[49] Kamikawa, K., Kinoshita, S., Furusyo, M., Takemoto, S., Matsuzaka, H., Uemura, M. (2007) *J. Org. Chem.*, **72**, 3394–3402.

[50] Rosillo, M., Domínguez, G., Pérez-Castells, J. (2007) *Chem. Soc. Rev.*, **36**, 1589–1604.

[51] Kamikawa, K., Uemura, M. (1996) *Tetrahedron Lett.*, **37**, 6359–6362.

[52] Davis, F. Higson, S. (2011) in *Macrocycles: Construction, Chemistry and Nanotechnology Applications*, John Wiley & Sons, Inc., Hoboken.

[53] Hattori, T., Koike, N., Okaishi, Y., Miyano, S. (1996) *Tetrahedron Lett.*, **37**, 2057–2060.

[54] Gulder, T., Baran, P. S. (2012) *Nat. Prod. Rep.*, **29**, 899–934.

[55] Pitsinos, E. N., Vidali, V. P., Couladouros, E. A. (2011) *Eur. J. Org. Chem.*, 1207–1222.

[56] Blankenstein, J., Zhu, J. (2005) *Eur. J. Org. Chem.*, 1949–1964.

[57] Beugelmans, R., Singh, G. P., Bois-Choussy, M., Chastanet, J., Zhu, J. (1994) *J. Org. Chem.*, **59**, 5535–5542.

[58] Jia, Y., Ma, N., Liu, Z., Bois-Choussy, M., Gonzalez-Zamora, E., Malabarba, A., Brunati, C., Zhu, J. (2006) *Chem. Eur. J.*, **12**, 5334–5351.

[59] Rama Rao, A. V., Gurjar, M. K., Lakshmipathi, P., Reddy, M. M., Nagarajan, M., Pal, S., Sarma, B., Tripathy, N. K. (1997) *Tetrahedron Lett.*, **38**, 7433–7436.

[60] Evans, D. A., Wood, M. R., Trotter, B. W., Richardson, T. I., Barrow, J. C., Katz, J. L. (1998) *Angew. Chem. Int. Ed.*, **37**, 2704–2708.

[61] Evans, D. A., Dinsmore, C. J., Watson, P. S., Wood, M. R., Richardson, T. I., Trotter, B. W., Katz, J. L. (1998) *Angew. Chem. Int. Ed.*, **37**, 2700–2704.

[62] Jones, E. P., Jones, P., Barrett, A. G. M. (2011) *Org. Lett.*, **13**, 1012–1015.

[63] Jones, E. P., Jones, P., White, A. J. P., Barrett, A. G. M. (2011) *Beilstein J. Org. Chem.*, **7**, 1570–1576.

[64] Pollegioni, L., Servi, S. (2011) in *Unnatural Amino Acids: Methods and Protocols*, Humana Press, New York.

[65] Soloshonok, V. A., Izawa, K. (2009) in *Asymmetric Synthesis and Application of α-Amino Acids*, ACS Symposium Series, American Chemical Society, Washington.

[66] Mąkosza, M., Sulikowski, D., Maltsev, O. (2008) *Synlett*, 1711–1713.

[67] Mąkosza, M. (2010) *Chem. Soc. Rev.*, **39**, 2855–2868.

[68] Seebach, D., Sting, A. R., Hoffmann, M. (1996) *Angew. Chem. Int. Ed. Engl.*, **35**, 2708–2748.

[69] Mąkosza, M., Chrominski, M., Sulikowski, D. (2011) *ARKIVOC*, **vi**, 82–91.

[70] Chaari, M., Jenhi, A., Lavergne, J. P., Viallefont, P. (1991) *J. Organomet. Chem.*, **401**, C10–C13.

[71] Sen, S., Potti, V. R., Surakanti, R., Murthy, Y. L., Pallepogu, R. (2011) *Org. Biomol. Chem.*, **9**, 358–360.

[72] Sulikowski, D., Mąkosza, M. (2010) *Eur. J. Org. Chem.*, 4218–4226.

[73] Barroso, S., Blay, G., Cardona, L., Fernández, I., García, B., Pedro, J. R. (2004) *J. Org. Chem.*, **69**, 6821–6829.

[74] Lotz, G. A., Palacios, S. M., Rossi, R. A. (1994) *Tetrahedron Lett.*, **35**, 7711–7714.

[75] Baumgartner, M. T., Lotz, G. A., Palacios, S. M. (2004) *Chirality*, **16**, 212–219.

[76] Shindo, M., Koga, K., Tomioka, K. (1992) *J. Am. Chem. Soc.*, **114**, 8732–8733.

[77] Shindo, M., Koga, K., Tomioka, K. (1998) *J. Org. Chem.*, **63**, 9351–9357.

[78] Tnay, Y. L., Chen, C., Chua, Y. Y., Zhang, L., Chiba, S. (2012) *Org. Lett.*, **14**, 3550–3553.

[79] Shindo, M., Yamamoto, Y., Yamada, K.-I., Tomioka, K. (2009) *Chem. Pharm. Bull.*, **57**, 752–754.

[80] Bella, M., Kobbelgaard, S., Jørgensen, K. A. (2005) *J. Am. Chem. Soc.*, **127**, 3670–3671.

[81] Kobbelgaard, S., Bella, M., Jørgensen, K. A. (2006) *J. Org. Chem.*, **71**, 4980–4987.

[82] Maruoka, K. (2008) in *Asymmetric Phase Transfer Catalysis*, Wiley-VCH, Weinheim

[83] Halpern, M. E. (1997) in *Phase-Transfer Catalysis. Mechanism and Synthesis*, ACS Symposium Series, American Chemical Society, Washington.

[84] Marti, C., Carreira, E. M. (2003) *Eur. J. Org. Chem.*, 2209–2219.

[85] Shirakawa, S., Koga, K., Tokuda, T., Yamamoto, K., Maruoka, K. (2014) *Angew. Chem. Int. Ed.*, **53**, 6220–6223.

[86] Shirakawa, S., Yamamoto, K., Tokuda, T., Maruoka, K. (2014) *Asian J. Org. Chem.*, **3**, 433–436.

[87] Armstrong, R. J., Smith, M. D. (2014) *Angew. Chem. Int. Ed.*, **53**, 12822–12826.

[88] Snyder, S. E., Carey, J. R., Shvets, A. B., Pirkle, W. H. (2005) *J. Org. Chem.*, **70**, 4073–4081.

[89] Feringa, B. L., van Delden, R. A. (1999) *Angew. Chem. Int. Ed.*, **38**, 3418–3438.

[90] Weissbuch, I., Lahav, M. (2011) *Chem. Rev.*, **111**, 3236–3267.

[91] *Spontaneous Chiral Crystallization of Achiral Materials and Absolute Asymmetric Transformation in the Chiral Crystalline Environment* by Sakamoto, M. (2007) in *Enantiomer Separation: Fundamentals and Practical Methods* (ed.: Toda, F.), Kluwer Academic, Dordrecht/Boston, pp. 103–133.

[92] *Asymmetric Reaction Using Molecular Chirality Controlled by Spontaneous Crystallization* by Sakamoto, M. Mino, T. (2012) in *Advances in Crystallization Processes* (ed.: Mastai, Y.), InTech, pp. 59–80.

[93] Sakamoto, M., Unosawa, A., Kobaru, S., Fujita, K., Mino, T., Fujita, T. (2007) *Chem. Commun.*, 3586–3588.

[94] Sakamoto, M., Fujita, K., Yagishita, F., Unosawa, A., Mino, T., Fujita, T. (2011) *Chem. Commun.*, **47**, 4267–4269.

9

HOMOLYTIC AROMATIC SUBSTITUTION

ROBERTO A. ROSSI, MARÍA E. BUDÉN, AND JAVIER F. GUASTAVINO
INFIQC, Departamento de Química Orgánica, Facultad de Ciencias Químicas, Universidad Nacional de Córdoba, Córdoba, Argentina

9.1 INTRODUCTION: SCOPE AND LIMITATIONS

Homolytic aromatic substitution (HAS) is defined as the replacement of a leaving group Y onto an aromatic ring by an attacking radical R• (Eq. 9.1) [1]:

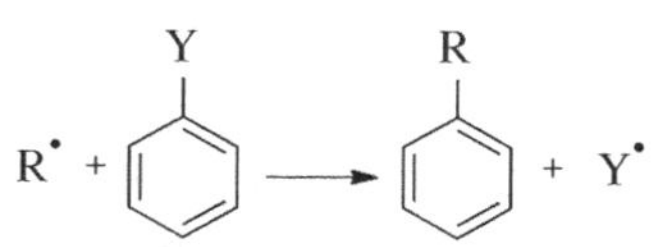

(9.1)

The HAS reaction proceeds via a sigma (σ) complex (**1**) with substitution being completed by the loss of the leaving group Y, which is usually hydrogen (Scheme 9.1, Y = H). Examples where the cyclohexadienyl radicals become trapped by fast reductants to form cyclohexadiene [2] and the detection of radical intermediates by ESR or CIDNP provide evidence that the cyclohexadienyl radicals are intermediates in this reaction [3]. In some systems, the addition of a radical onto the arene is the rate-determining step, because of the loss of aromaticity. For example, the rate constant for the addition of the *tert*-butyl radical to benzene at 79°C is $3.8 \times 10^2\,M^{-1}\,s^{-1}$ [4], which is clearly at the lower end of a useful radical reaction. The arene needs to be used at high concentration, or as the solvent, in order to compensate for poor rates. On the other hand, as the rate of addition of the phenyl radical to benzene is $4.5 \times 10^5\,M^{-1}\,s^{-1}$ [5], it is more useful in these kind of reactions.

Both electron-withdrawing group (EWG) and electron-donating group (EDG) activate the aromatic ring toward radical attack. Considering the nature of radicals, it is known that alkyl

Arene Chemistry: Reaction Mechanisms and Methods for Aromatic Compounds, First Edition. Edited by Jacques Mortier.

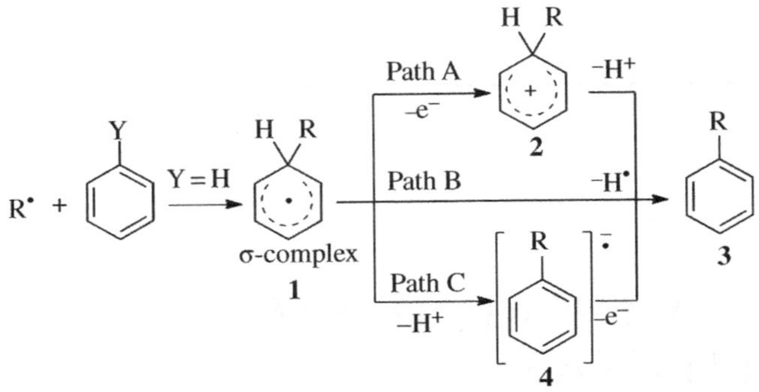

SCHEME 9.1 Homolytic aromatic substitution (HAS) mechanisms, where the leaving group is a hydrogen atom.

radicals (nucleophilic) react faster with electron-deficient arenes than with electron-rich arenes, with the opposite being true for electrophilic radicals such as the $^{\cdot}CF_3$ radical [6]. Although the regioselectivity of the substitution is generally poor for most substituted arenes, the regiochemistry can be improved by considering the electronic compatibility of the arene with the radical.

After the radical attack, **1** can mainly react by three different paths (Scheme 9.1). Path A denotes the oxidation to form a cyclohexadienyl cation **2**, followed by the rapid loss of the leaving group H^+ to give the substitution product **3**. This route is important under oxidative conditions such as in the presence of oxidizing metal ions (Mn^{3+}, Cu^{2+}, Ce^{4+}, Fe^{3+}). Path B is the direct loss of the hydrogen atom as a leaving group, which in general involves an abstraction process. Path C is a radical anion (**4**) formation by the loss of H^+, which provides **3** after an electron transfer (ET) reaction. The loss of the leaving group H^+ is assisted by a base. Thus, path C is usually called a base-promoted homolytic aromatic substitution (BHAS). Although hydrogen is the leaving group in most reported HAS reactions, *ipso* substitutions with NO_2, SO_2R, or C(O)R [7] as the leaving group ($Y^{\cdot}$) are also possible.

Many side reactions compete efficiently with the HAS process. Under non-oxidizing conditions, the cyclohexadienyl radical **1** is rather long lived and can dimerize to **5** (radical–radical coupling) or disproportionate to cyclohexadiene **6** (H-atom abstraction) and substitution product **3** (Scheme 9.2). Long-lived radical **1** may also couple with radicals derived from the radical initiator [8]. These reactions can be considered as termination steps [9]. In addition, radical $R^{\cdot}$ can be reduced by hydrogen atom abstraction from the solvent to yield R–H, before the attack on the benzene core preventing the HAS reaction.

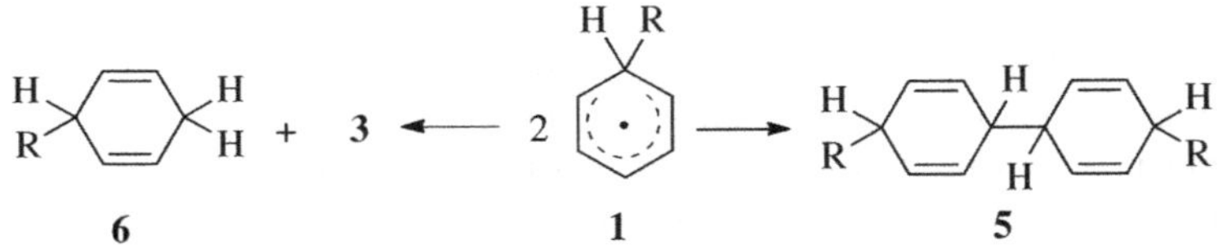

SCHEME 9.2 Termination steps in HAS reactions.

As intramolecular strategies minimize the problems of the poor regioselectivity obtained in intermolecular reactions, they are therefore more useful in synthesis. When the precursor radical is part of the chain connecting the arene, for example, substrate **7**, the radical intermediate **8** is formed by the loss of a leaving group (–Y•), with subsequent attack of this radical to arene forming cyclohexadienyl radical **9**, which finally gives the product **10** (Eq. 9.2):

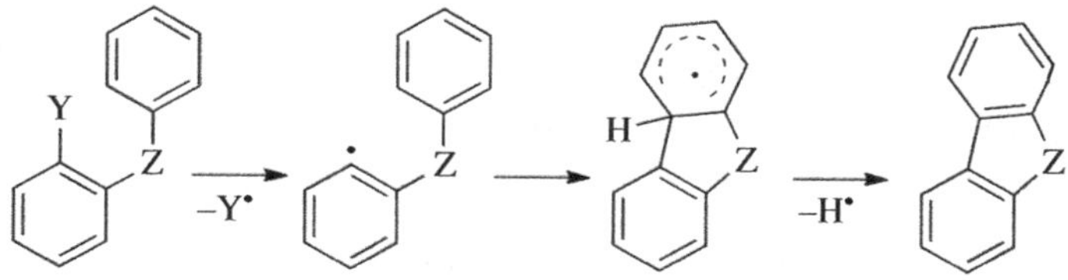

7 **8** **9** **10** (9.2)

Using this methodology in 1896, Pschorr first reported the synthesis of phenanthrenes from the corresponding (Z)-2-styrylbenzenediazonium salts promoted by Cu^+ [10]. Later, in 1924, Gomberg and Bachmann showed that biaryls could be prepared by intermolecular HAS from aryldiazonium salts and benzene [11]. However, yields were generally low (<40%), and many side reactions of diazonium salts were observed.

Another general procedure for intramolecular HAS reactions involves cyclization of pendant alkyl radicals onto arenes (Scheme 9.3), and the alkyl radical intermediates **11** can be generated in three different ways:

A. By oxidation of an anion in α-position to an EWG such as **12**
B. From **13**, (i) by ET and fragmentation of the C—X bond to give carbon-centered radical and the anion X^-, (ii) by homolysis of the C—X bond, or (iii) by X• abstraction
C. By radical translocation from aryl radical **14** (hydrogen migration)

Once again, the radical **11** adds to the arene, and the σ-complex intermediate **15** is afforded. Finally, the cyclized product **16** is obtained after loss of hydrogen atom or alternative pathways (Scheme 9.1).

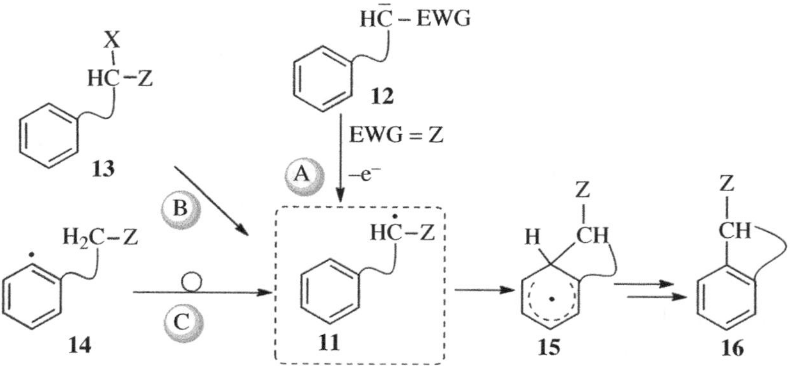

SCHEME 9.3 Different strategies of intramolecular HAS reactions by alkyl radicals.

The ring closure product can also be obtained by intermolecular–intramolecular tandem reactions. The first step is initiated by a bimolecular radical reaction, for example, the addition of radicals R• onto unsaturated bonds (generally alkenes and alkynes [12]) to generate radical **17**, with the second step involving intramolecular HAS from **17** to yield the cyclohexadienyl radical **18**, which finally affords **19** (Eq. 9.3). Another possibility is the addition of the radical R• to a carbene-type compound (e.g., isonitriles) to afford intermediate radical **20** that adds to the arene to afford the cyclohexadienyl radical **21** and finally yield **22** (Eq. 9.4):

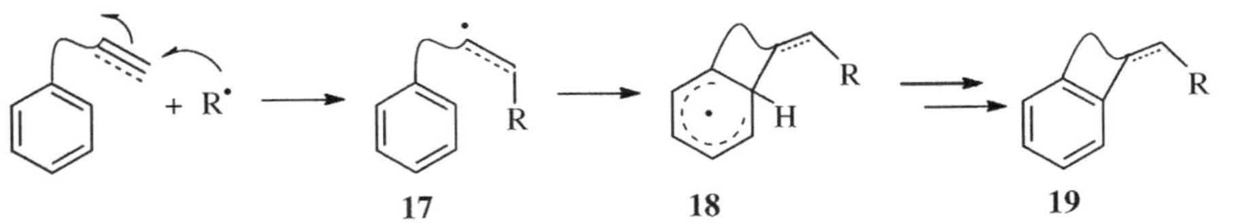

(9.3)

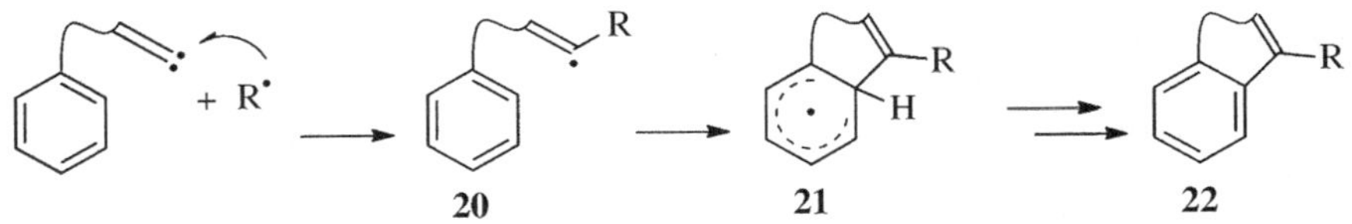

(9.4)

Regarding the regioselectivity of the cyclization, the 5-ring closure is generally favored over the 6-one (Scheme 9.4) [13]. Upon generation of radical **23**, kinetically favored 5-*ipso* attacks result in the formation of spirocyclohexadienyl radical **24** [13a, 14], which can undergo a concerted ring expansion [15] to furnish the thermodynamically more stable radical **25**. This ultimately yields product **26** after rearomatization. Alternatively, aryl radical **24** can form the fused cyclic system **27** via a 3-ring closure, followed by successive neophyl rearrangement [16] to give **28** and finally the regioisomer product **29**. The formation of the product **29** can also be explained through an initial 6-cyclization, which gives rise to the same aryl radical intermediate **28**.

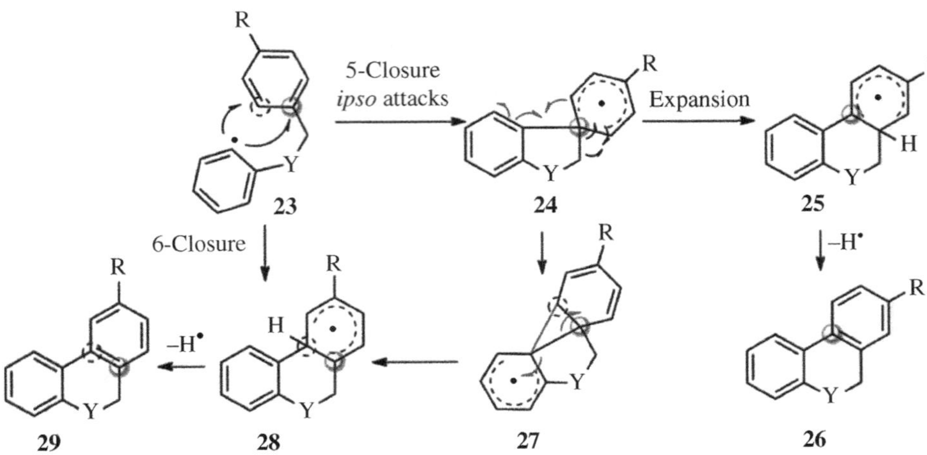

SCHEME 9.4 Regioselectivity of intramolecular HAS.

In some cases, aryl radical migration appeared as a side reaction [16a, 17], with the most extensively investigated one being the 1,2-aryl migration in β-aryl carbon-centered radicals, called the neophyl-type rearrangement (Eq. 9.5), where the neophyl radical **30** is converted to tertiary radical

32 via a cyclohexadienyl radical 31 [18]. This rearrangement has been used as radical clock in kinetic competition experiments. The rate of neophyl rearrangement is $k = 762\,s^{-1}$ at 25°C [19]:

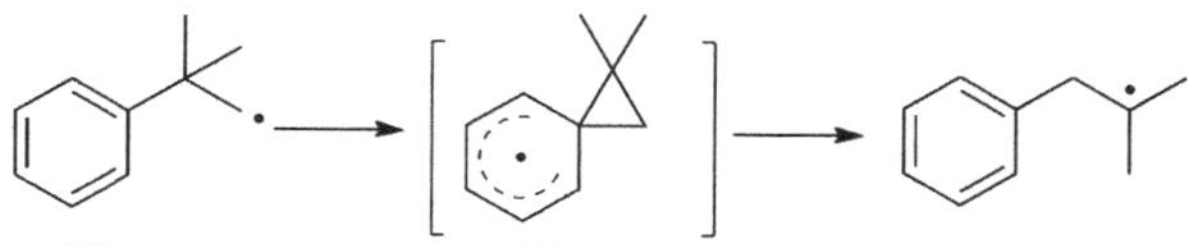

30 31 32

(9.5)

Aryl migrations in free radicals chemistry are not restricted to 1,2-migrations; 1,4- and 1,5-migrations of aryl groups are also an efficient method for C—C bond formation. Again, these migrations are usually considered to be side reactions, although some synthetically useful examples have been reported in the literature [17].

This chapter aims to provide an overview of the present state of the art of the HAS reactions and has been organized according to three different routes available to generate the initial radical intermediates: (i) homolysis, (ii) free radical initiators, and (iii) redox reactions (including BHAS reactions). However, owing to space limitations, only arylation reactions using carbon-centered radicals in combination with arenes (benzenoid) will be discussed. Heteroarenes can also act as radical acceptors in these processes; actually, due to marked polar effects, electron-poor heteroarenes can be used as radical traps for detecting the formation of nucleophilic alkyl radicals. Moreover, the alkylation of various heteroarenes, as radical acceptors, has been successfully achieved by Minisci reactions [20]. There are reports that highlighted the potential of xanthates as alkyl radical precursors in HAS [21], but these have not been included in this chapter.

9.2 RADICALS GENERATED BY HOMOLYTIC CLEAVAGE PROCESSES: THERMOLYSIS AND PHOTOLYSIS

Thermolysis and photolysis refer to the homolytic σ-bond cleavage upon heating or induced by light (particularly by UV light) to generate the corresponding radicals. In photolysis, cleavage occurs from the excited states (singlet or triplet).

An interesting example for radicals being generated by thermolysis is exemplified in Equation 9.6 [22]. The carbon-centered radical 34 generated from 33 via thermal C—O bond homolysis reacts by intramolecular HAS to yield the oxindole 35. It should be noted that, within these reactions, 2,2,6,6-tetramethylpiperidine-1-oxyl radical (TEMPO) acts as an oxidant in the rearomatization step:

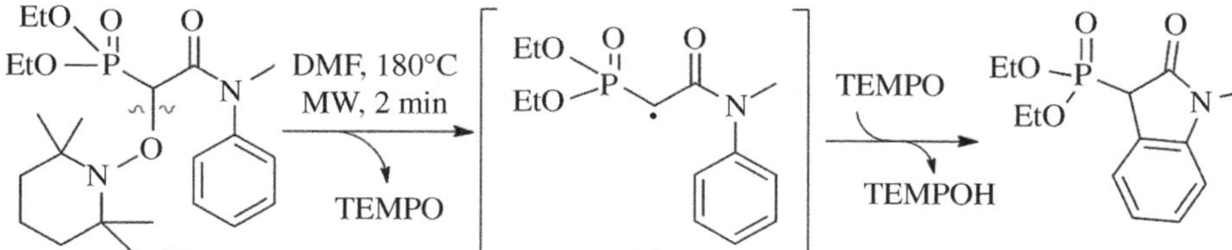

33 34 35 (81%)

(9.6)

In a similar approach, quinolines 40 were prepared starting from alkoxyamine 36 and isonitriles 37 induced by TEMPO using an inter-intramolecular sequence [23] (Eq. 9.7). The precursor 36 via thermal C—O homolysis and addition with aryl isonitrile 37 affords the corresponding imidoyl radical 38. Radical 38 can further react in a 5-*exo*-type cyclization to give the primary radical 39, which can undergo HAS and oxidation to provide the corresponding quinoline 40:

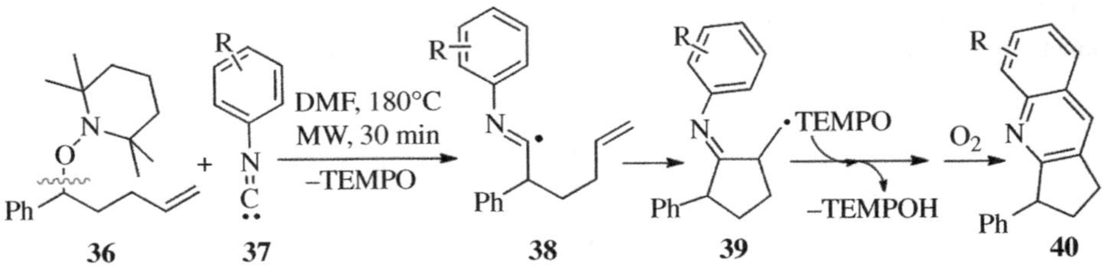

(9.7)

The initiating radicals (Init•) can also be generated from peroxides, azo compounds (Init–Init), etc., via thermolysis or photolysis (Eq. 9.8). A hydrogen abstraction reaction between radical Init• and a given H-donor (R–H) affords Init-H and a new radical (R•) that can initiate the radical chain reaction (Eq. 9.9):

$$\text{Init–Init} \xrightarrow{\Delta \text{ or } h\nu} 2\ \text{Init}^{\bullet} \tag{9.8}$$

$$\text{Init}^{\bullet} + \text{RH} \longrightarrow \text{Init-H} + \text{R}^{\bullet} \tag{9.9}$$

This procedure was employed for the synthesis of 6-substituted phenanthridine **43** from isonitriles **41** (Eq. 9.10). In this example, thermolysis of dibenzoyl peroxide (BPO) was used to generate two benzoyloxy radicals (PhCOO•), which abstracted the α-H of dioxane (Z=O) [24] or from cyclohexane ($Z=CH_2$) [25]. The radical **42** is produced and then intramolecular radical cyclization took place from it to finally form the 6- substituted phenanthridine **43** in good yields:

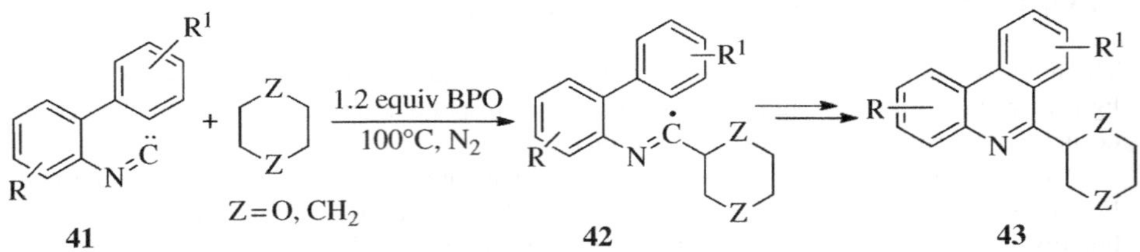

(9.10)

Photoinduced homolysis offers decisive advantages over thermolysis, because stronger bonds can be broken and fewer side reactions are generally observed with most reactions proceeding in the absence of radical initiators. The photochemical generation of vinyl radicals is actually restricted to the photolysis of vinyl halides, which have significant conjugation to enable absorption of UV light capable of this mode of radical initiation. The irradiation of a solution of iodoalkene **44** in MeCN led to the formation of 9-substituted phenanthrenes **45** with vinyl radicals as intermediates (Eq. 9.11) [26]:

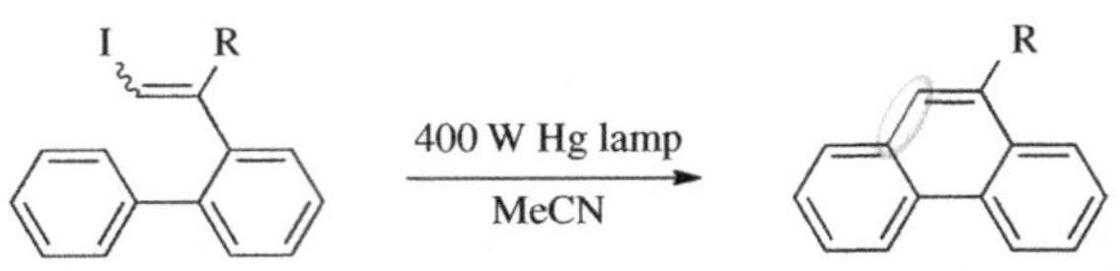

(9.11)

R = H (95%), OAc (80%), Cl (64%)

Photolysis of ArX generally involves radical intermediates through homolytic cleavage of the C–X bond [27], where the reactivity is ArI>ArBr>ArCl. For example, simple irradiation

($\lambda = 254$ nm) of *N*-(2-iodobenzyl)aniline **46** affords the dihydrophenanthridine **47**, which is spontaneously oxidized in the work-up to give the phenanthridine **48** (Eq. 9.12) [28]. Nevertheless, with bromide derivatives, the yields of such reactions were low. Photolysis has been found to be quite general and has been used to synthesize different heterocycles [29] and carbohelicenes [30]:

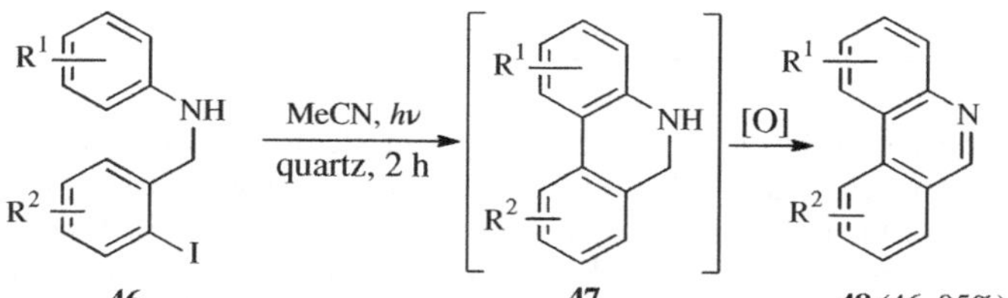

46 **47** **48** (46–95%) (9.12)

In a similar way, heteroaryl radicals can be generated by photolysis to synthesize a great number of different biaryls [31]. Alkyl radicals can be also generated by photolysis of the C—X bond (X = aryl selenide [32] or diazene [33]).

9.3 REACTIONS MEDIATED BY TIN AND SILICON HYDRIDES

Owing to space limitations, only a review of the most important concepts of the radical reactions mediated by tin and silicon hydrides will be discussed. However, these reactions have been excellently reviewed elsewhere [34].

Most radical reactions with reducing reagents can be outlined as shown in Scheme 9.5. Stannyl or silyl radicals are generated by a hydrogen abstraction reaction using a free radical initiator (Init•). Once $R_3Sn^•$ or $R_3Si^•$ is formed, carbon-centered radicals R• can be generated from suitable functional groups. In reduction reactions, R_3SnH or R_3SiH transfers hydrogen to R• to give R–H, thereby regenerating the $R_3Sn^•$ or $R_3Si^•$ radicals that propagate the reaction. It is worth noting that in HAS reactions R• adds to arene to yield the cyclohexadienyl radical **1**. Under these conditions, it seems that the initiator (or radicals derived from it) can abstract hydrogen from the cyclohexadienyl radical to give the aromatization product **3** [8, 34a, 35]. Importantly, the requirement of more than full equivalents of initiator and the dependence of aromatic product yields upon initiator half-lives are indicative of a nonchain reaction mechanism required to continually supplement the HAS with radicals [36].

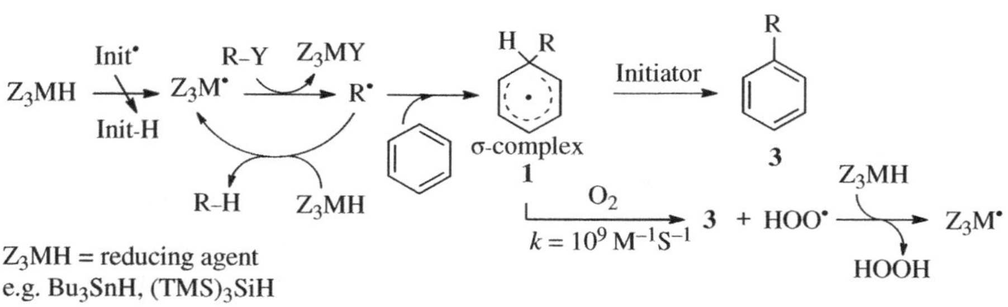

SCHEME 9.5 Reactions with reducing agent.

The O_2 can act as the chain carrier by abstracting hydrogen from **1** and forming the peroxyl radical HOO•, which abstracts hydrogen from Z_3MH to generate H_2O_2 and a radical $Z_3M^•$, thus propagating the chain reaction (Scheme 9.5) [37].

TABLE 9.1 Rate Constants for H-Atom Abstraction of Different Types of Carbon-Centered Radicals from $(TMS)_3SiH$ and Bu_3SnH [38]

Carbon-Centered Radical	$k_{(TMS)_3SiH}$ $(M^{-1} \times s^{-1})$	k_{Bu_3SnH} $(M^{-1} \times s^{-1})$
$Ph^{\bullet}$	3.0×10^8	7.8×10^8
$^{\bullet}CF_2(CF_2)_6CF_3$	5.1×10^7	2.0×10^8
$^{\bullet}CH_2$-Alkyl	3.8×10^5	2.4×10^6

Another factor to be considered is that the hydrogen transfer from Z_3MH to $R^{\bullet}$ is generally faster than $R^{\bullet}$ addition onto arene (Table 9.1) [38]. For example, as mentioned earlier (see Section 9.1), the rate of addition of the phenyl radical to benzene is $4.5 \times 10^5 M^{-1} s^{-1}$ [5]. Therefore, in order to avoid hydrogen transfer competition, the hydride is often added to the reaction mixture slowly (thereby lowering its concentration) by a syringe pump, together with the radical initiator.

As stannyl or silyl radicals have high rates of halogen abstraction, alkyl and aryl halides are the most common sources of radicals [39]. The reactivity order is R–I > R–Br > R–Cl, with R being either alkyl or aryl. Radical generation by chlorine abstraction occurs only in activated alkyl chlorides, such as α-chloroketones (or esters), and polyhalogenated carbons, whereas vinyl and aryl chlorides tend to be nonreactive to organostannyl radicals.

Intramolecular HAS induced by reducing reagents has been intensively applied to obtain a large number of compounds, where either alkyl, vinyl, aryl, or heteroaryl radicals successfully add onto arenes [31b, 40]. Different synthetic strategies, such as ring expansion [41], *ipso* substitution [42], 1,5-hydrogen translocation, and tandem cyclization [43], among others, have been applied to afford important cyclic compounds such as phenanthridines [44], polycyclic arenes [45], 6*H*-benzo[*c*] chromen-6-ones [13a], and strained helicenes [46].

To demonstrate the potential of these transformations, the synthesis of luotonin A (**54**, R = H) from **49** and aryl isonitrile **51** via a cascade radical annulation in the key step is presented (Scheme 9.6) [43b]. Radical addition of **50** onto **51** gave radical **52**, which after 5-*exo-dig* cyclization affords vinyl radical **53**. This radical adds to arene to yield a cyclohexadienyl radical. Then, after oxidation, it delivers the pentacyclic alkaloid luotonin A (**54**, R = H).

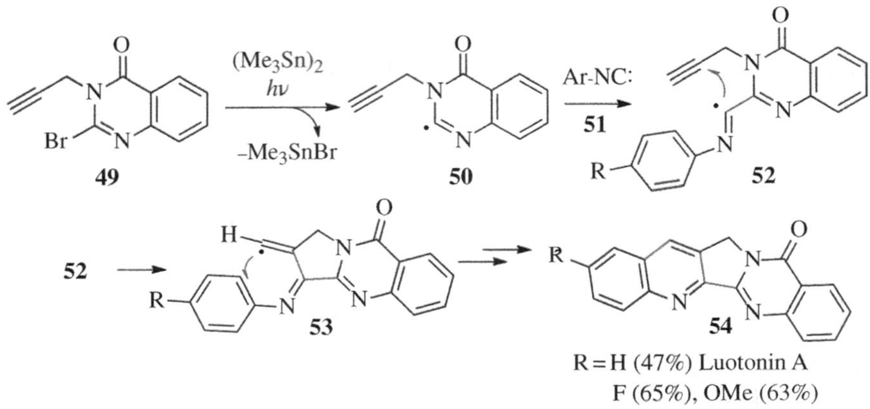

SCHEME 9.6 Synthesis of alkaloid luotonin A and **54**-ring substituted analogues.

As in the case of intramolecular HAS of arenes with aryl radicals (Scheme 9.4), also the vinyl radical intermediate can add to the arene in a 5-*ipso* attack or 6-cyclization, the product ratio dependent on the nature of the substituent on the arene under attack [47].

Aryl radicals have also been extensively used in intramolecular HAS mediated by reducing reagents. An example that shows the potential of this methodology is the synthesis of the strained 9,10-dimethoxy[7] helicene (**56**) by double cyclization from (*Z,Z*)-bis-halostilbene **55** using 1,1′-azobis-cyclohexanecarbonitrile (VAZO) as the initiator (Scheme 9.7) [46]. Furthermore, this intramolecular 6-radical cyclization was successfully applied to the synthesis of phenanthrene derivatives (**57**) [45b].

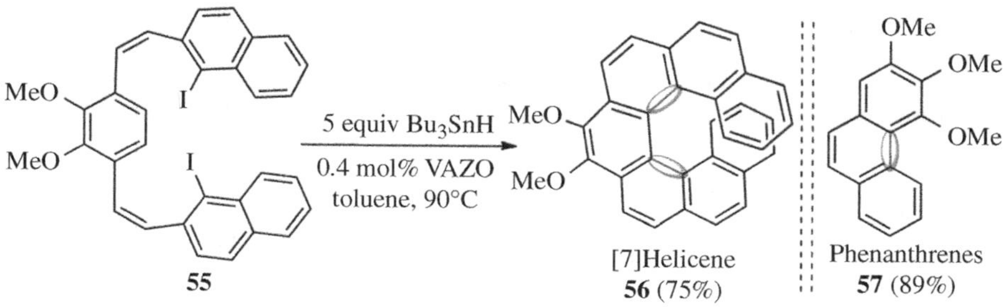

SCHEME 9.7 Synthesis of [7]helicene (**56**) and phenanthrene derivatives (**57**).

Aryl radical intermediates were also successfully used for the preparation of the strained polyarene cavicularin (**60**) from aryl iodide **58** (Eq. 9.13). In this transformation, the key step involved the generation of the σ-complex **59** with a sp^3 carbon that was able to alleviate the strain of the macrocycle and the cyclization proceeded without any significant distortion of arene **A**. The thermodynamic cost of bending the arene **A** was compensated by the rearomatization of arene **D** from the intermediate **59** [48]:

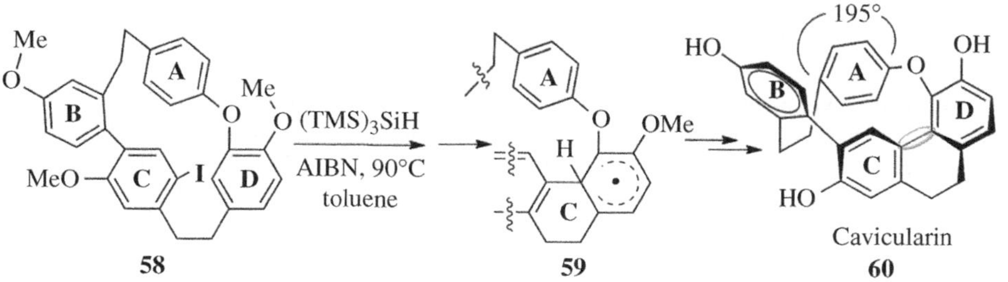

(9.13)

Another elegant example that shows the synthetic value of this type of reaction is the synthesis of the structurally unique alkaloid (+)-scholarisine A precursor **63** (Eq. 9.14) [49]. This synthetic sequence involves the formation of aryl radical from **61**, followed by 1,5-H atom transfer to generate regioselectively the tertiary alkyl radical **62**, which then adds onto the aryl group. This proposed mechanism is represented in a general way in Scheme 9.3 path C:

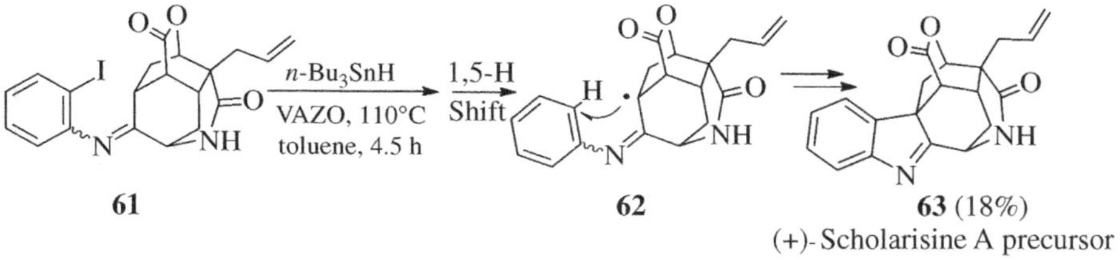

(9.14)

The intermolecular addition of radicals onto arenes normally results in reduced regioselectivity and moderate yields. An interesting approach to overcome these aspects is through an intramolecular *ipso* attack at a position with a good radical leaving group. Thus, aryl radical migrations have been exploited for the stereoselective generation of bonds [16a]. Both carbon and heteroatom (nitrogen, oxygen, phosphorous, silicon, sulfur, etc.) linkers can be used in aryl transfer reactions. An example illustrating this methodology is applied to the synthesis of biaryls and is shown in Scheme 9.8 [50]. Starting from phosphinate **64**, the radical **65** is generated. Then, intramolecular attack onto the arene in the substituted carbon gives the spirocyclohexadienyl radical intermediate **66** (*ipso* attack), and due to the generation of a stable radical **67**, β-scission of **66** is produced. After some steps (H abstraction to afford **68** and fragmentation), the biaryl **69** was obtained in 64% yield.

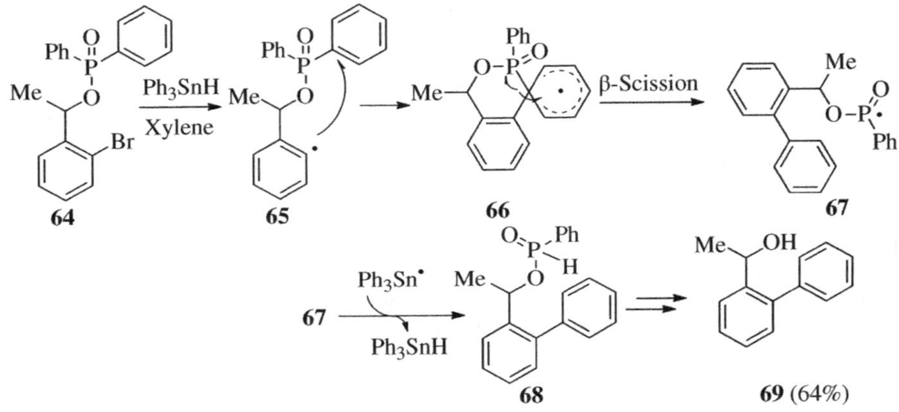

SCHEME 9.8 Synthesis of biaryl **69** by aryl radical migration.

When the cyclopentene derivative **70** was treated with Bu_3SnH and a trace of AIBN, the rearrangement compound **75** was obtained in high yields via an unprecedented radical cyclization cascade. This involves the addition of $Bu_3Sn^•$ to **70** to afford **71**, *ipso*-1,4-aryl migration to produce **74** via spirocyclohexadienyl radical **73**, which finally provides **75** as represented in Scheme 9.9 [51].

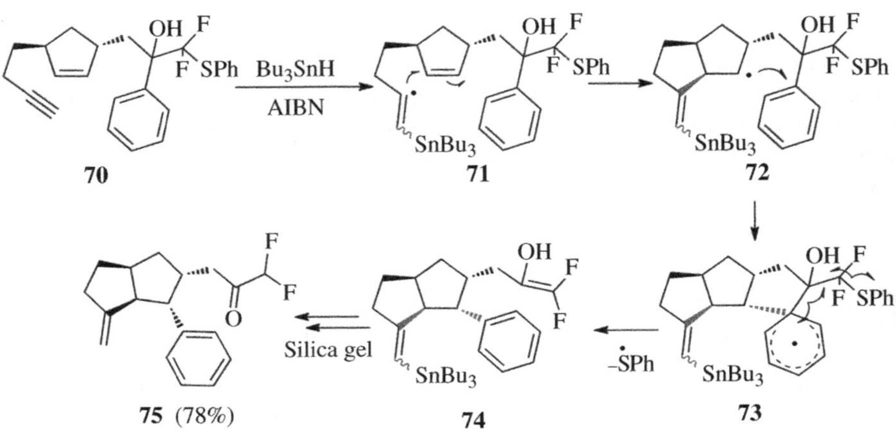

SCHEME 9.9 Radical cyclization/*ipso*-1,4-aryl migration cascade.

9.4 RADICALS GENERATED BY ET: REDOX REACTIONS

Radicals can also be generated via redox processes either by reduction or oxidation of an appropriate precursor, depending on the nature of the radical precursor.

9.4.1 Reducing Metals

9.4.1.1 Thermal Reactions The use of reducing metals (M^{n+}) allows the radical formation by ET from M^{n+} to RX to give $M^{(n+1)+}$ together with the corresponding radical $R^{\bullet}$ and anion X^- (Eq. 9.15):

$$RX + M^{n+} \longrightarrow M^{(n+1)+} + R^{\bullet} + X^- \quad (9.15)$$

Some typical M^{n+} are Co^{2+}, Cu^+, Sm^{2+}, Ni, and Fe salts. For example, intramolecular HAS has been induced by Fe^{2+} salts in the synthesis of carbazoles **77** from *N*-methyldiarylamines **76** (Eq. 9.16) [52]:

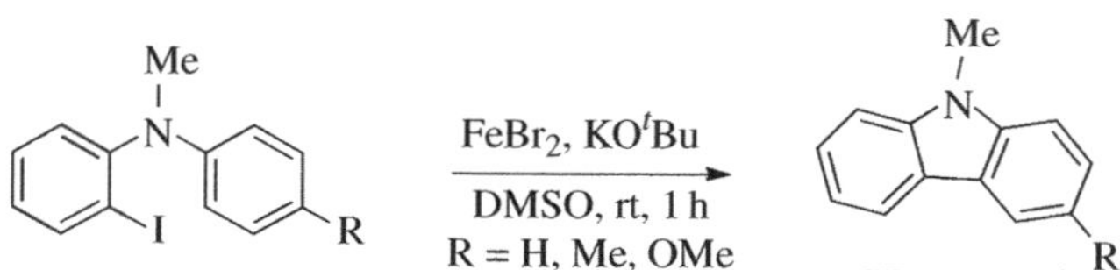

76 **77** (90–94%) (9.16)

The reaction of 2-bromo-*N,N*-diisopropylbenzamide derivatives **78** (X=Br) affords the 2,3-dihydro-1*H*-inden-1-one derivatives **79** and **80** by Ni-mediated intramolecular arylation (Scheme 9.10) [53]. A plausible mechanism could involve intramolecular H-atom abstraction from aryl radical **81** providing the stabilized tertiary alkyl radical **82**, which by attack to the benzene ring affords the radicals **83** and **84** to furnish ultimately the products observed. The same synthesis was

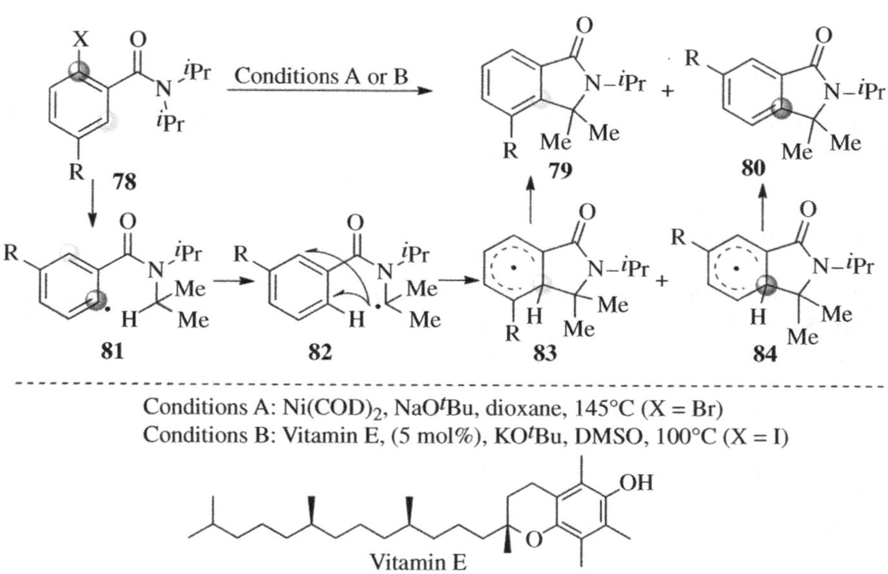

SCHEME 9.10 Synthesis of **79** and **80** derivatives by intramolecular H-atom migration.

successfully achieved using transition metal-free catalysis via BHAS from iodo derivatives **78** (X=I) [54]. In this case, only KO^{t}Bu and vitamin E as additives (5 mol%) in DMSO at 100°C yielded the cyclic products in 60–94%.

The reduction of ROOR via ET from M^{n+} produces the cleavage of the σ-bond (RO-OR), thereby liberating the corresponding alkoxide anion RO^-, the radical RO$^•$, and $M^{(n+1)+}$. A novel strategy for the synthesis of the oxindoles **87** from different *N*-substituted *N*-phenylacrylamides **85** was reported (Scheme 9.11) [55], which involves a selective sp^3 carbon functionalization by C-alkyl/C-aryl bond formation via a copper-catalyzed free radical cascade process. This efficient method utilized the ability of Cu^+ to promote the formation of the peroxide radical and oxidize the cyclohexadienyl radical. The peroxide radical (RO$^•$) fragments to acetophenone and Me$^•$ radical, which is responsible for the generation of the alkyl radical **88** (from alkane **86**) that reacts with **85** to initiate the chain propagation steps.

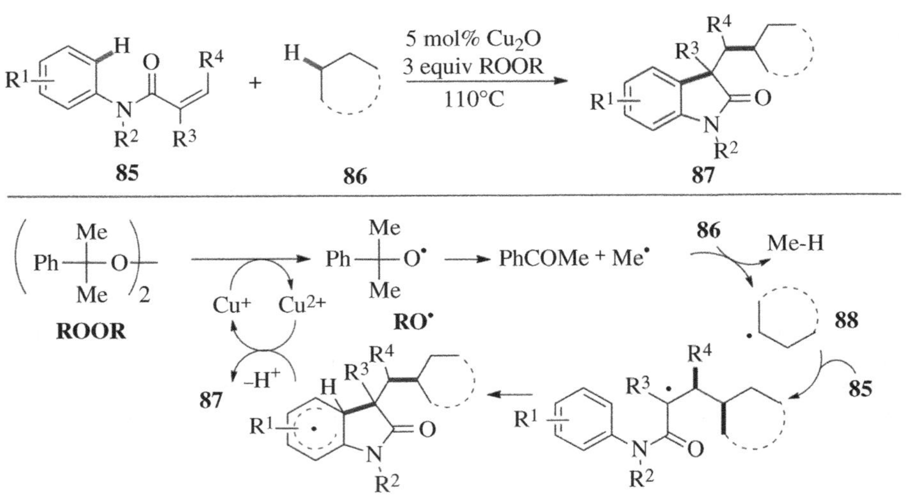

SCHEME 9.11 Radical reaction induced by Cu^+/ROOR.

A related approach used catalysis by $FeCl_3$, *t*-BuOOH (TBHP) as oxidant, and 1,8-diazabicycloundec-7-ene (DBU) as ligand. Several dialkyl ethers, sulfides, and amines have been used to generate alkyl radicals, which add to *N*-substituted *N*-phenylacrylamide [56]. This initiation methodology (M^{n+}/ROOR) was also used in the synthesis of 6-arylated phenanthridines [57, 58], biaryls [59], fluorenones and xanthones [60], etc.

9.4.1.2 Visible Light Photoredox Catalysis with Transition Metal Complexes: Ruthenium and Iridium Salts Redox reactions can also be promoted by light. In the past few years, visible light-induced ET (photoredox) catalysis has been a growing field of organic chemistry due to its low cost and easy availability. Very recently, photochemically induced ET from transition metals has also gained great attention. The Ru^{2+} and Ir^{3+} catalysts, especially tris-chelated complexes of Ru^{2+} with ligands such as 2,2′-bipyridyl [$(Ru(bpy)_3)Cl_2$] and 1,10-phenanthroline [$Ru(phen)_3Cl_2$], have been the most commonly used because of their photophysical and photochemical properties [61].

Ru: Many examples of radical arylations by photoredox catalysis have been recently reported [62–64]. For instance, when $ArN_2^+X^-$ salts are reduced, a C—N bond cleavage occurs to give N_2, an Ar$^•$ radical, and the anion X^-. An intramolecular photo-Pschorr reaction is exemplified by the conversion of $ArN_2^+X^-$ (**89**) into phenanthrene derivatives **92**, in quantitative yields, photocatalyzed by $[Ru(bpy)_3](BF_4)_2$ (Scheme 9.12). In this way, the excited state of this ruthenium complex was able to reduce the salt **89** to the corresponding aryl radical **90**. This in turn cyclized, thereby affording a radical tricyclic intermediate **91** that was oxidized to product **92** by $[Ru(bpy)]^{3+}$, which was formed in the initial ET step. Thus, the ruthenium complex was restored to the original oxidation state and proceeded as a photocatalyst [65].

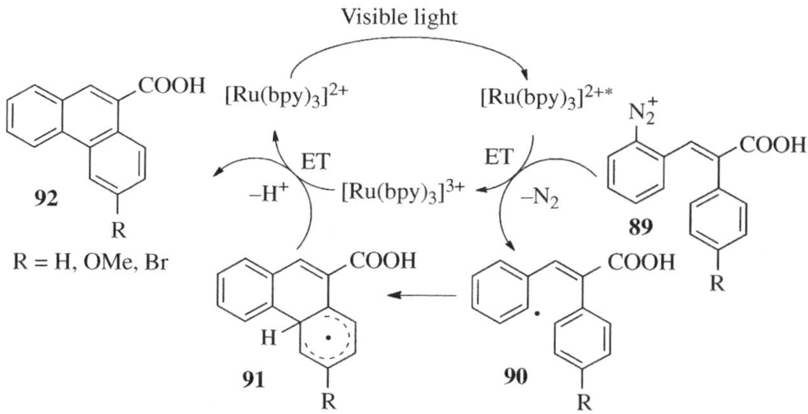

SCHEME 9.12 Aryl radical generated by excited $[Ru(bpy)_3]^{2+*}$.

Using a similar approach, a mild and efficient visible light-mediated diarylation of *N*-substituted *N*-phenylacrylamides **93** with $ArN_2^+X^-$ **94** was developed to afford 3,3-disubstituted oxindoles **95** by constructing two C—C bonds in one step via an inter-intramolecular sequence (Eq. 9.17) [66].

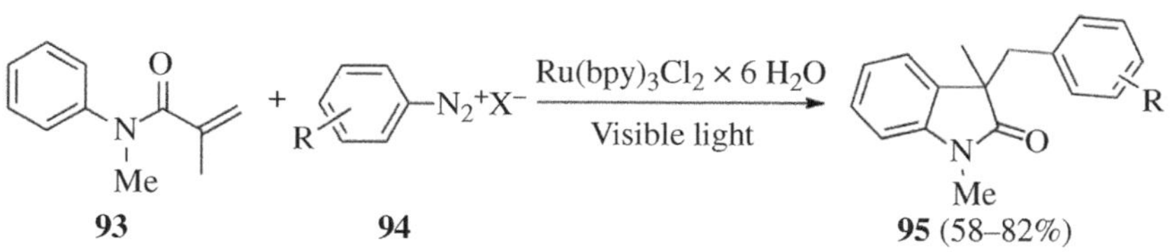

(9.17)

In the presence of a reducing agent, diaryliodonium salts (Ar_2IX) receive an electron and form the radical anion $[(Ar_2I)X]^{\bullet-}$, which then fragments to yield $Ar^{\bullet}$, ArI, and the anion X^-. This $Ar^{\bullet}$ formation also can be achieved by direct photodecomposition from Ar_2IX. Taking this into account, photoredox catalysis has been recently applied to the trifluoromethylation of arenes using Togni's reagent (**97** in Eq. 9.19) [67] or CF_3SO_2Cl as the source of $^{\bullet}CF_3$ radicals and $Ru(phen)_3Cl_2$ as catalysts. In the latter case, reduction of CF_3SO_2Cl by ET provides the $[CF_3SO_2Cl]^{\bullet-}$ radical anion, which fragments to Cl^- ions, SO_2, and $^{\bullet}CF_3$ radical [68]. By this methodology, heterocycles of five and six members as well as unactivated arenes were trifluoromethylated.

Ir: Aryl halides (X=I, Br) were believed to be unsuitable substrates for the visible light photoredox chemistry because of their high redox potentials. However, when the reaction with ArI was carried out in benzene in the presence of the Ir complex $[Cp^*IrHCl]_2$,($Cp^*=\eta^5$-pentamethylcyclopentadienyl) as catalyst and KO^{t}Bu at 80°C, the biaryls were obtained in a 43–73% yield [69]. In addition, in the photostimulated reaction with $Ir(ppy)_3$ (tris(2-phenylpyridine)iridium) as catalyst, biphenyls were obtained at room temperature (rt) from ArI or within the range rt to 90°C from ArBr (Eq. 9.18) [70]:

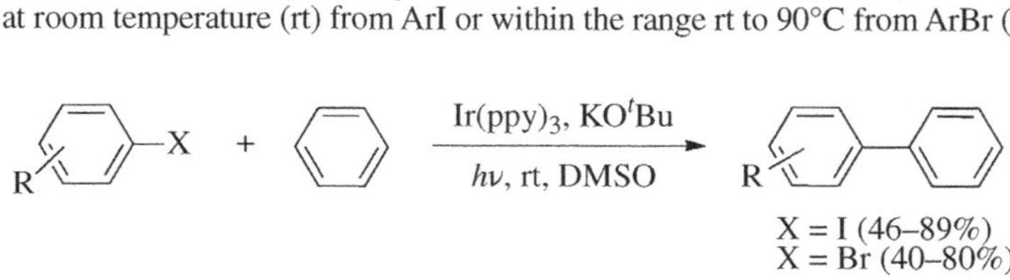

(9.18)

The Ir complexes are capable of affording trifluoromethylation of the different anilines **96** with Togni's reagent **97** as a source of the $^{\cdot}CF_3$ radical to afford **98** (Eq. 9.19) [71]. When the *para* position has no substituent, a mixture of products is obtained:

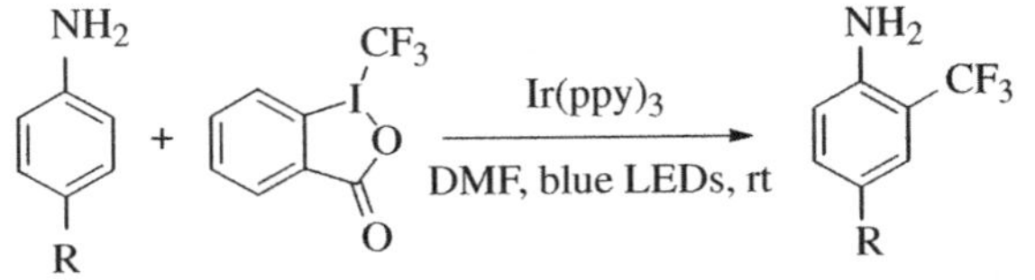

96 **97** **98** (38–81%) (9.19)

Alkyl radicals have been generated using *fac*-Ir(ppy)$_3$ under irradiation. This methodology was used for the synthesis of oxindoles **100** from **99** by an intramolecular HAS using a 40 W household fluorescent lamp at rt (Eq. 9.20) [72]:

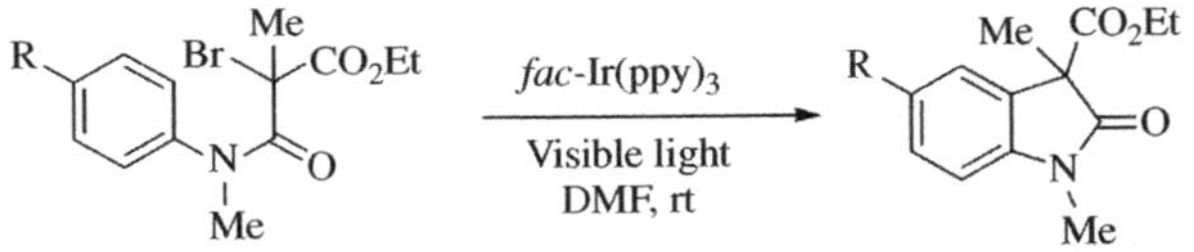

99 R = H, Me, OMe, Cl **100** (93–95%) (9.20)

Organic Photoredox Catalyst: The reduction of $ArN_2^+X^-$ to form $Ar^{\bullet}$ can be likewise accomplished by using the inexpensive visible light-absorbing dye eosin Y [73]. The $Ar^{\bullet}$ radical formed under these conditions has been used for the benzannulation of biaryldiazonium salts with alkynes, via a cascade radical addition and cyclization sequence (Scheme 9.13) [74]. In this approach, the photoexcitation of eosin Y by visible light generates excited eosin Y*, which reduces the diazonium salt **101** to form biaryl radical **102** and eosin Y$^{\bullet+}$. Once the cyclohexadienyl radical **104** is formed from **103**, it is oxidized to **105** by eosin Y$^{\bullet+}$ to afford **106** and regenerates the photocatalyst eosin Y, which continues the chain propagation.

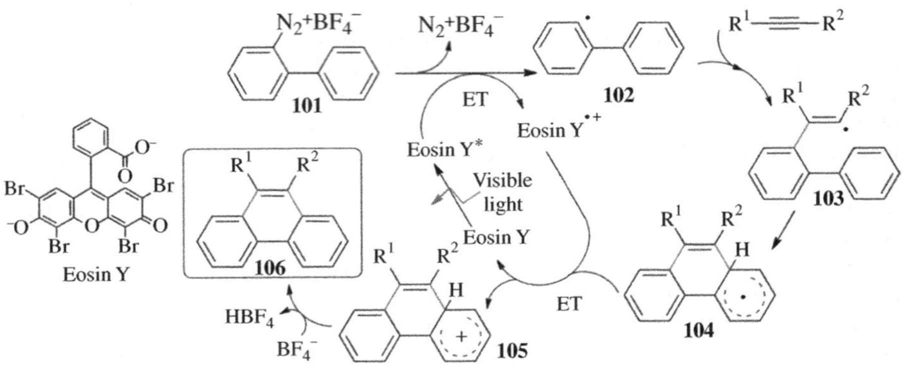

SCHEME 9.13 Benzannulation of biaryldiazonium and alkynes with eosin Y.

9.4.2 Other Reducing Agents

Examples of organic molecules [75] used as reducing agents include ascorbic acid, which has been successfully employed as a radical initiator in a C—H arylation of benzene with anilines **107** to afford the biaryl compounds **108** at rt [76] (Eq. 9.21):

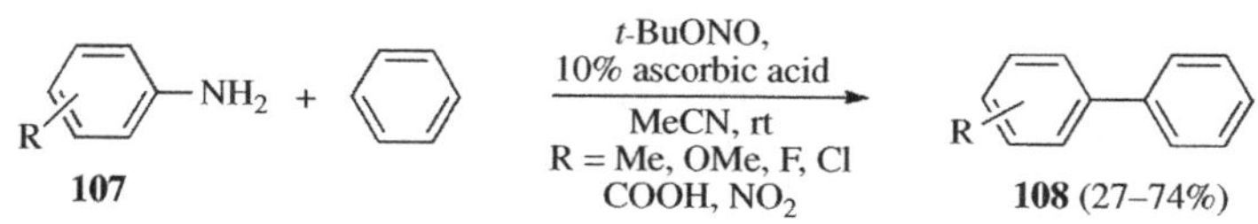

107 **108** (27–74%) (9.21)

The anilines **107** were transformed to the corresponding diazonium salts by *tert*-butyl nitrite (*t*-BuONO) *in situ*, and then ET through an inner-sphere mechanism generates N_2, an ascorbyl radical, and Ar$^{\bullet}$, which undergoes a HAS reaction with benzene as acceptor.

Similarly, the synthesis of 6-aryl-phenanthridines by aryl radical addition onto 2-isocyanobiphenyls was recently reported [77]. Aryl radicals were generated from diazonium salts, which were prepared *in situ* from anilines by diazotization. With the same approach, 3-benzyl-3-alkyloxindoles were prepared from diazotization of anilines to afford aryl radicals, which react with acrylamides, similar to compounds **85**, in one-pot procedure [78].

Sulfinatodehalogenation [79]: Different sources of sulfur dioxide radical anion, such as $Na_2S_2O_4$, $NaSO_2CH_2OH$ (Rongalite), or $Zn(SO_2CH_2OH)_2$ (Decroline), have been commonly used in perfluoroalkylation reactions [80]. The reaction of R_fX (X=I, Br) with $Na_2S_2O_4$ in MeCN/H_2O or DMF/H_2O, known as "sulfinatodehalogenation," provides a cheap, nontoxic, and simple method to generate perfluoroalkyl radicals ($R_f^{\bullet}$) from commercially available reagents [81]. Inert R_fCl can be activated in the presence of $Na_2S_2O_4$ in DMSO at 75–80°C [82, 83], whereas zinc sulfinate salts combined with TBHP can be used to transfer alkyl and $R_f^{\bullet}$ radicals to heterocycles [84].

In sulfinatodehalogenation, a reaction is induced by dissociation of $S_2O_4^{2-}$ to form $SO_2^{\bullet-}$ radical anions. Then, ET from $SO_2^{\bullet-}$ to R_fX after C—X bond fragmentation gives the electrophilic $R_f^{\bullet}$ radical, SO_2, and the anion X^- (Eq. 9.22):

$$R_fX + SO_2^{-\bullet} \longrightarrow R_f^{\bullet} + SO_2 + X^- \qquad (9.22)$$

According to this principle, perfluoroalkylation was performed with R_fX on aromatic nuclei bearing EDG [85, 86]. For example, aniline **109** is transformed into *ortho* and *para* perfluoroalkyl derivatives **110** and **111** using $R_fI/Na_2S_2O_4$ as the reagent system (Eq. 9.23) [87, 88]. Similar perfluoroalkylations have also been realized in aqueous DMF [89]:

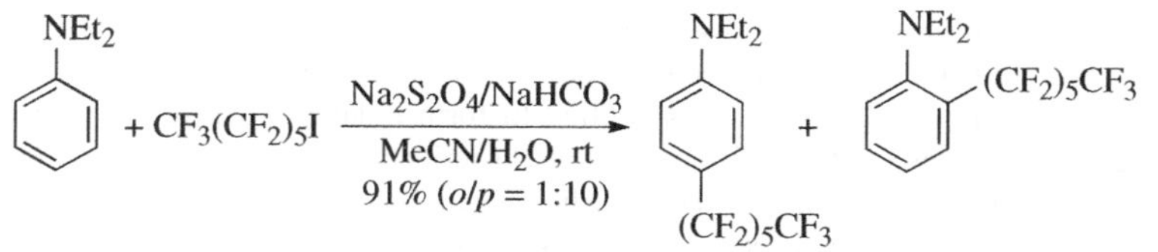

(9.23)

109 **110** **111**

9.4.3 Oxidizing Metals

Metal salts of Ce^{4+}, Mn^{3+}, Fe^{3+}, Cu^{2+}, and Ag^+ can be used for the oxidation of electron-rich species with concomitant formation of free radicals [90]. In this way, these oxidants are able to promote regio-, chemo-, and stereoselective free radical reactions.

The exposure of electron-rich species such as amines, hydrazines, or enolate anions to an oxidant results in ET, which generates a free (or metal-bonded) radical or radical cation [91]. For example, the oxidation of α-alkyl β-ketoester (**112**) with Mn^{3+} results in enolization to yield **113** in the rate-determining step, and this is followed by ET to form the electrophilic radical **114** [92] (Scheme 9.14). In the HAS reaction, cyclohexadienyl radicals **115** are oxidized by $Mn(OAc)_3$ to cation **116**, which by deprotonation regenerates the aromatic system **117**.

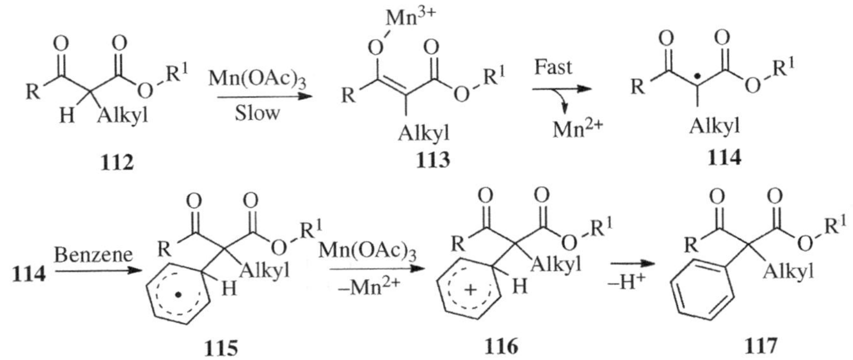

SCHEME 9.14 Carbon radicals generated by $Mn(OAc)_3$.

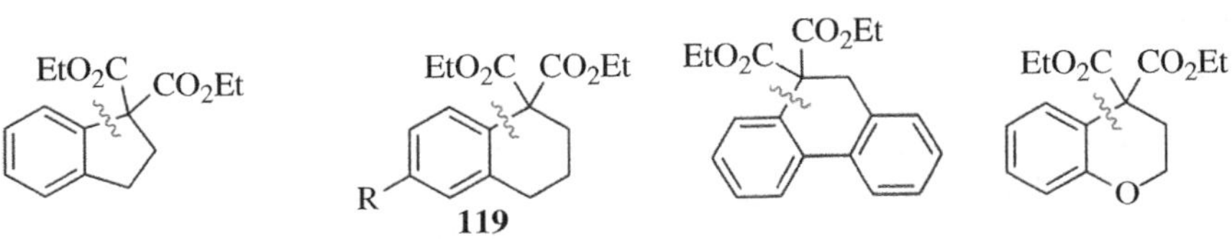

118 (39–51%) R = H (85%), NHCOMe (88%), NO_2 (85%), OMe (80), **120** (93%) **121** (90%)

FIGURE 9.1 Synthesis of polycyclic compounds using $Mn(OAc)_3$.

The intermolecular addition of electrophilic radicals to aromatic rings is more efficient with electron-rich substrates such as toluene, anisole, and heteroaromatics than with benzene. Intramolecular addition mediated by $Mn(OAc)_3$ [93], $(NH_4)_2Ce(NO_3)_6$ (CAN) [94], Fe^{3+} [95], and Cu^{2+} [96] salts has been reported [97].

The approach shown in Scheme 9.14 has further been used for ring closure reactions to yield cycles of different sizes as well as bridge structures [94, 98]. Cyclization of β-ketoesters, β-cyanoesters, and β-diketones has also been realized on electron-rich arenes [93, 99]. For instance, the cyclization of α-malonyl radicals was applied to form indane (**118**), tetrahydronaphthalene (**119**), dihydrophenanthrene (**120**), and chromane (**121**) derivatives using $Mn(OAc)_3$ (Fig. 9.1) [93].

Another important feature is the oxidative radical cyclization of a polyene chain to give a six-membered ring; this reaction proceeds with high stereoselectivity through chair transition states [100]. Thus, cyclization mediated by Mn^{3+} is a powerful approach for the synthesis of benzo-fused polycyclic natural products, such as **123** from **122** (Eq. 9.24):

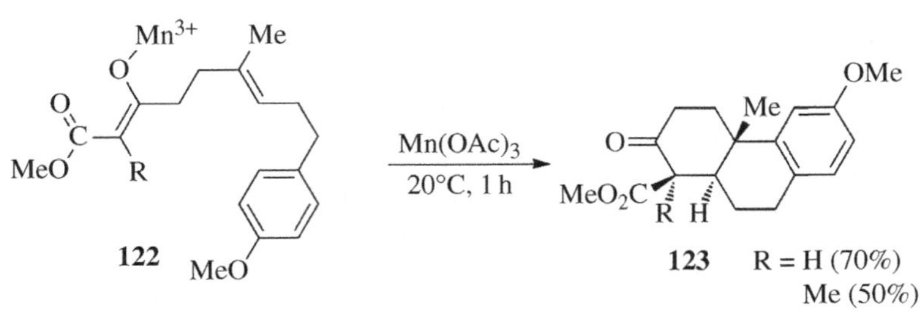

(9.24)

A new mode of oxidative organocatalytic activation has been reported, termed organo-SOMO catalysis, which has been successfully applied using an enantioselective intramolecular α-arylation of aldehydes via catalytic oxidative radical cyclization (Eq. 9.25) [101]. In this approach, the exposure of an aryl-tethered aldehyde (**124**) to a chiral secondary amine catalyst **125** and a suitable oxidant leads to enantio-enriched **126**:

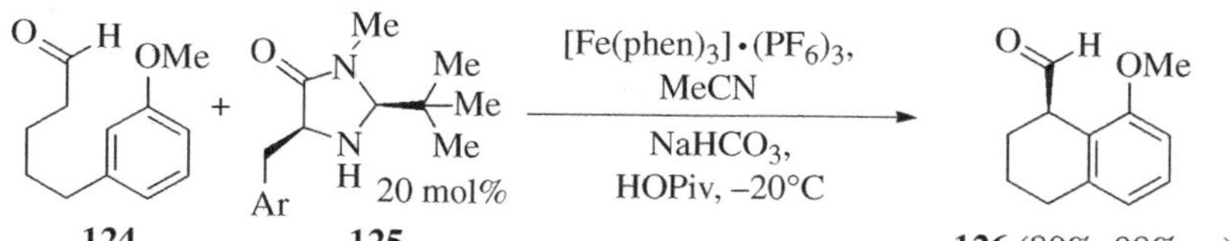

124 **125** **126** (80%, 98% ee) (9.25)

This approach has been extended to the enantioselective construction of multiple C—C bonds and contiguous stereocenters. In this way, an oxidative organocatalytic polycyclization reaction of suitable functionalized aldehydes such as **127** afforded di- to hexacyclization products in good yields and useful enantioselectivities (Eq. 9.26) [102]:

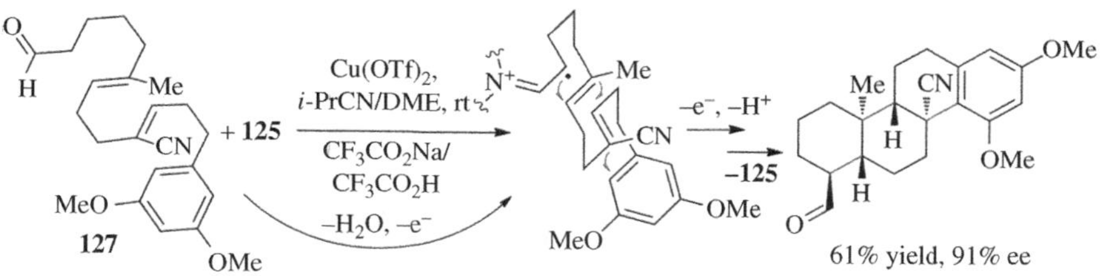

127

(9.26)

The mechanism of the reaction, the nature of the intermediates, and the generation of the enantioselectivity of this unprecedented transformation are outside the scope of the present chapter [103].

Arylhydrazines ($ArNHNH_2$) are oxidized by oxygen or by oxidizing metal ions to generate the corresponding $Ar^\bullet$ radical via the formation of unstable diazenes. The formation of biaryls was carried out by oxidizing $ArNHNH_2$ in the presence of $Mn(OAc)_3$ in benzene [104]. Oxidative coupling was observed in good yield with $ArNHNH_2$ substituted with either EDG or EWG, although the highest yields were attained with EDG at the *para* position. For example, oxidation of $ArNHNH_2$ was used in the regioselective arylation of 4-substituted anilines in the synthesis of 2-aminobiphenyls [105].

Oxidation of $RB(OH)_2$ with $Mn(OAc)_3$ also produces $R^\bullet$ that are subsequently added to aromatic rings to generate alkylated or arylated compounds [106], for example, in the synthesis of 6-alkyl- or 6-aryl-phenanthridines **129** from isonitriles **128** (Eq. 9.27) [107]:

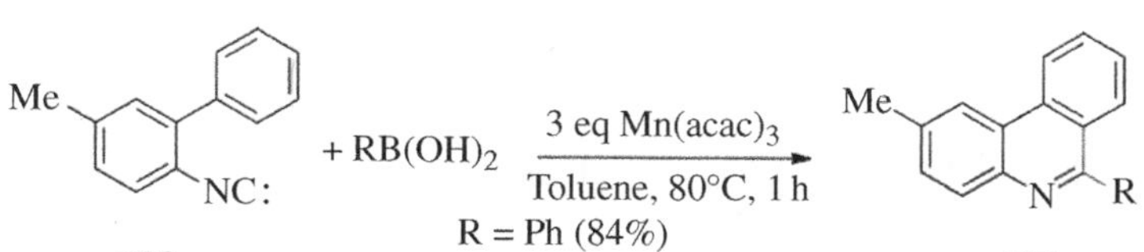

128 **129** (9.27)

9.4.4 Base-Promoted Homolytic Aromatic Substitution (BHAS)

Over the recent years, there have been many reported syntheses of biaryl compounds using BHAS [108], thereby avoiding the use of stoichiometric amounts of tin or silicon reagents, radical initiators, or transition metals. In 2008, Itami showed that KO^{t}Bu caused the addition of the aryl part of ArI or ArBr to pyrazine and other electron-poor arenes under elevated temperatures or MW irradiation (Eq. 9.28) [109]:

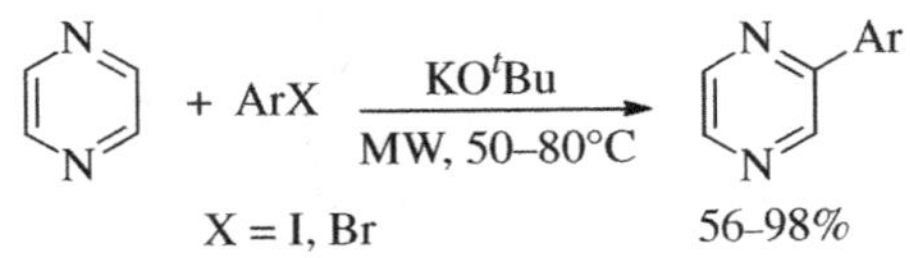

(9.28)

Furthermore, different research groups have reported independently the construction of biaryl compounds from unactivated aromatic compounds by direct C—H activation using NaO^{t}Bu or KO^{t}Bu and 1,10-phenanthroline [110, 111] or DMEDA [112] as ligands (Eqs. 9.29 and 9.30):

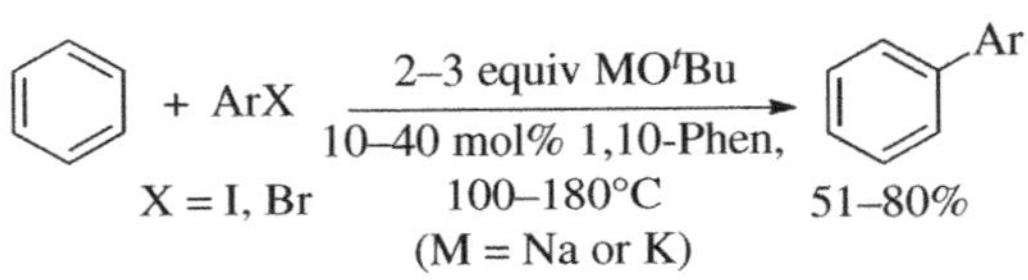

(9.29)

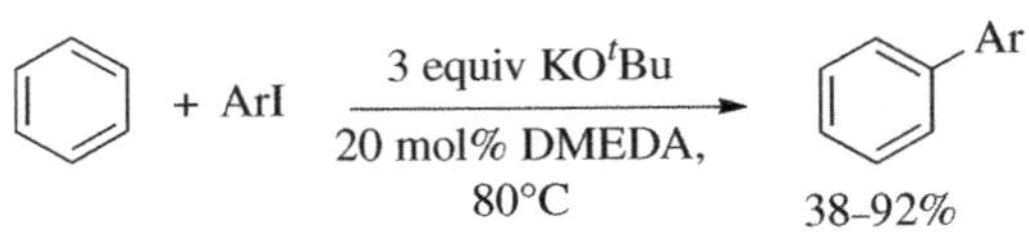

(9.30)

It was found that ArX gives Ar$^•$ radicals in the presence of different ligands, with t-BuO$^-$ anion at high temperatures [113]. This proposed BHAS mechanism is represented in Scheme 9.1 path C, where the formation of radical Ar$^•$ involved an ET to ArX as the initiation step to yield an ArX$^{•-}$, which then fragmented to produce an Ar$^•$ radical and an X$^-$ ion [108, 114].

The facts that the reactions require a large excess of acceptor, the appearance of the reduced haloarene with an H/D isotope effects, and inhibition by TEMPO or other radical scavengers are consistent with radicals being involved.

The presence of radical anions as intermediates has been proposed in the reactions of 1,2-, 1,3-, or 1,4-dihalobenzenes with benzene, which are then converted into disubstituted products (o-, m-, and p-terphenyl, **133**; Scheme 9.15) [110–112, 115]. Once the radical anion **130** is formed as the intermediate, it can follow two possible pathways. One of these is an intermolecular ET to give halobiphenyl **131** (monosubstituted product) (Path A). The other possibility is an intramolecular ET

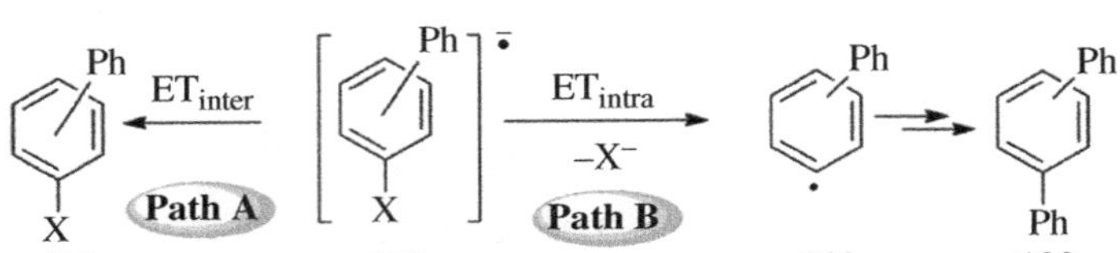

131 **130** **132** **133**

X = I, Br, Cl

SCHEME 9.15 Competing ET for the radical anions.

to the C—X bond that after fragmentation gives aryl radical **132**, which finally provides **133** (Path B). However, in all cases, only small amounts of monosubstituted product **131** were formed, even after very short reaction times.

Different kinds of ligands, such as quinoline-1-amino-2-carboxylic acid [116], amino acid proline [117], quinazoline alkaloid vasicine [118], simple alcohols [119], amino-linked nitrogen heterocyclic carbenes [120], porphyrins [121], 2,6-bis(imino)pyridine [122], and macrocyclic pyridine pentamer [123], among others, have been designed to promote biaryl synthesis from ArX and benzene in the presence of KO^{t}Bu. A different approach toward biaryl syntheses (ligand-free protocol) was reported using photoinduced direct C—H arylation of benzene in the presence of KO^{t}Bu at rt [115], which proceeds through photo- and BHAS.

A number of authors have suggested that ET from a complex **134** of NaO^{t}Bu or KO^{t}Bu with a "ligand" to PhI is the initiation step in the radical process to give the radical cation complexes **135** and the radical anion of PhI **136** (Eq. 9.31) [110, 111, 120, 124]:

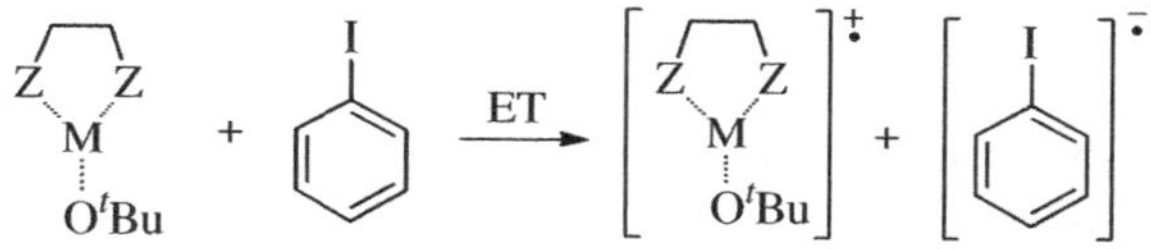

134 **135** **136** (9.31)

Z = NHMe, OH, etc.
M = Na, K

However, when Murphy *et al.* calculated the thermodynamics for the basic ET reaction between phenanthroline–MO^{t}Bu complexes **134** and PhI by DFT methods, they found that this step is highly endergonic both for NaO^{t}Bu (ΔG=63.9 kcal mol^{-1}) and for KO^{t}Bu (ΔG=59.5 kcal mol^{-1}) [125]. Therefore, they proposed that the initiation step might occur based on formation of heterocycle-derived organic electron donors generated *in situ* from phenanthroline (additive) or pyridine (solvent) without the requirement of KO^{t}Bu, or a simple derived complex, to act as organic super-electron donor in the initiation step for the generation of aryl radicals. When the coupling of simple PhI and benzene proceeds in the absence of additives (or ligands), a benzyne mechanism in the initiation is proposed. It is worth noting that KO^{t}Bu alone at high temperature (ca. 160°C) can promote the reaction without the addition of a ligand [126].

Intramolecular BHAS has been found to be an efficient method to synthesize different structures [127, 128], for example, phenanthridinones and dibenzoazepinones [129], pyrrolophenanthridone and pyridophenanthridone [130], phenanthridines [130, 131], and dibenzo[*d*,*f*][1,3]dioxepine [132], among others. Moreover, the total synthesis of trispheridine (a phenanthridine alkaloid), nornitidine, and norchelerythrine (benzo[*c*]phenanthridines) has been reported [130b].

9.5 SUMMARY AND OUTLOOK

Although the examples shown can only give a partial insight into the area of HAS reactions, with only carbon-centered radicals and arenes (benzenoid) as acceptors being discussed, we hope that we have nevertheless been able to illustrate that HAS is a powerful synthetic tool. The diversity of the chemistry and the wide effectiveness of the routes used for the generation of radicals from different sources, such as homolysis (thermolysis and photolysis), or the use of free radical initiators with tin hydrides or silanes, or redox and photoredox catalysis reactions, have revealed HAS to be a suitable alternative to gain access to complex scaffolds. Its utility in preparative chemistry is remarkable, with efficient methodologies now emerging, which should open up new perspectives and lead to more diverse applications.

ABBREVIATIONS

AIBN	α,α′-Azobisisobutyronitrile
BHAS	Base-promoted homolytic aromatic substitution
BPO	Benzoyl peroxide
bpy	2,2′-Bipyridyl
CIDNP	Chemically induced dynamic nuclear polarization
DBU	1,8-Diazabicycloundec-7-ene
DMEDA	*N,N′*-Dimethyl-1,2-ethanediamine
DMF	Dimethylformamide
DMSO	Dimethyl sulfoxide
EDG	Electron-donating group
ESR	Electron spin resonance
ET	Electron transfer
EWG	Electron-withdrawing group
fac	Facial
HAS	Homolytic aromatic substitution
M^n	Reducing metals
MW	Microwave
phen	1,10-Phenanthroline
SOMO	Singly occupied molecular orbital
TBHP	*tert*-Butyl hydroperoxide
TEMPO	2,2,6,6-Tetramethylpiperidine-1-oxyl radical
TMS	Trimethylsilyl
VAZO	1,1′-Azobis-(cyclohexanecarbonitrile)

REFERENCES

[1] (a) Studer, A. and Bossart, B. (2001) In *Radicals in Organic Synthesis*, ed. Renaud, P. and Sibi, M. P., Wiley-VCH, Weinheim, vol. **2**, pp. 62–80; (b) Fossey, J., Lefort, D., and Sorba, J. (1995) In *Free Radicals in Organic Chemistry*, John Wiley & Sons, Ltd, Chichester, United Kingdom, pp. 167–180.

[2] (a) Crich, D. and Hwang, J.-T. (1998) *J. Org. Chem.*, **63**, 2765–2770; (b) Crich, D., Grant, D., Krishnamurthy, V., and Patel, M. (2007) *Acc. Chem. Res.*, **40**, 453–463.

[3] (a) Lunazzi, L., Placucci, G., and Grossi, L. (1982) *J. Chem. Soc. Perkin Trans.* **2**, 875–880; (b) Fahrenholtz, S. R. and Trozzolo, A. M. (1972) *J. Am. Chem. Soc.*, **94**, 282–283.

[4] Citterio, A., Minisci, F., Porta, O., and Sesana, G. (1977) *J. Am. Chem. Soc.*, **99**, 7960–7968.

[5] Scaiano, J. C. and Stewart, L. C. (1983) *J. Am. Chem. Soc.*, **105**, 3609–3614.

[6] (a) Tiecco, M. (1981) *Pure Appl. Chem.*, **53**, 239–258; (b) Bravo, A., Bjørsvik, H.-R., Fontana, F., Liguori, L., Mele, A., and Minisci, F. (1997) *J. Org. Chem.*, **62**, 7128–7136.

[7] Fiorentino, M., Testaferri, L., Tiecco, M., and Troisi, L. (1977) *J. Chem. Soc. Chem. Commun.*, 316–317.

[8] McLoughlin, P. T. F., Clyne, M. A., and Aldabbagh, F. (2004) *Tetrahedron*, **60**, 8065–8071.

[9] Smith, M. B. and March, J. in *March's Advanced Organic Chemistry. Reactions, Mechanisms, and Structure*, John Wiley & Sons, Inc., Hoboken, New Jersey, 2007.

[10] Pschorr, R. (1896) *Ber. Dtsch. Chem. Ges.*, **29**, 496–501.

[11] Gomberg, M. and Bachmann, W. E. (1924) *J. Am. Chem. Soc.*, **42**, 2339–2343.

[12] Wille, U. (2013) *Chem. Rev.*, **113**, 813–853.

[13] (a) Bowman, W. R., Mann, E., and Parr, J. (2000) *J. Chem. Soc. Perkin Trans.* **1**, 2991–2999; (b) Harrowven, D. C., Nunn, M. I. T., Newman, N. A., and Fenwick, D. R. (2001) *Tetrahedron Lett.*, **42**, 961–964.

[14] Bowman, W. R., Heaney, H., and Jordan, B. M. (1991) *Tetrahedron*, **47**, 10119–10128.

[15] Benati, L., Calestani, G., Leardini, R., Minozzi, M., Nanni, D., Spagnolo, P., Strazzari, S., and Zanardi, G. (2003) *J. Org. Chem.*, **68**, 3454–3464.

[16] (a) Studer, A. and Bossart, M. (2001) *Tetrahedron*, **57**, 9649–9667; (b) Leardini, R., Nanni, D., Pareschi, P., Tundo, A., and Zanardi, G. (1997) *J. Org. Chem.*, **62**, 8394–8399.

[17] Sasaki, K., Cumpstey, I., and Crich, D. (2012) In *Encyclopedia of Radicals in Chemistry, Biology and Materials*, ed. Chatgilialoglu, C. and Studer, A., John Wiley & Sons, Ltd, Chichester, United Kingdom, vol. **1**, pp. 125–145.

[18] Urry, W. H. and Kharasch, M. S. (1944) *J. Am. Chem. Soc.*, **66**, 1438–1440.

[19] Lindsay, D. A., Lusztyk, J., and Ingold, K. U. (1984) *J. Am. Chem. Soc.*, **106**, 7087–7093.

[20] (a) Minisci, F., Vismara, E., Fontana, F., Morini, G., Serravalle, M., and Giordano, C. (1987) *J. Org. Chem.*, **52**, 730–736; (b) Punta, C. and Minisci, F. (2008) *Trends Heterocycl. Chem.*, **13**, 1–68; (c) Duncton, M. A. J. (2011) *Med. Chem. Commun.*, **2**, 1135–1161.

[21] (a) Coppa, F., Fontana, F., Minisci, F., Pianese, G., Tortoreto, P., and Zhao, L. (1992) *Tetrahedron Lett.*, **33**, 687–690; (b) Zard, S. Z., in *Radicals in Organic Synthesis*, ed. Renaud, P. and Sibi, M. P., Wiley-VCH, Weinheim, 2001, vol. **1**, pp. 90–106; (c) Binot, G. and Zard, S. Z. (2005) *Tetrahedron Lett.*, **46**, 7503–7506; (d) Bacque, E., Qacemi, M. E., and Zard, S. Z. (2004) *Org. Lett.*, **6**, 3671–3674.

[22] Teichert, A., Jantos, K., Harms, K., and Studer, A. (2004) *Org. Lett.*, **6**, 3477–3480.

[23] Janza, B. and Studer, A. (2006) *Org. Lett.*, **8**, 1875–1878.

[24] Wang, L., Sha, W., Dai, Q., Feng, X., Wu, W., Peng, H., Chen, B., and Cheng, J. (2014) *Org. Lett.*, **16**, 2088–2091.

[25] Sha, W., Yu, J.-T., Jiang, Y., Yang, H., and Cheng, J. (2014) *Chem. Commun.*, **50**, 9179–9181.

[26] Campos-Gómez, E., Campos, P. J., González, H. F., and Rodríguez, M. A. (2012) *Tetrahedron*, **68**, 4292–4295.

[27] (a) Wolf, W. and Kharasch, N. (1961) *J. Org. Chem.*, **26**, 283–284; (b) Sharma, R. K. and Kharasch, N. (1968) *Angew. Chem. Int. Ed.*, **7**, 36–44; (c) Ho, T., Ku, C., and Liu, S. H. (2001) *Tetrahedron Lett.*, **42**, 715–717.

[28] Linsenmeier, A. M., Williams, C. M., and Bräse, S. (2011) *J. Org. Chem.*, **76**, 9127–9132.

[29] Kessar, S. V., Gupta, Y. P., Dhingra, K., Sharma, G. S., and Narula, S. (1977) *Tetrahedron Lett.*, **18**, 1459–1462.

[30] Flammang-Barbieux, M., Nasielski, J., and Martin, R. H. (1967) *Tetrahedron Lett.*, **8**, 743–744.

[31] (a) Fozard, A. and Bradsher, C. K. (1967) *J. Org. Chem.*, **32**, 2966–2969; (b) Clyne, M. A. and Aldabbagh, F. (2006) *Org. Biomol. Chem.*, **4**, 268–277; (c) O'Connell, J. M., Moriarty, E., and Aldabbagh, F. (2012) *Synthesis*, **44**, 3371–3377; (d) Dai, Y., Zhang, W., Wang, K., Wang, W., and Zhang, W. (2013) *Tetrahedron*, **69**, 1912–1918.

[32] (a) Murakami, S., Ishii, H., Tajima, T., and Fuchigami, T. (2006) *Tetrahedron*, **62**, 3761–3769; (b) Murakami, S., Kim, S., Ishii, H., and Fuchigami, T. (2004) *Synlett*, 815–818.

[33] Nakamura, T. and Koga, Y. (1998) *J. Chem. Soc. Perkin Trans.* **2**, 659–662.

[34] (a) Beckwith, A. L. J., Bowry, V. W., Bowman, W. R., Mann, E., Parr, J., and Storey, J. M. D. (2004) *Angew. Chem. Int. Ed.*, **43**, 95–98; (b) Bowman, W. R. and Storey, J. M. D. (2007) *Chem. Soc. Rev.*, **36**, 1803–1822; (c) Chatgilialoglu, C. and Timokhin, V. I. (2008) *Adv. Organomet. Chem.*, **57**, 117–181; (d) Jasch, H. and Heinrich, M. R. (2012) In *Encyclopedia of Radicals in Chemistry, Biology & Materials*, ed. Chatgilialoglu, C. and Studer, A., John Wiley & Sons, Ltd, Chichester, United Kingdom, vol. **2**, pp. 1–32.

[35] Minisci, F., Fontana, F., Pianese, G., and Yan, Y. M. (1993) *J. Org. Chem.*, **58**, 4207–4211.

[36] Lynch, M., Hehir, S., Kavanagh, P., Leech, D., O'Shaughnessy, J., Carty, M. P., and Aldabbagh, F. (2007) *Chem. Eur. J.*, **13**, 3218–3226.

[37] Curran, D. P. and Keller, A. I. (2006) *J. Am. Chem. Soc.*, **128**, 13706–13707.

[38] Chatgilialoglu, C. and Newcomb, M. (1999) *Adv. Organomet. Chem.*, **44**, 67–112.

[39] Curran, D. P., Jasperse, C. P., and Collins, M. J. (1991) *J. Org. Chem.*, **56**, 7169–7172.

[40] (a) Markgraf, J. H., Dowst, A. A., Hensley, L. A., Jalobsche, C. E., Kaltner, C. J., Webb, P. J., and Zimmerman, P. W. (2005) *Tetrahedron*, **61**, 9102–9110; (b) Nuñez, A., Sánchez, A., Burgos, C., Alvarez-Builla, J. (2004) *Tetrahedron*, **60**, 6217–6224.

[41] Harrowven, D. C., L'Helias, N., Moseley, J. D., Blumire, N. J., and Flanagan, R. R. (2003) *Chem. Commun.*, 2658–2659.

[42] Clive D. L. J. and Kang, S. (2001) *J. Org. Chem.*, **66**, 6083–6091.

[43] (a) Curran, D. P., Liu, H., Josien, H., and Ko, S.-B. (1996) *Tetrahedron*, **52**, 11385–11404; (b) Tangirala, R., Smith, A., Agama, K., Pommier, Y., and Curran, D. P. (2005) *Synlett*, 2843–2846; (c) Reynolds, A. J., Scott, A. J., Turner, C. I., and Sherburn, M. S. (2003) *J. Am. Chem. Soc.*, **125**, 12108–12109.

[44] Rosa, A. M., Lobo, A. M., Branco, P. S., Prabhakar, S., and Pereira, A. M. D. L. (1997) *Tetrahedron*, **53**, 269–284.

[45] (a) Orito, K., Uchilto, S., Satoh, Y., Tatsuzawa, T., Harada, R., and Tokuda, M. (2000) *Org. Lett.*, **2**, 307–310; (b) Harrowven, D. C., Nunn, M. I. T., and Fenwick, D. R. (2002) *Tetrahedron Lett.*, **43**, 3185–3187.

[46] Harrowven, D. C., Guy, I. L., and Nanson, L. (2006) *Angew. Chem. Int. Ed.*, **45**, 2242–2245.

[47] Benati, L., Leardini, R., Minozzi, M., Nanni, D., Spagnolo, P., and Zanardi, G. (2000) *J. Org. Chem.*, **65**, 8669–8674.

[48] Harrowven, D. C., Woodcock, T., and Howes, P. D. (2005) *Angew. Chem. Int. Ed.*, **44**, 3899–3901.

[49] Smith, M. W. and Snyder, S. A. (2013) *J. Am. Chem. Soc.*, **135**, 12964–12967.

[50] (a) da Mata, M. L. E. N., Motherwell, W. B., and Ujjainwalla, F. (1997) *Tetrahedron Lett.*, **38**, 137–140; (b) Clive, D. L. J., and Kang, S. (2000) *Tetrahedron Lett.*, **41**, 1315–1319; (c) Studer, A., Bossart, M., and Vasella, T. (2000) *Org. Lett.*, **2**, 985–988.

[51] Thaharn, W., Soorukram, D., Kuhakarn, C., Tuchinda, P., Reutrakul, V., and Pohmakotr, M. (2014) *Angew. Chem. Int. Ed.*, **53**, 2212–2215.

[52] Melnika, I., Bringis, K., and Katkevics, M. (2013) *Chem. Heterocycl. Compd.*, **49**, 529–539.

[53] Wertjes, W. C., Wolfe, L. C., Waller, P. J., and Kalyani, D. (2013) *Org. Lett.*, **15**, 5986–5989.

[54] Bhakuni, B. S., Yadav, A., Kumar, S., Patel, S., Sharma, S., and Kumar, S. (2014) *J. Org. Chem.*, **79**, 2944–2954.

[55] Li, Z., Zhang, Y., Zhang, L., and Liu, Z. Q. (2014) *Org. Lett.*, **16**, 382–385.

[56] Wei, W.-T., Zhou, M.-B., Fan, J.-H., Liu, W., Song, R.-J., Liu, Y., Hu, M., Xie, P., and Li, J.-H. (2013) *Angew. Chem. Int. Ed.*, **52**, 3638–3641.

[57] Leifert, D., Daniliuc, C. G., and Studer, A. (2013) *Org. Lett.*, **15**, 6286–6289.

[58] Zhu, Z.-Q., Wang, T.-T., Bai, P., and Huang, Z.-Z. (2014) *Org. Biomol. Chem.*, **12**, 5839–5842.

[59] Uchiyama, N., Shirakawa, E., Nishikawa, R., and Hayashi, T. (2011) *Chem. Commun.*, **47**, 11671–11673.

[60] Wertz, S., Leifert, D., and Studer, A. (2013) *Org. Lett.*, **15**, 928–931.

[61] Xi, Y., Yia, H., and Lei, A. (2013) *Org. Biomol. Chem.*, **11**, 2387–2403.

[62] For reviews on recent developments in photoredox catalysis in organic synthesis, see: (a) Prier, C. K., Rankic, D. A., and MacMillan, D. W. C. (2013) *Chem. Rev.*, **113**, 5322–5363; (b) Xuan, J. and Xiao, W-J. (2012) *Angew. Chem. Int. Ed.*, **51**, 6828–6838; (c) Shi, L. and Xia, W.-J. (2012) *Chem. Soc. Rev.*, **41**, 7687–7697; (d) Narayanam, J. M. R. and Stephenson, C. R. J. (2011) *Chem. Soc. Rev.*, **40**, 102–113; (e) Yoon, T. P., Ischay, M. A., and Du, J. (2010) *Nat. Chem.*, **2**, 527–532; (f) Ravelli, D. and Fagnoni, M. (2012) *ChemCatChem*, **4**, 169–171.

[63] Juris, A., Balzani, V., Barigelletti, F., Campagna, S., Belser, P., and von Zelewsky, A. (1988) *Coord. Chem. Rev.*, **84**, 85–277.

[64] For reviews, see: (a) Hari, D. P. and König, B. (2013) *Angew. Chem. Int. Ed.*, **52**, 4734–4743; (b) Mo, F., Dong, G., Zhang, Y., and Wang, J. (2013) *Org. Biomol. Chem.*, **11**, 1582–1593; (c) Xuan, J., Lu, L.-Q., Chen, J.-R., and Xiao, W.-J. (2013) *Eur. J. Org. Chem.*, 2071–2075.

[65] Cano-Yelo, H. and Deronzier, A. (1984) *J. Chem. Soc. Perkin Trans. 2*, 1093–1098.

[66] Fu, W., Xu, F., Fu, Y., Zhu, M., Yu, J., Xu, C., and Zou, D. (2013) *J. Org. Chem.*, **78**, 12202–12206.

[67] Xu, P., Xie, J., Xue, Q., Pan, C., Cheng, Y., and Zhu, C. (2013) *Chem. Eur. J.*, **19**, 14039–14042.

[68] Nagib, D. A. and MacMillan, D. W. C. (2011) *Nature*, **480**, 224–228.

[69] Fujita, K., Nonogawa, M., and Yamaguchi, R. (2004) *Chem. Commun.*, 1926–1927.

[70] Cheng, Y., Gu, X., and Li, P. (2013) *Org. Lett.*, **15**, 2664–2667.

[71] Xie, J., Yuan, X., Abdukader, A., Zhu, C., and Ma, J. (2014) *Org. Lett.*, **16**, 1768–1771.

[72] Ju, X., Liang, Y., Jia, P., Li, W., and Yu, W. (2012) *Org. Biomol. Chem.*, **10**, 498–501.

[73] (a) Hari, D. P., Schroll, P., and König, B. (2012) *J. Am. Chem. Soc.*, **134**, 2958–2961; (b) Schroll, P., Hari, D. P., and König, B. (2012) *ChemistryOpen*, **1**, 130–133; (c) Hering, T., Hari, D. P., and König, B. (2012) *J. Org. Chem.*, **77**, 10347–10352.

[74] Xiao, T., Dong, X., Tang, Y., and Zhou, L. (2012) *Adv. Synth. Catal.*, **354**, 3195–3199.

[75] Farwaha, H. S., Bucher, G., and Murphy, J. A. (2013) *Org. Biomol. Chem.*, **11**, 8073–8081.

[76] Crisóstomo, F. P., Martín, T., and Carrillo, R. (2014) *Angew. Chem. Int. Ed.*, **53**, 2181–2185.

[77] Xia, Z., Huang, J., He, Y., Zhao, J., Lei, J., and Zhu, Q. (2014) *Org. Lett.*, **16**, 2546–2549.

[78] Tang, S., Zhou, D., Wang, Y.-C. (2014) *Eur. J. Org. Chem.*, 3656–3661.

[79] (a) Postigo, A. (2012) *Can. J. Chem.*, **90**, 493–497; (b) Barata-Vallejo, S. and Postigo, A. (2012) *Eur. J. Org. Chem.*, 1889–1899; (c) Zhang, C.-P., Chen, Q.-Y., Guo, Y., Xiao, J.-C., and Gu, Y.-C. (2012) *Chem. Soc. Rev.*, **41**, 4536–4559.

[80] Tordeux, M., Langlois, B., and Wakselman, C. (1989) *J. Org. Chem.*, **54**, 2452–2453.

[81] Dolbier, W. R. (1996) *Chem. Rev.*, **96**, 1557–1584.

[82] Long, Z.-Y. and Chen, Q.-Y. (1999) *J. Org. Chem.*, **64**, 4775–4782.

[83] Huang, X.-T., Long, Z.-Y., and Chen, Q.-Y. (2001) *J. Fluor. Chem.*, **111**, 107–113.

[84] (a) Fujiwara, Y., Dixon, J. A., O'Hara, F., Funder, E. D., Dixon, D. D., Rodriguez, R. A., Baxter, R. D., Herle, B., Sach, N., Collins, M. R., Ishihara, Y., and Baran, P. S. (2012) *Nature*, **492**, 95–99; (b) Fujiwara, Y., Dixon, J. A., Rodriguez, R. A., Baxter, R. D., Dixon, D. D., Collins, M. R., Blackmond, D. G., and Baran, P. S. (2012) *J. Am. Chem. Soc.*, **134**, 1494–1497; (c) Yiningji, Y., Brueckl, T., Baxter, R. D., Fujiwara, Y., Seiple, I. B., Su, S., Blackmond, D. G., and Baran, P. S. (2011) *Proc. Natl. Acad. Sci. U. S. A.*, **108**, 14411–14415.

[85] Dmowski, W. and Piasecka-Maciejewska, K. (2010) *J. Fluor. Chem.*, **131**, 746–750.

[86] Dmowski, W., Urbańczyk-Lipkowska, Z., and Wójcik, D. (2009) *J. Fluor. Chem.*, **130**, 509–511.

[87] Huang, W.-Y. and Ma, W.-P. (1992) *Chin. J. Chem.*, **10**, 180–185.

[88] Cao, H.-P., Xiao, J.-C., and Chen, Q.-Y. (2006) *J. Fluor. Chem.*, **127**, 1079–1086.

[89] (a) Tordeux, M., Langlois, B., and Wakselman, C. (1990) *J. Chem. Soc. Perkin Trans. 1*, 2293–2299; (b) Huang, W.-Y., Ma, W.-P., and Wang, W. (1990) *Chin. J. Chem.*, **8**, 175–181.

[90] Mn(III): (a) Snider, B. B. (1996) *Chem. Rev.*, **96**, 339–363; (b) Burton, J. W. (2012) In *Encyclopedia of Radicals in Chemistry, Biology & Materials*, ed. Chatgilialoglu, C. and Studer, A., John Wiley & Sons, Ltd, Chichester, United Kingdom, vol. **2**, pp 901–941; Ce(IV): (c) Nair, V. and Deepthi, A. (2007) *Chem. Rev.*, **107**, 1862–1891; Fe(III): (d) Sarhan, A. A. O. and Bolm, C. (2009) *Chem. Soc. Rev.*, **38**, 2730–2744.

[91] Snider, B. B. (2009) *Tetrahedron*, **65**, 10738–10744.

[92] Curran, D. P., Morgan, T. M., Schwartz, C. E., Snider, B. B., and Dombroski, M. A. (1991) *J. Am. Chem. Soc.*, **113**, 6607–6617.

[93] Citterio, A., Fancelli, D., Finzi, C., Pesce, L., and Santi, R. (1989) *J. Org. Chem.*, **54**, 2713–2718.

[94] Holzgrabe, U., Reinhardt, J., and Stoll, E. (1993) *Arch. Pharm.*, **326**, 985–987.

[95] Citterio, A., Cerati, A., Sebastiano, R., Finzi, C., and Santi, R. (1989) *Tetrahedron Lett.*, **30**, 1289–1292.

[96] (a) Jia, Y.-X. and Kündig, E. P. (2009) *Angew. Chem. Int. Ed.*, **48**, 1636–1639; (b) Klein, J. E. M. N., Perry, A., Pugh, D. S., and Taylor, R. J. K. (2010) *Org. Lett.*, **12**, 3446–3449.

[97] Citterio, A., Nicolini, M., and Sebastiano, R. (1993) *Gazz. Chim. Ital.*, **123**, 189–195.

[98] (a) Ji, N., Rosen, B. M., and Myers, A. G. (2004) *Org. Lett.*, **6**, 4551–4553; (b) Tsubusaki, T. and Nishino, H. (2009) *Tetrahedron*, **65**, 9448–9459.

[99] Jamie, J. F. and Rickards, R. W. (1996) *J. Chem. Soc. Perkin Trans. 1*, 2603–2613.

[100] (a) Snider, B. B., Mohan, R., and Kates, S. A. (1985) *J. Org. Chem.*, **50**, 3659–3661; (b) Snider, B. B. and Kwon, T. (1990) *J. Org. Chem.*, **55**, 4786–4788; (c) Yang, D., Ye, X.-Y., Gu, S., and Xu, M. (1999) *J. Am. Chem. Soc.*, **121**, 5579–5580.

[101] (a) Conrad, J. C., Kong, J., Laforteza, B. N., and MacMillan, D. W. C. (2009) *J. Am. Chem. Soc.*, **131**, 11640–11641; (b) Nicolaou, K. C., Reingruber, R. D., Sarlah, D., and Bräse, S. (2009) *J. Am. Chem. Soc.*, **131**, 2086–2087.

[102] Rendler, S. and MacMillan, D. W. C. (2010) *J. Am. Chem. Soc.*, **132**, 5027–5029.

[103] (a) Um, J. M., Gutierrez, O., Schoenebeck, F., Houk, K. N., and MacMillan, D. W. C. (2010) *J. Am. Chem. Soc.*, **132**, 6001–6005; (b) Macmillan, D. W. C. and Rendler, S., in *Asymmetric Synthesis II: More Methods and Applications*, ed. Christmann, M. and Bräse, S., Wiley-VCH, Weinheim, 2012, pp. 87–94.

[104] (a) Demir, A. S., Reis, Ö., and Özgül-Karaaslan, E. (2001) *J. Chem. Soc. Perkin Trans. 1*, 3042–3045; (b) Demir, A. S. and Findik, H. (2008) *Tetrahedron*, **64**, 6196–6201.

[105] Jasch, H., Scheumann, J., and Heinrich, M. R. (2012) *J. Org. Chem.*, **77**, 10699–10706.

[106] (a) Demir, A. S., Reis, Ö., and Emrullahoğlu, M. (2003) *J. Org. Chem.*, **68**, 578–580; (b) Demir, A. S., Findik, H., Saygili, N., and Tuna Subasi, N. (2010) *Tetrahedron*, **66**, 1308–1312.

[107] Tobisu, M., Koh, K., Furukawa, T., and Chatani, N. (2012) *Angew. Chem. Int. Ed.*, **51**, 11363–11366.

[108] (a) Studer, A. and Curran, D. P. (2011) *Angew. Chem. Int. Ed.*, **50**, 5018–5022; (b) Shirakawa, E. and Hayashi, T. (2012) *Chem. Lett.*, **41**, 130–134.

[109] Yanagisawa, S., Ueda, K., Taniguchi, T., and Itami, K. (2008) *Org. Lett.*, **10**, 4673–4676.

[110] Sun, C. L., Li, H., Yu, D. G., Yu, M., Zhou, X., Lu, X. Y., Huang, K., Zheng, S. F., Li, B. J., and Shi, Z. J. (2010) *Nat. Chem.*, **2**, 1044–1049.

[111] Shirakawa, E., Itoh, K., Higashino, T., and Hayashi, T. (2010) *J. Am. Chem. Soc.*, **132**, 15537–15539.

[112] Liu, W., Cao, H., Zhang, H., Zhang, H., Chung, K. H., He, C., Wang, H., Kwong, F. Y., and Lei, A. (2010) *J. Am. Chem. Soc.*, **132**, 16737–16740.

[113] Yanagisawa, S. and Itami, K. (2011) *ChemCatChem*, **3**, 827–829.

[114] Cohen, T. and Bhupathy, M. (1989) *Acc. Chem. Res.*, **22**, 152–161.

[115] Budén, M. E., Guastavino, J. F., Rossi, R. A. (2013) *Org. Lett.*, **15**, 1174–1177.

[116] Qiu, Y., Liu, Y., Yang, K., Hong, W., Li, Z., Wang, Z., Yao, Z., and Jiang, S. (2011) *Org. Lett.*, **13**, 3556–3559.

[117] Tanimoro, K., Ueno, M., Takeda, K., Kirihata, M., and Tanimori, S. (2012) *J. Org. Chem.*, **77**, 7844–7849.

[118] Sharma, S., Kumar, M., Kumar, V., and Kumar, N. (2013) *Tetrahedron Lett.*, **54**, 4868–4871.

[119] Liu, W., Tian, F., Wang, X., Yu, H., and Bi, Y. (2013) *Chem. Commun.*, **49**, 2983–2985.

[120] Chen, W.-C., Hsu, Y.-C., Shih, W.-C., Lee, C.-Y., Chuang, W.-H., Tsai, Y.-F., Chen, P. P.-Y., and Ong, T.-G. (2012) *Chem. Commun.*, **48**, 6702–6704.

[121] Ng, Y. S., Chan, C. S., and Chan, K. S. (2012) *Tetrahedron Lett.*, **53**, 3911–3914.

[122] A, S., Liu, X., Li, H., He, C., and Mu, Y. (2013) *Asian J. Org. Chem.*, **2**, 857–861.

[123] Zhao, H., Shen, J., Guo, J., Ye, R., and Zeng, H. (2013) *Chem. Commun.*, **49**, 2323–2325.

[124] Sun, C. L., Li, H., Gu, Y.-F., Wang, B., and Shi, Z.-J. (2011) *Chem. Eur. J.*, **17**, 10844–10847.

[125] Zhou, S., Anderson, G. M., Mondal, B., Doni, E., Ironmonger, V., Kranz, M., Tuttle, T., and Murphy, J. A. (2014) *Chem. Sci.*, **5**, 476–482.

[126] Cuthbertson, J., Gray, V., and Wilden, J. D. (2014) *Chem. Commun.*, **50**, 2575–2578.

[127] Roman, D. S., Takahashi, Y., and Charette, A. B. (2011) *Org. Lett.*, **13**, 3242–3245.

[128] Sun, C.-L., Gu, Y.-F., Huanga, W.-P., and Shi, Z.-J. (2011) *Chem. Commun.*, **47**, 9813–9815.

[129] Bhakuni, B. S., Kumar, A., Balkrishna, S. J., Sheikh, J. A., Konar, S., and Kumar, S. (2012) *Org. Lett.*, **14**, 2838–2841.

[130] (a) De, S., Ghosh, S., Bhunia, S., Sheikh, J. A., and Bisai, A. (2012) *Org. Lett.*, **14**, 4466–4469; (b) De, S., Mishra, S., Kakde, B. N., Dey, D., and Bisai, A. (2013) *J. Org. Chem.*, **78**, 7823–7844.

[131] Wu, Y., Wong, S. M., Mao, F., Chan, T. L., and Kwong, F. Q. (2012) *Org. Lett.*, **14**, 5306–5309.

[132] (a) Masters, K. S. and Bräse, S. (2013) *Angew. Chem. Int. Ed.*, **52**, 866–869; (b) Masters, K. S., Bihlmeier, A., Klopper, W., and Bräse, S. (2013) *Chem. Eur. J.*, **19**, 17827–17835.

10

RADICAL-NUCLEOPHILIC AROMATIC SUBSTITUTION

ROBERTO A. ROSSI, JAVIER F. GUASTAVINO, AND MARÍA E. BUDÉN
INFIQC, Departamento de Química Orgánica, Facultad de Ciencias Químicas, Universidad Nacional de Córdoba, Córdoba, Argentina

10.1 INTRODUCTION: SCOPE AND LIMITATIONS—BACKGROUND

Since its discovery from Bunnett in 1970 [1], the radical-nucleophilic aromatic substitution, unimolecular (the $S_{RN}1$ reaction), has been widely used to achieve new C—C or C–heteroatom bonds. In these reactions, a compound bearing an adequate leaving group is substituted at the *ipso* position by a nucleophile (Nu^-), and this process involves electron transfer (ET) steps. The global reaction is depicted in Equation 10.1:

$$ArX + Nu^- \rightarrow ArNu + X^- \quad (10.1)$$

The aromatic $S_{RN}1$ reaction is a chain process (Scheme 10.1). In the initiation step, the radical anion of the substrate $ArX^{\cdot-}$ is formed; this radical anion fragments to afford the radical $Ar^{\cdot}$ and the X^- ion. In some system, the ET and fragmentation occur simultaneously (dissociative ET (DET)) [2]. The radical thus formed can react with the Nu^- to give the radical anion $ArNu^{\cdot-}$, which by ET to the substrate affords the substitution product ArNu and the radical anion $ArX^{\cdot-}$ required to continue the propagation cycle.

Not many initiation events are needed, but the propagation cycle must be fast and efficient to afford a successful process. Radical probes have been used to demonstrate the formation of radicals along the propagation cycle and also as a synthetic tool. Thus, products from ring closure or ring opening or from rearrangement of the radicals were taken as evidence of the presence of these intermediates. Inhibition by radical traps with stable free radicals: 2,2,6,6-tetramethylpiperidine-1-oxyl (TEMPO), di-*t*-butyl nitroxide (DTBN), etc.; has been extensively used to provide mechanistic

Arene Chemistry: Reaction Mechanisms and Methods for Aromatic Compounds,
First Edition. Edited by Jacques Mortier.

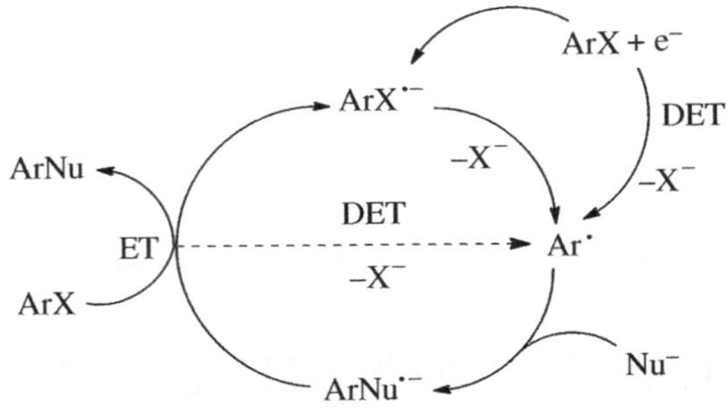

SCHEME 10.1 The $S_{RN}1$ mechanism.

evidence. Inhibition with good electron acceptors such as *p*-dinitrobenzene is taken as evidence of an ET process.

The process has considerable width of scope, and many substituents such as alkyl groups, OR, SAr, CF_3, CO_2R, NH_2, SO_2R, CN, COR, $CONH_2$, and F are compatible with the reaction. The reactions are not inhibited by nitro groups or negatively charged substituents, for example carboxylate groups, or oxyanions, but as a general rule they make slower the reactions. In addition to the aromatic and heteroaromatic substrates discussed in this chapter, the $S_{RN}1$ reaction has been applied to alkyl halides with electron-withdrawing groups (EWG) [3], vinyl [4], perfluoroalkyl [5], neopentyl [6], cycloalkyl [7], and bridgehead halides [8].

Besides to halides, other leaving groups (RS, PhSO, $PhSO_2$, PhSe, Ph_2S^+, N_2BF_4, N_2SR) are known. It is noteworthy that hydroxyl and amine groups are leaving groups through their phosphate esters [$(EtO)_2P(O)O$] and ammonium salts [R_3N^+], respectively.

Carbanions are some of the most common nucleophiles through which a new C—C bond can be formed. C—C instead of C—O or C—N bond formation can be achieved with oxygen and nitrogen bidentate nucleophiles in intermolecular reactions. However, in the intramolecular processes, C—O and C—N bonds could be formed. Nucleophiles derived from other heteroatoms react to form a new C–heteroatom bond.

Several reviews have been published in relation to the $S_{RN}1$ mechanism [9, 10], to aromatic photoinitiated substitutions [10a, b], to reactions performed under electrochemical catalysis [11], and to the synthetic applications of the process, which have become one of the common methodologies in modern synthesis [12]. This chapter aims to describe the basic research in the area, the mechanistic studies, and the recent synthetic strategies in $S_{RN}1$ reactions. The chapter is divided in five principal sections:

1. Introduction.
2. Mechanistic consideration.
3. Intermolecular $S_{RN}1$ reactions, divided according to the most utilized nucleophiles of the periodic table group: group 14 (C and Sn), group 15 (N, P, As), and group 16 (O, S, Se). The reactions with carbanions as Nu^- followed by a polar ring closure reaction have a subsection.
4. Intramolecular $S_{RN}1$ reactions to afford ring closure products.
5. Miscellaneous ring closure reactions.

We expect to cover recent development in this area, with an emphasis on the scope of the process in terms of synthetic capability and target applications.

10.2 MECHANISTIC CONSIDERATIONS

10.2.1 Initiation Step

In the initiation step, an ET from the Nu^- (or from a suitable electron source) to the ArX takes place. In some systems, spontaneous or thermal ET is a possible initiation step depending on the relationship between the electron affinity (EA) of the substrate and the oxidation potential (OP) of the Nu^- [13]. In this system, the ArX must be a good electron acceptor, and the Nu^- must be a good electron donor. In others words, if E^0 ($ArX/ArX^{\cdot-}$) > $E^0(Nu^\cdot/Nu^-)$, the spontaneous ET can occur. The ET involves an electronic transition from HOMO of the electron donor Nu^- to the LUMO of the ArX acceptor (Scheme 10.2).

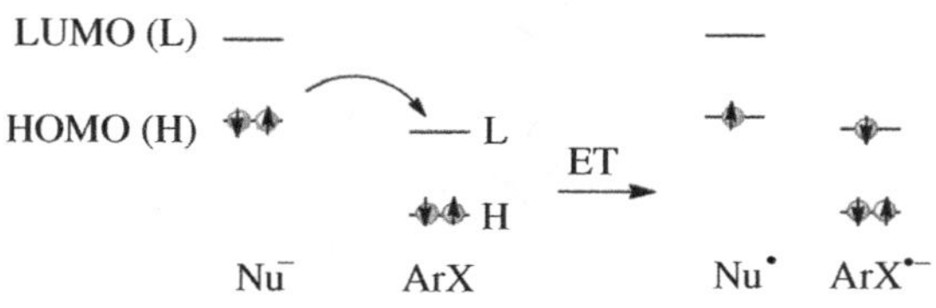

SCHEME 10.2 Spontaneous or thermal ET.

In "thermally" induced $S_{RN}1$ processes, the Nu^- itself serves as the electron donor without external stimulation. However, in most cases, induced ET is necessary to initiate $S_{RN}1$ reactions. Different methodologies are used to initiate $S_{RN}1$ reactions, such as (i) photostimulated reaction, (ii) electrochemical initiation, (iii) inorganic salts (usually Fe^{2+}), and (iv) solvated electrons from alkali metal or sodium amalgam in liquid ammonia, among others.

By far, photostimulation is the most used method to initiate $S_{RN}1$ reactions. The initiation step by photoinduced ET can conceivably be accomplished in one of the followings ways: (i) homolytic cleavage of the C—X bond in excited ArX, (ii) ET from the Nu^- to the excited ArX (or from the excited Nu^- to the ArX), and (iii) ET within an excited charge-transfer complex (CTC), among others (Scheme 10.3). Depending on the nature of the ArX, the Nu^-, and the experimental conditions, any of these mechanisms could probably be considered as an initiation step. It is noteworthy that the *t*-BuOK can form $[ArX]^{\cdot-}$ by ET under irradiation [14, 15].

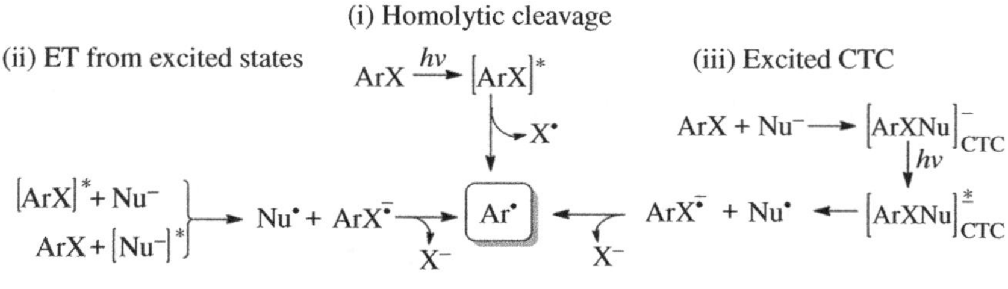

SCHEME 10.3 Initiation step by photoinduced ET.

Electrochemical initiation is an approach that has been successful in a considerable number of cases with aromatic and heteroaromatic substrates. The Savéant group has studied most of these reactions [11]. According to the difference of the standard potentials between the $ArX/ArX^{\cdot-}$ and $ArNu/ArNu^{\cdot-}$ couples, two different situations may arise. If $E^0(ArX/ArX^{\cdot-}) > E^0(ArNu/ArNu^{\cdot-})$ and $ArX^{\cdot-}$ fragments slowly, $Ar^\cdot$ and $ArNu^{\cdot-}$ are formed far from the electrode surface. The $ArNu^{\cdot-}$ formed can be oxidized at the electrode surface or in the solution (Eq. 10.2) to continue the propagation cycle. Under this situation, complete conversion of ArX into ArNu can be obtained with a catalytic amount of electrons whenever the radical-nucleophile coupling is fast enough to overcome side reactions:

$$ArX + [ArNu]^{\cdot-} \rightleftharpoons [ArX]^{\cdot-} + ArNu \quad (10.2)$$

When $ArX^{\cdot-}$ fragments very fast, $Ar^{\cdot}$ is formed close to the electrode surface and will be reduced to the anion before reacting with the Nu^- in the solution. However, this situation can be overcome by the presence of a redox mediator (RM) that is reduced to $RM^{\cdot-}$ at a more positive potential than ArX. The $RM^{\cdot-}$ lives long enough to reduce ArX far of the electrode, and then $Ar^{\cdot}$ is formed in the solution away from the electrode (Scheme 10.4).

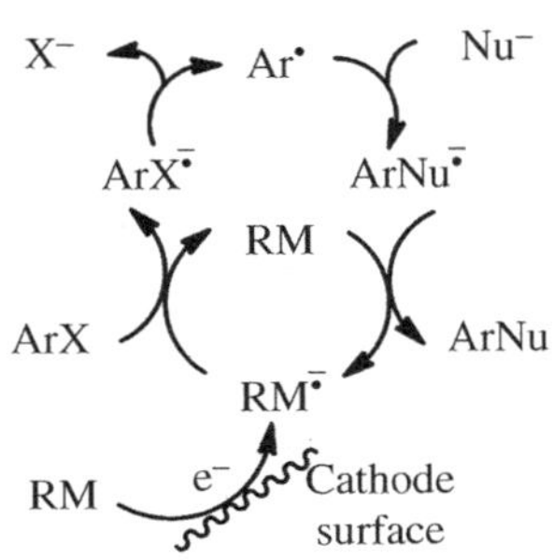

SCHEME 10.4 $S_{RN}1$ reaction mediated by a redox mediator (RM).

Another alternative for initiation is using inorganic salts. Although several inorganic salts have been tested so far, Fe^{2+} salts especially $FeSO_4$ in liquid ammonia and $FeCl_2$ or $FeBr_2$ in DMSO have provided the best results of substitution [9]. The role of the Fe^{2+} ion in the reaction remains intriguing. Its presence is required from catalytic to equimolecular amounts, which depends on the leaving group and the Nu^- used.

The process allows the possibility of affording the substitution of an unreactive Nu_1^- through *entrainment* conditions (Scheme 10.5) [16]. *Entrainment* is useful when the Nu_1^- is unreactive at initiation but quite reactive at propagation. Under these conditions, the addition of a second Nu_2^-, more reactive at initiation, increases the generation of radical intermediates and allows Nu_1^-, less reactive at initiation, to establish its own propagation [17].

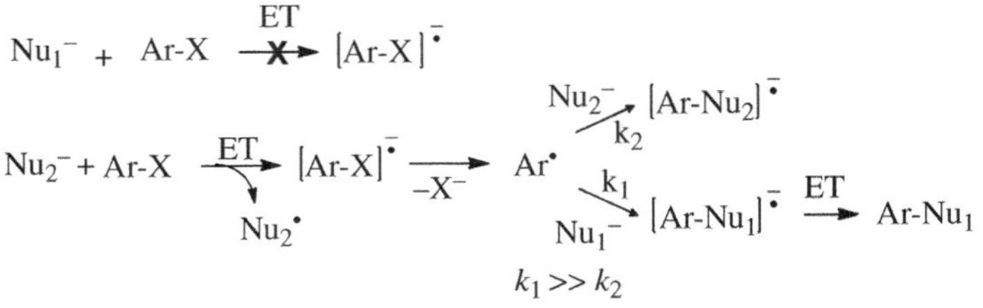

SCHEME 10.5 Reaction under *entrainment* conditions.

10.2.2 Propagation Steps

In the propagation steps, the fragmentation of the radical anion $ArX^{\cdot-}$ occurs through an intramolecular ET from the π^* molecular orbital (MO) to the σ^* MO of the C–X bond. The π^* and σ^* systems are adjacent and orthogonal, and the main reaction coordinates for the intramolecular ET from the π* MO to the σ* MO, which dissociates into Ar· radical and X^- ions, are C—X bond elongation and bending with respect to the Ar plane [18].

When aromatic substrates with two leaving groups such as **1** react with nucleophiles, monosubstitution with halogen retention **5** and disubstitution **7** products can be obtained, depending on the nature and position of the leaving groups and the nature of the Nu^- (Scheme 10.6).

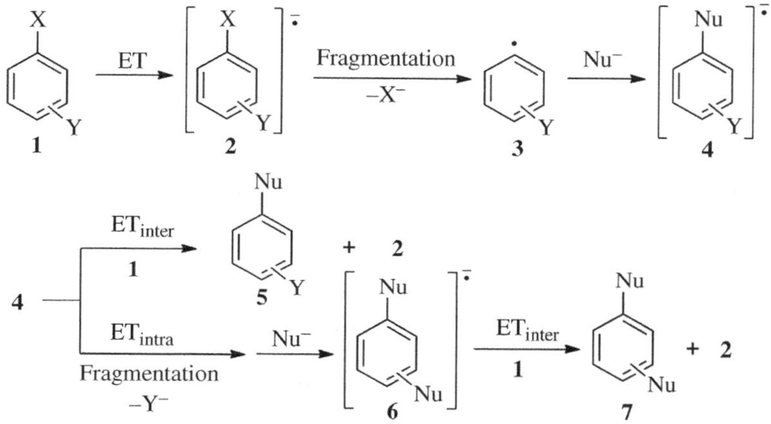

SCHEME 10.6 Reaction of a substrate with two leaving groups and nucleophiles.

When **1** receives an electron, it forms the radical anion **2**, which fragments at the more labile C—X bond to give radical **3**. This radical reacts with the Nu^- to form radical anion **4**, which by intermolecular ET to **1** (ET_{inter}) furnishes radical anion **2** and product **5**. Another possibility is that radical anion **4**, by intramolecular ET (ET_{intra}) to the second C—Y bond, fragments to afford a radical that by coupling with the Nu^- gives the radical anion **6**. This radical anion by an ET_{inter} reaction to **1** affords the disubstitution product **7** and **2**, which continued the chain propagation [9].

Another factor to be considered is the stability of the radical anion intermediate of the substitution product ($ArNu^{\cdot-}$). For example, the reaction of Ph· radical with $^-CH_2CN$ anion affords the radical anion **8** that by intramolecular ET fragments into $PhCH_2^{\cdot}$ radical and CN^- ions (Scheme 10.7), PhMe being the principal product observed by H abstraction from the solvent (S-H) [19].

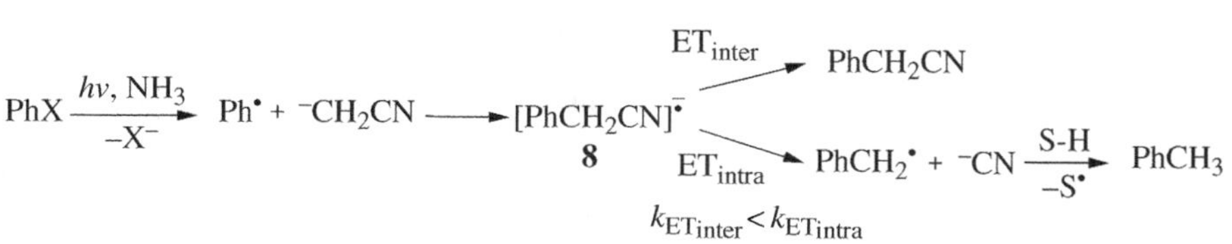

SCHEME 10.7 Reaction of Ph· radical with $^-CH_2CN$ anion.

In this reaction, the rate of the intermolecular ET is slower than the rate of intramolecular ET/fragmentation $\left(k_{ET_{inter}} < k_{ET_{intra}}\right)$. However, with polycyclic aromatic halides, haloarenes substituted with EWG, substitution products $ArCH_2CN$ are formed in good yields [20]. With these substrates, the extra electron of the radical anion intermediate resides in the low-energy π* MO of the aryl moiety, and fragmentation of the $(ArCH_2–CN)^{\cdot-}$ radical anion is avoided $\left(k_{ET_{inter}} > k_{ET_{intra}}\right)$.

10.2.3 Termination Steps

The termination steps depends on the ArX, the Nu⁻, and the experimental conditions. For instance, radical Ar˙ can react with S-H through a hydrogen atom abstraction reaction to yield Ar–H (Eq. 10.3) or by ET to yield the aryl anion (Ar⁻), which finally is protonated by the S-H to afford Ar–H (Eq. 10.4). When the coupling reaction between Ar˙ and the Nu⁻ is not too fast, the reduction product ArH increases the yield:

$$\mathrm{Ar}^{\bullet} \xrightarrow[-\mathrm{S}^{\bullet}]{\text{S-H}} \mathrm{Ar{-}H} \qquad (10.3)$$

$$\mathrm{Ar}^{\bullet} \xrightarrow{\text{ET}} \mathrm{Ar}^{-} \xrightarrow[-\mathrm{S}^{-}]{\text{S-H}} \mathrm{Ar{-}H} \qquad (10.4)$$

Poor hydrogen donor S-H, such as liquid ammonia or DMSO, are used to diminish the reduction reactions.

10.3 INTERMOLECULAR $S_{RN}1$ REACTIONS

10.3.1 Nucleophiles from Group 14: C and Sn

The reaction with carbanions and anions derived from stannanes to afford C–C and C–Sn bonds constitutes one of the more representative examples.

Carbanions: With carbanions derived from hydrocarbons, such as 1,3-pentadienyl, 1-(4-anisyl) propenyl, fluorenyl, and indenyl anions, it has been observed that the regiochemistry of the coupling favors the formation of the more stable radical anion intermediate [21]. For instance, the reaction of 1,3-pentadienyl anion (**9**) with PhBr induced by Na metal yields the 1-phenylpentane (**10**), the 1,1-diphenylpentane (**11**), and the 1,5-diphenylpentane (**12**) after catalytic hydrogenation (Eq. 10.5). In this transformation, PhBr receives an electron and fragments to give a Ph˙. This radical could react with **9** in position 1 or 3, but the mixture of regioisomeric products obtained (mono- and diarylation in 1,1- and 1,5- positions) indicates that the generation of the radical anion that has the most extended π MO system seem to be the principal way. In order to avoid the di- and triarylation, the carbanions are generally used in a 3 or 4 equiv excess:

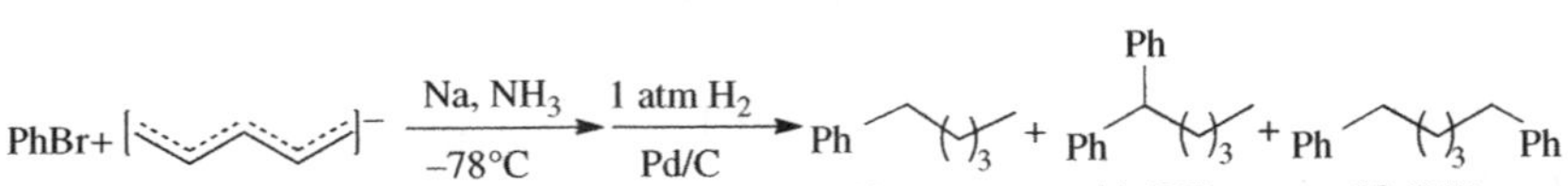

9 **10** (57%) **11** (9%) **12** (7%) (10.5)

Recently, the $S_{RN}1$ mechanism was used to promote the coupling of aryl Grignard (ArMgX) [22] or arylzinc (ArZnI) [23] reagents with ArX in the biaryl synthesis. This cross-coupling reaction, exemplified by the reaction of PhMgBr **13** with 2-iodo-1,1′-biphenyl **14** (Eq. 10.6), is initiated by ET to give the radical anion of **14**, which is stabilized by the Grignard reagent (as a π-Lewis acid). After C–I fragmentation, the aryl radical intermediate reacts very fast, and the nucleophilic combination is facilitated by magnesium [22b]:

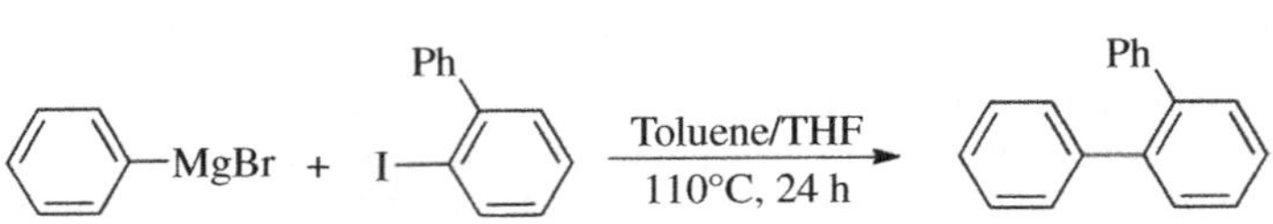

13 **14** **15** (93%) (10.6)

This approach using ArMgX bearing a leaving group was applied in radical/anionic $S_{RN}1$-type polymerizations for preparation of oligoarenes [24]. The reaction of **16** with TEMPO as initiator afforded the poly(*m*-phenylene) **17** (Eq. 10.7):

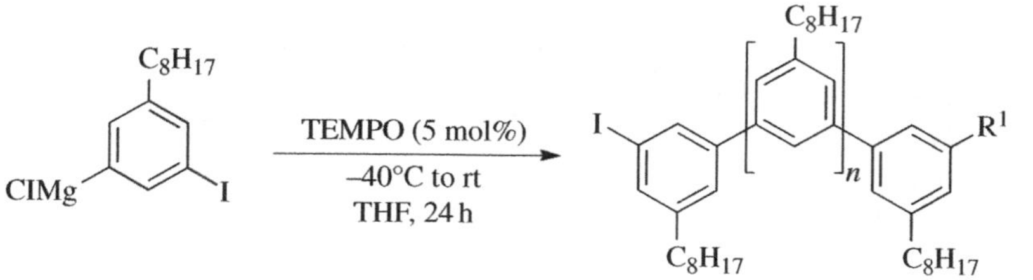

16 **17** R^1 = H, I (66%) (10.7)

In this transformation, the homocoupling of the ArMgX reagents to biaryls (Ar–Ar) by oxidation with TEMPO was applied as initiation step. This homocoupling proceeds via a biaryl radical anion intermediate **18**, which fragments at the C—I bond to give the radical **19** that initiates polymerization (Eq. 10.8) to yield **17** [25]:

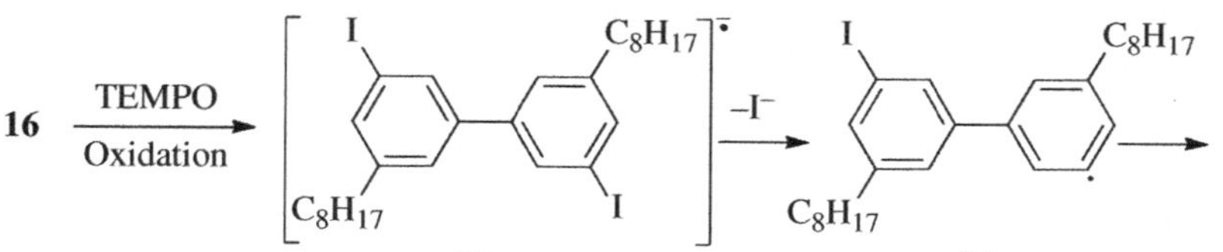

18 **19** (10.8)

Also, carbanions derived from esters, *N,N*-disubstituted amides, thioamides, imides, and *β*-cyanocarbonyl and *β*-dicarboxylic compounds have been successfully used as nucleophiles to form new C—C bonds with aromatic substrates [9].

The photostimulated reaction of 1-iodonaphthalene **20** with a chiral-assisted imide enolate anion **21** in liquid ammonia is an interesting example of reaction from carbanion α to an EWG. This provides a stereoselective coupling of an aromatic radical with a nucleophile. In this reaction, the diastereomeric isomers of the substitution compound (**22** and **23**) are formed (43–64%), and the selectivity observed is highly dependent on the metal counterion used (Li^+, Na^+, K^+, Cs^+, Ti^{4+}). The highest stereoselectivity is found with Li^+ at −78°C (d.r.: 78/22%, S/R) and with Ti^{4+} at −33°C (d.r.>98%, S/R) (Eq. 10.9) [26]:

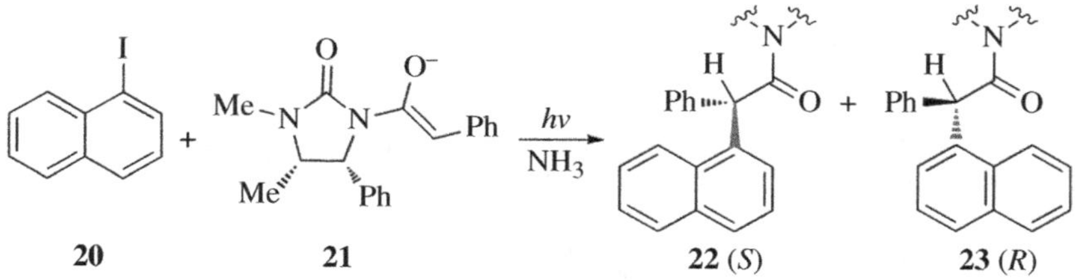

20 **21** **22** (*S*) **23** (*R*) (10.9)

Other anions that can be used in this type of reaction include anions from compounds with α-hydrogen to nitrile group such as phenylacetonitrile, 2-phenylbutyronitrile, and 2-cyclohexylidene-acetonitrile [9]. For example, the reaction of 2-iodoanisole or 1-iodo-2-methoxynaphthalene with 2-naphthylacetonitrile anion (**24**) gave good yields of α,α-diarylated acetonitriles (**25**) (Eq. 10.10). However, the regioselectivity decreases when the same anion reacts with $Ar^{\cdot}$ with less steric hindrance and substitution at the C_1 of the naphthyl **26** moiety begins to be observed ($4\text{-}MeOC_6H_4^{\cdot}$ and

$1\text{-}C_{10}H_7^{\cdot}$) [27]. This outcome has been explained based on the kinetic control, where the observed regioselectivity resulted mainly from the steric factors in the transition states:

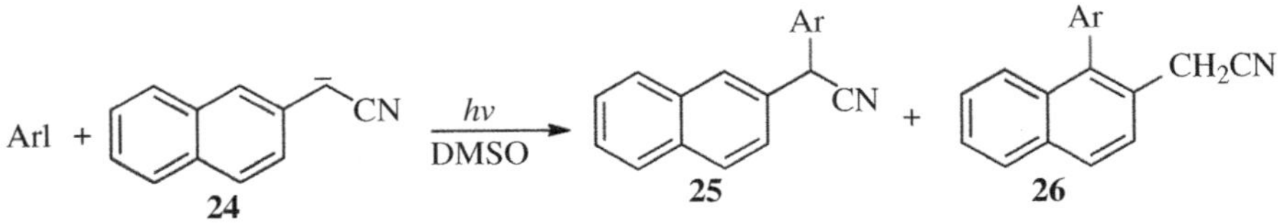

24 **25** **26**

Ar	25	26
Ar = 2-$MeOC_6H_4$	76%	—
2-MeO-1-$C_{10}H_6$	84%	—
4-$MeOC_6H_4$	47%	21%
1-$C_{10}H_7$	68%	16%

(10.10)

On the other hand, the enolate ions of ketones are the carbanions most widely studied. The enolate ions of acyclic and cyclic ketones react with ArX under photostimulation in liquid ammonia or DMSO to afford α-arylation products in good yields [9, 28]. The anions of aromatic ketones, such as acetophenone, methyl 2-naphthyl ketone, 2-acetylfuran, 2- and 3-acetylthiophene, 2- and 3-acetyl-*N*-methylpyrrole, anthrone, and tetralone, among others, have been efficiently α-arylated. Some examples are shown in Equation 10.11 [6a, 17a]:

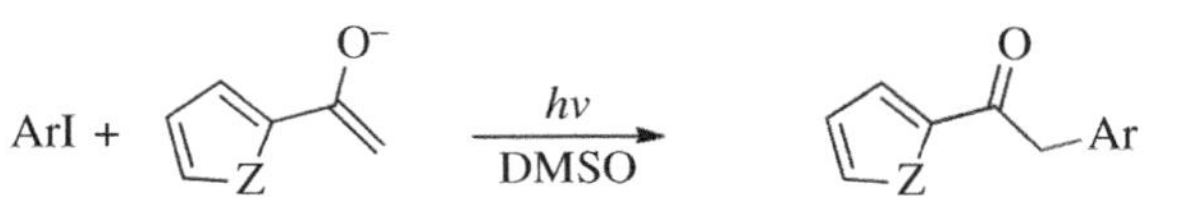

(10.11)

Ar = 1-naphthyl, Z = O (69%)
9-anthracenyl, Z = O (96%)
phenyl, Z = N-Me (89%)

Aromatic substrates with two leaving groups such as *o*-dibromobenzene react with $^{-}CH_2CO^{t}Bu$ anion to give only disubstitution product in the sense of Scheme 10.6 where $k_{ET_{intra}} > k_{ET_{inter}}$ [29]. On the other hand, *o*-iodohalobenzene (X = I; Y = Cl, Br, or I) reacts with several enolate ions from aromatic ketones to give the monosubstitution products with retention of one halogen $\left(k_{ET_{inter}} > k_{ET_{intra}}\right)$ [30]. With nucleophiles of aromatic ketones **27**, the extra electron in the radical anions **28**, formed by the coupling of radical **3** with **27**, is in the π* MO of the ArCO moiety and not in the ring with the labile C—Y bond. Because of this, the ET_{inter} is favored over ET_{intra} and the C—Y bond is conserved (Eq. 10.12):

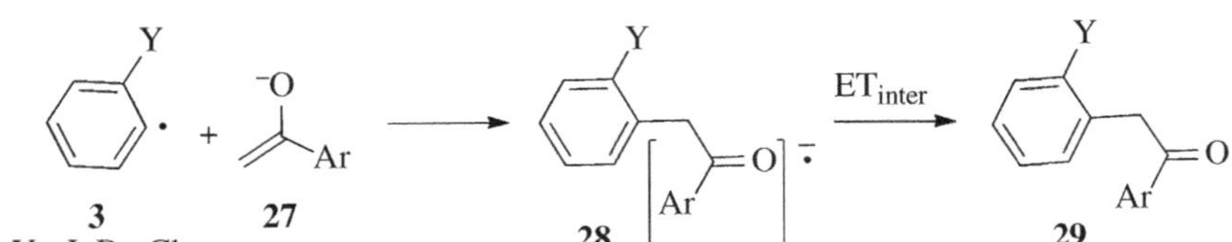

3 Y = I, Br, Cl; **27**; **28**; **29**

(10.12)

In unsymmetrical dialkyl ketones, isomeric enolate ions can be formed. The equilibrium concentration of two possible enolate ions and the selectivity of the attacking $Ar^{\cdot}$ radical largely determine the distribution of two possible α-arylated products [9]. For example, $Ph^{\cdot}$ radical coupled to the most substituted position of the enolate ion of the 2-butanone gives mainly product **30** [21, 31].

However, the increase of the steric hindrance in the enolate ion, such as 3-methyl-2-butanone, the radical coupled principally at the least substituted carbon to afford mainly product **31** (Eq. 10.13) [21, 32]:

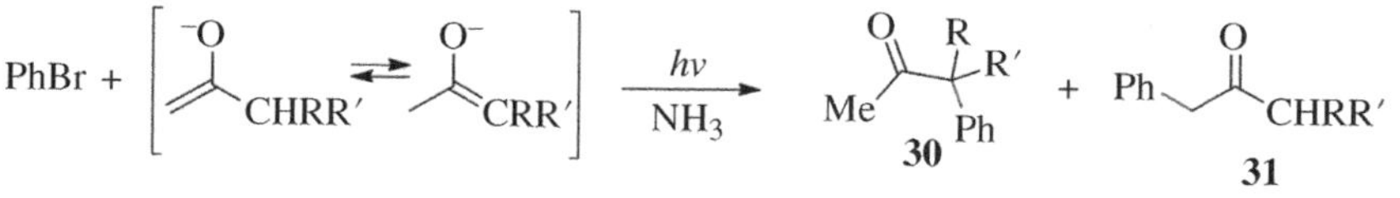

R = H; R′ = Me	50 min	61%	19%
R = R′ = Me	120 min	9%	81%

(10.13)

10.3.1.1 Intermolecular $S_{RN}1$ Reactions Followed by a Polar Ring Closure Reaction In this area, one of the most widely studied approaches to synthesize heterocycles is the $S_{RN}1$ substitution, followed by a polar ring closure. In this way, aromatic compounds that have an appropriate *ortho* substituent (Z) to the leaving group (X), such as **32**, react with a Nu^- by the $S_{RN}1$ mechanism to afford product **33**. The ring closure product **34** is achieved by the reaction between the Nu group and Z (Eq. 10.14):

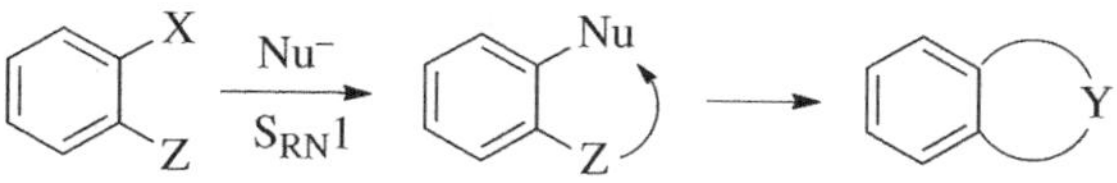

32 **33** **34** (10.14)

For instance, the photostimulated reaction of *o*-iodoaniline **35** with methyl ketone anions ($^-CH_2COR$) affords intermediate **36**, which by spontaneous dehydration gives 2-aryl/alkyl indoles **37** in 64–93% (Eq. 10.15) [33–36]. The reactions of **35** with enolate anions of cyclic ketones produced the fused indoles **38** and **39** in good yields [35b]:

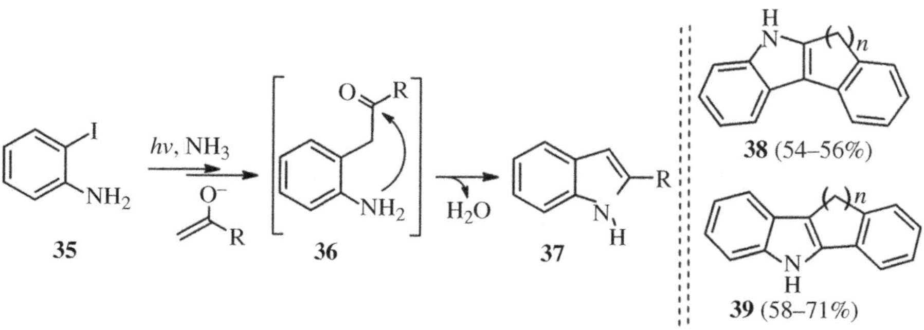

(10.15)

The synthesis of 3-substituted isoquinolinones can be performed by the photoinduced $S_{RN}1$ reactions in DMSO of *o*-iodobenzamide **40** with enolate ions of aromatic and aliphatic ketones in 68–87% yields. With cyclic ketones, the reactions afforded fused isoquinolinones **41** and **42** with good yields (Scheme 10.8) [37].

2-Iodobenzenesulfonamide **43** undergoes photostimulated $S_{RN}1$ reactions in liquid ammonia with alkyl and cycloalkyl enolate ions and gives fair to good yields of 3-alkyl 2*H*-1,2-benzothiazine-1,1-dioxides **44** and fused 2*H*-1,2-benzothiazine-1,1-dioxides **45** (Scheme 10.9) [38].

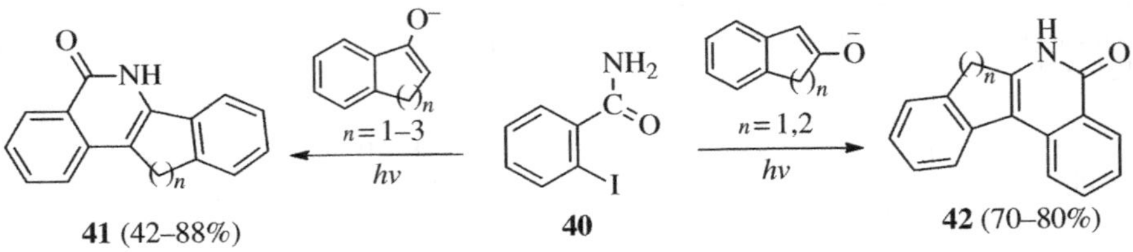

SCHEME 10.8 Synthesis of fused isoquinolinones.

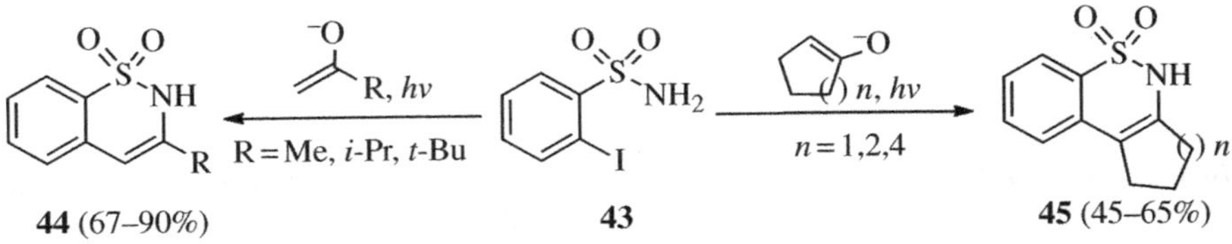

SCHEME 10.9 Synthesis of 2*H*-1,2-benzothiazine-1,1-dioxides.

Tin Nucleophiles: The $S_{RN}1$ reactions with anions derivate from stannanes have been recently reviewed [39]. ArBr and ArI react with tin nucleophiles by a halogen/metal exchange (HME) reaction (see Chapter 28) [40]. On the other hand, ArCl undergoes $S_{RN}1$ substitution by Me_3Sn^- ions in liquid ammonia under irradiation to afford $ArSnMe_3$ in high yields (88–100%). Also, disubstitution products are formed in the reaction of *o*-, *m*-, and *p*-$C_6H_4Cl_2$ under photostimulation to give 58, 90, and 88% yields, respectively. The photostimulated reaction of 1,3,5-trichlorobenzene with Me_3Sn^- ions afforded the trisubstituted product 1,3,5-tris(trimethylstannyl)benzene with 71% yield [41]. Importantly, these substitutions are carried out with Cl as leaving groups. This fact and the functional group tolerance make the $S_{RN}1$ reaction a successful synthetic alternative.

In a recent application, it was found that methyl 2,5-dichlorobenzoate **46** reacted with Me_3Sn^- ions in liquid ammonia under irradiation to afford disubstitution product **47** in 99% (Eq. 10.16) [42]:

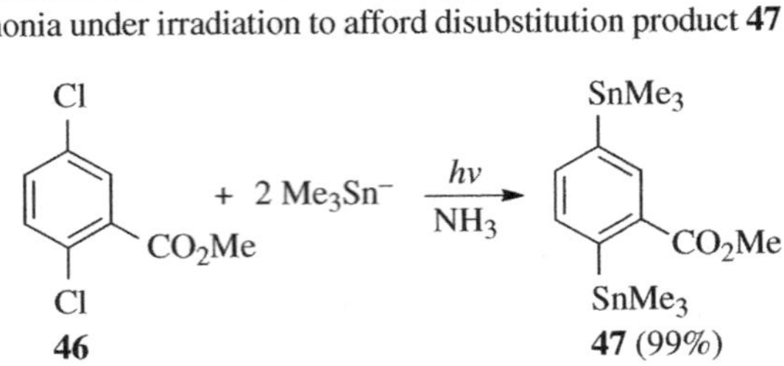

(10.16)

ArOH could be converted into $ArOP(O)(OEt_2)$, which by the photoinduced reaction with Me_3Sn^- anions affords $ArSnMe_3$ in good yields. This methodology was used for the synthesis of mono-, di-, and tristannylated products [43]. For instances, the diol **48** gave the diphosphate ester **49**, which affords the disubstitution product **50** with good yields (Eq. 10.17) [43b]. On the other hand, $ArNH_2$ can be converted into $ArSnMe_3$ through its $ArNMe_3^+$ salts [44, 45]:

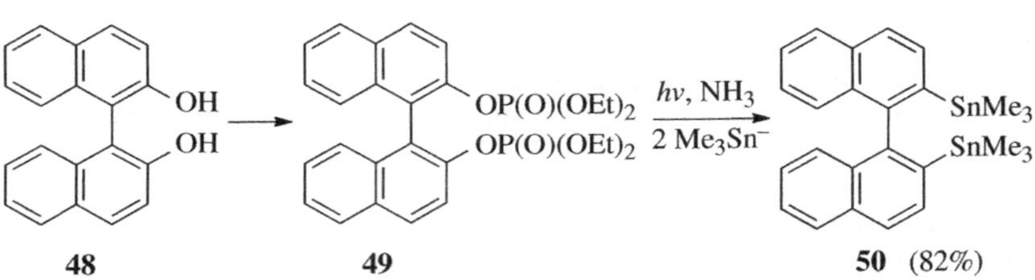

(10.17)

These reactions are employed to obtain a variety of stannanes, which are then used in the Stille cross-coupling reactions (see Chapter 19). Thus, the sequence $S_{RN}1$–Stille reactions can be successfully applied to build complex molecules [45, 46].This approach was used for the synthesis of 6-substituted uracils from **51**. By performing a one-pot three-step reaction: $S_{RN}1$–Stille–hydrolysis, it was possible to obtain 6-aryl- and 6-acyl-substituted uracils **52** in good yields (Eq. 10.18) [47]:

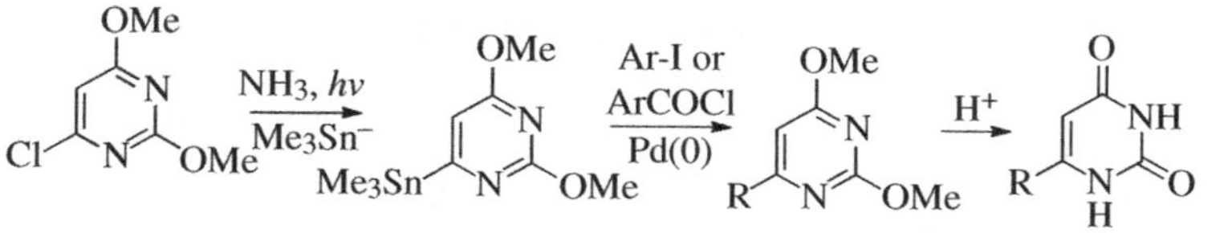

51 R = Ar, ArCO **52** (43–57%)

(10.18)

In Stille reactions, a substrate bearing two leaving groups I–Ar^1–Cl will react faster by the C–I bond. Chemoselective cross-coupling reaction of I–Ar^1–Cl with a stannane Ar–$SnMe_3$ (synthesized by an $S_{RN}1$ reaction) affords the product Ar–Ar^1–Cl. This chloroarene by another $S_{RN}1$ reaction with Me_3Sn^- anions gives stannane Ar–Ar^1–$SnMe_3$, which can furnish asymmetric product Ar–Ar^1–Ar^2 by a cross-coupling reaction with Ar^2–X (Eq. 10.19) [45]:

$$\text{Ar-Cl} \xrightarrow[S_{RN}1]{Me_3Sn^-} \text{Ar-SnMe}_3 \xrightarrow[Pd(0)]{\text{I-Ar}^1\text{-Cl}} \text{Ar-Ar}^1\text{-Cl} \xrightarrow[S_{RN}1]{Me_3Sn^-} \text{Ar-Ar}^1\text{-SnMe}_3 \xrightarrow[Pd(0)]{\text{Ar}^2\text{-X}} \text{Ar-Ar}^1\text{-Ar}^2 \quad (10.19)$$

The iterative sequence $S_{RN}1$ reaction Pd-catalyzed process was investigated in order to foster a methodology to build complex molecules [45]. In this way, when *m*-bis(trimethylstannyl)benzene **53** obtained through an $S_{RN}1$ reaction (from *m*-$Cl_2C_6H_4$) is allowed to react under Pd catalysis with *p*-iodochlorobenzene, the product **54** is obtained. Compound **54** reacted with Me_3Sn^- ions by the $S_{RN}1$ mechanism to afford compound **55**. In turn, **55** can suffer, under Pd catalysis, coupling with 1-iodonaphthalene to afford the hydrocarbon **56** in high yield (Scheme 10.10) [45].

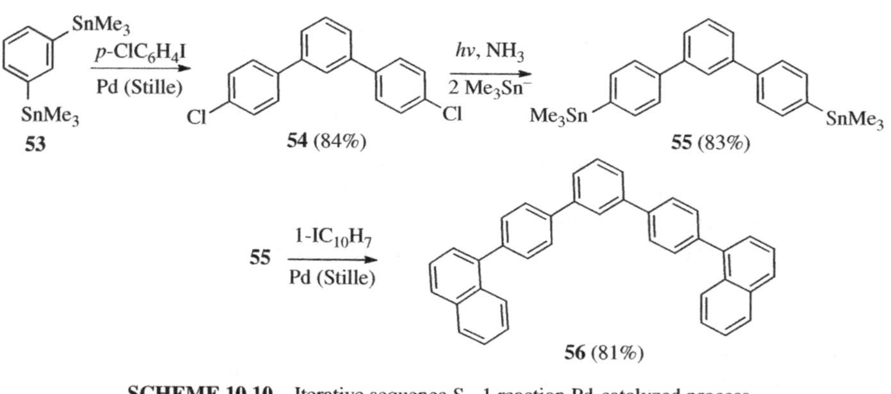

SCHEME 10.10 Iterative sequence $S_{RN}1$ reaction Pd-catalyzed process.

The leaving group ability of Me_3Sn group in electrophilic substitutions allowed the synthesis of diaryl ketones in good yields (40–78%) through the reaction of Me_3SnAr (synthesized by $S_{RN}1$ reactions) with ArCOCl [48]. These reactions were regioselective, allowing the synthesis of diaryl ketones not usually available under Friedel–Crafts reactions.

10.3.2 Nucleophiles Derived from Group 15: N, P, As, and Sb

The anions derived from anilines are bidentate nucleophiles, and they can react through the N- or the C-atoms. For instances, 2-naphthylamide **57** anion reacted under photoinitiation with ArI in liquid ammonia to give 1-aryl-2-naphthylamines **58** (C-arylation) in good yields and traces of N-arylated products **59** (Eq. 10.20) [49]. Double C-arylation was possible using *p*-bromoiodobenzene as substrate and **57** to afford the disubstitution product **60** [50]:

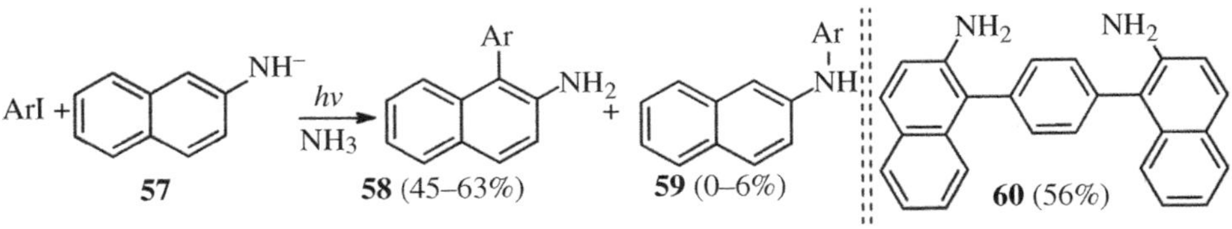

57 **58** (45–63%) **59** (0–6%) **60** (56%)

(10.20)

Following the same methodology, the photostimulated reaction of the anion of 9-phenanthryl amide with ArX gave 10-aryl phenanthren-9-amines (75–100%) [51].The double arylation of 9-phenanthryl amide with *p*-bromoiodobenzene afforded 40% yield of the disubstitution product [50].

The regiochemistry of the coupling of Ar· with different bidentate nitrogen heteroaryl anions has been investigated. For example, diverse arylpyrroles, arylindoles, and arylimidazoles are synthesized electrochemically by reaction of pyrrolyl [52], indolyl [53], and imidazolyl [52] anions in liquid ammonia with ArX with a C—C bond formation. The most reactive positions of nitrogen heteroaryl carbanions under $S_{RN}1$ conditions are the same as those observed in electrophilic substitution. For pyrrolyl ion, substitution at C-2 on the heterocycle is the main product [52]. For indolyl ions, substitution at C-3 is the major product observed [53]. The imidazolyl anion is the least reactive and leads to a mixture of arylation products at the C-2 and C-4(5), where the 4-position was estimated to be about four times more reactive (Eq. 10.21) [53]:

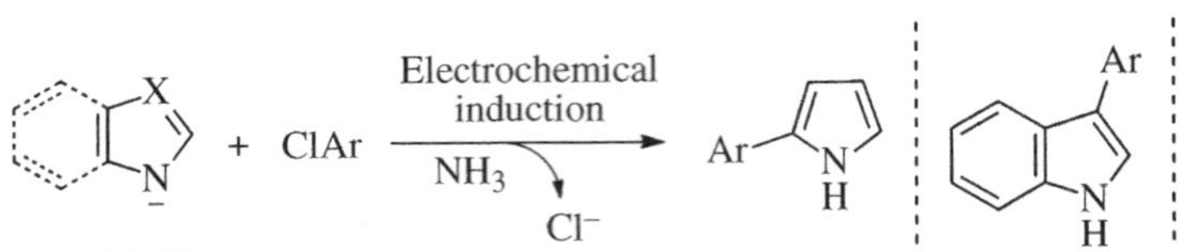

X = CH, N

Main products observed

(10.21)

In a similar procedure, electrochemically induced $S_{RN}1$ reactions have been employed for the synthesis of aryl uracils from ArX and the uracil anion **61** in DMSO (Eq. 10.22) [54], where the substitution always took place at the carbon leading to C-5-arylated uracils **62**. This approach offers a mild method to obtain pyrimidine nucleosides substituted at the 5-position without requiring the preparation of specific reagents:

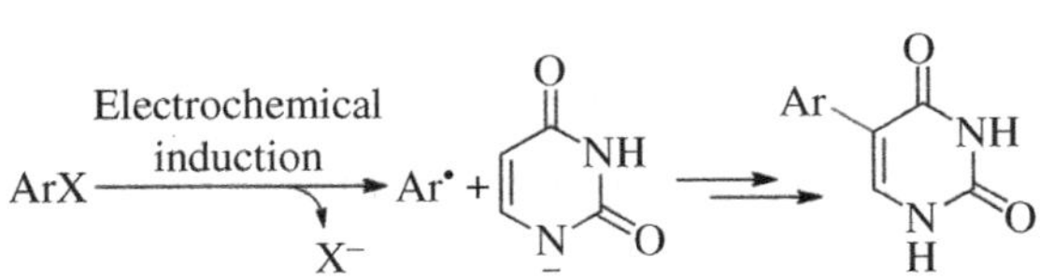

61 **62**

Ar = 4-PhCOC$_6$H$_4^-$ (55%)
4-CNC$_6$H$_4^-$ (50%)
4-NO$_2$C$_6$H$_4^-$ (55%)
2-CF$_3$C$_6$H$_4^-$ (30%)

(10.22)

The $(EtO)_2PO^-$ ions react with ArX to yield the C—P bond (Ar–P(O)(OEt)$_2$) under irradiation either in liquid ammonia, MeCN/THF, DMF, or DMSO as solvent [9, 55]. With substrates with two leaving groups, such as **63** (*o*-, *m*-, and *p*-dihalobenzenes) afforded mainly disubstitution products **64** in good yields (59–89%) (Eq. 10.23) [55a–c]:

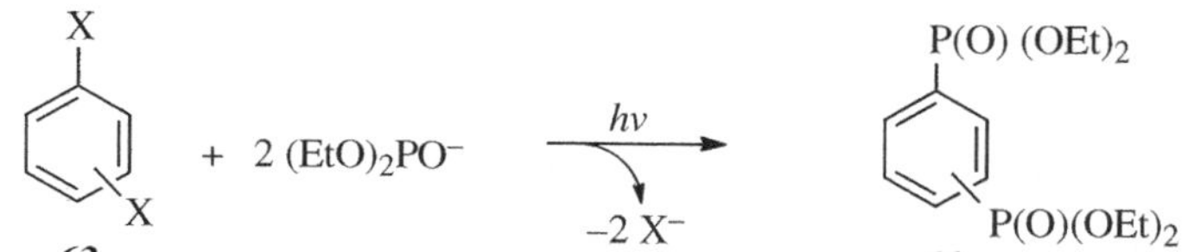

63 **64** (10.23)

Similar behavior was observed with other nucleophiles from phosphorus compounds. For example, Ph_2P^-, $Ph(OBu)PO^-$, Ph_2PO^-, $(Me_2N)_2PO^-$, and $(C_6H_5CH_2)_2PO^-$ react with ArX in liquid ammonia under photostimulation to give C—P bond formation in nearly quantitative yields [56, 57].

A different behavior was observed with nucleophiles derived from arsines. For example, Ph_2As^- ions react under irradiation with 4-halotoluenes, 4-haloanisoles, 1-bromonaphthalene, and 9-bromophenanthrene to yield substitution product $ArAsPh_2$ together to arsines with exchanged aryl ring (Ar_2AsPh, Ar_3As, and $AsPh_3$). These products are known as *scrambled* products [58]. For instance, the reaction with 4-chlorotoluene gave the scrambled arsines (Eq. 10.24):

$$\underset{\text{To = 4-tolyl}}{\text{ClTo} + {}^{-}\text{AsPh}_2} \xrightarrow{h\nu} \underset{15\%}{\text{AsPh}_3} + \underset{61\%}{\text{ToAsPh}_2} + \underset{15\%}{\text{To}_2\text{AsPh}} + \underset{2\%}{\text{To}_3\text{As}} \qquad (10.24)$$

The $S_{RN}1$ mechanism, involving the formation of arsine radical anions, adequately explains the formation of *scrambled* products by competition between ET_{inter} and C—As bond fragmentation (via ET_{intra}) (Scheme 10.11) [59]. In this system, the σ* MO of the C—As bonds have similar energy to the low-lying antibonding π* MO of the aromatic system, and the orbital crossing can occur. In this case, the extra electron in $[ArAsPh_2]^{\cdot-}$ could occupy the σ* MO of any C—As bonds, and the radical anion decomposes into different arsine anions and aromatic σ radicals ($Ar^{\cdot}$ and $Ph^{\cdot}$).

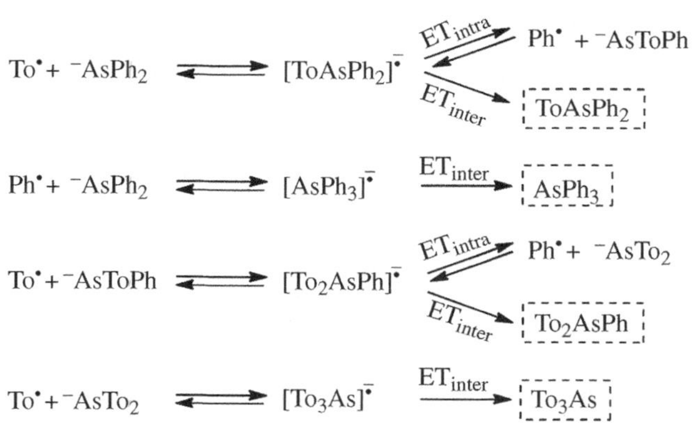

To = 4-tolyl

SCHEME 10.11 Representation of the scrambling reaction.

However, when the substrate ArX has a low-energy LUMO, such as 4-chlorobenzophenone [58], 2-chloroquinoline [58], and 6-chloro-2,4-dimethoxypyrimidine [60], the extra electron in the radical anion $(ArAsPh_2)^{\cdot-}$ intermediates is properly stabilized. In these case, the gap between π* MO and the σ* MO of the C—As bonds is large enough, and the transfer of the odd electron to the σ* MO is unfavorable. As consequence, the C—As bond fragmentation does not occur, and the straightforward arsine substitution products $ArAsPh_2$ are obtained in 100, 76, and 70% yields, respectively. On the other hand, when Ph_2Sb^- ions react under irradiation with ArX, the *scrambled* products are always formed [58]. These results suggest that the difference of the energy between the σ* MO of the weak C—Sb bonds and the π* MO is small. Consequently, the ET_{intra} to σ* MO and fragmentation occur even when the $(ArSbPh_2)^{\cdot-}$ radical anion intermediate has a low-energy π* MO.

10.3.3 Nucleophiles Derived from Group 16: O, S, Se, and Te

The anions derived from phenols are bidentate nucleophiles and can react through of the O- or the C-atoms. For example, the 2-naphthoxide ions **65** react with ArX to give substitution only at C-1 of the naphthalene ring [61]. In this reaction, the intermediated $Ar^{\cdot}$ radical couples in C-1 of **65** to yield the radical anion **66**, which by ET gives the product **67** that after tautomerization yields the more stable products **68** (Eq. 10.25). The rates of reactions of 2- and 4-anisyl and 2-methoxy-1-naphthyl radicals with **65** were determined (10^8–$10^9\,M^{-1}\,s^{-1}$) using an indirect method, a competition of the coupling reaction with the H-atom abstraction from the DMSO [62]:

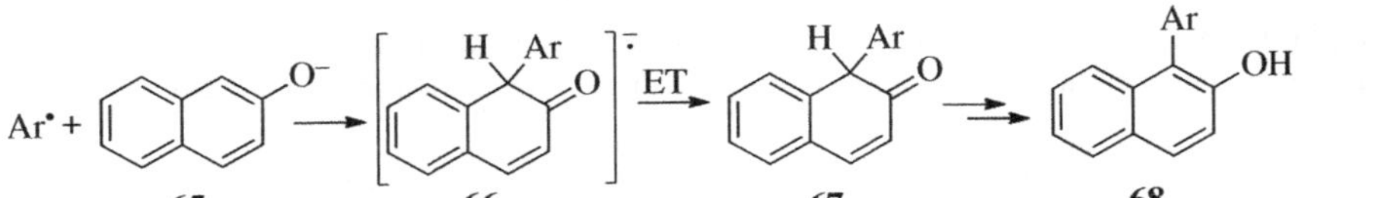

65 66 67 68

(10.25)

Other examples of this approach are the photostimulated reaction of 1-iodonaphthalene with ions **65** that afford the substitution product 1,1′-binaphthyl-2-ol (53%) [61] and the photostimulated reaction of PhI with the anion of 9-phenanthrol that afforded 10-phenylphenanthren-9-ol in 53% yield [51].

The photostimulated reaction of 4-hydroxycoumarin **69** with ArI (Ar=Ph, 4-anisyl, 2-anisyl, and 1-naphthyl) and *t*-BuOK as base in DMSO has been investigated, and 3-aryl-4-hydroxycoumarins **70** were obtained in good yields (Eq. 10.26) [63]:

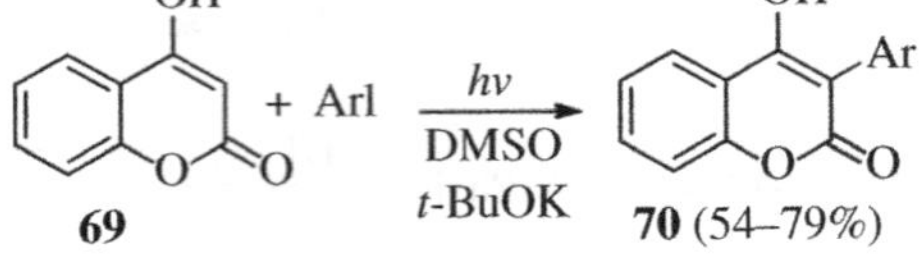

69 70 (54–79%)

(10.26)

Sulfur-centered anions were used in $S_{RN}1$ reactions, and the results observed depend not only on ArX but also on the thiolate anion used. For alkane thiolate ions, the fragmentation reactions of the radical anion intermediate $(ArSR)^{\cdot-}$ to afford ArS^- ion and radical $R^{\cdot}$ may compete with the substitution process. Nevertheless, the straightforward substitution product ArSR is obtained in the reaction of RS^- with ArX that has a low-energy LUMO, such as compounds bearing EWG or polycyclic halides (Scheme 10.12). These aromatic moieties stabilize the radical anion intermediates

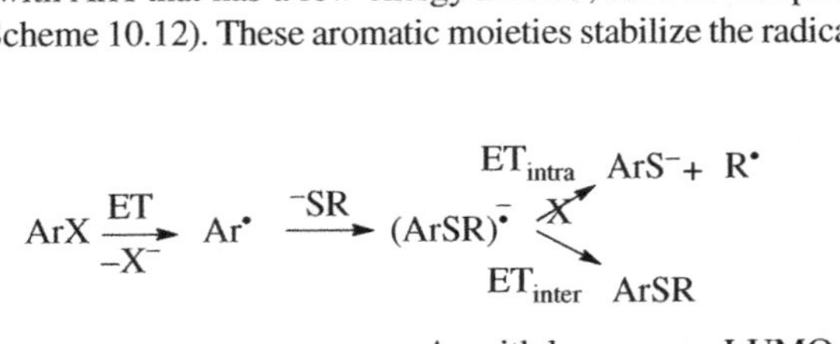

SCHEME 10.12 $S_{RN}1$ reactions with sulfur-centered anions.

$(ArSR)^{\cdot-}$ and prevent the S—R bond fragmentation [9]. In contrast, aromatic Ar^1S^- ions react with ArX in liquid ammonia under irradiation to form good yields of Ar^1SAr [9].

When 2-pyrimidinethiolate ions are allowed to react with *o*-, *m*-, and *p*-$C_6H_4Cl_2$, the monosubstitution product with chlorine retention is the principal product, which can be explained on the basis of Scheme 10.6 $\left(k_{ET_{inter}} > k_{ET_{intra}}\right)$ [64].

In a recent investigation, the photostimulated reaction of methyl 2,5-dichlorobenzoate **46** (R = H) or methyl 3,6-dichloro-2-methoxybenzoate **71** (R = OMe) with PhS^- ions in excess in liquid ammonia was studied (Eq. 10.27). Compound **46** afforded mainly the product **72** (R = H, 88%) with retention of the chlorine, whereas **71** yielded the disubstitution product **73** (R = OMe) with 80% yield. The different reactivity observed was explained on the basis of the activation energy necessary for the ET_{intra} from the π* to the σ* of the C—Cl bond and the fragmentation of the radical anion intermediates [65]:

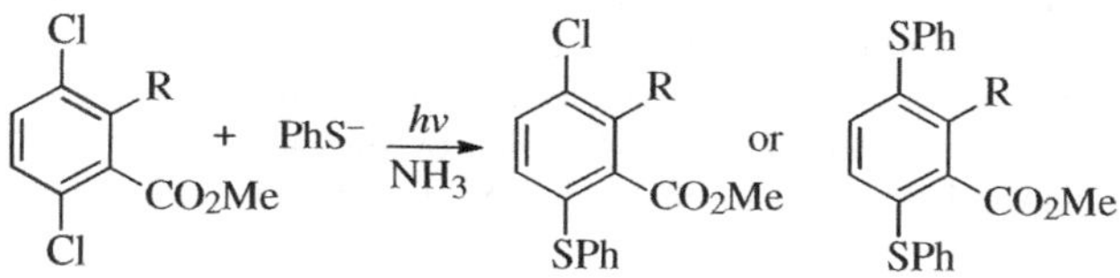

46 or **71** **72** **73** (10.27)

The photostimulated $S_{RN}1$ reactions of other sulfur nucleophiles such as thiourea anion [66] and thioacetate anion with ArX in DMSO have been reported as a one-pot two-step methodology for the synthesis of several sulfur aromatic compounds in moderate to good yields [14, 67]. Coupling of the radical $Ar^{\cdot}$ with thiourea anion **74** afforded the radical anion intermediate **75** that fragments into ArS^- ion and the radical **76** (Eq. 10.28). The deprotonation of the latter gave a new radical anion intermediate **77**, which by ET to the ArX yielded methanediimine **78** and $Ar^{\cdot}$ radicals to continue the propagation steps (Eq. 10.29) [14]:

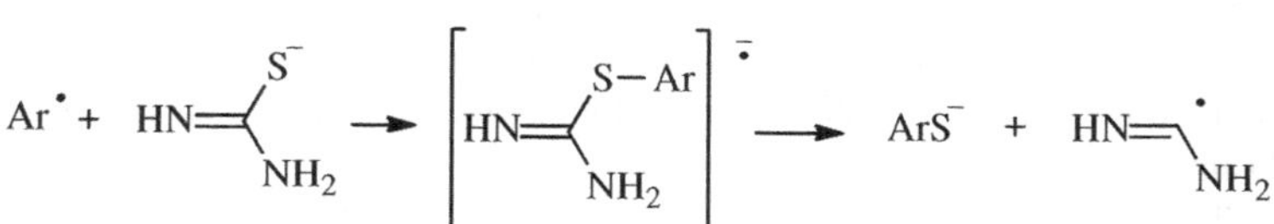

74 **75** **76** (10.28)

$$\mathbf{76} \xrightarrow[-t\text{-BuOH}]{t\text{-BuO}^-} \left[HN{=}C{=}NH\right]^{\cdot-} \xrightarrow[-X^-]{ArX} HN{=}C{=}NH + Ar^{\cdot} \quad (10.29)$$

77 **78**

The ArS^- ions thus obtained are quenched with MeI to yield ArSMe in a one-pot procedure, together with Ar_2S by the reaction of ArS^- anion and $Ar^{\cdot}$ radicals. With substrates with EWG, the yields of ArSMe are good (49–87%) [66].

Similar to the reaction with urea anion, the photostimulated reaction of 1-bromonaphthalene with selenobenzamide and selenourea anions afforded methyl-1-naphthyl selenide after quenching the reaction with MeI [68]. The Se^{-2} nucleophile can be formed by reaction of Se and Na metals in liquid ammonia and reacts under irradiation with PhI to give $(PhSe)_2$ (78%) after oxidation of the $PhSe^-$ ions formed [69]. The reaction of $PhSe^-$ or $PhTe^-$ with ArX under irradiation; straightforward and scrambled products are obtained and depends on the ArX employed [59, 70].

10.4 INTRAMOLECULAR $S_{RN}1$ REACTIONS

When a substrate has both the leaving group and the nucleophilic center, such as **79**, the intramolecular $S_{RN}1$ reaction affords the ring closure product **83** (Eq. 10.30):

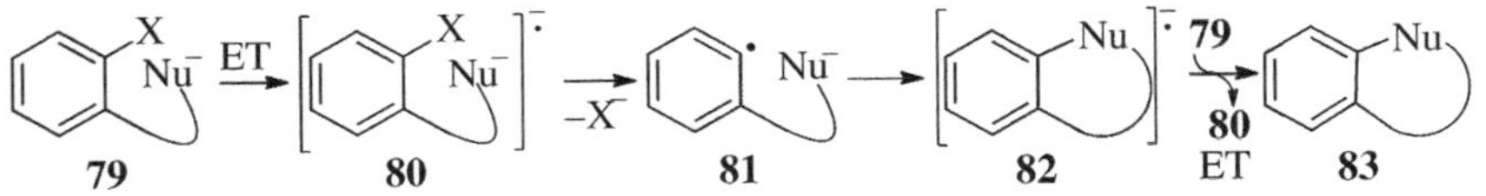
(10.30)

Substrate **79** receives an electron by ET and forms the radical dianion **80**, which by fragmentation of the C—X bond gives the distonic radical anion **81**. This is a high-energy intermediate, and the driving force to afford the ring closure product is the lowest energy of the conjugated radical anion **82**. Finally, an ET reaction furnishes the ring closure product **83**.

Another mechanistic possibility is an intramolecular ET from the nucleophilic center to the pendant aryl halide to give the diradical anion **84**, which by fragmentation of C—X bond gives the diradical **85** that collapses in the ring closure product **83** (Eq. 10.31). In this case, there are no chain propagation steps:

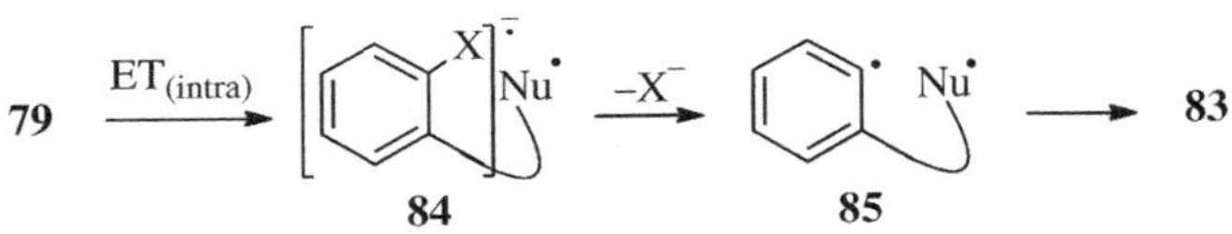
(10.31)

The carbanions have been investigated in intramolecular $S_{RN}1$ reactions to synthesize carbocycles of different size. Its usefulness was demonstrated in the key steps of synthesis of the complex alkaloids eupolauramine **86** [71], (±)-tortuosamine **87** [72], and the alkaloid-type ergot **88** (Fig. 10.1) [73].

Intramolecular hydrogen atom abstraction is an important reaction that competes with the intramolecular reaction from enolate ions of ketones [32, 74] and amides [75] to obtain six-membered rings.

In a recent example, ω-(2-bromophenyl)alkyl-2-oxazolines **89** were treated with LDA in THF, and the anions **90** were formed. These anions under irradiation afforded the ring closure products 1-phenyl-indanes **91a** and tetralin derivatives **91b** (Eq. 10.32). With similar substrate **89** ($n = 3$), the intramolecular $S_{RN}1$ reaction gave the benzocycloheptane **91c**, in which the oxazoline group had undergone a novel areneotropic migration from the end of the spacer to the benzo ring [76]:

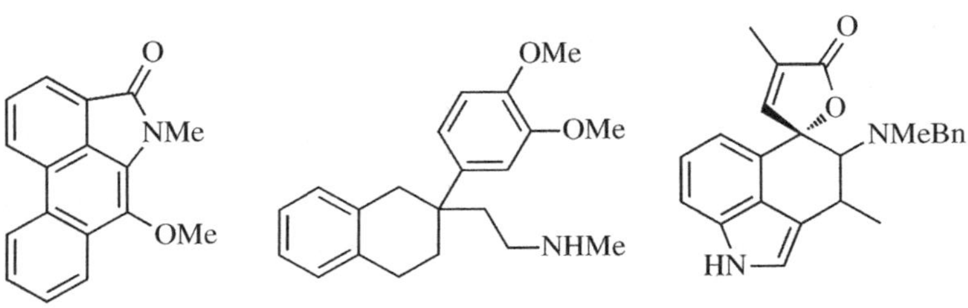

86 **87** **88**

FIGURE 10.1 Synthesis of alkaloid through $S_{RN}1$ reactions.

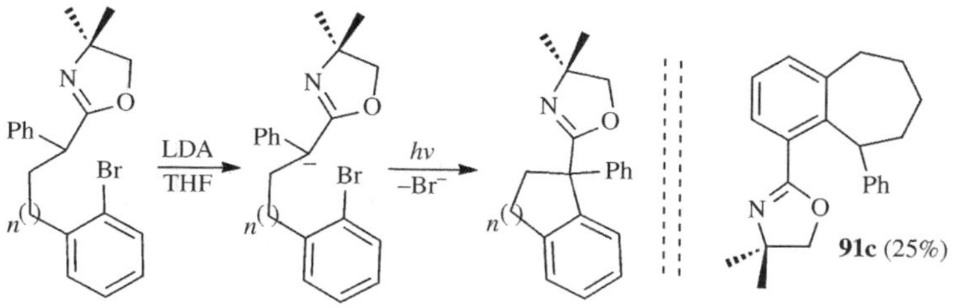

89 90 91a ($n=1$) (57%) 91b ($n=1$) (50%) 91c (25%)

(10.32)

Following a similar approach, the anion of **93**, prepared from **92**, afforded the fused tetracyclic derivative **94** when it was treated with an excess of LDA in THF for 48 h via an $S_{RN}1$ reaction, followed by 1,3-areneotropic migration (Eq. 10.33) [77]:

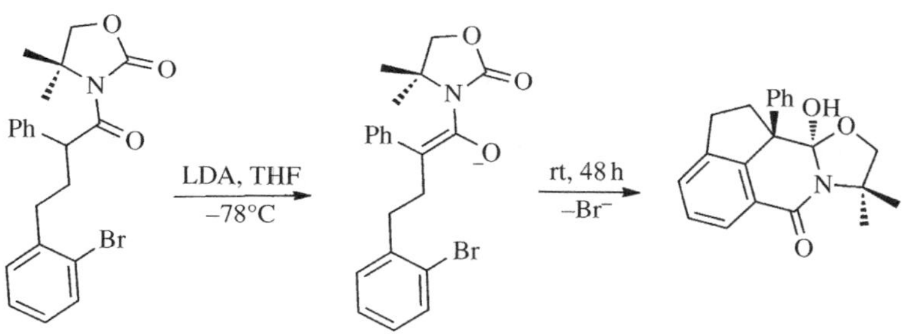

92 93 94 (75%)

(10.33)

Recently, the syntheses of six- to nine-membered benzo-fused heterocycles **98** in good to excellent yields were reported [78]. The synthetic approach involves the photostimulated $S_{RN}1$ reaction of ketone enolate anions **96**, prepared from ketones **95**, linked by a bridge Z to a pendant haloarene (Scheme 10.13).

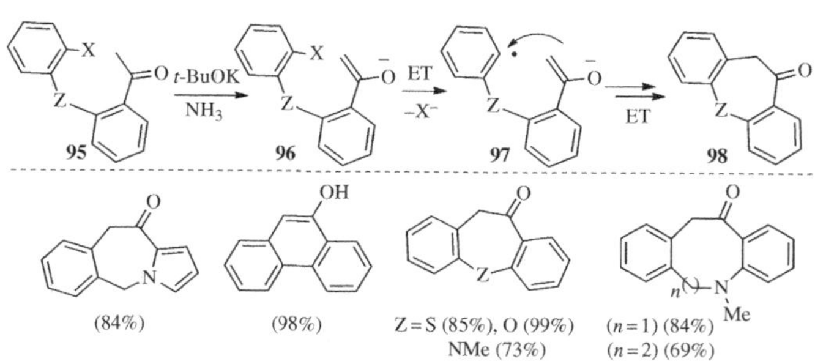

SCHEME 10.13 Fused heterocycles synthesized by $S_{RN}1$ reactions.

The cyclization reaction of *N*-(2-halophenyl)-*N*-methyl-2-phenylacetamide **99** affords 1-methyl-3-phenylindolin-2-one **100** under microwave (MW) heating with good yields and short reaction times (Eq. 10.34) [79]:

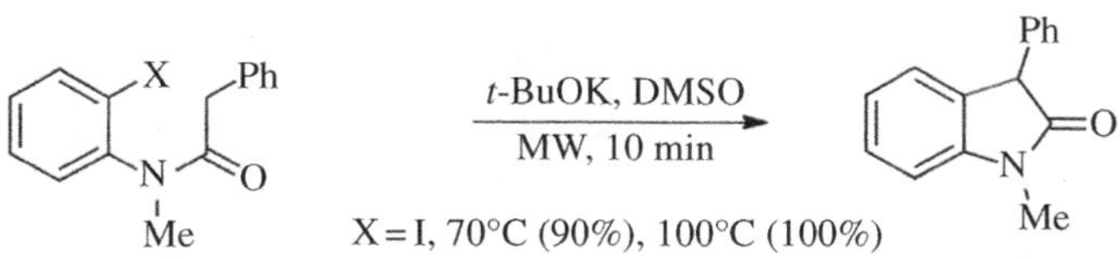

99 100

X = I, 70°C (90%), 100°C (100%)
X = Br, 100°C (80%)

(10.34)

Similar to the carbanions, the anions derived from anilines have been used to obtain new heterocycles through intramolecular C—C bond formation. For instance, the intramolecular arylation of iodobenzyl phenyl amide ions **101** afforded phenanthridines **103** in very good yields [80]. These reactions were carried out under irradiation in liquid ammonia or DMSO. Dihydrophenanthridines **102** were the ring closure products; however, they oxidized in the work-up to afford finally **103** (Eq. 10.35). Benzo[*a*]phenanthridine (98%), benzo[*c*]phenanthridine (84%), and the novel naphtho[2,3-*a*]phenanthridine (72%) were obtained following this synthetic strategy [80]:

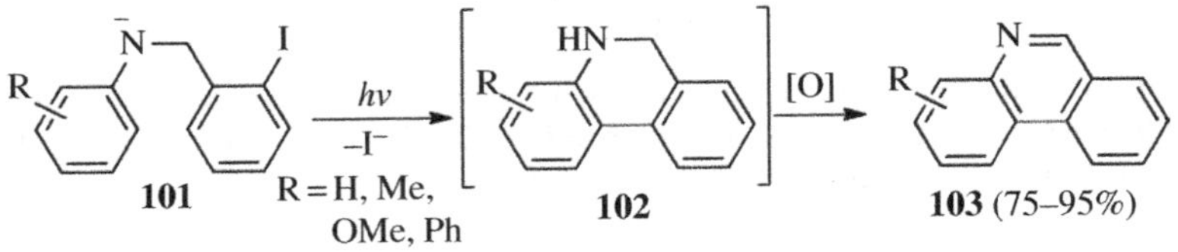

101 **102** **103** (75–95%) (10.35)

Intramolecular $S_{RN}1$ reactions have been successfully applied to the syntheses of five-membered rings through biaryl C—C bond formation as the key step. In this sense, the synthesis of a series of substituted 9*H*-carbazoles **105** by the photostimulated $S_{RN}1$ reaction of diarylamine anions **104** as starting substrates was performed in liquid ammonia or DMSO (Eq. 10.36) [81]. Furthermore, when N^4,$N^{4\prime}$-bis(2-bromophenyl)biphenyl-4,4′-diamine was treated with *t*-BuOK in DMSO and irradiated, 3,3′-bi(9*H*-carbazole) **106** was achieved in 67% isolated yield (Eq. 10.36) [81]:

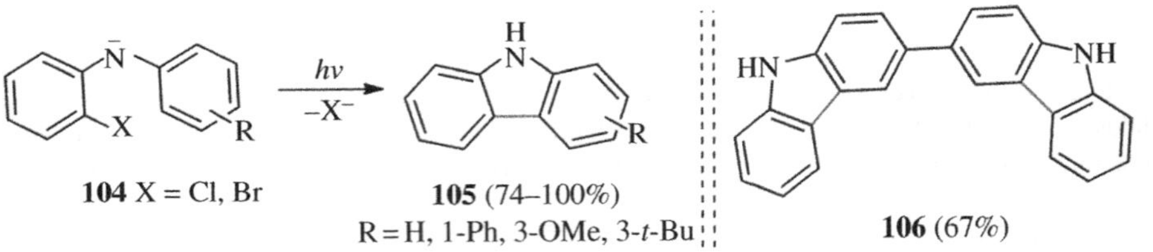

104 X = Cl, Br **105** (74–100%) R = H, 1-Ph, 3-OMe, 3-*t*-Bu **106** (67%)

(10.36)

When in the diarylamine one aryl group is replaced by a pyridine ring and the pyridyl moiety has the leaving group (like aniline halopyridines **107**), carbolines **108** were obtained through a C—C bond formation with good yields. This is the only synthetic route known to prepare all four carboline regioisomers (Scheme 10.14).

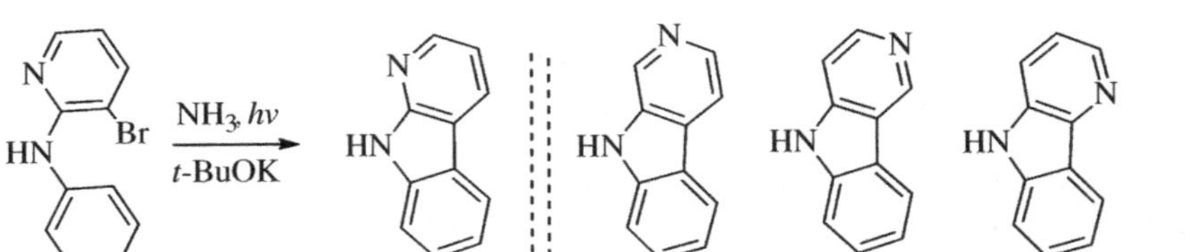

107 **108** *α* (75%) **108** *β* (70%) **108** *γ* (72%) **108** *δ* (80%)

SCHEME 10.14 Synthesis of four carboline regioisomers.

When 2-halophenyl pyridyl amide anions such as **109** are irradiated in liquid ammonia, pyrido[1,2-*a*]benzimidazoles **110** were obtained with good yields through a new C—N bond formation (Eq. 10.37) [82]:

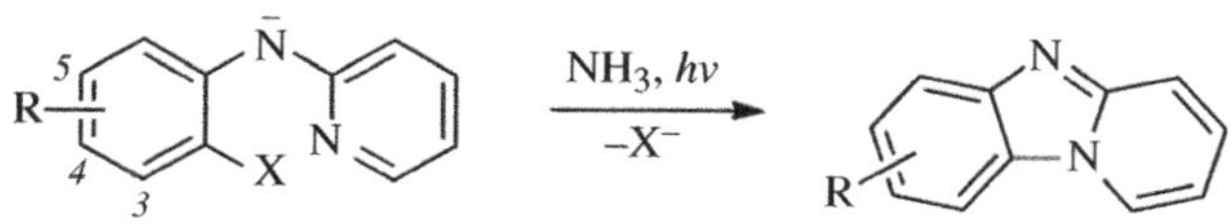

109 X = Cl, Br   R = 4-Me, 5-Me, 4-CF_3, 5-F   **110** (62–94%)

(10.37)

Another system where C—N bond formation was observed is in the synthesis of indazoles **112**. The synthesis of heterocycles **112** from (*Z*)-2-bromoacetophenone tosylhydrazones **111** involves a catalytic system composed of a diamine as ligand, such as N^1,N^2-dimethylethane-1,2-diamine and K_2CO_3 at rt (Eq. 10.38) [83]. The mechanism proposed is the intramolecular ET from the nucleophilic center to the pendant aryl bromide, fragmentation, and radical–radical coupling (see Eq. 10.31):

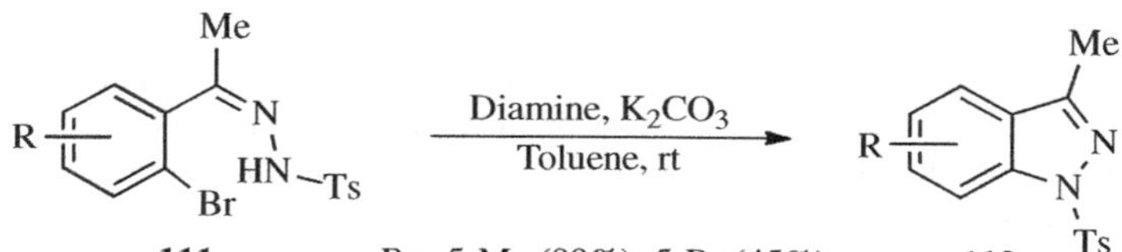

111   R = 5-Me (99%); 5-Br (45%); 5-Cl (40%); 5,6-$(OMe)_2$ (98%)   **112**

(Ts = *p*-Toluenesulfonyl)

(10.38)

Similar to the intermolecular version, the anions **113** (R = H, Me) derived from pyrazole, formed by *t*-BuOK in liquid ammonia, gave under irradiation the ring closure products **114** by a C—C bond formation, which were oxidized in the work-up to yield finally **115** (Eq. 10.39) [84]:

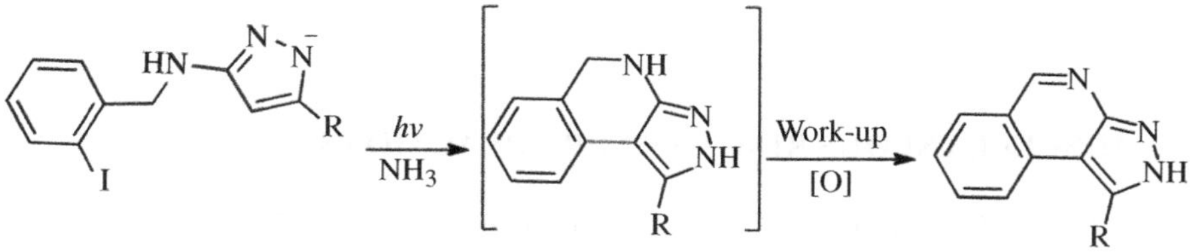

113 R = H, Me   **114**   **115** (54–56%)

(10.39)

The synthesis of a series of 6-substituted 2-pyrrolyl and 2-indolyl benzoxazoles **117** was obtained by photostimulated cyclization of anions from 2-pyrrole-carboxamides, 2-indole-carboxamides, or 3-indole-carboxamides **116**. It has been found that these reactions proceed with a C—O bond formation in good to excellent yields (Eq. 10.40) [85]:

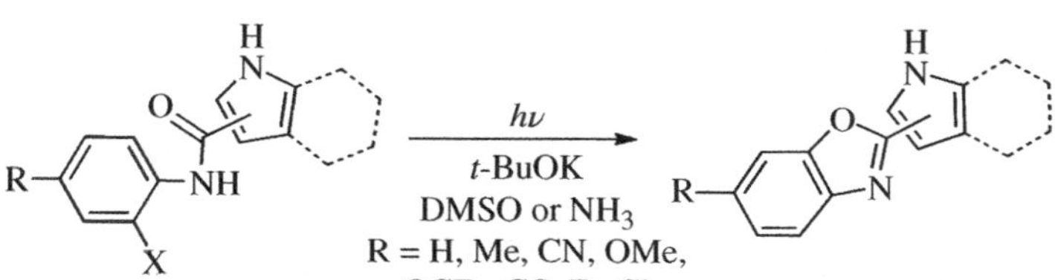

116 X = Cl, Br, I   **117** (41–100%)

(10.40)

On the basis of photochemical and photophysical experiments, it was proposed that 2-indolyl benzoxazole product **121** was obtained from the anion **118** by an intramolecular ET in the sense of

Equation 10.31. Once formed, the diradical **119** collapses to give the more stable tautomer **121** (Eq. 10.41) [86]:

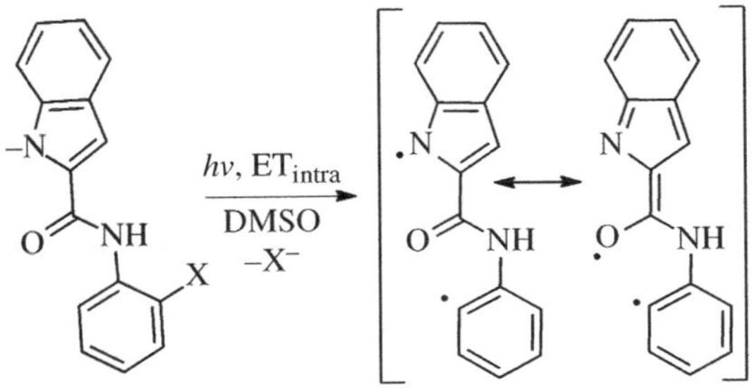

118 X = Cl, Br, I **119** **120** **121**

(10.41)

Like carbanions and anions from aryl amines, the anions from phenols have been used in intramolecular $S_{RN}1$ to obtain heterocycles by C—C bond formation [87]. One example of this approach is the photostimulated reaction of the phenoxide ion linked with a pendant bromoarene by *N*-substituted tetrahydroisoquinoline bridge such as **122**. Under this reaction condition, aporphine alkaloid derivatives **123a,b** ($n=1$) were obtained in good yields (Eq. 10.42). This approach was extended for the first time to the synthesis of a homoaporphine alkaloid **123c** ($n=2$) [88].

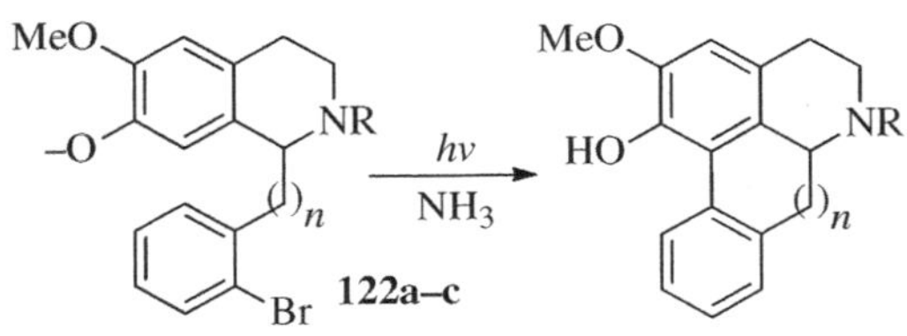

123a R: CO_2Me $n=1$ (75%)
123b R: COMe $n=1$ (73%)
123c R: CO_2Me $n=2$ (40%)

(10.42)

10.5 MISCELLANEOUS RING CLOSURE REACTIONS

10.5.1 *Exo* or *Endo* Radical Cyclization Followed by an $S_{RN}1$ Reaction

When an aromatic substrate has both the leaving group and a double bond at an appropriate distance, the radical formed after the ET and C—X bond fragmentation can be trapped by the double bond in an *exo* or *endo* ring closure mode. For instance, with substrates **124**, the radical **125** is formed and trapped by the double bond in a *5-exo-trig* ring closure mode to afford **126**. The resulting alkyl radical **126** is able to react with Nu^- to afford finally the ring closure substitution product **127** (Eq. 10.43). In this way, 3-substituted-2,3-dihydro-1-*H*-indoles **127** ($n=1$, Z=N-R) and 2,3-dihydrobenzofurans **127** ($n=1$, Z=O) [89] were obtained in very good yields. Moreover, the tandem 6-*exo-trig* cyclization–$S_{RN}1$ reactions with 1-(but-3-enyloxy)-2-halobenzenes afford 4-substituted chromanes **127** ($n=2$, Z=O) in good to excellent yields (Eq. 10.43) [90]:

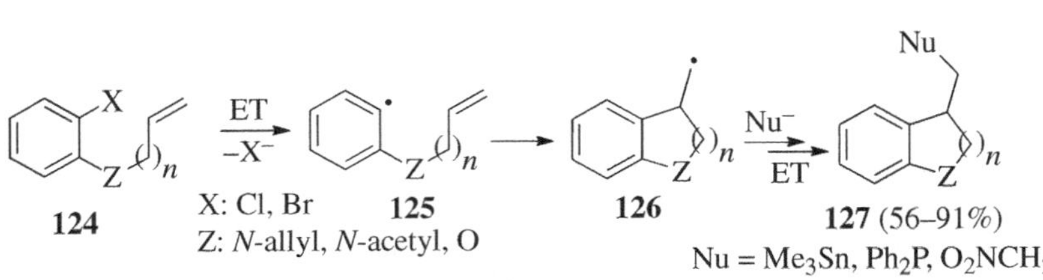

124 X: Cl, Br Z: *N*-allyl, *N*-acetyl, O **125** **126** **127** (56–91%) Nu = Me_3Sn, Ph_2P, O_2NCH_2

(10.43)

In a similar way, *N*-allyl-*N*-(2-chlorobenzyl)acetamide reacted with nucleophiles to furnish the cyclized-substituted compounds 6-*exo-trig* (35–50%) and 7-*endo-trig* (22–28%). The reactions were modeled with DFT method and relate the product distribution to the structure of the aliphatic radical intermediates [91]. In this sense, the acetamide **128** reacts with nucleophiles to yield only the 6-*exo-trig* ring closure product **129** (Eq. 10.44). However, changing the *N*-cyclohexene moiety of **128** with an *N*-(2-methylallyl)acetamide moiety afforded only the 7-*endo-trig* ring closure product **130** with good yields [91].

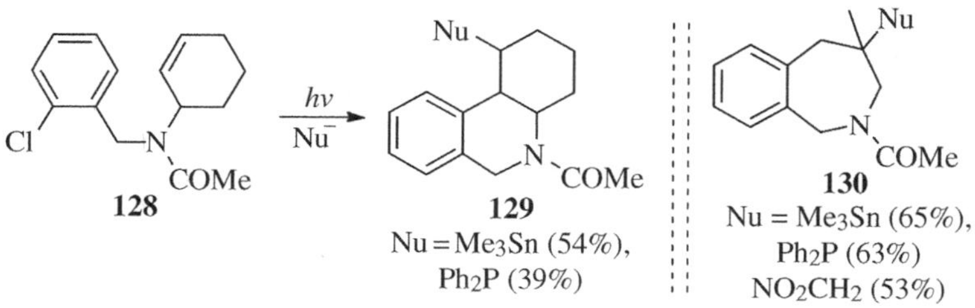

(10.44)

10.5.2 Intermolecular $S_{RN}1$ Reaction Followed by Intramolecular $S_{RN}1$ or BHAS Reaction

The photostimulated reaction of *o*-diiodobenzene **131** with 2-naphthoxide ion **65** in liquid ammonia afforded the tetracyclic compound **132** (Eq. 10.45) [92]:

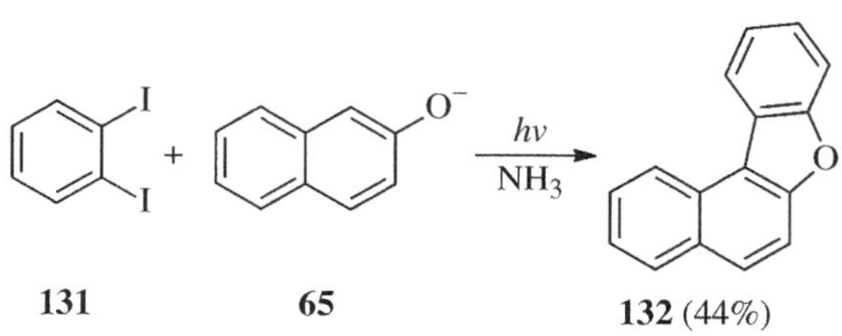

(10.45)

By ET to **131** and after C—I bond fragmentation, the radical **133** is formed, which by coupling with **65** gave finally **134**. By ET to **134** and after C—I bond fragmentation, the distonic radical anion **135** was formed, which by a coupling reaction yielded the conjugated radical anion **136**, which by ET gave ultimately the product **132** (Eq. 10.46):

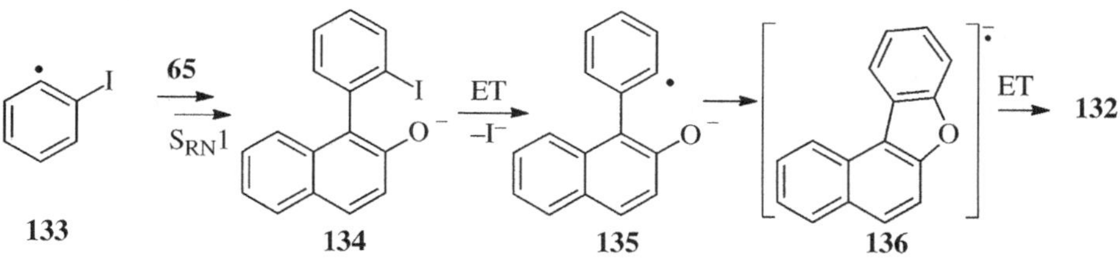

(10.46)

The aryl radical intermediate in $S_{RN}1$ reaction can be trapped by an aromatic ring. For example, the reaction of **131** with naphthalene-2-thiolate ions **137** afforded the ring closure product **141** (62% yield) [92]. In this case, by ET to **131** and after C—I bond fragmentation, the radical **133** is formed. This radical reacts with **137** and the radical anion is produced. After C—I bond fragmentation the radical **138** is generated. This radical by intramolecular HAS (see Chapter 9) affords the cyclohexadienyl radical **139**, which is deprotonated to yield the radical anion **140**. Finally, ET from **140** to **131** affords the product **141** and **133**, which continue the chain propagation (Eq. 10.47) [92]:

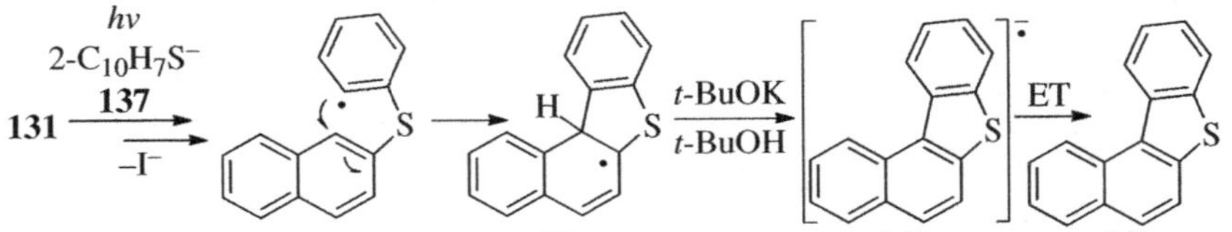

138 **139** **140** **141**

(10.47)

Finally, the reaction of 1,8-diiodonaphthalene **142** with 4-methylbenzenethiolate ions **143** gave under irradiation the ring closure product **144**. The mechanism of its formation is by a consecutive $S_{RN}1$ reaction followed by a HAS reaction (Eq. 10.48) [93], similar to Equation 10.47:

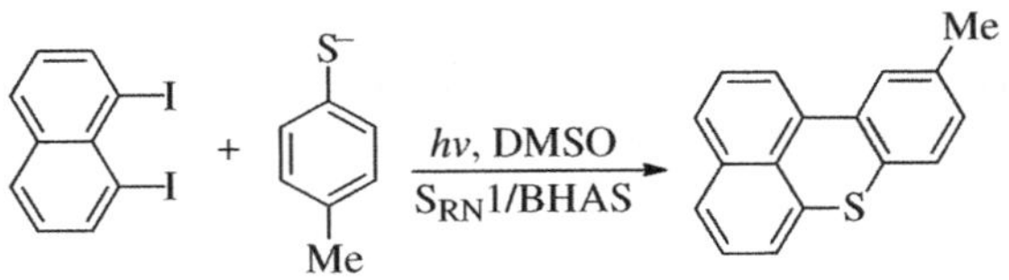

142 **143** **144** (55%)

(10.48)

10.6 SUMMARY AND OUTLOOK

The coupling between radical Ar· and a Nu⁻ is a broad and general reaction. The Nu⁻ that are able to react are carbanions, heteroatoms such as derivatives of Sn, and elements from the 15 and 16 groups. Di- and trisubstitution with substrates bearing more than one leaving group is also feasible.

The $S_{RN}1$ reaction of R_3Sn^- ions is quite versatile and affords triorganylstannyl compounds. The sequence of $S_{RN}1$ reactions to synthesize stannanes followed by a Pd-catalyzed process (Stille reaction) with electrophiles is a powerful synthetic tool; this sequence can be iteratively repeated to synthesize complex molecules.

One of the most widely studied approaches to heterocyclic synthesis is the intermolecular $S_{RN}1$ substitution of aromatic compounds that have an appropriate substituent Z *ortho* to the leaving group, followed by ring closure reactions between the Nu and a Z group.

Also, the intramolecular $S_{RN}1$ has proven to be highly efficient for the syntheses of heterocycles like carbazoles, carbolines, phenanthridines, etc. The Ar· radical intermediates can also be trapped by an appropriate double bond, and the resulting alkyl radical intermediates can react with a Nu⁻ to give substituted heterocycles.

Although the ring closure reactions to obtain heterocycles have been partially studied, due to the diversity of different nucleophiles and substrates that can be used, this is a field of vast possibilities of new developments.

ABBREVIATIONS

CTC	Charge-transfer complex
DET	Dissociative electron transfer
DTBN	di-*t*-Butyl nitroxide
EA	Electron affinity
ET	Electron transfer
EWG	Electron-withdrawing group
HAS	Homolytic aromatic substitution
Het	Heterocycle
HME	Halogen/metal exchange
HOMO	Highest occupied molecular orbital
LDA	Lithium diisopropylamide
LUMO	Lowest unoccupied molecular orbital
MO	Molecular orbital
MW	Microwave
Nu^-	Nucleophile
OP	Oxidation potential
RM	Redox mediator
rt	Room temperature
S-H	Solvent
TEMPO	2,2,6,6-Tetramethylpiperidine-1-oxyl

REFERENCES

[1] (a) Kim, J. K. and Bunnett, J. F. (1970) *J. Am. Chem. Soc.*, **92**, 7463–7464; (b) Kim, J. K. and Bunnett, J. F. (1970) *J. Am. Chem. Soc.*, **92**, 7464–7466.

[2] (a) Costentin, C., Donati, L., and Robert, M. (2009) *Chem. Eur. J.*, **15**, 785–792; (b) Costentin, C., Robert, M., and Savéant, J.-M. (2006) *Chem. Phys.*, **324**, 40–56.

[3] (a) Russell, G. A. and Danen, W. C. (1966) *J. Am. Chem. Soc.*, **88**, 5563–5565; (b) Kornblum, N., Michel, R. E., and Kerber, R. C. (1966) *J. Am. Chem. Soc.*, **88**, 5662–5663.

[4] (a) Bunnett, J. F., Creary, X., and Sundberg, J. E. (1976) *J. Org. Chem.*, **41**, 1707–1709; (b) Galli, C., Gentili, P., and Rappoport, Z. (1994) *J. Org. Chem.*, **59**, 6786–6795; (c) Santiago, A. N., Lassaga, G., Rappoport, Z., and Rossi, R. A. (1996) *J. Org. Chem.*, **61**, 1125–1128; (d) Chopa, A. B., Dorn, V. B., Badajoz, M. A., and Lockhart, M. T. (2004) *J. Org. Chem.*, **69**, 3801–3805.

[5] (a) Feiring, A. E. (1995) *J. Org. Chem.*, **50**, 3269–3274; (b) Vaillard, S. E., Postigo, A., and Rossi, R. A. (2004) *Organometallics*, **23**, 3003–3007.

[6] (a) Baumgartner, M. T., Pierini, A. B., and Rossi, R. A. (1999) *J. Org. Chem.*, **64**, 6487–6489; (b) Duca, J. S., Gallego, M. H., Pierini, A. B., and Rossi, R. A. (1999) *J. Org. Chem.*, **64**, 2626–2629; (c) Nazareno, M. A. and Rossi, R. A. (1996) *J. Org. Chem.*, **61**, 1645–1649.

[7] (a) Guastavino, J. F. and Rossi, R. A. (2009) *Organometallics*, **28**, 2646–2649; (b) Zhou, D. Y., Dou, H. Y., Zhao, C. X., and Chen, Q. Y. (2006) *J. Fluor. Chem.*, **127**, 740–745; (c) Nazareno, M. A. and Rossi, R. A. (1994) *Tetrahedron*, **50**, 9267–9274.

[8] (a) Santiago, A. N., Basso, S. M., Toledo, C. A., and Rossi, R. A. (2005) *New J. Chem.*, **29**, 875–880; (b) Lee, G. S., Bashara, J. N., Sabih, G., Oganesyan, A., Godjoian, G., Duong, H. M., Marinez, E. R., and Gutierrez, C. G. (2004) *Org. Lett.*, **6**, 1705–1707; (c) Santiago, A. N., Toledo, C. A., and Rossi, R. A. (2002) *J. Org. Chem.*, **67**, 2494–2500; (d) Lukach, A. E. and Rossi, R. A. (1999) *J. Org. Chem.*, **64**, 5826–5831; (e) Borosky, G. L., Pierini, A. B., and Rossi, R. A. (1990) *J. Org. Chem.*, **55**, 3705–3707.

[9] Rossi, R. A., Pierini, A. B., and Peñéñory, A. B. (2003) *Chem. Rev.*, **103**, 71–167.

[10] (a) Budén, M. E., Martín, S. E., and Rossi, R. A. (2012) Recent Advances in the Photoinduced Radical Nucleophilic Substitution Reactions, in *CRC Handbook of Organic Photochemistry and Photobiology*, 3th ed., Griesbeck, A. G., Oelgemöller, M., and Ghetti, F., Eds., CRC Press Inc., Boca Raton, pp. 347–368; (b) Rossi, R. A. (2005) Photoinduced Aromatic Nucleophilic Substitution Reactions, in *Synthetic Organic Photochemistry*, Griesbeck, A. G. and Mattay, J., Eds., Marcel Dekker: New York, pp. 495–527; (c) Bardagí, J. I., Vaillard, V. A., and Rossi, R. A., (2012) The $S_{RN}1$ Reaction, in *Encyclopedia of Radicals in Chemistry, Biology & Materials*, Chatgilialoglu, C., Studer, A., Eds., John Wiley & Sons Ltd, Chichester, pp. 333–364.

[11] (a) Savéant, J.-M. (1994) *Tetrahedron*, **50**, 10117–10165; (b) Andrieux, C. P., Hapiot, P., and Savéant, J.-M. (1990) *Chem. Rev.*, **90**, 723–738; (c) Savéant, J.-M. (1980) *Acc. Chem. Res.*, **13**, 323–329; (d) Savéant, J.-M. (1993) *Acc. Chem. Res.*, **26**, 455–461; (e) Savéant, J.-M. (1990) *Adv. Phys. Org. Chem.*, **26**, 1–130; (f) Savéant, J.-M. (1992) *New J. Chem.*, **16**, 131–150.

[12] Rossi, R. A., Pierini, A. B., and Santiago A. N. (1999) Aromatic Substitution by the $S_{RN}1$ Reaction, in *Organic Reactions*, Vol. **54**, Paquette, L. A., and Bittman, R., Eds., John Wiley & Sons, Inc., New York, pp. 1–271.

[13] Costentin, C., Hapiot, P., Médebielle, M., and Savéant, J.-M. (1999) *J. Am. Chem. Soc.*, **121**, 4451–4460.

[14] Schmidt, L. C., Argüello, J. E., and Peñéñory, A. B. (2007) *J. Org. Chem.*, **72**, 2936–2944.

[15] Budén, M. E., Guastavino, J. F., and Rossi, R. A. (2013) *Org. Lett.*, **15**, 1174–1177.

[16] Braslavsky, S. E. (2007) *Pure Appl. Chem.*, **79**, 293–465.

[17] See for example (a) Baumgartner, M. T., Gallego, M. H., and Pierini, A. B. (1998) *J. Org. Chem.*, **63**, 6394–6397; (b) Borosky, G. L., Pierini, A. B., and Rossi, R. A. (1992) *J. Org. Chem.*, **57**, 247–252.

[18] (a) Pierini, A. B. and Vera, D. M. A. (2003) *J. Org. Chem.*, **68**, 9191–9199; (b) Houmam, A. (2008) *Chem. Rev.*, **108**, 2180–2237.

[19] (a) Bunnett, J. F. and Gloor, B. F. (1975) *J. Org. Chem.*, **38**, 4156–4163; (b) Rossi, R. A., de Rossi, R. H., and Pierini, A. B. (1979) *J. Org. Chem.*, **44**, 2662–2667.

[20] Rossi, R. A., de Rossi, R. H., and López, A. F. (1976) *J. Org. Chem.*, **41**, 3371–3373.

[21] Rossi, R. A. and Bunnett, J. F. (1973) *J. Org. Chem.*, **38**, 3020–3025.

[22] (a) Shirakawa, E., Hayashi, Y., Itoh, K., Watabe, R., Uchiyama, N., Konagaya, W., Masui, S., and Hayashi, T. (2012) *Angew. Chem. Int. Ed.*, **51**, 218–221; (b) Haines, B. E. and Wiest, O. (2014) *J. Org. Chem.*, **79**, 2771–2774.

[23] (a) Minami, H., Wang, X., Wang, C., and Uchiyama, M. (2013) *Eur. J. Org. Chem.*, 7891–7894; (b) Shirakawa, E., Tamakuni, F., Kusano, E., Uchiyama, N., Konagaya, W., Watabe, R., and Hayashi, T. (2014) *Angew. Chem. Int. Ed.*, **53**, 521–525.

[24] Murarka, S. and Studer, A. (2012) *Angew. Chem. Int. Ed.*, **51**, 12362–12366.

[25] Tebben, L. and Studer, A. (2011) *Angew. Chem. Int. Ed.*, **50**, 5034–5068.

[26] Baumgartner, M. T., Lotz, G. A., and Palacios, S. M. (2004) *Chirality*, **16**, 212–219.

[27] Tempesti, T. C., Pierini, A. B., and Baumgartner, M.T. (2012) *New J. Chem.*, **36**, 597–602.

[28] Bunnett, J. F. and Sundberg, J. E. (1976) *J. Org. Chem.*, **41**, 1702–1706.

[29] Bunnett, J. F. and Singh, P. (1981) *J. Org. Chem.*, **46**, 5022–5025.

[30] Baumgartner, M. T., Jiménez, L. B., Pierini, A. B., and Rossi, R. A. (2002) *J. Chem. Soc. Perkin Trans.* **2**, 1092–1097.

[31] Scamehorn, R. G., Hardacre, J. M., Lukanich, J. M., and Sharpe, L. R. (1984) *J. Org. Chem.*, **49**, 4881–4883.

[32] Semmelhack, M. F. and Bargar, T. (1980) *J. Am. Chem. Soc.*, **102**, 7765–7776.

[33] Beugelmans, R. and Roussi, G. (1981) *Tetrahedron*, **37**, 393–397.

[34] Bard, R. R. and Bunnett, J. F. (1980) *J. Org. Chem.*, **45**, 1546–1547.

[35] (a) Baumgartner, M. T., Nazareno, M. A., Murguía, M. C., Pierini, A. B., and Rossi, R. A. (1999) *Synthesis*, 2053–2056; (b) Barolo, S. M., Lukach, A. E., and Rossi, R. A. (2003) *J. Org. Chem.*, **68**, 2807–2811.

[36] Beugelmans, R. and Chbani, M. (1995) *Bull. Soc. Chim. Fr.*, **132**, 729–733.

[37] Guastavino, J. F., Barolo, S. M., and Rossi, R. A. (2006) *Eur. J. Org. Chem.*, 3898–3902.

[38] Layman, W. J., Greenwood, T. D., Downey, A. L., and Wolfe, J. F. (2005) *J. Org. Chem.*, **70**, 9147–9155.

[39] Rossi, R. A. (2014) *J. Organomet. Chem.*, **751**, 201–212.

[40] Yammal, C. C., Podestá, J. C., and Rossi, R. A. (1982) *J. Org. Chem.*, **57**, 5720–5725.

[41] Córsico, E. F. and Rossi, R. A. (2000) *Synlett*, 227–229.

[42] Santiago, A. N., Basso, S. M., Montañez, J. P., and Rossi, R. A. (2006) *J. Phys. Org. Chem.*, **19**, 829–835.

[43] (a) Chopa, A. B., Lockhart, M. T., and Dorn, V. B. (2002) *Organometallics*, **21**, 1425–1429; (b) Chopa, A. B., Lockhart, M. T., and Silbestri, G. F. (2002) *Organometallics*, **21**, 5874–5878; (c) Silbestri, G. F., Lo Fiego, M. L., Lockhart, M. T., and Chopa, A. B. (2010) *J. Organomet. Chem.*, **695**, 2578–2585.

[44] (a) Chopa, A. B., Lockhart, M. T., and Silbestri, G. F. (2001) *Organometallics*, **20**, 3358–3360; (b) Silbestri, G. F., Lockhart, M. T., and Chopa, A. B. (2011) *Arkivoc*, **VII**, 210–220.

[45] Córsico, E. F. and Rossi, R. A. (2002) *J. Org. Chem.*, **67**, 3311–3316.

[46] Córsico, E. F. and Rossi, R. A. (2000) *Synlett*, 230–232.

[47] Bardagí, J. I. and Rossi, R. A. (2008) *J. Org. Chem.*, **73**, 4491–4495.

[48] (a) Silbestri, G. F., Masson, R. B., Lockhart, M. T., and Chopa, A. B. (2006) *J. Organomet. Chem.*, **691**, 1520–1524; (b) LoFiego, M. J., Badajoz, M. A., Silbestri, G. F., Lockhart, M. T., and Chopa, A. B. (2008) *J. Org. Chem.*, **73**, 9184–9187; (c) Lo Fiego, M. J., Silbestri, G. F., Chopa, A. B., and Lockhart, M. T. (2011) *J. Org. Chem.*, **76**, 1707–1714.

[49] Pierini, A. B., Baumgartner, M. T., and Rossi, R. A. (1987) *Tetrahedron Lett.*, **28**, 4653–4656.

[50] Jimenez, L. B., Torres, N. V., Borioni, J. L., and Pierini, A. B. (2014), *Tetrahedron*, **70**, 3614–3620.

[51] Tempesti, T. C., Pierini, A. B., and Baumgartner, M. T. (2005) *J. Org. Chem.*, **70**, 6508–6511.

[52] Chahma, M., Combellas, C., and Thiébault, A. (1994) *Synthesis*, 366–368.

[53] Chahma, M., Combellas, C., and Thiébault, A. (1995) *J. Org. Chem.*, **60**, 8015–8022.

[54] Médebielle, M., Oturan, M. A., Pinson, J., and Savéant, J.-M. (1993) *Tetrahedron Lett.*, **34**, 3409–3412.

[55] (a) Bunnett, J. F. and Creary, X. (1974) *J. Org. Chem.*, **39**, 3612–3614; (b) Bunnett, J. F. and Shafer, S. J. (1978) *J. Org. Chem.*, **43**, 1873–1877; (c) Bunnett, J. F. and Shafer, S. J. (1978) *J. Org. Chem.*, **43**, 1877–1879; (d) Bunnett, J. F. and Weiss, R. H. (1978) *Org. Synth.*, **58**, 134–137; (e) Bard, R. R., Bunnett, J. F., and Traber, R. P. (1979) *J. Org. Chem.*, **44**, 4918–4924.

[56] (a) Swartz, J. E. and Bunnett, J. F. (1979) *J. Org. Chem.*, **44**, 340–346; (b) Barolo, S. M., Martin, S. E., and Rossi, R. A. (2012) *Arkivoc*, **VIII**, 98–106.

[57] Swartz, J. E. and Bunnett, J. F. (1979) *J. Org. Chem.*, **44**, 4673–4677.

[58] Rossi, R. A., Alonso, R. A., and Palacios, S. M. (1981) *J. Org. Chem.*, **46**, 2498–2502.

[59] Rossi, R. A. (1982) *Acc. Chem. Res.*, **15**, 164–170.

[60] Bardagí, J. I. and Rossi, R. A. (2010) *J. Org. Chem.*, **75**, 5271–5277.

[61] Pierini, A. B., Baumgartner, M. T., and Rossi, R. A. (1988) *Tetrahedron Lett.*, **29**, 3429–3432.

[62] Tempesti, T. C., Pierini, A. B., and Baumgartner, M. T. (2009) *New J. Chem.*, **33**, 1523–1528.

[63] Rodriguez, S. A. and Baumgartner, M. T. (2010) *Tetrahedron Lett.*, **51**, 5322–5324.

[64] Amatore, C., Beugelmans, R., Bois-Choussy, M., Combellas, C., and Thiébault, A. (1989) *J. Org. Chem.*, **54**, 5688–5695.

[65] Uranga, J. G., Montañez, J. P., and Santiago, A. N. (2012) *Tetrahedron*, **68**, 584–589.

[66] Argüello, J. E., Schmidt, L. C., and Peñéñory, A. B. (2003) *Org. Lett.*, **5**, 4133–4136.

[67] Schmidt, L. C., Rey, V., and Peñéñory, A. B. (2006) *Eur. J. Org. Chem.*, 2210–2214.

[68] Bouchet, L. M., Peñéñory, A. B., and Argüello, J. E. (2011) *Tetrahedron Lett.*, **52**, 969–972.

[69] Rossi, R. A. and Peñéñory, A. B. (1981) *J. Org. Chem.*, **46**, 4580–4582.

[70] Pierini, A. B. and Rossi, R. A. (1979) *J. Org. Chem.*, **44**, 4667–4673.

[71] Goehring, R. R. (1992) *Tetrahedron Lett.*, **33**, 6045–6048.

[72] Goehring, R. R. (1994) *Tetrahedron Lett.*, **35**, 8145–8146.

[73] (a) Martin, S. F. and Liras, S. (1993) *J. Am. Chem. Soc.*, **115**, 10450–10451; (b) Liras, S., Lynch, C. L., Fryer, A. M., Vu, B. T., and Martin, S. F. (2001) *J. Am. Chem. Soc.*, **123**, 5918–5924.

[74] Semmelhack, M. F. and Bargar, T. M. (1977) *J. Org. Chem.*, **42**, 1481–1482.

[75] Dandekar, S. A., Greenwood, S. N., Greenwood T. D., Mabic S., Merola, J. S., Tanko, J. M., and Wolfe, J. F. (1999) *J. Org. Chem.*, **64**, 1543–1553.

[76] (a) Roydhouse, M. D. and Walton, J. C. (2005) *Chem. Comm.*, 4453–4455; (b) Marshall, L. J., Roydhouse, M. D., Slawin, A. M. Z., and Walton, J. C. (2007) *J. Org. Chem.*, **72**, 898–911.

[77] Roydhouse, M. D. and Walton, J. C. (2007) *Eur. J. Org. Chem.*, 1059–1063.

[78] Guastavino, J. F. and Rossi, R. A (2012) *J. Org. Chem.*, **77**, 460–472.

[79] Soria-Castro, S., Caminos, D. A., and Peñéñory, A. B. (2014) *RSC Adv.*, **4**, 17490–17497.

[80] Budén, M. E., Dorn, V. B., Gamba, M., Pierini, A. B., and Rossi, R. A. (2010) *J. Org. Chem.*, **75**, 2206–2218.

[81] Budén, M. E., Vaillard, V. A., Martín, S. E., and Rossi, R. A. (2009) *J. Org. Chem.*, **74**, 4490–4498.

[82] Barolo, S. M., Wang, Y., Rossi, R. A., and Cuny, G. D. (2013) *Tetrahedron*, **69**, 5487–5494.

[83] Thomé, I., Besson, C., Kleine, T., and Bolm, C. (2013) *Angew. Chem. Int. Ed.*, **52**, 7509–7513.

[84] Vaillard, V. A., Budén, M. E., Martín, S. E., and Rossi R. A. (2009) *Tetrahedron Lett.*, **50**, 3829–3832.

[85] Vaillard, V. A., Guastavino, J. F., Budén, M. E., Bardagi, J. I., Barolo, S. M., and Rossi R. A. (2012) *J. Org. Chem.*, **77**, 1507–1519.

[86] Vaillard, V. A., Rossi, R. A., and Argüello, J. E. (2012) *Org. Biomol. Chem.*, **10**, 9255–9261.

[87] (a) Theuns, H. G., Lenting, H. B. M., Salemink, C. A., Tanaka, H., Shibata, M., Ito, K., and Lousberg, R. J. J. Ch. (1984) *Heterocycles*, **22**, 2007–2011; (b) Wiegand, S. and Schaefer, H. J. (1995) *Tetrahedron*, **51**, 5341–5350.

[88] Barolo, S. M., Teng, X., Cuny, G. D., and Rossi, R. A. (2006) *J. Org. Chem.*, **71**, 8493–8499.

[89] Vaillard, S. E., Postigo, A., and Rossi, R. A. (2002) *J. Org. Chem.*, **67**, 8500–8506.

[90] Bardagí, J. I., Vaillard, S. E., and Rossi, R. A. (2007) *Arkivoc*, **IV**, 73–83.

[91] Peisino, L. E. and Pierini, A. B. (2013) *J. Org. Chem.*, **78**, 4719–4729.

[92] Baumgartner, M. T., Pierini, A. B., and Rossi, R. A. (1993) *J. Org. Chem.*, **58**, 2593–2598.

[93] Norris, R. K. and McMahon, J. A. (2003) *Arkivoc*, **X**, 139–155.

11

NUCLEOPHILIC SUBSTITUTION OF HYDROGEN IN ELECTRON-DEFICIENT ARENES

Mieczysław Mąkosza

Institute of Organic Chemistry, Polish Academy of Sciences, Warsaw, Poland

11.1 INTRODUCTION

Conventional nucleophilic substitution of halogens and other nucleofugal groups X in electron-deficient arenes, particularly nitroarenes, proceeds via addition of nucleophiles at positions occupied by X to form σ^X-adducts. The addition is connected with dearomatization—thus, the adducts, which are in fact nitronate anions of nitro cyclohexadienes, undergo rapid rearomatization via spontaneous departure of X^- to form products of nucleophilic aromatic substitution (S_NAr) reaction. Detailed discussion of these processes is presented in Chapter 6.

It was unambiguously shown that addition of nucleophiles to nitroarenes, also halonitroarenes, proceeds faster at positions occupied by hydrogen to form σ^H-adducts [1, 2]. However, due to the high energy of CH bond and hydride anion itself, it is unable to depart spontaneously from the σ^H-adducts. Thus, in order to recover aromaticity, conversion of the σ^H-adducts proceeds usually via departure of the nucleophile; hence, the formation of the σ^H-adducts is a fast and reversible process. Due to the reversibility of the addition at positions occupied by hydrogen to produce σ^H-adducts, they dissociate, and the slower but usually irreversible formation of σ^X-adducts can proceed. Subsequent fast departure of X^- gives products of S_NAr; thus, it can be the main observed process, whereas the initial formation of σ^H-adducts is usually overlooked. This situation is presented in Scheme 11.1. It can be therefore expected that the formation of σ^H-adducts is the initial process between nucleophiles and nitroarenes and that the further fast conversion of the σ^H-adducts should result in the formation of products of nucleophilic substitution of hydrogen in electron-deficient arenes, S_NArH. This reaction should therefore proceed faster than conventional S_NAr reaction (Scheme 11.1).

Arene Chemistry: Reaction Mechanisms and Methods for Aromatic Compounds,
First Edition. Edited by Jacques Mortier.

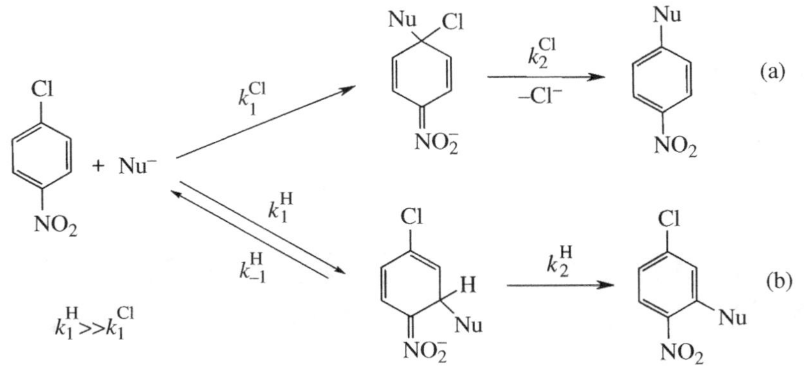

SCHEME 11.1 Relation of rates of nucleophilic addition at positions occupied by (a) chlorine and (b) hydrogen.

It should be stressed that in most textbooks in chapters on S_NAr, the nucleophilic substitution of hydrogen is not mentioned [3, 4]. On the other hand, in a recent monograph by Terrier, S_NArH is widely discussed [5]. Early examples of S_NArH reaction are thoroughly discussed in monograph by Chupakhin [6].

From Scheme 11.1, it is evident that S_NArH can be observed, provided further conversion of σ^H-adducts is a fast process. An important parameter is the state of equilibrium 11.1b; hence, factors that affect this equilibrium should decide about possibility of S_NArH. For a given combination of the educts and solvents, the addition equilibrium is governed by the temperature. Due to the entropy factor, at low temperature, the equilibrium is shifted toward adducts; thus, as it will be shown in the forthcoming parts, reactions of nucleophilic substitution of hydrogen proceed preferentially at low temperatures.

Up to now, a few ways of fast conversion of σ^H-adducts into products of S_NArH were developed such as oxidation by external oxidants, conversion into substituted nitrosoarenes according to intramolecular redox stoichiometry, elimination of HL when nucleophiles contain nucleofugal groups L at the nucleophilic center, *cine*- and *tele*-substitution, and addition of the nucleophile, ring opening, ring closure (ANRORC) process [1, 2].

11.2 OXIDATIVE NUCLEOPHILIC SUBSTITUTION OF HYDROGEN

Hydride anions cannot depart spontaneously from the σ^H-adducts; thus, they should be removed by external agents-oxidants. Oxidation of the σ^H-adducts by external oxidants seems to be the obvious way to products of oxidative nucleophilic substitution of hydrogen (ONSH). However, taking into account the reversibility of the formation of the σ^H-adducts and susceptibility of majority of nucleophiles to oxidation, the ONSH is limited to three variants: (i) nucleophiles are insensitive to the oxidants used, (ii) the equilibrium of the addition is shifted toward σ^H-adducts, and (iii) the addition is an irreversible process:

1. The most common nucleophiles that are resistant toward oxidation are ammonia and hydroxide anion. Indeed, such strong oxidant as $KMnO_4$ is soluble in liquid ammonia and forms solutions that are stable even at room temperature (under pressure). These solutions are efficient agents for amination of electron-deficient nitroarenes and azines that proceeds via addition of ammonia to the ring, followed by oxidation of the σ^H-adducts with $KMnO_4$. This amination process of azines is often referred to as the oxidative Chichibabin reaction [7]. For instance, the reaction

of 2,4-dinitrochlorobenzene with a solution of $KMnO_4$ in liquid ammonia results exclusively in ONSH with ammonia to give 2,4-dinitro-5-chloroaniline (Scheme 11.2).

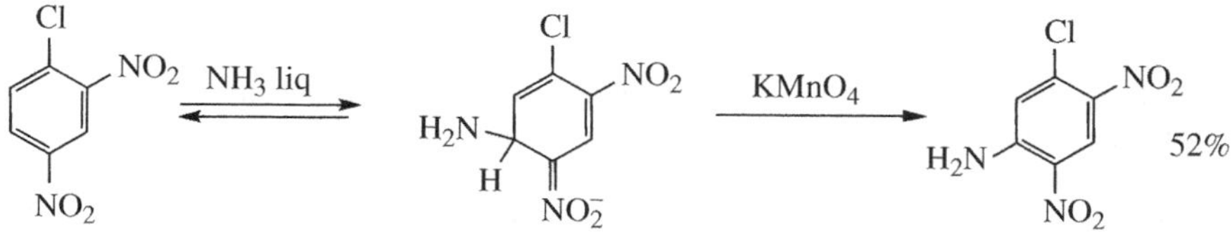

SCHEME 11.2 Oxidative amination of 2,4-dinitrochlorobenzene [7b].

Hydrolysis of *p*-chloronitrobenzene with KOH to *p*-nitrophenol is presented in almost every textbook as an example of the S_NAr reaction [3, 4]. However, nowhere there is an information that this is a slow secondary reaction, which proceeds at elevated temperatures, whereas at low temperatures in liquid ammonia solution, hydroxide anions supplied by KOH add rapidly at positions occupied by hydrogen. Subsequent oxidation of the resulting σ^H-adducts with oxygen gives 2-nitro-5-chlorophenol, the product of ONSH [8]. Thus, ONSH that proceeds below −35°C is obviously a much faster reaction than S_NAr of chlorine because the latter takes place only at an elevated temperature of, for example, +140°C (Scheme 11.3).

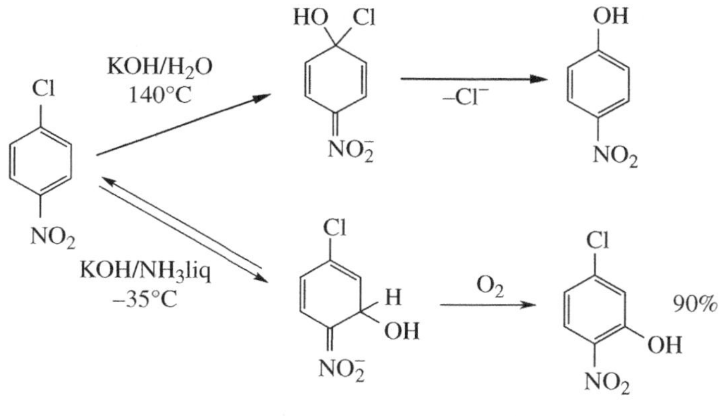

SCHEME 11.3 Reaction of *p*-chloronitrobenzene with KOH: fast primary ONSH and slow secondary S_NAr [8].

These examples of ONSH amination and hydroxylation of electron-deficient arenes provide unambiguous proofs of the relation of rates of nucleophilic addition at positions occupied by hydrogen and halogens presented in Scheme 11.1.

Oxidative amination of nitrobenzene with aniline and benzamide that proceeds efficiently in the presence of tetramethylammonium hydroxide also belongs to this category of ONSH reaction. It appears that in these processes, which are of substantial practical value, nitrobenzene and atmospheric oxygen afford the oxidation (Scheme 11.4) [9].

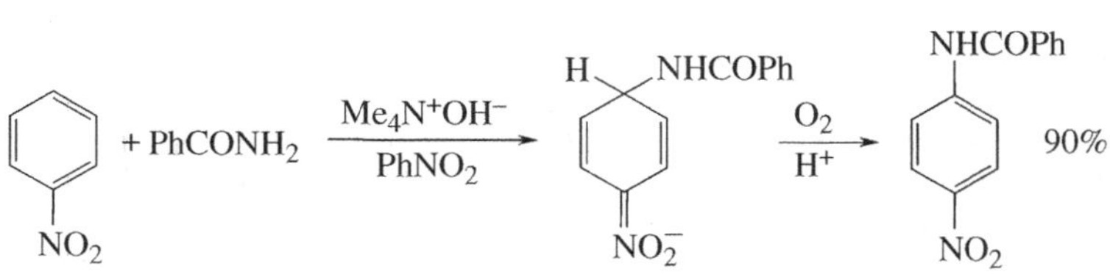

SCHEME 11.4 Synthesis of N-4-nitrophenylbenzamide via ONSH [9a].

2. Carbanions form an important group of nucleophiles that are sensitive to majority of oxidants; thus, ONSH with carbanions can proceed only when the addition equilibrium is shifted toward σ^H-adducts. For many years, it was a common opinion that mononitroarenes such as nitrobenzene, halonitrobenzenes, etc. are not sufficiently electrophilic to form long-lived σ^H-adducts with carbanions. We have shown that moderately stabilized carbanions of nitriles and esters of phenylalkanoic acids, esters of benzylphosphonic acids, etc. generated in liquid ammonia at temperature below −65°C add to nitrobenzene and other nitroarenes to form σ^H-adducts that under these conditions are long-lived intermediates [10–12]. The addition of carbanions to nitroarenes to form σ^H-adducts (and also σ^X-adducts) proceeds satisfactorily only when they are in the form of loose ion pairs with their countercations. Thus, the addition proceeds efficiently in liquid ammonia, solvent that assures solvation of sodium and potassium cations; hence, carbanions exhibit sufficient activity. Similar situation is in aprotic dipolar solvents; however, due to their high melting points and viscosity, they cannot be used at low temperature. On the other hand, mixtures of THF/DMF are very convenient systems for reactions of nitroarenes with carbanions, which should be carried out at low temperature. Lithium or sodium salts of carbanions in THF at low temperature do not add to nitroarenes to form σ^H-adducts; usually, the single-electron transfer (SET) dominates, leading to redox processes. However, addition of tetraalkylammonium (TAA) halides to lithium salts of carbanions results in an ion exchange and formation of loose ion pairs of the carbanions with TAA cations. In such form, the addition of carbanions to nitroarenes producing σ^H-adducts proceeds efficiently as it was shown in the reactions of methyl isobutyrate [13]. The addition of carbanions to nitroarenes in liquid ammonia or THF/DMF at temperatures below −65°C is a fast process, completed within minutes, and the equilibrium is shifted toward σ^H-adducts. Secondary carbanions of diethyl benzylphosphonate, ethyl phenylacetate, phenylacetonitrile, etc. add in liquid ammonia to nitroarenes at positions *ortho* and *para* to the nitro group. The produced σ^H-adducts upon subsequent oxidation with $KMnO_4$ form two isomeric products of ONSH. When the *para* position is occupied, even by fluorine atom, the addition at low temperature proceeds exclusively *ortho* to the nitro group, giving σ^H-adducts that upon oxidation form products of ONSH. No traces of products of S_NAr of fluorine are formed under these conditions (Scheme 11.5) [10, 11].

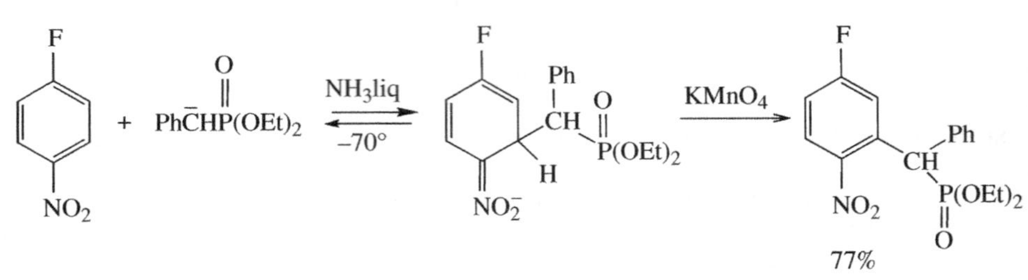

SCHEME 11.5 ONSH in p-fluoronitrobenzene with carbanion of diethyl benzylphosphonate.

It should be noted that at room or higher temperatures in DMF or DMSO, these carbanions react with *p*-fluoronitrobenzene via formation of σ^F-adducts to give products of conventional S_NAr of fluoride. On the other hand, due to steric effects, tertiary carbanions of similar esters and nitriles add to nitroarenes selectively at the *para* positions to the nitro group. Treatment of solutions of the formed σ^H-adducts with $KMnO_4$ results in rapid oxidation to give products of ONSH usually in high yields [12a].The $KMnO_4$ oxidation of the produced σ^H-adducts is very sensitive to steric hindrances created by substituents *ortho* to the addition site (i.e., *meta* to the nitro group). For instance, σ^H-adducts of 2-phenylpropionitrile to 3,5-dichloronitrobenzene, although formed quantitatively, are not oxidized by $KMnO_4$ under the reaction conditions (Scheme 11.6). In THF/DMF solvent, σ^H-adducts formed at low temperature can be also oxidized by dichlorodicyanobenzoquinone (DDQ), giving the expected ONSH products [10]. On the other hand, σ^H-adducts generated under these conditions can be oxidized by dimethyldioxirane (DMD) to give substituted phenols [14]. Oxidation of the σ^H-adducts by DMD is insensitive to steric hindrances at the addition site; thus, the adducts of tertiary carbanions to 3,5-dichloronitrobenzene oxidized with DMD give expected substituted phenols in high yields (Scheme 11.6).

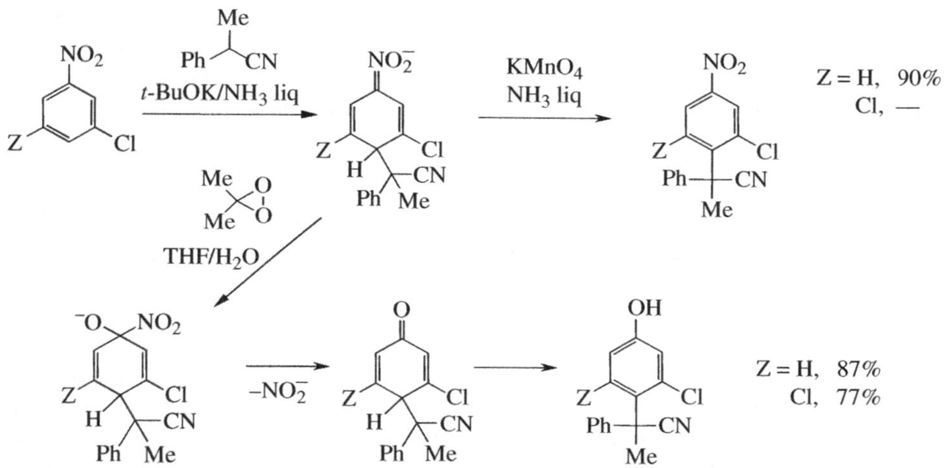

SCHEME 11.6 ONSH in nitroarenes with carbanion of 2-phenylpropionitrile, $KMnO_4$ [12a] and dimethyldioxirane DMD [14] oxidants.

These results indicate that DMD directs its action on the negatively charged nitro group of the σ^H-adducts to form intermediate cyclohexadienone and subsequent aromatization. Sensitivity of $KMnO_4$ oxidation of the σ^H-adducts to the steric effects at the addition site and high value of kinetic isotope effect (KIE) $k_H/k_D \approx 10$ at −70°C [12b] suggests that the oxidation proceeds via direct interaction of the oxidant with the oxidized site, hence probably via abstraction of the hydride anion.

The ONSH in nitroarenes with carbanions carried out at low temperature and $KMnO_4$ in liquid ammonia or DDQ and DMD in THF/DMF is a general process. For instance, this reaction has found valuable application for introduction of nitroaryl and hydroxyaryl substituents into molecules of α-amino acids. Thus, carbanions of esters of protected amino acids upon addition to nitroarenes form σ^H-adducts that are oxidized by $KMnO_4$ in liquid ammonia or DDQ in THF/DMF. The subsequent hydrolysis produces α-nitroaryl α-amino acids (Scheme 11.7) [15].

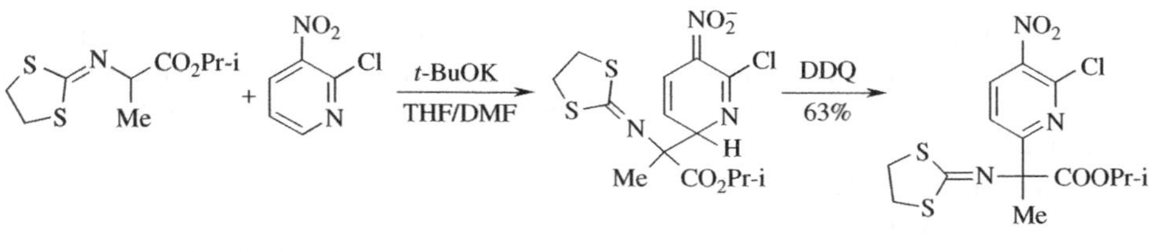

SCHEME 11.7 Oxidative nitroarylation of protected alanine ester [15a].

Oxidation of these σ^H-adducts with DMD results in an introduction of *p*-hydroxyaryl substituents into α-positions of amino acids [15]. The formation of σ^H-adducts is connected with creation of a stereogenic center. When the carbanions of protected amino acids contain a chiral center in the vicinity of the prochiral carbanion moiety, the formation of the σ^H-adducts proceeds with high stereoselectivity. Since the oxidation of the σ^H-adducts does not affect the newly formed stereogenic center, the ONSH in such cases proceeds with high diastereoselectivity and, as a consequence, enantioselectivity (Scheme 11.8) [16].

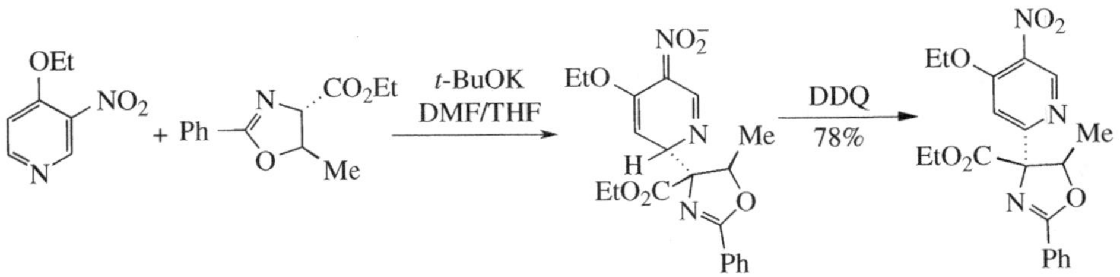

SCHEME 11.8 Diastereoselective ONSH with carbanion of protected threonine ester [16].

Use of oxygen, the most common oxidant, for oxidation of the σ^H-adducts is limited to the adducts of primary and secondary carbanions. Since the oxidation of these σ^H-adducts by oxygen proceeds efficiently only in the presence of an excess of base, we suppose that the dianions produced via further deprotonation of the σ^H-adducts are oxidized by oxygen [17]. This process is illustrated by direct ONSH in nitroarenes with enolates of ketones that proceeds selectively *para* to the nitro group [18, 19]. On the other hand, reaction of such enolates with *m*-nitroanilines—an efficient synthesis of nitroindoles proceeds mostly *ortho* to the amino and nitro groups [18]. Both of these reactions proceed in the presence of atmospheric oxygen and do not require another oxidant (Scheme 11.9).

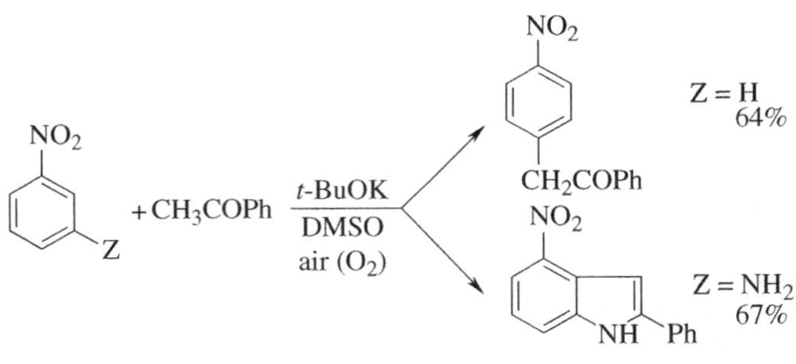

SCHEME 11.9 ONSH in nitroarenes with enolate of acetophenone. Direct synthesis of nitroindoles [18, 19].

ONSH is an efficient tool for introduction of perfluoroalkyl groups into electron-deficient arenes. Due to moderate nucleophilicity and stability of perfluoroalkyl carbanions, the reaction proceeds only with highly electrophilic arenes such as azinium salts (Scheme 11.10) [20, 21].

Similarly to ONSH reaction in nitroarenes with carbanions proceeds ONSH with diphenylphosphine anion. At low temperature in liquid ammonia, this anion adds to *p*-fluoronitrobenzene at the *ortho* position; subsequent oxidation of the produced σ^H-adduct with $KMnO_4$ gives

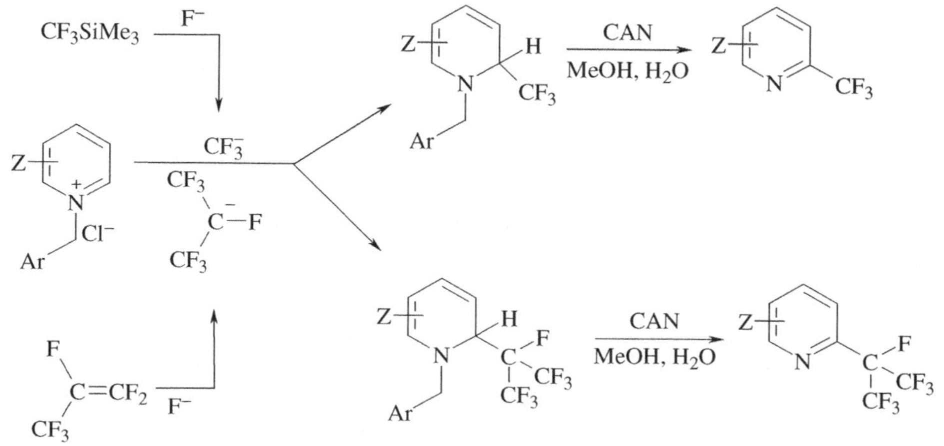

SCHEME 11.10 Synthesis of perfluoroalkylpyridines via ONSH in pyridinium salts with cerium ammonium nitrate (CAN) oxidant [20, 21].

the respective diphenyl-2-nitro-5-fluorophenyl phosphine oxide (Scheme 11.11) [22]. The formation of the σ^H-adduct is a reversible reaction; thus, at room temperature, the main process between these reagents is S_NAr of fluorine.

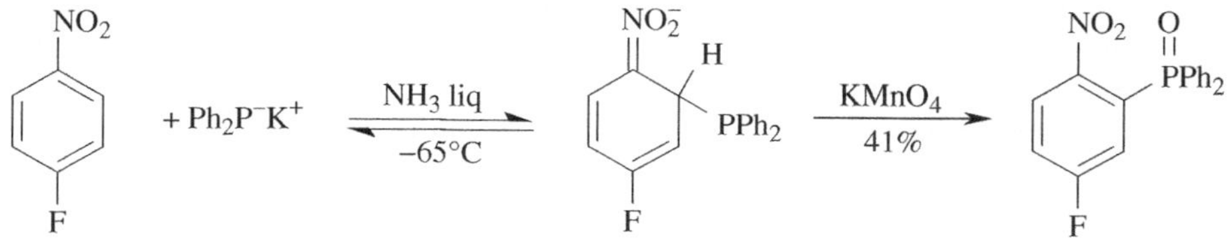

SCHEME 11.11 ONSH in *p*-fluoronitrobenzene with diphenylphosphine anion [22].

3. Addition of primary alkylmagnesium halides and alkyllithium compounds to nitroarenes leading to σ^H-adducts is an irreversible process, and because of that, the S_NAr of halogens with these nucleophiles does not proceed. The σ^H-adducts of these nucleophiles can be subsequently oxidized with a variety of oxidants [23]. The most convenient and efficient oxidant of these σ^H-adducts is a solution of $KMnO_4$ in liquid ammonia added after formation of the adducts [24]. The reaction is a valuable method of nucleophilic alkylation of nitroarenes (Scheme 11.12).

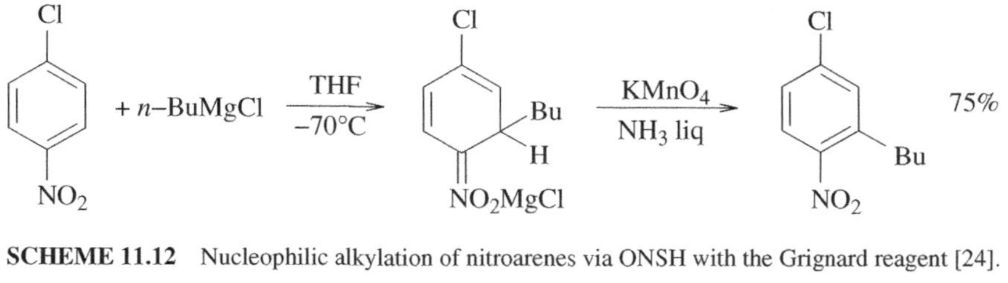

SCHEME 11.12 Nucleophilic alkylation of nitroarenes via ONSH with the Grignard reagent [24].

It should be noted that allylic, aromatic, and vinylic Grignard reagents do not form σ^H-adducts [23].

The σ^H-adducts of a variety of nucleophiles to nitroarenes can be also oxidized electrochemically. Mechanism of the anodic oxidation, that is, the consecutive removal of electrons, is different from the oxidation by chemical oxidants; however, as a rule, the products of ONSH induced by chemical oxidants and electrochemical oxidation are identical [25]. Similarly to oxidation by chemical oxidants, anodic oxidation of the σ^H-adducts can proceed, provided the addition equilibrium is shifted toward the adducts.

11.3 CONVERSION OF THE σ^H-ADDUCTS OF NUCLEOPHILES TO NITROARENES INTO SUBSTITUTED NITROSOARENES

Under proper conditions, σ^H-adducts of some nucleophiles to nitroarenes can be converted into substituted nitrosoarenes. This conversion, which proceeds according to intramolecular redox stoichiometry, thus is somewhat related to ONSH, takes place when the reactions between nucleophiles and nitroarenes are carried out in protic media or the σ^H-adducts are exposed to silylating agents or Lewis acids. Protonation, silylation, or complexation with the Lewis acids of the negatively charged oxygen atoms of the nitro group is followed by elimination of water or a silanol to form substituted nitrosoarenes. This process is of general character, because σ^H-adducts of carbon, nitrogen, and phosphorus nucleophiles can react in this way. Nitrosoarenes are more electrophilic than nitroarenes; hence, in the presence of bases, nucleophiles, or Lewis acids, they react further to form products, often of substantial interest [26]. For instance, carbanions of arylacetonitriles react in protic media (KOH, MeOH) with *p*-chloronitrobenzene and other *p*-substituted nitrobenzenes to form benzisoxazoles. The initially produced σ^H-adducts in the *ortho* position to the nitro group upon O-protonation followed by elimination of water form *o*-nitrosoaryl arylacetonitriles that being strong CH acids are immediately deprotonated. Subsequent cyclization of the produced carbanions via intramolecular addition–elimination gives benzisoxazoles (Scheme 11.13) [27a].

On the other hand, when σ^H-adducts of arylacetonitrile carbanions to *para*-substituted nitroarenes are converted into substituted nitrosoarenes in aprotic media via silylation in the presence of trialkylamines, the produced *o*-nitrosoaryl acetonitriles can undergo further heterocyclization on alternative pathway to produce substituted acridines [28]. Reactions of arylacetonitriles with nitroarenes unsubstituted in the *para* position under identical protic conditions result in the formation of *p*-nitrosoaryl arylacetonitriles that can be isolated in the tautomeric form of *p*-arylcyanomethylene quinone oximes (Scheme 11.13) [27b].

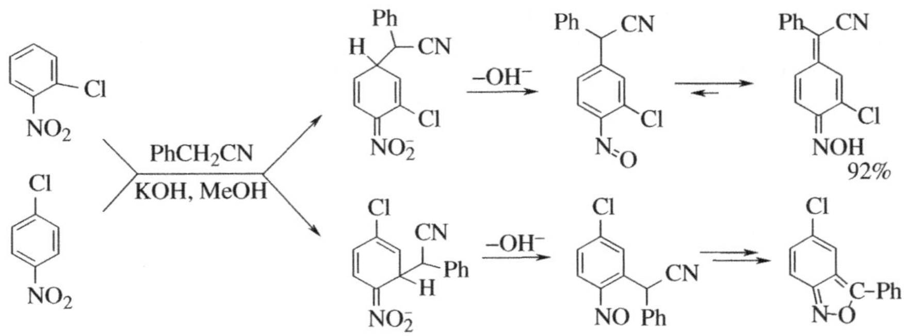

SCHEME 11.13 Reaction of carbanion of phenylacetonitrile with *o*- and *p*-chloronitrobenzene in protic media [27].

Allylic carbanions, under conditions that promote conversion of the σ^H-adducts into nitrosoarenes, react with nitroarenes to form a variety of heterocyclic systems. For instance, carbanion of allyl tolyl sulfone in the reaction with 2-nitrothiophene forms substituted pyridothiophene (Scheme 11.14) [29].

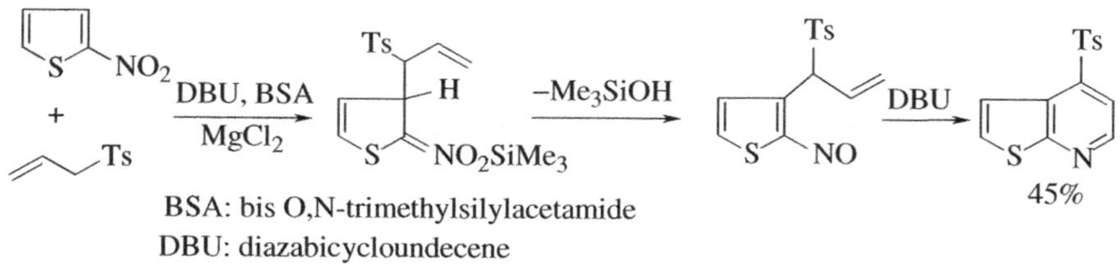

SCHEME 11.14 Synthesis of thieno[2,3-b]pyridine via reaction of 2-nitrothiophene with carbanion of allyl tolyl sulfone [29].

The reaction proceeds via conversion of the intermediate σ^H-adduct into substituted nitrosothiophene and subsequent deprotonation and cyclization of the produced nitrosoaryl allylic carbanion [29]. Of substantial value and interest are reactions of nitroarenes with anilines. Whereas simple heating of solutions of anilines with *p*-halonitrobenzenes results in conventional nucleophilic substitution of halogens (S_NAr) to give *p*-nitrodiarylamines, the reaction carried out at low temperature in the presence of strong base (*t*-BuOK) results in the formation of *o*-nitrosodiarylamines. The reaction proceeds via addition of the nitrogen nucleophiles (anilide anions) to nitroarenes at positions occupied by hydrogen, followed by conversion of the produced σ^H-adducts into substituted nitrosoarenes according to intramolecular redox stoichiometry (Scheme 11.15) [30].

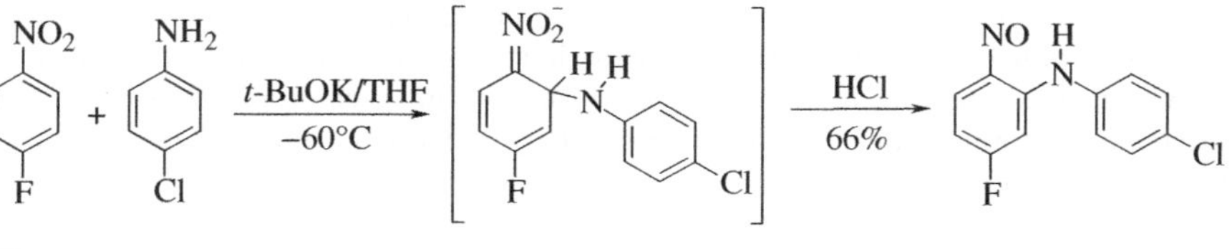

SCHEME 11.15 Synthesis of *o*-nitrosodiarylamine via S_NArH in *p*-fluoronitrobenzene with *p*-chloroaniline [30].

Of substantial interest are reactions between bicyclic nitroarenes, for example, 1-nitronaphthalenes and dimethylphosphite carried out in the presence of sodium methoxide in methanol. The initial step of these multistep processes is the formation of the σ^H-adducts. Further conversion of the σ^H-adducts according to the intramolecular redox stoichiometry gives dimethyl 1-nitrosonaphthalene-2-phosphonates that are deoxygenated with the excess of phosphite to produce highly active nitrenes. Intramolecular addition of the nitrenes to form azirine followed by addition of a nucleophile and ring opening of the aziridine gives substituted benzazepines [31a]. On the other hand, the reaction of 1-nitro-4-aminonaphthalene with dimethylphosphite under such conditions gives directly tetramethyl-1,4-diamino-2,3-naphthalene diphosphonate (Scheme 11.16) [31b].

When the σ^H-adducts of nucleophiles to nitroarenes are converted into substituted nitrosoarenes in the absence of nucleophiles or bases, they can be isolated often in good yields [12a, 23].

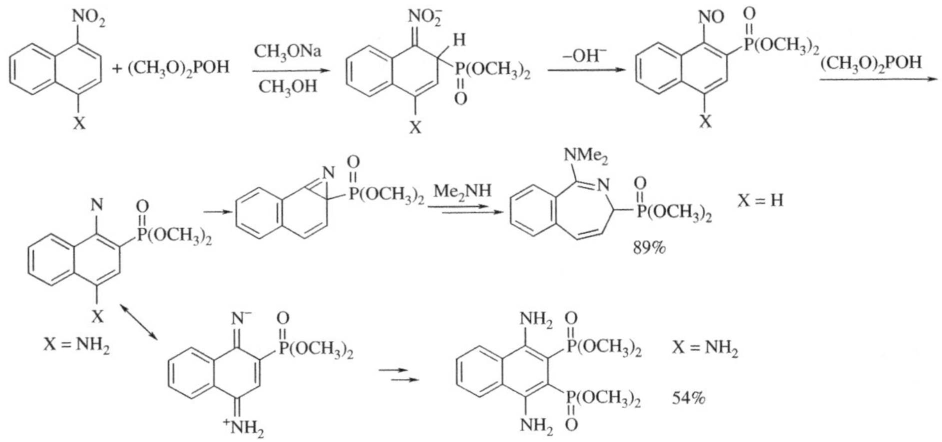

SCHEME 11.16 Reaction of 1-nitronaphthalenes with dimethylphosphite anion proceeds via dimethyl 1-nitrosonaphthalene-2-phosphonates [31].

11.4 VICARIOUS NUCLEOPHILIC SUBSTITUTION OF HYDROGEN

11.4.1 Introduction

When nucleophiles contain leaving groups L at the nucleophilic centers, as it is in the case of α-halocarbanions, the σ^H-adducts to nitroarenes can react further via base-induced β-elimination of HL at the expense of the ring hydrogen. The elimination of HL results in the formation of nitrobenzylic type carbanions, which upon protonation give products of nucleophilic substitution of hydrogen with the carbanion moieties. For instance, carbanion of chloromethyl phenyl sulfone reacts with nitrobenzene, *p*-fluoro- and *p*-chloronitrobenzenes, and other nitroarenes to form products of substitution of hydrogen *para* and *ortho* to the nitro group (Scheme 11.17) [32].

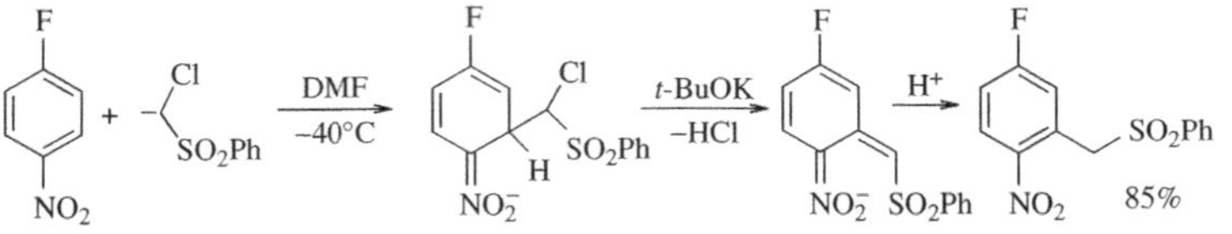

SCHEME 11.17 Vicarious nucleophilic substitution in *p*-halonitrobenzenes with carbanion of chloromethyl phenyl sulfone [32].

This reaction was named vicarious nucleophilic substitution of hydrogen (VNS), because the chlorine atom of the carbanion departs from the σ^H-adducts in the form of anion instead of the hydride anion. The most important feature of VNS is the strong preference for the substitution of hydrogen over conventional nucleophilic substitution of halogens, when present in the nitroaromatic rings. This strong preference of VNS over S_NAr is particularly evident in the reaction of 2,4-dinitrofluorobenzene, the Sanger reagent. In spite of well-known very fast S_NAr of fluorine, α-halocarbanions react with the Sanger reagent exclusively following the VNS of hydrogen pathway (Scheme 11.18) [33].

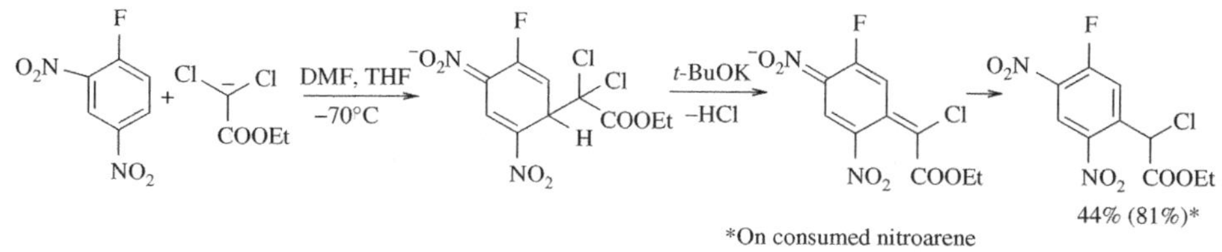

SCHEME 11.18 VNS reaction in the Sanger reagent [33].

Products of VNS are generated and exist in the reaction mixtures in the form of nitrobenzylic carbanions that are not electrophilic anymore; thus, the reaction proceeds selectively as monosubstitution. Nevertheless, the carbanion of chloromethyl phenyl sulfone can react with *m*-dinitrobenzene to form products of mono- and disubstitution (Scheme 11.19). Obviously, the second nitro group in the nitrobenzylic carbanion of the monosubstitution product imposes sufficient electrophilicity on the ring; thus, this carbanion behaves as the electrophilic partner, a kind of Michael acceptor that is able to react with the carbanion of chloromethyl phenyl sulfone [34]. Of course, the reaction of this carbanion with dinitrobenzene is much faster than with the nitrobenzylic carbanion of the monosubstitution product, so the rates of monosubstitution and disubstitution differ substantially, and these reactions proceed with high selectivity.

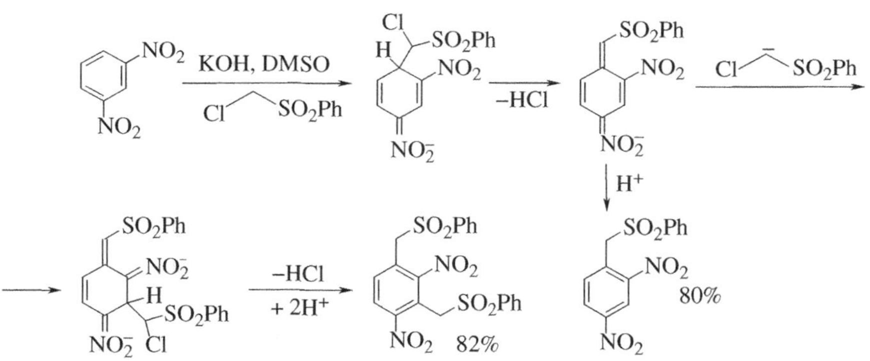

SCHEME 11.19 Mono- and di-VNS in *m*-dinitrobenzene [34].

For the VNS reaction to proceed satisfactorily, the carbanions should be in the form of loose ion pairs; when they form tight ion pairs or aggregates with countercations, the addition does not proceed. The most convenient base–solvent systems are therefore *t*-BuOK, NaOH, or NaH in dipolar aprotic solvents, such as DMF, DMSO, or their mixtures with THF. An excellent solvent for this reaction is liquid ammonia. According to stoichiometry of the reaction, the base should be used in an excess. In majority of cases, the reaction is carried out preferably at low temperatures of −20 to −70°C [35].

11.4.2 Mechanism of VNS Reaction

VNS proceeds in two distinct steps: formation of the σ^H-adducts of α-halocarbanions to nitroarenes and conversion of these σ^H-adducts into products of substitution of hydrogen. In order to clarify the mechanism, a few questions should be answered: How the addition proceeds? Is it really a reversible process? Does conversion of the σ^H-adducts proceeds as base-induced β-elimination? What factors affect relation of rates of these processes? Can σ^H-adducts, which are the postulated intermediates, be observed directly?

Formation of the σ^H-adducts of nucleophiles to nitroarenes is a common initial step for all S_NArH reactions: ONSH, VNS, *cine*- and *tele*-substitution, etc. The key differences between these reactions are pathways of conversion of the σ^H-adducts into final products. Nitroarenes can act not only as electrophilic partners but also as active electron acceptors. Since carbanions are good electron donors, the reaction between nitroarenes and carbanions can proceed via a simple direct addition or via an SET to produce the corresponding anion radicals of nitroarenes and radicals from the carbanions. Combination of these paramagnetic species can also produce σ^H-adducts. Indeed, in such systems, ESR spectra of the anion radicals of nitroarenes are often observed. On this basis, SET is often postulated as the way of formation of the σ-adducts (Scheme 11.20) [36].

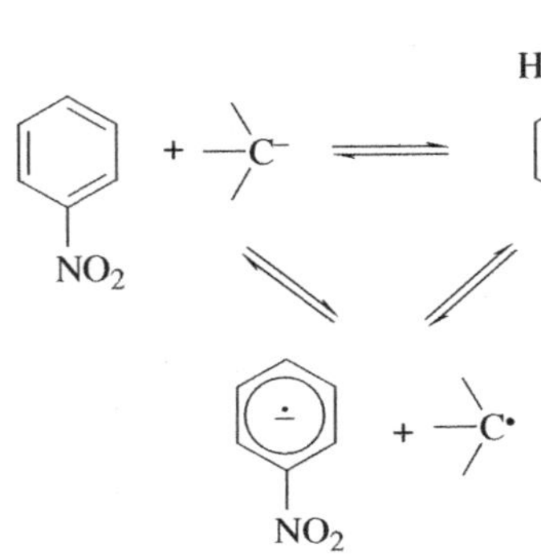

SCHEME 11.20 Formation of σ^H-adducts of carbanions to nitrobenzene via direct addition or SET pathway.

However, due to very high sensitivity of the ESR method, observation of the signal is often misleading, and differentiation between formation of the σ^H-adducts via direct addition or a two-step SET pathway should be made using more reliable criteria. It was shown by Bartoli that addition of primary alkylmagnesium halides to nitroarenes proceeds via SET, followed by combination of the nitroaromatic anion radicals with alkyl radicals [37]. This conclusion came from experiments with radical clock and is additionally supported by the observation that the ratio of *ortho* to *para* adducts to nitrobenzene is close to statistical, hence not affected by solvation effects and association of Mg cation with the nitro group [23, 37]. On the other hand, orientation (ratio of *ortho* to *para* isomers) of VNS with α-halocarbanions is strongly affected by solvation effects [38]. This observation indicates that addition of the charged species to the electron-deficient rings takes place. Reversibility and the observed secondary KIE of the addition of carbanions to nitroarenes also contradict the hypothetical SET pathway. The most convincing piece of evidence for the direct addition as the reaction pathway was provided by experiments with a very fast radical clock—carbanion of a chlorosulfone structure A^- (in Scheme 11.21). The radical $A^\cdot$ generated in separate experiments via abstraction of chlorine atom from dichlorosulfone A-Cl by the tributyltin radical undergoes a fast rearrangement (rate constant of 10^9 s^{-1}) [39a], whereas carbanion A^- enters into VNS with nitroarenes without formation of any rearranged products. One can therefore conclude that the carbanion is not converted into the corresponding radical $A^\cdot$ on the reaction path; thus, the formation of σ^H-adducts proceeds via direct nucleophilic addition (Scheme 11.21) [39b].

Supposition that the conversion of σ^H-adducts of α-halocarbanions into products of VNS proceeds as a base-induced β-elimination was based on the observation that base concentration often controls the rate of the reaction. Thus, it was observed that the competition between VNS and

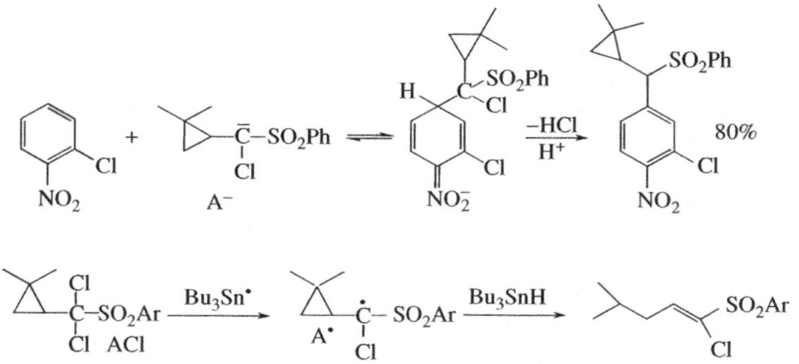

SCHEME 11.21 VNS with carbanion that can behave as a fast radical clock [39].

S_NAr of fluorine in *p*-fluoronitrobenzene with carbanion of chloromethyl phenyl sulfone was controlled by the strength and the concentration of the base [40].

Detailed mechanistic picture of VNS came from studies of the reaction of the carbanion of chloromethyl *p*-tolyl sulfone with 2-fluoro-4-bromo-, 2-fluoro-4-bromo-6-deutero-, and 2-deutero-4-bromonitrobenzenes (Scheme 11.22) [41].

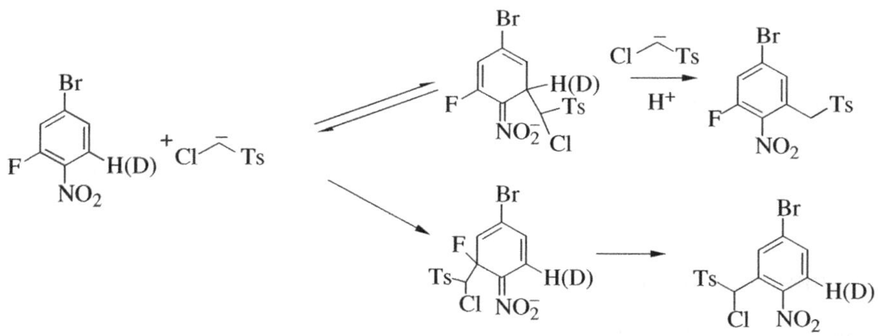

SCHEME 11.22 Mechanistic studies of VNS: relation of rates of VNS to S_NAr and kinetic isotope effects as function of base [41].

Bromine atom in these nitroarenes protects the *para* position to the nitro group against addition of the carbanion; thus, the reactions can proceed only at the *ortho* positions. The competition between VNS and S_NAr of fluorine and VNS of hydrogen and deuterium as a function of concentration of the carbanion (that acts also as the base) revealed that in the region of low concentrations of the carbanion, the ratio of VNS to S_NAr is a function of the concentration and the primary KIE ($VNS_{(H)}$:$VNS_{(D)}$) ≈ 4 is observed; hence, β-elimination is the rate-limiting step. On the other hand, in the region of high carbanion concentration, the ratio VNS:S_NAr is constant, and KIE ≈ 0.9. These results suggest the following general mechanistic picture of VNS. At a high carbanion concentration, σ^H-adducts, once formed, undergo rapid base-induced β-elimination of HCl, faster than their dissociation; thus, VNS:S_NAr ratio does not depend on the base concentration and reflects the relation of rates of the addition at positions 2- and 6- thus rates of formation of σ^H- and σ^F-adducts. Since under such conditions the addition is the rate-limiting step, the secondary KIE equal ≈ 0.9 is observed, typical for the reactions that proceed at carbon atom that undergoes rehybridization $sp^2 \rightarrow sp^3$. On the other hand, at low concentration of the carbanion, the β-elimination is the rate-limiting step; thus, VNS: S_NAr ratio is a function of concentration, and primary KIE equal ≈ 4 is observed. Under the latter conditions, there is an equilibrium between σ^H-adducts and educts [41]. These results suggest that for this reaction terms kinetic and thermodynamic control can be used but applied not for the final products but key intermediates. Under conditions that disfavor dissociation of the σ^H-adducts (low temperature) and assure high rate of the β-elimination-high concentration of strong base, the reaction is kinetically controlled. The σ^H-adducts once formed are converted into the products faster than they dissociate (equilibrate with educts). On the other hand at higher temperature that accelerates dissociation of the σ^H-adducts and at low base concentration the system can equilibrate and the reaction is under thermodynamic control. This can affect orientation of the substitution and also promote S_NAr when halogens are present in activated positions. Some results that illustrate this situation are presented in Scheme 11.23 [42a].

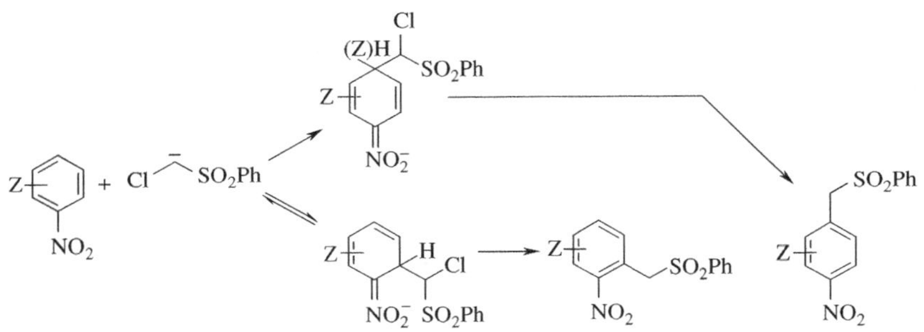

Kin: *t*-BuOK, THF-DMF, –50°C
Therm: KOH, DMSO, +20°C

	Kin,	Therm	Kin,	Therm
Z = H	74	0	26	100
4-F	100	0	0	100*
3-Cl	85*	0	15	100

* 2-Isomer

*Product of S_NAr of fluorine

SCHEME 11.23 Ratio of products of VNS reaction under kinetic and thermodynamic control [42a].

Since VNS can proceed under kinetic control, namely, initially formed σ^H-adducts can be converted into the products faster than they dissociate, the reaction can serve as a proper tool for determination of electrophilic activities of nitroarenes. Effects of substituents on rates of S_NAr was subject of thorough studies [43]; however, the results, although useful in practice of synthesis, cannot be considered as a reliable measure of electrophilic activities of nitroarenes because S_NAr of halogens is a slow secondary process preceded by a reversible formation of the σ^H-adducts. On the other hand, the rate of VNS reaction under kinetic control reflects the rates of the initial nucleophilic addition of carbanions to nitroaromatic rings, thus can be used as measure of electrophilic activities of these compounds. Particularly convenient and reliable way to determine such effects is the competitive experiments in which two nitroarenes compete for the VNS reaction with carbanion of chloromethyl phenyl sulfone under conditions that assure faster β-elimination of HCl from the σ^H-adducts than their dissociation [42]. Relative rate constants of the addition of this carbanion to some nitroarenes in relation to nitrobenzene are given in Scheme 11.24.

From these data, it is evident that electrophilic activity of nitroarenes is a superposition of two factors—electronic effects of additional substituents in the ring and aromatization energy, since the addition process is connected with dearomatization of nitroarenes.

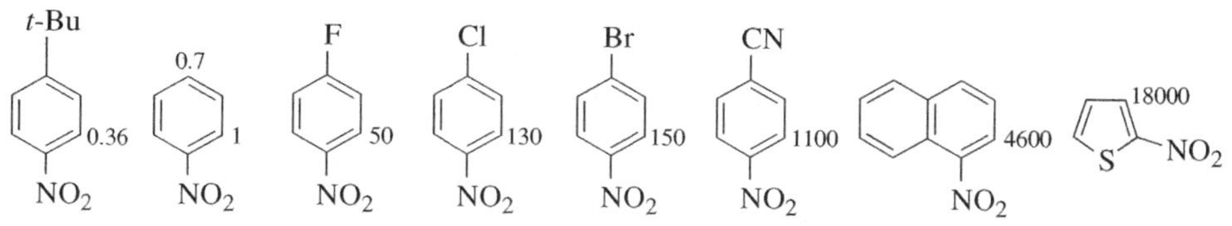

SCHEME 11.24 Electrophilic activity of nitroarenes expressed by relative rate constants of the nucleophilic addition [42].

The mechanistic picture elaborated for the VNS with carbanion of chloromethyl phenyl sulfone as a convenient model nucleophile is valid for other carbanions.

On the basis of the previously presented results and discussion, we can conclude that the VNS proceeds via direct addition of α-halocarbanions to nitroarenes. The produced σ^H-adducts react with base according to base-induced β-elimination of HL pathway to form products in the form of nitrobenzylic carbanions. Depending on the nature of the reaction partners and the reaction conditions—particularly strength and concentration of the base and temperature—the reaction can be controlled kinetically or thermodynamically. At low temperature in the presence of an excess of strong base, the elimination proceeds faster than dissociation of the σ^H-adducts; thus, the reaction proceeds under kinetic control. On the other hand, when the reaction is carried out at higher temperature in the presence of a weak base at low concentration, the dissociation of the σ^H-adducts is faster than β-elimination; thus, the addition is really a reversible process, and the reaction proceeds under thermodynamic control.

It should be stressed that these conclusions are supported by direct observation by NMR and UV-VIS spectroscopy of σ^H-adducts of carbanion of chloromethyl phenyl sulfone to nitroarenes and conversion of such σ^H-adducts into the VNS products [44].

11.4.3 Scope and Limitation of VNS

The VNS reaction between carbanions and nitroarenes that is presented in general Scheme 11.25 is a process of wide scope in respect to the both partners: electron-deficient arenes, particularly nitroarenes, and carbanions that should contain nucleofugal group L able to be eliminated as HL from the σ^H-adducts (Scheme 11.25) [35].

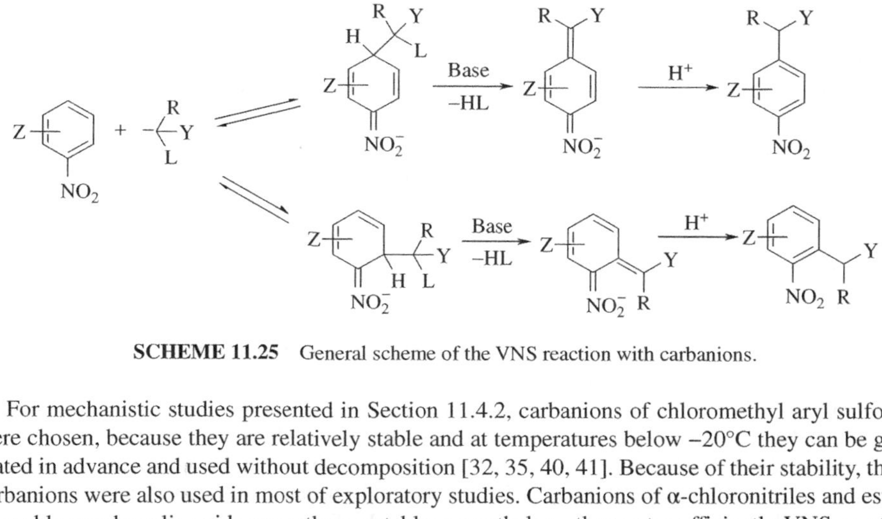

SCHEME 11.25 General scheme of the VNS reaction with carbanions.

For mechanistic studies presented in Section 11.4.2, carbanions of chloromethyl aryl sulfones were chosen, because they are relatively stable and at temperatures below −20°C they can be generated in advance and used without decomposition [32, 35, 40, 41]. Because of their stability, these carbanions were also used in most of exploratory studies. Carbanions of α-chloronitriles and esters of α-chlorocarboxylic acids are rather unstable; nevertheless, they enter efficiently VNS reaction when generated in the presence of electron-deficient arenes. Even very unstable trihalomethyl carbanions, generated by deprotonation of chloroform and bromoform enter VNS when generated in the presence of nitroarenes at low temperature –70°C. Due to instability of the majority of α-halocarbanions, better results were often obtained when for VNS reaction more stable carbanions that contain other leaving groups L such as RO and RS (R = alkyl, aryl) were used. Of particular interest are RS substituents, because they are not only good leaving groups but also exert carbanion-stabilizing effects; hence, dithioacetals, benzyl sulfides, etc. are convenient precursors of carbanions that are able to enter VNS. In conclusion, carbanions $RLYC^-$ formed by combination of any carbanion-stabilizing group Y = CN, COOR, COR, SO_2Ph, etc., with any leaving groups, Cl, Br, RO, RS, etc., and any substituents R that can be H, alkyl, aryl, halogens, RO, etc. are able to enter VNS reaction; thus, the process is of general character in respect to the carbanions.

VNS is also of general character in respect to electron-deficient arenes. Nitrobenzene, the model representative of these arenes, can enter into VNS with any carbanion presented earlier. The reaction of nitrobenzene with secondary carbanions such as chloromethyl aryl sulfones or aryloxyacetonitriles can proceed at positions *ortho* and *para*; the orientation is controlled by the conditions. At low temperature and excess of strong base (kinetic control), *ortho* substitution dominates, whereas thermodynamic control favors *para* substitution (Scheme 11.23) [42]. The reaction proceeds mainly in the *ortho* position also when carried out in *t*-BuOK/THF system due to specific solvation effect [38]. Sterically demanding tertiary carbanions, R ≠ H, react preferentially in *para* positions.

The nitrobenzene ring can contain unlimited variety of substituents: halogens and electron-withdrawing and electron-donating groups; they affect rates and orientation of the reaction, but do not disturb the reaction course [35, 45]. When the *para* position in the nitroarenes is occupied, even by halogens, the reaction proceeds exclusively *ortho* to the nitro group even with sterically demanding tertiary carbanions [46]. *Meta* substituted 3-X-nitrobenzenes can in principle form three isomeric products. The observed orientation of the reaction is interplay of electronic and steric effects. Interestingly, when X = F, Cl, Br, Me, MeO, etc., the reaction with secondary carbanions under kinetic control proceeds mainly in the most sterically hindered position 2 [38, 45]. Nitrobenzene ring can also contain substituents that under strongly basic reaction conditions are deprotonated, provided that the negative charge of the produced anions is not conjugated with the nitroaromatic rings. Thus, the reaction proceeds without problems with nitrobenzoic and nitroarenesulfonic acids that react in the form of anions [47], but nitrophenols [34], for obvious reasons, do not enter VNS. On the other hand, dinitrophenols and picric acid, which enter the reaction as di- and trinitrophenolates, react with α-halocarbanions satisfactorily. Similarly, in di- and trinitrobenzene, disubstitution and even trisubstitution are possible. [34].

The scope and versatility of the VNS reaction in respect to nitroarenes, nitroheteroarenes, and other electron-deficient arenes as well as carbanions are illustrated by selected examples presented in Schemes 11.26–11.40.

Nitroheteroarenes such as nitropyridines [49], nitroquinolines [50], nitroisoquinolines [51], nitrothiophenes, nitrofurans, nitropyrroles [52], nitrothiazoles [53], nitroimidazoles [54], nitropyrazoles [55] etc. are active electrophilic partners in the VNS reaction (Schemes 11.27–11.33).

VNS proceeds satisfactorily also with nitro derivatives of nonbenzenoic carbocyclic aromatics such as 1,6-methano[10]annulenes (Scheme 11.34) [56].

The VNS reaction proceeds also with arenes that do not contain a nitro group, but are electron deficient due to specific electronic configuration such as azulene [57] and a plethora of azines: 1,2,4-triazines [58], pteridines [59], pyridazines [60], etc. (Scheme 11.35–11.38).

When electron-deficient azines do not contain a nitro group, their reaction with α-halocarbanions can follow an alternative route. Due to high negative charge density on the vicinal nitrogen atom in the σ^H-adduct of carbanion of chloromethyl phenyl sulfone to quinoxaline, instead of β-elimination, intramolecular 1,3-substitution can proceed to form aziridines. On the other hand, σ^H-adducts of α-halocarbanions to N-oxides of such azines, in which the negative charge is efficiently delocalized, react according to the β-elimination pathway to give VNS products [61]. In general, N-oxides of azines are more active than azines in VNS reaction (Scheme 11.39) [62].

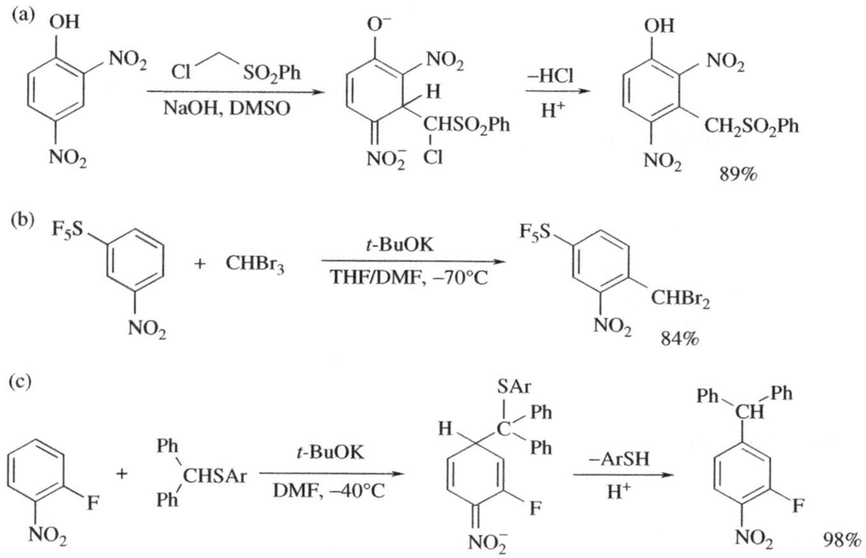

SCHEME 11.26 Examples of VNS in substituted nitrobenzenes: (a) in 2,4-dinitrophenol [47], (b) in *m*-pentafluorosulfenylnitrobenzene [48a], and (c) in synthesis of triarylmethanes via VNS [48b].

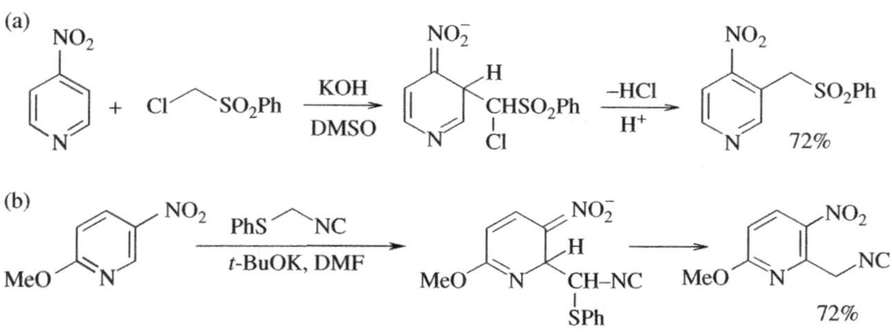

SCHEME 11.27 VNS reaction (a) in 4-nitropyridine [49a] and (b) in 2-methoxy-5-nitropyridine [49b].

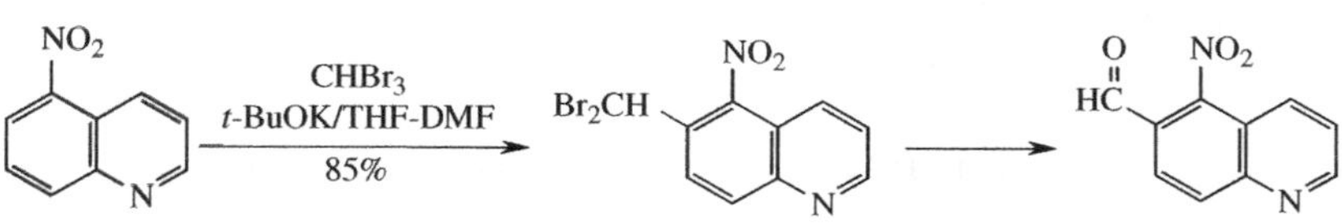

SCHEME 11.28 VNS reaction in 5-nitroquinoline used for nucleophilic formylation [50].

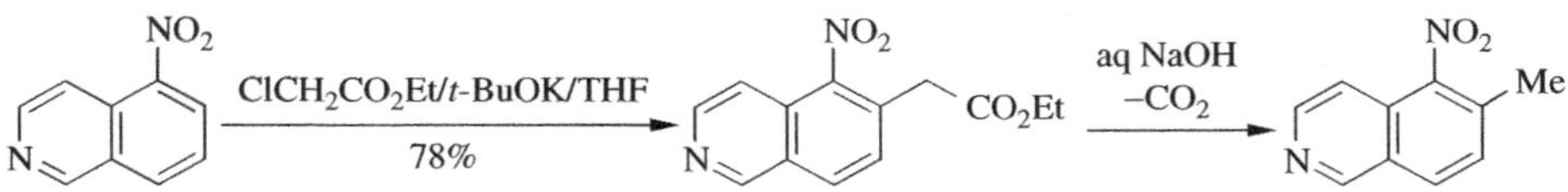

SCHEME 11.29 VNS reaction in 5-nitroisoquinoline used for nucleophilic methylation [51].

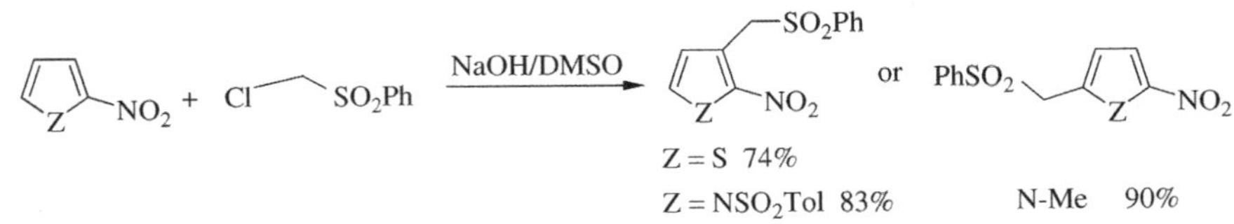

SCHEME 11.30 VNS reaction in 2-nitrothiophene and 2-nitropyrroles [52].

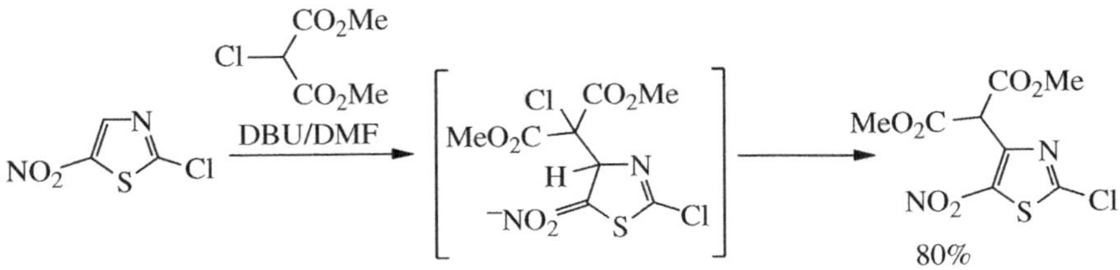

SCHEME 11.31 VNS reaction of carbanion of dimethyl chloromalonate with nitrothiazole [53].

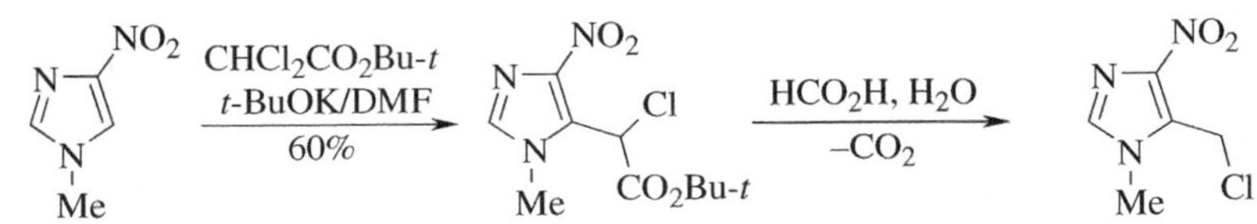

SCHEME 11.32 Chloromethylation of nitroimidazole via VNS with *t*-Bu dichloroacetate [54].

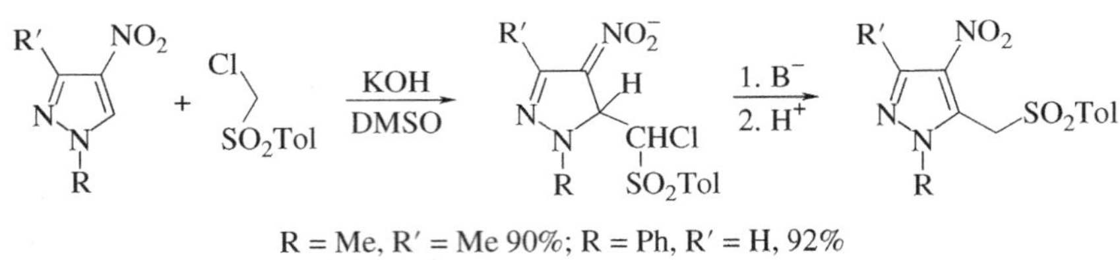

SCHEME 11.33 VNS in nitropyrazole [55].

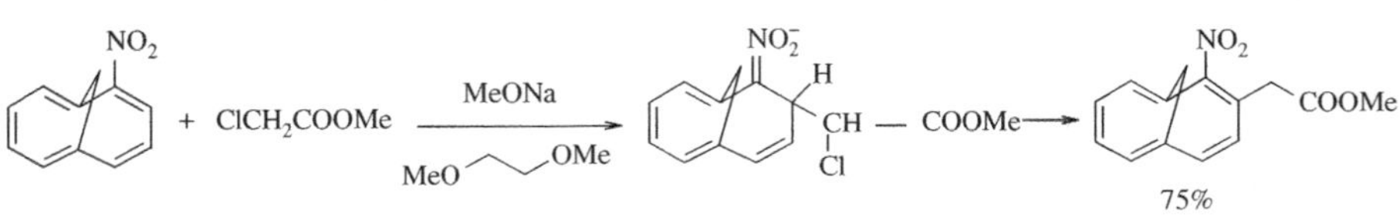

SCHEME 11.34 VNS does proceed with nitroannulenes [56].

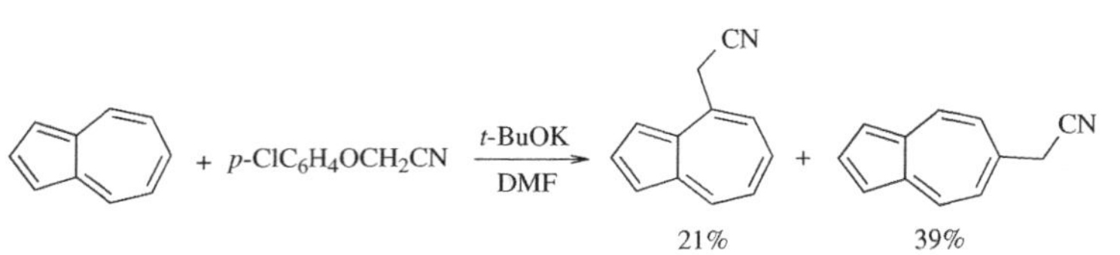

SCHEME 11.35 VNS cyanomethylation of azulene [57].

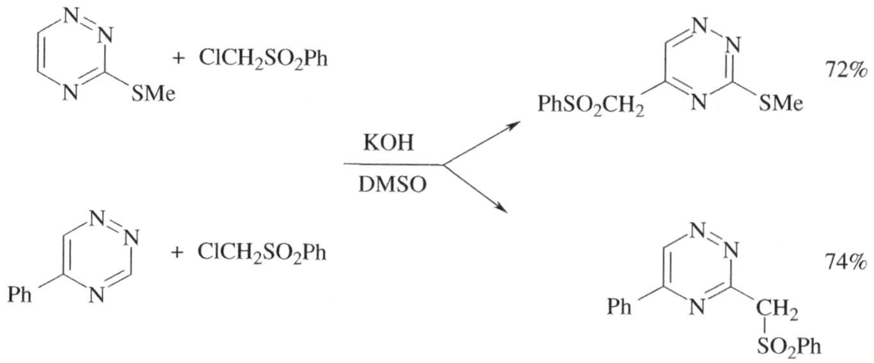

SCHEME 11.36 VNS can proceed in various positions of 1,2,4-triazines [58].

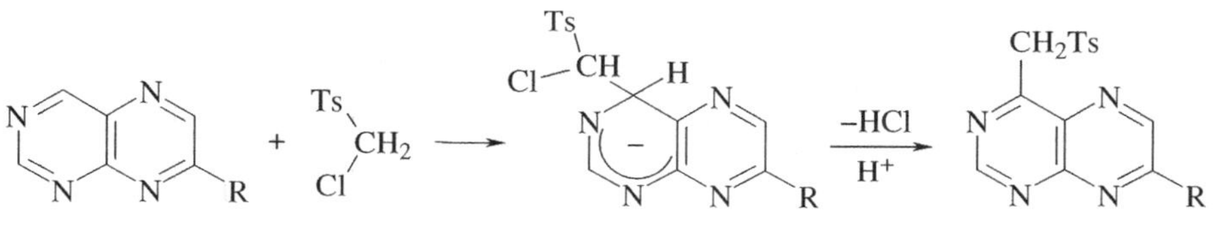

R = Ph, 72%; Me, 78%

SCHEME 11.37 VNS in pteridines [59].

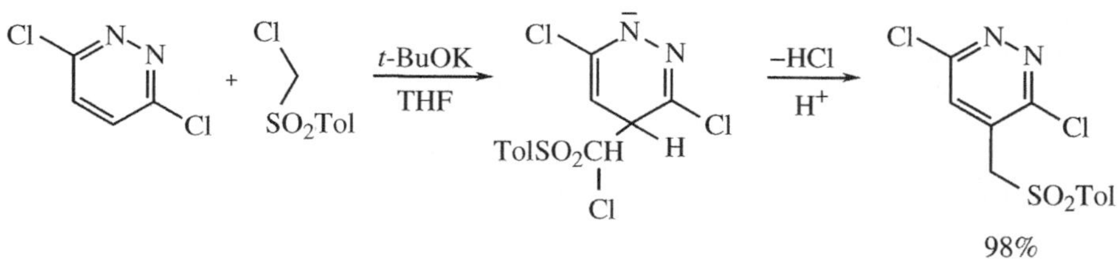

98%

SCHEME 11.38 VNS proceeds faster than S_NAr of chlorine in 3,6-dichloropyridazine [60].

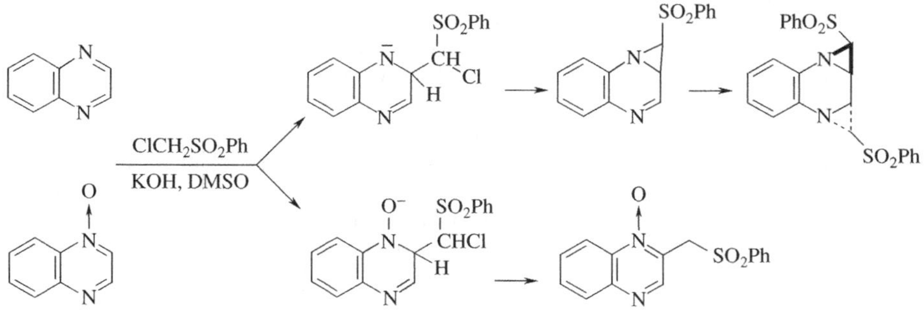

SCHEME 11.39 VNS versus aziridination of quinoxaline in the reaction with carbanion of chloromethyl phenyl sulfone [61].

On the other hand, highly electron-deficient hexahapto complexes of arenes with transition metals such as arene chromium tricarbonyl or cationic arene Fe(II)Cp complexes do not enter VNS [63]. Although they are able to form σ^H-adducts with α-halocarbanions and these σ^H-adducts can be oxidized to form ONSH products, the second step, the base-induced β-elimination of HL from these σ^H-adducts, does not proceed, perhaps due to steric reasons [63].

Products of VNS are generated in the form of carbanions and are isolated upon protonation. Instead of protonation, they can be introduced directly in subsequent reactions with electrophilic partners such as alkylating agents [64a] or aldehydes [64b]. For instance, the VNS in nitrobenzene with carbanion of 2-dichloromethyloxazoline produces nitrobenzylic α-chlorocarbanion that enters the Darzens reaction with aromatic aldehydes to give oxiranes (Scheme 11.40) [64b] whereas the reaction with carbanion of *t*-butyl dichloroacetate followed by Michael addition gives *t*-butyl nitroarylcyclopropane carboxylates [64c].

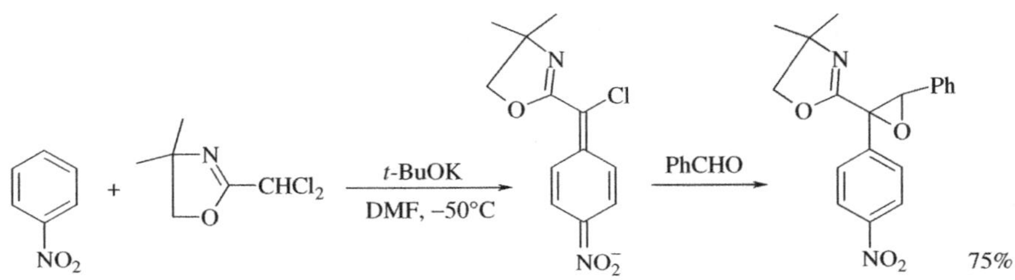

SCHEME 11.40 One-pot subsequent VNS and Darzens reactions [64b].

This short discussion of the scope of VNS indicates that the reaction is of general character in respect to the carbanions and electron-deficient arenes. In fact, almost any such arene that contains hydrogen atom in activated position can react according to the VNS pathway with any carbanion of the general structure $RLYC^-$ to form products in which this hydrogen is replaced with the RYCH substituent. Of course, there are some limitations. Since the second step of the VNS is a bimolecular reaction of base-induced β-elimination of HL from the intermediate σ^H-adducts, the reaction can proceed only when the σ^H-adducts are produced in the reaction mixture in a reasonable concentration. Thus, α-halocarbanions that are weak nucleophiles do not react with moderately electrophilic nitroarenes, but react with that of high electrophilicity. For instance, the carbanion of dimethyl chloromalonate does not react with nitrobenzene, but reacts with *m*-dinitrobenzene and nitrothiazoles (Scheme 11.31) [53].

The VNS reaction with carbanions is an efficient and general tool for the replacement of hydrogen atom in electron-deficient arenes with functionalized carbon substituents; hence, it can be considered as a process analogous to the Friedel–Crafts reaction, but proceeding according to opposite polarity. For instance, Friedel–Crafts reaction of chloroform with benzene and other arenes carried out in the presence of $AlCl_3$ gives dichloromethyl arenes (benzylidene chlorides) that usually react further to give triarylmethanes. The reaction proceeds via addition of dichloromethyl carbocations to nucleophilic aromatic rings and formation of cationic intermediate adducts.

On the other hand, the VNS reaction of chloroform with nitroarenes carried out in the presence of *t*-BuOK that gives dichloromethyl nitroarenes [65] proceeds via addition of trichloromethyl carbanions to electrophilic aromatic rings of nitroarenes and formation of anionic intermediates [65]. One can therefore consider VNS as an umpolung of the Friedel–Crafts reaction. It is also a process complementary to the Friedel–Crafts reaction, because it proceeds with nitroarenes that usually do not enter the Friedel–Crafts reaction (Scheme 11.41).

Versatility and wide scope of VNS in respect to electron-deficient arenes and carbanions discussed earlier are illustrated by series of examples presented in Schemes 11.17–11.40.

VNS is not limited to carbon nucleophiles–carbanions containing nucleofugal groups L. Anions of alkylhydroperoxides such as *t*-butyl or cumyl hydroperoxides contain at the nucleophilic

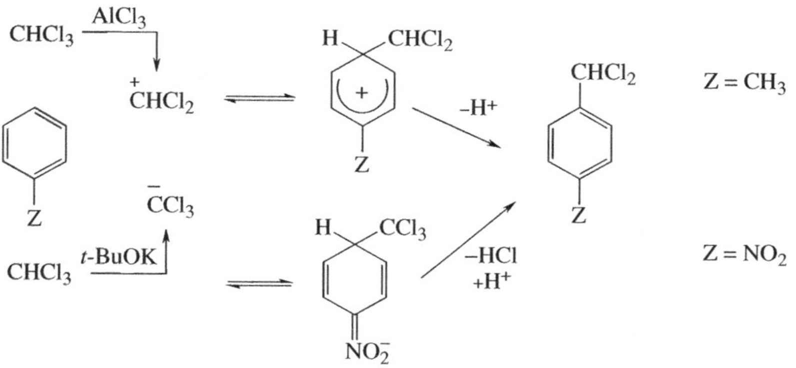

SCHEME 11.41 Similarity of the Friedel–Crafts and VNS reactions of chloroform with arenes.

centers nucleofugal alkoxy groups and thus can behave similarly to α-halocarbanions. This similarity is evident in the known reactions of these species with Michael acceptors. Addition of α-halocarbanions to the Michael acceptors produces intermediate γ-halocarbanions that enter fast intramolecular 1,3-substitution to produce substituted cyclopropanes. Similarly, addition of anions of these hydroperoxides to the Michael acceptors followed by intramolecular 1,3-substitution at the oxygen gives oxiranes. Indeed, similarly to the reactions of α-halocarbanions, *t*-butyl and cumyl hydroperoxides react with nitroarenes in the presence of an excess of base to form nitrophenols (Scheme 11.42) [66, 67]. Due to moderate nucleophilicity of these anions, the reaction proceeds satisfactorily with nitroarenes of high electrophilic activity: nitrobenzene that contains electron-withdrawing substituents, nitronaphthalene, etc. The VNS with these hydroperoxide anions in nitrobenzene and nitroanisoles proceeds sluggishly (Scheme 11.42).

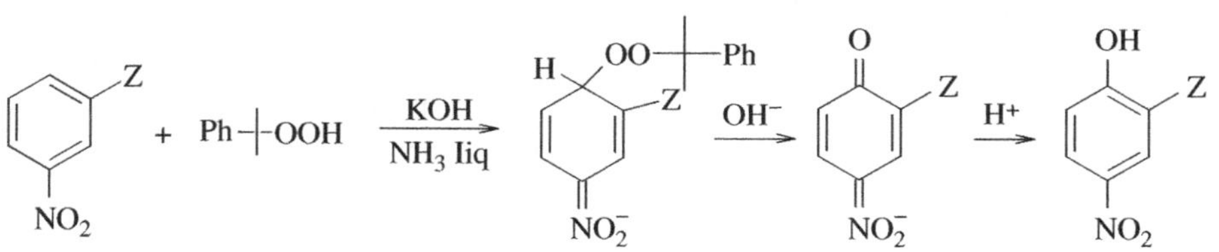

Z = F, 76%; Cl, 84%; CF_3, 86%; CN, 87%

SCHEME 11.42 VNS hydroxylation of nitroarenes [66].

Similarly, as in the case of carbanion nucleophiles, addition of these hydroperoxide anions proceeds faster at positions occupied by hydrogen that at those occupied by halogens.

VNS hydroxylation of nitroarenes with alkylhydroperoxides proceeds according to mechanism similar to that of the reaction with α-halocarbanions—reversible nucleophilic addition, followed by the base-induced β-elimination of alcohols from the intermediate σ^H-adducts. These mechanistic features were established by determination of effects of base on the rate of the reaction, namely, on the competition between VNS hydroxylation and S_NAr of chlorine in 4-chloronitrobenzene and 2,4-dinitrochlorobenzene. For instance, the reaction of these hydroperoxides with 2,4-dinitrochlorobenzene carried out in the presence of an excess of *t*-BuOK proceeds exclusively as VNS to give 2,4-dinitro-5-chlorophenol, whereas in the presence of equimolar amount of *t*-BuOK substantial quantity of 2,4-dinitrophenol, product of S_NAr is formed (Scheme 11.43) [66].

Effect of base on the rate of VNS hydroxylation is clearly observed in the reaction of *t*-butyl hydroperoxide with 1-nitronaphthalene [66]. In the presence of an excess of *t*-BuOK,

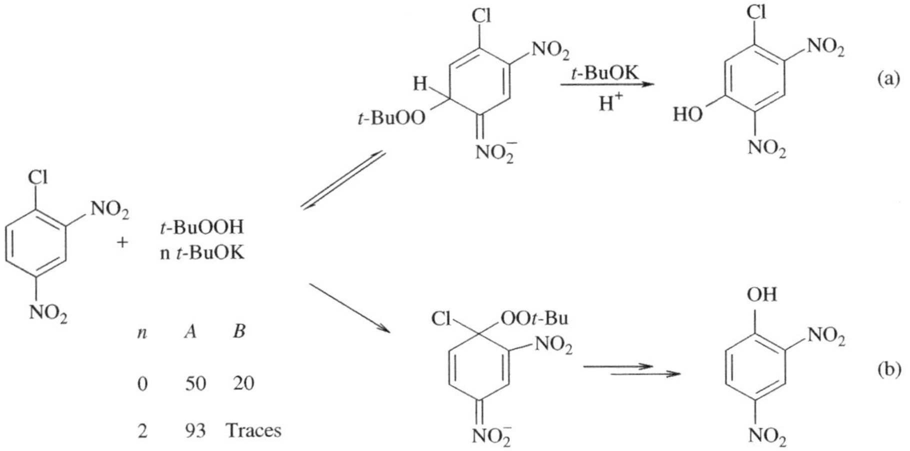

SCHEME 11.43 VNS versus S_NAr reaction of *t*-butyl hydroperoxide with 2,4-dinitrochlorobenzene [66].

1-nitro-2-naphthol was formed, whereas when an insoluble base NaOH was used 1-nitro-4-naphthol was the major product. These reactions can be considered as textbook examples of the kinetic and thermodynamic control of the reaction course (Scheme 11.44).

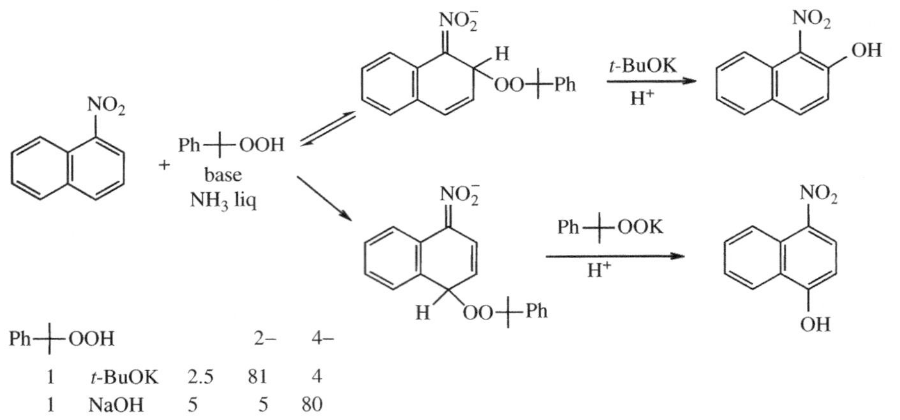

SCHEME 11.44 Kinetic versus thermodynamic control of the VNS hydroxylation of 1-nitronaphthalene [66].

VNS is also an efficient tool for introduction of amino groups into electron-deficient arenes, particularly nitroarenes. Amination of nitroarenes with hydroxylamine in basic media known for more than 100 years can be considered as early example of VNS. It is however limited to highly active nitroarenes: 1-nitronaphthalene, nitroquinolines, etc. [68], and its mechanism was not recognized or was misinterpreted. In our opinion, this reaction proceeds undoubtedly according to the VNS mechanism. Upon introduction of the concept of the VNS reaction with α-halocarbanions, a few aminating agents were designed: derivatives of hydroxylamine and hydrazine such as 4-amino-1,2,4-triazole [69], trimethylhydrazonium iodide [70], methoxyamine [71], and sulfenamides [72]. The N-anions generated by deprotonation of these compounds add to nitroarenes at positions occupied by hydrogen to produce σ^H-adducts that enter base-induced β-elimination of triazole, trimethylamine, methanol, or arene thiols, respectively, to form nitroanilines (Scheme 11.45).

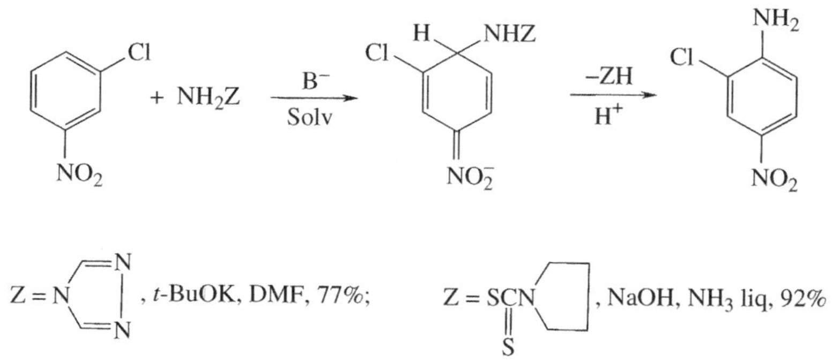

SCHEME 11.45 VNS amination of nitroarenes with 4-amino-1,2,4-triazole and sulfenamide [69, 72].

This direct amination process has found wide application in synthesis. For instance, double VNS amination of 2,4,6-trinitroaniline with an excess of trimethylhydrazinium iodide is an efficient method of manufacturing of 2,4,6-trinitro-1,3,5-triaminobenzene, a highly energetic material (Scheme 11.46) [70b].

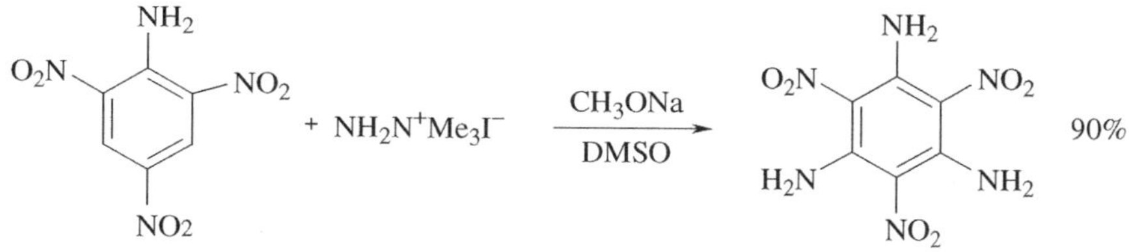

SCHEME 11.46 Synthesis of 1,3,5-triamino-2,4,6-trinitrobenzene via double VNS amination of 2,4,6-trinitroaniline [70b].

It should be stressed that similarly to the VNS hydroxylation with anions of hydroperoxides, VNS amination proceeds according to the addition– β-elimination mechanism. As in the case of other nucleophiles, N-anions of the aminating agents add faster at positions occupied by hydrogen than at those occupied by halogens.

11.5 OTHER WAYS OF CONVERSION OF THE σ^H-ADDUCTS

Besides the three major ways of conversion of the σ^H-adducts into products of S_NArH presented in Sections 11.2–11.4, there are a few other processes, less general, but nevertheless of substantial interest. There are many cases of conversion of the σ^H-adducts of nucleophiles to nitroarenes that proceed by departure of nucleofugal groups located in vicinity of the addition sites. This type of reactions is termed *cine*-substitution. Most frequently, the *cine*-substitution proceeds via departure of the nitro group in the form of nitrite anion from the σ^H-adducts in positions *ortho* to the nitro group (Scheme 11.47) [73].

Interestingly, some σ^H-adducts of α-halocarbanions to nitroarenes can react further in two ways. When generated in the presence of an excess of base, β-elimination of HCl results in VNS, whereas when additional base is absent, immediate protonation with a strong acid results in *cine*-substitution (Scheme 11.48) [74].

On the other hand, when the σ^H-adducts are converted via departure of a nucleofugal group from a position remote in the relation to the addition site, the process is termed *tele*-substitution. Such process is frequently observed when electron-deficient arenes contain trichloromethyl substituents (Scheme 11.49).

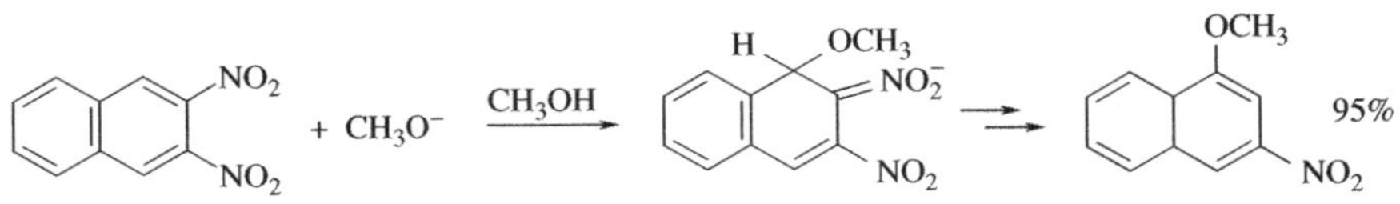

SCHEME 11.47 *Cine*-substitution in 2,3-dinitronaphthalene [73].

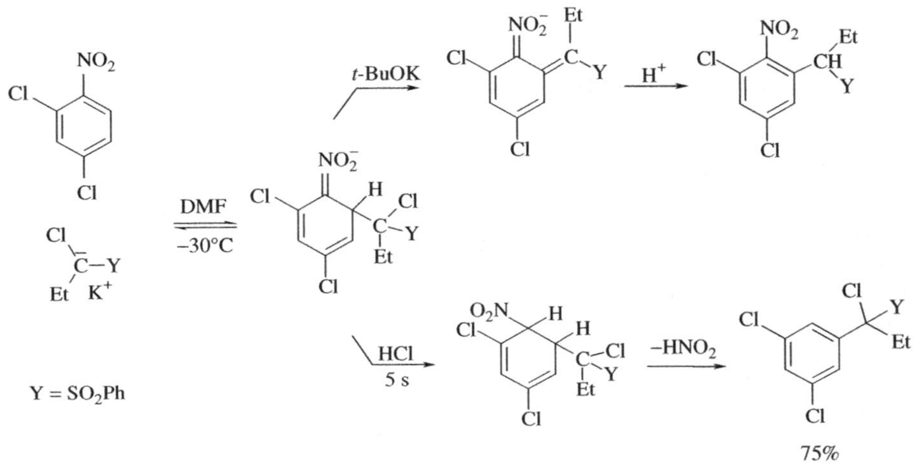

SCHEME 11.48 *Cine* versus VNS in the reaction of α-chloropropylphenyl sulfone carbanion with 2,4-dichloronitrobenzene [74].

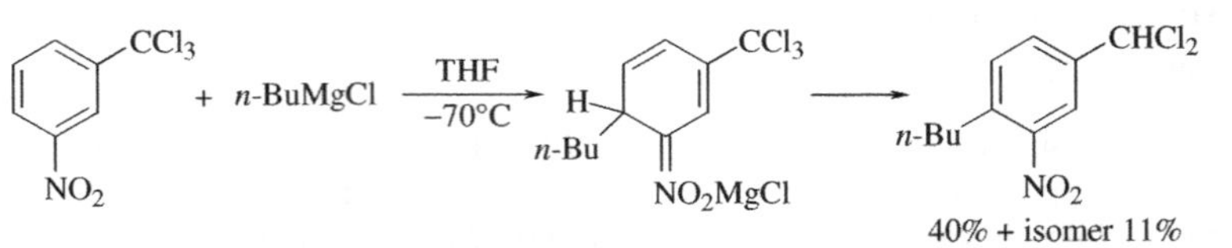

SCHEME 11.49 *Tele*-substitution in the reaction of *n*-butylmagnesium chloride with *m*-trichloromethylnitrobenzene [75].

The *cine*- and *tele*-substitution reactions are discussed in a review paper [76].

Somewhat related to the *tele*-substitution are reactions of nucleophiles with halogenated azines that proceed via addition of nucleophiles in positions occupied by hydrogen and subsequent conversion of the σ^H-adducts via ring opening and ring closure. For instance, reaction of 2-bromo-4-phenylpyrimidine with potassium amide gives 2-amino-4-phenylpyrimidine; however, the reaction does not proceed via a direct addition of the amide anion at position 2 that is occupied and protected against nucleophilic addition by the halogen. Instead, the addition takes place at position 6. Ring opening in the produced σ^H-adduct followed by elimination of bromide anion and ring closure gives the final product (Scheme 11.50). This kind of S_NAr reaction termed ANRORC is quite common in heterocyclic chemistry [77]. This reaction not only serves as a tool for introduction of substituents into azine rings but also for interesting ring transformations (Scheme 11.50) [77].

It should be stressed that although outcome of these reactions is identical to that of classical S_NAr of halogens, the substitution proceeds at the carbon atom of the ring connected with hydrogen; thus, it is fully justified to consider *cine*-, *tele*-, and ANRORC substitutions as processes of nucleophilic substitution of hydrogen S_NArH.

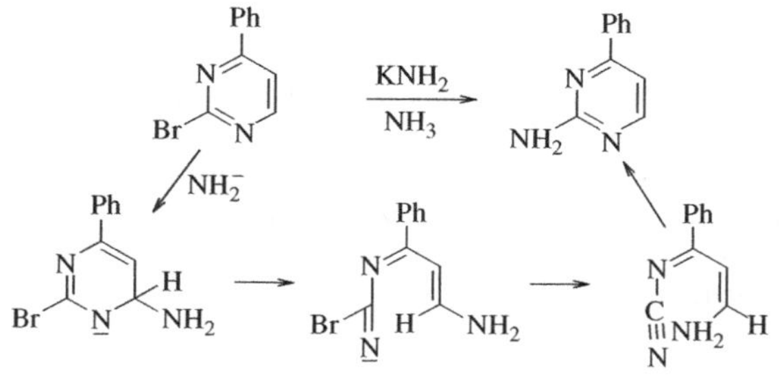

SCHEME 11.50 Reaction of 2-bromo-4-phenylpyrimidine with KNH_2 proceeds via ANRORC mechanism.

11.6 CONCLUDING REMARKS

From a variety of examples of a few variants of S_NArH reactions such as ONSH, VNS, etc. presented in this chapter, it is evident that these reactions are powerful and versatile tools in organic synthesis, particularly in the synthesis of aromatic compounds. There are two major applications of S_NArH: introduction of substituents into electron-deficient aromatic rings and construction of hetero- and carbocyclic systems. Of particular value is the introduction of carbon substituents—alkyl and functionalized alkyl groups. Thus, direct or indirect introduction of alkyl [23, 24], benzyl, benzhydryl [48b], chloromethyl [54], dihalomethyl [48a, 65], and perfluoroalkyl [21, 22] groups as well as alkyl groups functionalized with CN [56–58], COOR [51, 78], COR [18, 19], SPh [79], SO_2Ph [32], $PO(OMe)_2$ [11, 80], NC [49b], and other substituents into electron-deficient arenes can be efficiently realized via VNS or ONSH reactions. Introduction of heteroarylmethyl substituents into nitroarene rings proceeds efficiently via VNS with α-chlorocarbanions of chloromethylpyridines, benzothiazole, etc. [81]. S_NArH reactions are also an efficient tool for introduction of amino and substituted amino groups [6, 7, 9, 69–72] as well as hydroxylation of electron-deficient arenes [66, 67].

A variety of heterocyclic systems can be efficiently constructed directly in the course of S_NArH reactions or via simple transformations of S_NArH products [82]. Perhaps the most important is the synthesis of substituted indoles via direct reaction of ketone enolates with *m*-nitroanilines (Scheme 11.8) [18] and a variety of transformations of *o*-cyanomethyl [83], *o*-arylsulfonylmethyl [84], and other nitroarenes readily available via VNS reaction (Scheme 11.51).

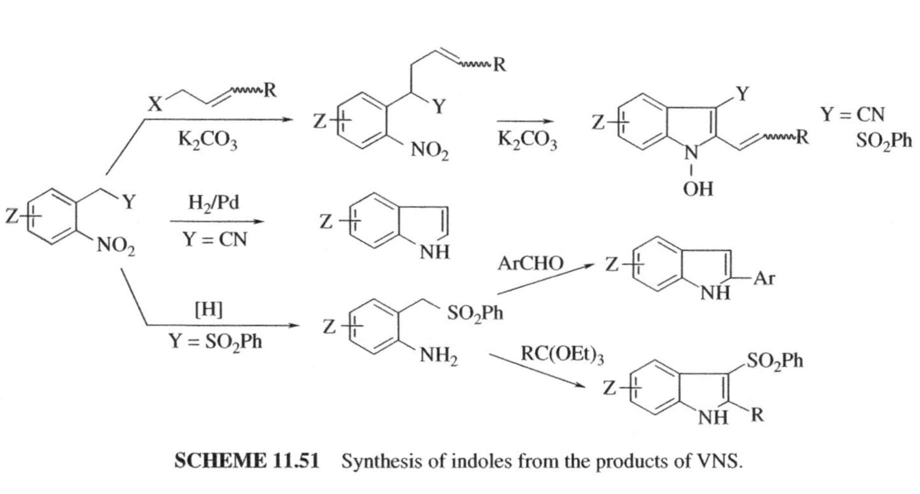

SCHEME 11.51 Synthesis of indoles from the products of VNS.

Contrary to majority of modern methods of synthesis of indoles, these reactions do not need transition metal catalysis, so the produced indoles can by directly used in pharmaceutical and other industries. A plethora of other heterocyclic systems such as quinolines [28], benzisoxazoles [26, 27], acridines [27], phenazines, [85] etc. can be similarly constructed via S_NArH reactions.

Products of VNS are versatile starting materials also for the synthesis of carbocyclic compounds. For instance, the reaction of nitrobenzylic sulfones—products of VNS reaction of *meta* functionalized nitrobenzenes—react with Michael acceptors to produce a variety of substituted nitronaphthalenes (Scheme 11.52) [86].

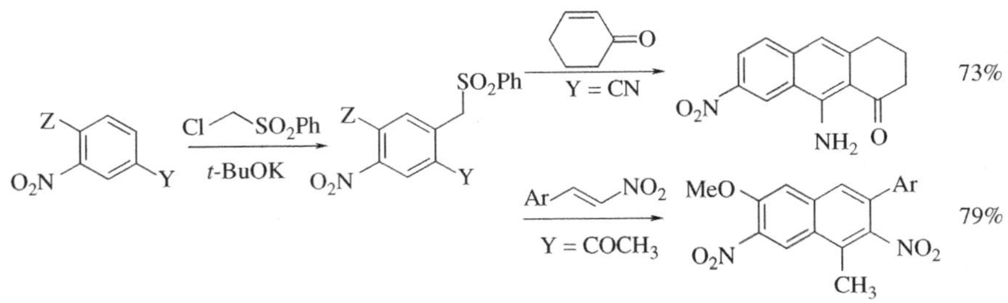

SCHEME 11.52 Synthesis of substituted naphthalenes via VNS reaction [86].

Besides of great practical values, the reactions of S_NArH discussed in this chapter provide important contribution to general knowledge of organic chemistry. First of all, it is unambiguously documented that addition of nucleophiles to electron-deficient arenes proceeds much faster at positions occupied by hydrogen to form σ^H-adducts than in those, equally activated, occupied by halogens to form the σ^X-adducts. When nature of arenes, nucleophiles, and conditions is such that fast further conversion of σ^H-adducts into products S_NArH can proceed, conventional S_NAr reaction does not occur because the competing substitution of hydrogen proceeds faster (Scheme 11.53).

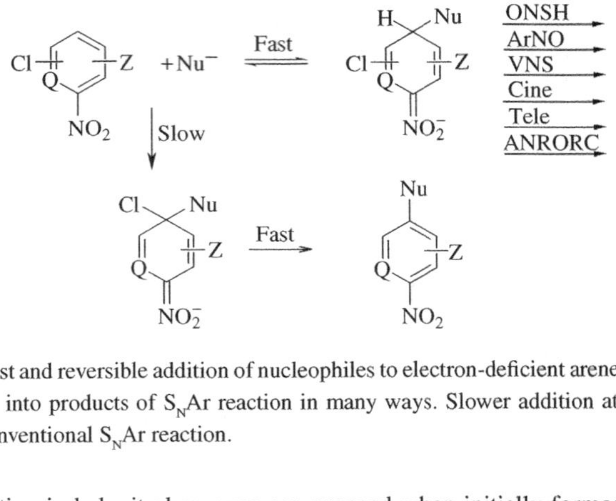

SCHEME 11.53 Fast and reversible addition of nucleophiles to electron-deficient arenes produces σ^H-adducts that can be converted into products of S_NAr reaction in many ways. Slower addition at position occupied by halogens results in conventional S_NAr reaction.

Thus, S_NAr reaction in halonitrobenzenes can proceed when initially formed σ^H-adducts are not converted into products of S_NArH but rearomatize via dissociation to starting nucleophiles and arenes. This equilibration followed by slower but irreversible (as a rule) addition at positions occupied by halogens X and subsequent departure of X^- anions from the σ^X-adducts results in S_NAr reaction. Since in halonitroarenes initial nucleophilic addition proceeds at positions occupied by hydrogen, whereas the addition in positions occupied by halogens is much slower, hence S_NAr does not proceed, one can formulate a paradoxical statement: halogens in electron-deficient arenes protect positions they occupy against nucleophilic attack. It should be also mentioned that formation of σ^X adducts, hence S_NAr reaction, can be partially or totally inhibited by faster formation of σ^H-adducts. These adducts at low temperature are long-living species; thus, S_NAr cannot proceed simply, because the reactants (Nu and ArX) are engaged in σ^H-adducts. In some papers on S_NAr reactions, fast formation of σ^H adducts is acknowledged; however, they are considered as "unproductive intermediates." Taking into consideration a variety of transformations of σ^H-adducts into a plethora of S_NArH products, it is obvious that σ^H-adducts are very productive key and versatile short-lived intermediates. It is therefore necessary to introduce proper correction in modern textbooks on organic chemistry. In chapters dealing with S_NAr, it should be mentioned that S_NAr in electron-deficient arenes proceeding via addition–elimination mechanism is a secondary process, preceded by reversible formation of σ^H-adducts that under proper conditions can be converted into products of S_NArH.

It should be stressed that the term "secondary" is used to indicate timing of these reactions. Great importance, value, and generality of S_NAr reactions discussed in Chapter 6 are unquestionable. Although the first example of nucleophilic substitution of hydrogen in nitroarenes, the von Richter reaction [87], is known for almost 150 years and a few other examples were published in the middle of the last century [27, 88], the general concept of these reactions was formulated only about 30 years ago. In spite of their generality and great value in synthesis as well as new mechanistic concepts formulated in arene chemistry, they are not sufficiently recognized, perhaps because in traditional thinking nucleophiles replace halogens and other nucleofugal groups, not hydrogen. I certainly hope that this chapter will help to change this situation.

ABBREVIATIONS

ANROCR	Addition nucleophile ring opening ring closure
CAN	Cerium ammonium nitrate
DDQ	Dichlorodicyanoquinone
DMF	Dimethylformamide
DMD	Dimethyldioxirane
DMSO	Dimethylsulfoxide
ESR	Electron spin resonance
KIE	Kinetic isotope effect
ONSH	Oxidative nucleophilic substitution of hydrogen
SET	Single-electron transfer
S_NAr	Nucleophilic aromatic substitution (of halogens)
S_NArH	Nucleophilic aromatic substitution of hydrogen
TAA	Tetra alkyl ammonium cation
THF	Tetrahydrofuran
TS	*p*-Toluenosulfonyl
VNS	Vicarious nucleophilic substitution

REFERENCES

[1] (a) Mąkosza, M. (**2011**) *Synthesis*, 2341–2356; (b) Mąkosza, M. (**2014**) *Chem. Eur. J.*, **20**, 5536–5546.

[2] Mąkosza, M. (**2010**) *Chem. Soc. Rev.*, **39**, 2855–2868.

[3] (a) Clayden, J., Greeves, N., Warren, S., and Wothers, P. (**2011**) *Organic Chemistry*, Oxford University Press, New York; (b) Anslyn, E.V. and Daugherty, W.A. (**2006**) *Modern Physical Organic Chemistry*, University Science Books, Sausalito.

[4] (a) Maloney, M.B. (2008) *Structure and Reactivity in Organic Chemistry*, Blackwell Publishing, Oxford; (b) Klein, D. (2012) *Organic Chemistry*, Wiley, Weinheim.

[5] Terrier, F. (2013) *Modern Nucleophilic Aromatic Substitution*, Wiley-VCH Verlag, Weinheim.

[6] Chupakhin, O.N., Charushin, V.N., and van der Plas, H.C. (1994) *Nucleophilic Aromatic Substitution of Hydrogen*, Academic Press, San Diego.

[7] (a) Van der Plas, H.C. and Woźniak, M. (1986) *Croat. Chem. Acta*, **59**, 33–49; (b) Szpakiewicz, B. and Grzegozek, M. (2004) *Russ. J. Org. Chem.*, **40**, 829–833.

[8] Malykhin, E.V., Kolesnichenko, G.A., and Shteingarts, V.D. (1985) *Zh. Org. Khim.*, **21**, 1150–1159.

[9] (a) Stern, M.K. and Cheng, B.K. (1993) *J. Org. Chem.*, **58**, 6883–6888; (b) Stern, M.K., Hileman, F.D., and Bashkin, J.K. (1992) *J. Am. Chem. Soc.*, **114**, 9237–9238.

[10] Mąkosza, M. and Sulikowski, D. (2009) *J. Org. Chem.*, **74**, 3827–3832.

[11] Mąkosza, M., Kamieńska-Trela, K., Paszewski, M., and Bechcicka, M. (2005) *Tetrahedron*, **61**, 11952–11964.

[12] (a) Mąkosza, M. and Staliński, K. (1997) *Chem. Eur. J.*, **3**, 2025–2031; (b) Mąkosza, M. and Staliński, K. (1998) *Tetrahedron Lett.*, **39**, 3575–3576.

[13] Mąkosza, M. and Surowiec, M. (2003) *Tetrahedron*, **59**, 6261–6266.

[14] (a) Adam, W., Mąkosza, M., Staliński, K., and Zhao, C-G. (1998) *J. Org. Chem.*, **63**, 4390–4391; (b) Adam, w., Mąkosza, M., Zhao, C-G., and Surowiec, M. (2001) *J. Org. Chem.*, **65**, 1099–1101.

[15] (a) Mąkosza, M., Surowiec, M., Szczepańska, A., Sulikowski, D., and Maltsev, O. (2007) *Synlett*, 470–474; (b) Mąkosza, M., Sulikowski, D., and Maltsev, O. (2008) *Synlett*, 1711–1713.

[16] Sulikowski, D. and Mąkosza, M. (2010) *Eur. J. Org. Chem.*, 4218–4226.

[17] Mąkosza, M. and Sypniewski, M. (1994) *Tetrahedron*, **50**, 4913–4920.

[18] Moskalev, N., Barbasiewicz, M., and Mąkosza, M. (2004) *Tetrahedron*, **60**, 347–358.

[19] Xu, Q.-L., Gao, H., Yousufuddin, M., Ess, D., and Kürti, L. (2013) *J. Am. Chem. Soc.*, **135**, 14048–14054.

[20] Loska, R. and Mąkosza, M. (2007) *J. Org. Chem.*, **72**, 1354–1365.

[21] Loska, R., Majcher, M., and Mąkosza, M. (2007) *J. Org. Chem.*, **72**, 5574–5580.

[22] Mąkosza, M., Paszewski, M., and Sulikowski, D. (2008) *Synlett*, 2938–2940.

[23] Bartoli, G. (1984) *Acc. Chem. Res.*, **17**, 109–115.

[24] Mąkosza, M. and Surowiec, M. (2001) *J. Organomet. Chem.*, **624**, 167–171.

[25] (a) Gallardo, I., Guirado, C., and Marquet, J. (2001) *Chem. Eur. J.*, **7**, 1759–1765; (b) Cruz, H., Gallardo, I., and Guirado, G. (2011) *Eur. J. Org. Chem.*, 7378–7389.

[26] Yamamoto, H. and Momiyama, N. (2005) *Chem. Comm.*, 3514–3525;

[27] (a) Davis, R.B. and Pizzini, L.C. (1960) *J. Org. Chem.*, **25**, 1884–1888; (b) Davis, R.B., Pizzini, L.C., and Bara, F.J. (1961) *J. Org. Chem.*, **26**, 4270–4274.

[28] Bobin, M., Kwast, A., and Wróbel, Z. (2007) *Tetrahedron*, **63**, 11048–11054.

[29] Wróbel, Z. (2000) *Eur. J. Org. Chem.*, 521–525.

[30] (a) Wróbel, Z. and Kwast, A. (2010) *Synthesis*, 3865–3872; (b) Wróbel, Z., Stachowska, K., Grudzień, K., and Kwast, A. (2011) *Synlett*, 1439–1443.

[31] (a) Danikiewicz, W. and Mąkosza, M. (1991) *J. Org. Chem.*, **56**, 1283–1286; (b) Danikiewicz, W. and Mąkosza, M. (1987) *Tetrahedron Lett.*, **28**, 1707–1710.

[32] (a) Goliński, J. and Mąkosza, M. (1978) *Tetrahedron Lett.*, **19**, 3495–3498; (b) Mąkosza, M., Goliński, J., and Baran, J. (1984) *J. Org. Chem.*, **49**, 1488–1494.

[33] Mąkosza, M. and Stalewski, J. (1991) *Liebigs Ann. Chem.*, 605–606.

[34] Mąkosza, M. and Ludwiczak, S. (1984) *J. Org. Chem.*, **49**, 4562–4563.

[35] Mąkosza, M. and Winiarski, J. (1987) *Acc. Chem. Res.*, **20**, 282–289.

[36] Bilkis, I.I. and Shein, S.M. (1975) *Tetrahedron*, **31**, 969–971.

[37] Dalpozzo, R. and Bartoli, G. (2005) *Curr. Org. Chem.*, **9**, 163–178.

[38] Mąkosza, M., Glinka, T., and Kinowski, J. (1984) *Tetrahedron*, **40**, 1863–1868.

[39] (a) Mąkosza, M. and Kwast, A. (1994) *Bull. Soc. Chim. Belg.*, **103**, 445–448; (b) Mąkosza, M. and Kwast, A. (2004) *Eur. J. Org. Chem.*, 2125–2130.

[40] Glinka, T. and Mąkosza, M. (1983) *J. Org. Chem.*, **48**, 3860–3861.

[41] Mąkosza, M., Lemek, T., Kwast, A., and Terrier, F. (2002) *J. Org. Chem.*, **67**, 394–400.

[42] (a) Błażej, S. and Mąkosza, M. (2008) *Chem. Eur. J.*, **14**, 11113–11122; (b) Seeliger, F., Błażej, S., Bernhardt, S., Mąkosza, M., and Mayr, H. (2008) *Chem. Eur. J.*, **14**, 6108–6118.

[43] Miller, J. (1968) *Aromatic Nucleophilic Substitution*, Elsevier, Amsterdam.

[44] Lemek, T., Mąkosza, M., Stephenson, D.S., and Mayr, H. (2003) *Angew. Chem. Int. Ed.*, **42**, 2793–2795.

[45] Mąkosza, M. and Kwast, A. (1998) *J. Phys. Org. Chem.*, **11**, 341–349.

[46] Mudryk, B. and Mąkosza, M. (1988) *Tetrahedron*, **44**, 209–213.

[47] Mąkosza, M. and Ludwiczak, S. (1986), *Synthesis*, 50–52.

[48] (a) Beier, P., Pastyrikova, T., and Iakobson, G. (2011) *J. Org. Chem.*, **76**, 4781–4786; (b) Mąkosza, M., Surowiec, M., and Voskresenski, S. (2000) *Synthesis*, 1237–1240.

[49] (a) Mąkosza, M., Chylińska, B., and Mudryk, B. (1984) *Liebigs Ann. Chem.*, 8; (b) Kinowski, A. and Ostrowski, S. (1993) *Synthesis*, 1215–1217.

[50] Couch, G.D., Burke, P.J., Knox, R.J., and Moody, C.J. (2008) *Tetrahedron*, **64**, 2816–2823.

[51] Achmatowicz, M., Thiel, O.R., Gorins, G., Goldstein, C., Affouard, C., Jensen, R., and Larsen, R.D. (2008) *J. Org. Chem.*, **73**, 6793–6799.

[52] Mąkosza, M. and Kwast, E. (1995) *Tetrahedron*, **51**, 8339–8354.

[53] (a) Mąkosza, M., Rydz, A., and Wróbel, Z. (1995) *Pol. J. Chem.*, **69**, 918–921; (b) Haglund, O. and Nilsson, M. (1993) *Synthesis*, 209–211.

[54] Tercel, M., Lee, A.E., Hogg, A., Anderson, R.F., Lee, H.H., Siim, B.G., Denny, W.A., and Wilson, W.R. (2001) *J. Med. Chem.*, **44**, 3511–3522.

[55] Bernard, M.K., Mąkosza, M., Szafran, B., and Wrzeciono, U. (1989) *Liebigs Ann. Chem.*, 545–549.

[56] Neidlein, R. and Lautenschlager, G. (1989) *Chem. Ber.*, **122**, 493–498.

[57] Mąkosza, M., Kuciak, R., and Wojciechowski, K. (1994) *Liebigs Ann. Chem.*, 615–618.

[58] Mąkosza, M., Goliński J., and Rykowski, A. (1983) *Tetrahedron Lett.*, **24**, 3277–3278.

[59] Mąkosza, M. and Ostrowski, S. (1988) *J. Prakt. Chem.*, **330**, 789–794.

[60] Ostrowicz, A., Bałoniak, S., Mąkosza, M., and Rykowski, A. (1992) *Tetrahedron Lett.*, **33**, 4787–4790.

[61] (a) Goliński, J., Mąkosza, M., and Rykowski, A. (1983) *Tetrahedron Lett.*, **24**, 3279–3280; (b) Mąkosza, M., Glinka, T., Ostrowski, S., and Rykowski, A. (1987) *Chem. Lett.*, 61–64.

[62] Hamana, M., Fujimura, Y., and Haradahira, T. (1987) *Heterocycles*, **25**, 229–233.

[63] (a) Ostrowski, S. and Mąkosza, M. (1989) *J. Organomet. Chem.*, **367**, 95–100; (b) Müller, P., Bernardinelli, G., and Motallebi, S. (1990) *Helv. Chim. Acta*, **73**, 1242–1249; (c) Balssa, F., Gagliardini, V., Le Corre-Sussanne, C., Rose-Munch, F., Rose, E., and Vaissermann, J. (1997) *Bull. Soc. Chim. Fr.*, **134**, 537–541.

[64] (a) Jackson, D.A., Lawrence, N.J., and Liddle, J. (1996) *Synlett*, 55–58; (b) Florio, S., Lorusso, R., Granito, C., and Luisi, R. (2004), *J. Org. Chem.*, **69**, 4961–4965; (c) Mąkosza, M., Bester, K., and Cmoch, P. (2015) *J. Org. Chem.*, **80**, 5436–5443.

[65] Mąkosza, M. and Owczarczyk, Z. (1989) *J. Org. Chem.*, **54**, 5094–5100.

[66] Mąkosza, M. and Sienkiewicz, K. (1998) *J. Org. Chem.*, **63**, 4199–4208.

[67] (a) Pews, R.G., Lysenko, Z., and Vosejpka, P.C. (1997) *J. Org. Chem.*, **62**, 8255–8256; (b) Beier, P., Pastyrikova, T., and Iakobson, G. (2011) *Tetrahedron Lett.*, **52**, 4392–4394.

[68] (a) Angeli, A. and Angelico, F. (1901) *Gazz. Chim. Ital.*, **31**, 27; (b) Price, C.G. and Voong, S.-T. (1948) *Org. Synth.*, **28**, 80.

[69] Katritzky, A.R. and Laurenzo, K.S. (1988) *J. Org. Chem.*, **53**, 3978–3982.

[70] (a) Pagoria, P.F., Mitchell, A.R., and Schmidt, R.D. (1996) *J. Org. Chem.*, **61**, 2934–2935; (b) Mitchell, A.R., Coburn, M.D., Schmidt, R.D., Pagoria, P.F., and Lee, G.S. (2002) *Thermochim. Acta*, **384**, 205–217.

[71] Seko, S., Miyake, K., and Kawamura, N. (1999) *J. Chem. Soc, Perkin Trans.* **1**, 1437–1444.

[72] Mąkosza, M. and Białecki, M. (1998) *J. Org. Chem.*, **63**, 4878–4888.

[73] Morrison, D.C. (1962) *J. Org. Chem.*, **27**, 296–297.

[74] Błażej, S., Kwast, A., and Mąkosza, M. (2004) *Tetrahedron Lett.*, **45**, 3193–3195.

[75] Mąkosza, M., Varvounis, G., Surowiec, M., and Giannopoulos, T. (2003) *Eur. J. Org. Chem.*, 3791–3797.

[76] Suwiński, J. and Świerczek, K. (2001) *Tetrahedron*, **57**, 1639–1662.

[77] van der Plas, H.C. (1978) *Acc. Chem. Res.*, **11**, 462–468.

[78] (a) Mąkosza, M. and Winiarski, J. (1984) *J. Org. Chem.*, **49**, 1494–1499; (b) Stahly, G.P., Stahly, B.C., and Lilje, K.C. (1984) *J. Org. Chem.*, **49**, 578–579.

[79] Mąkosza, M. and Winiarski, J. (1984) *J. Org. Chem.*, **49**, 5272–5274.

[80] (a) Mąkosza, M. and Goliński, J. (1982) *J. Angew. Chem.*, **21**, 451–452; (b) Lawrence, N.J., Liddle, J., and Jackson, D.A. (1995) *Tetrahedron Lett.*, **36**, 8477–8480.

[81] Florio, S., Lorusso, P., Luisi, R., Granito, C., Ronzini, L., and Troisi, L. (2004) *Eur. J. Org. Chem.*, 2118–2124;

[82] (a) Mąkosza, M. and Wojciechowski, K. (2004) *Chem. Rev.*, **104**, 2631–2666; (b) Mąkosza, M. and Wojciechowski, K. (2014) *Heterocycles*, **88**, 75–101; (c) Mąkosza, M. and Wojciechowski, K. (2014) *Top. Heterocycl. Chem.*, **37**, 51–105.

[83] (a) Wróbel, Z. and Mąkosza, M. (1993) *Synlett*, 597–598; (b) Mąkosza, M., Danikiewicz, W., and Wojciechowski, K. (1998) *Liebigs Ann. Chem.*, 203–208; (c) Iakobson, G., Posta, M., and Beier, P. (2013) *Synlett*, **24**, 855–859.

[84] (a) Wojciechowski, K. and Mąkosza, M. (1986) *Synthesis*, 97–99; (b) Wojciechowski, K. and Mąkosza, M. (1986) *Bull. Soc. Chim. Belg.*, **95**, 117–119.

[85] Kwast, A. Stachowska, K., Trawczyński, A., and Wróbel, Z. (2011), *Tetrahedron Lett.*, **52**, 6484–6488.

[86] Tyrała, A. and Mąkosza, M. (1994) *Synthesis*, 264–266.

[87] von Richter, V. (1871) *Chem. Ber.*, **4**, 21.

[88] (a) Traynelis, V.J. and Mc Sweeney, J.V. (1966) *J. Org. Chem.*, **31**, 243–247; (b) Nozaki, H., Yamamoto, R., and Noyori, R. (1966) *Tetrahedron Lett.*, 1123–1126.

PART III

ARYNE CHEMISTRY

12

THE CHEMISTRY OF ARYNES: AN OVERVIEW

Roberto Sanz and Anisley Suárez

Área de Química Orgánica, Facultad de Ciencias, Universidad de Burgos, Burgos, Spain

12.1 INTRODUCTION

Arynes are neutral intermediates but highly strained and kinetically unstable molecules. Although they are known for more than 60 years, it was not until recently when their use has experienced a rapid expansion. The interest in aryne chemistry has grown in parallel with the increasing available methods that allow their easy generation from new precursors under softer conditions. Due to their low-lying LUMO, arynes exhibit remarkable electrophilicity and have been shown to react with a wide variety of nucleophiles, ranging from very soft ones to organolithium reagents, as well as to participate in several pericyclic reactions, including cycloaddition and ene reactions. In addition, they are also able to undergo transition metal–catalyzed transformations. In this chapter, arynes will be shown as highly valuable intermediates for the regioselective construction of polysubstituted aromatic compounds. By covering the main types of aryne reactivity along with an overview of the mechanistic pathways involved, this chapter will only focus on the most significant and recent results in each topic as several comprehensive reviews have been published in previous years [1].

12.2 STRUCTURE AND REPRESENTATIVE REACTIONS OF ARYNES

Although the first suggestion about the existence of *o*-benzyne (1,2-didehydrobenzene) (**1**) dates from 1927, the isotopic labeling experiments carried out by Roberts *et al.* in 1953 provided the first compelling experimental evidence for the intermediacy of **1** in the reaction of chlorobenzene with KNH_2 [2].

Arene Chemistry: Reaction Mechanisms and Methods for Aromatic Compounds,
First Edition. Edited by Jacques Mortier.

As expected, the formal C—C triple bond in **1** is significantly weaker than in unstrained alkynes (as shown by the C≡C stretching vibrations: 1846 cm^{-1} for **1** [3] vs. 2150 cm^{-1} for typical alkynes), due to the distortion of the unhybridized p-orbitals that are no longer parallel to each other resulting in reduced orbital overlap and increased reactivity. Nevertheless, **1** is better described as a strained alkyne rather than a biradical, as further supported by the experimentally found C≡C bond length (1.24 Å) [4], which is more similar to a typical C—C triple bond rather than a C—C double bond. Due to the high strain created by the triple bond, arynes have low-lying LUMOs favoring their participation in a range of pericyclic reactions. This dienophilic nature of arynes has been exploited in a wide variety of cycloadditions, including ene reactions (Scheme 12.1).

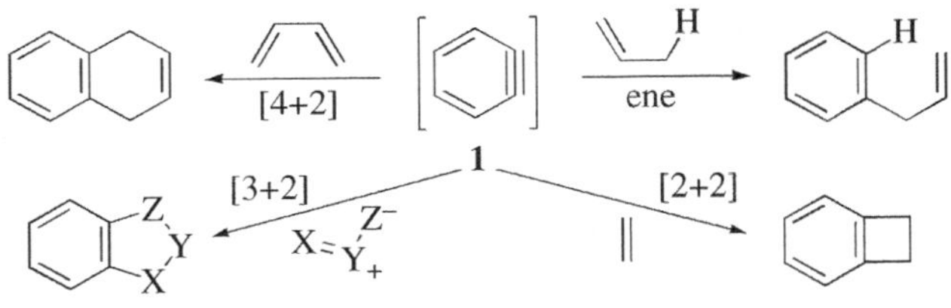

SCHEME 12.1 Pericyclic reactions.

Arynes also exhibit remarkable electrophilicity that makes the formal C≡C bond prone to nucleophilic attack. The resulting aryl anions **2** (a formal carbanion when using anionic nucleophiles, or an alternative zwitterion for uncharged nucleophiles) can be trapped by external or internal electrophiles (including a proton). Depending on the subsequent evolution of the aryne postnucleophilic addition, these reactions can be divided into four types. Trapping of the aryl anion by proton as the electrophile leads to monosubstitution of the aryne (Scheme 12.2, *path A*). This path also includes the use of uncharged nucleophiles bearing acidic protons, like alcohols or amines. If the nucleophilic addition is followed by intermolecular quenching by an electrophile, then a three-component reaction leads to disubstitution of the aryne. Using uncharged nucleophiles that lead to a zwitterion intermediate, further trapping with an electrophilic partner usually takes place with subsequent

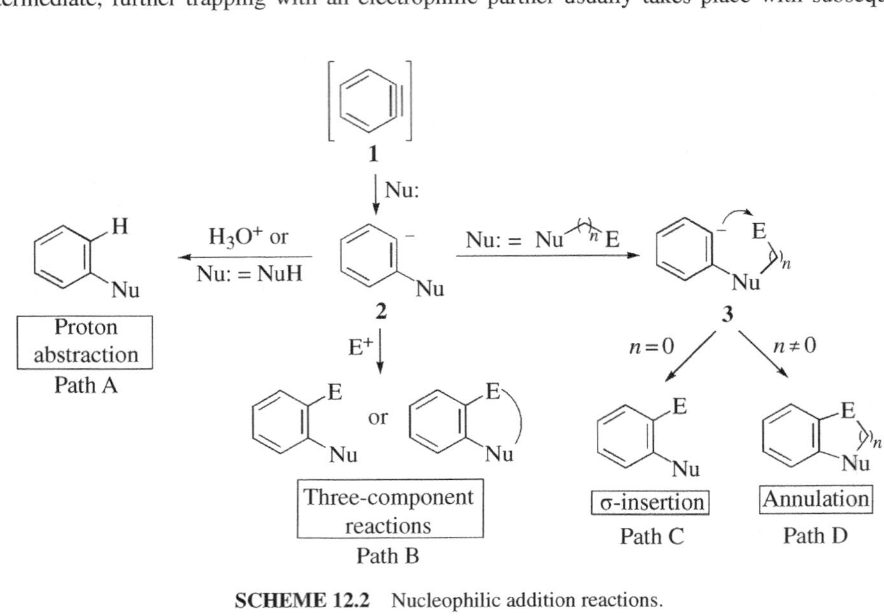

SCHEME 12.2 Nucleophilic addition reactions.

cyclization (Scheme 12.2, *path B*). In addition, by employing nucleophiles attached to a suitable electrophilic moiety with a σ-bond (Nu—E), arynes can be formally inserted into the Nu—E bond through a subsequent intramolecular reaction with the electrophile in the initially formed intermediate **3** ($n = 0$) (Scheme 12.2, *path C*). When the nucleophilic and electrophilic positions are separated by a longer tether ($n \neq 0$), annulation reactions can take place depending on both the reaction conditions and the substrate structures (Scheme 12.2, *path D*).

Finally, transition metal–catalyzed reactions of arynes have been explored as a useful method for the construction of a wide variety of carbo- and heterocycles. These reactions include cyclotrimerization of arynes, cocyclization of arynes with alkynes or alkenes, and carbopalladation of arynes with Pd-complexes. Moreover, some insertion reactions of arynes into σ-bonds are also catalyzed by metal complexes.

12.3 ARYNE GENERATION

Several methods have been developed for generating *o*-benzyne (**1**) [5]. As shown in Scheme 12.3, main strategies for accessing **1** can be considered as formal elimination reactions in which a precursor arene loses one or more groups, involving aryl anions, cations, radicals, zwitterions, and by fragmentation reactions. More recently, Hoye *et al.* have described a completely different strategy for generating arynes based on the intramolecular "hexadehydro–Diels–Alder" (HDDA) reaction of triynes [6].

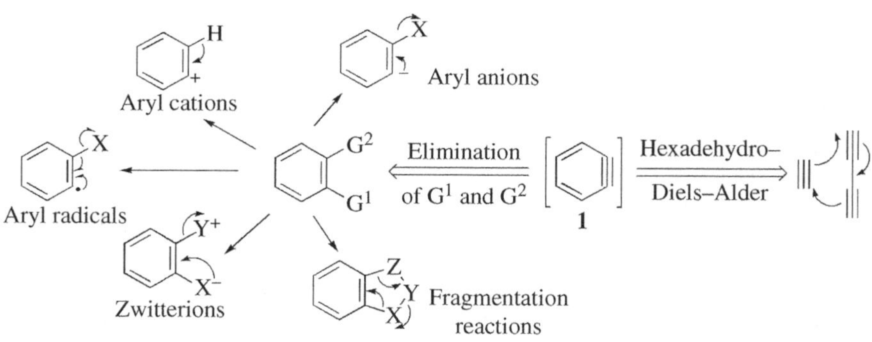

SCHEME 12.3 Methodologies for aryne generation.

12.3.1 Elimination Methods

Among the wide variety of methods available for the generation of benzyne (**1**) the formation of arylanions that undergo further elimination of a suitable leaving group is the most general procedure to access to these compounds. Traditionally, strong bases such as organolithiums or Grignard reagents, as well as alkali metal amides, were used for the generation of the required arylanion through deprotonation or halogen metal–exchange reactions. However, the limited functional group tolerance associated to these highly basic conditions is the main problem of these methods. Remarkably, in recent years the introduction of *o*-(silyl)phenyl triflates as benzyne precursors has supposed an important advance in this field because the simple treatment of these reagents with fluoride allows the easy generation of arynes under mild conditions.

Deprotonation. The *o*-metalation of aromatic halides is a well-established route to arynes. In cases where the starting material contains two different halogens in 1,3-positions, the deprotonation reaction occurs at 2-position and then the elimination may lead to two different aryne derivatives depending on the halide anion that suffers the elimination reaction. In this context, the solvent of the reaction plays a crucial role. Thus, for example, in the case of 3-fluoro-1-chlorobenzene, the trapping of the corresponding benzyne with a nonacidic diene showed that in THF the generation of 3-fluorobenzyne is preferred whereas in hydrocarbons 3-chlorobenzyne is formed. Rate studies reveal that LiF eliminates via monomer-based pathways requiring THF dissociation (Scheme 12.4) [7]. In this context, 3-functionalized arynes can be prepared by a chemo- and regioselective directed zincation of *m*-functionalized haloaromatics. The use of a lithium dialkyl(2,2,6,6-tetramethylpiperidino)zincate, $R_2Zn(TMP)Li$, bearing a methyl group (R = Me) is crucial for the smooth generation of the corresponding arynes (Scheme 12.4) [8].

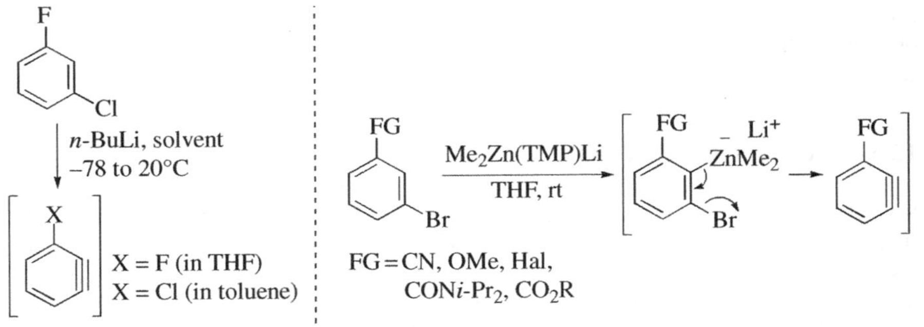

SCHEME 12.4 *ortho*-Metalation for aryne generation.

Halogen–Metal Exchange. Metalation of 1,2-dihalobenzenes and related systems, such as *o*-halotriflates, and *o*-halotosylates, through halogen–metal exchange, is a useful method to generate arynes. In the case of halogen–lithium exchange, this reaction takes place at low temperature and the relative stability of the intermediate *o*-functionalized aryllithium toward elimination is highly dependent on the nature of the group being eliminated as well as the rest of substituents. Scheme 12.5 shows how *o*-haloaryl triflate **4** behaves as a synthetic equivalent of 3-methoxy-1,4-benzdiyne allowing symmetrical or successive aryne trapping reactions in a one-pot manner [9]. As previously stated, the main limitation of the halogen–metal exchange procedure is the limited functional group tolerance. However, some advances in this regard have arisen in the last years. Thus, Knochel *et al.* have developed a useful method for accessing arynes bearing functional groups like ester, nitrile, and nitro, by the readily tunable elimination of 2-magnesiated aryl sulfonates prepared by I–Mg exchange from sulfonates **5** (Scheme 12.5) [10].

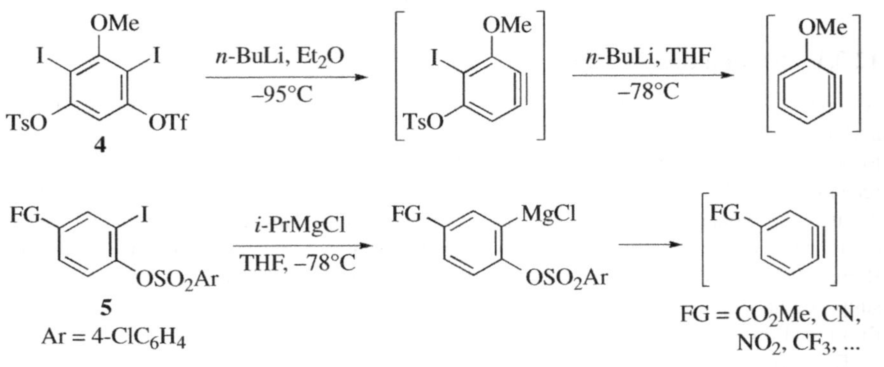

SCHEME 12.5 Halogen-metal exchange for aryne generation.

Fluoride-Induced Elimination. 2-(Trimethylsilyl)phenyl triflate (**6**), first reported by Kobayashi *et al.* in 1983 [11], and now commercially available, has turned into the most widely used benzyne precursor and has enabled the development of a wide variety of synthetic applications of arynes, which will be discussed along this chapter. Under very mild conditions, benzyne (**1**) is smoothly generated by a fluoride-induced 1,2-elimination. Apart from Kobayashi's original synthesis from *o*-chlorophenol, more efficient procedures have been later developed starting also from *o*-bromophenol or even phenol (Scheme 12.6) [12]. The most used route for synthesizing *o*-(TMS)phenyl triflates consists of a one-pot process involving *O*-silylation of the corresponding *o*-bromophenol, halogen–metal exchange, *O*- to *C*-silyl migration, and trapping of the phenoxide with Tf_2O [12a]. Different fluoride sources, mainly CsF, TBAF, and TBAT, have been used for generating the corresponding substituted benzynes.

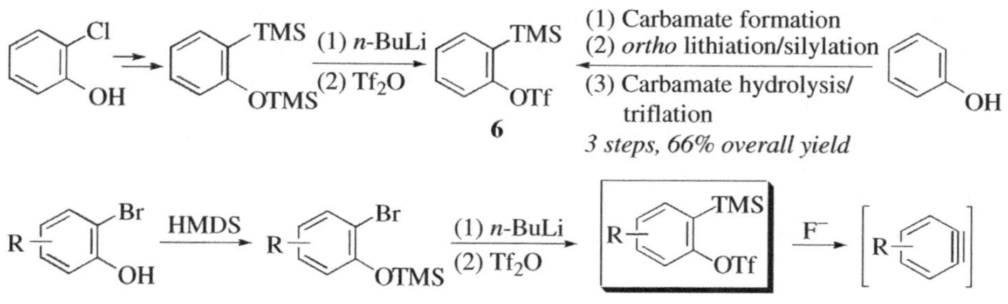

SCHEME 12.6 Routes for the synthesis of *o*-(TMS)phenyl triflates.

Kitamura *et al.* have developed (phenyl)[2-(trimethylsilyl)phenyl]iodonium triflate (**7**) as an excellent benzyne precursor [13], possessing the hypervalent iodine group with a salient leaving ability (~10^6 times greater than that of the OTf group). Treatment of **7**, which is prepared in two steps from 1,2-dichlorobenzene, with TBAF efficiently generates benzyne (**1**) (Scheme 12.7).

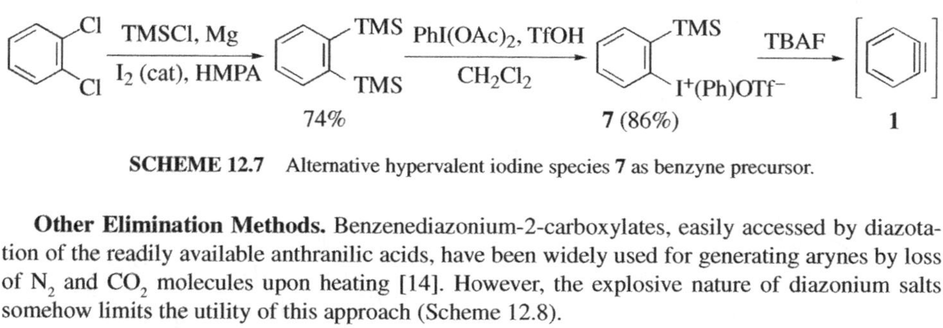

SCHEME 12.7 Alternative hypervalent iodine species **7** as benzyne precursor.

Other Elimination Methods. Benzenediazonium-2-carboxylates, easily accessed by diazotation of the readily available anthranilic acids, have been widely used for generating arynes by loss of N_2 and CO_2 molecules upon heating [14]. However, the explosive nature of diazonium salts somehow limits the utility of this approach (Scheme 12.8).

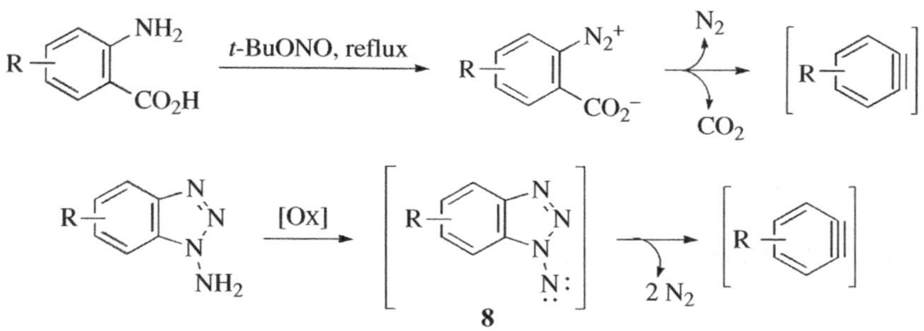

SCHEME 12.8 Other methods to generate arynes.

Fragmentation Reactions. Oxidation of 1-aminobenzotriazoles with $Pb(OAc)_4$ or NIS is used to generate arynes through the intermediacy of an aminonitrene **8**. The subsequent fragmentation with loss of two molecules of N_2 leads to the corresponding aryne derivative (Scheme 12.8) [15].

12.3.2 By Hexadehydro-Diels–Alder Reaction

More recently, Hoye *et al.* have described how an intramolecular HDDA reaction, the [4+2] cycloisomerization of a 1,3-diyne with a suitable "diynophile", produces arynes [6]. Moreover, this thermal transformation is metal-, reagent-, and by-product-free that is actually leading to find unprecedented reactivities for arynes. The authors discovered this surprising reaction by serendipity, when a trivial oxidation of alkynol **9** took an unexpected course involving the formation of the aryne derivative **10** in a [4+2] cycloaddition between a diyne and an electronically activated alkyne. This reaction cascade is terminated by an insertion of the aryne into the Si—O bond affording tricyclic product **11** that also constitutes a new pathway for benzyne trapping (Scheme 12.9). This type of "energized" arynes, like **12**, is also capable of the concerted removal of two vicinal hydrogen atoms from a hydrocarbon. This is the first example of a single-step, bimolecular reaction in which two hydrogen atoms are simultaneously transferred from a saturated alkane (Scheme 12.9) [16].

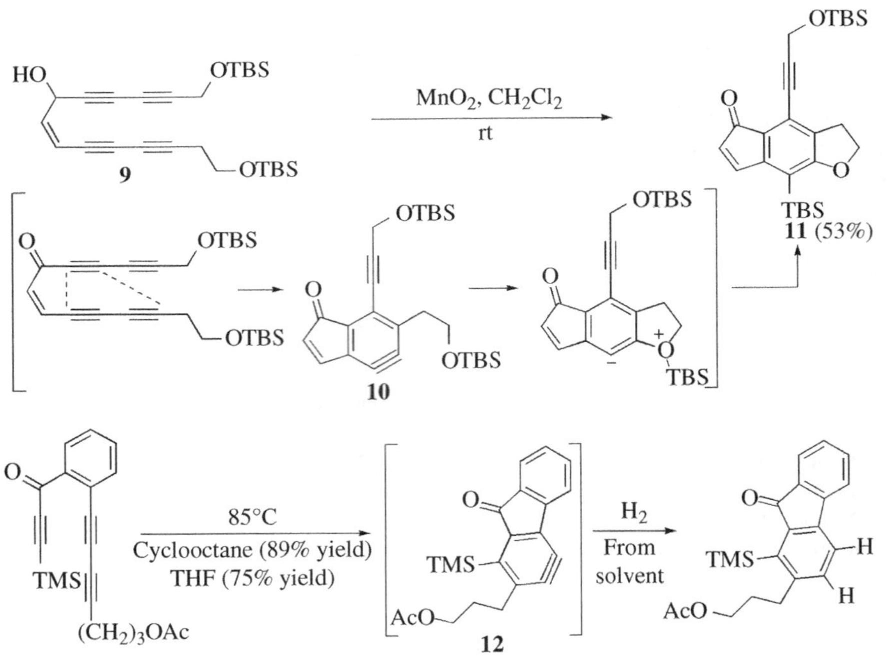

SCHEME 12.9 Hexadehydro-Diels–Alder reaction for accessing policyclic arynes.

12.4 PERICYCLIC REACTIONS

12.4.1 Diels–Alder Cycloadditions

[4+2] Cycloaddition reactions of arynes are known since the discovery of this reactive intermediate and several examples have been reported in the literature. Mechanistically, concerted or stepwise mechanisms could be considered, and substitution on both the aryne and the diene partners can influence the reaction pathway. In this section, only selected examples are highlighted in order to show the potential of this methodology. These examples have been organized according to the type of the diene partner.

Cyclic Dienes: Furans. *Cisoid* dienes like furans and isobenzofurans have been extensively studied in their cycloadditions with arynes. The benzyne–furan Diels–Alder reaction has been thoroughly employed as a method for accessing naphthol derivatives. For instance, trimethoxy-functionalized naphthol derivative **14** has been synthesized in three steps in nearly quantitative yield from 1,2,4-trimethoxybenzene and furan, involving a regioselective ring opening of tricyclic compound **13** (Scheme 12.10) [17]. When using unsymmetrically substituted arynes and furans like 3-fluorobenzyne and 2-trimethylsilylfuran, a mixture of regioisomeric 1,4-epoxy-1,4-dihydronaphthalenes is obtained. A partial solution to this problem of positional selectivity has been provided by using bulky trialkylsilyl groups in both components. In this situation the cycloaddition reaction proceeds in a regioselective way as shown in the formation of cycloadduct **15** (Scheme 12.10) [18].

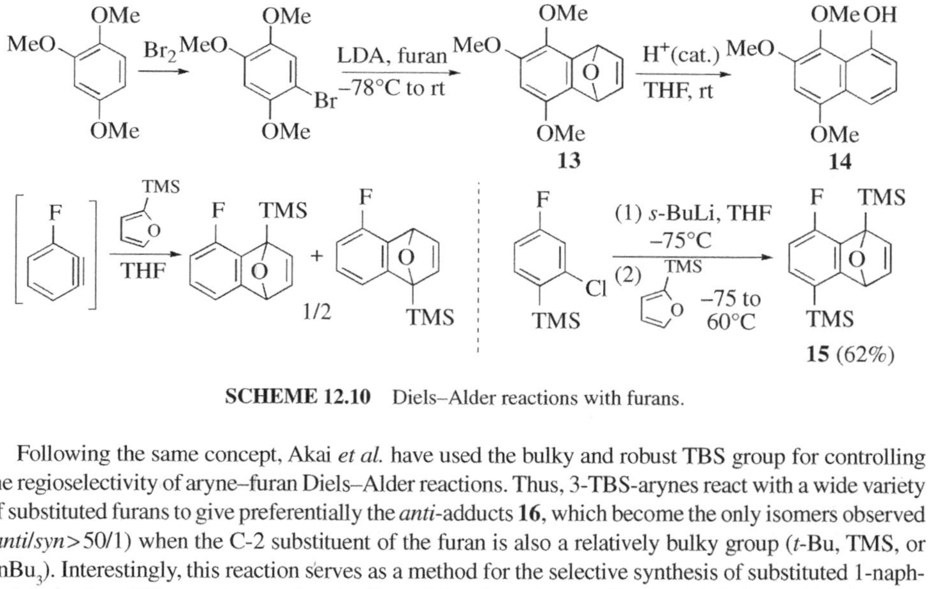

SCHEME 12.10 Diels–Alder reactions with furans.

Following the same concept, Akai *et al.* have used the bulky and robust TBS group for controlling the regioselectivity of aryne–furan Diels–Alder reactions. Thus, 3-TBS-arynes react with a wide variety of substituted furans to give preferentially the *anti*-adducts **16**, which become the only isomers observed (*anti*/*syn*>50/1) when the C-2 substituent of the furan is also a relatively bulky group (*t*-Bu, TMS, or $SnBu_3$). Interestingly, this reaction serves as a method for the selective synthesis of substituted 1-naphthol derivatives **17** by treatment of *anti*-adducts **16** with *p*-toluenesulfonic acid (Scheme 12.11) [19].

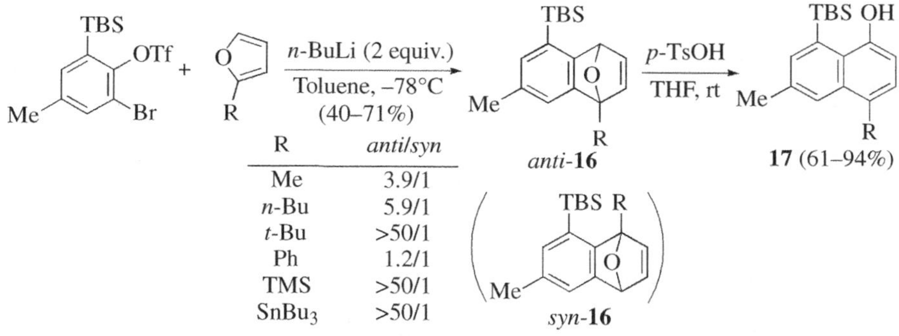

R	*anti*/*syn*
Me	3.9/1
n-Bu	5.9/1
t-Bu	>50/1
Ph	1.2/1
TMS	>50/1
$SnBu_3$	>50/1

SCHEME 12.11 Controlling regiochemistry in aryne–furan Diels–Alder reactions.

In order to control the regiochemistry of the aryne-furan cycloadditions, removable silicon tethers have been used and this methodology has been applied in total synthesis. This concept is exemplified with the regioselective preparation of naphthol **18** (Scheme 12.12) [20].

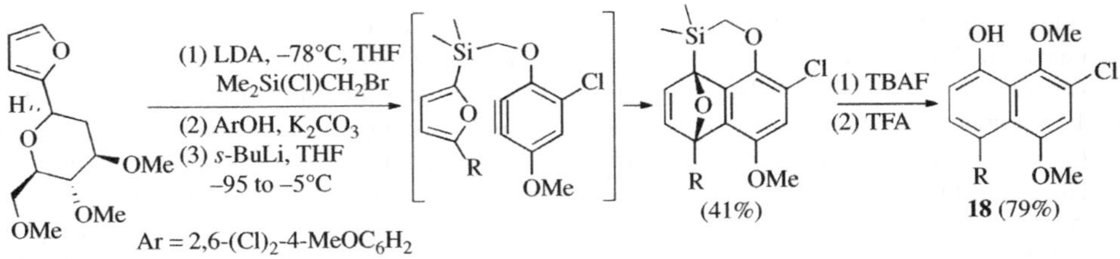

SCHEME 12.12 Intramolecular aryne–furan cycloadditions.

Other Cyclic Dienes. Besides furan, the Diels–Alder reaction of arynes with other cyclic dienes has been also reported. Lautens *et al.* have prepared azabicycle **20** from *N*-Boc-pyrrole and an aryne intermediate derived from dibromide **19**. The prepared cycloadduct has been used in the total synthesis of (+)-homochelidonine (Scheme 12.13) [21].

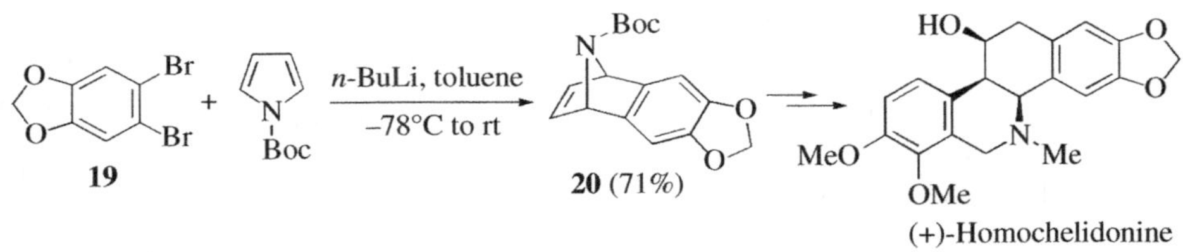

SCHEME 12.13 Aryne–pyrrole cycloadditions.

A general cycloaddition of 6-substituted and 6,6-disubstituted pentafulvenes with arynes has been developed by Biju *et al.*, leading to the formation of benzonorbornadiene derivatives **21** in high yields (Scheme 12.14) [22]. Other challenging dienes such as 1,2-benzoquinones have been studied by the same research group, reporting that these dienes show carbodiene-type reactivity in their [4+2] cycloadditions with arynes. In this way, the corresponding dioxobenzobicyclooctadienes **22** are obtained in usually high yields (Scheme 12.14) [23].

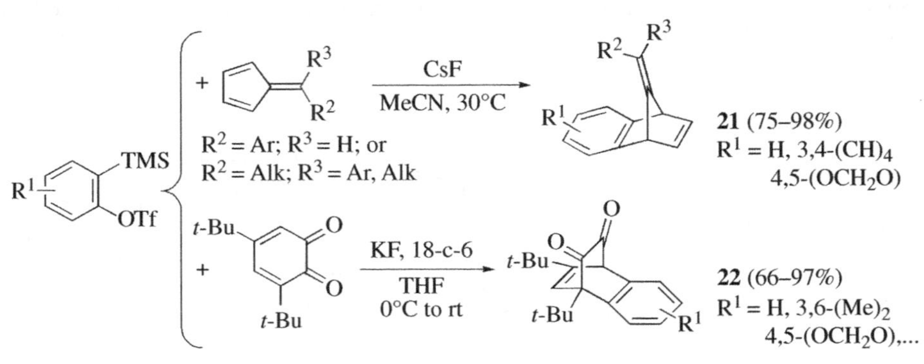

SCHEME 12.14 Diels–Alder cycloadditions with all-carbon cyclic dienes.

Other Dienes. Naphtho-fused oxindole derivatives **23** have been accessed by an aryne Diels–Alder reaction using methyleneindolines as dienes. The proposed mechanism involves a [4+2] cycloaddition followed by isomerization and further dehydrogenation. These results represent the first example of a styrene-type moiety employed as the 4π-component in aryne Diels–Alder reactions (Scheme 12.15) [24].

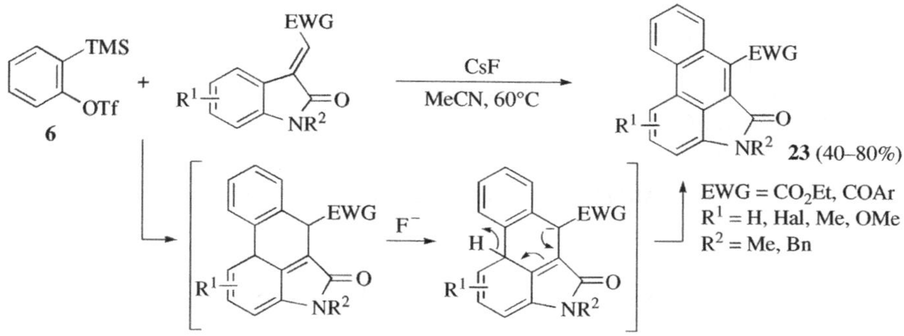

SCHEME 12.15 Diels–Alder cycloadditions with methyleneindolines.

Biju *et al.* have recently reported the synthesis of 9,10-dihydrophenanthrenes involving the use of styrenes as dienes. 9-Aryl-9,10-dihydrophenanthrenes **25** (2:1 adducts) are generally formed, probably through a concerted ene reaction of the initially generated [4+2] cycloadduct with another aryne molecule. However, when starting from styrenes bearing a strong electron-withdrawing group at the 4-position, the 1:1 adducts **24** derived from the initial Diels–Alder reaction and further isomerization were selectively obtained along with trace amounts of the corresponding cascade products **25** (Scheme 12.16) [25].

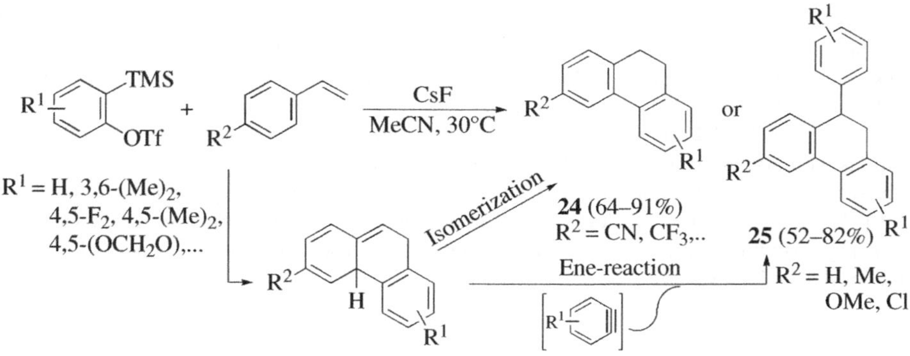

SCHEME 12.16 Styrenes as dienes for aryne [4+2] cycloadditions.

Lautens *et al.* have reported the synthesis of 1,4-dihydronaphthalenes **26** through the intermolecular benzyne Diels–Alder reaction with acyclic dienes possessing electron-donating or electron-withdrawing substituents. In the last case, an inverse electron-demand Diels–Alder pathway might be operating. In these reactions, benzyne (**1**) is generated from **6** by fluoride-induced elimination, or from benzenediazonium-2-carboxylate by thermal fragmentation (Scheme 12.17) [26].

12.4.2 [3+2] Cycloadditions

A wide variety of reagents have been used in 1,3-dipolar cycloadditions with arynes. For example, a general method for the preparation of *N*-unsubstituted indazoles **27** and 1-arylindazoles **28** by the [3+2] cycloaddition of arynes generated from *o*-(silyl)aryl triflates and diazomethane derivatives has been developed. The selective formation of **27** or **28** depends on the relative amounts of the

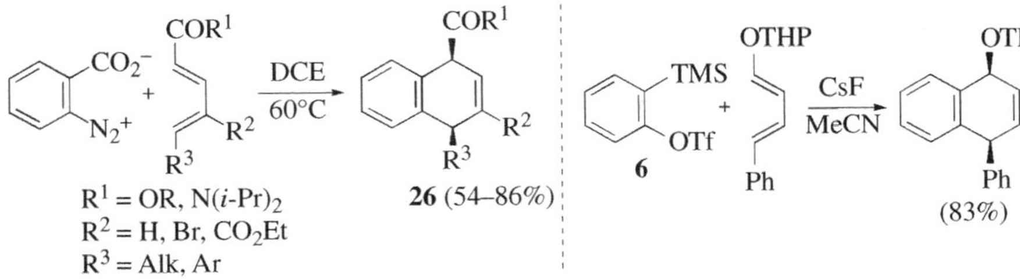

SCHEME 12.17 Acyclic dienes in aryne [4+2] cycloadditions.

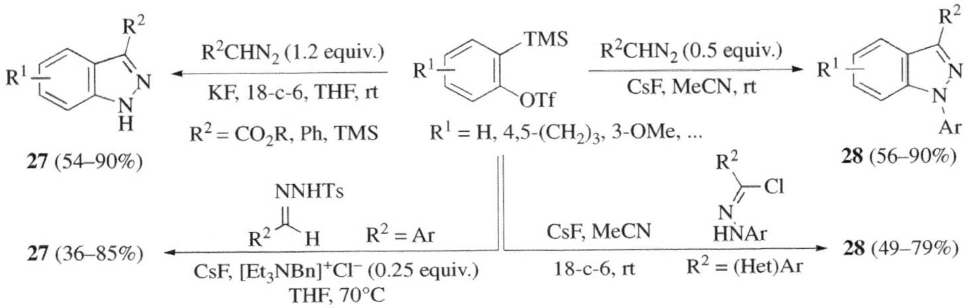

SCHEME 12.18 [3+2] Cycloadditions with diazoalkanes and nitrile imines.

reagents and the reaction conditions (Scheme 12.18) [27]. Later on, *N*-tosylhydrazones have been used as surrogates for diazo compounds, in order to obtain the same indazoles **27** with a wider range of substituents at C-3 [28]. In the same way, the synthesis of 1-aryl-1*H*-indazoles **28** has been described via the 1,3-dipolar cycloaddition of nitrile imines, generated *in situ* from hydrozonyl chlorides, with arynes (Scheme 12.18) [29].

Also in this field, Larock *et al.* have described that sydnones **29** undergo [3+2] dipolar cycloadditions with arynes affording 2*H*-indazoles **30** under mild reaction conditions. These heterocyclic derivatives are formed by a spontaneous extrusion of CO_2 from the initial cycloadduct (Scheme 12.19) [30]. The same authors have developed a simple route to pyrido[1,2-*b*]indazoles **31** via an aryne [3+2] cycloaddition with *N*-tosylpyridinium imides and subsequent fluoride-induced elimination of the tosylate anion (Scheme 12.19) [31].

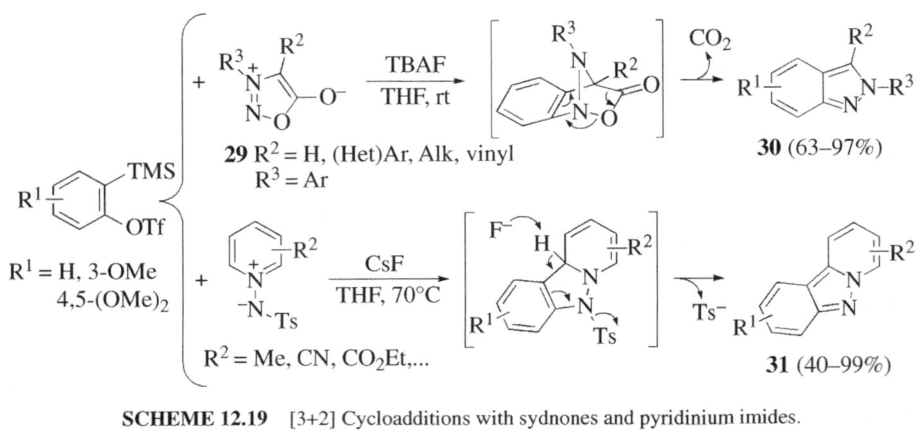

SCHEME 12.19 [3+2] Cycloadditions with sydnones and pyridinium imides.

12.4.3 [2+2] Cycloadditions with Alkenes

The [2+2] cycloaddition reactions of arynes with alkenes are highly interesting processes because they provide an alternative method for the construction of benzocyclobutenes. Due to the electrophilic nature of arynes, the reaction proceeds better with electron-rich alkenes, such as vinyl ethers, ketene acetals, and enamines. It has been reported that ketene silyl acetals (KSAs) are efficient counterparts for arynes in the regioselective synthesis of benzocyclobutenone derivatives. Suzuki *et al.* have developed a route to highly oxygenated tricyclobutabenzenes **33** based on the repeated regioselective [2+2] cycloaddition of arynes and KSA **32**. The observed regioselectivities are due to the directing effect of the methoxy group or by virtue of the ring strain (see Section 12.5.1) (Scheme 12.20) [32].

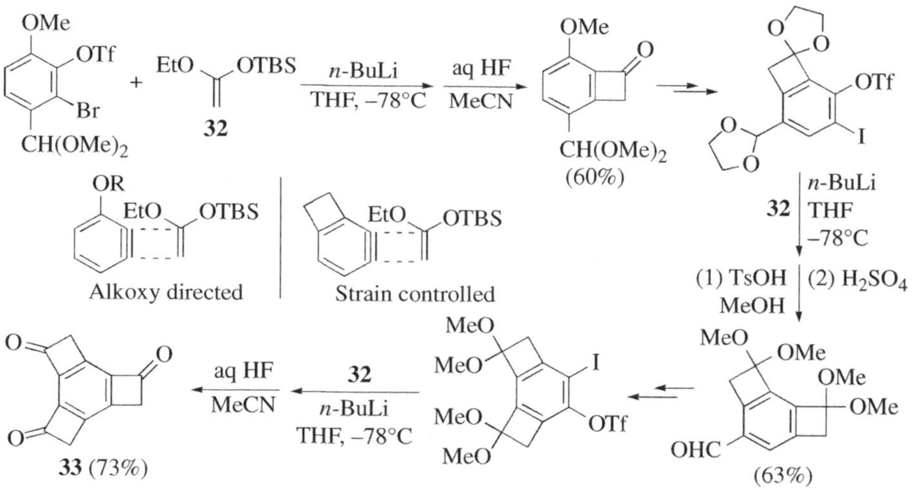

SCHEME 12.20 [2+2] Cycloadditions with ketene silyl acetals.

Enamides **34** have been also described as useful partners for [2+2] cycloaddition reactions with benzyne (**1**) to afford amido-functionalized benzocyclobutane derivatives **35**. By engaging this process with a pericyclic ring-opening intramolecular [4+2] cycloaddition, the total syntheses of chelidonine **37a** and norchelidonine **37b** from the corresponding aryne precursor and enamide **36** have been reported (Scheme 12.21) [33].

Limited success has been achieved for the insertion reactions of arynes into C–heteroatom double bonds that initially leads to the formal [2+2] cycloaddition-type adduct [34]. In these cases, the benzoxete intermediate tends to undergo ring opening at lower temperature, due to its high ring strain, leading to an *o*-quinone methide that evolves through different pathways. The first example of a synthetically useful insertion into the π-bond of carbonyl compounds was reported by Yoshida *et al.* who described the trapping of the aryne–aldehyde [2+2] adduct with an additional aryne giving rise to 9-arylxanthenes **38** (Scheme 12.22) [35].

However, most of the reported examples in this field refer to the insertion into the C=O π-bond of formamides. Miyabe *et al.* pioneered the use of DMF as a useful component in aryne-mediated reactions reporting the synthesis of 6-methoxysalicylaldehyde from 3-methoxybenzyne. Interestingly, they were also able to trap the four-membered ring intermediate **39** with dialkylzincs to afford *o*-aminoalkylphenols **40** (Scheme 12.23) [36].

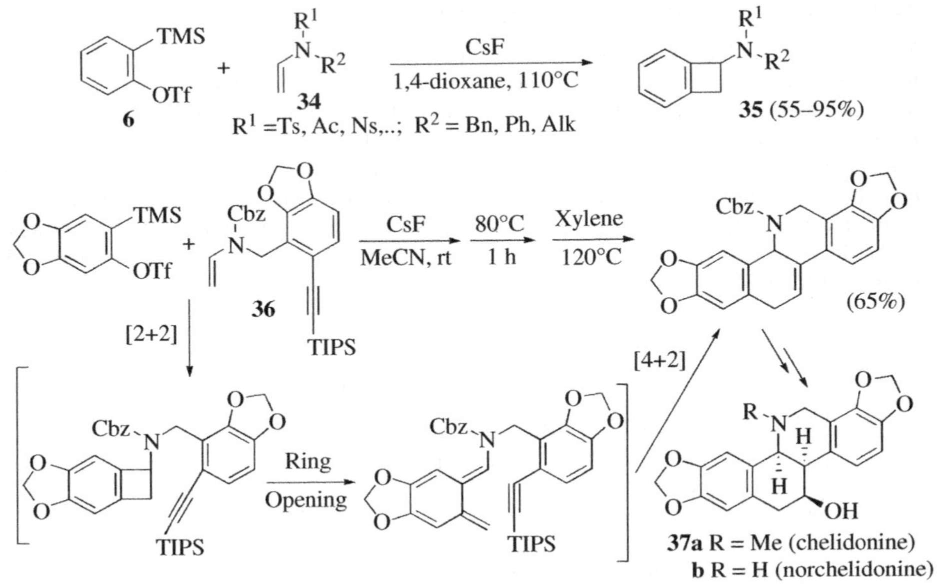

SCHEME 12.21 Enamide–aryne reactions.

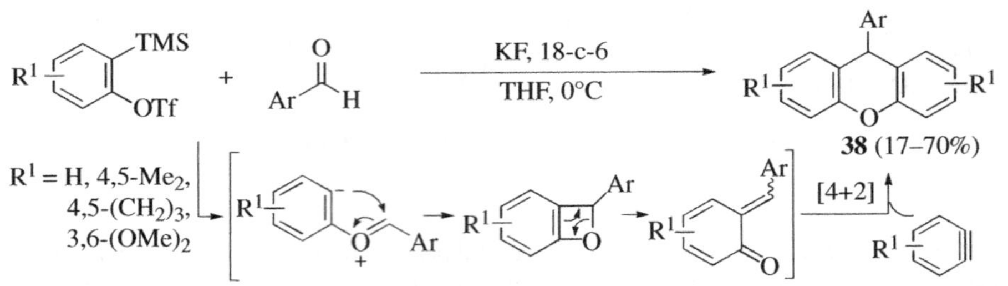

SCHEME 12.22 Aldehyde–aryne reactions.

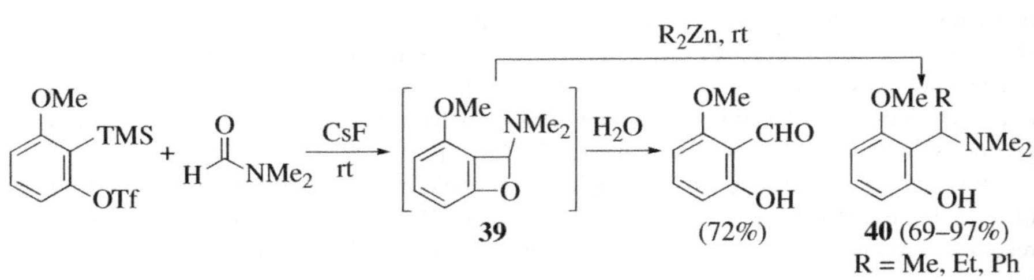

SCHEME 12.23 [2+2] Cycloadditions with DMF.

12.4.4 Ene Reactions

Few synthetically useful reports for aryne–ene reactions had been reported till Cheng *et al.* described the intermolecular reaction of arynes and alkynes to afford allene derivatives **41** (Scheme 12.24) [37]. More recently, the intermolecular ene reaction based on aryne and olefins has been also developed, tolerating the use of substituted chain olefins as well as cyclic olefins and affording allyl-substituted benzenes **42** in high yields (Scheme 12.24) [38]. Interestingly, a transition metal–free direct and regioselective *o*-arylation of anilines has been developed by Greaney *et al.* involving an aryne–heteroene reaction that accounts for the complete *ortho* selectivity, excluding *para* regioisomers. A possible second *o*-arylation is suppressed by using bulky *N*-tritylanilines **43** as starting materials (Scheme 12.24) [39].

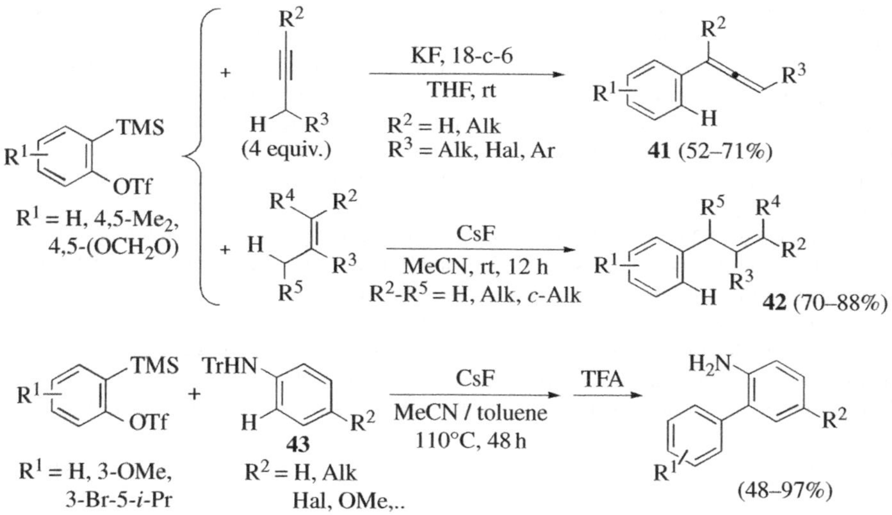

SCHEME 12.24 Intermolecular ene reactions of arynes.

Lautens *et al.* have developed the intramolecular version of this reaction, a highly stereo- and regioselective aryne–ene reaction, by using olefins as the ene component. The deprotonation strategy has been used to generate the intermediate arynes, which through an ene reaction evolve to a variety of carbo- and heterocycles **44**. By DFT calculations the authors proposed a concerted asynchronous transition state, and the reaction can be understood as a nucleophilic attack of the olefin onto the electrophilic aryne. It was also found that the geometry of the olefin (exclusive migration of the *trans*-allylic-H) as well as the electronic nature of the aryne play a crucial role in the success of these reactions (Scheme 12.25) [40].

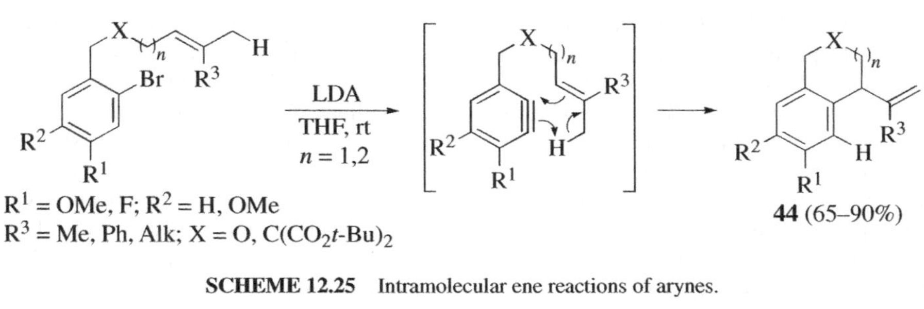

SCHEME 12.25 Intramolecular ene reactions of arynes.

12.5 NUCLEOPHILIC ADDITION REACTIONS TO ARYNES

12.5.1 Regioselectivity Issues for Functionalized Arynes

The addition reactions to unsymmetrically substituted arynes could lead to regioisomers, and the observed regiochemistry depends on the electronic and steric properties of the substituents, as well as their relative position to the aryne functionality. Reactions with one of the most employed functionalized aryne, 3-methoxybenzyne, usually occurred with high regiocontrol giving rise to products derived from attack at the C-1 position of the aryne, probably due to the higher electrophilicity of this carbon compared with the C-2 position. In addition, the steric repulsion between the methoxy group and the incoming nucleophile also favors the addition reaction at C-1. Other 3-substituted benzynes such as 1,2-naphthalyne, 3-phenylbenzyne, 3-chloro(fluoro)benzyne, and 3-halo-4-methoxybenzynes also show high regioselectivities in their addition reactions (Scheme 12.26). Moreover, reactions of an aryne possessing a fused four-membered ring are also highly regioselective due to the ring strain that causes a higher s-character of the C-3 orbital bound to C-2, rendering C-1 more electrophilic [41]. However, for 3-methylbenzyne, the steric effect is against the electron-donating inductive effect of the methyl group leading to moderate regioselectivity. On the other hand, 4-substituted arynes usually give mixtures of regioisomeric products. Stoltz *et al.* have prepared four highly substituted aryne precursors bearing several alkoxy groups in different relative positions, and have studied their reactivity in some aryne reactions showing complete regioselectivities (Scheme 12.26) [42]. The origin of the regioselectivity in the nucleophilic addition to substituted benzynes has been addressed by Garg and Houk, whose calculations show that polar effects can influence the distortion of an aryne and the transition state for a nucleophilic attack, but that ring fusion can also be a powerful directing effect. As under nucleophilic attack the transition state of an aryne is distorted to accommodate the developing negative charge (increasing s-character leads to decreased bond angle), attack is favored at the site that minimizes the change in distortion. This distortion model successfully explains the regioselectivities discussed earlier [43].

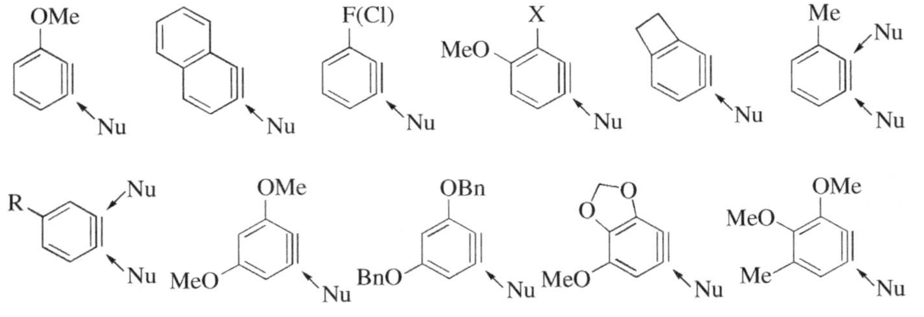

SCHEME 12.26 Regioselectivity in nucleophilic additions to arynes.

As established for 3-methylbenzyne, the repulsive steric interaction between the substituent and the nucleophile significantly reduces the expected regioselectivity of the addition based on the electronic effect. However, Akai *et al.* have reported the *ortho*-selective addition of primary alkyl amines to 3-trimethylsilylbenzynes **45** to afford 2-silylanilines **46**. The strong inductive effect of the silicate generated by the fluoride ion used for the aryne generation explains the unexpected high selectivity observed (Scheme 12.27) [44]. However, when an aromatic or secondary amine was used as the nucleophilic partner in these reactions, the addition products were obtained with low selectivities. Later on, Houk and Garg found, both experimentally and computationally, that if the incoming nucleophile is not bulky, the aryne distortion model works and the more electrophilic site of the aryne is attacked. However, using bulkier nucleophiles the attack on the more accessible

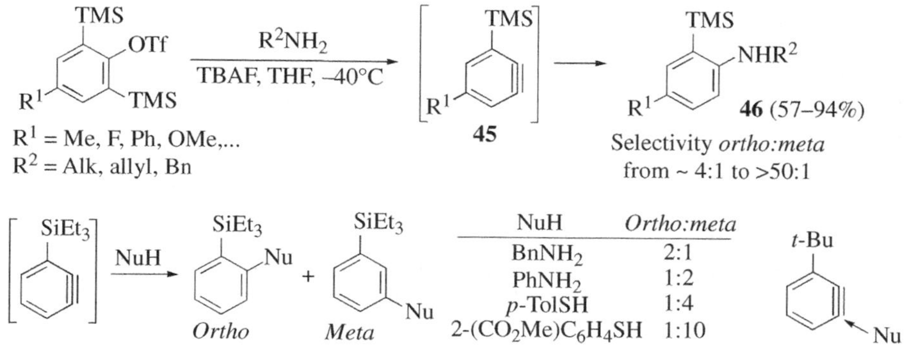

NuH	*Ortho:meta*
$BnNH_2$	2:1
$PhNH_2$	1:2
p-TolSH	1:4
$2\text{-}(CO_2Me)C_6H_4SH$	1:10

SCHEME 12.27 Fluoride-induced regioselective additions to arynes.

carbon is favored. These authors also reported that 3-*t*-butylbenzyne, which is sterically similar to 3-triethylsilylbenzyne but varies electronically, presents a minimal distortion, and so it undergoes sterically guided attack at C-1 to give *meta*-substituted products (Scheme 12.27) [45].

12.5.2 Proton Abstraction: Monosubstitution of the Aryne

Formally, the addition of anionic nucleophiles could be considered as the insertion of the aryne into the carbon- or heteroatom-metal bond whereas the addition of uncharged nucleophiles to arynes gives rise to a zwitterion. In this section, processes that involve reaction of the intermediate aryl anion or zwitterion with proton as electrophile, leading to monosubstitution of the aryne, will be described. Beller *et al.* had reported the formal cross-coupling reaction of aryl chlorides and amines, leading to arylamines by means of aryne chemistry. This transition metal–free reaction could be considered, at least in some cases, an efficient alternative to the well-established Pd-catalyzed Buchwald–Hartwig amination reaction. These reactions proceed efficiently with primary or secondary amines, including aromatic and aliphatic ones. As expected, *m*-anisidines were obtained from 2- or 3-chloroanisole. Interestingly, the reaction of 2- or 3-vinyl-substituted chlorobenzenes and anilines leads to indolines **47** through a tandem hydroamination–aryne amination reaction (Scheme 12.28) [46].

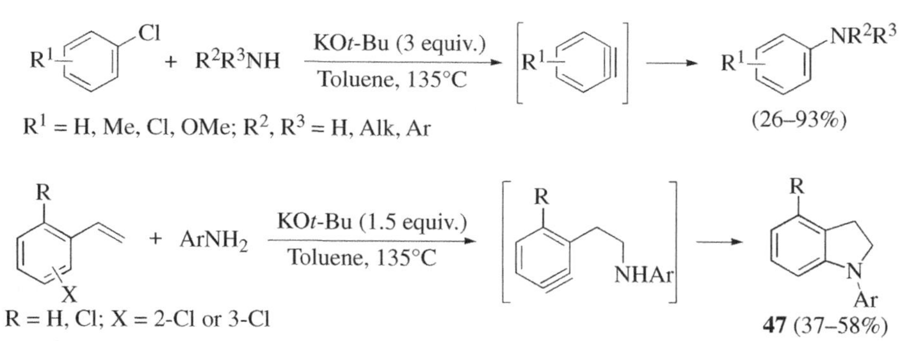

SCHEME 12.28 Synthesis of arylamines via arynes.

Based on this idea and looking for efficient transition metal–free procedures for the direct arylation of sp^2 C—H bonds, several years later Daugulis *et al.* developed the intramolecular arylation of phenols with aryl halides involving the initial formation of a benzyne intermediate followed by attack of the phenolate. The process implies a formal aromatic sp^2 C—H activation and leads to 6*H*-benzo[*c*]chromenes **48** as a mixture of regioisomers due to competitive cyclization via *ortho*- and/or *para*-positions (Scheme 12.29) [47]. This reaction clearly competes with the related Pd-catalyzed one. These authors have also developed the intermolecular trapping of arynes, generated from aryl chlorides, with phenols. When LTMP is used helicenes **49** are formed through the intermediacy of a benzocyclobutene, which upon ring opening and further addition to benzyne generates a tetracyclic intermediate. Subsequent opening of the four-membered ring gives a ketone that undergoes intramolecular nucleophilic attack followed by dehydration (Scheme 12.29) [48]. The desired *o*-arylation of phenols with aryl chlorides was achieved by employing NaO*t*-Bu as base in the presence of AgOAc. Related to this, the *ortho*-arylation of anilines by aryl halides and triflates has also been described using this strategy and providing a metal–free methodology, without protecting or directing groups on nitrogen, for accessing 2-arylanilines **50** (Scheme 12.29) [49].

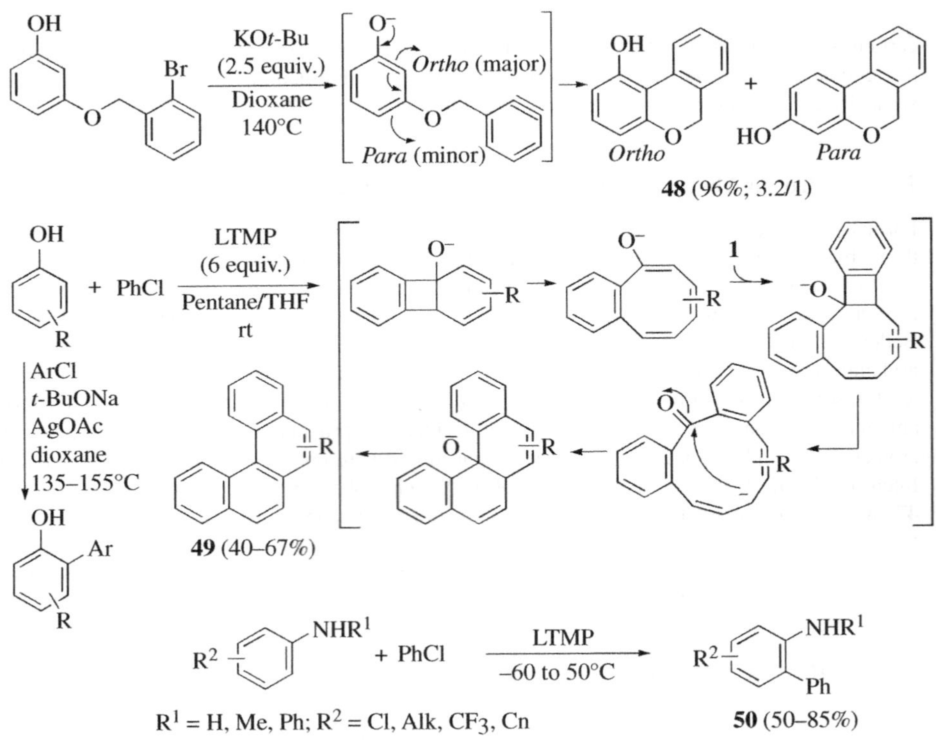

SCHEME 12.29 Arynes in transition metal–free arylation reactions.

Two arylation reactions via benzyne of *in situ* lithiated (hetero)aromatics or alkynes have also been developed. The base-promoted arylation of heterocycles and arenes by aryl chlorides and fluorides takes place at the most acidic C—H bond giving rise to arylated products **51** and **52**. For the transition metal–free arylation of terminal alkynes two set of conditions were developed tolerating different functional groups and leading to the corresponding alkynes **53** (Scheme 12.30) [50].

By using the fluoride-induced generation of arynes from *o*-(silyl)aryl triflates, Larock and Liu have developed the aryne-mediated *N*-arylation of amines, sulfonamides, and carbamates as well as the *O*-arylation of phenols and carboxylic acids. Remarkably, a wide variety of functional

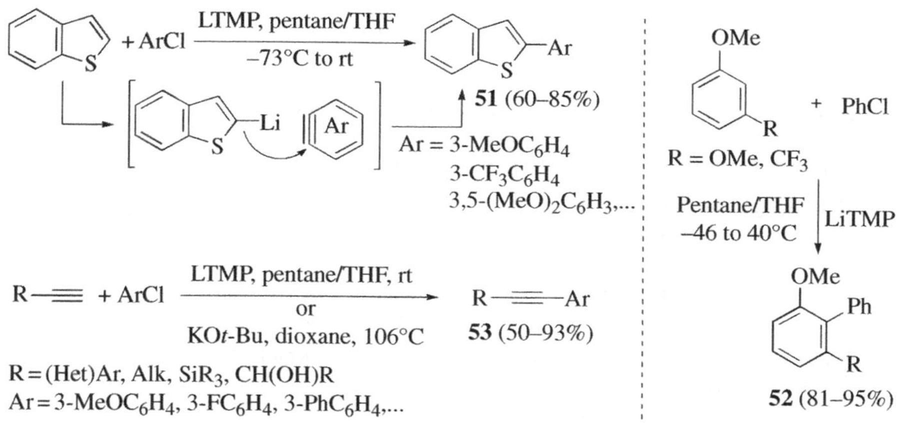

SCHEME 12.30 Aryne-mediated arylation of alkynes and (hetero)aromatics.

groups, including halides not compatible with Pd-catalyzed arylation procedures, are tolerated. In addition, monoarylated and diarylated amines can easily be obtained from primary amines by controlling the ratio of the reactants. High yields are usually obtained by applying this methodology that nicely complements well-established strategies based on transition metal-catalyzed processes (Scheme 12.31) [51].

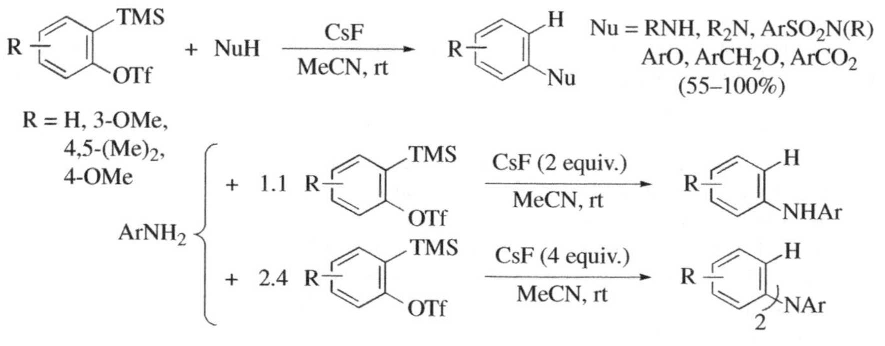

SCHEME 12.31 *N*- and *O*-arylation reactions.

12.5.3 Three-Component Reactions

When the nucleophilic addition to the aryne is followed by quenching of the resulting aryl anion with an external electrophile, an *o*-disubstituted arene derivative is obtained. The use of heteroatomic nucleophiles has lead to the development of a general procedure for the thio-(seleno-) and amino-magnesiation of arynes, being the resulting aryl magnesium species **54** capable of reacting with a variety of electrophiles and so, affording highly functionalized benzene derivatives such as **55–57** (Scheme 12.32) [52].

Although the intermolecular addition of organometallic species, such as organolithiums, to arynes is known since long, there are not many useful synthetic applications of this type of processes. During the total synthesis of clavilactone B, a three-component coupling reaction of an aryne, a Grignard reagent, and a carbonyl derivative has been developed. The reaction proceeds by addition of the Grignard reagent to the *in situ* generated aryne, followed by reaction of the newly formed aryl-magnesium species to the aldehyde leading to a key intermediate **58** (Scheme 12.33) [53].

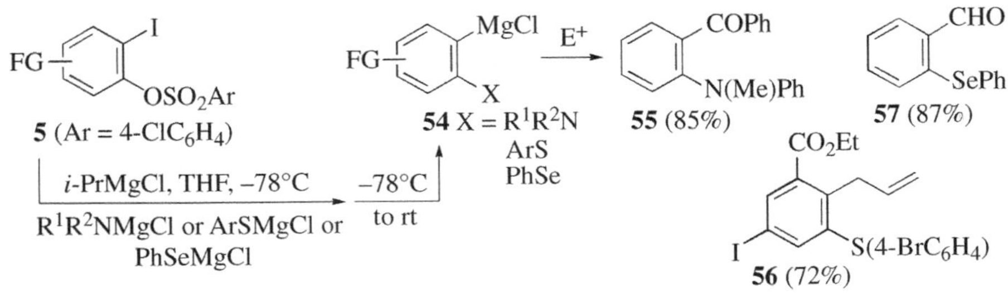

SCHEME 12.32 Addition of magnesium thiolates and amides to arynes.

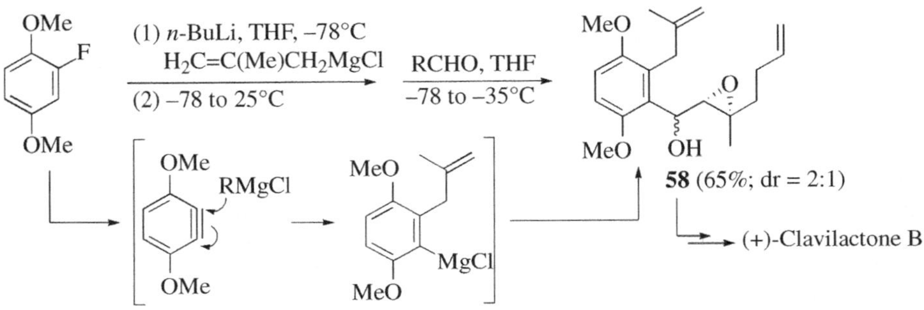

SCHEME 12.33 Three-component coupling with external electrophiles.

In this context, Leroux and Schlosser have elegantly developed a transition metal–free aryne–mediated aryl–aryl coupling methodology, based on the reaction of *in situ*–formed aryllithiums and arynes, with concomitant regeneration of a carbon–halogen bond (see Chapter 28) [54].

The intramolecular trapping of benzyne intermediates with tethered nucleophiles is a powerful strategy for the synthesis of benzo-fused heterocycles. In this field, 7-functionalized benzoxazoles **59** and benzothiazoles **60** have been prepared by anionic cyclization of the corresponding lithiated benzyne (thio)amides and (thio)carbamates, which were generated from 3-haloaniline derivatives (Scheme 12.34) [55]. Related to this, substituted indolines **61** have also been accessed by benzyne–mediated cyclization–functionalization from Boc-protected (2-bromo-4-methoxy)phenetylamines using $Mg(TMP)_2{\cdot}2LiCl$ as a base. This methodology has also been applied to the preparation of functionalized carbazoles (Scheme 12.34) [56].

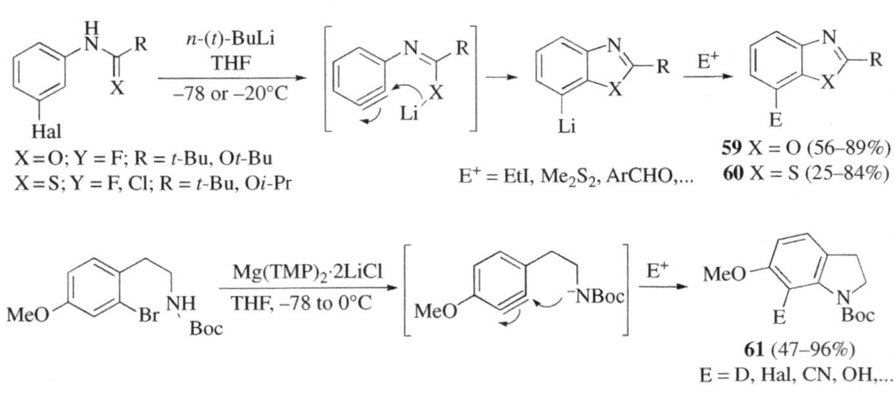

SCHEME 12.34 Intramolecular nucleophilic trapping of arynes.

Whereas the earlier examples about anionic cyclization reactions refer to heteroatom–centered nucleophiles, the formation and cyclization of aryne–tethered organolithiums have been independently explored by Bailey and Sanz and involves the formation of a new C—C bond in the cyclization step [57]. Sanz's group has developed a useful synthesis of C-4-functionalized indoles **62** by treatment of *N*-alkyl-*N*-(2-bromoallyl)-2-fluoroanilines with *t*-BuLi and further reaction with electrophiles. In addition, the authors took advantage from the intermediacy of a 3-methyleneindoline **63** to carry out subsequent Alder–ene reactions to prepare highly interesting 3,4-difunctionalized indoles **64** (Scheme 12.35) [57c]. The anionic cyclization of benzyne–tethered aryl-lithiums has been also reported to get, in a regioselective way, functionalized carbazoles **65**, dibenzofurans **66**, and dibenzothiophenes **67** (Scheme 12.35) [58]. The key step in all these reactions is a 5-*exo* addition of an aryl-lithium to the aryne. The viability of related 6-*exo* cyclizations was demonstrated with the synthesis of alkaloids trisphaeridine and *N*-methylcrinasiadine possessing a phenanthridine core [59].

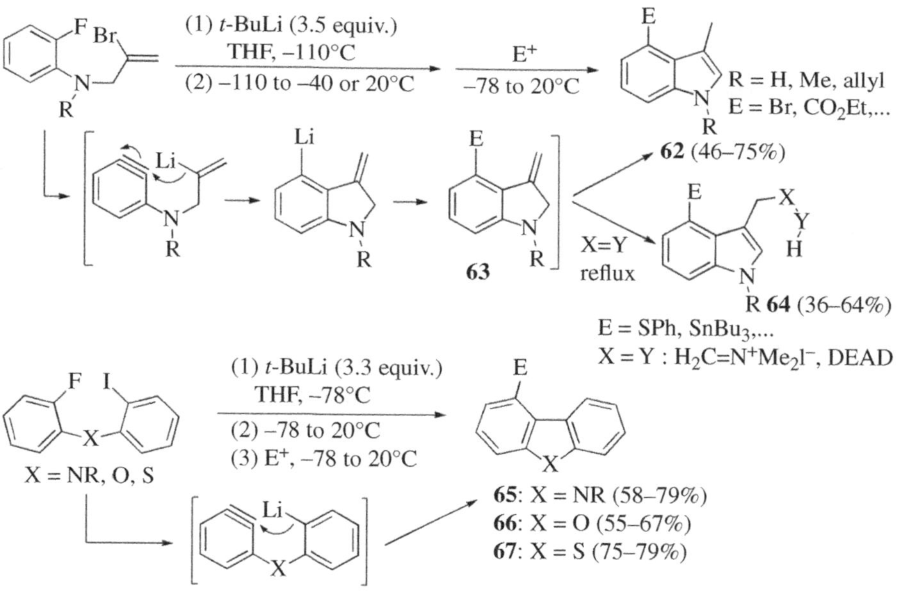

SCHEME 12.35 Indoles and dibenzoheteroles by cyclization of aryne–tethered organolithiums.

On the other hand, when using formally neutral nucleophiles for the addition onto the aryne, the subsequent trapping of the resulting zwitterions with an external electrophile leads to a disubstituted arene, which can further evolve through an intramolecular cyclization if the resulting cationic site is unsaturated. The first report on the use of isocyanide as a nucleophile in aryne MCR is by Yoshida *et al.* They described a three-component coupling reaction of arynes with isocyanides and aldehydes, leading to iminoisobenzofurans **68**. The reaction is initiated by nucleophilic addition of the isocyanide to the aryne and then, the resulting zwitterion is trapped by an aldehyde to give a new zwitterionic species that evolves by cyclization to yield the final product **68** (Scheme 12.36) [60]. The use of phenyl esters, as the electrophilic partner instead of aldehydes, has been more recently described and affords phenoxy iminoisobenzofurans **69**, which upon hydrolysis produces interesting *o*-ketobenzamides **70** (Scheme 12.36) [61].

Yoshida *et al.* have disclosed a MCR involving arynes, isocyanides, and alkynyl bromides, which affords *o*-iminobromoarenes **71** through the intermediacy of a bromine ate complex. Interestingly, these authors also found that cyclic ethers are able to participate in this process instead

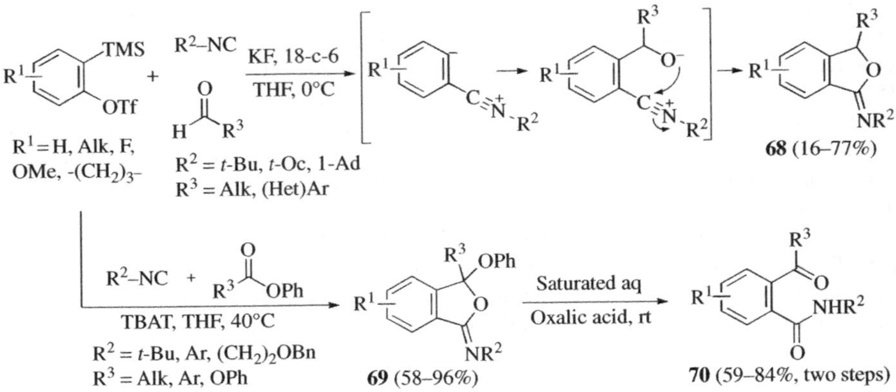

SCHEME 12.36 Isocyanides as nucleophiles in aryne MCR.

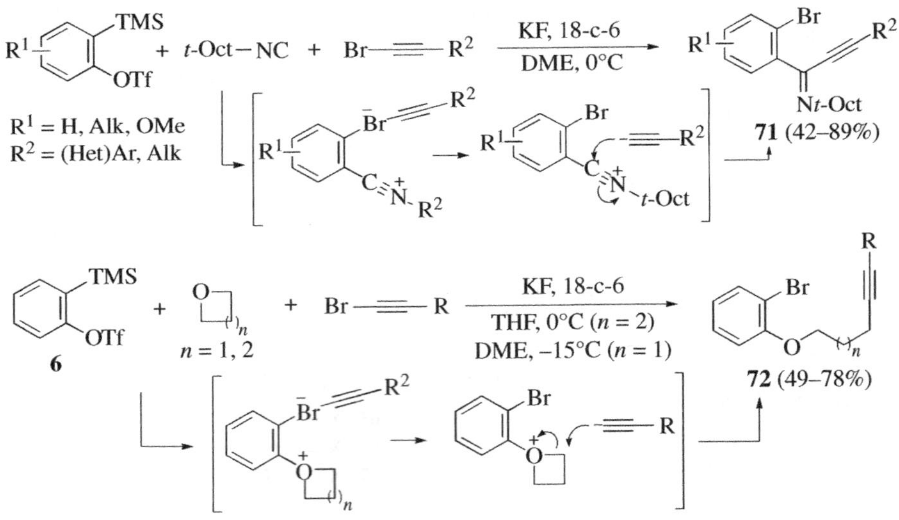

SCHEME 12.37 Bromoalkynes as electrophiles in aryne MCR.

of the isocyanide counterpart and so, *o*-bromophenyl ethers **72** can be obtained through a similar pathway (Scheme 12.37) [62].

Yoshida *et al.* have shown the potential of CO_2 as a C_1 source in aryne MCR developing a straightforward access to benzoxazinones **73** using imines as the nucleophilic component (Scheme 12.38) [63]. The same authors have used aminosilanes for the three-component coupling with arynes and aldehydes, including sulfonylimines, to provide 2-aminobenzhydrols and 2-aminobenzhydryl amines **74**. Benzoic acid plays a vital role in the reaction, probably favoring the formation of an amine from the aminosilane (Scheme 12.38) [64].

Cheng and Jeganmohan reported the first example of various *N*-heteroaromatic compounds, including pyridines, quinolines, and isoquinoline, as nucleophilic partners for the coupling with arynes and nitrile-containing solvents. The reported method allows the synthesis of *N*-arylated cyanomethyl–functionalized dihydro-*N*-heterocycles such as **75**. The negative charge of the zwitterionic species abstracts a proton from MeCN, and the resulting acetonitrile anion undergoes nucleophilic attack on the iminium intermediate (Scheme 12.39) [65]. In this field, Biju *et al.* have

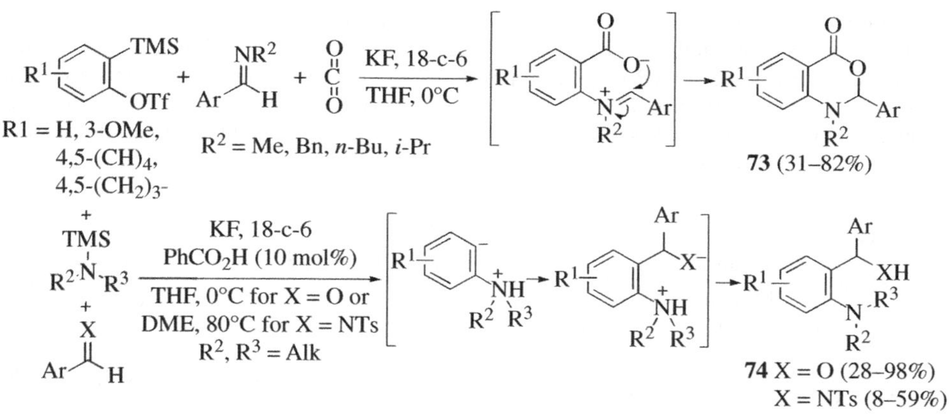

SCHEME 12.38 Three-component coupling of arynes with imines and aminosilanes as nucleophiles.

recently developed a new MCR involving arynes, *N*-heteroaromatics, and *N*-substituted isatins. When using isoquinoline as the nucleophilic partner, the reaction proceeds via the expected zwitterionic intermediate, leading to spiro-oxazino isoquinoline derivatives **76** with high diastereoselectivity (Scheme 12.39) [66].

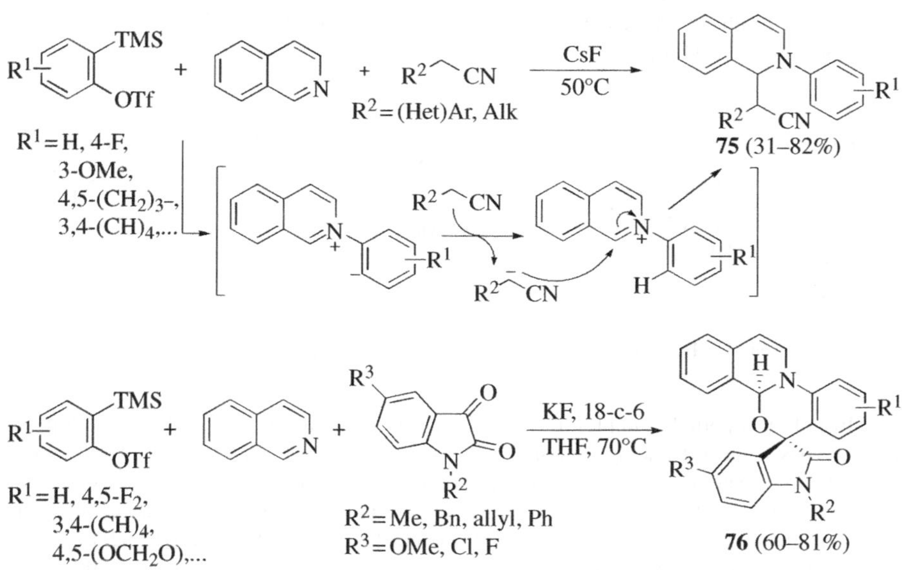

SCHEME 12.39 *N*-heteroaromatics as nucleophiles in aryne MCR.

12.5.4 Aryne Insertion Reactions into σ-Bonds

Different carbon- or heteroatom-centered nucleophilic positions σ-bonded to a suitable electrophilic functionality can undergo facile cleavage by adding to arynes and by further intramolecular nucleophilic substitution in the initially formed aryl anion or zwitterion intermediate. The overall process results in the insertion of an aryne into a nucleophilic–electrophilic σ-bond.

Insertion into Carbon–Carbon σ-Bonds. First examples of this concept were reported with anionic nucleophiles in the context of the addition of α-lithioalkyl and α-lithioarylacetonitriles to arynes. In 1984, Meyers and Pansegrau proposed a cyclization rearrangement mechanism to account for the formation of products **77** in the reaction of α-lithioacetonitriles to 3-oxazolylbenzyne. Initial attack of the nucleophile takes place at C-2 probably due to chelation of the lithium atom from the lithiated nitrile to the oxazoline moiety (Scheme 12.40) [67].

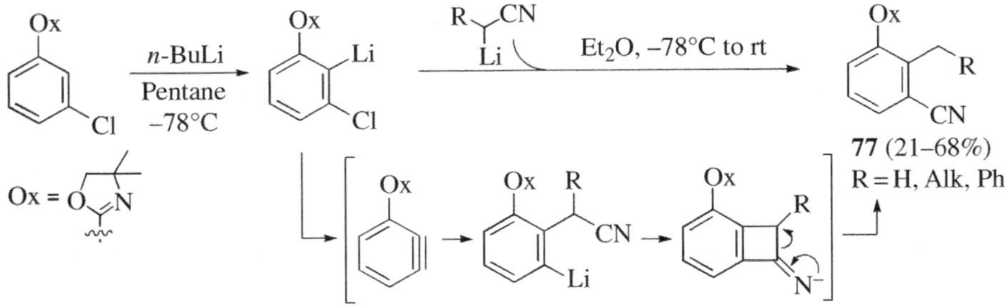

SCHEME 12.40 Aryne–mediated C—CN insertion reactions.

This σ-bond insertion reaction with anionic nucleophiles has been used in the context of the total synthesis of dynemicin A. The reaction of an aryne, generated from the LTMP metalation of a bromoarene, with lithiated dimethylmalonate leads to homophthalic ester **78** with a significant 71% yield (Scheme 12.41) [68].

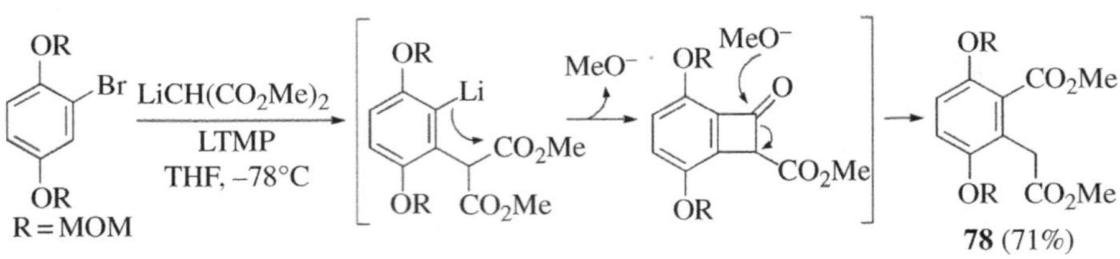

SCHEME 12.41 Synthesis of homophthalic esters via C—C σ-bond insertion.

Most of the reported examples for σ-bond aryne insertion reactions have been developed generating arynes from *o*-(trimethylsilyl)aryl triflates and using element–element σ-bond compounds bearing appropriate nucleophilic and electrophilic sites. Stoltz *et al.* have pioneered the insertion reaction of arynes into C—C σ-bonds using β-ketoesters to achieve the direct *o*-acyl–alkylation of arynes and providing a general access to new aromatic ketoesters **79** (Scheme 12.42) [69]. Although the formation of simple α-arylated products becomes competitive in the reaction with β-ketoesters bearing α-substitution, cyclic β-ketoesters efficiently undergo ring expansion to furnish medium-sized carbocycles. This strategy has been used by these authors for the synthesis of ketoester **80**—a key intermediate in the synthesis of (+)-amurensinine [70]— and also for the key step in the enantioselective synthesis of (−)-curvularin **81** (Scheme 12.42) [71].

In this field, Yoshida *et al.* have also made interesting contributions describing some insertion reaction of arynes into C—C σ-bonds. With *p*-toluenesulfonylacetonitrile or malononitrile two equivalents of the corresponding arynes react with the starting material. The overall process, which takes place through an initial C—C σ-bond insertion, involves the formation of three C—C and one C—H bonds, and affords diverse diarylmethane derivatives **82** in a straightforward manner (Scheme 12.43) [72].

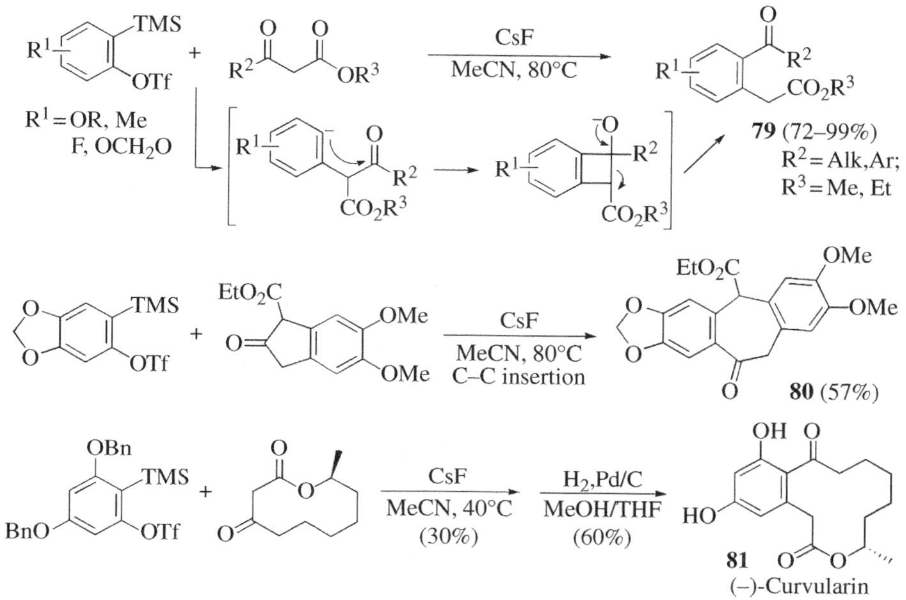

SCHEME 12.42 Insertion reactions into C—C σ-bonds. Stoltz's synthesis of (−)-*curvularin*.

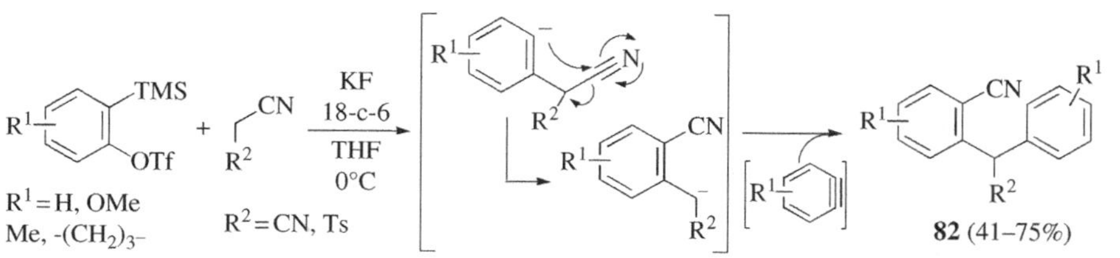

SCHEME 12.43 σ-bonds insertion reactions with nitriles.

Insertion into Heteroatom–Heteroatom σ-Bonds. Kunai *et al.* have demonstrated the addition of stannyl sulfides, aminosilanes, and tributyltin fluoride to arynes leading to thiostannylation [73], aminosilylation [74], and fluorostannylation [75] of arynes, respectively. So, a variety of 2-(arylthio) arylstannanes **83**, 2-silylanilines **84**, and 2-fluoroarylstannanes **85** can be obtained in modest to high yields. As expected, with unsymmetrical arynes only good regioselectivities were obtained with 3-methoxy, 3-halo, and 3-phenylbenzyne (Scheme 12.44).

Insertion into Carbon–Heteroatom σ-Bonds. Shirakawa *et al.* reported the first example of insertion of arynes into a C—N bond. Using ureas as substrates a wide variety of 1-amino-2-(aminocarbonyl)arenes **86**, including 1,4-benzodiazepine and 1,5-benzodiazocine derivatives, can be prepared (Scheme 12.45) [76]. Related to this work, Larock *et al.* developed the intermolecular C—N addition of amides across the aryne triple bond, providing access to (*o*-anilino)trifluoroacetophenones **87**. The presence of the CF_3 substituent resulted to be crucial for the insertion process, which were limited to secondary substrates (Scheme 12.45) [77].

Apart from C—N, other C—heteroatom σ-bond insertion reactions have been described in this field, such as the carbohalogenation of arynes with acylhalides, which enables simultaneous introduction of acyl- and halogen-functional groups into adjacent positions of arenes **88** (Scheme 12.46) [78]. Changing the solvent from acetonitrile to less acidic THF in order to prevent protonation of the

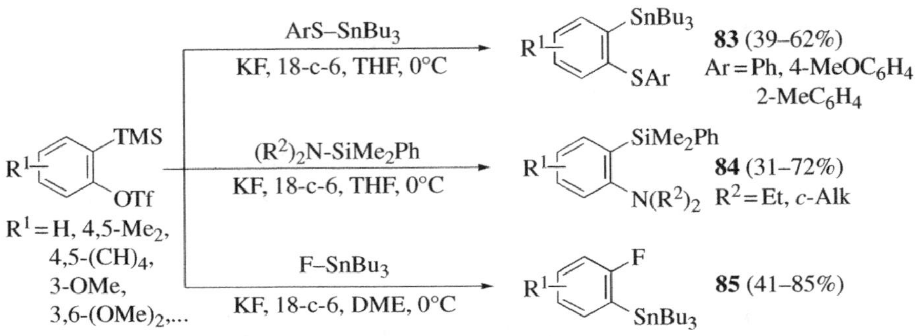

SCHEME 12.44 Aryne insertion reactions into heteroatom–heteroatom σ-bonds.

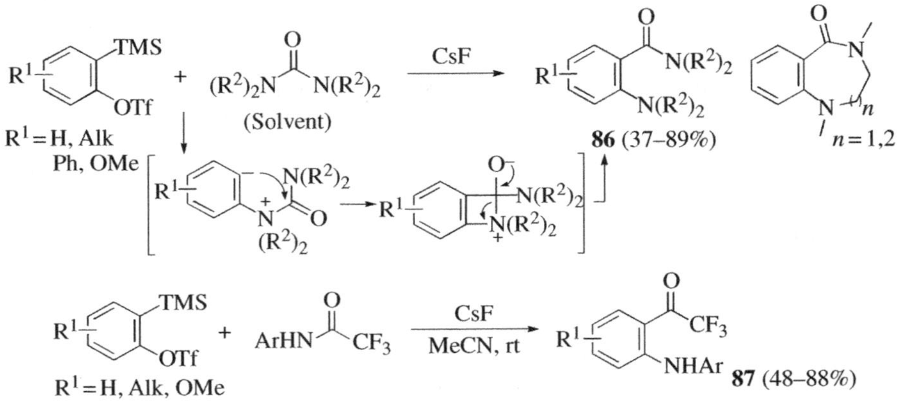

SCHEME 12.45 Insertion reactions into C—N σ-bonds.

zwitterionic intermediate, Larock and Dubrovskiy have achieved the C—O addition of nonconjugated carboxylic acids to arynes instead of the previously reported O—H addition. So, a selection of *o*-hydroxyaryl ketones **89** was prepared in moderate to good yields (Scheme 12.46) [79]. The reaction of *o*-(silyl)aryl triflates with ethoxyacetylene has also been described, affording 2-ethoxyethynylaryl derivatives **90** through a formal insertion of the aryne into the C(sp)—O(sp^3) bond of the alkyne. The observed regioselectivity suggests a mechanism that starts with the nucleophilic addition of the triple bond instead of the oxygen atom (Scheme 12.46) [80].

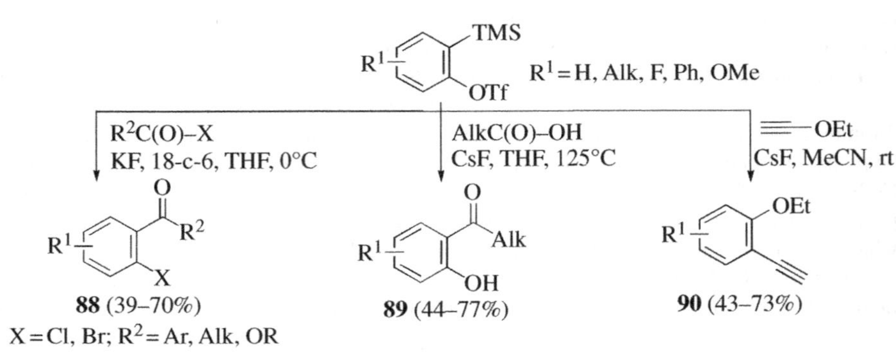

SCHEME 12.46 Other σ-bonds insertion reactions of arynes.

12.5.5 Aryne Annulation

When the nucleophilic and the electrophilic positions of the reagent confronted to the aryne are not σ-bonded, a cascade intermolecular nucleophilic addition–intramolecular electrophilic cyclization of arynes can take place. The fragmentation step, which is crucial for the insertion reaction of arynes into σ-bonds, is not involved in annulation processes because the intermediate obtained from the cyclization is usually a stable five- or six-membered ring system.

[2+2] Annulation Reactions. The trapping of arynes with ketone enolates has been investigated by the Caubère group over several years as a useful method for preparing strained aromatic polycyclic compounds. For instance, treatment of bromobenzene with enolates of five- to seven-membered ring ketones, and complex bases such as $NaNH_2$/NaO*t*-Bu leads to benzocyclobutenol derivatives like **91** (Scheme 12.47) [81]. In addition, the use of LTMP as a base for generating the aryne allows access to 1-arylbenzocyclobutenols **92** from simple ketones (Scheme 12.47) [82].

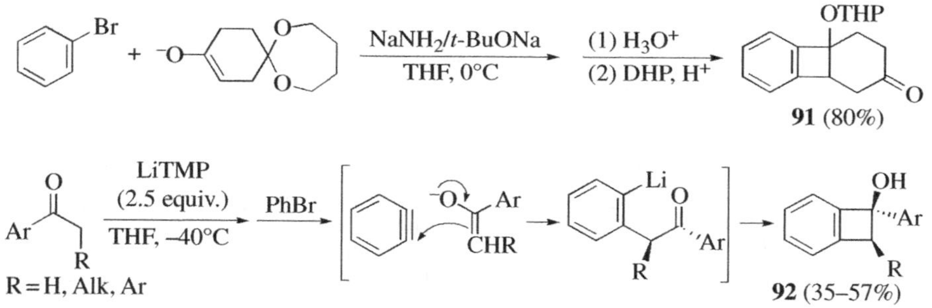

SCHEME 12.47 Trapping of arynes with ketone enolates.

[3+2] Annulation Reactions. Larock *et al.* have described the synthesis of different heterocyclic systems using a [3+2] annulation approach. For instance, pharmaceutically important pyrido[1,2-*a*]indole derivatives such as **93** are easily accessible from 2-substituted pyridines and aryne precursors (Scheme 12.48) [83]. More recently, the 1*H*-indazole skeleton has been accessed through a [3+2] annulation from arynes and hydrazones. The reaction with *N*-arylhydrazones leads to 1,3-disubstituted indazoles **94** through an annulation–oxidation process (Scheme 12.48). The use of *N*-tosylhydrazones also affords 3-substituted-*N*(H)-indazoles, although probably via a [3+2] cycloaddition (see Scheme 12.18) with *in situ* generated diazo compounds [84].

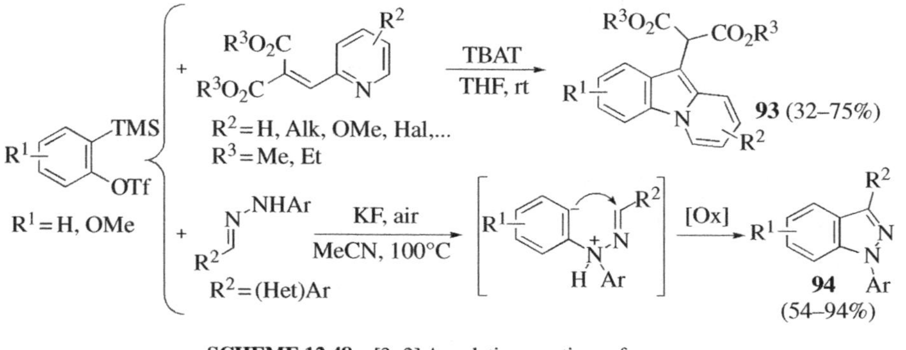

SCHEME 12.48 [3+2] Annulation reactions of arynes.

[4+2] Annulation Reactions. Larock *et al.* have developed a one-pot synthesis of xanthones, thioxanthones, and acridones **95a–c** employing salicylates, thiosalicylates, and *o*-aminobenzoates, respectively, as starting materials. In the last case, the reaction also proceeds with tertiary amines suggesting that the neutral amine itself is nucleophilic enough for this transformation (Scheme 12.49) [85]. The same authors have also developed a related methodology for the construction of acridines **96** in one step from commercially available *o*-aminoacetophenones and *o*-aminobenzophenones (Scheme 12.49) [86]. Similar heterocyclic systems had been previously obtained by Kim *et al.* but using 5-arylthianthrenium perchlorates as benzyne precursors, which are not commercially available and potentially explosives [41b].

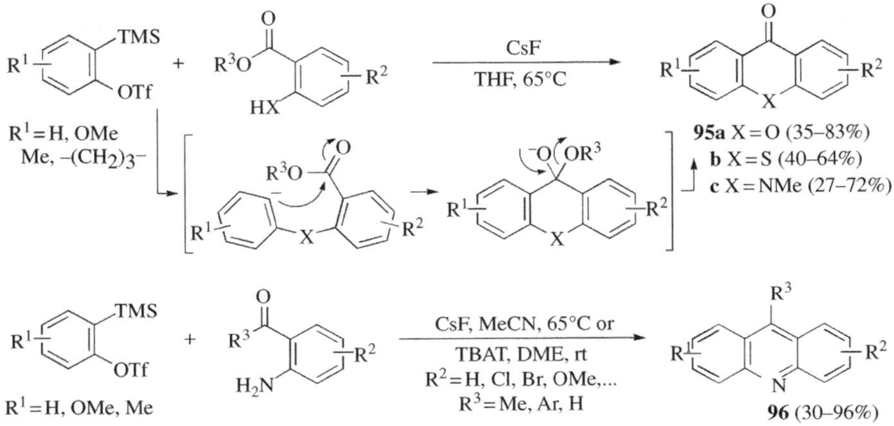

SCHEME 12.49 [4+2] Annulation reactions of arynes.

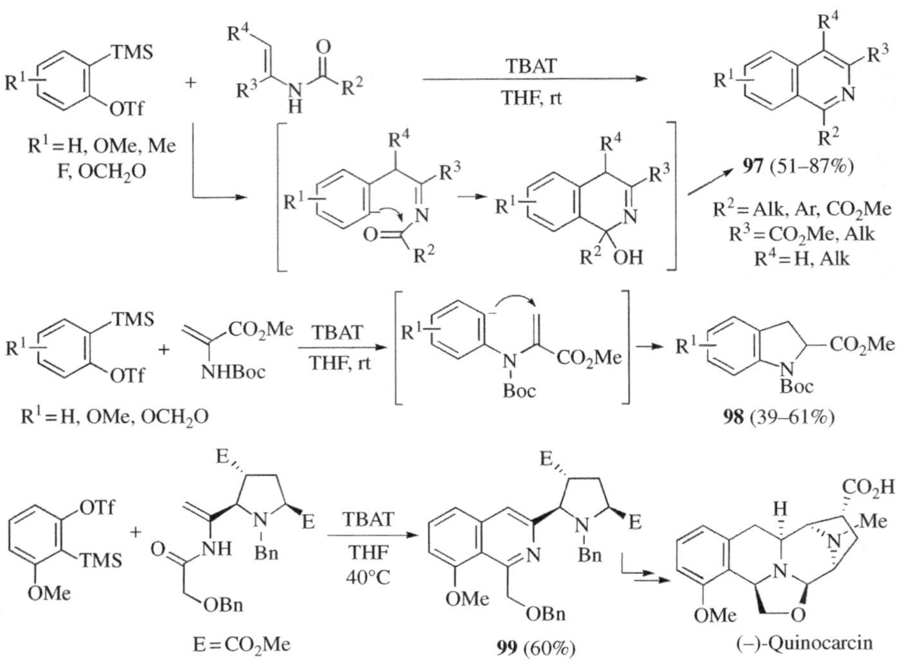

SCHEME 12.50 Annulation reactions of arynes with enamides.

Stoltz *et al.* have developed a method for the synthesis of isoquinolines **97** by the coupling reaction of *N*-acyl dehydroamino esters with arynes and have applied this strategy to a concise total synthesis of papaverine. They have also found that the corresponding carbamates ($R^2 = Ot$-Bu) display an orthogonal mode of reactivity leading to indolines **98** by initial nucleophilic attack of the nitrogen atom (Scheme 12.50) [87]. These authors have applied this interesting methodology to the development of an efficient total synthesis of (−)-quinocarcin. Thus, the reported aryne annulation reaction was used for accessing an elaborated intermediate **99** from the corresponding *N*-acyl enamine and 3-methoxybenzyne (Scheme 12.50) [88].

12.6 TRANSITION METAL–CATALYZED REACTIONS OF ARYNES

In 1998, Pérez, Guitián, *et al.* disclosed the Pd-catalyzed trimerization of arynes [89], demonstrating that these intermediates are able to undergo transition metal–catalyzed transformations and so, greatly expanding the potential of synthetic applications of arynes. The success of these processes depends on both the metal catalyst and the method of the aryne generation.

12.6.1 Cyclotrimerization of Arynes

The first metal–catalyzed reaction of arynes was described by Guitián in the late 1990s consisting in the cyclotrimerization of benzyne to yield triphenylene **100** (Scheme 12.51) [89]. These and other authors have extended this process to the trimerization of a variety of arynes, leading to polycyclic aromatic hydrocarbons (PAHs).

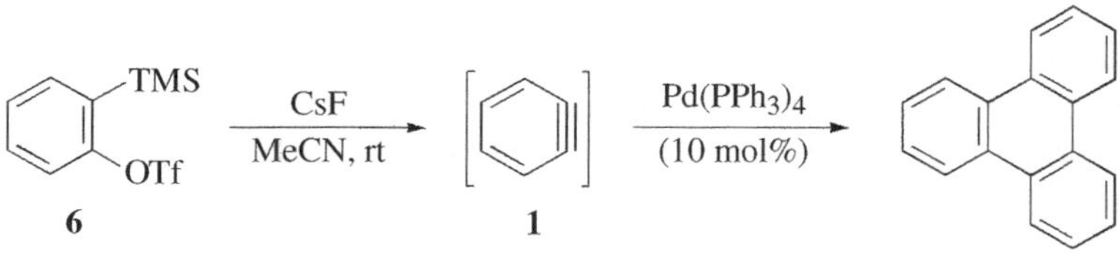

100 (83%)

SCHEME 12.51 Pd-catalyzed triphenylene synthesis.

12.6.2 Cocyclization of Arynes with Alkynes

Guitián *et al.* also reported the first cocyclization of arynes with alkynes such as dimethyl acetylenedicarboxylate (DMAD) leading to phenanthrenes **101** or naphthalenes **102**. The selectivity of the process can be tuned by an appropriate choice of the catalyst (Scheme 12.52) [90]. These authors have also described the first enantioselective version of this process, using a Pd-(BINAP) complex as catalyst, although the nonracemic 9,12-dimethoxypentahelicene **103** was isolated in a modest yield but with reasonable ee (Scheme 12.52) [91]. Diynes can also participate in [2+2+2] cycloaddition with arynes as demonstrated by Mori *et al.* in the total synthesis of taiwanins C and E **104a** and **b** (Scheme 12.52) [92].

12.6.3 Cocyclization of Arynes with Alkenes

Cheng *et al.* have demonstrated the [2+2+2] cocyclization of two arynes with bicyclic alkenes giving the corresponding norbornane anellated 9,10-dihydrophenanthrene derivatives **105**. The reaction probably proceeds by initial formation of a palladacycle from one aryne molecule and the

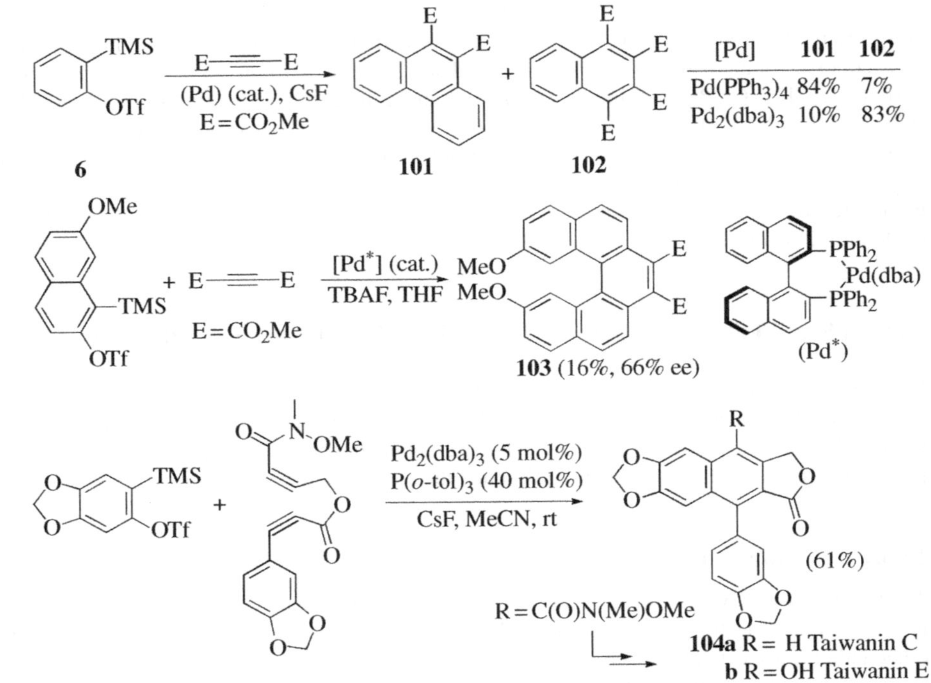

SCHEME 12.52 Aryne–alkyne [2+2+2] cocyclizations.

alkene, followed by a second aryne insertion and final reductive elimination (Scheme 12.53) [93]. Peña *et al.* showed for the first time that arynes can undergo Pd-catalyzed cocyclization with acyclic alkenes, selectively furnishing dihydrophenanthrenes **106** or *o*-olefinated biaryls **107** depending on the catalytic system. The most likely mechanism involves the formation of a palladacycle derived from two arynes (Scheme 12.53) [94].

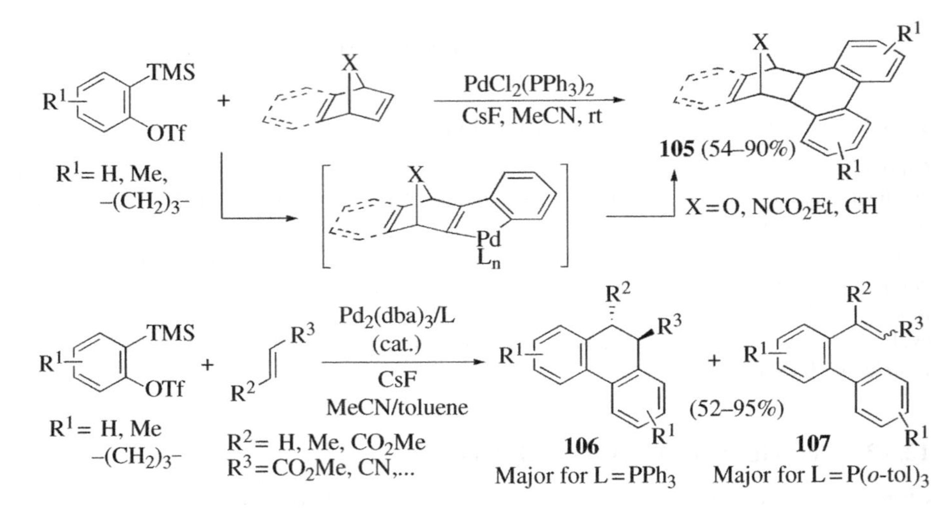

SCHEME 12.53 Aryne–alkene [2+2+2] cocyclizations.

12.6.4 Cocyclization of Arynes, Alkenes, and Alkynes

Xie and Qiu have reported the first example of the [2+2+2] carboannulation of arynes, activated alkenes, and arynes. Whereas Pd catalysts promote the two-component benzyne–alkene–benzyne cyclization, under Ni catalysis the three-component reaction is favored leading to 1,2-dihydronaphthalenes **108** from readily available materials. In this case, the catalytic cycle is likely initiated by oxidative coupling of aryne and alkene on Ni^0 to form a nickelacycle that undergoes subsequent insertion of the alkyne into the Ni–C(aryl) bond to give a seven-membered intermediate (Scheme 12.54) [95].

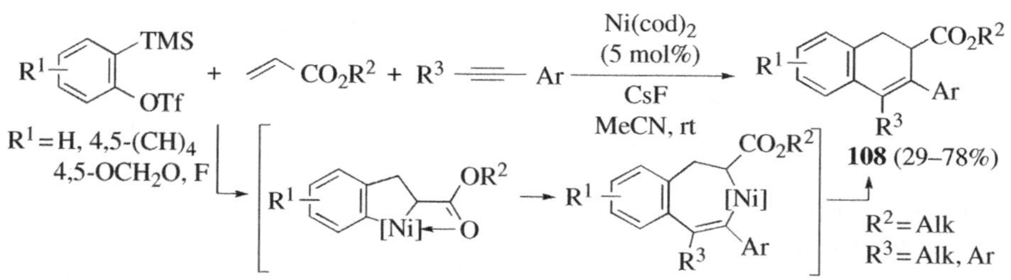

SCHEME 12.54 Ni-catalyzed aryne cocyclizations with alkynes and alkenes.

12.6.5 Intermolecular Carbopalladation of Arynes

The aryne C—C triple bond is highly reactive toward palladation processes, and the initially formed arylpalladium intermediate can undergo a variety of transformations with an appropriate reagent in an intermolecular, giving rise to 1,2-functionalized benzene derivatives, or in an intramolecular manner, leading to benzo-fused (hetero)carbocyclic systems.

Carbopalladation of Arynes by π-Allylpalladium and Related Complexes. Pioneering work from Yamamoto established that π-allylpalladium species are particularly effective for the carbopalladation of arynes leading to an arylpalladium intermediate **109**, which can be employed in three-component couplings with a second aryne to give phenanthrenes such as **110**, or with alkynes to produce naphthalenes such as **111**. In the first case, **109** evolves through insertion of another benzyne (**1**) molecule (carbopalladation of **1** by an arylpalladium) and further carbopalladation of the pendant alkene to afford a tricyclic species that undergoes β-hydride elimination (Scheme 12.55) [96].

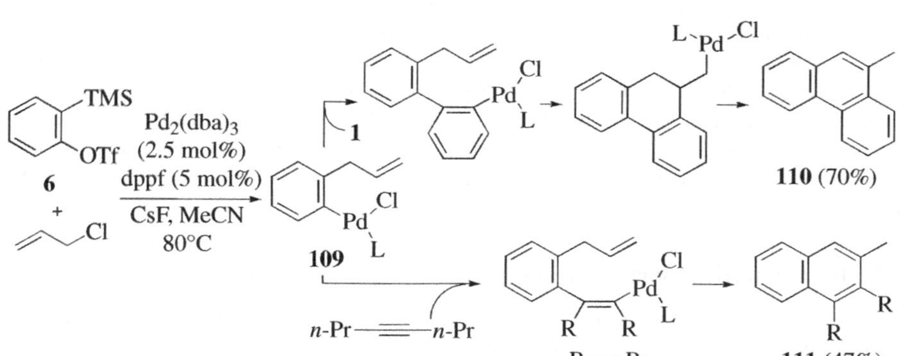

SCHEME 12.55 Carbopalladation of arynes by π-allylpalladium species.

Using this strategy, Cheng *et al.* demonstrated that initial carbopalladation with π-allylpalladium complexes could be engaged with a Stille coupling using alkynyl stannanes to afford 1-allyl-2-alkynylbenzenes **112** [97], or with a Suzuki coupling to produce 1-allyl-2-arylbenzenes **113** (Scheme 12.56) [98]. Benzyl bromides have been also used as the initial carbopalladation electrophile, as an alternative to allyl chlorides, with a Heck reaction used to trap the arylpalladium and leading to benzylphenyl derivatives **114** (Scheme 12.56) [99].

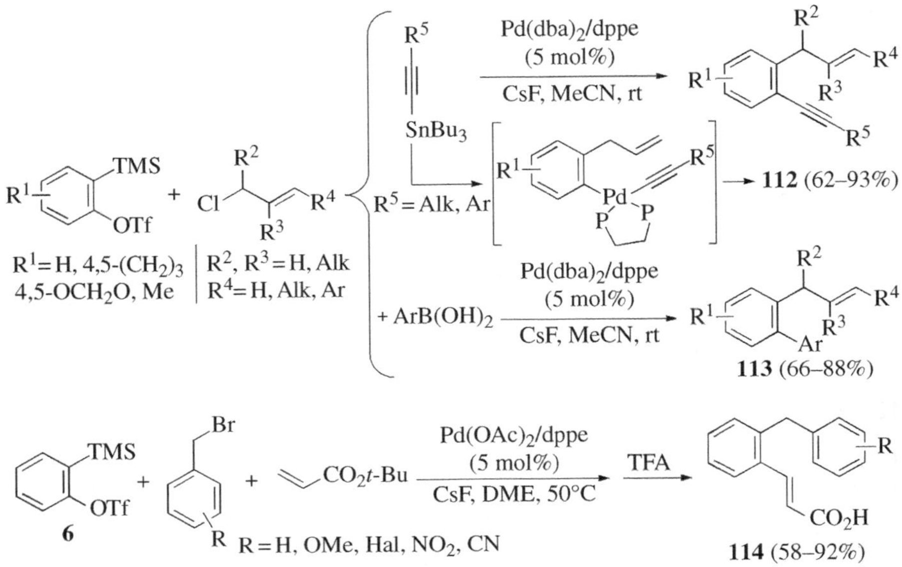

SCHEME 12.56 Combined Pd-catalyzed strategies initiated by carbopalladation of arynes.

Carbopalladation of Arynes by Arylpalladium Complexes: Pd-Catalyzed Annulation Reactions of Arynes. Whereas the Pd-catalyzed annulation of alkynes by aryl halides has proven to be an effective method for the construction of a wide variety of hetero- and carbocycles; similar carbopalladation of arynes were unprecedented until Larock and Zhang reported the synthesis of fluoren-9-ones **115** through the annulation of arynes by *o*-haloarenecarboxaldehydes (Scheme 12.57) [100]. Larock's group has also developed a Pd-catalyzed annulation of arynes by 2-halobiaryls, affording polycyclic aromatic and heteroaromatic compounds such as **116** from simple starting materials. Larock and Cheng have independently reported the related double annulation of arynes by simple aryl halides as an efficient route to functionalized triphenylenes (Scheme 12.57) [101].

In this field, Cheng *et al.* have also developed the carbocyclization reaction of aromatic iodides, bicyclic alkenes, and arynes to afford various annulated 9,10-dihydrophenanthrene derivatives **117** (Scheme 12.58) [102]. A related cross-coupling of arynes with aryl iodides but using alkynes as the third component has been described by Larock and Liu, furnishing substituted phenanthrenes **118** (Scheme 12.58) [103].

12.6.6 Catalytic Insertion Reactions of Arynes into σ-Bonds

Yoshida, Kunai *et al.* have reported some reactions of this type, complementing the transition metal–free aryne insertion into σ-bonds. Mechanistically, these reactions could proceed via carbopalladation of arynes by the palladium complex arising from the initial oxidative

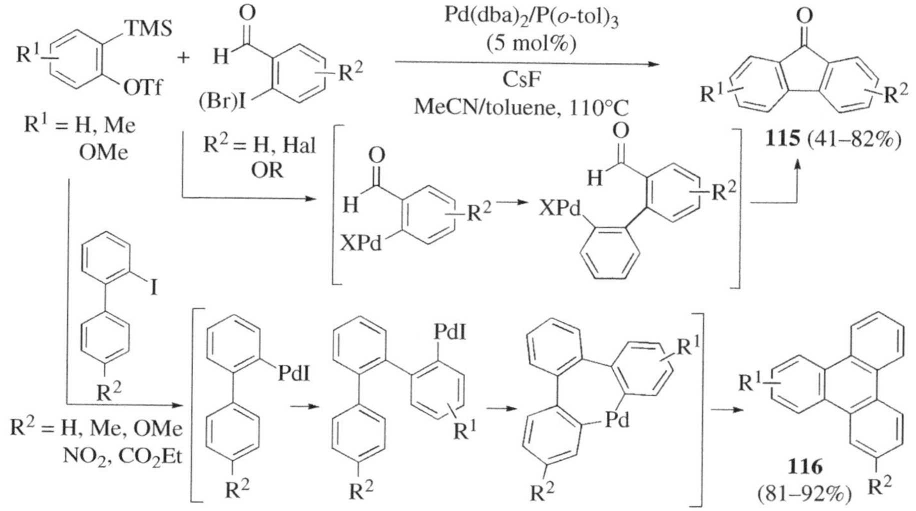

SCHEME 12.57 Carbopalladation of arynes by aryl halides.

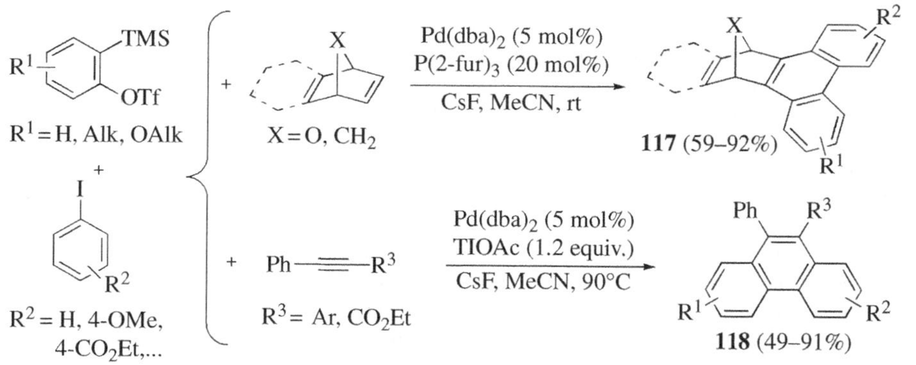

SCHEME 12.58 Three-component reactions initiated by carbopalladation of arynes.

addition of the corresponding σ-bond to the Pd^0 catalyst. However, an alternative prior interaction of the Pd^0 with the aryne to produce a three-membered palladacycle has been also proposed in some cases. Carbostannylation of arynes has been achieved using a Pd–iminophosphine complex that allows arynes to insert into the C—Sn bond of alkynyl and vinylstannanes, leading to *o*-functionalized arylstannanes **119** (Scheme 12.59) [104]. The distannylation of arynes has also been accomplished in the presence of a Pd-*t*-OctNC complex furnishing 1,2-distannylarenes **120** (Scheme 12.59) [105]. Interestingly, the same catalytic system that catalyzes the distannylation reaction also promotes the addition reaction of Si—Si σ-bonds to arynes. Thus, benzo-annulated disilacarbocycles **121** have been synthesized from cyclic disilanes (Scheme 12.59) [106].

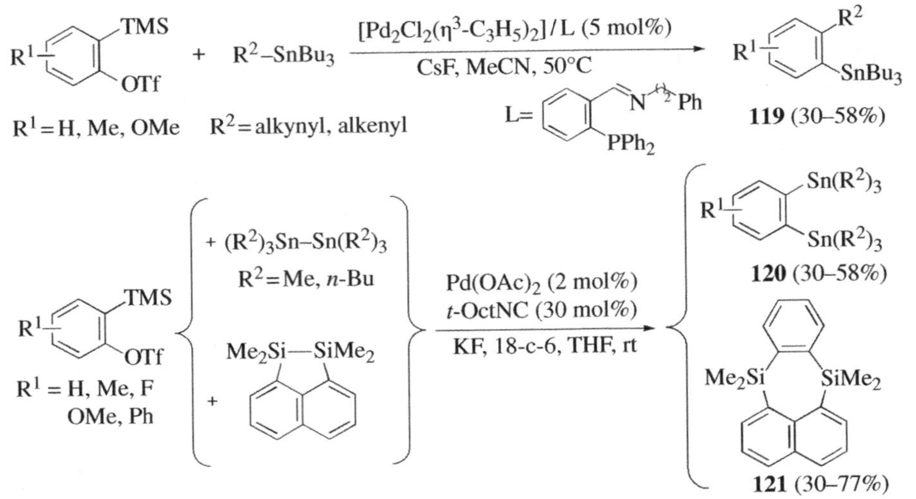

SCHEME 12.59 Pd-catalyzed Sn–C_{sp}, Sn–C_{sp^2}, and Sn–Sn aryne insertion reactions.

12.7 CONCLUSION

Although arynes have been known for more than half a century, the last few years have brought about a renaissance in their chemistry. The discovery of new methods for the easy and selective generation of arynes under mild reaction conditions is probably the main reason for this recent dramatic increase in aryne-based synthetic methodologies. The introduction of *o*-(trimethylsilyl)aryl triflates as aryne precursors, using the high affinity of fluoride for silicon and the leaving group ability of triflates, allows the generation of arynes under conditions compatible with the presence of a wide variety of nucleophilic reagents as well as with transition metal complexes. Apart from *o*-elimination processes, the intramolecular [4+2] cyclization of a diyne with an alkyne, the hexadehydro-Diels–Alder reaction has been recently described as a new reagent-free method for accessing substituted polycyclic arynes.

The strained nature of the aryne triple bond results in an easily accessible LUMO and so, arynes are excellent electrophiles that react with different anionic or neutral nucleophilic species. This low-lying LUMO also makes arynes prone to participate in a variety of pericyclic reactions. In addition, arynes are also able to undergo interesting transition metal–catalyzed processes. The increased availability of functionalized aryne precursors as well as the possibility of generating arynes under reagent-free conditions is allowing the development of new aryne-mediated synthetic strategies and the discovery of new reactivity patterns, which will expand the repertoire of chemistry that is actually possible with arynes.

We hope that the representative and recent developments selected for this chapter have provided an appropriate background for the present topic, and exciting new transformations are expected to appear in the near future.

ABBREVIATIONS

Boc *t*-Butoxycarbonyl
HDDA Hexadehydro-Diels–Alder
HMPA Hexamethylphosphoramide

KSA	Ketene silyl acetal
LTMP	Lithium 2,2,6,6-tetramethylpiperidide
LUMO	Lowest unoccupied molecular orbital
MCR	Multicomponent coupling reaction
NIS	*N*-Iodosuccinimide
TBAF	Tetrabutylammonium fluoride
TBAT	Tetrabutylammonium difluorotriphenylsilicate
TBS	*t*-Butyldimethylsilyl
Tf_2O	Trifluoromethanesulfonic anhydride
TMS	Trimethylsilyl

REFERENCES

[1] (a) Hoffmann, R. W. (1967) *Dehydrobenzene and Cycloalkynes*; Academic Press: New York; (b) Pellissier, H. and Santelli, M. (2003) *Tetrahedron*, **59**, 701–730. Recent reviews:(c) Sanz, R. (2008) *Org. Prep. Proced. Int.*, **40**, 215–291; (d) Chen, Y. and Larock, R. C. (2009) In *Modern Arylation Methods*; Ackermann, L., Ed.; Wiley-VCH: Weinheim; Ch. 12, pp 401–473; (e) Bhunia, A., Reddy Yetra, S., and Biju, A. T. (2012) *Chem. Soc. Rev.*, **41**, 3140–3152; (f) Gampe, C. M. and Carreira, E. M. (2012) *Angew. Chem. Int. Ed.*, **51**, 3766–3778; (g) Tadross, P. M. and Stoltz, B. M. (2012) *Chem. Rev.*, **112**, 3550–3577; (h) Dubrovskiy, A. V., Markina, N. A., and Larock, R. C. (2013) *Org. Biomol. Chem.*, **11**, 191–218; (i) Wu, C. and Shi, F. (2013) *Asian J. Org. Chem.*, **2**, 116–125; (j) Pérez, D., Peña, D., and Guitián, E. (2013) *Eur. J. Org. Chem.*, 5981–6013.

[2] Roberts, J. D., Simmons, H. E., Carlsmith, L. A., and Vaughan, C. W. (1953) *J. Am. Chem. Soc.*, **75**, 3290–3291.

[3] Radziszewski, J. G., Hess, B. A., Jr., and Zahradnik, R. (1992) *J. Am. Chem. Soc.*, **114**, 52–57.

[4] Orendt, A. M., Facelli, J. C., Radziszewski, J. G., Horton, W. J., Grant, D. M., and Michl, J. (1996) *J. Am. Chem. Soc.*, **118**, 846–852.

[5] Kitamura, T. (2010) *Aust. J. Chem.*, **63**, 987–1001.

[6] Hoye, T. R., Baire, B., Niu, D., Willoughby, P. H., and Woods, B. P. (2012) *Nature*, **490**, 208–212.

[7] Riggs, J. C., Ramírez, A., Cremeens, M. E., Bashore, C. G., Candler, J., Wirtz, M. C., Coe, J. W., and Collum, D. B. (2008) *J. Am. Chem. Soc.*, **130**, 3406–3412.

[8] Uchiyama, M., Miyoshi, T., Kajihara, Y., Sakamoto, T., Otani, Y., Ohwada, T., and Kondo, Y. (2002) *J. Am. Chem. Soc.*, **124**, 8514–8515.

[9] Hamura, T., Arisawa, T., Matsumoto, T., and Suzuki, K. (2006) *Angew. Chem. Int. Ed.*, **45**, 6842–6844.

[10] Sapountzis, I., Lin, W., Fischer, M., and Knochel, P. (2004) *Angew. Chem. Int. Ed.*, **43**, 4364–4366.

[11] Himeshima, Y., Sonoda, T., and Kobayashi, H. (1983) *Chem. Lett.*, **12**, 1211–1214.

[12] (a) Peña, D., Cobas, A., Pérez, D., and Guitián, E. (2002) *Synthesis*, 1454–1458; (b) Bronner, S. M. and Garg, N. K. (2009) *J. Org. Chem.*, **74**, 8842–8843.

[13] Kitamura, T., Yamane, M., Inoue, K., Todaka, M., Fukatsu, N., Meng, Z., and Fujiwara, Y. (1999) *J. Am. Chem. Soc.*, **121**, 11674–11679.

[14] Friedman, L. and Logullo, F. M. (1963) *J. Am. Chem. Soc.*, **85**, 1549.

[15] Birkett, M. A., Knight, D. W., and Mitchell, M. B. (1994) *Synlett*, 253–254.

[16] Niu, D., Willoughby, P. H., Woods, B. P., Baire, B., and Hoye, T. R. (2013) *Nature*, **501**, 531–534.

[17] Sörgel, S., Azap, C., and Reibig, H.-U. (2006) *Eur. J. Org. Chem.*, 4405–4418.

[18] Masson, E. and Schlosser, M. (2005) *Eur. J. Org. Chem.*, 4401–4405.

[19] Akai, S., Ikawa, T., Takayanagi, S-.i., Morikawa, Y., Mohri, S., Tsubakiyama, M., Egi, M., Wada, Y., and Kita, Y. (2008) *Angew. Chem. Int. Ed.*, **47**, 7673–7676.

[20] Kaelin, D. E., Jr., Sparks, S. M., Plake, H. R., and Martin, S. F. (2003) *J. Am. Chem. Soc.*, **125**, 12994–12995.

[21] McManus, H. A., Fleming, M. J., and Lautens, M. (2007) *Angew. Chem. Int. Ed.*, **46**, 433–436.

[22] Bhojgude, S. S., Kaicharla, T., Bhunia, A., and Biju, A. T. (2012) *Org. Lett.*, **14**, 4098–4101.

[23] (a) Kaicharla, T., Bhojgude, S. S., and Biju, A. T. (2012) *Org. Lett.*, **14**, 6238–6241; (b) For previous work using substituted cyclohexadienes as diene counterparts, see: Buszek, K. R. and Bixby, D. L. (1995) *Tetrahedron Lett.*, **50**, 9129–9132.

[24] Li, J., Wang, N., Li, C., and Jia, X. (2012) *Org. Lett.*, **14**, 4994–4997.

[25] Bhojgude, S. S., Bhunia, A., Gonnade, R. G., and Biju, A. T. (2014) *Org. Lett.*, **16**, 676–679.

[26] Dockendorff, C., Sahli, S., Olsen, M., Milhau, L., and Lautens, M. (2005) *J. Am. Chem. Soc.*, **127**, 15028–15029.

[27] Jin, T. and Yamamoto, Y. (2007) *Angew. Chem. Int. Ed.*, **46**, 3323–3325.

[28] Li, P., Zhao, J., Wu, C., Larock, R. C., and Shi, F. (2011) *Org. Lett.*, **13**, 3340–3343.

[29] Spiteri, C., Keeling, S., and Moses, J. E. (2010) *Org. Lett.*, **12**, 3368–3371.

[30] Wu, C., Fang, Y., Larock, R. C., and Shi, F. (2010) *Org. Lett.*, **12**, 2234–2237.

[31] Zhao, J., Wu, C., Li, P., Ai, W., Chen, H., Wang, C., Larock, R. C., and Shi, F. (2011) *J. Org. Chem.*, **76**, 6837–6843.

[32] Hamura, T., Ibusuki, Y., Uekusa, H., Matsumoto, T., and Suzuki, K. (2006) *J. Am. Chem. Soc.*, **128**, 3534–3535.

[33] (a) Feltenberger, J. B., Hayashi, R., Tang, Y., Babiash, E. S. C., and Hsung, R. P. (2009) *Org. Lett.*, **11**, 3666–3669; (b) Ma, Z.-X., Feltenberger, J. B., and Hsung, R. P. (2012) *Org. Lett.*, **14**, 2742–2745.

[34] Okuma, K. (2012) *Heterocycles*, **85**, 515–543.

[35] Yoshida, H., Watanabe, M., Fukushima, H., Ohshita, J., and Kunai, A. (2004) *Org. Lett.*, **6**, 4049–4051.

[36] Yoshioka, E., Kohtani, S., and Miyabe, H. (2010) *Org. Lett.*, **12**, 1956–1959.

[37] Jayanth, T. T., Jeganmohan, M., Cheng, M.-J., Chu, S.-Y., and Cheng, C.-H. (2006) *J. Am. Chem. Soc.*, **128**, 2232–2233.

[38] Chen, Z., Liang, J., Yin, J., Yu, G.-A., and Liu S. H. (2013) *Tetrahedron Lett.*, **54**, 5785–5787; See also ref. [25].

[39] Pirali, T., Zhang, F., Miller, A. H., Head, J. L., McAusland, D., and Greaney, M. F. (2012) *Angew. Chem. Int. Ed.*, **51**, 1006–1009.

[40] Candito, D. A., Dobrovolsky, D., and Lautens, M. (2012) *J. Am. Chem. Soc.*, **134**, 15572–15580.

[41] (a) Hamura, T., Ibusuki, Y., Sato, K., Matsumoto, T., Osamura, Y., and Suzuki, K. (2003) *Org. Lett.*, **5**, 3551–3554; (b) Yoon, K., Ha, S. M., and Kim, K. (2005) *J. Org. Chem.*, **70**, 5741–5744.

[42] Tadross, P. M., Gilmore, C. D., Bugga, P., Virgil, S. C., and Stoltz, B. M. (2010), *Org. Lett.*, **12**, 1224–1227.

[43] Im, G-Y. J., Bronner, S. M., Goetz, A. E., Paton, R. S., Cheong, P. H.-Y., Houk, K. N., and Garg, N. K. (2010) *J. Am. Chem. Soc.*, **132**, 17933–17944.

[44] Ikawa, T., Nishiyama, T., Shigeta, T., Mohri, S., Morita, S., Takayanagi, S.-i., Terauchi, Y., Morikawa, Y., Takagi, A., Ishikawa, Y., Fujii, S., Kita, Y., and Akai, S. (2011) *Angew. Chem. Int. Ed.*, **50**, 5674–5677.

[45] Bronner, S. M., Mackey, J. L., Houk, K. N., and Garg, N. K. (2012) *J. Am. Chem. Soc.*, **134**, 13966–13969.

[46] Beller, M., Breindl, C., Riermeier, T. H., and Tillack, A. (2001) *J. Org. Chem.*, **66**, 1403–1412.

[47] Bajracharya, G. B. and Daugulis, O. (2008) *Org. Lett.*, **10**, 4625–4628.

[48] Truong, T. and Daugulis, O. (2013) *Chem. Sci.*, **4**, 531–535.

[49] Truong, T. and Daugulis, O. (2012) *Org. Lett.*, **14**, 5964–5967.

[50] (a) Truong, T. and Daugulis, O. (2011) *J. Am. Chem. Soc.*, **133**, 4243–4245; (b) Truong, T. and Daugulis, O. (2011) *Org. Lett.*, **13**, 4172–4175.

[51] (a) Liu, Z. and Larock, R. C. (2003) *Org. Lett.*, **5**, 4673–4675; (b) Liu, Z. and Larock, R. C. (2004) *Org. Lett.*, **6**, 99–102; (c) Liu, Z. and Larock, R. C. (2006) *J. Org. Chem.*, **71**, 3198–3209.

[52] Lin, W., Sapountzis, I., and Knochel, P. (2005) *Angew. Chem. Int. Ed.*, **44**, 4258–4261.

[53] Larrosa, I., Da Silva, M. I., Gómez, P. M., Hannen, P., Ko, E., Lenger, S. R., Linke, S. R., White, A. J. P., Wilton, D., and Barrett, A. G. M. (2006) *J. Am. Chem. Soc.*, **128**, 14042–14043.

[54] For pioneering work, see: Leroux, R. and Schlosser, M. (2002) *Angew. Chem. Int. Ed.*, **41**, 4272–4274.

[55] (a) Clark, R. D. and Caroon, J. M. (1982) *J. Org. Chem.*, **47**, 2804–2806; (b) Stanetty, P. and Krumpak, B. (1996) *J. Org. Chem.*, **61**, 5130–5133.

[56] Noji, T., Fujiwara, H., Okano, K., and Tokuyama, H. (2013) *Org. Lett.*, **15**, 1946–1949.

[57] (a) Bailey, W. F. and Longstaff, S. C. (1998) *J. Org. Chem.*, **63**, 432–433; (b) Barluenga, J., Fañanás, F. J., Sanz, R., and Fernández, Y. (1999) *Tetrahedron Lett.*, **40**, 4865–4868; (c) Barluenga, J., Fañanás, F. J., Sanz, R., and Fernández, Y. (2002) *Chem. Eur. J.*, **8**, 2034–2046.

[58] Sanz, R., Fernández, Y., Castroviejo, M. P., Pérez, A., and Fañanás, F. J. (2006) *J. Org. Chem.*, **71**, 6291–6294.

[59] Sanz, R., Fernández, Y., Castroviejo, M. P., Pérez, A., and Fañanás, F. J. (2007) *Eur. J. Org. Chem.*, 62–69.

[60] Yoshida, H., Fukushima, H., Ohshita, J., and Kunai, A. (2004) *Angew. Chem. Int. Ed.*, **43**, 3935–3938.

[61] Allan, K. M., Gilmore, C. D., and Stoltz, B. M. (2011) *Angew. Chem. Int. Ed.*, **50**, 4488–4491.

[62] Yoshida, H., Asatsu, Y., Mimura, Y., Ito, Y. Ohshita, J., and Takaki, K. (2011) *Angew. Chem. Int. Ed.*, **50**, 9676–9679.

[63] Yoshida, H., Fukushima, H., Ohshita, J., and Kunai, A. (2006) *J. Am. Chem. Soc.*, **128**, 11040–11041.

[64] Morishita, T., Fukushima, H., Yoshida, H., Ohshita, J., and Kunai, A. (2008) *J. Org. Chem.*, **73**, 5452–5457.

[65] Jeganmohan, M. and Cheng, C.-H. (2006) *Chem. Commun.*, 2454–2456.

[66] Bhunia, A., Roy, T., Pachfule, P., Rajamohanan, P. R., and Biju, A. T. (2013) *Angew. Chem. Int. Ed.*, **52**, 10040–10043.

[67] Meyers, A. I. and Pansegrau, P. D. (1984) *Tetrahedron Lett.*, **35**, 2941–2944.

[68] Shair, M. D., Yoon, T. Y., Mosny, K. K., Chou, T. C., and Danishefsky, S. J. (1996) *J. Am. Chem. Soc.*, **118**, 9509–9525.

[69] Tambar, U. K. and Stoltz, B. M. (2005) *J. Am. Chem. Soc.*, **127**, 5340–5341.

[70] Tambar, U. K., Ebner, D. C., and Stoltz, B. M. (2006) *J. Am. Chem. Soc.*, **128**, 11752–11753.

[71] Tadross, P. M., Virgil, S. C., and Stoltz, B. M. (2010) *Org. Lett.*, **12**, 1612–1614.

[72] Yoshida, H., Watanabe, M., Morishita, T., Ohshita, J., and Kunai, A. (2007) *Chem. Commun.*, 1505–1507.

[73] Yoshida, H., Terayama, T., Oshita, J., and Kunai, A. (2004) *Chem. Commun.*, 1980–1981.

[74] Yoshida, H., Minabe, T., Oshita, J., and Kunai, A. (2005) *Chem. Commun.*, 3454–3456.

[75] Yoshida, H., Yoshida, R., and Takaki, K. (2013) *Angew. Chem. Int. Ed.*, **52**, 8629–8632.

[76] Yoshida, H., Shirakawa, E., Honda, Y., and Hiyama, T. (2002) *Angew. Chem. Int. Ed.*, **41**, 3247–3249.

[77] Liu, Z. and Larock, R. C. (2005) *J. Am. Chem. Soc.*, **127**, 13112–13113.

[78] Yoshida, H., Mimura, Y., Ohshita, J., and Kunai, A. (2007) *Chem. Commun.*, 2405–2407.

[79] Dubrovskiy, A. V. and Larock, R. C. (2010) *Org. Lett.*, **12**, 3117–3119.

[80] Laczkowski, K. Z., García, D., Peña, D., Cobas, A., Pérez, D., and Guitián, E. (2011) *Org. Lett.*, **13**, 960–963.

[81] Zouaoui, M. A., Mouaddib, A., Jamart-Gregoire, B., Ianelli, S., Nardelli, M., and Caubère, P. (1991) *J. Org. Chem.*, **56**, 4078–4081.

[82] Tripathy, S., Reddy, R., and Durst, T. (2003) *Can. J. Chem.*, **81**, 997–1002.

[83] Rogness, D. C., Markina, N. A., Waldo, J. P., and Larock, R. C. (2012) *J. Org. Chem.*, **77**, 2743–2755.

[84] Li, P., Wu, C., Zhao, J., Rogness, D. C., and Shi, F. (2012) *J. Org. Chem.*, **77**, 3149–3158.

[85] Zhao, J. and Larock, R. C. (2007) *J. Org. Chem.*, **72**, 583–588.

[86] Rogness, D. C. and Larock, R. C. (2010) *J. Org. Chem.*, **75**, 2289–2295.

[87] Gilmore, C. D., Allan, K. M., and Stoltz, B. M. (2008) *J. Am. Chem. Soc.*, **130**, 1558–1559.

[88] Allan, K. M. and Stoltz, B. M. (2008) *J. Am. Chem. Soc.*, **130**, 17270–17271.

[89] Peña, D., Escudero, S., Pérez, D., Guitián, E., and Castedo, L. (1998) *Angew. Chem. Int. Ed.*, **37**, 2659–2661.

[90] Peña, D., Pérez, D., Guitián, E., and Castedo, L. (1999) *J. Am. Chem. Soc.*, **121**, 5827–5828.

[91] Caeiro, J., Peña, D., Cobas, A., Pérez, D., and Guitián, E. (2006) *Adv. Synth. Catal.*, **348**, 2466–2474.

[92] Sato, Y., Tamura, T., and Mori, M. (2004) *Angew. Chem. Int. Ed.*, **43**, 2436–2440.

[93] Jayanth, T. T., Jenganmohan, M., and Cheng, C.-H. (2004) *J. Org. Chem.*, **69**, 8445–8450.

[94] Quintana, I., Boersma, A. J., Peña, D., Pérez, D., and Guitián, E. (2006) *Org. Lett.*, **8**, 3347–3349.

[95] Qiu, Z. and Xie, Z. (2009) *Angew. Chem. Int. Ed.*, **48**, 5729–5732.

[96] Yoshikawa, E. and Yamamoto, Y. (2000) *Angew. Chem. Int. Ed.*, **39**, 173–175.

[97] Jeganmohan, M. and Cheng, C.-H. (2004) *Org. Lett.*, **6**, 2821–2824.

[98] Jayanth, T. T., Jeganmohan, M., and Cheng, C.-H. (2005) *Org. Lett.*, **7**, 2921–2924.

[99] Henderson, J. L., Edwards, A. S., and Greaney, M. F. (2006) *J. Am. Chem. Soc.*, **128**, 7426–7427.

[100] Waldo, J. P., Zhang, X., Shi, F., and Larock, R. C. (2008) *J. Org. Chem.*, **73**, 6679–6685.

[101] (a) Liu, Z., Zhang, X., and Larock, R. C. (2005) *J. Am. Chem. Soc.*, **127**, 15716–15717; (b) Jayanth, T. T. and Cheng, C.-H. (2006) *Chem. Commun.*, 894–896.

[102] Bhuvaneswari, S., Jeganmohan, M., and Cheng, C.-H. (2006) *Org. Lett.*, **8**, 5581–5584.

[103] Liu, Z. and Larock R. C. (2007) *Angew. Chem. Int. Ed.*, **46**, 2535–2538.

[104] Yoshida, H., Konda, Y., Shirakawa, E., and Hiyama, T. (2001) *Chem. Commun.*, 1880–1881.

[105] Yoshida, H., Tanino, K, Ohshita, J., and Kunai, A. (2004) *Angew. Chem. Int. Ed.*, **43**, 5052–5055.

[106] Yoshida, H., Ikadai, J., Shudo, M., Ohshita, J., and Kunai, A. (2003) *J. Am. Chem. Soc.*, **125**, 6638–6639.

PART IV

REDUCTION, OXIDATION, AND DEAROMATIZATION REACTIONS

13

REDUCTION/HYDROGENATION OF AROMATIC RINGS

FRANCISCO FOUBELO AND MIGUEL YUS

Departamento de Química Orgánica, Facultad de Ciencias, Universidad de Alicante, Alicante, Spain

13.1 INTRODUCTION

The partial or total hydrogenation (or reduction) of substituted aromatic compounds is an important transformation in synthetic organic chemistry due to the abundance of aromatic molecules found in Nature. In the first case, a diene or an olefin is produced, which is much more reactive than the aromatic precursor, so a new wide range of possible reactivities can be explored. The total saturation of the aromatic ring is of special interest because this is a new entry to functionalized saturated carbocycles that generally show a different reactivity than the parent aromatics.

In this chapter, the famous Birch reaction will be initially studied, paying special attention not only to the dissolving metals typical version but also to the less developed enzymatic mode. The second part will be devoted to the catalytic hydrogenation with molecular hydrogen using different organometallics or metals under homogeneous or heterogeneous conditions, respectively. Other less used methodologies, including electrochemical procedures, will be finally considered.

13.2 THE BIRCH REACTION

The genuine Birch reaction was reported to take part using dissolving metals, and this has been the main application of this process. However, in the last decade, enzymatic methods have been reported for the same process that will be considered separately.

Arene Chemistry: Reaction Mechanisms and Methods for Aromatic Compounds,
First Edition. Edited by Jacques Mortier.

13.2.1 Dissolving Metals

Since the first account [1], the Birch reduction [2] of aromatic compounds has emerged as a useful and versatile tool to generate hydroaromatic derivatives. Initially, the reaction was performed with sodium metal dissolved in liquid ammonia and in the presence of a proton source such as ethanol. In the simplest case of benzene (**1**), 1,4-cyclohexadiene (**2**) was obtained with good yield (Scheme 13.1).

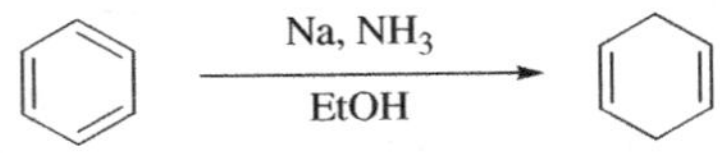

1 2

SCHEME 13.1 Preparation of 1,4-cyclohexadiene.

The proposed mechanism for this process involves initially an electron transfer from the metal to the LUMO of the aromatic ring of a molecule, giving an anion-radical **I**, which prefers to locate the two highly unstable carbon atoms (the radical and the carbanion) as far apart as possible, as in the intermediate **II**. Then, a proton from the solvent is abstracted giving the corresponding radical **III** that after capturing a second electron from the metal affords the anion **IV**, which finally takes a second proton from the solvent yielding the nonconjugated diene **2**, according to the principle of least motion, which states that reactions involving the least change in electronic configuration (and/or atomic positions) are favored [3] (Scheme 13.2).

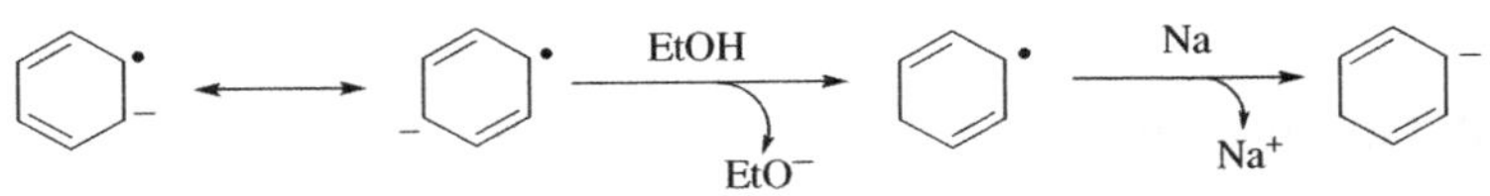

I II III IV

SCHEME 13.2 Proposed mechanism for the Birch reduction.

For substituted benzenes, the electronic nature of the substituent is responsible for the stability of the radical anion intermediate initially formed. Thus, in the case of anisole (**3**), the electron transfer is slower due to the presence of the electron-donating methoxy group, and from the two possible anion-radicals **V** and **VI**, the first one is less stable due to electron repulsion of vicinal carbanion and the electron-rich oxygen substituent. After taking a proton, radical **VII** is generated, which after successive capture of an electron and a proton yields the final product 2,5-dihydroderivative **4** (Scheme 13.3).

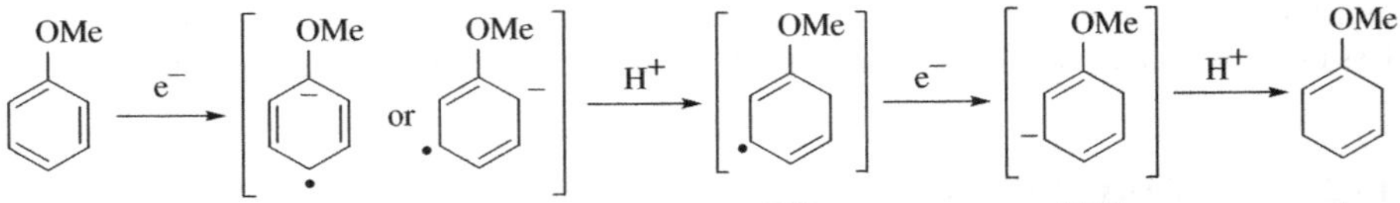

3 V VI VII VIII 4

SCHEME 13.3 Partial reduction of anisole.

However, when the substituent is an electron-withdrawing group, such as in the case of benzoic acid (**5**), the same sequence of steps shown in Scheme 13.3 applies. Initially, intermediates **IX** and **X** are formed, the first one being more stable due to the ability of the carboxyl group to stabilize the adjacent negative charge. Then, a proton is abstracted to give the intermediate **XI** and followed by addition of an electron (to give **XII**), and finally, a proton is abstracted resulting in the corresponding 1,4-dihydrodienic carboxylic acid **6**. As expected, the whole process involves the corresponding carboxylate, which is generated in the first step by reaction of the carboxylic acid with the metal (Scheme 13.4) [2].

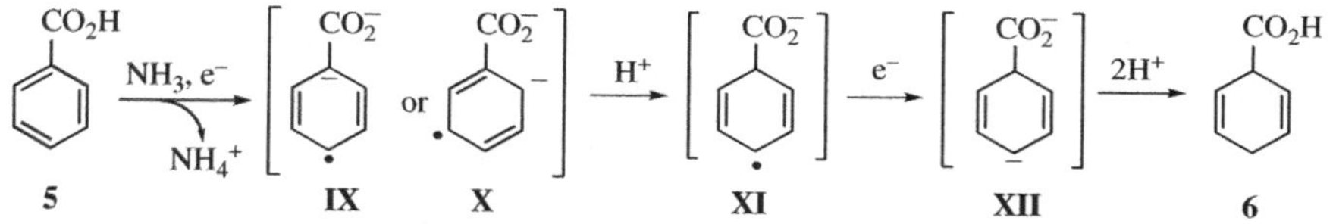

SCHEME 13.4 Partial reduction of benzoic acid.

The regiochemistry observed in the case of anisole (**3**, Scheme 13.3) and benzoic acid (**5**, Scheme 13.4) is general for electron-donating or electron-withdrawing groups, respectively.

Soon after the first report from Birch with sodium as the metal component, the employment of lithium metal was reported [4]. Scheme 13.5 shows a couple of examples using this combination. The amine **7** gives the expected vinyl ether **8** [5], and the dicarboxylic acid **9** yields the doubly reduced product **10** (Scheme 13.5) [6].

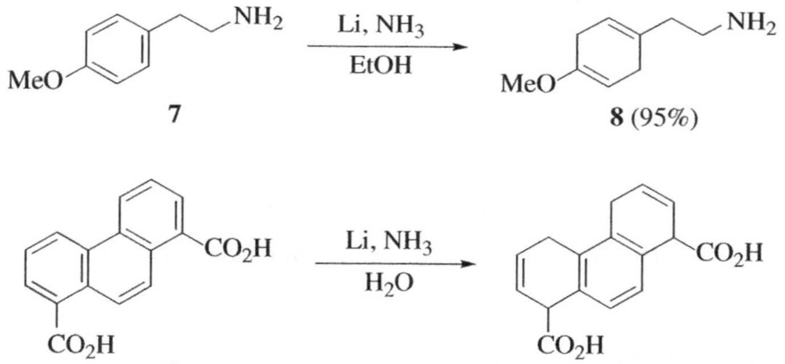

SCHEME 13.5 Partial reduction of compounds **7** and **9**.

From a synthetic point of view, methoxy arenes (in general alkoxy arenes) are of special interest because the final hydrolysis of the obtained vinyl ethers under acidic conditions produces enones. This chemistry has been successfully applied to the synthesis of steroids. Scheme 13.6 shows a simple example of this approach for the preparation of the enone **13** starting from the ether **11** and involving the corresponding vinyl ether **12**. Note that the conjugated double bond in **11** is also reduced under the conditions employed (isolated olefins are inert with dissolving metals). On the other hand, the final conjugated enone is the isomerized product coming from the initially formed β,γ-unsaturated ketone under strong reaction conditions (pH = 1) [7].

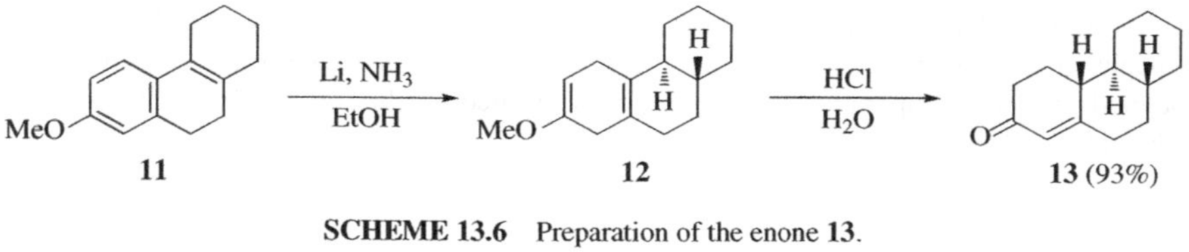

SCHEME 13.6 Preparation of the enone **13**.

An interesting example of the synthetic utility is the partial reduction of compound **14** under standard Birch conditions to give the diene **15**, and after acid hydrolysis the *exo*-enone **16**, of difficult access by other methodologies (Scheme 13.7) [8].

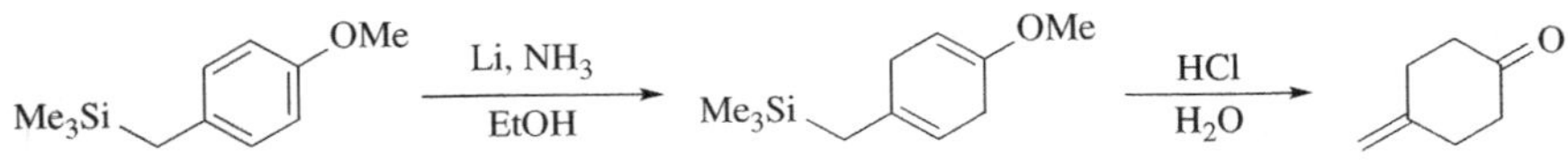

14 15 (84%) 16 (80%)

SCHEME 13.7 Preparation of the enone **16**.

Another interesting synthetic application of the Birch reduction is the capture of the *in situ*-generated carbanion with a carbon electrophile (usually an alkyl halide) forming a new carbon–carbon bond [9]; this process has also been studied in an asymmetric manner [10].

It is important to mention the pivotal role of the alcohol component in the typical Birch reduction. On one hand, the relatively acidic alcohol rapidly protonates the intermediate **IV** (Scheme 13.2) to give the 1,4 diene: in the absence of alcohol, an isomerization of this intermediate to the most stable 1,3-diene anion **XIII** takes place, and its final protonation affords the corresponding 1,3-diene (**17**), which is susceptible to reduction by the reaction media to give the final olefin cyclohexene (**18**) (Scheme 13.8). It is well known that 1,3-dienes are reduced by dissolving metals to the corresponding olefins in quantitative yields [11].

IV XIII 17 18

SCHEME 13.8 Isomerization of 1,4- to 1,3-dienic intermediates.

The second role of the alcohol is to avoid the reversion of the final 1,4-cyclohexadiene into the intermediate **IV** by means of the strong base (metal amide) generated in the reaction medium. This base is protonated by the alcohol giving a less strong base, the alkoxide, unable to deprotonate the 1,4-diene and to isomerize it to the most stable 1,3-system. In the absence of an alcohol, the reaction works at higher temperature, and usually, an amine is employed instead of ammonia.

Benzene (**1**) itself is reduced to cyclohexene (**18**) with lithium and ethylamine or ethylenediamine (Scheme 13.9) [12].

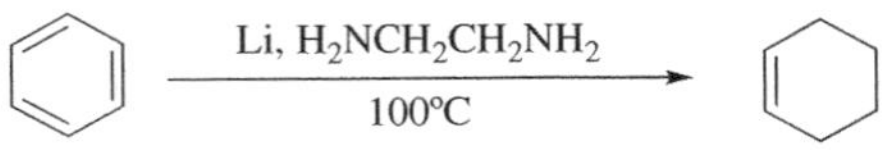

1 18 (51%)

SCHEME 13.9 Preparation of cyclohexene **18**.

An example that illustrates the role of the alcohol is depicted in Scheme 13.10 for the reduction of naphthalene (**19**). Whereas at low temperature and in the presence of ethanol compound **20** is obtained, at higher temperature and in absence of the alcohol, the corresponding olefin **21** is mainly isolated [13, 14].

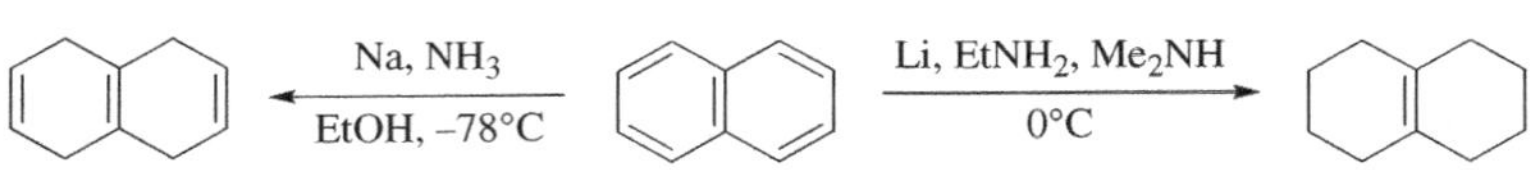

20 (80%) 19 21 (74%)

SCHEME 13.10 Reduction of naphthalene to compounds **20** and **21**.

Another simple case is the partial reduction of ethylbenzene (**22**) as shown in Scheme 13.11: depending on the presence or absence of propanol, compounds **23** or **24** are respectively isolated [15].

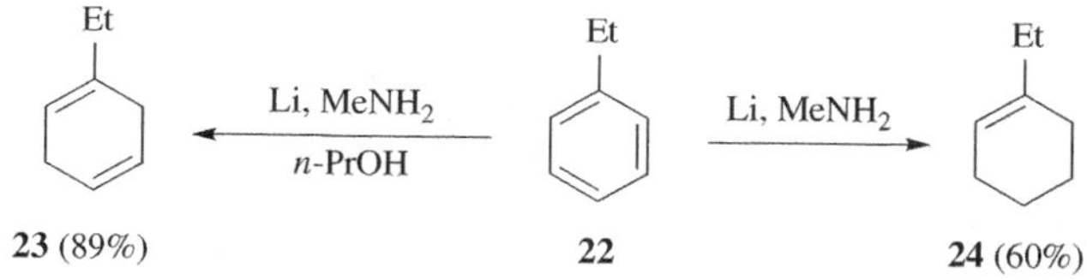

SCHEME 13.11 Preparation of compounds **23** and **24**.

The reduction of 1-naphthol and 2-naphthol offers an interesting regiocontrol contrast. Whereas 1-naphthol (**25**) gives the hydrogenation at the unsubstituted ring giving compound **26**, the 2-naphthol (**27**) affords saturation at the substituted one yielding 2-tetralone (**28**). The result can be easily explained considering that high-energy intermediate **XIV** (Scheme 13.12) is required for reduction of the substituted ring. This does not take place in intermediate **XV** resulting from 2-naphthol, wherein after protonation, the initially formed enol tautomerizes to 2-tetralone (**28**) [1, 16, 17].

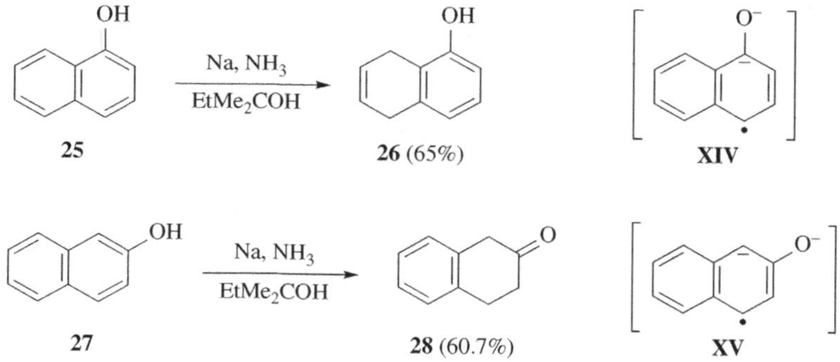

SCHEME 13.12 Partial reduction of compounds **25** and **27**.

For differently substituted aromatic rings, a mixture of the corresponding regioisomers is obtained. For instance, the steroid derivative **29** under standard Birch reduction conditions gives a mixture of both compounds **30** and **31**, which are formed from two possible regioisomeric radical anions **XVI** and **XVII**, respectively, of similar stability (Scheme 13.13) [18]. Regarding the

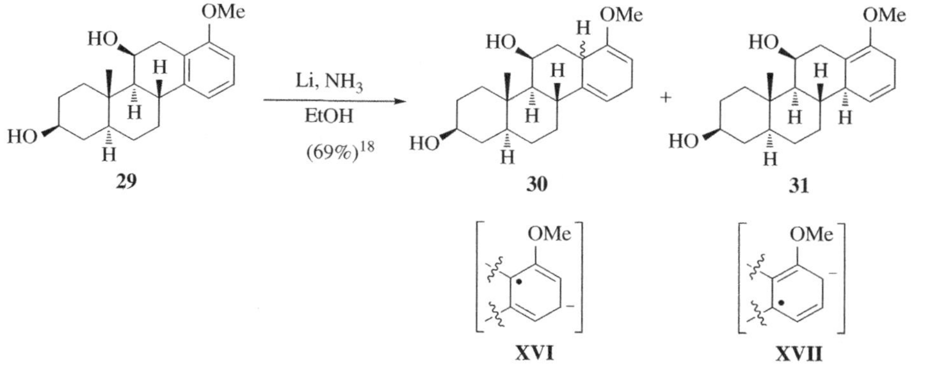

SCHEME 13.13 Partial reduction of compounds **29**.

stereochemistry of the Birch reduction in substituted aromatic compounds, since the conjugated anions and radical anion intermediates are planar systems, the configuration of the resulting hydrogenated products is determined by the presence of other stereogenic elements in the substrates, such as in the case of compound **31**.

Finally, apart from lithium and sodium, commonly used in the Birch reduction, calcium has shown to be effective for the partial reduction of aromatics, this procedure being especially interesting for scaling-up the reaction with industrial purposes [19]. Another interesting recent finding, regarding the Birch reduction of polynuclear aromatic compounds, was the use of Na_2K alloy in silica gel as reducing reagent. These reactions are performed in THF at room temperature and represent an alternative to the more tedious classical liquid ammonia conditions [20].

13.2.2 Enzymatic Reactions

In Nature, there is a biological Birch reduction that is promoted by benzoyl-coenzyme A reductase (BCR) under anaerobic conditions. The enzyme catalyzes the reduction of benzoyl-CoA (BCoA, **32**) to the corresponding 1,3-diene **33** rather than the 1,4-diene that is kinetically favored and uses ATP in connection with reduced ferredoxin (Fd). The reaction mechanism (involving intermediates **XVIII–XX**), which has been studied using deuterated reagents, was investigated using kinetic measurements and theoretical calculations (ab initio) and seems to be similar to the typical chemical Birch reduction, although, in this case, as for almost all enzymatic reactions, the hydrogenation takes place with higher levels of diastereocontrol. Thus, it was found, based mainly in NOESY experiments, that BCR-catalyzed reductions are *trans*-selective (Scheme 13.14) [21, 22]. More complex aromatic substrates undergo also selective dearomatization under enzymatic Birch reduction conditions. Thus, the nonactivated ring of 2-naphthoyl-CoA undergoes a four-electron reduction to 5,6,7,8-tetrahydro-2-naphthoyl-CoA by means of an oxygen-tolerant enzyme. This 2-naphthoyl-CoA reductase (NCR) contained FAD, FMN, and also a FeS cluster as cofactors [23].

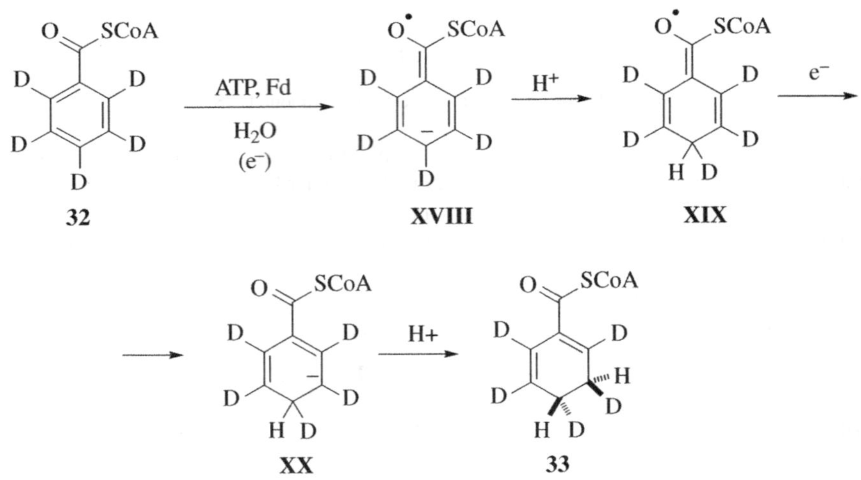

SCHEME 13.14 Enzymatic Birch reduction of BCoA (**32**).

The process shown in Scheme 13.14 can be reversible using a tungsten-containing class II BCR that catalyzes the fully reversible ATP-independent dearomatization of BCoA to give the corresponding 1,3-dienyl derivative at an extremely low redox potential. This reversible metalloenzyme catalysis is not possible using chemical reactions, which always work under irreversible conditions [24].

13.3 METAL-CATALYZED HYDROGENATIONS

Normally, a transition metal is the best catalyst to carry out the hydrogenation of aromatic compounds under both homogeneous and heterogeneous conditions that means in solution or in two-phase mode, respectively. Since the type of metal catalyst and its behavior strongly depend on the reaction conditions, both classes of catalysis will be considered separately in the following sections.

13.3.1 Homogeneous Conditions

13.3.1.1 Organometallic Complexes The hydrogenation of arenes has been traditionally performed under heterogeneous catalysis, whereas pure homogeneous catalysis was successfully applied in the hydrogenation of alkenes and, to a lesser extent, in the hydrogenation of arenes [25]. The majority of these homogeneous catalytic systems are based upon middle and later transition metals, and in many cases, successful reactivity is limited to polynuclear aromatic hydrocarbon substrates. The allyl–cobalt catalyst **34** [26] showed to be effective in the hydrogenation of different arenes under mild reaction conditions, namely, one atmosphere of hydrogen at 20°C. The reaction proved to be unequivocally homogeneous and showed to exhibit a remarkably high *cis* stereoselectivity [27]. In addition, the reactivity of **34** toward arenes was almost equivalent to that toward terminal olefins. Representative examples with conversion values are collected in Scheme 13.15.

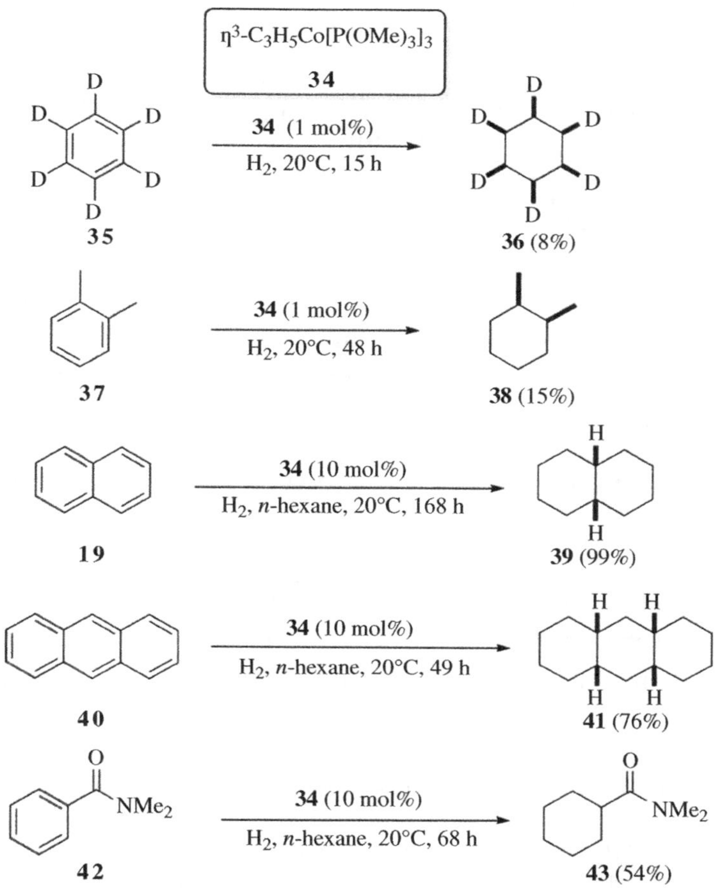

SCHEME 13.15 Homogeneous hydrogenation of arenes with η^3-$C_3H_5Co[P(OMe)_3]_3$ (**34**).

Thus, hydrogenation of perdeuterated benzene (**35**) yielded greater than 95% of the *cis* isomer **36** and *o*-xylene (**37**) gave only *cis*-1,2-dimethylcyclohexane (**38**) using only 1 mol% of **34**. In these two cases, the hydrogenation was performed without an additional solvent. On the other hand, naphthalene (**19**), anthracene (**40**), and *N,N*-dimethylbenzamide (**42**) led to *cis*-decalin (**39**), *cis-syn-cis*-perhydroanthracene (**41**), and *N,N*-dimethylcyclohexane carboxamide (**43**), respectively, as the major reaction products but using a larger amount of the catalyst (10 mol%) for longer reaction times in hexane. In these processes, the π donor capability of the arene must be relatively high for the hydrogenation cycle to take place, because electron-withdrawing groups deactivate the arenes toward the hydrogenation cycle with **34**. Thus, fluorine, cyano, and nitro groups rendered the benzene nucleus unreactive with this catalytic system [28]. More reactive analogues of allyl–cobalt catalyst **34** have been found, such as the trimethylphosphite derivative. However, they are also less resistant to the side reaction involving loss of propene leading to catalyst deactivation.

A mechanism has been proposed in order to explain the high *cis* stereoselectivity in the hydrogenation of arenes with allyl–cobalt compound **34**. In the presence of hydrogen, a dihydride was formed and interacted with the arene to give η^1-$C_3H_5CoH_2P(OMe)_3\eta^4$-$C_6H_5$ (**XXI**). The sequence shown in Scheme 13.16 suggests that no dissociation of the ring occurs after initial reductions to successively form intermediates **XXII–XXVIII**, making the last step irreversible because cyclohexane does not react with the cobalt complex. The proposed mechanism also accounts for the all *cis* stereochemistry of the cyclohexanes derived from substituted benzenes, such as *o*-xylene (**37**), as well as polycyclic aromatic hydrocarbons (PAHs) [29].

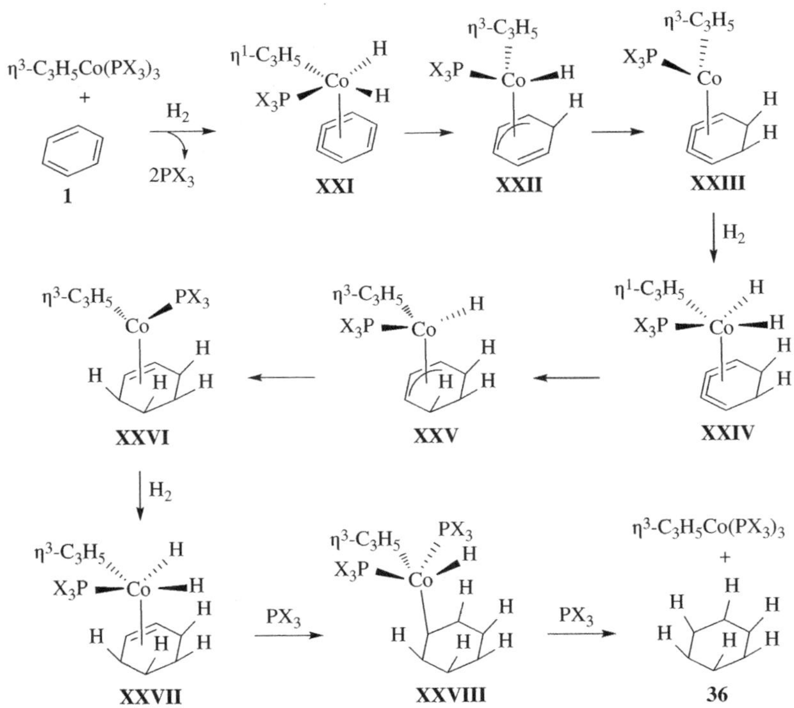

SCHEME 13.16 Mechanism of the homogeneous hydrogenation of arenes with η^3-$C_3H_5Co[P(OMe)_3]_3$ (**34**).

A variety of PAHs were also homogeneously hydrogenated in a highly regioselective manner but lacking stereoselectivity in the presence of $Co_2(CO)_8$ and synthesis gas ($CO+H_2$) at elevated temperature. For instance, 9,10-dimethylanthracene (**44**) led to nearly equal amounts of *cis*- and *trans*-9,10-dihydro-9,10-dimethylanthracene (**45** and **46**, respectively, Scheme 13.17) [30]. The operating mechanism should be different to that proposed for the hydrogenation with cobalt catalyst **34**. In this case, the active catalyst $HCo(CO)_4$ initially formed reacted with **44** to give free radical intermediate **XXIX**, which reacted with a second equivalent of $HCo(CO)_4$ to produce a mixture of *cis* and *trans* isomers **45** and **46**, respectively, and regenerate $Co_2(CO)_8$. This last step proceeded by an approximately random approach of $HCo(CO)_4$ to the radical center from either side of the nearly symmetrical molecular plane of **XXIX** (Scheme 13.17) [31].

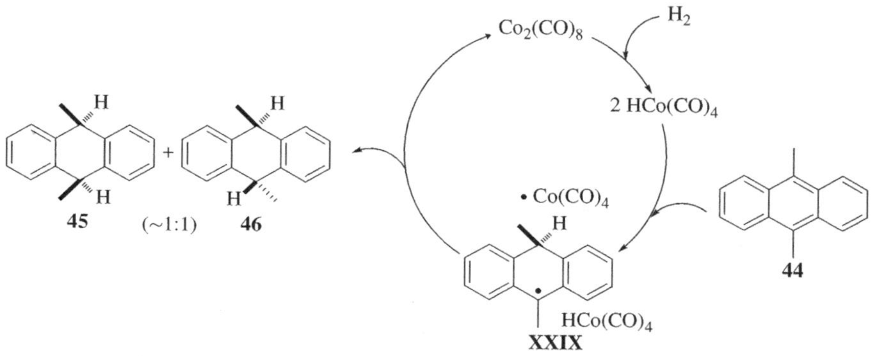

SCHEME 13.17 Proposed mechanism for the homogeneous hydrogenation of 9,10-dimethylanthracene with $Co_2(CO)_8$ and synthesis gas.

More recently, high-valent niobium and tantalum aryloxy hydrides were used as catalysts in the homogeneous hydrogenation of benzenes and polynuclear aromatic hydrocarbons. The hydrogenations occurred with high regioselectivity and *syn*-stereoselectivity. The reaction of a solution of niobium compound **47** [32] in cyclohexane with 20 equivalents of naphthalene (**19**) at 80°C under 1200psi of hydrogen gave 95.5% of tetralin (**49**) and 4.5% of *cis*-decalin (**39**). Under the same reaction conditions, hydrogenation of anthracene (**40**) led exclusively to 1,2,3,4,5,6,7,8-octahydroanthracene (**50**). Conversely, phenanthrene (**52**) was hydrogenated by **47** at a slower rate, because only 90% conversion was observed after 24h to give a mixture of 1,2,3,4,5,6,7,8-octahydrophenanthrene (**53**) and 9,10-dihydrophenanthrene (**54**). On the other hand, hydrogenation under similar conditions of naphthalene (**19**) and anthracene (**40**) with tantalum hydride **48** [33] produced tetralin (**49**) and 1,2,3,4-tetrahydroanthracene (**51**) with a small amount of 1,2,3,4,5,6,7,8-octahydroanthracene (**50**) in the latter case (Scheme 13.18) [34]. The tantalum complex **48** with the 2,6-dicyclohexylphenoxide ligand is generated by intramolecular hydrogenation of the *ortho*-phenyl rings in the 2,6-diphenylphenoxide precursor [35], whereas the niobium compounds similar to **47** are able to hydrogenate rapidly arylphosphine ligands to produce cyclohexylphosphine ligands [34]. High-valent ruthenium hydride complexes, such as $RuH_4(PPh_3)_3$, have also been found effective in the homogeneous hydrogenation of polynuclear aromatic hydrocarbons [36].

Asymmetric hydrogenation of monocyclic and bicyclic heteroaromatic compounds has been successfully achieved under homogeneous catalysis, taking place almost exclusively by reduction of the heteroatom-containing ring in the case of the bicyclic derivatives [37]. Conversely, poor enantioselectivity was obtained in the stereoselective hydrogenation of arenes, probably due to their lower ability to coordinate to the metal center and also the general difficulty of discriminating between the enantiotopic faces in these systems. However, a report described the enantioselective hydrogenation of the aromatic carbocyclic ring of substituted quinoxalines **56** and **58** by means of a homogeneous chiral ruthenium *N*-heterocyclic carbene (NHC) complex. The ruthenium complex was formed *in situ* from [Ru(cod)(2-methylallyl)$_2$] and a monodentate NHC. Importantly, the regioselectivity of the hydrogenation was mainly controlled by the right choice of the NHC ligand. Thus, hydrogenation product resulting from the reduction of the nitrogen-containing ring was obtained when the ruthenium complex with N-aryl-substituted NHC ligands was

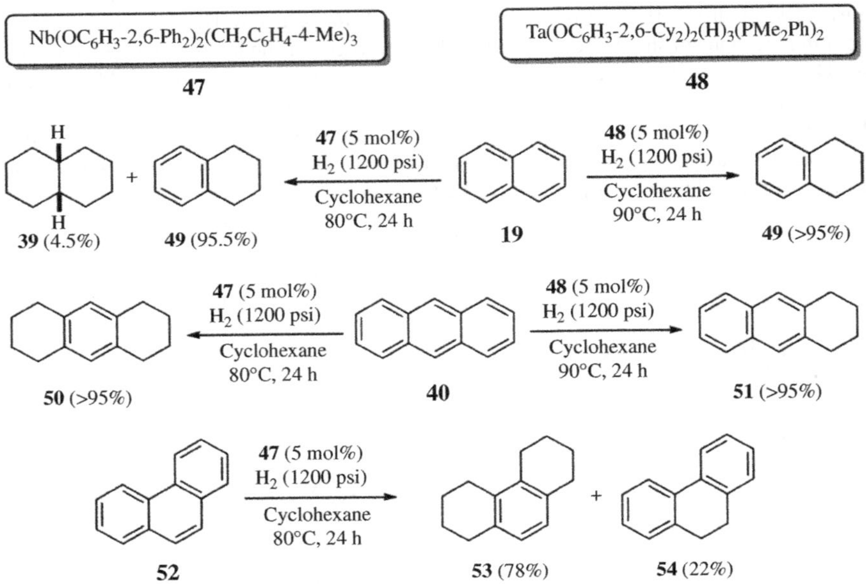

SCHEME 13.18 Homogeneous hydrogenation of arenes with niobium and tantalum compounds **47** and **48**.

used as the catalyst. Surprisingly, aromatic carbocyclic ring hydrogenation occurred with N-alkyl-substituted NHC ligands that represent the first example of homogeneous catalytic asymmetric hydrogenation of the carbocyclic ring for azoaromatic compounds. Optimal results were obtained working with the chiral ligand **55** under the reaction conditions depicted in Scheme 13.19. In general, higher enantiomeric ratios were obtained in the regioselective hydrogenation of 6-substituted quinoxalines **56** to yield **57**, whereas 5-methyl quinoxaline **58** gave **59** performing poorly considering stereoselectivity [38].

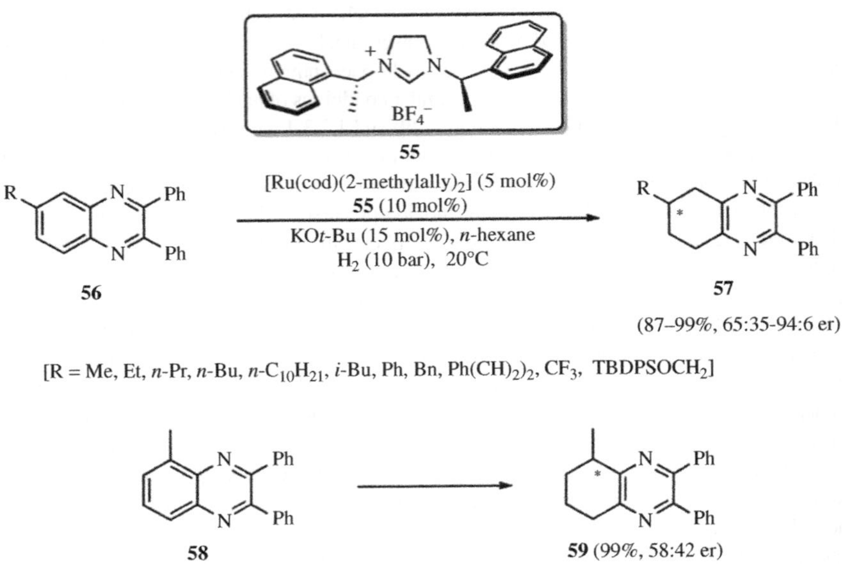

SCHEME 13.19 Asymmetric homogeneous hydrogenation of quinoxalines catalyzed by ruthenium *N*-heterocyclic carbene complexes.

13.3.1.2 Soluble Nanoparticles There is a controversy in the study of the arene hydrogenation catalysis regarding the true nature of the catalyst. Sometimes, it is not easy to distinguish homogeneous, single-metal-complex catalysts from soluble nanoparticle (NP, also called nanocluster) or colloid catalysts. Different protocols can be applied in order to solve the problem including the use of transmission electron microscopy (TEM) to detect metal aggregation, kinetic studies, catalyst poison experiments, and the important concept that the identity of the true catalyst will be consistent with all the data. There are examples in the literature of arene hydrogenation catalysts that were initially believed to be heterogeneous, but later evidence suggested that a soluble nanocluster was the true catalyst [39]. Over the last decade, there has been an increasing interest in the use of soluble metallic NPs as catalysts for arene hydrogenation, because these processes can be carried out under milder conditions. In addition, these soluble metallic NPs could be considered a borderline case between homogeneous and heterogeneous catalysts, combining the advantages of both types of systems in terms of activity, selectivity, and recovery [40]. In all processes involving metallic NPs, the stabilizing agent that avoids agglomeration plays an important role concerning activity and selectivity. Polymers, surfactants, and ionic liquids are the most commonly used stabilizing agents of NPs. Tetrabutylammonium polyoxoanions, amines, aromatic imines, phosphine, and phosphites have also been used as stabilizing agents of NPs, although to a lesser extent. The metals of choice in hydrogenation reactions are Ru, Rh, Pt, and Ir; showing also promising results are new nanocatalysts based on Co, Pd, Ni, and Fe.

Poly(vinylpyrrolidone) (PVP) is a nontoxic polymer that is soluble in many polar solvents and has been used as a stabilizing agent for metallic NPs. The polymer molecular weight is important because it determines many of its physical properties. In the case of PVP K-90, this polymer has an average molecular weight of 360,000. A nanocatalyst of Ru NPs was prepared from $RuCl_3{\cdot}H_2O$ and PVP K-90 in water and demonstrated extremely high activity in the hydrogenation of arenes. This could be explained by considering that the substrates are confined in the hydrophobic pocket of the biphasic system, which enhances the contact between them. Optimal results were obtained using a 1:10 metal/polymer ratio (considering the pyrrolidone ring), and 1:3000 metal/arene ratio, with cyclohexane being the optimal solvent for arenes. The highest TOF value was obtained with benzene (**1**) as a substrate. Also, relatively high TOF values were achieved with benzene derivatives bearing electron-donating substituents, such as anisole (**3**) and ethylbenzene (**22**). However, a lower TOF value was measured in the case of acetophenone (**63**) with an electron-withdrawing substituent on the arene ring (Scheme 13.20) [41].

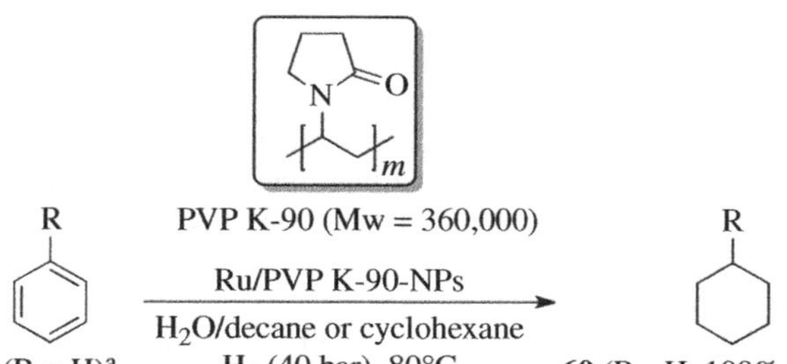

1 (R = H)[a]
3 (R = OMe)[a]
22 (R = Et)[b]
63 (R = COMe)[b]

60 (R = H, 100%, 45,000[c])
61 (R = OMe, 70%,[d] 16,000[c])
62 (R = Et, 100%, 30,000[c])
64 (R = CH(OH)Me, 83%,[e] 29,000[c])

[a] Solvent: H_2O/decane. [b] Solvent: cyclohexane. [c] Turnover frequency (TOF, h^{-1}) defined as number of moles of consumed H_2 per mole of Ru per hour. [d] Cyclohexanone (16%) and cyclohexanol (14%) were also isolated. [e] Ethylcyclohexane (17%) was also isolated.

SCHEME 13.20 Hydrogenation of aromatic rings with the Ru/PVP K-90-NPs catalyst.

Tetraalkylammonium salts were also used as stabilizing agents for the synthesis of metallic NPs, especially *N,N*-dimethyl-*N*-acetyl-*N*-(2-hydroxyethyl)ammonium halides (HEA-16-X). The surfactant counterion is important in the hydrogenation of monosubstituted benzene derivatives. For instance, the Rh/HEA-16-Cl catalytic system prepared under biphasic conditions using $[RhCl(1,5\text{-hexadiene})]_2$ and the ammonium salt gave the highest turnover frequencies compared with the HEA-16-Br one [42]. A HEA-16-Cl/M molar ratio of 2 was necessary to stabilize these NPs, and optimal results were obtained with substrate/metal ratio of 100. It was found that Rh was the most active metal, the hydrogenation taking place at room temperature under 1 bar of hydrogen pressure [42]. On the other hand, the hydrogenation with Ru/HEA-16-Cl-NPs occurred under *ca.* 30 bar of hydrogen pressure with moderate to high activities [43]. Steric and electronic factors affect also the catalytic activity in these processes. The presence of bulky substituents on the aromatic ring resulted in a decrease in catalytic activity, whereas arenes bearing electron-donating groups reacted faster than those with electron-withdrawing groups (Scheme 13.21).

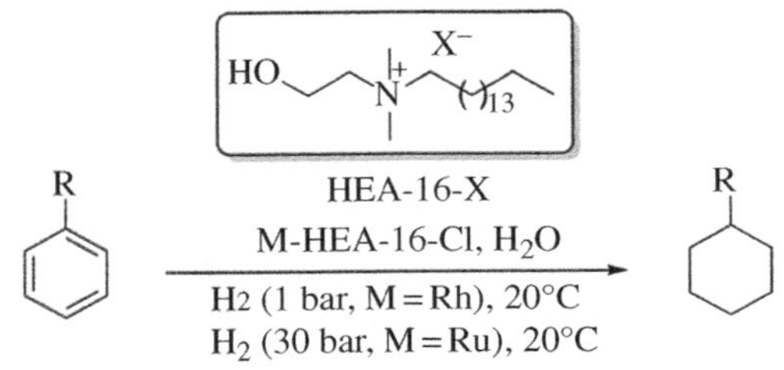

1 (R = H) **60** (R = H)
3 (R = OMe) **61** (R = OMe)
22 (R = Et) **62** (R = Et)
63 (R = COMe) **65** (R = CHMe)

	Rh		Ru	
	t (h)	TOF (h^{-1})	*t* (h)	TOF (h^{-1})
60	3.6	83	0.5	600
61	3.6	83	1	300
62	3.7	81	1	288
65	8.8	36	3.5	115

SCHEME 13.21 Hydrogenation of aromatic rings with Rh/- and Ru/HEA-16-Cl catalysts.

Stable and redispersible metal NPs of Rh, Ir, Ru, and Pt were obtained by reduction of different transition metal compounds dissolved in 1-*n*-butyl-3-methylimidazolium hexafluorophosphate ($BMI{\cdot}PF_6$), using molecular hydrogen as the reducing agent. The hydrogenation of benzene in a biphasic system catalyzed by these stabilized NPs dispersed in the ionic liquid was also studied under mild reaction conditions (75°C and 4 bar) [44]. Surprisingly, the hydrogenations were found to be faster when they were performed under heterogeneous neat conditions (Scheme 13.22). The influence of mass-transfer processes typical of a multiphase system could be a possible explanation for these results, although more recently it was found that imidazolium ionic liquids result in poisoning of Ir-NPs via the formation of Ir *N*-heterocyclic carbene complexes [45].

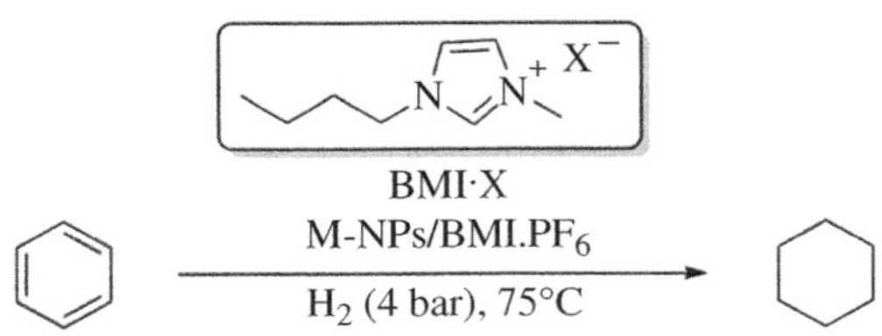

1 **60**

M	*t* (h)	TOF (h^{-1})
Rh	22 (12)	11 (21)
Ir	5 (2)	50 (125)
Ru	18.5 (5.5)	20 (82)
Pt	10 (9)	11 (28)

Values under solventless conditions are given in parenthesis

SCHEME 13.22 Hydrogenation of benzene with metal NPs in $BMI{\cdot}PF_6$.

The hydrogenation of disubstituted benzene derivatives using stabilized NP catalysts leads preferentially to *cis*-cyclohexanes and could be rationalized if the addition of hydrogen to each carbon atom of the aromatic ring takes place during a single period of adsorption onto the catalyst surface. On the other hand, the *trans* isomers could be formed when desorption of any partially reduced species occurs and, after reabsorption, is finally hydrogenated. There are only a few reports on the stereoselective hydrogenation of prochiral arenes. For instance, Rh NPs stabilized by the chiral amine (*R*)-*N*,*N*-dioctyl-1-cyclohexylethylamine (DOCEA) were used as catalysts in the hydrogenation of *o*-methylanisole (**66**) and *o*-methyl-*O*-trimethylsiloxybenzene (**67**). The resulting cyclohexane derivatives **68** and **69** were obtained with high *cis*-selectivity but extremely poor enantiomeric excesses (Scheme 13.23) [46].

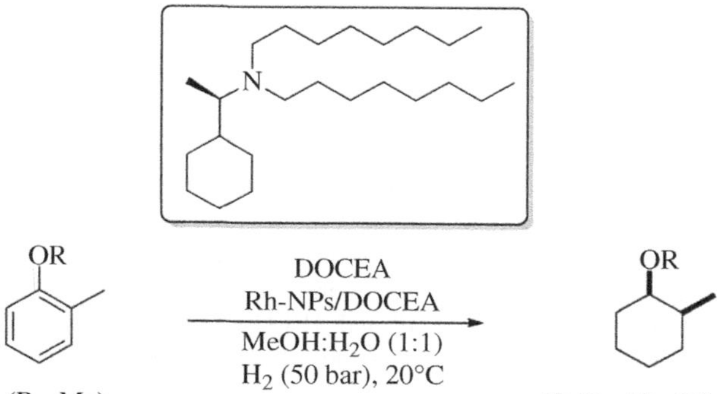

66 (R = Me)
67 (R = $SiMe_3$)

68 (R = Me, 3% ee)
69 (R = $SiMe_3$, 6% ee)
(97% *cis* selectivity)

SCHEME 13.23 Stereoselective hydrogenation of disubstituted benzene derivatives with Rh NPs stabilized by the chiral amine DOCEA.

13.3.2 Heterogeneous Conditions

The use of transition metal catalysts in the hydrogenation of arenes is a sustainable method that enables access to cyclohexane derivatives. The process can be carried out in an efficient manner using homogeneous catalysts. However, the presence of trace metals in the products makes this methodology of scarce industrial interest. In contrast, hydrogenation of arenes by means of heterogeneous catalysts is advantageous due to the easy handling. In addition, the reaction products can be obtained with less metal contamination. A major drawback of heterogeneous catalysis using conventional methodologies is that strongly acidic or basic conditions are required, along with high temperatures and also pressures over 10 atm. More recently, active supported transition metal nanoparticles have been used under milder reaction conditions in the hydrogenation of arenes [47]. A mechanism widely accepted common to cluster, nanoparticle, or bulk metal surface has been proposed, taking place a μ_3-η^2:η^2:η^2 coordination mode of the arene on the surface of the metal. In this way, the addition of hydrogen to one double bond would lead to a μ_3-η^2:η^2 cyclohexadiene. A second hydrogenation would produce a μ_3-η^2 cyclohexene, and a subsequent hydrogenation would give the cyclohexane that would be released from the metal surface. Partial hydrogenation of the arene could also be explained if replacement by a new arene substrate takes place in the hydrogenated products (Scheme 13.24). This stepwise hydrogenation mechanism is based on experimental evidences with some of the intermediates having been isolated and characterized by X-ray crystallography and also confirmed by DFT calculations [48]. The influence of the substituents on the reaction rate and on the stereochemical pathway of the hydrogenation was also studied, taking place mainly by the formation of the kinetically favored *cis* isomer.

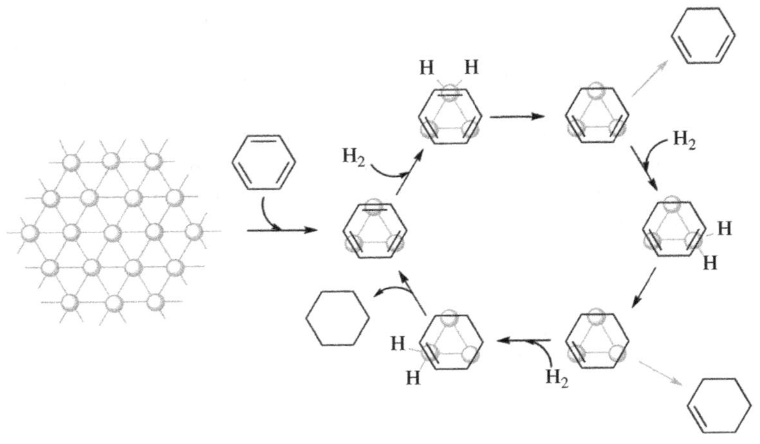

SCHEME 13.24 Proposed arene hydrogenation mechanism for heterogeneous catalysts.

13.3.2.1 Supported Materials Different metal sulfides, such as Ni–W and Ni–Mo sulfides, supported on alumina, silica, zeolite, and other oxides are able to catalyze the hydrogenation of arenes. However, temperatures ranging between 310 and 410°C and high pressures (80–140 atm) were needed for hydrogenation proceeds, and under these harsh operating conditions, the thermodynamic equilibrium toward saturated rings was inhibited. Much more attractive are supported transition metals of group VIII, which operate at much lower temperatures than the metal sulfides and with higher turnover frequencies. The catalytic activity of these metals is mainly determined by the chemical bond delocalization among the metal atoms and by their valence electronic configuration. In addition, Ru, Rh, Pd, Ir, and Pt are the catalysts of choice for arene hydrogenation since their lattice distances (from 0.36 to 0.38 nm) greatly facilitate the activation of adsorbed arenes [47].

A heterogeneous catalyzed hydrogenation of substituted arenes was carried out with commercial Rh on charcoal (Rh/C) under mild reaction conditions. The reaction rate was greatly enhanced using *i*-PrOH as solvent instead of the more environmentally friendly H_2O. It was also found that the reaction in *i*-PrOH takes place also at a significant extent at room temperature under 1 atm H_2. On the other hand, the cheaper and in some cases more practical heterogeneous catalyst Ru/C was found to be also effective in the hydrogenation of arenes, including phenols, although higher H_2 pressures were required for achieving high conversions. It is important to mention, taking into account environmental considerations, that these catalytic systems can be reused five times without any reactivation, although a gradual increase in reaction time was necessary. For instance, the hydrogenation of diphenylmethane (**70**), benzoic acid (**5**), and *N*-phenylacetamide (**73**) to give cyclohexane derivatives **71**, **72**, and **74**, respectively, proceeded in high yields at 60°C under 1 atm of H_2 with Rh/C as the catalyst. However, higher pressure (5 atm) was required with Ru/C in order to achieve similar conversions. Interestingly, partial reduction of binaphthol (**75**) occurred at 5 atm of H_2 under Rh/C catalysis, leading to the octahydro derivative **76**. The hydrogenation of phenols is of special synthetic interest because the cyclohexanol products are useful structural elements of many compounds such as pharmaceuticals and functional materials. Although it has been reported that the most active Rh/C catalysts may cause hydroxyl group elimination, phenol and alkyl-substituted phenols were completely hydrogenated with this Ru/C catalyst under optimized conditions to yield the corresponding cyclohexanols. In the case of 4-*tert*-butylphenol (**77**), a nearly 1:1 mixture of *cis*/*trans* stereoisomers **78** was obtained (Scheme 13.25) [49]. The hydrogenation of benzoic acid (**5**) to cyclohexanecarboxylic acid (**72**) was also successfully carried out in supercritical CO_2 medium with Rh/C catalyst at 50°C, under H_2 (40 atm) and CO_2 (100 atm) in near-quantitative yield. Under these reaction conditions, the catalyst was easily separated from the product without producing any waste [50].

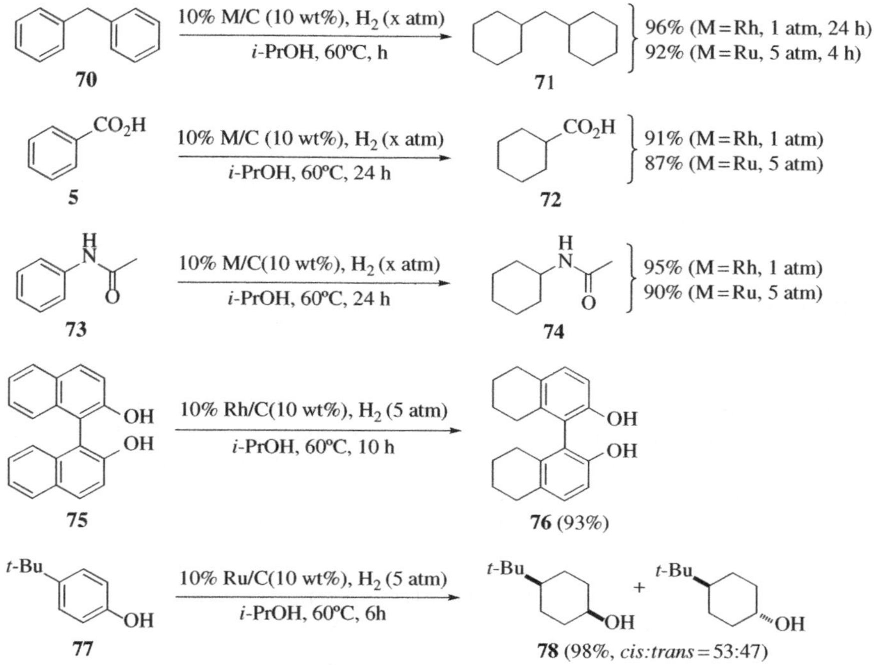

SCHEME 13.25 Hydrogenation of arenes with Rh and Ru heterogeneous catalysts.

13.3.2.2 Supported Nanoparticles Commercially available Rh/C shows excellent catalytic activity in a broad range of organic reactions. In spite of that, considerable effort has been aimed at enhancing the efficiency of this reagent. It was found that nanoporous carbon, carbon nanotubes and nanofibers, and layered clays are better supports than charcoal. On the other hand, the activity of these systems is also greatly determined by the size and shape-control of Rh-metal particles. For instance, surface area is increased by reducing the size of the particles. As a consequence, smaller particle size results in higher reactivity. Regarding the shape of the particles, low-temperature, kinetically controlled synthesis is one of the most common strategies for the shape evolution of nanoparticles. The catalytic activities of commercial Rh/C and tetrahedral (▲Rh/C) and spherical (•Rh/C) Rh NPs on activated charcoal for the hydrogenation of anthracene (**40**) were compared. Three different reaction products were obtained in these processes: the one resulting from the hydrogenation of the central ring (**79**), another from the hydrogenation of one side ring (**51**), and the third one from the hydrogenation of both side rings (**50**). It was found that the tetrahedral Rh NPs on charcoal (▲Rh/C, particle size 4.5–5.3 nm) showed excellent activity and selectivity in this reaction, being more efficient than either commercial Rh/C or spherical Rh NPs (•Rh/C, particle size 4.4–5.2 nm) (Scheme 13.26) [51]. The tetrahedral Rh NPs were also effective in the hydrogenation of a wide range of monoarenes. For instance, hydrogenation of anisole (**3**), benzene (**1**), toluene (**80**), and phenol (**82**) was complete within 1 h.

A new Rh catalyst [Rh/AlO(OH)] that is recyclable (can be recovered simply by filtration and reused 10 times without activity loss) and highly active in the hydrogenation of arenes was synthesized from $RhCl_3{\cdot}H_2O$, 2-butanol, and Al(O-*sec*-Bu)$_3$ and then entrapped in a highly porous and fibrous boehmite matrix. The particle size of Rh was estimated to be 2.5–3.0 nm. The catalyst was highly active at room temperature under 1 atm H_2 for monosubstituted arenes such as toluene (**80**), phenol (**82**), and ethyl benzoate (**84**). Methylcyclohexane (**81**) and cyclohexanol (**83**) were

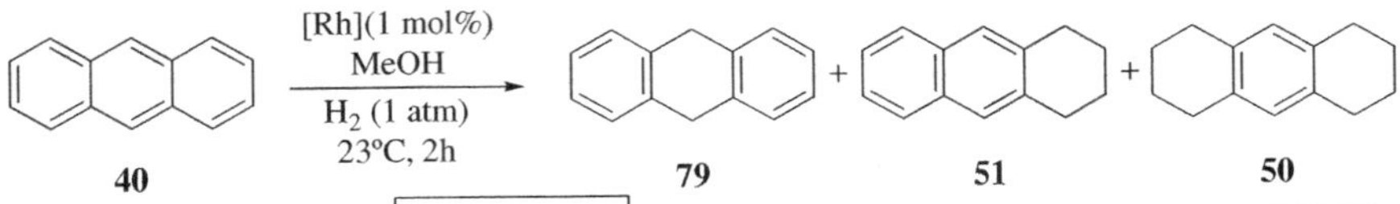

	Conversion (%)	79	51	50
(▲Rh/C)	100	2%	0%	98%
(●Rh/C)	99.5	6,7%	76.3%	16.0%
(Rh/C)	4.8	35.4%	45.8%	18.8%

SCHEME 13.26 Comparison of catalytic activities and selectivities of Rh on charcoal for the hydrogenation of anthracene (**40**).

quantitatively obtained after 1 h from toluene (**80**) and phenol (**82**), respectively. Hydrogenation of ethyl benzoate (**84**) to give ethyl cyclohexanecarboxylate (**85**) proceeded more slowly, and the process needed 4.5 h to be completed. Disubstituted arenes were also hydrogenated successfully with stereoselectivities between 94:6 in the case of *o*-xylene (**37**) and 52:48 for *o*-cresol (**86**). The *cis* isomer was always the major isomer of the reaction mixture (Scheme 13.27) [52].

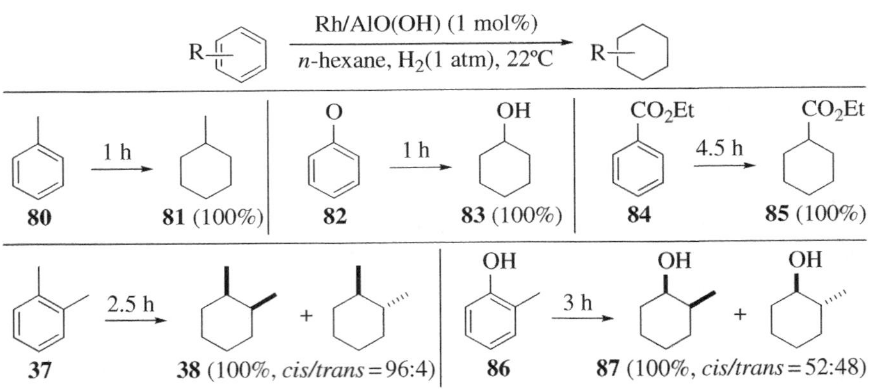

SCHEME 13.27 Hydrogenation of arenes with Rh NPs entrapped in a boehmite matrix [Rh/AlO(OH)].

Supported Ru NPs have also been used in catalytic hydrogenations of arenes. For instance, Ru NPs supported on platelet carbon nanofibers (CNF-P) displayed excellent catalytic activity. The catalyst was prepared by reaction of CNF-P with $Ru_3(CO)_{12}$ in toluene at 110°C. After TEM studies, the NPs size of Ru was determined to be 2–4 nm, and they were located on the edge of the graphite layers. Hydrogenation of monosubstituted aromatic compounds with supported Ru NPs took place at 100°C under 30 atm H_2 in excellent yields under solventless conditions. Toluene (**80**) and ethyl benzoate (**84**) were quantitatively converted into the corresponding cyclohexyl derivatives **81** and **85** after 2.5 and 5 h, respectively. However, longer reaction times were required for the nitrogen-containing compounds aniline (**88**) and (*R*)-1-phenylethylamine (**90**). Importantly, (*R*)-1-cyclohexylethylamine (**91**) was obtained without loss of optical purity with regard to the starting material **90** (Scheme 13.28) [53]. These hydrogenations required harsher reaction conditions than in the case of Rh NPs. On the other hand, Ru/CNF-P is a powerful and robust catalyst for the hydrogenation of various aromatic compounds, being also useful as a reusable catalyst.

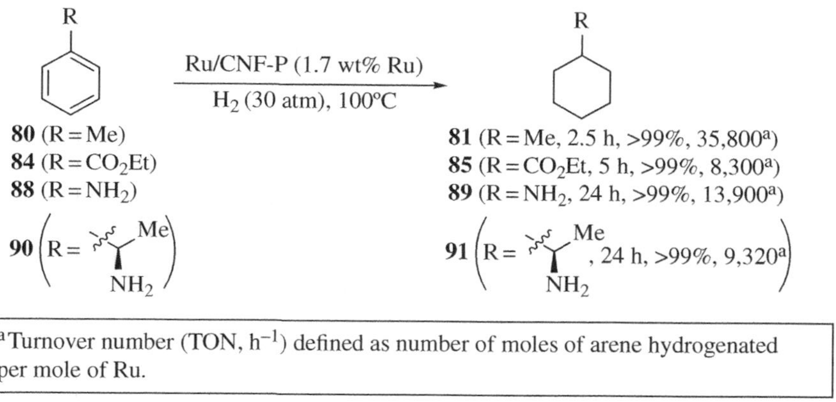

SCHEME 13.28 Hydrogenation of arenes with Ru NPs supported on platelet carbon nanofibers [Ru/CNF-P].

13.3.2.3 Organometallic Complexes Tethered on Silica-Supported Metal Catalysts Although homogeneous metal catalysts are generally more selective than heterogeneous metal catalysts, the separation and recovery of homogeneous catalysts from reaction solutions represents an important technical challenge. This could be overcome by tethering homogeneous metal catalysts to insoluble supports, combining the advantages of the high activity and selectivity of soluble complexes with the ease of catalyst recovery. Inorganic oxides are ideal supports for tethering metal complex catalysts due to their rigid structure and tolerance to a wide range of reaction conditions. Among these inorganic oxides, silica is the most commonly used. A silica-bound Rh(III) complex was reported in 1981 and used in the hydrogenation of aromatic compounds [54]. More recently, Rh isocyanide complexes $RhCl(CO)[CN(CH_2)_3Si(OEt)_3]_2$ ($Rh\text{-}CNR_2$) and $RhCl[CN(CH_2)_3Si(OEt)_3]_3$ ($Rh\text{-}CNR_3$) were tethered to the silica-supported metal heterogeneous catalysts $M\text{-}SiO_2$ (M = Pd, Pt, Ru) to give the tethered complex on supported metal (TCSM) catalysts $Rh\text{-}CNR_2/Pd\text{-}SiO_2$ and $Rh\text{-}CNR_3/M\text{-}SiO_2$ (M = Pd, Pt, Ru) (Fig. 13.1). It was found that some of these systems displayed a synergetic action of both the tethered complex and the supported metal [55].

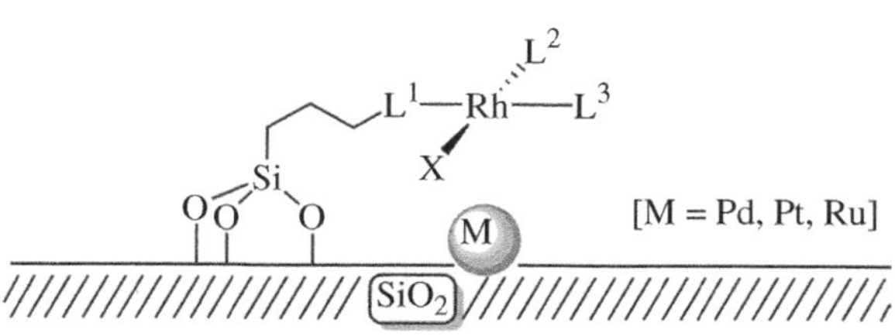

FIGURE 13.1 Illustration of a tethered homogeneous $[RhClL^1L^2L^3]$ complex catalyst on a silica-supported metal (M) heterogeneous catalyst.

These TCSM catalysts were used to catalyze the hydrogenation of arenes under, 40°C and 1 atm, relatively mild conditions. They exhibited activities that were higher than those of the separate homogeneous rhodium isocyanide complex, the separate silica-supported metal heterogeneous catalyst, or the Rh complex catalyst tethered on just silica. The activities of the TCSM catalysts were strongly affected by the nature and loading of the supported metal in the catalyst. Among the three silica-supported metal $M\text{-}SiO_2$ (M = Pd, Pt, Ru) catalysts, the Rh complex $Rh\text{-}CNR_3$ tethered on $Pd\text{-}SiO_2$ displayed the highest activity for the hydrogenation of toluene (**80**) [TOF = 5.5 mol H_2/(mol Rh min)] during 8.5 h to give methylcyclohexane (**81**). Importantly, after extended use, there was no observed Rh leaching into solution (Scheme 13.29) [55].

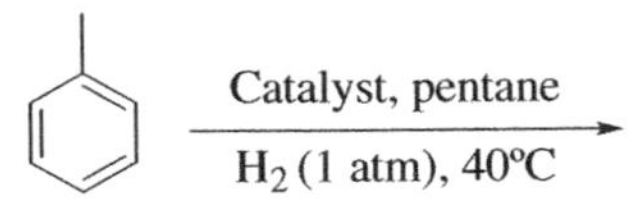

80 81

Catalyst	*t* (h)	TOF (min^{-1})
Pd-SiO_2	23	–
Rh-CNR_2	18	0.08
Rh-CNR_2/SiO_2	8.5	–
Rh-CNR_2/Pd-SiO_2	8.5	4.8
Rh-CNR_2	6.0	–
Rh-CNR_2/SiO_2	8.5	0.7
Rh-CNR_2/Pd-SiO_2	8.5	5.5

SCHEME 13.29 Hydrogenation of toluene with TCSM catalysts Rh-CNR_2/Pd–SiO_2 and Rh-CNR_3/Pd–SiO_2.

Other tethered complex catalysts, such as Rh–S–P/Pd–SiO_2, Rh–S/P–Pd–SiO_2, Rh(N–P)/Pd–SiO_2, and Rh(N–N)/Pd-SiO_2, were prepared from Rh complexes $Rh_2[\mu\text{-}S(CH_2)_3Si(OCH_3)_3]_2(CO)_4$ and $Rh_2[\mu\text{-}S(CH_2)_3Si(OCH_3)_3]_2[Ph_2P(CH_2)_3Si(OEt)_3]_2(CO)_2$, **92** and **93**, respectively. These catalytic systems were also active for the hydrogenation of toluene (**80**). The process was carried out at 40°C

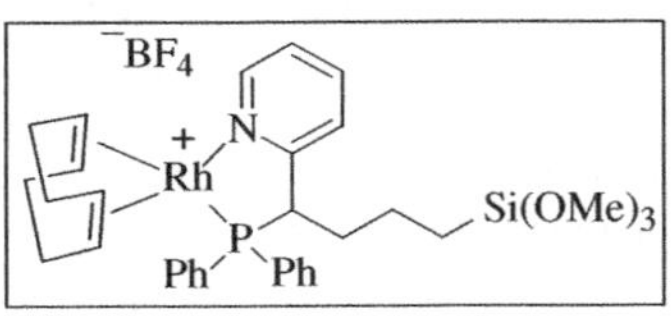

92

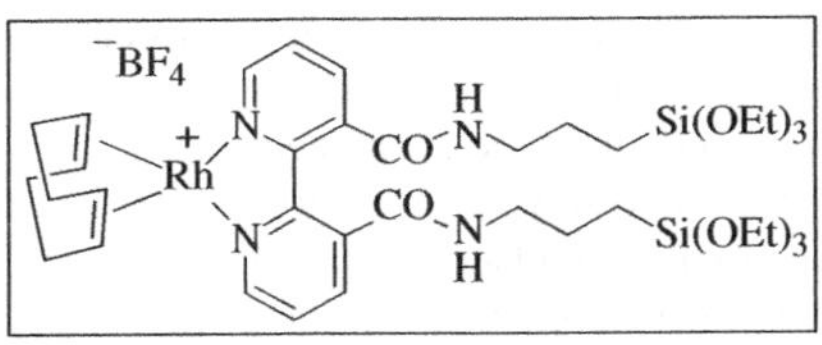

93

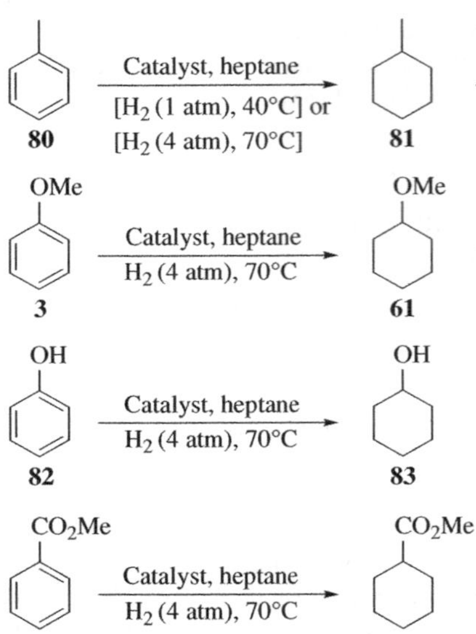

Catalyst	*t* (h)	TOF (min^{-1})
Rh-S-P/Pd-SiO_2[a]	21	0.95
Rh-S/P-(Pd-SiO_2)[a]	23.5	0.91
Rh-(N-P)/Pd-SiO_2[b]	6	48
Rh-(N-N)/Pd-SiO_2[b]	6	47
Rh-(N-P)/Pd-SiO_2	6	30
Rh-(N-N)/Pd-SiO_2	6	51
Rh-(N-P)/Pd-SiO_2[c]	6	11.8
Rh-(N-N)/Pd-SiO_2[d]	7	15.8
Rh-(N-P)/Pd-SiO_2	25	9.3
Rh-(N-N)/Pd-SiO_2	25	18.6

[a] H_2 (1 atm), 40°C. [b] H_2 (4 atm), 70°C. [c] 48% of cyclohexanone was also obtained. [d] 59% of cyclohexanone was also obtained.

SCHEME 13.30 Hydrogenation of arenes with TCSM catalysts Rh–S–P/Pd–SiO_2, Rh–S/P–Pd–SiO_2, Rh(N–P)/Pd–SiO_2, and Rh(N–N)/Pd–SiO_2.

and 1 atm H_2 with Rh–S–P/Pd–SiO_2 and Rh–S/P–Pd–SiO_2 [56] and at 70°C and 4 atm H_2 with the complexes containing the new pyridylphosphine and bipyridyl ligands **92** and **93** [57], respectively. In addition, TCSM catalysts Rh(N–P)/Pd–SiO_2 and Rh(N–N)/Pd–SiO_2 were also very active for the hydrogenation of other arenes at 70°C and 4 atm H_2. The higher rates of hydrogenation of toluene (**80**) and anisole (**3**), as compared with that of methyl benzoate (**94**) in the presence of either catalyst, suggested that electron-donating substituents in the arene accelerate the rate. Regarding the selectivity of the process, the hydrogenation of toluene (**80**), anisole (**3**), and methyl benzoate (**94**) provided methyl cyclohexane (**81**), methoxycyclohexane (**61**), and methyl cyclohexanecarboxylate (**95**), respectively, as the only reaction products. On the other hand, phenol (**82**), under the same reaction conditions, led to a mixture of cyclohexanol (**83**) and cyclohexanone (Scheme 13.30). The synergistic advantages of the tethered complex and supported metal in these TCSM catalysts were also observed for the hydrogenation of toluene.

13.3.2.4 Diastereoselective Hydrogenation The hydrogenation of substituted aromatic compounds bearing chiral auxiliaries could proceed in a stereoselective way. For instance, 2-oxazolidinone-substituted quinolines **96** were transformed into 5,6,7,8-tetrahydroquinolines **97** in the presence of PtO_2 as catalyst and trifluoroacetic acid as solvent at room temperature and 20 bar H_2. These were optimal conditions for achieving high levels of chemo- and diastereoselectivity, since hydrogenation took place only on the carbocyclic quinoline ring with diastereomeric ratios of up to 89:11. The relative position of the newly formed stereocenter and the chiral auxiliary is important in order to get good diastereoselectivities. Thus, the highest dr was obtained in the case of 8-methyl-substituted tetrahydroquinoline derivative **97c** (5 bonds) and the lowest for the 6-methyl derivative **97a** (7 bonds). In order to explain the stereochemical pathway of the hydrogenation, a model was proposed. Under acidic conditions and due to an intramolecular hydrogen bond, a rigid structure with the top face shielded by the isopropyl group of the auxiliary is formed. Adsorption on the surface of the heterogeneous catalyst and subsequent hydrogenation would take place from the less hindered bottom face of the system (Scheme 13.31) [58].

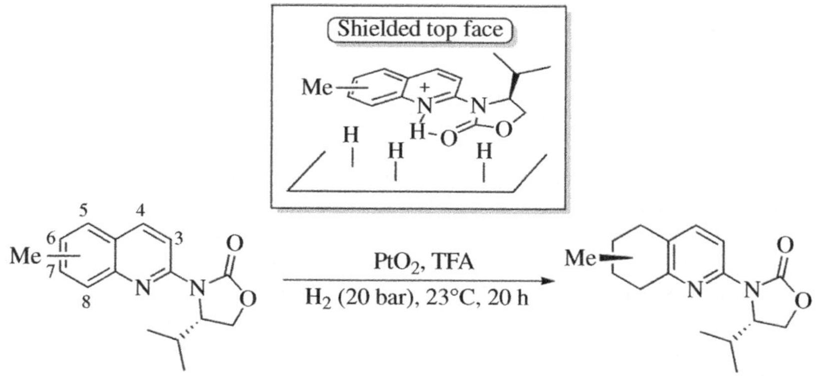

96

97a (6-Me, 99%, 81:19 dr)
97b (7-Me, 91%, 86:14 dr)
97c (8-Me, 95%, 89:11 dr)

SCHEME 13.31 Diastereoselective hydrogenation of methyl-substituted quinolines **96**.

13.4 ELECTROCHEMICAL REDUCTIONS

Large-scale catalytic hydrogenation of arenes to produce cycloalkanes requires the use of hydrogen plants, and many of these processes are carried out under high pressures, making them both economically and environmentally costly. The possibility of hydrogenation under mild conditions normally employed in electrosynthesis (room temperature and atmospheric pressure) makes it an attractive alternative to the high-pressure hydrogen treatment [59]. Furthermore, the electrochemical production of hydrogen directly at the surface of the catalyst would circumvent the compression, transportation, and storage of hydrogen. Electrochemical reduction of hydrated protons gives reactive atomic hydrogen species adsorbed on the surface of a negatively charged cathode material. When the active hydrogen species are formed on a catalytic surface, they can hydrogenate aromatic compounds. However, a competing reaction is the recombination of hydrogen on the surface to form hydrogen gas. The steps include the production of chemisorbed hydrogen, adsorption of the arene, and desorption of the hydrogenated compound, as illustrated in Scheme 13.32. In order to perform efficient hydrogenations, the arene should be soluble in the electrolytic medium. In addition, the nature of the electrode materials and their adsorption properties play an important role [60]. The electrochemical hydrogenations are usually developed in proton-donor solvents, such as alcohols and water, with cathodes of Ni, Rh, Pd, and Pt on different supports and concentrations.

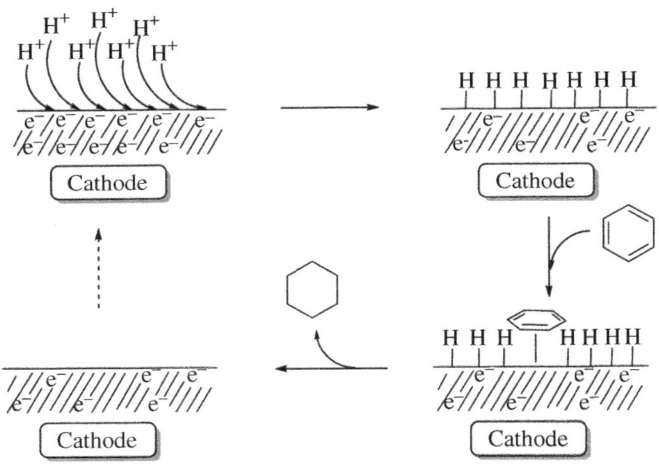

SCHEME 13.32 Simple model for the electrocatalytic hydrogenation of arenes.

The most widely used electrode in electrocatalytic hydrogenation has been a Pt electrode covered with different kinds of metal blacks. It was found that the hydrogenation of phenol (**82**) in dilute H_2SO_4 on a Pt/C electrode after passage of 12 F/mol gave cyclohexanol (**83**) as the only isolated reaction product, although in 54% yield. However, the same process on the Rh/C electrode gave cyclohexanol (**83**) in 92% yield. The observed difference between Pt and Rh is similar to the results obtained in other catalytic hydrogenations where Pt is known to give considerable hydrogenolysis of phenols and phenyl ethers as compared to an Rh catalyst. On the other hand, Pd/C gave only low electrocatalytic conversion, and the transformation did not take place at all in the case of Ni/C. It is important to point out that a variation of the current density from 0.2 to 1.0 A only had a small effect on the current efficiency. The hydrogenation of *o*-cresol (**86**) with Rh/C electrode gave 2-methylcyclohexanol (**87**) also in high yield and stereoselectively with a *cis/trans* ratio of 70:30. Electrocatalytic hydrogenation was also performed with the Rh/C electrode of other aromatic compounds bearing different substituents, such as aniline (**88**) and benzoic acid (**5**), although in lower yields of compounds **89** and **72**, respectively (Scheme 13.33) [61].

The effectiveness of electrocatalytic hydrogenation depends upon the nature of the support and the metal of the electrodes. For that reason, a variety of materials have been developed, including Raney Ni/Ni, Raney Ni/Hg pool, Pt/Pt, Pt/C, Rh/C [62], Pt/Pt, and Pt/C with different concentrations of Pt [63] and different types of composite materials incorporated in reticulated vitreous carbon like fractal Ni, crystalline Ni_2B, and metallic nanoaggregates deposited on nonconductive

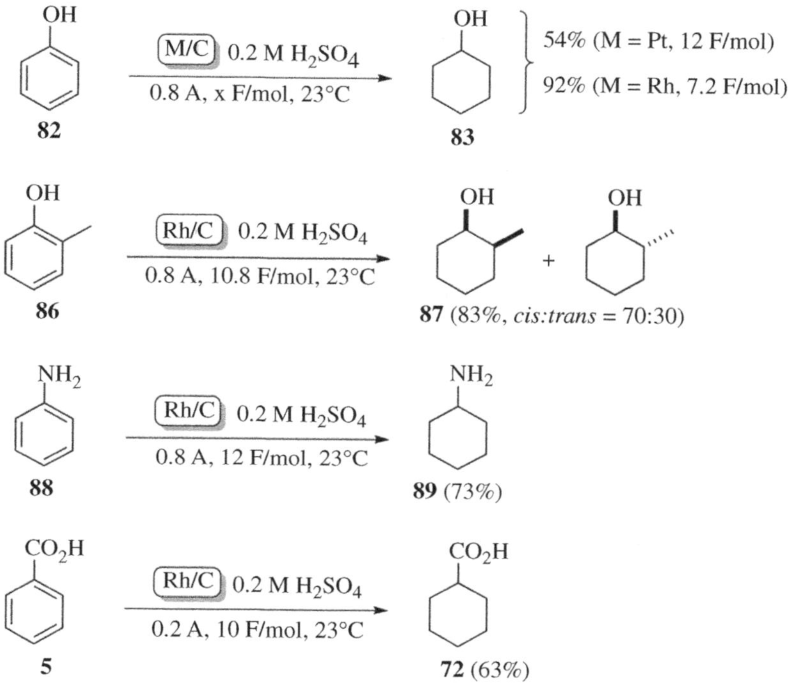

SCHEME 13.33 Electrocatalytic hydrogenation of substituted arenes.

powders like Al_2O_3, $Al(OH)_3$, $BaCO_3$, and $BaSO_4$ [61, 64]. More recently, the electrocatalytic hydrogenation of aromatic compounds in ionic liquid and water solutions over WS_2-on-glassy carbon and Raney nickel cathodes was also reported. The ionic liquid and water solutions provide high solubility of reactant, high conductivity, and protons for electrochemical hydrogenation [65].

13.5 OTHER METHODOLOGIES

Reduction of aromatic rings was also performed with the Raney Ni–Al alloy [66] in 1% aqueous KOH, using H_2O as solvent, under atmospheric pressure. Compared to catalytic hydrogenations with transition metals, this procedure is operationally simpler and safer, avoiding hydrogen gas manipulation, H_2O being the source of hydrogen in this case. The reduction of naphthalene (**19**) and acenaphthylene (**98**) at 90°C gave tetrahydronaphthalene (**49**) and hexahydroacenaphthylene (**99**), respectively, in high yields, maintaining the aromaticity of one of the rings. In the case of phenol (**82**), the reduction took place at lower temperatures (60°C) to give cyclohexanol (**83**), and 1-naphthol (**25**) led to 5,6,7,8-tetrahydro-1-naphthol (**100**) also in high yield, with the hydrogenation of the unsubstituted aromatic ring taking place exclusively (Scheme 13.34) [67].

Hydrogenation of aromatic compounds has also been achieved under metal-free conditions by activation of the molecule of dihydrogen through the use of frustrated Lewis pairs (FLPs), which are combinations of relatively bulky Lewis acids and bases where the formation of a Lewis acid–base adduct is wholly or partially inhibited [68]. This methodology has been successfully applied to the hydrogenation of N-substituted aromatic amines. Thus, the treatment of different aniline derivatives **101–104** with an equivalent of $B(C_6F_5)_3$ in toluene at 110°C and 4 atm H_2 for 36 h led to the corresponding cyclohexylammonium salts **105–108** in high yields (Scheme 13.35) [69]. Mechanistically, the formation of the salt $[R^1NH_2Ar][HB(C_6F_5)_3]$ is proposed previous the hydrogenation of the aromatic ring.

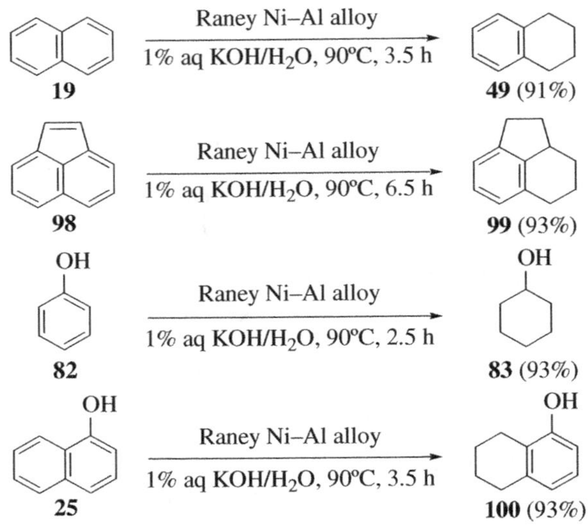

SCHEME 13.34 Hydrogenation of arenes with Raney Ni–Al alloy in aqueous media.

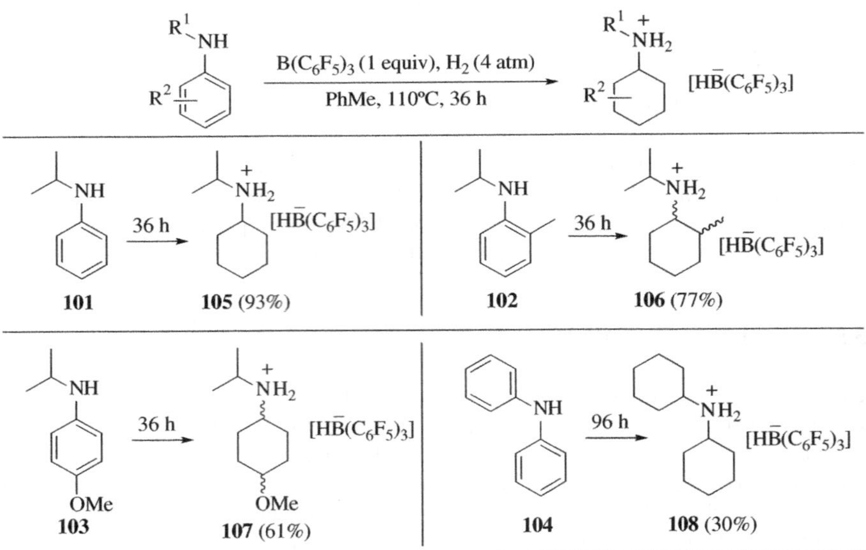

SCHEME 13.35 Metal-free hydrogenation of aromatic amines.

PAHs were also partially hydrogenated using a catalytic amount of the FLP resulting from the combination of $B(C_6F_5)_3$ and $Ph_2PC_6F_5$. The reactions were performed in dichloroethane at 80°C and 102 atm H_2. Anthracene (**40**) produced 9,10-dihydroanthracene (**109**) after 10 h in almost quantitative yield. Under identical reaction conditions, tetracene (**110**) gave 5,12-dihydrotetracene (**111**) in 90% yield, whereas dihydrogenation of tetraphene (**112**) was more difficult, leading to 7,12-tetraphene (**113**) in 30% yield after 48 h. In the proposed mechanism depicted on Scheme 13.36, the FLP $B(C_6F_5)_3/Ph_2PC_6F_5$ (**XXX**) produces the heterolysis of H_2 affording the

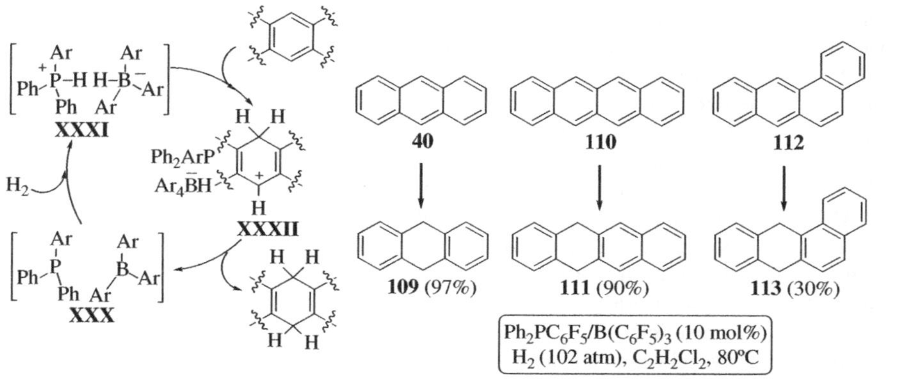

SCHEME 13.36 Metal-free partial hydrogenation of polycyclic aromatic hydrocarbons catalyzed by FLAs.

phosphonium borate salt **XXXI**, which protonates first the PAH. An irreversible hydride transfer to the resulting carbocation intermediate **XXXII** gives the dihydroderivatives, regenerating the FLP **XXX** [70].

13.6 SUMMARY AND OUTLOOK

The synthetic value of cycloalkenes and cycloalkanes as building blocks of complex molecules and in the fine chemical and pharmaceutical industries has greatly encouraged the development of efficient methods for arene hydrogenation over the last decades. Partial hydrogenation to 1,4-cyclohexadienes was achieved using dissolving metals under the so-called Birch reduction conditions, whereas catalytic hydrogenation constitutes the method of choice to produce total hydrogenation of aromatic systems, both in the presence of homogeneous and heterogeneous catalysts. In spite of being more selective than the heterogeneous catalysts, separation and recovery of the homogeneous catalysts from the reaction products is a major drawback. In addition, high pressures and costly equipments are required to carry out the hydrogenation of aromatic compounds under heterogeneous catalysis. Nonetheless, most of the large-scale industrial arene hydrogenation processes are performed with heterogeneous catalysts. Important progress has been made in the development of new catalysts combining the advantages of the high activity and selectivity of soluble complexes with the easy catalyst recovery of the heterogeneous catalysts, such as soluble nanoparticles and organometallic complexes tethered to solid supports. Electrocatalytic hydrogenation has also emerged as an attractive alternative to the previously mentioned methods due to the mild reaction conditions required for these processes to take place. Given the progress of this field, we believe that new catalysts will emerge in the coming years. Furthermore, the development of efficient catalytic stereoselective hydrogenation processes of polysubstituted aromatic compounds remains as a very interesting challenge.

ABBREVIATIONS

ATP	Adenosine triphosphate
BCoA	Benzoyl-CoA
BCR	Benzoyl-coenzyme A reductase
BMI·PF_6	1-*n*-Butyl-3-methylimidazolium hexafluorophosphate
CNF-P	Platelet carbon nanofibers

DFT	Density functional theory
DOCEA	(*R*)-*N*,*N*-Dioctyl-1-cyclohexylethylamine
FAD	Flavin adenine dinucleotide
Fd	Ferredoxin
FLPs	Frustrated Lewis pairs
FMN	Flavin mononucleotide
HEA-16-X	*N*,*N*-Dimethyl-*N*-acetyl-*N*-(2-hydroxyethyl)ammonium halides
LUMO	Lowest unoccupied molecular orbital
NCR	2-Naphthoyl-CoA reductase
NHC	*N*-Heterocyclic carbene
NOESY	Nuclear Overhauser effect spectroscopy
NP	Nanoparticle
PAHs	Polycyclic aromatic hydrocarbons
PVP	Poly(vinylpyrrolidone)
TCSM	Tethered complex on supported metal
TEM	Transmission electron microscopy
TOF	Turnover frequency

REFERENCES

[1] Birch, A. J. (1944) *J. Chem. Soc.*, 430–436.

[2] (a) Birch, A. J. and Smith, H. (1958) *Q. Rev.*, **12**, 17–33; (b) Rabideau, P. W. and Marcinow, Z. (1992) *Org. React.*, **42**, 1–334.

[3] Hine, J. (1966) *J. Org. Chem.*, **31**, 1236–1244.

[4] Wilds, A. L. and Nelson, N. A. (1953) *J. Am. Chem. Soc.*, **75**, 5360–5365.

[5] Bonjoch, J., Catena, J., and Valls, N. (1996) *J. Org. Chem.*, **61**, 7106–7115.

[6] Caluwe, P. and Pepper, T. (1988) *J. Org. Chem.*, **53**, 1786–1790.

[7] Chen, L., Gill, G. B., Pattenden, G., and Simonian, H. (1996) *J. Chem. Soc. Perkin Trans.* **1**, 31–43.

[8] Coughlin, D. J. and Salomon, R. G. (1979) *J. Org. Chem.*, **44**, 3784–3790.

[9] (a) Bachi, M. D., Epstein, J. W., Herzberg-Minzly, Y., and Loewenthal, H. J. E. (1969) *J. Org. Chem.*, **34**, 126–135; (b) Hook, J. M. and Mander, L. N. (1986) *Nat. Prod. Rep.*, **3**, 35–85.

[10] Schultz, A. G. (1999) *Chem. Commun.*, 1263–1271.

[11] Rabideau, P. W. and Huser, D. L. (1983) *J. Org. Chem.*, **48**, 4266–4271.

[12] Reggel, L., Friedel, R. A., and Wender, I. (1957) *J. Org. Chem.*, **22**, 891–894.

[13] Kaiser, E. M. and Benkeser, R. A. (1988) *Org. Synth. Coll.*, **6**, 852–855.

[14] Vogel, E., Klug, W., and Breuer, A. (1988) *Org. Synth. Coll.*, **6**, 731–736.

[15] (a) Benkeser, R. A., Burrous, M. I., Hazdra, J. J., and Kaiser, E. M. (1963) *J. Org. Chem.*, **28**, 1094–1097; (b) Benkeser, R. A., Robinson, R. E., Sauve, D. M., and Thomas, O. H. (1955) *J. Am. Chem. Soc.*, **77**, 3230–3233.

[16] Fried, J., Abraham, N. A., and Santhanakrishnan, T. S. (1967) *J. Am. Chem. Soc.*, **89**, 1044–1045.

[17] Dauben, W. G. and Teranishi, R. (1951) *J. Org. Chem.*, **16**, 550–555.

[18] Johnson, W. J., Pappo, R., and Johns, W. F. (1956) *J. Am. Chem. Soc.*, **78**, 6339–6347.

[19] Benkeser, R. A., Belmonte, F. G., and Yang, M. H. (1983) *Synth. Commun.*, **13**, 1103–1116.

[20] Nandi, P., Dye, J. L., and Jackson, J. E. (2009) *J. Org. Chem.*, **74**, 5790–5792.

[21] Thiele, B., Rieder, O., Golding, B. T., Müller, M., and Boll, M. (2008) *J. Am. Chem. Soc.*, **130**, 14050–14051.

[22] Möbitz, H. and Boll, M. (2002) *Biochemistry*, **41**, 1752–1758.

[23] Eberlein, C., Estelmann, S., Seifert, J., von Bergen, M., Müller, M., Meckenstock, R. U., and Boll, M. (2013) *Mol. Microbiol.*, **88**, 1032–1039.

[24] Kung, J. W., Baumann, S., von Bergen, M., Müller, M., Hagedoorn, P.-L., Hagen, W. R., and Boll, M. (2010) *J. Am. Chem. Soc.*, **132**, 9850–9856.

[25] Dyson, P. J. (2003) *Dalton Trans.*, 2964–2974.

[26] Rakowski, M. C., Hirsekorn, F. J., Stuhl, L. S., and Muetterties, E. L. (1976) *Inorg. Chem.*, **15**, 2379–2382.

[27] (a) Muetterties, E. L., Rakowski, M. C., Hirsekorn, F. J., Larson, W. D., Basus, V. J., and Anet, F. A. L. (1975) *J. Am. Chem. Soc.*, **97**, 1266–1267; (b) Bleeke, J. R. and Muetterties, E. L. (1081) *J. Am. Chem. Soc.*, **103**, 556–572.

[28] Stuhl, L. S., Rakowski DuBois, M., Hirsekorn, F. J., Bleeke, J. R., Stevens, A. E., and Muetterties, E. L. (1978) *J. Am. Chem. Soc.*, **100**, 2405–2410.

[29] Hirsekorn, F. J., Rakowski, M. C., and Muetterties, E. L. (1975) *J. Am. Chem. Soc.*, **97**, 237–238.

[30] Taylor, P. D. and Orchin, M. (1972) *J. Org. Chem.*, **37**, 3913–3915.

[31] Feder, H. M. and Halpern, J. (1975) *J. Am. Chem. Soc.*, **97**, 7186–7188.

[32] Chesnut, R. W., Jacob, G. G., Yu, J. S., Fanwick, P. E., and Rothwell, I. P. (1991) *Organometallics*, **10**, 321–328.

[33] Ankianiec, B. C., Fanwick, P. E., and Rothwell, I. P. (1991) *J. Am. Chem. Soc.*, **113**, 4710–4712.

[34] (a) Yu, J. S., Ankianiec, B. C., Rothwell, I. P., and Nguyen, M. T. (1992) *J. Am. Chem. Soc.*, **114**, 1927–1929; (b) Rothwell, I. P. (1997) *Chem. Commun.*, 1331–1338.

[35] Parkin, B. C., Clark, J. R., Visciglio, V. M., Fanwick, P. E., and Rothwell, I. P. (1995) *Organometallics*, **14**, 3002–3013.

[36] (a) Grey, R. A., Pez, G. P., and Wallo, A. (1980) *J. Am. Chem. Soc.*, **102**, 5948–5949; (b) Linn, D. E. and Halpern, J. (1987) *J. Am. Chem. Soc.*, **109**, 2969–2974.

[37] (a) Glorius, F. (2005) *Org. Biomol. Chem.*, **3**, 4171–4175; (b) Zhou, Y.-G. (2007) *Acc. Chem. Res.*, **40**, 1357–1366; (c) Wang, D.-S., Chen, Q.-A., Lu, S.-M., and Zhou, Y.-G. (2012) *Chem. Rev.*, **112**, 2557–2590.

[38] Urban, S., Ortega, N., and Glorius, F. (2011) *Angew. Chem. Int. Ed.*, **50**, 3803–3806.

[39] Widegren, J. A. and Finke, R. G. (2003) *J. Mol. Catal. A Chem.*, **191**, 187–207.

[40] Gual, A., Godard, C., Castillon, S., and Claver, C. (2010) *Dalton Trans.*, **39**, 11499–11512.

[41] (a) Lu, F., Liu, J., and Xu, J. (2006) *Adv. Synth. Catal.*, **348**, 857–861; (b) Lu, F., Liu, J., and Xu, J. (2007) *J. Mol. Catal. A Chem.*, **271**, 6–13.

[42] Roucoux, A., Schulz, J., and Patin, H. (2003) *Adv. Synth. Catal.*, **345**, 222–229.

[43] Nowicki, A., Le Boulaire, V., and Roucoux, A. (2007) *Adv. Synth. Catal.*, **349**, 2326–2330.

[44] Migowski, P. and Dupont, J. (2007) *Chem. Eur. J.*, **13**, 32–39.

[45] Ott, L. S., Cline, M. L., Deetlefs, M., Seddon, K. R., and Finke, R. G. (2005) *J. Am. Chem. Soc.*, **127**, 5758–5759.

[46] Nasar, K., Fache, F., Lemaire, M., Béziat, J. C., Besson, M., and Gallezot, P. (1994) *J. Mol. Catal.*, **87**, 107–115.

[47] Qi, S.-C., Wei, X.-Y., Zong, Z.-M., and Wang, Y.-K. (2013) *RSC Adv.*, **3**, 14219–14232.

[48] Morin, C., Simon, D., and Sautet, P. (2006) *Surf. Sci.*, **600**, 1339–1350.

[49] Maegawa, T., Akashi, A., Yaguchi, K., Iwasaki, Y., Shigetsura, M., Monguchi, Y., and Sajiki, H. (2009) *Chem. Eur. J.*, **15**, 6953–6963.

[50] Wang, H. and Zhao, F. (2007) *Int. J. Mol. Sci.*, **8**, 628–634.

[51] Park, K. H., Jang, K., Kim, H. J., and Son, S. U. (2007) *Angew. Chem. Int. Ed.*, **46**, 1152–1155.

[52] Park, I. S., Kwon, M. S., Kim, N., Lee, J. S., Kang, K. Y., and Park, J. (2005) *Chem. Commun.*, 5667–5669.

[53] Takasaki, M., Motoyama, Y., Higashi, K., Yoon, S.-H., Mochida, I., and Nagashima, H. (2007) *Chem. Asian J.*, **2**, 1524–1533.

[54] Ward, M. D. and Schwartz, J. (1981) *J. Am. Chem. Soc.*, **103**, 5253–5255.

[55] Gao, H. and Angelici, R. J. (1999) *Organometallics*, **18**, 989–995.

[56] Gao, H. and Angelici, R. J. (1999) *J. Mol. Catal. A Chem.*, **145**, 83–94.

[57] Yang, H., Gao, H., and Angelici, R. J. (2000) *Organometallics*, **19**, 622–629.

[58] Heitbaum, M., Fröhlich, R., and Glorius, F. (2010) *Adv. Synth. Catal.*, **352**, 357–362.

[59] Dubé, P., Kerdouss, F., Laplante, F., Proulx, P., Brossard, L., and Ménard, H. (2003) *J. Appl. Electrochem.*, **33**, 541–547.

[60] Wismeijer, A. A., Kieboom, A. P.G., and Van Bekkum, H. (1986) *Recl. Trav. Chim. Pays-Bas*, **105**, 129–136.

[61] Miller, L. L. and Christensen, L. (1978) *J. Org. Chem.*, **43**, 2059–2061.

[62] Casadei, M. A. and Pletcher, D. (1988) *Electrochim. Acta*, **33**, 117–120.

[63] Moutet, J. C. (1992) *Org. Prep. Proc. Int.*, **24**, 309–325.

[64] (a) Tountian, D., Brisach-Wittmeyer, A., Nkeng, P., Poillerat, G., and Menard, H. (2009) *J. Appl. Electrochem.*, **39**, 411–419; (b) Tountian, D., Brisach-Wittmeyer, A., Nkeng, P., Poillerat, G., and Menard, H. (2009) *Langmuir*, **25**, 11105–11111.

[65] Tsyganok, A., Holt, C. M., Murphy, S., Mitlin, D., and Gray, M. R. (2012) *Fuel*, **93**, 415–422.

[66] For a review on the use of nickel-aluminum alloy as a reducing reagent, see: Keefer, L. K. and Lunn, G. (1989) *Chem. Rev.*, **89**, 459–502.

[67] Tsukinoki, T., Kanda, T., Liu, G.-B., Tsuzuki, H., and Tashiro, M. (2000) *Tetrahedron Lett.*, **41**, 5865–5868.

[68] Stephan, D. W. and Erker, G. (2010) *Angew. Chem. Int. Ed.*, **49**, 46–76.

[69] Mahdi, T., Heiden, Z. M., Grimme, S., and Stephan, D. W. (2012) *J. Am. Chem. Soc.*, **134**, 4088–4091.

[70] Segawa, Y. and Stephan, D. W. (2012) *Chem. Commun.*, **48**, 11963–11965.

14

SELECTIVE OXIDATION OF AROMATIC RINGS

OXANA A. KHOLDEEVA
Boreskov Institute of Catalysis, Novosibirsk State University, Novosibirsk, Russia

14.1 INTRODUCTION

The selective oxidation of aromatic rings plays a central role in organic synthesis [1, 2] and biological systems [3]. Phenols are important antioxidants and intermediates in the production of resins, plastics, fine chemicals, and pharmaceuticals [1, 4]. Quinones serve as versatile building blocks en route to many biologically active compounds [2, 5–7]. Scheme 14.1 presents examples demonstrating utility of nuclear aromatic oxidation in the production of vital fine chemicals.

The insertion of oxygen atom into aromatic nucleus is one of the most difficult transformations in the organic synthesis. Until recently, oxyfunctionalization of arenes was accomplished through multistage technologies, employing environmentally unfriendly reagents [1, 4]. However, the ever-increasing attention toward the economical and environmental sustainability is prompting the scientific community to develop greener chemical processes, which eliminate the use of hazardous reactants and reduce generation of waste [8–10]. Therefore, the direct catalytic oxidation of arenes to the synthetically relevant scaffolds, phenols and quinones, with environmentally benign oxidants is a demanding task of the modern organic synthesis.

While numerous literature deals with oxygenation of sp^3 C—H bonds, reviews that focus, at least partially, on the oxidation of aromatic sp^2 C—H bonds are scarce [11–14]. The aim of this chapter is to give the reader an overview on the current state of the art in the field of the selective aromatic oxidation. The scope of this review is limited to aromatic C—H bond oxygenation and oxidative dehydrogenation, that is, to the formation of C—O and C=O bonds with retention of ring C—C

Arene Chemistry: Reaction Mechanisms and Methods for Aromatic Compounds,
First Edition. Edited by Jacques Mortier.

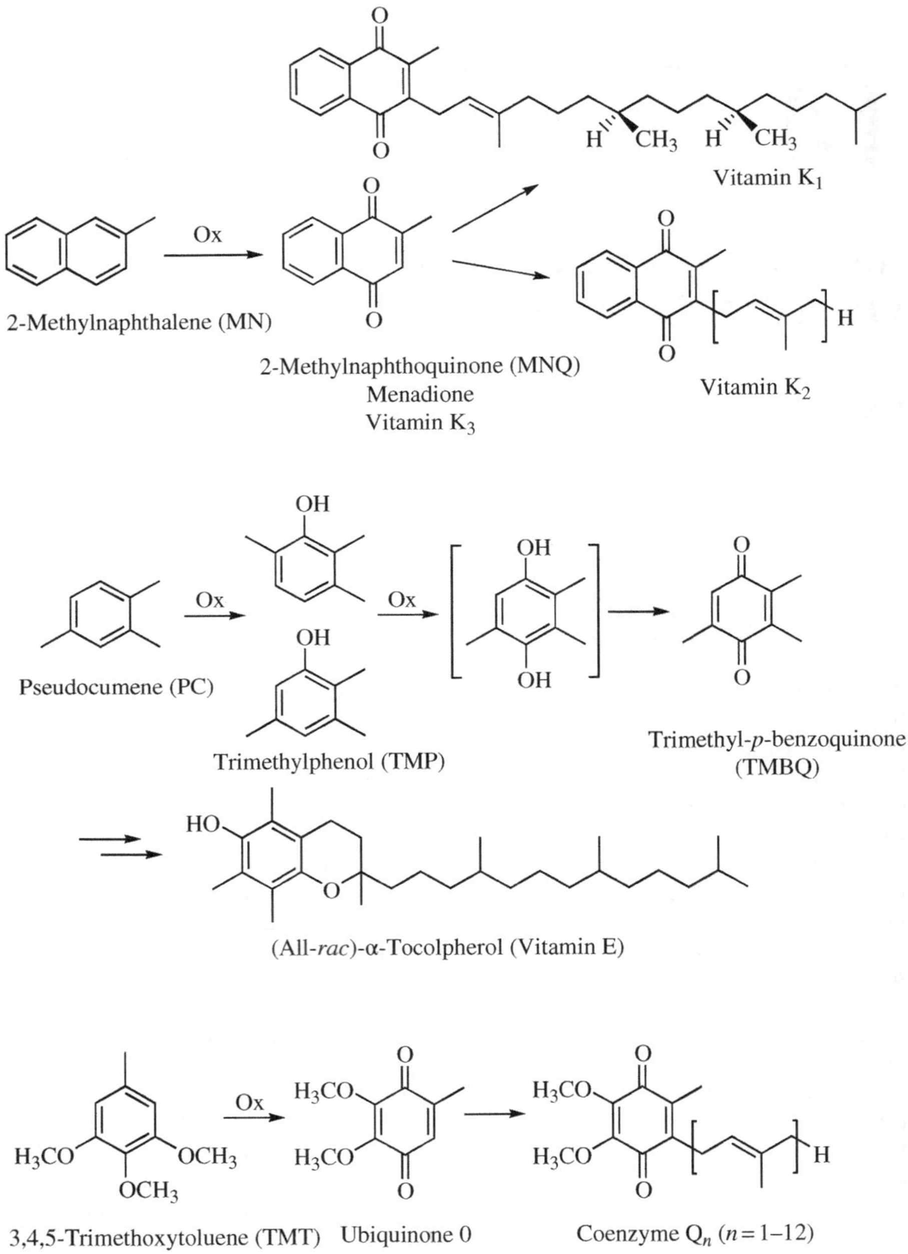

SCHEME 14.1 Aromatic oxidations as key steps in the synthesis of bioactive compounds.

bonding. Since Part VI of this book treats coupling reactions of arenes, here, we address oxidative C—C and C—O couplings only when such processes compete with the formation of phenolic and quinoid products.

The chapter starts with a brief description of the main mechanistic principles operating in the nuclear oxidation of arenes. The following sections cover different synthetic methodologies that allow realization of this demanding type of organic transformations. After a short discussion on stoichiometric oxidations, we will focus on catalytic methods, with particular attention drawn to environmentally benign ones that employ green oxidants—O_2 and H_2O_2. Special sections briefly treat photochemical and electrochemical methodologies as well as enzymatic hydroxylations.

14.2 MECHANISTIC PRINCIPLES

The majority of selective oxidation mechanisms can be divided into two fundamentally different types: homolytic and heterolytic ones [15]. Homolytic mechanisms involve one-electron elementary steps, such as hydrogen atom transfer (HAT), single electron transfer (SET), addition of a radical species to aromatic nuclear, etc. Heterolytic mechanisms do not engage radical species and merge a range of two-electron processes, that is, oxygen atom transfer or hydride transfer. In this section, we discuss some fundamental features of the mechanisms relevant for the selective oxidation of aromatic rings.

14.2.1 Autoxidation

A direct reaction of O_2 in its ground triplet state ($^3O_2$) with singlet organic molecules to give singlet products is a spin-forbidden process. The spin-conservation obstacle can be overcome by (i) involvement of $^3O_2$ in a radical chain process, (ii) interaction with paramagnetic transition metal ions, and (iii) photochemical activation of either $^3O_2$ to give singlet oxygen, $^1O_2$, or an organic substrate to transform it into triplet state.

Conventional free radical chain oxidation (referred to as autoxidation) involves chain initiation, propagation, and termination steps [15–17]. The possibility of hydrogen atom abstraction by radical species in arenes without alkyl side chains is small because of the strong sp^2 C—H bonds (the dissociation energies of the sp^2 Ar—H bonds and sp^3 $ArCR_2$—H are 112 and ca. 90 kcal/mol, respectively [18]). Phenols readily react by direct abstraction of $H^{\cdot}$ from the hydroxyl group to form phenoxyl radical $ArO^{\cdot}$, and this pathway is implicated in their antioxidant properties [16, 19]. Numerous literature addresses structure–reactivity relationships in the chemistry of phenols and phenoxyl radicals (see Refs. 1, 16, 19 and references therein).

In the absence of light, metal ions, or bases, autoxidation of phenols (Scheme 14.2) is typically a slow process.

Once $ArO^{\cdot}$ is formed (Eq. 14.1), it can be entrapped by O_2 producing peroxy radical $ArO_3^{\cdot}$ (Eq. 14.2), which abstracts H atom from another ArOH molecule to give $ArO^{\cdot}$ and hydroperoxide (HP) as the primary oxidation product (Eq. 14.3). At low oxygen pressure, chain termination via recombination of two $ArO^{\cdot}$ radicals results in the formation of a labile dimer that after enolization yields bisphenol (BP) that can be further oxidized to diphenoquinone (DPQ) (Eq. 14.4). Benzoquinone (BQ) can derive from chain termination steps (Eqs. 14.5 and 14.6) or appear in the course of HP transformations, for example, dehydration (Eq. 14.7), which is favored by the formation of conjugate C=O and C=C bonds.

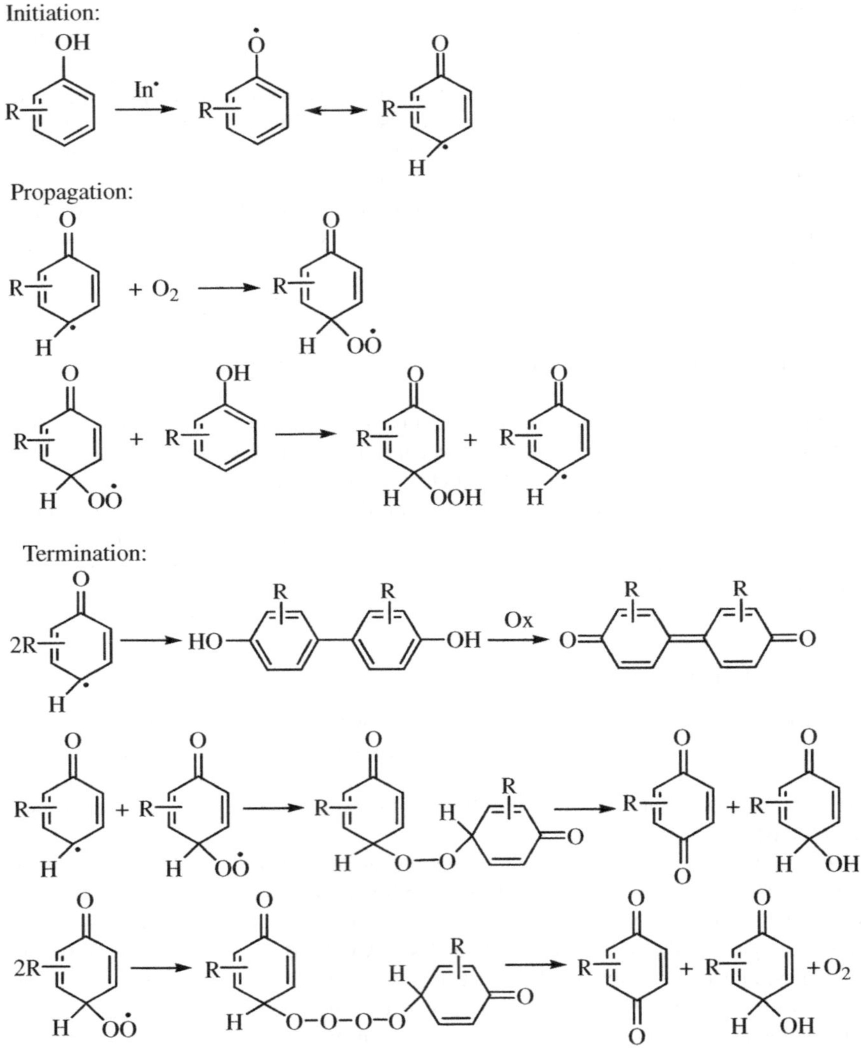

(14.1)

(14.2)

(14.3)

(14.4)

(14.5)

(14.6)

SCHEME 14.2 Main steps in free radical chain oxidation of phenols.

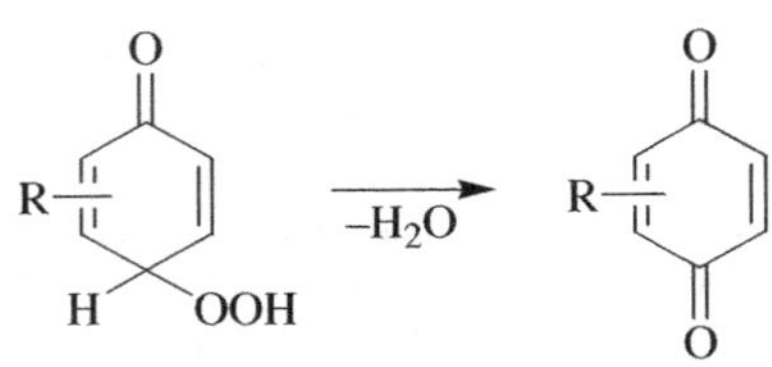

(14.7)

Various transformation routes of ArO$^{\cdot}$ result in a complex mixture of products [1]. However, some hindered phenols and naphthols, for which the radical coupling is sterically restricted, afford HP (Eq. 14.3) as stable major products [20].

Hydrogen bond accepting (HBA) solvents (alcohols, ketones, nitriles, etc.) form hydrogen bonds with the phenolic hydroxyl group, thus retarding the rate of HAT from ArOH to radical species [21]. Under basic conditions, a mechanism of sequential proton loss electron transfer (SPLET) becomes operative. This mechanism is favored with respect to HAT in solvents that are able to solvate and stabilize phenoxide anions ArO^- and is suppressed by nonionizing HBA solvents [21]. Details on phenol autoxidation under basic conditions can be found in the book chapter of Steenken and Neta [19b].

Diphenols are more reactive compounds than parent phenols, and their autoxidation easily proceeds under mild conditions (atmospheric pressure, 30–60°C) through the formation of semiquinone radicals that dissociate to a proton and semiquinone radical anion in polar solvents and can disproportionate via Equation 14.8 [16].

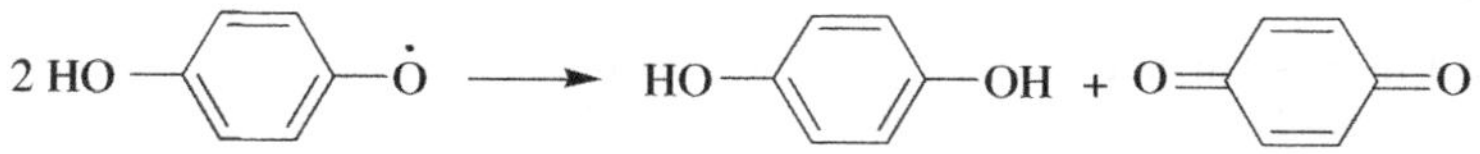

(14.8)

The capability of anthrahydroquinone derivatives to afford corresponding anthraquinones and H_2O_2 in the course of autoxidation, through decomposition of intermediate HP (Eq. 14.9), constitutes a basis for the main industrial process of the hydrogen peroxide production (the *Riedl–Pfleiderer* process) [22].

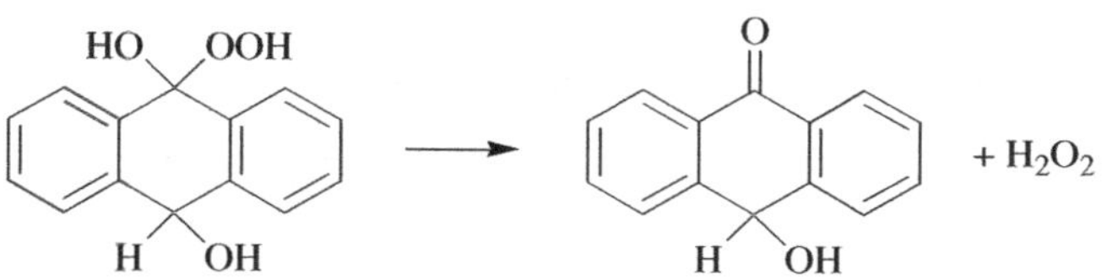

(14.9)

14.2.2 Spin-Forbidden Reactions with Triplet Oxygen

A limited number of organic substrates are able to react spontaneously with triplet oxygen in the absence of any catalyst or light at mild conditions [17]. Seip and Brauer have discovered that helianthrene (HEL) interacts with $^3O_2$ in nonpolar solvents in the dark to give endoperoxide (HELPO) (Eq. 14.10) [23].

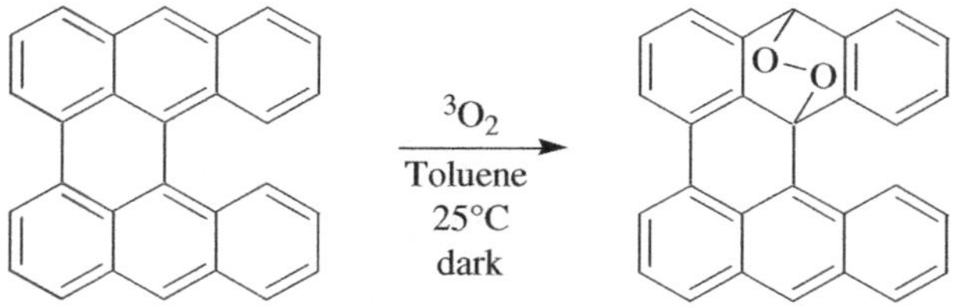

(14.10)

HEL HELPO

Solvent effects, in particular, a pronounced external heavy atom effect [24] (rate acceleration upon replacement of toluene for iodobenzene) found for the dark HEL oxidation suggested spin inversion (thermal intersystem crossing (ISC)) in the rate-limiting step, leading to the formation of a polar zwitterionic intermediate (^{1}Z).

$$\text{HEL} + {}^3\text{O}_2 \rightleftharpoons \text{HEL} \cdots {}^3\text{O}_2 \xrightleftharpoons{\text{ISC}} \underset{{}^1\text{Z}}{{}^+\text{HEL–O–O}^-} \rightleftharpoons \text{HELPO} \qquad (14.11)$$

Kholdeeva and Rossi have found that oxidation of 2-methyl-1-naphthol (MNL) with molecular oxygen smoothly proceeds in nonpolar organic solvents under mild conditions (3 atm O_2, 20–80°C) in the absence of any catalyst, in dark to give 2-methyl-1,4-naphthoquinone (MNQ, menadione, or vitamin K_3) in 80–85% yields [25]. This reaction revealed features similar to those found for the dark oxidation of HEL (external heavy atom effect, lack of selectivity in methanol, similar activation parameters, etc.), pointing out thermal ISC mechanism [26]. Radical initiators, transition metals, and low oxygen pressure push the reaction toward radical chain autoxidation.

The driving force for the spin-forbidden reactions is supposed to be either a strong strain in the substrate molecule, which reduces significantly in the product (the case of HEL and HELPO) (Ref. 23 and references therein), or resonance effects [26, 27].

14.2.3 Radical Hydroxylation (Addition–Elimination)

While radical species can hardly abstract H atom from benzene rings, some of them, in particular $HO^{\cdot}$ radical, are easily trapped by aromatic molecules. Radical hydroxylation most often follows Fenton-type chemistry that involves generation of $HO^{\cdot}$ in the course of homolytic decomposition of H_2O_2 induced by an iron(II) salt and addition of $HO^{\cdot}$ to an aromatic nuclear to give hydroxycyclohexadienyl radical **I**. This radical may dimerize, be oxidized to phenols (Cu(II) is one of the most effective oxidants for this), or undergo an acid-catalyzed collapse to radical cation [15, 28]. Scheme 14.3 shows the classical mechanism suggested by Walling [28].

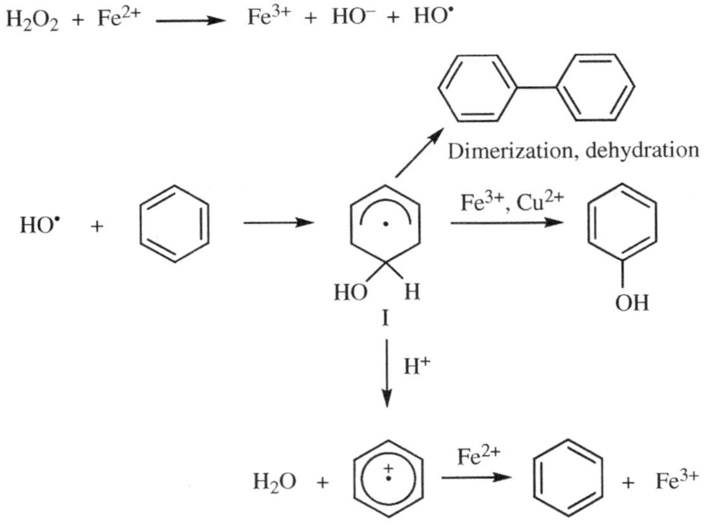

SCHEME 14.3 Radical hydroxylation of benzene with $HO^{\cdot}$ radical (Adapted from Ref. [28]).

In addition, both **I** and the radical cation can trap O_2, leading to new radical species and complication of the overall mechanistic picture [29]. In case of alkylaromatics, radical cations undergo side-chain fragmentation to benzyl radicals, giving rise to side-chain oxidation products [15, 28] (see Section 14.2.4).

Hydroxyl radicals possess week electrophilic properties as indicated by the order of reactivity of substituted benzenes and distribution of phenolic isomers, although the latter depends on the reaction conditions [28, 29]. The Fenton hydroxylation in aqueous solution reveals small (<5%) values of the "NIH shift" (i.e., migration of hydrogen atom from the site of hydroxylation to the adjacent carbon [30]). The reaction in CH_3CN demonstrated remarkably high shift values (30–40%) [31], which is typical of enzymatic processes [30]. Sawyer and coworkers proposed that the change in solvent might favor a mechanistic shift from $HO^{\cdot}$ to a metal-centered oxidant [32].

Radical addition–elimination mechanism with participation of an electrophilic V(IV)-OO· species was proposed for aromatic hydroxylation by a peroxo complex $[VO(O_2)(pic)(H_2O)_2]$ (pic = picolinate) that oxidizes benzene at room temperature in CH_3CN to produce phenol in a yield of 55%, without any coupling products (Scheme 14.4) [33]. A radical anion was suggested as an alternative hydroxylating species [34].

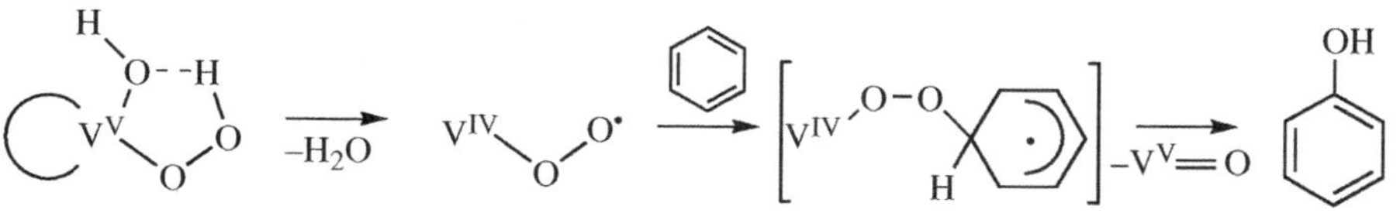

SCHEME 14.4 Benzene hydroxylation with $[VO(O_2)(pic)(H_2O)_2]$ (Adapted from Ref. [33]).

14.2.4 Electron Transfer Mechanisms

A range of aromatic oxidations involve direct SET from an organic substrate to the oxidant (catalyst, anode), leading to a radical cation [35]. Radical cations are much stronger acids than the parent hydrocarbon molecules [35a, b]. For example, the pK_a of toluene drops from 41 to ca. −13 with the removal of one electron, which makes deprotonation the predominant process in the transformation of the radical cation. Benzyl radicals formed in this way dimerize and participate in the side-chain oxidation (Scheme 14.5). On the other hand, radical cation can undergo attack by nucleophiles (H_2O, AcOH, etc.) followed by the second ET leading to the ring oxidation products [36].

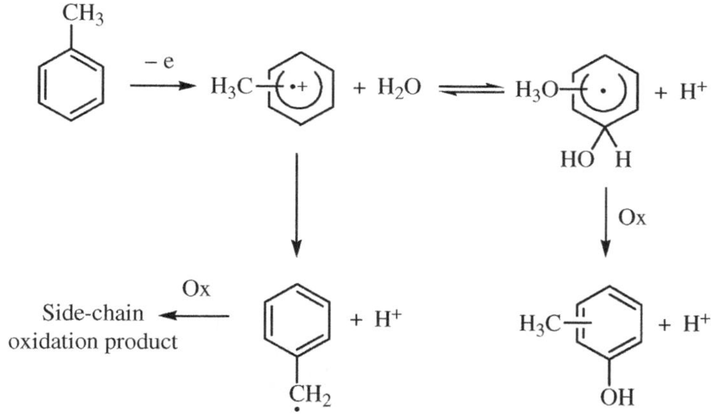

SCHEME 14.5 Possible transformations of toluene radical cation.

While the side-chain deprotonation of radical cations prevails for alkylbenzenes, for polycyclic aromatics, attack by nucleophiles can dominate [35b, 37, 38], resulting in oxidative nucleophilic substitution (Scheme 14.6).

ET mechanisms coupled with addition of nucleophiles operate, for example, in acetoxylation of aromatic compounds with the well-known ET agent, $K_2S_2O_8$, in acetic acid [36] or aerobic Cu-catalyzed *ortho*-acetoxylation of aryl C—H bonds [39].

Radical cations can also disproportionate (Eq. 14.12) [35a, 37] or undergo fast one-electron oxidation to dications that easily add nucleophiles to give oxygenated products (Eq. 14.13) [37, 40].

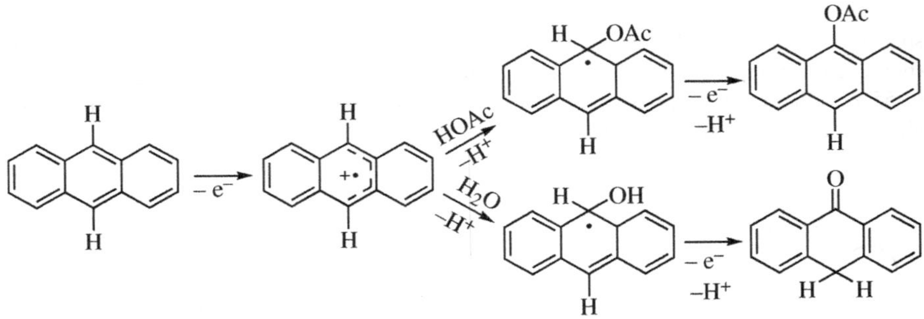

SCHEME 14.6 Oxidation of anthracene via electron transfer–nucleophilic addition.

$$2\,\mathrm{ArH}^{\bullet +} \rightleftharpoons \mathrm{ArH}^{2+} + \mathrm{ArH} \tag{14.12}$$

$$\mathrm{ArH} \xrightarrow[-e^-,\,-e^-]{} \mathrm{ArH}^{2+} \xrightarrow{\mathrm{OAc}^-} \overset{+}{\mathrm{Ar}}\langle^{\mathrm{H}}_{\mathrm{OAc}} \xrightarrow[-\mathrm{H}^+]{} \mathrm{ArOAc} \tag{14.13}$$

Khenkin and Neumann discovered that SET from polycyclic arenes to a mixed-addenda heteropoly acid (HPA), $H_5[PV_2Mo_{10}O_{40}]$, can be followed by oxygen atom transfer from the HPA to the radical cation (Scheme 14.7) [38]. This type of reactivity is well known in the area of heterogeneous gas-phase oxidation as a Mars–van Krevelen mechanism, whereby a metal oxide at high temperature transfers oxygen atom from the lattice to an activated hydrocarbon substrate, but in homogeneous catalysis, it is very rare.

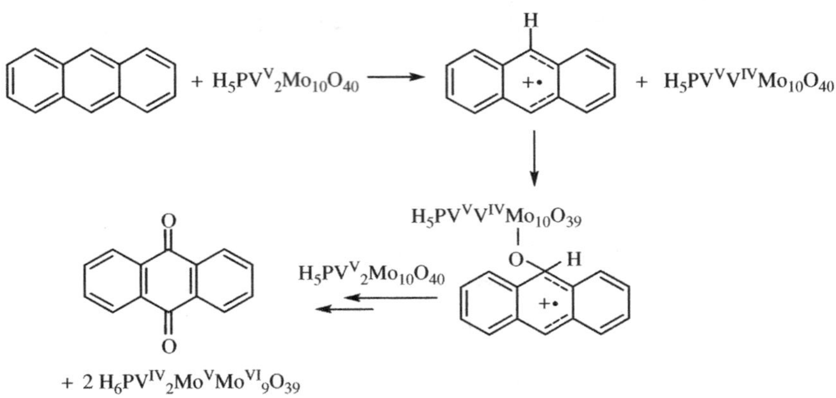

$$2\,H_6PV^{IV}{}_2Mo^VMo^{VI}{}_9O_{39} + 2\,H_2O \longrightarrow 2\,H_8PV^{IV}{}_2Mo^VMo^{VI}{}_9O_{40} \xrightarrow{1.5\,O_2} H_5PV^V{}_2Mo_{10}O_{40} + 3\,H_2O$$

$$2\,H_6PV^{IV}{}_2Mo^VMo^{VI}{}_9O_{39} + 1.5\,O_2 \longrightarrow 2\,H_5PV^V{}_2Mo_{10}O_{40} + H_2O$$

SCHEME 14.7 Anthracene oxidation via electron transfer–oxygenation mechanism (Elaborated from Ref. [38]).

Generation and various transformations of aromatic radical cations in superacidic media have been studied by Rudenko and Pragst [35c, d].

14.2.5 Electrophilic Hydroxylation via Oxygen Atom Transfer

In 1950, Derbyshire and Waters first considered the possibility of electrophilic aromatic hydroxylation with electrophilic hydroxyl (OH^+) [41]. While in gas phase heterolytic cleavage of the O—O bond in H_2O_2 molecule requires much more energy than homolytic cleavage (ca. 300 kcal/mol vs. 51 kcal/mol at 298 K), it becomes more favorable in polar solvents in the presence of Lewis or Brønsted acids [42]. Similarly, peroxy acids (organic or inorganic) can provide electrophilic oxidizing species [42, 43].

A range of enzymatic and biomimetic systems generate high-valent metal-oxo species, $L^{+\cdot}Fe^{IV}{=}O$ or $LFe^{V}{=}O$, one of the most efficient oxygen transfer reagents [44].

Electrophilic aromatic hydroxylations feature high values of the NIH shift [30, 44a]. Cationic cyclohexadiene intermediates or arene oxides are supposed to account for the 1,2-hydrogen migration. In addition, strong electrophilic agents are capable of attacking arenes at already substituted positions (*ipso* attack) to cause 1,2 shift of the substituent [30, 43, 45]. While radical hydroxylations may show a hydrogen migration, shifting of methyl groups is observed only for electrophilic hydroxylations. A stepwise mechanism leading to the substituent migration is shown in Scheme 14.8. The charge distribution in the intermediate determines the extent to which migration occurs [30].

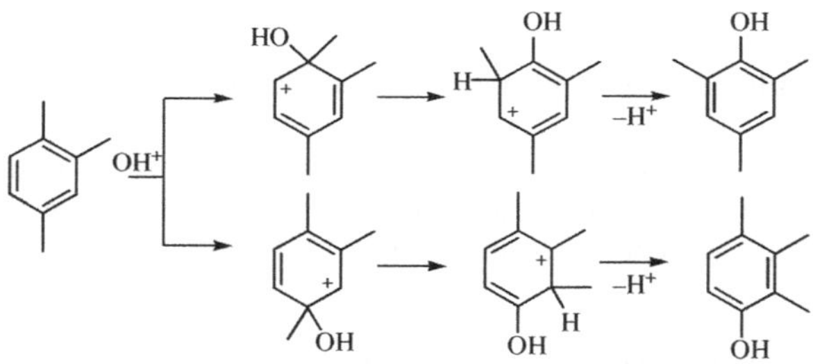

SCHEME 14.8 Migrations of methyl group in electrophilic hydroxylation of pseudocumene.

Arene oxide intermediates were postulated for the acid-catalyzed oxidation of arenes by dimethyldioxirane [46] and aromatic hydroxylations with hydrogen peroxide activated by methylrhenium trioxide (MTO, CH_3ReO_3) [47, 48]. Scheme 14.9 shows a proposed mechanism for pseudocumene (PC) oxidation with the MTO/H_2O_2 system.

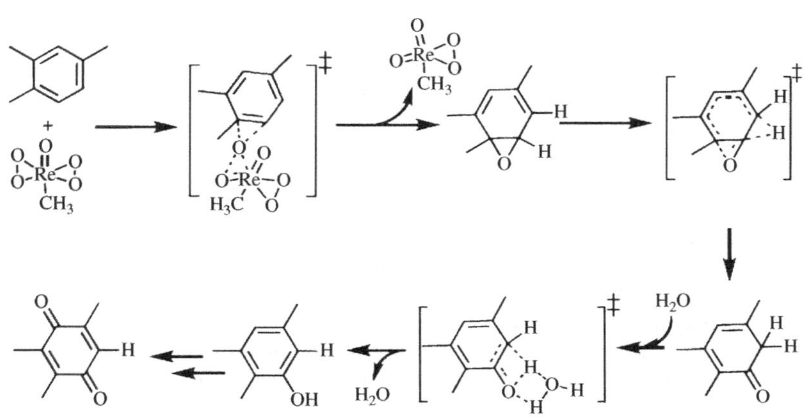

SCHEME 14.9 Proposed mechanism of pseudocumene oxidation with CH_3ReO_3/H_2O_2 (Reproduced with permission from Ref. [48]. Copyright (2011) Elsevier).

14.2.6 Heterolytic Activation of Substrate

Reactions involving electrophilic substitution of hydrogen in arenes are known for both nontransition [Hg(II), Tl(III), Pb(IV)] and transition metals [Au(III), Pd(II), Pt(IV)] [49]. Pd(II)-catalyzed acetoxylation involves arene activation via formation of an organometallic aryl-Pd σ-complex followed by oxidative addition of oxidant and reductive elimination to restore Pd(II) and release the product [11, 50]. Without oxidant, coupling reactions predominate, suggesting arylpalladium(IV) and arylpalladium(II) intermediates in the routes leading to aryl acetates and biaryls, respectively (Scheme 14.10).

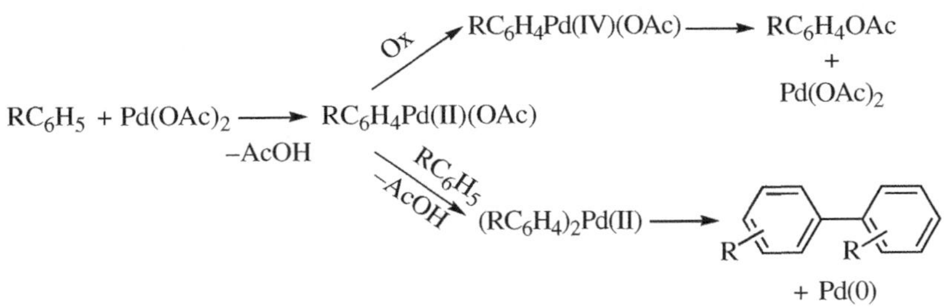

SCHEME 14.10 Pathways in Pd(II)-catalyzed nuclear oxidation (Elaborated from Ref. [50]).

Pd-catalyzed *ortho*-acetoxylations involve coordination of an arene functional group followed by cyclopalladation, two-electron oxidation to Pd(IV) intermediate, reductive C—O elimination, and, finally, ligand exchange to release the product [12, 13].

14.3 STOICHIOMETRIC OXIDATIONS

Until recently, noncatalytic stoichiometric oxidations of arenes with toxic inorganic reagents, such as CrO_3 and $K_2Cr_2O_7$, MnO_2 and $KMnO_4$, OsO_4, $Pb(OAc)_4$, $Tl(NO_3)_3$, nitric acid, ceric ammonium nitrate, and some others, were the main route for the production of oxygenated aromatic compounds [1, 2, 6, 51]. A classic example is the manufacture (~1500 t/year) of vitamin K_3 via stoichiometric oxidation of 2-methylnaphthalene (MN) with carcinogenic CrO_3 in sulfuric acid (Eq. 14.14) [52].

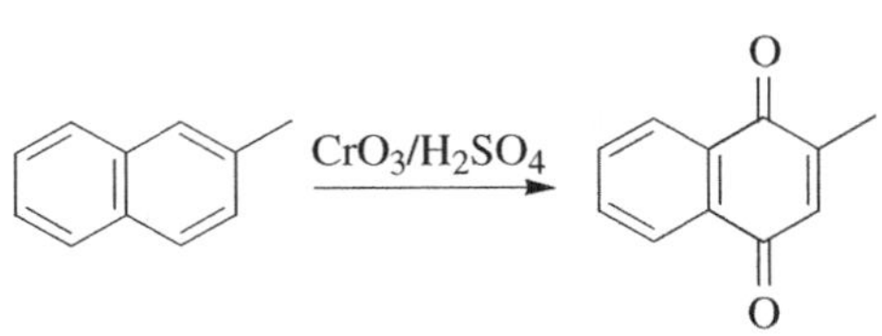

(14.14)

In this method, about 18 kg of toxic inorganic waste is produced per 1 kg of the target product (the so-called E-factor is ca. 18) [52]. Chromic acid is also used in the production of 9,10-anthraquinone (AQ) from anthracene (AN) [53].

Organic peroxy acids are well-known reagents for aromatic hydroxylation of arenes [42, 43]. *m*-Chloroperbenzoic acid was applied to prepare quinones in low to moderate yields [54]. Peroxy acids can be generated *in situ* from H_2O_2 and carboxylic acids or via aldehyde autoxidation [42]. Arnold et al. reported the synthesis of menadione in ca. 30% yield by using 30% H_2O_2 in acetic acid [55]. Following this strategy, Orita et al. oxidized a range of arenes by an excess of H_2O_2 in formic acid [56]. PC and 1,3,5-trimethoxybenzene were converted to 2,3,5-trimethyl-*p*-benzoquinone (TMBQ) and 2,6-dimethoxy-*p*-benzoquinone in 16 and 75% yields, respectively.

Yamamura comprehensively reviewed stoichiometric oxidations of phenols with both metal (V_2O_5, $VOCl_3$, MnO_2, CrO_3, $MoOCl_4$, Tl_2O_3, Ag_2O) and nonmetal compounds (hypervalent iodobenzenes, 2,3-dichloro-5,6-dicyano-*p*-benzoquinone, dioxirane, di-*tert*-butyl peroxide, di-*tert*-butyl peroxyoxalate) [57]. Some of these reagents, for example, metal-free $PhI(OAc)_2$ and $PhI(OCOCF_3)_2$, seem to be rather safe and are especially useful in the low-scale synthesis of complex molecules of natural products. However, one should remember that none of the stoichiometric oxidations is compatible with the concept of green chemistry because such processes produce vast amounts of effluents, which are difficult to dispose of. For example, the active oxygen content in bis(trifluoroacetoxy)iodobenzene is only 3.8%, and a typical E-factor for oxidations with this oxidant lies in the range of 15–25.

14.4 CATALYTIC OXIDATIONS

The development of catalytic methods for selective oxidations is aimed at the use of atom-efficient (i.e., containing high amount of active oxygen), clean, and cheap oxidants under mild reaction conditions [8–10]. The oxidants of choice are O_2 and H_2O_2 that contain 50(100) and 47% of active oxygen, respectively, and produce water as the sole by-product. Stable alkylhydroperoxides, for example, *tert*-butyl hydroperoxide (TBHP), are also viewed as environmentally acceptable oxidants because the alcohol by-products can be recycled or profitably converted [22a]. Nitrous oxide is potentially attractive as oxidant because it contains 36% of active oxygen and produces N_2 as the only by-product.

14.4.1 Benzene

Worldwide, the production of phenol (~8 megatons/year) is almost exclusively based on the three-step cumene process that generates acetone as a by-product (Eq. 14.15).

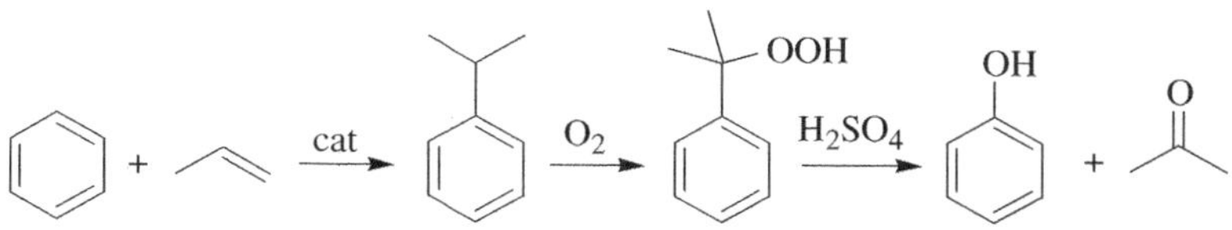

(14.15)

To satisfy environmental protection requirements and to meet the increasing demand for phenol, a considerable effort has been devoted to the development of effective one-step processes for catalytic hydroxylation of benzene to phenol [11, 58, 59].

14.4.1.1 Oxidation with Molecular Oxygen Molecular oxygen is the most attractive oxidant from both economic and environmental viewpoints [8–10]. The direct synthesis of phenol from benzene using O_2 is one of the most difficult challenges for catalysis [58].

Based on the finding that CuCl promotes benzene hydroxylation with O_2 [60], Cu-containing zeolites were intensively studied as catalysts for this reaction in both gas phase and liquid phase [61]. Hydroxyl radical generated from *in situ* formed H_2O_2 (Eqs. 14.16 and 14.17) has been identified as the active species operating on Cu/HY zeolites in aqueous solution [62]:

$$2Cu^I + O_2 + 2H^+ \rightarrow 2Cu^{II} + H_2O_2 \quad (14.16)$$

$$Cu^I + H_2O_2 + H^+ \rightarrow Cu^{II} + HO^\cdot + H_2O \quad (14.17)$$

Many research groups attempted to mimic a monooxygenase-type chemistry and used sacrificial reductants (SH_2), such as ascorbic acid, zinc powder, aldehydes, carbon oxide, or hydrogen, as a source of electrons to reductively activate O_2 (Eq. 14.18):

$$ArH + O_2 + SH_2 \rightarrow ArOH + S + H_2O \qquad (14.18)$$

Aerobic oxidation of aromatics (benzene, toluene, chlorobenzene) at room temperature in the presence of zinc powder as reducing agent and a μ-oxo binuclear iron complex as catalyst in CH_2Cl_2 led to phenols as major products [63]. Co-oxidation of benzene and crotonaldehyde catalyzed by $VO(dpm)_2$ [dmp = 1,3-bis(*p*-methoxyphenyl)-1,3-propanedionato] gave phenol in a 21% yield (Eq. 14.19) [64].

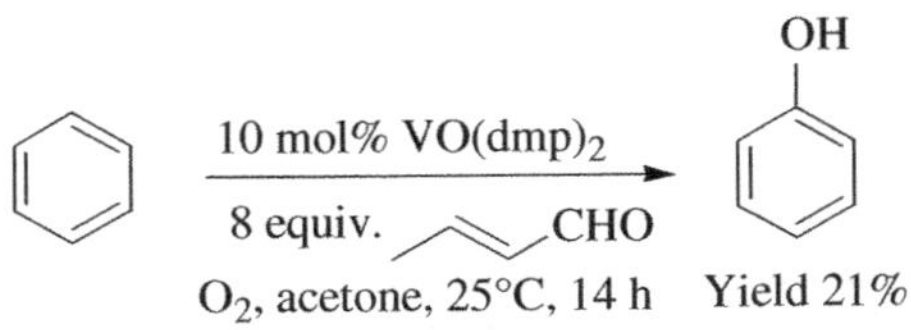

(14.19)

Vanadium phthalocyanine encapsulated in mesopores of SBA-15 mediated aerobic oxidation of benzene to phenol with 100% selectivity and 11.6% yield in the presence of ascorbic acid [65].

A large scientific effort was directed to employment of cheaper reducing agents such as carbon monoxide [66] and hydrogen [67]. A phenol yield of 27.3% was achieved by using air (15 atm), CO (5 atm), and HPA-*n* (Eq. 14.20) [66b].

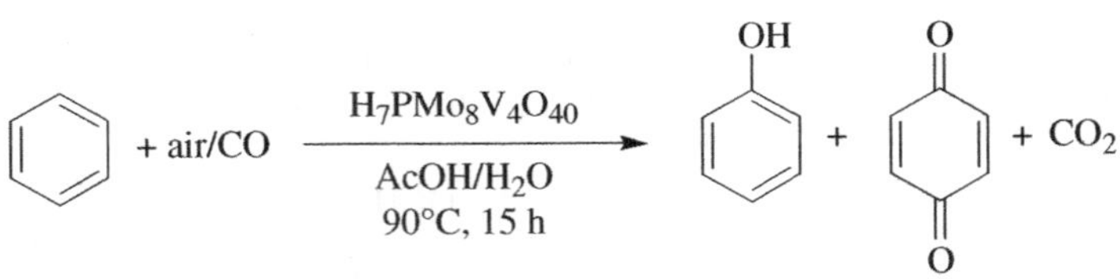

(14.20)

Benzene was selectively oxidized to phenol with O_2/H_2 on a Pt-V_2O_5/SiO_2 in acetic acid at 20–60°C, consuming 7–8 mol of hydrogen to form 1 mol of phenol [67a]. Gas-phase oxidation with a mixture of O_2 and H_2 over Pt-VO_x/SiO_2 catalysts at 413 K gave phenol with 97% selectivity at benzene conversion of 1.0% [67b]. The use of a palladium membrane resulted in encouraging results for the direct hydroxylation of benzene with O_2 and H_2 [67c]. Hydrogen penetrates through the membrane, dissociates on the Pd surface, and reacts with O_2 to give active species, which oxidize benzene to phenol with selectivities of 80–90% at conversions of 2–16%. This approach solves safety problems associated with the use of O_2/H_2 mixtures, but high cost of the membrane restricts its industrial applications.

Zeolite-supported Re_{10} clusters prepared by chemical vapor deposition (CVD) of CH_3ReO_3 onto zeolite HZSM-5 catalyzed efficiently the gas-phase oxidation of benzene with O_2 in the presence of ammonia [68]. Phenol selectivities of 91.6–93.9% at 1.7–9.9% conversions and 82.4–87.7% at 0.8–5.8% conversions were achieved in pulse reactions and steady-state reactions, respectively. The process involves structural transformation between Re_{10} clusters and Re monomers (Scheme 14.11). EXAFS confirmed the formation of the Re clusters after catalyst treatment with NH_3 [68].

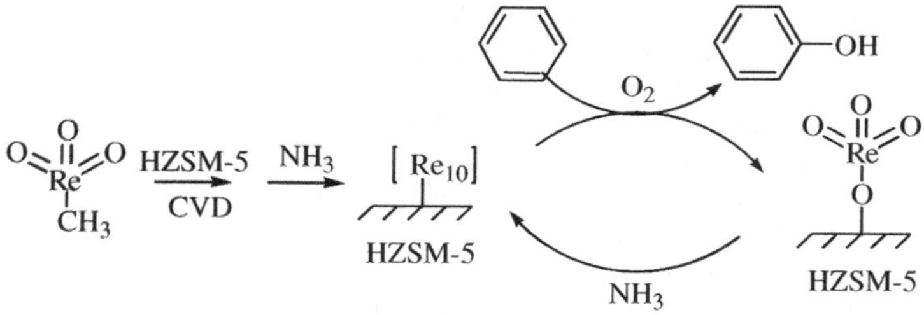

SCHEME 14.11 Benzene hydroxylation with O_2/NH_3 over Re_{10} clusters (Adapted from Ref. [68]).

14.4.1.2 Oxidation with Hydrogen Peroxide Aqueous hydrogen peroxide is a green, atom-efficient, and relatively cheap oxidant, and its use in the organic synthesis is expected to be permanently increased [22]. Benzene hydroxylation with H_2O_2 was extensively studied in the presence of both homogeneous and heterogeneous catalysts [11, 58]. One of the major difficulties is the ease of further oxidation of initially formed phenols (Eq. 14.21).

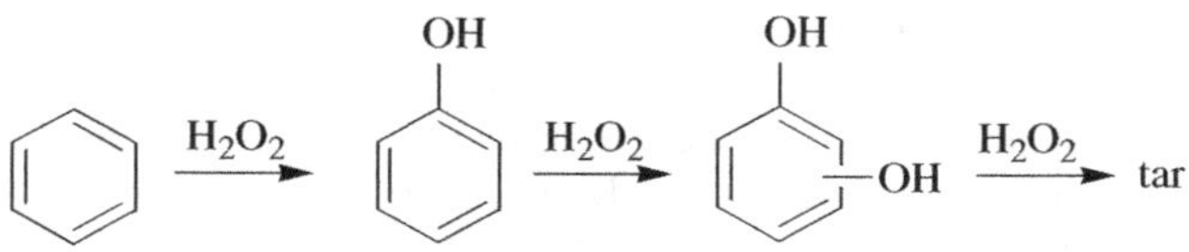

(14.21)

Olah et al. suggested an original approach to solve this problem by using a superacidic medium that protonates phenol and thus prevents it from overoxidation [45]. With highly concentrated (98%) H_2O_2 in $FSO_3H–SbF_5$ (1:1) at −78°C, phenol was obtained in a 54% isolated yield, based on both benzene and hydrogen peroxide [45]. With more practical 30% H_2O_2 combined with HF/BF_3, the yield was only 37% [69].

The Fenton-type system, Fe(II)/Cu(II)/H_2O_2, produced phenol with a moderate selectivity [28b]. In the absence of Cu(II) or Fe(III), biphenyl was the main product. Quinones, in particular 1,2-naphthoquinone-4-sulfonate, played a vital role in the Fe(III)-promoted hydroxylation of benzene with H_2O_2, giving moderate yields of phenol [70]. The addition of polyethyleneglycol to Fe(III)/H_2O_2 in aqueous acetonitrile enhanced the phenol yield and selectivity [71]. A heterophase reaction in the presence of surfactants (quaternary ammonium salts, crown ethers, and polymers) has led to a further improvement. Biphasic benzene hydroxylation in the system composed of aqueous H_2O_2, Fe_2SO_4 modified with 2-pyrazinecarboxylic acid (Hpca) derivatives, and CF_3COOH produced phenol with a selectivity of 97% (based on benzene) at substrate conversion of 8.6% (Eq. 14.22) [72]. The efficiency of H_2O_2 utilization attained 88% if the oxidant was charged slowly. The catalytic system (aqueous phase) could potentially be reused.

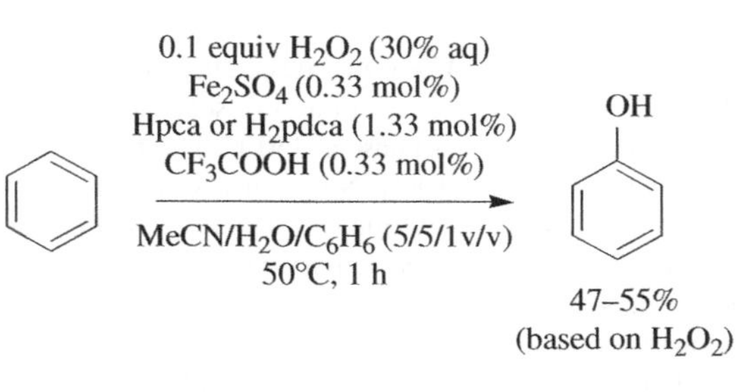

(14.22)

A green aqueous–ionic liquid biphasic hydroxylation of benzene was performed using metal dodecanesulfonate salts as catalysts [73]. With 1 equiv. of H_2O_2, the phenol selectivity reached 90% at 54 and 60% conversion of benzene and oxidant, respectively.

N-bridged diiron phthalocyanine $(FePctBu_4)_2N$ was capable of oxidizing benzene with H_2O_2 in MeCN under very mild conditions (20–40°C) to afford phenol as the main product [74]. Detection of benzene oxide among the minor products and a NIH shift supported a mechanism via the formation of high-valent diiron-oxo species.

A rhodium carbonyl cluster, $Rh_6(CO)_{16}$, catalyzed benzene hydroxylation in acetonitrile via the formation of an arene π-complex containing also an activated form of H_2O_2 [75]. A decamethylosmocene-based system, $(Me_5C_5)_2Os/H_2O_2$/py, that effectively generates $HO^{\cdot}$ radicals, produced phenol in a yield of 22% [76].

Vanadium compounds have been widely explored as catalysts for H_2O_2-based hydroxylation of benzene and other arenes [11, 77, 78]. A vanadyl complex grafted on to periodic mesoporous organosilicas, $[VO(acac)_2]$/PMO, was used as recyclable heterogeneous catalyst for benzene hydroxylation, producing phenol with nearly 100% selectivity at 27% benzene conversion but with a low oxidant utilization efficiency [79].

Numerous studies focused on the use of titanium-containing heterogeneous catalysts for benzene hydroxylation [10, 11, 58]. The best results were achieved for titanium-silicalite TS-1, the catalyst that had been developed by the ENI group in the early 1980s [80]. The active centers in TS-1 are Ti(IV) ions isolated in a hydrophobic microporous silicate matrix [80b, 81]. Benzene hydroxylation over TS-1, including synthetic, mechanistic, and economic aspects, has been comprehensively reviewed [58, 81]. With common solvents, low H_2O_2 to benzene ratios, and small-sized TS-1, the selectivity to phenol was below 50% at already 3–4% benzene conversion, while oxidant efficiency hardly surpassed 20% [82]. Overoxidation products comprised diphenols (30%), BQ (<4%), and tars (25–35%). Use of dimethyl and tetramethylene sulfones as solvents allowed a significant improvement of the phenol selectivity and oxidant efficiency up to 83 and 67%, respectively, at nearly doubled benzene conversions. Post-treatment of TS-1 with $(NH_4)HF_2$ and H_2O_2 raised the selectivities to phenol and oxidant to 94 and 83%, respectively, with negligible production of tars. These findings have created a basis of the *Polimeri Europa* TS-1/H_2O_2 process for the direct synthesis of phenol, which may compete in the future with the traditional cumene process [58].

14.4.1.3 Oxidation with Nitrous Oxide In 1983, Iwamoto et al. first employed N_2O for the oxidation of benzene to phenol using a vanadia catalyst [83]. However, the selectivity toward phenol was too low. In 1988, three groups independently discovered benzene hydroxylation with N_2O over ZSM-5 zeolites (Eq. 14.23) [84].

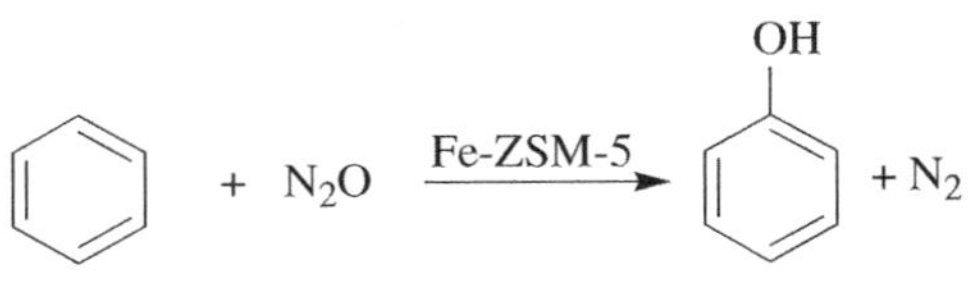

(14.23)

Panov's group in collaboration with *Monsanto* researchers contributed significantly to the development and implementation of this process in a pilot scale (nowadays, the *Solutia* process) [59, 85]. The reaction is performed in the gas phase at 350°C. The selectivity toward phenol attains 97–98% at 27% benzene conversion and 100% N_2O conversion. Dihydroxy benzenes (ca. 1%, mainly hydroquinone (HQ)) and carbon monoxide (0.2–0.3%) are main by-products. The catalyst half-life is restricted by a few days, but catalytic activity can be easily restored by burning-off coke deposits. The catalyst can thus sustain more than 100 regeneration cycles. The manufacture of adipic acid provides a cheap technical access to N_2O. More information about the *Solutia* process can be found in the two book chapters [58, 59]. Unfortunately, its commercialization has been postponed on the score of economics [9b].

The nature of the iron sites responsible of the unique reactivity is still under debate. Panov et al. suggested that ZSM-5 at elevated temperatures yields defect sites (the so-called α-sites) that interact with N_2O to produce on the zeolite surface iron-oxo species (the so-called α-oxygen, presumably, anion radical species $O^{\cdot-}$ bound to the iron), responsible for inserting oxygen atom into C—H bonds [59]. α-Oxygen is highly reactive and can readily oxidize various organic molecules, including benzene and methane, even at room temperature. This allows one to conduct the synthesis of phenol according to the scheme (Eqs. 14.24–14.26) that involves α-oxygen generation via interaction of N_2O with α-sites, its interaction with benzene at room temperature, and then phenol extraction from the catalyst surface [59].

$$N_2O + (\ )_\alpha \xrightarrow{470\text{–}520\ K} (O^-)_\alpha + N_2 \quad (14.24)$$

$$C_6H_6 + (O^-)_\alpha \xrightarrow{295\ K} (C_6H_5OH)_\alpha \quad (14.25)$$

$$(C_6H_5OH)_\alpha \xrightarrow[\text{Extraction}]{295\ K} C_6H_5OH + (\ )_\alpha \quad (14.26)$$

Reactions (14.24)–(14.26) represent the main mechanistic steps not only of the stoichiometric process but also of the steady-state catalytic oxidation of benzene.

14.4.2 Polycyclic Arenes

Polycyclic arenes are more easily oxidized than benzene owing to less strong sp^2 C—H bonds (ca. 105 vs. 112 kcal/mol). Some of them can undergo transformations that are not typical of other aromatic compounds, for example, catalytic oxygenation with dioxygen through the electron transfer-oxygenation sequence [38] (see Section 14.2.4) or noncatalytic spin-forbidden interaction with triplet oxygen [23] (see Section 14.2.2).

The oxidation selectivity strongly depends on the specific polyaromatic substrate. Oxidation of naphthalene with H_2O_2 in the liquid phase over a large pore zeolite, V-NCL-1 [86a], and a mesoporous vanadium-silicate, V-MCM-41 [86b], afforded 1,4-naphthoquinone (NQ) as the main product and minor amounts of 1- and 2-naphthols at the initial stage of reaction but, at conversions greater than 5%, phthalic anhydride predominated because of susceptibility of NQ to overoxidation. In contrast to NQ, AQ is rather stable to overoxidation and high selectivities can be attained at high substrate conversions.

Significant progress has been achieved in the development of heterogeneous catalysts for the selective oxidation of AN in the liquid-phase using TBHP as oxidant [87]. HPA-2 immobilized onto an amine-functionalized mesoporous silica SBA-15 showed 100% selectivity to AQ at 60% AN conversion [87a]. A mesoporous chromium-silicate, Cr-MCM-41 (0.46 wt% Cr), produced AQ with ca. 98% selectivity at 79% conversion [87b]. Even better results (ca. 90% yield) were reported for mesoporous metal-silicates Cr-SBA-15 [87c] and Ti-HMS [87d] as well as SiO_2-supported FePcS [87e]. A nearly quantitative AQ yield was obtained after 1.5 h at 100°C in chlorobenzene as solvent over a metal-organic framework (MOF), Cr-MIL-101 [88]. No leaching of active metal occurred at optimal reaction conditions, and the catalyst could be recycled several times without deterioration of the catalytic properties. The superior catalytic performance was attributed to the high affinity of the MOF toward aromatic molecules.

14.4.3 Alkylarenes

Reports on the nuclear oxidation of alkylated arenes with molecular oxygen are scarce because most of catalytic systems, in the absence of a sacrificial reductant, work with O_2 via HAT or ET mechanisms (see Section 14.2) and, therefore, tend to oxidize alkyl substituents. On the other hand, a range of two-electron electrophilic hydroxylations can be accomplished with alkylarenes using H_2O_2 as oxidant. The more substituted arene, the lower susceptibility of the oxygenated product toward overoxidation. Mesitylene is an especially "lucky substrate" because in the phenolic product, mesitol, all positions *ortho* and *para* to the hydroxyl group are occupied, thereby retarding further oxidation [43].

The hydroxylation of alkylbenzenes with 98% H_2O_2 in FSO_3H or $FSO_3H–SbF_5$ at −78°C afforded up to greater than 90% yields of oxygenated alkylbenzenes [45]. Aqueous 30% H_2O_2 with superacidic HF/BF_3 at −78°C produced hydroxylated products in satisfactory to good yields [69]. For monosubstituted alkylbenzenes, the *ortho*/*para* isomer ratio followed the predicted trend (from 3.33 for toluene to 0.54 for *tert*-butylbenzene) on consideration of the increasing steric hindrance to *ortho* substitution. Polyhydroxylation was practically suppressed owing to protonation of the phenols formed [45, 69].

Mitsubishi Gas Chemical Co. patented a method for the synthesis of menadione from MN using 60% H_2O_2 in MeOH in the presence of a large excess (12 equiv.) of HCl [89]. Oxidation of MN by H_2O_2 (50–80%) in the presence of carboxylic acids, respective anhydrides and strong mineral acids led to MNQ in a yield of 40–80% [90]. However, the use of concentrated H_2O_2 is considered as an inherent safety risk, and the strong acidic conditions cause potential corrosion problems for production on a larger scale.

Transition metal-catalyzed oxidations of alkylarenes may result in a wide variety of products. For example, MN gives not only target MNQ but also isomeric 6-MNQ as well as products of C–C coupling, side-chain oxidation, and overoxidation (Scheme 14.12).

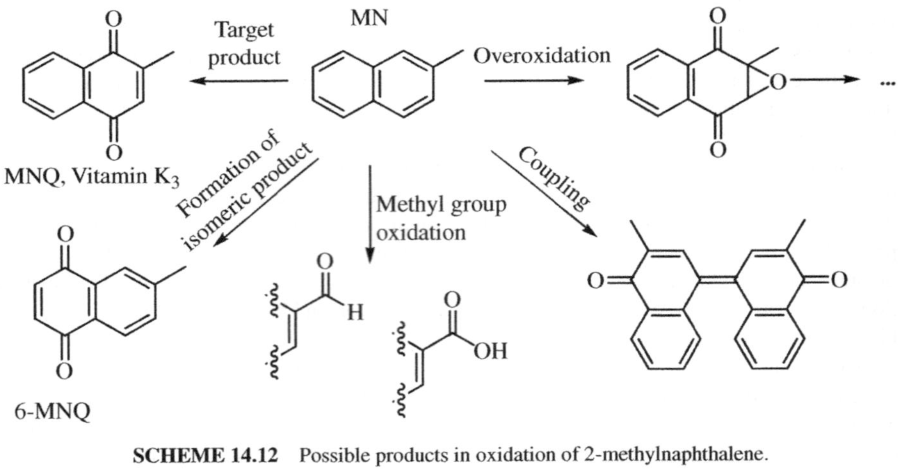

SCHEME 14.12 Possible products in oxidation of 2-methylnaphthalene.

Yamaguchi et al. reported the first metal-catalyzed synthesis of quinones where oxidation of methylbenzenes and naphthalenes was performed in acetic acid using 60% H_2O_2 in the presence of Pd(II)-sulfonated polystyrene-type resin [91]. The selectivities to quinones were significantly higher with naphthalenes than with methylbenzenes. While MNQ was obtained in a 60% yield at MN conversion of 93%, only 3.3% yield of TMBQ was attained in the oxidation of PC at 78% conversion.

Aromatic oxidation of substituted naphthalenes to corresponding quinones without oxidation of the methyl side chains was accomplished with 85% H_2O_2 in acetic acid in the presence of CH_3ReO_3 [92]. MN produced menadione in a yield of 47% at 81% substrate conversion after 4 h at 40°C [92a]. Alternatively, 35% H_2O_2 in acetic anhydride could be employed [92b]. Surprisingly, this system has led to a higher regioselectivity: the ratio of 2-MNQ and isomeric 6-MNQ was 11:1 versus 7:1 with 85% H_2O_2.

Jacob and Espenson oxidized various alkylarenes to *para*-BQ by 30% H_2O_2 (ca. 20-fold excess) in the presence of CH_3ReO_3 in acetic acid (Eq. 14.27) [47c].

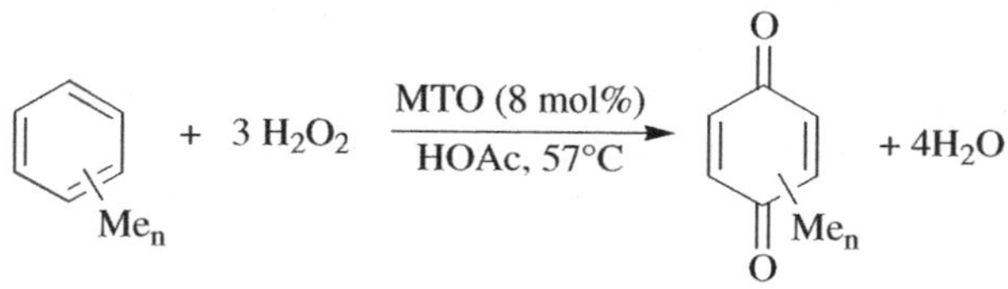

(14.27)

The more highly substituted substrates afforded greater yields of *para*-quinones. PC gave TMBQ with 67% selectivity at 75% substrate conversion. Recently, the MTO-based oxidation of PC was revisited to offer a more economically beneficial and environmentally friendly approach to the important industrial product, TMBQ [48]. In nitromethane, 73% TMBQ selectivity was attained at 30% conversion using just 4 equiv. of 50% H_2O_2 and 2 mol% of MTO modified with a Schiff base ligand. A mild procedure for the oxidation of PC based on the use of MTO and neutral amphiphile Brij 30 has led to a selectivity of 80% at 27% conversion [93].

Several vanadium-based catalytic systems have been reported for the nuclear hydroxylation of alkylaromatics [77, 78, 94, 95]. Mizuno and coworkers have discovered one of the most efficient system, which is based on the use of a divanadium-substituted phosphotungstate, $TBA_4H[\gamma\text{-}PW_{10}V_2O_{40}]$ (shortly, $PW_{10}V_2$; TBA = tetra-*n*-butylammonium) as homogeneous catalyst, mineral acid ($HClO_4$) as cocatalyst, 30% H_2O_2 as oxidant, and CH_3CN/*t*BuOH (1:1, v/v) as solvent [95]. In the oxidation of toluene, the ratio of ring-oxygenated products to side-chain-oxygenated products was 98:2. The system chemoselectively hydroxylated alkylarenes with highly reactive secondary and tertiary C—H bonds (ethylbenzene, cumene, xanthene, and diphenylmethane) without significant oxidation of the side chain and revealed unique regioselectivity for the formation of *para*-phenols. Anisole gave methoxyphenols in a 70% yield based on H_2O_2 with the *o*:*m*:*p* ratio of 3:<1:96 (Eq. 14.28), while *meta*-xylene produced dimethyl phenols in the ratio of 2,4:3,5:2,6 = 90:9:<1.

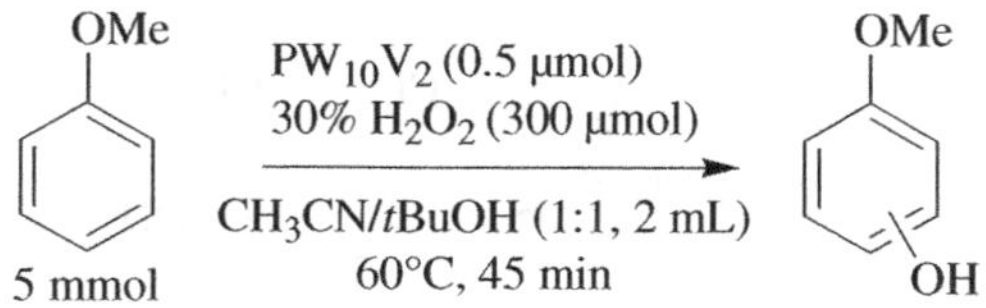

(14.28)

A high NIH shift (67%) was found for hydroxylation of 4-deuterotoluene. The authors suggested that the bis-μ-hydroxo core {OV-(μ-OH)$_2$-VO} in the phosphotungstate interacts with H_2O_2 to form a highly electrophilic peroxo vanadium species responsible for the observed catalysis. The high regioselectivity for *para*-phenols is likely due to the *ortho*-hydroxylation being suppressed by the steric crowding between the active site and the substituents [95].

Beller and coworkers suggested preparation of MNQ based on the use of easily available iron complexes (formed upon interaction of $FeCl_3 \cdot 6H_2O$, pyridine-2,6-dicarboxylic acid, and benzylamine derivatives) as catalysts, hydrogen peroxide as oxidant, and acetonitrile/water as solvent [96]. Up to 55% MNQ yield could be obtained by a simple and convenient procedure (room temperature, alcoholic solvents). They also oxidized naphthalene derivatives to corresponding quinones with good to excellent yields and high regioselectivity using 30% H_2O_2 and ruthenium-(2,2′,6′:2″-terpyridine)(2,6-pyridinedicarboxylate), [Ru(tpy)(pydic)], as catalyst [97]. The oxidation of MN gave 74% of the two quinones and 60% of menadione. Methods for the synthesis of the industrially important quinones, MNQ and TMBQ, have been recently surveyed [14]. In 2013, Sorokin reviewed production of quinones employing metal phthalocyanines [98].

Microporous TS-1 catalyzes heterogeneously aromatic hydroxylations, but alkyl substituents produce rate-retarding effects because steric hindrance prevails over electron donation [81].

Toluene reacts at a similar rate to that of benzene and produces ring-hydroxylated compounds (56% 4-methyl- and 41% 2-methylphenol) as the sole products [99]. In contrast, mesoporous titanium-silicates favor oxidation of the toluene side-chain [100]. The rate of hydroxylation of xylene isomers over TS-1 was sterically controlled, *ortho*<*meta*≪*para*, while over the larger pores zeolite Ti-MOR, it increased as expected for an electrophilic attack, *para*<*ortho*<*meta* [81]. Unfortunately, the microporous structure of zeolites does not enable oxidation of more sterically hindered aromatics because of pore size limitation. The development of an efficient heterogeneous catalyst for the selective oxidation of alkylarenes remains a great challenge [100].

14.4.4 Electron-Poor Aromatic Compounds

Reports on the ring hydroxylation of electron-poor aromatic compounds are relatively scarce. Chlorobenzene, nitrobenzene, benzonitrile, benzaldehyde, and benzoic acid are inert toward hydroxylation with H_2O_2 over TS-1 [81]. The hydroxylation of benzene derivatives XC_6H_5 (X=F, Cl, Br, NO_2) by H_2O_2 and in the presence of catalytic amounts of $[VO(O_2)(pic)(H_2O)_2]$ produced corresponding phenols in moderate yields [94]. The reaction was proved to be a radical chain process where initiation produces a radical anion of the peroxovanadium complex as the actual oxidizing species [34].

Hydroxylation of benzonitrile and fluoro- and chlorobenzenes was realized with N_2O over ZSM-5 zeolite [101]. In this system, chlorobenzene produced chlorophenol (*o*:*m*=28:72) with 58% selectivity at 23% substrate conversion [101a], while benzonitrile gave hydroxybenzonitrile (*o*:*m*:*p*=1:6:3) with 73% selectivity at 10% conversion [101b].

Even more rare are oxidations with dioxygen [102]. Khenkin et al. found that heating a 0.01 M solution of $H_5[PV_2Mo_{10}O_{40}]$ in neat nitrobenzene at 140°C under 2 bar O_2 selectively (>99%) yielded 2-nitrophenol at a 5% maximum yield [102a]. Liu et al. have developed a Cu-catalyzed "oxygenase-type" aerobic oxidation of arenes [102b]. The yield of 2,3,5,6-tetrachlorophenol reached 89% (Eq. 14.29).

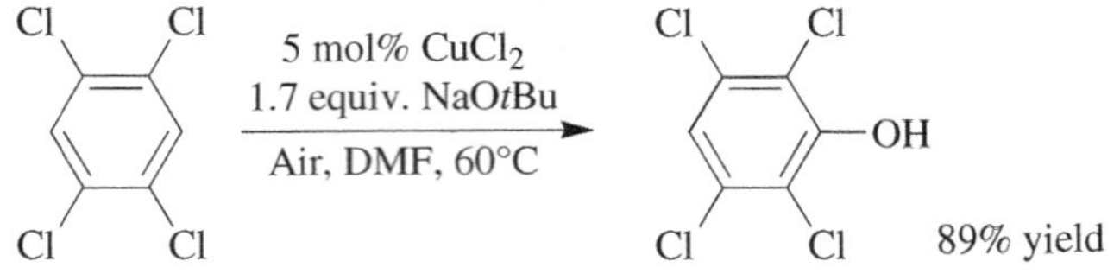

(14.29)

14.4.5 *ortho*-Hydroxylation Driven by Arene Functional Group

Que and coworkers first reported that the carboxylic group of benzoic acid can serve as a useful tool for implementation of a direct *ortho*-hydroxylation with H_2O_2 in the presence of a stoichiometric amount of a nonheme iron(II) complex [103]. Based on this pioneering work, Zhang and Yu demonstrated an *ortho*-hydroxylation of potassium benzoates with carboxylic acids as directing groups using air or O_2 (1–5 atm) as oxidant providing synthetically useful yields [104]. Additives of BQ significantly accelerated the reaction. Electron-rich arenes were readily hydroxylated in 60–82% yields using atmospheric oxygen (Eq. 14.30), while arenes with electron-withdrawing groups needed a higher pressure of O_2.

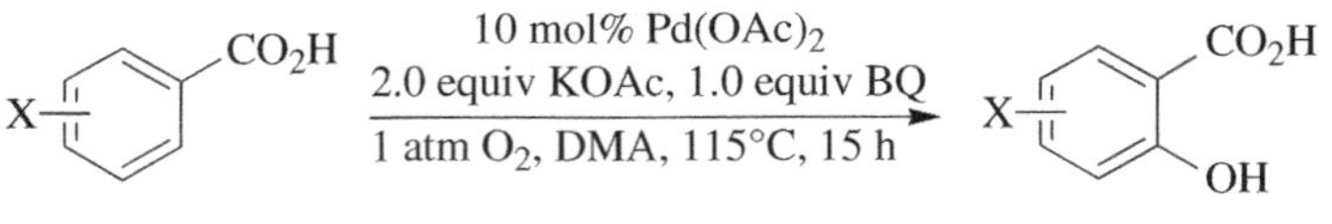

(14.30)

No reaction was observed with stoichiometric amounts of $Pd(OAc)_2$ under argon, suggesting that O_2 is likely involved in the product formation step rather than reoxidation of Pd(0). Labeling studies using $^{18}O_2$ and $H_2{}^{18}O$ supported a direct oxygenation of the arylpalladium intermediates with O_2 instead of an acetoxylation/hydrolysis sequence. Pyridyl group also enabled direct Cu-catalyzed *ortho*-selective acetoxylation of aryl C—H bonds with O_2 in $AcOH/Ac_2O$ [39].

A range of *ortho*-hydroxylations driven by weak coordination of an arene functional group have been reported using nontraditional oxidants [105]. *ortho*-Hydroxylated ethyl benzoates were produced using potassium persulfate, Selectfluor (1-chloromethyl-4-fluoro-1,4-diazoniabicyclo [2.2.2]octane bis(tetrafluoroborate)), or iodic acid as oxidants in the trifluoroacetic acid/trifluoroacetic anhydride (TFA/TFAA) cosolvent system catalyzed by a $[RuCl_2(p\text{-cymene})]_2$ complex (Eq. 14.31) [105a].

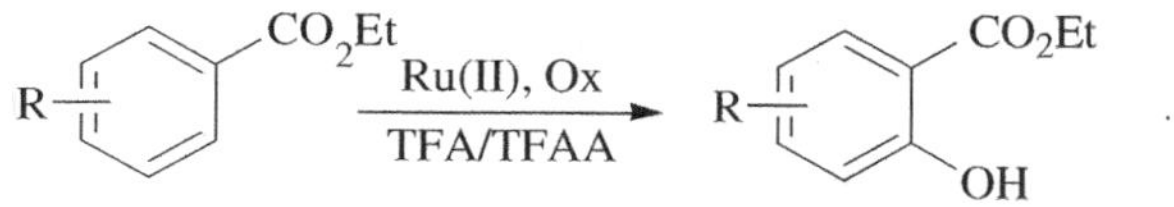

(14.31)

Ru-catalyzed hydroxylations of arenes bearing weakly coordinating amides were implemented with $PhI(OAc)_2$ as the oxidant [105b]. Mechanistic studies provided support for a reversible C—H bond metalation step. A ketone-directed highly *ortho*-selective hydroxylation of arylketones to produce *o*-acylphenols, versatile building blocks for the synthesis of fine chemicals and pharmaceuticals (e.g., warfarin), has been elaborated using $Pd(TFA)_2$ or $Pd(OAc)_2$ (5 mol%) in combination with $PhI(TFA)_2$ or $K_2S_2O_8$ (2 equiv) [105c]. Both aryl- and alkyl-substituted ketones with different steric properties gave satisfying yields of the corresponding *o*-acylphenols.

14.4.6 Phenol

Several catalytic technologies have been developed for direct hydroxylation of phenol to HQ and catechol (CAT) using H_2O_2 as oxidant [106]. More than 25,000 tons/year of these important chemicals is produced using three main processes (Table 14.1), a common feature of which is the use of an excess of substrate to avoid overoxidation.

The *Rhône-Poulenc* process is a typical acid-catalyzed electrophilic hydroxylation, suffering overoxidation at substrate conversions of greater than 5%. Phosphoric acid is required to sequester metal ions that could promote homolytic decomposition of H_2O_2. Oppositely, the *Brichima* process is based on Fenton chemistry and a low H_2O_2 utilization efficiency is characteristic of this process. The CAT/HQ ratio can be tuned by changing the nature of transition metal and/or employing chelate ligands. Copper(II) complexes with 2,6-dihydroxypyridine showed 93-95% selectivity of *ortho*-hydroxylation (CAT/HQ = 15–16) at 9–30% conversions [107]. Unusually high *para*-selectivity (HQ/CAT = 3.3–3.7 at 9–21% conversion) was achieved in phenol oxidation with H_2O_2 over MOFs, $[M^{II}(H_2O)_6]\cdot[Mn^{II}(phen)_2(H_2O)_2]_2\cdot 2BTC\cdot nH_2O$ (M = Cu, Mn; BTC = 1,3,5-benzenetricarboxylate) [108]. Silica-supported copper nanoparticles afforded phenol conversion of 69% after 15 min at 70°C with

TABLE 14.1 Comparison of Phenol Hydroxylation Processes[a]

Process/Catalyst	Rhône-Poulenc $HClO_4/H_3PO_4$	Brichima Fe^{II}, Co^{II}	EniChem/Borregaard TS-1
Phenol conversion	5	9–10	20–30
Selectivity on phenol	90	79–80	90–95
H_2O_2 efficiency	80–90	50–66	80–90
CAT/HQ	1.2–1.5	1.5–4	1.1–1.2

[a] Elaborated from Ref. [106].

selectivity greater than 95% (CAT/HQ~1) [109]. Gold nanoparticles deposited on diamond promoted Fenton chemistry at room temperature, converting phenol to dihydroxybenzenes (CAT/HQ=1.8) with a selectivity of 79% and oxidant efficiency of 91% [110].

A different type of chemistry has been realized in the process commercialized by *EniChem* in 1986, which is based on employment of H_2O_2 and the titanium-silicalite TS-1 as heterogeneous catalyst [106, 111]. The hydroxylation mechanism involves activation of hydrogen peroxide via the formation of a titanium hydroperoxo complex, TiOOH, followed by electrophilic oxygen atom transfer to phenol. Methanol and acetone are the solvents of choice to achieve high selectivity. The nature of solvent, phenol concentration, reaction time, and size of catalyst particles affect the CAT/HQ ratio. The TS-1-based process offers clear advantages in terms of conversion, selectivity, H_2O_2 efficiency (see Table 14.1), catalyst separation/recycling, and, hence, environmental impact.

Gas-phase oxidation of phenol in the mixture with benzene has been implemented using N_2O as oxidant and Fe-containing ZSM-5 zeolite as heterogeneous catalyst [112].

14.4.7 Alkylphenols and Alkoxyarenes

The presence of alkyl or alkoxy substituents in the phenol molecule facilitates the oxidation process and stabilizes the resulting quinones with regard to further oxidation, which enables high selectivity at high substrate conversions.

14.4.7.1 Oxidation with Molecular Oxygen Co(II)-Schiff base complexes have long been used as catalysts for selective oxidation of substituted phenols, and this subject has been comprehensively reviewed [15, 113]. Electron-releasing groups in the salen ligand favor the formation of BQ. Oxidation of 2,3,6-trimethylphenol (TMP) to TMBQ was accomplished with a 88% yield in DMF [114], while 2-methylnaphthol produced MNQ with a nearly quantitative yield in toluene or MeCN [115]. In CO_2-expanded CH_2Cl_2 (62 bar, 70°C), 2,6-di-*tert*-butyl-1,4-benzoquinone was obtained in a yield of 97% [113b]. *p*-Substituted phenols, for example, syringyl alcohol, could be oxidized to *p*-BQ in the presence of 5-coordinate Co(II)salen complexes (Scheme 14.13) [116].

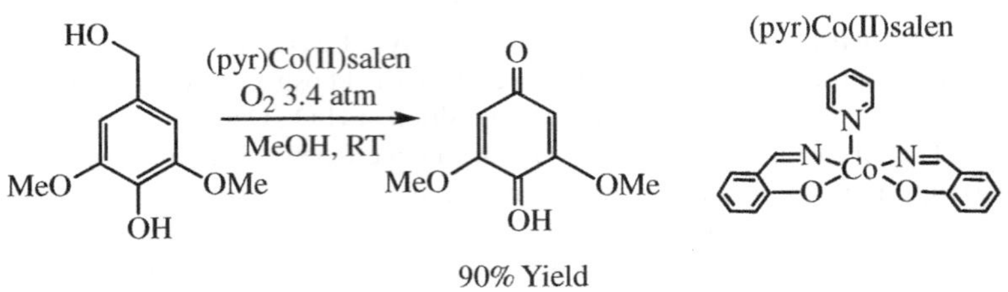

SCHEME 14.13 Oxidation of syringyl alcohol with O_2 catalyzed by (pyr)Co(salen).

A drawback of Schiff base complexes is their low productivity in oxidation catalysis because the salen ligand is highly prone to oxidative destruction. Kinetically more stable phthalocyanine complexes have been widely explored as catalysts for oxidation of alkylphenols. This subject has been recently reviewed by Sorokin [98].

Substituted phenols undergo a variety of oxidative transformations in the presence of copper compounds [15, 117–119]. The nature of both phenol and ligands and the reaction conditions strongly affect the reaction selectivity. Hay and coworkers discovered the oxidative C—O coupling of 2,6-dimethylphenol to polyphenylene ether, an important industrial polymer, in the presence of a homogeneous Cu(I) catalyst and amines [117a]. The presence of large substituents (*i*-Pr or *t*-Bu) at *ortho* positions of phenol and elevated reaction temperature shift the oxidation process toward the formation of the C—C coupling product, DPQ. High (10:1) molar ratios of pyridine to Cu favor C—O coupling while low ratios facilitate C—C coupling of intermediate phenoxyl radicals [117b].

Copper(II) salts, taken in quasi stoichiometric amounts, favor oxidation of alkylated phenols to *p*-BQ as the major products in DMF at 30 bar of O_2 [118a]. The vitamin E precursor, TMBQ, is currently produced in industry in yields of 95–98% via oxidation of TMP with O_2 (or air) in the presence of copper halides [7, 119a]. The use of a two-phase system (catalyst is in aqueous solution while substrate and product are dissolved in a heavy alcohol) enables an easy catalyst recovery and recycling [118b, c]. Since large amounts of $CuCl_2$ are used to ensure the high yields of TMBQ, special apparatus, resistant to corrosion, is required. The formation of Cl-containing by-products and product contamination with copper are other drawbacks of this process. The amount of $CuCl_2$ was substantially reduced using an ionic liquid, 1-butyl-3-methylimidazolium chloride ([BMIm]Cl), with *n*-butanol as cosolvent [120]. A 86% yield of TMBQ was achieved with 2.5 mol% of $[BMIm]CuCl_3$.

Alkylphenols can be converted to *p*-BQ using dioxygen and HPA-*n* (n=2–6) [121]. Oxidation of TMP in AcOH-H_2O (95/5) produced TMBQ with a selectivity of 86% at complete TMP conversion [121a]. The main reaction by-product was 2,2′,3,3′,5,5′-hexamethyl-4,4′-biphenol. Based on kinetic and spectroscopic studies, a stepwise mechanism that involves dissociation of HPA-*n* in the acidic medium generating the active species VO_2^+, oxidation of TMP by VO_2^+ to produce phenoxyl radical and VO^{2+} followed by reoxidation of V(IV) to V(V) with O_2 in the coordination sphere of the HPA-*n* has been implicated [121a]. Matveev et al. used of a two-phase system, that is, aqueous solution of HPA-*n* (0.1–0.5 M) and organic solvent, to convert MNL to MNQ in 85–88% yields [122]. Reoxidation of the reduced HPA-*n* with O_2 was performed in a separate step.

14.4.7.2 Oxidation with Hydrogen Peroxide Shimizu et al. have developed a method for the production of TMBQ and ubiquinone 0 based on the oxidation of TMP and 3,4,5-trimethoxytoluene (TMT), respectively, with H_2O_2 in the presence of HPAs $H_3PM_{12}O_{40}$ and $H_4SiM_{12}O_{40}$ (M = W, Mo) [123].

Methylrhenium trioxide was employed as catalyst for H_2O_2-based oxidation of phenols and methoxy derivatives [47, 124]. Alkylated phenols and methoxytoluenes were oxidized by the MTO/H_2O_2 system in a neutral ionic liquid, $[BMIM]BF_4$, leading to quinone yields varied from 70% to nearly quantitative [124a]. TMBQ and ubiquinone 0 were obtained from TMP and TMT, respectively, in yields of greater than 98% at more than 98% conversion.

Ruthenium and iron compounds were widely explored as catalysts for oxidation of substituted phenols and alkoxyarenes with H_2O_2 [14, 97, 98, 125]. Ito et al. reported 89 and 71% yields of TMBQ for TMP oxidation in acetic acid in the presence of $RuCl_3$ and $FeCl_3$, respectively [125a]. Hexacyanoferrate-catalyzed oxidation of trimethoxybenzenes with H_2O_2 furnished dimethoxy-*p*-benzoquinones with moderate yields [125b]. A system consisting of $FeCl_3{\cdot}6H_2O$, pyridine-2,6-dicarboxylic acid (H_2Pydic), and benzylamines ($FeCl_3$/H_2Pydic/amine = 1/1/2.2) afforded TMBQ and MNQ in 79 and 55% yields, respectively [14]. Finally, [Ru(tpy)(pydic)]-catalyzed oxidation led to *p*-BQ in 78–83% yields using 4.4 equiv. of H_2O_2 in MeOH [97].

Heterogeneous oxidation of alkyl-substituted phenols readily occurs with H_2O_2 over mesoporous titanium-silicates [100, 126]. Pinnavaia and coworkers first reported oxidation of 2,6-di-*tert*-butylphenol to a mixture of *p*-BQ and DPQ using Ti-HMS and Ti-MCM-41 [127]. Two groups independently reported the oxidation of TMP to TMBQ with 77–82% yields over MCM-41-type catalysts [128]. EPR and by-product studies implicated homolytic oxidation mechanism [129]. Nearly quantitative yields of TMBQ and other methylquinones were attained using a range of mesoporous Ti,Si catalysts [130]. The key point to achieve the high selectivity is the presence of Ti(IV) dimers or small oligomers on the silica surface (Scheme 14.14), which enables fast oxidation of ArO· and thus prevents formation of dimeric C—C coupling products [131].

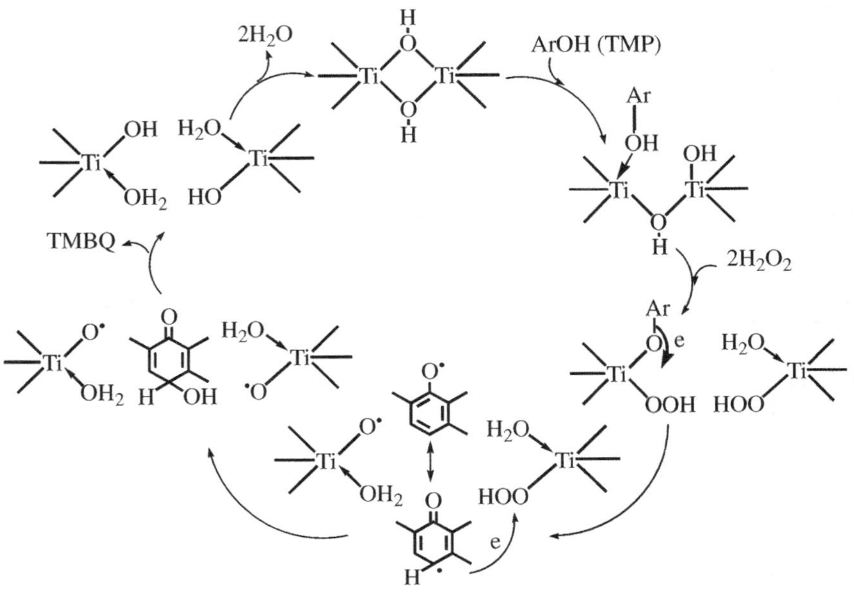

SCHEME 14.14 Oxidation of alkylphenols with H_2O_2 over Ti(IV) dimeric sites [131].

A Ti-based MOF, MIL-125, afforded alkylated *p*-BQ with a selectivity of greater than 99% but showed significantly lower H_2O_2 efficiencies as compared to the mesoporous Ti,Si catalysts [132].

The oxidation of 2-allylphenol and phenols bearing alcohol functional groups over a mesoporous titanium-silicate catalyst produces BQ with good to moderate yields keeping the other oxidizable sites intact [133]. 2-Hydroxybenzyl alcohol, which can alternatively be converted to salicylaldehyde, gave 2-hydroxymethyl-*p*-benzoquinone in a 73% yield.

A heterogeneous haloperoxidase-like catalyst based on WO_4^{2-}-exchanged layered double hydroxide has been developed by Sels et al. for the bromide-assisted oxidation of substituted phenols with H_2O_2 to corresponding *p*-quinols and *p*-quinol ethers [134].

14.4.7.3 Oxidation with tert-Butyl Hydroperoxide (TBHP) While catalytic amounts of $CuCl_2$ resulted in poor yields of TMBQ with O_2 as oxidant, the use of TBHP allowed for up to 80% yield with only 1.5 wt% of $CuCl_2$ and 2.8 wt% of NH_2OH—HCl cocatalyst under ambient conditions [135]. The Ru-catalyzed oxidation of *p*-substituted phenols with TBHP gave corresponding (*tert*-butyldioxy) cyclohexadienones that can be converted to 2-substituted *p*-BQ in the presence of a Lewis acid [136].

Iron tetrasulfophthalocyanine covalently bound to amino-modified silica (FePcS–SiO_2) catalyzed oxidation of a range of phenolic compounds with TBHP [98]. TMBQ was obtained in 87% yield at TMP conversion of 97% [98]. Phenols having other easily oxidizable sites, for example, alcoholic, double bond, or benzylic/allylic function, gave *p*-BQ with good to moderate yields. The FePcS–SiO_2 catalyst and FePcS immobilized within cages of the MOF MIL-101 demonstrated a better catalytic performance than homogeneous FePcS [137].

14.5 PHOTOCHEMICAL OXIDATIONS

Photochemical reactions of arenes are covered by Part IX of this book, so we give here just few notes to demonstrate the potential utility of photochemistry in aromatic oxidations.

Singlet oxygen reacts with polyacenes and electron-rich benzenes via [4+2] cycloaddition to afford endoperoxides (EP) that release $^1O_2$ upon heating (Eq. 14.32) [138].

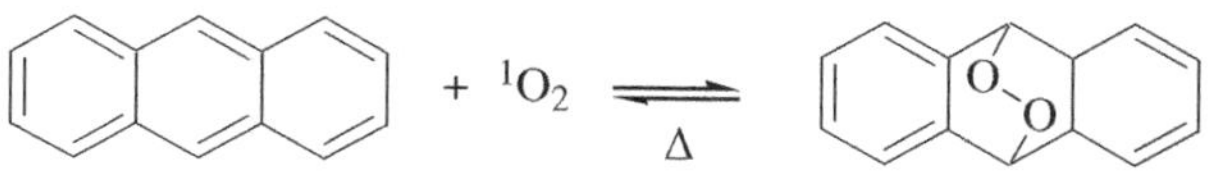

(14.32)

Some EPs give more stable 4-hydroperoxy-2,5-cyclohexadienones (Eq. 14.33) [139].

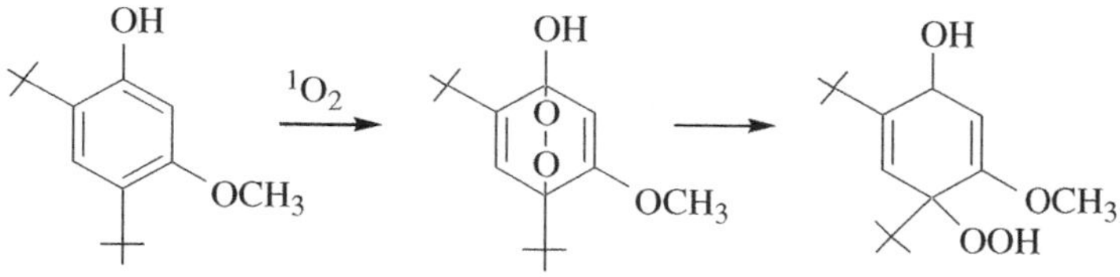

(14.33)

The development of efficient heterogeneous catalysts for generation of $^1O_2$ is an area of intensive research. Fullerene-coated aminomethylated poly(styrene-*co*-divinylbenzene) beads were used as recyclable catalysts in the photosensitized oxidation of 1-naphthol to naphthoquinone [140] (Scheme 14.15)

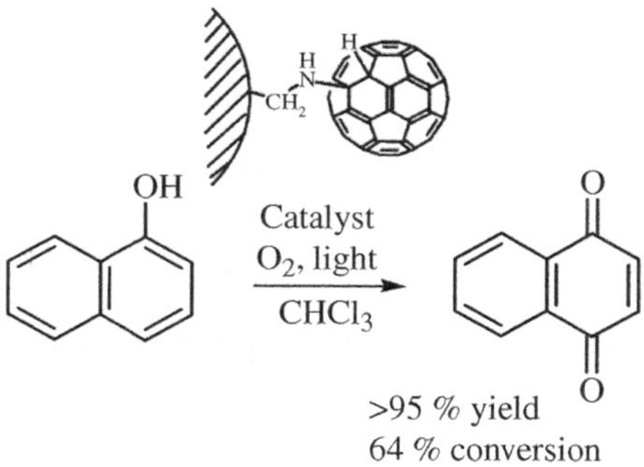

SCHEME 14.15 Fullerene-catalyzed oxidation of 1-naphthol using oxygen and light.

The selective photooxygenation of benzene has been accomplished in the presence of oxygen and water through photoinduced ET oxidation under ambient conditions using 3-cyano-1-methylquinolinium as homogeneous photocatalyst, producing phenol in a yield of 30% yield at benzene conversion of 31% [141].

Photocatalysis is particularly relevant to "sustainable and green chemistry" by benefit of the possibility to obtain fine chemicals with a low environmental impact. Maldotti and coworkers have recently reviewed application of heterogeneous photocatalysts for the selective oxidation of organic, including aromatic, compounds with molecular oxygen, so we send the interested reader directly to their book chapter [142].

14.6 ELECTROCHEMICAL OXIDATIONS

Electrochemistry is of significant importance in the practice of organic chemistry. Anodic oxidation of arenes has long been known [40, 57, 143]. Arenes with electron-withdrawing substituents were monohydroxylated in high yields by anodic oxidation in TFA/dichloromethane [144]. Hydroxylation of activated aromatic rings (polyaromatic or carrying electron-donating substituent) was achieved by anodic substitution with anions of TFA and *p*-toluenesulfonic acid, which result in ring deactivation upon substitution, followed by hydrolysis [145]. High yields at high substrate conversion for the electrochemical step were obtained in dry MeCN by constant potential electrolysis at graphite or platinum anodes.

In recent years, electrocatalysis has been widely employed to reductively activate dioxygen [22b, 146, 147]. The reduction of O_2 proceeds through two pathways, which are mainly determined by the electrocatalyst and electrode potential: two-electron reduction into H_2O_2 (Eq. 14.34) and four-electron reduction into H_2O (Eq. 14.35).

$$\text{Cathode}: O_2 + 2H^+ + 2e^- \rightarrow H_2O_2 \quad (14.34)$$

$$O_2 + 4H^+ + 4e^- \rightarrow 2H_2O \quad (14.35)$$

Electrochemical membrane reactors (Scheme 14.16 schematically shows the principle of their operation) may offer a less expensive alternative to catalytic membrane reactors [22b].

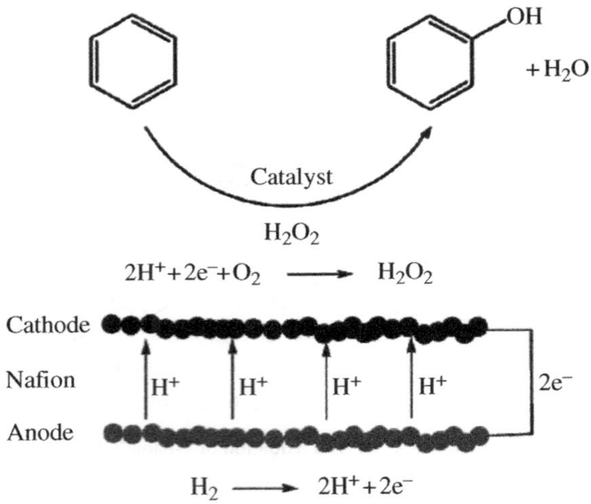

SCHEME 14.16 Phenol synthesis with *in situ* generated H_2O_2 in a proton-exchange membrane fuel cell Reproduced with permission from Ref. [22]. Copyright (2006) WILEY-VCH Verlag GmbH & Co. KGaA, Weinheim).

The selective hydroxylation of benzene and toluene has been realized by generation of an active oxygen species at a $SmCl_3$/graphite cathode in a phosphoric acid fuel cell [146a]. Benzene gave phenol as the only product while toluene produced nearly equal amounts of cresols and benzaldehyde. However, the current efficiency was only 5.5%, because oxygen reduction to water (Eq. 14.35) predominated. Importantly, the H_2–O_2 fuel cell system can cogenerate electricity and oxygenates [146a].

Hydroxylation of benzene to phenol was implemented at room temperature by applying H_2–O_2 fuel cells that continuously accumulated H_2O_2 at room temperature by using an Au cathode attached to a Nafion-H membrane as a protonic electrolyte [146b]. When an iron salt and benzene were loaded into the aqueous solution of HCl in the cathode compartment, phenol was formed due to generation of Fenton reagent.

A carbon whisker pretreated by hot aqueous HNO_3 turned out to be the most active carbon cathode for hydroxylation of benzene to phenol and HQ among other carbon materials [146b].

Without oxidative pretreatment, it was not active, suggesting that oxygenated groups (carboxyls and BQ–HQ ones) participate in the reductive activation of O_2. The addition of Pd-black enhanced the current remarkably, while the addition of Fe_2O_3 improved the oxidation efficiency for the formation of phenol and HQ. Their coaddition produced a synergistic effect.

An approach to overcoming the problem of the low current efficiency involves generation of an active oxygen species (most likely HO· radical) at the anode (Eq. 14.36):

$$\text{Anode}: H_2O \rightarrow O^* + 2H + 2e^- \quad (14.36)$$

$$C_6H_6 + O^* \rightarrow C_6H_5OH \quad (14.37)$$

A V_2O_5 anode was found to be the most efficient among anodes tested: the current efficiency for phenol production and selectivity were 41.7 and 100%, respectively [148].

However, for commercial application of the electrochemical methods, the current efficiency and the rate of oxygenations still need further improvements through the use of more efficient electrocatalysts and appropriate electrolytes.

14.7 ENZYMATIC HYDROXYLATION

The problem of the selective oxygenation of aromatic C—H bonds under mild conditions has been solved by Nature through the use of enzymes. To achieve high chemo-, regio-, and stereoselectivities, enzymes use two main strategies, that is, substrate orientation and size/shape selectivity. Overoxidation problems are efficiently solved in biological systems by separating catalysts and products into different environments. Active sites of many enzymes are placed into hydrophobic pockets where lipophilic substrates are readily oxidized, while the more hydrophilic oxygenated products are released into the aqueous environment and do not approach the catalytic site. Excellent monograph [44a], book chapters [149], and journal reviews [3, 150] with comprehensive coverage on the literature devoted to enzyme-mediated oxidations are available. The book chapter by Leak and coworkers [149b] and the review paper by Ullrich and Hofrichter [150c] are devoted to enzymatic aromatic oxidations.

Several types of enzymes, for example, heme-containing monooxygenases of the cytochrome P450 family, nonheme iron mono- and dioxygenases, are able to perform hydroxylation of aromatic compounds [44a, 149–151]. In general, mechanisms proposed for O_2 activation by various metallo-oxygenases involve the formation of an initial O_2 adduct (superoxo), conversion to a metal-peroxide (peroxo), and subsequent O—O bond cleavage to yield a high-valent oxidant (oxo) [3d].

Monooxygenases mediate hydroxylation by a process similar to that shown in Equation 14.15 where the role of electron supplier is played by a cofactor, NAD(P)H. Benzene hydroxylation with P450 leads mainly to phenol but also produces ketone and arene oxide as side products (Eq. 14.38). While phenol is less toxic than benzene, arene oxide is carcinogenic and mutagenic.

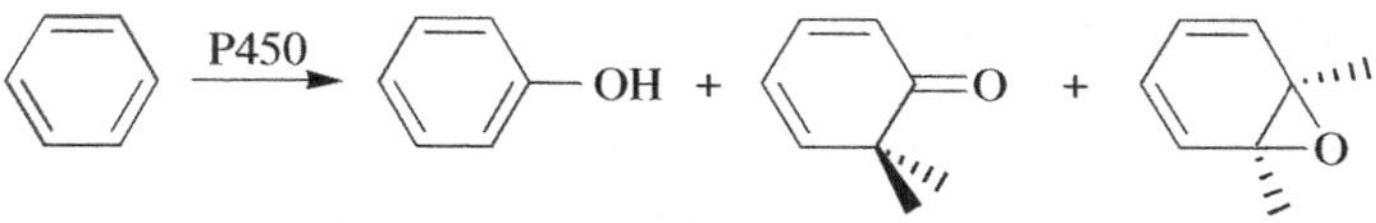

(14.38)

The aromatic hydroxylation by P450s is less well understood than aliphatic P450-mediated hydroxylation [152, 153]. The hydroxylation of arenes with enzymes features high values of the NIH shift [30, 44a, 151–153]. High-valent iron-oxo species were proposed to be involved in the oxygenation process with arene oxide as intermediate [44a, 149, 152]. Alternative mechanisms leading to the reaction products and responsible for the NIH shift have been discussed by Shaik and coworkers [152c, d]. DFT investigations revealed that electrophilic pathway dominates and involves an initial attack on the π-system of the benzene to produce a cationic σ-complex [153a]. de Visser and Shaik discovered computationally a mechanism of a porphyrin proton shuttle and suggested existence of an enzymatic pathway that converts benzene directly to a phenol and ketone, in addition to nonenzymatic production of these species by conversion of arene oxide to phenol and ketone [153a]. Safari et al. [154] found experimentally that both electron-withdrawing and electron-donating groups increase reactivity in the oxidation of several aromatics. This finding was rationalized by DFT computations under the assumption of a dual radical and cationic nature of the transition states [152c, 153b].

In spite of the great synthetic potential of P450s, their industrial implementation remains a challenge because of low activity, lack of stability, narrow substrate specificity, expensive cofactor requirements, and other reasons [150b, 155]. A range of molecular engineering techniques are nowadays available to solve these critical issues. A highly efficient hydroxylation of benzene and toluene catalyzed by wild-type cytochrome P450BM3 with the assistance of decoy molecules (linear perfluorinated carboxylic acids) has been reported [156]. This system demonstrated higher reaction rates and NADPH efficiencies than engineered P450BM3 mutants prepared by directed evolution [157].

While chemical methods for the controlled dihydroxylation of aromatic compounds are still missing, *Rieske* nonheme iron dioxygenases, are capable of regio- and stereoselective *cis*-dihydroxylation of aromatic compounds [149, 150]. The mechanism of this demanding oxidation where both atoms of dioxygen are incorporated into the *cis*-diol product has been carefully investigated [3d, 150a, 151].

The enzymatic hydroxylations play a crucial role in detoxification and excretion of toxic arenes and in drug metabolism. The enzymatic degradation of aromatics occurs via two different pathways: (i) monooxygenase-mediated formation of arene oxide followed by enzymatic hydrolysis to yield *trans*-1,2-diol and its further metabolism and (ii) dioxygenase-catalyzed arene *cis*-dihydroxylation and oxidative ring cleavage.

Enzymatic oxidation of phenols has been comprehensively reviewed [3, 57, 149, 150]. Several types of oxidoreductases enable *ortho*- and *para*-hydroxylations of phenols although the latter are relatively scarce. Metal-free enzymes, flavin-dependent monooxygenases, perform selective hydroxylation of phenols through the formation of a 4-*a*-(hydro)peroxoflavin [150b, c]. The binuclear copper enzymes tyrosinase and CAT oxidase activate O_2 via the formation of a (peroxo)dicopper(II) core that features an unusual side-on (μ-η^2:η^2) binding mode [3]. This core is supposed to be responsible for the *ortho*-hydroxylation of phenols, occurring by an electrophilic aromatic substitution pathway [3c, d]. Alternatively, a bis(μ-oxo)dicopper(III) species may act as a real oxidizing species for the *ortho*-hydroxylation [158]. The tyrosinase-catalyzed preparation of L-DOPA from tyrosine and molecular oxygen is considered as a potential method for the synthesis of this important anti-Parkinsonian drug. *Myko Tech's* original process uses extracellular tyrosinase produced by a microorganism and which can then be used to synthesize L-DOPA in an economical manner [159].

Finding inspiration in nature, chemists have designed synthetic systems to mimic the catalytic action of enzymes. Excellent book [44a] and review papers [3, 44b, 63, 150a, 160] on this topic have been published.

14.8 SUMMARY AND OUTLOOK

While the first examples of aromatic hydroxylation date to the middle of the twentieth century, over the last decades, there have been significant advances in the development of catalytic methods for the production of phenols, quinones, and other ring-oxygenated products. Enormous progress has been achieved in the implementation of sustainable and energy- and resource-efficient catalytic processes for the large-scale oxidation of benzene to phenol and phenol to dihydroxybenzenes. The pressure of increasingly strict environmental regulation has provided a stimulus for the replacement of stoichiometric fine chemicals processes that employ polluted and hazardous oxidants, such as Cr(VI) and Mn(IV) compounds, peroxy acids, and others, with cleaner, catalytic alternatives based on the use of the environmentally friendly oxidants, O_2, H_2O_2, TBHP, and N_2O. Many efficient systems employing both homogeneous and heterogeneous catalysts have been discovered for the selective nuclear oxidation of substituted arenes to produce important intermediates for the synthesis of fine chemicals and pharmaceuticals. The majority of them operate in the liquid phase because both substrates and products are thermally unstable and mild reaction conditions are required to achieve a high level of selectivity at reasonable conversions. However, these catalytic developments still need improvements in order industrial applications would become possible. Studies on the mechanisms of the nuclear aromatic oxidation are indispensable on the way toward 100% product selectivity and high oxidant efficiency. One of the major concerns is catalyst stability that is often not enough high to get a good productivity (catalyst turnover numbers) and recyclability. True heterogeneous catalysts, stable to metal leaching under the conditions of liquid-phase oxidation, still remain a great challenge. The employment of solvent free conditions or use of aqueous media is highly desirable, when possible. Therefore, further research is encouraged in order to provide truly green and sustainable processes for the production of aromatic ring oxidation products. Photo- and electrochemical methods as well as biotechnologies may become a promising alternative to traditional catalytic systems in the near future, provided cost and stability issues will be solved successfully.

ACKNOWLEDGMENTS

The author is grateful to Dr. Olga Zalomaeva for assistance in the preparation of this chapter and to the Russian Foundation for Basic Research (grant 13-03-12042) for funding research on arene oxidation.

ABBREVIATIONS

[Ru(tpy)(pydic)]	Ruthenium-(2,2′,6′:2″-terpyridine)(2,6-pyridinedicarboxylate)
AN	Anthracene
AQ	Anthraquinone
BP	Bisphenol
BQ	Benzoquinone
BTC	1,3,5-Benzenetricarboxylate
CAT	Catechol
CVD	Chemical vapor deposition
DMF	Dimethylformamide
DPQ	Diphenoquinone
EP	Endoperoxide
ET	Electron transfer
EXAFS	Extended X-ray absorption fine structure
FePcS	Iron tetrasulfophthalocyanine
H_2Pydic	Pyridine-2,6-dicarboxylic acid
HAT	Hydrogen atom transfer
HBA	Hydrogen bond accepting (solvent)
HEL	Helianthrene
HELPO	Corresponding endoperoxide
HP	Hydroperoxide
HPA	Heteropoly acid
HPA-*n*	Heteropoly acids of the general formula $H_{3+n}[PMo_{12-n}V_nO_{40}]$
Hpca	2-Pyrazinecarboxylic acid

HQ	Hydroquinone
ISC	Intersystem crossing
MN	2-Methylnaphthalene
MNL	2-Methyl-1-naphthol
MNQ	2-Methyl-1,4-naphthoquinone (menadione, vitamin K_3)
MOF	Metal-organic framework
MTO	Methylrhenium trioxide (CH_3ReO_3)
NQ	1,4-Naphthoquinone
PC	Pseudocumene (1,2,4-trimethylbenzene)
pic	Picolinate
SET	Single electron transfer
SPLET	Sequential proton loss electron transfer
TBHP	*tert*-Butyl hydroperoxide
TFA	Trifluoroacetic acid
TFAA	Trifluoroacetic anhydride
TMBQ	2,3,5-Trimethyl-*p*-benzoquinone
TMP	2,3,6-Trimethylphenol
TMT	3,4,5-Trimethoxytoluene
TS-1	Titanium-silicalite-1

REFERENCES

[1] (a) Rappoport, Z. ed. *The Chemistry of Phenols*, John Wiley & Sons, Ltd, Chichester, 2003; (b) Mihailović, M.L. and Čeković, Z. (1971) in *The Chemistry of the Hydroxyl Group, part 1* (ed. S. Patai) John Wiley & Sons, Ltd, London, p. 505; (c) Tyman, J.H.P. *Synthetic and Natural Phenols*; Studies in Organic Chemistry **52**, Elsevier, the Netherlands, 1996.

[2] Patai, S. and Rappoport, Z. eds. *The Chemistry of the Quinonoid Compounds*, John Wiley & Sons, Inc., New York, 1988.

[3] (a) Kitajima, N. and Morooka, Y. (1994) *Chem. Rev.*, **94**, 737–757; (b) Solomon, E.I., Sundaram, U.M., and Machonkin, T.E. (1996) *Chem. Rev.*, **96**, 2563–2605; (c) Itoh, S. and Fukuzumi, S. (2007) *Acc. Chem. Res.*, **40**, 592–600; (d) Que L., Jr. and Tolman, W.B. (2008) *Nature*, **455**, 333–340.

[4] (a) Lorenc, J.F., Lambeth, G., and Scheffer, W. (2003) in *Kirk-Othmer Encyclopedia of Chemical Technology*, Vol. **2**, Wiley-VCH, Weinheim, pp. 203–233; (b) Fiegel, H., Voges, H.-W., Hamamoto, T., Umemura, S., Iwata, T., Miki, H., Fujita, Y., Buysch, H.-J., Garbe, D., and Paulus, W. (2002) in *Ullmann's Encyclopedia of Industrial Chemistry*, Wiley-VCH, Weinheim.

[5] (a) Schudel, P., Mayer, H., and Isler, O. (1972) in *The Vitamins*, Vol. **5** (ed. W.H. Sebrell and R.S. Harris), Academic Press, New York, pp. 168–218; (b) Rucker, R.B., Suttie, J.W., McCormick, D.B., and Machlin, L.J. eds. *Handbook of Vitamins*, Marcel Dekker, New York, 2001.

[6] Nicolaou, K.C., Chen, J.S., Edmonds, D.J., and Estrada, A.A. (2009) *Angew. Chem. Int. Ed.*, **48**, 660–719.

[7] (a) Bonrath, W. and Netscher, T. (2005) *Appl. Catal. A Gen.*, **280**, 55–73; (b) Eggersdorfer, M., Laudert, D., Létinois, U., McClymont, T., Medlock, J., Netscher, T., and Bonrath, W. (2012) *Angew. Chem. Int. Ed.*, **51**, 12960–12990.

[8] Sheldon, R., Arends, I.W.C.E., and Hanefeld U. *Green Chemistry and Catalysis*, Wiley-VCH, Weinheim, 2007.

[9] (a) Cavani, F., Centi, G., Perathoner, S., and Trifiro, F. eds. *Sustainable Industrial Processes*, Wiley-VCH, Weinheim, 2009; (b) Cavani, F. and Teles J.H. (2009) *ChemSusChem*, **2**, 508–534.

[10] Clerici, M.G. and Kholdeeva, O.A. eds. *Liquid Phase Oxidation via Heterogeneous Catalysis: Organic Synthesis and Industrial Applications*, John Wiley & Sons, Inc., Hoboken, 2013.

[11] Lücke, B., Narayana, K.V., Martin, A., and Jähnisch, K. (2004) *Adv. Synth. Catal.*, **346**, 1407–1424.

[12] Alonso, D.A., Nájera, C., Pastor, I.M., and Yus, M. (2010) *Chem. Eur. J.*, **16**, 5274–5284.

[13] Enthaler, S. and Company, A. (2011) *Chem. Soc. Rev.*, **40**, 4912–4924.

[14] Möller, K., Wienhöfer, G., Westerhaus, F., Junge, K., and Beller, M. (2011) *Catal. Today*, **173** 68–75.

[15] Sheldon, R.A. and Kochi, J.K. *Metal-Catalyzed Oxidations of Organic Compounds*, Academic Press, New York, 1981.

[16] (a) Denisov, E.T., Mitskevich, N.I., and Agabekov, V.E. *Liquid-Phase Oxidation of Oxygen-Containing Compounds*, Consultants Press, New York, 1977; (b) Denisov, E.T. and Afanas'ev, I.B. *Oxidation and Antioxidants in Organic Chemistry and Biology*, Taylor & Francis, Boca Raton, 2005.

[17] Arpentinier, P., Cavani, F., and Trifiro, F. *The Technology of Catalytic Oxidations*, TECHNIP, Paris, 2001, pp. 67–119.

[18] (a) Blanksby, S.J. and Ellison, G.B. (2003) *Acc. Chem. Res.*, **36**, 255–263; (b) Warren, J.J., Tronic, T.A., and Mayer, J.M. (2010) *Chem. Rev.*, **110**, 6961–7001.

[19] (a) Mayer, J.M. (2011) *Acc. Chem. Res.*, **44**, 36–46; (b) Steenken, S. and Neta P. (2003) in *The Chemistry of Phenols* (ed. Z. Rappoport), John Wiley & Sons, Ltd, Chichester, pp. 1107–1152.

[20] Berkessel, A. and Vogl, N. (2006) in *The Chemistry of Peroxides*, Vol. 2, Part 1, (ed. Z. Rappoport), John Wiley & Sons, Ltd, Chichester, pp. 307–586.

[21] Litwinienko, G. and Ingold, K.U. (2007) *Acc. Chem. Res.*, **40**, 222–230.

[22] (a) Strukul, G. and Scarso, A. (2013)in *Liquid Phase Oxidation via Heterogeneous Catalysis: Organic Synthesis and Industrial Applications* (eds. M.G. Clerici and O.A. Kholdeeva), John Wiley & Sons, Inc., Hoboken, pp. 1–20; (b) Campos-Martin, J.M., Blanco-Brieva, G., and Fierro, J.L.G. (2006) *Angew. Chem. Int. Ed.*, **45**, 6962–6984.

[23] Seip, M. and Brauer, H.-D. (1992) *J. Am. Chem. Soc.*, **114**, 4486–4490.

[24] Birks, J.B. *Photophysics of Aromatic Molecules*, John Wiley & Sons, Ltd, London, 1970, pp. 209–211.

[25] (a) Kholdeeva, O.A. and Rossi, M., Patent No. WO 2,006,104,411, 2006; (b) Kholdeeva, O.A., Zalomaeva, O.V., Sorokin, A.B., Ivanchikova, I.D., Della Pina, C.D., and Rossi, M. (2007) *Catal. Today*, **121**, 58–64.

[26] Kholdeeva, O.A., Ivanchikova, I.D., Zalomaeva, O.V., Sorokin, A.B., Skobelev, I.Y., and Talsi, E.P. (2011) *J. Phys. Chem. B*, **115**, 11971–11983.

[27] (a) Barton, D.H.R., Haynes, R.K., Leclerc, G., Magnus, P.D., and Menzies, I.D.J. (1975) *Chem. Soc. Perkin Trans.* **1**, 2055–2065; (b) Clark, K.B., Howard, J.A., and Oyler, A.R. (1997) *J. Am. Chem. Soc.*, **119**, 9560–9561.

[28] (a) Walling, C. (1975) *Acc. Chem. Res.*, **8**, 125–131; (b) Walling, C. (1998) *Acc. Chem. Res.*, **31**, 155–157.

[29] (a) Kunai, A., Hata, S., Ito, S., and Sasaki, K. (1986) *J. Am. Chem. Soc.*, **108**, 6012–6016; (b) Karakhanov, E.A., Narin, S.Y., and Dedov, A.G. (1991) *Appl. Organomet. Chem.* **5**, 445–461.

[30] Guroff, G., Daly, J.W., Jerina, D.M., Renson, J., Witkop, B., and Udenfriend, S. (1967) *Science*, **157**, 1524–1530.

[31] Castle, L., Lindsay Smith, J.R., and Buxton, G.V. (1980) *J. Mol. Catal.*, **7**, 235–243.

[32] Sawyer, D.T., Sobkowiak, A., and Matsushita, T. (1996) *Acc. Chem. Res.*, **29**, 409–416.

[33] Mimoun, H., Saussine, L., Daire, E., Postel, M., Fischer, J., and Weiss, R. (1983) *J. Am. Chem. Soc.*, **105**, 3101–3110.

[34] Bonchio, M., Conte, V., Di Furia, F., Modena, G., and Moro, S. (1994) *J. Org. Chem.*, **59**, 6262–6267.

[35] (a) Fukuzumi, S., Nakanishi, I., and Tanaka, K. (1999) *J. Phys. Chem. A*, **103**, 11212–11220; (b) Tanko, J.M. and Wang, Y. (1997) Chem. Comm., 2387–2388; (c) Rudenko, A.P. (1994) *Zh. Org. Khim. (Russ. J. Org. Chem.)*, **30**, 1847–1881; (d) Rudenko, A.P. and Pragst, F. (1998) *Zh. Org. Khim.*, **34**, 1660–1696.

[36] Minisci, F., Citterio, A., and Giordano, C. (1983) *Acc. Chem. Res.*, **16**, 27–32.

[37] Parker, V.D. (1984) *Acc. Chem. Res.*, **17**, 243–250.

[38] (a) Khenkin, A.M. and Neumann, R. (2000) *Angew. Chem. Int. Ed.*, **39**, 4088–4090; (b) Khenkin, A., Weiner, L., Wang, Y., and Neumann, R. (2001) *J. Am. Chem. Soc.*, **123**, 8531–8542; (c) Neumann, R. and Khenkin, A.M. (2006) Chem. Commun., 2529–2538.

[39] Chen, X., Hao, X.-S., Goodhue, C.E., and Yu, J.-Q. (2006) *J. Am. Chem. Soc.*, **128**, 6790–6791.

[40] Eberson, L. and Nyberg, K. (1973) *Acc. Chem. Res.*, **6**, 106–112.

[41] Derbyshire, D.H. and Waters, W.A. (1950) *Nature*, **4191**, 401.

[42] Swern, D. ed. *Organic Peroxides*, Vol. 1 and 2, Wiley-Interscience, New York, 1970.

[43] Hart, H. (1971) *Acc. Chem. Res.*, **4**, 337–343.

[44] (a) Ortiz de Montellano, P.R. ed. *Cytochrome P-450: Structure, Mechanism and Biochemistry*, 3rd ed., Kluwer Academic/Plenum, New York, 2005; (b) Bryliakov, K.P. and Talsi, E.P. (2014) *Coord. Chem. Rev.*, **276**, 73–96.

[45] Olah, G.A. and Ohnishi, R. (1978) *J. Org. Chem.*, **43**, 865–867.

[46] Adam, W. and Shimizu, M. (1994) *Synthesis*, **4**, 560–562.

[47] (a) Adam, W., Herrmann, W.A., Lin, W., and Saha-Möller, C.R. (1994) *J. Org. Chem.*, **59**, 8281–8283; (b) Adam, W., Herrmann, W.A., Saha-Möller, C.R., and Shimizu, M. (1995) *J. Mol. Catal. A*, **97**, 15–20; (c) Jacob, J. and Espenson, J.H. (1998) *Inorg. Chim. Acta*, **270**, 55–59.

[48] Carril, M., Altmann, P., Drees, M., Bonrath, W., Netscher, T., Schütz, J., and Kühn, F.E. (2011) *J. Catal.* **283**, 55–67.

[49] Shilov, A.E. and Shul'pin, G.B. (1997) *Chem. Rev.*, **97**, 2879–2932.

[50] (a) Stock, L.M., Tse, K., Vorvick, L.J., and Wolstrum, S.A. (1981) *J. Org. Chem.*, **46** 1757–1759; (b) Benazzi, E., Cameron, C.J., and Mimoun, H. (1991) *J. Mol. Catal.*, **69**, 299–321.

[51] Haines, A.H. *Methods for the Oxidation of Organic Compounds: Alkanes, Alkenes, Alkynes and Arenes*, Academic Press, London, 1985.

[52] Sheldon, R.A. (1993) *Top. Curr. Chem.* **164**, 21–43.

[53] Vogel, A. and Bayer, A.G. (1985) in *Ullmann's Encyclopedia of Industrial Chemistry*, vol. **2**, Wiley-VCH, Weinheim, pp. 347–354.

[54] Asakawa, Y., Matsuda, R., Tori, M., and Sono, M. (1988) *Org. Chem.* **53**, 5453–5457.

[55] Arnold, R.T. and Larson, R. (1940) *J. Org. Chem.* **5**, 250–252.

[56] Orita, H., Shimizu, M., Hayakawa, T., and Takehira, K. (1989) *Bull. Chem. Soc. Jpn.*, **62**, 1652–1657.

[57] Yamamura, S. (2003) in *The Chemistry of Phenols* (ed. Z. Rappoport) John Wiley & Sons, Ltd, Chichester, pp. 1153–1346.

[58] Ricci, M., Bianchi, D., and Bortolo, R. (2009) in *Sustainable Industrial Processes* (eds. F. Cavani, G. Centi, S. Perathoner, and F. Trifiro) Wiley-VCH, Weinheim, pp. 453–476.

[59] Panov, G.I., Dubkov, K.A., and Kharitonov, A.S. (2009) in *Modern Heterogeneous Oxidation Catalysis: Design, Reactions and Characterization* (ed. N. Mizuno), Wiley-VCH, Weinheim, pp. 217–252.

[60] Ito, S., Yamasaki, T., Okada, H., Okino, S., and Sasaki, K. (1988) *J. Chem. Soc. Perkin Trans.*, **2**, 285–293.

[61] (a) Hamada, R., Shibata, Y., Nishiyama, S., and Tsuruya, S. (2003) *Phys.Chem.Chem.Phys.*, **5**, 956–965; (b) Kubacka, A., Wang, Z., Sulikowski, B., and Corberan, V.C. (2007) *J. Catal.*, **250**, 184–189.

[62] Häusser, A., Trautmann, M., and Roduner, E. (2011) *Chem. Commun.*, **47**, 6954–6956.

[63] Moro-oka, Y. and Akita, M. (1998) *Catal. Today*, **41**, 327–338.

[64] Hata, E., Takai, T., Yamada, T., and Mukaiyama, T. (1994) Chem. Lett., 1849–1852.

[65] Guo, Z.-W., Gu, Y.-Y., Zhou, S.-L., and Ren, C.-H. (2011) *Adv. Mater. Res.*, **233–235**, 1575–1580.

[66] (a) Jintoku, T., Takaki, K., Fujiwara, Y., Fuchita, Y., and Hiraki, K. (1990) *Bull. Chem. Soc. Jpn.*, **63**, 438–441; (b) Tani, M., Sakamoto, T., Mita, S., Sakaguchi, S., and Ishii Y. (2005) *Angew. Chem. Int. Ed.*, **44**, 2586–2588.

[67] (a) Miyake, T., Hamada, M., Sasaki, Y., and Oguri, M. (1995) *Appl. Catal. A*, **131**, 33–42; (b) Ehrich, H., Berndt, H., Pohl, M.-M., Jähnisch, K., and Baerns, M. (2002) *Appl. Catal. A*, **230**, 271–280; (c) Niwa, S., Eswaramoorthy, M., Nair, J., Raj, A., Itoh, N., Shoji, H., Namba, T., and Mizukami, F. (2002) *Science*, **295**, 105–107.

[68] (a) Bal, R., Tada, M., Sasaki, T., and Iwasawa, Y. (2006) *Angew. Chem. Int. Ed.* **45**, 448–452; (b) Tada, M., Bal, R., Sasaki, T., Uemura, Y., Inada, Y., Tanaka, S., Nomura, M., and Iwasawa, Y. (2007) *J. Phys. Chem. C*, **111**, 10095–10104.

[69] Olah, G.A., Fung, A.P., and Keumi, T. (1981) *J. Org. Chem.*, **46**, 4306–4307.

[70] Tamagaki, S., Suzuki, K., Okamoto, H., and Tagaki, W. (1983) *Tetrahedron Lett.*, **24**, 4847–4850.

[71] Karakhanov, E.A., Narin, S.Y., and Dedov, A.G. (1989) *Catal. Lett.* **3**, 31–36.

[72] Bianchi, D., Bortolo, R., Tassinari, R., Ricci, M., and Vignola, R. (2000) *Angew. Chem. Int. Ed.*, **39**, 4321–4323.

[73] Peng, J.J., Shi, F., Gu, Y.L., and Deng, Y.Q. (2003) *Green Chem.*, **5**, 224–226.

[74] Kudrik, E.V. and Sorokin, A.B. (2008) *Chem. Eur. J.*, **14**, 7123–7126.

[75] Shul'pin, G.B., Muratov, D.V., Shul'pina, L.S., Kudinov, A.R., Strelkova, T.V., and Petrovskiy, P.V. (2008) *Appl. Organomet. Chem.*, **22**, 684–688.

[76] Shul'pin, G.B., Kirillova, M.V., Kozlov, Y.N., Shul'pina, L.S., Kudinov, A.R., and Pombeiro, A.J.L. (2011) *J. Catal.*, **277**, 164–172.

[77] Kirillov, A.M. and Shul'pin, G.B. (2013) *Coord. Chem. Rev.*, **257**, 732–754.

[78] Mizuno, N. and Kamata, K. (2011) *Coord. Chem. Rev.*, **255**, 2358–2370.

[79] Borah, P., Ma, X., Truc Nguyen, K., and Zhao, Y. (2012) *Angew. Chem. Int. Ed.*, **51**, 7756–7761.

[80] (a) Taramasso, M., Perego, C., and Notari, B. US Patent 4,410,501, 1983; (b) Notari, B. (1996) *Adv. Catal.*, **41**, 253–334.

[81] Clerici, M.G. and Domine, M.E. (2013) in *Liquid Phase Oxidation via Heterogeneous Catalysis: Organic Synthesis and Industrial Applications* (eds. M.G. Clerici and O.A. Kholdeeva), John Wiley & Sons, Inc., Hoboken, pp. 21–93.

[82] Balducci, L., Bianchi, D., Bortolo, R., D'Aloisio, R., Ricci, M., Tassinari, R., and Ungarelli, R. (2003) *Angew. Chem. Int. Ed.*, **42**, 4937–4940.

[83] Iwamoto, M., Hirata, J., Matsukami, K., and Kagawa, S. (1983) *J. Phys. Chem.*, **87**, 903–905.

[84] (a) Suzuki, E., Nakashiro, K., and Ono, Y. (1988) *Chem. Lett.*, **17**, 953–956; (b) Gubelmann, M. and Tirel, P., France Patent 2 630 735, 1988; (c) Kharitonov, A.S., Aleksandrova, T.N., Vostrikova, L.A., Sobolev, V.I., Ione, K.G., and Panov G.I., USSR Patent 1 805 127, 1988.

[85] (a) Panov, G.I., Uriate, A., Rodkin, M.A., and Sobolev, V.I. (1998) *Catal. Today*, **41**, 365–385; (b) Panov, G.I. (2000) *Cattech*, **4**, 18–32.

[86] (a) Ramesh Reddy, K., Ramaswamy, A.V., and Ratnasamy, P. (1993). *J. Catal.* **143**, 275–285; (b) Shylesh, S. and Singh, A.P. (2004) *J. Catal.* **228**, 333–346.

[87] (a) Bordoloi, A., Lefebvre, F., and Halligudi, S.B. (2007) *J. Catal.* **247**, 166–175; (b) Srinivas, N., Radha Rani, V., Kulkarni, S.J., and Raghavan, K.V. (2002) *J. Mol. Catal. A*, **179**, 221–231; (c) Selvaraj, M. and Kawi, S. (2007) *Microporous Mesoporous Mater.*, **101**, 240–249; (d) Araújo, R.S., Azevedo, D.C.S., Rodríguez-Castellón, E., Jiménez-López, A., and Cavalcante, C.L., Jr. (2008) *J. Mol. Catal. A*, **281**, 154–163; (e) Pérollier, C., Pergrale-Mejean, C., and Sorokin, A.B. (2005) *New J. Chem.*, **29**, 1400–1403.

[88] Kholdeeva, O.A., Skobelev, I.Y., Ivanchikova, I.D., Kovalenko, K.A., Fedin, V.P., and Sorokin, A.B. (2014) *Catal. Today*, **238**, pp. 54–61.

[89] Sugano, J., Kuriyama, Y., Ishiuchi, Y., and Minamikawa, Y., US Patent 3953482, 1976.

[90] (a) Thiel, W., Sun, Y., Schubert, A., Bohle, A., and Schindler, G., World Patent 123644, 2005; (b) Bohle, A., Schubert, A., Sun, Y., and Thiel, W.R. (2006) *Adv. Synth. Catal.*, **348**, 1011–1015.

[91] Yamaguchi, S., Inone, M., and Enomoto, S. (1986) *Chem. Pharm. Bull.*, **59**, 2881–2884.

[92] (a) Adam, W., Herrmann, W.A., Lin, W., Saha-Möller, C.R., Fischer, R.W., and Correia, J.D.G. (1994) *Angew. Chem. Int. Ed.*, **33**, 2475–2476; (b) Herrmann, W.A., Haider, J.J., and Fischer, R.W. (1999) *J. Mol. Catal. A*, **138**, 115–121.

[93] Carril, M., Altmann, P., Bonrath, W., Netscher, T., Schütz, J., and Kühn, F.E. (2012) *Catal. Sci. Technol.*, **2**, 722–724.

[94] (a) Bianchi, M., Bonchio, M., Conte, V., Coppa, F., Di Furia, F., Modena, G., Moro, S., and Standen, S. (1993) *J. Mol. Catal. A*, **83**, 107–116; (b) Arichi, H.J., Eternot, M., and Louis, B. (2008) *Catal. Today*, **138**, 117–122; (c) Joseph, J.K., Singhal, S., Jain, S.L., Sivakumaran, R., Kumar, B., and Sain, B. (2009) *Catal. Today*, **141**, 211–214.

[95] Kamata, K., Yamaura, T., and Mizuno, N. (2012) *Angew. Chem. Int. Ed.*, **51**, 7275–7278.

[96] Möller, K., Wienhöfer, G., Schröder, K., Join, B., Junge, K., and Beller, M. (2010) *Chem. Eur. J.*, **16**, 10300–10303.

[97] Wienhöfer, G., Schröder, K., Möller, K., Junge, K., and Beller, M. (2010) *Adv. Synth. Catal.*, **352**, 1615–1620.

[98] Sorokin, A.B. (2013) *Chem. Rev.*, **113**, 8152–8191.

[99] Kumar, R. and Bhaumik, A. (1998) *Microporous Mesoporous Mater.*, **21**, 497–504.

[100] Kholdeeva, O.A. (2013) in *Liquid Phase Oxidation via Heterogeneous Catalysis: Organic Synthesis and Industrial Applications* (eds. M.G. Clerici and O.A. Kholdeeva), John Wiley & Sons, Inc., Hoboken, pp. 127–219.

[101] (a) Motz, J.L., Heinichen, H., and Hölderich, W.F. (1998) *J. Mol. Catal. A*, **136**, 175–184; (b) Ehrich, H., Schwieger, W., and Jähnisch, K. (2004) *Appl. Catal. A*, **272**, 311–319.

[102] (a) Khenkin, A., Weiner, L., and Neumann, R. (2005) *J. Am. Chem. Soc.*, **127**, 9988–9989; (b) Liu, Q., Wu, P., Yang, Y., Zeng, Z., Liu, J., Yi, H., and Lei, A. (2012) *Angew. Chem. Int. Ed.*, **124**, 4744–4748.

[103] Taktak, S., Flook, M., Foxman, B.M., Que, L., Rybak-Akimova, E.V., and Akimova, R. (2005) Chem. Commun., 5301–5303.

[104] Zhang, Y.H. and Yu, J.-Q. (2009) *J. Am. Chem. Soc.*, **131**, 14654–14655.

[105] (a) Yang, Y., Lin, Y., and Rao, Y. (2012) *Org. Lett.*, **14**, 2874–2877; (b) Thirunavukkarasu, V.S., Hubrich, J., and Ackermann, L. (2012) *Org. Lett.*, **14**, 4210–4213; (c) Mo, F., Trzepkowski, L.J., and Dong, G. (2012) *Angew. Chem. Int. Ed.*, **51**, 13075–13079; (d) Rao, Y. (2013) *Synlett*, **24**, 2472–2476.

[106] (a) Romano, U. and Ricci, M. (2013) in *Liquid Phase Oxidation via Heterogeneous Catalysis: Organic Synthesis and Industrial Applications* (eds. M.G. Clerici and O.A. Kholdeeva), John Wiley & Sons, Inc., Hoboken, pp. 541–462; (b) Clerici, M. (2001) in *Fine Chemicals through Heterogeneous Catalysis* (eds. R.A. Sheldon and H. van Bekkum) Wiley-VCH, Weinheim, pp. 538–551.

[107] Karakhanov, E.A., Maximov, A.L., Kardasheva, Y.S., Skorkin, V.A., Kardashev, S.V., Ivanova, E.A., Lurie-Luke, E., Seeley, J.A., and Cron, S.L. (2010) *Ind. Eng. Chem. Res.*, **49**, 4607–4613.

[108] Qui, L.-G., Xie, A.-J., and Zhang, L.-D. (2005) *Adv. Mater.*, **17**, 689–692.

[109] Karakhanov, E.A., Maximov, A.L., Kardasheva, Y.S., Skorkin, V.A., Kardashev, S.V., Predeina, V.V., Talanova, M.Y., Lurie-Luke, E., Seeley, J.A., and Cron, S.L. (2010) *Appl. Catal. A*, **385**, 62–72.

[110] Navalon, S., Martin, R., Alvaro, M., and Garcia, H. (2010) *Angew. Chem. Int. Ed.*, **49**, 8403–8407.

[111] Perego, C., Carati, A., Ingalina, P., Mantegazza, M.A., and Bellussi, G. (2001) *Appl. Catal. A*, **221**, 63–72.

[112] Ivanov, D.P., Sobolev, V.I., Pirutko, L.V., and Panov, G.I. (2002) *Adv. Synth. Catal.*, **344**, 986–995.

[113] (a) Simándi, L.I. (2003) in *Advances in Catalytic Activation of Dioxygen by Metal Complexes* (ed. L.I. Simándi), Kluwer Academic, Boston; (b) Gupta, K.C. and Sutar, A.K. (2008) *Coord. Chem. Rev.*, **252**, 1420–1450.

[114] de Jong, A.G. and van Helden, R., German Patents 2,460,665 and 2,517,870, 1975.

[115] Frostin-Rio, M., Pujol, D., Bied-Charreton, C., Perree-Fauvet, M., and Gaudemer, A. (1984) *J. Chem. Soc., Perkin Trans.*, **1**, 1971–1979.

[116] Bozell, J.J., Hames, B.R., and Dimmel, D.R. (1995) *J. Org. Chem.*, **60**, 2398–2404.

[117] (a) Hay, A.S., Blanchard, H.S., Endres, G.F., and Eunstance, J.W. (1959) *J. Am. Chem. Soc.*, **81**, 6335–6336; (b) Endres, G.F., Hay, A.S., and Eunstance, J.W. (1963) *J. Org. Chem.*, **28**, 1300–1305.

[118] (a) Brenner, W., German Patent 2,221,624, 1972; (b) Isshiki, T., Yui, T., Uno, H., and Abe, M., European Patent 0,127,888, 1984; (c) Hoercher, U., Jessel, B., Bockstiegel, B., Grafen, P., and Laas, H. US Patent 5041572, 1991.

[119] (a) Mercier, C. and Chabardes, P. (1994) *Pure Appl. Chem.*, **66**, 1509–1518; (b) Punniyamurthy, T. and Rout, L. (2008) *Coord. Chem. Rev.*, **252**, 134–154.

[120] Sun, H., Harms, K., and Sundermeyer, J. (2004) *J. Am. Chem. Soc.*, **126**, 9550–9551.

[121] (a) Kholdeeva, O.A., Golovin, A.V., Maksimovskaya, R.I., and Kozhevnikov, I.V. (1992) *J. Mol. Catal.*, **75**, 235–244; (b) Lissel, H., Jansen in de Wal, H., and Neumann, R. (1992) *Tetrahedron Lett.*, **33**, 1795–1798.

[122] Matveev, K.I., Odjakov, V.F., and Zhizhina, E.G. (1996) *J. Mol. Catal. A Chem.*, **114**, 151–160.

[123] (a) Shimizu, M., Orita, H., Hayakawa, T., and Takehira, K. (1989) *Tetrahedron Lett.*, **30**, 471–474; (b) Orita, H., Shimizu, M., Hayakawa, T., and Takehira, K., European Patent Appl. 0347021, 1988.

[124] (a) Bernini, R., Mincione, E., Barontini, M., Fabrizi, G., Pasqualetti, M., and Tempesta, S. (2006) *Tetrahedron*, **62**, 7733–7737; (b) Bernini, R., Mincione, E., Barontini, M., Crisante, F., Fabrizi, G., and Gambacorta, A. (2007) *Tetrahedron*, **63**, 6895–6900.

[125] (a) Ito, S., Aihara, K., and Matsumoto, M. (1983) *Tetrahedron Lett.*, **24**, 5249–5252; (b) Matsumoto, M. and Kobayashi, H. (1985) *J. Org. Chem.*, **50**, 1766–1768.

[126] See for review: Kholdeeva, O.A. (2014) *Catal. Sci. Technol.*, **4**(7), 1869–1889.

[127] Tanev, P.T., Chibwe, M., and Pinnavaia, T.J. (1994) *Nature*, **368**, 321–323.

[128] (a) Sorokin, A. and Tuel, A. (2000) *Catal. Today*, **57**, 45–59; (b) Trukhan, N.N., Romannikov, V.N., Paukshtis, E.A., Shmakov, A.N., and Kholdeeva, O.A. (2001) *J. Catal.*, **202**, 110–117.

[129] Zalomaeva, O.V., Trukhan, N.N., Ivanchikova, I.D., Panchenko, A.A., Roduner, E., Talsi, E.P., Sorokin, A.B., Rogov, V.A., and Kholdeeva, O.A. (2007) *J. Mol. Catal. A Chem.*, **277**, 185–192.

[130] (a) Kholdeeva, O.A., Trukhan, N.N., Vanina, M.P., Romannikov, V.N., Parmon, V.N., Mrowiec-Bialon, J., and Jarzebski, A.B. (2002) *Catal. Today*, **75**, 203–209; (b) Kholdeeva, O.A., Ivanchikova, I.D., Guidotti, M., and Ravasio, N. (2007) *Green Chem.*, **9**, 731–733; (c) Ivanchikova, I.D., Kovalev, M.K., Mel'gunov, M.S., Shmakov, A.N., and Kholdeeva, O.A. (2014) *Catal. Sci. Technol.*, **4**(1), 200–207.

[131] (a) Kholdeeva, O.A., Ivanchikova, I.D., Guidotti, M., Pirovano, C., Ravasio, N., Barmatova, M.V., and Chesalov, Y.A. (2009) *Adv. Synth. Catal.*, **351**, 1877–1889; (b) Kholdeeva, O.A., Ivanchikova, I.D., Guidotti, M., Ravasio, N., Sgobba, M., and Barmatova, M.V. (2009) *Catal. Today*, **141**, 330–336.

[132] Ivanchikova, I.D., Lee, J.S., Maksimchuk, N.V., Shmakov, A.N., Chesalov, Y.A., Ayupov, A.B., Hwang, Y.K., Jun, C.-H., Chang, J.-S., and Kholdeeva, O.A. (2014) Eur. J. Inorg. Chem., 132–139.

[133] Zalomaeva, O.V., Kholdeeva, O.A., and Sorokin, A.B. (2006) *Green Chem.*, **8**, 883–886.

[134] Sels, B.F., De Vos, D.E., and Jacobs, P.A. (2005) *Angew. Chem. Int. Ed.*, **44**, 310–313.

[135] Bodnar, Z., Mallat, T., and Baiker, A. (1996) *J. Mol. Catal. A Chem.*, **110**, 55–63.

[136] Murahashi, S.-I., Naota, T., Miyaguchi, N., and Noda, S. (1996) *J. Am. Chem. Soc.*, **118**, 2509–2510.

[137] Zalomaeva, O.V., Kovalenko, K.A., Chesalov, Yu.A., Mel'gunov, M.S., Zaikovskii, V.I., Kaichev, V.V., Sorokin, A.B., Kholdeeva, O.A., and Fedin, V.P. (2011) *Dalton Trans.*, **40**, 1441–1444.

[138] Adam, W. and Prein, M. (1996) *Acc. Chem. Res.*, **29**, 275–283.

[139] Korshin, E.E. and Bachi, M.D. (2006) in *The Chemistry of Peroxides*, Vol. 2, Part 1 (ed. Z. Rappoport), John Wiley & Sons, Ltd, Chichester, pp. 189–305.

[140] Jensen, A.W. and Daniels, C. (2003) *J. Org. Chem.*, **68**, 207–210.

[141] Ohkubo, K., Kobayashi, T., and Fukuzumi, S. (2011) *Angew. Chem. Int. Ed.*, **50**, 8652–8655.

[142] Maldotti, A., Amadelli, R., and Molinari, A. (2013) in *Liquid Phase Oxidation via Heterogeneous Catalysis: Organic Synthesis and Industrial Applications* (eds. M.G. Clerici and O.A. Kholdeeva), John Wiley & Sons, Inc., Hoboken, pp. 411–450.

[143] Fyfe, C.A. (1971) in *The Chemistry of the Hydroxyl Group*, Vol. 1 (ed. S. Patai), Wiley-Interscience, New York, pp. 83–127.

[144] (a) Fujimoto, K., Tokuda, Y., Maekawa, H., Matsubara, Y., Mizuno T., and Nishiguchi, I. (1996) *Tetrahedron*, **52**, 3889–3896; (b) George, T., Mabon, R., Sweeney, G., Sweeney, J.B., and Tavassoli, A. (2000) *J. Chem. Soc. Perkin Trans.* **1**, 2529–2574.

[145] Kreh, R.P., Tadros, M.E., Hand, H.M., Cockerham, M.P., and Smith, E.K. (1986) *J. Appl. Electrochem.*, **16**, 440–446.

[146] (a) Otsuka, K., Yamanaka, I., and Hosokawa, K. (1990) *Nature*, **345**, 697–698; (b) Otsuka, K. and Yamanaka, I. (1998) *Catal. Today*, **41**, 311–325.

[147] (a) Cai, R., Song, S., Ji, B., Yang, W., Sun, G., and Xin, Q. (2005) *Catal. Today*, **104**, 200–204; (b) Kornienko, V.L. and Kolyagin, G.A. (2003) *Russ. J. Electrochem.*, **38**, 1308–1316.

[148] Lee, B., Naito, H., and Hibino, T. (2012) *Angew. Chem. Int. Ed.*, **51**, 440–444.

[149] (a) Flitsch, S., Grogan, G., and Ashcroft D. (2002) in *Enzyme Catalysis in Organic Synthesis*, 2nd ed., Vol. 2 (eds. K. Drauz and H. Waldmann), Wiley-VCH, Weinheim, pp. 1065–1108; (b) Leak, D.J., Yin, Y., Zhang, J.-J., and Zhou, N.-Y. (2012) in *Enzyme Catalysis in Organic Synthesis*, 3rd ed., Vol. 3 (eds. K. Drauz, H. Gröger, and O. May), Wiley-VCH, Weinheim, pp. 1487–1534.

[150] (a) Costas, M., Mehn, M.P., Jensen, M.P., and Que, L., Jr. (2004) *Chem. Rev.*, **104**, 939–986; (b) Hollmann, F., Arends, I.W.C.E., Buehler, K., Schallmey A., and Bühler, B. (2011) *Green Chem.*, **13**, 226–265; (c) Ullrich, R. and Hofrichter, M. (2007) *Cell. Mol. Life Sci.* **64**, 271–293.

[151] (a) Mitchell, K.H., Rogge, C.E., Gierahn, T., and Fox, B.G. (2003) *Proc. Natl. Acad. Sci. U. S. A.*, **100**, 3784–3789; (b) Abu-Omar, M.M., Loaiza, A., and Hontzeas N. (2005) *Chem. Rev.*, **105**, 2227–2252.

[152] (a) Sono, M., Roach, M.P., Coulter, E.D., and Dawson, J.H. (1996) *Chem. Rev.*, **96**, 2841–2887; (b) Meunier, B., de Visser, S.P., and Shaik, S. (2004) *Chem. Rev.*, **104**, 3947–3980; (c) Shaik, S., Kumar, D., de Visser, S.P., Altun, A., and Thiel, W. (2005) *Chem. Rev.*, **105**, 2279–2328; (d) Shaik, S. and de Visser, S.P. in Shilov, A.E. and Shul'pin, G.B. (1997) *Chem. Rev.*, **97**, 2879–2932, pp. 45–85.

[153] (a) de Visser, S.P. and Shaik, S. (2003) *J. Am. Chem. Soc.*, **125**, 7413–7424; (b) Bathelt, C.M., Ridder, L., Mulholland, A.J., and Harvey, J.N. (2003) *J. Am. Chem. Soc.*, **125**, 15004–15005.

[154] Safari, N., Bahadoran, F., Mohammed, R.H., and Ghiasi, R. (2000) *J. Porphyrins Phthalocyanines*, **4**, 285–291.

[155] O'Reilly, E., Köhler, V., Flitsch, S.L., and Turner, N.J. (2011) *Chem. Commun.*, **47**, 2490–2501.

[156] Shoji, O., Kunimatsu, T., Kawakami, N., and Watanabe, Y. (2013) *Angew. Chem. Int. Ed.*, **52**, 6606–6610.

[157] Farinas, E.T., Alcalde, M., and Arnold, F. (2004) *Tetrahedron*, **60**, 525–528.

[158] Mirica, L.M., Vance, M., Rudd, D.J., Hedman, B., Hodgson, K.O., Solomon, E.I., and Stack, T.D.P. (2005) *Science*, **308**, 1890–1892.

[159] http://www.ideaconnection.com/inventions/11872-In-vitro-production-of-Ldihydroxyphenylalanine-L-DOPA.html

[160] (a) Korendovych, I.V., Kryatov, S.V., and Rybak-Akimova, E.V. (2007) *Acc. Chem. Res.*, **40** 510–521; (b) Friedle, S., Reisner, E., and Lippard, S.J. (2010) *Chem. Soc. Rev.*, **39**, 2768–2779; (c) Tinberg, C.E. and Lippard, S.J. (2011) *Acc. Chem. Res.*, **44**, 280–288; (d) Bryliakov, K.P. and Talsi, E.P. (2012) *Coord. Chem. Rev.*, **256**, 1418–1434; (e) Shteinman, A. (2008) *Russ. Chem. Rev.*, **77**, 945–966.

15

DEARORMATIZATION REACTIONS: AN OVERVIEW

F. Christopher Pigge

Department of Chemistry, University of Iowa, Iowa City, IA, USA

15.1 INTRODUCTION

Arenes and arene derivatives are widely distributed in nature, providing a convenient and abundant source of synthetic building blocks for the pharmaceutical and materials chemistry industries. There exist numerous well-established synthetic methods suitable for arene functionalization (e.g., electrophilic aromatic substitution) that lead to substituted arenes in which the aromaticity of the ring systems remains intact. In comparison, synthetic methods that result in permanent loss of aromaticity such that arene substrates are converted to nonaromatic (alicyclic) products are underdeveloped. Efficient dearomatization of arene derivatives, however, can afford structurally sophisticated ring systems, often possessing multiple stereogenic centers. Thus, dearomatization reactions may render even relatively simple arenes viable substrates for asymmetric transformations leading to enantio- and/or diastereoenriched synthetic building blocks [1].

In this chapter, an overview of dearomatization tactics employed in organic synthesis is presented. The material is organized according to mechanistic considerations and includes discussion of conventional as well as transition metal-mediated dearomatization reactions, with particular emphasis accorded to asymmetric processes. This chapter highlights the most common dearomatization reactions encountered in synthesis (with the exclusion of the Birch reduction—the topic of another chapter in this volume) and is not meant to provide a comprehensive treatise on the subject. Only dearomatization reactions of carbocyclic arenes are discussed.

Arene Chemistry: Reaction Mechanisms and Methods for Aromatic Compounds,
First Edition. Edited by Jacques Mortier.

15.2 ALKYLATIVE DEAROMATIZATION

15.2.1 C-Alkylation of Phenolate Anions

Phenols can be viewed as stable forms of enol tautomers, and phenolate anions display ambident nucleophilicity at oxygen as well as C2/C6 and C4 (*ortho*/*para* positions). Consequently, phenolate anions are susceptible to C—C bond formation upon reaction with appropriate organic electrophiles (e.g., alkyl halides and sulfonates). When bond formation occurs at a substituted arene carbon, a quaternary center is generated, which may lead to isolation of stable cyclohexadienone products and complete a net alkylative dearomatization (Scheme 15.1) [2].

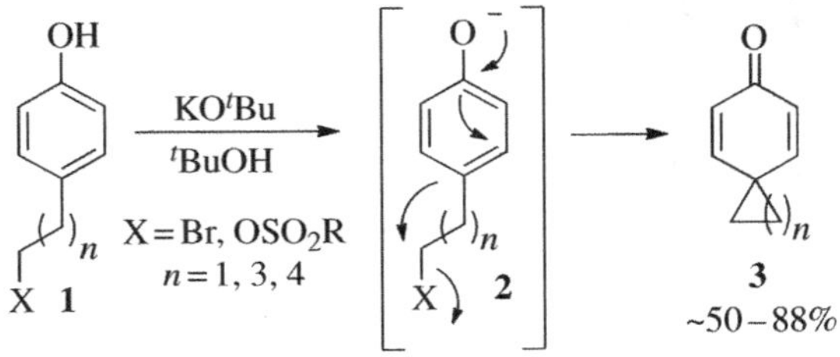

SCHEME 15.1 Intramolecular alkylative dearomatization.

Early examples of this reaction involved cyclizations of 4-substituted phenols tethered to alkyl sulfonates and halides [3]. Cyclizations involving carbonyl electrophiles (aldehydes, ketones) and imines have been reported as well, but esters are not sufficiently electrophilic to react [2]. Subsequent studies established that the facility of these so-called "Ar_1-n" cyclizations was strongly affected by the size of the newly formed ring in the order 3>5>6>>>4. Since the vast majority of alkylative dearomatizations involve intramolecular cyclizations (thereby avoiding competitive O-alkylation reactions), stereoelectronic effects operative in the transition states (resembling the TS of an S_N2 reaction) are crucially important. These sometimes subtle effects can result in differential reactivity of structurally similar substrates [4].

Intramolecular alkylative dearomatization offers a convenient means of constructing spirocyclohexadienone ring systems (Scheme 15.1). This reaction has been utilized in a number of total syntheses targeting construction of terpenoid ring systems found in natural products such as cedrene, hinesol, kaurene, β-vetivone, and isolongifolene [5]. Subsequently, alkylative dearomatizations have been employed as key steps in approaches toward more complex natural products such as *inter alia* duocarmycin SA [6], cortistatin A [7], and platensimycin (Scheme 15.2) [8]. This last example is notable as it illustrates the ability to employ silyl-protected phenols as masked phenolate anions in high-yielding alkylations.

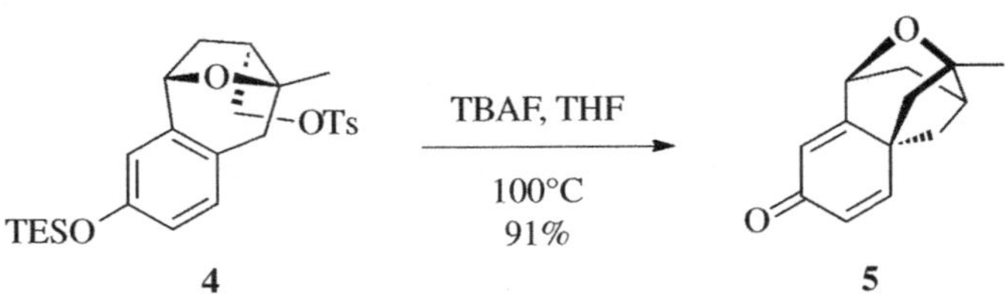

4 5

SCHEME 15.2 Approach to the core structure of platensimycin.

Many alkylative dearomatizations proceed with complete stereoselectivity due to conformational rigidity of the cyclization substrates. Diastereoselective dearomatization of more flexible substrates is possible provided nonbonding interactions encountered in the S_N2-like transition states are sufficient to differentiate prochiral faces of the phenolate anions. For example, benzoate saponification in **6** generates the corresponding phenolate anion that then undergoes alkylative dearomatization to afford **8** as a single diastereomer (Scheme 15.3). The stereoselectivity was attributed to cyclization through a single TS (**7**) that minimizes steric interactions [9].

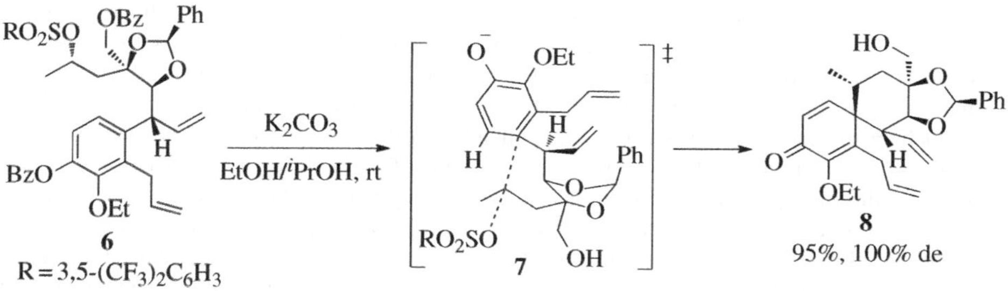

SCHEME 15.3 Diastereoselective alkylative dearomatization.

Phloroglucinols (1,3,5-trihydroxybenzenes) are a particularly reactive class of phenol derivative that are especially prone to dearomatization reactions due to their propensity to undergo regioselective C-alkylation in the presence of reactive electrophiles [10a]. This property extends to C-acylated phloroglucinols and has been exploited in the synthesis of phloroglucinol natural products featuring bicyclo[3.3.1]nonane core frameworks through Michael addition/elimination/Michael addition dearomatization cascades. The process has been rendered asymmetric through the use of chiral bis(*cinchona*)-alkaloid-derived phase-transfer catalysts (Scheme 15.4) [10b]. A binaphthylphosphonic acid phase-transfer catalyst has been found to mediate the intermolecular electrophilic fluorination of substituted phenols in high ee [11].

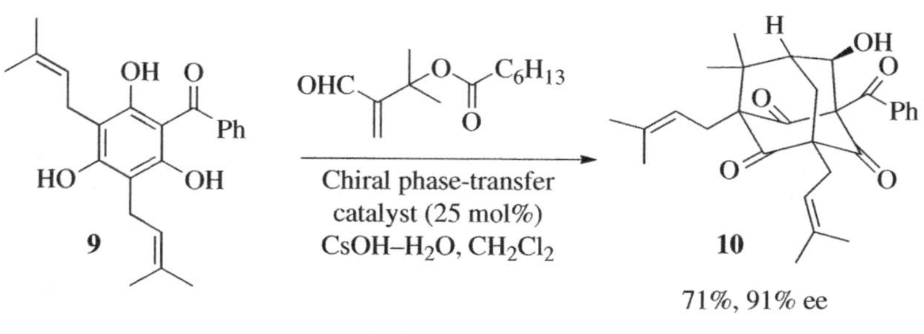

SCHEME 15.4 Asymmetric dearomatization of a phloroglucinol-based substrate.

15.2.2 Anionic Dearomatization

Anionic dearomatization entails the addition of nucleophiles to arenes or polycyclic arenes leading to transient formation of cyclohexadienyl anions that can be trapped with electrophiles (H^+ or E^+). Since carbocyclic aromatic ring systems are generally electron rich, activating electron-withdrawing substituents are usually required to promote these reactions. Examples of activating substituents include carboxylic acids, sterically hindered carboxylic esters and amides, nitriles, nitro groups, sulfones, sulfonamides, phosphinamides, and oxazolines (*vide infra*). Strong nucleophiles are typically used in these reactions, such as organolithium reagents, Grignard reagents, unstabilized carbanions, lithium amides, and hydrides. Tandem nucleophilic addition–electrophile trapping sequences offer a means of directly converting arenes to highly substituted alicyclic products in a stereoselective fashion, and anionic dearomatization has emerged as a valuable synthetic tool [12].

The most widely investigated substrates are substituted naphthalenes as polyaromatic ring systems are more susceptible than benzene derivatives to dearomatization because such reactions disrupt the aromatic character of only one ring while leaving the remaining arene(s) intact. Nucleophilic addition to naphthalenes has been used to access substituted dihydro- and tetrahydronaphthalenes, often with high stereoselectivity. Oxazoline rings are particularly effective activating groups as these moieties are electron withdrawing, easily prepared from carboxylic esters and amino acids, and resist direct reaction with nucleophiles. Chiral nonracemic oxazolines can be obtained from chiral amino acids, thereby offering a chiral auxiliary approach to influence absolute stereochemistry in addition reactions. An application of this methodology in the asymmetric synthesis of podophyllotoxin (**13**) is illustrated in Scheme 15.5a [13]. Regio- and stereoselective aryl lithium addition to β-oxazolinyl naphthalene **11** gave key intermediate **12** after diastereoselective protonation. Even monocyclic oxazolinyl arenes undergo nucleophilic addition under appropriate conditions (^{i}PrLi, DMPU solvent) to afford cyclohexadienes with complete diastereoselectivity (Scheme 15.5b) [14]. In this example, the anionic intermediate **15** formed from nucleophilic addition to **14** is alkylated with MeI to give **16** as a single diastereomer, which was then converted to novel carbasugar derivatives such as **17**.

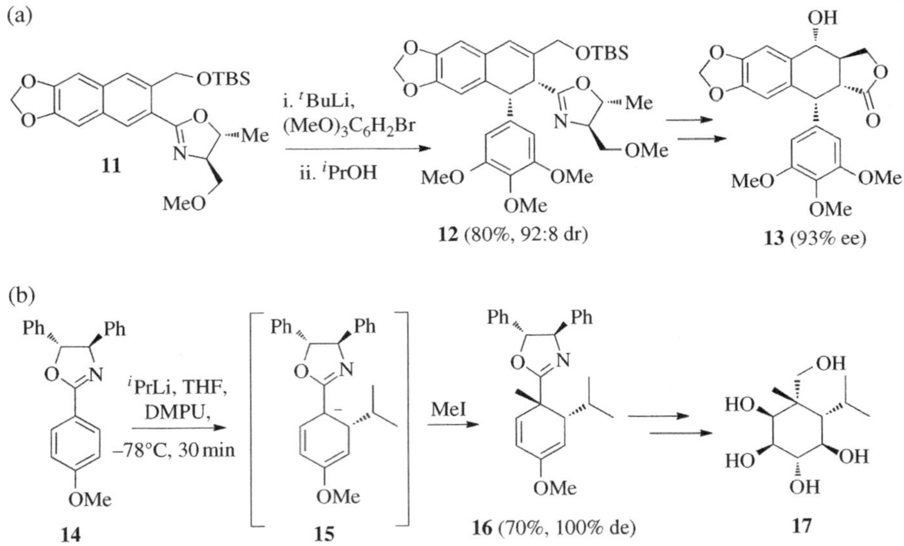

SCHEME 15.5 (a) Synthesis of podophyllotoxin (**13**) via anionic oxidative dearomatization. (b) Anionic dearomatization–alkylation of a simple oxazolinyl arene en route to carbocyclic sugar analogues.

The need for activating arene substituents that are compatible with strong organic nucleophiles presents a limitation of this process and precludes incorporation of many common carbonyl-containing functional groups. If carbonyl derivatives are sufficiently sterically bulky, however, then 1,4- or 1,6-addition to an arene may result. Steric bulk can be introduced transiently using extremely bulky Lewis acids such as ATPH (Scheme 15.6) [15]. The aluminum complex not only shields the carbonyl group to prevent 1,2-addition of the nucleophile but also increases the electrophilicity of the arene nucleus. Besides simple alkyl lithiums, ester enolates have been reported to give the reaction in good yield.

Intramolecular anionic dearomatization reactions have been developed starting from substituted naphthalenes and benzenes. Carbanions generated from deprotonation of *N*-benzyl aryl amides

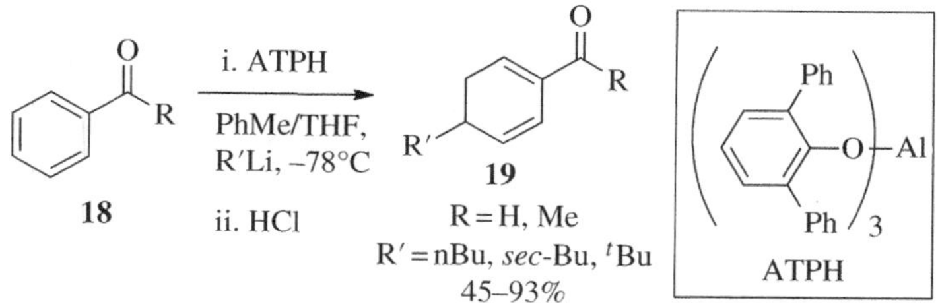

SCHEME 15.6 Anionic dearomatization in the presence of bulky Lewis acids.

participate in intramolecular cyclizations to deliver *cis*-fused isoindolones after quenching with mild acid [16]. Enantioselective dearomatizations have been developed using chiral amide bases in the deprotonation step. Thus, exposure of **20** to chiral amide base **21** followed by acidic hydrolysis gives optically pure (−)-kainic acid precursor **22** in good yield (Scheme 15.7) [17]. Other functional groups capable of activating benzene derivatives toward intramolecular nucleophilic addition include phosphinamides and sulfonamides [18].

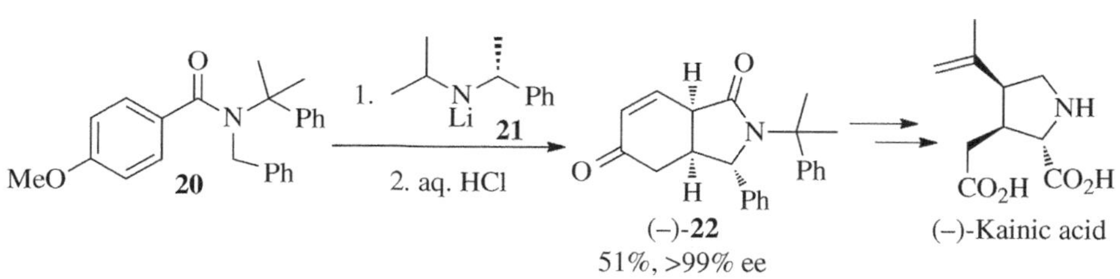

SCHEME 15.7 Enantioselective intramolecular anionic dearomatization.

15.2.3 Radical Dearomatization

Examples of radical additions to arenes leading to dearomatization are less common than other dearomatization processes. While various organic radicals are capable of reacting with arenes to form intermediate cyclohexadienyl radicals, these additions are often reversible, thus allowing radical precursors to undergo transformations involving alternative reaction manifolds. Moreover, cyclohexadienyl radicals exhibit limited reactivity in radical chain reactions and are frequently observed to suffer oxidation to cations followed by rearomatization. Nonetheless, several notable examples of radical dearomatization have been reported, and this type of dearomatization possesses several synthetically attractive features.

Intermolecular additions of aryl radicals to both carbocyclic and heterocyclic arenes have been reported [19]. These additions proceed efficiently only in the presence of phenylselenol (generated *in situ*) as this species reacts with cyclohexadienyl radicals significantly faster than other potential hydrogen atom donors, thereby allowing preferential formation of cyclohexa-1,4-dienes. This process has been applied in a formal synthesis of (±)-pancratistatin, with radical dearomatization of benzene providing key intermediate **24** (Scheme 15.8) [20].

Intramolecular radical dearomatization has been examined as a route to spirocyclic ring systems. Spiro[4.5]decane derivatives have been obtained from a tandem radical addition process initiated by $Mn(OAc)_3$ (Scheme 15.9) [21]. Yields of spirocyclic products were highest when substituents on the intermediate cyclohexadienyl radical (e.g., OMe in **26**) could promote subsequent oxidation to stable cyclohexadiene products via cation **27**. Related ring systems have been prepared under

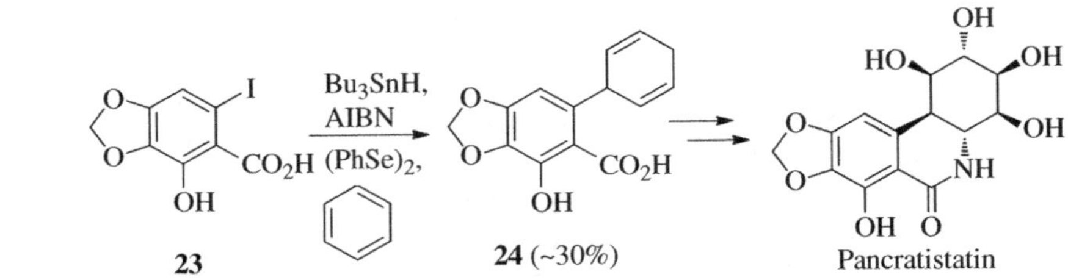

SCHEME 15.8 Intermolecular radical addition/dearomatization of benzene.

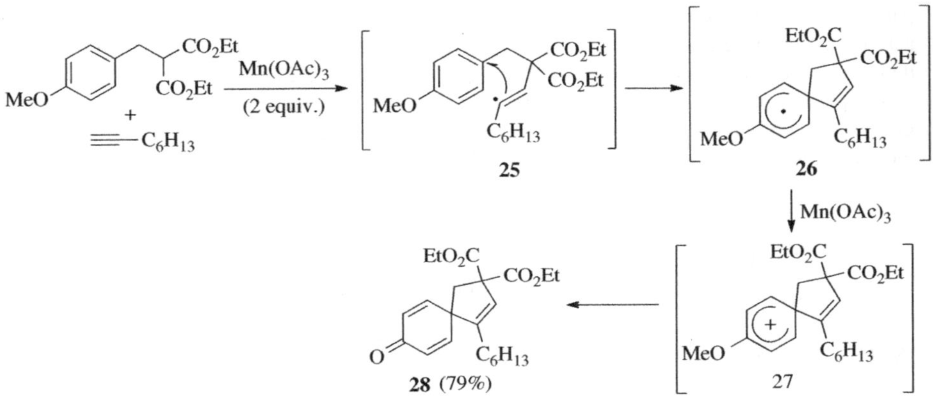

SCHEME 15.9 Spirocyclic derivatives from tandem radical addition/dearomatization.

reducing conditions via SmI_2-promoted addition of ketyl radicals to arenes possessing electron-withdrawing *para*-carbomethoxy groups [22].

Azaspirocyclic derivatives have been prepared via intramolecular atom transfer radical dearomatization of *N*-benzyltrichloroacetamides performed in the presence of nickel powder and acetic acid [23]. Similar transformations mediated by catalytic amounts of CuCl also have been disclosed [24]. Spirolactams have been prepared from *N*-benzylacetamide radicals generated via peroxide-initiated decomposition of xanthate esters. Similar to the example shown in Scheme 15.9, electron-donating substituents (OMe) present at the 2- and/or 4-position of the arene partner facilitate subsequent peroxide-mediated oxidation of cyclohexadienyl radicals to the corresponding cations, which then convert to the observed dienone products (Scheme 15.10) [25]. Intramolecular *ipso* addition of aryl radicals generated from *N*-halophenyl benzamides affords spirocyclohexadienone derivatives under oxidative conditions through an analogous mechanism [26].

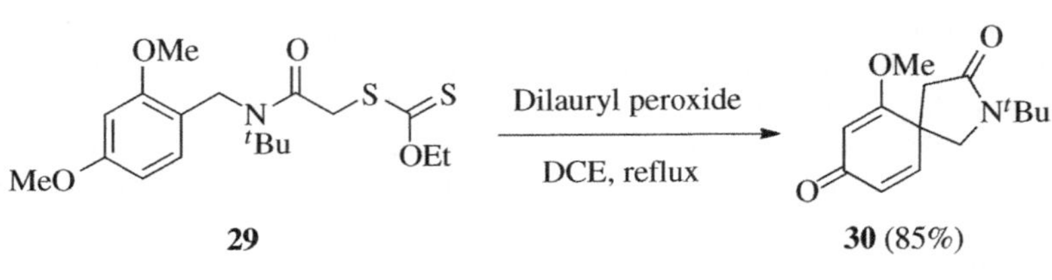

SCHEME 15.10 Azaspirocyclic products from *N*-benzylacetamide radicals.

15.3 PHOTOCHEMICAL AND THERMAL DEAROMATIZATION

15.3.1 Dearomatization by Photocycloaddition

Substituted benzenes display a fascinating array of photochemical reaction manifolds, several of which possess demonstrated synthetic utility via dearomatization [27]. Arenes and alkenes participate in photocycloaddition reactions that rapidly deliver intriguing and sophisticated polycyclic molecular frameworks in an atom economical fashion. Commonly observed modes of photocycloaddition between benzene and an alkene are illustrated in Scheme 15.11. Mechanistically, these cycloadditions are believed to proceed through formation of an exciplex between the photoexcited arene and the alkene. Selectivity among the [2+2], [3+2], and [4+2] modes of cycloaddition is governed by the excited state of the arene, the electronic features of the reacting components, and the degree of charge transfer in the exciplex. The [3+2] photocycloaddition (*meta*-photocycloaddition) is most common, followed by the [2+2] mode. *Para*-photocycloadditions ([4+2] mode) are rarely observed. Even if a single mode of photocycloaddition is operative for a particular substituted arene–alkene pair, multiple regio- and stereoisomeric (i.e., *endo*/*exo*) photoproducts are often observed. Hence, from a synthetic standpoint, the most useful photo-dearomatizations involve intramolecular cycloadditions in which the alkene and arene are tethered by an appropriate linker so that the mode, regiochemistry, and stereochemistry of photocycloaddition can be better controlled [28].

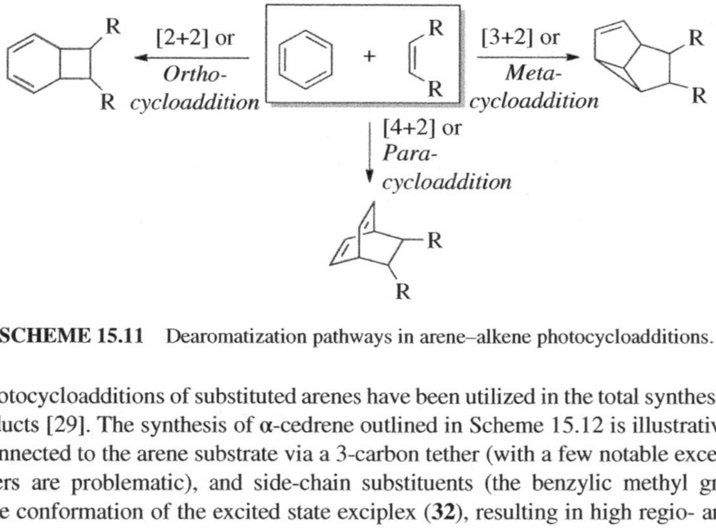

SCHEME 15.11 Dearomatization pathways in arene–alkene photocycloadditions.

meta-Photocycloadditions of substituted arenes have been utilized in the total synthesis of several natural products [29]. The synthesis of α-cedrene outlined in Scheme 15.12 is illustrative [30]. The alkene is connected to the arene substrate via a 3-carbon tether (with a few notable exceptions [31], longer tethers are problematic), and side-chain substituents (the benzylic methyl group in **31**) influence the conformation of the excited state exciplex (**32**), resulting in high regio- and stereoselectivity in formation of the initial cycloadduct (**33**). Conformational constraints of the tether result in preferential formation of *exo* cycloadducts. Intermediate **33** is most often envisioned as a biradical species possessing an electrophilic radical at C1 and a delocalized nucleophilic radical at C3/C5. The fate of biradical intermediates such as **33** is frequently a complicating feature of arene–alkene [3+2] photocycloadditions as radical recombination usually produces two regioisomeric cyclopropanes (**34** and **35**). In Wender's synthesis of α-cedrene, both **34** and **35** could be converted to the same intermediate en route to the target; however, this is not always possible, and so efficient elaboration of cyclopropane products is an important issue that impacts the overall synthetic utility of the process. In some instances, thermal equilibration of the vinylcyclopropane isomers is possible [32].

Related dearomatizations by [2+2] and [4+2] photocycloadditions are less common. It has been observed that cycloadditions featuring high degrees of charge separation in the reactive exciplex favor the [2+2] pathway [27a]. Consequently, examples of this reaction often involve arenes and/or alkenes substituted with electron-withdrawing groups. Benzocyclobutane products formed in [2+2]

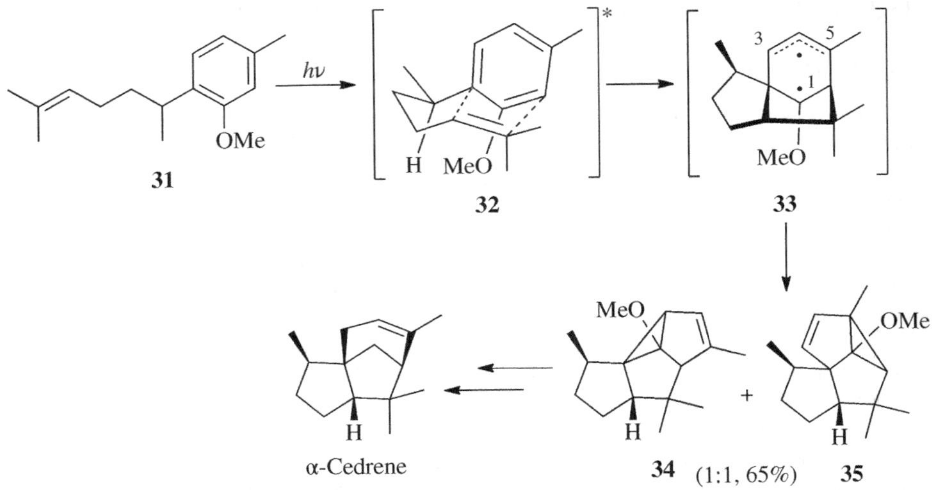

SCHEME 15.12 Synthesis of α-cedrene by intramolecular arene [3+2] photocycloaddition.

cycloadditions are typically quite reactive and can possess limited thermal and photochemical stability. Asymmetry in *ortho*-photocycloadditions has been achieved through application of chiral auxiliary technology [33].

Synthetic applications of [4+2] photocycloadditions have not been extensively developed. It has been observed, however, that intramolecular photocycloaddition of arenes and allenes proceeds preferentially via the [4+2] reaction mode. This process appears to be reasonably general across a diverse range of allenes attached to aromatic aldehydes and ketones with a variety of tethers. Thus, the protected aniline **36** undergoes *para*-cycloaddition to give the intriguing bridged polycyclic product **37** (Scheme 15.13) [34]. Mechanistic details of this and related transformations have not yet been reported.

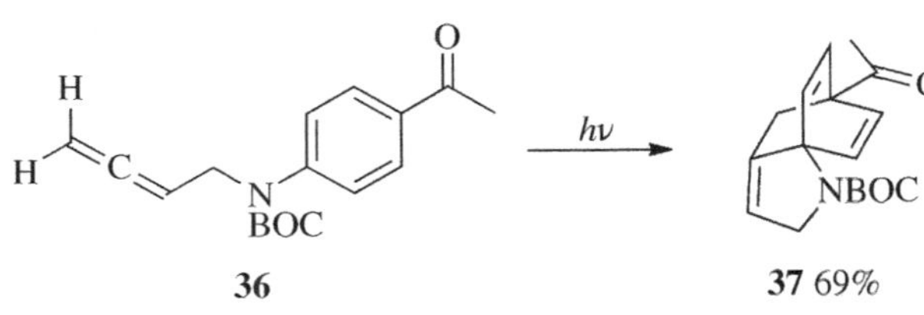

SCHEME 15.13 Dearomatization via arene–allene [4+2] photocycloaddition.

15.3.2 Dearomatization by Thermally Induced Rearrangement

Aromatic π systems are well-known participants in various pericyclic reactions, and aromaticity of substrates is often transiently disrupted in these transformations (e.g., Claisen rearrangement of *O*-allyl phenol). In most cases, a pathway for rearomatization exists (such as tautomerization) that ultimately leads to substituted aromatic products. In other instances, dearomatization upon pericyclic rearrangement results in formation of stable alicyclic products, and this tactic has been successfully employed in several synthetic studies.

Certain functionalized allyl phenyl ethers undergo Claisen rearrangement to afford stable cyclohexadienones. Steric congestion in the substrates and products is believed to facilitate these transformations [35]. For example, dimethylallyl ether **38** undergoes regioselective thermal rearrangement at relatively low temperature and in excellent yield to afford cyclohexadienone **39**, a precursor to the neurotrophic sesquiterpene tricycloillicinone (Scheme 15.14) [36]. Bulky Lewis acids have been shown to promote similar sigmatropic rearrangements as part of an improved synthetic approach to the same target [36b].

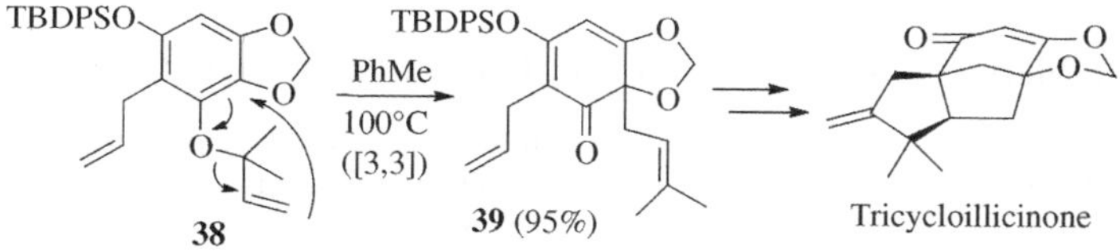

SCHEME 15.14 Dearomatization by thermal sigmatropic rearrangement.

Steric crowding along with remote electronic effects appear to be important factors contributing to regioselective Claisen rearrangements observed in several total syntheses of *Garcinia*-derived bioactive natural products. Initial dearomatized 2,4-cyclohexadienones obtained upon [3, 3]-rearrangement are immediately trapped via intramolecular Diels–Alder cycloaddition to afford highly sterically congested oxatricyclic decane ring systems (**41**) that are characteristic of these natural products (Scheme 15.15) [37].

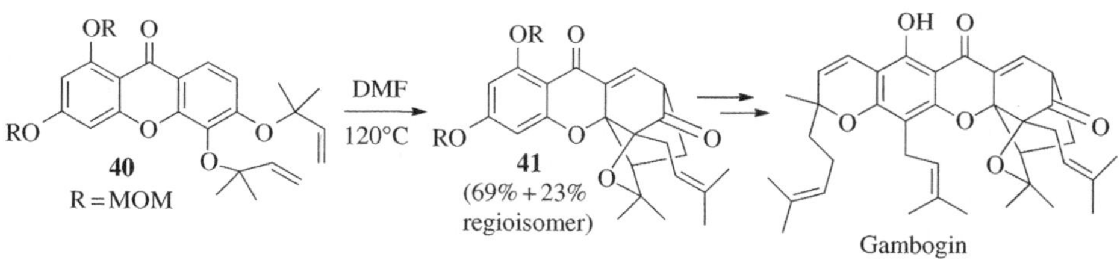

SCHEME 15.15 Dearomatization via tandem Claisen rearrangement/[4+2] cycloaddition.

Polycyclic aromatic compounds, such as naphthalene and anthracene, are well known to participate in Diels–Alder and related transformations as these reactions typically result in loss of aromaticity in only one of several arene rings [38]. Such cycloadditions are rarely observed in simpler monocyclic arenes. Several examples of intramolecular [3+2] cycloaddition between an arene and an attached nitrile oxide moiety have been reported [39]. Additionally, 4+2 cycloadditions (also intramolecular) between phenyl groups and attached allenes have been examined [40]. Allene substrates can be conveniently generated by base-induced isomerization of alkynyl amides under conditions also suitable for cycloaddition, thereby affording a one-pot method to convert substrates such as **42** to tricyclic products **43** (Scheme 15.16) [41]. Elaboration of **43** and structurally related

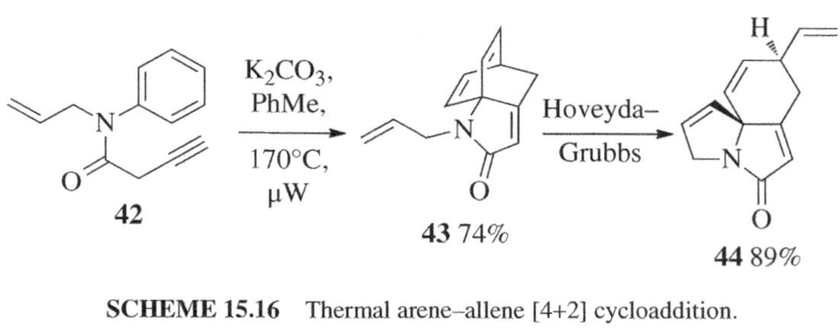

SCHEME 15.16 Thermal arene–allene [4+2] cycloaddition.

analogues using metathesis chemistry provides rapid access to complex heterocyclic ring systems (e.g., **44**) that should find utility in alkaloid synthesis. Notably, **43** is similar to products obtained from allene–arene photocycloadditions described earlier (**37**, Scheme 15.13).

15.4 OXIDATIVE DEAROMATIZATION

15.4.1 Oxidative Dearomatization with Formation of Carbon–Heteroatom Bonds

Of all the dearomatization processes discussed in this chapter, oxidative dearomatization reactions are among the most widely used and synthetically versatile. Provided a carbocyclic arene substrate is amenable to oxidation (i.e., electron rich), a number of mild and selective oxidizing agents can be employed to generate putative arene radical cations that are then susceptible to nucleophilic addition, ultimately leading to dearomatization. Examples of oxidants used in this regard include metal-based reagents (Fe(III), Cu(II), and Pb(IV) complexes) along with hypervalent iodine reagents [42]. The majority of oxidative dearomatizations feature arene substrates possessing at least one electron-donating substituent such as a hydroxyl, ether, or amine group. This substituent reduces the oxidation potential of the arene and provides a functional group handle for directing subsequent nucleophilic addition.

In many cases, oxidative dearomatization of arenes is accompanied by addition of heteroatomic nucleophiles, especially alcohols, carboxylic acids, and amides (the latter generally reacting through the carbonyl oxygen). Addition of alcohols to oxidized phenols provides convenient access to 1,3- and 1,4-cyclohexadienones as illustrated in Scheme 15.17 [43]. The exact structure of intermediates **45** varies as a function of oxidant, but in all cases, reactivity is consistent with a cyclohexadienyl cation generated from formal two-electron oxidation of the starting phenol. The site of nucleophilic addition can be influenced by the oxidant, but addition usually occurs at the dienyl carbon bearing the largest LUMO coefficient (a substituted *ortho* or *para* carbon when *R* is an electron-releasing alkyl or alkoxy group). The protected *o*- and *p*-quinones or quinols (**46**, **47**) produced by this process can be quite reactive, especially **46**, which show a tendency to spontaneously dimerize via 4+2 cycloaddition.

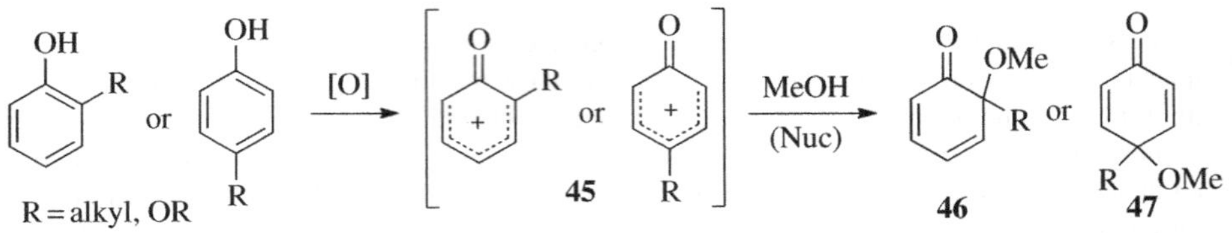

SCHEME 15.17 Overview of oxidative dearomatization pathways.

Cycloadditions involving compounds of general structure **46** have been exploited in several synthetic approaches to natural product targets [43]. A dienophile fragment is often tethered to the arene substrate prior to dearomatization so that intramolecular 4+2 cycloaddition can ensue at the expense of competing cyclohexadiene dimerization [44]. For example, treatment of phenol **48** with phenyl iodine diacetate (PIDA) in the presence of MeOH (CF_3CH_2OH is employed as a nonnucleophilic solvent) affords cyclohexadienone intermediate **49** that undergoes intramolecular [4+2] cycloaddition upon moderate heating (Scheme 15.18) [44a]. Cage-like structure **50** was subsequently converted to the bioactive diterpenoid vinigrol. The regioselectivity of MeOH addition to **48** is controlled by deactivation of one *ortho*-alkyl ether by a trifluoroethyl group.

Intramolecular [4+2] cycloaddition can also be initiated via incorporation of a dienophile into the nucleophilic participant in an oxidative dearomatization sequence (e.g., an allylic alcohol) [45]. Additionally, dienophilic alcohol moieties can be tethered to phenol substrates

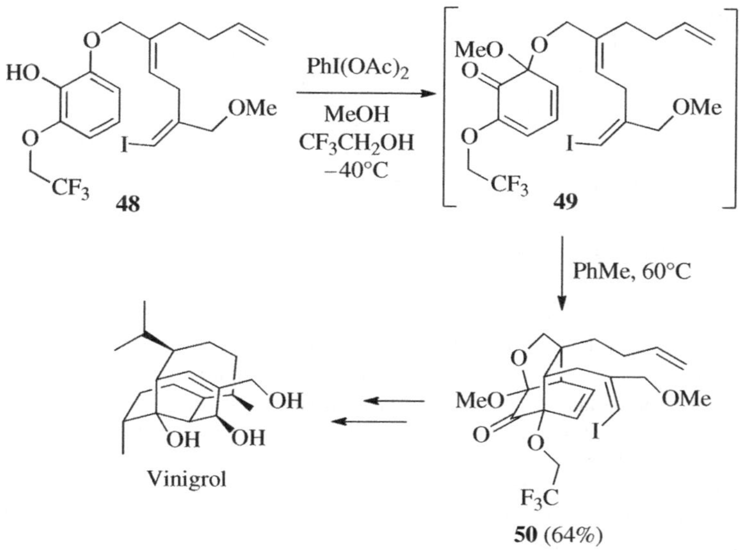

SCHEME 15.18 Tandem oxidative dearomatization–intramolecular 4+2 cycloaddition.

so that both nucleophilic addition to the oxidized arene and [4+2] cycloaddition are rendered intramolecular [46].

Synthetically useful applications stemming from oxidative dearomatization of 4-substituted phenols also have been reported. Often products obtained from these transformations are analogous to **47** (Scheme 15.17) and feature an achiral 1,4-cyclohexadienone fragment. Desymmetrization of this moiety offers convenient entry to chiral nonracemic materials [47a]. For example, Gaunt and coworkers developed a tandem oxidative dearomatization—organocatalyzed intramolecular Michael addition sequence for conversion of *para*-substituted phenols to enantioenriched bicyclic building blocks (Scheme 15.19) [47b]. Similar desymmetrization reactions of 4,4-disubstituted cyclohexadienones via organocatalyzed Stetter reaction and Michael additions catalyzed by chiral hydrogen bonding urea catalysts have also been reported [48, 49].

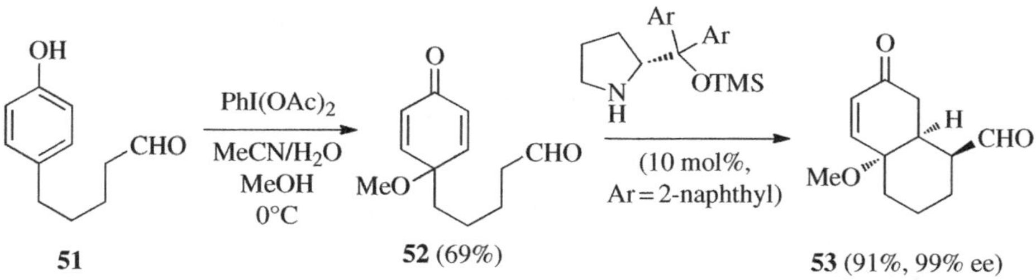

SCHEME 15.19 Desymmetrization of a *p*-quinol obtained from oxidative dearomatization.

Substrate-controlled routes to optically enriched materials via diastereoselective oxidative dearomatizations constitute a second strategy for harnessing this process in asymmetric synthesis. Best results are obtained in intramolecular dearomatizations in which a preexisting stereogenic center is present on the side-chain nucleophile of a prochiral arene substrate [50]. For example, intramolecular oxidative dearomatization of **54** proceeds diastereoselectively as a consequence of conformational effects operative in the course of acetal formation (Scheme 15.20) [51].

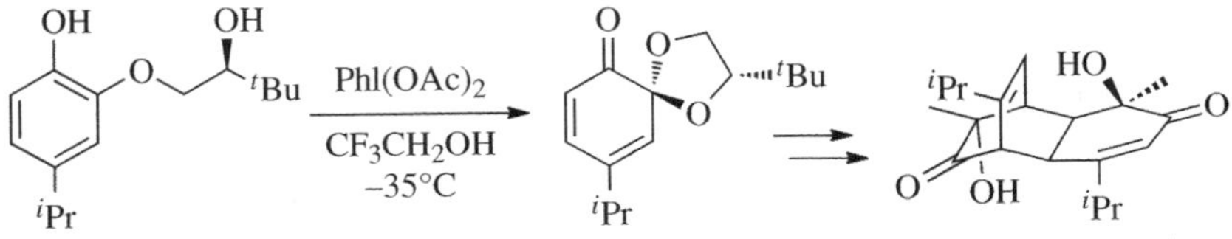

54 (Optically pure) **55** (>95:5 dr) (+)-Biscarvacrol 86% ee

SCHEME 15.20 Substrate-controlled diastereoselective oxidative dearomatization.

Methods to directly convert achiral phenols into chiral nonracemic *o*- and *p*-quinols via asymmetric oxidation have proven difficult to achieve. Some success has been reported using stoichiometric reagents (e.g., a sparteine–Cu(II)-oxo complex and chiral iodinanes) [52, 53], and approaches predicated on the use of hypervalent iodine *catalysts* appear to be quite promising [54]. Iodinane-catalyzed oxidative dearomatizations use reagents generated *in situ* from catalytic amounts of an iodoarene and a stoichiometric oxidant such as *meta*-chloroperbenzoic acid (MCPBA) [55]. Two chiral iodoarene precatalysts capable of mediating enantioselective oxidative lactonization of *ortho*-substituted naphthols and phenols have been identified (**56** and **57**, respectively, Scheme 15.21) [56]. Treatment of **58** with 10 mol% **57** in the presence of 1.2 equiv. MCPBA affords **59** in excellent yield and high enantiomeric excess [56b]. The success of these asymmetric dearomatization reactions suggests a mechanism involving ligation of the iodinane reagent by the phenolic residue of the substrate via ligand exchange. The positioning of the chiral iodoarene framework within proximity of the substrate facilitates enantioselective oxidation/lactonization. Generation of reactive phenoxonium cations by outer-sphere electron transfer is also possible, but such a mechanistic manifold is expected to result in little to no enantioselectivity. Attempts to extend this approach to encompass generation of chiral nonracemic *p*-quinols have resulted in only modest levels of enantioselection [57].

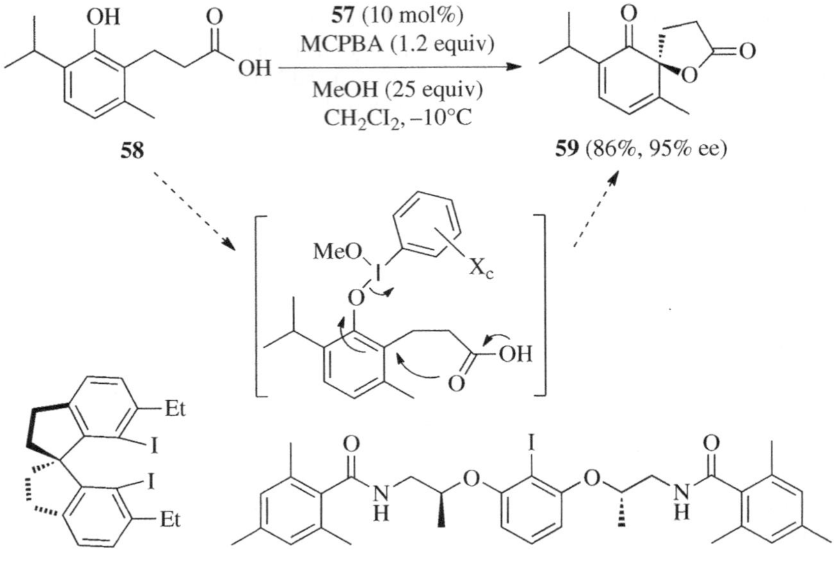

58 **59** (86%, 95% ee)

56 **57**

SCHEME 15.21 Iodinane-catalyzed asymmetric oxidative dearomatization.

Oxidative dearomatization leading to formation of C—N bonds is an important synthetic transformation, although the range of acceptable N-nucleophiles is somewhat limited. Free amines are problematic nucleophiles in oxidative dearomatizations, although exceptions have been noted [58]. Primary and secondary amides participate in oxidative dearomatization but tend to react through the carbonyl oxygen rather than the nitrogen atom. Secondary *N*-methoxy amides present an exception, and azaspirocyclic frameworks have been assembled from hypervalent iodine-mediated oxidative spirolactamization. Oxidation of the *N*-methoxy amide group to an acylnitrenium moiety that is then trapped by the attached electron-rich arene is postulated to be the operative mechanism [59]. Phenolic sulfonamides and phenolic oxazolines are viable N-nucleophiles, and intramolecular reaction of the latter species has been used to construct several azaspirane natural products (Scheme 15.22) [60]. Intermolecular addition of N-nucleophiles during the course of oxidative dearomatization has been achieved through development of Ritter-like reactions between acetonitrile and substituted phenols in the presence of I^{III} oxidants (e.g., PIDA) and trifluoroacetic acid [61].

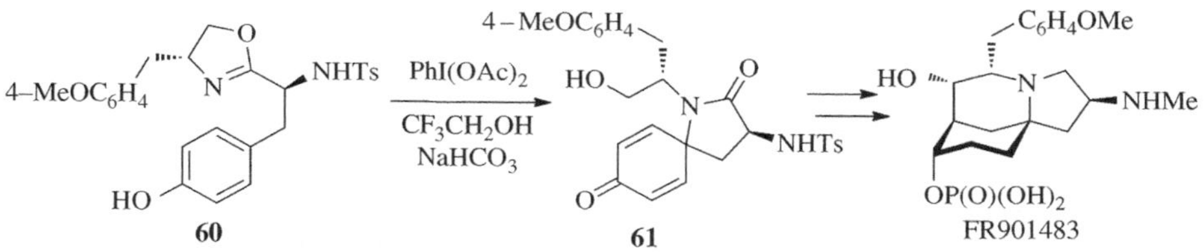

SCHEME 15.22 Oxidative dearomatization with C–N bond formation.

15.4.2 Oxidative Dearomatization with Formation of Carbon–Carbon Bonds

Arene oxidation leading to direct C—C bond formation allows rapid assembly of complex and stereochemically rich carbocyclic ring systems. Crucial to the success of this approach is the identification of carbon nucleophiles that are stable in the presence of oxidation agents typically used to effect arene dearomatization. Enolates and enol ethers are problematic as these species undergo rapid oxidation under mild conditions [62]. Stabilized enolates (such as those derived from activated methylenes) exhibit greater compatibility with oxidation conditions and have been used as nucleophilic participants in intramolecular oxidative dearomatizations initiated by $[Fe(CN)_6]^-$ and PIDA to afford spirocyclic cyclohexadienones [63, 64]. Detailed mechanisms for these reactions have not been defined so it is unclear whether bond formation occurs through ionic or radical intermediates.

Allyl silanes are a species of carbon nucleophile that has performed well in combination with oxidative dearomatization induced by hypervalent iodine reagents [65]. Inter- and intramolecular additions of allyl silanes to putative phenoxonium cations have been reported, and the latter reaction type provided a key transformation in the synthesis of the potent antibiotic (–)-platensimycin (Scheme 15.23) [66].

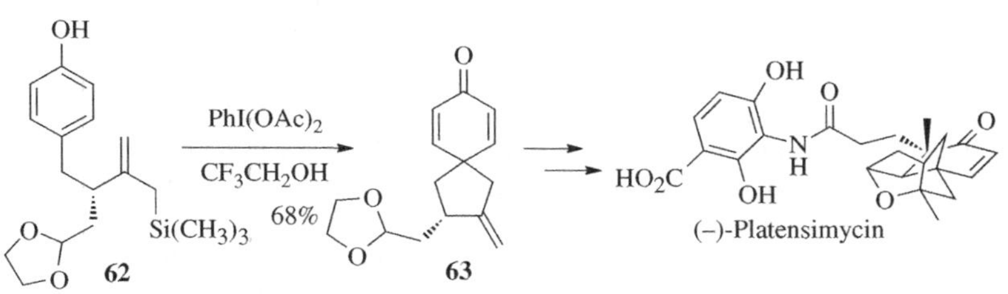

SCHEME 15.23 Oxidative dearomatization route to (–)-platensimycin.

Arenes, alkenes, and alkynes also participate as carbon nucleophiles in oxidative dearomatization sequences. These reactions can be thought of as variations on conventional electrophilic arene–alkene addition reactions, with intermediates similar to **45** (Scheme 15.17) serving as electrophiles [67]. When alkenes function as nucleophiles, carbocation-like intermediates are formed, the fate of which depends on the exact reaction and reaction conditions [68]. Interesting examples of alkenes participating in diastereoselective formal [5+2] cycloaddition cascades initiated by arene oxidation have been reported in the synthesis of several terpenoid natural products [69a]. Dearomatizations resulting from oxidative rearrangements (e.g., alkyl migrations and Prins–pinacol-type rearrangements) have been reported as well (Scheme 15.24) [69b, c].

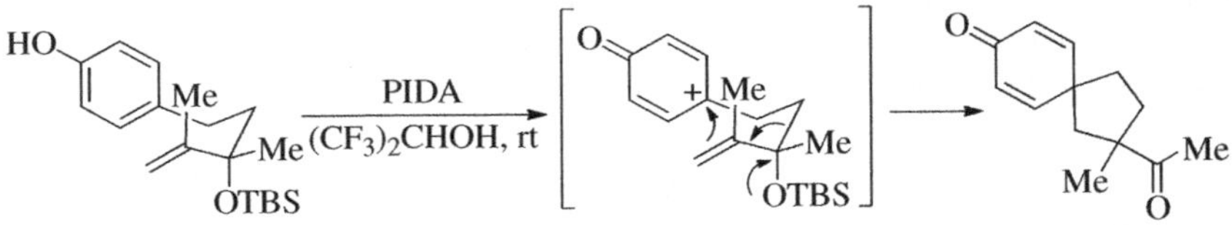

64 **65** **66** (68%)

SCHEME 15.24 Oxidative dearomatization via Prins–pinacol rearrangement.

Only a few examples of substrate-controlled diastereoselective intramolecular oxidative dearomatization have appeared. These transformations are analogous to the reaction shown in Scheme 15.20 in that stereogenic centers present on the linker connecting the carbon nucleophile to the arene substrate are responsible for diastereoselectivity [69, 70]. A rare example of a true metal-*catalyzed* oxidative dearomatization of substituted 2-naphthols is shown in Scheme 15.25. Treatment of naphthol **67** with Fe(salen) catalyst **68** in the presence of nitromethane (as a carbon nucleophile) and air (O_2 as stoichiometric oxidant) gave naphthone **69** possessing a quaternary carbon center in 90% ee [71].

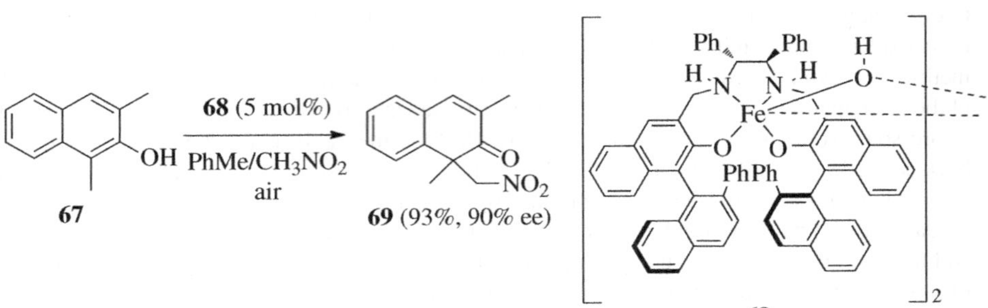

67 **69** (93%, 90% ee) **68**

SCHEME 15.25 Fe-catalyzed oxidative dearomatization of β-naphthol.

A final type of oxidative carbon–carbon bond forming dearomatization process involves electrophile-induced dearomatization. The most common variant of this reaction entails activation of an alkyne or alkene moiety with an electrophilic halide source to initiate intramolecular dearomatization accompanied by formal arene oxidation (Scheme 15.26) [72]. Proper positioning of an electron-donating methoxy group is crucial for success of this transformation. Other examples of halocyclization–dearomatization reactions involving appropriately substituted arenes tethered to alkynes and alkenes have been reported, along with an intramolecular Pummerer-type dearomatization initiated by an electrophilic thionium ion [73, 74].

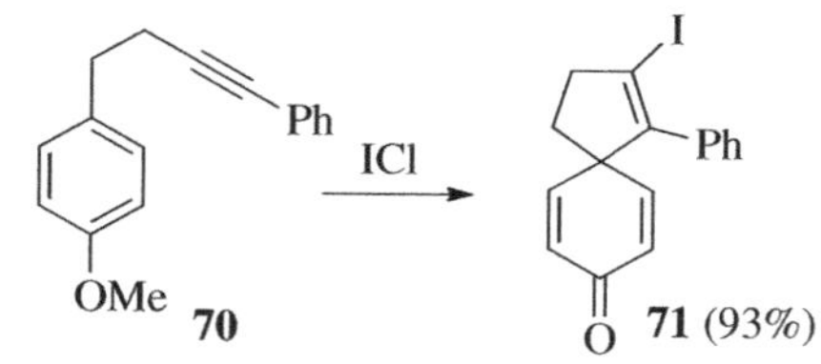

SCHEME 15.26 Dearomatization by halocyclization.

15.5 TRANSITION METAL-ASSISTED DEAROMATIZATION

Organotransition metal reagents have been used in both catalytic and stoichiometric dearomatization reaction sequences and provide mechanistically distinct pathways to alicyclic products. Additionally, metal-catalyzed asymmetric dearomatization is an attractive strategy for enantioselective synthesis [75].

15.5.1 Dearomatization Reactions of Metal Carbenoids

Arenes suffer dearomatization via cyclopropanation upon reaction with α-diazocarbonyl compounds (Büchner reaction) [76]. Initially formed norcaradiene products are usually present in equilibrium with cycloheptatrienes formed via electrocyclic cyclopropane ring opening. The reaction is dramatically promoted by transition metal catalysts (usually Cu(I) or Rh(II) complexes) that give metal-stabilized carbenoids upon reaction with diazo compounds. Inter- and intramolecular manifolds are known, and asymmetric variants employing substrate control and chiral transition metal catalysts have been developed [77]. Effective chiral catalysts for intramolecular Büchner reactions include $Rh_2(mandelate)_4$, rhodium carboxamidates, and Cu(I)-bis(oxazolines). While enantioselectivities as high as 95% have been reported, more modest levels of asymmetric induction are typically observed.

Substrate-controlled stereoselective dearomatizations provide cycloheptatriene derivatives in high diastereomeric excess, and the reaction has been used to prepare 7-membered ring systems found in several natural products. Scheme 15.27a illustrates the Rh(II)-catalyzed conversion of diazo derivative **72** to polycyclic cycloheptatriene **73**, which was subsequently converted to harringtonolide [78]. Note that the initial cycloheptatriene product of the Büchner reaction is converted to a more stable isomer by the action of DBU. In some instances, intramolecular Büchner reactions afford norcaradiene products that are not in equilibrium with the corresponding cycloheptatrienes. These examples arise as a consequence of conformational constraints inherent in the substrates. Cu-catalyzed Büchner reactions have been employed to access derivatives of stable norcaradiene fragments found in several natural products (e.g., gibberellin GA_{103} and (+)-salvileucalin B, Scheme 15.27b and c, respectively) [79].

15.5.2 Dearomatization Catalyzed by Palladium, Iridium, and Related Complexes

Several metal-catalyzed dearomatization sequences have been developed that rely upon the electrophilicity of π-allyl metal complexes to initiate carbon–carbon bond formation with arene nucleophiles. When alkylation of π-allyl metal intermediates takes place at a substituted arene carbon, then a quaternary center is generated, thereby preventing subsequent rearomatization. Remote electron-donating substituents (such as phenolic residues) help to increase and direct the reactivity of arene substrates in these transformations.

Allyl complexes of palladium and iridium are conveniently generated by oxidative addition between Pd(0) and Ir(I) catalysts and allylic acetates and carbonates. Chiral allyl complexes

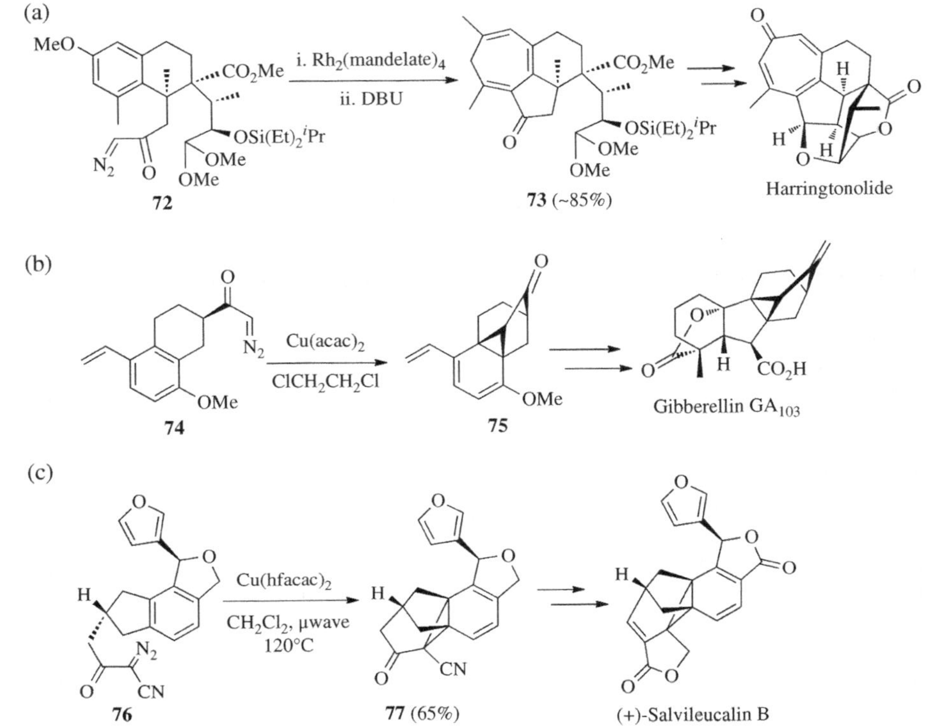

SCHEME 15.27 Examples of natural product synthesis featuring metal carbenoid dearomatization. (a) Conversion of arene to cycloheptatriene en route to harringtonolide [78]. (b) and (c) Copper carbenoid cyclopropanation of arenes to give stable norcaradiene products [79a,b].

can be accessed through placement of chiral ancillary phosphine ligands on the metal center. Asymmetric Pd- and Ir-catalyzed intramolecular dearomatization reactions involving phenol derivatives tethered to allylic carbonates have been developed (Scheme 15.28) [80]. The Ir-mediated process appears to exhibit a wider scope and a variety of carbo- and heterocyclic

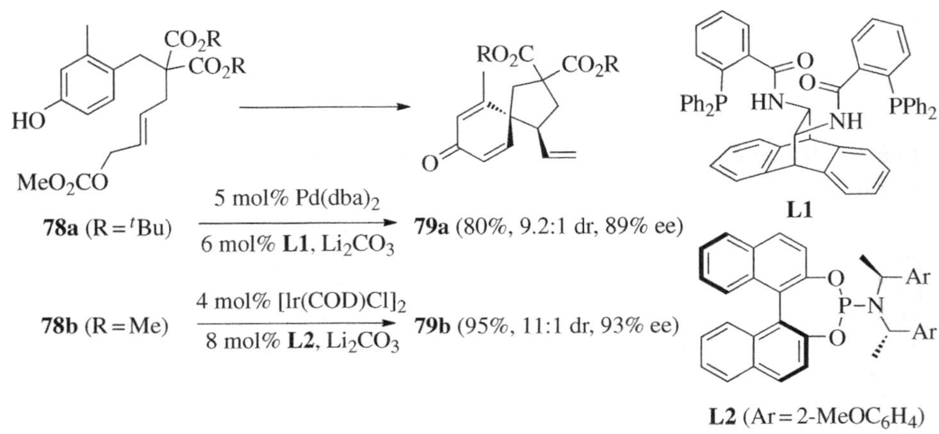

SCHEME 15.28 Pd- and Ir-catalyzed asymmetric dearomatization via (allyl)metal intermediates.

spirocyclohexadienones have been prepared using this procedure [80b]. Pd–allenyl electrophiles also give the reaction to afford more highly substituted spirocyclohexadienones [81]. Asymmetric Pd-catalyzed *intermolecular* dearomatizing allylations of substituted 2-naphthols have been reported with ee's as high as 97% in the presence of chelating phosphines analogous to **L1** [82]. While π-allyl-catalyzed dearomatization has yet to be extensively employed in synthesis, this reaction did constitute a key transformation in the total synthesis of tatanans B and C—two atropisomeric natural products reported to possess potent glucokinase activating effects (Scheme 15.29) [83].

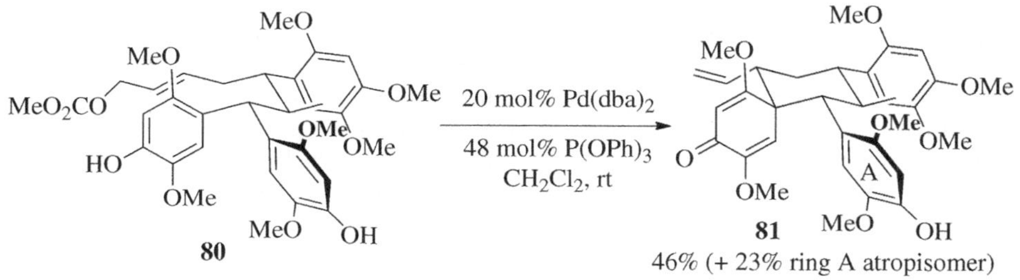

SCHEME 15.29 Synthesis of tatanans via Pd-catalyzed dearomatization.

Dearomatization of η^3-benzylpalladium complexes represents an electronically reversed variation on the reactions described earlier. Initial examples utilized allyl and allenyl stannanes as nucleophilic components in combination with benzyl halides and invoked mechanisms involving aryl–alkyl Pd(II) intermediates [84]. Subsequently, direct addition of stabilized nucleophiles (e.g., malonate anions) to η^3-naphthylpalladium complexes has been achieved [85].

Several other mechanistically distinct metal-catalyzed dearomatization procedures have been reported, and almost all involve phenol or naphthol derivatives undergoing dearomatization via intramolecular transformations. Intramolecular Pd- and Rh-catalyzed C4-arylation and alkylation of *para*-substituted phenols has been used to construct compounds of general structure **82** (Fig. 15.1) [86]. These reactions rely on generation of electrophilic aryl or alkyl σ-metal complex intermediates that participate in tandem C4 metalation–reductive elimination with an attached phenol. Ruthenium- and Pt-catalyzed reactions of naphthalenes and alkynes deliver spirocyclic products such as **83** [87, 88]. An asymmetric intramolecular naphthalene dearomatization catalyzed by Pd(0)–phosphine complexes has been used to prepare carbazole derivatives **84** in good enantiomeric excess from 1-(*N*-2-bromophenyl)aminonaphthalene precursors [89].

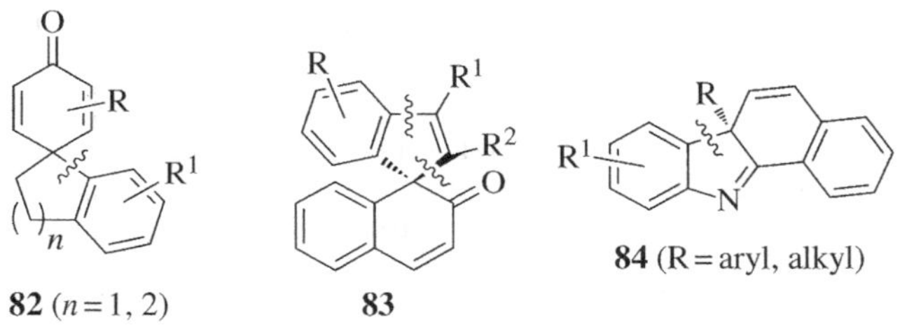

FIGURE 15.1 Survey of ring systems constructed via metal-catalyzed dearomatization (wavy lines indicate bonds formed during dearomatization).

15.5.3 Dearomatization of η^2-Arene Metal Complexes

Transient and reversible η^2-coordination of transition metal fragments to arene ligands is widely implicated in various catalytic processes. While this mode of coordination does not typically result in formation of isolable materials, Harman and coworkers have developed a series of tractable (η^2-arene)metal complexes. Regioselective binding of these metal fragments to arenes (e.g., benzene, naphthalene, phenol, *N,N*-dimethylaniline, anisole) results in perturbation of the aromatic π system such that the uncomplexed diene fragment exhibits reactivity toward dienophiles and electrophiles, thus providing a means to effect arene dearomatization [90]. Initially, arene complexes featuring the $(NH_3)_5Os(II)$ fragment were investigated [91]. Arenes bind to the metal in an η^2 fashion through the single vacant coordination site on Os, while, most importantly, the electron density surrounding the Os(II) center supplied by the strong σ-donating amine ligands facilitates pi-backbonding from the metal to the π^* orbitals of the arene. The combination of these binding interactions serves to isolate the coordinated π bond from the remaining uncoordinated π bonds in the arene ligand and provides a means to effect stereoselective dearomatization via tandem reaction with electrophiles and nucleophiles (Scheme 15.30) [92]. Metal-free organic products are isolated after mild oxidative demetalation (**87** to **88** in Scheme 15.30).

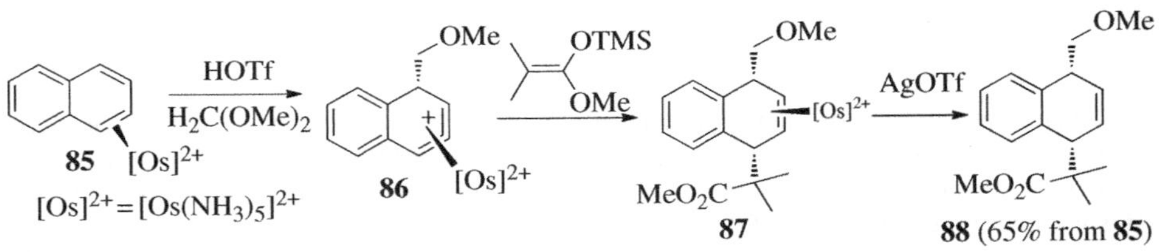

SCHEME 15.30 Os-promoted naphthalene dearomatization [92b].

Recognition of the importance of metal backbonding to the arene ligand facilitated subsequent identification of alternative metal fragments capable of η^2-arene coordination/dearomatization that circumvent issues of cost and toxicity associated with organoosmium chemistry [93]. In particular, a series of Re(I) complexes of the general formula {TpReL(CO)} (Tp = tris(pyrazolyl) borate) where L is a variable ligand capable of fine-tuning the electronic and steric properties of the complex (e.g., L = pyridine, 1-methylimidazole, NH_3, or PMe_3) was prepared. These complexes also possess a chiral metal center that can be harnessed in asymmetric dearomatizations via diastereoselective coordination of facially prochiral arenes [94]. The tungsten fragment $TpW(NO)PMe_3$ has subsequently emerged as a particularly useful dearomatization agent in terms of cost, synthetic accessibility, and ability to obtain diastereomerically pure η^2-arene complexes [95]. Once again, dearomatization is achieved via tandem reaction of coordinated arenes with electrophiles followed by nucleophiles. For example, protonation of the η^2-anisole complex **89** with diphenylammonium triflate (DPhAT) affords the anisolium complex **90** as a single diastereomer (Scheme 15.31) [96]. Stereoselective reaction of the uncomplexed alkene with acrolein gives **91**. A dearomatized oxadecalin ring system (**93**) is then obtained upon tandem nucleophilic addition/conjugate addition of the aldehyde carbonyl, protonation of the resulting enol ether (**92**), stereoselective reduction of the enolium complex, and oxidative demetalation. Regio- and diastereoselective W(0)-mediated dearomatizations of aniline and phenol derivatives have also been established [97, 98].

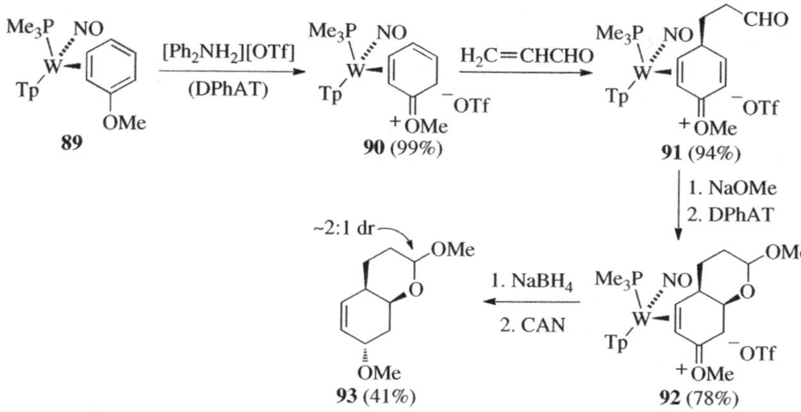

SCHEME 15.31 Representative dearomatization via an (η^2-arene)W(0) complex.

15.5.4 Dearomatization of η^6-Arene Metal Complexes

Numerous transition metal fragments form tractable η^6-arene complexes, but from an organic synthesis standpoint complexes incorporating $Cr(CO)_3$, $Mn(CO)_3^+$, $CpFe^+$, and $CpRu^+$ moieties are the most useful (Cp = cyclopentadienyl) [99]. The metal fragment can be thought of as a large electron-withdrawing group that renders the arene susceptible to stereoselective nucleophilic addition from the uncoordinated face of the ligand. Nucleophilic additions to (η^6-arene) metal complexes result in initial formation of η^5-cyclohexadienyl derivatives, the fate of which depends upon the identity of the metal fragment, the nucleophile, and the substituents on the starting arene. If S_NAr reactions are not possible and/or nucleophilic addition is irreversible, then dearomatization manifolds may become operative. Intermolecular chiral auxiliary-mediated diastereoselective nucleophilic additions to (η^6-arene)$Cr(CO)_3$ complexes are well established, and the resulting (cyclohexadienyl)Cr anion intermediates are also subject to stereocontrolled manipulation via reaction of electrophiles, often initially at the metal center as illustrated in Scheme 15.32 [100a, b]. A similar dearomatization sequence has been coupled with ring-closing metathesis to provide a range of *cis*-fused bicyclic ring systems [100c]. Intramolecular dearomatization of (η^6-arene)$Cr(CO)_3$ complexes also has been reported, particularly as a means to construct spirocyclic ring systems found in various terpene natural products [101].

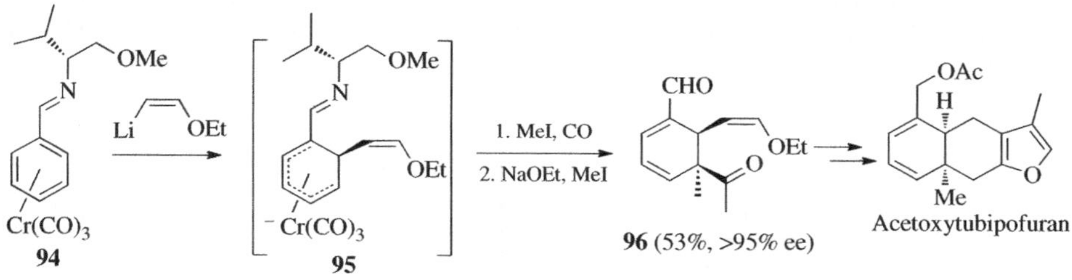

SCHEME 15.32 (Arene)$Cr(CO)_3$ dearomatization in the asymmetric synthesis of acetoxytubipofuran.

More recently, spirocyclizations utilizing (η^6-arene)$RuCp^+$ complexes have been reported [102]. Treatment of Ru(II) complex **97** with NaH in the presence of an aldehyde initiates tandem stereoselective nucleophilic aromatic addition–Horner–Wadsworth–Emmons olefination to deliver stable cyclohexadienyl complex **98** (Scheme 15.33) [103]. Dearomatization was completed via formal oxidation of the cyclohexadienyl ligand accompanied by addition of an external nucleophile. The stereodirecting effects of the metal center and the preexisting stereocenter completely control diastereoselectivity, resulting in formation of optically pure **99** with recovery of the CpRu fragment.

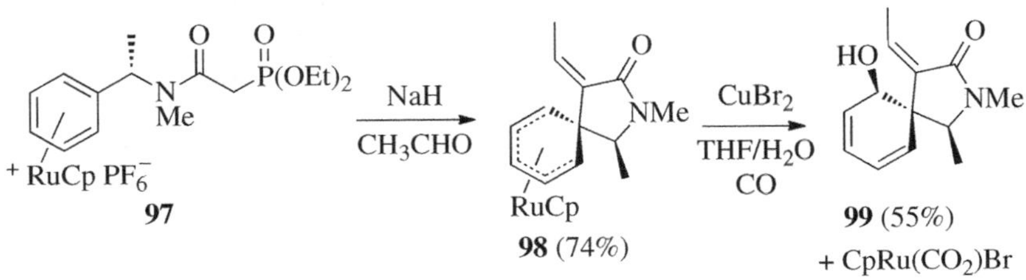

SCHEME 15.33 Asymmetric (η^6-arene)Ru(II)-mediated dearomatization.

An obvious drawback to widespread use of η^2 and η^6-arene complexes in synthesis is the need to employ stoichiometric amounts of transition metal. Nonetheless, these methods offer the ability to effect rapid stereocontrolled dearomatization of both simple and functionalized arenes through unique reaction manifolds. Uncovering procedures that would enable these transformations to be performed in the presence of transition metal catalysts would be a significant advance in this area.

15.6 ENZYMATIC DEAROMATIZATION

Several classes of microbial enzymes responsible for oxidative arene degradation have been identified, and synthetic applications of biocatalyzed dearomatization are growing in importance. These enzymes (e.g., toluene dioxygenase, naphthalene dioxygenase) are capable of converting simple arenes and naphthalenes into *cis*-dihydrodiols (**100**) that possess considerable potential in enantioselective synthesis and as precursors to carbocyclic sugar analogues (Scheme 15.34) [104, 105]. Benzoate dioxygenase displays altered regioselectivity toward benzoic acid substrates, affording *ipso ortho* oxidation products **101**, also enantioselectively [106]. Dearomatization via the action of reducing enzymes (benzoyl and naphthoyl CoA reductases) has also been noted, although synthetic applications have not been extensively explored [107].

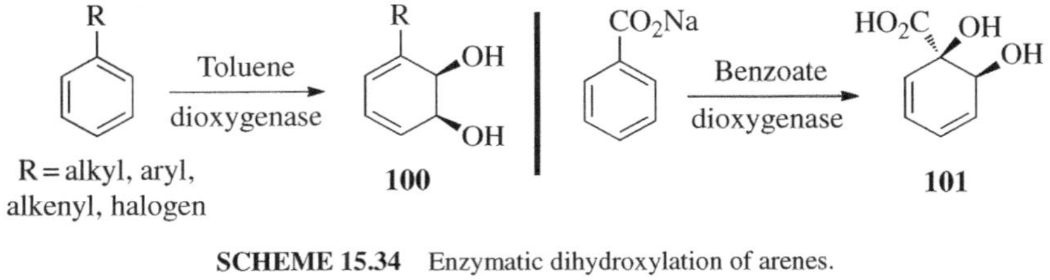

SCHEME 15.34 Enzymatic dihydroxylation of arenes.

Enzymatic dearomatization is typically performed on relatively simple arene substrates and so is usually conducted in the early stages of multistep synthetic sequences to access small-molecule chiral nonracemic building blocks. The recent example shown in Scheme 15.35 involves toluene dioxygenase-mediated dearomatization of **102** to give the chiral diol building block **103** as the first step in an asymmetric synthesis of morphinan derivatives [108]. Interestingly, **103** was ultimately converted to *ent*-hydromorphone through a sequence of transformations that included a tandem oxidative dearomatization–cycloaddition (**104–106**). Arenes that are suitable substrates for toluene dioxygenase and related enzymes number into the hundreds; thus, the study of biocatalyzed dearomatization is poised for significant growth as the value and power of these transformations to deliver enantiomerically enriched synthetic intermediates gain greater recognition.

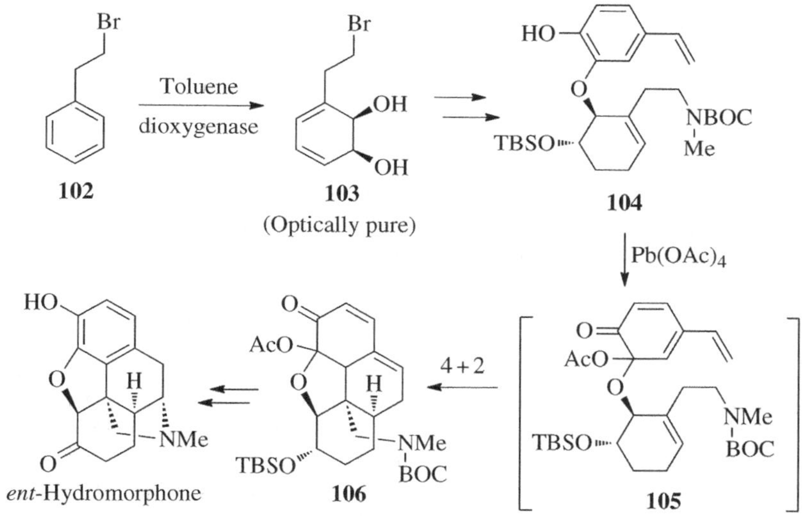

SCHEME 15.35 Enzymatic and oxidative dearomatization in the asymmetric synthesis of morphinan derivatives.

15.7 CONCLUSIONS AND FUTURE DIRECTIONS

The concept of dearomatization has emerged as an important and powerful strategy in organic synthesis, and its use is certain to gain increasing popularity as new methods for stereocontrolled dearomatization are discovered. The value of dearomatization tactics in delivering concise routes to architecturally complex alicyclic ring systems from readily available arene substrates has been amply demonstrated. While a wide range of carbocyclic arenes (both electron rich and electron deficient) participate in various dearomatization reactions, catalytic enantioselective dearomatization remains underdeveloped as few effective procedures currently exist. Additional efforts directed toward uncovering new organo- and transition metal-catalyzed dearomatization transformations are needed in order to achieve this important objective.

ABBREVIATIONS

AIBN	Azobis(isobutyronitrile)
ATPH	Aluminum tris(2,6-diphenylphenol)
CAN	Ceric ammonium nitrate

Cp	Cyclopentadienyl
de	Diastereomeric excess
DMPU	Dimethyltetrahydropyrimidinone
DPhAT	Diphenylammonium triflate
dr	Diastereomeric ratio
ee	Enantiomeric excess
LUMO	Lowest unoccupied molecular orbital
MCPBA	*meta*-Chloroperbenzoic acid
MVK	Methyl vinyl ketone
PIDA	Phenyl iodine diacetate
PTC	Phase-transfer catalyst
TBAF	Tetrabutylammonium fluoride
Tp	Hydrotris(pyrazolyl)borate
TS	Transition state

REFERENCES

[1] Roche, S.P. and Porco, J.A. (2011) *Angew. Chem. Int. Ed. Engl.*, **50**, 4068–4093.

[2] Murphy, W.S. and Wattanasin, S. (1983) *Chem. Soc. Rev.*, **12**, 213–250.

[3] (a) Winstein, S. and Baird, R. (1957) *J. Am. Chem. Soc.*, **79**, 756–757; (b) Dreiding, A.S. (1957) *Helv. Chim. Acta*, **40**, 1812–1814.

[4] (a) Masamune, S. (1961) *J. Am. Chem. Soc.*, **83**, 1009–1010; (b) Masamune, S. (1964) *J. Am. Chem. Soc.*, **86**, 288–289.

[5] (a) Corey, E.J., Girotra, N.N. and Mathew, C.T. (1969) *J. Am. Chem. Soc.*, **91**, 1557–1559; (b) Marshall, J.A. and Brady, S.F. (1970) *J. Org. Chem.*, **35**, 4068–4077; (c) Paul, T. and Mukherjee, D. (2003) *Tetrahedron Lett.*, **44**, 4985–4988; (d) Kraus, G.A. and Kesavan, S. (2005) *Tetrahedron Lett.*, **46**, 1111–1113.

[6] Boger, D.L. and Machiya, K. (1992) *J. Am. Chem. Soc.*, **114**, 10056–10058.

[7] Wang, Z., Dai, M., Park, P.K. and Danishefsky, S.J. (2011) *Tetrahedron*, **67**, 10249–10260.

[8] (a) Lalic, G. and Corey, E.J. (2007) *Org. Lett.*, **9**, 4921–4923; (b) McGrath, N.A., Bartlett, E.S., Sittihan, S. and Njardarson, J.T. (2009) *Angew. Chem. Int. Ed. Engl.*, **48**, 8543–8546.

[9] Ritter, T., Zarotti, P. and Carreira, E.M. (2004) *Org. Lett.*, **6**, 4371–4374.

[10] (a) Raikar, S.B., Nuhant, P., Delpech, B. and Marazano, C. (2008) *Eur. J. Org. Chem.*, **8**, 1358–1369; (b) Qi, J., Beeler, A.B., Zhang, Q. and Porco, J.A. (2010) *J. Am. Chem. Soc.*, **132**, 13642–13644.

[11] Phipps, R.J. and Toste, F.D. (2013) *J. Am. Chem. Soc.*, **135**, 1268–1271.

[12] Ortiz, F.L., Iglesias, M.J., Fernández, I., Sánchez, C.M.A. and Gómez, G.R. (2007) *Chem. Rev.*, **107**, 1580–1691.

[13] Andrews, R.C., Teague, S.J. and Meyers, A.I. (1988) *J. Am. Chem. Soc.*, **110**, 7854–7858.

[14] Clayden, J., Parris, S., Cabedo, N. and Payne, A.H. (2008) *Angew. Chem. Int. Ed. Engl.*, **47**, 5060–5062.

[15] Maruoka, K., Ito, M. and Yamamoto, H. (1995) *J. Am. Chem. Soc.*, **117**, 9091–9092.

[16] (a) Ahmed, A., Clayden, J. and Yasin, S.A. (1999) *Chem. Commun.*, 231–232; (b) Clayden, J. and Kenworthy, M.N. (2004) *Synthesis*, 1721–1736.

[17] (a) Clayden, J., Menet, C.J. and Mansfield, D.J. (2002) *Chem. Commun.*, 38–39; (b) Lemière, G., Sedehizadeh, S., Toueg, J., Fleary-Roberts, N. and Clayden, J. (2011) *Chem. Commun.*, **47**, 3745–3747.

[18] (a) Ruiz-Gómez, G., Iglesias, M.J., Serano-Ruiz, M., Garcia-Granda, S., Francesch, A., López-Ortiz, F. and Cuevas, C. (2007) *J. Org. Chem.*, **72**, 3790–3799; (b) Aggarwal, V.K., Alonso, E., Ferrara, M. and Spey, S.E. (2002) *J. Org. Chem.*, **67**, 2335–2344.

[19] Crich, D. and Patel, M. (2006) *Tetrahedron*, **62**, 7824–7837.

[20] Crich, D. and Krishnamurthy, V. (2006) *Tetrahedron*, **62**, 6830–6840.

[21] Citterio, A., Sebastiano, R., Maronati, A., Santi, R. and Bergamini, F. (1994) *J. Chem. Soc., Chem. Commun.*, 1517–1518.

[22] Ohno, H., Maeda, S., Okumura, M., Wakayama, R. and Tanaka, T. (2002) *Chem. Commun.*, 316–317.

[23] Boivin, J., Yousfi, M. and Zard, S. (1997) *Tetrahedron Lett.*, **38**, 5985–5988.

[24] Diaba, F., Montiel, J.A., Martínez-Laporta, A. and Bonjoch, J. (2013) *Tetrahedron Lett.*, **54**, 2619–2622.

[25] Martínez-Laporta, A., Ibarra-Rivera, T.R., El Kaïm, L. and Miranda, L.D. (2010) *Synthesis*, 1285–1290.

[26] (a) de Turiso, F.G.-L. and Curran, D.P. (2005) *Org. Lett.*, **7**, 151–154; (b) Lanza, T., Leardini, R., Minozzi, M., Nanni, D., Spagnolo, P. and Zanardi, G. (2008) *Angew. Chem. Int. Ed. Engl.*, **47**, 9439–9442.

[27] (a) Hoffmann, N. (2012) *Photochem. Photobiol. Sci.*, **11**, 1613–1641; (b) Cornelisse, J. (1993) *Chem. Rev.*, **93**, 615–669.

[28] (a) Streit, U. and Bochet, C.G. (2011) *Beilstein J. Org. Chem.*, **7**, 525–542; (b) Mattay, J. (2007) *Angew. Chem. Int. Ed. Engl.*, **46**, 663–665.

[29] (a) Chappell, D. and Russell, A.T. (2006) *Org. Biomol. Chem.*, **4**, 4409–4430; (b) Hoffmann, N. (2004) *Synthesis*, 481–495.

[30] Wender, P.A. and Howbert, J.J. (1981) *J. Am. Chem. Soc.*, **103**, 688–690.

[31] Boyd, J.W., Greaves, N., Kettle, J., Russell, A.T. and Steed, J.W. (2005) *Angew. Chem. Int. Ed. Engl.*, **44**, 944–946.

[32] Gaich, T. and Mulzer, J. (2010) *Org. Lett.*, **12**, 272–275.

[33] Wagner, P.J. and McMahon, K. (1994) *J. Am. Chem. Soc.*, **116**, 10827–10828.

[34] Streit, U., Birbaum, F., Quattropani, A. and Bochet, C.G. (2013) *J. Org. Chem.*, **78**, 6890–6910.

[35] Murray, R.D.H., Sutcliffe, M. and Hasegawa, M. (1975) *Tetrahedron*, **31**, 2966–2971.

[36] (a) Pettus, T.R.R., Inoue, M., Chen, X.-T. and Danishefsky, S.J. (2000) *J. Am. Chem. Soc.*, **122**, 6160–6168; (b) Lei, X., Dai, M., Hua, Z. and Danishefsky, S.J. (2008) *Tetrahedron Lett.*, **49**, 6383–6385.

[37] (a) Tisdale, E.J., Slobodov, I. and Theodorakis, E.A. (2004) *Proc. Natl. Acad. Sci. U. S. A.*, **101**, 12030–12035; (b) Nicolaou, K.C., Xu, H. and Wartmann, M. (2005) *Angew. Chem. Int. Ed. Engl.*, **44**, 756–761.

[38] Atherton, J.C.C. and Jones, S. (2003) *Tetrahedron*, **59**, 9030–9057.

[39] (a) Yonekawa, M., Koyama, Y., Kuwata, S. and Takata, T. (2012) *Org. Lett.*, **14**, 1164–1167; (b) Soro, Y., Bamba, F., Saika, S. and Coustard, J.-M. (2006) *Tetrahedron Lett.*, **47**, 3315–3319.

[40] Himbert, G. and Fink, D. (1985) *Tetrahedron Lett.*, **26**, 4363–4366.

[41] Lam, J.K., Schmidt, Y. and Vanderwal, C.D. (2012) *Org. Lett.*, **14**, 5566–5569.

[42] (a) Ding, Q., Ye, Y. and Fan, R. (2013) *Synthesis*, **45**, 1–16; (b) Pouységu, L., Deffieux, D. and Quideau, S. (2010) *Tetrahedron*, **66**, 2235–2261; (c) Tohma, H. and Kita, Y. (2003) *Top. Curr. Chem.*, **224**, 209–248.

[43] Magdziak, D., Meek, S.J. and Pettus, T.R.R. (2004) *Chem. Rev.*, **104**, 1383–1429.

[44] (a) Yang, Q., Draghici, C., Njardarson, J.T., Li, F., Smith, B.R. and Das, P. (2014) *Org. Biomol. Chem.*, **12**, 330–344; (b) Gong, J., Lin, G., Sun, W., Li, C.-C. and Yang, Z. (2010) *J. Am. Chem. Soc.*, **132**, 16745–16746.

[45] Zhao, C., Zheng, H., Jing, P., Fang, B., Xie, X. and She, X. (2012) *Org. Lett.*, **14**, 2293–2295.

[46] Cook, S.P., Polara, A. and Danishefsky, S.J. (2006) *J. Am. Chem. Soc.*, **128**, 16440–16441.

[47] (a) For a review, see Maertens, G., Ménard, M.-A. and Canesi, S. (2014) *Synthesis*, **46**, 1573–1582; (b) Vo, N.T., Pace, R.D.M., O'Hara, F. and Gaunt, M.J. (2008) *J. Am. Chem. Soc.*, **130**, 404–405.

[48] Liu, Q. and Rovis, T. (2006) *J. Am. Chem. Soc.*, **128**, 2552–2553.

[49] Gu, Q. and You, S.-L. (2011) *Org. Lett.*, **13**, 5192–5195.

[50] Cha, J.Y., Burnett, L., Huang, Y., Davidson, J.B. and Pettus, T.R.R. (2011) *J. Org. Chem.*, **76**, 1361–1371.

[51] Pouységu, L., Sylla, T., Garnier, T., Rojas, L.B., Charris, J., Deffieux, D. and Quideau, S. (2010) *Tetrahedron*, **66**, 5908–5917.

[52] Dong, S., Zhu, J. and Porco, J.A. (2008) *J. Am. Chem. Soc.*, **130**, 2738–2739.

[53] Boppisetti, J.K. and Birman, V.B. (2009) *Org. Lett.*, **11**, 1221–1223.

[54] Liang, H. and Ciufolini, M.A. (2011) *Angew. Chem. Int. Ed. Engl.*, **50**, 11849–11851.

[55] Dohi, T., Maruyama, A., Yoshimura, M., Morimoto, K., Tohma, H. and Kita, Y. (2005) *Angew. Chem. Int. Ed. Engl.*, **44**, 6193–6196.

[56] (a) Dohi, T., Takenaga, N., Nakae, T., Toyoda, Y., Yamasaki, M., Shiro, M., Fujioka, H., Maruyama, A. and Kita, Y. (2013) *J. Am. Chem. Soc.*, **135**, 4558–4566; (b) Uyanik, M., Yasui, T. and Ishihara, K. (2013) *Angew. Chem. Int. Ed. Engl.*, **52**, 9215–9218.

[57] Volp, K.A. and Harned, A.M. (2013) *Chem. Commun.*, **49**, 3001–3003.

[58] Scheffler, G., Seike, H. and Sorensen, E.J. (2000) *Angew. Chem. Int. Ed. Engl.*, **39**, 4593–4596.

[59] Wardrop, D.J. and Burge, M.S. (2005) *J. Org. Chem.*, **70**, 10271–10284.

[60] (a) Ousmer, M., Braun, N.A., Bavoux, C., Perrin, M. and Ciufolini, M.A. (2001) *J. Am. Chem. Soc.*, **123**, 7534–7538; (b) Canesi, S., Belmont, P., Bouchu, D., Rousset, L. and Ciufolini, M.A. (2002) *Tetrahedron Lett.*, **43**, 5193–5195.

[61] (a) Liang, H. and Ciufolini, M.A. (2008) *J. Org. Chem.*, **73**, 4299–4301; (b) Mendelsohn, B.A. and Ciufolini, M.A. (2009) *Org. Lett.*, **11**, 4736–4739.

[62] Dalko, P.I. (1995) *Tetrahedron*, **51**, 7579–7653.

[63] Kende, A.S., Koch, K. and Smith, C.A. (1988) *J. Am. Chem. Soc.*, **110**, 2210–2218.

[64] Liang, J., Chen, J., Du, F., Zeng, X., Li, L. and Zhang, H. (2009) *Org. Lett.*, **11**, 2820–2823.

[65] (a) Zheng, C., Wang, L., Li, W., Wang, L. and Wang, D.Z. (2013) *Org. Lett.*, **15**, 4046–4049; (b) Sabot, C., Guérard, K.C. and Canesi, S. (2009) *Chem. Commun.*, 2941–2943; (c) Quideau, S., Looney, M.A. and Pouységu, L. (1999) *Org. Lett.*, **1**, 1651–1654.

[66] Nicolaou, K.C., Edmonds, D.J., Li, A. and Tria, G.S. (2007) *Angew. Chem. Int. Ed. Engl.*, **46**, 3942–3945.

[67] (a) Hamamoto, H., Shiozaki, Y., Nambu, H., Hata, K., Tohma, H. and Kita, Y. (2004) *Chem. Eur. J.*, **10**, 4977–4982; (b) Jacquemot, G. and Canesi, S. (2012) *J. Org. Chem.*, **77**, 7588–7594.

[68] (a) Desjardins, S., Andrez, J.-C. and Canesi, S. (2011) *Org. Lett.*, **13**, 3406–3409; (b) Swenton, J.S., Callinan, A., Chen, Y., Rohde, J.J., Kerns, M.J. and Morrow, G.W. (1996) *J. Org. Chem.*, **61**, 1267–1274.

[69] (a) Green, J.C. and Pettus, T.R.R. (2011) *J. Am. Chem. Soc.*, **133**, 1603–1608; (b) Guérard, K.C., Guérinot, A., Bouchard-Aubin, C., Ménard, M.-A., Lepage, M., Beaulieu, M.A. and Canesi, S. (2012) *J. Org. Chem.*, **77**, 2121–2133; (c) Beaulieu, M.A., Guérard, K.C., Maertens, G., Sabot, C. and Canesi, S. (2011) *J. Org. Chem.*, **76**, 9460–9471.

[70] Rudolph, A., Bos, P.H., Meetsma, A., Minnaard, A.J. and Feringa, B.L. (2011) *Angew. Chem. Int. Ed. Engl.*, **50**, 5834–5838.

[71] Oguma, T. and Katsuki, T. (2012) *J. Am. Chem. Soc.*, **134**, 20017–20020.

[72] Zhang, X. and Larock, R.C. (2005) *J. Am. Chem. Soc.*, **127**, 12230–12231.

[73] (a) Tang, B.-X., Zhang, Y.-H., Song, R.-J., Tang, D.-J., Deng, G.-B., Wang, Z.-Q., Xie, Y.-X., Xia, Y.-Z and Li, J.-H. (2012) *J. Org. Chem.*, **77**, 2837–2849; (b) Yin, Q. and You, S.-L. (2012) *Org. Lett.*, **14**, 3526–3529; (c) Leon, R., Jawalekar, A., Redert, T. and Gaunt, M.J. (2011) *Chem. Sci.*, **2**, 1487–1490.

[74] Ovens, C., Martin, N.G. and Procter, D.J. (2008) *Org. Lett.*, **10**, 1441–1444.

[75] Zhuo, C.-X., Zhang, W. and You, S.-L. (2012) *Angew. Chem. Int. Ed. Engl.*, **51**, 12662–12686.

[76] Doyle, M.P., McKervey, M.A. and Ye, H. (1998) *Modern Catalytic Methods for Organic Synthesis with Diazo Compounds*, John Wiley & Sons, Inc., New York, pp. 289–354.

[77] (a) Sugimura, T., Sato, Y., Im, C.Y. and Okuyama, T. (2004) *Org. Lett.*, **6**, 4439–4442; (b) Maguire, A.R., Buckley, N.R., O'Leary, P. and Ferguson, G. (1998) *J. Chem. Soc. Perkin Trans.*, **1**, 4077–4091; (c) O'Neill, S., O'Keefe, S., Harrington, F. and Maguire, A.R. (2009) *Synlett*, 2312–2314.

[78] Frey, B., Wells, A.P., Rogers, D.H. and Mander, L.N. (1998) *J. Am. Chem. Soc.*, **120**, 1914–1915.

[79] (a) King, G.R., Mander, L.N., Monck, N.J.T., Morris, J.C. and Zhang, H. (1997) *J. Am. Chem. Soc.*, **119**, 3828–3829; (b) Levin, S., Nani, R.R. and Reisman, S.E. (2011) *J. Am. Chem. Soc.*, **133**, 774–776; (c) Reisman, S.E., Nani, R.R. and Levin, S. (2011) *Synlett*, 2437–2442.

[80] (a) Nemoto, T., Ishige, Y., Yoshida, M., Kohno, Y., Kanematsu, M. and Hamada, Y. (2010) *Org. Lett.*, **12**, 5020–5023; (b) Wu, Q.-F., Liu, W.-B., Zhuo, C.-X., Rong, Z.-Q., Ye, K.-Y. and You, S.-L. (2011) *Angew. Chem. Int. Ed. Engl.*, **50**, 4455–4458.

[81] (a) Nemoto, T., Zhao, Z., Yokosaka, T., Suzuki, Y., Wu, R. and Hamada, Y. (2013) *Angew. Chem. Int. Ed. Engl.*, **52**, 2217–2220; (b) Nemoto, T., Nozaki, T., Yoshida, M. and Hamada, Y. (2013) *Adv. Synth. Catal.*, **355**, 2693–2700.

[82] Zhuo, C.-X. and You, S.-L. (2013) *Angew. Chem. Int. Ed. Engl.*, **52**, 10056–10059.

[83] Xiao, Q., Jackson, J.J., Basak, A., Bowler, J.M., Miller, B.G. and Zakarian, A. (2013) *Nat. Chem.*, **5**, 410–416.

[84] Bao, M., Nakamura, H. and Yamamoto, Y. (2001) *J. Am. Chem. Soc.*, **123**, 759–760.

[85] Peng, B., Zhang, S., Yu, X., Feng, X. and Bao, M. (2011) *Org. Lett.*, **13**, 5402–5405.

[86] (a) Li, Y., Zhang, L., Zhang, L., Wu, Y. and Gong, Y. (2013) Eur. J. Org. Chem., 8039–8047; (b) Rousseaux, S., García-Fortanet, J., Aguila, M.A.D. and Buchwald, S.L. (2011) *J. Am. Chem. Soc.*, **133**, 9282–9285.

[87] Nan, J., Zuo, Z., Luo, L., Bai, L., Zheng, H., Yuan, Y., Liu, J., Luan, X. and Wang, Y. (2013) *J. Am. Chem. Soc.*, **135**, 17306–17309.

[88] Shibuya, T., Noguchi, K. and Tanaka, K. (2012) *Angew. Chem. Int. Ed. Engl.*, **51**, 6219–6222.

[89] García-Fortanet, J., Kessler, F. and Buchwald, S.L. (2009) *J. Am. Chem. Soc.*, **131**, 6676–6677.

[90] Harrison, D.P. and Harman, W.D. (2012) *Aldrichem. Acta*, **45**, 45–55.

[91] Harman, W.D. (1997) *Chem. Rev.*, **97**, 1953–1978.

[92] (a) Smith P.L, Chordia, M.D. and Harman, W.D. (2001) *Tetrahedron*, **57**, 8203–8225; (b) Winemiller, M.D. and Harman, W.D. (1998) *J. Am. Chem. Soc.*, **120**, 7835–7840.

[93] (a) Surendranath, Y. and Harman, W.D. (2006) Dalton Trans., 3957–3965; (b) Keane, J.M. and Harman, W.D. (2005) *Organometallics*, **24**, 1786–1798; (c) Brooks, B.C., Gunnoe, T.B. and Harman, W.D. (2000) *Coord. Chem. Rev.*, **206–207**, 3–61.

[94] Ding, F., Valahovic, M.T., Keane, J.M., Anstey, M.R., Sabat, M., Trindle, C.O. and Harman, W.D. (2004) *J. Org. Chem.*, **69**, 2257–2267.

[95] Pienkos, J.A., Zottig, V.E., Iovan, D.A., Li, M., Harrison, D.P., Sabat, M., Salomon, R.J., Strausberg, L., Teran, V.A., Myers, W.H. and Harman, W.D. (2013) *Organometallics*, **32**, 691–703.

[96] Lis, E.C., Salomon, R.J., Sabat, M., Myers, W.H. and Harman, W.D. (2008) *J. Am. Chem. Soc.*, **130**, 12472–12476.

[97] Salomon, R.J., Todd, M.A., Sabat, M., Myers, W.H. and Harman, W.D. (2010) *Organometallics*, **29**, 707–709.

[98] Todd, M.A., Sabat, M., Myers, W.H. and Harman, W.D. (2007) *J. Am. Chem. Soc.*, **129**, 11010–11011.

[99] (a) Kündig, E.P. and Pape, A.R. (2004) *Top. Organomet. Chem.*, **7**, 71–94; (b) Pape, A.R., Kaliappan, K.P. and Kündig, E.P. (2000) *Chem. Rev.*, **100**, 2917–2940.

[100] (a) Kündig, E.P., Cannas, R., Laxmisha, M., Ronggang, L. and Tchertchian, S. (2003) *J. Am. Chem. Soc.*, **125**, 5642–5643; (b) Kündig, E.P., Laxmisha, M.S., Cannas, R., Tchertchian, S. and Ronggang, L. (2005) *Helv. Chim. Acta*, **88**, 1063–1080; (c) Kündig, E.P., Bellido, A., Kaliappan, K.P., Pape, A.R. and Radix, S. (2006) *Org. Biomol. Chem.*, **4**, 342–351.

[101] Semmelhack, M.F. and Yamashita, A. (1980) *J. Am. Chem. Soc.*, **102**, 5924–5926.

[102] (a) Pigge, F.C., Dhanya, R. and Swenson, D.C. (2009) *Organometallics*, **28**, 3869–3875; (b) Pigge, F.C. and Dalvi, R. (2008) *Tetrahedron*, **64**, 10123–10131; (c) Pigge, F.C., Dhanya, R. and Hoefgen, E.R. (2007) *Angew. Chem. Int. Ed. Engl.*, **46**, 2887–2890.

[103] Pigge, F.C., Coniglio, J.J. and Dalvi, R. (2006) *J. Am. Chem. Soc.*, **128**, 3498–3499.

[104] (a) Froese, J., Endoma-Arias, M.A.A. and Hudlicky, T. (2014) *Org. Process Res. Dev.*, **18**. 801–809; (b) Trant, J.F., Froese, J. and Hudlicky, T. (2013) *Tetrahedron Asymmetry*, **24**, 184–190; (c) Khan, M.A., Mahon, M.F., Lowe, J.P., Stewart, A.J.W. and Lewis, S.E. (2012) *Chem. Eur. J.*, **18**, 13480–13493; (d) Hudlicky, T. and Reed, J. W. (2009) *Chem. Soc. Rev.*, **38**, 3117–3132; (e) Gibson, D.T., Hensley, M., Yoshioka, H. and Mabry, T.J. (1970) *Biochemistry*, **9**, 1626–1630.

[105] (a) Boyd, D.R., Sharma, N.D., Bowers, N.I., Coen, G.B., Malone, J.F., O'Dowd, C.R., Stevenson, P.J. and Allen, C.C.R. (2010) *Org. Biomol. Chem.*, **8**, 1415–1423; (b) Ley, S.V. and Sternfeld, F. (1988) *Tetrahedron*, **45**, 3463–3476.

[106] (a) Griffen, J.A., Kenwright, S.J., Abou-Shehada, S., Wharry, S., Moody, T.S. and Lewis, S.E. (2014) *Org. Chem. Front.*, **1**, 79–90; (b) Lewis, S.E. (2014) *Chem. Commun.*, **50**, 2821–2830.

[107] (a) Eberlein, C., Estelmann, S., Seifert, J., von Bergen, M., Müller, M., Meckenstock, R.U. and Boll, M. (2013) *Mol. Microbiol.*, **88**, 1032–1039; (b) Boll, M. (2005) *J. Mol. Microbiol. Biotechnol.*, **10**, 132–142.

[108] Varghese, V. and Hudlicky, T. (2014) *Angew. Chem. Int. Ed. Engl.*, **53**, 4355–4358.

PART V

AROMATIC REARRANGEMENTS

16

AROMATIC COMPOUNDS VIA PERICYCLIC REACTIONS

SETHURAMAN SANKARARAMAN
Department of Chemistry, Indian Institute of Technology Madras, Chennai, India

16.1 INTRODUCTION

Reactions that proceed through a cyclic transition state and in a concerted manner are defined as pericyclic reactions [1]. There are no intermediates formed in these single-step reactions. Being concerted, the bond-breaking and bond-making processes contribute to the transition state structure, although not necessarily to the same degree since these processes are not always synchronous, depending on the substitution pattern on the substrates. There are five types of pericyclic reactions known, namely, (i) electrocyclic ring closure and ring-opening reactions, (ii) cycloaddition and cycloreversion reactions, (iii) sigmatropic rearrangements, (iv) chelotropic reactions, and (v) group transfer reactions [2]. Although all these types of reactions proceed through cyclic transition states, only electrocyclic ring closure and cycloaddition reactions lead to cyclic compounds as products. Therefore, among these types, only electrocyclic reactions and cycloaddition reactions are considered in this chapter for the synthesis of aromatic compounds. Some examples from the chelotropic reactions are considered as part of the cycloaddition reaction, in those cases where the initially formed cycloadduct undergoes chelotropic elimination leading to the formation of aromatic compounds.

An acyclic conjugated polyene undergoing a concerted ring-closing reaction is known as electrocyclic ring-closing reaction, and the reverse process is known as electrocyclic ring-opening reaction. Addition reactions that proceed in a concerted manner and result in the formation of a cyclic compound are termed as cycloaddition reactions. The selection rules that govern these reactions are called the Woodward–Hoffmann rules [3].

Arene Chemistry: Reaction Mechanisms and Methods for Aromatic Compounds,
First Edition. Edited by Jacques Mortier.

16.2 ELECTROCYCLIC RING CLOSURE REACTION

From the point of view of synthesis of aromatic compounds, a 6π electron cyclization of either a 1,3,5-hexatriene or 1,2-divinylarene derivative followed by oxidation would be an appropriate strategy (Scheme 16.1). The intermediate dihydroaromatic derivatives can be oxidized to the fully aromatic compounds by a variety of oxidizing and dehydrogenating agents.

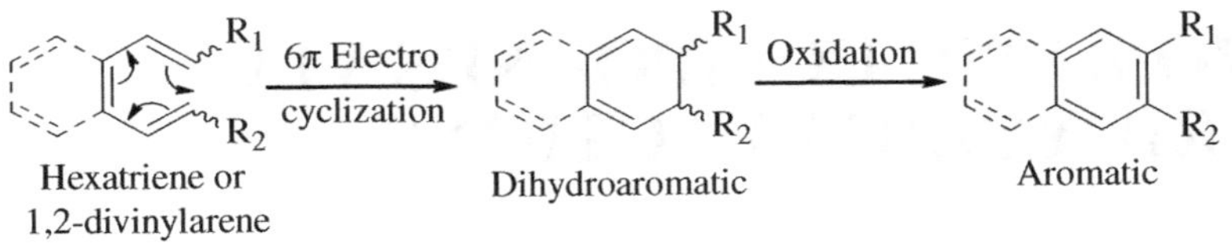

SCHEME 16.1 6π Electrocyclization/oxidation strategy for synthesis of aromatics.

According to the Woodward–Hoffmann rules for electrocyclic reactions, a 6π electrocyclization is thermally allowed in a disrotatory manner and photochemically allowed in a conrotatory manner. However, in the present context of synthesis of aromatic compounds as final products, which are devoid of any stereocenters, the stereochemical aspects of the substituents in the intermediate dihydroaromatic compound should not matter. The photochemical 6π electrocyclization of *cis*-stilbene derivatives followed by oxidation of the dihydroaromatic intermediate provides access to angularly fused polycyclic aromatic compounds (Scheme 16.2) [4].

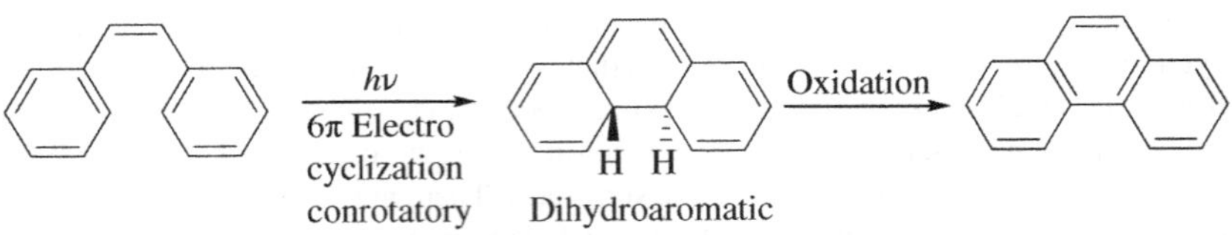

SCHEME 16.2 Photochemical 6π electrocyclization/oxidation method.

A general strategy for the synthesis of 1,2-divinylethene (1,3,5-hexatriene) involves Heck coupling of 1,2-dihaloethene or 1,2-dihaloarene derivatives. The electrocyclization of the divinyl derivative followed by dehydrogenation step completes the synthesis of aromatic compounds (Scheme 16.3). A review has appeared on this topic recently [5].

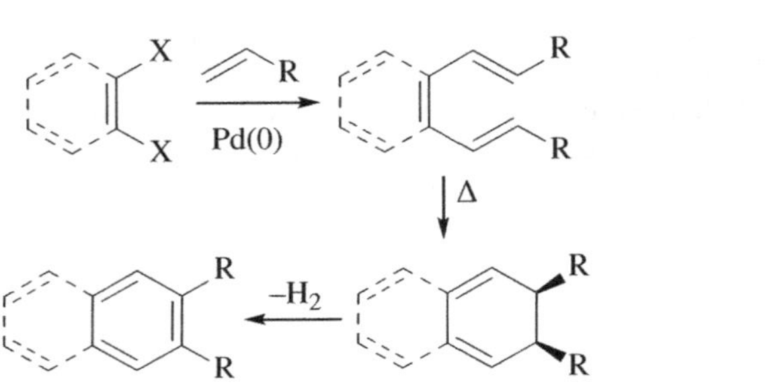

SCHEME 16.3 Domino Heck/electrocyclization/dehydrogenation methodology.

16.2.1 Application of Electrocyclic Ring Closure in Aromatic Synthesis

The general strategy illustrated in Scheme 16.3 is used in the synthesis of indane and tetralin derivatives (Scheme 16.4) [6]. Pd(0)-catalyzed double Heck reaction yielded the hexatriene framework, which on heating underwent disrotatory electrocyclization to yield the dihydroaromatic derivative. Dehydrogenation was carried out either with Pd/C or elemental sulfur to yield the aromatic compound.

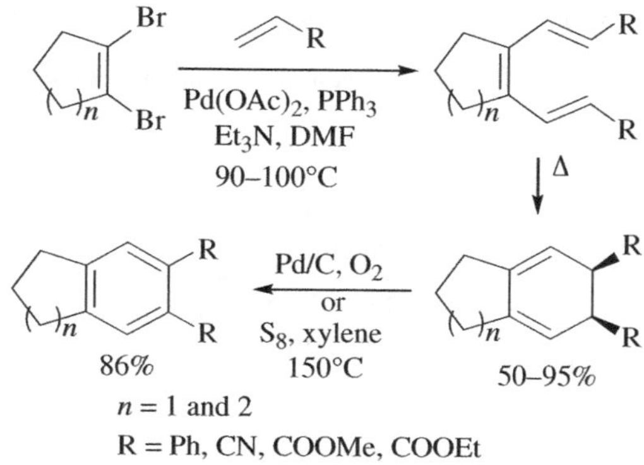

SCHEME 16.4 Synthesis of indane and tetralin derivatives.

The Heck reaction followed by domino electrocyclization/dehydrogenation strategy works well for the benzo fusion of [2.2]paracyclophanediene as shown in Scheme 16.5 [7]. The *in situ* formed electrocyclization product was aromatized by dehydrogenation using either Pd/C or elemental sulfur.

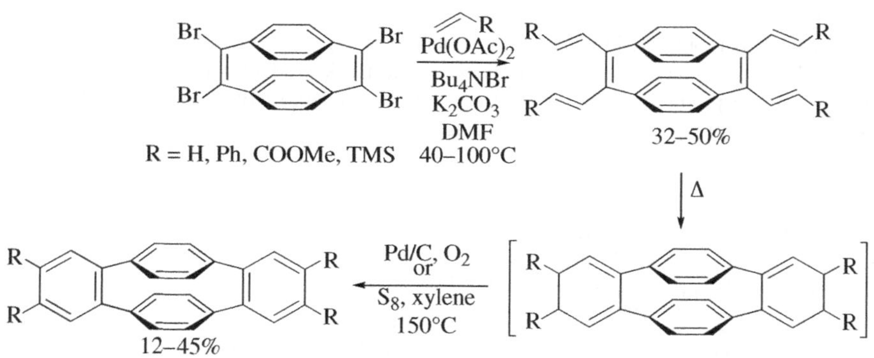

SCHEME 16.5 Benzannulation of [2.2]paracyclophane.

Substituted fluoranthenes were prepared from 2,3-dibromoacenaphthalene by Heck reaction with acrylates and styrene followed by electrocyclization and dehydrogenation that occurred in a one-pot sequence in good yields (Scheme 16.6) [8].

Similarly, 2,3-disubstituted anthraquinones were prepared from 2,3-dibromonaphthaquinone by a domino twofold Heck coupling/6π electrocyclization/dehydrogenation sequence that occurred in one pot (Scheme 16.7) [9]. Pd(0) present in these reactions might be responsible for the observed dehydrogenation of the product of electrocyclization.

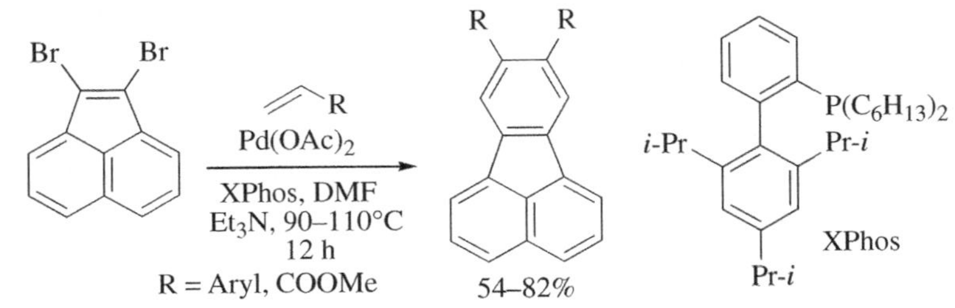

SCHEME 16.6 Synthesis of fluoranthene derivatives.

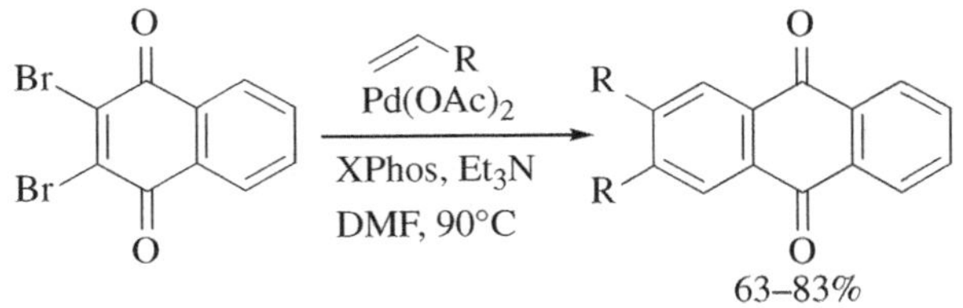

SCHEME 16.7 Synthesis of substituted anthraquinones by domino twofold Heck/electrocyclization/dehydrogenation in one pot.

In all of the previous examples, the central double bond of the hexatriene unit was not part of an aromatic sextet. Therefore, the electrocyclization reaction proceeded smoothly. The domino Heck/electrocyclization/dehydrogenation strategy has been widely applied for benzannulation of heterocyclic compounds. Examples of double benzannulation of thiophene and pyrrole are illustrated in Scheme 16.8 [10]. A sequence involving fourfold Heck coupling followed by electrocyclization and dehydrogenation has been adopted to effect these conversions. In case of the thiophene derivative, the electrocyclization/dehydrogenation took place in a single step at 140°C, whereas in case of the pyrrole derivative, reaction occurred only when the Heck product was heated at 200°C [11]. In these cases, the products of initial electrocyclization undergo isomerization by [1,5]-H shift to restore the aromaticity of the heterocyclic ring.

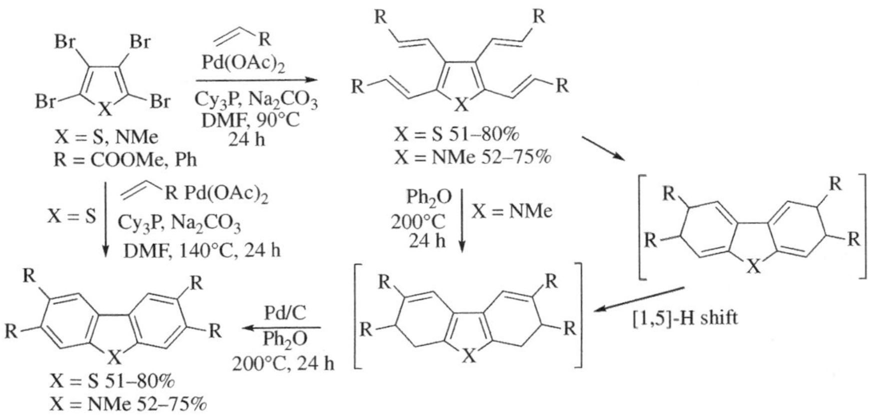

SCHEME 16.8 Double benzannulation of thiophene and *N*-methylpyrrole.

Although 2,3-dialkenylpyridine, pyrazine, and quinoxaline derivatives were synthesized from the corresponding 2,3-dichloro derivatives, the subsequent electrocyclization failed to occur under a variety of conditions (Scheme 16.9) [12].

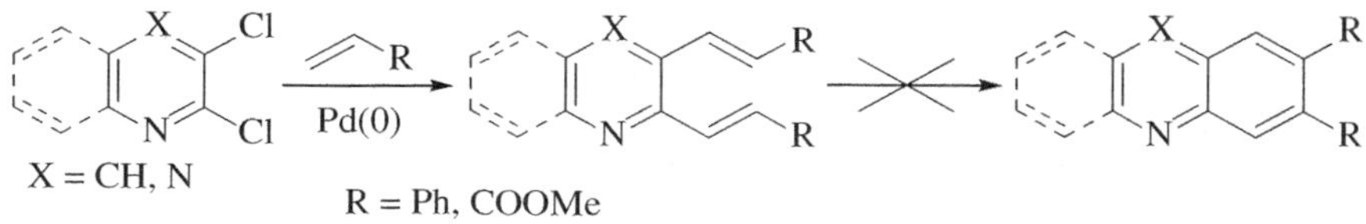

SCHEME 16.9 Double Heck reaction of six-membered *N*-heterocycles.

Benzannulation of bridged bicyclic systems has been reported [13]. The reaction involves *in situ* reduction of bridged bicyclic hexa-3-yn-1,5-diene to 1,3,5-hexatriene and electrocyclization to dihydroaromatic intermediate. Lewis acid-mediated dehydrogenation gave the benzannulated bridged bicyclic compound as the product (Scheme 16.10) [13]. Although the mechanism of Lewis acid-mediated dehydrogenation in the presence of benzaldehyde is not fully understood, it is presumed to occur by the abstraction of hydride ion from the dihydroaromatic intermediate by the Lewis acid-activated aldehyde.

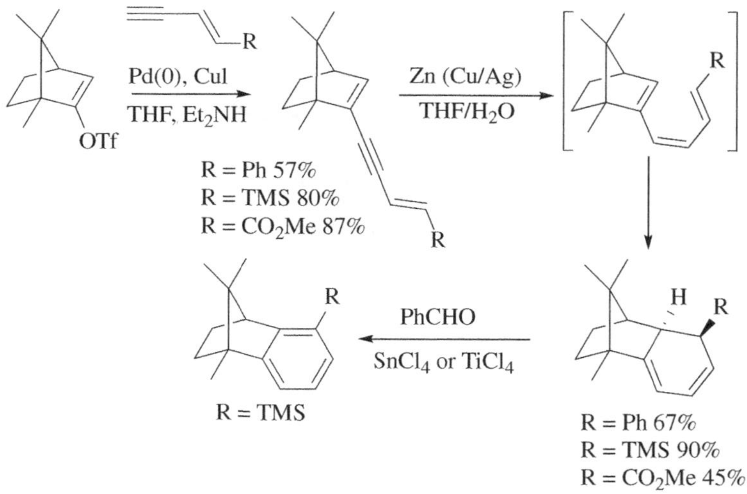

SCHEME 16.10 Benzannulation of bridged bicyclic derivatives through electrocyclization of 1,3,5-hexatriene unit.

Photochemical *trans–cis* isomerization of diarylethenes followed by photochemical 6π electrocyclization and oxidation is a standard protocol, and this protocol is widely applied in the synthesis of angularly fused aromatic compounds [4]. In particular, this strategy is widely used in the synthesis of helicenes, an important class of angularly fused polycyclic aromatic compounds (Scheme 16.11) [14].

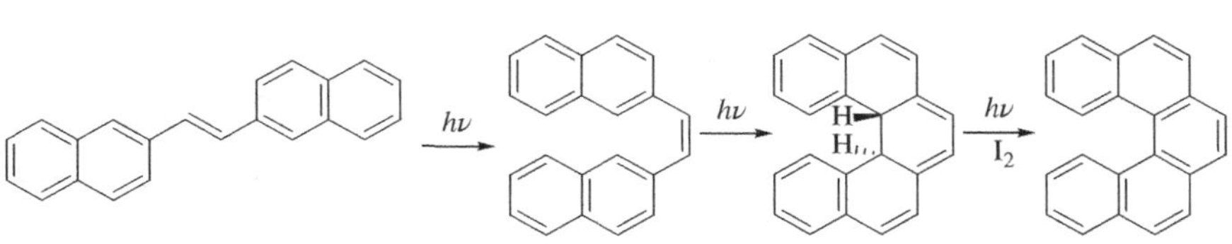

SCHEME 16.11 Synthesis of [5]helicene by *cis–trans* isomerization/electrocyclization/oxidation sequence.

With unsymmetrical olefins, formation of multiple products is possible due to the availability of multiple sites for cyclization as illustrated in the synthesis of [6]helicene (Scheme 16.12) [15]. Out of the four possible products (bond formation between a–c, a–d, b–c, and b–d), two are formed (bond formation between a–c and b–c) of which one of them is [6]helicene (Scheme 16.12). When a more symmetrical phenyl-substituted derivative was used, [6]helicene was formed exclusively (Scheme 16.12) [15]. Synthesis of the higher homologue, namely, [11]helicene, is illustrated in Scheme 16.13. From both the reactions given in Scheme 16.13, it is evident that the bay region carbons of phenanthrene and benzophenanthrene are more reactive for electrocyclization [16].

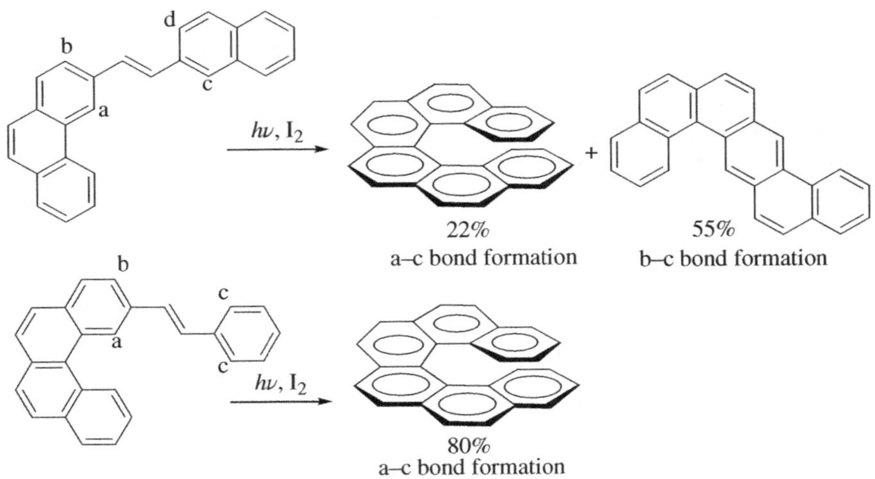

SCHEME 16.12 Synthesis of [6]helicene *via* electrocyclization/oxidation.

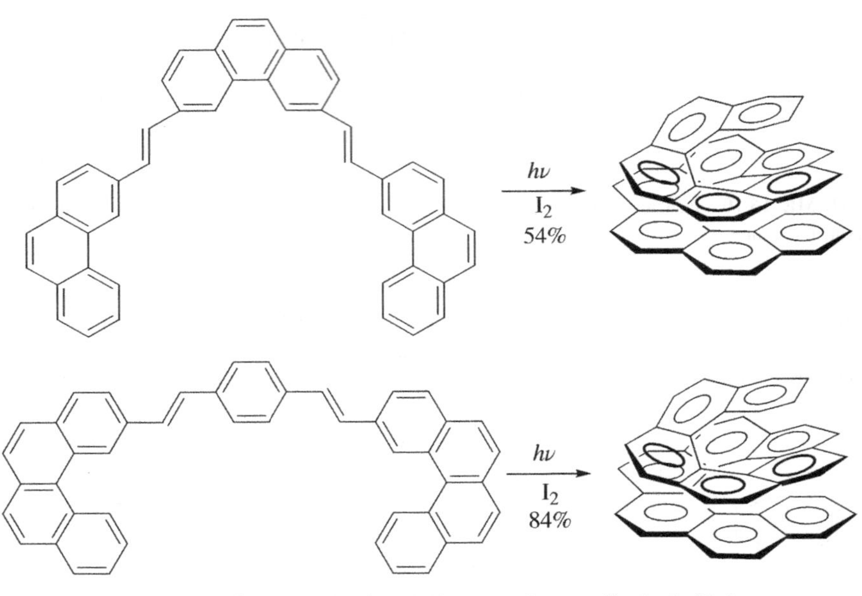

SCHEME 16.13 Synthesis of [11]helicene *via* electrocyclization/oxidation.

Synthesis of helicenes by electrocyclization using circularly polarized light has met with limited success due to poor enantioselectivities observed (Scheme 16.14) [17]. Although helicenes exhibit high optical rotation and hence it is possible to detect even small enantiomeric excess by optical rotational method, it is considered to be an unreliable method. Enantiomeric ratio determined from chiral HPLC analysis is more accurate and reliable, and it has made the optical rotation method obsolete [14b].

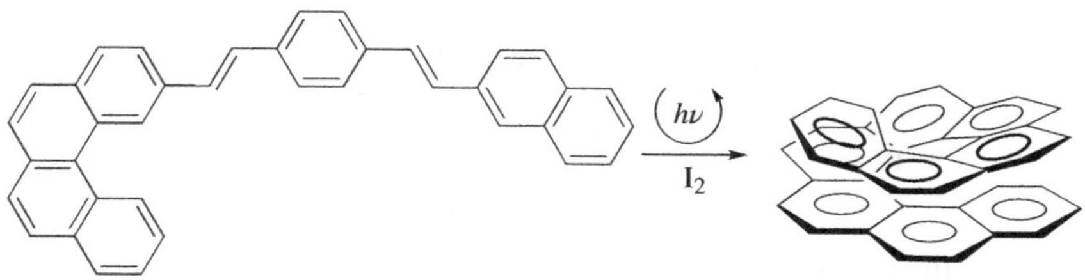

ee 0.42%

SCHEME 16.14 Asymmetric synthesis of [9]helicene using left circularly polarized light.

A one-pot procedure for the synthesis of functionalized *p*-terphenyl derivatives has been reported that involves tandem aldol condensation/electrocyclization/oxidative dehydrogenation (Scheme 16.15) [18]. The intermediate electrocyclization product, namely, the dihydroaromatic derivative, is presumably oxidized by oxygen under basic conditions.

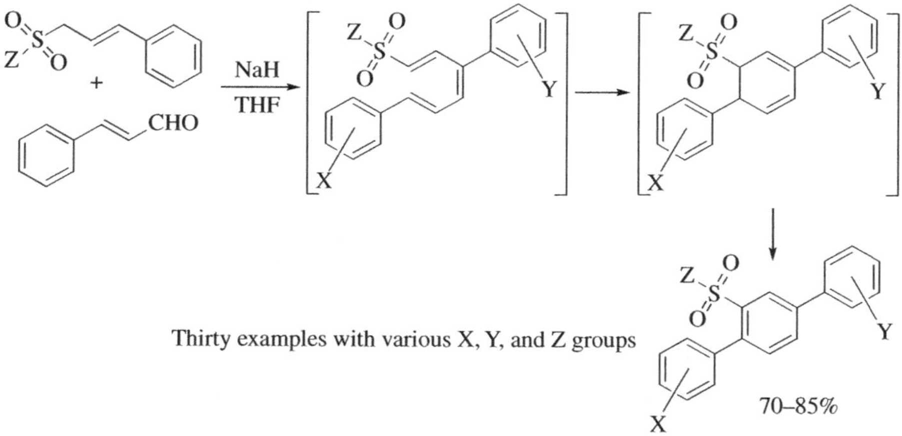

SCHEME 16.15 Tandem aldol/electrocyclization/oxidative dehydrogenation method toward *p*-terphenyl derivatives.

16.3 INTRODUCTION TO CYCLOADDITION REACTIONS

[4+2] Cycloaddition reactions involving 6π electrons result in the formation of either dihydroaromatic or tetrahydroaromatic derivatives. It depends upon whether the dienophile is an olefin or an acetylene. Dehydrogenation of the cycloadducts results in the formation of aromatic compounds (Scheme 16.16) [19]. Cycloaddition reactions are thermally allowed under suprafacial–suprafacial overlap mode $[4\pi_s+2\pi_s]$ [3].

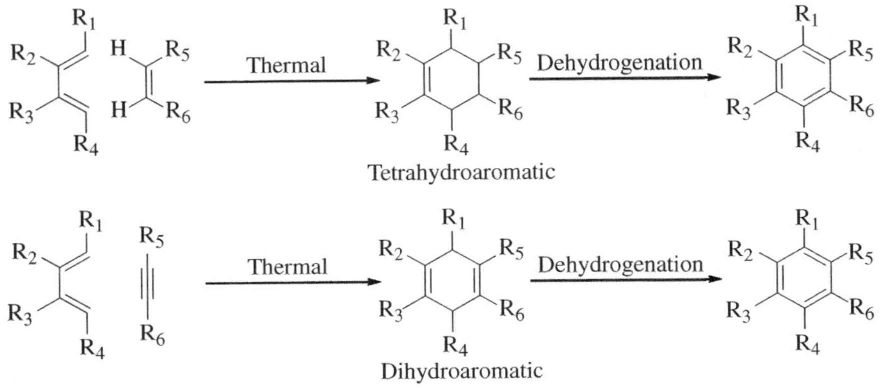

SCHEME 16.16 General strategy for synthesis of aromatics by [4+2] cycloaddition.

In addition, suitably substituted butadienes undergo Diels–Alder reaction followed by either 1,2- or 1,4-elimination of the substituents leading to the formation of aromatic compounds (Scheme 16.17). For the use of catalysts in cycloaddition reactions and their role in the mechanism of the reaction, see Refs. 2 and 3.

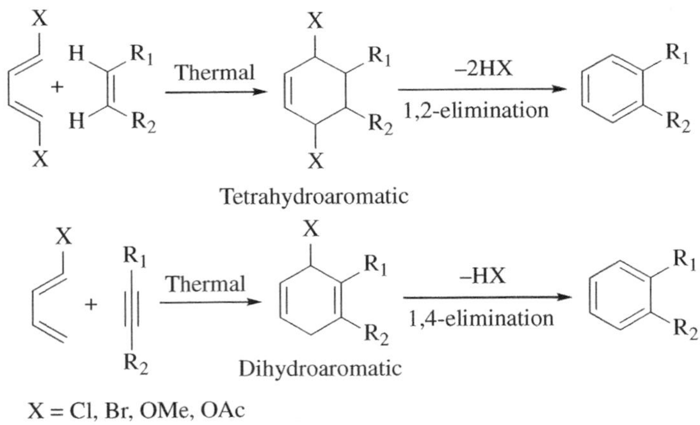

SCHEME 16.17 Domino [4+2] cycloaddition/elimination method.

The third methodology leading to the formation of aromatic compounds is domino [4+2] cycloaddition reaction followed by the chelotropic elimination from the cycloadduct (Scheme 16.18). The intermediate bridged bicyclic cycloadducts are strained, and they undergo elimination leading to aromatization usually under the reaction conditions of the Diels–Alder reaction.

16.3.1 Application of [4+2] Cycloaddition Method for Synthesis of Aromatic Compounds

A two-step sequence involving thermal [4+2] cycloaddition followed by dehydrogenation using 2,3-dichloro-5,6-dicyanobenzoquinone (DDQ) has been used for the synthesis of diethyl esters of naphthalene and anthracene dicarboxylates in good yields (Scheme 16.19) [20].

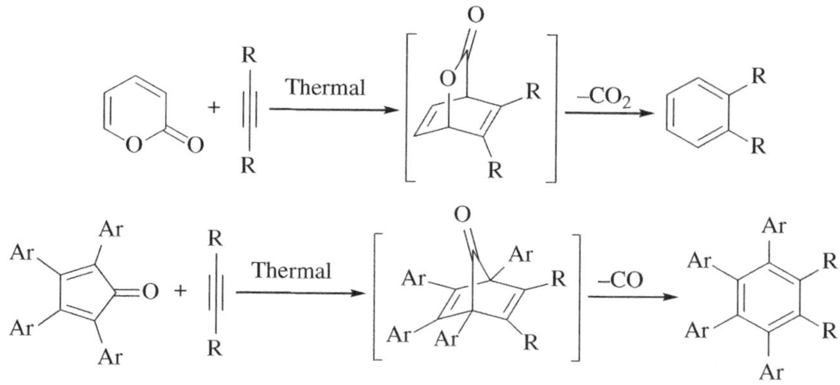

SCHEME 16.18 Domino [4+2] cycloaddition/chelotropic elimination to aromatics.

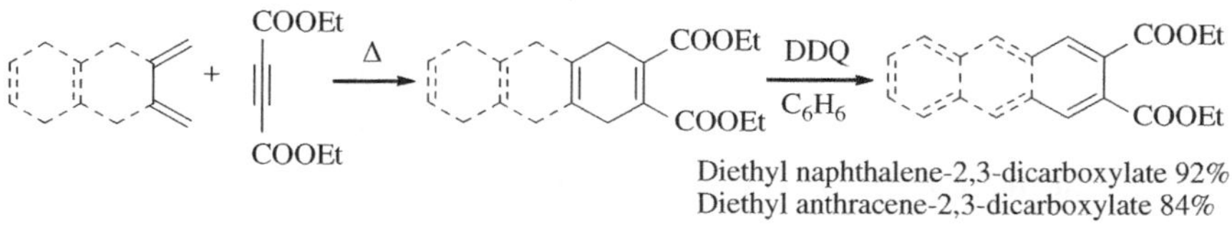

SCHEME 16.19 Cycloaddition/aromatization strategy for synthesis of 2,3-disubstituted naphthalene and anthracene derivatives.

Diels–Alder reaction followed by dehydrogenation of the cycloadduct using $KMnO_4$ supported on alumina under mild conditions has been reported (Scheme 16.20) [21]. Two other dihydroaromatic substrates (cycloadducts shown in the box) that have been oxidized by this method are also shown in Scheme 16.20.

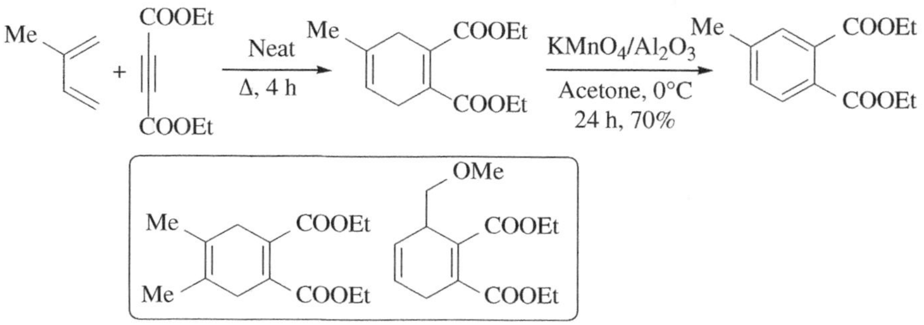

SCHEME 16.20 Cycloaddition/dehydrogenation using $KMnO_4$ on alumina.

Synthesis of substituted stilbenes has been reported through a tandem process involving Diels–Alder reaction/Wittig reaction/DDQ oxidation sequence (Scheme 16.21) [22]. The Diels–Alder step is catalyzed by cobalt(II) and Zn(II) salts.

Diels–Alder reaction of alkynyl sulfides followed by oxidation has been used for the synthesis of diaryl sulfides (Scheme 16.22) [23].

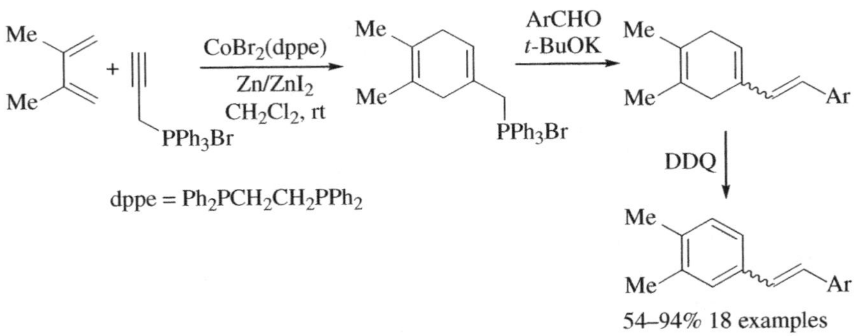

SCHEME 16.21 Stilbene synthesis via Diels–Alder/Wittig/oxidation sequence.

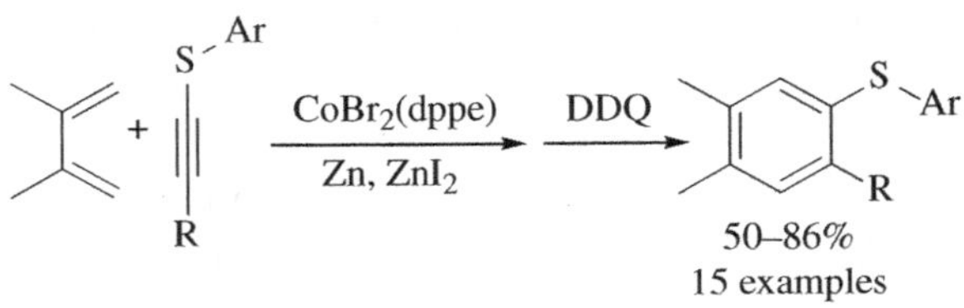

SCHEME 16.22 Diels–Alder reaction of alkynyl sulfide toward diaryl sulfides.

Dendralenes [24] (acyclic cross-conjugated polyenes) have been used as dienes in tandem Diels–Alder reactions, and a methodology for the synthesis of highly functionalized angularly anelated aromatic compounds has been developed (Scheme 16.23) [25]. A tandem double Diels–Alder reaction of DMAD with [3]dendralene followed by oxidation with DDQ gave the tetramethyl ester

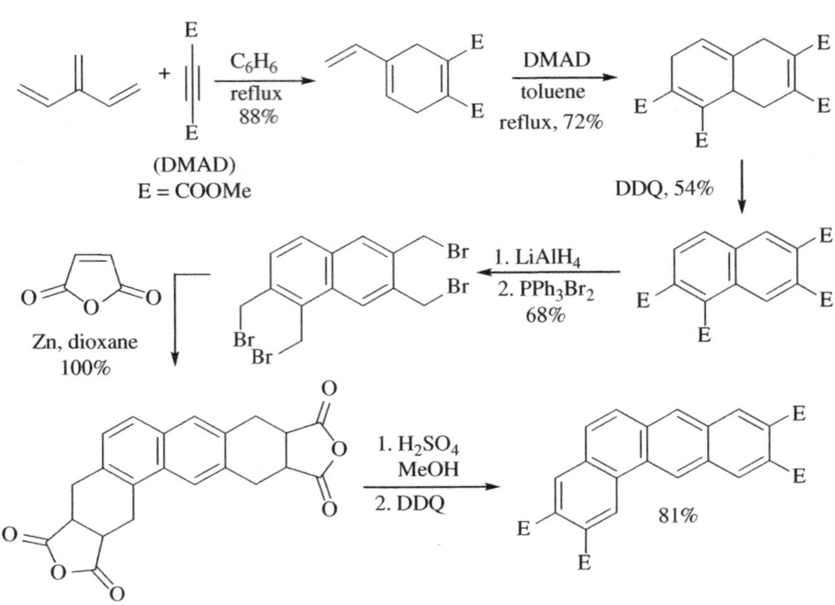

SCHEME 16.23 Diels–Alder reaction of [3]dendralene enroute to angular aromatic compounds.

of 2,3,5,6-naphthalenetetracarboxylic acid, which on reduction followed by treatment with PPh_3Br_2 gave the corresponding tetrabromide. The tetrabromide on treatment with Zn underwent double 1,4-elimination to form the corresponding bis(orthoquinodimethide), which was trapped *in situ* in the presence of maleic anhydride to yield the corresponding dianhydride. Esterification of the dianhydride followed by oxidation with DDQ afforded tetrasubstituted tetraphene in 81% yield.

Synthesis of a hexasubstituted phenanthrene derivative from [4]dendralene is illustrated in Scheme 16.24 [25]. The poor yields observed in the cycloaddition steps might be due to the thermal instability of [4]dendralene and the initially formed cycloadduct.

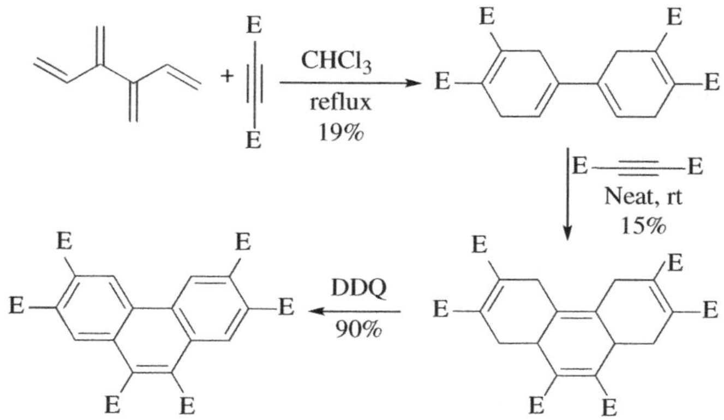

SCHEME 16.24 Synthesis of hexasubstituted phenanthrene derivative from [4]dendralene.

Isobenzofurans are reactive Diels–Alder dienes, and they have been used for the construction of aromatic skeletons extensively [26]. A mixture of *endo/exo* cycloadducts obtained from the Diels–Alder reaction of isobenzofuran with a variety of dienophiles underwent dehydration under acidic conditions to give aromatic compounds. Isobenzofuran was generated *in situ* in the reaction by the loss of ethanol from the starting material cyclic acetal (Scheme 16.25) [27]. Using the approach in Scheme 16.25, a variety of naphthalimides have been synthesized that find application in DNA detection and white light-emitting OLEDs [28].

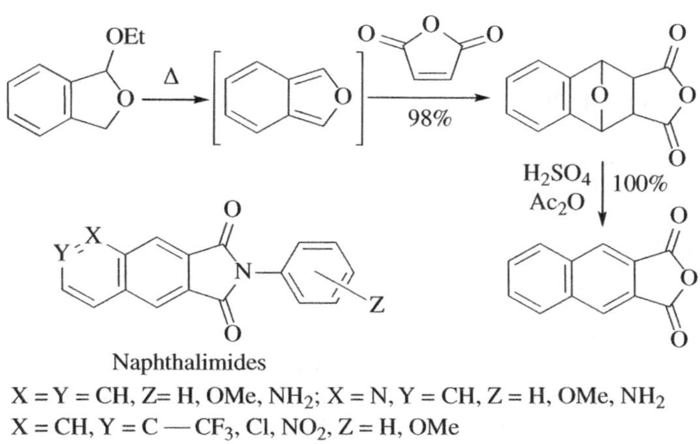

SCHEME 16.25 Synthesis of aromatic compounds from isobenzofuran cycloaddition.

Higher homologues of linear acenes have been synthesized using the isobenzofuran cycloaddition method as illustrated in the following schemes (Schemes 16.26 and 16.27) [29]. The cycloadduct of tetracyclone shown in Scheme 16.26 (in the box) is a synthetic equivalent of benzodifuran, and on heating, it releases 1,2,3,4-tetraphenylbenzene by a retro Diels–Alder reaction generating isobenzofuran *in situ*, probably in a sequential manner.

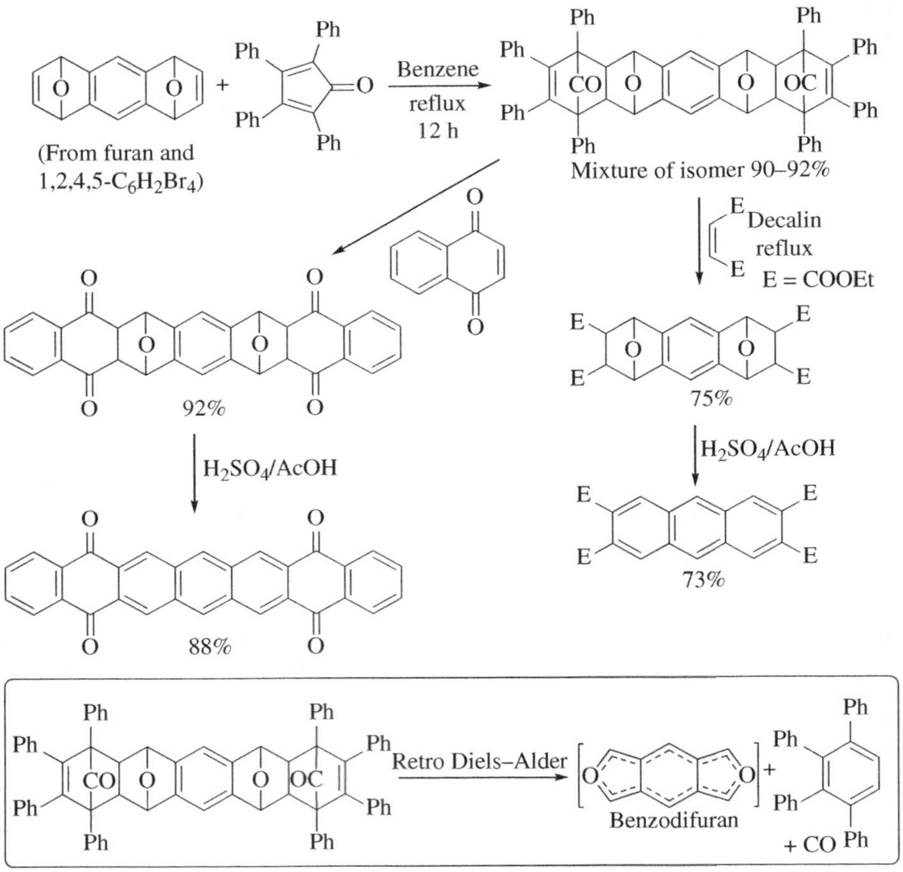

SCHEME 16.26 Synthesis of linear acenes from benzodifuran equivalent.

Synthesis of substituted rubrene has been reported through the cycloaddition of substituted isobenzofuran and naphthyne generated *in situ* (Scheme 16.27) [30]. Dehydration leading to aromatization of the Diels–Alder adduct was initiated by $AlBr_3$ as the Lewis acid in the presence of CsI.

This method has gained importance in recent time in view of the importance of linear acenes in molecular electronics applications. The following synthesis of substituted heptacene demonstrates the utility of this methodology (Scheme 16.28) [31].

Synthesis of phenylene-containing oligoacenes has been accomplished [32]. The starting material 3,4-bis-*exo*-methylenecyclobutene was obtained from the flash vacuum pyrolysis of hexa-1,5-diyne. Although the *exo*-methylene groups are oriented as *cisoid* diene in this molecule, it failed to undergo Diels–Alder reaction as a diene, presumably because the resulting Diels–Alder

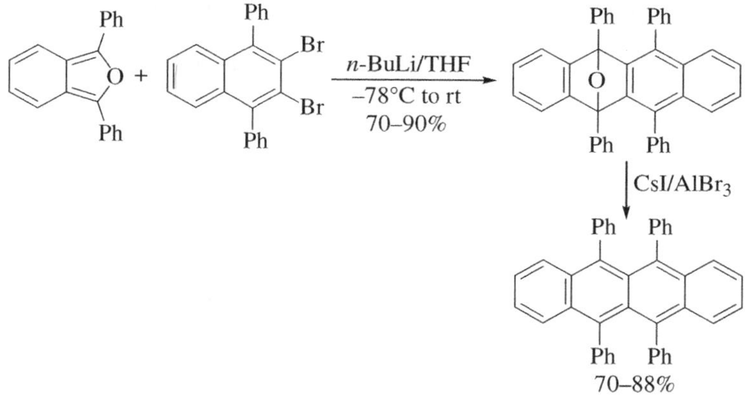

SCHEME 16.27 Synthesis of substituted rubrene through cycloaddition/aromatization method.

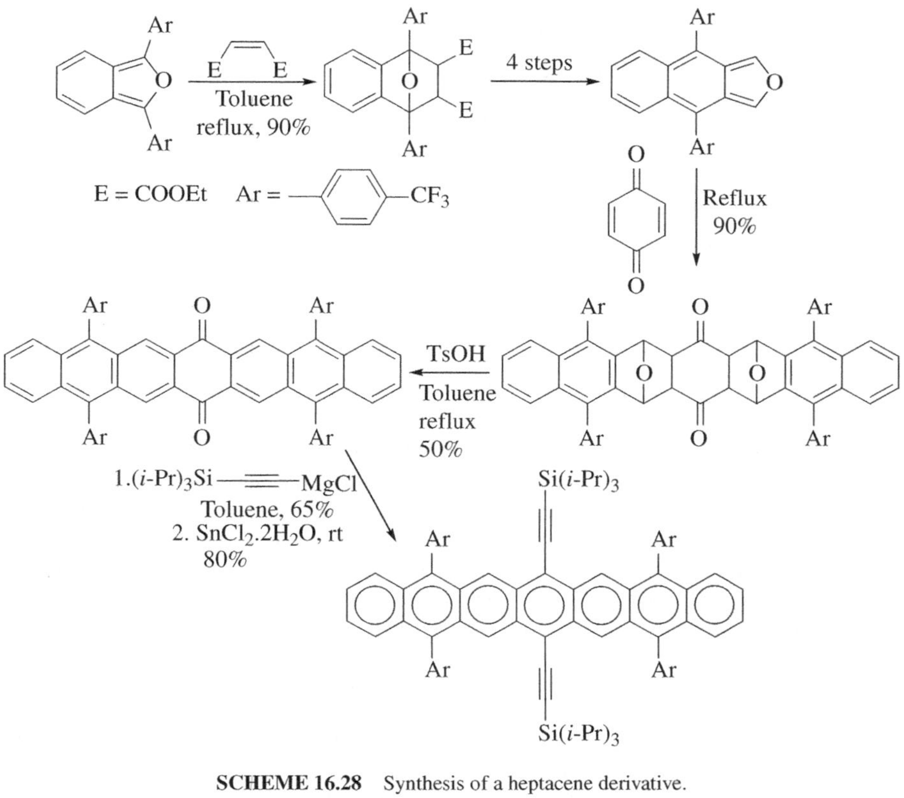

SCHEME 16.28 Synthesis of a heptacene derivative.

adduct would be a cyclobutadiene. Once the endo double bond has reacted as a dienophile with another diene, then the resulting cycloadduct was subjected to undergo further cycloaddition reaction through the bis-*exo*-methylene groups as a diene (Scheme 16.29) [32]. The last step in Scheme 16.28 involves addition of the acetylenic Grignard to the quinone moiety followed by reductive removal of the oxygens using $SnCl_2$.

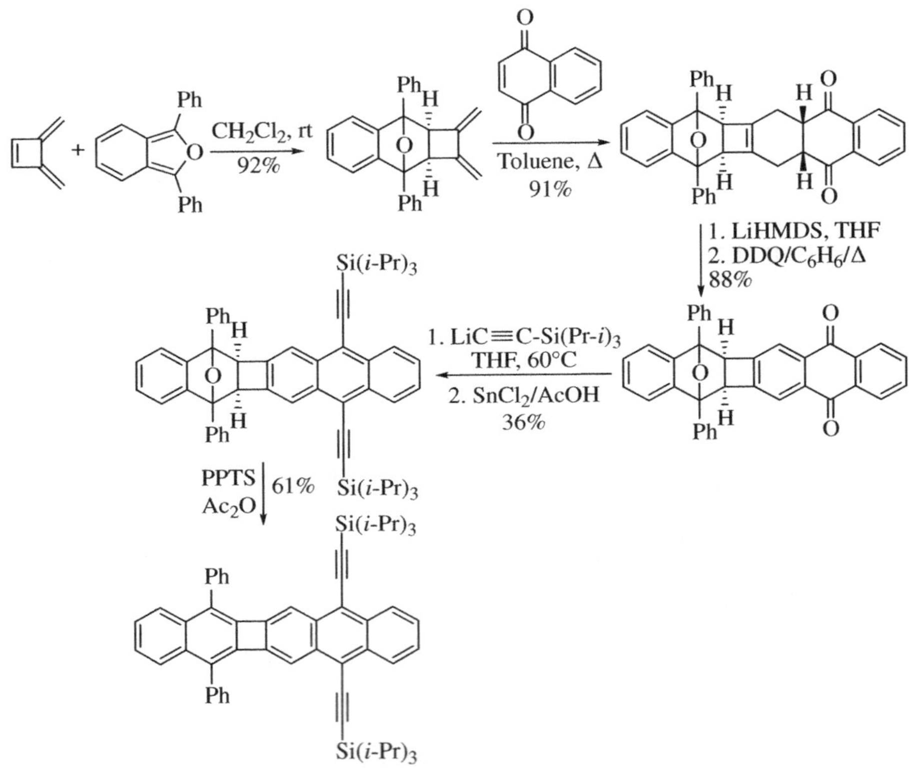

SCHEME 16.29 Synthesis of phenylene oligoacenes.

The tetrabromo derivative shown in Scheme 16.30 serves as a bis-aryne precursor, and the *in situ* generated aryne underwent double Diels–Alder reaction with the diene [32].

In Schemes 16.28 and 16.29, readily available quinones were used as the dienophiles, and the yield of the cycloadducts was excellent. The cycloadducts had to be further converted into aromatic compounds in subsequent steps. In Scheme 16.30, the bis-aryne generated from the tetrabromo derivative served as the dienophile. Although the yield of the cycloaddition step is only 36%, this strategy gave the aromatic compound directly in a single step.

2,5-Dimethylfuran and acrolein are products of degradation of biomass such as cellulose. They were used as starting materials in the synthesis of *p*-xylene, demonstrating the utilization of materials obtainable from biomass toward the synthesis of polyethylene terephthalate (Scheme 16.31) [33]. The key steps involve Lewis acid-catalyzed cycloaddition followed by oxidation and further aromatization through acid-mediated dehydration.

A strategy consisting of cycloaddition followed by elimination is common when alkoxy-substituted dienes are used. Upon heating the cycloadduct in pyridine, double elimination of MeOH and HCl leading to aromatization occurred (Scheme 16.32) [34].

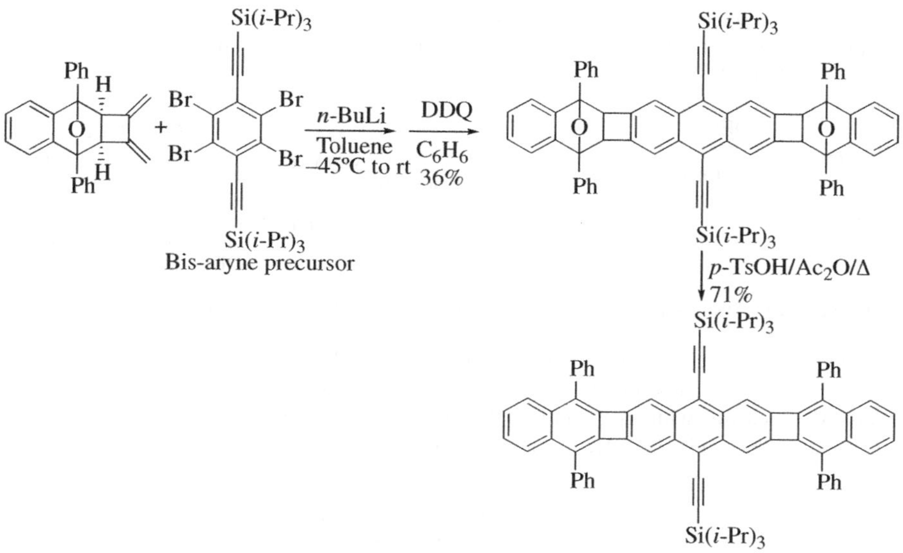

SCHEME 16.30 Synthesis of phenylene oligoacenes.

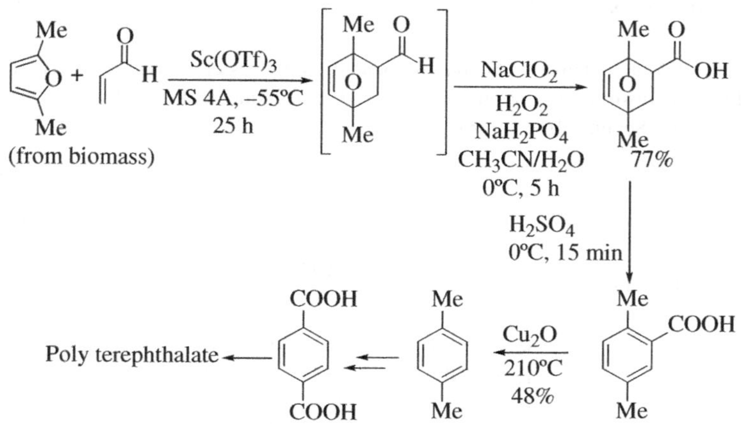

SCHEME 16.31 Raw material from biomass to aromatics toward polymers.

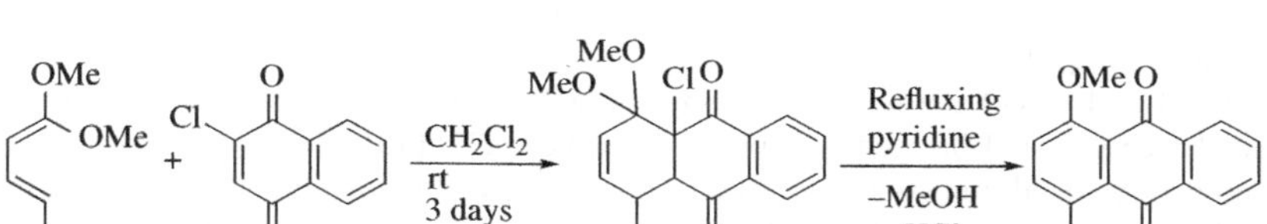

SCHEME 16.32 Diels–Alder/elimination sequence for the synthesis of anthraquinone derivative.

Highly regioselective double cycloaddition to 2,5-dichlorobenzoquinone and subsequent elimination has been used in the manufacturing of highly substituted anthraquinone derivative shown in Scheme 16.33 [35].

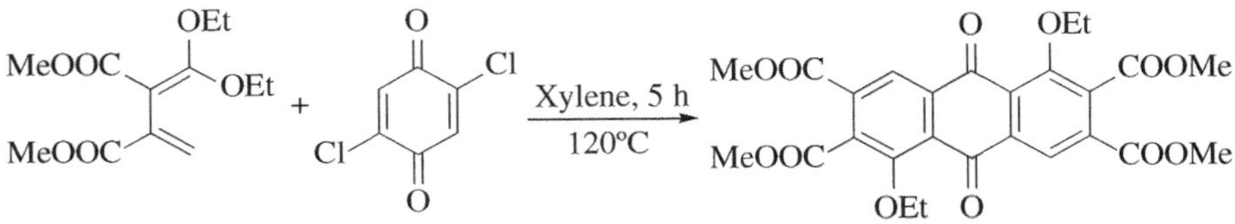

SCHEME 16.33 Double Diels–Alder/elimination reaction toward highly substituted anthraquinone synthesis.

2-Pyrone is a reactive diene for Diels–Alder reactions. The resulting cycloadducts are generally unstable and undergo loss of CO_2 to yield aromatic products [36]. 2-Pyrone is known to undergo Diels–Alder reaction with *bis*(trimethylstannyl)acetylene (Scheme 16.34) [37]. This reaction offers a method for the synthesis of 1,2-distannyl aromatic compounds.

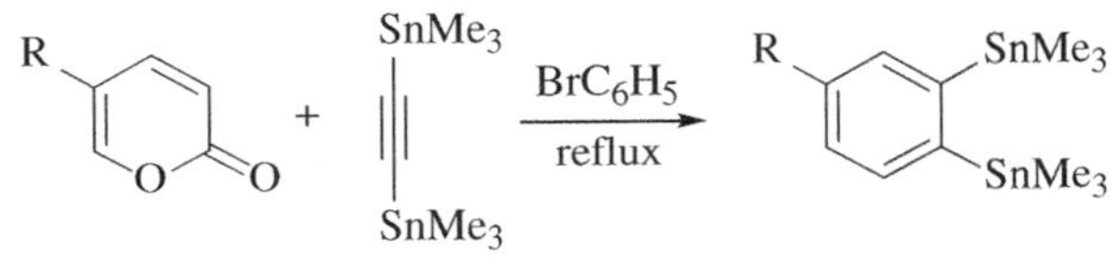

R = H (51%), Me (39%), COOMe (55%)

SCHEME 16.34 Diels–Alder reaction of 2-pyrone with bis(trimethylstannyl)acetylene.

Similarly, 1,2-phenylenebis(phosphonic acid dimethyl ester) derivatives have been obtained from differently substituted 2-pyrones and 1,2-bis(phosphonic acid dimethyl ester)acetylene as the reaction partners (Scheme 16.35) [38].

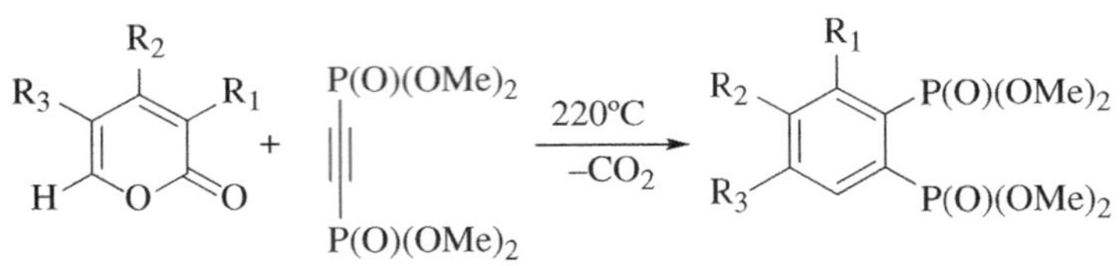

R_2 = COOMe, $R_1 = R_3$ = H 74%
R_1 = OMe, $R_2 = R_3$ = H 90%
R_2 = OMe, R_3 = Me, R_1 = H 62%
R_2 = OMe, R_1 = Me, R_3 = H 79%

SCHEME 16.35 Synthesis of 1,2-phenylenebis(phosphonic acid dimethyl esters).

A strategy for the synthesis of highly substituted anilines and *o*-phenylenediamine derivatives has been developed based on pyrone cycloaddition with acetylenic dienophiles (Scheme 16.36) [39].

A threefold Diels–Alder reaction of 2-pyrone with cyclic triacetylene shown in Scheme 16.37 provided access to silyl-bridged cyclophane (Scheme 16.37) [40].

Another cycloaddition strategy for the synthesis of symmetrically substituted [2.2]paracyclophane derivatives involves the Diels–Alder reaction between bis-allene and disubstituted acetylenes (Scheme 16.38) [41]. The intermediate paraquinodimethide dimerized to form the cyclophane product, presumably through a diradical pathway.

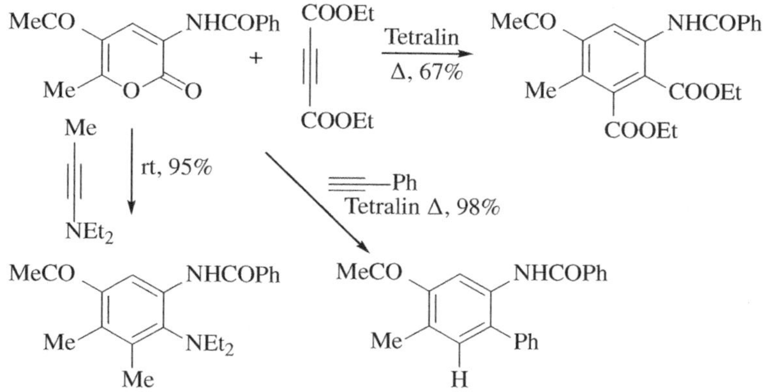

SCHEME 16.36 Synthesis of highly substituted anilines from 2-pyrone cycloaddition.

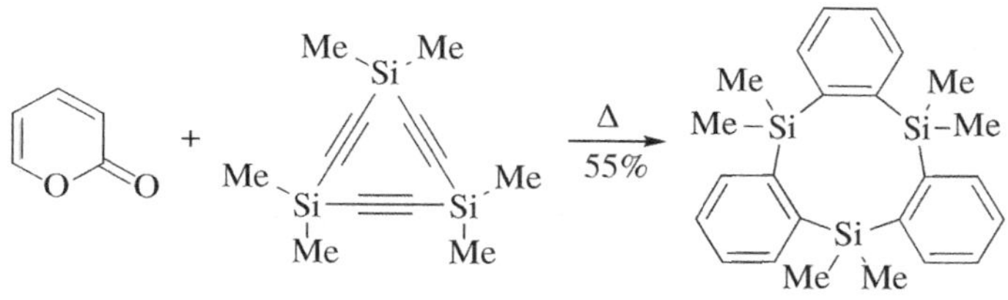

SCHEME 16.37 2-Pyrone cycloaddition to silicon-bridged cyclophane.

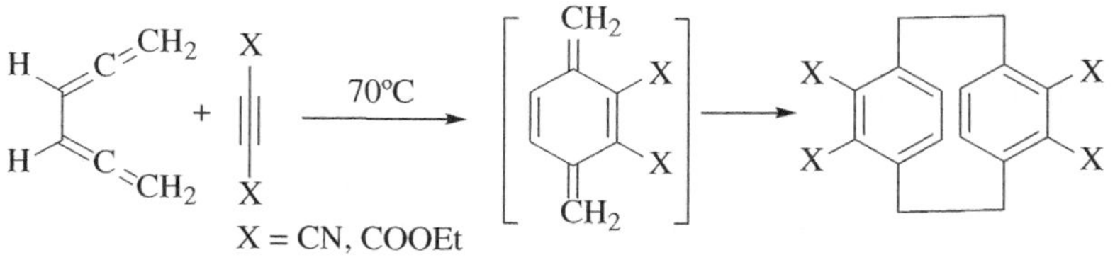

SCHEME 16.38 Synthesis of [2.2]paracyclophane derivatives through Diels–Alder reaction of bis-allene.

Diels–Alder reaction of tetraarylcyclopentadienone with diarylacetylene with concomitant elimination of carbon monoxide results in the formation of hexaaryl-substituted benzene [42]. The hexaarylbenzenes serve as precursors for the synthesis of hexabenzocoronene derivatives (Scheme 16.39) [43]. The Diels–Alder step generally gives excellent yields of the hexaarylbenzenes.

This strategy has been used for the synthesis of a wide variety of molecular shapes and sizes, (super acenes) molecules that are segments of graphene. The cycloaddition step is generally carried out in refluxing diphenyl ether (200–230°C) and the oxidative dehydrogenation using a variety of salts such as $FeCl_3$, $AlCl_3$, $Cu(OTf)_2$, etc. in nitromethane or dichloromethane as solvent [43]. Synthesis of polycyclic aromatic hydrocarbon (PAH) $C_{78}H_{26}$, characterized by LD-TOFMS, from simple starting materials such as tetracyclone and 1,4-bis(phenylethynyl)benzene is illustrated (Scheme 16.40) [44].

Using suitably substituted acetylenes as starting materials, super naphthalene and super triphenylene have been synthesized [45]. The unsubstituted super acenes are insoluble solids. In order to obtain soluble derivatives, *t*-butyl groups were added to the periphery of the super acenes. Synthesis of super triphenylene is illustrated in Scheme 16.41 [45].

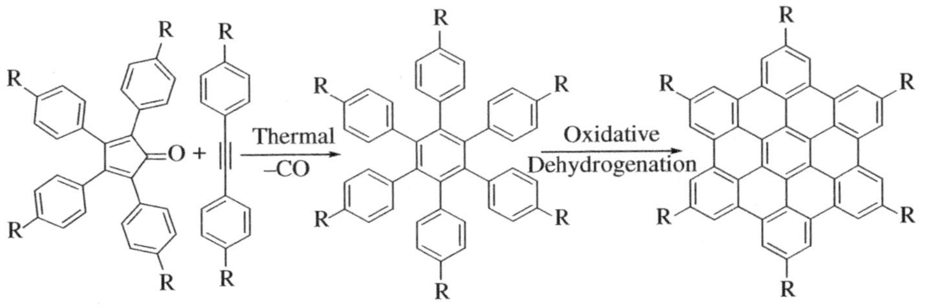

SCHEME 16.39 Synthesis of hexaarylbenzene and hexabenzocoronene via cycloaddition/dehydrogenation sequence.

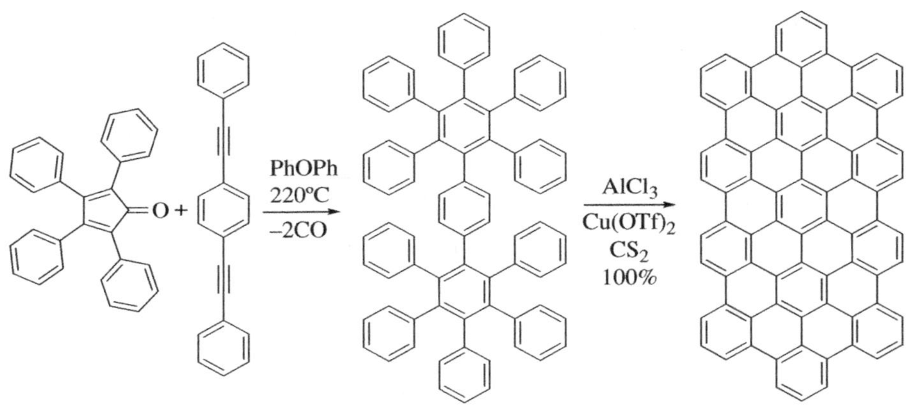

SCHEME 16.40 Synthesis of PAH $C_{78}H_{26}$ through cycloaddition/dehydrogenation method.

All of the PAH and super acenes are planar molecules. However, incorporation of a seven-membered ring results in nonplanar (curved) molecules that are of interest in the context of curved π surfaces and nonplanar aromatic compounds. Cycloaddition/dehydrogenation strategy shown in Scheme 16.42 results in the formation of seven-membered ring fused curved polycyclic aromatic molecule [46].

Synthesis of twisted hexabenzoperylene derivative is shown in Scheme 16.43 [47]. In this reaction, the dehydrogenation is carried out using DDQ and CH_3SO_3H, a method that has been reported earlier [48]. This reaction did not proceed all the way to form the planar PAH; instead, it is controlled by strongly electron donating alkoxy group that directed the cyclization only *ortho* to the alkoxy groups leading to a twisted structure that failed to undergo further oxidative cyclization to planar PAH. The twisted hexabenzoperylene has been isomerized at 110°C to the *anti* isomer with which it exists in an equilibrium with equilibrium constant of 0.35 in toluene [47].

Iron(III)-catalyzed benzannulation of 2-phenacylbenzaldehyde has been reported (Scheme 16.44) [49]. This reaction provides high substrate scope and high yields of the benzannulated naphthalene derivatives. The mechanism of the reaction is shown in Scheme 16.45. It involves enolization of 2-phenacylbenzaldehyde to give an *ortho* quinodimethide derivative that undergoes Diels–Alder reaction and elimination of a carboxylic acid from the adduct to yield the naphthalene derivative.

A sequence involving double Diels–Alder reaction followed by double retro Diels–Alder reaction that is reiterated again to obtain tetrabenzocyclophane has been reported (Scheme 16.46) [50]. The diene is a heterodiene, and the cycloadduct obtained from it is capable of losing PhNCO in the retro Diels–Alder step. The dienophile is a strained reactive cyclic diacetylene (cyclophyne).

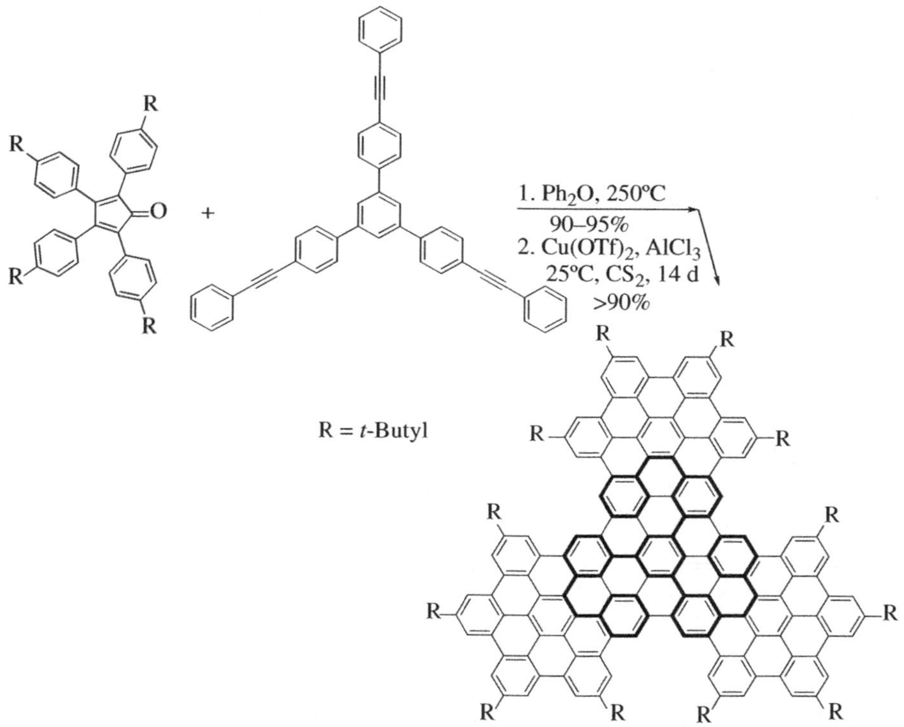

SCHEME 16.41 Synthesis of super triphenylene through cycloaddition/dehydrogenation. The starting triacetylene carbon skeleton is highlighted with thick bonds.

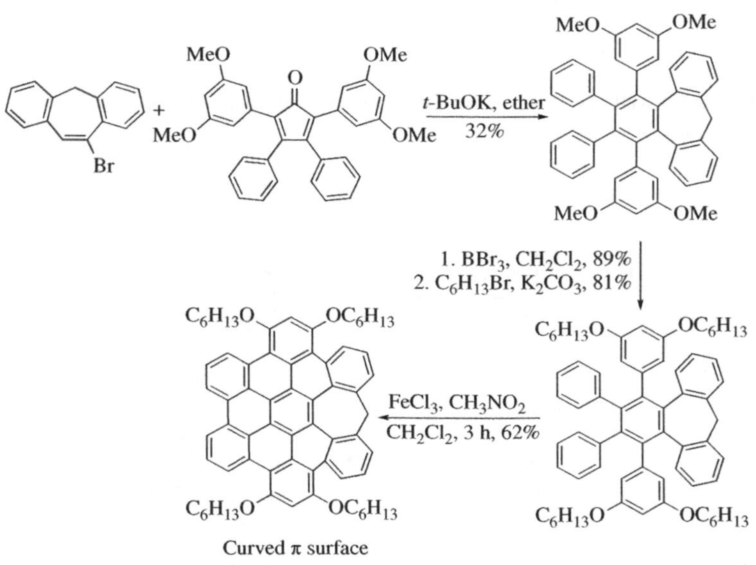

SCHEME 16.42 Synthesis of nonplanar (curved π surface) PAH.

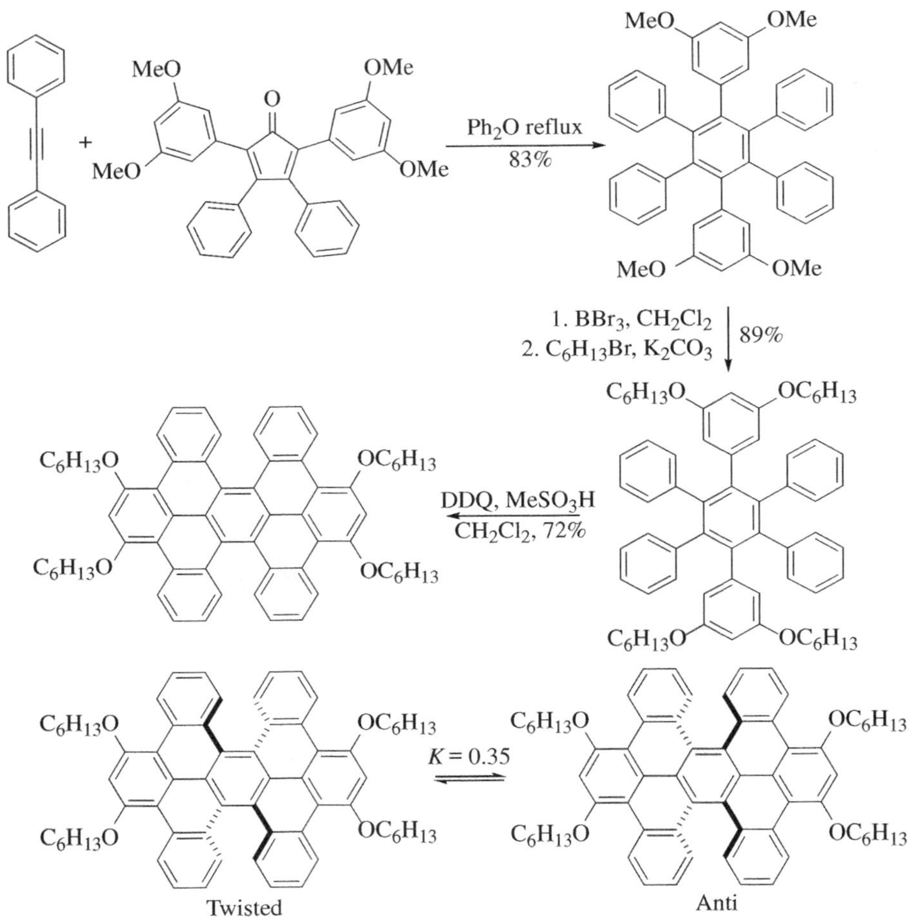

SCHEME 16.43 Synthesis of twisted hexabenzoperylene and equilibrium between the twisted and the *anti* forms at 110°C in toluene.

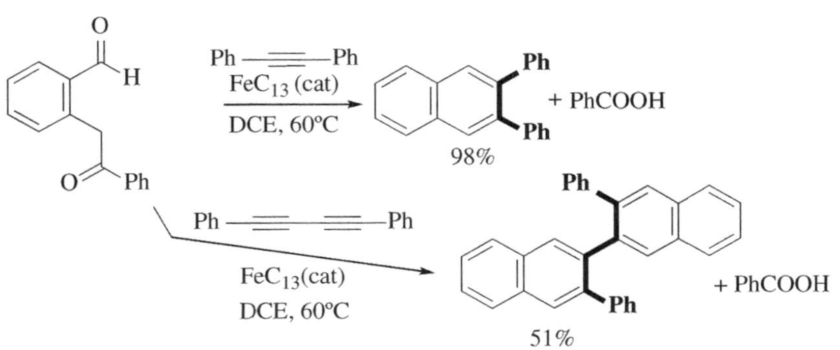

SCHEME 16.44 Ferric ion-catalyzed benzannulation to naphthalene derivatives. The starting acetylene carbon skeletons are highlighted in the products.

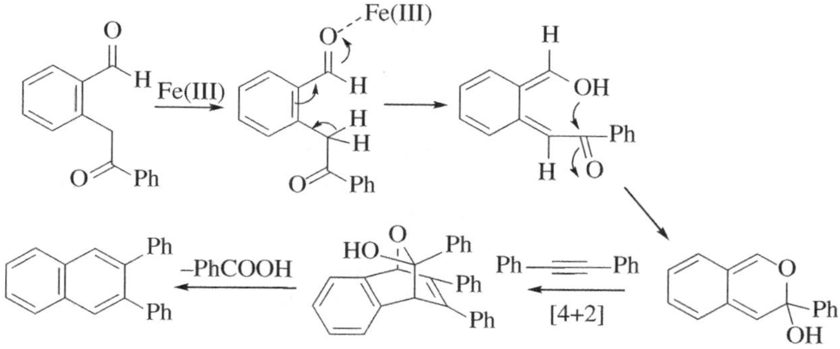

SCHEME 16.45 Mechanism of Fe(III)-catalyzed benzannulation.

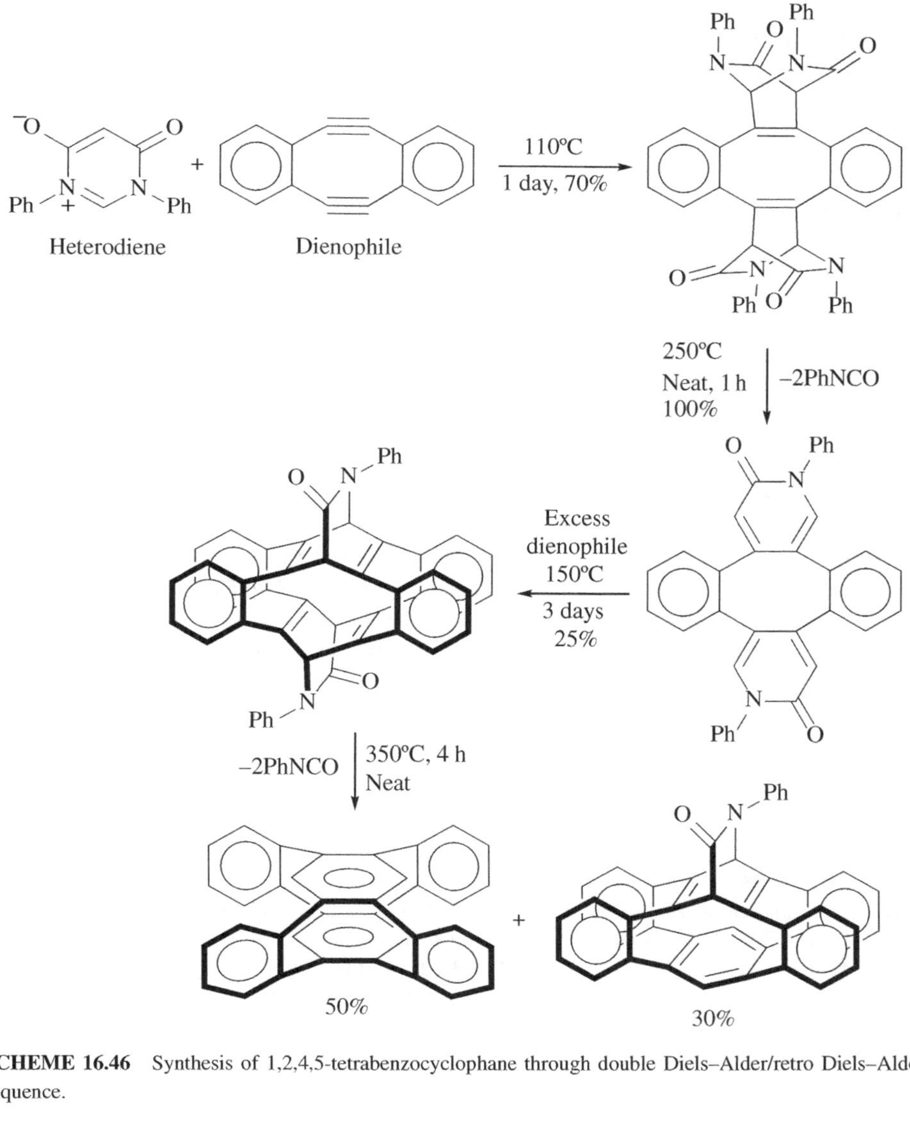

SCHEME 16.46 Synthesis of 1,2,4,5-tetrabenzocyclophane through double Diels–Alder/retro Diels–Alder sequence.

16.4 CONCLUSIONS

From the forgoing examples, it is evident that the two categories of pericyclic reactions, namely, 6π-electrocyclic ring closure and [4+2] cycloaddition reactions, have played significant roles in the synthesis of novel aromatic compounds. Both these classes of reactions yield cyclohexene/cyclohexadiene derivatives as products of initial pericyclic step that are oxidized to the fully aromatic compounds by a variety of oxidizing agents. Many examples are chosen deliberately so as to illustrate the complexity of molecular architecture of aromatic planar and nonplanar compounds. Some examples are also chosen to highlight the use of aromatic compounds in the area of organic materials. Although not exhaustive in literature review, it is hoped that this chapter has brought out salient and noteworthy contributions in this area.

ABBREVIATIONS

DDQ 2,3-Dichloro-5,6-dicyanobenzoquinone
PAH Polycyclic aromatic hydrocarbon

REFERENCES

[1] (a) Woodward, R. B., Hoffmann, R. (1970), *The Conservation of Orbital Symmetry*, Academic Press, New York; (b) Woodward, R. B., Hoffmann, R. (1969), *Angew. Chem. Int. Ed. Engl.*, **8**, 781–853.

[2] Sankararaman, S. (2005), *Pericyclic Reactions—A Textbook*, Wiley-VCH, Weinheim.

[3] (a) Fleming, I. (1999), *Pericyclic Reactions*, Oxford University Press, Oxford, pp 31–56; (b) Woodward, R. B., Hoffman, R. (1965), *J. Am. Chem. Soc.*, **87**, 395–397.

[4] Turro, N. J., Ramamurthy, V., Scaiano, J. C. (2010), *Modern Molecular Photochemistry of Organic Molecules*, University Science Books, Sausalito, pp 740–742.

[5] Hussain, M., Sung, T. V., Langer, P. (2012), *Synlett*, **23**, 2735–2744.

[6] (a) Lansky, A., Reiser, O., de Meijere, A. (1990), *Synlett*, 405–407; (b) Voigt, K., von Zezschwitz, P., Rosauer, K., Lansky, A., Adams, A., Reiser, O., de Meijere, A. (1998), *Eur. J. Org. Chem.*, 1521–1534.

[7] Reiser, O., Reichow, S., de Meijere, A. (1987), *Angew. Chem. Int. Ed. Engl.*, **26**, 1277–1278.

[8] Ullah, I., Nawaz, M., Villinger, A., Langer, P. (2011), *Tetrahedron Lett.*, **52**, 1888–1890.

[9] Hussain, M., Zinad, D. S., Salman, G. A., Sharif, M., Villinger, A., Langer, P. (2010), Synlett, 276–278.

[10] Toguem, S.-M. T., Knepper, I., Ehlers, P., Dang, T. T., Patonay, T., Langer, P. (2012), *Adv. Synth. Catal.*, **354**, 1819–1826.

[11] Toguem, S.-M. T., Langer, P. (2011) *Synlett*, 513–516.

[12] Malik, I., Hussain, M., Ali, A., Toguem, S.-M. T., Basha, F. Z., Fischer, C., Langer, P. (2010), *Tetrahedron*, **66**, 1637–1642.

[13] Benson, C. L., West, F. G. (2007), *Org. Lett.*, **9**, 2545–2548.

[14] (a) Martin, R. H. (1974), *Angew. Chem. Int. Ed. Engl.*, **13**, 649–660; (b) Hoffmann, N. (2014), *J. Photochem. Photobiol. C. Photochem. Rev.* **19**, 1–19; (c) Lefebvre, Q., Jentsch, M. Reuping, M. (2013), *Beilstein J. Org. Chem.*, **9**, 1883–1890.

[15] (a) Laarhoven, W. H., Prinsen, W. J. C. (1984), *Top. Curr. Chem.*, **125**, 63–130; (b) Laarhoven, W. H., Cuppen, Th. J. H. M., Nivard, R. J. F. (1970), *Tetrahedron*, **26**, 4865–4881.

[16] Martin, R. H., Baes, M. (1975), *Tetrahedron*, **31**, 2135–2137.

[17] Moradpour, A., Kagan, H., Baes, M., Morren, G., Martin, R. H. (1975), *Tetrahedron*, **31**, 2139–2143.

[18] Cheng, M.-Y., Chan, C.-K., Wu, M.-H. (2013), *Tetrahedron*, **69**, 7916–7924.

[19] Smith, M. B. (1994), *Organic Synthesis*, McGraw Hill, New York, pp 1113–1147.

[20] Angus, Jr, R. O., Johnson, R. P. (1983), *J. Org. Chem.*, **48**, 273–276.

[21] McBride, C. M., Chrisman, W., Harris, C. E., Singaram, B. (1999), *Tetrahedron Lett.*, **40**, 45–48.

[22] Hilt, G., Hengst, C. (2007), *J. Org. Chem.*, **72**, 7337–7342.

[23] Hilt, G., Lüers, S., Harms, K. (2004), *J. Org. Chem.*, **69**, 624–630.

[24] Hopf, H. (1984), *Angew. Chem. Int. Ed. Engl.*, **23**, 948–960.

[25] Hopf, H., Yildizhan, S. (2011), *Eur. J. Org. Chem.*, 2029–2034.

[26] (a) Wittig, G., Krebs, A. (1961), *Chem. Ber.*, **94**, 3260; (b) Wong, H. N. C. (1989), *Acc. Chem. Res.*, **22**, 145–152.

[27] Fier, S., Sullivan, R. W., Rickborn, B. (1988), *J. Org. Chem.*, **53**, 2353–2354.

[28] Nandhikonda, P., Heagy, M. D. (2010), *Org. Lett.*, **12**, 4796–4799.

[29] Luo, J., Hart, H. (1988), *J. Org. Chem.*, **53**, 1341–1343.

[30] Dodge, J. A., Bain, J. D., Chamberlin, A. R. (1990), *J. Org. Chem.*, **55**, 4190–4198.

[31] Qu, H., Chi, C. (2010), *Org. Lett.*, **12**, 3360–3363.

[32] Parkhurst, R. R., Swager, T. M. (2012), *J. Am. Chem. Soc.*, **134**, 15351–15356.

[33] Shiramizu, M., Toste, F. D. (2011), *Chem. Eur. J.*, **17**, 12452–12457.

[34] (a) Cameron, D. W., Feutrill, G. I., McKay, P. G. (1982), *Aust. J. Chem.*, **35**, 2095–2109; (b) Cameron, D. W., Feutrill, G. I., McKay, P. G. (1981), *Tetrahedron Lett.*, **22**, 701–702.

[35] Okada, H. (2011), *Jpn. Kokai Tokkyo Koho*, Patent No.2011136954.

[36] Reed, J. A., Schilling, Jr, C. L., Travin, R. F., Retting, T. A., Stille, J. K. (1969), *J. Org. Chem.*, **34**, 2188–2192.

[37] Evnin, A. B., Seyferth, D. (1967), *J. Am. Chem. Soc.*, **89**, 952–959.

[38] Ziegler, T., Layh, M., Effenberger, F. (1987), *Chem. Ber.*, **120**, 1347–1355.

[39] Kranjc, K., Stefane, B., Polanc, S., Kocevar, M. (2004), *J. Org. Chem.*, **69**, 3190–3193.

[40] Sakurai, H., Eriyama, Y., Hosomi, A., Nakadaira, Y., Kabuto, C. (1982), *Chem. Lett.*, 1971–1974.

[41] (a) Hopf, H., Lenich, Th. (1974), *Chem. Ber.*, **107**, 1891–1902; (b) Böhm, I., Herrmann, H., Menke, K., Hopf, H. (1978), *Chem. Ber.*, **111**, 523–537; (c) Hopf, H., Böhm, I., Kleinschroth, J. (1981), *Org. Syn.*, **60**, 41–48.

[42] (a) Dilthey, W., Hurtig, G. (1934), *Ber. Dtsch. Chem. Ges.*, **67**, 495–496; (b) Dilthey, W., Schommer, W., Hoschen, W., Dierichs, H. (1935), *Ber. Dtsch. Chem. Ges.*, **68**, 1159–1162.

[43] (a) Grzybowski, M., Skonieczny, K., Butenschön, H., Gryko, D. T. (2013), *Angew. Chem. Int. Ed.*, **52**, 9900–9930; (b) Ito, S., Wehmeier, M., Brand, D. J., Kübel, C., Epsch, R., Rabe, J. P., Müllen, K. *Chem. Eur. J.*, **6**, 4327–4342.

[44] Müller, M., Iyer, S. V., Kübel, C., Enkelmann, V., Müllen, K. (1997), *Angew. Chem. Int. Ed.*, **36**, 1607–1610.

[45] Iyer, S. V., Wehmeier, M., Brand, D. J., Keegstra, M. A., Müllen, K. (1997), *Angew. Chem. Int. Ed.*, **36**, 1602–1607.

[46] Yamamoto, K., Sonobe, H., Matsubara, H., Sato, M., Okamoto, S., Kitaura, K. (1996), *Angew. Chem. Int. Ed.*, **35**, 69–70.

[47] Luo, J., Xu, X., Mao, R., Miao, Q. (2012), *J. Am. Chem. Soc.*, **134**, 13796–13803.

[48] Zhai, L., Shukla, R., Rathore, R. (2009), *Org. Lett.*, **11**, 3474–3477.

[49] Zhu, S., Xiao, Y., Guo, Z., Jiang, H. (2013), *Org. Lett.*, **15**, 898–901.

[50] Brettreich, M., Bendikov, M., Chaffins, S., Perepichka, D. F., Dautel, O., Duong, H. Helgeson, R., Wudl, F. (2002), *Angew. Chem. Int. Ed.*, **114**, 3840–3843.

17

RING-CLOSING METATHESIS: SYNTHETIC ROUTES TO CARBOCYCLIC AROMATIC COMPOUNDS USING RING-CLOSING ALKENE AND ENYNE METATHESIS

Charles B. de Koning[1] and Willem A. L. van Otterlo[2]

[1] *School of Chemistry, University of the Witwatersrand, Johannesburg, South Africa*
[2] *Department of Chemistry and Polymer Science, Stellenbosch University, Matieland, South Africa*

17.1 INTRODUCTION

Arguably, one of the most important synthetic methods to have been discovered and developed over the last four to five decades is the ene–ene metathesis reaction, an efficient process that results in the shuffling of unsaturated carbons and their substituents, as shown in Figure 17.1a.

Along with its less utilized relative, the enyne metathesis (Fig. 17.1b), it would not be unjust to state that these reactions have made a significant impact on the field of carbon–carbon bond formation, be it in ring-opening metathesis (ROM), ring-opening metathesis polymerization (ROMP), cross metathesis (CM), or, as in this chapter, the forming of new aromatic rings, where the reaction could be the ring-closing metathesis (RCM) reaction (Fig. 17.2).

The history of the metathesis reaction is rich and varied, with scientific observations in the 1950s (Ziegler–Natta catalysis) [1] and advances in industrial processes in the 1960s (such as the Du Pont polynorbornene synthesis [2], Phillips olefin disproportionations [3], and Goodyear transalkylidenations [4]) playing their role. Coupled to this work were significant advances in the understanding of the structure of metallocarbenes by the research groups of Fischer, Grubbs, Chauvin, Pettit, Katz, Casey, Tebbe, and Schrock [5]. The culmination of this plethora of research resulted in the combination of catalysis and metal (pre)catalysts, eventually leading to the deserved

Arene Chemistry: Reaction Mechanisms and Methods for Aromatic Compounds,
First Edition. Edited by Jacques Mortier.

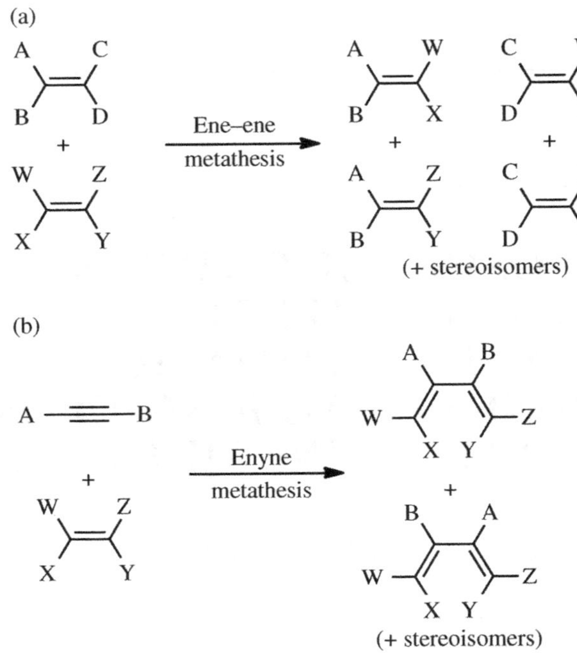

FIGURE 17.1 (a and b) Generic representation of ene–ene and enyne metathesis reactions.

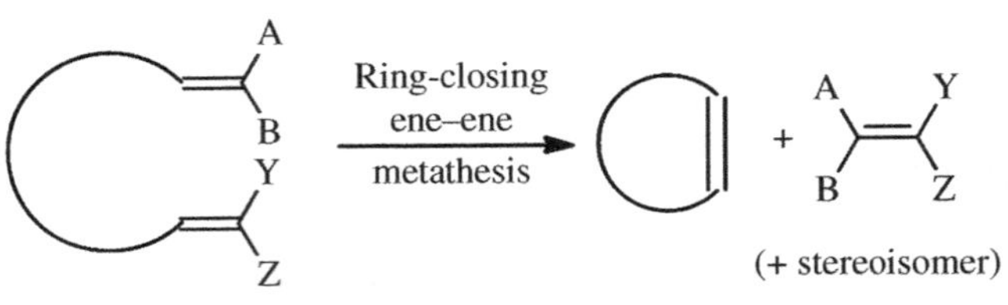

FIGURE 17.2 Generic representation of ring-closing ene–ene metathesis reactions.

awarding of the Nobel Prize in 2005 to Schrock, Grubbs, and Chauvin [6]. It became clear that the versatility of these metathetic processes had broad impact, with applications in synthetic to medicinal chemistry, to polymer science and nanotechnology. Hence, it should come as no surprise that this area has been reviewed extensively, with many recent reviews focusing on specific areas of impact [7].

In terms of the mechanism of the reaction, the commonly accepted view is that the metathesis reactions occur as postulated by Chauvin in 1971 (Scheme 17.1) [8]. Essentially, the metathesis cascade is promulgated by way of a series of [2+2] cycloadditions and cycloreversions, which center on a number of metallacyclobutane intermediates (Scheme 17.1). It is believed that these [2+2] processes are thermally allowed at the reaction temperatures due to overlap of the metal d-orbitals with the π systems involved. In addition, the driving force for the reaction to go to completion is often the liberation of gaseous ethylene that will escape from the reaction medium.

The regioselective synthesis of complex aromatic compounds has been an important and active research area for chemists over decades. Various strategies and documented synthetic methods have found favor for the synthesis of aromatic compounds. These would include the use of Diels–Alder reactions, Friedel–Crafts methodology, and a range of annulation methods such as those developed by Hauser, Kraus, Staunton, and Weinreb, as well as a range of transition

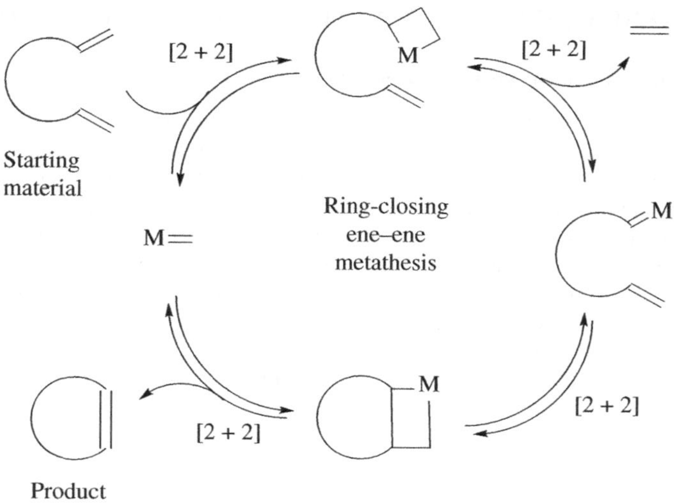

SCHEME 17.1 The ring-closing ene–ene metathesis reaction catalytic cycle.

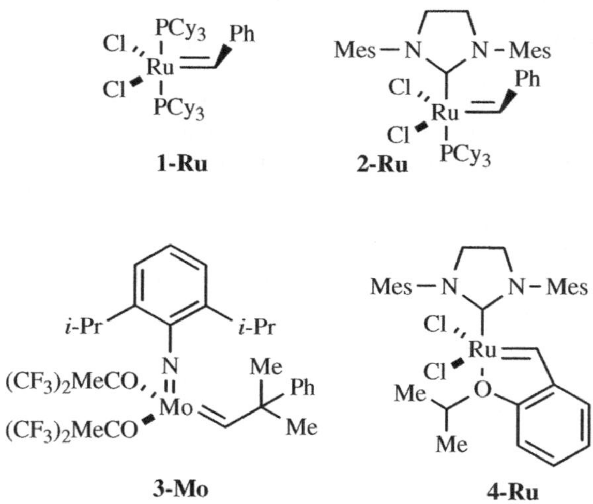

FIGURE 17.3 Commonly utilized metathesis catalysts (Reprinted with permission from Ref. [10]. Copyright (2009) American Chemical Society).

metal-mediated processes [9]. Therefore, it should come as no surprise that use of the transition metal-mediated metathesis reaction has also been utilized in the assembly of aromatic rings. In this review, we will strive to demonstrate the impact that RCM synthetic strategies have had on the de novo assembly of carbocyclic *aromatic* ring systems.

A small group of catalysts have become popular reagents of choice for effecting the RCM reaction. The most widely used catalysts are shown in Figure 17.3 and include the Grubbs first- and second-generation catalysts **1-Ru** and **2-Ru**, respectively [11], the Schrock catalyst **3-Mo** [12], and the second-generation Hoveyda–Grubbs catalyst **4-Ru** [13]. When required, other catalysts utilized for the synthesis of carbocyclic aromatic substrates will be highlighted in the appropriate sections where the work is described.

17.2 ALKENE RCM FOR THE SYNTHESIS OF AROMATIC COMPOUNDS

17.2.1 Synthesis of Substituted Benzenes

Among a plethora of reactions, the Friedel–Crafts reaction has historically been an important traditional method for preparing substituted benzenes. However, harsh reaction conditions and problems associated with regioselectivity often limit the use of this reaction in preparing substituted benzenes. In addition, the Friedel–Crafts methodology relies on the fact that the benzene ring-containing compounds are commercially available and the subsequent Friedel–Crafts reaction is performed on the preformed aromatic ring. As a result of the discovery of the RCM reaction, new methodology has become available to prepare substituted benzenes, where the metathesis reaction is used to assemble the benzene ring. For example, from a retrosynthetic point of view (shown in Fig. 17.4), phenol **1** can be considered to be a tautomer of the conjugated ketone **2**, and if this was to be prepared using RCM, triene **3** could be considered to be a reasonable precursor.

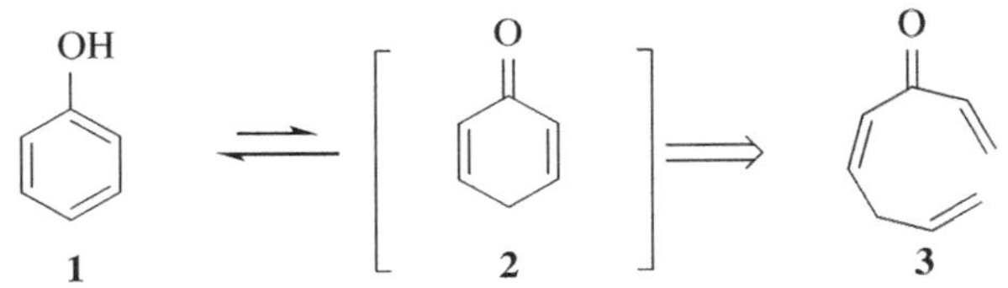

FIGURE 17.4 Synthesis of phenol from acyclic (Z)-octa-1,4,7-trien-3-one.

This type of retrosynthetic analysis has been utilized by Yoshida and Imamoto to prepare an impressive range of substituted phenols [14]. For example, as shown in Scheme 17.2, triene **4** afforded phenol **5** in 92% yield on exposure to **2-Ru**. In another example, exposure of triene **6**, with a different arrangement of the three double bonds, to the same catalyst resulted in the formation of phenol **7** via the presumed ketone intermediate **8**. It is also evident from these two examples that a biaryl system is formed during the RCM–aromatization process. This approach has thus added a new synthetic tool to the chemist's repertoire, in addition to traditional methodologies such as the Suzuki–Miyaura reaction.

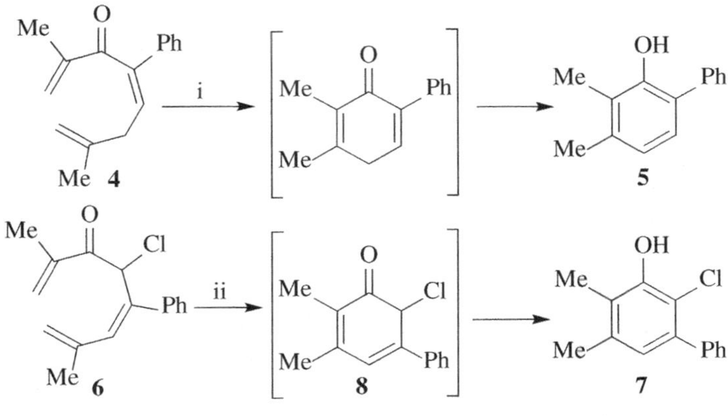

Reagents and conditions: (i) **2-Ru** (7.5 mol%), CH_2Cl_2, 40°C, 2 h (92%); (ii) **2-Ru** (7.5 mol%), toluene, 80°C, 12 h (84%).

SCHEME 17.2 Synthesis of substituted phenols from substituted (Z)-octa-1,4,7-trien-3-one and (Z)-octa-1,5,7-trien-3-one. (Adapted with permission from Ref. [10]. Copyright (2009) American Chemical Society).

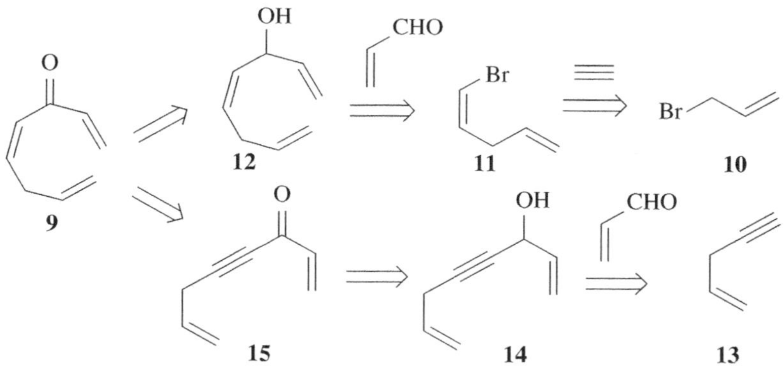

FIGURE 17.5 Retrosynthetic evaluation of (*Z*)-octa-1,4,7-trien-3-one.

Of course, the utility of this reaction would depend on the ease of synthesis of the trienes. Fortunately, Yoshida and Imamoto discovered that a range of the trienes with substitution on all the carbon atoms were easy to prepare [14]. The retrosynthetic rationale for two routes leading to the synthesis of 1,4,7-trien-3-one **9** is shown in Figure 17.5. The top route illustrates how allyl bromide **10** being subjected to a palladium-catalyzed reaction with acetylene furnishes **11**. This diene can be coupled to an unsaturated aldehyde to give alcohol **12**, a compound that can be oxidized to 1,4,7-trien-3-one **9**. Alternatively, enyne **13** can be coupled to an unsaturated aldehyde to give **15**, after the alcohol intermediate **14** has been oxidized. Treatment of **15** with Lindlar catalyst would then give the same 1,4,7-trien-3-one **9**. It should be noted that Yoshida and Imamoto synthesized a range of 1,4,7-trien-3-ones using both these synthetic routes [14].

The same group has also taken intermediate alcohols such as **12**, formed as shown in Figure 17.5, to synthesize numerous benzenes by means of a metathesis–dehydration procedure. In the one example shown in Scheme 17.3, the cyclic alcohol intermediate **17** was formed from triene **16** by reaction with catalyst **2-Ru**. The functionalized benzene **18** was then formed by the elimination of water upon the addition of *p*TsOH [15]. The authors were also able to synthesize an aniline derivative **20** from triene **19** using this approach [15b]. The group also developed a method by which compound **21** (and various others that contained a ketone and an alcohol functional group) underwent an RCM reaction to furnish the metathesis product **22**. Compounds of this type then underwent dehydration, oxidation, and tautomerization using *p*TsOH (as before). Alternatively, as shown in the following example, application of *t*-BuOK and MsCl afforded the substituted phenol **23** in an excellent yield as shown in Scheme 17.3 [16]. For other related strategies, a paper from the group of Yoshida and Imamoto can be consulted [17].

Other innovative approaches have been used to promote aromatization, once the RCM reaction involving the two alkenes has been accomplished. These two approaches are shown in Scheme 17.4. In the first one, triene **24** underwent an RCM reaction to afford the exocyclic double bond-containing compound **25** [18]. Use of the rhodium catalyst $[RhCl(cod)]_2$ then facilitated formation of the desired phenol **26**. Another neat way to accomplish formation of the aromatic benzene ring was to allow the intermediate cyclic Michael acceptor **28** (formed by RCM of **27**) to undergo a Heck–Mizoroki reaction with *p*-methoxybenzenediazonium tetrafluoroborate **30** to afford the second phenol **29** [18].

Clive and coworkers have recently disclosed a method for the general synthesis of *para*-disubstituted benzenes, represented by **34**, which involved conversion of 1,4-diketones **31** into the corresponding divinyl derivatives **32** after reaction with vinyllithium [19]. Subsequent RCM of these dienes afforded compounds **33**, which was then followed by aromatization to afford the substituted benzenes **34** (Scheme 17.5). Substituted aromatics **35** and **36**, two of the nine examples described in the paper, were synthesized in good yields from their respective diketone precursors.

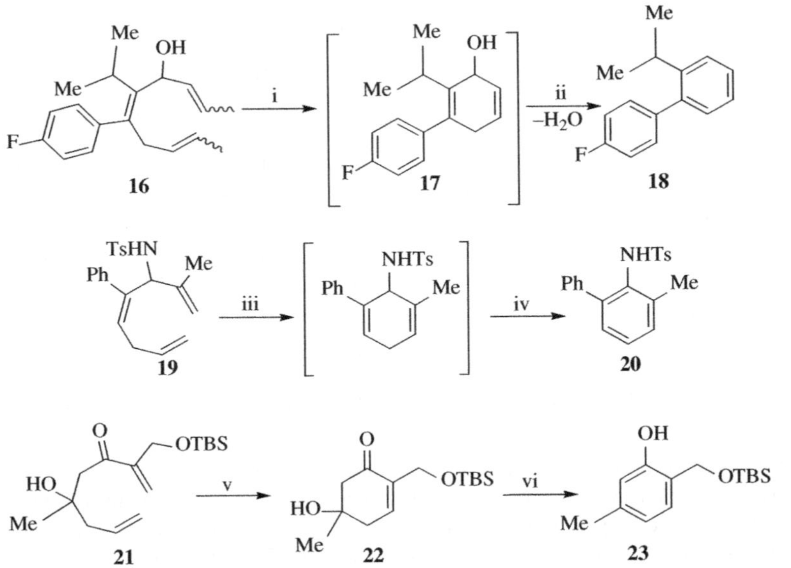

Reagents and conditions: (i) **2-Ru** (7.5 mol%), CH_2Cl_2, 40°C, 2 h; (ii) *p*-TsOH (10 mol%) rt, 1 h (98% over 2 steps); (iii) **2-Ru** (7.5 mol%), CH_2Cl_2, rt—40°C, 2 h (99%); (iv) MnO_2 (excess), rt (77%); (v) **2-Ru** (7.5 mol%), CH_2Cl_2, 40°C, 2 h (99%); (vi) *t*-BuOK (4.4 mol equiv.), MsCl (3.2 mol equiv.), CH_2Cl_2, rt, 2.5 h (88%).

SCHEME 17.3 Synthesis of substituted benzenes by way of a metathesis–dehydration approach.

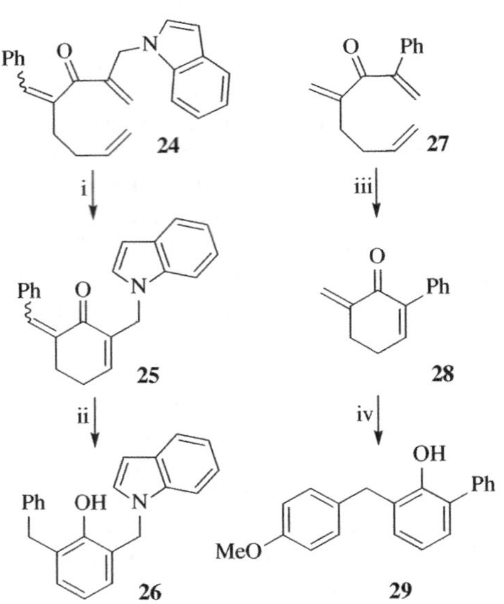

Reagents and conditions: (i) **2-Ru** (1.5 mol%), toluene, 40°C, 2 h (81%); (ii) $[RhCl(cod)]_2$ (1 mol%), Cs_2CO_3 (1 equiv.), dioxane-H_2O, 60°C, 5 h (69%); (iii) **2-Ru** (1.5 mol%), toluene, 40°C, 2 h (92%); (iv) *p*-$MeOC_6H_4N_2BF_4$ **30**, $Pd(OAc)_2$ (5 mol%), MeCN-H_2O (1:1), 60°C, 12 h (60%).

SCHEME 17.4 Further examples of the synthesis of substituted benzenes from acyclic precursors. (Reprinted with permission from Ref. [10]. Copyright (2009) American Chemical Society).

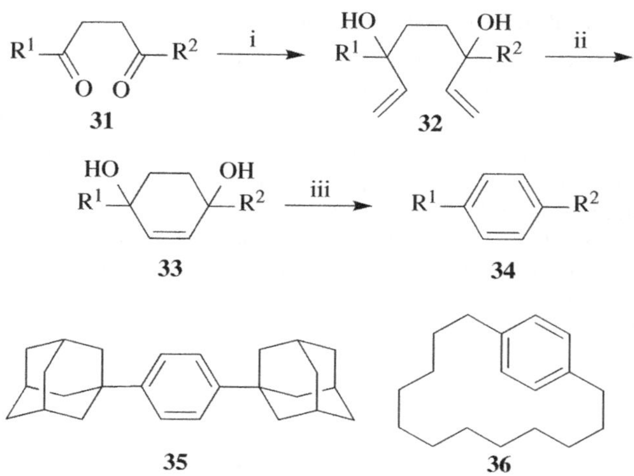

Reagents and conditions: (i) vinyllithium (from MeLi, tetravinyltin, Et_2O, 0°C), Et_2O, −78°C (72–93%); (ii) **1-Ru** (1.5 mol%) or **3-Mo** (20 mol%), CH_2Cl_2 (or C_6H_6), rt (or 80°C), various times (>90%); (iii) $TsOH.H_2O$, C_6H_6, reflux, 7 h (66%—quantitative).

SCHEME 17.5 Synthesis of *para*-substituted benzenes from 1,4-diketone precursors.

Another synthesis of benzene-containing compounds using alkene metathesis, in which a biaryl linkage is also constructed using the Suzuki–Miyaura coupling reaction, has recently been reported by Nishihara and coworkers [20]. The researchers commenced with, for example, the commercially available phenyl substituted 1-alkynylboronate **37** (Scheme 17.6). Treatment of **37** with Negishi's reagent (Cp_2ZrCl_2/*n*-BuLi) in the presence of P(*t*-Bu)$_3$ afforded the stabilized zirconacyclopropene **38**. This was followed by the addition of TMS-protected allyl alcohol **39** and then allyl chloride in the presence of CuCl, which allowed for the sequential double allylation to furnish **40** exclusively as the (*Z*)-alkenylboronate. With the double allylated alkenylboronate **40** in hand, the RCM reaction was attempted. While the **2-Ru** catalyst provided diene **41** in good yield (82%), it was found that the **1-Ru** catalyst provided the product **41** in a better yield of 95%. The cyclic diene **41** was then subjected to a coupling reaction in the presence of $PdCl_2$(dppf), followed by aromatization with DDQ to furnish the desired benzene ring-containing compound **42**.

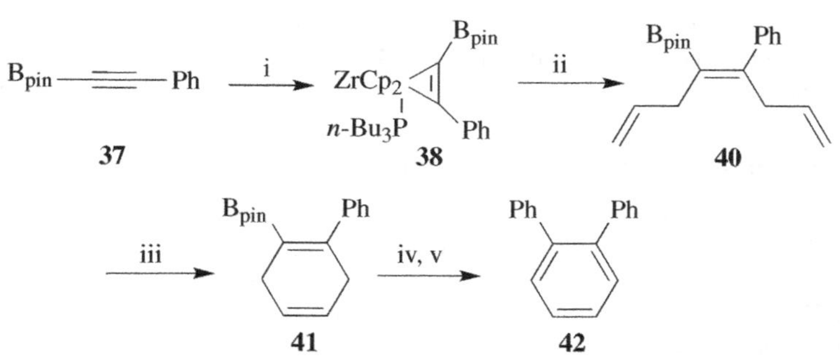

Reagents and conditions: (i) (a) Cp_2ZrCl_2/*n*-BuLi, (b) P(*t*-Bu)$_3$, THF, −78°C to rt, 1 h; (ii) CH_2=$CHCH_2OTMS$ **39**, then CH_2=$CHCH_2Cl$, CuCl, DMPU, 50°C, 1 h, (60%); (iii) **1-Ru** (1 mol%), toluene, 25°C, 1 h, (95%); (iv) $PdCl_2$(dppf) (5 mol%), PhI, NaOH, THF, rt, 24 h; (v) DDQ, CH_2Cl_2, rt, 30 min, (93%).

SCHEME 17.6 Synthesis of substituted aromatics by way of a zirconacyclopropene intermediate.

17.2.2 Synthesis of Substituted Naphthalenes

It should come as no surprise that the same sort of approach used for the synthesis of benzenes could also be used for the synthesis of substituted naphthalenes [21]. The difference in this approach is that the two alkenes would be stitched to a preexisting benzene nucleus, and hence, this would result in the formation of a second aromatic ring giving rise to naphthalene derivatives. In the example shown in Scheme 17.7, the diene precursor **43** could easily be synthesized from isovanillin in a few steps [22]. As with the benzene precursors, RCM on benzylic alcohol **43** resulted in the formation of the desired six-membered ring, which underwent dehydration to yield naphthalene **44**. Alternatively, the benzylic alcohol could be oxidized to afford the corresponding conjugated ketone **45** containing the two alkene side chains required for RCM. Subjecting substrate **45** to the **2-Ru** catalyst then afforded phenol **46**, presumably via the intermediate ketone **47**.

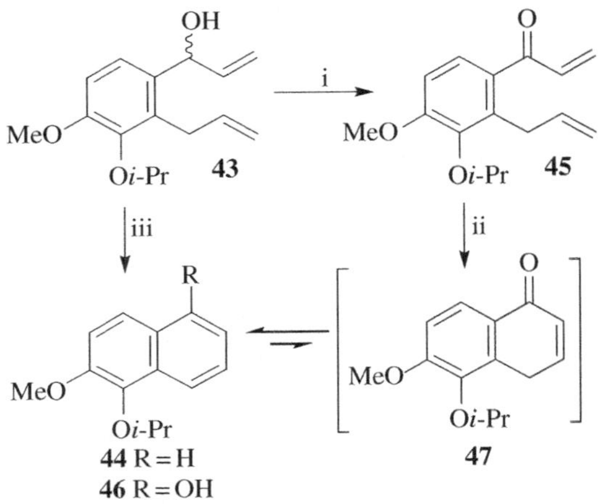

Reagents and conditions: (i) MnO_2, C_6H_6 (54%); (ii) **2-Ru** (5 mol%), CH_2Cl_2, reflux, (69%); (iii) **2-Ru** (5 mol%), CH_2Cl_2, reflux (98%).

SCHEME 17.7 Synthesis of substituted naphthalenes from 1-(2-allyl-3-isopropoxy-4-methoxyphenyl) prop-2-en-1-ol. (Reprinted with permission from Ref. [10]. Copyright (2009) American Chemical Society).

The desired diene-containing compounds such as **48** can also be synthesized using the Suzuki–Miyaura cross-coupling reaction. For example, cross-coupling of the aromatic diiodide **49** with allylboronic acid pinacol ester **50** readily furnished **48** [23]. After exposure of this diene **48** to an RCM reaction, followed by DDQ oxidation, the naphthalene **51** was produced in good yield (Scheme 17.8).

17.2.3 Synthesis of Substituted Phenanthrenes

The King group has used the same principle for the synthesis of phenanthrene **52** from **53**, using both **2-Ru** and **3-Mo** as catalysts for the RCM reaction. Initially, a Suzuki–Miyaura coupling strategy, as shown in Scheme 17.9, was utilized between the substituted styrenes **54** and **55** to synthesize the biphenyl **53**, which was subsequently cyclized [24] (see Ref. [25] for a related strategy).

Iuliano *et al.* utilized the same strategy in their synthesis of the substituted phenanthrene **56** (Scheme 17.10) from diene **57**. Of course, the synthesis of the precursor biaryl **57** may not be trivial to accomplish. In this case, the precursor **57** was obtained in two steps (*n*-BuLi and then DMF, followed by a Wittig reaction) from the corresponding biaryl dibromide **58** [26]. However, the dibromide **58** was synthesized from the related nonhalogenated biaryl precursor **59**. On

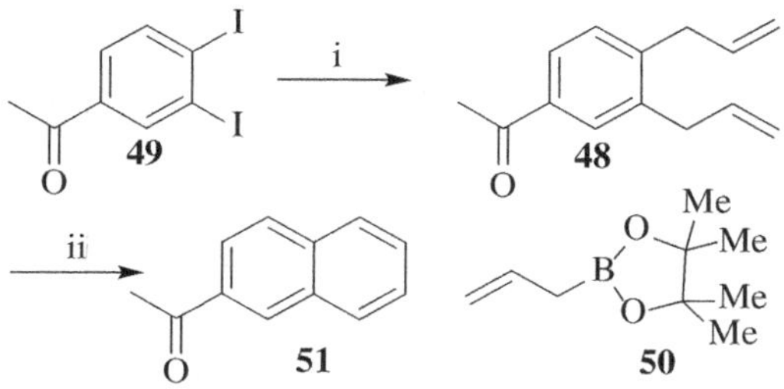

Reagents and conditions: (i) **50**, CsF, $Pd(PPh_3)_4$, THF, reflux (89%); (ii) **1-Ru** (3 mol%), CH_2Cl_2, rt, 20 min., then DDQ, C_6H_6, reflux (82% over 2 steps).

SCHEME 17.8 Synthesis of substituted naphthalenes from 1,2-diiodobenzenes—application of allylboronic acid pinacol ester. (Adapted with permission from Ref. [10]. Copyright (2009) American Chemical Society).

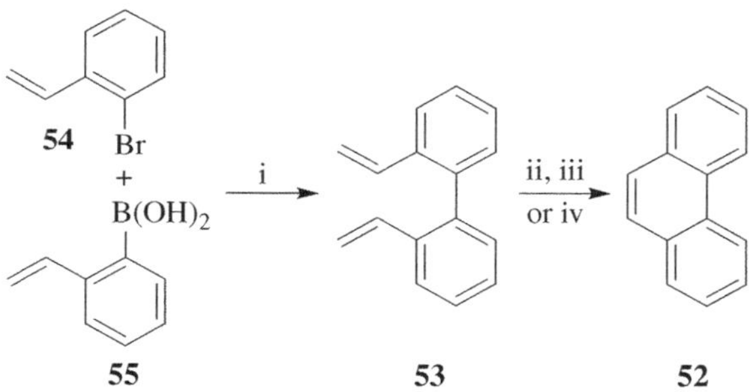

Reagents and conditions: (i) $Pd_2(dba)_3$, *t*-$Bu_3Ph.BF_4$, KF, THF, 20°C, 18 h (40%) [24]; (ii) **2-Ru** (10 mol%), CH_2Cl_2, 20°C, 24 h (50%) [24]; (iii) **2-Ru** (10 mol%), CD_2Cl_2, 25°C, 8 h, (87% by NMR) [23]; (iv) **3-Mo** (10 mol%), CS_2, rt, 1 h, then silica gel (71%) [23].

SCHEME 17.9 Synthesis of phenanthrene involving a Suzuki–Miyaura coupling followed by RCM. (Adapted with permission from Ref. [10]. Copyright (2009) American Chemical Society).

performing the bromination reaction on 2,2′,3,3′-tetramethoxybiphenyl **59** with BTEA·Br_3 at best, a separable mixture of three products—**58**, **60**, and **61**—was formed in respective yields of 45, 50, and 5%, somewhat limiting the use of RCM as a profitable procedure for preparing the substituted phenanthrenes by this methodology.

17.2.4 Synthesis of Anthraquinones and Benzo-Fused Anthraquinones

Since quinones are important natural products as a result of their biological activities, new methods for their synthesis have been an area of continued development. As far as RCM reactions are concerned, developments have often been centered on the addition of a further aromatic ring onto a preexisting quinone, thus illustrating that the catalyst and reaction conditions used for the assembly of another aromatic ring are tolerant of the quinone functional group.

The example shown in Scheme 17.11 outlines how naphthoquinones **62**, possessing two adjacent alkene chains in the form of allyl substituents, can undergo RCM in the presence of the **1-Ru** catalyst. Subsequent exposure of the product from this reaction to heat in the presence of Pd/C in toluene then afforded anthraquinones **63** [27].

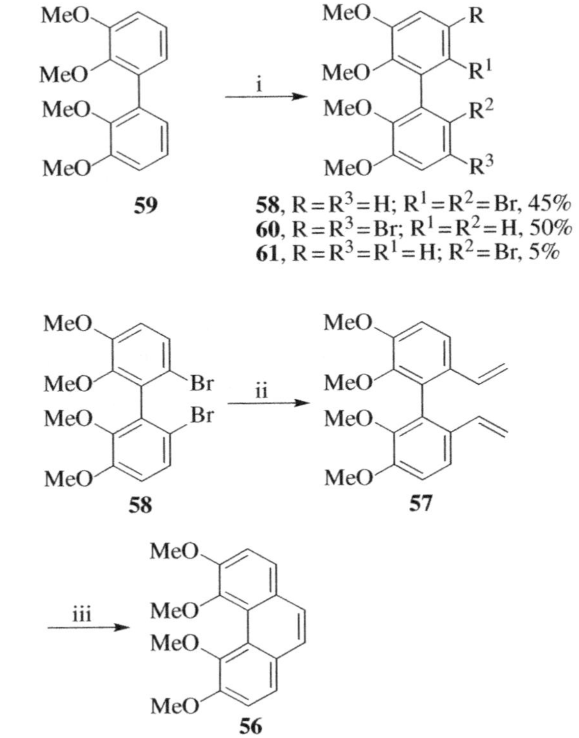

Reagents and conditions: (i) [BTEA·Br_3], CH_2Cl_2, MeOH, rt, (45%, **58**); (ii) (a) *n*-BuLi, –78°C, THF, (b) DMF, rt, 1 h; (c) CH_2=PPh_3, THF, rt; (iii) **2-Ru** (5 mol%), CH_2Cl_2, 40°C, N_2 (99%).

SCHEME 17.10 Synthesis of substituted phenanthrenes from the corresponding 6,6′-divinyl-1,1′-biphenyl. (Adapted with permission from Ref. [10]. Copyright (2009) American Chemical Society).

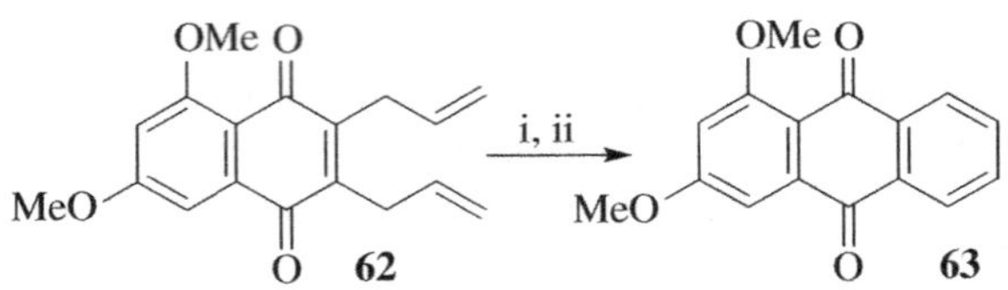

Reagents and conditions: (i) **1-Ru** (7 mol%), toluene, rt, 12 h; (ii) Pd/C, toluene, heat, (83% over two steps)

SCHEME 17.11 Synthesis of an anthraquinone. (Reprinted with permission from Ref. [10]. Copyright (2009) American Chemical Society).

Kotha and Mandal have taken this methodology one step further by introducing an extra ring onto a preexisting anthraquinone nucleus (Scheme 17.12). Commencing with anthraquinone **64**, bis-*O*-allyation, followed by a double Claisen rearrangement and protection of the resultant hydroquinone with acetic anhydride, furnished the bisacetate **65** [28]. At this stage, the RCM reaction was successfully performed to afford **66**. Treatment of **66** with DDQ then afforded the desired target **67**, where an additional aromatic ring had been stitched onto the anthraquinone nucleus.

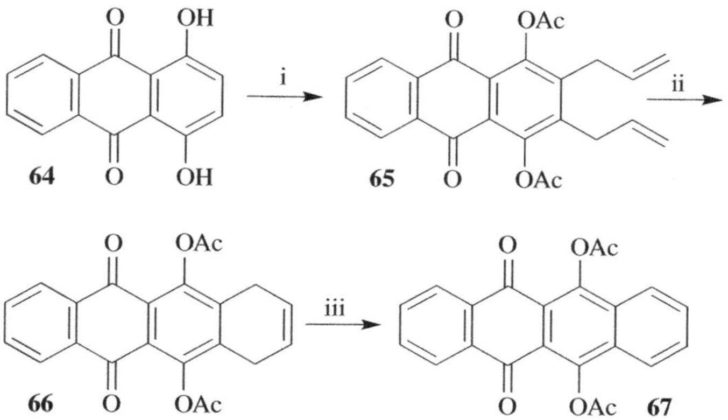

Reagents and conditions: (i) (a) allyl bromide, K_2CO_3, acetone, reflux (81%); (b) $Na_2S_2O_4$, DMF-H_2O (1:1), 130°C (71%); (c) Ac_2O, pyridine (95%); (ii) **2-Ru** (5 mol%), CH_2Cl_2, rt, 24 h; (iii) DDQ, C_6H_6, reflux (51% over two steps).

SCHEME 17.12 Synthesis of a substituted benzo-fused anthraquinone. (Adapted with permission from Ref. [10]. Copyright (2009) American Chemical Society).

17.2.5 Applications in the Synthesis of Polyarenes

The assembly of polyarenes has also been achieved using the same principles that were applied for the synthesis of phenanthrenes. Collins has demonstrated that a series of compounds known as the helicenes could be synthesized by olefin metathesis. Treatment of substrates such as **68** with catalyst **2-Ru** (microwave conditions, 100°C) or catalyst **4-Ru** (sealed tube, 40°C) resulted in the formation of [5]helicene **69** in excellent yields (Scheme 17.13) [29]. The group expanded their methodology to other substrates, including **70–72**, to furnish [6]- and [7]-membered helicenes, in excellent yields. Using the chiral ruthenium catalyst **5-Ru** shown in Scheme 17.13, their methodology was extended to the asymmetric synthesis of [7]helicene *M*-**73**, obtained in a maximum of 80% enantiomeric excess (ee) (38% conversion). The use of vinylcyclohexane and hexafluorobenzene as solvents also proved critical in the synthesis of *M*-[7]helicene **73** [30]. Collins and coworkers speculated that the olefin additive may allow for the *reversible* binding of the helicene precursor **72** to the catalyst, which could aid in the enantioselection of the catalyst. It could also be that the vinylcyclohexane changes the propagating carbene species of the catalytic cycle, making it more stable, hence resulting in more conversions. In addition, it may also have an influence on the ee of the process.

Use of the ROM–RCM reaction has been applied in an elegant manner for the synthesis of other complex aromatic molecular-bowl hydrocarbons and related compounds [31]. The asymmetric synthesis of trimethylsumanene [32], as well as the synthesis of sumanene **74** [33], has been accomplished using this methodology. Outlined in Scheme 17.14 are some of the key steps in the synthesis of sumanene **74** [33]. Trimerization of norbornadiene using the reagents shown in Scheme 17.14 resulted in the formation of both *syn*- and *anti*-**75** in 7% yield. More recently, a two-step procedure involving a tin–norbornadiene intermediate improved the yield of the transformation over two steps to 47%. Next, in the utilization of an ROM–RCM reaction, the *syn*-isomer of **75** was subjected to **1-Ru** (under an ethylene atmosphere), which resulted in the formation of **76** via the ROM intermediate **77** in a 30% yield. The product **76** was then oxidized with DDQ to give sumanene **74**, which is a bowl-shaped symmetric subunit of fullerene (C_{60}). An asymmetric synthesis of trimethylsumanene (not shown) was also accomplished by Higashibayashi and Sukurai, who used this ROM–RCM–aromatization approach for the synthesis of a chiral "buckybowl" [32].

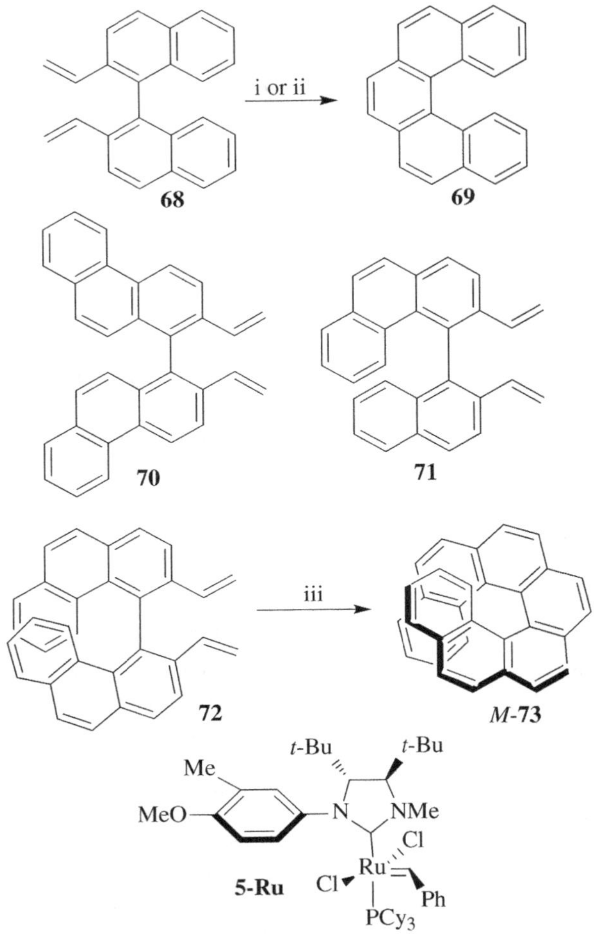

Reagents and conditions: (i) **2-Ru** (10 mol%), CH_2Cl_2, microwave, 100°C, 25 min (90%) [28]; (ii) **4-Ru** (10 mol%), C_6H_6, sealed tube, 40°C, 24 h (78–93%) [28]; (iii) **5-Ru** (5 mol%), C_6F_6, vinylcyclohexane (10 mol equiv.), rt, 2 h, (38% conversion, 80% ee) [29].

SCHEME 17.13 Synthesis of helicenes. (Adapted with permission from Ref. [10]. Copyright (2009) American Chemical Society).

17.2.6 Applications in the Synthesis of Natural Products

Use of RCM for the assembly of natural products has also been implemented with some success. Two examples will be presented here to illustrate the innovative use of the RCM reaction. The first will illustrate the construction of the benzene ring of the natural product (±)-cuparene **78** [34], while the second route to the natural product (±)-hasubanonine **79** uses RCM to assemble a phenanthrene-containing compound *en route* to this natural product [35].

The benzene ring of (±)-cuparene **78** has been assembled using RCM as a key step (Scheme 17.15) [34]. In this work, the advanced 1,4-dicarbonyl intermediate **80** was converted into diene **81**, which then underwent an RCM reaction to afford cyclohexene **82**. This compound was then aromatized under acidic conditions to afford (±)-cuparene **78** without any major problems, yielding the natural product as a racemate in very few steps.

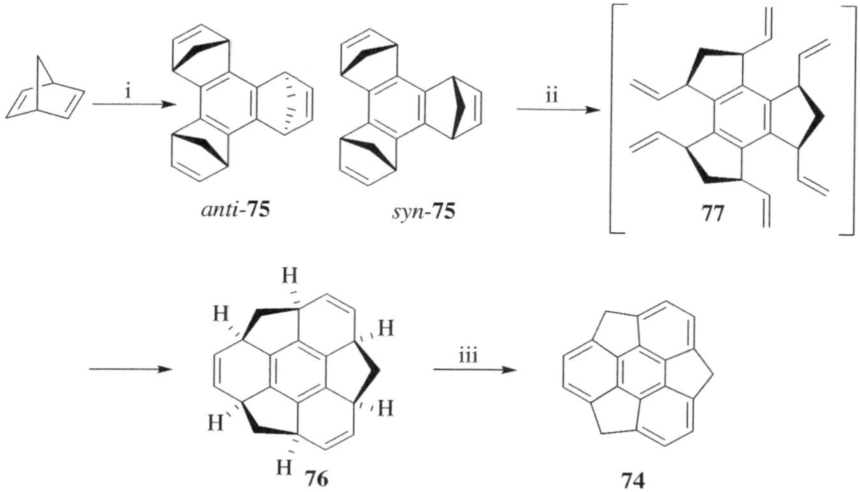

Reagents and conditions: (i) *n*-BuLi, *t*-BuOK, $BrCH_2CH_2Br$, THF, –78°C—rt, then CuI, rt (7%, *syn*:*anti* = 1:3); (ii) **1-Ru** (10 mol%), ethylene, toluene, –78°C—rt, 24 h (30%); (iii) DDQ, toluene, 110°C, 3 h, (70%).

SCHEME 17.14 Synthesis of sumanene. (Adapted with permission from Ref. [10]. Copyright (2009) American Chemical Society).

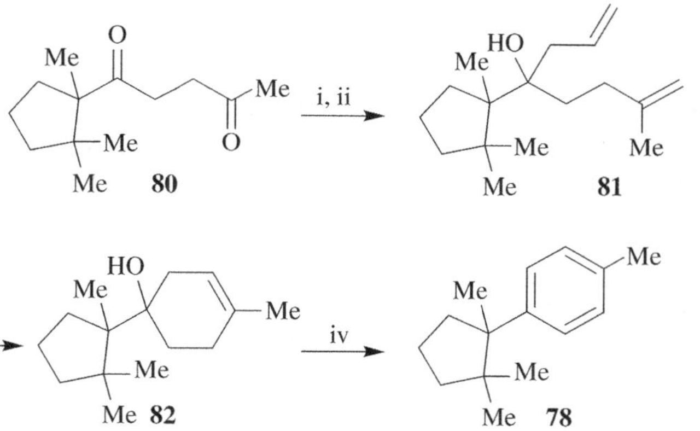

Reagents and conditions: (i) $Ph_3P{=}CH_2$, C_6H_6, 0°C, 30 min. (90%); (ii) allyl bromide, Li, THF, 15–20°C, 1 h, sonochemical irradiation (92%); (iii) **2-Ru** (3 mol%), C_6H_6, reflux, 30 min. (quantitative); (iv) *p*-TsOH, C_6H_6, air, reflux, 4 h (77%).

SCHEME 17.15 Synthesis of (±)-cuparene. (Adapted with permission from Ref. [10]. Copyright (2009) American Chemical Society).

The phenanthrene nucleus required for (±)-hasubanonine **79** (Scheme 17.16) was also assembled using RCM [35], in a similar way to that illustrated by Iuliano [26]. Initially, the biaryl compound **83** had to be assembled. This was done by synthesizing the aromatic iodide **84** and the aromatic pinacol ester **85**. These compounds were then subjected to a palladium-mediated coupling, followed by conversion of the aldehydes to alkenes, which proceeded smoothly to furnish **83**. Compound **83** was then subjected to **2-Ru** catalyst affording the required phenanthrene **86**, which was then converted into (±)-hasubanonine **79** over a number of further steps.

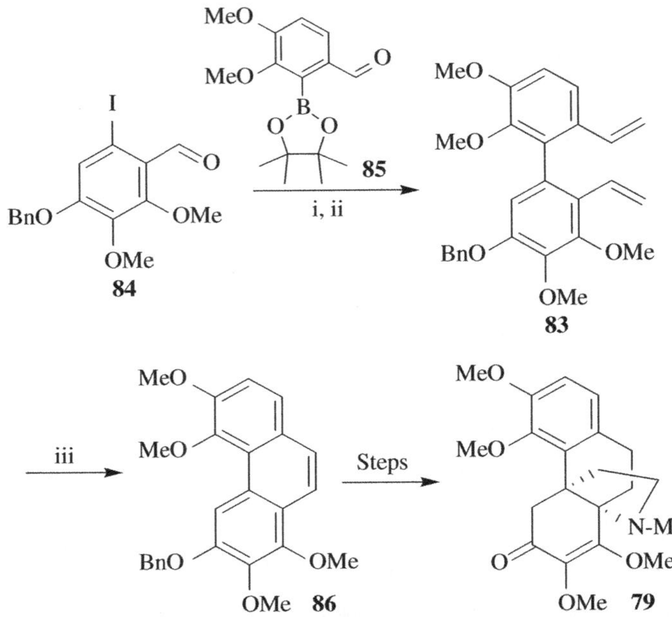

Reagents and conditions: (i) $Pd(dppf)_2Cl_2$, K_2CO_3, DMSO, 80°C, (79%); (ii) Ph_3PCH_3Br, *n*-BuLi, THF, 40°C, (93%); (iii) **2-Ru** (5 mol%), CH_2Cl_2, 40°C (99%).

SCHEME 17.16 Synthesis of (±)-hasubanonine. (Adapted with permission from Ref. [10]. Copyright (2009) American Chemical Society).

17.3 ENYNE METATHESIS FOLLOWED BY THE DIELS–ALDER REACTION FOR THE SYNTHESIS OF BENZENE RINGS IN COMPLEX AROMATIC COMPOUNDS

The strategy of using the enyne metathesis reaction, followed by a Diels–Alder reaction, has been used extensively for the synthesis of aromatic compounds. The enyne metathesis reaction allows for the construction of a diene, which will undergo reactions with dienophiles to afford the desired aromatic compounds after aromatization of the intermediate Diels–Alder adduct.

17.3.1 Synthesis of Substituted Benzenes

A number of research groups have utilized the strategy outlined above to synthesize benzene ring-containing compounds. For example, Kotha successfully employed it for the synthesis of phenylalanine derivatives [36]. Starting with alkyne **87** and the alkene allyl acetate, cross-enyne metathesis afforded diene **88**. This readily underwent a Diels–Alder reaction with dimethyl acetylenedicarboxylate (DMAD) to afford cyclohexene **89**. Aromatization of these compounds with DDQ subsequently furnished the phenylalanine **90** in good yield (Scheme 17.17).

Yoshida and Imamoto have also described a variation of the previous strategy, where the alkene and alkyne undergoing the enyne metathesis reaction are intramolecular variants [37]. In this work, a range of styrene-containing compounds was synthesized, and two specific examples are shown in Scheme 17.18. In the first example (**91**→**92**), the triene intermediate **93** was constructed from the enyne substrate **91** using **2-Ru** as a catalyst. This compound underwent aromatization to yield the desired styrene **92**. In this example, the reaction was only successful when the hydroxyl group of **91** was protected as the *O*-acetate. For the second reaction (**94**→**95**) shown in Scheme 17.18, it is interesting to note that the enyne metathesis reaction of **94** resulted in the substituted (1-phenylvinyl)benzene **95**.

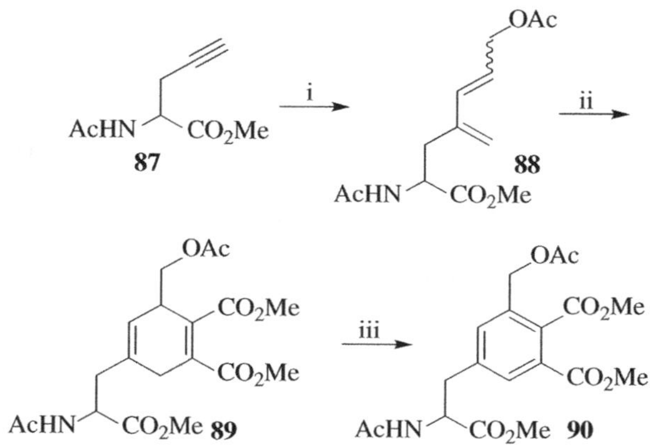

Reagents and conditions: (i) **1-Ru** (10–12 mol%), allyl acetate, C_6H_6, reflux, 40–50 h (45%, *E*/*Z* 1:1); (ii) DMAD, toluene, reflux; (iii) DDQ, C_6H_6, reflux (32% over two steps).

SCHEME 17.17 Synthesis of substituted benzene by enyne reaction between an alkyne and a 1,3-diene. (Adapted with permission from Ref. [10]. Copyright (2009) American Chemical Society).

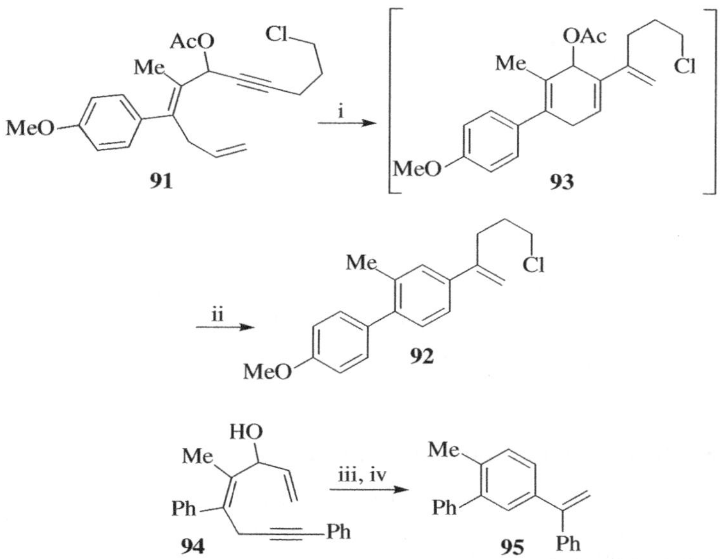

Reagents and conditions: (i) **2-Ru** (2.5–10 mol%), toluene, 80°C, 2 h; (ii) *p*-TsOH (15 mol%), rt, 1 h, (86% over 2 steps); (iii) **2-Ru** (7.5 mol%), toluene, 80°C, 2 h; (iv) *p*-TsOH (15 mol%), rt, 1 h (74% over 2 steps).

SCHEME 17.18 Synthesis of substituted styrenes by enyne RCM.

More recently, Wang, Zhao, and Shi succeeded in developing a synthesis of polysubstituted benzenes using an enyne RCM [38]. The authors indicated that in order for a metathesis reaction to occur, the initial dissociation of the PCy_3 ligand of the **2-Ru** catalyst needed to take place. They believed that this would occur in the presence of a catalytic amount of CuI [39]. Hence, exposure of a wide range of acetylenes (of general structure **96**) to 10 mol% of the **2-Ru** catalyst and 5 mol% of CuI under an atmosphere of ethylene gas when heated in toluene at 80°C for 24 h gave rise to the desired dienes **97**. These diene intermediates **97** were then heated in the presence of dienophiles such as DMAD to afford after oxidation with DDQ, an impressive range of tetrasubstituted benzenes **98** (Scheme 17.19).

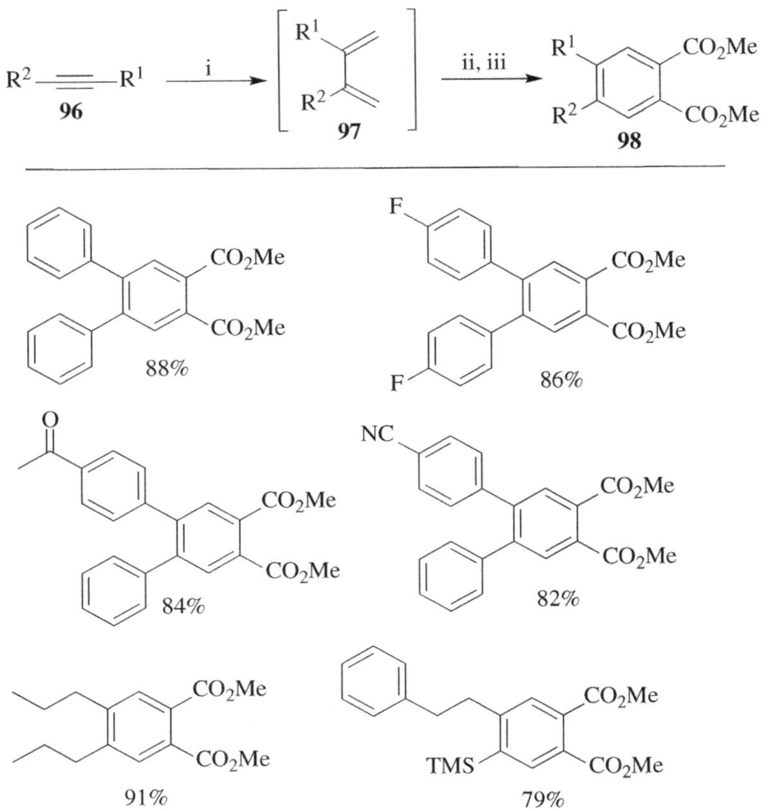

Reagents and conditions: (i) **2-Ru** (10 mol%), CuI (5 mol%), $CH_2{=}CH_2$ (1 atm), toluene, 80°C, 24 h; (ii) 5 equiv. DMAD, 100°C, 24 h; (iii) 1.2 equiv DDQ, 80°C, 4 h (yields in Scheme are for conversion of **96** to **98**).

SCHEME 17.19 Synthesis of polysubstituted benzenes by way of an enyne metathesis followed by a Diels–Alder reaction.

17.3.2 Synthesis of Substituted Phenanthrenes

Use of the enyne approach to assemble phenanthrenes has seen little activity. Katz has reported the use of a tungsten carbene complex 6-W to convert biaryl precursor **99** into 9-vinyl phenanthrene **100**, albeit in a mediocre yield (Scheme 17.20). Other examples are also outlined in this work but are generally low yielding [40].

17.3.3 Synthesis of Complex Naphthoquinones and Anthraquinones

Complex anthraquinones are often associated with biological activity, such as the class of quinones known as the angucyclinones. The synthesis of these types of quinones is possible by using a simple quinone as a dienophile in a Diels–Alder reaction. The diene for the reaction would then be assembled using the enyne metathesis reaction. In an example of the elegant synthesis of (+)-ochromycinone **101**, shown in Scheme 17.21, the enyne metathesis reaction of chiral **102** afforded the diene **103** [41]. After conversion into the diene **104**, this compound was reacted with the dienophile 5-acetoxy-2-bromo-1,4-naphthoquinone in a Diels–Alder fashion to establish the basic skeleton of the angucyclinones **105** in 45% yield over three steps (which included the removal of the acetyl group). The product was obtained as a single regioisomer and was then easily converted into (+)-ochromycinone **101**.

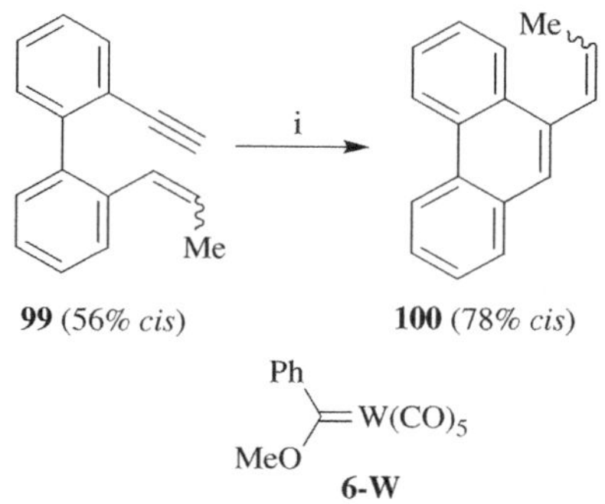

Reagents and conditions: (i) **6-W** (10 mol%), toluene, sealed tube, 75°C, 18 h (26%).

SCHEME 17.20 Synthesis of vinylphenanthridine using a tungsten carbene complex. (Adapted with permission from Ref. [10]. Copyright (2009) American Chemical Society).

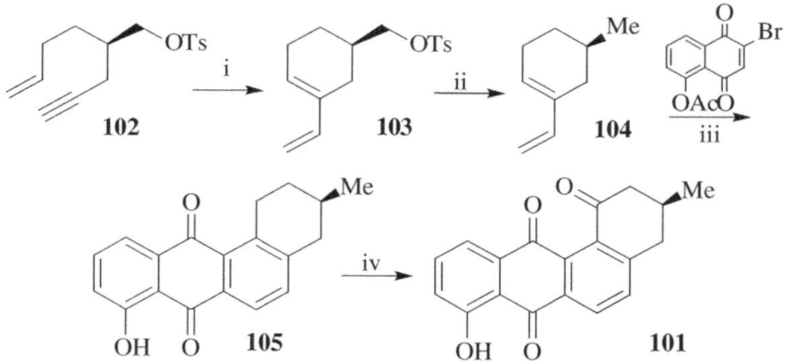

Reagents and conditions: (i) $CH_2{=}CH_2$, **1-Ru** (10 mol%), CH_2Cl_2, reflux, 12 h (quantitative); (ii) $LiAlH_4$, THF, 0°C—rt, 12 h; (iii) (a) toluene, 80°C, 16 h, (b) K_2CO_3, MeOH (45% over 3 steps); (iv) *hν*, O_2, C_6H_6, 20 h (82%).

SCHEME 17.21 Angucyclinone skeleton formation by enyne metathesis followed by a Diels–Alder reaction. (Adapted with permission from Ref. [10]. Copyright (2009) American Chemical Society).

Two other examples (**106**→**107** and **108**→**109**) of this elegant approach are shown in Figure 17.6. Using differently substituted enynes, Vanga and Kaliappan have synthesized (−)-zenkequinone B **110** (from enyne **111**) [42], as well as the naturally occurring compounds tetrangulol, kanglemycin M, X-14881-E, and anhydrolandomycinone (last 4 not shown) [43].

Reddy and coworkers proved that a sequential enyne metathesis/CM/Diels–Alder/aromatization strategy could be accomplished in a one-pot quadruple reaction sequence [44]. In this work, enyne **112** and alkene **113** were subjected to catalytic amounts of **2-Ru**, after which DMAD was added (Scheme 17.22). Finally, DDQ was used to aromatize the ring system into the corresponding benzene **114**, obtained in a respectable yield of 42% over the four synthetic steps involved. This compound was then readily converted into the natural product isofregenedadiol **115**.

A similar strategy has been used by Kotha to synthesize naphthoquinones and anthraquinones containing amino acid functionalities [45]. In this case, the racemic amino acid alkyne hybrid **116** was treated with ethylene in an enyne metathesis fashion to yield the diene **117**, which was then subjected to a Diels–Alder reaction with 2,3-dimethylbenzoquinone **118** (as shown in Scheme 17.23), to afford

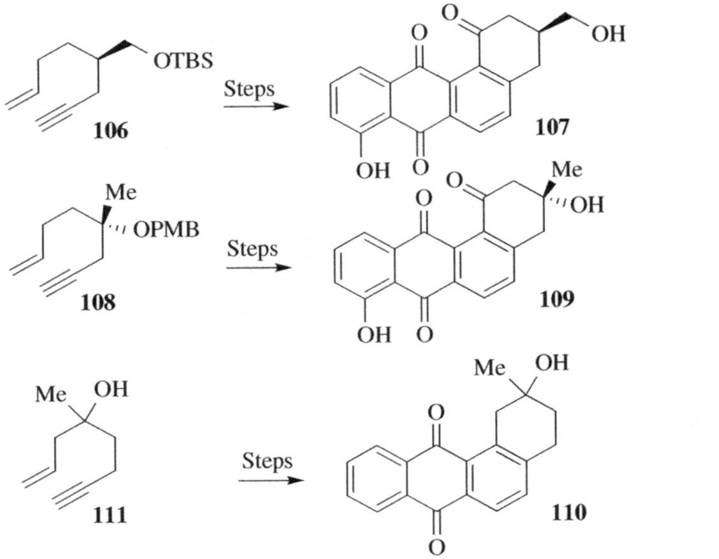

FIGURE 17.6 Further examples of the angucyclinone skeleton formed by enyne metathesis followed by a Diels–Alder reaction (Adapted with permission from Ref. [10]. Copyright (2009) American Chemical Society).

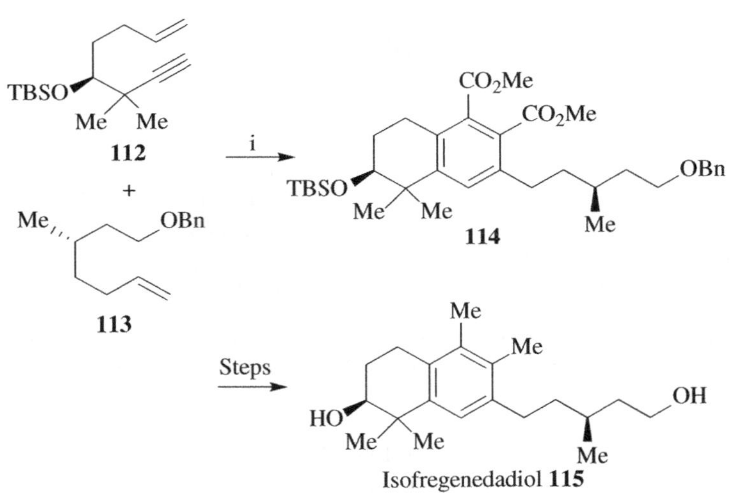

Reagents and conditions: (i) **2-Ru** (10 mol%), toluene, 50°C 18 h; then DMAD, 110°C 10 h; then DDQ, rt, 16 h (42% over 4 steps).

SCHEME 17.22 Synthesis of isofregenedadiol.

after aromatization the naphthoquinone **119**. The same approach was used to make the anthraquinone **120**, starting with **121** to prepare diene **122**, after which a Diels–Alder reaction with naphthoquinone **123** afforded the desired product **120** in good yield following aromatization with MnO_2.

Roulland and coworkers utilized a [2+2+2] cycloaddition and an RCM–aromatization strategy in an innovative approach to the central core of the landomycinone natural products [46]. As shown in Scheme 17.24, the triyne **124** was successfully cyclized with the catalyst $RhCl(PPh_3)_3$ to afford compound **125**, after which vinyllithium addition resulted in the hemiketal diene **126**. Immediate

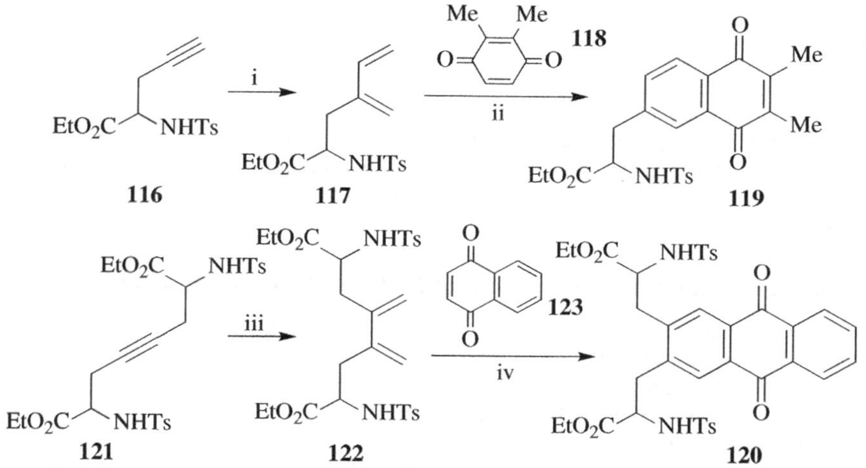

Reagents and conditions: (i) $CH_2{=}CH_2$ (1 atm), **1-Ru** (6 mol%), CH_2Cl_2, rt, 24 h (68%); (ii) (a) **121**, toluene, 90°C, 24 h, (b) MnO_2, dioxane, reflux, 30 h (81% over two steps);(iii) $CH_2{=}CH_2$ (sealed tube), **4-Ru** (7 mol%), C_6H_6, 80°C, 24 h (96%); (iv) (a) **123**, toluene, 110° C, 72 h, (b) MnO_2, dioxane, rt, 12 h (86% over two steps).

SCHEME 17.23 Enyne metathesis with ethylene to form dienes for Diels–Alder reactions.

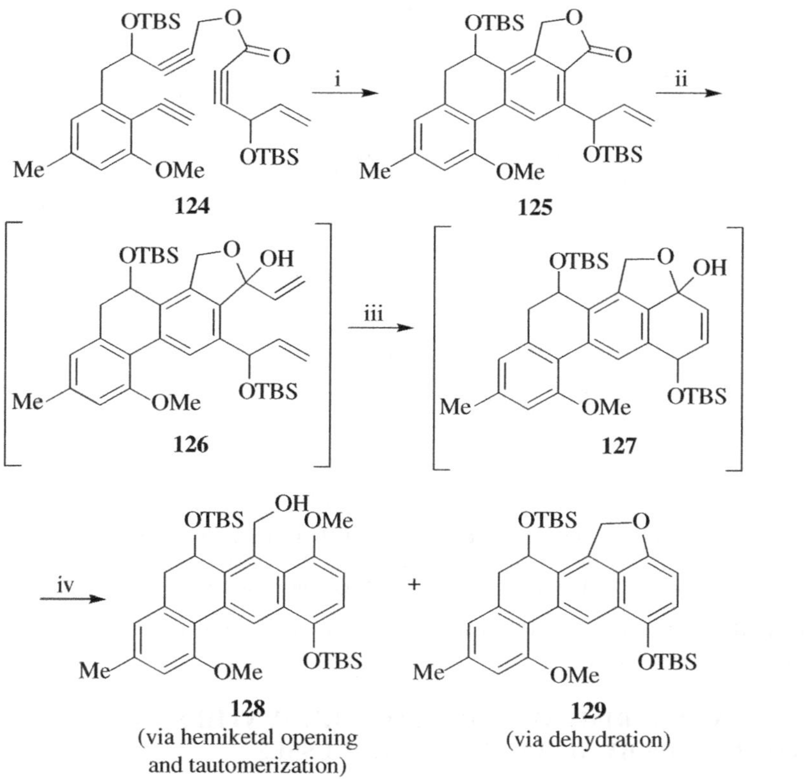

Reagents and conditions: (i) $RhCl(PPh_3)_3$ (5 mol%), CH_2Cl_2, reflux, 25 h (93%); (ii) vinyl bromide, *t*-BuLi, THF/Et_2O, –78°C, 24 h (96%); (iii) **2-Ru** (3 mol%), CH_2Cl_2, reflux, 1 h; (iv) Me_2SO_4, K_2CO_3, acetone, reflux, 24 h (**128** 33% and **129** 28% over three steps). Note: use of 2,6-lutidine (20 mol%) gave the following results: **128** 54% and **129** 13% over three steps.

SCHEME 17.24 Synthesis of the landomycinone skeleton.

treatment of this compound with **2-Ru** yielded **127**, followed by an *in situ* methylation to afford a binary mixture of compounds **128** and **129**. Subsequent optimization of the reaction conditions by the researchers indicated that addition of the base 2,6-lutidine afforded the desired product **128** in a much higher yield. The authors pointed out that the use of these key reactions allowed access to the landomycinone skeleton in an overall yield of 19% over 15 steps.

17.3.4 Applications to the Synthesis of Biologically Active Products

Botta and coworkers have used the enyne metathesis in the synthesis of the enantiopure antifungal agent (*S*)-bifonazole **130** [47]. In this work, the diene for the Diels–Alder reaction had to be synthesized from an alkyne. Reaction of alkyne (*R*)-**131** with ethyl vinyl ether, in the presence of catalyst **2-Ru**, thus afforded the desired diene **132** (Scheme 17.25) in an excellent yield of 88%. Next, the diene **132** was reacted with methyl vinyl ketone in a Diels–Alder reaction to afford compound **133**. Exposure of compound **133** to acid and then DDQ yielded the aromatic product **134**, where a newly formed benzene ring had been assembled. Further manipulations, including the formation of the required imidazole ring, allowed for the enantioselective synthesis of (*S*)-bifonazole **130**.

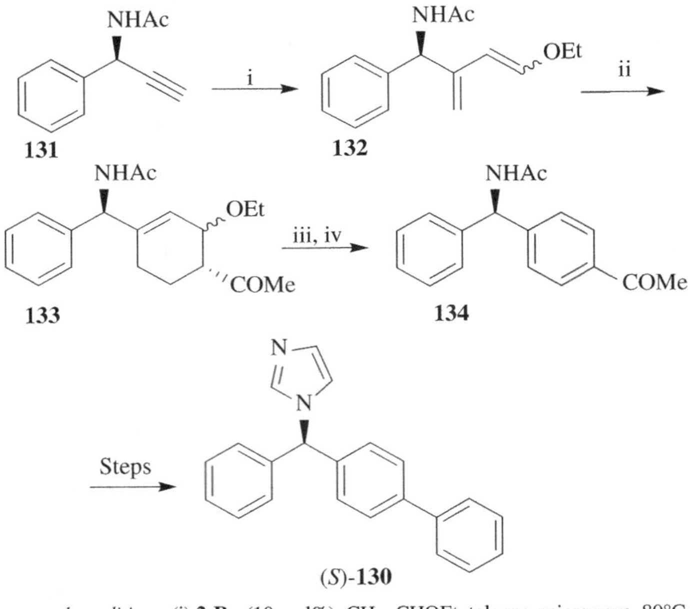

Reagents and conditions: (i) **2-Ru** (10 mol%), CH_2=CHOEt, toluene, microwave, 80°C (88%, ratio *E*/*Z* 2:1); (ii) CH_2=CHCOMe, $BF_3{\cdot}OEt_2$, CH_2Cl_2, –78°C (96%); (iii) H_2SO_4 (20%), THF, rt, 24 h (80%); (iv) DDQ, toluene, reflux, 3 h (48%).

SCHEME 17.25 Synthesis of (*S*)-bifonazole. (Adapted with permission from Ref. [10]. Copyright (2009) American Chemical Society).

17.4 CYCLOTRIMERIZATION FOR THE SYNTHESIS OF AROMATIC COMPOUNDS BY METATHETIC PROCESSES

Transition metal-catalyzed cyclotrimerizations [48] are a very important method for the synthesis of polysubstituted aromatic systems [49]. In this section, we will illustrate the use of this methodology for the *de novo* construction of a benzene ring, although the final products produced in many cases may be considered to be heteroaromatic.

In 1997, Peters and Blechert published an application of what is described as a "metathesis cascade" to afford substituted benzene-containing compounds from starting materials that had three embedded alkynes [50]. For example, as shown in Scheme 17.26, the tricyclic compound **135** could be assembled from the triyne precursor **136** upon exposure to 0.5 mol% of the catalyst **1-Ru**.

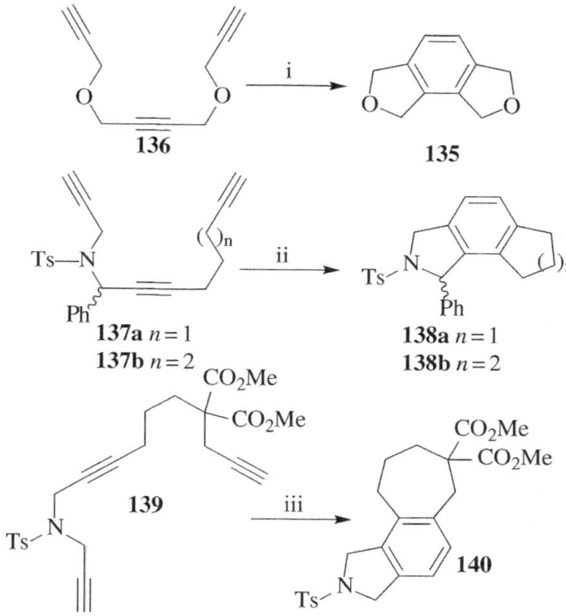

Reagents and conditions: (i) **1-Ru** (0.5 mol%), CH_2Cl_2 (1 M), rt, 2 h (88%); (ii) For $n = 1$, **1-Ru** (5 mol%), CH_2Cl_2 (1 M), rt, 12 h (74%), For $n = 2$, **1-Ru** (5 mol%), CH_2Cl_2 (1 M), rt, 48 h (35%); (iii) **1-Ru** (5 mol%), CH_2Cl_2 (1 M), rt, 48 h (15%).

SCHEME 17.26 The synthesis of substituted benzenes from triynes mediated by 1-Ru. (Adapted with permission from Ref. [10]. Copyright (2009) American Chemical Society).

Other compounds, where the central benzene ring was formed from three alkynes, were also made by this strategy as shown in Scheme 17.26, that is, **137a,b** → **138a,b** and **139** → **140**, although in some cases the yields of the reactions were mediocre. This was particularly noticeable when the size of the fused rings increased. In addition, these particular reactions necessitated larger amounts of catalyst and longer reaction times. It is highly probable that polymerization side reactions are taking place as well.

A mechanism has been proposed by Blechert for this "metathesis cascade," which involved the formation of a number of carbon–carbon bonds (in principle, a ruthenium-mediated [2+2+2] cycloaddition is also plausible for this transformation [49]). This postulated mechanism, as shown for the conversion of triyne **141** into the substituted aromatic system **142**, is depicted in Scheme 17.27 [50]. Initially, complex **1-Ru** adds to the less hindered acetylene of **141** to afford the vinyl carbene complex **143**, which then undergoes an intramolecular metathesis reaction to afford **144** via **145**. The conjugated complex **144** can then undergo a further RCM reaction to yield the product **142**.

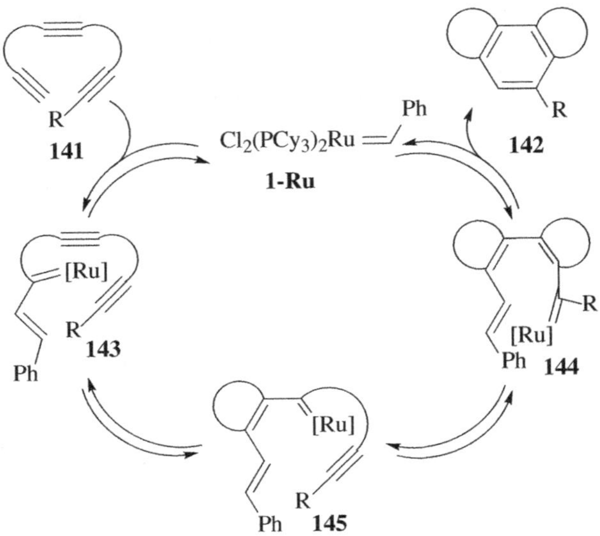

SCHEME 17.27 The proposed mechanism for the "metathesis cascade" leading to substituted benzenes (Adapted with permission from Ref. [10]. Copyright (2009) American Chemical Society).

Other examples of how this transformation has been utilized are shown in Scheme 17.28. In these examples, the transformations of **146** into **147a** (via **148**), as well as **149** into **147b**, employing the **1-Ru** catalyst, are illustrated. Use of microwave irradiation was also successful and the transformations were complete in 20 min with only 5% catalyst loading [51].

An intermolecular approach has also been found to be successful by Das and Roy, albeit resulting in the formation of a mixture of regioisomers (Scheme 17.29) [52]. These researchers found that treatment of the starting alkyne **150**, containing a carbohydrate moiety, again with the **1-Ru** catalyst, gave a mixture the 1,2,4-trisubstituted benzenes **151** and 1,3,5-trisubstituted benzenes **152**, with the former regioisomer as the major product.

Diynes with terminal alkynes have also been used by Witulski in an intermolecular cyclotrimerization reaction, together with the **1-Ru** catalyst [53]. Representative examples are shown in Scheme 17.30 where the final products are indolines and isoindolines, although the actual reaction that takes place results in the formation of a benzene ring. Commencing with **153** and **154**, the isomeric isoindolines **155** and **156** were formed, while with precursors **157** and **158**, the indolines **159** and **160** were furnished in acceptable yields.

17.5 STRATEGIES FOR THE SYNTHESIS OF AROMATIC CARBOCYCLES FUSED TO HETEROCYCLES BY THE RCM REACTION

17.5.1 Alkene RCM for the Synthesis of Benzene Rings in Indoles, Carbazoles, Benzo-Fused Pyridines and Pyridones, and Benzo-Fused Imidazoles

The indole nucleus is an important scaffold found in many biologically active compounds, and as a result, many methods have been developed for its construction [54]. Most of these rely on the assembly of the pyrrole ring on a preformed benzene nucleus. In principle, the indole nucleus could also be assembled by constructing a benzene ring on a preformed pyrrole ring. However, of the methods reported to date, the second method is much rarer. Recently, RCM has been utilized for

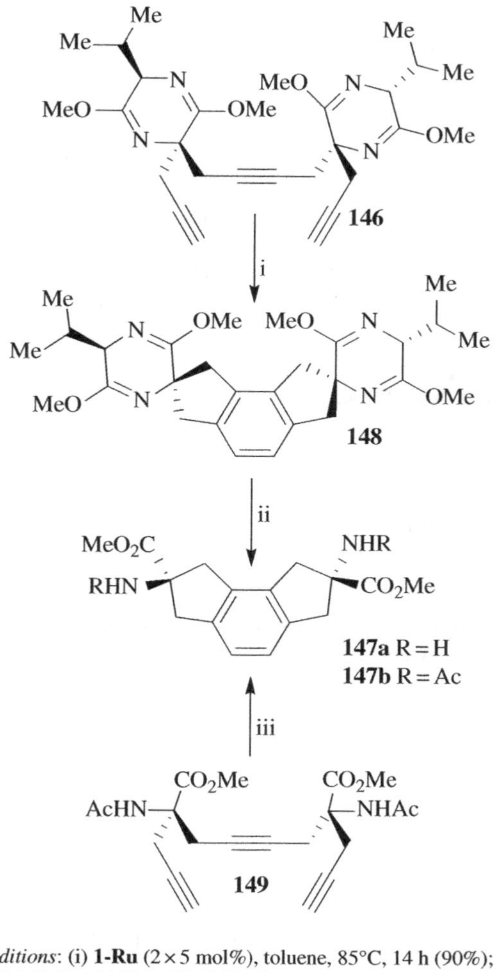

Reagents and conditions: (i) **1-Ru** (2 × 5 mol%), toluene, 85°C, 14 h (90%); (ii) TFA (0.1 M), $MeCN/H_2O$ (1:1), rt, 4 d (35% of **147a**); (iii) **1-Ru** (2 × 5 mol%), toluene, 85°C, 14 h (58% of **147b**).

SCHEME 17.28 Metathesis cascade on triynes leading to substituted benzenes. (Adapted with permission from Ref. [10]. Copyright (2009) American Chemical Society).

this purpose by Yoshida and coworkers [55]. As shown in Scheme 17.31, commencing from the easily synthesized pyrrole **161**, a cross-coupling reaction with a metal vinyl reagent, and the addition of an allylic metal reagent to the aldehyde of **161**, furnished diene **162**, albeit in poor yield (Scheme 17.31). Exposure of **162** to the **2-Ru** catalyst, followed by *p*TsOH to facilitate the subsequent dehydration reaction, allowed for the formation of the substituted indole **163**. An interesting variation allowed for the formation of a substituted phenol ring. Thus, starting from the same pyrrole **161**, a related ketone **164** was synthesized, and when this compound was subjected to the **2-Ru** catalyst, 7-hydroxyindole **165** was obtained in 93% yield.

As a result of the biological importance of a class of benzo-fused indole-containing compounds known as the carbazoles, many methods have been developed for their synthesis [56]. Therefore, it will come as no surprise that alkene RCM has been used as a strategy for the assembly of the benzene ring moieties that form part of the carbazole nucleus.

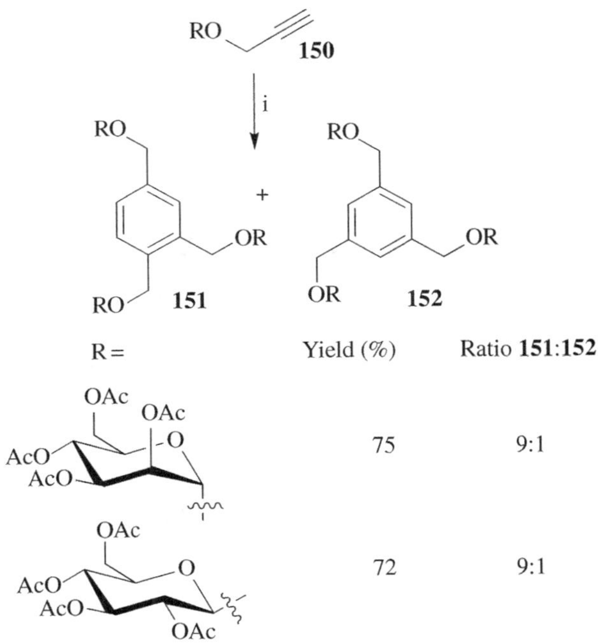

Reagents and conditions: (i) **1-Ru** (1.5 mol%), CH_2Cl_2 (1 M), 12 h, rt, for yields see above.

SCHEME 17.29 Intermolecular metathesis cascade on triynes leading to substituted benzenes. (Adapted with permission from Ref. [10]. Copyright (2009) American Chemical Society).

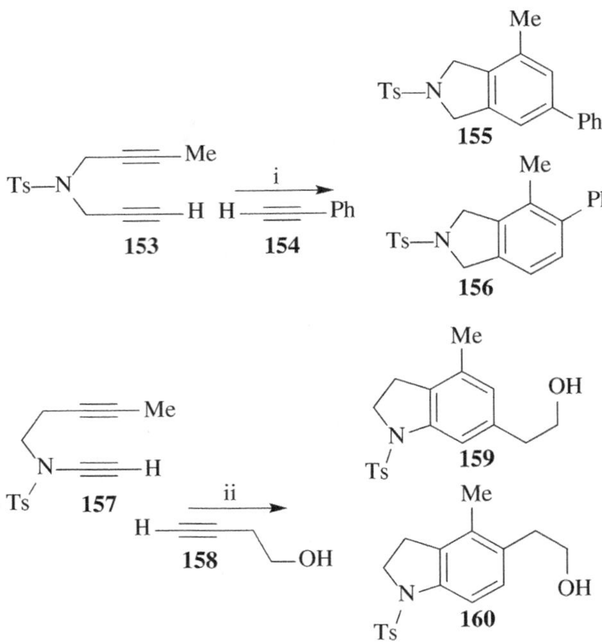

Reagents and conditions: (i) **1-Ru** (5–10 mol%), CH_2Cl_2, **154** (5 mol equiv.), 40°C, sealed tube, 10 h, (ratio **155**:**156**, 5:1), (82%); (ii) **1-Ru** (5–10 mol%), CH_2Cl_2, **158** (5 mol equiv.), 40°C, sealed tube, 10 h, (ratio **159**:**160**, 9:1), (51%).

SCHEME 17.30 Synthesis of indolines and isoindolines via the metathesis cascade. (Adapted with permission from Ref. [10]. Copyright (2009) American Chemical Society).

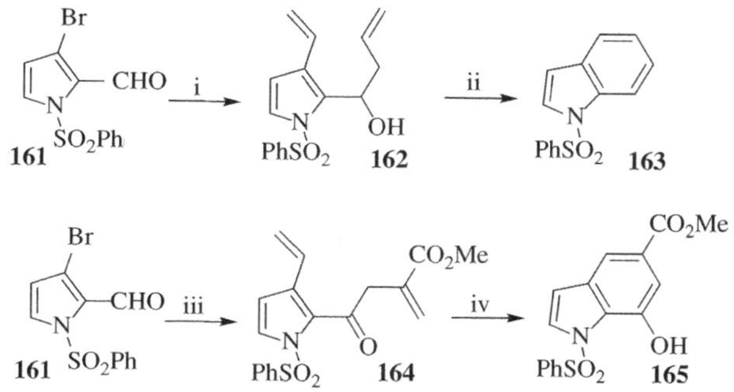

Reagents and conditions: (i) (a) vinyltrifluoroborate, $Pd(OAc)_2$, (5 mol%), PPh_3 (10 mol%), Cs_2CO_3, THF/H_2O; (b) $CH_2{=}CHCH_2BPin$, $ClCH_2CH_2Cl$, reflux, (17% over two steps); (ii) (a) **2-Ru**, (7.5 mol%), toluene, rt, 12 h; (b) *p*-TsOH.H_2O, (10 mol%), (98%); (iii) (a) vinyltrifluoroborate, $Pd(OAc)_2$, (5 mol%) PPh_3 (10 mol%), Cs_2CO_3 (3 equiv), THF/H_2O; (b) $CH_2{=}CHCH_2Bpin$, $ClCH_2CH_2Cl$, reflux, (79% over two steps); (c) DMP, CH_2Cl_2, 0°C, (70%); (iv) **2-Ru**, (7.5 mol%), toluene, rt, 12 h, (95%).

SCHEME 17.31 The synthesis of indoles from pyrroles via a metathesis–dehydration approach.

For the synthesis of the simple carbazole **166**, a similar strategy to that for the synthesis of naphthalenes has been used. The difference in this case is that the researchers started with a 2,3-disubstituted indole nucleus such as **167**, rather than an *ortho*-disubstituted benzene ring-containing compound. In the first example, the readily available 2-allyl indole **167**, containing an aldehyde at the 3-position, was treated with vinylmagnesium bromide to afford the indole **168** containing two alkenes (Scheme 17.32) [57]. An alkene RCM reaction mediated by the **1-Ru** catalyst, followed by an *in situ* dehydration, then led to the formation of **166** by forming a benzene ring in the last step. A closely related transformation commenced with the vinyl indole **169**. Addition of allylmagnesium bromide to the aldehyde of **169** provided the diene **170**, which on treatment with the **1-Ru** catalyst afforded carbazole **171** in a similar manner [58].

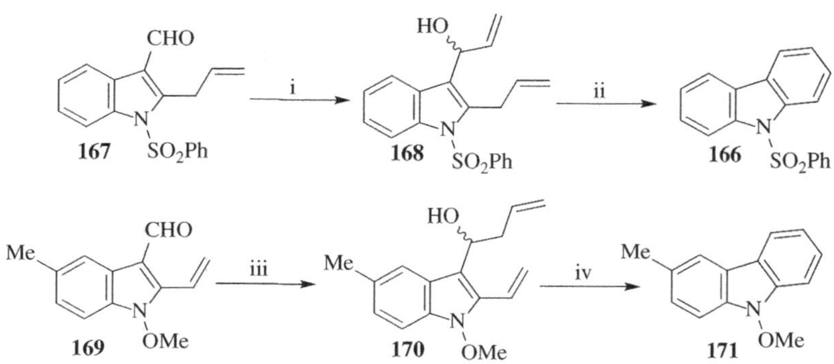

Reagents and conditions: (i) $BrMgCH{=}CH_2$, THF, –78 to 0°C, 3 h (75%); (ii) **1-Ru** (5 mol%), rt, CH_2Cl_2 (62%); (iii) Allyltributyltin, *n*-BuLi, –78°C (83%); (iv) **1-Ru** (5 mol%), CH_2Cl_2, rt to reflux (72%).

SCHEME 17.32 The synthesis of carbazoles from indoles via a metathesis–dehydration approach. (Adapted with permission from Ref. [10]. Copyright (2009) American Chemical Society).

The use of alkene RCM has also been employed by de Koning and coworkers to construct the central benzene ring of the indolo[2,3-c]carbazole nucleus **172** found in carbazoles such as rebeccamycin **173** [59]. Initially, conversion of the bisaldehyde **174** to **175** by means of the Wittig reaction was successfully accomplished. Subsequently, an RCM reaction performed on **175** afforded the desired carbazole **172** by means of constructing the central benzene ring in the last step, as shown in Scheme 17.33.

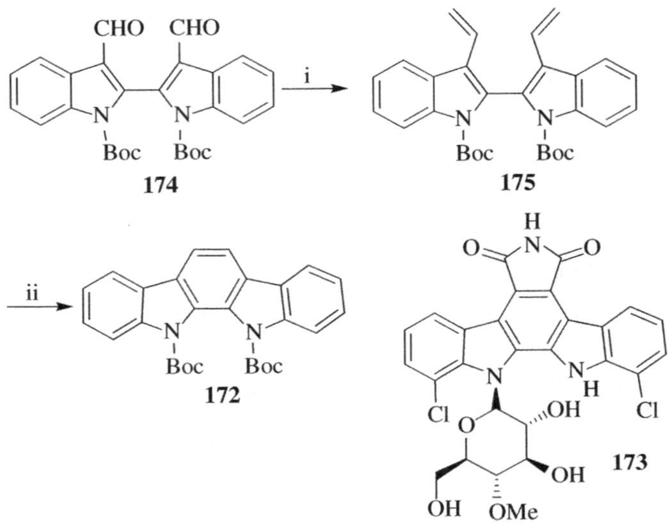

Reagents and conditions: (i) $MePPh_2Br$, *n*-BuLi, Et_2O, 0°C; (ii) **2-Ru** (8 + 4 mol%), toluene, 80°C, 20 + 4 h (64% over two steps).

SCHEME 17.33 The synthesis of the indolo[2,3-C]carbazole nucleus from bisindole aldehydes by means of ene–ene metathesis. (Adapted with permission from Ref. [10]. Copyright (2009) American Chemical Society).

The same type of disconnection (Scheme 17.34) was utilized for the synthesis of the thio analogue **176** of furostifoline **177**, where the central toluene ring was constructed by subjecting diene intermediate **178**, formed from **179**, to the **2-Ru** catalyst [59]. For a related example resulting in the formation of benzodithiophenes, see the following paper by Stephenson and coworkers [60].

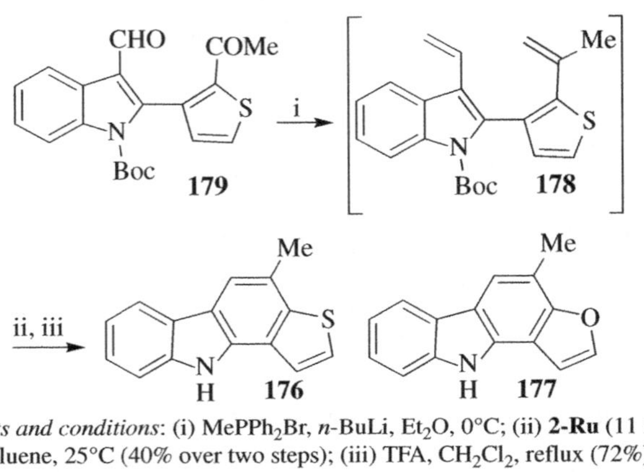

Reagents and conditions: (i) $MePPh_2Br$, *n*-BuLi, Et_2O, 0°C; (ii) **2-Ru** (11 mol%), toluene, 25°C (40% over two steps); (iii) TFA, CH_2Cl_2, reflux (72%).

SCHEME 17.34 The synthesis of the thio analogue of furostifoline via ene–ene metathesis. (Adapted with permission from Ref. [10]. Copyright (2009) American Chemical Society).

Similar ideas have also been used for the assembly of benzo-fused pyridones and pyridines. For example, the pyridone **180** was initially transformed into the desired diene **181** required for the RCM reaction. Exposure of **181** to the **1-Ru** catalyst allowed for the formation of intermediate **182**, which upon dehydrosulfonation readily yielded the isoquinolinone **183** (Scheme 17.35) [61].

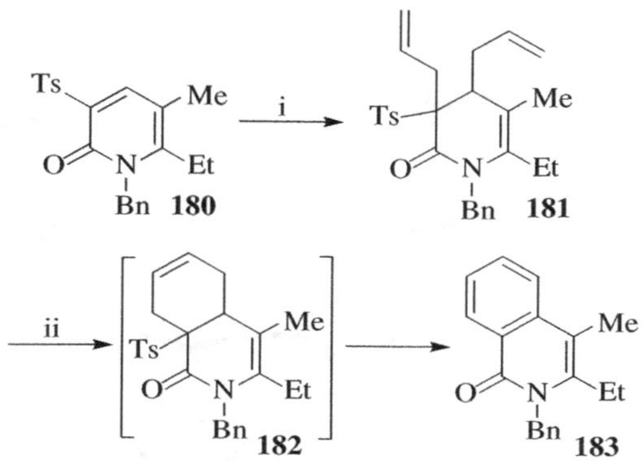

Reagents and conditions: (i) (a) allylmagnesium bromide, THF, rt; (b) allyl bromide, NaH, THF, rt (67% over 2 steps); (ii) (a) **1-Ru** (10 mol%), CH_2Cl_2, rt, 12 h; (b) *t*-BuOK, *t*-BuOH, reflux, 24 h (81% over 2 steps).

SCHEME 17.35 Ene–ene metathesis for the synthesis of the benzene ring of benzo-fused pyridones. (Adapted with permission from Ref. [10]. Copyright (2009) American Chemical Society).

The same type of transformations was used in the conversion of **184** to **185** via the intermediary compound **186**, as illustrated in Scheme 17.36 [61].

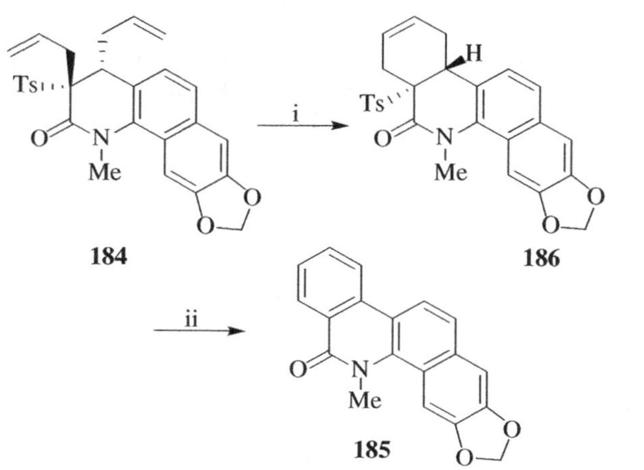

Reagents and conditions: (i) **1-Ru** (10 mol%), CH_2Cl_2, rt, 7.5 h (96%); (ii) (a) DBU, THF, 60°C, 20 h (93%); (b) DDQ, THF, reflux, 2 h (99%).

SCHEME 17.36 Ene–ene metathesis for the synthesis of the benzene ring of a complex benzo-fused pyridone. (Adapted with permission from Ref. [10]. Copyright (2009) American Chemical Society).

An example of the synthesis of a benzo-fused pyridine (i.e., a quinoline), containing an embedded ferrocene unit shown in structure **187**, has been reported by Mamane and Fort [62]. Commencing with the bisaldehyde **188** and subjecting it to a Wittig one-carbon homologation at each aldehyde group, it was converted into the bisvinyl complex **189**, setting this up for an alkene RCM reaction to form the benzene ring of the quinoline nucleus. In this event, exposure of **189** to the **2-Ru** catalyst satisfyingly afforded the aromatic ferroceno[*h*]quinoline **187** in an acceptable yield (Scheme 17.37). This demonstrated that the **2-Ru** catalyst is able to tolerate the ferrocene moiety.

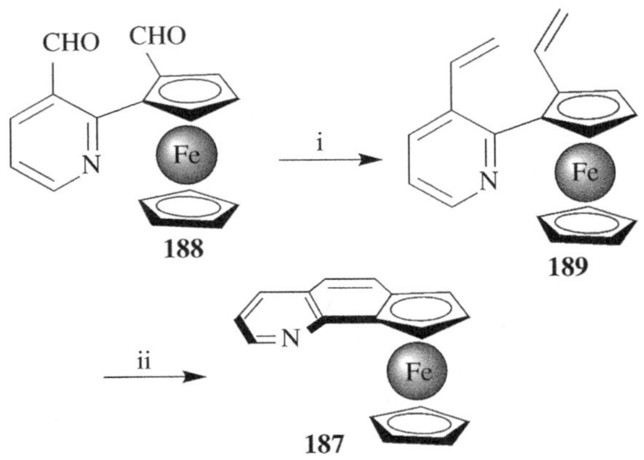

Reagents and conditions: (i) $Ph_3P{=}CH_2$, (3 mol equiv.), THF, –40°C to rt (69%); (ii) **2-Ru** (20 mol%), toluene, reflux, 30 min. (66%).

SCHEME 17.37 An alkene RCM reaction to form the benzene ring of the ferrocene-containing quinoline. (Adapted with permission from Ref. [10]. Copyright (2009) American Chemical Society).

Benzo-fused imidazoles can also be made in the same manner [63], provided that two alkene substituents are positioned to allow for the formation of a benzene ring fused to an imidazole nucleus. For example, Lovely has been able to do this by performing a metalation reaction on **190**, followed by quenching with acrolein to furnish the bisallyl imidazole **191** [64]. Conversion of **191** into an appropriate salt, followed by exposure to the **2-Ru** catalyst, resulted in the formation of the benzene ring of the benzo-fused imidazole **192** (Scheme 17.38).

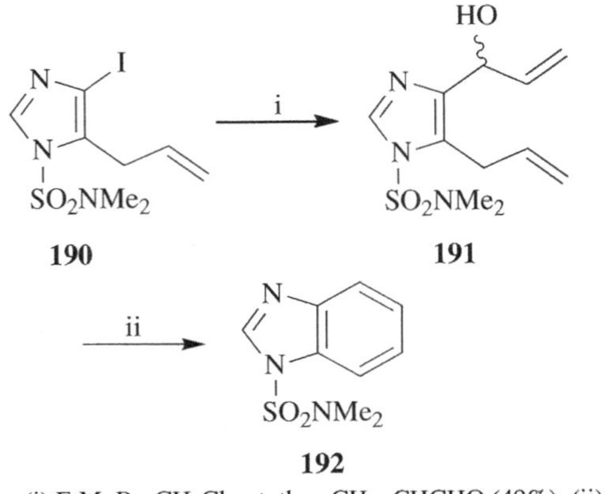

Reagents and conditions: (i) EtMgBr, CH_2Cl_2, rt, then $CH_2{=}CHCHO$ (49%); (ii) *p*-TsOH (1.1 equiv.), CH_2Cl_2 (0.1 M), reflux, 30 min., then **2-Ru** (5 mol%), reflux, 20 min., then rt, 1.5 h, (45%).

SCHEME 17.38 An ene–ene RCM reaction to form the benzene ring of a benzo-fused imidazole. (Adapted with permission from Ref. [10]. Copyright (2009) American Chemical Society).

17.5.2 Enyne RCM for the Synthesis of Benzene Rings in Tetrahydroisoquinolines, Annulated 1,2-Oxaza- and 1,2-Bisazacycles, and Indoles

As discussed in Section 17.3, the enyne RCM reaction has been innovatively used to construct dienes to be used in Diels–Alder reactions as illustrated in Scheme 17.17 (and others).

However, if the starting material for the enyne metathesis reaction is changed strategically by inserting a nitrogen atom in place of one of the carbon atoms, it is possible to synthesize related nitrogen-containing dienes. These can then be elaborated into relatively complex aromatic compounds where a benzene ring is synthesized utilizing the enyne RCM reaction, followed by a Diels–Alder reaction. Examples of this are illustrated in Scheme 17.39, where both the isomeric enynes **193** and **194**, containing a nitrogen atom, were subjected to an RCM reaction to yield the regioisomeric dienes **195** and **196**, respectively. Diene **195** was then treated with 1,4-naphthoquinone, followed by DDQ, to allow for aromatization to yield the tetrahydroisoquinoline **197** containing a newly formed benzene ring. Analogously, the isomeric diene **196** was exposed to DMAD, followed by aromatization with DDQ, to afford the tetrahydroisoquinoline **198** [65].

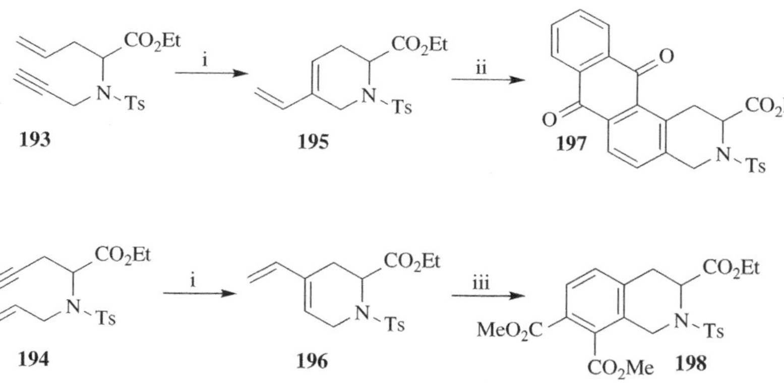

Reagents and conditions: (i) **1-Ru** (mol% not given in paper), toluene, reflux, 36 h, **195** (65%), **196** (70%); (ii) 1,4-naphthoquinone, then DDQ (52% over two steps); (iii) DMAD, then DDQ (85% over two steps).

SCHEME 17.39 Enyne RCM reactions and a Diels–Alder reaction for the synthesis of the benzene ring of isoquinolines. (Adapted with permission from Ref. [10]. Copyright (2009) American Chemical Society).

The process has been elaborated and extended to other enyne-containing starting materials where an extra oxygen has been inserted. For example, as shown in Scheme 17.40, compound **199** underwent the same type of RCM reaction to give diene **200**. Again, through the use of DMAD, the aromatic compound **201** could be accessed via **202** [66].

Alternatively, as shown in Scheme 17.41, the enyne **203**, now containing two protected nitrogen atoms, could go through the same sequence of transformations (**203** → **204** → **205** → **206**) to furnish the benzo-fused compound **206** [67].

The synthesis of 4-vinylindoles has also been accomplished using an enyne metathesis approach. Using the same ideas as those shown in Scheme 17.18, Yoshida and Yanagisawa were able to easily synthesize a range of the precursor pyrroles containing an acetylene and alkene functionality on adjacent carbons, namely, compound **207** [68]. When **207** was subjected to the **2-Ru** catalyst under an atmosphere of ethylene gas, the intermediate **208** was formed. When treated with *p*TsOH, the benzene ring of the 4-vinylindole **209** was subsequently formed in an excellent yield of 99% (Scheme 17.42). Other examples of this useful approach are illustrated in this paper [68].

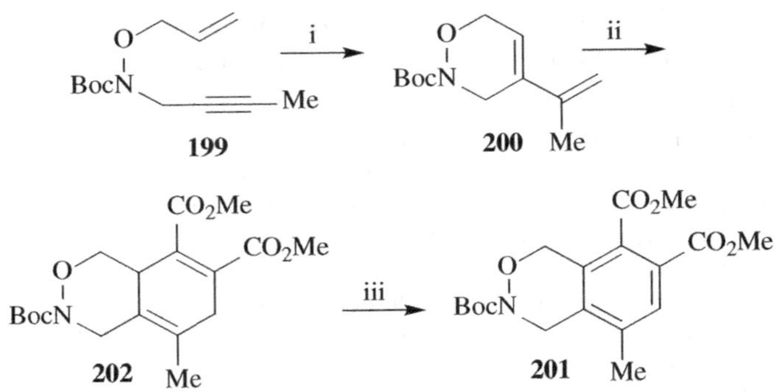

Reagents and conditions: (i) **1-Ru** (10 mol%), CH_2Cl_2 (0.007–0.02 M), 45°C, 4–34 h, (94%); (ii) DMAD, toluene, reflux, 6–9 h, (96%); (iii) DDQ (1 mol equiv.), C_6H_6, reflux, (94%).

SCHEME 17.40 An enyne RCM reaction and a Diels–Alder reaction for the synthesis of the benzene ring of **201**. (Adapted with permission from Ref. [10]. Copyright (2009) American Chemical Society).

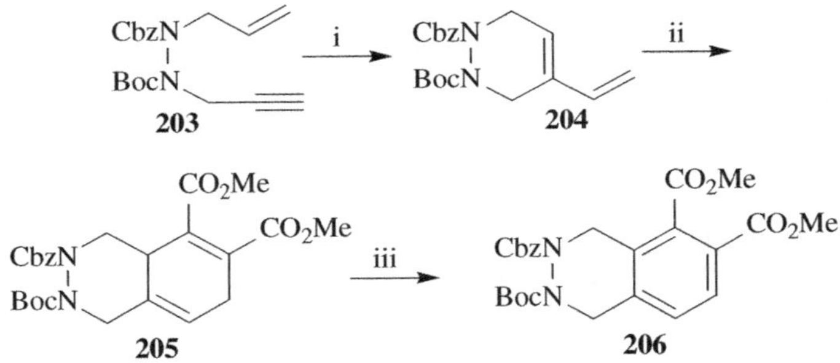

Reagents and conditions: (i) **1-Ru** (10 mol%), CH_2Cl_2 (0.02 M), reflux, 4–10 h, (99%); (ii) DMAD (1.2 mol equiv.), toluene, reflux, 6 h, (88%); (iii) DDQ (2 mol equiv.), toluene, reflux, (82%).

SCHEME 17.41 An enyne RCM reaction and a Diels–Alder reaction for the synthesis of the benzene ring of **206**. (Adapted with permission from Ref. [10]. Copyright (2009) American Chemical Society).

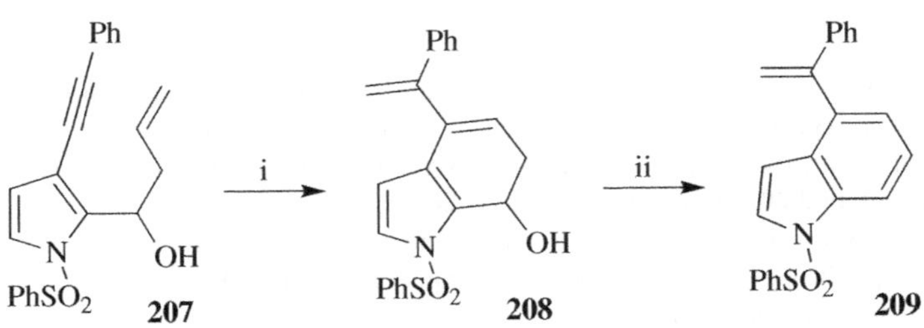

Reagents and conditions: (i) **2-Ru** (7.5 mol%), toluene, CH_2CH_2, 80°C, 12 h (ii) *p*-TsOH.H_2O (10 mol%), rt 1 h, (99%).

SCHEME 17.42 An enyne metathesis reaction for the synthesis of the benzene ring of 4-vinylindoles.

Finally, an unusual synthesis of indoles has been described using the enyne metathesis/Diels–Alder approach. The authors of this work reported that when (*Z*)-**210** was exposed to catalyst **2-Ru** for 15 min, it presumably formed the intermediate diene **211** by way of an enyne metathesis reaction (Scheme 17.43) [69]. This process was then followed by an intramolecular CM resulting in a

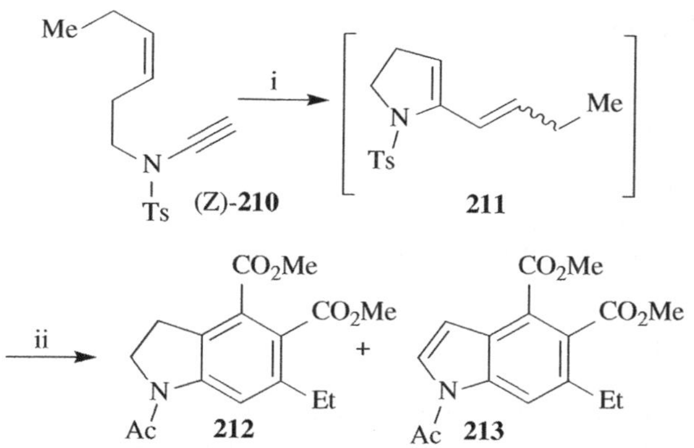

Reagents and conditions: For example (i) **2-Ru** (10 mol%), toluene, 80°C, 15 min.; (ii) (a) DMAD, toluene, 100°C, 3 h, (b) DDQ, toluene, 80°C, 20 h, **212** (20%) + **213** (12%), over three steps.

SCHEME 17.43 An enyne RCM reaction and a Diels–Alder reaction resulting in the synthesis of **212** and **213**. (Reprinted with permission from Ref. [10]. Copyright (2009) American Chemical Society).

metathetic transfer of the ethyl group. In this process, the 5-membered pyrrole ring of the indole moiety was constructed. However, since **211** is also a diene, it could undergo Diels–Alder reactions to allow for the construction of a benzene ring to form indoles. DDQ was therefore added with DMAD while heating to allow for aromatization to occur. Chromatography then resulted in the isolation of two compounds, namely, the indoline **212** in 20% yield and the substituted indole **213** in a yield of only 12%. Unfortunately, the authors of this work did not optimize this particular reaction, nor investigate the applicability of the enyne metathesis/CM/Diels–Alder/aromatization sequence to other indole systems.

17.6 FUTURE CHALLENGES

Of course, challenges exist in the use of alkene and enyne RCM for the assembly of aromatic rings. Firstly, while the methodology may not rely on the use of preformed aromatic rings, the metathesis precursors still need to be synthesized, and this synthesis can still be particularly challenging. Secondly, the experimental procedures, yields associated with the reactions, and the costs associated with the use of precious metal catalysts for the metathesis reaction may limit their use, particularly in an industrial setting. However, if cheaper, more air- and water-tolerant catalysts can be developed, this method for the synthesis of substituted benzene rings may find wider application.

17.7 CONCLUSIONS

In summary, as demonstrated by this chapter, the use of alkene and enyne RCM has seen a wide range of applications in the synthesis of aromatic compounds, where the construction of benzene rings or an additional fused benzene ring onto preexisting scaffolds has become possible. A distinct difference and possible advantage of the judicious use of this methodology, as compared to other available synthetic methods for assembling substituted benzene rings, is that it does not necessarily rely on preformed aromatic compounds.

ABBREVIATIONS

CM	Cross-metathesis
DMAD	Dimethyl acetylenedicarboxylate
DMG	Directing metalation group
RCM	Ring-closing metathesis
ROM	Ring-opening metathesis
ROMP	Ring-opening metathesis polymerization

REFERENCES

[1] Claverie, J. R. and Schaper, F. (2013) *MRS Bull.*, **38**, 213–218.

[2] Truett, W. L., Johnson, D. R., Robinson, I. M. and Montague, B. A. (1960) *J. Am. Chem. Soc.*, **82**, 2337–2340.

[3] Banks, R. L. and Bailey, G. C. (1964) *Ind. Eng. Chem. Prod. Res. Dev.*, **3**, 170–173.

[4] Calderon, N., Chen, H. Y. and Scott, K. W. (1967) *Tetrahedron Lett.*, **8**, 3327–3329.

[5] Astruc, D. (2005) *New J. Chem.*, **29**, 42–56.

[6] (a) Chauvin, Y. (2006) *Angew. Chem. Int. Ed.*, **45**, 3740–3747; (b) Grubbs, R. H. (2006) *Angew. Chem. Int. Ed.*, **45**, 3760–3765; (c) Schrock, R. R. (2006) *Angew. Chem. Int. Ed.*, **45**, 3748–3759.

[7] For topical reviews just in 2013, see the following references: (a) Wojtkielewicz, A. (2013) *Curr. Org. Synth.*, **10**, 43–66; (b) Vaddula, B. R., Varma, R. S. and Gonzalez, M. A. (2013) *Curr. Org. Chem.*, **17**, 2268–2278; (c) Tomasek, J. and Schatz, J. (2013) *Green Chem.*, **15**, 2317–2338; (d) Skowerski, K., Wierzbicka, C. and Grela, K. (2013) *Curr. Org. Chem.*, **17**, 2740–2748; (e) Simocko, C., Atallah, P. and Wagener, K. B. (2013) *Curr. Org. Chem.*, **17**, 2749–2763; (f) Shahane, S., Bruneau, C. and Fischmeister, C. (2013) *ChemCatChem*, **5**, 3436–3459; (g) Schrock, R. R. (2013) *Chem. Commun.*, **49**, 5529–5531; (h) Schanz, H.-J. (2013) *Curr. Org. Chem.*, **17**, 2575–2591; (i) Ranganath, K. V. S., Onitsuka, S., Kumar, A. K. and Inanaga, J. (2013) *Catal. Sci. Technol.*, **3**, 2161–2181; (j) Queval, P., Jahier, C., Basle, O. and Mauduit, M. (2013) *Curr. Org. Chem.*, **17**, 2560–2574; (k) Oehninger, L., Rubbiani, R. and Ott, I. (2013) *Dalton Trans.*, **42**, 3269–3284; (l) Hamad, F. B., Sun, T., Xiao, S. and Verpoort, F. (2013) *Coord. Chem. Rev.*, **257**, 2274–2292; (m) Hamad, F. B., Kai, C., Cai, Y., Xie, Y., Lu, Y., Ding, F., Sun, Y. and Verpoort, F. (2013) *Curr. Org. Chem.*, **17**, 2592–2608; (n) Fürstner, A. (2013) *Angew. Chem. Int. Ed.*, **52**, 2794–2819; (o) Fürstner, A. (2013) *Science*, **341**, 1357; (p) Dragutan, I., Dragutan, V., Demonceau, A. and Delaude, L. (2013) *Curr. Org. Chem.*, **17**, 2678–2720; (q) Deraedt, C., d'Halluin, M. and Astruc, D. (2013) *Eur. J. Inorg. Chem.*, **2013**, 4881–4908; (r) Dassonneville, B., Delaude, L., Demonceau, A., Dragutan, I., Dragutan, V., Etse, K. S. and Hans, M. (2013) *Curr. Org. Chem.*, **17**, 2609–2653; (s) Czaban, J., Torborg, C. and Grela, K. (2013) *Sustainable Catalysis: Challenges and Practices for the Pharmaceutical and Fine Chemical Industries*, edited by Dunn, P. J., Hii, K. K., Krische, M. J., Williams, M. T., John Wiley & Sons, Inc., Hoboken, NJ, pp. 163–214; (t) Buchmeiser, M. R. (2013) *Curr. Org. Chem.*, **17**, 2764–2775; (u) Atallah, P., Wagener, K. B. and Schulz, M. D. (2013) *Macromolecules*, **46**, 4735–4741.

[8] Hérisson, J.-L. and Chauvin, Y. (1971) *Makromol. Chem.*, **141**, 161–176.

[9] (a) Donner, C. D. (2013) *Tetrahedron*, **69**, 3747–3773; (b) Mal, D. and Pahari, P. (2007) *Chem. Rev.*, **107**, 1892–1918; (c) Rathwell, K. and Brimble, M. A. (2007) *Synthesis*, 643–662.

[10] van Otterlo, W. A. L., and de Koning, C. B. (2009) *Chem. Rev.*, **109**, 3743–3782.

[11] Trnka, T. M. and Grubbs, R. H. (2001) *Acc. Chem. Res.*, **34**, 18–29.

[12] Schrock, R. R. (1990) *Acc. Chem. Res.*, **23**, 158–165.

[13] Kingsbury, J. S., Harrity, J. P. A., Bonitatebus, P. J. and Hoveyda, A. H. (1999) *J. Am. Chem. Soc.*, **121**, 791–799.

[14] Yoshida, K. and Imamoto, T. (2005) *J. Am. Chem. Soc.*, **127**, 10470–10471.

[15] (a) Yoshida, K., Horiuchi, S., Iwadate, N., Kawagoe, F. and Imamoto, T. (2007) *Synlett*, 1561–1564; (b) Yoshida, K., Kawagoe, F., Iwadate, N., Takahashi, H. and Imamoto, T. (2006) *Chem. Asian J.*, **1**, 611–613.

[16] Yoshida, K., Toyoshima, T. and Imamoto, T. (2007) *Chem. Commun.*, 3774–3776.

[17] Yoshida, K., Takahashi, H. and Imamoto, T. (2008) *Chem. Eur. J.*, **14**, 8246–8261.

[18] Yoshida, K., Narui, R. and Imamoto, T. (2008) *Chem. Eur. J.*, **14**, 9706–9713.

[19] Ziffle, V. E., Cheng, P. and Clive, D. L. J. (2010) *J. Org. Chem.*, **75**, 8024–8038.

[20] Nishihara, Y., Suetsugu, M., Saito, D., Kinoshita, M. and Iwasaki, M. (2013) *Org. Lett.*, **15**, 2418–2421.

[21] de Koning, C. B., Rousseau, A. L. and van Otterlo, W. A. L. (2003) *Tetrahedron*, **59**, 7–36.

[22] van Otterlo, W. A. L., Ngidi, E. L., Coyanis, E. M. and de Koning, C. B. (2003) *Tetrahedron Lett.*, **44**, 311–313.

[23] Kotha, S. and Mandal, K. (2009) *Chem. Asian J.*, **4**, 354–362.

[24] Bonifacio, M. C., Robertson, C. R., Jung, J.-Y. and King, B. T. (2005) *J. Org. Chem.*, **70**, 8522–8526.

[25] Walker, E. R., Leung, S. Y. and Barrett, A. G. M. (2005) *Tetrahedron Lett.*, **46**, 6537–6540.

[26] Iuliano, A., Piccioli, P. and Fabbri, D. (2004) *Org. Lett.*, **6**, 3711–3714.

[27] Nguyen Van, T., D'hooghe, M., Pattyn, S. and De Kimpe, N. (2004) *Synlett*, 1913–1916.

[28] Kotha, S. and Mandal, K. (2004) *Tetrahedron Lett.*, **45**, 2585–2588.

[29] Collins, S. K., Grandbois, A., Vachon, M. P. and Côté, J. (2006) *Angew. Chem. Int. Ed. Engl.*, **45**, 2923–2926.

[30] Grandbois, A. and Collins, S. K. (2008) *Chem. Eur. J.*, **14**, 9323–9329.

[31] (a) Scott, L. T. (2004) *Angew. Chem. Int. Ed. Engl.*, **43**, 4994–5007; (b) Wu, Y.-T. and Siegel, J. S. (2006) *Chem. Rev.*, **106**, 4843–4867.

[32] Higashibayashi, S. and Sakurai, H. (2008) *J. Am. Chem. Soc.*, **130**, 8592–8593.

[33] Sakurai, H., Daiko, T. and Hirao, T. (2003) *Science*, **301**, 1878.

[34] Srikrishna, A., Satyanarayana, G. and Prasad, K. R. (2007) *Synth. Commun.*, **37**, 1511–1516.

[35] Jones, S. B., He, L. and Castle, S. L. (2006) *Org. Lett.*, **8**, 3757–3760.

[36] (a) Kotha, S., Halder, S. and Brahmachary, E. (2002) *Tetrahedron*, **58**, 9203–9208; (b) Kotha, S., Halder, S., Brahmachary, E. and Ganesh, T. (2000) *Synlett*, 853–855.

[37] Yoshida, K., Shishikura, Y., Takahashi, H. and Imamoto, T. (2008) *Org. Lett.*, **10**, 2777–2780.

[38] Feng, C., Wang, X., Wang, B.-Q., Zhao, K.-Q., Hu, P. and Shi, Z.-J. (2012) *Chem. Commun.*, **48**, 356–358.

[39] Voigtritter, K., Ghorai, S. and Lipshutz, B. H. (2011) *J. Org. Chem.*, **76**, 4697–4702.

[40] (a) Katz, T. J. and Sivavec, T. M. (1985) *J. Am. Chem. Soc.*, **107**, 737–738; (b) Sivavec, T. M., Katz, T. J., Chiang, M. J. and Yang, G. X.-Q. (1989) *Organometallics*, **8**, 1620–1625.

[41] (a) Kaliappan, K. P. and Ravikumar, V. (2007) *J. Org. Chem.*, **72**, 6116–6126; (b) Kaliappan, K. P. and Ravikumar, V. (2007) *Synlett*, 977–979.

[42] Vanga, D. G. and Kaliappan, K. P. (2012) *Synlett*, 2931–2934.

[43] Vanga, D. G. and Kaliappan, K. P. (2012) *Eur. J. Org. Chem.*, 2250–2259.

[44] Kurhade, S. E., Sanchawala, A. I., Ravikumar, V., Bhuniya, D. and Reddy, D. S. (2011) *Org. Lett.*, **13**, 3690–3693.

[45] Kotha, S., Mandal, K., Banerjee, S. and Mobin, S. M. (2007) *Eur. J. Org. Chem.*, 1244–1255.

[46] Bugaut, X., Guinchard, X. and Roulland, E. (2010) *J. Org. Chem.*, **75**, 8190–8198.

[47] Castagnolo, D., Giorgi, G., Spinosa, R., Corelli, F. and Botta, M. (2007) *Eur. J. Org. Chem.*, 3676–3686.

[48] (a) Chopade, P. R. and Louie, J. (2006) *Adv. Synth. Catal.*, **348**, 2307–2327; (b) Kotha, S., Brahmachary, E. and Lahiri, K. (2005) *Eur. J. Org. Chem.*, 4741–4767; (c) Yamamoto, Y. (2005) *Curr. Org. Chem.*, **9**, 503–519.

[49] Saito, S. and Yamamoto, Y. (2000) *Chem. Rev.*, **100**, 2901–2915.

[50] Peters, J.-U. and Blechert, S. (1997) *Chem. Commun.*, 1983–1984.

[51] (a) Hoven, G. B., Efskind, J., Rømming, C. and Undheim, K. (2002) *J. Org. Chem.*, **67**, 2459–2463; (b) Undheim, K., Efskind, J. and Hoven, G. B. (2003) *Pure Appl. Chem.*, **75**, 279–292.

[52] Das, S. K. and Roy, R. (1999) *Tetrahedron Lett.*, **40**, 4015–4018.

[53] Witulski, B., Stengel, T. and Fernández-Hernández, J. M. (2000) *Chem. Commun.*, 1965–1966.

[54] Platon, M., Amardeil, R., Djakovitch, L. and Hierso, J.-C. (2012) *Chem. Soc. Rev.*, **41**, 3929–3968.

[55] Yoshida, K., Hayashi, K. and Yanagisawa, A. (2011) *Org. Lett.*, **13**, 4762–4765.

[56] Schmidt, A. W., Reddy, K. R. and Knölker, H.-J. (2012) *Chem. Rev.*, **112**, 3193–3328.

[57] Bennasar, M.-L., Zulaica, E. and Tummers, S. (2004) *Tetrahedron Lett.*, **45**, 6283–6285.

[58] Selvakumar, N., Khera, M. K., Reddy, B. Y., Srinivas, D., Azhagan, A. M. and Iqbal, J. (2003) *Tetrahedron Lett.*, **44**, 7071–7074.

[59] Pelly, S. C., Parkinson, C. J., van Otterlo, W. A. L. and de Koning, C. B. (2005) *J. Org. Chem.*, **70**, 10474–10481.

[60] Stephenson, G. R., Cauteruccio, S. and Doulceta, J. (2014) *Synlett*, 701–707.

[61] Tsai, T.-H., Chung, W.-H., Chang, J.-K., Hsu, R.-T. and Chang, N.-C. (2007) *Tetrahedron*, **63**, 9825–9835.

[62] Mamane, V. and Fort, Y. (2005) *J. Org. Chem.*, **70**, 8220–8223.

[63] Du, H., He, Y., Sivappa, R. and Lovely, C. J. (2006) *Synlett*, 965–992.

[64] (a) Chen, Y., Dias, H. V. R. and Lovely, C. J. (2003) *Tetrahedron Lett.*, **44**, 1379–1382; (b) Lovely, C. J., Chen, Y. and Ekanayake, E. V. (2007) *Heterocycles*, **74**, 873–894.

[65] (a) Kotha, S. and Sreenivasachary, N. (2000) *Chem. Commun.*, 503–504; (b) Kotha, S. and Sreenivasachary, N. (2001) *Eur. J. Org. Chem.*, 3375–3383.

[66] Yang, Y.-K. and Tae, J. (2003) *Synlett*, 2017–2020.

[67] Tae, J. and Hahn, D.-W. (2004) *Tetrahedron Lett.*, **45**, 3757–3760.

[68] Hayashi, K., Yoshida, K. and Yanagisawa, A. (2013) *J. Org. Chem.*, **78**, 3464–3469.

[69] Mori, M., Wakamatsu, H., Saito, N., Sato, Y., Narita, R., Sato, Y. and Fujita, R. (2006) *Tetrahedron*, **62**, 3872–3881.

18

AROMATIC REARRANGEMENTS IN WHICH THE MIGRATING GROUP MIGRATES TO THE AROMATIC NUCLEUS: AN OVERVIEW

TIMOTHY J. SNAPE
School of Pharmacy and Biomedical Sciences, University of Central Lancashire, Preston, UK

18.1 INTRODUCTION

Aromatic rearrangements in which the migrating group migrates to the aromatic nucleus are widespread throughout organic chemistry, since they provide useful approaches to molecules that can otherwise be difficult to prepare. The mechanisms of such rearrangements can be largely divided into two main subsections—intramolecular and intermolecular—in which the aromatic ring in question either acts as an electrophile, and "accepts" the migrating group, or it acts as a nucleophile through a dissociative mechanism. As such, this chapter will cover examples of both possibilities (categorized accordingly) ranging from state-of-the-art organic synthesis to mechanistic studies and the application of such rearrangements to problems of academic and industrial interest.

In some cases, the mechanism may be contentious, or it does not fit neatly into the above categorization, in which case(s), the reaction has been included in the section that maintains the best flow within the text, but every effort has been made to include references relating to other possible mechanisms.

Arene Chemistry: Reaction Mechanisms and Methods for Aromatic Compounds,
First Edition. Edited by Jacques Mortier.

18.2 MECHANISMS BY CLASSIFICATION

18.2.1 Intramolecular Reactions: Nucleophilic Aromatic Substitution

18.2.1.1 Heteroatom→Heteroatom Aromatic Rearrangements (Smiles Rearrangement) The Smiles rearrangement encompasses molecular reorganizations that undergo an intramolecular nucleophilic aromatic substitution (S_NAr) process between substrates of type **1**, where X and Y are both heteroatoms capable of carrying a negative charge and where Y is sufficiently nucleophilic to displace X from the aromatic ring. Typically, the reaction proceeds via an addition–elimination mechanism and goes via the charged Meisenheimer complex. In the example shown in Scheme 18.1, the nucleophile (Y) attacks the electron-deficient *ipso*-position, before elimination of X occurs, and aromaticity is regained. The benzene ring can be further substituted (R), while R^1 usually needs to be electron withdrawing (i.e., $-NO_2$) to facilitate the nucleophilic addition step of the mechanism (Scheme 18.1). The carbon framework connecting X and Y (shown here as $-CH_2-CH_2-$) can also be widely varied and is not limited to sp^3-hybridized carbons [1].

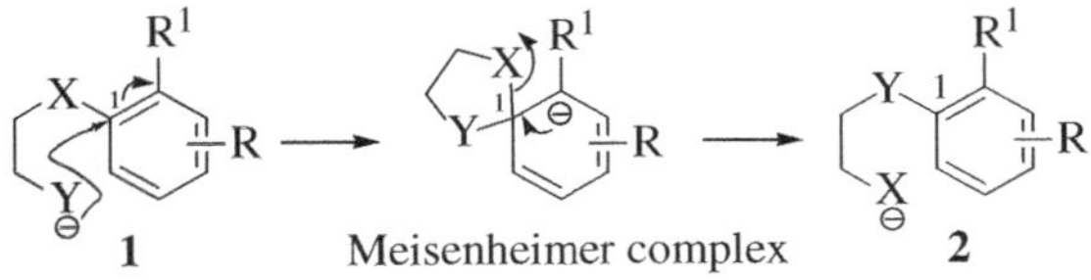

R = various, R^1 = electron withdrawing
X = O, S, CO_2, SO_2
YH = OH, NH_2, $CONH_2$, NHAc, SH

SCHEME 18.1 The Smiles rearrangement.

A range of substrates have been deemed successful in the Smiles rearrangement under basic conditions, but notable examples revealing its versatility are shown in Figure 18.1, all of which proceed via the same mechanism to that shown in Scheme 18.1.

Matsui et al. were interested in what kind of photochemical reactions would occur in molecules such as **1**, and their work describes the photochemical reactions of such aromatic molecules. Figure 18.1 exemplifies such a reaction (**3**→**4**) using a mercury lamp as the radiation source [2]. Wadia and Patil have also utilized the Smiles rearrangement (e.g., **5**→**6**) toward the synthesis of diclofenac analogues [3].

In studies toward the synthesis of glycyrol, Kim and coworkers employed a novel precursor, iodium acetate salt **7**, which was successfully incorporated into a novel Smiles rearrangement, ultimately leading to the first total synthesis of glycyrol (Figure 18.1) [4].

A more recent example sees the conversion of an aryl ether (**9**) to an *N*-aryl-2-hydroxypropionamide (**10**) in high overall yields from the starting materials, ultimately leading to a two-step Smiles–Sonogashira synthesis of indoles from 2,3-dihalophenols [5].

The Smiles rearrangement also proceeds under microwave conditions and has been shown to provide a facile C–N bond formation and a one-pot synthesis of arylamines (**13**) and derivatives (Figure 18.1) [6].

The Smiles rearrangement is also widely used on heterocyclic systems; however, due to the nature of this review, these will not be discussed here.

18.2.1.2 Heteroatom→Carbon Aromatic Rearrangements (Truce–Smiles Rearrangement) A simple variant of the Smiles rearrangement is the Truce–Smiles modification, a reaction that has been little used in recent times, despite it arguably being more synthetically useful due to the selective formation of carbon–carbon bonds (Scheme 18.2) [7]. As shown, a carbon-based nucleophile replaces that of the heteroatom nucleophile (Y, in Scheme 18.1).

Traditionally, the definition of this particular variant follows two important differences to the more common Smiles rearrangement above; these are as follows: (i) a carbanion is the nucleophile

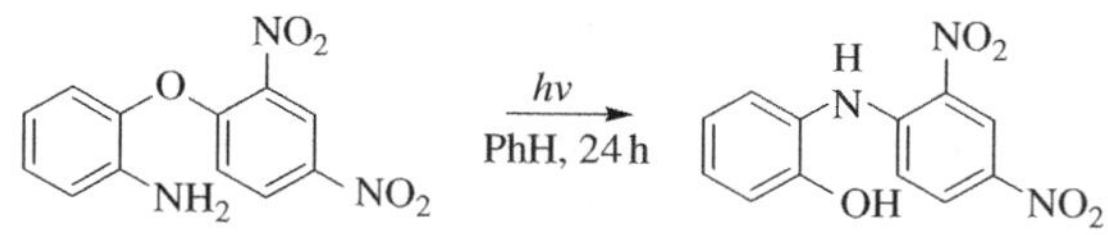

3 **4**

A photo-Smiles rearrangement

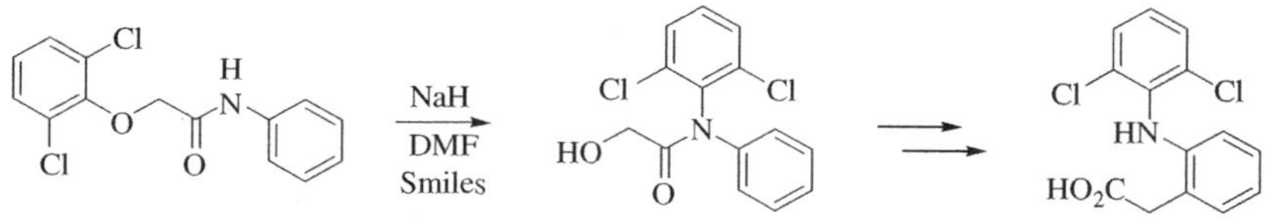

5 **6** Diclofenac

The Smiles rearrangement toward the synthesis of diclofenac

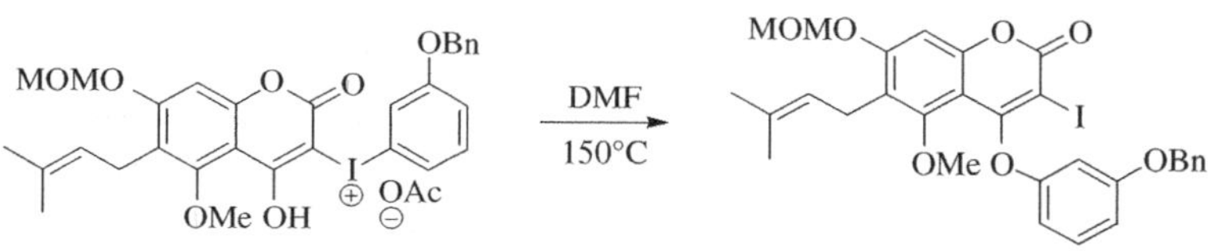

7 **8**

The Smiles rearrangement incorporated into the total synthesis of glycyrol

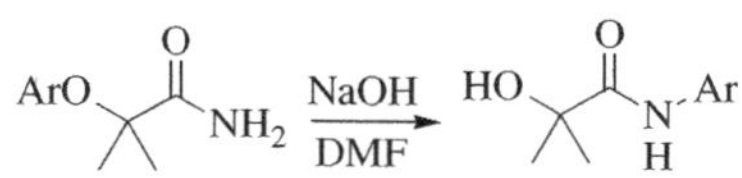

9 **10**

Synthesis of *N*-aryl-2-hydroxypropionamides

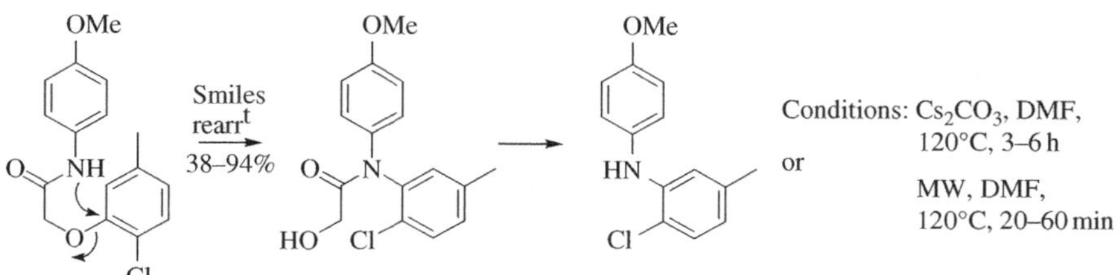

11 **12** **13**

Smiles rearrangement toward the synthesis of arylamines

FIGURE 18.1 Notable examples of the versatility of the Smiles rearrangement.

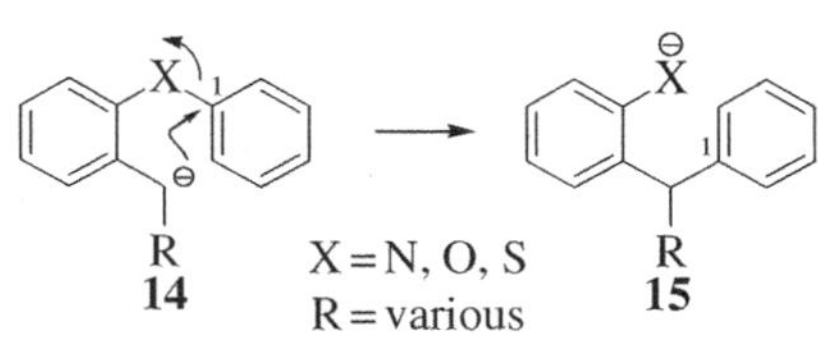

14 **15**

SCHEME 18.2 The Truce–Smiles rearrangement.

rather than a heteroatom, and (ii) whereas the Smiles rearrangement requires an activating substituent in the migrating aryl unit (e.g., *ortho-* or *para*-nitro groups), such activation is not needed, and perhaps would not be tolerated, in the Truce–Smiles rearrangement due to the occasional need for strong alkyl lithium bases for deprotonation [8]. Nevertheless, despite this apparent restriction, more and more examples have been, and continue to be, developed based on carbon nucleophiles that do make use of an activating group and therefore rearrange with weaker, more compatible bases. As such, due to this precedent, more recent reference to the Truce–Smiles rearrangement has been extended to mean any variation of the Smiles rearrangement, which utilizes carbanions in general as the nucleophile [9].

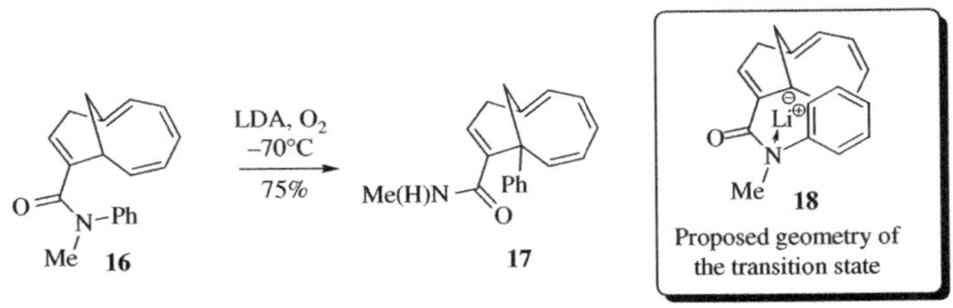

Ito's Truce–Smiles rearrangement of a substituted anilide

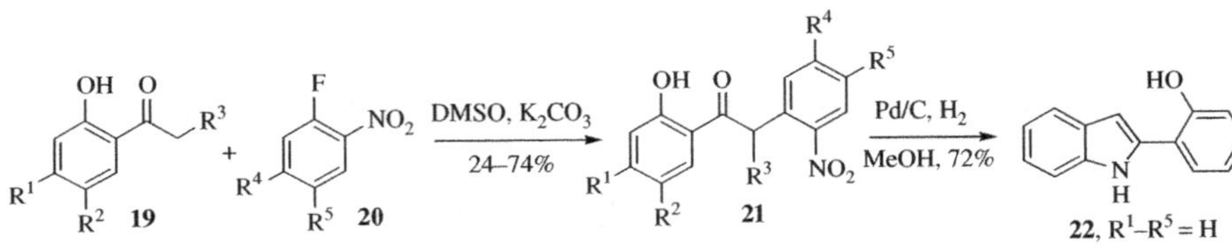

R^1–R^5 = H, Me, Br, CO_2Me, NO_2, OMe

Snape's work on expanding the substrate scope of the Truce–Smiles rearrangement

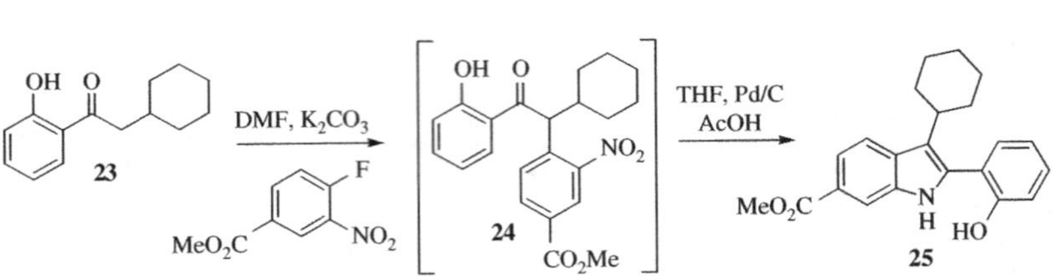

Merck's efficient process for the large-scale synthesis of a 2,3,6-trisubstituted indole

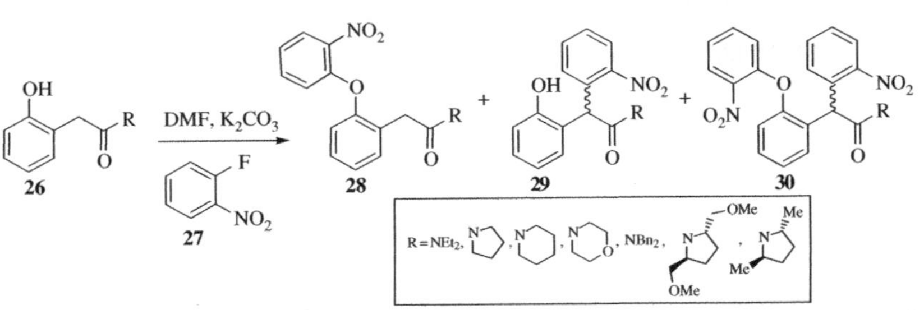

Snape's work developing the scope of the Truce–Smiles reaction toward new asymmetric reactions

FIGURE 18.2 Worthy examples of the versatility of the Truce–Smiles rearrangement.

W. E. Truce developed this variant in the late 1950s [7a] and went on to show that it was a reliable method for the synthesis of a variety of substituted aromatic sulfinic acids. Since then, a variety of substrate types have been developed by others. This particular rearrangement was subject to review by us in 2008 [10], and worthy examples are outlined in Figure 18.2.

Itô and coworkers discovered a unique Truce–Smiles rearrangement of the substituted anilide intermediate **16** upon exposure to LDA. When **16** was subjected to an oxidative process (LDA, THF, O_2, −70°C), the rearranged product **17** was obtained in 75% yield instead of the desired oxidation product. Their proposed transition state geometry (**18**) is depicted in the box in Figure 18.2 [11].

Furthermore, in 2008, we further developed the Truce–Smiles rearrangement when we applied it to the synthesis of substituted α-aryl aryl ketones (**21**, Figure 18.2) and the synthetic utility of the products was further demonstrated by the conversion of one of them to the substituted indole 2-(1*H*-indol-2-yl)phenol (**22**) via concomitant nitro reduction, cyclization, and aromatization [12]. The reaction proceeds via a two-step process, which involves (i) formation of a diaryl ether (akin to **28**), followed by (ii) Truce–Smiles rearrangement of the diaryl ether and migration of the aromatic ring from O→C. This two-step process has been proven, since attempted direct intermolecular arylation of an acetophenone derivative that lacks the phenolic —OH group fails to produce the C-arylated product. Our work has since been the inspiration for an efficient large-scale synthesis of a 2,3,6-trisubstituted indole, **25**, by researchers at Merck Sharp and Dohme [13]. In total, greater than 50 kg of the target indole was synthesized in 55% overall yield over five steps using this new Truce–Smiles route [13]. Furthermore, our efforts continued by attempting to produce the first asymmetric Truce–Smiles rearrangement by exploring amide substrates as potential precursors for new asymmetric reactions (Figure 18.2) [14]. During that work, the production of diastereomerically enriched α-aryl carbonyl compounds (**29** and **30**) was achieved. While we could only report modest diastereoselectivities (~1.6:1), the work demonstrated that the concept has potential for further optimization [14].

In related work, Ma and coworkers recently reported an efficient C—C bond forming reaction for the synthesis of a wide range of substituted of α-arylated ketones (**32**), under mild conditions, in moderate to excellent yields (41–97%; Scheme 18.3), reactions that proceed via intermediate **33** [15]. While most include heteroaromatic examples, there were a number of nonheteroaromatic cases exemplified.

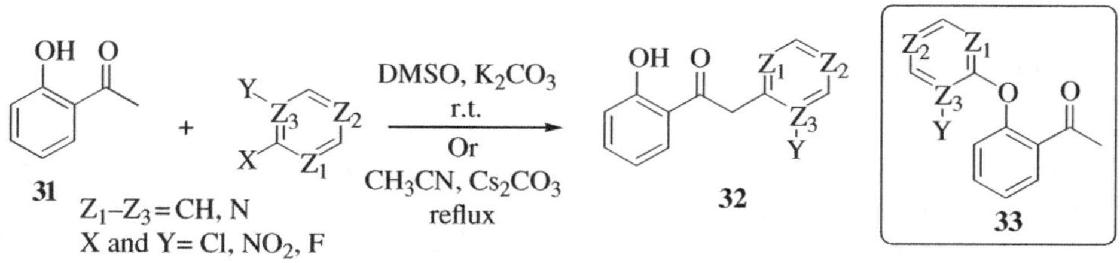

SCHEME 18.3 The synthesis of medicinally relevant scaffolds via the Truce–Smiles rearrangement.

With the advantages of cascade reactions in mind, El Kaim and coworkers developed a Smiles/Truce–Smiles cascade, which demonstrated that two different Smiles rearrangements can be combined to afford a multicomponent formation of isoquinolinones (**36**) and isoindolinones (**37**) from nitro methyl salicylate (**34**) (Scheme 18.4) [16]. After an Ugi–Smiles four-component coupling, the base-triggered cyclization of the resulting adduct (**35**) is followed by a Truce–Smiles rearrangement leading to therapeutically useful products [16].

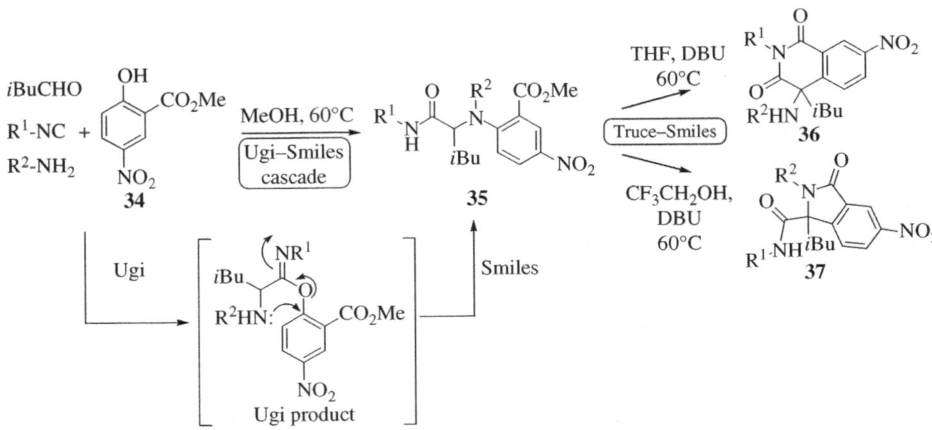

SCHEME 18.4 An Ugi–Smiles/Truce–Smiles cascade.

The value of such heteroatom→C rearrangements has been expertly demonstrated by Clayden and coworkers when it was shown that substituted diarylmethylamines could be prepared by a stereospecific intramolecular electrophilic arylation of lithiated ureas [17]. This work was soon followed-up and developed further to produce α-arylated cyclic amines [18], α-arylated benzylic alcohols [19], an asymmetric synthesis of clemastine (Scheme 18.5) [20], a sequential double α-arylation of *N*-allylureas by asymmetric deprotonation and N→C aryl migration [21], and a simple method for the synthesis of tertiary thiols and thioethers [22], and tertiary alcohols [23], all involving a Truce–Smiles rearrangement.

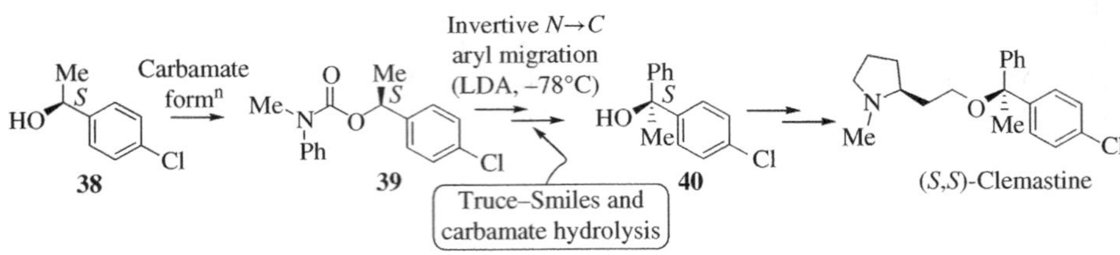

SCHEME 18.5 Clayden's rearrangement, illustrated using a key step in the synthesis of clemastine.

18.2.1.3 Rearrangement of Imidates to Amides (Chapman Rearrangement) and a Related Amidine Analogue Mechanistically related to the Smiles and Truce–Smiles rearrangements, the Chapman rearrangement effects the transformation of imidates, of general structure **41** (Scheme 18.6),

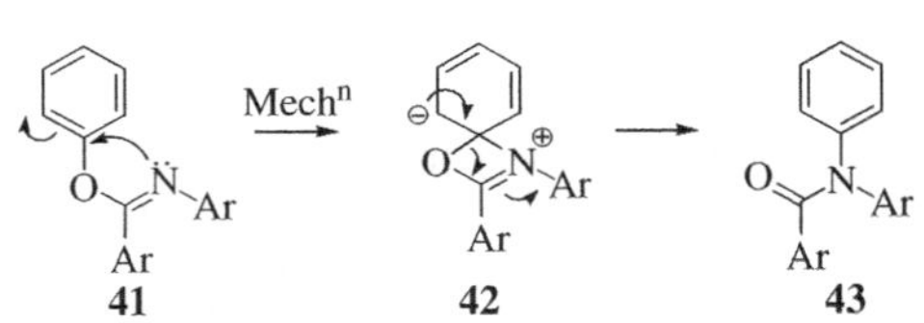

SCHEME 18.6 Chapman rearrangement of an imidate to an amide; a range of substituted rings are tolerated.

to amides (**43**). A reaction that proceeds under harsh reaction conditions (above 200°C), via an intramolecular mechanism, confirmed through crossover studies, in which a zwitterionic spirocyclic intermediate, **42**, can be postulated [24].

The Chapman rearrangement has been exploited in the synthesis of compounds that possess potent antiasthmatic activity. In particular, imidazo[4,5-*c*]quinolin-4(5*H*)-one derivatives (**44**) underwent reactions that were shown to proceed via an unusual thermally promoted rearrangement, wherein the methyl group at N-1 migrates to N-3 with concomitant Chapman rearrangement (Scheme 18.7) [25].

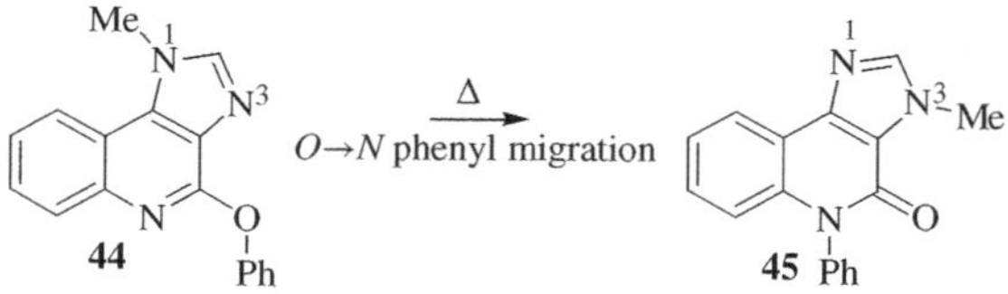

44 **45**

SCHEME 18.7 A Chapman rearrangement with concomitant methyl migration.

Application of this rearrangement in materials synthesis has been observed during the preparation of aromatic polyamides (**47**), a reaction that has been shown to occur either in the melt or in solution, at temperatures greater than 200°C. The resulting polymers are of interest due to their solubility in organic solvents, their high viscosity characteristics and good film-forming properties, as well as their high thermooxidative stability (Scheme 18.8) [26].

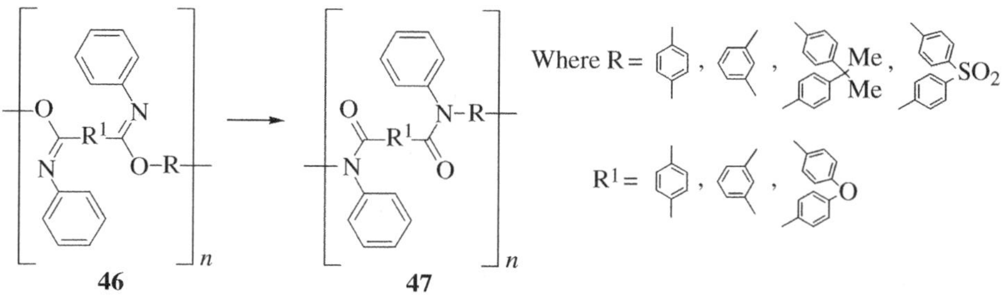

46 **47**

SCHEME 18.8 The Chapman rearrangement applied to the synthesis of aromatic polyamides.

A detailed theoretical study on the Chapman rearrangement has been conducted, and to the researchers, it was clear that, in all cases studied, the products were shown to be thermodynamically more stable than the reactants [27].

Despite both intramolecular and intermolecular mechanisms for the Chapman rearrangement having been postulated [28], it is the intramolecular one that has been confirmed by crossover experiments and isotopic labeling studies [29]. It has also been observed that the rearrangement follows first-order kinetics involving a nucleophilic aromatic substitution step [30].

A mechanistically related reaction is the analogous amidine rearrangement (Scheme 18.9). As with the other rearrangements, the aryl migration step can be facilitated by electron-withdrawing groups on the migrating ring. However, due to the equilibrium of this alternative reaction, its synthetic application has been limited.

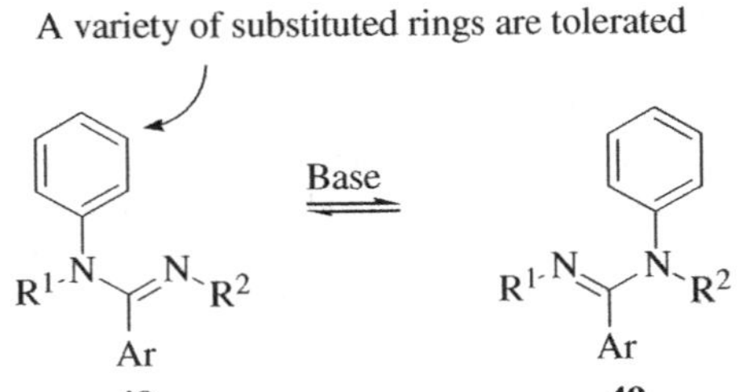

48 49

SCHEME 18.9 A Chapman-related rearrangement of amidines.

Despite this limitation, Esser and coworkers have managed to demonstrate the applicability of this 1,3-aryl migration, under neutral or basic conditions, in a synthesis of 2-arylamino-4,5,6,7-tetrahydro-1*H*-1,3-diazepines (**51**) (Scheme 18.10) [31]. Mechanistic studies on this amidine rearrangement in which a 1:1 mixture of two different rearrangement precursors (i.e., **50**) was reacted showed no evidence of intermolecular aryl transfer, thus providing evidence for an intramolecular transformation that, in comparison to the Chapman rearrangement, should proceed via a spirocyclic intermediate such as **52**.

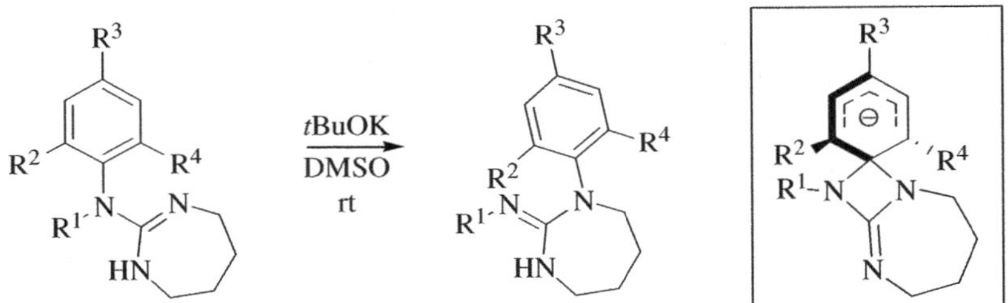

50 51 52

SCHEME 18.10 A Chapman-related rearrangement toward the synthesis of 1,3-diazepines (**51**).

18.2.1.4 Rearrangement of O-Thiocarbamates to S-Thiocarbamates (Newman–Kwart Rearrangement) Another mechanistically related reaction to the Chapman rearrangement is that named after Newman [32] and Kwart [33]. This particular variant involves the 1,3-migration of an aryl group from oxygen to sulfur in a thiocarbamate (Scheme 18.11), and the reaction has recently been reviewed [34].

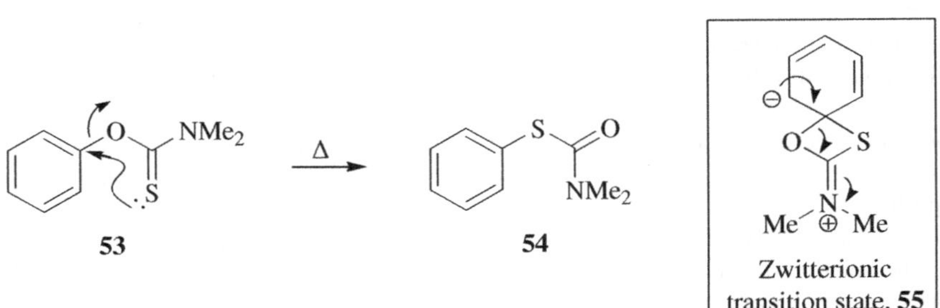

53 54

Zwitterionic transition state, **55**

SCHEME 18.11 The Newman–Kwart rearrangement of an *O*-thiocarbamate to an *S*-thiocarbamate.

The mechanism for the Newman–Kwart rearrangement follows a similar pathway to the Chapman and related reactions. For example, attack of the sulfur atom on the *ipso* carbon of **53** results in a transition state **55** consisting of a cyclohexadienyl moiety bearing a negative charge, a feature that explains the accelerating effect of electron-withdrawing aromatic substituents. Interestingly, computational analysis of the mechanism has suggested that, unlike traditional nucleophilic aromatic substitution reactions, the Newman–Kwart reaction proceeds via a transition state **55**, not an intermediate [34]. The variety of applications of the Newman–Kwart rearrangement is relatively large and has been reviewed [34], and a few examples of interest are shown in Figure 18.3.

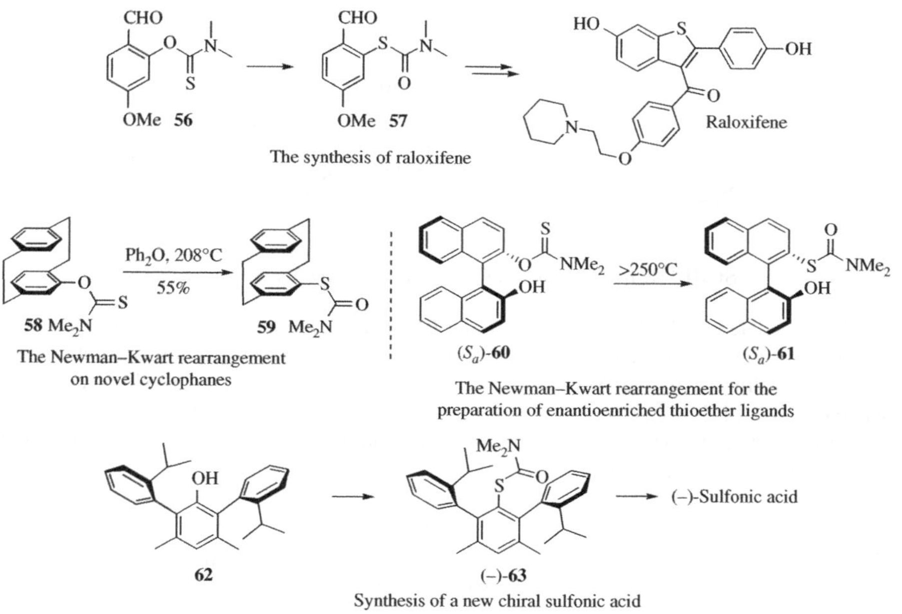

FIGURE 18.3 Interesting examples of the Newman–Kwart rearrangement.

Since its initial development, the rearrangement has been applied in a number of areas, such as a general synthesis of thiophenols (see Scheme 18.11, after hydrolysis of the *S*-thiocarbamate, **54**) [33,35], to the total synthesis of raloxifene (Figure 18.3) [36]. In addition, it has been used for the synthesis of novel cyclophane derivatives (**59**) [37] and to the synthesis of enantioenriched thioether ligands (**61**), which exhibit reduced levels of racemization compared to the equivalent BINOL-derived species [38]. Moreover, the synthesis of a novel chiral sulfonic acid, a class of compound that is paving the way for new opportunities to develop original catalysts, has also been obtained using this rearrangement [39].

Moreover, the Newman–Kwart rearrangement has also been shown to work under microwave radiation [39a], in scale-up procedures [40], as well as being made more practicable with the incorporation of a palladium catalysis step enabling a reduction in reaction temperatures to be obtained [41].

Inspired by the broad potential of the Newman–Kwart rearrangement, Jacobsen undertook a computational study to explore whether the scope of the rearrangement might be expanded and how that goal may be achieved [42]. The authors studied the rearrangement using a variety of thiocarbamates and density functional (B3LYP) and *ab initio* (MP2) methodologies, and their results confirmed the generally accepted mechanism shown in Scheme 18.11.

Scheme 18.12 highlights some of the interesting kinetic effects of this (and related) rearrangements that arise from steric factors caused by proximal substituents. For instance, there is an accelerating effect arising from *ortho*-substituents, which can be attributed to the restriction of rotation around the Ar—O bond. This results in a decrease in the loss of entropy on approach to the transition state geometry required for rearrangement [28a].

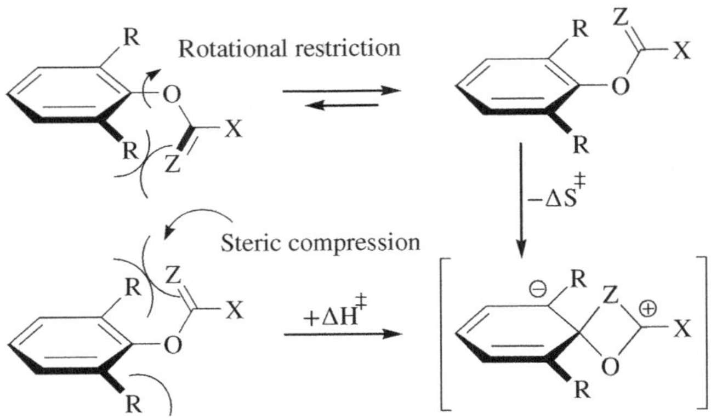

SCHEME 18.12 Kinetic effects of the Newman–Kwart rearrangement.

Closely related to the Newman–Kwart and Chapman reactions, the Schönberg rearrangement also involves the intramolecular rearrangement of an aryl group between nonadjacent atoms, from oxygen to sulfur as in the Newman–Kwart rearrangement, but in a diarylthiocarbonate (**64**; Scheme 18.13) [44]. It has been established that aryl thiocarbonates that lack a β-hydrogen atom undergo this kind of rearrangement under thermal conditions, and as such, it has wide application in the synthesis of substituted thiophenols.

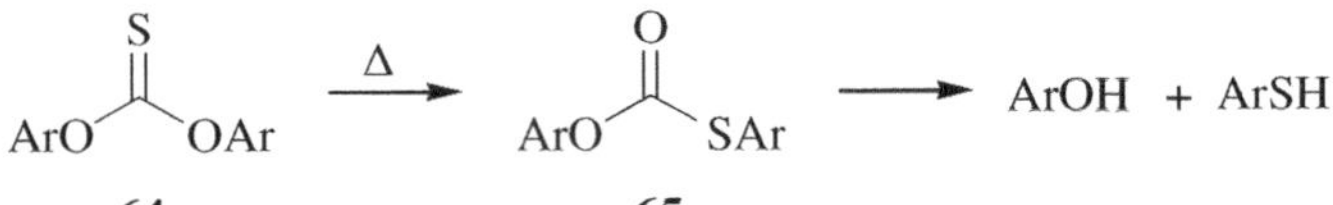

64 **65**

SCHEME 18.13 The Schönberg rearrangement [43].

18.2.2 Intramolecular: Sigmatropic Rearrangements

18.2.2.1 Aromatic Hydrazo Compounds and the Fischer Indole Synthesis The aromatic hydrazo rearrangement is best exemplified by the Fischer indole synthesis, since a cyclic [3,3]-sigmatropic aromatic rearrangement occurs (**66**→**67**) producing an intermediate imine (**67**), which ultimately goes on to produce the energetically favorable aromatic indole (**69**; Scheme 18.14) [45].

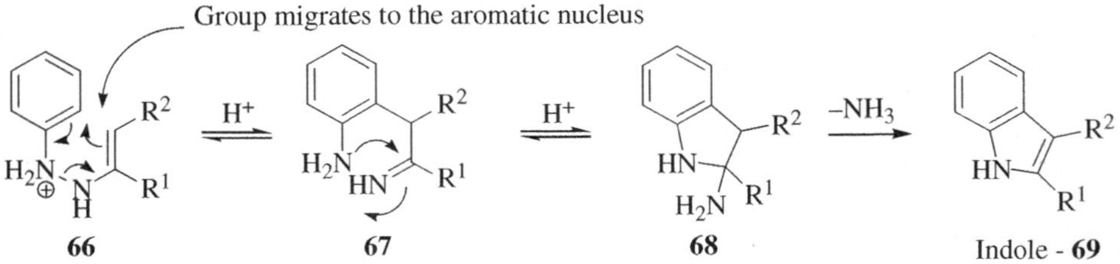

66 **67** **68** Indole - **69**

(3,3)-Sigmatropic rearr[t]

SCHEME 18.14 The Fischer indole synthesis by way of a [3,3]-sigmatropic aromatic hydrazo rearrangement.

2,4,6-Trichloro-1,3,5-triazine (TCT) is a mild, inexpensive, and readily available catalyst, and its widespread use has shown that product isolation is easy, and that when used, yields typically range from good to excellent. Spurred on by this, in 2013, Siddalingamurthy et al. developed a mild and efficient Fischer indole synthesis, which utilized TCT to drive the reaction (Figure 18.4) [46]. Under their conditions, several functional groups (ester, cyano, sulfone, amides, and ethers) are tolerated. The authors demonstrated the preparation of many functionalized analytically pure indoles without the need of purification using this method.

In efforts to increase the complexity of the Fischer indole products obtained using this aromatic rearrangement, List's group used a range of chiral phosphoric acid-based Brønsted acids to access chiral indolines (**75** and **76**) bearing a quaternary stereogenic center (Figure 18.4) [47]. They extended the method to include tethered substrates, which then enabled ring-closing cascades to be achieved, thus providing access to complex propellanes with two vicinal quaternary stereocenters. A proposed catalytic cycle was offered, which is in agreement with the accepted mechanism of the Fischer indolization. Briefly, the chiral phosphonic acid catalyzes the formation of a hydrazone, which, after tautomerization, is primed for the key [3,3]-sigmatropic rearrangement, directed by the chiral phosphonate counterion [47].

Moreover, Gutmann and coworkers have recently reported on the Fischer indole synthesis of 7-ethyltryptophol (**79**), under continuous flow conditions (Figure 18.4), a key intermediate in the synthesis of etodolac, and they outline both their mechanistic and process intensification efforts [48].

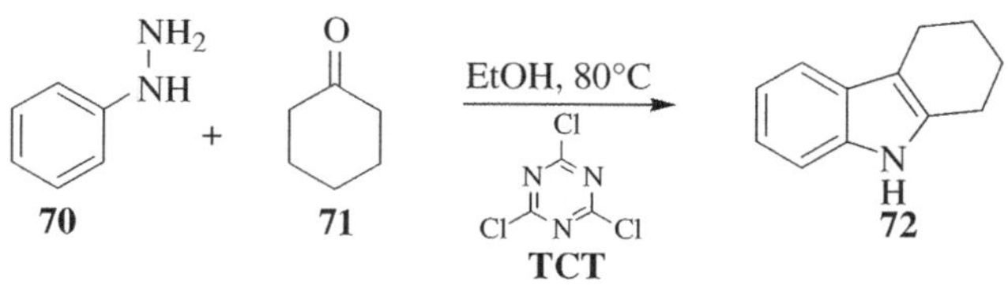

Reaction of phenylhydrazine with cyclohexanone in the presence of TCT by Siddalingamurthy

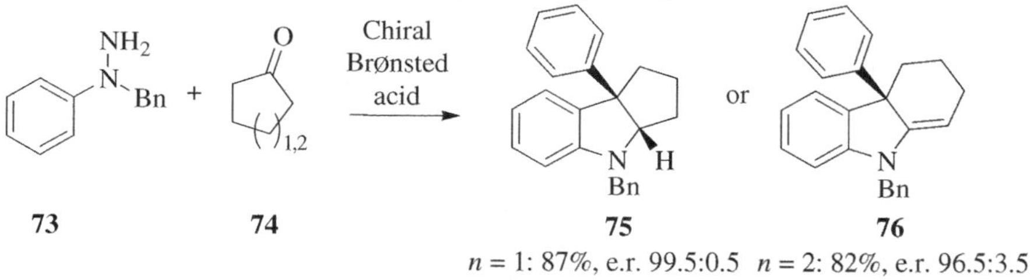

List's catalytic asymmetric Fischer indolization

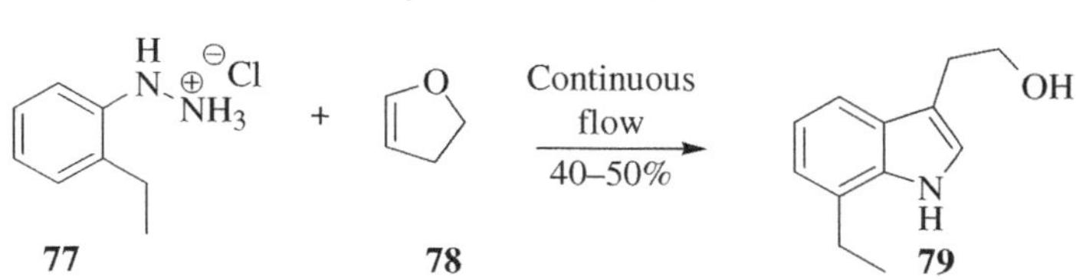

The Fischer indole synthesis under the continuous flow conditions of Gutmann

FIGURE 18.4 Significant illustrations of the Fischer indole synthesis.

18.2.2.2 Benzidine Rearrangement 4,4′-Diaminobiphenyl (aka benzidine, **81**) is commonly prepared from the rearrangement of 1,2-diphenylhydrazine (**80**) in an acidic environment (Scheme 18.15).

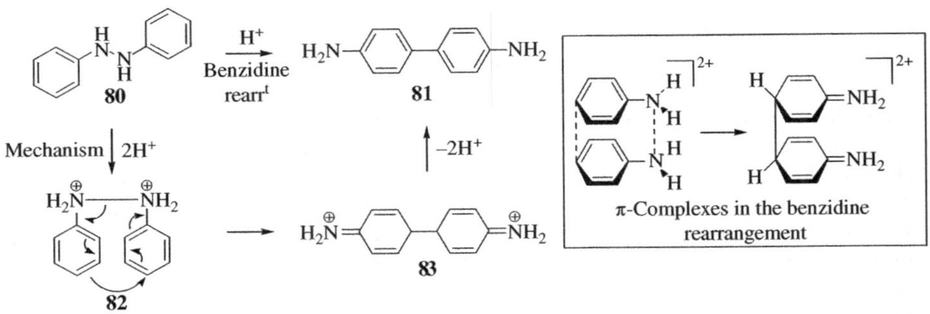

SCHEME 18.15 Diprotonated mechanism of the benzidine rearrangement (see text for details).

Kang and coworkers neatly employed a benzidine rearrangement for the regioselective [5,5]-sigmatropic rearrangement of aryl hydrazides **84**, in tandem with a palladium-coupling procedure (Figure 18.5). In their work, they demonstrated that the palladium-catalyzed formation of *N,N′*-aryl hydrazides, followed by a benzidine rearrangement, occurs in good to excellent yields on substrates

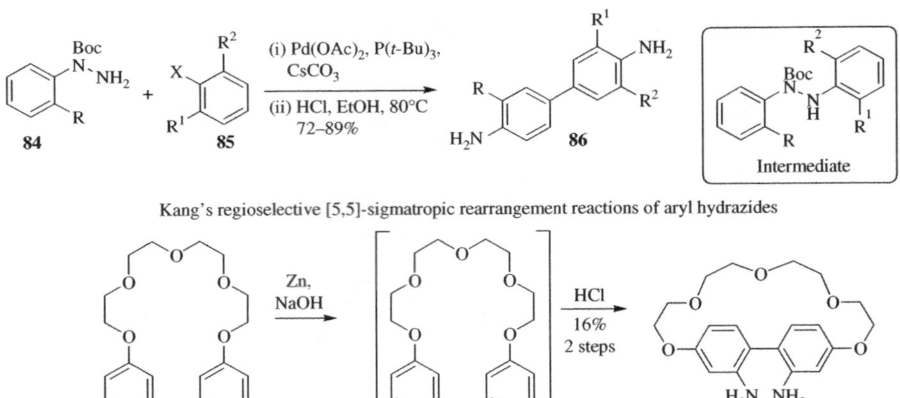

Kang's regioselective [5,5]-sigmatropic rearrangement reactions of aryl hydrazides

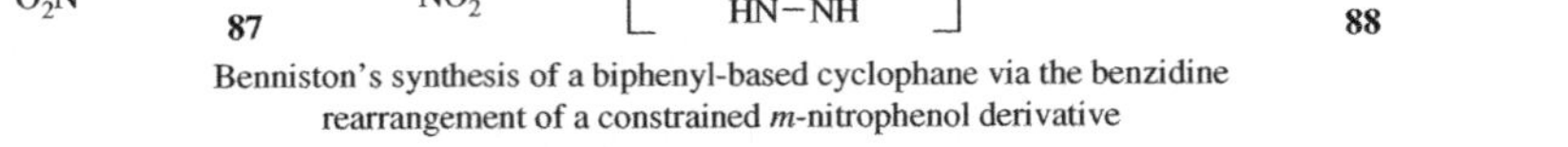

Benniston's synthesis of a biphenyl-based cyclophane via the benzidine rearrangement of a constrained *m*-nitrophenol derivative

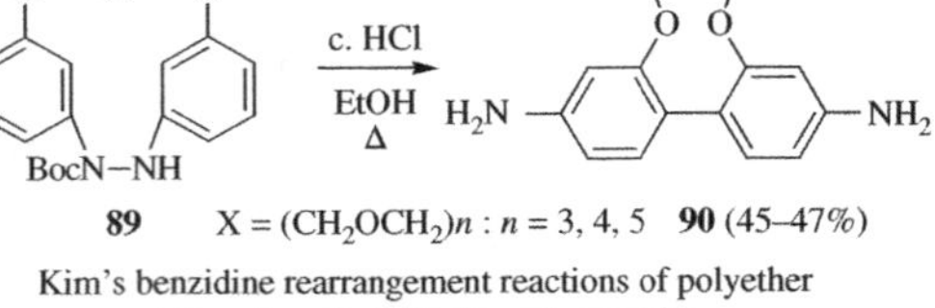

89 X = $(CH_2OCH_2)n$: *n* = 3, 4, 5 **90** (45–47%)

Kim's benzidine rearrangement reactions of polyether tethered cyclic *N,N′*-diaryl hydrazides

FIGURE 18.5 Noteworthy examples of the benzidine rearrangement.

containing *ortho-* or *meta*-substituents. They propose that the substituents suppress the formation of diphenylene, the major by-product of the conventional benzidine rearrangement (Figure 18.5) [49]. Incorporation of the benzidine rearrangement in the synthesis of supramolecular scaffolds has also been achieved by Benniston et al. (Figure 18.5), wherein they were able to successfully prepare a biphenyl-based cyclophane (**88**) via a benzidine rearrangement of a constrained *m*-nitrophenol derivative (**87**) [50]. Their success prompted them to suggest that the scope of the benzidine rearrangement on such substrates could be tested by employing other extended polyether, alkyl, and polyaza derivatives, a worthwhile achievement since such compounds could lead to an alternative strategy for producing new cyclophane hosts [50].

In this regard, in 2007, Kim et al. demonstrated the preparation of 4,4′-diamino-biphenyls (**90**) strapped to a polyether unit (X) at the 2,2′-positions using a polymer-bound benzidine rearrangement of polyether tethered cyclic *N,N*′-diaryl hydrazides (**89**) (Figure 18.5) [51].

From recent accounts, it seems as though the mechanism of the benzidine rearrangement is still contentious. For example, in a theoretical study in 2012, the monoprotonated mechanism of the benzidine acid-catalyzed rearrangement of hydrazobenzene (corresponding to a second-order kinetics) was studied and compared with the diprotonated mechanism (Scheme 18.15, corresponding to a third-order reaction). The nature of the two mechanisms was found to be completely different, wherein a concerted closed-shell sigmatropic shift takes place in the monoprotonated mechanism, and a stepwise radical cation recoupling occurs in the diprotonated case. The energetics of the two reaction mechanisms makes the third-order diprotonated mechanism favored (Scheme 18.15) with respect to the second-order monoprotonated mechanism for the rearrangement of hydrazobenzene, at typical experimental acid concentrations [52]. That said, it has also recently been shown that the acid-catalyzed benzidine mechanism may proceed via cation radicals formed by electron transfer to a proton from a hydrazyl nitrogen [53].

18.2.2.3 Claisen Rearrangement and Its Derivatives The Claisen rearrangement is a heavily utilized sigmatropic rearrangement in organic chemistry and a reaction that has been widely reviewed [54]. While the fundamental reaction has been expanded from the original substrate scope and conditions developed in 1912, the aromatic variant sees the [3,3]-sigmatropic rearrangement of an allyl phenyl ether (**91**) to an intermediate **92**, which tautomerizes to the *ortho*-substituted phenol (**93**) (Scheme 18.16).

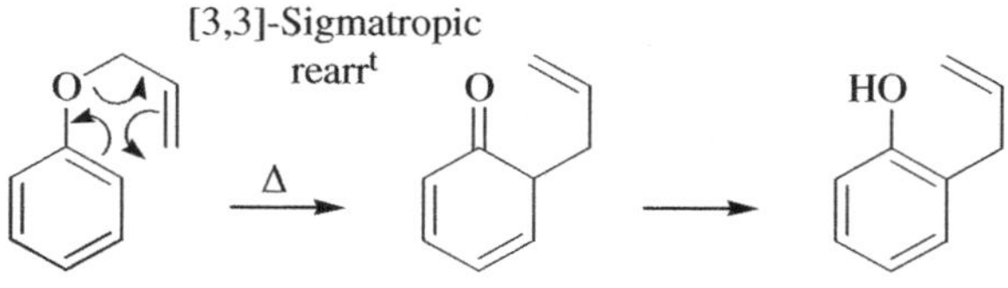

91 **92** **93**

SCHEME 18.16 The aromatic Claisen rearrangement.

After a century, a vast range of examples exist, which have been developed from the original variant, all of which undergo a mechanism related to that shown in Scheme 18.16 wherein a nucleophilic component migrates to the aromatic nucleus [54].

Despite the plethora of examples to choose from, Figure 18.6 outlines those diverse cases, which are aromatic sigmatropic Claisen rearrangements in which the migrating group migrates to an aromatic nucleus, applied to the synthesis of natural products.

For example, an aromatic Claisen rearrangement was successfully employed by Vyvyan and coworkers in their total synthesis of (±)-heliannuol C and E [55]. In their work, the authors prepared the natural products in 15 and 9% overall yields, respectively, over seven linear steps from

2-methylanisole. One of the key steps includes the regioselective aromatic Claisen rearrangement (**94–95**), which was used to install the requisite diene side chain. Further elaboration of **95** led to a common intermediate, which could be converted to both natural compounds in short order (Figure 18.6) [55].

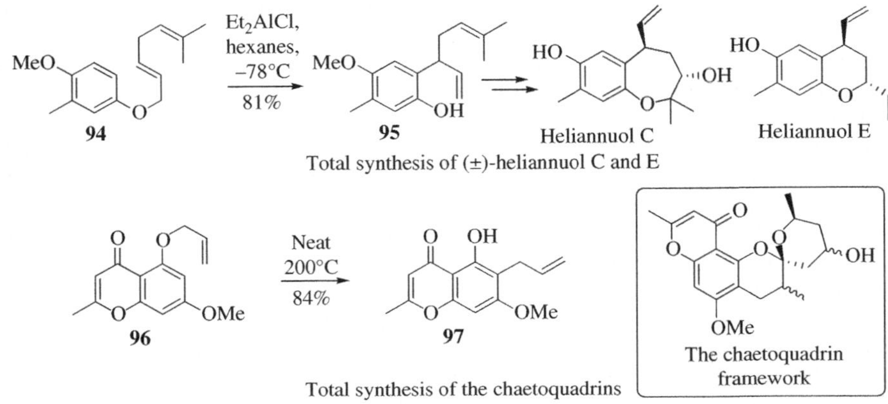

FIGURE 18.6 The aromatic Claisen rearrangement in the synthesis of natural products.

In a similar fashion, Kim and coworkers were also able to demonstrate the first total synthesis of the monoamine oxidase inhibitors, chaetoquadrins A–C (Figure 18.6), by exploiting an aromatic Claisen rearrangement as a key step (**96** → **97**), an intermediate that was developed over a number of steps into the chaetoquadrins [56].

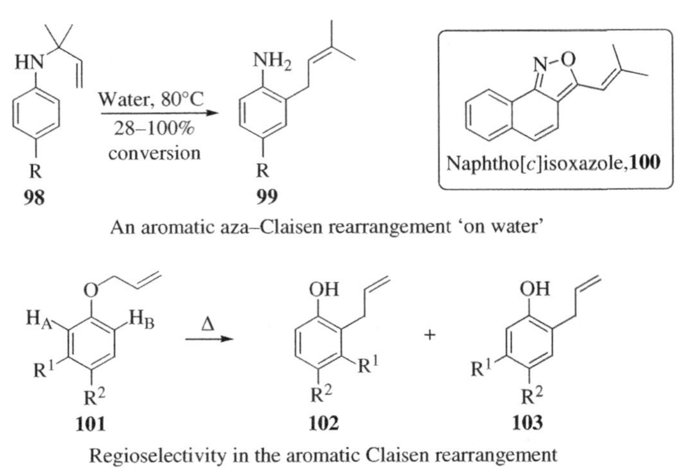

FIGURE 18.7 Aromatic Claisen rearrangements.

An example of an aromatic aza-Claisen rearrangement "on water" (a term defined due to the lack of water solubility of the reactants) has recently been achieved by McErlean and Beare [57]. They demonstrated the effect by transforming *N*-prenylated arylamines (**98**) into synthetically useful products (**99**) (Figure 18.7). The developed process was then used to probe the mechanism of the reaction and persuasive evidence in support of the acid-catalysis hypothesis for the "on-water" effect was obtained. The authors went on to demonstrate the applicability of the reaction with the synthesis of a previously inaccessible heterocyclic naphtho[*c*]isoxazole scaffold (**100**) in high yield.

Theoretical and mechanistic studies using QM/MM simulations have also looked at the solvent effects and "on-water" reactivity of the aromatic Claisen rearrangements of allyl *p*-R-phenyl ethers ($R = CH_3$, Br, and OCH_3) and allyl naphthyl ethers and showed that such aqueous systems can provide increased rate accelerations, yields, and specificity for several types of organic reaction classes compared to organic solvents [58]. Biologically relevant aromatic Claisen rearrangements have also been explored, which utilize the "on-water" effect [59].

Furthermore, efforts to predict the regioselectivity in aromatic Claisen rearrangements has also been attempted by Eberlin et al. in a series of eight *meta*-substituted allyl aryl ethers (**101**) (Figure 18.7) [60]. In doing so, their method was able to indorse an existing ^{1}H NMR protocol that can be used to predict aromatic Claisen regioselectivity from the ground-state conformational preference of the reactant allyloxy group.

18.2.2.4 Rearrangement of Ammonium Ylides (Sommelet–Hauser Rearrangement) The Sommelet–Hauser rearrangement portrays a sigmatropic rearrangement using charged species as outlined in Scheme 18.17 [61].

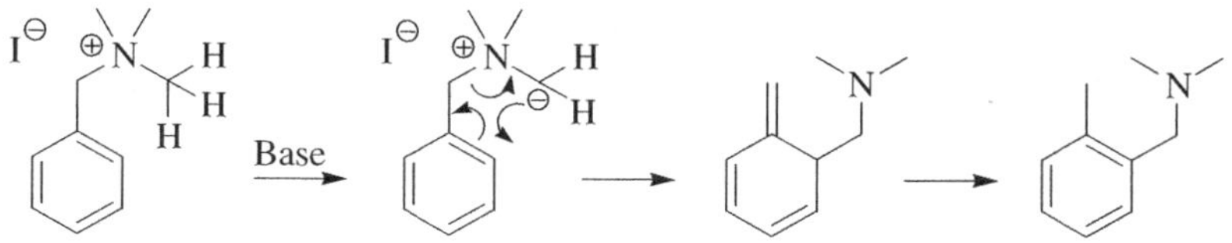

104 **105** **106** **107**

SCHEME 18.17 The Sommelet–Hauser rearrangement of ammonium ylides.

In 2007, Tayama and Kimura reported that an asymmetric Sommelet–Hauser rearrangement of enantiopure *N*-benzylic ammonium salts takes place when reaction conditions employing *t*BuOK in THF were used [62]. A surprising result, since when alternative conditions employing CsOH in 1,2-DCE were used, the reaction products were the result of a [1,2]-Stevens rearrangement. As a result of this serendipitous finding, optimization of the reaction to prepare a number of diastereomerically pure compounds, such as **109**, utilizing an (–)-8-phenylmenthol ester (**108**), were carried out. These reactions proceed with remarkably high levels of stereoselectivity to yield optically active α-aryl amino acid derivatives.

The rearrangement has also been shown to proceed catalytically, as exemplified by Wang's work developing a Rh(II)-catalyzed Sommelet–Hauser rearrangement [63]. The Rh(II)-catalyzed reaction of aryldiazoacetates (**110**) with ethyl benzylthioacetate (**111**) gave the Sommelet–Hauser products (**112**) in good to excellent yields (fourteen were prepared) and provides a way to introduce a substituent to the *ortho* position of such arylacetates. The required ylide is formed by the reaction between the Rh(II) carbene (formed *in situ* between **110** and the Rh(II) catalyst) and sulfide (**111**). Once formed, the ylide rearranges via a sigmatropic rearrangement (reminiscent of that shown in Scheme 18.17) prior to rearomatization to give **112** (Figure 18.8) [63].

108 109, d.r. up to >98%

The asymmetric Sommelet–Hauser rearrangement

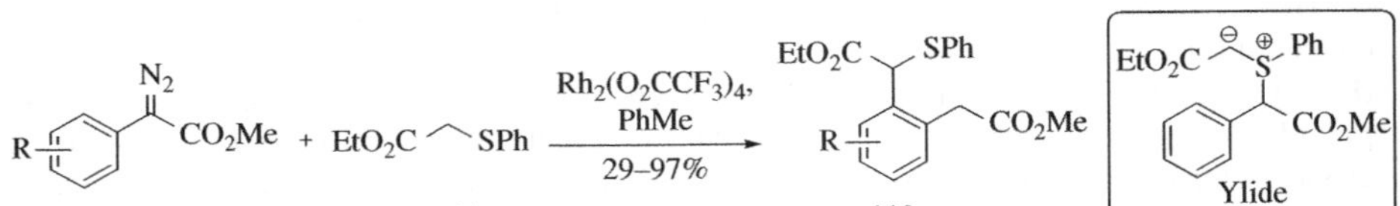

110 111 112

A Rh(II)-catalyzed Sommelet–Hauser rearrangement

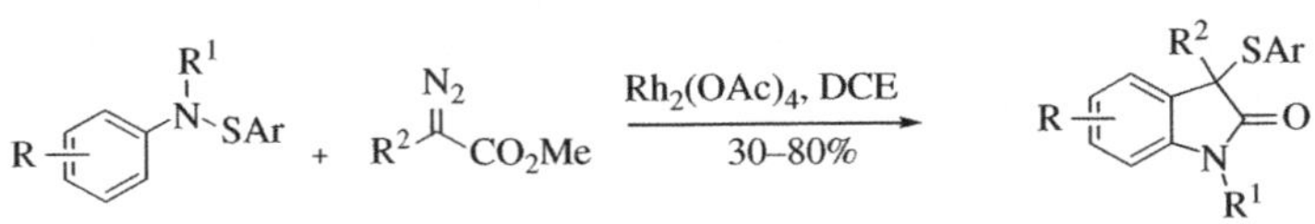

113 114 115

Catalytic thia-Sommelet–Hauser rearrangement toward the synthesis of oxindoles

FIGURE 18.8 Notable examples of the Sommelet–Hauser rearrangement.

Wang's group has further developed this variant of the rearrangement to the preparation of oxindoles (**115**) [64], whereby a similar sequence of events takes place between **113** and **114** but finishes with a lactam forming ring closure.

18.2.3 Intermolecular Rearrangements

18.2.3.1 Rearrangement of Aniline Derivatives (Bamberger, Orton, Fischer–Hepp, and Hofmann–Martius Rearrangements) The following molecular rearrangements are all transformations of various aniline derivatives, which share a common overall transformation. The general transposition for these reactions is shown in Scheme 18.18, whereby a substituent migrates from nitrogen to the aromatic nucleus.

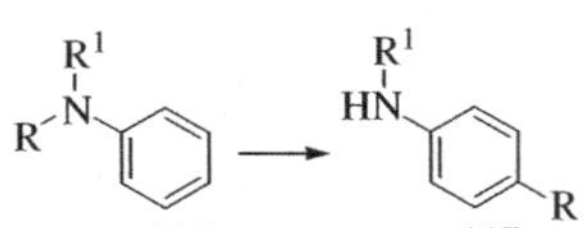

116 117

R = OH, R^1 = H, **Bamberger**
R = Cl or Br, R^1 = acyl, **Orton**
R = NO, R^1 = various (see examples below), **Fischer–Hepp**
R = various (see examples below), R^1 = H, **Hofmann–Martius**
Ortho-substituted products are also formed

SCHEME 18.18 The aromatic rearrangement of aniline derivatives.

*The Bamberger Rearrangement, **116** (R = OH, R′ = H) → p-Aminophenol* The Bamberger rearrangement is summarized by the transformation of *N*-phenylhydroxylamine (**118**) to *p*-aminophenol (**122**), and its *para*-selective mechanism is shown in Scheme 18.19.

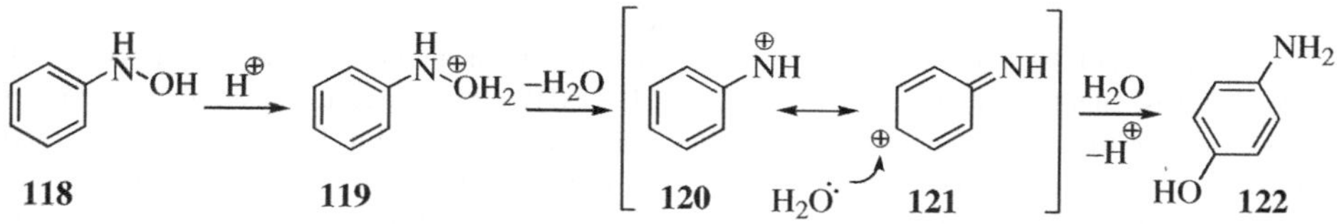

SCHEME 18.19 The mechanism of the Bamberger rearrangement.

As well as these traditional aqueous acid rearrangement conditions, Ratnam and coworkers have shown that the reaction can take place on solid acid catalysts, such as beta zeolite, K10 clay, sulfonated silica, and sulfated zirconias [65]. In their work, the rearrangement of *N*-phenylhydroxylamine to *para*-aminophenol was investigated, in water at 80°C, using a series of these solid acid catalysts, and it was found that both the activity and selectivity are affected by the choice of the catalyst used, and the rate constant shows a linear dependence on the number of acid sites [65].

Its use in synthesis has also been demonstrated, as highlighted in Scheme 18.20 [66]. In their work, the authors demonstrated a process that appears to involve a 1,2-Bamberger-type rearrangement for the annulation of enals (**123**) and nitroso compounds (**124**) by an unexpected *N*-heterocyclic carbene-catalyzed process, a process that generated a range of useful chiral seven-membered 4-azalactones (**125**) [66].

Furthermore, Zhang and Ying have utilized the same catalyst in an organocatalytic approach to secure a route to *N*-phenylisoxazolidin-5-ones, which can be converted to *N*-*p*-methoxyphenyl (N-PMP) protected beta-amino acid esters (**128**) using a similar Bamberger-like rearrangement in a mild one-pot protocol (Scheme 18.20) [67].

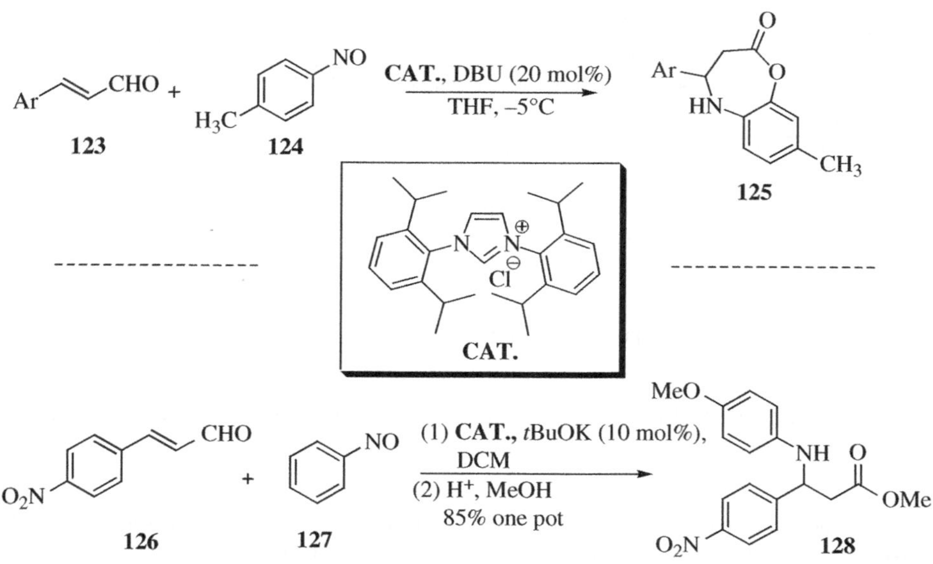

SCHEME 18.20 Bamberger-like rearrangements in synthesis.

As well as these more traditional synthetic uses, the Bamberger rearrangement has also been observed *in vitro*. When studying the growth of a *Pseudomonas putida* strain in media enriched with 4-nitrotoluene, or its metabolite 4-nitrobenzoate, an enzymatic cascade process, which includes a Bamberger-like rearrangement, was demonstrated [68].

Recently, the classical nitrenium-based mechanism for this reaction (Scheme 18.19) was contested when DFT studies showed a dication-like transition state may be taking part [69]. In this work, it was calculated that the nitrenium ion, C_6H_5—NH^+, often suggested as an intermediate, was absent owing to the high nucleophilicity of the water cluster around it. Furthermore, investigation showed that a diprotonated system, Ph—N(OH)H + $(H_3O^+)_2(H_2O)_{13}$, was evidenced, a system that possesses an activation energy similar to the experimental one. Based on this evidence, a new mechanism for the rearrangement was proposed [69].

Changing the —OH group in the Bamberger rearrangement to a halogen results in the Orton rearrangement (**116**, R = Cl or Br, R¹ = acyl; Scheme 18.18), a reaction that sees the migration of a halogen from the *N*-atom of an aniline derivative to the aromatic nucleus upon treatment with acid. The main product is the *para*-isomer, and in general, the amine must be acylated. Observed cross-halogenation products and labeling studies suggest an intermolecular mechanism, whereby, in the case of **116** (R = Cl), dechlorination of the starting material happens first and then the liberated Cl_2 halogenates the ring.

*The Fischer–Hepp Rearrangement, **116** (R = NO, R¹ = various) → p-Nitrosoamines* The Fischer–Hepp rearrangement describes the established transformation that *N*-nitroso amines undergo in their conversion into the corresponding *p*-nitrosoamines. There appear to be relatively few examples of the rearrangement in the synthetic literature, despite its similarity to the other rearrangements in this section (Scheme 18.18); however, the rearrangement's mechanism and kinetics were studied in depth in the 1970s by Williams [70] and Biggs [71].

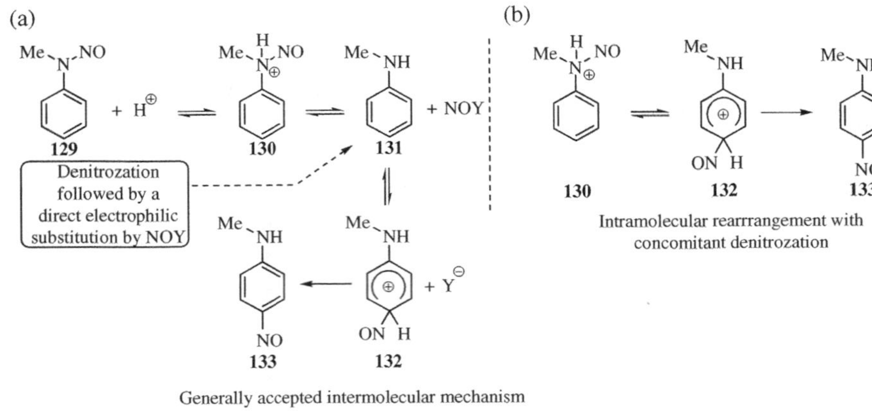

SCHEME 18.21 Mechanism of the Fischer–Hepp rearrangement (a) intermolecular mechanism and (b) intramolecular mechanism.

The mechanism of the rearrangement using aromatic *N*-nitroso-amines was studied in 1975, by Williams [70], whereby the rate equations were deduced for two of the possible mechanisms in acid solution. The mechanisms under examination were (a) the commonly assumed intermolecular process involving denitrosation to the secondary amine (**129** → **131**) and a free nitrosating agent, followed by a direct C-nitrosation of the secondary amine by the liberated nitrosating agent (Scheme 18.21), and (b) a mechanism whereby the rearrangement and denitrosation steps occur concomitantly (**130** → **132**), by two separate reactions of the protonated *N*-nitroso-amine (Scheme 18.21). The experimental evidence accrued supports the second mechanism while also being incompatible with the first [70].

Studies by Biggs examined the effects of methyl group substituents on both the kinetics and mechanism of the rearrangement as well as the competing denitrosation (Figure 18.9) [71b]. In such cases, a very large rate reduction was observed in the 2,6-disubstituted case (**136**), compared to the unsubstituted nitroso-amine, for both reactions; an effect that was attributed to steric hindrance toward protonation of the amino nitrogen atom. The other methyl substitution patterns gave rise to an activation of the system to denitrosation, by relatively small amounts in hydrochloric acid, but by greater amounts in sulfuric acid. Together, these results are taken to support the suggestion above (Scheme 18.21) that at high acidities, and in the absence of added nucleophiles, the alternative mechanism of denitrosation becomes important. Both 2- and 3-methyl substitution also increases the reactivity of the nitroso-amines toward rearrangement, with the 3-methyl group having the greater effect. Unexpectedly, the 3,5-dimethyl derivative (**138**) is significantly less reactive than the 3-methyl-nitroso-amine, suggesting that steric factors play a part in the intramolecular rearrangement process [71b].

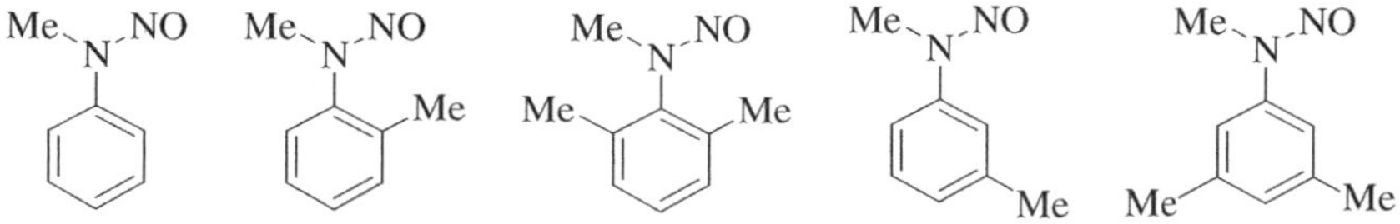

FIGURE 18.9 The effect of methyl substituents on the Fischer–Hepp rearrangement.

In contrast to the other mechanistically related mechanisms in this section, there do not appear to be any examples of the Fischer–Hepp reaction in synthesis on all-carbon aryl systems (i.e., benzene). That said, there is a recent heterocyclic example that claims to be the first example of the Fischer–Hepp-type rearrangement in pyrimidines, although its details will not be included here [72].

*The Hofmann–Martius Rearrangement, **116** (R = various alkyls, R^1 = H)→*C-*Alkylanilines* The conversion of *N*-alkylanilines into C-alkylanilines under acidic conditions is known as the Hofmann–Martius rearrangement, and there are relatively few reported examples of the rearrangement. However, in 2006, Magnus described a thermal, acid-catalyzed example of the Hofmann–Martius rearrangement using 3-*N*-aryl-2-oxindoles (**139** and **142**) into 3-(arylamino)-2-oxindoles (**140–144**) (Scheme 18.22) [73].

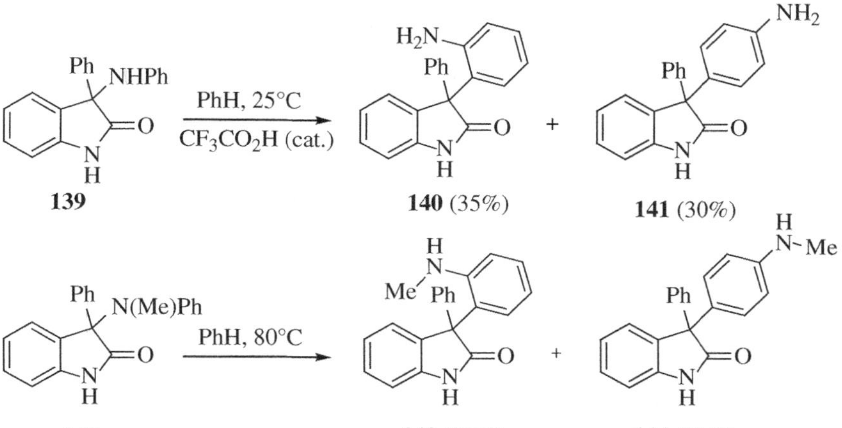

SCHEME 18.22 The Hofmann–Martius rearrangement.

In their work, the authors were able to demonstrate the rearrangement on a small number of derivatives and show that the reaction proceeded through a dissociative pathway, a mechanism that mirrors that of the phenol version of the same rearrangement. It is proposed that under the reaction conditions studied, the starting material dissociates, and the aniline derivative that forms is reattacked in a Friedel–Crafts manner to give the rearranged products. They were also able to show the synthetic utility of the products by converting **143** into the carbon framework relating to diazonamide and physostigmine analogues.

18.2.3.2 The Stieglitz Rearrangement The Stieglitz rearrangement is the reorganization of a trityl hydroxylamine (such as **145**; Scheme 18.23) to the corresponding triaryl imine (**146**), and the rearrangement has been observed under a variety of experimental conditions. In 1974, it was shown that lead tetraacetate could be used to promote this rearrangement and various mono-*para*-substituted triarylmethylamines have been studied using this reagent, and the corresponding migratory aptitudes determined [74]. Following trapping experiments in these reactions, a concerted mechanism was postulated that was consistent with all the data [74].

SCHEME 18.23 The Stieglitz rearrangement.

Alternatively, the Stieglitz rearrangement of tritylamines, benzhydrylamines, and benzylamines has also been induced by *para*-nitrobenzenesulfonyl peroxide [75]. In their work, the authors attempted to elucidate the intermediate in the lead tetraacetate-induced Stieglitz rearrangement from the results of migratory aptitude studies and trapping experiments, and the results indicated that a concerted cationic aryl migration takes place on the *O*-sulfonylhydroxylamine intermediate. When compared with other known Stieglitz rearrangements, the results suggest that the charge development on the migrating aryl group is structure and leaving-group dependent [75].

Moreover, Theodorou and coworkers observed the rearrangement occurring, albeit unexpectedly, during the treatment of tritylamine (**147**) with *n*-butyllithium (Scheme 18.24). The unexpected

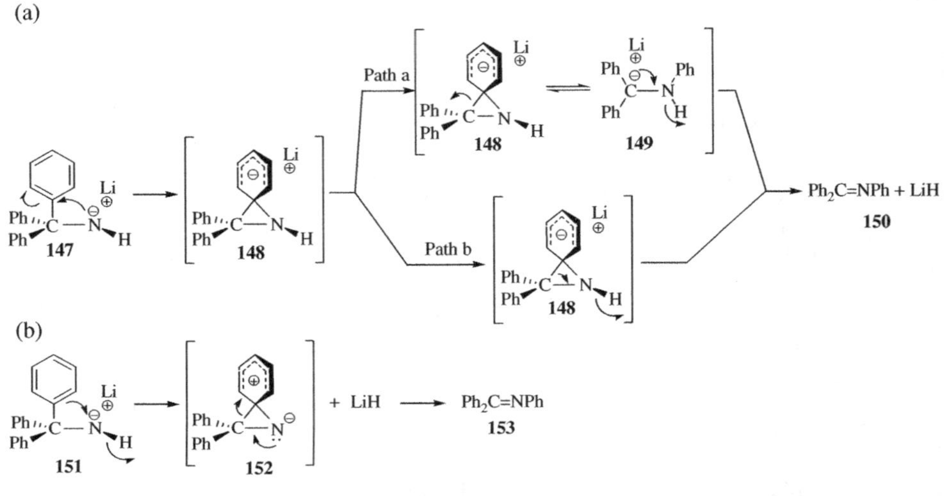

SCHEME 18.24 The Stieglitz rearrangement: proposed mechanisms for imine formation via (a) electrophilic phenyl rearrangement and (b) nucleophilic phenyl rearrangement.

rearrangement led to the formation of imine **150**, and two proposed mechanisms for its formation, via either an electrophilic phenyl rearrangement (a) or a nucleophilic phenyl rearrangement (b), have been postulated [76].

18.2.3.3 Acyl Migration (the Fries Rearrangement and Its Variants) The traditional Fries rearrangement (**154**→**155**; Scheme 18.25) proceeds via ionic intermediates but whether the exact mechanism is intermolecular or intramolecular is still under debate. Reports in the literature appear to present evidence to support either mechanism, and thus, the exact route depends on the structure of the substrates and the reaction conditions used. The photo-Fries rearrangement (Figure 18.10) proceeds via radical intermediates.

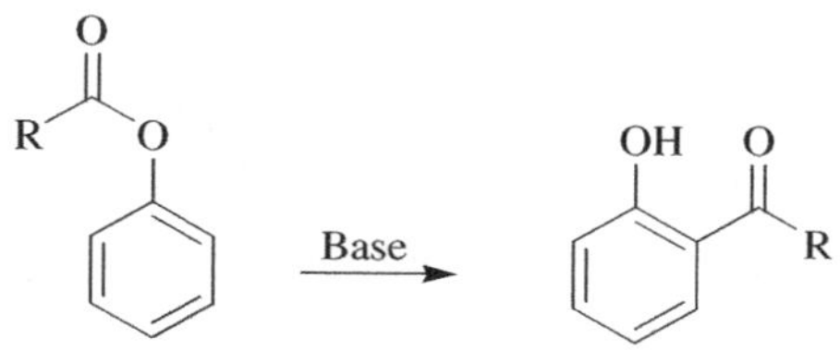

154 **155**

SCHEME 18.25 The Fries rearrangement.

Using a homologous anionic *ortho*-Fries variant of the reaction, Snieckus and coworkers demonstrated the first lateral functionalization of calix[4]arenes (Figure 18.10) [77]. Treatment of calix[4]arene bis-*O*-carbamates (**156**) with lithium diisopropylamide (LDA) results in the regio- and stereoselective introduction of both axial and equatorial carboxamide groups at the methylene bridges to give the products (**157**). The stereochemical outcome of the rearrangement appears to be dependent on both the conditions of the reaction and on the conformation of the starting material.

Other, nonclassical, Fries rearrangements have seen a thia-alternative that combines the rearrangement of aryl sulfonates (**158**→**159**), in solvent-free conditions under microwave activation, using an $AlCl_3$–$ZnCl_2$ mixture supported on silica gel (Figure 18.10) [78].

Alternatively, the Fries rearrangement has also been observed in ionic melts [79] when phenyl benzoates are shown to rearrange using 1-butyl-3-methylimidazolium chloroaluminate ($[BMIm]^+Al_2Cl_7^-$) as both a solvent and Lewis acid catalyst. Good yields and high selectivity are the features observed in this unconventional but interesting aprotic solvent; the rate of consumption of phenyl benzoate obeys first-order kinetics [79].

Furthermore, the use of light has also been used, wherein Magnus was able to effect the rearrangement and develop a photo-Fries rearrangement for the synthesis of the diazonamide macrocycle (**160**→**161**) [80]. Irradiation of **160** in benzene (0.001 M) with a medium pressure mercury vapor lamp at 23°C gave **161** (39%) along with about 20% of the *para*-isomer.

The mild conditions of the photo-Fries rearrangement in natural product synthesis were further demonstrated by Mulzer and colleagues with the total synthesis of the antibiotic kendomycin by a macrocyclization reaction using a combined photo-Fries rearrangement, ring-closing metathesis approach (**162**→**163**) (Figure 18.10) [81].

An unusual sodium hydride-mediated remote anionic 1,5-thia-Fries rearrangement reaction was developed by Xu and coworkers recently [82]. Their method provides an efficient approach for the regioselective synthesis of not only 2-(2-hydroxyphenyl)-3-indole triflones but also the related 3-sulfonylindoles (**164**→**165**) (Figure 18.10) [82].

Unsurprisingly, with such a useful and heavily utilized reaction, many variants have been developed, and one such example is the anionic phospho-Fries rearrangement (**166**→**167**). Jayasundera

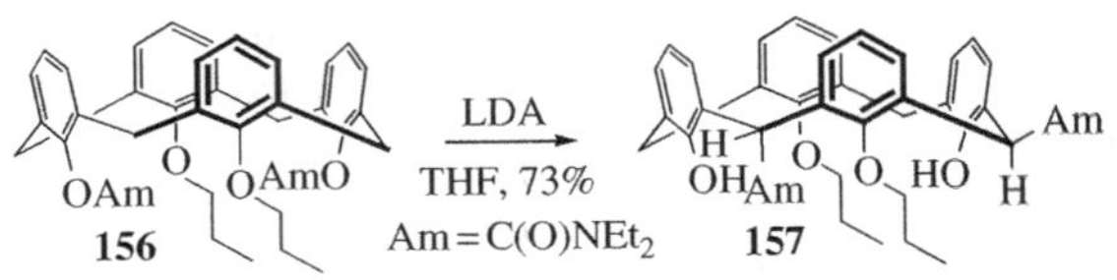

Lateral functionalization of calix[4]arenes using a homologous anionic *ortho*-Fries rearrangement

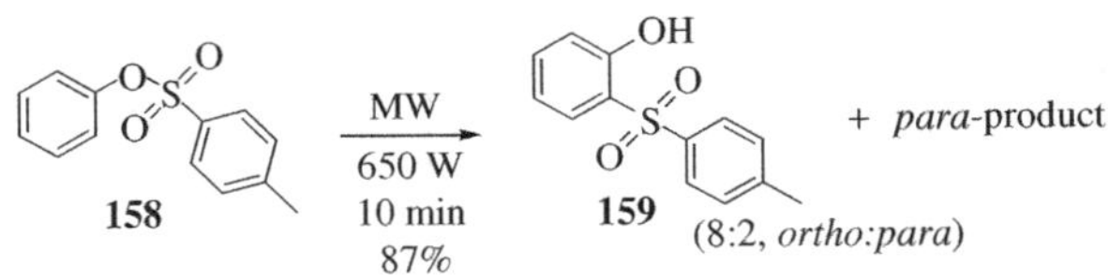

Thia-Fries rearrangement of aryl sulfonates under microwave irridiation

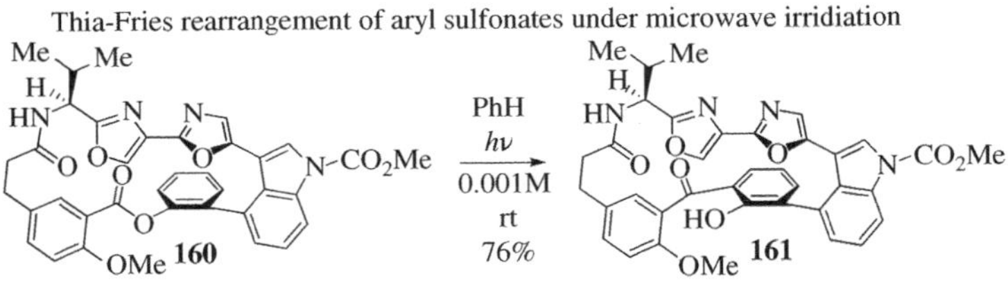

Photo-Fries used in the synthesis of the diazonamide macrocycle

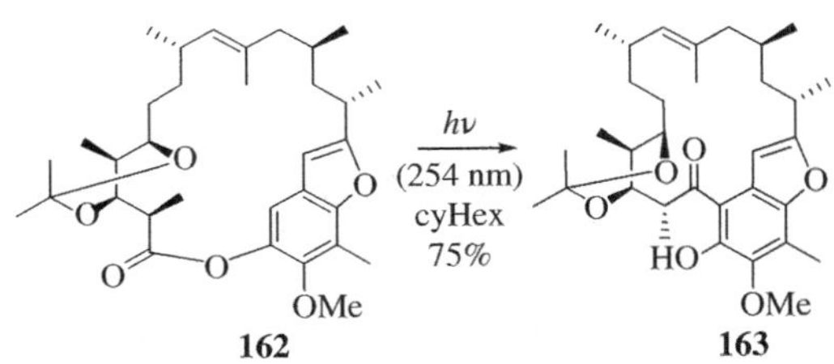

Photo-Fries rearrangement toward the synthesis of kendomycin

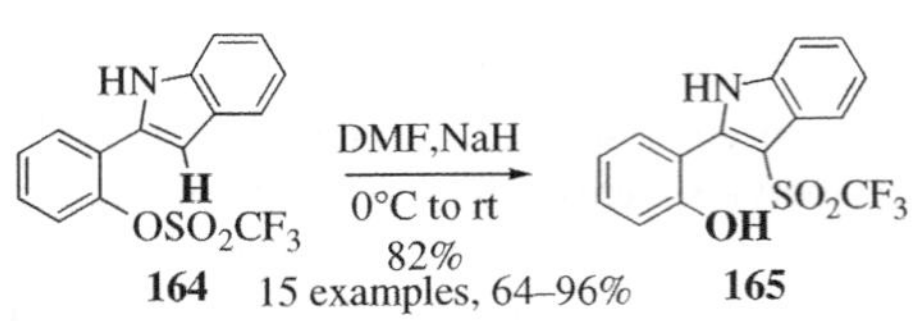

Remote anionic Fries rearrangement of sulfonates

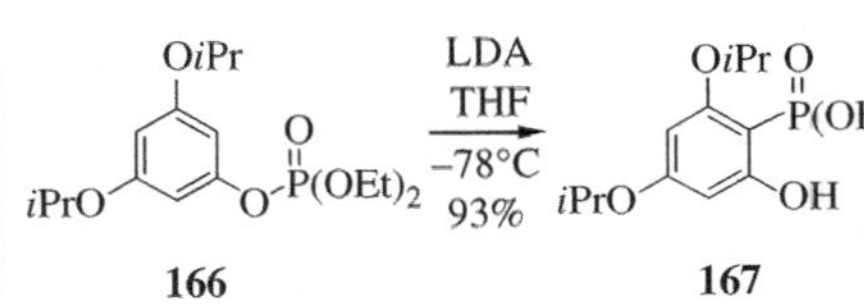

Anionic phospho-Fries rearrangement

FIGURE 18.10 Notable examples of the Fries rearrangement and its variants.

et al. used this modified version toward the synthesis of a tetrasubstituted arylphosphonate [83], whereby the rearrangement of phosphoric acid (3,5-di-isopropoxy)phenyl ester diethyl ester (**166**) gave rise to (2-hydroxy-4,6-di-isopropoxy-phenyl)phosphonic acid diethyl ester (**167**) in excellent yield [83]. The anionic phospho-Fries rearrangement has been reviewed [84].

18.2.3.4 Alkyl Migration (Jacobsen Rearrangement) The Jacobsen rearrangement is the migration of an alkyl group within a polyalkyl- or polyhalo-arylsulfonic acid (Scheme 18.26, **168** → **169**).

Evidence suggests that the rearrangement occurs intermolecularly and that the migrating group is transferred to a polyalkylbenzene, not to the sulfonic acid (i.e., migration of the alkyl group takes place first under the reaction conditions, followed by sulfonation *after* migration), and some support

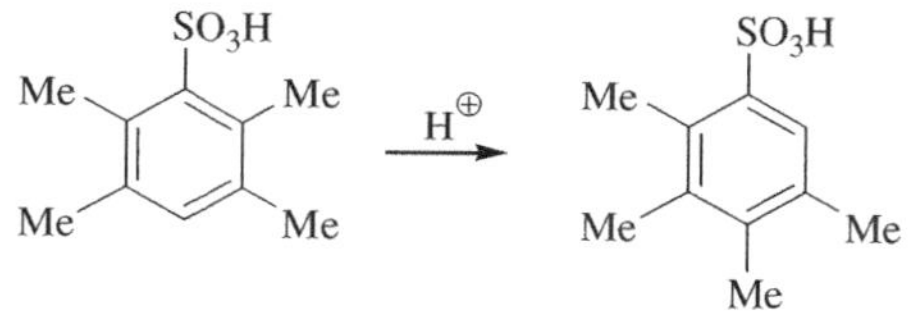

168 169

SCHEME 18.26 Examples of the Jacobsen rearrangement.

for the mechanism has been obtained. The reaction appears to be limited to benzene rings with at least four substituents (alkyl and/or halogen groups).

Applications of the reaction have been seen in the rearrangement of tetraethylbenzenesulfonic acids [85], wherein the reaction of *meta*- and *para*-di-, 1,3,5-tri- and 1,2,3,5- and 1,2,4,5-tetraethylbenzenes with concentrated aqueous sulfuric acid leads to clean sulfodeprotonation. Recent examples by Solari et al. witnessed a catalytic variation of the Jacobsen rearrangement using $ZrCl_4$ [86]. The rearrangement, previously only observed in concentrated sulfuric acid, appeared to be extremely mild under $ZrCl_4$-mediated conditions on permethylated arenes at room temperature in halogenated solvents.

18.2.3.5 Rearrangement of Azoxy Compounds (Wallach Rearrangement) The Wallach rearrangement is a reaction that converts an aromatic azoxy compound (**170**) into an azo compound (**174**) with sulfuric acid, wherein one ring is substituted by a hydroxyl group in the *para* position (Scheme 18.27).

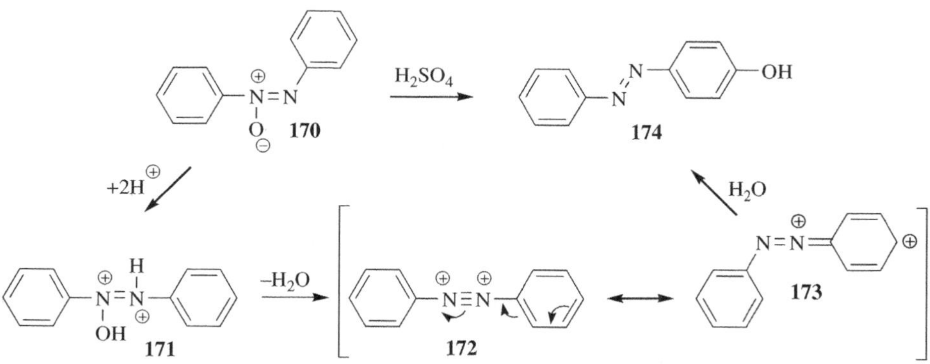

170 174 171 172 173

SCHEME 18.27 The Wallach rearrangement.

Perhaps the greatest contribution to understanding the Wallach rearrangement has come from the work of Buncel. Over the course of almost four decades, his contribution to the rearrangement has been instrumental in understanding its capabilities. From his early investigations in 1970 [87], right up until 2009 [88], his career's work had come "full circle" after a mechanism, which he proposed back in 1963, that had never actually been observed for any substrate before was validated.

There are only a relatively small number of examples of this rearrangement in the chemical literature; nevertheless, both a thermal and photochemical Wallach rearrangement of azoxybenzene has been achieved in zeolite cages [89] and an *ortho*-selective rearrangement has been developed on azoxybenzene-$SbCl_5$ complexes [90].

A computational study of the Wallach rearrangement has also been carried out, whereby molecular orbital theory employing the semiempirical AM1 method has been used to determine the energetics of the intermediates and the transition states for this rearrangement. This study was undertaken in order to try and confirm the mechanism of the reaction, since several plausible mechanisms for the rearrangement have been proposed [91].

18.3 SUMMARY AND OUTLOOK

As exemplified throughout this chapter, aromatic rearrangements in which the migrating group migrates to the aromatic nucleus are well established and embedded within the synthetic chemists' toolbox for the synthesis of numerous products. From their initial discovery to the range of variants and alternatives developed through time, such rearrangements continue to inspire chemists and throw up new challenges in determining their mechanisms of action. Nevertheless, that they are firmly embedded in the synthesis of natural products is testament to their importance, and new variants of these reactions offer unique opportunities to prepare synthetically useful reaction products, which are not easily accessible by other synthetic methods. With this in mind, these rearrangements could feature more heavily in synthesis and materials manufacture in the future.

ABBREVIATIONS

LDA	Lithium diisopropylamide
N-PMP	*N*-*p*-methoxyphenyl
TCT	2,4,6-Trichloro-1,3,5-triazine

REFERENCES

[1] W. E. Truce, E. M. Kreider, W. W. Brand, *Org. React.* 1970, **18**, 99.

[2] K. Matsui, N. Maeno, S. Suzuki, H. Shizuka, T. Morita, *Tetrahedron Lett.* 1970, **11**, 1467–1469.

[3] M. S. Wadia, D. V. Patil, *Synth. Commun.* 2003, **33**, 2725–2736.

[4] Y. L. Jin, S. Kim, Y. S. Kim, S.-A. Kim, H. S. Kim, *Tetrahedron Lett.* 2008, **49**, 6835–6837.

[5] R. Sanz, V. Guilarte, N. Garcia, *Org. Biomol. Chem.* 2010, **8**, 3860–3864.

[6] H. Yang, Z.-B. Li, D.-S. Shin, L.-Y. Wang, J.-Z. Zhou, H.-B. Qiao, X. Tian, X.-Y. Ma, H. Zuo, *Synlett* 2010, 483–487.

[7] (a) W. E. Truce, W. J. Ray Jr., O. L. Norman, D. B. Eickemeyer, *J. Am. Chem. Soc.* 1958, **80**, 3625–3629; (b) W. E. Truce, W. J. Ray Jr., *J. Am. Chem. Soc.* 1959, **81**, 481–484; (c) W. E. Truce, D. C. Hampton, *J. Org. Chem.* 1963, **28**, 2276–2279; (d) W. E. Truce, C. R. Robbins, E. M. Kreider, *J. Am. Chem. Soc.* 1966, **88**, 4027–4033; (e) W. E. Truce, B. VanGemert, W. W. Brand, *J. Org. Chem.* 1978, **43**, 101–105; (f) D. M. Snyder, W. E. Truce, *J. Am. Chem. Soc.* 1979, **101**, 5432–5433; (g) E. J. Madaj Jr., D. M. Snyder, W. E. Truce, *J. Am. Chem. Soc.* 1986, **108**, 3466–3469.

[8] W. E. Truce, E. J. Madaj Jr., *J. Sulfur Chem.* 1983, **3**, 259–287.

[9] L. H. Mitchell, N. C. Barvian, *Tetrahedron Lett.* 2004, **45**, 5669–5671.

[10] T. J. Snape, *Chem. Soc. Rev.* 2008, **37**, 2452–2458.

[11] Y. Fukazawa, N. Kato, S. Ito, *Tetrahedron Lett.* 1982, **23**, 437–438.

[12] T. J. Snape, *Synlett* 2008, 2689–2691.

[13] A. D. Alorati, A. D. Gibb, P. R. Mullens, G. W. Stewart, *Org. Process Res. Dev.* 2012, **16**, 1947–1952.

[14] D. Ameen, T. J. Snape, *Eur. J. Org. Chem.* 2014, **2014**, 1925–1934.

[15] Y. L. Liu, X. H. Zhang, Y. Ma, C. Ma, *Tetrahedron Lett.* 2013, **54**, 402–405.

[16] L. El Kaim, L. Grimaud, X. F. Le Goff, A. Schiltz, *Org. Lett.* 2011, **13**, 534–536.

[17] J. Clayden, J. Dufour, D. M. Grainger, M. Helliwell, *J. Am. Chem. Soc.* 2007, **129**, 7488–7489.

[18] R. Bach, J. Clayden, U. Hennecke, *Synlett* 2009, 421–424.

[19] J. Clayden, W. Farnaby, D. M. Grainger, U. Hennecke, M. Mancinelli, D. J. Tetlow, I. H. Hillier, M. A. Vincent, *J. Am. Chem. Soc.* 2009, **131**, 3410–3411.

[20] A. M. Fournier, R. A. Brown, W. Farnaby, H. Miyatake-Ondozabal, J. Clayden, *Org. Lett.* 2010, **12**, 2222–2225.

[21] D. J. Tetlow, U. Hennecke, J. Raftery, M. J. Waring, D. S. Clarke, J. Clayden, *Org. Lett.* 2010, **12**, 5442–5445.

[22] J. Clayden, P. MacLellan, *Beilstein J. Org. Chem.* 2011, **7**, 582–595.

[23] A. M. Fournier, J. Clayden, *Org. Lett.* 2012, **14**, 142–145.

[24] A. W. Chapman, *J. Chem. Soc.* 1925, **127**, 1992.

[25] T. Kuroda, F. Suzuki, *Tetrahedron Lett.* 1992, **33**, 2027–2028.

[26] V. F. Burdukovskiy, D. M. Mognonov, I. A. Farion, *J. Polym. Sci. Part A Polym. Chem.* 2007, **45**, 4656–4660.

[27] S. Noorizadeh, A. Ozhand, *Chin. J. Chem.* 2010, **28**, 1876–1884.

[28] (a) H. M. Relles, *J. Org. Chem.* 1968, **33**, 2245–2249; (b) O. H. Wheeler, F. Roman, O. Rosado, *J. Org. Chem.* 1969, **34**, 966–968; (c) D. J. Brown, R. V. Foster, *J. Chem. Soc.* 1965, 4911–4915.

[29] (a) H. Hettler, *Tetrahedron Lett.* 1968, **9**, 1793–1796; (b) J. Sarma, J. D. Dunitz, *Acta Cryst.* 1990, **46**, 780–784; (c) H. Taycher, M. Botoshansky, V. Shteiman, M. Kaftory, *Supramol. Chem.* 2001, **13**, 181–192.

[30] (a) A. W. Chapman, *J. Chem. Soc. Trans.* 1925, **127**, 1992–1998; (b) K. B. Wiberg, B. I. Rowland, *J. Am. Chem. Soc.* 1955, **77**, 2205–2209.

[31] F. Esser, K. Brandt, K. H. Pook, H. J. Forster, H. Koppen, J. M. Leger, A. Carpy, *J. Chem. Soc. Perkin Trans.* 1988, **1**, 3311–3316.

[32] M. S. Newman, H. A. Karnes, *J. Org. Chem.* 1966, **31**, 3980.

[33] H. Kwart, E. R. Evans, *J. Org. Chem.* 1966, **31**, 410.

[34] G. C. Lloyd-Jones, J. D. Moseley, J. S. Renny, *Synthesis* 2008, 661–689.

[35] M. S. Newman, H. A. Karnes, *J. Org. Chem.* 1966, **31**, 3980–3984.

[36] J. A. Aikins, T. Zhang, *Abs. Pap. Am. Chem. Soc.* 2000, **220**, U112–U113.

[37] V. V. Kane, A. Gerdes, W. Grahn, L. Ernst, I. Dix, P. G. Jones, H. Hopf, *Tetrahedron Lett.* 2001, **42**, 373–376.

[38] V. Albrow, K. Biswas, A. Crane, N. Chaplin, T. Easun, S. Gladiali, B. Lygo, S. Woodward, *Tetrahedron-Asymmetry* 2003, **14**, 2813–2819.

[39] (a) J. D. Moseley, R. F. Sankey, O. N. Tang, J. P. Gilday, *Tetrahedron* 2006, **62**, 4685–4689; (b) J. Das, F. Le Cavelier, J. Rouden, J. Blanchet, *Synthesis* 2012, **44**, 1349–1352.

[40] J. D. Moseley, P. Lenden, M. Lockwood, K. Ruda, J.-P. Sherlock, A. D. Thomson, J. P. Gilday, *Org. Process Res. Dev.* 2008, **12**, 30–40.

[41] J. N. Harvey, J. Jover, G. C. Lloyd-Jones, J. D. Moseley, P. Murray, J. S. Renny, *Angew. Chem. Int. Ed.* 2009, **48**, 7612–7615.

[42] H. Jacobsen, J. P. Donahue, *Can. J. Chem. Rev.* 2006, **84**, 1567–1574.

[43] H. R. Al-Kazimi, D. S. Tarbell, D. Plant, *J. Am. Chem. Soc.* 1955, **77**, 2479–2482.

[44] Z. Wang, in *Comprehensive Organic Name Reactions and Reagents*, John Wiley & Sons, Inc., Hoboken, 2010.

[45] S. Y. Ambekar, *Curr. Sci.* 1983, **52**, 578–582.

[46] E. Siddalingamurthy, K. M. Mahadevan, J. N. Masagalli, H. N. Harishkumar, *Tetrahedron Lett.* 2013, **54**, 5591–5596.

[47] A. Martinez, M. J. Webber, S. Mueller, B. List, *Angew. Chem. Int. Ed.* 2013, **52**, 9486–9490.

[48] B. Gutmann, M. Gottsponer, P. Elsner, D. Cantillo, D. M. Roberge, C. O. Kappe, *Org. Process Res. Dev.* 2013, **17**, 294–302.

[49] H. M. Kang, Y. K. Lim, I. J. Shin, H. Y. Kim, C. G. Cho, *Org. Lett.* 2006, **8**, 2047–2050.

[50] A. C. Benniston, W. Clegg, A. Harriman, R. W. Harrington, P. Y. Li, C. Sams, *Tetrahedron Lett.* 2003, **44**, 2665–2667.

[51] H.-Y. Kim, W.-J. Lee, H.-M. Kang, C.-G. Cho, *Org. Lett.* 2007, **9**, 3185–3186.

[52] G. Ghigo, A. Maranzana, G. Tonachini, *Tetrahedron* 2012, **68**, 2161–2165.

[53] A. Mamantov, *Prog. React. Kinet. Mech.* 2013, **38**, 1–31.

[54] (a) J. Rehbein, M. Hiersemann, *Synthesis* 2013, **45**, 1121–1159; (b) A. M. M. Castro, *Chem. Rev.* 2004, **104**, 2939–3002.

[55] J. R. Vyvyan, J. M. Oaksmith, B. W. Parks, E. M. Peterson, *Tetrahedron Lett.* 2005, **46**, 2457–2460.

[56] U. B. Kim, D. P. Furkert, M. A. Brimble, *Org. Lett.* 2013, **15**, 658–661.

[57] K. D. Beare, C. S. P. McErlean, *Org. Biomol. Chem.* 2013, **11**, 2452–2459.

[58] O. Acevedo, K. Armacost, *J. Am. Chem. Soc.* 2010, **132**, 1966–1975.

[59] S. Osuna, S. Kim, G. Bollot, K. N. Houk, *Eur. J. Org. Chem.* 2013, 2823–2831.

[60] F. C. Gozzo, S. A. Fernandes, D. C. Rodrigues, M. N. Eberlin, A. J. Marsaioli, *J. Org. Chem.* 2003, **68**, 5493–5499.

[61] J. B. Sweeney, *Chem. Soc. Rev.* 2009, **38**, 1027–1038.

[62] E. Tayama, H. Kimura, *Angew. Chem. Int. Ed.* 2007, **46**, 8869–8871.

[63] M. Liao, L. Peng, J. Wang, *Org. Lett.* 2008, **10**, 693–696.

[64] Y. Li, Y. Shi, Z. Huang, X. Wu, P. Xu, J. Wang, Y. Zhang, *Org. Lett.* 2011, **13**, 1210–1213.

[65] K. J. Ratnam, R. S. Reddy, N. S. Sekhar, M. L. Kantam, A. Deshpande, F. Figueras, *Appl. Catal. A Gen.* 2008, **348**, 26–29.

[66] L. Yang, B. Tan, F. Wang, G. Zhong, *J. Org. Chem.* 2009, **74**, 1744–1746.

[67] J. Seayad, P. K. Patra, Y. Zhang, J. Y. Ying, *Org. Lett.* 2008, **10**, 953–956.

[68] M. A. Hughes, M. J. Baggs, J. al-Dulayymi, M. S. Baird, P. A. Williams, *Appl. Environ. Microbiol.* 2002, **68**, 4965–4970.

[69] S. Yamabe, G. Zeng, W. Guan, S. Sakaki, *Beilstein J. Org. Chem.* 2013, **9**, 1073–1082.

[70] D. L. H. Williams, *Tetrahedron* 1975, **31**, 1343–1349.

[71] (a) I. D. Biggs, D. L. H. Williams, *J. Chem. Soc. Perkin Trans. 2* 1976, 601–605; (b) I. D. Biggs, D. L. H. Williams, *J. Chem. Soc. Perkin Trans. 2* 1977, 44–47.

[72] I. Cikotiene, M. Jonusis, V. Jakubkiene, *Beilstein J. Org. Chem.* 2013, **9**, 1819–1825.

[73] P. Magnus, R. Turnbull, *Org. Lett.* 2006, **8**, 3497–3499.

[74] A. J. Sisti, S. R. Milstein, *J. Org. Chem.* 1974, **39**, 3932–3936.

[75] R. V. Hoffman, D. J. Poelker, *J. Org. Chem.* 1979, **44**, 2364–2369.

[76] V. Theodorou, K. Skobridis, *Tetrahedron Lett.* 2005, **46**, 5021–5023.

[77] O. Middel, Z. Greff, N. J. Taylor, W. Verboom, D. N. Reinhoudt, V. Snieckus, *J. Org. Chem.* 2000, **65**, 667–675.

[78] F. M. Moghaddam, M. G. Dakamin, *Tetrahedron Lett.* 2000, **41**, 3479–3481.

[79] J. R. Harjani, S. J. Nara, M. M. Salunkhe, *Tetrahedron Lett.* 2001, **42**, 1979–1981.

[80] P. Magnus, C. Lescop, *Tetrahedron Lett.* 2001, **42**, 7193–7196.

[81] T. Magauer, H. J. Martin, J. Mulzer, *Angew. Chem. Int. Ed.* 2009, **48**, 6032–6036.

[82] X. H. Xu, M. Taniguchi, A. Azuma, G. K. Liu, E. Tokunaga, N. Shibata, *Org. Lett.* 2013, **15**, 686–689.

[83] K. P. Jayasundera, A. J. Watson, C. M. Taylor, *Tetrahedron Lett.* 2005, **46**, 4311–4313.

[84] C. M. Taylor, A. J. Watson, *Curr. Org. Chem.* 2004, **8**, 623–636.

[85] A. Koebergtelder, H. Cerfontain, *J. Chem. Soc. Perkin Trans. 2* 1977, 717–719.

[86] E. Solari, F. Musso, R. Ferguson, C. Floriani, A. Chiesivilla, C. Rizzoli, *Angew. Chem. Int. Ed.* 1995, **34**, 1510–1512.

[87] E. Buncel, W. M. Strachan, *Can. J. Chem.* 1970, **48**, 377–382.

[88] E. Buncel, S. R. Keum, S. Rajagopal, R. A. Cox, *Can. J. Chem.* 2009, **87**, 1127–1134.

[89] A. Lalitha, K. Pitchumani, C. Srinivasan, *J. Mol. Catal. A Chem.* 2000, **160**, 429–435.

[90] J. Yamamoto, Y. Nishigaki, M. Umezu, T. Matsuura, *Tetrahedron* 1980, **36**, 3177–3180.

[91] A. S. Ozen, S. S. Erdem, V. Aviyente, *Struct. Chem.* 1998, **9**, 15–25.

PART VI

TRANSITION METAL-MEDIATED COUPLING

19

TRANSITION METAL-CATALYZED CARBON–CARBON CROSS-COUPLING

ANNY JUTAND[1] AND GUILLAUME LEFÈVRE[2]

[1] *Ecole Normale Supérieure-PSL Research University, Département de Chimie, Sorbonne Universités, UPMC Univ Paris 06, CNRS UMR 8640 PASTEUR, Paris, France*

[2] *CEA–Saclay, IRAMIS Institute, SIS2M/LCCEF, Gif-sur-Yvette, Cedex, France*

19.1 INTRODUCTION

Before 1970, the reactions of aryl halides (ArX) with nucleophiles were restricted to electron-deficient aryl halides *via* S_NAr mechanisms. The reactivity of aryl halides in S_NAr reactions decreases in the order ArF > ArCl > ArBr >> ArI [1]. The range of reactive nucleophiles was narrow, and reactions of alkenes or alkynes were not known. The discovery in 1968 of oxidative additions of palladium(0) (Fitton [2a]) and then nickel(0) (Kochi [2b]) with aryl halides (ArI > ArBr > ArCl) that generate $ArM^{II}XL_n$ (M = Pd, Ni) paved the way to new palladium- and nickel-catalyzed reactions of aryl halides with C-nucleophiles, leading to the formation of C—C bonds. Palladium- or nickel-catalyzed cross-coupling reactions are described herein, following the historical order of discovery.

19.2 THE MIZOROKI–HECK REACTION

19.2.1 General Considerations and Mechanisms

The Mizoroki–Heck reaction between aryl halides and alkenes is catalyzed by palladium (Scheme 19.1). It is one of the most efficient routes for the arylation of alkenes in which a new Csp^2—Csp^2 bond is formed. The reactions are performed in the presence of a base and may generate the linear or/and the branched product [3]. If reactions of aryl iodides can be conducted in the absence of ligand L, the latter is required for reactions involving less reactive aryl bromides and chlorides.

Arene Chemistry: Reaction Mechanisms and Methods for Aromatic Compounds, First Edition. Edited by Jacques Mortier.

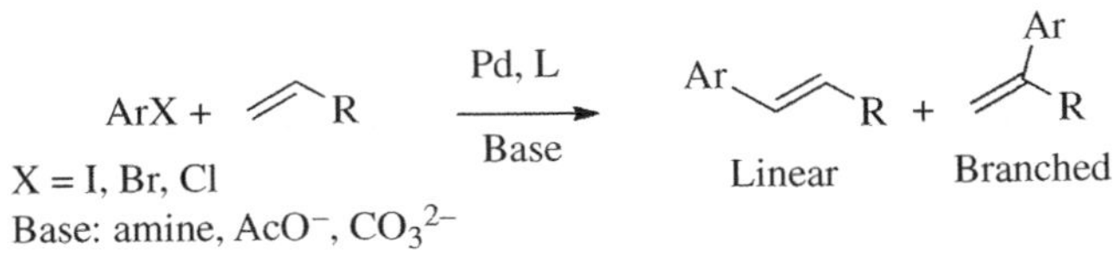

SCHEME 19.1 The Mizoroki–Heck reaction (1971–1972).

In 1968, Heck reported stoichiometric reactions of alkenes with [ArPd^{II}Cl] (generated *in situ* by reacting ArHgCl with Li_2PdCl_4) leading to ArCH=CHR [4a]. Mizoroki reported in 1971 the first catalytic reactions between PhI and alkenes, in the presence of catalytic $PdCl_2$ and KOAc as base [5a]. In 1972, the catalytic reactions were improved by Heck upon using $Pd(OAc)_2$ and $n\text{Bu}_3N$ as a base [4b]. In 1973, Mizoroki in a last contribution extended the scope of its preliminary work to PhBr with the reactivity order: PhI ≫ PhBr [5b]. In 1974, Heck developed the use of PPh_3 in association with $Pd(OAc)_2$ to allow reactions of ArBr at 100–135°C [4c]. In 1978, Heck introduced the tri-*o*-tolylphosphine $P(o\text{-Tol})_3$ in association with $Pd(OAc)_2$ that let the reaction with ArBr proceed at lower temperatures (75°C) [4d]. The foundation for what is called the Mizoroki–Heck reaction was established. Heck received the Nobel Prize in 2010.

In 1974, Heck proposed a mechanism for reactions catalyzed by $Pd(OAc)_2$ associated with monophosphine ligands L that involves as main steps: *oxidative addition*, *carbopalladation*, *β-hydride elimination*, and *reductive elimination* (Scheme 19.2) [3a, b, 4c, d].

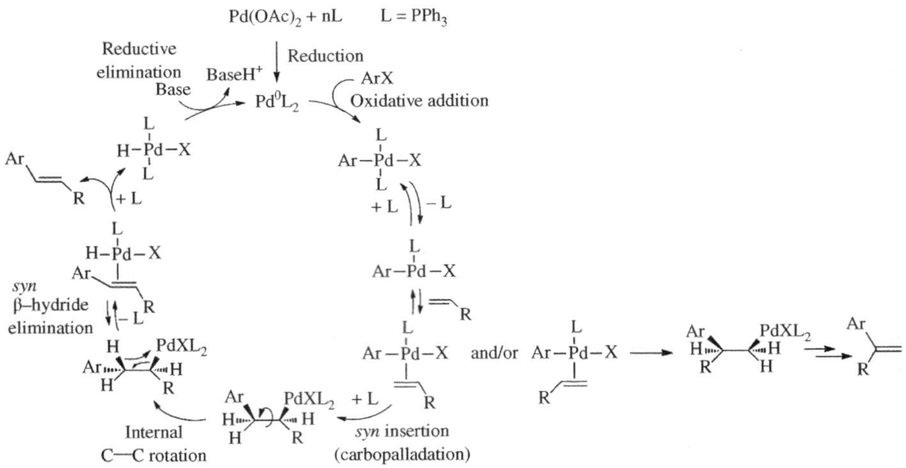

SCHEME 19.2 Mechanism proposed by Heck.

The main steps of the mechanism proposed by Heck have been further on confirmed. New ligands (diphosphines, carbenes, bulky monophosphines, polyphosphines) and new precatalysts (P,C-palladacycles) were introduced all along the last 50 years. Mechanistic investigations revealed that depending on the experimental conditions, the catalytic cycle may involve intermediate palladium complexes whose structure differs from the original ones proposed by Heck.

19.2.1.1 Precatalyst: $Pd(OAc)_2$ in Ligand-Free Heck Reactions Heck reactions from aryl iodides are catalyzed by $Pd(OAc)_2$ without any phosphine ligand [4b]. Their efficiency is greatly improved by additives such as ammonium salts $R_4N^+X^-$ (X = Br, Cl) (Jeffery [6a]). The thermolytic decomposition of $Pd(OAc)_2$ generates Pd^0 nanoparticles stabilized by $R_4N^+X^-$ (Scheme 19.3a, Reetz [6b]).

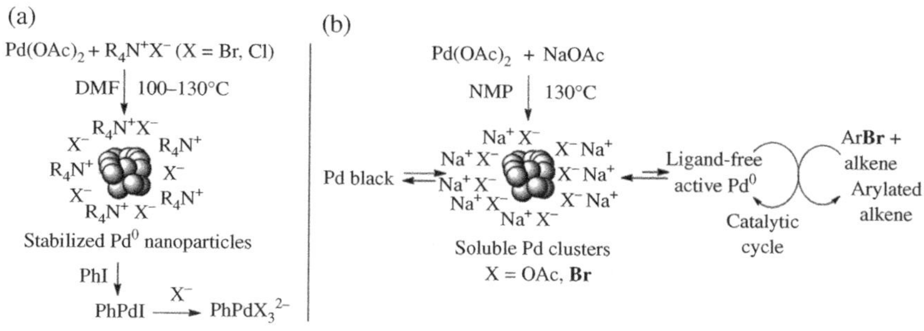

SCHEME 19.3 (a) Nanoparticles or (b) clusters in ligand-free Heck reactions.

De Vries has developed ligand-free Heck reactions involving ArBr in the presence of low loading of $Pd(OAc)_2$ (0.01–0.1 mol%) and NaOAc as a base [6c]. Soluble Pd^0 clusters stabilized by Na^+X^- ($X^-=AcO^-$, Br^-) are formed that inhibit the reversible formation of inactive palladium black and slowly deliver the active ligand-free Pd^0 (Scheme 19.3b) [6d]. Lower the $Pd(OAc)_2$ loading, higher the TOF. Aryl chlorides are unreactive.

19.2.1.2 *Precatalyst: $Pd(OAc)_2$ Associated with Phosphine Ligands*

Generation of Pd^0 Complexes The precatalyst $Pd(OAc)_2$ associated with PPh_3 is often used in Heck reactions [3, 4]. In 1991, Jutand discovered that a Pd^0 complex was generated *in situ* in DMF or THF at rt upon mixing $Pd(OAc)_2$ and *n* equiv. of PPh_3 ($n \geq 2$) [7a]. Kinetic data by means of electrochemical techniques revealed that a Pd^0 complex is formed by reductive elimination in $Pd(OAc)_2(PPh_3)_2$ (0th-order reaction for PPh_3, 1st-order reaction for Pd^{II}) (Scheme 19.4) [7a, b]. The intramolecular reduction of Pd^{II} to Pd^0 by the ligated PPh_3 delivers a phosphonium salt that is hydrolyzed to phosphine oxide in a faster step (0th-order reaction for H_2O) [7a]. Shortly after, Osawa/Hayashi reported the same process [8].

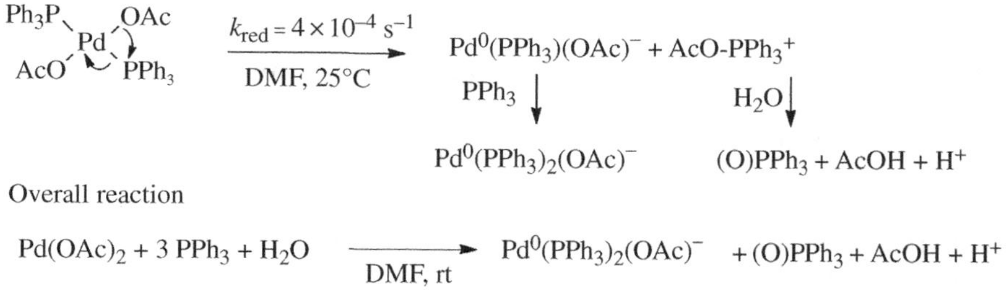

SCHEME 19.4 Mechanism of the formation *in situ* of a Pd^0 complex from $Pd(OAc)_2$ and PPh_3.

The Pd^0 complex generated from $Pd(OAc)_2$ and 3 equiv. PPh_3 is an anionic species $[Pd^0(PPh_3)_2(OAc)]^-$ ligated by AcO^- (Jutand/Amatore [7c]) whose structure is supported by DFT ($d_{Pd\text{-}O}=2.369$ Å, Shaik [7d]).

The formation of Pd^0 by reduction of Pd^{II} by a ligated phosphine takes place in all complexes possessing a Pd^{II}—OR bond (OH [7e, f], OAc [7a, b, 8], $OCOCF_3$ [7g]). The monophosphines might be aromatic, aliphatic [7a, b], or water soluble (TPPTS [7h], TPPTC [7i]). There is one exception; $P(o\text{-Tol})_3$ cannot reduce $Pd(OAc)_2$ to a Pd^0 complex because an activation of one Csp^3—H bond of the ligated $P(o\text{-Tol})_3$ by $Pd(OAc)_2$ occurs, affording a dimeric P,C-palladacycle (Scheme 19.5) that is used as precatalyst in Heck reactions involving ArBr and ArCl (Herrmann, [9, 3g, h]). Jutand has established by means of electrochemical techniques that *P,C*-palladacycles are a reservoir of Pd^0

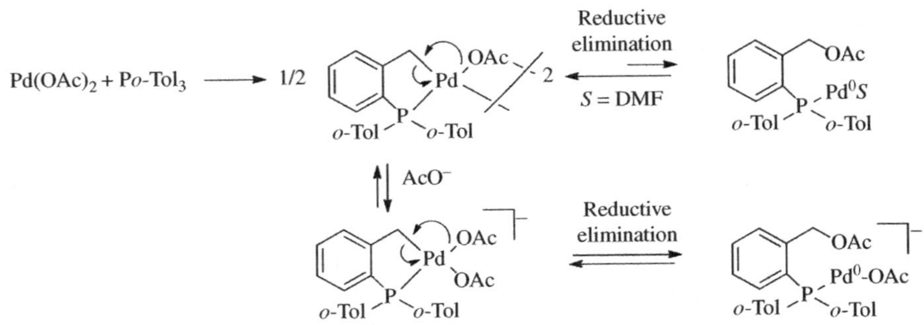

SCHEME 19.5 Formation of Pd^0 complexes from a P,C-palladacycle.

formed in an endergonic reductive elimination between the benzylic and *cis*-acetate ligands (Scheme 19.5) [10]. This reaction is favored in the presence of added acetates (often used as bases in Heck reactions) and by alkenes [31, s, 10].

The phosphine associated with $Pd(OAc)_2$ might be a diphosphine P^P [11]. In that case, water was found to play a crucial role (Binap, Hayashi [8]; dppp, Jutand/Amatore [7j]). Indeed, the reaction of $Pd(OAc)_2$ with dppp (1 equiv.) does not generate any detectable Pd^0 formed at low concentration due to the reversibility of the reductive elimination in $Pd(OAc)_2$(dppp) (Scheme 19.6). A Pd^0 complex is formed upon addition of dppp that substituted the monoligated Ph_2P-$(CH_2)_3$-$PPh_2(OAc)^+$ whose hydrolysis generates the hemioxide dppp(O), causing a shift of the successive equilibria toward the anionic $[Pd^0(dppp)(OAc)]^-$ (Scheme 19.6 [7j]) (DFT structure, Shaik [7d]).

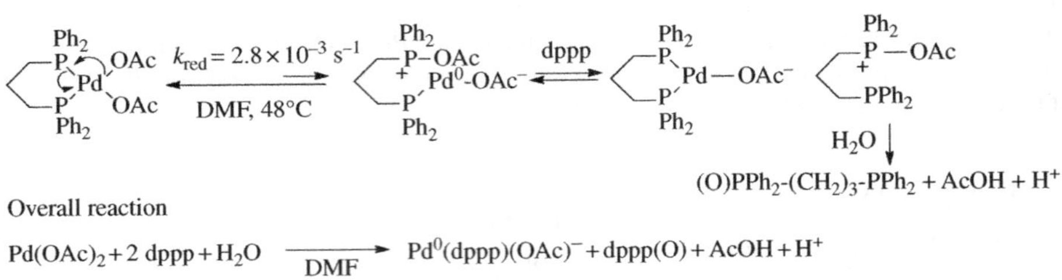

SCHEME 19.6 Formation of Pd^0 complexes from $Pd(OAc)_2$ and dppp.

Oxidative Addition of Pd^0 to ArX Kinetic data on the oxidative addition of PhI by means of electrochemical techniques revealed that $[Pd^0(PPh_3)_2(OAc)]^-$ (generated from $Pd(OAc)_2$ and PPh_3) is the reactive species (Scheme 19.7) [7c].

Carbopalladation (Complexation/Insertion) of the Alkene In the mechanism postulated by Heck (Scheme 19.2), *trans*-$PhPdIL_2$ was supposed to be formed in the oxidative addition of PhI and react with the alkene. The expected *trans*-$PhPdI(PPh_3)_2$ is not produced but instead *trans*-PhPd(OAc)$(PPh_3)_2$ [7c] in equilibrium with the cationic *trans*-$[PhPd(PPh_3)_2S]^+$ (Scheme 19.7) [7k].

Comparative reactivity of *trans*-$PhPdX(PPh_3)_2$ (X = I, OAc, BF_4) shows that *trans*-PhPd(OAc)$(PPh_3)_2$ reacts with styrene at 25°C in contrast to unreactive *trans*-$PhPdI(PPh_3)_2$ (Scheme 19.8a) [7c, l]. The reaction of the alkene with *trans*-$PhPdI(PPh_3)_2$ is limited by the dissociation of PPh_3 whereas that in *trans*-PhPd(OAc)$(PPh_3)_2$ is assisted by the bidentate character of the acetate ligand (Scheme 19.8a). This favors the approach of the alkene in a *cis*-position relative to the Ph group and facilitates the *syn*-insertion. The cationic *trans*-$[PhPd(PPh_3)_2(DMF)]^+$ is more reactive toward styrene than *trans*-$PhPdI(PPh_3)_2$ but less reactive than *trans*-PhPd(OAc)$(PPh_3)_2$ [7c, l] because the

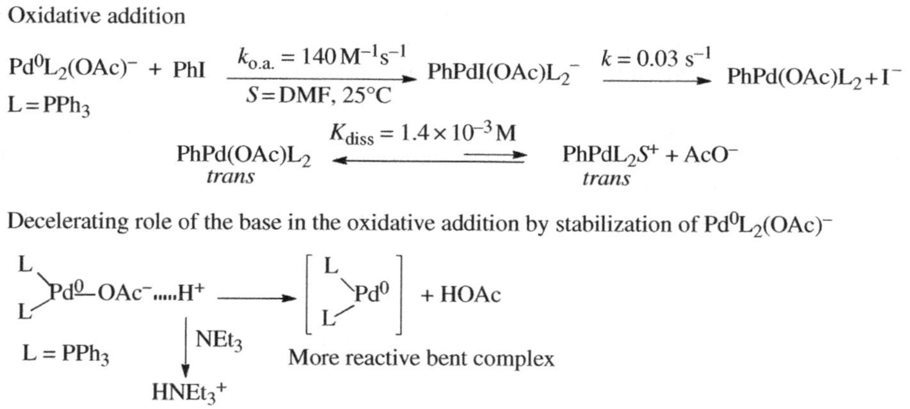

SCHEME 19.7 Oxidative addition of $[Pd^0(PPh_3)_2(OAc)]^-$ to PhI and role of NEt_3.

coordination of the cationic complex by the alkene gives a *trans*-complex (Scheme 19.8b). The insertion of the alkene is thus inhibited by the endergonic *trans/cis* isomerization of the alkene-ligated cationic complex, making route (b) slower than route (a). The following decreasing reactivity order is observed in DMF at rt:

$$trans\text{-PhPd(OAc)(PPh}_3)_2 > trans\text{-PhPd(PPh}_3)_2(\text{DMF})^+ \gg trans\text{-PhPdI(PPh}_3)_2$$

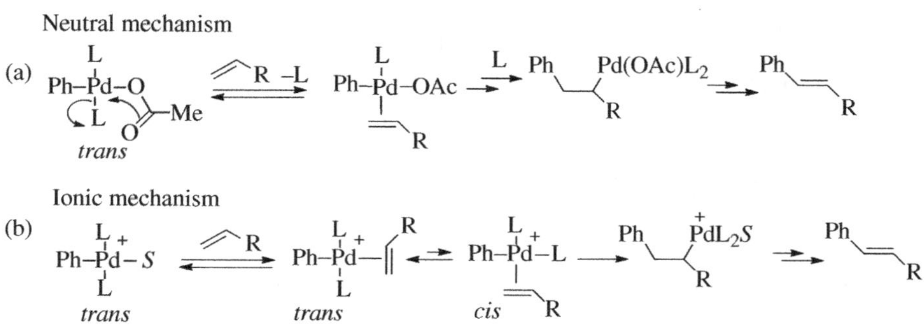

SCHEME 19.8 (a) Neutral or (b) ionic mechanism of the *carbopalladation*.

Consequently, *trans*-PhPd(OAc)$(PPh_3)_2$ is a key intermediate in the *carbopalladation*.

Amines and Alkenes as Factors Affecting the Rates of Both the Oxidative Addition and Carbopalladation The base (NEt_3) plays a multiple role. It stabilizes $Pd^0(PPh_3)_2(OAc)^-$ versus its decomposition to $Pd^0(PPh_3)_2$ by protons and consequently slows down the oxidative addition (Scheme 19.7) [7c, l, p]. The base accelerates the overall carbopalladation by shifting the equilibrium toward PhPd(OAc)L_2 upon quenching the proton (Scheme 19.9) [7l]. It favors the recycling of the Pd^0 complex from the hydrido-Pd^{II}. The formation of $HPdI(PPh_3)_2$ was proposed by Heck (Scheme 19.2). In DMF, the cationic $[HPd(PPh_3)_2S]^+$ must be formed with acetate as the counter anion (Amatore/Jutand [7m]).

The oxidative addition is slower when performed in the presence of alkenes due to their complexation to the reactive $[Pd^0(PPh_3)_2(OAc)]^-$, which generates nonreactive $[(\eta^2\text{–}CH_2{=}CHR)Pd^0(PPh_3)_2(OAc)]^-$ (Scheme 19.9) [7n].

Catalytic Cycle The mechanism of Heck reactions in Scheme 19.9 points out how the precatalyst, base, ligand, and alkene affect the structure and reactivity of intermediate Pd^0 or Pd^{II} complexes and consequently affect the efficiency of the catalytic reaction. From the rate constants of the main steps, it appears that for comparable PhI and styrene concentrations, the overall *carbopalladation* (complexation/insertion of the alkene) from $PhPd(OAc)(PPh_3)_2$ is the slowest step of the catalytic cycle. This mechanism highlights the crucial role of acetates delivered by the precatalyst $Pd(OAc)_2$. Indeed, AcO^- is a ligand of all Pd^0 and Pd^{II} complexes present in the catalytic cycle, mainly in $ArPd(OAc)L_2$ involved in the rate-determining *carbopalladation*.

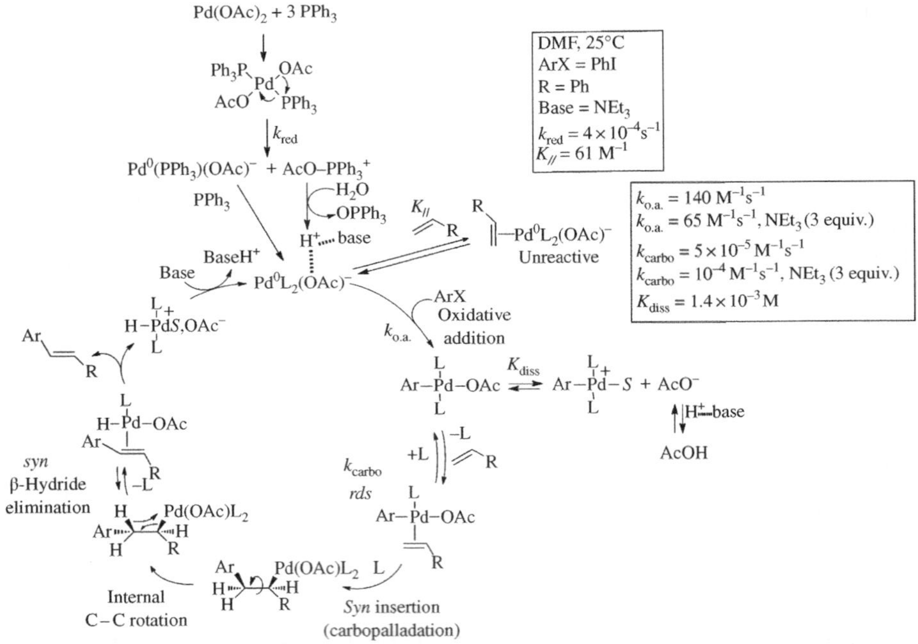

SCHEME 19.9 Mechanism of the Heck reaction with $Pd(OAc)_2$ as precatalyst associated with PPh_3, including the multiple role of the base and alkene.

The base, the alkene, and AcO^- play the same dual role in Heck reactions: deceleration of the fast *oxidative addition* and acceleration of the slow *carbopalladation*. This dual effect favors the efficiency of the catalytic reaction by making the rates of the *oxidative addition* and *carbopalladation* closer to each other [7o, p].

This mechanism is also valid for other precatalysts: $\{Pd^0(dba)_2$ and n $PPh_3\}$, $PdCl_2(PPh_3)_2$, or $Pd^0(PPh_3)_4$, if AcO^- is used as base. Indeed, AcO^- coordinates Pd^0L_2 to give $[Pd^0L_2(OAc)^-]$ or reacts with $ArPdIL_2$ to generate the reactive $ArPd(OAc)L_2$ [7l]. When considering less reactive ArBr or ArCl involved in the *rds oxidative addition*, the situation is worse since the alkene slows down the *oxidative addition* by complexation of the reactive $[Pd^0L_2(OAc)]^-$ (Scheme 19.9). To solve this problem, new ligands must be designed to make the Pd^0 more reactive.

19.2.1.3 Activation of Aryl Bromides and Chlorides in Heck Reactions

Electron-Rich Bulky Phosphines Electron-rich bulky phosphines were introduced by Fu in 1999 [12a]. Among them, $PtBu_3$ is the best ligand for reacting aryl chlorides under mild conditions [12d].

Phenyl chloride does not react with $Pd^0(PtBu_3)_2$ (Hartwig [12b]). This suggests that the concentration of the reactive $Pd^0(PtBu_3)$ in its equilibrium with $Pd^0(PtBu_3)_2$ is very low (Scheme 19.10a). Decreasing the ratio $PtBu_3$/Pd should favor the oxidative addition. This was observed with the precatalyst $Pd^0_2(dba)_3$ and Pd/$PtBu_3$ = 1 [12c]. $Pd^0(dba)(PtBu_3)$ dissociates to the reactive $Pd^0(PtBu_3)$ more easily than $Pd^0(PtBu_3)_2$ does.

Investigation by Fu on the recycling of the Pd^0 from the hydridopalladium(II) (formed in the β-hydride elimination) provides mechanistic insights on its key involvement in Heck reactions performed from ArCl [12c]. With Cs_2CO_3 as base, $HPd^{II}Cl(PtBu_3)_2$ was detected by ^{31}P NMR whereas when Cy_2NMe was employed, $HPd^{II}Cl(PtBu_3)_2$ was not observed but $Pd^0(PtBu_3)_2$ was detected instead. This is why Cy_2NMe is more efficient than Cs_2CO_3 because more efficient in the regeneration of Pd^0 from $HPdCl(PtBu_3)_2$. Fu has investigated the mechanism of formation of $Pd^0(PtBu_3)_2$ from *trans*-$HPdCl(PtBu_3)_2$ in the presence of Cy_2NMe. The reaction is inhibited by $PtBu_3$, which evidences a predissociation of one $PtBu_3$ prior reductive elimination (Scheme 19.10b) [12c]. Therefore, the last step of the catalytic cycle in which the Pd^0 is regenerated might be rate-determining, which is often underestimated.

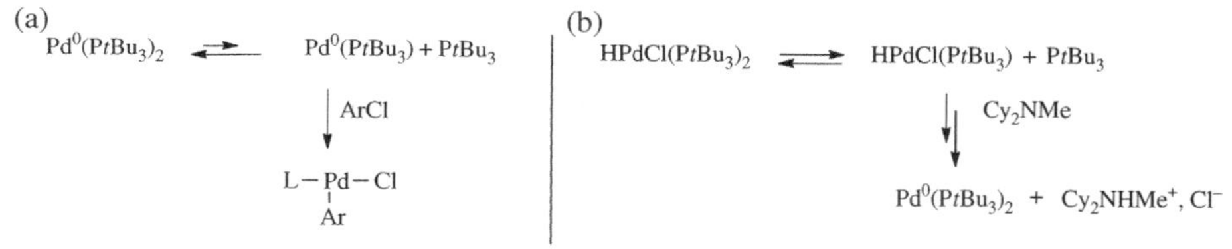

SCHEME 19.10 Two roles (a) and (b) of $PtBu_3$ in Heck reactions performed from ArCl.

N-Heterocyclic Carbene Ligands ***N*-heterocyclic carbene ligands** (NHC, named Cb in the following) were introduced by Herrmann *via* the precatalyst $PdI_2(Cb)_2$ to activate poorly reactive aryl bromides and chlorides in Heck reactions [13, 3h, k]. NHCs are indeed strong σ-donor [3k] that make Pd^0 complexes more reactive in the oxidative addition of aryl halides. The Heck reaction is accelerated upon addition of hydrazine, a reducing agent. This establishes that a Pd^0 is the active species in the oxidative addition [13].

In 2003, Roland/Jutand used $PdI_2(Cb^1)_2$ (Scheme 19.11) as precatalyst for Heck reactions performed from PhBr at moderate temperatures [14a]. Since $Pd^0(Cb^1)_2$ could not be isolated, its

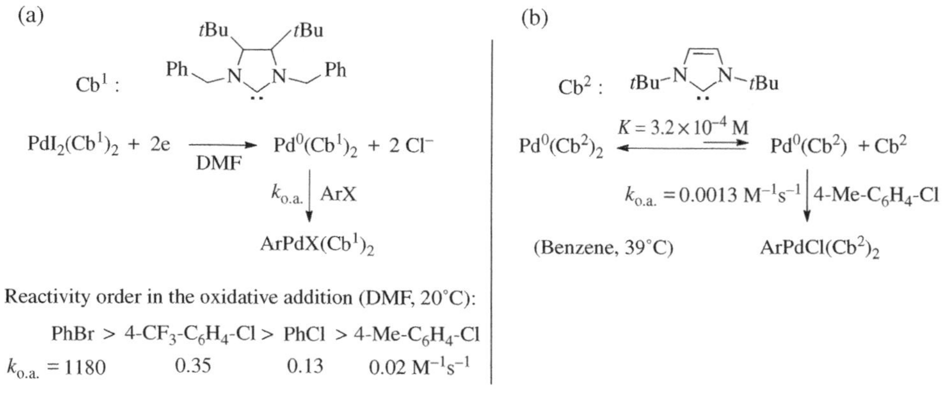

SCHEME 19.11 (a) *Associative* mechanism for the oxidative addition of ArX to $Pd^0(Cb^1)_2$. (b) *Dissociative* mechanism for the oxidative addition of ArX to $Pd^0(Cb^2)_2$.

reactivity with aryl halides was followed by cyclic voltammetry, the transient $Pd^0(Cb^1)_2$ being generated in the electrochemical reduction of $PdI_2(Cb^1)_2$ (Scheme 19.11a). The rate constants *k* of the oxidative addition of aryl halides (X=I, Br, Cl) to $Pd^0(Cb^1)_2$ were determined (*associative* mechanism) [14a, b]. Caddick/Cloke reported that the isolated $Pd^0(Cb^2)_2$ reacts with 4-CH_3—C_6H_4—Cl *via* $Pd^0(Cb^2)$ in a *dissociative* mechanism (Scheme 19.11b) [14c].

The stable Cb^2 carbene is bulky and prompted to dissociation from $Pd^0(Cb^2)_2$. Conversely, Cb^1 is less bulky than Cb^2 and less incited to dissociation from $Pd^0(Cb^1)_2$. Interestingly, comparison of the reactivities of $Pd^0(Cb^1)_2$ and $Pd^0(Cb^2)_2$ with 4-chlorotoluene shows that $Pd^0(Cb^1)_2$ that reacts via an *associative* mechanism (Scheme 19.11a) [14b] is more reactive than $Pd^0(Cb^2)_2$ that reacts via $Pd^0(Cb^2)$ in a *dissociative* mechanism (Scheme 19.11b) [14c]. The decreasing reactivity order was established: $Pd^0(Cb^1)_2 > Pd^0(Cb^2)$.

Consequently, the involvement of a monoligated $Pd^0(Cb)$ as the active species is not a guarantee for a fast oxidative addition, because $Pd^0(Cb)$ is always generated at low concentration in its endergonic equilibrium with the nonreactive $Pd^0(Cb)_2$.

Polyphosphines Ligands Polyphosphines ligands with CH_2-PPh_2 groups on a cyclopentyl moiety (e.g., *cis,cis,cis*-1,2,3,4-tetrakis(diphenylmethyl)cyclopentane, Santelli [15a]) or PPh_2 groups on the cyclopentadienyl ligands of the dppf ligand (e.g., 1,1′,2,2′-tetrakis(diphenylphosphino)-4,4′-di-*tert*-butylferrocene, Hierso [15b]) also proved to be good electron-rich ligands for Pd^0 complexes and thus favored the oxidative addition or aryl bromides (determination of the rate constants of the oxidative addition *via* electrochemical techniques, e.g., $k = 0.48\,M^{-1}s^{-1}$ for PhBr in THF, Lucas/Hierso) [15c, d]). In addition, the high number and flexibility of the PPh_2 groups assure the stability of the Pd^0 and intermediate Pd^{II} complexes, leading to the highest TONs reported for Heck reactions [15a, b, e].

All ligands that make Pd^0 complexes able to activate aryl bromides and chlorides in oxidative additions, such as palladacycles, NHC carbenes, bulky monophosphines, or polyphosphines, have also been successfully used for most cross-coupling reactions reported in the following that all involve oxidative addition as the first step of the catalytic cycle [3h, k, 9, 12d, 15e].

19.2.2 Scope of the Reaction

19.2.2.1 Regioselectivity of the Mizoroki–Heck Reaction Regioselectivity is a major problem in Heck reactions that could generate the linear and/or branched arylated alkenes (Scheme 19.1). The regioselectivity has been studied using $Pd(OAc)_2$ associated with a diphosphine ligand (dppp) and an electron-rich alkene, CH_2=CH—OR (R=alkyl) [11]. As observed by Cabri, linear and branched products are formed according to a nonselective complexation of CH_2=CH—OR to ArPdX(dppp) (*neutral* mechanism) (Scheme 19.12a) [11a]. In contrast, the branched product CH_2=CH(Ar)—OR is always formed as the major product in reactions performed in the presence of halide scavengers, under the conditions of an *ionic* mechanism triggered by a selective complexation of CH_2=CH—OR to $[ArPd(dppp)S]^+$ (Scheme 19.12b) [11a–d]. Xia made the same observation by working at high ionic strength, that is, when the cationic $[ArPd(dppp)S]^+$ is major (Scheme 19.12b) [11e, f].

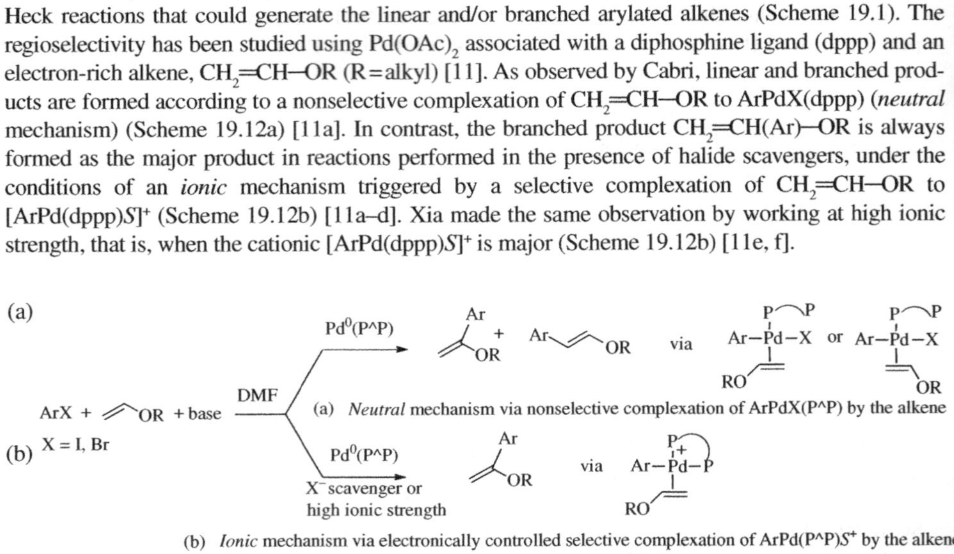

SCHEME 19.12 Experimentally observed regioselectivity and proposed mechanisms.

However, three main complexes are formed in the oxidative addition of ArI to $[Pd^0(dppp)(OAc)]^-$ (generated from $Pd(OAc)_2$ and 2 dppp): ArPdI(dppp), ArPd(OAc)(dppp) in equilibrium with the cationic $[ArPd(dppp)S]^+$ (Scheme 19.13) [7j].

The kinetics of the reactions of isolated PhPdX(dppp) (X=I, OAc) with CH_2=CH–Ph, electron-deficient CH_2=CH–CO_2Me or electron-rich CH_2=CH–O*i*Bu has been investigated in DMF (Jutand/Amatore [16a, b]). All reactions proceed from the *cationic* $[PhPd(dppp)S]^+$. Indeed, all reactions are retarded by added I^- and AcO^- and accelerated upon increasing the ionic strength (which favors the cationic complex), in agreement with an *ionic* mechanism but in contradiction with Cabri hypothesis (Scheme 19.12a). Moreover, Åkermark has reported reactions of PhPdX(dppp) (X=I, OAc, BF_4) with styrene [16c]. The ratio PhCH=CHPh/CH_2=CPh_2 decreases in the order OAc>I>BF_4. According to the kinetic results [16a] that establish the exclusive reactivity of styrene with the *cationic* $[PhPd(dppp)S]^+$ generated from PhPdX(dppp) (X=I, AcO), all complexes PhPdX(dppp) (X=I, AcO, BF_4) should afford the same regioselectivity, that given by $[PhPd(dppp)S]^+BF_4^-$, which is not observed experimentally [16c].

This suggests that I^- and AcO^- interfere after the reaction of the alkene with $[PhPd(dppp)S]^+$. A mechanism is proposed by Jutand/Amatore, which rationalizes the regioselectivity of Heck reactions performed in DMF (Scheme 19.13) [16b, 3t].

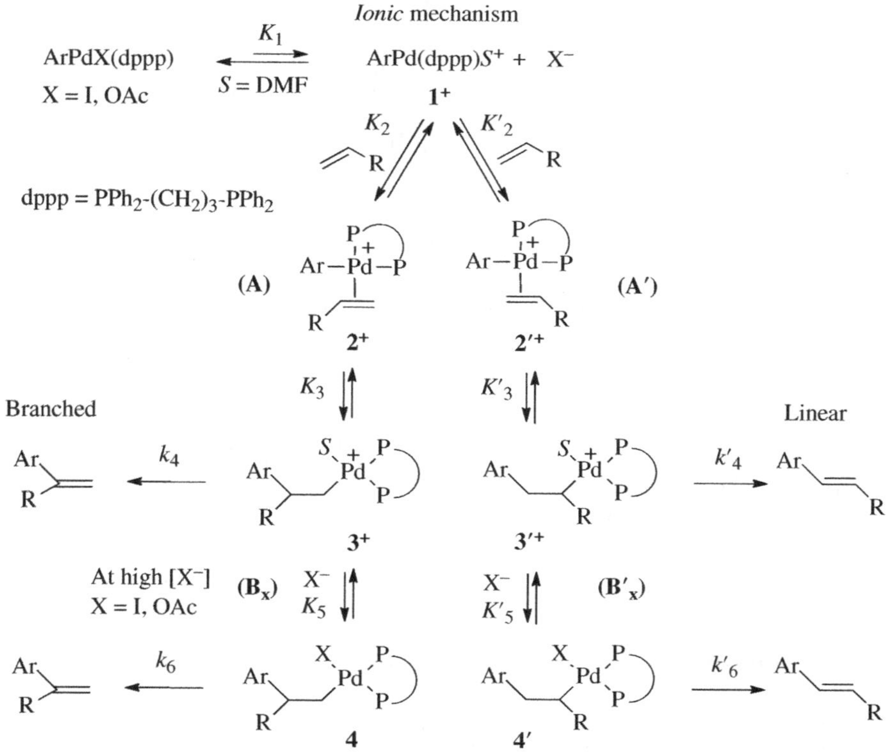

SCHEME 19.13 Regioselectivity of Heck reactions with dppp as ligand in DMF.

The complexation of the alkene to the cationic $[PhPd(dppp)S]^+$ (1^+) may generate the two isomers 2^+ and $2'^+$ and then complexes 3^+ and $3'^+$ in a reversible insertion step. If I^- or AcO^- is present at high concentrations, the cationic complexes 3^+ and $3'^+$ can be reversibly quenched by the anions to form the neutral **4** and **4′**, respectively.

When considering electron-rich alkenes, the branched product is formed as a major product via route **A** ($K_2K_3k_4 \gg K_2'K_3'k_4'$) as observed when reactions from aryl halides are performed in the presence of a halide scavenger [11a–d] or at high ionic strength [11e, f] (Scheme 19.12b). A mixture of linear and branched products will be formed if the reaction is performed in the absence of halide scavengers, that is, in the presence of a large excess of anions (I^-, AcO^-), as it often occurs in real catalytic reactions (competition **A′**+ $\mathbf{B_X}'$ vs. **A**+$\mathbf{B_X}$). Since all complexes are in equilibrium, the most stable **4′** or **4**, formed in the presence of X^- or AcO^-, may drive the regioselectivity toward the linear or branched product accordingly (Scheme 19.13). This is also valid for the less polarized styrene or electron-deficient acrylate.

Extension of Heck Reactions via β′-H Elimination In the case of cyclic alkenes (Scheme 19.14a) or intramolecular reactions (Scheme 19.14c) in which no *syn*-β-hydride is available because the C—C bond of the complex formed in the carbopalladation cannot rotate, a *syn*-β′-hydride elimination takes place, leading to nonconjugated alkenes [3a, c, d, m].

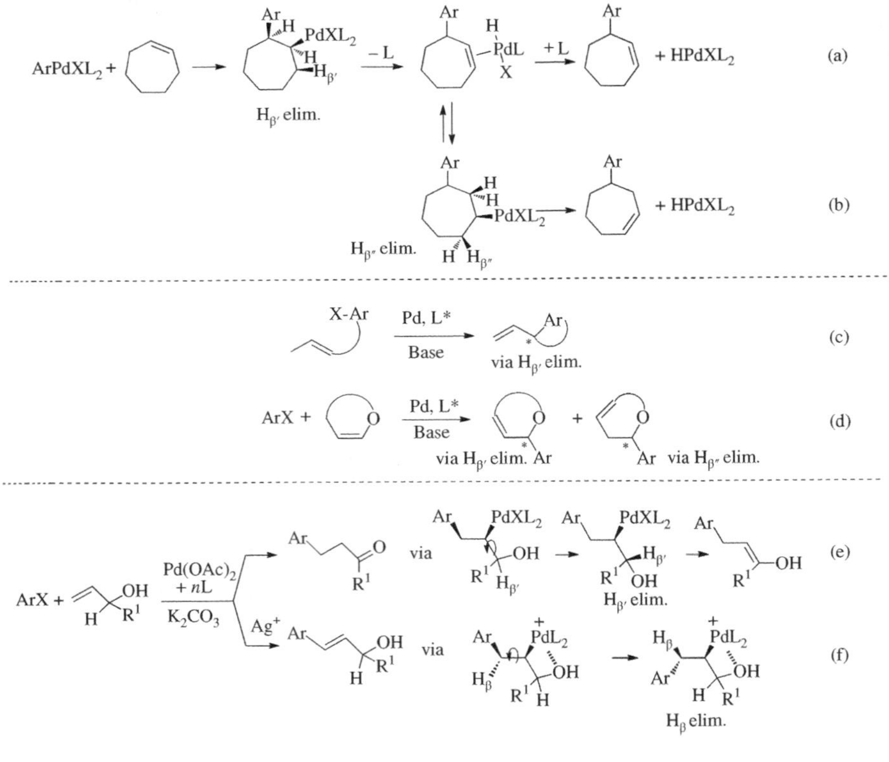

SCHEME 19.14 β′- and β″-hydride elimination and enantioselective Heck reactions.

Isomerization of the C=C bond may occur by a *syn*-readdition of HPdX with the reverse regioselectivity, followed by a *syn*-β″-hydride elimination (Scheme 19.14b and d).

When the alkene is an allylic alcohol, a nonclassical Heck reaction takes place due to a more favored *syn*-β′-hydride elimination, leading to a ketone (Scheme 19.14e). The classical Heck reaction involving the *syn*-β-hydride elimination occurs only in the presence of Ag^+ that generates the cationic $[ArPdL_2S]^+$. The coordination of the OH onto the Pd^+ center (Scheme 19.14f) prevents the C—C bond rotation observed previously [3c, m].

Enantioselective Reactions Due to *syn*-β′- or β″-hydride hydride elimination, stereogenic centers may be created in the presence of chiral ligands [3n–s, 17, 18]. The asymmetric intramolecular version (Scheme 19.14c) (Shibasaki, Overman) [17] is more developed than the asymmetric intermolecular version (Scheme 19.14d) (Hayashi) [18].

19.2.3 Synthetic Application

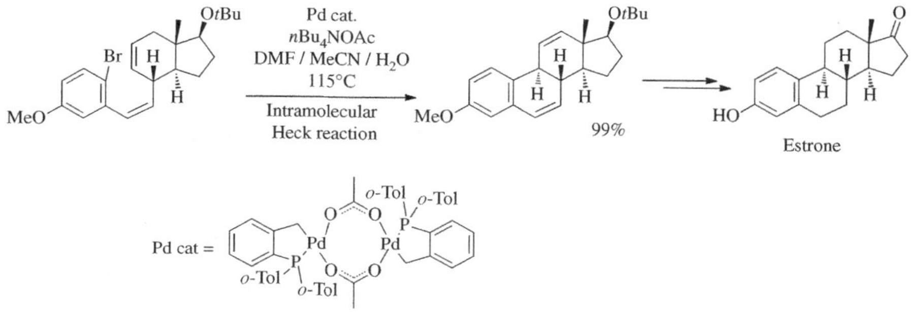

SCHEME 19.15 A Heck-type reaction in the synthesis of enantiopure estrone [19].

19.3 CROSS-COUPLING OF ARYL HALIDES WITH ANIONIC C-NUCLEOPHILES

19.3.1 The Kumada Reactions: Nickel-Catalyzed Cross-Coupling with Grignard Reagents

19.3.1.1 General Consideration In 1972, Kumada reported the first nickel-catalyzed cross-coupling of PhCl with secondary Grignard reagents [20a, b]. A by-product is formed *via* isomerization of the alkyl group, accompanied by formation of ArH (Scheme 19.16). The β-hydride elimination requires a free coordination site on the Ni center. This is why it is inhibited by diphosphines. The reaction was extended to EtMgBr and *n*BuMgBr [20b, c].

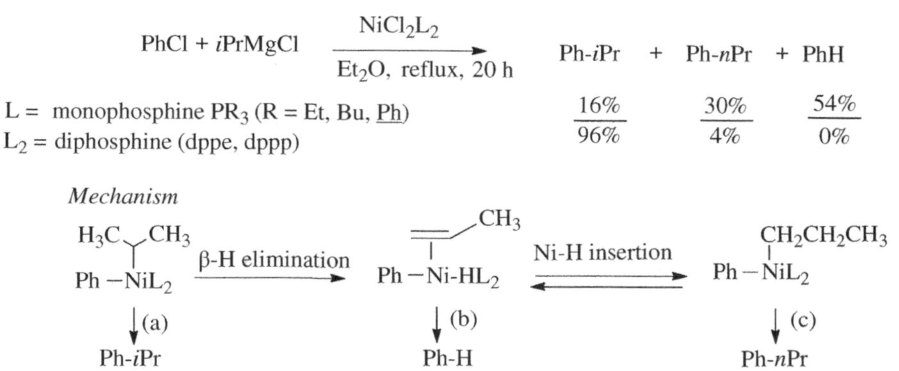

SCHEME 19.16 The Kumada reaction (1972).

19.3.1.2 Synthetic Application

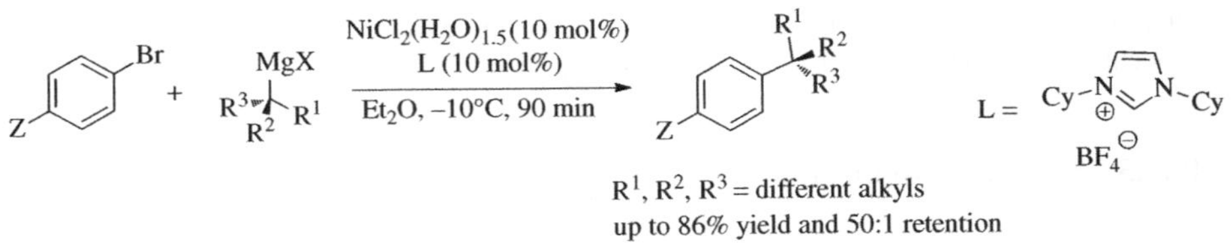

SCHEME 19.17 Bulky quaternary centers with absolute retention configuration [21].

19.3.2 Palladium-Catalyzed Cross-Coupling with Grignard Reagents

19.3.2.1 General Considerations In 1976, Fauvarque/Jutand reported the first palladium-catalyzed cross-coupling of aryl iodides and bromides with Grignard reagents (Scheme 19.18) [22]. The reduction product ArH is formed when a β-hydride elimination can occur in the intermediate complex $ArPdR(PPh_3)_2$ (R = Et, Bu) (Scheme 19.18), as established in reactions performed from isolated *trans*-$ArPdI(PPh_3)_2$ [22]. The same year, Ichikawa used $PhPdI(PPh_3)_2$ as precatalyst [23].

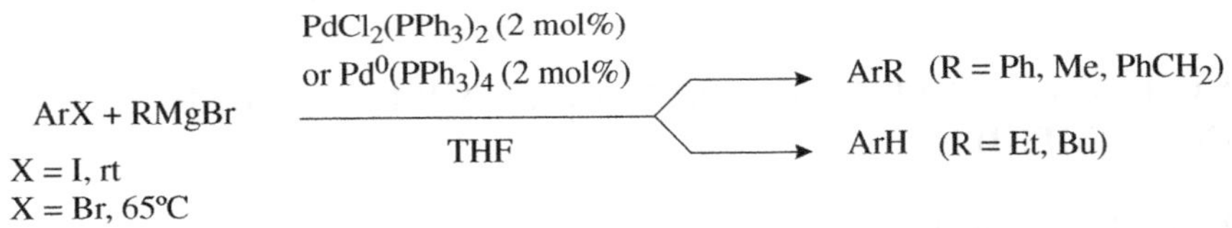

SCHEME 19.18 Palladium-catalyzed cross-coupling of Grignard reagents (1976).

In 1979, Kumada/Hayashi extended the scope of the palladium-catalyzed reaction using the diphosphine dppf to react alkyl Grignard reagents without any isomerization (Scheme 19.19) [24a, b]. Aryl bromides react at rt, which means that the dppf ligand accelerates the oxidative addition (electron-rich ligand) when compared to PPh_3. Remarkably, dppf also accelerates the reductive elimination (*cis* and bulky ligand) and suppresses the β-hydride elimination.

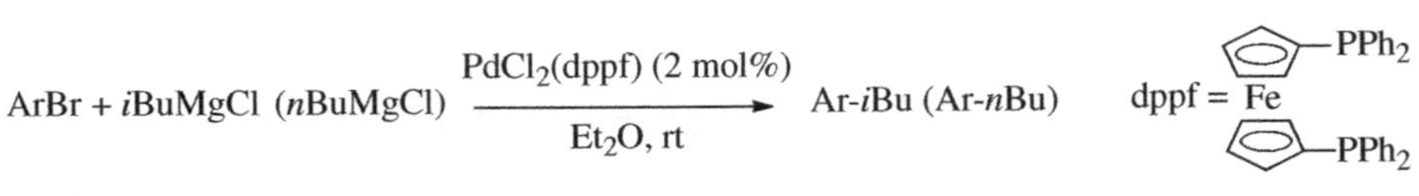

SCHEME 19.19 First use of the dppf ligand (1979).

19.3.2.2 Synthetic Application

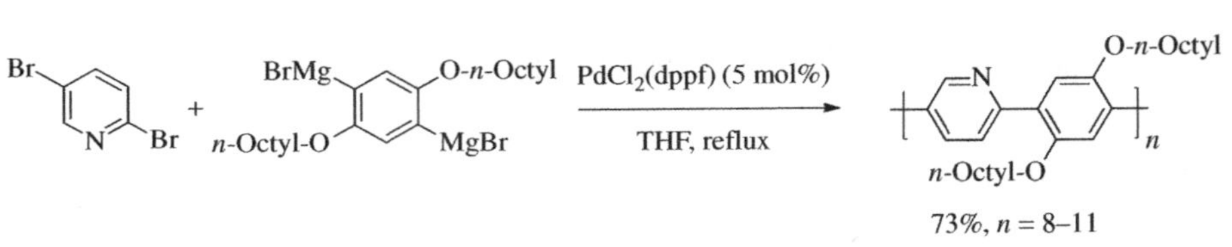

SCHEME 19.20 Synthesis of oligomers [25].

19.3.3 The Negishi Reaction: Palladium-Catalyzed Cross-Coupling with Organozinc Reagents

19.3.3.1 General Considerations In 1977, Negishi extended the palladium-catalyzed cross-coupling of aryl halides to organozinc reagents (Scheme 19.21) [26a]. The reactions are performed under mild conditions that are compatible with many functional groups in contrast to Grignard or organolithium reagents. The same year, Fauvarque/Jutand reported the cross-coupling of a zinc enolate, the Reformatsky reagent (vide infra, Scheme 19.29) [26b]. The coupling of organozinc reagents has been widely developed [26c–h], and the reaction became called the Negishi reaction. Negishi received the Nobel Prize in 2010 [26f]. In the early works, the precatalyst $PdCl_2(PPh_3)_2$ was reduced *in situ* into an active Pd^0 by $(i\text{Bu})_2AlH$. This is required when the organozinc reagent cannot reduce the Pd^{II} precatalyst.

$$\mathrm{ArX + RZnY} \xrightarrow[\text{or } \mathrm{Pd^0(PPh_3)_4}]{\mathrm{Pd^{II}Cl_2(PPh_3)_2} + (i\mathrm{Bu})_2\mathrm{AlH}\ (1:2)} \mathrm{ArR + XZnY}$$

X = I, Br
R = aryl, vinyl, alkynyl, alkyl

SCHEME 19.21 The Negishi Reaction (1977).

19.3.3.2 Synthetic Application

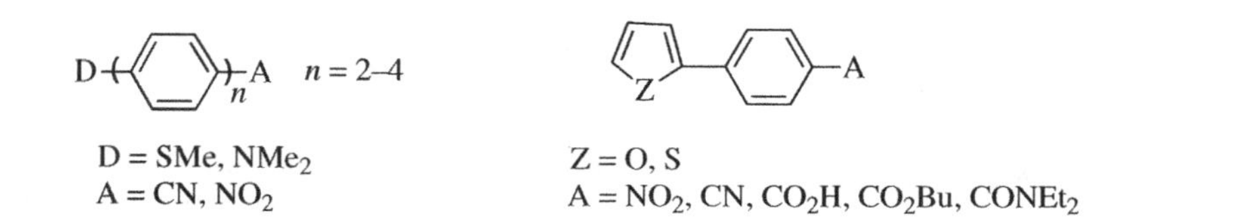

SCHEME 19.22 Synthesis of donor–acceptor bi- or polyaryls for nonlinear optics [27].

19.3.4 Palladium-Catalyzed Cross-Coupling with Organolithium Reagents

19.3.4.1 General Considerations The palladium-catalyzed cross-coupling between aryl iodides and organolithium reagents was pioneered by Murahashi in 1979 (Scheme 19.23a) [28a]. In 1981, Linstrumelle reported cross-coupling of allenyllithium reagents (Scheme 19.23b) [28b]. The active catalyst $Pd^0(PPh_3)_2LiCl$ is formed by reduction of $PdCl_2(PPh_3)_2$ by MeLi or $AlH(i\text{Bu})_2$ [28c].

In 2013, Feringa reported the palladium-catalyzed cross-coupling of organolithium reagents with aryl bromides and chlorides [29]. Pd^{II} or Pd^0 complexes are used associated with bulky electron-rich monophosphines ($Pt\text{Bu}_3$, Xphos) or electron-rich NHC ligands (PEPPSI-Ipent) (Scheme 19.24) that favor oxidative additions.

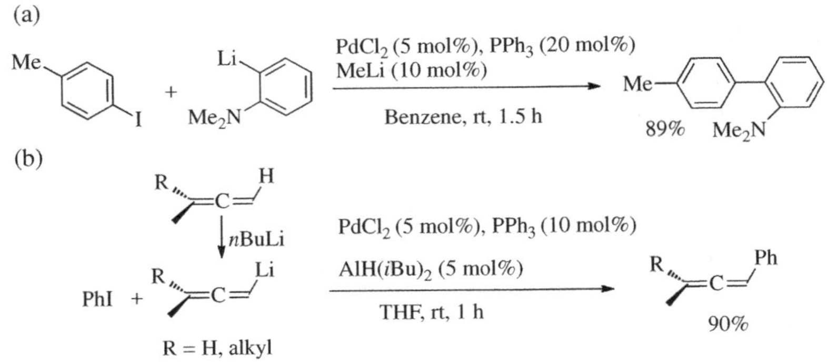

SCHEME 19.23 (a) and (b): Palladium-catalyzed cross-coupling of organolithium reagents (1979).

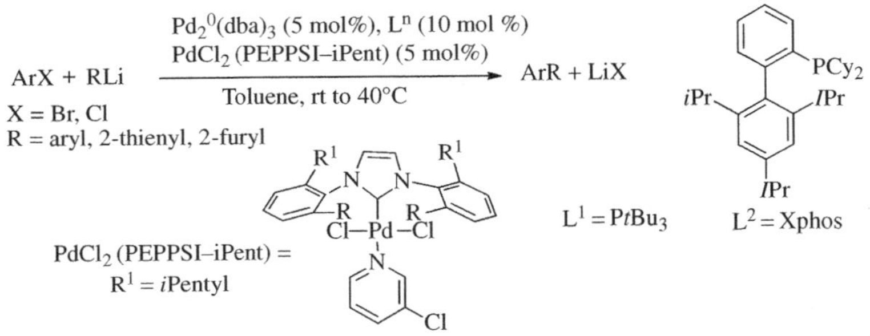

SCHEME 19.24 Palladium-catalyzed cross-coupling of organolithium reagents (2013).

19.3.4.2 *Synthetic Applications*

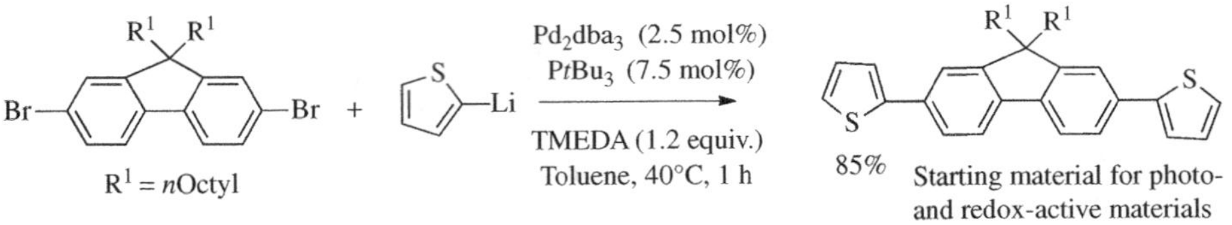

SCHEME 19.25 Cross-coupling of ArBr [29a].

19.3.5 Mechanism of Palladium-Catalyzed Cross-Couplings with Rm (m = Li, MgY, ZnY)

A general mechanism was proposed for cross-coupling reactions of aryl halides with Grignard [20, 22], organozinc [26] reagents (Scheme 19.26a). The first step is an *oxidative addition* of aryl halides to M^0L_2 that gives *trans*-$ArM^{II}XL_2$. A *transmetallation* by the nucleophile generates *trans*-$ArM^{II}RL_2$ that must isomerize in an endergonic process to the *cis*-$ArM^{II}RL_2$ from which a *reductive elimination* gives the cross-coupling product ArR and again the M^0 catalyst that enters into a second catalytic cycle. In the case of diphosphine ligands P^P, all complexes have the *cis*-stereochemistry (Scheme 19.26b). In the case of bulky monophosphines L, all Pd^0 and Ar-Pd^{II} may be ligated by only one ligand L, which facilitates both oxidative additions and reductive eliminations [12b, d].

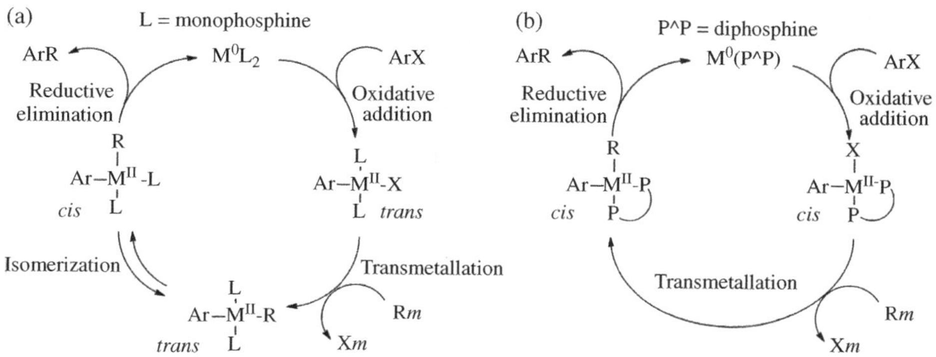

SCHEME 19.26 Proposed mechanisms for metal-catalyzed cross-coupling reactions: (a) monophosphine and (b) diphosphine ligands.

Isolated *trans*-PhPdI(PPh_3)$_2$ were found to react with RLi, RMgBr [22], and BrZnCO_2Et [26b] but with a rate lower than that of the overall catalytic reaction, suggesting that *trans*-PhPdI(PPh_3)$_2$ was not the reactive intermediate in the catalytic reactions, in contrast to what was proposed in Scheme 19.26a.

Murahashi has isolated $Pd^0(PPh_3)_2ClLi$ in the reduction of $PdCl_2(PPh_3)_3$ by MeLi [28c]. Negishi has observed that the ^{31}P NMR signal of the Pd^0 complex formed by reduction of $PdX_2(PPh_3)_2$ (X = I, Br, Cl) by RLi or HAl(tBu)$_2$) depends on X and the countercation m of the reductant. Anionic Pd^0 complexes were proposed $[Pd^0(PPh_3)X_n]^{n-}n(m^+)$ (n = 1 or 2) [30]. Such complexes react with PhI but the reactions are too fast to be monitored by ^{31}P NMR.

Electrochemical techniques have been used by Amatore/Jutand to mimic the chemical reduction of $PdCl_2(PPh_3)_2$ [31a, b]. The anionic $[Pd^0(PPh_3)_2Cl]^-$ is formed [31b] supported by DFT calculations (Shaik) [7d]. The rate constant $k_{o.a}$ of its oxidative addition to PhI was determined (Scheme 19.27a) [31b]. The cations Li^+ delivered in the chemical reduction of $PdCl_2(PPh_3)_2$ by RLi [30] accelerate the oxidative addition by interacting with $[Pd^0(PPh_3)_2Cl]^-$ [7l], which generates a more reactive bent $Pd^0(PPh_3)_2$ (Scheme 19.27b and c, DFT by Shaik [7p]).

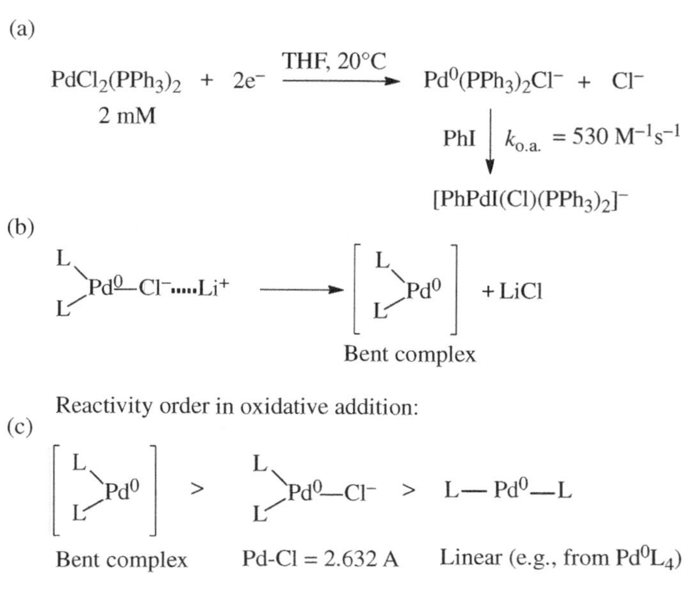

SCHEME 19.27 Oxidative addition of PhI to Pd^0 complexes: role of chloride ions (a–c) and cations (b).

Due to the anionic character of $[Pd^0(PPh_3)_2Cl]^-$, an anionic intermediate $[PhPdI(Cl)(PPh_3)_2]^-$ is first formed in the oxidative addition (Scheme 19.27a) on the way to *trans*-$PhPdI(PPh_3)_2$ (formed at longer times, ≈hs) via an equilibrium involving the neutral $PhPdIS(PPh_3)_2$ and Cl^- (Scheme 19.28) [31c]. $PhPdIS(PPh_3)_2$ reacts with the nucleophile R^- to generate a transient anionic pentacoordinated $[PhPdIR(PPh_3)_2]^-$ in which the two ligands Ph and R sit in adjacent positions as is required for the reductive elimination (Scheme 19.28) [71, 31c].

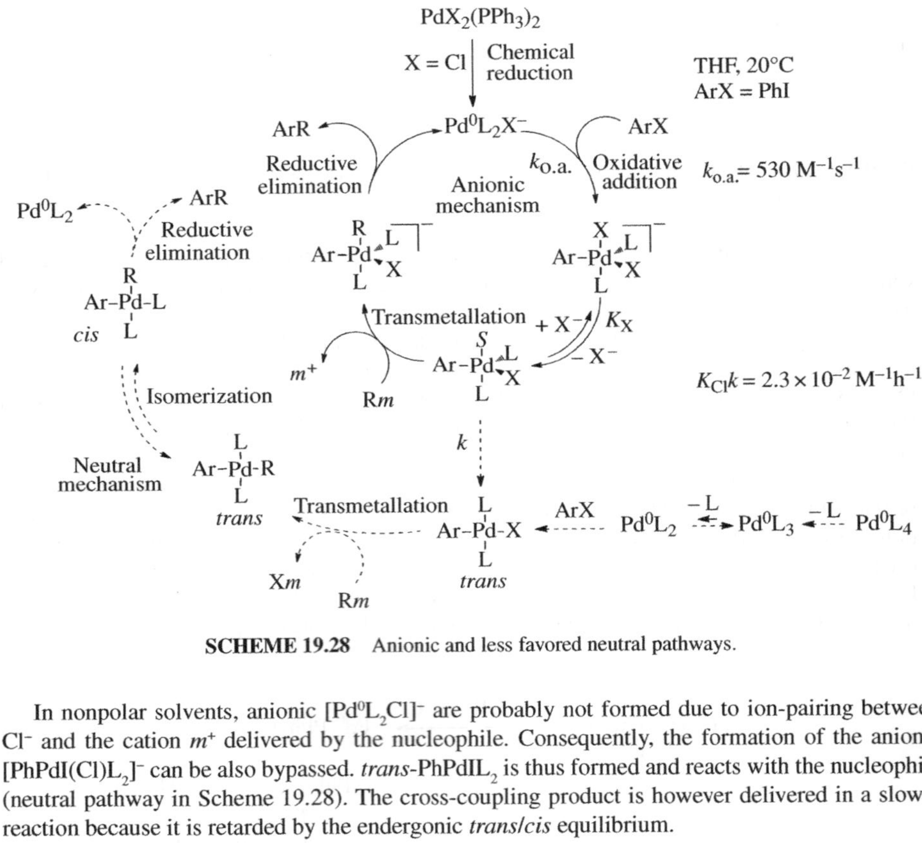

SCHEME 19.28 Anionic and less favored neutral pathways.

In nonpolar solvents, anionic $[Pd^0L_2Cl]^-$ are probably not formed due to ion-pairing between Cl^- and the cation m^+ delivered by the nucleophile. Consequently, the formation of the anionic $[PhPdI(Cl)L_2]^-$ can be also bypassed. *trans*-$PhPdIL_2$ is thus formed and reacts with the nucleophile (neutral pathway in Scheme 19.28). The cross-coupling product is however delivered in a slower reaction because it is retarded by the endergonic *trans*/*cis* equilibrium.

19.3.6 Nickel- and Palladium-Catalyzed Arylation of Ketone, Ester, and Amide Enolates

19.3.6.1 General Considerations and Mechanism In 1977, Fauvarque/Jutand reported the first palladium- and nickel-catalyzed cross-coupling reaction of aryl halides with a zinc ester enolate: the Reformatsky reagent (Scheme 19.29) [26b, 32]. When $PdCl_2(PPh_3)_2$ and $NiCl_2(PPh_3)_2$ are used

$$ArX + BrZnCH_2CO_2Et \xrightarrow[\text{Methylal/HMPA (1/1)}]{\substack{Ni^{II}Cl_2L_2\ (10\ mol\%) + L\ (20\ mol\%) + EtMgBr\ (20\ mol\%) \\ \text{or } Pd^{II}Cl_2L_2\ (10\ mol\%) + L\ (20\ mol\%) + EtMgBr\ (20\ mol\%) \\ \text{or } Pd^0L_4\ (L = PPh_3)}} ArCH_2CO_2Et + XZnBr$$

Ni : X = I, Br, Cl; 45°C, 3 h
Pd : X = I, 45°C, 6 h

SCHEME 19.29 Nickel- and Palladium-catalyzed arylation of zinc ester enolate (1977).

as precatalysts, addition of 2 equiv. of EtMgBr per precatalyst is required to reduce the M^{II} precatalysts to the active M^0, the Reformatsky reagent being not a reductant. Reactions of $BrZnCH_2CO_2Et$ with isolated *trans*-$ArPdI(PPh_3)_2$ deliver the cross-coupling product [26b, 32].

Twenty years later in 1997, Buchwald [33a] and Hartwig [33b] developed arylations of ketones using Pd^0 precatalysts associated with electron-rich mono- or diphosphines. The enolate is generated *in situ* by a base (Scheme 19.30a) [33, 34]. The *transmetallation* of $ArPdBrL_n$ with preformed enolates generates ArPd(enolate)L_n (Scheme 19.30d). Depending on the structure and electronic properties of the enolates and ligands, the Pd^{II} center is ligated by a C-enolate or by an O-enolate (Hartwig [34a]). Both complexes deliver the α-arylated ketone by thermolysis, due to isomerization of the C- and O-Pd^{II} complexes, leading to the mechanism in Scheme 19.30d.

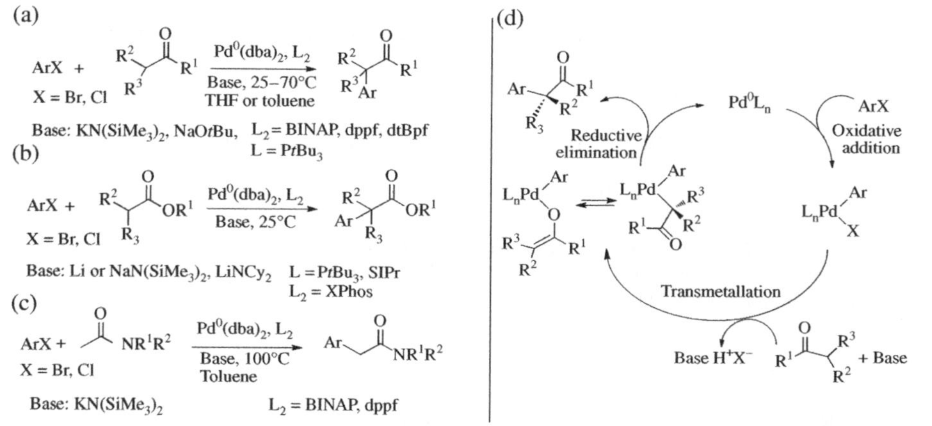

SCHEME 19.30 Palladium-catalyzed arylation of (a) ketones (1997), (b) esters, and (c) amides; (d) mechanism for ketone enolates.

Pd^0 complexes catalyze the arylation of ester enolates generated *in situ* in the presence of a base (Buchwald, Hartwig, Scheme 19.30b) [35]. Arylation of malonates proceeds under similar conditions [35b].

The arylation of amide enolates generated *in situ* by a base is catalyzed by Pd^0 but in harsher conditions because amides are less acidic (Scheme 19.30c, Hartwig [36a]). The intramolecular version with ArBr as R^1 delivers oxindoles (L=P*t*Bu_3 or carbene precursor SIPr, base=NaO*t*Bu). An enantioselective reaction takes place in the presence of a chiral ligand [36b]. Palladium catalyzes the arylation of zinc amide enolates (Hartwig, Cossy [36c, d]).

19.3.6.2 Synthetic Application

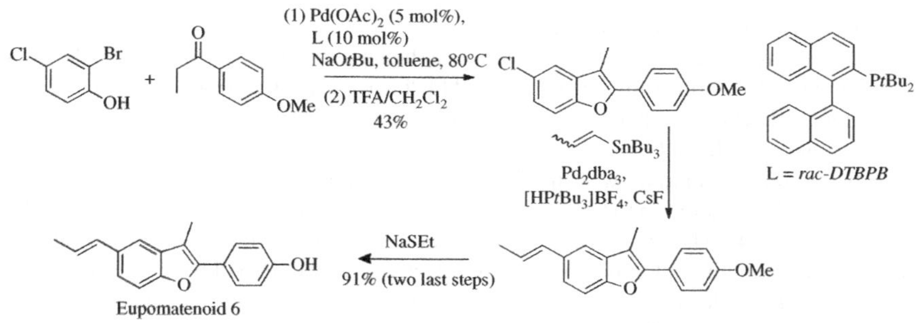

SCHEME 19.31 Construction of a benzofuran ring via a Pd-catalyzed arylation of a ketone followed by a Stille reaction (vide infra) in the synthesis of eupomatenoid 6 [37].

19.4 THE SONOGASHIRA REACTION

19.4.1 General Considerations and Mechanism

The Pd-catalyzed arylation of terminal alkynes was independently reported by Cassar [38] and Heck [39] in 1975 (Scheme 19.32a), then improved by Sonogashira [40] upon addition of a Cu^I cocatalyst (Scheme 19.32b). Both are now called the Sonogashira reactions [41]. They are performed in the presence of a base, usually an amine.

(a)

Cassar–Heck (1975)

$$ArI + RC{\equiv}CH + base \xrightarrow{Pd} RC{\equiv}CAr + baseH^+, X^-$$

(b)

Sonogashira (1975)

$$ArI + RC{\equiv}CH + base \xrightarrow{Pd/Cu} RC{\equiv}CAr + baseH^+, X^-$$

SCHEME 19.32 Arylation of terminal alkynes.

The Copper(I)-cocatalyzed Sonogashira reactions (b)

The role of the cocatalyst (CuI, CuCN) is the formation of an alkynyl-copper in the presence of the base (Scheme 19.33) [41].

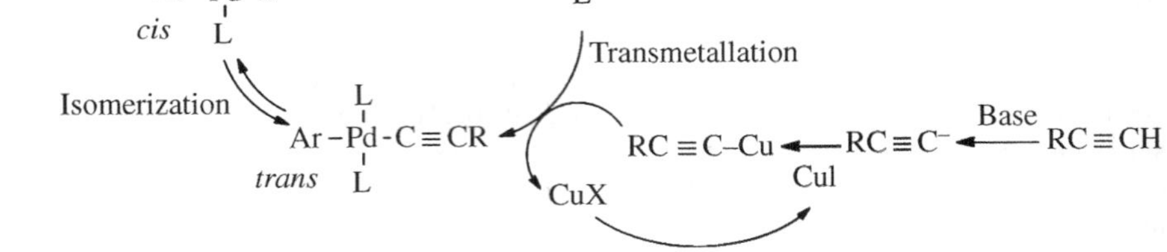

SCHEME 19.33 Postulated mechanism of Pd/Cu-catalyzed arylation of alkynes.

The copper-free Sonogashira reaction (a)

The copper-free reactions are very sensitive to the base, usually an amine used also as solvent, as was pioneered by Linstrumelle (catalyst: $Pd^0(PPh_3)_4$ [42]). The reaction in Scheme 19.34 is indeed more efficient when performed in piperidine than in morpholine [43a]. Its mechanism has been investigated by Jutand [43a].

Electrochemical techniques were used to characterize Pd^0L_2 generated as the minor but reactive complex in oxidative addition to PhI, from the major but nonreactive $Pd^0(dba)L_2$ or Pd^0L_3 ($L{=}PPh_3$) (Scheme 19.34) [44]. The oxidative addition is not the rate-determining step of the catalytic reaction since the oxidative addition follows the reactivity order: $Pd^0(PPh_3)_4 > \{Pd^0(dba)_2 + 2\ PPh_3\}$ whereas a reverse order is found for the catalytic reactions [42, 43a]. One additional role of the amine was discovered via NMR spectroscopy: the reversible substitution of one phosphine in *trans*-$PhPdI(PPh_3)_2$ to generate $PhPdI(PPh_3)$(amine) (Scheme 19.34) [43b].

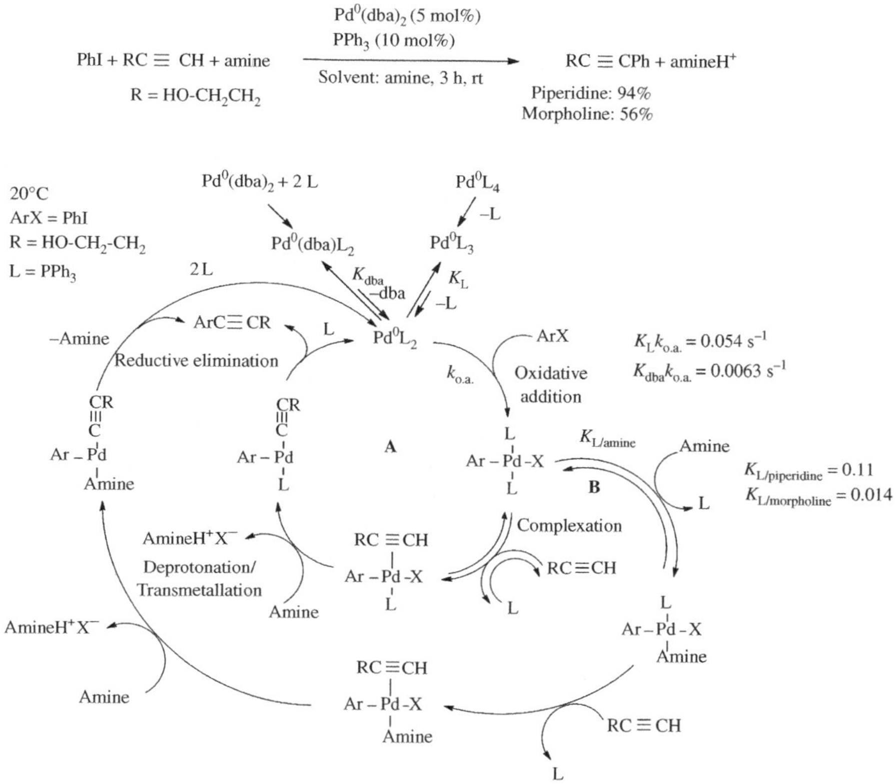

SCHEME 19.34 Cu-free, Pd-catalyzed arylation of alkynes and mechanism.

Two alternative mechanisms are proposed for the copper-free Sonogashira reactions, performed from PhI and HO—CH_2CH_2—C≡CH (Scheme 19.34): path **A** when the alkyne is a better ligand than the amine for the Pd^{II} center in $PhPdIL_2$ [43a, c] or path **B** when the amine is a better ligand than the alkyne [43a]. This explains why the catalytic reactions are very sensitive to the base (path **B** more efficient than path **A**) [43a]. Consequently, the amine does not react as a simple base in copper-free Sonogashira reactions but may also be involved as ligand for aryl–Pd^{II} complexes [43a, b].

19.4.2 Synthetic Applications

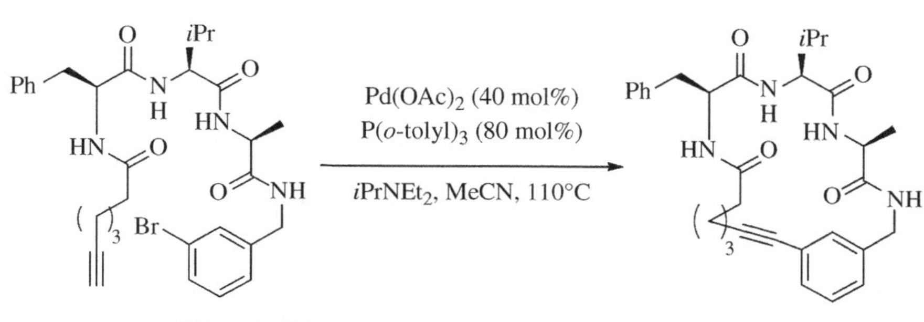

SCHEME 19.35 An intramolecular Sonogashira reaction [45].

19.5 THE STILLE REACTION

19.5.1 General Considerations and Mechanism

The palladium-catalyzed cross-coupling of aryl halides with Me_4Sn was discovered by Milstein/Stille in 1979 [46a]. $R^1Sn(nBu)_3$ were later on used as substrates due to the lack of reactivity of the *n*Bu group compared to other more reactive R^1 groups (Scheme 19.36) [46b–d]. $PdCl_2(PPh_3)_2$, $Pd^0(PPh_3)_4$, and {$Pd^0(dba)_2$ or $Pd^0_2(dba)_3$ associated with PAr_3 or $AsPh_3$} can be used as precatalysts [46b–d].

$$ArX + R^1Sn(R^2)_3 \xrightarrow{\text{Pd}} ArR^1 + XSn(R^2)_3$$

R^1 = vinyl, aryl, alkynyl, 2-furyl, 2-thienyl
$R^2 = n$Bu

SCHEME 19.36 The Stille reaction (1979).

The Stille reaction of PhI with $CH_2{=}CH{-}SnBu_3$ is more efficient with $AsPh_3$ than with PPh_3 associated with $Pd^0(dba)_2$ (Scheme 19.37, Farina [46b, 47]). According to Farina, the higher efficiency of $AsPh_3$ comes from an easier dissociation of $AsPh_3$ from $PhPdI(AsPh_3)_2$, which allows coordination of $CH_2{=}CH{-}SnBu_3$ *via* its C=C bond [46b, 47]. According to Liebeskind, this is also why CuI cocatalyzes the reaction by quenching one PPh_3 [46e]. The mechanism of this Stille reaction has been established by Amatore/Jutand in DMF (Scheme 19.37).

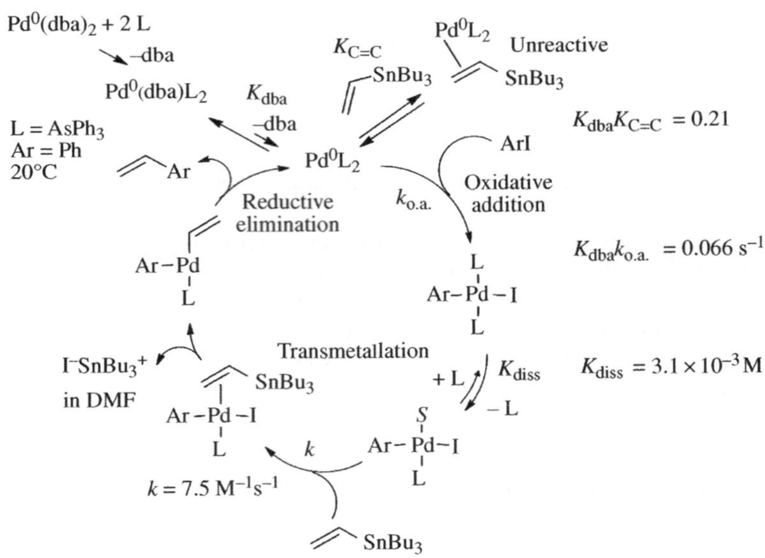

SCHEME 19.37 A Stille reaction with $AsPh_3$ as ligand and mechanism.

Electrochemical techniques have been used (i) to investigate the mechanism of the oxidative addition of PhI with the characterization of the reactive $Pd^0(AsPh_3)_2$ generated from the major but nonreactive $Pd^0(dba)(AsPh_3)_2$ (Scheme 19.37) [48a], (ii) to characterize the nonreactive $Pd^0(\eta^2{-}CH_2{=}CH{-}SnBu_3)(AsPh_3)_2$ generated in the presence of $CH_2{=}CH{-}SnBu_3$ (Scheme 19.37 [56a]), (iii) to characterize $PhPdI(AsPh_3)(DMF)$ formed by endergonic dissociation of $AsPh_3$ from $PhPdI(AsPh_3)_2$ [48b], and (iv) to establish and thus confirm the *dissociative* mechanism proposed by Farina [47] for the reaction of $CH_2{=}CH{-}SnBu_3$ with $PhPdI(AsPh_3)_2$ in the rate-determining transmetallation (Scheme 19.37) [48b]. Taking the opportunity that $ISnBu_3$ (formed at the same rate as $PhCH{=}CH_2$) is fully dissociated into the free ions I^- and Bu_3Sn^+ in DMF, the kinetics of formation of the ionic species, that is, the kinetics of the transmetallation was investigated by *conductivity measurements* [48b, 49]. From the retarding effect of $AsPh_3$ on the transmetallation, it is established that the *transmetallation* proceeds by reaction of $CH_2{=}CH{-}SnBu_3$ with $PhPdI(AsPh_3)S$ generated by dissociation of $AsPh_3$ from $PhPdI(AsPh_3)_2$ (Scheme 19.37) [48b]. The *dissociative* mechanism in DMF was later on supported by DFT calculations (Alvarez/de Lera [50], Espinet [51]).

The common decelerating effect of dba and $CH_2{=}CH{-}SnBu_3$ on the rate of the oxidative addition of PhI [48a] favors the efficiency of the catalytic reaction by bringing the rate of the fast *oxidative addition* of PhI closer to the rate of the slower *transmetallation* by $CH_2{=}CH{-}SnBu_3$ [7o, p].

Beneficial Role of Chloride Ions Stille reactions are often accelerated by additives such as LiCl. A dual role of chloride ions has been established (Scheme 19.38, Maes/Jutand [52]). The *rds* transmetallation of the Stille reactions is faster in the presence of Cl^- via *in situ* halide metathesis in $ArPdX(PPh_3)_2$ (X = I, Br) leading to more reactive $ArPdCl(PPh_3)_2$. The second role of Cl^- (evidenced by means of electrochemical techniques) is the stabilization of $Pd^0(PPh_3)_2$ (generated in the reduction of $PdCl_2(PPh_3)_2$ by $2\text{-}ThSnBu_3$) by formation of anionic $[Pd^0(PPh_3)_2Cl]^-$ preventing the Pd^0 decomposition, as was observed in the absence of Cl^-. The catalyst loading is maintained high in the presence of Cl^-; this why the catalytic reaction becomes faster [52].

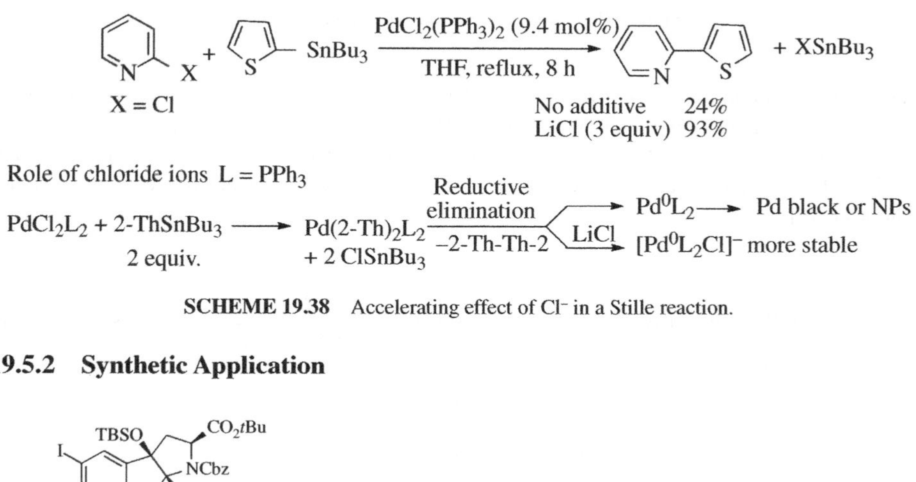

SCHEME 19.38 Accelerating effect of Cl^- in a Stille reaction.

19.5.2 Synthetic Application

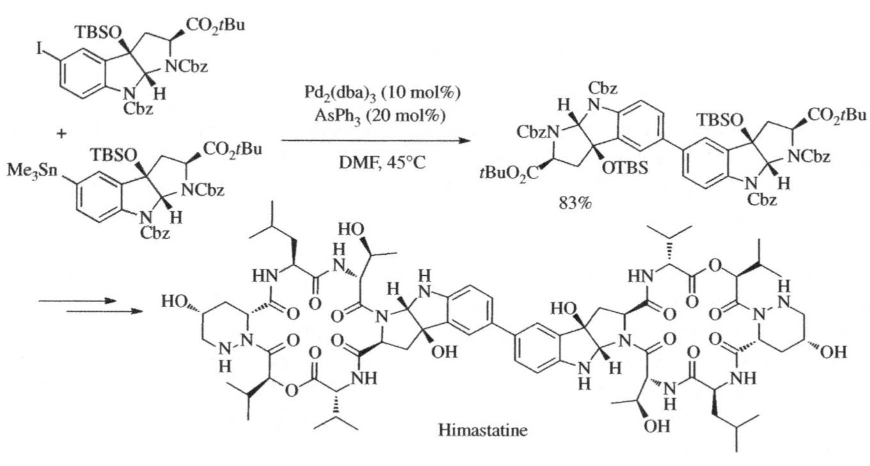

SCHEME 19.39 Synthesis of the symmetric biaryl unit of Himastatine via a Stille reaction [53].

19.6 THE SUZUKI–MIYAURA REACTION

19.6.1 General Considerations and Mechanism

The palladium-catalyzed Suzuki–Miyaura reaction is a cross-coupling reaction between aryl halides and boron derivatives (Scheme 19.40) [54]. Suzuki received the Nobel Prize in 2010 [54h]. Reactions are performed under mild conditions and many functional groups are tolerated. Water is not a problem and may be even beneficial (vide infra). The first reported reactions by Suzuki in 1979 insisted about the presence of a base, even if no proton is exchanged [54a, b]. Various bases have been used: EtO^-, MeO^-, OH^-, CO_3^{2-}, and F^- associated with countercations such as Na^+, K^+, Cs^+, Li^+, nBu_4N^+, Ag^+, or Tl^+ [54, 55]. The role of the base has led to many mechanistic interpretations involving among them anionic arylborates as reagents [54g, 56].

$$\text{ArX} + \text{R}B \xrightarrow[\text{Base}]{\text{Pd}} \text{ArR} + \text{X}B$$

R = vinyl, aryl, alkyl

$B = B(OH)_2, B(OR)_2$

SCHEME 19.40 The Suzuki–Miyaura reaction (1979).

19.6.1.1 Three Roles for Hydroxide Ions The mechanistic role of OH^- was investigated in the reactions of $Ar'B(OH)_2$ with isolated *trans*-$ArPdX(PPh_3)_2$ (X = I, Br, Cl) (generated in the oxidative addition of aryl halides to $Pd^0(PPh_3)_4$) (Scheme 19.41, Jutand/Amatore [57a]). The rate of formation of $Pd^0(PPh_3)_3$ generated at the same rate as ArAr′ was followed by means of electrochemical techniques [57a].

$$\underset{C_0}{trans\text{-}ArPdX(PPh_3)_2} + \underset{\beta C_0}{Ar'B(OH)_2} + \underset{\alpha C_0}{\text{base}} + \underset{2 C_0}{PPh_3} \xrightarrow[\text{DMF, 25°C}]{k_{obs}} Ar\text{-}Ar' + Pd^0(PPh_3)_3$$

Base = OH^-, CO_3^{2-}, F^-

SCHEME 19.41 Model reaction.

The variations of the observed rate constant k_{obs} versus the concentration of OH^- (at $Ar'B(OH)_2$ constant, Fig. 19.1a) or versus the concentration of $Ar'B(OH)_2$ (at constant OH^- concentration) always exhibit a maximum. This reveals that OH^- ions are required for the reaction but at high $Ar'B(OH)_2$ concentrations they are quenched to form the unreactive arylborate $Ar'B(OH)_3^-$ (Eq. 1, Scheme 19.42).

In contrast to *trans*-$ArPdBr(PPh_3)_2$ that does not react with $Ar'B(OH)_2$ in the absence of OH^-, *trans*-$ArPd(OH)(PPh_3)_2$ (formed by X/OH exchange in Eq. 2, Scheme 19.42 [7e, 58a]) reacts with $Ar'B(OH)_2$ [57a]. Therefore, OH^- ions are involved into two kinetically antagonistic effects, because involved into two competitive equilibria, one delivers the unreactive arylborate $Ar'B(OH)_3^-$ and the second one the reactive *trans*-$ArPd(OH)(PPh_3)_2$ that reacts with $Ar'B(OH)_2$ in the *rds* transmetallation to form the intermediate *trans*-$Ar\text{-}Pd\text{-}Ar'(PPh_3)_2$ (Scheme 19.42). The transmetallation is triggered by the oxophilicity of the boron center in $Ar'B(OH)_2$ that allows a precomplexation of the hydroxo of $ArPd(OH)L_2$ (Scheme 19.42) [57a].

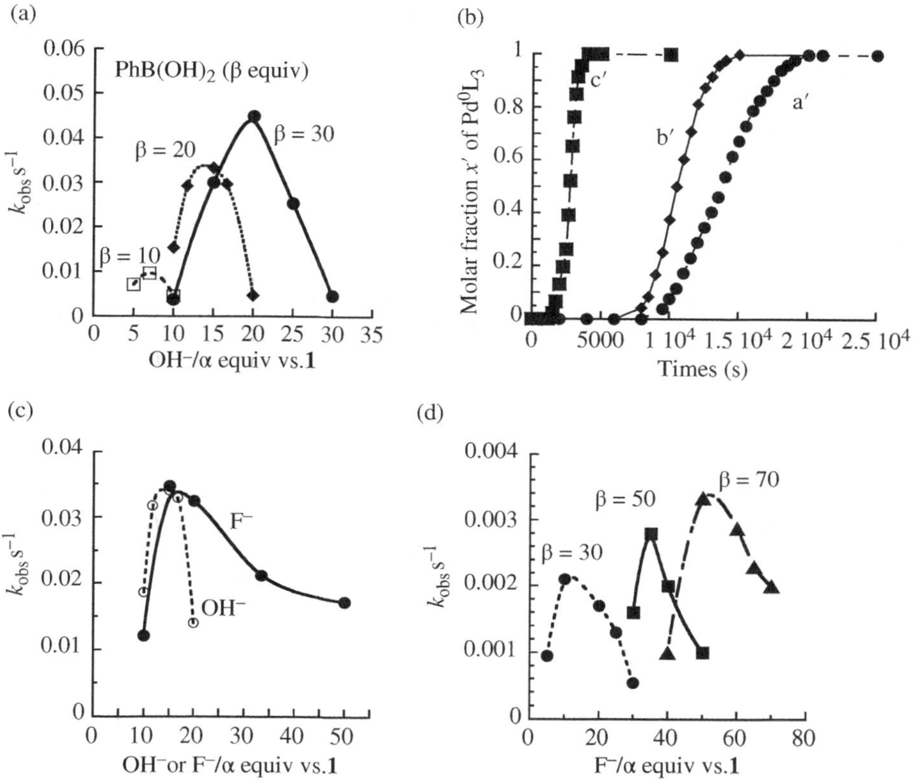

FIGURE 19.1 (a–d) Kinetics of the transmetallation performed from p—CN—C_6H_4—$PdBr(PPh_3)_2$ **1** (C_0 = 1.9 mM) in the presence of PPh_3 (2 equiv.) in DMF, 25°C. (a) Plot of k_{obs} versus α (equiv. of OH^-) for the reaction of $PhB(OH)_2$ (β equiv.). (b) Molar fraction x' of $Pd^0(PPh_3)_3$ versus time formed in the reaction of **1** (1.9 mM) with $PhB(OH)_2$ (50 equiv.) in the presence of Cs_2CO_3 (25 equiv.) and (a′) 0 equiv., (b′) 25 equiv., and (c′) 1750 equiv of water. (c) Plot of k_{obs} versus α equiv. of F^- (●) or OH^- (O) for the reaction of $PhB(OH)_2$ (20 equiv.). (d) Reaction of **1** (2 mM) with $PhSi(OMe)_3$ (β equiv.) in the presence of F^- (α equiv.) at 70°C. Plot of k_{obs} versus α.

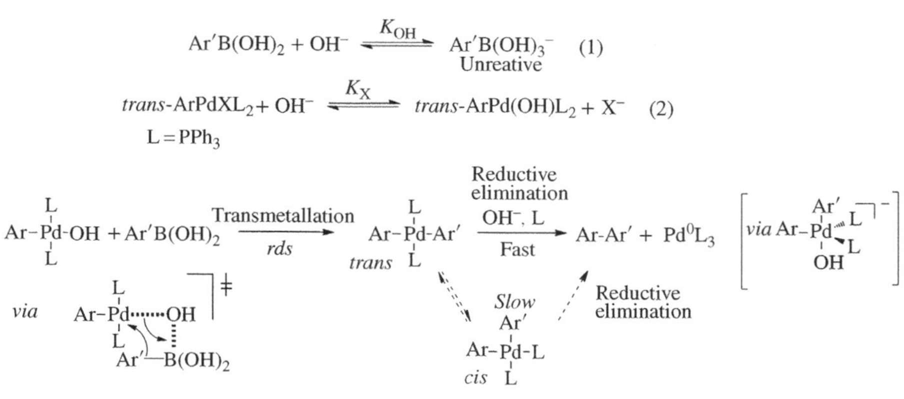

SCHEME 19.42 Mechanism of the transmetallation/reductive elimination.

Trans-Ar-Pd-Ar′$(PPh_3)_2$ could be quite stable and characterized [57a, e]. It is only after addition of OH^- that they leads to ArAr′ and $Pd^0(PPh_3)_3$. Therefore, OH^- catalyzes the reductive elimination from the *trans*-complex *via* a pentacoordinated complex [57a]. OH^- plays the well-known role of a fifth ligand on the Pd^{II} that favors the reductive elimination [31c, 59]. This bypasses the classical reductive elimination from the *cis*-complex, which is slow due to the endergonic *trans/cis* isomerization (Scheme 19.42 [57a]).

Other hypothetical pathways [56] have been kinetically ruled out [57a]. The preformed arylborate Ar′$B(OH)_3^-$ does not react with *trans*-ArPd(OH)$(PPh_3)_2$. The intrinsic reactivity of Ar′$B(OH)_3^-$ with *trans*-ArPdX$(PPh_3)_2$ was tested in the presence of excess X^- (as present in catalytic reactions) to avoid the X/OH exchange due to the generation of OH^- from Ar′$B(OH)_3^-$ (Eq. 1). No reaction was observed. In the absence of X^-, a reaction from *trans*-ArPdX$(PPh_3)_2$ may occur but via reacting ArPd(OH)$(PPh_3)_2$ (formed in Eq. 2) and Ar′$B(OH)_2$ but that reaction will be very slow because of the low thermodynamic concentrations of OH^- and Ar′$B(OH)_2$ generated from Ar′$B(OH)_3^-$ (Eq. 1). Consequently, the arylborate Ar′$B(OH)_3^-$ is intrinsically unreactive, and the major pathway is the reaction of ArPd(OH)$(PPh_3)_2$ with Ar′$B(OH)_2$. This was later on confirmed by Hartwig using ^{31}P NMR performed at −40°C [60a]. The nonreactivity of arylborates was established by Schmidt in ligand-less reactions [60b].

The mechanism of the Suzuki–Miyaura reaction is presented in Scheme 19.43 [57a]. The hydroxides OH^- plays three roles: (i) formation of the reactive *trans*-ArPd(OH)L_2 in the *rds* transmetallation with Ar′$B(OH)_2$, (ii) catalysis of the reductive elimination from *trans*-ArPdAr′L_2, and (iii) formation of unreactive Ar′$B(OH)_3^-$, leading to two kinetically antagonistic role in the rate of the *transmetallation* controlled by the concentration ratio $[OH^-]/[Ar'B(OH)_2]$, which must be smaller than 1 (Fig. 19.1a).

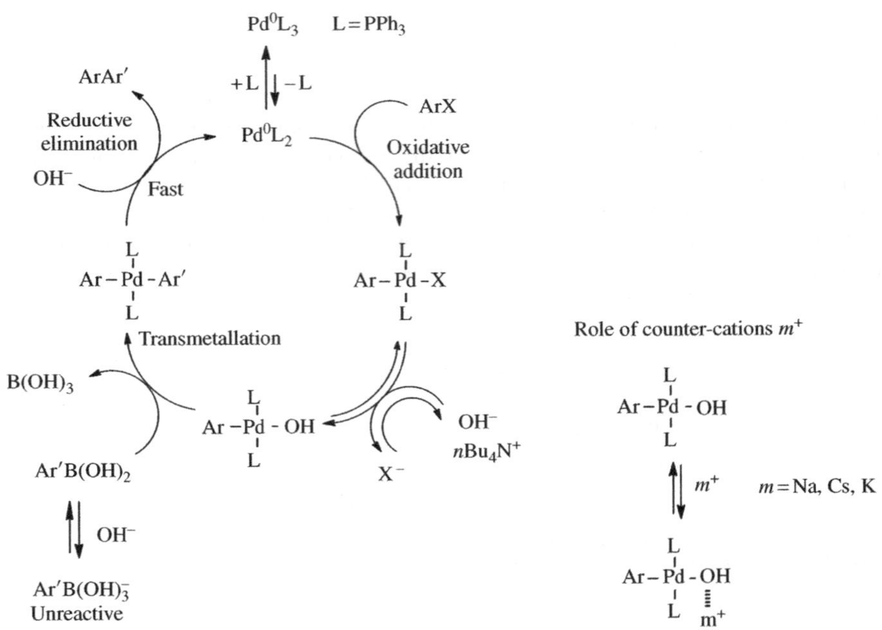

SCHEME 19.43 Mechanism of the Suzuki–Miyaura reaction with nBu_4NOH as the "base" and the inhibiting role of countercations m^+.

19.6.1.2 Fourth Antagonistic Kinetic Role Due to the Countercation m^+ of Hydroxide Some countercations of the base, Ag^+, Tl^+ accelerate Suzuki–Miyaura reactions in THF (Kishi [55]). This is rationalized by a shift of the equilibrium *trans*-$ArPdXL_2$/*trans*-$ArPd(OH)L_2$ by quenching X^- to form insoluble AgX or TlX in THF. Consequently, the concentration of the *trans*-$ArPd(OH)L_2$ increases, making the *transmetallation* faster. In contrast, some countercations that do not have any affinity for X^- ($m^+ = Na^+$, Cs^+, K^+) in the polar DMF solvent slow down the *transmetallation* in the order [57b] $n Bu_4N^+ > K^+ > Cs^+ > Na^+$. The concentration of *trans*-$ArPd(OH)(PPh_3)_2$ decreases via interaction of its OH group with m^+, and the competitive *transmetallation* becomes slower (Scheme 19.43) [57b].

19.6.1.3 Role of Carbonate Ions Carbonates, m_2CO_3 ($m^+ = K^+$, Cs^+, Ag^+, Tl^+), are often used as bases [54, 55]. The effect of carbonate on the rate of formation of $Pd^0(PPh_3)_3$ in the reaction in Scheme 19.41 has been tested. The reaction performed in supposedly "dry" DMF and "dry" Cs_2CO_3 is considerably slower than that performed in the presence of $n Bu_4NOH$ under similar conditions. Moreover, a long induction period was observed (Fig. 19.1b) due to accumulation of *trans*-$ArPdAr'L_2$. The reaction is faster and faster in the presence of increasing amounts of water, whereas the induction period is shorter and shorter [57b]. CO_3^{2-} does not react with *trans*-$ArPdX(PPh_3)_2$ in dry DMF, but the introduction of a small amount of water leads to the formation of *trans*-$ArPd(OH)(PPh_3)_2$ [57b], due to the formation of OH^- by reaction of carbonate with water.

Therefore, water favors the Suzuki–Miyaura reaction involving carbonates by formation of OH^-. The mechanism is thus similar to that reported in Scheme 19.43. However, a small amount of OH^- is generated, controlled by the amount of water. Consequently, compared to pure OH^- at the same concentration as CO_3^{2-}, the *transmetallation* is slower because of the low concentration of $ArPd(OH)L_2$. For the same reason, the *reductive elimination* is also slower; this is why an induction period is observed for the formation of the Pd^0 (Fig. 19.1b) [56b].

19.6.1.4 Three Antagonist Roles for Fluoride Ions Fluoride ions, F^- ions, are also used as "bases" in Suzuki–Miyaura reactions performed from $Ar'B(OH)_2$ [61]. The influence of F^- (α equiv., added as $n Bu_4NF$ to minimize the role of the countercation) on the rate of the reaction in Scheme 19.41 reveals two kinetically antagonistic roles of F^- in the *transmetallation* (Fig. 19.1c) [57c]. Fluoride ions are required but the reaction becomes slower when their concentration increases, which attests that anionic arylfluoroborates $Ar'B(OH)_{3-n}F_n^-$, formed at high fluoride concentrations are less reactive than $Ar'B(OH)_2$ in the *transmetallation* (Scheme 19.44) as proposed in literature [62].

trans-$ArPdF(PPh_3)_2$ are generated by reversible exchange of the halide of *trans*-$ArPdXL_2$ (Scheme 19.44) [57c, 58b, c]. Isolated $ArPdF(PPh_3)_2$ reacts with $Ar'B(OH)_2$ to generate the stable intermediate *trans*-$ArPdAr'(PPh_3)_2$ [57c]. The *rds* transmetallation is triggered by the fluorophilicity of the boron center (Scheme 19.44). $Pd^0(PPh_3)_3$, and ArAr′ were formed after addition of F^- that catalyzes the reductive elimination from *trans*-$ArPdAr'L_2$. The mechanism of the Suzuki–Miyaura performed in the presence of $n Bu_4NF$ is given in Scheme 19.44.

The fluoride F^- plays three kinetically antagonistic roles: (i) formation of *trans*-$ArPdFL_2$ that reacts with $Ar'B(OH)_2$ in the *rds* transmetallation, (ii) catalysis of the reductive elimination from *trans*-$ArPdAr'L_2$, and (iii) formation of less or unreactive arylfluoroborates. The rate of the *rds* transmetallation is controlled by the ratio $[F^-]/[Ar'B(OH)_2]$.

Therefore, anionic bases, F^-, OH^- (when used as it or generated from CO_3^{2-} in the presence of water), do not play the role of bases but serve as ligands for aryl–Pd^{II} complexes. In Suzuki reactions involving a fast oxidative addition (ArI and EAG-substituted ArBr), one needs to increase the rate of the rate-determining transmetallation. This requires (i) to select the best base: $OH^- > CO_3^{2-}$,

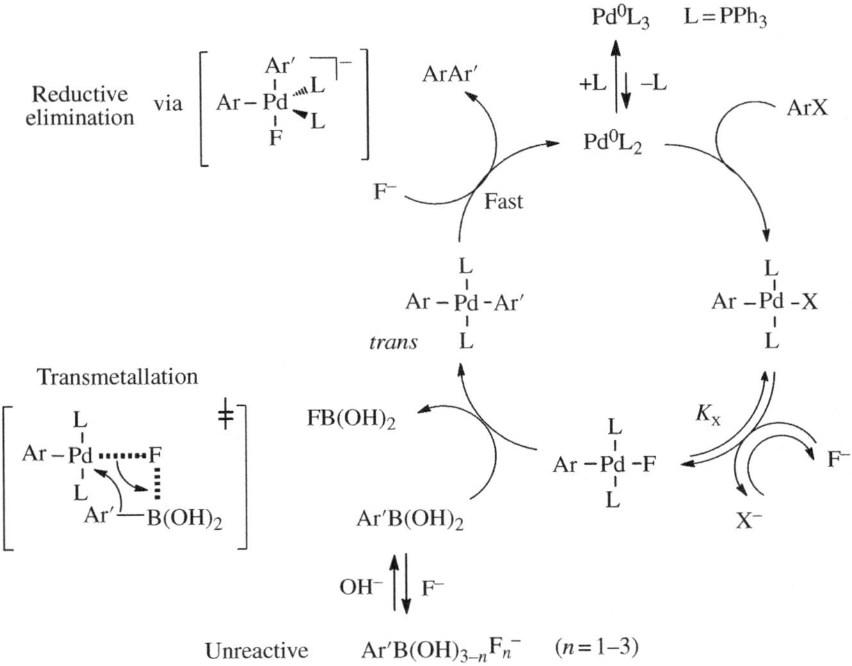

SCHEME 19.44 Mechanism of the Suzuki–Miyaura reaction in the presence of $n\text{Bu}_4\text{NF}$.

(ii) to optimize the ratio ($[OH^-]$ or $[F^-]$)/$[Ar'B(OH)_2]$, and (iii) to select the best countercation of the anionic base: Tl^+ (Ag^+)>$n\text{Bu}_4\text{N}^+$>K^+>Cs^+>Na^+. This optimization is of course less crucial when the oxidative addition is *rds* (for ArCl). However, the bell-shaped curves observed for the transmetallation (Fig. 19.1a and c) show that the latter can be very slow (at very low or very high concentrations of OH^- or F^-) and may even be slower than the oxidative addition of ArCl. Therefore, the mechanism of the transmetallation/reductive elimination elucidated for OH^- and F^- remains always valid [57d].

19.6.1.5 Suzuki–Miyaura Reactions from Trifluoroborates The convenient use of stable RBF_3^- as an alternative to $RB(OH)_2$ was developed in 1999 by Genêt [63a] and Molander [63b] (Scheme 19.45).

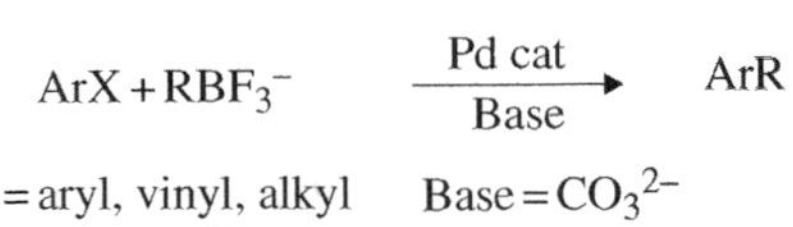

$$ArX + RBF_3^- \xrightarrow[\text{Base}]{\text{Pd cat}} ArR$$

R = aryl, vinyl, alkyl Base = CO_3^{2-}

SCHEME 19.45 Suzuki reactions from potassium aryltrifluoroborates (1999).

Batey [62a] and Molander [62b] came to the conclusion that RBF_3^- were not reactive due to reaction of the base that generates $RB(OH)_{3-n}F_n^-$. This was later on confirmed by Lloyd-Jones who establishes that $Ar'B(OH)_2$ generated by reaction of OH^- with $Ar'BF_3^-$ was the reactive species [62c]. The rate of hydrolysis of RBF_3^- to $RB(OH)_2$ however depends on the structure of R [62d].

19.6.2 Synthetic Application

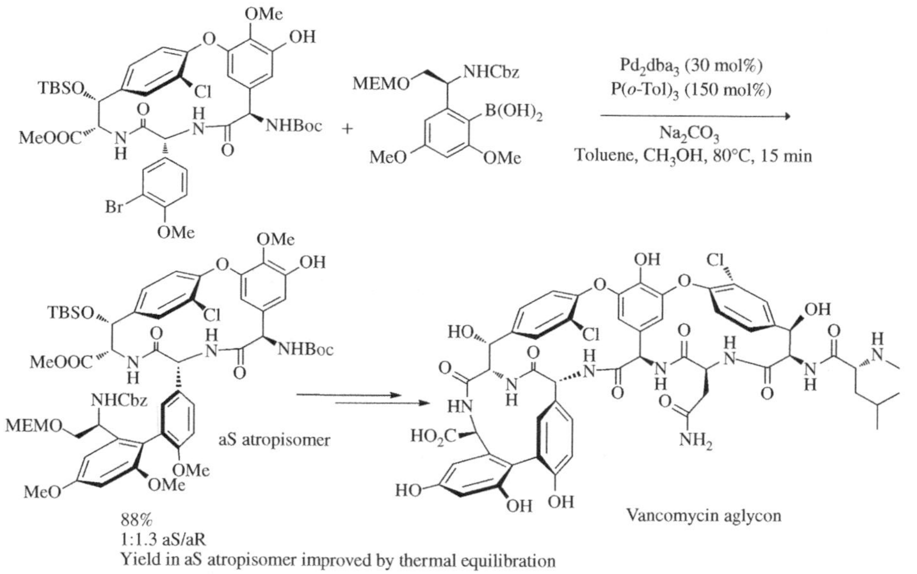

SCHEME 19.46 Construction of the biaryl unit in vancomycin aglycon via a Suzuki–Miyaura reaction [64].

19.7 THE HIYAMA REACTION

19.7.1 General Considerations and Mechanism

The palladium-catalyzed Hiyama reaction is a cross-coupling between aryl halides and silanes (Scheme 19.47) [65]. Silane derivatives are easily available. They are stable with a low toxicity. However, they exhibit a low reactivity, and the reactions must be performed at high temperatures. Reactions are promoted by fluoride ions [65].

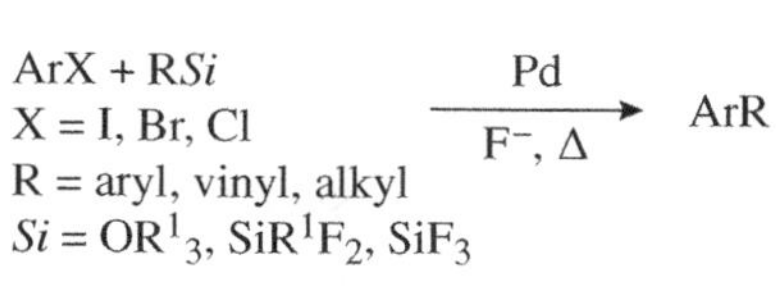

SCHEME 19.47 The Hiyama reaction (1988).

Various silanes may be used: $RSi(OR^1)_3$, $RSiR^1F_2$, $RSiF_3$ [65] or silanols ($R^1CH{=}CH{-}Si(OH)Me_2$) developed by Denmark in 2000 [66a–c]. All reactions are accelerated in the presence of F^-. Fluoride-free reactions have been developed by Denmark with silanolates $[R^1CH{=}CH{-}Si(Me)_2{-}O]^-$

[66d–g]. The mechanism of the cross-coupling between aryl halides and silanolates was investigated by Denmark [66e, f] and the role of F^- in the cross-coupling of silanols as well [66b].

The mechanism of the cross-coupling of aryl halides and $Ar'Si(OMe)_3$ catalyzed by $Pd^0(PPh_3)_4$ has been investigated in the presence of F^- (Scheme 19.48, Jutand/Grimaud [67]). Hypotheses in the literature propose arylsilicates as reagents [65b] and/or a Pd—F bond in the transition state [65c].

Electrochemical techniques have been used to investigate the influence of F^- on the kinetics on the model reaction in Scheme 19.48 [67]. Reactions were performed at 70°C (Fig. 19.1d).

$$\underset{C_0}{trans\text{-}ArPdX(PPh_3)_2} + \underset{\beta C_0}{PhSi(OMe)_3} + \underset{\alpha C_0}{F^-} + \underset{2 C_0}{PPh_3} \xrightarrow[\text{DMF, 70°C}]{k_{obs}} ArPh + Pd^0(PPh_3)_3 + FSi(OMe)_3$$

SCHEME 19.48 Model reaction.

The mechanism of the Hiyama reaction displayed in Scheme 19.49 shows three roles for fluorides F^- that favor the reaction (i) by formation of *trans*-$ArPdF(PPh_3)_2$, which reacts with $Ar'Si(OMe)_3$ in the *rds transmetallation*, and (ii) by promoting the *reductive elimination* from intermediate *trans*-$ArPdAr'(PPh_3)_2$. Conversely, F^- disfavors the reaction by formation of unreactive arylsilicate $Ar'Si(OMe)_3F^-$, leading to two antagonistic effects of F^- involved in two competitive equilibria. The rate of the transmetallation is controlled by the concentration ratio $[F^-]/[Ar'Si(OMe)_3]$, which must be less than 1 (Fig. 19.1d) [67].

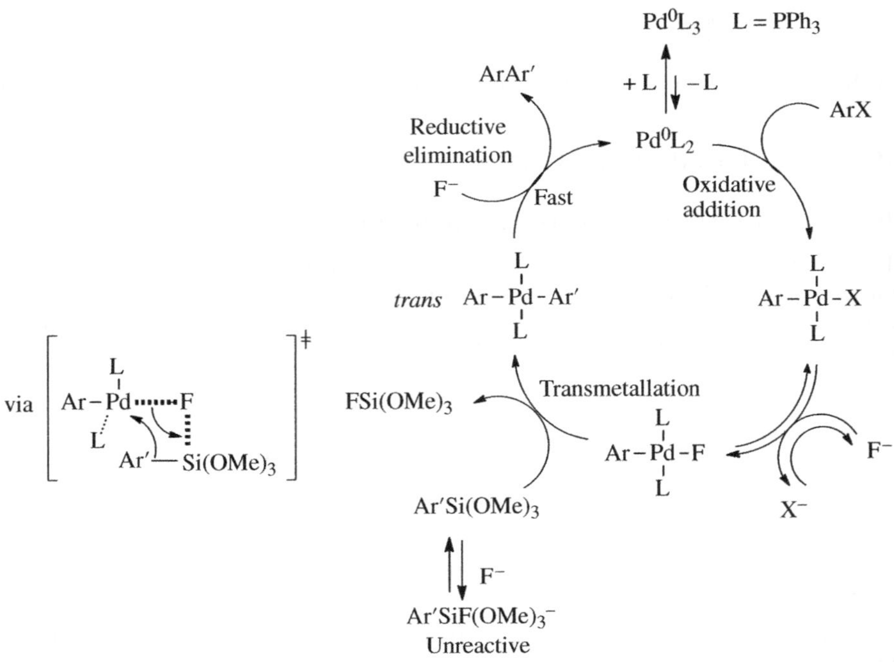

SCHEME 19.49 Mechanism of the Hiyama reaction performed in the presence of $n$$Bu_4NF$.

19.7.2 Synthetic Applications

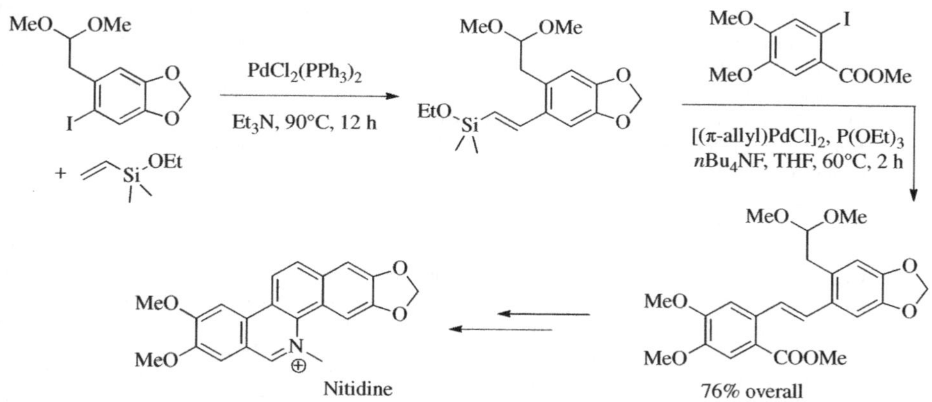

SCHEME 19.50 Heck followed by Hiyama reactions in the synthesis of nitidine [68].

19.8 SUMMARY AND OUTLOOK

Much work has been done in the last 50 years to perform nickel- or palladium- catalyzed C—C cross-coupling reactions under mild conditions with high turnover numbers (TON) and frequency (TOF), to react aryl chlorides, to improve the efficiency, the regio- and enantioselectivity of the reactions. Meanwhile, the mechanisms have been investigated to explain the high dependence of the reaction efficiency on the nature of the precatalysts, ligands, additives, and bases when required.

ABBREVIATIONS

Binap	2,2′-bis(diphenylphosphino)-1,1′-binaphthyl
DMF	Dimethylformamide
Dppp	1,3-bis(diphenylphosphino)propane
P^P	Diphosphine
rt	Room temperature
Rds	Rate-determining step
THF	Tetrahydrofuran
Tol-Binap	2,2′-bis(di-4-tolylphosphino)-1,1′-binaphthyl
TPPTC	Trisodium tri(3-carboxyphenyl)phosphane
TPPTS	Trisodium tri(3-sulfonatophenyl)phosphane

REFERENCES

[1] March, J. (1985) *Advanced Organic Chemistry, Reactions, Mechanisms and Structure*, 3ed. John Wiley & Sons, Inc., New York.

[2] (a) Fitton, P., Johnson, M.P., and McKeon, J.E. (1968) *J. Chem. Soc. Chem. Commun.*, 6–7; (b) Tsou, T.T. and Kochi, J.K. (1979) *J. Am. Chem. Soc.*, **101**, 6319–6332.

[3] For reviews and books on Heck reactions; see: (a) Heck, R.F. (1979) *Acc. Chem. Res.*, **12**, 146–151; (b) Heck, R.F. (1982) *Org. Reat.*, **27**, 345–390; (c) de Meijere, A. and Meyer, F.E. (1994) *Angew. Chem. Int. Ed. Engl.*, **33**, 2379–2411; (d) Beletskaya, I.P. and Cheprakov, A.V. (2000) *Chem. Rev.*, **100**, 3009–3066; (e) Bräse, S. and de Meijere, A. (2002) in Handbook of Organopalladium Chemistry for Organic Synthesis, A. Palladium-Catalyzed Reactions Involving Carbopalladation, vol. **I**, (ed. E.I. Negishi) Wiley-Interscience, New York, pp. 1123–1132; (f) Alonso, F., Beletskaya, I.P., and Yus, M. (2005) *Tetrahedron*, **61**, 11771–11835; (g) Herrmann, W.A., Böhm, V.P.W., and Reisinger, C.-P. (1999) *J. Organomet. Chem.*, **576**, 23–41; (h) Herrmann, W.A., Ölefe, K., von Preysing, D., and Schneider, S.K. (2003) *J. Organomet. Chem.*, **687**, 229–248; (i) Beletskaya, I.P. and Cheprakov, A.V. (2004) *J. Organomet. Chem.*, **689**, 4055–4082; (j) Littke, A.F. and Fu, G.C. (2002) *Angew. Chem. Int. Ed.*, **41**, 4176–4211; (k) Herrmann, W.A. (2002) *Angew. Chem. Int. Ed.*, **41**, 1290–1309; (l) Oestreich, M. (2009) The Mizoroki-Heck Reaction, John Wiley & Sons, Ltd, Hoboken; (m) Larhed, A. and Hallberg, A. (2002) in Handbook of Organopalladium Chemistry for Organic Synthesis-Alkene Substitution via Carbopalladation-Dehydropalladation and Related Carbopalladation Reactions, vol. **1**, (ed. E.I. Negishi) Wiley-Interscience, New York, pp. 1133–1178; (n) Hayashi, T., Kubo, A., and Ozawa, F. (1992) *Pure & Appl. Chem.*, **64**, 421–427; (o) Shibasaki, M., Boden, C.D.J., and Kojima, A. (1997) *Tetrahedron*, **53**, 7371–7395; (p) Loiseleur, O., Hayashi, M., Schmees, N., and Pfaltz, A. (1997) Synthesis, 1338–1345; (q) Shibasaki, M. and Vogl, E.M. (1999) *J. Organomet. Chem.*, **576**, 1–15; (r) Link, J.T. (2002) The Intramolecular Heck Reaction, (ed. L. Overman), John Wiley & Sons, Inc., New York; Org. React., 60, 157–534; (s) Oestreich, M. (2005) Eur. J. Org. Chem., 783–792; (t) Jutand A. (2008) *Chem. Rev.*, **108**, 2300–2347.

[4] (a) Heck, R.F. (1968) *J. Am. Chem. Soc.*, **90**, 5518–5526; (b) Heck, R.F. and Nolley, J.P., Jr. (1972) *J. Org. Chem.*, **37**, 2320–2322; (c) Dieck, H.A. and Heck, R.F. (1974) *J. Am. Chem. Soc.*, **96**, 1133–1136; (d) Ziegler, C.B. and Heck, R.F. (1978) *J. Org. Chem.*, **43**, 2941–2946.

[5] (a) Mizoroki, T., Mori, K., and Ozaki, A. (1971) *Bull. Soc. Chim. Jpn.*, **44**, 581–581; (b) Mori, K., Mizoroki, T. and Ozaki, A. (1973) *Bull. Soc. Chim. Jpn.*, **46**, 1505–1508.

[6] (a) Jeffery, T. (1996) *Tetrahedron*, **52**, 10113–10130; (b) Reetz, M.T. and Westermann, E. (2000) *Angew. Chem. Int. Ed.*, **39**, 165–168; (c) de Vries, A.H.M., Mulders, J.M.C.A., Mommers, J.H.M., Hendrickx, H.J.W., and de Vries, J.G. (2003) *Org. Lett.*, **5**, 3285–3288; (d) de Vries, A.H.M., Parlevliet, F.J., Schmieder-van de Vonderwoort, L., Mommers, J.H.M., Hendrickx, H.J.W., Walet, M.A.M., and de Vries, J.G. (2002) *Adv. Synth. Catal.*, **344**, 996–1002.

[7] (a) Amatore, C., Jutand, A., and M'Barki, M.A. (1992) *Organometallics*, **11**, 3009–3013; (b) Amatore, C., Carré, E., Jutand, A., M'Barki, M.A., and Meyer, G. (1995) *Organometallics*, **14**, 1818–1826; (c) Amatore, C., Carré, E., Jutand, A., M'Barki, M.A., and Meyer, G. (1995) *Organometallics*, **14**, 5605–5614; (d) Kozuch, S., Shaik, S., Jutand, A., and Amatore, C. (2004) *Chem. Eur. J.*, **10**, 3072–3080; (e) Grushin, V.V. and Alper, H. (1993) *Organometallics*, **12**, 1890–1901; (f) Amatore, C., Jutand, A., and Medeiros, M.J. (1996) *New J. Chem.*, **20**, 1143–1148; (g) Amatore, C., Jutand, A., Lemaître, F., Ricard, J.L., Kozuch, S., and Shaik, S. (2004) *J. Organomet. Chem.*, **689**, 3728–3734; (h) Amatore, C., Blart, E., Genêt, J.P., Jutand, A., Lemaire-Audoire, S., and Savignac, M. (1995) *J. Org. Chem.*, **60**, 6829–6839; (i) Gélin, E., Amengual, R., Michelet, V., Savignac, M., Jutand, A., and Genêt, J.P. (2004) *Adv. Synth. Catal.*, **346**, 1733–1741; (j) Amatore, C., Jutand, A., and Thuilliez, A. (2001) *Organometallics*, **20**, 3241–3249; (k) Amatore, C., Carré, E., and Jutand, A. (1998) *Acta Chem. Scand.*, **52**, 100–106; (l) Amatore, C. and Jutand, A. (2000) *Acc. Chem. Res.*, **33**, 314–321; (m) Amatore, C., Jutand, A., Meyer, G., Carelli, I., and Chiarotto, I. (2000) *Eur. J. Inorg. Chem.*, 1855–1859; (n) Amatore, C., Carré, E., Jutand, A., and Medjour, Y. (2002) *Organometallics*, **21**, 4540–4545; (o) Amatore, C. and Jutand, A. (1999) *J. Organomet. Chem.* **576**, 254–278; (p) Kozuch, S., Amatore, C., Jutand, A., and Shaik, S. (2005) *Organometallics*, **24**, 2319–2330.

[8] Osawa, F., Kubo, A., and Hayashi, T. (1992) Chem. Lett., 2177–2180.

[9] Herrmann, W.A., Brossmer, C., Ölefe, K., Reisinger, C.-P., Priermeier, T., Beller, M., and Fischer, H. (1995) *Angew. Chem. Int. Ed.*, **34**, 1844–1848.

[10] d'Orlyé, F. and Jutand, A. (2005) *Tetrahedron*, **61**, 9670–9678.

[11] For Heck reactions catalyzed by $Pd(OAc)_2$ associated with diphosphines, see (a) Cabri, W. and Candiani, I. (**1995**) *Acc. Chem. Res.*, **28**, 2–7; (b) Vallin, K.S.A., Larhed, M., and Hallberg, A. (2001) *J. Org. Chem.*, **66**, 4340–4343; (c) Karabelas, K., Westerlund, C., and Hallberg, A. (1985) *J. Org. Chem.*, **50**, 3896–3900; (d) Karabelas, K. and Hallberg, A. (1986) *J. Org. Chem.*, **51**, 5286–5290; (e) Xu, L.J., Chen, W.P., Ross, J., and Xiao, J. (2001) *Org. Lett.*, **3**, 295–297; (f) Mo, J., Xu, L., and Xiao, J. (2005) *J. Am. Chem. Soc.* **127**, 751–760.

[12] (a) Littke, A.F. and Fu, G.C. (1999) *J. Org. Chem.*, **64**, 10–11; (b) Stambuli, J.P., Incarvito, C.D., Bühl, M., and Hartwig, J.F. (2004) *J. Am. Chem. Soc.*, **126**, 1184–1194; (c) Littke, A.F. and Fu, G.C. (2004) *Org. Synth.* **126**, 13178–13179; (d) Littke, A.F. and Fu, G.C. (2002) *Angew. Chem. Int. Ed*, **41**, 4176-4211.

[13] Herrmann, W.A., Elison, M., Fischer, J., Köcher, C., and Artus, G.R.J. (1995) *Angew. Chem. Int. Ed. Engl.*, **34**, 2371–2374.

[14] (a) Pytkowicz, J., Roland, S., Mangeney, P., Meyer, G., and Jutand, A. (2003) *J. Organomet. Chem.*, **678**, 166–179; (b) Roland, S., Mangeney, P., and Jutand, A. (2006) Synlett, 3088–3094; (c) Lewis, A.K.D., Caddick, S., Cloke, F.G.N., Billingham, N.C., Hitchcock, P.B., and Leonard, J. (2003) *J. Am. Chem. Soc.*, **125**, 10066–10073.

[15] (a) Feuerstein, M., Doucet, H., and Santelli, M. (2001) *J. Org. Chem.*, **66**, 5923–5925; (b) Hierso, J.C., Firhi, A., Amardeil, R., and Meunier, P. (2003) *Organometallics*, **22**, 4490–4499; (c) Evrard, D., Lucas, D., Mugnier, Y., Meunier, P., and Hierso, J.-C. (2008) *Organometallics*, **27**, 2643–2653; (d) Zinovyeva, V.A., Monn, S., Fournier, S., Devilliers, C., Cattey, H., Doucet, H., Hierso, J.-C., and Lucas, D. (2013) *Inorg. Chem.*, **52**, 11923–11933; (e) Hierso, J.-C., Beaupérin, M., and Meunier, P. (2007) *Eur. J. Inorg. Chem.*, 3767–3780.

[16] (a) Amatore, C., Godin, B., Jutand, A., and Lemaitre, F. (2007) *Chem. Eur. J.*, **13**, 2002–2011; (b) Amatore, C., Godin, B., Jutand, A., and Lemaitre, F. (2007) *Organometallics*, **26**, 1757–1761; (c) Ludwig, M., Strömberg, S., Svensson, M., and Åkermark, B. (1999) *Organometallics*, **18**, 970–975.

[17] (a) Sato, Y., Sodeoka, M., and Shibasaki, M. (1989) *J. Org. Chem.*, **54**, 4738–4739; (b) Carpenter, N.E., Kucera, D.J., and Overman, L.E. (1989) *J. Org. Chem.*, **54**, 5846–5848; (c) Ashimori, A. and Overman, L.E. (1992) *J. Org. Chem.*, **57**, 4571–4572.

[18] (a) Ozawa, F., Kubo, A., and Hayashi, T. (1991) *J. Am. Chem. Soc.*, **113**, 1417–1419; (b) Ozawa, F., Kubo, A., Matsumoto, Y., and Hayashi, T. (1993) *Organometallics*, **12**, 4188–4196; (c) Worste, T.H. and Oestreich, M. (2011) *Chem. Eur. J.*, **17**, 11914–11918.

[19] Tietze, L.F., Nöbel, T., and Spescha, M. (1998) *J. Am. Chem. Soc.*, **120**, 8971–8977.

[20] (a) Tamao, K., Sumitani, K., and Kumada, K. (1972) *J. Am. Chem. Soc.* **94**, 4374–4376; (b) *idem* (1972) *J. Am. Chem. Soc.*, **94**, 9268–9269; (c) Tamao, K., Sumitani, K., Niso, Y., and Kumada, M. (1976) Bull. Soc. Chim. Jpn., **48**, 1958–1969; (d) Kumada, M., (1980) *Pure & Appl. Chem.*, **52**, 669–679.

[21] Joshi-Pangu, A., Wang, C.-Y., and Biscoe, M.R. (2011). *J. Am. Chem. Soc.*, **133**, 8478–8481.

[22] Fauvarque, J.-F. and Jutand, A. (1976) *Bull. Soc. Chim. Fr.*, 765–770.

[23] Sekiya, A. and Ishikawa, N. (1976) *J. Organomet. Chem.*, **118**, 349–354.

[24] (a) Hayashi, T, Konishi, M., and Kumada, M. (1979) *Tetrahedron Lett.*, **22**, 1871–1874; (b) Minato, A., Tamao, K.T., Hayashi, T., Suzuki, K., and Kumada, K. (1981) *Tetrahedron Lett.*, **22**, 5319–5322; (c) Hayashi, T., Konishi, M., Kobori, Y., Tamao, K., Kumada, M., Higuchi, T., and Hirotsu, K. (1984) *J. Am. Chem. Soc.*, **106**, 158–163.

[25] Babudri, F., Colangiuli, D., Farinola G.M., and Naso, F. (2002) *Eur. J. Org. Chem.*, **16**, 2785–2791.

[26] (a) Negishi, E.-I., King, A.O., and Okukado, N. (**1977**) *J. Org. Chem.*, **42**, 1821–1823; (b) Fauvarque, J.-F. and Jutand, A. (**1977**) *J. Organomet. Chem*, **132**, C17–C19; (c) King, A.O., Negishi, E.-I., Villani, F.J., and Silveira, A. (**1978**) *J. Org. Chem.*, **43**, 358–360; (d) Negishi, E.-I. (**1982**) *Acc. Chem. Res.*, **15**, 340–348; (e) Negishi, E.-I. and Luigi, A. (**2003**) *Chem. Rev.*, **103**, 1979–2017; (f) Negishi, E.-I. (**2011**) *Angew. Chem. Int. Ed.*, **50**, 6738–6764; (g) Knochel, P., Calaza, M.I., and Hupe, E. (**2004**) Carbon-Carbon bond forming reactions mediated by organozinc reagents in *Metal-catalyzed Cross-Coupling reactions*, (eds. A. de Mejeire and F. Diederich), Wiley-VCH, Weinheim, pp. 619–670; (h) Manolikales, G., Schade, M.A., Munoz Hernandez, C., Mayr, H., and Knochel, P. (2008) *Org. Lett.*, **10**, 2765–2768.

[27] (a) Amatore, C. Fauvarque, J.F., Jutand, A., and Negri, S. (1990) *J. Organomet. Chem.*, **390**, 389–398; (b) Pucetti, G., Ledoux, I., Zyss, J., Jutand, A., and Amatore, C. (1992) *Chem. Phys.*, **160**, 467–475.

[28] (a) Murahashi, S.-I., Yamamura, M., Yanagisawa, K.-I., Mita, N., and Kundo, K. (1979) *J. Org. Chem.*, **44**, 2408–2417; (b) Jeffery-Luong, T. and Linstrumelle, G. (1982) *Synthesis*, **9**, 738–740; (c) Murahashi, S.I. (2002) *J. Organomet. Chem.*, **653**, 27–33.

[29] (a) Giannerini, M., Fañanàs-Mastral, M., and Feringa, B.L. (2013) *Nat. Chem.*, **5**, 667–672; (b) Hornillos, V., Giannerini, M., Villa, C., Fañanàs-Mastral, M., and Feringa, B.L. (2013) *Org. Lett.*, **15**, 5114–5117; (c) Giannerini, M., Hornillos, V., Villa, C., Fañanàs-Mastral, M., and Feringa, B.L. (2013) *Angew. Chem. Int. Ed.*, **52**, 13329–13333.

[30] Negishi, E.-I., Takahashi, T., and Akiyoshi, K. (1986) J. Chem. Soc. Chem. Commun., 1338–1339.

[31] (a) Amatore, C., Azzabi, M., Calas, P., Jutand, A., Lefrou, C., and Rollin, Y. (1990) *J. Electroanal. Chem.*, **288**, 45–63; (b) Amatore, C., Azzabi, M., and Jutand, A. (1991) *J. Am. Chem. Soc.*, **113**, 8375–8384; (c) Amatore, C. Jutand, A., and Suarez, A. (1993) *J. Am. Chem. Soc.*, **115**, 9531–9541.

[32] Fauvarque, J.F. and Jutand, A. (1979) *J. Organomet. Chem.*, **177**, 273–281.

[33] (a) Palucki, M. and Buchwald, S.L. (1997) *J. Am. Chem. Soc.*, **119**, 11108–11109; (b) Hamann, B.C. and Hartwig, J.F. (1997) *J. Am. Chem. Soc.*, **119**, 12382–12383.

[34] (a) Culkin, D.A. and Hartwig, J.F. (2001) *J. Am. Chem. Soc.*, **123**, 5816–5817; (b) Culkin, D.A. and Hartwig, J.F. (2003) *Acc. Chem. Res.*, **36**, 234–2450.

[35] (a) Moradi, W.A. and Buchwald, S.L, (2001) *J. Am. Chem. Soc.*, **123**, 7996–8002; (b) Lee, S., Beare, N.A., and Hartwig, J.F. (2001) *J. Am. Chem. Soc.*, **123**, 8410–8411; (c) Jorgenssen, M., Lee, S., Beare, N.A., and Hartwig, J.F. (2002) *J. Am. Chem. Soc.*, **124**, 12557–12565; (d) Hama, T. and Hartwig, J.F. (2008) *Org. Lett.*, **10**, 1549–1552.

[36] (a) Shaughnessy, K.H., Hamann, B.C., and Hartwig, J.F. (1998) *J. Org. Chem.*, **63**, 6546–6553; (b) Lee, S. and Hartwig, J.F. (2001) *J. Org. Chem.*, **66**, 3402–3415; (c) Hama, T., Liu, X., Culkin, A.A., and Hartwig, J.F. (2003) *J. Am. Chem. Soc.*, **124**, 11176–11177; (d) de Filippis, A., Pardo, D.G., and Cossy, J. (2004) *Tetrahedron*, **60**, 9757–9767.

[37] Eidamshaus, C. and Burch, J.D. (2008) *Org. Lett.*, **10**, 4211–4214.

[38] Cassar, L. (1975) *J. Organomet. Chem.*, **93**, 253–257.

[39] Dieck, H.A. and Heck, R.F. (1975) *J. Organomet. Chem.*, **93**, 259–263.

[40] Sonogashira, K., Tohda, Y., and Hagihara, N. (1975) *Tetrahedron Lett.* **50**, 4467–4470.

[41] (a) Chinchilla, R., and Najera, C. (2007) *Chem. Rev.*, **107**, 874–922; (b) Doucet, H., and Hierso, J.-C. (2007) *Angew. Chem. Int. Ed.*, **46**, 834–871.

[42] Alami, M., Ferri, F., and Linstrumelle, G. (1993) *Tetrahedron Lett.*, **34**, 6403–6406.

[43] (a) Tougerti, A., Negri, S., and Jutand, A. (2007) *Chem. Eur. J.*, **13**, 666–673; (b) Jutand, A., Negri, S., and Principaud, A. (2005) *Eur. J. Inorg. Chem.*, 631–635; (c) for DFT calculations, see: Garcia-Melchor, M., Pacheco, M.C., Najera, C., Lledos, A., and Ujaque, G. (2012) *ACS Catal.*, **2**, 135–144.

[44] Amatore, C., Jutand, A., Khalil, F., M'Barki, M.A., and Mottier, L. (1993) *Organometallics*, **12**, 3168–3178.

[45] Balraju, V., Srinivasa Reddy, D., Periasamy, M., and Iqbal, J. (2005) *J. Org. Chem.*, **70**, 9626–9628.

[46] (a) Milstein, D. and Stille, J.K. (1979) *J. Am. Chem. Soc.*, **101**, 4992–4998; (b) Farina, V. and Roth, G.P. (1996) Recent advances in the Stille reaction in advances in Metal-Organic Chemistry, vol. **5**, (ed. L.S. Liebeskind), JAI Press Inc., Greenwich, pp. 1; (c) Farina, V., Krishnamurthy, V., and Scott, W.J. (1997) The Stille Reaction in Organic Reactions, John Wiley & Sons, Inc., New York, vol. **50**, pp. 1; (d) Espinet, P. and Echavarren, A.M. (2004) *Angew. Chem. Int. Ed.*, **43**, 4704–4734; (e) Farina, V., Kapadia, S., Krishnan, B., Wang, C., and Liebeskind, L.S. (1994) *J. Org. Chem.*, **59**, 5905–5911.

[47] Farina, V. and Krishnan, B. (1991) *J. Am. Chem. Soc.*, **113**, 9585–9595.

[48] (a) Amatore, C., Bucaille, A., Fuxa, A., Jutand, A., Meyer, G., and Ndedi Ntepe, A. (2001) *Chem. Eur. J.*, **7**, 2134–2142; (b) Amatore, C., Bahsoun, A.A., Jutand, A., Meyer, G., Ndedi Ntepe, A., and Ricard, L.(2003) *J. Am. Chem. Soc.*, **125**, 4212–4222.

[49] Jutand, A. (2003) *Eur. J. Inorg. Chem.*, 2017–2040.

[50] Alvarez, R., Faza, O.N., Lopez, C.S., and de Lera, A.R. (2006) *Org. Lett.*, **8**, 35–38.

[51] Nova, A., Ujaque, G., Maseras, F., Lledos, A., and Espinet, P. (2006) *J. Am. Chem. Soc.*, **128**, 14571–14578.

[52] Verbeeck, S., Meyers, C., Franck, P., Jutand, A., and Maes, B. (2010) *Chem. Eur. J.*, **16**, 12831–12837.

[53] Kamenecka, T.M. and Danishefsky, S.J. (1998) *Angew. Chem. Int. Ed.*, **37**, 2995–2998.

[54] (a) Miyaura, N. and Suzuki, A. (1979) *J. Chem. Soc.*, 866–867; (b) Miyaura, N., Yanagi, T., and Suzuki, A. (1981) *Synth. Commun.*, **11**, 513–519; (c) For reviews, see: Miyaura, N. and Suzuki, A. (1995) *Chem. Rev.*, **95**, 2457–2483; (d) Miyaura, N. (1998) Synthesis of biaryls via the cross-coupling reaction of arylboronic acids in Advances in Metal-Organic Synthesis. JAI Press Inc., Stanford, vol. **6**, pp. 187–243; (e) Suzuki, A. (1999) *J. Organomet. Chem.*, **576**, 147–168; (f) Nicolaou, K.C., Bulger, P.G., and Sarlah, D.

(2005) *Angew. Chem. Int. Ed.*, **44**, 4442–4489; (g) Suzuki, A. (2005) *Chem. Commun.*, 4759–4763; (h) Suzuki, A. (2011) *Angew. Chem. Int. Ed.*, **50**, 6723–6737.

[55] Uenishi, J.-I., Beau, J.-M., Armstrong, R.W., and Kishi, Y. (1987) *J. Am. Chem. Soc.*, **109**, 4756–4758.

[56] (a) Smith, G.B., Dezeny, G.C., Hughes, D.L., King, A.O., and Verhoeven, T.R. (1994) *J. Org. Chem.*, **59**, 8151–8156; (b) Matos, K. and Soderquist, J.A. (1998) *J. Org. Chem.*, **63**, 461–470; (c) Moriya, T., Miyaura, N., and Suzuki, A. (1994) Synlett, 149–151; (d) Braga, A.A.C., Morgon, N.H., Ujaque, G., Llédos, A., and Maseras, F. (2006) *J. Organomet. Chem.*, **691**, 4459–4466; (e) Glaser, R. and N. Knotts, (2006) *J. Phys. Chem. (A)*, **110**, 1295–1304; (f)Jover, J., FeyPurdie, M., Lloyd-Jones, G.C., and Harvey, J.N. (2010) *J. Mol. Cat A*, **324**, 39–47.

[57] (a) Amatore, C., Jutand, A., and Le Duc, G. (2011) *Chem. Eur. J.*, **17**, 2492–2503; (b) Amatore, C., Jutand, A., and Le Duc, G. (2012) *Chem. Eur. J.*, **18**, 6616–6625; (c) Amatore, C., Jutand, A., and Le Duc, G. (2012) *Angew. Chem. Int. Ed.*, **51**, 1379–1382; (d) Amatore, C., Jutand, A., and Le Duc, G. (2013) *Chem. Eur. J.*, **19**, 10082–10093; (e) Adamo, C., Amatore, C., Ciofini, I., Jutand, A., and Lakmini, H. (2006) *J. Am. Chem. Soc.*, **128**, 6829–6836.

[58] (a) Grushin, V.V. and Alper, H. (1996) *Organometallics*, **15**, 5242–5245; (b) Fraser, S.L., Antipin, M.Y., Khroustalyov, V.N., and Grushin, V.V. (1997) *J. Am. Chem. Soc.*, **119**, 4769–4770; (c) Pilon, M.C. and Grushin, V.V. (1998) *Organometallics*, **17**, 1774–1781.

[59] (a) For reductive elimination induced by a fifth ligand on square planar d^{10} complexes, see: Giovannini, R. and Knochel, P. (1998) *J. Am. Chem. Soc.*, **120**, 11186–11187; (b) Johnson, J.B. and Rovis, T. (2008) *Angew. Chem. Int. Ed.*, **47**, 840–871; (c) Jarvis Amanda, A.G. and Fairlamb, I.J.S. (2011) *Curr. Org. Chem.*, **15**, 3175–3196; (d) Gramage-Doria, R., Armspach, D., Matt, D., and Toupet, L. (2012) *Chem. Eur. J.*, **18**, 10813–10816.

[60] (a) Carrow, B.P. and Hartwig, J.F., (2011) *J. Am. Chem. Soc.*, **133**, 2116–42119; (b) Schmidt, A.F., Kurokhtina, A.A., and Larina, E.V. (2011) *Russ. J. Gen. Chem.*, **81**, 1573–1574.

[61] (a) Wright, S.W., Hageman, D.L., and McClure, L.D. (1994) *J. Org. Chem.*, **59**, 6095–6097; (b) Old, D.W., Wolfe, J.P., and Buchwald, S.T. (1998) *J. Am. Chem. Soc.*, **120**, 9722–9723; (c) Littke, A.F. and Fu, G.C. (1998) *Angew. Chem. Int. Ed.*, **37**, 3387–3388; (d) Littke, A.F., Dai, C., and Fu, G.C. (2000) *J. Am. Chem. Soc.*, **122**, 4020–4028.

[62] (a) Batey, R.A., Quach, T.D. (2001) *Tetrahedron Lett.*, **42**, 9099–9103; (b) Molander, G.A. and Ito, T. (2001) *Org. Lett.*, **3**, 393–396; (c) Butters, M., Harvey, J.N., Jover, J., Lennox, A.J.J., Lloyd-Jones, G.C., and Murray, P.M. (2010) *Angew. Chem. Int. Ed.*, **49**, 5156–5160; (d) Lloyd-Jones, G.C. (2012) *J. Am. Chem. Soc.*, **134**, 7431–7431.

[63] (a) Darses, S. and Genêt, J.P. (1999) *Eur. J. Org. Chem.*, **8**, 1875–1883; (b) Molander, G.A., and Rivero, M.R. (2002) *Org. Lett.*, **4**, 107–109.

[64] Boger, D.L., Miyazaki, S., Heon Kim, S., and Wu, J.H., Loiseleur, O., and Castle, S.L. (1999), *J. Am. Chem. Soc.*, **121**, 3226–3227.

[65] (a) Hatanaka, Y. and Hiyama, T. (1988) *Tetrahedron Lett.*, **29**, 97–98; (b) Hatanaka, Y., Goda, K.-I., Okahara, Y., and Hiyama, T. (1994) *Tetrahedron*, **50**, 8301–8316; (c) Hatanaka, Y. and Hiyama, T. (1990) *J. Am. Chem. Soc.*, **112**, 7793–7794; (d) Gouda, K.-I., Hagiwara, E., Hatanaka, Y., and Hiyama, T. (1996) *J. Org. Chem.*, **61**, 7232–7233; (e) Shibata, K., Miyazawa, K., and Goto, Y. (1997) Chem. Commun., 1309–1310; (f) Mowery, E. and DeShong, P. (1999) *Org. Lett.*, **1**, 2137–2140; (g) Molander, G.A. and Iannazzo, L. (2011) *J. Org. Chem.*, **76**, 9182–9187; (h) Hatanaka, Y. and Hiyama, T. (1991) Synlett, 845–853; (i) Hiyama, T. and Shirakawa, E. (2002) Organosilicon Compounds in Topics in Current Chemistry, vol. **219**, Springer-Verlag, Berlin, Heidelberg, pp. 61–85.

[66] (a) Denmark, S.E. and Wehri, D. (2000) *Org. Lett.*, **2**, 565–568; (b) Denmark, S.E., Sweis, R.F., and Wehri, D. (2004) *J. Am. Chem. Soc.*, **126**, 4865–4875; (c) Denmark, S.E. and Liu, J.H.C. (2010) *Angew. Chem. Int. Ed.*, **49**, 2978–2986; (d) Denmark, S.E. and Sweis, R.F. (2001) *J. Am. Chem. Soc.*, **123**, 6439–6440; (e) Denmark, S.E. and Sweis, R.F. (2004) *J. Am. Chem. Soc.*, **126**, 4876–4882; (f) Denmark, S.E. and Baird, J.D. (2006) *Chem. Eur. J.*, **12**, 4954–4963; (g) Denmark, S.E., Smith, R.C., and Chang, W-T. (2011) *Tetrahedron*, **67**, 4391–4396.

[67] Amatore, C., Grimaud, L., Le Duc, G., and Jutand, A. (2014) *Angew. Chem. Int. Ed.*, **53**, 6982–6985.

[68] Hanaoka, M., Yamagishi, H., Marutani, M., Mukai, C. (1986) *Chem. Pharm. Bull.*, **35**, 2348–2355.

20

TRANSITION METAL-MEDIATED CARBON–HETEROATOM CROSS-COUPLING (C—N, C—O, C—S, C—Se, C—Te, C—P, C—As, C—Sb, AND C—B BOND FORMING REACTIONS): AN OVERVIEW

Masanam Kannan, Mani Sengoden, and Tharmalingam Punniyamurthy

Department of Chemistry, Indian Institute of Technology Guwahati, Guwahati, India

20.1 INTRODUCTION

During the past few decades, transition metal-catalyzed cross-coupling reactions have become a powerful tool for the construction of C—C and C–heteroatom bonds [1]. This strategy allows the conceptually simple and yet powerful and reliable approach for synthesizing structurally complex pharmaceuticals and biologically active molecules. The two most vastly used transition metal catalysts in carbon–heteroatom bond formation are palladium (mainly depends on its ancillary ligands) and copper (depends on the optimization of the catalytic system as a whole copper source, solvent, base, concentrations, etc.). Besides, considerable developments have also been made with other transition metal catalysts such as nickel, iron, etc.

In general, transition metal-catalyzed cross-coupling reactions can be classified into four types based on the coupling partners (Scheme 20.1): (i) regular cross-coupling, (ii) oxidative cross-coupling, (iii) reductive cross-coupling, and (iv) umpolung cross-coupling. The most common one is the regular cross-coupling of carbon electrophile with heteroatom nucleophile. The next fast growing one is oxidative cross-coupling of carbon nucleophile with heteroatom nucleophile, and the other two types are least known in the literature.

Arene Chemistry: Reaction Mechanisms and Methods for Aromatic Compounds,
First Edition. Edited by Jacques Mortier.

Entry	Name	Coupling partners	Example				
1	Regular cross-coupling	Electrophile with nucleophile	PhI	+ $PhNH_2$	→	Ph-NHPh
2	Oxidative cross-coupling	Nucleophile with nucleophile	$PhB(OH)_2$	+ $PhNH_2$	→	Ph-NHPh
3	Reductive cross-coupling	Electrophile with electrophile	ArOTf	+ Ph_2PCl	→	$Ar\text{-}PPh_2$
4	Inverse (or) umpolung cross-coupling	Nucleophile with electrophile	Ph_2Zn	+ Ar_2NHOBz	→	Ar_2NPh

SCHEME 20.1 General classification of carbon–heteroatom cross-coupling reactions.

Typical mechanism of transition metal-catalyzed regular cross-coupling involves three main steps (Scheme 20.2) [2]: (i) oxidative addition of preactivated M (M = Pd, Cu, or Ni) complex **I** with aryl electrophile to generate an aryl-M complex **II**, which is facilitated by the use of electron-rich ancillary ligands; (ii) ligand exchange—the coordination of nucleophile followed by base-promoted proton abstraction results in a σ-aryl-M-nucleophile intermediate **III**; and (iii) the reductive elimination to give the coupling product and the active M-species from intermediate **III** that is the rate-determining step. Typically, bulky ancillary ligands increase the steric strain in **III** that facilitates the reductive elimination. On the other hand, oxidative cross-coupling involves transmetalation followed by oxidative addition and reductive elimination (Scheme 20.3) [3]. The active catalytic species rely on high-valent metal complexes Pd(II)/Pd(IV) pair or Cu(I)/Cu(III) pair instead of Pd(0)/Pd(II) pair that is involved in regular cross-coupling reactions, while the reductive and umpolung cross-couplings follow oxidative addition followed by transmetalation and reductive elimination. The main difference between these two couplings is the former involves aryl electrophile and the later utilizes heteroatom electrophile (Schemes 20.4 and 20.5) [4, 5].

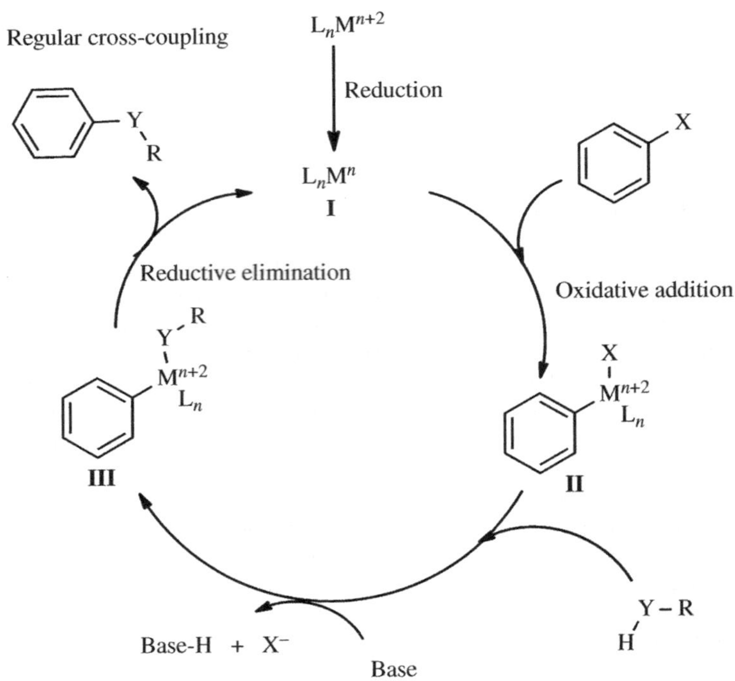

M = Pd(0), Ni(0) or Cu(I); L_n = ligand; Y = NH, O, S, etc.

X = halides or OTf

SCHEME 20.2 Catalytic cycle for regular cross-coupling reactions.

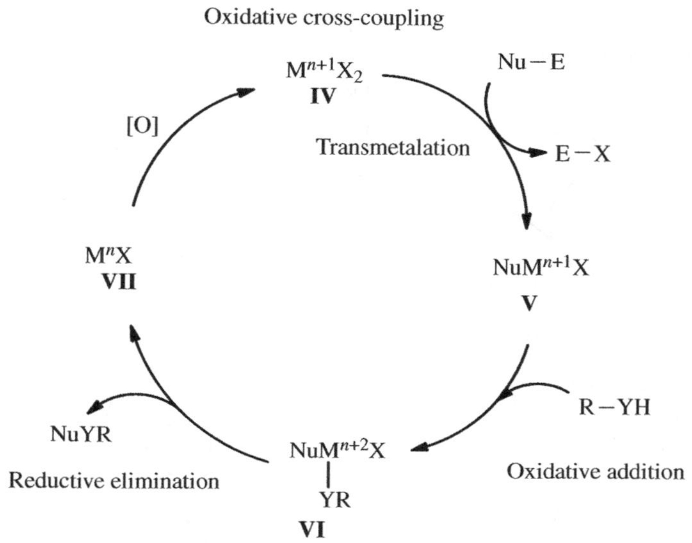

M = Pd(II), Cu(I); Y = NH, O, S, etc.; E. = Bi^{3+}, B^{3+}, Sn^{4+}, etc.

SCHEME 20.3 Catalytic cycle for oxidative cross-coupling reactions.

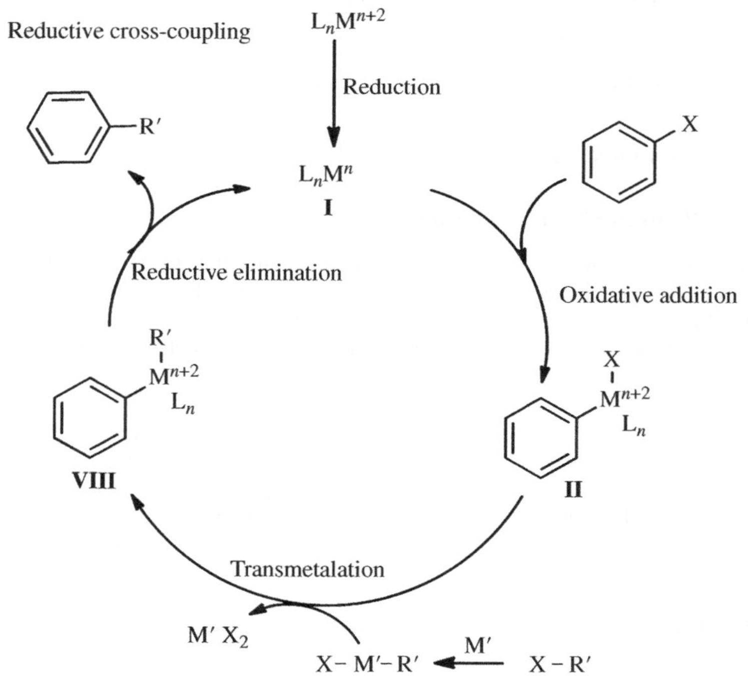

M = Pd(0), Ni(0); M′ = Zn, Mg; X = halides and triflates

SCHEME 20.4 Catalytic cycle for reductive cross-coupling reactions.

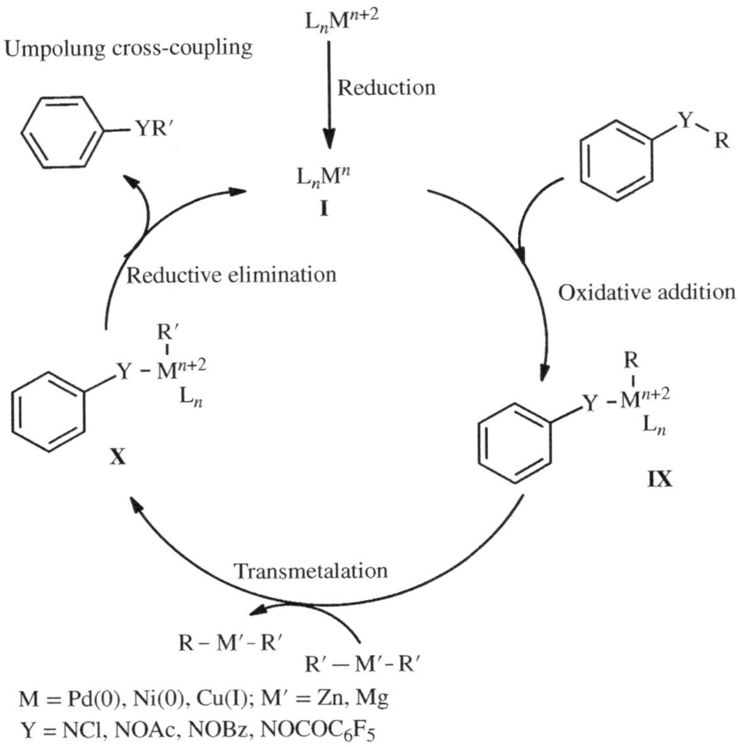

SCHEME 20.5 Catalytic cycle for umpolung cross-coupling reactions.

20.2 C—N CROSS-COUPLING

20.2.1 Palladium-Catalyzed Reactions

In 1983, Migita and coworkers reported the first palladium-catalyzed C—N coupling of organotin derivatives as nitrogen nucleophile using P(*o*-tol)$_3$ as ligand (Scheme 20.6) [6]. The limitation of this method is the formation of low amount of desired anilines over the β-hydride-eliminated by-product arene. This could be attributed to the labile nature of the bulky monophosphine ligand.

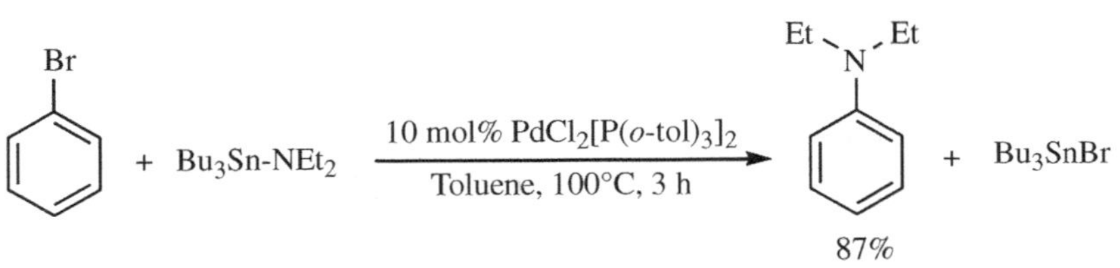

SCHEME 20.6 Pd-catalyzed C—N cross-coupling of aryl halides with organotin compound.

Buchwald and Hartwig groups subsequently explored bidentate ligands **L1–L2** for the amination of aryl bromides and aryl iodides (Scheme 20.7) [7, 8]. The steric, geometric, and electronic effects of various bidentate ligands **L3–L12** have been demonstrated (Fig. 20.1) [9]. The bidentate ligands could cause difficulty in accessing three-coordinate monophosphine complex responsible

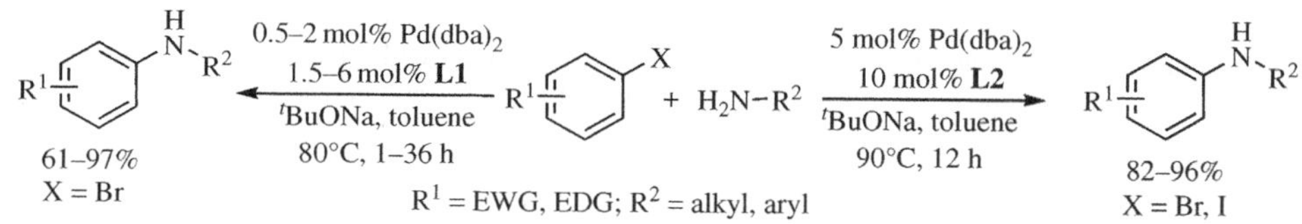

SCHEME 20.7 Examples using bidentate phosphorous ligands.

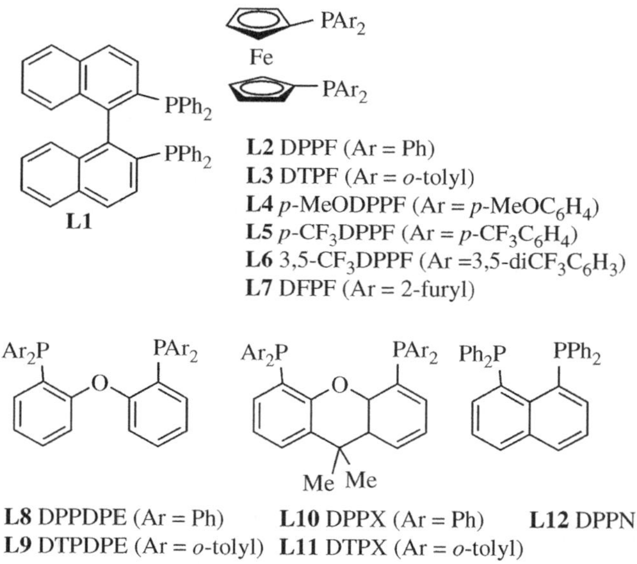

FIGURE 20.1 Bidentate phosphorous ligands.

for β-hydride elimination. Moreover, the product selectivity depends on the other characteristics of a chelating ligand such as electron-donating ability and "bite angle" (P-M-P angle). Generally, large bite angle ligand increases the reductive elimination, but they observed small bite angle ligands (BINAP and DPPN) to facilitate the monoarylation compared to diarylation (Scheme 20.8).

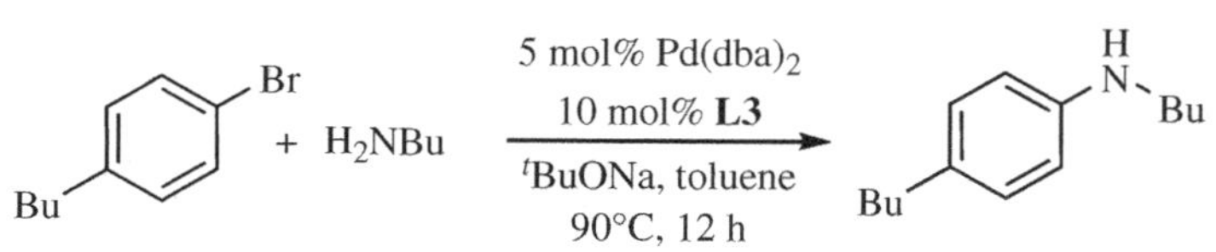

Ligand	Bite angle (deg)	*p*-BuC$_6$H$_4$NHBu	(*p*-BuC$_6$H$_4$)$_2$NHBu
DPPN	82	78	6
BINAP	85	91	3
DPPF	99	52	22
DPPR	101	12	5
DPPDPE	101	11	4
DPPX	109	47	7

SCHEME 20.8 Effect of bite angle in C—N cross-coupling reaction.

Beller and coworkers developed di(1-adamantyl)-*n*-butylphosphines **L13** and **L14** that facilitate the coupling of less reactive sterically hindered aryl chlorides with bulky amines (Scheme 20.9) [10]. Subsequently, the base effect on the amination of aryl chloride is reported [11].

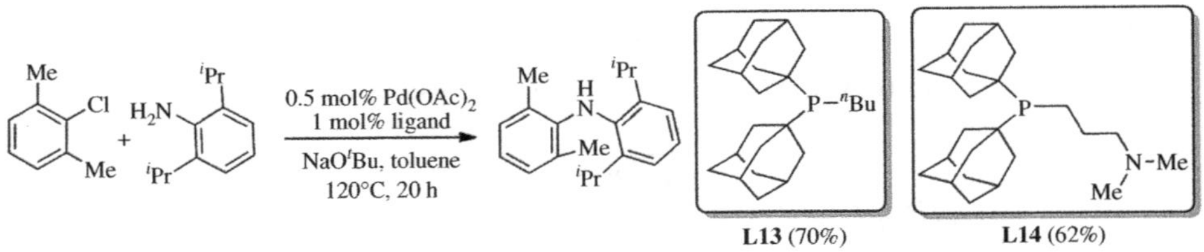

SCHEME 20.9 Pd-catalyzed cross-coupling of hindered aryl chlorides with aryl amines.

A versatile family of structurally related dialkylbiaryl phosphines **L15–L25** has been subsequently developed for the coupling of aryl halides with variety of amines (Fig. 20.2) [12–21]. BrettPhos and RuPhos found to be the most effective for coupling of 1° and 2° amines (Scheme 20.10). The main advantages of this class of ligands are its air stability, easy handling, crystallinity, and commercial availability.

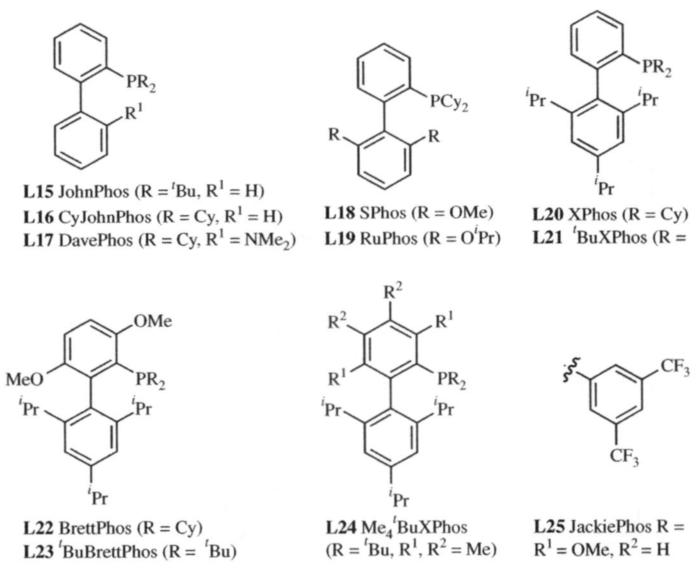

FIGURE 20.2 Examples of dialkylbiaryl phosphorous ligands for amination.

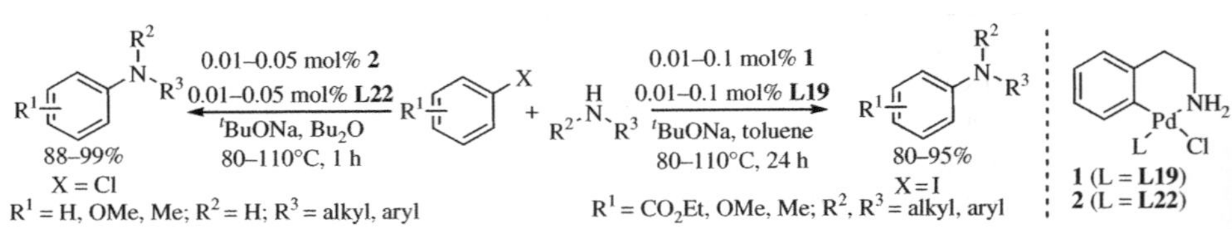

SCHEME 20.10 Arylation of primary and secondary amines.

The development of bulky bicyclic triaminophosphines **L26–L35** has been shown for the amination of a wide array of aryl bromides (Scheme 20.11) [22–25]. The salient features of these ligands are as follows: commercial availability, electron-rich phosphorus atom, and basicity enhancement by N → P bridgehead transannulation.

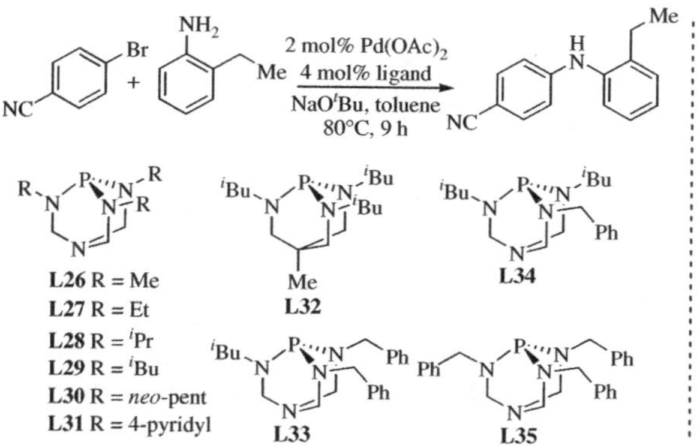

Entry	Ligand	Yield (%)
1	**L26**	15
2	**L27**	5
3	**L28**	0
4	**L29**	97
5	**L30**	30
6	**L31**	0
7	$P(NMe_2)_3$	10
8	$P(NEt_2)_3$	11
9	$P[N(^iBu)_2]_3$	15

SCHEME 20.11 The use of triaminophosphines in C—N cross-coupling reaction.

Relatively stable and highly active monodentate *N*-substituted heteroaryl phosphines **L36–L45** have been screened for the amination of aryl chlorides [26]. The catalyst with dialkylphosphino-*N*-arylindoles **L44** exhibit the best results (TON 8000 and TOF = 14,000 h^{-1} at 75% conversion) (Scheme 20.12).

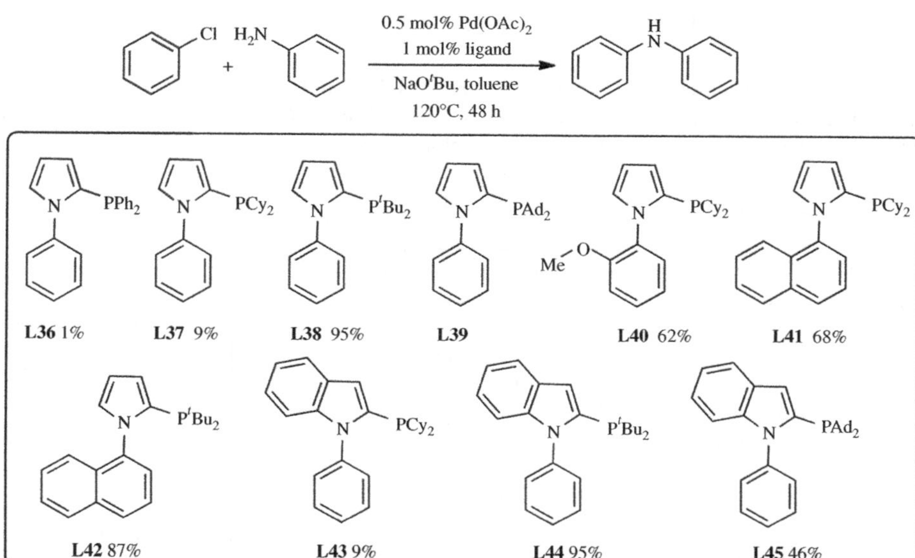

SCHEME 20.12 Examples of the use of N-substituted heteroaryl phosphine ligands.

Air- and moisture-stable palladium(II)-*N*-heterocyclic carbene complexes **3–6** have been used for the reaction of less reactive aryl chlorides with amines at room temperature (Scheme 20.13) [27].

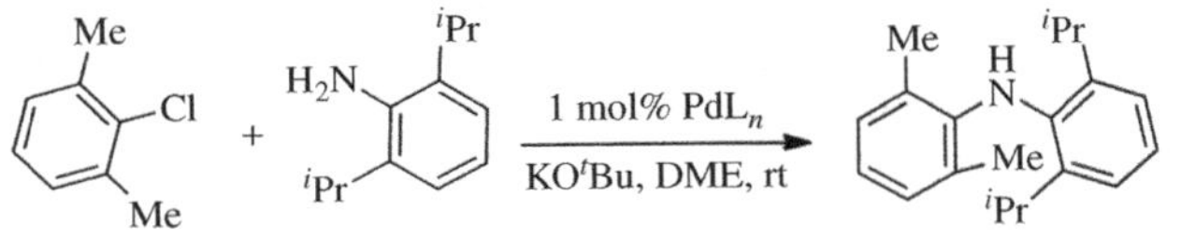

PdL$_n$	Time (h)	GC Conv. (%)
3	20	73
4	15	90
5	6	98
6	1.5	97

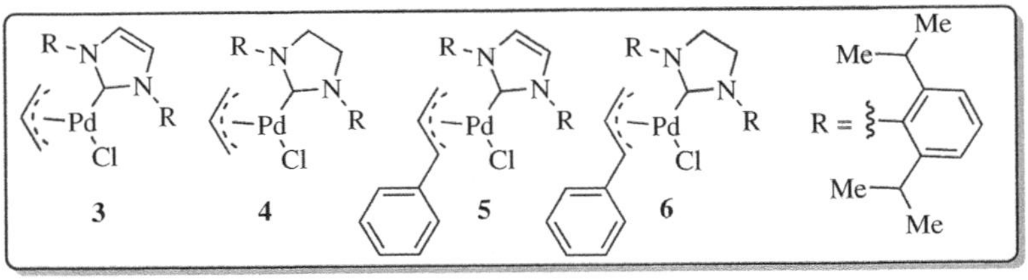

SCHEME 20.13 Pd-NHC-catalyzed room temperature C—N cross-coupling reaction.

The significance of structurally rigid and electron-rich Josiphos **L46** has been explored for coupling of aryl halides with aliphatic amines (Scheme 20.14) [28]. Better product selectivity, long lifetime of the catalyst, and low catalyst loading have been demonstrated.

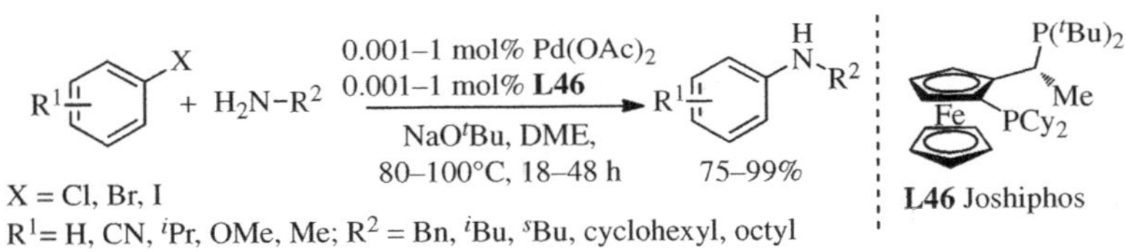

SCHEME 20.14 Coupling of aryl halide with alkyl amines.

Kwong and coworkers utilized amino phosphines **L47–L48** for the amination of aryl mesylates [29]. The use of air-stable phenyl-bridged P,N-ligands **L49–L56** has been subsequently demonstrated for the arylation of ammonia with aryl chlorides (Scheme 20.15) [30].

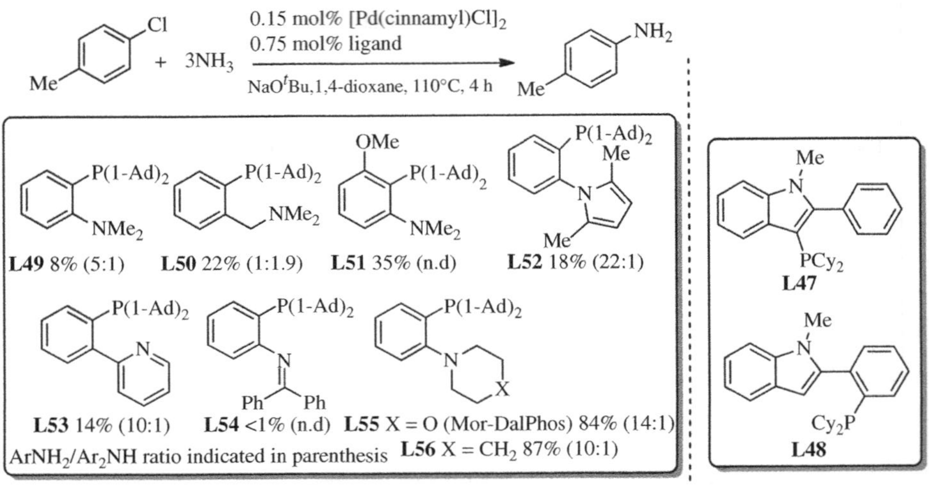

SCHEME 20.15 Application of P,N-ligands in aniline synthesis.

Buchwald and coworkers reported the synthesis of nitroarenes via coupling of aryl chlorides, triflates, and nonaflates with sodium nitrate [31]. Subsequently, palladium complex bearing Mor-DalPhos **L57** has been utilized for the coupling of aryl chlorides and tosylates with hydrazines (Scheme 20.16) [32].

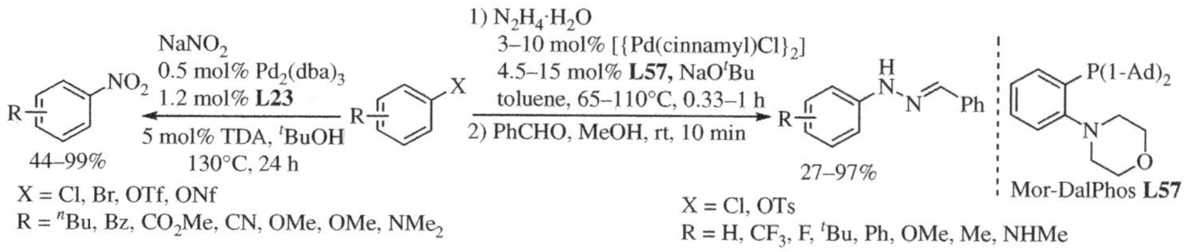

SCHEME 20.16 Pd-catalyzed nitration and hydrazination of aryl halides.

20.2.2 Copper-Catalyzed Reactions

20.2.2.1 Regular Cross-Coupling

Reaction of Arylamines Copper-catalyzed C—N coupling affords powerful tool for the synthesis of nitrogenated compounds [33]. In 1987, Paine reported soluble cuprous ion as the active catalytic species in Ullmann coupling [34]. Soluble air-stable copper(I) complex, $Cu(PPh_3)_3Br$, has been used for the synthesis of functionalized diaryl and triaryl amines (Scheme 20.17) [35]. Copper(I) complexes **7–8** and CuI–PBu_3 have been employed for the coupling of aryl halides with aryl amines [36, 37]. The catalyst with PBu_3 could be used for the coupling of less reactive aryl chlorides in the presence of KO^{t}Bu.

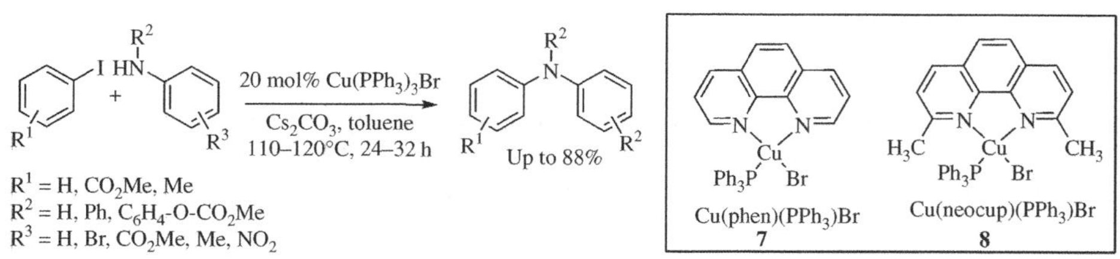

SCHEME 20.17 Cu(I)-catalyzed C—N cross-coupling.

Heterogeneous catalytic systems, Cu–Cu_2O [38, 39], Cu-Fe-hydrotalcite [40], and Cu-fluorapatite [41], have been employed for the coupling of aryl halides with amines (Scheme 20.18). These reactions provide the advantages of simplified product isolation and easy recovery and recyclability of the catalysts.

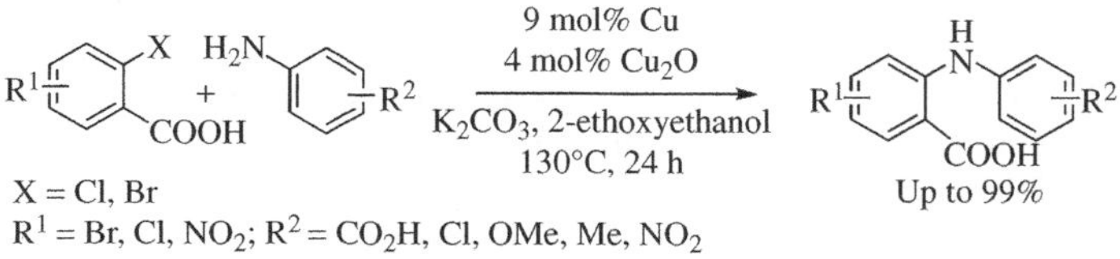

SCHEME 20.18 Heterogeneous Cu-catalyzed C—N cross-coupling.

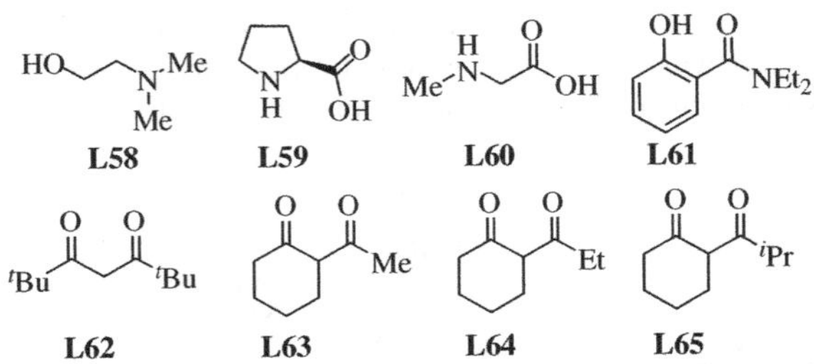

FIGURE 20.3 Ligands of the Cu-catalyzed cross-coupling of alkyl amines with aryl halides.

Reaction of Alkyl Amines Copper-catalyzed cross-coupling of aliphatic amines with aryl halides has made considerable progress. Several methods have used CuI with β-amino alcohols **L58** [42], amino acids **L59**, **L60** [43], *N,N*-diethyl salicylaldehyde **L61** [44], or β-diketones **L62–L65** [45] (Fig. 20.3). Among these, the protocol using β-diketones exhibited the superior results catalyzing the reaction even at room temperature (Scheme 20.19).

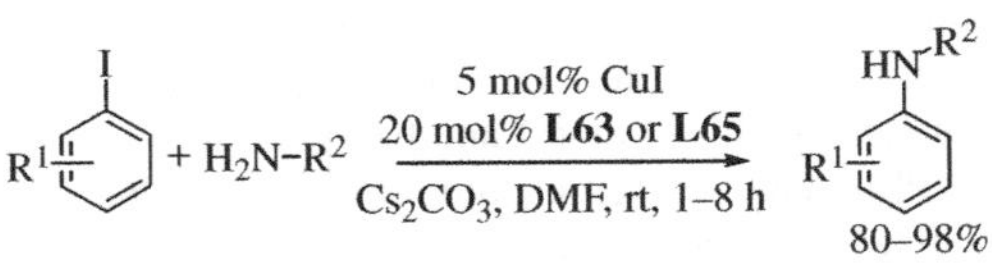

R^1= 4-Br, 4-OH, 3-NH_2, 3-Me, 3-CH_2OH, 3,5-Me_2, 3,4-F_2
R^2 = allyl, alkyl, benzyl

SCHEME 20.19 Cu-catalyzed room temperature arylation of aliphatic amines.

The usefulness of ancillary ligands **L66–L78** to promote the copper-catalyzed N-arylation of aryl halides with aliphatic amines has been demonstrated (Schemes 20.20 and 20.21) [46, 47]. In absence of the ligands, the coupling of aryl halides with alkyl amines requires higher temperature [48].

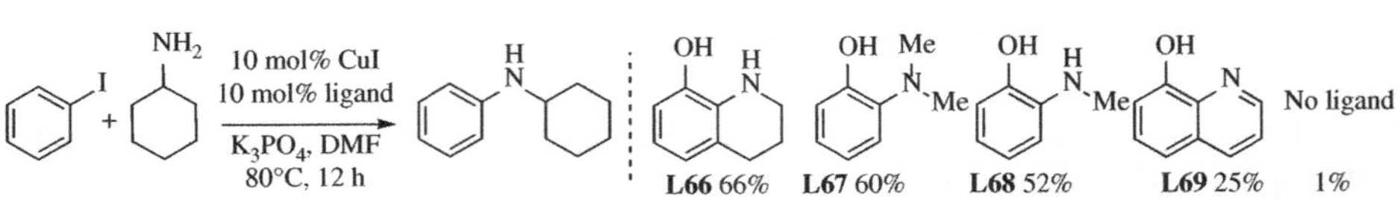

SCHEME 20.20 Ligand-assisted Cu-catalyzed C—N cross-coupling.

Lie and coworkers developed various quaternary ammonium salts **9** and phosphorus-based organic ionic bases **10** that promote the coupling of aryl halides with aliphatic amines, which are difficult to achieve with traditional alkaline bases (Fig. 20.4) [49]. In a set of screened promoters, TBAA, TBPE, and TBPM exhibited the best results (Scheme 20.22).

Further extending the study of the ligand effect in determining the product selectivity, β-diketone **L65** has been investigated for the N-arylation of amino alcohols (Scheme 20.23). Computational studies using density functional theory suggest that the reaction takes place via single-electron transfer (SET) [50, 51]. In this reaction, copper(I) undergoes reaction with amine

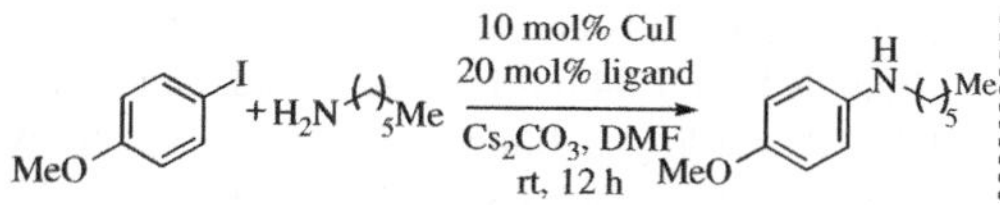

Entry	Ligand	Yield (%)	Entry	Ligand	Yield (%)
1	—	N.R.	6	**L74**	21
2	**L70**	45	7	**L75**	<10
3	**L71**	43	8	**L76**	97
4	**L72**	N.R.	9	**L77**	88
5	**L73**	60	10	**L78**	85

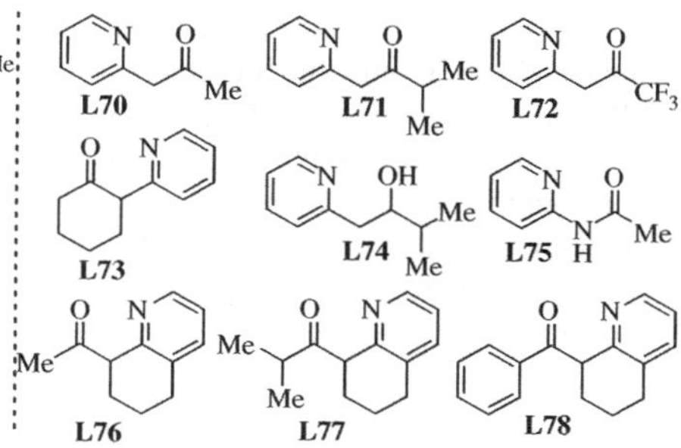

SCHEME 20.21 Room temperature cross-coupling of alkyl amine with aryl halides.

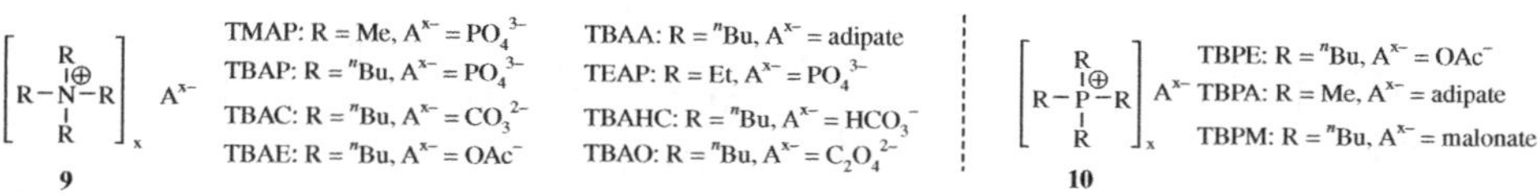

FIGURE 20.4 Examples of quaternary ammonium and phosphonium salts.

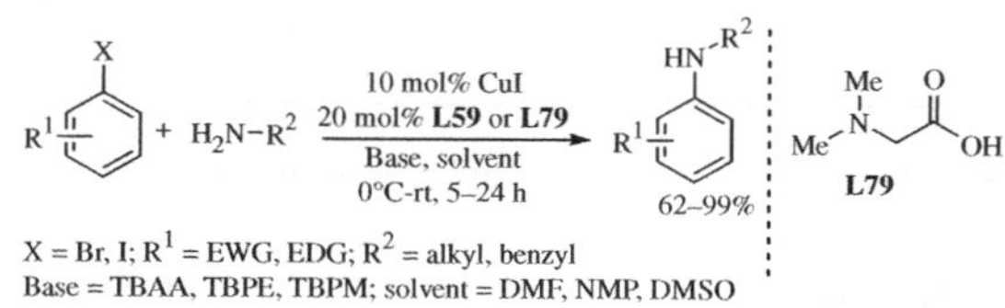

SCHEME 20.22 The C—N coupling of alkyl amines promoted by organic ionic base.

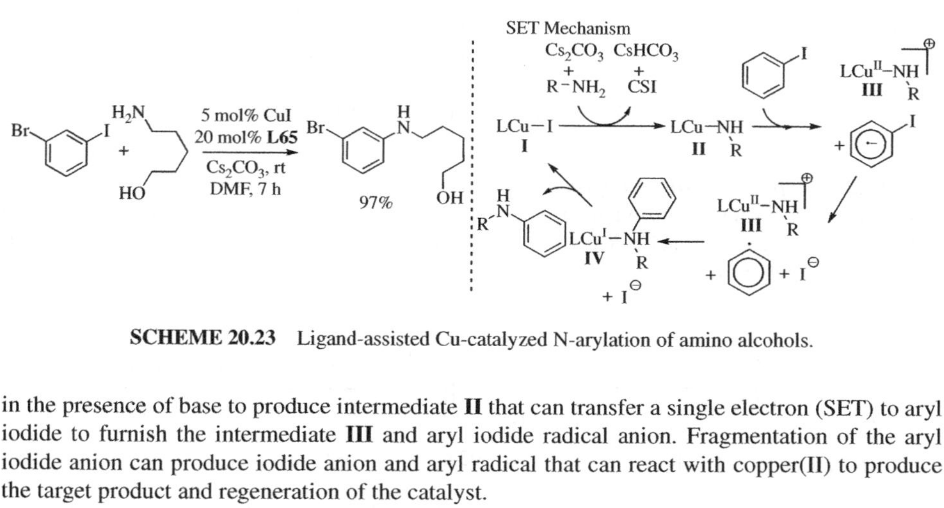

SCHEME 20.23 Ligand-assisted Cu-catalyzed N-arylation of amino alcohols.

in the presence of base to produce intermediate **II** that can transfer a single electron (SET) to aryl iodide to furnish the intermediate **III** and aryl iodide radical anion. Fragmentation of the aryl iodide anion can produce iodide anion and aryl radical that can react with copper(II) to produce the target product and regeneration of the catalyst.

The coupling of bulky secondary amines remains synthetically challenging compared to primary amines. To overcome this limitation, a series of ligands **L79–L84** has been screened, and DMPAO **L83** afforded the best results [52] (Scheme 20.24). Accordingly, Jiang and coworkers employed CuBr-8-hydroxyquinoline-*N*-oxide **L85** for the coupling of aryl halides with aliphatic amines [53]. This method provides the advantage of low catalyst loading, mild reaction conditions, and the use of inexpensive aryl chlorides for large-scale synthesis. Later, Cu/L-proline [54] and Cu(acac)$_2$ [55] have been utilized for the amination of aryl iodides and bromides using NH_4Cl and aqueous NH_3 (Scheme 20.25).

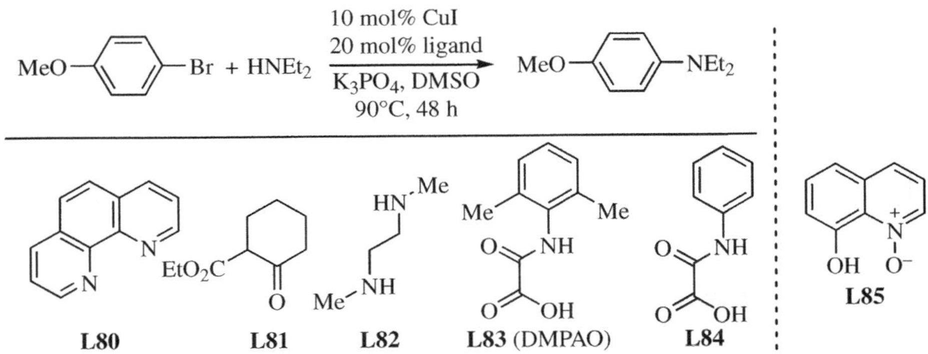

Ligand	L79	L80	L81	L82	L83	L84
Yield (%)	20	0	0	0	95	22

SCHEME 20.24 Cu-catalyzed C—N cross-coupling of secondary alkyl amines with aryl halides.

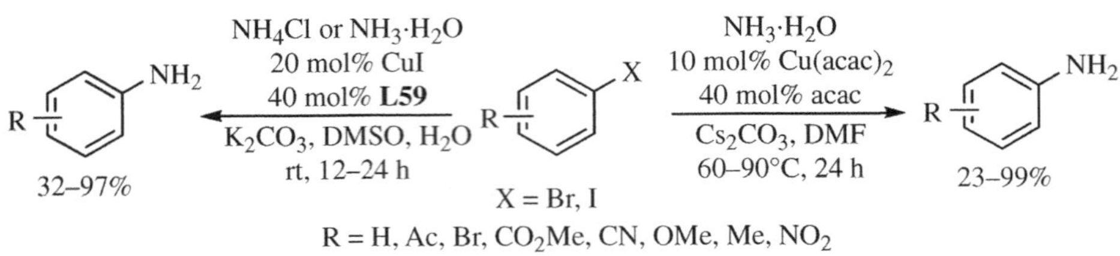

SCHEME 20.25 Cu-catalyzed aniline synthesis.

20.2.2.2 Umpolung Cross-Coupling Johnson and coworkers reported electrophilic amination of organozinc using *o*-benzoyl hydroxylamine as electrophilic nitrogen (Scheme 20.26) [56]. This process allows the facile synthesis of sterically hindered amines via *ortho*-metalation/*trans*-metalation/catalytic amination.

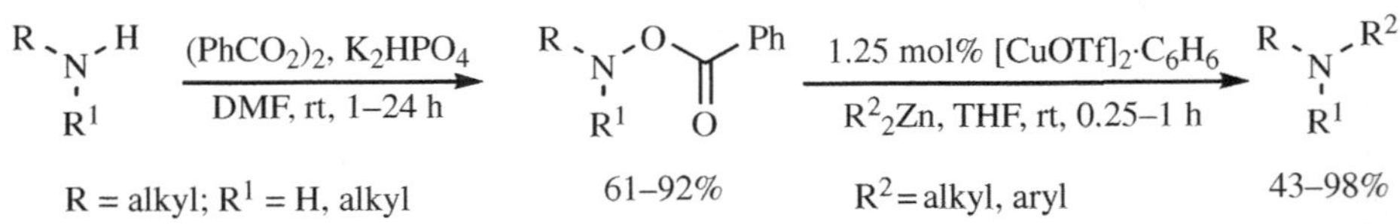

SCHEME 20.26 Umpolung cross-coupling of secondary amines with organozinc reagents.

20.2.2.3 Oxidative Cross-Coupling Chan and coworkers reported the coupling of organobismuth [57] or phenylboronic acid [58] as electrophilic source with equivalent amount of $Cu(OAc)_2$ (Scheme 20.27). The use of myristic acid as an additive led the system to be catalytic at room temperature [59].

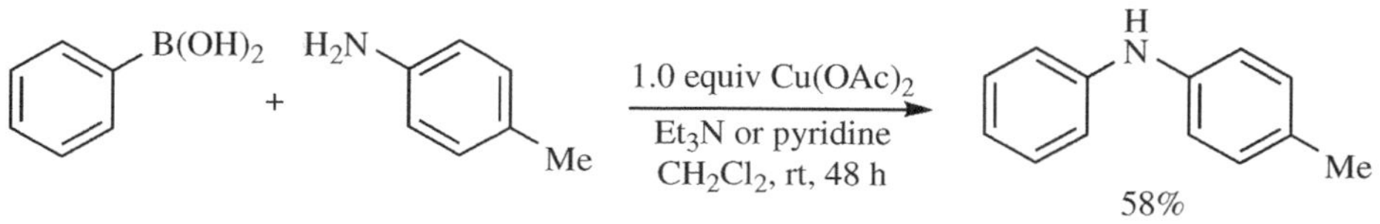

SCHEME 20.27 The cross-coupling of aryl amine with aryl boronic acid.

The use of oxidant such as molecular oxygen [60] or di-*tert*-butylperoxide [61] with catalytic amount of copper led to eliminate the use of base in the boronic acid oxidative coupling (Scheme 20.28). Fu and coworkers used atmospheric oxygen as oxidant with Cu_2O for the selective synthesis of primary amines at room temperature using aqueous ammonia as an amine source [62]. Yamamoto and coworkers developed air- and water-stable cyclic triolborate complex to increase the nucleophilicity of the attached group in boronic acid derivative [63]. Copper-mediated selective coupling of methylboronic acids with primary amines has been subsequently developed [64].

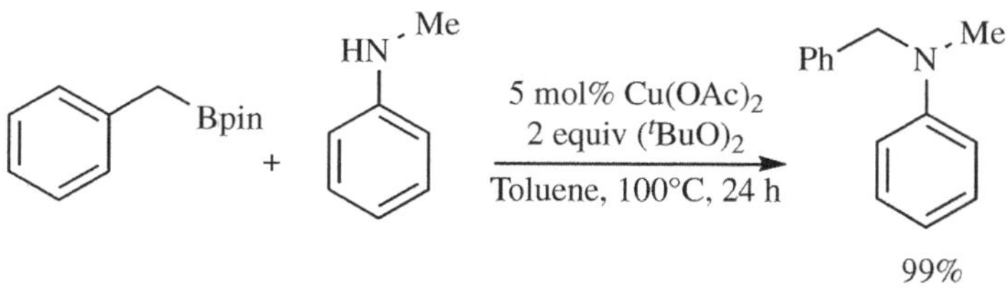

SCHEME 20.28 The cross-coupling of secondary amines with boronic acid derivatives.

20.2.3 Other Transition Metal-Catalyzed Reactions

Nickel plays a prominent role in C—N bond formation. The first nickel-catalyzed C—N coupling has been reported using $Ni(cod)_2$-dppf and $Ni(cod)_2$-1,10-phenanthroline for the synthesis of aryl amine (Scheme 20.29) [65]. The use of Ni/bipyridine [66], Ni/C-dppf [67], and Ni(II)-(NHC) [68, 69] has been subsequently described for the coupling of aryl halides with amines in the presence of sodium alkoxide. Ackermann and coworkers used $Ni(cod)_2$-dppf for

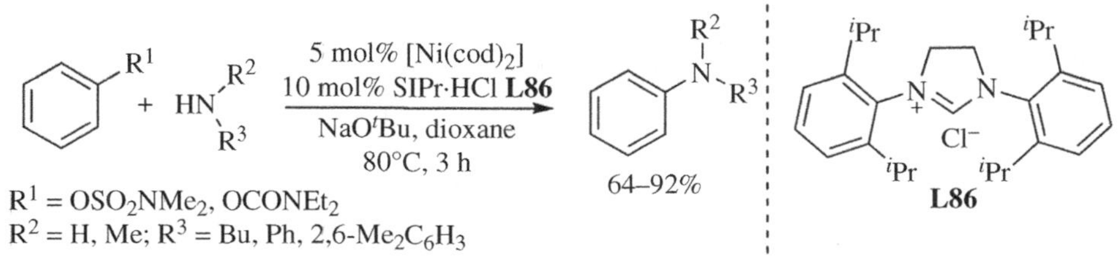

SCHEME 20.29 Cross-coupling of amines with aryl sulfamates and carbamates.

the amination of aryl sulfonates and sulfamates [70]. Later, the coupling of alkyl and aryl amines has been demonstrated with aryl sulfamates [71] and aryl carbamates [72] using Ni(NHC) **L86** (Scheme 20.29), while the coupling of alkyl amines could be accomplished with aryl pivalates utilizing Ni(NHC) [73]. Ni-bipyridyl has been utilized for the coupling of boronic acids with amines in the presence of organic base DBU [74] and microwave-assisted coupling of aryl halides with amines under ligand-free conditions [75]. Few studies are focused on the use of Fe_2O_3-L-proline [76], Fe/C_g [77], and Rh-NHC [78] for the coupling of amines with aryl halides.

20.2.4 Synthetic Applications

N-Aryl amines are important class of compounds that are widely employed in material and biological sciences. Several light-emitting carbazoles has been synthesized using palladium-catalyzed C—N cross-coupling [79]. Scheme 20.30 shows examples of some hole transporting and blue light-emitting materials.

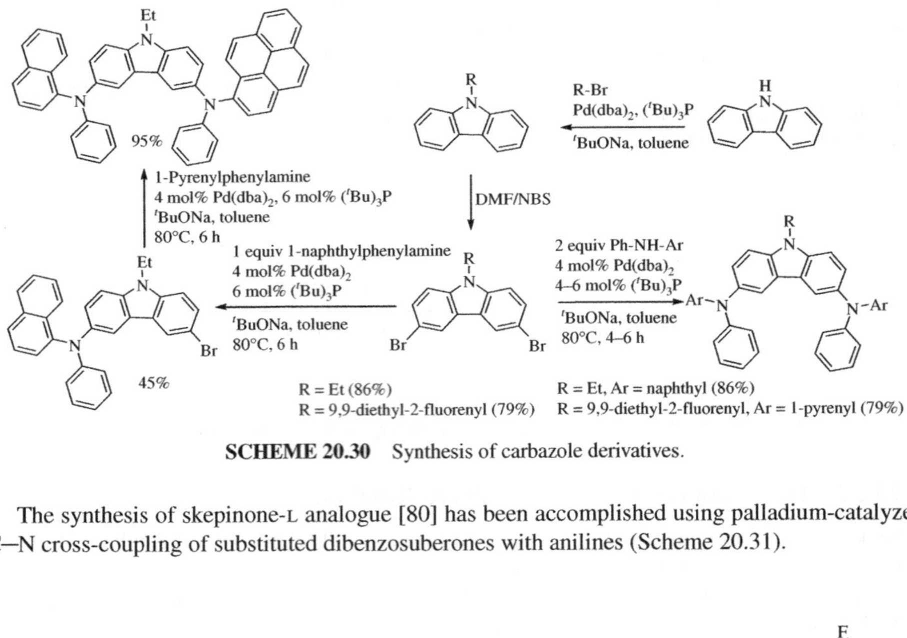

SCHEME 20.30 Synthesis of carbazole derivatives.

The synthesis of skepinone-L analogue [80] has been accomplished using palladium-catalyzed C—N cross-coupling of substituted dibenzosuberones with anilines (Scheme 20.31).

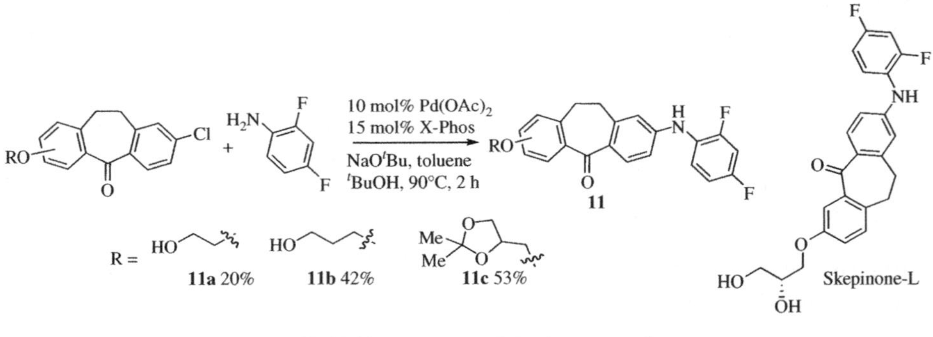

SCHEME 20.31 Synthesis of skepinone-L analogous.

20.3 C—O CROSS-COUPLING

20.3.1 Reactions with Aromatic Alcohols

20.3.1.1 Palladium-Catalyzed Reactions Palladium-catalyzed C—O cross-coupling has been well established with the discovery of a wide variety of phosphorous-based ligands (Scheme 20.32). Sterically hindered alkylphosphine **L87** found to accelerate the reductive elimination [81]. Subsequently, phosphine ligands, **L15**, **L17**, and **L88–L91**, have been explored for the coupling of electron-rich and electron-deficient aryl halides and sulfonates [82, 83]. Beller and coworkers reported the coupling of aryl chlorides with phenols using 2-phosphino-*N*-arylpyrroles **L37–L39** and -indoles **L43–L45** (cataCXium P ligands; see Scheme 20.12) [84]. Increase of the bulkiness of phosphorus substitution in cataCXium led to facilitate the reductive elimination. However, these ligand systems are inefficient to catalyze the coupling of *ortho*-substituted and electron-deficient phenols. Modified phosphorus ligands such as $Me_4{}^tBu$X-Phos **L24** [85] and cata*CX*ium phosphines **L92** [86] have thus been developed. Further increase of the steric hindrance in *ortho* position of ^{t}BuBrett-Phos **L93** [87] led to effect the target coupling at room temperature.

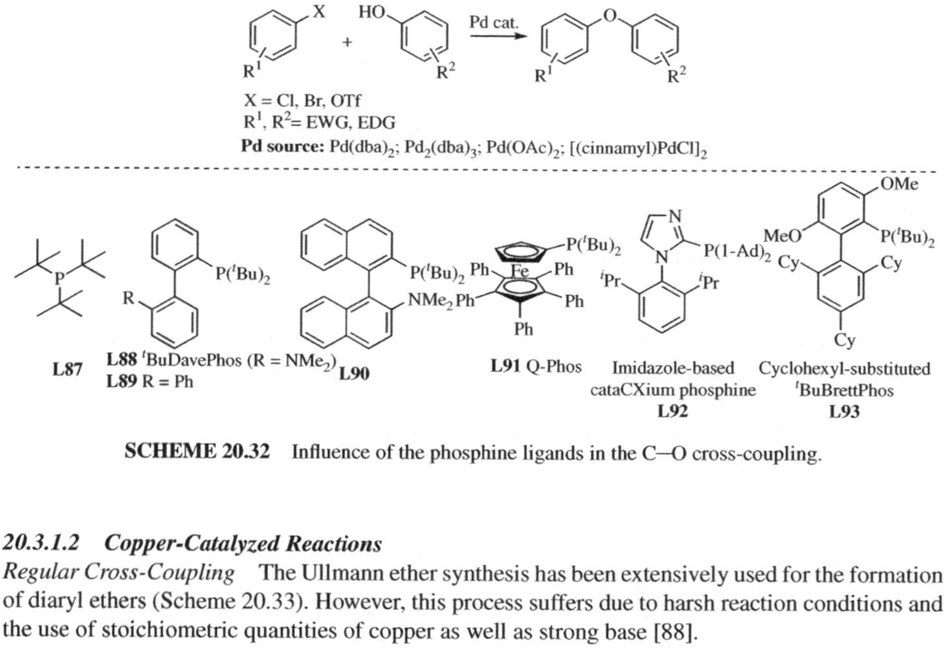

SCHEME 20.32 Influence of the phosphine ligands in the C—O cross-coupling.

20.3.1.2 Copper-Catalyzed Reactions

Regular Cross-Coupling The Ullmann ether synthesis has been extensively used for the formation of diaryl ethers (Scheme 20.33). However, this process suffers due to harsh reaction conditions and the use of stoichiometric quantities of copper as well as strong base [88].

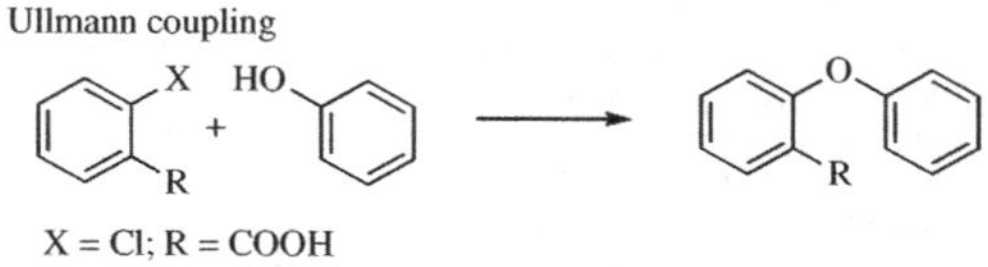

SCHEME 20.33 Ullmann synthesis of diaryl ethers.

Some of these drawbacks have been overcome by the recent developments in copper-catalyzed cross-coupling reactions. Buchwald and coworkers reported the coupling of variety of phenols with aryl bromides in the presence of Cs_2CO_3 (Scheme 20.34) [89]. The less reactive phenols required an equivalent amount of 1-naphthoic acid and molecular sieves to facilitate the transformation. Under these conditions, the reaction of unactivated aryl halides and less reactive or hindered phenols can be achieved.

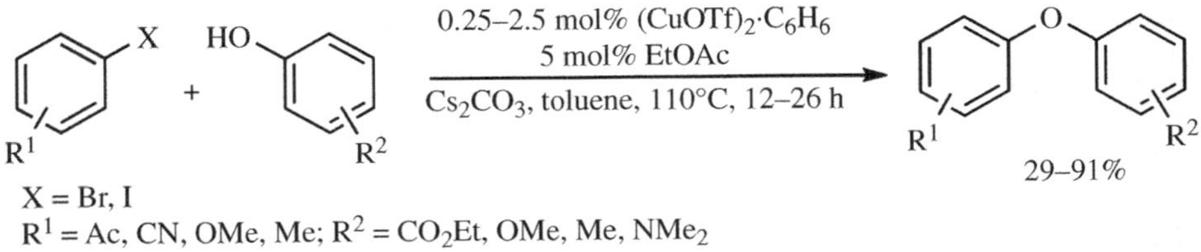

SCHEME 20.34 Copper(I)-catalyzed synthesis of diaryl ethers.

Copper complexes **7**, **8** [36] and $Cu(PPh_3)_3Br$ have been explored for the coupling of aryl bromides with phenols in the presence of Cs_2CO_3 or NMP (Scheme 20.35) [90].

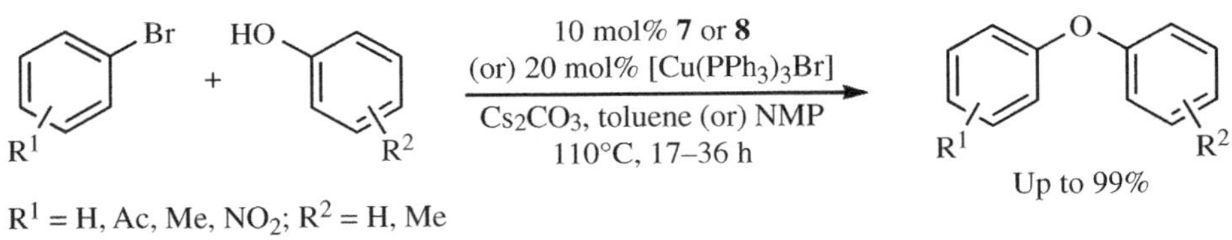

SCHEME 20.35 Coupling of aryl bromides with phenols.

Enhancement in the rate of the reaction (10–15-fold) has been reported using 2,2,6,6-tetramethylheptane-3,5-dione (TMHD) **L62** [91] (Scheme 20.36). In this reaction, TMHD exists as tautomer and forms a cocomplex with the copper(I)–phenolate species that equilibrates rapidly and accelerates the rate-determining step. *N,N*-Dimethylglycine **L79** [92, 93] and pyridine-derived Schiff bases **L94** and **L95** have been employed for the coupling of a wide range of aryl bromides with phenols [94, 95]. Pyrrolidine-2-phosphonic acid phenyl monoester (PPAPM) **L96** has been

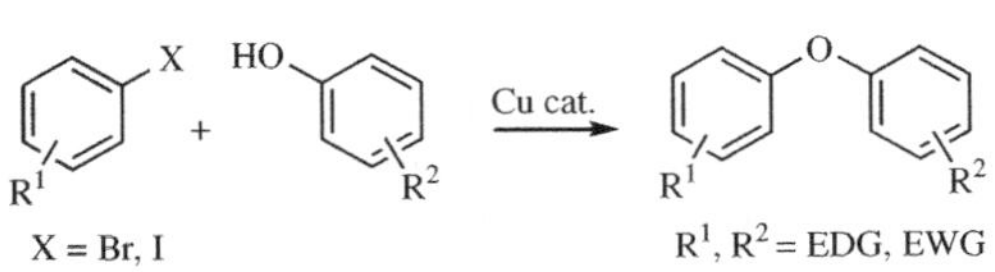

Reaction conditions	Yield (%)	Ref.
a) 50 mol% CuCl, 10 mol% **L62**, Cs_2CO_3, 120°C, 4–47 h	51–85	91
b) 2–10 mol% CuI, 7.5–30 mol% **L79**, Cs_2CO_3, 90°C, 4–24 h	64–97	92, 93
c) 5 mol% Cu_2O, 20 mol% **L95**, Cs_2CO_3, 82–110°C, 24–118 h	80–98	94
d) 10 mol% CuI, 10 mol% **L94**, **L95**, K_3PO_4, 60–80°C, 24 h	21–97	95
e) 10 mol% CuI, 20 mol% **L96**, K_3PO_4, 110°C, 13–36 h	69–98	96
f) 10 mol% CuBr, 20 mol% **L81**, Cs_2CO_3, 60–80°C, 22–36 h	72–97	97
g) 10 mol% CuI, 20 mol% **L80**, KF/Al_2O_3, 110°C, 11–15 h	30–98	98
h) 5 mol% Cu/C, 0.5 equiv **L80**, Cs_2CO_3, MW, 180–220°C, 0.5–3 h	79–89	99

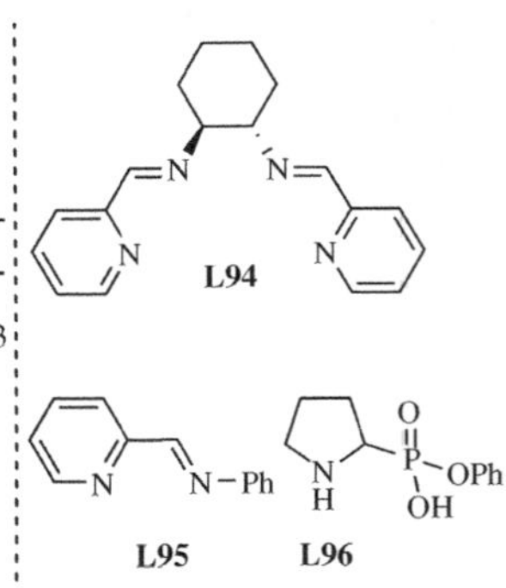

SCHEME 20.36 Cu-catalyzed C—O cross-coupling of aryl halides with phenols.

used for the coupling of phenols with aryl iodides or bromides [96]. The coupling of phenol derivatives with aryl bromides has been subsequently accomplished using CuBr–β-ketoester **L81** [97], CuI–1,10-phenanthroline **L80** [98], and Cu/C under microwave irradiation [99]. Xia and Taillefer utilized CuBr/TMHD for the coupling of less reactive aryl chlorides with phenols [100].

Effort has also been made on the development of recyclable catalytic systems. For example, CuO nanoparticles [101, 102], Cu-SiO_2 [103], Cu-SiO_2-bipyridyl [104], and $CuFe_2O_4$ nanoparticles [105] have been investigated for the coupling of aryl halides with phenols.

Oxidative Cross-Coupling (Chan–Evans–Lam Reaction) The coupling of phenols with aryl boronic acids has been accomplished (Scheme 20.37). Initially, Chan and coworkers studied an equivalent amount of $Cu(OAc)_2$ for the coupling reaction [58]. Subsequently, Evans and coworkers found the use of MS 4 Å led to improvement in the yield of the target product [106]. Later, Lam and coworkers made the process catalytic using oxygen as an oxidant [107]. Their mechanistic aspects have been recently demonstrated [108].

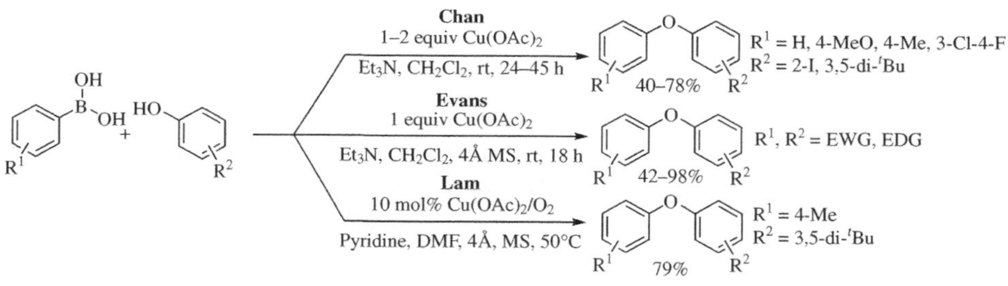

SCHEME 20.37 Chan–Evans–Lam reaction: oxidative cross-coupling.

20.3.2 Reactions with Aliphatic Alcohols

20.3.2.1 Palladium-Catalyzed Reactions The coupling of tertiary alcohols with electron-deficient aryl bromides has been reported using Pd(0)/DPPF in the presence of *tert*-butoxide [109]. Palladium-$^t Bu_3P$ **L87** is also effective and the coupling of aryl halides (electron-deficient and -rich) can be achieved [110]. Parrish and Buchwald reported the synthesis of aryl *tert*-butyl ethers using biphenyl-based phosphine ligands **L15**, **L17**, and **L97** [111].

The C—O cross-coupling of primary and secondary alcohols with aryl halides has been synthetically challenging due to β-hydride elimination (Scheme 20.38, see Scheme 20.2). This has been overcome using binaphthyl-based phosphines **L98** and **L90** [112]. Buchwald and coworkers have explored sterically hindered phosphines for the alkoxylation of aryl substrates with secondary alcohols, and ligands **L99–L100** with 1-phenyl and naphthyl backbone exhibited superior results (Scheme 20.39) [113].

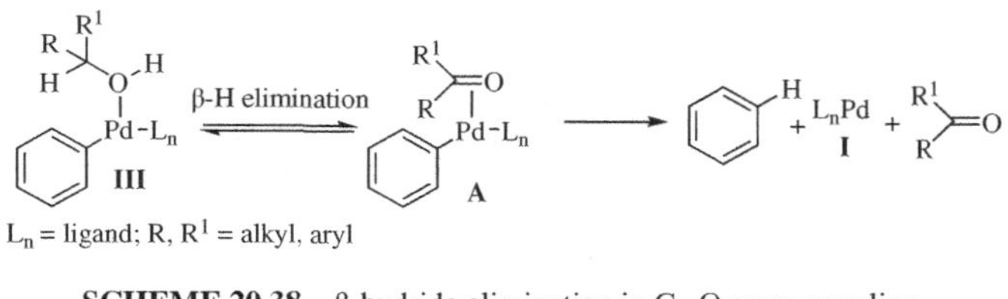

SCHEME 20.38 β-hydride elimination in C—O cross-coupling.

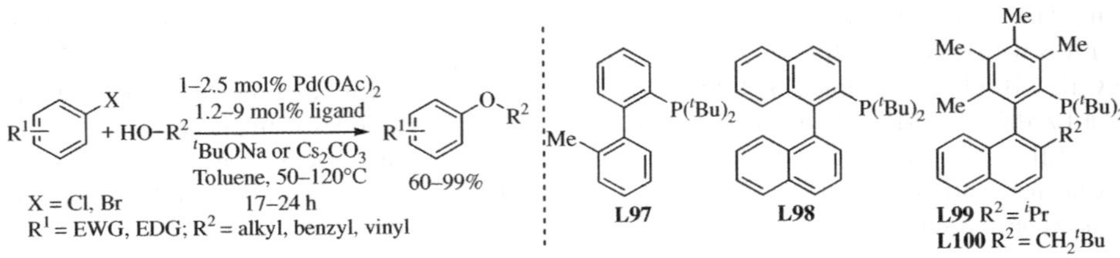

SCHEME 20.39 Cross-coupling of aryl halides with aliphatic alcohols.

Beller and coworkers reported the coupling of primary alcohols with aryl halides using $Pd(OAc)_2$ with a series of ligands **L2**, **L13**, **L21**, **L101–L110**, and **L105** (Scheme 20.40) [114].

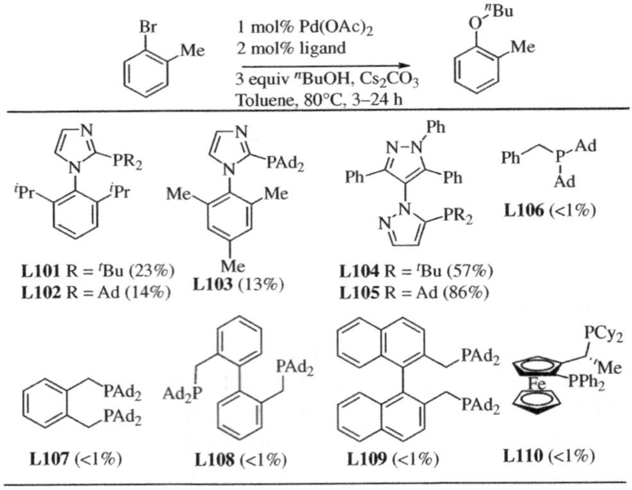

SCHEME 20.40 Ligand effect in Pd-catalyzed C—O coupling of 2-bromotoluene with ^{n}BuOH.

The use of RockPhos **L111** has been demonstrated for the coupling of substituted aryl halides with primary and secondary alcohols (Scheme 20.41) [115]. The reaction of aryl halide with smallest alcohol (MeOH) had long been considered as challenging both because of easy β-elimination, and a very small nucleophile reluctant to undergo reductive elimination, but it has been resolved by using **L105** [116].

Microwave-assisted coupling of aryl halides with vinyl trialkoxysilanes has been performed using $Pd(OAc)_2$-**L112** in the presence of NaOH [117]. The coupling of aryl chlorides with $K[B(OMe)_4]$ is reported using $Pd_2(dba)_3$-**L21** [118] (Scheme 20.42).

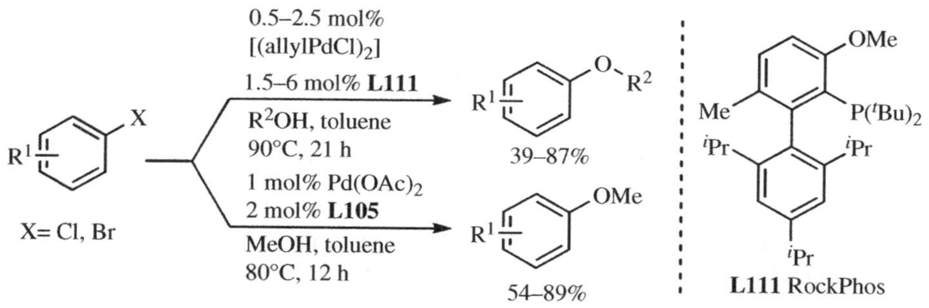

SCHEME 20.41 Coupling of primary and secondary alcohols with aryl bromides/chlorides.

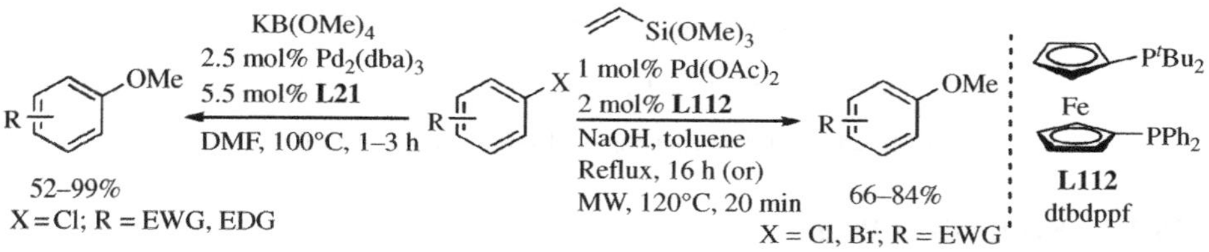

SCHEME 20.42 Methoxylation of aryl halides with alkoxysilanes or borate salts.

20.3.2.2 Copper-Catalyzed Reactions Copper-catalyzed C—O cross-coupling of alcohols with aryl halides made a remarkable progress. CuI–1,10-phenanthroline **L80** has been employed for the coupling of aryl iodides with aliphatic alcohols in the presence of Cs_2CO_3 [119]. The utilization of CuI–3,4,7,8-tetramethyl-1,10-phenanthroline (Me_4-phen) **L113** has been demonstrated for the coupling of aryl bromides with alcohols [120]. Air-stable copper(I)–bipyridyl complex **13** [121], Cu_8 clusters ($[Cu_8\{S_2P(OR)_2\}_6(\mu_8\text{-}Cl)]PF_6$) [122], CuI–*N,N*-dimethylglycine **L79** [123], CuI–1,1′-binaphthyl-2,2′-diamine (BINAM) **L114** [124], and CuI–8-hydroxyquinoline **L69** [125] have been subsequently used for the cross-coupling of aryl halides with aliphatic alcohols (Scheme 20.43). Interestingly, the OH group can be chemoselectively coupled with aryl halides without affecting NH_2 group using **L113** (Scheme 20.44) [50].

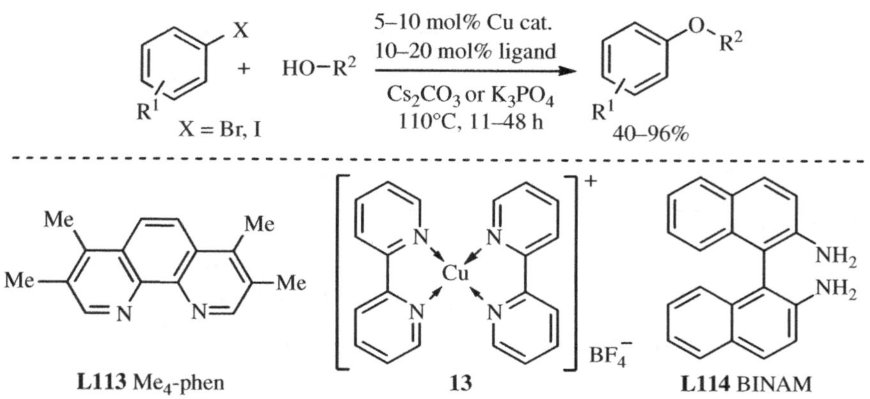

SCHEME 20.43 Cu-catalyzed synthesis of aryl alkyl ethers.

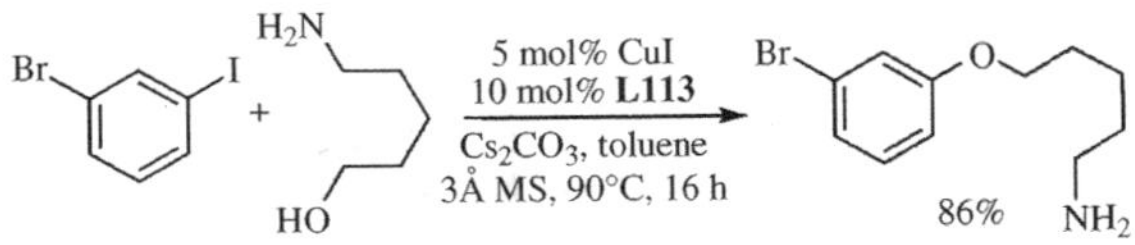

SCHEME 20.44 Cu(I)-catalyzed O-arylation of amino alcohol.

The coupling of phenols/aliphatic alcohols with potassium organotrifluoroborate could be accomplished using monohydrated $Cu(OAc)_2$-DMAP at room temperature (Scheme 20.45) [126].

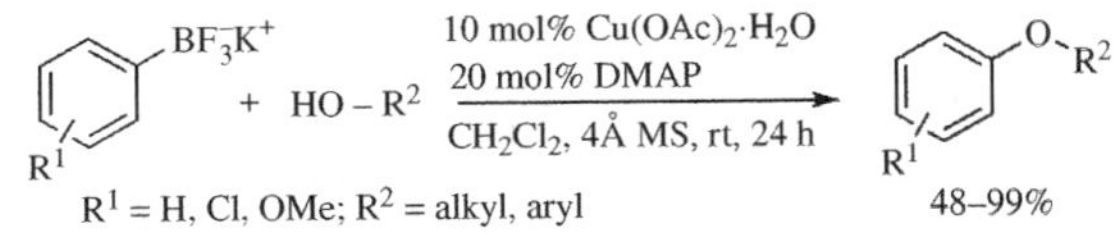

SCHEME 20.45 Cross-coupling of alcohols with organotrifluoroborates.

20.3.3 Synthesis of Phenols

20.3.3.1 Palladium-Catalyzed Reactions Few studies are focused on the synthesis of phenols via cross-coupling aryl halides and hydroxy nucleophiles. The cross-coupling of aryl halides with KOH has been accomplished using the bulky phosphine ligands **L21** and **L24** [127] and **L92** [128] at 100°C. The use of CsOH as a nucleophile led to effect this transformation at room temperature, in which intermediate **14** that could form during oxidation addition has been isolated (Scheme 20.46) [129]. Later, the use of K_3PO_4 with **L87** [130] and K_2CO_3 [131] using Herrmann's palladacycle **15–L21** has been reported for the hydroxylation of aryl halides. The hydroxylation of aryl chloride has been carried out using $Pd_2(dba)_3$-**L115** and precatalyst **16–L23** at room temperature [132].

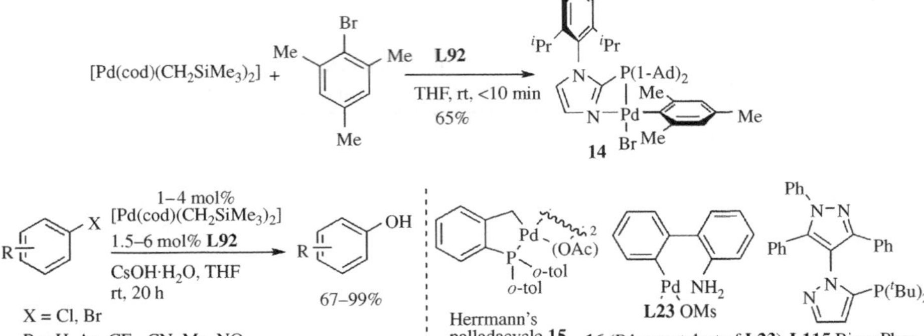

SCHEME 20.46 Pd-catalyzed synthesis of phenols at room temperature.

20.3.3.2 Copper-Catalyzed Reactions The use of CuI was explored in combination with several ligands for the coupling of aryl halides with hydroxide nucleophiles. In 2006, Leadbeater reported the coupling aryl halides with NaOH using CuI/L-proline in water under microwave irradiation (200–300°C) [133]. Subsequently, CuI–1,3-diketone **L62** [134], CuI–1,10-phenanthroline **L80** [135], CuI–8-hydroxyquinalidine [136], CuI–lithium pipecolinate [137], CuI–8-hydroxyquinoline **L69** [138], and CuI–8-hydroxyquinoline-*N*-oxide **L85** [139] have been screened for the coupling of aryl halides with KOH or CsOH. Later, Cu_2O-pyridine-2-aldoxime [140], $Cu(OAc)_2$-D-glucose [141], CuCl-NHC [142], and $Cu(OH)_2$-glycolic acid [143] have been examined for the hydroxylation of aryl bromides, iodides, and activated aryl chlorides. Effort has also been made on the use of heterogeneous catalytic systems employing CuI nanoparticles [144], CuO-silica [145], and copper-graphitic carbon nitride (Cu—g—C_3N_4) (Scheme 20.47) [146].

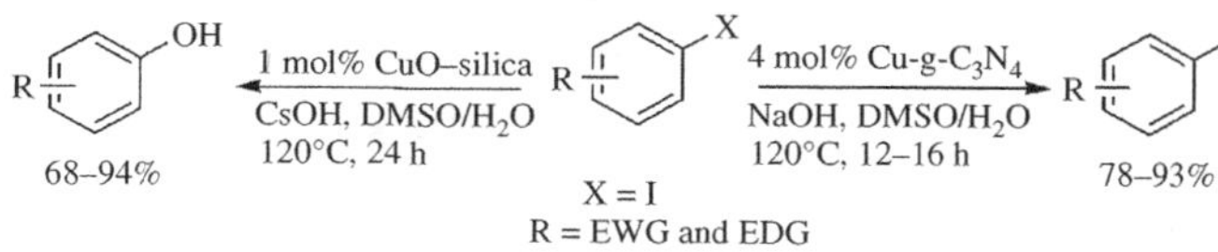

SCHEME 20.47 Cu-catalyzed hydroxylation of aryl bromides/iodides.

20.3.4 Synthetic Applications

The diaryl ether motif is abundant in a number of natural products and medicinally significant compounds. Evans and coworkers employed the C—O cross-coupling for the synthesis of tyrosine derivative **17** (Scheme 20.48) [58], which is an intermediate in Hems' synthesis of L-thyroxine [147].

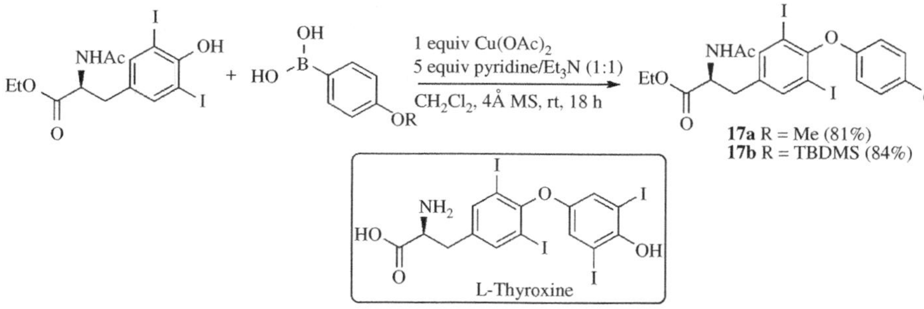

SCHEME 20.48 Synthesis of tyrosine derivative.

The convergent total synthesis of (*S*,*S*)-isodityrosine has been accomplished from natural α-amino acids, L-tyrosine, and L-phenylalanine (Scheme 20.49). In this synthesis, boronic acid is used as the aryl donor, and tyrosine derivatives are used as the phenolic partner [148].

In the synthesis of piperazinomycin, key ring closure step has been performed using the intramolecular O-arylation of phenol with arylboronic acid [149]. Reaction smoothly proceeded using $Cu(OAc)_2$ and Et_3N in presence of 4 Å MS at room temperature (Scheme 20.50).

The synthesis of 2-hydroxy-2′-methoxydiphenyl ether present in verbenachalcone has been described using the coupling of 2-benzyloxybromobenzene with 2-methoxyphenol (Scheme 20.51) [150].

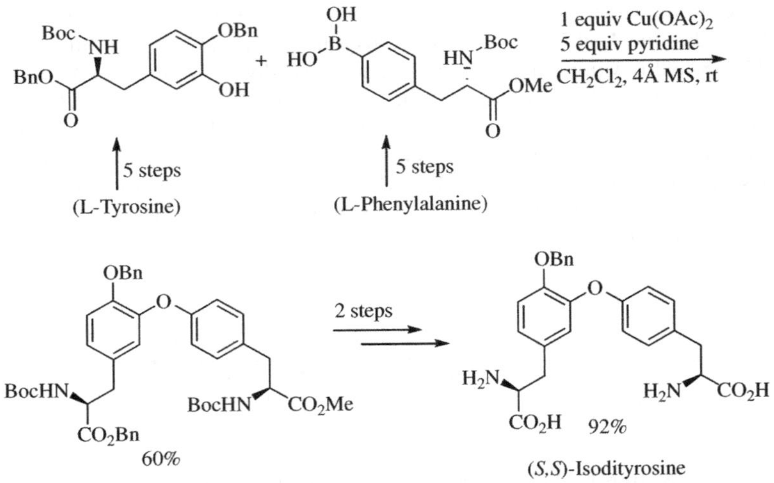

SCHEME 20.49 Synthesis of (*S*,*S*)-isodityrosine from natural amino acids.

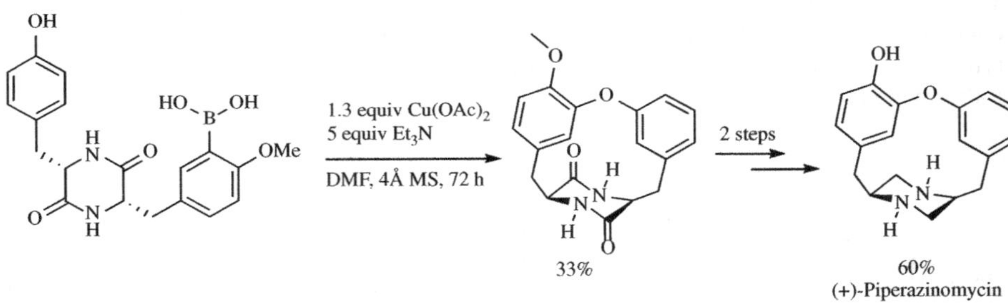

SCHEME 20.50 Synthesis of (+)-piperazinomycin.

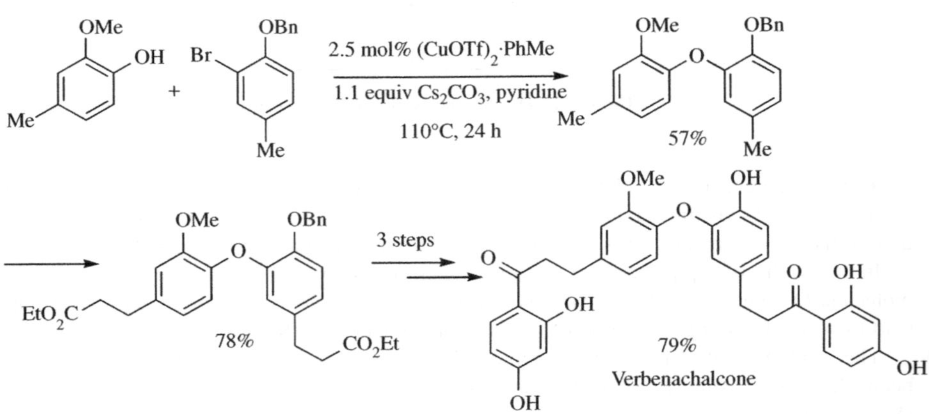

SCHEME 20.51 Synthesis of verbenachalcone.

20.4 C—S CROSS-COUPLING

20.4.1 Palladium-Catalyzed Reactions

Transition metal-catalyzed C—S cross-coupling reaction remains a challenging task in synthetic chemistry because of catalytic poisoning by sulfur [151]. In 1980, Migita and coworkers utilized $Pd(Ph_3)_4$ for the first palladium-catalyzed coupling of aryl halides with thiols [152]. Hartwig and coworkers studied the electronic, steric, and ligand hybridization effect of palladium-catalyzed C—S bond formation [153]. Ligands with large bite angles, electron-deficient carbon-bound ligands, and electron-rich thiolate ligands facilitate the reductive elimination. The relative rates of the reductive elimination with alkenyl and aryl substituents are found to be high compared to that of alkynyl and alkyl substituents. Phosphine oxide [154], bidentate phosphines **L8** [155], **L116** [156], and **L46** [157, 158] have been employed for the palladium-catalyzed coupling of aryl halides with aryl thiols (Scheme 20.52). For mechanism, see Scheme 20.2.

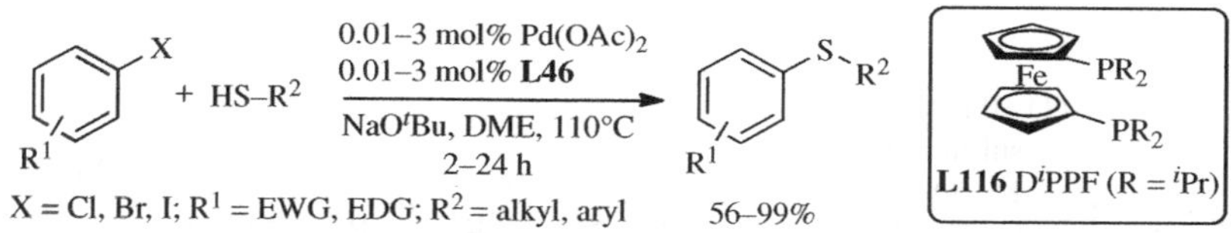

SCHEME 20.52 Pd-catalyzed aryl thioether synthesis.

Further invention of pyridine-enhanced precatalyst preparation stabilization and initiation (PEPPSI) ligands has led to a remarkable improvement in palladium-catalyzed C—S coupling [159, 160]. This catalyst effectively couples the less reactive aryl chlorides even at room temperature [161]. Recently, NHC-based catalyst, $[Pd(IPr^{*OMe})(cin)Cl]$ **18**, has been developed for the coupling of aryl halides with aliphatic or aromatic thiols (Scheme 20.53) [162].

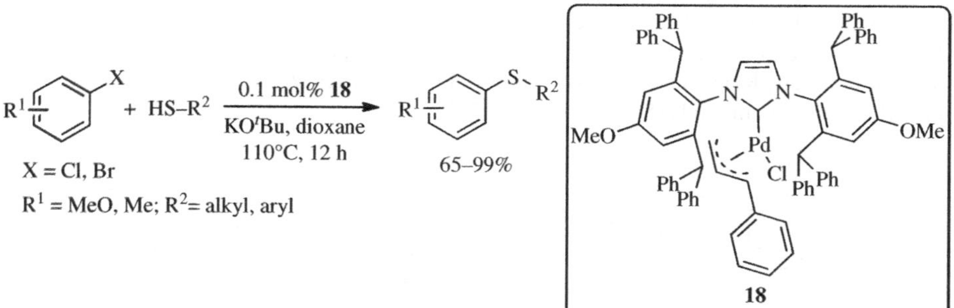

18
$[Pd(IPr^{*OMe})(cin)Cl]$

SCHEME 20.53 Pd-NHC catalyzed aryl thioether synthesis.

20.4.2 Copper-Catalyzed Reactions

The first copper-mediated C—S coupling of aryl boronic acids with alkyl thiols was reported in the presence of an organic base, pyridine (Scheme 20.54) [163]. Copper complexes, CuI–neocuproine [164], CuI–ethylene glycol [165], CuI–bipyridine/air [166], and $CuSO_4$–1,10-phenanthroline/O_2 [167], have been subsequently explored for the coupling of aryl halides and boronic acids with thiols. Oxidizable and foul-smelling nature of the arene thiols led to the use of alternative sulfur sources such as ethyl xanthogenate [168] and sulfur powder [169] for the coupling reactions (Scheme 20.55). In sulfur powder, diaryl disulfide intermediates are formed at first that is reduced to aryl thiol and further coupled with alkyl halides. The nature of base plays a crucial role in controlling the selectivity of the formation of diaryl disulfide and diaryl thioether [170]. Besides, Cu(I)–thiolate complex has been utilized for the synthesis of diaryl thioether by coupling of aryl iodide with disulfides [171].

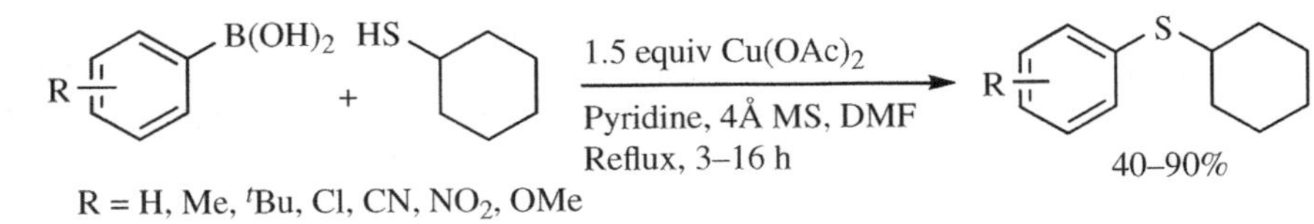

SCHEME 20.54 Cu-catalyzed oxidative C—S cross-coupling.

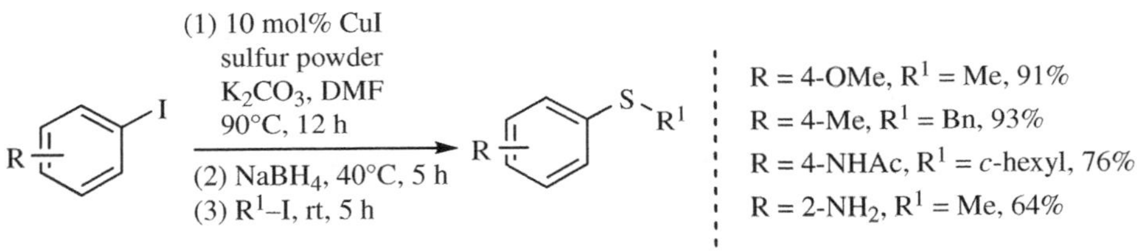

SCHEME 20.55 Cu-catalyzed coupling of aryl iodides with sulfur powder.

The use of nanoparticle for the C—S cross-coupling has been reported under ligand-free conditions. Punniyamurthy and coworkers introduced air-stable CuO nanoparticles for the coupling of thiols with aryl iodides [102, 172]. Karvembu [173] and Kakulapati [174] subsequently employed CuO nanoparticles for the coupling of aryl halides with thiophenol and ethyl potassium xanthogenate, respectively. Kamal and coworkers used CuO nanoparticles/graphene oxide for the coupling of aryl chlorides with thiols (Scheme 20.56) [175].

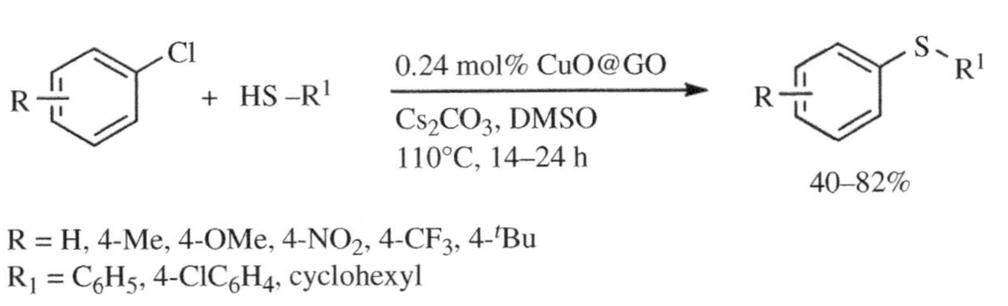

SCHEME 20.56 CuO nanoparticle graphene oxide-catalyzed aryl thioethers synthesis.

20.4.3 Other Transition Metal-Catalyzed Reactions

Few studies are focused on the use of CoI_2(dppe)/Zn [176], $NiCl_2{\cdot}6H_2O$/TBAB [177], [(NHC)-Ni(allyl)Cl] [178], and Ni/NHC [179] for the coupling of aryl halides with aryl and alkyl thiols.

20.5 C—Se CROSS-COUPLING

Diaryl selenides are important class of compounds in medicinal and pharmaceutical sciences [180]. Traditional synthesis of diaryl selenides employs cross-coupling of aryl halides with aryl selenols in the presence of transition metal catalysts. Suzuki first reported CuI-mediated diaryl selenide synthesis [181]. Catalytic methods have been subsequently developed using $(byp)_2NiBr_2$, $Pd(PPh_3)_4$, and $(PPh_3)Cu(phen)I$ [182–184]. The use of diselenide as a selenium source with transition metal catalysts, CuI–neocuproine [185], Cu_2O–byp/Mg [186], $(PPh_3)_2PdCl_2$ [187], and CuI [188], has been demonstrated for the synthesis of diaryl selenides. Domino C—Se coupling was reported using CuI–1,10-phenanthroline **L80** via intra- and intermolecular couplings (Scheme 20.57) [189]. The coupling of aryl halides with diselenides has been reported using CuO nanoparticle [190, 191], CuS/Fe [192], and $CuFe_2O_4$ nanoparticle [193] as recyclable catalysts.

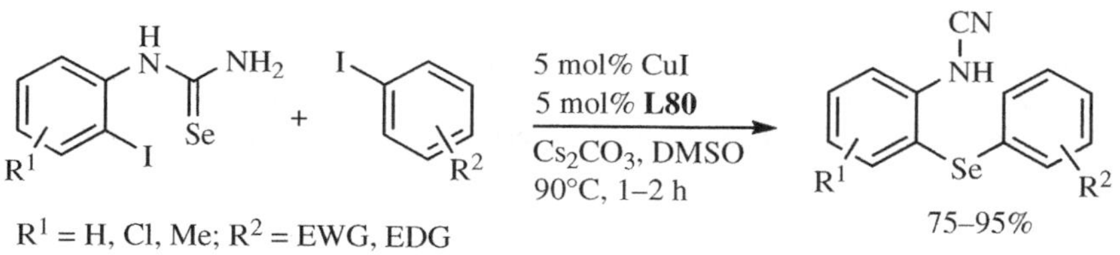

SCHEME 20.57 Cu-catalyzed domino C—Se cross-coupling.

20.6 C—Te CROSS-COUPLING

First, Suzuki and coworkers reported CuI-mediated synthesis of diaryl tellurides via C—Te cross-coupling of aryl iodides with benzene tellurolate ions [194]. Catalytic protocol has been subsequently developed using CuI/1,10-phenanthroline **L80** for the coupling of aryl iodides with organotellurolates [195] (Scheme 20.58). Further improvement was achieved in the coupling of ditellurides with boronic acid derivatives using copper-based catalytic systems, CuI–bpy [166], CuCl–bpy [196], and $CuSO_4{\cdot}5H_2O$/1,10-phenanthroline [197].

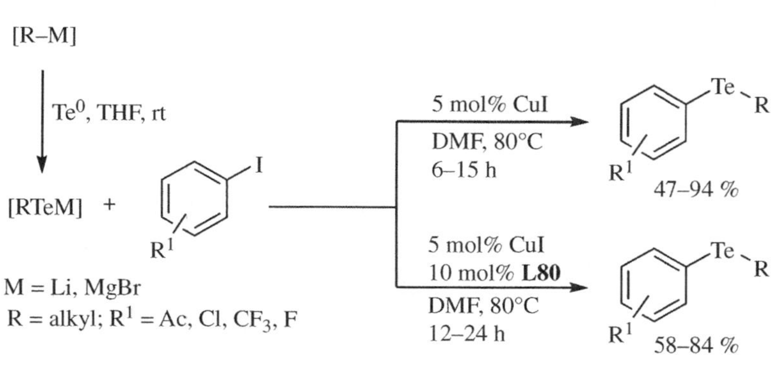

SCHEME 20.58 Cu-catalyzed synthesis of aryl alkyl tellurides.

20.7 C—P CROSS-COUPLING

Transition metal-catalyzed C—P cross-coupling affords powerful tool for the preparation of aryl phosphorus compounds [198]. The scope of the phosphorus-based nucleophiles used in cross-coupling with aryl electrophiles in the presence of transition metal (Pd-, Cu-, and Ni-based) catalysts is very broad. Figure 20.5 summarizes the common phosphorus nucleophiles used for cross-coupling reactions (for mechanism, see Scheme 20.2).

20.7.1 Palladium-Catalyzed Reactions

20.7.1.1 Reaction of* H*-Phosphonate Ester Hirao and coworkers reported the coupling of aryl halides with *H*-phosphonate ester using $Pd(PPh_3)_4$ in the presence of Et_3N [199] (Scheme 20.59). The scope of the protocol has been extended to aryl triflates [200] and several phosphine ligands (dppp, dppb, $[3\text{-}(SO_3M)C_6H_4]PPh_2$, and Xantphos) [201], and microwave conditions have been developed for the mild coupling processes [202]. Mechanistic studies suggest that the rate of ligand substitution depends on the nature of the anionic additives [203, 204]. $Pd(OAc)_2$ has been found to be effective for the coupling of aryl imidazolylsulfonate with *H*-phosphonate ester [205].

Montchamp and coworkers reported the coupling of electron-deficient aryl chlorides using $Pd(OAc)_2$-dppf [206]. The reaction of inactive electron-rich aryl chlorides has been accomplished

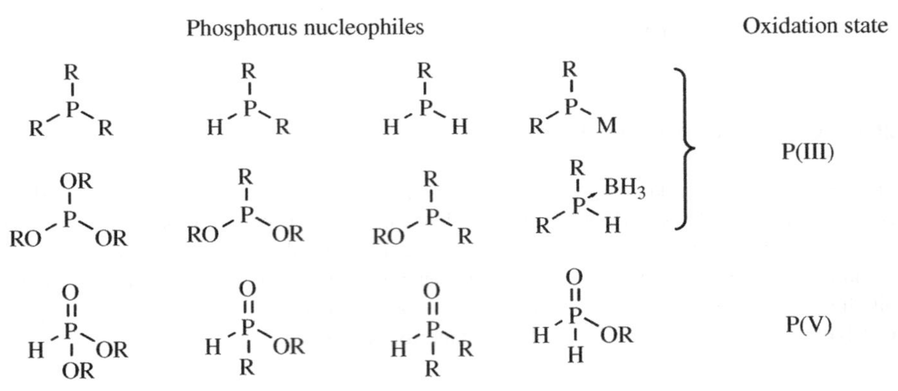

FIGURE 20.5 Commonly used phosphorus nucleophiles for cross-coupling reactions.

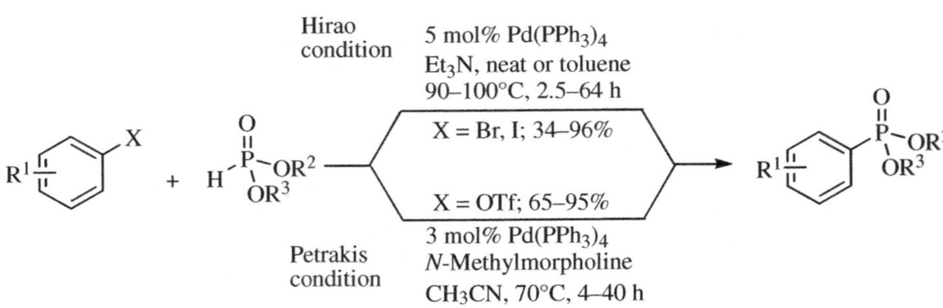

SCHEME 20.59 Synthesis of aryl phosphonates.

using palladacycle **19** associated with *X*-Phos (Scheme 20.60) [207, 208]. Using $PdCl_2$/bipyridine, the coupling of triarylbismuth with *H*-phosphonate ester has been reported [209].

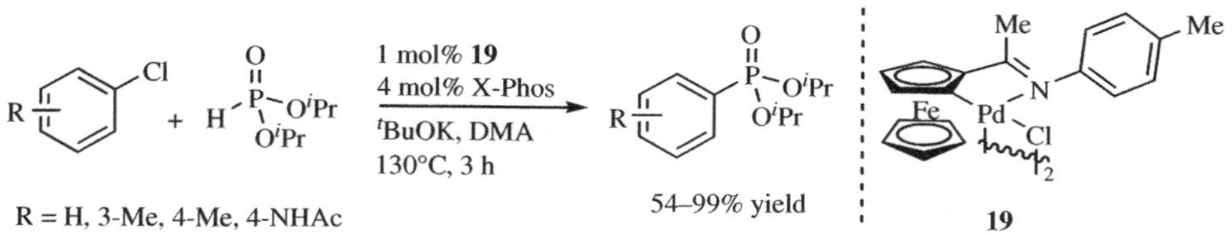

SCHEME 20.60 Reaction of aryl chlorides with diisopropyl *H*-phosphonate.

Larhed and coworkers reported the coupling of arylboronic acids and trifluoroborates with diethyl phosphites in the presence of $Pd(OAc)_2$-2,9-dimethyl-1,10-phenanthroline (dmphen) under microwave irradiation, which uses *p*-benzoquinone as an oxidant [210] (Scheme 20.61).

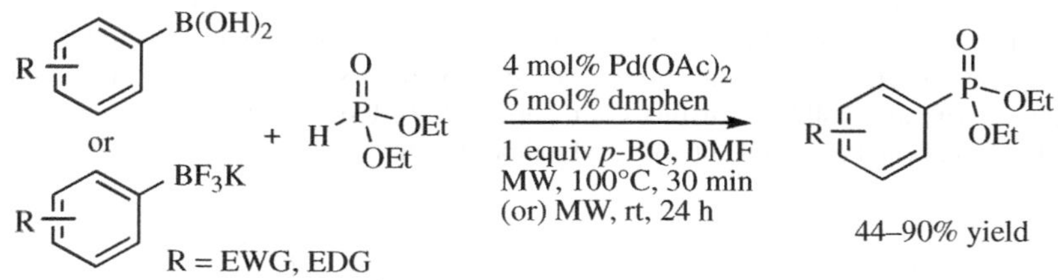

SCHEME 20.61 Pd(II)-catalyzed synthesis of aryl phosphonate diesters.

20.7.1.2 Reaction of* H*-Phosphinate Ester The coupling of *H*-phosphinate ester with aryl bromides or iodides has been explored using $Pd(PPh_3)_4$ in the presence of Et_3N [211] (Scheme 20.62). The coupling of optically active phosphonites can be accomplished with aryl bromides in retention configuration [212]. The utilization of $Pd(OAc)_2$-Xantphos has been demonstrated for the coupling of *H*-phosphinate ester with aryl chlorides [213]. In this reaction, the additive helps to tautomerize the phosphonites into less nucleophilic P(V) to the more reactive P(III).

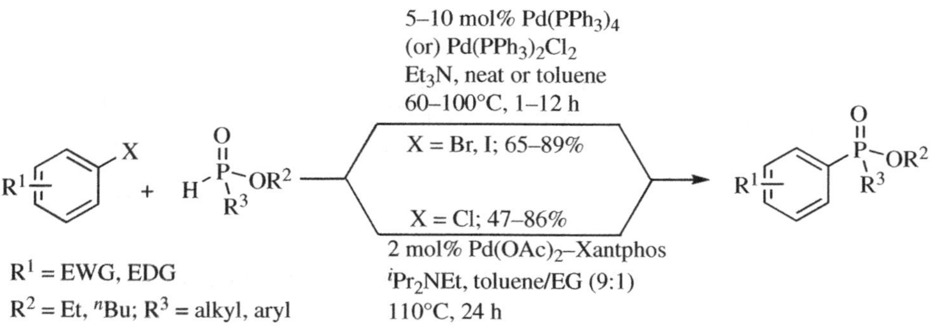

SCHEME 20.62 Reaction of aryl halides with H-phosphinate ester.

20.7.1.3 Reactions of Secondary Phosphine Oxides The coupling of secondary phosphine oxides was reported with aryl bromides using $Pd(PPh_3)_4$ in the presence of Et_3N [214]. Under these conditions, electron-donating aryl halides yielded poor results. A similar system is reported for the coupling of electron-rich aryl halides using $Pd_2(dba_3)$-(*S*)-BINAP [215] (Scheme 20.63). The use of water as solvent has been demonstrated under microwave irradiation in the presence of Pd/C [216]. The oxidative coupling of aryl boronic acids has been performed with phosphine oxide using $Pd(OAc)_2$-dppb in air [217].

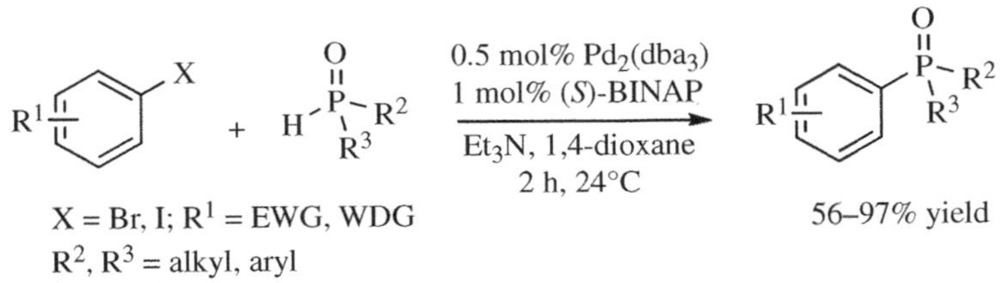

SCHEME 20.63 Room temperature P-arylation of secondary phosphine oxides.

20.7.1.4 Reactions of Phosphinic Acids Mono- and diarylation of hypophosphites could be performed with aryl iodides in the presence of $Pd(PPh_3)_2Cl_2$ (Scheme 20.64) [218]. The product selectivity can be controlled by varying the ratio of hypophosphites with aryl iodides. Propylene oxide is used as HI scavenger to inhibit the ester hydrolysis. The coupling of anilinium hypophosphites with aryl halides or triflates is reported [219]. Using $Pd(OAc)_2$-dppp, the coupling of phosphinic acid can be carried out with aryl iodides [220].

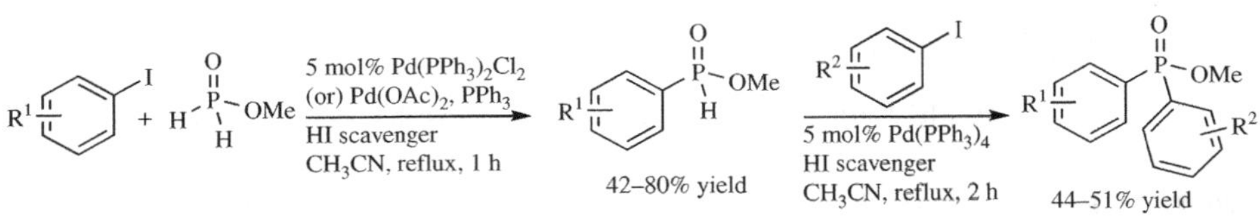

SCHEME 20.64 Mono- and diarylation of methyl phosphinate with aryl iodides.

The coupling of trimethylphosphite with *p*-halo-toluenetricarbonylchromium can be accomplished using $Pd(dba)_2$ and $PdCl_2$ (Scheme 20.65) [221].

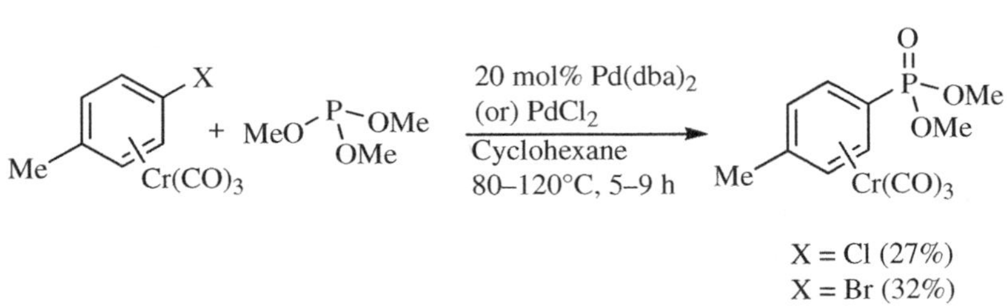

SCHEME 20.65 Pd-catalyzed arylation of trimethylphosphite.

20.7.1.5 Reactions of Phosphines Palladium-catalyzed coupling of triarylphosphine can be accomplished with aryl bromides and triflates [222] (Scheme 20.66). A solvent-free approach has been demonstrated with improved yield [223]. The use of excess of triarylphosphine is crucial for second oxidative addition. The substrate scope can be extended to aryl chlorides employing Pd/C in the presence of NaI [224]. Scheme 20.66 shows the catalytic cycle. The oxidative addition of Pd(0) **I** with aryl halides/triflates can give the intermediate Pd(II) **II** that can lead to the formation

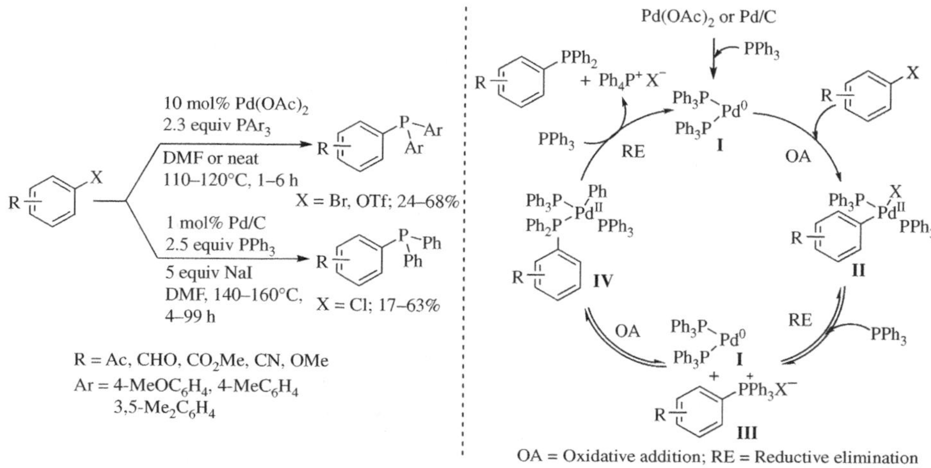

SCHEME 20.66 Phosphination of aryl halides/triflates using triarylphosphines.

of phosphonium salt **III** via the reductive elimination. The oxidative addition of **III** with Pd(0)-complex **I** can generate the Pd-complex **IV** that can undergo reductive elimination to give the target product and complete the catalytic cycle.

Limited studies are focused on the coupling of primary and secondary phosphines with aryl electrophiles due to palladium-catalyzed aerobic oxidation of R_2PH to $R_2P(O)H$ [225]. Palladium-catalyzed cross-coupling of primary or secondary phosphines has been reported with aryl iodides (Scheme 20.67) [226, 227]. Buchwald and coworkers utilized $Pd(OAc)_2$-D^{i}PPF for the coupling of secondary aryl phosphines with aryl halides [256]. This system is also effective for the reaction of phosphines with aryl chlorides.

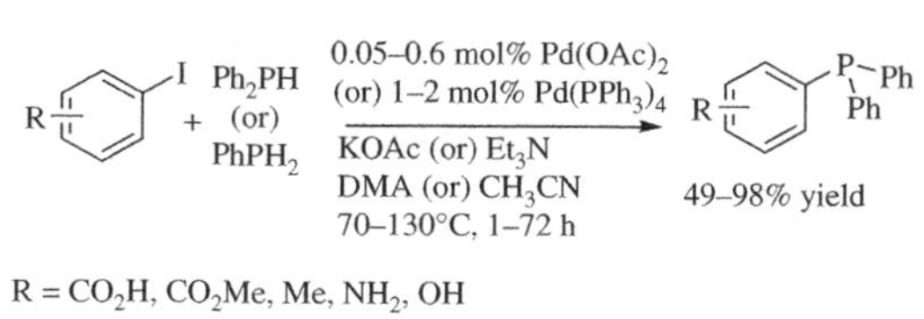

SCHEME 20.67 Cross-coupling of aryl iodides with primary or secondary phosphines.

Imamoto and coworkers described phosphine–borane complexes as a coupling partner for the synthesis of aryl phosphines with aryl electrophiles using $Pd(PPh_3)_4$ at room temperature (Scheme 20.68) [228, 229]. The borane moiety can be easily removed by excess use of diethyl amine or morpholine. Gaumont and coworkers demonstrated the palladium-catalyzed C—P cross-coupling in imidazolium-based ionic liquid and that the catalyst can be recycled up to six cycles [230].

Tinney and Stille reported the first C—P bond formation using silyl and stannyl phosphines with aryl iodides (Scheme 20.69) [231]. Under these conditions, aryl chlorides and aryl bromides are ineffective. The coupling of *tert*-butyl(trimethylsilyl)phosphine has been subsequently accomplished with aryl halides [232]. Rossi and coworkers reported the coupling of (trialkylstannyl)diphenylphosphines with aryl iodides [233]. The reaction conditions can be extended for the coupling of aryl triflates using CuI as cocatalyst [234].

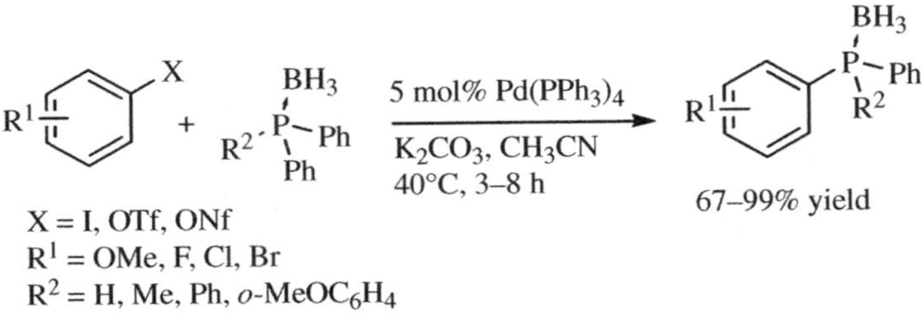

SCHEME 20.68 Pd(0)-catalyzed synthesis of phosphine–boranes.

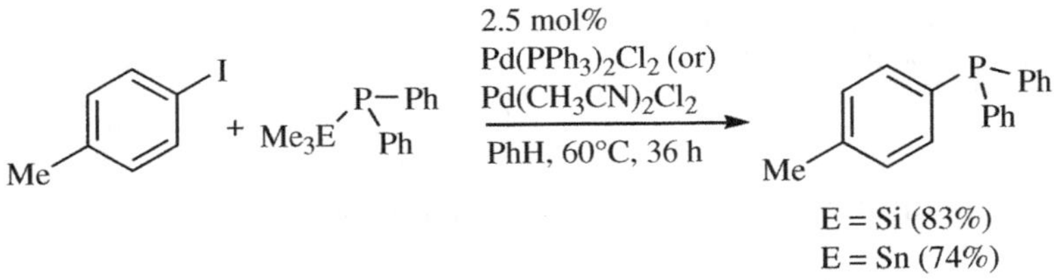

SCHEME 20.69 Phosphination of aryl iodides with silyl/stannyl phosphines.

20.7.2 Copper-Catalyzed Reactions

Copper-catalyzed C—P cross-coupling is less explored. In 1983, Osuka and coworkers reported the arylation of phosphite using stoichiometric amount of CuI [235]. The catalytic version of the reaction has been reported using inorganic bases, K_2CO_3, K_3PO_4, and Cs_2CO_3 [236]. Buchwald and coworkers utilized CuI/DMEDA for the coupling of the secondary phosphines and phosphites with aryl halides (Scheme 20.70) [237]. The coupling of varies phosphine oxides can be accomplished with aryl bromides using CuI and ligands such as pyrrolidine-2-phosphonic acid phenyl monoester (PPAPM) [96], L-proline or picolinic acid in the presence of Cs_2CO_3 or DMAP [238]. The coupling of secondary phosphine oxides with aryl bromides can be accomplished in the presence of CuI–(*S*)-α-phenylethylamine [239].

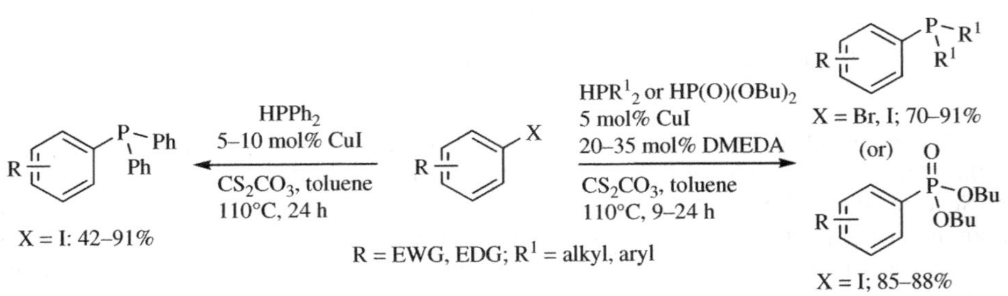

SCHEME 20.70 Cu(I)-catalyzed C—P cross-coupling.

The coupling of phosphine oxides with aryl boronic acids [240] and diaryliodonium salts [241] has been reported at room temperature (Scheme 20.71).

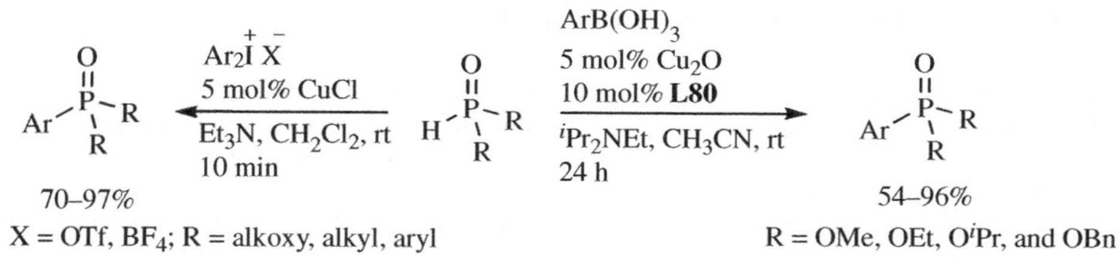

SCHEME 20.71 Coupling of arylboronic acid/iodonium salts with phosphorus nucleophiles.

20.7.3 Nickel-Catalyzed Reactions

In 1970, Tavs reported the first nickel-catalyzed arylation of trialkyl phosphites or dialkyl aryl phosphonites with aryl bromides [242] (Scheme 20.72). In this reaction, trialkyl phosphite, a phosphorylating agent, plays a crucial role in the reduction of Ni(II) to active Ni(0) complex [243].

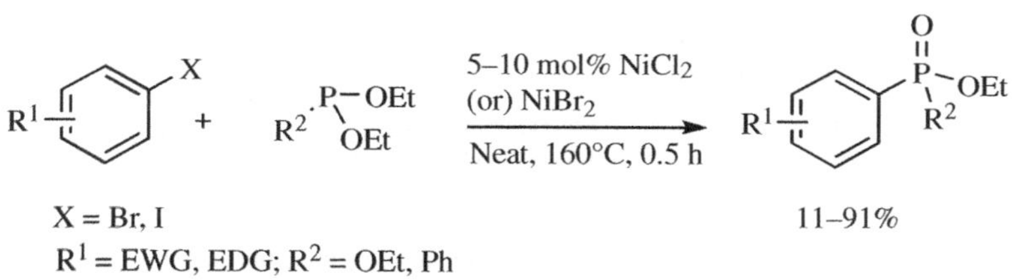

SCHEME 20.72 The cross-coupling of trialkyl phosphite with aryl halides.

Using $NiCl_2{\cdot}6H_2O$/bipyridine, the coupling of $Ph_2P(O)H$ has been carried out with aryl halides in the presence of Zn in water [244] (Scheme 20.73). The coupling of aryl chlorides and sulfonates can be carried out with $Ph_2P(O)H$ in the presence of $NiCl_2$-1,2-dimethoxyethane (DME) or dppp [245]. The coupling of phenols could be accomplished using $NiCl_2$/dppp, in which phenol C—O bond is activated by bromo-tris-pyrrolidino-phosphonium hexafluorophosphate [246]. The coupling of aryl nitriles can be performed with $TMSPPh_2$ via C—CN bond cleavage using $[NiCl_2(PPh_3)_2]$ in the presence of ^{t}BuOK [247]. Gao and coworkers reported the coupling of aryl boronic acid with phosphine oxides using $NiBr_2$-pyridine [248].

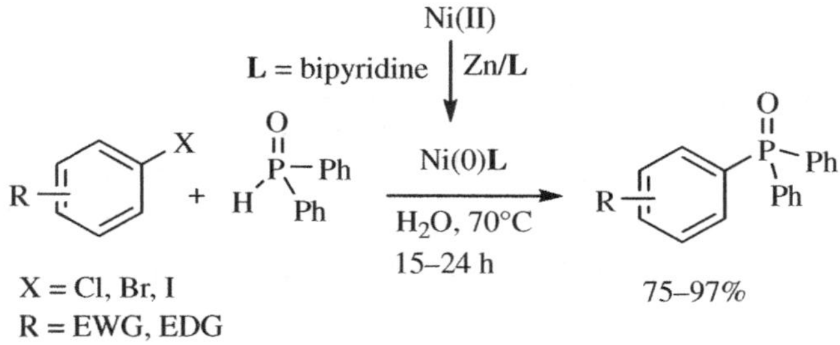

SCHEME 20.73 Reductive coupling of aryl halides with diphenylphosphine oxide.

20.8 C—As AND C—Sb CROSS-COUPLING

Organoarsanes and organostibanes are an important class of compounds in organic synthesis as intermediates as well as ligands in transition metal-catalyzed reactions [249]. Palladium-catalyzed cross-coupling of triphenyl-arsines and stibines with aryl iodides has been reported. The reaction of Ph_2M^- anion with Bu_3SnCl affords the stannanes Bu_3SnMPh_2. The palladium-catalyzed coupling has been carried out with this stannanes and *para*-substituted aryl iodides (Scheme 20.74) [250]. The scope of the reaction can be extended to *ortho*-substituted aryl iodides [251] and aryl triflates [234] using additive such as LiCl, CuI, and CsF. The synthesis of biphenyl-based arsine **20** [252] and chiral bis(arsine) **21** has been accomplished using C—As cross-coupling [253].

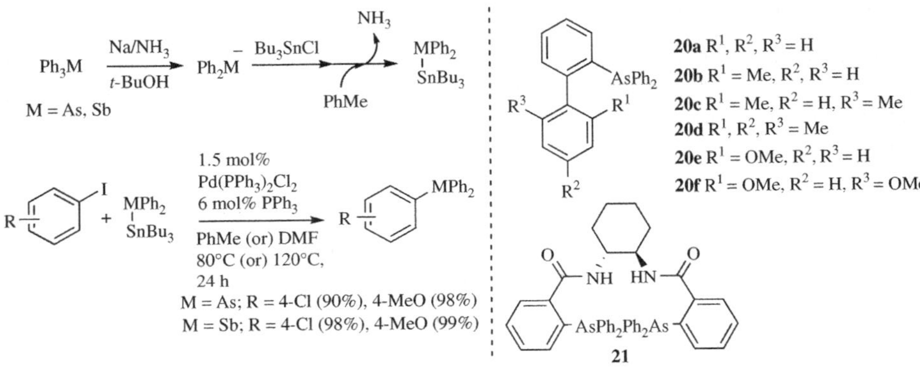

SCHEME 20.74 Pd-catalyzed arsination or stibanation.

20.9 C—B CROSS-COUPLING

Arylboronic acids are important class of reagents [254] that are extensively employed as a coupling partner in C—C and C–heteroatom bond formation [255]. Miyaura and coworkers first reported $PdCl_2$(dppf) for coupling of bis(pinacolato)diborane (B_2pin_2) with aryl halides [256] and aryl triflates [257] (Scheme 20.75). The same catalytic system has been applied for borylation of aryldiazonium tetrafluoroborate salts [258]. Dialkoxy borane (HBpin) can be employed as a boron nucleophile for coupling of aryl halides [259] and triflates [260]. The introduction of PCy_3 [261], NHC [262], DPEphos [263], and biarylmonophoshines [264] has led to effect the palladium catalysis for the coupling of boranes with aryl halides and diazonium salts. Diisopropylaminoborane can be used as a borylating reagent for palladium-catalyzed borylation of aryl chloride [265] (for mechanism, see Scheme 20.2).

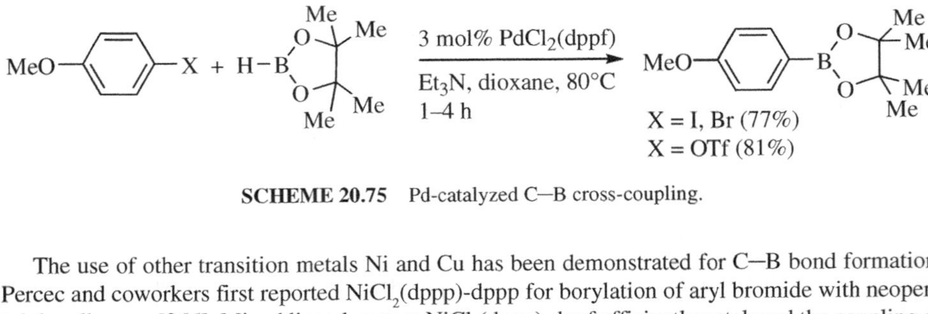

SCHEME 20.75 Pd-catalyzed C—B cross-coupling.

The use of other transition metals Ni and Cu has been demonstrated for C—B bond formation. Percec and coworkers first reported $NiCl_2$(dppp)-dppp for borylation of aryl bromide with neopentylglycolborane [266]. Mixed ligand system $NiCl_2$(dppp)-dppf efficiently catalyzed the coupling of less reactive [267] and *ortho*-substituted aryl halides [268, 269]. Trialkyl phosphorous-based ligands, PMe_3 and PCy_3, have been subsequently reported for coupling of aryl halides [270] and carbamates [271] with diboranes. $NiCl_2$(dppp)-PPh_3 has been employed for borylation of aryl halides, mesylates, triflates, and sulfamates with bis-boronic acid at room temperature [272]. CuI has been used for the coupling of HBpin [273] and B_2pin_2 [274] with aryl halides. Using Fe/Cu cooperative catalytic system, the borylation aryl bromides has been reported [275].

20.10 SUMMARY AND OUTLOOK

Transition metal-catalyzed cross-coupling methodology has become a powerful tool for carbon–heteroatom bond formation, and its application in industry and academic research is limitless. The design, synthesis, and development of bulky and electron-rich ancillary ligands, mostly phosphines, as well as the optimization of the reaction conditions have made palladium and copper catalysts a premier choice due to atom economy. Further progress in the past two decades has cracked out the challenges in "contemporary methods" including shorter lifetime of the catalyst, accessing less reactive halides and sterically hindered substrates. The discovery of alternate metal source in coupling reactions has not been completely disclosed yet. The search for a new catalyst and more convenient process to access wide variety of substrates is a future goal of research in this area.

ABBREVIATIONS

BINAM	1,1′-Binaphthyl-2,2′-diamine
DME	1,2-Dimethoxyethane
dmphen	2,9-Dimethyl-1,10-phenanthroline
Me_4-phen	3,4,7,8-Tetramethyl-1,10-phenanthroline
PEPPSI	Pyridine-enhanced precatalyst preparation stabilization and initiation
PPAPM	Pyrrolidine-2-phosphonic acid phenyl monoester
SET	Single-electron transfer
TMHD	2,2,6,6-Tetramethylheptane-3,5-dione

REFERENCES

[1] (a) Evano, G., Blanchard, N. eds. (2013) *Copper-Mediated Cross-Coupling Reactions*, John Wiley & Sons, Inc., Hoboken; (b) Ribas, X. ed. (2013) in *C-H and C-X Bond Functionalization: Transition Metal Mediation, RSC Catalysis Series 11*, RSC Publishing, Cambridge.

[2] Corbet, J.-P., Mignani, G. (2006) *Chem. Rev.*, **106**, 2651–2710.

[3] (a) Beletskaya, I. P., Cheprakov, A. V. (2012) *Organometallics*, **31**, 7753–7808; (b) Casitas, A., Ribas, X. (2013) *Chem. Sci.*, **4**, 2301–2318.

[4] (a) Liu, L., Wu, H.-C., Yu, J.-Q. (2011) *Chem. Eur. J.*, **17**, 10828–10831; (b) Ager, D. J., Laneman, S. A. (1997) *Chem. Commun.*, 2359–2360.

[5] (a) Rucker, R. P., Whittaker, A. M., Dang, H., Lalic, G. (2012) *Angew. Chem. Int. Ed.*, **51**, 3953–3956; (b) He, C., Chen, C., Cheng, J., Liu, C., Liu, W., Li, Q., Lei, A. (2008) *Angew. Chem. Int. Ed.*, **47**, 6414–6417.

[6] Kosugi, M., Kameyama, M., Migita, T. (1983) *Chem. Lett.*, 927–928.

[7] Wolfe, J. P., Wagaw, S., Buchwald, S. L. (1996) *J. Am. Chem. Soc.*, **118**, 7215–7216.

[8] Driver, M. S., Hartwig, J. F. (1996) *J. Am. Chem. Soc.*, **118**, 7217–7218.

[9] Hamann, B. C., Hartwig, J. F. (1998) *J. Am. Chem. Soc.*, **120**, 3694–3703.

[10] Ehrentraut, A., Zapf, A., Beller, M. (2002) *J. Mol. Catal. A Chem.*, **182–183**, 515–523.

[11] Alcazar-Roman, L. M., Hartwig, J. F. (2001) *J. Am. Chem. Soc.*, **123**, 12905–12906.

[12] Wolfe, J. P., Tomori, H., Sadighi, J. P., Yin, J., Buchwald, S. L. (2000) *J. Org. Chem.*, **65**, 1158–1174.

[13] Strieter, E. R., Blackmond, D. G., Buchwald, S. L. (2003) *J. Am. Chem. Soc.*, **125**, 13978–13980.

[14] Barder, T. E., Buchwald, S. L. (2007) *J. Am. Chem. Soc.*, **129**, 5096–5101.

[15] Barder, T. E., Buchwald, S. L. (2007) *J. Am. Chem. Soc.*, **129**, 12003–12010.

[16] Surry, D. S., Buchwald, S. L. (2008) *Angew. Chem. Int. Ed.*, **47**, 6338–6361.

[17] Fors, B. P., Watson, D. A., Biscoe, M. R., Buchwald, S. L. (2008) *J. Am. Chem. Soc.*, **130**, 13552–13554.

[18] Biscoe, M. R., Fors, B. P., Buchwald S. L. (2008) *J. Am. Chem. Soc.*, **130**, 6686–6687.

[19] Fors, B. P., Davis, N. R., Buchwald, S. L. (2009) *J. Am. Chem. Soc.*, **131**, 5766–5768.

[20] Surry, D. S., Buchwald, S. L. (2011). *Chem. Sci.*, **2**, 27–50.

[21] Maiti, D., Fors, B. P., Henderson, J. L., Nakamura Y., Buchwald S. L. (2011) *Chem. Sci.*, **2**, 57–68.

[22] Urgaonkar, S., Nagarajan, M., Verkade, J. G. (2003) *Org. Lett.*, **5**, 815–818.

[23] Urgaonkar, S., Xu, J.-H., Verkade, J. G. (2003) *J. Org. Chem.*, **68**, 8416–8423.

[24] Urgaonkar, S., Nagarajan, M., Verkade, J. G. (2003) *J. Org. Chem.*, **68**, 452–459.

[25] Urgaonkar, S., Verkade, J. G. (2004) *J. Org. Chem.*, **69**, 9135–9142.

[26] Rataboul, F., Zapf, A., Jackstell, R., Harkal, S., Riermeier, T., Monsees, A., Dingerdissen, U., Beller, M. (2004) *Chem. Eur. J.*, **10**, 2983–2990.

[27] Marion, N., Navarro, O., Mei, J., Stevens, E. D., Scott, N. M., Nolan, S. P. (2006) *J. Am. Chem. Soc.*, **128**, 4101–4111.

[28] Shen, Q., Ogata, T., Hartwig, J. F. (2008) *J. Am. Chem. Soc.*, **130**, 6586–6596.

[29] So, C. M., Zhou, Z., Lau, C. P., Kwong, F. Y. (2008) *Angew. Chem. Int. Ed.*, **47**, 6402–6406.

[30] Lundgren, R. J., Peters, B. D., Alsabeh, P. G., Stradiotto, M. (2010) *Angew. Chem. Int. Ed.*, **49**, 4071–4074.

[31] Fors, B. P., Buchwald, S. L. (2009) *J. Am. Chem. Soc.*, **131**, 12898–12899.

[32] Lundgren, R. J., Stradiotto, M. (2010) *Angew. Chem. Int. Ed.*, **49**, 8686–8690.

[33] (a) Ley, S. V., Thomas, A. W. (2003) *Angew. Chem. Int. Ed.*, **42**, 5400–5449; (b) Evano, G., Blanchard, N., Toumi, M. (2008) *Chem. Rev.*, **108**, 3054–3131; (c) Evano, G., Toumi, M., Coste, A. (2009) *Chem. Commun.*, 4166–4175.

[34] Paine, A. P. (1987) *J. Am. Chem. Soc.*, **109**, 1496–1502.

[35] Gujadhur, R., Venkataraman, D., Kintigh, J. T. (2001) *Tetrahedron Lett.*, **42**, 4791–4793.

[36] Gujadhur, R. K., Bates, C. G., Venkataraman, D. (2001) *Org. Lett.*, **3**, 4315–4317.

[37] Patil, N. M., Kelkar, A. A., Nabi, Z., Chaudhari, R. V. (2003) *Chem. Commun.*, 2460–2461.

[38] Mei, X., August, A. T., Wolf, C. (2006) *J. Org. Chem.*, **71**, 143–149.

[39] Wolf, C., Liu, S., Mei, X., August, A. T., Casimir, M. D. (2006) *J. Org. Chem.*, **71**, 3270–3273.

[40] Jadhav, V. H., Dumbre, D. K., Phapale, V. B., Borate, H. B., Wakharkar, R. D. (2007) *Catal. Commun.*, **8**, 65–68.

[41] Kantam, M. L., Venkanna, G. T., Sridhar, C., Sreedhar, B., Choudary, B. M. (2006) *J. Org. Chem.*, **71**, 9522–9524.

[42] Lu, Z., Twieg, R. J., Huang, S. D. (2003) *Tetrahedron Lett.*, **44**, 6289–6292.

[43] Ma, D., Cai, Q., Zhang, H. (2003) *Org. Lett.*, **5**, 2453–2455.

[44] Kwong, F. Y., Buchwald, S. L. (2003) *Org. Lett.*, **5**, 793–796.

[45] Shafir, A., Buchwald. S. L. (2006) *J. Am. Chem. Soc.*, **128**, 8742–8743.

[46] Wang, H., Li, Y., Sun, F., Feng, Y., Jin, K., Wang, X. (2008) *J. Org. Chem.*, **73**, 8639–8642.

[47] Wang, D., Ding, K. (2009) *Chem. Commun.*, 1891–1893.

[48] (a) Jiao, J., Zhang, X.-R., Chang, N.-H., Wang, J., Wei, J.-F., Shi, X.-Y., Chen, Z.-G. (2011) *J. Org. Chem.*, **76**, 1180–1183; (b) Güell, I., Ribas, X. (2014) *Eur. J. Org. Chem.*, 3188–3195.

[49] Yang, C.-T., Fu, Y., Huang, Y.-B., Yi, J., Guo, Q.-X., Liu, L. (2009) *Angew. Chem. Int. Ed.*, **48**, 7398–7401.

[50] Shafir, A., Lichtor, P. A., Buchwald, S. L. (2007) *J. Am. Chem. Soc.*, **129**, 3490–3491.

[51] Jones, G. O., Liu, P., Houk, K. N., Buchwald, S. L. (2010) *J. Am. Chem. Soc.*, **132**, 6205–6213.

[52] Zhang, Y., Yang, X., Yao, Q., Ma, D. (2012) *Org. Lett.*, **14**, 3056–3059.

[53] Yang, K., Qiu, Y., Li, Z., Wang, Z., Jiang, S. (2011) *J. Org. Chem.*, **76**, 3151–3159.

[54] Kim, J., Chang, S. (2008) *Chem. Commun.*, 3052–3054.

[55] Xia, N., Taillefer, M. (2009) *Angew. Chem. Int. Ed.*, **48**, 337–339.

[56] Berman, A. M., Johnson, J. S. (2006) *J. Org. Chem.*, **71**, 219–224.

[57] Chan, D. M. T. (1996) *Tetrahedron Lett.*, **37**, 9013–9016.

[58] Chan, D. M. T., Monaco, K. L., Wang, R.-P., Winters, M. P. (1998) *Tetrahedron Lett.*, **39**, 2933–2936.

[59] Antilla, J. C., Buchwald, S. L. (2001) *Org. Lett.*, **3**, 2077–2079.

[60] Quach, T. D., Batey, R. A. (2003) *Org. Lett.*, **5**, 4397–4400.

[61] Sueki, S., Kuninobu, Y. (2013) *Org. Lett.*, **15**, 1544–1547.

[62] Rao, H., Fu, H., Jiang, Y., Zhao, Y. (2009) *Angew. Chem. Int. Ed.*, **48**, 1114–1116.

[63] Yamamoto, Y., Takizawa, M., Yu, X.-Q., Miyaura, N. (2008) *Angew. Chem. Int. Ed.*, **47**, 928–931.

[64] González, I., Mosquera, J., Guerrero, C., Rodríguez, R., Cruces, J. (2009) *Org. Lett.*, **11**, 1677–1680.

[65] Wolfe, J. P., Buchwald, S. L. (1997) *J. Am. Chem. Soc.*, **119**, 6054–6058.

[66] (a) Brenner, E., Fort, Y. (1998) *Tetrahedron Lett.*, **39**, 5359–5362; (b) Brenner, E., Schneider, R., Fort, Y. (1999) *Tetrahedron*, **55**, 12829–12842.

[67] Lipshutz, B. H., Ueda, H. (2000) *Angew. Chem. Int. Ed.*, **39**, 4492–4494.

[68] Desmarets, C., Schneider, R., Fort, Y. (2002) *J. Org. Chem.*, **67**, 3029–3036.

[69] Chen, C., Yang, L.-M. (2007) *J. Org. Chem.*, **72**, 6324–6327.

[70] Ackermann, L., Sandmann, R., Song, W. (2011) *Org. Lett.*, **13**, 1784–1786.

[71] (a) Ramgren, S. D., Silberstein, A. L., Yang, Y., Garg, N. K. (2011) *Angew. Chem. Int. Ed.*, **50**, 2171–2173; (b) Nathel, N. F. F., Kim, J., Hie, L., Jiang, X., Garg, N. K. (2014) *ACS Catal.*, **4**, 3289–3293.

[72] Mesganaw, T., Silberstein, A. L., Ramgren, S. D., Nathel, N. F. F., Hong, X., Liu, P., Garg, N. K. (2011) *Chem. Sci.*, 1766–1771.

[73] Shimasaki, T., Tobisu, M., Chatani, N. (2010) *Angew. Chem. Int. Ed.*, **49**, 2929–2932.

[74] Raghuvanshi, D. S., Gupta, A. K., Singh, K. N. (2012) *Org. Lett.*, **14**, 4326–4329.

[75] Gupta, A. K., Rao, G. T., Singh, K. N. (2012) *Tetrahedron Lett.*, **53**, 2218–2221.

[76] Guo, D., Huang, H., Xu, J., Jiang, H., Liu, H. (2008) *Org. Lett.*, **10**, 4513–4516.

[77] Swapna, K., Kumar, A. V., Reddy, V. P., Rao, K. R. (2009) *J. Org. Chem.*, **74**, 7514–7517.

[78] Kim, M., Chang, S. (2010) *Org. Lett.*, **12**, 1640–1643.

[79] Thomas, K. R. J., Lin, J. T., Tao, Y.-T., Ko, C.-W. (2001) *J. Am. Chem. Soc.*, **123**, 9404–9411.

[80] (a) Koeberle, S. C., Fischer, S., Schollmeyer, D., Schattel, V., Grütter, C., Rauh, D., Laufer, S. A. (2012) *J. Med. Chem.*, **55**, 5868–5877; (b) Fischer, S., Wentsch, H. K., Mayer-Wrangowski, S. C., Zimmermann, M., Bauer, S. M., Storch, K., Niess, R., Koeberle, S. C., Grütter, C., Boeckler, F. M., Rauh, D., Laufer, S. F. (2013) *J. Med. Chem.*, **56**, 241–253.

[81] Mann, G., Incarvito, C., Rheingold, A. L., Hartwig, J. F. (1999) *J. Am. Chem. Soc.*, **121**, 3224–3225.

[82] Aranyos, A., Old, D. W., Kiyomori, A., Wolfe, J. P., Sadighi, J. P., Buchwald, S. L. (1999) *J. Am. Chem. Soc.*, **121**, 4369–4378.

[83] Shelby, Q., Kataoka, N., Mann, G., Hartwig, J. (2000) *J. Am. Chem. Soc.*, **122**, 10718–10719.

[84] Harkal, S., Kumar, K., Michalik, D., Zapf, A., Jackstell, R., Rataboul, F., Riermeier, F., Monsees, A., Beller, M. (2005) *Tetrahedron Lett.*, **46**, 3237–3240.

[85] Burgos, C. H., Barder, T. E., Huang, X., Buchwald, S. L. (2006) *Angew. Chem. Int. Ed.*, **45**, 4321–4326.

[86] Hu, T., Schulz, T., Torborg, C., Chen, X., Wang, J., Beller, M., Huang, J. (2009) *Chem. Commun.*, 7330–7332.

[87] Salvi, L., Davis, N. R., Ali, S. Z., Buchwald, S. L. (2012) *Org. Lett.*, **14**, 170–173.

[88] (a) Ullmann, F. (1904) *Ber. Dtsch. Chem. Ges.*, **37**, 853–854; (b) Ullmann, F., Sponagel, P. (1905) *Ber. Dtsch. Chem. Ges.*, **38**, 2211–2212.

[89] Marcoux, J.-F., Doye, S., Buchwald, S. L. (1997) *J. Am. Chem. Soc.*, **119**, 10539–10540.

[90] Gujadhur, R., Venkataraman, D. (2001) *Synth. Commun.*, **31**, 2865–2879.

[91] Buck, E., Song, Z. J., Tschaen, D., Dormer, P. G., Volante, R. P., Reider, P. J. (2002) *Org. Lett.*, **4**, 1623–1626.

[92] Ma, D., Cai, Q. (2003) *Org. Lett.*, **5**, 3799–3802.

[93] Cai, Q., Zou, B., Ma, D. (2006) *Angew. Chem. Int. Ed.*, **45**, 1276–1279.

[94] Cristau, H.-J., Cellier, P. P., Hamada, S., Spindler, J.-F., Taillefer, M. (2004) *Org. Lett.*, **6**, 913–916.

[95] (a) Ouali, A., Spindler, J.-F., Cristau, H.-J., Taillefer, M. (2006) *Adv. Synth. Catal.*, **348**, 499–505; (b) Ouali, A., Spindler, J.-F., Jutand, A., Taillefer, M. (2007) *Adv. Synth. Catal.*, **349**, 1906–1916; (c) Ouali, A., Taillefer, M. (2007) *Organometallics*, **26**, 65–74.

[96] Rao, H., Jin, Y., Fu, H., Jiang, Y., Zhao, Y. (2006) *Chem. Eur. J.*, **12**, 3636–3646.

[97] Lv, X., Bao, W. (2007) *J. Org. Chem.*, **72**, 3863–3867.

[98] Hosseinzadeh, R., Tajbakhsh, M., Mohadjerani, M., Alikarami, M. (2005) *Synlett*, 1101–1104.

[99] Lipshutz, B. H., Unger, J. B., Taft, B. R. (2007) *Org. Lett.*, **9**, 1089–1092.

[100] Xia, N., Taillefer, M. (2008) *Chem. Eur. J.*, **14**, 6037–6039.

[101] Kidwai, M., Mishra, N. K., Bansal, V., Kumar, A., Mozumdar, S. (2007) *Tetrahedron Lett.*, **48**, 8883–8887.

[102] Jammi, S., Sakthivel, S., Rout, L., Mukherjee, T., Mandal, S., Mitra, R., Saha, P., Punniyamurthy, T. (2009) *J. Org. Chem.*, **74**, 1971–1976.

[103] Miao, T., Wang, L. (2007) *Tetrahedron Lett.*, **48**, 95–99.

[104] Benyahya, S., Monnier, F., Taillefer, M., Man, M. W. C., Bied, C., Ouazzani, F. (2008) *Adv. Synth. Catal.*, **350**, 2205–2208.

[105] Zhang, R., Liu, J., Wang, S., Niu, J., Xia, C., Sun, W. (2011) *ChemCatChem*, **3**, 146–149.

[106] Evans, D. A., Katz, J. L., West, T. R. (1998) *Tetrahedron Lett.*, **39**, 2937–2940.

[107] Lam, P. Y. S., Vincent, G., Clark, C. G., Deudon, S., Jadhav, P. K. (2001) *Tetrahedron Lett.*, **42**, 3415–3418.

[108] Anderson, K. W., Ikawa, T., Tundel, R. E., Buchwald, S. L. (2006) *J. Am. Chem. Soc.*, **128**, 10694–10695.

[109] Mann, G., Hartwig, J. F. (1996) *J. Am. Chem. Soc.*, **118**, 13109–13110.

[110] Watanabe, M., Nishiyama, M., Koie, Y. (1999) *Tetrahedron Lett.*, **40**, 8837–8840.

[111] Parrish, C. A., Buchwald, S. L. (2001) *J. Org. Chem.*, **66**, 2498–2500.

[112] Torraca, K. E., Huang, X., Parrish, C. A., Buchwald, S. L. (2001) *J. Am. Chem. Soc.*, **123**, 10770–10771.

[113] Vorogushin, A. V., Huang, X., Buchwald, S. L. (2005) *J. Am. Chem. Soc.*, **127**, 8146–8149.

[114] Gowrisankar, S., Sergeev, A. G., Anbarasan, P., Spannenberg, A., Neumann, H., Beller, M. (2010) *J. Am. Chem. Soc.*, **132**, 11592–11598.

[115] Wu, X., Fors, B. P., Buchwald, S. L. (2011) *Angew. Chem. Int. Ed.*, **50**, 9943–9947.

[116] Gowrisankar, S., Neumann, H., Beller, M. (2012) *Chem. Eur. J.*, **18**, 2498–2502.

[117] Milton, E. J., Fuentes, J. A., Clarke, M. L. (2009) *Org. Biomol. Chem.*, **7**, 2645–2648.

[118] Tolnai, G. L., Pethő, B., Králl, P., Novák, Z. (2014) *Adv. Synth. Catal.*, **356**, 125–129.

[119] Wolter, M., Nordmann, G., Job, G. E., Buchwald, S. L. (2002) *Org. Lett.*, **4**, 973–976.

[120] Altman, R. A., Shafir, A., Choi, A., Lichtor, P. A., Buchwald, S. L. (2008) *J. Org. Chem.*, **73**, 284–286.

[121] Niu, J., Zhou, H., Li, Z., Xu, J., Hu, S. (2008) *J. Org. Chem.*, **73**, 7814–7817.

[122] Manbeck, G. F., Lipman, A. J., Stockland, R. A. Jr., Freidl, A. L., Hasler, A. F., Stone, J. J., Guzei, I. A. (2005) *J. Org. Chem.*, **70**, 244–250.

[123] Zhang, H., Ma, D., Cao, W. (2007) *Synlett*, 243–246.

[124] Naidu, A. B., Sekar, G. (2008) *Tetrahedron Lett.*, **49**, 3147–3151.

[125] Niu, J., Guo, P., Kang, J., Li, Z., Xu, J., Hu, S. (2009) *J. Org. Chem.*, **74**, 5075–5078.

[126] Quach, T. D., Batey, R. A. (2003) *Org. Lett.*, **5**, 1381–1384.

[127] (a) King, A. E., Brunold, T. C., Stahl, S. S. (2009) *J. Am. Chem. Soc.*, **131**, 5044–5045; (b) King, A. E., Ryland, B. L., Brunold, T. C., Stahl, S. S. (2012) *Organometallics*, **31**, 7948–7957.

[128] Schulz, T., Torborg, C., Schäffner, B., Huang, J., Zapf, A., Kadyrov, R., Börner, A., Beller, M. (2009) *Angew. Chem. Int. Ed.*, **48**, 918–921.

[129] Sergeev, A. G., Schulz, T., Torborg, C., Spannenberg, A., Neumann, H., Beller, M. (2009) *Angew. Chem. Int. Ed.*, **48**, 7595–7599.

[130] Chen, G., Chan, A. S. C., Kwong, F. Y. (2007) *Tetrahedron Lett.*, **48**, 473–476.

[131] Yu, C.-W., Chen, G. S., Huang, C.-W., Chern, J.-W. (2012) *Org. Lett.*, **14**, 3688–3691.

[132] (a) Lavery, C. B., Rotta-Loria, N. L., McDonald, R., Stradiotto, M. (2013) *Adv. Synth. Catal.*, **355**, 981–987; (b) Cheung, C. W., Buchwald, S. L. (2014) *J. Org. Chem.*, **79**, 5351–5358.

[133] Kormos, C. M., Leadbeater, N. E. (2006) *Tetrahedron*, **62**, 4728–4732.

[134] Tlili, A., Xia, N., Monnier, F., Taillefer, M. (2009) *Angew. Chem. Int. Ed.*, **48**, 8725–8728.

[135] Zhao, D., Wu, N., Zhang, S., Xi, P., Su, X., Lan, J., You, J. (2009) *Angew. Chem. Int. Ed.*, **48**, 8729–8732.

[136] Paul, R., Ali, M. A., Punniyamurthy, T. (2010) *Synthesis*, 4268–4272.

[137] Jing, L., Wei, J., Zhou, L., Huang, Z., Li, Z., Zhou, X. (2010) *Chem. Commun.*, **46**, 4767–4769.

[138] Maurer, S., Liu, W., Zhang, X., Jiang, Y., Ma, D. (2010) *Synlett*, 976–978.

[139] Yang, K., Li, Z., Wang, Z., Yao, Z., Jiang, S. (2011) *Org. Lett.*, **13**, 4340–4343.

[140] Yang, D., Fu, H. (2010) *Chem. Eur. J.*, **16**, 2366–2370.

[141] Thakur, K. G., Sekar, G. (2011) *Chem. Commun.*, **47**, 6692–6694.

[142] Songis, O., Boulens, P., Benson, C. G. M., Cazin, C. S. J. (2012) *RSC Adv.*, **2**, 11675–11677.

[143] Xiao, Y., Xu, Y., Cheon, H.-S., Chae, J. (2013) *J. Org. Chem.*, **78**, 5804–5809.

[144] Xu, H.-J., Liang, Y.-F., Cai, Z.-Y., Qi, H.-X., Yang, C.-Y., Feng, Y.-S. (2011) *J. Org. Chem.*, **76**, 2296–2300.

[145] Chan, C.-C., Chen, Y.-W., Su, C.-S., Lin, H.-P., Lee, C.-F. (2011) *Eur. J. Org. Chem.*, 7288–7293.

[146] Ding, G., Han, H., Jiang, T., Wu, T., Han, B. (2014) *Chem. Commun.*, **50**, 9072–9075.

[147] Chalmers, J. R., Dickson, G. T., Elks, J., Hems, B. A. (1949) *J. Chem. Soc.*, 3424–3433.

[148] Jung, M. E., Lazarova, T. I. (1999) *J. Org. Chem.*, **64**, 2976–2977.

[149] Ghosh, S., Kumar, A. S., Mehta, G. N., Soundararajan, R., Sen, S. (2009) *ARKIVOC*, 72–79.

[150] Xing, X., Padmanaban, D., Yeh, L.-A., Cuny, G. D. (2002) *Tetrahedron*, **58**, 7903–7910.

[151] (a) Beletskaya, I. P., Ananikov, V. P. (2011) *Chem. Rev.*, **111**, 1596-1636; (b) Font, M., Parella, T., Costas, M., Ribas, X. (2012) *Organometallics*, **31**, 7976–7982.

[152] Migita, T., Shimizu, T., Asami, Y., Shiobara, J.-I., Kato, Y., Kosugi, M. (1980) *Bull. Chem. Soc. Jpn.*, **53**, 1385–1389.

[153] Mann, G., Baranano, D., Hartwig, J. F., Rheingold, A. L., Guzei, I. A. (1998) *J. Am. Chem. Soc.*, **120**, 9205–9219.

[154] Li, G. Y., Zheng, G., Noonan, A. F. (2001) *J. Org. Chem.*, **66**, 8677–8681.

[155] Schopfer, U., Schlapbach, A. (2001) *Tetrahedron*, **57**, 3069–3073.

[156] Murata, M., Buchwald, S. L. (2004) *Tetrahedron*, **60**, 7397–7403.

[157] Fernández-Rodríguez, M. A., Shen, Q., Hartwig, J. F. (2006) *Chem. Eur. J.*, **12**, 7782–7796.

[158] Fernández-Rodríguez, M. A., Hartwig, J. F. (2009) *J. Org. Chem.*, **74**, 1663–1672.

[159] Sayah, M., Organ, M. G. (2011) *Chem. Eur. J.*, **17**, 11719–11722.

[160] Valente, C., Pompeo, M., Sayah, M., Organ, M. G. (2014) *Org. Process Res. Dev.*, **18**, 180–190.

[161] Sayah, M., Organ, M. G. (2013) *Chem. Eur. J.*, **19**, 16196–16199.

[162] Bastug, G., Nolan, S. P. (2013) *J. Org. Chem.*, **78**, 9303–9308.

[163] Herradura, P. S., Pendola, K. A., Guy, R. K. (2000) *Org. Lett.*, **2**, 2019–2022.

[164] Bates, C. G., Gujadhur, R. K., Venkataraman, D. (2002) *Org. Lett.*, **4**, 2803–2806.

[165] Kwong, F. Y., Buchwald, S. L. (2002) *Org. Lett.*, **4**, 3517–3520.

[166] Taniguchi, N. (2007) *J. Org. Chem.*, **72**, 1241–1245.

[167] Xu, H.-J., Zhao, Y.-Q., Feng, T., Feng, Y.-S. (2012) *J. Org. Chem.*, **77**, 2878–2884.

[168] Prasad, D. J. C., Sekar, G., (2011) *Org. Lett.*, **13**, 1008–1011.

[169] Jiang, Y., Qin, Y., Xie, S., Zhang, X., Dong, J., Ma, D. (2009) *Org. Lett.*, **11**, 5250–5253.

[170] Chen, H.-Y., Peng, W.-T., Lee, Y.-H., Chang, Y.-L., Chen, Y.-J., Lai, Y.-C., Jheng, N.-Y. Chen, H.-Y. (2013) *Organometallics*, **32**, 5514–5522.

[171] Martinek, M., Korf, M., Srogl, J. (2010) *Chem. Commun.*, **46**, 4387–4389.

[172] Rout, L., Sen, T. K., Punniyamurthy, T. (2007) *Angew. Chem. Int. Ed.*, **46**, 5583–5586.

[173] Babu, S. G., Karvembu, R. (2013) *Tetrahedron Lett.*, **54**, 1677–1680.

[174] Akkilagunta, V. K., Kakulapati, R. R. (2011) *J. Org. Chem.*, **76**, 6819–6824.

[175] Kamal, A., Srinivasulu, V., Murty, J. N. S. R. C., Shankaraiah, N., Nagesh, N., Reddy, T. S., Rao, A. V. S. (2013) *Adv. Synth. Catal.*, **355**, 2297–2307.

[176] Wong, Y.-C., Jayanth, T. T., Cheng, C.-H. (2006) *Org. Lett.*, **8**, 5613–5616.

[177] Jammi, S., Barua, P., Rout, L., Saha, P., Punniyamurthy, T. (2008) *Tetrahedron Lett.*, **49**, 1484–1487.

[178] Iglesias, M. J., Prieto, A., Nicasio, M. C. (2010) *Adv. Synth. Catal.*, **352**, 1949–1954.

[179] Guan, P., Cao, C., Liu, Y., Li, Y., He, P., Chen, Q., Liu, G., Shi, Y. (2012) *Tetrahedron Lett.*, **53**, 5987–5992.

[180] Nogueira, C. W., Zeni, G., Rocha, J. B. T. (2004) *Chem. Rev.*, **104**, 6255–6286.

[181] Suzuki, H., Abe, H., Osuka, A. (1981) *Chem. Lett.*, 151–152.

[182] Cristau, H. J., Chabaud, B., Labaudiniere, R., Christol, H. (1985) *Organometallics*, **4**, 657–661.

[183] Nishiyama, Y., Tokunaga, K., Sonoda, N. (1999) *Org. Lett.*, **1**, 1725–1727.

[184] Beletskaya, I. P., Sigeev, A. S., Peregudov, A. S., Petrovskii, P. V. (2003) *Tetrahedron Lett.*, **44**, 7039–7041.

[185] Gujadhur, R. K., Venkataraman, D. (2003) *Tetrahedron Lett.*, **44**, 81–84.

[186] Taniguchi, N., Onami, T. (2004) *J. Org. Chem.*, **69**, 915–920.

[187] Bonaterra, M., Martín, S. E., Rossi, R. A. (2006) *Tetrahedron Lett.*, **47**, 3511–3515.

[188] Dandapat, A., Korupalli, C., Prasad, D. J. C., Singh, R., Sekar, G. (2011) *Synthesis*, 2297–2302.

[189] Ramana, T., Punniyamurthy, T. (2011) *Eur. J. Org. Chem.*, 4757–4759.

[190] Reddy, V. P., Kumar, A. V., Swapna, K., Rao, K. R. (2009) *Org. Lett.*, **11**, 951–953.

[191] Alves, D., Santos, C. G., Paixão, M. W., Soares, L. C., de Souza, D., Rodrigues, O. E. D., Braga, A. L. (2009) *Tetrahedron Lett.*, **50**, 6635–6638.

[192] Li, Y., Wang, H., Li, X., Chen, T., Zhao, D. (2010) *Tetrahedron*, **66**, 8583–8586.

[193] Swapna, K., Murthy, S. N., Nageswar, Y. V. D. (2011) *Eur. J. Org. Chem.*, 1940–1946.

[194] Suzuki, H., Abe, H., Ohmasa, N., Osuka, A. (1981) *Chem. Lett.*, 1115–1116.

[195] Silva, M. S., Comasseto, J. V. (2011) *Tetrahedron*, **67**, 8763–8768.

[196] Cheng, J.-H., Yi, C.-L., Liu, T.-J. Lee, C.-F. (2012) *Chem. Commun.*, **48**, 8440–8442.

[197] Kumar, A., Kumar, S. (2014) *Tetrahedron*, **70**, 1763–1772.

[198] (a) Beletskaya, I. P., Kazankova, M. A. (2002) *Russ. J. Org. Chem.*, **38**, 1391–1430; (b) Tappe, F. M. J., Trepohl, V. T., Oestreich, M. (2010) *Synthesis*, 3037–3062; (c) Glueck, D. S. (2010) *Top. Organomet. Chem.*, **31**, 65–100.

[199] (a) Hirao, T., Masunaga, T., Ohshiro, Y., Agawa, T. (1981) *Synthesis*, 56–57; (b) Hirao, T., Masunaga, T., Yamada, N., Ohshiro, Y., Agawa, T. (1982) *Bull. Chem. Soc. Jpn.*, **55**, 909–913.

[200] Petrakis, K. S., Nagabhushan, T. L. (1987) *J. Am. Chem. Soc.*, **109**, 2831–2833.

[201] (a) Kurz, L., Lee, G., Morgans, D., Waldyke, M. J., Ward, T. (1990) *Tetrahedron Lett.*, **31**, 6321–6324; (b) Yan, Y.-Y., RajanBabu, T. V. (2000) *J. Org. Chem.*, **65**, 900–906; (c) Casalnuovo, A. L., Calabrese, J. C. (1990) *J. Am. Chem. Soc.*, **112**, 4324–4330; (d) Lavén, G., Kalek, M., Jezowska, M., Stawinski, J. (2010) *New J. Chem.*, **34**, 967–975.

[202] Kalek, M., Ziadi, A., Stawinski, J. (2008) *Org. Lett.*, **10**, 4637–4640.

[203] Kalek, M., Stawinski, J. (2007) *Organometallics*, **26**, 5840–5847.

[204] (a) Kalek, M., Stawinski, J. (2008) *Organometallics*, **27**, 5876–5888; (b) Kalek, M., Jezowska, J., Stawinski, J. (2009) *Adv. Synth. Catal.*, **351**, 3207–3216.

[205] Luo, Y., Wu, J. (2009) *Organometallics*, **28**, 6823–6826.

[206] Belabassi, Y., Alzghari, S., Montchamp, J.-L. (2008) *J. Organomet. Chem.*, **693**, 3171–3178.

[207] Xu, K., Hu, H., Yang, F., Wu, Y. (2013) *Eur. J. Org. Chem.*, 319–325.

[208] Xu, K., Yang, F., Zhang, G., Wu, Y. (2013) *Green Chem.*, **15**, 1055–1060.

[209] Wang, T., Sang, S., Liu, L., Qiao, H., Gao, Y., Zhao, Y. (2014) *J. Org. Chem.*, **79**, 608–617.

[210] Andaloussi, M., Lindh, J., Sävmarker, J., Sjöberg, P. J. R., Larhed, M. (2009) *Chem. Eur. J.*, **15**, 13069–13074.

[211] (a) Xu, Y., Li, Z., Xia, J., Guo, H., Huang, Y. (1983) *Synthesis*, 377–378; (b) Xu, Y., Zhang, J. (1984) *Synthesis*, 778–780; (c) Bennett, S. N. L., Hall, R. G. (1995) *J. Chem. Soc. Perkin Trans.*, **1**, 1145–1151.

[212] (a) Xu, Y., Zhang, J. (1986) *J. Chem. Soc. Chem. Commun.*, **1606**; (b) Zhang, J., Xu, Y., Huang, G., Guo, H. (1988) *Tetrahedron Lett.*, **29**, 1955–1958.

[213] Deal, E. L., Petit, C., Montchamp, J.-L. (2011) *Org. Lett.*, **13**, 3270–3273.

[214] Xu, Y., Li, Z., Xia, J., Guo, H., Huang, Y. (1984) *Synthesis*, 781–782.

[215] Bloomfield, A. J., Herzon, S. B. (2012) *Org. Lett.*, **14**, 4370–4373.

[216] Rummelt, S. M., Ranocchiari, M., van Bokhoven, J. A. (2012) *Org. Lett.*, **14**, 2188–2190.

[217] Fu, T., Qiao, H., Peng, Z., Hu, G., Wu, X., Gao, Y., Zhao, Y. (2014) *Org. Biomol. Chem.*, **12**, 2895–2902.

[218] (a) Schwabacher, A. W., Stefanescu, A. D. (1996) *Tetrahedron Lett.*, **37**, 425–428; (b) Lei, H., Stoakes, M. S., Schwabacher, A. W. (1992) *Synthesis*, 1255–1260.

[219] Montchamp, J.-L., Dumond, Y. R. (2001) *J. Am. Chem. Soc.*, **123**, 510–511.

[220] Bravo-Altamirano, K., Huang, Z., Montchamp, J.-L. (2005) *Tetrahedron*, **61**, 6315–6329.

[221] Chauvin, R. (1990) *J. Organomet. Chem.*, **387**, C1–C4.

[222] (a) Kwong, F. Y., Lai, C. W., Tian, Y., Chan, K. S. (2000) *Tetrahedron Lett.*, **41**, 10285–10289; (b) Kwong, F. Y., Lai, C. W., Yu, M., Tian, Y., Chan, K. S. (2003) *Tetrahedron*, **59**, 10295–10305.

[223] Kwong, F. Y., Lai, C. W., Chan, K. S. (2002) *Tetrahedron Lett.*, **43**, 3537–3539.

[224] Wang, Y. Lai, C. W., Kwong, F. Y., Jia, W., Chan, K. S. (2004) *Tetrahedron*, **60**, 9433–9439.

[225] Martorell, G., Garcías, X., Janura, M., Saá, J. M. (1998) *J. Org. Chem.*, **63**, 3463–3467.

[226] Herd, O., Heßler, A., Hingst, M., Tepper, M., Stelzer, O. (1996) *J. Organomet. Chem.*, **522**, 69–76.

[227] Stadler, A., Kappe, C. O. (2002) *Org. Lett.*, **4**, 3541–3543.

[228] (a) Imamoto, T., Oshiki, T., Onozawa, T., Kusumoto, T., Sato, K. (1990) *J. Am. Chem. Soc.*, **112**, 5244–5252; (b) Imamoto, T. (1993) *Pure Appl. Chem.*, **65**, 655–660.

[229] Lipshutz, B. H., Buzard, D. J., Yun, C. S. (1999) *Tetrahedron Lett.*, **40**, 201–204.

[230] Vallette, H., Pican, S., Boudou, C., Levillain, J., Plaquevent, J.-C., Gaumont, A.-C. (2006) *Tetrahedron Lett.*, **47**, 5191–5193.

[231] Tunney, S. E., Stille, J. K. (1987) *J. Org. Chem.*, **52**, 748–753.

[232] Beletskaya, I. P., Veits, Y. A., Leksunkin, V. A., Foss, V. L. (1992) *Russ. Chem. Bull.*, **41**, 1272–1274.

[233] Martín, S. E., Bonaterra, M., Rossi, R. A. (2002) *J. Organomet. Chem.*, **664**, 223–227.

[234] Bonaterra, M., Rossi, R. A., Martín, S. E. (2009) *Organometallics*, **28**, 933–936.

[235] Osuka, A., Ohmasa, N., Yoshida, Y., Suzuki, H. (1983) *Synthesis*, 69–71.

[236] Allen, D. V., Venkataraman, D. (2003) *J. Org. Chem.*, **68**, 4590–4593.

[237] Gelman, D., Jiang, L., Buchwald, S. L. (2003) *Org. Lett.*, **5**, 2315–2318.

[238] Huang, C., Tang, X., Fu, H., Jiang, Y., Zhao, Y. (2006) *J. Org. Chem.*, **71**, 5020–5022.

[239] Stankevič, M., Włodarczyk, A. (2013) *Tetrahedron*, **69**, 73–81.

[240] Zhuang, R., Xu, J., Cai, Z., Tang, G., Fang, M., Zhao, Y. (2011) *Org. Lett.*, **13**, 2110–2113.

[241] Xu, J., Zhang, P., Gao, Y., Chen, Y., Tang, G., Zhao, Y. (2013) *J. Org. Chem.*, **78**, 8176–8183.

[242] Tavs, P. (1970) *Chem. Ber.*, **103**, 2428–2436.

[243] Balthazor, T. M., Grabiak, R. C. (1980) *J. Org. Chem.*, **45**, 5425–5426.

[244] Zhang, X., Liu, H., Hu, X., Tang, G., Zhu, J., Zhao, Y. (2011) *Org. Lett.*, **13**, 3478–3481.

[245] (a) Zhang, H.-Y., Sun, M., Ma, Y.-N., Tian, Q.-P., Yang, S.-D. (2012) *Org. Biomol. Chem.*, **10**, 9627–9633; (b) Shen, C., Yang, G., Zhang, W. (2012) *Org. Biomol. Chem.*, **10**, 3500–3505.

[246] (a) Zhao, Y.-L., Wu, G.-J., Li, Y., Gao, L.-X. Han, F.-S. (2012) *Chem. Eur. J.*, **18**, 9622–9627; (b) Zhao, Y.-L., Wu, G.-J., Han, F.-S. (2012) *Chem. Commun.*, **48**, 5868–5870.

[247] Sun, M., Zhang, H.-Y., Han, Q., Yang, K., Yang, S.-D. (2011) *Chem. Eur. J.*, **17**, 9566–9570.

[248] Hu, G., Chen, W., Fu, T., Peng, Z., Qiao, H., Gao, Y., Zhao, Y. (2013) *Org. Lett.*, **15**, 5362–5365.

[249] Norman, N. C. ed. (1998) *Chemistry of Arsenic, Antimony and Bismuth*, Blackie Academic and Professional, London.

[250] Bonaterra, M., Martín, S. E., Rossi, R. A. (2003) *Org. Lett.*, **5**, 2731–2734.

[251] Uberman, P. M., Lanteri, M. N., Martín, S. E. (2009) *Organometallics*, **28**, 6927–6934.

[252] Uberman, P. M., Lanteri, M. N., Parajón Puenzo, S. C., Martín, S. E. (2011) *Dalton Trans.*, **40**, 9229–9237.

[253] Uberman, P. M., Caira, M. R., Martín, S. E. (2013) *Organometallics*, **32**, 3220–3226.

[254] Hall, D. G. ed. (2011) *Boronic Acids: Preparation and Applications in Organic Synthesis Medicine and Materials*, Wiley-VCH, Weinheim.

[255] Miyaura, N., Suzuki, A. (1995) *Chem. Rev.*, **95**, 2457–2483.

[256] Ishiyama, T., Murata, M., Miyaura, N. (1995) *J. Org. Chem.*, **60**, 7508–7510.

[257] Ishiyama, T., Itoh, Y., Kitano, T., Miyaura, N. (1997) *Tetrahedron Lett.*, **38**, 3447–3450.

[258] Willis, D. M., Strongin, R. M. (2000) *Tetrahedron Lett.*, **41**, 8683–8686.

[259] Murata, M., Watanabe, S., Masuda, Y. (1997) *J. Org. Chem.*, **62**, 6458–6459.

[260] Murata, M., Oyama, T., Watanabe, S., Masuda, Y. (2000) *J. Org. Chem.*, **65**, 164–168.

[261] Ishiyama, T., Ishida, K., Miyaura, N. (2001) *Tetrahedron*, **57**, 9813–9816.

[262] (a) Fürstner, A., Seidel, G. (2002) *Org. Lett.*, **4**, 541–543; (b) Ma, Y., Song, C., Jiang, W., Xue, G., Cannon, J. F., Wang, X., Andrus, M. B. (2003) *Org. Lett.*, **5**, 4635–4638.

[263] (a) Broutin, P.-E., Čerňa, I., Campaniello, M., Leroux, F., Colobert, F. (2004) *Org. Lett.*, **6**, 4419–4422; (b) Murata, M., Sambommatsu, T., Watanabe, S., Masuda, Y. (2006) *Synlett*, 1867–1870.

[264] (a) Billingsley, K. L., Barder, T. E., Buchwald, S. L. (2007) *Angew. Chem. Int. Ed.*, **46**, 5359–5363; (b) Billingsley, K. L., Buchwald, S. L. (2008) *J. Org. Chem.*, **73**, 5589–5591.

[265] Guerrand, H. D. S., Marciasini, L. D., Jousseaume, M., Vaultier, M., Pucheault, M. (2014) *Chem. Eur. J.*, **20**, 5573–5579.

[266] Rosen, B. M., Huang, C., Percec, V. (2008) *Org. Lett.*, **10**, 2597–2600.

[267] Moldoveanu, C., Wilson, D. A., Wilson, C. J., Corcoran, P., Rosen, B. M., Percec, V. (2009) *Org. Lett.*, **11**, 4974–4977.

[268] Moldoveanu, C., Wilson, D. A., Wilson, C. J., Leowanawat, P., Resmerita, A.-M., Liu, C., Rosen, B. M., Percec, V. (2010) *J. Org. Chem.*, **75**, 5438–5452.

[269] (a) Leowanawat, P., Resmerita, A.-M., Moldoveanu, C., Liu, C., Zhang, N., Wilson, D. A., Hoang, L. M., Rosen, B. M., Percec, V. (2010) *J. Org. Chem.*, **75**, 7822–7828; (b) Wilson, D. A., Wilson, C. J., Moldoveanu, C., Resmerita, A.-M., Corcoran, P., Hoang, L. M., Rosen, B. M., Percec, V. (2010) *J. Am. Chem. Soc.*, **132**, 1800–1801.

[270] Yamamoto, T., Morita, T., Takagi, J., Yamakawa, T. (2011) *Org. Lett.*, **13**, 5766–5769.

[271] Huang, K., Yu, D.-G., Zheng, S.-F., Wu, Z.-H., Shi, Z.-J. (2011) *Chem. Eur. J.*, **17**, 786–791.

[272] Molander, G. A., Cavalcanti, L. N., García-García, C. (2013) *J. Org. Chem.*, **78**, 6427–6439.

[273] Zhu, W., Ma, D. (2006) *Org. Lett.*, **8**, 261–263.

[274] Kleeberg, C., Dang, L., Lin, Z., Marder, T. B. (2009) *Angew. Chem. Int. Ed.*, **48**, 5350–5354.

[275] Labre, F., Gimbert, Y., Bannwarth, P., Olivero, S., Duñach, E., Chavant, P. Y. (2014) *Org. Lett.*, **16**, 2366–2369.

21

TRANSITION METAL-MEDIATED AROMATIC RING CONSTRUCTION

KEN TANAKA

Department of Applied Chemistry, Graduate School of Science and Engineering, Tokyo Institute of Technology, Meguro-ku, Tokyo, Japan

21.1 INTRODUCTION

Recent significant advance in the area of the transition metal-mediated aromatic ring construction reactions enables the construction of substituted benzenoid aromatic rings with efficient and convenient manners [1]. These aromatic ring construction reactions provide promising new routes to the complex benzenoid aromatic compounds. In this chapter, I summarize the intermolecular multicomponent reactions and the intramolecular single-component reactions, which are able to construct benzenoid aromatic rings.

As the intermolecular multicomponent reactions, three-component cycloaddition reactions (**21.2** [2+2+2] cycloaddition and **21.3** [3+2+1] cycloaddition) and two-component cycloaddition reactions (**21.4** [4+2] cycloaddition) are described. As the intramolecular single-component reactions, cycloaromatization reactions (**21.5** intramolecular hydroarylation of alkynes and cyclization via transition metal vinylidenes) are described. Aromatic ring construction reactions involving aryne reactions (Chapter 12), rearrangement reactions (Chapters 16 and 18), metathesis reactions (Chapter 17), and coupling reactions (Chapters 19 and 20) are described in these different chapters.

21.2 [2+2+2] CYCLOADDITION

In 1948, Reppe reported the first example of the transition metal-catalyzed [2+2+2] cycloaddition of alkynes by using a nickel complex [2]. Vollhardt extensively studied utilization of the cobalt-mediated [2+2+2] cycloaddition in organic synthesis [3a, b], and Yamazaki studied general reactivity of

---

Arene Chemistry: Reaction Mechanisms and Methods for Aromatic Compounds, First Edition. Edited by Jacques Mortier.

cobaltacycles [3c]. After these pioneering works, many transition metal catalysts have been developed and utilized in organic synthesis [4]. The most frequently employed catalysts are cobalt [$CpCoL_2$, CpCo(alkene)(L), and CoX_2/M/L], nickel [Ni(0)-phosphine], ruthenium [Cp^*Ru(cod)Cl], rhodium [$RhCl(PPh_3)_3$ and $Rh(I)^+$-diphosphine], and iridium [Ir(I)-diphosphine] complexes [4]. These complexes catalyze the [2+2+2] cycloaddition through metallacycle intermediates. In this subchapter, I summarized the transition metal-catalyzed [2+2+2] cycloaddition reactions of alkynes for the construction of benzenoid aromatic rings by classifying the reaction patterns.

21.2.1 Mechanism

Generally proposed mechanisms of the transition metal-catalyzed [2+2+2] cycloaddition of alkynes are shown in Schemes 19.1 and 19.2. Two alkynes react with the transition metal complex to generate metallacyclopentadiene **A**. The subsequent [4+2] cycloaddition of **A** with the alkyne affords metallabicyclo[2.2.0]heptadiene **B**. Reductive elimination affords the desired benzene (Schemes 21.1). Alternatively, insertion of the alkyne into **A** leading to metallacycloheptatriene **C** followed by reductive elimination also affords the desired benzene (Schemes 19.1). The mechanism via the intermediates **A** and **B** is confirmed in the CpCo(I)-phosphine complex-catalyzed [2+2+2] cycloaddition of alkynes by the theoretical calculation [5].

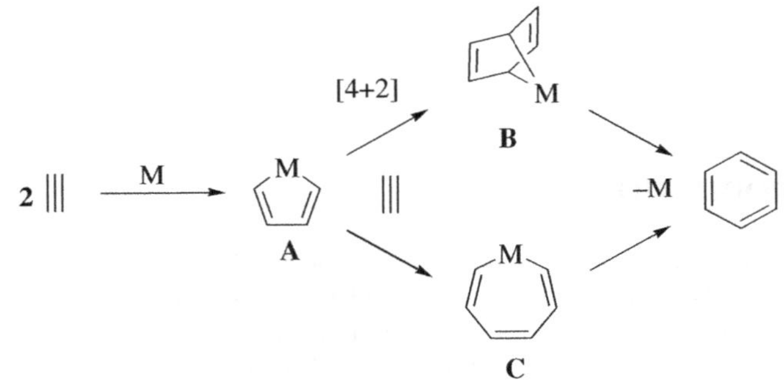

SCHEME 21.1 Mechanism of transition metal-catalyzed [2+2+2] cycloaddition (1).

When the initially formed five-membered metallacycle exists as not metallacyclopentadiene **A** but metallacyclopentatriene (biscarbene) **D**, the [2+2+2] cycloaddition proceeds via a mechanism shown in Scheme 21.2. The [2+2] cycloaddition of metallacyclopentatriene **D** with the alkyne affords metallabicyclo[3.2.0]heptatriene **E**. Skeletal rearrangement leading to metallacycloheptatetraene **F** followed by

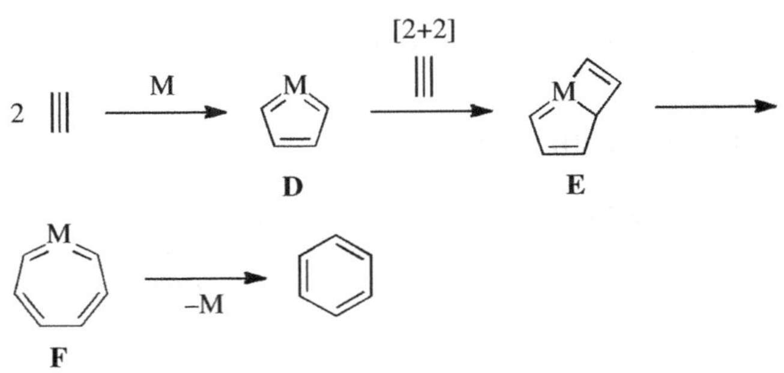

SCHEME 21.2 Mechanism of transition metal-catalyzed [2+2+2] cycloaddition (2).

reductive elimination affords the desired benzene. This mechanism (Schemes 21.1) is confirmed in the Cp*Ru(cod)Cl complex-catalyzed [2+2+2] cycloaddition of alkynes by the theoretical calculation [6].

21.2.2 [2+2+2] Cycloaddition of Monoynes

In the intermolecular [2+2+2] cycloaddition of unsymmetric monoynes, it is difficult to control its regioselectivity. Cationic rhodium(I)/biaryl bisphosphine complexes are highly active and selective catalysts for the 1,2,4-selective [2+2+2] cycloaddition of terminal alkynes (Scheme 21.3) [7]. However, unfortunately, a general and practical catalyst that enables the 1,3,5-selective [2+2+2] cycloaddition of terminal alkynes has not been developed.

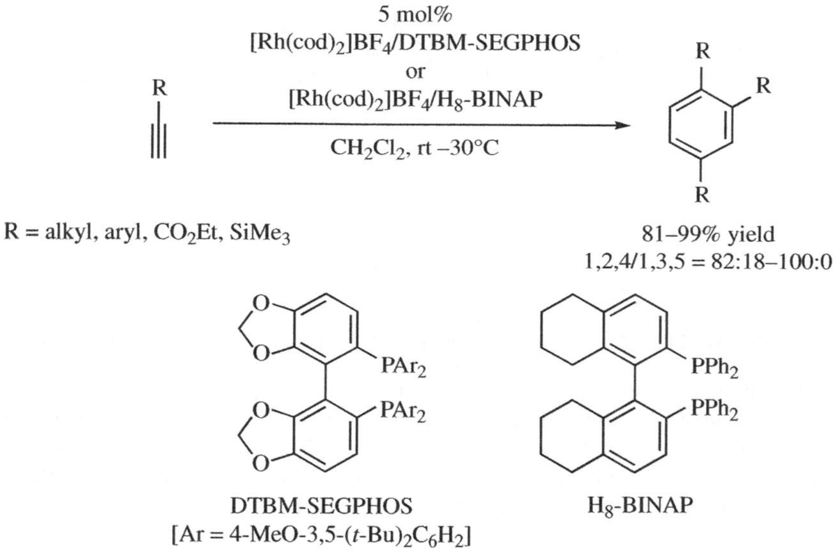

SCHEME 21.3 Rh-catalyzed regioselective intermolecular [2+2+2] cycloaddition.

The cationic rhodium(I)/H_8-BINAP complex also catalyzed the chemo- and regioselective cross-[2+2+2] cycloaddition of dialkyl acetylenedicarboxylate with two molecules of electron-rich terminal alkynes (Scheme 21.4) [7].

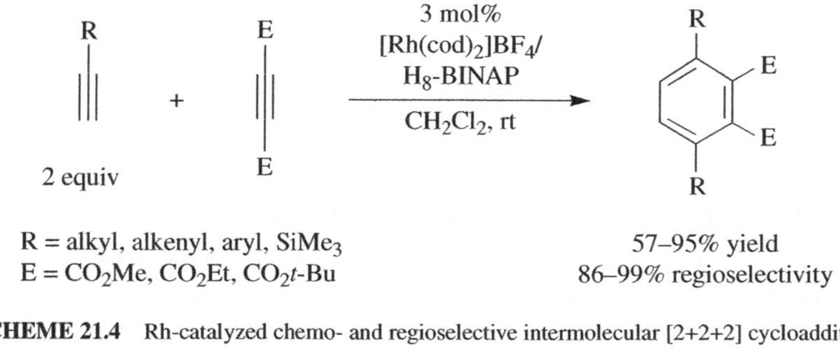

SCHEME 21.4 Rh-catalyzed chemo- and regioselective intermolecular [2+2+2] cycloaddition.

The cobalt catalyst is highly effective for the [2+2+2] cycloaddition of internal monoynes. For example, the [2+2+2] cycloaddition of a diaryl acetylene proceeded to give an oligoaryl, which is used as an electrochemical switch (Scheme 21.5) [8].

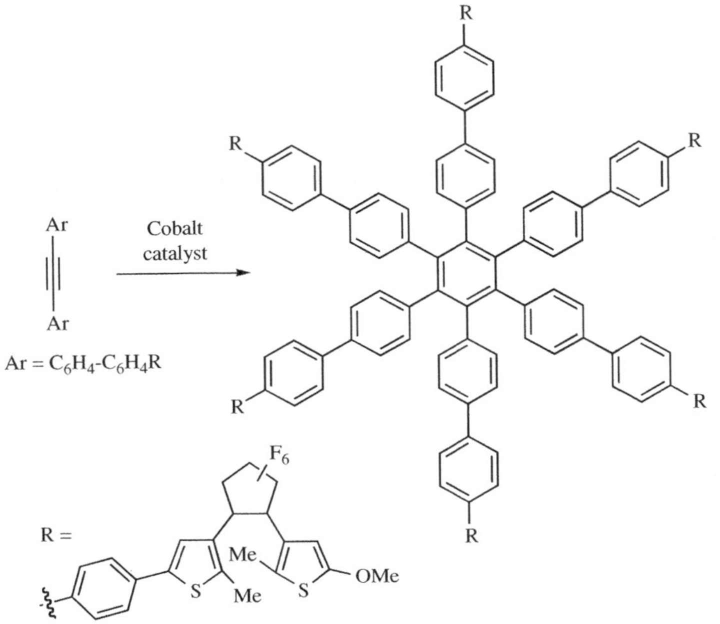

SCHEME 21.5 Cocatalyzed intermolecular [2+2+2] cycloaddition of internal monoynes.

Although stoichiometric amounts of transition metal complexes were employed, the cross-[2+2+2] cycloaddition of three different alkynes has been achieved using zirconium and nickel complexes. The reaction of 2-butyne, 4-octyne, and Cp_2ZrEt_2 selectively afforded unsymmetrical zirconacyclopentadiene. Subsequently, it reacted with 3-hexyne in the presence of $NiBr_2(PPh_3)_2$ to give the desired hexasubstituted benzene (Scheme 21.6) [9].

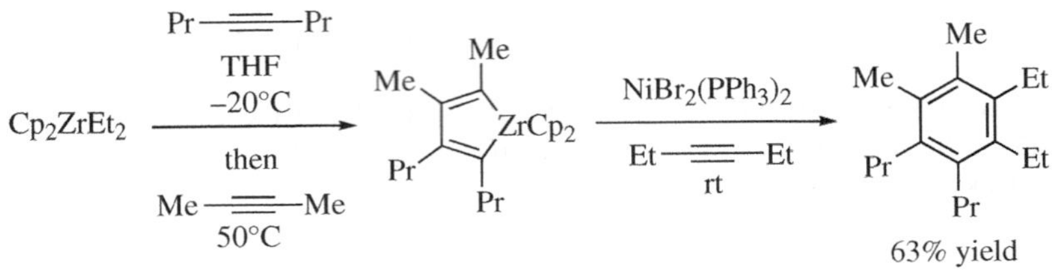

SCHEME 21.6 Zr/Ni-mediated chemo- and regioselective intermolecular [2+2+2] cycloaddition.

This method has been applied to the synthesis of substituted pentacenes by homologation (Scheme 21.7) [10].

21.2.3 [2+2+2] Cycloaddition of Diynes with Monoynes

The chemo- and regioselectivity problems can be solved by the [2+2+2] cycloaddition of diynes with monoynes, although the accessible products are limited to bicyclic compounds. The [2+2+2] cycloaddition of diynes with monoynes has been applied to the synthesis of biologically active molecules and functional materials. The first application in the natural product synthesis was the *dl*-estrone synthesis using $CpCo(CO)_2$ as a catalyst. The [2+2+2] cycloaddition followed by the benzocyclobutane to *ortho*-quinodimethane rearrangement and intramolecular Diels–Alder reaction afforded a *dl*-estrone precursor (Scheme 21.8) [11].

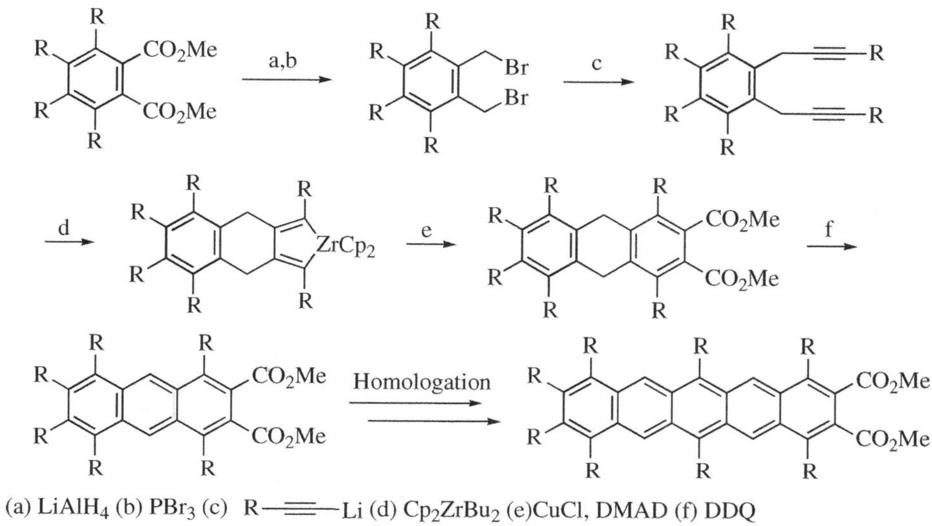

SCHEME 21.7 Synthesis of substituted pentacenes.

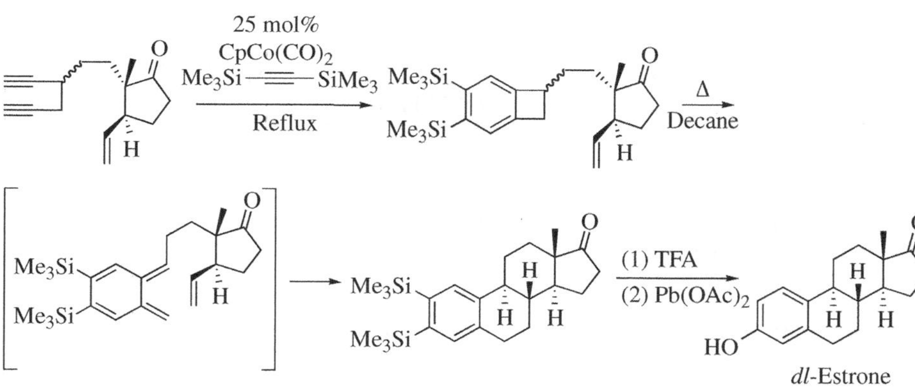

SCHEME 21.8 Co-catalyzed [2+2+2] cycloaddition of diyne with monoyne for synthesis of *dl*-estrone.

The ruthenium-catalyzed [2+2+2] cycloaddition of a functionalized chloro-diyne with a functionalized propargylic alcohol furnished the corresponding chlorinated indene derivative as a single regioisomer. This product was successfully transformed into the natural product, sporolide B (Scheme 21.9) [12].

The use of heteroatom-linked diynes enables the efficient synthesis of substituted heterofluorenes. For example, substituted carbazoles were synthesized by the $RhCl(PPh_3)_3$-catalyzed [2+2+2] cycloaddition of tosylamide-linked 1,6-diynes with terminal alkynes (Scheme 21.10) [13].

Substituted dibenzofurans were also synthesized by the cationic rhodium(I)/H_8-BINAP complex-catalyzed [2+2+2] cycloaddition of phenol-linked 1,6-diynes with alkynes (Scheme 21.11) [14].

The [2+2+2] cycloaddition of silicon-linked diynes with monoynes afforded 9-silafluorene (silole) [15]. This reaction was applied to the synthesis of a densely substituted ladder-type silafluorene (Scheme 21.12) [15].

Iodoalkynes could be employed in the ruthenium catalysis. The double [2+2+2] cycloaddition of diiodotetrayne with acetylene afforded 4,4′-diiodobiaryl [16a]. This biaryl was subjected to the

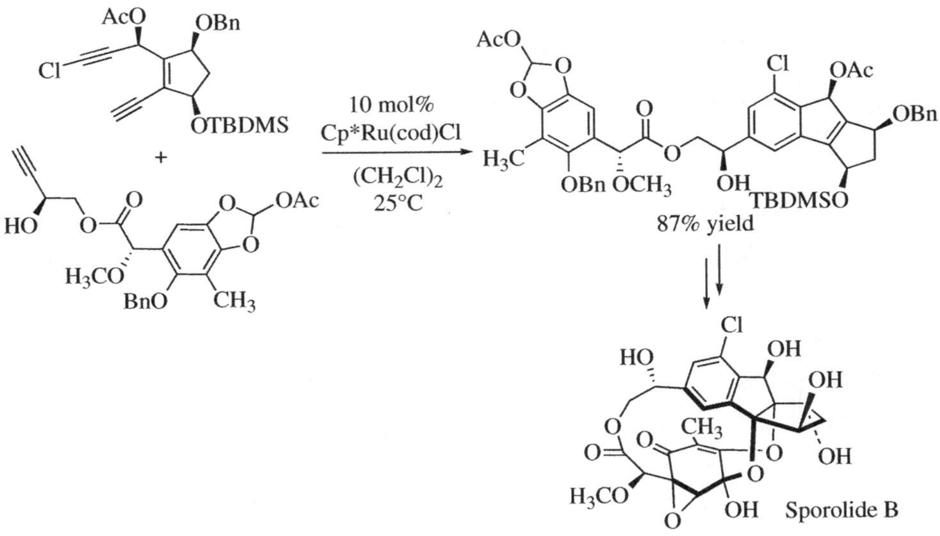

SCHEME 21.9 Ru-catalyzed [2+2+2] cycloaddition of chloro-diyne with monoyne for synthesis of sporolide B.

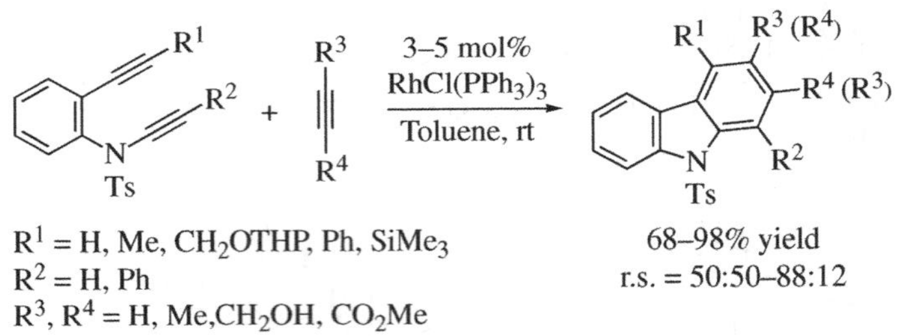

SCHEME 21.10 Rh-catalyzed [2+2+2] cycloaddition of tosylamide-linked diynes with alkynes.

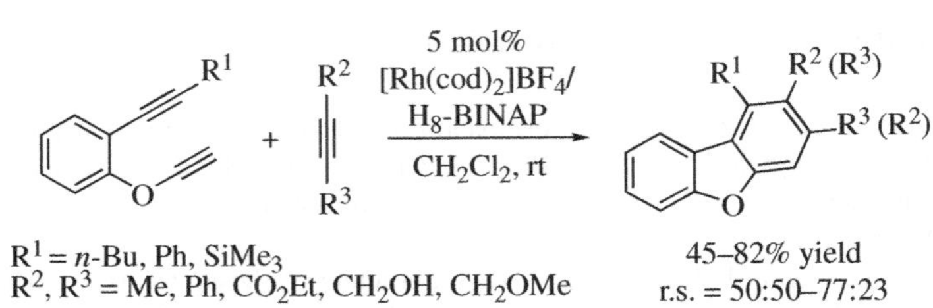

SCHEME 21.11 Rh-catalyzed [2+2+2] cycloaddition of phenol-linked diynes with alkynes.

palladium-catalyzed double Suzuki–Miyaura cross-coupling with *p*-biarylboronate, prepared from diynylboronate and acetylene [16b], to give the corresponding hexaphenylene (Scheme 21.13) [16a].

A π-conjugated long ladder molecule, containing triphenylene and fluorene fragments, was successfully synthesized via the rhodium-catalyzed [2+2+2] cycloaddition followed by dehydration (Scheme 21.14) [17].

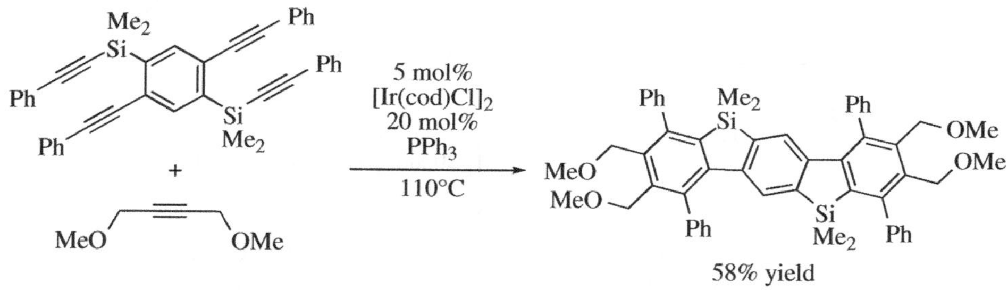

SCHEME 21.12 Ir-catalyzed [2+2+2] cycloaddition of silicon-linked diyne with alkyne.

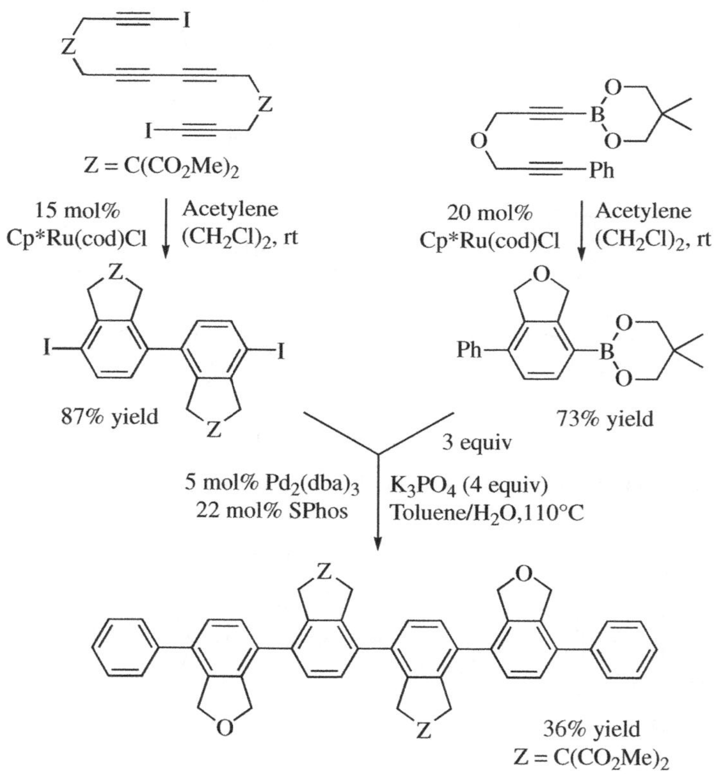

SCHEME 21.13 Ru-catalyzed [2+2+2] cycloaddition of diiododiyne with acetylene.

Not only a planar π-conjugated molecule but also nonplanar π-conjugated molecules, highly curved buckybowls containing a corannulene fragment, were also synthesized via the rhodium-catalyzed [2+2+2] cycloaddition followed by dehydrochlorination (Scheme 21.15) [18].

The transition metal-catalyzed [2+2+2] cycloaddition has been applied to the atroposelective biaryl synthesis. The first example is the chiral cobalt(I) complex-catalyzed [2+2+2] cycloaddition of aryl-diynes with nitriles to produce axially chiral arylpyridines [19a]. Subsequently, the enantioselective synthesis of axially chiral biaryl phosphine oxides has also been achieved (Scheme 21.16) [19b].

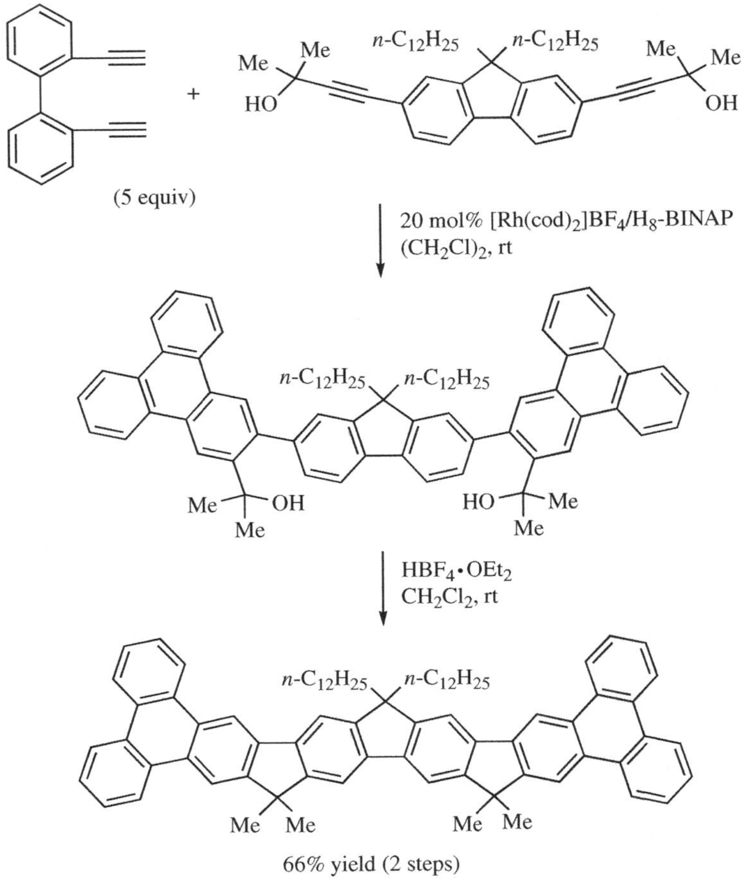

SCHEME 21.14 Synthesis of long ladder molecule containing triphenylene and fluorene fragments.

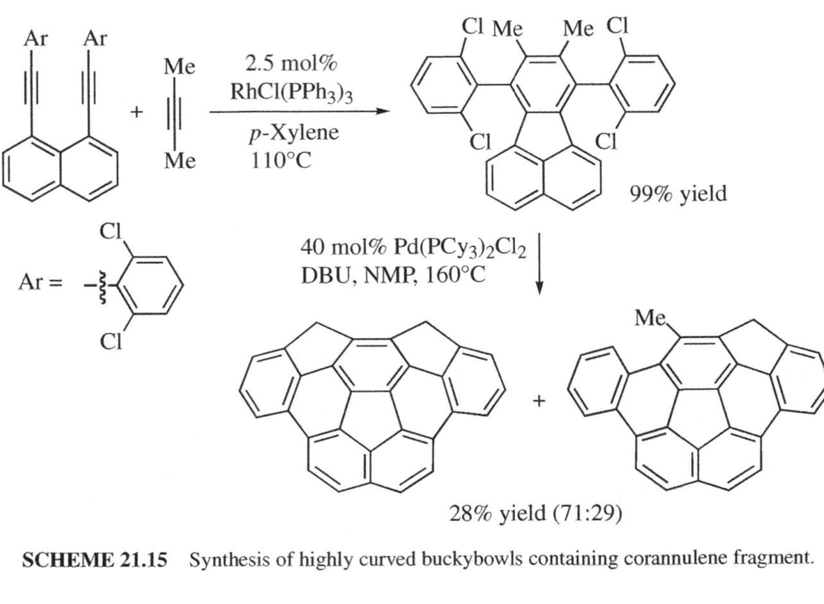

SCHEME 21.15 Synthesis of highly curved buckybowls containing corannulene fragment.

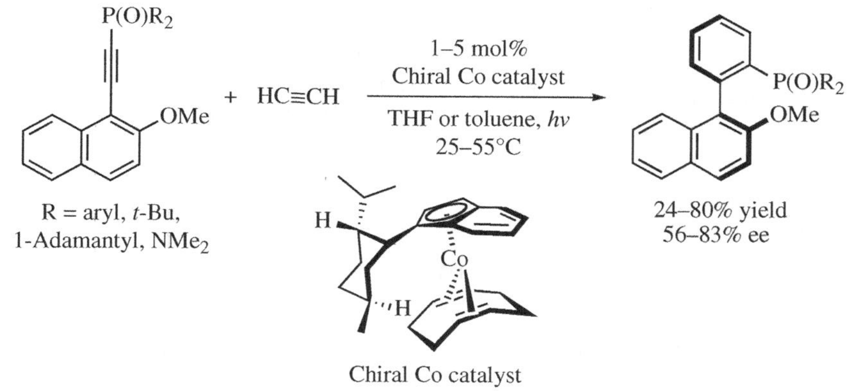

SCHEME 21.16 Co-catalyzed atroposelective [2+2+2] cycloaddition.

The second example is the neutral iridium(I)/Me-Duphos complex-catalyzed [2+2+2] cycloaddition [20]. This reaction afforded 1,4-teraryls with two atropisomeric chiralities with excellent enantio- and diastereoselectivity (Scheme 21.17) [20].

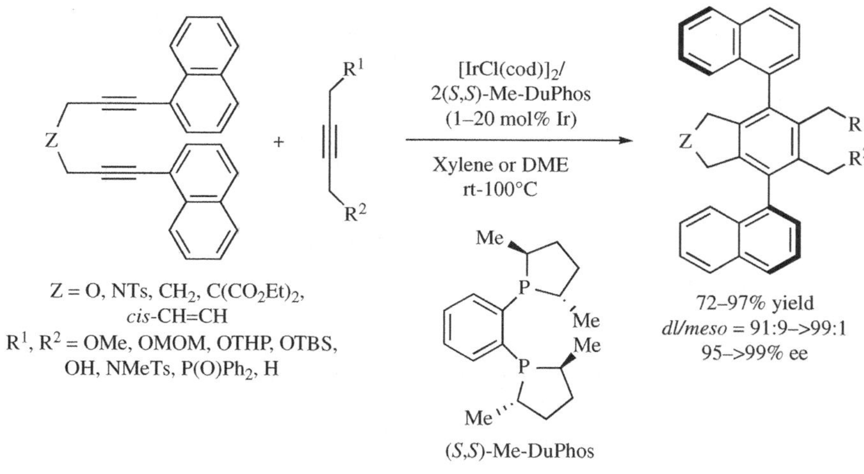

SCHEME 21.17 Ir-catalyzed atroposelective [2+2+2] cycloaddition.

The third example is the cationic rhodium(I)/H_8-BINAP complex-catalyzed [2+2+2] cycloaddition of aryl-diynes with monoynes [21]. This catalyst has been applied to the enantioselective synthesis of axially chiral biaryl phosphine oxides (Scheme 21.18) [22a]. The atroposelective synthesis of axially chiral biaryl esters was also reported [22b].

The application of the rhodium(I)-catalyzed double [2+2+2] cycloaddition approach to the synthesis of symmetric biaryl diphosphorus compounds was first accomplished in the reactions of 1,4-bis(diphenylphosphinoyl)buta-1,3-diyne with terminal diynes to give achiral biaryl bisphosphine oxides [23a]. C_2-Symmetric axially chiral biaryl diphosphonates were obtained with perfect enantioselectivity by using a phosphonate-substituted 1,3-butadiyne and internal 1,6-diynes (Scheme 21.19) [23b]. Axially chiral biaryl dicarboxylates were also obtained by this method [23b].

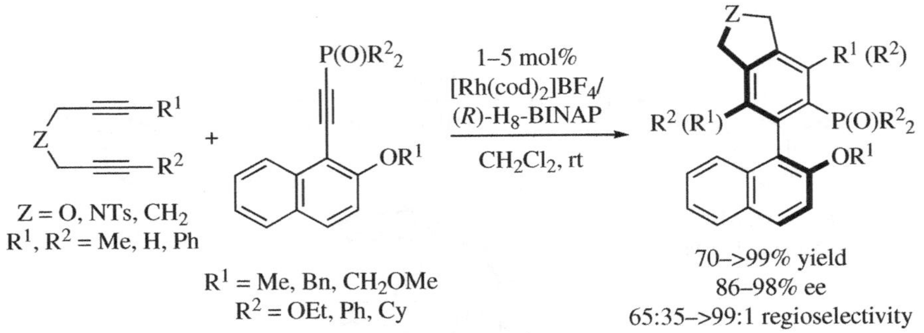

SCHEME 21.18 Rh-catalyzed atroposelective [2+2+2] cycloaddition.

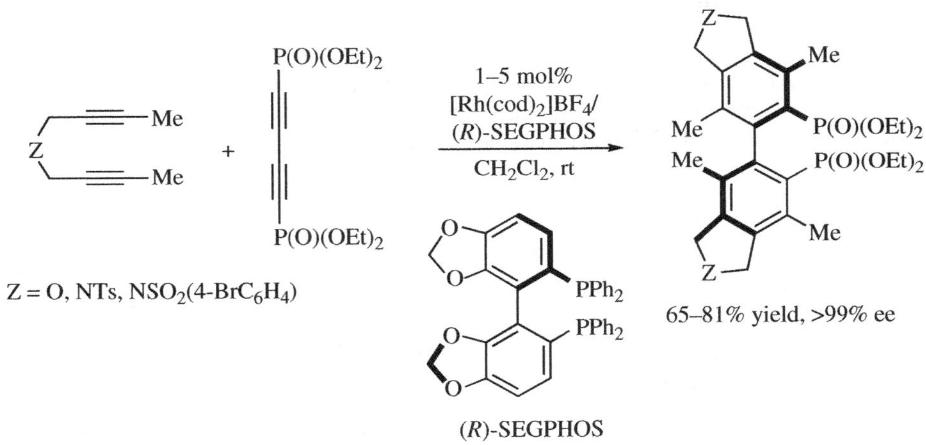

SCHEME 21.19 Rh-catalyzed atroposelective double [2+2+2] cycloaddition.

Substituted tetraphenylenes are known as interesting biaryl-based chiral cyclic scaffolds. The cationic rhodium(I)/Cy-BINAP or QuinoxP* complex-catalyzed enantioselective double homo-[2+2+2] cycloaddition of triynes afforded chiral tetraphenylenes with high enantioselectivity (Scheme 21.20) [24].

Anilides, bearing a sterically demanding *ortho*-substituent, are known to exist as atropisomers due to a high rotational barrier around aryl-nitrogen single bond. The cationic rhodium(I)/xyl-BINAP complex-catalyzed enantioselective [2+2+2] cycloaddition of 1,6-diynes with trimethylsilylynamides afforded axially chiral anilides with good to excellent enantioselectivity (Scheme 21.21) [25].

The high-yielding and highly enantioselective synthesis of carba[10]paracyclophanes has been achieved by the cationic rhodium(I)/(*S,S*)-BDPP complex-catalyzed [2+2+2] cycloaddition of cyclic diynes with terminal monoynes (Scheme 21.22) [26].

The cationic rhodium(I)/axially chiral biaryl bisphosphine complex-catalyzed [2+2+2] cycloaddition of biaryl-linked tetraynes with dialkynylketones or dialkynylphosphine oxides afforded helically chiral 1,1′-bitriphenylenes, containing a densely substituted fluorenone or phosphafluorene core (Scheme 21.23) [27].

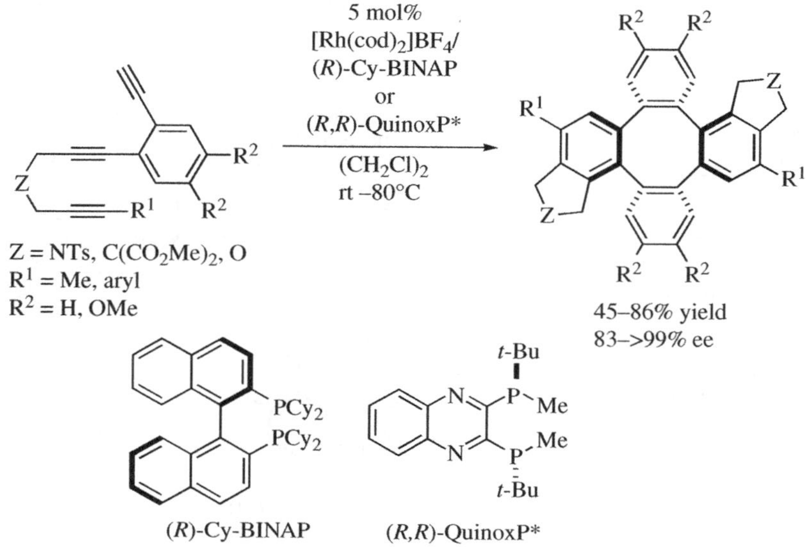

SCHEME 21.20 Enantioselective synthesis of chiral tetraphenylenes.

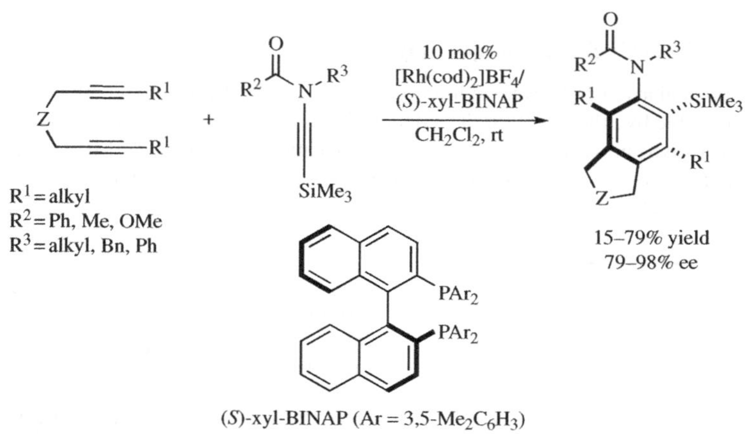

SCHEME 21.21 Enantioselective synthesis of axially chiral anilides.

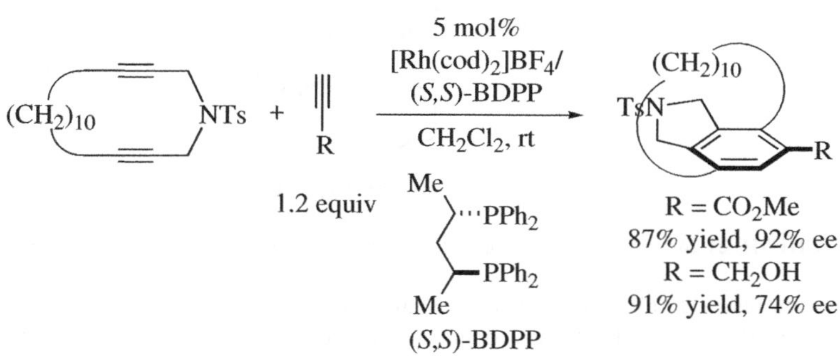

SCHEME 21.22 Enantioselective synthesis of planar chiral cyclophanes.

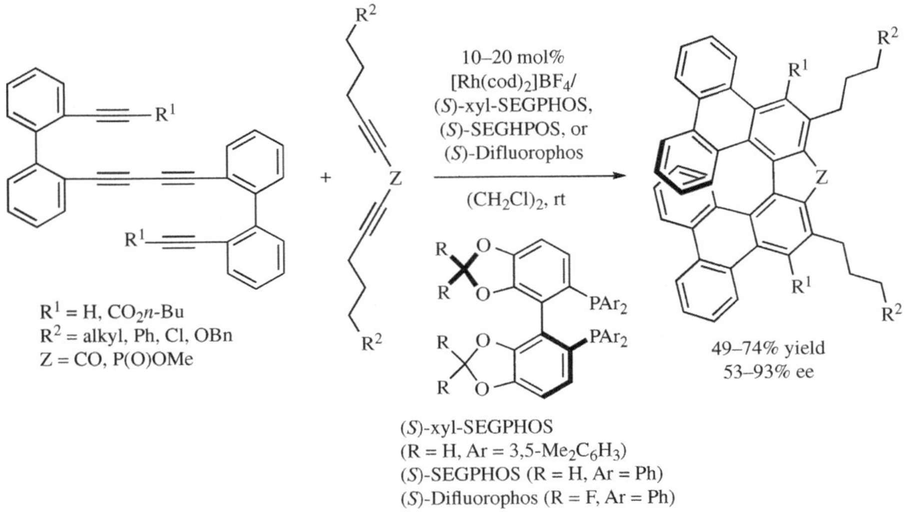

SCHEME 21.23 Enantioselective synthesis of helically chiral 1,1′-bitriphenylenes.

21.2.4 [2+2+2] Cycloaddition of Triynes

The intramolecular [2+2+2] cycloaddition of triynes affords tricyclic compounds, which are not readily accessible by other methods. The double [2+2+2] cycloaddition of a diphenylphosphinoyl-substituted hexayne proceeded in the presence of the cationic rhodium(I)/tol-BINAP catalyst to give the corresponding C_2-symmetric axially chiral biaryl bisphosphine oxide with high enantioselectivity (Scheme 21.24) [28].

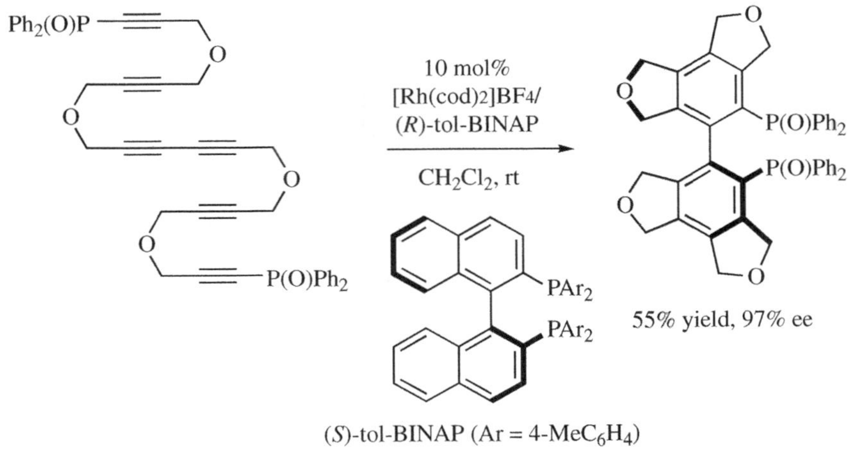

SCHEME 21.24 Rh-catalyzed atroposelective intramolecular double [2+2+2] cycloaddition.

Interesting planar chiral tripodal cyclophanes have also been synthesized with high enantioselectivity by the cationic rhodium(I)/Me-Duphos complex-catalyzed [2+2+2] cycloaddition of branched triynes (Scheme 21.25) [29].

The intramolecular [2+2+2] cycloaddition of triynes is particularly effective for the synthesis of helicenes and helicene-like molecules. For example, the novel direct synthesis of fully

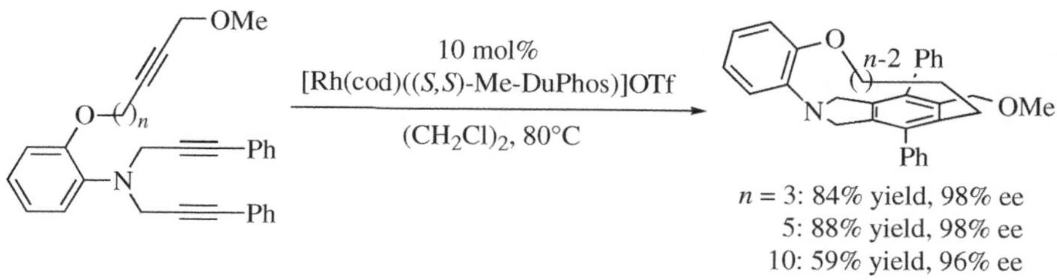

SCHEME 21.25 Enantioselective synthesis of planar chiral tripodal cyclophanes.

aromatic [5]–[7]helicenes was achieved in the nickel-catalyzed [2+2+2] cycloaddition of *cis,cis*-dienetriynes (Scheme 21.26) [30].

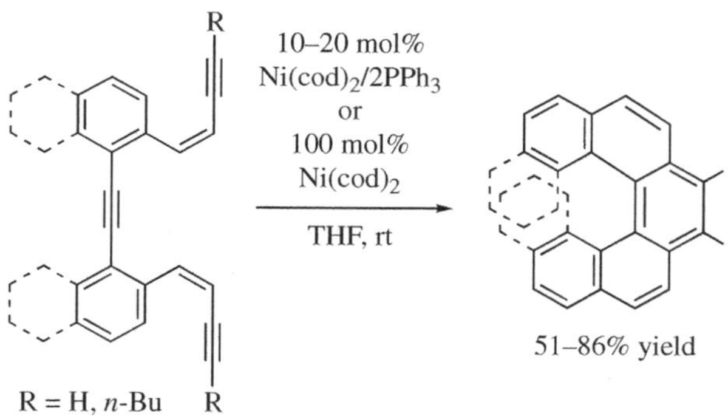

SCHEME 21.26 Synthesis of fully aromatic [5]–[7]helicenes.

The highly enantioselective synthesis of [7]helicene-like molecules has been achieved in the cationic rhodium(I)/Me-Duphos complex-catalyzed [2+2+2] cycloaddition of triynes (Scheme 21.27) [31].

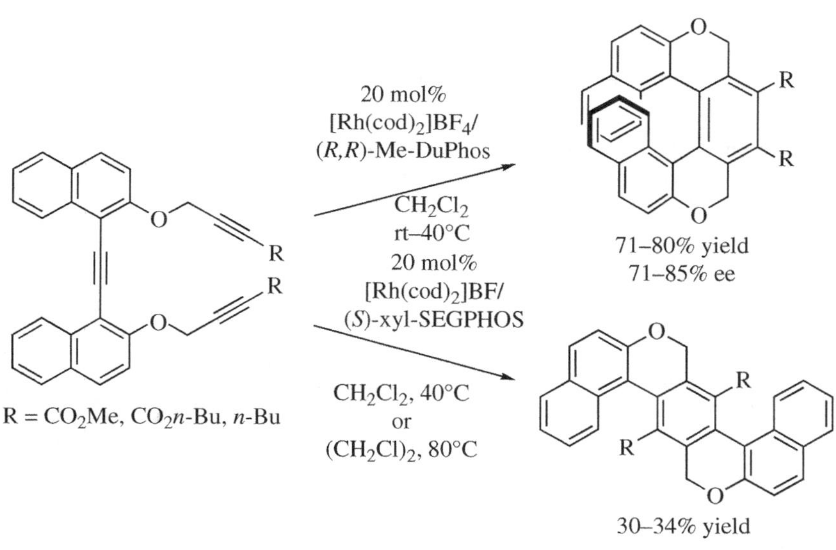

SCHEME 21.27 Enantioselective synthesis of [7]helicene-like molecules.

Interestingly, the use of xyl-Segphos as a ligand afforded unexpected [2+1+2+1] products through the C≡C triple-bond cleavage (Scheme 21.27) [31].

The sequential [2+2+2] cycloaddition enabled the synthesis of long helicenes. The total syntheses of [7]heliphenes and [9]heliphenes were achieved via the cobalt-mediated double and triple [2+2+2] cycloadditions (Scheme 21.28) [32].

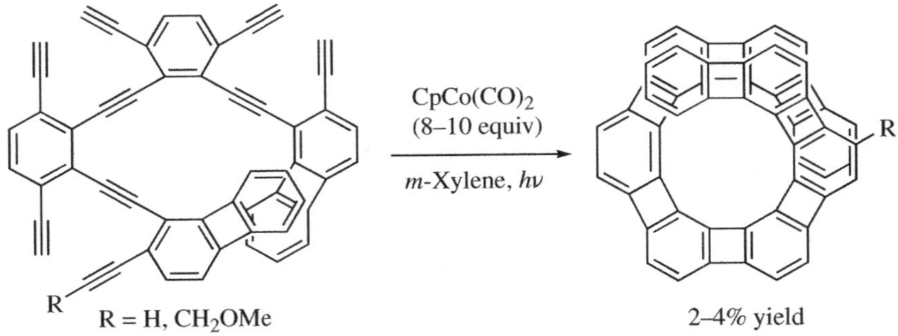

SCHEME 21.28 Syntheses of [7]heliphenes and [9]heliphenes.

A racemic anthra[11]helicene was also prepared via the cobalt-mediated double [2+2+2] cycloaddition of a hexayne followed by the acetic acid elimination and dehydrogenation (Scheme 21.29) [33].

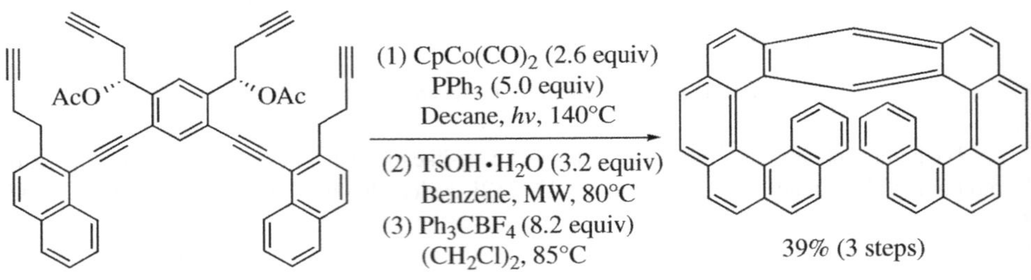

SCHEME 21.29 Syntheses of anthra[11]helicene.

Very recently, the enantioselective synthesis of a completely *ortho*-fused [11]helicene-like molecule has been achieved via the rhodium-mediated intramolecular double [2+2+2] cycloadditions (Scheme 21.30) [34].

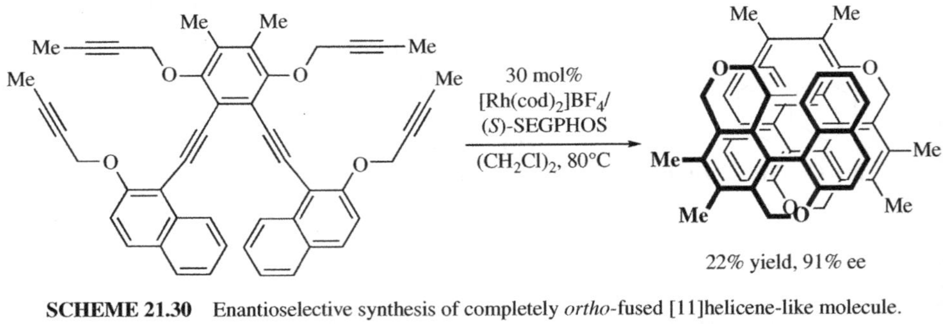

SCHEME 21.30 Enantioselective synthesis of completely *ortho*-fused [11]helicene-like molecule.

21.3 [3+2+1] CYCLOADDITION

The [3+2+1] cycloaddition of an α,β-unsaturated or aryl carbene complex of chromium, an alkyne, and carbon monoxide, that is named the Dötz benzannulation, is a useful method for the synthesis of a phenol or naphthol derivative, although this reaction requires a stoichiometric amount of the chromium carbene complex of chromium [35]. A mechanism of this reaction is shown in Scheme 21.31. When unsymmetrical alkynes are used, the regioselectivity is determined in the alkyne insertion step by the steric effect.

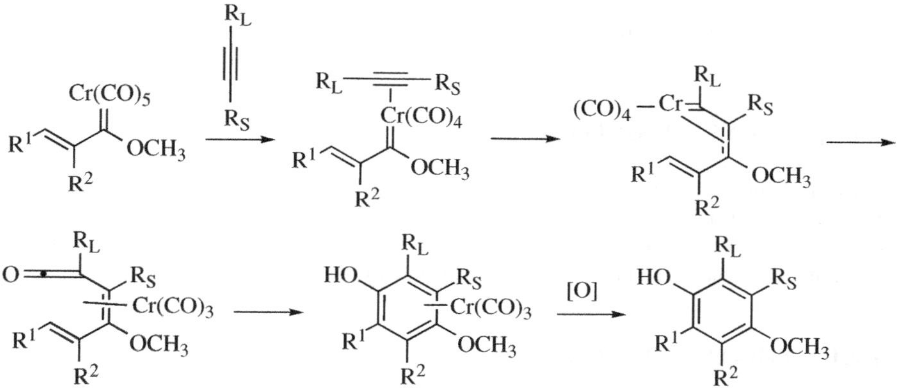

SCHEME 21.31 Mechanism of Dötz benzannulation.

Several total syntheses employed the Dötz benzannulation as key steps. For example, the Dötz benzannulation was employed in the total syntheses of Vitamin E (Scheme 21.32) [36].

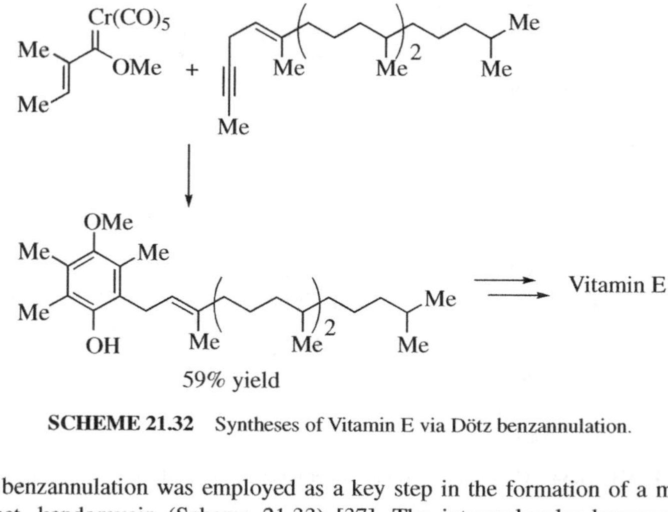

SCHEME 21.32 Syntheses of Vitamin E via Dötz benzannulation.

The Dötz benzannulation was employed as a key step in the formation of a metacyclophane natural product, kendomycin (Scheme 21.33) [37]. The intramolecular benzannulation of the vinylcarbene and alkyne, linked with the highly functionalized tether, afforded the corresponding metacyclophane in good yield.

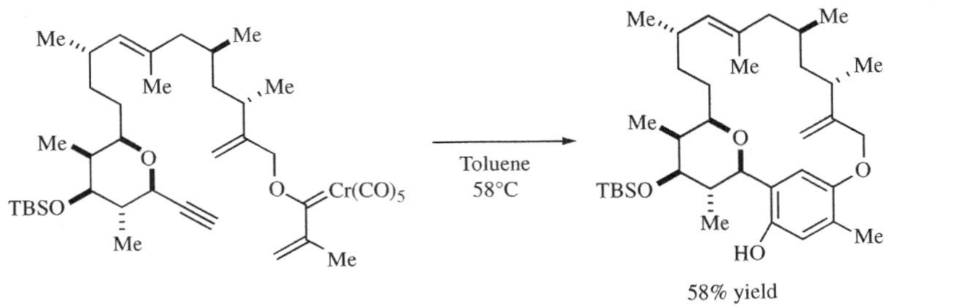

SCHEME 21.33 Syntheses of metacyclophane by Dötz benzannulation.

21.4 [4+2] CYCLOADDITION

In this subchapter, I summarized the intermolecular two-component cycloaddition reactions, which are able to construct benzenoid aromatic rings. Four types of transition metal-catalyzed [4+2] cycloaddition reactions (Diels–Alder reactions, reactions of enynes with alkynes, reactions via pyrylium intermediates, and reactions via acylmetallacycles) are described.

21.4.1 Diels–Alder Reactions

The thermal Diels–Alder reaction of electron-rich 1,3-dienes with electron-deficient alkynes affords the corresponding cyclohexa-1,4-dienes, aromatization of which gives the corresponding benzenes. As the thermal Diels–Alder reaction proceeds in a concerted fashion that is best described by the Woodward–Hoffmann rules, it is difficult to use electronically neutral substrates. On the other hand, as the transition metal-catalyzed Diels–Alder reaction may proceed in a stepwise fashion, electronically neutral substrates can be employed. For the transition metal-catalyzed Diels–Alder reaction, cobalt-based complexes are the most frequently employed catalysts [38], although some rhodium-based complexes have also been used [39].

For example, the reaction of a 1,3-diene with alkynyl sulfides in the presence of an *in situ* generated active cobalt catalyst followed by dehydrogenation with DDQ afforded the corresponding trisubstituted benzene with good regioselectivity (Scheme 21.34) [40].

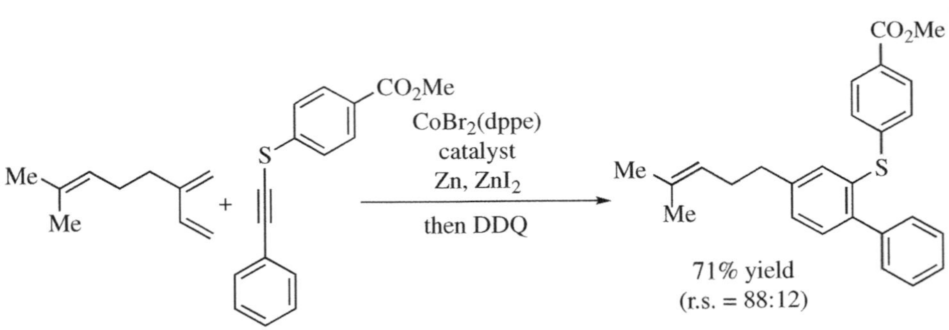

SCHEME 21.34 Co-catalyzed Diels–Alder reaction of 1,3-diene with alkynyl sulfides.

A proposed mechanism of the cobalt-catalyzed Diels–Alder reaction is shown in Schemes 21.35 [41]. Reduction of $CoBr_2$(dppe) generates cationic Co(I)-dppe complex **A**. This active catalyst **A** reacts with a diene and alkyne to give Co(I)-diene complex **B**. This complex **B** may be transformed into Co(I)-cyclohexadiene complex **C**. Although a precise mechanism is not understood, this transformation may proceed in a stepwise fashion. Releasing a cyclohexadiene regenerates the active catalyst **A**.

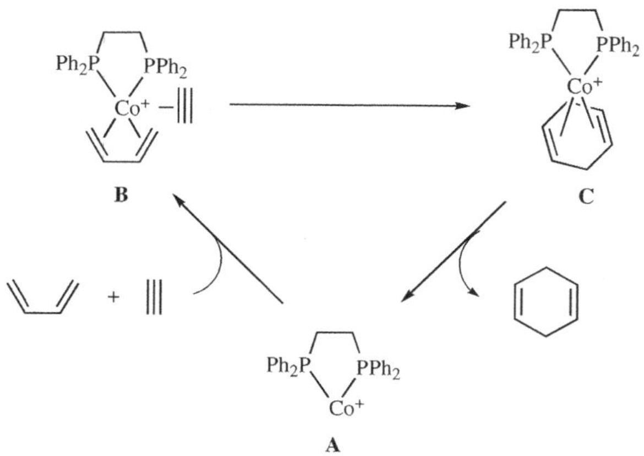

SCHEME 21.35 Mechanism of Co-catalyzed Diels–Alder reaction.

On the other hand, the reactions of a dienyl ether or sulfide with an alkynylborane in the presence of the same cobalt catalyst afforded the corresponding disubstituted benzenes through elimination of phenol or benzenethiol (Scheme 21.36) [42].

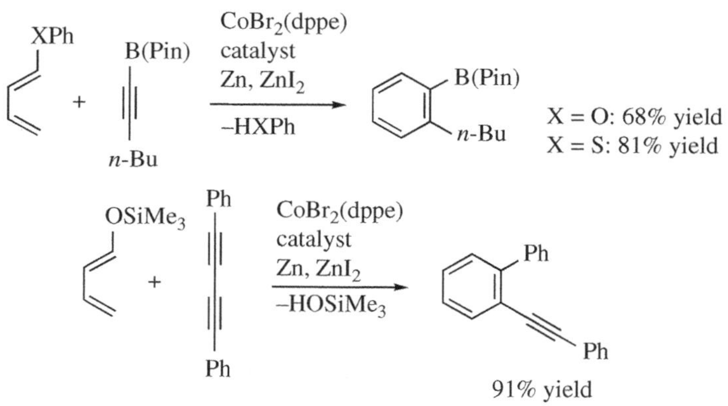

SCHEME 21.36 Co-catalyzed Diels–Alder reaction of dienyl ether or sulfide with alkynylborane.

An interesting ligand effect was observed in the cobalt-catalyzed regioselective Diels–Alder reaction of 2-aryl-1,3-butadiene with an alkynylborane. The use of dppe as a ligand predominately afforded the *para*-substituted product. On the contrary, the use of a pyridine–imine ligand predominately afforded the *meta*-substituted product (Scheme 21.37) [43].

21.4.2 Reactions of Enynes with Alkynes

The Lewis acid-mediated [4+2] cycloaddition of conjugated enynes with alkynes affords arenes; however, it is limited to the intramolecular cycloaddition [44]. The use of transition metal complexes realizes the catalytic intermolecular variants. The first example was reported in the palladium(0)-catalyzed homo-[4+2] cycloaddition of monosubstituted conjugated enynes giving substituted styrenes (Scheme 21.38) [45]. Importantly, this reaction proceeded with definite regioselectivity.

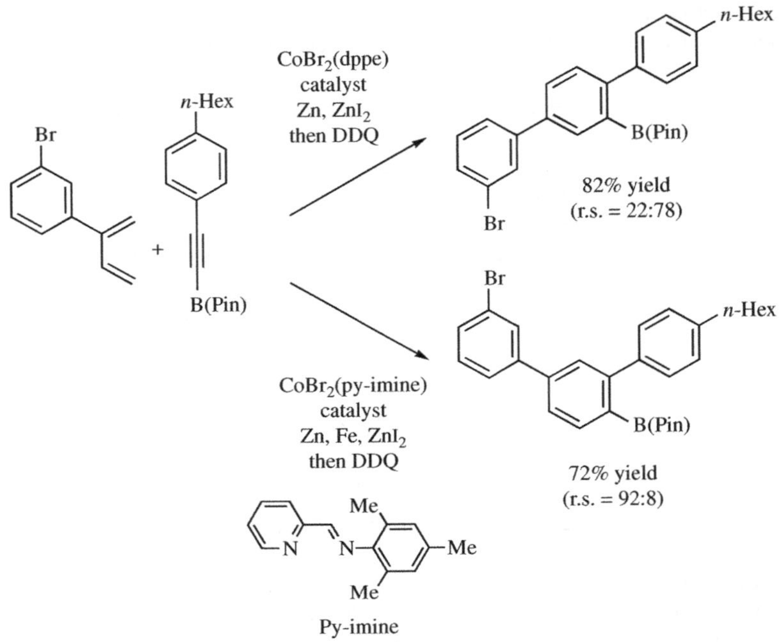

SCHEME 21.37 Co-catalyzed Diels–Alder reaction of 2-aryl-1,3-butadiene with alkynylborane.

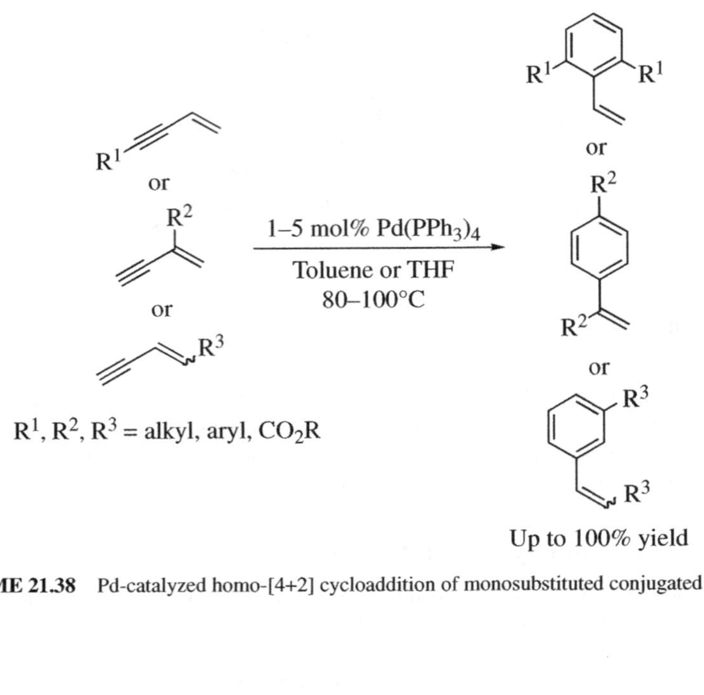

SCHEME 21.38 Pd-catalyzed homo-[4+2] cycloaddition of monosubstituted conjugated enynes.

When electron-withdrawing groups exist in the substrates, 1,2- or 2,4-disubstituted enynes can participate in this cycloaddition (Scheme 21.39) [45c, d].

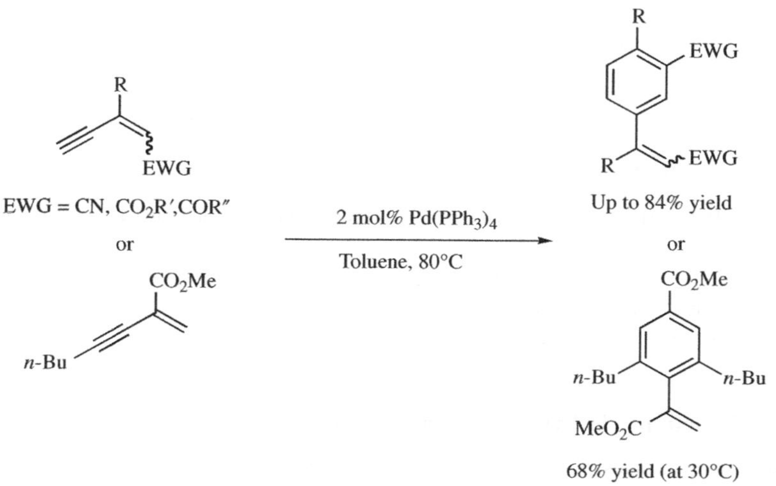

SCHEME 21.39 Pd-catalyzed homo-[4+2] cycloaddition of electron-deficient disubstituted enynes.

Dimerization through the [4+2] cycloaddition of electron-rich 2,4-disubstituted enynes was realized by using a palladium(II)/*N*-heterocyclic carbene/electron-rich phosphine catalyst system (Scheme 21.40) [46].

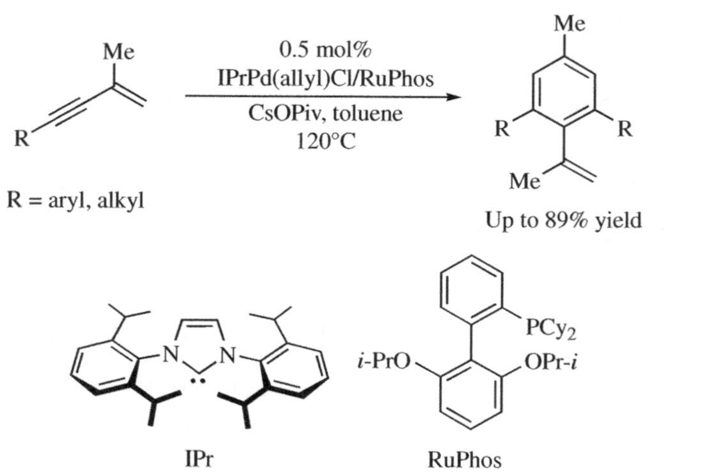

SCHEME 21.40 Pd-catalyzed homo-[4+2] cycloaddition of electron-rich disubstituted enynes.

The palladium-catalyzed cross-[4+2] cycloaddition proceeded between conjugated enynes and diynes with excellent chemo- and regioselectivity to give densely substituted aryl acetylenes (Scheme 21.41) [47].

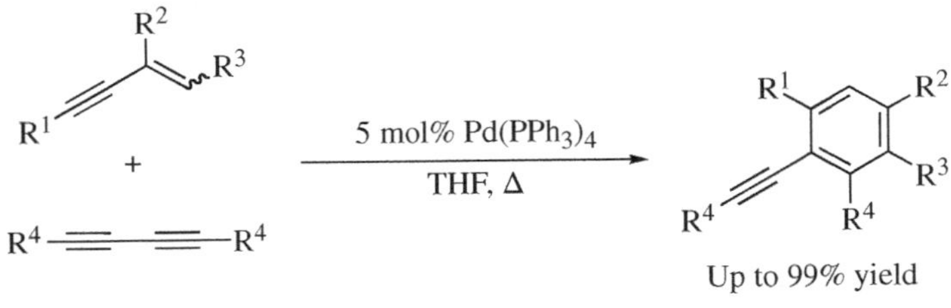

R^1, R^2, R^3 = H, alkyl, aryl, TMS, CO_2R

SCHEME 21.41 Pd-catalyzed cross-[4+2] cycloaddition of conjugated enynes and diynes.

Although a mechanism of the above palladium-catalyzed cross-benzannulation between conjugated enynes and diynes is not perfectly understood, the reaction may proceed via not a concerted mechanism but a stepwise mechanism [48].

21.4.3 Reactions via Pyrylium Intermediates

Pyrylium intermediates have been widely used as a four carbon atoms unit in various [4+2] cycloaddition reactions. The *in situ* gold(III)-catalyzed formation of pyrylium intermediates from a 2-alkynylbenzaldehyde followed by the [4+2] cycloaddition with alkynes and bond shift afforded trisubstituted naphthalenes (Scheme 21.42) [49].

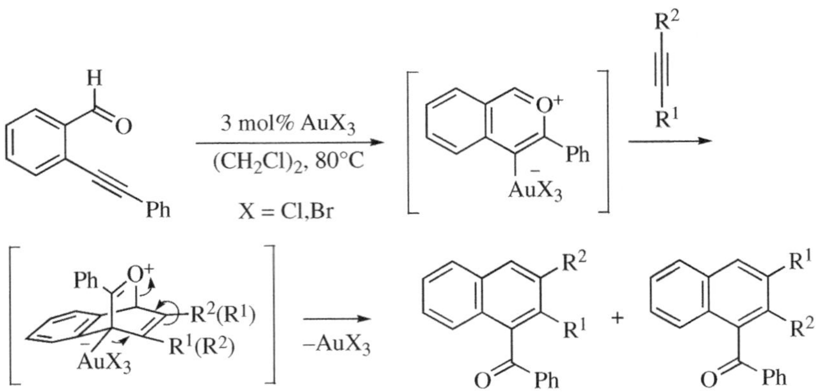

R^1 = Ph, R^2 = H: 100% yield (r.s. = >99:1)
R^1 = Ph, R^2 = TMS: 92% yield (r.s. = >99:1)
R^1 = Ac, R^2 = H: 75% yield (r.s. = >99:1)

SCHEME 21.42 Au-catalyzed [4+2] cycloaddition of 2-alkynylbenzaldehyde with alkynes.

Enolizable carbonyl compounds could also be employed in the gold(III)-catalyzed [4+2] cycloaddition with 2-alkynylbenzaldehydes to give the corresponding naphthyl ketones (Scheme 21.43) [50]. This reaction may proceed via the inverse electron demand Diels–Alder reaction of benzopyrylium intermediates with the enol form of carbonyl compounds followed by dehydration.

Interestingly, the reactions of the 2-alkynylbenzaldehyde with alkynes were conducted in the presence of $Cu(OTf)_2$ (5 mol %) and a Brønsted acid (1 equiv) afforded decarbonylated naphthalenes

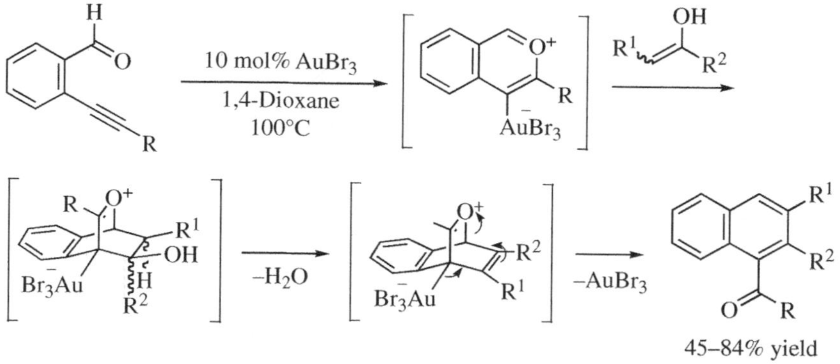

SCHEME 21.43 Au-catalyzed [4+2] cycloaddition of 2-alkynylbenzaldehyde with enolizable carbonyl compounds.

(Scheme 21.44) [49]. This reaction would proceed via the protonolysis of the C—Cu bond of the [4+2] cycloaddition intermediate followed by the bond rearrangement.

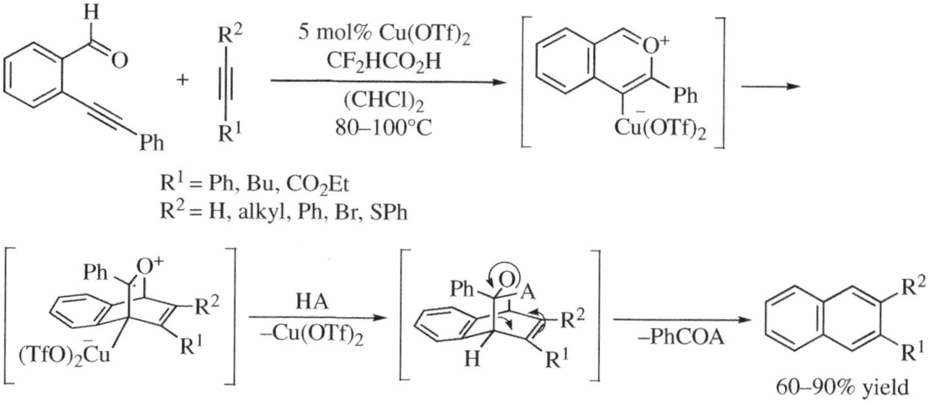

SCHEME 21.44 Cu-catalyzed [4+2] cycloaddition of 2-alkynylbenzaldehyde with alkynes.

Platinum(II) chloride catalyzed the [4+2] cycloaddition of 2-alkynylbenzoates with vinyl ethers to give the corresponding naphthyl ketones (Scheme 21.45) [51]. This reaction may proceed via the [3+2] cycloaddition of platinum carbonyl ylides with the vinyl ethers followed by 1,2-alkyl migration.

21.4.4 Reactions via Acylmetallacycles

The [4+2] cycloaddition reactions between acylmetallacycles and alkynes afford substituted phenols. For example, the nickel-catalyzed [4+2] cycloaddition of 3-phenylcyclobutenone with 3-hexyne afforded the corresponding tetrasubstituted phenol (Scheme 21.46) [52]. This cycloaddition proceeds via an acylnickelacycle generated through carbon–carbon bond cleavage with the nickel(0) complex.

A rhodium(I) carbonyl complex also catalyzed the [4+2] cycloaddition of 2,3-dipropylcyclobutenone with 3-hexyne via an acylrhodacycle at elevated temperature to give the corresponding pentasubstituted phenol (Scheme 21.47) [53].

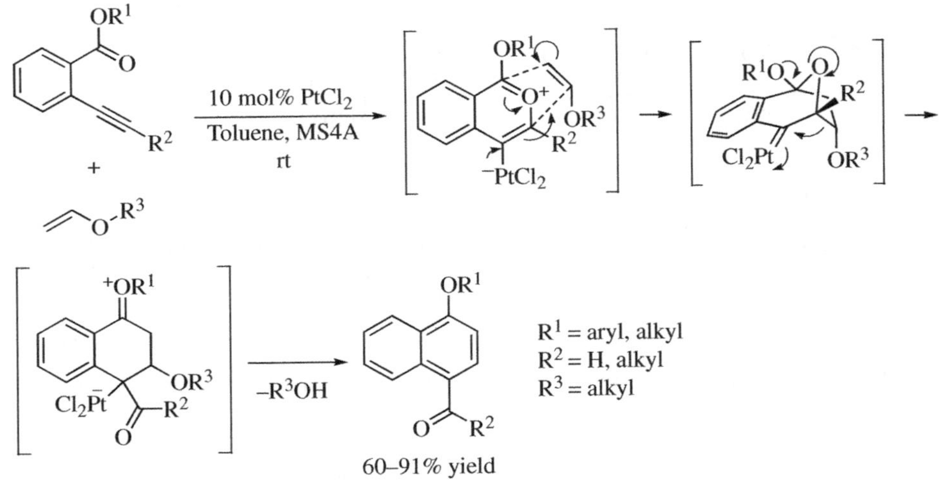

SCHEME 21.45 Pt-catalyzed [4+2] cycloaddition of 2-alkynylbenzaldehyde with vinyl ethers.

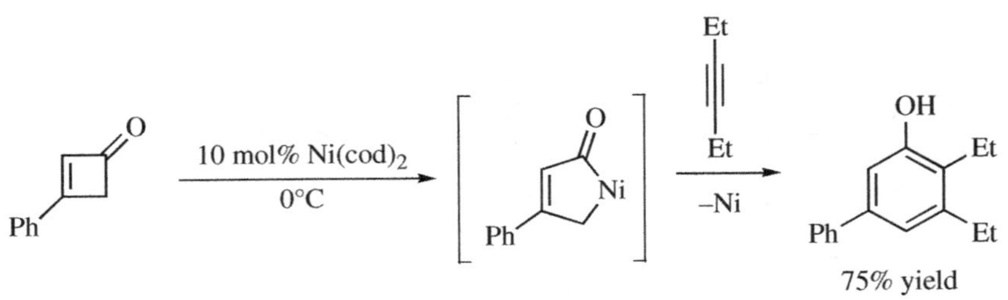

SCHEME 21.46 Ni-catalyzed [4+2] cycloaddition of 3-phenylcyclobutenone with 3-hexyne.

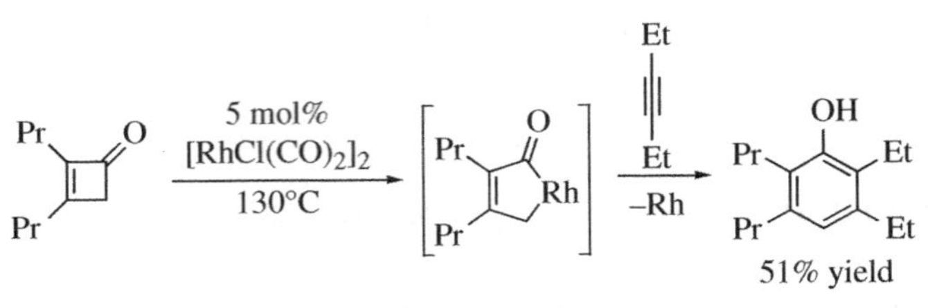

SCHEME 21.47 Ni-catalyzed [4+2] cycloaddition of 2,3-dipropylcyclobutenone with 3-hexyne.

21.5 INTRAMOLECULAR CYCLOAROMATIZATION

In this subchapter, I summarized the intramolecular single-component reactions, which are able to construct benzenoid aromatic rings. Two types of cycloaromatization reactions (intramolecular hydroarylation of alkynes and cyclization via transition metal vinylidenes) are described.

21.5.1 Intramolecular Hydroarylation of Alkynes

The intramolecular hydroarylation of alkynes is a useful method for the synthesis of fused arenes. For example, biphenyl derivatives, bearing an alkyne unit at the *ortho* position, were converted into substituted phenanthrene derivatives in the presence of various π-electrophilic transition metal catalysts such as platinum, palladium, gold, gallium, indium, iron, and so on (Scheme 21.48) [54].

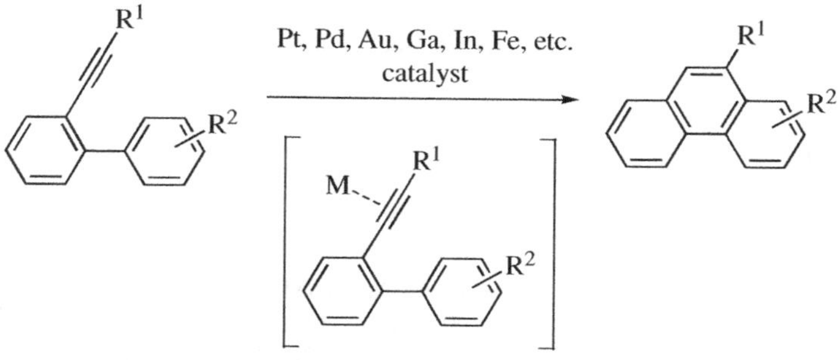

SCHEME 21.48 Transition metal-catalyzed intramolecular hydroarylation of alkynes.

The sequential intramolecular hydroarylation of alkynes is applied to the synthesis of structurally complex extended π-systems. The gold-catalyzed sequential intramolecular hydroarylation of triynes followed by aromatization with DDQ proceeded to give triaryl-substituted diacenaphtho[1,2-*j*:1′,2′-l] fluoranthenes, which can be used for organic light-emitting devices (Scheme 21.49) [55].

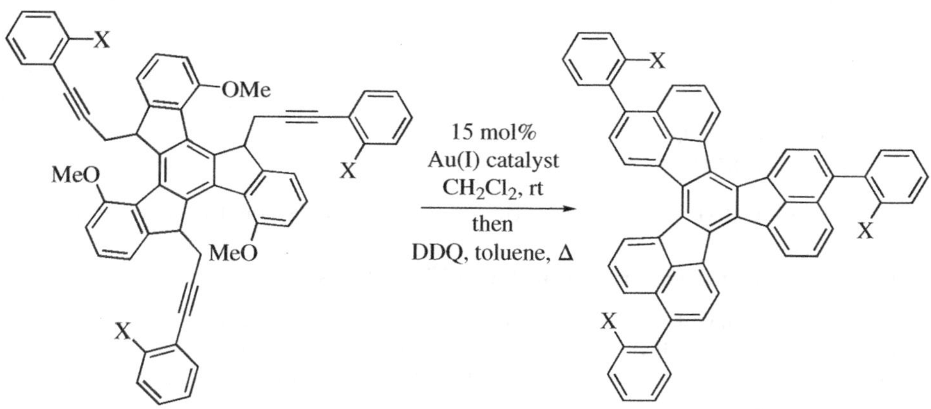

X = H, Br

72–77% yield

SCHEME 21.49 Au-catalyzed sequential intramolecular hydroarylation of triynes.

The sequential intramolecular hydroarylation of alkynes has been applied to the catalytic helicene synthesis. The platinum-catalyzed sequential intramolecular hydroarylation of diynes proceeded to give substituted [6]helicenes (Scheme 21.50) [56].

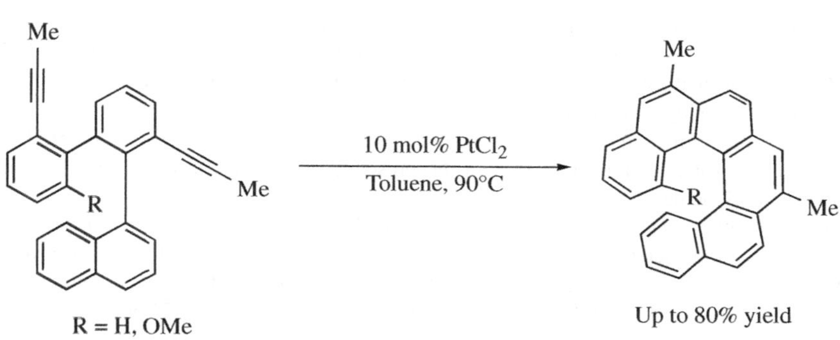

R = H, OMe

Up to 80% yield

SCHEME 21.50 Synthesis of substituted [6]helicenes.

Very recently, the enantio- and diastereoselective synthesis of an S-shaped double azahelicene has been achieved via the gold-catalyzed sequential intramolecular alkyne hydroarylation (Scheme 21.51) [57].

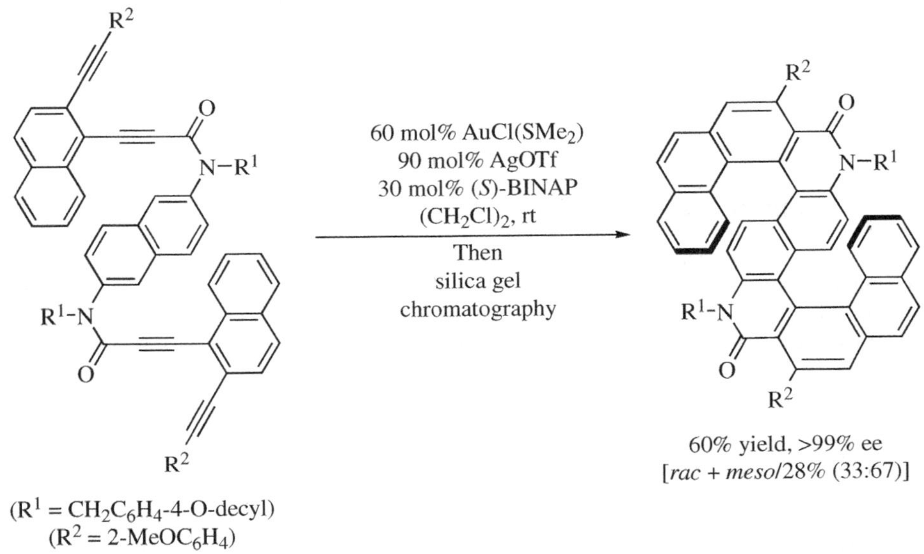

SCHEME 21.51 Enantio- and diastereoselective synthesis of S-shaped double azahelicene.

21.5.2 Cyclization via Transition Metal Vinylidenes

Transition metal–vinylidene complexes [58] are generated through the reaction of transition metal complexes and terminal alkynes via 1,2-hydrogen transfer (Scheme 21.52). The reactive transition metal–vinylidene complexes thus generated have been used in organic synthesis including the synthesis of substituted arenes [59].

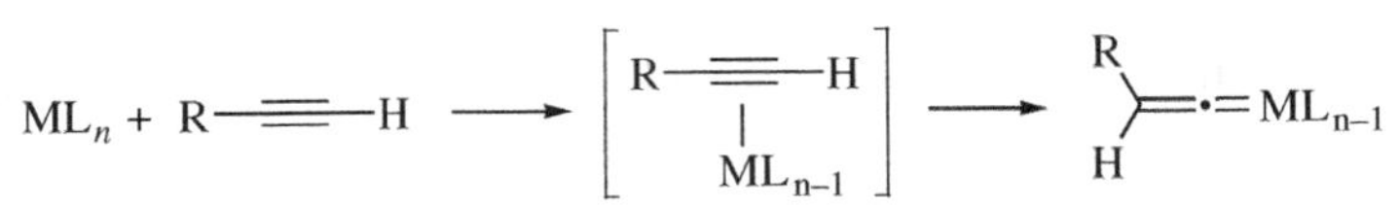

SCHEME 21.52 Formation of transition metal–vinylidene complexes from terminal alkynes.

For example, the cycloaromatization of tungsten–vinylidene complexes, generated from 2-ethynylstyrenes, proceeded to give the corresponding substituted naphthalenes (Scheme 21.53) [60].

The cycloaromatization of a conjugated dienyne afforded the corresponding tricyclic product via a ruthenium–vinylidene complex (Scheme 21.54) [61].

In the ruthenium-catalyzed cycloaromatization of a benzene-linked diyne, double C—C bond forming cycloaromatization proceeded in the presence of an external carbon nucleophile, ethyl acetoacetate (Scheme 21.55) [62].

Not only ruthenium but also rhodium was used in the transition metal-catalyzed cycloaromatization via transition metal–vinylidene complexes. For example, the reaction of an acyclic (*Z*)-3-ene-1,5-diyne in the presence of $RhCl(P(i\text{-}Pr)_3)_2$ (5 mol%) afforded the corresponding allylic benzene presumably through the rhodium–vinylidene complex (Scheme 21.56) [63].

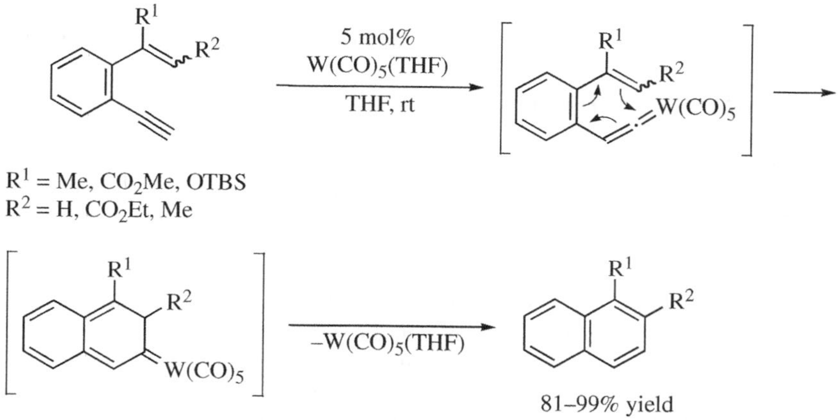

SCHEME 21.53 Cycloaromatization via W–vinylidene complex.

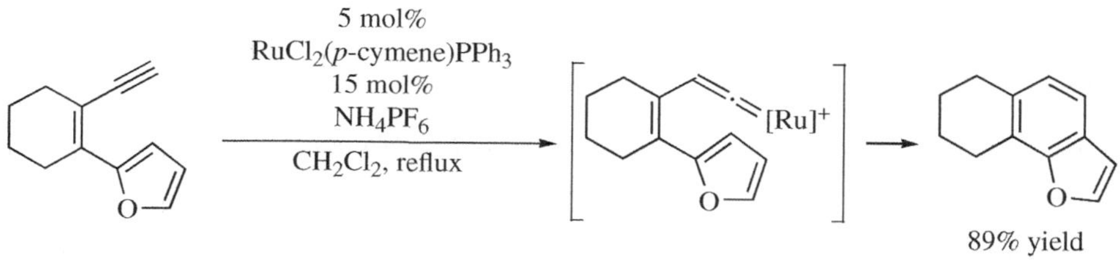

SCHEME 21.54 Cycloaromatization via Ru–vinylidene complex.

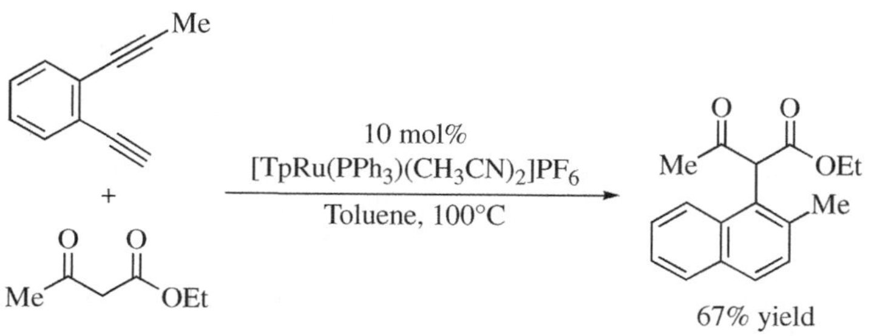

SCHEME 21.55 Ru-catalyzed double C—C bond forming cycloaromatization.

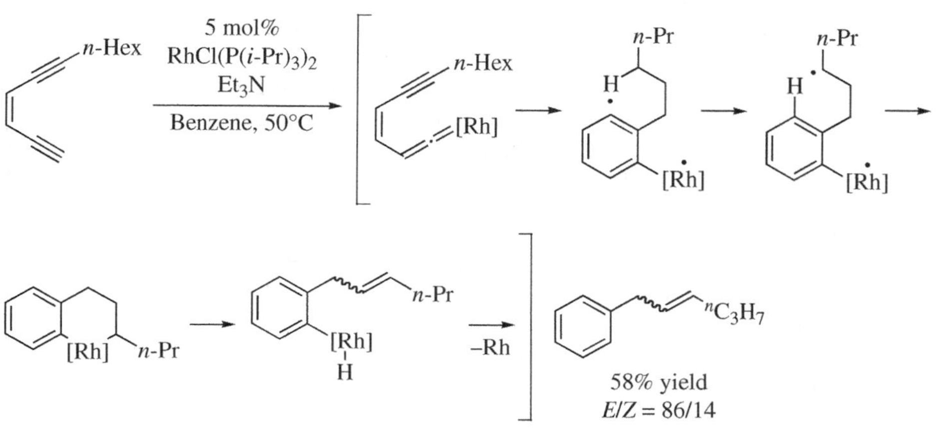

SCHEME 21.56 Cycloaromatization via Rh–vinylidene complex.

21.6 SUMMARY AND OUTLOOK

In this chapter, the intermolecular multicomponent aromatic ring construction reactions and intramolecular single-component aromatic ring construction reactions are described. Among them, the [2+2+2] cycloaddition and intramolecular hydroarylation reactions are the most widely employed and reliable method. Various polycyclic and sterically hindered aromatic compounds have been synthesized by this method. In the past 10 years, the asymmetric [2+2+2] cycloaddition and intramolecular hydroarylation reactions have been developed, which enabled the enantioselective synthesis of sterically hindered chiral aromatic compounds, such as axially chiral biaryls, planar chiral cyclophanes, and helically chiral helicenes. Details of the transition metal-mediated aromatic ring construction reactions are comprehensively covered in the recently published book [1]. Further development of efficient and powerful transition metal-mediated aromatic ring construction reactions would enable the more facile and diverse synthesis of complex benzenoid aromatic compounds.

REFERENCES

[1] For a review, see: *Transition-Metal-Mediated Aromatic Ring Construction*, (Ed.: K. Tanaka), John Wiley & Sons, Inc., Hoboken, 2013.

[2] W. Reppe, W. J. Schweckendiek, *Justus Liebigs Ann. Chem.* 1948, **560**, 104.

[3] (a) K. P. C. Vollhardt, *Acc. Chem. Res.* 1977, **10**, 1; (b) K. P. C. Vollhardt, *Angew. Chem. Int. Ed. Engl.* 1984, **23**, 539; (c) H. Yamazaki, *J. Synth. Org. Chem. Jpn.* 1987, **45**, 244.

[4] For selected recent reviews, see: (a) S. Okamoto, Y. Sugiyama, *Synlett* 2013, **24**, 1044; (b) D. L. J. Broere, E. Ruijter, *Synthesis* 2012, **44**, 2639; (c) Y. Shibata, K. Tanaka, *Synthesis* 2012, **44**, 323; (d) N. Weding, M. Hapke, *Chem. Soc. Rev.* 2011, **40**, 4525; (e) G. Domínguez, J. Pérez-Castells, *Chem. Soc. Rev.* 2011, **40**, 3430; (f) P. A. Inglesby, P. A. Evans, *Chem. Soc. Rev.* 2010, **39**, 2791; (g) B. R. Galan, T. Rovis, *Angew. Chem. Int. Ed.* 2009, **48**, 2830; (h) K. Tanaka, *Chem. Asian J.* 2009, **4**, 508; (i) T. Shibata, K. Tsuchikama, *Org. Biomol. Chem.* 2008, **6**, 1317; (j) K. Tanaka, *Synlett* 2007, 1977; (k) B. Heller, M. Hapke, *Chem. Soc. Rev.* 2007, **36**, 1085; (l) N. Agenet, O. Buisine, F. Slowinski, V. Gandon, C. Aubert, *M. Malacria in Organic Reactions* (Ed.: L. E. Overman), John Wiley & Sons, Inc., Hoboken, 2007, vol. **68**, p. 1; (m) P. R. Chopad, J. Louie, *Adv. Synth. Catal.* 2006, **348**, 2307; (n) V. Gandon, C. Aubert, M. Malacria, *Chem. Commun.* 2006, 2209; (o) S. Kotha, E. Brahmachary, K. Lahiri, *Eur. J. Org. Chem.* 2005, 4741; (p) V. Gandon, C. Aubert, M. Malacria, *Curr. Org. Chem.* 2005, **9**, 1699; (q) Y. Yamamoto, *Curr. Org. Chem.* 2005, **9**, 503.

[5] N. Agenet, V. Gandon, K. P. C. Vollhardt, M. Malacria, C. Aubert, *J. Am. Chem. Soc.* 2007, **129**, 8860.

[6] Y. Yamamoto, T. Arakawa, R. Ogawa, K. Itoh, *J. Am. Chem. Soc.* 2003, **125**, 12143.

[7] (a) K. Tanaka, K. Shirasaka, *Org. Lett.* 2003, **5**, 4697; (b) K. Tanaka, K. Toyoda, A. Wada, K. Shirasaka, M. Hirano, *Chem. Eur. J.* 2005, **11**, 1145.

[8] J. Areephong, H. Logtenberg, W. R. Browne, B. L. Feringa, *Org. Lett.* 2010, **12**, 2132.

[9] T. Takahashi, F. Tsai, Y. Li, K. Nakajima, M. Kotora, *J. Am. Chem. Soc.* 1999, **121**, 11093.

[10] (a) T. Takahashi, M. Kitamura, B. Shen, K. Nakajima, *J. Am. Chem. Soc.* 2000, **122**, 12876; (b) T. Takahashi, S. Li, W. Huang, F. Kong, K. Nakajima, B. Shen, T. Ohe, K. Kanno, *J. Org. Chem.* 2006, **71**, 7967.

[11] R. L. Funk, K. P. C. Vollhardt, *J. Am. Chem. Soc.* 1980, **102**, 5253.

[12] (a) K. C. Nicolaou, Y. Tang, J. Wang, *Angew. Chem. Int. Ed.* 2009, **48**, 3449; (b) P. Li, D. Menche, *Angew. Chem. Int. Ed.* 2009, **48**, 5078.

[13] B. Witulski, C. Alayrac, *Angew. Chem. Int. Ed.* 2002, **41**, 3281.

[14] Y. Komine, A. Kamisawa, K. Tanaka, *Org. Lett.* 2009, **11**, 2361.

[15] T. Matsuda, S. Kadowaki, T. Goya, M. Murakami, *Org. Lett.* 2007, **9**, 133.

[16] (a) Y. Yamamoto, K. Hattori, H. Nishiyama, *J. Am. Chem. Soc.* 2006, **128**, 8336; (b) Y. Yamamoto, K. Hattori, J. Ishii, H. Nishiyama, *Tetrahedron* 2006, **62**, 4294.

[17] K. Murayama, Y. Sawada, K. Noguchi, K. Tanaka, *J. Org. Chem.* 2013, **78**, 6202.

[18] T.-C. Wu, H.-J. Hsin, M.-Y. Kuo, C.-H. Li, Y.-T. Wu, J. Am, *Chem. Soc.* 2011, **133**, 16319.

[19] (a) A. Gutnov, B. Heller, C. Fischer, H.-J. Drexler, A. Spannenberg, B. Sundermann, C. Sundermann, *Angew. Chem. Int. Ed.* 2004, **43**, 3795; (b) B. Heller, A. Gutnov, C. Fischer, H.-J. Drexler, A. Spannenberg, D. Redkin, C. Sundermann, B. Sundermann, *Chem. Eur. J.* 2007, **13**, 1117.

[20] (a) T. Shibata, T. Fujimoto, K. Yokota, K. Takagi, *J. Am. Chem. Soc.* 2004, **126**, 8382; (b) T. Shibata, K. Tsuchikama, *Chem. Commun.* 2005, **48**, 6017; (c) T. Shibata, Y. Arai, K. Takami, K. Tsuchikama, T. Fujimoto, S. Takebayashi, K. Takagi, *Adv. Synth. Catal.* 2006, **348**, 2475.

[21] K. Tanaka, G. Nishida, A. Wada, K. Noguchi, *Angew. Chem. Int. Ed.* 2004, **43**, 6510.

[22] (a) G. Nishida, K. Noguchi, M. Hirano, K. Tanaka, *Angew. Chem. Int. Ed.* 2007, **46**, 3951; (b) S. Ogaki, Y. Shibata, K. Noguchi, K. Tanaka, *J. Org. Chem.* 2011, **76**, 1926.

[23] (a) S. Doherty, J. G. Knight, C. H. Smyth, R. W. Harrington, W. Clegg, *Org. Lett.* 2007, **9**, 4925; (b) G. Nishida, S. Ogaki, Y. Yusa, T. Yokozawa, K. Noguchi, K. Tanaka, *Org. Lett.* 2008, **10**, 2849.

[24] T. Shibata, T. Chiba, H. Hirashima, Y. Ueno, K. Endo, *Angew. Chem. Int. Ed.* 2009, **48**, 8066.

[25] (a) K. Tanaka, K. Takeishi, K. Noguchi, *J. Am. Chem. Soc.* 2006, **128**, 4586; (b) T. Suda, K. Noguchi, M. Hirano, K. Tanaka, *Chem. Eur. J.* 2008, **14**, 6593.

[26] T. Araki, K. Noguchi, K. Tanaka, *Angew. Chem. Int. Ed.* 2013, **52**, 5617.

[27] Y. Sawada, S. Furumi, A. Takai, M. Takeuchi, K. Noguchi, K. Tanaka, *J. Am. Chem. Soc.* 2012, **134**, 4080.

[28] F. Mori, N. Fukawa, K. Noguchi, K. Tanaka, *Org. Lett.* 2011, **13**, 362.

[29] T. Shibata, T. Uchiyama, K. Endo, *Org. Lett.* 2009, **11**, 3906.

[30] F. Teplý, I. G. Stará, I. Starý, A. Kollárovič, D. Šaman, L. Rulíšek, P. Fiedler, *J. Am. Chem. Soc.* 2002, **124**, 9175.

[31] K. Tanaka, A. Kamisawa, T. Suda, K. Noguchi, M. Hirano, *J. Am. Chem. Soc.* 2007, **129**, 12078.

[32] (a) S. D. Han, D. R. Anderson, A. D. Bond, H. V. Chu, R. L. Disch, D. Holmes, J. M. Schulman, S. J. Teat, K. P. C. Vollhardt, G. D. Whitener, *Angew. Chem. Int. Ed.* 2002, **41**, 3227; (b) S. D. Han, A. D. Bond, R. L. Disch, D. Holmes, J. M. Schulman, S. J. Teat, K. P. C. Vollhardt, G. D. Whitener, *Angew. Chem. Int. Ed.* 2002, **41**, 3223.

[33] P. Sehnal, I. G. Stará, D. Šaman, M. Tichý, J. Míšek, J. Cvačka, L. Rulíšek, J. Chocholousova, J. Vacek, G. Goryl, M. Szymonski, I. Císařová, I. Starý, *Proc. Natl. Acad. Sci. U. S. A.* 2009, **106**, 13169.

[34] Y. Kimura, N. Fukawa, Y. Miyauchi, K. Noguchi, K. Tanaka, *Angew. Chem. Int. Ed.* 2014, **53**, 8480–8483.

[35] K. H. Dötz, *Angew. Chem. Int. Ed. Engl.* 1975, **14**, 644.

[36] K. H. Dötz, W. Kuhn, *Angew. Chem. Int. Ed. Engl.* 1983, **22**, 732.

[37] K. Tanaka, M. Watanabe, K. Ishibayashi, H. Matsuyama, Y. Saikawa, M. Nakata, *Org. Lett.* 2010, **12**, 1700.

[38] G. Hilt, *Chem. Rec.* 2014, **14**, 386.

[39] (a) I. Matsuda, M. Shibata, S. Sato, Y. Izumi, *Tetrahedron Lett.* 1987, **28**, 3361; (b) S.-J. Paik, S. U. Son, Y. K. Chung, *Org. Lett.* 1999, **1**, 2045; (c) S. I. Lee, S. Y. Park, J. H. Park, I. G. Jung, S. Y. Choi, Y. K. Chung, *J. Org. Chem.* 2006, **71**, 91; (d) T. Shibata, D. Fujiwara, K. Endo, *Org. Biomol. Chem.* 2008, **6**, 464.

[40] (a) G. Hilt, S. Lüers, *Synthesis* 2003, 1784; (b) G. Hilt, S. Lüers, K. Harms, *J. Org. Chem.* 2004, **69**, 624.

[41] L. Fiebig, J. Kuttner, G. Hilt, M. Schwarzer, G. Frenking, H.-G. Schmalz, M. Schäfer *J. Org. Chem.* 2013, **78**, 10485.

[42] A.-L. Auvinet, J. P. A. Harrity, G. Hilt, *J. Org. Chem.* 2010, **75**, 3893.

[43] M. Danz, G. Hilt, *Adv. Synth. Catal.* 2011, **353**, 303.

[44] R. L. Danheiser, A. E. Gould, R. Fernandez de la Pradilla, A. L. Helgason, *J. Org. Chem.* 1994, **59**, 5514.

[45] (a) S. Saito, M. M. Salter, V. Gevorgyan, N. Tsuboya, K. Tando, Y. Yamamoto, *J. Am. Chem. Soc.* 1996, **118**, 3970; (b) V. Gevorgyan, K. Tando, N. Uchiyama, Y. Yamamoto, *J. Org. Chem.* 1998, **63**, 7022; (c) S. Saito, N. Tsuboya, Y. Chounan, T. Nogami. Y. Yamamoto, *Tetrahedron Lett.*, 1999, **40**, 7529; (d) S. Saito, Y. Chounan, T. Nogami, T. Fukushi, N. Tsuboya, Y. Yamada, H. Kitahara, Y. Yamamoto, *J. Org. Chem.* 2000, **65**, 5350.

[46] O. V. Zatolochnaya, A. V. Galenko, V. Gevorgyan, *Adv. Synth. Catal.* 2012, **354**, 1149.

[47] (a) V. Gevorgyan, A. Takeda, Y. Yamamoto, *J. Am. Chem. Soc.* 1997, **119**, 11313; (b) V. Gevorgyan, N. Sadayori, Y. Yamamoto, *Tetrahedron Lett.* 1997, **38**, 8603–8604; (c) V. Gevorgyan, A. Takeda, M. Homma, N. Sadayori, U. Radhakrishnan, Y. Yamamoto, *J. Am. Chem. Soc.* 1999, **121**, 6391.

[48] M. Rubina, M. Conley, V. Gevorgyan, *J. Am. Chem. Soc.* 2006, **128**, 5818.

[49] (a) N. Asao, K. Takahashi, S. Lee, T. Kasahara, Y. Yamamoto, *J. Am. Chem. Soc.* 2002, **124**, 12650; (b) N. Asao, T. Nogami, S. Lee, Y. Yamamoto, *J. Am. Chem. Soc.* 2003, **125**, 10921.

[50] N. Asao, H. Aikawa, Y. Yamamoto, *J. Am. Chem. Soc.* 2004, **126**, 7458.

[51] (a) H. Kusama, H. Funami, J. Takaya, N. Iwasawa, *Org. Lett.* 2004, **6**, 605; (b) H. Kusama, H. Funami, N. Iwasawa, *Synthesis* 2007, 2014.

[52] M. A. Huffman, L. S. Liebeskind, *J. Am. Chem. Soc.* 1991, **113**, 2771.

[53] (a) T. Kondo, Y. Taguchi, Y. Kaneko, M. Niimi, T. Mitsudo, *Angew. Chem. Int. Ed.* 2004, **43**, 5369; (b) T. Kondo, M. Niimi, M. Nomura, K. Wada, T. Mitsudo, *Tetrahedron Lett.* 2007, **48**, 2837.

[54] For recent reviews, see: (a) T. Kitamura, *Eur. J. Org. Chem.* 2009, 1111; (b) C. Nevado, A. Echavarren, *Synthesis*, 2005, 167.

[55] S. Pascual, C. Bour, P. de Mendoza, A. M. Echavarren, *Beilstein J. Org. Chem.* 2011, **7**, 1520.

[56] (a) J. Storch, J. Sýkora, J. Čermák, J. Karban, I. Cisařová, A. Růžička, *J. Org. Chem.* 2009, **74**, 3090; (b) J. Storch, M. Bernard, J. Sýkora, J. Karvan, J. Čermák, *Eur. J. Org. Chem.* 2013, 260.

[57] K. Nakamura, S. Furumi, M. Takeuchi, T. Shibuya, K. Tanaka, *J. Am. Chem. Soc.* 2014, **136**, 5555.

[58] (a) O. S. Mills, A. D. Redhouse, *Chem. Commun.* 1966, 444; (b) O. S. Mills, A. D. Redhouse, *J. Chem. Soc. A* 1968, 1282.

[59] For a recent review, see: *Metal Vinylidenes and Allenylidenes in Catalysis*, (Eds.: C. Bruneau, P. H. Dixneuf), Wiley-VCH, Weinheim, 2008.

[60] K. Maeyama, N. Iwasawa, *J. Org. Chem.* 1999, **64**, 1344.

[61] C. A. Merlic, M. E. Pauly, *J. Am. Chem. Soc.* 1996, **118**, 11319.

[62] A. Odedra, C.-J. Wu, T. B. Pratap, C.-W. Huang, Y.-F. Ran, R.-S. Liu, *J. Am. Chem. Soc.* 2005, **127**, 3406.

[63] K. Ohe, M. Kojima, K. Yonehara, S. Uemura, *Angew. Chem. Int. Ed. Engl.* 1996, **35**, 1823.

22

Ar–C BOND FORMATION BY AROMATIC CARBON–CARBON IPSO-SUBSTITUTION REACTION

Maurizio Fagnoni[1] and Sergio M. Bonesi[2]

[1] *PhotoGreen Lab, Department of Chemistry, University of Pavia, Pavia, Italy*
[2] *CIHIDECAR CONICET, Departamento de Química Orgánica, Facultad de Ciencias Exactas y Naturales, Universidad de Buenos Aires, Buenos Aires, Argentina*

22.1 INTRODUCTION

The use of metal-mediated reactions is nowadays recognized as the best approach for the formation of an Ar–C bond, a key reaction for the synthesis of valuable compounds (please see also Chapter 17 of this book) [1]. This process involves the metal-catalyzed cross-coupling reaction between an electrophilic aromatic (Ar–X, mainly an aryl halide or an aryl sulfonate) and a carbon-based nucleophile (Nu–Y; Scheme 22.1). The driving force of these *ipso*-substitution reactions is the formation of an Ar–C bond at the expense of a weaker Ar–X bond (X ≠ H). However, in the last few years, various C-based leaving groups, including carbinol, CN, COOH (COOR), CO-X-CO (X = O, NR, S), and carbonyl groups, have been developed and used in arylation reactions [2, 3]. We designate here such a process as the ARomatic Carbon–carbon Ipso-Substitution (ARCIS) reaction.

Various ARCIS reactions are described here, organized according to the Ar–C bond formed, with emphasis on the preparation of alkyl-substituted, vinyl, and alkynyl aromatics as well as biphenyls.

Arene Chemistry: Reaction Mechanisms and Methods for Aromatic Compounds,
First Edition. Edited by Jacques Mortier.

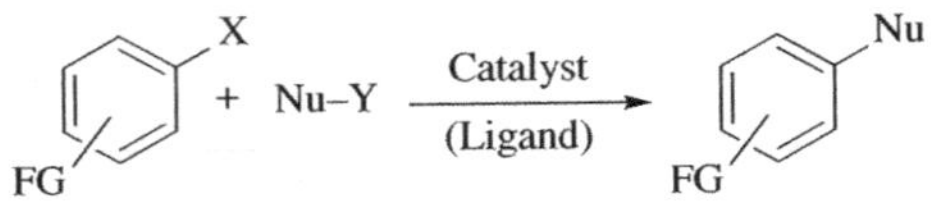

Classical condition: X = mainly I, Br, OSO_2R
ARCIS reaction: X = carbon-based leaving group
Nu–Y = carbon-based nucleophile
Y = B, Sn, Si, Zn, Mg, and others

SCHEME 22.1 Metal-catalyzed Ar–C bond formation.

22.2 FORMATION OF Ar–C(sp^3) BONDS

22.2.1 Ni-Catalyzed Reactions

The activation of the Ar–CN bond by the catalyst $NiCl_2(PMe_3)_2$ has been exploited in cross-coupling reactions with strong nucleophiles, such as Grignard reagents, resulting in an efficient method for the preparation of alkylarene derivatives. The Grignard reagents have typically been modified by the addition of either *t*-BuOLi or PhSLi, in order to reduce the amount of unwanted nucleophilic addition to the nitrile group during the cross-coupling event (Scheme 22.2) [4].

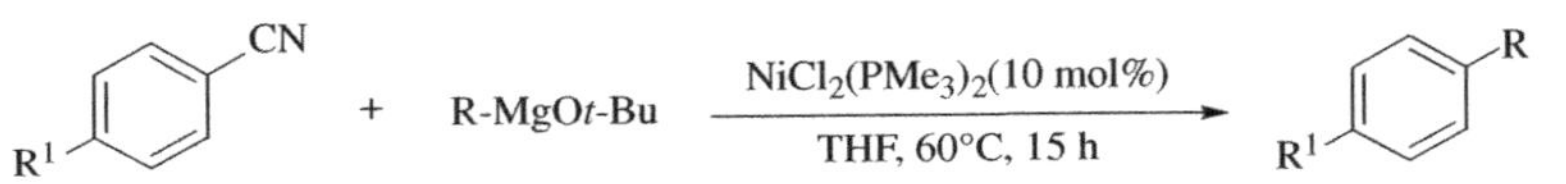

R^1 = H, Me, MeO, F

R = Me, *n*-Bu, cyclopropyl

R^1 = R = Me, 66%
R^1 = Me, R = *n*-Bu, 43%

SCHEME 22.2 Alkylated benzenes from the Ni-catalyzed cleavage of the Ar–CN bond.

The use of milder nucleophiles (e.g., olefins), however, is feasible if C–C bond activation is coupled with alkene insertion as in the arylcyanation of unactivated alkenes [5]. This cleavage followed by addition of both Ar and CN groups to a C–C double bond provides ready access to highly functionalized nitriles. The process can be carried out intermolecularly, a typical example being the arylcyanation of norbornene and norbornadiene to give bicyclo[2.2.1]heptane and bicyclo[2.2.1]heptene derivatives, respectively [6]. However, the intramolecular version of this reaction gives access to indane derivatives in highly enantioenriched form as depicted in Scheme 22.3 [7].

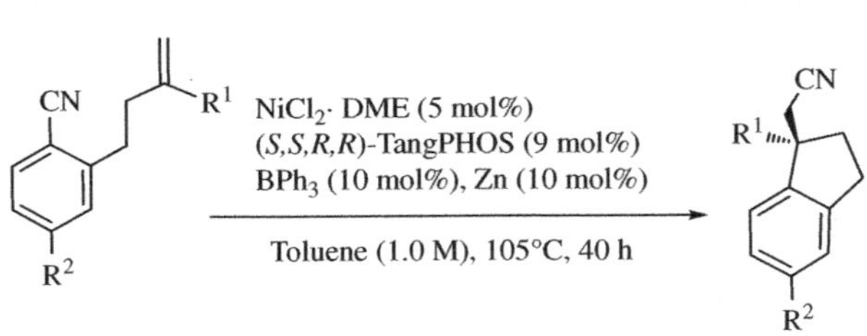

R^1	R^2	Yield (%)
Me	H	85 (93% ee)
Me	F	84 (92% ee)
p-$CF_3C_6H_4$	H	49 (92% ee)

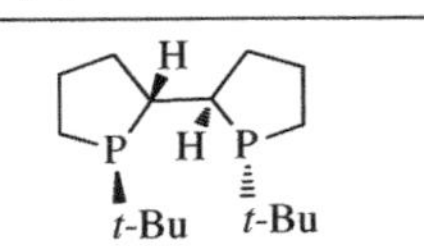

(*S,S,R,R*)-TangPHOS

R^1 = Me, *n*-Pr, *i*-Bu, Ph, *p*-$CF_3C_6H_4$, *p*-$MeOC_6H_4$
R^2 = H, F, MeO

SCHEME 22.3 Nickel-catalyzed intramolecular arylcyanation of alkenes.

Bidentate chiral ligands such as (*S,S,R,R*)-TangPHOS afford the corresponding indanes in moderate to high enantioselectivities under $Ni(cod)_2$ catalysis. However, substrates bearing sterically demanding or electron-deficient alkene substituents were found to require elevated catalyst loadings (10 mol% $NiCl_2$.DME, 18 mol% TangPHOS, 20 mol% BPh_3, and 20 mol% Zn) and extended reaction times to give useful product yields. A related protocol was later applied to the synthesis of natural compounds, namely, (−)-eptazocine and (−)-esermethole [8].

Aryl imides are interesting electrophilic partners that undergo a Ni-mediated decarbonylative cross-coupling reaction in the presence of organozinc reagents such as nucleophiles [9]. This ARCIS reaction has been successful for a range of imide nitrogen substituents, showing a significant functional group tolerance and proceeding with yields generally greater than 75% (Scheme 22.4a) [9]. The mechanism proposed for the reaction is shown in Scheme 22.4b for the case of **1** and involves the initial oxidative addition of a Ni(0) species to form intermediate **2** followed by a transmetalation/migration mechanism upon addition of the organometallic species.

(a)

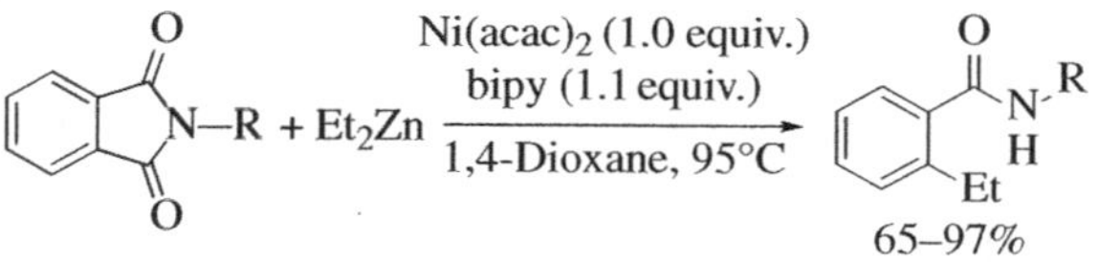

R = Ph, 4-EtC_6H_4, 4-iPrC_6H_4, 4-EtOOCC_6H_4, 3,5-$(CF_3)_2C_6H_3$
2,5-$(MeO)_2C_6H_3$, Me, Bu, CH_2CH_2Ph, $(CH_2)_3NMe_2$

(b)

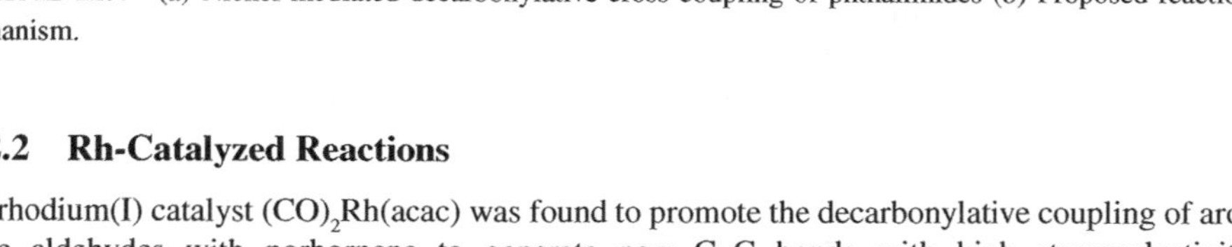

SCHEME 22.4 (a) Nickel-mediated decarbonylative cross-coupling of phthalimides (b) Proposed reaction mechanism.

22.2.2 Rh-Catalyzed Reactions

The rhodium(I) catalyst $(CO)_2Rh(acac)$ was found to promote the decarbonylative coupling of aromatic aldehydes with norbornene to generate new C–C bonds with high stereoselectivity (Scheme 22.5) [10]. The reaction took place through an oxidative addition of the aldehyde C–H

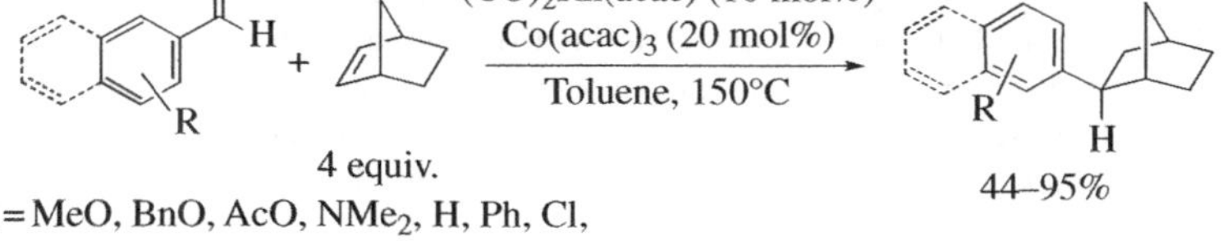

R = MeO, BnO, AcO, NMe_2, H, Ph, Cl,
COOMe, Me, α-naphthyl, β-naphthyl

SCHEME 22.5 Decarbonylative coupling of aromatic aldehydes and norbornenes catalyzed by rhodium.

bond to the metal catalyst and required the presence of $Co(acac)_3$ (20 mol%) as the cocatalyst to avoid the reduction of the Rh(I) species. The exo-product was formed exclusively, demonstrating the excellent stereoselectivity of this reaction [10].

A related approach involves the direct decarbonylation of stable ketones. Daugulis and Brookhart demonstrated that the rhodium-catalyzed decarbonylation of diaryl ketones was feasible [11]. Efficient extrusion of CO from alkyl aryl ketones to form alkylarenes was easily achieved by rhodium(I) catalysis directed by a pyridyl *ortho* to the RCO group (Scheme 22.6) [12]. $(CO)_2Rh(acac)$ was found to be the optimal catalyst and the methodology had a broad substrate scope. This method offers an alternative way to synthesize alkyl benzenes through an ARCIS reaction, complementary to the known Friedel–Crafts alkylation reaction of arenes.

Mechanistically, the decarbonylation is initiated by the oxidative cleavage of the C–C bond in the square planar Rh(I) complex, with the assistance of the pyridine group to form the octahedral acyl/Rh(III) species **3** or **4**. After the reverse migratory insertion, reductive elimination afforded the desired alkylarene and the Rh(I) catalyst that reentered the catalytic cycle (Scheme 22.6b).

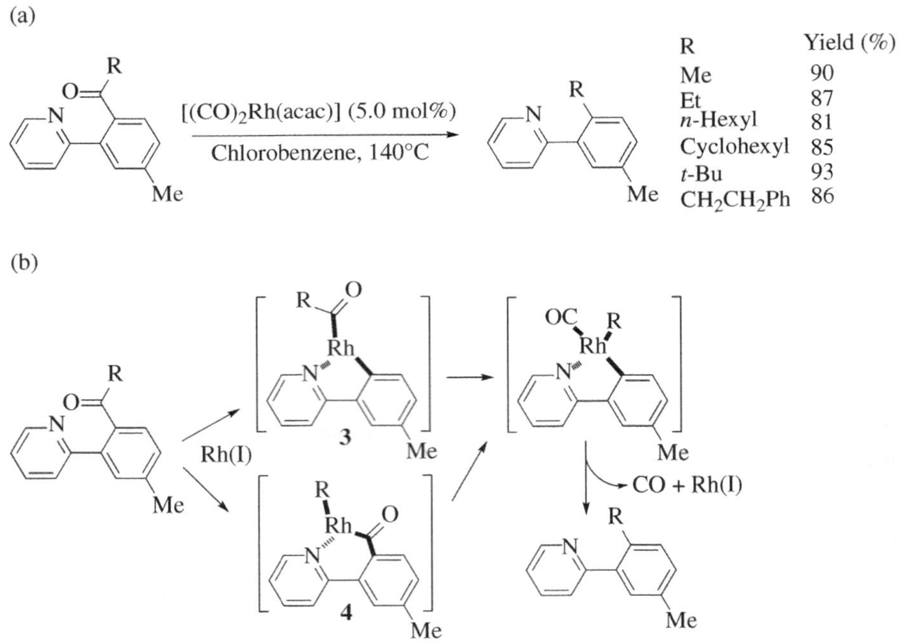

SCHEME 22.6 (a) Rhodium(I)-catalyzed extrusion of CO from aryl ketones and (b) proposed reaction mechanism.

By contrast, the COOH group is lost in the Rh(I)-catalyzed conjugate addition of 2,6-difluorinated benzoic acids to electron-poor olefins (Michael acceptors; Scheme 22.7) [13]. Alkyl acrylates and *N,N*-dimethylacrylamide have been found to give higher yields than methyl vinyl ketone [13]. The use of aqueous toluene as the solvent avoids the formation of the Mizoroki–Heck-type product, whereby the reaction has thus far only been applied to fluorinated benzoic acids **5** [13].

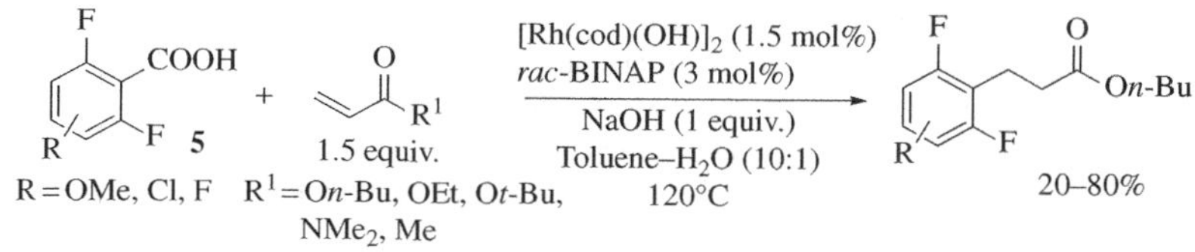

SCHEME 22.7 Rhodium-mediated decarboxylative conjugate addition of fluorinated benzoic acids.

22.2.3 Pd-Catalyzed Reactions

There are only sparse reports of Pd-catalyzed reactions in which an Ar–C(sp^3) bond is formed by an ARCIS reaction [14], one example being the versatile preparation of 2-deoxy-C-aryl glycosides [15]. The strategy was based on a palladium-catalyzed decarboxylative coupling reaction between substituted benzoic acids and glycals in the presence of stoichiometric amounts of Ag_2CO_3. The reaction afforded the desired C-aryl glycosides in moderate to good yields with exclusive regio- and excellent anomeric stereoselectivity (Scheme 22.8). This preparation was inspired by the pioneering work of Myers [16]. Glycals, such as glucals, galactals, glycals derived from rhamnose and ribose, and even a disaccharide, afforded the desired arylated C-glycosides in good yields with complete stereocontrol. The reaction was found to be particularly well suited for benzoic acids substituted with strong electron-donating groups at the 2,4- or 2,6-positions.

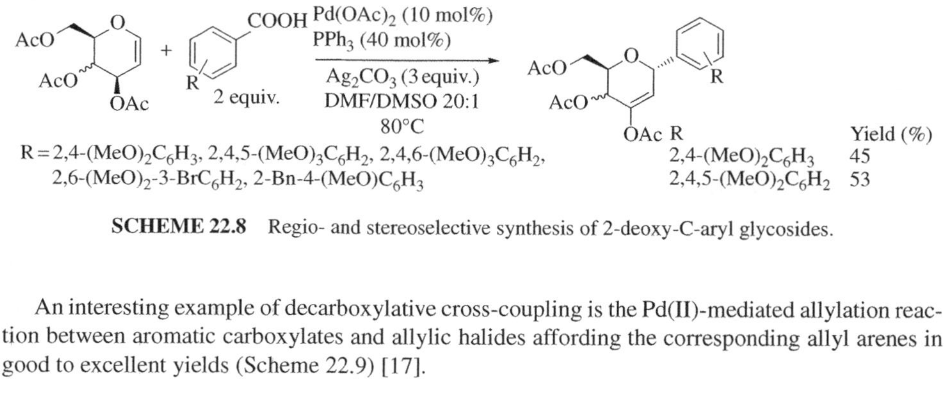

SCHEME 22.8 Regio- and stereoselective synthesis of 2-deoxy-C-aryl glycosides.

An interesting example of decarboxylative cross-coupling is the Pd(II)-mediated allylation reaction between aromatic carboxylates and allylic halides affording the corresponding allyl arenes in good to excellent yields (Scheme 22.9) [17].

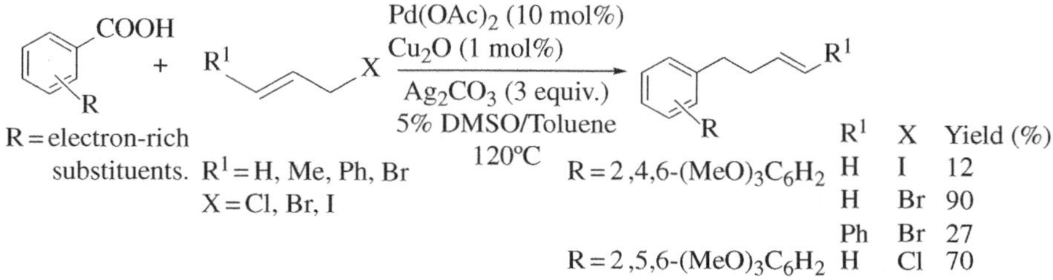

SCHEME 22.9 Pd(II)-catalyzed decarboxylative allylation of arene carboxylates with allylic halides.

Moreover, the Pd(II)/Ag-catalyzed decarboxylative addition of benzoic acids to allyl alcohols readily yields the corresponding β-aryl ketones and aldehydes [18].

Wu and coworkers have demonstrated the feasibility of decarboxylative 1,2-addition of carboxylic acids to aldehydes or imines for the efficient preparation of diaryl methanols and diaryl *N*-tosyl amines, respectively (Scheme 22.10) [19]. Aliphatic aldehydes and ketones did not react under the conditions optimized for the aryl reagents.

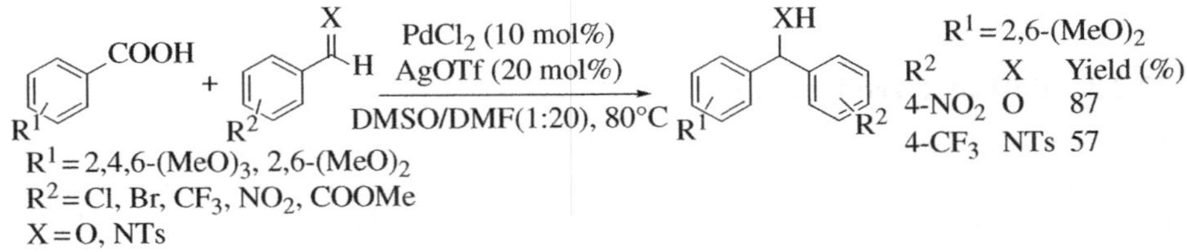

SCHEME 22.10 Palladium-catalyzed decarboxylative 1,2-addition of carboxylic acids to aldehydes or imines.

22.3 FORMATION OF Ar–C(sp²) BONDS

22.3.1 Synthesis of Aryl Ketones and Amidines

The preparation of aryl ketones can be achieved starting from benzoic acids and nitriles (as the reaction medium) under Pd catalysis via the ketimine intermediates (Scheme 22.11) [20a]. A similar protocol has been developed for the synthesis of aryl amidines replacing the nitrile with a cyanamide [20b].

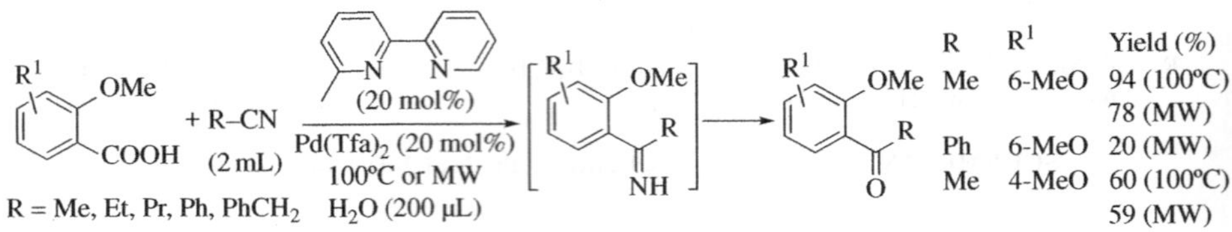

SCHEME 22.11 Synthesis of aryl ketones by palladium(II)-catalyzed decarboxylative reactions of benzoic acids with nitriles.

22.3.2 Formation of Ar–Vinyl Bonds

For the most part, studies on the formation of Ar–vinyl bonds have involved Ar–COOH (or their derivatives) as the arylating agents in Pd-catalyzed reactions with substituted olefins.

22.3.2.1 Ni-Catalyzed Reactions The nickel-catalyzed cross-coupling of alkenyl Grignard reagents with aryl nitrile derivatives has been exploited by Miller and coworkers [4]. The conditions are similar to those described in Scheme 22.2. In general, alkenyl Grignards were *E*/*Z* mixtures and therefore provided styrene derivatives as a mixture of *E* and *Z* isomers in modest to good yields (43–80%). Milder conditions were later developed in the Ni-catalyzed arylcyanation of alkynes, in which the insertion of alkynes into an aromatic C–CN bond took place (Scheme 22.12a) [5, 21a, b]. The intramolecular version of the reaction was likewise reported, and cyclization in a 5-exo-dig fashion occurred (Scheme 22.12b) [21c]. The addition of Lewis acids, such as $AlMe_2Cl$, $AlMe_3$, and

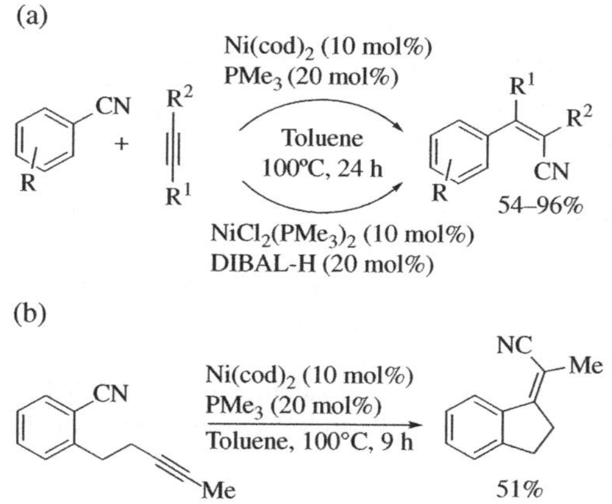

SCHEME 22.12 Nickel-catalyzed arylcyanation of alkynes: (a) intermolecular reaction and (b) intramolecular reaction.

BPh_3, as cocatalysts was found to promote the reaction significantly, giving (*Z*)-adducts in good to excellent yields. Recently, Hiyama and coworkers reported the perfluoroarylcyanation of alkynes in good yields and with notable regioselectivity [21d].

The nickel-catalyzed activation of the Ar–CN bond has been extended to *o*-arylcarboxybenzonitriles, and substituted coumarins are easily formed in the reaction with alkynes (Scheme 22.13) [22]. The nitrile group is not incorporated in the end products, and the formation of a new Ar–CN bond completes the sequence. In the specific case of the reaction of *o*-{[(4-(dimethylamino)benzoyl]oxy} benzonitrile (**6**) with unsymmetrical alkynes, regioisomers (**7a** and **7b**) were isolated in different ratios, depending on the steric bulk of the alkyne substituent. Terminal alkynes failed to participate in the reaction.

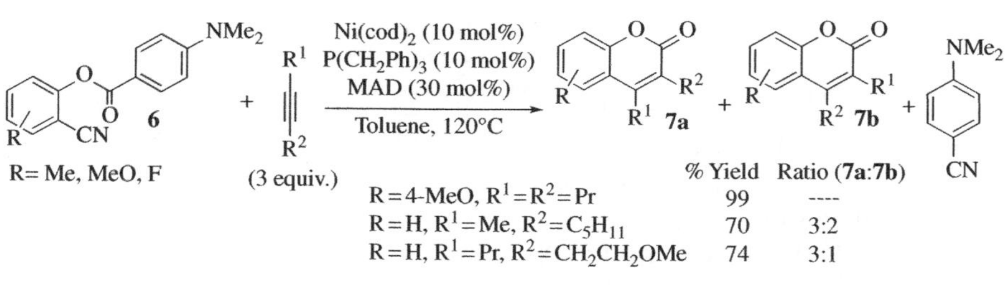

MAD = methylaluminum bis(2,6-di-tert-butyl-4-methylphenoxide)

SCHEME 22.13 Ni-catalyzed cycloaddition of *o*-arylcarboxybenzonitriles and alkynes.

The decarbonylative addition of (thio)phthalic anhydrides to alkynes for the preparation of valuable (thio)isocoumarins is another application of nickel(0) catalysts for the formation of Ar–C(sp^2) bonds (Scheme 22.14) [23a, b]. When the reaction is applied to thiophthalic anhydrides, benzothiophenes or thiochromanones can be prepared selectively in place of thioisocoumarins, depending on the reaction conditions employed [23b]. In related reactions, the nickel-catalyzed

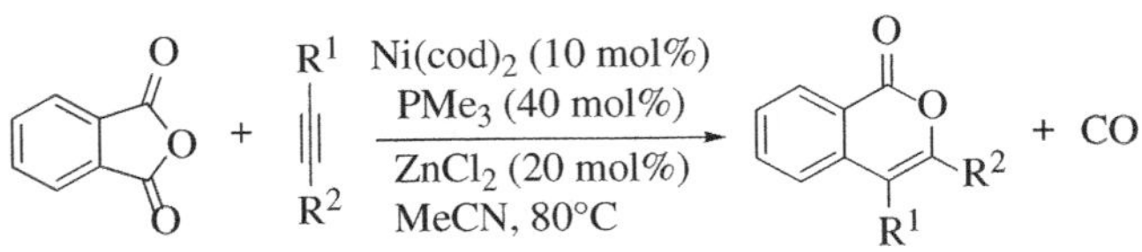

R^1	R^2	Yield (%)
Pr	Pr	87
Me_3Si	Me	93
Me_3Si	Ph	86

SCHEME 22.14 Ni-catalyzed decarbonylative addition of anhydrides to alkynes.

cycloaddition between benzoxazinones and alkynes has been found to yield substituted quinolines [23c], whereas the decarbonylative alkylidenation of phthalimides in the presence of the catalyst Ni(0)/PMe_3/MAD furnished isoindolinones upon reaction with trimethylsilyl-substituted alkynes [23d]. A regioselective decarbonylative cycloaddition of (thio)phthalic anhydrides with allenes in the presence of a Ni(0) catalyst provided δ-(thio)lactones in good yields (Scheme 22.15) [23e].

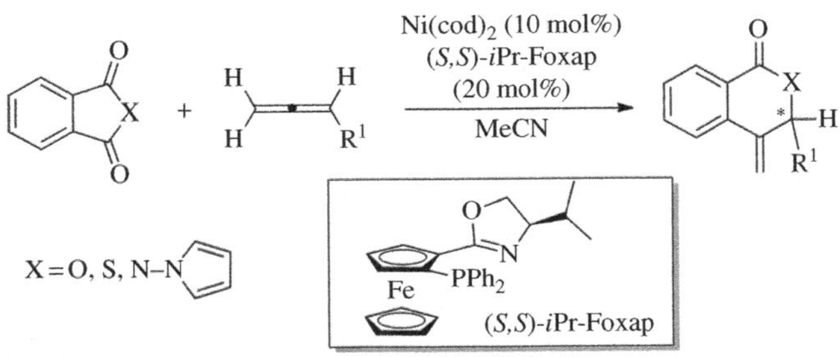

X=S	
R^1	Yield (%)
C_5H_{11}	99 (82% ee)
Cy	87 (87% ee)
Ph	90 (82% ee)
X=O	
R^1	Yield (%)
C_5H_{11}	70 (39% ee)
Cy	64 (81% ee)

SCHEME 22.15 Decarbonylative cycloaddition of (thio)phthalic anhydrides with allenes.

Matsubara and coworkers have recently developed a nickel-catalyzed cycloaddition of isatoic anhydrides with alkynes to afford 2,3-disubstituted indoles in good yields (Scheme 22.16). The cycloaddition reaction proceeds via a decarbonylation, decarboxylation, and alkyne insertion sequence [24].

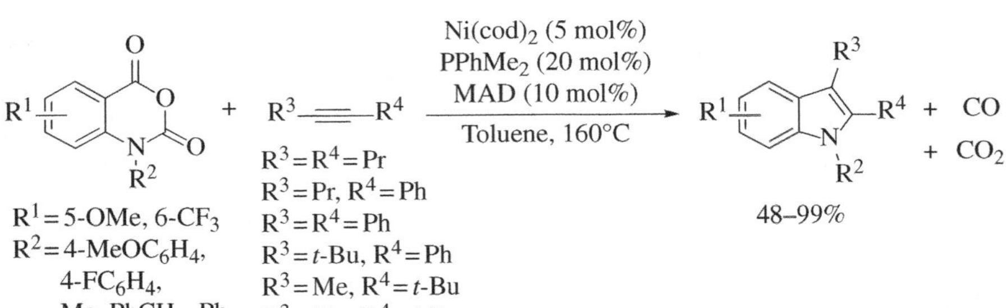

SCHEME 22.16 Ni-catalyzed decarbonylative and decarboxylative cycloaddition of isatoic anhydrides with alkynes.

22.3.2.2 Ru-Catalyzed Reactions Arylalkenes can also be formed by the Ru-catalyzed decarbonylative coupling of aromatic aldehydes with terminal alkynes in a Wittig-like reaction [25]. The reaction is primarily applicable to aromatic aldehydes bearing electron-rich substituents, since electron-poor benzaldehydes or aliphatic aldehydes give poor results (Scheme 22.17).

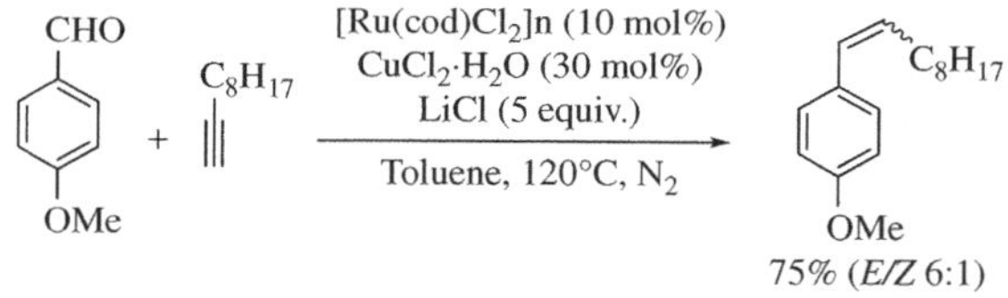

SCHEME 22.17 Olefin formation by ruthenium-catalyzed decarbonylative addition of aldehydes to terminal alkynes.

22.3.2.3 Rh-Catalyzed Reactions Zhao and coworkers have shown that the reaction of *ortho*-substituted arene carboxylic acids with electron-deficient olefins in the presence of a hydrorhodium(I) complex as the catalyst gave the Mizoroki–Heck products (**8**) and the conjugate addition products (**9**) with high selectivity (Scheme 22.18a) [13, 26]. The selectivity depends on the content of water in the reaction mixture: higher water content favored the hydrolysis of the enolato intermediate **10**, while lower water content favored β-H elimination. This reaction works only for *ortho*-substituted arene carboxylic acids and with electron-deficient olefins (Scheme 22.18b).

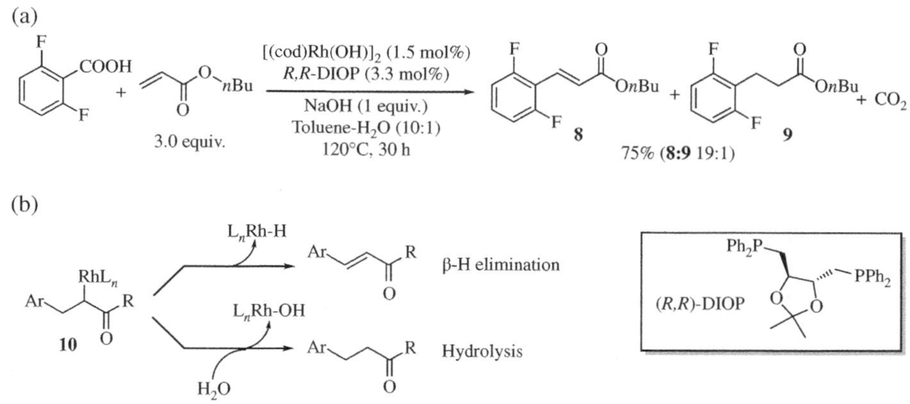

SCHEME 22.18 (a) Heck–Mizoroki products from Rh(I)-catalyzed decarboxylative reactions of arenecarboxylic acids and (b) competitive pathways depending on the water content in the reaction.

Extrusion of CO from acyl-2-phenylpyridines is another way to form an Ar–C(sp²) bond. Rh(I) catalysis was found to be effective in the conversion of styryl ketones into the corresponding stilbenes [12]. Aroyl chlorides react with acyclic alkenes in the presence of a rhodium–ethylene complex, [{RhCl(C_2H_4)$_2$}$_2$], in refluxing *o*-xylene under N_2 to give Mizoroki–Heck-type products [27a]. This ARCIS reaction proved useful in the synthesis of a series of stilbenes and butyl cinnamates. By contrast, [RhCl(cod)]$_2$ with PPh_3 was found to catalyze the decarbonylative addition of aroyl chlorides onto terminal alkynes to afford the corresponding vinyl chlorides [27b].

A Rh(III) complex was found to catalyze selective C–C bond activation in secondary aryl alcohols, which in the presence of olefins afforded C–C coupling Mizoroki–Heck products [28]. This unprecedented strategy was directed by a pyridinyl group that favored β-C elimination. Secondary alcohols (but not primary) bearing both aryl and aliphatic groups underwent β-C elimination and further alkenylation smoothly, generating aldehydes as by-products (Scheme 22.19). Styrene derivatives, vinylnaphthalenes, and aliphatic terminal alkenes were suitable coupling partners in this process [28].

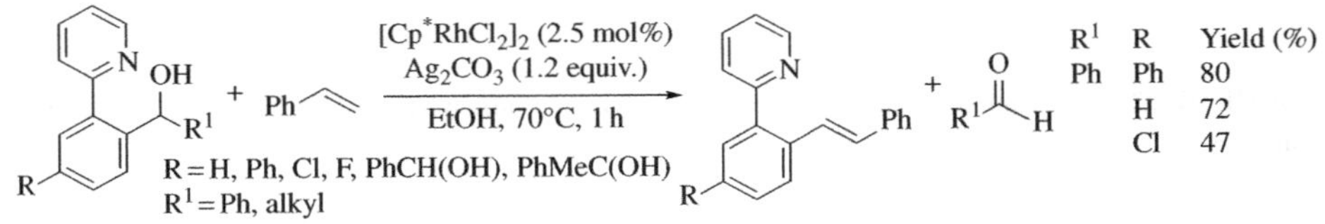

SCHEME 22.19 Rh(III)-catalyzed C–C bond cleavage of secondary arylmethanols.

As a final example, Scheme 22.20 shows a Mizoroki–Heck-type reaction between substituted aryl nitriles and vinylsilanes to give alkenylsilanes in good yields. The process is initiated by the reaction of the catalyst with a disilane (e.g., $Me_3SiSiMe_3$) to form a reactive silylrhodium species [29].

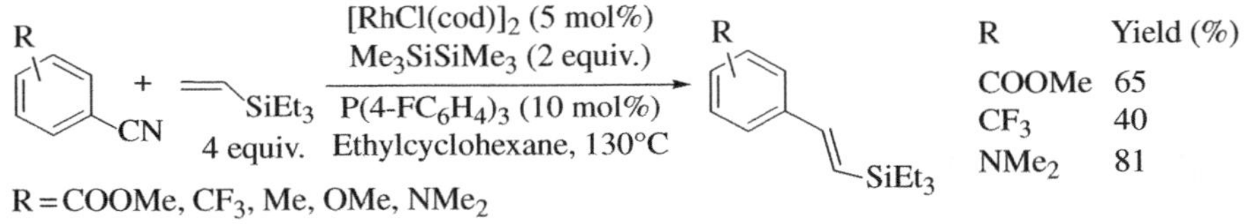

SCHEME 22.20 Rh-catalyzed alkenylation of benzonitriles.

Benzocyclobutenols have also been exploited as substrates for a rhodium-catalyzed ARCIS reaction (Scheme 22.21). The reaction is based on exclusive ring opening from the proximal C(sp²)–C(sp³) bond upon $[Rh(OH)(cod)]_2$ catalysis, followed by alkynes insertion. Dihydronaphthalene derivatives have been prepared with noticeable regioselectivity (Scheme 22.21) [30]. Thus, the ring-opened arylrhodium(I) intermediate **11** undergoes 1,2-addition across the carbon–carbon triple bond to furnish an alkenylrhodium(I) species (**12**) that adds back onto the carbonyl group in a 6-exo mode to afford the alkoxyrhodium(I) intermediate **13** and subsequently dihydronaphthalene after protonation.

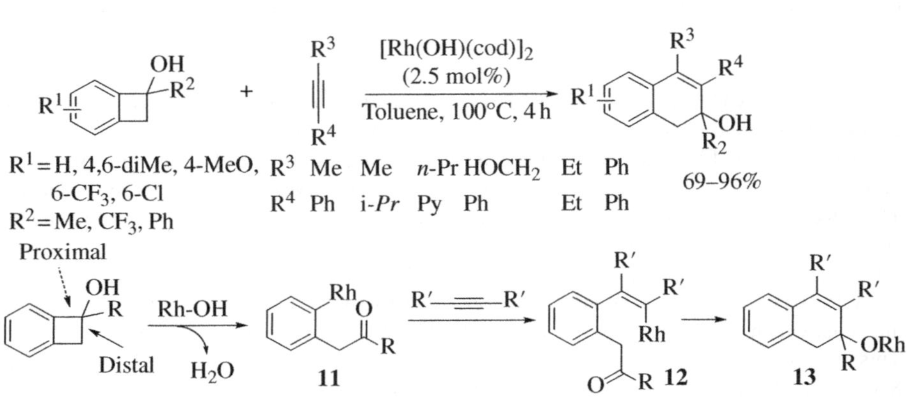

SCHEME 22.21 Rh-catalyzed ring opening of benzocyclobutenols followed by alkyne insertion.

22.3.2.4 Ir-Catalyzed Reactions (*Z*)-Vinyl chlorides are obtained selectively by the iridium-catalyzed decarbonylative addition of aroyl chlorides to terminal alkynes, using RuPhos as a phosphine ligand in refluxing toluene (20 h) (Scheme 22.22) [31].

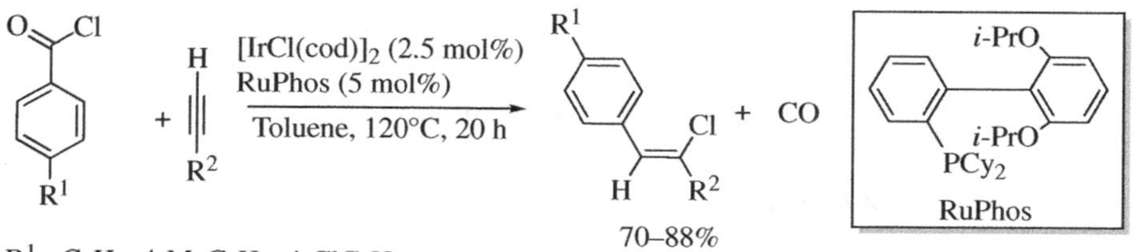

$R^1 = C_6H_5$, 4-MeC_6H_4, 4-ClC_6H_4
$R^2 = C_6H_5$, *n*-C_6H_{13}

SCHEME 22.22 Ir-catalyzed addition of acid chlorides to terminal alkynes.

22.3.2.5 Pd-Catalyzed Reactions Aromatic carboxylic acids can serve as arylating reagents in reactions with alkynes, olefins, and even saturated ketones and nitroalkanes, forming Ar–vinyl bonds [14]. As an example, Glorius and coworkers have developed the palladium-catalyzed formal [4+2] annulation of 2-phenylbenzoic acids with alkynes via successive cleavage of both C–H and C–C bonds, to give the corresponding polycyclic aromatic hydrocarbons (e.g., phenanthrenes; Scheme 22.23) [32]. This ARCIS reaction makes possible the synthesis of benzo[*c*]phenanthrene and benzo[*c*]anthracene derivatives, with only negligible amounts of 1-phenylnaphthalenes.

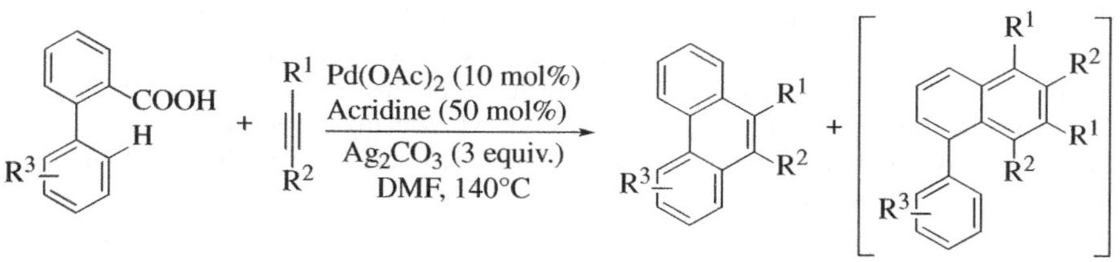

R^1	R^2	R^3	Yield (%)	
Ph	Ph	H	67	9
Et	Ph	H	52	—
Ph	Ph	4-Me	65	4
CH_2OMe	Ph	4-Me	51	—

SCHEME 22.23 Synthesis of phenanthrenes by Pd-catalyzed decarboxylative coupling of 2-phenylbenzoic acids with alkynes.

Su and coworkers established a Pd-catalyzed method for decarboxylative Mizoroki–Heck coupling, in which 1.2 equiv. of *p*-benzoquinone is used in place of the Ag(I) salt. This method met with some success only with electron-rich (hetero)aromatic carboxylic acids [33]. Subsequently, the same authors reported that the Pd catalyst itself can induce decarboxylative Mizoroki–Heck coupling of aromatic carboxylic acids when dioxygen is used as the terminal oxidant completely replacing the Ag salt [34]. Depending on the structure of the acids, two different Pd catalysts were required for the Mizoroki–Heck coupling to occur: $Pd(OAc)_2$ worked efficiently for electron-rich aromatic carboxylic acids, while the $Pd(OAc)_2$/SIPr system (SIPr: 1,3-bis(2,6-diisopropylphenyl)-4,5-dihydroimidazol-2-ylidene) enabled the use of electron-deficient substituents (Scheme 22.24) [34].

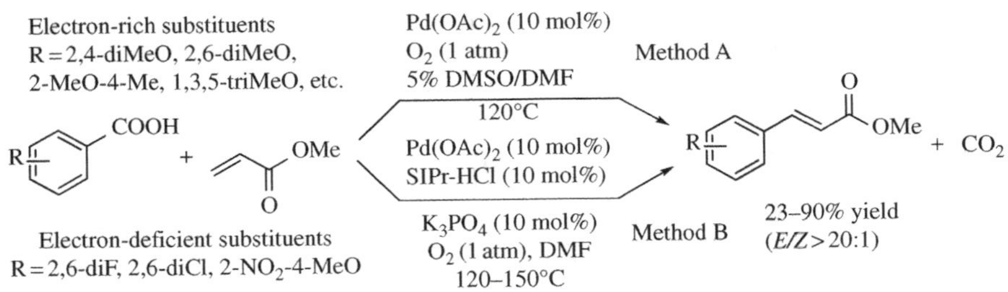

SCHEME 22.24 Pd-catalyzed decarboxylative Heck coupling.

A broad range of both mono- and disubstituted olefins (mainly α,β-unsaturated esters and styrenes) can be used as coupling partners in this ARCIS reaction. It is noteworthy that this methodology overcomes the generally lower reactivity of 1,2-disubstituted olefins in Mizoroki–Heck coupling reactions.

An interesting protocol for the decarboxylative Mizoroki–Heck vinylation of benzoic acids has been disclosed by Gooßen and coworkers [14, 35]. Various 2-nitrobenzoates react with electron-poor olefins or styrene derivatives to afford the corresponding vinyl arenes with high selectivity (the (*E*)-configuration of the double bond is formed exclusively). A cooperative Pd(II)/Ag catalytic system was applied for the decarboxylative cross-coupling between arene carboxylic acids and 2-cycloalken-1-ones to yield the corresponding 3-aryl derivatives [36]. Gooßen and coworkers have developed a new strategy that involves the salt-free synthesis of vinyl arenes from enol esters (isoprenyl carboxylates) through a Pd-mediated decarbonylative Mizoroki–Heck reaction in 1-methyl-2-pyrrolidone (NMP) at 160°C [37]. The general applicability was investigated with a series of isoprenyl aryl carboxylates (prepared from the corresponding carboxylic acids and propyne gas [37]) with different olefins (mainly styrenes; Scheme 22.25). The hydroxyl-substituted ammonium salts *N*-dodecyl-*N*-methylephedrinium bromide and tri-*n*-butyl(2-hydroxyethyl)ammonium bromide were found to stabilize the palladium in solution for a sufficient time to give quantitative conversion along with high selectivity. *Trans*-stilbenes (**14**) were formed as the main products along with variable amounts of the 1,1-arylphenylethylene isomers (**15**; Scheme 22.25). The reaction tolerated many functionalities, including esters, ethers, nitro, keto, trifluoromethyl, and even formyl groups. In analogy to traditional Mizoroki–Heck olefination, even higher selectivities were obtained with activated olefins such as acrylates.

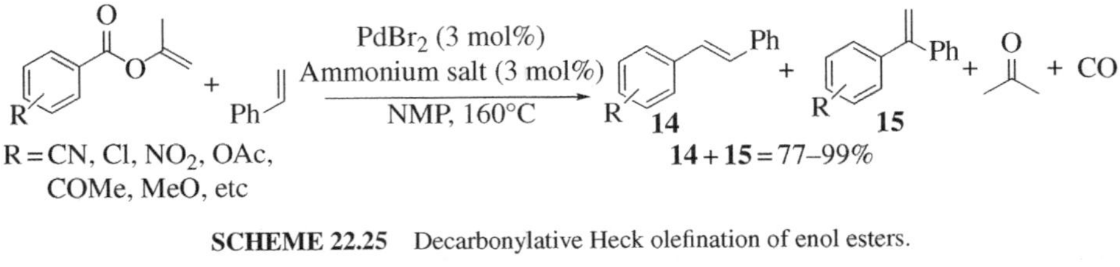

SCHEME 22.25 Decarbonylative Heck olefination of enol esters.

Aromatic carboxylic anhydrides have been found to be suitable substrates as aryl sources in the Pd-mediated decarboxylative Mizoroki–Heck cross-coupling reaction (Scheme 22.26) [38]. Olefins with electron-withdrawing groups generally give high yields of the (*E*)-β-arylated products. With terminal olefins, a very rapid isomerization occurs and the arylated products are obtained as a mixture of isomers.

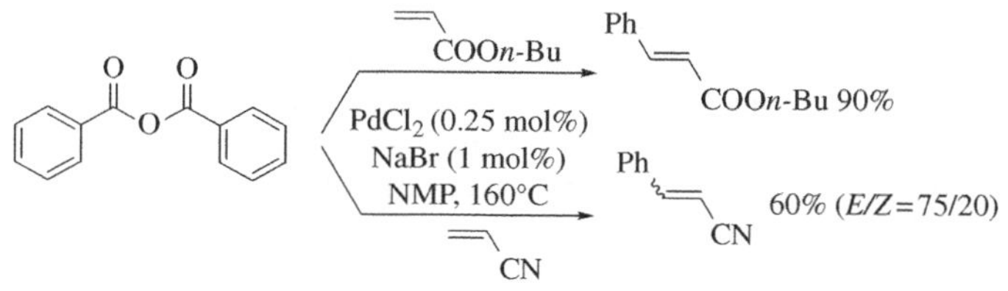

SCHEME 22.26 Aromatic carboxylic anhydrides as arylating agents in Heck reactions.

A similar Pd-mediated decarbonylation reaction involves the conversion of aromatic carboxylic acids *in situ* to mixed anhydrides by treatment with di-*tert*-butyl dicarbonate, which are then vinylated with extrusion of CO. All by-products were found to be volatile, which makes product isolation particularly easy (Scheme 22.27) [39].

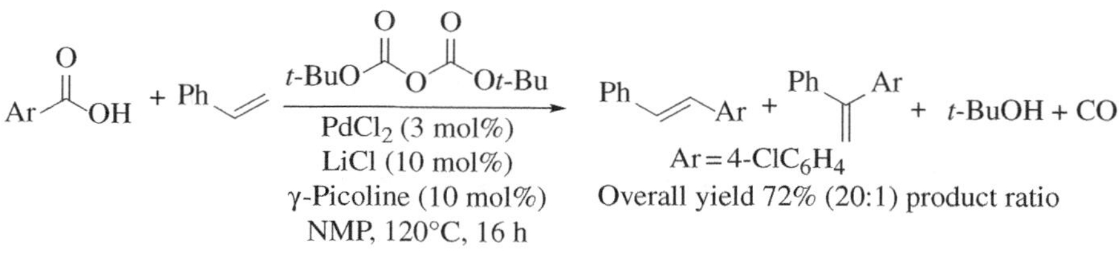

SCHEME 22.27 *In situ* activation of benzoic acids by di-*tert*-butyl dicarbonate for olefination reactions.

Pd catalysts also promote the decarbonylative Mizoroki–Heck reactions of even poorly reactive *p*-nitrophenyl aryl carboxylates with olefins to give the vinyl arenes in good to excellent yields, along with CO and the corresponding phenols [40]. Mizoroki–Heck-type arylation of styrene and acrylate esters by use of aroyl chlorides can also be performed in the presence of $PdCl_2(PhCN)_2$/$(PhCH_2)Bu_3NCl$ as the catalytic system without adding base [41].

Allylic esters act as effective coupling partners (Scheme 22.28). Electron-rich substituents on the aromatic moiety of the carboxylic acids provide the expected Mizoroki–Heck products in high yields. On the other hand, 2,6-difluoro-, 2,4,6-trifluoro-, and perfluorobenzoic acids afford only very low yields of the Mizoroki–Heck products. Substituted allylic esters react efficiently in this ARCIS reaction [17].

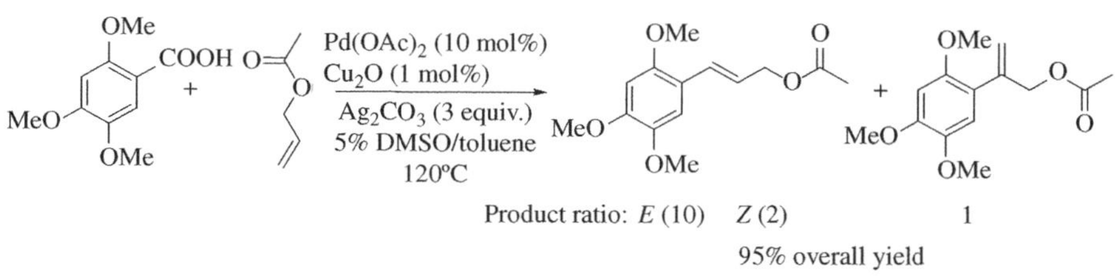

SCHEME 22.28 Allylic esters as effective coupling partners in Pd-catalyzed Mizoroki–Heck reactions.

An unexpected Pd-mediated coupling reaction of arene carboxylic acids with nitroethane via a combination of decarboxylation and dehydrogenation has been reported by Su and coworkers [42]. This method provides exclusively (*E*)-β-nitrostyrenes. Su et al. varied the reaction conditions according to the nature of the substituents tethered to the aromatic ring of the benzoic acid, as summarized in Scheme 22.29 [42]. The addition of pivalic acid optimizes the ARCIS process in electron-rich derivatives, whereas it is detrimental with electron-deficient benzoic acids, for which it leads to undesired protodecarboxylation [42]. The use of 1-nitropropane instead of nitroethane was tested in the reaction with 2,4-dimethoxybenzoic acid, but the corresponding β-nitrostyrene was obtained in poor yield (24%, as a 3:2 mixture of *E*/*Z* isomers). More sterically demanding nitro compounds, such as 2-nitropropane, nitrocyclohexane, and α- and β-nitroethylbenzene, were found to be ineffective.

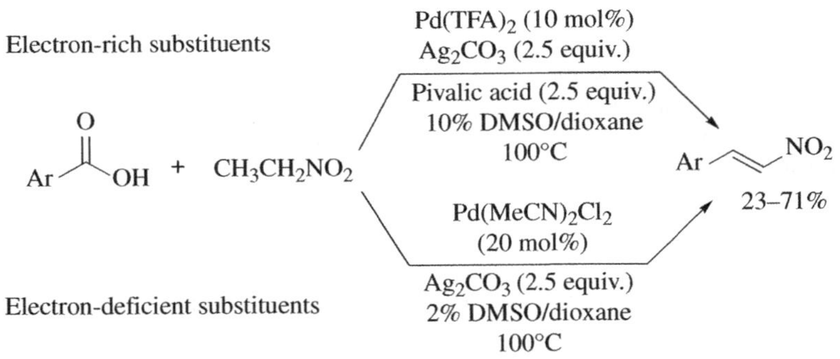

SCHEME 22.29 Pd-catalyzed cross-coupling of carboxylic acids with nitroethane.

Recently, Su and coworkers developed an interesting strategy to generate a new C–C bond. The method involves an *in situ* dehydrogenation of saturated ketones to form enones, which can undergo decarboxylative cross-coupling to form the Mizoroki–Heck-type product [43]. This tandem reaction with Pd(II)/Ag(I) catalysis offers a straightforward and highly efficient synthetic method for the preparation of chalcone derivatives. In some cases, $Cu(OAc)_2$ or CuF_2 were found to be more efficient than Ag(I) when electron-rich carboxylic acids were used.

22.3.3 Formation of Ar–Ar Bonds

A large number of known ARCIS processes are devoted to the construction of Ar–Ar bonds. Compounds of several metals, notably Ni, Cu, Rh, Pd, and bimetallic systems (Pd/Ag, Pd/Cu), are commonly employed for this purpose. Most of the reactions of this type employ arenes with carbon-based leaving groups, such as Ar–COR, Ar–CN, and $Ar–CR_3$. These undergo cross-coupling with aryl halides, esters, or boronic esters or with arylzinc or arylmagnesium reagents to form Ar–Ar bonds. The direct substitution of H in Ar–Ar bond formation is discussed in detail in the following sections.

22.3.3.1 Ni-Catalyzed Reactions The nickel-mediated decarbonylative cross-coupling of diarylzinc reagents with phthalimides, in the role of electrophilic coupling partner, has been reported, using reaction conditions similar to those described in Scheme 22.4 [9]. This reaction provides a highly efficient means to construct *ortho*-substituted benzamides (Scheme 22.30).

Unsymmetrical biaryls can easily be prepared in good to excellent yields by the nickel-mediated cross-coupling of arenecarbonitriles with aryl Grignard reagents in the presence of PhSLi or *t*-BuOLi under mild conditions (Scheme 22.31) [44]. These processes tolerate the presence of many functionalities on both the aryl nitriles and the aryl Grignard reagents. Aryl manganese reagents (prepared from the corresponding organolithium reagents and Li_2MnCl_4) have likewise been used for the Ni-catalyzed synthesis of unsymmetrical biaryls, requiring no additives [45].

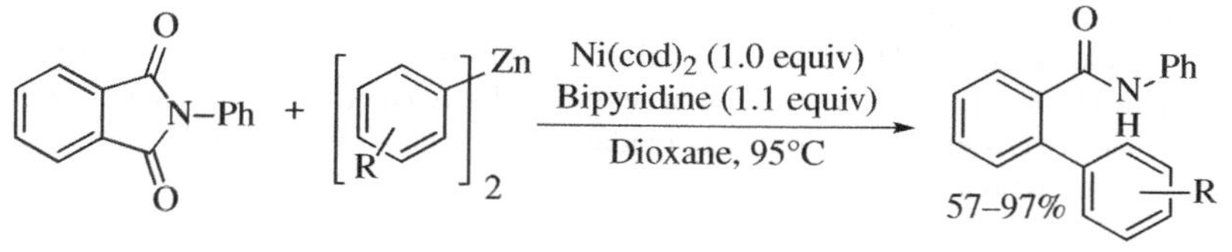

R = H, 4-MeO, 3-MeO, 2-MeO, 4-MeS, 4-CF_3, 4-F, 4-Me, 4-Me_2N, 3-Me, 3,5-$(CF_3)_2$

SCHEME 22.30 Ar–Ar bond formation via Ni-mediated decarbonylative cross-coupling of phthalimides with diorganozinc reagents.

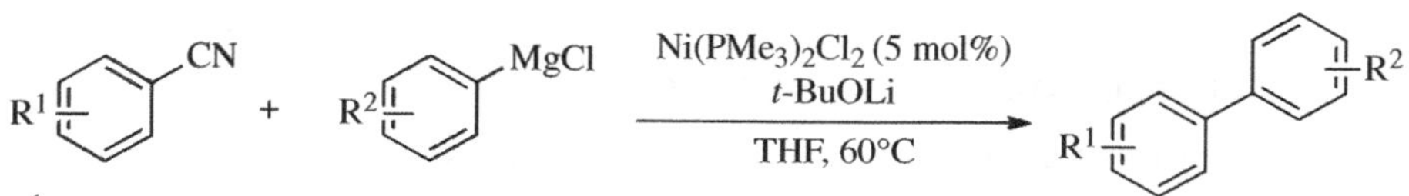

R^1 = 4-F, 4-NMe_2, H, 4-Me, 4-MeO, 3-COO*t*-Bu
R^2 = H, 4-Me, 3-MeO

82–97%

SCHEME 22.31 Biaryls from Ni-catalyzed cross-coupling of aryl nitriles.

*22.3.3.2 **Cu-Catalyzed Reactions*** The COOH group in arene carboxylic acids can be activated under Cu catalysis with no need of other transition metals. The process involves a Cu-catalyzed decarboxylative cross-coupling of potassium polyfluorobenzoates with aryl iodides and bromides [46]. This ARCIS reaction was found to afford the desired polyfluorobiaryls from aryl iodides, with diglyme as the solvent at 130°C (Scheme 22.32, *path a*), whereas the addition of 1,10-phenanthroline as a ligand was mandatory for the success of the reaction with aryl bromides (Scheme 22.32, *path b*).

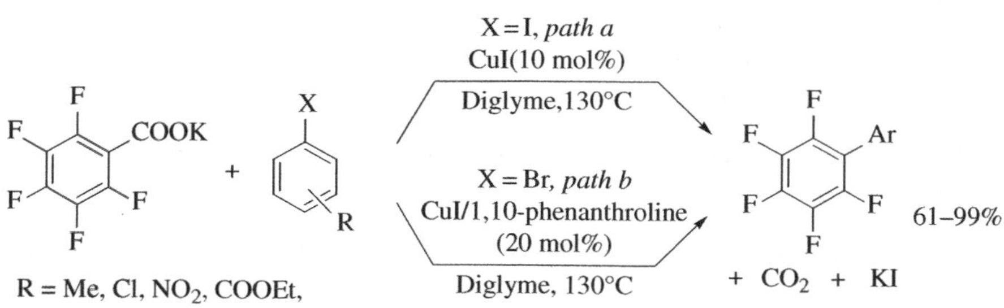

R = Me, Cl, NO_2, COOEt, MeO, CN, CF_3, etc.

SCHEME 22.32 Cu-catalyzed decarboxylative cross-coupling of polyfluorobenzoates with aryl iodides and bromides.

*22.3.3.3 **Rh-Catalyzed Reactions*** Various Ar–C bonds can be activated by Rh catalysis in order to synthesize biaryls. As an example, a decarbonylative Suzuki coupling, catalyzed by a $[Rh(ethylene)_2Cl]_2$/KF system, has been applied to the preparation of unsymmetrical biaryls [47a]. This ARCIS reaction involves the use of aromatic carboxylic anhydrides or acid chlorides with triarylboroxines as the coupling partners. The desired adducts have generally been accompanied by a nonnegligible amount of the corresponding diaryl ketone. Better results can be obtained in the preparation of arylated arenes by reacting 2-arylpyridines with benzoic anhydride in the presence of a Rh(I) catalyst by a phosphine ligand-free decarbonylative C–H activation cross-coupling reaction (Scheme 22.33) [47b]. The synthesis of biaryls can also be accomplished by Rh(I) catalysis by exploiting the extrusion of CO from aromatic aldehydes [48]. This has been demonstrated by the decarbonylative C–H activation cross-coupling between substituted benzaldehydes and 2-arylpyridines in the presence of *tert*-butyl peroxide as an oxidant. This reaction was found to form a mixture of mono- and bis-arylation products in good yields.

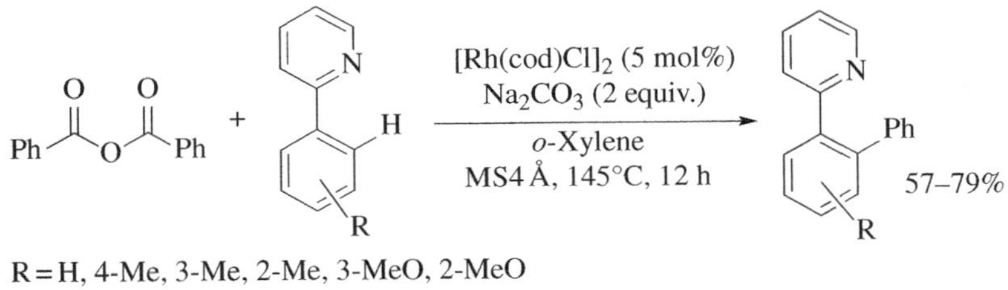

SCHEME 22.33 Rh(I)-catalyzed direct arylations by using benzoic anhydrides.

A Rh(I) catalyst has likewise been found to promote cross-coupling between readily available benzoic acids as arylating agents and arenes bearing N-based directing groups such as pyridinyl, heteroaryl, and iminyl moieties (Scheme 22.34). The process involves the intermediacy of a mixed anhydride formed *in situ* by the reaction between the benzoic acid and $(t\text{-BuCO})_2O$ [49].

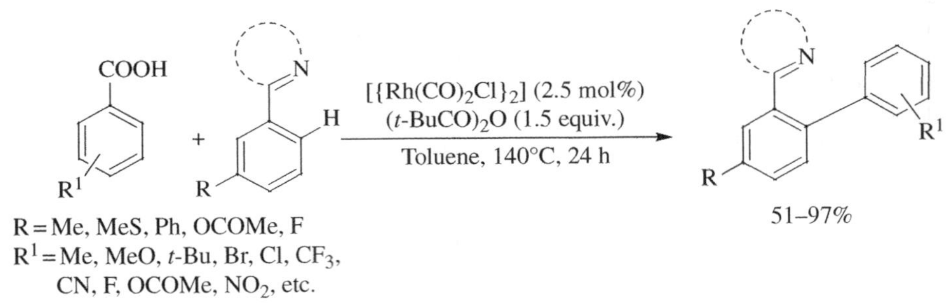

SCHEME 22.34 Rh(I)-catalyzed cross-coupling of carboxylic acids with arenes containing N-based directing groups.

Another example of an ARCIS reaction is the efficient Rh(I)-catalyzed decarboxylative coupling of ethyl benzo[*h*]quinoline-10-carboxylate with organoboron compounds, which proceeds through C–C and C–B cleavage with the assistance of an N-containing directing group (Scheme 22.35) [50].

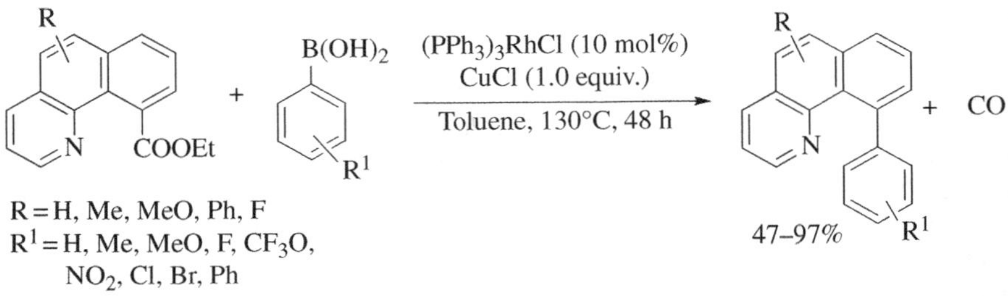

SCHEME 22.35 Rh-catalyzed reactions of aryl esters with organoboron compounds.

An intramolecular version of the decarbonylative C–H arylation of 2-aryloxybenzoic acids has been reported for the synthesis of dibenzofurans [51].

Another example of intramolecular Rh(I)-mediated biaryl coupling has been reported in the efficient cross-coupling between aryl cyanides and aryl chlorides in the presence of hexamethyldisilane [52]. Dibenzofuran, carbazole, and fluorene derivatives are readily obtained, albeit pyridine-containing substrates fail to react under the standard conditions, due to catalyst deactivation (Scheme 22.36). However, addition of catalytic amounts of $InCl_3$ in place of phosphine ligand dramatically increases the product yield.

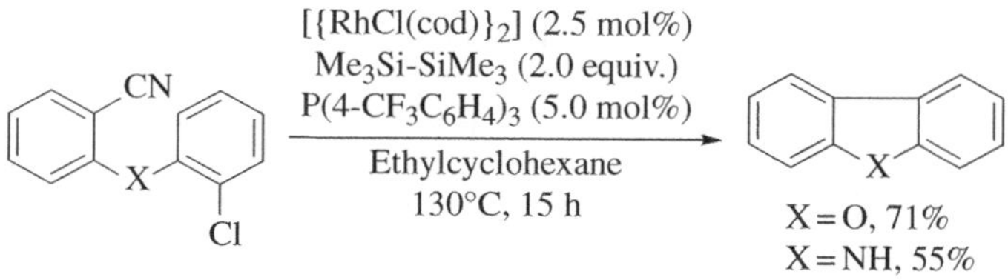

SCHEME 22.36 Rh-catalyzed synthesis of dibenzofuran derivatives.

22.3.3.4 Pd-Catalyzed Reactions The decarboxylative cross-coupling reaction of potassium polyfluorobenzoates with aryl bromides, chlorides, and triflates was achieved by use of Pd catalysis, with diglyme as the solvent [53a]. In several cases, aryl triflates (but not aryl tosylates) can also be successfully converted. The reaction proceeds smoothly when both the *ortho* positions in the aryl moiety are substituted with F, Cl, or CF_3 groups. The $Pd(OAc)_2/PPh_3$ catalytic system has been applied to achieve a concise and efficient protocol for the intramolecular arylation via decarboxylative cross-coupling of arene carboxylic acids with aryl halides [53b]. Substituted 6*H*-benzo[*c*]chromenes have been formed in this manner, starting from 2-[(2-bromobenzyl)oxy] benzoic acids (Scheme 22.37). Esters can likewise be adopted in place of the free acids, since ester hydrolysis and the cross-coupling sequence proceed efficiently in a one-pot fashion. The formation of biaryls is likewise efficient when the etheric O is replaced with an N-Me or an N-Ac group, but when the nitrogen atom is substituted by CH_2 or S, no cross-coupling reaction has been observed, even at 150°C and for longer reaction times (24 or 48 h).

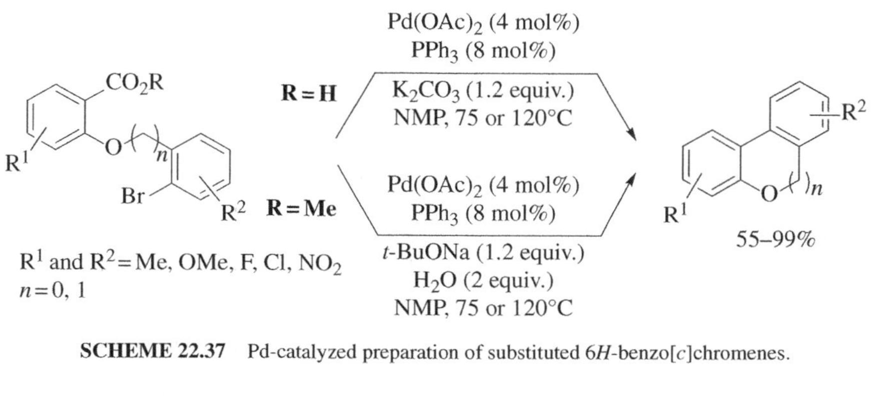

SCHEME 22.37 Pd-catalyzed preparation of substituted 6*H*-benzo[*c*]chromenes.

Aromatic alcohols such as α,α-disubstituted arylmethanols have been found to be viable substitutes for benzoic acids in the Pd-mediated cross-coupling reaction with aryl halides [54]. However, the required β-carbon elimination sometimes competes with *ortho* C–H bond cleavage, depending on the structure of the carbinol (Scheme 22.38a). Indeed, blocking the *ortho* position with an appropriate substituent on 2-phenyl-2-propanol (other than a phenyl group) can selectively induce β-carbon elimination. For example, the reaction of 2-(2-methylphenyl)-2-propanol with phenyl bromide, under conditions similar to those used for unsubstituted 2-phenyl-2-propanol, was found to give selectively 2-methylbiphenyl in 77% yield through β-carbon elimination with extrusion of acetone (Scheme 22.38b). The reaction of triphenylcarbinol with 2-methylphenyl bromide under Pd(II) catalysis gives the corresponding biaryl in excellent yield and selectivity, thanks to the use of bulky phosphines such as PCy_3, $P(t\text{-}Bu)_3$, and $P(o\text{-}tolyl)_3$.

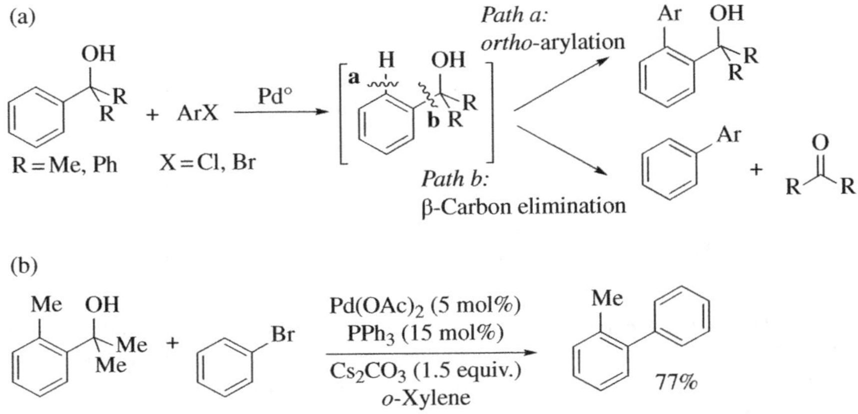

SCHEME 22.38 (a) Possible Pd-catalyzed C-H bond cleavage pathways of α,α-disubstituted arylmethanols and (b) selective preparation of 2-methylbiphenyl.

The use of PCy_3 as a ligand also favors the cross-coupling of triphenylcarbinol with a variety of substituted chlorobenzenes (Scheme 22.39) [54].

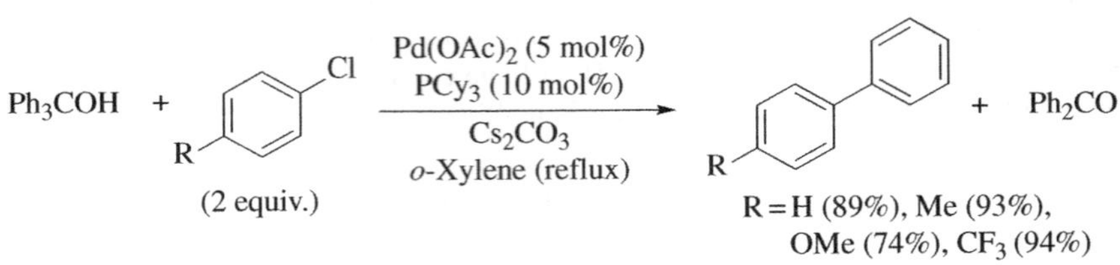

SCHEME 22.39 Biaryls from the Pd-catalyzed activation of the Ar–C bond in triphenylcarbinols.

22.3.3.5 Pd-Catalyzed Reactions in the Presence of Ag Salts A bimetallic system containing a Pd catalyst has been used in some cases to achieve Ar–Ar bond formation. Palladium-mediated decarboxylative cross-coupling of aromatic acids and aryl iodides takes place efficiently under microwave heating (MW), affording the corresponding biaryls, as shown in Scheme 22.40 (*path a*) [55].

The use of microwave heating allows a reduction in the reaction times from many hours to a few minutes. However, the biaryl product is always accompanied by the arene generated from decarboxylation of the benzoic acid. Similarly, substituted benzoic acids react with aryl iodides in the presence of $PdCl_2$ as the catalyst, BINAP as the ligand, and 3 equivalents of Ag_2CO_3 as an additive in dimethylacetamide (DMA) (Scheme 22.40, *path b*) [56]. Unsymmetrical biaryls have been obtained in 58–90% yield by the reaction of aryl iodides with arene carboxylic acids by use of a $PdCl_2/AsPh_3$ catalytic system in the presence of Ag_2CO_3 [57].

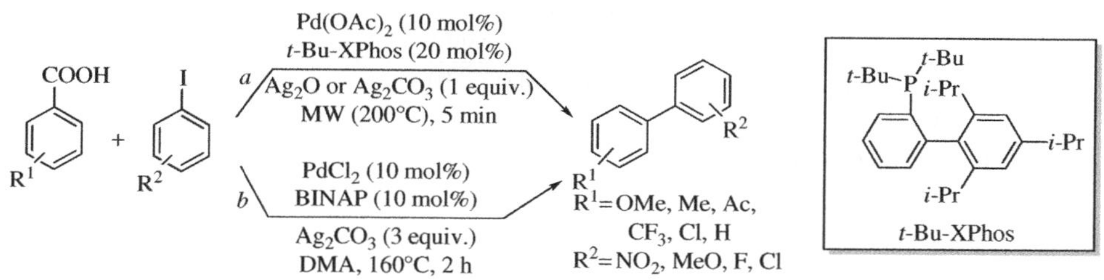

SCHEME 22.40 Pd-catalyzed decarboxylative coupling of aromatic acids with aryl iodides.

Alternatively, biaryls (including sterically hindered biaryls) can be synthesized by the reaction of arene carboxylic acids with diaryliodonium salts by means of a $PdCl_2$/DEPhos catalytic system at 150°C in the presence of Ag_2CO_3 [58].

The use of aryl sulfonates (mainly triflates) allows slightly lower temperature (130°C) for the decarboxylative cross-coupling with potassium carboxylates, with a Pd/Ag catalytic system [59]. The corresponding biaryls are obtained with catalytic loads of $PdCl_2$ and Ag_2CO_3 in the presence of inexpensive PPh_3 and catalytic amounts of 2,6-lutidine. This ARCIS reaction proceeds efficiently under microwave irradiation in only 5 min. The reaction also meets with some success with aromatic carboxylates functionalized in the *ortho* position.

Decarboxylative Suzuki cross-coupling has been observed when benzoic acids react with aryl boronic esters of 2,2-dimethylpropane-1,3-diol in the presence of catalytic amounts of $Pd(TFA)_2$ and 3 equivalents of Ag_2CO_3 [60]. The latter reaction tolerates coupling partners with diverse electronic properties and a variety of functional groups (Scheme 22.41). Other Pd catalysts, including $Pd(OAc)_2$, $PdCl_2$, and $Pd(CH_3CN)_4(BF_4)_2$, and other Ag salts, such as AgOAc, Ag_2O, Ag_3PO_4, and AgF_2, also induce the reaction but in a lower yield.

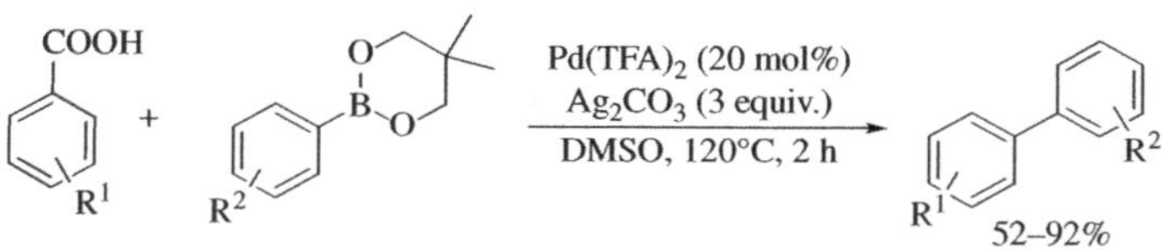

R^1=MeO, F, Br, NO_2, CH_3
R^2=Me, Cl, Br, MeO, *t*-Bu, NO_2, COMe, CN, OCOMe

SCHEME 22.41 Pd-catalyzed decarboxylative Suzuki reactions.

Pd/Ag bimetallic catalytic systems can likewise promote direct decarboxylative C–H arylation, allowing the intermolecular coupling of a variety of electron-poor and electron-rich benzoic acids with simple (hetero)arenes and polyfluoroarenes [61a–d]. This ARCIS methodology was found to proceed efficiently with high regio- and chemoselectivity when electron-poor benzoic acids were reacted with a series of indole derivatives to give C-3 arylated adducts, thereby representing an excellent alternative to double C–H activation oxidative couplings (Scheme 22.42a) [61a].

Recently, Luo and coworkers reported the arylation reaction of simple arenes with perfluorobenzoic acids by use of a Pd/Ag bimetallic catalyst system to give perfluorobiaryls, in which aryl halides and triflates were directly replaced by simple arenes (Scheme 22.42b) [61b].

The arylation of polyfluoroarenes has also been achieved with Pd/Ag catalysis (Scheme 22.43a) [61c]. Mechanistic investigations have demonstrated that the Pd–phosphine complex and the silver salts are both involved in the decarboxylation of electron-rich aromatic acids (Scheme 22.43b;

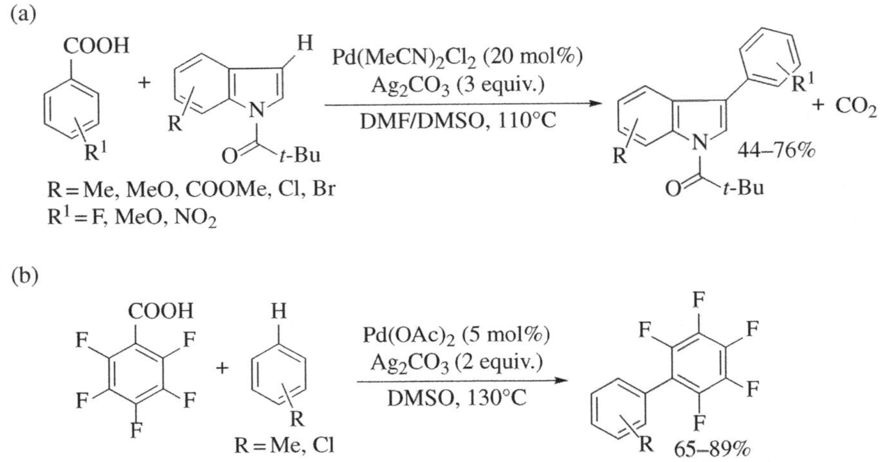

SCHEME 22.42 Ar–Ar bond formation via decarboxylative cross-coupling of (a) substituted benzoic acids and (b) perfluorobenzoic acid.

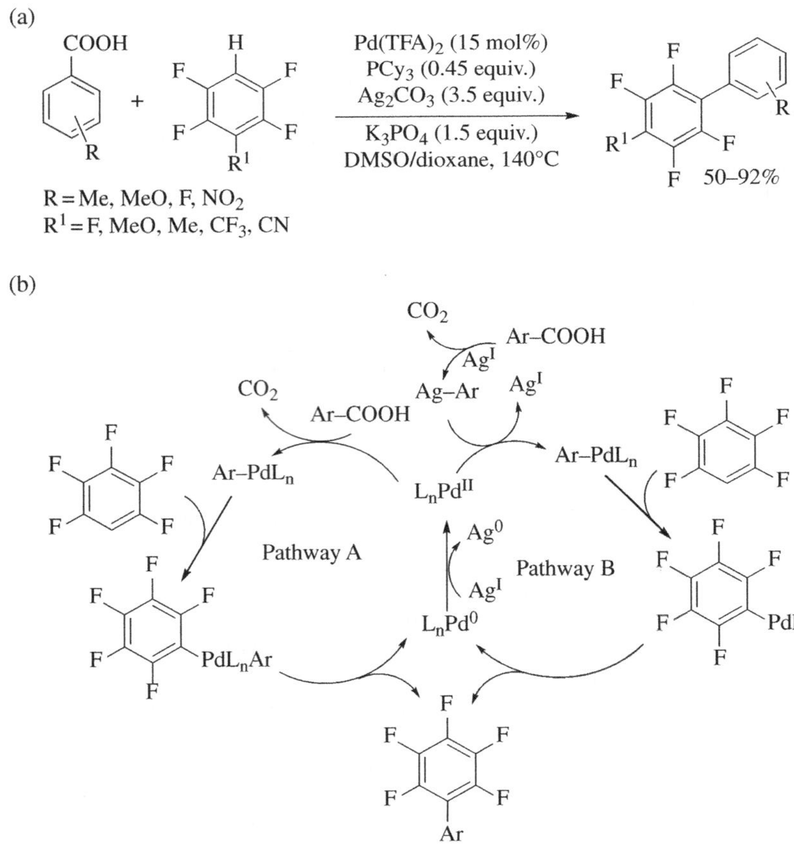

SCHEME 22.43 (a) Pd/PR$_3$-catalyzed direct arylation of electron-deficient polyfluoroarenes and (b) proposed reaction mechanism.

pathway A), whereas only silver salts are required in the decarboxylation of electron-poor benzoic acids (Scheme 22.43b, pathway B). The $Pd(OAc)_2/PCy_3$ or $Pd(TFA)_2$ systems, in combination with Ag_2CO_3, were found to be suitable catalysts in promoting the decarboxylative C–H bond arylation of thiophenes, using inexpensive and readily available benzoic acids as arylating agents [61d].

An intramolecular direct decarboxylative C–H arylation of substituted 2-phenoxybenzoic acids has been reported by Glorius and coworkers by use of Pd-mediated catalysis in the presence of stoichiometric amounts of Ag_2CO_3, affording a variety of dibenzofurans [61e].

In a handful of cases, two COOH groups have been activated for the synthesis of biaryls. Larrosa and coworkers reported for the first time the decarboxylative homocoupling of aromatic acids mediated by Pd and Ag [62a]. The reaction makes use of $Pd(TFA)_2$ as a catalyst and Ag_2CO_3 as an additive to afford the desired biaryls in 76–95% yields. The only by-products observed were due to the proto-decarboxylation of the aryl carboxylic acid. Both metals are essential for the reaction, and the role of the Ag salt is not only as the terminal oxidant but also as a mediator of the decarboxylation process. The method is subject to some limitations on the substituents on the benzoic acids. Thus, *m*- and *p*-nitrobenzoic acids as well as benzoic acids *ortho* substituted with F, Br, or MeO failed to give decarboxylative homocoupling products. In all cases, protodecarboxylations to the corresponding arenes were the main products observed. The same problem was reported in the protocol developed by Deng and coworkers, where the best results were obtained with $PdCl_2$ and PPh_3 in the presence of Ag_2CO_3 [62b].

A general method for Pd-mediated cross-coupling of a wide range of aryl carboxylic acids has been recently developed (Scheme 22.44) [62c]. The reaction proceeds efficiently by using $Pd(TFA)_2$ as the catalyst in the presence of Ag_2CO_3 as an additive. Among the ligands tested, PCy_3 has been found to yield the best results, and a DMSO/DME (3:17) mixture has proved to be most effective as the reaction medium. This process is based on the ability of Ag(I) salts to promote the decarboxylation of both aryl carboxylic acid reactants, generating the respective aryl-Ag(I) species. Transmetalation of both aryl groups from Ag(I) to Pd(II) then generates a single Pd complex bearing both aryl moieties. Finally, reductive elimination yields the desired biaryls (Scheme 22.44b).

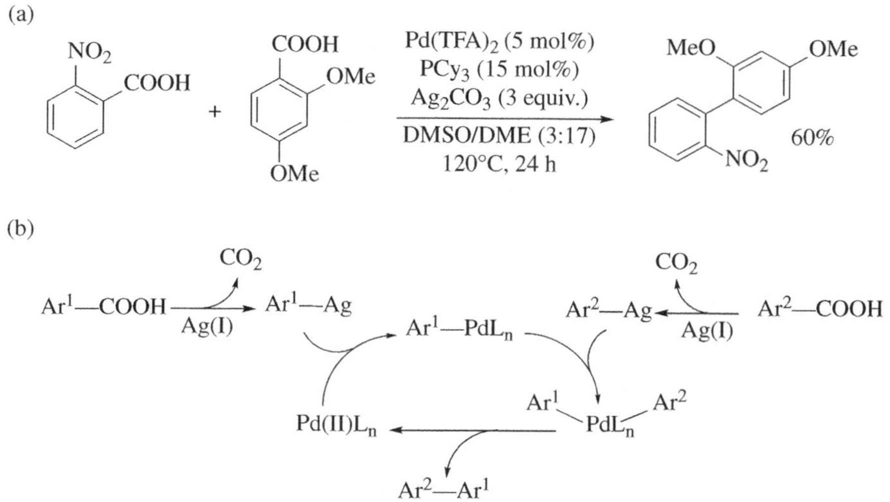

SCHEME 22.44 (a) Synthesis of biaryl compounds from Pd-catalyzed decarboxylative cross-coupling reaction between aryl carboxylic acids and (b) proposed reaction mechanism.

An efficient protocol for the Pd-mediated cross-coupling reaction of benzocyclobutenols with aryl bromides has been developed, which affords selectively unsymmetrical biaryl derivatives (Scheme 22.45) [63]. The cross-coupling reaction fails, however, with aryl chlorides, because of a prohibitively difficult oxidative addition step, while aryl iodides give the expected biaryl derivatives, along with a significant amount of the ring-opened product **16**. Aryl triflates give predominantly the undesired **16**. This chemistry has been successfully extended to the synthesis of functionalized phenanthrenes and cyclic imines, using a cross-coupling and condensation sequence.

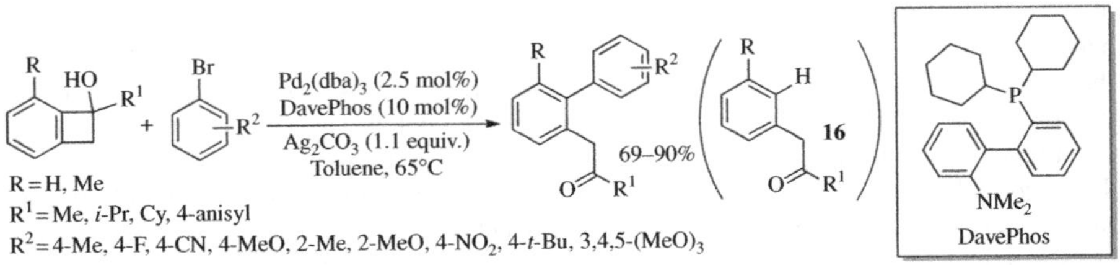

SCHEME 22.45 Pd-catalyzed reactions between benzocyclobutenols and aryl bromides.

22.3.3.6 Pd-Catalyzed Reactions in the Presence of Cu Salts Gooßen's group has developed a straightforward catalytic dual metal-mediated decarboxylative cross-coupling reaction between aryl carboxylic acids (or their potassium salts) and haloarenes to afford unsymmetrical biaryls [14, 64]. The coupling reaction proceeds efficiently with two distinct protocols: Pd(acac)$_2$/P(*i*-Pr)Ph$_2$ catalyst system with stoichiometric amounts of both $CuCO_3$ (as the base) and KF (as the additive) and, in alternative, made use of substoichiometric amounts of CuI, Pd(acac)$_2$, and 1,10-phenanthroline. Scheme 22.46 shows the typical experimental conditions and the proposed reaction mechanism.

Gooßen's group subsequently developed a different catalytic system (CuI, 1,10-phenanthroline, PdI_2, and di(*tert*-butyl)biphenylphosphane at 160°C) in order to use nonactivated aryl chlorides (electron-rich derivatives) as electrophilic substrates for cross-coupling with benzoates [14, 65]. An interesting application of the Pd/Cu substoichiometric-mediated cross-coupling reaction has been reported in the synthesis of a key synthon in the preparation of the angiotensin II receptor antagonist telmisartan (Boehringer Ingelheim, Micardis) [66].

An alternative solution to the limitations of the aryl carboxylate salts mentioned previously is a new bimetallic catalyst system that allows the replacement of aryl halides by aryl triflates [14, 67]. The reaction proceeds efficiently with PdI_2, Tol-BINAP, Cu_2O, and 1,10-phenanthroline, affording the corresponding unsymmetrical biaryls (Scheme 22.47). It is noteworthy that the reaction is generally applicable to aromatic carboxylic acid salts, regardless of their substitution pattern. The use of microwave irradiation has proved beneficial for this ARCIS reaction, since under these conditions the Pd(II) catalyst is almost instantaneously reduced to Pd(0) [68]. Similarly, substituted aryl tosylates have been found to be suitable substrates in decarboxylative coupling reactions with a full range of benzoic acids (Scheme 22.47) [69].

Recently, an example of intramolecular formation of Ar–Ar bonds has been reported, consisting in the preparation of dibenzopyranone derivatives by a highly selective, Pd(acac)$_2$/CuCl-mediated tandem reaction [70]. Substituted 2-nitrobenzoic acids and methyl 2-bromo benzoate derivatives are used as starting materials.

22.3.4 Formation of Benzocondensed Derivatives

A particular case of an ARCIS reaction leading to a benzocondensed derivative is represented by the formation of two Ar–Ar bonds in the same step. Triarylmethanols are generally used as the electrophilic partner in the reaction with internal alkynes, catalyzed by a Rh/Cu-based system

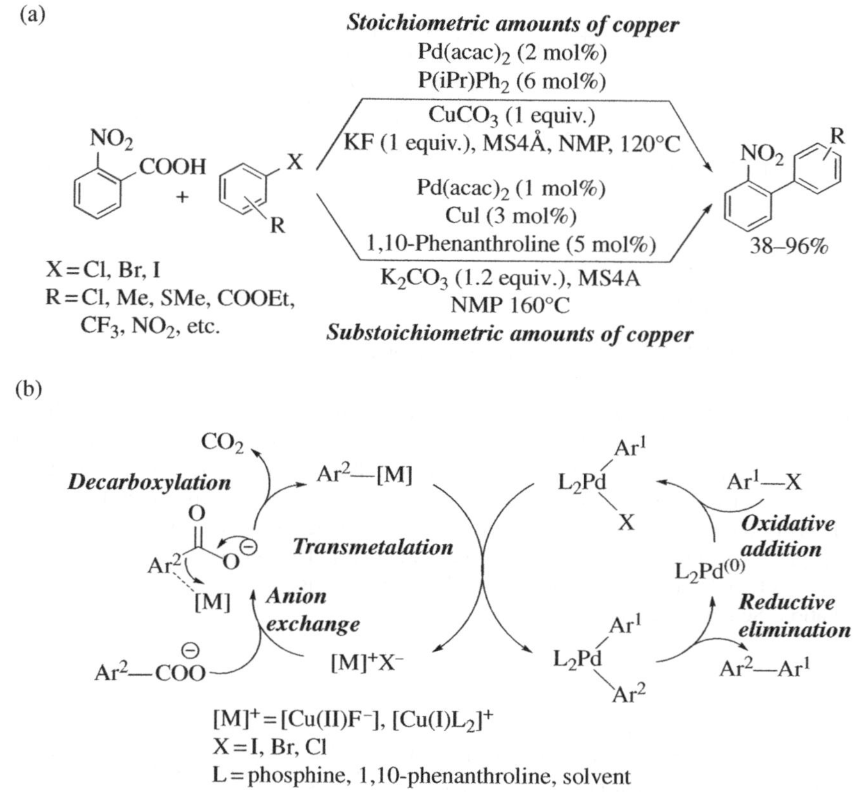

SCHEME 22.46 (a) Unsymmetrical biaryls via Pd-catalyzed activation of the Ar–C bond in benzoic acids and (b) mechanism of the Pd-mediated cross-coupling reaction.

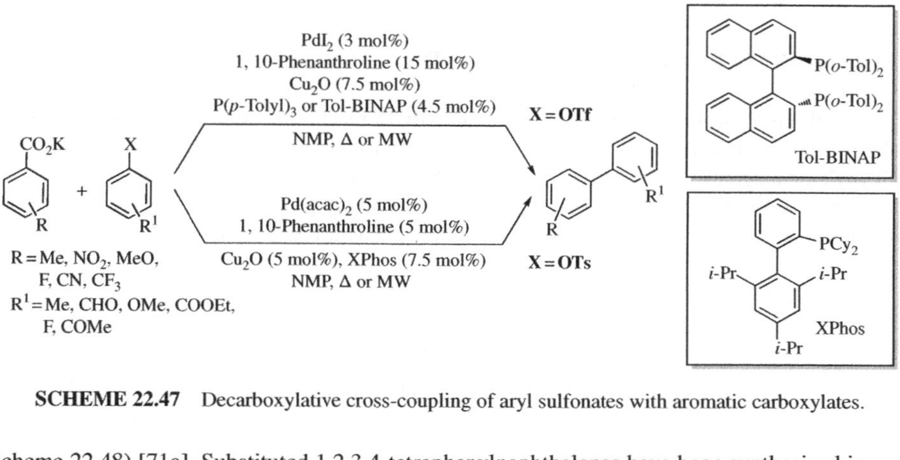

SCHEME 22.47 Decarboxylative cross-coupling of aryl sulfonates with aromatic carboxylates.

(Scheme 22.48) [71a]. Substituted 1,2,3,4-tetraphenylnaphthalenes have been synthesized in good yields in this manner, the reaction proceeding by means of consecutive Ar–H and Ar–C bond cleavages, starting from diphenylacetylenes.

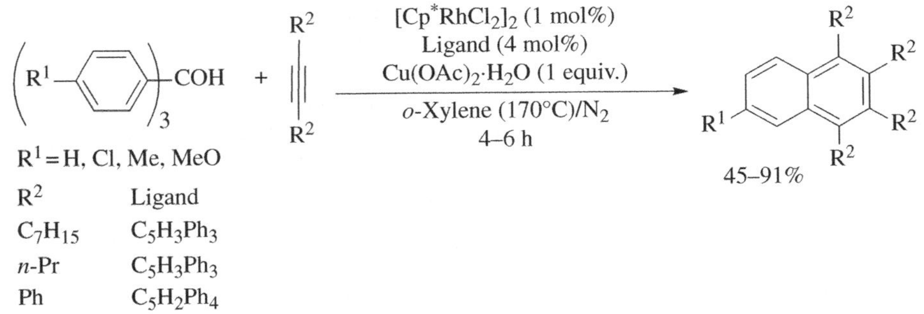

SCHEME 22.48 Rh-catalyzed reactions of triarylmethanols with internal alkynes.

In a related ARCIS process, the Pd-based cross-coupling of *o*-bromobenzyl alcohols with *o*-iodobiphenyls under basic conditions (cesium carbonate) with palladium catalysis allows the preparation of several highly substituted triphenylenes. The use of an electron-deficient phosphine ligand is sufficient to assure the formation of two C–C bonds at the expense of a C–C and a C–H bond [71b]. Another approach for the synthesis of triphenylenes involves the [2+2+2] trimerization of benzynes, generated *in situ* from benzoic acids [71c]. The cooperative Pd(II)/Cu(II) catalytic system was found to be suitable for *ortho* C–H bond activation in benzoic acids followed by decarbonylation (Scheme 22.49).

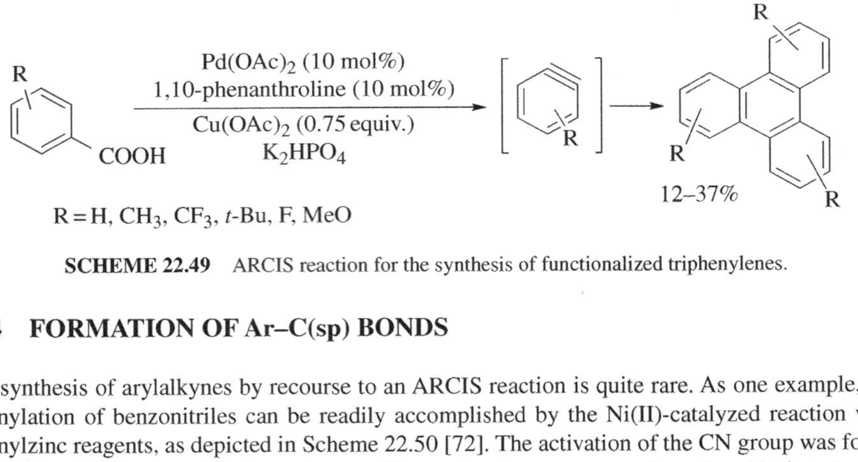

SCHEME 22.49 ARCIS reaction for the synthesis of functionalized triphenylenes.

22.4 FORMATION OF Ar–C(sp) BONDS

The synthesis of arylalkynes by recourse to an ARCIS reaction is quite rare. As one example, the alkynylation of benzonitriles can be readily accomplished by the Ni(II)-catalyzed reaction with alkynylzinc reagents, as depicted in Scheme 22.50 [72]. The activation of the CN group was found to be very selective, even in the presence of better leaving groups, such as the chlorine atom in 4-chlorobenzonitrile, which remains virtually untouched during the reaction.

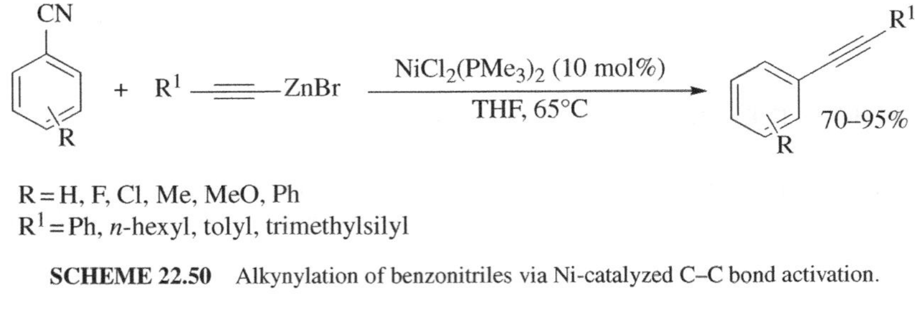

SCHEME 22.50 Alkynylation of benzonitriles via Ni-catalyzed C–C bond activation.

22.5 SUMMARY AND OUTLOOK

The reaction types described in this chapter are summarized in Table 22.1. ARCIS reactions are very promising for the formation of Ar–C(sp^2) bonds, both Ar–vinyl and Ar–Ar. Arylalkynes are virtually unobtainable by an ARCIS reaction, whereas Ar–C(sp^3) bonds were formed with some success in some cases. It is apparent that Ni and Rh compounds constitute the most versatile catalysts for ARCIS reactions, because they are able to activate Ar–C bonds with different carbon leaving groups.

TABLE 22.1 Ar–C Formation by a Metal-Catalyzed Activation of Another Ar–C Bond

Metal Catalyst	Carbon Leaving Group	Bond Formed
Ni	CN, CO(NR)CO	Ar-sp^3
Ni	CN, CO(O)CO, CO(S)CO	Ar-sp^2 (Ar–vinyl)
Ni	CN, CO(NR)CO	Ar-sp^2 (Ar–Ar)
Ni	CN	Ar-sp
Cu	COOH	Ar-sp^2 (Ar–Ar)
Rh	CHO, COR, COOH	Ar-sp^3
Rh	COOH, C(OH)R_2, CN	Ar-sp^2 (Ar–vinyl)
Rh	CN, CHO, COOH(R), CO(O)CO	Ar-sp^2 (Ar–Ar)
Pd	COOH	Ar-sp^2 (ArCO)
Pd	COOH(R), CO(O)CO	Ar-sp^2 (Ar–vinyl)
Pd	COOH(R), C(OH)R_2	Ar-sp^2 (Ar–Ar)
Ru	CHO	Ar-sp^2 (Ar–vinyl)
Cu/Rh	COOR	Ar-sp^2 (Ar–Ar)
Cu/Rh	C(OH)R_2	Ar-sp^2 (benzocondensed)
Pd/Ag	COOH	Ar-sp^3
Pd/Ag	COOH	Ar-sp^2 (Ar–vinyl)
Pd/Ag	COOH, C(OH)R_2	Ar-sp^2 (Ar–Ar)
Pd/Cu	COOH	Ar-sp^3
Pd/Cu	COOH	Ar-sp^2 (Ar–vinyl)
Pd/Cu	COOH	Ar-sp^2 (Ar–Ar)
Pd/Cu	COOH	Ar-sp^2 (benzocondensed)

Pd catalysts alone or accompanied by other metals (Ag or Cu) are mostly employed in the cross-coupling reaction of benzoic acids. The contribution of other metals, such as Cu and Ru by themselves, is quite limited. Most of these cross-coupling reactions are very recent, reported in the last 15 years, so there is still ample room to improve the reactions and achieve milder reaction conditions. Although in many examples the reactions require high temperature, less prohibitive conditions have generally been possible when using a strong nucleophile as the reaction partner (e.g., arylzinc or Grignard reagents) or when employing microwave irradiation. Despite much progress in the use of aromatics bearing carbon-based leaving groups as the electrophilic partner in cross-coupling reactions, they have not yet attained the utility of aryl halides and aryl sulfonate esters. This calls for improvements in catalytic systems and for the design of new, more versatile C-based leaving groups.

ABBREVIATIONS

ARCIS Aromatic carbon–carbon ipso-substitution
DMA Dimethylacetamide
MW Microwave heating
NMP 1-Methyl-2-pyrrolidone

REFERENCES

[1] (a) Beller, M., Bolm, C. Eds, *Transition Metals for Organic Synthesis*, Vol. **1**, 2nd Edition. Wiley-VCH, Weinheim, 2004; (b) Crawley, M. L., Trost, B. M., *Applications of Transition Metal Catalysis in Drug Discovery and Development*. John Wiley & Sons, Inc., Hoboken, 2012.

[2] (a) Bonesi, S. M., Fagnoni, M., Albini, A. (2008). Biaryl formation involving carbon-based leaving groups: why not? *Angewandte Chemie International Edition*, **47**, 10022–10025; (b) Bonesi, S., Fagnoni, M. (2010). The aromatic carbon–carbon ipso-substitution reaction. *Chemistry-An European Journal*, **16**, 13572–13589.

[3] Murakami, M., Matsuda, T. (2011). Metal-catalysed cleavage of carbon–carbon bonds. *Chemical Communications*, **47**, 1100–1105.

[4] Miller, J. A., Dankwardt, J. W. (2003). Nickel catalyzed cross-coupling of modified alkyl and alkenyl Grignard reagents with aryl- and heteroaryl nitriles: activation of the C—CN bond. *Tetrahedron Letters*, **44**, 1907–1910.

[5] (a) Nakao, Y. (2012). Nickel/Lewis acid-catalyzed carbocyanation of unsaturated compounds. *Bulletin of the Chemical Society of Japan*, **85**, 731–745; (b) Tobisu, M., Chatani, N. (2008). Catalytic reactions involving the cleavage of carbon–cyano and carbon–carbon triple bonds. *Chemical Society Reviews*, **37**, 300–307.

[6] Nakao, Y., Yada, A., Satoh, J., Ebata, S., Oda, S., Hiyama, T. (2006). Arylcyanation of norbornene and norbornadiene catalyzed by nickel. *Chemistry Letters*, **35**, 790–791.

[7] Watson, M. P., Jacobsen, E. N. (2008). Asymmetric intramolecular arylcyanation of unactivated olefins via C-CN bond activation. *Journal of the American Chemical Society*, **130**, 12594–12595.

[8] Nakao, Y., Ebata, S., Yada, A., Hiyama, T., Ikawa, M., Ogoshi, S. (2008). Intramolecular arylcyanation of alkenes catalyzed by nickel/$AlMe_2Cl$. *Journal of the American Chemical Society*, **130**, 12874–12875.

[9] Havlik, S. E., Simmons, J. M., Winton, V. J., Johnson, J. B. (2011). Nickel-mediated decarbonylative cross-coupling of phthalimides with in situ generated diorganozinc reagents. *The Journal of Organic Chemistry*, **76**, 3588–3593.

[10] Yang, L., Guo, X., Li, C.-J. (2010). The first decarbonylative coupling of aldehydes and norbornenes catalyzed by rhodium. *Advanced Synthesis and Catalysis*, **352**, 2899–2904.

[11] Daugulis, O., Brookhart, M. (2004). Decarbonylation of aryl ketones mediated by bulky cyclopentadienylrhodium bis(ethylene) complexes. *Organometallics*, **23**, 527–534.

[12] Lei, Z.-Q., Li, H., Li, Y., Zhang, X.-S., Chen, K., Wang, X., Sun, J., Shi, Z.-J. (2012). Extrusion of CO from aryl ketones: rhodium(I)-catalyzed C—C bond cleavage directed by a pyridine group. *Angewandte Chemie International Edition*, **51**, 2690–2694.

[13] Sun, Z.-M., Zhao, P. (2009). Rhodium-mediated decarboxylative conjugate addition of fluorinated benzoic acids: stoichiometric and catalytic transformations. *Angewandte Chemie International Edition*, **48**, 6726–6730.

[14] (a) Gooßen, L. J., Rodriguez, N., Gooßen, K. (2008). Carboxylic acids as substrates in homogeneous catalysis. *Angewandte Chemie International Edition*, **47**, 3100–3120; (b) Rodríguez, N., Gooßen, L. J. (2011). Decarboxylative coupling reactions: a modern strategy for C–C-bond formation. *Chemical Society Reviews*, **40**, 5030–5048; (c) Dzik, W. I., Lange, P. P., Gooßen, L. J. (2012). Carboxylates as sources of carbon nucleophiles and electrophiles: comparison of decarboxylative and decarbonylative pathways. *Chemical Science*, **3**, 2671–2678.

[15] Xiang, S., Cai, S., Zeng, J., Liu, X.-W. (2011). Regio- and stereoselective synthesis of 2-deoxy-*C*-aryl glycosides via palladium catalyzed decarboxylative reactions. *Organic Letters*, **13**, 4608–4611.

[16] Myers, A. G., Tanaka, D., Mannion, M. R. (2002). Development of a decarboxylative palladation reaction and its use in a Heck-type olefination of arene carboxylates. *Journal of the American Chemical Society*, **124**, 11250–11251.

[17] Wang, J., Cui, Z., Zhang, Y., Li, H., Wu, L.-M., Liu, Z. (2011). Pd(II)-catalyzed decarboxylative allylation and Heck-coupling of arene carboxylates with allylic halides and esters. *Organic and Biomolecular Chemistry*, **9**, 663–666.

[18] Huang, L., Qi, J., Wu, X., Huang, K., Jiang, H. (2013). Highly selective β-hydride elimination in Pd-catalyzed decarboxylative Heck-type reaction. *Organic Letters*, **15**, 2330–2333.

[19] Luo, Y., Wu, J. (2010). Palladium-catalyzed decarboxylative 1,2-addition of carboxylic acids to aldehydes or imines. *Chemical Communications*, **46**, 3785–3787.

[20] (a) Lindh, J., Sjöberg, P. J. R., Larhed, M. (2010). Synthesis of aryl ketones by palladium(II)-catalyzed decarboxylative addition of benzoic acids to nitriles. *Angewandte Chemie International Edition*, **49**, 7733–7737; (b) Rydfjord, J., Svensson, F., Trejos, A., Sjöberg, P. J. R., Sköld, C., Sävmarker, J., Odell, L. R., Larhed, M. (2013). Decarboxylative palladium(II)-catalyzed synthesis of aryl amidines from aryl carboxylic acids: development and mechanistic investigation. *Chemistry-An European Journal*, **19**, 13803–13810.

[21] (a) Nakao, Y., Oda, S., Hiyama, T. (2004). Nickel-catalyzed arylcyanation of alkynes. *Journal of the American Chemical Society*, **126**, 13904–13905; (b) Nakao, Y., Yada, A., Ebata, S., Hiyama, T. (2007). A dramatic effect of Lewis-acid catalysts on nickel-catalyzed carbocyanation of alkynes. *Journal of the American Chemical Society*, **129**, 2428–2429; (c) Nakao, Y., Oda, S., Yada, A., Hiyama, T. (2006). Arylcyanation of alkynes catalyzed by nickel. *Tetrahedron*, **62**, 7567–7576; (d) Minami, Y., Yoshiyasu, H., Nakao, Y., Hiyama, T. (2013). Highly chemoselective carbon-carbon σ-bond activation: nickel/Lewis acid catalyzed polyfluoroarylcyanation of alkynes. *Angewandte Chemie International Edition*, **52**, 883–887.

[22] Nakai, K., Kurahashi, T., Matsubara, S. (2011). Nickel-catalyzed cycloaddition of o-arylcarboxybenzonitriles and alkynes via cleavage of two carbon-carbon σ bonds. *Journal of the American Chemical Society*, **133**, 11066–11068.

[23] (a) Kajita, Y., Kurahashi, T., Matsubara, S. (2008). Nickel-catalyzed decarbonylative addition of anhydrides to alkynes. *Journal of the American Chemical Society*, **130**, 17226–17227; (b) Inami, T., Baba, Y., Kurahashi, T., Matsubara, S. (2011). Nickel-catalyzed cycloadditions of thiophthalic anhydrides with alkynes. *Organic Letters*, **13**, 1912–1915; (c) Maizuru, N., Inami, T., Kurahashi, T., Matsubara, S. (2011). Nickel-catalyzed cycloadditions of benzoxazinones with alkynes: synthesis of quinolines and quinolones. *Chemistry Letters*, **40**, 375–377; (d) Shiba, T., Kurahashi, T., Matsubara, S. (2013). Nickel-catalyzed decarbonylative alkylidenation of phthalimides with trimethylsilyl-substituted alkynes. *Journal of the American Chemical Society*, **135**, 13636–13639; (e) Ochi, Y., Kurahashi, T., Matsubara, S. (2011). Decarbonylative cycloaddition of phthalic anhydrides with allenes. *Organic Letters*, **13**, 1374–1377.

[24] Nakai, K., Kurahashi, T., Matsubara, S. (2013). Nickel-catalyzed decarbonylative and decarboxylative cycloaddition of isatoic anhydrides with alkynes. *Chemistry Letters*, **42**, 1238–1240.

[25] Guo, X., Wang, J., Li, C.-J. (2009). An olefination via ruthenium-catalyzed decarbonylative addition of aldehydes to terminal alkynes. *Journal of the American Chemical Society*, **131**, 15092–15093.

[26] Sun, Z.-M., Zhang, J., Zhao, P. (2010). Rh(I)-catalyzed decarboxylative transformations of arenecarboxylic acids: ligand- and reagent-controlled selectivity toward hydrodecarboxylation or Heck-Mizoroki products. *Organic Letters*, **12**, 992–995.

[27] (a) Sugihara, T., Satoh, T., Miura, M., Nomura, M. (2004). Rhodium-catalyzed coupling reaction of aroyl chlorides with alkenes. *Advanced Synthesis and Catalysis*, **346**, 1765–1772; (b) Kokubo, K., Matsumasa, K., Miura, M., Nomura, M. (1996). Rhodium-catalyzed reaction of aroyl chlorides with alkynes. *The Journal of Organic Chemistry*, **61**, 6941–6946.

[28] Li, H., Li, Y., Zhang, X.-S., Chen, K., Wang, X., Shi, Z.-J. (2011). Pyridinyl directed alkenylation with olefins via Rh(III)-catalyzed C-C bond cleavage of secondary arylmethanols. *Journal of the American Chemical Society*, **133**, 15244–15247.

[29] Kita, Y., Tobisu, M., Chatani, N. (2010). Rhodium-catalyzed alkenylation of nitriles via silicon-assisted C–CN bond cleavage. *Organic Letters*, **12**, 1864–1867.

[30] Ishida, N., Sawano, S., Masuda, Y., Murakami, M. (2012). Rhodium-catalyzed ring opening of benzocyclobutenols with site-selectivity complementary to thermal ring opening. *Journal of the American Chemical Society*, **134**, 17502–17504.

[31] Iwai, T., Fujihara, T., Terao, J., Tsuji, Y. (2009). Iridium-catalyzed addition of acid chlorides to terminal alkynes. *Journal of the American Chemical Society*, **131**, 6668–6669.

[32] Wang, C., Rakshit, S., Glorius, F. (2010). Palladium-catalyzed intermolecular decarboxylative coupling of 2-phenylbenzoic acids with alkynes via C-H and C-C bond activation. *Journal of the American Chemical Society*, **132**, 14006–14008.

[33] Hu, P., Kan, J., Su, W., Hong, M. (2009). $Pd(O_2CCF_3)_2$/benzoquinone: a versatile catalyst system for the decarboxylative olefination of arene carboxylic acids. *Organic Letters*, **11**, 2341–2344.

[34] Fu, Z., Huang, S., Su, W., Hong, M. (2010). Pd-catalyzed dearboxylative Heck coupling with dioxygen as the terminal oxidant. *Organic Letters*, **12**, 4992–4995.

[35] Gooßen, L. J., Zimmermann, B., Knauber, T. (2010). Pd-catalyzed decarboxylative Heck vinylation of 2-nitrobenzoates in the presence of CuF_2. *Beilstein Journal of Organic Chemistry*, **6**, 43.

[36] Tanaka, D., Myers, A. G. (2004). Heck-type arylation of 2-cycloalken-1-ones with arylpalladium intermediates formed by decarboxylative palladation and by aryl iodide insertion. *Organic Letters*, **6**, 433–436.

[37] Gooßen, L. J., Paetzold, J. (2004). Decarbonylative Heck olefination of enol esters: salt-free and environmentally friendly access to vinyl arenes. *Angewandte Chemie International Edition*, **43**, 1095–1098.

[38] Stephan, M. S., Teunissen, A. J. J. M., Verzijl, G. K. M., de Vries, J. G. (1998). Heck reactions without salt formation: aromatic carboxylic anhydrides as arylating agents. *Angewandte Chemie International Edition*, **37**, 662–664.

[39] Gooßen, L. J., Paetzold, J., Winkel, L. (2002). Pd-catalyzed decarbonylative Heck olefination of aromatic carboxylic acids activated in situ with di-*tert*-butyl dicarbonate. *Synlett*, 1721–1723.

[40] Gooßen, L. J., Plaetzold, J. (2002). Pd-catalyzed decarbonylative olefination of aryl esters: towards a waste-free Heck reaction. *Angewandte Chemie International Edition*, **41**, 1237–1241.

[41] Sugihara, T., Satoh, T., Miura, M. (2005). Mizoroki–Heck type arylation of alkenes using aroyl chlorides under base-free conditions. *Tetrahedron Letters*, **46**, 8269–8271.

[42] Zhang, M., Zhou, J., Kan, J., Wang, M., Su, W., Hong, M. (2010). Pd-catalyzed cross-coupling of carboxylic acids with nitroethane via combination of decarboxylation and dehydrogenation. *Chemical Communications*, **46**, 5455–5457.

[43] Zhou, J., Wu, G., Zhang, M., Jie, X., Su, W. (2012). Pd-catalyzed cross-coupling of aryl carboxylic acids with propiophenones through a combination of decarboxylation and dehydrogenation. *Chemistry-An European Journal*, **18**, 8032–8036.

[44] Miller, J. A., Dankwardt, J. W., Penny, J. M. (2003). Nickel catalyzed cross-coupling and amination reactions of aryl nitriles. *Synthesis*, 1643–1648.

[45] Liu, N., Wang, Z.-X. (2012). Nickel-catalyzed cross-coupling of arene- or heteroarenecarbonitriles with aryl- or heteroarylmanganese reagents through C-CN bond activation. *Advanced Synthesis and Catalysis*, **354**, 1641–1645.

[46] Shang, R., Fu, Y., Wang, Y., Xu, Q., Yu, H.-Z., Liu, L. (2009). Copper-catalyzed decarboxylative cross-coupling of potassium polyfluorobenzoates with aryl iodides and bromides. *Angewandte Chemie International Edition*, **48**, 9350–9354.

[47] (a) Gooßen, L. J., Paetzold, J. (2004). New synthesis of biaryls via Rh-catalyzed decarbonylative Suzuki-coupling of carboxylic anhydrides with arylboroxines. *Advanced Synthesis and Catalysis*, **346**, 1665–1668; (b) Jin, W., Yu, Z., He, W., Ye, W., Xiao, W.-J. (2009). Efficient Rh(I)-catalyzed direct arylation and alkenylation of arene C-H bonds via decarbonylation of benzoic and cinnamic anhydrides. *Organic Letters*, **11**, 1317–1320.

[48] Shuai, Q., Yang, L., Guo, X., Basle, O., Li, C.-J. (2010). Rhodium-catalyzed oxidative C-H arylation of 2-arylpyridine derivatives via decarbonylation of aromatic aldehydes. *Journal of the American Chemical Society*, **132**, 12212–12213.

[49] Pan, F., Lei, Z.-Q., Wang, H., Li, H., Sun, J., Shi, Z.-J. (2013). Rhodium(I)-catalyzed redox-economic cross-coupling of carboxylic acids with arenes directed by N-containing groups. *Angewandte Chemie International Edition*, **52**, 2063–2067.

[50] Wang, J., Liu, B., Zhao, H., Wang, J. (2012). Rhodium-catalyzed cross-coupling reactions of carboxylate and organoboron compounds via chelation-assisted C-C bond activation. *Organometallics*, **31**, 8598–8607.

[51] Maetani, S., Fukuyama, T., Ryu, I. (2013). Rhodium-catalyzed decarbonylative C-H arylation of 2-aryloxybenzoic acids leading to dibenzofuran derivatives. *Organic Letters*, **15**, 2754–2757.

[52] Tobisu, M., Kita, Y., Ano, Y., Chatani, N. (2008). Rhodium-catalyzed silylation and intramolecular arylation of nitriles via the silicon-assisted cleavage of carbon-cyano bonds. *Journal of the American Chemical Society*, **130**, 15982–15989.

[53] (a) Shang, R., Xu, Q., Jiang, Y.-Y., Wang, Y., Liu, L. (2010). Pd-catalyzed decarboxylative cross coupling of potassium polyfluorobenzoates with aryl bromides, chlorides, and triflates. *Organic Letters*, **12**, 1000–1003; (b) Shen, Z., Ni, Z., Mo, S., Wang, J., Zhu, Y. (2012). Palladium-catalyzed intramolecular decarboxylative coupling of arene carboxylic acids/esters with aryl bromides. *Chemistry-An European Journal*, **18**, 4859–4865.

[54] Terao, Y., Wakui, H., Nomoto, M., Satoh, T., Miura, M., Nomura, M. (2003). Palladium-catalyzed arylation of α,α-disubstituted arylmethanols via cleavage of a C–C or a C–H bond to give biaryls. *The Journal of Organic Chemistry*, **68**, 5236–5243.

[55] Voutchkova, A., Coplin, A., Leadbeater, N. E., Crabtree, R. H. (2008). Palladium-catalyzed decarboxylative coupling of aromatic acids with aryl halides or unactivated arenes using microwave heating. *Chemical Communications*, 6312–6314.

[56] Wang, Z., Ding, Q., He, X., Wu, J. (2009). Pd-catalyzed decarboxylative couplings of arenecarboxylic acids with aryl iodides. *Tetrahedron*, **65**, 4635–4638.

[57] Becht, J. M., Catala, C., Le Drian, C., Wagner, A. (2007). Synthesis of biaryls via decarboxylative Pd-catalyzed cross-coupling reaction. *Organic Letters*, **9**, 1781–1783.

[58] Becht, J. M., Le Drian, C., (2008). Biaryl synthesis via decarboxylative Pd-catalyzed reactions of arenecarboxylic acids and diaryliodonium triflates. *Organic Letters*, **10**, 3161–3164.

[59] Gooßen, L. J., Lange, P. P., Rodriguez, N., Linder, C. (2010). Low-temperature Ag/Pd-catalyzed decarboxylative cross-coupling of aryl triflates with aromatic carboxylate salts. *Chemistry-An European Journal*, **16**, 3906–3909.

[60] Dai, J.-J., Liu, J.-H., Luo, D.-F., Liu, L. (2011). Pd-catalysed decarboxylative Suzuki reactions and orthogonal Cu-based O-arylation of aromatic carboxylic acids. *Chemical Communications*, 677–679.

[61] (a) Cornella, J., Lu, P., Larrosa, I. (2009). Intermolecular decarboxylative direct C-3 arylation of indoles with benzoic acids. *Organic Letters*, **11**, 5506–5509; (b) Luo, H.-Q., Dong, W., Loh, T.-P. (2013). Pd-catalyzed decarboxylative cross-coupling of perfluorobenzoic acids with simple arenes. *Tetrahedron Letters*, **54**, 2833–2836; (c) Zhao, H., Ye, W., Xu, J., Kan, J., Su, W. (2011). Pd/PR_3-catalyzed cross-coupling of aromatic carboxylic acids with electron-deficient polyfluoroarenes via combination of decarboxylation with sp^2 C-H cleavage. *The Journal of Organic Chemistry*, **76**, 882–893; (d) Hu, P., Zhang, M., Jie, X., Su, W. (2012). Palladium-catalyzed decarboxylative C-H bond arylation of thiophenes. *Angewandte Chemie International Edition*, **51**, 227–231; (e) Wang, C., Piel, I., Glorius, F. (2009). Palladium-catalyzed intramolecular direct arylation of benzoic acids by tandem decarboxylation/C-H activation. *Journal of the American Chemical Society*, **131**, 4194–4195.

[62] (a) Cornella, J., Lahlali, H., Larrosa, I. (2010). Decarboxylative homocoupling of (hetero)aromatic carboxylic acids. *Chemical Communications*, 8276–8278; (b) Xie, K., Wang, S., Yang, Z., Liu, J., Wang, A., Li, X., Tan, Z., Guo, C.-C., Deng, W. (2011). Synthesis of biaryls by Pd-catalyzed decarboxylative homo- and heterocoupling of substituted benzoic acids. *European Journal of Organic Chemistry*, 5787–5790; (c) Hu, P., Shang, Y., Su, W. (2012). A general Pd-catalyzed decarboxylative cross-coupling reaction between aryl carboxylic acids: synthesis of biaryl compounds. *Angewandte Chemie International Edition*, **51**, 5945–5949.

[63] Chtchemelinine, A., Rosa, D., Orellana, A. (2011). Palladium-catalyzed selective carboelimination and cross-coupling reactions of benzocyclobutenols with aryl bromides. *The Journal of Organic Chemistry*, **76**, 9157–9162

[64] (a) Gooßen, L. J., Deng, G., Levy, M. L. (2006). Synthesis of biaryls via catalytic decarboxylative coupling. *Science*, **313**, 662–664; (b) Gooßen, L. J., Rodriguez, N., Melzer, B., Linder, C., Deng, G., Levy, L. M. (2007). Biaryl synthesis via Pd-catalyzed decarboxylative coupling of aromatic carboxylates with aryl halides. *Journal of the American Chemical Society*, **129**, 4824–4833.

[65] Gooßen, L. J., Zimmermann, B., Knauber, T. (2008). Palladium/copper-catalyzed decarboxylative cross-coupling of aryl chlorides with potassium carboxylates. *Angewandte Chemie International Edition*, **47**, 7103–7106.

[66] Gooßen, L. J., Knauber, T. (2008). Concise synthesis of telmisartan via decarboxylative cross-coupling. *The Journal of Organic Chemistry*, **73**, 8631–8634.

[67] Gooßen, L. J., Rodriguez, N., Linder, C. (2008). Decarboxylative biaryl synthesis from aromatic carboxylates and aryl triflates. *Journal of the American Chemical Society*, **130**, 15248–15249.

[68] Gooßen, L. J., Linder, C., Rodriguez, N., Lange, P. P. (2009). Biaryl and aryl ketone synthesis via Pd-catalyzed decarboxylative coupling of carboxylate salts with aryl triflates. *Chemistry-An European Journal*, **15**, 9336–9349.

[69] Gooßen, L. J., Rodriguez, N., Lange, P. P., Linder, C. (2010). Decarboxylative cross-coupling of aryl tosylates with aromatic carboxylate salts. *Angewandte Chemie International Edition*, **49**, 1111–1114.

[70] Luo, J., Lu, Y., Liu, S., Liu, J., Deng, G.-J. (2011). Efficient one-pot synthesis of dibenzopyranones via a decarboxylative cross-coupling and lactonization sequence. *Advanced Synthesis and Catalysis*, **353**, 2604–2608.

[71] (a) Uto, T., Shimizu, M., Ueura, K., Tsurugi, T., Satoh, T., Miura, M. (2008). Rhodium-catalyzed oxidative coupling of triarylmethanols with internal alkynes via successive C-H and C-C bond cleavages. *The Journal of Organic Chemistry*, **73**, 298–300; (b) Iwasaki, M., Iino, S., Nishihara, Y. (2013). Palladium-catalyzed annulation of o-iodobiphenyls with o-bromobenzyl alcohols: synthesis of functionalized triphenylenes via C-C and C-H bond cleavages. *Organic Letters*, **15**, 5326–5329; (c) Cant, A. A., Roberts, L., Greaney, M. F. (2010). Generation of benzyne from benzoic acid using C-H activation. *Chemical Communications*, **46**, 8671–8673.

[72] Penney, J. M., Miller, J. A. (2004). Alkynylation of benzonitriles via nickel catalyzed C-C bond activation. *Tetrahedron Letters*, **45**, 4989–4992.

PART VII

C–H FUNCTIONALIZATION

23

CHELATE-ASSISTED ARENE C–H BOND FUNCTIONALIZATION

Marion H. Emmert and Christopher J. Legacy
Department of Chemistry and Biochemistry, Worcester Polytechnic Institute, Worcester, MA, USA

23.1 INTRODUCTION

During the last two decades, transition metal-catalyzed, direct C–H functionalizations have become a widely used strategy for the synthesis of arenes bearing substituents. The strategy of using directing groups has been of particular use for the synthesis of complex molecules; these groups coordinate to the catalyst during the reaction and thus enable C–H cleavage as well as new bond formations in proximity to the catalyst coordination site. Since most of these reactions are believed to proceed through cyclometalated chelate complexes as intermediates, the term "chelate assisted" has been introduced to describe the relevant reactivity pattern [1]. Other researchers refer to this type of reactivity as "directed" or "ligand-directed" C–H functionalizations [2]; however, this terminology might be misleading for substrates in which directing substituent effects on site selectivity—similar to effects in electrophilic substitution [3]—as well catalyst directing effects are possible. Similarly, several authors refer to functionalizations at C–H bonds in α-positions of heteroaromatics such as pyridine as heteroatom-directed reactions [4]. In order to avoid confusion, this review will primarily use the terminology "chelate-assisted" instead of "directed" C–H functionalizations (Scheme 23.1).

The strategy of using directing groups for C–H activation and subsequent functionalization has enabled the introduction of many different functional groups on arenes without intermediate steps, which has the potential to be particularly useful for the diversification of complex molecules at late stages of syntheses and thus drug discovery applications [5–7]. Conceptual challenges of this approach are the introduction and/or removal of the used directing group, which is not always desired in the target molecule. Various creative approaches have been taken to overcome this limitation (addressed in Sections 23.1.2 and 23.1.3). Nondirected C–H functionalizations of arenes have also been realized and are addressed in the next chapter of this book. Due to page limitations and

Arene Chemistry: Reaction Mechanisms and Methods for Aromatic Compounds,
First Edition. Edited by Jacques Mortier.

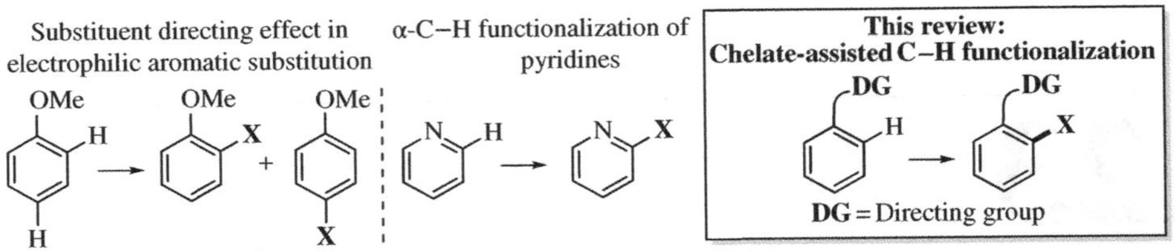

SCHEME 23.1 Illustration of different "directed" reactivity in organic chemistry.

in order to stay within the scope of this book, this chapter exclusively reviews chelate-assisted functionalizations of nonheteroaromatic C–H bonds.

23.1.1 Mechanisms of Chelate-Assisted C–H Bond Functionalization and Activation

C–H functionalizations can generally proceed through three types of mechanisms [5, 8]: (i) radical rebound mechanisms, which proceed through radical intermediates; (ii) direct insertions into C–H bonds; and (iii) mechanisms through an organometallic intermediate with a metal–carbon bond. These three different pathways are outlined in Scheme 23.2.

Radical rebound mechanisms (*1*) are typically proposed for metal-oxo catalysts, which are common in enzyme catalysis and beyond [9]. Due to the high bond dissociation energy of aromatic C–H bonds [10], radical rebound pathways are rarely proposed for the herein reviewed chelate-assisted C–H bond functionalizations. In analogy, direct insertion mechanisms (*2*) are most often proposed for functionalizations of weak C–H bonds, as these pathways also require a complete breaking of the C–H bond during the rate determining step. Thus, among the three possible functionalization mechanisms, C–H functionalization through organometallic intermediates (*3*) is the most commonly proposed and investigated mechanism for chelate-assisted modes of reactivity [2, 11–13] and will be reviewed in detail in the following paragraphs.

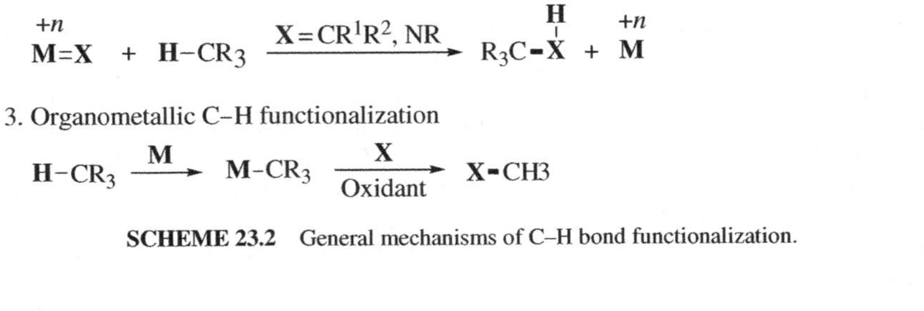

SCHEME 23.2 General mechanisms of C–H bond functionalization.

Chelate-assisted C–H bond functionalizations through C–H activation are typically initiated by catalyst precoordination, which situates the aromatic C–H bond in the proximity of the catalyst (Scheme 23.3). The C–H bond is then cleaved, which results in formation of an organometallic intermediate **1**.

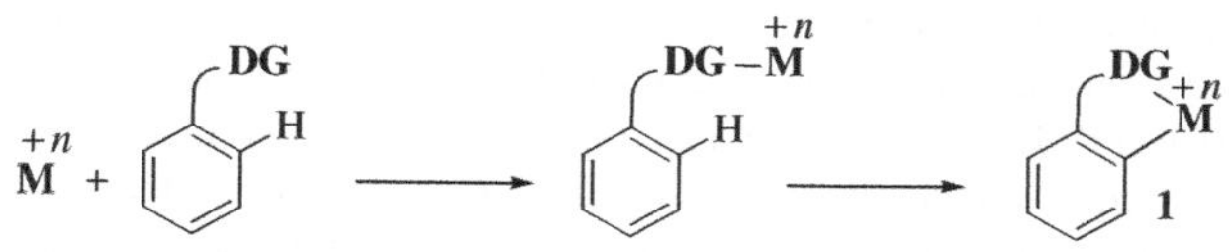

SCHEME 23.3 Chelate-assisted C–H activation. DG, directing group.

Various reactivity modes for this first step of chelate-assisted C–H functionalization—transition metal-mediated C–H activation—are reported in the literature [8, 14]. The exact mechanism for C–H activation in a particular system depends on the identity of the metal and its oxidation state as well as the ancillary ligands around the metal center. For example, in the presence of carboxylate or carbonate ligands at the metal center, C–H cleavage through a 6-membered transition state is often proposed (concerted metalation–deprotonation; Scheme 23.4, *1A*). Furthermore, both heteroatom- and carbon-bound ligands such as methoxy or methyl can be involved, which is referred to as concerted metalation–deprotonation through a 4-membered transition state (*1B*) or sigma bond metathesis (*2*). Computations suggest that these two mechanisms are distinctly different with respect to the involved orbitals: In CMD through a 4-membered transition state, the forming X–H bond is not based on the same orbital as the breaking M–X bond; therefore, these steps are not necessarily happening in a concerted fashion. In contrast, M–R bond breaking and R–H bond formation are concerted processes in the SBM mechanism. An additional difference between these mechanisms is the nature of the involved ligands on the metal center: X is typically a heteroatom with a lone pair (O, N), while R in the SBM mechanism stands for a C-based group (alkyl, aryl). Finally, oxidative addition mechanisms (*3*) are possible for various transition metal complexes of the third period (Ir^{III} [15], Pt^{II} [16]).

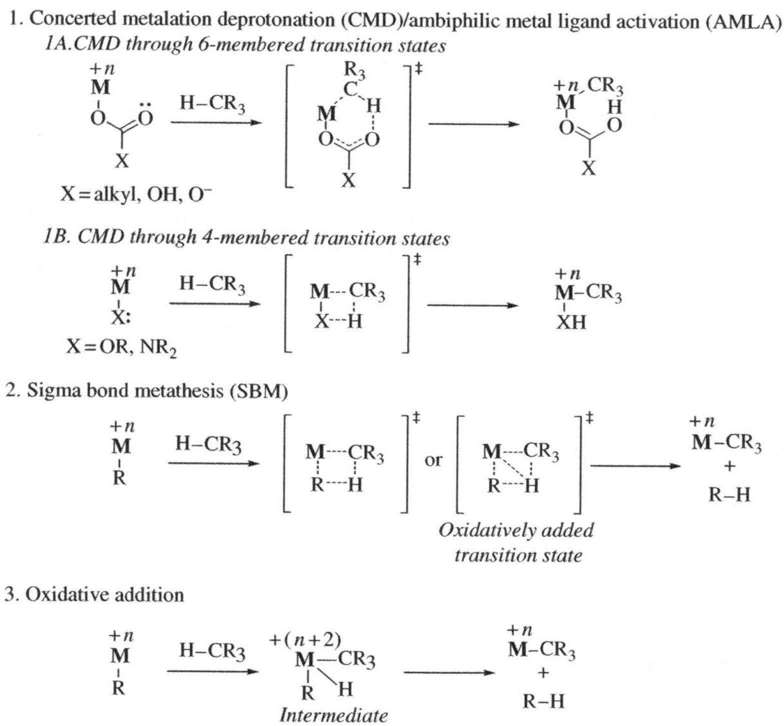

SCHEME 23.4 Potential mechanistic manifolds of C–H activation.

Once C–H activation has been achieved and the organometallic intermediate **1** is formed, different pathways can be taken that all result in C–H functionalization, depending on the used metal catalyst (Cu, Pd, Rh, Ir, Ru), oxidant (one-electron vs. two-electron oxidants, soluble vs. gaseous vs. salt-based oxidants), the formed bond (C–C vs. C–X), and the reaction conditions (pH, temperature, solvent). Two potential pathways for the steps following C–H activation are shown in Scheme 23.5: Path A proceeds through initial oxidation of intermediate **1** with subsequent bond formation, while path B forms the new bond first before the catalyst is regenerated by oxidation. Polarized bonds (e.g., C–O) are frequently proposed to be formed through initial oxidation, when product formation by reductive elimination is challenging from a catalyst with a lower oxidation state [12, 17]. On the other hand, thermodynamically favorable C–C bond reductive eliminations tend to occur readily before catalyst oxidation [18].

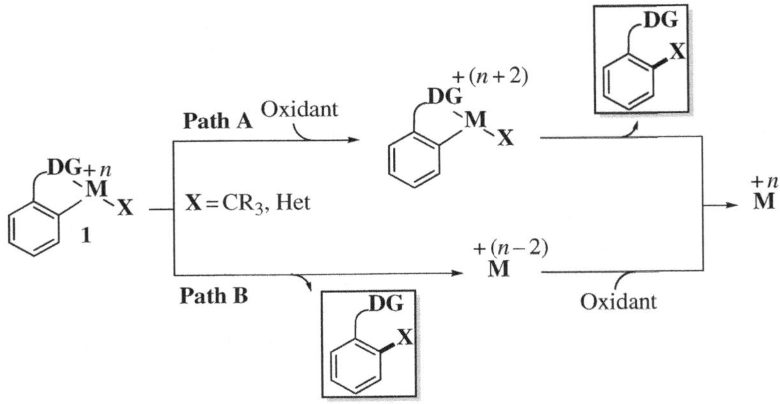

SCHEME 23.5 C–H functionalization of organometallic intermediate **1**.

The mechanism of chelate-assisted C–H activation has fundamentally important consequences for the site selectivity of substitution. Since C–H activation takes place in *ortho* position to the site of the directing group in order to form energetically favorable 5- or 6-membered chelate complexes as intermediates, formation of a new C–C or C–X bond takes place at the same site. Thus, *meta*- and *para*-products can typically not be obtained. However, a modification of this strategy with longer tethers as directing group allows exclusive *meta*-C–H functionalizations through *meta*-C–H activation (Scheme 23.6) [19, 20].

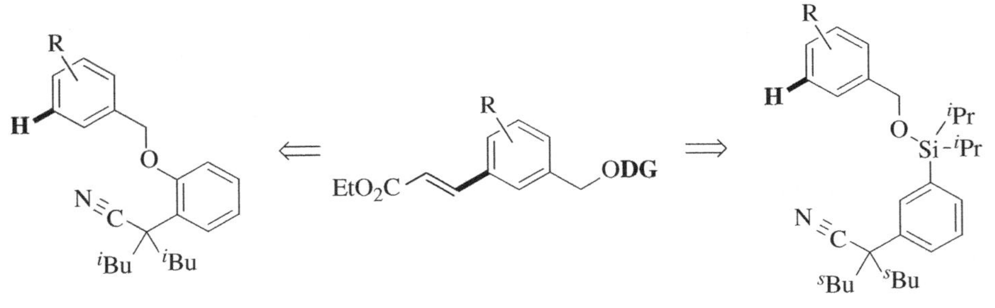

SCHEME 23.6 Directing groups (DGs) for *meta*-C–H functionalization.

23.1.2 Weakly and Strongly Coordinating Directing Groups

Potential directing groups can possess a large variety of structure, with coordinating atoms being O, N, S, C, and P (see Scheme 23.7). Since commonly used late transition metals tend to form weaker bonds with O-donor atoms than with N- or P-donor atoms, carboxylate-derived directing groups have been referred to as weakly coordinating groups in the literature. However, the strength of the formed coordination bond is dependent on the identity and the oxidation state of the transition metal catalyst [21], and as such, a general classification of directing groups as strong or weak might be somewhat misleading. Interestingly, recent results with Pd catalysts suggest that the strength of the coordinative bond may influence the rate of catalytic C–H functionalization [22].

23.1.3 Common Directing Groups

Scheme 23.7 depicts various common directing groups for chelate-assisted C–H functionalizations. Directing groups that coordinate through their N-atom such as pyridines and other heterocycles (**1a–r**) have been most widely used [2, 4, 23].

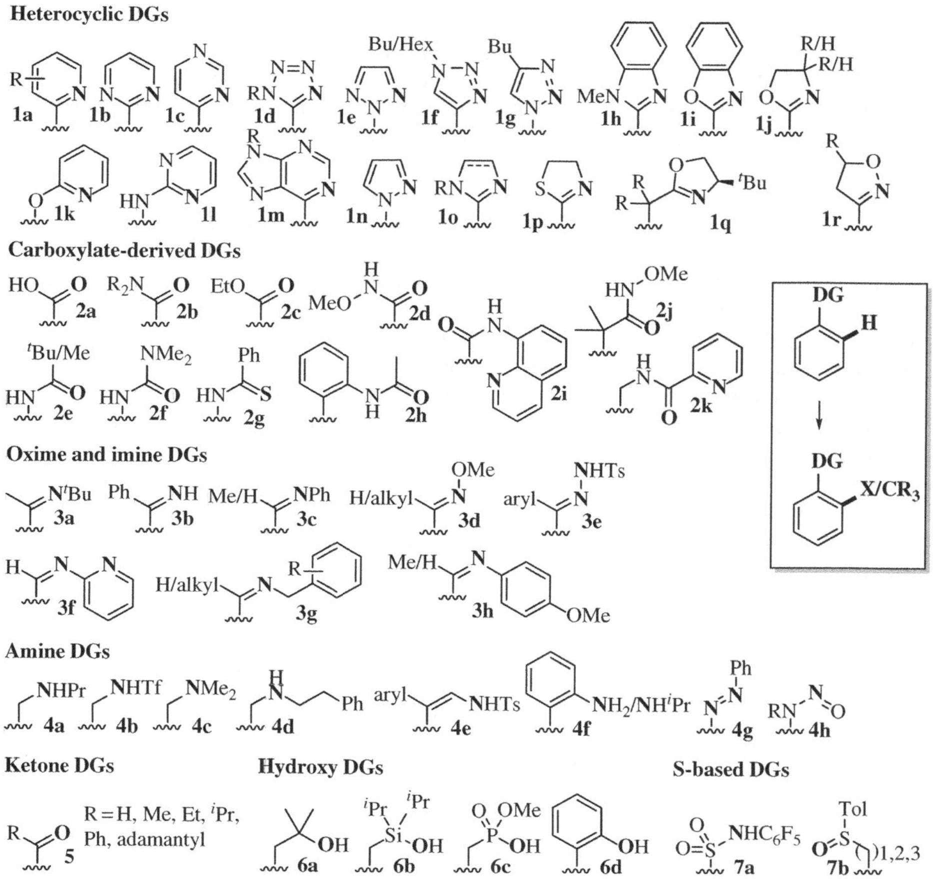

SCHEME 23.7 Directing groups (DGs) for chelate-assisted C–H functionalization.

Carboxylates and carboxylic amides are another important class of directing groups (**2a–k**) [24]. These functional groups are perceived to be more synthetically useful, due to the ease of converting them into other substructures. Based on carboxylic amide functionalities, chelating directing groups (**2i**, **2k**) have been developed, which presumably occupy two coordination sites at the metal center of the catalyst. These structures have enabled a broad range of applications, in particular in diastereoselective C–H functionalization [25, 26]. Furthermore, imines and oximes (**3a–h**) have been widely used as directing groups, both of which coordinate through their N-atom to the transition metal catalyst. Furthermore, various reactions are known to proceed through coordination of a ketone functionality on the substrate (**5**) [4, 23, 27]. A more unusual example is directing groups that coordinate to the catalyst through hydroxy groups (**6a–d**), and a few examples of this type have been reported [28–31]. Amine directing groups are slightly more common and are found as primary, secondary, and tertiary amines (**4a–d**, **f**), as well as amine-derived functional groups (**4e**, **h**) [2, 23, 32]. Additionally, tailored sulfonamide (**7a**) and sulfoxide functionalities (**7b**) have been shown to act as directing groups for aromatic C–H functionalizations [6, 33].

23.1.4 Transformable and *In Situ* Generated Directing Groups

From a synthetic perspective, direct and intermolecular C–H functionalizations are more favorable than sequential functional group conversions, as the former approaches approach can minimize the number of linear steps in a synthetic sequence [7, 34]. However, if additional steps like the installation and removal of directing groups have to be added to the sequence, the overall step count might not decrease, which often reduces the usefulness of C–H functionalization protocols. Several directing groups can address this challenge: Groups based on carboxylate structures fall into this category, as the further modification of carboxylic acids and amides is well established; one example is shown in Scheme 23.8 [35].

Furthermore, several functional groups have been developed specifically for the purpose of subsequent synthetic modification. These directing groups are prepared in prior reactions and are readily converted into other functional groups after C–H functionalization. One example of this strategy is the use of *N*-acetyl oximes as directing groups by Neufeldt and Sanford (Scheme 23.9) [36]. After C–H acetoxylation, halogenation, or arylation, *N*-acetyl oximes can be transformed into a variety of functional groups that are useful for subsequent synthesis. Scheme 23.10 depicts an example of a transformable directing group developed by Kornienko and coworkers. The directing group used is a transformable heterocycle that can be reacted with alkynes after C–H functionalization to form two new C–C bonds and eliminate N_2 through an inverse electron demand Diels–Alder sequence [37].

A slightly different strategy takes advantage of labile Si–O or Si–N bonds that can be formed quantitatively by initial reaction of NH and OH groups with $^tBu_2SiClH$, $^tBu_2SiBr_2$, or Et_2SiH_2. This

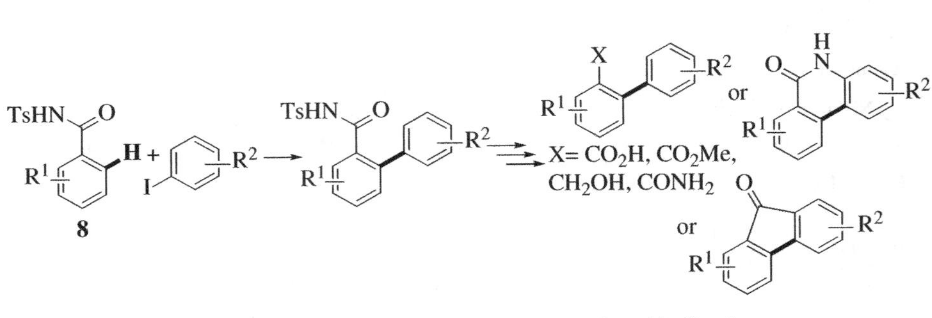

SCHEME 23.8 Carboxylic amide **8** as substrate transformable directing group.

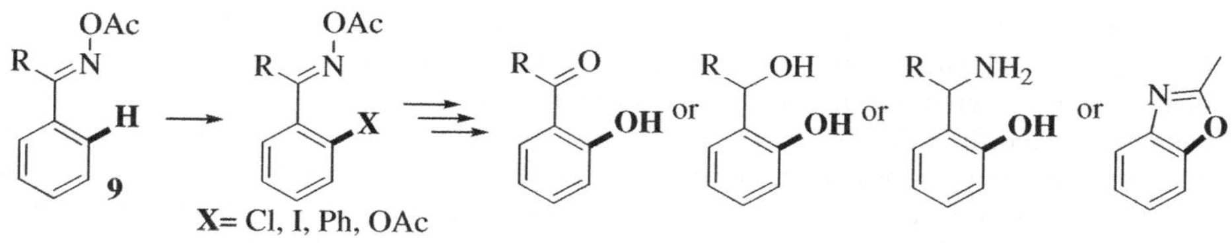

SCHEME 23.9 *N*-acetyl oximes **9** as precursors to difunctionalized arenes.

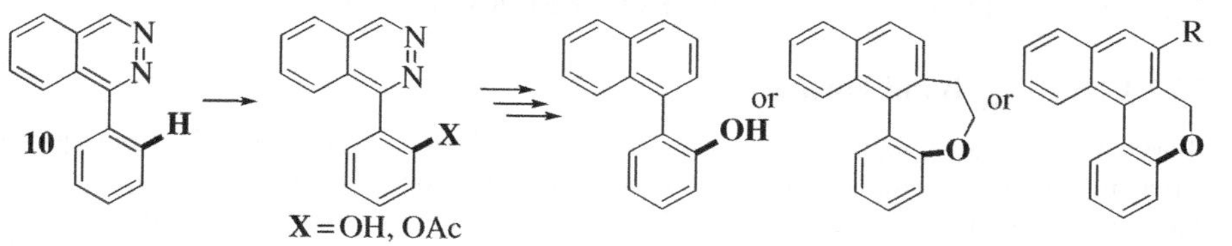

SCHEME 23.10 Transformable heterocycle strategy.

leads to the quantitative formation of a directing group that then catalyzes the highly selective subsequent C–H functionalization [28, 38–40]. One example of this kind is provided later in this chapter (Scheme 23.31), depicting a silane-directed C–H borylation protocol.

Another approach toward traceless directing groups prepares them in the same pot in which C–H functionalization occurs. This strategy is commonly referred to as the use of *in situ* generated directing groups. Work by Bedford and Limmert (Scheme 23.11) is an elegant example of this approach, as it exploits the power of both organocatalysis and transition metal catalysis [41, 42]. In the shown reaction, C–H arylation takes place in *ortho* position to a hydroxy group on the arene **11**. However, this functional group is not acting as a directing group; instead, a phosphonate group that can be transferred *in situ* between phenol moieties of different molecules directs the Rh catalyst to the functionalization site. Another good example for the use of this strategy is Catellani-type reactions, in which norbornene insertion into metal–C bonds generates a directing group for *ortho*-C–H functionalization *in situ* (Scheme 23.12). This type of reactivity has found manifold applications such as in Suzuki–Miyaura couplings and direct arylations and has been reviewed elsewhere [43, 44].

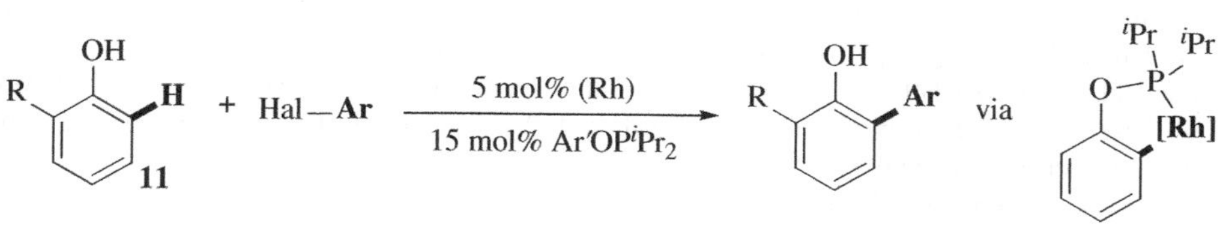

SCHEME 23.11 Rhodium-catalyzed *ortho*-arylation with an *in situ* generated DG.

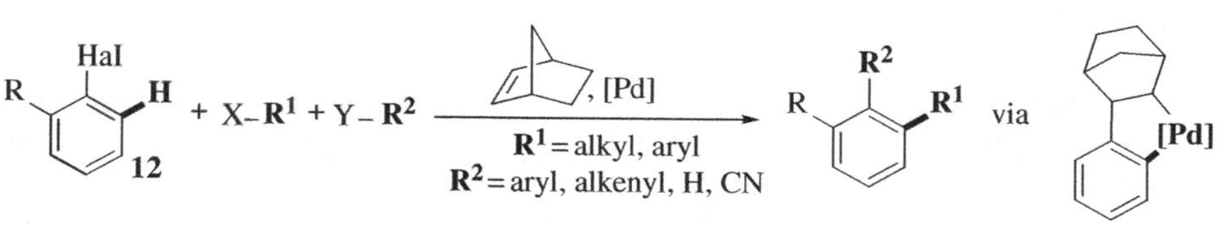

SCHEME 23.12 *In situ* generated DGs in Catellani-type reactions.

In summary, the use of directing groups has been highly successful in enabling chelate-assisted C–H cleavage and functionalization. The selectivity of these reactions is determined by the structure of the employed groups, with *ortho*-C–H functionalization being the most common reactivity pattern observed. In order to lower the step count for these reactions, transformable and *in situ* generated directing groups have been developed. The following chapters will provide an overview of the types of C–C and C–X bonds that can be formed using the previously outlined strategies and will highlight examples of common catalyst systems and reaction conditions.

23.2 CARBON–CARBON (C–C) BOND FORMATIONS

23.2.1 C–C_{Aryl} Bond Formations

Biaryl moieties are ubiquitous in natural products and pharmaceuticals [45]. Consequentially, efficient and direct synthetic protocols for the formation of biaryl molecules are highly desirable. The earliest report of chelate-assisted C–H arylation from 1998 describes the use of Wilkinson's catalyst [$RhCl(PPh_3)_3$] for the *ortho*-C–H functionalization of 2-arylpyridines with tetraphenylstannane (Scheme 23.13) [46]. Monoarylated products can be obtained by using a blocking group on one of the *ortho* positions on the pyridine moiety of the substrate (**13**) that prevents arylation of the second *ortho*-C–H bond. In the absence of a blocking substituent in the substrate **14**, diarylated products are formed.

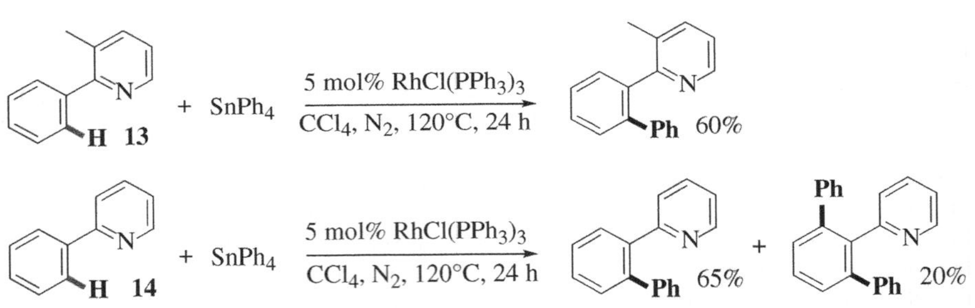

SCHEME 23.13 Rh-catalyzed *ortho*-arylation of 2-aryl pyridines.

Since this first report, a large variety of catalysts (e.g., $Pd(OAc)_2$, $RhCl(PPh_3)_3$, $RuCl_2(\eta^6\text{-}C_6H_6)_2$, $Ni(cod)_2$), directing groups (heterocycles, carboxylate based, phenols), and aryl transfer reagents (Ar-Hal, Ar-OR, Ar-$B(OR)_2$, Ar-H) has been employed to achieve direct arylations [47–51]. Many of these reactions have been reviewed elsewhere [52–55], and only few representative examples will be discussed in detail in this chapter.

As one example of this large body of literature, we depict in Scheme 23.14 the redox-neutral *ortho*-C–H arylation of anilides with simple $Pd(OAc)_2$ as efficient catalyst. Daugulis and coworkers reported very high yields for this reaction, forming both mono- and diarylated products under mild reaction conditions [56]. Interestingly, the protocol tolerates various functional groups on the anilide including halogens.

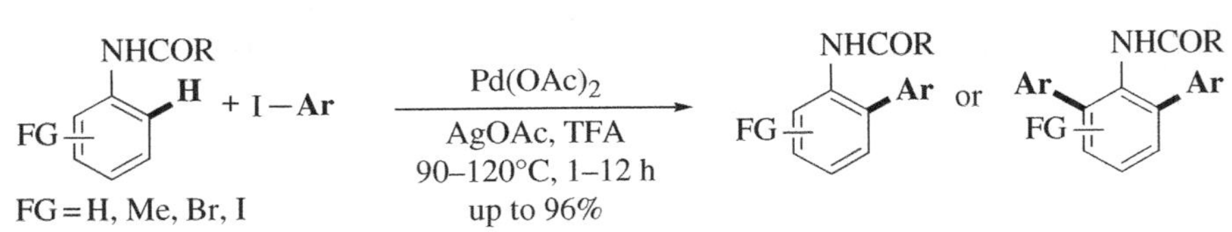

SCHEME 23.14 Pd-catalyzed *ortho*-arylation of anilides.

In an effort to move away from precious metal catalysts, various reports in recent years have focused on the use of first-row metal catalysts for direct arylations [57–60]. As a representative example of these new developments, we illustrate in Scheme 23.15 the chelate-assisted *ortho*-C–H arylation of arenes with Fe catalysts [61]. With iron being cheap, nontoxic, and ubiquitous, this protocol is highly attractive for pharmaceutical syntheses. Using the catalyst precursor $Fe(acac)_3$ in conjunction with bidentate pyridine ligands, Zn-aryl reagents as aryl transfer reagents and 1,2-dichloroisobutane as the oxidant, excellent yields of the arylated product were obtained. An interesting feature of this reaction is the hydrolysis of the imine moiety after work-up. The reaction conditions tolerate additional functionalities such as cyanides, chlorides, triflates, tosylates, and thiophenes.

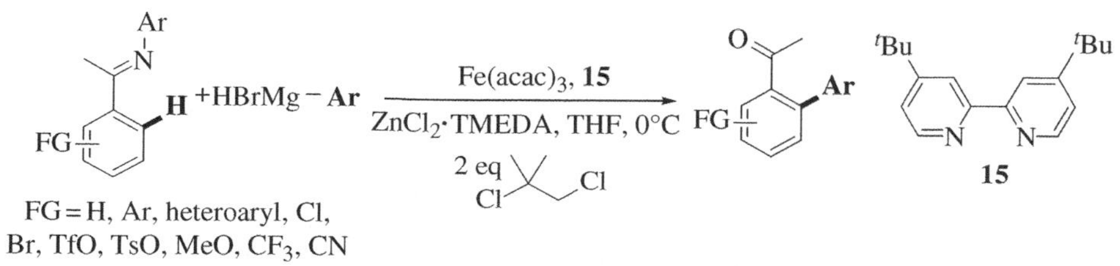

SCHEME 23.15 Fe-catalyzed *ortho*-arylation of aromatic imines.

23.2.2 C–C$_{Alkenyl}$ Bond Formations

Aromatics with alkenyl substituents are central substructures in synthetic chemistry and can be found in vast numbers of materials such as dyes [62] and optical and electronic polymers [63]. Transition metal catalysis has played an invaluable role in the direct synthesis of these important materials [64].

Even though not (yet) among the typically employed reactions for the production of materials, C–H olefinations of arenes are a valuable, more sustainable alternative for the synthesis of olefin-substituted arenes. C–H olefinations are among the first C–H functionalizations ever reported [65], and the originally stoichiometric modifications of simple arenes with olefins in the presence of Pd^{II} have initiated broad research efforts. To this day, C–H alkenylations are more common using aromatic substrates without directing groups [18], and these reactions will be addressed in the next chapter of this book. However, several examples of chelate-assisted C–H olefinations are known [30, 66–74], which exhibit excellent *ortho*-selectivity.

Ru complexes catalyze the reaction between 2-aryl pyridines and alkenyl esters (Scheme 23.16) [75]. These types of olefins are often only slowly reacting substrates in oxidative Heck (Fujiwara–Moritani) reactions [18]; thus, their use in Ru-catalyzed C–H olefinations is a remarkable advance. The shown reaction is overall redox neutral and produces 1 equivalent of alcohol or acid as side product. A variety of heterocyclic directing groups can be employed, and many functional groups are tolerated. However, the olefin scope of the reaction is somewhat limited to alkyl- and aryl-substituted alkenes.

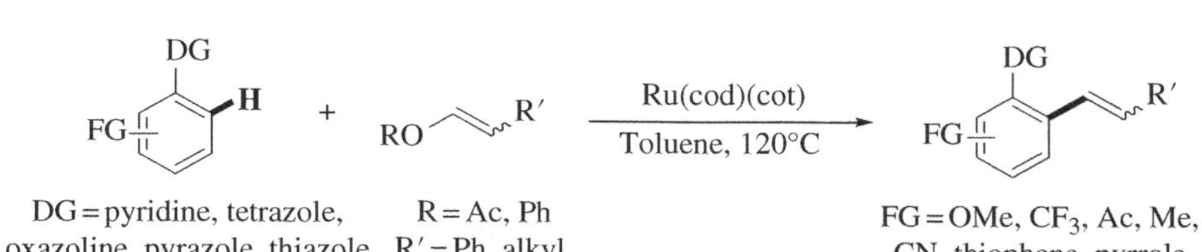

SCHEME 23.16 Ru-catalyzed C–H alkenylation with alkenyl ethers and esters.

Phosphine oxide directing groups enable both redox-neutral and oxidative C–H olefinations (Scheme 23.17) [76, 77]. In combination with [Ru(*p*-cymene)Cl_2]$_2$, *cis*-selective hydroarylations of alkynes can be achieved. Furthermore, in the presence of [Cp*$RhCl_2$]$_2$/$AgSbF_6$, $Cu(OAc)_2$, and Ag_2CO_3, oxidative C–H olefinations are possible with electron-poor olefins as coupling partners. The latter reactions proceed with complete selectivity for the *trans*-product.

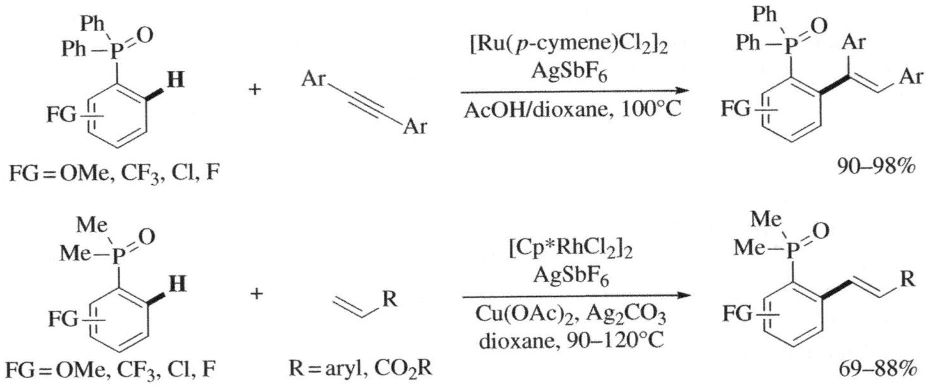

SCHEME 23.17 C–H olefinations with phosphine oxide directing groups.

23.2.3 C–C_{Alkyl} Bond Formations

In contrast to C–H arylations and olefinations, C–H alkylations are a rather recent development. Interestingly, these reactions can proceed under oxidative or redox-neutral conditions and use different alkyl precursors [78–80]. One of the earliest examples for this reaction was published by Murai and coworkers, using olefins as reactants [81]. The overall hydroarylation of the olefin with a Ru catalyst leads to *ortho*-alkyl-substituted aryl ketones, which are difficult to synthesize selectively with classical methods (Scheme 23.18).

A more recent example from the Yu group uses alkylboronic acids as alkyl precursors, which enables a broader product scope and even the introduction of strained cyclopropyl groups [82]. Pyridine directing groups enable the complete *ortho*-selectivity in the presence of Ag^I salts, benzoquinone, and air as oxidants (Scheme 23.19). Interestingly, the protocol can be employed for C–H alkylations of both sp^2 and sp^3 C–H bonds.

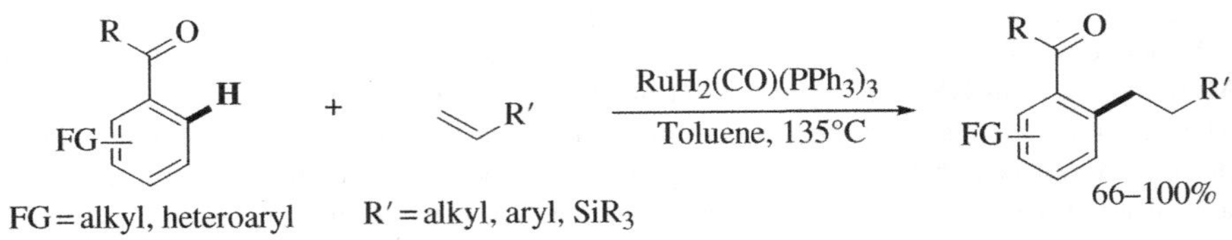

SCHEME 23.18 Ru-catalyzed hydroarylation of olefins.

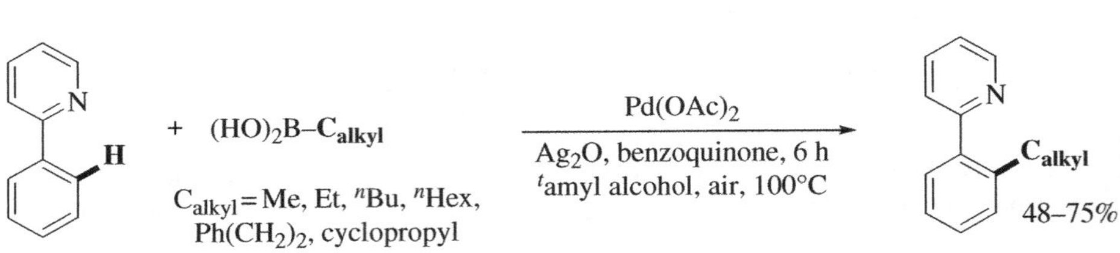

SCHEME 23.19 Pd-catalyzed oxidative alkylation with alkylboronic acids.

Redox-neutral *ortho*-C–H alkylations of phenols can be achieved with simple alcohols in the presence of a Ru catalyst (Scheme 23.20) [83]. Using this method, several complex molecules (e.g., **16**, **17**, and **18**) have been synthesized. Mechanistic experiments show the loss of deuterium labeling in α-position to the phenol functionality and no primary kinetic isotope effect ($k_H/k_D = 1.1 \pm 0.1$). Both findings are consistent with rapid and reversible C–H bond activation before C–C bond formation. However, the mechanism of the reaction has not been studied in detail, and the source of the unusual directing group effect of the phenol moiety is unclear.

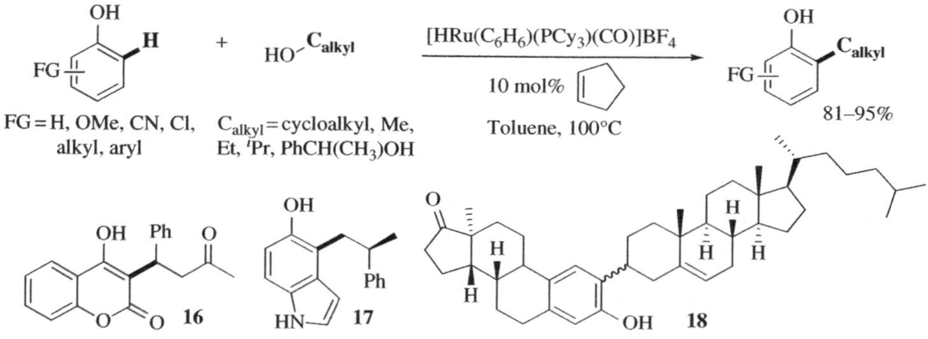

SCHEME 23.20 Dehydrative C–H alkylation of phenols with alcohols.

23.2.4 C–C$_{Acyl}$ Bond Formations

C–H bond acylations are a relatively new area of C–H functionalizations. These reactions have the potential to provide alternatives to Friedel–Crafts acylations, which are used to produce aryl ketones on a multiton scale per year [84]. Using directing groups, only *ortho*-C–H acylations are possible, which might limit their usefulness on a large scale.

C–H acylations can proceed using various acyl precursors and most reactions proceed under oxidative conditions. The Rh-catalyzed protocol in Scheme 23.21 is an exemption from this trend, as the carboxylic anhydride reagent acts as oxidant and acyl source at once.

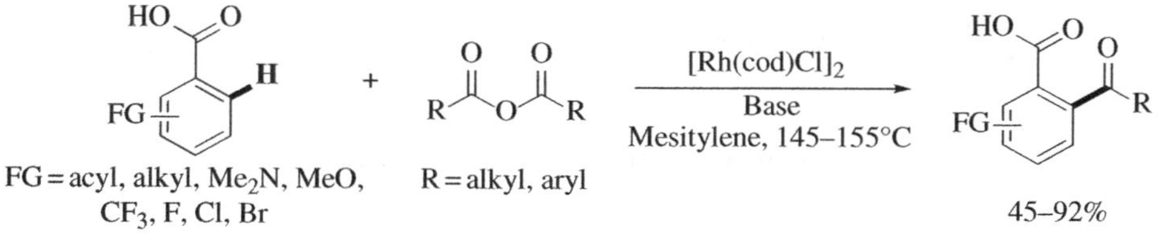

SCHEME 23.21 Rh-catalyzed *ortho*-C–H acylation of aromatic carboxylic acids.

When using aldehydes as acyl synthons, C–H bond acylations require the presence of an oxidant. Commonly employed oxidants are Ag salts (Scheme 23.22) [85] in Rh-catalyzed protocols or *tert*-butyl hydroperoxide for Pd-catalyzed methodologies [86].

Another approach to achieve C–H acylations relies on synthesizing the carbonyl group *in situ* from seemingly unrelated precursors. Following this strategy, *tert*-butyl hydroperoxide can be used to oxidize benzylic methyl groups, which leads to C–H acylation of acetanilides in the presence of a Pd catalyst (Scheme 23.23). Remarkably, these reactions tolerate the presence of other benzylic methyl groups as well as aryl bromides [87].

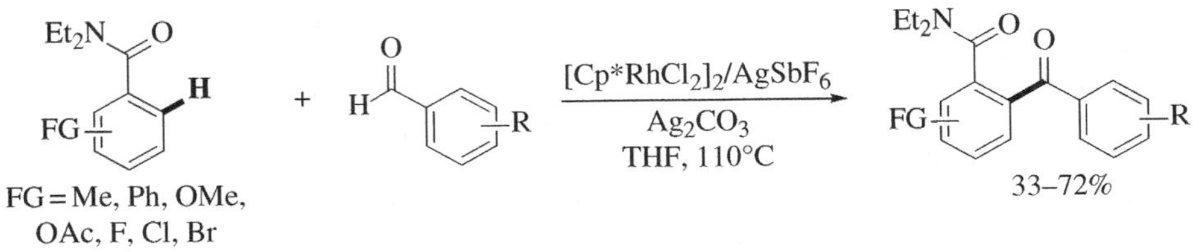

SCHEME 23.22 Rh-catalyzed oxidative *ortho*-C–H acylation with aldehydes.

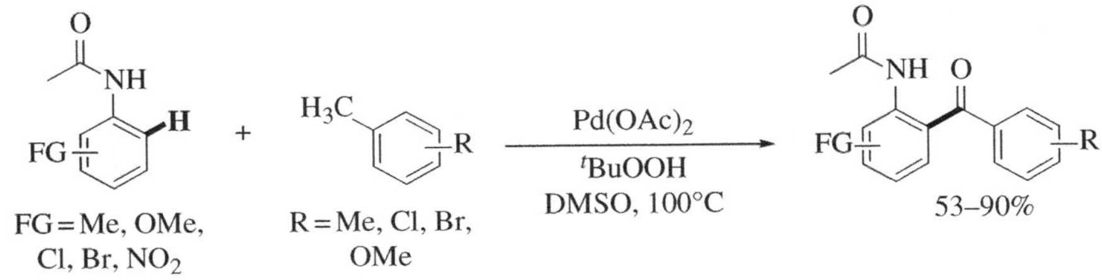

SCHEME 23.23 Pd-catalyzed C–H acylation by benzylic oxidation.

23.2.5 C–CN Bond Formations

In contrast to all the C–C bond formations described previously, C–H cyanations are much rarer. This might be due to the possible strong coordination of free CN ligands to transition metal centers, which would preclude C–H activation as the first step of the catalytic cycle for C–H functionalization. The first C–CN bond formation was reported in 2006 by Yu and coworkers along with a series of Cu-catalyzed and Cu-mediated C–H functionalizations of 2-phenyl pyridine (Scheme 23.24) [88]. C–CN bond formations can thus be achieved with Me_3SiCN or CH_3NO_2 as cyanating reagents in the presence of 1 equivalent of $Cu(OAc)_2$ and air.

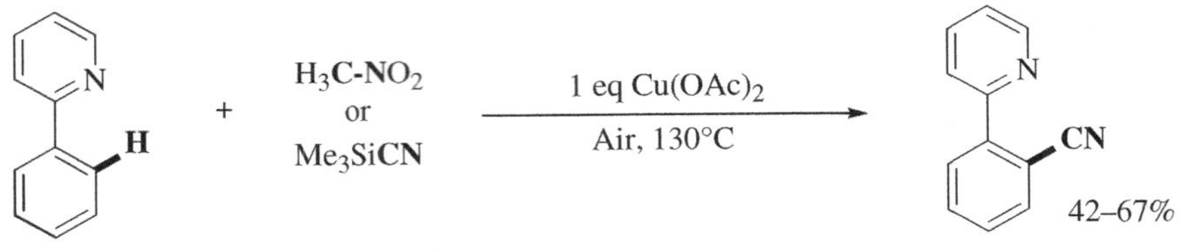

SCHEME 23.24 Cu-mediated C–H cyanation.

Since then, more recent developments have focused on using less toxic reagents (Scheme 23.25) [89], on broadening the scope of the used directing group (Scheme 23.26) [90], and on establishing functional group tolerant protocols. A great example for the use of cyanide sources with low toxicity is the example shown in Scheme 23.25, which employs $K_3Fe(CN)_6$ as reagent of choice. A variety of functional groups are tolerated under these conditions and pyrazole directing groups are effective for this transformation in addition to pyridine-based directing groups, [89].

A Rh-catalyzed protocol with PhTsN(CN) as cyanide source exhibits a remarkably broad scope of functionalized substrates, leaving even unprotected phenol groups, boronates, epoxides, and reactive aliphatic C–Cl bonds untouched under the reaction conditions (Scheme 23.26) [90].

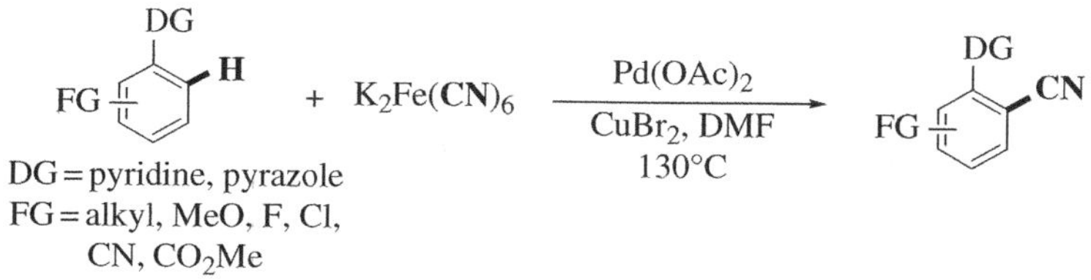

SCHEME 23.25 Pd-catalyzed C–H cyanation with nontoxic $K_3Fe(CN)_6$ as cyanide source.

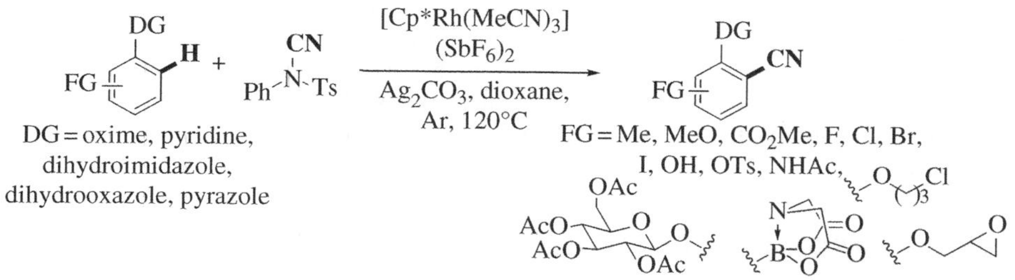

SCHEME 23.26 C–H cyanation with broad functional group tolerance.

23.2.6 C–CF_3 Bond Formations

Trifluoromethyl (CF_3) groups are electron-withdrawing substituents that increase the lipophilicity of molecules at the same time. Furthermore, CF_3 substitution can enhance the bioavailability of a drug by increasing its metabolic stability. Therefore, the selective installation of CF_3 groups is of substantial interest for pharmaceutical applications [91].

Chelate-assisted C–H bond trifluoromethylations are rare, and so far, only two reagents have been used successfully in these reactions. The first reagent **19** is an electrophilic source of CF_3^+, which reacts under Pd catalysis with substrates bearing directing groups (Scheme 23.27) [92, 93]. The reaction also requires the presence of $Cu(OAc)_2$ and trifluoroacetic acid as promoters; both amide-based and heterocyclic directing groups can be employed.

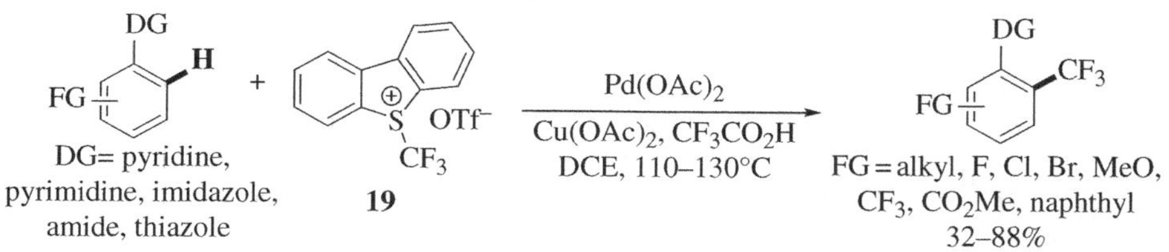

SCHEME 23.27 Pd-catalyzed *ortho*-C–H trifluoromethylation.

Interestingly, the rather unusual triazene directing group also promotes chelate-assisted C–H bond trifluoromethylations in combination with super-stoichiometric amounts of AgF (Scheme 23.28) [94]. The reagent used in this process, Me_3Si-CF_3 (**20**), is considerably less expensive than reagent **19**, which makes the protocol highly attractive. However, no other directing groups have been shown to promote analogous C–H trifluoromethylations. Interestingly, very similar conditions to those shown in Scheme 23.28 also promote *ortho*-C–H pentafluoroethylation, heptafluoropropylation, and ethoxycarbonyldifluoromethylation of aromatic triazenes using the respective Me_3Si-R_f reagents [95].

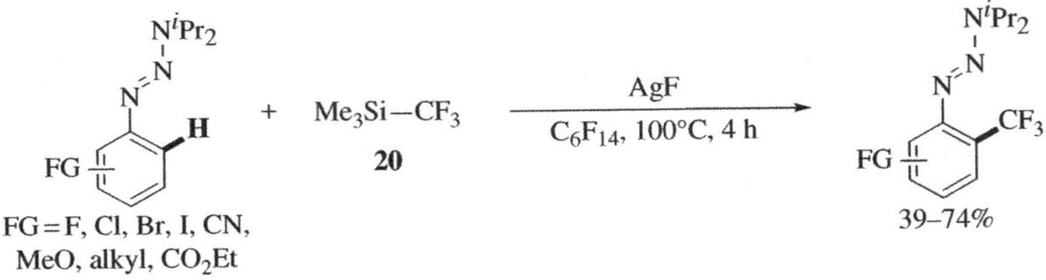

SCHEME 23.28 Triazene/Ag-mediated C–H bond trifluoromethylation.

In summary, a wide variety of chelate-assisted C–C bond formations is possible using various transition metal catalysts. These protocols introduce $C–C_{aryl}$, $C–C_{alkenyl}$, $C–C_{alkyl}$, $C–C_{acyl}$, C–CN, and $C–CF_3$ bonds in *ortho* positions of suitable directing groups. As such, these methods can be employed in order to build carbon frameworks. With their increasing popularity in the synthetic community, many more applications of these protocols can be expected in the years to come.

23.3 CARBON–HETEROATOM (C–X) BOND FORMATIONS

23.3.1 C–B Bond Formations

C–H borylations as methods to prepare synthetically valuable aryl boronic acids and esters have been intensively studied for substrates without directing groups since their discovery by Smith and coworkers in 1999 [96, 97]. Continuing this work, several groups have pursued ways to control the selectivity of these reactions. Among the applied strategies, the use of directing groups to achieve *ortho*-selective C–H borylations has been realized with late transition metal catalysts such as Pd, Rh, and Ir. Interestingly, some of the created protocols proceed under oxidative conditions (Scheme 23.29), resulting in stoichiometric by-products derived from the used oxidant [98]. The presence of the ligand **21** is crucial for high yields in the shown protocol. In contrast, early examples of C–H borylations as well as most Rh- and Ir-catalyzed protocols proceed under mild conditions in the absence of added oxidants [96, 97, 99, 100].

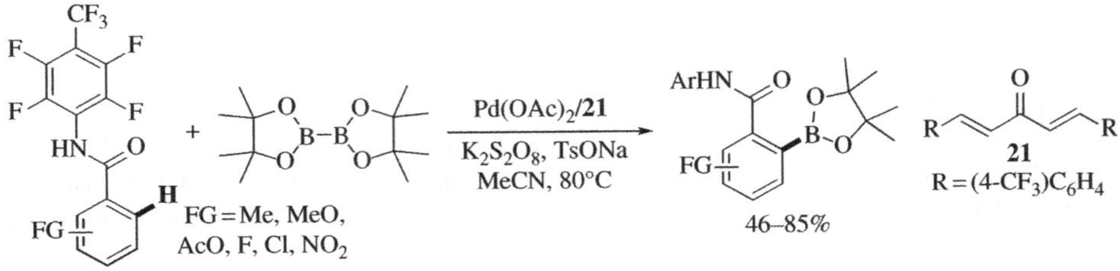

SCHEME 23.29 Pd-catalyzed oxidative *ortho*-C–H borylation.

In order to make the developed catalysts more synthetically useful, anchoring the ligands to surfaces has been a successful strategy, resulting in a protocol with significant tolerance of steric hindrance around the cleaved C–H bond (Scheme 23.30) [101, 102]. An additional feature of this protocol is its capability to employ various and unusual directing groups, including esters, amides, sulfonic esters, and even Cl.

A second interesting development uses traceless, silyl-based directing groups as shown in Scheme 23.31. This strategy allows *ortho*-C–H borylation of phenols, arylamines, and alkylarenes in one pot without additional steps, which are often required to install and remove directing groups [103].

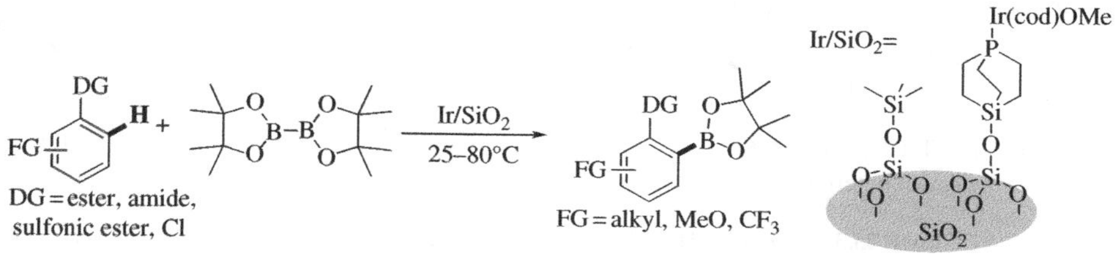

SCHEME 23.30 Directed *ortho*-C–H borylation with supported Ir catalyst.

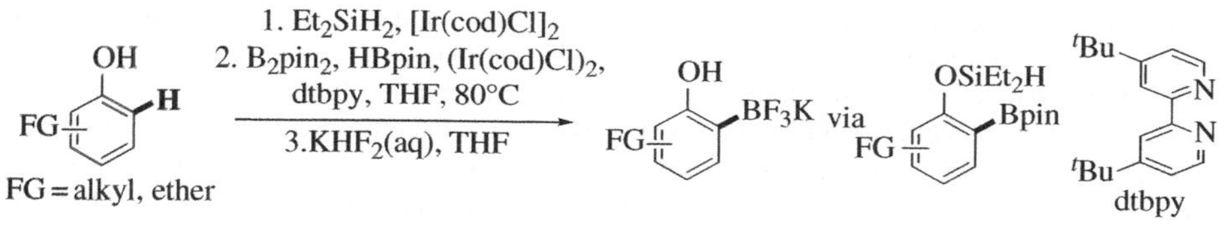

SCHEME 23.31 One-pot *ortho*-C–H borylation of phenols.

23.3.2 C–Si Bond Formations

Similar to C–H borylations, C–H silylations form new C–X bonds with small electronegativity differences between the bonding partners (ΔEN[C–B] = 0.51; ΔEN[C–Si] = 0.65). This leads to organosilicon and organoboron reagents possessing a relatively lower reactivity compared with other organometallic compounds (e.g., for organolithium reagents, ΔEN[C–Li] = 1.57) and a better functional group tolerance [104]. Therefore, both C–H borylations and silylations lead to valuable and versatile organometallic intermediates, which have been used for very diverse subsequent transformations [105].

C–H silylations are possible without an intramolecularly tethered directing group using Pt, Ru, or Ir catalysts [106–108]. In these cases, the site selectivity of C–Si bond formation is mainly controlled by the electronic and steric properties of the substrates. Silylations using a directing group circumvent this inherent substrate control of selectivity and enforce *ortho*-C–H silylation. Directing groups promoting these reactions include silanes (Scheme 23.32) [108, 109] and transformable protecting–directing groups (PG/DGs) of arylboronic acids (Scheme 23.33) [110, 111]. In the latter case, the directing group enables C–H functionalizations in the *ortho* position of a C_{aryl}–B bond, which can be used for further transformations after reaction with pinacol under acidic conditions [110]. Both directing–protecting groups shown in Scheme 23.33 can be recovered from the reaction mixtures and are thus reusable.

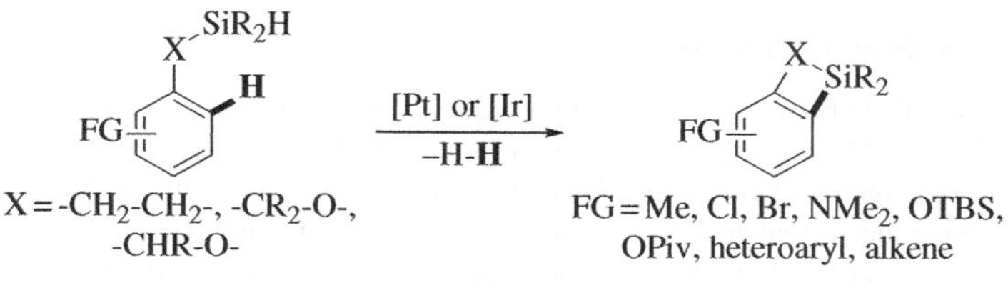

SCHEME 23.32 Silyl-directed intramolecular C–H bond silylations.

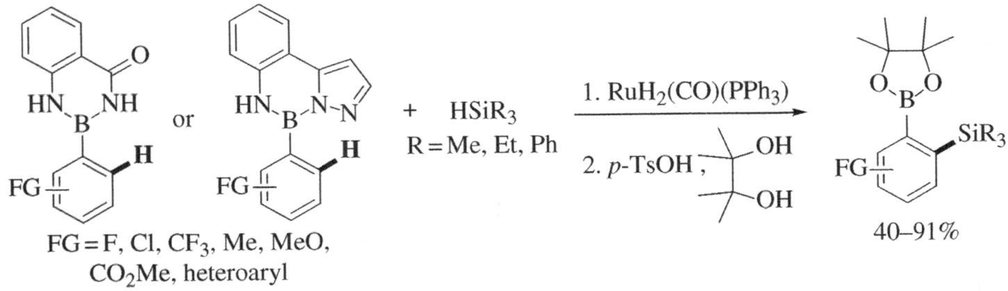

SCHEME 23.33 *Ortho*-C–H silylation of arylboronic acids with transformable PG/DGs.

23.3.3 C–O Bond Formations

Transition metal-catalyzed hydroxylations, acetoxylations, and analogous C–H oxygenations of simple arenes have been studied by the catalysis community for several decades [112–114]. However, *ortho*-selective C–H oxygenations are a more recent development with the first example being reported by Sanford and coworkers in 2004 [115]. Since then, a large variety of directing groups (heterocycles, carboxylates, carboxylic amides, oximes) and catalysts ($Pd(OAc)_2$, $PdCl_2$, $[RuCl_2(p\text{-cymene})_2]_2$, $Cu(OAc)_2$) enabling C–H oxygenations has been established [2, 11, 40, 116–125]. Typically, these reactions require the presence of a strong oxidant (O_2, I^{III} reagents, oxone, $K_2S_2O_8$), which is likely needed to promote C–O reductive elimination through oxidation of the catalyst [12, 17].

An interesting recent development in the field is the use of directing groups that are not predicted to coordinate strongly to the catalyst, for example, phosphoric and phosphonic acids as shown in Scheme 23.34 [126].

Other remarkable developments include the use of oxygen as the terminal oxidant to directly form C_{aryl}–OH functionalities without the intermediacy of phenolic ester structures. Several types of directing groups, among them pyridyl and carboxylate groups (Scheme 23.35), have been reported to enable these reactions. Notably, each of these protocols requires the presence of a cocatalyst (benzoquinone or *N*-hydroxyphthalimide) [127, 128].

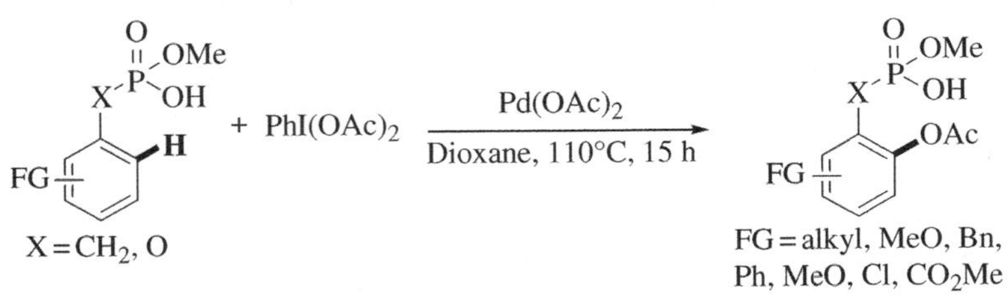

SCHEME 23.34 C–H bond acetoxylation of phosphonic and phosphoric acids.

23.3.4 C–N Bond Formations

Chelate-assisted C–H bond aminations often require the presence of an oxidant in analogy to the previously discussed C–H oxygenations. Therefore, many protocols use a mixture of amine source and oxidant as reagents in order to achieve C–N bond formation. Intermolecular directed C–H aminations have been realized with this approach by Che and coworkers (Scheme 23.36). Remarkably, this methodology is capable of employing a variety of different directing groups as well as primary amides as amine sources [129].

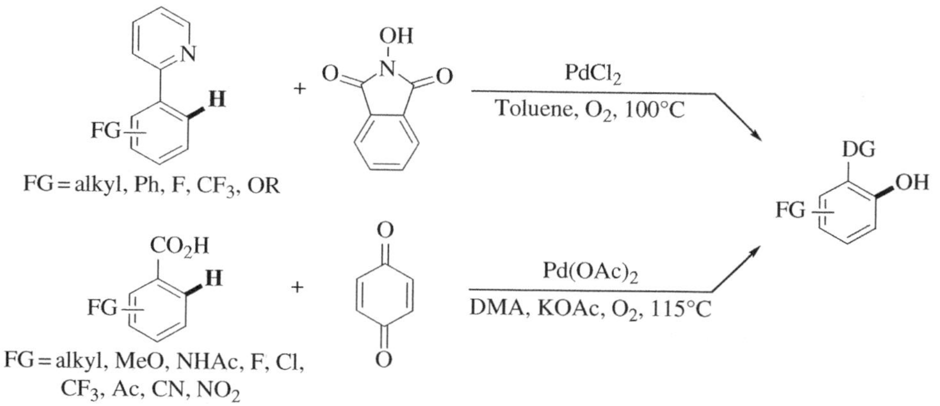

SCHEME 23.35 Pd-catalyzed aerobic C–H hydroxylations.

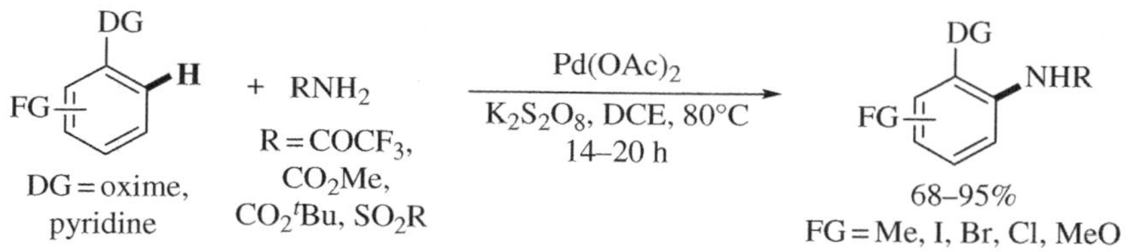

SCHEME 23.36 Intermolecular, Pd-catalyzed C–H amination with $K_2S_2O_8$ as oxidant.

However, the approach of using mixtures of oxidant and amine source in C–H amination protocols poses the challenge of reagent compatibility, in particular with easily oxidized amine sources. In order to address this issue, several protocols have been developed, which combine the oxidant and amine source in one reagent. Both organic azides and protected hydroxylamine derivatives have been used with this approach in mind to result in high yields of aminated products (Scheme 23.37) [130–134].

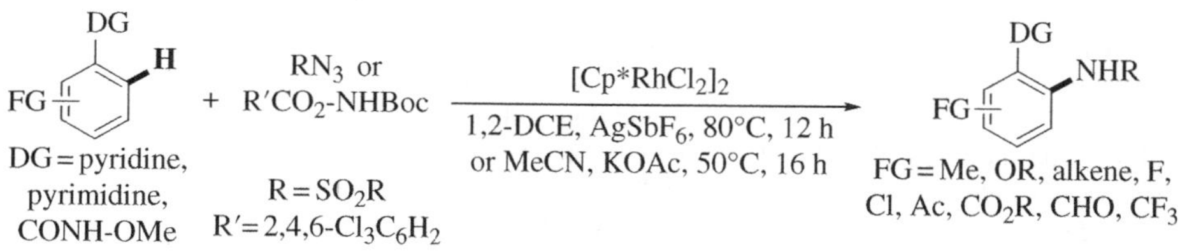

SCHEME 23.37 Rh-catalyzed C–H aminations with electrophilic amination reagents.

A second strategy to broaden the functional group tolerance of C–H aminations consists of employing inert amine sources [27] or relatively inert oxidants [135]. Using the latter approach in the protocol shown in Scheme 23.38, the presence of various heterocyclic and functionalized structures in arene substrates can be tolerated with $Cu(OAc)_2$ and O_2 as oxidants. The elaborate directing group that is required for high reactivities can be readily cleaved from the product and recovered for reuse.

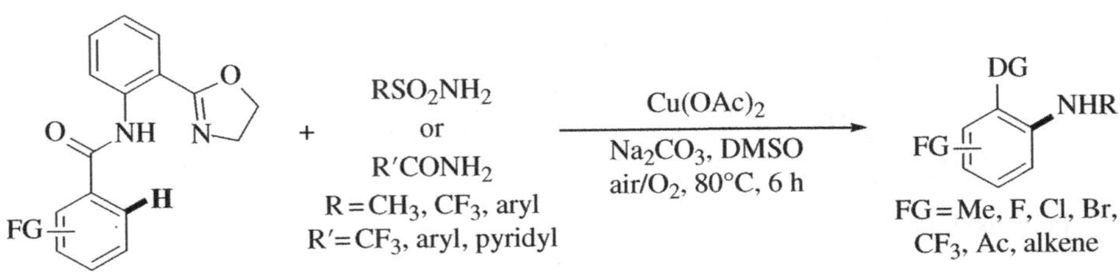

SCHEME 23.38 Cu-mediated, aerobic C–H amination of arenes.

23.3.5 C–P Bond Formations

Functional groups containing phosphorous in the oxidation state +III are typically very easily oxidized, even by relatively inert oxidants such as O_2. Therefore, strategies for oxidative C–H bond phosphorylations typically avoid the presence of P^{III} in the presence of other oxidants. Intramolecularly, this issue can be addressed by using P^V reagents [136] or by performing the reaction without external oxidants (Scheme 23.39) [137]. The latter example possibly proceeds through C–P bond activation as well as C–H bond activation. Since the reaction products are susceptible to air oxidation, they are isolated as oxides **22** after treatment with H_2O_2.

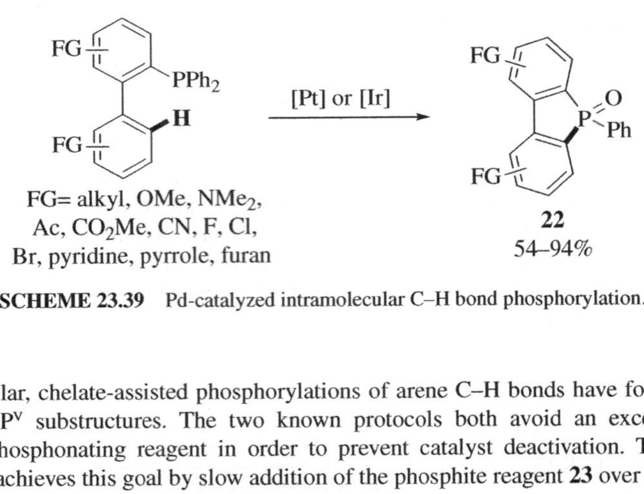

SCHEME 23.39 Pd-catalyzed intramolecular C–H bond phosphorylation.

Intermolecular, chelate-assisted phosphorylations of arene C–H bonds have focused so far on reagents with P^V substructures. The two known protocols both avoid an excess of strongly coordinating phosphonating reagent in order to prevent catalyst deactivation. The example in Scheme 23.40 achieves this goal by slow addition of the phosphite reagent **23** over a period of 13 h [138]. The modified phosphorylation reagent **24** (Scheme 23.41) achieves a similar effect: the active phosphite reactant (HP(=O)Bu_2 or [P(=O)Bu_2]$^-$) is slowly liberated from the masked reagent **24** under the reaction conditions by base-promoted elimination of acetone [139].

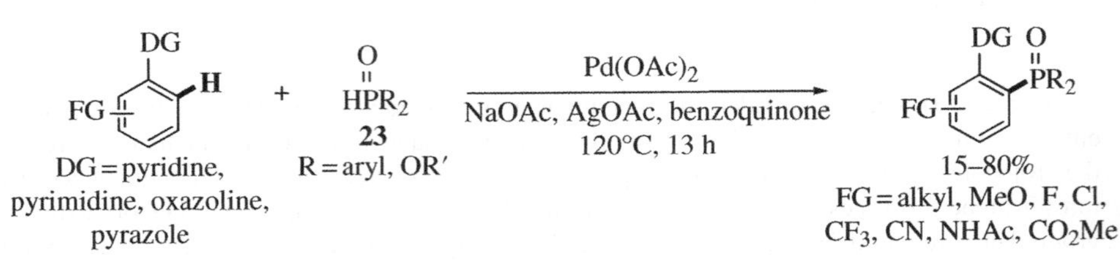

SCHEME 23.40 Pd-catalyzed C–H bond phosphorylation.

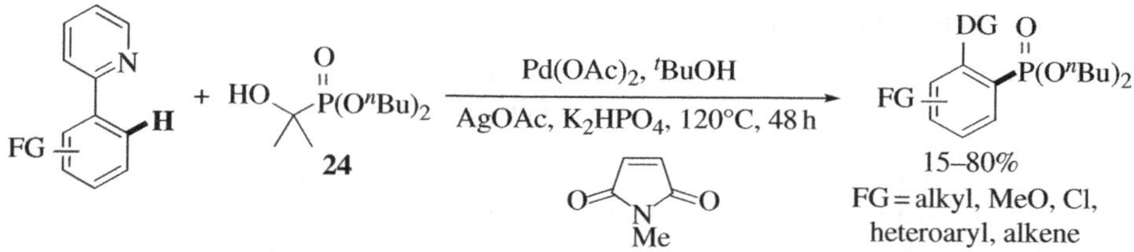

SCHEME 23.41 Pd-catalyzed C–H bond phosphorylation with masked reagent 24.

23.3.6 C–S Bond Formations

Together with C–H phosphorylations, C–H sulfenylations are rather rare among the large body of literature covering chelate-assisted C–H functionalizations. An initial example of this type of reactivity was reported by Yu and coworkers, using PhSH or MeSSMe as reagents for the sulfenylation of 2-phenyl pyridine (Scheme 23.42) [88].

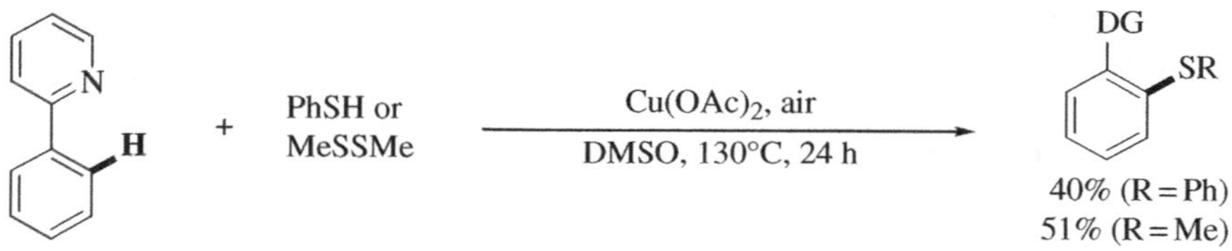

SCHEME 23.42 First chelate-assisted Cu-catalyzed C–H sulfenylation.

The general concept shown in Scheme 23.42 has later been extended by Daugulis and coworkers to arrive at a general auxiliary-promoted Cu-catalyzed C–H sulfenylation protocol for benzoic acids (Scheme 23.43). The reaction exclusively affords disulfenylated products when both *ortho*-C–H bonds are available; otherwise, monosulfenylated products can be obtained [140].

Notably, a C–H sulfonylation protocol by Dong and coworkers provides a quite different approach to forming C–S bonds (Scheme 23.44) [141]. Using pyridine- or oxime-based directing groups, *ortho*-C–H bonds can be directly sulfonylated through Pd-catalyzed reaction with arylsulfonyl chlorides.

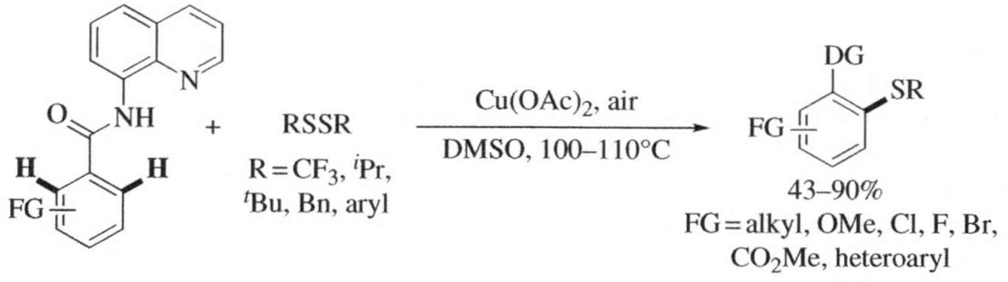

SCHEME 23.43 Auxiliary-promoted, Cu-catalyzed C–H sulfenylation.

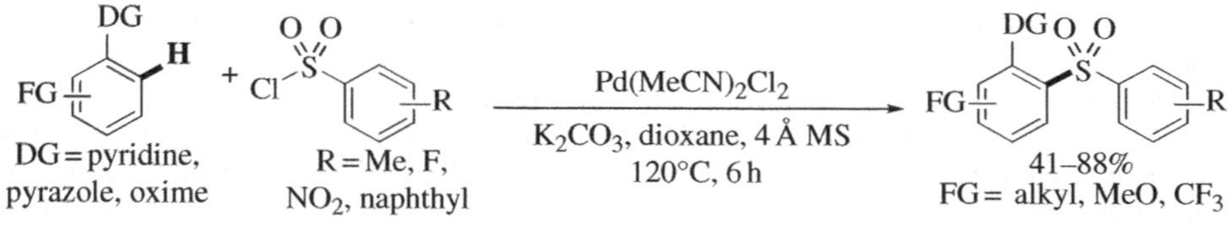

SCHEME 23.44 Pd-catalyzed, chelate-assisted C–H sulfonylation.

23.3.7 C–Halogen Bond Formations

Chelate-assisted C–H bond halogenations are among the more developed C–H bond functionalizations and can generally employ a large variety of directing groups such as pyridines, pyrimidines, oximes, carboxylates, and amides [88, 142, 143]. Some protocols use electrophilic halogen sources such as NCS, NBS, or NIS (Scheme 23.45) [143–147]; furthermore, C–I bonds can be introduced by reacting the substrates with elementary I_2 [88] or IOAc [148].

Furthermore, both C–H iodinations and brominations can be achieved using the respective halide salts in combination with oxidants [148]. Scheme 23.46 shows one example of this approach, in which calcium halides act as halide sources for C–H bond halogenations with $Cu(O_2CCF_3)_2$ as stoichiometric oxidant [142].

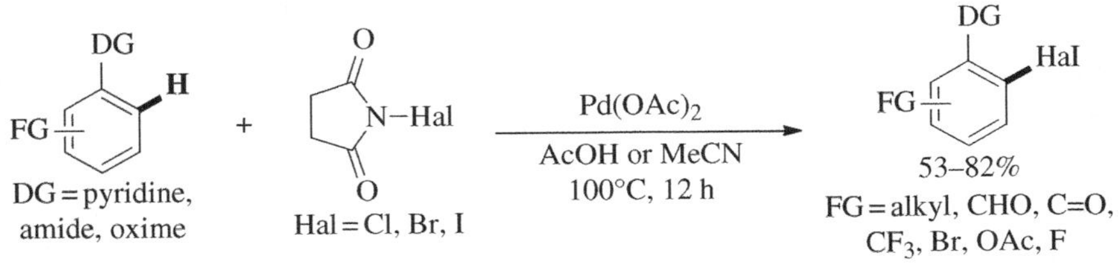

SCHEME 23.45 Pd-catalyzed C–H bond halogenations with NXS (X = Cl, Br, I).

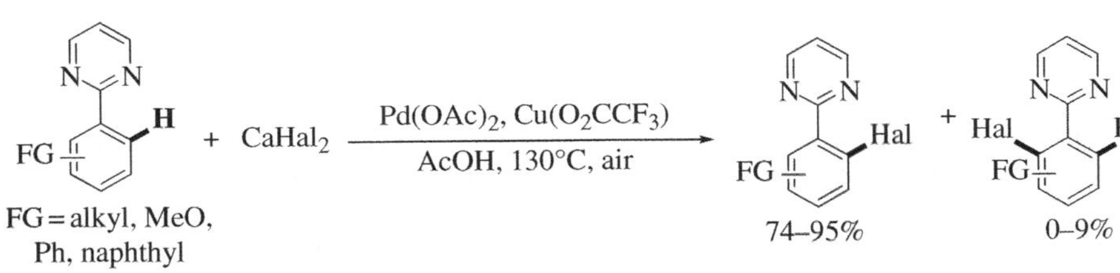

SCHEME 23.46 Pd-catalyzed selective monohalogenation of aryl pyrimidines.

Clearly, various methods are available to achieve C–I, C–Br, and C–Cl bond formations following chelate-assisted C–H activation. However, none of these methods are applicable to C–H fluorinations. Several possible challenges for establishing this reaction have been described in the literature: (i) The C–F coupling step through reductive elimination from transition metal aryl complexes is challenging. As such, relatively few examples of this reaction have been reported so far [149, 150]. (ii) C–H bond fluorination protocols are sensitive to the directing groups used in the transformation [151]. (iii) Only few electrophilic F^+ sources are available, which can be used conveniently in methodology development. The alternative pathway, fluoride salts in combination with an oxidant has shown to be successful for C–F bond formation, but so far only for the fluorination of benzylic C–H bonds [152].

Despite these challenges, several groups have developed chelate-assisted fluorinations of C–H bonds. The earliest example from 2006 uses **25** as F^+ source and oxidant in combination with microwave irradiation (Scheme 23.47) [149]. Using a pyridine directing group, difluorination is observed when the arene has no additional substituents in *ortho* or *meta* position.

Similar selectivity issues can be found in triflamide-directed reactions (Scheme 23.48) [153]. Using a modified fluorinating reagent **26**, mono- and difluorinated products can be obtained. A more recent development focuses on addressing this selectivity issue with an auxiliary-based approach [154]. Finally, quinoxaline-based directing groups are also able to promote mono-selective C–H bond fluorinations in combination with *N*-fluorobenzenesulfonimide as F^+ source [151].

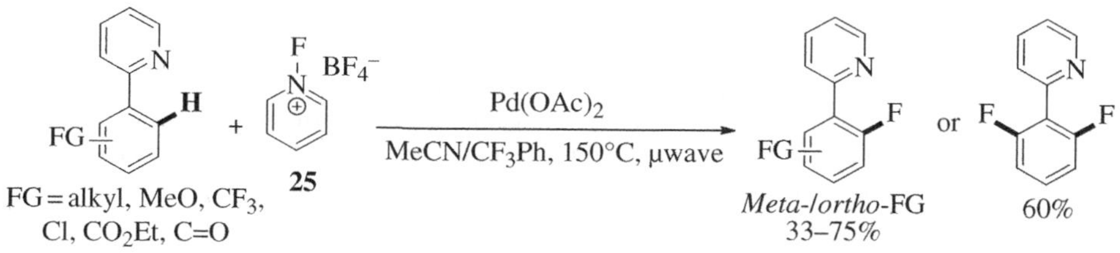

SCHEME 23.47 Pd-catalyzed C–H fluorination of 2-phenyl pyridines.

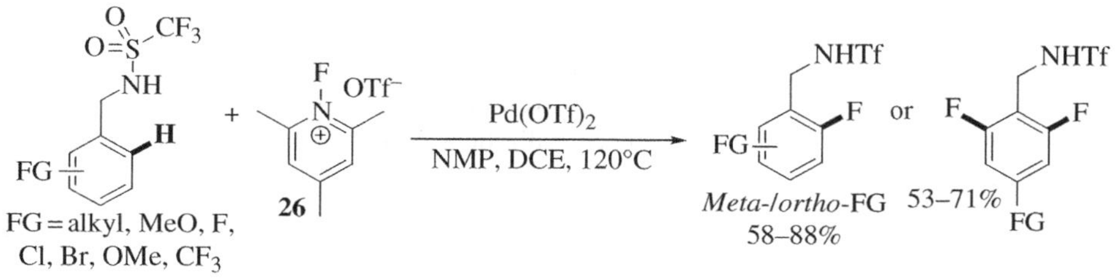

SCHEME 23.48 Pd-catalyzed C–H bond fluorination of benzylic triflamides.

23.3.8 C–D Bond Formations

Incorporating deuterium into organic molecules is highly useful for the detection and identification of metabolites and decomposition products of bioactive molecules. Therefore, deuterating reactions such as H/D exchange have been used for many environmental and pharmaceutical applications [155]. Moreover, molecules with C–D bonds are helpful for analyzing transition metal-catalyzed reactions [156, 157].

In order to achieve selective deuterium incorporation in *ortho* positions to a directing group, chelate-assisted C–H deuterations can be performed with cationic Ir complexes such as Crabtree's catalyst **27** (Scheme 23.49) [155, 158, 159]. Moreover, Pd complexes catalyze the same type of reactivity, even with weakly directing groups (Scheme 23.50) [22]. The proposed mechanisms for

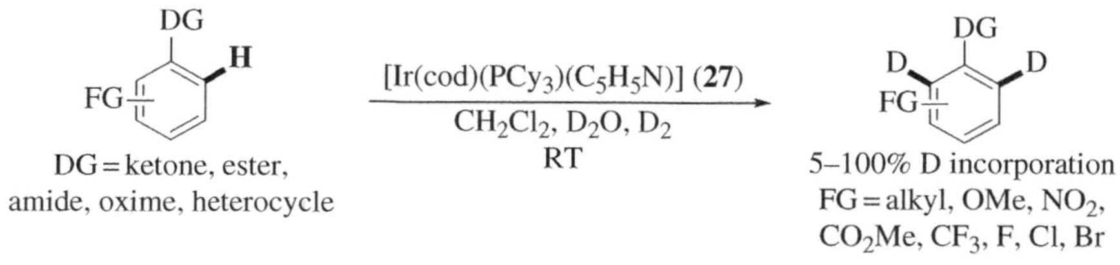

SCHEME 23.49 C–H deuteration of C–H bonds with Crabtree's catalyst **27**.

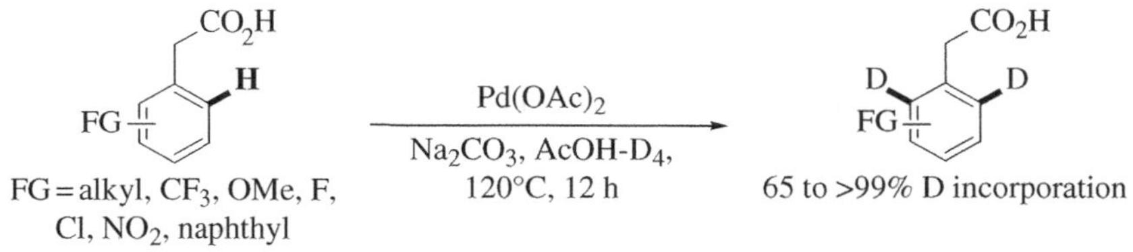

SCHEME 23.50 H/D exchange with weakly coordinating directing groups.

these processes essentially consist of C–H activation, followed by the microscopic reverse of C–H activation to form the new C–D bond [22, 160].

In summary, various functional groups can be introduced in *ortho* positions of arene substrates by using suitable directing groups. The scope of directing groups is extremely broad, as documented in this chapter. The large variety of introducible functional groups promises high versatility of these protocols for organic syntheses and will very likely find broad applications, in particular for the late-stage modification of complex structures [6, 161, 162].

23.4 STEREOSELECTIVE C–H FUNCTIONALIZATIONS

As aromatic structures are typically planar—and therefore often not the focus of method development in enantioselective synthesis—this chapter contains only few examples of stereoselective arene C–H functionalizations. Despite this restriction, stereoselective syntheses can be performed through chelate-assisted C–H functionalizations of arene moieties in specially designed substrates.

One such example is biaryl substrates with low rotation barriers around the biaryl C–C bond. Rotation around this bond converts one axial enantiomer into the other [163]. The naphthyl pyridine **28** falls into this class of substrates (Scheme 23.51). Chelate-assisted C–H alkylation of **28** results in generation of two enantiomeric products with 49% ee, since the rotation around the biaryl C–C bond is frozen in the product.

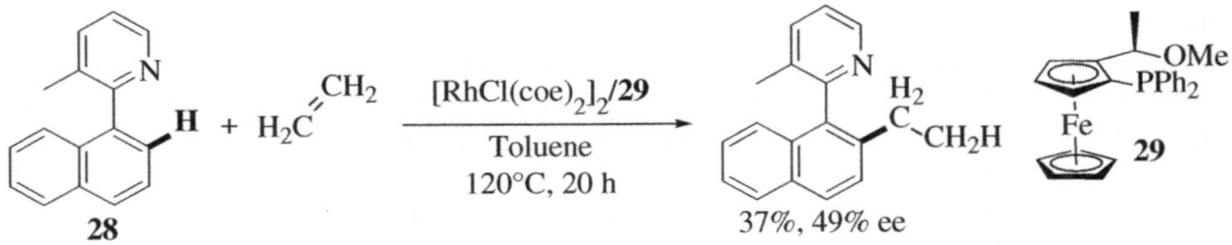

SCHEME 23.51 Rh-catalyzed C–H alkylation inducing axial chirality.

Notably, diastereoselective iodinations of aryl and other substituents can be achieved through an auxiliary-based approach using enantiomerically pure oxazolines as directing groups (Scheme 23.52) [164]. Diastereomeric ratios as well as yields are high using this protocol, providing mono-iodinated aryl products.

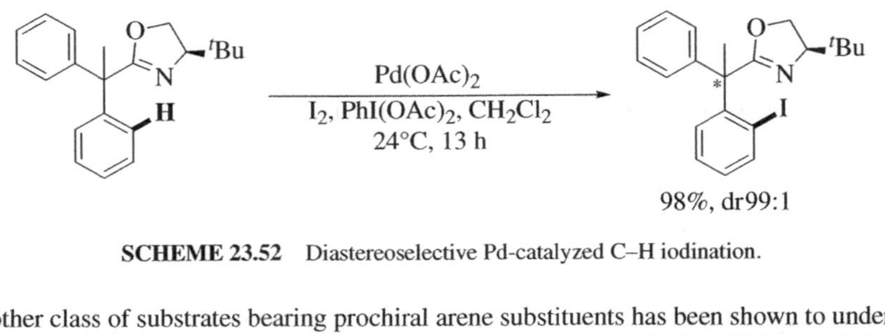

SCHEME 23.52 Diastereoselective Pd-catalyzed C–H iodination.

Another class of substrates bearing prochiral arene substituents has been shown to undergo Pd-catalyzed C–C couplings in high yields and enantioselectivities up to 95% ee (Scheme 23.53) [165]. Key to this reactivity is a broad screening of carboxylate coligands such as **30** whose design takes advantage of the chiral pool of amino acids.

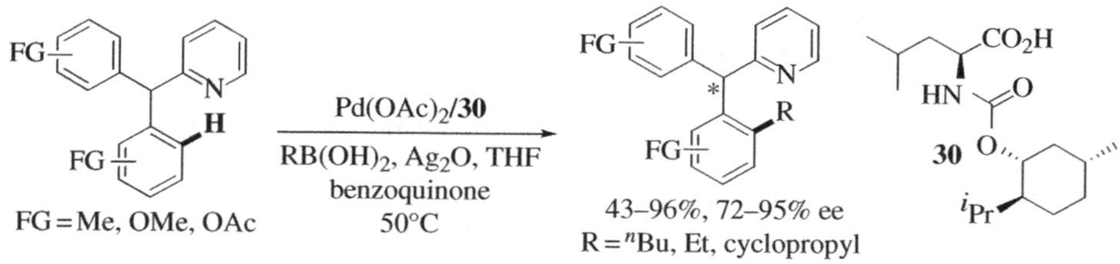

SCHEME 23.53 Enantioselective Pd-catalyzed C–H alkylation with alkyl boronic acids.

Interestingly, none of the stereoselective reactions described previously assembles a new chiral center; instead, the applied strategies rely on previously existing, special chiral characteristics of the substrates (prochiral moieties or fluxional axial chirality). In contrast to these approaches, a recently published report relies solely on catalyst design to generate a new stereocenter in the product through chelate-assisted hydroarylation (Scheme 23.54) [166]. The most selective Rh catalyst **31** uses a biaryl-substituted Cp ring as the structural element to induce enantioselectivity. After the reaction, the hydroarylation product with a new quaternary stereocenter can be isolated in 87% yield and 91% ee.

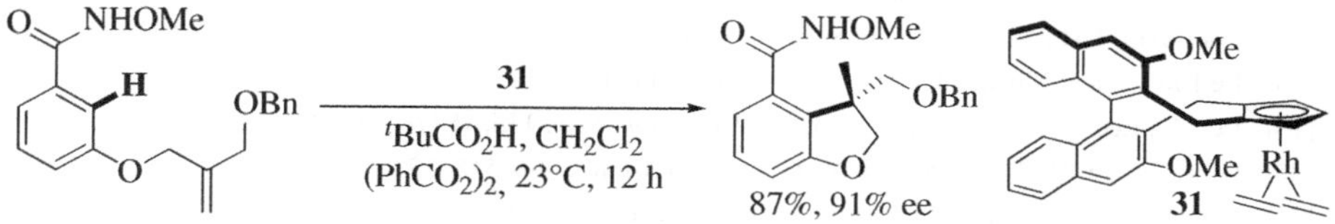

SCHEME 23.54 Enantioselective hydroarylation generating a quaternary stereocenter.

In conclusion, several approaches have been taken in order to enable stereoselective C–H functionalizations of arene substrates. Most currently known examples take advantage of chiral features in the substrate, but recent developments aim at establishing catalyst-controlled enantioselectivity.

Overall, the field of chelate-assisted C–H bond functionalization has evolved dramatically over the last 10 years and is expected to contribute even more to streamlining organic synthesis and enabling efficient synthetic methodologies in the years to come. The following chapter will cover developments in the area of nonchelate-assisted C–H bond functionalizations, some of which have been mentioned briefly in this chapter.

ABBREVIATIONS

AMLA	Ambipilic metal ligand activation
CF_3	Trifluoromethyl
CMD	Concerted metalation deprotonation
PG/DGs	Protecting-directing groups
SBM	Sigma bond metathesis

REFERENCES

[1] Kuhl, N., Hopkinson, M.N., Wencel-Delord, J. and Glorius, F. (2012) *Angew. Chem. Int. Ed.*, **51**, 10236–10254.

[2] Lyons, T.W. and Sanford, M.S. (2010) *Chem. Rev.*, **110**, 1147–1169.

[3] Perrin, C.L. and Skinner, G.A. (1971) *J. Am. Chem. Soc.*, **93**, 3389–3394.

[4] Colby, D.A., Bergman, R.G. and Ellman, J.A. (2010) *Chem. Rev.*, **110**, 624–655.

[5] Davies, H.M. and Manning, J.R. (2008) *Nature*, **451**, 417–424.

[6] Dai, H.X., Stepan, A.F., Plummer, M.S., Zhang, Y.H. and Yu, J.Q. (2011) *J. Am. Chem. Soc.*, **133**, 7222–7228.

[7] Bergman, R.G. (2007) *Nature*, **446**, 391–393.

[8] Balcells, D., Clot, E. and Eisenstein, O. (2010) *Chem. Rev.*, **110**, 749–823.

[9] Borovik, A.S. (2011) *Chem. Soc. Rev.*, **40**, 1870–1874.

[10] Blanksby, S.J. and Ellison, G.B. (2003) *Acc. Chem. Res.*, **36**, 255–263.

[11] Suess, A.M., Ertem, M.Z., Cramer, C.J. and Stahl, S.S. (2013) *J. Am. Chem. Soc.*, **135**, 9797–9804.

[12] Powers, D.C., Xiao, D.Y., Geibel, M.A. and Ritter, T. (2010) *J. Am. Chem. Soc.*, **132**, 14530–14536.

[13] Tauchert, M.E., Incarvito, C.D., Rheingold, A.L., Bergman, R.G. and Ellman, J.A. (2012) *J. Am. Chem. Soc.*, **134**, 1482–1485.

[14] Boutadla, Y., Davies, D.L., Macgregor, S.A. and Poblador-Bahamonde, A.I. (2009) *Dalton Trans.*, 5820–5831.

[15] Arndtsen, B.A., Bergman, R.G., Mobley, T.A. and Peterson, T.H. (1995) *Acc. Chem. Res.*, **28**, 154–162.

[16] Heyduk, A.F., Zhong, H.A., Labinger, J.A. and Bercaw, J.E. (2004) in *Activation and Functionalization of C-H Bonds*, vol. **885**, American Chemical Society, Washington, DC, pp. 250–263.

[17] Racowski, J.M., Dick, A.R. and Sanford, M.S. (2009) *J. Am. Chem. Soc.*, **131**, 10974–10983.

[18] Le Bras, J. and Muzart, J. (2011) *Chem. Rev.*, **111**, 1170–1214.

[19] Lee, S., Lee, H. and Tan, K.L. (2013) *J. Am. Chem. Soc.*, **135**, 18778–18781.

[20] Leow, D., Li, G., Mei, T.S. and Yu, J.Q. (2012) *Nature*, **486**, 518–522.

[21] Kruppa, M. and König, B. (2006) *Chem. Rev.*, **106**, 3520–3560.

[22] Ma, S., Villa, G., Thuy-Boun, P.S., Homs, A. and Yu, J.Q. (2014) *Angew. Chem. Int. Ed.*, **53**, 734–737.

[23] Arockiam, P.B., Bruneau, C. and Dixneuf, P.H. (2012) *Chem. Rev.*, **112**, 5879–5918.

[24] Park, S.H., Kim, J.Y. and Chang, S. (2011) *Org. Lett.*, **13**, 2372–2375.

[25] He, G., Zhang, S.-Y., Nack, W.A., Li, Q. and Chen, G. (2013) *Angew. Chem. Int. Ed.*, **52**, 11124–11128.

[26] He, G., Zhao, Y., Zhang, S., Lu, C. and Chen, G. (2011) *J. Am. Chem. Soc.*, **134**, 3–6.

[27] Xiao, B., Gong, T.J., Xu, J., Liu, Z.J. and Liu, L. (2011) *J. Am. Chem. Soc.*, **133**, 1466–1474.

[28] Mewald, M., Schiffner, J.A. and Oestreich, M. (2012) *Angew. Chem. Int. Ed.*, **51**, 1763–1765.

[29] Meng, X. and Kim, S. (2013) *Org. Lett.*, **15**, 1910–1913.

[30] Lu, Y., Wang, D.-H., Engle, K.M. and Yu, J.-Q. (2010) *J. Am. Chem. Soc.*, **132**, 5916–5921.

[31] Miura, M., Tsuda, T., Satoh, T. and Nomura, M. (1997) *Chem. Lett.*, **26**, 1103.

[32] Liu, B., Fan, Y., Gao, Y., Sun, C., Xu, C. and Zhu, J. (2013) *J. Am. Chem. Soc.*, **135**, 468–473.

[33] Wang, B., Shen, C., Yao, J., Yin, H. and Zhang, Y. (2014) *Org. Lett.*, **16**, 46–49.

[34] Covell, D.J. and White, M.C. (2013) *Tetrahedron*, **69**, 7771–7778.

[35] Péron, F., Fossey, C., Cailly, T. and Fabis, F. (2012) *Org. Lett.*, **14**, 1827–1829.

[36] Neufeldt, S.R. and Sanford, M.S. (2009) *Org. Lett.*, **12**, 532–535.

[37] Rastogi, S.K., Medellin, D.C. and Kornienko, A. (2014) *Org. Biomol. Chem.*, **12**, 410–413.

[38] Robbins, D.W., Boebel, T.A. and Hartwig, J.F. (2010) *J. Am. Chem. Soc.*, **132**, 4068–4069.

[39] Huang, C., Chattopadhyay, B. and Gevorgyan, V. (2011) *J. Am. Chem. Soc.*, **133**, 12406–12409.

[40] Huang, C., Ghavtadze, N., Chattopadhyay, B. and Gevorgyan, V. (2011) *J. Am. Chem. Soc.*, **133**, 17630–17633.

[41] Bedford, R.B., Coles, S.J., Hursthouse, M.B. and Limmert, M.E. (2003) *Angew. Chem. Int. Ed.*, **42**, 112–114.

[42] Bedford, R.B. and Limmert, M.E. (2003) *J. Org. Chem.*, **68**, 8669–8682.

[43] Catellani, M. (2005) in *Palladium in Organic Synthesis*, vol. **14** (ed. Tsuji, J.), Springer, Berlin/Heidelberg, pp. 21–53.

[44] Martins, A., Mariampillai, B. and Lautens, M. (2010) in *C-H Activation*, vol. **292** (ed. Yu, J.-Q. and Shi, Z.), Springer, Berlin Heidelberg, pp. 1–33.

[45] Welsch, M.E., Snyder, S.A. and Stockwell, B.R. (2010) *Curr. Opin. Chem. Biol.*, **14**, 347–361.

[46] Oi, S., Fukita, S. and Inoue, Y. (1998) *Chem. Commun.*, 2439–2440.

[47] Satoh, T., Kametani, Y., Terao, Y., Miura, M. and Nomura, M. (1999) *Tetrahedron Lett.*, **40**, 5345–5348.

[48] Kakiuchi, F., Kan, S., Igi, K., Chatani, N. and Murai, S. (2003) *J. Am. Chem. Soc.*, **125**, 1698–1699.

[49] Xiao, B., Fu, Y., Xu, J., Gong, T.-J., Dai, J.-J., Yi, J. and Liu, L. (2009) *J. Am. Chem. Soc.*, **132**, 468–469.

[50] Zhao, X., Yeung, C.S. and Dong, V.M. (2010) *J. Am. Chem. Soc.*, **132**, 5837–5844.

[51] Wencel-Delord, J., Nimphius, C., Wang, H. and Glorius, F. (2012) *Angew. Chem. Int. Ed.*, **51**, 13001–13005.

[52] Alberico, D., Scott, M.E. and Lautens, M. (2007) *Chem. Rev.*, **107**, 174–238.

[53] Song, G., Wang, F. and Li, X. (2012) *Chem. Soc. Rev.*, **41**, 3651–3678.

[54] Liu, C., Zhang, H., Shi, W. and Lei, A. (2011) *Chem. Rev.*, **111**, 1780–1824.

[55] Yeung, C.S. and Dong, V.M. (2011) *Chem. Rev.*, **111**, 1215–1292.

[56] Daugulis, O. and Zaitsev, V.G. (2005) *Angew. Chem. Int. Ed.*, **44**, 4046–4048.

[57] Tasker, S.Z., Standley, E.A. and Jamison, T.F. (2014) *Nature*, **509**, 299–309.

[58] Pieber, B., Cantillo, D. and Kappe, C.O. (2012) *Chem. Eur. J.*, **18**, 5047–5055.

[59] Wang, J., Ferguson, D.M. and Kalyani, D. (2013) *Tetrahedron*, **69**, 5780–5790.

[60] Sun, C.-L., Li, B.-J. and Shi, Z.-J. (2010) *Chem. Rev.*, **111**, 1293–1314.

[61] Yoshikai, N., Matsumoto, A., Norinder, J. and Nakamura, E. (2009) *Angew. Chem. Int. Ed.*, **48**, 2925–2928.

[62] Hara, K., Sato, T., Katoh, R., Furube, A., Yoshihara, T., Murai, M., Kurashige, M., Ito, S., Shinpo, A., Suga, S. and Arakawa, H. (2005) *Adv. Funct. Mater.*, **15**, 246–252.

[63] Grimsdale, A.C., Leok Chan, K., Martin, R.E., Jokisz, P.G. and Holmes, A.B. (2009) *Chem. Rev.*, **109**, 897–1091.

[64] Babudri, F., Farinola, G.M. and Naso, F. (2004) *J. Mater. Chem.*, **14**, 11–34.

[65] Jia, C., Kitamura, T. and Fujiwara, Y. (2001) *Acc. Chem. Res.*, **34**, 633–639.

[66] Yu, M., Xie, Y., Xie, C. and Zhang, Y. (2012) *Org. Lett.*, **14**, 2164–2167.

[67] Kakiuchi, F., Yamamoto, Y., Chatani, N. and Murai, S. (1995) *Chem. Lett.*, **24**, 681–682.

[68] Graczyk, K., Ma, W. and Ackermann, L. (2012) *Org. Lett.*, **14**, 4110–4113.

[69] Muralirajan, K., Parthasarathy, K. and Cheng, C.-H. (2011) *Angew. Chem. Int. Ed.*, **50**, 4169–4172.

[70] Patureau, F.W., Besset, T., Kuhl, N. and Glorius, F. (2011) *J. Am. Chem. Soc.*, **133**, 2154–2156.

[71] Zhang, X.-S., Zhu, Q.-L., Zhang, Y.-F., Li, Y.-B. and Shi, Z.-J. (2013) *Chem. Eur. J.*, **19**, 11898–11903.

[72] Gandeepan, P. and Cheng, C.-H. (2012) *J. Am. Chem. Soc.*, **134**, 5738–5741.

[73] Minami, Y., Shiraishi, Y., Yamada, K. and Hiyama, T. (2012) *J. Am. Chem. Soc.*, **134**, 6124–6127.

[74] Li, G., Leow, D., Wan, L. and Yu, J.-Q. (2013) *Angew. Chem. Int. Ed.*, **52**, 1245–1247.

[75] Ogiwara, Y., Tamura, M., Kochi, T., Matsuura, Y., Chatani, N. and Kakiuchi, F. (2013) *Organometallics*, **33**, 402–420.

[76] Mo, J., Lim, S., Park, S., Ryu, T., Kim, S. and Lee, P.H. (2013) *RSC Adv.*, **3**, 18296–18299.

[77] Itoh, M., Hashimoto, Y., Hirano, K., Satoh, T. and Miura, M. (2013) *J. Org. Chem.*, **78**, 8098–8104.

[78] Sonoda, M., Kakiuchi, F., Kamatani, A., Chatani, N. and Murai, S. (1996) *Chem. Lett.*, **25**, 109–110.

[79] Martinez, R., Chevalier, R., Darses, S. and Genet, J.-P. (2006) *Angew. Chem. Int. Ed.*, **45**, 8232–8235.

[80] Martinez, R., Simon, M.-O., Chevalier, R., Pautigny, C., Genet, J.-P. and Darses, S. (2009) *J. Am. Chem. Soc.*, **131**, 7887–7895.

[81] Murai, S., Kakiuchi, F., Sekine, S., Tanaka, Y., Kamatani, A., Sonoda, M. and Chatani, N. (1993) *Nature*, **366**, 529–531.

[82] Chen, X., Goodhue, C.E. and Yu, J.-Q. (2006) *J. Am. Chem. Soc.*, **128**, 12634–12635.

[83] Lee, D.-H., Kwon, K.-H. and Yi, C.S. (2012) *J. Am. Chem. Soc.*, **134**, 7325–7328.

[84] Mamone, P., Danoun, G. and Gooßen, L.J. (2013) *Angew. Chem. Int. Ed.*, **52**, 6704–6708.

[85] Park, J., Park, E., Kim, A., Lee, Y., Chi, K.-W., Kwak, J.H., Jung, Y.H. and Kim, I.S. (2011) *Org. Lett.*, **13**, 4390–4393.

[86] Xiao, F., Shuai, Q., Zhao, F., Basle, O., Deng, G. and Li, C.J. (2011) *Org. Lett.*, **13**, 1614–1617.

[87] Yin, Z. and Sun, P. (2012) *J. Org. Chem.*, **77**, 11339–11344.

[88] Chen, X., Hao, X.-S., Goodhue, C.E. and Yu, J.-Q. (2006) *J. Am. Chem. Soc.*, **128**, 6790–6791.

[89] Jia, X., Yang, D., Wang, W., Luo, F. and Cheng, J. (2009) *J. Org. Chem.*, **74**, 9470–9474.

[90] Gong, T.-J., Xiao, B., Cheng, W.-M., Su, W., Xu, J., Liu, Z.-J., Liu, L. and Fu, Y. (2013) *J. Am. Chem. Soc.*, **135**, 10630–10633.

[91] Liang, T., Neumann, C.N. and Ritter, T. (2013) *Angew. Chem. Int. Ed.*, **52**, 8214–8264.

[92] Wang, X., Truesdale, L. and Yu, J.-Q. (2010) *J. Am. Chem. Soc.*, **132**, 3648–3649.

[93] Zhang, X.-G., Dai, H.-X., Wasa, M. and Yu, J.-Q. (2012) *J. Am. Chem. Soc.*, **134**, 11948–11951.

[94] Hafner, A. and Bräse, S. (2012) *Angew. Chem. Int. Ed.*, **51**, 3713–3715.

[95] Hafner, A., Bihlmeier, A., Nieger, M., Klopper, W. and Bräse, S. (2013) *J. Org. Chem.*, **78**, 7938–7948.

[96] Iverson, C.N. and Smith, M.R. (1999) *J. Am. Chem. Soc.*, **121**, 7696–7697.

[97] Cho, J.-Y., Tse, M.K., Holmes, D., Maleczka, R.E. and Smith, M.R. (2002) *Science*, **295**, 305–308.

[98] Dai, H.-X. and Yu, J.-Q. (2011) *J. Am. Chem. Soc.*, **134**, 134–137.

[99] Ishiyama, T., Isou, H., Kikuchi, T. and Miyaura, N. (2010) *Chem. Commun.*, **46**, 159–161.

[100] Ros, A., Estepa, B., López-Rodríguez, R., Álvarez, E., Fernández, R. and Lassaletta, J.M. (2011) *Angew. Chem. Int. Ed.*, **50**, 11724–11728.

[101] Kawamorita, S., Ohmiya, H., Hara, K., Fukuoka, A. and Sawamura, M. (2009) *J. Am. Chem. Soc.*, **131**, 5058–5059.

[102] Kawamorita, S., Miyazaki, T., Ohmiya, H., Iwai, T. and Sawamura, M. (2011) *J. Am. Chem. Soc.*, **133**, 19310–19313.

[103] Boebel, T.A. and Hartwig, J.F. (2008) *J. Am. Chem. Soc.*, **130**, 7534–7535.

[104] Sun, C.-L., Li, B.-J. and Shi, Z.-J. (2010) *Chem. Commun.*, **46**, 677–685.

[105] Hartwig, J.F. (2011) *Acc. Chem. Res.*, **45**, 864–873.

[106] Ishiyama, T., Sato, K., Nishio, Y. and Miyaura, N. (2003) *Angew. Chem. Int. Ed.*, **42**, 5346–5348.

[107] Klare, H.F.T., Oestreich, M., Ito, J.-I., Nishiyama, H., Ohki, Y. and Tatsumi, K. (2011) *J. Am. Chem. Soc.*, **133**, 3312–3315.

[108] Tsukada, N. and Hartwig, J.F. (2005) *J. Am. Chem. Soc.*, **127**, 5022–5023.

[109] Simmons, E.M. and Hartwig, J.F. (2010) *J. Am. Chem. Soc.*, **132**, 17092–17095.

[110] Ihara, H., Koyanagi, M. and Suginome, M. (2011) *Org. Lett.*, **13**, 2662–2665.

[111] Ihara, H. and Suginome, M. (2009) *J. Am. Chem. Soc.*, **131**, 7502–7503.

[112] Rachkovskaya, L.N., Matveev, K.I., Il'lnich, G.N. and Eremenko, E.K. (1977) *Kinet. Katal.*, **18**, 854–856.

[113] Yoneyama, T. and Crabtree, R.H. (1996) *J. Mol. Catal. A Chem.*, **108**, 35–40.

[114] Burton, H.A. and Kozhevnikov, I.V. (2002) *J. Mol. Catal. A Chem.*, **185**, 285–290.

[115] Dick, A.R., Hull, K.L. and Sanford, M.S. (2004) *J. Am. Chem. Soc.*, **126**, 2300–2301.

[116] Desai, L.V., Stowers, K.J. and Sanford, M.S. (2008) *J. Am. Chem. Soc.*, **130**, 13285–13293.

[117] Wang, G.-W., Yuan, T.-T. and Wu, X.-L. (2008) *J. Org. Chem.*, **73**, 4717–4720.

[118] Xiao, B., Gong, T.-J., Liu, Z.-J., Liu, J.-H., Luo, D.-F., Xu, J. and Liu, L. (2011) *J. Am. Chem. Soc.*, **133**, 9250–9253.

[119] Mo, F., Trzepkowski, L.J. and Dong, G. (2012) *Angew. Chem. Int. Ed.*, **51**, 13075–13079.

[120] Yang, Y., Lin, Y. and Rao, Y. (2012) *Org. Lett.*, **14**, 2874–2877.

[121] Shan, G., Yang, X., Ma, L. and Rao, Y. (2012) *Angew. Chem. Int. Ed.*, **51**, 13070–13074.

[122] Thirunavukkarasu, V.S. and Ackermann, L. (2012) *Org. Lett.*, **14**, 6206–6209.

[123] Shan, G., Han, X., Lin, Y., Yu, S. and Rao, Y. (2013) *Org. Biomol. Chem.*, **11**, 2318–2322.

[124] Yang, X., Shan, G. and Rao, Y. (2013) *Org. Lett.*, **15**, 2334–2337.

[125] Yang, F. and Ackermann, L. (2013) *Org. Lett.*, **15**, 718–720.

[126] Chan, L.Y., Meng, X. and Kim, S. (2013) *J. Org. Chem.*, **78**, 8826–8832.

[127] Zhang, Y.-H. and Yu, J.-Q. (2009) *J. Am. Chem. Soc.*, **131**, 14654–14655.

[128] Yan, Y., Feng, P., Zheng, Q.Z., Liang, Y.F., Lu, J.F., Cui, Y. and Jiao, N. (2013) *Angew. Chem. Int. Ed.*, **52**, 5827–5831.

[129] Thu, H.Y., Yu, W.Y. and Che, C.M. (2006) *J. Am. Chem. Soc.*, **128**, 9048–9049.

[130] Kim, J.Y., Park, S.H., Ryu, J., Cho, S.H., Kim, S.H. and Chang, S. (2012) *J. Am. Chem. Soc.*, **134**, 9110–9113.

[131] Shin, K., Baek, Y. and Chang, S. (2013) *Angew. Chem. Int. Ed.*, **52**, 8031–8036.

[132] Zheng, Q.-Z., Liang, Y.-F., Qin, C. and Jiao, N. (2013) *Chem. Commun.*, **49**, 5654–5656.

[133] Grohmann, C., Wang, H. and Glorius, F. (2013) *Org. Lett.*, **15**, 3014–3017.

[134] Kim, J., Kim, J. and Chang, S. (2013) *Chem. Eur. J.*, **19**, 7328–7333.

[135] Shang, M., Sun, S.-Z., Dai, H.-X. and Yu, J.-Q. (2014) *J. Am. Chem. Soc.*, **136**, 3354–3357.

[136] Kuninobu, Y., Yoshida, T. and Takai, K. (2011) *J. Org. Chem.*, **76**, 7370–7376.

[137] Baba, K., Tobisu, M. and Chatani, N. (2013) *Angew. Chem. Int. Ed.*, **52**, 11892–11895.

[138] Feng, C.-G., Ye, M., Xiao, K.-J., Li, S. and Yu, J.-Q. (2013) *J. Am. Chem. Soc.*, **135**, 9322–9325.

[139] Li, C., Yano, T., Ishida, N. and Murakami, M. (2013) *Angew. Chem. Int. Ed.*, **52**, 9801–9804.

[140] Tran, L.D., Popov, I. and Daugulis, O. (2012) *J. Am. Chem. Soc.*, **134**, 18237–18240.

[141] Zhao, X., Dimitrijević, E. and Dong, V.M. (2009) *J. Am. Chem. Soc.*, **131**, 3466–3467.

[142] Song, B., Zheng, X., Mo, J. and Xu, B. (2010) *Adv. Synth. Catal.*, **352**, 329–335.

[143] Kalyani, D., Dick, A.R., Anani, W.Q. and Sanford, M.S. (2006) *Org. Lett.*, **8**, 2523–2526.

[144] Schröder, N., Wencel-Delord, J. and Glorius, F. (2012) *J. Am. Chem. Soc.*, **134**, 8298–8301.

[145] John, A. and Nicholas, K.M. (2012) *J. Org. Chem.*, **77**, 5600–5605.

[146] Sun, X., Shan, G., Sun, Y. and Rao, Y. (2013) *Angew. Chem. Int. Ed.*, **52**, 4440–4444.

[147] Sun, X., Sun, Y., Zhang, C. and Rao, Y. (2014) *Chem. Commun.*, **50**, 1262–1264.

[148] Mei, T.-S., Giri, R., Maugel, N. and Yu, J.-Q. (2008) *Angew. Chem. Int. Ed.*, **47**, 5215–5219.

[149] Hull, K.L., Anani, W.Q. and Sanford, M.S. (2006) *J. Am. Chem. Soc.*, **128**, 7134–7135.

[150] Furuya, T. and Ritter, T. (2008) *J. Am. Chem. Soc.*, **130**, 10060–10061.

[151] Lou, S.-J., Xu, D.-Q., Xia, A.-B., Wang, Y.-F., Liu, Y.-K., Du, X.-H. and Xu, Z.-Y. (2013) *Chem. Commun.*, **49**, 6218–6220.

[152] McMurtrey, K.B., Racowski, J.M. and Sanford, M.S. (2012) *Org. Lett.*, **14**, 4094–4097.

[153] Wang, X., Mei, T.-S. and Yu, J.-Q. (2009) *J. Am. Chem. Soc.*, **131**, 7520–7521.

[154] Chan, K.S.L., Wasa, M., Wang, X. and Yu, J.-Q. (2011) *Angew. Chem. Int. Ed.*, **50**, 9081–9084.

[155] Atzrodt, J., Derdau, V., Fey, T. and Zimmermann, J. (2007) *Angew. Chem. Int. Ed.*, **46**, 7744–7765.

[156] Crabtree, R.H. (2004) *J. Organomet. Chem.*, **689**, 4083–4091.

[157] Marcus, D.M., Mclachlan, K.A., Wildman, M.A., Ehresmann, J.O., Kletnieks, P.W. and Haw, J.F. (2006) *Angew. Chem. Int. Ed.*, **45**, 3133–3136.

[158] Ellames, G.J., Gibson, J.S., Herbert, J.M. and Mcneill, A.H. (2001) *Tetrahedron*, **57**, 9487–9497.

[159] Cross, P.W.C., Ellames, G.J., Gibson, J.S., Herbert, J.M., Kerr, W.J., Mcneill, A.H. and Mathers, T.W. (2003) *Tetrahedron*, **59**, 3349–3358.

[160] Junk, T. and Catallo, W.J. (1997) *Chem. Soc. Rev.*, **26**, 401–406.

[161] Hale, J.J., Mills, S.G., Maccoss, M., Finke, P.E., Cascieri, M.A., Sadowski, S., Ber, E., Chicchi, G.G., Kurtz, M., Metzger, J., Eiermann, G., Tsou, N.N., Tattersall, F.D., Rupniak, N.M.J., Williams, A.R., Rycroft, W., Hargreaves, R. and Macintyre, D.E. (1998) *J. Med. Chem.*, **41**, 4607–4614.

[162] Nadin, A., Hattotuwagama, C. and Churcher, I. (2012) *Angew. Chem. Int. Ed.*, **51**, 1114–1122.

[163] Kakiuchi, F., Le Gendre, P., Yamada, A., Ohtaki, H. and Murai, S. (2000) *Tetrahedron: Asymmetry*, **11**, 2647–2651.

[164] Giri, R., Chen, X. and Yu, J.-Q. (2005) *Angew. Chem. Int. Ed.*, **44**, 2112–2115.

[165] Shi, B.-F., Maugel, N., Zhang, Y.-H. and Yu, J.-Q. (2008) *Angew. Chem. Int. Ed.*, **47**, 4882–4886.

[166] Ye, B., Donets, P.A. and Cramer, N. (2014) *Angew. Chem. Int. Ed.*, **53**, 507–511.

24

REACTIVITY AND SELECTIVITY IN TRANSITION METAL-CATALYZED, NONDIRECTED ARENE FUNCTIONALIZATIONS

Dipannita Kalyani and Elodie E. Marlier
St. Olaf College, Northfield, MN, USA

24.1 INTRODUCTION

Transition metal-catalyzed direct conversions of aromatic C—H bonds to C—C or C—heteroatom bonds have emerged as powerful strategies for the synthesis of structurally diverse arenes [1]. These methods serve as environmentally friendly and atom economical alternatives to traditional cross-coupling reactions (e.g., Suzuki, Stille, Negishi, Kumada) by obviating the requirement for prefunctionalized arene substrates (e.g., aryl halides or pseudohalides, arylboronic acids) [2]. The key challenge toward realizing synthetically useful and industrially applicable arene C—H functionalizations is the selective transformation of one specific C(Ar)—H moiety in the presence of other C(Ar)—H bonds in the same substrate. In the absence of such selectivity, several undesired by-products might result, thereby necessitating tedious purification procedures toward obtaining the desired product.

The vast majority of current transition metal-catalyzed arene C—H functionalizations achieve selectivity by employing substrates with appropriate directing groups (DG) [3]. As discussed in Chapter 23, this strategy precludes the formation of undesired product mixtures because the DG directs the functionalization of the adjacent C—H bond, thereby affording the *ortho*-functionalized products selectively. Despite this advantage, the DG approach has a number of limitations. First, substrates bearing DGs might not be readily available in the context of complex molecule synthesis. Second, additional synthetic steps are often required to both install the DG in the substrate and to remove or manipulate it after the desired C—H functionalization, thus reducing the step economy of the overall process. Finally, in most cases, the DG strategy affords *ortho*-functionalized products selectively, restricting the scope of accessible products.

Arene Chemistry: Reaction Mechanisms and Methods for Aromatic Compounds,
First Edition. Edited by Jacques Mortier.

A more appealing strategy for arene C—H functionalizations involves the use of simple benzene derivatives lacking DGs. The overall pathway for such transformations entails metal mediated C—H activation to afford the metal-aryl intermediate **Ar—M** followed by subsequent functionalization of **Ar—M** to release the desired product (Scheme 24.1). The C—H activation step can proceed via oxidative addition, sigma bond metathesis, or concerted metalation deprotonation pathway. The exact mechanism of C—H cleavage is dependent on the nature of the metal and the ancillary ligands [1, 3].

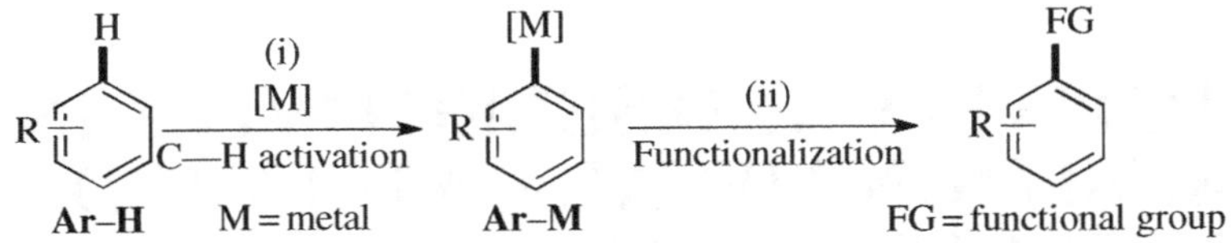

SCHEME 24.1 A general pathway for arene C—H functionalization.

While this route is an attractive approach, the lack of DG in substrates presents challenges pertaining to both selectivity and reactivity. As mentioned earlier, mixtures of products might result from activation of chemically distinct C—H bonds (step i, Scheme 24.1). Furthermore, the lack of directing groups renders the C—H activation step to be kinetically slow thereby impacting the overall efficiency of these transformations [1]. Despite these challenges, exciting progress has been made toward the direct conversion of arene C—H bonds to C—X (X=C, O, N, B, Si) bonds, and many of these accomplishments have been guided by careful mechanistic explorations.

This chapter overviews the state of the art in this field by highlighting the scope and limitations as well as the mechanistic understanding of nondirected intermolecular arene C—H functionalizations reported until the beginning of 2014. In general, only homogeneous reactions proceeding *via* aryl metal intermediates will be detailed, excluding transformations involving nonorganometallic or radical pathways [4]. Furthermore, since nondirected transition metal-catalyzed arene functionalizations have been reviewed very recently [5], this chapter predominantly focuses on Pd-catalyzed C—H transformations [6] since they encompass the vast majority of reports in this field to date. However, a brief synopsis of methods using transition metals other than palladium is discussed toward the end. In the context of this chapter, the phrase "arene C—H functionalization" strictly refers to transformations of unactivated relatively nonacidic C(Ar)—H bonds in simple benzene derivatives. Discussions on the functionalization of activated relatively acidic C—H bonds adjacent to electronegative atoms in heteroarenes (e.g., indoles, benzofurans, azoles, etc.) or in perfluorobenzenes will not be presented [5, 7]. The chapter is divided into eight main sections with the first seven sections detailing the advances in Pd-catalyzed aromatic C—H arylations, alkenylations, alkylations, carboxylation, oxygenation, thiolation, and amination. The final section highlights transition metal-catalyzed arene functionalizations that have not been accomplished using palladium.

24.2 ARYLATION

The biaryl structural motif is prevalent in many functional materials including pharmaceuticals, agrochemicals, and materials applications [1]. As such, the development of efficient and economically viable methods for the construction of biaryl structures is an important challenge. An emerging strategy to achieve this goal involves the direct replacement of C—H bonds in arenes by aromatic groups in aryl halides (or pseudohalides), organometallic reagents, or other arenes. Numerous reports for the arylation of arenes using Pd catalysis have been published. These can be divided into two broad classes: (i) direct arylation [8] and (ii) cross-dehydrogenative coupling (CDC) [9]. The former involves the coupling of an aromatic C—H bond with Ar—X or Ar—M, while the latter entails the direct coupling of two aromatic C—H bonds. As will be discussed later, both direct arylations and CDC reactions can proceed *via* either a $Pd^{0/II}$ or a $Pd^{II/IV}$ catalytic cycle.

24.2.1 Direct Arylations

24.2.1.1 Mechanisms of Direct Arylations A plausible mechanism for $Pd^{0/II}$-catalyzed intermolecular direct arylations with ArX (X=Cl, Br, I) involves: (i) oxidative addition of a Pd^0 catalyst into an aryl halide, (ii) intermolecular C—H activation, and (iii) reductive elimination to release the arylated product and regenerate the catalyst (Scheme 24.2a) [1, 5, 6, 8].

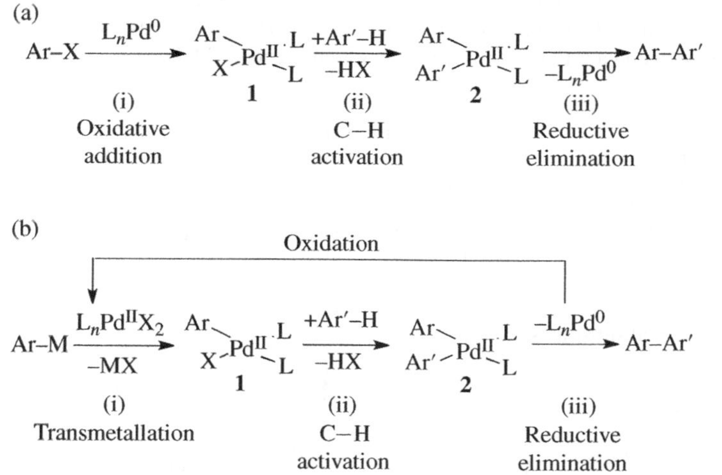

SCHEME 24.2 General mechanisms for $Pd^{0/II}$-catalyzed direct arylation.

Alternatively, these arylations can be achieved using transmetallating reagents ArM (e.g., boron reagents) in place of ArX (X=halide; Scheme 24.2b) [8]. The mechanism of these transformations differs from those using aryl halides in two ways. First, the Pd^{II}–aryl intermediate **1** is generated *via* transmetallation instead of oxidative addition. Secondly, an external oxidant (e.g., benzoquinone, Ag_2CO_3, O_2) is required to regenerate the Pd^{II} catalyst by oxidizing the Pd^0 species formed upon reductive elimination.

A general mechanistic manifold for $Pd^{II/IV}$-catalyzed direct arylations involves (i) Pd^{II}-mediated C—H activation to afford **1**, (ii) oxidation of **1** to Pd^{IV}–aryl complex **3**, and (iii) C—C bond formation *via* reductive elimination to release the product and regenerate the Pd^{II} catalyst (Scheme 24.3). Aryl iodides (Scheme 24.3a) or aryl iodonium salts (Scheme 24.3b) are the most commonly employed

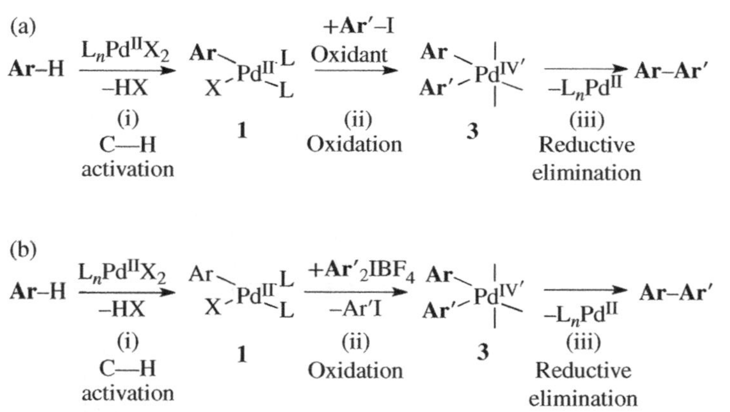

SCHEME 24.3 General mechanisms for $Pd^{II/IV}$-catalyzed direct arylation.

oxidants for these arylations. Copper(II) or silver(I) salts are generally used as cooxidants for $Pd^{II/IV}$-arylations using aryl iodides. Aryl iodonium salts are strong oxidants and hence serve as both the aryl source and the terminal oxidants [1, 5, 6, 8].

24.2.1.2 $Pd^{0/II}$-Catalyzed Direct Arylations: Synthetic Scope Several examples of $Pd^{0/II}$-catalyzed direct arylations with aryl halides have been demonstrated. The Fagnou group reported the first efficient Pd-catalyzed coupling of arenes with aryl bromides in 2006 [10]. The reaction of electron-neutral and electron-deficient aryl halides with excess benzene affords the desired biaryl products in good yields using a Pd/Davephos catalyst system (Scheme 24.4).

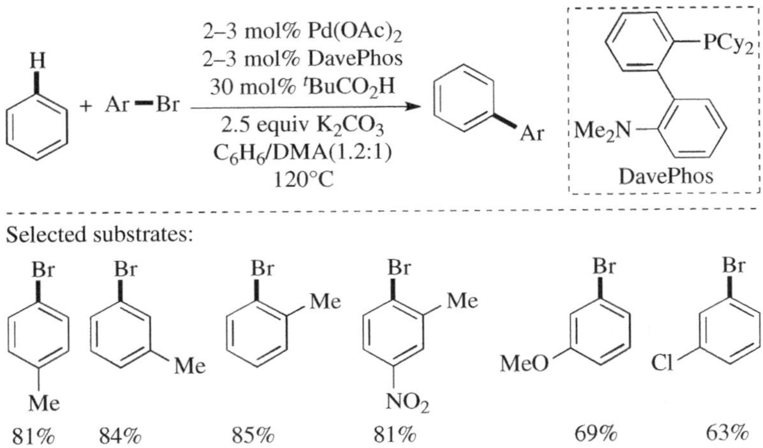

SCHEME 24.4 Representative examples for direct arylation using aryl bromides.

The use of carbonate bases in conjunction with substoichiometric amounts of a pivalate source (e.g., ^{t}BuCO$_2$H) is critical to obtain the products in high yields. Density functional theory (DFT) studies suggest that the pivalate anion serves to decrease the transition state energy for C—H cleavage. In competition studies, electron-deficient arenes are arylated preferentially. For example, the reaction of *p*-tolyl bromide with equimolar amounts of benzene and fluorobenzene leads to the formation of **A** and **B** in a 1:11 ratio; product **B** is formed as a 22:3:1 mixture of *o*/*m*/*p* isomers (Scheme 24.5). These results implicate the importance of C—H acidity in these systems. Notably,

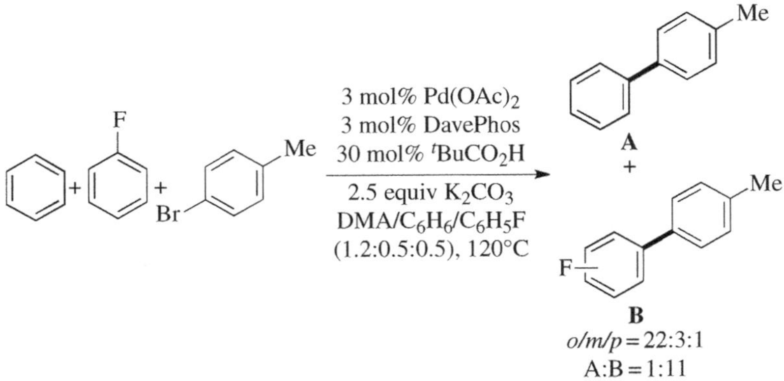

SCHEME 24.5 Electronic effect for direct arylations using *p*-tolyl bromide.

these reactions represent the first high-yielding arylations of an unactivated arenes. However, they are limited to the use of aryl bromides since aryl chlorides and iodides are not effective in these arylations and afford the desired products in trace yields.

Inspired by the preferential arylation of the most acidic C–H bonds in the aforementioned reactions, Fagnou group pioneered the use of nitroarenes for direct arylations [11]. Interestingly, a directing effect by the NO_2 group is observed since the *ortho*-arylated isomer is the major product of these reactions. Particularly noteworthy is that the use of pivalic acid, and a carbonate base is critical for enhancing both the efficiency and *ortho*-selectivity in these transformations (Scheme 24.6).

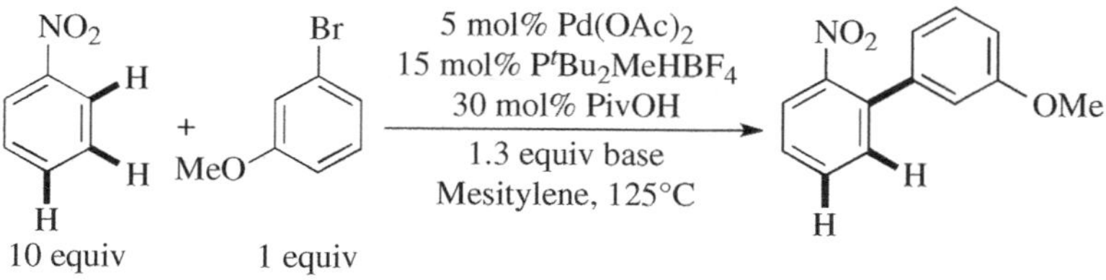

Conditions tested:

Entry	Base	PivOH	Yield	*o*/(*m*+*p*)
1	K_3PO_4	Yes	29%	6:1
2	Cs_2CO_3	Yes	69%	36:1
3	K_2CO_3	Yes	77%	34:1
4	K_2CO_3	No	43%	5.1:1

SCHEME 24.6 Base and additive effects on direct arylation of nitrobenzene.

Under the optimal reaction conditions, a variety of nitro aromatics are coupled with aryl bromides to afford the corresponding products in modest to good yields (Scheme 24.7). This transformation is applicable toward the synthesis of Boscalid, which is a commercially available agrochemical.

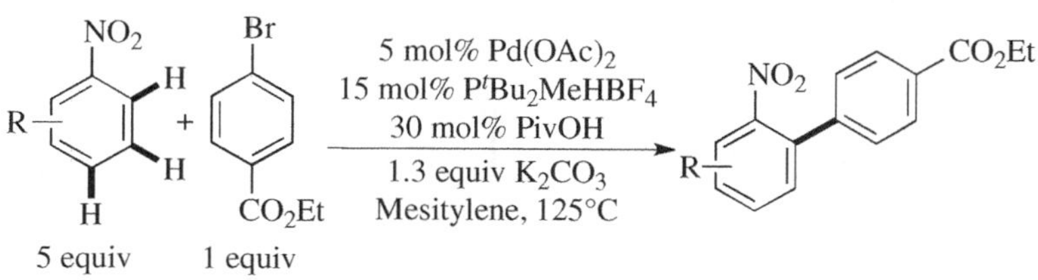

Selected aryl nitro substrates:

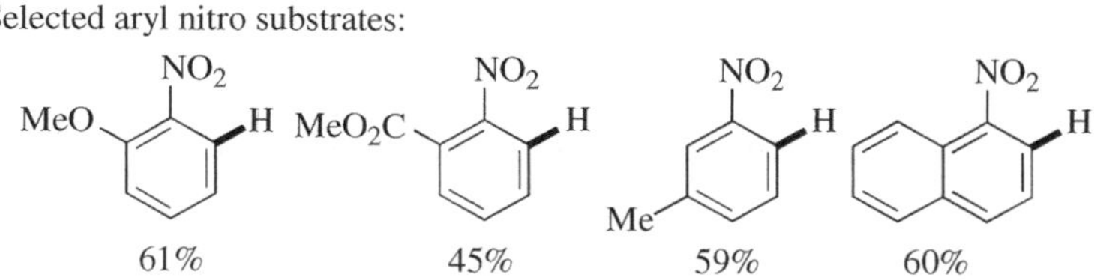

SCHEME 24.7 Representative examples for direct arylations of nitro aromatics.

$Pd^{0/II}$-catalyzed direct arylations have also been accomplished using arylboronic acids. As shown in Scheme 24.8, the $Pd(OAc)_2$-catalyzed reaction of electron-rich arenes with $PhB(OH)_2$ affords the phenylated products in modest to good yields using $Cu(OAc)_2$ as the oxidant under an O_2 atmosphere [12].

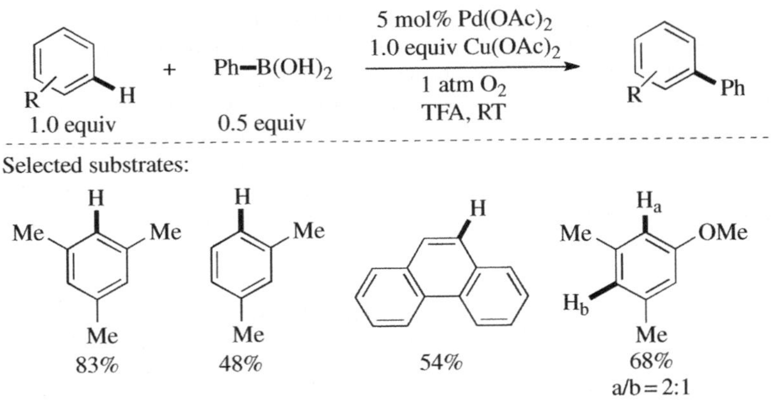

SCHEME 24.8 Direct arylation using phenylboronic acids.

24.2.1.3 Pd$^{II/IV}$-Catalyzed Direct Arylations: Synthetic Scope Aryl iodides, aryl stannanes, and aryl iodonium reagents have all been successfully employed for the direct conversion of C_{Ar-H} to C_{Ar-Ar} *via* Pd$^{II/IV}$ catalytic cycles. For example, the Pd-catalyzed reaction of naphthalene with $PhSnCl_3$ in the presence of $CuCl_2$ affords the phenylated product in modest yield as a mixture of the α/β isomers (Scheme 24.9) [13]. This transformation can be applied toward the phenylation of other electron-neutral arenes, albeit, with poor efficiencies. Interestingly, reactions of naphthalene and xylenes yield products *via* preferential arylation of the more sterically hindered C—H bond. Preliminary mechanistic studies suggest a Pd$^{II/IV}$ catalytic cycle for these reactions.

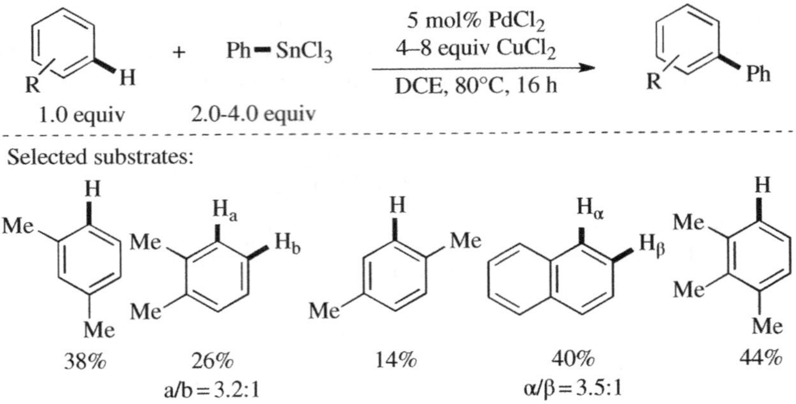

SCHEME 24.9 Direct arylations using $PhSnCl_3$.

The Lu group demonstrated the Pd-catalyzed arylation of naphthalene using aryl iodides in the presence of stoichiometric $AgOCOCF_3$ (Scheme 24.10) [14]. Electronically diverse aryl iodides effectively participate in these reactions to afford the corresponding products in good yields. The α-isomer is the major product in all cases regardless of the electronic nature of the aryl iodides. The Pd$^{II/IV}$ pathway is proposed to be operative under these reaction conditions, primarily because Pd^0 catalysts do not afford the desired biaryl products in the absence of the $AgOCOCF_3$.

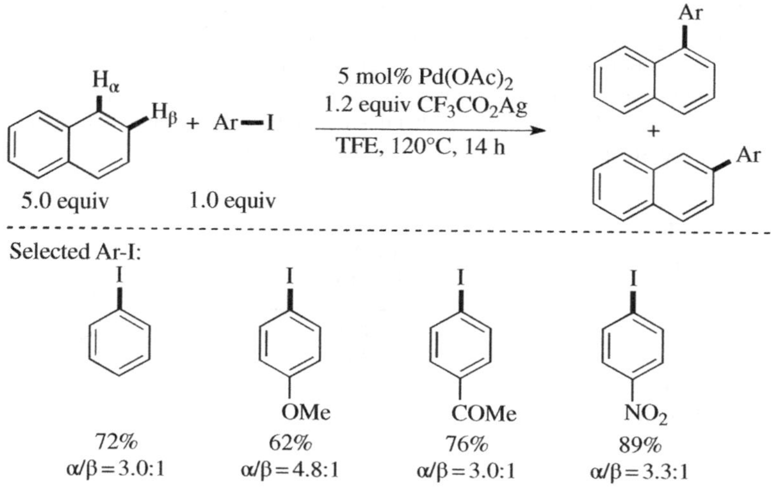

SCHEME 24.10 Direct arylation using aryl iodides.

Similar direct arylations have been accomplished using hypervalent iodine reagents and the Herrmann–Beller catalyst **4** (Scheme 24.11) [15]. This method is applicable toward the phenylation of electron-neutral and electron-rich arene substrates and provides the products in modest to good yields. Increasing steric hindrance on the arene moiety leads to decreased product yields. An isomeric mixture of products is obtained, and the selectivity for phenylation of xylenes is similar to the transformations in Scheme 24.9.

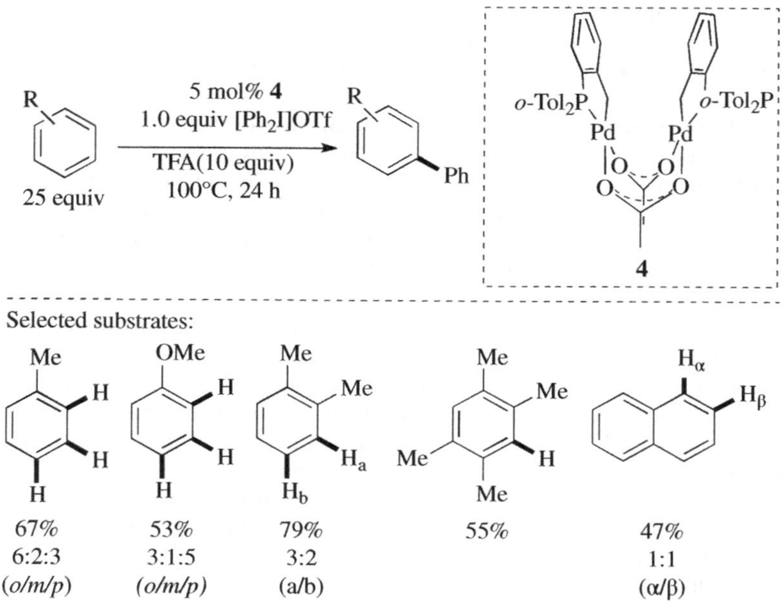

SCHEME 24.11 Direct arylation using catalyst 4.

The palladacyclic complex **4**-catalyzed reactions can be applied toward arylation of *para*-xylene using electronically and sterically varied iodonium salts (Scheme 24.12). The yields of the arylated products are generally higher with electron-deficient oxidants than electron-rich ones. The use of iodonium reagents bearing sterically hindered *ortho*-substituted aryl groups lead to diminished yields of the products. Most interestingly, halide-containing oxidants are well tolerated under the reaction conditions suggesting against a $Pd^{0/II}$ mechanism for these arylations.

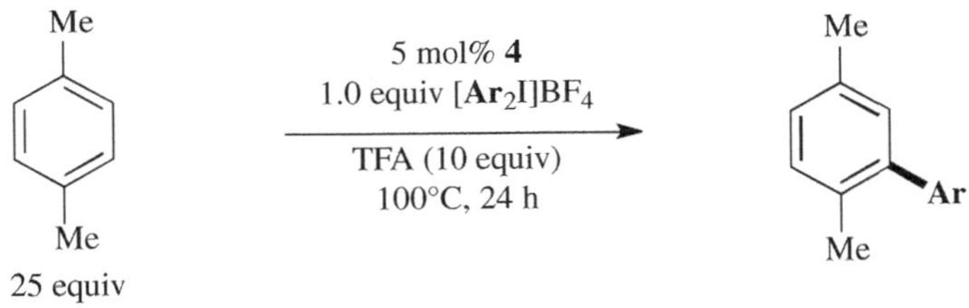

Selected Ar:

p-MeOC_6H_4	o-MeC_6H_4	p-ClC_6H_4	p-BrC_6H_4	p-CO_2MeC_6H_4
42%	19%	99%	56%	53%

SCHEME 24.12 Scope of aryl iodonium salts for 4-catalyzed direct arylation.

One critical drawback of the aforementioned direct arylations is the inability to predictably control the site-selectivity of these reactions. For example, the Pd-catalyzed arylation of naphthalene affords varying ratios of α:β arylated isomers depending on the aryl source and the reaction conditions (Schemes 24.9, 24.10, and 24.11). Catalyst-based control serves as a powerful alternative to systematically modulate the site-selectivities of organometallic transformations. Modification of the steric and electronic nature of the ancillary ligands has the potential to allow the preferential functionalization of the desired C—H bond. However, catalyst-based control remains challenging for Pd-catalyzed C—H functionalizations since these transformations generally employ simple Pd salts that do not contain readily tunable ligands.

Sanford and coworkers initiated the first systematic investigation of the catalyst-controlled site and chemoselectivity for the Pd-catalyzed arylation of unactivated arenes in 2011 [16]. The effect of the electronic and steric nature of the ligands on the site-selectivity of naphthalene phenylation using $[Ph_2I]BF_4$ as the oxidant was examined. Simple Pd^{II} salts are effective for catalyzing the phenylation of naphthalene, with $PdCl_2$ providing the best balance of reactivity and selectivity. As such, $PdCl_2$ complexes bearing different bidentate ligands were further explored. As shown in Scheme 24.13, the diimine-bearing complexes **6–12** lead to higher yield of the desired products than the bipyridine complex **5**. Diimine complexes **6–12** were chosen to explore the effect of the electronic and steric nature of the ligands on the site-selectivity of phenylation. However, as illustrated in Scheme 24.13, no direct correlation between the ligand electronics and the selectivity is observed, suggesting that the ligand effects are not predominantly electronic in nature. Ultimately, the best balance of yield and selectivity is obtained with the 2,6-Cl diimine complex **12** that affords the products in 70% yield as a 71:1 (α:β) mixture of isomers.

The chemoselectivity of these transformations was explored through competition experiments by subjecting a 1:1 mixture of naphthalene and other arenes to substoichiometric Ph_2IBF_4 (0.4 equiv) (Scheme 24.14). The transformation is extremely selective for the phenylation of naphthalene regardless of the electronic nature of the competing arene.

These results suggest that the chemoselectivity-determining step does not proceed *via* a classical electrophilic aromatic substitution. These data together with other mechanistic studies including an intermolecular kinetic isotope effect (KIE) ($k_H/k_D = 1.07 \pm 0.03$) as well as the kinetic order of the C—H

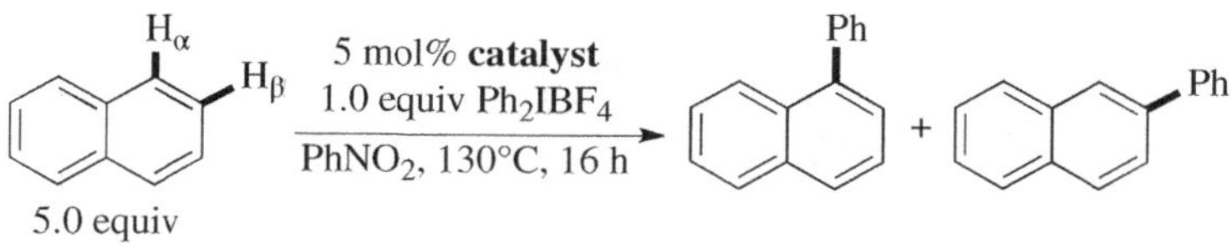

Selected catalysts:

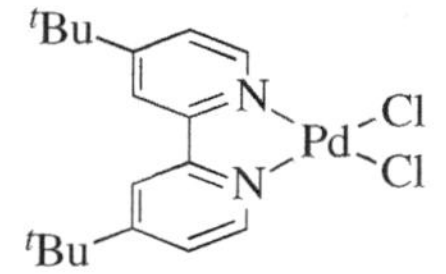

5, 22%, α/β = 78:1

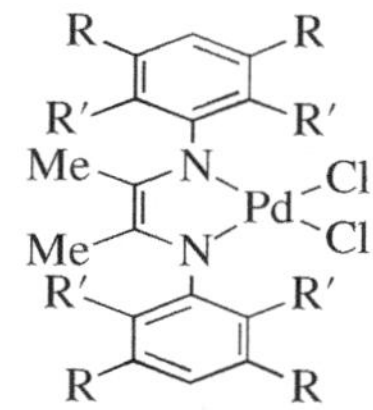

Complex	R	R′	Yield	α/β
6	H	H	54%	9:1
7	Me	H	47%	9:1
8	CF_3	H	50%	10:1
9	^{t}Bu	H	32%	18:1
10	Cl	H	42%	22:1
11	H	Me	43%	27:1
12	H	Cl	70%	71:1

SCHEME 24.13 Ligand investigation for site-selective phenylation of naphthalene.

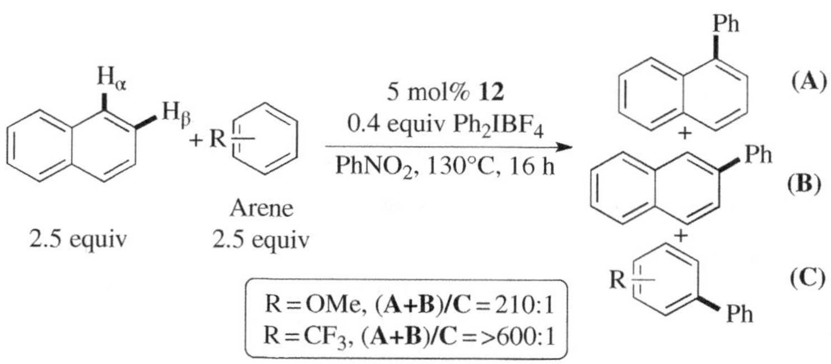

SCHEME 24.14 Electronic effect for phenylation of naphthalene.

arylation in naphthalene (zero order) and Ph_2IBF_4 (first order) are consistent with the mechanism involving a rate-determining oxidation of Pd^{II} to Pd^{IV} prior to C—H activation (Scheme 24.15). The small magnitude of k_H/k_D is proposed to arise due to the π-coordination of the arene to Pd^{IV} prior to C—H bond breaking. This π-complexation is also proposed to be the chemoselectivity-determining step because it explains the high preference for naphthalene functionalization over other arenes since naphthalene is well known to form more stable metal complexes than benzene derivatives.

Consistent with a turnover limiting oxidation, electron-deficient and electron-neutral iodonium salts provide the desired arylated products in higher yields than corresponding electron-rich iodonium salts (Scheme 24.16). Collectively, these results exemplify that both reactivity and selectivity in Pd^{IV}-mediated C—H arylation can be tuned through modification of the supporting ligand structures.

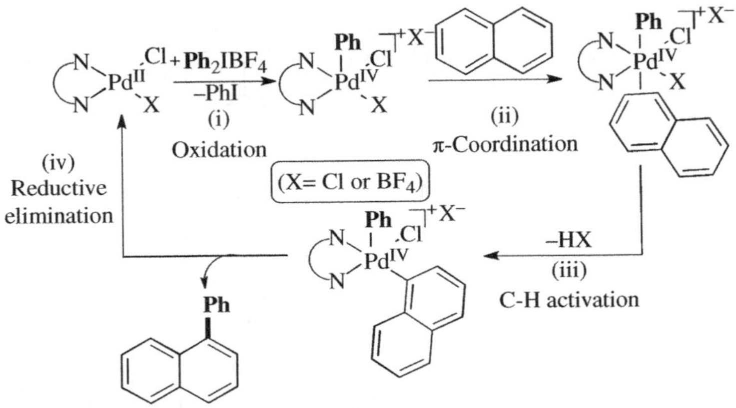

SCHEME 24.15 Mechanistic proposal for direct arylation using aryliodonium salts.

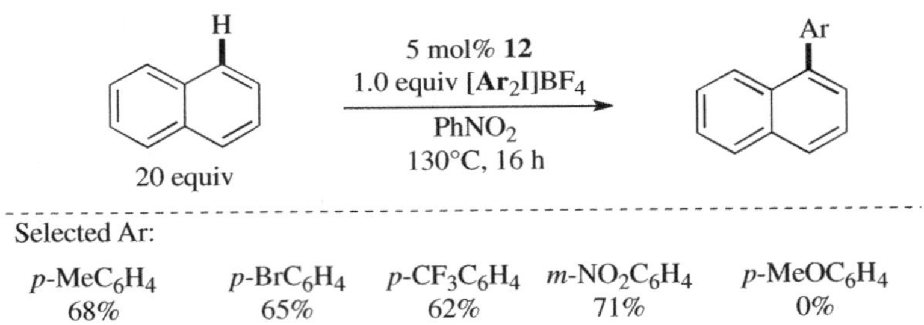

SCHEME 24.16 Scope of aryl iodonium salts for direct arylation of naphthalene.

24.2.2 Cross-Dehydrogenative Arylations

The aforementioned direct arylations involve the coupling of an arene with a prefunctionalized reagent (e.g., Ar—X, Ar—M etc.). A more attractive strategy for biaryl synthesis involves the direct coupling of two aromatic C—H bonds because it precludes the formation of stoichiometric amounts of halogenated or organometallic by-products [9]. The CDC reactions are particularly challenging because both the chemo- and site-selectivity of arene C—H activation and C—C couplings need to be controlled to avoid forming mixtures of isomeric homo- and cross-coupled biaryl products. Despite this challenge, early reports by Fujiwara and Bercaw have inspired the development of several Pd-catalyzed CDC reactions [17]. In 2006, an example of the $Pd(OAc)_2$-catalyzed CDC of two simple arenes using $K_2S_2O_8$ as the terminal oxidant was reported (Scheme 24.17) [18]. The concentration of TFA impacts both the turnover number (TON) and the selectivity of these arylations. For example, the reaction of anisole with excess benzene affords a 1.3:1 ratio of the homo-coupled and cross-coupled products in the presence of 6.3 equivalents of TFA. Significantly higher selectivity for the cross-coupled product is observed with decreased amounts of TFA albeit with diminished efficiency (lower TON). The increased TON for **B** at higher [TFA] is in part due to the enhanced electrophilicity of the Pd^{II} species (presumably bearing a trifluoroacetate ligand) involved in the C—H activation of the limiting arene substrate. The formation of homo- and cross-coupled products is attributed to the similar reactivity of the two arenes (benzene and anisole) in the C—H activation step.

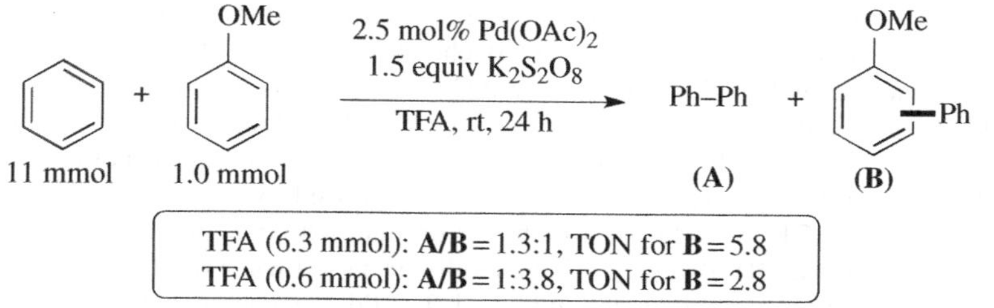

SCHEME 24.17 Effect of TFA in CDC reactions.

24.2.2.1 Strategies and Mechanisms for Cross-Dehydrogenative Arylations In order to suppress homo-coupling, two different strategies have been used in CDC reactions. The first involves the use of two electronically differentiated arenes (**Strategy I**, Scheme 24.18). The second strategy involves the use of a directing group on one of the two arenes (**Strategy II**, Scheme 24.19) [9, 19]. Biaryl syntheses using **Strategy I** generally involve the coupling of a heteroaromatic arene with a nonheteroaromatic arene. In these systems, the selective formation of the cross-coupled products requires the catalyst to react with one arene (e.g., heteroarene, Ar—H in Scheme 24.18) in the first step of the mechanism and then invert its selectivity in the second step to react exclusively with the other arene (Ar′—H, Scheme 24.18).

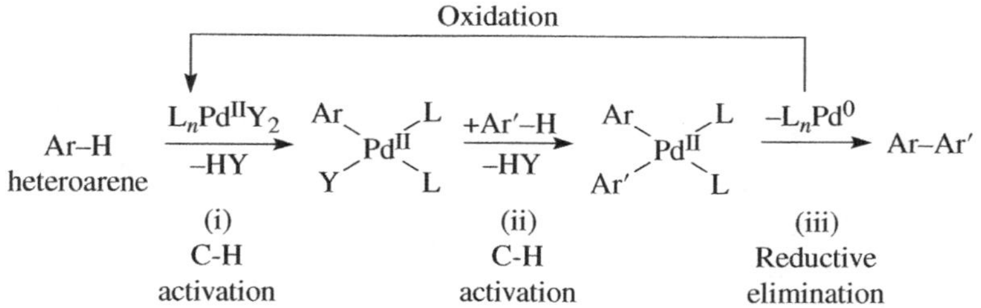

SCHEME 24.18 General mechanism for CDC using **Strategy I**.

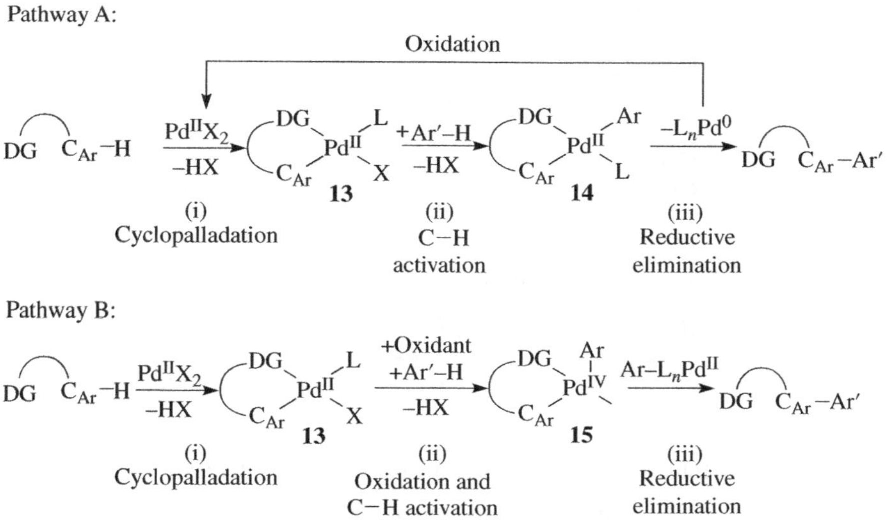

SCHEME 24.19 General mechanism for CDC reactions using **Strategy II**.

A general mechanism for the CDC of two arenes using **Strategy II** is depicted in Scheme 24.19 (pathway A). It involves (i) chelate-directed C—H activation to afford the cyclopalladated complex **13**, (ii) nondirected C—H activation of arene Ar′—H by the Pd complex **13**, (iii) reductive elimination to release the biaryl product, and (iv) oxidation of Pd^0 to regenerate the catalyst. Alternatively, intermediate **13** can undergo oxidation to a Pd^{IV} complex. Subsequent C—H activation of Ar′H at Pd^{IV} followed by C—C bond forming reductive elimination from **15** leads to the desired biaryl product with concomitant regeneration of the Pd^{II} catalyst (pathway B). Pathway A is generally proposed when oxidants such as benzoquinone, Ag_2CO_3, $Cu(OAc)_2$, and O_2 are employed. The $Pd^{II/IV}$ mechanism is generally thought to be operative in the presence of persulfate, electrophilic F^+, or hypervalent iodine reagents as the terminal oxidants (pathway B) [9, 19].

24.2.2.2 Cross-Dehydrogenative Arylations Using Strategy I The successful implementation of **Strategy I** was first demonstrated toward the cross-coupling of indoles with simple arenes [20]. The products are obtained as a mixture of isomers with modest to good selectivity for the preferential functionalization at the C-3 position of the indole (Scheme 24.20). Interestingly, the oxidants play an instrumental role in controlling the site-selectivities of these arylations [17].

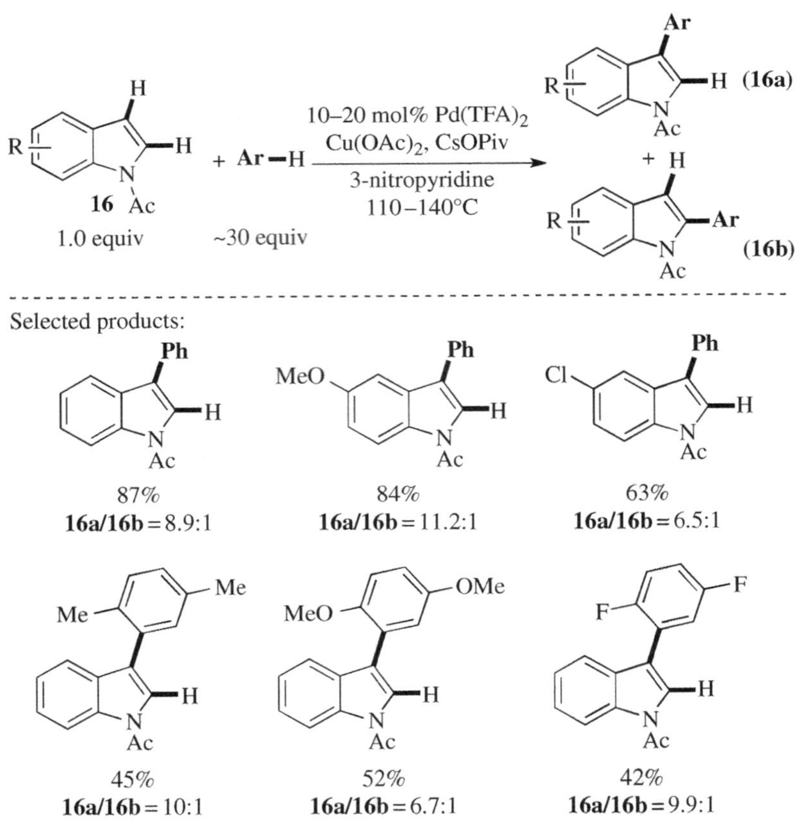

SCHEME 24.20 CDC of indoles with arenes.

The use of $Cu(OAc)_2$ and AgOAc as oxidants affords C-3 and C-2 arylated isomers, respectively, as the major products (Schemes 24.20 and 24.21). The *N*-protecting group also influences the site-selectivity of indole arylations. While C-3 arylated products are preferentially formed using *N*-acetyl indoles (Scheme 24.21), the C-2 isomer is the major product using indoles bearing SEM, benzyl, MOM, Me, and tosyl protecting groups (Scheme 24.22) [21].

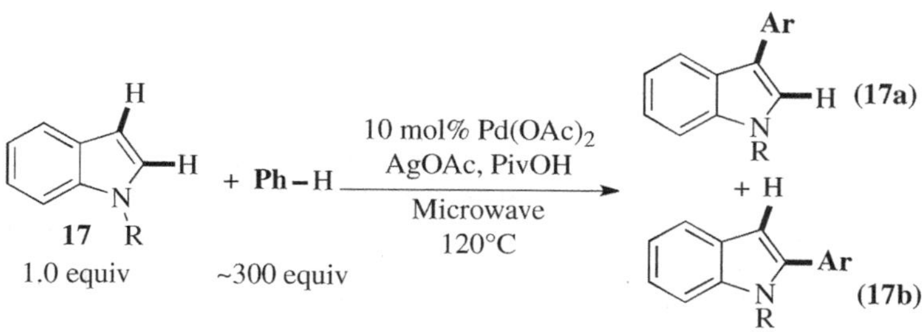

Results with selected *N*-protecting groups, R:

R = SEM
69%
17a/17b = 1:3

R = Bn
57%
17a/17b = 1:4

R = MOM
56%
17a/17b = 1:4

R = Me
31%
17a/17b = 1:4

R = Ts
49%
only isomer **17b**

SCHEME 24.21 Site-selectivity of CDC between indoles and arenes.

The Stahl group demonstrated the catalyst-controlled reactivity and selectivity for the arylation of indoles using O_2 as the oxidant (Scheme 24.22) [22]. A number of nitrogen ligands were evaluated for the reaction of *N*-pivaloyl indole **18** with benzene. The yield and the C-2/C-3 selectivity are both affected by the nature of the ligands. The diazafluorene ligand **22** affords the highest yield of the

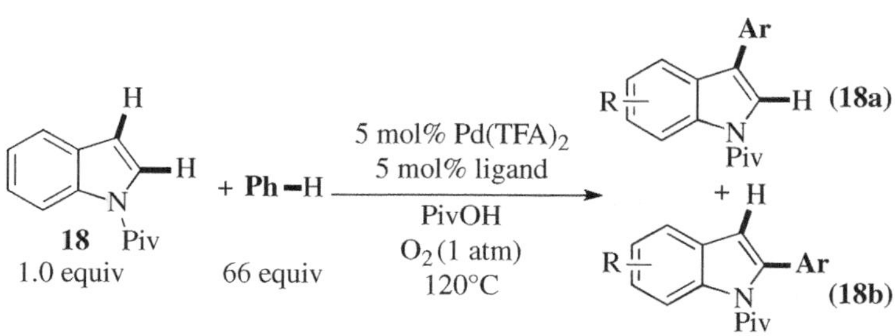

Results with selected ligands:

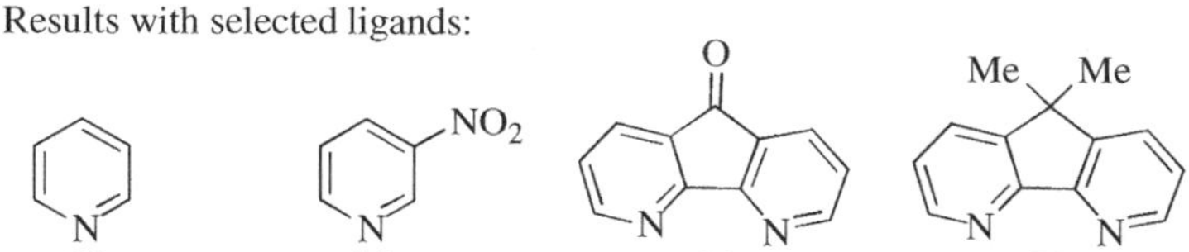

19
22%
18a/18b = 1:1

20
31%
18a/18b = 1:2.4

21
45%
18a/18b = 1.7:1

22
66%
18a/18b = 3.8:1

SCHEME 24.22 Ligand effect on the CDC between indoles and arenes.

products *via* preferential functionalization at the C-3 position of **18**. Since these initial reports on the CDC of indoles with arenes, several other heteroaromatic substrates (e.g., benzofuran, pyrrole, thiophene, imidazo[1,2-a]pyridine, and xanthines) have been effectively coupled with simple arenes [23].

24.2.2.3 Cross-Dehydrogenative Arylations Using Strategy II CDCs using **Strategy II** have been accomplished most commonly using substrates bearing amide-directing groups (Scheme 24.23). In 2008, the Shi group disclosed an example of the Pd-catalyzed coupling of acetanilides with arenes using O_2 as the oxidant [24]. Mono-, di-, and trisubstituted arenes couple with a variety of acetanilides to afford the biaryl products in good yields. In general, the least sterically hindered C—H bond of the arene couples with the *ortho* C—H bond on the anilide to afford the biaryl products. For example, the reaction of disubstituted arenes affords a single isomer of the product. However, mixtures of *meta*- and *para*-functionalized biaryl products are obtained from reactions using electron-rich and electron-deficient monosubstituted arenes. Interestingly, *ortho*-isomers of the product are not observed for these CDCs with monosubstituted arenes suggesting that these arylations are very sensitive to the steric environment of the C—H bond in the nondirected C—H activation step (Scheme 24.19, pathway A, step ii).

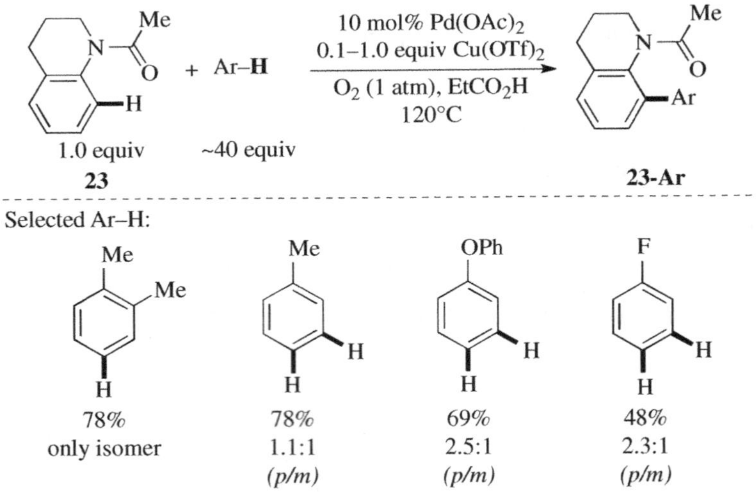

SCHEME 24.23 Representative scope for CDC between 23 and arenes.

Following the report described in Scheme 24.23, Buchwald group published an example for the aerobic CDC of anilides with arenes (Scheme 24.24) [25]. The use of TFA is critical for the efficiency of these arylations. Unlike the reactions in Scheme 24.23, the CDC of anilides **24** with monosubstituted arenes leads to small amounts of the *ortho*-biaryl product along with the *meta*- and *para*-isomers.

The CDC of *O*-phenyl carbamates with arenes using $Na_2S_2O_8$ in place of O_2 was described in 2011 (Scheme 24.25) [26]. The scope and selectivity of these arylations are similar to those described for transformations in Scheme 24.24. Preliminary studies were conducted to gain insight into the mechanism of the undirected C—H activation (step ii, pathway A, Scheme 24.19). A competition reaction of **26** with an equimolar amount of C_6H_6 and 1,2-dichlorobenzene leads to biaryl **26a** in 90% yield (Scheme 24.26). These results suggest that electron-rich arenes undergo arylation at a faster rate than electron-neutral arenes. The authors propose an electrophilic aromatic

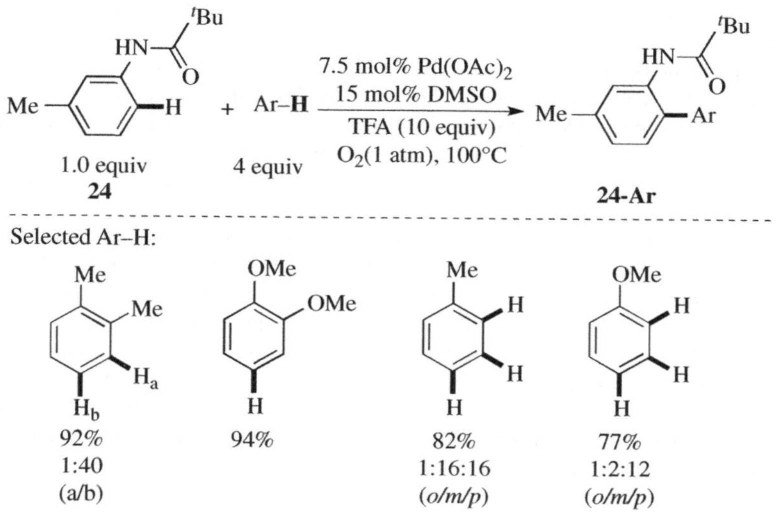

SCHEME 24.24 Representative scope for CDC between 24 and arenes.

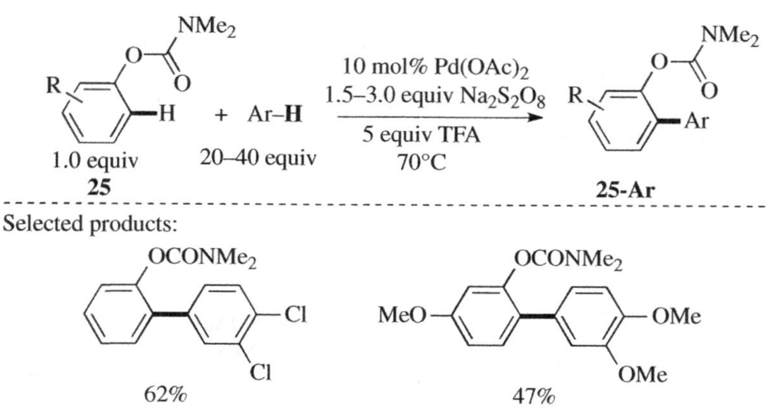

SCHEME 24.25 Representative scope for CDC between 25 and arenes.

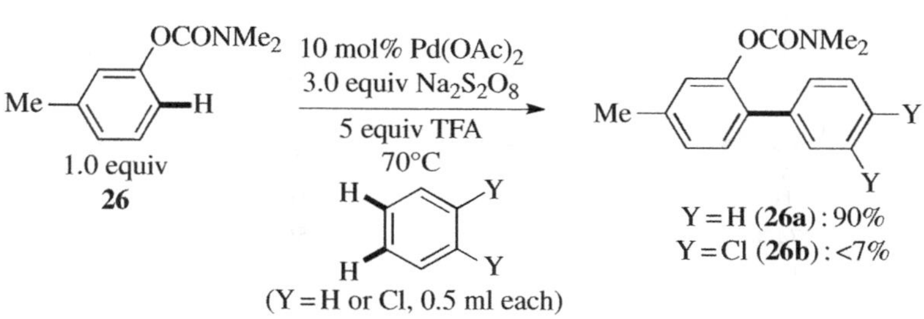

SCHEME 24.26 Electronic effect for CDC between 26 and arenes.

substitution mechanism for the nondirected C—H activation of the arene (Scheme 24.19, pathway A, step ii).

Yu group demonstrated the first example of highly *para*-selective CDC with monosubstituted arenes as coupling partners (Scheme 24.27) [27]. The Pd-catalyzed reaction of benzamides with arenes using *N*-fluorobenzenesulfonimide (NFSI) as the oxidant affords the desired products in good yields and high *para*-selectivity. The use of a F^+ oxidant is critical for the observed *para*-selectivity. The use of other oxidants such as $K_2S_2O_8$ affords the products with significantly attenuated preference for the *para*-isomer.

The abovementioned CDC reactions employ amide substrates as the coupling partner undergoing directed C—H activation. Sanford group demonstrated the use of pyridine directing groups for CDC reactions (Scheme 24.28). Notably, these transformations were the first successful application of **Strategy II** toward CDCs. The Pd-catalyzed coupling of benzo[*h*]quinoline (bzq) with arenes (~65–100 equiv) proceeds to afford the biaryl products in good yields using Ag_2CO_3 as the oxidant. The use of both Ag_2CO_3 and benzoquinone (BQ) is important for the efficiency of these CDCs. Analogous to the amide-directed reactions described previously, these transformations also generally exhibit high selectivity for the arylation of less sterically hindered C—H bonds of arenes [28]. However, reaction of nitrobenzene affords the product via functionalization of the *meta*-C—H bond.

As described earlier, significant progress has been made toward CDC reactions of two arenes. Remarkably, these transformations exhibit high selectivities for cross-coupling, and little or no homo-coupling of arenes is observed. Despite these advances, the synthetic applicability of these CDCs is limited due to the inability to predict and control the site-selectivity of these reactions. The functionalization of the same arene affords the biaryl product in significantly different *o*/*m*/*p* ratios depending on the reaction conditions employed. For example, the Pd-catalyzed coupling of anilide **27** with anisole leads to *para*-functionalized product selectively (Scheme 24.27). In contrast, the Pd-catalyzed arylation of benzo[*h*]quinoline **28** with anisole affords a 1:2.6:3.3 mixture of *o*/*m*/*p* isomers of the product (Scheme 24.28).

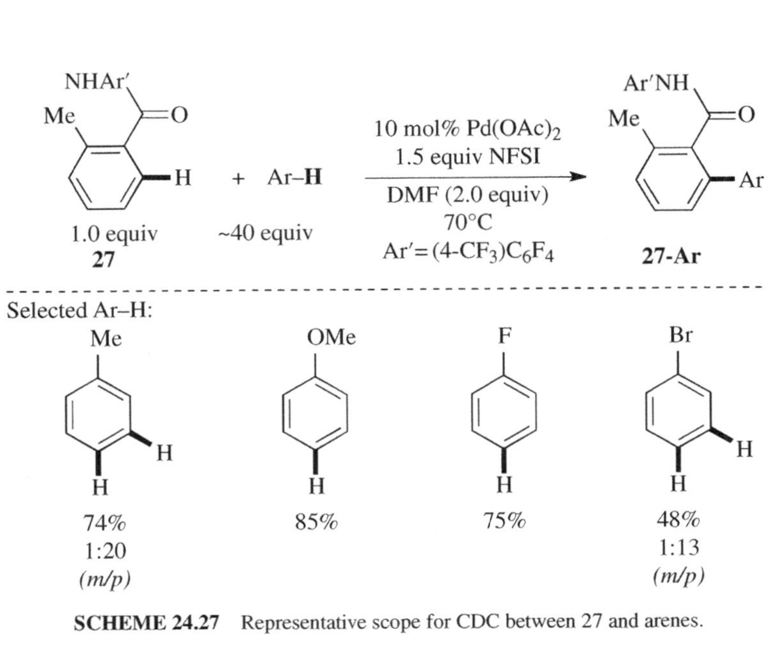

SCHEME 24.27 Representative scope for CDC between 27 and arenes.

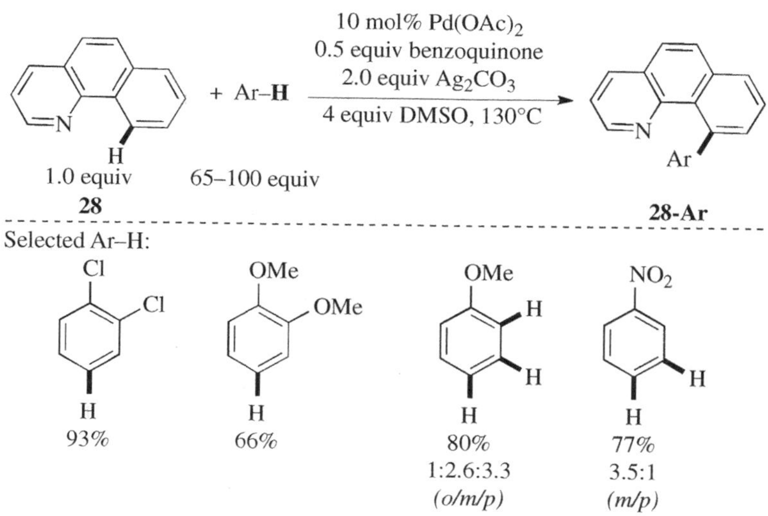

SCHEME 24.28 Representative scope for CDC between 28 and arenes.

Sanford and coworkers reported a systematic study toward modulating the site-selectivity of cross-dehydrogenative arylation by tuning the electronics and sterics of the ancillary ligands at the Pd center [25a–c]. These investigations are guided by careful consideration of the mechanism of these reactions. A thorough investigation of the Pd-catalyzed oxidative cross-coupling of benzo[*h*] quinoline (bzq) with arenes promoted by BQ in the presence of a Ag(I) oxidant was conducted. Based on kinetic and stoichiometric experiments, the proposed mechanism of the reaction involves (i) cyclopalladation of bzq to generate **29**, (ii) reversible activation of the Ar–H bond, (iii) BQ complexation, (iv) C–C bond forming reductive elimination to release the product, and (iv) finally reoxidation of Pd^0 by Ag(I) to regenerate the Pd(II) catalyst. Interestingly, the rate-determining step (r.d.s) is dependent upon the concentration of the quinone promoter. While at low BQ concentrations quinone complexation is rate determining (**regime** 1), at high BQ concentrations the C–H activation of ArH is turnover limiting (**regime** 2) (Scheme 24.29).

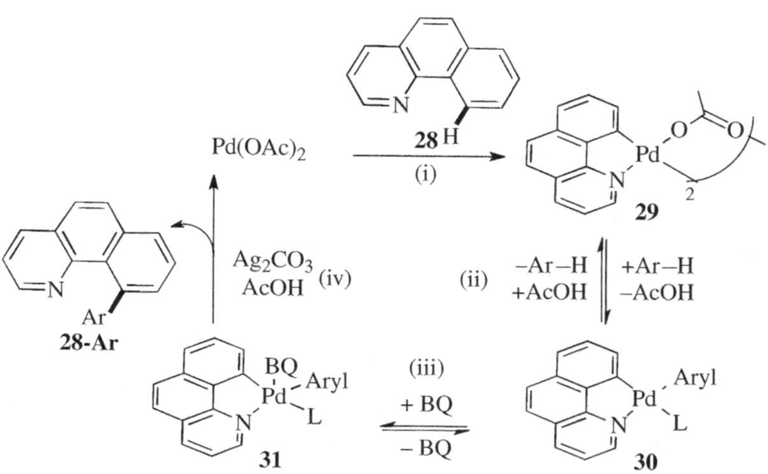

SCHEME 24.29 Mechanistic proposal for CDC between 28 and arenes.

Based on these data, the authors hypothesize that the change in the r.d.s might also lead to a change in the selectivity. It is reasoned that the modification of the ancillary ligands at Pd intermediates **29–31** might allow catalyst-controlled selectivity for the functionalization of C(Ar)–H. To corroborate this hypothesis, a systematic investigation of the site-selectivity for the stoichiometric arylation of palladacycle **29** with 1,3-dimethoxybenzene was conducted for both **regimes 1** and **2** as a function of the quinone and the anionic ligand at Pd. As shown in Scheme 24.30, the selectivity of the reaction is highly dependent on both the concentration and the nature of the quinone. The ratio of isomeric coupling products **28a**:**28b** range from 5:1 to 1:1.7 with 1 and 40 equivalences of BQ respectively. Furthermore, in **regime** 1 (low [BQ]), the quinone substitution has a dramatic effect on the selectivity with **28a**:**28b** increasing from 5:1 with BQ (**32**) to 14:1 with 2,3,5,6-tetramethylbenzoquinone (**33**). The reaction is more selective with *para*-substituted 2,5-diarylquinones bearing electron-donating substituents. For example, the reaction using **35** leads to a 1.1:1 ratio of **28a**:**28b**, while the use of **34** affords a 5.7:1 ratio of **28a**:**28b**. Significantly decreased selectivities are observed for regime 2 (high [BQ]) with all quinones investigated.

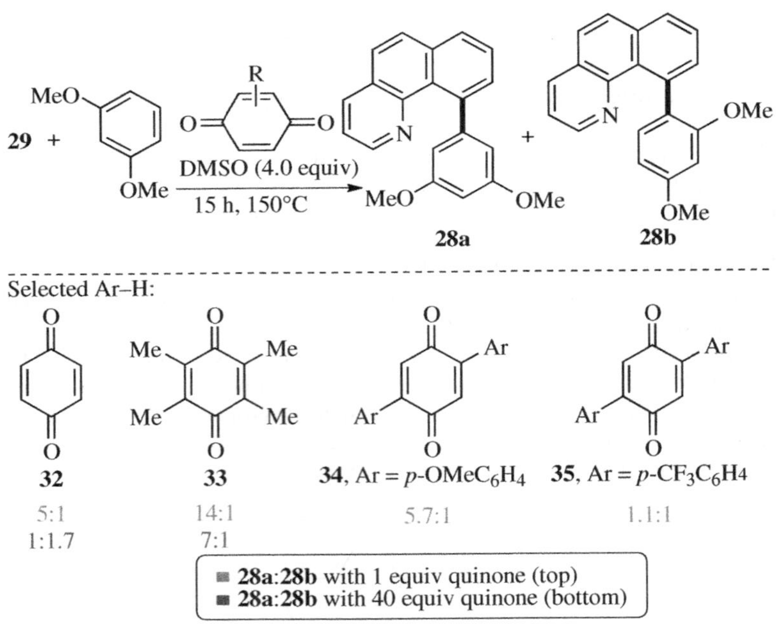

SCHEME 24.30 Effect of benzoquinone structure on CDC between 28 and 1,3-dimethoxybenzene.

The influence of the electronic and steric characteristics of the anionic ligands at Pd on the ratio of **28a**:**28b** was also explored (Scheme 24.31). Modest steric (entries 1 and 2) and electronic (entries 3–5) effects on **28a**:**28b** are observed in both regimes 1 and 2. Interestingly, changing the anionic ligand from carboxylate to carbonate leads to a complete reversal in selectivity **for regime** 1 (entry 1 vs. 6). This reversal of selectivity is general for the coupling of the cyclometallated complex with various simple arenes. Computational studies have been undertaken to understand this selectivity change between the carbonate and the acetate systems [25d]. Remarkably, the insights gained from these investigations have enabled the rational design of experiments to predictably modulate the site-selectivity in these transformations. Overall, these studies demonstrate that detailed mechanistic explorations can serve as a powerful platform to predictably control the site-selectivities of arene functionalizations.

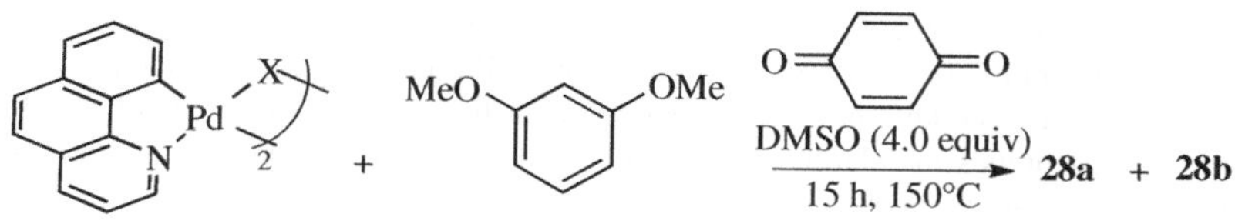

Selected anionic ligands, X:

Entry	X	28a:28b regime 1	28a:28b regime 2
1	OAc	5:1	1:1.5
2	OPiv	9:1	1.8:1
3	$OCO(p\text{-}OMeC_6H_4)$	14:1	1:1.7
4	$OCO(C_6H_5)$	10:1	1:2
5	$OCO(p\text{-}CF_3C_6H_4)$	11:1	1:2.9
6	CO_3^{2-}	1:6	n.d.

28a:**28b** with 1 equiv benzoquinone (regime 1)
28a:**28b** with 20 equiv benzoquinone (regime 2)

SCHEME 24.31 Effect of anionic ligand structure on CDC between **28** and 1,3-dimethoxybenzene.

24.3 ALKENYLATION

Since its initial discovery by Fujiwara [29], several reports on the Pd-catalyzed olefination of simple arenes have been published [30]. Most early examples were limited to the reaction between alkenes and electron-rich arenes and resulted in a mixture of *ortho*-, *meta*-, and *para*-olefinated products with monosubstituted arene substrates. In 2009, the Yu group reported the first example of a *meta*-selective olefination of highly electron-deficient arenes using O_2 as the terminal oxidant (Scheme 24.32) [31]. This report is a significant advance over previous alkenylations that were

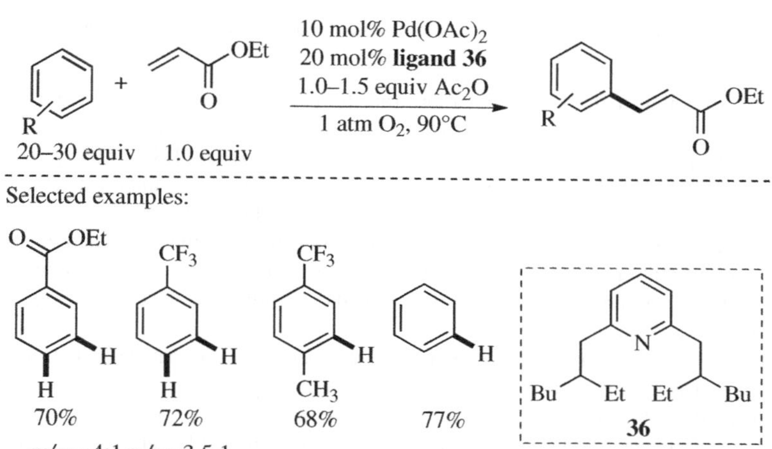

SCHEME 24.32 Representative examples for aerobic alkenylation of arenes.

primarily limited to the functionalization of electron-neutral and electron-rich arenes. The key to this success is the employment of **36** as a ligand for the Pd^{II} catalyst. As depicted in Scheme 24.32, the optimized reaction conditions can be applied toward the coupling of a variety of monosubstituted electron-deficient arenes and olefin substrates to afford the products as a mixture of *meta*- and *para*-isomers. The *meta*-isomer is the major product in all cases.

Detailed computational studies were conducted to gain further insight into the overall mechanism of these alkenylations [32]. Additionally, these investigations assisted in elucidating the role of the pyridine ligand (**36**) and the origin of the observed *meta*-selectivity. The proposed mechanism of the olefination involves (i) rate-determining arene C—H activation via a concerted metallation/deprotonation (CMD) pathway to afford **37**, (ii) displacement of the pyridine ligand by the olefin, (iii) insertion of the alkene into the Pd—C bond, (iv) β-hydride elimination, (v) product dissociation, and (vi) finally catalyst regeneration (Scheme 24.33).

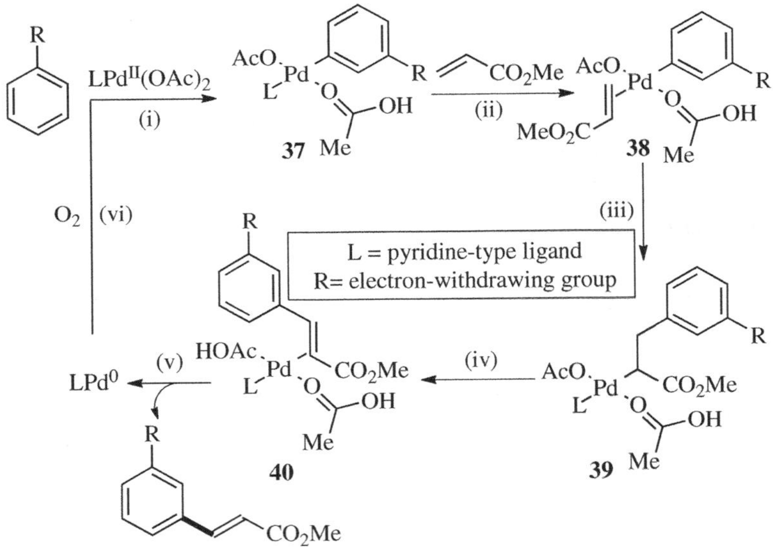

SCHEME 24.33 Mechanistic proposal for aerobic alkenylation of arenes.

An investigation of the C—H activation transition states for electron-deficient aromatic substrates revealed that the transition state ($T.S._{meta}$) for *meta*-C—H bond activation has a lower energy than the corresponding transition states for activation of the *ortho*- ($T.S._{ortho}$) and *para*-C—H bonds ($T.S._{para}$). The higher energy of $T.S._{ortho}$ (vs. $T.S._{meta}$ and $T.S._{para}$) is manifested in the catalytic reactions, which only afford *meta*- and *para*-alkenylated products. The absence of *ortho*-substituted products is attributed to the unfavorable steric interactions between the ligand and the substituent on the arene ring. Additionally, the predominant formation of the *meta*-olefinated products is proposed to arise from the preferential functionalization of the slightly more electron-rich *meta* (vs. *para)* C—H bond in the electron-poor substrates.

As mentioned previously, the use of **36** is critical to obtain the olefinated products in synthetically useful yields. The use of other structurally similar pyridine ligands (e.g., 41–43) lead to significantly diminished yields of the desired products (Scheme 24.34).

The greater efficiency of **36** than **41–43** is attributed to a number of ligand characteristics. First, the steric bulkiness of the ligand should be sufficient to allow the formation of mono-ligated ligand~Pd complexes, yet attenuated enough to ensure the binding of the ligand to the Pd center. Importantly, the

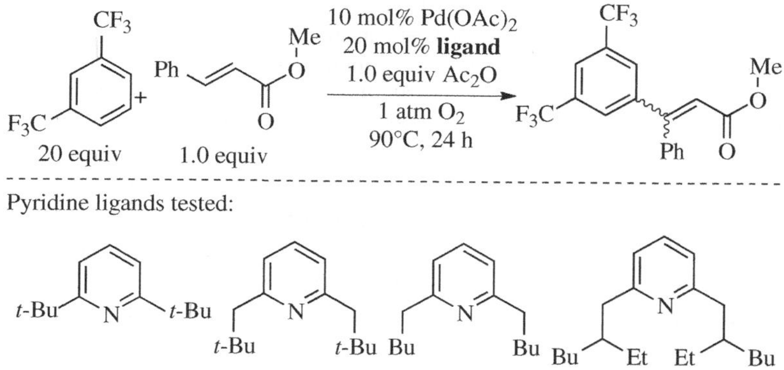

41	42	43	36
<2%	10%	24%	52%
		$E/Z = 7.8$	$E/Z = 5.9$

SCHEME 24.34 Ligand effects for the aerobic alkenylation of arenes.

formation of mono- (vs. poly-) ligated complexes is critical toward providing a coordination site for the arene, a step that is necessary for the C—H activation step of the proposed mechanism. Second, the electron-deficient character of the pyridine-type ligands is desirable to render the Pd center sufficiently electrophilic for arene C—H activation. Third, the labile nature of the ligand allows for the replacement of itself with the olefin substrate. Finally, the pyridine ligand promotes the release of the product and the regeneration of the catalyst. The result in Scheme 24.34 suggests that **36** is most effective in achieving the optimal balance of sterics and electronics to affect the desired arene olefination.

Subsequent to the report by the Yu group, Sanford and coworkers demonstrated the Pd-catalyzed alkenylation of arenes using *tert*-butyl peroxy benzoate as the terminal oxidant [33]. The coupling reaction proceeds to afford the desired products using commercially available pyridine ligands. The Pd/pyr ratio affects both the rate and efficiency of the transformation. A 1:1 Pd/pyr ratio accelerates the rate of the reaction, while a 1:2 Pd/pyr retards the rate of the reaction. Further detailed studies revealed that Pd/pyr ratios of 1:0.5 to ~1:1 afford the highest yields of the product. A systematic screening of other pyridine ligands revealed that 3,5-dichloropyridine (**45**) leads to the highest yields of the product (Scheme 24.35). Interestingly, the most efficient ligand **36** (Scheme 24.34)

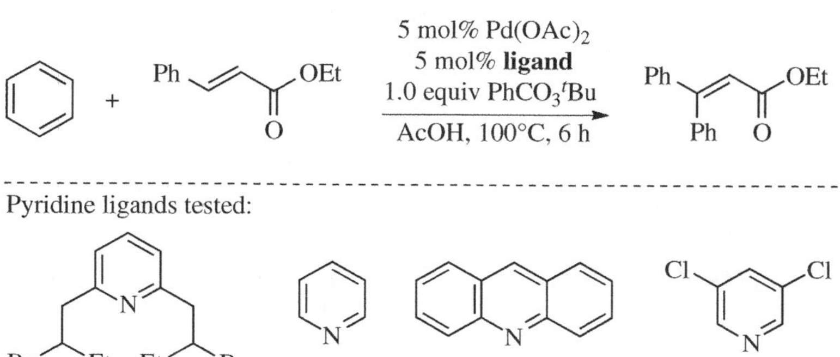

36	19	44	45
61%	76%	80%	81%

SCHEME 24.35 Ligand effects for arene alkenylation using $PhCO_3{}^tBu$.

used for aerobic olefinations by Yu is less effective than pyridine for the reactions in Scheme 24.35, suggesting that the ligand effects in arene C—H functionalization can be oxidant dependent.

The optimal reaction conditions enable the coupling of both electron-rich and electron-deficient arenes (Scheme 24.36). In contrast to Yu report, the olefination of electron-deficient arenes affords small amounts of the *ortho*-functionalized products along with the *meta* and *para*-isomers.

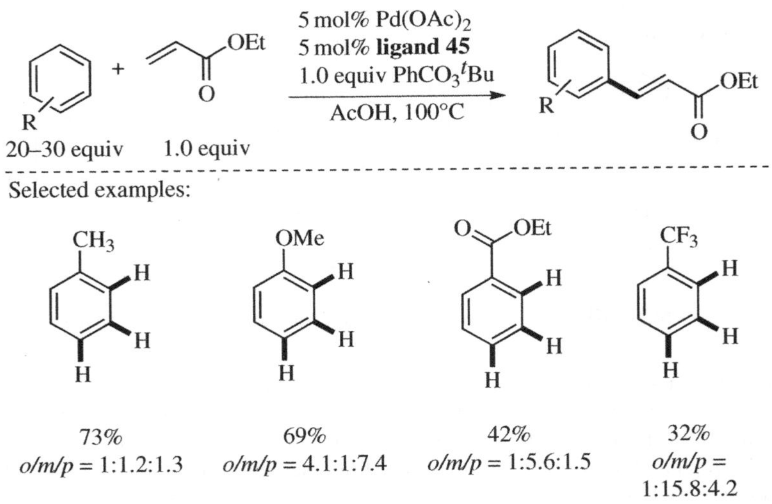

SCHEME 24.36 Representative scope for arene alkenylations using 45 as ligand.

Duan and coworkers recently reported the use of a bidentate ligand for Pd-catalyzed alkenylation of arenes using $Cu(OAc)_2$ as the stoichiometric oxidant (Scheme 24.37) [34]. Specifically, the $Pd(OAc)_2$/**46** catalyst system is effective for the coupling of arenes with mono- and disubstituted alkenes. A variety of electronically differentiated monosubstituted arenes couple with *n*-butyl acrylate to afford the products in good to excellent yields. In analogy to the aforementioned alkenylations, the *meta*-substituted alkenes are

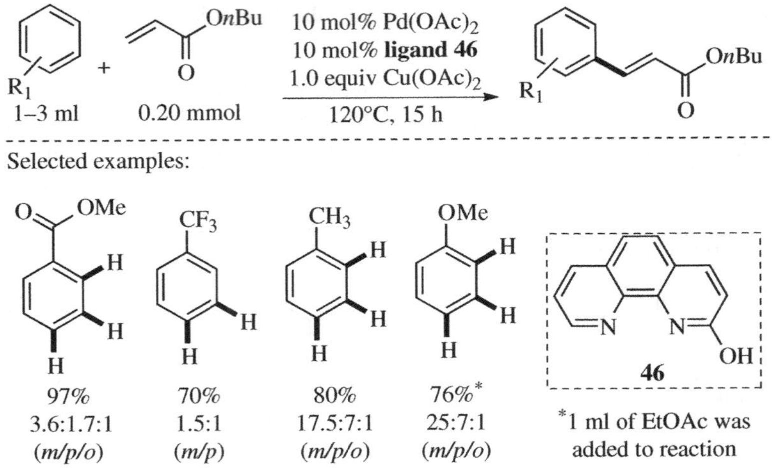

SCHEME 24.37 Representative scope of arenes for Pd-catalyzed alkenylation using ligand **46**.

the major products in the reaction of electron-deficient arene substrates. Interestingly however, in contrast to previous reports, the reaction of electron-rich arenes, such as toluene and anisole, also proceed *via* the preferential functionalization of the *meta*-C—H bond. These alkenylations can also be conducted using substoichiometric $Cu(OAc)_2$ in the presence of O_2 as the terminal oxidant.

The site-selectivity of these transformations is highly dependent on the ligand (Scheme 24.38). For example, the use of **46** and **47** leads to the *meta*-isomer of the product predominantly. In contrast, the *ortho*-isomer of the product is formed preferentially when **49–51** are employed as ligands.

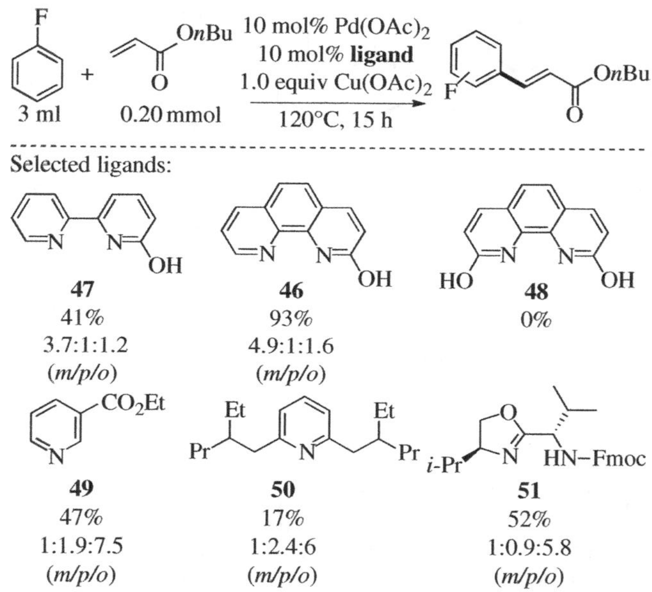

SCHEME 24.38 Ligand survey for Pd-catalyzed alkenylation of fluorobenzene.

Bäckvall group accomplished the Pd-catalyzed aerobic synthesis of cinnamyl derivatives *via* dehydrogenative coupling of arenes with allyl esters (Scheme 24.39) [35]. The use of substoichiometric benzoquinone (BQ) and iron phthalocyanine (Fe(Pc)) cooxidants is essential for the facile reoxidation of Pd^0 to Pd^{II} (similar to step vi, Scheme 24.33), thereby enabling catalyst turnover. The coupling of various mono- and disubstituted arenes with allyl acetate proceeds to afford the cinnamyl product in modest yields. The reaction exhibits high *E*/*Z* and linear/branched selectivity favoring the *E*-isomer of the linear product. However, modest site-selectivity is observed for the C—H activation of the arene substrate as exemplified by the formation of mixtures of isomeric products. Electron-deficient and electron-neutral arenes are less reactive than electron-rich arenes and require higher Pd catalyst loadings suggesting an electrophilic palladation pathway for the C—H activation step. Additionally, the significant primary K.I.E (k_H/k_D=4.2) implicates that C—H activation is the rate-determining step in these transformations.

Pd-catalyzed oxidative cross-coupling of arenes with chromanones was reported by the Hong group [36]. These reactions proceed through *in situ* formation of the chromones via Pd-catalyzed dehydrogenation of the starting chromanones (Scheme 24.40, step i). Subsequent coupling of the alkene moiety in the chromones with the arenes affords the final products.

The transformation is selective for the formation of C-2 substituted chromones and is tolerant toward a number of functional groups including nitro, free hydroxyl, acetyl, triflates, trifluoromethyl, and halogens (Scheme 24.41). The Bäckvall group subsequently reported a similar transformation that enabled the Pd-catalyzed coupling of arenes with cyclic saturated ketones [37].

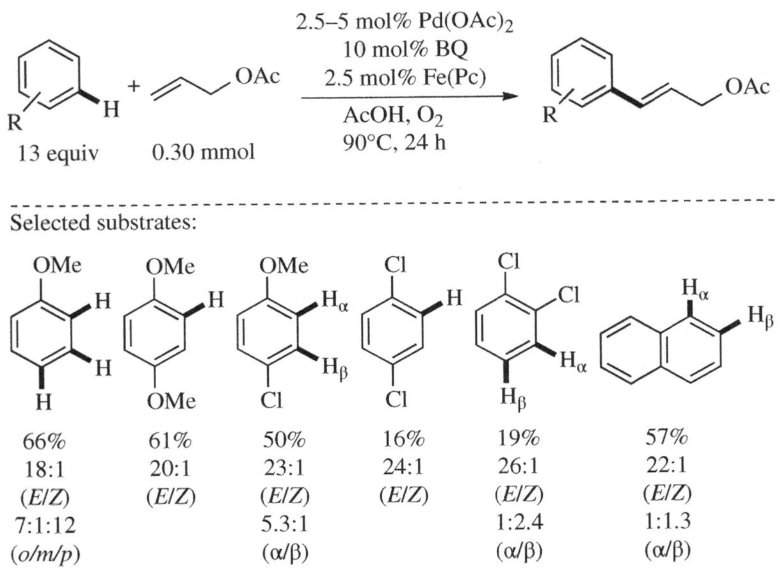

SCHEME 24.39 Representative scope for coupling of arenes with allyl acetate.

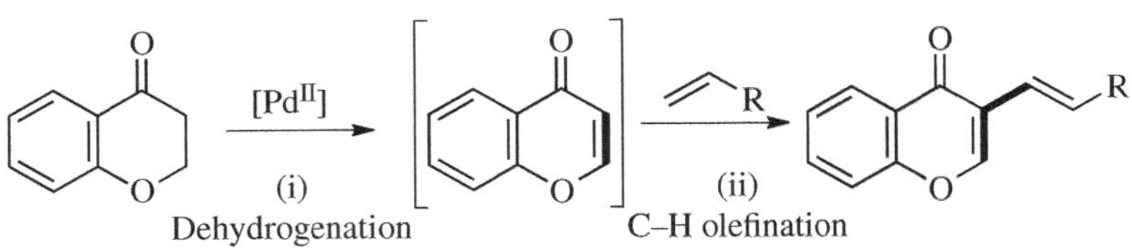

SCHEME 24.40 General pathway for coupling of arenes with chromanones.

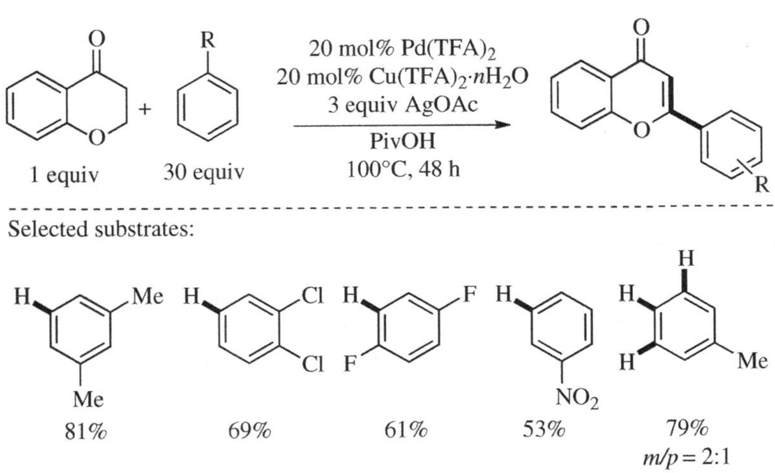

SCHEME 24.41 Pd-catalyzed C-2 arylation of chromones.

24.4 ALKYLATION

Significant progress has been made toward C—H arylations under Pd catalysis. In sharp contrast, methods for the Pd-catalyzed nondirected alkylation of arenes remain sporadically studied. The paucity of methods for alkylation could be partly due to the facile, undesired β-hydride elimination from the Pd–alkyl intermediate formed upon oxidative addition of the alkyl halide into Pd^0 catalysts. In 2009, Wang and coworkers reported the first example of palladium-catalyzed alkylation of highly electron-deficient perylene-3,4:9,10-tetracarboxylic acid bisimides (PBIs) with alkyl bromides and iodides (Scheme 24.42) [38]. The products are obtained in at best modest yields, partly due to the competitive formation of dialkylated PBIs. Most interestingly, these transformations exhibit high selectivity for the alkylation of the *meta*-C—H bonds in PBI. The authors attribute the *meta*-selectivity to the greater reactivity of the *meta*-C—H bond in electron-deficient arenes such as PBI. The proposed mechanism of these alkylations is similar to that of direct arylations using aryl halides.

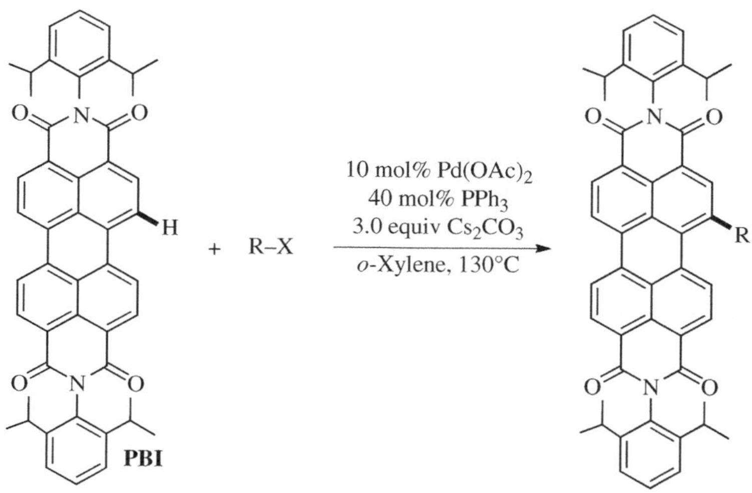

Selected RX:

n-$C_6H_{13}Br$	n-$C_6H_{13}I$	$CF_3CF_2C_2H_4I$	n-$C_4H_9(OCH_2CH_2)_3Br$
41%	55%	22%	28%

SCHEME 24.42 Representative example of arene alkylation.

Michael *et al.* demonstrated an example of alkylation of arenes in the context of Pd-catalyzed carboamination of alkenes [39]. In these systems, the Pd–alkyl intermediate generated upon intramolecular aminopalladation undergoes a subsequent coupling with an arene to afford the final products. The authors propose a $Pd^{II/IV}$ catalytic cycle for these reactions since the carboamination products are not observed in the absence of the electrophilic fluorine oxidant, *N*-fluorobenzenesulfonimide. Furthermore, the carboaminations are tolerant toward aryl bromides suggesting against the intermediacy of Pd^0 species under the reaction conditions. Although mechanistic studies were not conducted, the authors propose two different pathways for the coupling of the arene with the Pd–alkyl intermediate. These pathways are consistent with the higher reactivity of nucleophilic arenes than their electron-deficient counterparts (Scheme 24.43).

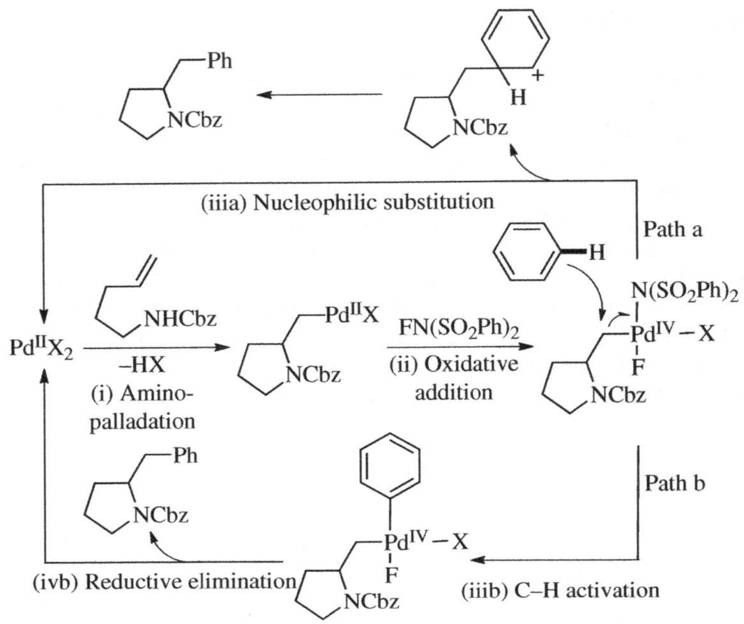

SCHEME 24.43 Mechanistic proposal for carboamination.

Interestingly, the reactions involving monosubstituted arene substrates exhibit high selectivity for the formation of *para*-functionalized products (Scheme 24.44). *Ortho*- and *meta*-isomers of the product are not reported under the optimal conditions. The *para*-selectivity is similar to that observed for the C—H arylations depicted in Scheme 24.27. The electrophilic fluorine oxidant used for both the arylations in Scheme 24.27 and the carboaminations (Scheme 24.44) might be responsible for the high *para*-selectivity.

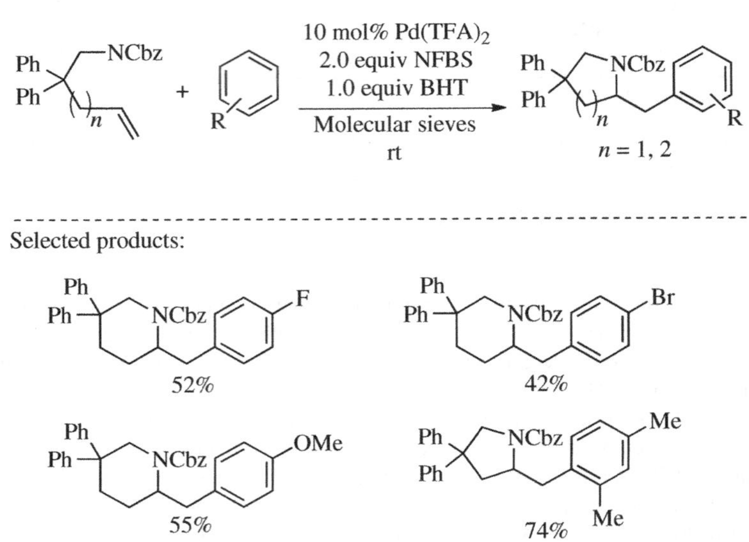

SCHEME 24.44 Representative scope of carboamination.

24.5 CARBOXYLATION

Palladium-catalyzed carboxylation of aromatic C—H bonds is an attractive atom economical route for the conversion of arenes to benzoic acid derivatives. Early reports for arene C—H carboxylations under palladium catalysis employed CO and CO_2 as the carbon source [40]. More recently, alternative carboxylation methods using formic acid in place of CO or CO_2 have been developed to obviate the difficulty associated with handling gaseous reactants. Nozaki and coworkers developed a Pd-catalyzed carboxylation of arenes utilizing formic acid as the carbonyl source and $K_2S_2O_8$ as the oxidant (Scheme 24.45) [41]. The transformation allows the carboxylation of electron-neutral arenes affording benzoic acid products in good yields *via* preferential functionalization of less sterically hindered C—H bonds. However, electron-deficient arenes are inert under these reaction conditions. Notably, the use of **52** as an additive is critical toward obtaining the benzoic acids in good yields. Under otherwise analogous conditions, benzoic acid is formed in 84% and 18% yields with and without **52**, respectively. The authors propose that the phosphenium salt serves as an electron-withdrawing ligand and promotes the Pd^{II}-mediated cleavage of the arene C—H bonds by enhancing the electrophilicity of the Pd catalyst.

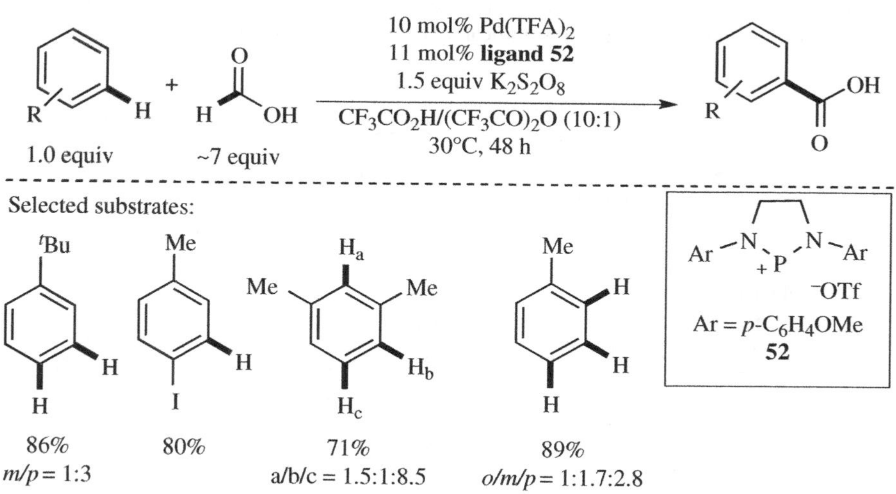

SCHEME 24.45 Representative example of arene carboxylation.

24.6 OXYGENATION

Direct oxygenations of simple arenes have the potential to provide a direct route from arenes to phenols, which serve as commodity chemicals and building blocks for the synthesis of functional materials. The Pd-catalyzed oxygenation of simple arenes has been known since 1971 and is generally thought to proceed *via* $Pd^{II/IV}$ catalytic cycles (Scheme 24.46) [42]. The vast majority of C—H oxygenations afford acetoxylated arenes and are accomplished using Pd^{II} salts in conjunction with strong oxidants (e.g., $PhI(OAc)_2$, $Pb(OAc)_4$, K_2CrO_4, $KMnO_4$, heteropoly acids/O_2). Reagents such as $PhI(OAc)_2$ and $Pb(OAc)_4$ serve as both the oxidant and the acetate source. However, the use of other oxidants (e.g., heteropoly acids/O_2, $K_2S_2O_8$) generally requires the use of alkali metal salts (e.g., KOAc) or solvents (e.g., acetic acid) as external acetate sources.

As mentioned earlier, Pd-catalyzed arene oxygenations have been known for over four decades. However, the vast majority of the early reports are plagued by low TONs and competing biphenyl formation, which often leads to catalyst decomposition [42]. Furthermore, oxygenations of substituted arenes generally afford the desired products as mixtures of *o*/*m*/*p* isomers, which decrease the synthetic

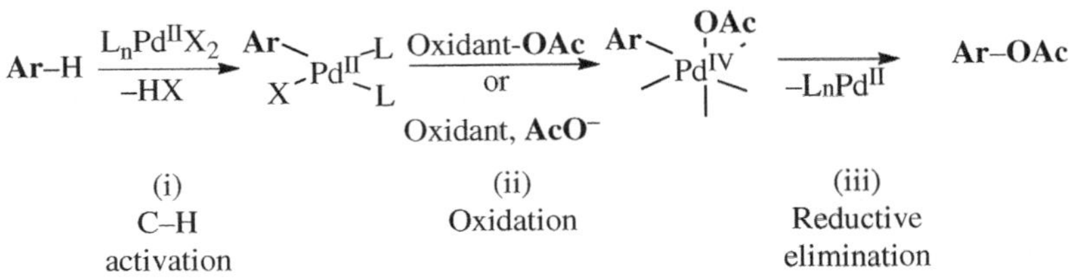

SCHEME 24.46 General mechanism for arene acetoxylation.

applicability of these reactions. In 2011, Sanford and coworkers evaluated the effect of ligands on the reactivity and site-selectivity of the C—H acetoxylation of arenes (Scheme 24.47) [43]. The effect of ligand/metal ratio on the efficiency of $Pd(OAc)_2$-catalyzed acetoxylation of arenes using $PhI(OAc)_2$ as the terminal oxidant was explored. These studies revealed that Pd–ligand ratios of 1:0.5 to 1:1.3 lead to the highest rates of these oxygenations. A TON of 4756 is achieved for the acetoxylation of benzene using 0.01 mol% $Pd(OAc)_2$/0.009 mol% pyridine after 306 h at 100°C. The increased reactivity of the ~1:1 Pd/pyr system is general for a variety of arenes (Scheme 24.47). Notably, the absence of pyridine or the use of 1:2 Pd/pyridine leads to significantly reduced yields of the desired products.

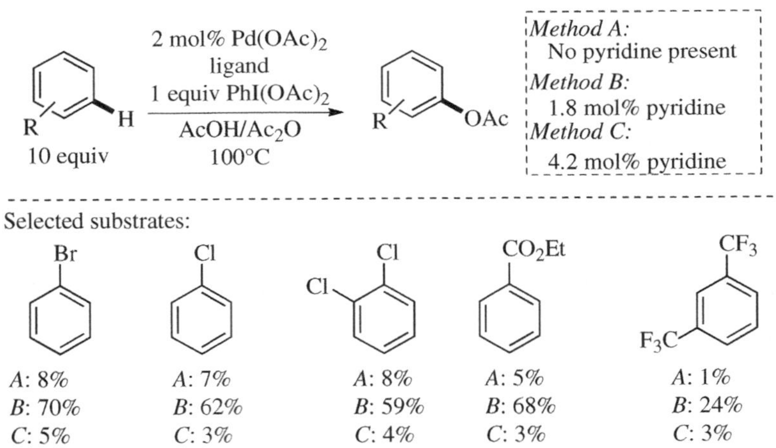

SCHEME 24.47 Effect of pyridine equivalence on efficiency of arene acetoxylation.

In order to probe the role of added pyridine in accelerating these reactions, the primary KIE for the reaction was determined by comparing the initial rates of the reactions of benzene and C_6D_6. The obtained k_H/k_D of about 4.8 implicates C—H activation to be the rate-determining step. Based on this observation, the authors attribute the rate enhancement upon addition of pyridine to the acceleration of the rate-determining C—H activation step. Further, they hypothesize that the increased rate of C—H activation is due to the generation of an open coordination site at Pd (by deaggregation of trimeric/polymeric $Pd(OAc)_2$) in the presence of 1 equivalent of pyridine. The addition of pyridine also influences the site-selectivity of these reactions. For example, the reaction of 1,2-dichlorobenzene with $PhI(OAc)_2$ in the presence of substoichiometric Pd/pyr (1:0.9) leads to a 1:2.4 mixture of α- and β-product isomers versus the 1:1.4 ratio obtained in the absence of pyridine. Interestingly, the use of a bulkier oxidant $MesI(OAc)_2$ further improves the selectivity for the β-isomer (Scheme 24.48).

Prompted by the encouraging results for the modulation of site-selectivity for arene functionalization by the use of pyridine (Scheme 24.48), a more extensive exploration of ligand effects on the site-selectivity for arene acetoxylation was undertaken [44]. Several ligands lead to increased selectivities for the acetoxylation of 1,2-dichlorobenzene than those obtained in the absence of ligand, and the highest

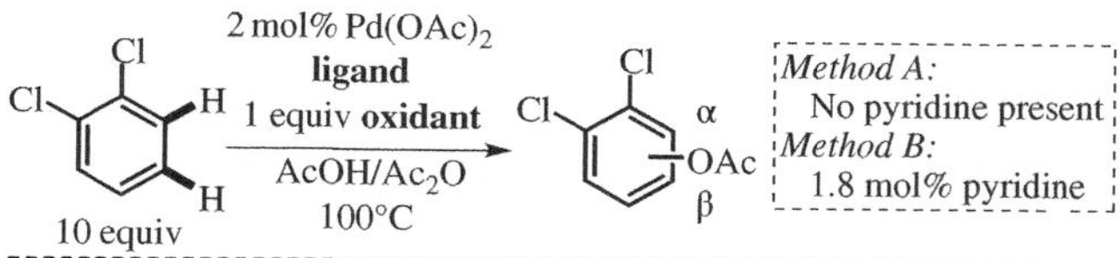

Conditions tested:

Entry	Method	Oxidant	Yield	α/β
1	A	$PhI(OAc)_2$	8%	1:1.4
2	B	$PhI(OAc)_2$	59%	1:2.4
3	A	$MesI(OAc)_2$	8%	1:1.7
4	B	$MesI(OAc)_2$	64%	1:8.1

SCHEME 24.48 Site-selectivity for the acetoxylation of 1,2-dichlorobenzene.

selectivity is observed using acridine as the ligand. Interestingly, exploration of the influence of Pd/acridine (**44**) ratios revealed that the highest selectivity (α:β >1:99) is obtained using a Pd/**44** ratio of 1:10 albeit with significantly diminished product yields (29%). The best balance of yield and selectivity is obtained using 1:3 Pd–**44** (76% yield, 1:49 selectivity). Taken together, the results in Schemes 24.47 and 24.49 suggest that optimal Pd–ligand ratios are dependent on the nature of the ligand.

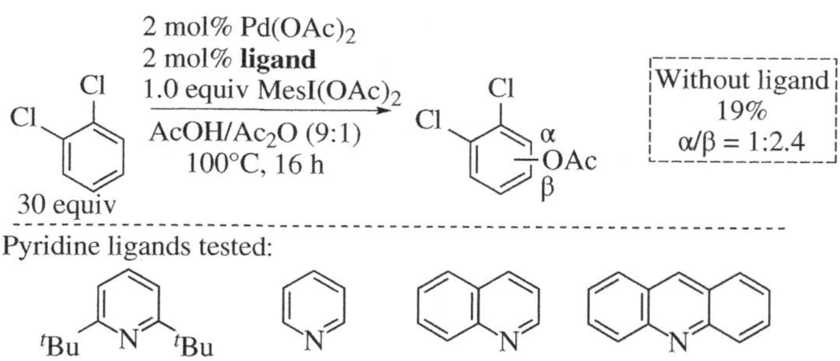

41 28% α/β = 1:2.7 | **19** 71% α/β = 1:16 | **53** 82% α/β = 1:13 | **44** 78% α/β = 1:19

SCHEME 24.49 Ligand effects on the site-selectivity of 1,2-dichlorobenzene.

The optimal conditions are applied toward the acetoxylation of diverse mono-, di-, and trisubstituted arenes (Scheme 24.50). The influence of added acridine on the reactivity and selectivity of functionalization is general across a wide array of substituted arenes.

While this method represents an important advance in the field of arene oxygenations, the use of $PhI(OAc)_2$ as a terminal oxidant is a significant limitation. $PhI(OAc)_2$ is expensive and leads to the formation of undesired iodobenzene as a stoichiometric by-product. In order to address this drawback, the use of $K_2S_2O_8$ in place of $PhI(OAc)_2$ has been explored [45]. Potassium persulfate ($K_2S_2O_8$) is more than an order of magnitude less expensive than $PhI(OAc)_2$ and leads to easily separable water soluble by-products. Based on previous results using $PhI(OAc)_2$ (Scheme 24.50), the use of 1:1 Pd/pyr catalytic system was investigated for the acetoxylation of benzene with $K_2S_2O_8$ as the oxidant (Scheme 24.51). Analogous to the results in Scheme 24.50 earlier, enhanced yields of PhOAc are obtained in the presence of pyridine. However, the yields are significantly diminished than those obtained using $PhI(OAc)_2$ as the oxidant. Gratifyingly, the use of cationic ligands **54** and **55** in place of pyridine lead to significantly higher yields of PhOAc under otherwise identical conditions. Additionally, the selectivity for PhOAc versus PhPh was high in both cases (~36:1 with **54**

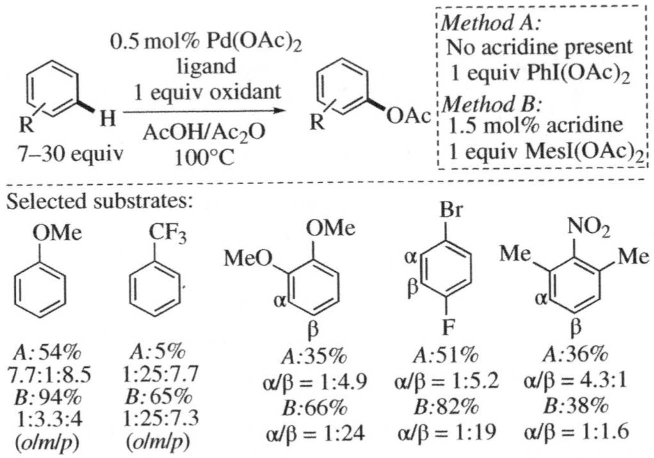

SCHEME 24.50 Representative examples of arene acetoxylations using acridine as ligand.

and 65:1 with **55**). Systematic studies were conducted to understand the greater effectiveness of **54** and **55** versus pyridine in these oxygenations. Based on these investigations, the authors conclude that the primary reason for the enhanced efficiencies of **54** and **55** is due to the ability of these cationic ligands to serve as a phase transfer catalyst to facilitate interactions between the poorly soluble $[S_2O_8]^{2-}$ and the Pd catalyst. Overall, the results discussed in this section exemplify the potential for the ligand-modulated control over the reactivity and site-selectivity of arene oxygenations.

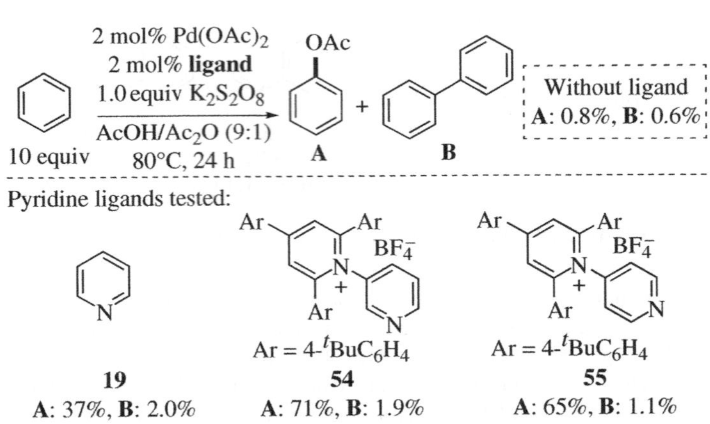

SCHEME 24.51 Ligand effects for arene acetoxylations using $K_2S_2O_8$ as the oxidant.

24.7 THIOLATION

Aryl sulfide scaffolds find important applications in the synthesis of bioactive molecules and organic semiconductors. Anbarasan and coworkers recently reported the first example of Pd-catalyzed direct phenylthiolation of simple arenes (Scheme 24.52) [46]. $Pd(OAc)_2$ serves as an efficient catalyst for the thiolation of various arenes using the succinimide reagent **56** as the source of the phenyl thiol moiety. Mono-, di-, and trisubstituted arenes participate in these reactions to afford the products in good to

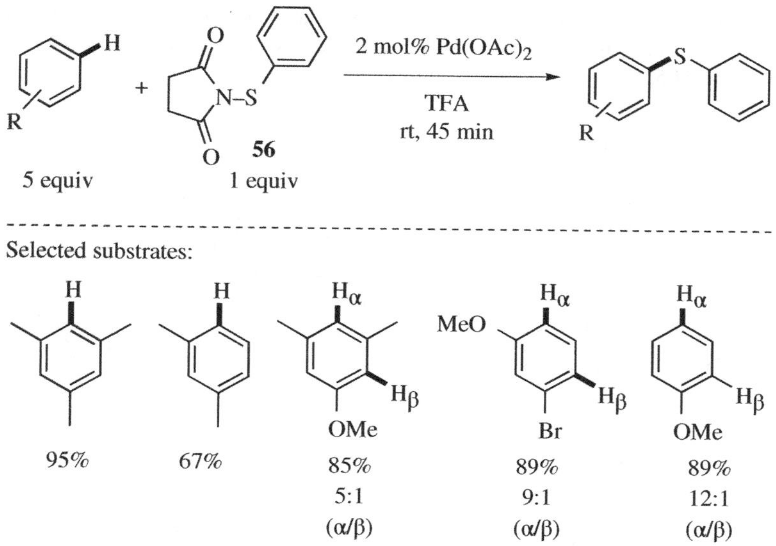

SCHEME 24.52 Representative scope of arenes for thiolation.

excellent yields. Furthermore, a number of functional groups including ethers, benzylic hydrogens, free hydroxyl, ester, and aromatic bromides are well tolerated under the reaction conditions.

The method can be applied toward the arylthiolation of mesitylene and trimethoxybenzene. Electron-rich, electron-deficient, *ortho*-substituted, and heterocyclic arylthiol-bearing products (**57–60**) can be obtained. Furthermore, alkylthiol reagents can also be effectively utilized to afford the corresponding products (**61** and **62**) in good yields (Scheme 24.53).

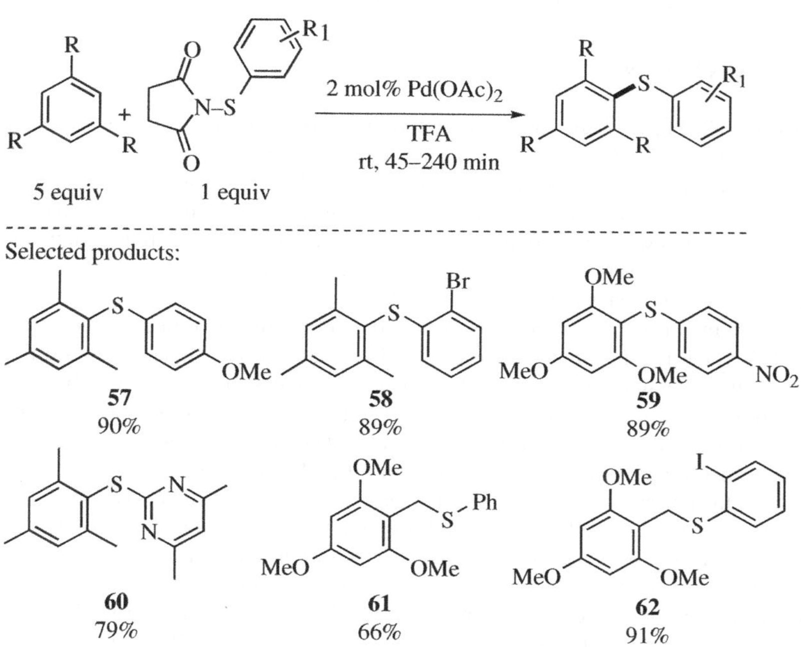

SCHEME 24.53 Representative scope of aryl and alkyl thiolation.

The authors propose a $Pd^{II/IV}$ mechanism for these aryl and alkyl thiolations. Importantly, this proposal is consistent with the compatibility of aryl iodides and bromides under the reaction conditions (Scheme 24.54).

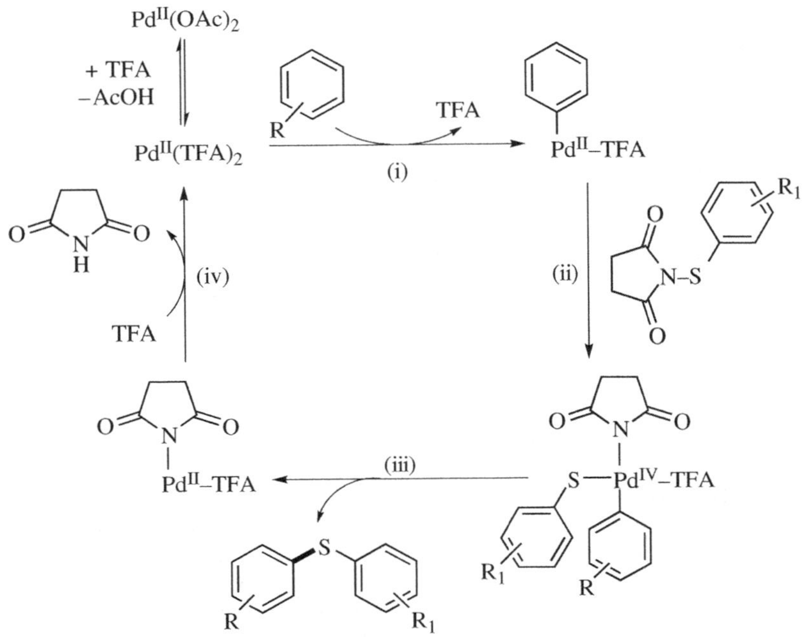

SCHEME 24.54 Mechanistic proposal for Pd-catalyzed C—H thiolation.

24.8 AMINATION

Amines are valuable synthetic targets because of their abundance in medicine, agrochemicals, and materials science [47]. Transition metal-catalyzed direct conversions of aromatic C—H bonds to C—N bonds are attractive synthetic routes toward accessing aryl amines [48]. These reactions are more advantageous over the conventional amination methods that either requires harsh conditions (e.g., nitration followed by reduction) [49] or prefunctionalized substrates (amination of aryl halides) [50, 51]. The Hartwig group recently reported the first example of Pd-catalyzed intermolecular amination of arenes (Scheme 24.55) [52a].

The coupling of mono-, di-, and trisubstituted arenes with phthalimide is accomplished using the $Pd(OAc)_2/P^tBu_3$ catalyst system. Electron-rich and electron-deficient arenes efficiently participate in these aminations to afford the products in modest to good yields. In analogy to the oxygenations described previously, halogenated arenes undergo C—H amination and by-products resulting from the amination of the C—X (X = halide) bond are not observed. The site-selectivity of these transformations is predominantly controlled by the sterics of the arene counterpart, resulting in the preferential amination of the least sterically hindered C—H bonds. For example, the reaction of 1,2,3-trimethylbenzene affords the 1,2,3,5-functionalized product exclusively. Additionally, the use of monosubstituted arenes leads to the *meta*- and *para*-aminated isomers in greater amounts than the corresponding *ortho*-isomer of the product.

Following the Hartwig report, the Emmert group disclosed a method for the Pd-catalyzed amination of acetamide using $K_2S_2O_8$ as the oxidant (Scheme 24.56) [52b]. Importantly, the optimal

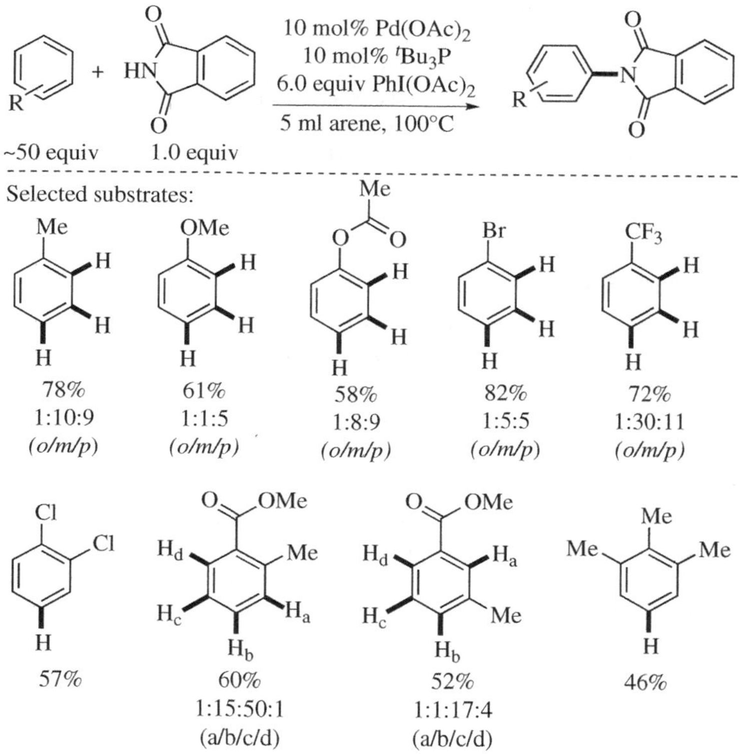

SCHEME 24.55 Representative examples of arene C—H amination.

conditions for amination were obtained upon a detailed screening of catalysts, ligands, Pd–ligand ratios, Pd loading, and additives. As shown in Scheme 24.56, the use of acridine (**44**) in conjunction with TFA and 18-crown-6 is critical for these aminations.

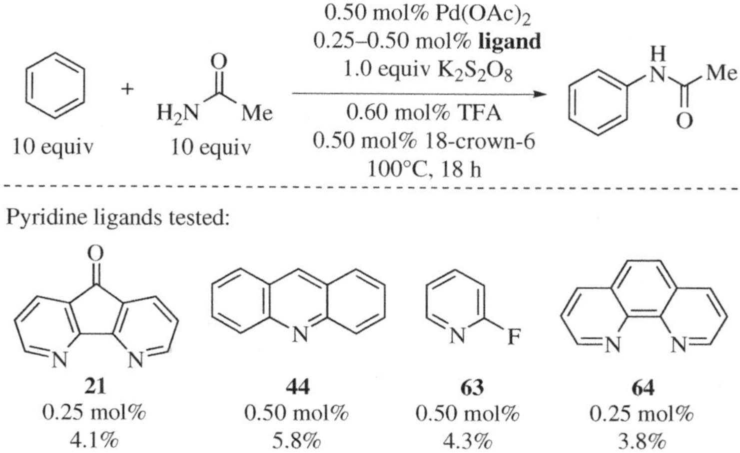

SCHEME 24.56 Ligand survey for Pd-catalyzed synthesis of acetanilides.

24.9 MISCELLANEOUS

As mentioned in the introduction, the majority of currently known transition metal-catalyzed arene C—H functionalizations employ Pd catalysts. Many of these transformations have served as an inspiration for the development of arene functionalizations using other transition metal catalysts (e.g., Ni [53], Pt [54], Rh [55], Ru [56], Ir [57], Au [58], Cu [59], Fe [60]) [5]. Herein, we showcase some of these examples, highlighting bond formations (e.g., C—alkynyl, C—Hal, C—Si, C—B) that are either unknown or less efficient using Pd catalysts.

24.9.1 Halogenation

Aromatic halides serve as important precursors for a plethora of synthetically useful reagents and intermediates (e.g., organometallics and benzynes). Furthermore, aryl halides are potential substrates for numerous transformations including cross-coupling reactions and nucleophilic aromatic substitution. As such, extensive research efforts have been devoted to the development of general and efficient direct halogenation of arenes [61]. However, the synthetic utility of most of the reported transformations is limited due to the requirement of strong acids and/or a limited substrate scope. Recently, a method for the Au-catalyzed bromination of a wide range of electron-neutral and electron-rich arenes was reported (Scheme 24.57) [62a]. Importantly, these halogenations do not require the use of strong acids and are tolerant toward benzylic alkyl groups. The high efficiency of these reactions is attributed to the dual role of the Au catalyst in activating both the succinimide component (in NBS) and the arene. The site-selectivities of these reactions are consistent with the proposed electrophilic aromatic metallation mechanism for the C—H activation step. Specifically, the reactions exhibit *ortho/para*-selectivity for electron-rich substrates and *meta*-selectivity for electron-poor arenes. Subsequent to this report, this method was expanded toward the Au-catalyzed halogenation of aromatic boronates [62b].

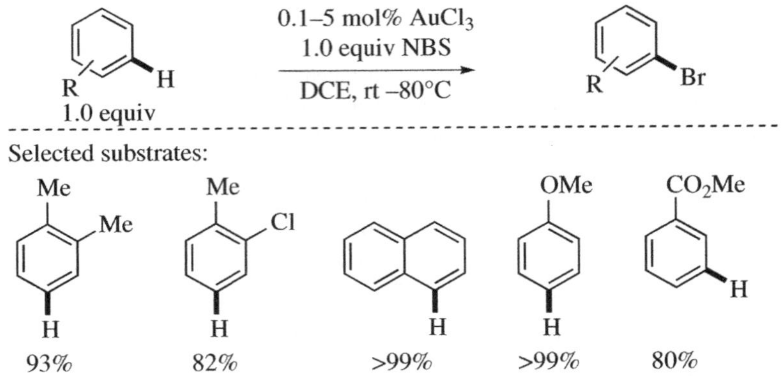

SCHEME 24.57 Representative example of arene halogenation.

24.9.2 Silylation

Curtis and coworkers reported the first example of nondirected arene C—H silylation using an iridium catalyst in 1982 [63]. Since then, a number of metal catalysts (e.g., Ni, Pt, Rh) [64] have been employed for the direct coupling of arenes with silanes [65]. While these examples represent significant contributions to the field of arene functionalizations, many of them suffer from low yields or a limited arene scope. Recently, however, an example of a high-yielding and general Ir-catalyzed protocol for arene silylations was disclosed [66]. The use of the Ir[I]/dipphen catalyst system affects the halosilylation of electron-rich and electron-deficient arenes to afford the desired products in excellent yields (Scheme 24.58). The site-selectivity of these reactions is predominantly controlled by steric factors with statistical mixture of *meta*- and *para*-products being observed for monosubstituted arenes;1,3-disubstituted arenes lead exclusively to 1,3,5-trisubstituted products. Importantly, the arylhalosilane products and their hypervalent derivatives serve as versatile reagents for carbon–carbon and carbon–heteroatom bond forming organic reactions.

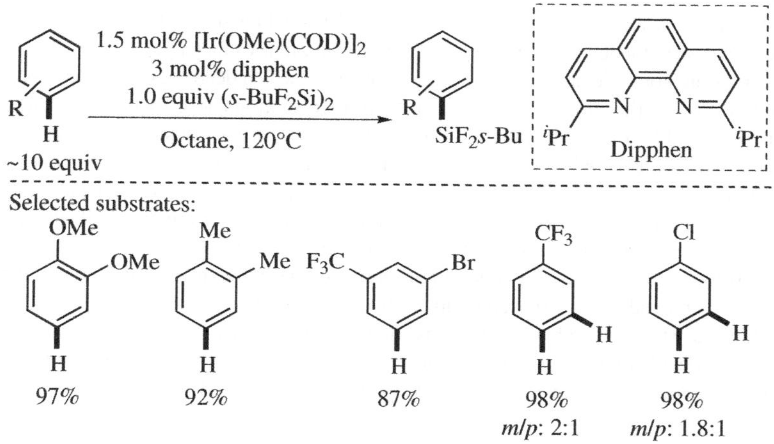

SCHEME 24.58 Representative example of arene silylation.

24.9.3 Borylation

The direct borylation of arenes is an attractive strategy for accessing synthetically useful arylboron reagents. Iridium complexes have emerged as the catalyst of choice for the selective borylation of arenes using HB(pin) or $B_2(pin)_2$ [67]. Extensive studies by the Hartwig, Ishiyama, and Miyaura groups have led to the identification of the $Ir^IOMe(cod)_2$/dtbpy as the optimal catalyst system for these transformations [68–70]. As illustrated in Scheme 24.59, electronically diverse arenes are borylated at room temperature to afford the products in excellent yields. The site-selectivity of these

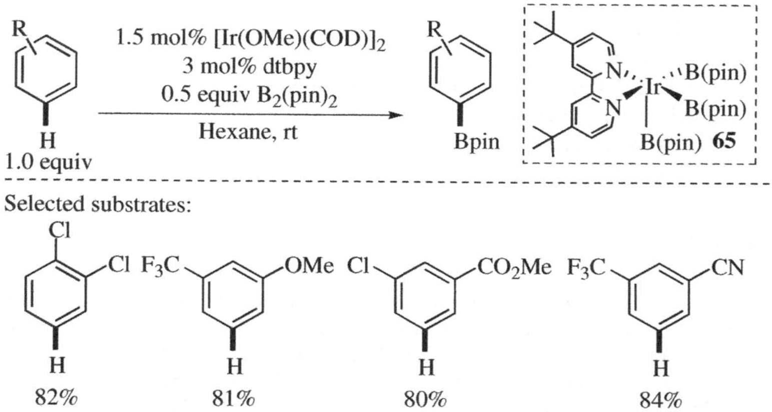

SCHEME 24.59 Representative example of arene borylation.

reactions is similar to that of the C—H silylations described earlier. Mechanistic studies show that the triborylated Ir^{III} complex **65** bearing one bidentate dtbpy ligand is the catalytically active species in these reactions.

24.10 SUMMARY AND OUTLOOK

Transition metal-catalyzed functionalization of simple arenes is an elegant, atom-economic strategy to access structurally diverse aromatic compounds. While this field is still in its infancy, numerous exciting advances have been made over the past decade. However, many challenges remain. These include the use of super stoichiometric amounts of the arene substrate, which decreases the atom economy of these transformations. Furthermore, in majority of the transformations described herein, the selectivity is primarily dictated by the steric and electronic nature of the arene substrate leading to a mixture of isomeric products. Furthermore, methods affording high selectivity for a single isomer are plagued with the inability to selectively yield the other isomeric products, thereby limiting the range of accessible aromatics. A more powerful approach toward modulating the site-selectivity entails the rational design of catalysts or reaction conditions to override substrate control and predictably allow the selective functionalization of a specific C—H bond in the arene. As described in this chapter, a number of recent reports have achieved catalyst-based control for C—H arylations, alkenylations, and oxygenations. These efforts have been guided by careful consideration of the mechanism of these reactions. As such, we anticipate that detailed mechanistic investigations will continue to inform the development of selective and efficient strategies for the burgeoning field of versatile nondirected arene C—H functionalizations.

ABBREVIATIONS

BQ	Benzoquinone
bzq	benzo[*h*]quinoline
CDC	Cross-dehydrogenative coupling
CMD	Concerted metallation/deprotonation
DFT	Density functional theory
DG	Directing group
Dipphen	Diisopropyl-1,10-phenanthroline
Fe(Pc)	Iron phthalocyanine
KIE	Kinetic isotope effect
Mes	Mesityl
MOM	Methoxymethyl ether
NFSI	*N*-fluorobenzenesulfonimide
PBI	Perylene-3,4:9,10-tetracarboxylic acid bisimide
SEM	2-(trimethylsilyl)ethoxymethyl
TFA	Trifluoroacetic acid
TON	Turnover number

REFERENCES

[1] For recent reviews on C—H activation, see: (a) Kakiuchi, F., Kochi, T. (2008) *Synthesis*, **19**, 3013–3039; (b) Giri, R., Shi, B.-F., Engle, K. M., Maugel, N., Yu, J.-Q. (2009) *Chem. Soc. Rev.*, **38**, 3242–3272; (c) Thansandote, P., Lautens, M. (2009) *Chem. Eur. J.*, **15**, 5874–5883; (d) Chiusoli, G. P., Catellani, M., Costa, M., Motti, E., Della Ca, N., Maestri, G. (2010) *Coord. Chem. Rev.*, **254**, 456–469; (e) Wencel-Delord, J., Dröge, T., Liu, F., Glorius, F. (2011) *Chem. Soc. Rev.*, **40**, 4740–4761; (f) Ackermann, L. (2011) *Chem. Rev.*, **111**, 1315–1345; (g) Li, B., Dixneuf, P. H. (2013) *Chem. Soc. Rev.*, **42**, 5744–5767; (h) Mousseau, J. J., Charette, A. B. (2013) *Acc. Chem. Res.*, **46**, 412–424; (i) Zhang, X.-S., Chen, K., Shi, Z.-J. (2014) *Chem. Sci.*, **5**, 2146–2159.

[2] (a) de Meijere, A., Diederich, F. Eds. (2004) *Metal-Catalyzed Cross-Coupling Reactions*, 2nd ed., Wiley-VCH, Weinheim; (b) Negishi, E. (2011) *Angew. Chem. Int. Ed.*, **50**, 6738–6764; (c) Suzuki, A. (2011) *Angew. Chem. Int. Ed.*, **50**, 6722–6737.

[3] See Chapter 23 of this book for references and details.

[4] Selected examples for transition-metal-catalyzed biaryl coupling through radical mechanisms: (a) Fujita, K., Nonogawa, M., Yamaguchi, R. (2004) *Chem. Commun.*, 1926–1927; (b) Proch, S., Kempe, R. (2007) *Angew. Chem. Int. Ed.*, **46**, 3135–3138; (c) Xie, G., Li, T., Qu, X., Mao, J. (2011) *J. Mol. Catal. A*, **340**, 48–52; (d) Liu, W., Cao, H., Lei, A. (2010) *Angew. Chem. Int. Ed.*, **49**, 2004–2008; (e) Vallée, F., Mousseau, J. J., Charette, A. B. (2010) *J. Am. Chem. Soc.*, **132**, 1514–1516; (f) Li, H., Sun, C.-L., Yu, M., Yu, D.-G., Li, B.-J., Shi, Z.-J. (2011) *Chem. Eur. J.*, **17**, 3593–3597; (g) Liu, W., Cao, H., Xin, J., Jin, L., Lei, A. (2011) *Chem. Eur. J.*, **17**, 3588–3592.

[5] Kuhl, N., Hopkinson, M. N., Wencel-Delord, J., Glorius, F. (2012) *Angew. Chem. Int. Ed.*, **51**, 10236–10254.

[6] (a) Chen, X., Engle, K. M., Wang, D.-H., Yu, J.-Q. (2009) *Angew. Chem. Int. Ed.*, **48**, 5094–5115; (b) Lyons, T. W., Sanford, M. S. (2010) *Chem. Rev.*, **110**, 1147–1169; (c) Daugulis, O. (2010) *Top. Curr. Chem.*, **292**, 57–84; (d) Hickman, A. J., Sanford M. S. (2012) *Nature*, **484**, 177–185.

[7] (a) Ackermann, L., Vincente, R., Kapdi, A. R. (2009) *Angew. Chem. Int. Ed.*, **48**, 9792–9826; (b) Rossi, R., Bellina, F., Lessi, M., Manzini, C. (2014) *Adv. Synth. Catal.*, **356**, 17–117.

[8] (a) Alberico, D., Scott, M. E., Lautens, M. (2007) *Chem. Rev.*, **107**, 174–238; (b) McGlacken, G. P., Bateman, L. M. (2009) *Chem. Soc. Rev.*, **38**, 2447–2464.

[9] (a) Yeung, C. S., Dong, V. M. (2011) *Chem. Rev.*, **111**, 1215–1292; (b) Liu, C., Zhang, H., Shi, W., Lei, A. (2011) *Chem. Rev.*, **111**, 1780–1824.

[10] (a) Campeau, L.-C., Parisien, M., Jean, A., Fagnou, K. (2006) *J. Am. Chem. Soc.*, **128**, 581–590; (b) Lafrance, M., Fagnou, K. (2006) *J. Am. Chem. Soc.*, **128**, 16496–16497.

[11] Caron, L., Campeau, L.-C., Fagnou, K. (2008) *Org. Lett.*, **10**, 4533–4536.

[12] Yang, S.-D., Sun, C.-L., Fang, Z., Li, B.-H., Li, Y.-Z., Shi, Z.-J. (2008) *Angew. Chem. Int. Ed.*, **47**, 1473–1476.

[13] Kawai, H., Kobayashi, Y., Oi, S., Inoue, Y. (2008) *Chem. Commun.*, 1464–1466.

[14] Qin, C., Lu, W. (2008) *J. Org. Chem.*, **73**, 7424–7427.

[15] Storr, T. E., Greaney, M. F. (2013) *Org. Lett.*, **15**, 1410–1413.

[16] Hickman, A. J., Sanford, M. S. (2011) *ACS Catal.*, **1**, 170–174.

[17] (a) Fujiwara, Y., Moritani, I., Ikegami, K., Tanaka, R., Teranishi, S. (1970) *Bull. Chem. Soc. Jpn.*, **43**, 863–867; (b) Ackerman, L. J., Sadighi, J. P., Kurtz, D. M., Labinger, J. A., Bercaw, J. E. (2003) *Organometallics*, **22**, 3884–3890.

[18] (a) Li, R., Jiang, L., Lu, W. (2006) *Organometallics*, **25**, 5973–5975; (b) Zhou, L., Lu, W. (2012) *Organometallics*, **31**, 2124–2127.

[19] (a) Li, C.-J. (2009) *Acc. Chem. Res.*, **42**, 335–344; (b) Scheuermann, C. J. (2010) *Chem. Asian J.*, **5**, 436–451.

[20] (a) Stuart, D. R., Fagnou, K. (2007) *Science*, **316**, 1172–1175; (b) Stuart, D. R., Villemure, E., Fagnou, K. (2007) *J. Am. Chem. Soc.*, **129**, 12072–12073.

[21] (a) Dwight, T. A., Rue, N. R., Charyk, D., Josselyn, R., DeBoef, B. (2007) *Org. Lett.*, **9**, 3137–3139; (b) Potavathri, S., Dumas, A. S., Dwight, T. A., Naumiec, G. R., Hammann, J. M., DeBoef, B. (2008) *Tetrahedron Lett.*, **49**, 4050–4053; (c) Potavathri, S., Pereira, K. C., Gorelsky, S. I., Pike, A., LeBris, A. P., DeBoef, B. (2010) *J. Am. Chem. Soc.*, **132**, 14676–14681.

[22] Campbell, A. N., Meyer, E. B., Stahl, S. S. (2011) *Chem. Commun.*, **47**, 10257–10259.

[23] (a) Malakar, C. C., Schmidt, D., Conrad, J., Beifuss, U. (2011) *Org. Lett.*, **13**, 1378–1381; (b) Wang S., Liu, W., Cen, J., Liao, J., Huang, J., Zhan, H. (2014) *Tetrahedron Lett.*, **55**, 1589–1592.

[24] Li, B.-H., Tian, S.-L., Fang, Z., Shi, Z.-J. (2008) *Angew. Chem. Int. Ed.*, **47**, 1115–1118.

[25] Brasche, G., Garcia-Fortanet, J., Buchwald, S. L. (2008) *Org. Lett.*, **10**, 2207–2210.

[26] (a) Zhao, X., Yeung, C. S., Dong, V. M. (2010) *J. Am. Chem. Soc.*, **132**, 5837–5844; (b) Yeung, C. S., Zhao, X., Borduas, N., Dong, V. M. (2010) *Chem. Sci.*, **1**, 331–336.

[27] Wang, X., Leow, D., Yu, J.-Q. (2011) *J. Am. Chem. Soc.*, **133**, 13864–13867.

[28] (a) Hull, K. L., Sanford, M. S. (2007) *J. Am. Chem. Soc.*, **129**, 11904–11905; (b) Hull, K. L., Sanford, M. S. (2009) *J. Am. Chem. Soc.*, **131**, 9651–9653; (c) Lyons, T. W., Hull, K. L., Sanford, M. S. (2011) *J. Am. Chem. Soc.*, **133**, 4455–4464; (d) Sanhueza, I. A., Wagner, A. M., Sanford, M. S., Schoenebeck, F. (2013) *Chem. Sci.*, **4**, 2767–2775.

[29] Fujiwara, Y., Moratini, I., Danno, S., Asano, R., Teranish, S. (1969) *J. Am. Chem. Soc.*, **91**, 7166–7169.

[30] Zhou, L., Lu, W. (2014) *Chem. Eur. J.*, **20**, 634–642.

[31] Zhang, Y.-H., Shi, B.-F., Yu, J.-Q. (2009) *J. Am. Chem. Soc.*, **131**, 5072–5074.

[32] Zhang, S., Shi, L., Ding, Y. (2011) *J. Am. Chem. Soc.*, **133**, 20218–20229.

[33] Kubota, A., Emmert, M. H., Sanford, M. S. (2012) *Org. Lett.*, **14**, 1760–1763.

[34] Ying, C.-H., Yan, S.-B., Duan, W.-L. (2014) *Org. Lett.*, **16**, 500–503.

[35] Gigant, N., Bäckvall, J.-E. (2014) *Org. Lett.*, **16**, 1664–1667.

[36] Moon, Y., Kwon, D., Hong, S. (2012) *Angew. Chem. Int. Ed.*, **51**, 11333–11336.

[37] Gigant, N., Bäckvall, J.-E. (2014) *Chem. Eur. J.*, **20**, 5890–5894.

[38] Yue, W., Li, Y., Jiang, W., Zhen, Y., Wang, Z. (2009) *Org. Lett.*, **11**, 5430–5433.

[39] Rosewall, C. F., Sibbald, P. A., Liskin, D. V., Michael, F. E. (2009) *J. Am. Chem. Soc.*, **131**, 9488–9489.

[40] Liu, Q., Zhang, H., Lei, A. (2011) *Angew. Chem. Int. Ed.*, **50**, 10788–10799.

[41] (a) Shibahara, F., Kinoshita, S., Nozaki, K. (2004) *Org. Lett.*, **6**, 2437–2439; (b) Sakakibara, K., Yamashita, M., Nozaki, K. (2005) *Tetrahedron Lett.*, **46**, 959–962.

[42] (a) Henry, P. M. (1971) *J. Org. Chem.*, **36**, 1886–1890; (b) Eberson, L., Jönsson, L. (1974) *J. Chem. Soc. Chem. Commun.*, 885–886; (c) Eberson, L., Jönsson, E. (1974) *Acta Chem. Scand. B*, **28**, 771–776; (d) Eberson, L., Jönsson, L. (1976) *Acta Chem. Scand. B*, **30**, 361–364; (e) Eberson, L., Jönsson, L. (1977) *Justus Liebigs Ann. Chem.*, 233–241; (f) Stock, L. M., Tse, K.-T., Vorvick, L. J., Walstrum, S. A. (1981) *J. Org. Chem.*, **46**, 1757–1759; (g) Yoneyama, T., Crabtree, R. H. (1996) *J. Mol. Catal. A*, **108**, 35–40.

[43] Emmert, M. H., Cook, A. K., Xie, Y. J., Sanford, M. S. (2011) *Angew. Chem. Int. Ed.*, **50**, 9409–9412.

[44] Cook, A. K., Emmert, M. H., Sanford, M. S. (2013) *Org. Lett.*, **15**, 5428–5431.

[45] Gary, J. B., Cook, A. K., Sanford, M. S. (2013) *ACS Catal.*, **3**, 700–703.

[46] Saravanan P., Anbarasan, P. (2014) *Org. Lett.*, **16**, 848–851.

[47] (a) Cheng, J., Kamiya, K., Kodama, I. (2001) *Cardiovasc. Drug Rev.*, **19**, 152–171; (b) Horton, D. A., Bourne, G. T., Smythe, M. L. (2003) *Chem. Rev.*, **103**, 893–930; (c) Sańchez, C., Mećdez, C., Salas, J. A. (2006) *Nat. Prod. Rep.*, **23**, 1007–1045; (d) Bhattacharya, S., Chaudhuri, P. (2008) *Curr. Med. Chem.*, **15**, 1762–1777.

[48] Louillat, M.-L., Patureau, F. W. (2014) *Chem. Soc. Rev.*, **43**, 901–910.

[49] (a) Larock, R. C. (1989) In *Comprehensive Organic Transformations*, Wiley-VCH, New York, Weinhein, Cambridge; (b) March, J. (1992) *Advanced Organic Chemistry*, 4th ed., John Wiley & Sons, Inc., New York.

[50] (a) Beletskaya, I. P., Cheprakov, A. V. (2004) *Coord. Chem. Rev.*, **248**, 2337–2364; (b) Evano, G., Blanchard, N., Toumi, T. (2008) *Chem. Rev.*, **108**, 3054–3131; (c) Monguchi, D., Fujiwara, T., Furukawa, H., Mori, A. (2009) *Org. Lett.*, **11**, 1607–1610; (d) Monnier, F., Taillefer, M. (2009) *Angew. Chem. Int. Ed.*, **48**, 6954–6971; (e) Sperotto, E., van Klink, G. P. M., van Koten, G., de Vries, J. G. (2010) *Dalton Trans.*, **39**, 10338–10351.

[51] (a) Hartwig, J. F. (2002) In *Handbook of Organopalladium Chemistry for Organic Synthesis*, (Negishi, E., Ed.), Wiley-Interscience, New York, p 1051; (b) Jiang, L., Buchwald, S. L. (2004) *Metal-Catalyzed Cross-Coupling Reactions*, 2nd ed., Wiley-VCH, Weinheim, Vol. **2**, p 699.

[52] (a) Shrestha, R., Mukherjee, P., Tan, Y., Litman, Z. C., Hartwig, J. F. (2013) *J. Am. Chem. Soc.*, **135**, 8480–8483; (b) Bandara, H. M. D., Sosin, M. H., McKeogh, B. J., Emmert, M. H. (2013) *GSTF J. Chem.*, **1**, 17–31.

[53] Yamaguchi, J., Muto, K., Itami, K. (2013) *Eur. J. Org. Chem.*, 19–30.

[54] Wagner, A. M., Hickman, A. J., Sanford, M. S. (2013) *J. Am. Chem. Soc.*, **135**, 15710–15713.

[55] (a) Colby, D. A., Bergman, R. G., Ellman, J. A. (2010) *Chem. Rev.*, **110**, 624–655; (b) Bouffard, J., Itami, K. (2010) *Top. Curr. Chem.*, **292**, 231–280; (c) Song, G., Wang, F., Li, X. (2012) *Chem. Soc. Rev.*, **41**, 3651–3678; (d) Kuhl, N., Schroder, N., Glorius, F. (2014) *Adv. Synth. Catal.*, **356**, 1443–1460; (e) Pham,

M. V., Cramer, N. (2014) *Angew. Chem. Int. Ed.*, **53**, 3484–3487; (f) Vora, H. U., Silvestri, A. P., Engelin, C. J., Yu, J.-Q. (2014) *Angew. Chem. Int. Ed.*, **53**, 2683–2686.

[56] (a) Ackermann, L., Vicente, R. (2010) *Top. Curr. Chem.*, **292**, 211–229; (b) Arockiam, P. B., Bruneau, C., Dixneuf, P. H. (2012) *Chem. Rev.*, **112**, 5879–5918; (c) Ackermann, L. (2014) *Acc. Chem. Res.*, **47**, 281–295.

[57] Choi, J., Goldman, A. S. (2011) *Top. Organomet. Chem.*, **34**, 139–167.

[58] (a) Boorman, T. C., Larrosa, I. (2011) *Chem. Soc. Rev.*, **40**, 1910–1925; (b) Hopkinson, M. N., Gee, A. D., Gouverneur, V. (2011) *Chem. Eur. J.*, **17**, 8248–8262; (c) de Haro, T., Nevado, C. (2011) *Synthesis*, **16**, 2530–2539.

[59] (a) Díaz-Requejo, M. M., Pérez, P. J. (2008) *Chem. Rev.*, **108**, 3379–3394; (b) Philips, R. J., Gaunt, M. J. (2009) *Science*, **323**, 1593–1597; (c) Zhang, M. (2010) *Appl. Organomet. Chem.*, **24**, 269–284; (d) Liu, W., Bi, Y. (2012) *Chin. J. Org. Chem.*, **32**, 1041–1050; (e) Wendlandt, A. E., Suess, A. M., Stahl, S. S. (2011) *Angew. Chem. Int. Ed.*, **50**, 11062–11087.

[60] (a) Norinder, J., Matsumoto, A., Yoshikai, N., Nakamura, E. (2008) *J. Am. Chem. Soc.*, **130**, 5858–5859; (b) Sun, C.-L., Li, B.-J., Shi, Z.-J. (2011) *Chem. Rev.*, **111**, 1293–1314.

[61] (a) Yadav, J. S., Reddy, B. V. S., Reddy, R. S. R., Basak, A. K., Narsaiah, A. V. (2004) *Adv. Synth. Catal.*, **346**, 77–82; (b) Prakash, G. K. S., Mathew, T., Hoole, D., Esteves, P. M., Wang, Q., Rasul, G., Olah, G. A. (2004) *J. Am. Chem. Soc.*, **126**, 15770–15776; (c) Zhang, Y., Shibatomi, K., Yamamoto, H. (2005) *Synlett*, **18**, 2837–2842; (d) Bagheri, M., Azizi, N., Saidi, M. R. (2005) *Can. J. Chem.*, **83**, 146–149; (e) Ganguly, N. C., De, P., Dutta, S. (2005) *Synthesis*, 1103–1108; (f) Zysman-Colman, E., Arias, K., Siegel, J. S. (2009) *Can. J. Chem.*, **87**, 440–447; (g) Al-Zoubi, R. M., Hall, D. G. (2010) *Org. Lett.*, **12**, 2480–2483.

[62] (a) Mo, F., Yan, J. M., Qiu, D., Li, F., Zhang, Y., Wang, J. (2010) *Angew. Chem. Int. Ed.*, **49**, 2028–2032; (b) Qiu, D., Mo, F., Zheng, Z., Zhang, Y., Wang, J. (2010) *Org. Lett.*, **12**, 5474–5477.

[63] Gustavson, W. A., Epstein, P. S., Curtis, M. D. (1982) *Organometallics*, **1**, 884–885.

[64] (a) Sakakura, T., Tokunaga, Y., Sodeyama, T., Tanaka, M. (1987) *Chem. Lett.*, **12**, 2375–2378; (b) Ishikawa, M., Okazaki, S., Naka, A., Sakamoto, H. (1992) *Organometallics*, **11**, 4135–4139; (c) Uchimaru, Y., Mouneer, A., Elsayed, M., Tanaka, M. (1993) *Organometallics*, **12**, 2065–2069; (d) Ishikawa, M., Naka, A., Ohshita, J. (1993) *Organometallics*, **12**, 4987–4992; (e) Ezbiansky, K., Djurovich, P. I., LaForest, M., Sinning, D. J., Zayes, R., Berry, D. H. (1998) *Organometallics*, **17**, 1455–1457; (f) Tsukada, N., Hartwig, J. F. (2005) *J. Am. Chem. Soc.*, **127**, 5022–5023; (g) Murata, M., Fukuyama, N., Wada, J., Watanabe, S., Masuda, Y. (2007) *Chem. Lett.*, **36**, 910–911.

[65] Hartwig, J. F. (2012) *Acc. Chem. Res.*, **45**, 864–873.

[66] (a) Ishiyama, T., Sato, K., Nishio, Y., Miyaura, N. (2003) *Angew. Chem. Int. Ed.*, **42**, 5346–5348; (b) Saiki, T., Nishio, Y., Ishiyama, T., Miyaura, N. (2006) *Organometallics*, **25**, 6068–6073.

[67] (a) Miyaura, N. (2008) *Bull. Chem. Soc. Jpn.*, **81**, 1535–1553; (b) Mkhalid, I. A. I., Barnard, J. H., Marder, T. B., Murphy, J. M., Hartwig, J. F. (2010) *Chem. Rev.*, **110**, 890–931; (c) Hartwig, J. F. (2011) *Chem. Soc. Rev.*, **40**, 1992–2002.

[68] (a) Ishiyama, T., Takagi, J., Ishida, K., Miyaura, N., Anastasi, N. R., Hartwig, J. F. (2002) *J. Am. Chem. Soc.*, **124**, 390–391; (b) Ishiyama, T., Takagi, J., Hartwig, J. F., Miyaura, N. (2002) *Angew. Chem. Int. Ed.*, **41**, 3056–3058; (c) Ishiyama, T., Nobuta, Y., Hartwig, J. F.; Miyaura, N. (2003) *Chem. Commun.*, 2924–2925; (d) Boller, T. M., Murphy, J. M., Hapke, M., Ishiyama, T., Miyaura, N., Hartwig, J. F. (2005) *J. Am. Chem. Soc.*, **127**, 14263–14278.

[69] (a) Tamura, H., Yamazaki, H., Sato, H., Sakaki, S. (2003) *J. Am. Chem. Soc.*, **125**, 16114–16126; (b) Liskey, C. W., Wei, C. S., Pahls, D. R., Hartwig, J. F. (2009) *Chem. Commun.*, **37**, 5603–5605; (c) Chotana, G. A., Vanchura, B. A., II, Tse, M. K., Staples, R. J., Meleczka, R. E., Smith, M. R., III, Jr., (2009) *Chem. Commun.*, 5731–5733.

[70] (a) Chotana, G. A., Rak, M. A., Smith, M. R. (2005) *J. Am. Chem. Soc.*, **127**, 10539–10544; (b) Hurst, T. E., Macklin, T. K., Becker, M., Hartmann, E., Kügel, W., Parisienne-La Salle, J.-C., Batsanov, A. S., Marder, T. B., Snieckus, V. (2010) *Chem. Eur. J.*, **16**, 8155–8161.

25

FUNCTIONALIZATION OF ARENES VIA C—H BOND ACTIVATION CATALYSED BY TRANSITION METAL COMPLEXES: SYNERGY BETWEEN EXPERIMENT AND THEORY

Amalia Isabel Poblador-Bahamonde
Department of Organic Chemistry, University of Geneva, Geneva, Switzerland

25.1 INTRODUCTION

C—H bond activation is an extremely important process due to its potential for producing functionalized hydrocarbons and its fundamental scientific interest. The use of transition metals to catalyze the formation of a new C—C bond by the C—H activation of a hydrocarbon is a very interesting alternative in terms of atom economy [1]. The improvement of the catalytic C—H bond activation step has been the common denominator in many investigations, mainly intramolecular [2, 3], as intermolecular functionalization still remains challenging [4].

Determining the precise nature of a C—H activation step experimentally can be especially difficult due to the fact that the dividing line between the various mechanistic possibilities can be unclear and the isolation of key intermediates very challenging. As consequence, the use of a theoretical approach has become ideal to provide insight into such mechanistic issues. Improvements in methodology [5, 6] and computer power have helped to broaden the range of systems that can be studied and the nature of the problems that can be tackled.

This chapter will highlight the powerful synergy that the combination of experiment and computational studies provide for the understanding of the mechanisms of arene C—H activation by transition metal complexes as shown in several previous reviews [7]. First of all, a schematic summary of the mechanistic possibilities from classical to recent classifications will be described (Section 25.2). The following will address different strategies to promote arene functionalization emphasizing how theory can give important insight into experiment (Section 25.3.1) and the use of

Arene Chemistry: Reaction Mechanisms and Methods for Aromatic Compounds,
First Edition. Edited by Jacques Mortier.

carboxylate or carbonate bases for an easy C—H activation (Section 25.3.2). Finally, computational studies of catalytic cycles for the formation of new C—C bonds are assessed (Section 25.4).

The methodology used will not be described here, unless stated, and all the computational work cited was carried out using density functional theory (DFT) [7]. All energies are given in kilocalories per mole.

25.2 MECHANISMS OF C—H BOND ACTIVATION

The mechanisms describing C—H activation by organometallic complexes were traditionally divided into (I) oxidative addition (OA), (II) σ-bond metathesis (SBM) and (III) electrophilic activation (EA) (Fig. 25.1). The classification was generally identified with the nature of the metal center; therefore, electron-rich low-valence metal complexes usually promote oxidative addition, early complexes undergo σ-bond metathesis, and electron-deficient late transition metal complexes react through electrophilic activation.

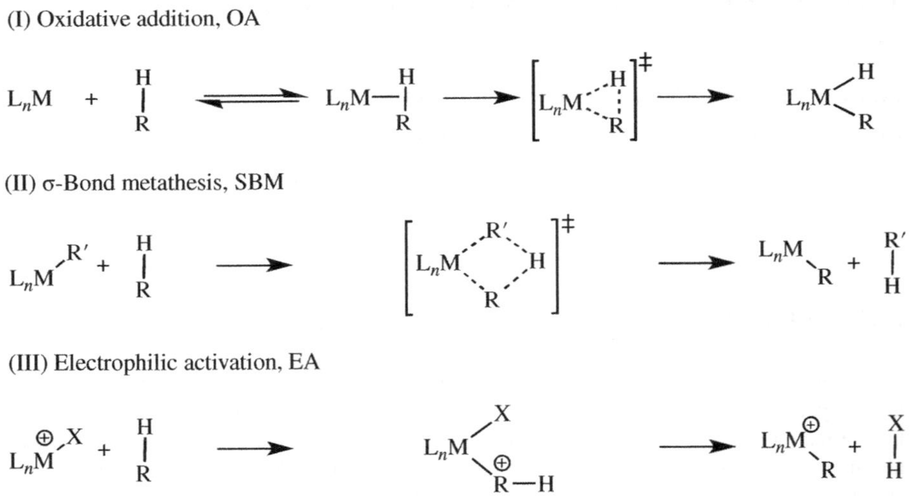

FIGURE 25.1 Schematic representation of the different mechanism for C—H bond activation.

The use of computational methods has shed further light onto the mechanism of C—H activation step by showing the existence of electrophilic, ambiphilic, and nucleophilic key interactions as well as key ligands/substrate functionalization [8]. To mechanistically distinguish all these metalations, a handful of acronyms are now usually found in to the literature, such a concerted metalation–deprotonation (CMD), internal electrophilic substitution (IES), ambiphilic metal ligand activation (AMLA), oxidative hydrogen migration (OHM), oxydatively added transition state (OATS), and σ-complex-assisted metathesis (σ-CAM) among others.

The major part of these mechanisms invokes the assistance of bifunctional ligands bearing an additional Lewis-basic heteroatom, such a secondary phosphine oxide or carboxylates during the C—H cleavage process. Both ligands will undergo through a transition state analogous to that depicted in Figure 25.2, but different acronyms will be used depending on the role of the metal, ligand, or substrate on the C—H activation step. A more detailed description of each mechanism will be discussed further in the chapter (Sections 25.3 and 25.4).

Hall and coworkers [9] described this mechanistic spectrum by an analysis of the geometry computed for the transition state. To do so, they used the reaction coordinate middle point defined by Bader's atoms in molecules (AIM) approach [10], which describes the bond critical points

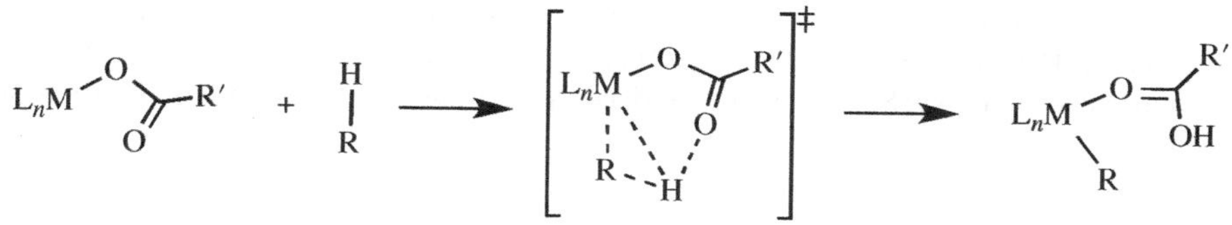

FIGURE 25.2 Schematic representation for an assisted C—H bond activation mechanism.

(BCP) and ring critical points (RCP) present in the system. Through this study, it was proposed that some of the mechanisms previously described could be differentiated by variations in the position and number of BCP and RCP and suggested the assignments depicted in Figure 25.3.

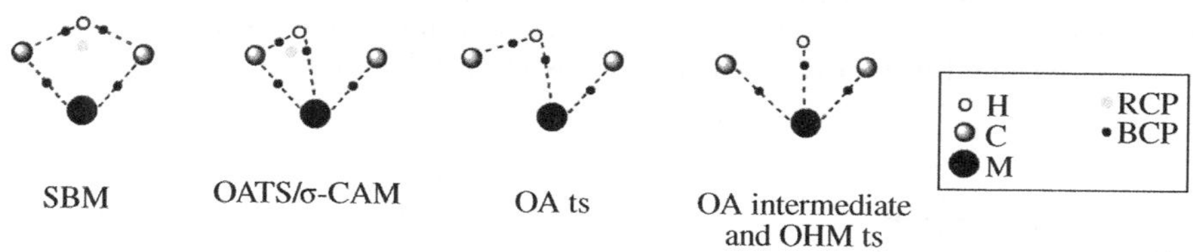

FIGURE 25.3 Spectrum of mechanisms for C—H activation assigned via AIM analysis [9].

Recently, Vidossich and Lledós [11] proposed the analysis of the chemical bonding by the use of the localized orbitals approach. The outcome of this approach provides the centroid of the charge and the area covered by each orbital. The molecular structure together with the orbital centroids will help on the assignation of an oxidation state to the metal center, as we could visualize "where" the electrons are and to "whom" they belong.

The authors extended the study to the calculation of the intrinsic reaction coordinates (IRC) of a reaction in order to use this approach for the visualization of the reaction path by the superposition of all IRCs computed. An example of the visualization is shown in Figure 25.4. The representation

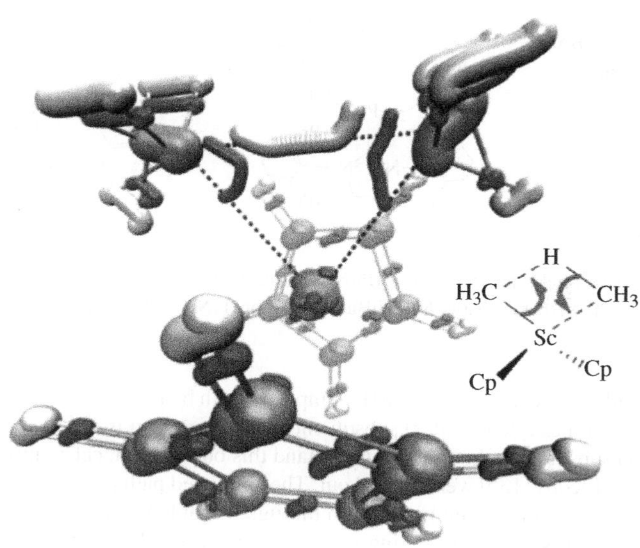

FIGURE 25.4 Superposition of configurations from the intrinsic reaction coordinates (IRC) of the C—H bond activation for $Cp_2Sc(CH_4)(CH_3)$ [11]. Reproduced by permission from Ref. [11]. Copyright (2014) The Royal Society of Chemistry.

shows the C—H bond activation for the $Cp_2Sc(CH_4)(CH_3)$ complex via a σ-bond metathesis mechanism. The transformation of the Sc—C bond to the new C—H bond and that of the C—H bond from methane to the new Sc—C bond is described by the white and grey delocalized lines between the atoms.

25.3 DEVELOPMENT OF STOICHIOMETRIC C—H BOND ACTIVATION

Bergman et al. [12] reported one of the first studies of C—H bond activation with a transition metal system capable of intermolecular oxidative addition. The reaction involved the photolysis of $[Ir(\eta\text{-}Cp^*)(PMe_3)(H)_2]$ in different hydrocarbon solvents. Figure 25.5 shows that the reaction likely proceeds via H_2 loss to form a very reactive Ir^I 16 electron metal intermediate. The C—H activation proceeds via a 3-centered transition state leads to an Ir^{III} hydrido–alkyl complex in a high yield at room temperature. The process is well described as an oxidative addition of the alkane.

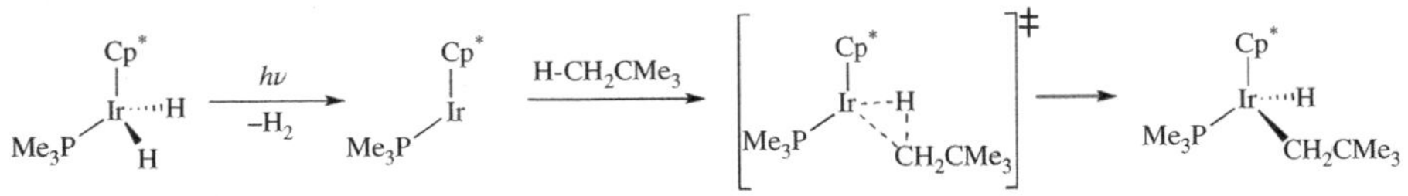

FIGURE 25.5 C—H activation upon photolysis of $[Ir(\eta\text{-}Cp^*)(PMe_3)(H)_2]$.

Oxidative addition is still one of the key processes observed in the literature. In fact, a better understanding of the steric and electronic properties of complexes that can promote C—H bond activation throughout this mechanism has been widely studied. The recent work of Conejero and coworkers [13] highlights this interest by the alteration of the environment of the *N*-heterocyclic carbene (NHC) ligands on a series of stable T-shaped $[Pt(NHC')(NHC)][BAr^f_4]$, where NHC′ represents the cyclometallated ligand, species (Fig. 25.6).

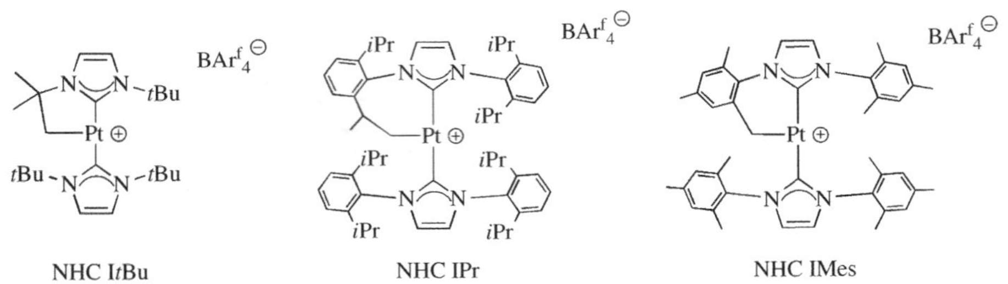

FIGURE 25.6 Structural representation of $[Pt(NHC')(NCH)][BAr^f_4]$ complexes. NHC′ refers to the cyclometallated ligand. The NHC for *t*-butyl (NHC I*t*Bu, left), *i*-propyl (NHC IPr, center), and mesytilen (NHC IMes, right) are depicted.

Experimentally, the reaction of the Pt(II) complexes with benzene only occurs when the NHC is functionalized with mesitylene (IMes) substituents and no reaction is observed with *t*-butyl (I*t*Bu) or *i*-propyl (IPr) substituents. In order to understand this observation, calculations in collaboration with the group of Lledós [13b] were carried out. The computed path proceeds through an oxidative addition and reductive elimination mechanism through unstable Pt(IV) intermediates. In all cases, the oxidative addition step is the limiting step of the process although the overall energy profile differs notably depending on the NHC ligand used. For NHC I*t*Bu and IPr, the energy barrier required is over 40 kcal/mol, while the functionalization by IMes lowers it by at least 10 kcal/mol, which is good agreement with experimental results. The analysis of the computed geometry for each system reveals the inability of the IMes ligand to establish internal agostic interactions with the metal center. This feature is key for favoring the C—H bond activation, as the Pt center is more accessible for reaction.

In 1985, Watson and Parshall [14] reported the first example of a σ-bond metathesis. The interchange between $[Cp^*_2MCH_3]$ (M=Y and Lu) and ^{13}C-labeled methane was observed. They suggested a concerted 4-centered transition state, as depicted in Figure 25.7.

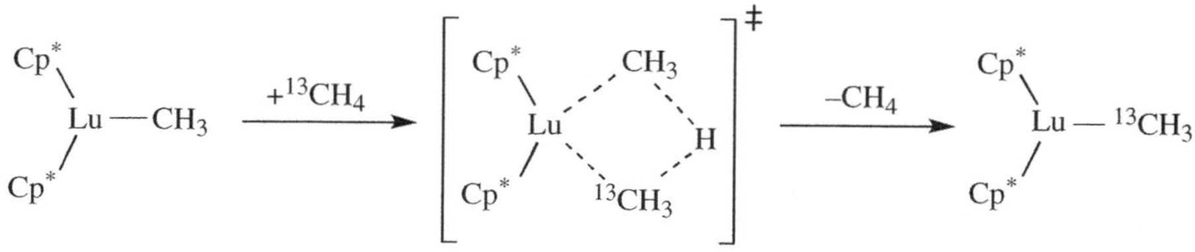

FIGURE 25.7 Methyl exchange reaction of $Cp^*_2LuCH_3$ with $^{13}CH_4$ via a σ-bond metathesis mechanism [14].

The use of computational methods has also been extended to understand the factors that make σ-bond metathesis favorable. Ziegler et al. [15] investigated the ability of $[Cp_2Sc{-}H]$ and $[Cp_2Sc{-}CH_3]$ to undergo σ-bond metathesis with the C—H bonds of methane, ethene, and ethyne. A barrier of 10.8 kcal/mol was computed for the exchange of the methyl substituent in $[Cp_2ScCH_3]$ complex, which correlates with the experimental process of Lu described earlier (11.7 kcal/mol). Lower barriers were computed for ethene and ethyne. The authors concluded that the σ-bond metathesis in this system was favored due to greater *s* character and lower *p* character at the carbon atom included in the {Sc···R···H···R′} moiety, which favors the directionality of the new σ-bond formed.

Following these previous examples, Bergman [16] reported a C—H and Si—H bond activation employing an Ir^{III} complex $[Ir(\eta^5\text{-}Cp^*)(PMe_3)(CH_3)(OTf)]$ (where $OTf{=}OSO_2CF_3$). Their proposed mechanism is shown in Figure 25.8. After ionization of the $[Ir(\eta^5\text{-}Cp^*)(PMe_3)(CH_3)(OTf)]$ complex by elimination of OTf^-, the active species promotes the C—H activation step. Interestingly, both σ-bond metathesis and oxidative addition mechanism were considered as, experimentally, it was not clear whether an Ir(V) intermediate would be accessible.

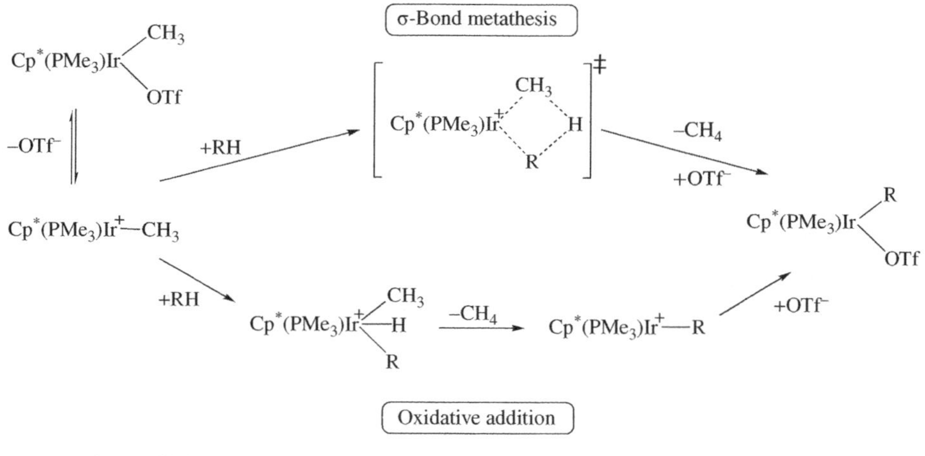

FIGURE 25.8 Proposed mechanisms for C—H activation at $[Ir(\eta^5\text{-}Cp^*)(PMe_3)(CH_3)]^+$ [16].

Computational strategies have reached such a level of accuracy that DFT calculations can be used to distinguish these type of issues. Calculations carried out by Strout et al. [17] on the model complex $[Ir(\eta^5\text{-Cp})(PH_3)(CH_3)]^+$ reacting with methane suggested that the oxidative addition mechanism is favored. The activation energy required is only 11.5 kcal/mol and involves the initial formation of an agostic intermediate more stable than the separated reactants. All attempts to locate a σ-bond metathesis mechanism failed. This result was confirmed by the calculations carried out by Niu and Hall [18] on the analogous $[Ir(\eta^5\text{-Cp}^*)(PMe_3)(CH_3)]^+$ complex. Experimentally, these two pathways are hard to differentiate as σ-bond metathesis leads to the same product as oxidative addition followed by reductive elimination.

The most commonly observed examples of C—H bond activation by electrophilic activation occur with Pd(II), Rh(III) and Au(III). The most detailed experimental study of electrophilic activation appeared in 1985 by Ryabov et al. [19]. A detailed kinetic and mechanistic study of activation of *N,N*-dimethylbenzylamine (dmba-H) by palladium acetate in chloroform solution was reported. The formation of a cyclopalladated acetate-bridged dimer, $[Pd(OAc)(dmba)]_2$, is shown in Figure 25.9. It is important to highlight that the reaction occurs at room temperature. The analysis of the H/D kinetic isotope effects and the observation of large negative entropy allowed the proposal of a highly ordered transition state similar to a Wheland intermediate. In this transition state, the approach of the metal toward the aryl group leads to the delocalization of a positive charge around the ring. Additionally, the stabilization of this transition state by an interaction between the leaving hydrogen and the acetate is suggested. The electrophilic nature of the reaction was experimentally analyzed by a comparison of the rate of the cyclopalladation step for a series of *para*-substituted amines. Thus, electron-withdrawing substituents result in higher rates of reaction.

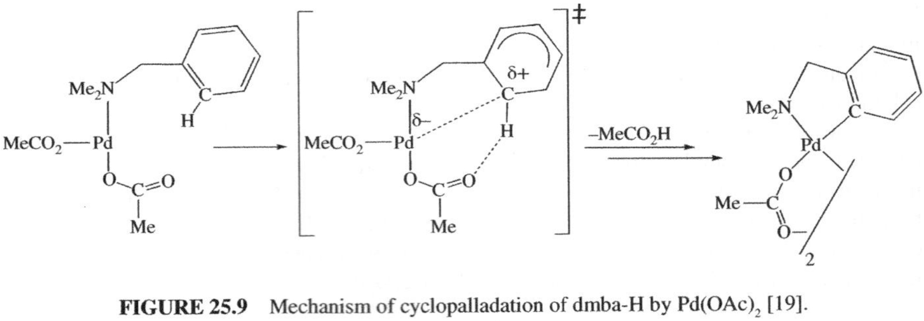

FIGURE 25.9 Mechanism of cyclopalladation of dmba-H by $Pd(OAc)_2$ [19].

Macgregor et al. [20] computed the mechanism of this cyclometalation. Three possible transitions states were modeled for the C—H activation step: (i) a 6-membered acetate-assisted process, where the proton is transferred to the outer oxygen of acetate; (ii) a 4-membered process via a transfer to the inner oxygen; and (iii) oxidative addition (see Fig. 25.10). The acetate-assisted

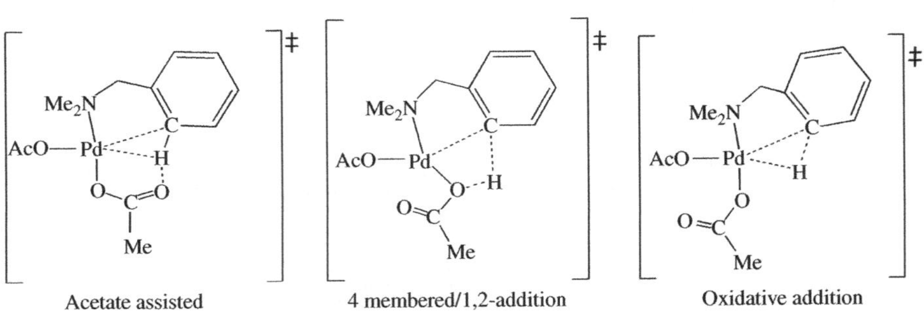

FIGURE 25.10 Suggested transition states for the C—H activation of dmba-H by $[Pd(OAc)_2]$ [20].

6-membered transition state was found to be the most favorable pathway for the C—H activation step and required an energy barrier of 11.1 kcal/mol. The other two processes were computed to be associated with very high computed barriers of 34.3 kcal/mol and 25.7 kcal/mol for the 4-membered and oxidative addition mechanisms, respectively. The 4-membered transition state could be described as a σ-bond metathesis due to its geometry.

A detailed reaction path for the respective cyclometalation through a 6-membered transition state is shown in Figure 25.11. The displacement of one arm of the κ²-acetate from [Pd(κ²-OAc)(κ¹-OAc)(dmba-H)] (**1a**) allows the approach of the *ortho* C—H bond via the $\mathbf{TS}_{1a-2a}$ transition state, which leads to an agostic intermediate **2a**. The second step describes an H-transfer through a 6-membered transition state ($\mathbf{TS}_{2a-3a}$), which involves the C—H bond from the aryl moiety, the metal, and the free oxygen arm of the acetate ligand. This final step occurs via a minimal activation barrier and leads to the final cyclometallated product **3a**. In contrast to the mechanism proposed by Ryabov [19] via a Wheland intermediate, the reaction proceeds via a C—H agostic intermediate **2a**. The formation of the agostic intermediate is not only the rate-determining step but also the key stage of the process as it enhances the acidity of the C—H bond and facilitates the following proton transfer.

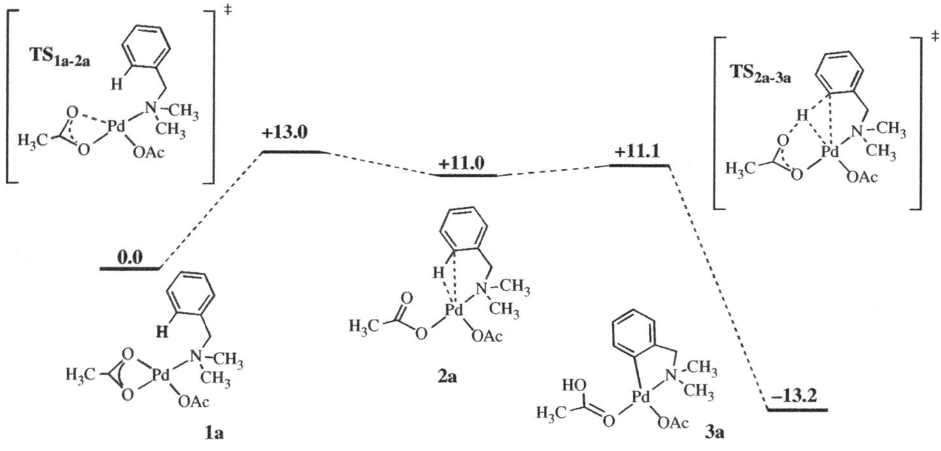

FIGURE 25.11 Computed reaction profile (kcal/mol) for the cyclometalation of dmba-H by $Pd(OAc)_2$ via a 6-membered C—H activation transition state [20].

25.3.1 Mechanistic Ambiguity: The Power of Theory

Once the classics of C—H activation mechanisms have been described, our interest will focus to more up-to-date examples were the choice of the mechanism is more ambiguous. Lin, Lau, and Eisenstein [4c] studied the C—H activation of methane by [M(η-Tp)(PH_3)(CH_3)] (M = Fe, Ru, and Os; Tp = hydridotris(3,5-dimethylpyrazolyl)borate). As seen in Figure 25.12, depending on the nature of the metal center, the computed activation barriers increase, following the trend Os < Ru < Fe. An oxidative addition path was computed for the respective Os catalyst. The two-step mechanism identifies an Os^{IV} intermediate, which is typical for an oxidative addition mechanism. In the case of the Fe system, no intermediate is located, and the transition state features a very short Fe···H distance (1.53 Å). Finally, the Ru system exhibits an intermediate behavior, as a very shallow Ru^{IV} intermediate was identified. After a structural study of each of the transition states, the authors concluded that the Fe system undergoes reaction through a transition state that resembles an oxidative addition mechanism. Therefore, transition states which feature short M···H contacts have been named "oxidatively added transition state" (OATS) by Lin and coworkers [22].

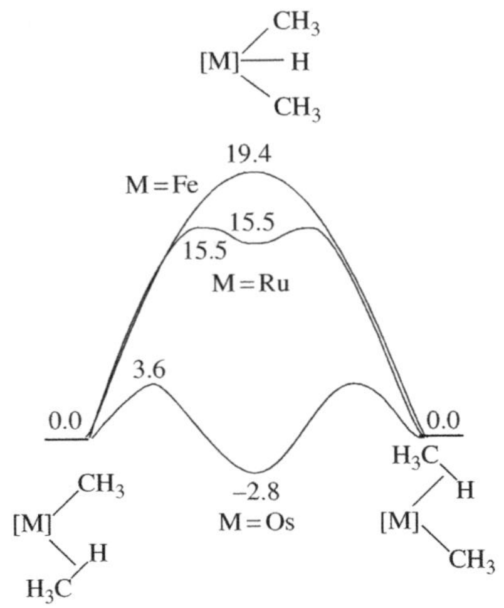

FIGURE 25.12 Computed C—H activation energy profiles (kcal/mol) for species ([M(η-Tp)(PH_3)] [21] where (M=Fe, Ru and Os; Tp=hydridotris(3,5-dimethylpyrazolyl)borate) [4c].

The computational analysis of Goddard and Periana [23] on the hydroarylation of ethene by benzene catalyzed by the [Ir(acac′)Ph(H_2O)] complex (**A**) locate the intermediate [Ir(acac′)(η^2-C_6H_6)C_2H_4Ph] (**B**) as product for the C—H activation (see Fig. 25.13). The structure of the computed transition state exhibits a seven-coordinate Ir^V species and a short Ir—H bond (1.58Å). In this case, the mechanism was named OHM. Different than previous analysis, they suggested that by reducing the electron density in the Ir^{III} center by electron-donating ligands, the following C—H activation could change from an oxidative addition to an OHM mechanism.

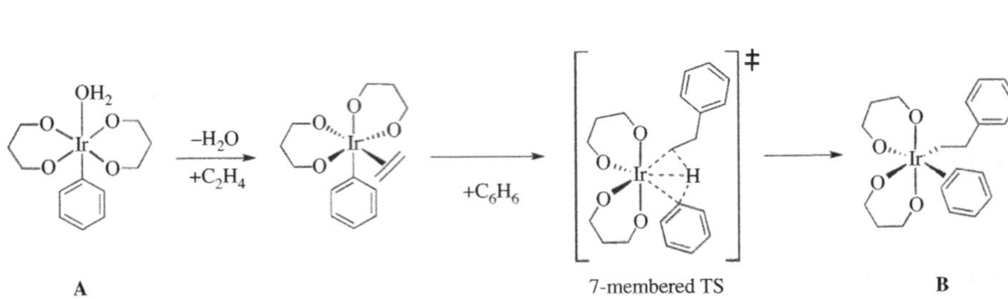

FIGURE 25.13 Mechanism of C—H activation in [Ir(acac′)(η^2-C_6H_6)C_2H_4Ph] [23b].

The importance of steric hindrance was studied by Bergman et al. [21] in the computational analysis of the C—H activation of methane by [CpRe$(CO)_2$] and [TpRe$(CO)_2$] (where Cp=C_5H_5 and Tp=tris(1-pyrazolyl)borate). The seven-coordinate product [TpRe$(CO)_2$(H)(Me)] is not favorable as it suffers from closer intramolecular contacts, while the [CpRe$(CO)_2$(H)(Me)] complex is less crowded. The author's observations suggest that nonbulky ligands will favor a less crowded coordination sphere and promote an exothermic C—H activation via an oxidative addition intermediate.

Changes in the metal environment can be also generated by a prior coordination of a H—R ligand, which acts as a two electron donor to the metal center in a σ-complex. A detailed study by Perutz and Sabo-Etienne [24] based on the experimental observation of reactant and product σ-complexes lead to the proposal of σ-CAM. This idea allows σ-bond metathesis to be adapted for C—H activation of σ-complexes at late transition metals. Figure 25.14 shows the mechanism that describes dynamic rearrangement of σ-complexes. To date, this process has been only observed experimentally for silanes and boranes; however, in principle this can be extended to hydrocarbon substrates.

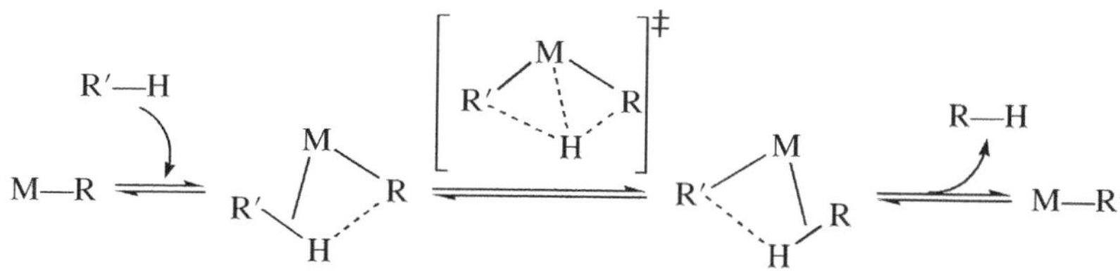

FIGURE 25.14 σ-Complex-assisted metathesis [24].

25.3.2 C—H Activation Assisted by Carboxylate or Carbonate Bases

Recently, interesting new systems that include carboxylates or carbonate in their structure have been broadly studied. One of the most common complexes used is palladium acetate, [$Pd(OAc)_2$]. With this catalyst precursor, the acetate is thought to play an active role in the C—H activation step [7d]. Therefore, a number of computational studies oriented toward the understanding of the mechanism involved in the C—H activation step have appeared in recent years, and these will be the focus of the following section.

The first computational study on such a systems was carried out by Sakaki [25] and compared the C—H activation of benzene and methane by [$M(\kappa^2\text{-}O_2CH)_2$] (M=Pd and Pt). The reaction profiles computed for both systems with [$Pd(\kappa^2\text{-}O_2CH)_2$] are shown in Figure 25.15. The coordination of the benzene (Fig. 25.15, right side) leads to the formation of a more stable complex at −2.3 kcal/mol. The partial dissociation of the one oxygen atom of one acetate ligand leads to an η^2-benzene intermediate that features a short H···O contact. This interaction facilitates an extra stabilization of 9.8 kcal/mol in the η^2-benzene intermediate. The following C—H cleavage from this intermediate requires an activation energy of 16.1 kcal/mol. Interestingly, methane intermediates have a different behavior and feature a weaker C—H···M σ-interaction (+8.8 kcal/mol). Consequently, a much higher-energy intermediate is computed for the C—H cleavage, and the following proton transfer is the rate-limiting step of the process with an overall energy barrier of 21.5 kcal/mol. Exchanging the acetate ligand by PH_3 showed that C—H activation results in an oxidative addition mechanism with high-energy barriers (for $Pd(PH_3)_2$: $\Delta E^{\ddagger}(CH_4)=34.0$ kcal/mol and $\Delta E^{\ddagger}(C_6H_6)=22.7$ kcal/mol). Overall, the difficulty of the processes is overtaken by the assistance of the free oxygen of the $\kappa^1\text{-}O_2CH$ of the [$Pd(\kappa^2\text{-}O_2CH)_2$] complex as a less constrained 6-membered transition state is achieved.

Lledós [26] reported on a similar mechanism, also with [$Pd(OAc)_2$], for the selective cyclopalladation of iminophosphoranes $R_3P{=}NCH_2Ar$. The energy profile found for the exo cyclometalation (Fig. 25.16) exhibited an agostic intermediate (**5**) in analogy to the mechanism described by Macgregor and coworkers (see Fig. 25.11).

Experimental studies by Davies and coworkers [27] showed that dmba-H reacts with [$IrCl_2Cp^*$]$_2$ in presence of sodium acetate under mild conditions to form [Ir(η-Cp*)(dmba)Cl] as the final cyclometalated product. To observe the final product, the use of acetate as a base was a must, while the use of other bases such as triethylamine remained unsuccessful. This methodology is also used for the cyclometalation of imines, and similar reactivity is seen with [$RhCl_2Cp^*$]$_2$ and [$RuCl_2$(*p*-cymene)]$_2$ [28].

Based on these experimental results, Macgregor et al. [29] reported a computational study of this cyclometalation reaction focusing on the mechanism involved in the C—H activation step (Fig. 25.17). The model system considered was [$Ir(\eta^5\text{-}Cp)(\kappa^2\text{-}OAc)(dmba\text{-}H)$]$^+$ (**7**). The lowest-energy pathway involves a 6-membered transition state, **TS(7-9)**, with an activation barrier of

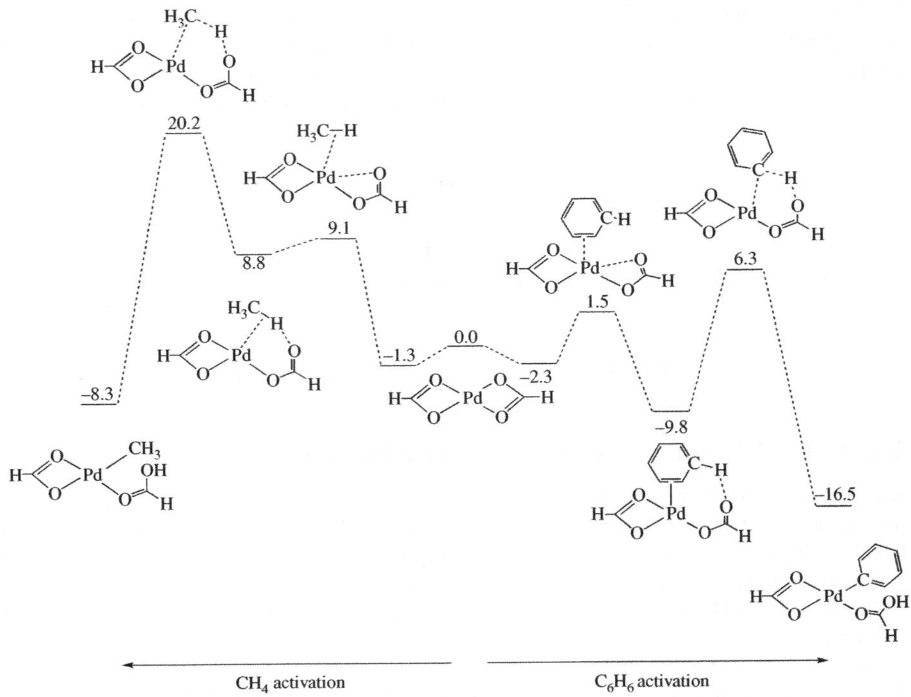

FIGURE 25.15 C—H bond activation of benzene (right) and methane (left) by [Pd(κ^2-$O_2CH)_2$] [25].

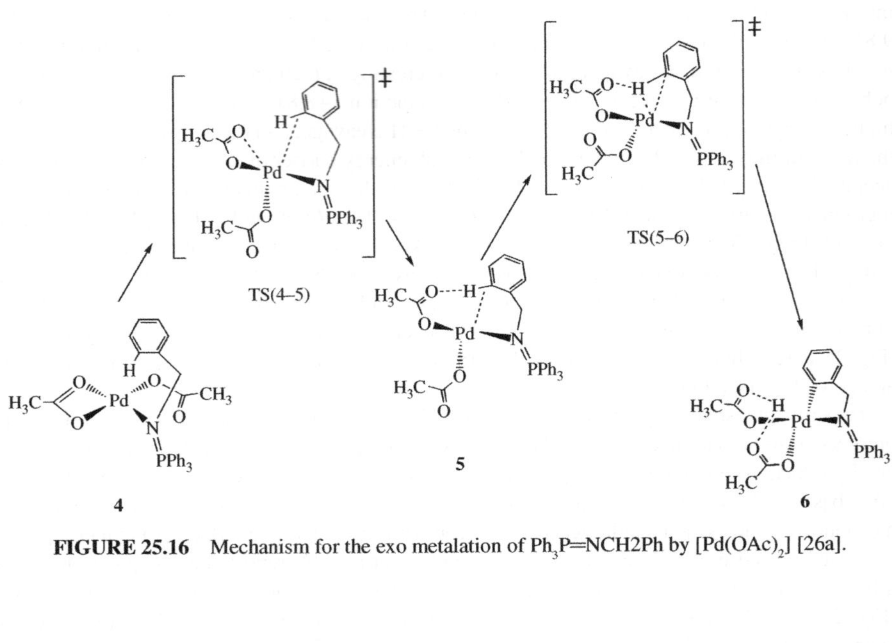

FIGURE 25.16 Mechanism for the exo metalation of Ph_3P=NCH2Ph by [Pd(OAc)$_2$] [26a].

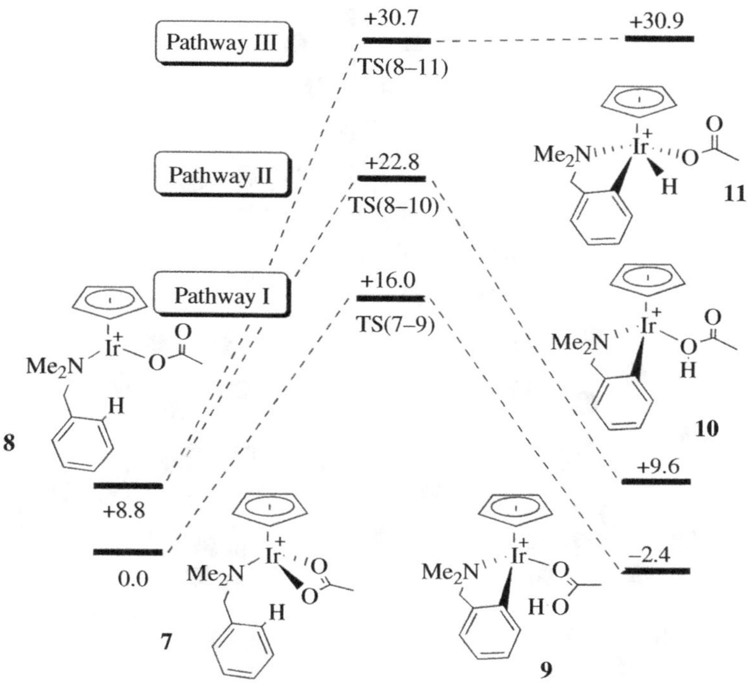

FIGURE 25.17 Computed reaction profiles (kcal/mol) for C—H activation in [Ir(η^5-Cp)(dmba-H)(κ^2-OAc)]$^+$ [29].

16.0 kcal/mol, which is similar to the activation barrier computed with [Pd(OAc)$_2$(dmba-H)] (+13.0 kcal/mol, see Fig. 25.11). Both the 1,2 addition and oxidative addition pathways are higher in energy (**TS(8–10)** = 22.8 and **TS(8–11)**= 30.7 kcal/mol) and follow the same pattern as seen for the palladium system as a κ^2–κ^1 displacement (**8**) is required before the following C—H activation.

A comparison between the key distances in the three C—H activation transition states is shown in Figure 25.18. The availability of the free oxygen of the acetate ligand is key for the formation of the cyclometallated product in all three pathways. Pathway I correspond to a one-step process involving a 6-membered transition state, **TS(7–9)**, where dissociation of one arm of the acetate is coupled to the hydrogen transfer from the activated C—H bond of dmba. This transition state bears two key features at once: (i) an agostic C—H···M interaction and (ii) an intramolecular C—H···O—H bond. Pathway II presents a 4-membered transition state where a rotation around the Ir—O$_1$ bond permits the hydrogen to be transferred to the metal-bound oxygen, via **TS(8–10)**. For Pathway III, the free oxygen is twisted away from the hydrogen, and the metal must become fully involved in the C—H cleavage (**TS(8–11)**). Calculations with Cp* instead of Cp did not result in any significant change to the reaction energies profiles.

Further studies from the same group [30], considering the use of different chelating bases, allow a better insight into the factors that control this type of C—H activation process. In this paper, the authors report a stabilization of all the stationary points along the three computed pathways. Pathway I is still the most favored although now, it has been described by a two-step process. The first and rate-limiting step is the dissociation of one arm of the acetate ligand, and the second step, almost barrierless, is the C—H activation (Fig. 25.19).

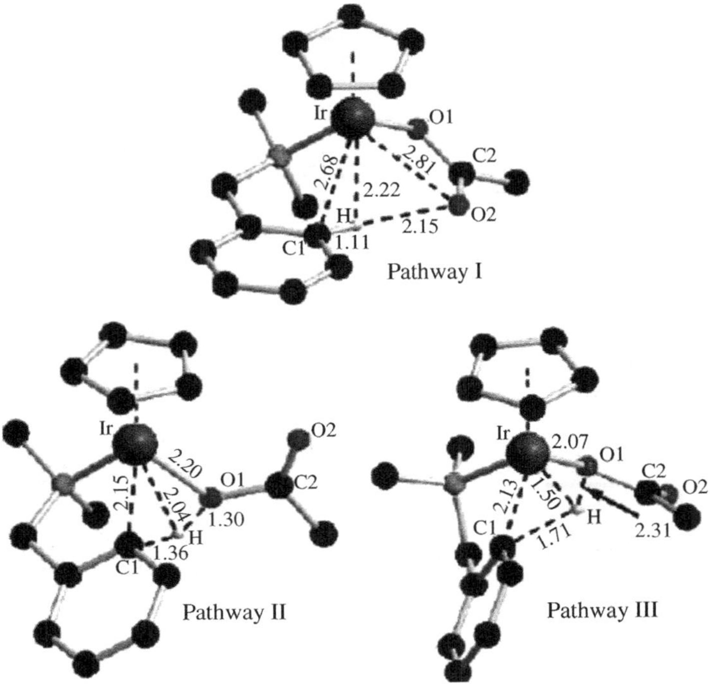

FIGURE 25.18 Computed C—H activation transition states along Pathways I–III described in Figure 25.17. Non participating H atoms are omitted for clarity and distances in Å [29] Adapted with permission from Ref. [29]. Copyright (2006) American Chemical Society.

Changes in the chelating base, from stronger to weaker, were also investigated. The displacement of one arm of the base becomes easier for weaker bases, while the energy barrier of the C—H activation step increases its energy barrier although it always remains feasible. Experimentally, the chelating base affects the cyclometalation process, but no simple correlations between the calculations and reactivity has been observed yet. These studies have highlighted the important roles of both the metal and the acetate in the C—H activation process. The agostic interaction between the metal center and the C—H bond that is to be cleaved enhances the proton acidity and facilitates the transfer. Therefore, an ambiphilic behavior, in which the metal and the internal base work synergistically in order to facilitate the C—H cleavage, leads to the definition of an AMLA mechanism. This type of mechanism could promote C—H activation through a 4- or 6-membered transition state named AMLA-4 and AMLA-6, respectively.

A subsequent study by Goddard and Periana [31] in the intermolecular C—H activation of benzene and methane by $[Ir(acac')_2(X)]$ complexes ($X = CH_3COO$ and OH) also facilitates an easy bond activation process and demonstrates a similar ambiphilic bonding character. An acetate-assisted deprotonation via a 6-membered transition state was computed as the lowest-energy process in both cases. This underlines again the high potential of such ambiphilic character for an easier C—H activation process.

Maseras and Echavarren [32] have observed similar behavior in intramolecular arylation promoted by $[Pd(OAc)_2]$ (Fig. 25.20). Interestingly, only the use of an excess of K_2CO_3 leads to satisfactory results, whereas Et_3N or DBU (1,8-diazabicycloundec-7-ene) led to starting material. This experimental observation suggests that the mechanism should occur via an assisted C—H activation process.

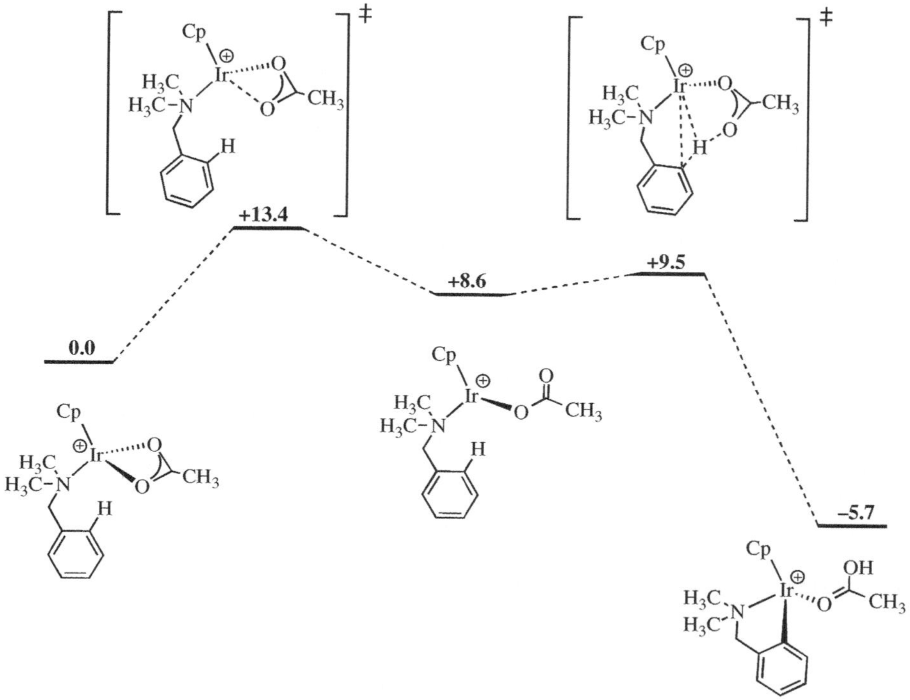

FIGURE 25.19 Computed two steps reaction profile (kcal/mol) for the cyclometalation of dmba-H by [Ir(η[5]-Cp)(dmba-H)(κ[2]-OAc)]$^+$ via a 6-membered C—H activation (AMLA-6) transition state [30].

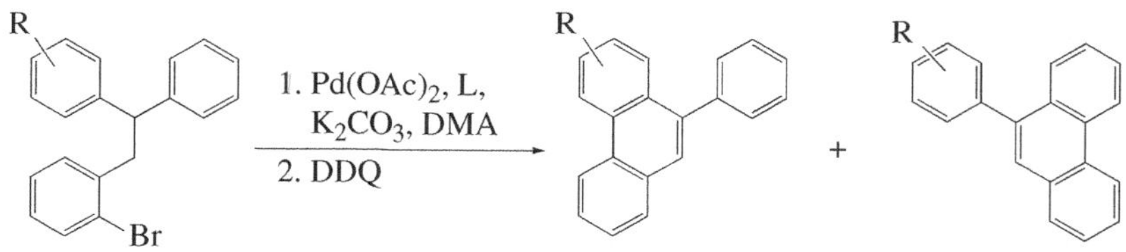

12 13 14

FIGURE 25.20 Pd-catalyzed intramolecular arylation [32].

The coordination of the Pd complex to the reactant via C—Br oxidative addition of the substrate gives intermediate **15**. The substitution of Br^- by an intramolecular base, HCO_3^-, leads to species **16**. The energy barrier for the C—H activation is as low as 23.5 kcal/mol, consistent with the experimental temperature of 100–135°C, and occurs via a 6-membered transition state. The authors named this process as "proton abstraction mechanism" analogous to the acetate-assisted C—H bond activation (AMLA), although in this process bicarbonate instead of acetate is bonded to the metal (see Fig. 25.21).

One year later, Echavarren and Maseras [33] reported new variants on the "proton abstraction mechanism" described earlier. In this case, special attention was paid to the substituents on the

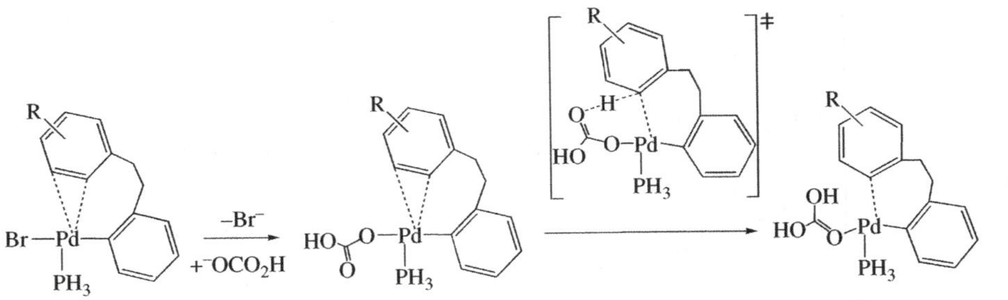

15 **16** **17**

FIGURE 25.21 Mechanism computed for the formation of a palladacycle via a 6-membered transition state using [Pd(PH_3)(κ^1-OCO_2H)(*o*-$(CH_2)_2$-C_6H_4R)Ph)] (R = H) as model system [32].

aromatic ring to probe their directing effect. As before, [Pd(PH_3)Br(*o*-((CH_2)$_2$-C_6H_4R)Ph)], intermediate **15**, was used as a model system. Three possible pathways were considered and are depicted in Figure 25.22. The highest-energy barrier is computed for the path where nonassistance from the bicarbonate is present and the proton transfer occurs directly onto Br. This correlates with the need of a base in the reaction. The activation barriers found for the two assisted pathways are 17.4 kcal/mol for the intermolecular assisted route and 23.5 kcal/mol for the intramolecular mechanism. Interestingly, changes in the R group on the aryl moiety, from CH_3 to CF_3, invert this pattern

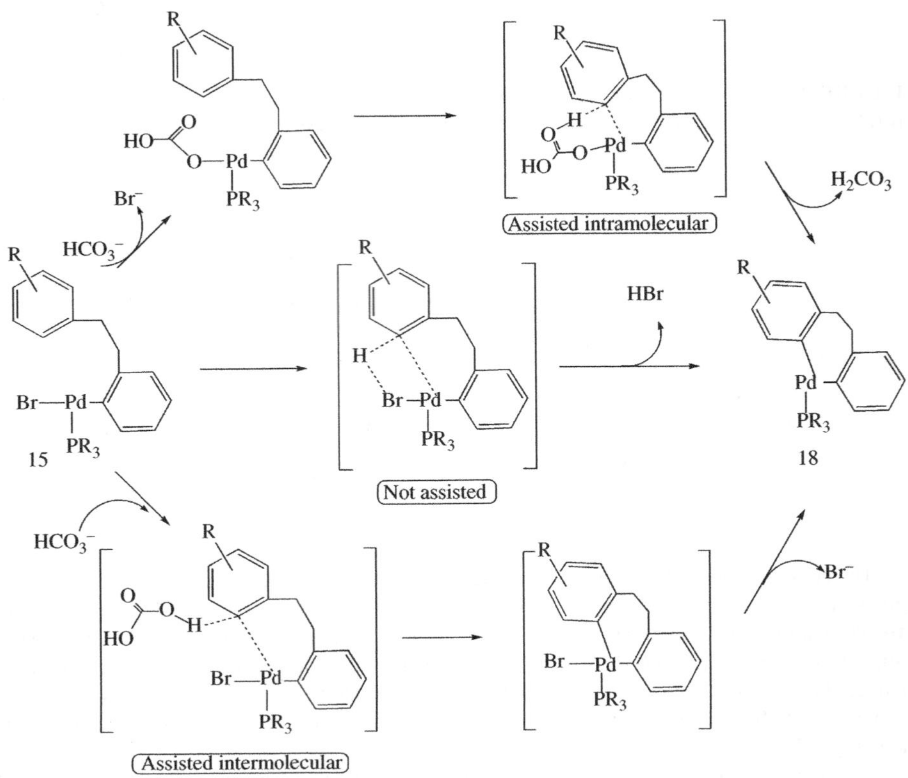

FIGURE 25.22 Three possible mechanisms proposed for the formation of a palladacycle [33].

(14.4 kcal/mol vs. 13.2 kcal/mol, for inter- vs. intramolecular pathway). This change on behavior emphasizes that both mechanisms can be operative and the preference for one over another may depend on the substrate.

Following this study, Fagnou et al. [34] reported the direct arylation of different aromatic substrates promoted by $[Pd(OAc)_2]$. This process followed a CMD mechanism, which describes simultaneous making–breaking bonds. A closer look to the transition state highlights that the elongation of the C—H bond is not in relation with the acidity of the arene as electron-rich and electron-deficient arenes promote the C—H activation by similar activation barriers (see numbers in bold in Fig. 25.23). This observation highlights that the metal and the base are key to promote the arylation reaction with no dependency on the nature of the substrate.

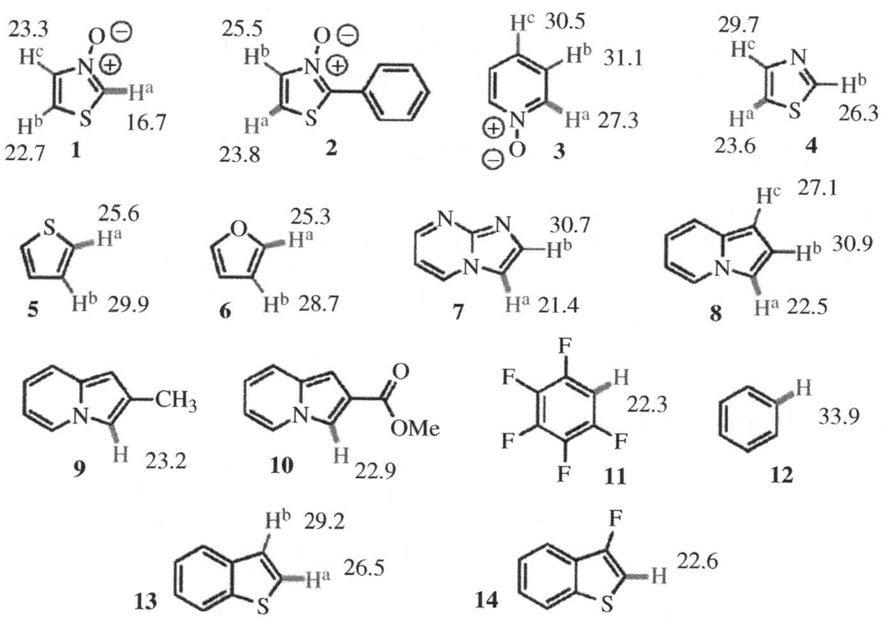

FIGURE 25.23 Aromatic aryl substrates functionalized by benzene on the presence of $[Pd(OAc)_2]$. Numbers in bold are free energy of activation in kilocalories per mole. Bold bonds indicate the experimental observed sites of arylation [34]. Adapted with permission from Ref. [34]. Copyright (2008) American Chemical Society.

Recently, the mechanistic analysis of iridium-catalyzed direct arylation of heteroarenes by Gorelsky and Woo [35] embraces the analysis of different mechanisms (oxidative addition, CMD, and electrophilic activation) and the comparison of the C—H bond activation step by Ir(III) and Pd(II). Both metals promote this step through a concerted metalation–deprotonation (CMD) mechanism. Interestingly, the involvement of the metal center is different, and it is related to the larger electrophilic character of the Ir(III) vs. Pd(II). The iridium center models a stronger metal-carbon and weaker base-proton interactions than the Pd species as it is depicted in Figure 25.24. The arrows in the figure indicate charge donation from the C atom of the arene to the metal and from the base to the arene. The size of the arrow is directly related to the magnitude of the charge transfer interactions. The involvement of the metal in the arylation shows that both metal and ligand are key for the C—H bond activation step.

In conclusion, the presence of carboxylate or carbonate bases in the functionalization of different substrates has defined a new class of mechanisms in which metal and base work synergistically

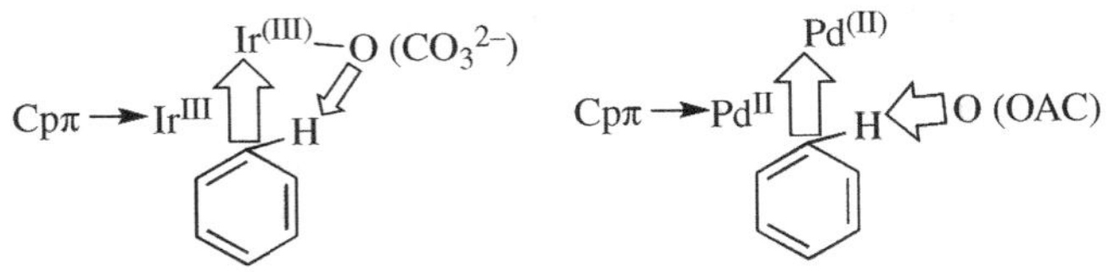

FIGURE 25.24 Schematic representation of the donor–acceptor interactions for the CMD transition states with Ir(III)-carbonate (left) and Pd(II)-acetate catalysts (right) [35].

to facilitate the C—H bond activation step. Even if can be challenging to differentiate between the discussed mechanisms, the cooperative behavior between these two features in the transition metal complexes during the C—H activation step is the key. Overall, the proton abstraction mechanism typically models a weaker involvement of the metal center than the CDM, or the AMLA mechanisms require a stronger involvement of the metal center for the C—H bond activation.

25.4 CATALYTIC C—H ACTIVATION AND FUNCTIONALIZATION

The real challenge in organometallic chemistry is the integration of the C—H activation processes described previously into catalytic transformations. Such reactions are very promising and powerful for the construction of C—C bond frameworks. A wide variety of methodologies have been developed on the basis of the understanding of the mechanism. Combination of experimental and computational chemistry has aimed to the development of new catalysts. The following will describe the study of the C—H activation of arenes into successful catalytic cycles.

25.4.1 Hydroarylation of Alkenes

The groups of Goddard and Periana [23, 36] have focused some of their efforts on the study of the mechanism of Ir^{III} catalyzed hydroarylation of alkenes with $[Ir(acac)_2(Ph)(H_2O)]$ as catalyst [37]. The dissociation of the water ligand from the catalyst $[Ir(acac)_2(Ph)(H_2O)]$ (**49**) is followed by a *cis* rearrangement of the acac ligands, allowing alkene coordination to form $[Ir(acac)_2(Ph)(\eta^2\text{-}C_2H_4)]$ (**50**). Ethylene and phenyl ligands need to be in a *cis* fashion in order to promote the insertion step, which requires an activation barrier of $\Delta H^{\ddagger} = 35.1$ kcal/mol and $\Delta G^{\ddagger} = 26.1$ kcal/mol, in good agreement with the free energy of activation determined experimentally (28.7 kcal/mol). The catalytic cycle starts from $[Ir(acac)_2(Ph)(\eta^2\text{-}C_2H_4)]$ (**50**), and its reaction profile is shown in Figure 25.25 [23b]. The first intermediate $[Ir(acac)_2(C_2H_4Ph)]$ (**51**) is obtained by migratory insertion via a 4-membered transition state, which requires an activation energy of $\Delta H^{\ddagger} = 27.0$ kcal/mol. Its geometry emphasizes an agostic interaction with one *ortho* C—H bond of the phenyl group. Therefore, the coordination of a second benzene requires the rotation of this ligand away from the metal to give **53**. Finally, an oxygen hydrogen migration transition state is computed (12.0 kcal/mol) to give **54**. To complete the catalytic cycle, this complex dissociates ethyl benzene (uphill by 2.9 kcal/mol) and coordinates a new ethene molecule ($\Delta H = -19.2$ kcal/mol). The limiting step of the process is the alkene insertion ($\Delta H^{\ddagger} = 27.0$ kcal/mol).

The authors also consider two main side reactions from structure **51**: (i) β-H elimination and (ii) the binding of a second molecule of ethene. They assessed both possibilities by computing the set of reaction depicted in Figures 25.26 and 25.27. β-H-transfer was computed to be more favorable than the coordination of a benzene molecule. However, the energy required for the elimination of styrene was slightly higher than the insertion of an ethylene molecule. The authors concluded that this side reaction is possible and facile, but reversible, that it will still allow the catalytic cycle.

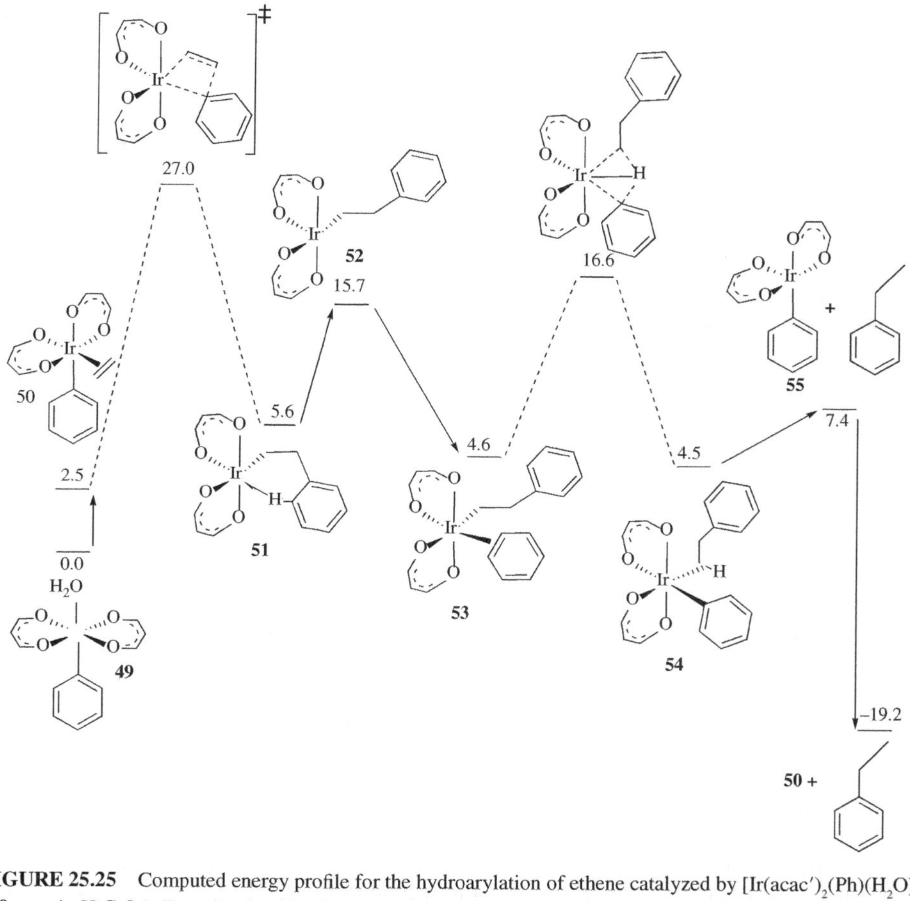

FIGURE 25.25 Computed energy profile for the hydroarylation of ethene catalyzed by $[Ir(acac')_2(Ph)(H_2O)]$ (**49**, acac′ = $H_3C_3O_2$). Energies (kcal/mol) are reported relative to free ethene and **49** [23b].

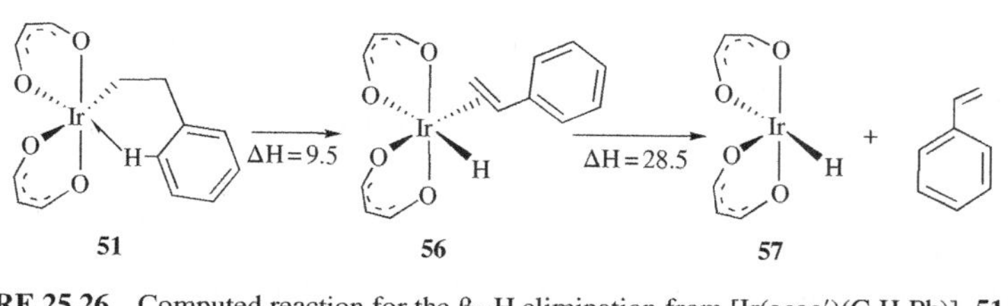

FIGURE 25.26 Computed reaction for the β—H elimination from $[Ir(acac')(C_2H_3Ph)]$, **51** [23b].

The second side reaction, the addition of a second ethylene molecule to **51**, was also favorable (Fig. 25.27). Complex **58** can promote an insertion of the second ethylene molecule (**59**) or the C—H activation to form the vinyl complex **60**. Both reactions were computed to be competitive with ethylene insertion.

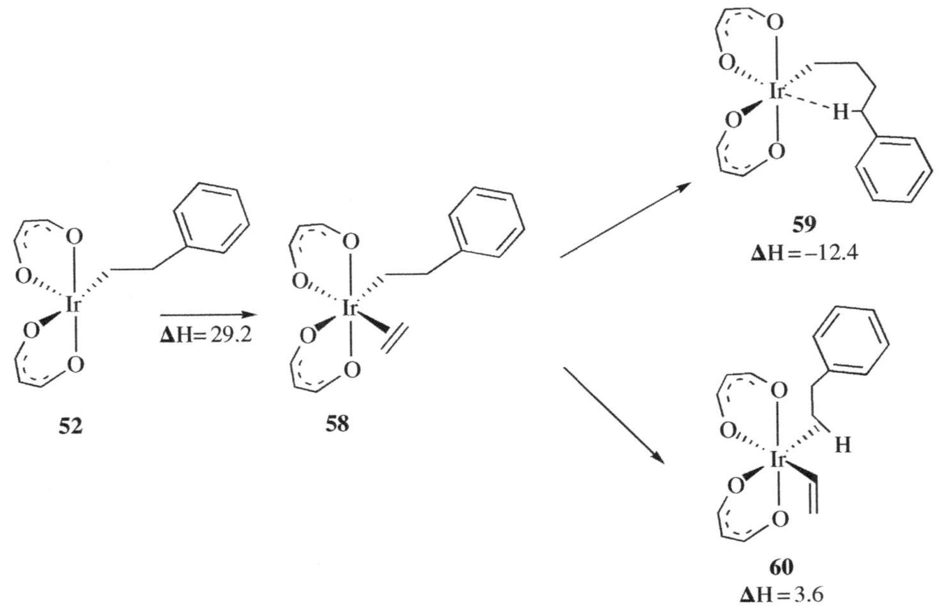

FIGURE 25.27 Computed sides reactions of [Ir(acac′)(C_2H_4Ph)], **52**, with ethene [23b].

Another system that catalyzes the hydroarylation of alkenes is the [RuTp(CO)(NCMe)(Ph)] catalyst reported by Gunnoe and coworkers [38]. Figure 25.28 shows the computational analysis of the catalytic system. There are two key steps: (i) the alkene insertion and (ii) the C—H activation step via an OHM process. As before, alkene insertion into the metal–aryl bond is the limiting step of the process. The overall cycle is exothermic by 18.0 kcal/mol.

Some years later, Gunnoe and Cundari [39] reported how the nature of the *para*-substituents could affect the C—H activation step. Their efforts were focused on comparing two different complexes, [RuTp(CO)(Me)(C_6H_5X)] and [RuTp(PMe_3)(Me)(C_6H_5X)] (X=CN, H, NH_2, NO_2, Br, Cl, F, or OCH_3). The analysis of the structures of the transitions states led to longer Ru—H bond distances for electron-donating substituents. The metal center coordinates to the C—H bond and promotes an intramolecular transfer (Fig. 25.29) similar to a 1,2-addition across Ru—X (X=OH or NHPh).

Following these results, the authors constructed the computational Hammett plots using the Gibb's free energy to calculate the rate constants of the reaction for each complex. Their calculations afford more positive values in these systems than those usually found for σ-bond metathesis processes.

The comparison of these systems with the previously described Ir complexes highlights that the mechanism for the hydroarylation is similar in both species. The limiting step of the process is, in both cases, the alkene insertion. Interestingly, ruthenium and iridium species suggested opposing trends, as alkene insertion is favored for the ruthenium systems, while C—H activation is easier for the iridium complexes.

This observation reveals that a potential new catalyst could be achieved by analysis of both processes by trying to balance both steps. Therefore, Periana and Goddard [4b] carried out a detail mechanistic analysis in order to investigate the generality of this trend. Different metals (Rh, Pd, Os, and Pt) and ligands ({acac} and {Tp(CO)}) were studied. Figure 25.30 shows a correlation between the computed ΔH of the C—H activation and the alkene insertion step. An inverse relationship between both processes is observed, which complicates the optimization of the catalytic process.

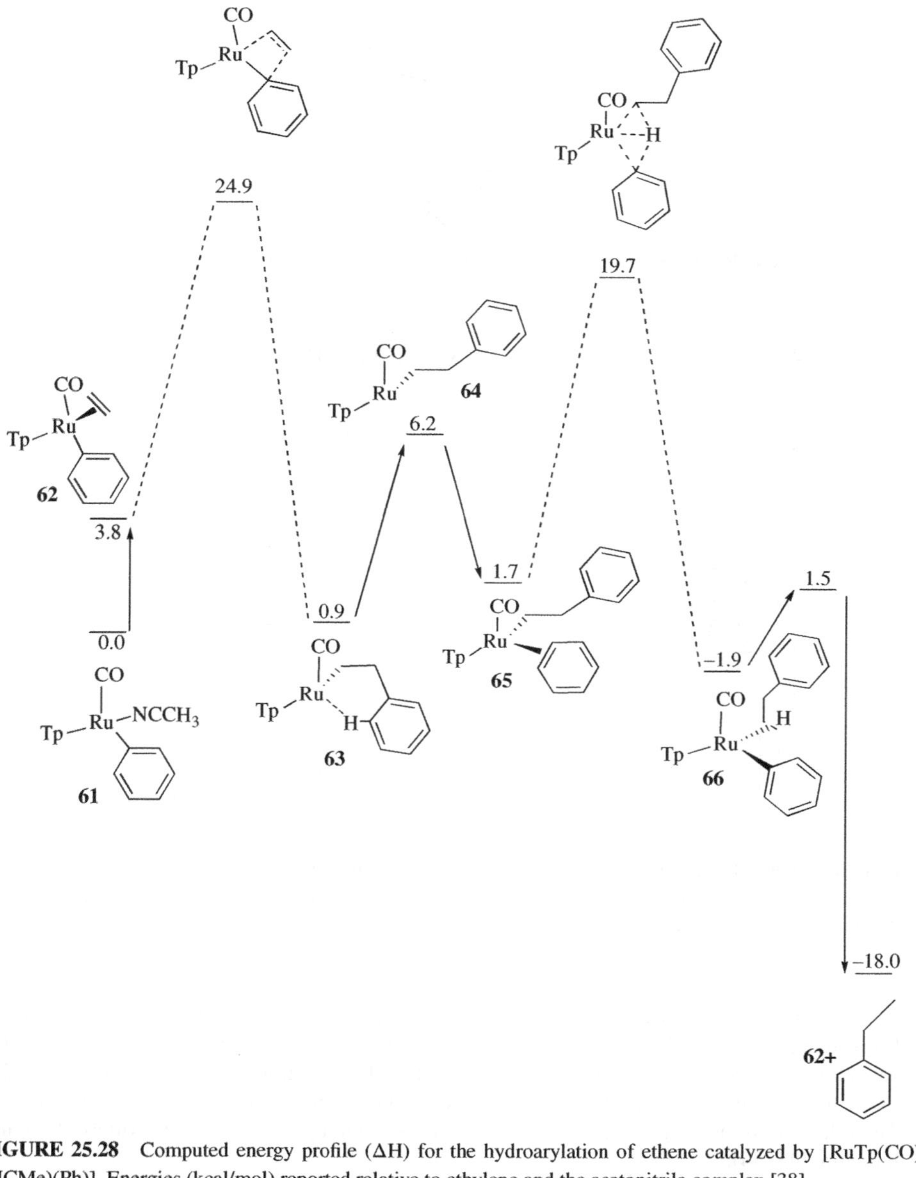

FIGURE 25.28 Computed energy profile (ΔH) for the hydroarylation of ethene catalyzed by [RuTp(CO)(NCMe)(Ph)]. Energies (kcal/mol) reported relative to ethylene and the acetonitrile complex [38].

Recently, Eisenstein and Perutz [40] explored computationally the functionalization of electron-deficient arenes by nickel phosphine complexes. They based the study on the experimental observation of the reaction by Nakao and coworkers [41]. Experimentally, either the C—H or C—F bond activated products could be observed; however, the authors only proposed a mechanism where the C—H activated product was obtained via an oxidative addition mechanism (see scheme on the left on Fig. 25.31). The aim of Eisenstein and Perutz was to explore both possibilities, C—H or C—F bond activation mechanisms, as previously they demonstrated that the C—H oxidative addition could compete with the C—F oxidative addition [42].

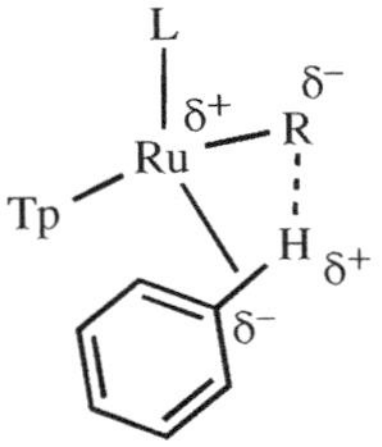

FIGURE 25.29 Model for C—H activation in [RuTp(L)(R)(C_6H_5X)] systems [39].

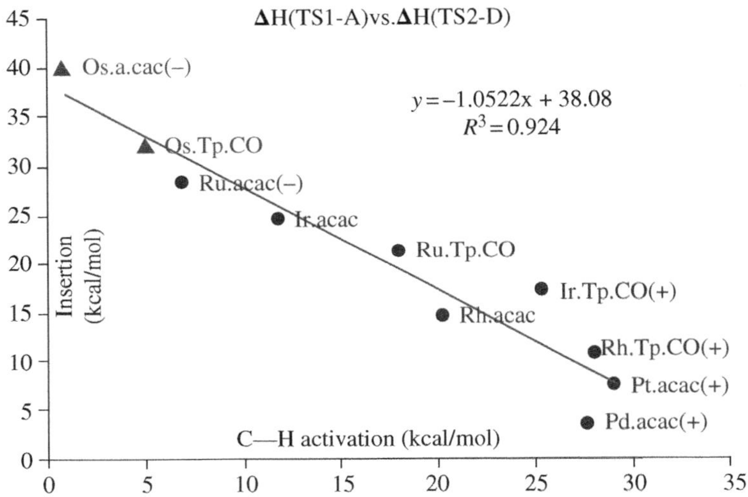

FIGURE 25.30 The correlation of the activation energy for insertion versus the activation energy for C—H activation. The linear trend fits the points with a correlation of $R^2 = 0.924$ [4b]. Adapted with permission from Ref. [4b]. Copyright (2004) American Chemical Society.

Both proposed catalytic cycles are depicted in Figure 25.31. The mechanisms proposed by Eisenstein and Perutz (right) and those described by Nakao and coworkers (left) differ on the C—H activation path. Nakao and coworkers proposed a C—H activation step via an oxidative addition mechanism on the bases that the C—H activated product could carry out the hydrofluoroarylation. The role of fluorine in the selectivity of the C—H activation process has been widely study in the literature [43, 44].

Thus, Eisenstein and Perutz proposed a new pathway via a Ni(phosphine)(alkyne) complex allowing the coordination of the arene moiety. From here, the C—H activation step transfers the proton from the σ-C—H bond of the coordinated arene to the alkyne by a process named by the authors as ligand-to-ligand hydrogen transfer (LLHT). Following a *cis* rearrangement of the aryl and vinyl groups, a reductive elimination leads to the formation of the arylalkene product. This mechanism is in full agreement with the experimental lack of kinetic isotopic effect observed.

The new proposed mechanism differs from others (σ-CAM, CMD, or AMLA processes) as it is the alkyne moiety that promotes the proton transfer and not a strong Lewis base as the acetate or carbonate ligands and, one similarity, the importance of the metal center.

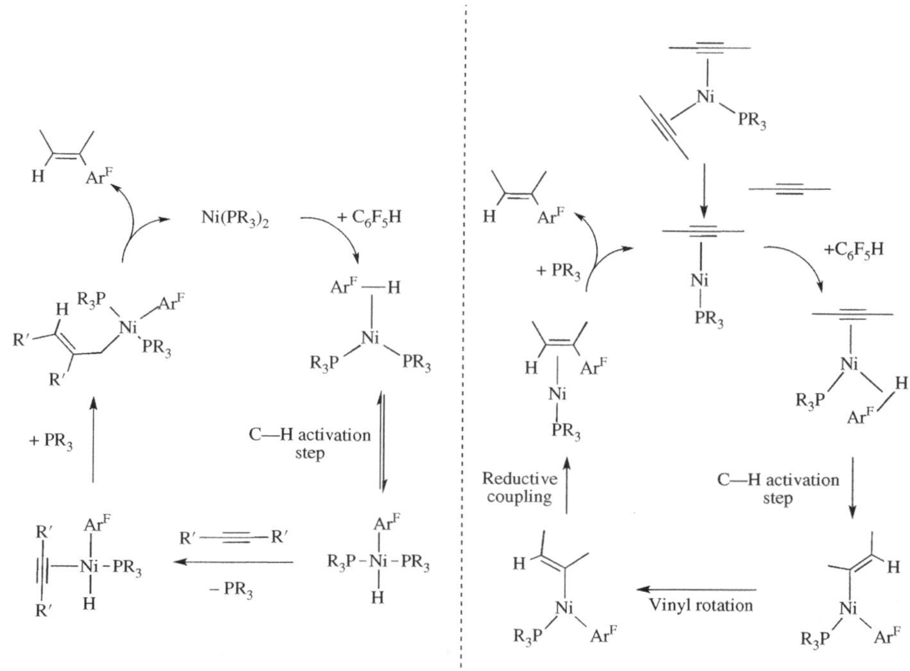

FIGURE 25.31 On the left, proposed catalytic cycle by Nakao and coworkers and on the right, computed catalytic cycle by Eisenstein and Perutz for the hydroarylation of alkynes by Ni complexes [42].

25.4.2 Arene Functionalization via a Base-Assisted Mechanism

The use of an inter- or intramolecular base to promote C—H activation has been previously described. This section will summarize the use of this promising gadget into a catalytic cycle. Only few computational studies have tackled the overall cycle, as they have been mainly focused in the understanding of the C—H activation step.

One of the first breakthroughs in this area was presented by Fagnou and coworkers in 2006 [44] on the arylation reaction of electron-deficient benzenes in excellent yields (Fig. 25.32).

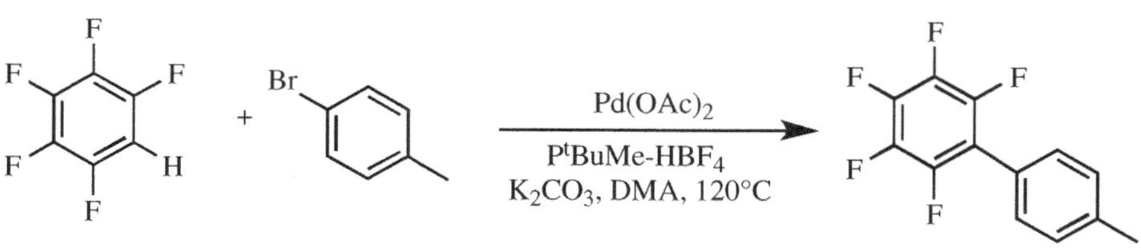

67 **68** **69**

FIGURE 25.32 Reaction scheme for the catalytic intermolecular arylation of pentafluorobenzene by $Pd(OAc)_2$ [44].

They observed that the C—H activation of the pentafluorobenzene was preferred. Several mechanisms were studies computationally, ranging from OA to electrophilic activation, and the lowest-energy pathway computed revealed the base-assisted transition state shown in Figure 25.33.

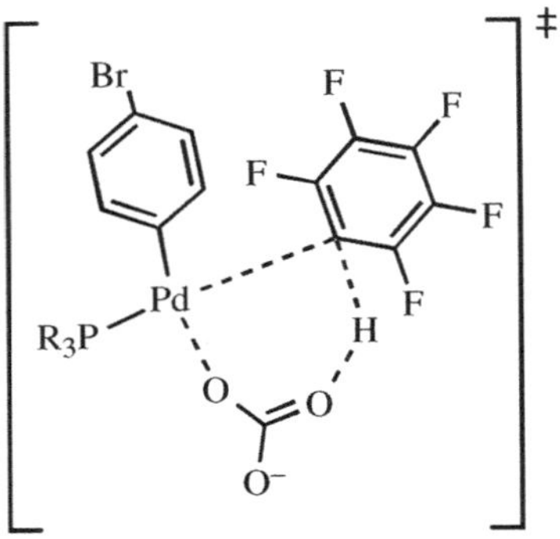

FIGURE 25.33 Transition state computed for the C—H activation step of the arylation reaction of pentafluorobenzene by $Pd(OAc)_2$ [44].

This transition state is analogous to the AMLA-6 or CMD transition states, previously described (see Section 25.3.2).

Davies and coworkers [45] studied computationally the ethylene functionalization of benzene by $[Ir(\kappa^2\text{-}OAc)(PMe_3)Cp]^+$ complex. The choice of the catalyst was based on the promising presence of acetate as an intramolecular base. Figure 25.34 shows the computed reaction path for the process. The C—H activation step undergoes an AMLA-6 transition state, and the rate-limiting step of the reaction is the insertion of ethylene. The C—H activation and the protonolysis have similar barriers of 15.8 and 15.3 kcal/mol, respectively. The authors also investigated a number of competing processes, from protonolysis of benzene to double insertion of ethylene among others.

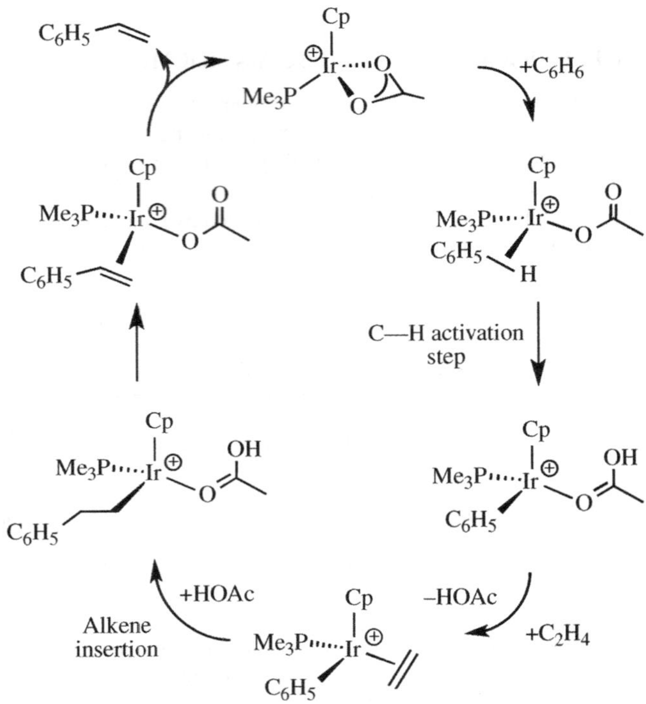

FIGURE 25.34 Computed catalytic cycle for ethene hydroarylation catalyzed by $[Ir(\kappa^2\text{-}OAc)(PMe_3)Cp]^+$ [45].

The analysis of the C—H activation involves the deprotonation of the acetate and the formation of a vacant site at the metal center. The authors suggest that this process could be improved by the use of a more weakly coordinated base. The final step returns the proton to the phenylethyl ligand by an accessible energy barrier. However, the β-H elimination from the $[Ir(\kappa^1\text{-}HOAc)(PMe_3)(CH_2CH_2Ph)Cp]^+$ complex was computed to be a competing process.

Jutand and Dixneuf [46] reported experimental evidence on the nature of the intra- or intermolecular acetate-assisted mechanism for the monoarylation of 2-phenylpyridine catalyzed by $[Ru(OAc)_2$ (*p*-cymene)] and $[Pd(OAc)_2]$ complexes. The kinetic data presented shows that the reaction on $[Ru(OAc)_2$ (*p*-cymene)] complex is faster in the presence of added acetates, which means that the acetate is involved in the rate-determining step of the processes. On the contrary, $[Pd(OAc)_2]$ systems do not present such behavior. These evidences suggest different C—H activation mechanisms for each complex.

The mechanism proposed for the monoarylation of 2-phenylpyridine by $[Ru(OAc)_2$(*p*-cymene)] is shown in Figure 25.35. It highlights three different roles of the acetic acid: (i) the autocatalytic process in the C—H activation step, (ii) its importance for the formation of the cationic B complex, and (iii) the regeneration of the starting arene from the cationic B complex. The authors propose an intermolecular deprotonation for the C—H activation step (S_E3 mechanism) as the reaction only occurs by the presence of added acetates and its deuterated spectroscopy analysis evidence its involvement on the formation of the cationic B complex.

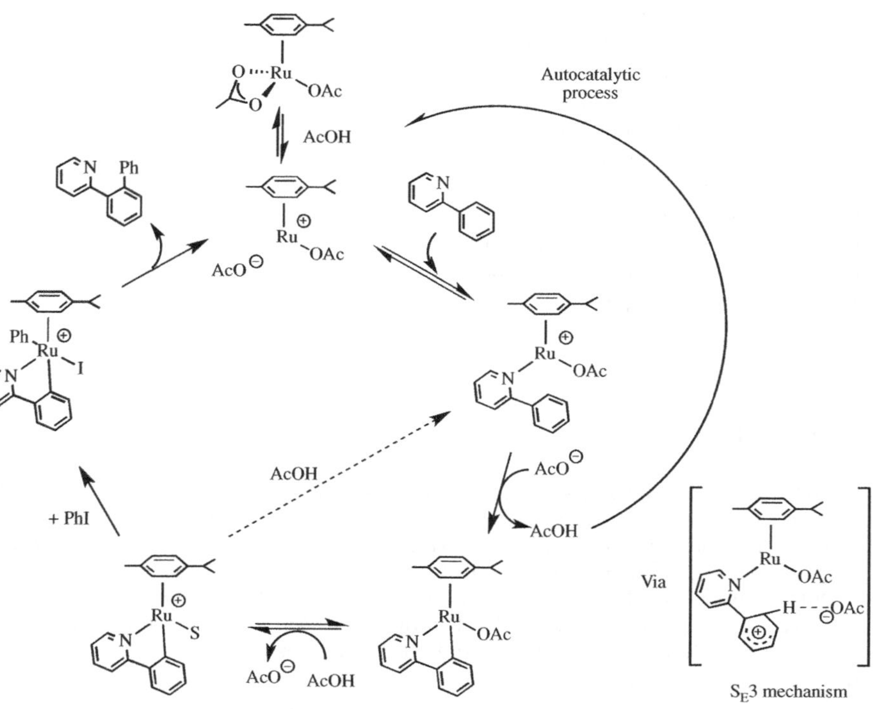

FIGURE 25.35 Proposed catalytic cycle of the phenylation of arenes catalyzed by $[Ru(OAc)_2$(*p*-cymene)] complex [46].

On the other hand, the same reaction catalyzed by $Pd(OAc)_2$ complex was faster, and no affect of additives on the rate of the reaction was observed. Therefore, the authors proposed a CMD mechanism for the C—H activation step. However, they recognize that their kinetic data cannot discriminate between mechanisms. Thus, AMLA, LLHT, or OHM mechanisms could also be proposed, which would only be distinguishable by a computational approach.

25.5 SUMMARY

Computational methods are ideally suited to gain further insight into the overall mechanisms of arene C—H activation by transition metals complexes. The collaborations that have been formed over the last years between theoreticians and experimentalists have facilitated the understanding of the factors that affect and control the C—H activation step. Along this chapter, the cooperative behavior between metal and ligand leads to an easier C—H activation, which can be extrapolated as milder reaction conditions from an experimental point of view. The comprehension of this step has as goal the synthesis of new catalysts where C—H activation can be exploited without being an issue.

ABBREVIATIONS

σ-CAM	σ-Complex-assisted metathesis
AIM	Atoms in molecules
AMLA	Ambiphilic metal ligand activation
BCP	Bond critical points
CMD	Concerted metalation–deprotonation
DFT	Density functional theory
EA	Electrophilic activation
IES	Internal electrophilic substitution
IRC	Intrinsic reaction coordinates
LLHT	Ligand-to-ligand hydrogen transfer
NHC	*N*-Heterocyclic carbene
OA	Oxidative addition
OATS	Oxidatively added transition state
OHM	Oxidative hydrogen migration
RCP	Ring critical points
SBM	σ-Bond metathesis

REFERENCES

[1] (a) R. G. Bergman., *Nature*, 2007, **446**, 391; (b) R. H. Crabtree., *J. Chem. Soc. Dalton Trans.*, 2001, 2951; (c) Y. Guari, S. Sabo-Etienne, B. Chaudret., *Eur. J. Inorg. Chem.*, 1999, 1047; (d) B. Li, T. Roisnel, C. Darcel, P. H. Dixneuf., *Dalton Trans.*, 2012, **41**, 10934; (e) C. Wang, S. Rakshit, F. Glorius., *J. Am. Chem. Soc.*, 2010, 132, 14006; (f) M. H. Emmert, A. K. Cook, Y. J. Xie, M. Sandford., *Angew. Chem. Int. Ed.*, 2011, 50, 9409; (g) Y. Tan, J. F. Hartwig., *J. Am. Chem. Soc.*, 2011, 133, 3308; (h) I. A. I. Mkhalid, J. H. Barnard, T. B. Marder, J. M. Murphy, J. F. Hartwig., *Chem. Rev.*, 2010, 110, 890; (i) G. E. Dobereiner, R. H. Crabtree., *Chem. Rev.*, 2010, 110, 681; (j) K. Godula, D. Sames., *Science*, 2006, 312, 67.

[2] R. B. Bedford., *Chem. Comm.*, 2003, 1787.

[3] J. C. Lewis, R. G. Bergman, J. A. Ellman., *Acc. Chem. Res.*, 2008, 41, 1013.

[4] (a) J. F. Hartwig, K. S. Cook, M. Hapke, C. D. Incarvito, Y. B. Fan, C. E. Webster, M. B. Hall., *J. Am. Chem. Soc.*, 2005, 127, 2538; (b) J. Oxgaard, R. A. Periana, W. A. Goddard., *J. Am. Chem. Soc.*, 2004, 126, 11658; (c) W. H. Lam, G. C. Jia, Z. Y. Lin, C. P. Lau, O. Eisenstein., *Chem. Eur. J.*, 2003, **9**, 2775.

[5] (a) C. J. Cramer, D. G. Truhlar., *Phys. Chem. Chem. Phys.*, 2009, **11**, 10757; (b) P. Comba., *Modelling of Molecular Properties*, Wiley-VCH, Weinheim, 2011.

[6] W. Koch, M. C. Holthausen., *A Chemist's Gide to Density Functional Theory*, Wiley-VCH, Weinheim, 2001.

[7] (a) A. Dedieu., *Chem. Rev.*, 2000, **100**, 543; (b) S. Q. Niu, M. B. Hall., *Chem. Rev.*, 2000, **100**, 353; c(c) Y. Boutadla, D. L. Davies, S. A. Macgregor, A. I. Poblador-Bahamonde., *Dalton Trans.*, 2009, 5820; (d) L. Ackerman., *Chem. Rev.*, 2011, **111**, 1315; (e) D. Balcells, E. Clot, O. Eisenstein., *Chem. Rev.*, 2010, **110**, 749; (f) D. Lupp, N. J. Christensen, P. Fristrup., *Dalton Trans.*, 2014, **43**, 11093.

[8] D. H. Ess, R. J. Nielsen, W. A. Goddard, R. A. Periana., *J. Am. Chem. Soc.*, 2009, **131**, 11686.

[9] B. A. Vastine, M. B. Hall., *J. Am. Chem. Soc.*, 2007, **129**, 12068.

[10] R. F. W. Bader., *Acc. Chem. Res.*, 1985, **18**, 9.

[11] P. Vidossich, A. Lledós., *Dalton Trans.*, 2014, **43**, 11145.

[12] A. H. Janowicz, R. G. Bergman., *J. Am. Chem. Soc.*, 1982, **104**, 352.

[13] (a) O. Rivada-Wheelaghan, B. Donnadieu, C. Maya, S. Conejero., *Chem. Eur. J.*, 2010, **16**, 10323; (b) O. Rivada-Wheelaghan, M. A. Ortuño, J. Díez, A. Lledós, S. Conejero., *Angew. Chem. Int. Ed.*, 2012, **51**, 3936.

[14] P. L. Watson, G. W. Parshall., *Acc. Chem. Res.*, 1985, **18**, 51.

[15] T. Ziegler, E. Folga, A. Berces., *J. Am. Chem. Soc.*, 1993, **115**, 636.

[16] P. Burger, R. G. Bergman., *J. Am. Chem. Soc.*, 1993, **115**, 10462.

[17] D. L. Strout, S. Zaric, S. Q. Niu, M. B. Hall., *J. Am. Chem. Soc.*, 1996, **118**, 6068.

[18] S. Q. Niu, M. B. Hall., *J. Am. Chem. Soc.*, 1998, **120**, 6169.

[19] A. D. Ryabov, I. K. Sakodinskaya, A. K. Yatsimirsky., *J. Chem. Soc. Dalton Trans.*, 1985, **12**, 2629.

[20] D. L. Davies, S. M. A. Donald, S. A. Macgregor., *J. Am. Chem. Soc.*, 2005, **127**, 13754.

[21] R. G. Bergman, T. R. Cundari, A. M. Gillespie, T. B. Gunnoe, W. D. Harman, T. R. Klinckman, M. D. Temple, D. P. White., *Organometallics*, 2003, **22**, 2331.

[22] Z. Y. Lin., *Coord. Chem. Rev.*, 2007, **251**, 2280.

[23] (a) G. Bhalla, J. Oxgaard, W. A. Goddard, R. A. Periana., *Organometallics*, 2005, **24**, 5499; (b) J. Oxgaard, R. P. Muller, W. A. Goddard, R. A. Periana., *J. Am. Chem. Soc.*, 2004, **126**, 352.

[24] R. N. Perutz, S. Sabo-Etienne., *Angew. Chem. Int. Ed.*, 2007, **46**, 2578.

[25] B. Biswas, M. Sugimoto, S. Sakaki., *Organometallics*, 2000, **19**, 3895.

[26] (a) R. Bielsa, R. Navarro, E. P. Urriolabeitia, A. Lledós., *Inorg. Chem.*, 2007, **46**, 10133; (b) D. Aguilar, R. Bielsa, M. Contel, A. Lledós, R. Navarro, T. Soler, E. P. Urriolabeitia., *Organometallics*, 2008, **27**, 2929.

[27] D. L. Davies, O. Al-Duaij, J. Fawcett, M. Giardiello, S. T. Hilton, D. R. Russell., *Dalton Trans.*, 2003, **41**, 4132.

[28] (a) L. Ling, W. W. Brennessel, W. D. Jones., *Organometallics.*, 2009, **12**, 3492; (b) J. Kovach, M. Peralta, W. D. Jones., *J. Mol. Stru.*, 2011, **992**, 33; (c) A. P. Walsh, W. W. Brennessel, W. D. Jones., *Inorg. Chim. Acta*, 2013, **407**, 131.

[29] D. L. Davies, S. M. A. Donald, O. Al-Duaij, S. A. Macgregor, M. Polleth., *J. Am. Chem. Soc.*, 2006, **128**, 4210.

[30] Y. Boutadla, D. L. Davies, S. A. Macgregor, A. I. Poblador-Bahamonde., *Dalton Trans.*, 2009, 5887.

[31] D. H. Ess, S. M. Bischof, J. Oxgaard, R. A. Periana, W. A. Goddard., *Organometallics*, 2008, **27**, 6440.

[32] D. Garcia-Cuadrado, A. A. C. Braga, F. Maseras, A. M. Echavarren., *J. Am. Chem. Soc.*, 2006, **128**, 1066.

[33] D. Garcia-Cuadrado, P. de Mendoza, A. A. C. Braga, F. Maseras, A. M. Echavarren., *J. Am. Chem. Soc.*, 2007, **129**, 6880.

[34] S. I. Gorelsky, D. Lapointe, K. Fagnou., *J. Am. Chem. Soc.*, 2008, **130**, 10848.

[35] M. García-Melchor, S. I. Gorelsky, T. K. Woo., *Chem. Eur. J.*, 2011, **17**, 13847.

[36] G. Bhalla, X. Y. Liu, J. Oxgaard, W. A. Goddard, R. A. Periana., *J. Am. Chem. Soc.*, 2005, **127**, 11372.

[37] R. A. Periana, X. Y. Liu, G. Bhalla., *Chem. Comm.*, 2002, 3000.

[38] M. Lail, B. N. Arrowood, T. B. Gunnoe., *J. Am. Chem. Soc.*, 2003, **125**, 7506.

[39] N. J. DeYonker, N. A. Foley, T. R. Cundari, T. B. Gunnoe, J. L. Petersen., *Organometallics*, 2007, **26**, 6604.

[40] J. Guihaumé, S. Halbert, O. Eisenstein, R. N. Perutz., *Organometallics*, 2012, **31**, 1300.

[41] K. S. Kanyiva, N. Kashihara, Y. Nakao, T. Hiyama, M. Ohashi, S. Ogoshi., *Dalton Trans.*, 2010, **31**, 10483.

[42] A. Nova, M. Reinhold, R. N. Perutz, S. A. Macgregor, J. E. McGrady., *Organometallics*, 2010, **29**, 1824.

[43] (a) A. Nova, S. Erhardt, N. A. Jasim, R. N. Perutz, S. A. Macgregor, J. E. MaGrady, A. C. Whitwood., *J. Am. Chem. Soc.*, 2008, **130**, 15499; (b) E. Clot, C. Mégret, O. Eisenstein, R. N. Perutz., *J. Am. Chem. Soc.*, 2006, **128**, 8350; (c) E. Clot, C. Mégret, O. Eisenstein, R. N. Perutz., *J. Am. Chem. Soc.*, 2009, **131**, 7817.

[44] M. Lafrance, C. N. Rowley, T. K. Woo, K. Fagnou., *J. Am. Chem. Soc.*, 2006, **128**, 8754.

[45] D. L. Davies, S. A. Macgregor, A. I. Poblador-Bahamonde., *Dalton Trans.*, 2010, **39**, 10520.

[46] (a) I. Fabre, N. von Wolff, G. Le Duc, E. Ferrer Flegeau, C. Bruneau, P. H. Dixneuf, A. Jutand., *Chem. Eur. J.*, 2013, **19**, 7595; (b) E. Ferrer Flegeau, C. Bruneau, P. H. Dixneuf, A. Jutand., *J. Am. Chem. Soc.*, 2013, **133**, 10161.

PART VIII

DIRECTED METALATION REACTIONS

26

DIRECTED METALATION OF ARENES WITH ORGANOLITHIUMS, LITHIUM AMIDES, AND SUPERBASES

FRÉDÉRIC R. LEROUX[1] AND JACQUES MORTIER[2]

[1] *Laboratoire de Chimie Moléculaire, CNRS and University of Strasbourg, UMR CNRS 7509, ECPM, Strasbourg Cedex 2, France*

[2] *Institut des Molécules et Matériaux du Mans, Faculté des Sciences et Techniques, UMR CNRS 6283, Université du Maine and CNRS, Le Mans Cedex, France*

26.1 INTRODUCTION

The metalation of arenes in the vicinity of an appropriate directing metalation group (DMG) by organolithiums, lithium amides, and superbases is one of the most powerful methods for the regioselective preparation of synthetically useful polyfunctional arenes. The term "metalation" was coined by Gilman in 1934 for "reactions involving replacement of hydrogen by a metal to give a true organometallic compound" [1].

Research efforts in the field have been devoted to the discovery of new DMGs, recognition of new types of metalating agents, elucidation of metalation mechanism, and the development of new synthetic methods. Numerous papers have witnessed the constant interest, the increasing mechanistic comprehension, and the industrial applications of metalation reactions.

This chapter aims to provide a broad understanding of the theory and practices of aromatic metalation reactions, which allow direct functionalization of unactivated aromatic C—H bonds by the stoichiometric formation of aryllithium(potassium) intermediates, namely (Fig. 26.1), (i) directed *ortho* metalation (DoM) of arenes bearing one or more DMGs, (ii) directed remote

Dedicated to the memory of Prof. Manfred Schlosser for his outstanding contribution in the field of polar organometallic chemistry.

Arene Chemistry: Reaction Mechanisms and Methods for Aromatic Compounds, First Edition. Edited by Jacques Mortier.

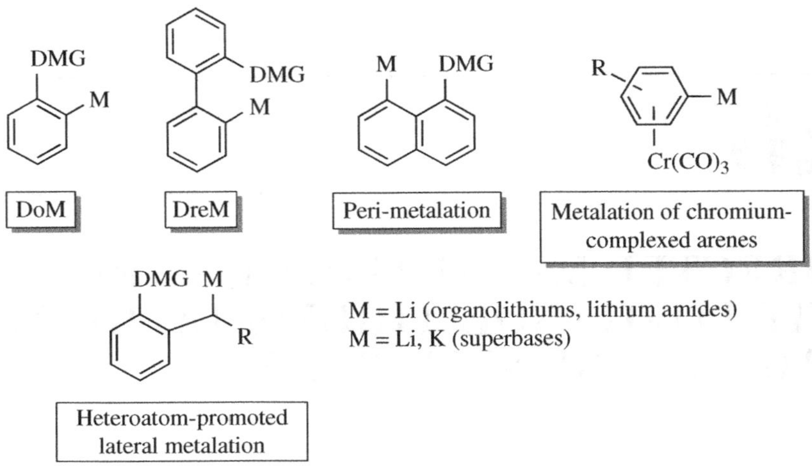

FIGURE 26.1 Aromatic metalation reactions.

metalation (DreM) of polyarenes, (iii) peri metalation of naphthalenes, and (iv) metalation of metal-complexed arenes (mainly chromium complexes).

Reactivity differences between aryl and benzyllithium species may suggest the use of lateral lithiation over the DoM, and vice versa. This is why it seems appropriate to discuss heteroatom-promoted lateral metalation reactions in this chapter.

The main analytical methods and techniques (crystallography, NMR spectroscopy, quantitative analysis) and important synthetic applications are presented to highlight the recent advances of these techniques and their impact for improving the design of pharmaceutical compounds and organic materials. Metalations with alkali metal–nonalkali metal organic combinations (ate compounds, turbo-Grignard reagents, and synergistic pairs of metal amides) are discussed in Chapter 27.

26.2 PREPARATION AND REACTIVITY OF ORGANOLITHIUM COMPOUNDS

26.2.1 Bases and Complexing Agents

Organolithiums (RLi) are organometallic compounds that contain C—Li bonds. In solution, organolithium compounds exist as aggregates (oligomers) or polymers [2, 3]. Aggregates consist of equilibrium mixtures, which differ by their state of aggregation, degree of solvation, and electronic structure [4, 5]. The formation of aggregates is influenced by steric effects and by the coordination between lithium and solvent molecules or additives. The knowledge of the structure of these reactive species is crucial for the elucidation of reaction mechanisms [6].

The nature of the aggregates is dependent on both the solvent and the organolithium concentration [7, 8]. The aggregates of *n*-BuLi are tetramers in ether and hexamers in pentane. Phenyllithium is constituted of dimer–tetramer aggregates in ether solution and of monomer–dimer aggregates in THF. At −96°C in THF, *s*-BuLi is a monomer–dimer mixture at a concentration of 1.2*M* [9], whereas it is mainly monomeric in more diluted medium (0.2*M*) [5]. The extremely polar nature of the C—Li bond makes organolithiums strong bases ($pKa_{butane} \approx 50$). Use of high concentrations of *n*-BuLi enhances the solubility of the metalated species by generating mixed-soluble aggregates [10]. Organolithiums can also react as nucleophiles and as reducing agents, depending on the other reactants. Details on the preparation and properties of organolithiums and information for safe handling are provided by recent references [11, 12].

The structure of organolithium aggregates is also affected by the presence of aprotic Lewis base additives such as tetramethylethylenediamine (TMEDA), tridentate *N,N,N′,N″,N″*-pentamethyldiethylenetriamine (PMDTA), tetradentate *N,N,N′,N″,N‴,N‴*-hexamethyltriethylenetetramine (HMTTA), 1,4-diazabicyclo[2,2,2]octane (DABCO), and hexamethylphosphoramide (HMPA) (Fig. 26.2) [3, 8]. The increase in reactivity observed can be attributed to the higher reactivity of small aggregates compared to larger aggregates. Reactivity increases only if the new aggregation state of the base allows to lower the activation energy of the reaction [13, 14].

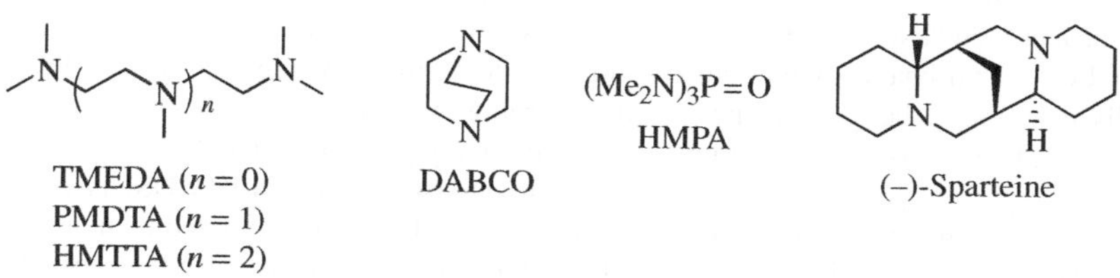

FIGURE 26.2 Aprotic Lewis base additives.

In the presence of *n*-BuLi in hexane, the benzene molecule gives phenyllithium in low yield (<1%) (Scheme 26.1). TMEDA that manifests a highly substrate-dependent affinity for lithium [15] increases the reactivity of the base, allowing effective metalation (92%).

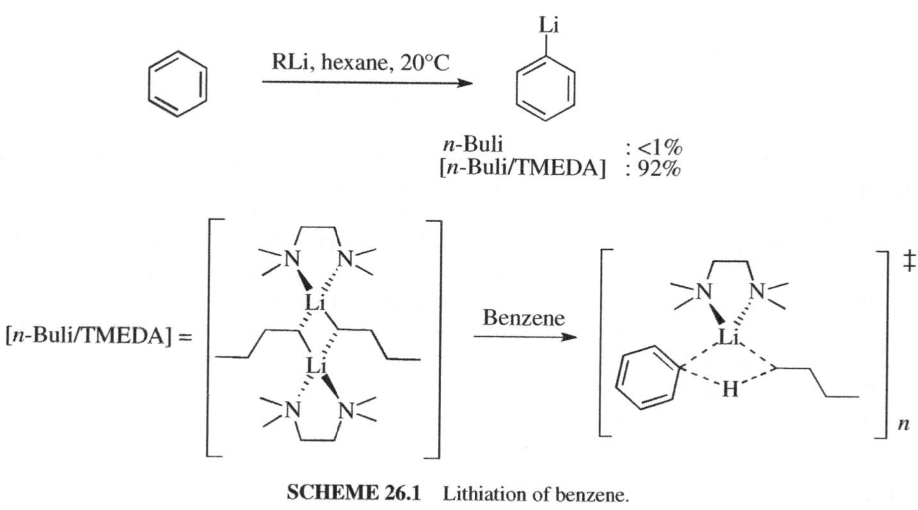

SCHEME 26.1 Lithiation of benzene.

TMEDA breaks down the alkyllithium aggregates with concomitant increase in basicity. However, in THF, TMEDA drives the PhLi monomer–dimer equilibrium completely to dimer [15]. Even incremental amounts of TMEDA (≈0.2 equiv per alkyllithium) in ether can cause a significant rate acceleration of the metalation reaction [16, 17]. Reactivity of butyllithiums further increases when TMEDA is replaced by PMDTA or HMTTA chelating agents. The aggregation state of organolithiums can also be modified by adding inorganic lithium salts (usually lithium halides) [18–20].

The 1:1 complex *s*-BuLi/TMEDA is probably the most potent metalating organolithium reagent. With (−)-sparteine (Fig. 26.2), metalation leads to asymmetric complexes allowing the formation of enantiomerically enriched functionalized compounds [21].

Transmetalation reactions for preparing organolithium reagents by tin–lithium exchange is an effective and well-established methodology [22]. The halogen–metal exchange is also an important tool for preparing organolithium reagents (see Chapter 28) [23]. Organosodium bases (RNa), which exhibit a much less convenient reactivity than organolithium compounds [24, 25], are more difficult to prepare and thermically more unstable.

Nonnucleophilic (sterically hindered) lithium amides can be prepared by the simple reaction of the corresponding amines with *n*-BuLi in nonpolar organic solvents (Fig. 26.3). Lithium amides are more soluble in hydrocarbons than their heavier element congeners (Na, K). LiN(*i*-Pr)$_2$ (LDA) is also cheaper than KN(*i*-Pr)$_2$ (KDA) and is more widely used. Lithium amides, which have a much lower Lewis acid character than alkyllithiums, also form aggregates in solution [26–28]. They usually react under thermodynamic control according to a classic acid–base mechanism. The p*K*a of diisopropylamine is 36 [8]. Lithium 2,2,6,6-tetramethylpiperidide (LTMP) is slightly more basic ($pKa_{2,2,6,6\text{-tetramethylpiperidine}} = 37$) [29]. Lithium bis(trimethylsilyl)amide (LiHMDS) is a weaker base ($pKa_{HMDS} = 30$). The importance of lithium amides in synthesis has prompted many studies of their mechanisms of action [30]. A number of interesting chiral amides have also been tested [4, 31].

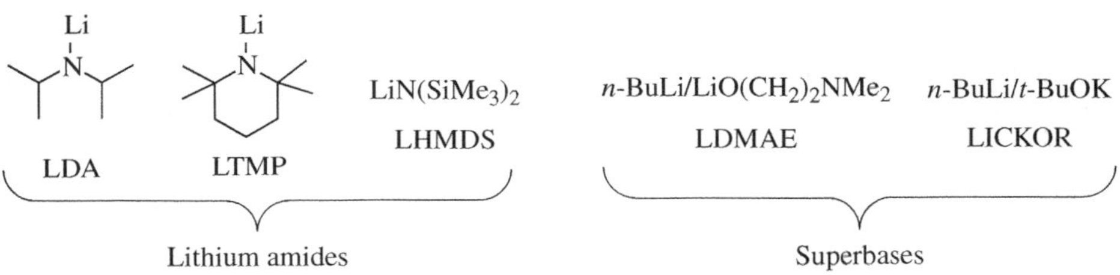

FIGURE 26.3 Lithium amides and superbases.

The term superbases is applied to those bases resulting from a mixing of two (or more) bases leading to new basic species possessing inherent new properties. The basicity and the nucleophilicity of the superbases differ from those of the individual parent bases. There are two main classes of superbases: unimetallic superbases (such as *n*-BuLi/LiO(CH$_2$)$_2$NMe$_2$) [32] and bimetallic superbases (Li-K) [33, 34]. The 1:1 mixture of an alkyllithium (LiC) and a hindered potassium alkoxide (KOR, usually KO*t*-Bu), popularly known as "LICKOR" or Schlosser–Lochmann superbase, is particularly efficient in the deprotonation of arenes [33, 34]. The main advantage of these bases is their low complexing ability combined with high basicity and low nucleophilicity [35].

Reactions of organolithium compounds are generally carried out under argon or dry nitrogen at low to very low temperatures (−20 → −90°C).

26.2.2 Solvents

The choice of the solvent imposes its straightforward purification from water and peroxides, low-enough freezing point, and its stability against organolithium compounds [7]. Strong bases are not infinitely stable in ethereal solvents. The stability order is hydrocarbons > Et$_2$O > THF.

In the laboratory, THF is the solvent of choice for generating metalated aromatic species. Diethyl ether is highly flammable, more toxic, and prone to the formation of peroxides [7]. Aromatic hydrocarbons can be used; however, toluene is relatively easily lithiated, and benzene is toxic and cannot be used at low temperatures. The obvious trade-off of using a hydrocarbon as solvent is the lower solubility of starting materials as well as of the organolithium intermediate. The presence of excess *n*-BuLi can assist the overall solubility by inclusion of *n*-BuLi into the solubilized aggregates [10]. Industrial optimizations cut back on THF or even eliminated it [10]. THF is completely water soluble, which is what renders the water-quenching step (used to remove metal) problematic. 2-Methyltetrahydrofuran (MeTHF) is a commercially available solvent produced from a renewable resource (furfural) through hydrogenation, which provides clean phase separation during work-up, with little tendency to form emulsions or rag layers [36]. However, MeTHF has the drawback of easy peroxide formation, it is instable under acidic conditions, and it has a low flash point (−12°C). Cyclopentyl methyl ether (CPME) is a commercial, highly hydrophobic and thus easy to dry ethereal solvent with low formation of peroxides and high boiling point (106°C) useful for both bench-size and plant-scale applications [37].

26.2.3 Electrophiles

Many electrophiles react with lithiated arenes (Table 26.1) [38]. Particular emphasis is put on the introduction of carbon, oxygen, nitrogen, halogen, sulfur, phosphorus, and silicon substituents, and we refer here only to typical trapping reactions. Electrophiles are potentially hazardous to humans (because our bodies are largely composed of nucleophiles) and should be handled with care. For instance, breathing iodomethane fumes can cause lung, liver, kidney, and central nervous system damage [39].

C—C Bond Formation: Treatment of an organometallic intermediate with dry ice and acidification with aqueous HCl affords a carboxylic acid derivative. With paraformaldehyde, one obtains an alcohol, and with oxiranes, oxetanes, and 1-iodo-4-(methoxymethoxy)butane, assisted by copper(I) iodide or other Lewis acids, alcohols with different chain lengths are accessible.

With MeI or dimethylsulfate, the methyl derivatives are prepared; with acetyl chloride-CuI or the Weinreb amides, the ketones; with methyl chloroformate, the methyl esters; and with dialkyl oxalate, the α-oxo esters. Alkyl long-chain derivatives are better prepared by lateral (benzylic) metalation (see Section 26.7). Besides, aryllithiums may undergo transmetalation to afford a large panel of organometallic species (Li → Zn, Cu, Sn, Si, B, etc.), key reactants for transition metal-catalyzed reactions. The cross-coupling of aryllithium reagents with alkenyl triflates in the presence of the commercially available catalytic system [$Pd_2(dba)_3$]/DavePhos] was recently reported [40].

C—N Bond Formation: The introduction of a NH_2-group can be achieved either by reaction with *N*-lithiated *O*-methylhydroxylamine or by azidation followed by reduction. With imines, the aminomethyl functionality is introduced. Arylhydrazines and benzamides are obtained by reaction with di-*tert*-butyl azodicarboxylate and isocyanates, respectively.

C—O Bond Formation: Trapping of ArLi intermediates can be achieved with trimethyl borate or triisopropyl borate [41]. The intermediate lithium borate complexes, when treated with water, afford oxidizable boronates. With fluorodimethoxyborane, spontaneous elimination of fluoride leads directly to boronates, which are precursors of phenols by oxidation. Chemoselective metal–heteroatom exchange or deprotonation generates carbanions in the presence of dialkyl peroxides, and intramolecular trapping provides an efficient route to dihydrobenzofuranes and dihydrobenzopyrans [42].

C—S Bond Formation: Dialkyl and diaryl sulfides are effective trapping reagents for aryllithium species.

C—Halogen Bond Formation: Fluorination of organometallics can be achieved with perchloryl fluoride (F-ClO_3) [43]. However, violent explosions have been reported when this reagent accumulates in the reaction mixture. In recent years, fluoro-aza reagents such as Selectfluor®, NFSI, and *N*-fluoropyridinium have been conveniently used as fluorine donors [44–46].

Elemental chlorine that is too aggressive to be employed directly can be substituted effectively by hexachloroethane, NCS, or 1,1,2-trichloro-1,2,2-trifluoroethane. For bromination, elemental bromine can be used below −78°C. Otherwise, NBS, CBr_4, 1,2-dibromoethane, and 1,2-dibromo-1,1,2,2-tetrafluoroethane give good results. Iodination can be performed with elemental iodine or alternatively with 1,2-diodoethane.

TABLE 26.1 Selected Electrophile (for Further and More Detailed Information, See Ref. [24])

Reagent	Substituent	Reagent	Substituent
CO_2	$-CO_2H$	$MeONH_2Li$	$-NH_2$
$(CH_2O)_n$	$-CH_2OH$	$PhSO_2N_3$, TsN_3 (then LAH reduction)	$-N_3$ $-NH_2$
Oxiranes, CuI	$-(CH_2)_2OH$	RCH=NR′	—CHR-NHR′
Oxetane, BF_3-OEt_2	$-(CH_2)_3OH$	*t*-BuO_2C—N=N—CO_2*t*-Bu	$-NHNH_2$
$I(CH_2)_4OCH_2OMe$	$-(CH_2)_4OH$	RNCO	—CONHR
D_2O, DCl, MeOD	—D	$B(OR)_3$ and then OH^-/H_2O_2 $FB(OMe)_2$ and then oxidation	—OH
MeI, Me_2SO_4	—Me	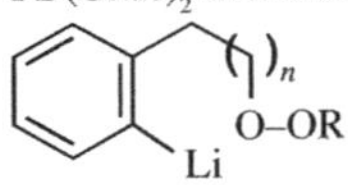 $n = 1, 2$	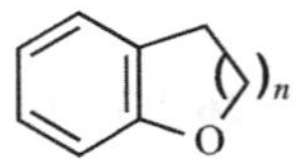
RCHO	—CH(R)OH	RSSR	—SR
DMF (HCO_2Et)	—CHO	Selectfluor®,[a] NFSI[b] *N*-Fluoropyridinium	—F
RCOCl, CuI	—COR	C_2Cl_6, NCS	—Cl
RCONMe(OMe)		Cl_2FC-CF_2Cl	
$ClCO_2Me$	$-CO_2Me$	Br_2, NBS, CBr_4, $BrCH_2CH_2Br$, CF_2Br-CF_2Br	—Br
$(CO_2R)_2$	$-COCO_2R$	I_2, ICH_2CH_2I	—I
TfOCR=CHR′ cat[$Pd_2(dba)_3$]/DavePhos][c]	—CR=CHR′	R_3SiCl	$-SiR_3$
		Bu_3SnCl	$-SnBu_3$
		R_2PCl	$-PR_2$

[a] Selectfluor®: 1-chloromethyl-4-fluoro-1,4-diazoniabicyclo[2.2.2]octane bis(tetrafluoroborate).
[b] NFSI, *N*-fluorodibenzenesulfonimide.
[c] dba, dibenzylideneacetone; DavePhos, 2-dicyclohexylphosphino-2′-(*N*,*N*-dimethylamino)biphenyl.

C—Si and C—P Bond Formation: Finally, chlorotrialkylsilanes generate arylsilanes. TMSCl can be used for *in situ* trapping experiments (see Section 26.3.4) [47]. Chloro dialkyl- and diarylphosphines provide the corresponding phosphines.

26.3 DIRECTED *ortho*-METALATION (DoM)

DoM is a very powerful method for the preparation of *ortho*-substituted aromatic systems, which are difficult to obtain by classic substitution routes. The DoM reaction is widely applied in academic and industrial fields. To date, a large number of reviews have been published on this valuable strategy [48–53]. The discussion that follows is limited to mechanistic aspects and key features of the reaction.

26.3.1 Mechanisms: Complex-Induced Proximity Effect Process, Kinetically Enhanced Metalation, and Overriding Base Mechanism

The *ortho*-lithiation reaction proceeds according to three limiting mechanisms. The so-called complex-induced proximity effect (CIPE) process [54], advanced by Beak and Meyers [55], accounts for regioselective lithiation of aromatic compounds **1** bearing a Lewis basic heteroatom (generally O and N) on the directing group (Scheme 26.2). The lithium of the base, acting as a Lewis acid, coordinates with the lone pairs of the heteroatom forming a prelithiation complex (PLC). In the transition state (TS), the basic R group can remove the hydrogen in *ortho* position,

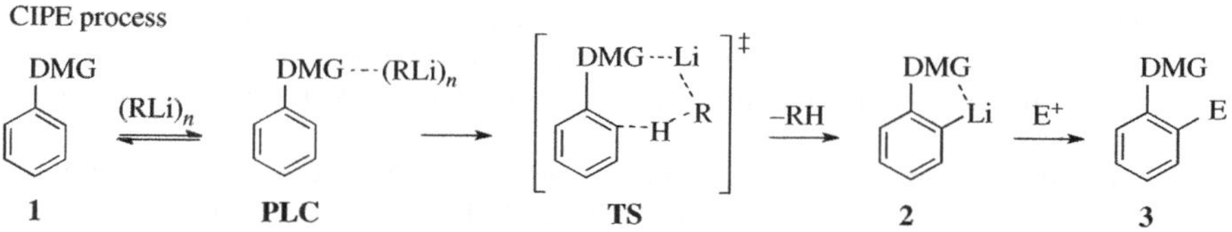

SCHEME 26.2 Complex-induced proximity effect (CIPE) process.

leading to *ortho*-lithiated species **2**. The deprotonation is the limiting step [13, 14, 56]. The driving force originates from the stabilization of the *ortho*-lithiated aromatic compound by intramolecular complexation of Li by the DMG and/or from the inductive stabilization provided by the DMG. An electrophile can then replace the lithium atom to give **3**.

In agreement with this mechanism, a stopped-flow IR spectroscopic study on the metalation of *N,N*-dialkylbenzamides with *s*-BuLi/TMEDA [57] suggested the presence of amide-Li complex intermediates before the formation of the lithiated product. Theoretical (MNDO) evidence was also found for the formation of weakly chelated complexes suggesting that a coordination event leads to a deprotonation process [58, 59]. Inter- and intramolecular kinetic isotope effect (KIE) studies also support a CIPE mechanism [60, 61].

An alternative scenario in which complexation of the base and proton abstraction occur simultaneously in a kinetically controlled one-step reaction was postulated by Schleyer (Fig. 26.4a) [13, 14]. In this *kinetically enhanced metalation* (KEM) model, "this is not precomplexation that is important but the existence of a stabilizing metal-substituent interaction at the rate-limiting transition structures" [13]. The fact that KIEs of inter- and intramolecular lithiation of anisole by *n*-BuLi in Et_2O are identical ($k_H/k_D = 2.5 \pm 0.2$) [63] supports this one-step mechanism.

Both theories furnish a complete picture if adequately put together (Fig. 26.4b) [62]. The directing and accelerating effect of the substituents is likely due to the stabilization of *both* the initial complex and the transition structure. Coordination by the DMG would be stronger in the TS than in the initial complex. As a result, complexation would increase the rate of reaction by providing a new mechanism with lower activation energy (E_a).

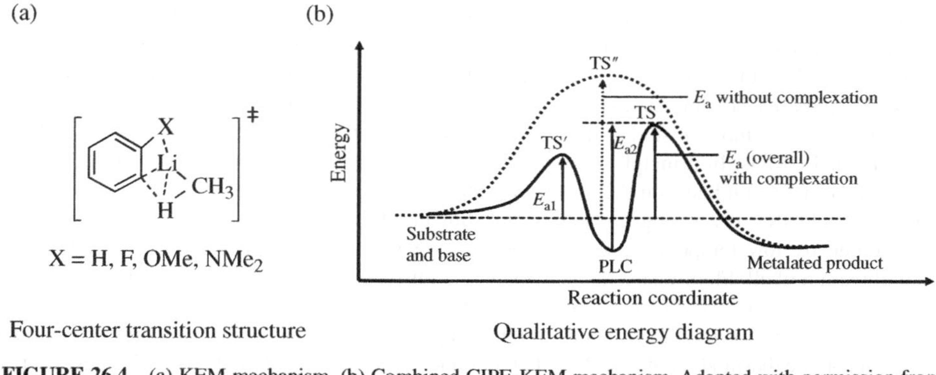

FIGURE 26.4 (a) KEM mechanism. (b) Combined CIPE-KEM mechanism. Adapted with permission from Ref. [62]. Copyright (2005) American Chemical Society.

The *overriding base* mechanism suggests that metalation is driven by the acidity of *ortho* hydrogen atoms resulting from strong electronegativities of DMGs such as halo, trifluoromethyl, and cyano groups [38, 49, 64, 65]. Therefore, the stronger the electron-withdrawing group, the greater the acidity of the *ortho* hydrogen. Superbases are not significantly influenced by *ortho*-directing groups and preferentially attack the inductively activated aromatic position next to the most electronegative heteroatom and/or the most acidic position available [66, 67]. The importance of inductive effects in DoM chemistry is further supported by kinetic studies and ab initio calculations [56].

In conclusion, no single mechanism can be applied to all DoM reactions. Most DMGs direct metalation by a combination of factors where CIPE/KEM/Overriding-base mechanisms are operative. Weakly solvated polar organometallics preferentially coordinate with the DMG, whereas fully complexed bases deprotonate positions where the resulting negative charge can be efficiently stabilized [38, 49, 64, 65]. DMGs do not enable to entirely control the site selectivity and other factors such as steric effects and the presence of other functional groups need also to be considered.

26.3.2 Directing Metalation Groups (DMGs)

A list of common DMGs classified as carbon-based and heteroatom-based DMGs is shown in Table 26.2. DMGs generally provide useful functionality for further transformation. A good DMG must be a good coordination site but a poor electrophile to prevent nucleophilic attack by the base.

The $CONEt_2$ (tertiary amide) and $OCONEt_2$ (*O*-carbamate) discovered by Beak and Snieckus are probably the most popular DMGs. Secondary amides (CONHR), carboxylic acids (CO_2H), methylalcohols (CH_2OH), secondary thioamides (CSNHR), secondary amines (NHR), carbamates (NCO_2H), alcohols (OH), and thiols (SH) are deprotonated themselves prior to ring *ortho* metalation, leading to dimetalated species. The carboxylate group (CO_2Li) which might be thought to be susceptible to nucleophilic addition by the organolithium bases (affording ketones and alcohols) [100] retains its integrity by low temperature kinetic control (−78°C) [81, 101]. For more detailed information, the reader may refer to the publications cited in the Table 26.2 of the research groups most concerned on a specific DMG.

26.3.3 Optional Site Selectivity: Selected Examples

Since a number of biological compounds contain multisubstituted arenes, a knowledge of the relative *ortho*-directing abilities of DMGs toward metalation is crucial. This aspect has been assessed thanks to inter- and intramolecular competition experiments. Competition studies using two DMGs like OMe, $CONEt_2$, oxazoline, $OCONEt_2$, and CO_2Li as anchoring groups allow a concrete picture of the metalation scenario to be drawn (Fig. 26.5).

There are a number of studies dealing with lithiations of arenes carrying *ortho- meta* and *para*-interrelated DMGs. In principle, two *meta*-interrelated DMGs function in concert to direct introduction of the metal between them, however there are exceptions (vide infra). The most powerful DMG ($OCONR_2$) is a strong Lewis base substituent whose strong electrophilic character can be overcome by low temperature kinetic control (−78°C), steric effect, charge deactivation, or a combination of all. Fluorine exerts the strongest acidifying properties on *ortho* positions among all halogen atoms, whereas Cl and Br show little difference [102, 103]. There are numerous examples in the literature showing the influence of long-range inductive effects of substituents (including Cl and Br) on the thermodynamic acidity of arenes and kinetics of metalation [104]. The ease of metalation, as well as the stability of the metalated species, is somewhat greater with polyfluorinated benzenes. Bromine and iodine have been used in conjunction with DoM for benzyne reactions (Chapter 12) and metal–halogen exchange and halogen dance rearrangements (Chapter 28).

Schlosser introduced the concept of *optional site selectivity through mechanism-based substrate reagent matching* to describe the metalation of arenes carrying two different DMGs

TABLE 26.2 Selected Directed Metalation Groups (DMGs)

Carbon-Based DMGs		Heteroatom-Based DMGs	
$CONEt_2$	Beak [68]	F	Gilman [69]
$CONMeCH(TMS)_2$	Snieckus [70]	Cl	Iwao [71]
CONLiR	Hauser [72]	Br	Schlosser [73]
$CONLiC(Me)_2Ph$	Snieckus [74]	NLiBoc	Gschwend [75]
$C(OLi)NR_2$ (→ CHO)	Comins [76]	NLiCO*t*-Bu	Muchowski [77]
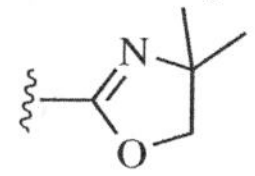	Meyers [78]	NCO_2Li	Lee [79]
	Gschwend [80]		
CO_2Li (K)	Mortier, Bennetau [81]	N=C(OLi)–Nu (→ NHCONu)	Lloyd-Jones, Booker-Milburn [82]
CH_2OLi	Uemura [83]	$OCONEt_2$	Snieckus [84]
$C(OLi)_2$ (→ –CO–)	Mortier [85]	OCH_2OMe	Christensen [86]
CSNLiR	Gschwend [87]	OLi	Posner [88]
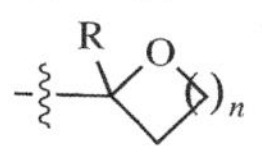 $n = 1,2$	Capriati [89, 90]	$O(CH_2)_nOLi$	Screttas [91]
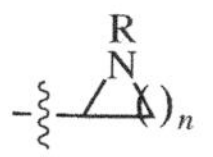 $n = 1,2$	Florio, Luisi [92, 93]	$OCON(Me)C(Me)_2Ph$	Snieckus [74]
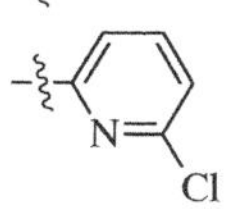	Fort [94]	OSO_2NR_2	Snieckus [95]
		SLi	Martin [96]
		SO_2NLiR, SO_2NR_2	Hauser [97]
		$SO_2NLiC(Me)_2Ph$	Snieckus [74]
		$P(O)(OLi)_2$	Lopez-Ortiz [98, 99]

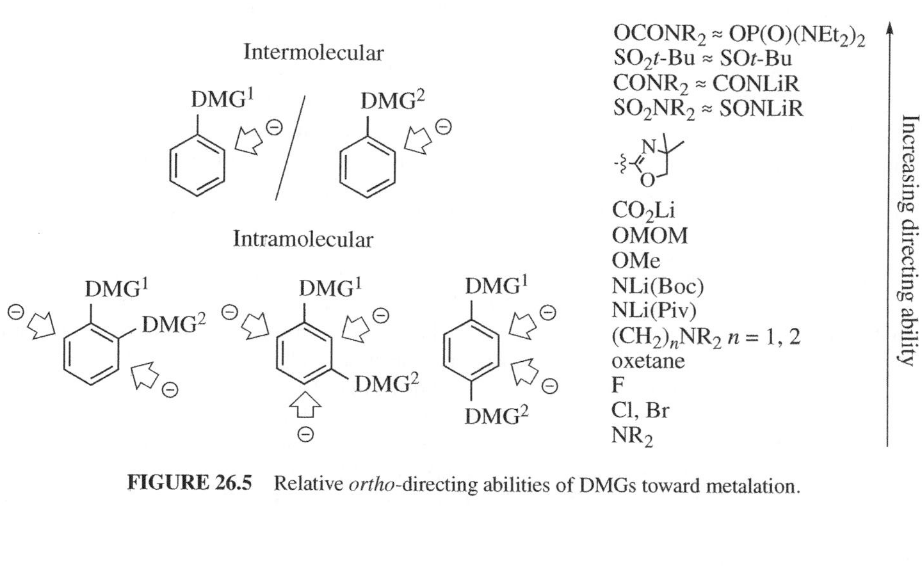

FIGURE 26.5 Relative *ortho*-directing abilities of DMGs toward metalation.

(Fig. 26.6) [35]. Superbases preferentially attack the inductively activated aromatic position next to the most electronegative substituent (overriding base mechanism), while organolithiums, which require precoordination, deprotonate in the vicinity of the most powerful electron-donor substituent (CIPE mechanism).

N-protected *ortho*- and *para*-anisidines **4** and **5** undergo hydrogen–metal exchange in the position adjacent to the oxygen with LICKOR and *ortho* to the nitrogen atom with *t*-BuLi [105]. On the other hand, the metalation of 2- and 4-fluoroanisoles **6** and **7** occurs at the oxygen-adjacent position with *n*-BuLi and *ortho* to the fluorine with LICKOR or *n*-BuLi/PMDTA [105]. The CO_2Li DMG permits an excellent degree of regiocontrol between nonequivalent *ortho* centers. Metalation of *ortho*-anisic acid **8** with *s*-BuLi/TMEDA proceeds exclusively in the position adjacent to the carboxylate, while a complete reversal of regioselectivity is observed with LICKOR [106].

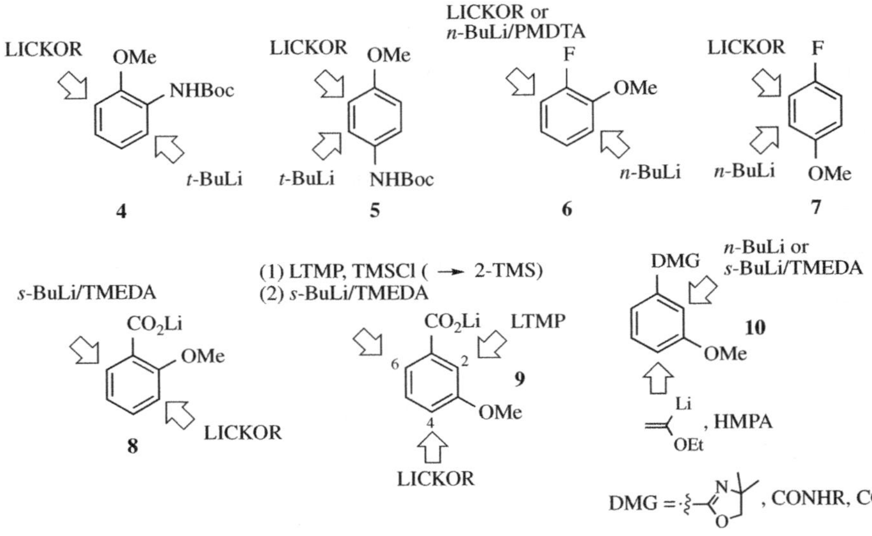

FIGURE 26.6 Optional site selectivity through mechanism-based substrate reagent matching.

The influence of two *meta*-interrelated DMGs can be concerted to direct the metalation in between them. LTMP metalates *meta*-anisic acid **9** in THF at 0°C at the doubly activated position (C2) [62]. The regiochemistry of this lithiation is truly thermodynamically controlled: resonance and inductive effects favor removal of the H2 proton. LICKOR deprotonates preferentially the C4 position. To prepare 6-substituted benzoates, one has to (i) protect the C2 site by introducing a trimethylsilyl group with LTMP, (ii) lithiate with *s*-BuLi/TMEDA, (iii) quench with an electrophile, and (iv) remove the protecting group in C2.

Meta-methoxy phenyloxazoline and secondary and tertiary benzamides **10** are deprotonated at the C2 position by *n*-BuLi and *s*-BuLi/TMEDA [107, 108], whereas metalation occurs exclusively *ortho* to the methoxy with *α*-ethoxyvinyllithium/HMPA [109]. These observations indicate that the directing effect of the DMGs result from both kinetic (coordination) and thermodynamic (acidity) factors.

Iterative metalation using introduced electrophiles as DMGs was conceptualized by Snieckus [110]. This "walk-around-the-ring" metalation sequence, which provides multisubstituted arenes, is exemplified in Scheme 26.3. After TMS protection of the 2-position of the benzamide **11**, a metalation is achieved adjacent to the carboxamide DMG of **12**, and a second one next to the freshly introduced DMG of **13**. Protodesilylation of **14** can be realized with CsF to give **15**. In the presence of benzaldehyde in refluxing dry DMF, the reaction gives amide carbinols, which cyclize upon treatment with TsOH to give phthalides **16** [111].

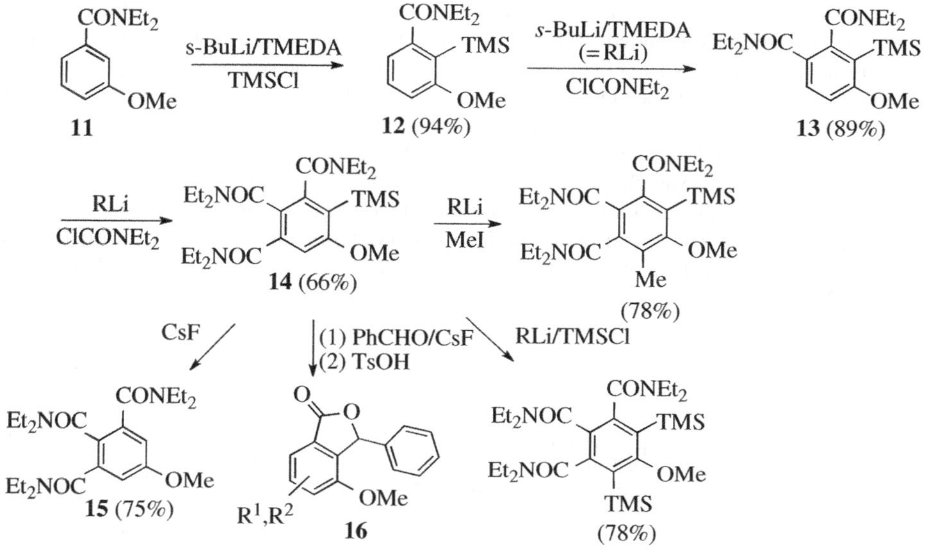

SCHEME 26.3 *Walk-around-the-ring* metalation sequence.

Some stereogenic–chiral DMGs have also been studied: oxazolines [112], masked aldehydes [113], amides [114, 115], sulfonamides [116], and sulfoxides [117–119]. Aldehydes, ketones (leading to chiral alcohols), and imines (leading to chiral amines) are standard prochiral electrophiles. With chiral arene sulfoxides, enantiopure aromatic phenyl and naphthyl sulfoxides **17** can be prepared by reaction of (*S*)-*t*-butyl *t*-butanethiosulfinate with aryllithium derivatives (Scheme 26.4) [120]. The DoM reaction is performed with *n*-BuLi followed by addition of the lithiated intermediates to *N*-tosylimines, affording the chiral arene **18**.

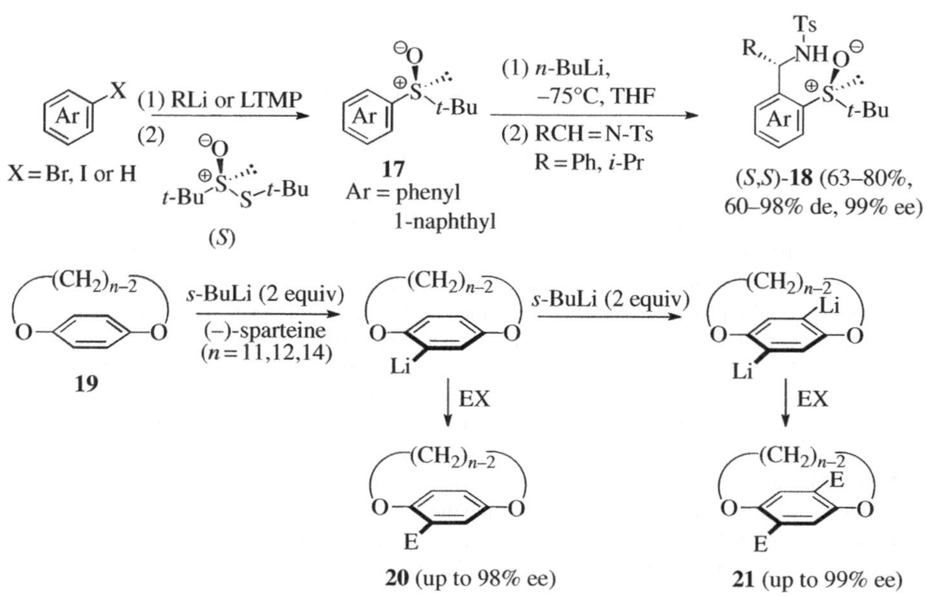

SCHEME 26.4 Chiral *ortho*-metalation.

The first example of catalytic and enantioselective *ortho*-lithiation for the generation of planar chirality was recently reported [121]. Enantioselective monolithiation and dilithiation of 1,*n*-dioxa[*n*]paracyclophanes **19** are performed with *s*-BuLi in the presence of a catalytic or stoichiometric amount of sparteine. Quenching with various electrophiles provides access to chiral mono- and disubstituted paracyclophanes **20** and **21** with excellent ee's.

Despite the general interest in the transposition of solution-phase organic reactions to the solid phase [122], only few examples of DoM chemistry applied on a solid support have been described (Scheme 26.5). A phthalide library was prepared by directed *ortho*-lithiation of resin-bound benzamides **22** [123, 124]. The lithiation substitution is followed by cyclization–resin release induced by simply warming of the reaction mixture in toluene or dioxane, yielding the desired phthalide compounds **23** in high purity.

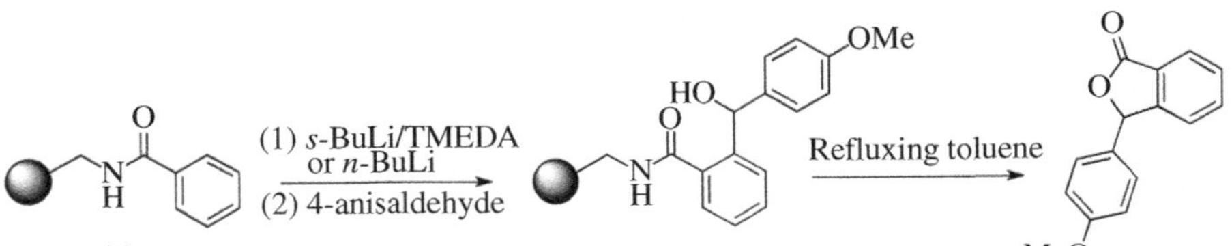

SCHEME 26.5 Solid-phase DoM reactions.

26.3.4 External and *In Situ* Quench Conditions

Under conventional external quench (EQ) conditions, the arene and the base are premixed prior to addition of the electrophile. In general, the thermodynamic equilibrium existing between the anions intermediately formed is displaced toward the most stable (less basic) anion owing to its stabilization by the substituents (Fig. 26.7).

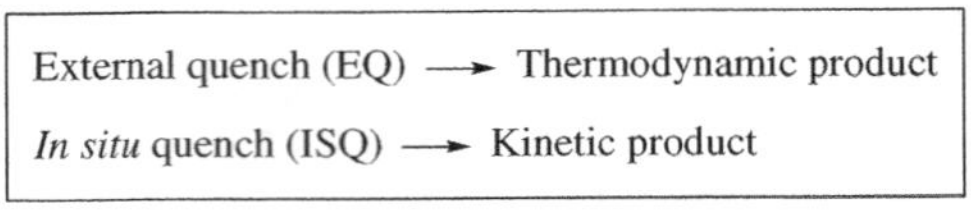

FIGURE 26.7 EQ/ISQ conditions.

The *in situ* quench (ISQ) technique [47] involves premixing of a lithium amide base (usually LDA or LTMP) and the electrophile at low temperature before addition of the arene. As soon as the *ortho*-lithio anion forms, it can immediately react with the electrophile. The inverse addition protocol is equally productive, that is, a mixture of the arene and the electrophile is treated with a lithium amide base. The electrophile must be either unreactive to or react nondestructively with the lithium amide base, which therefore drastically limits useful base–electrophile combinations. This concept was introduced by Martin for cyanobenzene deprotonation–silylation sequences [47]. Low concentrations of aryllithiums lead to increased functional group tolerance. The ISQ technique was extended to a number of electrophiles that are compatible with lithium amide bases, including TMSCl, Me_3SnCl, $B(OiPr)_3$ [125, 126], benzaldehyde, MeI, EtI, and Me_2S_2.

The ISQ sequence may be illustrated through the example of trisilylation of 1,3,5-trifluorobenzene **24** (Scheme 26.6). In contrast to what was stated in an early report published 40 years ago and regularly quoted in the literature since then [127], the reaction between 1,3,5-trifluorobenzene **24** and an excess of *t*-BuLi and subsequent treatment with an excess of TMSCl at −115°C affording 1,3,5-trifluoro-2,4,6-tris(trimethylsilyl)benzene **26** does not occur via the trilithiated species **25**.

A silylation step is interposed between the first, second, and third hydrogen–lithium exchange; in other words, the threefold electrophilic substitution is mediated by monometalated intermediates [128].

A change of the addition mode can cause complete reversal of regioselectivity of the metalation (Scheme 26.7) [129]. Under ISQ conditions, veratric acid **27** and LTMP/TMSCl give predominantly

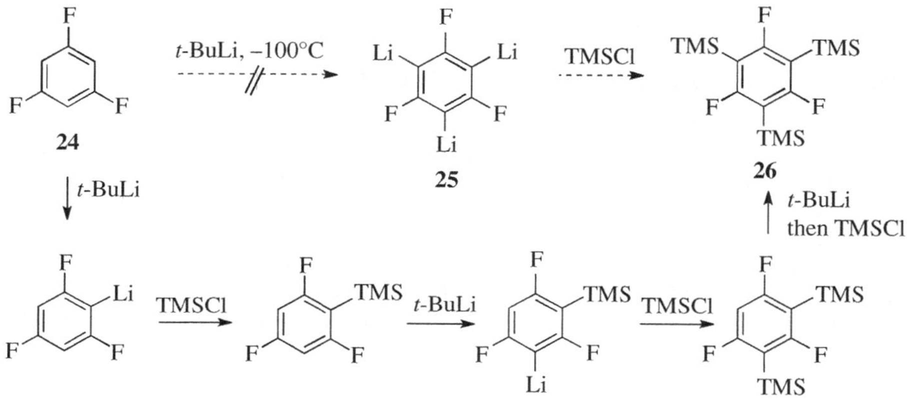

SCHEME 26.6 Trisilylation of 1,3,5-trifluorobenzene.

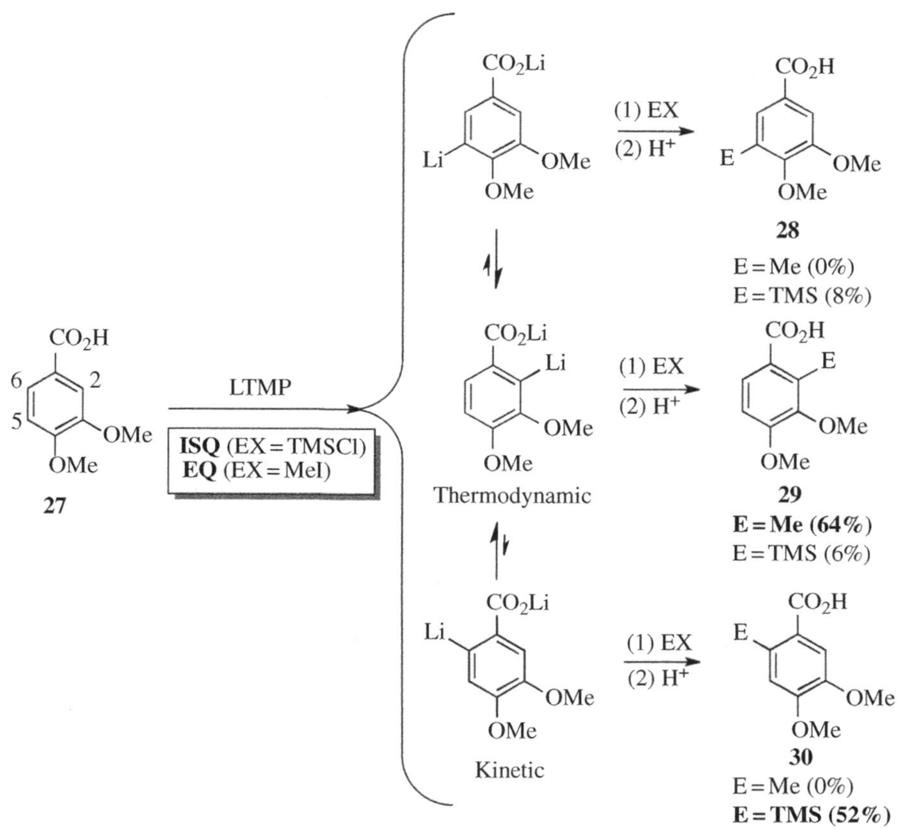

SCHEME 26.7 Reversal of regioselectivity (EQ/ISQ conditions).

the kinetic regioisomer **30** (E=TMS) with small amounts of isomers **28** and **29**, suggesting the intermediate formation of three dianions. Due to the *ortho*-directing effect of the 1,3-interrelated CO_2Li and MeO, the hydrogen H2 is thermodynamically more acidic. Under EQ conditions, equilibrium between the *ortho*-lithiated species toward the thermodynamically more stable lithium 2-lithio-3,4-dimethoxybenzoate is followed by reaction with iodomethane leading to the 2-methyl regioisomer **29** as the sole product.

26.3.5 Apparent Anomalies in the Reactivity of Certain Electrophiles

Apparent anomalies in reactivity have appeared to exist for a number of reactions. It is not generally known that under conventional EQ conditions, an electrophile can promote aromatic metalation *during* the quench step. It was stated in an early report [130] that the reaction of phenothiazine **31** with 2 equivalents of *n*-BuLi proceeds with formation of dilithio intermediate **33** via monolithio amide **32** (Scheme 26.8). The fact that the regioselectivity of the reaction is dependent on the electrophile used was not properly analyzed by the authors: whereas trapping experiments with RCOX=DMF, $PhCONMe_2$, $MeCO_2Li$, $PhCO_2Li$, and CO_2 give *C(1)*-acylation products **34**, the reaction of RX=MeCOCl, MeI, and ethylene oxide provides *N(10)*-substitution products **35**.

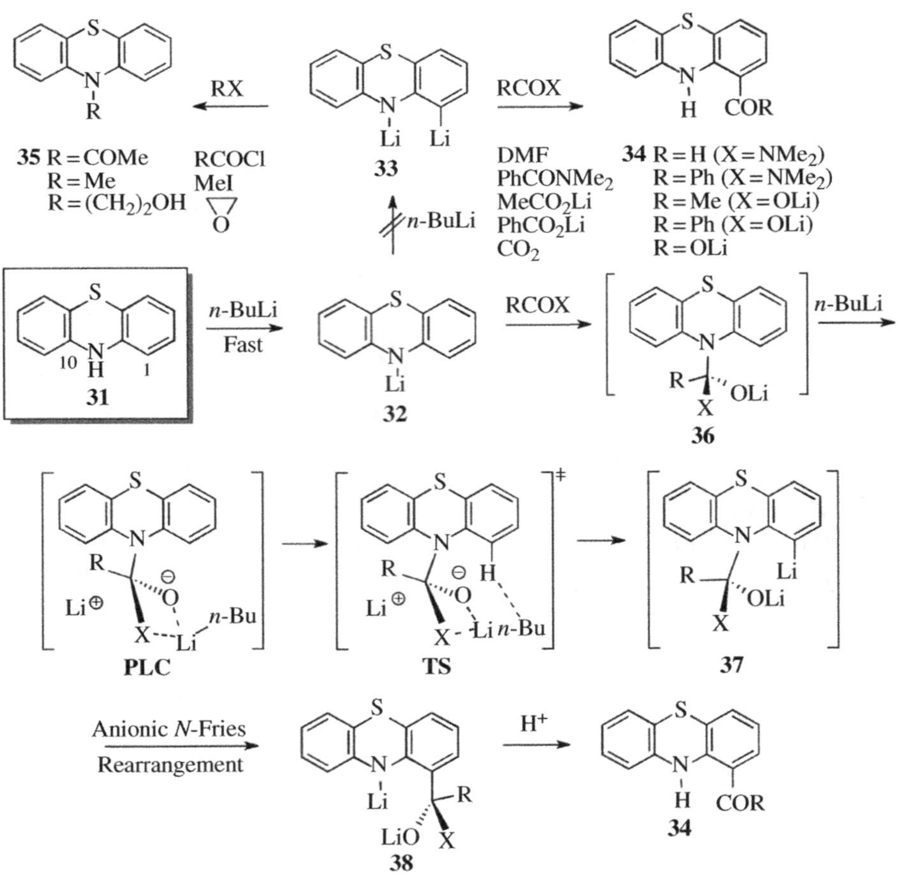

SCHEME 26.8 Lithiation of phenothiazine.

After careful analysis of the results obtained by different authors [131], it is possible to unravel the pathway of this reaction. Metalation does not occur via the dilithio intermediate **33**. DMF, $PhCONMe_2$, $MeCO_2Li$, $PhCO_2Li$, and CO_2 are responsible for a second deprotonation, which occurs *during the quench*. Nucleophilic addition of the monolithium amide **32** to these electrophiles, which affords tetrahedral species **36**, is faster than the trap of *n*-BuLi by the electrophile. *Ortho*-lithiation directed by the *in situ* formed lithium aminoalkoxide DMG gives the dianionic intermediate **37**. This deprotonation presumably proceeds via a PLC (CIPE Process), which immediately precedes the formation of an eight-membered ring TS. Internal deprotonations have been suggested to be optimal for eight-membered rings [132]. This deprotonation is followed by a rapid, irreversible anionic *N*-Fries rearrangement leading to the thermodynamically more stable (less basic) lithium amide **38**. Quenching with water gives the observed 1-acyl products **34**. In agreement with the proposed mechanism, *N*-substitution products **35** are formed exclusively with MeCOCl, MeI, and ethylene oxide.

26.4 DIRECTED remote METALATION (DreM)

The mechanistic concept of CIPE, introduced in association with DoM reactions, is also utilized to rationalize metalations at remote positions relative to the DMG group or DreM. However, recent mechanistic studies have provided an in-depth comprehension of these transformations that extend beyond the scope of the CIPE model.

DreM reactions can be classified into four categories (Fig. 26.8) [133]. Directed remote aromatic and lateral metalations on flexible biaryl structures bearing an electrophilic coordinating

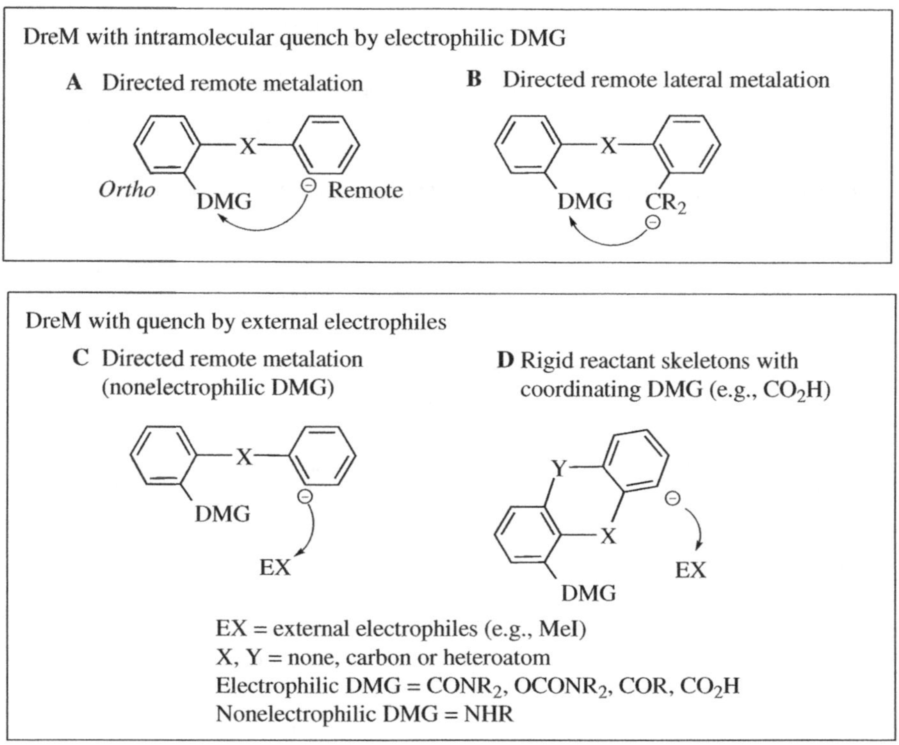

FIGURE 26.8 Directed remote metalation (DreM) categories.

DMG are followed by intramolecular trapping of the remote anion by the DMG (categories A and B). Besides, quenching of remote anions with external electrophiles has been reported for nonelectrophilic DMGs on flexible biaryl structures and for electrophilic coordinating DMGs when the rigidity of structures does not allow for intramolecular electrophilic trapping (categories C and D).

Two relevant examples for categories A and D are presented (Schemes 26.9 and 26.10). Reaction of *N,N*-diethyl 2-biphenyl carboxamide **39** with *s*-BuLi/TMEDA (1.1 equiv) at −78°C followed by the addition of an electrophile affords C3-substituted derivatives **40** (DoM process, CIPE mechanism). In contrast, cyclization to give fluorenone **45** occurs with LDA at 0°C-rt (84%, DreM process) [134]. When an excess of LDA (4 equiv) and TMSCl (4 equiv) are premixed in THF at −78°C prior to addition of *N,N*-diethyl 2-biphenyl carboxamide **39** (ISQ conditions), the 3-trimethylsilyl derivative **42** is formed exclusively (65%).

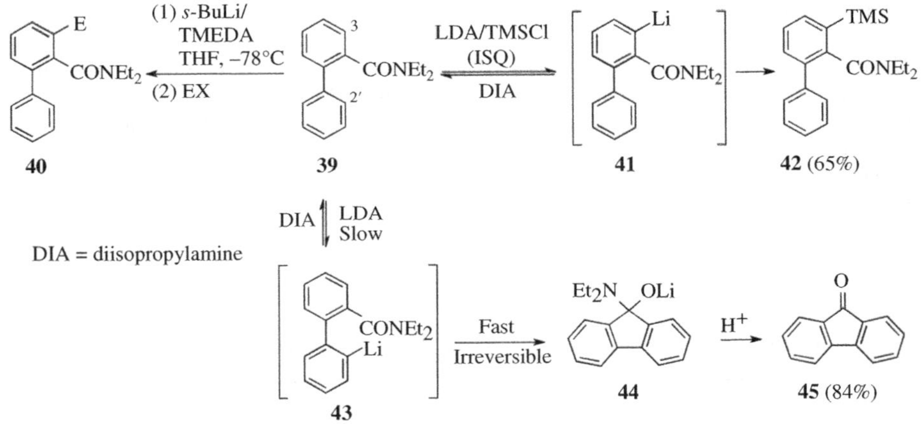

SCHEME 26.9 Mechanism of the DoM and DreM reactions of *N,N*-dialkylbiphenyl 2-carboxamides.

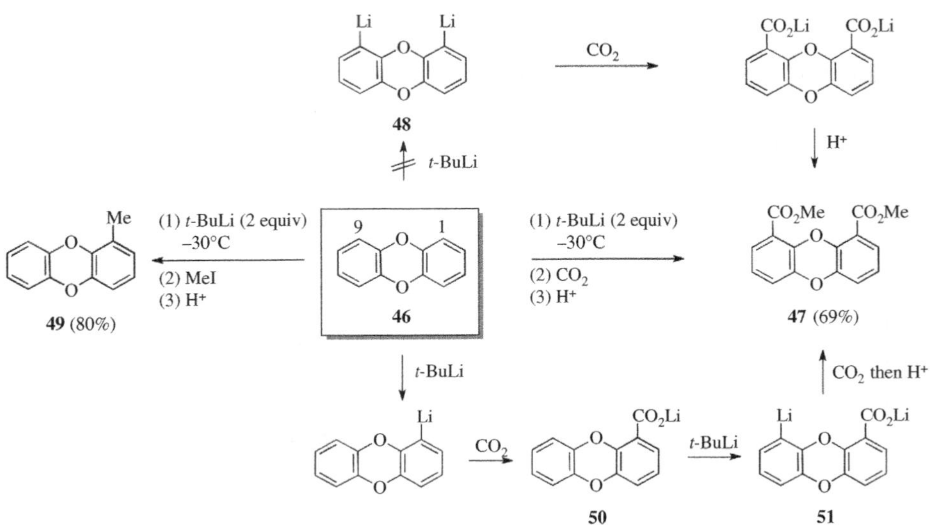

SCHEME 26.10 DreM reaction of dibenzodioxin.

The LDA metalation is proposed to proceed via initial amide base complexation and equilibrium formation of the 3-lithio anion **41**, whose fast reaction with an *in situ* electrophile (TMSCl) to afford the 3-trimethylsilylated product **42** prevents its equilibration with the 2′-lithio anion **43**. In the absence of TMSCl, the 3-lithio anion undergoes equilibration with **43**, which cyclizes to a stable tetrahedral carbinolamine oxide which, upon hydrolysis, affords fluorenone **45**.

In the case of rigid substrates (category D), when the lack of flexibility of the molecule does not allow for intramolecular quenching, reaction of remote metalated species by external electrophiles becomes possible (Scheme 26.10). Treatment of dibenzodioxin **46** with *t*-BuLi followed by quench with CO_2 and esterification provides a 1,9-diester **47**, suggesting that 1,9-dianion **48** is generated [135]. However, using iodomethane as an electrophile, only 1-methyldibenzodioxin **49** is isolated, indicating that at −30°C only monometalation occurs: the initially formed 1-carboxylate salt **50** is able to direct a remote metalation at the 9-position to give **51** *during* the quench. Carbon dioxide reacts presumably faster with mixed aggregates than with *n*-BuLi in solution.

26.5 PERI LITHIATION OF SUBSTITUTED NAPHTHALENES

The peri lithiation reaction, in which a DMG at position C1 directs deprotonation at C8 position of a naphthalene derivative, has received so far less attention (Scheme 26.11) [136]. The best DMGs for peri metalation are those which coordinate to the base *and* do not acidify—hence activate the *ortho* (C2) position [137]. Naphthalenes carrying F, Cl, CF_3, CF_3O, $CONR_2$ at position 1 are kinetically and thermodynamically deprotonated by organometallic reagents at position 2. In contrast, peri lithiation is favored if the DMG is OLi, NLi_2, NLiMe, NMe_2, and CH_2NMe_2.

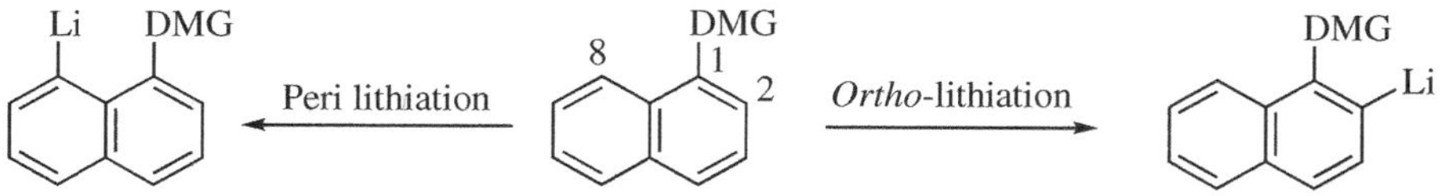

SCHEME 26.11 Peri lithiation of substituted naphthalenes.

The methoxy group shows a hybrid behavior: *t*-BuLi reacts both at the *ortho* and peri positions of 1-methoxynaphthalene (1:7 ratio), whereas reversed values (7:1) are obtained with *n*-BuLi/TMEDA [138]. This regiodivergent reactivity of MeO can be rationalized by the ease of coordination of this group to the organometallic reagent (CIPE), and its electron-withdrawing properties resulting in an increase in CH-acidity of the adjacent protons (KIE) [136]. DFT calculations (B3LYP/6-31+G*) reveal that peri lithiation of 1-naphthol model is a slow process taking place preferentially through an open-dimer route [139].

DMGs like CO_2H, CO_2R, COR, CH=NR, NO_2, SO_2R, and BR_2 groups at position 1 or 2 may enhance conjugate addition of the metalating agents to the naphthalene π-system (Scheme 26.12) [136].

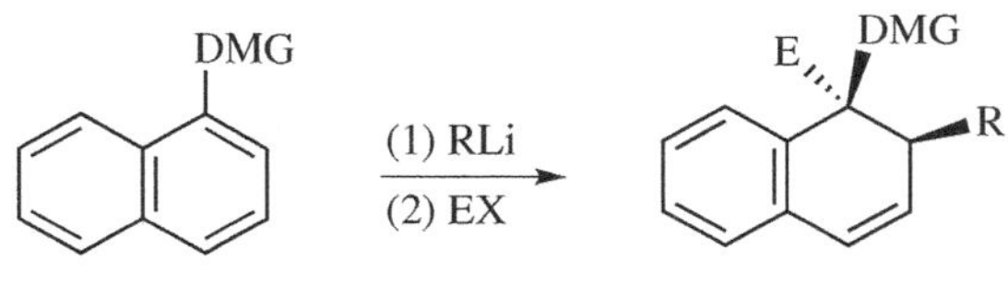

DMG = CO_2H, CO_2R, COR, $CONR_2$,
CH=NR, NO_2, SO_2R, and BR_2

SCHEME 26.12 1,4-Addition of organolithium reagents to the naphthalene π-system.

26.6 LITHIATION OF METAL ARENE COMPLEXES

Coordination of an arene with a transition metal renders the ring hydrogens more acidic compared to the free arene (Scheme 26.13). Metal-complexed aryllithiums react like a stabilized version of the free aryllithiums. Most of the general methodologies related to this process concern the $Cr(CO)_3$ system, and this topic has been addressed in several reviews [140–142].

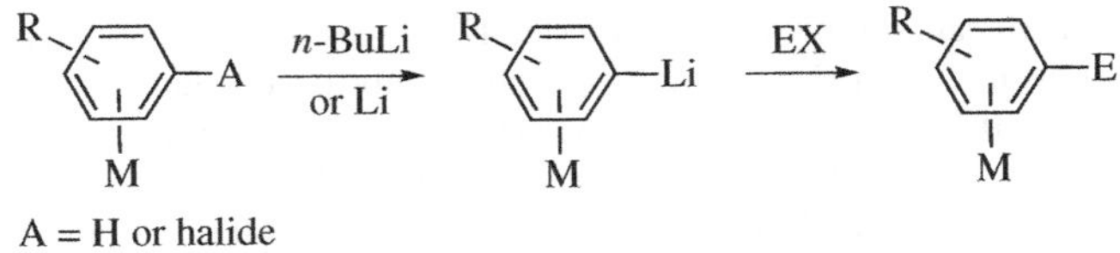

SCHEME 26.13 Lithiation of metal arene complexes.

For monosubstituted arenes, kinetically controlled discrimination between the two enantiotopic *ortho* hydrogens of the planar chiral benzene chromium tricarbonyl complex leads to nonracemic products. Asymmetric lithiation is more efficient when one or more oxygen atoms, such as ether linkages, are present in the starting prochiral complex (Scheme 26.14). Treatment of $Cr(CO)_3$-anisole complex **52** with the chiral lithium amide **53**, in the presence of TMSCl under ISQ conditions, affords (+)-*ortho*-silylated complex **54** with good chemical yield and ee value [143–145]. The isobenzofuran system **55** reacts as well to give α-silylated product **56** [146].

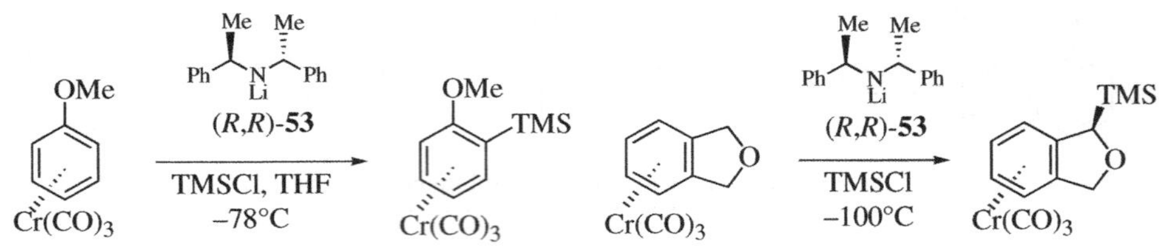

52 **54** (83%, 84% ee) **55** **56** (82%, 76% ee)

SCHEME 26.14 Asymmetric lithiation of chromium arene complexes.

Sugar residues can also be efficiently employed as chiral auxiliaries for the asymmetric formation of *ortho*-substituted tricarbonylchromium complexes. Properly selected carbohydrate auxiliaries enable therefore to generate planar chiral $Cr(CO)_3$-aromatic derivatives **57** (Scheme 26.15) [147]. The chiral sugar auxiliary is easily removed to give **58**.

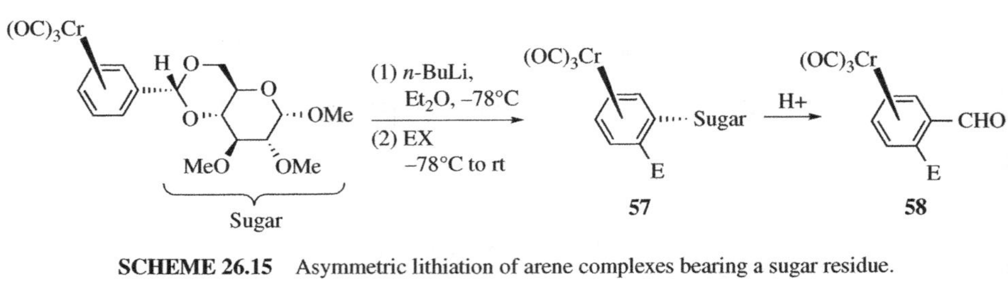

57 **58**

SCHEME 26.15 Asymmetric lithiation of arene complexes bearing a sugar residue.

26.7 LATERAL LITHIATION

Lateral lithiation is the site-selective proton removal at benzylic position by alkyllithiums or lithium dialkylamides (Scheme 26.16) [53, 148]. As for DoM and DreM reactions, DMGs increase the acidity by inductive effect and/or coordination. Useful DMGs include CO_2H, $CONR_2$, CONHR, CH_2NR_2, $OCONR_2$, NHCOR, $NHCO_2R$, dialkylamines, aziridines, ketone enolates, and sulfonates. Benzylic anions demonstrate good reactivity, and conventional electrophiles that react with aryllithium reagents can be used (see Table 26.2). Primary, allylic, and benzylic halides usually give good yields of laterally alkylated products. Successful reaction with these substrates is noteworthy since aryllithiums arising from DoM reactions do not alkylate efficiently or give poor yields with alkyl halides other than iodomethane. Secondary and acetylenic halides have also been used. The CH_2NR_2 and CH_2OR groups are more efficient DMGs of lateral lithiation than NR_2 and OR.

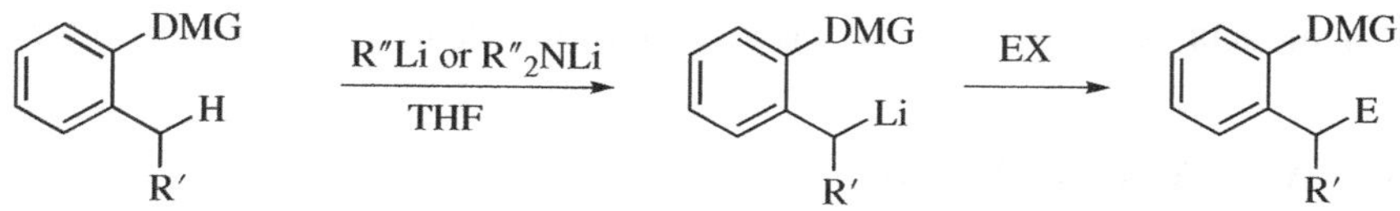

DMGs = CO_2H, $CONR_2$, CONHR, CH_2NR_2, $OCONR_2$, NHCOR, $NHCO_2R$, NR_2, OR, etc.
R′ = H, alkyl, aryl

$CH_2NR_2 > NR_2$
$CH_2OR > OR$

SCHEME 26.16 Lateral lithiation.

Depending on the base used, there is possible competition between *ortho* and lateral metalation. *s*-BuLi/TMEDA reacts irreversibly with 4-methyl-*N,N*-diisopropylbenzamide **59** *ortho* to the amide DMG under kinetic control (CIPE mechanism) to give **60** by trapping with MeOD [108], whereas LDA affords **61** via the thermodynamically more stable benzylic anion (Scheme 26.17). The regioselectivity observed is dependent on many factors (nature of the base, presence of additives, temperature, solvent, etc.).

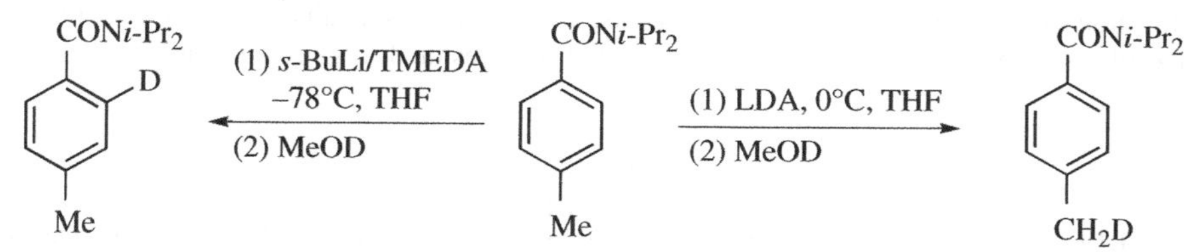

60 (57% d_1) **59** **61** (50% d_1)

SCHEME 26.17 Competition between *ortho* and lateral metalation.

The benzylic anion **63** derived from *O*-aryl carbamate **62** undergoes rearrangement to give benzylic amide **64** upon warming (Snieckus–Fries rearrangement) via migration of the carbonyl carbon from oxygen to carbon (Scheme 26.18) [84]. On the other hand, deracemization of racemic diarylmethanes can be effected via lateral lithiation–protonation sequence with a chiral ligand such as sparteine [149]. *Ortho*-lithiation of *N,N*-diisopropyl-1-naphthamide **65** with *s*-BuLi followed by quench with iodoethane gives **66** whose subsequent lateral lithiation and electrophilic quench provide **67** with good yield and high level of diastereoselectivity [150].

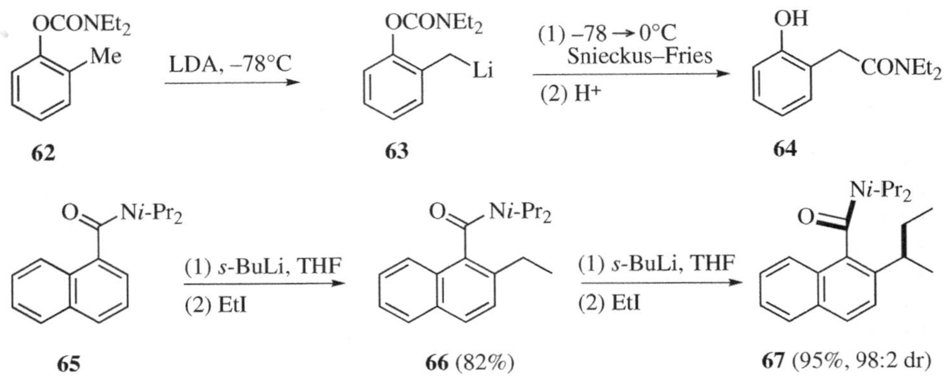

SCHEME 26.18 Snieckus–Fries rearrangement. Atroposelective lithiation of *N,N*-diisopropyl-1-naphthamide.

26.8 ANALYTICAL METHODS

26.8.1 Quantitative Determination of Organolithiums

Organolithium compounds are highly sensitive to air and humidity. The commercial RLi usually contains varying amounts of alkoxides as impurities. The concentration of the organolithium solution can be easily checked by several methods, which can be divided into two categories: (i) direct titration methods, based usually on a color change of self-indicators, and (ii) double titration methods. Such different methods have been reviewed in detail [12, 151]. The traditional double titration method provides the content of active organolithium and of lithium alkoxides/hydroxide (Fig. 26.9).

First, the total base is determined by acid–base titration of LiOH formed by the hydrolysis of a sample of organolithium solution. Therefore, a known amount of the organolithium solution is quenched with excess of water and titrated with 0.1N aq. H_2SO_4 or HCl using phenolphthalein as color indicator.

In the second step, the organolithium reagent is destroyed with an organic halide (e.g., benzyl chloride or 1,2-dibromoethane). It is supposed that the organolithium reagent reacts quantitatively affording LiCl or LiBr and that the residual base (LiOH and lithium alkoxides) does not react. After hydrolysis with an excess of water, the amount of free (residual) base is determined.

Total basicity

$$\text{RLi} + \text{Alkaline components} \xrightarrow{H_2O} \text{LiOH} + \text{Alkaline components}$$

Residual basicity

$$\text{RLi} + \text{Alkaline components} \xrightarrow[\text{(2) } H_2O]{\text{(1) BnCl or } BrCH_2CH_2Br} \text{LiCl or LiBr} + \text{Alkaline components}$$

Active base RLi = Total base – Residual base

FIGURE 26.9 Quantitative determination of organolithiums.

Among the direct titration methods, one can perform (i) the titration of colored charge–transfer complexes between organolithiums and polycyclic aromatic bases, which requires a standard solution of 2-butanol [152]; (ii) the back-titration of a colored anion with a standard solution of benzoic acid or 2-butanol; and (iii) the colored direct titration based on the deprotonation of an organic reagent where the endpoint becomes visible via the formation of an intensively colored dianion [153]. Whereas the first two methods require the use of standardized solutions, the latter one is rather simple as long as strongly basic organolithiums are concerned. For example, when the organolithium solution is added to diphenylacetic acid, the latter is progressively deprotonated until all acid is consumed. One drop more will lead to the formation of a yellow-colored dianion. Several other methods based on the same principle have been developed. Each method has certainly its advantages and limitations.

The classical double titration method has certainly the advantage that it can also be employed to titrate organomagnesium intermediates, metal dialkylamides, and even radical anions. Moreover, self-indicators and anhydrous and oxygen-free conditions are not required.

26.8.2 Qualitative Determination of Organolithiums

In addition to these quantitative tests, qualitative methods allow to reveal the presence of organolithium intermediates and to differentiate between alkyl- and aryllithium reagents. The Gilman–Schulze test (Color Test I) [151] responds on all-polar organometallic intermediates and reagents. When a sample containing the organolithium compounds is added to a solution of 4,4′-bis(dimethylamino)benzophenone **68** (Michler's ketone) in toluene, upon addition of water and a few drops of an iodine solution in glacial acetic acid leads, when positive, to an intense malachite green–blue color of a benzhydryl cation **69** (Scheme 26.19).

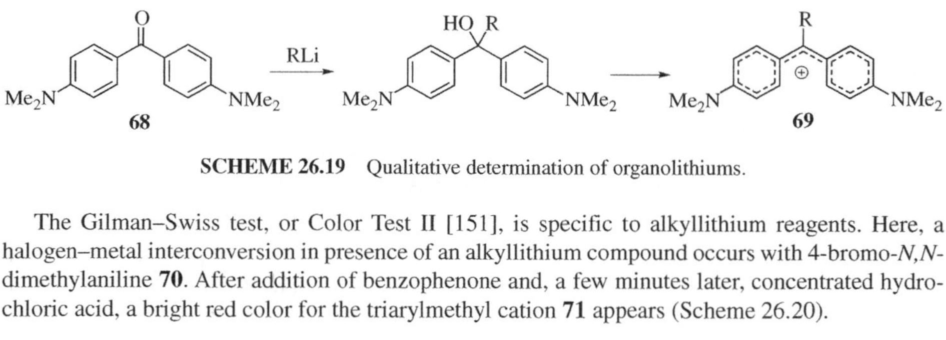

SCHEME 26.19 Qualitative determination of organolithiums.

The Gilman–Swiss test, or Color Test II [151], is specific to alkyllithium reagents. Here, a halogen–metal interconversion in presence of an alkyllithium compound occurs with 4-bromo-*N,N*-dimethylaniline **70**. After addition of benzophenone and, a few minutes later, concentrated hydrochloric acid, a bright red color for the triarylmethyl cation **71** appears (Scheme 26.20).

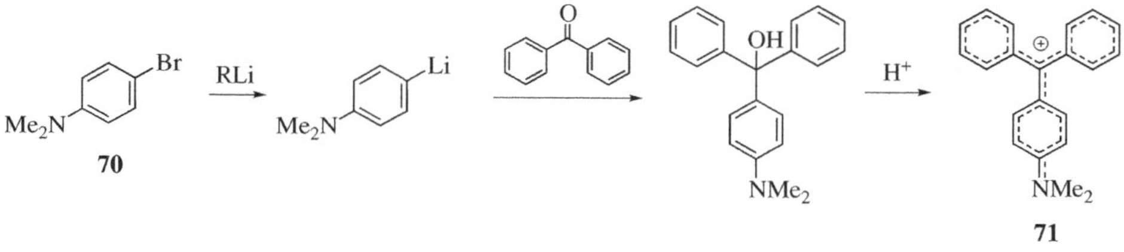

SCHEME 26.20 Gilman–Swiss test (Color Test II).

26.8.3 Crystallography

X-ray diffraction studies on monocrystals allow a much better understanding of the structure and reactivity of organolithium compounds. Solvent-free organolithium species and, most frequently, organolithium species coordinated with solvent molecules or additives have been reviewed at several occasions [4, 154].

The X-ray crystallographic study of an early DoM reaction of *N,N*-dimethylbenzylamine with BuLi in diethyl ether hexane has allowed the identification of a tetranuclear aryllithium cluster $Li_4[C_6H_4\text{-}2\text{-}(CH_2NMe_2)]_4$ where each Li atom is associated with three aryl carbanion carbons and one nitrogen atom [155]. When THF is added, a slow deaggregation affords a dinuclear $Li_2[C_6H_4\text{-}2\text{-}(CH_2NMe_2)]_2(THF)_4$ complex.

The X-ray structures of lithiated *N,N*-diisopropylbenzamide and 1-naphthamide revealed for the first time that the oxygen coordination of the lithium center is maintained in *ortho*-lithiated aromatic tertiary amides [156] like *N,N*-diisopropyl benzamide and naphthamide, giving species that have been characterized as solid-state dimers (Fig. 26.10). In a $(CLi)_2$ core, each Li-ion is further stabilized by one solvent molecule (Et_2O or THF) and an amide *O*-center [157]. These solid-state structures are coherent with a CIPE-induced lithiation of benzamides through amide(O)–Li coordination.

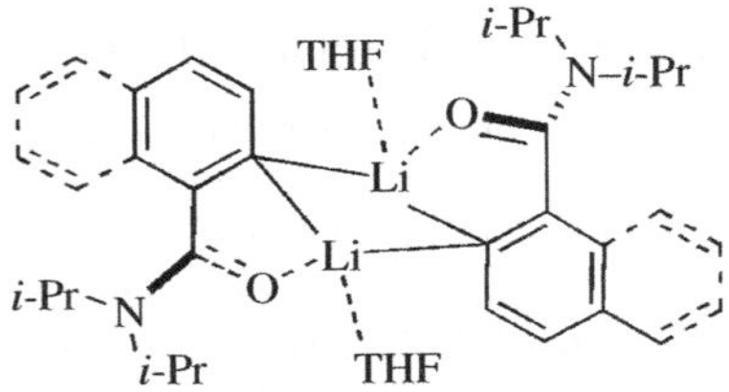

FIGURE 26.10 Structures of *ortho*-lithiated aromatic tertiary amides.

After metalation of 2-ethyl-*N,N*-diisopropyl benzamide **72** with *t*-BuLi in THF-toluene at −78°C, crystals can be obtained and identified as *N,N*-diisopropyl-2-ethyl-6-lithiobenzamide–THF **73** by X-ray crystallography (Scheme 26.21). The solid-state structure showed that each metal center is stabilized by an amide O-center and one THF molecule. In contrast, in presence of PMDTA, lateral lithiation occurs and leads to crystals that reveal by X-ray diffraction to be *α*-lithio-2-ethyl-*N,N*-diisopropyl-1-benzamide–PMDTA **74** [158, 159].

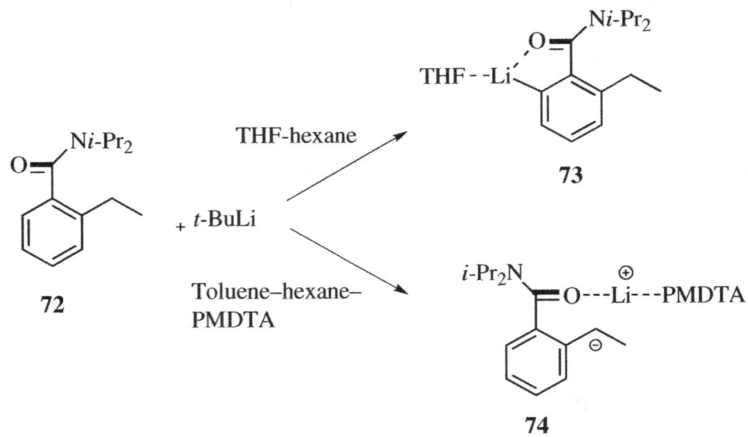

72

73

74

SCHEME 26.21 Structures of a laterally deprotonated aromatic tertiary amide.

26.8.4 NMR Spectroscopy

The use of ^{6}Li and ^{7}Li NMR based on the coupling of these nuclei with other nuclei like ^{13}C and ^{15}N allows the characterization of the structure of organolithium compounds and reaction dynamics based on chemical shift, quadrupolar interaction, and the study of relaxation times dependent on temperature. For more details on solution NMR spectroscopy, see Refs. [154, 160–162]. The lithium spectroscopy relies on the presence of two NMR-active lithium nuclei, ^{7}Li and ^{6}Li, whose natural abundance is 92.6% and 7.4%, respectively. Both isotopes are suitable for NMR spectroscopy as they have a nuclear spin of $I=3/2$ and 1. When the exchange process is slow, coupling information can be obtained based on the multiplicity according to $M=2nI+1$ (n is the number of coupled nuclei and I the spin quantum number). A quadrupolar relaxation, due to their nuclear spin greater than ½, may contribute to the general relaxation mechanism by decreasing the relaxation times T_1 and T_2. This results in a line broadening in the NMR spectra. Information on the aggregation state of organolithium species is determined by measuring the ^{6}Li–X coupling constant (X = ^{13}C or ^{15}N) and the multiplicity of the X nucleus. In cases where the multiplicity is not accessible for different reasons, the aggregation is difficult or impossible to determine via standard NMR experiments. Molecular diffusion study by high-resolution NMR has emerged as a powerful tool to overcome these difficulties and to obtain specific information on the physical behavior of molecules (distribution of molecular weight, hydrodynamic radiuses, solvation, mobility with respect to solvent/temperature/concentration). They are characterized with respect to their chemical shift and their molecular diffusion coefficient. In particular, this allows the separation and identification of different aggregates within a mixture [163].

The reactivity of organolithium compounds depends to a large extent on the nature of the solvent and on changes in aggregation states [4]. Higher aggregates are less reactive than lower ones. Rapid-injection nuclear magnetic resonance (RINMR) emerged as a powerful tool to determine the reactivity of aggregates and afford kinetic data, which are otherwise difficult to assess. Reich and coworkers published in 2007 the use of RINMR apparatus [164]. NMR experiments at very low temperatures (down to −130 to −135°C) were performed. As the absolute rates of interconversion between dimers and tetramers are at this temperature much lower, a much broader range of reactivity could be studied. For further details, refer to recent publications [160, 165].

In contrast to solution NMR spectroscopy, the contribution of the quadrupolar interaction in solid-state NMR is insignificant. Therefore, due to its greater natural abundance, greater sensitivity, and shorter T_1 relaxation time, ^{7}Li NMR is more used. Solid-state NMR provides additional information to X-ray diffraction and theoretical calculations (DFT), with which it is generally combined [163].

26.9 SYNTHETIC APPLICATIONS

Organolithium chemistry is nowadays well established in organic synthesis [166]. In the present section, we highlight some representative examples where selective metalation reactions play a key role. We focus on combinations between *ortho* metalations and subsequent functionalizations like cross-coupling strategies. DoM is the most employed approach, followed by DreM, lateral, and peri metalation [167].

26.9.1 DoM and C—C Cross-Coupling

Transition metal-catalyzed cross-coupling methodologies have become privileged approaches and are recognized as powerful, practical, and versatile methods for carbon–carbon and carbon–heteroatom bond formation (see Chapters 19 and 20) [168–170]. They play an important role in both modern drug discovery and the synthesis of agrochemical ingredients. The Corriu–Kumada, Negishi, Suzuki–Miyaura, and Stille coupling reactions have strongly impacted the efficient access of aromatics and heteroaromatics as building blocks for more complex target molecules. In such context, DoM allows for the regioselective preparation of aromatic $ArB(OH)_2$, ArMgX, ArZnX, and $ArSnR_3$ reagents and can be combined with cross-coupling strategies to afford substituted biaryls and heterobiaryls [171, 172]. Thus, this approach has been increasingly employed by academic and medicinal chemists for small-scale synthesis and by process chemists for multikilogram scale routes for clinical candidates and commercial pharmaceuticals and agrochemicals.

The DoM C—C cross-coupling was notably employed in the synthesis of telmisartan **75**, an angiotensin II receptor antagonist [173]. The reaction pathway is based on a one-pot protocol employing sequential metalation, transmetalation to zincate, and Negishi transition metal-catalyzed cross-coupling. At final step, the oxazoline DMG is removed by hydrolysis to afford the expected carboxylic acid product **75** (Scheme 26.22).

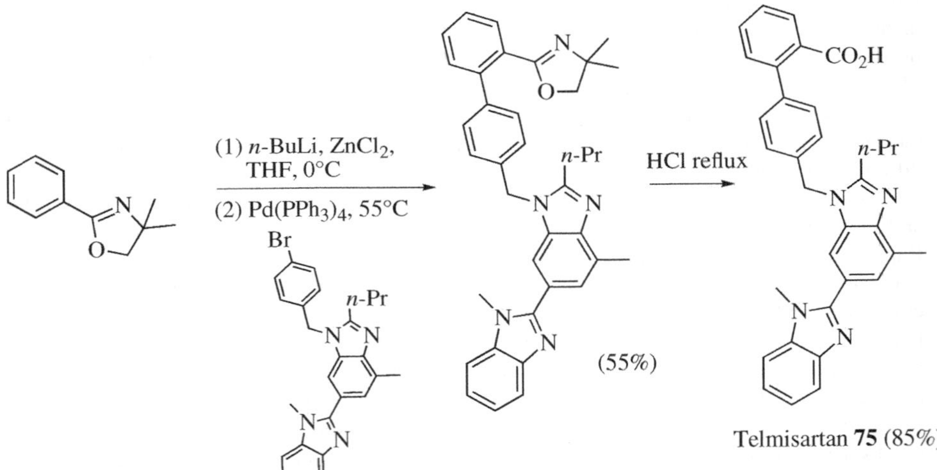

Telmisartan **75** (85%)

SCHEME 26.22 Synthesis of telmisartan (angiotensin II receptor antagonist).

Sanz and coworkers reported on the synthesis of valuable 4-fluoro-2-substituted-1*H*-indoles and benzoxazoles employing either transition metal-catalyzed C—C coupling or the use of benzynes as transition metal-free alternative (Scheme 26.23). DoM-mediated iodination of 3-fluorotrifluoroacetanilide and subsequent Sonogashira coupling–cyclization lead to the indole derivative **76**. In contrast, when the DoM reaction is carried out at temperatures higher than −60°C, benzyne formation occurs via lithium fluoride elimination (see Chapters 12 and 28) followed by intramolecular cyclization to afford the benzoxazole core **77** [174].

26.9.2 DoM, DreM, and Anionic Fries Rearrangement

The DreM of *O*-carbamates can be efficiently linked to transition metal-catalyzed cross-couplings like the Negishi and Suzuki–Miyaura couplings, eventually performed after DoM protocols [175]. For example, the synthesis of alkaloid schumanniophytine **79** has been accomplished involving DoM of *O*-carbamate **78** followed by borylation, Suzuki–Miyaura cross-coupling, and an *ortho*-silicon-induced *O*-carbamate remote anionic Snieckus–Fries rearrangement (Scheme 26.24).

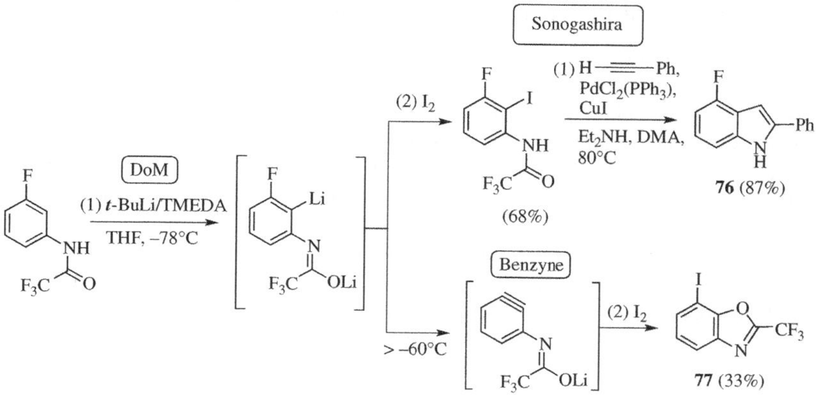

SCHEME 26.23 Synthesis of 4-fluoro-2-substituted-1H-indoles and benzoxazoles.

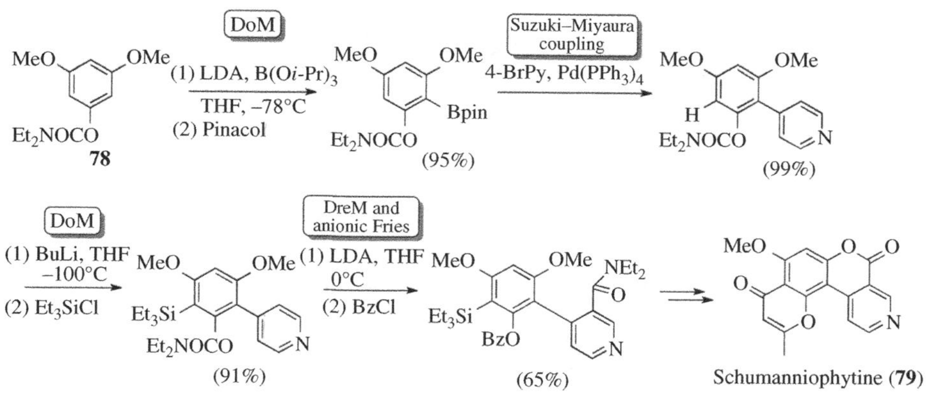

SCHEME 26.24 Synthesis of alkaloid schumanniophytine.

A general alkyllithium–LDA-mediated rearrangement of *N*-carbamoyl diarylamines **81** has been devised for the regioselective efficient construction of acridone alkaloid yukodine **83** [176]. The *N*-carbamoyl diarylamine precursor **80**, readily prepared using Buchwald–Hartwig C—N bond cross-coupling, is submitted to 3 equivalents LTMP. In a highly regioselective and quantitative manner, *N*-Fries rearrangement leads to anthranilamide **81**. Unfortunately, for the second cyclization, another anionic step is unsuccessful, and the authors had to rely on a more classical approach based on the treatment with Tf_2O to afford the acridone **82**. Selective deisopropylation [177] gives then yukodine **83** (Scheme 26.25).

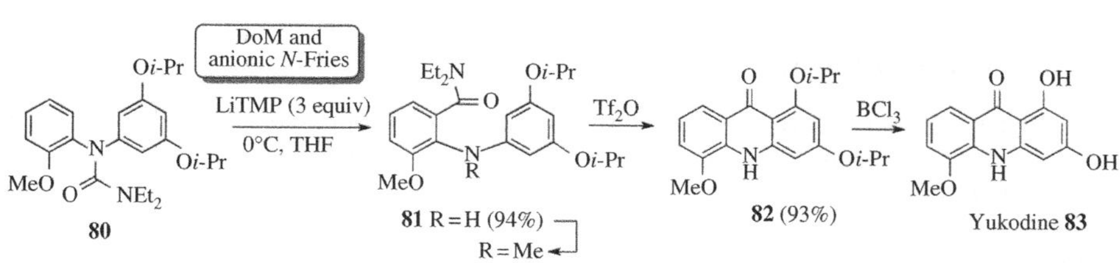

SCHEME 26.25 Synthesis of yukodine.

26.9.3 Industrial Scale-Up of *Ortho* Metalation Reactions

Nowadays, the scale-up of metalation chemistry can be effectively used on industrial large-scale synthesis [178]. For example, Merck developed an efficient multihundred gram synthesis of 3-bromo-6-chloro-phenanthrene-9,10-dione **88**, a useful building block for the modular synthesis of phenanthrenequinones or pharmaceutically important phenanthreneimidazoles employing a DoM–cross-coupling–DreM sequence (Scheme 26.26). In the first step, the amide **84** is subjected to DoM under ISQ conditions (with LDA in the presence of B(*i*-OPr)$_3$) to afford the corresponding boronic acid **85** in greater than 97:1 regioselectivity. The regioselectivity of the borylation is highly dependent on the solvent, with DME being the best. After the cross-coupling step, the anionic cyclization of **86** leading to **87** is realized with lithium diethylamide. Lithium diisopropylamide (LDA) can also be used. However, due to transamidation (Et$_2$N for *i*-Pr$_2$N) a somewhat less effective cyclization (90% vs. 95% assay yield) can be observed.

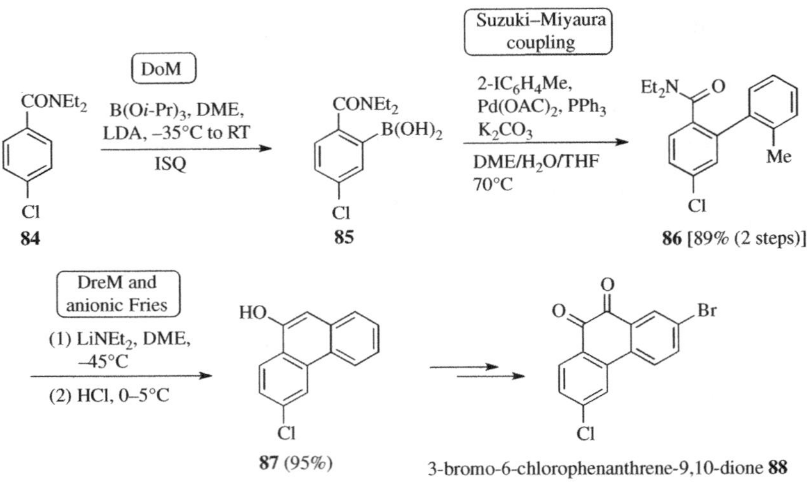

SCHEME 26.26 Synthesis of building blocks for pharmaceutically relevant phenanthrenequinones and phenanthreneimidazoles.

These compounds are useful building blocks for pharmaceutically relevant phenanthrenequinones and phenanthreneimidazoles [179].

The large-scale synthesis of mPGE synthase I inhibitor **93** achieved by Merck on kilogram scale employing DoM–cross-coupling–DreM strategies was reported [180]. When the amide **89** is subjected to metalation under ISQ conditions with B(O*i*-Pr)$_3$ as an *in situ* trap, regioisomerically pure borate **90** is obtained (Scheme 26.27). After successful Suzuki–Miyaura coupling leading to **91**, careful optimization of the reaction conditions allow for the anionic cyclization via DreM giving **92** using lithium diethylamide as a base.

26.9.4 Lateral Lithiation

The polyphenolic binaphthyl gossypol **94** displays interesting pharmacological applications like male oral contraceptive, treatment of bronchitis, total inhibitor of HIV 1, and *in vitro* antiviral activity against herpes type 2 virus and influenza virus (Scheme 26.28). Apogossypol **95**, a degradation derivative of gossypol that lacks the aldehyde functions, still has the potent pan-active inhibitor activity of antiapoptotic Bcl-2 family proteins. The naphthoic acid precursor can be obtained by

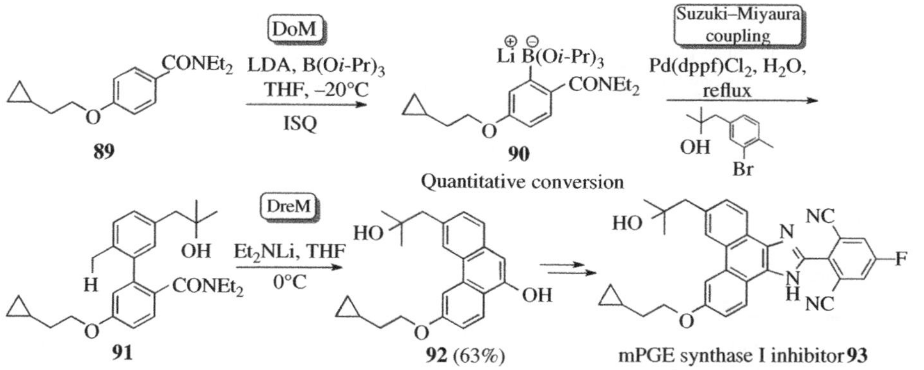

SCHEME 26.27 Large-scale synthesis of mPGE synthase I inhibitor **93**.

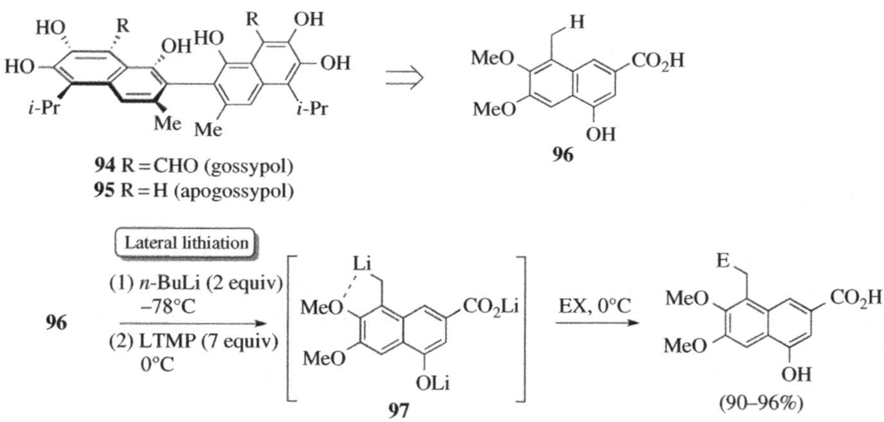

SCHEME 26.28 Synthesis of Apogossypol analogues.

lateral metalation of 4-hydroxy-6,7-dimethoxy-8-methyl-2-naphthoic acid **96** employing LTMP as the base [181]. Optimum reaction conditions revealed to deprotonate first the phenol and carboxylic acid function with 2 equivalents of *n*-BuLi followed by lateral metalation with excess of LTMP generating a trianionic species **97**. The use of 7 equivalents of LTMP is required as the p-electrons of the methoxy and lithium alkoxide coordinate strongly with the base.

26.9.5 Superbase Metalation

The nonsteroidal anti-inflammatory flurbiprofen **100** has been prepared via deprotonation of 3-fluorotoluene **98** with Schlosser–Lochmann superbase (Scheme 26.29) [182]. The selectivity improves significantly when a combination of *tert*-butyllithium and potassium *tert*-butoxide is used as the mixed-metal reagent. The deprotonation occurs at the least-hindered position adjacent to fluorine. Trapping of the organometallic intermediate with fluorodimethoxyborane-diethyletherate and hydrolysis affords the boronic acid, which is then employed in a Suzuki–Miyaura coupling reaction. Another superbase metalation of **99**, now with a combination of LDA and potassium *tert*-butoxide, allows the deprotonation of the benzylic position, followed by carboxylation and a second metalation, and trapping with MeI to afford flurbiprofen **100**.

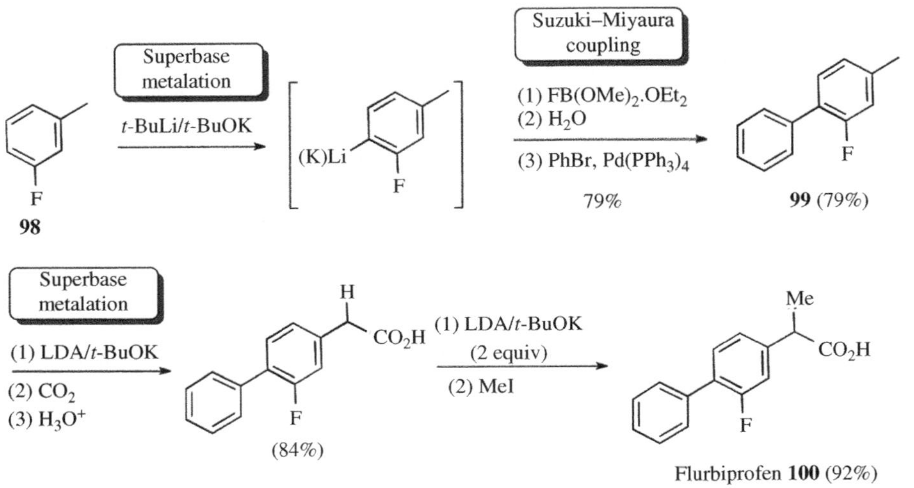

SCHEME 26.29 Synthesis of nonsteroidal anti-inflammatory flurbiprofen.

26.10 CONCLUSION

Polar organometallic chemistry, a term coined for the first time by Manfred Schlosser, and the various facets of metalation chemistry belong nowadays to the toolbox methods of any synthetic chemist. DoM has emerged as a powerful and general technique to construct polysubstituted aromatic and heteroaromatic compounds with perfect control of regioselectivity. Academic researchers and chemists working in the field of life sciences have nowadays methods in hand to assess all possible regioisomers of a given aromatic scaffold and to introduce at will numerous electrophiles in order to get novel bioactive targets for high-throughput screening and for understanding their structure–activity relationships. Moreover, the chemistry is often robust enough to be applied by process chemists on multikilogram scale with pharmaceutical and agrochemical commercialization.

The combination of metalation strategies in modern transition metal-catalyzed reactions offers the possibility to access a myriad of new compounds in a rapid and straightforward manner. Since a couple of years, transition metal-mediated direct C—H functionalization has emerged as a promising strategy. Some mechanistic principles like the CIPE process can be found also here. Although this approach is still in its infancy compared to the tremendous work accomplished in the field of polar organometallic reactions, both strategies are not in competition but complementary with the goal of functionalizing aromatic and heteroaromatic scaffolds to a high degree.

We have tried to summarize in this chapter the basic tendencies and mechanisms of directed metalation. DoM in presence of DMG(s) allows for the regioexhaustive functionalization of aromatics. DreM, frequently combined with migration of the DMG or rearrangement, gives a powerful tool for the synthesis of natural products. Peri and lateral lithiations have shown to further enable functionalization for aromatic scaffolds, and last but not least, when coordinated to and thus activated by a metal complex, aromatic ring systems can be subjected to enantioselective metalation reactions. Although such a chapter cannot cover the topic in an exhaustive manner, we hope to have found a compromise between scholarly presentation and citation of relevant literature.

ABBREVIATIONS

CIPE	Complex-induced proximity effect
DMG	Directing metalation group
DoM	Directed *ortho* metalation
DreM	Directed remote metalation
EQ	External quench conditions
HMPA	Hexamethylphosphoramide
HMTTA	Tetradentate *N,N,N′,N″,N‴,N‴*-hexamethyltriethylenetetramine
ISQ	*In situ* quench conditions
KEM	Kinetically enhanced metalation
LDA	LiN(*i*-Pr)$_2$
LICKOR	*n*-BuLi/*t*-BuOK
LiHMDS	Lithium bis(trimethylsilyl)amide
LTMP	Lithium 2,2,6,6-tetramethylpiperidide
PLC	Prelithiation complex
PMDTA	Tridentate *N,N,N′,N″,N″*-pentamethyldiethylenetriamine
RINMR	Rapid-injection nuclear magnetic resonance
TMEDA	Tetramethylethylenediamine
TMSCl	Chlorotrimethylsilane
TS	Transition state

REFERENCES

[1] Gilman, H., Young, R. V. (1934) *J. Am. Chem. Soc.*, **56**, 1415–1416.

[2] Wardell, J. L. (1982) Alkali Metals. in *Comprehensive Organometallic Chemistry* Vol. **1** (eds.: Stone, E. A., Abel, E. W.), Pergamon Press, Oxford, pp. 43–120.

[3] Carl, E., Stalke, D. (2014) Structure—Activity Relationship in Organolithium Compounds. in *Lithium Compounds in Organic Synthesis* (eds.: Luisi, R., Capriati, V.), Wiley-VCH, Weinheim, pp. 3–51.

[4] Reich, H. J. (2013) *Chem. Rev.*, **113**, 7130–7178.

[5] Bauer, W., Winchester, W. R., Schleyer, P. V. (1987) *Organometallics*, **6**, 2371–2379.

[6] Gessner, V. H., Daschlein, C., Strohmann, C. (2009) *Chem. Eur. J.*, **15**, 3320–3334.

[7] Wakefield, B. J. (1988) *Organolithium Methods*, 4th ed., Academic Press, London.

[8] Reich, H. J., Rigby, J. H. (1999) *Handbook of Reagents for Organic Synthesis Acidic and Basic Reagents*, John Wiley & Sons, Ltd, Chichester.

[9] Fraenkel, G., Henrichs, M., Hewitt, M., Su, B. M. (1984) *J. Am. Chem. Soc.*, **106**, 255–256.

[10] Rathman, T., Bailey, W. F. (2009) *Org. Process Res. Dev.*, **13**, 144–151.

[11] Rathman, T., Schwindeman, J. A. (2014) *Org. Process Res. Dev.*, **18**, 1192–1210.

[12] Deggenaro, L., Giovine, A., Caroccia, L., Luisi, R. (2014) Practical Aspects of Organolithium Chemistry. in *Lithium Compounds in Organic Synthesis* (eds.: Luisi, R., Capriati, V.), Wiley-VCH, Weinheim, pp. 513–538.

[13] Hommes, N. J. R. V., Schleyer, P. V. (1994) *Tetrahedron*, **50**, 5903–5916.

[14] Abbotto, A., Leung, S. S. W., Streitwieser, A., Kilway, K. V. (1998) *J. Am. Chem. Soc.*, **120**, 10807–10813.

[15] Collum, D. B. (1992) *Acc. Chem. Res.*, **25**, 448–454.

[16] Slocum, D. W., Moon, R., Thompson, J., Coffey, D. S., Li, J. D., Slocum, M. G., Siegel, A., Gaytongarcia, R. (1994) *Tetrahedron Lett.*, **35**, 385–388.

[17] Slocum, D. W., Thompson, J., Friesen, C. (1995) *Tetrahedron Lett.*, **36**, 8171–8174.

[18] Gupta, L., Hoepker, A. C., Singh, K. J., Collum, D. B. (2009) *J. Org. Chem.*, **74**, 2231–2233.

[19] Singh, K. J., Hoepker, A. C., Collum, D. B. (2008) *J. Am. Chem. Soc.*, **130**, 18008–18017.

[20] Hevia, E., Mulvey, R. E. (2011) *Angew. Chem. Int. Ed.*, **50**, 6448–6450.

[21] Kizirian, J.-C. (2010) Mechanism and Stereochemical Features in Asymmetric Deprotonation Using RLi/(–)-Sparteine Bases. in *Topics in Stereochemistry, Stereochemical Aspects of Organolithium Compounds* (eds.: Gawley, R. E., Siegel, J. S.), Verlag Helvetica Chimica Acta, Zurich, pp. 189–251.

[22] Guijarro, D., Pastor, I. M., Yus, M. (2011) *Curr. Org. Chem.*, **15**, 2362–2389.

[23] Leroux, F., Schlosser, M., Zohar, E., Marek, I. (2009) The Preparation of Organolithium Reagents and Intermediates. in *The Chemistry of Organolithium Compounds, Patai Series: The Chemistry of Functional Groups* Vol. **Part 1** (eds.: Rappoport, Z., Marek, I.), John Wiley & Sons, Ltd, Chichester, pp. 435–493.

[24] Schlosser, M. (2013) Organoalkali Reagents. in *Organometallics in Synthesis: A Manual*, 3rd ed. (ed.: Schlosser, M.), John Wiley & Sons, Inc., Hoboken, pp. 1–222.

[25] Gissot, A., Becht, J. M., Desmurs, J. R., Pevere, V., Wagner, A., Mioskowski, C. (2002) *Angew. Chem. Int. Ed.*, **41**, 340–343.

[26] Romesberg, F. E., Collum, D. B. (1994) *J. Am. Chem. Soc.*, **116**, 9198–9202.

[27] Remenar, J. F., Lucht, B. L., Kruglyak, D., Romesberg, F. E., Gilchirst, J. H., Collum, D. B. (1997) *J. Org. Chem.*, **62**, 5748–5754.

[28] Lappert, M., Protchenko, A., Power, P., Seeber, A. (2008) *Metal Amide Chemistry*, John Wiley & Sons, Ltd, Chichester.

[29] Kampmann, D., Stuhlmuller, G., Simon, R., Cottet, F., Leroux, F., Schlosser, M. (2005) *Synthesis*, 1028–1029.

[30] Collum, D. B., McNeil, A. J., Ramirez, A. (2007) *Angew. Chem. Int. Ed.*, **46**, 3002–3017.

[31] Harrison-Marchand, A., Maddaluno, J. (2014) Advances in the Chemistry of Chiral Lithium Amides. in *Lithium Compounds in Organic Synthesis* (eds.: Luisi, R., Capriati, V.), Wiley-VCH, Weinheim, pp. 297–328.

[32] Gros, P., Choppin, S., Mathieu, J., Fort, Y. (2002) *J. Org. Chem.*, **67**, 234–237.

[33] Schlosser, M. (1992) *Mod. Synth. Methods*, **6**, 227–271.

[34] Lochmann, L., Janata, M. (2014) *Cent. Eur. J. Chem.*, **12**, 537–548.

[35] Mongin, F., Maggi, R., Schlosser, M. (1996) *Chimia*, **50**, 650–652.

[36] Aycock, D. F. (2007) *Org. Process Res. Dev.*, **11**, 156–159.

[37] Watanabe, K., Yamagiwa, N., Torisawa, Y. (2007) *Org. Process. Res. Dev.*, **11**, 251–258.

[38] Schlosser, M. (1994) Organoalkali Reagents. in *Organometallics in Synthesis: A Manual*, 1st ed. (ed.: Schlosser, M.), John Wiley & Sons, Ltd, Chichester, pp. 1–166.

[39] Dikshith, T. S. S. (2013) *Hazardous Chemicals: Safety Management and Global Regulations*, CRC Press, Boca Raton.

[40] Vila, C., Hornillos, V., Giannerini, M., Fañanás-Mastral, M., Feringa, B. L. (2014) *Chem. Eur. J.*, **20**, 13078–13083.

[41] Brown, H. C., Cole, T. E. (1983) *Organometallics*, **2**, 1316–1319.

[42] Willand-Charnley, R., Puffer, B. W., Dussault, P. H. (2014) *J. Am. Chem. Soc.*, **136**, 5821–5823.

[43] Schuetz, R. D., Taft, D. D., O'Brien, J. P., Shea, J. L., Mork, H. M. (1963) *J. Org. Chem.*, **28**, 1420–1422.

[44] Anbarasan, P., Neumann, H., Beller, M. (2010) *Chem. Asian J.*, **5**, 1775–1778.

[45] Anbarasan, P., Neumann, H., Beller, M. (2010) *Angew. Chem. Int. Ed.*, **49**, 2219–2222.

[46] Yamada, S., Gavryushin, A., Knochel, P. (2010) *Angew. Chem. Int. Ed.*, **49**, 2215–2218.

[47] Krizan, T. D., Martin, J. C. (1983) *J. Am. Chem. Soc.*, **105**, 6155–6157.

[48] Snieckus, V. (1990) *Chem. Rev.*, **90**, 879–933.

[49] Schlosser, M. (2005) *Angew. Chem. Int. Ed.*, **44**, 376–393.

[50] Gschwend, H. W., Rodriguez, H. R. (1979) *Org. React.*, **26**, 1–360.

[51] Snieckus, V., Macklin, T. K. (2005) Directed *Ortho* and Remote Metalation (DoM and DreM). in *Handbook of CH Transformations* (ed.: Dyker, G.), Wiley-VCH, Weinheim, pp. 106–118.

[52] Hartung, C. G., Snieckus, V. (2002) The Directed *ortho* Metalation Reaction-A Point of Departure for New Synthetic Aromatic Chemistry. in *Modern Arene Chemistry* (ed.: Astruc, D.), Wiley-VCH, Weinheim, pp. 330–367.

[53] Clayden, J. (2004) Directed Metallation of Aromatic Compounds. in *The Chemistry of Organolithium Compounds the Chemistry of Functional Groups*, (eds.: Rappoport, Z., Marek, I.), John Wiley & Sons, Ltd, Chichester, pp. 495–646.

[54] Whisler, M. C., MacNeil, S., Snieckus, V., Beak, P. (2004) *Angew. Chem. Int. Ed.*, **43**, 2206–2225.

[55] Beak, P., Meyers, A. I. (1986) *Acc. Chem. Res.*, **19**, 356–363.

[56] Chadwick, S. T., Rennels, R. A., Rutherford, J. L., Collum, D. B. (2000) *J. Am. Chem. Soc.*, **122**, 8640–8647.

[57] Hay, D. R., Song, Z., Smith, S. G., Beak, P. (1988) *J. Am. Chem. Soc.*, **110**, 8145–8153.

[58] Saa, J. M., Morey, J., Frontera, A., Deya, P. M. (1995) *J. Am. Chem. Soc.*, **117**, 1105–1116.

[59] Saa, J. M., Deya, P. M., Suner, G. A., Frontera, A. (1992) *J. Am. Chem. Soc.*, **114**, 9093–9100.

[60] Anderson, D. R., Faibish, N. C., Beak, P. (1999) *J. Am. Chem. Soc.*, **121**, 7553–7558.

[61] Resek, J. E., Beak, P. (1994) *J. Am. Chem. Soc.*, **116**, 405–406.

[62] Nguyen, T. H., Chau, N. T. T., Castanet, A. S., Nguyen, K. P. P., Mortier, J. (2005) *Org. Lett.*, **7**, 2445–2448.

[63] Stratakis, M. (1997) *J. Org. Chem.*, **62**, 3024–3025.

[64] Maggi, R., Schlosser, M. (1999) *Tetrahedron Lett.*, **40**, 8797–8800.

[65] Schlosser, M., Mongin, F., Porwisiak, J., Dmowski, W., Buker, H. H., Nibbering, N. M. M. (1998) *Chem. Eur. J.*, **4**, 1281–1286.

[66] Bauer, W., Lochmann, L. (1992) *J. Am. Chem. Soc.*, **114**, 7482–7489.

[67] Shi, G. Q., Takagishi, S., Schlosser, M. (1994) *Tetrahedron*, **50**, 1129–1134.

[68] Beak, P., Brown, R. A. (1977) *J. Org. Chem.*, **42**, 1823–1824.

[69] Gilman, H., Soddy, T. (1957) *J. Org. Chem.*, **22**, 1715–1716.

[70] Cuevas, J. C., Patil, P., Snieckus, V. (1989) *Tetrahedron Lett.*, **30**, 5841–5844.

[71] Iwao, M. (1990) *J. Org. Chem.*, **55**, 3622–3627.

[72] Puterbaugh, W. H., Hauser, C. R. (1964) *J. Org. Chem.*, **29**, 853–856.

[73] Mongin, F., Desponds, O., Schlosser, M. (1996) *Tetrahedron Lett.*, **37**, 2767–2770.

[74] Metallinos, C., Nerdinger, S., Snieckus, V. (1999) *Org. Lett.*, **1**, 1183–1186.

[75] Fuhrer, W., Gschwend, H. W. (1979) *J. Org. Chem.*, **44**, 1133–1136.

[76] Comins, D. L., Brown, J. D. (1984) *J. Org. Chem.*, **49**, 1078–1083.

[77] Muchowski, J. M., Venuti, M. C. (1980) *J. Org. Chem.*, **45**, 4798–4801.

[78] Meyers, A. I., Mihelich, E. D. (1975) *J. Org. Chem.*, **40**, 3158–3159.

[79] Wu, C. J., Lee, S. H., Yu, S. T., Na, S. J., Yun, H., Lee, B. Y. (2008) *Organometallics*, **27**, 3907–3917.

[80] Gschwend, H. W., Hamdan, A. (1975) *J. Org. Chem.*, **40**, 2008–2009.

[81] Mortier, J., Moyroud, J., Bennetau, B., Cain, P. A. (1994) *J. Org. Chem.*, **59**, 4042–4044.

[82] Houlden, C. E., Lloyd-Jones, G. C., Booker-Milburn, K. I. (2010) *Org. Lett.*, **12**, 3090–3092.

[83] Uemura, M., Tokuyama, S., Sakan, T. (1975) *Chem. Lett.*, **11**, 1195–1198.

[84] Sibi, M. P., Snieckus, V. (1983) *J. Org. Chem.*, **48**, 1935–1937.

[85] Tilly, D., Samanta, S. S., De, A., Castanet, A. S., Mortier, J. (2005) *Org. Lett.*, **7**, 827–830.

[86] Christensen, H. (1975) *Synth. Commun.*, **5**, 65–78.

[87] Fitt, J. J., Gschwend, H. W. (1976) *J. Org. Chem.*, **41**, 4029–4031.

[88] Posner, G. H., Canella, K. A. (1985) *J. Am. Chem. Soc.*, **107**, 2571–2573.

[89] Coppi, D. I., Salomone, A., Perna, F. M., Capriati, V. (2012) *Angew. Chem. Int. Ed.*, **51**, 7532–7536.

[90] Mallardo, V., Rizzi, R., Sassone, F. C., Mansueto, R., Perna, F. M., Salomone, A., Capriati, V. (2014) *Chem. Commun.*, **50**, 8655–8658.

[91] Salteris, C. S., Kostas, I. D., Micha-Screttas, M., Heropoulos, G. A., Screttas, C. G. (1999) *J. Org. Chem.*, **64**, 5589–5592.

[92] Capriati, V., Florio, S., Luisi, R., Musio, B. (2005) *Org. Lett.*, **7**, 3749–3752.

[93] Affortunato, F., Florio, S., Luisi, R., Musio, B. (2008) *J. Org. Chem.*, **73**, 9214–9220.

[94] Fort, Y., Rodriguez, A. L. (2003) *J. Org. Chem.*, **68**, 4918–4922.

[95] Macklin, T. K., Snieckus, V. (2005) *Org. Lett.*, **7**, 2519–2522.

[96] Figuly, G. D., Loop, C. K., Martin, J. C. (1989) *J. Am. Chem. Soc.*, **111**, 654–658.

[97] Watanabe, H., Gay, R. L., Hauser, C. R. (1968) *J. Org. Chem.*, **33**, 900–903.

[98] Yañez Rodríguez, V., del Águila, M. Á., Iglesias, M. J., López Ortiz, F. (2012) *Tetrahedron*, **68**, 7355–7362.

[99] Fernandez, I., Burgos, P. O., Gomez, G. R., Bled, C., Garcia-Granda, S., Ortiz, F. L. (2007) *Synlett*, **4**, 611–614.

[100] Jorgenson, M. J. (1970) *Org. React.*, **18**, 1–97.

[101] Nguyen, T. H., Chau, N. T., Castanet, A. S., Nguyen, K. P., Mortier, J. (2007) *J. Org. Chem.*, **72**, 3419–3429.

[102] Moyroud, J., Guesnet, J. L., Bennetau, B., Mortier, J. (1995) *Tetrahedron Lett.*, **36**, 881–884.

[103] Mongin, F., Schlosser, M. (1996) *Tetrahedron Lett.*, **37**, 6551–6554.

[104] Klis, T., Lulinski, S., Serwatowski, J. (2008) *Curr. Org. Chem.*, **12**, 1479–1501.

[105] Maggi, R., Schlosser, M. (1996) *J. Org. Chem.*, **61**, 5430–5434.

[106] Nguyen, T. H., Castanet, A. S., Mortier, J. (2006) *Org. Lett.*, **8**, 765–768.

[107] Osmunddesilva, S., Reed, J. N., Snieckus, V. (1978) *Tetrahedron Lett.*, 5099–5102.

[108] Beak, P., Brown, R. A. (1982) *J. Org. Chem.*, **47**, 34–46.

[109] Shimano, M., Meyers, A. I. (1994) *J. Am. Chem. Soc.*, **116**, 10815–10816.

[110] Mills, R. J., Taylor, N. J., Snieckus, V. (1989) *J. Org. Chem.*, **54**, 4372–4385.

[111] Zhao, Z., Snieckus, V. (2005) *Org. Lett.*, **7**, 2523–2526.

[112] Meyers, A. I., Hanagan, M. A., Trefonas, L. M., Baker, R. J. (1983) *Tetrahedron*, **39**, 1991–1999.

[113] Juaristi, E., Beck, A. K., Hansen, J., Matt, T., Mukhopadhyay, T., Simson, M., Seebach, D. (1993) *Synthesis*, **12**, 1271–1290.

[114] Beak, P., Tse, A., Hawkins, J., Chen, C. W., Mills, S. (1983) *Tetrahedron*, **39**, 1983–1989.

[115] Matsui, S., Uejima, A., Suzuki, Y., Tanaka, K. (1993) *Perkin Trans.* **1**, 701–704.

[116] Takahashi, H., Tsubuki, T., Higashiyama, K. (1991) *Chem. Pharm. Bull.*, **39**, 260–265.

[117] Ogawa, S., Furukawa, N. (1991) *J. Org. Chem.*, **56**, 5723–5726.

[118] Pollet, P., Turck, A., Ple, N., Queguiner, G. (1999) *J. Org. Chem.*, **64**, 4512–4515.

[119] Snieckus, V., Beaulieu, F., Mohri, K., Han, W., Murphy, C. K., Davis, F. A. (1994) *Tetrahedron Lett.*, **35**, 3465–3468.

[120] Le Fur, N., Mojovic, L., Ple, N., Turck, A., Reboul, V., Metzner, P. (2006) *J. Org. Chem.*, **71**, 2609–2616.

[121] Kanda, K., Endo, K., Shibata, T. (2010) *Org. Lett.*, **12**, 1980–1983.

[122] Braese, S. (2007) *Combinatorial Chemistry on Solid Supports, Topics in Current Chemistry* Vol. **278**, Springer, New York.

[123] Garibay, P., Toy, P. H., Hoeg-Jensen, T., Janda, K. D. (1999) *Synlett*, **9**, 1438–1440.

[124] Garibay, P., Vedso, P., Begtrup, M., Hoeg-Jensen, T. (2001) *J. Comb. Chem.*, **3**, 332–340.

[125] Kristensen, J., Lysen, M., Vedso, P., Begtrup, M. (2001) *Org. Lett.*, **3**, 1435–1437.

[126] Li, W. J., Nelson, D. P., Jensen, M. S., Hoerrner, R. S., Cai, D. W., Larsen, R. D., Reider, P. J. (2002) *J. Org. Chem.*, **67**, 5394–5397.

[127] Dua, S. S., Gilman, H. (1974) *J. Organomet. Chem.*, **64**, C1–C2.

[128] Schlosser, M., Guio, L., Leroux, F. (2001) *J. Am. Chem. Soc.*, **123**, 3822–3823.

[129] Chau, N. T. T., Nguyen, T. H., Castanet, A.-S., Nguyen, K. P. P., Mortier, J. (2008) *Tetrahedron*, **64**, 10552–10557.

[130] Hallberg, A., Alshowaier, I., Martin, A. R. (1984) *J. Heterocycl. Chem.*, **21**, 197–199.

[131] Mortier, J., Nguyen, T. H., Tilly, D., Castanet, A. S. (2007) *Arkivoc*, **6**, 47–54.

[132] Bernardi, A., Capelli, A. M., Cassinari, A., Comotti, A., Gennari, C., Scolastico, C. (1992) *J. Org. Chem.*, **57**, 7029–7034.

[133] Tilly, D., Magolan, J., Mortier, J. (2012) *Chem. Eur. J.*, **18**, 3804–3820.

[134] Tilly, D., Fu, J. M., Zhao, B. P., Alessi, M., Castanet, A. S., Snieckus, V., Mortier, J. (2010) *Org. Lett.*, **12**, 68–71.

[135] Palmer, B. D., Boyd, M., Denny, W. A. (1990) *J. Org. Chem.*, **55**, 438–441.

[136] Pozharskii, A. F., Ryabtsova, O. V. (2006) *Russ. Chem. Rev.*, **75**, 709–736.

[137] Clayden, J., Frampton, C. S., McCarthy, C., Westlund, N. (1999) *Tetrahedron*, **55**, 14161–14184.

[138] Ruzziconi, R., Spizzichino, S., Giurg, M., Castagnetti, E., Schlosser, M. (2010) *Synthesis*, 1531–1535.

[139] Saá, J. M. (2002) *Helv. Chim. Acta*, **85**, 814–840.

[140] Semmelhack, M. F. (1995) Transition Metal Arene Complexes: Ring Lithiation. in *Comprehensive Organometallic Chemistry II* (eds.: Abel, E. W., Stone, F. G. A., Wilkinson, G.), Pergamon, New York, pp. 1017–1038.

[141] Berger, A., Djukic, J.-P., Michon, C. (2002) *Coord. Chem. Rev.*, **225**, 215–238.

[142] Semmelhack, M. F., Chlenov, A. (2004) (Arene)Cr(CO)3 Complexes: Arene Lithiation/Reaction with Electrophiles. in *New Aspects of Transition Metal Arene pi-Complexes in Organic Synthesis and Catalysis, Topics in Organometallic Chemistry* Vol. **7** (ed.: Kündig, E. P.), Springer, New York, pp. 21–42.

[143] Ariffin, A., Blake, A. J., Li, W. S., Simpkins, N. S. (1997) *Synlett*, 1453–1455.

[144] Simpkins, N. S. (1996) *Pure Appl. Chem.*, **68**, 691–694.

[145] Uemura, M., Hayashi, Y., Hayashi, Y. (1994) *Tetrahedron: Asymmetry*, **5**, 1427–1430.

[146] Ewin, R. A., MacLeod, A. M., Price, D. A., Simpkins, N. S., Watt, A. P. (1997) *Perkin Trans.*, **1**, 401–415.

[147] Han, J. W., Son, S. U., Chung, Y. K. (1997) *J. Org. Chem.*, **62**, 8264–8267.

[148] Clark, R. D., Jahangir, A. (1995) *Org. React.*, **47**, 1–314.

[149] Prat, L., Mojovic, L., Levacher, V., Dupas, G., Queguiner, G., Bourguignon, J. (1998) *Tetrahedron: Asymmetry*, **9**, 2509–2516.

[150] Clayden, J., Pink, J. H. (1997) *Tetrahedron Lett.*, **38**, 2561–2564.

[151] Schlosser, M. (2002) Organoalkali Chemistry. in *Organometallics in Synthesis: A Manual*, 2nd ed. (ed.: Schlosser, M.), John Wiley & Sons, Ltd, Chichester, pp. 1–352.

[152] Watson, S. C., Eastham, J. F. (1967) *J. Organomet. Chem.*, **9**, 165–168.

[153] Kofron, W. G., Baclawski, L. M. (1976) *J. Org. Chem.*, **41**, 1879–1880.

[154] Harrison-Marchand, A., Mongin, F. (2013) *Chem. Rev.*, **113**, 7470–7562.

[155] Jastrzebski, J. T. B. H., Vankoten, G., Konijn, M., Stam, C. H. (1982) *J. Am. Chem. Soc.*, **104**, 5490–5492.

[156] Clayden, J., Davies, R. P., Hendy, M. A., Snaith, R., Wheatley, A. E. H. (2001) *Angew. Chem. Int. Ed.*, **40**, 1238–1240.

[157] Bond, A. D., Clayden, J., Wheatley, A. E. H. (2001) *Acta Crystallogr. E*, **57**, o292–o294.

[158] Armstrong, D. R., Clayden, J., Haigh, R., Linton, D. J., Schooler, P., Wheatley, A. E. H. (2003) *Chem. Commun.*, 1694–1695.

[159] Armstrong, D. R., Boss, S. R., Clayden, J., Haigh, R., Kirmani, B. A., Linton, D. J., Schooler, P., Wheatley, A. E. H. (2004) *Angew. Chem. Int. Ed.*, **43**, 2135–2138.

[160] Reich, H. J. (2012) *J. Org. Chem.*, **77**, 5471–5491.

[161] Gunther, H., Moskau, D., Bast, P., Schmalz, D. (1987) *Angew. Chem. Int. Ed.*, **26**, 1212–1220.

[162] Bauer, W., Schleyer, P. v. R. (1992) Advances in Carbanion Chemistry. in *Advances in Carbanion Chemistry* Vol. **1** (ed.: Snieckus, V.), Jai Press, Greenwich, pp. 81–175.

[163] Sebban, M., Guilhaudis, L., Oulyadi, H. (2014) Spectroscopic Advances in Structural Lithium Chemistry: Diffusion-Ordered Spectroscopy and Solid-State NMR. in *Lithium Compounds in Organic Synthesis* (eds.: Luisi, R., Capriati, V.), Wiley-VCH, Weinheim, pp. 85–121.

[164] Jones, A. C., Sanders, A. W., Bevan, M. J., Reich, H. J. (2007) *J. Am. Chem. Soc.*, **129**, 3492–3493.

[165] Jones, A. C. (2014) Advances in the Chemistry of Chiral Lithium Amides. in *Lithium Compounds in Organic Synthesis* (eds.: Luisi, R., Capriati, V.), Wiley-VCH, Weinheim, pp. 53–84.

[166] Shen, Y., Chen, J., Liu, M., Ding, J., Gao, W., Huang, X., Wu, H. (2014) *Chem. Commun.*, **50**, 4292–4295.

[167] Snieckus, V. (2011) *Beilstein J. Org. Chem.*, **7**, 1215–1218.

[168] de Meijere, A., Diederich, F. (1998) *Metal-Catalyzed Cross-Coupling Reactions*, 1st ed., Wiley-VCH Verlag GmBH, Weinheim.

[169] de Meijere, A., Diederich, F. (2004) *Metal-Catalyzed Cross-Coupling Reactions*, 2nd ed., Wiley-VCH Verlag GmBH, Weinheim.

[170] Miyaura, N. (2002) *Cross-Coupling Reactions, Topics in Current Chemistry* Vol. **219**, Springer, Berlin.

[171] Board, J., Cosman, J. L., Rantanen, T., Singh, S. P., Snieckus, V. (2013) *Platin. Met. Rev.*, **57**, 234–258.

[172] Chauder, B., Green, L., Snieckus, V. (1999) *Pure Appl. Chem.*, **71**, 1521–1529.

[173] Kumar, A. S., Ghosh, S., Mehta, G. N. (2010) *J. Chem. Res.*, 95–97.

[174] Guilarte, V., Castroviejo, M. P., Garcia-Garcia, P., Fernandez-Rodriguez, M. A., Sanz, R. (2011) *J. Org. Chem.*, **76**, 3416–3437.

[175] Macklin, T. K., Reed, M. A., Snieckus, V. (2008) *Eur. J. Org. Chem.*, 1507–1509.

[176] MacNeil, S. L., Wilson, B. J., Snieckus, V. (2006) *Org. Lett.*, **8**, 1133–1136.

[177] Sala, T., Sargent, M. V. (1979) *Perkin Trans.*, **1**, 2593–2598.

[178] Han, Y.-H., Zhou, T., Sui, Y., Hua, R. (2014) *Org. Process. Res. Dev.*, **18**, 1229–1233.

[179] Limanto, J., Dorner, B. T., Hartner, F. W., Tan, L. (2008) *Org. Process. Res. Dev.*, **12**, 1269–1272.

[180] Gosselin, F., Lau, S., Nadeau, C., Trinh, T., O'Shea, P. D., Davies, I. W. (2009) *J. Org. Chem.*, **74**, 7790–7797.

[181] Le, T. T., Chau, N. T. T., Nguyen, T. T., Brien, J., Trieu, T. T., Nourry, A., Castanet, A. S., Kim, P. P. N., Mortier, J. (2011) *J. Org. Chem.*, **76**, 601–608.

[182] Schlosser, M., Geneste, H. (1998) *Chem. Eur. J.*, **4**, 1969–1973.

27

DEPROTONATIVE METALATION USING ALKALI METAL–NONALKALI METAL COMBINATIONS

FLORIS CHEVALLIER[1], FLORENCE MONGIN[1], RYO TAKITA[2,3], AND MASANOBU UCHIYAMA[2,3]

[1] *Chimie et Photonique Moléculaires, UMR 6226 CNRS-Université de Rennes 1, Rennes, France*
[2] *RIKEN Center for Sustainable Resource Science, Wako-shi, Saitama, Japan*
[3] *Graduate School of Pharmaceutical Sciences, The University of Tokyo, Bunkyo-ku, Tokyo, Japan*

27.1 INTRODUCTION

This chapter will describe arene deprotometalation [1–5] reactions performed by organic bases containing two metals, namely, an alkali metal and a nonalkali metal. In general, using very hindered amino group such as 2,2,6,6-tetramethylpiperidino (TMP) as ligand in these combinations leads to synergic bases as none of the corresponding monometallic compounds are capable of performing similar reactions [6–12]. Such a result can be reached using ate compounds, in which both anionic activation [13, 14] and reduced aggregation [15] are present, using suitable media. A synergy through both anionic activation and reduced aggregation can be similarly reached using salt-activated organometallic compounds (e.g., "turbo"-Grignard reagents) [9, 12]. Finally, synergic pairs of metal amides, which complement each other in deprotometalation reactions, are also efficient combinations to fulfill this purpose [9, 12].

The chapter is thus intended to depict the behavior toward arenes of organic bimetallic compounds for which efficient or/and chemoselective reactions have been described. Mixtures containing an alkali or alkaline earth metal (Li, Na, Mg), a group 2 (Mg), group 3 (La), group 4 (Zr), group 7 (Mn), group 8 (Fe), group 9 (Co), group 11 (Cu), group 12 (Zn, Cd), or group 13 (Al) metal, and ligands (alkyls with, in most cases, at least a hindered amino group) have thus been chosen to this purpose. The chapter is divided into five main parts. After this introduction (Part 1), generalities

Arene Chemistry: Reaction Mechanisms and Methods for Aromatic Compounds,
First Edition. Edited by Jacques Mortier.

about bimetallic combinations (among which, bases) and their behavior are presented in parts 2 and 3. Parts 4 and 5 illustrate through various examples the mechanisms involved and the scope and applications of the deprotometalation using alkali metal–nonalkali metal combinations.

The formulas used refer to the overall compositions, and do not necessarily correspond to the stoichiometries of the species actually present in solution.

27.2 PREPARATION OF THE BIMETALLIC COMBINATIONS AND THEIR STRUCTURAL FEATURES

27.2.1 Preparation of Alkali Metal–Nonalkali Metal Basic Combinations

- Formation by acid/base approach. LiCl-activated Grignard reagents such as *i*-PrMgCl·LiCl are basic enough to deprotonate hindered amines such as HTMP, and TMPMgCl·LiCl can be prepared by this way [16]. Note that it is also possible to prepare TMPZnCl·LiCl by zinc insertion into TMPCl [17].
- Formation by metathesis approach. Bimetallic (and trimetallic) combinations can be obtained by reaction of an inorganic salt (or a chelate) with (a) polar organo- or/and amidometallic compound(s) (see examples in Table 27.1) [12]. Salts such as LiCl are simultaneously formed using this metathesis approach and can influence the course of the reactions in which the compounds are employed [25].
- Formation by cocomplexation approach. The cocomplexation approach is an alternative way to bimetallic combinations using the corresponding monometallic compounds [12]. It is helpful for the synthesis of mixed-ligand bases. Examples are given in Table 27.2.

27.2.2 Ate Compounds

Ate compounds refer to the bimetallic combinations containing a metal as part of a complex anion; they correspond to an interaction between a species for which the metal is a Lewis acid and a second metallic species (e.g., an organometal, an amide) with an anionic fragment behaving as a Lewis

TABLE 27.1 Access to Polymetallic Combinations Using the Metathesis Approach (THF)

Entry	Base	Precursors
1. Ref. [18]	TMPMgCl·LiCl	$MgCl_2$, LiTMP
2. Ref. [19]	$Mg(TMP)_2·2LiCl$	$MgCl_2$, 2 LiTMP
3. Ref. [20]	$Al(TMP)_3·3LiCl$	$AlCl_3$, 3 LiTMP
4. Ref. [21]	$Mn(TMP)_2·2MgCl_2·4LiCl$	$MnCl_2·2LiCl$, 2 TMPMgCl·LiCl
5. Ref. [22]	$Fe(TMP)_2·2MgCl_2·4LiCl$	$FeCl_2·2LiCl$, 2 TMPMgCl·LiCl
6. Ref. [23]	$Zn(TMP)_2·2MgCl_2·2LiCl$	$ZnCl_2$, 2 TMPMgCl·LiCl
7. Ref. [24]	TMPZnCl·LiCl	$ZnCl_2$, LiTMP

TABLE 27.2 Access to Bimetallic Combinations Using the Cocomplexation Approach

Entry	Base	Precursors	Solvent
1. Ref. [26]	[(TMEDA)Na(TMP)BuMg(TMP)]	Bu_2Mg, BuNa, HTMP, TMEDA	Hexane
2. Ref. [27]	$Mg(TMP)_2·2LiCl$	TMPMgCl·LiCl, LiTMP	THF
3. Ref. [28]	Li(TMP)Al(*i*-Bu)$_3$	Al(*i*-Bu)$_3$, LiTMP	THF
4. Ref. [29]	Li(TMP)Zn(*t*-Bu)$_2$·2LiCl	Zn(*t*-Bu)$_2$·2LiCl, LiTMP	THF

base [14, 30]. Their stability, structure, and behavior depend on the metals involved, on the ligand size and charge stabilization [31, 32], and on the solvent/additive [33] in which they are generated or used.

Depending on the number of ligands around the central metal, either *lower- or higher-order ate compounds* can be generated. For example, merging a diorganomagnesium and an organolithium stoichiometrically results in the formation of a lithium trialkylmagnesate, whereas a 1:2 ratio leads to a dilithium tetraalkylmagnesate [34–38]. The higher-order (or highly coordinated) ate compounds are in general more reactive than the lower-order ones. For example, toluene can be deprotometalated using DABCO-activated Na_2MgBu_4 but not using DABCO-activated $NaMgBu_3$ [39]. The ^1H NMR field shift values can allow for comparing the anionic character of species, for example, with −1.08 and −1.44 ppm observed in THF at −20°C for the methyl signals in $LiZnMe_3$ (less anionic character) and Li_2ZnMe_4 (more anionic character), respectively [40].

27.2.3 Salt-Activated Compounds

The presence of lithium salts can modify the ionic strength of a system but can also affect the nature of the reacting species (and, thus, the course of the reactions) due to aggregation change [25]. For example, in the presence of nonalkali organometallic compounds, LiCl can participate in new structures and thus contribute to new reactivities. On the basis of electrospray ionization mass spectroscopy, electrical conductivity measurements, and NMR spectroscopy, the formation of organozincates (known as more reactive than the corresponding diorganozincs [41]) in the course of organozinc halide and (to a lesser extent) diorganozinc preparations was evidenced and proposed to rationalize the special reactivity exhibited by these compounds [42, 43].

27.2.4 Contacted and Solvent-Separated Ion Pairs

Depending on the metal and ligand components, but also on the solvent/additive(s) employed, bimetallic compounds can exist as contacted ion pairs or/and separated ion pairs [44–47]. For instance, the presence of HMPA can reduce the formation of aggregates and favor the formation of solvent-separated ion pairs [48, 49]. Contacted ion pairs with the deprotonating ligand shared between the two metals being in general the key to successful deprotometalation [12], taking into account these structural data is fundamental.

27.3 BEHAVIOR OF ALKALI METAL–NONALKALI METAL COMBINATIONS

27.3.1 One-Electron and Two-Electron Transfers

For bimetallic compounds in which one metal can increase its oxidation state (e.g., Mn, Fe, and Co; alkali or alkaline earth metals and some others are not concerned), one-electron and two-electron transfers are possible changes (Scheme 27.1) [50].

SCHEME 27.1 One-electron and two-electron transfers from bimetallic compounds.

27.3.2 Base and Nucleophile Ligand Transfers

By avoiding electron transfers through a suitable choice of the type and number of ligands, and of course from compounds in which both metals are at their highest possible oxidation state, ligand transfer is favored. It is thus possible to perform reactions with a substrate containing an acidic enough hydrogen (deprotometalation) or with an electrophile (nucleophilic transfer, e.g., to intercept an aryl bimetallic compound formed by deprotometalation) (Scheme 27.2) [50]. 1,2-Migration is a nucleophilic transfer within a bimetallic compound from the nonalkali metal; it is generally observed with unstable higher-order ate compounds [51]. From bimetallic compounds containing different ligands, selective transfers are often noted; in spite of numerous studies dedicated to this important point [12], no general conclusions have been drawn.

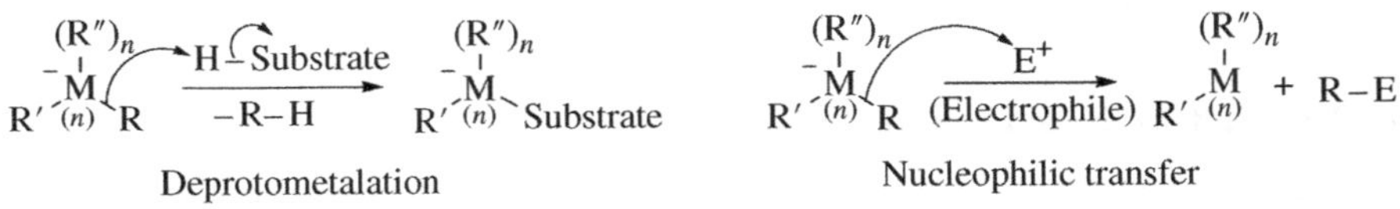

SCHEME 27.2 Deprotometalation and nucleophilic transfer from bimetallic compounds.

27.4 MECHANISTIC STUDIES ON THE DEPROTOMETALATION USING ALKALI METAL–NONALKALI METAL COMBINATIONS

27.4.1 Deprotometalation Using Alkali Metal–Amidozincate Complexes

The first report on the deprotometalation using lithium TMP-zincate (t-Bu$_2$Zn(TMP)Li) describes the high chemoselectivity of directed *ortho*-metalation (DoM) in the presence of ester or cyano groups, which are very sensitive toward the nucleophilic addition of polar organometallic reagents [29]. Since the commonly utilized strong bases, such as alkyllithiums or alkyl Grignard reagents, cause such competitive side reactions, TMP-zincate has paved the way in DoM strategy, allowing its application to various organic architectures (the feature of unique reactivity of TMP-zincates is more precisely discussed later in Section 27.5.7). Therefore, the benefit of TMP-zincate is quite remarkable in the regio- and chemoselective functionalization of aromatic compounds, and thus, the origin of such high selectivity is of great interest.

First, the structure of a representative TMP-zincate, t-Bu$_2$Zn(TMP)Li, was studied using DFT calculation, X-ray crystallography, and NMR spectroscopic technique [52]. Among two possible structures of t-Bu$_2$Zn(TMP)Li, DFT calculations in gas phase indicated that the "asymmetric" structure is more stable (−17.3 kcal/mol) than the other "symmetric" complex (Scheme 27.3). In accordance with the theoretical results, the X-ray crystal structure was found to possess the asymmetric architecture, with a trigonal planar geometry around the Zn center, and the lithium cation coordinated by the nitrogen atom of the TMP ligand, one of the t-Bu groups and THF. The NMR spectra also supported the "asymmetric" structure, with the separately observed α-carbons of two t-Bu groups at −50°C. In addition, t-Bu$_2$Zn(TMP)Li should keep this "asymmetric" structure in the reaction closely relevant to the calculated results.

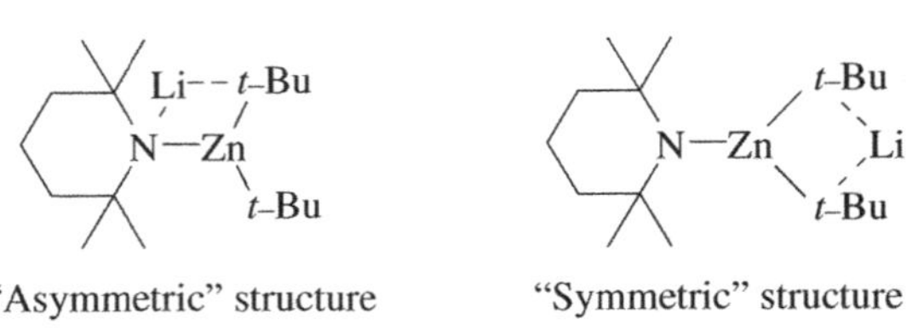

SCHEME 27.3 Two possible structures of t-Bu$_2$Zn(TMP)Li.

Thus, to get insight into the reaction pathway of the DoM process, DFT calculations were performed employing $Me_2Zn(NMe_2)Li$ as a model complex for TMP-zincates and the coordinating Me_2O as a model for THF. The reaction pathways of the DoM reaction of anisole mediated by $Me_2Zn(NMe_2)Li{\cdot}OMe_2$ thus obtained are summarized in Scheme 27.4. Given the possibility of deprotonation of anisole by the methyl ligand in the $Me_2Zn(NMe_2)Li$ complex, the reaction pathway through **TS1** was identified though the activation barrier of this step is found to be rather high (+32.4 kcal/mol). On the other hand, two transition state (TS) structures, **TS2** (closed form) and **TS3** (open form), were obtained for the DoM reaction from the complex **CP2**, where Me_2N ligand deprotonates anisole to afford Me_2NH coordinated to the resultant arylzincate species in **PD2**. The activation energies of both pathways are reasonably low because the DoM reaction is facilitated by the push–pull synergy of the Lewis acidic Li cation and the negatively charged $Me_2Zn(NMe_2)$ moiety. Detailed analysis revealed that in the case of the Me ligand on the zincate, the contracted and ill-directed sp^3-like orbital of the carbon atom needs to be used in the C—H bond scission. In contrast, the orientation of the lone pair orbital of the nitrogen atom in **TS2** or **TS3** appears to be ideal to interact with the breaking aromatic C—H bond, strongly suggesting the kinetic preference of *N*-ligand over *C*-ligand transfer.

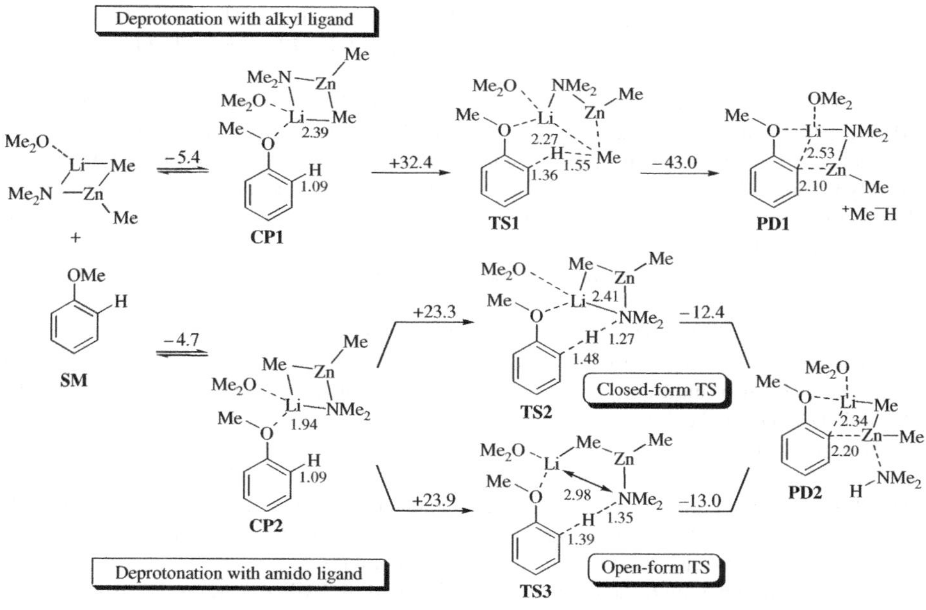

SCHEME 27.4 Reaction pathways of the DoM reaction of anisole using $Me_2Zn(NMe_2)Li$. Energy changes and bond lengths at the B3LYP/6-31+G* and SVP (Zn) level of theory are shown in kcal/mol and Å, respectively.

While the deprotonation with amido ligand is kinetically favored in the results of DFT calculations, **PD1** in Scheme 27.4, which is produced by the deprotonation of anisole with the alkyl ligand on the zincate complex, is suggested to be thermodynamically more stable than **PD2**. In fact, the structurally related zincate, *t*-BuZn(*μ*-*t*-Bu)(*μ*-TMP)Na·TMEDA, has been shown to deprotonate, giving the **PD1**-type intermediate with the elimination not of TMP-H but of *t*-Bu-H [53]. Further detailed study combining X-ray crystallography and DFT calculations brought comprehensive understanding of the whole deprotonation process [54].

When the DoM reaction of *N*,*N*-diisopropylnaphthamide was examined with EtZn(*μ*-Et)(*μ*-TMP) Li complex, the resulting crystalline deposit was identified as $EtZn[\mu\text{-}C_{10}H_6CON(i\text{-}Pr)_2]_2Li{\cdot}2THF$ by

X-ray diffraction, suggesting the possibility of dibasic properties of the parent zincate complex (Scheme 27.5).

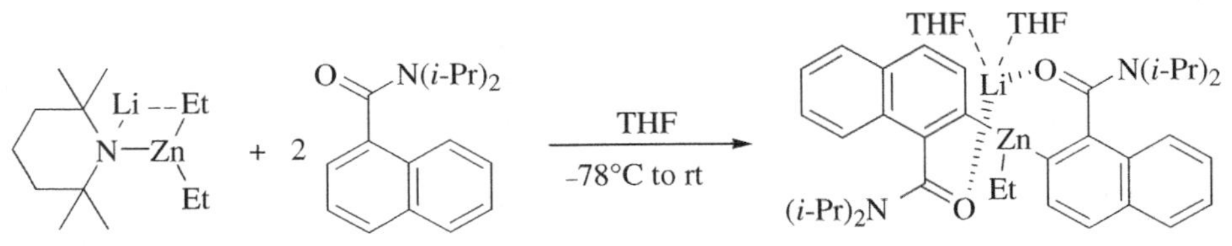

SCHEME 27.5 DoM reaction of *N,N*-diisopropylnaphthamide using EtZn(μ-Et)(μ-TMP)Li complex to afford EtZn[μ-$C_{10}H_6$CON(i-Pr)$_2$]$_2$Li·2THF.

Theoretical investigations with *N,N*-dimethylbenzamide as a model substrate were effective to elucidate the ligand transfer preferences of alkyl and amido to disconnect aromatic C—H bond (Scheme 27.6). Similar to the case with anisole, the deprotonation by the bridging-alkyl ligand, that is, the conversion of MeZn(μ-Me)(μ-NMe_2)Li·OMe_2 and Me_2NCOPh (**SM**) into MeZn(Ar′)(NMe_2) Li·OMe_2 and methane (**PD1**, Ar′ = $Me_2NCOC_6H_4$), is exothermic (23.3 kcal/mol) but proceeds via **TS1** accessed by a very high Gibbs energy of activation ($\Delta G^{\ddagger}$ = 37.8 kcal/mol). On the other hand, utilization of dimethylamido (NMe_2) ligand as a base in the *ortho*-deprotonation significantly lowers the activation barrier of **TS2** ($\Delta G^{\ddagger}$ = 24.0 kcal/mol), affording Me_2Zn(Ar′)Li·OMe_2 and Me_2NH (**PD2**) with a nominal endothermicity (ΔG = 2.2 kcal/mol). Following coordination of the *in situ*-generated amine at the alkali metal center in Me_2Zn(Ar′)Li·OMe_2 generates **CP3**, and **PD1** is readily produced by the subsequent methane loss from this intermediate. This reaction is found to be significantly exothermic (−25.5 kcal/mol) with a 21.6 kcal/mol activation barrier. Thus, this stepwise sequence (first deprotonation by amido ligand and second ligand exchange reaction to afford alkane loss) is crucial to rationalize the observation of **PD1**-type products in benzenoid deprotonation systems without the need to invoke direct alkyl elimination. Moreover, dibasic or tribasic properties of dialkyl(amido)zincate complexes were also revealed by further DFT calculations [54].

Moreover, additional information should be provided herein: (i) the stepwise DoM reaction sequence was also indicated with the structurally related complex *t*-BuZn(μ-*t*-Bu)(μ-TMP) Na·TMEDA, having the different countercation (Na) and also additional ligand (TMEDA), by comprehensive DFT calculations [55], and (ii) it was experimentally revealed that the prepared *t*-Bu_2(Ar)ZnLi complex (Ar = 2-MeO-C_6H_4) smoothly deprotonates TMP-H to afford *t*-Bu(Ar) Zn(TMP)Li, which replicates the second ligand exchange reaction of the stepwise sequence [56]. Thus, it was concluded that, in general, the stepwise reaction sequence should be the crucial reaction pathway of the DoM reaction with TMP-zincate complexes.

Lastly, the origin of high chemoselectivity in the DoM by TMP-zincates is discussed. Given that the reactions of benzonitrile with alkyllithiums or alkyl Grignard reagents preferentially afford the corresponding 1,2-adducts, as the results of nucleophilic attack to the cyano group, not the DoM products, the reaction pathways of 1,2-addition and DoM reactions using alkyllithiums or TMP-zincates were clarified to elucidate the origin of this unique chemoselectivity.

The reactions of benzonitrile with $(MeLi)_2$ and $Me_2Zn(NMe_2)$Li, as the simplest possible chemical models of the alkyllithium and TMP-zincate reagents, were investigated by DFT calculations (Scheme 27.7) [57]. After the formation of the intermediate complex $\mathbf{CP}_{RLi}$, the TS for the 1,2-addition of the Me ligand to the cyano group ($\mathbf{TS1}_{RLi}$) is only 4.0 kcal/mol higher in energy than $\mathbf{CP}_{RLi}$, and more favorable than $\mathbf{TS2}_{RLi}$ by 15.7 kcal/mol, because of double activation of the nitrogen atom of the cyano group by both lithium atoms of the alkyllithium dimer. On the other hand, the investigations with $Me_2Zn(NMe_2)$Li revealed that the activation energies from the prereaction complexes for $\mathbf{TS1}_{zincate}$ (1,2-addition of Me group) and $\mathbf{TS2}_{zincate}$ (deprotonation by NMe_2 ligand) are 29.8 and 27.0 kcal/mol, respectively. Though the energy difference between both TSs is only a

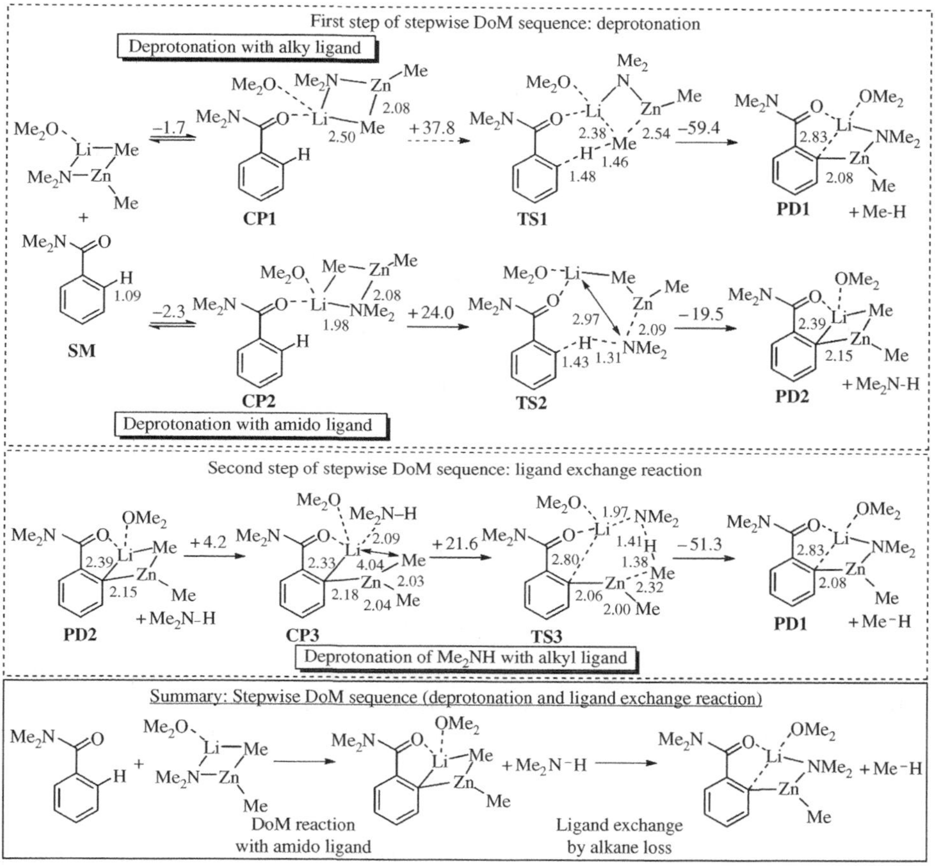

SCHEME 27.6 Reaction pathways of the DoM reaction of *N,N*-dimethylbenzamide using $Me_2Zn(NMe_2)Li$. Energy changes and bond lengths at the B3LYP/6-31+G* and SVP (Zn) level of theory are shown in kcal/mol and Å, respectively.

few kilocalories per mole, these results are consistent with the observed unique chemoselectivity. As seen in $\mathbf{TS2}_{zincate}$, the rearrangement of the Li position in the zincate proceeds smoothly so that the Me_2N ligand can interact with the *ortho*-hydrogen atom. In addition, for instance, flexible change of TS structures (open or close form) was observed in the case of DoM of anisole, as shown in Scheme 27.4. Thus, these results reflect that the zincate bases can change their TS structures flexibly to adapt to various substrates and thus show high substrate compatibility.

27.4.2 Deprotometalation Using Alkali Metal–Amidoaluminate Complexes

Organoaluminum compounds are widely used in organic synthesis, especially in aliphatic chemistry. Aromatic aluminum chemistry, however, has not been well developed, mainly due to the lack of efficient preparation methods compatible with various functional groups on aromatics. In 2004, an aluminum ate base, $i\text{-}Bu_3Al(TMP)Li$ (TMP-aluminate), was developed as effective reagent for the regio- and chemoselective DoM reaction to generate a wide range of functionalized aromatic aluminum compounds [28]. The typical results of DoM reactions using TMP-aluminates are presented later, in Section 27.5.2.

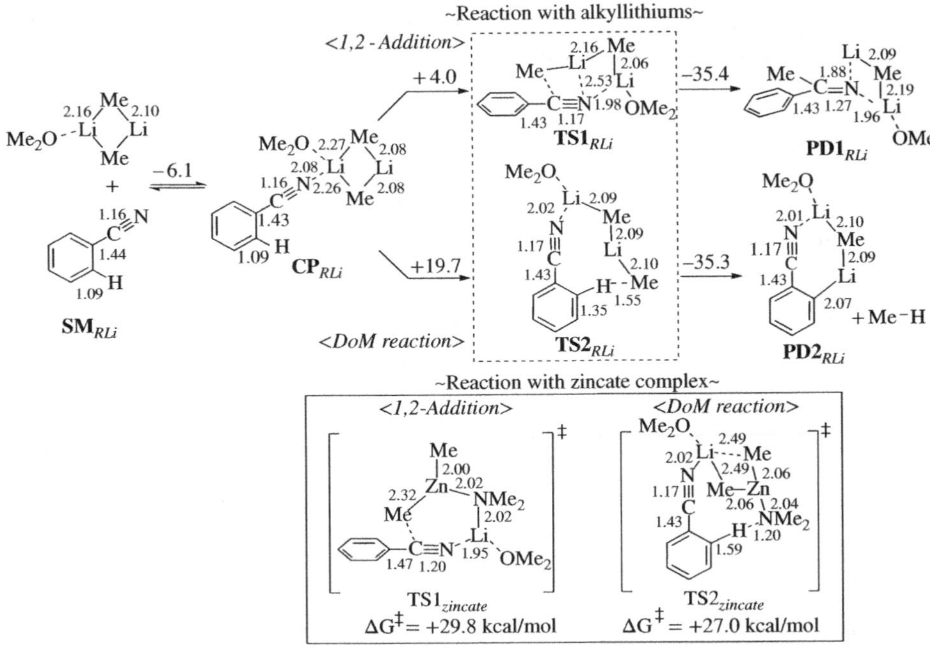

SCHEME 27.7 (Top) Reaction pathways of the 1,2-addition and DoM reaction of benzonitrile using $(MeLi)_2$. (Bottom box) TS structures in the 1,2-addition and DoM reaction of benzonitrile using $Me_2Zn(NMe_2)Li$. Energy changes and bond lengths at the B3LYP/6-31+G* and SVP (Zn) level of theory are shown in kcal/mol and Å, respectively.

The structure of the *i*-Bu_3Al(TMP)Li complex was firstly analyzed using NMR spectroscopy, and the ^{27}Al NMR chemical shift (153.4 ppm) showed clean formation of a tetrahedrally disposed aluminum center, indicating the formation of "ate" complex [58]. Since the three isobutyl groups are identical on the NMR time scale, it was difficult to decide among the two possible structures: "*C*-bridged" structure, in which Li^+ is connected to two alkyl groups, and "*N*-bridged" structure, in which Li^+ is connected to TMP and one alkyl group (Scheme 27.8). DFT calculation at B3LYP/6-31+G* level of theory indicated the preference of "*N*-bridged" structure (by 10–13 kcal/mol), and finally, X-ray crystal structure of the *i*-Bu_3Al(TMP)Li·THF complex unequivocally revealed the "*N*-bridged" structure. Furthermore, the aluminated arene intermediate, obtained from the DoM reaction of anisole with *i*-Bu_3Al(TMP)Li, was found to have three isobutyl groups on its aluminum center by ^{13}C NMR spectroscopic analysis, a result that might suggest deprotonation with amido ligand on aluminate complex.

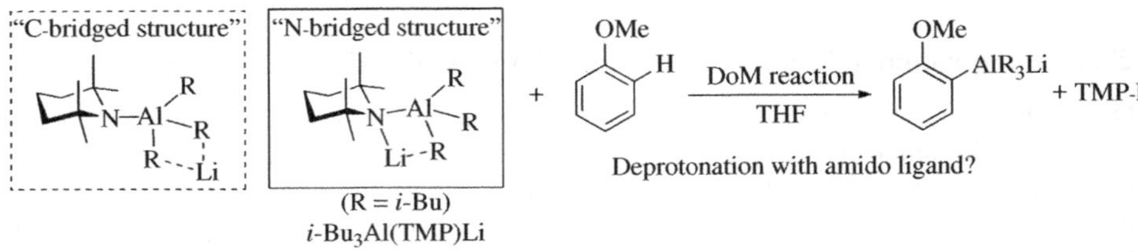

SCHEME 27.8 Two possible structures of *i*-Bu_3Al(TMP)Li and DoM reaction of anisole to afford aluminated arene complex having three alkyl groups on aluminum center.

It is also reported that the efficient isolation of TMP-free, triisobutyl aluminate complex from a benzamide substrate provides the rather strong evidence that the dominant active base is the (solvated) i-$Bu_3Al(TMP)Li$ complex, while the disproportionation between complexes occurs in the solution [59].

The possible reaction pathways for the deprotonative metalation of *N,N*-dimethylbenzamide using $Me_3Al(NMe_2)Li{\cdot}OMe_2$ have been modeled using DFT calculations (Scheme 27.9) [60]. Reaction of **SM** gives the relatively stable electrostatic complexes **CP1** and **CP2**, which resembles the experimentally observed adducts between *N,N*-diisopropylbenzamide and $R_3Al(TMP)Li$ (R = Me, *i*-Bu) [58, 61]. The following reaction step is the deprotonation with alkyl or amido ligand at the *ortho* position of the aromatic compound, similar to the case of TMP-zincates. An activation barrier of +52.6 kcal/mol strongly suggested a kinetically unfavorable deprotonation by a methyl ligand in $Me_3Al(NMe_2)Li{\cdot}OMe_2$ (**CP1–TS1–PD1**). On the other hand, calculating the Me_2N-mediated deprotonation process (**CP2–TS2–PD2**) revealed an activation barrier of +30.7 kcal/mol en route to forming aluminated aromatic compound **PD2**. The deprotonation of the resulting Me_2NH in **PD2** with a methyl ligand (ligand exchange reaction, **PD2–TS3–PD1**) requires an activation barrier of +37.7 kcal/mol, and the energy of **TS3** relative to **CP2** is +51.2 kcal/mol. These data strongly suggest that the deprotonation of Me_2NH by an Al-bonded alkyl ligand is very unlikely, a conclusion significantly at odds with TMP-zincate reactivity but consistent with crystallographic observation of trialkylaluminate intermediate.

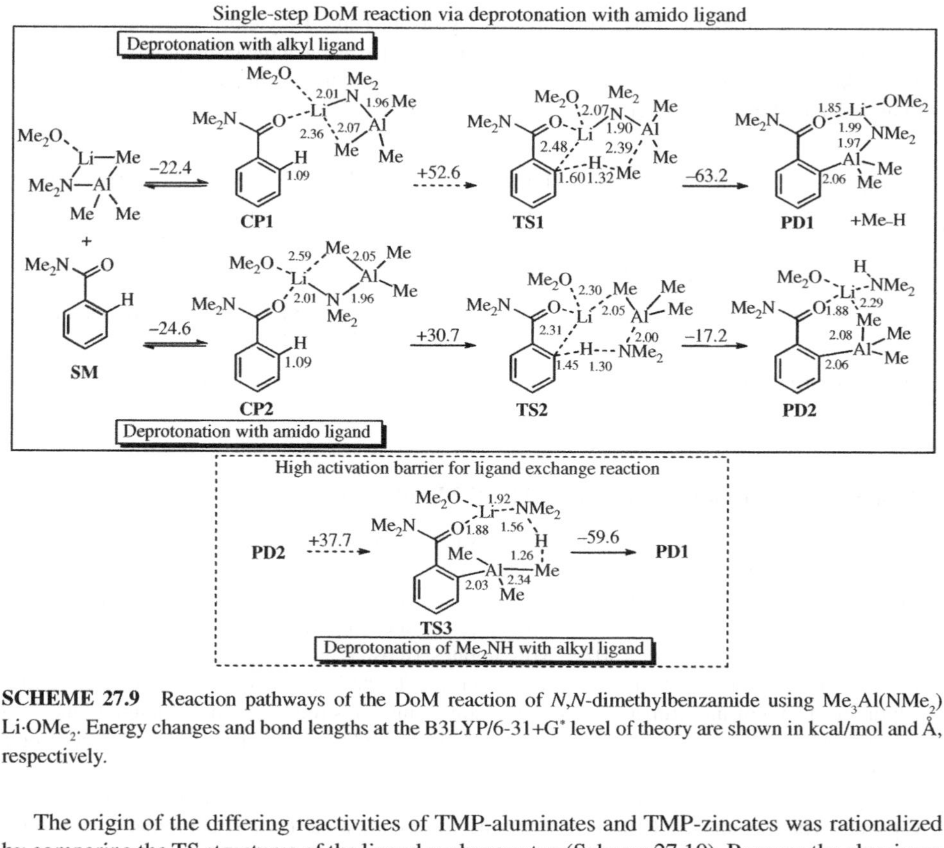

SCHEME 27.9 Reaction pathways of the DoM reaction of *N,N*-dimethylbenzamide using $Me_3Al(NMe_2)Li{\cdot}OMe_2$. Energy changes and bond lengths at the B3LYP/6-31+G* level of theory are shown in kcal/mol and Å, respectively.

The origin of the differing reactivities of TMP-aluminates and TMP-zincates was rationalized by comparing the TS structures of the ligand exchange step (Scheme 27.10). Because the aluminum center is coordinatively saturated, it is unable to stabilize the $[Me_2N]^-$ anion that is emerging in the

TS. In contrast, zinc is unsaturated (**16e**) and thus can stabilize the loosely associated $[Me_2N]^-$ anion, lowering the total activation energy for this quenching step.

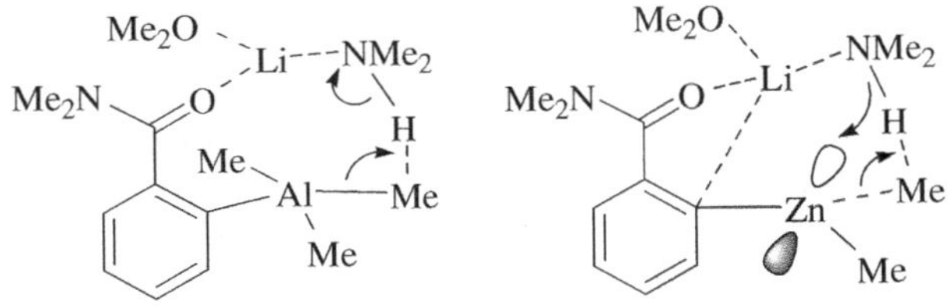

SCHEME 27.10 (Left) Disfavored operation of the TS in the deprotonation of Me_2NH by a methyl ligand with the aluminate complex and (right) favored operation of the TS with the zincate complex.

Therefore, TMP-aluminate is a kinetically controlled base with monobasicity, complementary to TMP-zincates; such stoichiometric reactivity clarifies the status of the dummy ligand (i.e., the *i*-Bu group) and ensures freedom from undesired and uncontrolled ligand scrambling. For instance, as an alternative application, such feature is highlighted in the regioselective deprotonation of an unsymmetrical ketone (Scheme 27.11). When 2-methylcyclohexanone was treated with *i*-$Bu_3Al(TMP)Li$ at 0°C for 0.5 h, electrophilic trapping of this enolate with benzaldehyde proceeded smoothly to give only the kinetically favored regioisomer in 69% yield, without any trace of the undesired regioisomer. The high selectivity was maintained, even at the reflux temperature for 18 h. Such a kinetically controlled behavior, even at high temperature, is unique to TMP-aluminates and in high demand from a synthetic viewpoint.

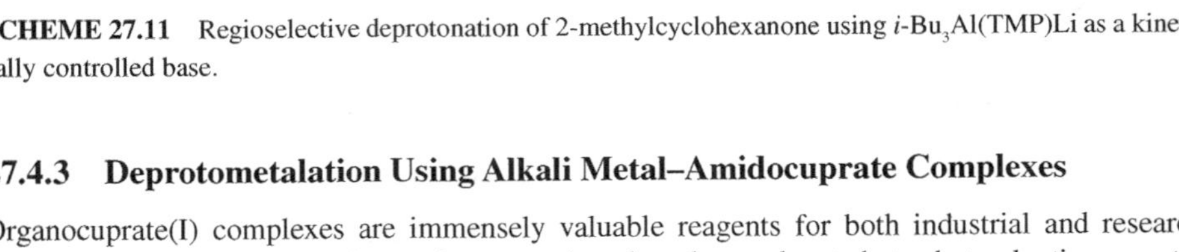

SCHEME 27.11 Regioselective deprotonation of 2-methylcyclohexanone using *i*-$Bu_3Al(TMP)Li$ as a kinetically controlled base.

27.4.3 Deprotometalation Using Alkali Metal–Amidocuprate Complexes

Organocuprate(I) complexes are immensely valuable reagents for both industrial and research chemistries. In particular, increasing attention has been devoted to heteroleptic cuprates ($[R—Cu—R']^-$) as well as the large body of homoleptic cuprates ($[R—Cu—R]^-$). Organo-amidocuprates also represent an important class of heteroleptic cuprates in organic transformations, especially in stereoselective syntheses. In 2007, TMP-cuprates (most typically, $[RCu(TMP)(CN)Li_2]$, where R = alkyl, phenyl, or TMP) have been found to promote the highly chemoselective directed *ortho*-cupration (DoC) of multifunctionalized aromatic compounds [62, 63]. The arylcuprate intermediates can be employed not only in the trapping of electrophiles but also in oxidative ligand coupling to form new C—C bonds with alkyl/aryl groups (as depicted later in Section 27.5.6).

Organocuprates(I) chemistry is dominated by two structure types: the Gilman-type and the Lipshutz-type (Scheme 27.12). The former one adopts a homodimeric structure, and the latter one exhibits heteroaggregate structures. With the help of X-ray crystallography, the structure–reactivity relationship study using TMP-cuprates revealed that (i) the cyclic dimer of the Gilman-type cuprate

$[(TMP)_2CuLi]$ is unreactive in DoC and (ii) a variety of Lipshutz-type amidocuprates, including $[(TMP)_2Cu(CN)Li_2]$, exhibit enhanced deprotonative reactivity [64].

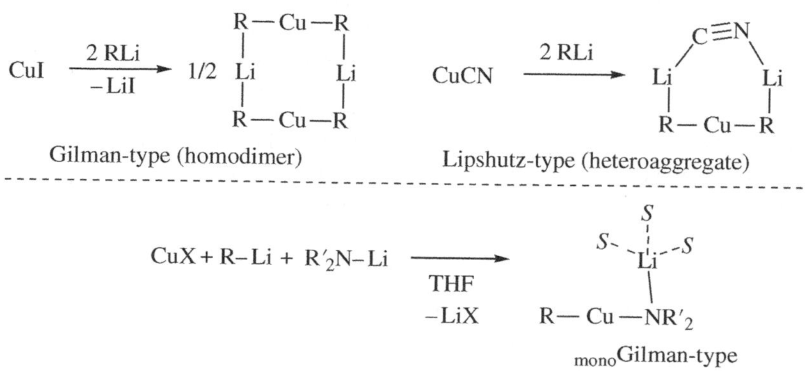

SCHEME 27.12 (Top) Structures of Gilman- and Lipshutz-type cuprates (bottom) structure of $_{mono}$Gilman-type cuprates with the coordination of solvent (*S*, solvent molecule).

To gain further insights into the reactive cuprate species and the mechanisms by which they react, a DFT study of the DoC of *N,N*-dimethylbenzamide was performed. A summary of the deprotonation process using a model Lipshutz-type cuprate $[MeCu(NMe_2)(CN)Li_2]\cdot 2OMe_2$ is shown in Scheme 27.13 [64].

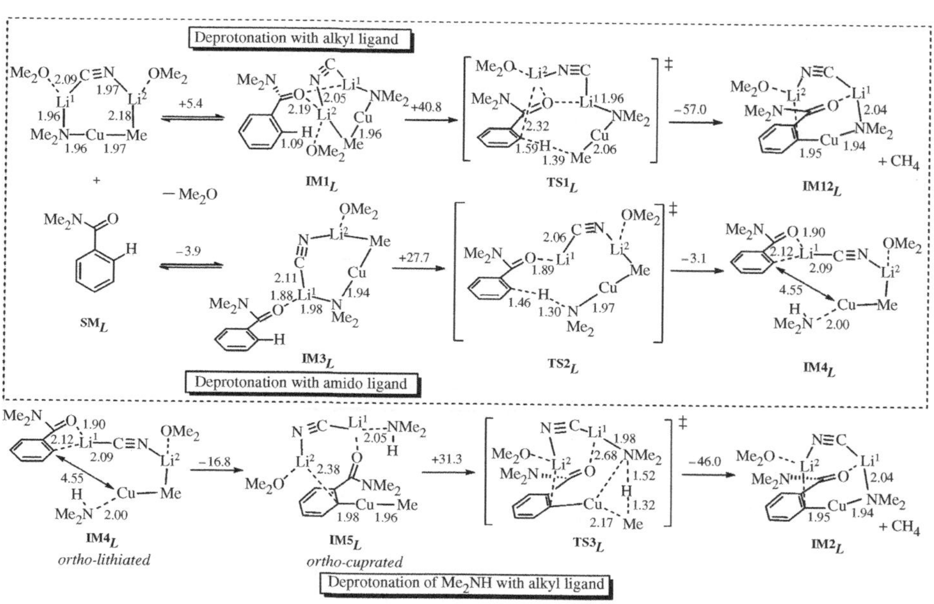

SCHEME 27.13 Reaction pathways of the DoM reaction of *N,N*-dimethylbenzamide using the Lipshutz-type $[MeCu(NMe_2)(CN)Li_2]\cdot 2OMe_2$. Energy changes and bond lengths at the B3LYP/6-31+G* and SVP (Cu) level of theory are shown in kcal/mol and Å, respectively.

In common with the previously noted reaction pathways for TMP-zincates and TMP-aluminates, two plausible pathways were discovered for the regioselective DoM, depending on if the deprotonation is triggered by the amido or alkyl ligand of the Lipshutz-type cuprate. Although the formation of $\mathbf{IM2}_L$ represents a thermodynamically favorable outcome, the pathway in which alkyl ligand promotes the *ortho*-deprotonation ($\mathbf{IM1}_L$–$\mathbf{TS1}_L$–$\mathbf{IM2}_L$) is kinetically unfavorable on account of the relatively high energy of $\mathbf{TS1}_L$ ($\Delta G^{\ddagger}$ = 40.8 kcal/mol). $\mathbf{TS2}_L$ is then attained via an amido ligand-promoted deprotonation ($\mathbf{IM3}_L$–$\mathbf{TS2}_L$–$\mathbf{IM4}_L$). While the activation barrier is slightly high (+27.7 kcal/mol) relative to $\mathbf{IM3}_L$, it is still relatively accessible, allowing metalated intermediate $\mathbf{IM4}_L$ to form. Ligand exchange step ($\mathbf{IM5}_L$–$\mathbf{TS3}_L$–$\mathbf{IM2}_L$) requires the high activation barrier (+31.3 kcal/mol), which implies a difficult transformation.

Further investigations into possible pathways using the Lipshutz-type reagent led to the identification of an *ortho*-cupration pathway based on the action of a monomeric Gilman-type complex (in the bottom of Scheme 27.12), the structure of which was previously clarified with TMP-cuprates having some coordinating molecules [65]. Calculations suggest that the $_{\text{mono}}$Gilman complexes $\mathbf{IM1}_G$ and $\mathbf{IM3}_G$ are formed by the elimination of solvated LiCN, with one of the solvent molecules coordinated to the lithium ion being substituted by *N,N*-dimethylbenzamide (Scheme 27.14). Although $\mathbf{IM1}_G$ is more stable than $\mathbf{IM3}_G$, it fails to promote the deprotonation with alkyl ligand ($\mathbf{IM1}_G$–$\mathbf{TS1}_G$–$\mathbf{IM2}_G$), a process that reveals very high activation energy (+40.3 kcal/mol). The pathway $\mathbf{IM3}_G$–$\mathbf{TS2}_G$–$\mathbf{IM4}_G$, in which deprotonation is promoted by the amido ligand of

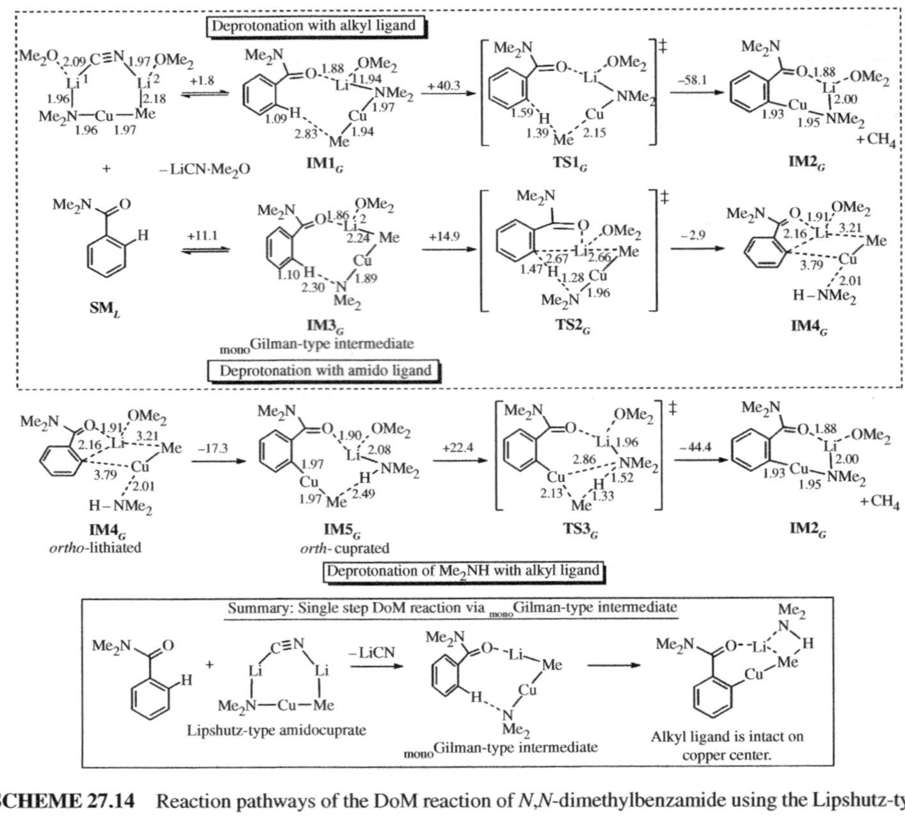

SCHEME 27.14 Reaction pathways of the DoM reaction of *N,N*-dimethylbenzamide using the Lipshutz-type $[MeCu(NMe_2)(CN)Li_2]\cdot 2OMe_2$ via $_{\text{mono}}$Gilman-type intermediate. Energy changes and bond lengths at the B3LYP/6-31+G* and SVP (Cu) level of theory are shown in kcal/mol and Å, respectively.

the ${}_{\mathrm{mono}}$Gilman-type base, has a very low activation barrier (+14.9 kcal/mol with respect to **IM3**$_G$) en route to **IM4**$_G$. Although the total activation energy (26.0 kcal/mol) from **SM**$_L$ to **TS2**$_G$ is comparable to that from **SM**$_L$ to **TS2**$_L$ (23.8 kcal/mol, in Scheme 27.13), the ${}_{\mathrm{mono}}$Gilman pathway does not need more than 15 kcal/mol in energy for any given step, and this contrasts with the Lipshutz path requiring over 27 kcal/mol for the deprotonation step. As with the Lipshutz system, quenching of the amine Me_2NH (**IM5**$_G$–**TS3**$_G$–**IM2**$_G$) was felt unlikely to proceed on account of the rather high energy difference between **SM**$_L$ and **TS3**$_G$ (more than 28 kcal/mol), while the activation barrier to reach **TS3**$_G$ is entropically permissible at 22.4 kcal/mol.

The oxidative ligand coupling in the arylcuprate intermediate, generated from the DoC of *N,N*-diisopropylbenzamide using [MeCu(TMP)(CN)Li_2], by nitrobenzene afforded 2-methyl-*N,N*-diisopropylbenzamide (Scheme 27.15) [62]. This result clearly shows that the methyl ligand is intact on the copper center of the arylcuprate intermediate. Moreover, a significant increase of the activation energy of deprotonation starting with the Gilman dimer was found, in good agreement with experimental results (described earlier). In summary, the DoM reaction using the Lipshutz-type amidocuprates proceeds via formation of the ${}_{\mathrm{mono}}$Gilman-type intermediate, and the alkyl ligand is intact on the central copper metal in the resultant arylcuprate species without ligand scrambling.

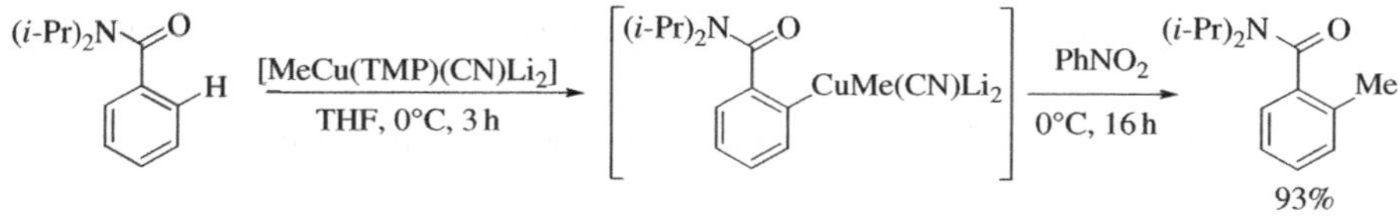

SCHEME 27.15 Oxidative ligand coupling of the arylcuprate intermediate.

27.4.4 Deprotometalation Using Alkali Metal–Amidocadmate Complexes

Given that amido ligands in amidozincates, aluminates, and cuprates preferentially act as an actual base, compared to alkyl ligands, poly(amido)metal ate complexes would be of interest as effective deprotonating reagents with polybasicity, especially in terms of catalytic use of metal reagents. The formation of $(TMP)_3ZnLi$ from LiTMP and $(TMP)_2Zn$, however, has been found to be a thermodynamically unfavorable process by DFT study, and the analysis of the ^{13}C spectra experimentally revealed that in THF solution containing 1:3 mixture of $ZnCl_2$·TMEDA and LiTMP, the main species are LiTMP and $(TMP)_2Zn$ [66]. In contrast, exothermicity was predicted for the generation of $(TMP)_3CdLi$ from LiTMP and $(TMP)_2Cd$ by DFT calculation (Scheme 27.16). Actually, such *in situ*-prepared poly(TMP)-cadmate was found to be quite effective for the regio- and chemoselective DoM reaction of (hetero)aromatics, even using only a substoichiometric amount of $CdCl_2$·TMEDA (0.5 equiv) (as described in Section 27.5.9) [67, 68].

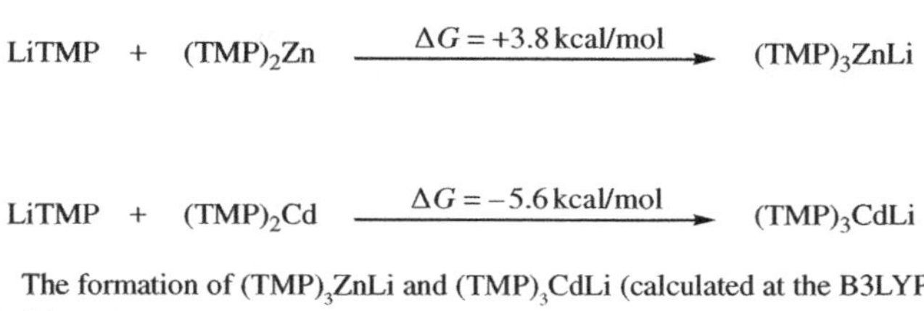

SCHEME 27.16 The formation of $(TMP)_3ZnLi$ and $(TMP)_3CdLi$ (calculated at the B3LYP/6-31G* and SVP (Zn and Cd) level of theory).

The reaction pathways of the DoM reaction promoted by poly(amido)cadmate were found to be different from those previously noted for amidozincates (Scheme 27.17) [69].

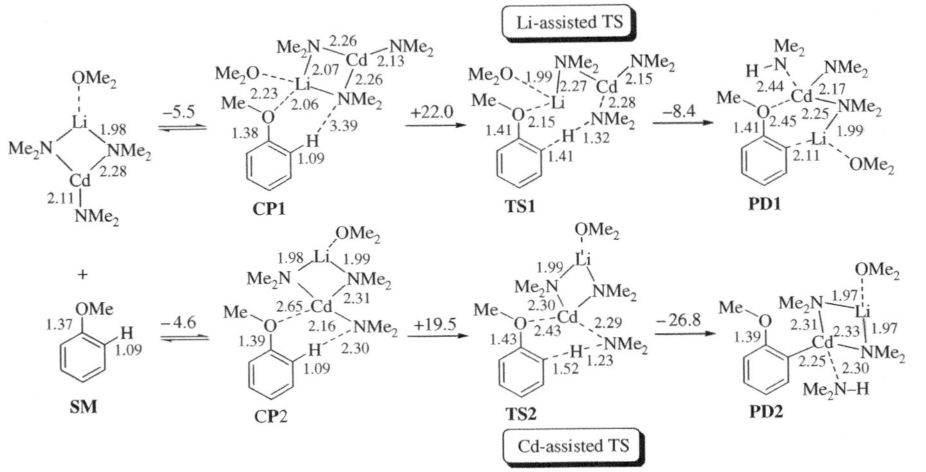

SCHEME 27.17 Reaction pathways of the DoM reaction of anisole using the $(Me_2N)_3CdLi$. Energy changes and bond lengths at the B3LYP/6-31G* and SVP (Cd) level of theory are shown in kcal/mol and Å, respectively.

Two plausible pathways for the deprotonation reaction using $(Me_2N)CdLi$ and anisole through "Li-assisted" and "Cd-assisted" deprotonation TSs were determined. In both cases, the reaction coordinates start with formation of a relatively stable initial complex (**CP1** or **CP2**) between anisole oxygen and the countercation Li or the central Cd metal. The Li-assisted deprotonation is a pathway similar to that seen in TMP-zincate-mediated DoM reactions and takes place via "open form" **TS1** with a reasonable activation energy (+22.0 kcal/mol). The Cd-assisted deprotonation, a unique pathway of this cadmate base, proceeds smoothly with a smaller activation energy (19.5 kcal/mol). This deprotonation is facilitated by the direct push–pull synergy of the Lewis acidic Cd metal and the negatively charged NMe_2 moiety to generate a stable product (**PD2**). Thus, the identified reaction mode is quite informative and beneficial for the further development of novel metal ate complexes and (catalytic) DoM reactions.

27.5 SCOPE AND APPLICATIONS OF THE DEPROTOMETALATION

27.5.1 Using Lithium– or Sodium–Magnesium Mixed-Metal Bases

The addition of cryptands or alkoxides to diorganomagnesiums in Et_2O or apolar solvents makes the organometallic reagents more basic (e.g., capable of deprotonating chloro- and fluorobenzenes), and triorganomagnesates were early proposed as possible deprotonating species [70–72]. In the same way, the example shown in Scheme 27.18 depicts a crown ether activation responsible for the deprotonation of the 5-bromo-1,3-xylylene using a diarylmagnesium [73–75].

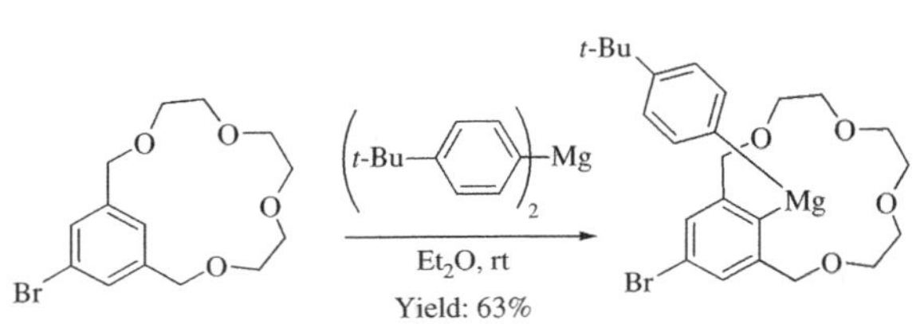

SCHEME 27.18 5-Bromo-1,3-xylylene-15-crown-4: deprotometalation using a diarylmagnesium.

A lithium magnesate able to deprotometalate benzenes, Bu_3MgLi, was for the first time patented in 1992. (Trifluoromethyl)benzenes 3-substituted by a CF_3 (Scheme 27.19), OMe, or NMe_2 group can be functionalized using this organo base in Et_2O at room temperature [76]. The presence of TMEDA, the replacement of butyl by TMP ligands, and the use of higher-order reagents in general result in improved conversions [77]. The method was applied to different substituted benzenes, and various functionalizations were achieved (Table 27.3) [78].

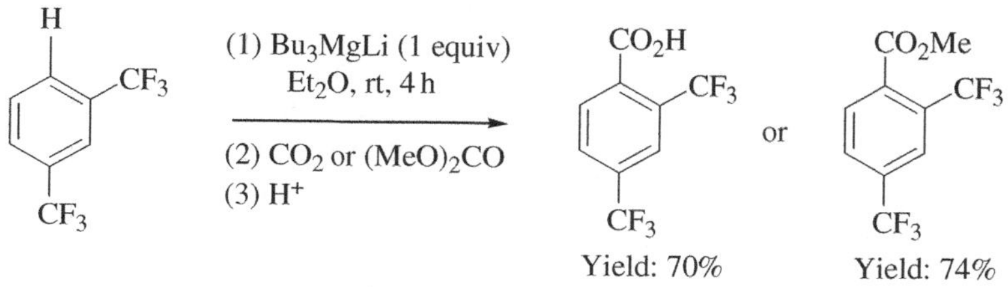

SCHEME 27.19 1,3-Bis(trifluoromethyl)benzene: deprotometalation using a lithium–magnesium base and subsequent electrophilic trapping.

TABLE 27.3 Substituted Benzenes: Deprotometalation Using a Lithium–Magnesium Base and Subsequent Interception with Different Electrophiles

Ar–H —[(1) Magnesate (n equiv), THF, conditions; (2) Electrophile; (3) Hydrolysis]→ Ar–E

Entry	Ar–H	Magnesate (n), Conditions	Electrophile (E)	Yield (%)
1	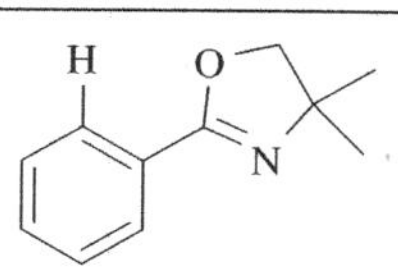	Bu_3MgLi (0.4), rt, 2 h	O_2 (OH)	55
2	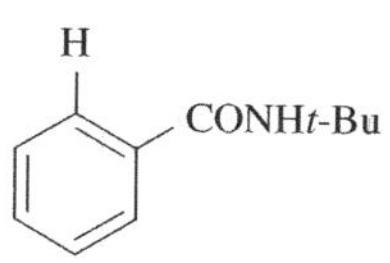	Bu_4MgLi_2 (0.55), rt, 2 h	$(PhSO_2)_2NF$ (F)	59
3	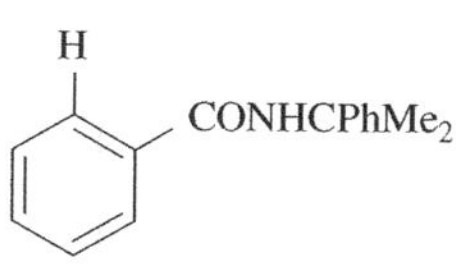	Bu_4MgLi_2 (0.55), rt, 2 h	$MeCH(OCH_2)$ ($CH_2CH(OH)Me$)	61
4	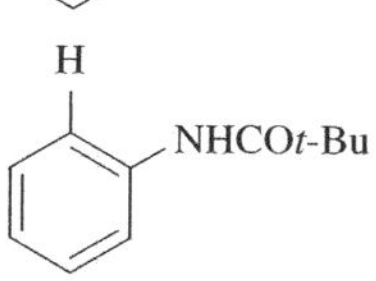	Bu_4MgLi_2 (1), reflux, 2 h	I_2 (I)	70
5	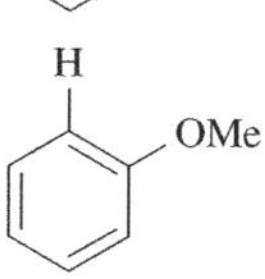	Bu_4MgLi_2 (0.3), rt, 2 h	$BrC(Me){=}CH_2$[a] ($C(Me){=}CH_2$)	66

[a] In the presence of catalytic $PdCl_2$·DPPF.

It is possible to deprotomagnesate bare benzene [26] and toluene [79], for example, by using [(TMEDA)Na(TMP)BuMg(TMP)] at hexane reflux, to afford [(TMEDA)Na(TMP)PhMg(TMP)] and [(TMEDA)Na(TMP)(3-tolyl)Mg(TMP)] (regioselective metalation of toluene at the 3 position) in 44% and 58% yield, respectively. Benzene and toluene dideprotometalation can also be observed using magnesates [80–83], for example, using $NaMg(TMP)_2Bu$ free of TMEDA to give the "inverse crown compounds" $[Na_4Mg_2(1,4\text{-}C_6H_4)(TMP)_6]$ and $[Na_4Mg_2(1\text{-Me-}2,5\text{-}C_6H_3)(TMP)_6]$, respectively [80].

In contrast with the other lithium-metal bases, the presence of bulky amino groups is not mandatory to observe deprotometalation using lithium magnesates. Nevertheless, reactions using amido-type bimetallic bases are in general more efficient. The reactivity of hindered magnesium amides can be strengthened in the presence of alkali metal salts. Salt-activated (or "turbo-Grignard" reagents) TMPMgCl·LiCl is suitable to deprotometalate benzenes bearing functional groups, provided that the temperature is adapted (Table 27.4) [16, 84–92]. The efficiency of the reactions could be in relation with the ability of LiCl to break up oligomeric aggregates of magnesium

TABLE 27.4 Substituted Benzenes: Deprotometalation Using a Lithium–Magnesium Base and Subsequent Interception with Different Electrophiles

$$\text{Ar–H} \xrightarrow[\text{(2) Electrophile; (3) Hydrolysis}]{\text{(1) TMPMgCl·LiCl (1 equiv), THF, conditions}} \text{Ar–E}$$

Entry	Ar–H	Conditions	Electrophile (E)	Yield (%)
1	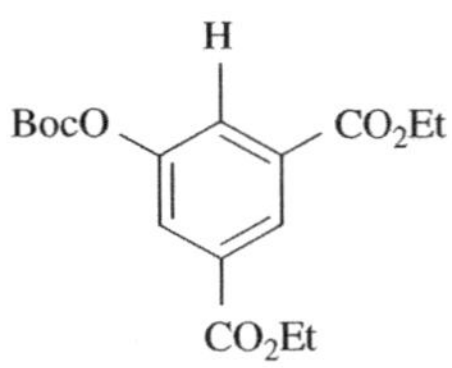	0°C, 1 h	$NCCO_2Et$ (CO_2Et)	85
2	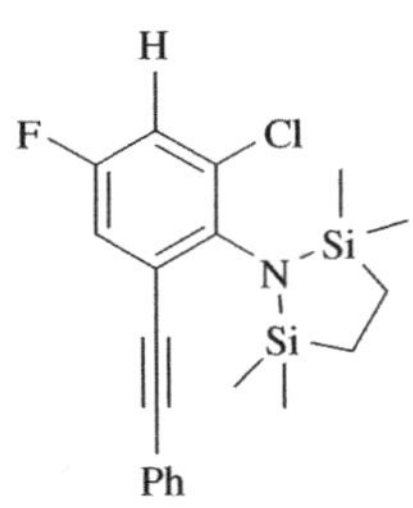	rt, 2 h	TsCN (CN)	80
3	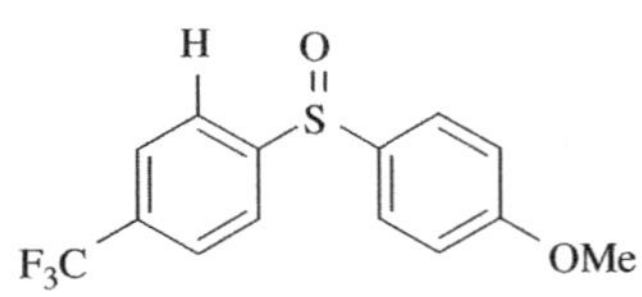	−30°C, 0.3 h	4-ClC_6H_4I[a] (C_6H_4-4-Cl)	75
4	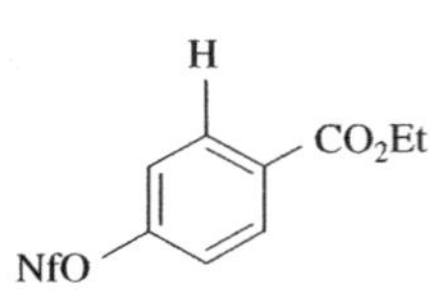	−20°C, 3 h	4-AnisylCHO (CH(OH)-4-anisyl)	72

[a] Under palladium catalysis after transmetalation with $ZnCl_2$.

TABLE 27.5 Substituted Benzenes and Naphthalenes: Deprotometalation Using a Lithium–Magnesium Base and Subsequent Interception with Different Electrophiles

$$\text{Ar–H} \xrightarrow[\text{(2) Electrophile}]{\text{(1) Mg(TMP)}_2\cdot\text{2LiCl (1 equiv), THF, conditions}} \text{Ar–E}$$

Entry	Ar–H	Conditions	Electrophile (E)	Yield (%)
1	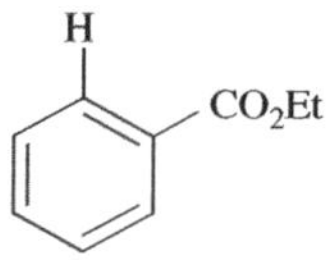	25°C, 0.75 h	4-MeC_6H_4Br[a] (C_6H_4-4-Me)	71
2	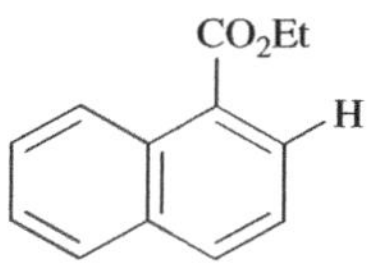	25°C, 0.75 h	Boc_2O (CO_2*t*-Bu)	69
3	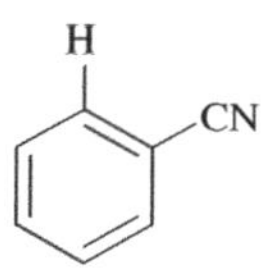	−30°C, 3 h	4-$EtOCOC_6H_4I$[b] (C_6H_4-4-CO_2Et)	70
4	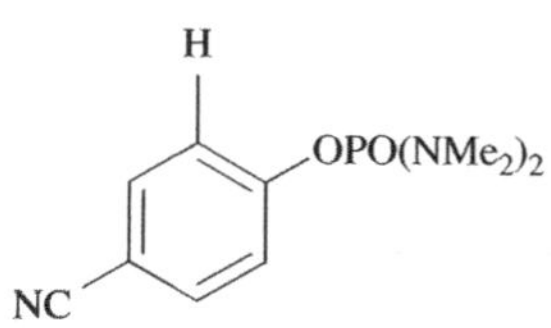	0°C, 1 h	*t*-BuCOCl[c] (CO*t*-Bu)	81

[a] Under palladium and ruthenium catalysis after transmetalation with $ZnCl_2$.
[b] Under palladium catalysis after transmetalation with $ZnCl_2$.
[c] In the presence of CuCN·2LiCl after transmetalation with $ZnCl_2$.

amides, a possibility supported by the isolation of crystals of [$(THF)_2LiMgCl_2(TMP)(THF)$] [18]. Nevertheless, this solid-state structure is not retained in THF solution, and the solvent-separated nature in this case distinguishes TMPMgCl·LiCl from most of the mixed alkyl-amido species, which are contact ion-pair arrangements, suggesting distinct reaction mechanisms [93].

As TMPMgCl·LiCl only works readily with well-activated substrates, $Mg(TMP)_2\cdot 2LiCl$ was developed to remedy this lack of efficiency. The bis-amide is more suitable to abstract protons from more moderately activated arenes, and various functionalizations can also be reached (Table 27.5) [19, 92, 94–97]. Other hindered RR'N groups (R = *i*-Pr, R' = *t*-Bu) can replace TMP in TMPMgCl·LiCl and $Mg(TMP)_2\cdot 2LiCl$ [98]. Starting from bromobenzenes, subsequent formation of benzynes through elimination of amidomagnesium bromide is not avoided; this possibility was elegantly applied to the synthesis of substituted indolines and carbazoles [99].

27.5.2 Using Lithium–Aluminum Mixed-Metal Bases

When used in THF, Li(TMP)Al(*i*-Bu)$_3$ (2 equiv) is capable of deprotonating substituted benzenes such as anisole and *N*,*N*-diisopropylbenzamide at room temperature and sensitive substrates such as benzonitrile and 1-chloro-4-iodobenzene at −78°C. Iodine is the electrophile of choice, but others can also be

TABLE 27.6 Substituted Benzenes: Deprotometalation Using a Lithium–Aluminum Base and Subsequent Interception with Electrophiles

$$\text{Ar–H} \xrightarrow[\text{(2) Electrophile}]{\text{(1) Li(TMP)Al}(i\text{-Bu})_3 \text{ (2 equiv)}\ \text{THF, conditions}} \text{Ar–E}$$

Entry	Ar–H	Conditions	Electrophile (E)	Yield (%)
1	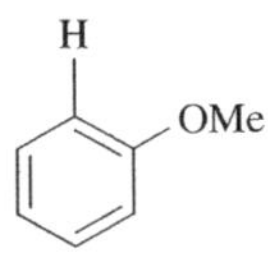	rt, 3h	I_2 (I)	99
2	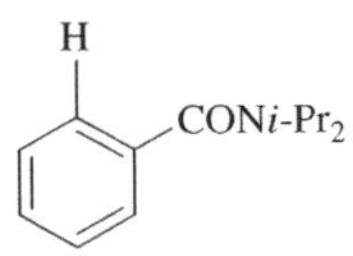	rt, 3h	AllylBr (allyl)	100
3	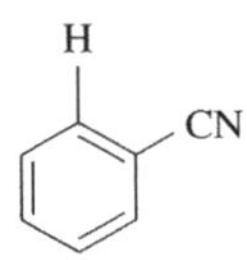	−78°C, 2h	I_2 (I)	100
4	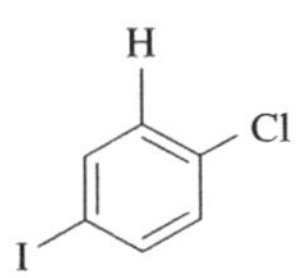	−78°C, 3h	I_2 (I)	100

employed (Table 27.6) [28, 58]. Crystals of the deprotometalated *N*,*N*-diisopropylbenzamide were isolated, allowing its structure to be identified as $[(THF)_3Li\{O(=C)N(i\text{-}Pr)_2C_6H_4\}Al(i\text{-}Bu)_3]$ using X-ray diffraction [59].

The good tolerance toward halogens of $Li(TMP)_2Al(i\text{-}Bu)_2$ allowed the 4-halo anisoles (chloro, bromo, iodo) to be deprotoaluminated next to the oxygen-containing group (due to the strong Lewis acidic character of aluminum) in hexane at room temperature. Iodo, bromo, and chloro derivatives were prepared, respectively, using iodine, *N*-bromosuccinimide (more efficient than Br_2, no reaction using $Cl_2BrCCBrCl_2$), and sulfuryl chloride (more efficient than *N*-chlorosuccinimide, no reaction using Cl_3CCCl_3) (Scheme 27.20) [100].

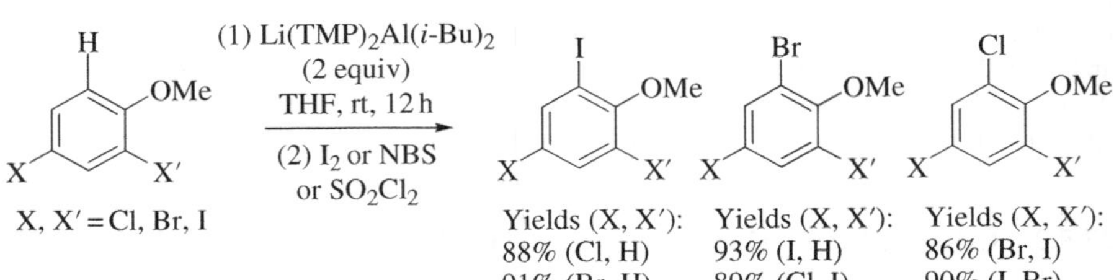

SCHEME 27.20 4-Halo anisoles: deprotometalation using a lithium–aluminum base and subsequent interception with iodine, *N*-bromosuccinimide, and sulfuryl chloride.

TABLE 27.7 Substituted Benzenes and Naphthalenes: Deprotometalation Using a Lithium–Aluminum Base and Subsequent Interception with Electrophiles

$$\text{Ar–H} \xrightarrow[\text{(2) CuCN·2LiCl, Electrophile}]{\text{(1) Al(TMP)}_3\text{·3LiCl (1 equiv), THF, conditions}} \text{Ar–E}$$

Entry	Ar–H	Conditions	Electrophile (E)	Yield (%)
1	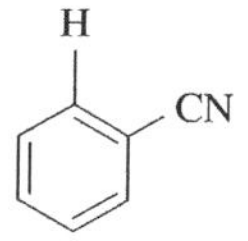	−10°C, 4h	2-FurylCOCl (CO-2-furyl)	70
2	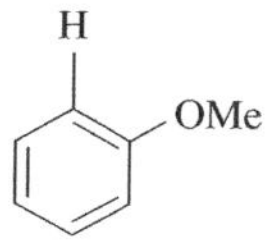	25°C, 10h	4-ClC$_6$H$_4$COCl (COC$_6$H$_4$-4-Cl)	78
3	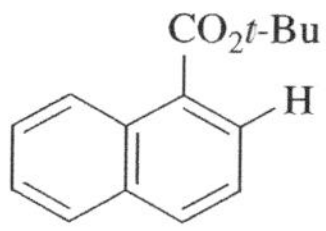	−5°C, 6h	PhCOCl (COPh)	74

Deprotoalumination next to halogen is possible using THF-solvated Li(TMP)Al(i-Bu)$_3$ in hexane. In contrast, using amide-rich Li(TMP)$_2$Al(i-Bu)$_2$ leads to benzyne formation, for example, from 3-iodoanisole after metalation between both substituents [101].

Salt-activated Al(TMP)$_3$·3LiCl can be used to achieve the functionalization of substituted benzenes and naphthalenes. Provided that a reduced temperature is used, the reaction tolerates the presence of functional groups (Table 27.7) [20]. Al(N(CH(i-Pr)t-Bu)t-Bu)$_3$·3LiCl is an efficient alternative, for example, to achieve subsequent palladium-catalyzed cross-coupling [102].

27.5.3 Using Lithium–, Sodium–, or Magnesium–Manganese Mixed-Metal Bases

When benzene is treated by [(TMEDA)Na(TMP)(CH$_2$SiMe$_3$)Mn(TMP)] (0.5 equiv) at hexane reflux, monodeprotonation takes place to afford [(TMEDA)Na(TMP)(Ph)Mn(TMP)] (61% yield) in which the CH$_2$SiMe$_3$ ligand, chosen because no thermal β-H decomposition can occur, is replaced. Without TMEDA, using Mn(CH$_2$SiMe$_3$)$_2$ (2 equiv), NaTMP (4 equiv), and HTMP (2 equiv) leads to benzene 1,4-dideprotonation, a result evidenced by the isolation of [Na$_4$Mn$_2$(TMP)$_6$(C$_6$H$_4$)] crystals (Scheme 27.21) [103].

Under similar reaction conditions, toluene is 3,5-dimetalated, a result demonstrated by the isolation of [(TMP)$_6$Na$_4$(3,5-Mn$_2$C$_6$H$_3$CH$_3$)] in 68% yield [83]. Again in hexane, but at room temperature, anisole and N,N-diisopropylbenzamide are *ortho*-metalated, respectively affording [(TMEDA)Na(TMP)(2-anisyl)Mn(TMP)] and [(TMEDA)Na(TMP){2-[C(O)Ni-Pr$_2$]C$_6$H$_4$}Mn(Me$_3$SiCH$_2$)], the former together with generation of tetramethylsilane and the latter with liberation of HTMP (Scheme 27.22) [104].

These reaction conditions were applied to the functionalization of naphthalene, 1-methoxynaphthalene, and 2-methoxynaphthalene. The three compounds were deprotometalated regioselectively, either at the 2-position, when available, or otherwise at the 3-position. Crystalline products were isolated in all cases, in 88%, 65%, and 85% yield, respectively (Scheme 27.23) [105].

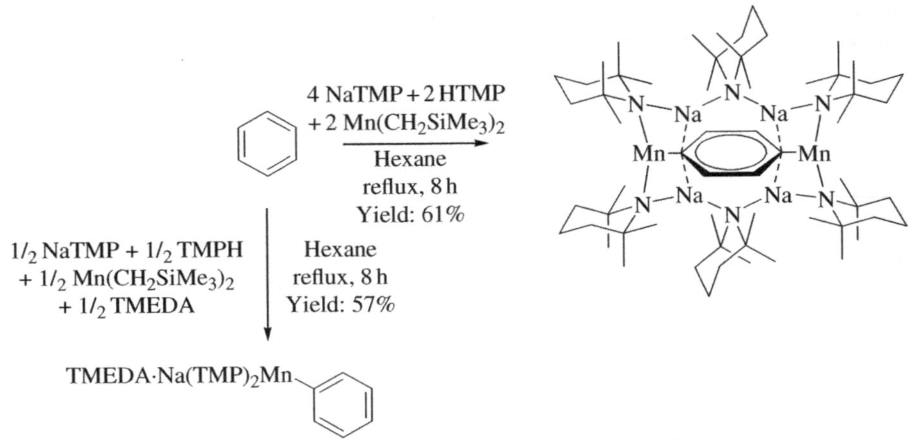

SCHEME 27.21 Benzene: dideprotometalation using a sodium–manganese base.

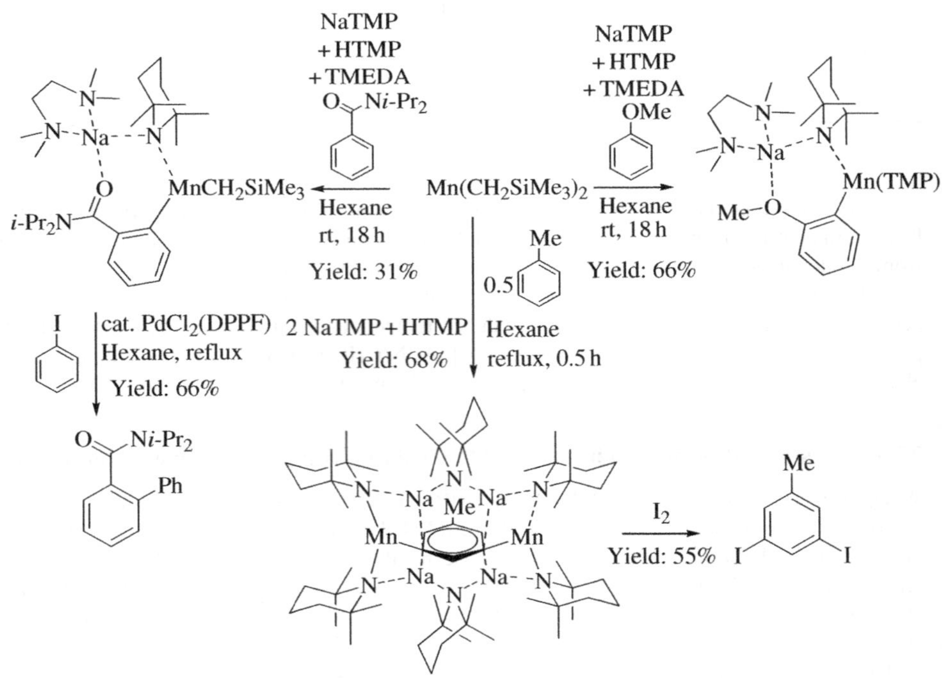

SCHEME 27.22 Toluene, anisole, and *N,N*-diisopropylbenzamide: deprotometalation using a sodium–manganese base.

Benzenes bearing electron-withdrawing groups, including reactive functions, can be deprotonated in THF at room temperature using salt-activated $Mn(TMP)_2{\cdot}2MgCl_2{\cdot}4LiCl$. Recourse to copper or palladium catalysis allows the generated diarylmanganese species to be functionalized in different ways (Table 27.8). Notably, arylamines were synthesized by conversion to copper(I) species, treatment with lithium amides, and oxidative amination [21, 106].

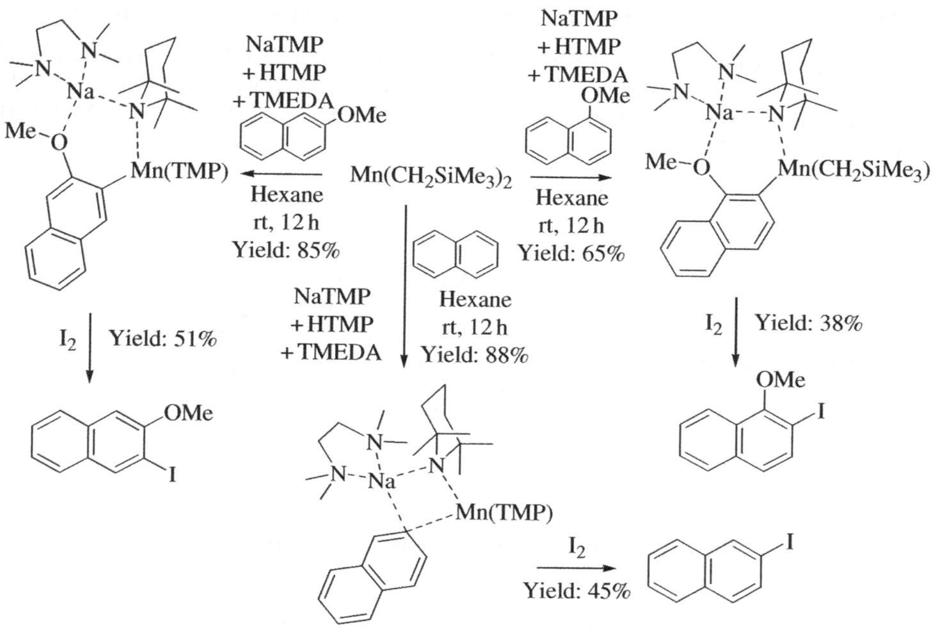

SCHEME 27.23 Naphthalene, 1-methoxynaphthalene, and 2-methoxynaphthalene: deprotometalation using a sodium–manganese base.

TABLE 27.8 Activated Benzenes: Deprotometalation Using a Lithium–Magnesium–Manganese Base and Subsequent Trapping

$$\text{Ar–H} \xrightarrow[\text{THF, rt, reaction time}]{\text{Mn(TMP)}_2\cdot 2\text{MgCl}_2\cdot 4\text{LiCl (0.6 equiv)}} \xrightarrow[\text{(2) Hydrolysis}]{\text{(1) Electrophile}} \text{Ar–E}$$

Entry	Ar–H	Reaction Time (h)	Electrophile (E)	Yield (%)
1	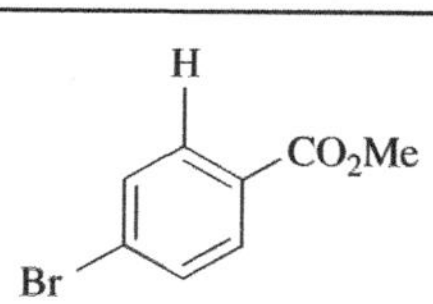	3.5	2-ThienylCOCl[a] (CO-2-thienyl)	77
2	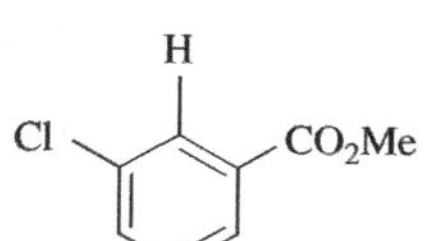	2	$3\text{-F}_3\text{CC}_6\text{H}_4\text{I}$[b] ($\text{C}_6\text{H}_4\text{-3-CF}_3$)	77
3	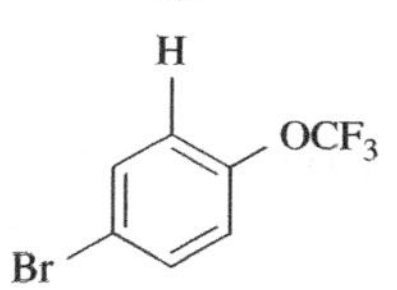	10	NCCO_2Et (CO_2Et)	77

[a] In the presence of catalytic $CuCN \cdot 2LiCl$.
[b] In the presence of catalytic $Pd(PPh_3)_4$.

27.5.4 Using Lithium–, Sodium–, or Magnesium–Iron Mixed-Metal Bases

The base prepared by mixing in hexane TMEDA·Fe(CH_2SiMe_3)$_2$, NaTMP, and HTMP in a 1:2:2 ratio is strong enough to dideprotonate benzene, as shown by the isolation (in 39% yield) and the identification by X-ray diffraction of an iron-host inverse crown complex, [$Na_4Fe_2(TMP)_6(C_6H_4)$] (Scheme 27.24) [107].

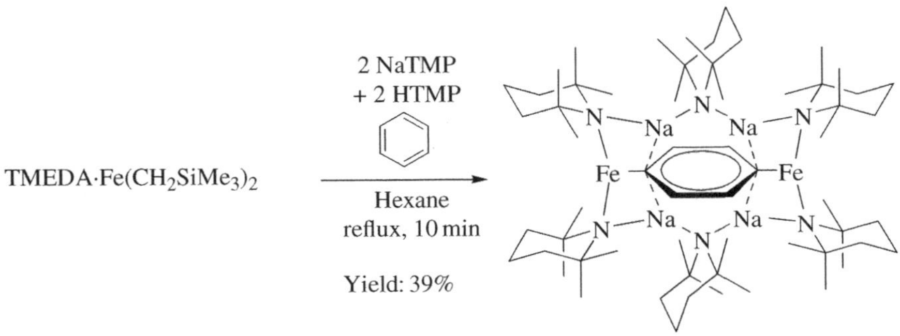

SCHEME 27.24 Benzene: dideprotometalation using a sodium–iron base.

Salt-activated Fe(TMP)$_2$·2$MgCl_2$·4LiCl was successfully used in THF at room temperature to convert activated 1,3- and 1,4-disubstituted arenes into the corresponding diaryliron(II) species. The latter were cross-coupled under nickel catalysis with alkyl iodides and bromides or even benzyl chloride. The reaction tolerates a large range of functional groups (Table 27.9) [22].

TABLE 27.9 Activated Benzenes: Deprotometalation Using a Lithium–Magnesium–Iron Base and Subsequent Nickel-Catalyzed Alkylation

$$\text{Ar–H} \xrightarrow[\text{THF, rt, reaction time}]{\text{Fe(TMP)}_2\cdot 2\text{MgCl}_2\cdot 4\text{LiCl (0.75 equiv)}} \xrightarrow[\text{NiCl}_2\text{ (0.5 mol.\%)}]{\text{R-X}} \text{Ar–R}$$

Entry	Ar–H	Reaction Time (h)	RX	Yield (%)
1[a]	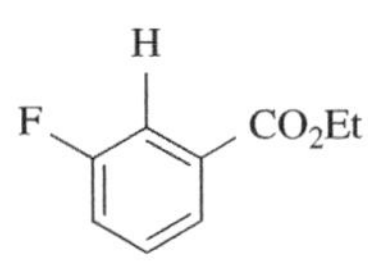	3	Oct-X	88 (X=I), 74 (X=Br)
2[a]			c-Hex-X	83 (X=I), 60 (X=Br)
3			Bn-Cl	88
4[a]	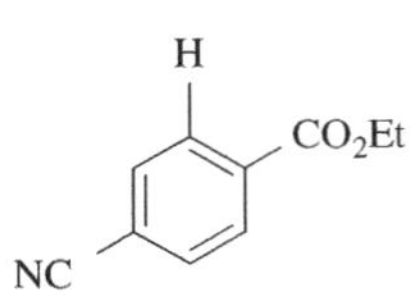	48	$F_3C(CH_2)_3$-I	65
5	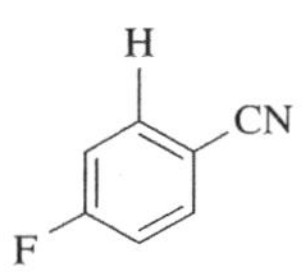	18	$(EtO)_2P(O)CH_2$-I	72

[a] In the presence of 4-fluorostyrene (4–10 mol%).

The base prepared from LiTMP and $FeBr_2$ in a 3:1 ratio was employed in THF for the deprotometalation of a less activated substrate, anisole. After 2 h contact at room temperature and the addition of an electrophile, iodine, or pivalaldehyde, the expected derivatives were isolated in 71% or 47% yield, respectively (Scheme 27.25) [108].

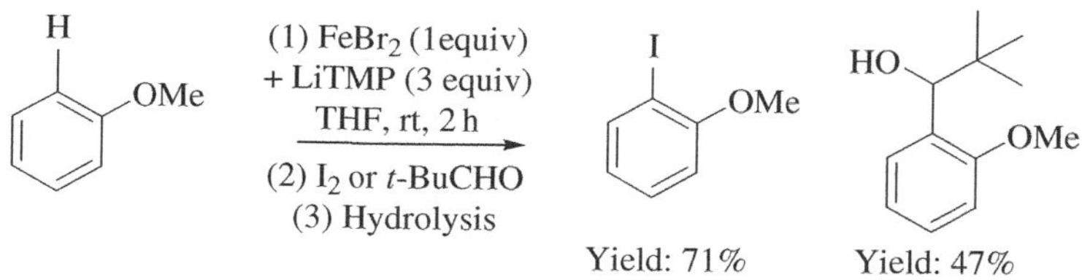

SCHEME 27.25 Anisole: deprotometalation using a lithium–iron base and subsequent interception with iodine and pivalaldehyde.

27.5.5 Using Lithium–Cobalt Mixed-Metal Bases

When used in THF at room temperature, the base prepared from LiTMP and $CoBr_2$ in a 3:1 ratio proves capable of deprotometalating different methoxybenzenes. The generated arylmetal species can be trapped by different electrophiles: iodine, anisaldehyde, and chlorodiphenylphosphine to respectively generate the corresponding iodide, alcohol, and phosphine in high yields and benzophenone, benzoyl chloride, and allyl bromide to afford the expected alcohol, ketone, and allylated compound in moderate yields (Table 27.10). The side formation of symmetrical dimers was observed in the course of these reactions, in particular using allyl bromide; such a side reaction (reduced by using the base in excess) could be initiated by one-electron transfer from a cobaltate species to the electrophile. This lithium–cobalt base is not suitable for the functionalization of sensitive substrates such as ethyl benzoate [109, 110].

27.5.6 Using Lithium–Copper Mixed-Metal Bases

Among several amidocuprates, Lipshutz-type cuprate $Li_2(CN)Cu(TMP)Me$ was identified as the best candidate to perform deprotocupration. When substituted benzenes are treated by 2 equiv of this base in THF at 0°C for 3 h, the corresponding arylcuprates form quantitatively and regioselectively, a result evidenced by subsequent quenching with different electrophiles (Table 27.11). The formation of symmetrical dimers is only observed using an oxidative agent (nitrobenzene) [62].

Albeit more powerful in azine series, the base prepared from LiTMP and CuCl in a 2:1 ratio ($LiCu(TMP)_2{\cdot}LiCl$) can abstract protons from activated benzenes such as 1,4-dimethoxybenzene. Using iodine, which is also an oxidative agent, to intercept the arylmetal derivative is not efficient, leading to mixtures including the symmetrical dimer; on the other hand, allyl bromide and aroyl chlorides are more suitable electrophiles, affording the allylated derivative and ketones, respectively (Scheme 27.26) [63, 111].

27.5.7 Using Lithium–, Sodium–, or Magnesium–Zinc Mixed-Metal Bases

Due to the preferred kinetic reactivity of the zinc–nitrogen bonds in comparison with the zinc–carbon bonds, amidozincates are more reactive bases than alkylzincates. When used in THF at room temperature, $(t\text{-Bu})_2Zn(TMP)Li$ is able to deprotonate functionalized substrates such as alkyl benzoates and benzonitrile; the generated arylzincates can be quenched by iodine, benzaldehyde (Table 27.12), and bromine [29, 112–116]. Crystals of the base were isolated and its structure identified by X-ray diffraction as an ion-contacted zincate [117]. In addition, different reaction outcomes being obtained by changing the nature of the alkali metal (sodium vs. lithium), a reactivity as solvent-separated ion pairs was found unlikely [118].

TABLE 27.10 Methoxybenzenes: Deprotometalation Using a Lithium–Cobalt Base and Subsequent Interception with Electrophiles

$$\text{Ar–H} \xrightarrow[\text{(2) Electrophile; (3) Hydrolysis}]{\text{(1) CoBr}_2\text{ (x equiv) + LiTMP (2x equiv), THF, rt, 0.5 h}} \text{Ar–E + Ar–Ar}$$

Entry	Ar–H	x	Electrophile (E)	Yields (%)	
1	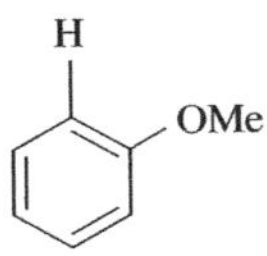	2	I_2 (I)	93[a]	5
2			4-AnisylCHO (CH(OH)-4-anisyl)	84	15
3			Ph_2PCl (PPh_2)	82	15
4			Ph_2CO ($C(OH)Ph_2$)	45	10
5			PhCOCl (COPh)	30	16
6	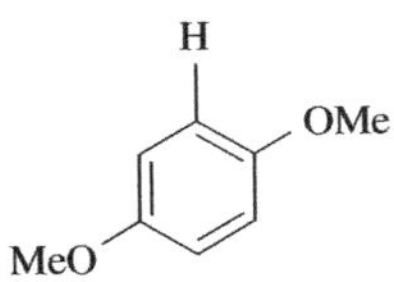	2	I_2 (I)	76	10
7	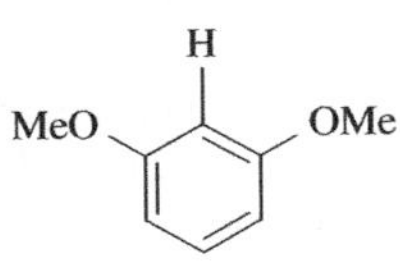	1	I_2 (I)	97	0
8	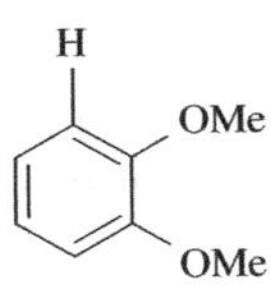	2	I_2 (I)	74	0
9			AllylBr (Allyl)	23	62

[a] Twenty-two percent yield from ethyl benzoate.

TABLE 27.11 Substituted Benzenes: Deprotometalation Using a Lithium–Copper Base and Subsequent Interception with Electrophiles

$$\text{Ar–H} \xrightarrow[\text{(2) Electrophile, conditions}]{\text{(1) Li}_2\text{(CN)Cu(TMP)Me (2 equiv), THF, 0°C, 3h}} \text{Ar–E}$$

Entry	Ar–H	Electrophile, Conditions (E)	Yield (%)
1	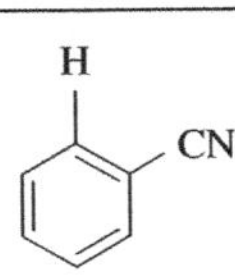	I_2, rt, 16h (I)	91
2	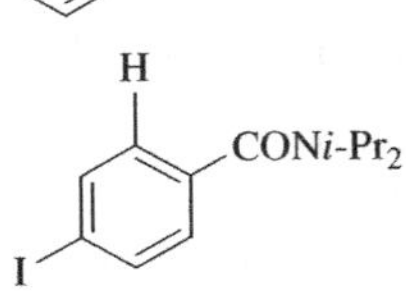	I_2, rt, 16h (I)	100
3	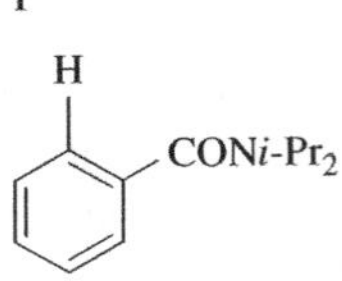	D_2O, rt, 0.5h (D)	100
4		MeI, rt, 2h (Me)	99
5		Me_3SiCl, 50°C, 16h ($SiMe_3$)	99

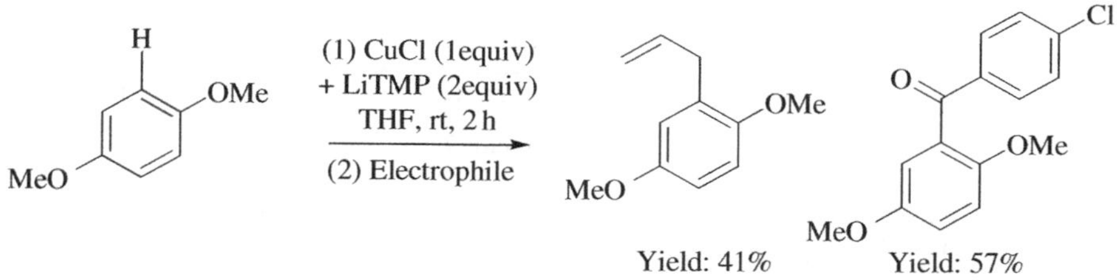

SCHEME 27.26 1,4-Dimethoxybenzene: deprotometalation using a lithium–copper base and subsequent interception with electrophiles.

The deprotometalation of 3-substituted bromobenzenes has been studied using two Li(TMP) ZnR_2 bases in THF. With different substituents (OMe, Cl, F, CN, CON*i*-Pr_2), subsequent iodination after treatment using Li(TMP)Zn(*t*-Bu)$_2$ (2 equiv) showed a regioselective attack of the site between both groups. Hindered CF_3 led to a reaction at the other position next to the bromo group, whereas CO_2Et and CO_2*t*-Bu ester functions did not allow the substrates to be functionalized regioselectively (competitive reaction at the other position next to CO_2R). The *tert*-butyl ligands are essential for the stability of the formed arylzincates; indeed, when Li(TMP)$ZnMe_2$ is employed to deprotonate the same substrates, formation of benzynes cannot be avoided (an example is given in Scheme 27.27) due to a lower activation energy for benzyne formation in the latter case (easier interaction between the dialkylzinc moiety and bromine). The method was used to prepare key intermediates in the synthesis of benzo[*b*]furans [119]. Similar reactions were described from 1,4-disubstituted benzenes for which one of the substituents is a bromo, a chloro, or a triflate group [120, 121].

For the reaction performed in THF at room temperature on *N*,*N*-diisopropylbenzamide using Li(TMP)$ZnEt_2$, a stepwise mechanism where HTMP liberated during the first step is recycled through elimination of ethane was proposed on the basis of theory and experiment [54]. Additional structural elements in favor of such a two-step mechanism were more recently obtained [56].

TABLE 27.12 Substituted Benzenes: Deprotometalation Using a Lithium–Zinc Base and Subsequent Interception with Electrophiles

$$\text{Ar–H} \xrightarrow[\text{(2) Electrophile, conditions; (3) Hydrolysis}]{\text{(1) }(t\text{-Bu})_2\text{Zn(TMP)Li (x equiv), THF, rt, 3 h}} \text{Ar–E}$$

Entry	Ar–H	x	Electrophile, Conditions (E)	Yield (%)
1	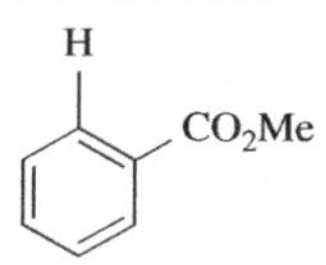	2	I_2, rt, 1 h (I)	73
2	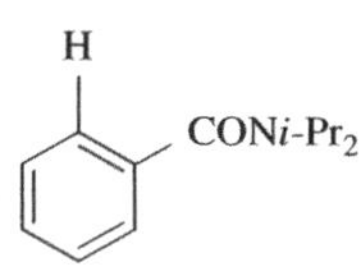	2	I_2, rt, 1 h (I)	95
3	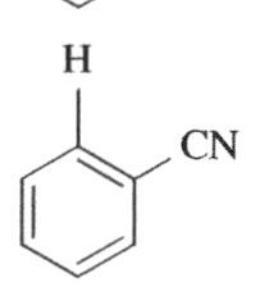	1	PhCHO, rt, 24 h (CH(OH)Ph)	96

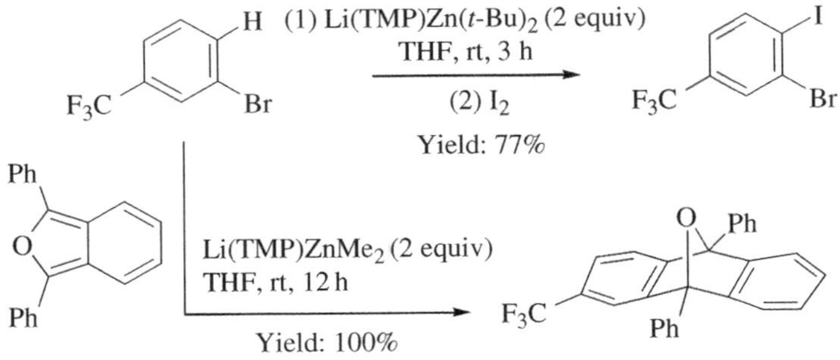

SCHEME 27.27 1-Bromo-3-(trifluoromethyl)benzene: outcomes of the deprotometalation using Li(TMP)Zn(t-Bu)$_2$ and Li(TMP)ZnMe$_2$.

Li(TMP)ZnEt$_2$ can be used in a similar protocol with phenyl *N,N*-diethylcarbamate (Scheme 27.28) [122], whereas the anionic Fries rearrangement is spontaneously observed using classical lithium bases. The method was later applied to the synthesis of IKK2 inhibitors [123]. In the course of the deprotometalation of anisole, it was similarly shown that [(THF)Li(TMP)(t-Bu)Zn(t-Bu)] functions on the whole as an alkyl base [124].

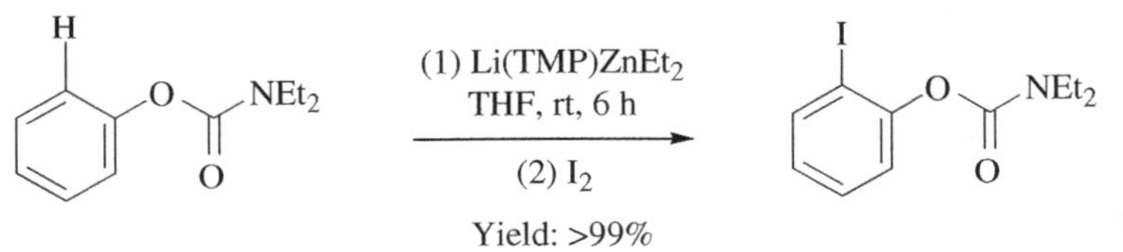

SCHEME 27.28 Phenyl *N,N*-diethylcarbamate: deprotometalation using a lithium–zinc base and subsequent iodolysis.

Changing the alkali metal of M(TMP)Zn(t-Bu)$_2$ from lithium to sodium implies other conditions to perform the deprotometalation [52]. Using [(TMEDA)Na(TMP)(t-Bu)Zn(t-Bu)] (1 equiv) in hexane at room temperature allowed benzene to be deprotometalated, a result evidenced by identification of [(TMEDA)Na(TMP)(Ph)Zn(t-Bu)] crystals, isolated in 51% yield after 0.5 h reaction time [53]. A stepwise mechanism, starting with a kinetic deprotonation involving the TMP ligand, was later shown [55]. The use of [(TMEDA)Na(TMP)(t-Bu)Zn(t-Bu)] in an apolar solvent is not restricted to bare benzene. (Methoxymethyl)benzene is similarly functionalized, at the *ortho* site [125]. The presence of nitrile functions is tolerated, as exemplified by the reaction of 3-tolunitrile and 1-cyanonaphthalene [126]. Whereas toluene [127] and (trifluoromethyl)benzene [128] lead to complex mixtures, *N,N*-dimethyl aniline is mainly deprotonated at its *meta*-position to provide [(TMEDA)Na(TMP)(3-Me$_2$NC$_6$H$_4$)Zn(t-Bu)], and its 3-methyl derivative can be similarly functionalized (Scheme 27.29) [129, 130]; this is a remarkable result, *ortho*-metalation being only observed using classical lithium bases (butyllithium at hexane reflux) [131].

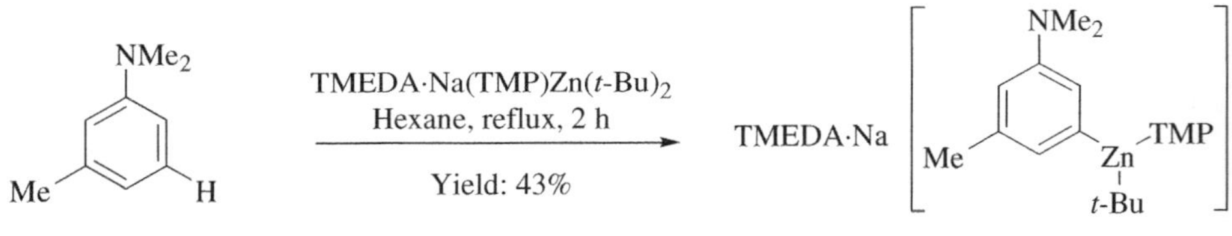

SCHEME 27.29 *N,N*-dimethyl 3-tolylamine: deprotometalation using a sodium–zinc base.

[(TMEDA)Na(TMP)(*t*-Bu)Zn(*t*-Bu)] is also capable of dideprotometalation. From benzene, delaying the addition of TMEDA results in its 1,4-dimetalation, as manifested in crystalline [C_6H_4-1,4-{(TMEDA)Na(TMP)Zn(*t*-Bu)}$_2$], isolated in 39% yield [82]. From naphthalene, either monodeprotonation (at C2) or dideprotonation (at C2 and C6) can be observed, the former using 1 equiv of base at room temperature and the latter using 2 equiv of base at hexane reflux (Scheme 27.30) [132].

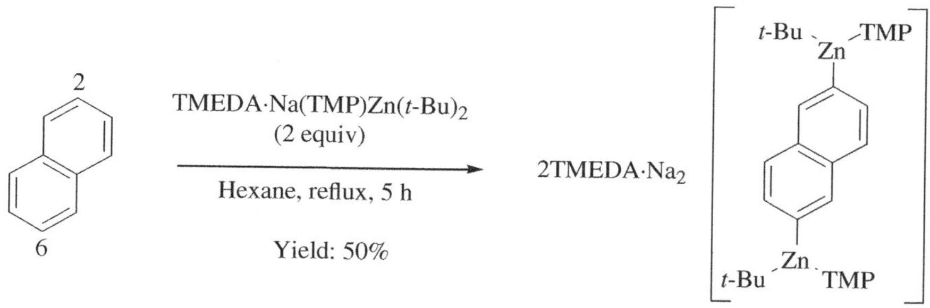

SCHEME 27.30 Naphthalene: dideprotometalation using a sodium–zinc base.

Salt-activated Zn(TMP)$_2$·2MgCl$_2$·2LiCl has been developed in order to tolerate substrates containing sensitive functions. Long reaction times are often required to reach complete deprotometalation in THF, even with activated substrates; they can be shortened upon recourse to microwave irradiation. Various functionalizations are possible by subsequent metal-catalyzed trapping (Table 27.13) [23, 96, 133, 134]. Albeit in a lower chemoselectivity, Zn(N(*i*-Pr)*t*-Bu)$_2$·2MgCl$_2$·2LiCl can be alternatively used [98].

More chemoselective zincations (e.g., compatible with the presence of nitro groups) can be reached employing salt-activated TMPZnCl·LiCl in THF, either under classical conditions or under microwave irradiation depending on the reactivity of the substrate (Table 27.14) [24, 88, 135, 136].

TABLE 27.13 Substituted Benzenes: Deprotometalation Using a Lithium–Zinc Base and Subsequent Interception with Electrophiles

$$\text{Ar–H} \xrightarrow[\text{(2) Electrophile}]{\text{(1) Zn(TMP)}_2\cdot 2\text{MgCl}_2\cdot 2\text{LiCl (0.6 equiv)}\ \ \text{THF, conditions}} \text{Ar–E}$$

Entry	Ar–H	Conditions	Electrophile (E)	Yield (%)
1	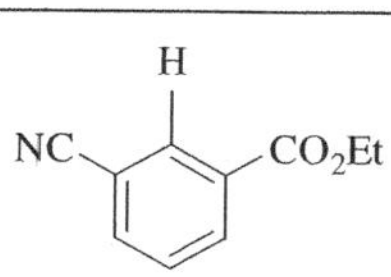	25°C, 30h	PhCOCl[a] (COPh)	83
2	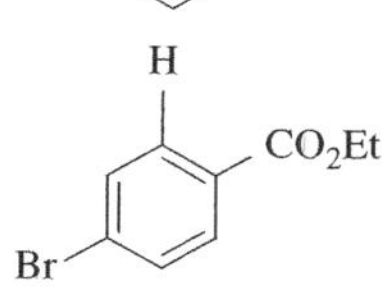	25°C, 10h	PhCOCl[a] (COPh)	83
3	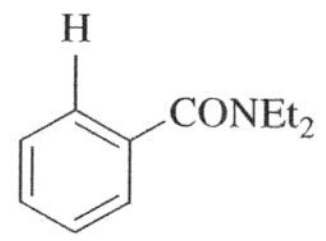	MW, 120°C, 5h	3-$F_3CC_6H_4I$[b] (C_6H_4-3-CF_3)	85

[a] In the presence of CuCN·2LiCl.
[b] Using catalytic Pd(dba)$_2$ and P(2-furyl)$_3$.

TABLE 27.14 Substituted Benzenes: Deprotometalation Using a Lithium–Zinc Base and Subsequent Interception with Electrophiles

$$\text{Ar–H} \xrightarrow[\text{(2) Electrophile}]{\text{(1) TMPZnCl·LiCl (1.1 equiv), THF, conditions}} \text{Ar–E}$$

Entry	Ar–H	Conditions	Electrophile (E)	Yield (%)
1	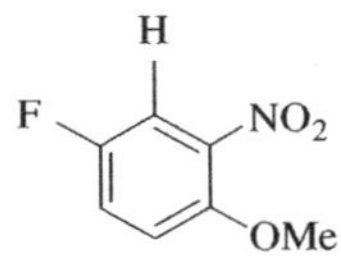	25°C, 6 h	$CH_2C(CO_2Et)CH_2Br$[a] ($CH_2C(CO_2Et){=}CH_2$)	67
2	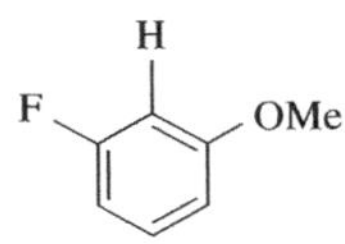	MW, 160°C, 2 h	PhCOCl[b] (COPh)	72
3	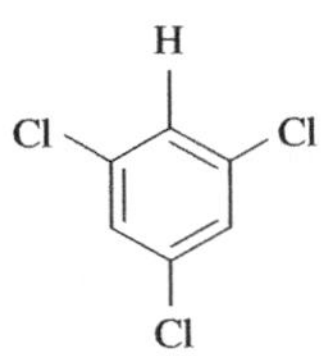	MW, 80°C, 0.75 h	$CH_2C(CO_2Et)CH_2Br$[a] ($CH_2C(CO_2Et){=}CH_2$)	75

[a] After transmetalation with substoichiometric CuCN·2LiCl.
[b] After transmetalation with stoichiometric CuCN·2LiCl.

Pairs of metal compounds that complement each other in deprotometalation reactions have been developed. The TMP-based lithium–zinc mixture, prepared *in situ* from $ZnCl_2$·TMEDA and LiTMP (3 equiv) and supposed to be 1:1 LiTMP·2LiCl(±TMEDA)-$Zn(TMP)_2$ [137, 138], was identified as a suitable reagent to functionalize both sensitive and inactivated aromatic compounds including naphthalenes (Table 27.15) [68, 139, 140]. Deprotolithiation followed by *in situ* trapping of the generated aryllithium by a zinc species is proposed as responsible for the synergy observed using this combination [66]. Similar approaches consisting in using $Mg(TMP)_2$·2LiCl in the presence of $ZnCl_2$ [134, 141] or CuCN·2LiCl [134] and TMPMgCl·LiCl in the presence of $ZnCl_2$·2LiCl [142] have since been developed.

27.5.8 Using Lithium– or Magnesium–Zirconium Mixed-Metal Bases

Zirconation of functionalized activated arenes can be achieved using salt-activated $Zr(TMP)_4$·4$MgCl_2$·6LiCl in THF. The resulting arylmetal species display a high reactivity toward a large range of electrophiles such as carbon dioxide (Table 27.16), aldehydes, epoxides, and chlorotrimethylstannane [143].

27.5.9 Using Lithium–Cadmium Mixed-Metal Bases

The TMP-based lithium–cadmium mixture, prepared *in situ* from $CdCl_2$·TMEDA and LiTMP (3 equiv), allows both sensitive and inactivated aromatic compounds to be functionalized at room temperature (Table 27.17). It is even more powerful than the corresponding lithium–zinc reagent [67–69]. In the absence of TMEDA, such a reagent proved to be 1:1 LiTMP·2LiCl-$Cd(TMP)_2$ [138].

TABLE 27.15 Benzenes and Naphthalenes: Deprotometalation Using a Lithium–Zinc Combination and Subsequent Iodolysis

$$\text{Ar–H} \xrightarrow[\text{(2) } I_2]{\text{(1) } ZnCl_2\cdot\text{TMEDA (x equiv)} + \text{LiTMP (3x equiv), THF, rt, 2 h}} \text{Ar–I}$$

Entry	Ar–H	x	Yield (%)
1	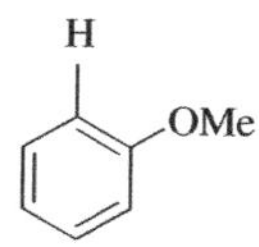	0.5	84
2	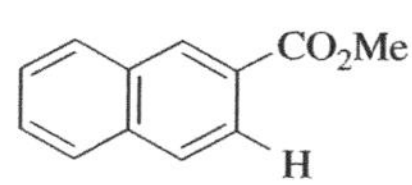	1	51
3	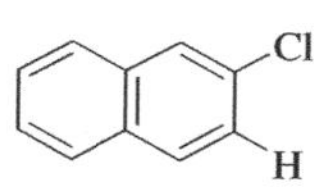	1	78

TABLE 27.16 Substituted Benzenes: Deprotometalation Using a Lithium–Zirconium Base and Subsequent Conversion to Carboxylic Acids

$$\text{Ar–H} \xrightarrow[\text{(2) } CO_2\text{, 1 atm, 50°C, 8 h}]{\text{(1) Zr(TMP)}_4\cdot 4MgCl_2\cdot 6LiCl \text{ (0.25 equiv), THF, conditions}} \text{Ar–}CO_2H$$

Entry	Ar–H	Conditions	Yield (%)
1	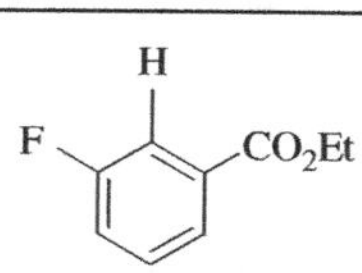	25°C, 0.75 h	89
2	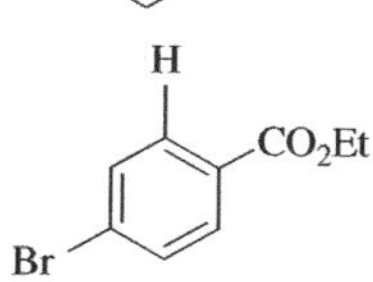	−15°C, 1.5 h	79
3	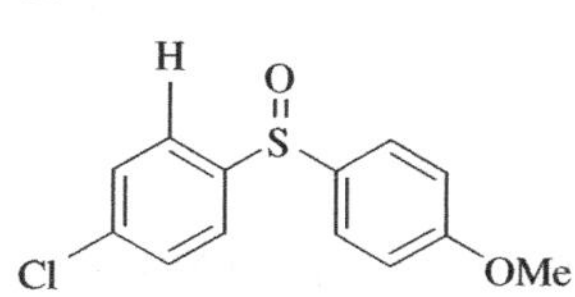	0°C, 0.75 h	86

27.5.10 Using Lithium– or Magnesium–Lanthanum Mixed-Metal Bases

Salt-activated $La(TMP)_3\cdot 3MgCl_2\cdot 5LiCl$ can be used to deprotometalate functionalized activated arenes in THF. As previously noted using $Zr(TMP)_4\cdot 4MgCl_2\cdot 6LiCl$, a stoichiometric amount of base can be employed. Palladium-catalyzed cross-coupling with aryl iodides or direct electrophilic

TABLE 27.17 Substituted Benzenes: Deprotometalation Using a Lithium–Cadmium Combination and Subsequent Iodolysis

$$\text{Ar–H} \xrightarrow[\text{(2) } I_2]{\text{(1) } CdCl_2\cdot\text{TMEDA (0.5 equiv)} + \text{LiTMP (1.5 equiv)}, \text{THF, rt, 2 h}} \text{Ar–I}$$

Entry	Ar–H	Yield (%)	Entry	Ar–H	Yield (%)
1	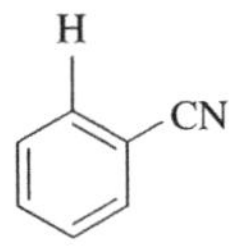	68	3	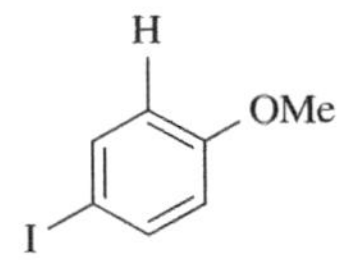	83
2	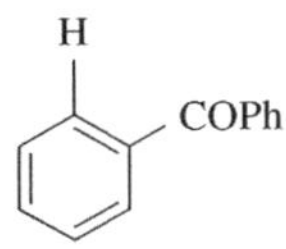	60	4	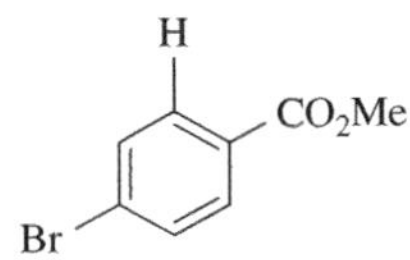	60

TABLE 27.18 Substituted Benzenes: Deprotometalation Using a Lithium–Lanthanum Base and Subsequent Interception with Electrophiles

$$\text{Ar–H} \xrightarrow[\text{(2) Electrophile; (3) Hydrolysis}]{\text{(1) } La(TMP)_3\cdot 3MgCl_2\cdot 5LiCl \text{ (0.35 equiv)}, \text{THF, conditions}} \text{Ar–E}$$

Entry	Ar–H	Conditions	Electrophile (E)	Yield (%)
1	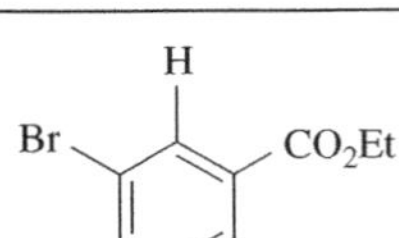	25°C, 2.5 h	$ClCOCO_2Et$ ($COCO_2Et$)	67
2	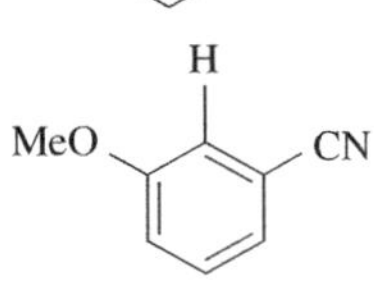	25°C, 1.5 h	*c*-HexCHO (CH(OH)-*c*-Hex)	74
3	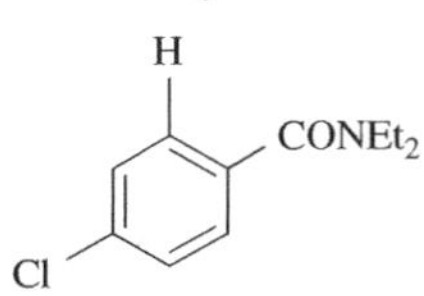	0°C, 2 h	4-NCC_6H_4I[a] (C_6H_4-4-CN)	80

[a] In the presence of $Pd(PPh_3)_4$ (2.5 mol%).

trapping with aldehydes and ketones to afford alcohols (Table 27.18) and with acid chlorides and anhydrides can be used with success.

27.6 CONCLUSION AND PERSPECTIVES

Our goal was to show how arenes can be functionalized using organic bases containing two metals, namely, an alkali metal and a softer nonalkali metal. Emphasis focused on the TMP-containing organic bimetallic combinations, which exhibit a special behavior because of unprecedented reactivity and/or tolerance toward sensitive substituents. Of course, the subject is not exhaustively covered because of the huge work done in this field, but the authors tried to choose examples that illustrate the various approaches developed.

Even if many studies have been devoted to a better understanding of the synergies exhibited in the course of these deprotometalation processes, much remains to be done due to the complexity of the reaction mixtures. A better command of all the parameters at the origin of the reactivities and chemoselectivities observed should lead to the design of more simple and catalytic systems, making the method even more in accordance with sustainable chemistry.

ACKNOWLEDGMENTS

F.M. acknowledges the Institut Universitaire de France.

ABBREVIATIONS

Bn	Benzyl
Boc	*tert*-Butyloxycarbonyl
Bu	*n*-Butyl
DABCO	1,4-Diazabicyclo[2.2.2]octane
DoC	Directed *ortho*-cupration
DoM	Directed *ortho*-metalation
DPPF	1,1′-Bis(diphenylphosphino)ferrocene
equiv	Equivalent
HMPA	Hexamethylphosphoramide
MW	Microwave
NMR	Nuclear magnetic resonance
THF	Tetrahydrofuran
TMEDA	*N,N,N′,N′*-Tetramethylethylenediamine
TMP	2,2,6,6-Tetramethylpiperidino
TS	Transition state

REFERENCES

[1] Gschwend, H. W., Rodriguez, H. R. (1979) *Org. React.*, **26**, 1–360.

[2] Beak, P., Snieckus, V. (1982) *Acc. Chem. Res.*, **15**, 306–312.

[3] Snieckus, V. (1990) *Chem. Rev.*, **90**, 879–933.

[4] Gant, T. G., Meyers, A. I. (1994) *Tetrahedron*, **50**, 2297–2360.

[5] Schlosser, M. (2002) *Organometallics in Synthesis*, 2nd ed. (Ed.: Schlosser, M.), John Wiley & Sons, Inc.: New York, Chapter I.

[6] Mulvey, R. E. (2006) *Organometallics*, **25**, 1060–1075.

[7] Mulvey, R. E., Mongin, F., Uchiyama, M., Kondo, Y. (2007) *Angew. Chem. Int. Ed.*, **46**, 3802–3824.

[8] Mulvey, R. E. (2009) *Acc. Chem. Res.*, **42**, 743–755.

[9] Haag, B., Mosrin, M., Ila, H., Malakhov, V., Knochel, P. (2011) *Angew. Chem. Int. Ed.*, **50**, 9794–9824.

[10] Mongin, F., Uchiyama, M. (2011) *Curr. Org. Chem.*, **15**, 2340–2361.

[11] Mulvey, R. E. (2013) *Dalton Trans.*, **42**, 6676–6693.

[12] Mongin, F., Harrison-Marchand, A. (2013) *Chem. Rev.*, **113**, 7563–7727.

[13] Wittig, G. (1958) *Angew. Chem.*, **70**, 65–71.

[14] Tochtermann, W. (1966) *Angew. Chem. Int. Ed. Engl.*, **5**, 351–371.

[15] Harrison-Marchand, A., Mongin, F. (2013) *Chem. Rev.*, **113**, 7470–7562.

[16] Krasovskiy, A., Krasovskaya, V., Knochel, P. (2006) *Angew. Chem. Int. Ed.*, **45**, 2958–2961.

[17] Unsinn, A., Ford, M. J., Knochel, P. (2013) *Org. Lett.*, **15**, 1128–1131.

[18] García-Álvarez, P., Graham, D. V., Hevia, E., Kennedy, A. R., Klett, J., Mulvey, R. E., O'Hara, C. T., Weatherstone, S. (2008) *Angew. Chem. Int. Ed.*, **47**, 8079–8081.

[19] Clososki, G. C., Rohbogner, C. J., Knochel, P. (2007) *Angew. Chem. Int. Ed.*, **46**, 7681–7684.

[20] Wunderlich, S. H., Knochel, P. (2009) *Angew. Chem. Int. Ed.*, **48**, 1501–1504.

[21] Wunderlich, S. H., Kienle, M., Knochel, P. (2009) *Angew. Chem. Int. Ed.*, **48**, 7256–7260.

[22] Wunderlich, S. H., Knochel, P. (2009) *Angew. Chem. Int. Ed.*, **48**, 9717–9720.

[23] Wunderlich, S. H., Knochel, P. (2007) *Angew. Chem. Int. Ed.*, **46**, 7685–7688.

[24] Mosrin, M., Monzon, G., Bresser, T., Knochel, P. (2009) *Chem. Commun.*, **37**, 5615–5617.

[25] Hevia, E., Mulvey, R. E. (2011) *Angew. Chem. Int. Ed.*, **50**, 6448–6450.

[26] Hevia, E., Gallagher, D. J., Kennedy, A. R., Mulvey, R. E., O'Hara, C. T., Talmard, C. (2004) *Chem. Commun.*, 2422–2423.

[27] Rohbogner, C. J., Wagner, A. J., Clososki, G. C., Knochel, P. (2009) *Org. Synth.*, **86**, 374–384.

[28] Uchiyama, M., Naka, H., Matsumoto, Y., Ohwada, T. (2004) *J. Am. Chem. Soc.*, **126**, 10526–10527.

[29] Kondo, Y., Shilai, M., Uchiyama, M., Sakamoto, T. (1999) *J. Am. Chem. Soc.*, **121**, 3539–3540.

[30] Wittig, G. (1966) *Q. Rev.*, **20**, 191–210.

[31] Ashby, E. C., Chao, L.-C., Laemmle, J. (1974) *J. Org. Chem.*, **39**, 3258–3263.

[32] Ashby, E. C., Laemmle, J. T. (1975) *Chem. Rev.*, **75**, 521–546.

[33] Caubère, P. (1993) *Chem. Rev.*, **93**, 2317–2334.

[34] Wittig, G., Meyer, F. J., Lange, G. (1951) *Justus Liebigs Ann. Chem.*, **571**, 167–201.

[35] Coates, G. E., Heslop, J. A. (1968) *J. Chem. Soc. A*, 514–518.

[36] Thoennes, D., Weiss, E. (1978) *Chem. Ber.*, **111**, 3726–3731.

[37] Greiser, T., Kopf, J., Thoennes, D., Weiss, E. (1981) *Chem. Ber.*, **114**, 209–213.

[38] Schubert, B., Weiss, E. (1984) *Chem. Ber.*, **117**, 366–375.

[39] Andrikopoulos, P. C., Armstrong, D. R., Hevia, E., Kennedy, A. R., Mulvey, R. E., O'Hara, C. T. (2005) *Chem. Commun.*, 1131–1133.

[40] Uchiyama, M., Kameda, M., Mishima, O., Yokoyama, N., Koike, M., Kondo, Y., Sakamoto, T. (1998) *J. Am. Chem. Soc.*, **120**, 4934–4946.

[41] Maclin, K. M., Richey, H. G., Jr. (2002) *J. Org. Chem.*, **67**, 4602–4604.

[42] Fleckenstein, J. E., Koszinowski, K. (2011) *Organometallics*, **30**, 5018–5026.

[43] McCann, L. C., Hunter, H. N., Clyburne, J. A. C., Organ, M. G. (2012) *Angew. Chem. Int. Ed.*, **51**, 7024–7027.

[44] Chabanel, M. (1990) *Pure Appl. Chem.*, **62**, 35–46.

[45] Lambert, C., Schleyer, P. v. R. (1994) *Angew. Chem. Int. Ed. Engl.*, **33**, 1129–1140.

[46] Wheatley, A. E. H. (2004) *New J. Chem.*, **28**, 435–443.

[47] Graham, D. V., Hevia, E., Kennedy, A. R., Mulvey, R. E. (2006) *Organometallics*, **25**, 3297–3300.

[48] Reich, H. J., Borst, J. P., Dykstra, R. R., Green, P. D. (1993) *J. Am. Chem. Soc.*, **115**, 8728–8741.

[49] Torvisco, A., Ruhlandt-Senge, K. (2011) *Organometallics*, **30**, 986–991.

[50] Uchiyama, M. (2002) *Yakugaku Zasshi*, **122**, 29–46.

[51] Negishi, E., Akiyoshi, K., O'Connor, B., Takagi, K., Wu, G. (1989) *J. Am. Chem. Soc.*, **111**, 3089–3091.

[52] Uchiyama, M., Matsumoto, Y., Nobuto, D., Furuyama, T., Yamaguchi, K., Morokuma, K. (2006) *J. Am. Chem. Soc.*, **128**, 8748–8750.

[53] Andrikopoulos, P. C., Armstrong, D. R., Barley, H. R. L., Clegg, W., Dale, S. H., Hevia, E., Honeyman, G. W., Kennedy, A. R., Mulvey, R. E. (2005) *J. Am. Chem. Soc.*, **127**, 6184–6185.

[54] Kondo, Y., Morey, J. V., Morgan, J. C., Naka, H., Nobuto, D., Raithby, P. R., Uchiyama, M., Wheatley, A. E. H. (2007) *J. Am. Chem. Soc.*, **129**, 12734–12738.

[55] Nobuto, D., Uchiyama, M. (2008) *J. Org. Chem.*, **73**, 1117–1120.

[56] Clegg, W., Conway, B., Hevia, E., McCall, M. D., Russo, L., Mulvey, R. E. (2009) *J. Am. Chem. Soc.*, **131**, 2375–2384.

[57] Uchiyama, M., Matsumoto, Y., Usui, S., Hashimoto, Y., Morokuma, K. (2007) *Angew. Chem. Int. Ed.*, **46**, 926–929.

[58] Naka, H., Uchiyama, M., Matsumoto, Y., Wheatley, A. E. H., McPartlin, M., Morey, J. V., Kondo, Y. (2007) *J. Am. Chem. Soc.*, **129**, 1921–1930.

[59] Conway, B., Hevia, E., García-Álvarez, J., Graham, D. V., Kennedy, A. R., Mulvey, R. E. (2007) *Chem. Commun.*, 5241–5243.

[60] Naka, H., Morey, J. V., Haywood, J., Eisler, D. J., McPartlin, M., Garcia, F., Kudo, H., Kondo, Y., Uchiyama, M., Wheatley, A. E. H. (2008) *J. Am. Chem. Soc.*, **130**, 16193–16200.

[61] García-Álvarez, J., Hevia, E., Kennedy, A. R., Klett, J., Mulvey, R. E. (2007) *Chem. Commun.*, 2402–2404.

[62] Usui, S., Hashimoto, Y., Morey, J. V., Wheatley, A. E. H., Uchiyama, M. (2007) *J. Am. Chem. Soc.*, **129**, 15102–15103.

[63] Nguyen, T. T., Marquise, N., Chevallier, F., Mongin, F. (2011) *Chem. Eur. J.*, **17**, 10405–10416.

[64] Komagawa, S., Usui, S., Haywood, J., Harford, P. J., Wheatley, A. E. H., Matsumoto, Y., Hirano, K., Takita, R., Uchiyama, M. (2012) *Angew. Chem. Int. Ed.*, **51**, 12081–12085.

[65] Haywood, J., Morey, J. V., Wheatley, A. E. H., Liu, C.-Y., Yasuike, S., Kurita, J., Uchiyama, M., Raithby, P. R. (2009) *Organometallics*, **28**, 38–41.

[66] L'Helgoual'ch, J. M., Seggio, A., Chevallier, F., Yonehara, M., Jeanneau, E., Uchiyama, M., Mongin, F. (2008) *J. Org. Chem.*, **73**, 177–183.

[67] L'Helgoual'ch, J. M., Bentabed-Ababsa, G., Chevallier, F., Yonehara, M., Uchiyama, M., Derdour, A., Mongin, F. (2008) *Chem. Commun.*, 5375–5377.

[68] Snégaroff, K., L'Helgoual'ch, J. M., Bentabed-Ababsa, G., Nguyen, T. T., Chevallier, F., Yonehara, M., Uchiyama, M., Derdour, A., Mongin, F. (2009) *Chem. Eur. J.*, **15**, 10280–10290.

[69] Snégaroff, K., Komagawa, S., Yonehara, M., Chevallier, F., Gros, P. C., Uchiyama, M., Mongin, F. (2010) *J. Org. Chem.*, **75**, 3117–3120.

[70] Richey, H. G., Jr., King, B. A. (1982) *J. Am. Chem. Soc.*, **104**, 4672–4674.

[71] Screttas, C. G., Micha-Screttas, M. (1985) *J. Organomet. Chem.*, **290**, 1–13.

[72] Farkas, J., Jr., Stoudt, S. J., Hanawalt, E. M., Pajerski, A. D., Richey, H. G., Jr. (2004) *Organometallics*, **23**, 423–427.

[73] Markies, P. R., Nomoto, T., Akkerman, O. S., Bickelhaupt, F., Smeets, W. J. J., Spek, A. L. (1988) *Angew. Chem.*, **100**, 1143–1144.

[74] Markies, P. R., Nomoto, T., Schat, G., Akkerman, O. S., Bickelhaupt, F., Smeets, W. J. J., Spek, A. L. (1991) *Organometallics*, **10**, 3826–3837.

[75] Bickelhaupt, F. (1992) *Acta Chem. Scand.*, **46**, 409–417.

[76] Castaldi, G., Borsotti, G. (1992) Eur. Patent 491326A2 (Chem. Abstr. 1992, 117, 150667).

[77] Awad, H., Mongin, F., Trécourt, F., Queguiner, G., Marsais, F., Blanco, F., Abarca, B., Ballesteros, R. (2004) *Tetrahedron Lett.*, **45**, 6697–6701.

[78] Bellamy, E., Bayh, O., Hoarau, C., Trécourt, F., Queguiner, G., Marsais, F. (2010) *Chem. Commun.*, **46**, 7043–7045.

[79] Andrikopoulos, P. C., Armstrong, D. R., Graham, D. V., Hevia, E., Kennedy, A. R., Mulvey, R. E., O'Hara, C. T., Talmard, C. (2005) *Angew. Chem. Int. Ed.*, **44**, 3459–3462.

[80] Armstrong, D. R., Kennedy, A. R., Mulvey, R. E., Rowlings, R. B. (1999) *Angew. Chem. Int. Ed.*, **38**, 131–133.

[81] Andrews, P. C., Kennedy, A. R., Mulvey, R. E., Raston, C. L., Roberts, B. A., Rowlings, R. B. (2000) *Angew. Chem. Int. Ed.*, **39**, 1960–1962.

[82] Armstrong, D. R., Clegg, W., Dale, S. H., Graham, D. V., Hevia, E., Hogg, L. M., Honeyman, G. W., Kennedy, A. R., Mulvey, R. E. (2007) *Chem. Commun.*, 598–600.

[83] Blair, V. L., Carrella, L. M., Clegg, W., Conway, B., Harrington, R. W., Hogg, L. M., Klett, J., Mulvey, R. E., Rentschler, E., Russo, L. (2008) *Angew. Chem. Int. Ed.*, **47**, 6208–6211.

[84] Lin, W., Baron, O., Knochel, P. (2006) *Org. Lett.*, **8**, 5673–5676.

[85] Stoll, A. H., Knochel, P. (2008) *Org. Lett.*, **10**, 113–116.

[86] Rauhut, C. B., Melzig, L., Knochel, P. (2008) *Org. Lett.*, **10**, 3891–3894.

[87] Melzig, L., Rauhut, C. B., Knochel, P. (2009) *Synthesis*, 1041–1048.

[88] Monzon, G., Knochel, P. (2010) *Synlett*, 304–308.

[89] Melzig, L., Rauhut, C. B., Naredi-Rainer, N., Knochel, P. (2011) *Chem. Eur. J.*, **17**, 5362–5372.

[90] Stathakis, C. I., Bernhardt, S., Quint, V., Knochel, P. (2012) *Angew. Chem. Int. Ed.*, **51**, 9428–9432.

[91] Monzon, G., Tirotta, I., Knochel, P. (2012) *Angew. Chem. Int. Ed.*, **51**, 10624–10627.

[92] Unsinn, A., Rohbogner, C. J., Knochel, P. (2013) *Adv. Synth. Catal.*, **355**, 1553–1560.

[93] Armstrong, D. R., García-Álvarez, P., Kennedy, A. R., Mulvey, R. E., Parkinson, J. A. (2010) *Angew. Chem. Int. Ed.*, **49**, 3185–3188.

[94] Rohbogner, C. J., Clososki, G., Knochel, P. (2008) *Angew. Chem. Int. Ed.*, **47**, 1503–1507.

[95] Rohbogner, C. J., Wirth, S., Knochel, P. (2010) *Org. Lett.*, **12**, 1984–1987.

[96] Wunderlich, S. H., Rohbogner, C. J., Unsinn, A., Knochel, P. (2010) *Org. Process Res. Dev.*, **14**, 339–345.

[97] Frischmuth, A., Unsinn, A., Groll, K., Stadtmueller, H., Knochel, P. (2012) *Chem. Eur. J.*, **18**, 10234–10238.

[98] Rohbogner, C. J., Wunderlich, S. H., Clososki, G. C., Knochel, P. (2009) *Eur. J. Org. Chem.*, 1781–1795.

[99] Noji, T., Fujiwara, H., Okano, K., Tokuyama, H. (2013) *Org. Lett.*, **15**, 1946–1949.

[100] Conway, B., Crosbie, E., Kennedy, A. R., Mulvey, R. E., Robertson, S. D. (2012) *Chem. Commun.*, **48**, 4674–4676.

[101] Crosbie, E., Kennedy, A. R., Mulvey, R. E., Robertson, S. D. (2012) *Dalton Trans.*, **41**, 1832–1839.

[102] Groll, K., Blümke, T. D., Unsinn, A., Haas, D., Knochel, P. (2012) *Angew. Chem. Int. Ed.*, **51**, 11157–11161.

[103] Carrella, L. M., Clegg, W., Graham, D. V., Hogg, L. M., Kennedy, A. R., Klett, J., Mulvey, R. E., Rentschler, E., Russo, L. (2007) *Angew. Chem. Int. Ed.*, **46**, 4662–4666.

[104] Blair, V. L., Clegg, W., Conway, B., Hevia, E., Kennedy, A., Klett, J., Mulvey, R. E., Russo, L. (2008) *Chem. Eur. J.*, **14**, 65–72.

[105] Blair, V. L., Clegg, W., Mulvey, R. E., Russo, L. (2009) *Inorg. Chem.*, **48**, 8863–8870.

[106] Wunderlich, S. H., Bresser, T., Dunst, C., Monzon, G., Knochel, P. (2010) *Synthesis*, 2670–2678.

[107] Alborés, P., Carrella, L. M., Clegg, W., García-Álvarez, P., Kennedy, A. R., Klett, J., Mulvey, R. E., Rentschler, E., Russo, L. (2009) *Angew. Chem. Int. Ed.*, **48**, 3317–3321.

[108] Nagaradja, E., Chevallier, F., Roisnel, T., Jouikov, V., Mongin, F. (2012) *Tetrahedron*, **68**, 3063–3073.

[109] Dayaker, G., Chevallier, F., Gros, P. C., Mongin, F. (2010) *Tetrahedron*, **66**, 8904–8910.

[110] Tilly, D., Snégaroff, K., Dayaker, G., Chevallier, F., Gros, P. C., Mongin, F. (2012) *Tetrahedron*, **68**, 8761–8766.

[111] Nguyen, T. T., Chevallier, F., Jouikov, V., Mongin, F. (2009) *Tetrahedron Lett.*, **50**, 6787–6790.

[112] Szumigala, R. H., Jr., Devine, P. N., Gauthier, D. R., Jr., Volante, R. P. (2004) *J. Org. Chem.*, **69**, 566–569.

[113] Xu, C., Yamada, H., Wakamiya, A., Yamaguchi, S., Tamao, K. (2004) *Macromolecules*, **37**, 8978–8983.

[114] Yamaguchi, S., Tamao, K. (2005) *Chem. Lett.*, **34**, 2–7.

[115] Gauthier, D. R., Jr., Limanto, J., Devine, P. N., Desmond, R. A., Szumigala, R. H., Jr., Foster, B. S., Volante, R. P. (2005) *J. Org. Chem.*, **70**, 5938–5945.

[116] Sanz, R., Castroviejo, M. P., Guilarte, V., Perez, A., Fananas, F. J. (2007) *J. Org. Chem.*, **72**, 5113–5118.

[117] Clegg, W., Dale, S. H., Hevia, E., Honeyman, G. W., Mulvey, R. E. (2006) *Angew. Chem. Int. Ed.*, **45**, 2370–2374.

[118] Clegg, W., Dale, S. H., Harrington, R. W., Hevia, E., Honeyman, G. W., Mulvey, R. E. (2006) *Angew. Chem. Int. Ed.*, **45**, 2374–2377.

[119] Guilarte, V., Castroviejo, M. P., Alvarez, E., Sanz, R. (2011) *Beilstein J. Org. Chem.*, **7**, 1255–1260.

[120] Uchiyama, M., Miyoshi, T., Kajihara, Y., Sakamoto, T., Otani, Y., Ohwada, T., Kondo, Y. (2002) *J. Am. Chem. Soc.*, **124**, 8514–8515.

[121] Uchiyama, M., Kobayashi, Y., Furuyama, T., Nakamura, S., Kajihara, Y., Miyoshi, T., Sakamoto, T., Kondo, Y., Morokuma, K. (2008) *J. Am. Chem. Soc.*, **130**, 472–480.

[122] García, F., McPartlin, M., Morey, J. V., Nobuto, D., Kondo, Y., Naka, H., Uchiyama, M., Wheatley, A. E. H. (2008) *Eur. J. Org. Chem.*, **4**, 644–647.

[123] Lin, X., Busch-Petersen, J., Deng, J., Edwards, C., Zhang, Z., Kerns, J. K. (2008) *Synlett*, 3216–3220.

[124] Clegg, W., Dale, S. H., Drummond, A. M., Hevia, E., Honeyman, G. W., Mulvey, R. E. (2006) *J. Am. Chem. Soc.*, **128**, 7434–7435.

[125] Balloch, L., Kennedy, A. R., Klett, J., Mulvey, R. E., O'Hara, C. T. (2010) *Chem. Commun.*, **46**, 2319–2321.

[126] Clegg, W., Dale, S. H., Hevia, E., Hogg, L. M., Honeyman, G. W., Mulvey, R. E., O'Hara, C. T., Russo, L. (2008) *Angew. Chem. Int. Ed.*, **47**, 731–734.

[127] Armstrong, D. R., Garcia-Alvarez, J., Graham, D. V., Honeyman, G. W., Hevia, E., Kennedy, A. R., Mulvey, R. E. (2009) *Chem. Eur. J.*, **15**, 3800–3807.

[128] Armstrong, D. R., Blair, V. L., Clegg, W., Dale, S. H., Garcia-Alvarez, J., Honeyman, G. W., Hevia, E., Mulvey, R. E., Russo, L. (2010) *J. Am. Chem. Soc.*, **132**, 9480–9487.

[129] Armstrong, D. R., Clegg, W., Dale, S. H., Hevia, E., Hogg, L. M., Honeyman, G. W., Mulvey, R. E. (2006) *Angew. Chem. Int. Ed.*, **45**, 3775–3778.

[130] Armstrong, D. R., Balloch, L., Hevia, E., Kennedy, A. R., Mulvey, R. E., O'Hara, C. T., Robertson, S. D. (2011) *Beilstein J. Org. Chem.*, **7**, 1234–1248.

[131] Lepley, A. R., Khan, W. A., Giumanini, A. B., Giumanini, A. G. (1966) *J. Org. Chem.*, **31**, 2047–2051.

[132] Clegg, W., Dale, S. H., Hevia, E., Hogg, L. M., Honeyman, G. W., Mulvey, R. E., O'Hara, C. T. (2006) *Angew. Chem. Int. Ed.*, **45**, 6548–6550.

[133] Wunderlich, S., Knochel, P. (2008) *Org. Lett.*, **10**, 4705–4707.

[134] Dong, Z., Clososki, G. C., Wunderlich, S. H., Unsinn, A., Li, J., Knochel, P. (2009) *Chem. Eur. J.*, **15**, 457–468.

[135] Mosrin, M., Knochel, P. (2009) *Org. Lett.*, **11**, 1837–1840.

[136] Bresser, T., Mosrin, M., Monzon, G., Knochel, P. (2010) *J. Org. Chem.*, **75**, 4686–4695.

[137] García-Álvarez, P., Mulvey, R. E., Parkinson, J. A. (2011) *Angew. Chem. Int. Ed.*, **50**, 9668–9671.

[138] Armstrong, D. R., Kennedy, A. R., Mulvey, R. E., Parkinson, J. A., Robertson, S. D. (2012) *Chem. Sci.*, **3**, 2700–2707.

[139] Snégaroff, K., Komagawa, S., Chevallier, F., Gros, P. C., Golhen, S., Roisnel, T., Uchiyama, M., Mongin, F. (2010) *Chem. Eur. J.*, **16**, 8191–8201.

[140] Kadiyala, R. R., Tilly, D., Nagaradja, E., Roisnel, T., Matulis, V. E., Ivashkevich, O. A., Halauko, Y. S., Chevallier, F., Gros, P. C., Mongin, F. (2013) *Chem. Eur. J.*, **19**, 7944–7960.

[141] Dong, Z.-B., Zhu, W.-H., Zhang, Z.-G., Li, M.-Z. (2010) *J. Organomet. Chem.*, **695**, 775–780.

[142] Unsinn, A., Wunderlich, S. H., Knochel, P. (2013) *Adv. Synth. Catal.*, **355**, 989–995.

[143] Jeganmohan, M., Knochel, P. (2010) *Angew. Chem. Int. Ed.*, **49**, 8520–8524.

28

THE HALOGEN/METAL INTERCONVERSION AND RELATED PROCESSES (M = Li, Mg)

ARMEN PANOSSIAN AND FRÉDÉRIC R. LEROUX

Laboratoire de Chimie Moléculaire, CNRS and University of Strasbourg, UMR CNRS 7509, ECPM, Strasbourg Cedex 2, France

28.1 INTRODUCTION

The bromine– or iodine–lithium permutation has been discovered almost simultaneously by G. Wittig [1] and H. Gilman [2]. It is one of the most important tools for preparing organolithium intermediates [3–6]. The utility of the halogen/metal interconversion in general made it a routine technique for many synthetic chemists. In the literature, more examples can be found of the halogen/lithium interconversion than of the halogen/magnesium one. Yet, the last decade has seen a burst of activity in the field of halogen/magnesium exchange, thanks, mainly, to the work of Knochel and coworkers. Recent years witnessed significant improvements like the development of new exchange reagents (e.g., mono- and bimetallic magnesium reagents), new strategies for the preparation of otherwise difficultly accessible haloarenes, the regioselective and/or chemoselective halogen/metal interconversion on polyhalogenated substrates, enantioselective halogen/metal interconversion allowing the preparation of enantioenriched or enantiopure compounds as well as the functionalization in presence of functional groups. In addition, related processes like the sulfoxide/metal exchange give access to enantiopure compounds in a modular way.

In Chapter 28, the most important reagents for halogen/metal interconversions will be presented and the mechanisms described. We will show which kinds of substituents are tolerated during the process with respect to the exchange reagents. Selectivity issues will be discussed and illustrated on representative examples. The related sulfoxide/metal and phosphorus/metal interconversions will be presented. Finally, the recent improvements in the aryl—aryl coupling based on the preparation of aryllithium or arylmagnesium intermediates *via* halogen/metal interconversions will be shown.

Arene Chemistry: Reaction Mechanisms and Methods for Aromatic Compounds,
First Edition. Edited by Jacques Mortier.

28.2 GENERALITIES

28.2.1 Monometallic Organolithium Reagents

In general, aryl bromides or iodides are treated with an alkyllithium reagent like butyllithium at low temperatures (typically −78°C). The reaction is very fast and is complete in a few seconds if conducted in tetrahydrofuran. For bromobenzenes, it is slower in diethyl ether, sometimes even at −50°C depending on the substituents of the aromatic ring, but rapid at 0°C. The reaction can also be conducted in toluene or other hydrocarbon solvents, but at higher temperatures and even then, the exchange may be slow [7–9]. Aryl iodides react instantaneously in any case. In contrast, aryl fluorides and chlorides are practically inert toward organolithium reagents [10, 11]:

$$\text{ArI} > \text{ArBr} \gg \text{ArCl} > \text{ArF}$$

As outlined in the following text, the halogen/metal exchange is an equilibrium, and thus the position of the equilibrium depends on the basicity of the exchange reagent. *tert*-butyllithium outperforms *sec*-butyllithium, which is superior to butyllithium. Methyllithium is the least reactive alkyllithium reagent but can still be more effective than phenyllithium in some cases:

$$t\text{BuLi} > s\text{BuLi} > \text{BuLi} > \text{MeLi} \approx \text{PhLi}$$

Hetero-substituents such as dialkylamino or bis(trialkylsilyl)amino, alkoxy and 2-tetrahydropyranyloxy, lithiooxy or lithiooxycarbonyl, fluoro or trifluoromethyl, and chloro are well tolerated. Cyano or nitro and bromo or iodo substituents are tolerated when the solvent and temperature are carefully chosen [3].

28.2.2 Monometallic Organomagnesium Reagents

The iodine/magnesium exchange, employing, for example, *i*PrMgCl or *i*PrMgBr, can also be realized in THF at temperatures below 0°C although it is slower than the exchange with lithium. In the presence of electron-donating groups, however, higher temperatures (25°C) are required. The reactivity of the carbon–magnesium bond is also very temperature dependent, and only reactive electrophiles like aldehydes react rapidly below 0°C [5]. The halogen/metal exchange using nonactivated Grignard reagents has been reviewed by Knochel and coworkers [12, 13].

28.2.3 Bimetallic Organolithium/Magnesium Reagents

The synthetic usefulness of lithium organomagnesates (R_3MgLi) for halogen/metal exchange has been demonstrated by Oshima and coworkers [14, 15]. The lithium trialkylmagnesates are prepared from 2 equivalents of organolithium reagent with 1 equivalent of alkylmagnesium halide in THF at 0°C. Two of the three alkyl groups are exchanged, and the halogen/magnesium exchange proceeds more readily and is less sensitive to electron density of the aromatic ring when compared to *i*PrMgBr, for example. Since then, the use of bimetallic reagents has been intensively studied and recently reviewed [16].

In order to promote faster exchanges, especially bromine/magnesium ones, as well as reactions at lower temperatures, Knochel and coworkers showed that LiCl activates efficiently *i*PrMgCl [17]. *i*PrMgCl·LiCl ("turbo-"Grignard reagents) is prepared by adding *i*PrCl to a mixture of Mg turnings and LiCl in THF and reacts efficiently with a wide range of aromatic and heteroaromatic bromides even at −78°C.

28.3 MECHANISM OF THE HALOGEN/METAL INTERCONVERSION

28.3.1 Reactivity

An important feature of the halogen/metal interconversion is that it is an equilibrium-controlled process, which is therefore shifted toward the most stable organolithium or organomagnesium reagent or, to put it in other words, toward the weakest base [18–20].

As a first consequence, the reaction of a halogenoarene and an alkyllithium or an alkyl Grignard reagent leads to the formation of the arylmetal species almost irreversibly. In the case of lithium, the reaction can become practically irreversible when 2 equivalents of *tert*-butyllithium are used [21–23], since, formally, the first equivalent produces *tert*-butyl halide, which reacts with the remaining equivalent of *t*BuLi to eliminate LiX and release volatile isobutylene and isobutane, constituting the driving force of the reaction. The use of 2 equivalents of *t*BuLi has, however, recently been questioned [24]. On the other hand, the other typical reagent for Li/X exchange, butyllithium, leads to less clean reaction mixtures, where side products of elimination or nucleophilic substitution (Wurtz coupling) are sometimes formed from the released 1-halobutane and the aryllithium species (Scheme 28.1), particularly in THF solution [25].

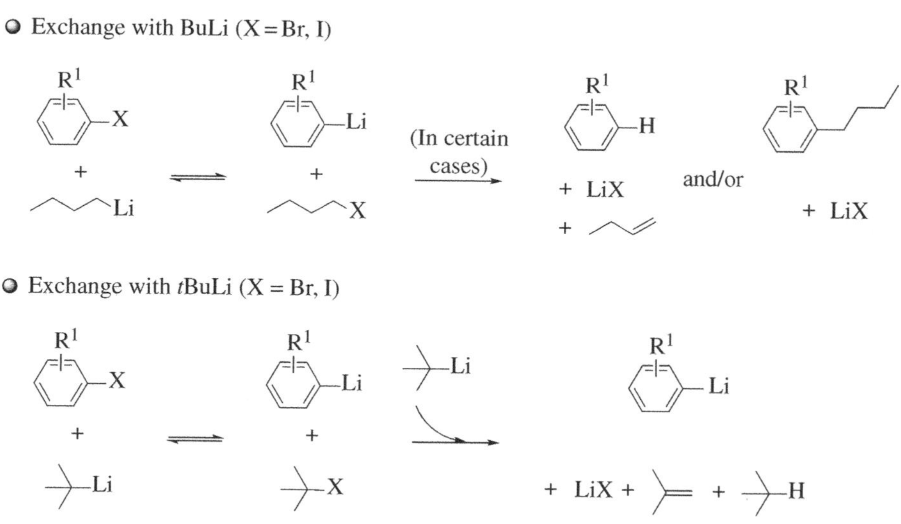

SCHEME 28.1 Halogen/lithium exchange using BuLi and *t*BuLi.

A second consequence concerns the exchange reaction between a haloarene and an arylmetal—and, as a corollary, the evolution of a halogen-substituted arylmetal over time. In those cases, the equilibrium is shifted toward the arylmetal bearing the metal closest to as many –I (inductively electron-withdrawing) and +M (mesomerically electron-donating) substituents as possible, the –I effect being the most influential [26].

Interestingly, the thermodynamically favored product of such a reaction is also the kinetically favored one [27–30], thanks to the same kind of stabilization by electron-donating or electron-withdrawing substituents of the aromatic ring in the "ate complex" transition state or intermediate (*vide infra*).

Another important feature of the halogen/metal interconversion is that it proceeds almost exclusively on reactants RX where X is iodine or bromine. In the case of haloarenes, a few examples of chlorine/metal exchange have been reported in very particular cases—strong electron depletion in the arene [31–33] or buttressing effect [34] (*vide infra*)—but fluorides are unreactive. Iodides are the most reactive toward exchange and are several orders of magnitude faster than bromides. When an alkyllithium is used as exchange reagent, the reaction is extremely fast in polar coordinating solvents (THF or diethyl ether), even at temperatures below −78°C. This allows chemoselective functionalization of haloarenes bearing even carbonyl, carboxyl, amide, etc. groups in some cases. This chemoselectivity can even hold true for milder exchange reagents such as standard or "turbo-" Grignard reagents. Noteworthily, the presence of lithium halide salts can have dramatic effects on the rate of exchange; for instance, in the interconversion with organomagnesium species, added LiCl leads to spectacular improvements [17] as it appeared in the development of turbo-Grignard reagents, while LiBr may be detrimental to the speed of the halogen/lithium exchange [35].

28.3.2 Mechanism

The mechanism of the halogen/metal exchange has been a subject of debate over several decades. Several mechanistic proposals were discussed and reviewed [36] (Scheme 28.2): (a) a concerted four-center transition state; (b) a single electron transfer (S.E.T.) pathway; (c) an S_N2 mechanism, with a concerted transition state; and (d) an S_N2-like pathway going through an "ate complex" intermediate. In the end, thanks to—conveniently summarized [37–39]—several elegant synthetic,

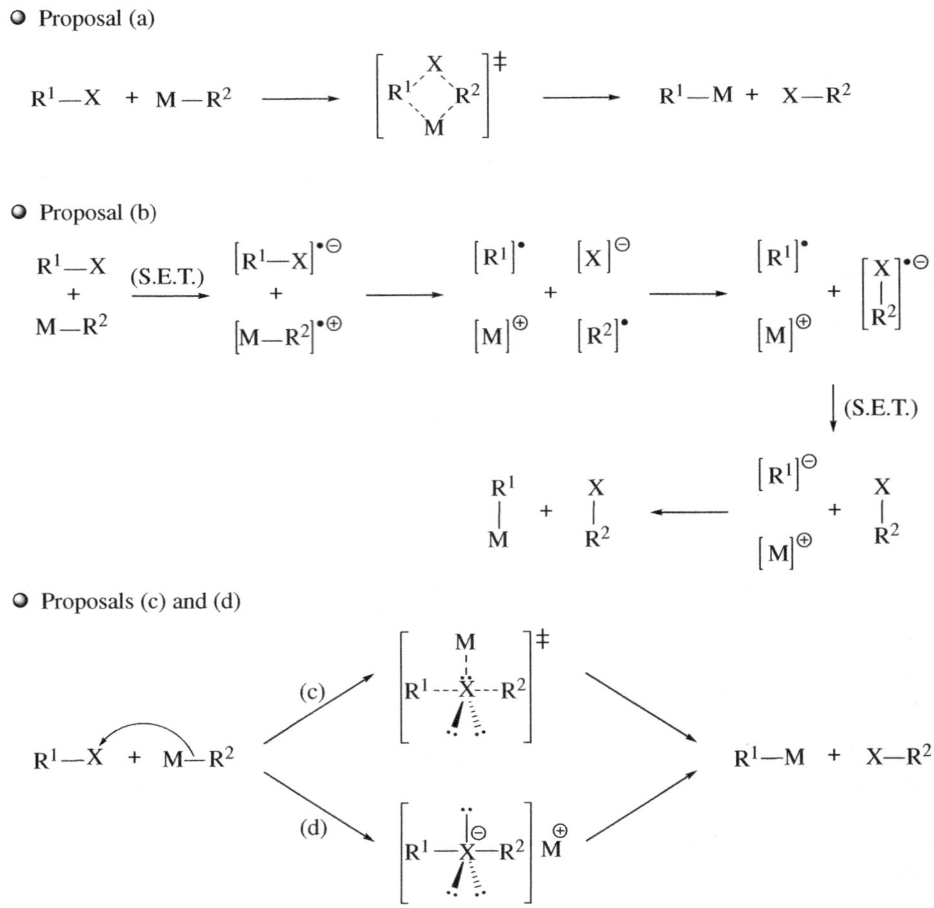

SCHEME 28.2 Mechanistic proposals for the halogen/metal exchange.

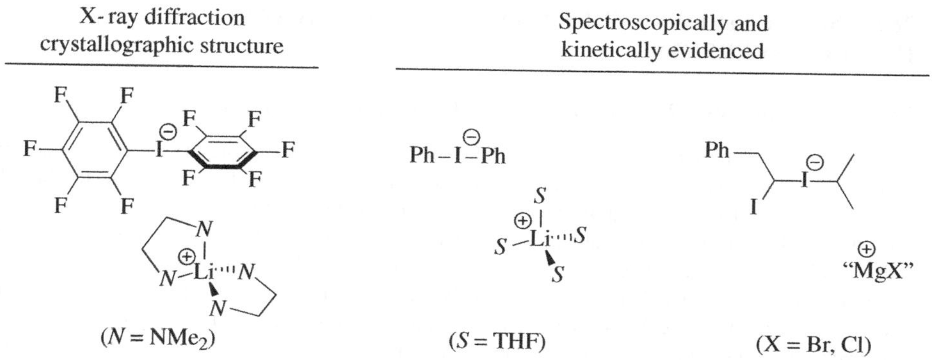

FIGURE 28.1 Observed halogen ate complexes.

kinetic, spectroscopic, crystallographic, and theoretical studies, it was showed, on the one hand, that for aryl halides proposals (a) and (b) could be ruled out and, on the other hand, that proposals (c) and (d) were actually two sides of the same coin, with ate complex intermediates being observable and even isolable [40] in the case of sufficiently but not exceedingly electron-poor carbon groups so as to stabilize the negative charge [41–43], and/or by addition of a ligand to sequester the lithium or magnesium cation (Fig. 28.1) [40, 44–46].

28.4 HALOGEN MIGRATION ON AROMATIC COMPOUNDS

The basicity-driven heavy-halogen migration is an important tool, mainly in heteroaromatic chemistry, for regiodivergent functionalization [34]. The first example in the aromatic series was the LiTMP promoted migration of bromine in 2-bromobenzotrifluoride (Scheme 28.3) [47]. After a first, rapid deprotonation at −75°C in THF at the position adjacent to bromine, the intermediate 2-bromo-3-(trifluoromethyl)phenyllithium converts immediately into the thermodynamically more stable 2-bromo-6-(trifluoromethyl)phenyllithium, which can be trapped with carbon dioxide as electrophile. It has been shown that the mechanism of this halogen migrations or "halogen dance" reactions implies a dibrominated species (2,3-dibromobenzotrifluoride). Thus, initiated by a metalation adjacent to a halogen atom, the basicity gradient-driven halogen migration implies a halogen/metal exchange as the final step. Iodine migrates with particular ease. The migration requires in general the use of LiTMP, whereas all reactions triggered by a metalation adjacent to a fluorine atom can be realized with LDA [34].

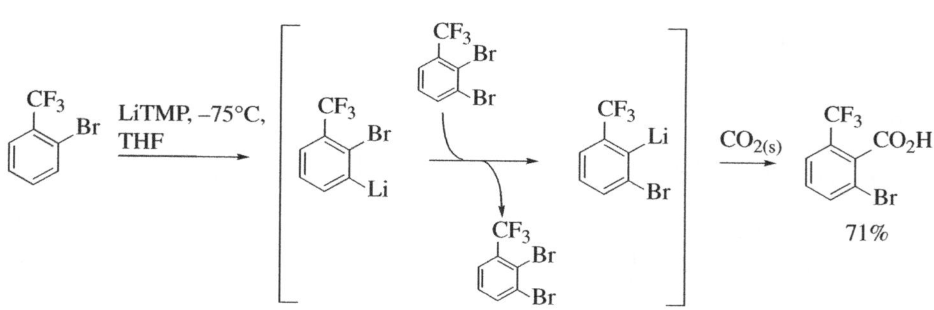

SCHEME 28.3 Halogen migration on 2-bromobenzotrifluoride.

28.5 SELECTIVE SYNTHESIS VIA HALOGEN/METAL INTERCONVERSION

28.5.1 Chemo and Regioselectivity of Halogen/Metal Interconversions

Selectivity aspects play a role when more than one halogen atom are present on an aromatic ring system. They can be sequentially replaced by a metal and then trapped with a suitable electrophile. The halogen/metal exchange is an equilibrium process, which favors the most stable, less basic organometallic intermediate. Thus, if potentially different organometallic intermediates are possible, the exchange will take place first at the intrinsically most acidic position and ending with the least acidic one.

If one has to distinguish between iodine and bromine, iodine is displaced faster than bromine (by several powers of 10) and chlorine is exchange resistant [34]. For example, bromoiodobenzenes afford selectively the corresponding bromophenyllithium reagents [14]. 2-Bromo-2′,6-diiodo-6′-methoxy-1,1′-biphenyl undergoes chemoselective iodine/lithium exchange (Fig. 28.2a) [48]. Similarly, 2-bromo-3,4,5-trifluoro-1-iodobenzene cleanly reacts with butyllithium by kinetically favored iodine/lithium exchange (Fig. 28.2a), whereas 1,2-dibromo-3,4,5-trifluorobenzene undergoes bromine/lithium exchange at the 2-position (flanked by two other halogens), which is energetically favored (Fig. 28.2b) [49].

If one heteroatom is adjacent to one of the halogens to be exchanged, it is always the one next to the electronegative substituent that will be exchanged first. For example, in 2,4-dibromoanisole, first the bromine atom next to the methoxy group is exchanged with butyllithium [50]. 5,7-Dibromo-2,3-dihydro-1-benzofuran exclusively undergoes the first bromine/lithium exchange at the 7-position and next at the 5-position [51]. Similarly, 2,4-dibromo-*N*-pivaloylaniline [52] and 2,4-dibromo-1-nitrobenzene [53] react with a perfect regioselectivity at the bromine atom adjacent to the nitrogen atom (Fig. 28.2b).

The symmetrical 2,2′,6,6′-tetrabromobiphenyl [54] can undergo, according to the reaction conditions, a selective either single or double halogen/metal exchange, and subsequently a variety of electrophiles, phosphinyl groups included, can be readily introduced (Scheme 28.4) [48, 55].

Regioselectivity issues are less obvious, when the halogen atoms to be exchanged are not adjacent to a heteroelement. Systematic studies by Leroux et al. revealed reactivity differences for the bromine/lithium exchange of *meta*-dibrominated substrates. The activating effect of an additional halogen substituent was demonstrated in an intramolecular competition experiment. In the presence of butyllithium, 1,3-dibromo-5-(((3-bromobenzyl)oxy)methyl)benzene exclusively underwent bromine/metal permutation at the higher substituted ring (Scheme 28.5) [55]. Even in conjugated systems like biphenyls, the reaction occurs exclusively at the doubly halogenated ring as shown in the case of 2,2′,6-tribromobiphenyl [55].

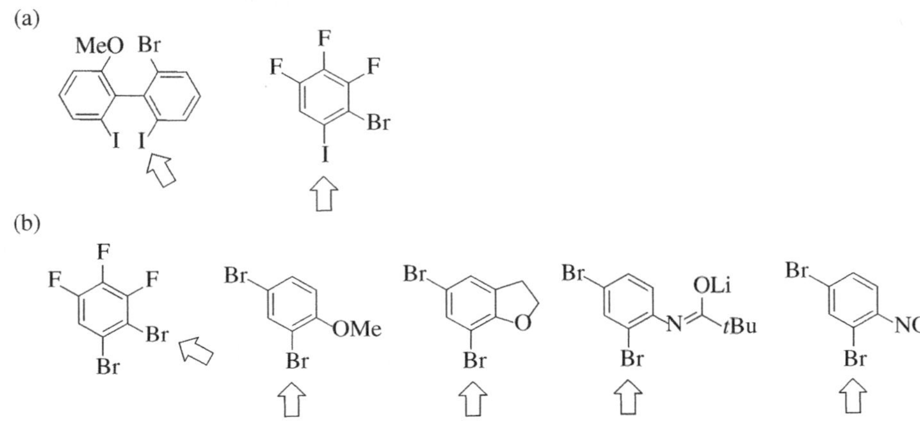

FIGURE 28.2 (a) Chemo- and (b) regioselectivity of the halogen/metal exchange.

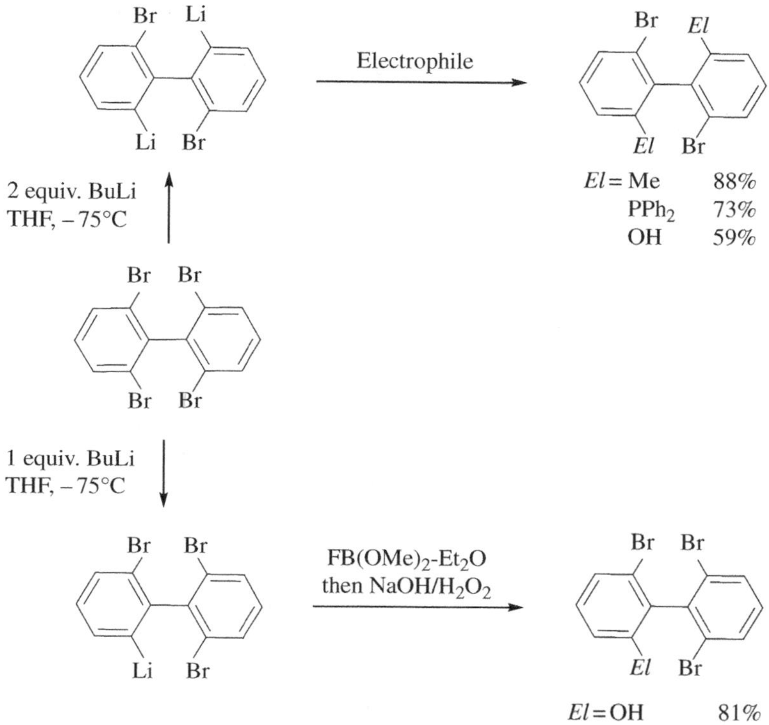

SCHEME 28.4 Regioselective bromine/lithium interconversions on equivalent positions.

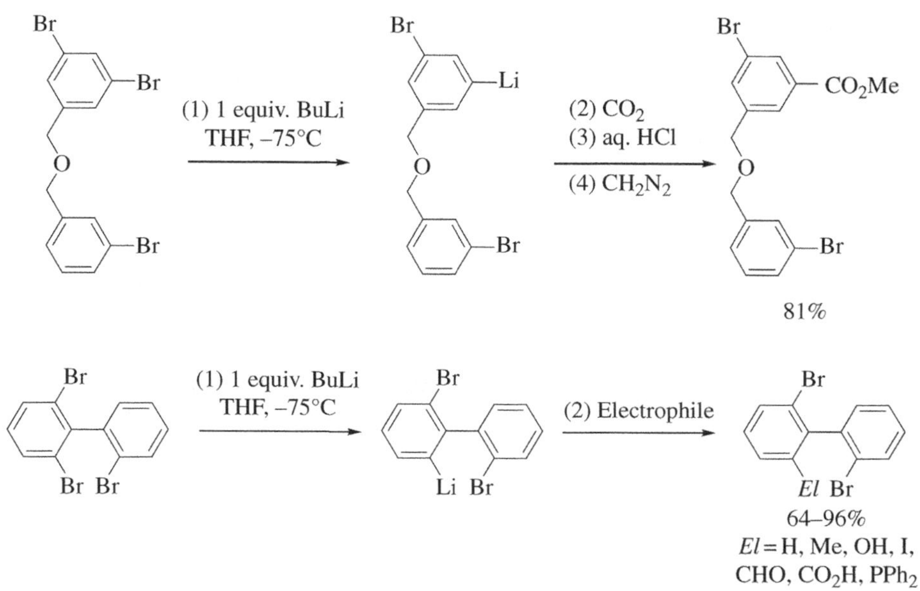

SCHEME 28.5 Regioselective bromine/lithium interconversions on seemingly equivalent positions.

These examples underline that the outcome of monolithiation in polybrominated starting materials are under thermodynamic control. When the chemical environment of the different bromine atoms is not identical, always the most stable organolithium intermediate is formed first. The relative basicities of aryllithiums bearing methoxy, chlorine, fluorine, trifluoromethyl, and trifluoromethoxy substituents at the *ortho*, *meta*, and *para* positions have been studied by Schlosser et al., and $\Delta\Delta G$ increments have been calculated from the equilibrium constants [26]. These thermodynamic values allow to explain these regioselectivity issues.

Schlosser and coworkers showed that Bu_3MgLi can be employed for a selective monodebromination of a tribromotrifluorobenzene. Only the bromine adjacent to fluorine was removed [49]. The magnesate could also be used for selective bromine/magnesium exchange on carboxylic acids (Scheme 28.6) [56].

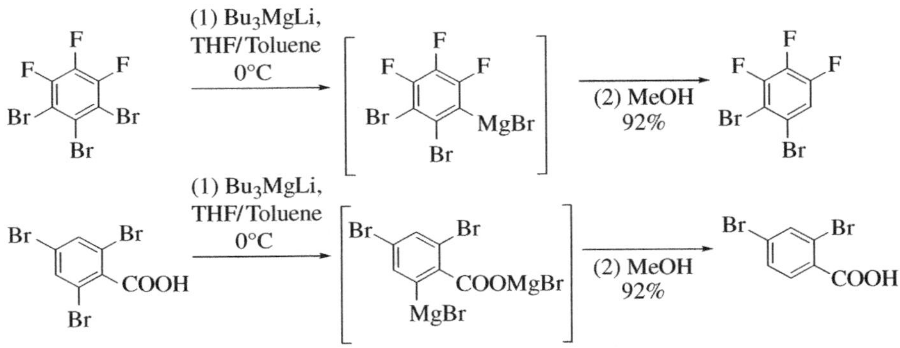

SCHEME 28.6 Selective single halogen/metal exchange with magnesates.

It has been shown that trialkylsilyl groups exert a strong influence on the exchange rate of halogen/metal permutations. In contrast to what has been expected, steric pressure overrides steric hindrance, and the departure of a voluminous heavy-halogen atom close to a bulky silyl substituent relieves strain. Thus, (2,4,6-tribromophenyl)trimethylsilane affords mainly the 2-lithio species when treated with phenyllithium, and (2,6-dichlorophenyl)trimethylsilane undergoes one of the rare chlorine/lithium exchanges in the presence of *sec*-butyllithium along with remote metalation at the 5-position (Scheme 28.7) [34].

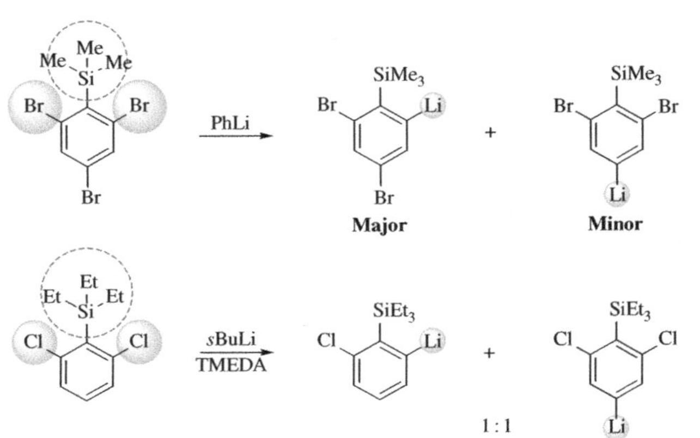

SCHEME 28.7 Strain relief-driven regioselective halogen/metal exchange.

28.5.2 Stereoselectivity of Halogen/Metal Interconversions

Kagan and Alexakis have reported independently the first enantioselective versions of the halogen/metal exchange reaction on arenes. Whereas Kagan succeeded in the monolithiation of 1,3-dibromo-5-methyl-2-(1′-naphthyl)benzene with BuLi in the presence of stoichiometric amounts of (-)-sparteine in 66% yield and 26% ee [57], Alexakis reported on the desymmetrization of 2,2′,6,6′-tetrabromobiphenyl and analogues in presence of stoichiometric amounts of (−)-sparteine or diamine-based ligands affording ee's up to 66% [58]. More recently, this approach has been optimized employing now catalytic amounts of the ligands [59]. Numerous diamines or diether derivatives have been evaluated, and a mechanism for the enantioselective bromine/lithium exchange has been proposed. Best results have been obtained with the Tomioka ligand. In this manner, axially chiral compounds in high yield and high enantioselectivity become accessible (Scheme 28.8).

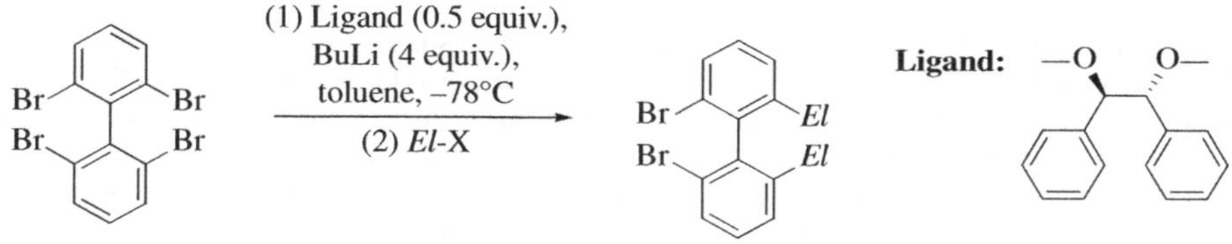

El-X: DMF; *El* = CHO: 89%, 80% ee.
El-X: CO_2; *El* = COOH: 85%, 80% ee.
El-X: MeI; *El* = Me: 85%, 81% ee.
El-X: I_2; *El* = I: 82%, 80% ee.
El-X: TMSCl; *El* = TMS: 77%, 80% ee.

SCHEME 28.8 Enantioselective bromine/lithium exchange-based desymmetrization of 2,2′,6,6′-tetrabromobiphenyl.

Brückner and coworkers reported on the asymmetric halogen/metal exchange to desymmetrize prochiral bis(bromoaryl)alcohols using $i\text{Pr}_2\text{Mg}$ in the presence of stoichiometric amounts of enantiopure lithium alkoxides or phenoxides. The desymmetrized arylmagnesium compounds were quenched with electrophiles. The best ee/yield combination was (52% ee, 58% yield) (Scheme 28.9) [60].

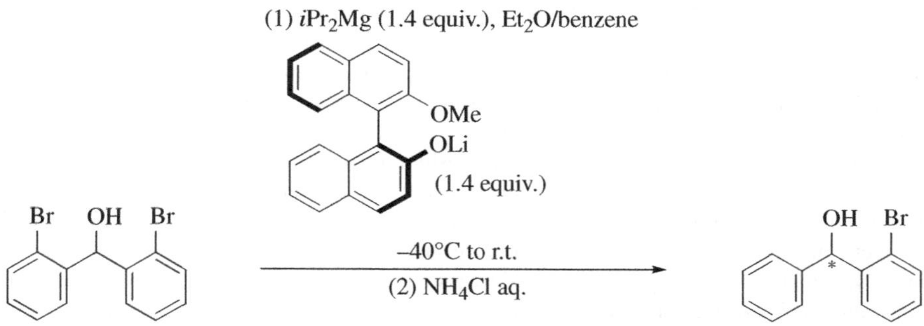

58%, 52% ee (*R*)

SCHEME 28.9 Enantioselective bromine/lithium exchange-based desymmetrization of bis(2-bromoaryl) carbinols.

28.6 THE SULFOXIDE/METAL AND PHOSPHORUS/METAL INTERCONVERSIONS

While the interconversion of halogens is the most used type of exchange with lithium or magnesium, several other elements can be interchanged, such as most of the elements belonging both to columns 14–16 and rows 3–6 (i.e., around the metalloid diagonal) [41]. Among those, we will focus on the sulfoxide/metal and the phosphorus/metal interconversions, due to the synthetical use of the former and the potential involvement of the latter along the preparation of the vast family of phosphorus ligands for catalysis.

28.6.1 The Sulfoxide/Metal Interconversion

Although the sulfoxide/metal interchange (Scheme 28.10) had been observed previously as a side reaction [61], Mislow was the first to actually study it, and he believed it to involve a sulfine intermediate [62]. Efforts to better understand the mechanism of the reaction were conducted by Durst, Lockard, and Furukawa [63–65]. Surprisingly, the powerful synthetic utility of this reaction seems to have eclipsed any further investigation on the mechanism of the reaction. The studies mentioned previously nevertheless showed that (i) the C—S bond that is cleaved is the one producing the most stable carbanion, and (ii) the interconversion reaction proceeded with inversion of the relative configuration at sulfur, by an S_N2-type pathway involving an oxasulfurane transition state or intermediate (Scheme 28.10). These features point at the close analogy between the halogen/metal and sulfoxide/metal interconversions.

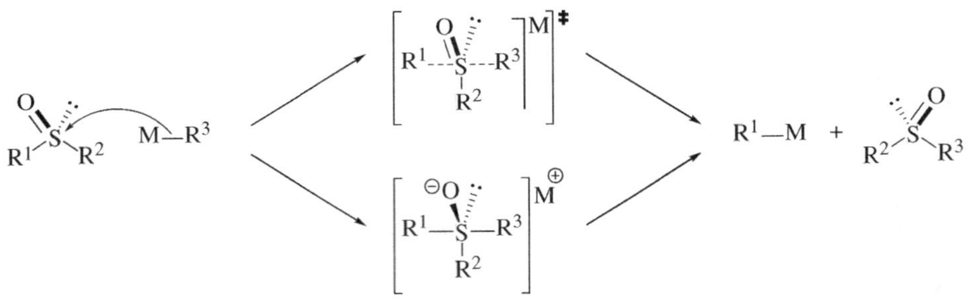

SCHEME 28.10 Mechanistic scenarios for the sulfoxide/metal exchange.

Notably, Andersen showed that the use of a four or fivefold excess of aryllithium reagent affords other products [66]. Yet the sulfoxide/metal interconversion usually leads to high yields of the desired product after trapping of the intermediate organometal by electrophiles, when the reaction is conducted at low temperature in THF and using 1–2 equivalents of exchange reagent. Also, one should note that unlike arylsulfinyl groups, *tert*-butylsulfinyl substituents on arenes cannot be cleaved by sulfoxide/lithium or sulfoxide/magnesium interconversion [64].

The peculiar properties of sulfoxides as chiral auxiliaries, and as directing groups for *ortho*-metalation, combined with the sulfoxide/metal exchange, make them a very powerful, yet underexploited tool for arene functionalization. Three main strategies have been described. The first one consists in the introduction of the sulfinyl group, followed by a stereoselective sulfoxide-oriented lateral functionalization (see Chapter 26) and completed by removal of the sulfinyl group by sulfoxide/metal interconversion and trapping with an electrophile (Scheme 28.11) [67–76]. A related strategy comprises a sulfoxide-oriented functionalization of the aromatic ring itself, that is, the use of the sulfinyl group as a directing group for *ortho*-metalation (see Chapter 26), which actually turns the arene ring into a masked *ortho*-dianion, as underlined by Knochel (Scheme 28.12) [77–80].

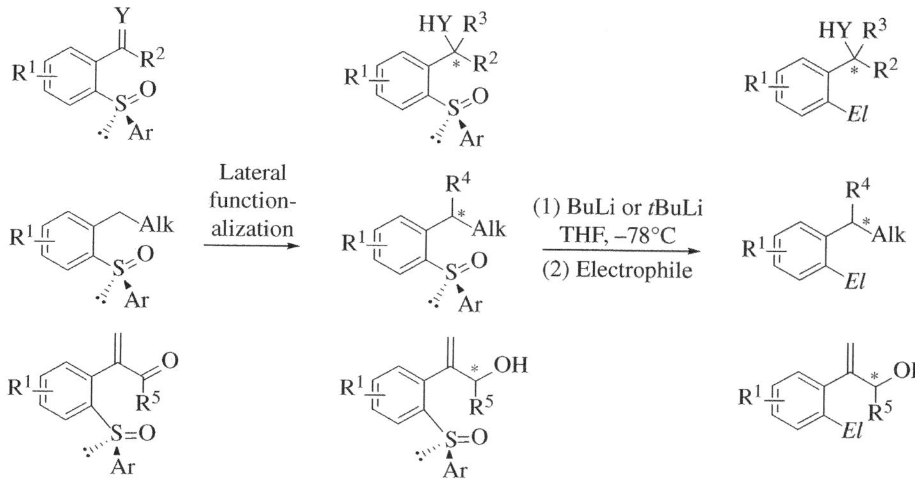

Ar = *p*Tol, 2,4,6-Me_3-C_6H_2, 2,4,6-*i*Pr_3-C_6H_2
El = H or other substituent

SCHEME 28.11 Asymmetric sulfoxide-directed lateral functionalization.

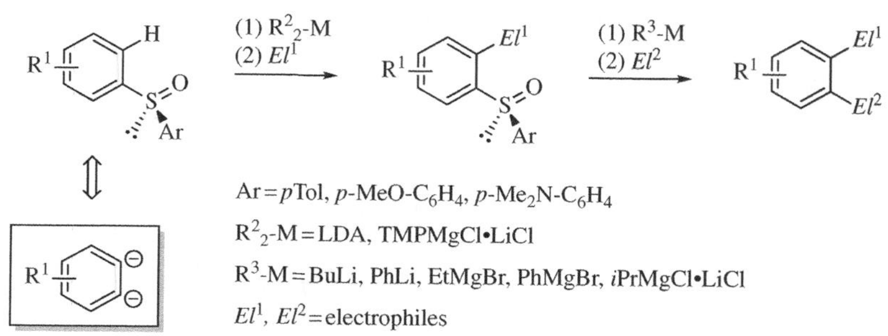

Ar = *p*Tol, *p*-MeO-C_6H_4, *p*-Me_2N-C_6H_4
$R^2{}_2$-M = LDA, TMPMgCl•LiCl
R^3-M = BuLi, PhLi, EtMgBr, PhMgBr, *i*PrMgCl•LiCl
El^1, El^2 = electrophiles

SCHEME 28.12 Sulfoxide-mediated *ortho*-difunctionalization.

In a third elegant use of the sulfoxide/metal interchange, sulfoxides were used as "traceless resolving agents," a term coined by Clayden, for the selective obtention of atropo-enriched compounds, either by dynamic thermodynamic resolution or by physical separation of nonequilibrating atropo-diastereomers (Schemes 28.13 and 28.14) [81–85]. In the latter case, the conversion of the sulfinyl moiety to various functional groups was reported by F. Leroux and coworkers to be not only stereo- but also chemoselective, with the exchange with sulfoxides being favored over the one with bromine and allowing controlled sequential functionalization of biaryls (Scheme 28.14) [82].

Similarly, planar chirality could be introduced on [2.2]paracyclophanes by means of the *p*-tolylsulfinyl group. After separation of diastereomers and sulfoxide/lithium exchange, various enantiopure, planar chiral, 4-substituted [2.2]paracyclophanes were obtained (Scheme 28.15) [86, 87].

An elegant approach, somewhat related to the strategies described earlier, was recently applied to the total synthesis of Cavicularin, a natural product. A sulfinyl group was first introduced in a nonasymmetric way to facilitate a subsequent S_NAr; it was then replaced by sulfoxide/lithium interchange, by the same sulfinyl group but, enantioselectively, for the desymmetrization of an advanced intermediate (Scheme 28.16) [88].

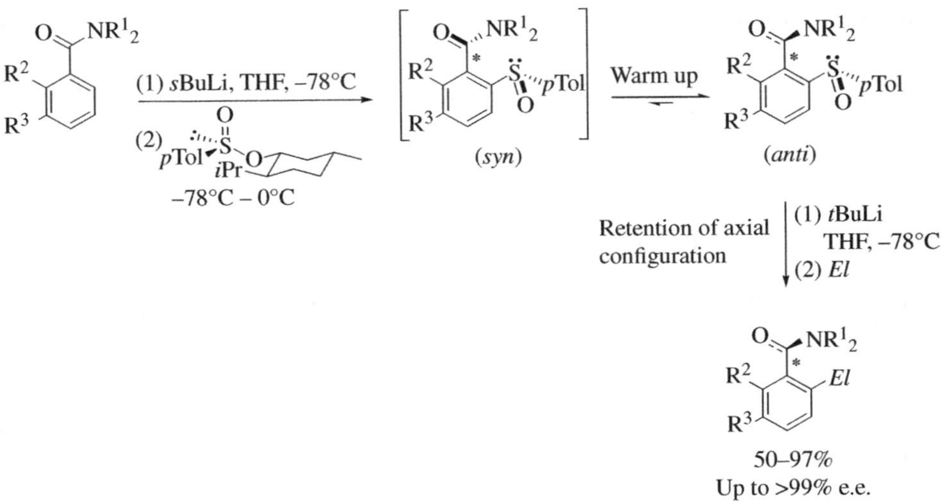

SCHEME 28.13 Resolution-sulfoxide/metal exchange sequence in the preparation of atropo-enriched benzamides.

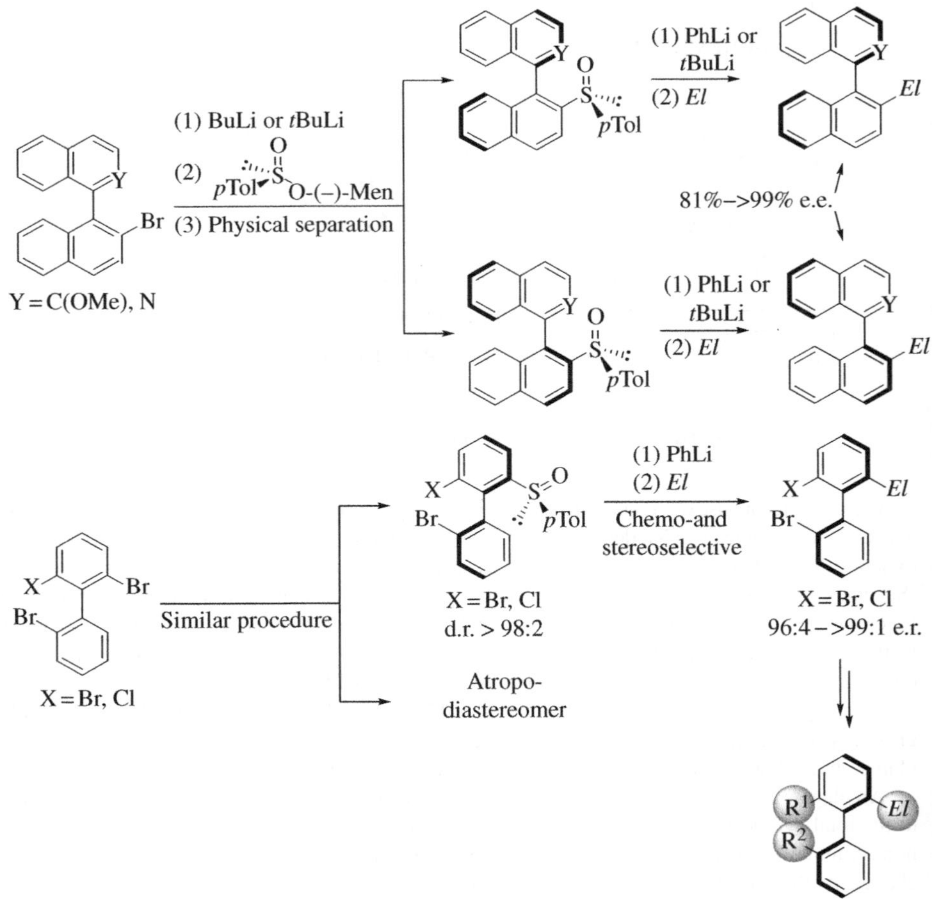

SCHEME 28.14 Sulfoxides as traceless auxiliaries in the resolution or desymmetrization of biaryls.

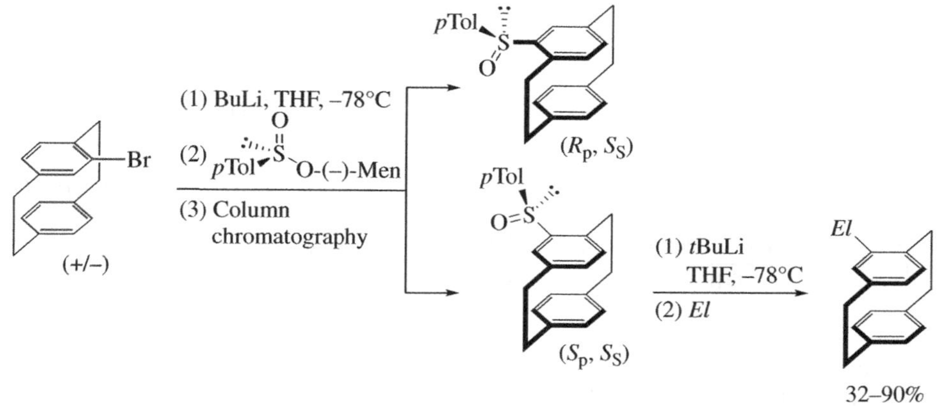

SCHEME 28.15 Sulfoxides as traceless auxiliaries in the resolution or desymmetrization of paracyclophanes.

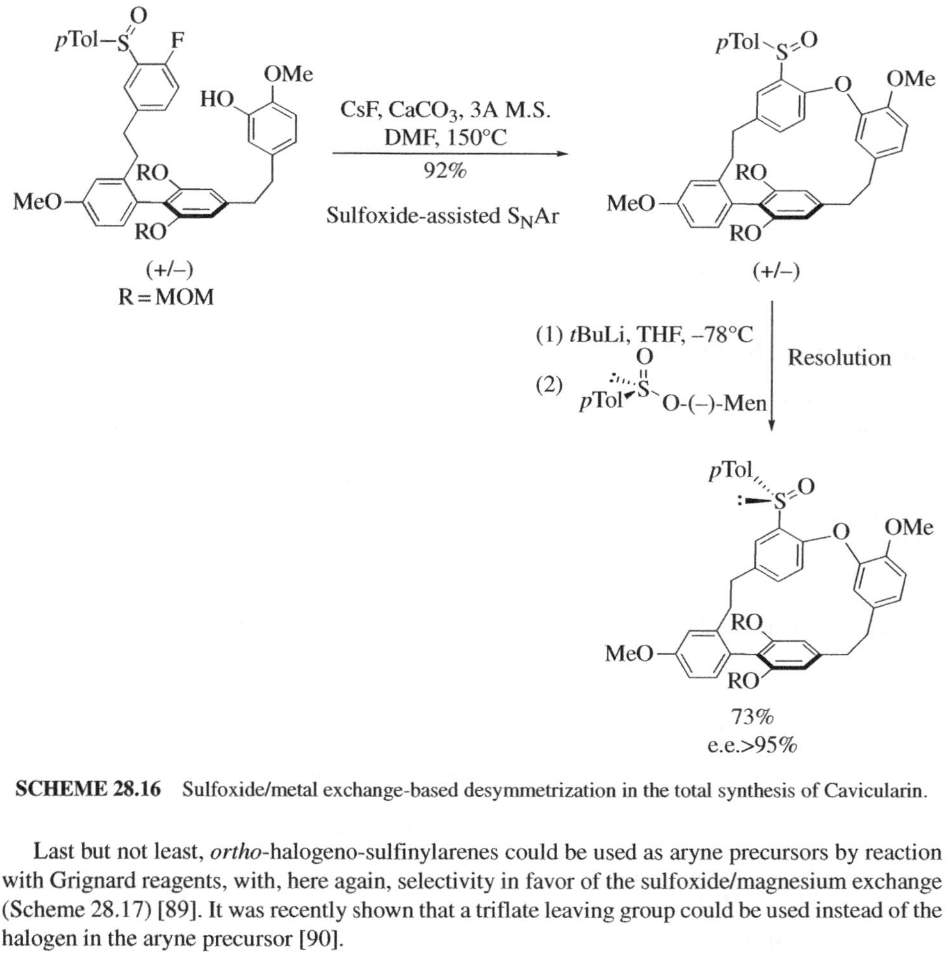

SCHEME 28.16 Sulfoxide/metal exchange-based desymmetrization in the total synthesis of Cavicularin.

Last but not least, *ortho*-halogeno-sulfinylarenes could be used as aryne precursors by reaction with Grignard reagents, with, here again, selectivity in favor of the sulfoxide/magnesium exchange (Scheme 28.17) [89]. It was recently shown that a triflate leaving group could be used instead of the halogen in the aryne precursor [90].

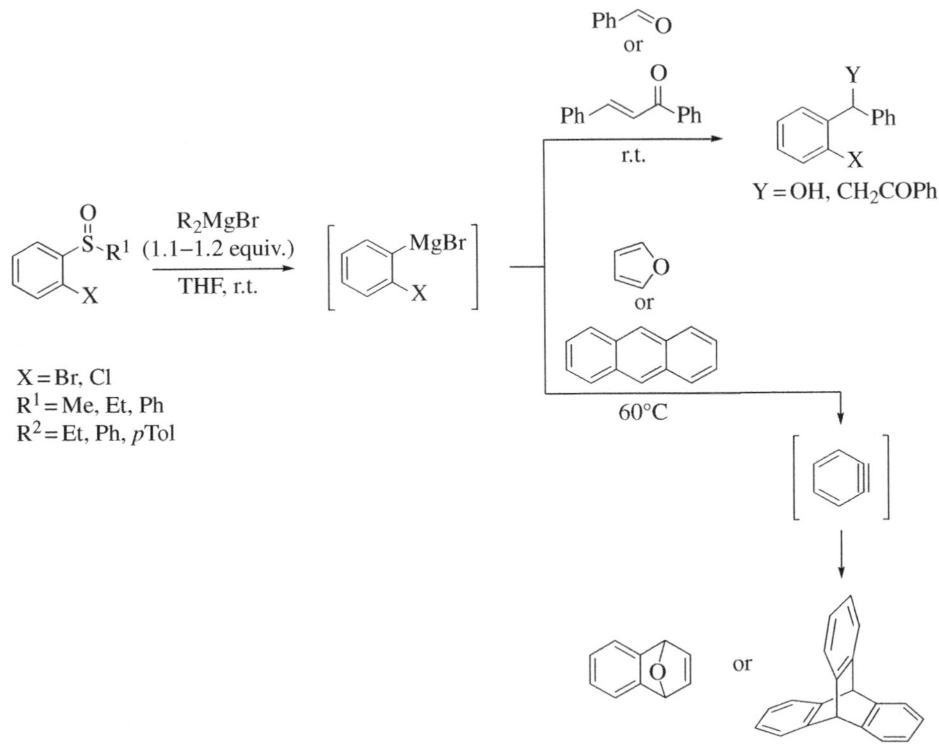

SCHEME 28.17 Sulfoxide/metal exchange for the generation of arynes.

28.6.2 The Phosphorus/Metal Interconversion

The phosphorus/metal exchange was first reported in the case of simple triarylphosphine oxides reacting with alkyllithium or aryllithium or arylmagnesium reagents [91]. Interestingly, it was shown to be a stereospecific reaction, with inversion of the relative configuration of the phosphine oxide, similarly to sulfoxides [92]. Departure of an aryllithium or arylmagnesium was possible as well on trivalent phosphines and also phosphine–borane complexes, but rather than a means to functionalize the aryl group, it has been employed as a useful way to access dibenzophospholes, as depicted in Scheme 28.18 (for leading references, see Ref. 93). In fact, one should take this exchange reaction into account during the synthesis of phosphorus ligands *via* biaryllithium or biarylmagnesium intermediates as it may lead to considerable loss of enantioenrichment [94, 95].

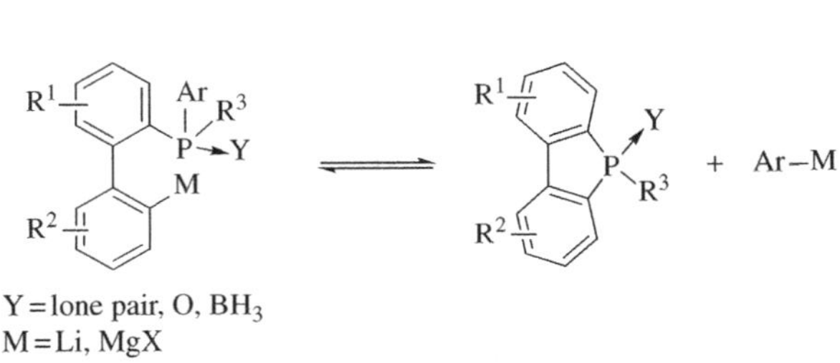

SCHEME 28.18 Phosphorus/metal interchange by cyclization of 2′-phosphinylbiarylmetals.

A single example of arene functionalization by means of a synthetically useful phosphorus/metal exchange was described by Hayashi, who accessed various MOP ligands by conversion of a diphenylphosphinoyl group into diverse substituents with retention of axial enantiopurity (Scheme 28.19) [94].

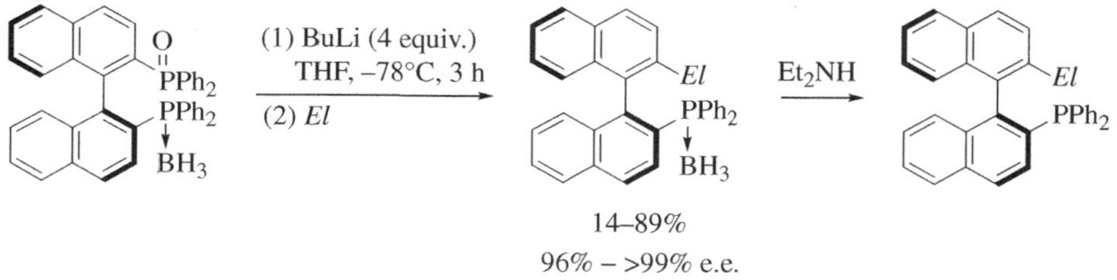

SCHEME 28.19 Synthesis of MOP ligands from the BINAPO-borane complex by phosphine oxide/metal exchange.

28.7 ARYL—ARYL COUPLING THROUGH HALOGEN/METAL INTERCONVERSION

Aryllithium or arylmagnesium reagents are used in several established aryl—aryl coupling reactions, such as the arylic S_NAr, the Kumada–Corriu coupling, the Negishi coupling with *in situ* generation of the arylzinc species, and their oxidative homo- or heterocouplings (see Chapters 6–10 and 19). Since the halogen/magnesium exchange has only been extensively studied and used in the last decade or so, mainly thanks to the work of Knochel, the number of examples of the abovementioned aryl—aryl couplings through the title interconversion remain limited. The literature on the use of the halogen/lithium interconversion is more abundant, although considerably limited in comparison to, for example, the Suzuki–Miyaura coupling, due to the high reactivity and less convenient handling of aryllithium species.

In this section, we will rather focus, on one hand, on the recent successful work on the discovery or development of transformations considered thus far as side reactions or inefficient processes and, on the other hand, on the preparation of biaryls by coupling of an aryllithium or arylmagnesium and an aryne, especially the so-called ARYNE coupling, which involves 3 successive halogen/metal interconversions in one step.

28.7.1 (Re)emerging Methods for Aryl—Aryl Coupling Through Halogen/Metal Interconversion

Due to the cost and environmental impact of transition metals, special emphasis has been given lately to "transition metal-free" coupling reactions [96, 97]. Knochel, Mayr, and coworkers reported on the homocoupling of Grignard reagents obtained by halogen/magnesium interconversion, in the presence of 3,3′,5,5′-tetra-*tert*-butyldiphenoquinone as an organic oxidant (Scheme 28.20) [98]. Although the oxidant is used in stoichiometric amount, it can be recycled at the end of the reaction. The absence of involvement of free aryl radicals was demonstrated. A similar methodology was described by Studer and coworkers, with TEMPO as organic oxidant (Scheme 28.20) [99, 100]. The participation of trace transition metal—present in TEMPO and the Grignard reagents—could not be ruled out.

On the other hand, the direct coupling, under palladium catalysis, of aryllithiums obtained by halogen/lithium exchange with haloarenes was recently reinvestigated. First reported by

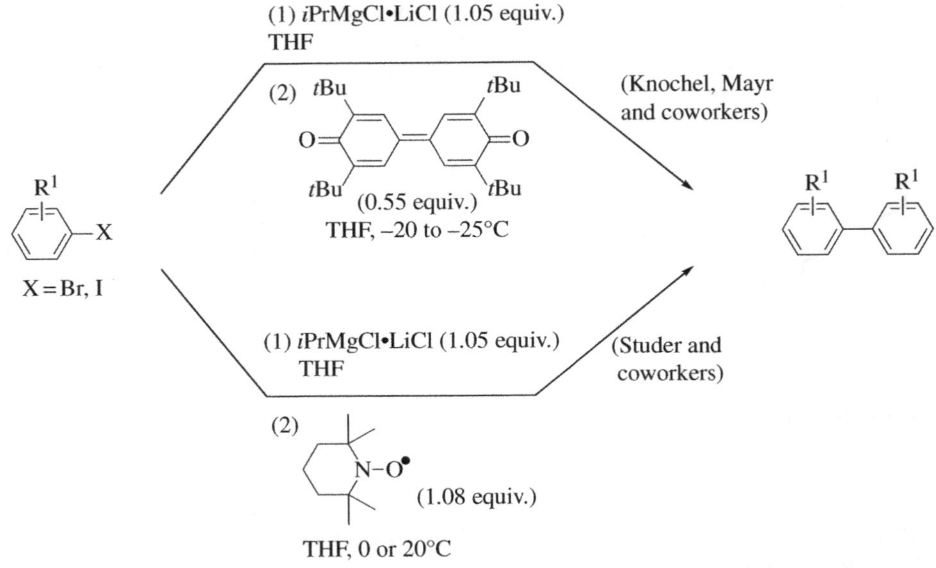

SCHEME 28.20 Oxidative aryl—aryl coupling *via* halogen/metal exchange.

Murahashi et al. [101], the reaction hibernated for three decades, until the Yoshida group managed to obtain synthetically useful yields in an integrated flow microreactor system [102]. Feringa, Fañanás-Mastral, and coworkers developed the equivalent batch process, using a syringe pump, thanks to a careful choice of catalytic system avoiding the formation of usual side products of the halogen/lithium exchange (Scheme 28.21) [103–105].

28.7.2 Aryne-Mediated Aryl—Aryl Coupling

The chemistry of *ortho*-arynes [106–109] has recently been undergoing a tremendous burst. Indeed, these versatile transient intermediates exhibit a unique reactivity, making them partners of choice for the functionalization of arenes by additions across the triple bond or for the synthesis of fused rings by pericyclic reactions.

The reaction of aryllithium compounds with *ortho*-dihaloarenes, proceeding *via* the *in situ* formation of arynes allows the preparation of *ortho*-halobiaryls. This "ARYNE coupling" has become a robust method for aryl—aryl coupling and combines several advantages: the use of cheap and/or easily accessible halogeno aromatic compounds, the access to biaryls bearing two distinct aromatic units and to polyhalogenated biaryls—which can be functionalized further—the use of lithium or magnesium reagents (i.e., in the absence of transition metals), and multigram reaction scales [110]. The mechanism (Scheme 28.22) is based on a subtle interplay of several organometallic species and their relative basicities. The chain reaction proceeds as follows: (1) A thermodynamically stable organolithium intermediate is formed, of which (2) a small amount performs a halogen/metal exchange on the coupling partner *via* an ate complex, generating a thermodynamically unstable *ortho*-halophenyllithium intermediate, which (3) eliminates spontaneously lithium halide affording an aryne. Then, (4) the transient aryne species undergoes nucleophilic addition of the organolithium precursor, and (5) the resulting 2-biaryl lithium intermediate is finally stabilized by *in situ* transfer of bromine or iodine from the starting material, affording again some *ortho*-halophenyllithium to pursue the chain reaction.

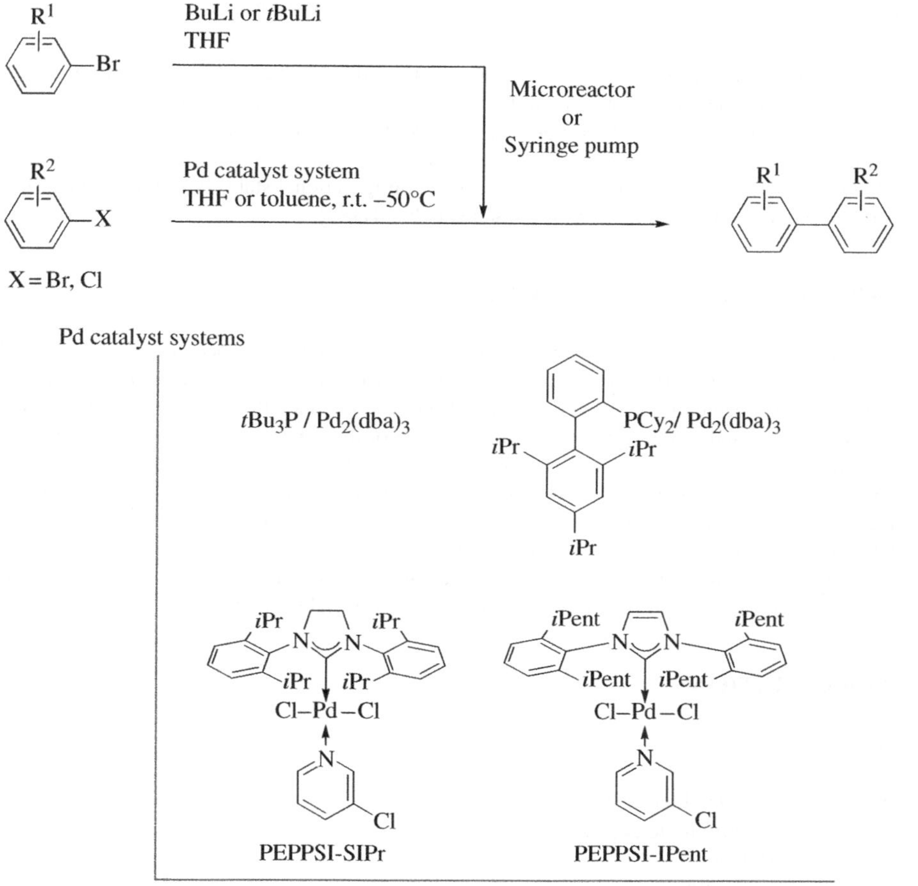

SCHEME 28.21 Pd-catalyzed coupling of haloarenes and aryllithiums obtained by halogen/lithium interconversion.

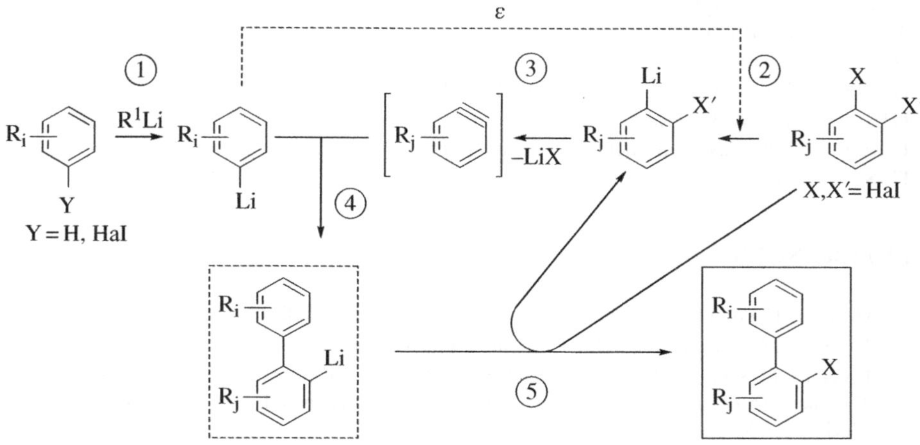

SCHEME 28.22 Mechanism of the ARYNE coupling.

It has been shown that the aryl—aryl bond formation in the case of dissymmetrical benzyne precursors can be conducted with complete control of regioselectivity [111]. Moreover, due to the regioselective introduction (see earlier) of a traceless chiral auxiliary (an enantiopure *p*-tolylsulfinyl group), the separation of atropo-diastereoisomers by simple crystallization becomes possible. Chemoselective sulfoxide/lithium exchange with configurational stability of the biaryl axis allows for the obtention of enantiopure biaryls (Scheme 28.14) [82].

28.8 SUMMARY AND OUTLOOK

The halogen/metal interconversion has matured over eight decades to become a very powerful tool for the structural decoration of arenes. Even if its mechanism has been elucidated only lately, extensive work has been carried out on the scope of the reaction and allowed for practically exhaustive chemo-, regio-, and stereoselective functionalization of the benzene ring. The development of new reagents for more selective transformations, milder reaction conditions, or increased reactivity is still underway and promises new vistas in the organic chemistry of arenes. The use of the halogen/metal exchange in aryl—aryl bond formation, among other transformations, shows the competitiveness of this method with regard to transition metal-mediated reactions. The development of a catalytic version of the title process—that is, using a substoichiometric amount of the lithium or magnesium reagent—will probably be the next endeavor in this effervescent field of research.

ABBREVIATIONS

Ar	Aryl
Cy	Cyclohexyl
dba	Dibenzylideneacetone
ee	Enantiomeric excess
Hal	Halogen
*i*Pent	*iso*-Pentyl
*i*Pr	*iso*-Propyl
LDA	Lithium *N*,*N*-di*iso*propylamide
LiTMP	Lithium 2,2,6,6-tetramethylpiperidide
Men	Menthyl
*p*Tol	*para*-Tolyl
*s*Bu	*sec*-Butyl
*t*Bu	*tert*-Butyl
TEMPO	2,2,6,6-Tetramethylpiperidine 1-oxyl
THF	Tetrahydrofuran
TMEDA	*N*,*N*,*N'*,*N'*-Tetramethylethylenediamine
TMS	Trimethylsilyl

REFERENCES

[1] G. Wittig, U. Pockels, H. Droge, *Ber. Dtsch. Chem. Ges. B* 1938, **71B**, 1903–1912.

[2] H. Gilman, W. Langham, A. L. Jacoby, *J. Am. Chem. Soc.* 1939, **61**, 106–109.

[3] F. Leroux, M. Schlosser, E. Zohar, I. Marek, in *The Chemistry of Organolithium Compounds*, Patai Series: The Chemistry of Functional Groups, Vol. **Part 1** (Eds.: Z. Rappoport, I. Marek), John Wiley & Sons, Ltd, Chichester, 2009, pp. 435–493.

[4] J. Clayden, in *The Chemistry of Organolithium Compounds*, The Chemistry of Functional Groups, Vol. **1**, (Eds.: Z. Rappopart, I. Marek), John Wiley & Sons, Ltd, Chichester, **2004**.

[5] P. Knochel, A. Krasovskiy, I. Sapountzis, in *Handbook of Functionalized Organometallics: Applications in Synthesis* **1** (Ed.: P. Knochel), Wiley-VCH, Weinheim, 2005.
[6] C. Nájera, J. M. Sansano, M. Yus, *Tetrahedron* 2003, **59**, 9255–9303.
[7] W. J. Trepka, R. J. Sonnenfeld, *J. Organomet. Chem.* 1969, **16**, 317–320.
[8] H. R. Rogers, J. Houk, *J. Am. Chem. Soc.* 1982, **104**, 522–525.
[9] W. F. Bailey, M. R. Luderer, K. P. Jordan, *J. Org. Chem.* 2006, **71**, 2825–2828.
[10] H. Gilman, F. W. Moore, *J. Am. Chem. Soc.* 1940, **62**, 1843–1846.
[11] W. Langham, R. Q. Brewster, H. Gilman, *J. Am. Chem. Soc.* 1941, **63**, 545–549.
[12] P. Knochel, W. Dohle, N. Gommermann, F. F. Kneisel, F. Kopp, T. Korn, I. Sapountzis, V. A. Vu, *Angew. Chem. Int. Ed.* 2003, **42**, 4302–4320.
[13] H. Ila, O. Baron, A. J. Wagner, P. Knochel, *Chem. Commun.* 2006, 583–593.
[14] A. Inoue, K. Kitagawa, H. Shinokubo, K. Oshima, *J. Org. Chem.* 2001, **66**, 4333–4339.
[15] K. Kitagawa, A. Inoue, H. Shinokubo, K. Oshima, *Angew. Chem. Int. Ed.* 2000, **39**, 2481–2483.
[16] D. Tilly, F. Chevallier, F. Mongin, P. C. Gros, *Chem. Rev.* 2014, **114**, 1207–1257.
[17] A. Krasovskiy, P. Knochel, *Angew. Chem. Int. Ed.* 2004, **43**, 3333–3336.
[18] D. E. Applequist, D. F. O'Brien, *J. Am. Chem. Soc.* 1963, **85**, 743–748.
[19] H. J. S. Winkler, H. Winkler, *J. Am. Chem. Soc.* 1966, **88**, 964–969.
[20] L. I. Zakharkin, O. Y. Okhlobystin, K. A. Bilevitch, *J. Organomet. Chem.* 1964, **2**, 309–313.
[21] D. Seebach, H. Neumann, *Chem. Ber.* 1974, **107**, 847–853.
[22] E. J. Corey, D. J. Beames, *J. Am. Chem. Soc.* 1972, **94**, 7210–7211.
[23] W. F. Bailey, E. R. Punzalan, *J. Org. Chem.* 1990, **55**, 5404–5406.
[24] C. Waldmann, O. Schober, G. Haufe, K. Kopka, *Org. Lett.* 2013, **15**, 2954–2957.
[25] R. E. Merrill, E. Negishi, *J. Org. Chem.* 1974, **39**, 3452–3453.
[26] J. Gorecka-Kobylinska, M. Schlosser, *J. Org. Chem.* 2009, **74**, 222–229.
[27] L. Shi, Y. Chu, P. Knochel, H. Mayr, *Angew. Chem. Int. Ed.* 2008, **47**, 202–204.
[28] L. Shi, Y. Chu, P. Knochel, H. Mayr, *J. Org. Chem.* 2009, **74**, 2760–2764.
[29] L. Shi, Y. Chu, P. Knochel, H. Mayr, *Org. Lett.* 2009, **11**, 3502–3505.
[30] L. Shi, Y. Chu, P. Knochel, H. Mayr, *Org. Lett.* 2012, **14**, 2602–2605.
[31] N. J. Foulger, B. J. Wakefield, *J. Organomet. Chem.* 1974, **69**, 321–326.
[32] E. T. McBee, R. A. Sanford, *J. Am. Chem. Soc.* 1950, **72**, 5574–5575.
[33] M. D. Rausch, F. E. Tibbetts, H. B. Gordon, *J. Organomet. Chem.* 1966, **5**, 493–500.
[34] M. Schlosser, *Angew. Chem. Int. Ed.* 2005, **44**, 376–393.
[35] H. J. S. Winkler, H. Winkler, *J. Am. Chem. Soc.* 1966, **88**, 969–974.
[36] W. F. Bailey, J. J. Patricia, *J. Organomet. Chem.* 1988, **352**, 1–46.
[37] J. Clayden, *Organolithiums: Selectivity for Synthesis*, Tetrahedron Organic Chemistry Series, Vol. **23**, Elsevier, Oxford, 2002.
[38] H. J. Reich, *J. Org. Chem.* 2012, **77**, 5471–5491.
[39] H. J. Reich, *Chem. Rev.* 2013, **113**, 7130–7178.
[40] W. B. Farnham, J. C. Calabrese, *J. Am. Chem. Soc.* 1986, **108**, 2449–2451.
[41] G. Boche, M. Schimeczek, J. Cioslowski, P. Piskorz, *Eur. J. Org. Chem.* 1998, 1851–1860.
[42] M. Müller, M. Brönstrup, O. Knopff, V. Schulze, R. W. Hoffmann, *Organometallics* 2003, **22**, 2931–2937.
[43] K. B. Wiberg, S. Sklenak, W. F. Bailey, *J. Org. Chem.* 2000, **65**, 2014–2021.
[44] R. W. Hoffmann, M. Brönstrup, M. Müller, *Org. Lett.* 2003, **5**, 313–316.
[45] B. Jedlicka, R. H. Crabtree, P. E. M. Siegbahn, *Organometallics* 1997, **16**, 6021–6023.
[46] A. Krasovskiy, B. F. Straub, P. Knochel, *Angew. Chem. Int. Ed.* 2005, **45**, 159–162.
[47] F. Mongin, O. Desponds, M. Schlosser, *Tetrahedron Lett.* 1996, **37**, 2767–2770.
[48] F. R. Leroux, H. Mettler, *Adv. Synth. Catal.* 2007, **349**, 323–336.

[49] C. Heiss, M. Schlosser, *Eur. J. Org. Chem.* 2003, 447–451.
[50] S. V. Sunthankar, H. Gilman, *J. Org. Chem.* 1951, **16**, 8–16.
[51] R. A. Murphy, H. F. Kung, M. P. Kung, J. Billings, *J. Med. Chem.* 1990, **33**, 171–178.
[52] P. Stanetty, B. Krumpak, I. K. Rodler, *J. Chem. Res. Synop.* 1995, **9**, 342–343.
[53] G. Voss, H. Gerlach, *Chem. Ber.* 1989, **122**, 1199–1201.
[54] F. R. Leroux, R. Simon, N. Nicod, *Lett. Org. Chem.* 2006, **3**, 948–954.
[55] F. Leroux, N. Nicod, L. Bonnafoux, B. Quissac, F. Colobert, *Lett. Org. Chem.* 2006, **3**, 165–169.
[56] C. Heiss, E. Marzi, M. Schlosser, *Eur. J. Org. Chem.* 2003, 4625–4629.
[57] C. A. Fan, B. Ferber, H. B. Kagan, O. Lafon, P. Lesot, *Tetrahedron: Asymmetry* 2008, **19**, 2666–2677.
[58] Q. Perron, J. Praz, A. Alexakis, *Tetrahedron: Asymmetry* 2009, **20**, 1004–1007.
[59] Q. Perron, A. Alexakis, *Adv. Synth. Catal.* 2010, **352**, 2611–2620.
[60] D. Sälinger, R. Brückner, *Chem. Eur. J.* 2009, **15**, 6688–6703.
[61] H. Gilman, S. H. Eidt, *J. Am. Chem. Soc.* 1956, **78**, 3848–3850.
[62] J. Jacobus, K. Mislow, *J. Am. Chem. Soc.* 1967, **89**, 5228–5234.
[63] J. P. Lockard, C. W. Schroeck, C. R. Johnson, *Synthesis* 1973, 485–486.
[64] T. Durst, M. J. LeBelle, R. Van den Elzen, K.-C. Tin, *Can. J. Chem.* 1974, **52**, 761–766.
[65] N. Furukawa, S. Ogawa, K. Matsumura, H. Fujihara, *J. Org. Chem.* 1991, **56**, 6341–6348.
[66] K. K. Andersen, S. A. Yeager, *J. Org. Chem.* 1963, **28**, 865–867.
[67] J. L. Garcia Ruano, L. Marzo, V. Marcos, C. Alvarado, J. Aleman, *Chem. Eur. J.* 2012, **18**, 9775–9779.
[68] S. Nakamura, H. Yasuda, Y. Watanabe, T. Toru, *J. Org. Chem.* 2000, **65**, 8640–8650.
[69] J. L. Garcia Ruano, M. A. Fernandez-Ibañez, J. A. Fernandez-Salas, M. C. Maestro, P. Marquez-Lopez, M. M. Rodriguez-Fernandez, *J. Org. Chem.* 2009, **74**, 1200–1204.
[70] M. I. Antczak, F. Cai, J. M. Ready, *Org. Lett.* 2011, **13**, 184–187.
[71] S. Nakamura, M. Oda, H. Yasuda, T. Toru, *Tetrahedron* 2001, **57**, 8469–8480.
[72] J. L. García Ruano, M. T. Aranda, M. Puente, *Tetrahedron* 2005, **61**, 10099–10104.
[73] S. Nakamura, H. Yasuda, T. Toru, *Tetrahedron: Asymmetry* 2002, **13**, 1509–1518.
[74] S. Nakamura, H. Yasuda, Y. Watanabe, T. Toru, *Tetrahedron Lett.* 2000, **41**, 4157–4160.
[75] A. Almorín, M. C. Carreño, Á. Somoza, A. Urbano, *Tetrahedron Lett.* 2003, **44**, 5597–5600.
[76] H. Sugimoto, S. Nakamura, Y. Shibata, N. Shibata, T. Toru, *Tetrahedron Lett.* 2006, **47**, 1337–1340.
[77] S. Ogawa, N. Furukawa, *J. Org. Chem.* 1991, **56**, 5723–5726.
[78] L. Melzig, C. B. Rauhut, N. Naredi-Rainer, P. Knochel, *Chem. Eur. J.* 2011, **17**, 5362–5372.
[79] C. B. Rauhut, L. Melzig, P. Knochel, *Org. Lett.* 2008, **10**, 3891–3894.
[80] L. Melzig, C. B. Rauhut, P. Knochel, *Synthesis* 2009, 1041–1048.
[81] J. Clayden, P. M. Kubinski, F. Sammiceli, M. Helliwell, L. Diorazio, *Tetrahedron* 2004, **60**, 4387–4397.
[82] F. R. Leroux, A. Berthelot, L. Bonnafoux, A. Panossian, F. Colobert, *Chem. Eur. J.* 2012, **18**, 14232–14236.
[83] J. Clayden, D. Mitjans, L. H. Youssef, *J. Am. Chem. Soc.* 2002, **124**, 5266–5267.
[84] J. Clayden, S. P. Fletcher, J. J. W. McDouall, S. J. M. Rowbottom, *J. Am. Chem. Soc.* 2009, **131**, 5331–5343.
[85] F. Geittner, T. Thaler, P. Knochel, *Synlett* 2007, **17**, 2655–2658.
[86] P. B. Hitchcock, G. J. Rowlands, R. Parmar, *Chem. Commun.* 2005, 4219–4221.
[87] R. Parmar, M. P. Coles, P. B. Hitchcock, G. J. Rowlands, *Synthesis* 2010, 4177–4187.
[88] H. Takiguchi, K. Ohmori, K. Suzuki, *Angew. Chem. Int. Ed.* 2013, **52**, 10472–10476.
[89] N. Furukawa, T. Shibutani, H. Fujihara, *Tetrahedron Lett.* 1987, **28**, 2727–2730.
[90] S. Yoshida, K. Uchida, T. Hosoya, *Chem. Lett.* 2014, **43**, 116–118.
[91] D. Seyferth, D. E. Welch, J. K. Heeren, *J. Am. Chem. Soc.* **1964**, **86**, 1100–1105.

[92] C. Cardellicchio, V. Fiandanese, F. Naso, S. Pacifico, M. Koprowski, K. M. Pietrusiewicz, *Tetrahedron Lett.* 1994, **35**, 6343–6346.

[93] A. Oukhrib, L. Bonnafoux, A. Panossian, S. Waifang, D. H. Nguyen, M. Urrutigoity, F. Colobert, M. Gouygou, F. R. Leroux, *Tetrahedron* 2014, **70**, 1431–1436.

[94] T. Shimada, H. Kurushima, Y.-H. Cho, T. Hayashi, *J. Org. Chem.* 2001, **66**, 8854–8858.

[95] M. Cereghetti, W. Arnold, E. A. Broger, A. Rageot, *Tetrahedron Lett.* 1996, **37**, 5347–5350.

[96] V. P. Mehta, B. Punji, *RSC Adv.* 2013, **3**, 11957.

[97] R. A. D. Arancon, C. S. K. Lin, C. Vargas, R. Luque, *Org. Biomol. Chem.* 2014, **12**, 10–35.

[98] A. Krasovskiy, A. Tishkov, V. del Amo, H. Mayr, P. Knochel, *Angew. Chem. Int. Ed.* 2006, **45**, 5010–5014.

[99] M. S. Maji, T. Pfeifer, A. Studer, *Angew. Chem. Int. Ed.* 2008, **47**, 9547–9550.

[100] A. Studer, M. Maji, *Synthesis* 2009, 2467–2470.

[101] S.-I. Murahashi, M. Yamamura, K.-I. Yanagisawa, N. Mita, K. Kondo, *J. Org. Chem.* 1979, **44**, 2408–2417.

[102] A. Nagaki, A. Kenmoku, Y. Moriwaki, A. Hayashi, J. Yoshida, *Angew. Chem. Int. Ed.* 2010, **49**, 7543–7547.

[103] M. Giannerini, V. Hornillos, C. Vila, M. Fañanás-Mastral, B. L. Feringa, *Angew. Chem. Int. Ed.* 2013, **52**, 13329–13333.

[104] M. Giannerini, M. Fañanás-Mastral, B. L. Feringa, *Nat. Chem.* 2013, **5**, 667–672.

[105] V. Hornillos, M. Giannerini, C. Vila, M. Fañanás-Mastral, B. L. Feringa, *Org. Lett.* 2013, **15**, 5114–5117.

[106] P. M. Tadross, B. M. Stoltz, *Chem. Rev.* 2012, **112**, 3550–3577.

[107] A. Bhunia, S. R. Yetra, A. T. Biju, *Chem. Soc. Rev.* 2012, **41**, 3140–3152.

[108] H. Yoshida, K. Takaki, *Synlett* 2012, **23**, 1725–1732.

[109] R. Sanz, *Org. Prep. Proced. Int.* 2008, **40**, 215–291.

[110] F. R. Leroux, L. Bonnafoux, C. Heiss, F. Colobert, D. A. Lanfranchi, *Adv. Synth. Catal.* 2007, **349**, 2705–2713.

[111] V. Diemer, M. Begaud, F. R. Leroux, F. Colobert, *Eur. J. Org. Chem.* 2011, 341–354.

PART IX

PHOTOCHEMICAL REACTIONS

29

AROMATIC PHOTOCHEMICAL REACTIONS

NORBERT HOFFMANN AND EMMANUEL RIGUET

Institut de Chimie Moléculaire de Reims, UMR 6229 CNRS et Université de Reims Champagne-Ardenne, UFR Sciences, Reims, France

29.1 INTRODUCTION

In photochemical reactions, molecules are electronically excited by absorption of light. Thus changing the electronic configuration, the chemical reactivity of a compound is considerably modified when compared to the ground state. Polarities in functional groups are often inverted leading to an umpolung of the reactivity. The redox potentials change, and redox reactions are facilitated. Due to these facts, photochemical reactions are now frequently applied to organic synthesis [1, 2]. They considerably enrich the methodology of organic chemistry. In the same context, photoredox catalysis is currently intensively investigated [3, 4]. As far as pure organic reactions are concerned, different spin multiplicities of the reactive excited state causes additional modifications of the chemical reactivity. Generally, these phenomena are explained by potential energy surface topology of the ground and the excited state [5].

In contrast to most of the ground state reactions, photochemical reactions often don't need activation reagents such as acids, bases, metals, or enzymes. In this context, the photon is considered as a traceless reagent [6, 7]. In this way, waste and side product formation is reduced. Photochemical reactions are thus an important part of sustainable chemistry methodology [8].

A lot of photochemical reactions of aromatic compounds have been described in the literature. In this context, photocycloadditions are typical examples [6, 9]. In such reactions involving $\pi\pi^*$ excitation of the chromophore, the aromatic character is not reestablished in the final products as is typical for ground state reactions. Very efficient methods based on this reactivity have been developed for the construction of polycyclic compounds. Thus, molecular complexity is generated from simple and easily available starting compounds in only one step. For these reasons, photochemical

Arene Chemistry: Reaction Mechanisms and Methods for Aromatic Compounds, First Edition. Edited by Jacques Mortier.

reactions of aromatic compounds have been successfully applied to both the synthesis of polycyclic natural products and the diversity oriented synthesis. In particular, the second topic is of high interest to find new biologically active compounds.

Nevertheless, an increasing number of photochemical reactions of aromatic substrates exist in which the aromaticity in the products is built up again. These reactions are interesting for application to the preparation of various biologically active compounds. They are also considered systematically to be applied to the synthesis of organic materials where aromatic moieties are of particular interest.

29.2 AROMATIC COMPOUNDS AS CHROMOPHORES

29.2.1 Photocycloaddition and Photochemical Electrocyclic Reactions Involving Aromatics

Among the photochemical reactions of aromatic compounds, the photocycloadditions are most frequently applied to the synthesis of complex polycyclic compounds [6, 9]. The [2+3] or *meta* photocycloaddition of aromatic compounds and alkenes is the most prominent example [10]. This transformation also demonstrates complementarities between photochemical and ground state reactions since such reactions are almost impossible using conventional activation. A [2+2] or *ortho* photocycloaddition between carbocyclic aromatic compounds and alkenes is observed as well. It is often competitive with other cycloaddition modes, in particular the [2+3] mode [11]. Many of these reactions are reversible, and photostationary equilibria are involved. This reaction was much less applied to organic synthesis. Recently, it was found that an acidic reaction medium may have an influence on the outcome of the reaction. The intramolecular photocycloaddition of resorcinol derivatives such as **1** is difficult due to its reversibility (Scheme 29.1). However, in an acidic reaction medium, the cycloadducts **2a,b** are protonated at the oxygen atom of the tetrahydrofuran moiety

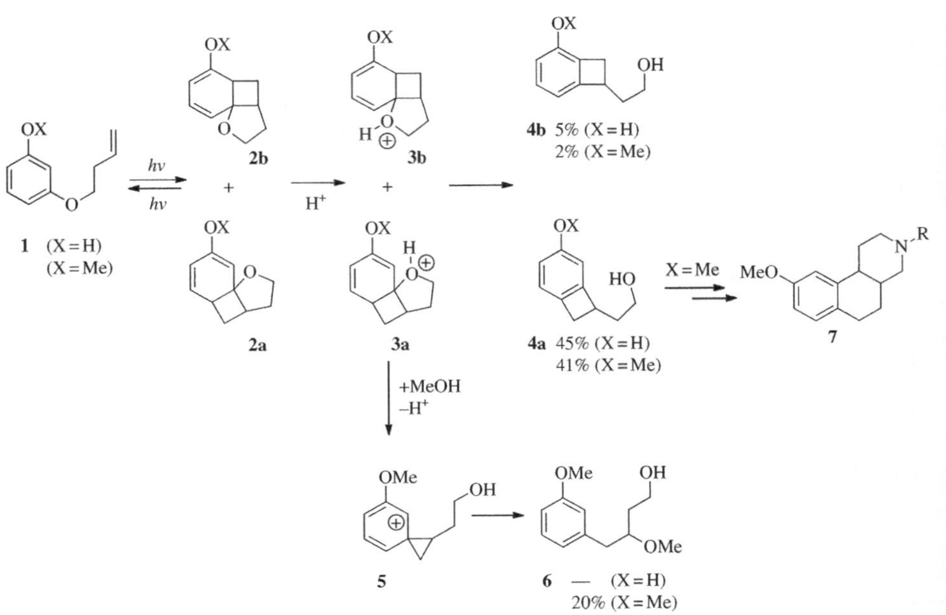

SCHEME 29.1 Acid-catalyzed [2+2] photocycloaddition of resorcinol derivatives. The irradiation was carried out at $\lambda = 254$ nm.

(**3a,b**), which leads to ring opening and to the formation benzocyclobutenes (**4a,b**) [12]. In a competing reaction, the protonated intermediates **3a,b** also rearrange to the spirocyclic intermediate **5**, which leads to the final product **6** when the reaction is carried out in methanol as solvent. Since the aromatic character is reestablished in the final products, compounds **4a,b** and **6** may be considered as products of an aromatic substitution reaction. The benzocyclobutene **4a** (X=Me) has been transformed into nitrogen-containing heterocycles such as **7** [13]. These compounds are rigidified analogues of dopamine and are therefore of interest for the treatment of Parkinson's disease.

A similar reaction has been performed with resorcinol derivatives possessing a naphthalene moiety on the side chain (Scheme 29.2) [14]. After excitation ($\lambda = 300$ nm) of the substrate **8**, an intramolecular exciplex is formed. Such steps are often reversible on the potential energy surface. The exciplex is also transformed into the [2+2] photocycloadduct **9**. As in the case of the transformation of **1** in Scheme 29.1, this photocycloaddition is reversible. The back reaction is induced by light absorption of **9**. Once again, protonation of the tetrahydrofuran moiety (**10**) leads to ring opening. Intramolecular trapping of the carbenium ion in **11** leads to the formation of the tetrahydrofuran derivative **12a,b**. In the overall reaction, again an aromatic substitution occurred on the resorcinol moiety.

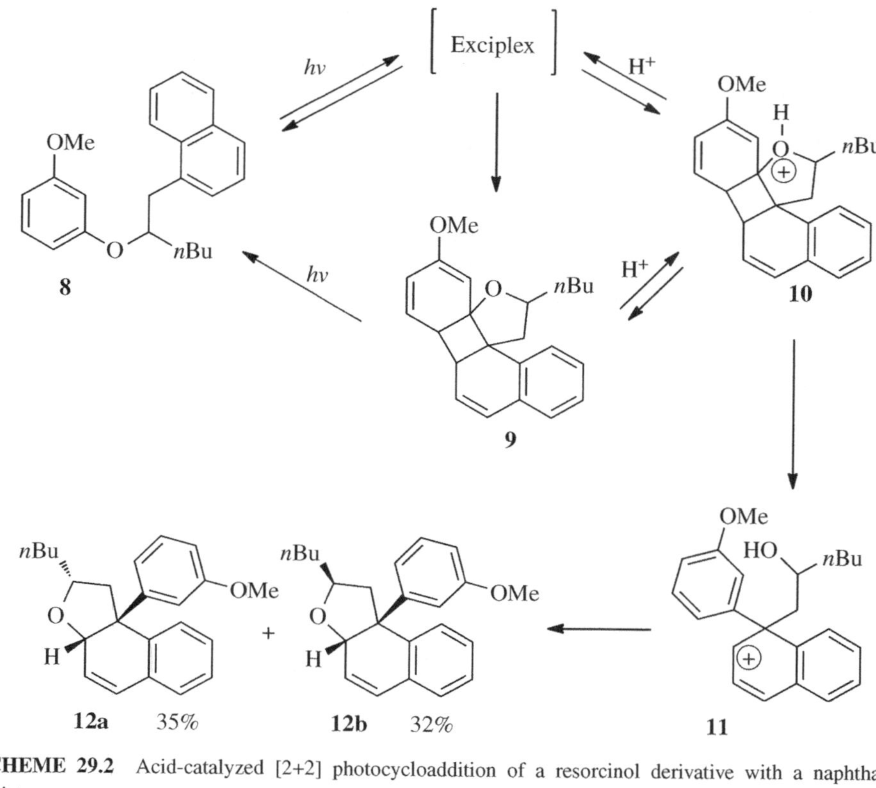

SCHEME 29.2 Acid-catalyzed [2+2] photocycloaddition of a resorcinol derivative with a naphthalene moiety.

Electrocyclic reactions of polyenes are controlled by the orbital symmetry [15]. Thermal or photochemical conditions lead to a complementary stereochemistry. Such reactions are also observed when aromatic moieties are involved. Thus, stilbene moieties undergo photocyclization leading intermediately to tricyclic compounds in which the aromatic character is suspended [16]. Often, an oxidation then leads to aromatic phenanthrene units. This reaction is a key step in the synthesis of helicenes [17]. Many helicene syntheses are based on this step. Helicenes are composed of *ortho* annellated benzene rings [18]. Steric repulsion due to face to face overlapping leads to the helix structure. Particular attention is paid to the chirality of these compounds and to their high optical activity [19]. Numerous applications are currently envisaged [20].

Stilbene derivatives such as compound **13** (Scheme 29.3) are easily obtained as *trans* isomers. In a photostationary equilibrium, the *cis* isomer **14** dominates. This isomer readily undergoes a photochemical conrotatory electrocyclization (compare Ref. [15]), and compound **15** is formed [21, 22]. Such products are easily oxidized due to an aromatization. Thus, the final heptahelicene **16** is obtained. In the present case, iodine was used. The resulting hydroiodic acid often causes decomposition of either the starting material or the final products. Such side reactions may be avoided when epoxides capable of trapping HI are added to the reaction mixture [23]. The reaction conditions are further optimized by using continuous flow reactors [24]. It must further be pointed out that the cyclization of compound **14** is regiospecific *via* C—C bond formation in positions 4 and 4′. Only compound **15** is formed. The alternative cyclization in position 4 and 2′ in conformer **17** leading to the final product **18** is not observed. The prediction of the regioselectivity of such reactions is possible using simple Hückel calculations and the determination of the sum of the free valence numbers for the excited state [17, 22, 25]. These parameters may also be calculated at a higher level. The sum of the free valence numbers for the excited state ΣF^*_{rs} for two positions r and s is defined as $F_r^* + F_s^* = (\sqrt{3} - \Sigma P_r^*) + (\sqrt{3} - \Sigma P_s^*)$. ΣP^* is the sum of the bond orders at the excited state in a particular position. The following tendencies are observed. The photocyclization does not occur when $\Sigma F^*_{r,s} < 1$. When two reactions are in competition, only one of them is observed when $\Delta\left(\Sigma F^*_{r,s}\right) \geq 0.1$ between the two positions. Furthermore, in the case when $\Sigma F^*_{r,s} > 1$ and when the formation of planar product is in competition with the formation of a nonplanar product, the first one is favored. In the case of compound **14**, $\Sigma F^*_{2,2'} = 0.927$. For the formation of compound **18** *via* conformer **17**, $\Sigma F^*_{4,2'} = 1.027$. Compound **18** or the resulting pentahelicene substructure is nonplanar. The formation of compound **15** is favored since $\Sigma F^*_{4,4'} = 1.126$. Generally, the regioselectivity leading to helicenes is favored. For this reason, the electrocyclization of stilbene derivatives is one of the standard methods for the synthesis of these compounds.

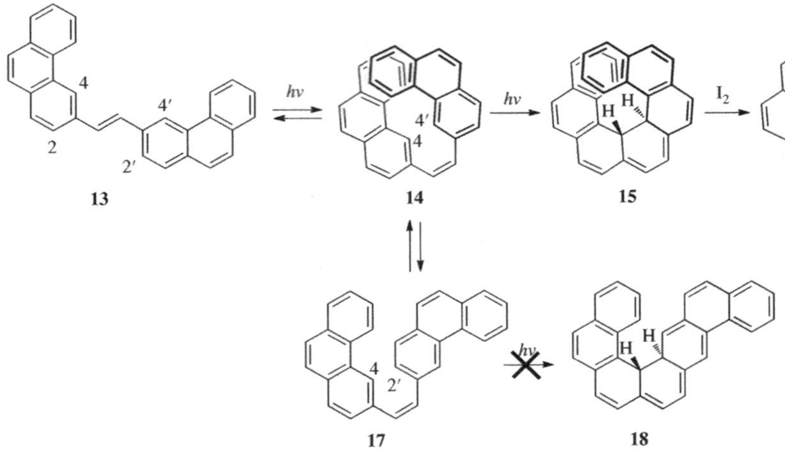

SCHEME 29.3 Photochemical electrocyclization applied to the synthesis of helicenes. A Hanovia 450 W medium-pressure mercury vapor lamp (quartz well) was used for UV irradiation.

Multiple electrocyclizations have been used for the synthesis of higher helicenes, and once again, the determination of the sum of the free valence numbers for the excited state may be used to optimize the synthesis. In compound **19**, three phenanthrene moieties are linked by two ethylene moieties (Scheme 29.4). Photocyclization leads the undecahelicene **20** in moderate yield. In compound **21**, the alkene moieties are arranged in a different way [26]. The corresponding cyclizations are much more efficient, and **20** is obtained in an excellent yield. The positions 1 of the benzophenanthrene units in **21** are particularly reactive. Generally, the presence of simple isolated benzene units in the precursors leads to a higher selectivity. In this way, helicenes containing up to 14 benzene rings have been synthesized [26].

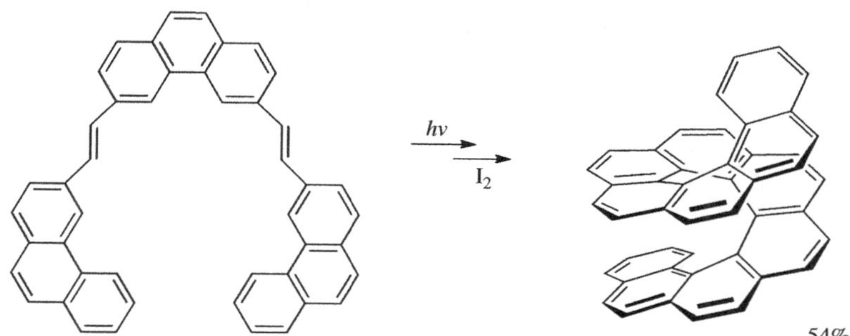

19 **20**

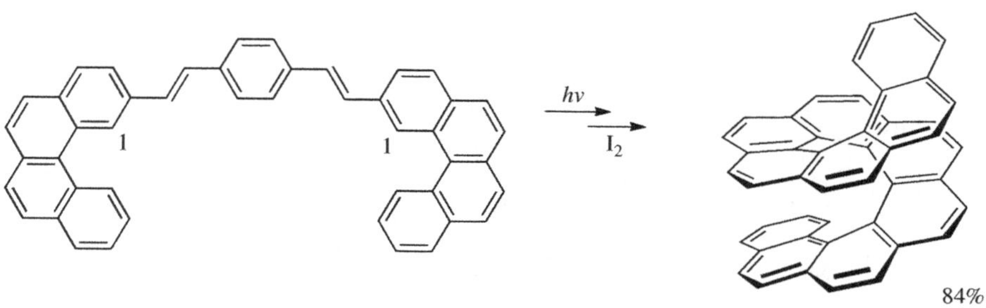

21 **20**

SCHEME 29.4 Photochemical electrocyclization applied to the optimized synthesis of higher helicenes. A Hanovia 450W high-pressure mercury vapor lamp (Pyrex well) was used for UV irradiation.

Photochemical electrocyclic reactions are also carried out with heteroatom-substituted precursors. Typical examples are depicted in Scheme 29.5. Amides are highly polar functional groups. Thus, the zwitterionic mesomeric structure **23** of the α,β-unsaturated amid **22** significantly contributes to the description of this functional group. Upon irradiation, a conrotatory electrocyclic reaction therefore takes place, and the intermediate **24** is formed. After 1,5-hydrogen shift, two diastereoisomers of the final products **25a,b** are obtained [27]. In this step, the aromaticity of the phenyl substituent is regenerated. The diastereomeric ratio **25a**:**25b** depends on the solvent used for the transformation and varies between 15:1 when the reaction is carried out in diethyl ether and 2:5 when carried out in methanol. The position of functional group with respect to the alkene and the phenyl moiety may be inverted as in the case of the enamine derivative **26** [28]. Once again, the photochemical electrocyclization *via* intermediate **28** may easily be explained by the mesomeric structure **27**. In this case, only one stereoisomer (**29**) was obtained. Both reactions have also been performed in the presence of oxidants such as iodine. Under such conditions, the alkene moiety is also reestablished in the final products. The reaction was also applied to the synthesis of natural products such as crinan [29], and a similar one was used as key step in the synthesis of aspidospermidine [30]. The photochemical electrocyclization can also be performed with highly functionalized enamine derivatives. A more recent report describes the transformation of **30** into the tricyclic compound **32** (Scheme 29.5) [31]. In this case, the negative partial charge at the oxygen atom of the amide function is further stabilized by a hydrogen bond as it is also depicted in the intermediate **31** with the polar mesomeric structure. It must further be pointed out that almost all substrates of such transformations must carry an additional substituent on the amide function. The privileged conformation of corresponding secondary amides is unfavorable for electrocyclization. Only few additional steps were needed to transform the photochemical product **32** into either (+)-pancratistatin or (+)-narciclasine. These natural products and several of their derivatives possess antitumor activities.

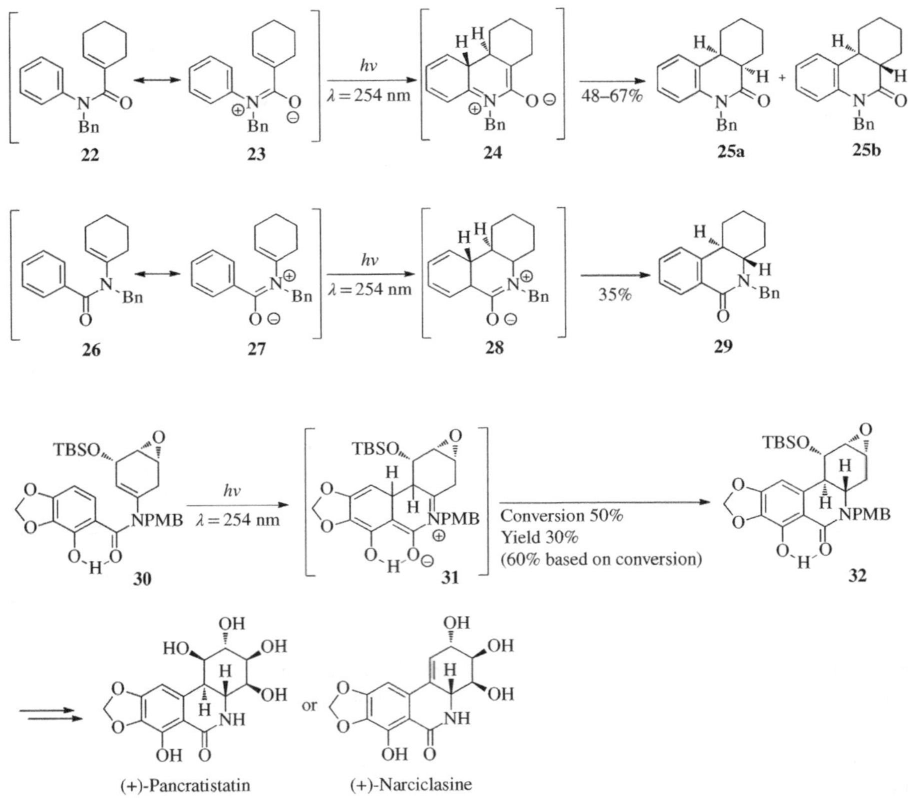

SCHEME 29.5 Photochemical electrocyclization of amides.

29.2.2 Photoinduced Radical Reactions

As it has already been pointed out, electronic excitation modifies the redox potentials of chemical compounds [32]. Generally, electron transfer is facilitated when electronic excitation is involved. The Rehm–Weller equation (Eq. 29.1) [33] enables an estimation of the exothermicity of a photochemical electron transfer (free enthalpy of electron transfer: ΔG_{ET}). Even when the electron transfer at the ground state ($E^0(D^+/D) - E^0(A/A^-)$) is endothermic, this may be compensated by the excitation energy (E_{0-0}). The attraction of the resulting ions is given by the term w, which is derived from Coulomb's law. It is reduced when reactions are carried out in a polar reaction medium, thus stabilizing the ions with respect to back electron transfer. Most frequently, in the case of organic compounds, radical ions are generated [34]. These intermediates may generate final products, or they may lead to neutral radical intermediates:

$$\Delta G_{\mathrm{ET}} = E^0(\mathrm{D}^+/\mathrm{D}) - E^0(\mathrm{A}/\mathrm{A}^-) - E_{0-0} - w \quad (29.1)$$

When compared to electrophilic aromatic substitution, nucleophilic aromatic substitution is not very common at the ground state. Under photochemical reaction conditions, however, such transformations become very efficient when an electron transfer is involved [35, 36]. Such transformations frequently occur according to a $S_{RN}1$ mechanism [37]. A classical example is depicted in Scheme 29.6 [38]. 1,4-Dicyanobenzene **33** is photochemically excited to the singlet state. An electron transfer from allyltrimethylsilane **34** to **33** is now exothermic, and the radical ion pair **35** and **36** is generated. In the radical cation **36**, the C—Si bond is elongated due to the delocalization of the positive charge toward the silicon atom. This is an element of stabilization of **36** and reduces back electron transfer leading to the starting compounds. Radical combination with the radical anion leads to the anion intermediate **37** and the silyl cation **38**. In the last step, cyanide is eliminated, and the final product **39** is formed in high yield. The released cyanide combines with **38**, which yields **40**. Acetonitrile is used in many cases of photochemical electron transfer-induced reactions. This polar solvent is capable of stabilizing the radical ion pair obtained after the electron transfer. Generally, this solvent is inert. In few cases, however, it traps cation intermediates [39]. When 1,4-dicyanobenzene **32** is transferred under the same conditions with isobutene **41** in acetonitrile as solvent, photochemical electron transfer leads to the corresponding radical ion pair **35** and **42** (Scheme 29.6) [40]. The cation **42** is trapped by acetonitrile, and the radical cation **43** is formed. This step also reduces back electron transfer. Radical combinations with the radical anion **35** lead to the zwitterionic intermediate **44**. The bicyclic compound **45** is generated by immediate charge combination. Finally, elimination of HCN generates the dihydroisoquinoline derivative **46**. As already pointed out several times, back electron transfer at the stage of the radical ion pair often considerably diminishes the efficiency of such reactions. When carried out using photochemical electron transfer sensitization, such side processes are inhibited, and the transformations are improved [36]. Recently, the substitution of a cyano group by attack of alkyl radicals was carried out with cyanopyridines [41]. The alkyl radicals were generated using triplet sensitization. A variety of pyridine derivatives has been obtained, some of them of particular interest due to biological activity.

Halogen-substituted aromatic compounds are transformed in a similar way [42]. Such transformations may be highly efficient. For this reason, they are applied as a key step in multistep syntheses. A macrocyclization in compound **47** was carried out in this way (Scheme 29.7) [43]. Compound **47** contains an indole and a dihydroindole moiety, which must be coupled to complete the construction of the carbon skeleton of the final product (−)-diazonamide A. After electronic excitation by UV light irradiation, intramolecular electron transfer from the more electron-rich acetoxylated indole moiety to the brominated dehydroindole occurs. After immediate deacylation, the radical anion **48** is formed. The presence of lithium ions in the reaction mixture increases the polarity of the medium and stabilizes the radical anion **48**, both inhibiting back electron transfer. Release of LiBr and radical combination then leads to the coupled intermediate **49**. The final product **50** of this coupling of a benzene ring and an indole moiety is generated by tautomerization. The carbon skeleton is built up, and only functional group transformations has to be carried out to obtain (−)-diazonamide A. Among them, the phenolic hydroxyl function was removed. In this context, it must be pointed out that the presence of the corresponding acetate group in the substrate **47** considerably favored the electron transfer step and thus improved the overall synthesis yield. The transformations of similar compounds without this acetoxy group were much less efficient [44]. The initial determination of the structure was wrong, and the total syntheses of this compound were also necessary to determine the correct molecular

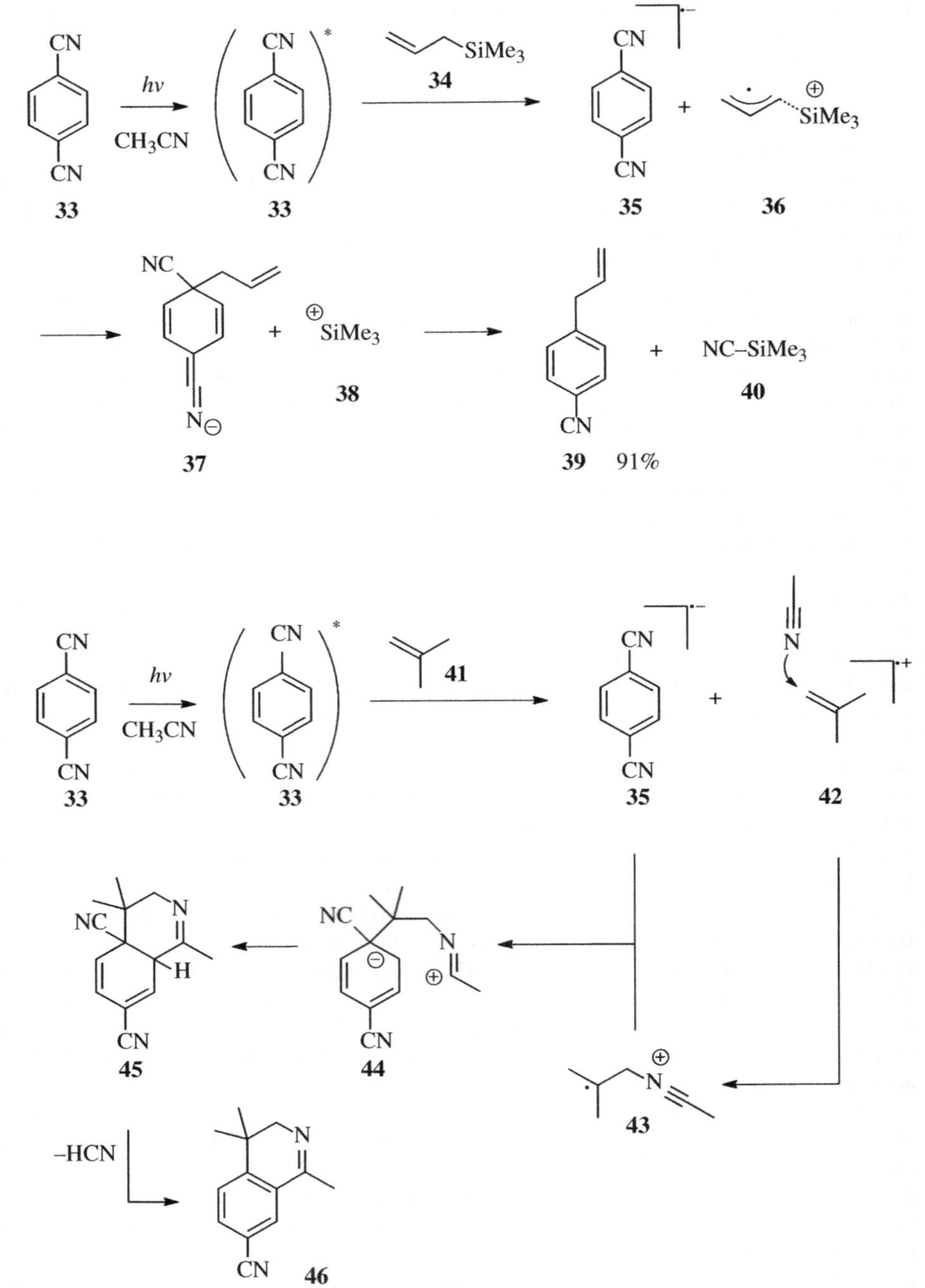

SCHEME 29.6 Photochemical electron transfer-mediated $S_{RN}1$ reactions. The irradiations were carried out with a high-pressure mercury lamp.

structure [43, 45]. Intensive synthesis studies have been performed on (−)-diazonamide A due to its cytostatic properties.

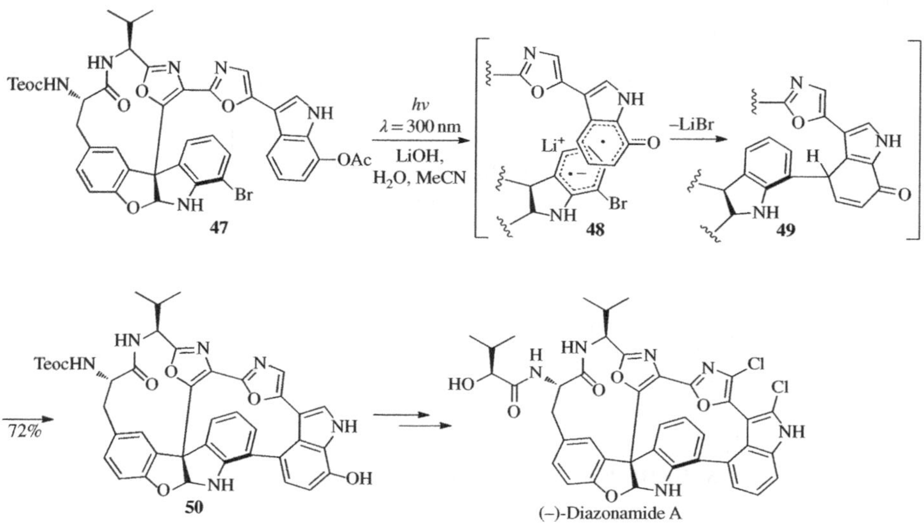

SCHEME 29.7 Photochemical electron transfer-mediated macrocyclization.

The previously reported reaction is close to the Witkop reaction, which has been widely applied to organic synthesis mainly to the synthesis of indole derivatives [46]. The reaction is most frequently carried out with α-chlorocaronic acid amides such as compound **51** (Scheme 29.8) [47]. According to the generally accepted mechanism, a photochemically induced electron transfer takes place from the indole moiety to the chloroacid amide leading to an intramolecular radical ion pair **52**. After elimination of chloride (**53**), radical combination often occurs *via* C—C fond formation on the benzene moiety in position 4 of the indole (**54**). When the reaction is carried out in dry acetonitrile as solvent, only this reaction is observed. After deprotonation and elimination of HCl, the tricyclic final product **55** possessing an eight-membered ring was obtained. The reaction was also carried out in acetonitrile containing water. In this case, a competition between a reaction in position 4 and 2 is observed. To a less extent, the radical combination step takes place in position 2 of the indole system leading *via* intermediate **56** to the final product **57**. In this case, HCl is eliminated by an intramolecular nucleophilic substitution generating a cyclic ether moiety in **57**. Under these reaction conditions, the intermediate **54** resulting from a radical combination in position 4 is transformed into compound **58**, which is isolated as a 4/1 mixture of epimers. This mixture of stereoisomers was transformed into (−)-indolactam V, a natural product that was isolated from *Streptomyces*. A further key operation in this total synthesis is the ring expansion from an eight-membered to a nine-membered ring. The same type of a Witkop reaction was also applied to the synthesis of dragmacidin E in which a ring contraction from the eight-membered ring to a seven-membered ring was carried out [48].

The Witkop reaction was also applied to the synthesis of (+)-decursivine (Scheme 29.9) [49]. Photochemical cyclization of the 5-hydroxytryptophane derivative **59** followed by elimination of HCl leads to the intermediate **60**. It was shown by a detailed mechanistic study that this intermediate undergoes photochemical cyclization in order to generate the benzotetrahydrofuran moiety. Two diastereoisomers **61** and **62** are obtained in a ratio of 1/6. The major isomer **62** doesn't possess correct configuration in the α-position of the lactam function. A base-catalyzed epimerization to **63** is therefore necessary to advance in the synthesis. (+)-Decursivine was isolated from leaves of

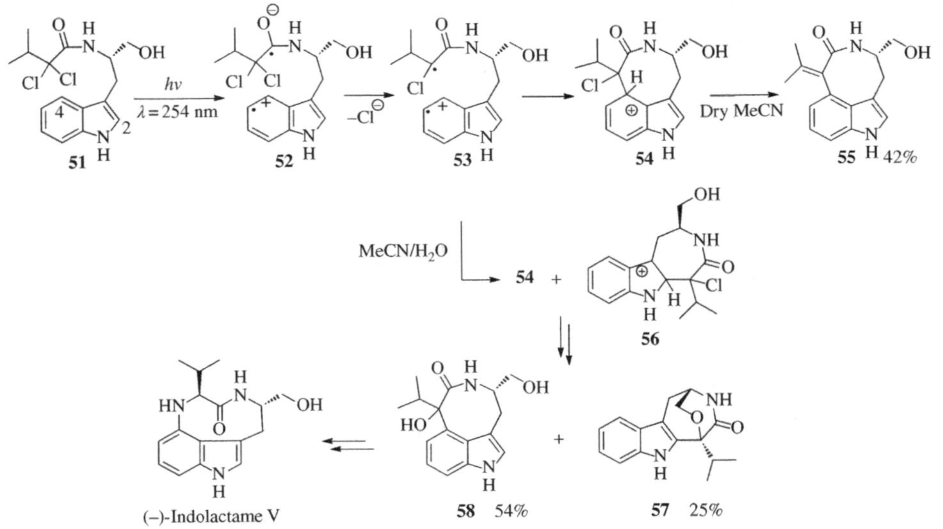

SCHEME 29.8 Various mechanistic steps in a Witkop reaction.

Rhaphidophora decursiva Schott (Araceae). It is active against the chloroquine-resistant malaria parasite *Plasmodium falciparum*. For this reason, such compounds are currently of high interest in pharmacy. The structural analogue serotobenine has been synthesized using the same strategy. For further synthesis studies, see Ref. [50].

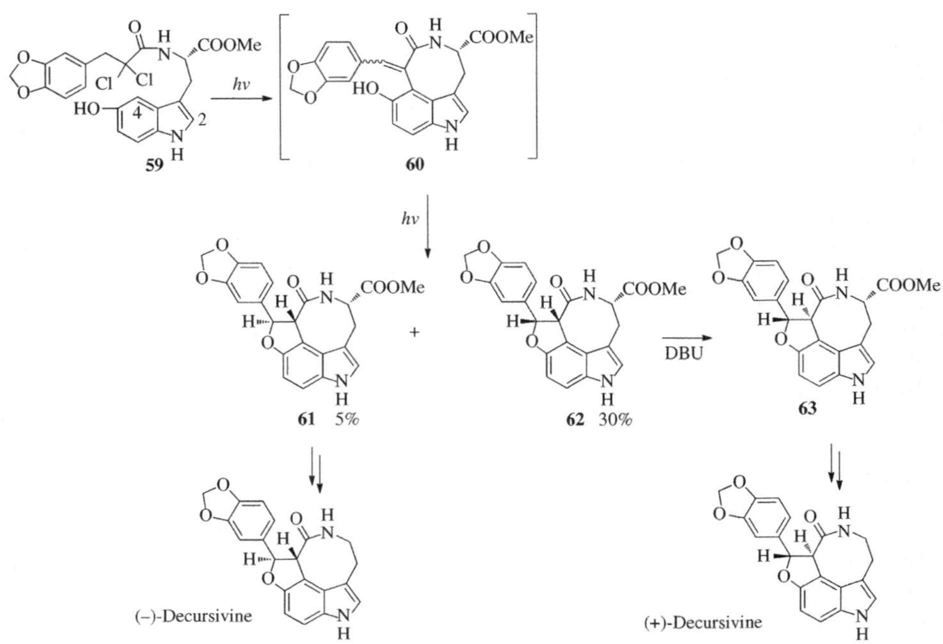

SCHEME 29.9 Witkop reaction applied to the synthesis of a polycyclic natural product. The irradiations were carried out with a high-pressure mercury lamp.

The aspects of photochemical electron transfer-mediated generation of radical intermediates are also well discussed for transformation of naphthylamine derivatives such as **64** with electron-deficient alkenes such as **65** (Scheme 29.10) [51]. The adducts **66** and **67** are obtained in high yield. The addition of nucleophilic species in the α position of such α,β-unsaturated lactones is unusual when compared to ground state reactivity of these compounds. Only few examples of photochemical electron transfer conditions have been reported for similar reactions [52]. Furthermore, the transformation is interesting from the sustainable chemistry point of view since a C—C bond is generated without chemical activation, which diminishes considerably the formation of side products. The photon is considered as a traceless reagent. In contrast to the previously discussed examples for aromatic nitriles or halogen-substituted aromatics, no leaving group is necessary, and the reaction can be discussed in the context of C—H activation [53]. Products such as **66** are particularly interesting due to their pharmaceutical activities [54]. An electron transfer from the naphthylamine derivative **64** to the α,β-unsaturated lactone **65** is endothermic at the electronic ground state. However, after excitation of **64**, electron transfer becomes exothermic, and a radical ion pair (**66**) is generated. From the results of a mechanistic investigation, it was concluded that the radical ions remain in contact, which favors back electron transfer regenerating the substrates. Thus, the quantum yield ($\Phi = 0.1$) is relatively low. In the presence of water, the radical anion is protonated in the β position leading to the neutral oxoallyl radical **70**. The radical cation **69** is released. An electron is transferred from the amine **64** to this electrophilic radical. The enolate **71** and the radical cation **69** are formed. Charge combination occurs in position 4 and to a lower extent in the position 5 (**72**, **73**). The formation of the minor isomer **67** is unusual, and it was explained by the overall electron distribution in the radical cation **69**.

The direct addition of electrophilic radicals such as fluoroalkyl radicals, without previous electron transfer, leads to substitution in positions 4 and 2 [55]. Recently, such a reaction was carried out using the photochemical electron transfer from an electronically excited naphthylamine derivative similar to that one shown in Scheme 29.10 to induce a radical chain reaction [56].

Radical ion species may also be generated by photochemically induced bond cleavage without previous inter- or intramolecular electron transfer. Upon irradiation, halogenated benzene derivatives react under heterolytic carbon halogen bond cleavage and release of a halogenide ion [57, 58]. Such reactions are particularly efficient when the aromatic compound possess an electron donor substituent in the *para* position such as in **74** (Scheme 29.11). The resulting cation **75** undergoes intersystem crossing to reach its triplet ground state **76**, which possesses less strain in the six-membered ring [59]. The vibrationally relaxed structure resembles that of a radical cation or a σ-radical (for the structure of radicals, see Ref. [60]). Such a cation generated from *p*-chloro-*N,N*-dimethylaminobenzene **77** readily adds to variety of electron-rich aromatic compounds such as 2-methylfuran **78**, which mainly leads to the adduct **79** (Scheme 29.11) [61]. The reaction is similar to an electrophilic aromatic substitution. Therefore, minor amounts of a coupling product **81** resulting from an addition with **80** are also formed. The reduction of the corresponding cation **75** or **76** leads to the formation of a second minor side product **80**. The main reaction can also be compared to a Suzuki reaction in which no palladium catalysis is needed and no borate leaving group is required on the nucleophilic reaction partner. In a similar reaction, the sesamol derivative **82** and related compounds are coupled with terminal alkynes such as **83** [62]. The reaction is induced by triplet sensitization with acetone. After electronic excitation and release of chloride, the resulting cation **84** reacts with the alkyne **83** leading to the spirocyclic intermediate **85**. Rearrangement and deprotonation generates the final product **86**. This transformation corresponds to a Sonogashira reaction in which no copper or palladium catalysts are needed. Such metal-free transformations are highly recommended in the context of sustainable chemistry. Phenol derivatives instead of corresponding halides have been transformed in the same way when the phenol group was transformed into a nonaflate substituent ($ArO{-}Nfs = ArO{-}SO_2C_4F_9$) as leaving group [63]. Reactions of various substituted aryl cations with a variety of nucleophiles have been reported. For recent examples, see Ref. [59] and citations therein.

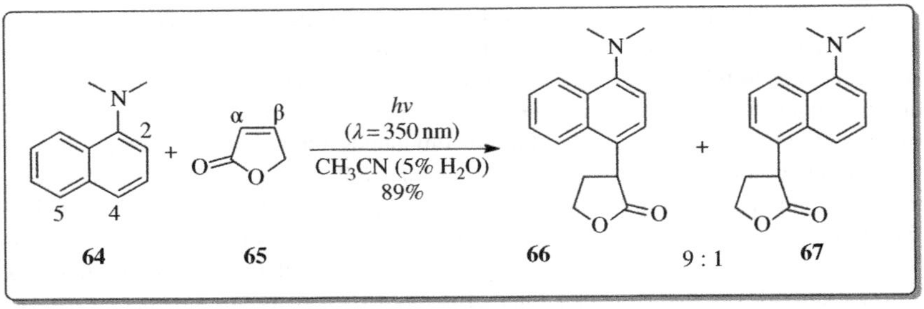

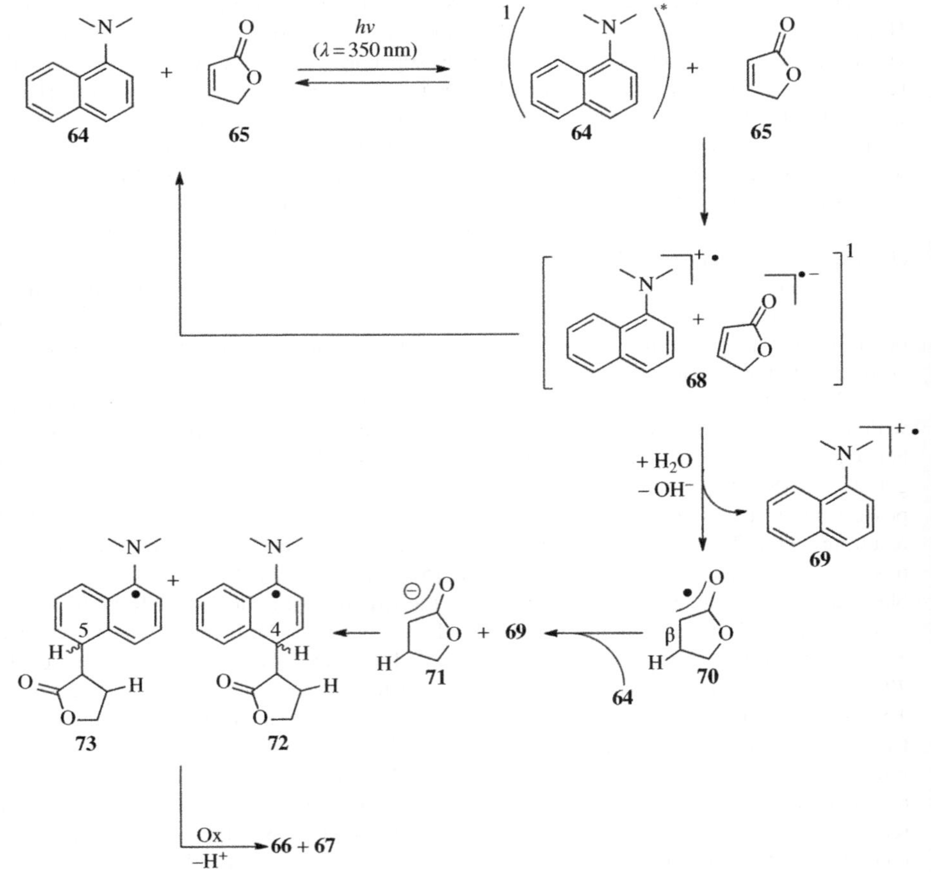

SCHEME 29.10 Photochemical electron transfer-mediated **α**-selective addition of a naphthylamine derivative to an **α**,β-unsaturated lactone.

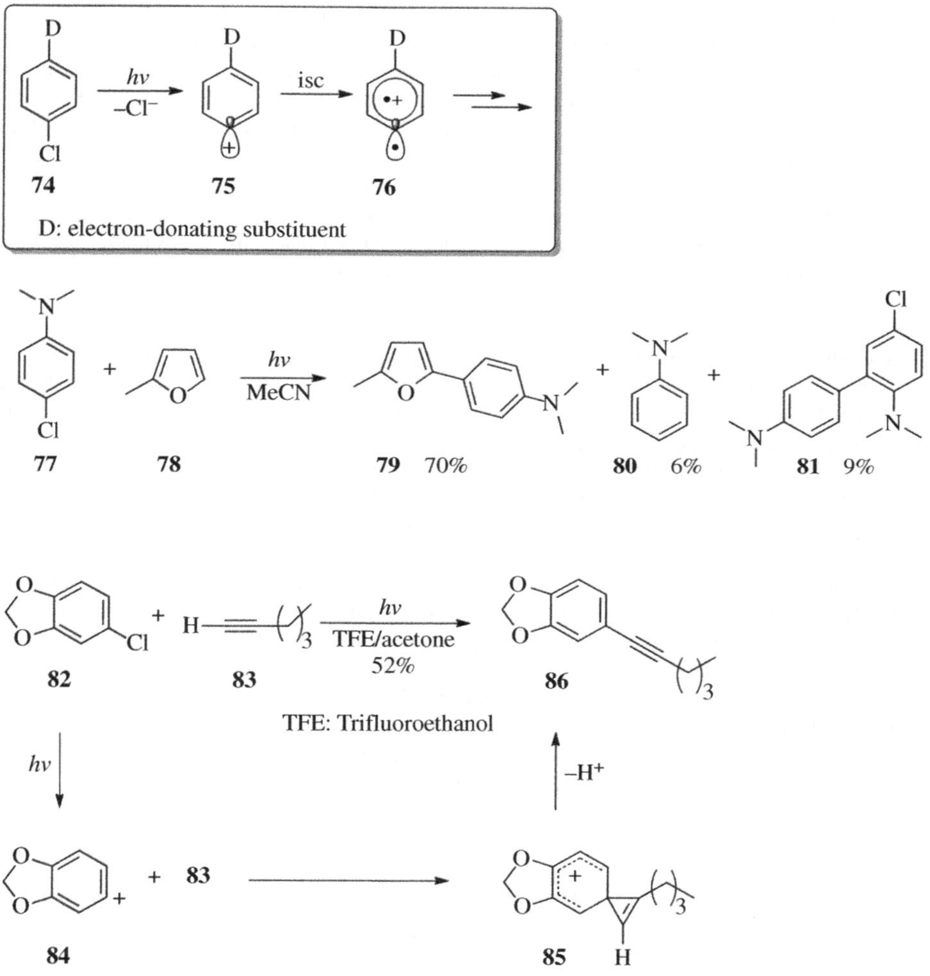

SCHEME 29.11 Metal-free Suzuki and Sonogashira analogous transformation. The irradiations were carried out with high-pressure mercury lamps.

29.3 PHOTOSENSITIZED AND PHOTOCATALYZED REACTIONS

29.3.1 Metal-Catalyzed Reactions

Historically, the photochemistry of aromatic compounds was mainly concerned by the reactivity of (π,π^*) electronic excited state arising from direct irradiation of these chromophoric compounds. On this respect, the photochemistry of aromatic compounds is a powerful illustration of how the electronic excited state significantly changes the reactivity of organic compounds. The photochemical reactions of aromatic compounds that lead to the formation of a new bond on the aromatic ring have not been subject to great attention. This lack of attention could be explained by the huge amount of methodologies involving transition metal catalysis allowing the formation of aromatic-carbon or aromatic heteroatom bonds. Notable exceptions involving radical intermediates and phenyl cation with triplet character are described in Section 27.2.2 of this chapter.

The use of photocatalysis for an efficient access to highly reactive radical intermediates by single-electron transfer has been extensively studied in recent years [3]. These studies provide a nice illustration of the expected benefits using light as activating reagent, such as soft activation of the organic compounds and increasing of the sustainability of the overall processes [8]. Selected examples of this powerful methodology based on the use of stable metal complexes as photocatalyst are discussed in this section.

A big part of these works involves the use of aryldiazonium salts as reaction partners. They are extremely popular reagents in organic synthesis as they significantly impacted the beginning of synthetic arenes chemistry through the emergence of named reactions, for example, Sandmeyer, Pschorr, Gomberg-Bachmann, and Meerwein reactions [64]. In such reactions, a stoichiometric or a catalytic amount of transition metal is used to generate the aryl radical intermediate. In 1984, a pioneer study [65] that takes advantage of photoredox properties of tris(2,2′-bipyridine) ruthenium(II) ion $[Ru(bpy)_3]^{2+}$ in organic synthesis [66] described the photocatalytic Pschorr reaction in phenanthrene series (Scheme 29.12). The arene diazonium **87** efficiently affords the phenanthrene **88** in quantitative yield upon selective excitation by visible light of an acetonitrile solution containing a catalytic amount of $[Ru(bpy)_3]^{2+}$. Based on a careful investigation, it was proposed that the mechanism involves single-electron reduction of arene diazonium salt **87** by the electronically excited state of ruthenium complex $^*[Ru(bpy)_3]^{2+}$(oxidative quenching). An intramolecular cyclization of the free aryl radical **89** then occurs through a radical pathway. The oxidation of the resulting cyclized radical **90** by $[Ru(bpy)_3]^{3+}$ leads to a Wheland-type cationic intermediate. The phenanthrene **88** is then obtained by loss of a proton. According to this reaction mechanism, it is interesting to note that both reductive and oxidative steps of the photocatalytic system are involved in the product formation. As the completion of catalytic system doesn't require a sacrificial electron donor or acceptor reagent, the reaction proceeds with high atom economy efficiency. This study highlights the synthetic utility of visible light photoinduced electron transfer to achieve selective reaction. Indeed, when a solution of arene diazonium **87** in acetonitrile is irradiated by UV light, the complete degradation of this compound is observed, while only a small amount (10–20%) of the phenanthrene **88** is formed. Furthermore, it was demonstrated that the main product observed in this reaction condition could be explained by the addition of acetonitrile on phenyl cation resulting from the photoinduced homolytic cleavage of diazonium salts **87**.

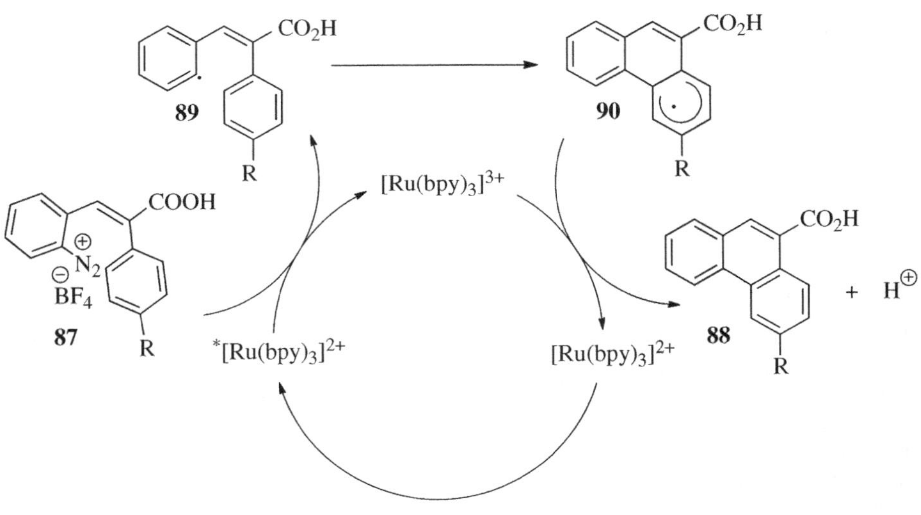

SCHEME 29.12 Synthesis of phenanthrene by photocatalytic Pschorr reaction.

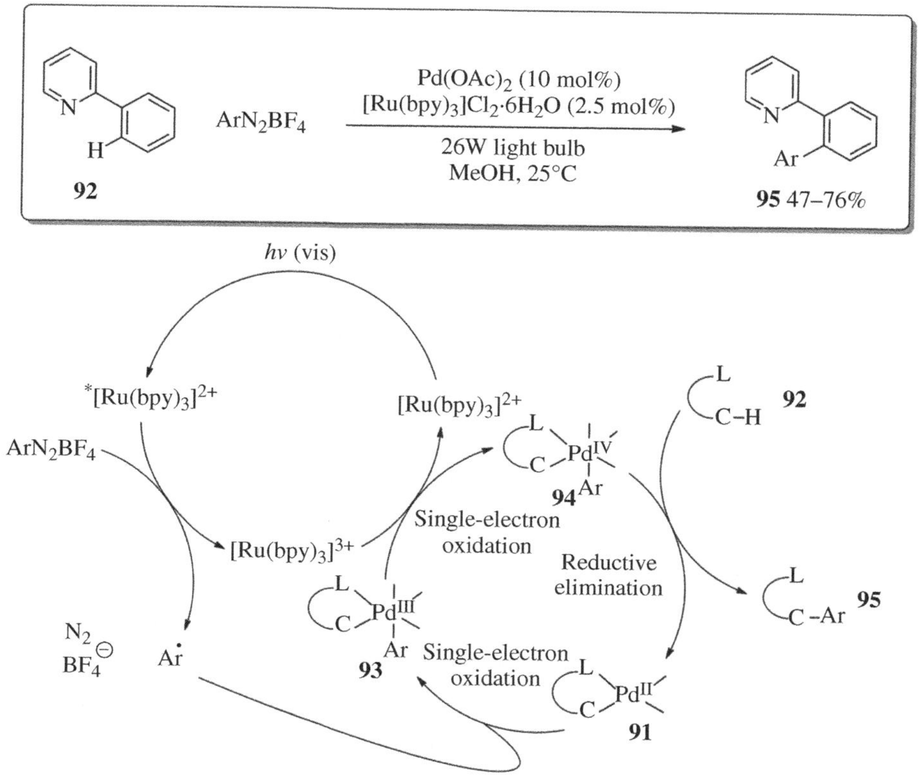

SCHEME 29.13 C—H functionalization at room temperature by merged photocatalysis and high-valence palladium catalysis.

Twenty years after this pioneer work, a huge amount of studies dealing with photocatalysts applied to organic synthesis appeared in the scientific literature. These recent studies have pushed the tris(2,2′-bipyridine)ruthenium(II) ion $[Ru(bpy)_3]^{2+}$ to the class of privileged photocatalyst. In this context, the synthetic utility of arene diazonium salts was also reevaluated through numerous reports on efficient direct functionalization of arenes derivatives [67]. The field of arene chemistry was recently strongly impacted by the development of metal-catalyzed methodologies allowing direct functionalization of the arene moiety, avoiding the use of activating groups, the so-called C—H activation [53]. The C—H functionalization strategies involving substrates that contain one or more functional groups able to chelate the metal catalyst and position it for selective C—H cleavage is well recognized [68]. According to this strategy, methods for the synthesis of biaryl compounds were described, using high-valent palladium catalysis (Scheme 29.13). The C—H arylation of amide and arylpyridine substrates was achieved using diaryliodonium salts as aryl donor and oxidative agent [69]. The overall process involved the formation of a Pd^{II} palladacycle, a two-electron oxidation of Pd^{II} to Pd^{IV} and reductive elimination affording the biaryl product and regenerating the Pd^{II}. Although the efficiency of the methodology was well established, notably in regard of the possibility to introduce selectively a wide range of substituted aryl ring at a C—H position, both high temperatures and use of acidic solvent were required to obtain the reaction products in high yields [70]. Based on mechanistic investigation, it was proposed to merge photocatalysis and high-valence palladium catalysis to perform this transformation at room temperature [71].

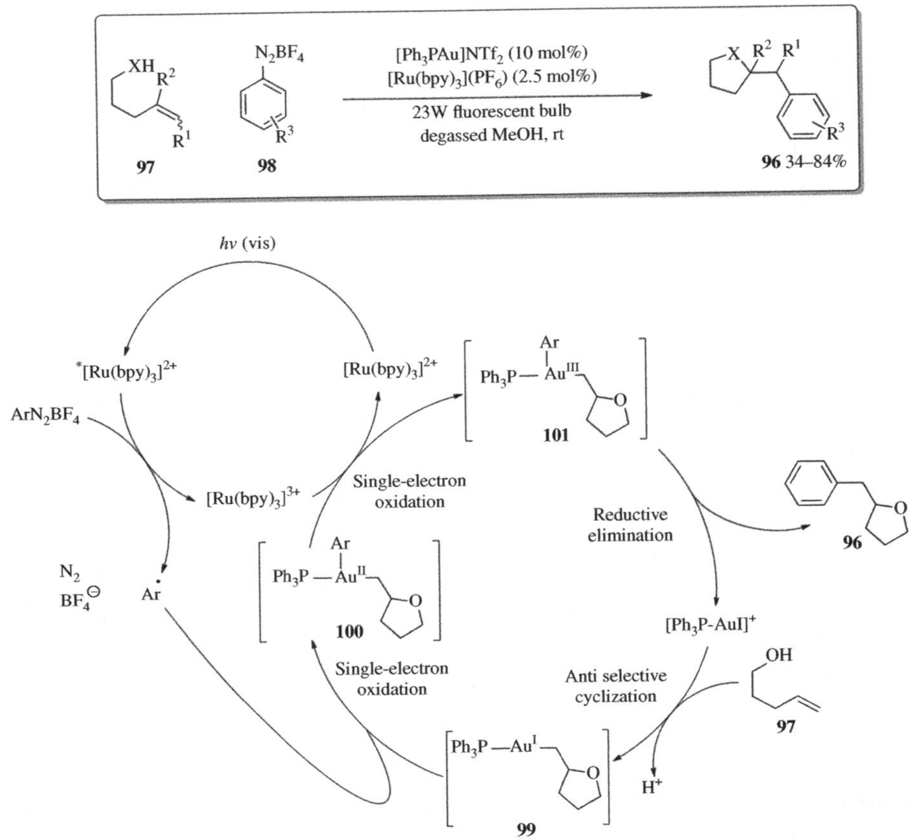

SCHEME 29.14 Combined photocatalysis and gold catalysis for oxy- and aminoarylation of alkenes.

The work hypothesis was that the aryl radical would be a more kinetically arylating reagents than diaryliodonium salts. Thus, a room temperature procedure, involving formation of aryl radical by photocatalysis, has been developed for the arylation of amide and arylpyridine derivatives leading the reaction product in high yields. According to the proposed mechanism, the palladium catalytic cycle is initiated by the formation of palladacycle **91** by C—H activation of the substrate **92**. The aryl radical generated through the photocatalytic cycle then undergoes an addition to palladium to afford the Pd^{III} intermediate **93**. A single-electron oxidation of this intermediate by [Ru(bpy)$_3$]$^{3+}$ regenerates the photocatalyst and afford Pd^{IV} intermediate **94**. The reaction product **95** is then obtained by C—C bond forming reductive elimination, and thus the Pd^{II} catalyst is regenerated.

A conceptually related strategy based on the combination of photocatalysis and gold catalysis was recently reported leading to an efficient oxy- and aminoarylation of alkenes (Scheme 29.14) [72]. Earlier works in gold catalysis mainly exploited the π-lewis acidity of gold and didn't imply a change of redox state of the Au^{I} or Au^{III} gold catalyst [73]. However, methodologies that involve Au^{I}/Au^{III} redox catalytic cycle have been recently described [74]. These methodologies extend the scope of gold catalysis by allowing the creation of a new C—X bond during the demetalation step in place of proto-deauration. In such processes, a strong external oxidant is required to achieve oxidation of Au^{I} to Au^{III}.

Compounds **96** are obtained in good yields starting from alkenes **97** bearing hydroxy or amino group and aryldiazonium salts **98** by combining gold and photoredox catalysis and thus avoiding the use of an external oxidant. According to the proposed mechanism, the gold catalytic cycle begins with alkene activation by Au^{I} catalyst followed by an *anti*-selective cyclization to form the Au^{I} intermediate **99**. The aryl radical formed through the photocatalytic cycle then adds to Au^{I} intermediate **99** entailing the formation of Au^{III} intermediate **100** by a single-electron oxidation process. A second single-electron transfer occurs from the intermediate **100** to $[Ru(bpy)_3]^{3+}$ to form the Au^{III} intermediate **101** and regenerates the $[Ru(bpy)_3]^{2+}$ photocatalyst. The new arene carbon bond is then formed by reductive elimination giving the reaction product **96**. The two synthetic methodologies described previously nicely illustrate how visible light photocatalysis can positively interfere with a metal catalytic cycle to access new reactive intermediates.

Based on success obtained with $[Ru(bpy)_3]^{2+}$ as photocatalyst to generate phenyl radicals by single-electron transfer, researchers reinvestigated the Meerwein-type reactions in such conditions [75]. The classic Meerwein reaction allows the carbon sp^2–sp^2 coupling of aryldiazonium and styrene (or electron-deficient alkenes). The accepted mechanism involves a radical process mediated by reversible oxidation of copper salts catalysts. Unfortunately, high catalyst loading is required (15–20 mol%), and low reaction yields are usually obtained (20–40%) due to the formation of side products, thus entailing limited applications of this reaction to organic synthesis. When a solution of aryldiazonium salts and styrene is irradiated at 455 nm in presence of 1 mol% $[Ru(bpy)_3]^{2+}$, substituted stilbenes **102** derivatives are obtained in moderate to good yields (55–87%) (Scheme 29.15). The reported procedure is efficient with a range of substituted aryl diazonium salts and tolerates a variety of functional groups. The proposed mechanism involves the attack of an aryl radical, formed by photocatalysis, to the olefinic double bond of styrene affording the benzyl radical **103**. A single-electron oxidation then occurs to form the carbenium ion **104**. This oxidation can be mediated either by $[Ru(bpy)_3]^{3+}$ complex regenerating the $[Ru(bpy)_3]^{2+}$ photocatalyst or by aryl diazonium salts initiating a radical chain mechanism. The stilbene **102** is then obtained after deprotonation.

The α-arylation of enol acetates was carried out following a related approach (Scheme 29.16) [76]. The reaction is especially efficient with aryldiazonium salts bearing a nitro substituent. The formation of the products can be explained by an analogous mechanism as described for the photocatalyzed coupling reaction between aryldiazonium salts and styrenes. The cationic species **105** is formed by aryl radical addition to enol acetate and single-electron oxidation. The α-arylated ketones are then obtained by elimination of an acyl cation, which is simultaneously trapped by the solvent. The synthetic interest of this methodology has been nicely illustrated by a two-step synthesis of a range of 2-substituted indoles starting from 2-nitrophenyldiazonium and easy accessible enol acetates (Scheme 29.16).

Recent studies have demonstrated the ability of photocatalysis based on $[Ru(bpy)_3]^{2+}$ and $Ir^{III}(ppy)_3$ to generate a great variety of synthetically valuable reactive radicals and allowed the development of innovative methods for the creation of aryl–carbon bonds. A methodology enabling a direct α C—H arylation of amines was developed using $Ir^{III}(ppy)_3$ as photocatalyst (Scheme 29.17) [77]. A key feature of the system is that completion of photoredox catalytic cycle by the activation of both reaction partners avoids the use of sacrificial reagents. Aromatic rings **106** bearing an electron-withdrawing group and a cyano group acting as leaving group are privileged aromatic reactant in this new reaction. The reactive radical anion **107** is obtained from aromatic compound **106** by single-electron reduction occurring from the iridium complex at the excited state $^*Ir^{III}(ppy)_3$. The resulting $Ir^{IV}(ppy)_3$ is then able to oxidize amine **108** leading to α-aminoalkyl radical **109** after loss of a proton and regenerating the photocatalyst. The key bond formation step then occurs by radical–radical coupling of compounds **107** and **109**. The elimination of the cyanide anion from species **110** then leads to the benzylic amine **111**. A wide range of benzyl amines was prepared in very good yield according to this highly efficient and operationally trivial reaction protocols.

The ability of $iridium^{III}$ photocatalytic cycle to induce concomitant reduction of cyanoaromatic and oxidation of amine was nicely illustrated further by the direct β-arylation of saturated ketones

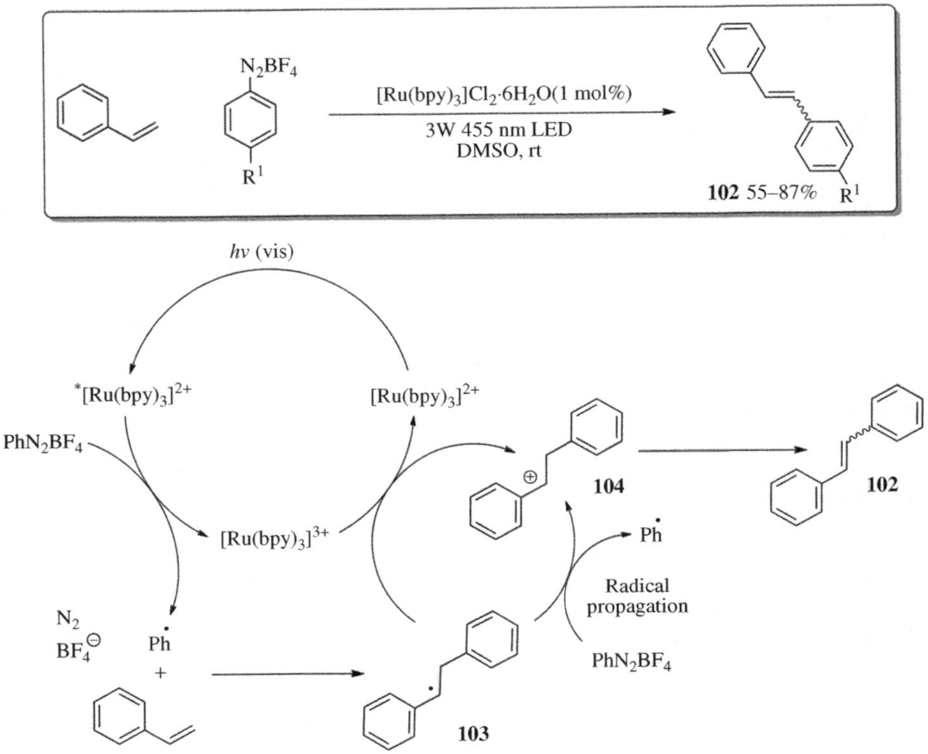

SCHEME 29.15 Photocatalytic Meerwein-type reaction.

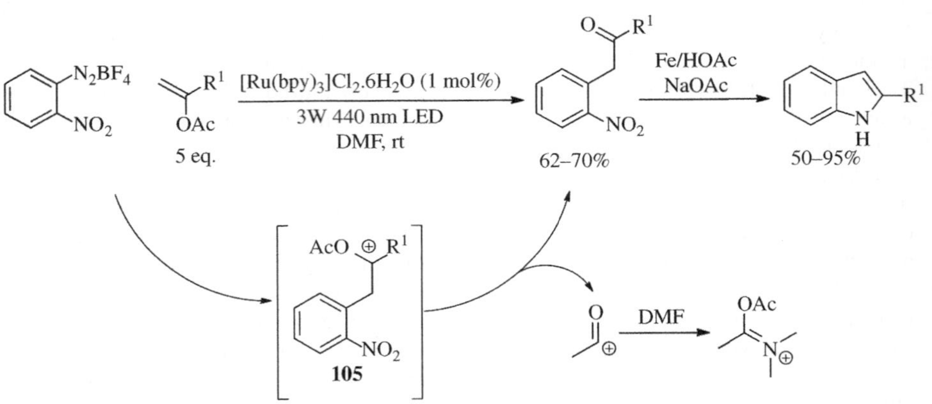

SCHEME 29.16 Photocatalytic α-arylation of enol acetates.

and aldehydes (Scheme 29.18) [78]. The chemistry of aliphatic aldehydes and ketones is dominated by nucleophilic addition to carbonyl group as well as a broad range of methodologies leading to α-functionalization. By contrast, methodology allowing direct β-functionalization of saturated aldehydes and ketones was previously unknown. By merging an organocatalytic cycle using secondary amine catalyst and photocatalysis, highly efficient β-arylation was developed. The two

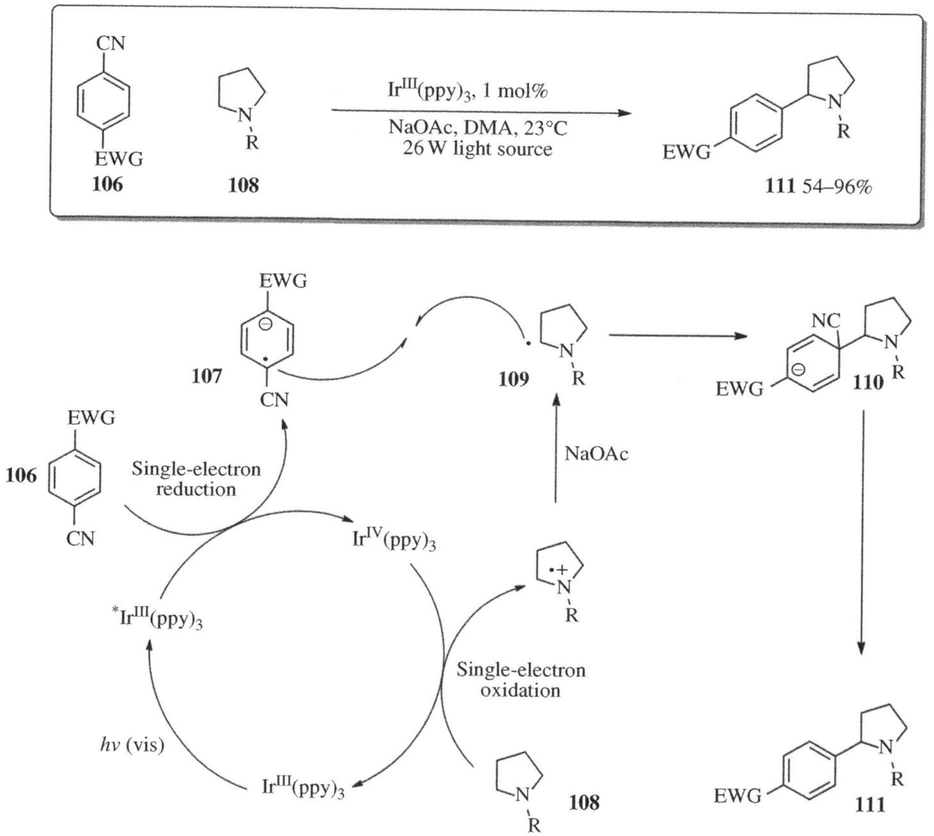

SCHEME 29.17 Photocatalytic α C—H arylation of amines.

reactive radical species are formed through the photocatalytic cycle. The single-electron reduction of aromatic reaction partner **112** by $^{*}Ir^{III}(ppy)_3$ affords the reactive radical anion **113** and the $Ir^{IV}(ppy)_3$ complex. The organocatalytic cycle begins with the condensation of the amine organocatalyst **114** and the aldehyde **115**. The resulting enamine is then oxidized by $Ir^{IV}(ppy)_3$ complex leading to the radical cation **116** and completing the photocatalytic cycle. The 5π electron intermediate **117** is then obtained by release of a proton. The radical coupling step between compounds **117** and **113** affords the anion **118**. The reaction product **119** is obtained after elimination of cyanide anion and hydrolysis of enamine regenerating the organocatalyst.

In strategies described earlier, the formation of reactive aromatic intermediate by photocatalysis is the key step. An alternative approach could be envisaged based on the photocatalytic generation of reactive intermediate capable of reacting with an inactivated arene. A typical example of this class of reactions is the direct photocatalytic trifluoromethylation of arenes (Scheme 29.19) [79]. The key issue of the methodology is the use of suitable trifluoromethyl reagent affording efficient formation of trifluoromethyl radical from ruthenium complex at the excited state ($^{*}[Ru(bpy)_3]^{2+}$). The single-electron transfer reduction of trifluoromethanesulfonyl chloride (CF_3SO_2Cl) by $^{*}[Ru(bpy)_3]^{2+}$ affords the CF_3SO_2Cl radical anion, which rapidly collapses to $^{\bullet}CF_3$, sulfur dioxide, and chloride anion. The addition of $^{\bullet}CF_3$ to arene derivatives yields the trifluoromethyl arene radical **120**. The Wheland cationic intermediate is then formed by a single-electron transfer to the strongly oxidizing $[Ru(bpy)_3]^{3+}$ to regenerate the photocatalyst. The deprotonation of the Wheland cationic

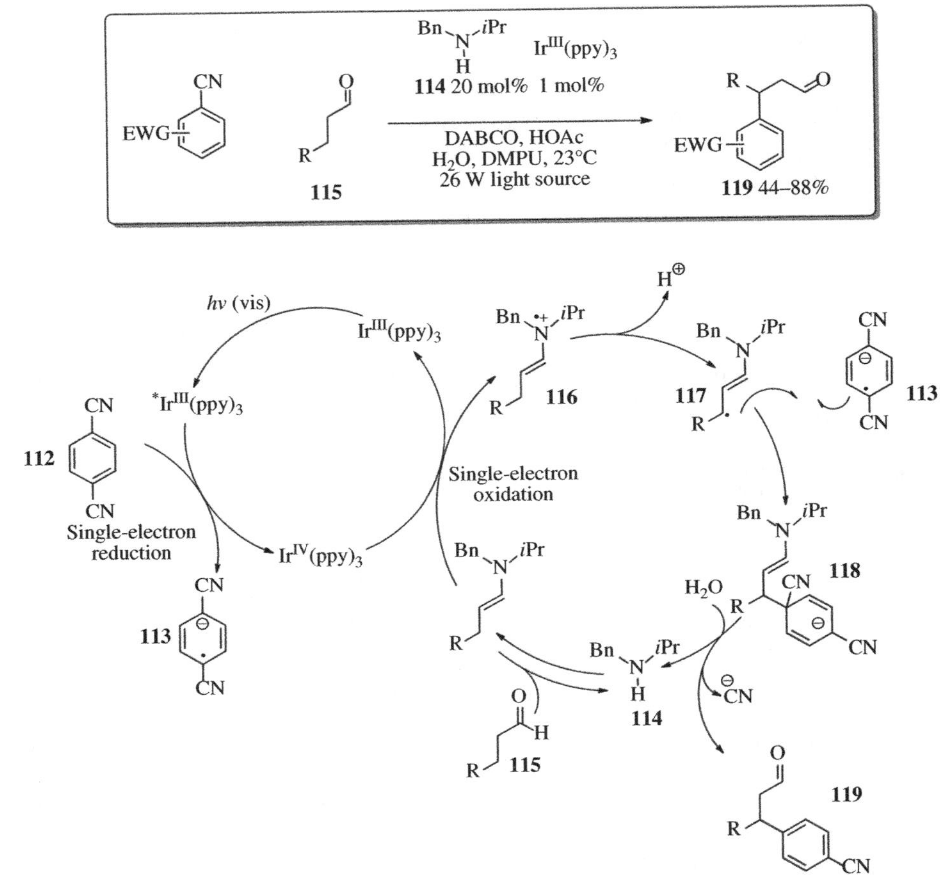

SCHEME 29.18 Photocatalytic direct β-arylation of saturated ketone and aldehydes.

intermediate **121** leads to the trifluoromethyl arenes in usually high yields. This simple and mild reaction conditions allow the trifluoromethylation of a wide range of arenes and heteroarenes without the need for arene prefunctionalization. Moreover, direct and regioselective trifluoromethylation of biologically active molecules is achieved, which provides new opportunities for pharmacomodulation by late-stage modification of pharmaceutical agents.

The mono-benzoyloxylation of electron-rich aromatic compounds was also achieved through the addition on aromatic of photocatalytically generated benzoyl radical [80].

29.3.2 Metal-Free Reactions

The use of substoichiometric amounts of organic molecules, which absorb light and induce photochemical transformations by sensitization either by energy or electron transfer to reaction partners, is a well-recognized strategy in organic photochemistry. Benzophenone derivatives have been widely used in this context.

A photochemically sensitized radical tandem addition cyclization reaction was carried out with aniline derivatives such as **122** (Scheme 29.20) [81]. Thus, tetrahydroquinoline derivatives such as **123a,b** are obtained in with diastereoselectivities around 90%. The reaction is highly efficient

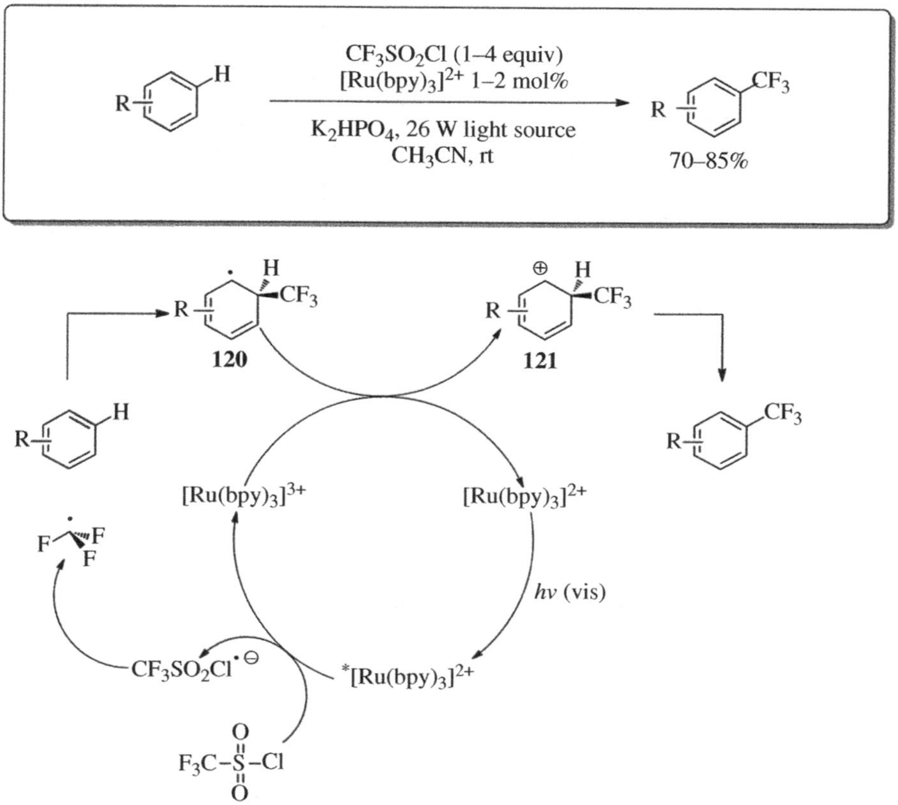

SCHEME 29.19 Photocatalytic direct trifluoromethylation of arene.

because electron donor-substituted aromatic ketones such as **124** are used as sensitizer [82]. These compounds are only needed in catalytic amounts and are recovered up to 90% after the reaction. Thus, these ketones become efficient photocatalysts. Due to high photoreduction and pinacolization rates [83], conventional sensitizers such as benzophenone are much less efficient. The reaction starts with a photochemical electron transfer from the tertiary amine **122** to the excited sensitizer. Proton exchange leads to the neutral radical intermediates **125** and **126**. α-Aminoalkyl radicals such as **126** are nucleophilic and therefore add easily to electron-deficient C=C double bonds as in the case of **127**. The resulting electrophilic oxoallyl radical **128** undergoes cyclization, and the tricyclic radical intermediate **129** is formed. In the presence of acetone as an oxidant, the latter intermediate is oxidized to the final product **123a,b**. Acetone may also act as a sensitizer. However, this compound is much less efficient as the aromatic ketones. The present transformation was carried out using homogeneous photocatalysis. The same reaction was also performed using heterogeneous photocatalysis with inorganic semiconductors as photoredox sensitizers [84]. The reaction has also been observed as a side process in the unsensitized photochemical transformation of α,β-unsaturated ketones with *N,N*-dimethylaniline **122** [85].

Since a few years, ruthenium- and iridium-based polypyridyl complexes are privileged photocatalysts in visible light photocatalysis. However, organic dyes should constitute a valuable alternative not only due to their relatively lower cost and wider availability but also by giving access to new transformations. The direct arylation of heteroarenes (e.g., furan, thiophene, and pyrrole)

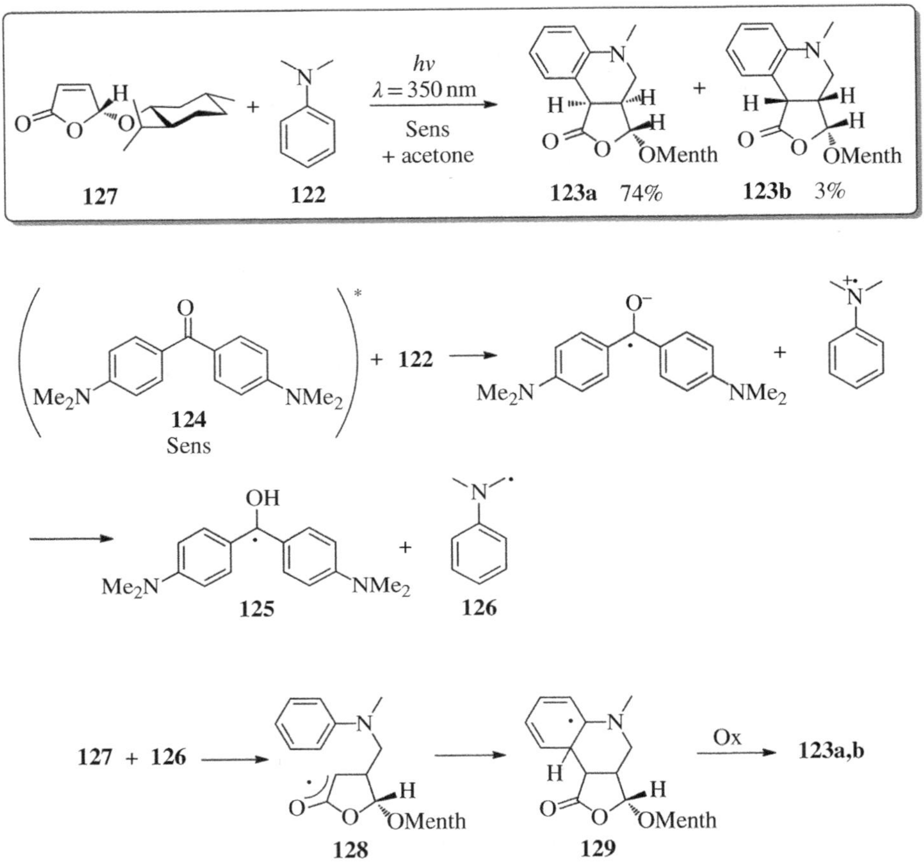

SCHEME 29.20 Photochemically sensitized radical tandem addition cyclization reaction.

with aryl diazonium salts using eosin as photocatalyst was reported (Scheme 29.21) [86]. According to the proposed mechanism, the eosin dye at the excited state is able to produce an aryl radical by single-electron reduction of the aryl diazonium salt leading to the formation of the eosin radical cation. The key bond formation step involves the addition of aryl radical to heteroarene affording the radical **130**. The formation of the Wheland-type cationic intermediate **131** occurs either by oxidation by eosin radical cation and thereby regenerating the photocatalyst or by electron transfer from aryl diazonium salts inducing radical chain propagation. The coupling product is then obtained after deprotonation. A wide range of substituted furans was obtained in moderate to good yields by applying this methodology.

The ability of eosin at the electronically excited state to promote the formation of aryl radical from aryl diazonium salts was further applied to the synthesis of benzothiophene avoiding the use of transition metal and harsh condition usually required in such synthesis (Scheme 29.22) [87]. The radical **132** is obtained by single-electron transfer reduction of *o*-methylthio-benzenediazonium salt **133**. The vinyl radical **134** is then formed by addition of intermediate **132** to alkyne **135**. The cyclization of the vinyl radical **134** followed by single-electron oxidation yields the benzothiophenium ion **136**, which after transfer of the methyl group to DMSO affords benzothiophene derivatives **137**. The proposed methodology allows the synthesis of a wide range of synthetically valuable benzothiophenes including a key intermediate in the synthesis of the drug raloxifene.

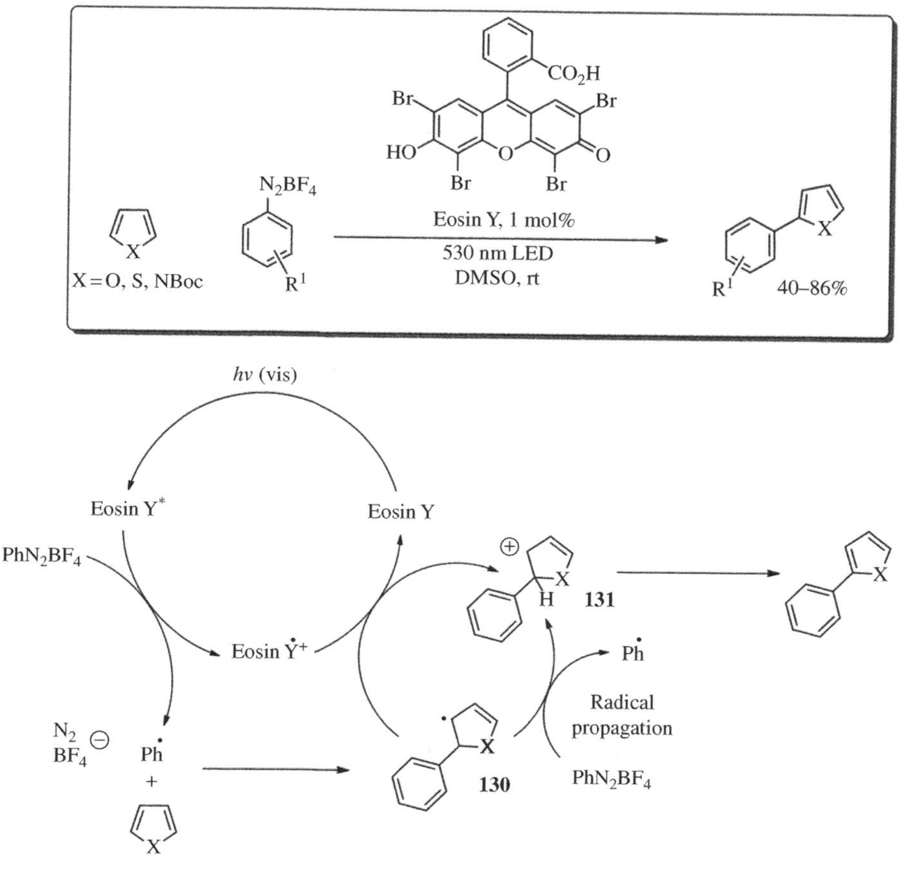

SCHEME 29.21 Photocatalytic direct arylation of heteroarenes.

In a related approach, the synthesis of several phenanthrene derivatives **138** was performed by coupling of alkyne derivatives **139** to biphenyl diazonium salts **140** (Scheme 29.23) [88]. The synthesis of a range of phenanthrene derivatives with various substituents was performed in moderate to good yields using this smooth reaction conditions.

The photocatalytic generation of aryl radicals was also successfully applied to the formation of carbon heteroatom bonds. The aryl pinacolboronates **141** can be easily achieved by visible light irradiation of a solution of aryl diazonium salts **142** and diboron pinacol ester **143** containing 5 mol% of eosin (Scheme 29.24) [89]. The proposed mechanism involves the addition of aryl radical **144** to the complex **145** that is generated by interaction of tetrafluoroborate anion and diboran pinacol ester **143**. This process leads to the formation of the aryl pinacolboronates **141** and the radical anion intermediate **146**. The oxidation of this intermediate by eosin radical cation completes the catalytic cycle.

Similar reaction conditions were applied for the synthesis of thioether derivatives using dialkylsulfide as sulfur reagent (Scheme 29.25) [90]. The key step of the process involves the oxidative cleavage of disulfide by the photogenerated aryl radical.

The formation of carbon oxygen bond was also investigated under visible light photocatalysis using an organic dye as photocatalyst. A strategy based on the generation of highly reactive oxygen

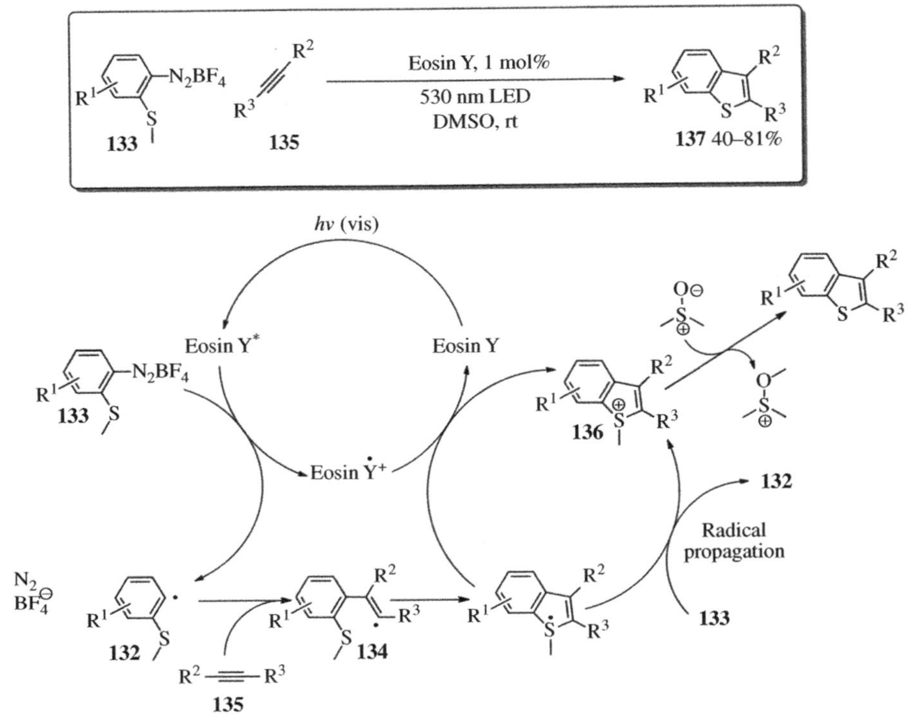

SCHEME 29.22 Synthesis of benzothiophene by photocatalysis.

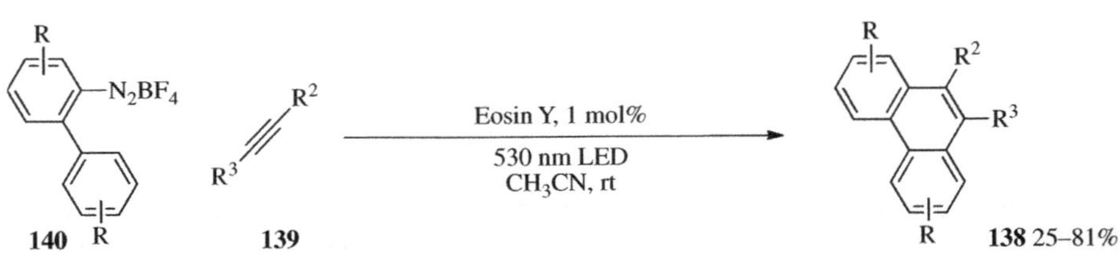

SCHEME 29.23 Photocatalytic synthesis of phenanthrenes.

species from molecular oxygen by electron transfer was reported. The conversion of aryl boronic acid into phenol was achieved under oxygen atmosphere using methylene blue as photocatalyst and Hünig's base as sacrificial electron donor (Scheme 29.26) [91]. Noteworthy, similar reaction conditions were previously reported for this transformation using $[Ru(bpy)_3]^{2+}$ as photocatalyst [92]. Mechanistic investigation and kinetic analysis showed that methylene blue is more efficient than $[Ru(bpy)_3]^{2+}$. After reaching the excited state, the methylene blue $^*(MB^+)$ is reduced by the Hünig's base affording methylene blue radical ($MB^•$). This species is able to act as electron donor to the molecular oxygen leading to its reduced form $O_2^{•-}$. The $O_2^{•-}$ nucleophilic attack on boron atom followed by hydrogen abstraction leads to the boronate intermediate **148**. The aryl transfer to an oxygen atom followed by hydrolysis affords the phenol derivatives in high yield. Due to the mild and environmentally benign conditions, this methodology should become a privileged strategy for the introduction of hydroxyl group on aromatic ring.

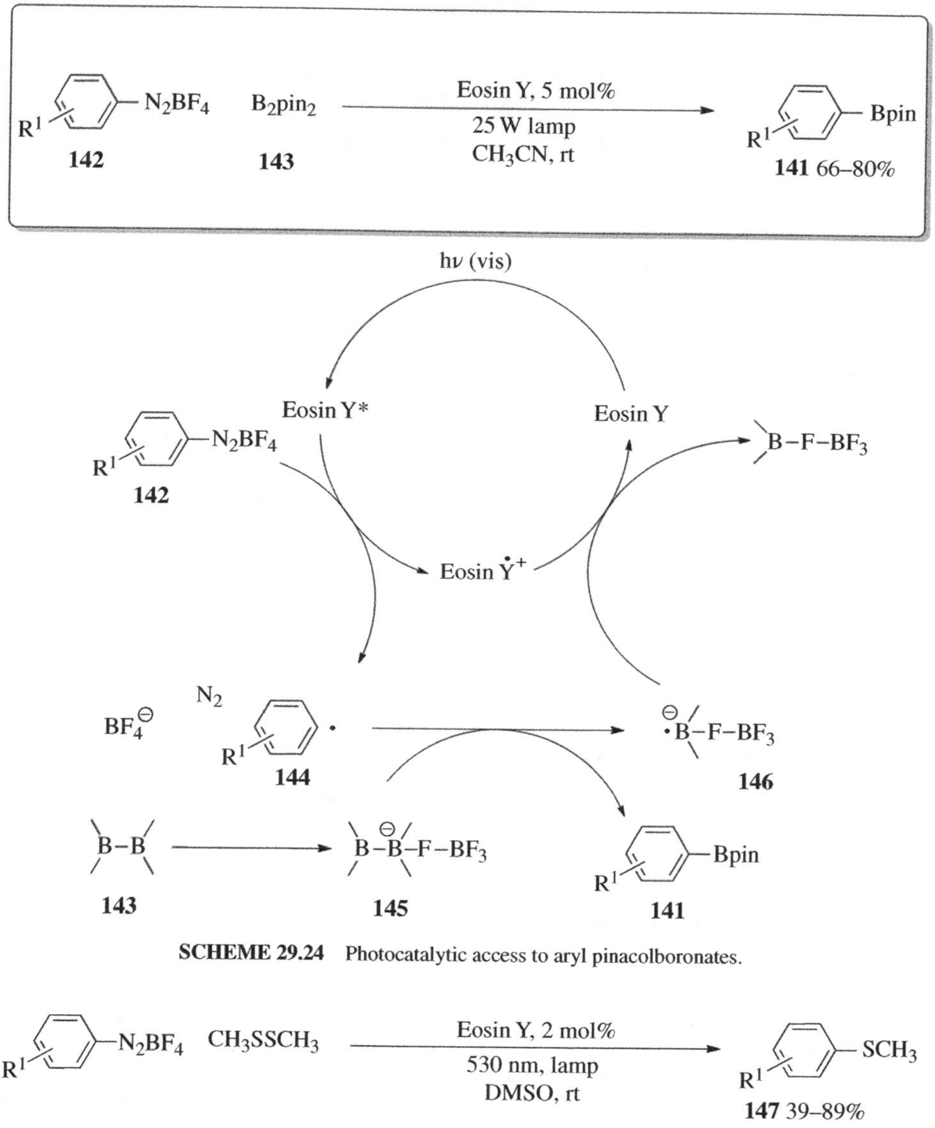

SCHEME 29.24 Photocatalytic access to aryl pinacolboronates.

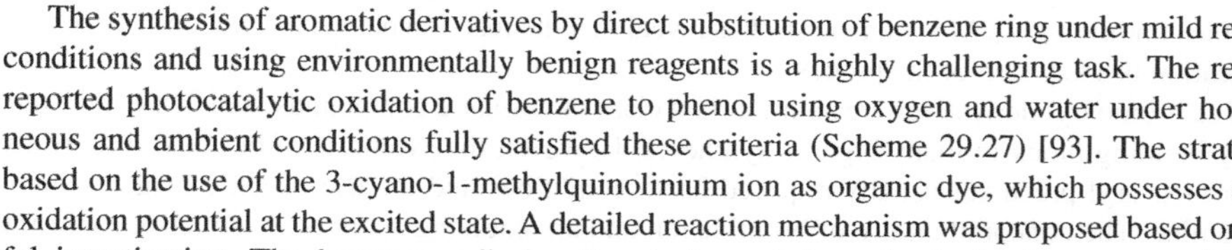

SCHEME 29.25 Photocatalytic synthesis of thioether.

The synthesis of aromatic derivatives by direct substitution of benzene ring under mild reaction conditions and using environmentally benign reagents is a highly challenging task. The recently reported photocatalytic oxidation of benzene to phenol using oxygen and water under homogeneous and ambient conditions fully satisfied these criteria (Scheme 29.27) [93]. The strategy is based on the use of the 3-cyano-1-methylquinolinium ion as organic dye, which possesses a high oxidation potential at the excited state. A detailed reaction mechanism was proposed based on careful investigation. The benzene radical cation is formed through single-electron transfer from benzene to the 3-cyano-1-methylquinolinium ion at its excited state. The reaction of water with the benzene radical cation produces the hydroxy benzene radical adduct **149**. The molecular oxygen is

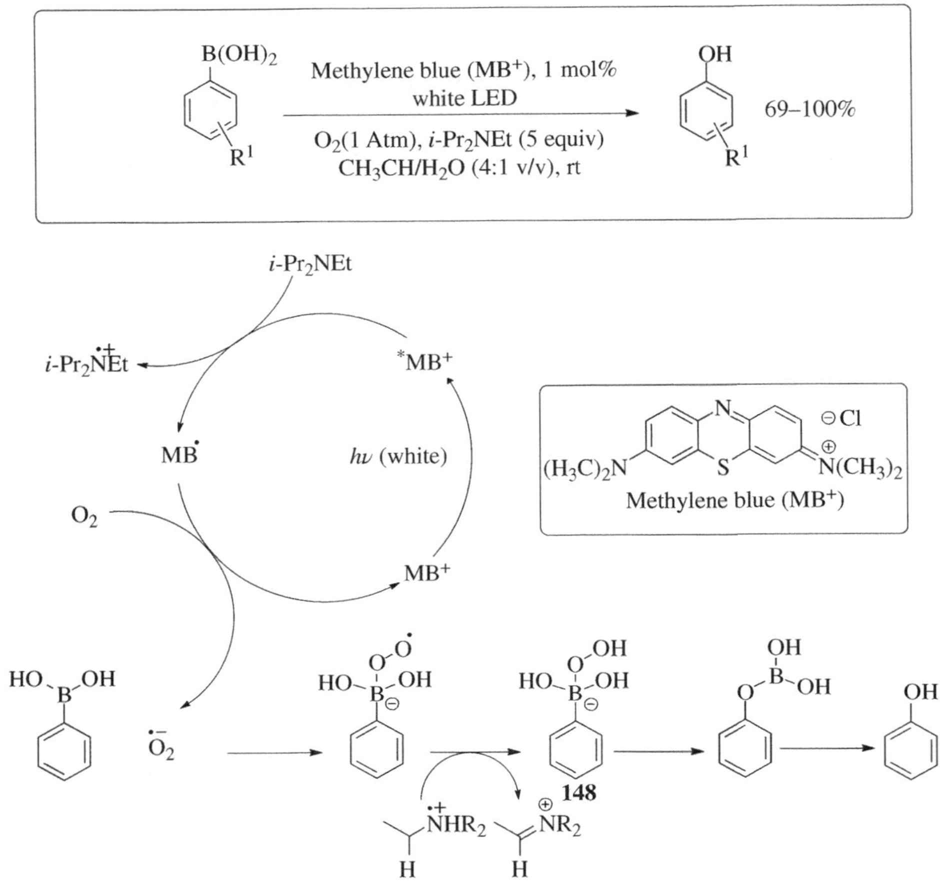

SCHEME 29.26 Photocatalytic access to phenol from boronic acid.

both involved in the completion of the photocatalytic cycle and the formation of phenol derivative. The regeneration of the photocatalyst occurs through the reduction of molecular oxygen leading to the $O_2^{\bullet-}$. The protonation of this species leads to the peroxide radical (HOO$^{\bullet}$), which performs hydrogen abstraction on hydroxy benzene radical adduct **149** to afford the phenol. These simple reaction conditions are applied to the transformation of 2.3 g of benzene into 1.1 g of phenol (41% yields). More recently, visible light-mediated oxygenation of benzene was carried out using 2,3-dichloro-5,6-dicyano-*p*-benzoquinone (DDQ) as photocatalyst. The phenol is obtained in high yield using catalytic amount of DDQ (33 mol%) and *tert*-butylnitrite as a recycle agent under aerobic condition [94]. Mechanistic studies reveal that the reaction between water and benzene radical cation is a key step of the process.

The bromoaryl derivatives are an important class of aromatic compounds, which are involved as a key building block in many organic syntheses. However, the synthesis of bromoaryl derivatives from their aryl precursor usually required harsh conditions and the use of toxic reagents such as dibromine. The concomitant use of bromide ion and an oxidizing agent is an attractive root for the bromination of aromatic compounds under environmentally friendly conditions. In this context, combination of hydrogen peroxide and bromine anion is efficient for bromination of arenes producing only water as side product. However, the use of molecular oxygen as oxidant remains more

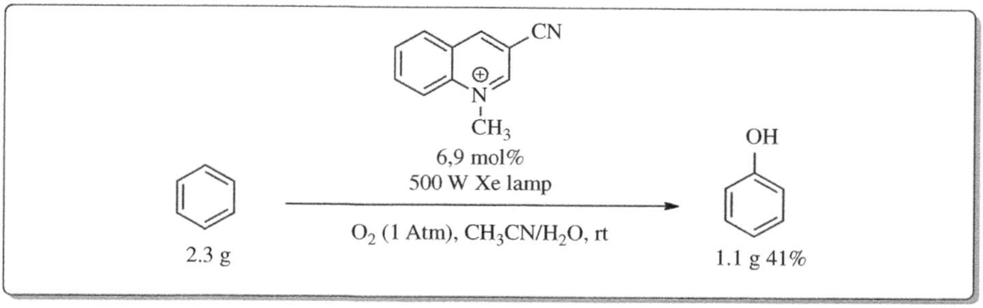

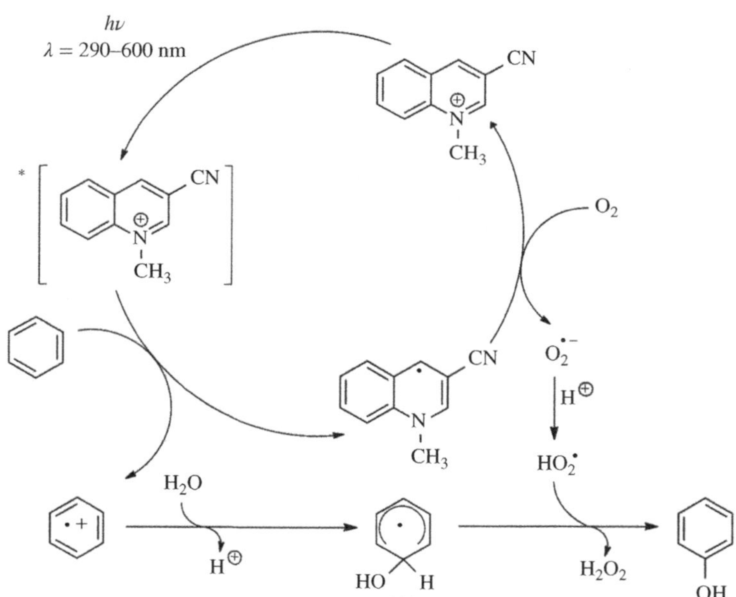

SCHEME 29.27 Photocatalytic oxidation of benzene to phenol.

attractive. An efficient bromination of electron-rich arene using aqueous HBr as bromine source and molecular oxygen as an oxidant was recently reported (Scheme 29.28) [95]. This method exploited the photoredox properties of the 9-mesityl-10-methylacridinium perchlorate (Acr^{+}–Mes). Upon irradiation, an intramolecular electron transfer occurs on this species (Acr^{+}–Mes) leading to an electronic excited state, which possesses radical character on the acridine moiety and a radical cation character on the mesitylene moiety ($Acr^{\bullet}$–$Mes^{\bullet+}$). At this electronic configuration ($Acr^{\bullet}$–$Mes^{\bullet+}$), the photocatalyst displays both oxidizing and reducing properties. Consequently, the oxidation of the electron-rich arene to the corresponding radical cation, and the reduction of oxygen to peroxide radicals $HOO^{\bullet}$ are performed by photoinduced electron transfer through the excited state of the 9-mesityl-10-methylacridinium. The reaction of an arene radical cation and bromide ion produces the arene bromine radical adduct **150**. The hydrogen abstraction by $HOO^{\bullet}$ leads to the expected bromoarene and hydrogen peroxide. As the mixture of hydrobromic acid and hydrogen peroxide is also capable of performing bromination of the arene, the yields of the expected products are further increased by this secondary reaction.

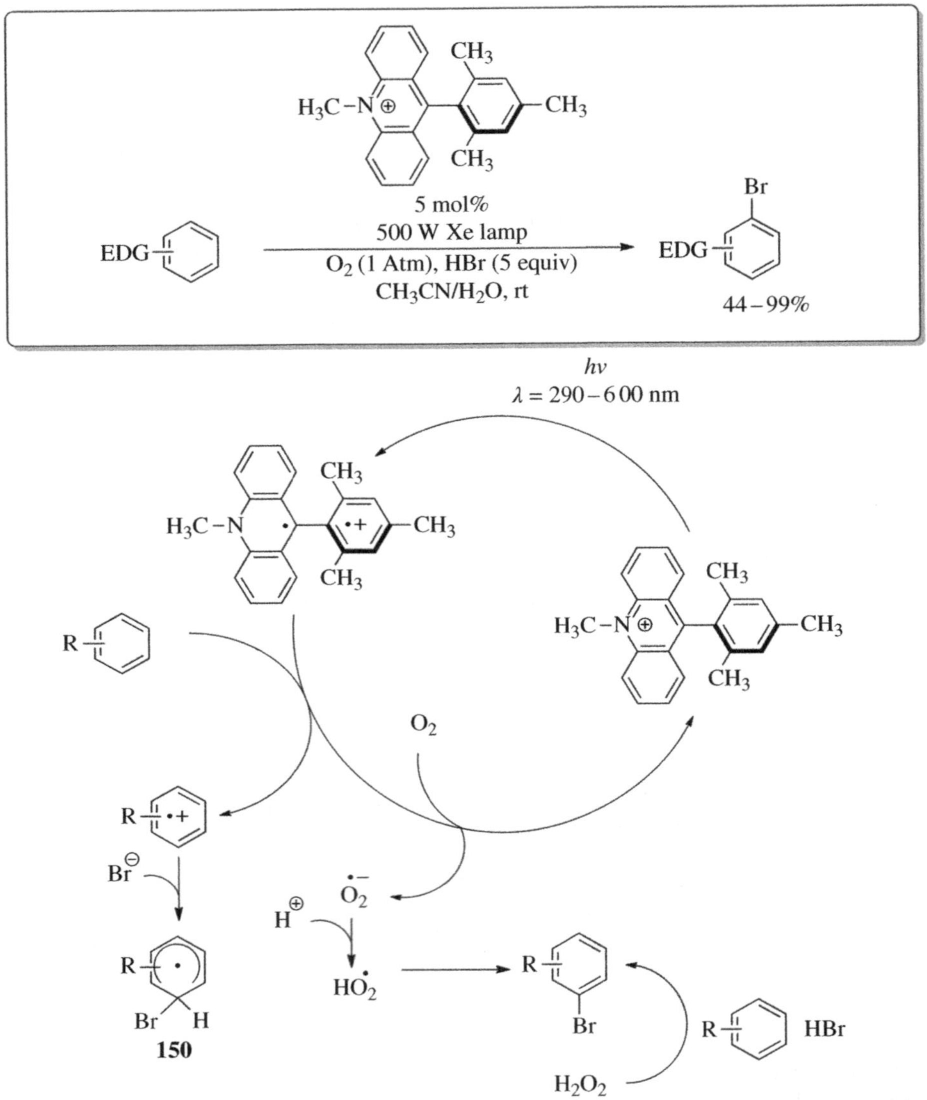

SCHEME 29.28 Photocatalytic bromination of arenes.

29.4 CONCLUSION

The photochemical reactivity of aromatic compounds is mainly characterized by a high tendency to suppress aromaticity in the final products. Numerous applications to organic synthesis have been reported and are currently investigated. Aside from these reactions, a lot of photochemical transformations enable coupling-type processes in which aromaticity is maintained. Several of these reactions such as photochemical electrocyclic reactions and nucleophilic aromatic substitution involving photochemical electron transfer in S_{RN} mechanisms have also been investigated in the past. With the recently increasing interest of photocatalytic reactions in organic chemistry, particularly the second topic as well as almost all kind of electron transfer-mediated reactions of aromatic compounds are now focused. Photoredox catalytic reactions, especially when combined with other forms of catalysis, provide particularly mild reaction conditions for the synthesis of a large variety of aromatic compounds. Beyond the interest in application to the synthesis of biologically active compounds, such reactions are also of high interest for the preparation of organic materials where π-conjugated molecules are needed.

ABBREVIATION

DMSO Dimethylsulfoxide

REFERENCES

[1] (a) Hoffmann, N. (2008) *Chem. Rev.*, **108**, 1052–1103; (b) Bach, T., Hehn, J. P. (2011) *Angew. Chem. Int. Ed.*, **50**, 1000–1045.

[2] (a) Klán, P., Wirz, J. *Photochemistry of Organic Compounds*, John Wiley & Sons, Ltd, Chichester, 2009; (b) Turro, N. J., Ramamurthy, V., Scaiano, J. C. *Modern Molecular Photochemistry of Organic Compounds*, University Science Books, Sausalito, 2010.

[3] (a) Hoffmann, N. (2008) *J. Photochem. Photobiol. C Photochem. Rev.*, **9**, 43–60; (b) Yoon, T. P., Ischay, M. A., Du, J. (2010) *Nat. Chem.*, **2**, 527–532; (c) Teplý, F. (2011) *Collect. Czech. Chem. Commun.*, **76**, 859–917; (d) Narayanam, J. M. R., Stephenson, C. R. J. (2011) *Chem. Soc. Rev.*, **40**, 101–113; (e) Shi, L., Xia, W. (2012) *Chem. Soc. Rev.*, **41**, 7687–7697; (f) Prier, C. K., Rankic, D. A., MacMillan, D. W. C. (2013) *Chem. Rev.*, **113**, 5322–5363; (g) Reckenthähler, M., Griesbeck, A. G. (2013) *Adv. Synth. Catal.*, **355**, 2727–2744; (h) König, B., Ed. *Chemical Photocatalysis*, Walter de Gruyter GmbH & Co. KG, Berlin, 2013.

[4] (a) Ravelli, D., Protti, S., Fagnoni, M., Albini, A. (2013) *Curr. Org. Chem.*, **17**, 2366–2373; (b) Hoffmann, N. (2012) *ChemSusChem*, **5**, 362–371.

[5] (a) Zimmerman, H. E. (1969) *Angew. Chem. Int. Ed.*, **8**, 1–11; (b) Turro, N. J. (1986) *Angew. Chem. Int. Ed.*, **25**, 882–901; (c) Olivucci, M., Santoro, F. (2008) *Angew. Chem. Int. Ed.*, **47**, 6322–6325.

[6] Hoffmann, N. (2012) *Photochem. Photobiol. Sci.*, **11**, 1613–1641.

[7] Givens, R. S., Rubina, M., Stensrud, K., Cope, E. D., Perera, C., Senadheera, S., Sebej, P., Anthony, L. A., Klan, P., Toscano, J. P., Anthony, E. S. (2010) Pacifichem 2010, International Chemical Congress of Pacific Basin Societies, Honolulu, HI, United States, December 15–20, ORGN-960.

[8] (a) Albini, A., Fagnoni, M. (2008) *ChemSusChem*, **1**, 63–66; (b) Oelgemöller, M., Jung, C., Mattay, J. (2007) *Pure Appl. Chem.*, **79**, 1939–1947.

[9] (a) Hoffmann, N. (2004) *Synthesis*, **4**, 481–495; (b) Streit, U., Bochet, C. G. (2011) *Beilstein J. Org. Chem.*, **7**, 525–542; (c) Döpp, D., Kruse, C., Erian, W. A., Pies, M. (2013) *ARKIVOC*, **(3)**, 317–337.

[10] (a) Cornelisse, J. (1993) *Chem. Rev.*, **93**, 615–669; (b) Wender, P. A., Ternansky, R., DeLong, M., Singh, S., Olivero, A., Rice, K. (1990) *Pure Appl. Chem.*, **62**, 1597–1602; (c) Chappell, D., Russell, A. T. (2006) *Org. Biomol. Chem.*, **4**, 4409–4430.

[11] De Keukeleire, D. (1994) *Aldrichim. Acta*, **27**, 59–69.

[12] Hoffmann, N., Pete, J.-P. (1997) *J. Org. Chem.*, **62**, 6952–6960.

[13] (a) Verrat, C., Hoffmann, N., Pete, J.-P. (2000) Synlett, 1166–1168; (b) Verrat, C. (2000) Photocycloadditions [2+2] intramoléculaires d'éthers de polyphénols: accès au squelette de produits naturels hétérocycliques, *PhD Dissertation*, Université de Reims Champagne-Ardenne.

[14] (a) Hoffmann, N., Pete, J.-P., Inoue, Y., Mori, T. (2002) *J. Org. Chem.*, **67**, 2315–2322; (b) Hoffmann, N. (2002) *Tetrahedron*, **58**, 7933–7941.

[15] (a) Woodward, R. B., Hoffmann, R. (1969) *Angew. Chem. Int. Ed.*, **8**, 781–853; (b) Geerlings, P., Ayers, P. W., Toro-Labbé, A., Chattaraj, P. K., De Proft, F. (2012) *Acc. Chem. Res.*, **45**, 683–695.

[16] (a) Mallory, F. B., Mallory, C. W. (1984) *Org. React.*, **30**, 1–456; (b) Laarhoven, W. H. (1989) *Org. Photochem.*, **10**, 163–308; (c) Arai, T., Tokumaru, K. (1993) *Chem. Rev.*, **93**, 23–39; (d) Mazzucato, U., Momicchioli, F. (1991) *Chem. Rev.*, **91**, 1679–1719.

[17] Hoffmann, N. (2014) *J. Photochem. Photobiol. C Photochem. Rev.*, **19**, 1–19.

[18] (a) Vögtle, F., *Reizvolle Moleküle in der Organischen Chemie*, B. G. Teubner, Stuttgart, 1989; (b) Martin, R. H. (1974) *Angew. Chem. Int. Ed.*, **13**, 649–660; (c) Gingras, M. (2013) *Chem. Soc. Rev.*, **42**, 968–1006; (d) Shen, Y., Chen, C. F. (2012) *Chem. Rev.*, **112**, 1463–1535.

[19] (a) Moradpour, A., Nicoud, J. F., Balavoine, G., Kagan, H., Tsoucaris, G. (1971) *J. Am. Chem. Soc.*, **93**, 2353–2354; (b) Grimme, S., Harren, J., Sobanski, A., Vögtle, F. (1998) Eur. J. Org. Chem., 1491–1509; (c) Urbano, A. Carreño, M. C. (2013) *Org. Biomol. Chem.*, **11**, 699–708; (d) Gingras, M., Félix, G., Peresutti, R. (2013) *Chem. Soc. Rev.*, **42**, 1007–1050.

[20] Gingras, M. (2013) *Chem. Soc. Rev.*, **42**, 1051–1095.

[21] Flammang-Barbieux, M., Nasielski, J., Martin, R. H. (1967) *Tetrahedron Lett.*, **8**, 743–744.

[22] Laarhoven, W. H., Cuppen, T. H. J. H. M., Nivard, R. J. F. (1970) *Tetrahedron*, **26**, 4865–4881.

[23] Sudhakar, A., Katz, T. J. (1986) *Tetrahedron Lett.*, **27**, 2231–2234.

[24] Lefevre, Q., Jentsch, M., Rueping, M. (2013) *Beilstein J. Org. Chem.*, **9**, 1883–1890.

[25] Dietz, F., Scholz, M. (1968) *Tetrahedron*, **24**, 6845–6849.

[26] Martin, R. H., Baes, M. (1975) *Tetrahedron*, **31**, 2135–2137.

[27] Ninomiya, I., Yamauchi, S., Kiguchi, T., Shinohara, A., Naito, T. (1974) *J. Chem. Soc.*, Perkin Trans **1**, 1747–1751.

[28] Ninomiya, I., Naito, T., Kiguchi, T. (1973) *J. Chem. Soc.*, Perkin Trans **1**, 2257–2261.

[29] Ninomiya, I., Naito, T., Kiguchi, T. (1973) *J. Chem. Soc.*, Perkin Trans **1**, 2261–2264.

[30] Benchekroun-Mounir, N., Dugat, D., Gramain, J.-C., Husson, H.-P. (1993) *J. Org. Chem.*, **58**, 6457–6465.

[31] Rigby, J. H., Maharoof, U. S. M., Mateo, M. E. (2000) *J. Am. Chem. Soc.*, **122**, 6624–6628.

[32] (a) Kavarnos, G. J. *Fundamentals of Photoinduced Electron Transfer*, VCH Publishers, Inc., New York, 1993; (b) Mattay, J. (1987) *Angew. Chem. Int. Ed.*, **26**, 825–845; (c) Julliard, M., Chanon, M. (1983) *Chem. Rev.*, **83**, 425–506.

[33] Rehm, D., Weller, A. (1970) *Isr. J. Chem.*, **8**, 259–271.

[34] Todres, Z. V. *Ion-Radical Organic Chemistry*, second edition, CRC Press Taylor & Francis Group, LLC, Boca Raton, 2009.

[35] (a) Cornelisse, J., Havinga, E. (1975) *Chem. Rev.*, **75**, 353–388; (b) Peñéñory, A. B., Argüello, J. E. *Handbook of Synthetic Photochemistry* (Albini, A., Fagnoni, M., Eds.), Wiley-VCH Verlag GmbH & Co KGaA, Weinheim, 2010; pp. 319–351; (c) Budén, M. E., Martin, S. E., Rossi, R. A. *CRC Handbook of Organic Photochemistry and Photobiology*, third edition, vol. **1** (Griesbeck, A., Oelgemöller, M., Ghetti, F., Eds.), CRC Press Taylor & Francis Group, Boca Raton, 2012, pp. 347–391; (d) Pretali, L., Albini, A. *CRC Handbook of Organic Photochemistry and Photobiology*, third edition, vol. **1** (Griesbeck, A., Oelgemöller, M., Ghetti, F., Eds.), CRC Press Taylor & Francis Group, Boca Raton, 2012, pp. 369–391; (e) Ravelli, D., Fagnoni, M. *CRC Handbook of Organic Photochemistry and Photobiology*, third edition, vol. **1** (Griesbeck, A., Oelgemöller, M., Ghetti, F., Eds.), CRC Press Taylor & Francis Group, Boca Raton, 2012, pp. 393–417.

[36] Mangion, D., Arnold, D. R. (2002) *Acc. Chem. Res.*, **35**, 297–304.

[37] Rossi, R. A., Pierini, A. B., Santiago, A. N. (1999) *Org. React.*, **54**, 1–271.

[38] (a) Mizuno, K., Ikeda, M., Otsuji, Y. (1985) *Tetrahedron Lett.*, **26**, 461–464; (b) Mizuno, K. *Photochemical Key Steps in Organic Synthesis* (Mattay, J., Griesbeck, A., Eds.), VCH Verlagsgesellschaft mbH, Weinheim, 1994, p. 196.

[39] Vanossi, M., Mella, M., Albini, A. (1994) *J. Am. Chem. Soc.*, **116**, 10070–10075.

[40] de Lijser, H. J. P., Arnold, D. R. (1997) *J. Org. Chem.*, **62**, 8432–8438.

[41] Hoshikawa, T., Inoue, M. (2013) *Chem. Sci.*, **4**, 3118.

[42] Rossi, R. A., Pierini, A. B., Peñéñory, A. B. (2003) *Chem. Rev.*, **103**, 71–168.

[43] Burgett, A. W. G., Li, Q., Wei, Q., Harran, P. G. (2003) *Angew. Chem. Int. Ed.*, **42**, 4961–4966.

[44] (a) Li, J., Jeong, S., Esser, L., Harran, P. G. (2001) *Angew. Chem. Int. Ed.*, **40**, 4765–4770; (b) Nicolaou, K. C., Bella, M., Chen, D. Y. K., Huang, X., Ling, T., Snyder, S. A. (2002) *Angew. Chem. Int. Ed.*, **41**, 3495–3499; (c) Nicolaou, K. C., Chen, D. Y. K., Huang, X., Ling, T., Bella, M., Snyder, S. A. (2004) *J. Am. Chem. Soc.*, **126**, 12888–12896.

[45] Nicolaou, K. C., Chen, J. S. (2008) *Pure Appl. Chem.*, **80**, 727–742.

[46] (a) Yonemitsu, O., Cerutti, P., Witkop, B. (1966) *J. Am. Chem. Soc.*, **88**, 3941–3945; (b) Gritsch, P. J., Leitner, C., Pfaffenbach, M., Gaich, T. (2014) *Angew. Chem. Int. Ed.*, **53**, 1208–1217.

[47] Mascal, M., Moody, C. J., Slawin, A. M. Z., Williams, D. J. (1992) *J. Chem. Soc.*, Perkin Trans. **1**, 829–830.

[48] Feldman, K. S., Ngernmeesri, P. (2012) *Synlett*, **23**, 1882–1892.

[49] Hu, W., Qin, H., Cui, Y., Jia, Y. (2013) *Chem. Eur. J.*, **19**, 3139–3147.

[50] (a) Mascal, M., Modes, K. V., Durmus, A. (2011) *Angew. Chem. Int. Ed.*, **50**, 4445–4446; (b) Qin, H., Xu, Z., Cui, Y., Jia, Y. (2011) *Angew. Chem. Int. Ed.*, **50**, 4447–4449.

[51] Jahjah, R., Gassama, A., Dumur, F., Marinković, S., Richert, S., Landgraf, S., Lebrun, A., Cadiou, C., Selles, P., Hoffmann, N. (2011) *J. Org. Chem.*, **76**, 7104–7118.

[52] (a) Bowman, R. M., Chamberlain, T. R., Huang, C.-W., McCullough, J. J. (1974) *J. Am. Chem. Soc.*, **96**, 692–700; (b) Ohashi, M., Tanaka, Y., Yamada, S. (1977) *Tetrahedron Lett.*, **18**, 3629–3632; (c) Mangion, D., Frizzle, M., Arnold, D. R., Cameron, T. S. (2001) *Synthesis*, **8**, 1215–1222.

[53] Sezen, B., Sames, D. *Handbook of C-H Transformations*, vol. **1** (Dyker, G., Ed.), Wiley-VCH Verlag GmbH & Co. KGaA, Weinheim, 2005, pp. 3–10.

[54] Borck, J., Dahm, H., Koppe, V., Krämer, J., Schorre, G., Hovy, J. W. H., Schorscher, E. (1971) German Patent, DE1695385.

[55] Strekowski, L., Hojjat, M., Patterson, S. E., Kiselyov, A. S. (1994) *J. Heterocycl. Chem.*, **31**, 1413–1416.

[56] Barata-Vallejo, S., Flesia, M. M., Lantaño, B., Argüello, J. E., Peñéñory, A. B., Postigo, A. (2013) Eur. J. Org. Chem., 998–1008.

[57] (a) Aschi, M., Harvey, J. N. (1999) *J. Chem. Soc.*, Perkin Trans. **2**, 1059–1062; (b) Lazzaroni, S., Dondi, D., Fagnoni, M., Albini, A. (2008) *J. Org. Chem.*, **73**, 206–211; (c) Laali, K. K., Rasul, G., Prakash, G. K. S., Olah, G. A. (2002) *J. Org. Chem.*, **67**, 2913–2918; (d) Dichiarante, V., Fagnoni, M., Albini, A. (2008) *J. Org. Chem.*, **73**, 1282–1289.

[58] (a) Dichiarante, V., Fagnoni, M. (2008) *Synlett*, **6**, 787–800; (b) Fagnoni, M., Albini, A. (2005) *Acc. Chem. Res.*, **38**, 713–721.

[59] Protti, S., Dichiarante, V., Dondi, D., Fagnoni, M., Albini, A. (2012) *Chem. Sci.*, **3**, 1330–1337

[60] (a) Fossey, J., Lefort, D., Sorba, J. *Free Radicals in Organic Chemistry*, John Wiley & Sons, Ltd, Chichester, 1995; (b) Giese, B. (1984) *Acc. Chem. Res.*, **17**, 438; (c) Giese, B., Beckhaus, H.-D. (1978) *Angew. Chem. Int. Ed.*, **17**, 594–595.

[61] Guizzardi, B., Mella, M., Fagnoni, M., Albini, A. (2000) *Tetrahedron*, **56**, 9383–9389.

[62] Protti, S., Fagnoni, M., Albini, A. (2005) *Angew. Chem. Int. Ed.*, **44**, 5675–5678.

[63] Raviola, C., Canevari, V., Protti, S., Albini, A., Fagnoni, M. (2013) *Green Chem.*, **15**, 2704–2708.

[64] Mo, F., Dong, G., Zhang, Y., Wang, J. (2013) *Org. Biomol. Chem.*, **11**, 1582–1593.

[65] Cano-Yelo, H., Deronzier, A. (1984) *J. Chem. Soc.*, Perkin Trans. **2**, 1093–1098.

[66] (a) Juris, A., Balzani, V., Barigelletti, F., Campagna, S., Belser, P., von Zelewsky, A. (1988) *Coord. Chem. Rev.*, **84**, 85–277; (b) Tucker, J. W., Stephenson, C. R. J. (2012) *J. Org. Chem.*, **77**, 1617–1622.

[67] Hari, D. P., König, B. (2013) *Angew. Chem. Int. Ed.*, **52**, 4734–4743.

[68] Lyons, T. W., Sanford, M. S. (2010) *Chem. Rev.*, **110**, 1147–1169.

[69] Kalyani, D., Deprez, N. R., Desai, L. V., Sanford, M. S. (2005) *J. Am. Chem. Soc.*, **127**, 7330–7331.

[70] Deprez, N. R., Sanford, M. S. (2009) *J. Am. Chem. Soc.*, **131**, 11234–11241.

[71] Kalyani, D., McMurtrey, K. B., Neufeldt, S. R., Sanford, M. S. (2011) *J. Am. Chem. Soc.*, **133**, 18566–18569.

[72] Sahoo, B., Hopkinson, M. N., Glorius, F. (2013) *J. Am. Chem. Soc.*, **135**, 5505–5508.

[73] (a) Hashmi, A. S. K. (2010) *Angew. Chem. Int. Ed.*, **49**, 5232–5241; (b) Krause, N., Winter, C. (2011) *Chem. Rev.*, **111**, 1994–2009; (c) Liu, L.-P., Hammond, G. B. (2012) *Chem. Soc. Rev.*, **41**, 3129–3139.

[74] (a) Hopkinson, M. N., Gee, A. D., Gouverneur, V. (2011) *Chem. Eur. J.*, **17**, 8248–8262; (b) Wegner, H. A., Auzias, M. (2011) *Angew. Chem. Int. Ed.*, **50**, 8236–8247.

[75] Schroll, P., Hari, D. P., König, B. (2012) *ChemistryOpen*, **1**, 130–133.

[76] Hering, T., Hari, D. P., König, B. (2012) *J. Org. Chem.*, **77**, 10347–10352.

[77] McNally, A., Prier, C. K., MacMillan, D. W. C. (2011) *Science*, **334**, 1114–1117.

[78] Pirnot, M. T., Rankic, D. A., Martin, D. B. C., MacMillan, D. W. C. (2013) *Science*, **339**, 1593–1596.

[79] Nagib, D. A. M., MacMillan, D. W. C. (2011) *Nature*, **480**, 224–228.

[80] Rao, H., Wang, P., Li, C.-J. (2012) *Eur. J. Org. Chem.*, **2012**, 6503–6507.

[81] (a) Bertrand, S., Hoffmann, N., Pete, J. P., Bulach, V. (1999) Chem. Commun., 2291–2292; (b) Bertrand, S., Hoffmann, N., Humbel, S., Pete, J. P. (2000) *J. Org. Chem.*, **65**, 8690–8703; (c) Marinković, S., Brulé, C., Hoffmann, N., Prost, E., Nuzillard, J.-M., Bulach, V. (2004) *J. Org. Chem.*, **69**, 1646–1651.

[82] (a) Bertrand, S., Hoffmann, N., Pete, J. P. (2000) Eur. J. Org. Chem., 2227–2238; (b) Hoffmann, N., Görner, H. (2004) *Chem. Phys. Lett.*, **383**, 451–455.

[83] Cohen, S. G., Parola, A., Parsons, G. H., Jr. (1973) *Chem. Rev.*, **73**, 141–161.

[84] Marinkovic, S., Hoffmann, N. (2004) Eur. J. Org. Chem., 3102–3107.

[85] Yoon, U. C., Mariano, P. S. (1992) *Acc. Chem. Res.*, **25**, 233–340.

[86] Hari, D. P., Schroll, P., König, B. (2012) *J. Am. Chem. Soc.*, **134**, 2958–2961.

[87] Hari, D. P., Hering, T., König, B. (2012) *Org. Lett.*, **14**, 5334–5337.

[88] Xiao, T., Dong, X., Tang, Y., Zhou, L. (2012) *Adv. Synth. Catal.*, **354**, 3195–3199.

[89] Yu, J., Zhang, L., Yan, G. (2012) *Adv. Synth. Catal.*, **354**, 2625–2628.

[90] Majek, M., von Wangelin, A. J. (2013) *Chem. Commun.*, **49**, 5507–5509.

[91] Pitre, S. P., McTiernan, C. D., Ismaili, H., Scaiano, J. C. (2013) *J. Am. Chem. Soc.*, **135**, 13286–13289.

[92] Zou, Y.-Q., Chen, J.-R., Liu, X.-P., Lu, L.-Q., Davis, R. L., Jørgensen, K. A., Xiao, W.-J. (2012) *Angew. Chem. Int. Ed.*, **51**, 784–788.

[93] Ohkubo, K., Kobayashi, T., Fukuzumi, S. (2011) *Angew. Chem. Int. Ed.*, **50**, 8652–8655.

[94] Ohkubo, K., Fujimoto, A., Fukuzumi, S. (2013) *J. Am. Chem. Soc.*, **135**, 5368–5371.

[95] Ohkubo, K., Mizushima, K., Iwata, R., Fukuzumi, S. (2011) *Chem. Sci.*, **2**, 715–722.

30

PHOTOCHEMICAL BERGMAN CYCLIZATION AND RELATED REACTIONS

RANA K. MOHAMED, KEMAL KAYA, AND IGOR V. ALABUGIN
Department of Chemistry & Biochemistry, Florida State University, Tallahassee, FL, USA

30.1 INTRODUCTION: THE DIVERSITY OF CYCLOAROMATIZATION REACTIONS

Photochemical cycloaromatization reactions have two attractive features: use of light as a reagent and formation of aromatic compounds as products. It is well accepted in the chemical community that light is a desirable energy source for inducing chemical transformations. From a practical point of view, light is an abundant and environmentally benign source of energy. More so, under the influence of electronic excitation, molecules behave in theoretically intriguing and synthetically useful ways, often producing unusual outcomes. The two quintessential cycloaromatization reactions, the Bergman and Myers–Saito cyclizations, shown in Figure 30.1, have been extensively studied. In these reactions, an acyclic conjugated reactant (enediyne or enyne allene) can be transformed into a cycle with simultaneous aromatization, providing a conceptually unique approach to the preparation of aromatic compounds. Certain features distinguish cycloaromatization reactions from other types of cycle-forming processes. Most obvious is that unstable reactive centers (e.g., diradicals) are created from a closed-shell neutral molecule [1]. In the Bergman (as well as in the related Myers–Saito cyclization), two radical centers are formed as *two* π-bonds are traded for only *one* σ-bond, resulting in a net loss of a chemical bond. Although this "trade-off" is clearly unfavorable, its thermodynamic cost is partially compensated for by the aromatic stabilization gained in the product. Photochemical excitation can assist in cycloaromatization processes by providing additional electronic energy to the reactant.

The discovery of the remarkable biological activity of natural enediynes and enyne allenes has further fueled interest in understanding the fundamentals that govern these transformations. Indeed, cycloaromatization reactions have served as a productive "playground" for testing numerous concepts related to structure and reactivity of organic and organometallic compounds [2]. Furthermore, the

Arene Chemistry: Reaction Mechanisms and Methods for Aromatic Compounds,
First Edition. Edited by Jacques Mortier.

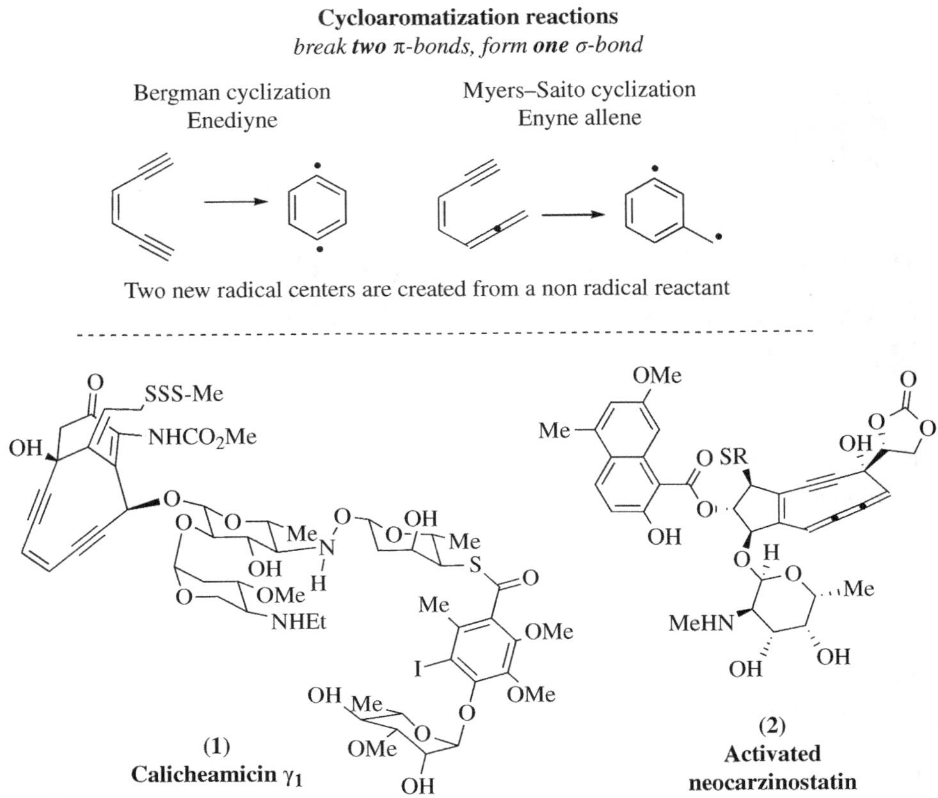

FIGURE 30.1 The two quintessential cycloaromatization reactions, the Bergman and Myers–Saito cyclizations (top) and natural products that display biological activity based on these cyclizations (bottom).

scope of this field has grown immensely, as illustrated in a recent comprehensive examination of cycloaromatization processes [3]. The early development of photochemical Bergman cyclization (BC) has been reviewed by Jones and Russell [4] and later by Alabugin, Yang, and Pal [5]. Basak and coworkers [6a] compared different approaches for the selective activation of enediyne prodrugs including photo- and pH activation. Unique properties of photoactivated enediyne–amino acid hybrids and their potential in phototherapy of cancer were recently discussed by Breiner *et al.* [7].

In this chapter, we narrow our analysis to outlining the electronic feature of the photo-BC, with respect to the introduction of different substituents and geometric strain in the starting material. The scope of this reaction is exemplified by a summary of metal-mediated photoinitiation strategies. Furthermore, the limitations of the photo-BC, with respect to direct excitation and reaction pathway divergence, are discussed. Finally, we present a concluding discussion of the potential of photoactivation to allow spatial and temporal control over enediyne reactivity in medicinal applications [6].

30.2 ELECTRONIC FACTORS IN PHOTO-BC

Although the seminal study of Bergman reported that no photochemical cyclization occurs during irradiation of the parent enediyne [8], a photochemical transformation that can be classified as a photo-Bergman reaction had been reported as early as 1968 by Campbell and Eglinton (Scheme 30.1) [9].

The substituted naphthalene structures for the products were assigned on the basis of mass-spectral analysis and their conversion to naphthalene upon hydrogenolysis. The yields for the cyclic products were low, most likely due to the lack of suitable H-donor in these experiments.

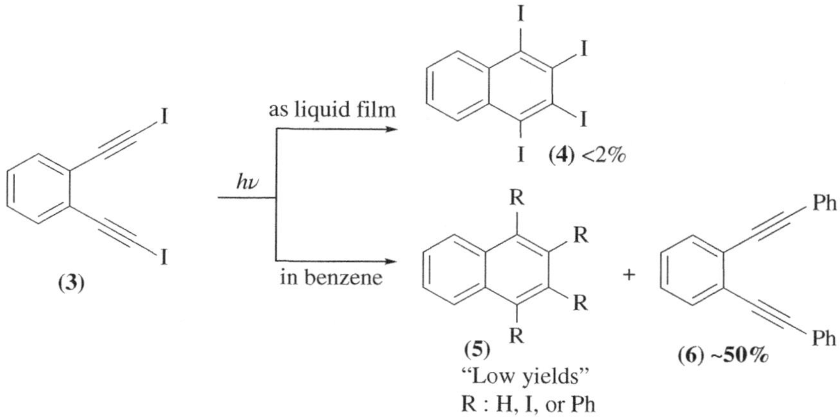

SCHEME 30.1 First report of photochemical Bergman cyclization.

Since the early examples were discouraging, it took about 25 years and the discovery of anticancer enediyne antibiotics for the idea of photochemical triggering of the BC to reemerge in the focus of active scientific pursuit. The first attempt to use enediynes for photochemical DNA cleavage was reported by Kagan *et al.* in 1993 [10]. However, irradiation of 1,6-diphenyl-3-hexene-1,5-diyne in THF did not yield any of the expected Bergman product, *o*-terphenyl. Instead, the enediyne underwent *cis/trans* bond isomerization along with the formation of a minor amount of polymeric product. This disappointing result notwithstanding, the authors found that the same enediyne causes extensive single-strand DNA photocleavage in supercoiled pBR322 DNA. The origin of the photodamage remained unclear until a recent report disclosed that efficient DNA cleavage can be achieved by monoalkynes incapable of the BC [11].

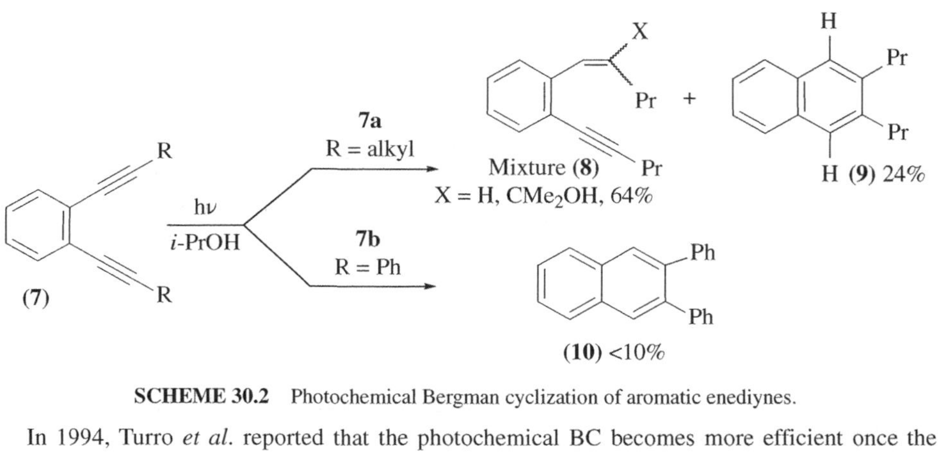

SCHEME 30.2 Photochemical Bergman cyclization of aromatic enediynes.

In 1994, Turro *et al.* reported that the photochemical BC becomes more efficient once the central bond isomerization is prevented through incorporation of this bond into a ring [12]. Although photochemical alkyne reduction remained a major reaction pathway and the yield of the cyclized product was rather low (24% in isopropanol), the result was conceptually important and encouraged a number of subsequent studies (Scheme 30.2).

Generally, reactions can be accelerated via transition state stabilization or reactant destabilization, which for enediynes can be investigated by varying the electronic nature of substituents at either the ene or alkynyl end or by inducing ring strain.

30.2.1 Substituent Effects

The significant effects of substituents framing the enediyne architecture were corroborated by Evenzahav and Turro [12b], as indicated by the less efficient photo-BC upon substitution of phenyl groups in enediyne **7b** (Scheme 30.2). Literature data for the photochemical reactivity of this specific molecule are mixed. More recent reports suggest that, despite the high (~75%) conversion, the yield of the naphthalene products is low (<10% [13], ~1% [14]). Spence [15] reported that naphthalene-fused derivatives of enediynes with electron-donating methoxy substituents were found to undergo BC upon excitation at 300 nm.

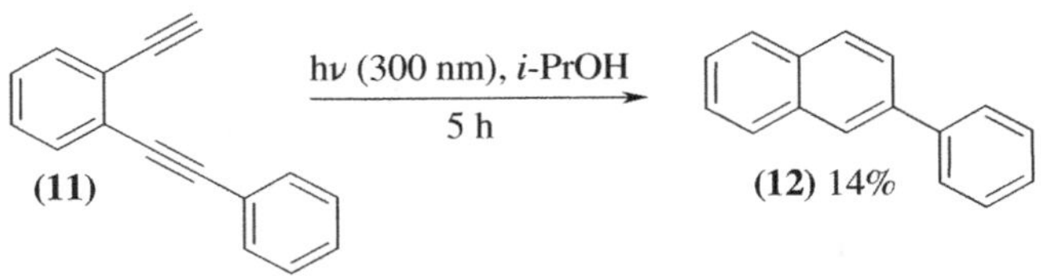

SCHEME 30.3 Photoreactivity of 1-ethynyl-2-phenylethynylbenzene.

A very interesting observation by Spence and coworkers [16] is that 1-ethynyl-2-phenylethynylbenzene requires shorter irradiation times and affords higher yields of the photo-BC product than 1,2-bis (2-phenylethynyl)benzene (Scheme 30.3).

30.2.2 Introducing Strain

An interesting experimental observation of Jones and coworkers illustrates the effect of strain at the ene core on the efficiency of photochemical BC [17]. Variations in the size of the annealed cycle strongly affect the yield of the cycloaromatized product (Scheme 30.4). Initially, an increase in the ring size attenuates strain and leads to an increase in yield, but after reaching the maximum at the 6-membered cycle, the yields drop again. Interestingly, this decrease in efficiency occurs despite the appreciable reduction in the C_1–C_6 distance between the terminal acetylenic carbons for the seven- and eight-membered systems (by 0.071 Å and 0.125 Å, respectively) relative to the six-membered analogue. This unusual trend in efficiency may be a function of how photochemical excitation is distributed in the reactive excited state. Comparison of reaction yields with enediyne geometries suggests that the cyclization is more efficient for those enediynes where rotation of the aromatic ring out of the enediyne plane forces the aromatic π-orbitals to overlap with the *in-plane* π-orbitals of the enediyne system. This involves the *in-plane* π-orbitals in a more extended conjugated system and decreases the energy gap between frontier *in-plane* orbitals.

Interestingly, when the enediyne part is coupled with a moderately bulky group attached at the *meta*-position through an alkyne linker, the photochemical BC became quite efficient (44%). This result still awaits a proper explanation.

Poor photoreactivity of cyclopentene diynes was also reported by Kaneko *et al.* [18] (Scheme 30.5) who found that only dimethyl substituted enediyne **15d** was reactive even under harsh irradiation conditions (254 nm). None of the substrates were reactive under Pyrex-filtered light. The latter result is not surprising because enediynes without terminal aryl substituents do not absorb above 280–310 nm. No explanation for the remarkable effect of the two methyl substituents has been offered to date.

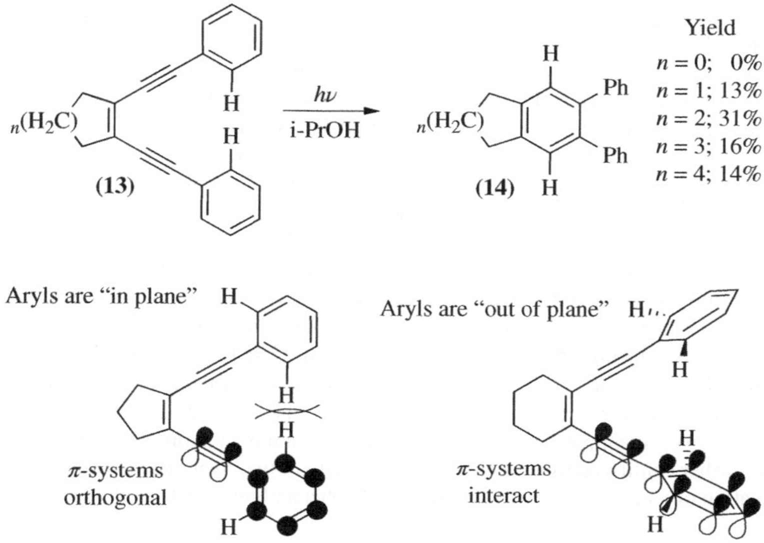

SCHEME 30.4 Effect of annealed ring size on the photochemical Bergman cyclization.

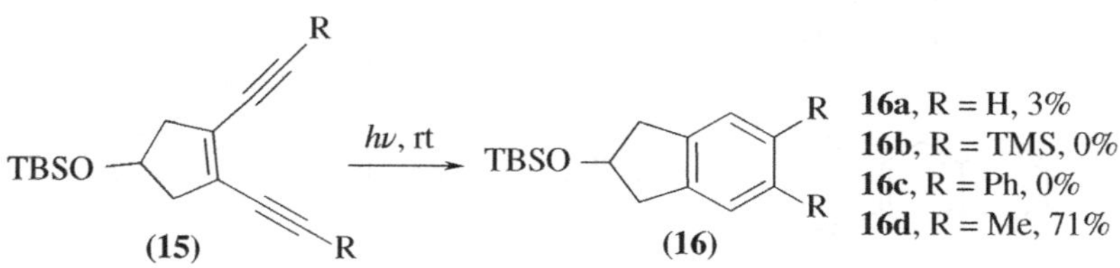

SCHEME 30.5 The photoreactivity of cyclopentene diynes is drastically increased upon methyl substitution at the alkyne termini.

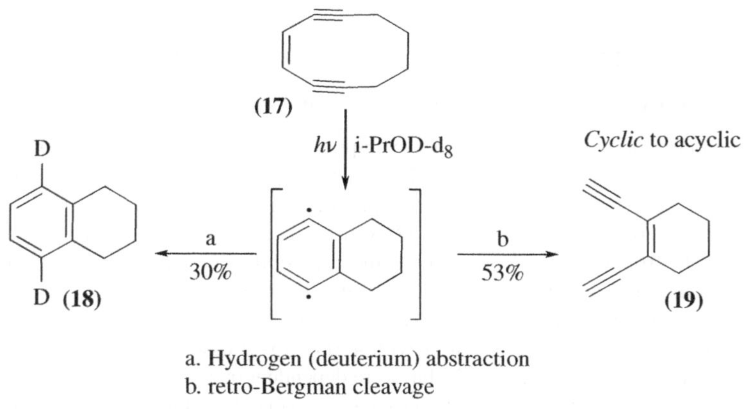

SCHEME 30.6 Competition between photo-Bergman cyclization and retro-Bergman cleavage in photochemistry of cyclodec-3-ene-1,5-diyne.

The rate of hydrogen abstraction by the intermediate 1,4-diradical becomes particularly significant when the enediyne is restricted in a 10-membered ring as shown in Scheme 30.6. In the case of cyclodec-3-ene-1,5-diyne, hydrogen abstraction by the *p*-benzyne intermediate is apparently slower than the rate of the retro-Bergman opening leading to the formation of 1,2-ethynylcyclohexene [19]. The overall transformation can be considered as a strain-driven photo-Cope rearrangement. This result suggests that the ratio of products is likely to depend on the nature and concentration of the hydrogen donor. Although the hydrogen donor concentration effects were not studied, the product ratio was found to differ from solvent to solvent.

In agreement with the results earlier, Funk *et al.* [20] found that strained, cyclic enediynes annealed to a polycyclic aromatic core undergo especially facile photochemical BC, while related terminal acetylenic compounds such as 4,5-diethynylpyrene are unreactive. Obviously, the retro-Bergman cleavage in this case is possible only in the direction of the starting material, and no isomerization is expected, unlike the case of cyclodecenediyne (**25**). However, the fact that the cyclization in Scheme 30.7 proceeded more slowly with less than 30 equivalents of 1,4-cyclohexadiene (1,4-CHD) suggests that the cyclization is reversible and that the retro-Bergman ring cleavage may provide a more efficient pathway for diradical deactivation than hydrogen abstraction. This observation is further confirmed by the contrasting reactivity of two naphthalene-fused enediynes—only the formation of a more thermodynamically stable phenanthrene cycle is observed.

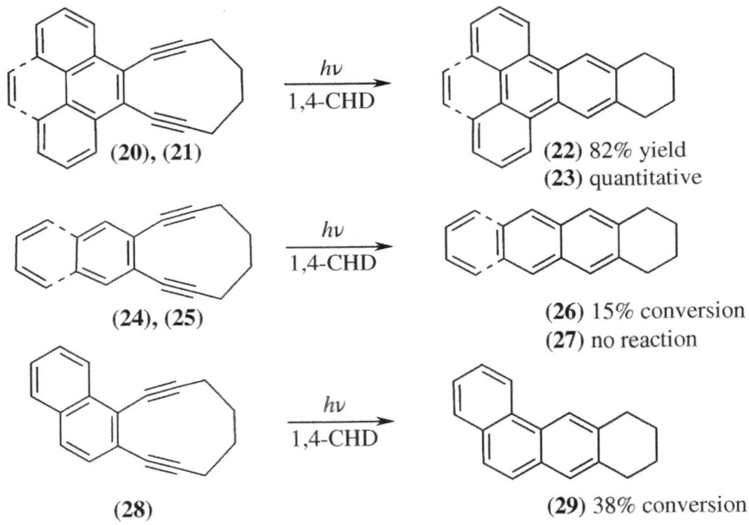

SCHEME 30.7 Photoreactivity of cyclic enediynes with polyaromatic core.

Branda and coworkers provided another example of an acyclic enediyne that undergoes a photochemical reaction different from a photo-Bergman ring closure [21]. In this case, photochemical Bergman in strained enediynes is slower than photochromic 6π-electrocyclic ring closures. Only the latter process is observed upon UV (365 nm) excitation of the strained enediyne shown in Scheme 30.8. Branda and coworkers used the reversibility of this process for an on-demand "uncaging" of thermally reactive enediynes by visible light (>525 nm).

Peterson *et al.* has reported photochemical reactivity of imidazole-fused cyclic enediynes [22]. Irradiation of degassed enediyne solutions with 450 W low-pressure Hg lamp with a quartz filter gave benzimidazole derivatives in yields that were dependent on the nature of solvent and substituents at the alkyne termini (Scheme 30.9). The cycloaromatized product was obtained in the highest yields (26–64%) for R = Ph, with the best yield obtained using THF as solvent. Yields for the two alkyl substituted enediynes (R = Me and Bu) depended very strongly on the solvent. The cyclic

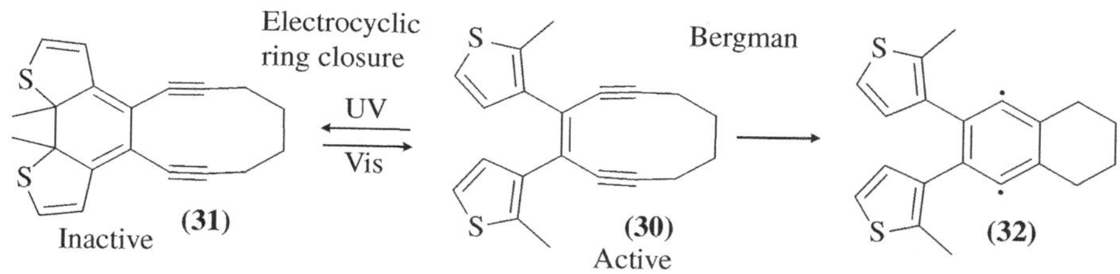

SCHEME 30.8 Competition of photo-Bergman ring closure with hexatriene electrocyclization in photochromic enediynes.

enediyne (**35**) afforded 45% of cycloaromatized product upon irradiation, even though related 11-membered ring enediynes were shown to be thermally unreactive [23].

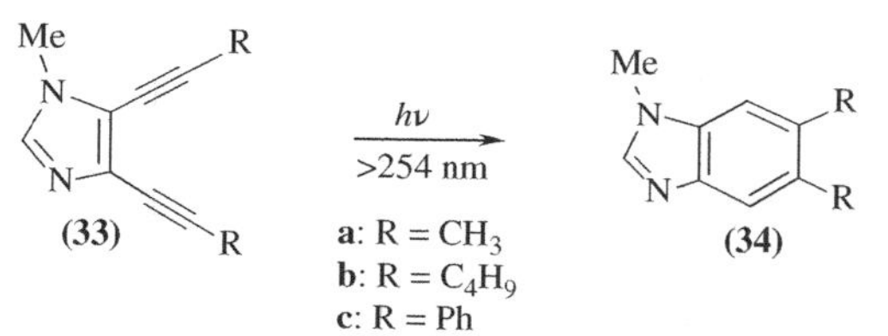

Solvent	a	b	c
THF	6%	31%	64%
i-PrOH	Trace	24%	58%
Hexane	7%	22%	47%
Cyclohexane	6%	16%	51%
CH_3CN	6%	Trace	46%
n-Hexane/ 1,4-CHD	6%	15%	44%
CH_2Cl_2	Trace	Trace	40%
t-Butanol	Trace	Trace	26%

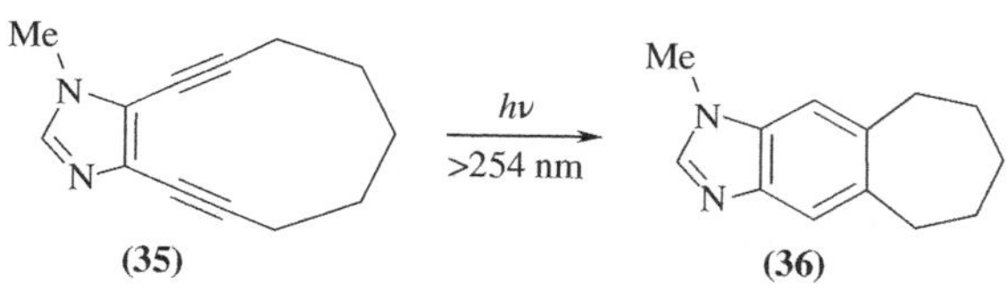

45% in hexane
45% in *i*-PrOH

SCHEME 30.9 Photochemical reactivity of imidazole-fused cyclic enediynes.

Russell and coworkers reported thermal and photochemical cyclizations of cyclic pyrimidine enediynol and enediynone (Scheme 30.10) [24]. Again, the cyclic structure favors the cyclization, and irradiation of the alcohol provides 83% of the cyclized product. On the other hand, irradiation of the corresponding ketone leads to a complex mixture with only 10% of cyclized products. No solvent or hydrogen donor concentration effects were studied.

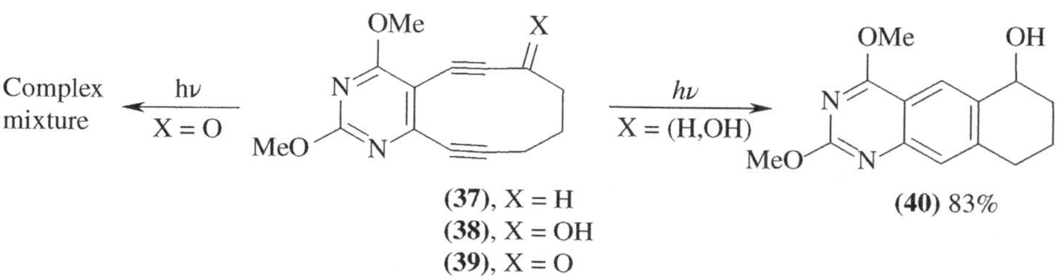

SCHEME 30.10 Photochemical reactivity of pyrimidine-fused cyclic enediynes.

Photochemical activation of bis-(phenylethynyl) sulfide in hexane in the presence of 1,4-CHD has been reported by the Matzger group to produce 3,4-diphenylthiophene through the presumed intermediacy of the 2,5-didehydrothiophene diradical (Scheme 30.11) [25]. Although reported, the thermal version of this process is inefficient as predicted by computations [26] and evidenced by the exclusive recovery of the starting bis-(phenylethynyl) sulfide after thermolysis below 240°C.

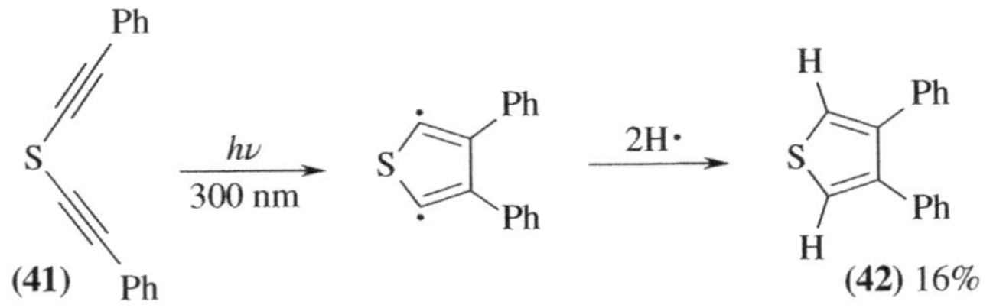

SCHEME 30.11 Photochemical transformation of bis-(phenylethynyl) sulfide in 3,4-diphenylthiophene.

30.3 SCOPE AND LIMITATIONS OF THE PHOTO-BC

The photochemistry of the BC has been largely expanded by metal-mediated strategies aiming to achieve control over enediyne reactivity. The effects of direct incorporation of metal centers in metalloenediynes, porphyrin-based enediynes, as well as inhibitory complexation of simple acyclic and cyclic enediynes are discussed in the following.

30.3.1 Metal-Mediated Photochemistry

Zaleski *et al.* explored metal–ligand charge transfer-mediated photo-BC in metal-coordinated enediynes. This approach utilizes d orbital overlap with low-energy σ or π orbitals of enediynes, thus enabling lower energy photon absorption at wavelengths that otherwise do not activate uncomplexed enediynes for photochemical BC. In particular, vanadium (V) metalloenediyne compounds (Scheme 30.12) exhibited strong ligand-to-metal charge transfer (LMCT) transitions in the near-IR region because of low redox potentials of the high-valent vanadium center and the easily oxidizable metal-binding motif [27]. These LMCT transitions were used to photothermally activate the metalloenediyne toward BC upon 785 nm laser excitation, despite the compound's relative inertness to photo-Bergman reactivity upon electronic excitation in the ultraviolet spectral region. Formation of black, sparingly soluble products with vibrational characteristics of poly(*p*-phenylene) together with the detection of high-molecular weight species (MW up to 274,000 daltons) suggests photopolymerization as the major process. Copper-complexed enediynes were photoactivated at wavelengths within the near-IR regions of the electromagnetic spectrum, while analogous enediynes that were not complexed were inert under UVA and UVB irradiations. This dramatic difference in the energies required to initiate photochemical BC establishes the incentive to develop more efficient LMCT complexes.

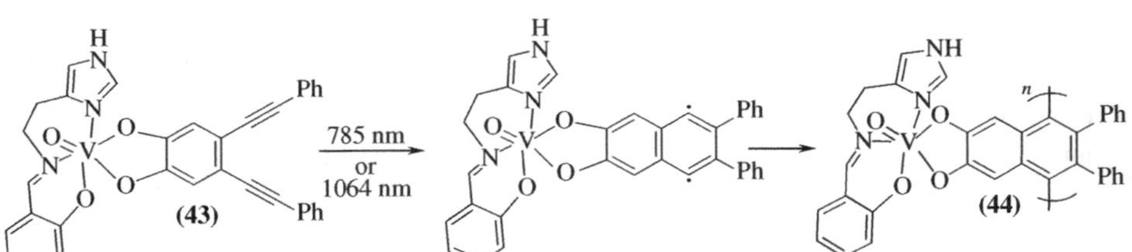

SCHEME 30.12 Photothermal cyclization of vanadium (V) metalloenediyne in the solid state.

Zaleski *et al.* [27a] suggested that MLCT from the metal dπ orbital to the empty π* orbital of the pyridine ring is responsible for the BC of the enediyne Cu(I) complexes in Scheme 30.13. Neither the uncomplexed ligand nor the analogous Zn complex was photochemically reactive. In the case of Cu(II) complex, the charge transfer from π-orbital of the pyridine to d orbital of the metal results in a partial π-hole in the pyridine ring that is transferred, to some extent, onto the enediyne unit. Thus, the photoreaction may involve an enediyne-centered state, despite initial population of the formal Cu–pyridine CT manifold.

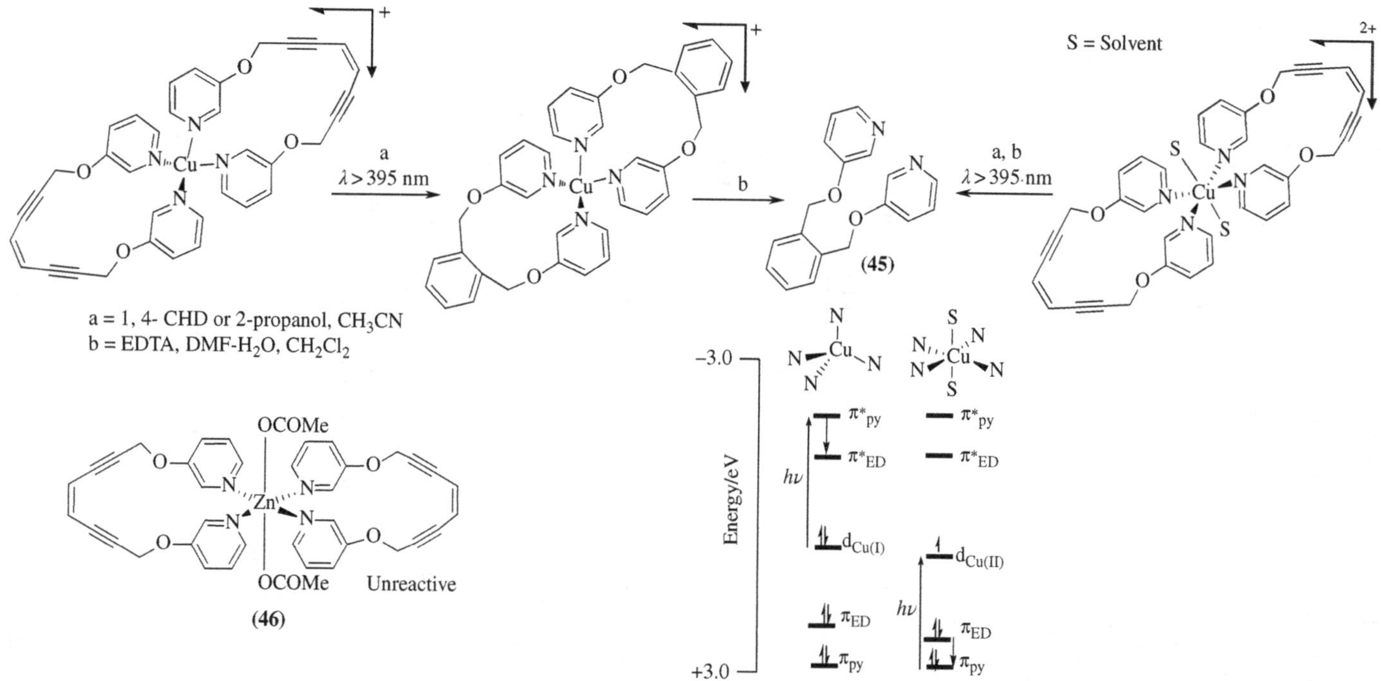

SCHEME 30.13 Top: photo-Bergman cyclization of copper metalloenediynes via metal–ligand charge transfer complex. Bottom: relative energies of copper and ligand MOs proposed to be involved in the photoelectronic Bergman cyclization of Cu(I) and Cu(II)-metalloenediynes. Adapted with permission from Benites *et al.* [27b]. © American Chemical Society.

A more detailed analysis with time-dependent density functional theory (TD-DFT) indicated that chelation to Cu(I) or Cu(II) creates several low-energy (<3.0 eV) charge transfer excited states. The multiconfigurational nature of the charge transfer excited states indicates the electronic similarities between the excited MLCT or LMCT configuration and the cyclization transition state's electronic structure, which drives the BC of metalloenediynes. In agreement with the experimental data, the Zn(II) complex does not exhibit low-energy charge transfer states because of the poor energetic overlap between the Zn(II) d orbitals and the pyridine orbitals. C=C cis/trans photoisomerization is observed instead.

Zaleski and coworkers also prepared an interesting family of enediynes based on the porphyrin core [28]. These porphyrinic enediynes (Scheme 30.14) show systematic red shifts in their electronic spectra as a function of the number of conjugated alkyne units (~13 nm/alkyne), which indicate involvement of the enediyne units in the electronic excited states. Photochemical activation of the free base of tetraphenyl octakis(phenylethynyl) porphyrin derivative only led to the reduction of the porphyrin backbone via H-atom addition at the opposing *meso*-positions, suggesting significant activation barriers for the photo-BC.

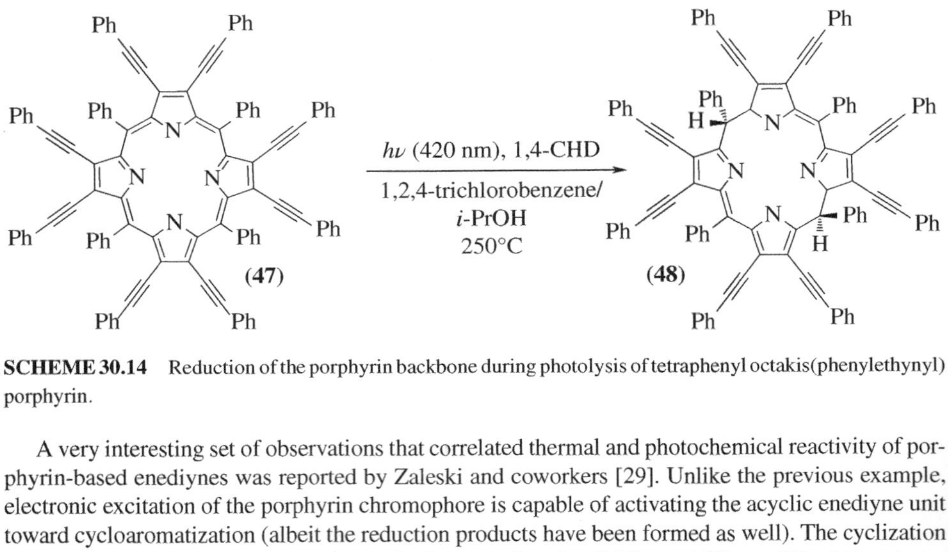

SCHEME 30.14 Reduction of the porphyrin backbone during photolysis of tetraphenyl octakis(phenylethynyl) porphyrin.

A very interesting set of observations that correlated thermal and photochemical reactivity of porphyrin-based enediynes was reported by Zaleski and coworkers [29]. Unlike the previous example, electronic excitation of the porphyrin chromophore is capable of activating the acyclic enediyne unit toward cycloaromatization (albeit the reduction products have been formed as well). The cyclization efficiency depends, however, on the excitation wavelengths (35% vs. 15% vs. 6% of the photoproduct for the λ ≥ 395 nm (Soret), ≥ 515 nm, and ≥590 nm (Q-band) excitation, respectively). The initially formed cyclic product is trapped via radical addition to the adjacent *meso*-phenyl rings and subsequent rearomatization with the formation of final picenoporphyrin products (Scheme 30.15).

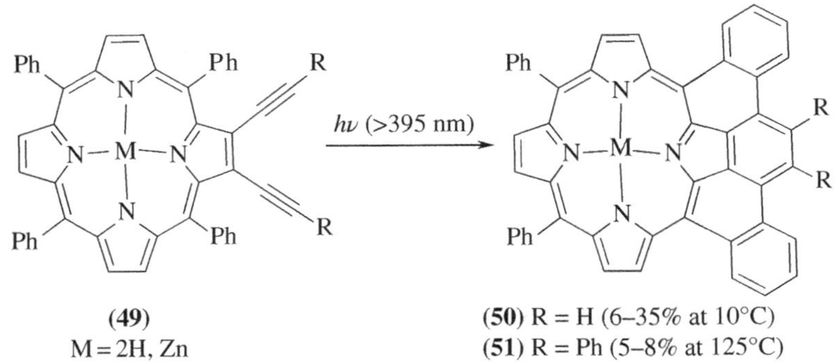

(49)
M = 2H, Zn

(50) R = H (6–35% at 10°C)
(51) R = Ph (5–8% at 125°C)

SCHEME 30.15 Formation of picenoporphyrins from dialkynylporphyrinic enediynes.

Barriers for the BC step in the excited state depend on the steric bulk of the substituents R at the alkyne termini. For R = H, the corresponding photoproduct is formed in up to 35% yield even at approximately 10°C, whereas photolysis of the diphenyl substituted enediyne produced only small amounts (5–8%) of picenoporphyrin even at 125°C. Because no thermal BC has been observed at 125°C with R = Ph, this result suggests that the excited-state cyclization barrier is still lower than the ground-state barrier. Based on the comparison of thermal and photochemical reactivity for the two enediynes, authors suggested qualitative potential energy surfaces summarized in Scheme 30.16. The presence of the substantial activation barriers at the excited-state hypersurface is consistent with the theoretical analysis of formally forbidden photo-BC.

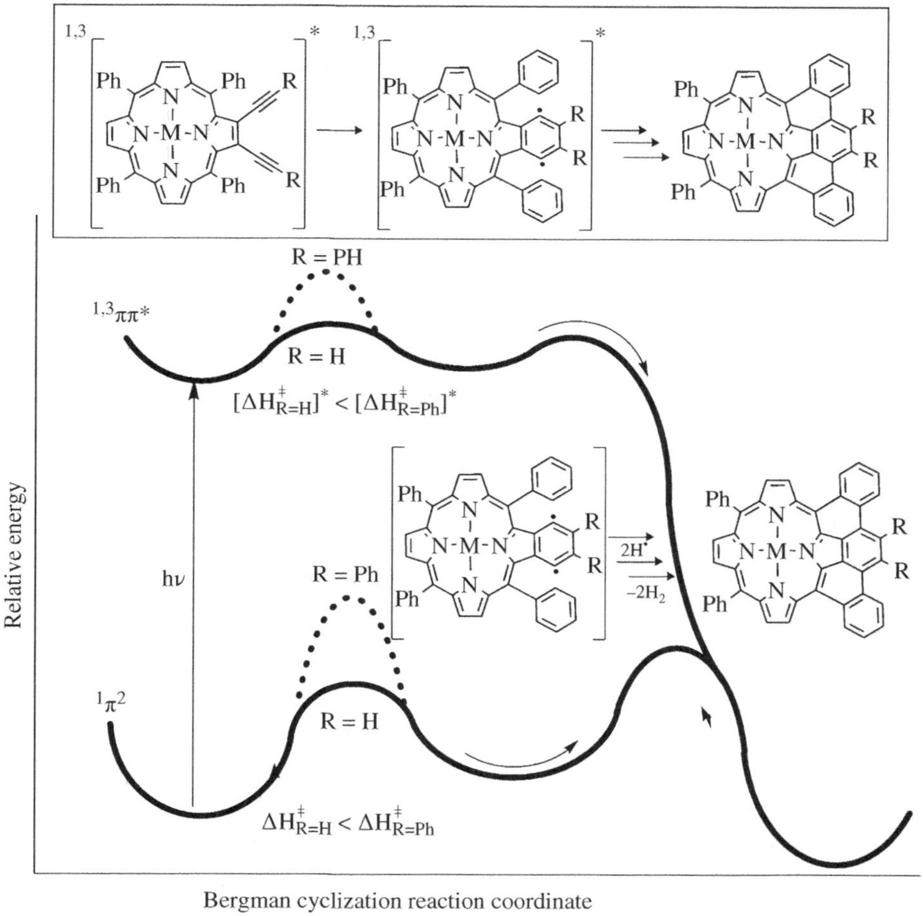

SCHEME 30.16 Suggested reaction energy profiles for ground- and excited-state Bergman cyclization of porphyrin-based enediynes. Reprinted with permission from Nath *et al.* [29a]. © American Chemical Society.

An interesting variation where the photo-BC is inhibited by metal complexation is described by O'Connor *et al.* [30] (Scheme 30.17). In contrast to the parent benzannelated cyclic enediyne, its ruthenium complex did not produce any cyclized product upon photoactivation. The reluctance of this compound to be involved in photochemical cycloaromatization is tentatively attributed to decreased aromaticity in the incipient 1,4-diradical, potentially providing an interesting corollary to the results summarized in Scheme 30.7.

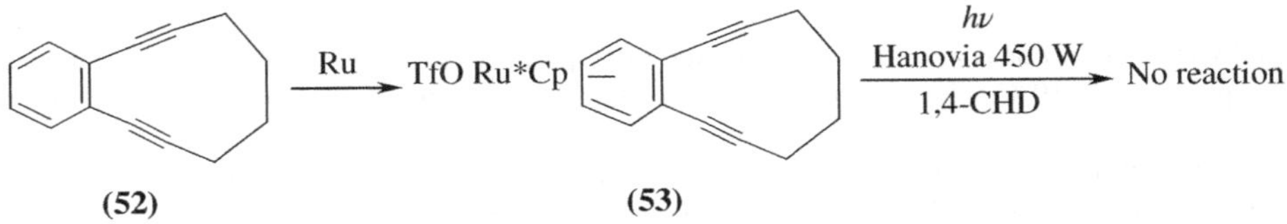

SCHEME 30.17 Inhibition of photo-Bergman cyclization upon Ru complexation.

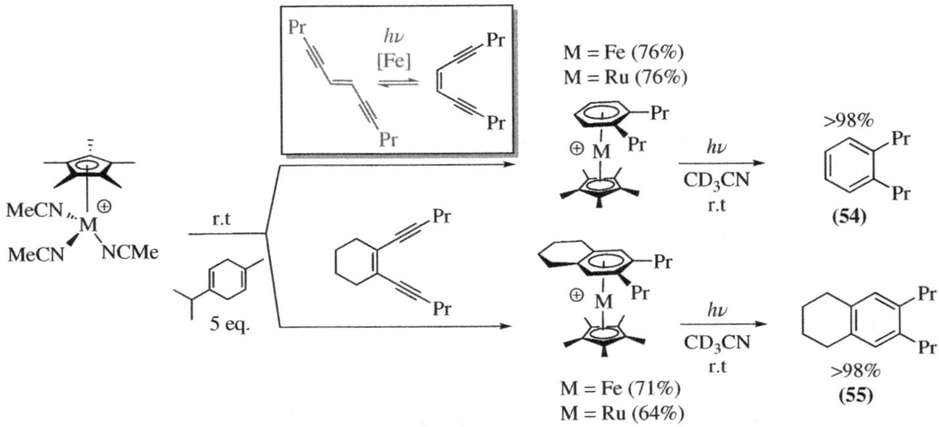

SCHEME 30.18 The Ru- and Fe-catalyzed cycloaromatization of nonbenzannelated enediynes.

In the absence of the competing coordination with arene, the Ru- and Fe-catalyzed reactions of nonbenzannelated enediynes proceed quickly and efficiently, although photochemical liberation of the arene ligand occurs more effectively with Fe due to its lower "arenophilicity" relative to ruthenium (Scheme 30.18). Facile photochemical dissociation of the product-metal complexes can be used for the creation of catalytic cycles. A fascinating feature of this approach is that it allows incorporation of trans/cis isomerization in the catalytic cycle. The possibility of using trans enediynes in cycloaromatization reactions has the potential of dramatically increasing the scope of these reactions.

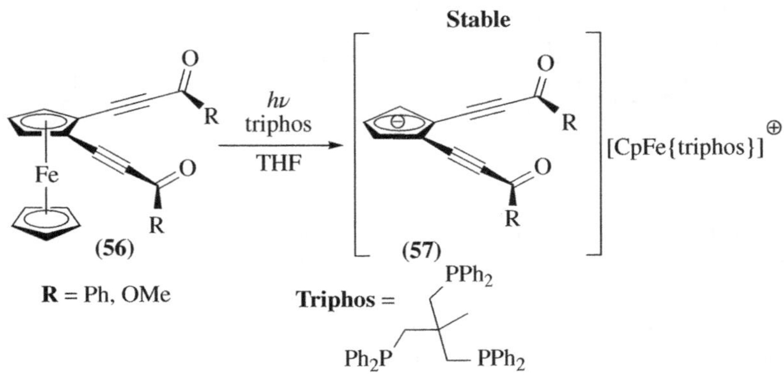

SCHEME 30.19 Photochemical liberation of the enediyne from the metal complex results in the formation of stable cyclopentadienido enediyne.

O'Connor and coworkers [31] also described photochemical liberation of enediynes from a metal complex to generate stable cyclopentadienido enediyne anions (Scheme 30.19). These compounds had not been previously reported and exhibit interesting features such as a long distance between the terminal alkyne carbons (4.6Å, R=OMe and 4.4Å, R=Ph), perhaps accounting for their low reactivity.

30.3.2 Diverting from BC Pathway: Direct Excitation and Photoinduced Electron Transfer

Although photosensitized reactions with various enediynes often yield cyclized products identical to those expected from thermal cycloaromatization (i.e., diradicals), the multichannel nature of alkyne/enediyne photoreactivity necessitates basic understanding of the factors controlling the nature of excitation in acetylenic systems.

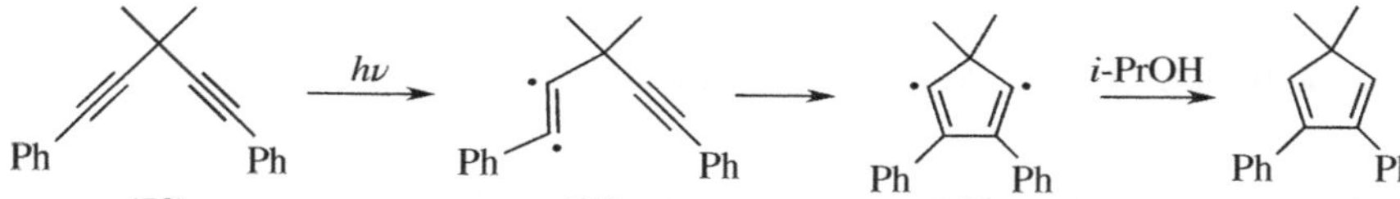

(58) (59) (60) (61)

SCHEME 30.20 Photochemical cyclization yielding a nonaromatic five-membered ring.

For example, the photochemical cyclization of diethynylmethanes, discovered by Zimmerman and Pincock [32] around the same time as the thermal BC was reported by Bergman, does not produce an aromatic product. The reaction can proceed upon direct excitation of a nonconjugated 1,4-diyne in isopropyl alcohol or upon triplet sensitization by acetophenone or xanthone. A moderately high quantum yield (0.25) upon sensitization with acetophenone indicates a relatively efficient triplet reaction (Scheme 30.20).

Furthermore, diversion from the Bergman pathway, yielding fulvenes, has been reported by Ramkumar *et al.* for enediyne under oxidative electron transfer conditions (Scheme 30.21) [33, 34].

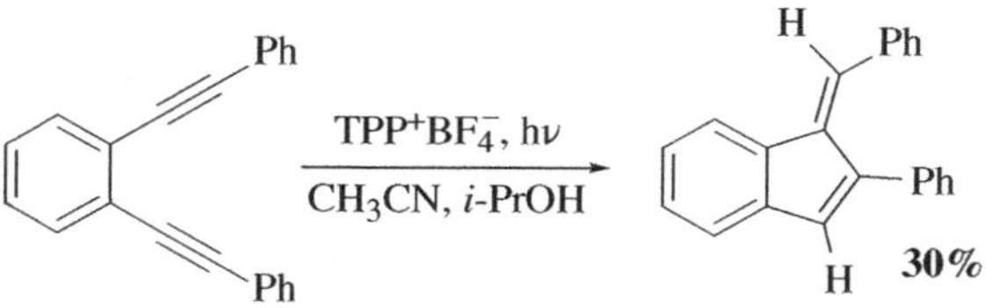

(62) (63)

SCHEME 30.21 Photochemical transformations of 1,2-bis(2-phenylethynyl)benzene under oxidative conditions facilitates formation of fulvenes.

Kovalenko and Alabugin reported that C_1–C_5 cyclization of benzannelated enediynes with tetrafluoropyridinyl (TFP) substituents at the terminal alkyne carbons forms indenes rather than fulvenes (Scheme 30.22). The radical/anionic C_1–C_5 cyclization of enediynes represents a new type of cycloaromatization reaction—the "cyclo*re*aromatization" process driven by rearomatization in the vicinity of the TS. This process is triggered via photoinduced electron transfer (PET) [35–38]. Experimental work unambiguously established PET as the triggering event for the C_1–C_5 cascade and the intermediacy of the second PET step in the indene-forming cascade [39]. Unlike the stable benzene product of the Bergman cycloaromatization, the fulvene intermediate of C_1–C_5 cyclization (prepared independently through Bu_3Sn radical-mediated 5-exo-cyclization of enediynes) [37, 38] is capable of further photoreduction to indenes with abstraction of two additional hydrogen atoms from the environment.

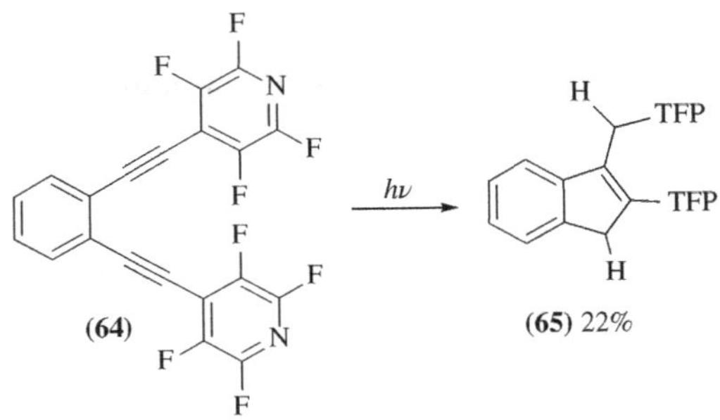

SCHEME 30.22 C_1–C_5 cyclization of benzannelated enediynes with TFP substituents form indenes.

When the core chromophore is changed to pyrazine (Scheme 30.23, bottom), interaction of enediyne moiety with 1,4-cyclohexadiene proceeds as a cycloaddition [40]. The latter change in reactivity is attributed to the particularly fast formation of an electrophilic triplet state that attacks the π-bond of 1,4-CHD. In the presence of an *ortho*-amide functionality, the same chromophore undergoes either a 7-endo-dig cyclization or photohydration [41].

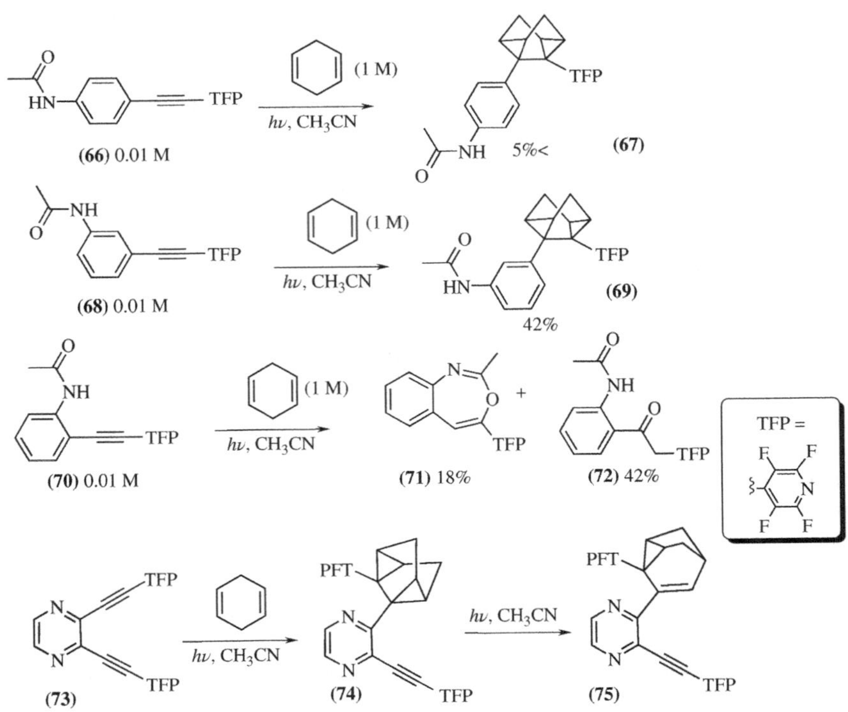

SCHEME 30.23 New photochemical reactions of enediynes promoted by variations at the core and terminus of the enediyne.

30.4 ENEDIYNE PHOTOCYCLIZATIONS: TOOL FOR CANCER THERAPY

Natural enediynes are the origin for the inspiration that has led to the vast amount of research covered in this chapter. These molecules utilize the sufficient lifetime of diradical intermediates as means for abstracting hydrogens from the DNA backbone of cells. Since these potent anti-cancer, antibiotic compounds do not differentiate between healthy cells and cancerous cells, much of research efforts have focused on tuning the reactivity and selectivity of enediynes toward DNA in cancer cells. Such a feat requires these compounds to be activated at physiological temperature at the correct time, that is, once the enediyne is in the correct vicinity to attack DNA. Because such a narrow temperature/time window is difficult to achieve, research has turned to light as a triggering source for this special reactivity [10]. Some natural enediynes such as esperamicin and neocarzinostatin also cause double-strand (ds) DNA cleavage after 254 nm irradiation [42].

Tissue-penetrating light enables prodrug transformation into therapeutic species at the right place and the right time. Once the tumor has become saturated with the prodrug, the tissue-penetrating light can be delivered selectively at the tumor site. Because tissue is not uniformly transparent but strongly absorbs in the UV-Vis range, drugs that go beyond activation on the surface of tissues need to be activated by light within approximately 650–900 nm range (the "therapeutic window of tissue"). Two-photon activation is a particularly attractive option for solving this problem, as this method can also be used for three-dimensional control over drug activation. It was already shown that simple enediyne chromophores can undergo two-photon activation [43].

Further efforts to promote the selectivity and reactivity of enediynes have been undertaken. For example, TFP enediynes and acetylenes cause a significantly increased amount of therapeutically important ds DNA photocleavage at the lower pH characteristic for hypoxic cancer tissues (Scheme 30.24). Not only do these compounds induce 100–1000 times more ds DNA cleavage than expected from a combination of random coincident single-stranded cleavages [44], but the ds/ss ratios for the DNA damage exceed that of calicheamicin, the most effective natural enediyne ds DNA cleaver known. Modular design of pH-regulated DNA-binding groups was particularly effective in the design of more reactive conjugates [45]. Further efforts of finding the right partner for the photoactivated enediyne warhead continue [46–51].

30.5 CONCLUSION

Selective photochemical initiation is a viable option for inducing cycloaromatizations of enediynes and related compounds. Although increasing the efficiency of photochemical cycloaromatizations has been shown to be feasible, further research is needed to fully elucidate the electronic factors that control these processes. The presence of two orthogonal π-systems in alkynes is important for the understanding and control of photochemical cycloaromatization reactions since the two π-systems play complementary and important roles.

Elucidation of the chemical mechanism of DNA damage and DNA binding of enediynes is crucial for developing light-activated drugs. General understanding of the key photophysical processes and barriers on the excited-state surfaces is still insufficient. Excited states that involve the *in-plane* MOs are likely to play a key role in the control of photochemical reactivity. Two-photon excitation of enediynes should open new avenues in the design of anticancer drugs activated by tissue-penetrating photons. Overall, the potential of photochemical cycloaromatizations as tools for efficient DNA photocleavage and design of light-activated drugs for cancer therapy continues to expand. However, a robust theoretical model that incorporates the disjointed experimental data in a coherent and logical framework is necessary for guiding the future development of this field.

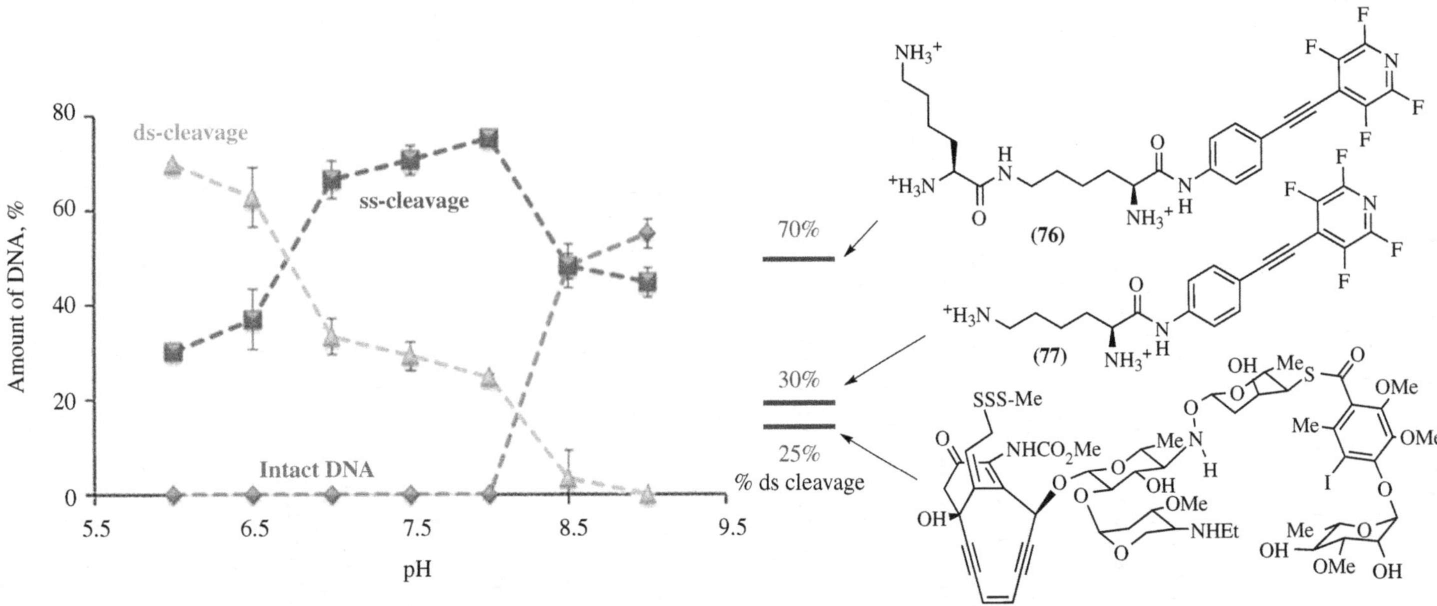

SCHEME 30.24 Left: comparison of DNA photocleavage with TFP-acetylene bis-lysine conjugates at pH range of 6–9 after 10 min. of UV (>300 nm) irradiation. Code: diamond; form I = intact supercoiled DNA, square; form II = relaxed form (ss cleavage), triangle; form III = linear form (ds cleavage). Right: comparison of maximum ds/ss ratios for two of the "designed" DNA photocleavers and calicheamicin.

ABBREVIATIONS

1,4-CHD	I,4-Cyclohexadiene
BC	Bergman cyclization
ds	Double strand
LMCT	Ligand-to-metal charge transfer
MLCT	Metal-to-ligand charge transfer
MW	Molecular weight
PET	Photoinduced electron transfer
ss	Single strand
TD-DFT	Time-dependent density functional theory
TFP	Tetrafluoropyridinyl
THF	Tetrahydrofuran

REFERENCES

[1] For another example where the borders are blurred, see anionic 5-endo cyclizations: Gilmore, K.; Manoharan, M.; Wu, J.; Schleyer, P. v. R.; Alabugin, I. V. *J. Am. Chem. Soc.* 2012, **134**, 10584.

[2] Alabugin, I. V.; Breiner, B.; Manoharan, M. *Advances in Physical Organic Chemistry*. Richard, J. P. (Ed.); Elsevier: London, 2008; Vol. **42**; p 1.

[3] Mohamed, R. K.; Peterson, P.; Alabugin, I. V. *Chem. Rev.* 2013, **113**, 7089.

[4] Jones, G. B.; Russell, K. C. *CRC Handbook of Organic Photochemistry and Photobiology*. 2nd ed.; CRC Press: Boca Raton, 2004; pp 29/1–29/21.

[5] Alabugin, I. V.; Yang, W.-Y.; Pal, R. *CRC Handbook of Organic Photochemistry and Photobiology*. Griesbeck, A.; Oelgemöller, M.; Ghetti, F. (Eds); Taylor & Francis: Boca Raton, 2012.

[6] (a) Kar, M.; Basak, A. *Chem. Rev.* 2007, **107**, 2861. (b) Jones, G. B.; Fouad, F. S. *Curr. Pharm. Des.* 2002, **8**, 2415.

[7] Breiner, B.; Kaya, K.; Roy, S.; Yang, W.-Y.; Alabugin, I. V. *Org. Biomol. Chem.* 2012, **10**, 3974.

[8] (a) Jones, R. R.; Bergman, R. G. *J. Am. Chem. Soc.* 1972, **94**, 660. (b) Bergman, R. G. *Acc. Chem. Res.* 1973, **6**, 25.

[9] Campbell, I. D.; Eglinton, G. *J. Chem. Soc. C* 1968, **16**, 2120.

[10] Kagan, J.; Wang, X.; Chen, X.; Lau, K. Y.; Batac, I. V.; Tuveson, R. W.; Hudson, J. B. *J. Photochem. Photobiol B Biol.* 1993, **21**, 135.

[11] Yang, W.-Y.; Breiner, B.; Kovalenko, S. V.; Ben, C.; Singh, M.; LeGrand, S. N.; Sang, Q.-X. A.; Strouse, G. F.; Copland, J. A.; Alabugin, I. V. *J. Am. Chem. Soc.* 2009, **131**, 11458.

[12] (a) Turro, N. J.; Evenzahav, A.; Nicolaou, K. C. *Tetrahedron Lett.* 1994, **35**, 8089. (b) Evenzahav, A.; Turro, N. *J. Am. Chem. Soc.* 1998, **120**, 1835.

[13] Spence, J. D.; Hargrove, A. E.; Crampton, H. L.; Thomas, D. W., *Tetrahedron Lett.* 2007, **48**, 725.

[14] Lewis, K. D.; Matzger, A. J. *J. Am. Chem. Soc.* 2005, **127**, 9968.

[15] (a) Korovina, N.; Gherman, B. F.; Spence, J. D. 238th ACS National Meeting, Washington, DC, August 2009. (b) Chang, M. L.; Spence, J. D. 237th ACS National Meeting, Salt Lake City, UT, March 2009.

[16] Korovina, N. V.; Chang, M. L.; Nguyen, T. T.; Fernandez, R.; Walker, H. J.; Olmstead, M. M.; Gherman, B. F.; Spence, J. D. *Org. Lett.* 2011, **13**, 3660–3663.

[17] Jones, G. B.; Wright, J. M.; Plourde II, G.; Purohit, A. D.; Wyatt, J. K.; Hynd, G.; Fouad, F. *J. Am. Chem. Soc.* 2000, **122**, 9872.

[18] Kaneko, T.; Takanashi, M.; Hirama, M. *Angew. Chem. Int. Ed. Engl.* 1999, **38**, 1267.

[19] Kaneko, T.; Takahashi, M.; Hirama, M. *Tetrahedron Lett.* 1999, **40**, 2015.

[20] Funk, R.L.; Young, E.R.R.; Williams, R. M.; Flanagan, M. F.; Cecil, T. L., *J. Am. Chem. Soc.* 1996, **118**, 3291.

[21] Sud, D.; Wigglesworth, T. J.; Branda, N. R. *Angew. Chem. Int. Ed.* 2007, **46**, 8017.

[22] Zhao, Z.; Peacock, J. G.; Gubler, D. A.; Peterson, M. A. *Tetrahedron Lett.* 2005, **46**, 1373.

[23] Wandel, H.; Wiest, O. *J. Org. Chem.* 2002, **67**, 388.

[24] Choy, N.; Blanco, B.; Wen, J.; Krishan, A.; Russell, K. C. *Org. Lett.* 2000, **2**, 3761.

[25] (a) Lewis, K. D.; Wenzler, D. L.; Matzger, A. J. *Org. Lett.* 2003, **5**, 2195. Related Thermal Cycloaromatization: (b) Lewis, K. D.; Rowe, M. P.; Matzger, A. *J. Tetrahedron* 2004, **60**, 7191.

[26] Kawatkar, S. P.; Schreiner, P. R. *Org. Lett.* 2002, **4**, 3643.

[27] (a) Kraft, B. J.; Coalter, N. L.; Nath, M.; Clark, A. E.; Siedle, A. R.; Huffman, J. C.; Zaleski, J. M. *Inorg. Chem.* 2003, **42**, 1663. (b) Benites, P. J.; Holmberg, R. C.; Rawat, D. S.; Kraft, B. J.; Klein, L. J.; Peters, D. G.; Thorp, H. H.; Zaleski, J. M. *J. Am. Chem. Soc.* 2003, **125**, 6434.

[28] Chandra, T.; Kraft, B. J.; Huffman, J. C.; Zaleski, J. M. *Inorg. Chem.* 2003, **42**, 5158.

[29] (a) Nath, M.; Pink, M.; Zaleski, J. M. *J. Am. Chem. Soc.* 2005, **127**, 478. (b) Boerner, L. J. K.; Pink, M.; Park, H.; LeSueur, A.; Zaleski, J. M. *Chem. Commun.* 2013, **49**, 2145.

[30] (a) O'Connor, J. M.; Lee, L. I.; Gantzel, P.; Rheingold, A. L.; Lam, K.-C. *J. Am. Chem. Soc.* 2000, **122**, 12057. For Fe- and Ru-mediated reactivity, see (b) O'Connor, J. M.; Friese, S. J.; Rodgers, B. L., *J. Am. Chem. Soc.* 2005, **127**, 16342. (c) O'Connor, J. M.; Friese, S. J.; Tichenor, M. *J. Am. Chem. Soc.* 2002, **124**, 3506. (d) O'Connor, J. M.; Friese, S. *J. Organometallics* 2008, **27**, 4280.

[31] O'Connor, J. M.; Baldridge, K. K.; Rodgers, B. L.; Aubrey, M.; Holland, R. L.; Kassel, W. S.; Rheingold, A. L. *J. Am. Chem. Soc.* 2010, **132** (32), 132.

[32] Zimmerman, H. E.; Pincock, J. A. Exploratory and Mechanistic Organic Photochemistry. LXXVII, *J. Am. Chem. Soc.* 1973, **95**, 3246.

[33] Ramkumar, D.; Kalpana, M.; Varghese, B.; Sankararaman, S. Jagadeesh, M. N.; Chandrasekhar, J. *J. Org. Chem.* 1996, **61**, 2247.

[34] For a follow-up mechanistic study, see: Schmittel, M.; Kiau, S. *Liebigs. Ann. Recl.* 1997, **7**, 1391.

[35] (a) Alabugin I. V.; Kovalenko, S. V. *J. Am. Chem. Soc.* 2002, **124**, 9052. (b) Alabugin I. V.; Manoharan, M. *J. Am. Chem. Soc.* 2003, **125**, 4495.

[36] For mechanistic studies of related radical cyclizations, see(a) Alabugin, I. V.; Manoharan, M. *J. Am. Chem. Soc.* 2005, **127**, 9534. (b) Alabugin, I. V.; Timokhin, V. I.; Abrams, J. N.; Manoharan, M.; Ghiviriga, I.; Abrams, R. *J. Am. Chem. Soc.* 2008, **130**, 10984. (c) Alabugin, I. V.; Gilmore, K.; Manoharan, M. *J. Am. Chem. Soc.* 2011, **133**, 12608. (d) Gilmore, K.; Alabugin, I. V. *Chem. Rev.* 2011, **111**, 6513.

[37] Kovalenko, S. V.; Peabody, S.; Manoharan, M.; Clark, R. J.; Alabugin, I. V. *Org. Lett.* 2004, **6**, 2457.

[38] Peabody, S.; Breiner, B.; Kovalenko, S. V.; Patil, S.; Alabugin I. V. *Org. Biomol. Chem.* 2005, **3**, 218.

[39] Breiner, B.; Schlatterer, J. C.; Kovalenko, S. V.; Greenbaum, N. L.; Alabugin I. V. *Angew. Chem. Int. Ed.* 2006, **45**, 3666.

[40] (a) Zeidan, T.; Clark, R. J.; Kovalenko, S. V.; Ghiviriga, I.; Alabugin I. V. *Chem. Eur. J.* 2005, **11**, 4953. (b) Zeidan, T.; Kovalenko, S. V.; Manoharan, M.; Clark, R. J.; Ghiviriga, I.; Alabugin I. V. *J. Am. Chem. Soc.* 2005, **127**, 4270.

[41] Yang, W.-Y.; Marrone, S. A.; Minors, N.; Zorio, D. A. R.; Alabugin, I. V. *Beilstein J. Org. Chem.* 2011, **7**, 813.

[42] Uesawa, Y.; Suguira, Y., *Biochem. Biophys. Res. Commun.* 1989, **164**, 903.

[43] (a) Kauffman, J. F.; Turner, J. M.; Alabugin, I. V.; Breiner, B.; Kovalenko, S. V.; Badaeva, E. A.; Masunov, A.; Tretiak, S. *J. Phys. Chem. A* 2006, **110**, 241. (b) Poloukhtine, A.; Popik, V. V. *J. Org. Chem.* 2006, **71**, 7417.

[44] (a) Kovalenko, S. V.; Alabugin, I. V. *Chem. Commun.* 2005, 1444.

[45] Yang, W-Y.; Roy, S.; Phrathep, B.; Rengert, Z.; Kenworthy, R.; Zorio, D. A. R.; Alabugin, I. V. *J. Med. Chem.* 2011, **54** (24), 8501.

[46] Du, Y.; Creighton, C. J.; Yan, Z.; Gauthier, D. A.; Dahl, J. P.; Zhao, B.; Belkowski, S. M.; Reitz, A. B. *Bioorg. Med. Chem.* 2005, **13**, 5936.

[47] Kaiser, J.; Esseveldt, B. C. J.; Segers, M. J. A.; van Delft, F. L.; Smits, J. M. M.; Butterworth, S.; Rutjes, F.P.J.T. *Org. Biomol. Chem.* 2009, **7**, 695.

[48] Basak, A.; Bag, S. S. *Bioorg. Med. Chem.* 2005, **13**, 4096.

[49] Jerić, I.; Chen, H-M. *Tetrahedron Lett.* 2007, **48**, 4687.

[50] (a) Fouad, F. S.; Wright, J. M.; Plourde II, G.; Purohit, A. D.; Wytat, J. K.; El-Shafey, A.; Hynd, G.; Crasto, C. F.; Lin, Y.; Jones, G. B. *J. Org. Chem.* 2005, **70**, 9789. (b) Plourde II, G.; El-Shafey, A.; Fouad, F. S.; Purohit A. S.; Jones, G. B. *Bioorg. Med. Lett.* 2002, **12** (20), 2985. For Cleavage of Chymotrypsin with neutral conjugates, see also: (c) Dutta, S.; Basak, A.; Dasgupta, S. *Bioorg. Med. Chem.* 2009, **17**, 3990.

[51] Hatial, I.; Addy, P. S.; Ghosh, A. K.; Basak, A. *Tetrahedron Lett.* 2013, **54**, 854.

31

PHOTO-FRIES REACTION AND RELATED PROCESSES

Francisco Galindo[1], M. Consuelo Jiménez[2], and Miguel Angel Miranda[2]

[1] *Departamento de Química Inorgánica y Orgánica, Universitat Jaume I de Castellón, Castellón de la Plana, Spain*

[2] *Departamento de Química/Instituto de Tecnología Química UPV-CSIC, Universitat Politècnica de València, València, Spain*

31.1 INTRODUCTION

More than half a century ago, Anderson and Reese reported the first example of a photo-Fries rearrangement (PFR), where 2-hydroxyphenylacetate was converted into 2,3- and 3,4-dihydroxyacetophenone upon UV irradiation [1]. Since then, hundreds of articles have appeared on this reaction, which has been extended from the initial aryl esters to amides, carbonates, carbamates, thioesters, sulfonates, etc. [2–4].

The present chapter covers the publications on PFR appearing until the end of 2013, with special emphasis on the developments of the last decade. In addition to recent synthetic applications, the accumulated mechanistic evidence based on direct spectroscopic studies and the peculiarities of the reaction in organized and constrained media are highlighted.

31.2 MECHANISTIC ASPECTS

31.2.1 General Scheme

The PFR of phenyl acetate, which affords *o*-hydroxyacetophenone, *p*-hydroxyacetophenone, and phenol, is a convenient model to illustrate the basic mechanistic aspects of the process. Initial homolysis of the carbonyl-oxygen bond from the singlet excited state gives an acetyl/phenoxyl radical pair within a solvent cage, where spin is delocalized through the oxygen atom and the *o*- and

Arene Chemistry: Reaction Mechanisms and Methods for Aromatic Compounds, First Edition. Edited by Jacques Mortier.

p-aromatic carbons [5]. Subsequent in-cage recombination yields the starting ester and two isomeric acylcyclohexadienones, which tautomerize to the final products [6–11]. Alternatively, escape of the phenoxyl radical from the solvent cage leads eventually to phenol (Scheme 31.1). In the case of substituted aromatic esters, alternative reaction pathways (photodecarboxylation [12, 13], photodecarbonylation [14–16], etc.) can compete.

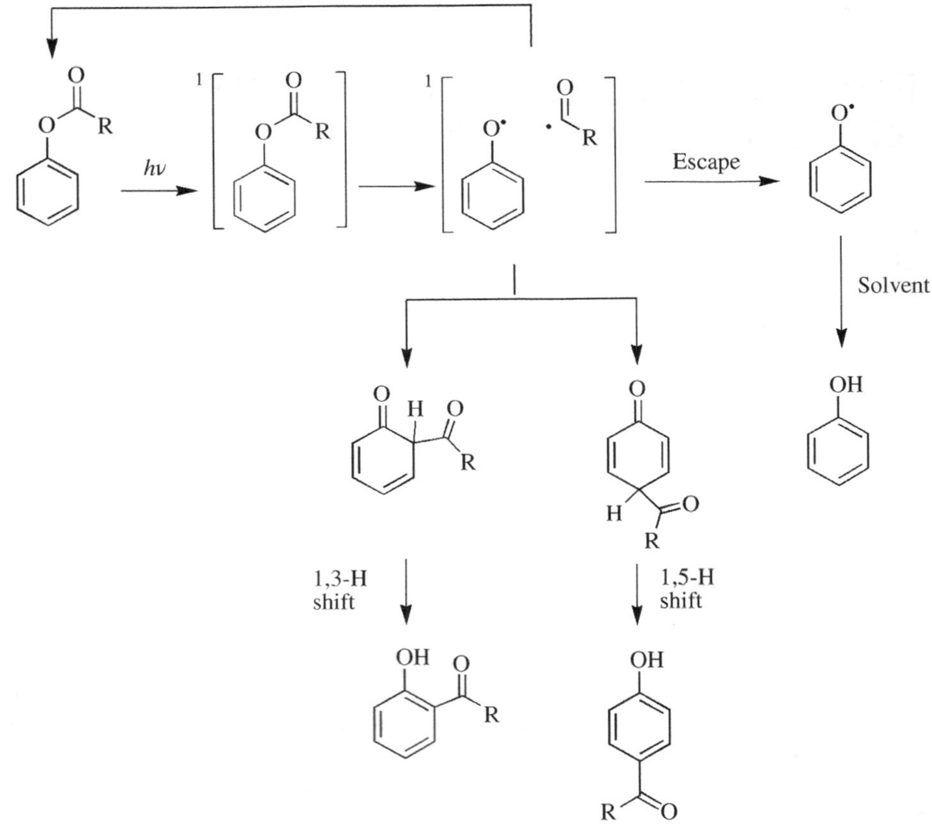

SCHEME 31.1 Mechanistic pathways in the PFR of phenyl acetate.

A similar mechanism is accepted for the photo-Claisen rearrangement of aromatic ethers, where allyl, benzyl, or other activated alkyl groups migrate from the oxygen to the *o*- and *p*-carbons of the aromatic ring [17].

31.2.2 Experimental Evidence: Steady-State Photolysis

It has been established that the PFR occurs from the lowest singlet excited state, on the basis of the conversions and product distributions obtained in photosensitization and quenching experiments [18, 19]; however, a minor contribution of upper triplet excited states has been suggested in some cases [20–22].

As regards the proposed caged radical pair model, it is consistent with the results obtained upon photolysis of phenyl acetate in the vapor phase, which gives almost exclusively the escape product (phenol) [10, 11]. Conversely, in rigid matrixes (where the cage effect is enhanced), only radical recombination products are obtained [23].

A direct proof supporting recombination of the acyl and phenoxyl radicals to give the PFR products has been provided by independent generation of both species. For instance, irradiation of pinacolone in the presence of *p*-trifluoromethylphenol leads to *o*-acetyl-*p*-trifluoromethylphenol (Scheme 31.2). Parallel formation of *p*-trifluoromethylphenyl acetate confirms the feasibility of O—C radical coupling under PFR conditions to give the starting phenyl esters [24].

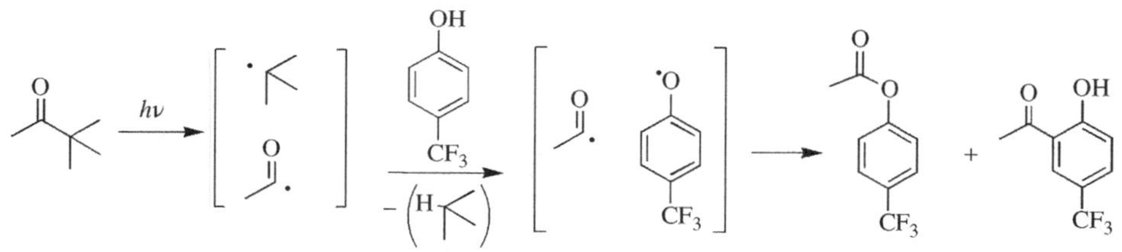

SCHEME 31.2 Intermolecular PFR between pinacolone and *p*-trifluoromethylphenol.

Unambiguous chemical evidence for the involvement of cyclohexadienones as PFR intermediates has been provided by the isolation of acylnaphthalenones in the irradiation of partially blocked naphthyl esters (Scheme 31.3) [25, 26].

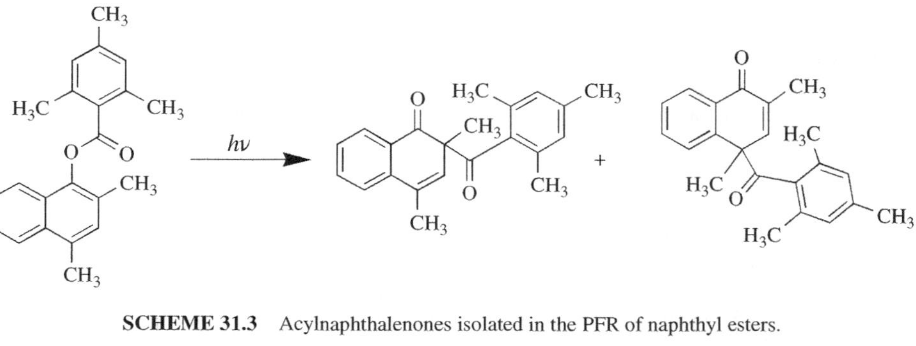

SCHEME 31.3 Acylnaphthalenones isolated in the PFR of naphthyl esters.

31.2.3 Experimental Evidence: Time-Resolved Studies

Investigation of the PFR of 2-naphthyl acetate in poly(methyl methacrylate) by means of time-resolved fluorescence has shown that the photoproducts act as long-range quenchers for the starting ester. This confirms singlet excited state involvement and suggests a nonrandom distribution of the generated chromophores [27].

The PFR of phenyl acetate with conventional flash lamp photolysis demonstrates the generation of long-lived (submillisecond) phenoxyl radicals. In addition, a cyclohexadienone transient is observed, decaying within *ca.* 1 s to *o*-hydroxyacetophenone [8, 9]. The kinetics of this process has been examined in more detail by laser flash photolysis. The rate constant for the sigmatropic 1,3-hydrogen shift is $3.6\,s^{-1}$, almost two orders of magnitude higher than that of the 1,5-shift. As expected, in the case of phenyl acetate-d_5, the corresponding deuterium shift is slower than in the nondeuterated ester [28, 29]. Likewise, naphthoxyl radicals peaking at 320–350 nm have been detected in the laser flash photolysis of naphthyl esters. Formation of the rearranged hydroxyketones is directly monitored through the enhanced absorbance at 360 nm, with a rate constant in the order of $10^4\,s^{-1}$ [22, 30].

A further step in the study of cyclohexadienones as PFR reaction intermediates has consisted in the secondary photolysis of these relatively long-lived species by using the two-laser–two-color technique, which gives rise to formation of a dienic ketene, absorbing at $\lambda_{max}=330$ nm (Fig. 31.1) [31].

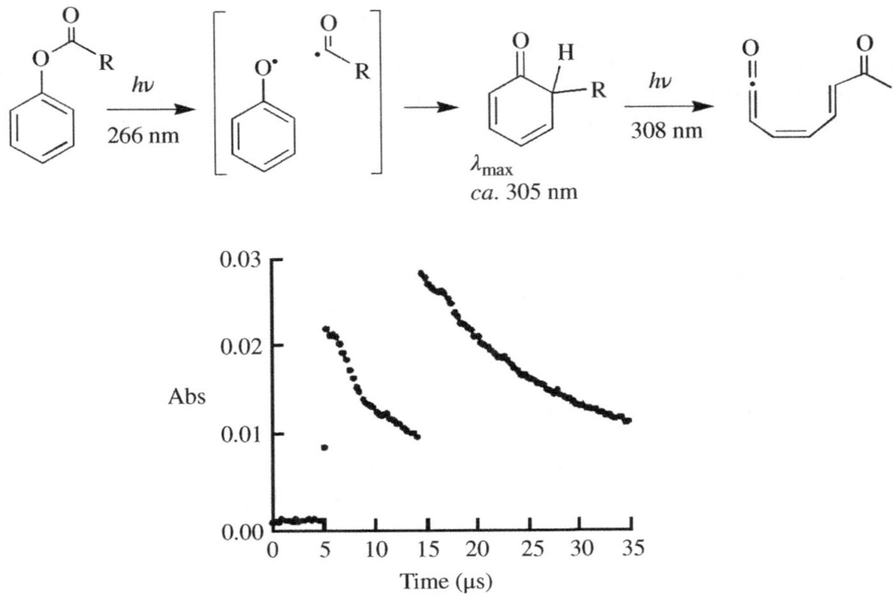

FIGURE 31.1 Two-laser–two-color photolysis of a phenyl ester (Abs at 330 nm vs. time). The jump at 5 μs of the kinetic trace corresponds to the first laser shot (266 nm) and the jump at 15 μs to the second one (308 nm), producing the cyclohexadienone ring opening. Reproduced from Jiménez et al. [31] with permission from the Royal Society of Chemistry.

The acetyl radical has been directly detected by time-resolved FTIR studies during the PFR of 1-naphthyl acetate adsorbed onto NaY zeolite as a transient species at 2127 cm^{-1}; its lifetime has been estimated as 75 μs. Labeling (^{13}C) of the substrate causes a 34 cm^{-1} red shift in the vibration frequency [32]. Spontaneous Raman spectroscopy has also been used for mechanistic studies on the PFR of phenyl esters, confirming the established model. Thus, three species with different lifetimes are detected after laser photolysis (266 nm) in water. The shortest-lived one corresponds to phenoxyl radical (assigned by comparison with the spectrum generated by direct irradiation of phenol), whereas the two longer-lived species are tentatively assigned to 6-acetyl- and 4-acetylcyclohexadienones [33].

Further insight into the early events involved in PFR has been provided by improving time resolution, using two-color femtosecond pump–probe spectroscopy. It has been reported that the primary carbonyl-oxygen cleavage of 4-*tert*-butylphenol acetate (in cyclohexane) occurs within 2 ps, while recombination in the radical cage takes 13 ps [34]. A somewhat higher value for the lifetime of the caged singlet radical pair has been inferred from the results of nanosecond laser flash photolysis studies on 1-naphthyl acetate under magnetic fields in the range of 0–7 T. Thus, the yield of escaped 1-naphthoxyl radical increases with magnetic field in the mT range, whereas the opposite trend is found at higher fields. Such inverted magnetic field effects have been taken as an indication of a lifetime in the nanosecond range for the spin-correlated radical pair [35, 36].

A full picture of the PFR of phenyl acetate has recently been obtained by UV-vis and IR transient spectroscopy in the (sub)picosecond range (Fig. 31.2). The combination of the two techniques provides accurate data of the involved ultrafast processes. The singlet lifetime is 28 ps, both from the decay of its $S_1 \rightarrow S_n$ absorption, monitored at 524 nm, and its vibration at 1807 cm^{-1}. The phenoxyl

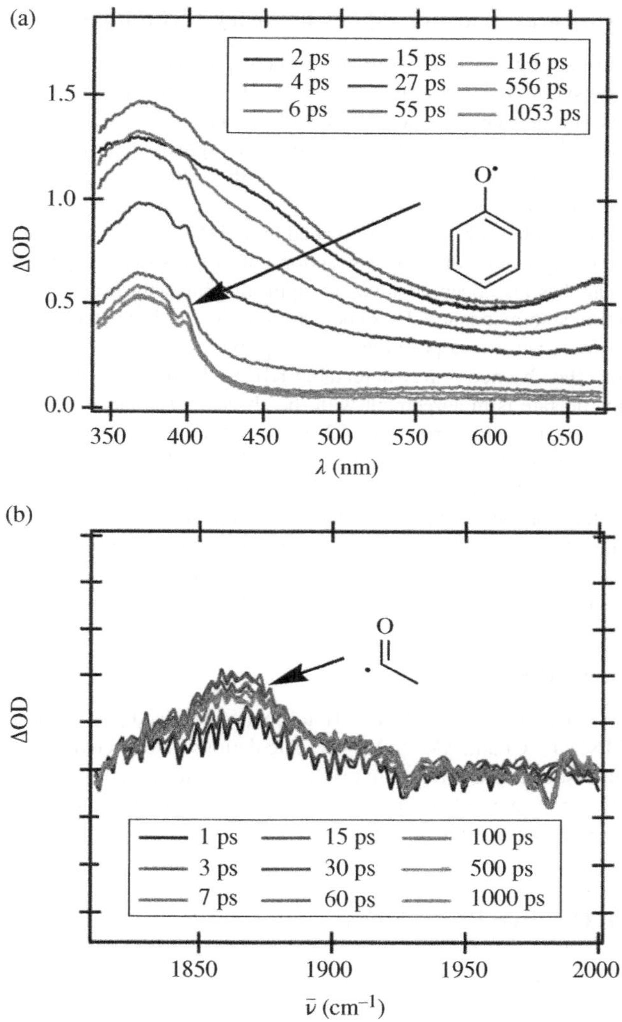

FIGURE 31.2 (a) Transient electronic absorption spectrum ($S_1 \rightarrow S_n$) of the phenyl acetate singlet excited state (4 ps after the laser pulse) and of the phenoxyl radical (*ca.* 1 ns later). (b) Transient vibrational absorption spectrum of the acetyl radical at several pump–probe delay times. Reproduced from Harris et al. [37] with permission from the Royal Society of Chemistry.

radical is detected through its absorption at 350–400 nm, whereas the acetyl radical is evidenced by its C═O stretching at 1866 cm^{-1}.

Formation of cyclohexadienones occurs in 42 ps and can be followed by the appearance of characteristic carbonyl stretching bands at 1697 and 1765 cm^{-1} from the enone and acetyl groups, respectively. Using appropriate kinetic models, the recombination quantum yield is estimated as 0.74, and hence, the corresponding value for the escape process is 0.26. Recombination includes recovery of S_0 and formation of cyclohexadienones (quantum yields of 0.54 and 0.20, respectively). Final tautomerization to the final PFR products is not detected in the maximum time scale (2.5 ns), confirming that carbon to oxygen H transfer is a very slow process [37].

31.2.4 Experimental Evidence: Spin Chemistry Techniques

The observation of strongly polarized signals in the NMR spectrum recorded during steady-state irradiation of *p*-cresyl *p*-chlorobenzoate supports the involvement of a radical pair mechanism and allows discriminating between "in-cage" and "out-of-cage" processes [38]. In the case of 1- and 2-naphthyl acetates, laser-induced CIDNP results are in agreement with O—CO homolysis taking place mainly from the first excited singlet state (S_1), with a minor contribution from a reactive upper triplet excited state (T_2) [22, 30].

Formation of acyl radicals has been confirmed by CIDNP through the signals corresponding to the aldehydes resulting from hydrogen abstraction out of the solvent cage [22]. Indirect detection of the acyl radicals has also been achieved by spin trapping with 2-methyl-2-nitrosopropane and subsequent electron spin resonance (ESR) detection of the resulting stable radicals [39].

Measurement of kinetic isotope effects (KIE) provides valuable insight into the mechanism of the PFR. Thus, the absence of any detectable KIE in the rearrangement of ^{14}C- or ^{18}O-labeled 4-methoxyphenyl acetate is in agreement with the absence of an activation energy barrier for the O—CO bond cleavage and with formation of 2-hydroxy-5-methoxyacetophenone by recombination of a caged radical pair produced from a singlet excited state [40, 41].

In the PFR of 1-naphthyl acetate, the yield of 2-acetyl-1-naphthol exhibits a positive external magnetic field effect for the ester labeled by the magnetically active isotope ^{13}C, but no effect for the ^{12}C ester. This effect is explained by considering hyperfine coupling between ^{1}H or ^{13}C of the naphthoxyl radical and the unpaired electron of the acetyl radical, which is in agreement with formation of the in-cage product from a singlet radical pair [42].

31.2.5 Theoretical Studies

A theoretical explanation using molecular orbital theory has been proposed for the PFR of phenyl acetate, using MNDOC-CI and AM1/AM1-HE calculations. Irrespective of the electronic configuration of the nondissociative excited state reached upon photoexcitation, surface crossing occurs to give a dissociative $\pi\sigma^*$state (where the σ orbital is located along the O—CO bond), followed by homolysis and subsequent rearrangement [43].

31.3 SCOPE OF THE REACTION

In this section, it is not intended to provide a compilation of all the reported PFR processes, but rather to present a number of selected examples illustrating the scope and limitations of this type of photorearrangement.

31.3.1 Esters

The archetypal PFR is represented by phenyl acetate. The reaction has been extended to a wide variety of aromatic esters, where a number of structural modifications have been introduced in both the phenolic and acyl moieties.

In particular, the reactivity of phenyl esters of α,β-unsaturated carboxylic acids is interesting because the resulting 2-alkenylphenols (Scheme 31.4) can undergo intramolecular cyclization, leading to benzopyranone derivatives [44–46]. Higher unsaturated esters including acyl side chains with acetylenic [47] or allenic [48] functionalities lead also to cyclizable PFR products.

The studies have been extended to esters that contain condensed polyaromatics in the phenolic substructure. For instance, 1-naphthyl acetate gives 2- and 4-acetyl-1-naphthols [22, 30, 42], whereas in the case of 2-naphthyl acetate, 1-, 3-, 6-, and (marginally) 8-acetyl-2-naphthols are obtained [27, 49–53]. The reaction also works with heterocyclic compounds, such as furan [54], pyridine [55–57], indole [58–60], or chromane [61] derivatives.

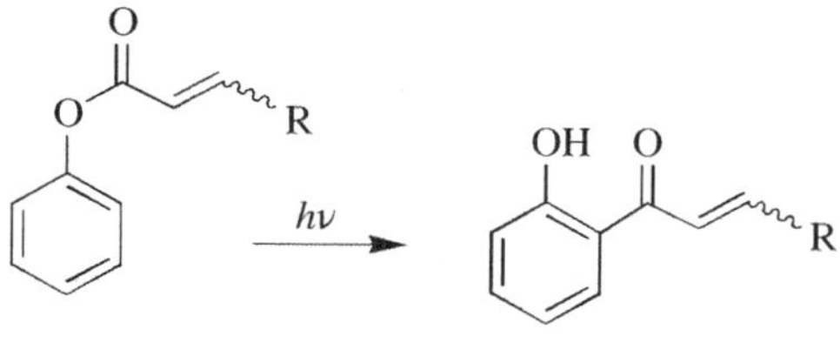

SCHEME 31.4 PFR of phenyl esters of α,β-unsaturated carboxylic acids.

31.3.2 Amides

Anilides are also substrates for the PFR, although their reactivity is lower than that of the analogous phenyl esters. Thus, irradiation of acetanilide leads to 2-aminoacetophenone, 4-aminoacetophenone, and aniline (Scheme 31.5) [62]; the quantum yields of product formation (<0.1) are markedly lower than those of phenyl acetate [63, 64] and decrease with solvent polarity, due to excited state quenching. Similar results are obtained with benzanilides [65, 66].

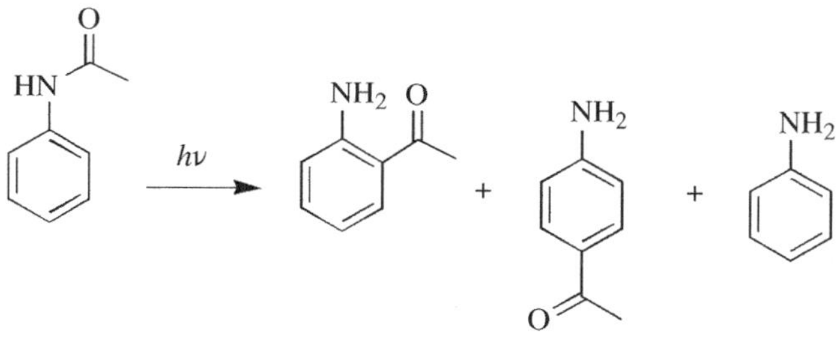

SCHEME 31.5 PFR of acetanilide.

In the case of 2-allylacetanilide, a normal PFR is observed (Scheme 31.6). Conversely, in the cinnamyl analog, rearrangement is suppressed, while photocyclization and *trans*/*cis*-isomerization are favored [67].

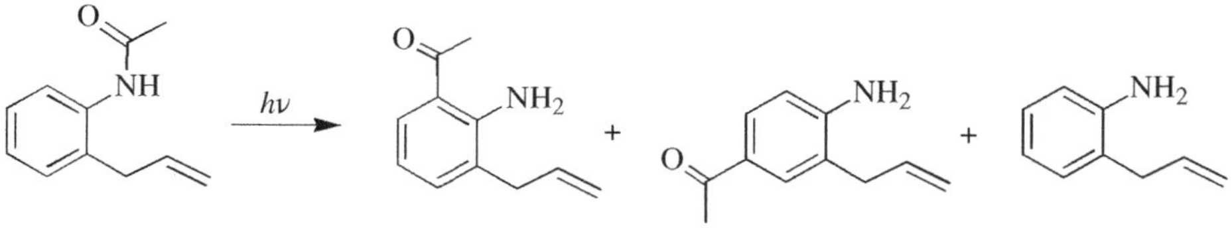

SCHEME 31.6 PFR of 2-allylacetanilide.

Photo-Fries products are also formed upon irradiation of pyridine [68, 69], indole [70, 71], and carbazole [72] derivatives. In the case of imides, only one of the acyl groups tends to migrate [73–76]; however, macrocyclic imides give rise to photoproducts diacylated at the aromatic ring [77].

31.3.3 Other

Aryl carbonates and carbamates are building blocks of polycarbonates and polyurethanes, respectively. Upon UV irradiation of diphenyl carbonate PFR is observed [78, 79]. Product studies reveal that the primary phenyl salicylate rearranges further to 2,2′-dihydroxybenzophenone and 2,4′-dihydroxybenzophenone (Scheme 31.7).

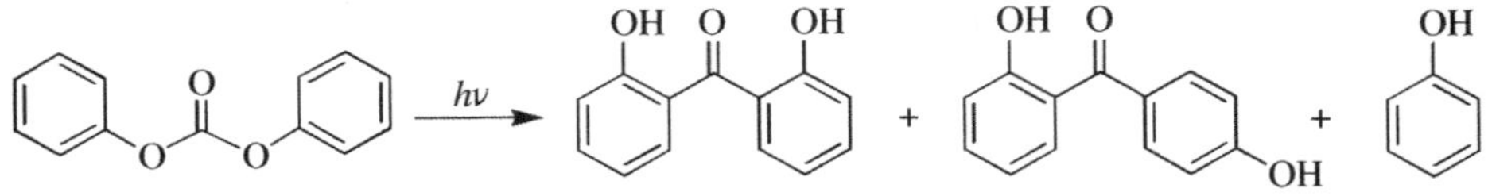

SCHEME 31.7 PFR of diphenyl carbonate.

Carbamates could in principle undergo cleavage of either the C—O [80] or the C—N [81–93] bond, although the latter process is much more common. This is illustrated later for *O*-ethyl-*N*-phenyl carbamate (Scheme 31.8).

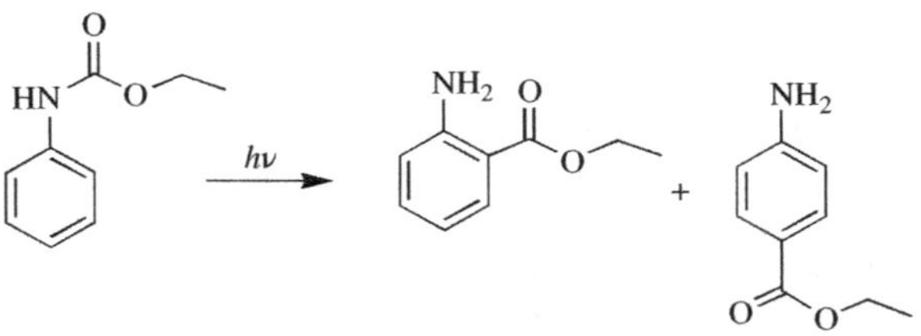

SCHEME 31.8 PFR of *O*-ethyl-*N*-phenyl carbamate.

Photochemical rearrangement of sulfonic acid esters affords sulfones in moderate yields [94–98]. For instance, irradiation of carbazolyl sulfonates affords C-sulfonated hydroxyl carbazoles (Scheme 31.9).

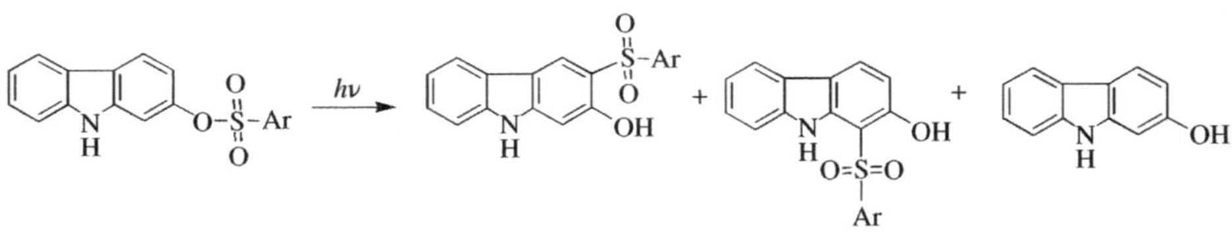

SCHEME 31.9 PFR of a carbazolyl sulfonate.

Likewise, sulfonamides undergo PFR to give the corresponding aminophenyl sulfones (Scheme 31.10) [70, 99–101].

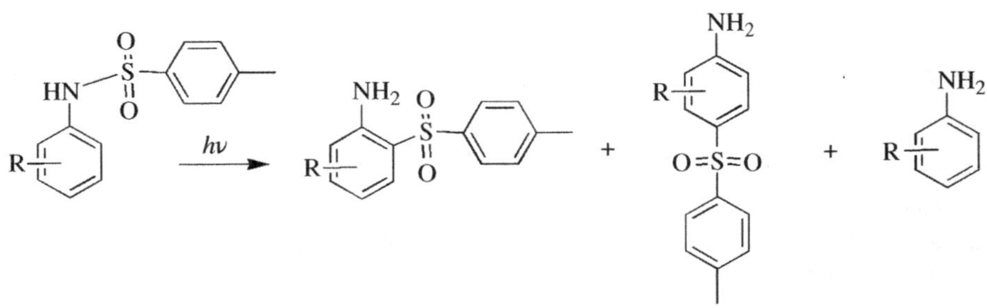

SCHEME 31.10 PFR of sulfonamides.

Although less common, some examples of PFR with thioesters, selenoesters, and telluroesters have also been reported [102–105]. Thus, irradiation of *S*-phenyl thioacetate gives minor amounts of *ortho*- and *para*-rearranged products; escape of the phenylthio radical out of the cage predominates, affording diphenyl disulfide as major product, together with thiophenol [102]. The rearrangement also occurs when the sulfur atom belongs to the acyl moiety [103].

31.4 (MICRO) HETEROGENEOUS SYSTEMS AS REACTION MEDIA

Being the prototype of a reaction involving in-cage recombination versus diffusion of radicals, the PFR constitutes an appropriate probe to disclose how the nature of the experienced microenvironment modulates photochemical reactivity and selectivity. Among the investigated hosts, special attention is devoted to cyclodextrins, micelles, zeolites, proteins, and polymeric matrixes.

31.4.1 Cyclodextrins

The PFR of phenyl esters [106–110], anilides [108, 111, 112], and sulfonyl anilides [113] has been thoroughly investigated in the presence of cyclodextrins. In general, complexation within this hydrophobic microenvironment favors formation of the *ortho*-rearranged isomer. Likewise, in the case of 1-naphthyl esters [114–116], the selectivity toward 2-acylnaphthols increases when the reaction is performed with the inclusion complexes.

31.4.2 Micelles

Enhanced *ortho*-selectivity has also been reported in anionic (sodium dodecyl sulfate) or cationic (cetyltrimethylammonium bromide) micelles [114, 117–119]. Singlet sensitization, in addition to micellar effect, is observed during the photolysis of 1-naphthyl acetate in aqueous solutions of an antenna polyelectrolyte containing fluorene chromophores. In this system, competition between in-cage and out-of-cage processes can be controlled by modifying fluorene content. Thus, the lower fluorene content, the higher yield of noncage product 1-naphthol is obtained [120].

31.4.3 Zeolites

The PFR of aromatic esters [121–123] and amides [124] has been studied within faujasite and pentasil zeolites. In the former, the predominating product is again the *ortho*-isomer, while in the latter, the *para*-isomer is favored. The observed differences have been attributed to the size and shape of the cavities and channels of the zeolites, which is consistent with the selectivity trends found in faujasites containing alkaline cation of diverse sizes. The same *ortho*-selectivity is found in Y zeolites [125].

31.4.4 Proteins

The influence of human and bovine serum albumins as hosts on the PFR has been investigated using 4-methoxy-1-naphthyl hydrogen glutarate and the corresponding acetate (Scheme 31.11). The photorearrangement does indeed occur in these microenvironments, and its efficiency depends on the nature of the higher-affinity binding sites for the specific substrate. The reaction efficiency seems to be higher within binding site I (acetate) than within site II (glutarate). As a practical advantage, the intraprotein PFR is convenient to work in aqueous media with hardly soluble aryl esters [126].

31.4.5 Other Organized Media

That the PFR is an ideal reaction probe for the study of the matrix morphology has been demonstrated by examining the photoreactivity of 2-naphthyl esters included in stretched or unstretched polyethylene films [127, 128]. The obtained selectivity data indicate that the average sites experienced by reaction intermediates in the polymer are nearly cylindrical, since the photoproducts

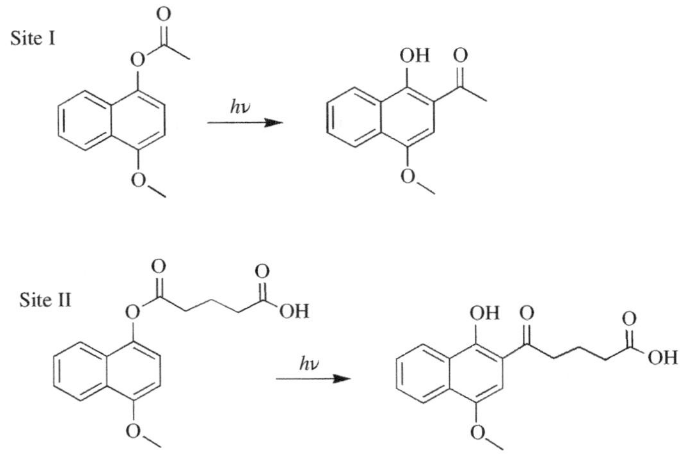

SCHEME 31.11 PFR of 4-methoxy-1-naphthyl hydrogen glutarate and the corresponding acetate within the binding sites of human and bovine serum albumin.

derived from coupling at C3 and C6 (with a more extended shape) are preferred. Similar results have been obtained for other aromatic esters [129–134] in the same media. Likewise, the PFR of 1-naphthyl esters (Scheme 31.12) has been examined in poly(vinyl acetate) films as alternative polymeric media, above and below their glass transition temperatures (T_g). As expected for a singlet radical pair mechanism, the distribution of PFR products is largely determined by the initial conformation of the included esters, due to the "templating" effect of the latter on the reaction cavities, especially below T_g. The radical pair recombination rates for formation of the dienone precursor of 2-(2-phenylpropanoyl)-1-naphthol upon irradiation of 1-naphthyl 2-phenylpropanoate in the polymer seem to be reduced with the decreasing temperature from above to below T_g [135].

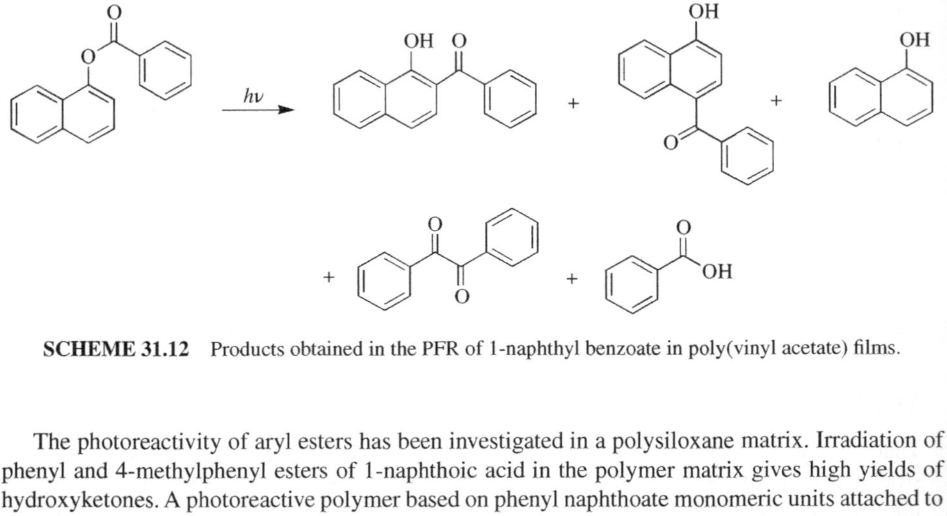

SCHEME 31.12 Products obtained in the PFR of 1-naphthyl benzoate in poly(vinyl acetate) films.

The photoreactivity of aryl esters has been investigated in a polysiloxane matrix. Irradiation of phenyl and 4-methylphenyl esters of 1-naphthoic acid in the polymer matrix gives high yields of hydroxyketones. A photoreactive polymer based on phenyl naphthoate monomeric units attached to a norbornene ring shows a high PFR yield that is associated with a large increase in the refractive index, making this polymer promising for optical applications, such as waveguiding and diffractive optics [136]. Photosensitive thiol-bearing aryl esters such as phenyl 16-mercaptohexadecanoate have been applied as a self-assembled monolayer on gold. Following PFR (Scheme 31.13), the adsorbates are selectively derivatized to yield amino-functionalized surfaces. Then, micrometer-scale patterned surfaces are produced, using a contact mask and an argon ion laser [137].

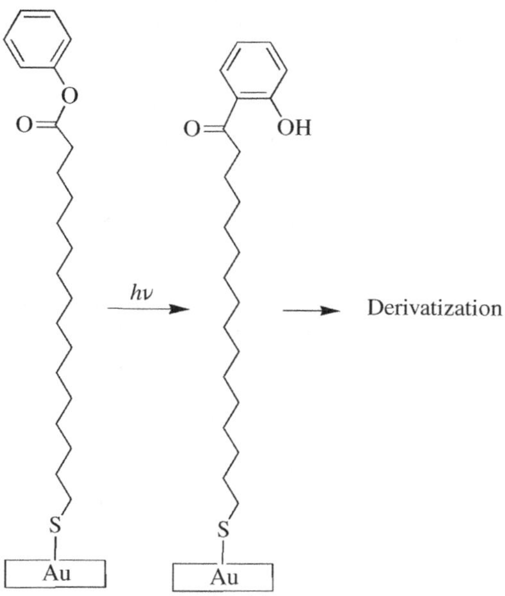

SCHEME 31.13 PFR of thiol-bearing aryl esters applied as a self-assembled monolayer on gold.

Trichloroorganosilanes bearing photoreactive aryl ester groups have been anchored to thin silane layers on silicon oxide surfaces. Upon PFR, a change in chemical reactivity of the surface is noticed. Photopatterned surfaces are then produced using a contact mask during illumination. Photolithographic modification provides a versatile and powerful tool for fabricating functionalized patterned surfaces (Scheme 31.14) [138].

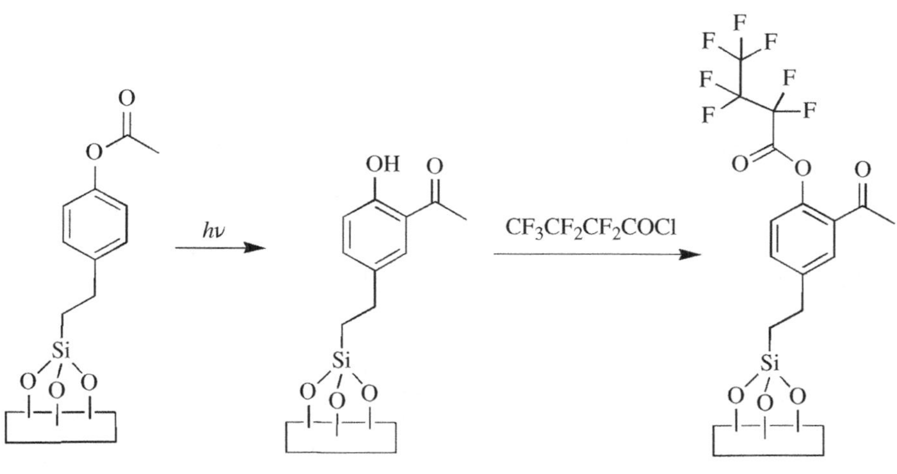

SCHEME 31.14 PFR and postexposure modification reaction of an aryl ester bearing a trichloroorganosilane moiety.

The PFR of 1-naphthyl esters has been examined in aqueous medium using styrene-based water-soluble polymer as lipophilic hosts. The intrapolymeric domains restrict the mobility of the starting esters, the generated intermediates, and the final products, resulting in a remarkable selectivity toward the 2-acylnaphthols [139]. A similar trend is observed using water-soluble poly(alkyl aryl ether) dendrimers [140], methanol-swollen Nafion beads as microreactors [141], or the capsule formed by a deep-cavity cavitand with eight carboxylic acid groups [142].

Finally, the PFR of 1-naphthyl acetate has been performed in supercritical carbon dioxide; the obtained results point to an enhanced solvent cage effect on the product ratios at lower supercritical fluid bulk densities, which is consistent with the existence of supercritical solvent–solute clusters with a lifetime similar to or longer than that of the primary singlet radical pair [143].

31.5 APPLICATIONS IN ORGANIC SYNTHESIS

The Lewis acid-catalyzed Fries rearrangement is performed under strong acidic conditions and therefore requires protection of sensitive functional groups. Conversely, the PFR can be achieved under milder conditions (organic solvents, neutral media, room temperature, etc.) and is therefore free from this type of problems. Moreover, polyacylation of aromatics is usually difficult to achieve. In the PFR, this problem can be solved by using acetals [144–146] or enol esters [147] as carbonyl protecting groups. Addition of anhydrous K_2CO_3 is convenient to avoid deprotection [148].

A possible problem for the use of PFR at preparative scale is the internal light filter effect of the *ortho*-rearranged products. This drawback has been circumvented in synthetic routes leading to chromanones by *in situ* cyclization of the α,β-unsaturated *ortho*-hydroxyphenones resulting from PFR. For that purpose, irradiation is performed in one pot by using a two-phase benzene/aqueous NaOH system [149]. On this basis, a photochemical approach based on PFR has been developed for the synthesis of chromenes, with biological activity as juvenile hormone inhibitors (see example in Scheme 31.15) [150–155]. Likewise, the PFR or aryl phenylpropinoates afford 2-hydroxyaryl phenylethynyl ketones, whose cyclization leads to flavones or aurones, depending on the base and solvent employed [47]. A similar PFR reaction can be performed with β-ketoesters; subsequent cyclization of the rearranged β-diketones provides a direct entry to the chromone skeleton [156].

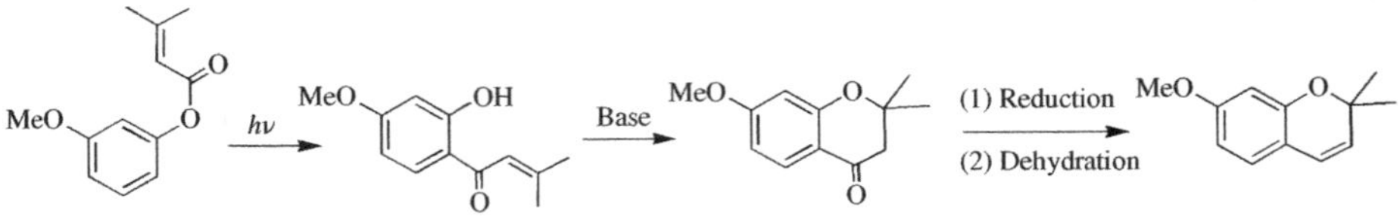

SCHEME 31.15 Application of the PFR to the synthesis of chromenes.

Xanthones can be obtained by PFR of salicylic and thiosalicylic acid derivatives (Scheme 31.16) [157, 158]. Starting from the anthranilic acid analogs, the corresponding acridones are formed [159]. In the case of aryl hydrogen succinates, the PFR products are 4-(2-hydroxyaryl)-4-oxobutanoic acids, which can be readily cyclized to 5-(2-acetoxyaryl)-2(3*H*)-furanones [160–164].

The PFR has also been used as a key step in the synthesis of more complex natural products or biologically active compounds, such as capillarol [165], bikaverin [166, 167], adriamycin [168, 169], rutaretin methyl ether [170], arizonine and caseadine [171], diazonamide [172], and *Strychnos* and *Aspidosperma* alkaloids [173]. A more recent example is the elegant synthesis of (−)-kendomycin involving the intramolecular PFR of a macrocyclic lactone (Scheme 31.17) [174, 175].

Substituted tetrahydroisoquinolines have been obtained by PFR of appropriate 4-methoxyphenylacetates, followed by tandem reduction/cyclization and further reduction. The

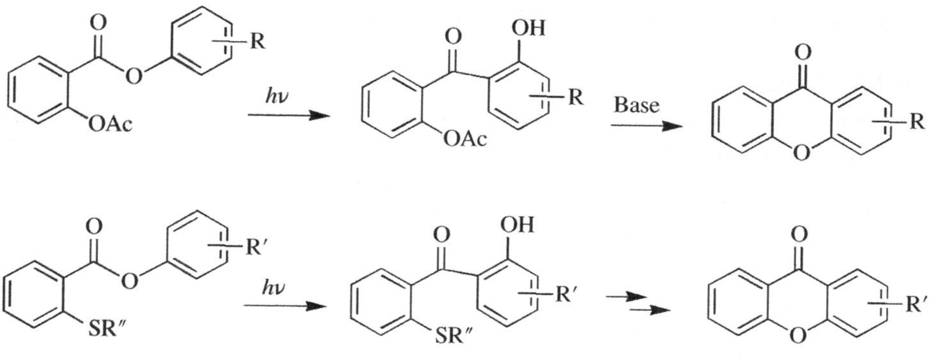

SCHEME 31.16 Synthesis of xanthones by PFR of salicylic or thiosalicylic acid phenyl esters.

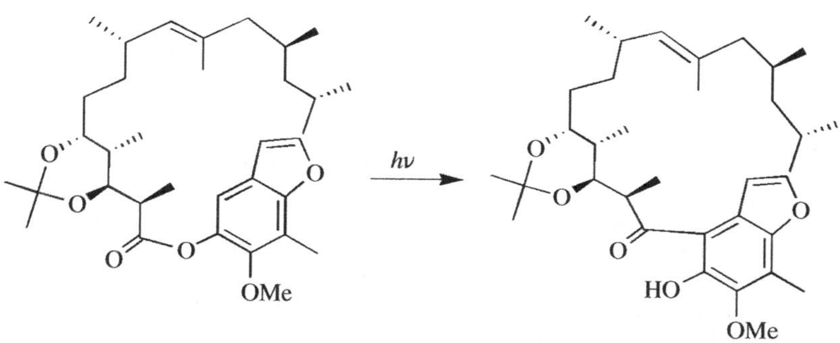

SCHEME 31.17 Photorearrangement of a macrocyclic lactone as a step in the total synthesis of (−)-kendomycin.

synthesized compounds displace D2 dopamine receptor from its specific binding site in the rat striatal membranes [176]. Related functionalized quinolines and tetrahydronaphthyridines have been prepared by PFR of *p*-substituted anilides, followed by trapping with acetylenic Michael acceptors [177].

The PFR of *N*-chloroacetylanthranylates afford benzophenones, whose direct reaction with ammonia gives 1,3-quinazolines. Alternatively, treatment with potassium iodide before addition of ammonia leads to 1,3-dihydro-2*H*-1,4-benzodiazepin-2-ones [178]. A related synthesis of benzodiazepines is based on an initial PFR of anilides derived from *N*-Boc-Ala-OH (Scheme 31.18) [179].

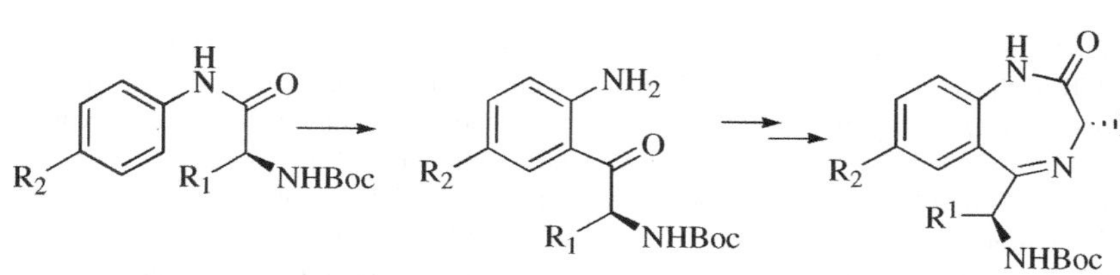

SCHEME 31.18 Synthesis of benzodiazepines by PFR of anilides.

31.6 BIOLOGICAL AND INDUSTRIAL APPLICATIONS

31.6.1 Drugs

The PFR has been observed in compounds with pharmacological activity possessing aromatic ester or amide moieties. Benorylate, employed in rheumatoid arthritis therapy, is obtained by esterification of paracetamol with aspirin. Its irradiation leads to PFR with initial cleavage of the central C—O bond to yield 5-acetamido-2′-acetoxy-2-hydroxybenzophenone, which undergoes transacetylation to 5′-acetamido-2′-acetoxy-2-hydroxybenzophenone (Scheme 31.19) [180].

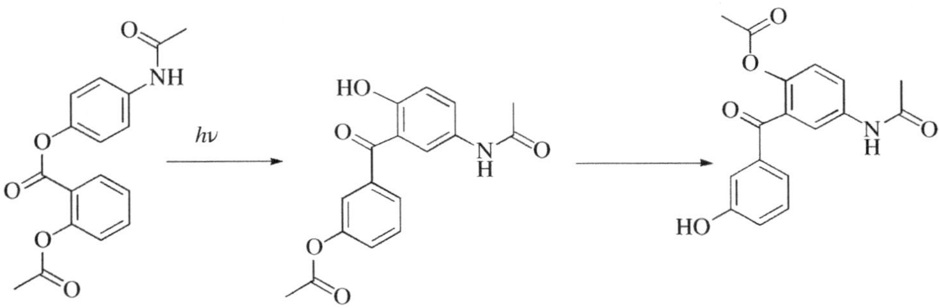

SCHEME 31.19 PFR of benorylate.

Photodegradation of the analgesic and antipyretic drug paracetamol under UV exposure at 254 nm leads to PFR, affording 2-amino-5-hydroxyacetophenone (quantum yield in the order of 10^{-3}). Formation of 4-aminophenol is observed as a minor competitive pathway (Scheme 31.20). In the presence of oxygen, a photooxygenation process takes place, leading to the formation of a peroxyester. The PFR product is more toxic than the parent drug, when examined by a luminescent bacteria test. This is interesting in connection with risk assessment associated with the presence of widely used pharmaceuticals in aquatic environments [181–183].

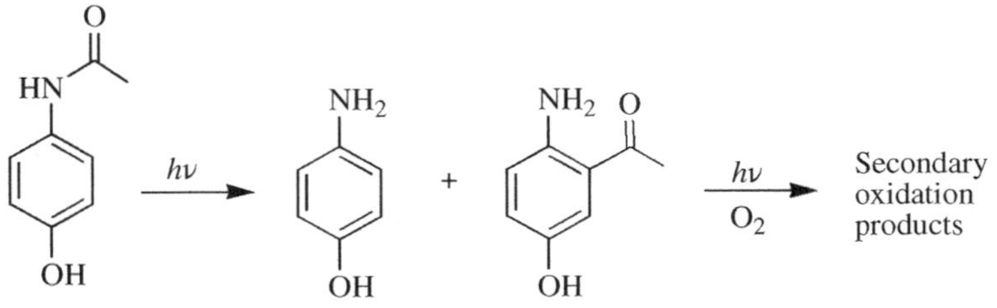

SCHEME 31.20 PFR of paracetamol.

31.6.2 Agrochemicals

Irradiation of the herbicide propanil gives rise to complex mixtures containing the products arising from PFR of the anilide moiety [184]. In the carbamate series, isoprocarb and promecarb give rise to the *ortho*- and *para*-hydroxybenzamides, whereas bendiocarb, thiobencarb, furathiocarb, fenoxycarb, and pirimicarb lead mainly to the corresponding phenols [185].

Carbaryl is photolyzed in aerated aqueous solution with a low quantum yield (2.1×10^{-3}) to give several naphthoquinone derivatives, whereas in organic solvents, it is mainly converted into 1-naphthol, and no PFR products were found (Scheme 31.21). Laser flash photolysis in water reveals formation of the triplet excited state (λ_{max} = 410 nm, τ_T *ca.* 3 μs), in addition to naphthoxyl radicals and solvated electrons [186].

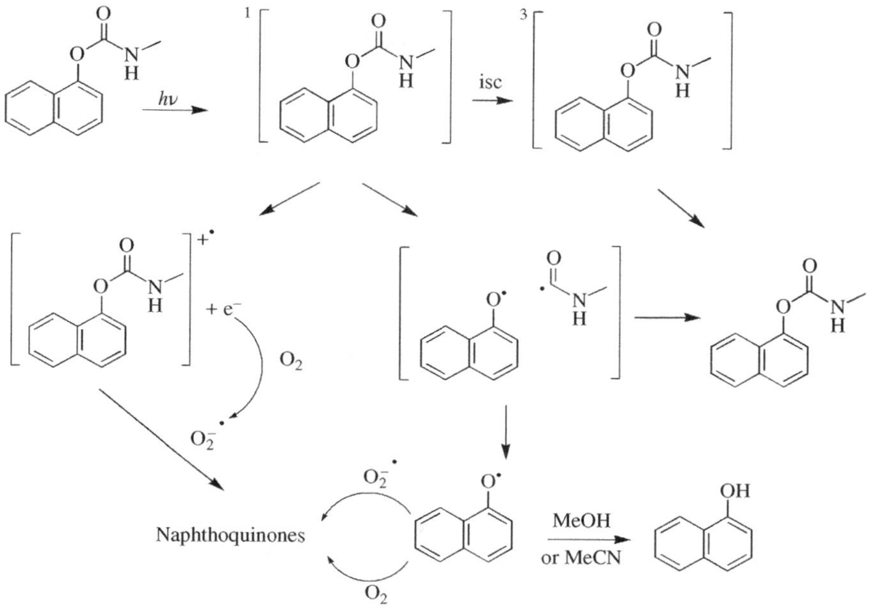

SCHEME 31.21 Photodegradation mechanism of carbaryl.

The photobehavior of phenylurea herbicides in aqueous solution depends on ring substitution. For nonhalogenated derivatives, the main reaction is PFR (Scheme 31.22), whereas in the case of halogenated phenylureas, photohydrolysis prevails. Photocatalytic oxidation can be induced by TiO_2, iron salts, or humic substances [187, 188].

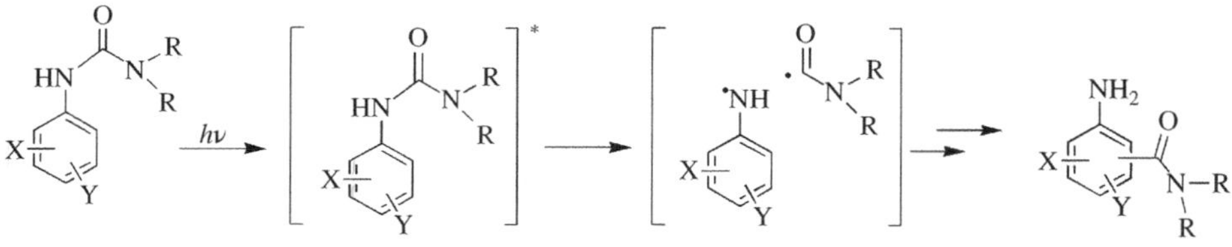

SCHEME 31.22 PFR of phenylurea herbicides.

Irradiation of cyanophos in aqueous solutions with UV light (254–313 nm) or solar light leads to various processes, such as hydrolysis, homolytic bond dissociations, and PFR (Scheme 31.23).

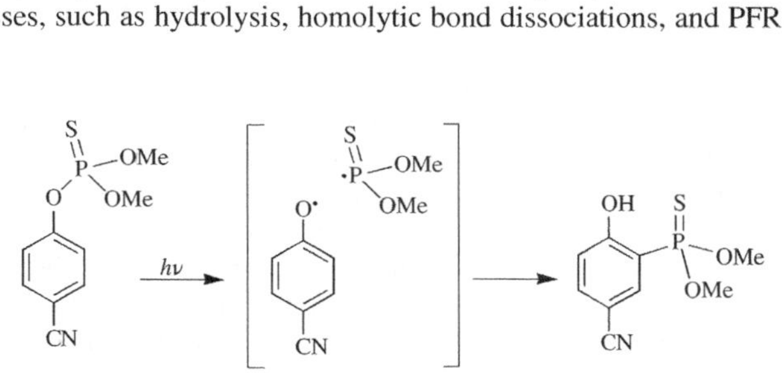

SCHEME 31.23 PFR of the pesticide cyanophos.

Combined photosensitization and quenching experiments suggest the involvement of both singlet and triplet excited states. The latter is indeed observed in laser flash photolysis [189].

31.6.3 Polymers

Aromatic polyesters [190–201], polyamides [202, 203], polycarbonates [204–211], and polyurethanes [212, 213] undergo PFR either in the polymeric backbone [214] (see example in Scheme 31.24) or in the pendant groups [215, 216].

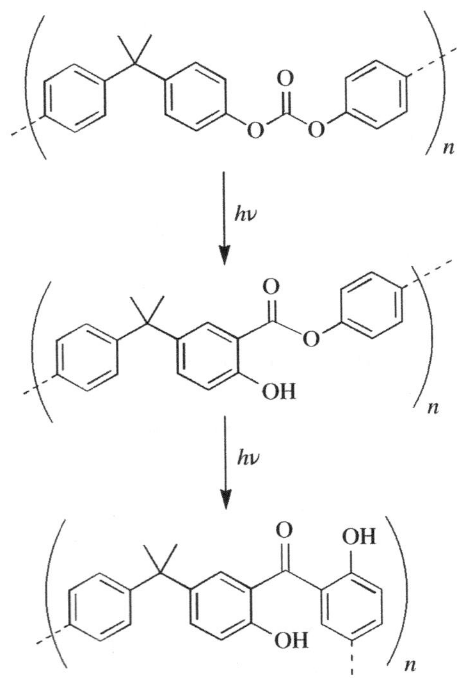

SCHEME 31.24 Example of PFR of polymers in the backbone.

An important application of the PFR in polymers is photostabilization [217]. This is associated with light absorption by the aromatic chromophores and by the internal filter effect of the *ortho*-rearranged products.

The photochemical behavior of semiflexible and rigid polyesters containing photoreactive mesogenic units derived from *p*-phenylenediacrylic acid and cinnamic acid has been studied both in solution and in films. In solution, PFR is observed, together with [2+2] photocycloaddition and E/Z isomerization. Conversely, in spin-coated films, [2+2] photocycloaddition predominates, leading to cross-linking. Liquid crystalline cells have been made with polarized irradiated films as aligning layers. The photoinduced anisotropy is stable at high temperatures, and the liquid crystalline molecules are insoluble in the irradiated polymer [218].

Thermally enhanced photoinduced molecular reorientation has been observed upon irradiation of liquid crystalline polymethacrylates with pendant aryl benzoate side groups (Scheme 31.25), using linearly polarized UV light [219, 220]. Annealing the irradiated films in the liquid crystalline temperature range of the material amplifies their photoinduced optical anisotropy. Double exposure using one photomask without alignment allows fabricating a patterned quarter-wave plate.

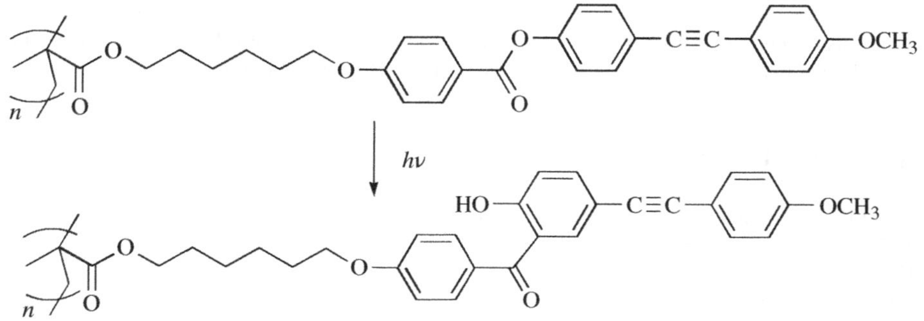

SCHEME 31.25 PFR in liquid crystalline polymethacrylates with pendant aryl benzoates.

Photosensitive polymers containing aryl esters or amides undergo PFR leading to aromatic hydroxyketones or aminoketones in the illuminated areas (Scheme 31.26). In thin films, the photoreaction leads to significant modifications of the refractive index. Appropriate postexposure reactions can be used for selective functionalization of the polymer. For example, the photogenerated aromatic hydroxyl groups readily react with acyl chlorides or sulfonyl chlorides to form new carboxylic esters or sulfonates. In combination with lithographic techniques, the postexposure reactions can be exploited to obtain patterned functionalized surfaces and to optical applications such as waveguiding or holographic data storage [221–223]. Related work has been reported on polynorbornene-based copolymers bearing *ortho*-nitrobenzyl and phenyl ester groups capable of undergoing the photoinduced cleavage reaction and in a subsequent step optionally the PFR upon UV irradiation [224, 225].

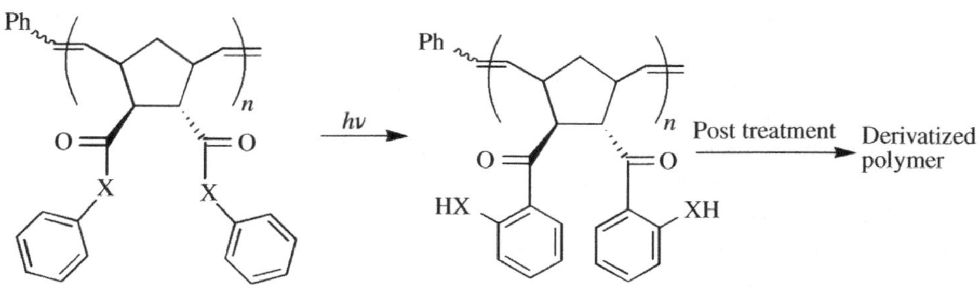

X = O, NH

SCHEME 31.26 PFR of photosensitive polymers containing aryl esters or amides in the side chains.

31.7 SUMMARY AND OUTLOOK

The PFR was discovered half a century ago as the light-induced conversion of phenyl esters into *o*- and *p*-acyl phenols. Since then, it has been extended to a wide variety of related substrates. A full mechanistic picture has recently been obtained by a combination of UV-vis and IR transient spectroscopy in the (sub)picosecond range, providing direct evidence for the involvement of the singlet excited states, caged singlet phenoxyl/acyl radical pairs, and cyclohexadienones. Tautomerization to the final rearranged products is a very slow process, easily detectable by transient absorption spectroscopy at longer time scales. The reaction constitutes the prototype of a caged radical pair process, which is initiated by homolytic photocleavage of the CO—O bond and is followed by in-cage radical recombination. Hence, it is a suitable probe for investigating the microenvironments provided by a wide variety of materials and media, from cyclodextrins to polymers or protein cavities. In addition to its mechanistic interest, the PFR has found application as the key step in the synthesis of a number of interesting compounds and plays a relevant role in the photodegradation of biologically active agents, as well as in the design of functional polymers.

ABBREVIATIONS

CIDNP	Chemically induced dynamic nuclear polarization
ESR	Electron spin resonance
FTIR	Fourier transform infrared
KIE	Kinetic isotope effect
NMR	Nuclear magnetic resonance
PFR	Photo-Fries rearrangement
UV	Ultraviolet

REFERENCES

[1] Anderson, J. C. and Reese C. B. (1960) *Proc. Chem. Soc. London*, 217.

[2] Miranda, M. A. and Galindo, F. (2003) in *Molecular and Supramolecular Photochemistry*, vol. **9** (ed. Ramamurthy, V. and Schanze, K. S.), Marcel Dekker, New York, pp. 43–131.

[3] Miranda, M. A. and Galindo F. (2004) in *Handbook of Organic Photochemistry and Photobiology*, 2nd ed. (ed. Horspool, W. M. and Lenci, F.), CRC Press, Boca Raton, FL, pp. 42/1–42/11.

[4] Bellus, D. (1971) *Adv. Photochem.*, **8**, 109–159.

[5] Stone, T. J. and Waters, W. A. (1964) *J. Chem. Soc.*, 213–218.

[6] Bellus, D., Hrdlovic, P., and Slama, P. (1968) *Collect. Czechoslov. Chem. Commun.*, **33**, 2646–2655.

[7] Bellus, D., Hrdlovic, P., and Slama, P. (1968) *Collect. Czechoslov. Chem. Commun.*, **33**, 3752–3760.

[8] Kalmus, C. E. and Hercules, D. M. (1972) *Tetrahedron Lett.*, **13**, 1575–1577.

[9] Kalmus, C. E. and Hercules, D. M. (1974) *J. Am. Chem. Soc.*, **96**, 449–456.

[10] Meyer, J. W. and Hammond, G. S. (1970) *J. Am. Chem. Soc.*, **92**, 2187–2189.

[11] Meyer, J. W. and Hammond, G. S. (1972) *J. Am. Chem. Soc.*, **94**, 2219–2228.

[12] Mori, T., Takamoto, M., Wada, T., and Inoue, Y. (2003) *Photochem. Photobiol. Sci.*, **2**, 1187–1199.

[13] Mori, T., Weiss, R. G., and Inoue, Y. (2004) *J. Am. Chem. Soc.*, **126**, 8961–8975.

[14] Xu, J. and Weiss R. G. (2005) *J. Org. Chem.*, **70**, 1243–1252.

[15] Xu, J. and Weiss, R. G. (2005) *Photochem. Photobiol. Sci.*, **4**, 210–215.

[16] Xu, J. and Weiss, R. G. (2005) *Photochem. Photobiol. Sci.*, **4**, 348–358.

[17] Galindo, F. (2005) *J. Photochem. Photobiol. C: Photochem. Rev.*, **6**, 123–138.

[18] Shizuka, H., Morita, T., Mori, Y., and Tanaka, I. (1969) *Bull. Chem. Soc. Jpn.*, **42**, 1831–1836.

[19] Finnegan, R. A. and Mattice, J. J. (1965) *Tetrahedron*, **21**, 1015–1026.

[20] Lally J. M. and Spillane, W. J. (1987) *J. Chem. Soc. Chem. Commun.*, 8–9.

[21] Lally J. M. and Spillane, W. J. (1991) *J. Chem. Soc. Perkin Trans.* **2**, 803–807.

[22] Gritsan, N. P., Tsentalovich, Y. P., Yurkovskaya, A. V., and Sagdeev, R. (1996) *J. Phys. Chem.*, **100**, 4448–4458.

[23] Coppinger, G. M. and Bell, E. R. (1966) *J. Phys. Chem.*, **70**, 3479–3489.

[24] Jiménez, M. C., Leal, P., Miranda, M. A., and Tormos, R. (1995) *J. Chem. Soc. Chem. Commun.*, 2009–2010.

[25] Mori, T., Takamoto, M., Saito, H., Wada, T., and Inoue Y. (2004) *Chem. Lett.*, **33**, 254–255.

[26] Mori, T., Takamoto, M., Saito, H., Furo, T., Wada, T., and Inoue, Y. (2004) *Chem. Lett.*, **33**, 256–257.

[27] Wang, Z., Holden, D. A., and McCourt, F. R. W. (1990) *Macromolecules*, **23**, 3773–3779.

[28] Arai, T., Tobita, S., and Shizuka, H. (1994) *Chem. Phys. Lett.*, **223**, 521–526.
[29] Arai, T., Tobita, S., and Shizuka, H. (1995) *J. Am. Chem. Soc.*, **117**, 3968–3975.
[30] Molokov, I. F., Tsentalovich, Y. P., Yurkovskaya, A. V., and Sagdeev, R. Z. (1997) *J. Photochem. Photobiol. A Chem.*, **110**, 159–165.
[31] Jiménez, M. C., Miranda, M. A., Scaiano, J. C., and Tormos, R. (1997), *J. Chem. Soc. Chem. Commun.*, 1487–1488.
[32] Vasenkov, S. and Frei, H. (1998) *J. Am. Chem. Soc.*, **120**, 4031–4032.
[33] Beck, S. M. and Brus, L. E. (1982) *J. Am. Chem. Soc.*, **104**, 1805–1808.
[34] Lochbrunner, S., Zissler, M., Piel, J., Riedle, E., Spiegel, A., and Bach, T. (2004) *J. Chem. Phys.*, **120**, 11634–11639.
[35] Gohdo, M. and Wakasa, M. (2010) *Chem. Lett.*, **39**, 106–107.
[36] Gohdo, M., Takamasu, T., and Wakasa, M. (2011) *Phys. Chem. Chem. Phys.*, **13**, 755–761.
[37] Harris, S. J., Murdock, D., Grubb, M. P., Greetham, G. M., Clark, I. P., Towrie, M., and Ashfold, M. N. R. (2014) *Chem. Sci.*, **5**, 707–714.
[38] Adam, W. and de Sanabia, J. A. (1973) *J. Org. Chem.*, **38**, 2571–2572.
[39] Rosenthal, I., Mossoba, M. M., and Riesz, P. (1982) *Can. J. Chem.*, **60**, 1486–1492.
[40] Schutte, L. and Havinga, E. (1967) *Tetrahedron*, **23**, 2281–2284.
[41] Shine, H. J. and Subotkowski, W. (1987) *J. Org. Chem.*, **52**, 3815–3821.
[42] Nakagaki, R., Hiramatsu, M., Watanabe, T., Tanimoto, Y., and Nagakura, S. (1985) *J. Phys. Chem.*, **89**, 3222–3226.
[43] Grimme, S. (1992) *Chem. Phys.*, **163**, 313–330.
[44] Alvaro, M., García, H., Miranda, M. A., and Primo, J. (1986) *Recl. Trav. Chim. Pays-Bas* **105**, 233–234.
[45] Obara, H. and Takahashi, H. (1967) *Bull. Chem. Soc. Jpn.*, **40**, 1012.
[46] Obara, H., Takahashi, H., and Hirano, H. (1969) *Bull. Chem. Soc. Jpn.*, **42**, 560–561.
[47] García, H., Iborra, S., Primo, J., and Miranda, M. A. (1986) *J. Org. Chem.*, **51**, 4432–4436.
[48] Trifonov, L. S., Orahovats, A. S., Prewo, R., Bieri, J. H., and Heimgartner, H. (1986) *J. Chem. Soc. Chem. Commun.*, 708–709.
[49] Bellus, D., Schaffner, K., and Hoigne, J. (1968) *Helv. Chim. Acta*, **51**, 1980–1989.
[50] Ohto, Y., Shizuka, H., Sekiguchi, S., and Matsui, K. (1974) *Bull. Chem. Soc. Jpn.*, **47**, 1209–1214.
[51] Holden, D. A., Jordan, K., and Safarzadeh-Amiri, A. (1986) *Macromolecules*, **19**, 895–901.
[52] Cui, C., Wang, X., and Weiss, R. G. (1996) *J. Org. Chem.*, **61**, 1962–1974.
[53] Baldvins, J. E., Cui, C., and Weiss, R. G. (1996) *Photochem. Photobiol.*, **63**, 726–738.
[54] Park, K. K., Kim, S.-H., and Park, J. W. (2004) *J. Photochem. Photobiol. A Chem.*, **163**, 241–247.
[55] Le Goff, M.-T. and Beugelmans, R. (1972) *Bull. Soc. Chim. Fr.*, 1115–1124.
[56] Le Goff, M.-T. and Beugelmans, R. (1974) *Bull. Soc. Chim. Fr.*, 2047–2055.
[57] Sakamoto, M., Yagi, T., Fujita, S., Mino, T., Karatsu, T., and Fujita, T. (2002) *J. Org. Chem.*, **67**, 1843–1847.
[58] Oldroyd D. L. and Weedon, A. C. (1974) *J. Org. Chem.*, **59**, 1333–1343.
[59] Schwartz, A., Pal, Z., and Szabo, L. (1987) *J. Heterocycl. Chem.*, **24**, 651–654.
[60] Chan, A. C. and Hilliard Jr., P. R. (1989) *Tetrahedron Lett.*, **30**, 6483–6486.
[61] Miranda, M. A., Primo, J., and Tormos, R. (1989) *Tetrahedron*, **45**, 7593–7600.
[62] Shizuka, H. and Tanaka, I. (1968) *Bull. Chem. Soc. Jpn.*, **41**, 2343–2349.
[63] Shizuka, H. (1969) *Bull. Chem. Soc. Jpn.*, **42**, 52–57.
[64] Shizuka, H. (1969) *Bull. Chem. Soc. Jpn.*, **42**, 57–65.
[65] Elad, V., Rao, D. V., and Stenberg, V. I. (1965) *J. Org. Chem.*, **30**, 3252–3254.
[66] Mayouf, A. M. and Park, Y.-T. (2002) *J. Photochem. Photobiol. A Chem.*, **150**, 115–123.
[67] Benali, O., Miranda, M. A., and Tormos, R. (2002) *Eur. J. Org. Chem.*, 2317–2322.

[68] Edward, J. T. and Mo L. Y. S. (1973) *J. Heterocycl. Chem.*, **6**, 1047–1049.
[69] Itoh, K. and Kanaoka, Y. (1974) *Chem. Pharm. Bull.*, **22**, 1431–1432.
[70] Somei, M. and Natsume, M. (1973) *Tetrahedron Lett.*, **14**, 2451–2454.
[71] Carruthers, W. and Evans, N. (1974) *J. Chem. Soc., Perkin Trans* **1**, 1523–1525.
[72] Ghosh, S., Das, T. K., Datta, D. B., and Mehta, S. (1987) *Tetrahedron Lett.*, **28**, 4611–4614.
[73] Erra-Balsells, R. and Frasca, A. R. (1984) *Tetrahedron Lett.*, **25**, 5363–5366.
[74] Bonesi, S. M. and Erra-Balsells, R. (1991) *J. Photochem. Photobiol. A Chem.*, **56**, 55–72.
[75] Katsuhara, Y., Maruyama, H., Shigemitsu, Y., and Odaira, Y. (1973) *Tetrahedron Lett.*, **14**, 1323–1326.
[76] Kan, R. O. and Furey, R. L. (1966) *Tetrahedron Lett.*, **7**, 2573–2578.
[77] Heerklotz, J. A., Fu, C., Linden, A., and Hesse, M. (2000) *Helv. Chim. Acta*, **83**, 1809–1824.
[78] Davis, A. and Golden, J. H. (1968) *J. Chem. Soc. B*, 425–427.
[79] Horspool, W. M. and Pauson, P. L. (1965) *J. Chem. Soc.*, 5162–5166.
[80] Trecker, D. J., Foote, C. S., and Osborn, C. L. (1968) *Chem. Commun. (London)*, 1034–1035.
[81] Bellus, D. and Schaffner, K. (1968) *Helv. Chim. Acta*, **51**, 221–224.
[82] Schwetlick, K., Noack, R., and Schmieder, G. (1972) *Z. Chem.*, **12**, 107–108.
[83] Schwetlick, K. and Noack, R. (1972) *Z. Chem.*, **12**, 109–110.
[84] Noack, R. and Schwetlick, K. (1972) *Z. Chem.*, **12**, 140–141.
[85] Noack, R. and Schwetlick, K. (1972) *Z. Chem.*, **12**, 143–144.
[86] Stumpe, J., Mehlhorn, A., and Schwetlick, K. (1978) *J. Photochem.*, **8**, 1–9.
[87] Schultze, H. (1973) *Z. Naturforsch.*, **B28**, 339–352.
[88] Hageman, H. J. (1972) *Recl. Trav. Chim. Pays-Bas*, **91**, 362–366.
[89] Mehlhorn, A., Schwenzer, B., and Schwetlick, K. (1977) *Tetrahedron*, **35**, 63–68.
[90] Masilamani, D., Hutchins, R. O., and Ohr, J. (1976) *J. Org. Chem.*, **41**, 3687–3691.
[91] Herweh, J. E. and Hoyle, C. E. (1980) *J. Org. Chem.*, **45**, 2195–2201.
[92] Noack, R. and Schwetlick, K. (1974) *Tetrahedron*, **30**, 3799–3805.
[93] Passananti, M., Cermola, F., DellaGreca, M., Iesce, M. R., Previtera, L., Sferruzza, R., and Temussi, F. (2014) *Int. J. Photoenergy*, Article ID 864361.
[94] Stratenus, J. L. and Havinga, E. (1966) *Recl. Trav. Chim. Pays-Bas*, **85**, 434–436.
[95] Ogata, Y., Takagi, K., and Yamada, S. (1977) *J. Chem. Soc. Perkin Trans* **2**, 1629–1632.
[96] Snell, B. K. (1968) *J. Chem. Soc. C*, 2367–2370.
[97] Nasielski-Hinkens, Y., Maeck, J., and Tenvoorde, M. (1972) *Tetrahedron*, **28**, 5025–5028.
[98] Crevatín, L. K., Bonesi, S. M., and Erra-Balsells, R. (2006) *Helv. Chim. Acta*, **89**, 1147–1157.
[99] Nozaki, H., Okada, T., Noyori, R., and Kawanishi, M. (1966) *Tetrahedron*, **22**, 2177–2180.
[100] Chakrabarti, A., Biswas, G. K., and Chakraborty, D. P. (1989) *Tetrahedron*, **45**, 5059–5064.
[101] Park, K. K., Lee, J. J., and Ryu, J. (2003) *Tetrahedron*, **59**, 7651–7659.
[102] Loveridge, E. L., Beck, B. R., and Bradshaw, J. S. (1971) *J. Org. Chem.*, **36**, 221–222.
[103] Rungwerth, D. and Schwetlick, K. (1974) *Z. Chem.*, **14**, 17–18.
[104] Höhne, G., Lohner, W., and Praefcke, K. (1978) *Tetrahedron Lett.*, **7**, 613–614.
[105] Lohner, W., Martens, J., Praefcke, K., and Simon, H. (1978) *J. Organomet. Chem.*, **154**, 263–271.
[106] Ohara, M. and Watanabe, K. (1975) *Angew. Chem. Int. Ed. Engl.*, **14**, 820.
[107] Chenevert, R. and Voyer, N. (1984) *Tetrahedron Lett.*, **25**, 5007–5008.
[108] Syamala, M. S., Rao, B. N., and Ramamurthy, V. (1988) *Tetrahedron*, **44**, 7234–7242.
[109] Veglia, A. V., Sanchez, A. M., and de Rossi, R. H. (1990) *J. Org. Chem.*, **55**, 4083–4086.
[110] Veglia, A. V. and de Rossi, R. H. (1993) *J. Org. Chem.*, **58**, 4941–4944.
[111] Nassetta, M., de Rossi, R. H., and Cosa, J. J. (1988) *Can. J. Chem.*, **66**, 2794–2798.
[112] Chenevert, R. and Plante, V. (1983) *Can. J. Chem.*, **61**, 1092–1095.
[113] Pitchumani, K., Durai, M. C., and Srinivasan, C. (1991) *Tetrahedron Lett.*, **32**, 2975–2978.

[114] Xie, R.-Q., Liu, Y.-C., and Lei, X.-G. (1992) *Res. Chem. Intermed.*, **18**, 61–69.

[115] Banu, H. S., Pitchumani, K., and Srinivasan, C. (1999) *Tetrahedron*, **55**, 9601–9610.

[116] Koodanjeri, S., Pradhan, A. R., Kaanumalle, L. S., and Ramamurthy, V. (2003) *Tetrahedron Lett.*, **44**, 3207–3210.

[117] Singh, A. K. and Raghuraman, T. S. (1985) *Tetrahedron Lett.*, **26**, 4125–4128.

[118] Suau, R., Torres, G., and Valpuesta, M. (1995) *Tetrahedron Lett.*, **36**, 1311–1314.

[119] Singh, A. K. and Raghuraman, T. S. (1986) *Synth. Commun.*, **16**, 485–490.

[120] Nowakowska, M., Storsberg, J., Zapotoczny, S., and Guillet, J. E. (1999) *New J. Chem.*, **23**, 617–623.

[121] Pitchumani, K., Warrier, M., and Ramamurthy, V. (1996) *J. Am. Chem. Soc.*, **118**, 9428–9429.

[122] Pitchumani, K., Warrier, M., and Ramamurthy, V. (1999) *Res. Chem. Intermed.*, **25**, 623–631.

[123] Pitchumani, K., Warrier, M., Cui, C., Weiss, R. G., and Ramamurthy, V. (1996) *Tetrahedron Lett.*, **37**, 6251–6254.

[124] Balkus, J. K., Khanmamedova, A. K., and Woo, R. (1998) *J. Mol. Catal. A Chem.*, **134**, 137–143.

[125] Gu, W., Warrier, M., Ramamurthy, V., and Weiss, R. G. (1999) *J. Am. Chem. Soc.*, **121**, 9467–9468.

[126] Marín, M., Lhiaubet-Vallet, V., and Miranda, M. A. (2011) *J. Phys. Chem. B*, **115**, 2910–2915.

[127] Weiss, R. G. and Cui, C. (1993) *J. Am. Chem. Soc.*, **115**, 9820–9821.

[128] Cui, C., Naciri, J., He, Z., Jenkins, R. M., Lu, L., Ramesh, V., Hammond, G. S., and Weiss, R. G. (1993) *Quim Nova*, **16**, 578–585.

[129] Gu, W., Hill, A. J., Wang, X., Cui, C., and Weiss, R. G. (2000) *Macromolecules*, **33**, 7801–7811.

[130] Gu, W. and Weiss, R. G. (2001) *J. Org. Chem.*, **66**, 1775–1780.

[131] Gu, W., Abdallah, D. J., and Weiss, R. G. (2001) *J. Photochem. Photobiol. A Chem.*, **139**, 79–87.

[132] Gu, W. and Weiss, R. G. (2000) *Tetrahedron*, **56**, 6913–6925.

[133] Gu, W. and Weiss, R. G. (2001) *J. Photochem. Photobiol. C Photochem. Rev.*, **2**, 117–137.

[134] Chen, Y.-Z. and Weiss, R. G. (2009) *Photochem. Photobiol. Sci.*, **8**, 916–925.

[135] Gu, W., Bi, S., and Weiss, R. G. (2002) *Photochem. Photobiol. Sci.*, **1**, 52–59.

[136] Höfler, T., Grießer, T., Gruber, M., Jakopic, G., Trimmel, G., and Kern, W. (2008) *Macromol. Chem. Phys.*, **209**, 488–498.

[137] Griesser, T., Adams, T., Wappel, J., Kern, W., Leggett, G. J., and Trimmel, G. (2008) *Langmuir*, **24**, 12420–12425.

[138] Höfler, T., Track, A. M., Pacer, P., Shen, Q., Flesch, H.-G., Hlawacek, G., Soller, G., Ramsey, M. G., Schennach, R., Resel, R., Teichert, C., Kern, W., Trimmel, G., and Griesser, T. (2010) *Mater. Chem. Phys.*, **119**, 287–293.

[139] Arumugam, S., Vutukuri, D. R., Thayumanavan, S., and Ramamurthy V. (2005) *J. Am. Chem. Soc.*, **127**, 13200–13206.

[140] Kaanumalle, L. S., Nithyanandhan, J., Pattabiraman, M., Jayaraman, N., and Ramamurthy, V. (2004) *J. Am. Chem. Soc.*, **126**, 8999–9006.

[141] Arumugam, S., Kaanumalle, L. S., and Ramamurthy, V. (2006) *Photochem. Photobiol.*, **82**, 139–145.

[142] Kaanumalle, L. S., Gibb, C. L. D., Gibb, B. C., and Ramamurthy, V. (2007) *Org. Biomol. Chem.*, **5**, 236–238.

[143] Andrew, D., Des Islet, B. T., Margaritis, A., and Weedon, A. C. (1995) *J. Am. Chem. Soc.*, **117**, 6132–6133.

[144] García, H., Martinez-Utrilla, R., and Miranda, M. A. (1985) *Tetrahedron Lett.*, **41**, 3131–3134.

[145] García, H., Iborra, S., Miranda, M. A., and Primo, J. (1986) *Heterocycles*, **24**, 2511–2517.

[146] García, H., Iborra, S., Miranda, M. A., and Primo, J. (1985) *Heterocycles*, **23**, 1983–1989.

[147] García, H., Miranda, M. A., Roquet-Jalmar, M. F., and Martínez-Utrilla, R. (1982) *Liebigs Ann. Chem.*, **12**, 2238–2243.

[148] García, H., Miranda, M. A., and Primo, J. (1986) *J. Chem. Research (S)*, 100–101.

[149] Primo, J., Tormos, R., and Miranda, M. A. (1982) *Heterocycles*, **19**, 1819–1822.

[150] Bowers, W. S., Ohta, T., Cleere, J. S., and Marsella, P. A. (1976) *Science*, **193**, 542–547.

[151] Pratt, G. E. and Bowers, W. S. (1977) *Nature*, **265**, 548–550.

[152] Miranda, M. A., Primo, J., and Tormos, R. (1988) *Heterocycles*, **27**, 673–681.

[153] Miranda, M. A., Primo, J., and Tormos, R. (1991) *Heterocycles*, **32**, 1159–1166.

[154] Miranda, M. A., Primo, J., and Tormos, R. (1987) *Tetrahedron*, **43**, 2323–2328.

[155] López, C. S., Erra-Balsells, R., and Bonesi, S. M. (2010) *Tetrahedron Lett.*, **51**, 4387–4390.

[156] Alvaro, M., García, H., Iborra, S., Miranda, M. A., and Primo, J. (1987) *Tetrahedron*, **43**, 143–148.

[157] Diaz-Mondejar, M. R., and Miranda, M. A. (1982) *Tetrahedron*, **38**, 1523–1526.

[158] Belled, C., Miranda, M. A., and Simon-Fuentes, A. (1989) *An. Quim.*, **85**, 39–47.

[159] Belled, C., Miranda, M. A., and Simon-Fuentes, A. (1990) *An. Quim.*, **86**, 431–435.

[160] Martinez-Utrilla, R. and Miranda, M. A. (1980) *Tetrahedron Lett.*, **21**, 2281–2282.

[161] Fillol, L., Martinez-Utrilla, R., Miranda, M. A., and Morera, I. M. (1989) *Monatsh. Chem.*, **120**, 863–870.

[162] Martinez-Utrilla, R. and Miranda, M. A. (1981) *Tetrahedron*, **37**, 2111–2114.

[163] García, H., Martínez-Utrilla, R., and Miranda, M. A. (1981) *Tetrahedron Lett.*, **22**, 1749–1750.

[164] García, H., Martínez-Utrilla, R., Miranda, M. A., and Roquet-Jalmar, M. F. (1982) *J. Chem. Res. (S)*, 350–351.

[165] Okada, K., Suzuki, R., and Yokota, T. (1999) *Biosci. Biotechnol. Biochem.*, **63**, 257–260.

[166] Katagiri, N., Nakano, J., and Kato, T. (1981) *J. Chem. Soc. Perkin Trans.* **1**, 2710–2716.

[167] Lewis, J. R. and Paul, J. G. (1981) *J. Chem. Soc. Perkin Trans. 1*, **3**, 770–775.

[168] Crouse, D. J., Hurlbut, S. L., and Wheeler, D. M. S. (1981) *J. Org. Chem.*, **46**, 374–378.

[169] Kende, A. S., Belletire, T. J., Hume, E., and Airey, J. (1975) *J. Am. Chem. Soc.*, **97**, 4425–4427.

[170] Ishii, H., Sekiguchi, F., and Ishikawa, T. (1981) *Tetrahedron*, **37**, 285–290.

[171] Suau, R., Valpuesta, M., and Torres, G. (1995) *Tetrahedron Lett.*, **36**, 1315–1318.

[172] Magnus, P. and Lescop, C. (2001) *Tetrahedron Lett.*, **42**, 7193–7196.

[173] Ban, Y., Yoshida, K., Goto, J., and Oishi, T. (1981) *J. Am. Chem. Soc.*, **103**, 6990–6992.

[174] Magauer, T., Martin, H. J., and Mulzer, J. (2009) *Angew. Chem. Int. Ed.*, **48**, 6032-6036.

[175] Magauer, T., Martin, H. J., and Mulzer, J. (2010) *Chem. Eur. J.*, **16**, 507–519.

[176] Andreu, I., Cabedo, N., Torres, G., Chagraoui, A., Ramírez de Arellano, M. C., Gil, S., Bermejo, A., Valpuesta, M., Protais, P., and Cortes, D. (2002) *Tetrahedron*, **58**, 10173–10179.

[177] Guerrini, G., Taddei, M., and Ponticelli, F. (2011) *J. Org. Chem.*, **76**, 7597–7601.

[178] Algarra, F. and Miranda, M. A. (1993) *Heterocycles*, **36**, 2335–2344.

[179] Ferrini, S., Ponticelli, F., and Taddei M. (2006) *J. Org. Chem.*, **71**, 9217–9220.

[180] Castell, J. V., Gómez-Lechón, M. J., Mirabet, V., Miranda, M. A., and Morera, I. M. (1987) *J. Pharm. Sci.*, **76**, 374–378.

[181] Kawabata, K., Sugihara, K., Sanoh, S., Kitamura, S., and Ohta, S. (2012) *J. Photochem. Photobiol. A Chem.*, **249**, 29–35.

[182] Martignac, M., Oliveros, E., Maurette, M.-T., Claparols, C., and Benoit-Marquié, F. (2013) *Photochem. Photobiol. Sci.*, **12**, 527–535.

[183] Pozdnyakov, I. P., Zhang, X., Maksimova, T. A., Yanshole, V. V., Wu, F., Grivin, V. P., and Plyusnin, V. F. (2014) *J. Photochem. Photobiol. A. Chem.*, **274**, 117–123.

[184] Sturini, M., Fasani, E., Prandi, C., and Albini, A. (1997) *Chemosphere*, **35**, 931–937.

[185] Climent, M. J. and Miranda, M. A. (1996) *J. Chromatogr. A*, **738**, 225–231.

[186] Brahmia, O. and Richard C. (2003) *J. Photochem. Photobiol. A. Chem.*, **156**, 9–14.

[187] Amine-Khodja, A., Boulkamh, A., and Boule, P. (2004) *Photochem. Photobiol. Sci.*, **3**, 145–156.

[188] López, M. C., Fernández, M. I., Rodríguez, S., Santaballa, J. A., Steenken, S., and Vulliet, E. (2005) *ChemPhysChem*, **6**, 2064–2074.

[189] Menager, M., Pilichowski, J. F., and Sarakha, M. (2010) *Photochem. Photobiol.*, **86**, 247–254.

[190] Bellus, D., Hrdlovic, P., and Manasek Z. (1966) *Polym. Lett.*, **4**, 1–5.

[191] Humphrey Jr., J. S., Schultz, A. R., and Jacquiss, D. B. G. (1973) *Macromolecules*, **6**, 305–314.

[192] Gupta, A., Liang, R., Moacanin, J., Goldbeck, K., and Kliger, D. (1980) *Macromolecules*, **13**, 262–267.

[193] Torikai, A., Murata, T., and Fueki, K. (1984) *Polym. Photochem.*, **4**, 255–269.

[194] Gupta, M. C. and Tahilyani, G. V. (1988) *Colloid Polym. Sci.*, **266**, 620–623.

[195] Webb, J. D. and Czanderna, A. W. (1986) *Macromolecules*, **19**, 2810–2825.

[196] Gupta, A., Rembaum, A., and Moacanin, J. (1978) *Macromolecules*, **11**, 1285–1288.

[197] Factor, A., Lynch, J. C., and Greenberg, F. H. (1987) *J. Polym. Sci. A. Polym. Chem.*, **25**, 3413–3422.

[198] Lemaire, J., Gardette, A., Rivaton, A., and Roger, A. (1986) *Polym. Degrad. Stab.*, **15**, 1–13.

[199] Rivaton, A., Sallet, D., and Lemaire, J. (1986) *Polym. Degrad. Stab.*, **14**, 23–40.

[200] Mullen, P. A. and Searle, N. Z. (1970) *J. Appl. Polym. Sci.*, **14**, 765–776.

[201] Lo, J., Lee, S. N., and Pearce, E. M. (1984) *J. Appl. Polym. Sci.*, **29**, 35–43.

[202] Beachell, H. and Chang, I. (1972) *J. Polym. Sci. Polym. Chem.*, **10**, 503–520.

[203] Schultze, H. (1973) *Makromol. Chem.*, **172**, 57–75.

[204] Rivaton, A., Sallet, D., and Lemaire, J. (1986) *Polym. Degrad. Stab.*, **14**, 1–22.

[205] Rivaton, A., Sallet, D., and Lemaire, J. (1983) *Polym. Photochem.*, **3**, 463–481.

[206] Factor, A. and Chu, M. L. (1980) *Polym. Degrad. Stab.*, **2**, 203–223.

[207] Rivaton, A. and Gardette, J.-L. (1998) *Angew. Makromol. Chem.*, **211–262**, 173–188.

[208] Factor, A., Ligon, W. V., and May, R. J. (1987) *Macromolecules*, **20**, 2461–2468.

[209] Rivaton, A., Mailhot, B., Soulestin, J., Varghese, H., and Gardette, J. L. (2002) *Polym. Degrad. Stab.*, **75**, 17–33.

[210] Clark, D. T. and Munro, H. S. (1982) *Polym. Degrad. Stab.*, **4**, 441–457.

[211] Hoyle, C. E., Shah, H., and Nelson, G. L. (1992) *J. Polym. Sci. A Polym. Chem.*, **30**, 1525–1533.

[212] Wilhelm, C., Rivaton, A., and Gardette, J.-L. (1998) *Polymer*, **39**, 1223–1232.

[213] Liaw, D. J., Lin, S. P., and Liaw, B. Y. (1999) *J. Polym. Sci. A Polym. Chem.*, **37**, 1331–1339.

[214] Cohen, S. M., Young, R. H., and Markhart, A. H. (1971) *J. Polym. Sci. A*, **9**, 3263–3299.

[215] Li, S. K. L. and Guillet, J. E. (1977) *Macromolecules*, **10**, 840–844.

[216] Bellus, D., Slama, P., Hrdlovic, P., Manasek, Z., and Durisinova, L. (1969) *J. Polym. Sci. C*, **22**, 629–643.

[217] Allen, N. S. (1983) *Polym. Photochem.*, **3**, 167–242.

[218] Tejedor, R. M., Oriol, L., Piñol, M., Serrano, J. L., Strehmel, V., Stiller, B., and Stumpe, J. (2005) *J. Polym. Sci. A Polym. Chem.*, **43**, 4907–4921.

[219] Kawatsuki, N., Neko, T., Kurita, M., Nishiyama, A., and Kondo, M. (2011) *Macromolecules*, **44**, 5736–5742.

[220] Kawatsuki, N., Matsushita, H., Washio, T., Kurita, M., and Kondo M. (2012) *Macromolecules*, **45**, 8547–8554.

[221] Griesser, T., Höfler, T., Temmel, S., Kern, W., and Trimmel, G. (2007) *Chem. Mater.*, **19**, 3011–3017.

[222] Griesser, T., Kuhlmann, J.-C., Wieser, M., Kern, W., and Trimmel, G. (2009) *Macromolecules*, **42**, 725–731.

[223] Köpplmayr, T., Cardinale, M., Jakopic, G., Trimmel, G., Kern, W., and Griesser, T. (2011) *J. Mater. Chem.*, **21**, 2965–2973.

[224] Griesser, T., Höfler, T., Jakopic, G., Belzik, M., Kern, W., and Trimmel, G. (2009) *J. Mater. Chem.*, **19**, 4557–4565.

[225] Edler, M., Mayrbrugger, S., Fian, A., Trimmel, G., Radl, S., Kern, W., and Griesser, T. (2013) *J. Mater. Chem. C*, **1**, 3931–3938.

PART X

BIOTRANSFORMATIONS

32

BIOTRANSFORMATIONS OF ARENES: AN OVERVIEW

SIMON E. LEWIS
Department of Chemistry, University of Bath, Bath, UK

32.1 INTRODUCTION

Nature deploys a vast array of enzymes to catalyze synthetically useful biotransformations of aromatic substrates. Increasingly, chemists are able to utilize these enzymes to effect biocatalytic transformations of arenes. There are several motivations for wishing to do so: sometimes, the desired transformation may be achieved by conventional chemical means involving forcing conditions and expensive or toxic reagents. Sometimes, by conventional chemical means, the desired transformation cannot be achieved at all. The use of enzyme-catalyzed methods can offer multiple advantages in terms of rate of reaction, chemoselectivity, regioselectivity, and stereoselectivity. Furthermore, enzymatic methods accord with many of the principles of green chemistry, such as the use of nontoxic reagents, environmentally benign solvents (including water), low energy demand (as reactions are usually performed at or near ambient temperature), and no generation of hazardous waste.

The purpose of this chapter is not to offer an exhaustive coverage of the known enzymatic transformations of arenes. Nor is it to describe in detail the mechanisms by which these transformations occur. Rather, it aims to provide an overview of biocatalytic methods that have been applied (or might plausibly be applied) to arenes on a preparatively useful scale, which may be of use to the reader in planning syntheses of their own. Transformations are categorized by reaction type, and references both to primary literature and to relevant specialized reviews are given.

32.2 DEAROMATIZING ARENE DIHYDROXYLATION

Among bacteria that are capable of metabolizing arenes, certain strains express *arene dioxygenase* enzymes capable of effecting the dearomatizing dihydroxylation of an arene **1** to give a *cis*-cyclohexa-3,5-diene-1,2-diol **2** (Scheme 32.1). Ordinarily, such diene diols are merely fleeting

Arene Chemistry: Reaction Mechanisms and Methods for Aromatic Compounds,
First Edition. Edited by Jacques Mortier.

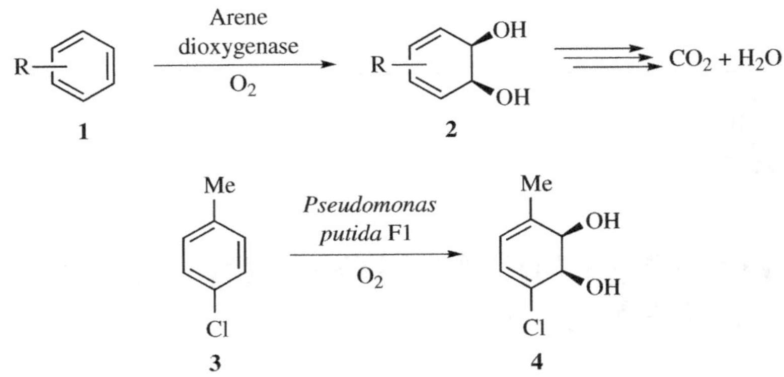

SCHEME 32.1 Biocatalytic dearomatizing dihydroxylation of an arene.

metabolic intermediates, as they undergo rapid oxidative rearomatization and, ultimately, mineralization. However, in 1968, a diene-*cis*-diol of type **2** was isolated for the first time [1]. Specifically, fermentation of *para*-chlorotoluene **3** by *Pseudomonas putida* F1 gave diol **4**.

The significance of this transformation stems from the paucity of nonbiocatalytic methods for carrying out such a dearomatization; to date, very few are known [2]. Shortly after the first report, the mutant strain *P. putida* F39/D was described, in which the enzyme responsible for the consumption of **2** by oxidative rearomatization was inactive [3]. This strain could therefore be used to produce *cis*-diols of type **2** on a multigram scale. Industrial interest in this biocatalytic technology led to reports of the synthesis of poly-*para*-phenylene **5** from benzene [4] and of indigo **6** from indole [5]. Applications of diols of type **2** as starting materials for total synthesis were reported from the late 1980s onward; Figure 32.1 depicts a selection of natural and nonnatural products accessed from building blocks of type **2**.

As regards the regio- and stereoselectivity of this biocatalytic dihydroxylation, most reported arene dioxygenase-mediated dearomatizing dihydroxylations of monosubstituted arenes of type **21** give products of type **22**. These bear hydroxyl groups "*ortho*" and "*meta*" to the original substituent, with the absolute stereochemistry shown (Scheme 32.2a) [20]. The products arising from monosubstituted arenes are generally single enantiomers, although a notable exception is when R = F, as the enzyme is no longer able to discriminate effectively between the two possible orientations of the substrate in the active site [21]. When disubstituted arenes are employed as substrates, the *ee* of the product varies from case to case. Broadly, in the case of *para*-disubstituted arenes, the larger the difference in the steric demand of the two substituents, the larger the enantiomeric excess. Thus, biocatalytic dihydroxylation of *para*-toluoyl iodide gives a product of 80% *ee*, whereas *para*-toluoyl chloride **3** gives a product **4** of only 15% *ee* [20]. A marked exception to the general trend in regioselectivity is observed upon using benzoate dioxygenase to effect the dihydroxylation of benzoic acid **23** (Scheme 32.2b). The product **24** has undergone dihydroxylation at the *ipso* and *ortho* positions, with the opposite absolute sense of enantioinduction to that observed for **22**. This complementary outcome and the presence of a quaternary center render **24** a versatile building block in its own right [22].

More than 400 arene *cis*-dihydrodiols derived from a wide variety of aromatic substrates have now been reported. For a more in-depth coverage, the reader is directed to several excellent reviews that have appeared recently [23]. Bicyclic and heterocyclic arenes are viable substrates for the transformation, although when 5-membered heteroarenes (furan, thiophene) undergo this transformation, the products are not always stable. Arene dioxygenases are membrane-bound proteins, and as such, this biotransformation is not carried out using isolated enzymes. Rather, whole-cell fermentation approaches are employed. While these are not as operationally simple as the use of isolated enzymes, they can nevertheless be carried out without recourse to any particularly unusual

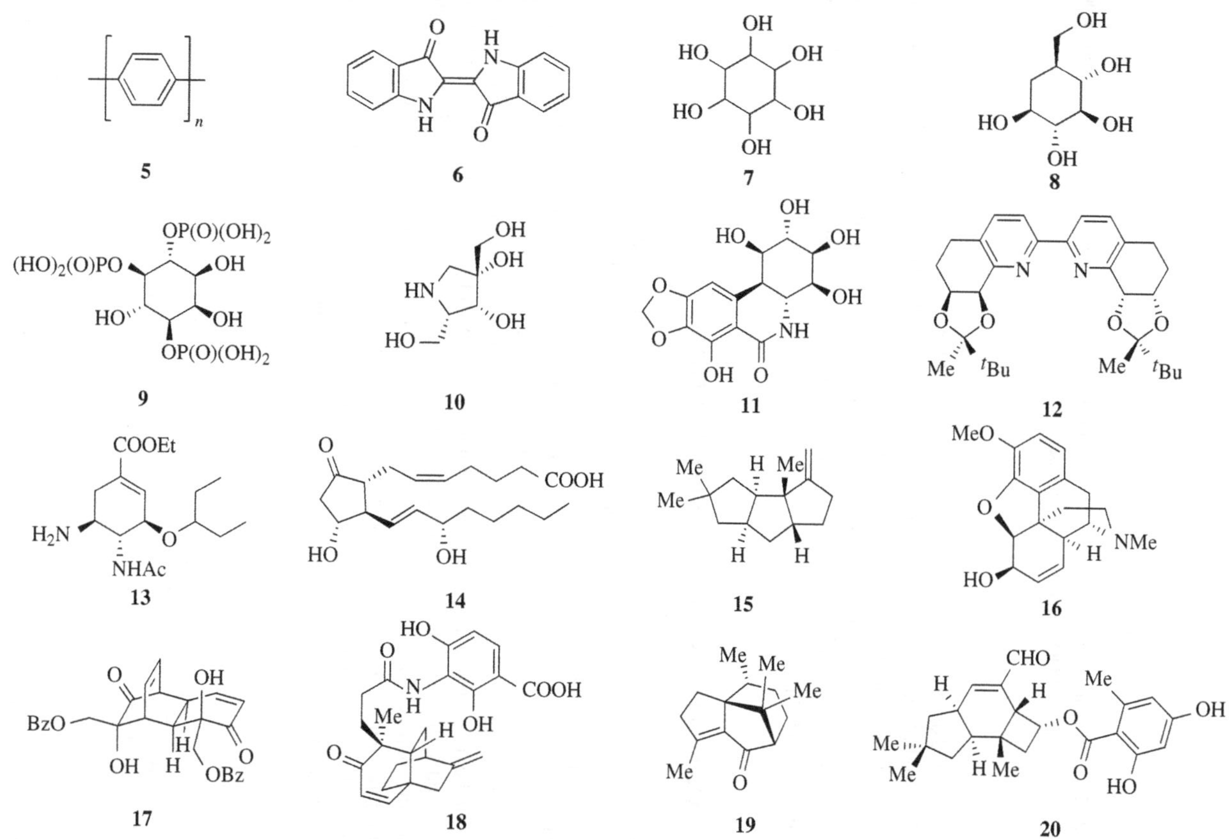

FIGURE 32.1 Natural and nonnatural products synthesized from *cis*-diols derived biocatalytically. (**5**, poly-*para*-phenylene [4]; **6**, indigo [5]; **7**, inositol isomers [6]; **8**, carba-β-L-glucopyranose [7]; **9**, (−)-(D)-*myo*-inositol 1,4,5-trisphosphate [8]; **10**, a furanose iminosugar [9]; **11**, (+)-pancratistatin [10]; **12**, an enantiopure 2,2′-bipyridyl ligand [11]; **13**, oseltamivir [12]; **14**, prostaglandin $E_2\alpha$ [13]; **15**, (−)-hirsutene [14]; **16**, (+) and (−)-codeine [15]; **17**, (+)-grandifloracin [16]; **18**, (−)-platencin [17]; **19**, (−)-patchoulenone [18]; **20**, (+)-armillarivin [19].)

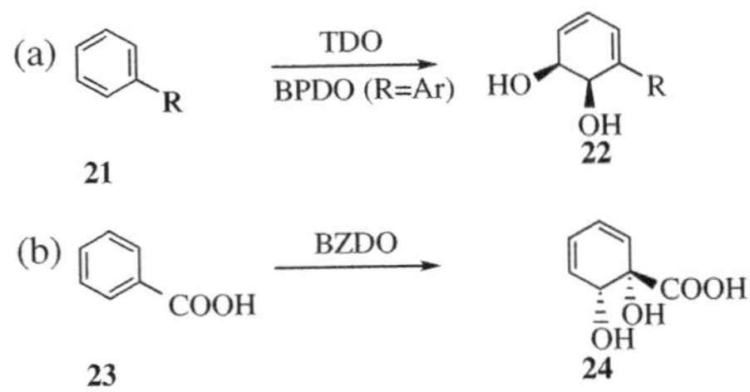

SCHEME 32.2 Regio- and stereoselectivity of arene dihydroxylation by arene dioxygenases. (a) For TDO (toluene dioxygenase) and BPDO (biphenyl dioxygenase). (b) For BZDO (benzoate dioxygenase).

equipment; detailed descriptions of the procedures have been published [23f, 24]. Importantly, most arene *cis*-dihydrodiols are unstable at elevated temperatures or low pH, due to their facile rearomatization to the corresponding phenols through dehydration, and must therefore be stored at low temperature. Finally, it should be noted that several of the more common arene *cis*-dihydrodiols are commercially available (e.g., **22**, R=Cl, Br, I, Me, CN, COOMe) from Almac Sciences and Sigma-Aldrich, among other suppliers.

32.3 DEAROMATIZING ARENE EPOXIDATION

Although far less developed for synthesis than the dearomatizing arene dihydroxylation, there are nevertheless examples of arene monooxygenase enzymes being used to effect the epoxidation of an arene. Upon metabolism by wood-rotting fungi of the *Phellinus* genus, methyl benzoate **25** underwent epoxidation to give **26**, as shown in Scheme 32.3 [25]. Such oxabicyclo[4.1.0]heptadienes derived from arenes are known simply as "arene oxides" and exist in tautomeric equilibrium with the corresponding oxepines, which form from the arene oxides through an electrocyclic ring opening. Crucially, all the ring carbons in the oxepine tautomer are sp^2 hybridized, and thus, any enantiomeric excess imparted in the formation of the arene oxide is eroded through equilibration with the oxepine until a racemic mixture is formed. In the case of arene oxide **26**, the equilibrium was found to favor the oxepine tautomer **27**. Among all methyl benzoates examined, methyl 2-trifluoromethyl benzoate **28** was found to be a unique case, insofar as the arene oxide formed upon its fungal metabolism, **29**, was the major component of the equilibrium. Indeed, oxepine tautomer **30** was present in such

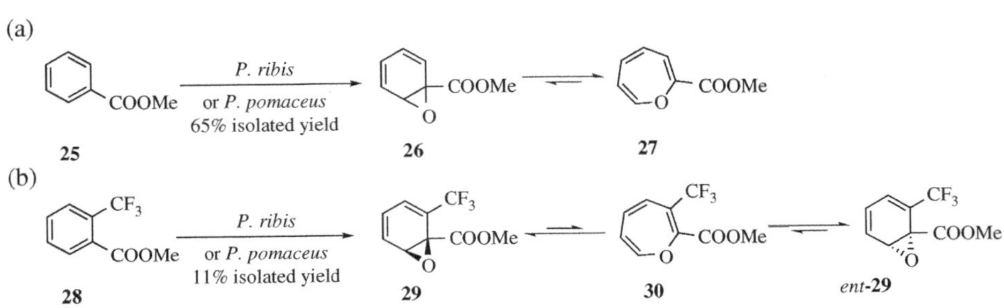

SCHEME 32.3 Biocatalytic formation of arene oxides and their oxepine tautomers. (a) For methyl benzoate. (b) For methyl 2-trifluoromethylbenzoate.

small amounts at equilibrium that the erosion in the *ee* of the initially formed **29** was rather slow, such that optical rotation data and circular dichroism spectra could be acquired, leading to the assignment of the absolute stereochemistry of **29** as shown. It must be stressed that the scarcity of reports of the isolation of arene oxides does not mean their formation is rare; rather, it is the case that in many instances, arene oxides are formed but undergo further transformations giving different products (e.g., phenols, *vide infra* Section 32.10).

32.4 ARENE ALKYLATION (BIOCATALYTIC FRIEDEL–CRAFTS)

In nature, the alkylation of aromatic rings most commonly proceeds using *S*-adenosylmethionine ("SAM," **31a**) as the source of an electrophilic methyl group. Such alkylations are catalyzed by C-methyltransferase enzymes and are analogous to Friedel–Crafts alkylations. Two C-methyltransferases, NovO and CouO (from *Streptomyces spheroides* and *Streptomyces rishiriensis*, respectively), have been expressed in recombinant *E. coli* and have been shown to have a synthetically useful substrate scope. Not only are they able to catalyze the regioselective alkylation of various coumarins **32** (X=O), 2-quinolones **32** (X=NH), and naphthalenediols **34–36**, but in addition, they are able to accept nonnatural cofactors **31b–f** in place of SAM and so can transfer various alkyl groups other than methyl to the substrates (Scheme 32.4) [26].

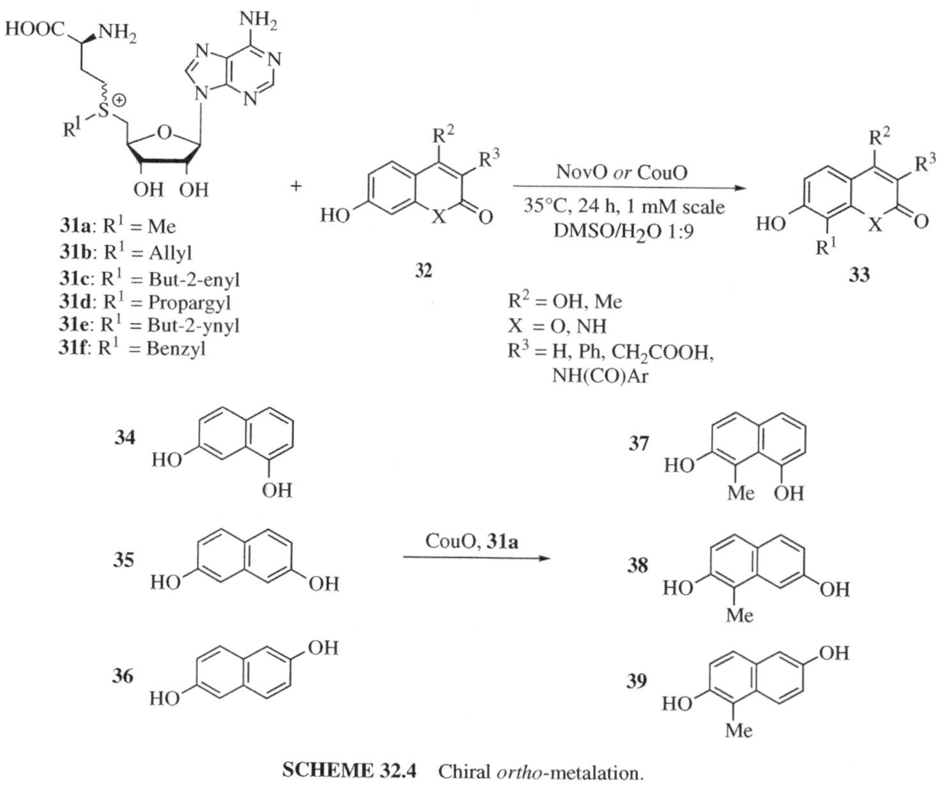

SCHEME 32.4 Chiral *ortho*-metalation.

Squalene–hopene cyclase (SHC) enzymes have been used to effect intramolecular "biocatalytic Friedel–Crafts" reactions of phenyl ethers [27]. The natural substrate for such enzymes, squalene, is a C_{30} hexaene, but it has been demonstrated that an SHC from *Zymomonas mobilis* can accept a truncated substrate bearing an arene (which is not present in squalene). As shown in Scheme 32.5, geranyl phenyl ether **40** (with a monoterpene C_{10} side chain) is cleanly cyclized by *Zmo*SHC1 to tricycle **41** by a cationic cyclization/Friedel–Crafts pathway. The fidelity of the transformation begins to break down with larger substrates, however: farnesyl phenyl ether **42** (with a sesquiterpene C_{15} side chain) cyclizes to give the desired tetracycle **43** as only the minor product, in a 1:2 ratio with by-product **44**. This by-product arises through successful aliphatic cation cyclization, but failure of the Friedel–Crafts step, with loss of a proton giving the alkene in **44** instead. For an even larger substrate, geranylgeranyl phenyl ether **45** (with a diterpene C_{20} side chain), no successful Friedel–Crafts cyclization to **46** is observed at all, with only isomeric by-products **47** and **48** being isolated.

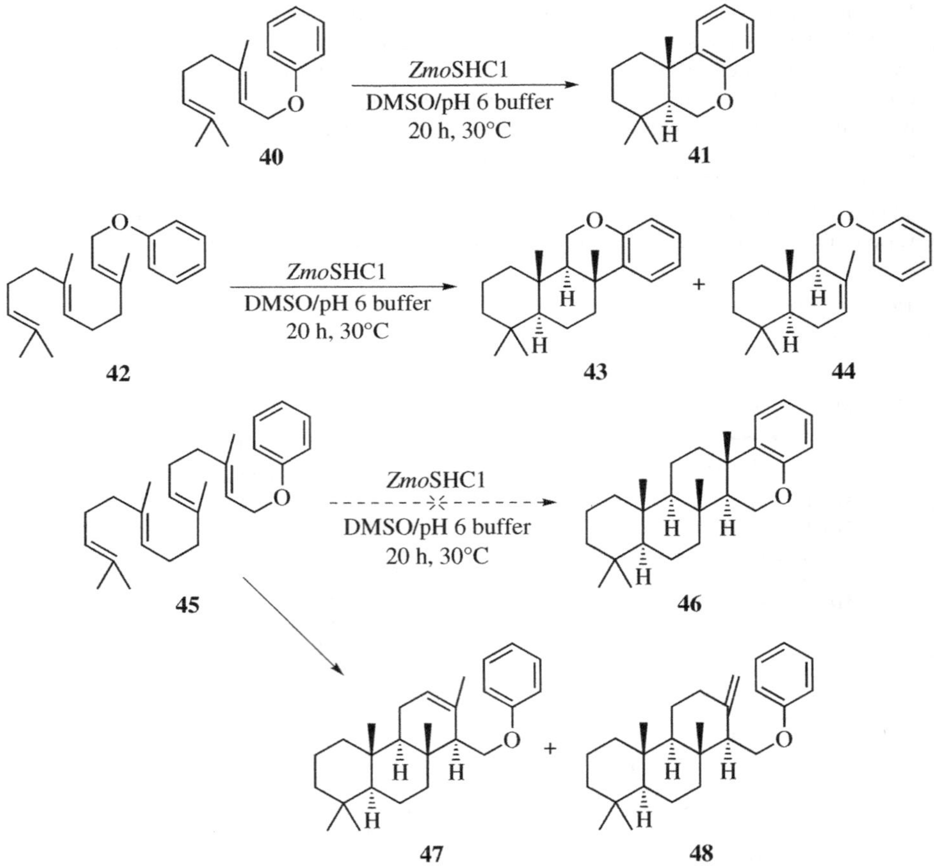

SCHEME 32.5 Tandem cyclization/aryl alkylation catalyzed by squalene hopene cyclase.

Other enzymes involved in terpene biosynthesis have also been harnessed for biocatalytic reactions of arenes. Prenyltransferase enzymes that can affect the addition of C_5 isoprenyl units both at carbon and at heteroatoms have been used for biocatalytic arene alkylations. For example, L-tryptophan **49** undergoes prenylation at various positions on the indole core in a wholly regioselective fashion, depending on the enzyme used [28] (Scheme 32.6).

In some cases, such prenyltransferase enzymes can process substituted tryptophans **53** and/or different alkylating agents other than DMAPP, the common requirement being that the alkylating agent **54** is able to form a stabilized (allylic/benzylic) carbocation, although the regioselectivity in

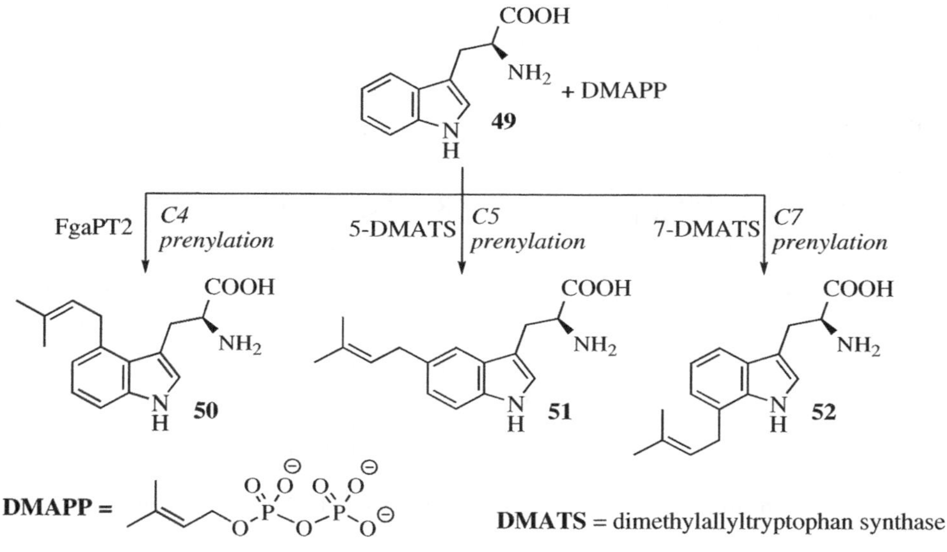

SCHEME 32.6 C-prenylation of L-tryptophan **49**.

formation of **55** may be lowered with such nonnatural alkylating agents [28, 29] (Scheme 32.7). Additionally, the substrate scope for such enzymes is not limited to tryptophan derivatives: other indole-containing substrates may also be alkylated [30, 31]. One particular case nicely illustrates the versatility of this approach: the same indole diketopiperazine substrate may be selectively alkylated at all seven different positions on the indole core, through choice of an appropriate biocatalyst [32]. Biocatalytic prenylations of flavonoids [33] as well as of naphthols and naphthalenediols have also been described [34].

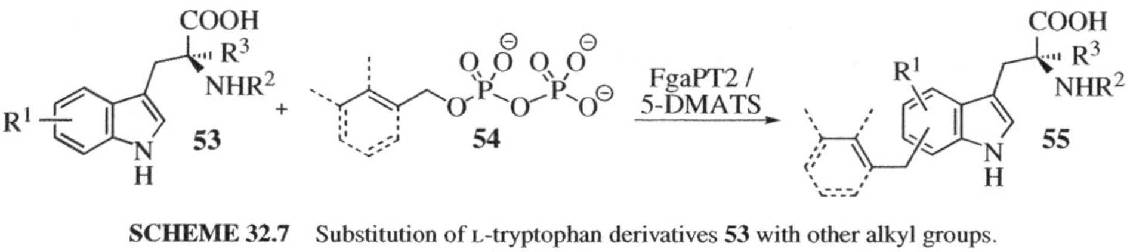

SCHEME 32.7 Substitution of L-tryptophan derivatives **53** with other alkyl groups.

A concerted biocatalytic cyclization of a phenol onto an *N*-methyl group has been demonstrated using berberine bridge enzyme (BBE) [35], as depicted in Scheme 32.8. When a racemic substrate (±)-**56** is used, two modes of biotransformation have been described. A kinetic resolution was reported first [35a], which gives the product **57** as the (*S*)-enantiomer (in up to 50% theoretical yield) and returns unreacted (*R*)-**56**. More recently, a dynamic kinetic resolution has been described [35b, c], in which a monoamine oxidase (MAO-N variant D-11) selectively oxidizes only the (*R*)-enantiomer of **56** to the corresponding iminium species, which is in turn reduced by an achiral reductant (borane–morpholine complex) to give (±)-**56**. By this process, all the starting material may be isomerized to (*S*)-**56** and cyclized as previously to give (*S*)-**57** in 79–97% yield. The reaction is also regioselective, favoring formation of *ortho*-product **57** over *para*-product **58**, although this can be influenced by choice of substrate [36]. Reaction conditions have been optimized, and BBE has been shown to have good tolerance for a high proportion of organic cosolvent in the reaction medium [37]. Berberine alkaloids have been shown to have diverse biological effects, for example, analgesia, sedation, and muscle relaxation.

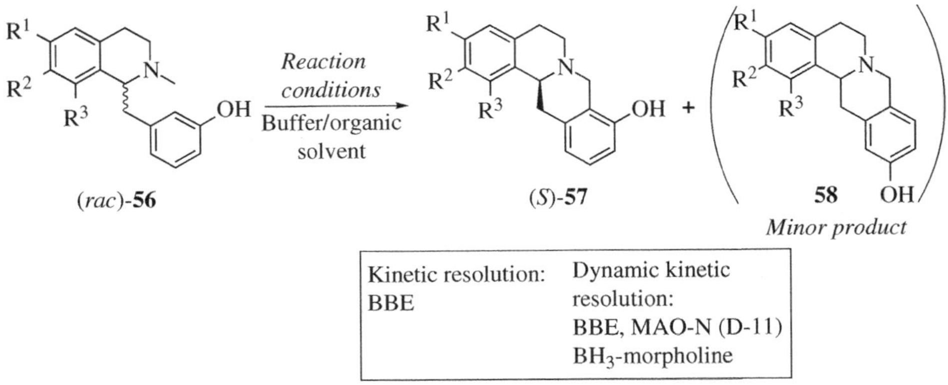

SCHEME 32.8 Formation of enantiopure tetracycle **57** from racemic **56**.

32.5 ARENE DEACYLATION (BIOCATALYTIC *RETRO* FRIEDEL–CRAFTS)

Certain enzymes are known to catalyze the deacylation of electron-rich arenes [38], and very recently, two of these have been exploited for preparative chemistry [39, 40]. As shown in Scheme 32.9, the enzyme PhlG (cloned from *Pseudomonas fluorescens*) catalyzes the monohydrolysis of diacetylphloroglucinol **59** to give **60**. The biotransformation has been optimized such that when diisopropyl ether is used in a biphasic mixture to help solubilize the substrate, a yield of 63% of **60** is obtained starting from a gram of **59**. Whereas the activity of PhlG is seemingly specific to substrate **59**, the enzyme Phy shows greater substrate promiscuity. Phy (cloned from *Eubacterium ramulus* or *Aspergillus niger*) is able to catalyze the deacylation of various monoacyl electron-rich arenes **62** to give products **63**.

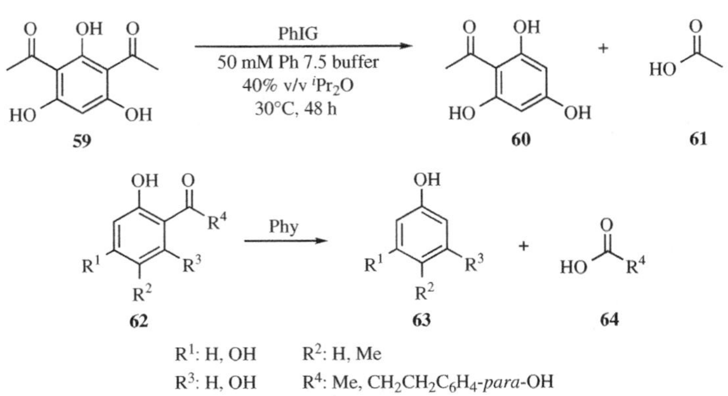

SCHEME 32.9 Biocatalytic deacylation of electron-rich arenes.

32.6 ARENE CARBOXYLATION (BIOCATALYTIC KOLBE–SCHMITT)

Various enzymes are known to catalyze arene carboxylation or decarboxylation in nature [41]. The first report [42] of a preparatively useful biocatalytic arene carboxylation was the *para*-carboxylation of phenol using the enzyme phenyl phosphate carboxylase. As the name implies, this requires the phenol to be phosphorylated prior to reaction. Subsequently, phenol *para*-carboxylation without prior phosphorylation was demonstrated using enzymes such as 4-hydroxybenzoate decarboxylase [43–45] and 3,4-dihydroxybenzoate decarboxylase [46, 47] (Scheme 32.10). Although the natural function of these latter enzymes is to catalyze *de*carboxylation reactions, it must be noted that these enzymes can catalyze reactions *in either direction*, dependent on the reaction conditions.

More extensively developed is the biocatalytic phenol *ortho*-carboxylation, for which multiple biocatalysts have been identified [48, 49]. Collectively, these various enzymes can affect *ortho*-carboxylation of a variety of substituted phenols (**68–70**, **74–81**). The availability of entirely regioselective carboxylations, both *ortho* and *para*, gives the biocatalytic approach a significant advantage over the traditional Kolbe–Schmidt process: the reaction of a metal phenoxide with CO_2 at high temperature and pressure typically gives a mixture of both regioisomers, although the choice of metal cation can influence the selectivity to a degree. Biocatalytic carboxylations of pyrrole [50] and indole [51] have also been reported, although the substrate scope is not broad in these cases.

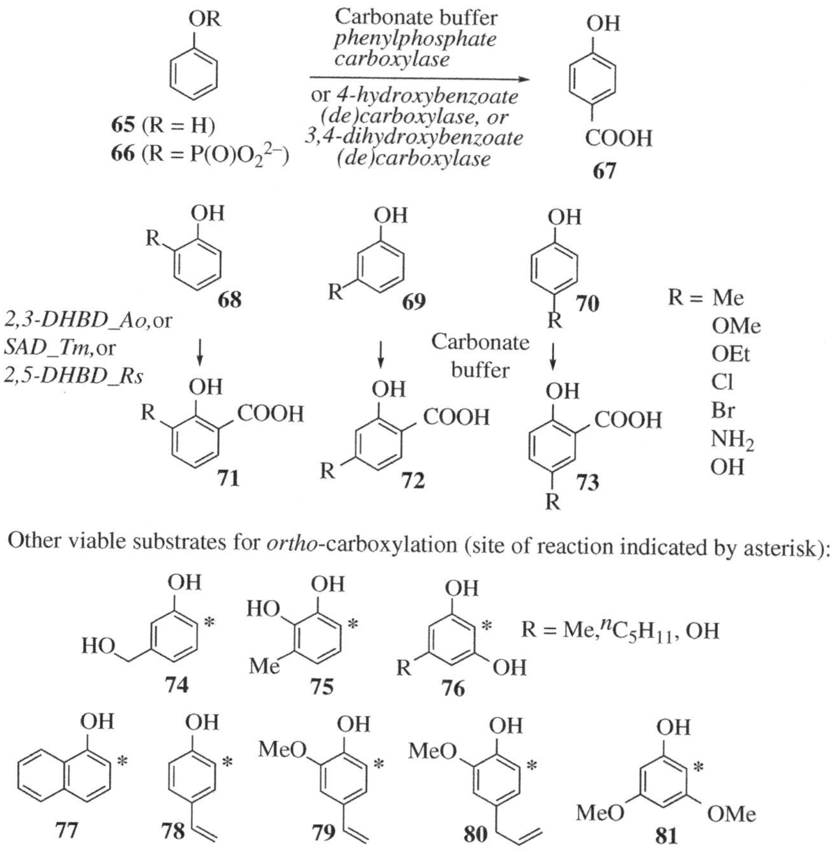

SCHEME 32.10 Biocatalytic arene carboxylation.

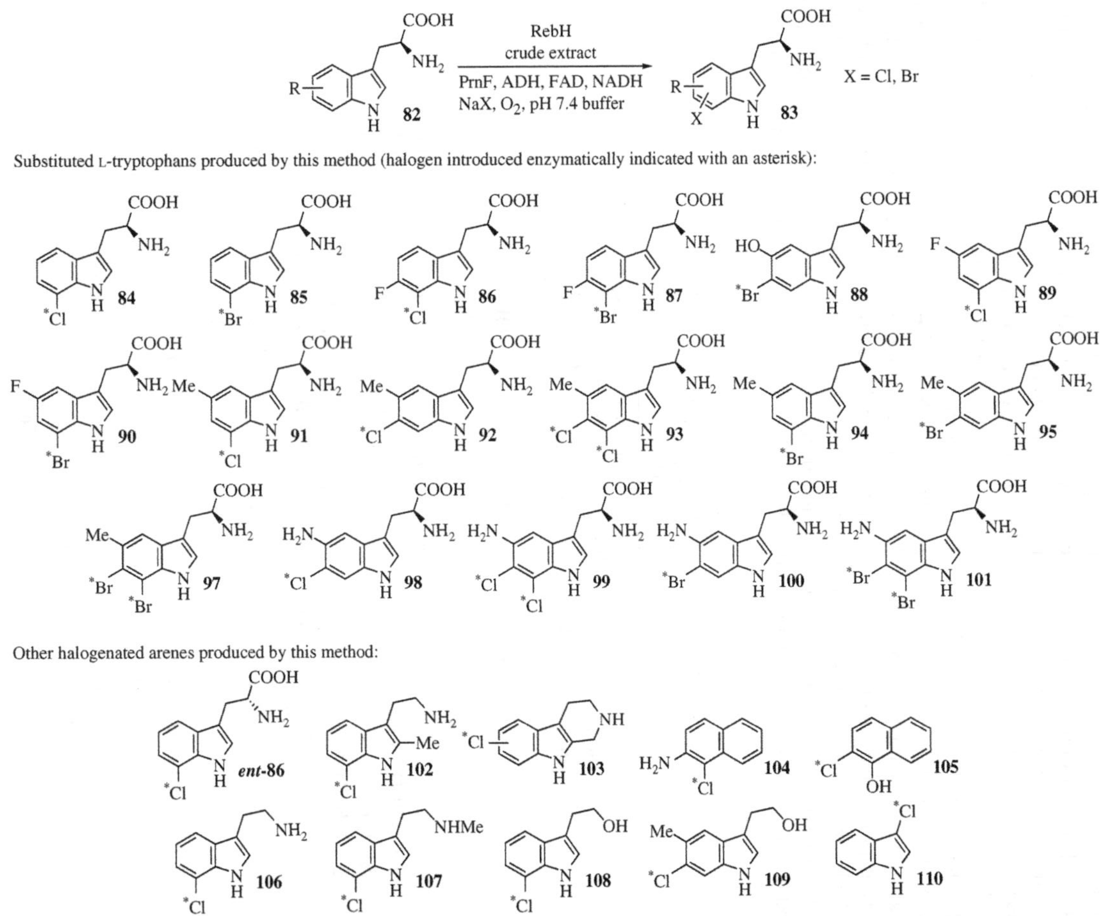

SCHEME 32.11 Biocatalytic arene halogenation.

32.7 ARENE HALOGENATION (HALOGENASES)

Halogenase enzymes are able to effect electrophilic halogenation of arenes and can do so in a regioselective fashion whereby the steric constraints of the enzyme's active site override the inherent regioselectivity of the substrate. By these means, halogenated arenes may be synthesized with substitution patterns that are difficult to access by conventional chemical methods. Such reactivity means that synthetically practicable biocatalytic arene halogenations are a desirable goal. Significant successes have been reported using halogenases for *in vivo* mutasyntheses of halogenated natural product analogues; that is to say that plants engineered to express certain arene halogenases have been able to produce halogenated variants of the alkaloids they ordinarily produce [52–55]. However, such techniques are not accessible to organic chemists at large. In contrast, isolated arene halogenase biocatalysts would be of greater preparative utility to the synthetic community, but their development has not been straightforward [56].

The tryptophan halogenase RebH from *Lechevalieria aerocolonigenes* has been investigated for the halogenation not just of tryptophan but of various indole derivatives [57–61], as shown in Scheme 32.11. The biotransformation requires the addition of various cofactors and supporting enzymes but proceeds under very mild conditions. A key point is that the halogen source is simply the relevant sodium halide salt, and the terminal oxidant is oxygen, with the necessary electrophilic halogenating species being generated *in situ*. On natural (unsubstituted) L-tryptophan, halogenation occurs at C7 selectivity, in contrast to the C3 (or C2) selectivity observed when indoles are treated with electrophilic halogen sources in the corresponding traditional chemical process. For substituted tryptophans, the preference for C7 functionalization persists, with reaction at C6 also being observed in some cases.

The first reports on the use of RebH describe problems of instability of the biocatalyst, but more recent reports describe strategies to overcome this, such as directed evolution of the RebH enzyme [58] or immobilization of all the necessary enzymes in a cross-linked aggregate [61]. This latter approach has allowed for the production of 7-bromo-L-Trp **85** on greater than 1.5 g scale. As an aside, it should be noted that in one of the studies on RebH [60], the substituted L-Trp substrates for halogenation were themselves prepared biocatalytically, by alkylation of the relevant indoles at the 3-position with L-serine, that is, another example of a biocatalytic Friedel–Crafts reaction [62, 63]. RebH is not the only enzyme to have been exploited for arene halogenation—PrnA also effects the C7 halogenation of L-Trp, and this latter enzyme has recently been engineered to effect C5 halogenation also [64]. Finally, it should be noted that another halogenase, Rdc2 from *Pochonia chlamydosporia*, has recently been reported to effect the biocatalytic chlorination of 4- and 6-hydroxyisoquinoline [65].

32.8 ARENE OXIDATION WITH LACCASES

Laccase enzymes are capable of oxidizing aromatic rings in multiple ways, leading to a variety of different product types. An advantage of this class of biotransformations is the ready commercial availability of a variety of laccases. They are thus of considerable synthetic value and their use has been reviewed [66, 67]. The substrates are typically phenols, catechols, or hydroquinones, and quinone intermediates are often formed. One reaction outcome is oxidative dimerization to give a biaryl [68], with selected examples shown in Scheme 32.12. Various salicylic acids **111** dimerize upon exposure to laccase C (from a *Trametes* species) to give biphenyls **112** [69], and tetrahydronaphthol **113** dimerizes upon exposure to laccase MtL (from *Myceliophtora thermophyla*) to give a mixture of symmetric and nonsymmetric dimers, **114** and **115**, respectively [70]. Notably, the biotransformation is far more selective for formation of **114** over **115** than the corresponding reaction carried out by a nonbiocatalytic method (with MnO_2). Given the importance of (homochiral) analogues of **114** in asymmetric catalysis, this work suggests a possible biocatalytic route to new axially chiral ligands and catalysts for enantioselective synthesis.

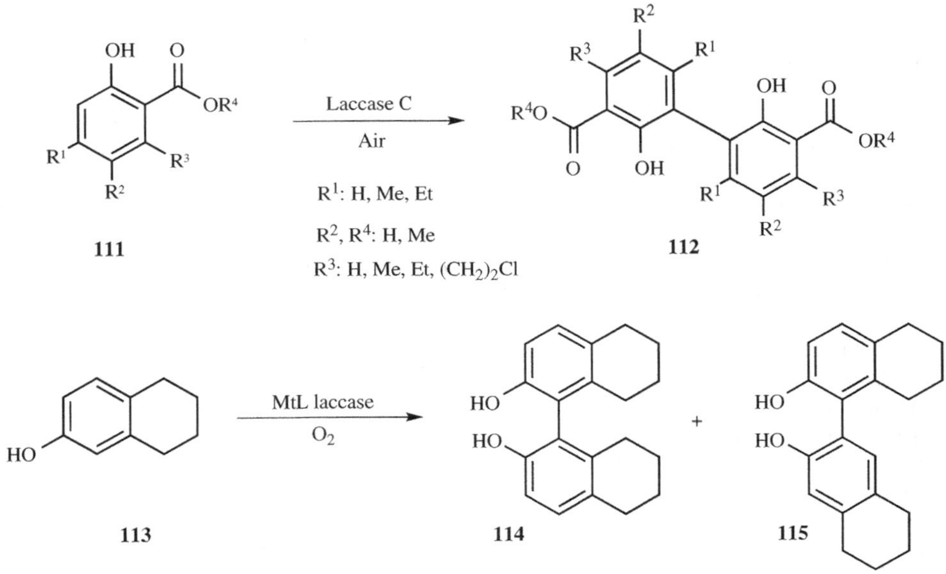

SCHEME 32.12 Biocatalytic oxidative dimerization mediated by laccases.

A common biocatalytic reaction cascade mediated by laccases is the oxidation of a catechol/hydroquinone to a quinone, Michael addition of an external nucleophile (giving a substituted catechol/hydroquinone), and then a further oxidation back to the quinone oxidation state. Representative examples are shown in Scheme 32.13: hydroquinones **116** are oxidatively coupled with one or two equivalents of anilines **117** to give monoamino- or diamino-*para*-benzoquinones **118** and/or **119**, respectively [71]. Analogously, naphthalene-1,4-diol **120** may be coupled with diverse anilines **121** to give naphthoquinone adducts **122** [72]. When a nucleophilic solvent (methanol) was used, 3-methylcatechol **123** gave an oxidized *o*-quinone product **124** in which a molecule of solvent had been incorporated; the methoxide group could be displaced by various aliphatic primary amines in a separate nonbiocatalytic step to give a range of products **125** [73]. As well as nitrogen nucleophiles, the attack of carbon and oxygen nucleophiles on quinones generated by laccases is known. For example, laccase-catalyzed oxidation of catechols **127** generates an *ortho*-quinone *in situ* to which *N*-Boc oxindoles of type **126** will add. Methylation then affords adducts of general structure **128** [74]. Interestingly, under these reaction conditions, the catechol formed upon addition of **126** does not undergo a second oxidation (i.e., the final product is a catechol bis(ether), as opposed to an *ortho*-quinone).

Biocatalytic annulation reactions mediated by laccases have been reported. For example, as shown in Scheme 32.14, barbituric acid derivatives **129** react with hydroquinone **116** or catechol **127** to give 5-deaza-10-oxaflavins **130** and **131**, respectively. The analogous reactions with dimedone derivatives **132** give **133** and **134** [75]. In the laccase-mediated annulation of functionalized hydroquinones **135** with heteroaromatic aminoamides **136**, the reaction outcome varies depending on the nature of the substitution on the hydroquinone (Scheme 32.15). When R is not a good leaving group, two substitutions occur on the hydroquinone ring, giving products **137**, comprising a 6-7-5 tricyclic ring system. On the other hand, when R is able to act as a leaving group, its displacement instead gives rise to products **138a** and **138b** comprising 6-8-5 tricyclic ring systems [76].

Several methods have been described for biocatalytic annulation of catechols with 1,3-dicarbonyls. One approach employs a laccase in conjunction with a lipase [77], and another employs a laccase in conjunction with a tyrosinase [78]. They are depicted in Scheme 32.16; in the former instance,

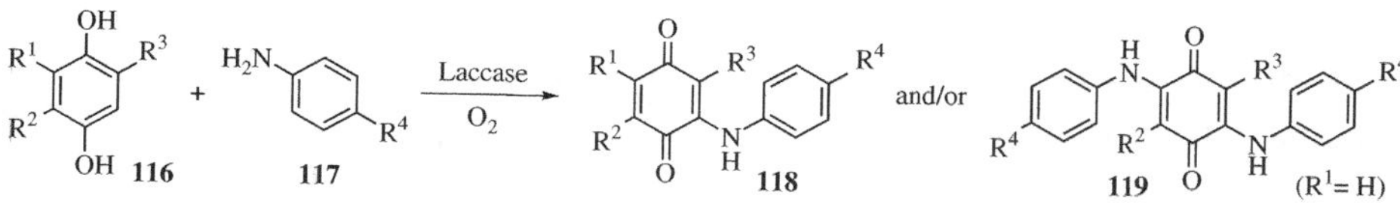

R^1, R^2, R^3 = H, Me, ^{t}Bu, Ac, COOMe, COOEt, $CONH_2$, $CONH(CH_2)_2OH$
R^4 = ^{i}Pr, COOH, Ac, $(CH_2)_2OH$, CH_2COOH, COOMe, $CONHCH_2COOH$, $N(CH_2CH_2)_2O$

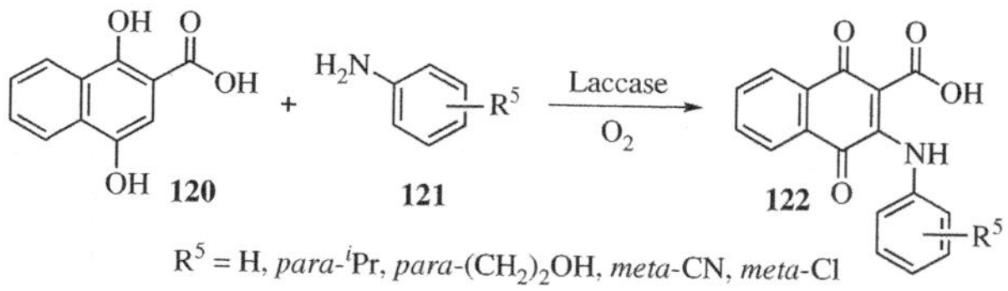

R^5 = H, *para*-^{i}Pr, *para*-$(CH_2)_2OH$, *meta*-CN, *meta*-Cl

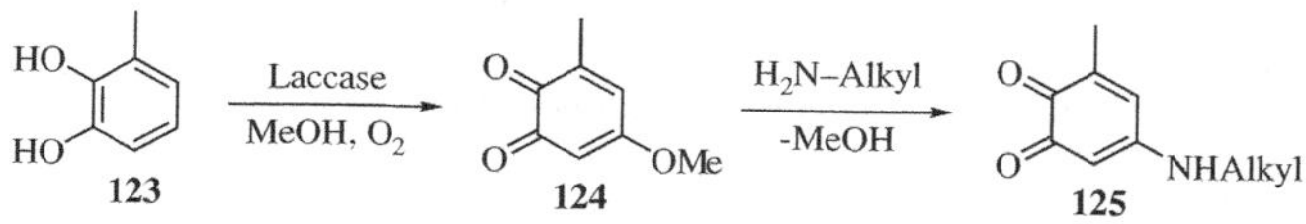

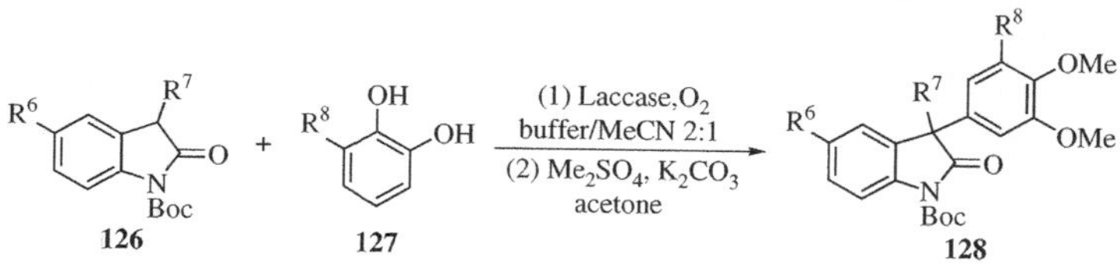

R^6 = H, Me, OMe R^7 = Ph, C_6H_4-*para*-OMe, Bn, $CH_2C_6C_4$-*meta*-F

R^8 = H, Me, OMe, F, Br

SCHEME 32.13 Oxidation/Michael addition/oxidation cascades mediated by laccases.

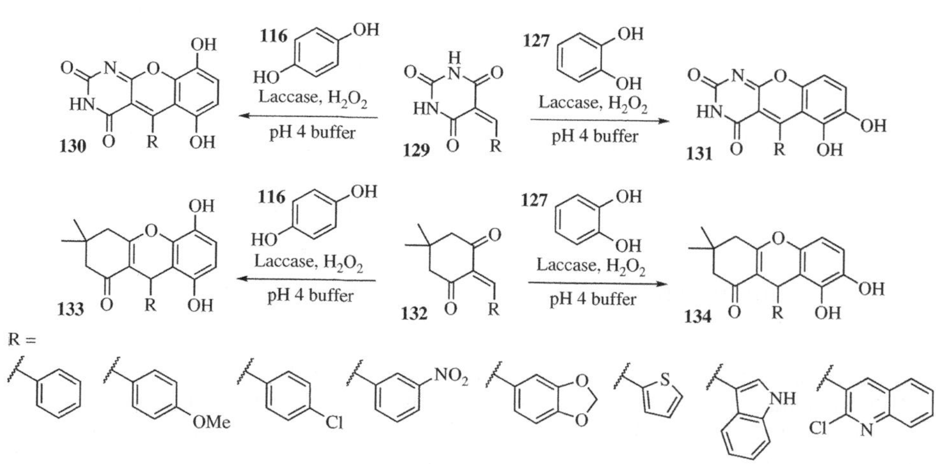

SCHEME 32.14 Laccase-mediated annulation of barbituric acid and dimedone derivatives.

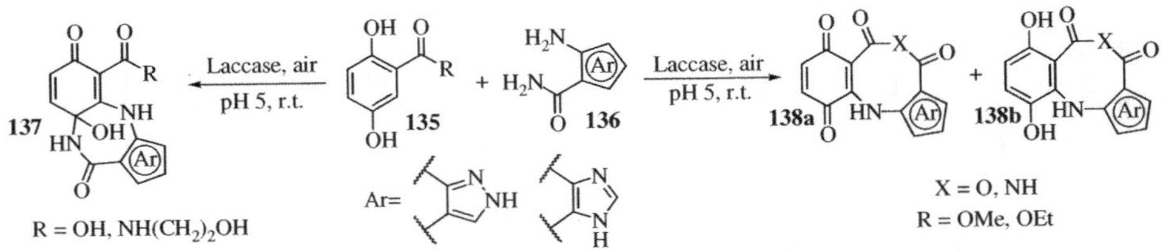

SCHEME 32.15 Laccase-mediated annulations of hydroquinones with aminoamides.

catechols **127** and β-diketones/β-ketoesters **139** give functionalized benzofurans **140**. In the latter instance, phenol **141** is used in place of catechol—the tyrosinase catalyzes its oxidation to a catechol, which is then the substrate for the laccase; using various cyclic β-dicarbonyls **142**, tri- and tetracyclic products **143** were accessible. A further extension of this methodology involved the use of heterocyclic β-dicarbonyls. Reaction of catechols **127** with heterocycles **144** gave 1:1 adducts **145**, analogous to **143**. In contrast, however, reaction of catechols **127** with (thio)barbituric acids **146** unexpectedly gave doubly spirocyclic 2:2 adducts **147**.

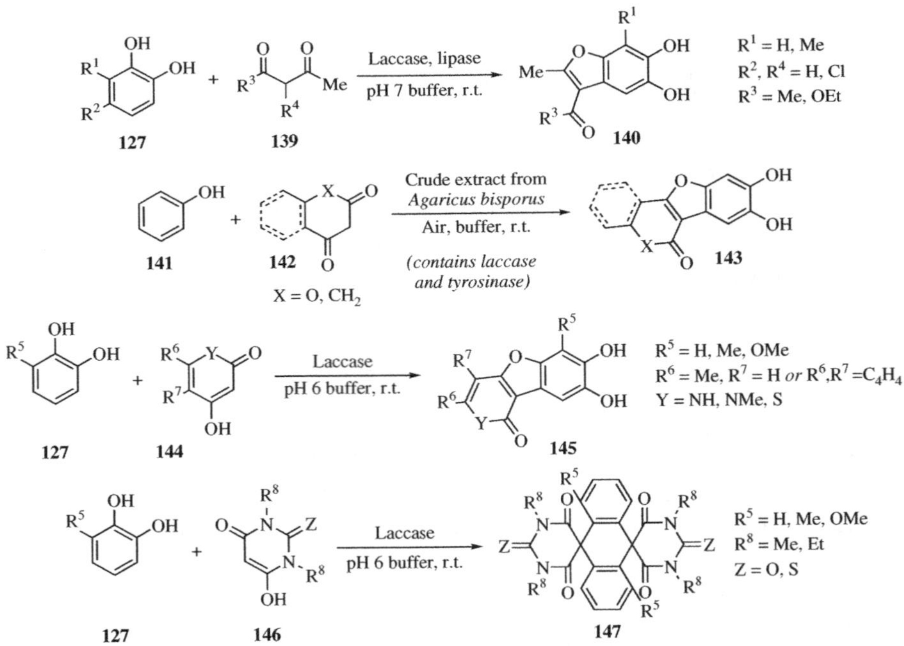

SCHEME 32.16 Laccase-mediated annulations of catechols with β-dicarbonyls.

Numerous biocatalytic arene dimerizations mediated by laccases are also described. 3-Hydroxyanthranilic acid **148** is a natural substrate for oxidative dimerization by fungal laccases to give cinnabarinic acid **149**, and the applicability of this transformation to other nonnatural substrates has been demonstrated (e.g., **150** to **151**) [79, 80]. Crossed dimers arising from the oxidative coupling of two different hydroxyaniline substrates have also been reported [81]. Some more unusual examples of laccase-mediated oligomerizations include the oxidative dimerization/cyclization of tyrosol **152** to give **153** [82] and the oxidative trimerization of indole **154** to give

155 [83]. Finally, mention should be made of other laccase-mediated biotransformations of synthetic utility such as oxidation of aryl methyl groups to aldehydes [84] and formation of benzimidazoles [85] (Scheme 32.17).

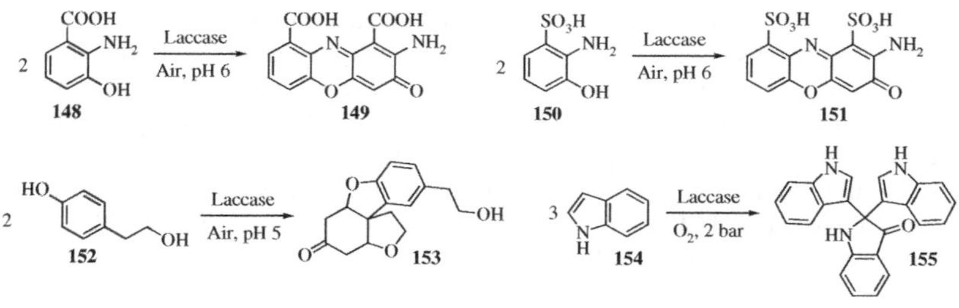

SCHEME 32.17 Some examples of laccase-mediated oligomerizations.

32.9 TETRAHYDROISOQUINOLINE SYNTHESIS (BIOCATALYTIC PICTET–SPENGLER)

Several enzymes are known to catalyze Pictet–Spengler cyclizations in nature. Norcoclaurine synthase catalyzes the reaction of dopamine **156** and 4-hydroxyphenylacetaldehyde **157** to give (*S*)-norcoclaurine **158**, a reaction that has been carried out biocatalytically on scale (10 mmol substrate, 80% yield, 93% *ee*) [86]. The use of this enzyme as a biocatalyst has been explored, and it has been determined that it has a wide substrate scope with respect to the aldehyde component. The phenethylamine component seemingly requires the hydroxyl group *para* to the site of substitution (Scheme 32.18) [87, 88]. Very recently, norcoclaurine synthase has also been used in a chemoenzymatic cascade synthesis of (*S*)-benzylisoquinoline and (*S*)-tetrahydroprotoberberine alkaloids [89]. Of course, the Pictet–Spengler cyclizations described could readily be carried out using nonbiocatalytic methods; the value of the biocatalyst in this context is in the *enantioselectivity* with which it forms the products. A second Pictet–Spenglerase, strictosidine synthase (for which the natural substrate is an indole as opposed to a catechol), has also been exploited to access novel "natural product-like" structures [90, 91].

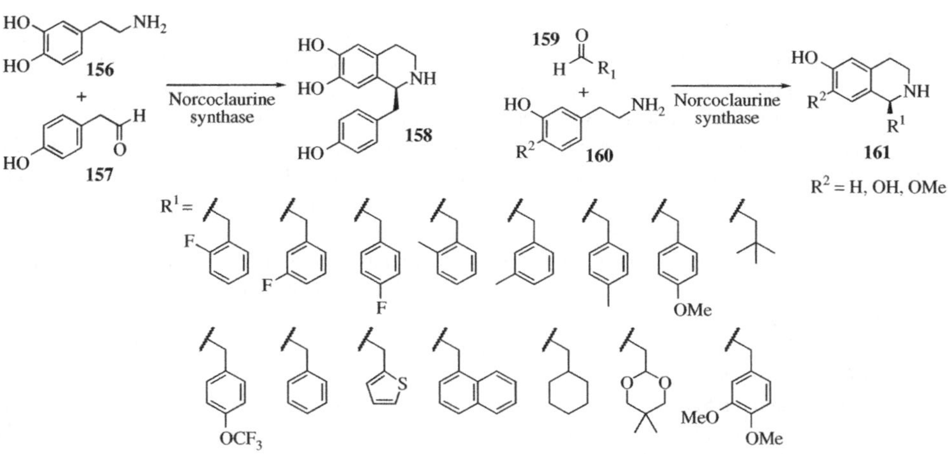

SCHEME 32.18 Enantioselective Pictet–Spengler reactions catalyzed by norcoclaurine synthase.

32.10 ARENE HYDROXYLATION

The oxidation of phenols to catechols or hydroquinones by tyrosinase enzymes has been developed for biocatalysis. For example, the *ortho*-hydroxylation of L-tyrosine **162** (and also substituted variants) to give L-DOPA **163** has been extensively studied due to the importance of L-DOPA in the treatment of Parkinson's disease [92, 93]. An arene hydroxylating enzyme having a broad substrate scope is 2-hydroxybiphenyl 3-monooxygenase from *Pseudomonas azelaica*, which is able to oxidize many *ortho*-substituted phenols **68** to the corresponding catechols **127** [94], as shown in Scheme 32.19. A notable example of an industrial biocatalytic arene hydroxylation that has been employed on very large scale (100 m^3 fermentation) is the *para*-hydroxylation of (*R*)-2-phenoxypropionic acid **164** by whole cells of *Beauveria bassiana* Lu 700 to give (*R*)-2-(4-hydroxyphenoxy)propionic acid **165**, an important intermediate in herbicide manufacture [95].

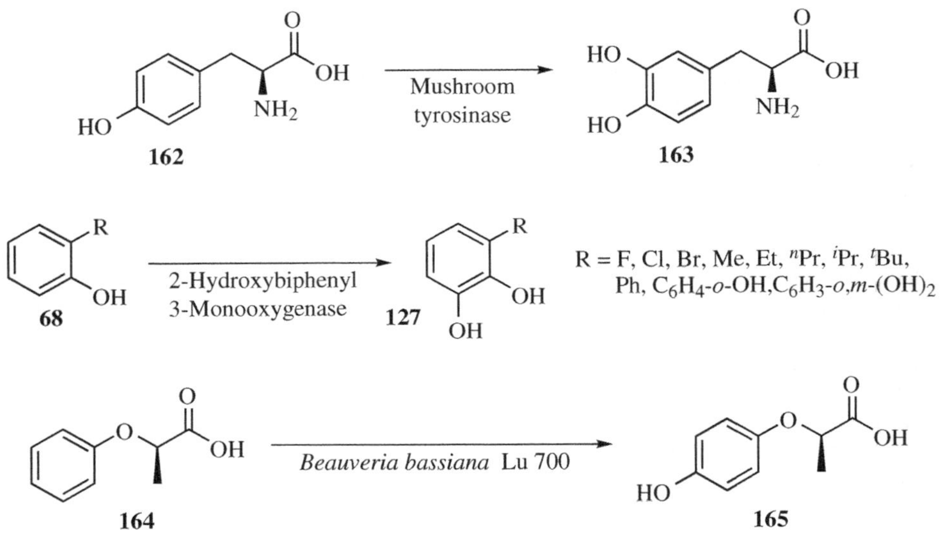

SCHEME 32.19 Selected examples of biocatalytic arene hydroxylation.

Cytochromes P450 are a group of enzymes that catalyze the oxidation of a very diverse range of both natural and xenobiotic substrates. A potential problem with the use of P450s to effect aromatic hydroxylation is that their substrate promiscuity may potentially lead to oxidation at multiple sites and hence mixtures of products. However, for certain specific categories of substrates, successes have been achieved with engineered P450s. An especially well studied P450 enzyme is P450 BM-3 from *Bacillus megaterium* (officially designated CYP102A1). The wild-type enzyme displays selectivity for hydroxylating long alkyl chains, but it has been shown that a single point mutation in this enzyme (F87V) results in a significant increase in activity with a variety of aromatic substrates **166** (Scheme 32.20) [96, 97]. Different engineered variants of P450 BM-3 were also reported to be able to hydroxylate polycyclic aromatic hydrocarbons **168** and **170** [98]. Such polycyclic aromatics are environmental pollutants, and their oxidation as shown is useful in the context of bioremediation. Preparative use of P450s to hydroxylate drug molecules such as diclofenac **175** and chlorzoxazone **177** has also been reported [99].

As mentioned in Section 32.2, arene *cis*-dihydrodiols of type **2** undergo facile dehydration reactions (usually catalyzed by Brønsted acid) that rearomatize the ring and give phenols **166** (Scheme 32.21). Such dehydrations are far more rapid than for the corresponding dehydration of arene *trans*-dihydrodiols (which may be accessed by hydrolysis of the epoxide ring in arene oxides

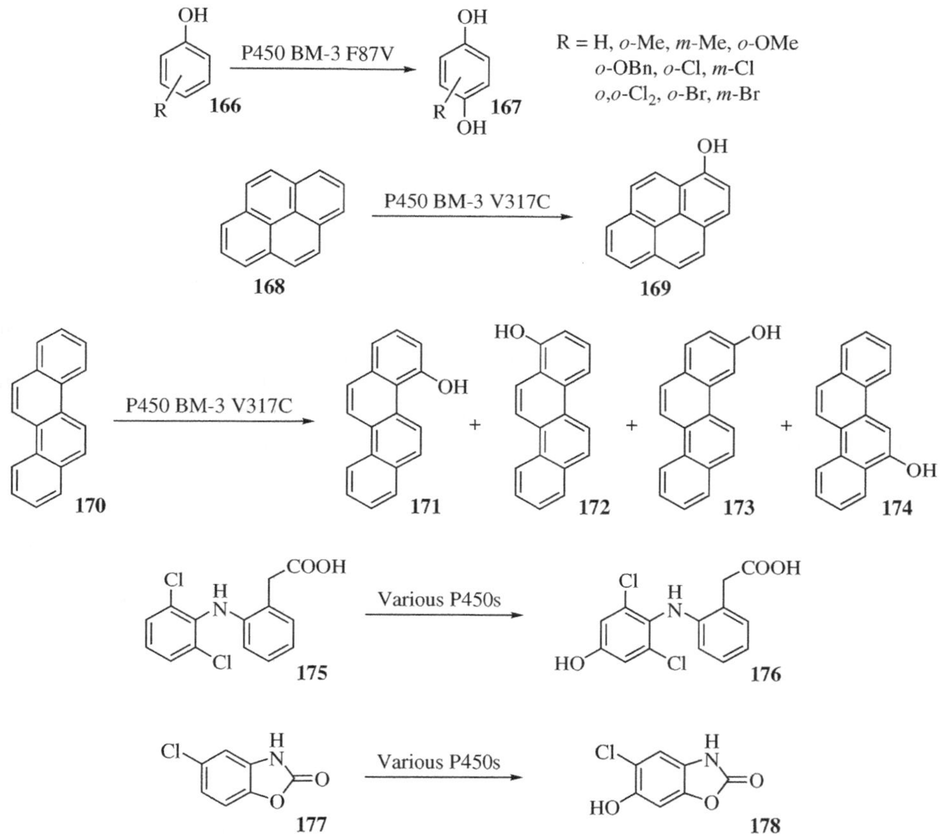

SCHEME 32.20 Selected examples of arene hydroxylation with cytochromes P450.

such as **26** or **29**). The mechanism of arene *cis*-dihydrodiol dehydration has been studied in some detail [100]. Overall, the transformation of **1** to **166** represents an appealing method for effecting biocatalytic arene hydroxylation.

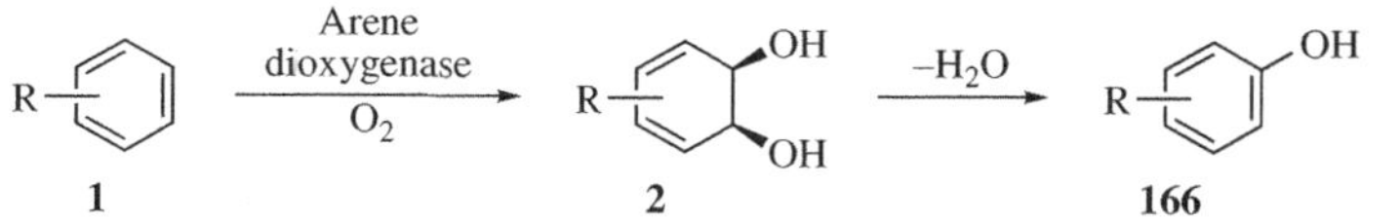

SCHEME 32.21 Arene hydroxylation via an intermediate *cis*-dihydrodiol.

It must be stressed that arene hydroxylation is often achieved in nature via multistep pathways, particularly via arene oxides. In contrast to the dehydration required to access phenols **166** from arene *cis*-diols **2**, the transformation of an arene oxide into a phenol is simply an isomerization. Several mechanisms may be proposed by which such an isomerization may occur, and extensive studies with isotopically labeled substrates have elucidated a process known as the "NIH shift" that is common to many arene oxide isomerizations [101]. As shown in Scheme 32.22, if a *para*-deuterated arene **179** is oxidized by an arene monooxygenase to the corresponding arene oxide **180**, this may undergo epoxide ring opening to give zwitterionic intermediate **181**, which contains a resonance-stabilized cation. Two potential pathways from **181** to the product phenol **182** may be envisaged. A direct deprotonation of **181** to rearomatize the ring (Scheme 32.22, top) will lead to **182**, in which no deuterium has been retained (since the phenolic proton is of course exchangeable). In contrast, **181** may undergo a hydride shift ("NIH shift") to give cyclohexadienone **183**, from which no deuterium has yet been lost. Tautomerization of **183** to the product **182** will then occur, but in this instance, whereas loss of deuteron in this process will give **182**, loss of proton instead will give **182-*d*$_1$**, in which deuterium has been retained (albeit now *meta* to the R substituent). When such experiments were in fact carried out, products in which a significant amount of deuterium was retained were isolated. Indeed, more **182-*d*$_1$** was isolated than **182**, which is to be expected given that the kinetic isotope effect for the enolization will favor loss of the proton over the deuteron. This conclusively demonstrated the operation of the hydride shift pathway, with further supporting evidence being obtained upon the use of tritiated substrates (the *para*-tritiated analogue of **179** gave a product distribution that even more heavily favored formation of **182-*t*$_1$** over **182**, due to the greater kinetic isotope effect with the heavier isotope of hydrogen). The overall transformation of an arene to the corresponding phenol via an arene oxide intermediate renders the use of monooxygenases a viable strategy for biocatalytic arene hydroxylation.

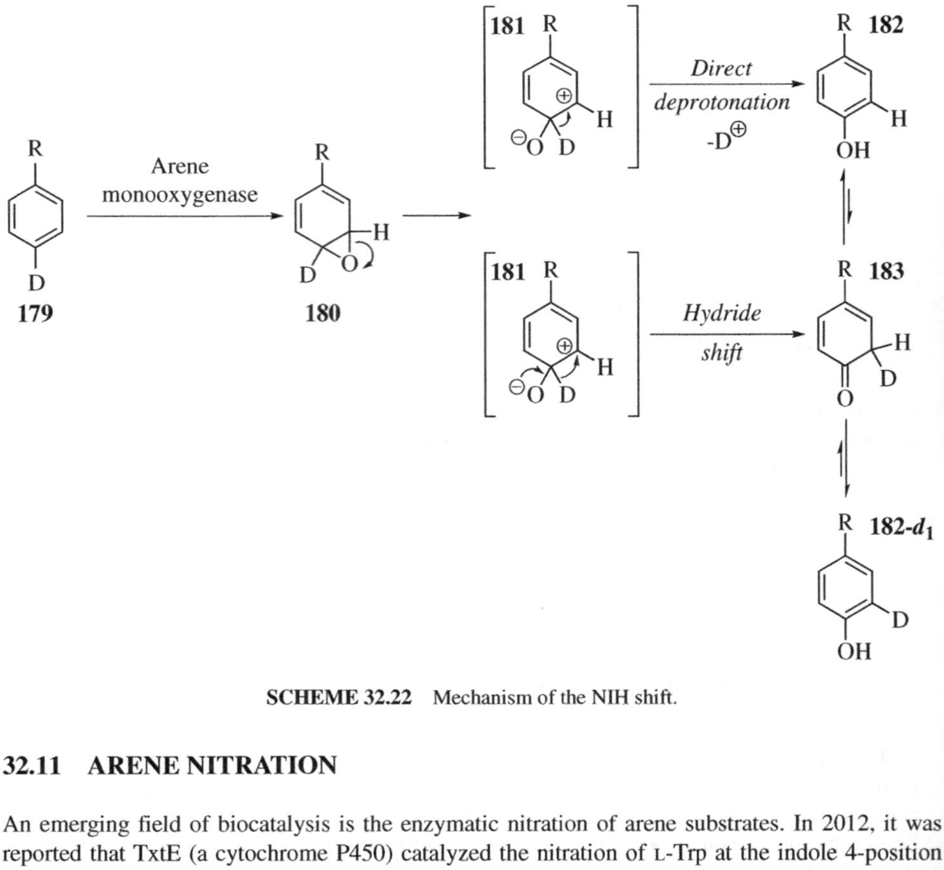

SCHEME 32.22 Mechanism of the NIH shift.

32.11 ARENE NITRATION

An emerging field of biocatalysis is the enzymatic nitration of arene substrates. In 2012, it was reported that TxtE (a cytochrome P450) catalyzed the nitration of L-Trp at the indole 4-position during the biosynthesis of thaxtomin A [102]. It was further determined that the source of nitrogen for this transformation is endogenously generated nitric oxide. A more recent report attempted to define the substrate scope, by delineating which substituted tryptophan analogues could be turned over by TxtE (Scheme 32.23) [103]. Given the unusual regioselectivity of this transformation, as well as the mildness of the reaction conditions, it seems likely that biocatalytic arene nitrations will find use in preparative chemistry before too long.

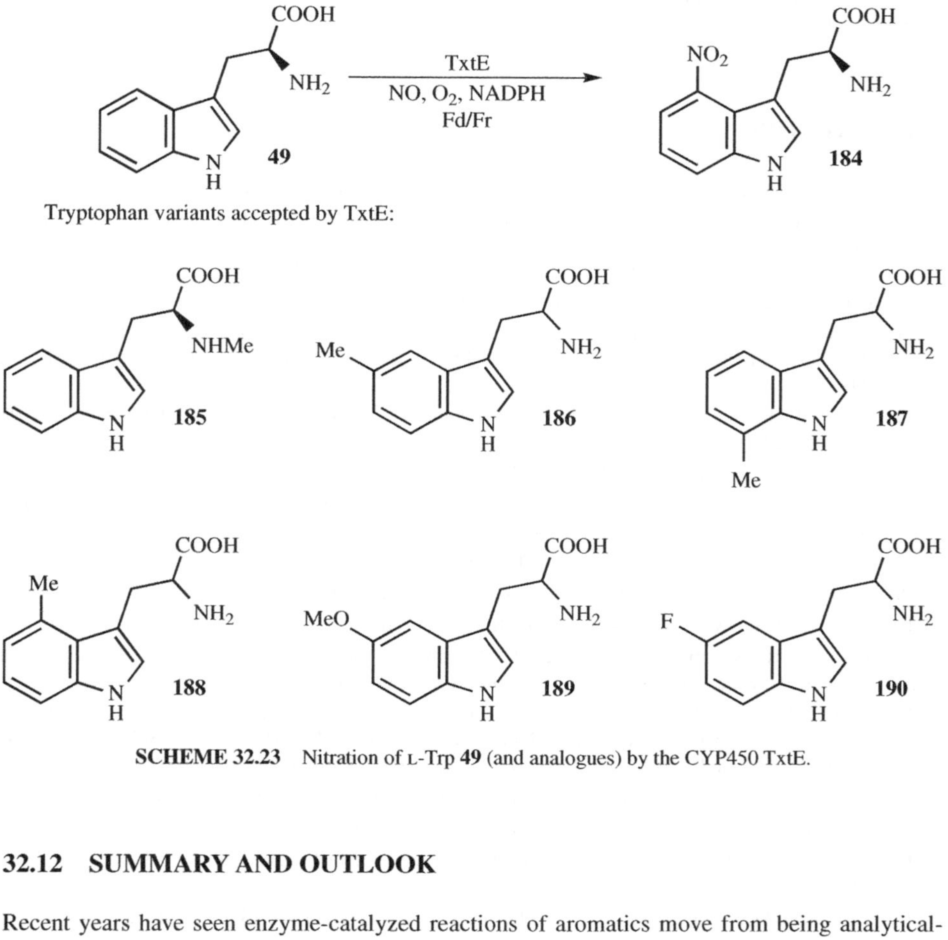

SCHEME 32.23 Nitration of L-Trp **49** (and analogues) by the CYP450 TxtE.

32.12 SUMMARY AND OUTLOOK

Recent years have seen enzyme-catalyzed reactions of aromatics move from being analytical-scale processes, of interest primarily to biologists, to being preparatively useful methods that are ever more accessible to synthetic organic chemists in general. Many of the biocatalysts described in this chapter are now commercially available or may be available on request from the relevant authors. Many biotransformations can be carried out without the need for specialized equipment. The advantages that biocatalysis can confer are significant, both from the perspective of selectivity and also from the perspective of sustainability. In view of the previous text, the reader is urged to consider biocatalytic processes as a plausible option in synthetic planning, especially bearing in mind that the repertoire of biocatalytic arene transformations will undoubtedly increase in the coming years.

ABBREVIATIONS

BBE	Berberine bridge enzyme
BPDO	Biphenyl dioxygenase
BZDO	Benzoate dioxygenase
MAO	Monoamine oxidase
SAM	*S*-adenosylmethionine
SHC	Squalene–hopene cyclase
TDO	Toluene dioxygenase

REFERENCES

[1] Gibson, D. T., Koch, J. R., Schuld, C. L., and Kallio, R. E. (1968) *Biochemistry*, **7**, 3795–3802.

[2] (a) Motherwell, W. B. and Williams, A. S. (1995) *Angew. Chem. Int. Ed.*, **34**, 2031–2033. (b) Jung, P. M. J., Motherwell, W. B., and Williams, A. S. (1997) *Chem. Commun.*, 1283–1284. (c) Feng, Y., Ke, C.-Y., Xue, G., and Que Jr., L. (2009) *Chem. Commun.*, 50–52, doi:10.1039/b817222f.

[3] Gibson, D. T., Hensley, M., Yoshioka, H., and Mabry, T. J. (1970) *Biochemistry*, **9**, 1626–1630.

[4] (a) Ballard, D. G. H., Courtis, A., Shirley, I. M., and Taylor, S. C. (1983) *J. Chem. Soc. Chem. Commun.*, 954–955. (b) Ballard, D. G. H., Courtis, A., Shirley, I. M., and Taylor, S. C. (1988) *Macromolecules*, **21**, 294–304.

[5] Ensley, B. D., Ratzkin, B. J., Osslund, T. D., Simon, M. J., Wackett, L. P., and Gibson, D. T. (1983) *Science*, **222**, 167–169.

[6] Carless, H. A. J., Busia, K., and Oak, O. Z. (1993) *Synlett*, 672–674.

[7] Boyd, D. R., Sharma, N. D., Llamas, N. M., Malone, J. F., O'Dowd, C. R., and Allen, C. C. R. (2005) *Org. Biomol. Chem.*, **3**, 1953–1963.

[8] Ley, S. V., Parra, M., Redgrave, A. J., and Sternfeld, F. (1990) *Tetrahedron*, **46**, 4995–5026.

[9] Adams, D. R., van Kempen, J., Hudlický, J. R., and Hudlický, T. (2014) *Heterocycles*, **88**, 1255–1274.

[10] Tian, X., Hudlický, T., and Koenigsberger, K. (1995) *J. Am. Chem. Soc.*, **117**, 3643–3644.

[11] Boyd, D. R., Sharma, N. D., Sbircea, L., Murphy, D., Belhocine, T., Malone, J. F., James, S. L., Allen, C. C. R., and Hamilton, J. T. G. (2008) *Chem. Commun.*, 5535–5537.

[12] (a) Shie, J.-J., Fang, J.-M., and Wong, C.-H. (2008) *Angew. Chem. Int. Ed.*, **47**, 5788–5791. (b) Matveenko, M., Willis, A. C., and Banwell, M. G. (2008) *Tetrahedron Lett.*, **49**, 7018–7020. (c) Sullivan, B., Carrera, I., Drouin, M., and Hudlický, T. (2009) *Angew. Chem. Int. Ed.*, **48**, 4229–4231.

[13] Hudlický, T., Luna, H., Barbieri, G., Kwart, L. D. (1988) *J. Am. Chem. Soc.*, **110**, 4735–4741.

[14] Banwell, M. G., Edwards, A. J., Harfoot, G. J., and Jolliffe, K. A. (2004) *Tetrahedron*, **60**, 535–547.

[15] Leisch, H., Omori, A. T., Finn, K. J., Gilmet, J., Bissett, T., Ilceski, D., and Hudlický, T. (2009) *Tetrahedron*, **65**, 9862–9875.

[16] Palframan, M. J., Kociok-Köhn, G., and Lewis, S. E. (2011) *Org. Lett.*, **13**, 3150–3153.

[17] Chang, E. L., Schwartz, B. D., Draffan, A. G., Banwell, M. G., and Willis, A. C. (2015) *Chem. Asian J.*, **10**, 427–439.

[18] Banwell, M. and McLeod, M. (1998) *Chem. Commun.*, 1851–1852.

[19] Schwartz, B. D., Matousova, E., White, R., Banwell, M. G., and Willis, A. C. (2013) *Org. Lett.*, **15**, 1934–1937.

[20] Boyd, D. R., Sharma, N. D., Hand, M. V., Groocock, M. R., Kerley, N. A., Dalton, H., Chima, J., and Sheldrake, G. N. (1993) *J. Chem. Soc. Chem. Commun.*, 974–976.

[21] Boyd, D. R., Sharma, N. D., Byrne, B., Hand, M. V., Malone, J. F., Sheldrake, G. N., Blacker, J., and Dalton, H. (1998) *J. Chem. Soc. Perkin Trans.* **1**, 1935–1944.

[22] Lewis, S. E. (2014) *Chem. Commun.*, **50**, 2821–2830.

[23] (a) Bon, D. J.-Y. D., Lee, B., Banwell, M. G., and Cade, I. A. (2012) *Chim. Oggi*, **30**, 22–27. (b) Hudlický, T. (2010) *Pure Appl. Chem.*, **82**, 1785–1796. (c) Hudlický, T. and Reed, J. W. (2009) *Synlett*, 685–703. (d) Austin, K. A. B., Matveenko, M., Reekie, T. A., and Banwell, M. G. (2008) *Chem. Aust.*, **75**, 3–7. (e) Boyd, D. R., and Bugg, T. D. H. (2006) *Org. Biomol. Chem.*, **4**, 181–192. (f) Johnson, R. A. (2004) *Org. React.*, **63**, 117–264. (g) Banwell, M. G., Edwards, A. J., Harfoot, G. J., Jolliffe, K. A., McLeod, M. D., McRae, K. J., Stewart, S. G., and Vögtle, M. (2003) *Pure Appl. Chem.*, **75**, 223–229. (h) Hudlický, T., Gonzalez, D., and Gibson, D. T. (1999) *Aldrichim. Acta*, **32**, 35–62. (i) Boyd, D. R. and Sheldrake, G. N. (1998) *Nat. Prod. Rep.*, 309–324.

[24] (a) Hudlický, T., Stabile, M. R., Gibson, D. T., and Whited, G. M. (1999) *Org. Synth.*, **76**, 77–85. (b) Vila, M. A., Brovetto, M., Gamenara, D., Bracco, P., Zinola, G., Seoane, G., Rodriguez, S., and Carrera, I. (2013) *J. Mol. Catal. B Enzym.*, **96**, 14–20.

[25] Boyd, D. R., Sharma, N. D., Harrison, J. S., Malone, J. F., McRoberts, W. C., Hamilton, J. T. G., and Harper, D. B. (2008) *Org. Biomol. Chem.*, **6**, 1251–1259.

[26] Stecher, H., Tengg, M., Ueberbacher, B. J., Remler, P., Schwab, H., Griengl, H., and Gruber-Khadjawi, M. (2009) *Angew. Chem. Int. Ed.*, **48**, 9546–9548.

[27] Hammer, S. C., Dominicus, J. M., Syren, P.-O., Nestl, B. M., and Hauer, B. (2012) *Tetrahedron*, **68**, 7624–7629.

[28] Liebhold, M., Xie, X., and Li, S.-M. (2012) *Org. Lett.*, **14**, 4882–4885.

[29] Liebhold, M. and Li, S.-M. (2013) *Org. Lett.*, **15**, 5834–5837.

[30] Liebhold, M., Xie, X., and Li, S.-M. (2013) *Org. Lett.*, **15**, 3062–3065.

[31] Tarcz, S., Ludwig, L., and Li, S.-M. (2014) *ChemBioChem*, **15**, 108–116.

[32] Fan, A.and Li, S.-M. (2013) *Adv. Synth. Catal.*, **355**, 2659–2666.

[33] Yu, X. and Li, S.-M. (2011) *ChemBioChem*, **12**, 2280–2283.

[34] Pockrandt, D., Sack, C., Kosiol, T., and Li, S.-M. (2014) *Appl. Microbiol. Biotechnol.*, **98**, 4987–4994.

[35] (a) Schrittwieser, J. H., Resch, V., Sattler, J. H., Lienhart, W.-D., Durchschein, K., Winkler, A., Gruber, K., Macheroux, P., and Kroutil, W. (2011) *Angew. Chem. Int. Ed.*, **50**, 1068–1071. (b) Schrittwieser, J. H., Groenendaal, B., Resch, V., Ghislieri, D., Wallner, S., Fischereder, E.-M., Fuchs, E., Grischek, B., Sattler, J. H., Macheroux, P., Turner, N. J., and Kroutil, W. (2014) *Angew. Chem. Int. Ed.*, **53**, 3731–3734. (c) Schrittwieser, J. H., Groenendaal, B., Willies, S. C., Ghislieri, D., Rowles, I., Resch, V., Sattler, J. H., Fischereder, E.-M., Grischek, B., Lienhart, W.-D., Turner, N. J., and Kroutil, W. (2014) *Catal. Sci. Technol.*, **4**, 3657–3664.

[36] Resch, V., Lechner, H., Schrittwieser, J. H., Wallner, S., Gruber, K., Macheroux, P., and Kroutil, W. (2012) *Chem. Eur. J.*, **18**, 13173–13179.

[37] Resch, V., Schrittwieser, J. H., Wallner, S., Macheroux, P., and Kroutil, W. (2011) *Adv. Synth. Catal.*, **353**, 2377–2383.

[38] Siirola, E., Frank, A., Grogan, G., and Kroutil, W. (2013) *Adv. Synth. Catal.*, **355**, 1677–1691.

[39] Frank, A., Siirola, E., Kroutil, W., and Grogan, G. (2014) *Top. Catal.*, **57**, 376–384.

[40] Siirola, E. and Kroutil, W. (2014) *Top. Catal.*, **57**, 392–400.

[41] Glueck, S. M., Gümüs, S., Fabian, W. M. F., and Faber, K. (2010) *Chem. Soc. Rev.*, **39**, 313–328.

[42] Aresta, M., Quaranta, E., Liberio, R., Dileo, C., and Tommasi, I. (1998) *Tetrahedron*, **54**, 8841–8846.

[43] Huang, J., He, Z., and Wiegel, J. (1999) *J. Bacteriol.*, **181**, 5119–5122.

[44] Liu, J., Zhang, X., Zhou, S., Tao, P., and Liu, J. (2007) *Curr. Microbiol.*, **54**, 102–107.

[45] Matsui, T., Yoshida, T., Hayashi, T., and Nagasawa, T. (2006) *Arch. Microbiol.*, **186**, 21–29.

[46] He, Z. and Wiegel, J. (1996) *J. Bacteriol.*, **178**, 3539–3543.

[47] Yoshida, T., Inami, Y., Matsui, T., and Nagasawa, T. (2010) *Biotechnol. Lett.*, **32**, 701–705.

[48] Wuensch, C., Gross, J., Steinkellner, G., Lyskowski, A., Gruber, K., Glueck, S. M., and Faber, K. (2014) *RSC Adv.*, **4**, 9673–9679.

[49] Wuensch, C., Glueck, S. M., Gross, J., Koszelewski, D., Schober, M., and Faber, K. (2012) *Org. Lett.*, **14**, 1974–1977.

[50] Wieser, M., Yoshida, T., and Nagasawa, T. (1998) *Tetrahedron Lett.*, **39**, 4309–4310.

[51] Yoshida, T., Fujita, K., and Nagasawa, T. (2002) *Biosci. Biotechnol. Biochem.*, **66**, 2388–2394.

[52] Runguphan, W., Qu, X. D., and O'Connor, S. E. (2010) *Nature*, **468**, 461–464.

[53] Roy, A. D., Grüschow, S., Cairns, N., and Goss, R. J. M. (2010) *J. Am. Chem. Soc.*, **132**, 12243–12245.

[54] Glenn, W. S., Nims, E., and O'Connor, S. E. (2011) *J. Am. Chem. Soc.*, **133**, 19346–19349.

[55] Runguphan, W. and O'Connor, S. E. (2013) *Org. Lett.*, **15**, 2850–2853.

[56] Smith, D. R. M., Grüschow, S., and Goss, R. J. M. (2013) *Curr. Opin. Chem. Biol.*, **17**, 276–283.

[57] Payne, J. T., Andorfer, M. C., and Lewis, J. C. (2013) *Angew. Chem. Int. Ed.*, **52**, 5271–5274.

[58] Poor, C. B., Andorfer, M. C., and Lewis, J. C. (2014) *ChemBioChem*, **15**, 1286–1289.

[59] Payne, J. T. and Lewis, J. C. (2014) *Synlett*, 1345–1349.

[60] Frese, M., Guzowska, P. H., Voß, H., and Sewald, N. (2014) *ChemCatChem*, **6**, 1270–1276.

[61] Frese, M. and Sewald, N. (2015) *Angew. Chem. Int. Ed.*, **54**, 298–301.

[62] Goss, R. J. M. and Newill, P. L. A. (2006) *Chem. Commun.*, 4924–4925.

[63] Smith, D. R. M., Willemse, T. Gkotsi, D. S., Schepens, W., Maes, B. U. W., Ballet, S., and Goss, R. J. M. (2014) *Org. Lett.*, **16**, 2622–2625.

[64] Lang, A., Polnick, S., Nicke, T., William, P., Patallo, E. P., Naismith, J. H., and van Pée, K.-H. (2011) *Angew. Chem. Int. Ed.*, **50**, 2951–2953.

[65] Zeng, J. Lytle, A. K., Gage, D., Johnson, S. J., and Zhan, J. (2013) *Bioorg. Med. Chem. Lett.*, **23**, 1001–1003.

[66] Witayakran, S. and Ragauskas, A. J. (2009) *Adv. Synth. Catal.*, **351**, 1187–1209.

[67] Burton, S. J. (2003) *Curr. Org. Chem.*, **7**, 1317–1331.

[68] Aldemir, H., Richarz, R., and Gulder, T. A. M. (2014) *Angew. Chem. Int. Ed.*, **53**, 8286–8293.

[69] Ciecholewski, S., Hammer, E., Manda, K., Bose, G., Nguyen, V. T. H., Langer, P., and Schauer, F. (2005) *Tetrahedron*, **61**, 4615–4619.

[70] Intra, A. Nicotra, S., Riva, S., and Danieli, B. (2005) *Adv. Synth. Catal.*, **347**, 973–977.

[71] Niedermeyer, T. H. J., Mikolasch, A., and Lalk, M. (2005) *J. Org. Chem.*, **70**, 2002–2008.

[72] Wellington, K. W. and Kolesnikova, N. I. (2012) *Bioorg. Med. Chem.*, **20**, 4472–4481.

[73] Herter, S., Mikolasch, A., Michalik, D., Hammer, E., Schauer, F., Bornscheuer, U., and Schmidt, M. (2011) *Tetrahedron*, **67**, 9311–9321.

[74] Pietruszka, J. and Wang, C. (2012) *ChemCatChem*, **4**, 782–785.

[75] Kidwai, M., Poddar, R., Diwaniyan, S., and Kuhad, R. C. (2009) *Adv. Synth. Catal.*, **351**, 589–595.

[76] Hahn, V., Davids, T., Lalk, M., Schauer, F., and Mikolasch, A. (2010) *Green Chem.*, **12**, 879–887.

[77] Witayakran, S. and Ragauskas, A. J. (2009) *Eur. J. Org. Chem.*, 358–363.

[78] Leutbecher, H., Hajdok, S., Braunberger, C., Neumann, M., Mika, S., Conrad, J., and Beifuss, U. (2009) *Green Chem.*, **11**, 676–679.

[79] Bruyneel, F., Enaud, E., Billottet, L., Vanhulle, S., and Marchand-Brynaert, J. (2008) *Eur. J. Org. Chem.*, 72–79.

[80] Bruyneel, F., Payen, O. Rescigno, A., Tinant, B., and Marchand-Brynaert, J. (2009) *Chem. Eur. J.*, **15**, 8283–8295.

[81] Bruyneel, F., Dive, G., and Marchand-Brynaert, J. (2012) *Org. Biomol. Chem.*, **10**, 1834–1846.

[82] Chakroun, H., Bouaziz, M., Yangui, T., Blibech, I., Dhouib, A., and Sayadi, S. (2013) *J. Mol. Catal. B Enzym.*, **87**, 11–17.

[83] Ganachaud, C., Garfagnoli, V., Tron, T., and Iacazio, G. (2008) *Tetrahedron Lett.*, **49**, 2476–2478.

[84] Potthast, A., Rosenau, T., Chen, C.-L., and Gratzl, J. S. (1995) *J. Org. Chem.*, **60**, 4320–4321.

[85] Leutbecher, H., Constantin, M.-A., Mika, S., Conrad, J., and Beifuss, U. (2011) *Tetrahedron Lett.*, **52**, 604–607.

[86] Bonamore, A., Rovardi, I., Gasparrini, F., Baiocco, P., Barba, M., Molinaro, C., Botta, B., Boffi A., and Macone, A. (2010) *Green Chem.*, **12**, 1623–1627.

[87] Ruff, B. M., Bräse, S., and O'Connor, S. E. (2012) *Tetrahedron Lett.*, **53**, 1071–1074.

[88] Nishihachijo, M., Hirai, Y., Kawano, S., Nishiyama, A., Minami, H., Katayama, T., Yasohara, Y., Sato, F., and Kumagai, H. (2014) *Biosci. Biotechnol. Biochem.*, **78**, 701–707.

[89] Lichman, B. R., Lamming, E. D., Pesnot, T., Smith, J. M., Hailes, H. C., and Ward, J. M. (2015) *Green Chem.*, **17**, 852–855.

[90] Wu, F., Zhu, H., Sun, L., Rajendran, C., Wang, M., Ren, X., Panjikar, S., Cherkasov, A., Zou, H., and Stöckigt, J. (2012) *J. Am. Chem. Soc.*, **134**, 1498–1500.

[91] Yan, W., Ge, H. M., Wang, G., Jiang, N., Mei, Y. N., Jiang, R, Li, S. J., Chen, C. J., Jiao, R. H., Xu, Q., Ng, S. W., and Tan, R. X. (2014) *Proc. Natl. Acad. Sci. U. S. A.*, **111**, 18138–18143.

[92] Phillips, R. S., Fletcher, J. G., Vontersch, R. L., and Kirk, K. L. (1990) *Arch. Biochem. Biophys.*, **276**, 65–69.

[93] Carvalho, G. M. J., Alves, T. L. M., and Freire, D. M. G. (2000) *Appl. Biochem. Biotechnol.*, **84–86**, 791–800.

[94] Schmid, A., Vereyken, I., Held, M., and Witholt, B. (2001) *J. Mol. Catal. B Enzym.*, **11**, 455–462.

[95] Dingler, C., Ladner, W., Krei, G. A., Cooper, B., and Hauer, B. (1996) *Pestic. Sci.*, **46**, 33–35.

[96] Sulistyaningdyah, W. T., Ogawa, J., Li, Q.-S., Maeda, C., Yano, Y., Schmid, R. D., and Shimizu, S. (2005) *Appl. Microbiol. Biotechnol.*, **67**, 556–562.

[97] Kitamura, E., Otomatsu, T., Maeda, C., Aoki, Y., Ota, C., Misawa, N., and Shindo, K. (2013) *Biosci. Biotechnol. Biochem.*, **77**, 1340–1343.

[98] Sideri, A., Goyal, A., Di Nardo, G., Tsotsou, G. E., and Gilardi, G. (2013) *J. Inorg. Biochem.*, **120**, 1–7.

[99] Weis, R., Winkler, M., Schittmayer, M., Kambourakis, S., Vink, M., Rozzell, J. D., and Glieder, A. (2009) *Adv. Synth. Catal.*, **351**, 2140–2146.

[100] (a) Kudavalli, J. S., Rao, S. N., Bean, D. E., Sharma, N. D., Boyd, D. R., Fowler, P. W., Gronert, S., Kamerlin, S. C. L., Keeffe, J. R., and More O'Ferrall, R. A. (2012) *J. Am. Chem. Soc.*, **134**, 14056–14069. (b) Lawlor, D. A., Kudavalli, J. S., MacCormac, A. C., Coyne, D. A., Boyd, D. R., and More O'Ferrall, R. A. (2011) *J. Am. Chem. Soc.*, **133**, 19718–19728. (c) Kudavalli, J. S., Boyd, D. R., Sharma, N. D., and More O'Ferrall, R. A. (2011) *J. Org. Chem.*, **76**, 9338–9343. (d) Kudavalli, J. S., Boyd, D. R., Coyne, D., Keeffe, J. R., Lawlor, D. A., MacCormac, A. C., More O'Ferrall, R. A., Rao, S. N., and Sharma, N. D., (2010) *Org. Lett.*, **12**, 5550–5553. (e) Boyd, D. R., Blacker, J., Byrne, B., Dalton, H., Hand, M. V., Kelly, S. C., More O'Ferrall, R. A., Rao, S. N., Sharma, N. D., and Sheldrake, G. N. (1994) *J. Chem. Soc. Chem. Commun.*, **313–314**.

[101] (a) Daly, J. W., Jerina, D. M., and Witkop, B. (1972) *Experientia*, **28**, 1129–1149. (b) Guroff, G., Daly, J. W., Jerina, D. M., Renson, J., Witkop, B., and Udenfriend, S. (1967) *Science*, **157**, 1524–1530.

[102] Barry, S. M., Kers, J. A., Johnson, E. G., Song, L., Aston, P. R., Patel, B., Krasnoff, S. B., Crane, B. R., Gibson, D. M., Loria, R., and Challis, G. L. (2012) *Nat. Chem. Biol.*, **8**, 814–816.

[103] Dodani, S. C., Cahn, J. K. B., Heinisch, T., Brinkmann-Chen, S., McIntosh, J. A., and Arnold, F. H. (2014) *ChemBioChem*, **15**, 2259–2267.

INDEX

Note: Page numbers in *italics* refer to Figures; those in **bold** to Tables.

Arene Chemistry: Reaction Mechanisms and Methods for Aromatic Compounds,
First Edition. Edited by Jacques Mortier.

d and bound by CPI Group (UK) Ltd, Croydon, CR0 4YY
16/09/2026
14963430-0001